A PEARSON AUSTRALIA CUSTOM BOOK

ENGG104 Electrical Systems

This custom book has been compiled from:

INTRODUCTORY CIRCUIT ANALYSIS

12TH EDITION

BOYLESTAD

UNIVERSITY OF WOLLONGONG

Pearson Australia
Suite 1001 Level 10
151 Castlereagh St
Sydney NSW 2000
Australia
Ph:02 9454 2200
www.pearson.com.au

Project Management Team Leader: Jill Gillies
Project Manager: Aida Reyes Cruz
Production Controller: Brad Smith
Custom Product Specialist: Nicki Barlow

ISBN: 978 1 4886 1153 7

A NOTE ABOUT THE CONTENTS OF THIS CUSTOM BOOK

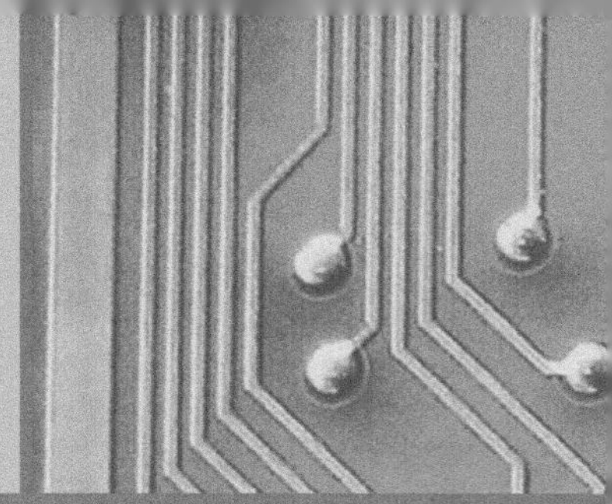

Welcome to *ENGG104 Electrical Systems*.

The chapters in this custom book have been chosen specifically to meet your course requirements by your lecturer.

In reading this book, please be aware that some chapters have not been included in this compilation.

Both the chapter numbers and page numbering in this Custom Publication are the same as in the original source book.

The index will then reference page numbers that are not included.

Please note that due to copyright restrictions some images have been removed, this will not impact your learning of this subject matter in any way but please keep in mind that you will see references to images that will not appear in this compilation.

We hope that you will find this text easy to follow and enjoyable.

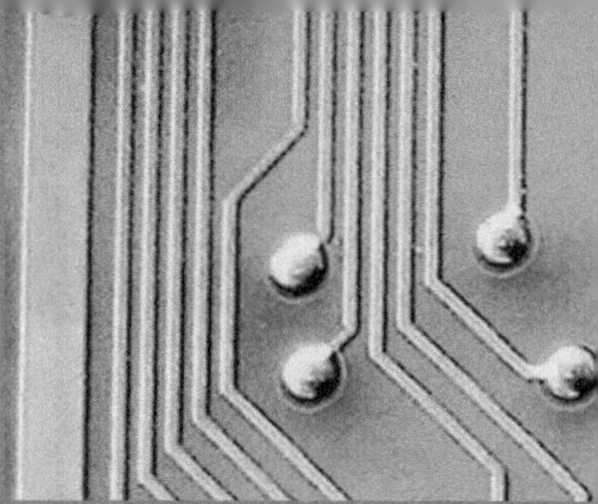

Brief Contents

Contents

1 Introduction 1

2 Voltage and Current 33

3 Resistance 63

4 Ohm's Law, Power, and Energy 101

INTRODUCTION

Objectives

- *Become aware of the rapid growth of the electrical/electronics industry over the past century.*
- *Understand the importance of applying a unit of measurement to a result or measurement and to ensuring that the numerical values substituted into an equation are consistent with the unit of measurement of the various quantities.*
- *Become familiar with the SI system of units used throughout the electrical/electronics industry.*
- *Understand the importance of powers of ten and how to work with them in any numerical calculation.*
- *Be able to convert any quantity, in any system of units, to another system with confidence.*

1.1 THE ELECTRICAL/ELECTRONICS INDUSTRY

Over the past few decades, technology has been changing at an ever-increasing rate. The pressure to develop new products, improve the performance of existing systems, and create new markets will only accelerate that rate. This pressure, however, is also what makes the field so exciting. New ways of storing information, constructing integrated circuits, and developing hardware that contains software components that can "think" on their own based on data input are only a few possibilities.

Change has always been part of the human experience, but it used to be gradual. This is no longer true. Just think, for example, that it was only a few years ago that TVs with wide, flat screens were introduced. Already, these have been eclipsed by high-definition TVs with images so crystal clear that they seem almost three-dimensional.

Miniaturization has also made possible huge advances in electronic systems. Cell phones that originally were the size of notebooks are now smaller than a deck of playing cards. In addition, these new versions record videos, transmit photos, send text messages, and have calendars, reminders, calculators, games, and lists of frequently called numbers. Boom boxes playing audio cassettes have been replaced by pocket-sized iPods® that can store 30,000 songs or 25,000 photos. Hearing aids with higher power levels that are almost invisible in the ear, TVs with 1-inch screens—the list of new or improved products continues to expand because significantly smaller electronic systems have been developed.

This reduction in size of electronic systems is due primarily to an important innovation introduced in 1958—the **integrated circuit (IC).** An integrated circuit can now contain features less than 50 nanometers across. The fact that measurements are now being made in nanometers has resulted in the terminology **nanotechnology** to refer to the production of integrated circuits called *nanochips.* To understand nanometers, consider drawing 100 lines within the boundaries of 1 inch. Then attempt drawing 1000 lines within the same length. Cutting 50-nanometer features would require drawing over 500,000 lines in 1 inch. The integrated circuit shown in Fig. 1.1 is an Intel® Core 2 Extreme quad-core processor that has 291 million

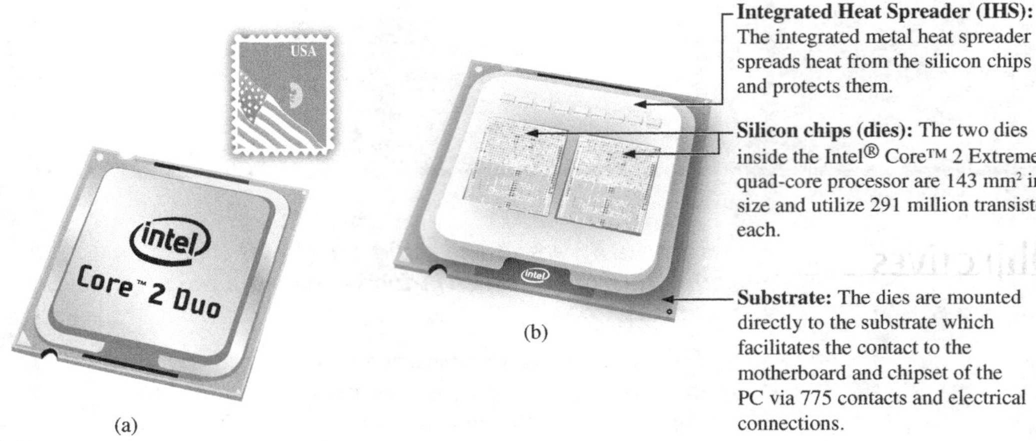

FIG. 1.1
Intel® Core™ 2 Extreme quad-core processer: (a) surface appearance, (b) internal chips.

transistors in each dual-core chip. The result is that the entire package, which is about the size of three postage stamps, has almost 600 million transistors—a number hard to comprehend.

However, before a decision is made on such dramatic reductions in size, the system must be designed and tested to determine if it is worth constructing as an integrated circuit. That design process requires engineers who know the characteristics of each device used in the system, including undesirable characteristics that are part of any electronic element. In other words, there are *no ideal (perfect) elements* in an electronic design. Considering the limitations of each component is necessary to ensure a reliable response under all conditions of temperature, vibration, and effects of the surrounding environment. To develop this awareness requires time and must begin with understanding the basic characteristics of the device, as covered in this text. One of the objectives of this text is to explain how ideal components work and their function in a network. Another is to explain conditions in which components may not be ideal.

One of the very positive aspects of the learning process associated with electric and electronic circuits is that once a concept or procedure is clearly and correctly understood, it will be useful throughout the career of the individual at any level in the industry. Once a law or equation is understood, it will not be replaced by another equation as the material becomes more advanced and complicated. For instance, one of the first laws to be introduced is Ohm's law. This law provides a relationship between forces and components that will always be true, no matter how complicated the system becomes. In fact, it is an equation that will be applied in various forms throughout the design of the entire system. The use of the basic laws may change, but the laws will not change and will always be applicable.

It is vitally important to understand that the learning process for circuit analysis is sequential. That is, the first few chapters establish the foundation for the remaining chapters. Failure to properly understand the opening chapters will only lead to difficulties understanding the material in the chapters to follow. This first chapter provides a brief history of the field followed by a review of mathematical concepts necessary to understand the rest of the material.

1.2 A BRIEF HISTORY

In the sciences, once a hypothesis is proven and accepted, it becomes one of the building blocks of that area of study, permitting additional investigation and development. Naturally, the more pieces of a puzzle available, the more obvious is the avenue toward a possible solution. In fact, history demonstrates that a single development may provide the key that will result in a mushrooming effect that brings the science to a new plateau of understanding and impact.

If the opportunity presents itself, read one of the many publications reviewing the history of this field. Space requirements are such that only a brief review can be provided here. There are many more contributors than could be listed, and their efforts have often provided important keys to the solution of some very important concepts.

Throughout history, some periods were characterized by what appeared to be an explosion of interest and development in particular areas. As you will see from the discussion of the late 1700s and the early 1800s, inventions, discoveries, and theories came fast and furiously. Each new concept broadens the possible areas of application until it becomes almost impossible to trace developments without picking a particular area of interest and following it through. In the review, as you read about the development of radio, television, and computers, keep in mind that similar progressive steps were occurring in the areas of the telegraph, the telephone, power generation, the phonograph, appliances, and so on.

There is a tendency when reading about the great scientists, inventors, and innovators to believe that their contribution was a totally individual effort. In many instances, this was not the case. In fact, many of the great contributors had friends or associates who provided support and encouragement in their efforts to investigate various theories. At the very least, they were aware of one another's efforts to the degree possible in the days when a letter was often the best form of communication. In particular, note the closeness of the dates during periods of rapid development. One contributor seemed to spur on the efforts of the others or possibly provided the key needed to continue with the area of interest.

In the early stages, the contributors were not electrical, electronic, or computer engineers as we know them today. In most cases, they were physicists, chemists, mathematicians, or even philosophers. In addition, they were not from one or two communities of the Old World. The home country of many of the major contributors introduced in the paragraphs to follow is provided to show that almost every established community had some impact on the development of the fundamental laws of electrical circuits.

As you proceed through the remaining chapters of the text, you will find that a number of the units of measurement bear the name of major contributors in those areas—*volt* after Count Alessandro Volta, *ampere* after André Ampère, *ohm* after Georg Ohm, and so forth—fitting recognition for their important contributions to the birth of a major field of study.

Time charts indicating a limited number of major developments are provided in Fig. 1.2, primarily to identify specific periods of rapid development and to reveal how far we have come in the last few decades. In essence, the current state of the art is a result of efforts that began in earnest some 250 years ago, with progress in the last 100 years being almost exponential.

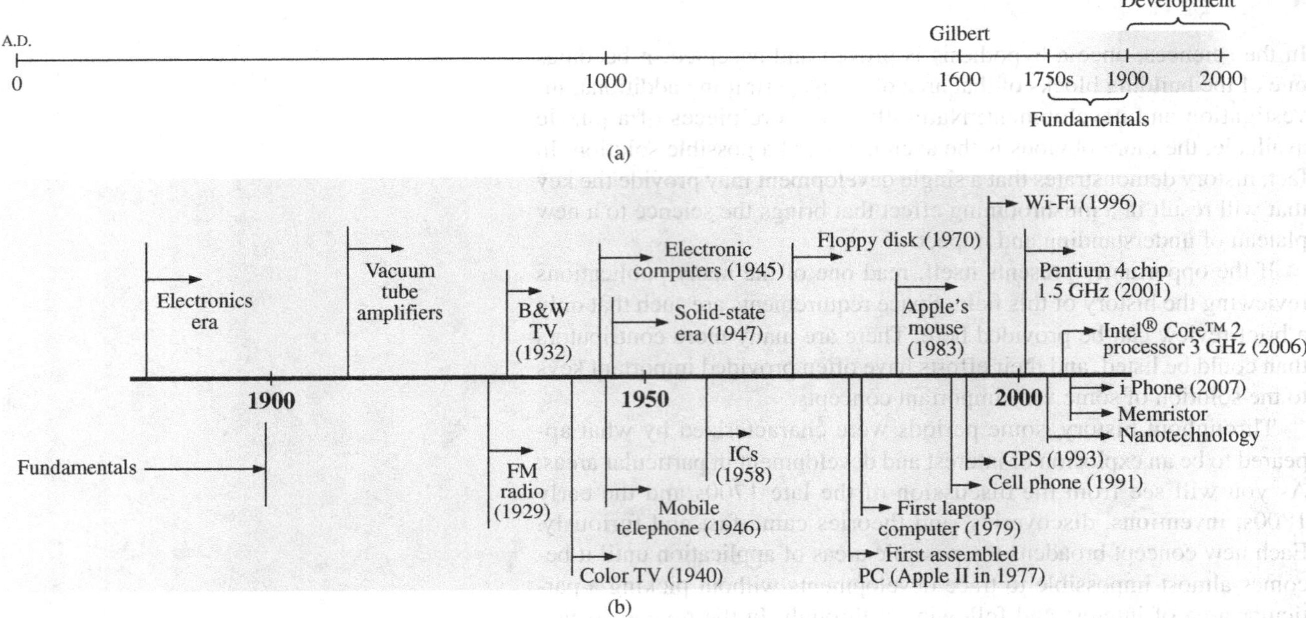

FIG. 1.2
Time charts: (a) long-range; (b) expanded.

As you read through the following brief review, try to sense the growing interest in the field and the enthusiasm and excitement that must have accompanied each new revelation. Although you may find some of the terms used in the review new and essentially meaningless, the remaining chapters will explain them thoroughly.

The Beginning

The phenomenon of **static electricity** has intrigued scholars throughout history. The Greeks called the fossil resin substance so often used to demonstrate the effects of static electricity *elektron,* but no extensive study was made of the subject until William Gilbert researched the phenomenon in 1600. In the years to follow, there was a continuing investigation of electrostatic charge by many individuals, such as Otto von Guericke, who developed the first machine to generate large amounts of charge, and Stephen Gray, who was able to transmit electrical charge over long distances on silk threads. Charles DuFay demonstrated that charges either attract or repel each other, leading him to believe that there were two types of charge—a theory we subscribe to today with our defined positive and negative charges.

There are many who believe that the true beginnings of the electrical era lie with the efforts of Pieter van Musschenbroek and Benjamin Franklin. In 1745, van Musschenbroek introduced the **Leyden jar** for the storage of electrical charge (the first capacitor) and demonstrated electrical shock (and therefore the power of this new form of energy). Franklin used the Leyden jar some 7 years later to establish that lightning is simply an electrical discharge, and he expanded on a number of other important theories, including the definition of the two types of charge as *positive* and *negative.* From this point on, new discoveries and theories seemed to occur at an increasing rate as the number of individuals performing research in the area grew.

In 1784, Charles Coulomb demonstrated in Paris that the force between charges is inversely related to the square of the distance between the charges. In 1791, Luigi Galvani, professor of anatomy at the University of Bologna, Italy, performed experiments on the effects of electricity on animal nerves and muscles. The first **voltaic cell,** with its ability to produce electricity through the chemical action of a metal dissolving in an acid, was developed by another Italian, Alessandro Volta, in 1799.

The fever pitch continued into the early 1800s, with Hans Christian Oersted, a Danish professor of physics, announcing in 1820 a relationship between magnetism and electricity that serves as the foundation for the theory of **electromagnetism** as we know it today. In the same year, a French physicist, André Ampère, demonstrated that there are magnetic effects around every current-carrying conductor and that current-carrying conductors can attract and repel each other just like magnets. In the period 1826 to 1827, a German physicist, Georg Ohm, introduced an important relationship between potential, current, and resistance that we now refer to as *Ohm's law.* In 1831, an English physicist, Michael Faraday, demonstrated his theory of *electromagnetic induction,* whereby a changing current in one coil can induce a changing current in another coil, even though the two coils are not directly connected. Faraday also did extensive work on a storage device he called the condenser, which we refer to today as a *capacitor.* He introduced the idea of adding a dielectric between the plates of a capacitor to increase the storage capacity (Chapter 10). James Clerk Maxwell, a Scottish professor of natural philosophy, performed extensive mathematical analyses to develop what are currently called *Maxwell's equations,* which support the efforts of Faraday linking electric and magnetic effects. Maxwell also developed the *electromagnetic theory of light* in 1862, which, among other things, revealed that electromagnetic waves travel through air at the velocity of light (186,000 miles per second or 3×10^8 meters per second). In 1888, a German physicist, Heinrich Rudolph Hertz, through experimentation with lower-frequency electromagnetic waves (microwaves), substantiated Maxwell's predictions and equations. In the mid-1800s, Gustav Robert Kirchhoff introduced a series of laws of voltages and currents that find application at every level and area of this field (Chapters 5 and 6). In 1895, another German physicist, Wilhelm Röntgen, discovered electromagnetic waves of high frequency, commonly called *X-rays* today.

By the end of the 1800s, a significant number of the fundamental equations, laws, and relationships had been established, and various fields of study, including electricity, electronics, power generation and distribution, and communication systems, started to develop in earnest.

The Age of Electronics

Radio The true beginning of the electronics era is open to debate and is sometimes attributed to efforts by early scientists in applying potentials across evacuated glass envelopes. However, many trace the beginning to Thomas Edison, who added a metallic electrode to the vacuum of the tube and discovered that a current was established between the metal electrode and the filament when a positive voltage was applied to the metal electrode. The phenomenon, demonstrated in 1883, was referred to as the **Edison effect.** In the period to follow, the transmission of radio waves and the development of the radio received widespread attention. In 1887, Heinrich Hertz, in his efforts to verify Maxwell's equations, transmitted radio waves for the first time in his laboratory. In 1896, an Italian scientist, Guglielmo Marconi (often called the father of the radio), demonstrated that telegraph signals could be sent through the air over long distances (2.5 kilometers) using a grounded antenna. In the same year, Aleksandr Popov sent what might have been the first radio message some 300 yards. The message was the name "*Heinrich Hertz*" in respect for Hertz's earlier contributions. In 1901, Marconi established radio communication across the Atlantic.

In 1904, John Ambrose Fleming expanded on the efforts of Edison to develop the first diode, commonly called **Fleming's valve**—actually the first of the *electronic devices.* The device had a profound impact on the design of detectors in the receiving section of radios. In 1906, Lee De Forest added a third element to the vacuum structure and created the first amplifier, the triode. Shortly thereafter, in 1912, Edwin Armstrong built the first regenerative circuit to improve receiver capabilities and then used the same contribution to develop the first nonmechanical oscillator. By 1915, radio signals were being transmitted across the United States, and in 1918 Armstrong applied for a patent for the superheterodyne circuit employed in virtually every television and radio to permit amplification at one frequency rather than at the full range of incoming signals. The major components of the modern-day radio were now in place, and sales in radios grew from a few million dollars in the early 1920s to over $1 billion by the 1930s. The 1930s were truly the golden years of radio, with a wide range of productions for the listening audience.

Television The 1930s were also the true beginnings of the television era, although development on the picture tube began in earlier years with Paul Nipkow and his *electrical telescope* in 1884 and John Baird and his long list of successes, including the transmission of television pictures over telephone lines in 1927 and over radio waves in 1928, and simultaneous transmission of pictures and sound in 1930. In 1932, NBC installed the first commercial television antenna on top of the Empire State Building in New York City, and RCA began regular broadcasting in 1939. World War 2 slowed development and sales, but in the mid-1940s the number of sets grew from a few thousand to a few million. Color television became popular in the early 1960s.

Computers The earliest computer system can be traced back to Blaise Pascal in 1642 with his mechanical machine for adding and subtracting numbers. In 1673, Gottfried Wilhelm von Leibniz used the *Leibniz wheel* to add multiplication and division to the range of operations, and in 1823 Charles Babbage developed the **difference engine** to add the mathematical operations of sine, cosine, logarithms, and several others. In the years to follow, improvements were made, but the system remained primarily mechanical until the 1930s when electromechanical systems using components such as relays were introduced. It was not until the 1940s that totally electronic systems became the new wave. It is interesting to note that, even though IBM was formed in 1924, it did not enter the computer industry until 1937. An entirely electronic system known as **ENIAC** was dedicated at the University of Pennsylvania in 1946. It contained 18,000 tubes and weighed 30 tons but was several times faster than most electromechanical systems. Although other vacuum tube systems were built, it was not until the birth of the solid-state era that computer systems experienced a major change in size, speed, and capability.

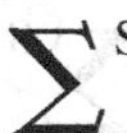

The Solid-State Era

In 1947, physicists William Shockley, John Bardeen, and Walter H. Brattain of Bell Telephone Laboratories demonstrated the point-contact **transistor** (Fig. 1.3), an amplifier constructed entirely of solid-state materials with no requirement for a vacuum, glass envelope, or heater voltage for the filament. Although reluctant at first due to the vast amount of material available on the design, analysis, and synthesis of tube networks, the industry eventually accepted this new technology as the wave of the future. In 1958, the first **integrated circuit (IC)** was developed at Texas Instruments, and in 1961 the first commercial integrated circuit was manufactured by the Fairchild Corporation.

It is impossible to review properly the entire history of the electrical/electronics field in a few pages. The effort here, both through the discussion and the time graphs in Fig. 1.2, was to reveal the amazing progress of this field in the last 50 years. The growth appears to be truly exponential since the early 1900s, raising the interesting question, Where do we go from here? The time chart suggests that the next few decades will probably contain many important innovative contributions that may cause an even faster growth curve than we are now experiencing.

FIG. 1.3
The first transistor.
(Used with permission of Lucent Technologies Inc./Bell Labs.)

1.3 UNITS OF MEASUREMENT

One of the most important rules to remember and apply when working in any field of technology is to use the correct units when substituting numbers into an equation. Too often we are so intent on obtaining a numerical solution that we overlook checking the units associated with the numbers being substituted into an equation. Results obtained, therefore, are often meaningless. Consider, for example, the following very fundamental physics equation:

$$\boxed{v = \frac{d}{t}} \qquad \begin{aligned} v &= \text{velocity} \\ d &= \text{distance} \\ t &= \text{time} \end{aligned} \qquad \textbf{(1.1)}$$

Assume, for the moment, that the following data are obtained for a moving object:

$$d = 4000 \text{ ft}$$
$$t = 1 \text{ min}$$

and v is desired in miles per hour. Often, without a second thought or consideration, the numerical values are simply substituted into the equation, with the result here that

$$v = \frac{d}{t} = \frac{4000 \text{ ft}}{1 \text{ min}} = \cancel{4000 \text{ mph}}$$

As indicated above, the solution is totally incorrect. If the result is desired in *miles per hour,* the unit of measurement for distance must be *miles,* and that for time, *hours.* In a moment, when the problem is analyzed properly, the extent of the error will demonstrate the importance of ensuring that

the numerical value substituted into an equation must have the unit of measurement specified by the equation.

The next question is normally, How do I convert the distance and time to the proper unit of measurement? A method is presented in Section 1.9 of this chapter, but for now it is given that

$$1 \text{ mi} = 5280 \text{ ft}$$
$$4000 \text{ ft} = 0.76 \text{ mi}$$
$$1 \text{ min} = \tfrac{1}{60} \text{ h} = 0.017 \text{ h}$$

Substituting into Eq. (1.1), we have

$$v = \frac{d}{t} = \frac{0.76 \text{ mi}}{0.017 \text{ h}} = \mathbf{44.71\ mph}$$

which is significantly different from the result obtained before.

To complicate the matter further, suppose the distance is given in kilometers, as is now the case on many road signs. First, we must realize that the prefix *kilo* stands for a multiplier of 1000 (to be introduced in Section 1.5), and then we must find the conversion factor between kilometers and miles. If this conversion factor is not readily available, we must be able to make the conversion between units using the conversion factors between meters and feet or inches, as described in Section 1.9.

Before substituting numerical values into an equation, try to mentally establish a reasonable range of solutions for comparison purposes. For instance, if a car travels 4000 ft in 1 min, does it seem reasonable that the speed would be 4000 mph? Obviously not! This self-checking procedure is particularly important in this day of the hand-held calculator, when ridiculous results may be accepted simply because they appear on the digital display of the instrument.

Finally,

if a unit of measurement is applicable to a result or piece of data, then it must be applied to the numerical value.

To state that $v = 44.71$ without including the unit of measurement *mph* is meaningless.

Eq. (1.1) is not a difficult one. A simple algebraic manipulation will result in the solution for any one of the three variables. However, in light of the number of questions arising from this equation, the reader may wonder if the difficulty associated with an equation will increase at the same rate as the number of terms in the equation. In the broad sense, this will not be the case. There is, of course, more room for a mathematical error with a more complex equation, but once the proper system of units is chosen and each term properly found in that system, there should be very little added difficulty associated with an equation requiring an increased number of mathematical calculations.

In review, before substituting numerical values into an equation, be absolutely sure of the following:

1. ***Each quantity has the proper unit of measurement as defined by the equation.***
2. ***The proper magnitude of each quantity as determined by the defining equation is substituted.***
3. ***Each quantity is in the same system of units (or as defined by the equation).***
4. ***The magnitude of the result is of a reasonable nature when compared to the level of the substituted quantities.***
5. ***The proper unit of measurement is applied to the result.***

1.4 SYSTEMS OF UNITS

In the past, the *systems of units* most commonly used were the English and metric, as outlined in Table 1.1. Note that while the English system is based on a single standard, the metric is subdivided into two interrelated standards: the **MKS** and the **CGS.** Fundamental quantities of these systems are compared in Table 1.1 along with their abbreviations. The MKS and CGS systems draw their names from the units of measurement used with each system; the MKS system uses *M*eters, *K*ilograms, and *S*econds, while the CGS system uses *C*entimeters, *G*rams, and *S*econds.

TABLE 1.1

Comparison of the English and metric systems of units.

ENGLISH	METRIC		SI
	MKS	CGS	
Length: Yard (yd) (0.914 m)	Meter (m) (39.37 in.) (100 cm)	Centimeter (cm) (2.54 cm = 1 in.)	**Meter (m)**
Mass: **Slug** (14.6 kg)	Kilogram (kg) (1000 g)	Gram (g)	**Kilogram (kg)**
Force: **Pound** (lb) (4.45 N)	Newton (N) (100,000 dynes)	Dyne	**Newton (N)**
Temperature: Fahrenheit (°F) $\left(= \frac{9}{5}\,°\text{C} + 32\right)$	Celsius or Centigrade (°C) $\left(= \frac{5}{9}(°\text{F} - 32)\right)$	Centigrade (°C)	**Kelvin (K)** K = 273.15 + °C
Energy: Foot-pound (ft-lb) (1.356 joules)	Newton-meter (N•m) or joule (J) (0.7376 ft-lb)	Dyne-centimeter or erg (1 joule = 10^7 ergs)	**Joule (J)**
Time: Second (s)	Second (s)	Second (s)	**Second (s)**

Understandably, the use of more than one system of units in a world that finds itself continually shrinking in size, due to advanced technical developments in communications and transportation, would introduce unnecessary complications to the basic understanding of any technical data. The need for a standard set of units to be adopted by all nations has become increasingly obvious. The International Bureau of Weights and Measures located at Sèvres, France, has been the host for the General Conference of Weights and Measures, attended by representatives from all nations of the world. In 1960, the General Conference adopted a system called Le Système International d'Unités (International System of Units), which has the international abbreviation **SI.** It was adopted by the Institute of Electrical and Electronic Engineers (IEEE) in 1965 and by the United States of America Standards Institute (USASI) in 1967 as a standard for all scientific and engineering literature.

For comparison, the SI units of measurement and their abbreviations appear in Table 1.1. These abbreviations are those usually applied to each unit of measurement, and they were carefully chosen to be the most effective. Therefore, it is important that they be used whenever applicable

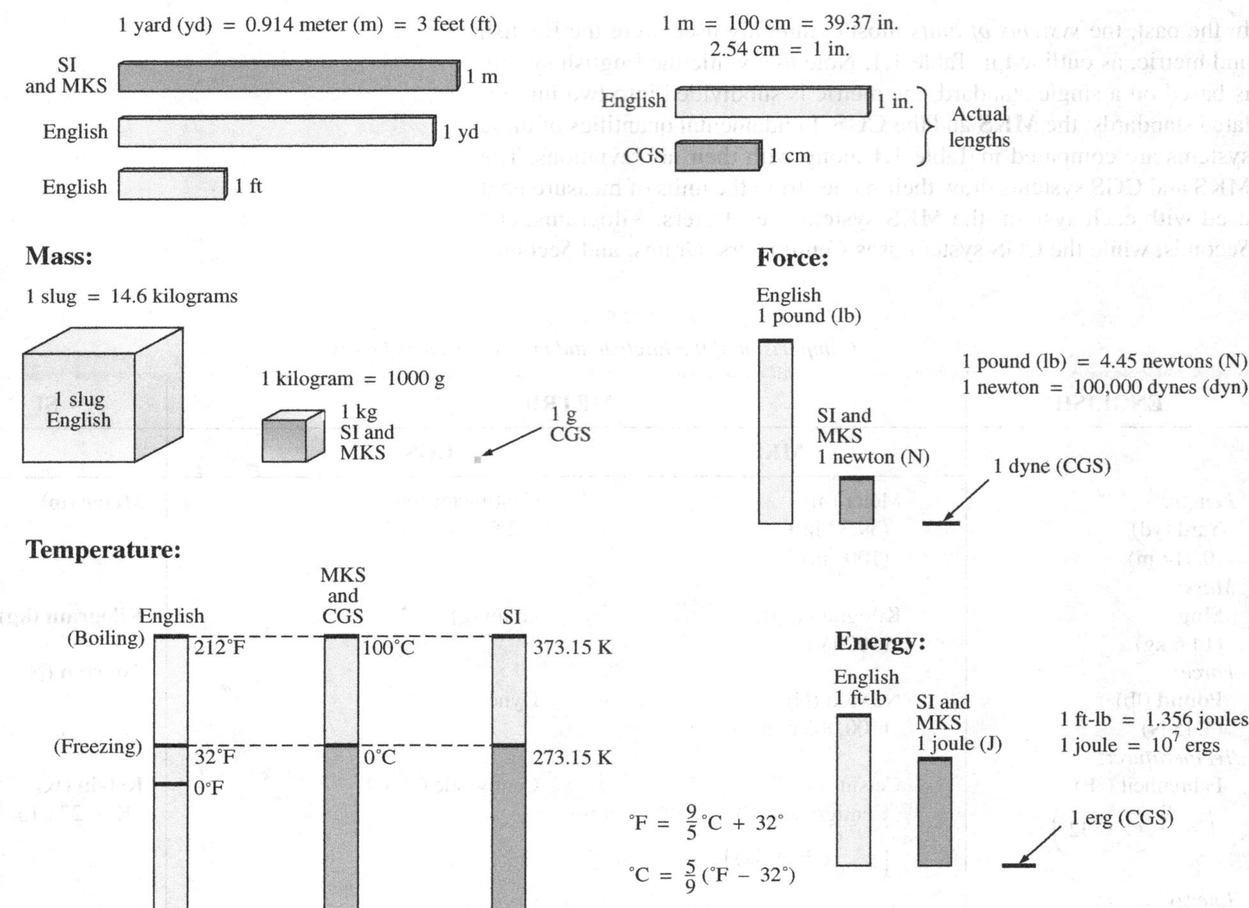

FIG. 1.4
Comparison of units of the various systems of units.

to ensure universal understanding. Note the similarities of the SI system to the MKS system. This text uses, whenever possible and practical, all of the major units and abbreviations of the SI system in an effort to support the need for a universal system. Those readers requiring additional information on the SI system should contact the information office of the American Society for Engineering Education (ASEE).*

Figure 1.4 should help you develop some feeling for the relative magnitudes of the units of measurement of each system of units. Note in the figure the relatively small magnitude of the units of measurement for the CGS system.

A standard exists for each unit of measurement of each system. The standards of some units are quite interesting.

The **meter** was originally defined in 1790 to be 1/10,000,000 the distance between the equator and either pole at sea level, a length preserved

*American Society for Engineering Education (ASEE), 1818 N Street N.W., Suite 600, Washington, D.C. 20036-2479; (202) 331-3500; http://www.asee.org/.

on a platinum–iridium bar at the International Bureau of Weights and Measures at Sèvres, France.

The meter is now defined with reference to the speed of light in a vacuum, which is 299,792,458 m/s.

The kilogram is defined as a mass equal to 1000 times the mass of 1 cubic centimeter of pure water at 4°C.

This standard is preserved in the form of a platinum–iridium cylinder in Sèvres.

The **second** was originally defined as 1/86,400 of the mean solar day. However, since Earth's rotation is slowing down by almost 1 second every 10 years,

the second was redefined in 1967 as 9,192,631,770 periods of the electromagnetic radiation emitted by a particular transition of the cesium atom.

1.5 SIGNIFICANT FIGURES, ACCURACY, AND ROUNDING OFF

This section emphasizes the importance of knowing the source of a piece of data, how a number appears, and how it should be treated. Too often we write numbers in various forms with little concern for the format used, the number of digits that should be included, and the unit of measurement to be applied.

For instance, measurements of 22.1 in. and 22.10 in. imply different levels of accuracy. The first suggests that the measurement was made by an instrument accurate only to the tenths place; the latter was obtained with instrumentation capable of reading to the hundredths place. The use of zeros in a number, therefore, must be treated with care, and the implications must be understood.

In general, there are two types of numbers: *exact* and *approximate.* Exact numbers are precise to the exact number of digits presented, just as we know that there are 12 apples in a dozen and not 12.1. Throughout the text, the numbers that appear in the descriptions, diagrams, and examples are considered *exact* so that a battery of 100 V can be written as 100.0 V, 100.00 V, and so on, since it is 100 V at any level of precision. The additional zeros were not included for purposes of clarity. However, in the laboratory environment, where measurements are continually being taken and the level of accuracy can vary from one instrument to another, it is important to understand how to work with the results. Any reading obtained in the laboratory should be considered *approximate.* The analog scales with their pointers may be difficult to read, and even though the digital meter provides only specific digits on its display, it is limited to the number of digits it can provide, leaving us to wonder about the less significant digits not appearing on the display.

The precision of a reading can be determined by the number of *significant figures (digits)* present. Significant digits are those integers (0 to 9) that can be assumed to be accurate for the measurement being made. The result is that all nonzero numbers are considered significant, with zeros being significant in only some cases. For instance, the zeros in 1005 are considered significant because they define the size of the number and are surrounded by nonzero digits. For the number 0.4020, the zero to the left of the decimal point is not significant, but the other two zeros are because they define the magnitude of the number and the fourth-place accuracy of the reading.

When adding approximate numbers, it is important to be sure that the accuracy of the readings is consistent throughout. To add a quantity accurate only to the tenths place to a number accurate to the thousandths place will result in a total having accuracy only to the tenths place. One cannot expect the reading with the higher level of accuracy to improve the reading with only tenths-place accuracy.

In the addition or subtraction of approximate numbers, the entry with the lowest level of accuracy determines the format of the solution.

For the multiplication and division of approximate numbers, the result has the same number of significant figures as the number with the least number of significant figures.

For approximate numbers (and exact numbers, for that matter), there is often a need to *round off* the result; that is, you must decide on the appropriate level of accuracy and alter the result accordingly. The accepted procedure is simply to note the digit following the last to appear in the rounded-off form, add a 1 to the last digit if it is greater than or equal to 5, and leave it alone if it is less than 5. For example, $3.186 \cong 3.19 \cong 3.2$, depending on the level of precision desired. The symbol $\cong$ means *approximately equal to*.

EXAMPLE 1.1 Perform the indicated operations with the following approximate numbers and round off to the appropriate level of accuracy.

a. $532.6 + 4.02 + 0.036 = 536.656 \cong$ **536.7** (as determined by 532.6)
b. $0.04 + 0.003 + 0.0064 = 0.0494 \cong$ **0.05** (as determined by 0.04)

EXAMPLE 1.2 Round off the following numbers to the hundredths place.

a. 32.419 = **32.42**
b. 0.05328 = **0.05**

EXAMPLE 1.3 Round off the result 5.8764 to

a. tenths-place accuracy.
b. hundredths-place accuracy.
c. thousands-place accuracy.

Solution:

a. **5.9**
b. **5.88**
c. **5.876**

1.6 POWERS OF TEN

It should be apparent from the relative magnitude of the various units of measurement that very large and very small numbers are frequently encountered in the sciences. To ease the difficulty of mathematical operations with numbers of such varying size, *powers of ten* are usually employed. This notation takes full advantage of the mathematical properties of powers

of ten. The notation used to represent numbers that are integer powers of ten is as follows:

$$1 = 10^0 \qquad 1/10 = 0.1 = 10^{-1}$$
$$10 = 10^1 \qquad 1/100 = 0.01 = 10^{-2}$$
$$100 = 10^2 \qquad 1/1000 = 0.001 = 10^{-3}$$
$$1000 = 10^3 \qquad 1/10{,}000 = 0.0001 = 10^{-4}$$

In particular, note that $10^0 = 1$ and, in fact, any quantity to the zero power is 1 ($x^0 = 1$, $1000^0 = 1$, and so on). Numbers in the list *greater than 1 are associated with positive powers of ten,* and numbers in the list *less than 1 are associated with negative powers of ten.*

A quick method of determining the proper power of ten is to place a caret mark to the right of the numeral 1 wherever it may occur; then count from this point to the number of places to the right or left before arriving at the decimal point. Moving to the right indicates a positive power of ten, whereas moving to the left indicates a negative power. For example,

$$10{,}000.0 = 1_\wedge \underbrace{0,000}_{1\,2\,3\,4}. = 10^{+4}$$

$$0.00001 = 0.\underbrace{0000}_{5\,4\,3\,2}1_\wedge{}_{1} = 10^{-5}$$

Some important mathematical equations and relationships pertaining to powers of ten are listed below, along with a few examples. In each case, n and m can be any positive or negative real number.

$$\boxed{\frac{1}{10^n} = 10^{-n} \qquad \frac{1}{10^{-n}} = 10^n} \tag{1.2}$$

Eq. (1.2) clearly reveals that shifting a power of ten from the denominator to the numerator, or the reverse, requires simply changing the sign of the power.

EXAMPLE 1.4

a. $\dfrac{1}{1000} = \dfrac{1}{10^{+3}} = \mathbf{10^{-3}}$

b. $\dfrac{1}{0.00001} = \dfrac{1}{10^{-5}} = \mathbf{10^{+5}}$

The product of powers of ten:

$$\boxed{(10^n)(10^m) = (10)^{(n+m)}} \tag{1.3}$$

EXAMPLE 1.5

a. $(1000)(10{,}000) = (10^3)(10^4) = 10^{(3+4)} = \mathbf{10^7}$
b. $(0.00001)(100) = (10^{-5})(10^2) = 10^{(-5+2)} = \mathbf{10^{-3}}$

The division of powers of ten:

$$\boxed{\frac{10^n}{10^m} = 10^{(n-m)}} \tag{1.4}$$

EXAMPLE 1.6

a. $\dfrac{100{,}000}{100} = \dfrac{10^5}{10^2} = 10^{(5-2)} = \mathbf{10^3}$

b. $\dfrac{1000}{0.0001} = \dfrac{10^3}{10^{-4}} = 10^{(3-(-4))} = 10^{(3+4)} = \mathbf{10^7}$

Note the use of parentheses in part (b) to ensure that the proper sign is established between operators.

The power of powers of ten:

$$(10^n)^m = 10^{nm} \tag{1.5}$$

EXAMPLE 1.7

a. $(100)^4 = (10^2)^4 = 10^{(2)(4)} = \mathbf{10^8}$
b. $(1000)^{-2} = (10^3)^{-2} = 10^{(3)(-2)} = \mathbf{10^{-6}}$
c. $(0.01)^{-3} = (10^{-2})^{-3} = 10^{(-2)(-3)} = \mathbf{10^6}$

Basic Arithmetic Operations

Let us now examine the use of powers of ten to perform some basic arithmetic operations using numbers that are not just powers of ten. The number 5000 can be written as $5 \times 1000 = 5 \times 10^3$, and the number 0.0004 can be written as $4 \times 0.0001 = 4 \times 10^{-4}$. Of course, 10^5 can also be written as 1×10^5 if it clarifies the operation to be performed.

Addition and Subtraction To perform addition or subtraction using powers of ten, the power of ten *must be the same for each term;* that is,

$$A \times 10^n \pm B \times 10^n = (A \pm B) \times 10^n \tag{1.6}$$

Eq. (1.6) covers all possibilities, but students often prefer to remember a verbal description of how to perform the operation.

Eq. (1.6) states

when adding or subtracting numbers in a power-of-ten format, be sure that the power of ten is the same for each number. Then separate the multipliers, perform the required operation, and apply the same power of ten to the result.

EXAMPLE 1.8

a. $6300 + 75{,}000 = (6.3)(1000) + (75)(1000)$
$= 6.3 \times 10^3 + 75 \times 10^3$
$= (6.3 + 75) \times 10^3$
$= \mathbf{81.3 \times 10^3}$

b. $0.00096 - 0.000086 = (96)(0.00001) - (8.6)(0.00001)$
$= 96 \times 10^{-5} - 8.6 \times 10^{-5}$
$= (96 - 8.6) \times 10^{-5}$
$= \mathbf{87.4 \times 10^{-5}}$

Multiplication In general,

$$(A \times 10^n)(B \times 10^m) = (A)(B) \times 10^{n+m} \quad \textbf{(1.7)}$$

revealing that the *operations with the power of ten can be separated from the operation with the multipliers.*

Eq. (1.7) states

when multiplying numbers in the power-of-ten format, first find the product of the multipliers and then determine the power of ten for the result by adding the power-of-ten exponents.

EXAMPLE 1.9

a. $(0.0002)(0.000007) = [(2)(0.0001)][(7)(0.000001)]$
$= (2 \times 10^{-4})(7 \times 10^{-6})$
$= (2)(7) \times (10^{-4})(10^{-6})$
$= \mathbf{14 \times 10^{-10}}$

b. $(340{,}000)(0.00061) = (3.4 \times 10^5)(61 \times 10^{-5})$
$= (3.4)(61) \times (10^5)(10^{-5})$
$= 207.4 \times 10^0$
$= \mathbf{207.4}$

Division In general,

$$\frac{A \times 10^n}{B \times 10^m} = \frac{A}{B} \times 10^{n-m} \quad \textbf{(1.8)}$$

revealing again that the *operations with the power of ten can be separated from the same operation with the multipliers.*

Eq. (1.8) states

when dividing numbers in the power-of-ten format, first find the result of dividing the mutipliers. Then determine the associated power for the result by subtracting the power of ten of the denominator from the power of ten of the numerator.

EXAMPLE 1.10

a. $\dfrac{0.00047}{0.002} = \dfrac{47 \times 10^{-5}}{2 \times 10^{-3}} = \left(\dfrac{47}{2}\right) \times \left(\dfrac{10^{-5}}{10^{-3}}\right) = \mathbf{23.5 \times 10^{-2}}$

b. $\dfrac{690{,}000}{0.00000013} = \dfrac{69 \times 10^4}{13 \times 10^{-8}} = \left(\dfrac{69}{13}\right) \times \left(\dfrac{10^4}{10^{-8}}\right) = \mathbf{5.31 \times 10^{12}}$

Powers In general,

$$(A \times 10^n)^m = A^m \times 10^{nm} \quad \textbf{(1.9)}$$

which again permits the separation of the *operation with the power of ten from the multiplier.*

Eq. (1.9) states

when finding the power of a number in the power-of-ten format, first separate the multiplier from the power of ten and determine each separately. Determine the power-of-ten component by multiplying the power of ten by the power to be determined.

EXAMPLE 1.11

a. $(0.00003)^3 = (3 \times 10^{-5})^3 = (3)^3 \times (10^{-5})^3$
$= \mathbf{27 \times 10^{-15}}$

b. $(90{,}800{,}000)^2 = (9.08 \times 10^7)^2 = (9.08)^2 \times (10^7)^2$
$= \mathbf{82.45 \times 10^{14}}$

In particular, remember that the following operations are not the same. One is the product of two numbers in the power-of-ten format, while the other is a number in the power-of-ten format taken to a power. As noted below, the results of each are quite different:

$$(10^3)(10^3) \neq (10^3)^3$$
$$(10^3)(10^3) = 10^6 = 1{,}000{,}000$$
$$(10^3)^3 = (10^3)(10^3)(10^3) = 10^9 = 1{,}000{,}000{,}000$$

1.7 FIXED-POINT, FLOATING-POINT, SCIENTIFIC, AND ENGINEERING NOTATION

When you are using a computer or a calculator, numbers generally appear in one of four ways. If powers of ten are not employed, numbers are written in the **fixed-point** or **floating-point notation.**

The fixed-point format requires that the decimal point appear in the same place each time. In the floating-point format, the decimal point appears in a location defined by the number to be displayed.

Most computers and calculators permit a choice of fixed- or floating-point notation. In the fixed format, the user can choose the level of accuracy for the output as tenths place, hundredths place, thousandths place, and so on. Every output will then fix the decimal point to one location, such as the following examples using thousandths-place accuracy:

$$\frac{1}{3} = \mathbf{0.333} \qquad \frac{1}{16} = \mathbf{0.063} \qquad \frac{2300}{2} = \mathbf{1150.000}$$

If left in the floating-point format, the results will appear as follows for the above operations:

$$\frac{1}{3} = \mathbf{0.333333333333} \qquad \frac{1}{16} = \mathbf{0.0625} \qquad \frac{2300}{2} = \mathbf{1150}$$

Powers of ten will creep into the fixed- or floating-point notation if the number is too small or too large to be displayed properly.

Scientific (also called *standard*) **notation** and **engineering notation** make use of powers of ten, with restrictions on the mantissa (multiplier) or scale factor (power of ten).

Scientific notation requires that the decimal point appear directly after the first digit greater than or equal to 1 but less than 10.

A power of ten will then appear with the number (usually following the power notation E), even if it has to be to the zero power. A few examples:

$$\frac{1}{3} = \mathbf{3.33333333333E{-}1} \qquad \frac{1}{16} = \mathbf{6.25E{-}2} \qquad \frac{2300}{2} = \mathbf{1.15E3}$$

Within scientific notation, the fixed- or floating-point format can be chosen. In the above examples, floating was employed. If fixed is chosen and set at the hundredths-point accuracy, the following will result for the above operations:

$$\frac{1}{3} = \mathbf{3.33E{-}1} \qquad \frac{1}{16} = \mathbf{6.25E{-}2} \qquad \frac{2300}{2} = \mathbf{1.15E3}$$

Engineering notation specifies that

all powers of ten must be 0 or multiples of 3, and the mantissa must be greater than or equal to 1 but less than 1000.

This restriction on the powers of ten is because specific powers of ten have been assigned prefixes that are introduced in the next few paragraphs. Using scientific notation in the floating-point mode results in the following for the above operations:

$$\frac{1}{3} = \mathbf{333.333333333E{-}3} \qquad \frac{1}{16} = \mathbf{62.5E{-}3} \qquad \frac{2300}{2} = \mathbf{1.15E3}$$

Using engineering notation with two-place accuracy will result in the following:

$$\frac{1}{3} = \mathbf{333.33E{-}3} \qquad \frac{1}{16} = \mathbf{62.50E{-}3} \qquad \frac{2300}{2} = \mathbf{1.15E3}$$

Prefixes

Specific powers of ten in engineering notation have been assigned prefixes and symbols, as appearing in Table 1.2. They permit easy recognition of the power of ten and an improved channel of communication between technologists.

TABLE 1.2

Multiplication Factors	SI Prefix	SI Symbol
1 000 000 000 000 000 000 = 10^{18}	exa	E
1 000 000 000 000 000 = 10^{15}	peta	P
1 000 000 000 000 = 10^{12}	tera	T
1 000 000 000 = 10^{9}	giga	G
1 000 000 = 10^{6}	mega	M
1 000 = 10^{3}	kilo	k
0.001 = 10^{-3}	milli	m
0.000 001 = 10^{-6}	micro	μ
0.000 000 001 = 10^{-9}	nano	n
0.000 000 000 001 = 10^{-12}	pico	p
0.000 000 000 000 001 = 10^{-15}	femto	f
0.000 000 000 000 000 001 = 10^{-18}	atto	a

EXAMPLE 1.12

a. 1,000,000 ohms $= 1 \times 10^6$ ohms
$= \mathbf{1\ megohm = 1\ M\Omega}$

b. 100,000 meters $= 100 \times 10^3$ meters
$= \mathbf{100\ kilometers = 100\ km}$

c. 0.0001 second $= 0.1 \times 10^{-3}$ second
$=$ **0.1 millisecond = 0.1 ms**

d. 0.000001 farad $= 1 \times 10^{-6}$ farad
$=$ **1 microfarad = 1 μF**

Here are a few examples with numbers that are not strictly powers of ten.

EXAMPLE 1.13

a. 41,200 m is equivalent to 41.2×10^3 m $=$ 41.2 kilometers $=$ **41.2 km.**

b. 0.00956 J is equivalent to 9.56×10^{-3} J $=$ 9.56 millijoules $=$ **9.56 mJ.**

c. 0.000768 s is equivalent to 768×10^{-6} s $=$ 768 microseconds $=$ **768 μs.**

d. $$\frac{8400 \text{ m}}{0.06} = \frac{8.4 \times 10^3 \text{ m}}{6 \times 10^{-2}} = \left(\frac{8.4}{6}\right) \times \left(\frac{10^3}{10^{-2}}\right) \text{m}$$
$$= 1.4 \times 10^5 \text{ m} = 140 \times 10^3 \text{ m} = 140 \text{ kilometers} = \mathbf{140\ km}$$

e. $$(0.0003)^4 \text{ s} = (3 \times 10^{-4})^4 \text{ s} = 81 \times 10^{-16} \text{ s}$$
$$= 0.0081 \times 10^{-12} \text{ s} = 0.0081 \text{ picosecond} = \mathbf{0.0081\ ps}$$

1.8 CONVERSION BETWEEN LEVELS OF POWERS OF TEN

It is often necessary to convert from one power of ten to another. For instance, if a meter measures kilohertz (kHz—a unit of measurement for the frequency of an ac waveform), it may be necessary to find the corresponding level in megahertz (MHz). If time is measured in milliseconds (ms), it may be necessary to find the corresponding time in microseconds (μs) for a graphical plot. The process is not difficult if we simply keep in mind that an increase or a decrease in the power of ten must be associated with the opposite effect on the multiplying factor. The procedure is best described by the following steps:

1. ***Replace the prefix by its corresponding power of ten.***
2. ***Rewrite the expression, and set it equal to an unknown multiplier and the new power of ten.***
3. ***Note the change in power of ten from the original to the new format. If it is an increase, move the decimal point of the original multiplier to the left (smaller value) by the same number. If it is a decrease, move the decimal point of the original multiplier to the right (larger value) by the same number.***

EXAMPLE 1.14 Convert 20 kHz to megahertz.

Solution: In the power-of-ten format:

$$20 \text{ kHz} = 20 \times 10^3 \text{ Hz}$$

The conversion requires that we find the multiplying factor to appear in the space below:

Increase by 3

$$20 \times 10^3 \text{ Hz} \Rightarrow \underline{\quad} \times 10^6 \text{ Hz}$$

Decrease by 3

Since the power of ten will be *increased* by a factor of *three*, the multiplying factor must be *decreased* by moving the decimal point *three* places to the left, as shown below:

$$\underbrace{020}_{3} = 0.02$$

and $$20 \times 10^3 \text{ Hz} = 0.02 \times 10^6 \text{ Hz} = \mathbf{0.02\ MHz}$$

EXAMPLE 1.15 Convert 0.01 ms to microseconds.

Solution: In the power-of-ten format:

$$0.01 \text{ ms} = 0.01 \times 10^{-3} \text{ s}$$

and $$0.01 \times 10^{-3} \text{ s} \Rightarrow \_\_ \times 10^{-6} \text{ s}$$

Decrease by 3

Increase by 3

Since the power of ten will be *reduced* by a factor of three, the multiplying factor must be *increased* by moving the decimal point three places to the right, as follows:

$$0.\underbrace{010}_{3} = 10$$

and $$0.01 \times 10^{-3} \text{ s} = 10 \times 10^{-6} \text{ s} = \mathbf{10\ \mu s}$$

There is a tendency when comparing -3 to -6 to think that the power of ten has increased, but keep in mind when making your judgment about increasing or decreasing the magnitude of the multiplier that 10^{-6} is a great deal smaller than 10^{-3}.

EXAMPLE 1.16 Convert 0.002 km to millimeters.

Solution:

$$0.002 \times 10^3 \text{ m} \Rightarrow \_\_ \times 10^{-3} \text{ m}$$

Decrease by 6

Increase by 6

In this example we have to be very careful because the difference between $+3$ and -3 is a factor of 6, requiring that the multiplying factor be modified as follows:

$$0.\underbrace{002000}_{6} = 2000$$

and $$0.002 \times 10^3 \text{ m} = 2000 \times 10^{-3} \text{ m} = \mathbf{2000\ mm}$$

1.9 CONVERSION WITHIN AND BETWEEN SYSTEMS OF UNITS

The conversion within and between systems of units is a process that cannot be avoided in the study of any technical field. It is an operation, however, that is performed incorrectly so often that this section was included to provide one approach that, if applied properly, will lead to the correct result.

There is more than one method of performing the conversion process. In fact, some people prefer to determine mentally whether the conversion factor is multiplied or divided. This approach is acceptable for some elementary conversions, but it is risky with more complex operations.

The procedure to be described here is best introduced by examining a relatively simple problem such as converting inches to meters. Specifically, let us convert 48 in. (4 ft) to meters.

If we multiply the 48 in. by a factor of **1,** the magnitude of the quantity remains the same:

$$\boxed{48 \text{ in.} = 48 \text{ in.}(\mathbf{1})} \qquad \mathbf{(1.10)}$$

Let us now look at the conversion factor for this example:

$$1 \text{ m} = 39.37 \text{ in.}$$

Dividing both sides of the conversion factor by 39.37 in. results in the following format:

$$\frac{1 \text{ m}}{39.37 \text{ in.}} = (\mathbf{1})$$

Note that the end result is that the ratio 1 m/39.37 in. equals 1, as it should since they are equal quantities. If we now substitute this factor (**1**) into Eq. (1.10), we obtain

$$48 \text{ in.}(\mathbf{1}) = 48 \cancel{\text{in.}}\left(\frac{1 \text{ m}}{39.37 \cancel{\text{in.}}}\right)$$

which results in the cancellation of inches as a unit of measure and leaves meters as the unit of measure. In addition, since the 39.37 is in the denominator, it must be divided into the 48 to complete the operation:

$$\frac{48}{39.37} \text{ m} = \mathbf{1.219 \text{ m}}$$

Let us now review the method:

1. ***Set up the conversion factor to form a numerical value of (1) with the unit of measurement to be removed from the original quantity in the denominator.***
2. ***Perform the required mathematics to obtain the proper magnitude for the remaining unit of measurement.***

EXAMPLE 1.17 Convert 6.8 min to seconds.

Solution: The conversion factor is

$$1 \text{ min} = 60 \text{ s}$$

Since the minute is to be removed as the unit of measurement, it must appear in the denominator of the (**1**) factor, as follows:

Step 1:

$$\left(\frac{60 \text{ s}}{1 \text{ min}}\right) = (\mathbf{1})$$

Step 2:

$$6.8 \text{ min } (\mathbf{1}) = 6.8 \cancel{\text{min}}\left(\frac{60 \text{ s}}{1 \cancel{\text{min}}}\right) = (6.8)(60) \text{ s}$$

$$= \mathbf{408 \text{ s}}$$

EXAMPLE 1.18 Convert 0.24 m to centimeters.

Solution: The conversion factor is

$$1\text{ m} = 100\text{ cm}$$

Since the meter is to be removed as the unit of measurement, it must appear in the denominator of the (**1**) factor as follows:

Step 1: $$\left(\frac{100\text{ cm}}{1\text{ m}}\right) = \mathbf{1}$$

Step 2: $$0.24\text{ m}(\mathbf{1}) = 0.24\,\cancel{\text{m}}\left(\frac{100\text{ cm}}{1\,\cancel{\text{m}}}\right) = (0.24)(100)\text{ cm} = \mathbf{24\text{ cm}}$$

The products (**1**)(**1**) and (**1**)(**1**)(**1**) are still **1**. Using this fact, we can perform a series of conversions in the same operation.

EXAMPLE 1.19 Determine the number of minutes in half a day.

Solution: Working our way through from days to hours to minutes, always ensuring that the unit of measurement to be removed is in the denominator, results in the following sequence:

$$0.5\,\cancel{\text{day}}\left(\frac{3\,\cancel{\text{h}}}{1\,\cancel{\text{day}}}\right)\left(\frac{60\text{ min}}{1\,\cancel{\text{h}}}\right) = (0.5)(24)(60)\text{ min} = \mathbf{720\text{ min}}$$

EXAMPLE 1.20 Convert 2.2 yards to meters.

Solution: Working our way through from yards to feet to inches to meters results in the following:

$$2.2\,\cancel{\text{yards}}\left(\frac{3\,\cancel{\text{ft}}}{1\,\cancel{\text{yard}}}\right)\left(\frac{12\,\cancel{\text{in.}}}{1\,\cancel{\text{ft}}}\right)\left(\frac{1\text{ m}}{39.37\,\cancel{\text{in.}}}\right) = \frac{(2.2)(3)(12)}{39.37}\text{ m} = \mathbf{2.012\text{ m}}$$

The following examples are variations of the above in practical situations.

EXAMPLE 1.21 In Europe, Canada, and many other countries, the speed limit is posted in kilometers per hour. How fast in miles per hour is 100 km/h?

Solution:

$$\left(\frac{100\text{ km}}{\text{h}}\right)(\mathbf{1})(\mathbf{1})(\mathbf{1})(\mathbf{1})$$

$$= \left(\frac{100\,\cancel{\text{km}}}{\text{h}}\right)\left(\frac{1000\,\cancel{\text{m}}}{1\,\cancel{\text{km}}}\right)\left(\frac{39.37\,\cancel{\text{in.}}}{1\,\cancel{\text{m}}}\right)\left(\frac{1\,\cancel{\text{ft}}}{12\,\cancel{\text{in.}}}\right)\left(\frac{1\text{ mi}}{5280\,\cancel{\text{ft}}}\right)$$

$$= \frac{(100)(1000)(39.37)}{(12)(5280)}\frac{\text{mi}}{\text{h}}$$

$$= \mathbf{62.14\text{ mph}}$$

TABLE 1.3

Symbol	Meaning
$\neq$	Not equal to $6.12 \neq 6.13$
$>$	Greater than $4.78 > 4.20$
$>>$	Much greater than $840 >> 16$
$<$	Less than $430 < 540$
$<<$	Much less than $0.002 << 46$
$\geq$	Greater than or equal to $x \geq y$ is satisfied for $y = 3$ and $x > 3$ or $x = 3$
$\leq$	Less than or equal to $x \leq y$ is satisfied for $y = 3$ and $x < 3$ or $x = 3$
$\cong$	Appoximately equal to $3.14159 \cong 3.14$
Σ	Sum of $\Sigma\,(4 + 6 + 8) = 18$
$\vert\;\vert$	Absolute magnitude of $\vert a\vert = 4$, where $a = -4$ or $+4$
$\therefore$	Therefore $x = \sqrt{4} \therefore x = \pm 2$
$\equiv$	By definition Establishes a relationship between two or more quantities
$a{:}b$	Ratio defined by $\frac{a}{b}$
$a{:}b = c{:}d$	Proportion defined by $\frac{a}{b} = \frac{c}{d}$

Many travelers use 0.6 as a conversion factor to simplify the math involved; that is,

$$(100 \text{ km/h})(0.6) \cong 60 \text{ mph}$$

and

$$(60 \text{ km/h})(0.6) \cong 36 \text{ mph}$$

EXAMPLE 1.22 Determine the speed in miles per hour of a competitor who can run a 4-min mile.

Solution: Inverting the factor 4 min/1 mi to 1 mi/4 min, we can proceed as follows:

$$\left(\frac{1 \text{ mi}}{4 \cancel{\text{min}}}\right)\left(\frac{60 \cancel{\text{min}}}{4}\right) = \frac{60}{4} \text{ mi/h} = \mathbf{15 \text{ mph}}$$

1.10 SYMBOLS

Throughout the text, various symbols will be employed that you may not have had occasion to use. Some are defined in Table 1.3, and others will be defined in the text as the need arises.

1.11 CONVERSION TABLES

Conversion tables such as those appearing in Appendix A can be very useful when time does not permit the application of methods described in this chapter. However, even though such tables appear easy to use, frequent errors occur because the operations appearing at the head of the table are not performed properly. In any case, when using such tables, try to establish mentally some order of magnitude for the quantity to be determined compared to the magnitude of the quantity in its original set of units. This simple operation should prevent several impossible results that may occur if the conversion operation is improperly applied.

For example, consider the following from such a conversion table:

To convert from	To	Multiply by
Miles	Meters	1.609×10^3

A conversion of 2.5 mi to meters would require that we multiply 2.5 by the conversion factor; that is,

$$2.5 \text{ mi}(1.609 \times 10^3) = \mathbf{4.023 \times 10^3 \text{ m}}$$

A conversion from 4000 m to miles would require a division process:

$$\frac{4000 \text{ m}}{1.609 \times 10^3} = 2486.02 \times 10^{-3} = \mathbf{2.486 \text{ mi}}$$

In each of the above, there should have been little difficulty realizing that 2.5 mi would convert to a few thousand meters and 4000 m would be only a few miles. As indicated above, this kind of anticipatory thinking will eliminate the possibility of ridiculous conversion results.

1.12 CALCULATORS

In most texts, the calculator is not discussed in detail. Instead, students are left with the general exercise of choosing an appropriate calculator and learning to use it properly on their own. However, some discussion about the use of the calculator is needed to eliminate some of

the impossible results obtained (and often strongly defended by the user—because the calculator says so) through a correct understanding of the process by which a calculator performs the various tasks. Time and space do not permit a detailed explanation of all the possible operations, but the following discussion explains why it is important to understand how a calculator proceeds with a calculation and that the unit cannot accept data in any form and still generate the correct answer.

When choosing a scientific calculator, be absolutely certain that it can operate on complex numbers and determinants, needed for the concepts introduced in this text. The simplest way to determine this is to look up the terms in the index of the operator's manual. Next, be aware that some calculators perform the required operations in a minimum number of steps, while others require a lengthy and complex series of steps. Speak to your instructor if you are unsure about your purchase.

The calculator examples in this text use the Texas Instruments TI-89 of Fig. 1.5.

When using any calculator for the first time, the unit must be set up to provide the answers in a desired format. Following are the steps needed to use a calculator correctly.

FIG. 1.5
Texas Instruments TI-89 calculator.
(Courtesy of Texas Instruments, Inc.)

Initial Settings

In the following sequences, the arrows within the square indicate the direction of the scrolling required to reach the desired location. The shape of the keys is a relatively close match of the actual keys on the TI-89.

Notation

The first sequence sets the **engineering notation** as the choice for all answers. It is particularly important to take note that you must select the ENTER key twice to ensure the process is set in memory.

ON HOME MODE ↓ Exponential Format → ↓

ENGINEERING ENTER ENTER

Accuracy Level Next, the accuracy level can be set to two places as follows:

MODE ↓ Display Digits → ↓ 3:FIX 2 ENTER ENTER

Approximate Mode For all solutions, the solution should be in decimal form to second-place accuracy. If this is not set, some answers will appear in fractional form to ensure the answer is EXACT (another option). This selection is made with the following sequence:

MODE F2 ↓ Exact/Approx → ↓ 3: APPROXIMATE ENTER ENTER

Clear Screen To clear the screen of all entries and results, use the following sequence:

F1 ↓ 8: Clear Home ENTER

Clear Current Entries To delete the sequence of current entries at the bottom of the screen, select the **CLEAR** key.

Turn Off To turn off the calculator, apply the following sequence:

2ND ON

Order of Operations

Although setting the correct format and accurate input is important, improper results occur primarily because users fail to realize that no matter how simple or complex an equation, the calculator performs the required operations in a specific order.

For instance, the operation

$$\frac{8}{3+1}$$

is often entered as

8 ÷ 3 + 1 ENTER $= \frac{8}{3} + 1 = 2.67 + 1 = \mathbf{3.67}$

which is incorrect (2 is the answer).

The calculator *will not* perform the addition first and then the division. In fact, addition and subtraction are the last operations to be performed in any equation. It is therefore very important that you carefully study and thoroughly understand the next few paragraphs in order to use the calculator properly.

1. *The first operations to be performed by a calculator can be set using parentheses ( ). It does not matter which operations are within the parentheses. The parentheses simply dictate that this part of the equation is to be determined first. There is no limit to the number of parentheses in each equation—all operations within parentheses will be performed first. For instance, for the example above, if parentheses are added as shown below, the addition will be performed first and the correct answer obtained:*

$\frac{8}{(3+1)} =$ 8 ÷ (3 + 1) ENTER = 2.00

2. *Next, powers and roots are performed, such as x^2, $\sqrt{x}$, and so on.*
3. *Negation (applying a negative sign to a quantity) and single-key operations such as sin, $\tan^{-1}$, and so on, are performed.*
4. *Multiplication and division are then performed.*
5. *Addition and subtraction are performed last.*

It may take a few moments and some repetition to remember the order, but at least you are now aware that there is an order to the operations and that ignoring them can result in meaningless results.

EXAMPLE 1.23 Determine

$$\sqrt{\frac{9}{3}}$$

Solution:

2ND √ (9 ÷ 3) ENTER = **1.73**

In this case, the left bracket is automatically entered after the square root sign. The right bracket must then be entered before performing the calculation.

For all calculator operations, the number of right brackets must always equal the number of left brackets.

EXAMPLE 1.24 Find

$$\frac{3+9}{4}$$

Solution: If the problem is entered as it appears, the *incorrect answer* of 5.25 will result.

[3] [+] [9] [÷] [4] [ENTER] $= 3 + \frac{9}{4} = 5.25$

Using brackets to ensure that the addition takes place before the division will result in the correct answer as shown below:

[(] [3] [+] [9] [)] [÷] [4] [ENTER]

$$= \frac{(3+9)}{4} = \frac{12}{4} = \mathbf{3.00}$$

EXAMPLE 1.25 Determine

$$\frac{1}{4} + \frac{1}{6} + \frac{2}{3}$$

Solution: Since the division will occur first, the correct result will be obtained by simply performing the operations as indicated. That is,

[1] [÷] [4] [+] [1] [÷] [6] [+] [2] [÷]

[3] [ENTER] $= \frac{1}{4} + \frac{1}{6} + \frac{2}{3} = \mathbf{1.08}$

Powers of Ten

The [EE] key is used to set the power of ten of a number. Setting up the number $2200 = 2.2 \times 10^3$ requires the following keypad selections:

[2] [•] [2] [EE] [3] [ENTER] = **2.20E3**

Setting up the number 8.2×10^{-6} requires the negative sign (−) from the *numerical keypad. Do not* use the negative sign from the mathematical listing of ÷, ×, −, and +. That is,

[8] [•] [2] [EE] [(−)] [6] [ENTER] = **8.20E−6**

EXAMPLE 1.26 Perform the addition $6.3 \times 10^3 + 75 \times 10^3$ and compare your answer with the longhand solution of Example 1.8(a).

Solution:

6 · 3 EE + 7 5 EE
3 ENTER = 81.30E3

which confirms the results of Example 1.8(a).

EXAMPLE 1.27 Perform the division $(69 \times 10^4)/(13 \times 10^{-8})$ and compare your answer with the longhand solution of Example 1.10(b).

Solution:

6 9 EE 4 ÷ 1 3 EE (−) 8
ENTER = 5.31E12

which confirms the results of Example 1.10(b).

EXAMPLE 1.28 Using the provided format of each number, perform the following calculation in one series of keypad entries:

$$\frac{(0.004)(6 \times 10^{-4})}{(2 \times 10^{-3})^2} = ?$$

Solution:

((0 · 0 0 4) × (

= 600.00E−3 = 0.6

Brackets were used to ensure that the calculations were performed in the correct order. Note also that the number of left brackets equals the number of right brackets.

1.13 COMPUTER ANALYSIS

The use of computers in the educational process has grown exponentially in the past decade. Very few texts at this introductory level fail to include some discussion of current popular computer techniques. In fact, the very accreditation of a technology program may be a function of the depth to which computer methods are incorporated in the program.

There is no question that a basic knowledge of computer methods is something that the graduating student must learn in a two-year or four-year program. Industry now requires students to be proficient in the use of a computer.

Two general directions can be taken to develop the necessary computer skills: the study of computer languages or the use of software packages.

Languages

There are several languages that provide a direct line of communication with the computer and the operations it can perform. A **language** is a set

of symbols, letters, words, or statements that the user can enter into the computer. The computer system will "understand" these entries and will perform them in the order established by a series of commands called a **program.** The program tells the computer what to do on a sequential, line-by-line basis in the same order a student would perform the calculations in longhand. The computer can respond only to the commands entered by the user. This requires that the programmer understand fully the sequence of operations and calculations required to obtain a particular solution. A lengthy analysis can result in a program having hundreds or thousands of lines. Once written, the program must be checked carefully to ensure that the results have meaning and are valid for an expected range of input variables. Some of the popular languages applied in the electrical/electronics field today include C++, QBASIC, Java, and FORTRAN. Each has its own set of commands and statements to communicate with the computer, but each can be used to perform the same type of analysis.

Software Packages

The second approach to computer analysis—**software packages**—avoids the need to know a particular language; in fact, the user may not be aware of which language was used to write the programs within the package. All that is required is a knowledge of how to input the network parameters, define the operations to be performed, and extract the results; the package will do the rest. However, there is a problem with using packaged programs without understanding the basic steps the program uses. You can obtain a solution without the faintest idea of either how the solution was obtained or whether the results are valid or way off base. It is imperative that you realize that the computer should be used as a tool to assist the user—it must not be allowed to control the scope and potential of the user! Therefore, as we progress through the chapters of the text, be sure that you clearly understand the concepts before turning to the computer for support and efficiency.

Each software package has a **menu,** which defines the range of application of the package. Once the software is entered into the computer, the system will perform all the functions appearing in the menu, as it was preprogrammed to do. Be aware, however, that if a particular type of analysis is requested that is not on the menu, the software package cannot provide the desired results. The package is limited solely to those maneuvers developed by the team of programmers who developed the software package. In such situations the user must turn to another software package or write a program using one of the languages listed above.

In broad terms, if a software package is available to perform a particular analysis, then that package should be used rather than developing new routines. Most popular software packages are the result of many hours of effort by teams of programmers with years of experience. However, if the results are not in the desired format, or if the software package does not provide all the desired results, then the user's innovative talents should be put to use to develop a software package. As noted above, any program the user writes that passes the tests of range and accuracy can be considered a software package of his or her authorship for future use.

Two software packages are used throughout this text: Cadence's OrCAD's PSpice Release 16.2 and Multisim Version 10.1. Although both PSpice and Multisim are designed to analyze electric circuits, there are sufficient differences between the two to warrant covering each approach. However, you are not required to obtain both programs in order to proceed with the content of this text. The primary reason for including the programs is simply to introduce each and demonstrate how each can support the learning process. In most cases, sufficient detail has been provided to allow use of the software package to perform the examples provided, although it would certainly be helpful to have someone to turn to if questions arise. In addition, the literature supporting the packages has improved dramatically in recent years and should be available through your bookstore or a major publisher. Cadence's OrCAD PSpice. Release 16.2 has been provided as an attachment to this text. However, Appendix B lists are the system requirements for each software package, including how to obtain the software.

PROBLEMS

Note: More difficult problems are denoted by an asterisk (*) throughout the text.

SECTION 1.2 A Brief History

1. Visit your local library (at school or home) and describe the extent to which it provides literature and computer support for the technologies—in particular, electricity, electronics, electromagnetics, and computers.
2. Choose an area of particular interest in this field and write a very brief report on the history of the subject.
3. Choose an individual of particular importance in this field and write a very brief review of his or her life and important contributions.

SECTION 1.3 Units of Measurement

4. What is the velocity of a rocket in mph if it travels 20,000 ft in 10 s?
5. In a recent Tour de France time trial, Lance Armstrong traveled 31 mi in a time trial in 1 h, 4 min. What was his average speed in mph?
6. How long in seconds will it take a car traveling at 60 mph to travel the length of a football field (100 yd)?

* 7. A pitcher has the ability to throw a baseball at 95 mph.
   - **a.** How fast is the speed in ft/s?
   - **b.** How long does the hitter have to make a decision about swinging at the ball if the plate and the mound are separated by 60 ft?
   - **c.** If the batter wanted a full second to make a decision, what would the speed in mph have to be?

SECTION 1.4 Systems of Units

8. Are there any relative advantages associated with the metric system compared to the English system with respect to length, mass, force, and temperature? If so, explain.
9. Which of the four systems of units appearing in Table 1.1 has the smallest units for length, mass, and force? When would this system be used most effectively?

*10. Which system of Table 1.1 is closest in definition to the SI system? How are the two systems different? Why do you think the units of measurement for the SI system were chosen as listed in Table 1.1? Give the best reasons you can without referencing additional literature.

11. What is room temperature (68°F) in the MKS, CGS, and SI systems?
12. How many foot-pounds of energy are associated with 1000 J?
13. How many centimeters are there in 1/2 yd?
14. In parts of the world outside the United States, the majority of countries use the centigrade scale rather than the Fahrenheit scale. This can cause problems for travelers not familiar with what to expect at certain temperature levels. To alleviate this problem, the following approximate conversion is typically used:

$$°F = 2(°C) + 30°$$

Comparing to the exact formula of $°F = \frac{9}{5}°C + 32°$, we find the ratio 9/5 is approximated to equal 2, and the temperature of 32° is changed to 30° simply to make the numbers easier to work with and slightly compensate for the fact that 2(°C) is more than 9/5(°C).
   - **a.** The temperature of 20°C is commonly accepted as normal room temperature. Using the approximate formula, determine (in your head) the equivalent Fahrenheit temperature.
   - **b.** Use the exact formula and determine the equivalent Fahrenheit temperature for 20°C.
   - **c.** How do the results of parts (a) and (b) compare? Is the approximation valid as a first estimate of the Fahrenheit temperature?
   - **d.** Repeat parts (a) and (b) for a high temperature of 30°C and a low temperature of 5°C.

SECTION 1.5 Significant Figures, Accuracy, and Rounding Off

15. Write the following numbers to tenths-place accuracy.
   - **a.** 14.6026
   - **b.** 056.0420
   - **c.** 1,046.06
   - **d.** 1/16
   - **e.** π
16. Repeat Problem 15 using hundredths-place accuracy.
17. Repeat Problem 15 using thousandths-place accuracy.

SECTION 1.6 Powers of Ten

18. Express the following numbers as powers of ten to hundredths-place accuracy:
a. 10,000 **b.** 1,000,000
c. 1000 **d.** 0.001
e. 1 **f.** 0.1

19. Using only those powers of ten listed in Table 1.2, express the following numbers in what seems to you the most logical form for future calculations:
a. 15,000 **b.** 0.005
c. 2,400,000 **d.** 60,000
e. 0.00040200 **f.** 0.0000000002

20. Perform the following operations:
a. $4200 + 48{,}000$
b. $9 \times 10^4 + 3.6 \times 10^5$
c. $0.5 \times 10^{-3} - 6 \times 10^{-5}$
d. $1.2 \times 10^3 + 50{,}000 \times 10^{-3} - 400$

21. Perform the following operations:
a. $(100)(1000)$ **b.** $(0.01)(1000)$
c. $(10^3)(10^6)$ **d.** $(100)(0.00001)$
e. $(10^{-6})(10{,}000{,}000)$ **f.** $(10{,}000)(10^{-8})(10^{28})$

22. Perform the following operations:
a. $(50{,}000)(0.002)$
b. 2200×0.002
c. $(0.000082)(1.2 \times 10^6)$
d. $(30 \times 10^{-4})(0.004)(7 \times 10^8)$

23. Perform the following operations:
a. $\dfrac{100}{10{,}000}$ **b.** $\dfrac{0.010}{1000}$
c. $\dfrac{10{,}000}{0.001}$ **d.** $\dfrac{0.0000001}{100}$
e. $\dfrac{10^{38}}{0.000100}$ **f.** $\dfrac{(100)^{1/2}}{0.01}$

24. Perform the following operations:
a. $\dfrac{2000}{0.00008}$ **b.** $\dfrac{0.004}{4 \times 10^6}$
c. $\dfrac{0.000220}{0.00005}$ **d.** $\dfrac{78 \times 10^{18}}{4 \times 10^{-6}}$

25. Perform the following operations:
a. $(100)^3$ **b.** $(0.0001)^{1/2}$
c. $(10{,}000)^8$ **d.** $(0.00000010)^9$

26. Perform the following operations:
a. $(200)^2$
b. $(5 \times 10^{-3})^3$
c. $(0.004)(3 \times 10^{-3})^2$
d. $((2 \times 10^{-3})(0.8 \times 10^4)(0.003 \times 10^5))^3$

27. Perform the following operations:
a. $(-0.001)^2$ **b.** $\dfrac{(100)(10^{-4})}{1000}$
c. $\dfrac{(0.001)^2(100)}{10{,}000}$ **d.** $\dfrac{(10^3)(10{,}000)}{1 \times 10^{-4}}$
e. $\dfrac{(0.0001)^3(100)}{1 \times 10^6}$ ***f.** $\dfrac{[(100)(0.01)]^{-3}}{[(100)^2][0.001]}$

28. Perform the following operations:
a. $\dfrac{(300)^2(100)}{3 \times 10^4}$ **b.** $[(40{,}000)^2][(20)^{-3}]$
c. $\dfrac{(60{,}000)^2}{(0.02)^2}$ **d.** $\dfrac{(0.000027)^{1/3}}{200{,}000}$
e. $\dfrac{[(4000)^2][300]}{2 \times 10^{-4}}$
f. $[(0.000016)^{1/2}][(100{,}000)^5][0.02]$
***g.** $\dfrac{[(0.003)^3][0.00007]^{-2}[(160)^2]}{[(200)(0.0008)]^{-1/2}}$ (a challenge)

SECTION 1.7 Fixed-Point, Floating-Point, Scientific, and Engineering Notation

29. Write the following numbers in scientific and engineering notation to hundredths place:
a. 20.46
b. 50,420
c. 0.000674
d. 000.0460

30. Write the following numbers in scientific and engineering notation to tenths place:
a. 5×10^{-2}
b. $0.45 \times 10^{+2}$
c. 1/32
d. π

SECTION 1.8 Conversion between Levels of Powers of Ten

31. Fill in the blanks of the following conversions:
a. $6 \times 10^4 =$ _____ $\times 10^6$
b. $0.4 \times 10^{-3} =$ _____ $\times 10^{-6}$
c. $50 \times 10^5 =$ _____ $\times 10^3 =$ _____ $\times 10^6$
$=$ _____ $\times 10^9$
d. $12 \times 10^{-7} =$ _____ $\times 10^{-3} =$ _____ $\times 10^{-6}$
$=$ _____ $\times 10^{-9}$

32. Perform the following conversions:
a. 0.05 s to milliseconds
b. 2000 μs to milliseconds
c. 0.04 ms to microseconds
d. 8400 ps to microseconds
e. 100×10^3 mm to kilometers

SECTION 1.9 Conversion within and between Systems of Units

33. Perform the following conversions to thousands-place accuracy:
a. 1.5 min to seconds
b. 2×10^{-2} h to seconds
c. 0.05 s to microseconds
d. 0.16 m to millimeters
e. 0.00000012 s to nanoseconds
f. 4×10^8 s to days

34. Perform the following metric conversions to tenths-place accuracy:
a. 80 mm to centimeters
b. 60 cm to kilometers
c. 12×10^{-3} m to micrometers
d. 60 sq cm (cm^2) to square meters (m^2)

35. Perform the following conversions between systems to hundredths-place accuracy:
a. 100 in. to meters
b. 4 ft to meters
c. 6 lb to newtons
d. 60,000 dyn to pounds
e. 150,000 cm to feet
f. 0.002 mi to meters (5280 ft = 1 mi)

36. What is a mile in feet, yards, meters, and kilometers?

37. Convert 60 mph to meters per second.

38. How long would it take a runner to complete a 10-km race if a pace of 6.5 min/mi were maintained?

39. Quarters are about 1 in. in diameter. How many would be required to stretch from one end of a football field to the other (100 yd)?

40. Compare the total time required to drive a long, tiring day of 500 mi at an average speed of 60 mph versus an average speed of 75 mph. Is the time saved for such a long trip worth the added risk of the higher speed?

***41.** Find the distance in meters that a mass traveling at 600 cm/s will cover in 0.016 h.

***42.** Each spring there is a race up 86 floors of the 102-story Empire State Building in New York City. If you were able to climb 2 steps/second, how long would it take in minutes to reach the 86th floor if each floor is 14 ft high and each step is about 9 in.?

***43.** The record for the race in Problem 42 is 10.22 min. What was the racer's speed in min/mi for the race?

***44.** If the race of Problem 42 were a horizontal distance, how long would it take a runner who can run 5-min miles to cover the distance? Compare this with the record speed of Problem 43. Did gravity have a significant effect on the overall time?

SECTION 1.11 Conversion Tables

45. Using Appendix A, determine the number of
a. Btu in 5 J of energy.
b. cubic meters in 24 oz of a liquid.
c. seconds in 1.4 days.
d. pints in 1 m^3 of a liquid.

SECTION 1.12 Calculators

Perform the following operations using a single sequence of calculator keys:

46. $6\,(4 \times 2 + 8) =$

47. $\dfrac{42 + \frac{6}{5}}{3} =$

48. $\sqrt{5^2 + \left(\dfrac{2}{3}\right)^2} =$

49. $\cos 21.87° =$

***50.** $\tan^{-1} \dfrac{3}{4} =$

***51.** $\sqrt{\dfrac{400}{6^2 + \frac{10}{5}}} =$

***52.** $\dfrac{8.2 \times 10^{-3}}{0.04 \times 10^{3}}$ (in engineering notation) =

***53.** $\dfrac{(0.06 \times 10^{5})(20 \times 10^{3})}{(0.01)^2}$ (in engineering notation) =

***54.** $\dfrac{4 \times 10^{4}}{2 \times 10^{-3} + 400 \times 10^{-5}} + \dfrac{1}{2 \times 10^{-6}}$
(in engineering notation) =

SECTION 1.13 Computer Analysis

55. Investigate the availability of computer courses and computer time in your curriculum. Which languages are commonly used, and which software packages are popular?

56. Develop a list of three popular computer languages, including a few characteristics of each. Why do you think some languages are better for the analysis of electric circuits than others?

GLOSSARY

Cathode-ray tube (CRT) A glass enclosure with a relatively flat face (screen) and a vacuum inside that will display the light generated from the bombardment of the screen by electrons.

CGS system The system of units employing the Centimeter, Gram, and Second as its fundamental units of measure.

Difference engine One of the first mechanical calculators.

Edison effect Establishing a flow of charge between two elements in an evacuated tube.

Electromagnetism The relationship between magnetic and electrical effects.

Engineering notation A method of notation that specifies that all powers of ten used to define a number be multiples of 3 with a mantissa greater than or equal to 1 but less than 1000.

ENIAC The first totally electronic computer.

Fixed-point notation Notation using a decimal point in a particular location to define the magnitude of a number.

Fleming's valve The first of the electronic devices, the diode.

Floating-point notation Notation that allows the magnitude of a number to define where the decimal point should be placed.

Integrated circuit (IC) A subminiature structure containing a vast number of electronic devices designed to perform a particular set of functions.

Joule (J) A unit of measurement for energy in the SI or MKS system. Equal to 0.7378 foot-pound in the English system and 10^7 ergs in the CGS system.

Kelvin (K) A unit of measurement for temperature in the SI system. Equal to 273.15 + °C in the MKS and CGS systems.

Kilogram (kg) A unit of measure for mass in the SI and MKS systems. Equal to 1000 grams in the CGS system.

Language A communication link between user and computer to define the operations to be performed and the results to be displayed or printed.

Leyden jar One of the first charge-storage devices.

Menu A computer-generated list of choices for the user to determine the next operation to be performed.

Meter (m) A unit of measure for length in the SI and MKS systems. Equal to 1.094 yards in the English system and 100 centimeters in the CGS system.

MKS system The system of units employing the *M*eter, *K*ilogram, and *S*econd as its fundamental units of measure.

Nanotechnology The production of integrated circuits in which the nanometer is the typical unit of measurement.

Newton (N) A unit of measurement for force in the SI and MKS systems. Equal to 100,000 dynes in the CGS system.

Pound (lb) A unit of measurement for force in the English system. Equal to 4.45 newtons in the SI or MKS system.

Program A sequential list of commands, instructions, and so on, to perform a specified task using a computer.

Scientific notation A method for describing very large and very small numbers through the use of powers of ten, which requires that the multiplier be a number between 1 and 10.

Second (s) A unit of measurement for time in the SI, MKS, English, and CGS systems.

SI system The system of units adopted by the IEEE in 1965 and the USASI in 1967 as the International System of Units (*S*ystème *I*nternational d'Unités).

Slug A unit of measure for mass in the English system. Equal to 14.6 kilograms in the SI or MKS system.

Software package A computer program designed to perform specific analysis and design operations or generate results in a particular format.

Static electricity Stationary charge in a state of equilibrium.

Transistor The first semiconductor amplifier.

Voltaic cell A storage device that converts chemical to electrical energy.

2

Voltage and Current

Objectives

- *Become aware of the basic atomic structure of conductors such as copper and aluminum and understand why they are used so extensively in the field.*
- *Understand how the terminal voltage of a battery or any dc supply is established and how it creates a flow of charge in the system.*
- *Understand how current is established in a circuit and how its magnitude is affected by the charge flowing in the system and the time involved.*
- *Become familiar with the factors that affect the terminal voltage of a battery and how long a battery will remain effective.*
- *Be able to apply a voltmeter and ammeter correctly to measure the voltage and current of a network.*

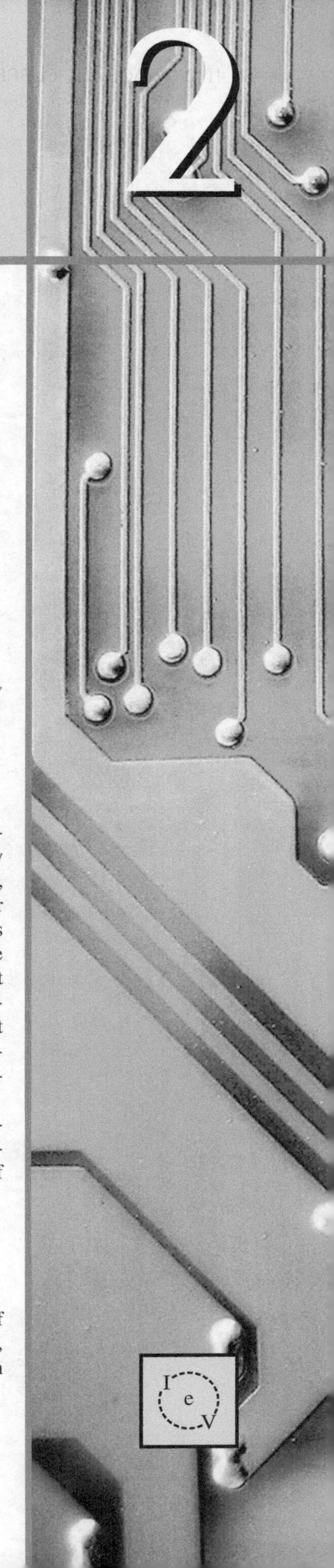

2.1 INTRODUCTION

Now that the foundation for the study of electricity/electronics has been established, the concepts of voltage and current can be investigated. The term **voltage** is encountered practically every day. We have all replaced batteries in our flashlights, answering machines, calculators, automobiles, and so on, that had specific voltage ratings. We are aware that most outlets in our homes are 120 volts. Although **current** may be a less familiar term, we know what happens when we place too many appliances on the same outlet—the circuit breaker opens due to the excessive current that results. It is fairly common knowledge that current is something that moves through the wires and causes sparks and possibly fire if there is a "short circuit." Current heats up the coils of an electric heater or the range of an electric stove; it generates light when passing through the filament of a bulb; it causes twists and kinks in the wire of an electric iron over time; and so on. All in all, the terms *voltage* and *current* are part of the vocabulary of most individuals.

In this chapter, the basic impact of current and voltage and the properties of each are introduced and discussed in some detail. Hopefully, any mysteries surrounding the general characteristics of each will be eliminated, and you will gain a clear understanding of the impact of each on an electric/electronics circuit.

2.2 ATOMS AND THEIR STRUCTURE

A basic understanding of the fundamental concepts of current and voltage requires a degree of familiarity with the atom and its structure. The simplest of all atoms is the hydrogen atom, made up of two basic particles, the **proton** and the **electron,** in the relative positions shown in Fig. 2.1(a). The **nucleus** of the hydrogen atom is the proton, a positively charged particle.

The orbiting electron carries a negative charge equal in magnitude to the positive charge of the proton.

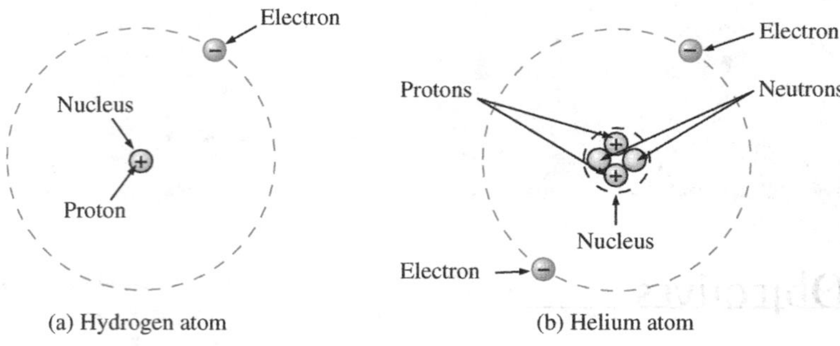

FIG. 2.1
Hydrogen and helium atoms.

In all other elements, the nucleus also contains **neutrons**, which are slightly heavier than protons and have no *electrical charge*. The helium atom, for example, has two neutrons in addition to two electrons and two protons, as shown in Fig. 2.1(b). In general,

the atomic structure of any stable atom has an equal number of electrons and protons.

Different atoms have various numbers of electrons in concentric orbits called *shells* around the nucleus. The first shell, which is closest to the nucleus, can contain only two electrons. If an atom has three electrons, the extra electron must be placed in the next shell. The number of electrons in each succeeding shell is determined by $2n^2$, where n is the shell number. Each shell is then broken down into subshells where the number of electrons is limited to 2, 6, 10, and 14 in that order as you move away from the nucleus.

Copper is the most commonly used metal in the electrical/electronics industry. An examination of its atomic structure will reveal why it has such widespread application. As shown in Fig. 2.2, it has 29 electrons in orbits around the nucleus, with the 29th electron appearing all by itself in the 4th shell. Note that the number of electrons in each shell and subshell is as

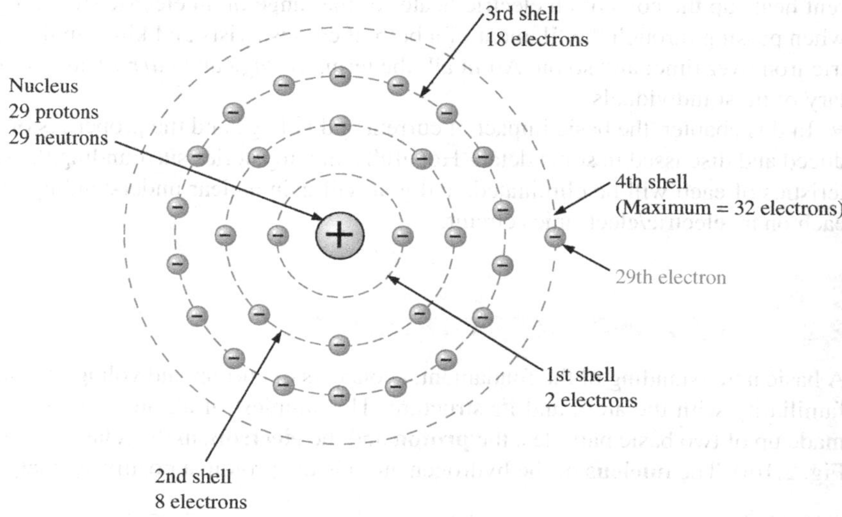

FIG. 2.2
The atomic structure of copper.

defined above. There are two important things to note in Fig. 2.2. First, the 4th shell, which can have a total of $2n^2 = 2(4)^2 = 32$ electrons, has only one electron. The outermost shell is incomplete and, in fact, is far from complete because it has only one electron. Atoms with complete shells (that is, a number of electrons equal to $2n^2$) are usually quite stable. Those atoms with a small percentage of the defined number for the outermost shell are normally considered somewhat unstable and volatile. Second, the 29th electron is the farthest electron from the nucleus. Opposite charges are attracted to each other, but the farther apart they are, the less the attraction. In fact, the force of attraction between the nucleus and the 29th electron of copper can be determined by **Coulomb's law** developed by Charles Augustin Coulomb (Fig. 2.3) in the late 18th century:

$$F = k\frac{Q_1Q_2}{r^2} \quad \text{(newtons, N)} \qquad \textbf{(2.1)}$$

where F is in newtons (N), k = a constant = 9.0×10^9 N·m²/C², Q_1 and Q_2 are the charges in coulombs (a unit of measure discussed in the next section), and r is the distance between the two charges in meters.

At this point, the most important thing to note is that the distance between the charges appears as a squared term in the denominator. First, the fact that this term is in the denominator clearly reveals that as it increases, the force will decrease. However, since it is a squared term, the force will drop dramatically with distance. For instance, if the distance is doubled, the force will drop to 1/4 because $(2)^2 = 4$. If the distance is increased by a factor of 4, it will drop by 1/16, and so on. The result, therefore, is that the force of attraction between the 29th electron and the nucleus is significantly less than that between an electron in the first shell and the nucleus. The result is that the 29th electron is loosely bound to the atomic structure and with a little bit of pressure from outside sources could be encouraged to leave the parent atom.

If this 29th electron gains sufficient energy from the surrounding medium to leave the parent atom, it is called a **free electron.** In 1 cubic in. of copper at room temperature, there are approximately 1.4×10^{24} free electrons. Expanded, that is 1,400,000,000,000,000,000,000,000 free electrons in a 1–in. square cube. The point is that we are dealing with enormous numbers of electrons when we talk about the number of free electrons in a copper wire—not just a few that you could leisurely count. Further, the numbers involved are clear evidence of the need to become proficient in the use of powers of ten to represent numbers and use them in mathematical calculations.

Other metals that exhibit the same properties as copper, but to a different degree, are silver, gold, and aluminum, and some rarer metals such as tungsten. Additional comments on the characteristics of conductors are in the following sections.

FIG. 2.3
Charles Augustin Coulomb.

French (Angoulème, Paris)
(1736–1806) Scientist and Inventor
Military Engineer, West Indies

Attended the engineering school at Mézières, the first such school of its kind. Formulated *Coulomb's law,* which defines the force between two electrical charges and is, in fact, one of the principal forces in atomic reactions. Performed extensive research on the friction encountered in machinery and windmills and the elasticity of metal and silk fibers.

2.3 VOLTAGE

If we separate the 29th electron in Fig. 2.2 from the rest of the atomic structure of copper by a dashed line as shown in Fig. 2.4(a), we create regions that have a net positive and negative charge as shown in Fig. 2.4(b) and (c). For the region inside the dashed boundary, the number of protons in the nucleus exceeds the number of orbiting electrons by 1, so the net charge is positive as shown in both figures. This positive region created by separating the free electron from the basic atomic structure is

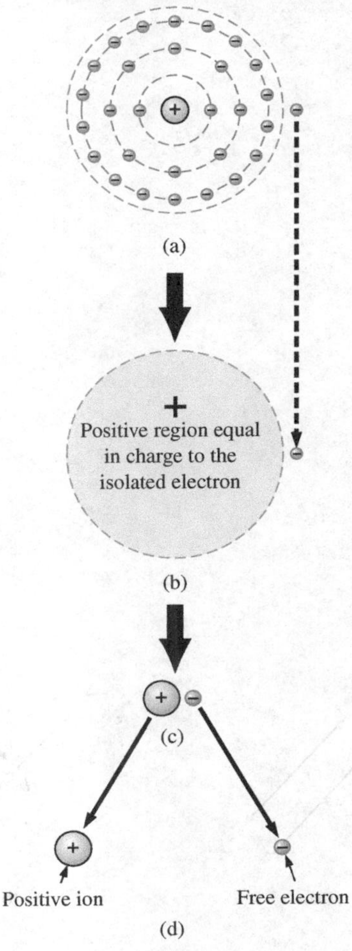

FIG. 2.4
Defining the positive ion.

called a **positive ion.** If the free electron then leaves the vicinity of the parent atom as shown in Fig. 2.4(d), regions of positive and negative charge have been established.

This separation of charge to establish regions of positive and negative charge is the action that occurs in every battery. Through chemical action, a heavy concentration of positive charge (positive ions) is established at the positive terminal, with an equally heavy concentration of negative charge (electrons) at the negative terminal.

In general,

every source of voltage is established by simply creating a separation of positive and negative charges.

It is that simple: If you want to create a voltage level of any magnitude, simply establish regions of positive and negative charge. The more the required voltage, the greater is the quantity of positive and negative charge.

In Fig. 2.5(a), for example, a region of positive charge has been established by a packaged number of positive ions, and a region of negative charge by a similar number of electrons, separated by a distance *r*. Since it would be inconsequential to talk about the voltage established by the separation of a single electron, a package of electrons called a **coulomb (C)** of charge was defined as follows:

One coulomb of charge is the total charge associated with 6.242×10^{18} electrons.

A coulomb of positive charge would have the same magnitude but opposite polarity.

In Fig. 2.5(b), if we take a coulomb of negative charge near the surface of the positive charge and move it toward the negative charge, we must expend energy to overcome the repulsive forces of the larger negative charge and the attractive forces of the positive charge. In the process of moving the charge from point *a* to point *b* in Fig. 2.5(b):

if a total of 1 joule (J) of energy is used to move the negative charge of 1 coulomb (C), there is a difference of 1 volt (V) between the two points.

The defining equation is

$$\boxed{V = \frac{W}{Q}} \qquad \begin{aligned} V &= \text{volts (V)} \\ W &= \text{joules (J)} \\ Q &= \text{coulombs (C)} \end{aligned} \qquad \textbf{(2.2)}$$

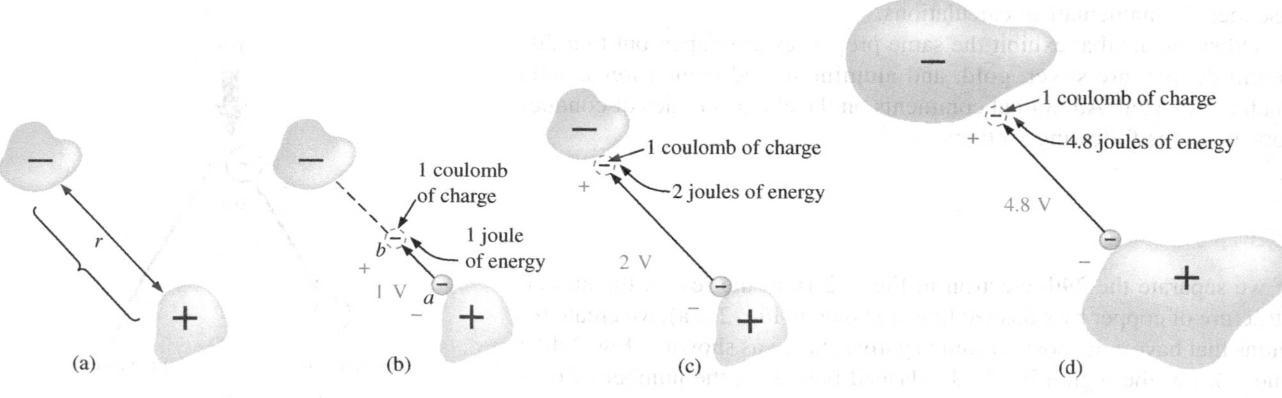

FIG. 2.5
Defining the voltage between two points.

Take particular note that the charge is measured in coulombs, the energy in joules, and the voltage in volts. The unit of measurement, **volt,** was chosen to honor the efforts of Alessandro Volta, who first demonstrated that a voltage could be established through chemical action (Fig. 2.6).

If the charge is now moved all the way to the surface of the larger negative charge as shown in Fig. 2.5(c), using 2 J of energy for the whole trip, there are 2 V between the two charged bodies. If the package of positive and negative charge is larger, as shown in Fig. 2.5(d), more energy will have to be expended to overcome the larger repulsive forces of the large negative charge and attractive forces of the large positive charge. As shown in Fig. 2.5(d), 4.8 J of energy were expended, resulting in a voltage of 4.8 V between the two points. We can therefore conclude that it would take 12 J of energy to move 1 C of negative charge from the positive terminal to the negative terminal of a 12 V car battery.

Through algebraic manipulations, we can define an equation to determine the energy required to move charge through a difference in voltage:

$$\boxed{W = QV} \quad \text{(joules, J)} \tag{2.3}$$

Finally, if we want to know how much charge was involved, we use

$$\boxed{Q = \frac{W}{V}} \quad \text{(coulombs, C)} \tag{2.4}$$

EXAMPLE 2.1 Find the voltage between two points if 60 J of energy are required to move a charge of 20 C between the two points.

Solution: Eq. (2.2): $V = \frac{W}{Q} = \frac{60\text{ J}}{20\text{ C}} = \mathbf{3\text{ V}}$

EXAMPLE 2.2 Determine the energy expended moving a charge of 50 μC between two points if the voltage between the points is 6 V.

Solution: Eq. (2.3):

$$W = QV = (50 \times 10^{-6}\text{ C})(6\text{ V}) = 300 \times 10^{-6}\text{ J} = \mathbf{300\ \mu J}$$

There are a variety of ways to separate charge to establish the desired voltage. The most common is the chemical action used in car batteries, flashlight batteries, and, in fact, all portable batteries. Other sources use mechanical methods such as car generators and steam power plants or alternative sources such as solar cells and windmills. In total, however, the sole purpose of the system is to create a separation of charge. In the future, therefore, when you see a positive and a negative terminal on any type of battery, you can think of it as a point where a large concentration of charge has gathered to create a voltage between the two points. More important is to recognize that a voltage exists between two points—for a battery between the positive and negative terminals. Hooking up just the positive or the negative terminal of a battery and not the other would be meaningless.

Both terminals must be connected to define the applied voltage.

FIG. 2.6
Count Alessandro Volta.
(© orion_eff / fotolia.com)

Italian (Como, Pavia)
(1745–1827)
Physicist
Professor of Physics,
Pavia, Italy

Began electrical experiments at the age of 18 working with other European investigators. Major contribution was the development of an electrical energy source from chemical action in 1800. For the first time, electrical energy was available on a continuous basis and could be used for practical purposes. Developed the first *condenser,* known today as the *capacitor.* Was invited to Paris to demonstrate the *voltaic cell* to Napoleon. The International Electrical Congress meeting in Paris in 1881 honored his efforts by choosing the *volt* as the unit of measure for electromotive force.

As we moved the 1 C of charge in Fig. 2.5(b), the energy expended would depend on where we were in the crossing. The *position* of the charge is therefore a factor in determining the voltage level at each point in the crossing. Since the **potential energy** associated with a body is defined by its position, the term *potential* is often applied to define voltage levels. For example, the difference in potential is 4 V between the two points, or the **potential difference** between a point and ground is 12 V, and so on.

2.4 CURRENT

The question, "Which came first—the chicken or the egg?" can be applied here also because the layperson has a tendency to use the terms *current* and *voltage* interchangeably as if both were sources of energy. It is time to set things straight:

The applied voltage is the starting mechanism—the current is a reaction to the applied voltage.

In Fig. 2.7(a), a copper wire sits isolated on a laboratory bench. If we cut the wire with an imaginary perpendicular plane, producing the circular cross section shown in Fig. 2.7(b), we would be amazed to find that there are free electrons crossing the surface in both directions. Those free electrons generated at room temperature are in constant motion in random directions. However, at any instant of time, the number of electrons crossing the imaginary plane in one direction is exactly equal to that crossing in the opposite direction, so the *net flow in any one direction is zero.* Even though the wire seems dead to the world sitting by itself on the bench, internally, it is quite active. The same would be true for any other good conductor.

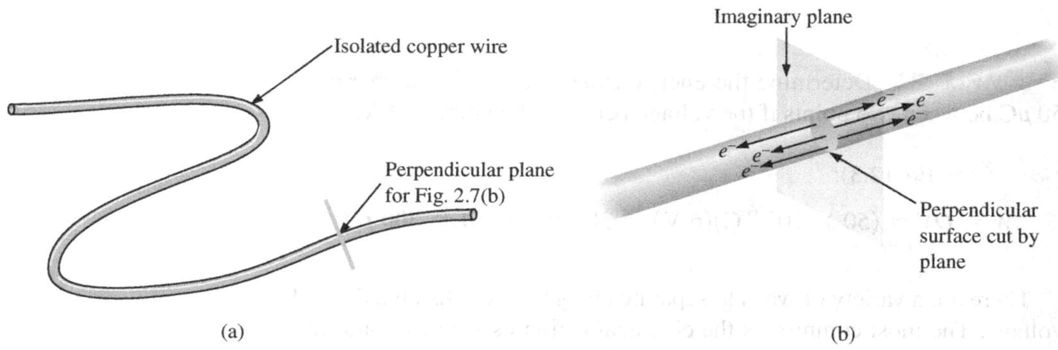

FIG. 2.7
There is motion of free carriers in an isolated piece of copper wire, but the flow of charge fails to have a particular direction.

Now, to make this electron flow do work for us, we need to give it a direction and be able to control its magnitude. This is accomplished by simply applying a voltage across the wire to force the electrons to move toward the positive terminal of the battery, as shown in Fig. 2.8. The instant the wire is placed across the terminals, the free electrons in the wire drift toward the positive terminal. The positive ions in the copper wire simply oscillate in a mean fixed position. As the electrons pass through the wire, the negative terminal of the battery acts as a supply of additional

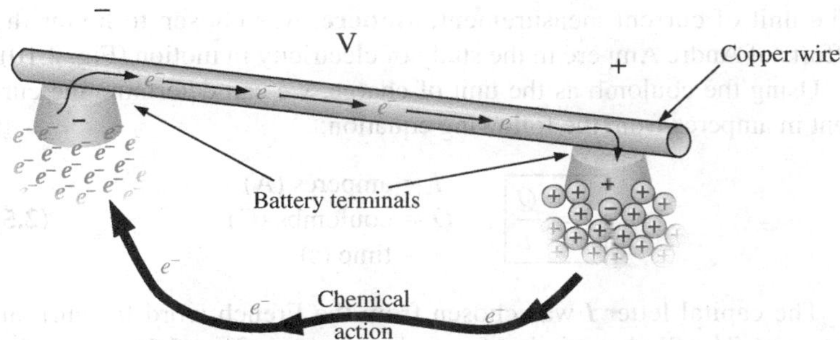

FIG. 2.8
Motion of negatively charged electrons in a copper wire when placed across battery terminals with a difference in potential of volts (V).

electrons to keep the process moving. The electrons arriving at the positive terminal are absorbed, and through the chemical action of the battery, additional electrons are deposited at the negative terminal to make up for those that left.

To take the process a step further, consider the configuration in Fig. 2.9, where a copper wire has been used to connect a light bulb to a battery to create the simplest of electric circuits. The instant the final connection is made, the free electrons of negative charge drift toward the positive terminal, while the positive ions left behind in the copper wire simply oscillate in a mean fixed position. The flow of charge (the electrons) through the bulb heats up the filament of the bulb through friction to the point that it glows red-hot and emits the desired light.

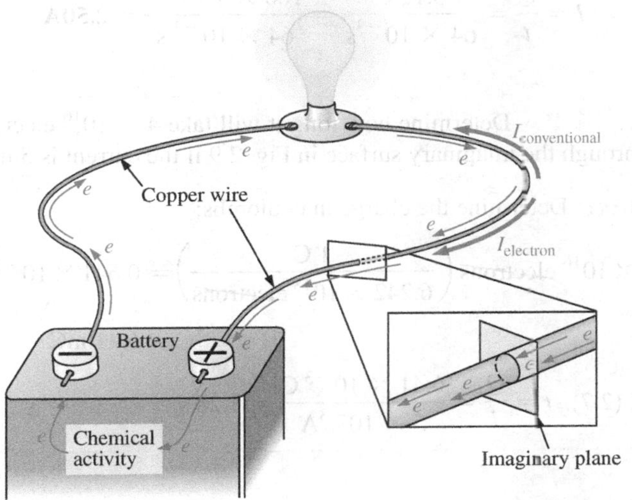

FIG. 2.9
Basic electric circuit.

In total, therefore, the applied voltage has established a flow of electrons in a particular direction. In fact, by definition,

if 6.242×10^{18} electrons (1 coulomb) pass through the imaginary plane in Fig. 2.9 in 1 second, the flow of charge, or current, is said to be 1 ampere (A).

FIG. 2.10
André Marie Ampère.
(© orion_eff / fotolia.com)

French (Lyon, Paris)
(1775–1836)
Mathematician and Physicist
Professor of Mathematics,
École Polytechnique, Paris

On September 18, 1820, introduced a new field of study, electrodynamics, devoted to the effect of electricity in motion, including the interaction between currents in adjoining conductors and the interplay of the surrounding magnetic fields. Constructed the first *solenoid* and demonstrated how it could behave like a magnet (the first *electromagnet*). Suggested the name *galvanometer* for an instrument designed to measure current levels.

The unit of current measurement, **ampere**, was chosen to honor the efforts of André Ampère in the study of electricity in motion (Fig. 2.10).

Using the coulomb as the unit of charge, we can determine the current in amperes from the following equation:

$$\boxed{I = \frac{Q}{t}} \qquad \begin{aligned} I &= \text{amperes (A)} \\ Q &= \text{coulombs (C)} \\ t &= \text{time (s)} \end{aligned} \qquad \textbf{(2.5)}$$

The capital letter I was chosen from the French word for current, *intensité*. The SI abbreviation for each quantity in Eq. (2.5) is provided to the right of the equation. The equation clearly reveals that for equal time intervals, the more charge that flows through the wire, the larger is the resulting current.

Through algebraic manipulations, the other two quantities can be determined as follows:

$$\boxed{Q = It} \qquad \text{(coulombs, C)} \qquad \textbf{(2.6)}$$

and

$$\boxed{t = \frac{Q}{I}} \qquad \text{(seconds, s)} \qquad \textbf{(2.7)}$$

EXAMPLE 2.3 The charge flowing through the imaginary surface in Fig. 2.9 is 0.16 C every 64 ms. Determine the current in amperes.

Solution: Eq. (2.5):

$$I = \frac{Q}{t} = \frac{0.16\text{ C}}{64 \times 10^{-3}\text{s}} = \frac{160 \times 10^{-3}\text{ C}}{64 \times 10^{-3}\text{ s}} = \mathbf{2.50A}$$

EXAMPLE 2.4 Determine how long it will take 4×10^{16} electrons to pass through the imaginary surface in Fig. 2.9 if the current is 5 mA.

Solution: Determine the charge in coulombs:

$$4 \times 10^{16} \cancel{\text{electrons}} \left(\frac{1\text{ C}}{6.242 \times 10^{18} \cancel{\text{electrons}}}\right) = 0.641 \times 10^{-2}\text{C}$$

$$= 6.41\text{ mC}$$

$$\text{Eq. (2.7): } t = \frac{Q}{I} = \frac{6.41 \times 10^{-3}\text{C}}{5 \times 10^{-3}\text{A}} = \mathbf{1.28\ s}$$

In summary, therefore,

the applied voltage (or potential difference) in an electrical/electronics system is the "pressure" to set the system in motion, and the current is the reaction to that pressure.

A mechanical analogy often used to explain this is the simple garden hose. In the absence of any pressure, the water sits quietly in the hose with no general direction, just as electrons do not have a net direction in the absence of an applied voltage. However, release the spigot, and the

applied pressure forces the water to flow through the hose. Similarly, apply a voltage to the circuit, and a flow of charge or current results.

A second glance at Fig. 2.9 reveals that two directions of charge flow have been indicated. One is called *conventional flow,* and the other is called *electron flow.* This text discusses only conventional flow for a variety of reasons; namely, it is the most widely used at educational institutions and in industry, it is employed in the design of all electronic device symbols, and it is the popular choice for all major computer software packages. The flow controversy is a result of an assumption made at the time electricity was discovered that the positive charge was the moving particle in metallic conductors. Be assured that the choice of conventional flow will not create great difficulty and confusion in the chapters to follow. Once the direction of I is established, the issue is dropped and the analysis can continue without confusion.

Safety Considerations

It is important to realize that even small levels of current through the human body can cause serious, dangerous side effects. Experimental results reveal that the human body begins to react to currents of only a few milliamperes. Although most individuals can withstand currents up to perhaps 10 mA for very short periods of time without serious side effects, any current over 10 mA should be considered dangerous. In fact, currents of 50 mA can cause severe shock, and currents of over 100 mA can be fatal. In most cases, the skin resistance of the body when dry is sufficiently high to limit the current through the body to relatively safe levels for voltage levels typically found in the home. However, if the skin is wet due to perspiration, bathing, and so on, or if the skin barrier is broken due to an injury, the skin resistance drops dramatically, and current levels could rise to dangerous levels for the same voltage shock. In general, therefore, simply remember that *water and electricity don't mix.* Granted, there are safety devices in the home today [such as the ground fault circuit interrupt (GFCI) breaker, discussed in Chapter 4] that are designed specifically for use in wet areas such as the bathroom and kitchen, but accidents happen. Treat electricity with respect—not fear.

2.5 VOLTAGE SOURCES

The term **dc,** used throughout this text, is an abbreviation for **direct current,** which encompasses all systems where there is a unidirectional (one direction) flow of charge. This section reviews dc voltage supplies that apply a fixed voltage to electrical/electronics systems.

The graphic symbol for all dc voltage sources is shown in Fig. 2.11. Note that the relative length of the bars at each end define the polarity of the supply. The long bar represents the positive side; the short bar, the negative. Note also the use of the letter E to denote *voltage source.* It comes from the fact that

an electromotive force (emf) is a force that establishes the flow of charge (or current) in a system due to the application of a difference in potential.

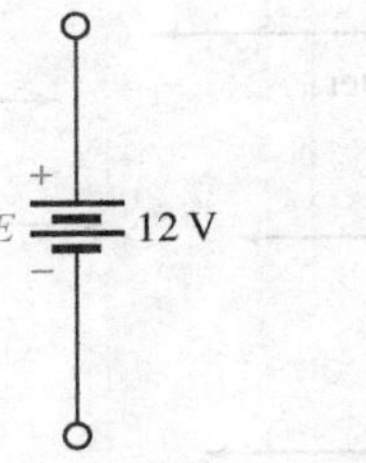

FIG. 2.11
Standard symbol for a dc voltage source.

In general, dc voltage sources can be divided into three basic types: (1) batteries (chemical action or solar energy), (2) generators (electromechanical), and (3) power supplies (rectification—a conversion process to be described in your electronics courses).

Batteries

General Information For the layperson, the battery is the most common of the dc sources. By definition, a battery (derived from the expression "battery of cells") consists of a combination of two or more similar **cells,** a cell being the fundamental source of electrical energy developed through the conversion of chemical or solar energy. All cells can be divided into the **primary** or **secondary** types. The secondary is rechargeable, whereas the primary is not. That is, the chemical reaction of the secondary cell can be reversed to restore its capacity. The two most common rechargeable batteries are the lead-acid unit (used primarily in automobiles) and the nickel–metal hydride (NiMH) battery (used in calculators, tools, photoflash units, shavers, and so on). The obvious advantages of rechargeable units are the savings in time and money of not continually replacing discharged primary cells.

All the cells discussed in this chapter (except the **solar cell,** which absorbs energy from incident light in the form of photons) establish a potential difference at the expense of chemical energy. In addition, each has a positive and a negative *electrode* and an **electrolyte** to complete the circuit between electrodes within the battery. The electrolyte is the contact element and the source of ions for conduction between the terminals.

Primary Cells (Non-rechargeable) The popular alkaline primary battery uses a powdered zinc anode (+); a potassium (alkali metal) hydroxide electrolyte; and a manganese dioxide/carbon cathode (−) as shown in Fig. 2.12(a). In Fig. 2.12(b), note that for the cylindrical types (AAA, AA, C, and D), the voltage is the same for each, but the ampere-hour (Ah) rating increases significantly with size. The ampere-hour rating is an indication of the level of current that the battery can provide for a specified period of time (to be discussed in detail in Section 2.6). In particular, note that for the large, lantern-type battery, the voltage is only 4 times that of the AAA battery, but the ampere-hour rating of 52 Ah is almost 42 times that of the AAA battery.

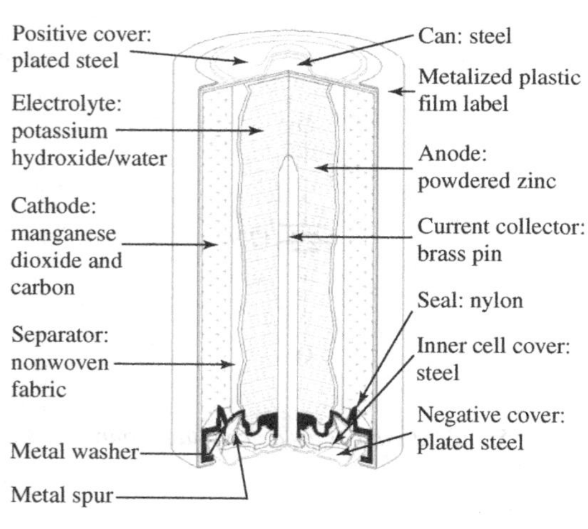

(a)

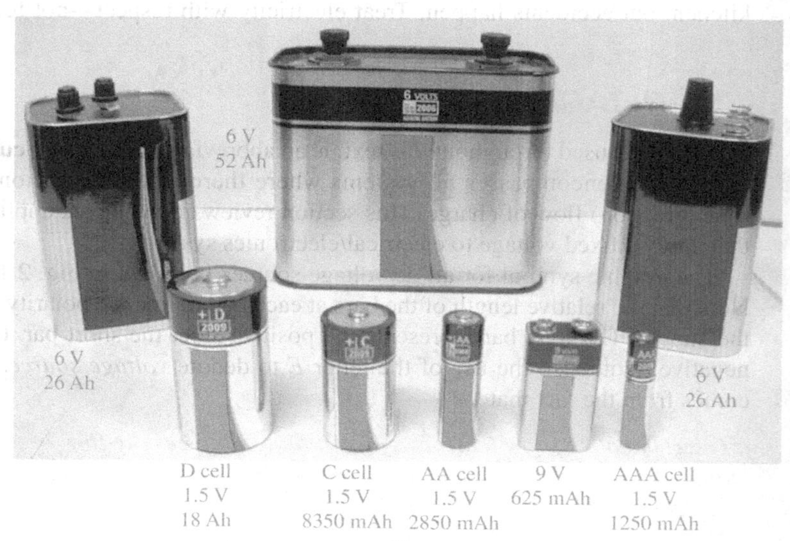

(b)

FIG. 2.12
Alkaline primary cell: (a) Cutaway of cylindrical Energizer® cell; (b) various types of primary cells.

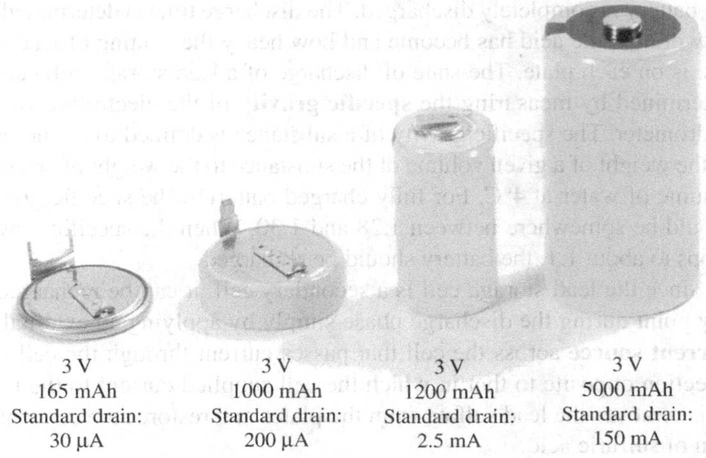

FIG. 2.13
Lithium primary batteries.

Another type of popular primary cell is the lithium battery, shown in Fig. 2.13. Again, note that the voltage is the same for each, but the size increases substantially with the ampere-hour rating and the rated drain current. It is particularly useful when low temperature is encountered.

In general, therefore,

for batteries of the same type, the size is dictated primarily by the standard drain current or ampere-hour rating, not by the terminal voltage rating.

Secondary Cells (Rechargeable)

Lead-Acid: The 12 V of Fig. 2.14, typically used in automobiles, has an electrolyte of sulfuric acid and electrodes of spongy lead (Pb) and lead peroxide (PbO_2). When a load is applied to the battery terminals, there is a transfer of electrons from the spongy lead electrode to the lead peroxide electrode through the load. This transfer of electrons will continue until

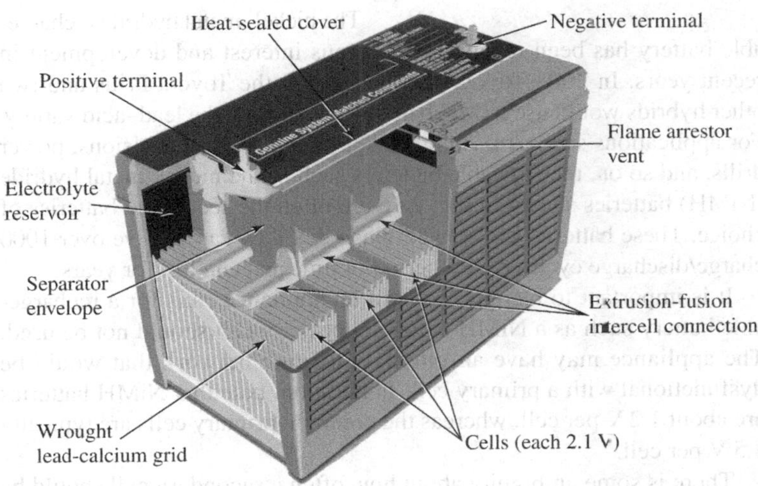

FIG. 2.14
Maintenance-free 12 V (actually 12.6 V) lead-acid battery.

the battery is completely discharged. The discharge time is determined by how diluted the acid has become and how heavy the coating of lead sulfate is on each plate. The state of discharge of a lead storage cell can be determined by measuring the **specific gravity** of the electrolyte with a hydrometer. The specific gravity of a substance is defined to be the ratio of the weight of a given volume of the substance to the weight of an equal volume of water at 4°C. For fully charged batteries, the specific gravity should be somewhere between 1.28 and 1.30. When the specific gravity drops to about 1.1, the battery should be recharged.

Since the lead storage cell is a secondary cell, it can be recharged at any point during the discharge phase simply by applying an external **dc current source** across the cell that passes current through the cell in a direction opposite to that in which the cell supplied current to the load. This removes the lead sulfate from the plates and restores the concentration of sulfuric acid.

The output of a lead storage cell over most of the discharge phase is about 2.1 V. In the commercial lead storage batteries used in automobiles, 12.6 V can be produced by six cells in series, as shown in Fig. 2.14. In general, lead-acid storage batteries are used in situations where a high current is required for relatively short periods of time. At one time, all lead-acid batteries were vented. Gases created during the discharge cycle could escape, and the vent plugs provided access to replace the water or electrolyte and to check the acid level with a hydrometer. The use of a grid made from a wrought lead–calcium alloy strip, rather than the lead–antimony cast grid commonly used, has resulted in maintenance-free batteries, shown in Fig. 2.14. The lead–antimony structure was susceptible to corrosion, overcharge, gasing, water usage, and self-discharge. Improved design with the lead–calcium grid has either eliminated or substantially reduced most of these problems.

It would seem that with so many advances in technology, the size and weight of the lead–acid battery would have decreased significantly in recent years, but even today it is used more than any other battery in automobiles and all forms of machinery. However, things are beginning to change with interest in nickel–metal hydride and lithium-ion batteries, which both pack more power per unit size than the lead–acid variety. Both will be described in the sections to follow.

Nickel–Metal Hydride (NiMH): The nickel–metal hydride rechargeable battery has been receiving enormous interest and development in recent years. In 2008 Toyota announced that the Toyota Prius and two other hybrids would use NiMH batteries rather than the lead–acid variety. For applications such as flashlights, shavers, portable televisions, power drills, and so on, rechargeable batteries such as the nickel–metal hydride (NiMH) batteries shown in Fig. 2.15 are often the secondary batteries of choice. These batteries are so well made that they can survive over 1000 charge/discharge cycles over a period of time and can last for years.

It is important to recognize that if an appliance calls for a rechargeable battery such as a NiMH battery, a primary cell should not be used. The appliance may have an internal charging network that would be dysfunctional with a primary cell. In addition, note that NiMH batteries are about 1.2 V per cell, whereas the common primary cells are typically 1.5 V per cell.

There is some ambiguity about how often a secondary cell should be recharged. Generally, the battery can be used until there is some indication that the energy level is low, such as a dimming light from a

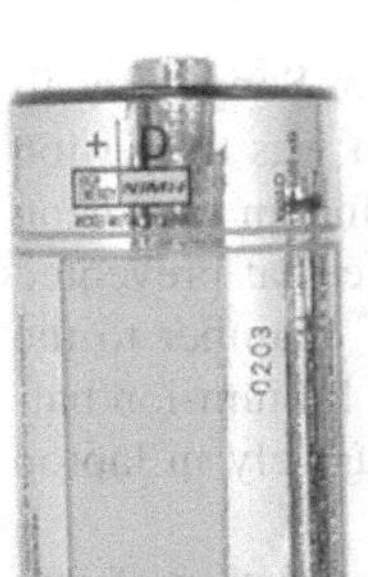
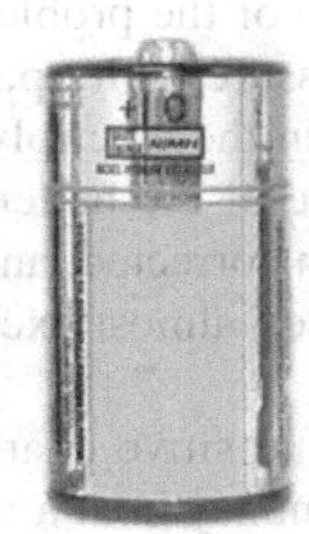
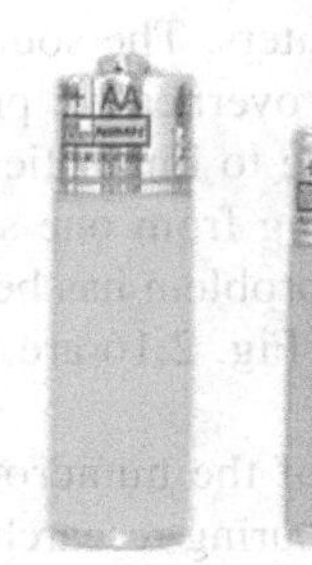

FIG. 2.15
Nickel–metal hydride (NiMH) rechargeable batteries.

flashlight, less power from a drill, or a signal from low-battery indicator. Keep in mind that secondary cells do have some "memory." If they are recharged continuously after being used for a short period of time, they may begin to believe they are short-term units and actually fail to hold the charge for the rated period of time. In any event, always try to avoid a "hard" discharge, which results when every bit of energy is drained from a cell. Too many hard-discharge cycles will reduce the cycle life of the battery. Finally, be aware that the charging mechanism for nickel–cadmium cells is quite different from that for lead–acid batteries. The nickel–cadmium battery is charged by a constant–current source, with the terminal voltage staying fairly steady through the entire charging cycle. The lead–acid battery is charged by a constant voltage source, permitting the current to vary as determined by the state of the battery. The capacity of the NiMH battery increases almost linearly throughout most of the charging cycle. Nickel–cadmium batteries become relatively warm when charging. The lower the capacity level of the battery when charging, the higher is the temperature of the cell. As the battery approaches rated capacity, the temperature of the cell approaches room temperature.

Lithium-ion (Li-ion): The battery receiving the most research and development in recent years is the lithium-ion battery. It carries more energy in a smaller space than either the lead–acid or NiMH rechargeable batteries. Its positive characteristics are such that once they can be properly channeled in a safe and efficient manner at a reasonable price, it may sweep the others off the scene. However, for the moment it is used extensively in smaller applications such as computers, a host of consumer products, and recently in power tools. It has a way to go before taking over the automobile market. The Chevrolet Volt, a General Motors plug-in hybrid concept car, uses a Li-ion battery but is limited to a 40-mile run before it needs to use a small gas engine. The sleek Tesla roadster with its battery of more than 6800 Li-ion cells can travel some 250 miles, but it carries a battery pack that costs between \$10,000 and \$15,000. Another problem is shelf life. Once manufactured, these batteries begin to slowly die even though they may go through normal charge/discharge cycles, which makes them similar to a normal primary cell, so lifetime is a major concern. You may remember

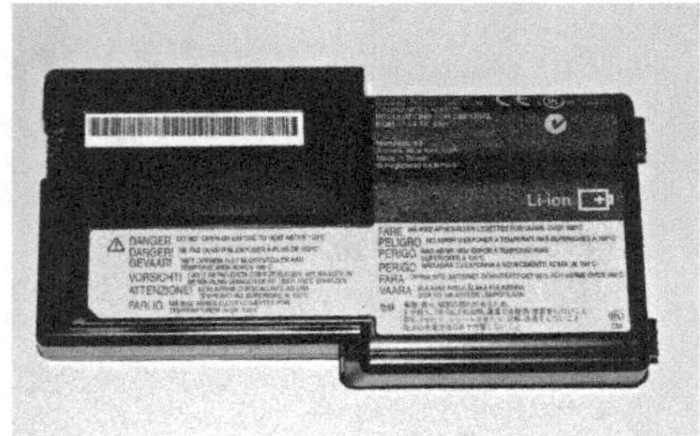

FIG. 2.16
Dell laptop lithium-ion battery: 11.1 V, 4400 mAh.
(© Herbert Rubens / Fotolia.com)

the laptops bursting into flames in 2006 and the need for Sony to recall some 6 million computers. The source of the problem was the Li-ion battery, which simply overheated; pressure built up, and an explosion occurred. This was due to impurities in the electrolyte that prevented the lithium ions moving from one side of the battery chamber to the other. Since then this problem has been corrected, and lithium-ion batteries as appearing in Fig. 2.16 are used almost exclusively in laptop computers.

Industry is aware of the numerous positive characteristics of this power source and is pouring research money in at a very high rate. Recent use of nanotechnology and microstructures has alleviated many of the concerns addressed here.

Solar Cell

The use of solar cells as part of the effort to generate "clean" energy has grown exponentially in the last few years. At one time the cost and the low conversion efficiencies were the main stumbling blocks to widespread use of the solar cell. However, the company Nanosolar has significantly reduced the cost of solar panels by using a printing process that uses a great deal less of the expensive silicon material in the manufacturing process. Whereas the cost of generating solar electricity is about 20 to 30 ¢/kWh, compared to an average of 11 ¢/kWh using a local utililty, this new printing process will have a significant impact on reducing the cost level. Another factor that will reduce costs is the improving level of efficiency being obtained by manufacturers. At one time the accepted efficiency level of conversion was between 10% and 14%. Recently, however, almost 20% has been obtained in the laboratory, and some feel that 30% to 60% efficiency is a possibility in the future. Given that the maximum available wattage on an average bright, sunlit day is 100 mW/cm^2, the efficiency is an important element in any future plans for the expansion of solar power. At 10% to 14% efficiency the maximum available power per cm^2 would only be 10 to 14 mW. For 1 m^2 the return would be 100 to 140 W. However, if the efficiency could be raised to 20%, the output would be significantly higher at 200 W for the 1-m^2 panel.

The relatively small three-panel solar unit appearing on the roof of the garage of the home of Fig. 2.17(a) can provide an energy source of 550 watt-hours (the watt-hour unit of measurement for energy will be discussed in detail in Chapter 4). Such a unit can provide sufficient electrical energy to run an energy-efficient refrigerator for 24 hours per day while simultaneously running a color TV for 7 hours, a microwave for 15 minutes, a 60 W bulb for 10 hours, and an electric clock for 10 hours. The basic system operates as shown in Fig. 2.17(b). The solar panels (1) convert sunlight into dc electric power. An inverter (2) converts the dc power into the standard ac power for use in the home (6). The batteries (3) can store energy from the sun for use if there is insufficient sunlight or a power failure. At night or on dark days when the demand exceeds the solar panel and battery supply, the local utility company (4) can provide power to the appliances (6) through a special hookup in the electrical panel (5). Although there is an initial expense to setting up the system, it is vitally important to realize that the source of energy is free—no monthly bill for sunlight to contend with—and will provide a significant amount of energy for a very long period of time.

FIG. 2.17
Solar System: (a) panels on roof of garage; (b) system operation.
(© majorosl66 / Fotolia.com)

Generators

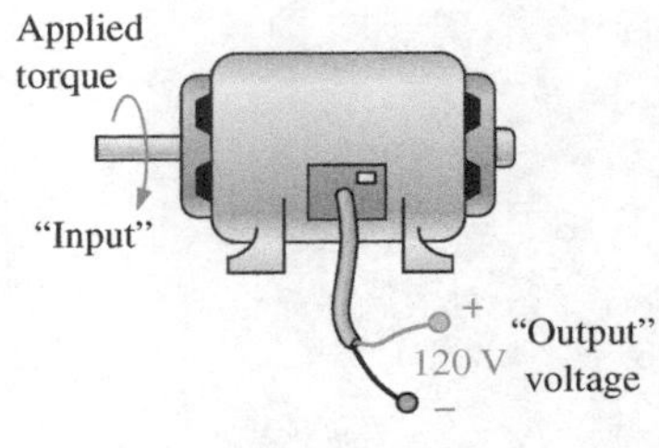

FIG. 2.18
dc generator.

The **dc generator** is quite different from the battery, both in construction (Fig. 2.18) and in mode of operation. When the shaft of the generator is rotating at the nameplate speed due to the applied torque of some external source of mechanical power, a voltage of rated value appears across the external terminals. The terminal voltage and power-handling capabilities of the dc generator are typically higher than those of most batteries, and its lifetime is determined only by its construction. Commercially used dc generators are typically 120 V or 240 V. For the purposes of this text, the same symbols are used for a battery and a generator.

Power Supplies

The dc supply encountered most frequently in the laboratory uses the **rectification** and *filtering* processes as its means toward obtaining a steady dc voltage. Both processes will be covered in detail in your basic electronics courses. In total, a time-varying voltage (such as ac voltage available from a home outlet) is converted to one of a fixed magnitude. A dc laboratory supply of this type is shown in Fig. 2.19.

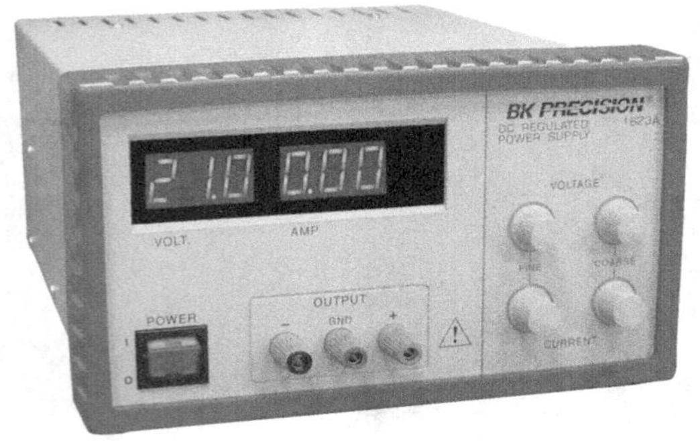

FIG. 2.19
A 0 V to 60 V, 0 to 1.5 A digital display dc power supply.
(Courtesy of B&K Precision.)

Most dc laboratory supplies have a regulated, adjustable voltage output with three available terminals, as indicated horizontally at the bottom of Fig 2.19 and vertically in Fig 2.20(a). The symbol for ground or zero potential (the reference) is also shown in Fig. 2.20(a). If 10 V above ground potential are required, the connections are made as shown in Fig. 2.20(b). If 15 V below ground potential are required, the connections are made as shown in Fig. 2.20(c). If connections are as shown in Fig. 2.20(d), we say we have a "floating" voltage of 5 V since the reference level is not included. Seldom is the configuration in Fig. 2.20(d) used since it fails to protect the operator by providing a direct low-resistance path to ground and to establish a common ground for the system. In any case, *the positive and negative terminals must be part of any circuit configuration.*

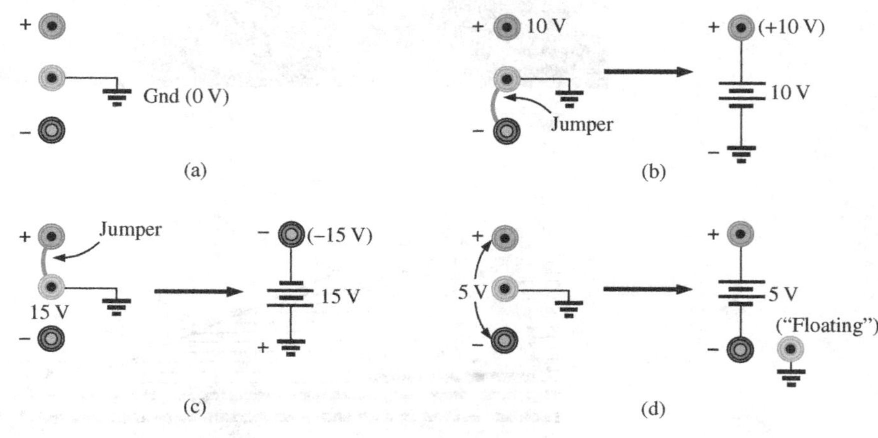

FIG. 2.20
dc laboratory supply: (a) available terminals; (b) positive voltage with respect to (w.r.t.) ground; (c) negative voltage w.r.t. ground; (d) floating supply.

Fuel Cells

One of the most exciting developments in recent years has been the steadily rising interest in **fuel cells** as an alternative energy source. Fuel cells are now being used in small stationary power plants, transportation (buses), and a wide variety of applications where portability is a major factor, such as the space shuttle. Millions are now being spent by major automobile manufacturers to build affordable fuel-cell vehicles.

Fuel cells have the distinct advantage of operating at efficiencies of 70% to 80% rather than the typical 20% to 25% efficiency of current internal combustion engine of today's automobiles. They also have no moving parts, produce little or no pollution, generate very little noise, and use fuels such as hydrogen and oxygen that are readily available. Fuel cells are considered primary cells (of the continuous-feed variety) because they cannot be recharged. They hold their characteristics as long as the fuel (hydrogen) and oxygen are supplied to the cell. The only byproducts of the conversion process are small amounts of heat (which is often used elsewhere in the system design), water (which may also be reused), and negligible levels of some oxides, depending on the components of the process. Overall, fuel cells are environmentally friendly.

The operation of the fuel cell is essentially opposite to that of the chemical process of electrolysis. Electrolysis is the process whereby electric current is passed through an electrolyte to break it down into its fundamental components. An electrolyte is any solution that will permit conduction through the movement of ions between adjoining electrodes. For instance, passing current through water results in a hydrogen gas by the cathode (negative terminal) and oxygen gas at the anode (positive terminal). In 1839, Sir William Grove believed this process could be reversed and demonstrated that the proper application of the hydrogen gas and oxygen results in a current through an applied load connected to the electrodes of the system. The first commercial unit was used in a tractor in 1959, followed by an energy pack in the 1965 Gemini program. In 1996, the first small power plant was designed, and today it is an important component of the shuttle program.

The basic components of a fuel cell are depicted in Fig. 2.21(a) with details of the construction in Fig. 2.21(b). Hydrogen gas (the fuel) is supplied to the system at a rate proportional to the current required by the load. At the opposite end of the cell, oxygen is supplied as needed. The net result is a flow of electrons through the load and a discharge of water with a release of some heat developed in the process. The amount of heat is minimal, although it can also be used as a component in the design to improve the efficiency of the cell. The water (very clean) can simply be discharged or used for other applications such as cooling in the overall application. If the source of hydrogen or oxygen is removed, the system breaks down. The flow diagram of the system is relatively simple, as shown in Fig. 2.21(a). In an actual cell, shown in Fig. 2.21(b), the hydrogen gas is applied to a porous electrode called the *anode* that is coated with a platinum catalyst. The catalyst on the anode serves to speed up the process of breaking down the hydrogen atom into positive hydrogen ions and free electrons. The electrolyte between the electrodes is a solution or membrane that permits the passage of positive hydrogen ions but not electrons. Facing this wall, the electrons choose to pass through the load and light up the bulb, while the positive hydrogen ions migrate toward the cathode. At the porous cathode (also coated with the catalyst), the incoming oxygen atoms combine with the arriving hydrogen

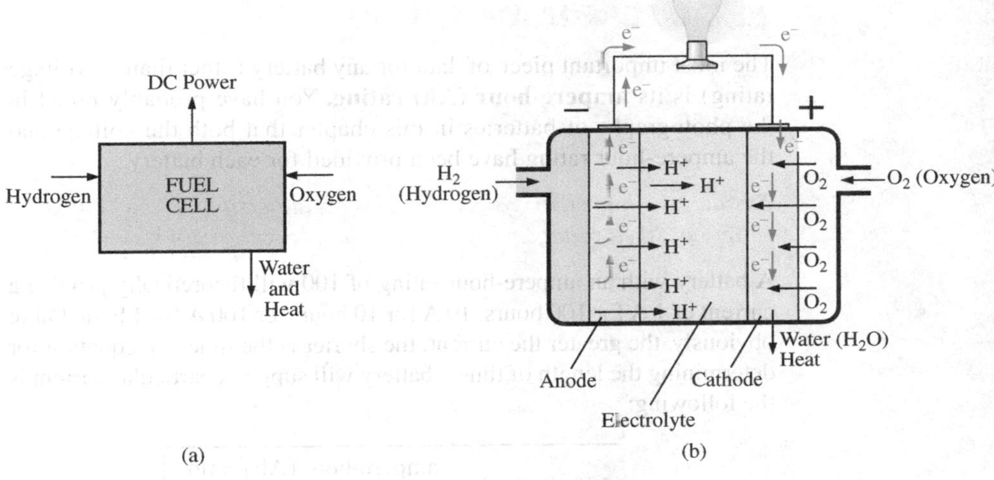

FIG. 2.21
Fuel cell (a) components; (b) basic construction.

ions and the electrons from the circuit to create water (H_2O) and heat. The circuit is, therefore, complete. The electrons are generated and then absorbed. If the hydrogen supply is cut off, the source of electrons is shut down, and the system is no longer an operating fuel cell.

In some fuel cells, either a liquid or molten electrolyte membrane is used. Depending on which the system uses, the chemical reactions will change slightly but not dramatically from that described above. The phosphoric acid fuel cell is a popular cell using a liquid electrolyte, while the PEM uses a polymer electrolyte membrane. The liquid or molten type is typically used in stationary power plants, while the membrane type is favored for vehicular use.

The output from a single fuel cell is a low-voltage, high-current dc output. Stacking the cells in series or parallel increases the output voltage or current level.

Fuel cells are receiving an enormous amount of attention and development effort. It is certainly possible that fuel cells may some day replace batteries in the vast majority of applications requiring a portable energy source. Fig. 2.22 shows the components of a hydrogen fuel-cell automobile.

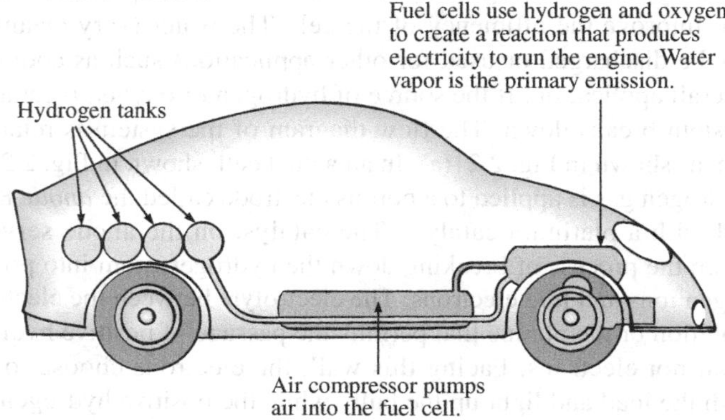

FIG. 2.22
Hydrogen fuel-cell automobile.

2.6 AMPERE-HOUR RATING

The most important piece of data for any battery (other than its voltage rating) is its **ampere-hour (Ah) rating.** You have probably noted in the photographs of batteries in this chapter that both the voltage and the ampere-hour rating have been provided for each battery.

The ampere-hour (Ah) rating provides an indication of how long a battery of fixed voltage will be able to supply a particular current.

A battery with an ampere-hour rating of 100 will theoretically provide a current of 1 A for 100 hours, 10 A for 10 hours, or 100 A for 1 hour. Quite obviously, the greater the current, the shorter is the time. An equation for determining the length of time a battery will supply a particular current is the following:

$$\text{Life (hours)} = \frac{\text{ampere-hour (Ah) rating}}{\text{amperes drawn (A)}} \tag{2.8}$$

EXAMPLE 2.5 How long will a 9 V transistor battery with an ampere-hour rating of 520 mAh provide a current of 20 mA?

Solution: Eq. (2.8): Life $= \dfrac{520 \text{ mAh}}{20 \text{ mA}} = \dfrac{520}{20} \text{ h} = \mathbf{26 \text{ h}}$

EXAMPLE 2.6 How long can a 1.5 V flashlight battery provide a current of 250 mA to light the bulb if the ampere-hour rating is 16 Ah?

Solution: Eq. (2.8): Life $= \dfrac{16 \text{ Ah}}{250 \text{ mA}} = \dfrac{16}{250 \times 10^{-3}} \text{ h} = \mathbf{64 \text{ h}}$

2.7 BATTERY LIFE FACTORS

The previous section made it clear that the life of a battery is directly related to the magnitude of the current drawn from the supply. However, there are factors that affect the given ampere-hour rating of a battery, so we may find that a battery with an ampere-hour rating of 100 can supply a current of 10 A for 10 hours but can supply a current of 100 A for only 20 minutes rather than the full 1 hour calculated using Eq. (2.8). In other words,

the capacity of a battery (in ampere-hours) will change with change in current demand.

This is not to say that Eq. (2.8) is totally invalid. It can always be used to gain some insight into how long a battery can supply a particular current. However, be aware that there are factors that affect the ampere-hour rating. Just as with most systems, including the human body, the more we demand, the shorter is the time that the output level can be maintained. This is clearly verified by the curves in Fig. 2.23 for the Eveready Energizer D cell. As the constant-current drain increased, the ampere-hour rating decreased from about 18 Ah at 25 mA to around 12 Ah at 300 mA.

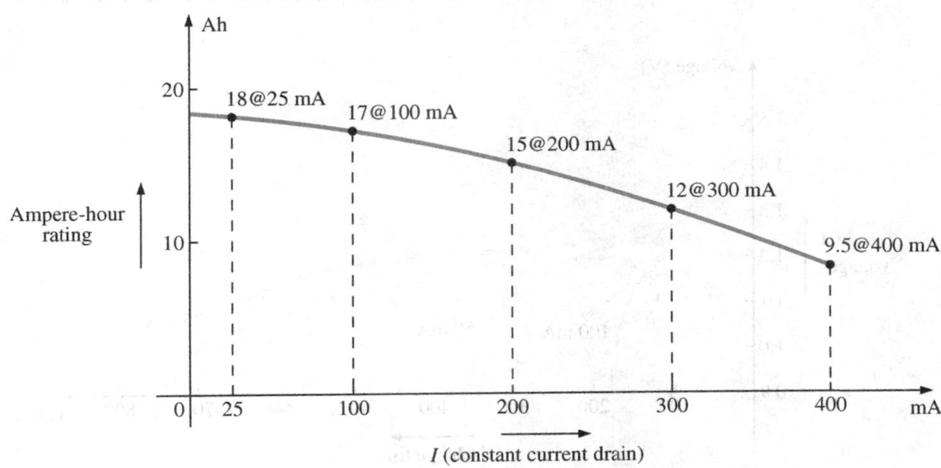

FIG. 2.23
Ampere-hour rating (capacity) versus drain current for an Energizer® D cell.

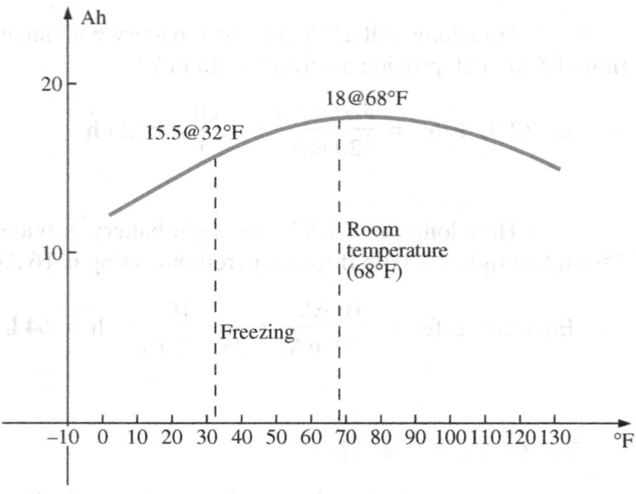

FIG. 2.24
Ampere-hour rating (capacity) versus temperature for an Energizer® D cell.

Another factor that affects the ampere-hour rating is the temperature of the unit and the surrounding medium. In Fig. 2.24, the capacity of the same battery plotted in Fig. 2.23 shows a peak value near the common room temperature of 68°F. At very cold temperatures and very warm temperatures, the capacity drops. Clearly, the ampere-hour rating will be provided at or near room temperature to give it a maximum value, but be aware that it will drop off with an increase or decrease in temperature. Most of us have noted that the battery in a car, radio, two-way radio, flashlight, and so on seems to have less power in really cold weather. It would seem, then, that the battery capacity would increase with higher temperatures—which, however, is not always the case. In general, therefore,

the ampere-hour rating of a battery will decrease from the room-temperature level with very cold and very warm temperatures.

Another interesting factor that affects the performance of a battery is how long it is asked to supply a particular voltage at a continuous drain current. Note the curves in Fig. 2.25, where the terminal voltage dropped at each level of drain current as the time period increased. The lower the

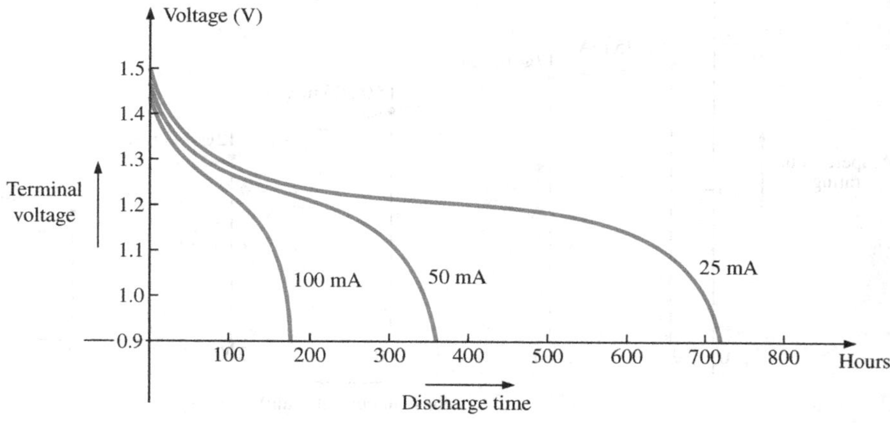

FIG. 2.25
Terminal voltage versus discharge time for specific drain currents for an Energizer® D cell.

current drain, the longer it could supply the desired current. At 100 mA, it was limited to about 100 hours near the rated voltage, but at 25 mA, it did not drop below 1.2 V until about 500 hours had passed. That is an increase in time of 5 : 1, which is significant. The result is that

the terminal voltage of a battery will eventually drop (at any level of current drain) if the time period of continuous discharge is too long.

2.8 CONDUCTORS AND INSULATORS

Different wires placed across the same two battery terminals allow different amounts of charge to flow between the terminals. Many factors, such as the density, mobility, and stability characteristics of a material, account for these variations in charge flow. In general, however,

conductors are those materials that permit a generous flow of electrons with very little external force (voltage) applied.

In addition,

good conductors typically have only one electron in the valence (most distant from the nucleus) ring.

Since **copper** is used most frequently, it serves as the standard of comparison for the relative conductivity in Table 2.1. Note that aluminum, which has seen some commercial use, has only 61% of the conductivity level of copper. The choice of material must be weighed against the cost and weight factors, however.

Insulators are those materials that have very few free electrons and require a large applied potential (voltage) to establish a measurable current level.

A common use of insulating material is for covering current-carrying wire, which, if uninsulated, could cause dangerous side effects. Power line workers wear rubber gloves and stand on rubber mats as safety measures when working on high-voltage transmission lines. A few different types of insulators and their applications appear in Fig. 2.26.

Be aware, however, that even the best insulator will break down (permit charge to flow through it) if a sufficiently large potential is applied across it. The breakdown strengths of some common insulators are listed

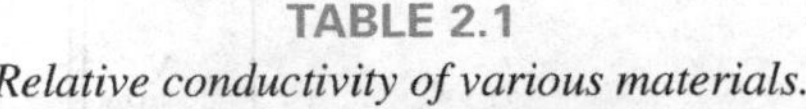

TABLE 2.1
Relative conductivity of various materials.

Metal	Relative Conductivity (%)
Silver	105
Copper	**100**
Gold	70.5
Aluminum	61
Tungsten	31.2
Nickel	22.1
Iron	14
Constantan	3.52
Nichrome	1.73
Calorite	1.44

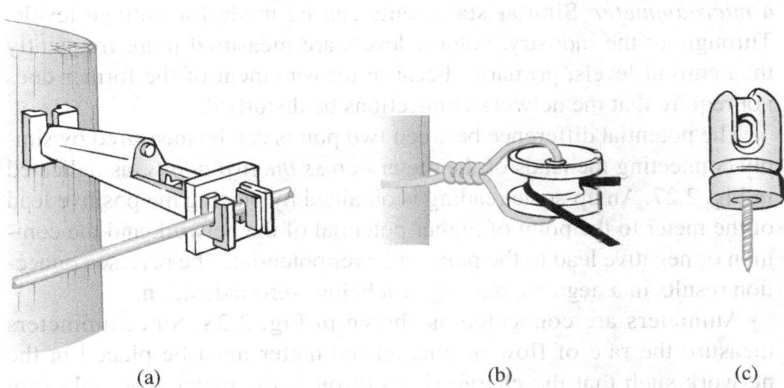

FIG. 2.26
Various types of insulators and their applications. (a) Fi-Shock extender insulator; (b) Fi-Shock corner insulator; (c) Fi-Shock screw-in post insulator.

TABLE 2.2
Breakdown strength of some common insulators.

Material	Average Breakdown Strength (kV/cm)
Air	30
Porcelain	70
Oils	140
Bakelite®	150
Rubber	270
Paper (paraffin-coated)	500
Teflon®	600
Glass	900
Mica	2000

in Table 2.2. According to this table, for insulators with the same geometric shape, it would require 270/30 = 9 times as much potential to pass current through rubber as through air and approximately 67 times as much voltage to pass current through mica as through air.

2.9 SEMICONDUCTORS

Semiconductors are a specific group of elements that exhibit characteristics between those of insulators and those of conductors.

The prefix *semi*, included in the terminology, has the dictionary definition of *half, partial*, or *between*, as defined by its use. The entire electronics industry is dependent on this class of materials since the electronic devices and integrated circuits (ICs) are constructed of semiconductor materials. Although *silicon* (Si) is the most extensively employed material, *germanium* (Ge) and *gallium arsenide* (GaAs) are also used in many important devices.

Semiconductor materials typically have four electrons in the outermost valence ring.

Semiconductors are further characterized as being photoconductive and having a negative temperature coefficient. Photoconductivity is a phenomenon in which the photons (small packages of energy) from incident light can increase the carrier density in the material and thereby the charge flow level. A negative temperature coefficient indicates that the resistance (a characteristic to be described in detail in the next chapter) decreases with an increase in temperature (opposite to that of most conductors). A great deal more will be said about semiconductors in the chapters to follow and in your basic electronics courses.

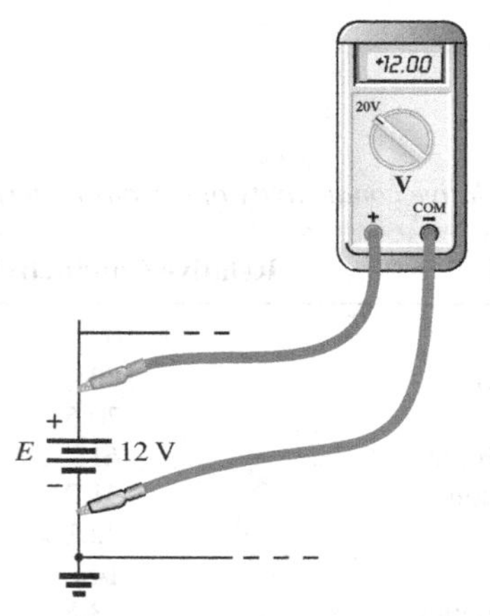

FIG. 2.27
Voltmeter connection for an up-scale (+) reading.

2.10 AMMETERS AND VOLTMETERS

It is important to be able to measure the current and voltage levels of an operating electrical system to check its operation, isolate malfunctions, and investigate effects impossible to predict on paper. As the names imply, **ammeters** are used to measure current levels; **voltmeters,** the potential difference between two points. If the current levels are usually of the order of milliamperes, the instrument will typically be referred to as a *milliammeter*, and if the current levels are in the microampere range, as a *microammeter*. Similar statements can be made for voltage levels. Throughout the industry, voltage levels are measured more frequently than current levels, primarily because measurement of the former does not require that the network connections be disturbed.

The potential difference between two points can be measured by simply connecting the leads of the meter *across the two points,* as indicated in Fig. 2.27. An up-scale reading is obtained by placing the positive lead of the meter to the point of higher potential of the network and the common or negative lead to the point of lower potential. The reverse connection results in a negative reading or a below-zero indication.

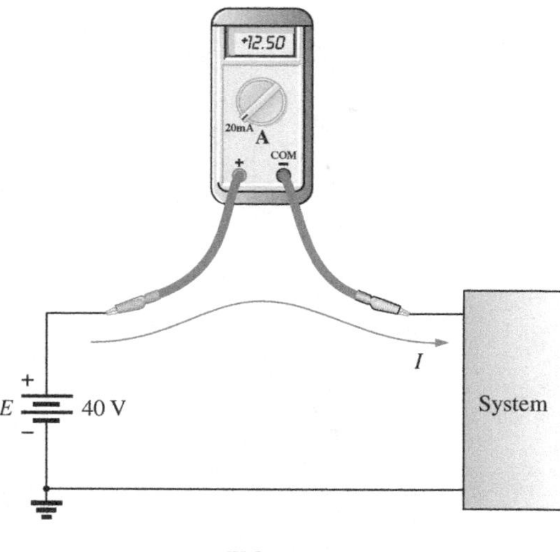

FIG. 2.28
Ammeter connection for an up-scale (+) reading.

Ammeters are connected as shown in Fig. 2.28. Since ammeters measure the rate of flow of charge, the meter must be placed in the network such that the charge flows through the meter. The only way this can be accomplished is to open the path in which the current is to be measured and place the meter between the two resulting terminals. For the configuration in Fig. 2.28, the voltage source lead (+) must be

disconnected from the system and the ammeter inserted as shown. An upscale reading will be obtained if the polarities on the terminals of the ammeter are such that the current of the system enters the positive terminal.

The introduction of any meter into an electrical/electronic system raises a concern about whether the meter will affect the behavior of the system. This question and others will be examined in Chapters 5 and 6 after additional terms and concepts have been introduced. For the moment, let it be said that since voltmeters and ammeters have internal components, they will affect the network when introduced for measurement purposes. The design of each, however, is such that the impact is minimized.

There are instruments designed to measure just current or just voltage levels. However, the most common laboratory meters include the *volt-ohm-milliammeter* (VOM) and the *digital multimeter* (DMM), shown in Figs. 2.29 and 2.30, respectively. Both instruments measure voltage and current and a third quantity, resistance (introduced in the next chapter). The VOM uses an analog scale, which requires interpreting the position of a pointer on a continuous scale, while the DMM provides a display of numbers with decimal-point accuracy determined by the chosen scale. Comments on the characteristics and use of various meters will be made throughout the text. However, the major study of meters will be left for the laboratory sessions.

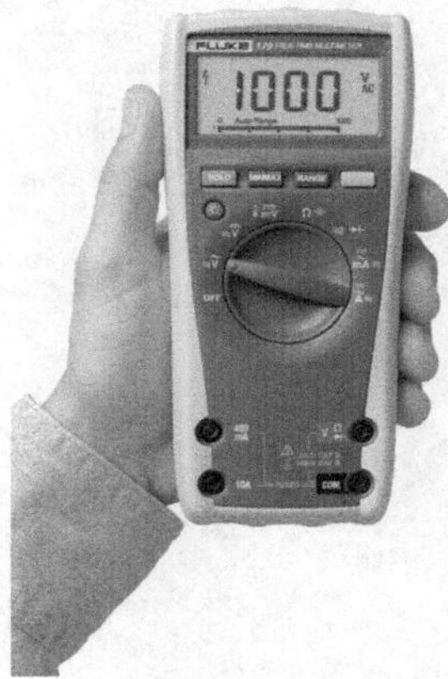

FIG. 2.30
Digital multimeter (DMM).
(Courtesy of Fluke Corporation. Reproduced with permission.)

2.11 APPLICATIONS

Throughout the text, Applications sections such as this one have been included to permit a further investigation of terms, quantities, or systems introduced in the chapter. The primary purpose of these Applications is to establish a link between the theoretical concepts of the text and the real, practical world. Although the majority of components that appear in a system may not have been introduced (and, in fact, some components will not be examined until more advanced studies), the topics were chosen very carefully and should be quite interesting to a new student of the subject matter. Sufficient comment is included to provide a surface understanding of the role of each part of the system with the understanding that the details will come at a later date. Since exercises on the subject matter of the Applications do not appear at the end of the chapter, the content is designed not to challenge the student but rather to stimulate his or her interest and answer some basic questions such as how the system looks inside, what role specific elements play in the system, and, of course, how the system works. In essence, therefore, each Applications section provides an opportunity to begin to establish a practical background beyond simply the content of the chapter. Do not be concerned if you do not understand every detail of each application. Understanding will come with time and experience. For now, take what you can from the examples and then proceed with the material.

Flashlight

Although the flashlight uses one of the simplest of electrical circuits, a few fundamentals about its operation do carry over to more sophisticated systems. First, and quite obviously, it is a dc system with a lifetime totally dependent on the state of the batteries and bulb. Unless it is the rechargeable type, each time you use it, you take some of the life out of

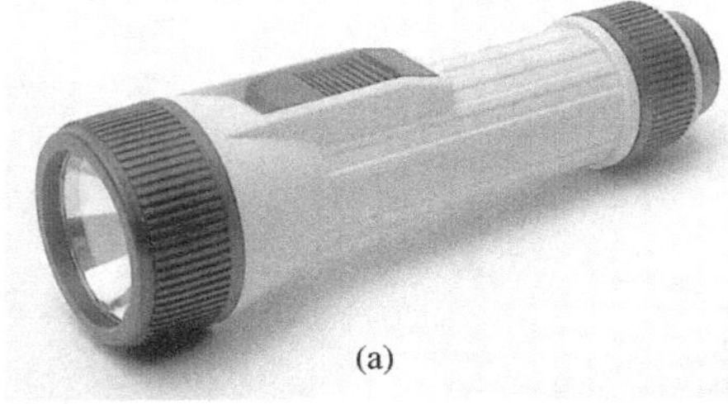

(a)

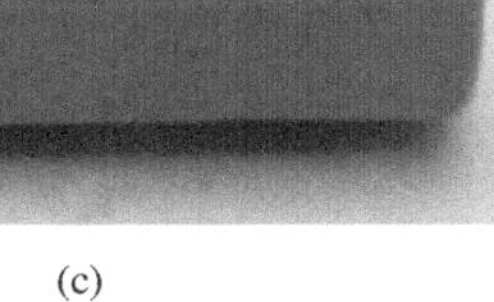

(c)

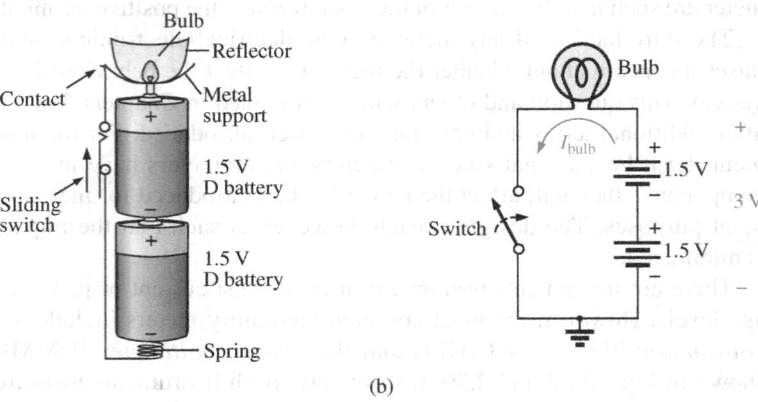

(b)

FIG. 2.31
(a) D cell flashlight; (b) electrical schematic of flashlight of part (a); (c) D cell battery.

it. For many hours, the brightness will not diminish noticeably. Then, however, as it reaches the end of its ampere-hour capacity, the light becomes dimmer at an increasingly rapid rate (almost exponentially). The standard two-battery flashlight is shown in Fig. 2.31(a) with its electrical schematic in Fig. 2.31(b). Each 1.5 V battery has an ampere-hour rating of about 18 as indicated in Fig. 2.12. The single-contact miniature flange-base bulb is rated at 2.5 V and 300 mA with good brightness and a lifetime of about 30 hours. Thirty hours may not seem like a long lifetime, but you have to consider how long you usually use a flashlight on each occasion. If we assume a 300 mA drain from the battery for the bulb when in use, the lifetime of the battery, by Eq. (2.8), is about 60 hours. Comparing the 60 hour lifetime of the battery to the 30 hour life expectancy of the bulb suggests that we normally have to replace bulbs more frequently than batteries.

However, most of us have experienced the opposite effect. We can change batteries two or three times before we need to replace the bulb. This is simply one example of the fact that one cannot be guided solely by the specifications of each component of an electrical design. The operating conditions, terminal characteristics, and details about the actual response of the system for short and long periods of time must be considered. As mentioned earlier, the battery loses some of its power each time it is used. Although the terminal voltage may not change much at first, its ability to provide the same level of current drops with each usage. Further, batteries slowly discharge due to "leakage currents" even if the switch is not on. The air surrounding the battery is not "clean" in the sense that moisture and other elements in the air can provide a conduction path for leakage currents through the air through the surface of the battery itself, or through other nearby surfaces, and the battery eventually discharges. How often have we left a flashlight with new batteries in a car for a long period of time only to find the light very dim or the batteries dead when we need the flashlight the most? An additional problem is acid leaks that appear as brown stains or corrosion on the casing of the battery. These leaks also affect the life of the battery. Further, when the flashlight is turned on, there is an initial surge in current that drains the battery more than continuous use for a period of time. In other words, continually turning the flashlight on and off has a very detrimental effect on its life. We must also realize that the 30 hour rating of the bulb is for continuous use, that is, 300 mA flowing through the bulb for a continuous 30 hours. Certainly, the filament in the bulb and the bulb itself will get hotter with time, and this heat has a detrimental effect on the filament wire. When the flashlight is turned on and off, it gives the bulb a chance to cool down and regain its normal characteristics, thereby avoiding any real damage. Therefore, with normal use we can expect the bulb to last longer than the 30 hours specified for continuous use.

Even though the bulb is rated for 2.5 V operation, it would appear that the two batteries would result in an applied voltage of 3 V, which suggests poor operating conditions. However, a bulb rated at 2.5 V can easily handle 2.5 V to 3 V. In addition, as was pointed out in this chapter, the terminal voltage drops with the current demand and usage. Under normal operating conditions, a 1.5 V battery is considered to be in good condition if the loaded terminal voltage is 1.3 V to 1.5 V. When it drops to the range from 1 V to 1.1 V, it is weak, and when it drops to the range from 0.8 V to 0.9 V, it has lost its effectiveness. The levels can be related directly to the test band now appearing on Duracell® batteries, such as on the one shown in Fig. 2.31(c). In the test band on this battery, the upper voltage area (green) is near 1.5 V (labeled 100%); the lighter area to the right, from about 1.3 V down to 1 V; and the replace area (red) on the far right, below 1 V.

Be aware that the total supplied voltage of 3 V will be obtained only if the batteries are connected as shown in Fig. 2.31(b). Accidentally placing the two positive terminals together will result in a total voltage of 0 V, and the bulb will not light at all. *For the vast majority of systems with more than one battery, the positive terminal of one battery will always be connected to the negative terminal of another. For all low-voltage batteries, the end with the nipple is the positive terminal, and the end with the flat end is the negative terminal. In addition, the flat or negative end of a battery is always connected to the battery casing with the helical coil to keep the batteries in place. The positive end of the battery is always connected to a flat spring connection or the element to be operated.* If you look carefully at the bulb, you will find that the nipple connected to the positive end of the battery is insulated from the jacket around the base of the bulb. The jacket is the second terminal of the battery used to complete the circuit through the on/off switch.

If a flashlight fails to operate properly, the first thing to check is the state of the batteries. It is best to replace both batteries at once. A system with one good battery and one nearing the end of its life will result in pressure on the good battery to supply the current demand, and, in fact, the bad battery will actually be a drain on the good battery. Next, check the condition of the bulb by checking the filament to see whether it has opened at some point because a long-term, continuous current level occurred or because the flashlight was dropped. If the battery and bulb seem to be in good shape, the next area of concern is the contacts between the positive terminal and the bulb and the switch. Cleaning both with emery cloth often eliminates this problem.

12 V Car Battery Charger

Battery chargers are a common household piece of equipment used to charge everything from small flashlight batteries to heavy-duty, marine, lead-acid batteries. Since all are plugged into a 120 V ac outlet

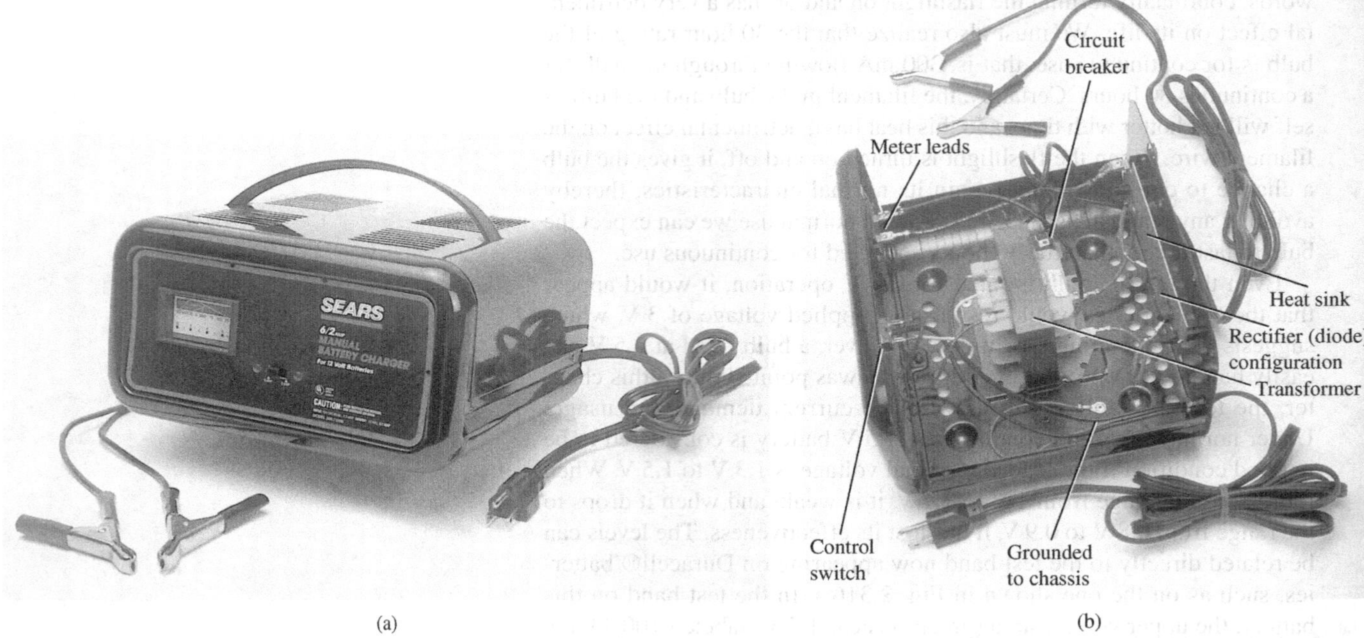

FIG. 2.32
Battery charger: (a) external appearance; (b) internal construction.

such as found in the home, the basic construction of each is quite similar. In every charging system, a *transformer* (Chapter 22) must be included to cut the ac voltage to a level appropriate for the dc level to be established. A *diode* (also called *rectifier*) arrangement must be included to convert the ac voltage, which varies with time, to an average dc level such as described in this chapter. Diodes and/or rectifiers will be discussed in detail in your first electronics course. Some dc chargers also include a *regulator* to provide an improved dc level (one that varies less with time or load). The car battery charger, one of the most common, is described here.

The outside appearance and the internal construction of a Sears 6/2 AMP Manual Battery Charger are provided in Fig. 2.32. Note in Fig. 2.32(b) that the transformer (as in most chargers) takes up most of the internal space. The additional air space and the holes in the casing are there to ensure an outlet for the heat that will develop due to the resulting current levels.

The schematic in Fig. 2.33 includes all the basic components of the charger. Note first that the 120 V from the outlet are applied directly across the primary of the transformer. The charging rate of 6 A or 2 A is determined by the switch, which simply controls how many turns of the primary will be in the circuit for the chosen charging rate. If the battery is charging at the 2 A level, the full primary will be in the circuit, and the ratio of the turns in the primary to the turns in the secondary will be a maximum. If it is charging at the 6 A level, fewer turns of the primary are in the circuit, and the ratio drops. When you study transformers, you will find that the voltage at the primary and secondary is directly related to the *turns ratio.* If the ratio from primary to secondary drops, the voltage drops also. The reverse effect occurs if the turns on the secondary exceed those on the primary.

The general appearance of the waveforms appears in Fig. 2.33 for the 6 A charging level. Note that so far, the ac voltage has the same wave

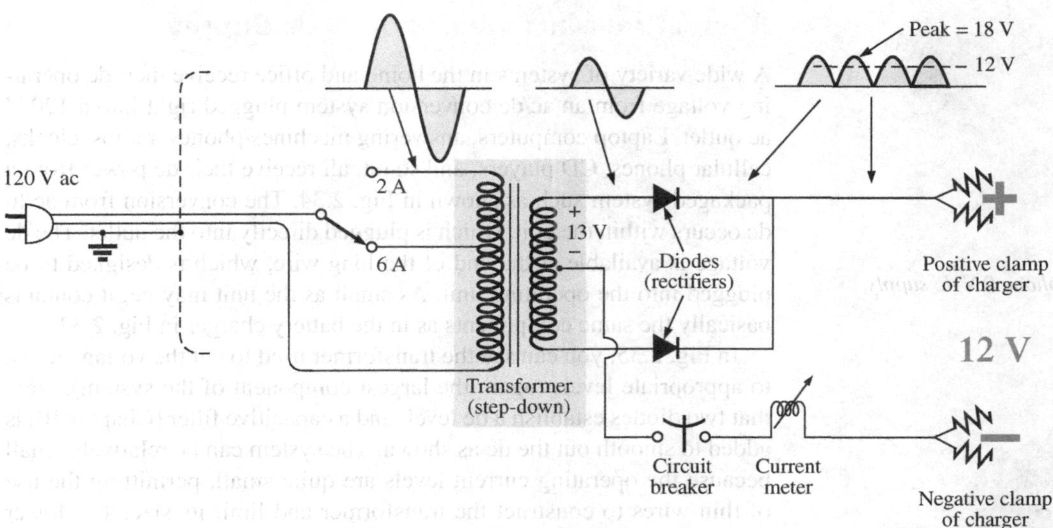

FIG. 2.33
Electrical schematic for the battery charger of Fig. 2.32.

shape across the primary and secondary. The only difference is in the peak value of the waveforms. Now the diodes take over and convert the ac waveform, which has zero average value (the waveform above equals the waveform below), to one that has an average value (all above the axis) as shown in the same figure. For the moment simply recognize that diodes are semiconductor electronic devices that permit only conventional current to flow through them in the direction indicated by the arrow in the symbol. Even though the waveform resulting from the diode action has a pulsing appearance with a peak value of about 18 V, it charges the 12 V battery whenever its voltage is greater than that of the battery, as shown by the shaded area. Below the 12 V level, the battery cannot discharge back into the charging network because the diodes permit current flow in only one direction.

In particular, note in Fig. 2.32(b) the large plate that carries the current from the rectifier (diode) configuration to the positive terminal of the battery. Its primary purpose is to provide a *heat sink* (a place for the heat to be distributed to the surrounding air) for the diode configuration. Otherwise, the diodes would eventually melt down and self-destruct due to the resulting current levels. Each component of Fig. 2.33 has been carefully labeled in Fig. 2.32(b) for reference.

When current is first applied to a battery at the 6 A charge rate, the current demand as indicated by the meter on the face of the instrument may rise to 7 A or almost 8 A. However, the level of current decreases as the battery charges until it drops to a level of 2 A or 3 A. For units such as this that do not have an automatic shutoff, it is important to disconnect the charger when the current drops to the fully charged level; otherwise, the battery becomes overcharged and may be damaged. A battery that is at its 50% level can take as long as 10 hours to charge, so don't expect it to be a 10-minute operation. In addition, if a battery is in very bad shape with a lower-than-normal voltage, the initial charging current may be too high for the design. To protect against such situations, the circuit breaker opens and stops the charging process. Because of the high current levels, it is important that the directions provided with the charger be carefully read and applied.

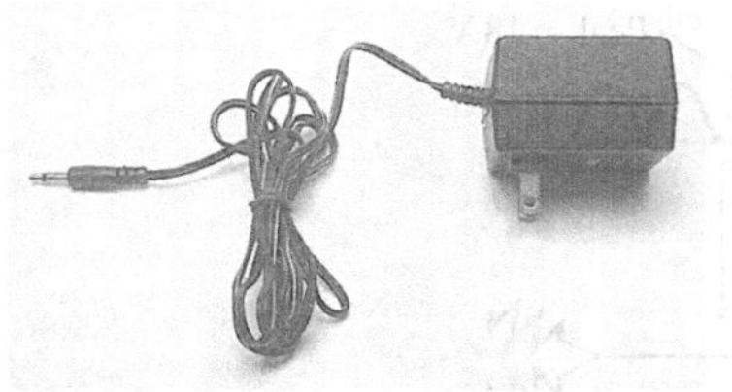

FIG. 2.34
Answering machine/phone 9 V dc supply.

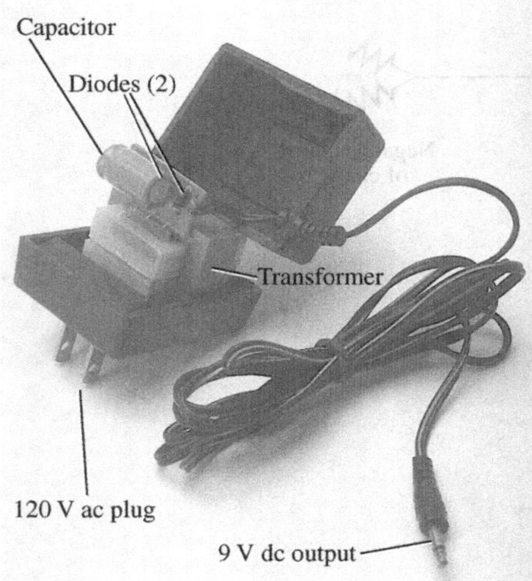

FIG. 2.35
Internal construction of the 9 V dc supply in Fig. 2.34.

Answering Machines/Phones dc Supply

A wide variety of systems in the home and office receive their dc operating voltage from an ac/dc conversion system plugged right into a 120 V ac outlet. Laptop computers, answering machines/phones, radios, clocks, cellular phones, CD players, and so on, all receive their dc power from a packaged system such as shown in Fig. 2.34. The conversion from ac to dc occurs within the unit, which is plugged directly into the outlet. The dc voltage is available at the end of the long wire, which is designed to be plugged into the operating unit. As small as the unit may be, it contains basically the same components as in the battery charger in Fig. 2.32.

In Fig. 2.35, you can see the transformer used to cut the voltage down to appropriate levels (again the largest component of the system). Note that two diodes establish a dc level, and a capacitive filter (Chapter 10) is added to smooth out the dc as shown. The system can be relatively small because the operating current levels are quite small, permitting the use of thin wires to construct the transformer and limit its size. The lower currents also reduce the concerns about heating effects, permitting a small housing structure. The unit in Fig. 2.35, rated at 9 V at 200 mA, is commonly used to provide power to answering machines/phones. Further smoothing of the dc voltage is accomplished by a regulator built into the receiving unit. The regulator is normally a small IC chip placed in the receiving unit to separate the heat that it generates from the heat generated by the transformer, thereby reducing the net heat at the outlet close to the wall. In addition, its placement in the receiving unit reduces the possibility of picking up noise and oscillations along the long wire from the conversion unit to the operating unit, and it ensures that the full rated voltage is available at the unit itself, not a lesser value due to losses along the line.

2.12 COMPUTER ANALYSIS

In some texts, the procedure for choosing a dc voltage source and placing it on the schematic using computer methods is introduced at this point. This approach, however, requires students to turn back to this chapter for the procedure when the first complete network is installed and examined. Therefore, the procedure is introduced in Chapter 4 when the first complete network is examined, thereby localizing the material and removing the need to reread this chapter and Chapter 3.

PROBLEMS

SECTION 2.2 Atoms and Their Structure

1. a. The numbers of orbiting electrons in aluminum and silver are 13 and 47, respectively. Draw the electronic configuration for each, and discuss briefly why each is a good conductor.
   b. Using the Internet, find the atomic structure of gold and explain why it is an excellent conductor of electricity.

2. Find the force of attraction in newtons between the charges Q_1 and Q_2 in Fig. 2.36 when
   a. $r = 1$ m
   b. $r = 3$ m
   c. $r = 10$ m
   d. Did the force drop off quickly with an increase in distance?

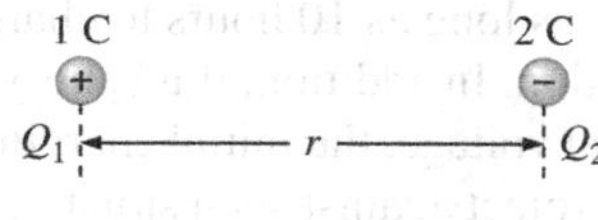

FIG. 2.36
Problem 2.

*3. Find the force of repulsion in newtons between Q_1 and Q_2 in Fig. 2.37 when
   a. $r = 1$ mi
   b. $r = 10$ ft
   c. $r = 1/16$ in.

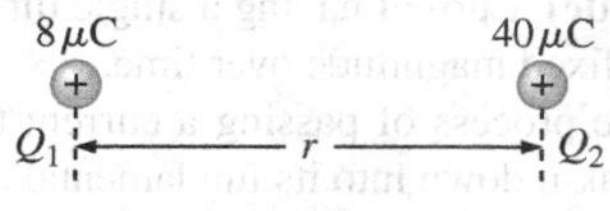

FIG. 2.37
Problem 3.

*4. a. Plot the force of attraction (in newtons) versus separation (in inches) between two unlike charges of 2 μC. Use a range of 1 in. to 10 in. in 1 in. increments. Comment on the shape of the curve. Is it linear or nonlinear? What does it tell you about plotting a function whose magnitude is affected by a squared term in the denominator of the equation?
   b. Using the plot of part (a), find the force of attraction at a 2.5 in. separation.
   c. Calculate the force of attraction with a 2.5 in. separation and compare with the result of part (b).

*5. For two similar charges the force F_1 exists for a separation of r_1 meters. If the distance is increased to r_2, find the new level of force F_2 in terms of the original force and the distance involved.

*6. Determine the distance between two charges of 20 μC if the force between the two charges is 3.6×10^4 N.

*7. Two charged bodies Q_1 and Q_2, when separated by a distance of 2 m, experience a force of repulsion equal to 1.8 N.
   a. What will the force of repulsion be when they are 10 m apart?
   b. If the ratio $Q_1/Q_2 = 1/2$, find Q_1 and Q_2 ($r = 10$ m).

SECTION 2.3 Voltage

8. What is the voltage between two points if 1.2 J of energy are required to move 20 μC between the two points?

9. If the potential difference between two points is 60 V, how much energy is expended to bring 8 mC from one point to the other?

10. Find the charge in electrons that requires 120 μJ of energy to be moved through a potential difference of 20 mV.

11. How much charge passes through a radio battery of 9 V if the energy expended is 72 J?

*12. a. How much energy in electron-volts is required to move 1 trillion (1 million million) electrons through a potential difference of 40 V?
   b. How many joules of energy does the result of part (a) represent?
   c. Compare results (a) and (b). What can you say about the use of joules and electron-volts as a unit of measure. Under what conditions should they be applied?

SECTION 2.4 Current

13. Find the current in amperes if 12 mC of charge pass through a wire in 2.8 s.

14. If 312 C of charge pass through a wire in 2 min, find the current in amperes.

15. If a current of 40 mA exists for 0.8 min, how many coulombs of charge have passed through the wire?

16. How many coulombs of charge pass through a lamp in 1.2 min if the current is constant at 250 mA?

17. If the current in a conductor is constant at 2 mA, how much time is required for 6 mC to pass through the conductor?

18. If $21.847 \times 10^{+18}$ electrons pass through a wire in 12 s, find the current.

19. How many electrons pass through a conductor in 1 min and 30 s if the current is 4 mA?

20. Will a fuse rated at 1 A "blow" if 86 C pass through it in 1.2 min?

*21. If $0.84 \times 10^{+16}$ electrons pass through a wire in 60 ms, find the current.

*22. Which would you prefer?
   a. A penny for every electron that passes through a wire in 0.01 μs at a current of 2 mA, or
   b. A dollar for every electron that passes through a wire in 1.5 ns if the current is 100 μA.

*23. If a conductor with a current of 200 mA passing through it converts 40 J of electrical energy into heat in 30 s, what is the potential drop across the conductor?

*24. Charge is flowing through a conductor at the rate of 420 C/min. If 742 J of electrical energy are converted to heat in 30 s, what is the potential drop across the conductor?

*25. The potential difference between two points in an electric circuit is 24 V. If 0.4 J of energy were dissipated in a period of 5 ms, what would the current be between the two points?

SECTION 2.6 Ampere-Hour Rating

26. What current will a battery with an Ah rating of 200 theoretically provide for 40 h?

27. What is the Ah rating of a battery that can provide 0.8 A for 75 h?

28. For how many hours will a battery with an Ah rating of 32 theoretically provide a current of 1.28 A?

29. A standard 12 V car battery has an ampere-hour rating of 40 Ah, whereas a heavy-duty battery has a rating of 60 Ah. How would you compare the energy levels of each and the available current for starting purposes?

30. At what current drain does the ampere-hour rating of the Energizer D Cell of Fig. 2.23 drop to 75% of its value at 25 mA?

31. What is the percentage loss in ampere-hour rating from room temperature to freezing for the Energizer D Cell of Fig. 2.24?

32. Using the graph of Fig. 2.25, how much longer can you maintain 1.2 V at a discharge rate of 25 mA compared to discharging at 100 mA?

*33. A portable television using a 12 V, 3 Ah rechargeable battery can operate for a period of about 6 h. What is the average current drawn during this period? What is the energy expended by the battery in joules?

SECTION 2.8 Conductors and Insulators

34. Discuss two properties of the atomic structure of copper that make it a good conductor.

35. Explain the terms *insulator* and *breakdown strength.*

36. List three uses of insulators not mentioned in Section 2.8.

37. **a.** Using Table 2.2, determine the level of applied voltage necessary to establish conduction through 1/2 inch of air.
b. Repeat part (a) for 1/2 in. of rubber.
c. Compare the results of parts (a) and (b).

SECTION 2.9 Semiconductors

38. What is a semiconductor? How does it compare with a conductor and an insulator?

39. Consult a semiconductor electronics text and note the extensive use of germanium and silicon semiconductor materials. Review the characteristics of each material.

SECTION 2.10 Ammeters and Voltmeters

40. What are the significant differences in the way ammeters and voltmeters are connected?

41. Compare analog and digital scales:
a. Which are you more comfortable with? Why?
b. Which can usually provide a higher degree of accuracy?
c. Can you think of any advantages of the analog scale over a digital scale? Be aware that the majority of scales in a plane's cockpit or in the control room of major power plants are analog.
d. Do you believe it is necessary to become proficient in reading analog scales? Why?

GLOSSARY

Ammeter An instrument designed to read the current through elements in series with the meter.

Ampere (A) The SI unit of measurement applied to the flow of charge through a conductor.

Ampere-hour (Ah) rating The rating applied to a source of energy that will reveal how long a particular level of current can be drawn from that source.

Cell A fundamental source of electrical energy developed through the conversion of chemical or solar energy.

Conductors Materials that permit a generous flow of electrons with very little voltage applied.

Copper A material possessing physical properties that make it particularly useful as a conductor of electricity.

Coulomb (C) The fundamental SI unit of measure for charge. It is equal to the charge carried by 6.242×10^{18} electrons.

Coulomb's law An equation defining the force of attraction or repulsion between two charges.

Current The flow of charge resulting from the application of a difference in potential between two points in an electrical system.

dc current source A source that will provide a fixed current level even though the load to which it is applied may cause its terminal voltage to change.

dc generator A source of dc voltage available through the turning of the shaft of the device by some external means.

Direct current (dc) Current having a single direction (unidirectional) and a fixed magnitude over time.

Electrolysis The process of passing a current through an electrolyte to break it down into its fundamental components.

Electrolytes The contact element and the source of ions between the electrodes of the battery.

Electron The particle with negative polarity that orbits the nucleus of an atom.

Free electron An electron unassociated with any particular atom, relatively free to move through a crystal lattice structure under the influence of external forces.

Fuel cell A nonpolluting source of energy that can establish current through a load by simply applying the correct levels of hydrogen and oxygen.

Insulators Materials in which a very high voltage must be applied to produce any measurable current flow.

Neutron The particle having no electrical charge found in the nucleus of the atom.

Nucleus The structural center of an atom that contains both protons and neutrons.

Positive ion An atom having a net positive charge due to the loss of one of its negatively charged electrons.

Potential difference The algebraic difference in potential (or voltage) between two points in an electrical system.

Potential energy The energy that a mass possesses by virtue of its position.

Primary cell Sources of voltage that cannot be recharged.

Proton The particle of positive polarity found in the nucleus of an atom.

Rectification The process by which an ac signal is converted to one that has an average dc level.

Secondary cell Sources of voltage that can be recharged.

Semiconductor A material having a conductance value between that of an insulator and that of a conductor. Of significant importance in the manufacture of electronic devices.

Solar cell Sources of voltage available through the conversion of light energy (photons) into electrical energy.

Specific gravity The ratio of the weight of a given volume of a substance to the weight of an equal volume of water at 4°C.

Volt (V) The unit of measurement applied to the difference in potential between two points. If 1 joule of energy is required to move 1 coulomb of charge between two points, the difference in potential is said to be 1 volt.

Voltage The term applied to the difference in potential between two points as established by a separation of opposite charges.

Voltmeter An instrument designed to read the voltage across an element or between any two points in a network.

RESISTANCE

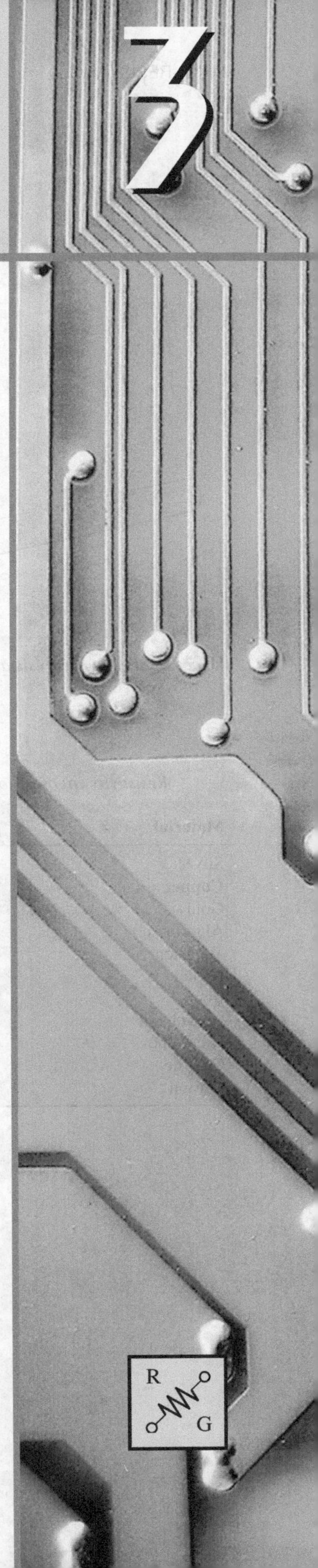

Objectives

- *Become familiar with the parameters that determine the resistance of an element and be able to calculate the resistance from the given dimensions and material characteristics.*
- *Understand the effects of temperature on the resistance of a material and how to calculate the change in resistance with temperature.*
- *Develop some understanding of superconductors and how they can affect future development in the industry.*
- *Become familiar with the broad range of commercially available resistors available today and how to read the value of each from the color code or labeling.*
- *Become aware of a variety of elements such as thermistors, photoconductive cells, and varistors and how their terminal resistance is controlled.*

3.1 INTRODUCTION

In the previous chapter, we found that placing a voltage across a wire or simple circuit results in a flow of charge or current through the wire or circuit. The question remains, however, What determines the level of current that results when a particular voltage is applied? Why is the current heavier in some circuits than in others? The answers lie in the fact that there is an opposition to the flow of charge in the system that depends on the components of the circuit. This opposition to the flow of charge through an electrical circuit, called **resistance,** has the units of **ohms** and uses the Greek letter *omega* (Ω) as its symbol. The graphic symbol for resistance, which resembles the cutting edge of a saw, is provided in Fig. 3.1.

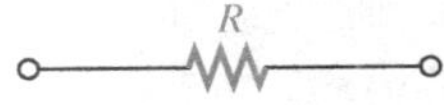

FIG. 3.1
Resistance symbol and notation.

This opposition, due primarily to collisions and friction between the free electrons and other electrons, ions, and atoms in the path of motion, converts the supplied electrical energy into **heat** that raises the temperature of the electrical component and surrounding medium. The heat you feel from an electrical heater is simply due to current passing through a high-resistance material.

Each material with its unique atomic structure reacts differently to pressures to establish current through its core. Conductors that permit a generous flow of charge with little external pressure have low resistance levels, while insulators have high resistance characteristics.

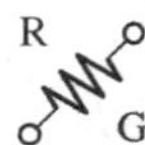

3.2 RESISTANCE: CIRCULAR WIRES

The resistance of any material is due primarily to four factors:

1. *Material*
2. *Length*
3. *Cross-sectional area*
4. *Temperature of the material*

As noted in Section 3.1, the atomic structure determines how easily a free electron will pass through a material. The longer the path through which the free electron must pass, the greater is the resistance factor. Free electrons pass more easily through conductors with larger cross-sectional areas. In addition, the higher the temperature of the conductive materials, the greater is the internal vibration and motion of the components that make up the atomic structure of the wire, and the more difficult it is for the free electrons to find a path through the material.

The first three elements are related by the following basic equation for resistance:

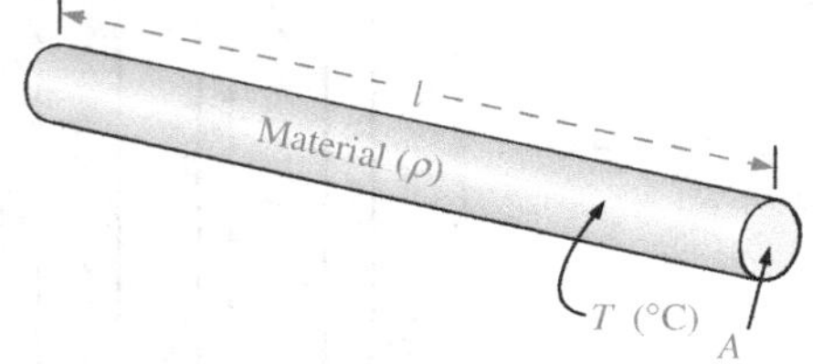

FIG. 3.2
Factors affecting the resistance of a conductor.

$$R = \rho \frac{l}{A} \qquad \begin{array}{l} \rho = \text{CM-}\Omega\text{/ft at } T = 20°\text{C} \\ l = \text{feet} \\ A = \text{area in circular mils (CM)} \end{array} \qquad \textbf{(3.1)}$$

with each component of the equation defined by Fig. 3.2.

The material is identified by a factor called the **resistivity,** which uses the Greek letter *rho* (ρ) as its symbol and is measured in CM-Ω/ft. Its value at a temperature of 20°C (room temperature = 68°F) is provided in Table 3.1 for a variety of common materials. Since the larger the resistivity, the greater is the resistance to setting up a flow of charge, it appears as a multiplying factor in Eq. (3.1); that is, it appears in the numerator of the equation. It is important to realize at this point that since the resistivity is provided at a particular temperature, *Eq. (3.1) is applicable only at room temperature.* The effect of higher and lower temperatures is considered in Section 3.4.

TABLE 3.1
Resistivity (ρ) of various materials.

Material	ρ (CM-Ω/ft)@20°C
Silver	9.9
Copper	**10.37**
Gold	14.7
Aluminum	17.0
Tungsten	33.0
Nickel	47.0
Iron	74.0
Constantan	295.0
Nichrome	600.0
Calorite	720.0
Carbon	21,000.0

Since the resistivity is in the numerator of Eq. (3.1),

the higher the resistivity, the greater is the resistance of a conductor

as shown for two conductors of the same length in Fig. 3.3(a).

Further,

the longer the conductor, the greater is the resistance

since the length also appears in the numerator of Eq. (3.1). Note Fig. 3.3(b).

Finally,

the greater the area of a conductor, the less is the resistance

because the area appears in the denominator of Eq. (3.1). Note Fig. 3.3(c).

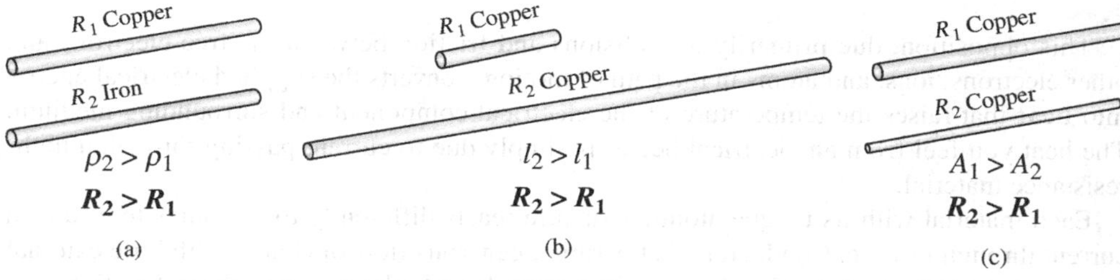

FIG. 3.3
Cases in which $R_2 > R_1$. For each case, all remaining parameters that control the resistance level are the same.

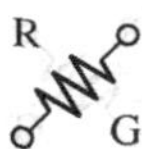

Circular Mils (CM)

In Eq. (3.1), the area is measured in a quantity called **circular mils** (CM). It is the quantity used in most commercial wire tables, and thus it needs to be carefully defined. The *mil* is a unit of measurement for length and is related to the inch by

$$\mathbf{1\ mil = \frac{1}{1000}\ in.}$$

or

$$\mathbf{1000\ mils = 1\ in.}$$

In general, therefore, the mil is a very small unit of measurement for length. There are 1000 mils in an inch, or 1 mil is only 1/1000 of an inch. It is a length that is not visible with the naked eye, although it can be measured with special instrumentation. The phrase *milling* used in steel factories is derived from the fact that a few mils of material are often removed by heavy machinery such as a lathe, and the thickness of steel is usually measured in mils.

By definition,

a wire with a diameter of 1 mil has an area of 1 CM.

as shown in Fig. 3.4.

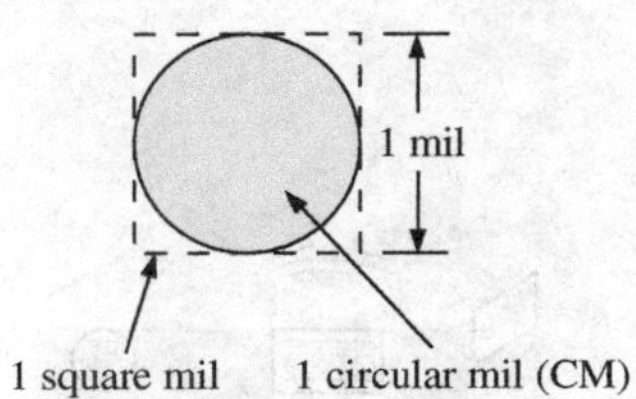

FIG. 3.4
Defining the circular mil (CM).

An interesting result of such a definition is that the area of a circular wire in circular mils can be defined by the following equation:

$$A_{CM} = (d_{mils})^2 \tag{3.2}$$

Verification of this equation appears in Fig. 3.5, which shows that a wire with a diameter of 2 mils has a total area of 4 CM, and a wire with a diameter of 3 mils has a total area of 9 CM.

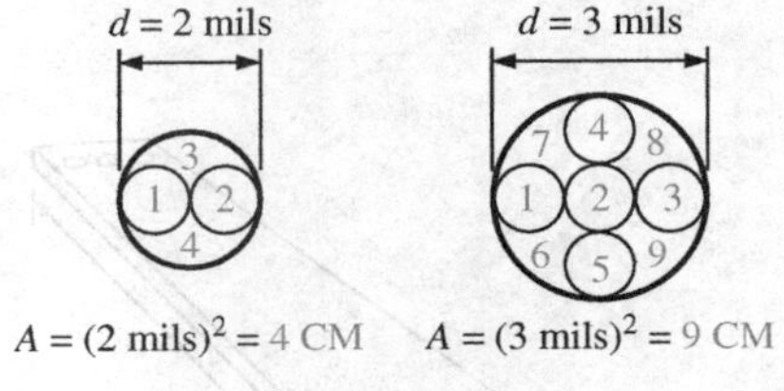

FIG. 3.5
Verification of Eq. (3.2): $A_{CM} = (d_{mils})^2$.

Remember, to compute the area of a wire in circular mils when the diameter is given in inches, first convert the diameter to mils by simply writing the diameter in decimal form and moving the decimal point three places to the right. For example,

$$\frac{1}{8}\text{ in.} = 0.125\text{ in.} = 125\text{ mils} \quad (\text{3 places})$$

Then the area is determined by

$$A_{CM} = (d_{mils})^2 = (125\text{ mils})^2 = \mathbf{15{,}625\ CM}$$

Sometimes when you are working with conductors that are not circular, you will need to convert square mils to circular mils, and vice versa. Applying the basic equation for the area of a circle and substituting a diameter of 1 mil results in

$$A = \frac{\pi d^2}{4} = \frac{\pi}{4}(1\text{ mil})^2 = \frac{\pi}{4}\text{ sq mils} \equiv 1\text{ CM} \quad (\text{by definition})$$

from which we can conclude the following:

$$1\text{ CM} = \frac{\pi}{4}\text{ sq mils} \tag{3.3}$$

or

$$1\text{ sq mil} = \frac{4}{\pi}\text{ CM} \tag{3.4}$$

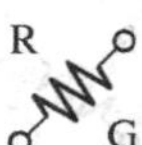

EXAMPLE 3.1 What is the resistance of a 100 ft length of copper wire with a diameter of 0.020 in. at 20°C?

Solution:

$$\rho = 10.37 \frac{\text{CM-}\Omega}{\text{ft}} \qquad 0.020 \text{ in.} = 20 \text{ mils}$$

$$A_{CM} = (d_{\text{mils}})^2 = (20 \text{ mils})^2 = 400 \text{ CM}$$

$$R = \rho \frac{l}{A} = \frac{(10.37 \text{ CM-}\Omega/\text{ft})(100 \text{ ft})}{400 \text{ CM}}$$

$$R = \mathbf{2.59\ \Omega}$$

EXAMPLE 3.2 An undetermined number of feet of wire have been used from the carton in Fig. 3.6. Find the length of the remaining copper wire if it has a diameter of 1/16 in. and a resistance of 0.5 Ω.

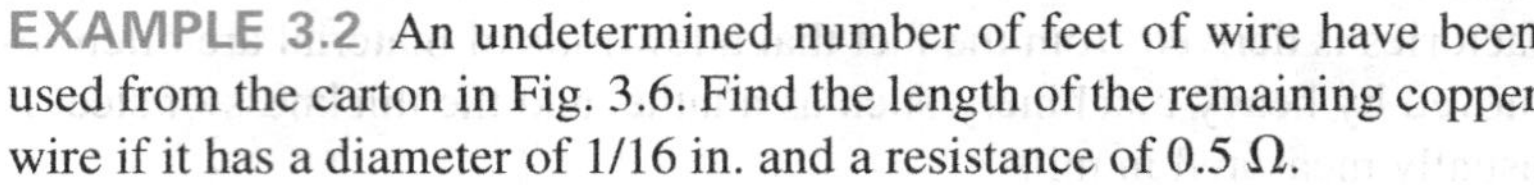

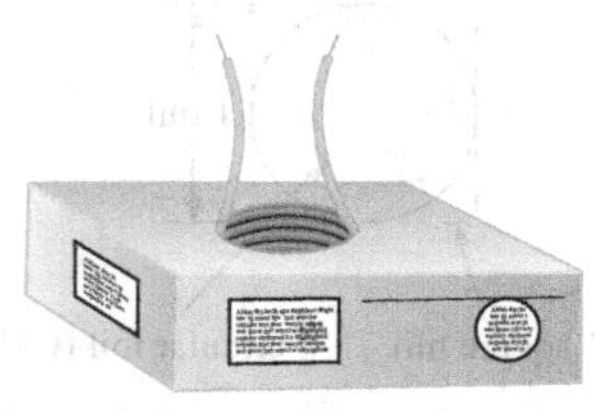

FIG. 3.6
Example 3.2.

Solution:

$$\rho = 10.37 \text{ CM-}\Omega/\text{ft} \qquad \frac{1}{16} \text{ in.} = 0.0625 \text{ in.} = 62.5 \text{ mils}$$

$$A_{CM} = (d_{\text{mils}})^2 = (62.5 \text{ mils})^2 = 3906.25 \text{ CM}$$

$$R = \rho \frac{l}{A} \Rightarrow l = \frac{RA}{\rho} = \frac{(0.5\ \Omega)(3906.25 \text{ CM})}{10.37 \dfrac{\text{CM-}\Omega}{\text{ft}}} = \frac{1953.125}{10.37}$$

$$l = \mathbf{188.34\ ft}$$

EXAMPLE 3.3 What is the resistance of a copper bus-bar, as used in the power distribution panel of a high-rise office building, with the dimensions indicated in Fig. 3.7?

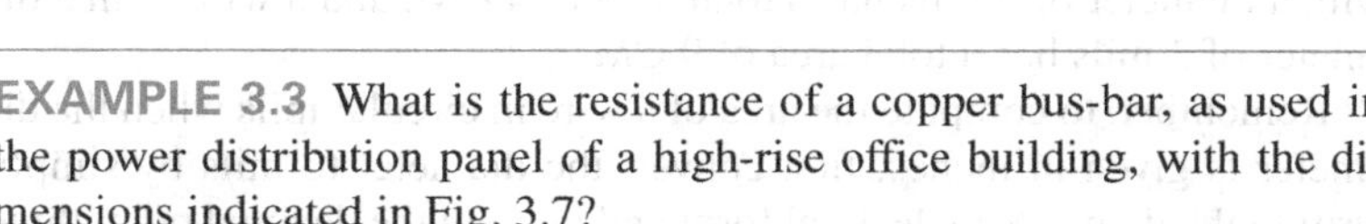

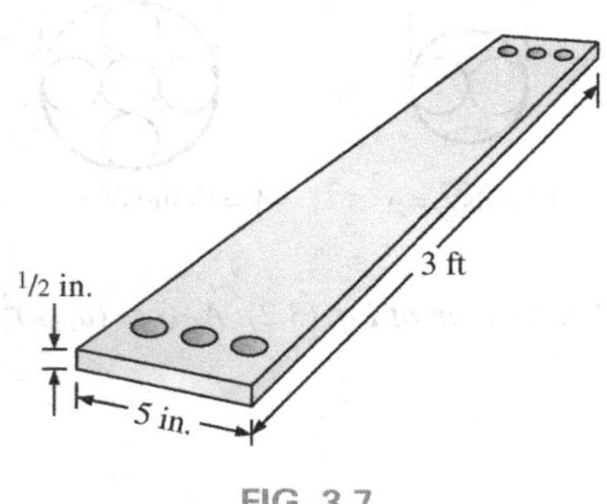

FIG. 3.7
Example 3.3.

Solution:

$$A_{CM} \begin{cases} 5.0 \text{ in.} = 5000 \text{ mils} \\ \dfrac{1}{2} \text{ in.} = 500 \text{ mils} \\ A = (5000 \text{ mils})(500 \text{ mils}) = 2.5 \times 10^6 \text{ sq mils} \\ \quad = 2.5 \times 10^6 \cancel{\text{sq mils}} \left(\dfrac{4/\pi \text{ CM}}{1 \cancel{\text{sq mil}}}\right) \\ A = 3.183 \times 10^6 \text{ CM} \end{cases}$$

$$R = \rho \frac{l}{A} = \frac{(10.37 \text{ CM-}\Omega/\text{ft})(3 \text{ ft})}{3.183 \times 10^6 \text{ CM}} = \frac{31.11}{3.183 \times 10^6}$$

$$R = \mathbf{9.774 \times 10^{-6}\ \Omega}$$

(quite small, 0.000009774 Ω ≅ 0 Ω)

You will learn in the following chapters that the lower the resistance of a conductor, the lower are the losses in conduction from the source to

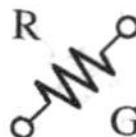

the load. Similarly, since resistivity is a major factor in determining the resistance of a conductor, the lower the resistivity, the lower is the resistance for the same-size conductor. It would appear from Table 3.1 that silver, copper, gold, and aluminum would be the best conductors and the most common. In general, there are other factors, however, such as **malleability** (ability of a material to be shaped), **ductility** (ability of a material to be drawn into long, thin wires), temperature sensitivity, resistance to abuse, and, of course, cost, that must all be weighed when choosing a conductor for a particular application.

In general, copper is the most widely used material because it is quite malleable, ductile, and available; has good thermal characteristics; and is less expensive than silver or gold. It is certainly not cheap, however. Contractors always ensure that the copper wiring has been removed before leveling a building because of its salvage value. Aluminum was once used for general wiring because it is cheaper than copper, but its thermal characteristics created some difficulties. The heating due to current flow and the cooling that occurred when the circuit was turned off resulted in expansion and contraction of the aluminum wire to the point where connections eventually loosened, resulting in dangerous side effects. Aluminum is still used today, however, in areas such as integrated circuit manufacturing and in situations where the connections can be made secure. Silver and gold are, of course, much more expensive than copper or aluminum, but the cost is justified for certain applications. Silver has excellent plating characteristics for surface preparations, and gold is used quite extensively in integrated circuits. Tungsten has a resistivity three times that of copper, but there are occasions when its physical characteristics (durability, hardness) are the overriding considerations.

3.3 WIRE TABLES

The wire table was designed primarily to standardize the size of wire produced by manufacturers. As a result, the manufacturer has a larger market, and the consumer knows that standard wire sizes will always be available. The table was designed to assist the user in every way possible; it usually includes data such as the cross-sectional area in circular mils, diameter in mils, ohms per 1000 feet at 20°C, and weight per 1000 feet.

The American Wire Gage (AWG) sizes are given in Table 3.2 for solid, round copper wire. A column indicating the maximum allowable current in amperes, as determined by the National Fire Protection Association, has also been included.

The chosen sizes have an interesting relationship:

The area is doubled for every drop in 3 gage numbers and increased by a factor of 10 for every drop of 10 gage numbers.

Examining Eq. (3.1), we note also that *doubling the area cuts the resistance in half, and increasing the area by a factor of 10 decreases the resistance of 1/10 the original,* everything else kept constant.

The actual sizes of some of the gage wires listed in Table 3.2 are shown in Fig. 3.8 with a few of their areas of application. A few examples using Table 3.2 follow.

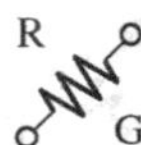

TABLE 3.2
American Wire Gage (AWG) sizes.

AWG #		Area (CM)	Ω/1000 ft at 20°C	Maximum Allowable Current for RHW Insulation (A)*
(4/0)	**0000**	211,600	0.0490	**230**
(3/0)	**000**	167,810	0.0618	**200**
(2/0)	**00**	133,080	0.0780	**175**
(1/0)	**0**	105,530	0.0983	**150**
	1	83,694	0.1240	**130**
	2	66,373	0.1563	**115**
	3	52,634	0.1970	**100**
	4	41,742	0.2485	**85**
	5	33,102	0.3133	—
	6	26,250	0.3951	**65**
	7	20,816	0.4982	—
	8	16,509	0.6282	**50**
	9	13,094	0.7921	—
	10	10,381	0.9989	**30**
	11	8,234.0	1.260	—
	12	6,529.9	1.588	**20**
	13	5,178.4	2.003	—
	14	4,106.8	2.525	**15**
	15	3,256.7	3.184	
	16	2,582.9	4.016	
	17	2,048.2	5.064	
	18	1,624.3	6.385	
	19	1,288.1	8.051	
	20	1,021.5	10.15	
	21	810.10	12.80	
	22	642.40	16.14	
	23	509.45	20.36	
	24	404.01	25.67	
	25	320.40	32.37	
	26	254.10	40.81	
	27	201.50	51.47	
	28	159.79	64.90	
	29	126.72	81.83	
	30	100.50	103.2	
	31	79.70	130.1	
	32	63.21	164.1	
	33	50.13	206.9	
	34	39.75	260.9	
	35	31.52	329.0	
	36	25.00	414.8	
	37	19.83	523.1	
	38	15.72	659.6	
	39	12.47	831.8	
	40	9.89	1049.0	

*Not more than three conductors in raceway, cable, or direct burial.

Source: Reprinted by permission from NFPA No. SPP-6C, National Electrical Code®, copyright © 1996, National Fire Protection Association, Quincy, MA 02269. This reprinted material is not the complete and official position of the NFPA on the referenced subject, which is represented only by the standard in its entirety. *National Electrical Code* is a registered trademark of the National Fire Protection Association, Inc., Quincy, MA, for a triennial electrical publication. The term *National Electrical Code,* as used herein, means the triennial publication constituting the National Electrical Code and is used with permission of the National Fire Protection Association.

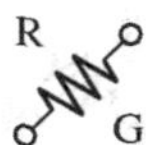

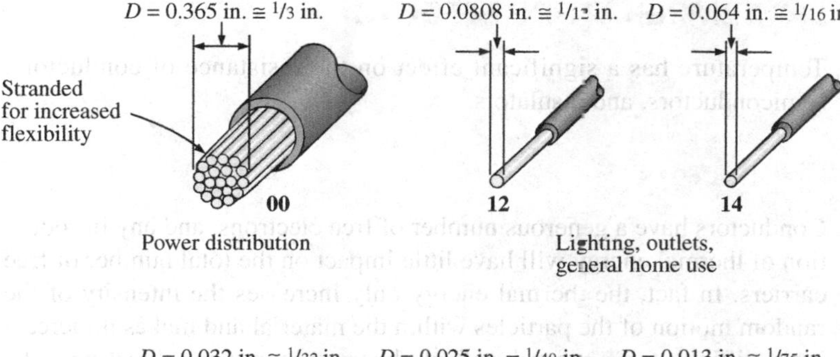

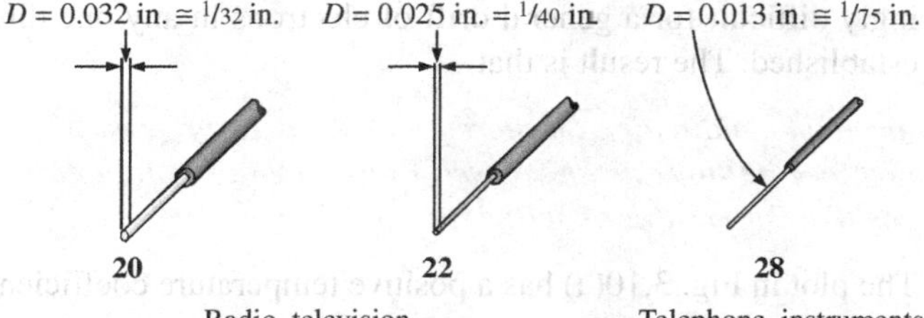

FIG. 3.8
Popular wire sizes and some of their areas of application.

EXAMPLE 3.4 Find the resistance of 650 ft of #8 copper wire (T = 20°C).

Solution: For #8 copper wire (solid), Ω/1000 ft at 20°C = 0.6282 Ω, and

$$650\,\cancel{\text{ft}}\left(\frac{0.6282\ \Omega}{1000\,\cancel{\text{ft}}}\right) = \mathbf{0.408\ \Omega}$$

EXAMPLE 3.5 What is the diameter, in inches, of a #12 copper wire?

Solution: For #12 copper wire (solid), A = 6529.9 CM, and

$$d_{\text{mils}} = \sqrt{A_{\text{CM}}} = \sqrt{6529.9\text{ CM}} \cong 80.81\text{ mils}$$

$$d = \mathbf{0.081\text{ in.}}\ \text{(or close to } 1/12\text{ in.)}$$

EXAMPLE 3.6 For the system in Fig. 3.9, the total resistance of *each* power line cannot exceed 0.025 Ω, and the maximum current to be drawn by the load is 95 A. What gage wire should be used?

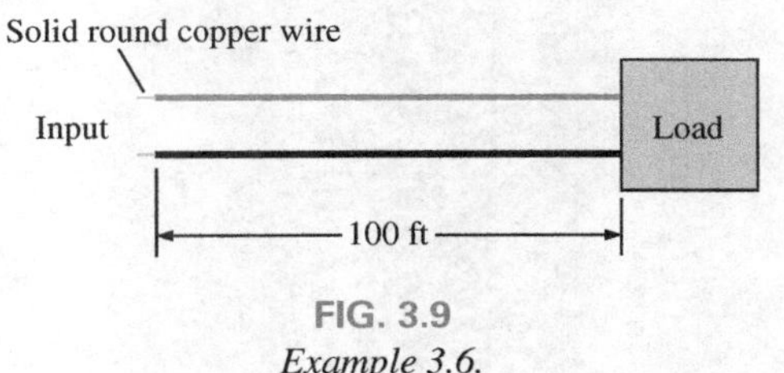

FIG. 3.9
Example 3.6.

Solution:

$$R = \rho\frac{l}{A} \Rightarrow A = \rho\frac{l}{R} = \frac{(10.37\text{ CM-}\Omega/\text{ft})(100\text{ ft})}{0.025\ \Omega} = \mathbf{41{,}480\text{ CM}}$$

Using the wire table, we choose the wire with the next largest area, which is #4, to satisfy the resistance requirement. We note, however, that 95 A must flow through the line. This specification requires that **#3 wire** be used since the #4 wire can carry a maximum current of only 85 A.

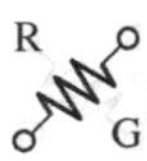

3.4 TEMPERATURE EFFECTS

Temperature has a significant effect on the resistance of conductors, semiconductors, and insulators.

Conductors

Conductors have a generous number of free electrons, and any introduction of thermal energy will have little impact on the total number of free carriers. In fact, the thermal energy only increases the intensity of the random motion of the particles within the material and makes it increasingly difficult for a general drift of electrons in any one direction to be established. The result is that

for good conductors, an increase in temperature results in an increase in the resistance level. Consequently, conductors have a positive temperature coefficient.

The plot in Fig. 3.10(a) has a positive temperature coefficient.

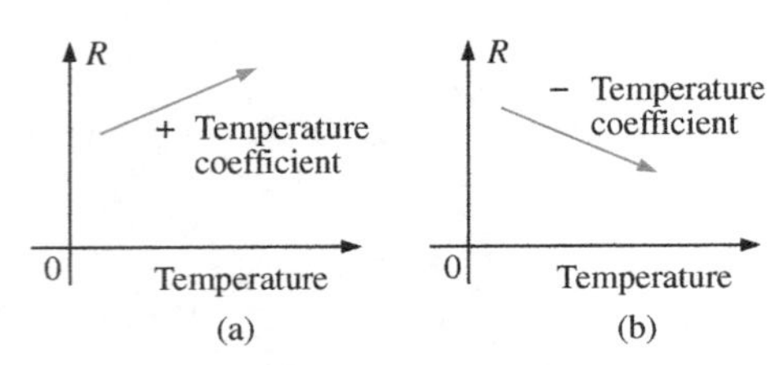

FIG. 3.10
Demonstrating the effect of a positive and a negative temperature coefficient on the resistance of a conductor.

Semiconductors

In semiconductors, an increase in temperature imparts a measure of thermal energy to the system that results in an increase in the number of free carriers in the material for conduction. The result is that

for semiconductor materials, an increase in temperature results in a decrease in the resistance level. Consequently, semiconductors have negative temperature coefficients.

The thermistor and photoconductive cell discussed in Sections 3.12 and 3.13, respectively, are excellent examples of semiconductor devices with negative temperature coefficients. The plot in Fig. 3.10(b) has a negative temperature coefficient.

Insulators

As with semiconductors, an increase in temperature results in a decrease in the resistance of an insulator. The result is a negative temperature coefficient.

Inferred Absolute Temperature

Fig. 3.11 reveals that for copper (and most other metallic conductors), the resistance increases almost linearly (in a straight-line relationship)

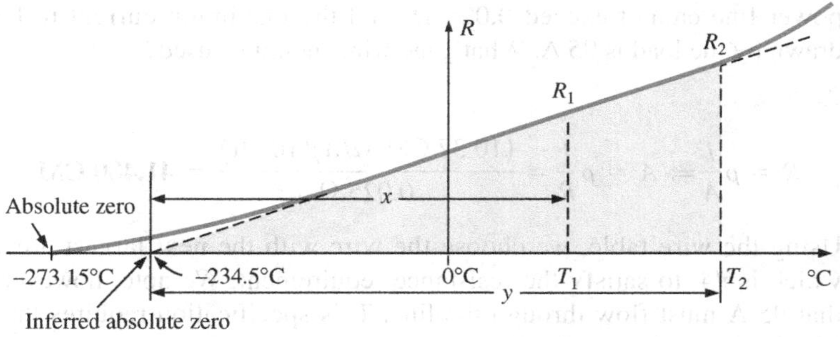

FIG. 3.11
Effect of temperature on the resistance of copper.

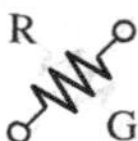

with an increase in temperature. Since temperature can have such a pronounced effect on the resistance of a conductor, it is important that we have some method of determining the resistance at any temperature within operating limits. An equation for this purpose can be obtained by approximating the curve in Fig. 3.11 by the straight dashed line that intersects the temperature scale at −234.5°C. Although the actual curve extends to **absolute zero** (−273.15°C, or 0 K), the straight-line approximation is quite accurate for the normal operating temperature range. At two temperatures T_1 and T_2, the resistance of copper is R_1 and R_2, respectively, as indicated on the curve. Using a property of similar triangles, we may develop a mathematical relationship between these values of resistance at different temperatures. Let x equal the distance from −234.5°C to T_1 and y the distance from −234.5°C to T_2, as shown in Fig. 3.11. From similar triangles,

$$\frac{x}{R_1} = \frac{y}{R_2}$$

or

$$\boxed{\frac{234.5 + T_1}{R_1} = \frac{234.5 + T_2}{R_2}} \qquad \textbf{(3.5)}$$

The temperature of −234.5°C is called the **inferred absolute temperature** of copper. For different conducting materials, the intersection of the straight-line approximation occurs at different temperatures. A few typical values are listed in Table 3.3.

TABLE 3.3
Inferred absolute temperatures (T_i).

Material	°C
Silver	−243
Copper	**−234.5**
Gold	−274
Aluminum	−236
Tungsten	−204
Nickel	−147
Iron	−162
Nichrome	−2,250
Constantan	−125,000

The minus sign does not appear with the inferred absolute temperature on either side of Eq. (3.5) because x and y are the *distances* from −234.5°C to T_1 and T_2, respectively, and therefore are simply magnitudes. For T_1 and T_2 less than zero, x and y are less than −234.5°C, and the distances are the differences between the inferred absolute temperature and the temperature of interest.

Eq. (3.5) can easily be adapted to any material by inserting the proper inferred absolute temperature. It may therefore be written as follows:

$$\boxed{\frac{|T_1| + T_1}{R_1} = \frac{|T_1| + T_2}{R_2}} \qquad \textbf{(3.6)}$$

where $|T_1|$ indicates that the inferred absolute temperature of the material involved is inserted as a positive value in the equation. In general, therefore, associate the sign only with T_1 and T_2.

EXAMPLE 3.7 If the resistance of a copper wire is 50 Ω at 20°C, what is its resistance at 100°C (boiling point of water)?

Solution: Eq. (3.5):

$$\frac{234.5°\text{C} + 20°\text{C}}{50\ \Omega} = \frac{234.5°\text{C} + 100°\text{C}}{R_2}$$

$$R_2 = \frac{(50\ \Omega)(334.5°\text{C})}{254.5°\text{C}} = \mathbf{65.72\ \Omega}$$

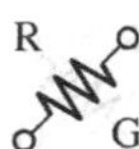

EXAMPLE 3.8 If the resistance of a copper wire at freezing (0°C) is 30 Ω, what is its resistance at −40°C?

Solution: Eq. (3.5):

$$\frac{234.5°\text{C} + 0}{30\ \Omega} = \frac{234.5°\text{C} - 40°\text{C}}{R_2}$$

$$R_2 = \frac{(30\ \Omega)(194.5°\text{C})}{234.5°\text{C}} = \mathbf{24.88\ \Omega}$$

EXAMPLE 3.9 If the resistance of an aluminum wire at room temperature (20°C) is 100 mΩ (measured by a milliohmmeter), at what temperature will its resistance increase to 120 mΩ?

Solution: Eq. (3.5):

$$\frac{236°\text{C} + 20°\text{C}}{100\ \text{m}\Omega} = \frac{236°\text{C} + T_2}{120\ \text{m}\Omega}$$

and

$$T_2 = 120\ \text{m}\Omega\left(\frac{256°\text{C}}{100\ \text{m}\Omega}\right) - 236°\text{C}$$

$$T_2 = \mathbf{71.2°C}$$

Temperature Coefficient of Resistance

There is a second popular equation for calculating the resistance of a conductor at different temperatures. Defining

$$\boxed{\alpha_{20} = \frac{1}{|T_1| + 20°\text{C}}} \quad (\Omega/°\text{C}/\Omega) \qquad \textbf{(3.7)}$$

as the **temperature coefficient of resistance** at a temperature of 20°C and R_{20} as the resistance of the sample at 20°C, we determine the resistance R_1 at a temperature T_1 by

$$\boxed{R_1 = R_{20}[1 + \alpha_{20}(T_1 - 20°\text{C})]} \qquad \textbf{(3.8)}$$

The values of α_{20} for different materials have been evaluated, and a few are listed in Table 3.4.

TABLE 3.4
Temperature coefficient of resistance for various conductors at 20°C.

Material	Temperature Coefficient (α_{20})
Silver	0.0038
Copper	**0.00393**
Gold	0.0034
Aluminum	0.00391
Tungsten	0.005
Nickel	0.006
Iron	0.0055
Constantan	0.000008
Nichrome	0.00044

Eq. (3.8) can be written in the following form:

$$\alpha_{20} = \frac{\left(\dfrac{R_1 - R_{20}}{T_1 - 20°\text{C}}\right)}{R_{20}} = \frac{\dfrac{\Delta R}{\Delta T}}{R_{20}}$$

from which the units of Ω/°C/Ω for α_{20} are defined.

Since $\Delta R/\Delta T$ is the slope of the curve in Fig. 3.11, we can conclude that

the higher the temperature coefficient of resistance for a material, the more sensitive is the resistance level to changes in temperature.

Referring to Table 3.3, we find that copper is more sensitive to temperature variations than is silver, gold, or aluminum, although the differences are quite small. The slope defined by α_{20} for constantan is so small that the curve is almost horizontal.

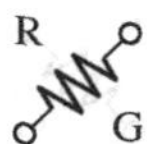

Since R_{20} of Eq. (3.8) is the resistance of the conductor at 20°C and $T_1 - 20°C$ is the change in temperature from 20°C, Eq. (3.8) can be written in the following form:

$$R = \rho\frac{l}{A}[1 + \alpha_{20}\,\Delta T] \tag{3.9}$$

providing an equation for resistance in terms of all the controlling parameters.

PPM/°C

For resistors, as for conductors, resistance changes with a change in temperature. The specification is normally provided in parts per million per degree Celsius (**PPM/°C**), providing an immediate indication of the sensitivity level of the resistor to temperature. For resistors, a 5000 PPM level is considered high, whereas 20 PPM is quite low. A 1000 PPM/°C characteristic reveals that a 1° change in temperature results in a change in resistance equal to 1000 PPM, or 1000/1,000,000 = 1/1000 of its nameplate value—not a significant change for most applications. However, a 10° change results in a change equal to 1/100 (1%) of its nameplate value, which is becoming significant. The concern, therefore, lies not only with the PPM level, but also with the range of expected temperature variation.

In equation form, the change in resistance is given by

$$\Delta R = \frac{R_{nominal}}{10^6}(\text{PPM})(\Delta T) \tag{3.10}$$

where $R_{nominal}$ is the nameplate value of the resistor at room temperature and ΔT is the change in temperature from the reference level of 20°C.

EXAMPLE 3.10 For a 1 kΩ carbon composition resistor with a PPM of 2500, determine the resistance at 60°C.

Solution:

$$\Delta R = \frac{1000\ \Omega}{10^6}(2500)(60°\text{C} - 20°\text{C}) = 100\ \Omega$$

and $$R = R_{nominal} + \Delta R = 1000\ \Omega + 100\ \Omega = \mathbf{1100\ \Omega}$$

3.5 TYPES OF RESISTORS

Fixed Resistors

Resistors are made in many forms, but all belong in either of two groups: fixed or variable. The most common of the low-wattage, fixed-type resistors is the film resistor shown in Fig. 3.12. It is constructed by depositing a thin layer of resistive material (typically carbon, metal, or metal oxide) on a ceramic rod. The desired resistance is then obtained by cutting away some of the resistive material in a helical manner to establish a long, continuous band of high-resistance material from one

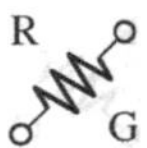

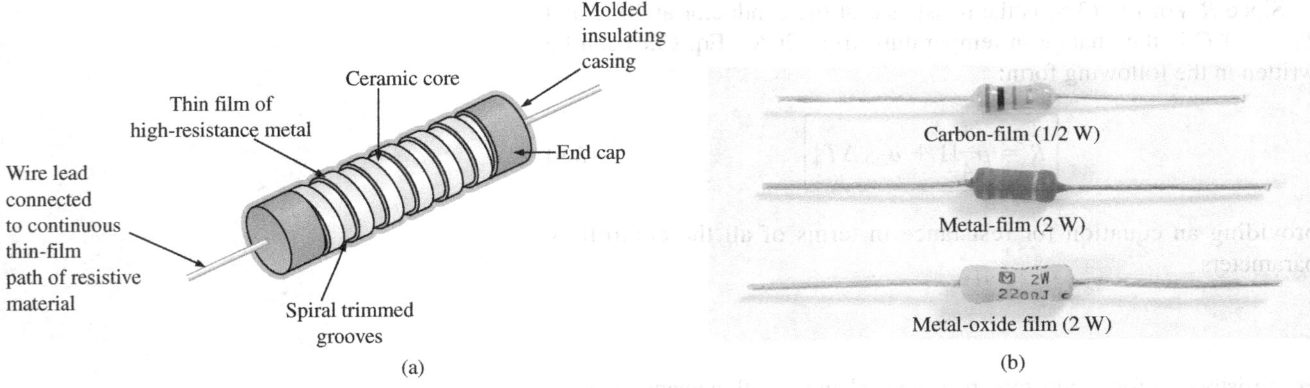

FIG. 3.12
Film resistors: (a) construction; (b) types.

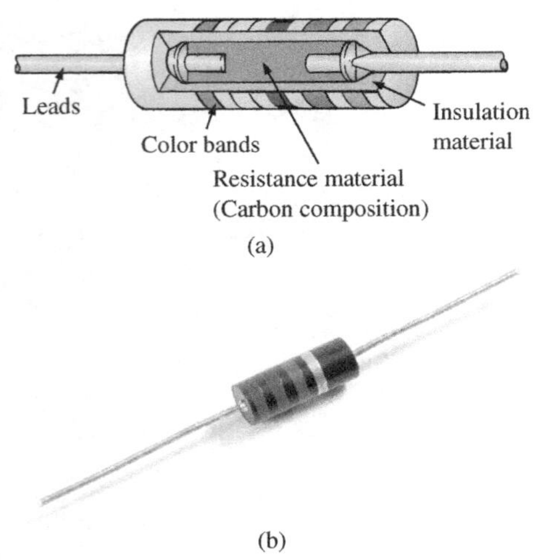

FIG. 3.13
Fixed-composition resistors: (a) construction; (b) appearance.

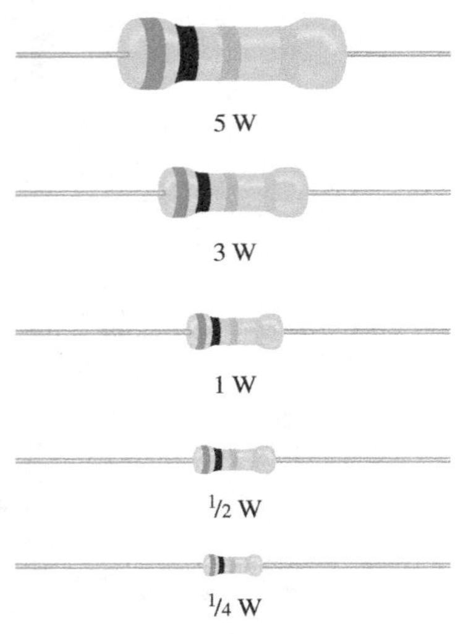

FIG. 3.14
Fixed metal-oxide resistors of different wattage ratings.

end of the resistor to the other. In general, carbon-film resistors have a beige body and a lower wattage rating. The metal-film resistor is typically a stronger color, such as brick red or dark green, with higher wattage ratings. The metal-oxide resistor is usually a softer pastel color, such as rating powder blue shown in Fig. 3.12(b), and has the highest wattage rating of the three.

When you search through most electronics catalogs or visit a local electronics dealer such as Radio Shack to purchase resistors, you will find that the most common resistor is the film resistor. In years past, the carbon composition resistor in Fig. 3.13 was the most common, but fewer and fewer companies are manufacturing this variety, with its range of applications reduced to applications in which very high temperatures and inductive effects (Chapter 11) can be a problem. Its resistance is determined by the carbon composition material molded directly to each end of the resistor. The high resistivity characteristics of carbon (ρ = 21,000 CM-Ω/ft) provide a high-resistance path for the current through the element.

For a particular style and manufacturer, the size of a resistor increases with the power or wattage rating.

The concept of power is covered in detail in Chapter 4, but for the moment recognize that increased power ratings are normally associated with the ability to handle higher current and temperature levels. Fig. 3.14 depicts the actual size of thin-film, metal-oxide resistors in the 1/4 W to 5 W rating range. All the resistors in Fig. 3.14 are 1 MΩ, revealing that

the size of a resistor does not define its resistance level.

A variety of other fixed resistors are depicted in Fig. 3.15. The wire-wound resistors of Fig. 3.15(a) are formed by winding a high-resistance wire around a ceramic core. The entire structure is then baked in a ceramic cement to provide a protective covering. Wire-wound resistors are typically used for larger power applications, although they are also available with very small wattage ratings and very high accuracy.

Fig. 3.15(c) and (g) are special types of wire-wound resistors with a low percent tolerance. Note, in particular, the high power ratings for the wire-wound resistors for their relatively small size. Figs. 3.15(b), (d), and (f) are power film resistors that use a thicker layer of film material than used in the variety shown in Fig. 3.12. The chip resistors in Fig. 3.15(f)

100 Ω, 25 W

2 kΩ, 8 W

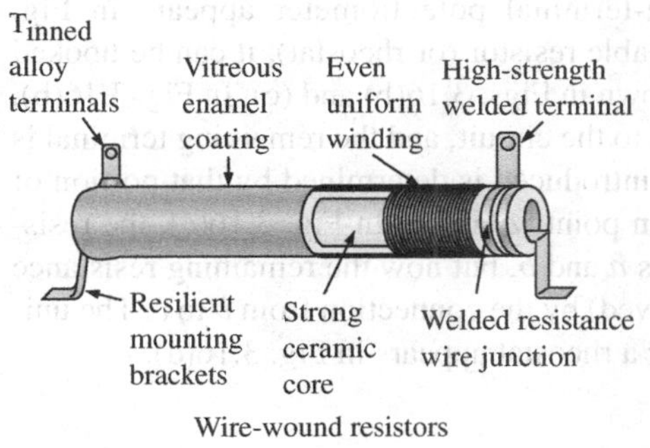

Wire-wound resistors

(a)

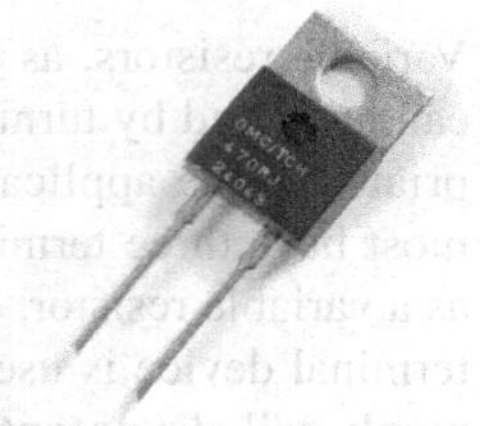

470 Ω, 35 W
Thick-film power resistor
(b)

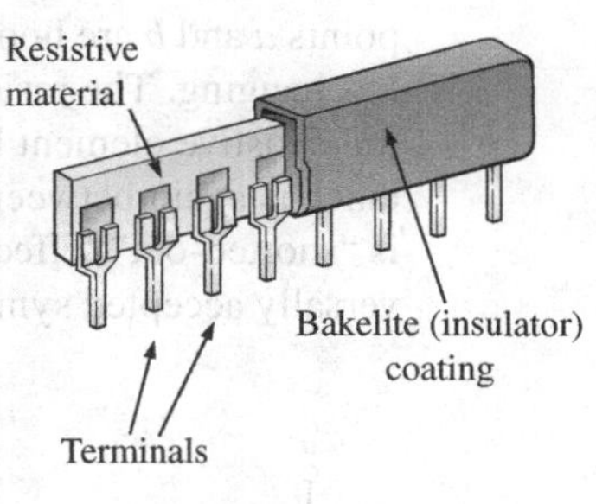

1 kΩ, 25 W

Aluminum-housed, chassis-mount resistor–precision wire-mount
(c)

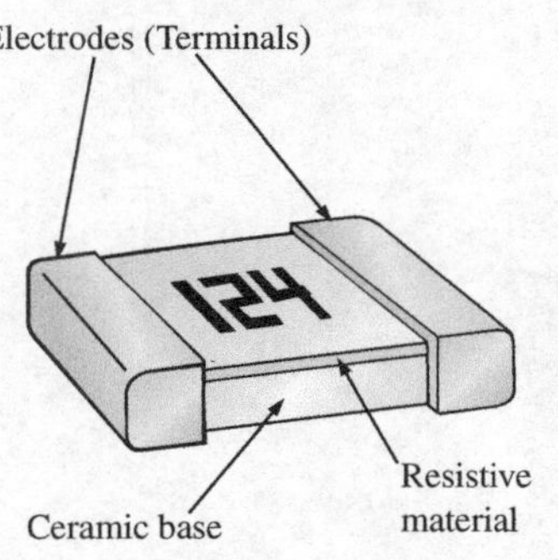

100 MΩ, 0.75 W
Precision power film resistor
(d)

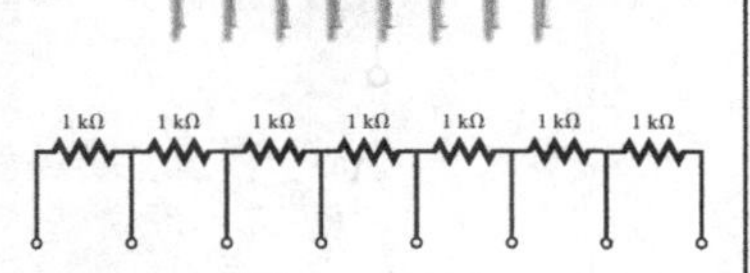

1 kΩ bussed (all connected on one side) single in-line resistor network
(e)

22 kΩ, 1W
Surface mount thick-film chip resistors with gold electrodes
(f)

25 kΩ, 5 W
Silicon-coated, wire-wound resistor
(g)

FIG. 3.15
Various types of fixed resistors.

are used where space is a priority, such as on the surface of circuit board. Units of this type can be less than 1/16 in. in length or width, with thickness as small as 1/30 in., yet they can still handle 0.5 W of power with resistance levels as high as 1000 MΩ—clear evidence that size does not determine the resistance level. The fixed resistor in Fig. 3.15(c) has terminals applied to a layer of resistor material, with the resistance between the terminals a function of the dimensions of the resistive material and the placement of the terminal pads.

Variable Resistors

Variable resistors, as the name implies, have a terminal resistance that can be varied by turning a dial, knob, screw, or whatever seems appropriate for the application. They can have two or three terminals, but most have three terminals. If the two- or three-terminal device is used as a variable resistor, it is usually referred to as a **rheostat.** If the three-terminal device is used for controlling potential levels, it is then commonly called a **potentiometer.** Even though a three-terminal device can be used as a rheostat or a potentiometer (depending on how it is connected), it is typically called a *potentiometer* when listed in trade magazines or requested for a particular application.

The symbol for a three-terminal potentiometer appears in Fig. 3.16(a). When used as a variable resistor (or rheostat), it can be hooked up in one of two ways, as shown in Figs. 3.16(b) and (c). In Fig. 3.16(b), points a and b are hooked up to the circuit, and the remaining terminal is left hanging. The resistance introduced is determined by that portion of the resistive element between points a and b. In Fig. 3.16(c), the resistance is again between points a and b, but now the remaining resistance is "shorted-out" (effect removed) by the connection from b to c. The universally accepted symbol for a rheostat appears in Fig. 3.16(d).

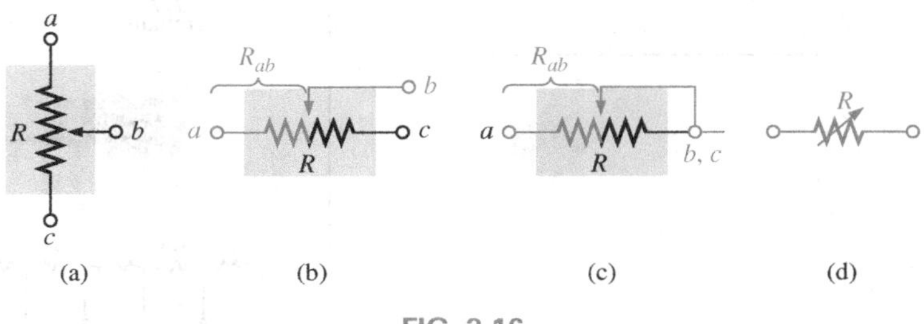

FIG. 3.16
Potentiometer: (a) symbol; (b) and (c) rheostat connections; (d) rheostat symbol.

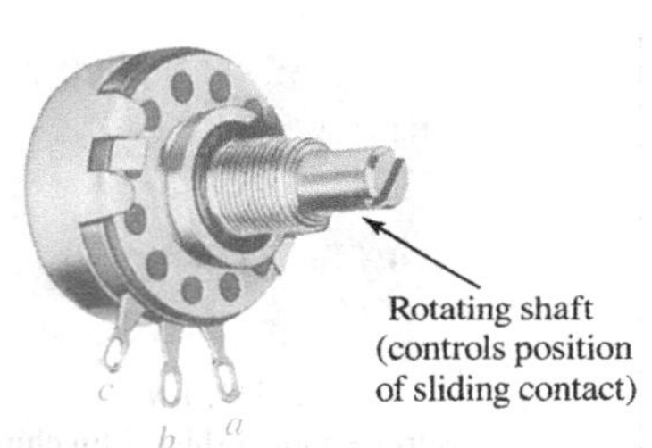

(a) External view

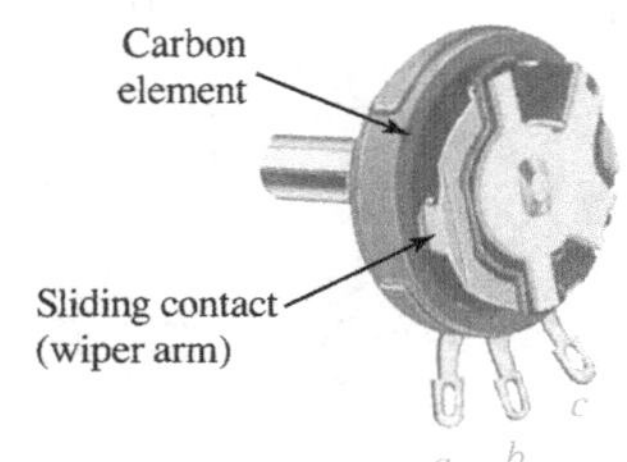

(b) Internal view

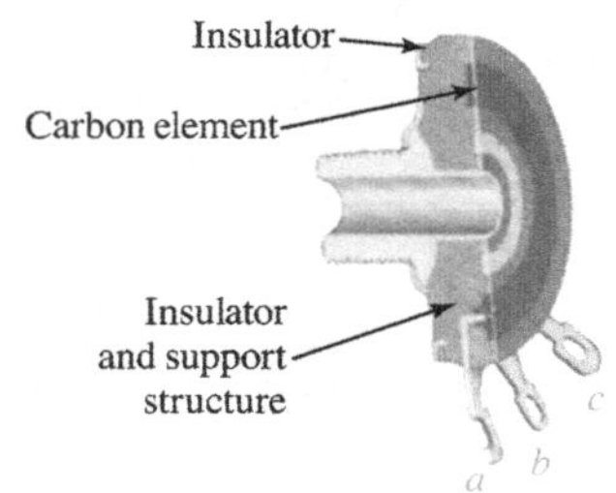

(c) Carbon element

FIG. 3.17
Molded composition-type potentiometer.

Most potentiometers have three terminals in the relative positions shown in Fig. 3.17. The knob, dial, or screw in the center of the housing controls the motion of a contact that can move along the resistive element connected between the outer two terminals. The contact is connected to the center terminal, establishing a resistance from movable contact to each outer terminal.

The resistance between the outside terminals a and c in Fig. 3.18(a) (and Fig. 3.17) is always fixed at the full rated value of the potentiometer, regardless of the position of the wiper arm b.

In other words, the resistance between terminals a and c in Fig. 3.18(a) for a 1 MΩ potentiometer will always be 1 MΩ, no matter how we turn the control element and move the contact. In Fig. 3.18(a), the center contact is not part of the network configuration.

The resistance between the wiper arm and either outside terminal can be varied from a minimum of 0 Ω to a maximum value equal to the full rated value of the potentiometer.

In Fig. 3.18(b), the wiper arm has been placed 1/4 of the way down from point a to point c. The resulting resistance between points a and b will therefore be 1/4 of the total, or 250 kΩ (for a 1 MΩ potentiometer), and the resistance between b and c will be 3/4 of the total, or 750 kΩ.

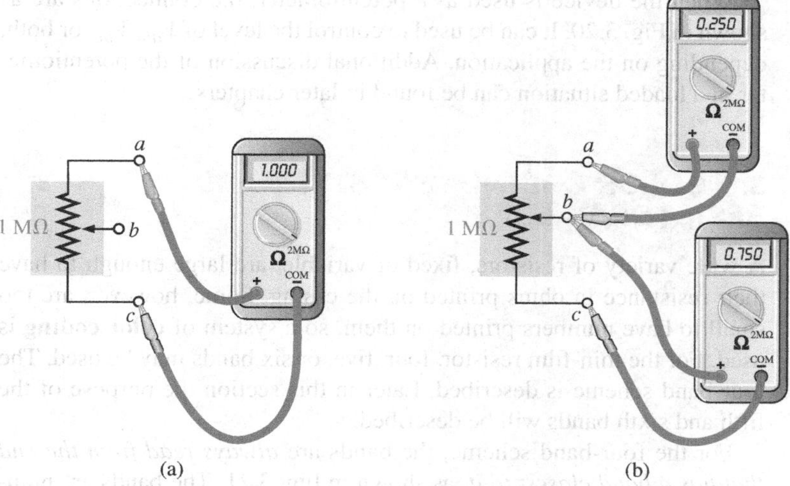

FIG. 3.18

Resistance components of a potentiometer: (a) between outside terminals; (b) between wiper arm and each outside terminal.

The sum of the resistances between the wiper arm and each outside terminal equals the full rated resistance of the potentiometer.

This is demonstrated in Fig. 3.18(b), where 250 kΩ + 750 kΩ = 1 MΩ. Specifically,

$$\boxed{R_{ac} = R_{ab} + R_{bc}} \tag{3.11}$$

Therefore, as the resistance from the wiper arm to one outside contact increases, the resistance between the wiper arm and the other outside terminal must decrease accordingly. For example, if R_{ab} of a 1 kΩ potentiometer is 200 Ω, then the resistance R_{bc} must be 800 Ω. If R_{ab} is further decreased to 50 Ω, then R_{bc} must increase to 950 Ω, and so on.

The molded carbon composition potentiometer is typically applied in networks with smaller power demands, and it ranges in size from 20 Ω to 22 MΩ (maximum values). A miniature trimmer (less than 1/4 in. in diameter) appears in Fig. 3.19(a), and a variety of potentiometers that use a cermet resistive material appear in Fig. 3.19(b). The contact point of the three-point wire-wound resistor in Fig. 3.19(c) can be moved to set the resistance between the three terminals.

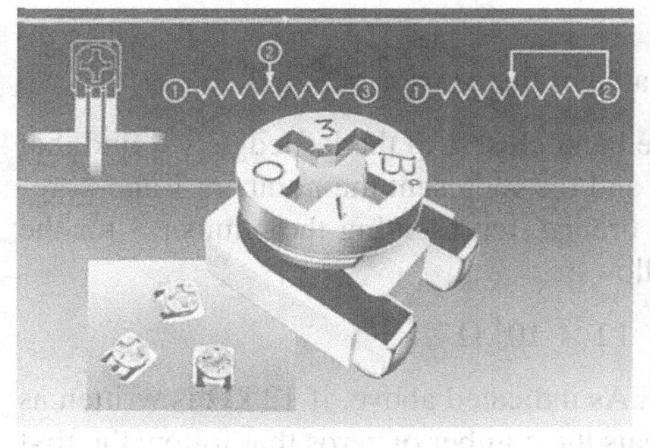

(a)

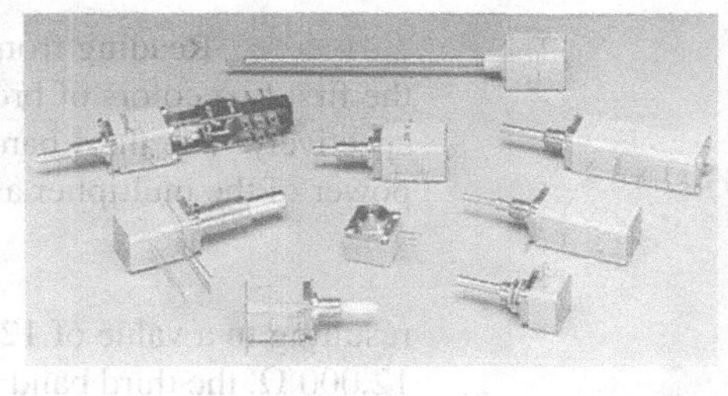

(b)

(c)

FIG. 3.19

Variable resistors: (a) 4 mm (≈ 5/32 in.) trimmer; (b) conductive plastic and cermet elements; (c) three-point wire-wound resistor. (Courtesy Honeywell International Inc.)

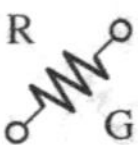

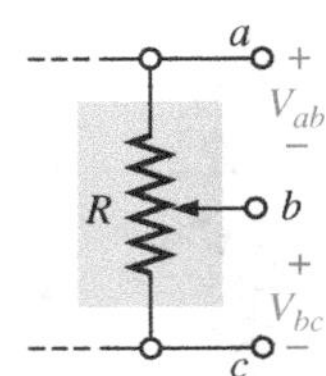

FIG. 3.20
Potentiometer control of voltage levels.

When the device is used as a potentiometer, the connections are as shown in Fig. 3.20. It can be used to control the level of V_{ab}, V_{bc}, or both, depending on the application. Additional discussion of the potentiometer in a loaded situation can be found in later chapters.

3.6 COLOR CODING AND STANDARD RESISTOR VALUES

A wide variety of resistors, fixed or variable, are large enough to have their resistance in ohms printed on the casing. Some, however, are too small to have numbers printed on them, so a system of **color coding** is used. For the thin-film resistor, four, five, or six bands may be used. The four-band scheme is described. Later in this section the purpose of the fifth and sixth bands will be described.

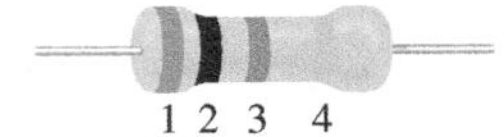

1 2 3 4

FIG. 3.21
Color coding for fixed resistors.

For the four-band scheme, the bands are *always read from the end that has a band closest to it,* as shown in Fig. 3.21. The bands are numbered as shown for reference in the discussion to follow.

The first two bands represent the first and second digits, respectively.

They are the actual first two numbers that define the numerical value of the resistor.

The third band determines the power-of-ten multiplier for the first two digits (actually the number of zeros that follow the second digit for resistors greater than 10 Ω).

The fourth band is the manufacturer's tolerance, which is an indication of the precision by which the resistor was made.

If the fourth band is omitted, the tolerance is assumed to be ±20%.

Number	Color
0	Black
1	Brown
2	Red
3	Orange
4	Yellow
5	Green
6	Blue
7	Violet
8	Gray
9	White
±5% (0.1 multiplier if 3rd band)	Gold
±10% (0.01 multiplier if 3rd band)	Silver

FIG. 3.22
Color coding.

The number corresponding to each color is defined in Fig. 3.22. The fourth band will be either ±5% or ±10% as defined by gold and silver, respectively. To remember which color goes with which percent, simply remember that ±5% resistors cost more and gold is more valuable than silver.

Remembering which color goes with each digit takes a bit of practice. In general, the colors start with the very dark shades and move toward the lighter shades. The best way to memorize is to simply repeat over and over that red is 2, yellow is 4, and so on. Simply practice with a friend or a fellow student, and you will learn most of the colors in short order.

FIG. 3.23
Example 3.11.

EXAMPLE 3.11 Find the value of the resistor in Fig. 3.23.

Solution: Reading from the band closest to the left edge, we find that the first two colors of brown and red represent the numbers 1 and 2, respectively. The third band is orange, representing the number 3 for the power of the multiplier as follows:

$$12 \times 10^3\ \Omega$$

resulting in a value of 12 kΩ. As indicated above, if 12 kΩ is written as 12,000 Ω, the third band reveals the number of zeros that follow the first two digits.

Now for the fourth band of gold, representing a tolerance of ±5%: To find the range into which the manufacturer has guaranteed the resistor

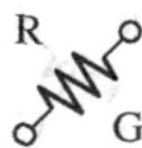

will fall, first convert the 5% to a decimal number by moving the decimal point two places to the left:

$$5\% \Rightarrow 0.05$$

Then multiply the resistor value by this decimal number:

$$0.05(12\text{ k}\Omega) = 600\ \Omega$$

Finally, add the resulting number to the resistor value to determine the maximum value, and subtract the number to find the minimum value. That is,

$$\text{Maximum} = 12{,}000\ \Omega + 600\ \Omega = 12.6\text{ k}\Omega$$
$$\text{Minimum} = 12{,}000\ \Omega - 600\ \Omega = 11.4\text{ k}\Omega$$
$$\text{Range} = \mathbf{11.4\text{ k}\Omega \text{ to } 12.6\text{ k}\Omega}$$

The result is that the manufacturer has guaranteed with the 5% gold band that the resistor will fall in the range just determined. In other words, the manufacturer does not guarantee that the resistor will be exactly 12 kΩ, but rather that it will fall in a range as defined above.

Using the above procedure, the smallest resistor that can be labeled with the color code is 10 Ω. However,

the range can be extended to include resistors from 0.1 Ω to 10 Ω by simply using gold as a multiplier color (third band) to represent 0.1 and using silver to represent 0.01.

This is demonstrated in the next example.

EXAMPLE 3.12 Find the value of the resistor in Fig. 3.24.

FIG. 3.24
Example 3.12.

Solution: The first two colors are gray and red, representing the numbers 8 and 2, respectively. The third color is gold, representing a multiplier of 0.1. Using the multiplier, we obtain a resistance of

$$(0.1)(82\ \Omega) = 8.2\ \Omega$$

The fourth band is silver, representing a tolerance of ±10%. Converting to a decimal number and multiplying through yields

$$10\% = 0.10 \quad \text{and} \quad (0.1)(8.2\ \Omega) = 0.82\ \Omega$$

$$\text{Maximum} = 8.2\ \Omega + 0.82\ \Omega = 9.02\ \Omega$$
$$\text{Minimum} = 8.2\ \Omega - 0.82\ \Omega = 7.38\ \Omega$$

so that

$$\text{Range} = \mathbf{7.38\ \Omega \text{ to } 9.02\ \Omega}$$

Although it will take some time to learn the numbers associated with each color, it is certainly encouraging to become aware that

the same color scheme to represent numbers is used for all the important elements of electrical circuits.

Later on, you will find that the numerical value associated with each color is the same for capacitors and inductors. Therefore, once learned, the scheme has repeated areas of application.

Some manufacturers prefer to use a **five-band color code.** In such cases, as shown in the top portion of Fig. 3.25, three digits are provided

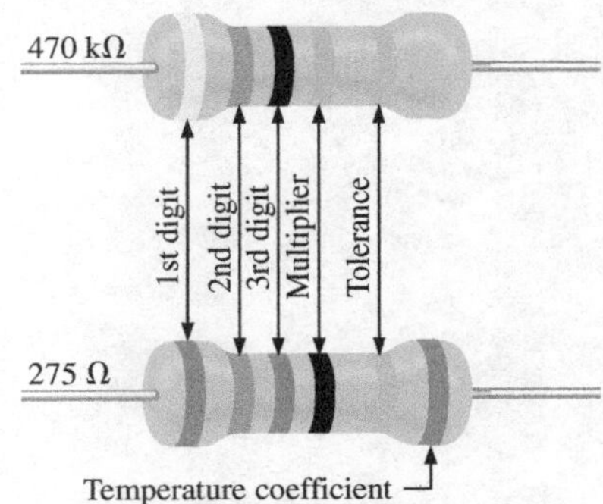

FIG. 3.25
Five-band color coding for fixed resistors.

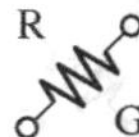

before the multiplier. The fifth band remains the tolerance indicator. If the manufacturer decides to include the temperature coefficient, a sixth band will appear as shown in the lower portion of Fig. 3.25, with the color indicating the PPM level.

For four, five, or six bands, if the tolerance is less than 5%, the following colors are used to reflect the % tolerances:

brown = ±1%, red = ±2%, green = ±0.5%, blue = ±0.25%, and violet = ±0.1%.

You might expect that resistors would be available for a full range of values such as 10 Ω, 20 Ω, 30 Ω, 40 Ω, 50 Ω, and so on. However, this is not the case, with some typical commercial values as 27 Ω, 56 Ω, and 68 Ω. There is a reason for the chosen values, which is best demonstrated by examining the list of standard values of commercially available resistors in Table 3.5. The values in boldface are the most common and typically available with 5%, 10%, and 20% tolerances.

TABLE 3.5

Standard values of commercially available resistors.

Ohms (Ω)					Kilohms (kΩ)		Megohms (MΩ)	
0.10	**1.0**	**10**	**100**	**1000**	**10**	**100**	**1.0**	**10.0**
0.11	1.1	11	110	1100	11	110	1.1	11.0
0.12	**1.2**	**12**	**120**	**1200**	**12**	**120**	**1.2**	**12.0**
0.13	1.3	13	130	1300	13	130	1.3	13.0
0.15	**1.5**	**15**	**150**	**1500**	**15**	**150**	**1.5**	**15.0**
0.16	1.6	16	160	1600	16	160	1.6	16.0
0.18	**1.8**	**18**	**180**	**1800**	**18**	**180**	**1.8**	**18.0**
0.20	2.0	20	200	2000	20	200	2.0	20.0
0.22	**2.2**	**22**	**220**	**2200**	**22**	**220**	**2.2**	**22.0**
0.24	2.4	24	240	2400	24	240	2.4	
0.27	**2.7**	**27**	**270**	**2700**	**27**	**270**	**2.7**	
0.30	3.0	30	300	3000	30	300	3.0	
0.33	**3.3**	**33**	**330**	**3300**	**33**	**330**	**3.3**	
0.36	3.6	36	360	3600	36	360	3.6	
0.39	**3.9**	**39**	**390**	**3900**	**39**	**390**	**3.9**	
0.43	4.3	43	430	4300	43	430	4.3	
0.47	**4.7**	**47**	**470**	**4700**	**47**	**470**	**4.7**	
0.51	5.1	51	510	5100	51	510	5.1	
0.56	**5.6**	**56**	**560**	**5600**	**56**	**560**	**5.6**	
0.62	6.2	62	620	6200	62	620	6.2	
0.68	**6.8**	**68**	**680**	**6800**	**68**	**680**	**6.8**	
0.75	7.5	75	750	7500	75	750	7.5	
0.82	**8.2**	**82**	**820**	**8200**	**82**	**820**	**8.2**	
0.91	9.1	91	910	9100	91	910	9.1	

Examining the impact of the tolerance level will help explain the choice of numbers for the commercial values. Take the sequence 47 Ω–68 Ω–100 Ω, which are all available with 20% tolerances. In Fig. 3.26(a), the tolerance band for each has been determined and plotted on a single axis. Note that with this tolerance (which is all that the manufacturer will guarantee), the full range of resistor values is available from 37.6 Ω to 120 Ω. In other words, the manufacturer is guaranteeing

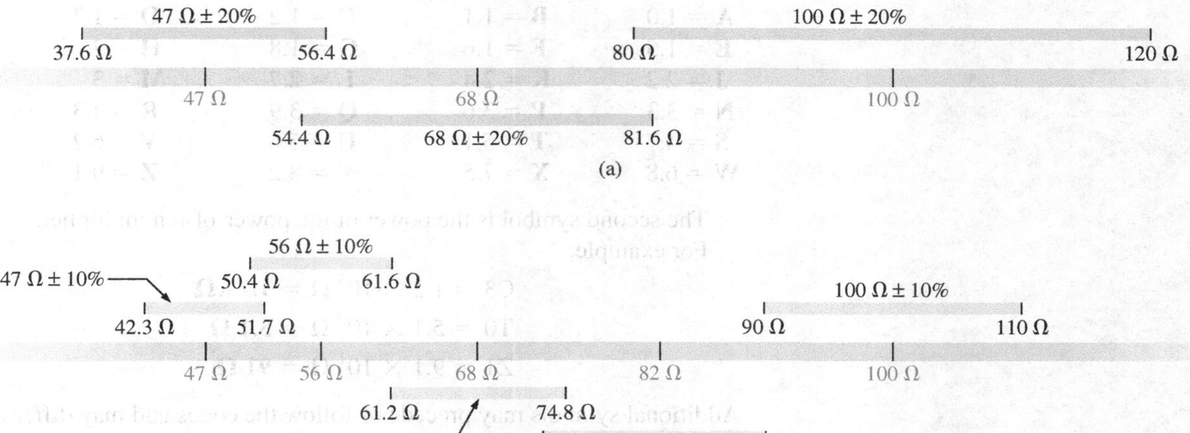

FIG. 3.26

Guaranteeing the full range of resistor values for the given tolerance: (a) 20%; (b) 10%.

the full range, using the tolerances to fill in the gaps. Dropping to the 10% level introduces the 56 Ω and 82 Ω resistors to fill in the gaps, as shown in Fig. 3.26(b). Dropping to the 5% level would require additional resistor values to fill in the gaps. In total, therefore, the resistor values were chosen to ensure that the full range was covered, as determined by the tolerances employed. Of course, if a specific value is desired but is not one of the standard values, combinations of standard values often result in a total resistance very close to the desired level. If this approach is still not satisfactory, a potentiometer can be set to the exact value and then inserted in the network.

Throughout the text, you will find that many of the resistor values are not standard values. This was done to reduce the mathematical complexity, which might interfere with the learning process. In the problem sections, however, standard values are frequently used to ensure that you start to become familiar with the commercial values available.

Surface Mount Resistors

In general, surface mount resistors are marked in three ways: color coding, three symbols, and two symbols.

The **color coding** is the same as just described earlier in this section for through-hole resistors.

The **three-symbol** approach uses three digits. The first two define the first two digits of the value; the last digit defines the power of the power-of-ten multiplier.

For instance:

$$820 \text{ is } 82 \times 10^0\ \Omega = \mathbf{82\ \Omega}$$
$$222 \text{ is } 22 \times 10^2\ \Omega = 2200\ \Omega = \mathbf{2.2\ k\Omega}$$
$$010 \text{ is } 1 \times 10^0\ \Omega = \mathbf{1\ \Omega}$$

The **two-symbol** marking uses a letter followed by a number. The letter defines the value as in the following list. Note that all the numbers of the commercially available list of Table 3.5 are included.

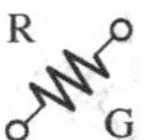

A = 1.0	B = 1.1	C = 1.2	D = 1.3
E = 1.5	F = 1.6	G = 1.8	H = 2
J = 2.2	K = 2.4	L = 2.7	M = 3
N = 3.3	P = 3.6	Q = 3.9	R = 4.3
S = 4.7	T = 5.1	U = 5.6	V = 6.2
W = 6.8	X = 7.5	Y = 8.2	Z = 9.1

The second symbol is the power of the power-of-ten multiplier. For example,

$$\text{C3} = 1.2 \times 10^3\ \Omega = \mathbf{1.2\ k\Omega}$$
$$\text{T0} = 5.1 \times 10^0\ \Omega = \mathbf{5.1\ \Omega}$$
$$\text{Z1} = 9.1 \times 10^1\ \Omega = \mathbf{91\ \Omega}$$

Additional symbols may precede or follow the codes and may differ depending on the manufacturer. These may provide information on the internal resistance structure, power rating, surface material, tapping, and tolerance.

3.7 CONDUCTANCE

By finding the reciprocal of the resistance of a material, we have a measure of how well the material conducts electricity. The quantity is called **conductance,** has the symbol G, and is measured in *siemens* (S) (note Fig. 3.27). In equation form, conductance is

$$\boxed{G = \frac{1}{R}} \qquad \text{(siemens, S)} \qquad \textbf{(3.12)}$$

A resistance of 1 MΩ is equivalent to a conductance of 10^{-6} S, and a resistance of 10 Ω is equivalent to a conductance of 10^{-1} S. The larger the conductance, therefore, the less is the resistance and the greater is the conductivity.

In equation form, the conductance is determined by

$$\boxed{G = \frac{A}{\rho l}} \qquad \text{(S)} \qquad \textbf{(3.13)}$$

indicating that increasing the area or decreasing either the length or the resistivity increases the conductance.

FIG. 3.27
Werner von Siemens.
(© orion_eff / fotolia.com)

German (Lenthe, Berlin)
(1816–92)
Electrical Engineer
Telegraph Manufacturer,
Siemens & Halske AG

Developed an *electroplating process* during a brief stay in prison for acting as a second in a duel between fellow officers of the Prussian army. Inspired by the electronic telegraph invented by Sir Charles Wheatstone in 1817, he improved on the design and with the help of his brother Carl proceeded to lay cable across the Mediterranean and from Europe to India. His inventions included the first *self-excited generator,* which depended on the *residual* magnetism of its electromagnet rather than an inefficient permanent magnet. In 1888 he was raised to the rank of nobility with the addition of *von* to his name. The current firm of Siemens AG has manufacturing outlets in some 35 countries and sales offices in some 125 countries.

EXAMPLE 3.13

a. Determine the conductance of a 1 Ω, a 50 kΩ, and a 10 MΩ resistor.
b. How does the conductance level change with increase in resistance?

Solution: Eq. (3.12):

a. $1\Omega\text{: } G = \dfrac{1}{R} = \dfrac{1}{1\ \Omega} = \mathbf{1\ S}$

$$50\ \text{k}\Omega\text{: } G = \frac{1}{R} = \frac{1}{50\ \text{k}\Omega} = \frac{1}{50 \times 10^3\Omega} = 0.02 \times 10^{-3}\text{S} = \mathbf{0.02\ mS}$$

$$10\ \text{M}\Omega\text{: } G = \frac{1}{R} = \frac{1}{10\ \text{M}\Omega} = \frac{1}{10 \times 10^6\Omega} = 0.1 \times 10^{-6}\text{S} = \mathbf{0.1\ \mu S}$$

b. The conductance level decreases rapidly with significant increase in resistance levels.

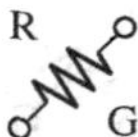

EXAMPLE 3.14 What is the relative increase or decrease in conductivity of a conductor if the area is reduced by 30% and the length is increased by 40%? The resistivity is fixed.

Solution: Eq. (3.13):

$$G_i = \frac{1}{R_i} = \frac{1}{\dfrac{\rho_i l_i}{A_i}} = \frac{A_i}{\rho_i l_i}$$

with the subscript i for the initial value. Using the subscript n for the new value, we obtain

$$G_n = \frac{A_n}{\rho_n l_n} = \frac{0.70A_i}{\rho_i(1.4l_i)} = \frac{0.70}{1.4}\frac{A_i}{\rho_i l_i} = \frac{0.70G_i}{1.4}$$

and $\quad G_n = \mathbf{0.5}\boldsymbol{G_i}$

3.8 OHMMETERS

The **ohmmeter** is an instrument used to perform the following tasks and several other useful functions:

1. *Measure the resistance of individual or combined elements.*
2. *Detect open-circuit (high-resistance) and short-circuit (low-resistance) situations.*
3. *Check the continuity of network connections and identify wires of a multilead cable.*
4. *Test some semiconductor (electronic) devices.*

For most applications, the ohmmeters used most frequently are the ohmmeter section of a VOM or DMM. The details of the internal circuitry and the method of using the meter will be left primarily for a laboratory exercise. In general, however, the resistance of a resistor can be measured by simply connecting the two leads of the meter across the resistor, as shown in Fig. 3.28. There is no need to be concerned about which lead goes on which end; the result is the same in either case since resistors offer the same resistance to the flow of charge (current) in either direction. If the VOM is used, a switch must be set to the proper resistance range, and a nonlinear scale (usually the top scale of the meter) must be properly read to obtain the resistance value. The DMM also requires choosing the best scale setting for the resistance to be measured, but the result appears as a numerical display, with the proper placement of the decimal point determined by the chosen scale. When measuring the resistance of a single resistor, it is usually best to remove the resistor from the network before making the measurement. If this is difficult or impossible, at least one end of the resistor must not be connected to the network, otherwise the reading may include the effects of the other elements of the system.

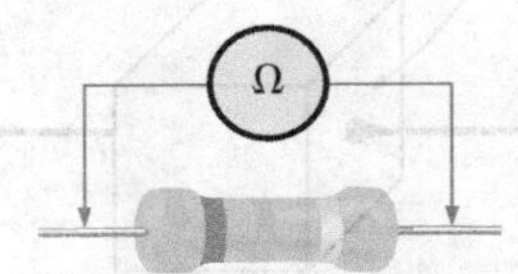

FIG. 3.28
Measuring the resistance of a single element.

If the two leads of the meter are touching in the ohmmeter mode, the resulting resistance is zero. A connection can be checked as shown in Fig. 3.29 by simply hooking up the meter to either side of the connection. If the resistance is zero, the connection is secure. If it is other than zero, the connection could be weak; if it is infinite, there is no connection at all.

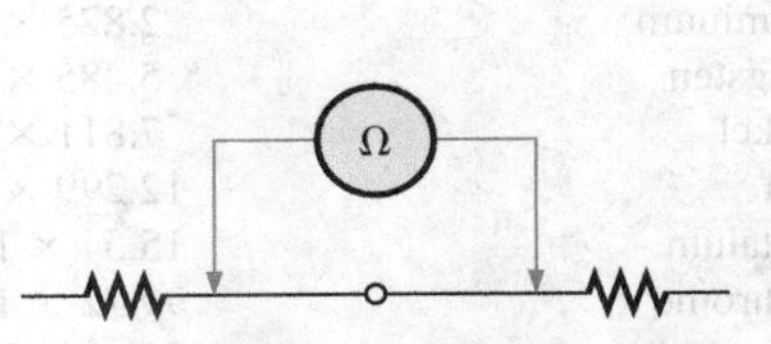

FIG. 3.29
Checking the continuity of a connection.

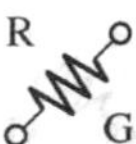

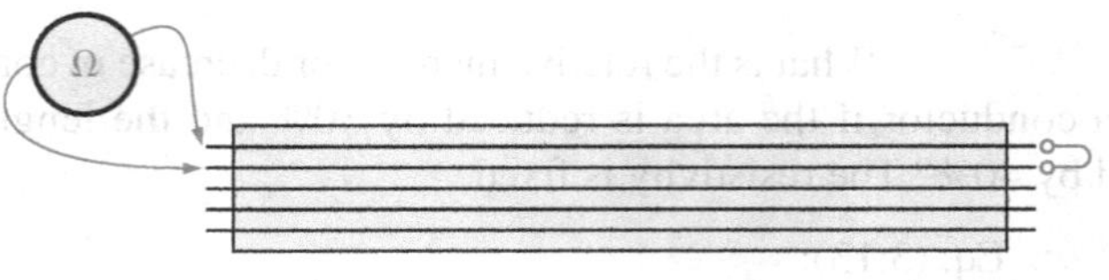

FIG. 3.30
Identifying the leads of a multilead cable.

If one wire of a harness is known, a second can be found as shown in Fig. 3.30. Simply connect the end of the known lead to the end of any other lead. When the ohmmeter indicates zero ohms (or very low resistance), the second lead has been identified. The above procedure can also be used to determine the first known lead by simply connecting the meter to any wire at one end and then touching all the leads at the other end until a zero ohm indication is obtained.

Preliminary measurements of the condition of some electronic devices such as the diode and the transistor can be made using the ohmmeter. The meter can also be used to identify the terminals of such devices.

One important note about the use of any ohmmeter:

Never hook up an ohmmeter to a live circuit!

The reading will be meaningless, and you may damage the instrument. The ohmmeter section of any meter is designed to pass a small sensing current through the resistance to be measured. A large external current could damage the movement and would certainly throw off the calibration of the instrument. In addition:

Never store a VOM or a DMM in the resistance mode.

If the two leads of the meter touch, the small sensing current could drain the internal battery. VOMs should be stored with the selector switch on the highest voltage range, and the selector switch of DMMs should be in the off position.

3.9 RESISTANCE: METRIC UNITS

The design of resistive elements for various areas of application, including thin-film resistors and integrated circuits, uses metric units for the quantities of Eq. (3.1) introduced in Section 3.2. In SI units, the resistivity would be measured in ohm-meters, the area in square meters, and the length in meters. However, the meter is generally too large a unit of measure for most applications, and so the centimeter is usually employed. The resulting dimensions for Eq. (3.1) are therefore

ρ:	ohm-centimeters
l:	centimeters
A:	square centimeters

The units for ρ can be derived from

$$\rho = \frac{RA}{l} = \frac{\Omega\text{-cm}^2}{\text{cm}} = \Omega\text{-cm}$$

The resistivity of a material is actually the resistance of a sample such as that appearing in Fig. 3.31. Table 3.6 provides a list of values of ρ in

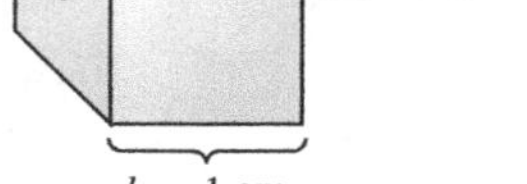

FIGURE 3.31
Defining ρ in ohm-centimeters.

TABLE 3.6
Resistivity (r) of various materials.

Material	Ω-cm
Silver	1.645×10^{-6}
Copper	$\mathbf{1.723 \times 10^{-6}}$
Gold	2.443×10^{-6}
Aluminum	2.825×10^{-6}
Tungsten	5.485×10^{-6}
Nickel	7.811×10^{-6}
Iron	12.299×10^{-6}
Tantalum	15.54×10^{-6}
Nichrome	99.72×10^{-6}
Tin oxide	250×10^{-6}
Carbon	3500×10^{-6}

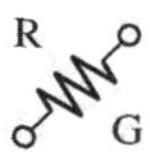

ohm-centimeters. Note that the area now is expressed in square centimeters, which can be determined using the basic equation $A = \pi d^2/4$, eliminating the need to work with circular mils, the special unit of measure associated with circular wires.

EXAMPLE 3.15 Determine the resistance of 100 ft of #28 copper telephone wire if the diameter is 0.0126 in.

Solution: Unit conversions:

$$l = 100\ \text{ft}\left(\frac{12\ \text{in.}}{1\ \text{ft}}\right)\left(\frac{2.54\ \text{cm}}{1\ \text{in.}}\right) = 3048\ \text{cm}$$

$$d = 0.0126\ \text{in.}\left(\frac{2.54\ \text{cm}}{1\ \text{in.}}\right) = 0.032\ \text{cm}$$

Therefore,

$$A = \frac{\pi d^2}{4} = \frac{(3.1416)(0.032\ \text{cm})^2}{4} = 8.04 \times 10^{-4}\ \text{cm}^2$$

$$R = \rho\frac{l}{A} = \frac{(1.723 \times 10^{-6}\Omega\text{-cm})(3048\ \text{cm})}{8.04 \times 10^{-4}\ \text{cm}^2} \cong \mathbf{6.5\ \Omega}$$

Using the units for circular wires and Table 3.2 for the area of a #28 wire, we find

$$R = \rho\frac{l}{A} = \frac{(10.37\ \text{CM-}\Omega/\text{ft})(100\ \text{ft})}{159.79\ \text{CM}} \cong \mathbf{6.5\ \Omega}$$

EXAMPLE 3.16 Determine the resistance of the thin-film resistor in Fig. 3.32 if the **sheet resistance** R_s (defined by $R_s = \rho/d$) is 100 Ω.

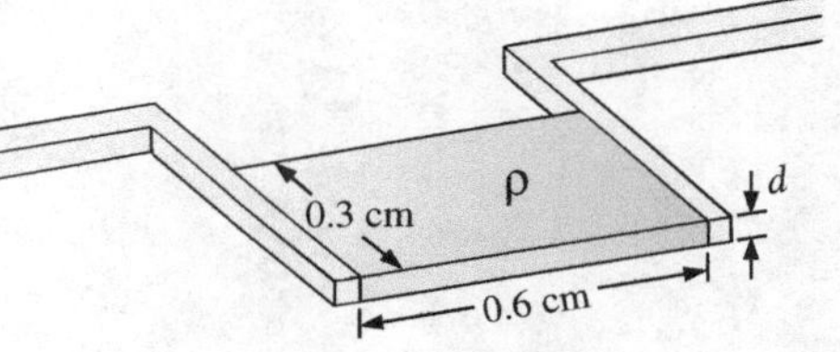

FIG. 3.32
Thin-film resistor. Example 3.16.

Solution: For deposited materials of the same thickness, the sheet resistance factor is usually employed in the design of thin-film resistors. Eq. (3.1) can be written

$$R = \rho\frac{l}{A} = \rho\frac{l}{dw} = \left(\frac{\rho}{d}\right)\left(\frac{l}{w}\right) = R_s\frac{l}{w}$$

where l is the length of the sample and w is the width. Substituting into the above equation yields

$$R = R_s\frac{l}{w} = \frac{(100\ \Omega)(0.6\ \text{cm})}{0.3\ \text{cm}} = \mathbf{200\ \Omega}$$

as one might expect since $l = 2w$.

The conversion factor between resistivity in circular mil-ohms per foot and ohm-centimeters is the following:

$$\boxed{\rho\,(\Omega\text{-cm}) = (1.662 \times 10^{-7}) \times (\text{value in CM-}\Omega/\text{ft})}$$

For example, for copper, $\rho = 10.37$ CM-Ω/ft:

$$\rho\,(\Omega\text{-cm}) = 1.662 \times 10^{-7}\,(10.37\ \text{CM-}\Omega/\text{ft}) = 1.723 \times 10^{-6}\ \Omega\text{-cm}$$

as indicated in Table 3.6.

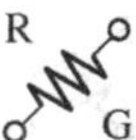

The resistivity in an integrated circuit design is typically in ohm-centimeter units, although tables often provide ρ in ohm-meters or microhm-centimeters. Using the conversion technique of Chapter 1, we find that the conversion factor between ohm-centimeters and ohm-meters is the following:

$$1.723 \times 10^{-6}\Omega\text{-}\cancel{\text{cm}}\left[\frac{1\text{ m}}{100\cancel{\text{cm}}}\right] = \frac{1}{100}[1.723 \times 10^{-6}]\Omega\text{-m}$$

or the value in ohm-meters is 1/100 the value in ohm-centimeters, and

$$\boxed{\rho(\Omega\text{-m}) = \left(\frac{1}{100}\right) \times (\text{value in } \Omega\text{-cm})} \quad \textbf{(3.14)}$$

Similarly,

$$\boxed{\rho(\mu\Omega\text{-cm}) = (10^6) \times (\text{value in } \Omega\text{-cm})} \quad \textbf{(3.15)}$$

For comparison purposes, typical values of ρ in ohm-centimeters for conductors, semiconductors, and insulators are provided in Table 3.7.

TABLE 3.7
Comparing levels of ρ in Ω-cm.

Conductor (Ω-cm)	Semiconductor (Ω-cm)		Insulator (Ω-cm)
Copper 1.723×10^{-6}	Ge	50	In general: 10^{15}
	Si	200×10^3	
	GaAs	70×10^6	

In particular, note the power-of-ten difference between conductors and insulators (10^{21})—a difference of huge proportions. There is a significant difference in levels of ρ for the list of semiconductors, but the power-of-ten difference between the conductor and insulator levels is at least 10^6 for each of the semiconductors listed.

3.10 THE FOURTH ELEMENT—THE MEMRISTOR

In May 2008 researchers at Hewlett Packard Laboratories led by Dr. Stanley Williams had an amazing announcement—the discovery of the "missing" link in basic electronic circuit theory called a **memristor,** shown in Fig. 3.33, and on the cover of this book. Up to this point the basic passive elements of circuit theory were the resistor, the capacitor, and the inductor, with the last two to be introduced later in this text. The presence of this fourth element was postulated in a seminal 1971 paper in the *IEEE Transactions on Circuit Theory* by Leon Chua of the University of California at Berkeley. However, it was not until this announcement that the device was actually constructed and found to function as predicted. Many attempts were made to build a memristor through the years, but it was not until work was done at the nanometer scale that success was obtained. It turns out that the smaller the structure, the more prominent is the *memristance* response. The level of memristance at the nanometer scale is a million times stronger than at the micrometer scale and is almost undetectable at the millimeter scale.

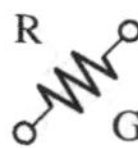

However, this property can work to the advantage of current IC designs that are already in the nanometer range.

The four basic circuit quantities of charge, current, voltage, and magnetic flux can be related in six ways. Three relations derive from the basic elements of the resistor, the capacitor, and the inductor. The resistor provides a direct relationship between current and voltage, the capacitor provides a relationship between charge and voltage, and the inductor provides a relationship between current and magnetic flux. That leaves the relationship between the magnetic field and the charge moving through an element. Chua sought a device that would define the relationship between magnetic flux and charge similar to that between the voltage and current of a resistor.

In general, Chua was looking for a device whose resistance would be a function of how much charge has passed through it. In Chapter 11 the relationship between the movement of charge and the surrounding magnetic field will be described in keeping with the need to find a device relating charge flow and the surrounding magnetic field.

The memristor is a device whose resistance increases with increase in the flow of charge in one direction and decreases as the flow of charge decreases in the reverse direction. Furthermore, and vastly important, it maintains its new resistance level when the excitation has been removed.

This behavior in the nanometer range was discovered using the semiconductor titanium dioxide (TiO_2), which is a highly resistive material but can be doped with other materials to make it very conductive. In this material the dopants move in response to an applied electric field and drift in the direction of the resulting current. Starting out with a memristor with dopants only one side and pure TiO_2 on the other, one can apply a biasing voltage to establish a current in the memristor. The resulting current will cause the dopants to move to the pure side and reduce the resistance of the element. The greater the flow of charge, the lower is the resulting resistance. In other words, as mentioned, TiO_2 has a high resistance, and on moving the dopants into the pure TiO_2, the resistance drops. The entire process of moving the dopants is due to the applied voltage and resulting motion of charge. Reversing the biasing voltage reverses the direction of current flow and brings the dopants back to the other side, thereby letting the TiO_2 return to its high-resistance state; on the surface this seems rather simple and direct.

An analog often applied to describe the action of a memristor is the flow of water (analogous to charge) through a pipe. In general, the resistance of a pipe to the flow of water is directly related to the diameter of the pipe: the smaller the pipe, the greater is the resistance, and the larger the diameter, the lower is the resistance. For the analogy to be appropriate in describing the action of a memristor, the diameter of the pipe must also be a function of the speed of the water and its direction. Water flowing in one direction will cause the pipe to expand and reduce the resistance. The faster the flow, the greater is the diameter. For water flowing in the opposite direction, the faster the flow, the smaller is the diameter and the greater is the resistance. The instant the flow of water is stopped in either direction, the pipe keeps its new diameter and resistance.

There are 17 memristors in Fig. 3.33 lined up in a row, each with a width of about 50 nm. Each has a bottom wire connected to one side of the device and a top wire connected to the opposite side through a network of wires. Each will then exhibit a resistance depending on the direction and magnitude of the charge through each one. The current

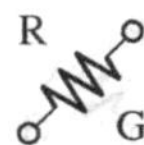

choice for the electronic symbol is also provided in Fig. 3.33. It is similar in design to the resistor symbol but also markedly different.

Thus, we have a memory device that will have a resistance dependent on the direction and level of charge flowing through it. Remove the flow of charge, and it maintains its new resistance level. The impact of such a device is enormous—computers would remember the last operation and display when they were turned off. Come back in a few hours or days and the display would be exactly as you left it. The same would be true for any system working through a range of activities and applications—instant startup exactly where you left off. It will be quite interesting to follow how this fourth element affects the electronics field in general.

Like the transistor, which was initially questioned and now is of such enormous importance, the memristor may stimulate the same awe-inspiring change in every electronic application.

3.11 SUPERCONDUCTORS

The field of electricity/electronics is one of the most exciting of our time. New developments appear almost weekly from extensive research and development activities. The research drive to develop a superconductor capable of operating at temperatures closer to room temperature has been receiving increasing attention in recent years due to the need to cut energy losses.

What are superconductors? Why is their development so important? In a nutshell,

superconductors are conductors of electric charge that, for all practical purposes, have zero resistance.

In a conventional conductor, electrons travel at average speeds of about 1000 mi/s (they can cross the United States in about 3 seconds), even though Einstein's theory of relativity suggests that the maximum speed of information transmission is the speed of light, or 186,000 mi/s. The relatively slow speed of conventional conduction is due to collisions with atoms in the material, repulsive forces between electrons (like charges repel), thermal agitation that results in indirect paths due to the increased motion of the neighboring atoms, impurities in the conductor, and so on. In the superconductive state, there is a pairing of electrons, denoted the **Cooper effect,** in which electrons travel in pairs and help each other maintain a significantly higher velocity through the medium. In some ways this is like "drafting" by competitive cyclists or runners. There is an oscillation of energy between partners or even "new" partners (as the need arises) to ensure passage through the conductor at the highest possible velocity with the least total expenditure of energy.

Even though the concept of superconductivity first surfaced in 1911, it was not until 1986 that the possibility of superconductivity at room temperature became a renewed goal of the research community. For over 70 years, superconductivity could be established only at temperatures colder than 23 K. (Kelvin temperature is universally accepted as the unit of measurement for temperature for superconductive effects. Recall that $K = 273.15° + °C$, so a temperature of 23 K is $-250°C$, or $-418°F$.) In 1986, however, physicists Alex Muller and George Bednorz of the IBM Zurich Research Center found a ceramic material—lanthanum barium copper oxide—that exhibited superconductivity at 30 K. This discovery introduced a new direction to the research effort and spurred others to

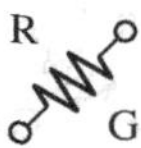

improve on the new standard. (In 1987, both scientists received the Nobel prize for their contribution to an important area of development.)

In just a few short months, Professors Paul Chu of the University of Houston and Man Kven Wu of the University of Alabama raised the temperature to 95 K using a superconductor of yttrium barium copper oxide. The result was a level of excitement in the scientific community that brought research in the area to a new level of effort and investment. The major impact of this discovery was that liquid nitrogen (boiling point of 77 K) rather than liquid helium (boiling point of 4 K) could now be used to bring the material down to the required temperature. The result is a tremendous saving in the cooling expense since liquid nitrogen is at least ten times less expensive than liquid helium. Pursuing the same direction, some success has been achieved at 125 K and 162 K using a thallium compound (unfortunately, however, thallium is a very poisonous substance).

Fig. 3.34 illustrates how the discovery in 1986 of using a ceramic material in superconductors led to rapid developments in the field. In 2008 a tin–copper oxide superconductor with a small amount of indium reached a new peak of 212 K—an enormous increase in temperature.

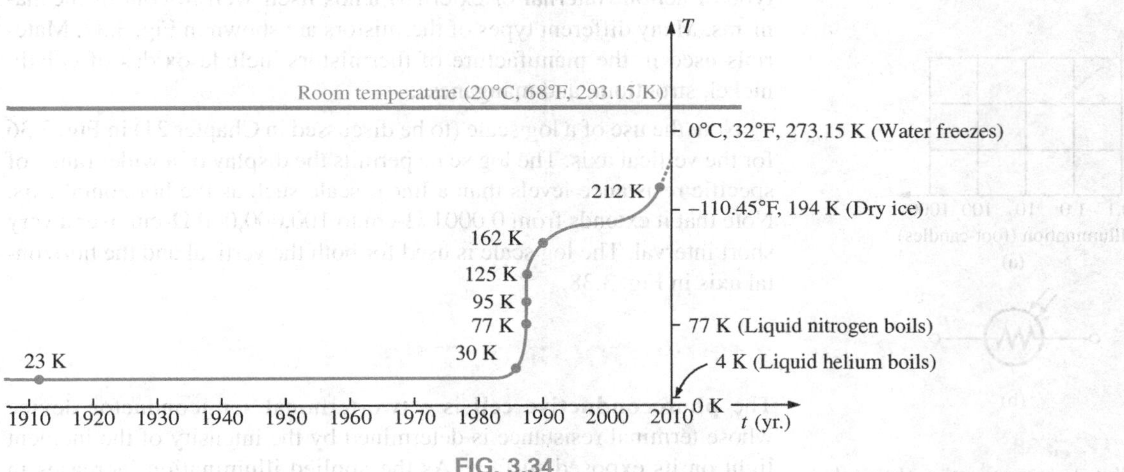

FIG. 3.34
Rising temperatures of superconductors.

The temperature at which a superconductor reverts back to the characteristics of a conventional conductor is called the *critical temperature,* denoted by T_c. Note in Fig. 3.35 that the resistivity level changes abruptly at T_c. The sharpness of the transition region is a function of the purity of the sample. Long listings of critical temperatures for a variety of tested compounds can be found in reference materials providing tables of a wide variety to support research in physics, chemistry, geology, and related fields. Two such publications include the CRC (Chemical Rubber Co.) *Handbook of Tables for Applied Engineering Science* and the CRC *Handbook of Chemistry and Physics.*

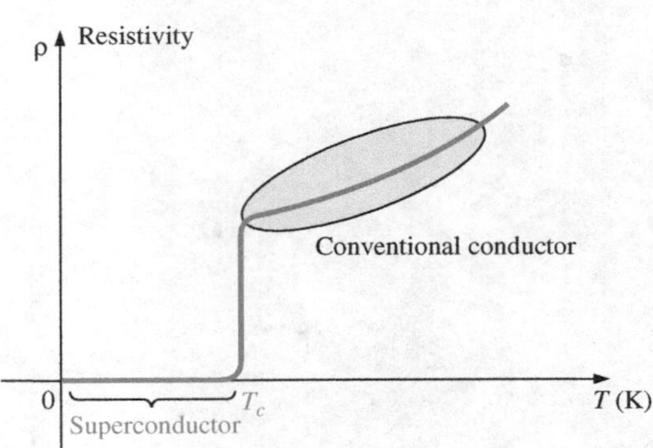

FIG. 3.35
Defining the critical temperature T_c.

Although room-temperature success has not been attained, numerous applications for some of the superconductors have been developed. It is simply a matter of balancing the additional cost against the results obtained or deciding whether any results at all can be obtained without the use of this zero-resistance state. Some research efforts require high-energy accelerators or strong magnets attainable only with superconductive materials. Superconductivity is currently applied in the design of Maglev trains (trains that ride on a cushion of air established by opposite magnetic poles) that exceed 300 mi/h, in powerful motors and generators,

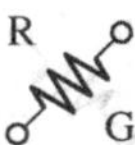

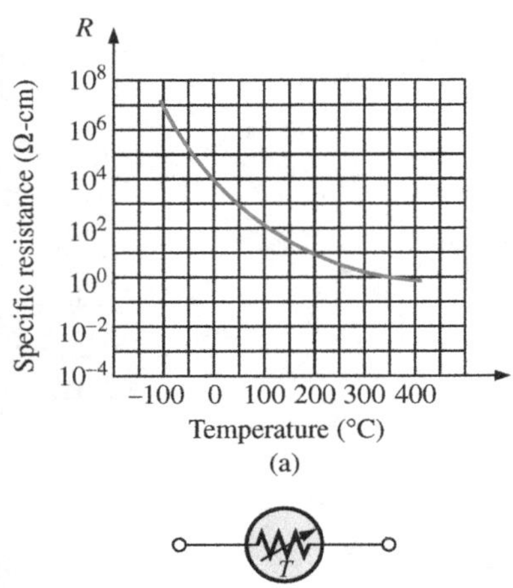

FIG. 3.36
Thermistor: (a) characteristics; (b) symbol.

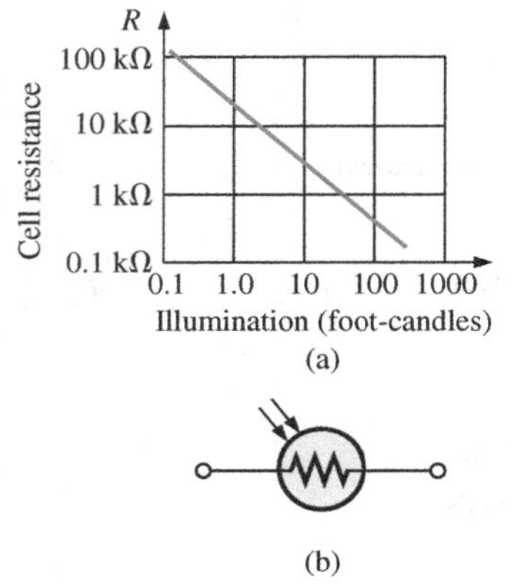

FIG. 3.38
Photoconductive cell: (a) characteristics. (b) symbol.

in nuclear magnetic resonance imaging (MRI) systems to obtain cross-sectional images of the brain (and other parts of the body), in the design of computers with operating speeds four times that of conventional systems, and in improved power distribution systems.

3.12 THERMISTORS

The **thermistor** is a two-terminal semiconductor device whose resistance, as the name suggests, is temperature sensitive. A representative characteristic appears in Fig. 3.36 with the graphic symbol for the device. Note the nonlinearity of the curve and the drop in resistance from about 5000Ω-cm to 100Ω-cm for an increase in temperature from 20°C to 100°C. The decrease in resistance with an increase in temperature indicates a negative temperature coefficient.

The temperature of the device can be changed internally or externally. An increase in current through the device raises its temperature, causing a drop in its terminal resistance. Any externally applied heat source results in an increase in its body temperature and a drop in resistance. This type of action (internal or external) lends itself well to control mechanisms. Many different types of thermistors are shown in Fig. 3.37. Materials used in the manufacture of thermistors include oxides of cobalt, nickel, strontium, and manganese.

Note the use of a log scale (to be discussed in Chapter 21) in Fig. 3.36 for the vertical axis. The log scale permits the display of a wider range of specific resistance levels than a linear scale such as the horizontal axis. Note that it extends from 0.0001 Ω-cm to 100,000,000 Ω-cm over a very short interval. The log scale is used for both the vertical and the horizontal axis in Fig. 3.38.

3.13 PHOTOCONDUCTIVE CELL

The **photoconductive cell** is a two-terminal semiconductor device whose terminal resistance is determined by the intensity of the incident light on its exposed surface. As the applied illumination increases in

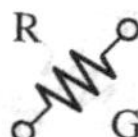

intensity, the energy state of the surface electrons and atoms increases, with a resultant increase in the number of "free carriers" and a corresponding drop in resistance. A typical set of characteristics and the photoconductive cell's graphic symbol appear in Fig. 3.38. Note the negative illumination coefficient. Several cadmium sulfide photoconductive cells appear in Fig. 3.39.

3.14 VARISTORS

Varistors are voltage-dependent, nonlinear resistors used to suppress high-voltage transients; that is, their characteristics enable them to limit the voltage that can appear across the terminals of a sensitive device or system. A typical set of characteristics appears in Fig. 3.40(a), along with a linear resistance characteristic for comparison purposes. Note that at a particular "firing voltage," the current rises rapidly, but the voltage is limited to a level just above this firing potential. In other words, the magnitude of the voltage that can appear across this device cannot exceed that level defined by its characteristics. Through proper design techniques, this device can therefore limit the voltage appearing across sensitive regions of a network. The current is simply limited by the network to which it is connected. A photograph of a number of commercial units appears in Fig. 3.40(b).

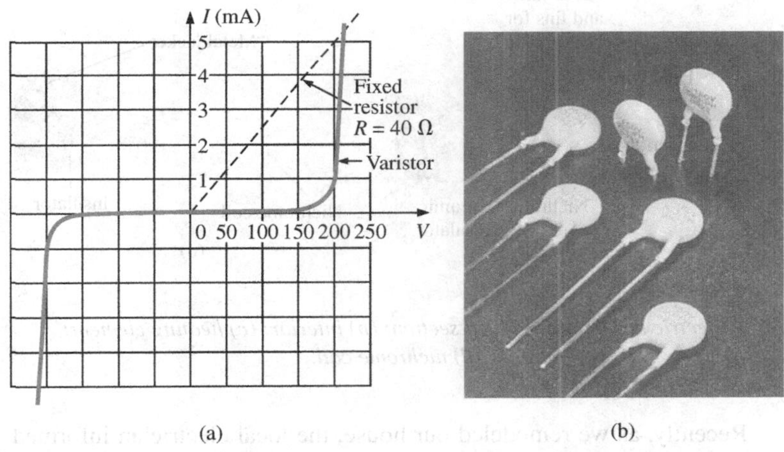

FIG. 3.40
Varistors available with maximum dc voltage ratings between 18 V and 615 V.
(Courtesy of Royal Philips.)

3.15 APPLICATIONS

The following are examples of how resistance can be used to perform a variety of tasks, from heating to measuring the stress or strain on a supporting member of a structure. In general, resistance is a component of every electrical or electronic application.

Electric Baseboard Heating Element

One of the most common applications of resistance is in household fixtures such as toasters and baseboard heating where the heat generated by current passing through a resistive element is employed to perform a useful function.

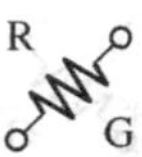

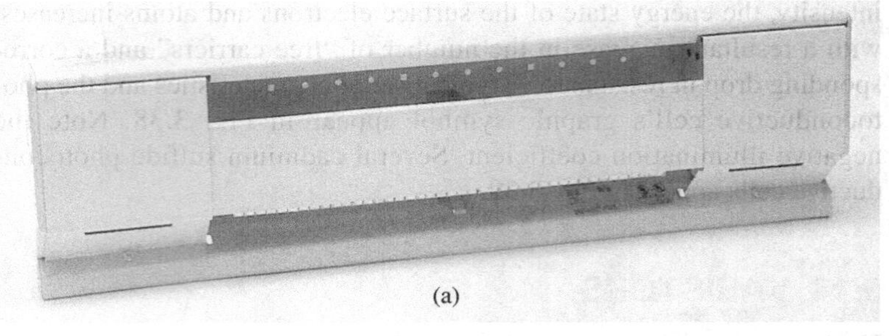

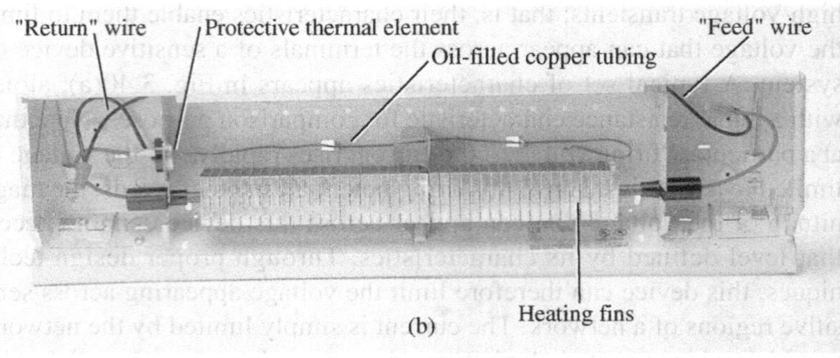

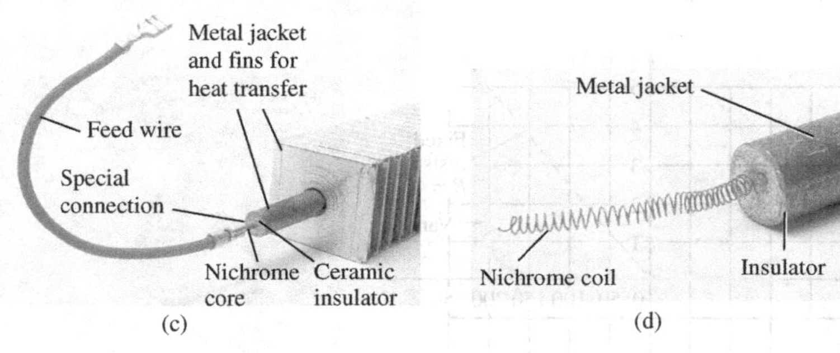

FIG. 3.41
Electric baseboard: (a) 2-ft section; (b) interior; (c) heating element; (d) nichrome coil.

Recently, as we remodeled our house, the local electrician informed us that we were limited to 16 ft of electric baseboard on a single circuit. That naturally had me wondering about the wattage per foot, the resulting current level, and whether the 16-ft limitation was a national standard. Reading the label on the 2-ft section appearing in Fig. 3.41(a). I found VOLTS AC 240/208, WATTS 750/575 (the power rating is described in Chapter 4), AMPS 3.2/2.8. Since my panel is rated 240 V (as are those in most residential homes), the wattage rating per foot is 750 W/2 or 375 W at a current of 3.2 A. The total wattage for the 16 ft is therefore 16 × 375 W, or 6000 W.

In Chapter 4, you will find that the power to a resistive load is related to the current and applied voltage by the equation $P = VI$. The total resulting current can then be determined using this equation in the following manner: $I = P/V = 6000\text{ W}/240\text{ V} = 25\text{ A}$. The result was that we needed a circuit breaker larger than 25 A; otherwise, the circuit breaker would trip every time we turned the heat on. In my case, the electrician used a 32 A breaker to meet the National Fire Code requirement that does not permit exceeding 80% of the rated current for a

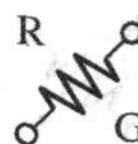

conductor or breaker. In most panels, a 30 A breaker takes two slots of the panel, whereas the more common 20 A breaker takes only one slot. If you have a moment, take a look in your own panel and note the rating of the breakers used for various circuits of your home.

Going back to Table 3.2, we find that the #12 wire commonly used for most circuits in the home has a maximum rating of 20 A and would not be suitable for the electric baseboard. Since #11 is usually not commercially available, a #10 wire with a maximum rating of 30 A was used. You might wonder why the current drawn from the supply is 19.17 A while that required for one unit was only 2.8 A. This difference is due to the parallel combination of sections of the heating elements, a configuration that will be described in Chapter 6. It is now clear why the requirement specifies a 16-ft limitation on a single circuit. Additional elements would raise the current to a level that would exceed the code level for #10 wire and would approach the maximum rating of the circuit breaker.

Fig. 3.41(b) shows a photo of the interior construction of the heating element. The red feed wire on the right is connected to the core of the heating element, and the black wire at the other end passes through a protective heater element and back to the terminal box of the unit (the place where the exterior wires are brought in and connected). If you look carefully at the end of the heating unit as shown in Fig. 3.41(c), you will find that the heating wire that runs through the core of the heater is not connected directly to the round jacket holding the fins in place. A ceramic material (insulator) separates the heating wire from the fins to remove any possibility of conduction between the current passing through the bare heating element and the outer fin structure. Ceramic materials are used because they are excellent conductors of heat. They also have a high retentivity for heat, so the surrounding area remains heated for a period of time even after the current has been turned off. As shown in Fig. 3.41(d), the heating wire that runs through the metal jacket is normally a nichrome composite (because pure nichrome is quite brittle) wound in the shape of a coil to compensate for expansion and contraction with heating and also to permit a longer heating element in standard-length baseboard. On opening the core, we found that the nichrome wire in the core of a 2-ft baseboard was actually 7 ft long, or a 3.5 : 1 ratio. The thinness of the wire was particularly noteworthy, measuring out at about 8 mils in diameter, not much thicker than a hair. Recall from this chapter that the longer the conductor and the thinner the wire, the greater is the resistance. We took a section of the nichrome wire and tried to heat it with a reasonable level of current and the application of a hair dryer. The change in resistance was almost unnoticeable. In other words, all our effort to increase the resistance with the basic elements available to us in the lab was fruitless. This was an excellent demonstration of the meaning of the temperature coefficient of resistance in Table 3.4. Since the coefficient is so small for nichrome, the resistance does not measurably change unless the change in temperature is truly significant. The curve in Fig. 3.11 would therefore be close to horizontal for nichrome. For baseboard heaters, this is an excellent characteristic because the heat developed, and the power dissipated, will not vary with time as the conductor heats up with time. The flow of heat from the unit will remain fairly constant.

The feed and return cannot be soldered to the nichrome heater wire for two reasons. First, you cannot solder nichrome wires to each other or to other types of wire. Second, if you could, there might be a problem because

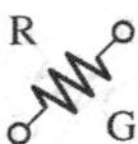

the heat of the unit could rise above 880°F at the point where the wires are connected, the solder could melt, and the connection could be broken. Nichrome must be spot welded or crimped onto the copper wires of the unit. Using Eq. (3.1) and the 8-mil measured diameter, and assuming pure nichrome for the moment, we find that the resistance of the 7-ft length is

$$R = \frac{\rho l}{A}$$

$$= \frac{(600)(7\text{ ft})}{(8\text{ mils})^2} = \frac{4200}{64}$$

$$R = \mathbf{65.6\ \Omega}$$

In Chapter 4, a power equation will be introduced in detail relating power, current, and resistance in the following manner: $P = I^2R$. Using the above data and solving for the resistance, we obtain

$$R = \frac{P}{I^2}$$

$$= \frac{575\text{ W}}{(2.8\text{ A})^2}$$

$$R = \mathbf{73.34\ \Omega}$$

which is very close to the value calculated above from the geometric shape since we cannot be absolutely sure about the resistivity value for the composite.

During normal operation, the wire heats up and passes that heat on to the fins, which in turn heat the room via the air flowing through them. The flow of air through the unit is enhanced by the fact that hot air rises, so when the heated air leaves the top of the unit, it draws cold air from the bottom to contribute to the convection effect. Closing off the top or bottom of the unit would effectively eliminate the convection effect, and the room would not heat up. A condition could occur in which the inside of the heater became too hot, causing the metal casing also to get too hot. This concern is the primary reason for the thermal protective element introduced above and appearing in Fig. 3.41(b). The long, thin copper tubing in Fig. 3.41 is actually filled with an oil-type fluid that expands when heated. If it is too hot, it expands, depresses a switch in the housing, and turns off the heater by cutting off the current to the heater wire.

Dimmer Control in an Automobile

A two-point rheostat is the primary element in the control of the light intensity on the dashboard and accessories of a car. The basic network appears in Fig. 3.42 with typical voltage and current levels. When the light switch is closed (usually by pulling the light control knob out from the dashboard), current is established through the 50 Ω rheostat and then to the various lights on the dashboard (including the panel lights, ashtray light, radio display, and glove compartment light). As the knob of the control switch is turned, it controls the amount of resistance between points *a* and *b* of the rheostat. The more resistance between points *a* and *b*, the less is the current and the less is the brightness of the various lights. Note the additional switch in the glove compartment light, which is activated by the opening of the door of the compartment. Aside from the glove compartment light, all the lights in Fig. 3.42 will be on at the same time when the light switch is activated. The first branch after the

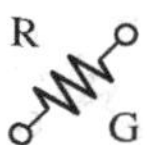

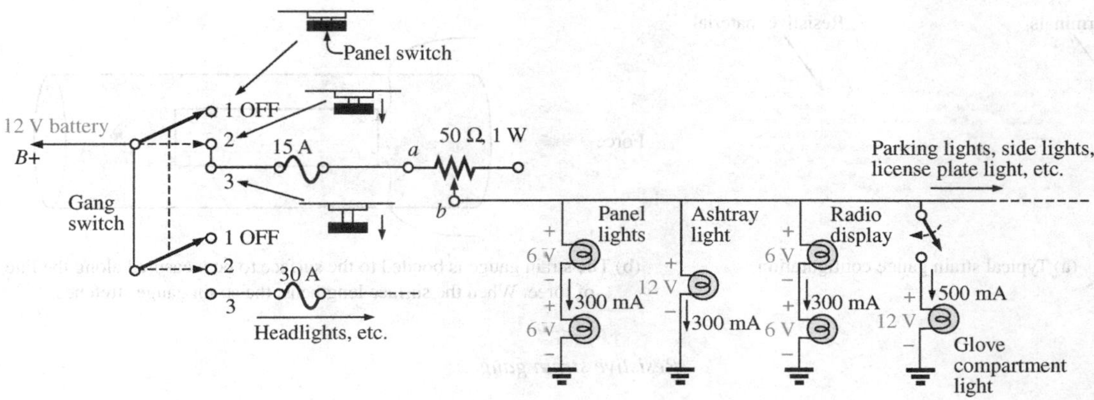

FIG. 3.42
Dashboard dimmer control in an automobile.

rheostat contains two bulbs of 6 V rating rather than the 12 V bulbs appearing in the other branches. The smaller bulbs of this branch produce a softer, more even light for specific areas of the panel. Note that the sum of the two bulbs (in series) is 12 V to match that across the other branches. The division of voltage in any network is covered in detail in Chapters 5 and 6.

Typical current levels for the various branches have also been provided in Fig. 3.42. You will learn in Chapter 6 that the current drain from the battery and through the fuse and rheostat approximately equals the sum of the currents in the branches of the network. The result is that the fuse must be able to handle current in amperes, so a 15 A fuse was used (even though the bulbs appear in Fig. 3.42 as 12 V bulbs to match the battery).

Whenever the operating voltage and current levels of a component are known, the internal "hot" resistance of the unit can be determined using Ohm's law, introduced in detail in Chapter 4. Basically this law relates voltage, current, and resistance by $I = V/R$. For the 12 V bulb at a rated current of 300 mA, the resistance is $R = V/I = 12\text{ V}/300\text{ mA} = 40\ \Omega$. For the 6 V bulbs, it is $6\text{ V}/300\text{ mA} = 20\ \Omega$. Additional information regarding the power levels and resistance levels is discussed in later chapters.

The preceding description assumed an ideal level of 12 V for the battery. In actuality, 6.3 V and 14 V bulbs are used to match the charging level of most automobiles.

Strain Gauges

Any change in the shape of a structure can be detected using strain gauges whose resistance changes with applied stress or flex. An example of a strain gauge is shown in Fig. 3.43. Metallic strain gauges are constructed of a fine wire or thin metallic foil in a grid pattern. The terminal resistance of the strain gauge will change when exposed to compression or extension. One simple example of the use of resistive strain gauges is to monitor earthquake activity. When the gauge is placed across an area of suspected earthquake activity, the slightest separation in the earth changes the terminal resistance, and the processor displays a result sensitive to the amount of separation. Another example is in alarm systems where the slightest change in the shape of a supporting beam when someone walks

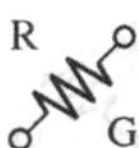

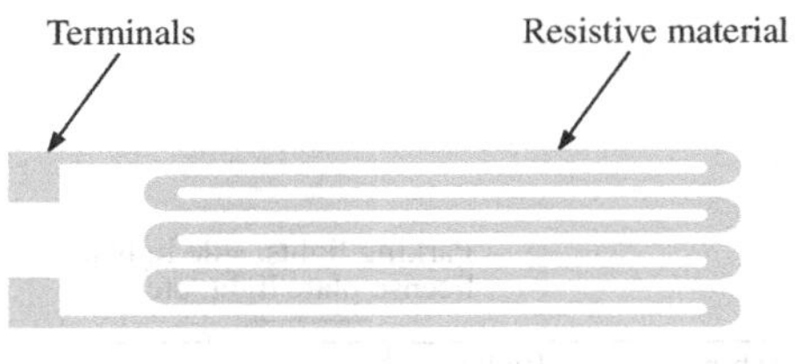

(a) Typical strain gauge configuration.

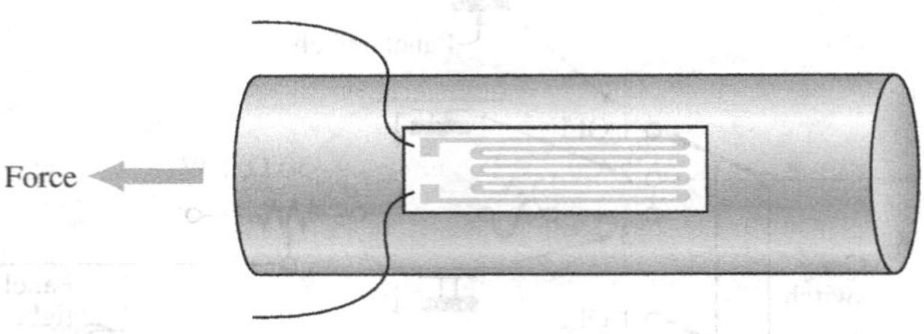

(b) The strain gauge is bonded to the surface to be measured along the line of force. When the surface lengthens, the strain gauge stretches.

FIG. 3.43
Resistive strain gauge.

overhead results in a change in terminal resistance, and an alarm sounds. Other examples include placing strain gauges on bridges to monitor their rigidity and on very large generators to check whether various moving components are beginning to separate because of a wearing of the bearings or spacers. The small mouse control within a computer keyboard can be a series of strain gauges that reveal the direction of compression or extension applied to the controlling element on the keyboard. Movement in one direction can extend or compress a resistance gauge, which can monitor and control the motion of the mouse on the screen.

PROBLEMS

SECTION 3.2 Resistance: Circular Wires

1. Convert the following to mils:
a. 0.5 in. **b.** 0.02 in.
c. 1/4 in. **d.** 10 mm
e. 0.01 ft **f.** 0.1 cm

2. Calculate the area in circular mils (CM) of wires having the following diameter:
a. 30 mils **b.** 0.016 in.
c. 1/8 in. **d.** 1 cm
e. 0.02 ft **f.** 4 mm

3. The area in circular mils is
a. 1600 CM **b.** 820 CM
c. 40,000 CM **d.** 625 CM
e. 6.25 CM **f.** 0.3×10^6 CM
What is the diameter of each wire in inches?

4. What is the resistance of a copper wire 200 ft long and 1/50 in. in diameter ($T = 20°C$)?

5. a. What is the area in circular mils of an aluminum conductor that is 80 ft long with a resistance of 2.5 Ω?
b. What is its diameter in inches?

6. A 2.2 Ω resistor is to be made of nichrome wire. If the available wire is 1/32 in. in diameter, how much wire is required?

7. a. What is the diameter in inches of a copper wire that has a resistance of 3.3 Ω and is as long as a football field (100 yd) ($T = 20°C$)?
b. Without working out the numerical solution, determine whether the area of an aluminum wire will be smaller or larger than that of the copper wire. Explain.
c. Repeat (b) for a silver wire.

8. A wire 1000 ft long has a resistance of 0.5 kΩ and an area of 94 CM. Of what material is the wire made ($T = 20°C$)?

9. a. A contractor is concerned about the length of copper hookup wire still on the reel of Fig. 3.44. He measured the resistance and found it to be 3.14 Ω. A tape measure indicated that the thickness of the stranded wire was about 1/32 in. What is the approximate length in feet?
b. What is the weight of the wire on the reel?
c. It is typical to see temperature ranges for materials listed in centigrade rather than Fahrenheit degrees. What is the range in Fahrenheit degrees? What is unique about the relationship between degrees Fahrenheit and degrees centigrade at −40°C?

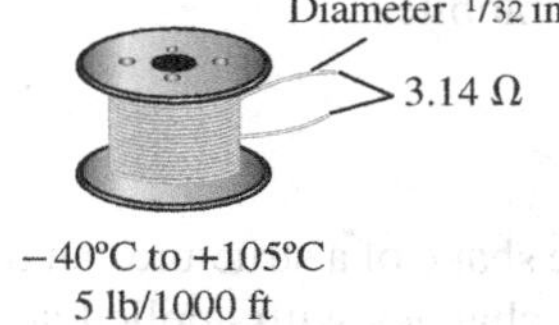

FIG. 3.44
Problem 9.

10. a. What is the cross-sectional area in circular mils of a rectangular copper bus bar if the dimensions are 3/8 in. by 4.8 in.?
b. If the area of the wire commonly used in house wiring has a diameter close to 1/12 in., how many wires would have to be combined to have the same area?

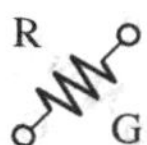

11. a. What is the resistance of a copper bus-bar for a high-rise building with the dimensions shown ($T = 20°C$) in Fig. 3.45?
   b. Repeat (a) for aluminum and compare the results.

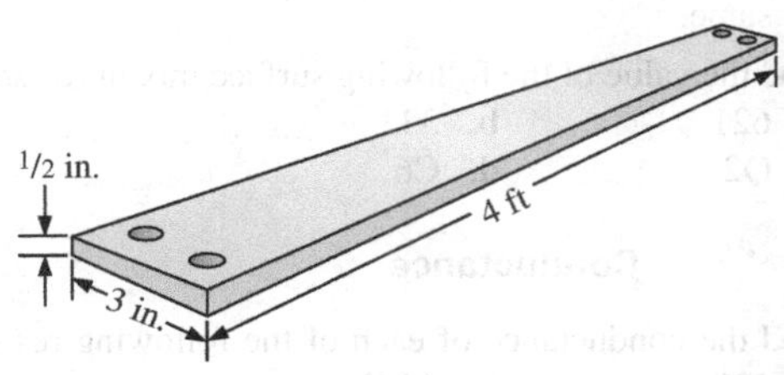

FIG. 3.45
Problem 11.

12. Determine the increase in resistance of a copper conductor if the area is reduced by a factor of 4 and the length is doubled. The original resistance was 0.2 Ω. The temperature remains fixed.

*13. What is the new resistance level of a copper wire if the length is changed from 200 ft to 100 yd, the area is changed from 40,000 CM to 0.04 in.^{2}, and the original resistance was 800 mΩ?

SECTION 3.3 Wire Tables

14. a. In construction the two most common wires employed in general house wiring are #12 and #14, although #12 wire is the most common because it is rated at 20 A. How much larger in area (by percent) is the #12 wire compared to the #14 wire?
   b. The maximum rated current for #14 wire is 15 A. How does the ratio of maximum current levels compare to the ratio of the areas of the two wires?

15. a. Compare the area of a #12 wire with the area of a #9 wire. Did the change in area substantiate the general rule that a drop of three gage numbers results in a doubling of the area?
   b. Compare the area of a #12 wire with that of a #0 wire. How many times larger in area is the #0 wire compared to the #12 wire? Is the result significant? Compare it to the change in maximum current rating for each.

16. a. Compare the area of a #20 hookup wire to a #10 house romax wire. Did the change in area substantiate the general rule that a drop of 10 gage numbers results in a tenfold increase in the area of the wire?
   b. Compare the area of a #20 wire with that of a #40 wire. How many times larger in area is the #20 wire than the #40 wire? Did the result support the rule of part (a)?

17. a. For the system in Fig. 3.46, the resistance of each line cannot exceed 6 mΩ, and the maximum current drawn by the load is 110 A. What minimum size gage wire should be used?
   b. Repeat (a) for a maximum resistance of 3 mΩ, $d = 30$ ft, and a maximum current of 110 A.

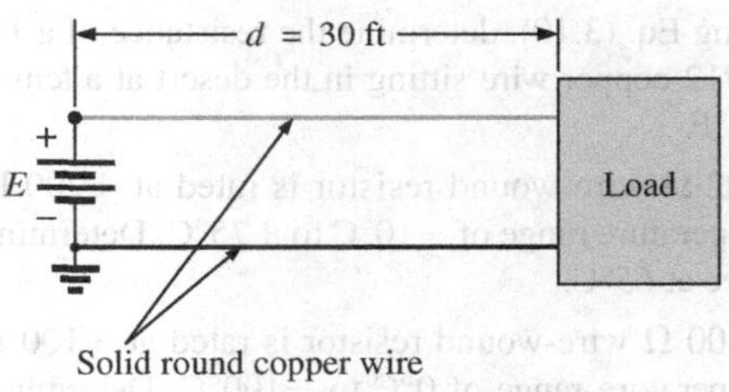

FIG. 3.46
Problem 17.

*18. a. From Table 3.2, determine the maximum permissible current density (A/CM) for an AWG #0000 wire.
   b. Convert the result of (a) to A/in.^{2}
   c. Using the result of (b), determine the cross-sectional area required to carry a current of 5000 A.

SECTION 3.4 Temperature Effects

19. The resistance of a copper wire is 2 Ω at 10°C. What is its resistance at 80°C?

20. The resistance of an aluminum bus-bar is 0.02 Ω at 0°C. What is its resistance at 100°C?

21. The resistance of a copper wire is 4 Ω at room temperature (68°F). What is its resistance at a freezing temperature of 32°F?

22. The resistance of a copper wire is 0.025 Ω at a temperature of 70°F.
   a. What is the resistance if the temperature drops 10° to 60°F?
   b. What is the resistance if it drops an additional 10° to 50°F?
   c. Noting the results of parts (a) and (b), what is the drop for each part in milliohms? Is the drop in resistance linear or nonlinear? Can you forecast the new resistance if it drops to 40°F, without using the basic temperature equation?
   d. If the temperature drops to −30°F in northern Maine, find the change in resistance from the room temperature level of part (a). Is the change significant?
   e. If the temperature increases to 120°F in Cairns, Australia, find the change in resistance from the room temperature of part (a). Is the change significant?

23. a. The resistance of a copper wire is 1 Ω at 4°C. At what temperature (°C) will it be 1.1 Ω?
   b. At what temperature will it be 0.1 Ω?

24. a. If the resistance of 1000 ft of wire is about 1 Ω at room temperature (68°F), at what temperature will it double in value?
   b. What gage wire was used?
   c. What is the approximate diameter in inches, using the closest fractional form?

25. a. Verify the value of α_{20} for copper in Table 3.6 by substituting the inferred absolute temperature into Eq. (3.9).
   b. Using Eq. (3.10), find the temperature at which the resistance of a copper conductor will increase to 1 Ω from a level of 0.8 Ω at 20°C.

26. Using Eq. (3.10), find the resistance of a copper wire at 16°C if its resistance at 20°C is 0.4 Ω.

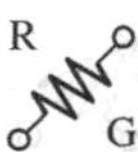

*27. Using Eq. (3.10), determine the resistance of a 1000-ft coil of #12 copper wire sitting in the desert at a temperature of 115°F.

28. A 22 Ω wire-wound resistor is rated at +200 PPM for a temperature range of −10°C to +75°C. Determine its resistance at 65°C.

29. A 100 Ω wire-wound resistor is rated at +100 PPM for a temperature range of 0°C to +100°C. Determine its resistance at 50°C.

SECTION 3.5 Types of Resistors

30. a. What is the approximate increase in size from a 1 W to a 2 W carbon resistor?
 b. What is the approximate increase in size from a 1/2 W to a 2 W carbon resistor?
 c. In general, can we conclude that for the same type of resistor, an increase in wattage rating requires an increase in size (volume)? Is it almost a linear relationship? That is, does twice the wattage require an increase in size of 2:1?

31. If the resistance between the outside terminals of a linear potentiometer is 10 kΩ, what is its resistance between the wiper (movable) arm and an outside terminal if the resistance between the wiper arm and the other outside terminal is 3.5 kΩ?

32. If the wiper arm of a linear potentiometer is one-fourth the way around the contact surface, what is the resistance between the wiper arm and each terminal if the total resistance is 2.5 kΩ?

*33. Show the connections required to establish 4 kΩ between the wiper arm and one outside terminal of a 10 kΩ potentiometer while having only zero ohms between the other outside terminal and the wiper arm.

SECTION 3.6 Color Coding and Standard Resistor Values

34. Find the range in which a resistor having the following color bands must exist to satisfy the manufacturer's tolerance:

	1st band	2nd band	3rd band	4th band
a.	gray	red	brown	gold
b.	red	red	brown	silver
c.	white	brown	orange	—
d.	white	brown	red	gold
e.	orange	white	green	—

35. Find the color code for the following 10% resistors:
 a. 68 Ω b. 0.33 Ω
 c. 22 kΩ d. 5.6 MΩ

36. a. Is there an overlap in coverage between 20% resistors? That is, determine the tolerance range for a 10 Ω 20% resistor and a 15 Ω 20% resistor, and note whether their tolerance ranges overlap.
 b. Repeat part (a) for 10% resistors of the same value.

37. Given a resistor coded yellow, violet, brown, silver that measures 492 Ω, is it within tolerance? What is the tolerance range?

38. a. How would Fig. 3.26(a) change if the resistors of 47 Ω, 68 Ω, and 100 Ω were changed to 4.7 kΩ, 6.8 kΩ, and 10 kΩ, respectively, if the tolerance remains the same.
 b. How would Fig. 3.26(a) change if the resistors of 47 Ω, 68 Ω, and 100 Ω were changed to 4.7 MΩ, 6.8 MΩ, and 10 MΩ, respectively, and the tolerance remained the same.

39. Find the value of the following surface mount resistors:
 a. 621 b. 333
 c. Q2 d. C6

SECTION 3.7 Conductance

40. Find the conductance of each of the following resistances:
 a. 120 Ω b. 4 kΩ
 c. 2.2 MΩ d. Compare the three results.

41. Find the conductance of 1000 ft of #12 AWG wire made of
 a. copper b. aluminum

42. a. Find the conductance of a 10 Ω, 20 Ω, and 100 Ω resistor in millisiemens.
 b. How do you compare the rate of change in resistance to the rate of change in conductance?
 c. Is the relationship between the change in resistance and change in associated conductance an inverse-linear relationship or an inverse nonlinear relationship?

*43. The conductance of a wire is 100 S. If the area of the wire is increased by two thirds and the length is reduced by the same amount, find the new conductance of the wire if the temperature remains fixed.

SECTION 3.8 Ohmmeters

44. Why do you **never** apply an ohmmeter to a live network?

45. How would you check the status of a fuse with an ohmmeter?

46. How would you determine the on and off states of a switch using an ohmmeter?

47. How would you use an ohmmeter to check the status of a light bulb?

SECTION 3.9 Resistance: Metric Units

48. Using metric units, determine the length of a copper wire that has a resistance of 0.2 Ω and a diameter of 1/12 in.

49. Repeat Problem 11 using metric units; that is, convert the given dimensions to metric units before determining the resistance.

50. If the sheet resistance of a tin oxide sample is 100 Ω, what is the thickness of the oxide layer?

51. Determine the width of a carbon resistor having a sheet resistance of 15 0 Ω if the length is 1/2 in. and the resistance is 500 Ω.

*52. Derive the conversion factor between ρ (CM-Ω/ft) and ρ (Ω-cm) by
 a. Solving for ρ for the wire in Fig. 3.47 in CM-Ω/ft.
 b. Solving for ρ for the same wire in Fig. 3.47 in Ω-cm by making the necessary conversions.
 c. Use the equation $\rho_2 = k\rho_1$ to determine the conversion factor k if ρ_1 is the solution of part (a) and ρ_2 the solution of part (b).

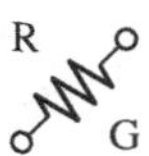

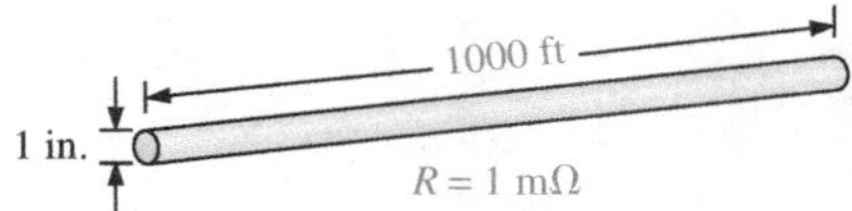

FIG. 3.47
Problem 52.

SECTION 3.11 Superconductors

53. In your own words, review what you have learned about superconductors. Do you feel it is an option that will have significant impact on the future of the electronics industry, or will its use be very limited? Explain why you feel the way you do. What could happen that would change your opinion?

54. Visit your local library and find a table listing the critical temperatures for a variety of materials. List at least five materials with critical temperatures that are not mentioned in this text. Choose a few materials that have relatively high critical temperatures.

55. Find at least one article on the application of superconductivity in the commercial sector, and write a short summary, including all interesting facts and figures.

***56.** Using the required 1 MA/cm^2 density level for integrated circuit manufacturing, determine what the resulting current would be through a #12 house wire. Compare the result obtained with the allowable limit of Table 3.2.

***57.** Research the SQUID magnetic field detector and review its basic mode of operation and an application or two.

SECTION 3.12 Thermistors

***58. a.** Find the resistance of the thermistor having the characteristics of Fig. 3.36 at −50°C, 50°C, and 200°C. Note that it is a log scale. If necessary, consult a reference with an expanded log scale.

b. Does the thermistor have a positive or a negative temperature coefficient?

c. Is the coefficient a fixed value for the range −100°C to 400°C? Why?

d. What is the approximate rate of change of ρ with temperature at 100°C?

SECTION 3.13 Photoconductive Cell

59. a. Using the characteristics of Fig. 3.38, determine the resistance of the photoconductive cell at 10 and 100 foot-candles of illumination. As in Problem 58, note that it is a log scale.

b. Does the cell have a positive or a negative illumination coefficient?

c. Is the coefficient a fixed value for the range 0.1 to 1000 foot-candles? Why?

d. What is the approximate rate of change of R with illumination at 10 foot-candles?

SECTION 3.14 Varistors

60. a. Referring to Fig. 3.40(a), find the terminal voltage of the device at 0.5 mA, 1 mA, 3 mA, and 5 mA.

b. What is the total change in voltage for the indicated range of current levels?

c. Compare the ratio of maximum to minimum current levels above to the corresponding ratio of voltage levels.

GLOSSARY

Absolute zero The temperature at which all molecular motion ceases; −273.15°C.

Circular mil (CM) The cross-sectional area of a wire having a diameter of 1 mil.

Color coding A technique using bands of color to indicate the resistance levels and tolerance of resistors.

Conductance (G) An indication of the relative ease with which current can be established in a material. It is measured in siemens (S).

Cooper effect The "pairing" of electrons as they travel through a medium.

Ductility The property of a material that allows it to be drawn into long, thin wires.

Inferred absolute temperature The temperature through which a straight-line approximation for the actual resistance-versus-temperature curve intersects the temperature axis.

Malleability The property of a material that allows it to be worked into many different shapes.

Memristor Resistor whose resistance is a function of the current through it; capable of remembering and retaining its last resistance value.

Negative temperature coefficient of resistance The value revealing that the resistance of a material will decrease with an increase in temperature.

Ohm (Ω) The unit of measurement applied to resistance.

Ohmmeter An instrument for measuring resistance levels.

Photoconductive cell A two-terminal semiconductor device whose terminal resistance is determined by the intensity of the incident light on its exposed surface.

Positive temperature coefficient of resistance The value revealing that the resistance of a material will increase with an increase in temperature.

Potentiometer A three-terminal device through which potential levels can be varied in a linear or nonlinear manner.

PPM/°C Temperature sensitivity of a resistor in parts per million per degree Celsius.

Resistance A measure of the opposition to the flow of charge through a material.

Resistivity (ρ) A constant of proportionality between the resistance of a material and its physical dimensions.

Rheostat An element whose terminal resistance can be varied in a linear or nonlinear manner.

Sheet resistance Defined by ρ/d for thin-film and integrated circuit design.

Superconductor Conductors of electric charge that have for all practical purposes zero ohms.

Thermistor A two-terminal semiconductor device whose resistance is temperature sensitive.

Varistor A voltage-dependent, nonlinear resistor used to suppress high-voltage transients.

Problem 52

Superconductors

*53. In your own words, review what you have learned about superconductors. Do you feel it is an option that will have significant impact on the future of the electronics industry, or will its use be very limited? Explain why you feel the way you do. What could happen that would change your opinion?

*54. Visit your local library and find a table listing the critical temperatures for a variety of materials. List at least five materials with critical temperatures that are not mentioned in this text. Choose a few materials that have a relatively high critical temperature.

*55. Find at least one article on the application of superconductivity in the commercial sector, and write a short summary, including all interesting facts and figures.

*56. Using the required 1 MA/cm² density level for integrated circuit manufacturing, determine what the resulting current would be through a #12 house wire. Compare the result obtained with the allowable limit of Table 3.2.

*57. Research the SQUID magnetic field detector and review its basic mode of operation and an application or two.

Thermistors

*58. a. Find the resistance of the thermistor having the characteristics of Fig. 3.26 at −50°C, 50°C, and 200°C. Note that it is a log scale. If necessary, consult a reference with an extended log scale.
b. Does the thermistor have a positive or a negative temperature coefficient?
c. Is the coefficient a fixed value for the range −100°C to 400°C? Why?
d. What is the approximate rate of change of ρ with temperature at 100°C?

Photoconductive Cell

59. a. Using the characteristics of Fig. 3.28, determine the resistance of the photoconductive cell at 10 and 100 foot-candles of illumination. As in Problem 58, note that it is a log scale.
b. Does the cell have a positive or a negative illumination coefficient?
c. Is the coefficient a fixed value for the range 0.1 to 1000 foot-candles? Why?
d. What is the approximate rate of change of *R* with illumination at 10 foot-candles?

Varistors

60. a. Referring to Fig. 3.30(a), find the terminal voltage of the device at 0.5 mA, 1 mA, 3 mA, and 5 mA.
b. What is the total change in voltage for the indicated range of current levels?
c. Compare the ratio of maximum to minimum current levels above to the corresponding ratio of voltage levels.

GLOSSARY

Absolute zero The temperature at which all molecular motion ceases; −273.15°C.

Circular mil (CM) The cross-sectional area of a wire having a diameter of 1 mil.

Color coding A technique using bands of color to indicate the resistance levels and tolerance of resistors.

Conductance (G) An indication of the relative ease with which current can be established in a material. It is measured in siemens (S).

Cooper effect The "pairing" of electrons as they travel through a medium.

Ductility The property of a material that allows it to be drawn into long, thin wires.

Inferred absolute temperature The temperature through which a straight-line approximation for the actual resistance-versus-temperature curve intersects the temperature axis.

Malleability The property of a material that allows it to be worked into many different shapes.

Memristor Resistor whose resistance is a function of the current through it, capable of remembering and retaining its last resistance value.

Negative temperature coefficient of resistance The value revealing that the resistance of a material will decrease with an increase in temperature.

Ohm (Ω) The unit of measurement applied to resistance.

Ohmmeter An instrument for measuring resistance levels.

Photoconductive cell A two-terminal semiconductor device whose terminal resistance is determined by the intensity of the incident light on its exposed surface.

Positive temperature coefficient of resistance The value revealing that the resistance of a material will increase with an increase in temperature.

Potentiometer A three-terminal device through which potential levels can be varied in a linear or nonlinear manner.

PPM/°C Temperature sensitivity of a resistor in parts per million per degree Celsius.

Resistance A measure of the opposition to the flow of charge through a material.

Resistivity (ρ) A constant of proportionality between the resistance of a material and its physical dimensions.

Rheostat An element whose terminal resistance can be varied in a linear or nonlinear manner.

Sheet resistance Defined by ρ/d for thin-film and integrated circuit design.

Superconductor Conductors of electric charge that have, for all practical purposes, zero ohms.

Thermistor A two-terminal semiconductor device whose resistance is temperature sensitive.

Varistor A voltage-dependent, nonlinear resistor used to suppress high-voltage transients.

Ohm's Law, Power, and Energy

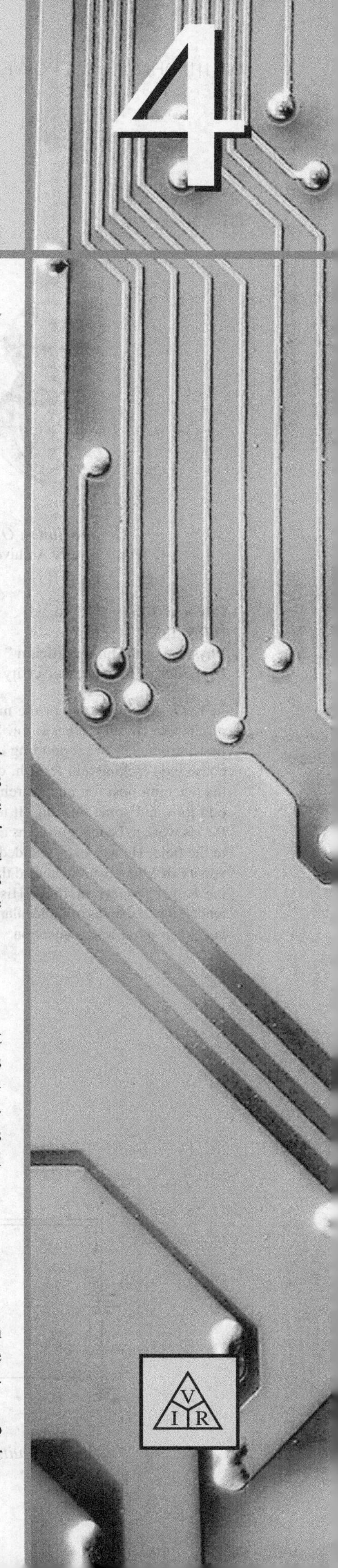

Objectives

- ***Understand the importance of Ohm's law and how to apply it to a variety of situations.***
- ***Be able to plot Ohm's law and understand how to "read" a graphical plot of voltage versus current.***
- ***Become aware of the differences between power and energy levels and how to solve for each.***
- ***Understand the power and energy flow of a system, including how the flow affects the efficiency of operation.***
- ***Become aware of the operation of a variety of fuses and circuit breakers and where each is employed.***

4.1 INTRODUCTION

Now that the three important quantities of an electric circuit have been introduced, this chapter reveals how they are interrelated. The most important equation in the study of electric circuits is introduced, and various other equations that allow us to find power and energy levels are discussed in detail. It is the first chapter where we tie things together and develop a feeling for the way an electric circuit behaves and what affects its response. For the first time, the data provided on the labels of household appliances and the manner in which your electric bill is calculated will have some meaning. It is indeed a chapter that should open your eyes to a wide array of past experiences with electrical systems.

4.2 OHM'S LAW

As mentioned above, the first equation to be described is without question one of the most important to be learned in this field. It is not particularly difficult mathematically, but it is very powerful because it can be applied to any network in any time frame. That is, it is applicable to dc circuits, ac circuits, digital and microwave circuits, and, in fact, any type of applied signal. In addition, it can be applied over a period of time or for instantaneous responses. The equation can be derived directly from the following basic equation for all physical systems:

$$\boxed{\text{Effect} = \frac{\text{cause}}{\text{opposition}}} \tag{4.1}$$

Every conversion of energy from one form to another can be related to this equation. In electric circuits, the *effect* we are trying to establish is the flow of charge, or *current*. The *potential difference*, or voltage, between two points is the *cause* ("pressure"), and the opposition is the *resistance* encountered.

An excellent analogy for the simplest of electrical circuits is the water in a hose connected to a pressure valve, as discussed in Chapter 2. Think of the electrons in the copper wire as the water

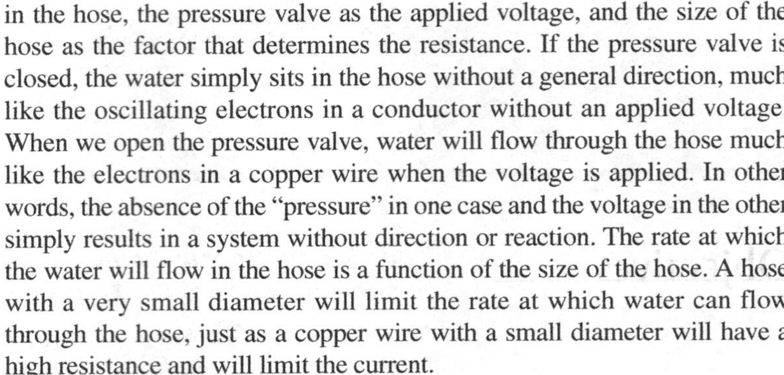

in the hose, the pressure valve as the applied voltage, and the size of the hose as the factor that determines the resistance. If the pressure valve is closed, the water simply sits in the hose without a general direction, much like the oscillating electrons in a conductor without an applied voltage. When we open the pressure valve, water will flow through the hose much like the electrons in a copper wire when the voltage is applied. In other words, the absence of the "pressure" in one case and the voltage in the other simply results in a system without direction or reaction. The rate at which the water will flow in the hose is a function of the size of the hose. A hose with a very small diameter will limit the rate at which water can flow through the hose, just as a copper wire with a small diameter will have a high resistance and will limit the current.

In summary, therefore, the absence of an applied "pressure" such as voltage in an electric circuit will result in no reaction in the system and no current in the electric circuit. Current is a reaction to the applied voltage and not the factor that gets the system in motion. To continue with the analogy, the greater the pressure at the spigot, the greater is the rate of water flow through the hose, just as applying a higher voltage to the same circuit results in a higher current.

Substituting the terms introduced above into Eq. (4.1) results in

$$\text{Current} = \frac{\text{potential difference}}{\text{resistance}}$$

and

$$\boxed{I = \frac{E}{R}} \qquad \text{(amperes, A)} \qquad \textbf{(4.2)}$$

Eq. (4.2) is known as **Ohm's law** in honor of Georg Simon Ohm (Fig. 4.1). The law states that for a fixed resistance, the greater the voltage (or pressure) across a resistor, the greater is the current; and the greater the resistance for the same voltage, the lower is the current. In other words, the current is proportional to the applied voltage and inversely proportional to the resistance.

By simple mathematical manipulations, the voltage and resistance can be found in terms of the other two quantities:

$$\boxed{E = IR} \qquad \text{(volts, V)} \qquad \textbf{(4.3)}$$

and

$$\boxed{R = \frac{E}{I}} \qquad (\text{ohms, } \Omega) \qquad \textbf{(4.4)}$$

All the quantities of Eq. (4.2) appear in the simple electrical circuit in Fig. 4.2. A resistor has been connected directly across a battery to establish a current through the resistor and supply. Note that

the symbol E is applied to all sources of voltage

and

the symbol V is applied to all voltage drops across components of the network.

Both are measured in volts and can be applied interchangeably in Eqs. (4.2) through (4.4).

FIG. 4.1
George Simon Ohm.

German (Erlangen, Cologne)
(1789–1854)
Physicist and Mathematician
Professor of Physics, University of Cologne

In 1827, developed one of the most important laws of electric circuits: *Ohm's law.* When the law was first introduced, the supporting documentation was considered lacking and foolish, causing him to lose his teaching position and search for a living doing odd jobs and some tutoring. It took some 22 years for his work to be recognized as a major contribution to the field. He was then awarded a chair at the University of Munich and received the Copley Medal of the Royal Society in 1841. His research also extended into the areas of molecular physics, acoustics, and telegraphic communication.

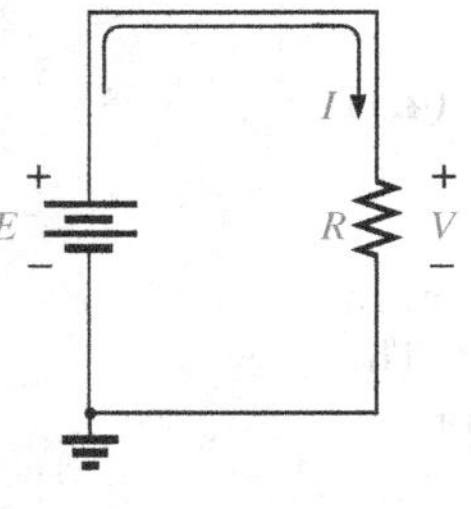

FIG. 4.2
Basic circuit.

Since the battery in Fig. 4.2 is connected directly across the resistor, the voltage V_R across the resistor must be equal to that of the supply. Applying Ohm's law, we obtain

$$I = \frac{V_R}{R} = \frac{E}{R}$$

Note in Fig. 4.2 that the voltage source "pressures" current (conventional current) in a direction that leaves the positive terminal of the supply and returns to the negative terminal of the battery. *This will always be the case for single-source networks.* (The effect of more than one source in the same network is investigated in a later chapter.) Note also that the current enters the positive terminal and leaves the negative terminal for the load resistor R.

For any resistor, in any network, the direction of current through a resistor will define the polarity of the voltage drop across the resistor

as shown in Fig. 4.3 for two directions of current. Polarities as established by current direction become increasingly important in the analyses to follow.

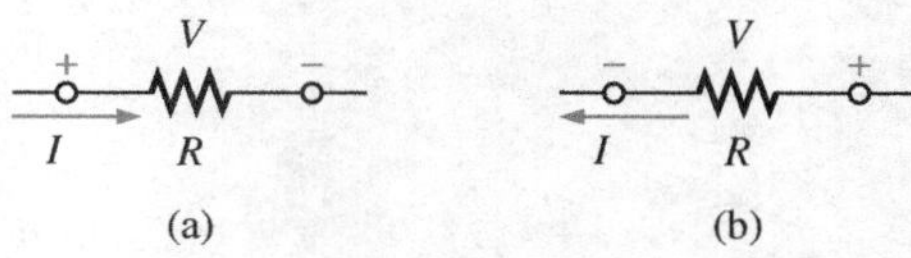

FIG. 4.3
Defining polarities.

EXAMPLE 4.1 Determine the current resulting from the application of a 9 V battery across a network with a resistance of 2.2 Ω.

Solution: Eq. (4.2):

$$I = \frac{V_R}{R} = \frac{E}{R} = \frac{9\text{ V}}{2.2\ \Omega} = \mathbf{4.09\ A}$$

EXAMPLE 4.2 Calculate the resistance of a 60 W bulb if a current of 500 mA results from an applied voltage of 120 V.

Solution: Eq. (4.4):

$$R = \frac{V_R}{I} = \frac{E}{I} = \frac{120\text{ V}}{500 \times 10^{-3}\text{ A}} = \mathbf{240\ \Omega}$$

EXAMPLE 4.3 Calculate the current through the 2 kΩ resistor in Fig. 4.4 if the voltage drop across it is 16 V.

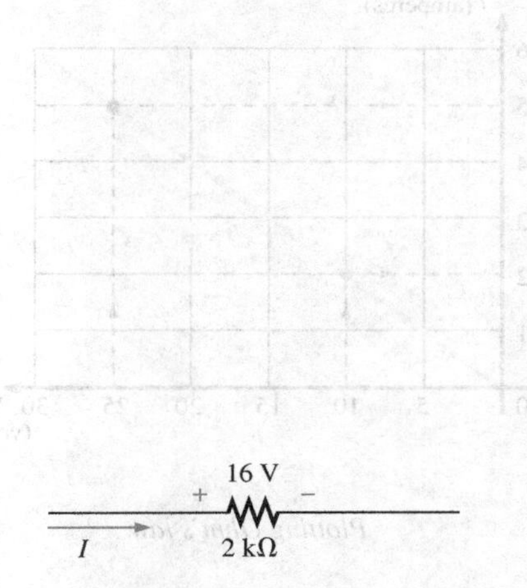

FIG. 4.4
Example 4.3.

Solution:

$$I = \frac{V}{R} = \frac{16\text{ V}}{2 \times 10^3\ \Omega} = \mathbf{8\ mA}$$

EXAMPLE 4.4 Calculate the voltage that must be applied across the soldering iron in Fig. 4.5 to establish a current of 1.5 A through the iron if its internal resistance is 80 Ω.

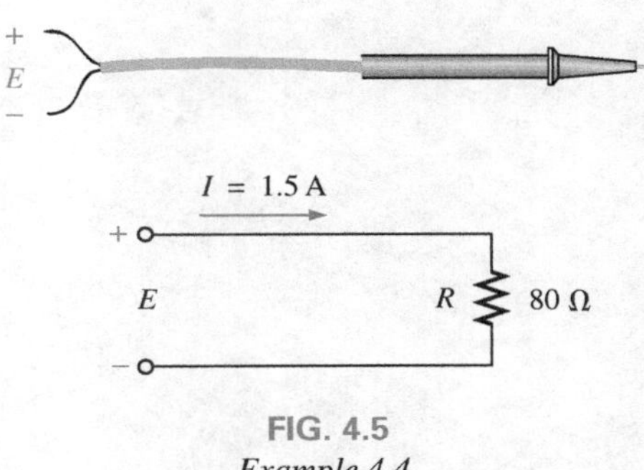

FIG. 4.5
Example 4.4.

Solution:

$$E = V_R = IR = (1.5\text{ A})(80\ \Omega) = \mathbf{120\ V}$$

In a number of the examples in this chapter, such as Example 4.4, the voltage applied is actually that obtained from an ac outlet in the home, office, or laboratory. This approach was used to provide an

opportunity for the student to relate to real-world situations as soon as possible and to demonstrate that a number of the equations derived in this chapter are applicable to ac networks also. Chapter 13 will provide a direct relationship between ac and dc voltages that permits the mathematical substitutions used in this chapter. In other words, don't be concerned that some of the voltages and currents appearing in the examples of this chapter are actually ac voltages, because the equations for dc networks have exactly the same format, and all the solutions will be correct.

4.3 PLOTTING OHM'S LAW

Graphs, characteristics, plots, and the like play an important role in every technical field as modes through which the broad picture of the behavior or response of a system can be conveniently displayed. It is therefore critical to develop the skills necessary both to read data and to plot them in such a manner that they can be interpreted easily.

For most sets of characteristics of electronic devices, the current is represented by the vertical axis (ordinate) and the voltage by the horizontal axis (abscissa), as shown in Fig. 4.6. First note that the vertical axis is in amperes and the horizontal axis is in volts. For some plots, I may be in milliamperes (mA), microamperes (μA), or whatever is appropriate for the range of interest. The same is true for the levels of voltage on the horizontal axis. Note also that the chosen parameters require that the spacing between numerical values of the vertical axis be different from that of the horizontal axis. The linear (straight-line) graph reveals that the resistance is not changing with current or voltage level; rather, it is a fixed quantity throughout. The current direction and the voltage polarity appearing at the top of Fig. 4.6 are the defined direction and polarity for the provided plot. If the current direction is opposite to the defined direction, the region below the horizontal axis is the region of interest for the current I. If the voltage polarity is opposite to that defined, the region to the left of the current axis is the region of interest. For the standard fixed resistor, the first quadrant, or region, of Fig. 4.6 is the only region of interest. However, you will encounter many devices in your electronics courses that use the other quadrants of a graph.

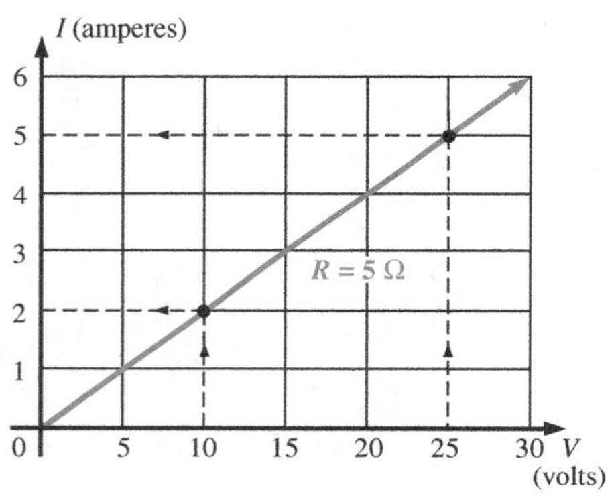

FIG. 4.6
Plotting Ohm's law.

Once a graph such as Fig. 4.6 is developed, the current or voltage at any level can be found from the other quantity by simply using the resulting plot. For instance, at $V = 25$ V, if a vertical line is drawn on Fig. 4.6 to the curve as shown, the resulting current can be found by drawing a horizontal line over to the current axis, where a result of 5 A is obtained. Similarly, at $V = 10$ V, drawing a vertical line to the plot and a horizontal line to the current axis results in a current of 2 A, as determined by Ohm's law.

If the resistance of a plot is unknown, it can be determined at any point on the plot since a straight line indicates a fixed resistance. At any point on the plot, find the resulting current and voltage, and simply substitute into the following equation:

$$\boxed{R_{dc} = \frac{V}{I}} \tag{4.5}$$

To test Eq. (4.5), consider a point on the plot where $V = 20$ V and $I = 4$ A. The resulting resistance is $R_{dc} = 20\text{ V}/I = 20\text{ V}/4\text{ A} = 5\ \Omega$. For comparison purposes, a 1 Ω and a 10 Ω resistor were plotted on the graph in Fig. 4.7. Note that the lower the resistance, the steeper is the slope (closer to the vertical axis) of the curve.

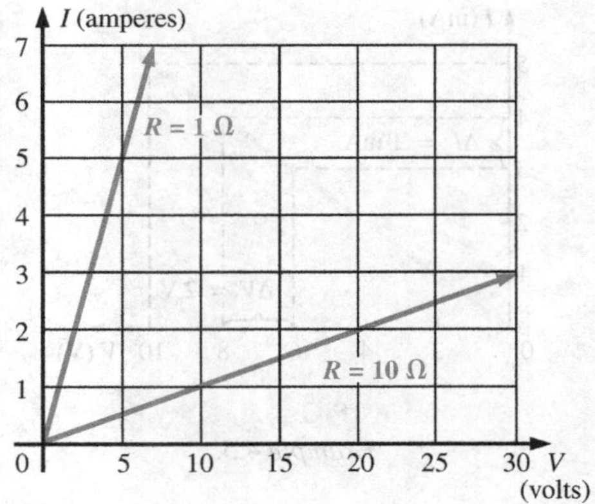

FIG. 4.7
Demonstrating on an I-V plot that the lower the resistance, the steeper is the slope.

If we write Ohm's law in the following manner and relate it to the basic straight-line equation

$$I = \frac{1}{R} \cdot E + 0$$
$$\downarrow \quad \downarrow \quad \downarrow \quad \downarrow$$
$$y = m \cdot x + b$$

we find that the slope is equal to 1 divided by the resistance value, as indicated by the following:

$$m = \text{slope} = \frac{\Delta y}{\Delta x} = \frac{\Delta I}{\Delta V} = \frac{1}{R} \tag{4.6}$$

where Δ signifies a small, finite change in the variable.

Eq. (4.6) reveals that the greater the resistance, the lower is the slope. If written in the following form, Eq. (4.6) can be used to determine the resistance from the linear curve:

$$R = \frac{\Delta V}{\Delta I} \quad \text{(ohms)} \tag{4.7}$$

The equation states that by choosing a particular ΔV (or ΔI), you can obtain the corresponding ΔI (or ΔV, respectively) from the graph, as shown in Fig. 4.8, and then determine the resistance. If the plot is a straight line, Eq. (4.7) will provide the same result no matter where the equation is applied. However, if the plot curves at all, the resistance will change.

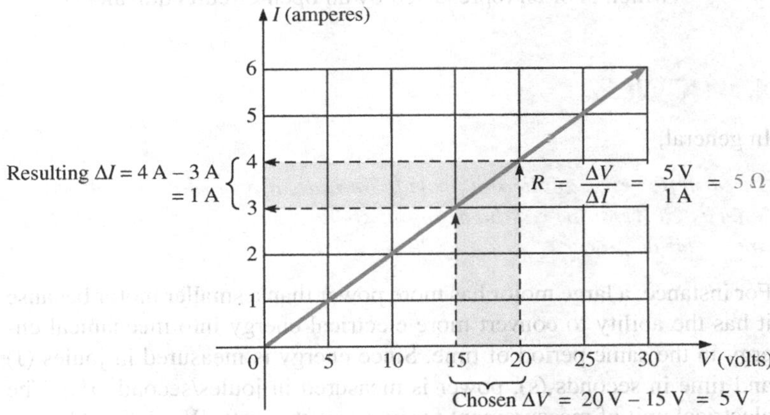

FIG. 4.8
Applying Eq. (4.7).

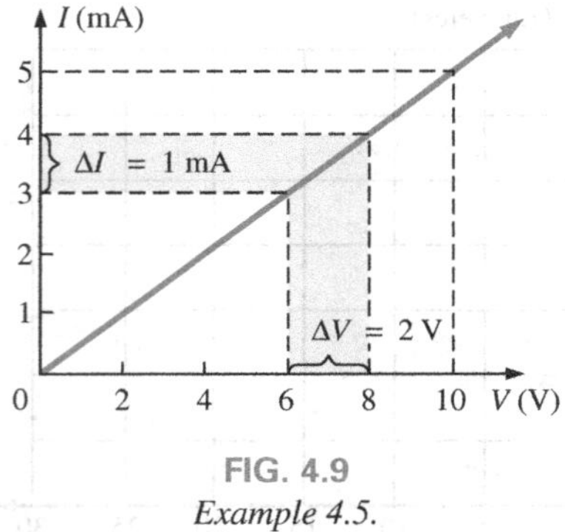

FIG. 4.9
Example 4.5.

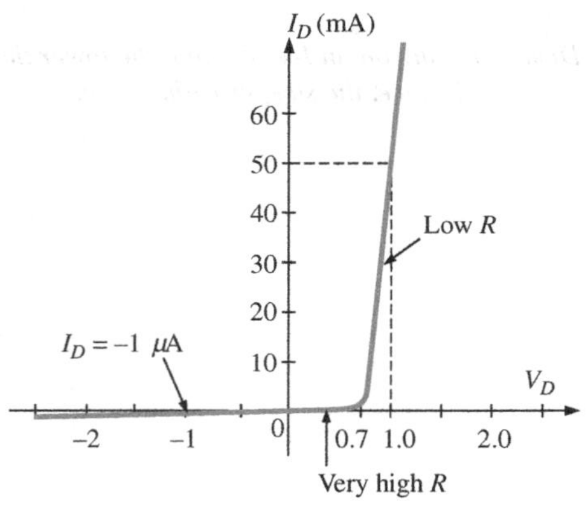

FIG. 4.10
Semiconductor diode characteristics.

EXAMPLE 4.5 Determine the resistance associated with the curve in Fig. 4.9 using Eqs. (4.5) and (4.7), and compare results.

Solution: At $V = 6$ V, $I = 3$ mA, and

$$R_{dc} = \frac{V}{I} = \frac{6\text{ V}}{3\text{ mA}} = \mathbf{2\text{ k}\Omega}$$

For the interval between 6 V and 8 V,

$$R = \frac{\Delta V}{\Delta I} = \frac{2\text{ V}}{1\text{ mA}} = \mathbf{2\text{ k}\Omega}$$

The results are equivalent.

Before leaving the subject, let us first investigate the characteristics of a very important semiconductor device called the **diode,** which will be examined in detail in basic electronics courses. This device ideally acts as a low-resistance path to current in one direction and a high-resistance path to current in the reverse direction, much like a switch that passes current in only one direction. A typical set of characteristics appears in Fig. 4.10. Without any mathematical calculations, the closeness of the characteristic to the voltage axis for negative values of applied voltage indicates that this is the low-conductance (high resistance, switch opened) region. Note that this region extends to approximately 0.7 V positive. However, for values of applied voltage greater than 0.7 V, the vertical rise in the characteristics indicates a high-conductivity (low resistance, switch closed) region. Application of Ohm's law will now verify the above conclusions.

At $V_D = +1$ V,

$$R_{\text{diode}} = \frac{V_D}{I_D} = \frac{1\text{ V}}{50\text{ mA}} = \frac{1\text{ V}}{50 \times 10^{-3}\text{ A}} = 20\ \Omega$$

(a relatively low value for most applications)

At $V_D = -1$ V,

$$R_{\text{diode}} = \frac{V_D}{I_D} = \frac{1\text{V}}{1\ \mu\text{A}} = 1\text{ M}\Omega$$

(which is often represented by an open-circuit equivalent)

4.4 POWER

In general,

the term power is applied to provide an indication of how much work (energy conversion) can be accomplished in a specified amount of time; that is, power is a rate of doing work.

For instance, a large motor has more power than a smaller motor because it has the ability to convert more electrical energy into mechanical energy in the same period of time. Since energy is measured in joules (J) and time in seconds (s), power is measured in joules/second (J/s). The electrical unit of measurement for power is the watt (W), defined by

$$\boxed{1\text{ watt (W)} = 1\text{ joule/second (J/s)}} \tag{4.8}$$

In equation form, power is determined by

$$\boxed{P = \frac{W}{t}} \quad \text{(watts, W, or joules/second, J/s)} \qquad \textbf{(4.9)}$$

with the **energy** (W) measured in joules and the time t in seconds.

The unit of measurement—the watt—is derived from the surname of James Watt (Fig. 4.11), who was instrumental in establishing the standards for power measurements. He introduced the **horsepower** (hp) as a measure of the average power of a strong dray horse over a full working day. It is approximately 50% more than can be expected from the average horse. The horsepower and watt are related in the following manner:

$$\boxed{1 \text{ horsepower} \cong 746 \text{ watts}}$$

The power delivered to, or absorbed by, an electrical device or system can be found in terms of the current and voltage by first substituting Eq. (2.5) into Eq. (4.9):

$$P = \frac{W}{t} = \frac{QV}{t} = V\frac{Q}{t}$$

But

$$I = \frac{Q}{t}$$

so that

$$\boxed{P = VI} \quad \text{(watts, W)} \qquad \textbf{(4.10)}$$

By direct substitution of Ohm's law, the equation for power can be obtained in two other forms:

$$P = VI = V\left(\frac{V}{R}\right)$$

and

$$\boxed{P = \frac{V^2}{R}} \quad \text{(watts, W)} \qquad \textbf{(4.11)}$$

or

$$P = VI = (IR)I$$

and

$$\boxed{P = I^2R} \quad \text{(watts, W)} \qquad \textbf{(4.12)}$$

The result is that the power absorbed by the resistor in Fig. 4.12 can be found directly, depending on the information available. In other words, if the current and resistance are known, it pays to use Eq. (4.12) directly, and if V and I are known, use of Eq. (4.10) is appropriate. It saves having to apply Ohm's law before determining the power.

The power supplied by a battery can be determined by simply inserting the supply voltage into Eq. (4.10) to produce

$$\boxed{P = EI} \quad \text{(watts, W)} \qquad \textbf{(4.13)}$$

The importance of Eq. (4.13) cannot be overstated. It clearly states the following:

The power associated with any supply is not simply a function of the supply voltage. It is determined by the product of the supply voltage and its maximum current rating.

The simplest example is the car battery—a battery that is large, difficult to handle, and relatively heavy. It is only 12 V, a voltage level that

FIG. 4.11
James Watt.
(© orion_eff / fotolia.com)

Scottish (Greenock, Birmingham)
(1736–1819)
Instrument Maker and Inventor
Elected Fellow of the Royal Society of London in 1785

In 1757, at the age of 21, used his innovative talents to design mathematical instruments such as the *quadrant, compass,* and various *scales.* In 1765, introduced the use of a separate *condenser* to increase the efficiency of steam engines. In the following years, he received a number of important patents on improved engine design, including a rotary motion for the steam engine (versus the reciprocating action) and a double-action engine, in which the piston pulled as well as pushed in its cyclic motion. Introduced the term **horsepower** as the average power of a strong dray (small cart) horse over a full working day.

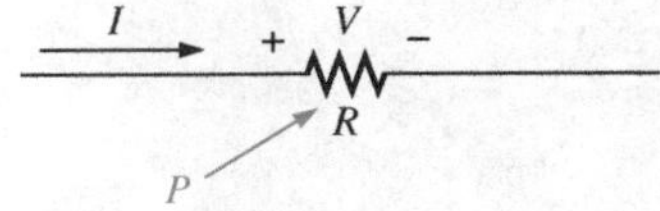

FIG. 4.12
Defining the power to a resistive element.

could be supplied by a battery slightly larger than the small 9 V portable radio battery. However, to provide the **power** necessary to start a car, the battery must be able to supply the high surge current required at starting—a component that requires size and mass. In total, therefore, it is not the voltage or current rating of a supply that determines its power capabilities; it is the product of the two.

Throughout the text, the abbreviation for energy (*W*) can be distinguished from that for the watt (W) because the one for energy is in italics while the one for watt is in roman. In fact, all variables in the dc section appear in italics, while the units appear in roman.

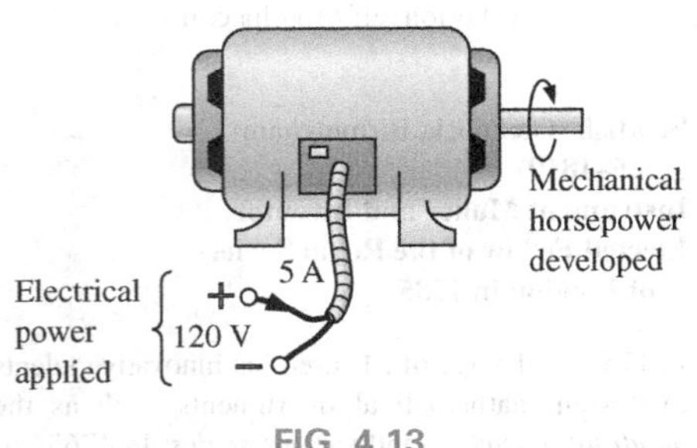

FIG. 4.13
Example 4.6.

EXAMPLE 4.6 Find the power delivered to the dc motor of Fig. 4.13.

Solution: $P = EI = (120\text{ V})(5\text{ A}) = 600\text{ W} = \mathbf{0.6\ kW}$

EXAMPLE 4.7 What is the power dissipated by a 5 Ω resistor if the current is 4 A?

Solution:

$$P = I^2R = (4\text{ A})^2(5\ \Omega) = \mathbf{80\ W}$$

EXAMPLE 4.8 The *I-V* characteristics of a light bulb are provided in Fig. 4.14. Note the nonlinearity of the curve, indicating a wide range in resistance of the bulb with applied voltage. If the rated voltage is 120 V, find the wattage rating of the bulb. Also calculate the resistance of the bulb under rated conditions.

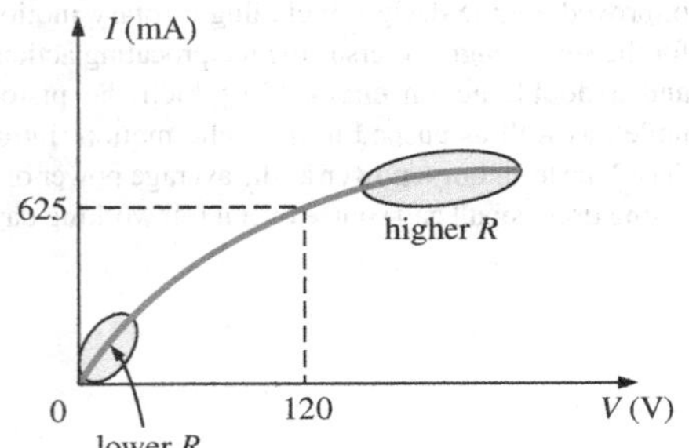

FIG. 4.14
The nonlinear I-V characteristics of a 75 W light bulb (Example 4.8).

Solution: At 120 V,

$$I = 0.625\text{ A}$$

and
$$P = VI = (120\text{ V})(0.625\text{ A}) = \mathbf{75\ W}$$

At 120 V,

$$R = \frac{V}{I} = \frac{120\text{ V}}{0.625\text{ A}} = \mathbf{192\ \Omega}$$

Sometimes the power is given and the current or voltage must be determined. Through algebraic manipulations, an equation for each variable is derived as follows:

$$P = I^2R \Rightarrow I^2 = \frac{P}{R}$$

and
$$\boxed{I = \sqrt{\frac{P}{R}}} \qquad \text{(amperes, A)} \qquad \textbf{(4.14)}$$

$$P = \frac{V^2}{R} \Rightarrow V^2 = PR$$

and
$$\boxed{V = \sqrt{PR}} \qquad \text{(volts, V)} \qquad \textbf{(4.15)}$$

EXAMPLE 4.9 Determine the current through a 5 kΩ resistor when the power dissipated by the element is 20 mW.

Solution: Eq. (4.14):

$$I = \sqrt{\frac{P}{R}} = \sqrt{\frac{20 \times 10^{-3}\ \text{W}}{5 \times 10^{3}\ \Omega}} = \sqrt{4 \times 10^{-6}} = 2 \times 10^{-3}\ \text{A}$$
$$= \mathbf{2\ mA}$$

FIG. 4.15
James Prescott Joule.
© World History Archive / Alamy

British (Salford, Manchester)
(1818–89)
Physicist
Honorary Doctorates from the Universities of Dublin and Oxford

Contributed to the important fundamental *law of conservation of energy* by establishing that various forms of energy, whether electrical, mechanical, or heat, are in the same family and can be exchanged from one form to another. In 1841 introduced *Joule's law,* which stated that the heat developed by electric current in a wire is proportional to the product of the current squared and the resistance of the wire (I^2R). He further determined that the heat emitted was equivalent to the power absorbed, and therefore heat is a form of energy.

4.5 ENERGY

For power, which is the rate of doing work, to produce an energy conversion of any form, it must be *used over a period of time.* For example, a motor may have the horsepower to run a heavy load, but unless the motor is *used* over a period of time, there will be no energy conversion. In addition, the longer the motor is used to drive the load, the greater will be the energy expended.

The **energy** (W) lost or gained by any system is therefore determined by

$$\boxed{W = Pt} \qquad \text{(wattseconds, Ws, or joules)} \qquad \textbf{(4.16)}$$

Since power is measured in watts (or joules per second) and time in seconds, the unit of energy is the *wattsecond* or *joule* (note Fig. 4.15). The wattsecond, however, is too small a quantity for most practical purposes, so the *watthour* (Wh) and the *kilowatthour* (kWh) are defined, as follows:

$$\boxed{\text{Energy (Wh)} = \text{power (W)} \times \text{time (h)}} \qquad \textbf{(4.17)}$$

$$\boxed{\text{Energy (kWh)} = \frac{\text{power (W)} \times \text{time (h)}}{1000}} \qquad \textbf{(4.18)}$$

Note that the energy in kilowatthours is simply the energy in watthours divided by 1000. To develop some sense for the kilowatthour energy level, consider that *1 kWh is the energy dissipated by a 100 W bulb in 10 h.*

The **kilowatthour meter** is an instrument for measuring the energy supplied to the residential or commercial user of electricity. It is normally connected directly to the lines at a point just prior to entering the power distribution panel of the building. A typical set of dials is shown in Fig. 4.16, along with a photograph of an analog kilowatthour meter. As indicated, each power of ten below a dial is in kilowatthours. The more rapidly the aluminum disc rotates, the greater is the energy demand. The dials are connected through a set of gears to the rotation of this disc. A solid-state digital meter with an extended range of capabilities also appears in Fig. 4.16.

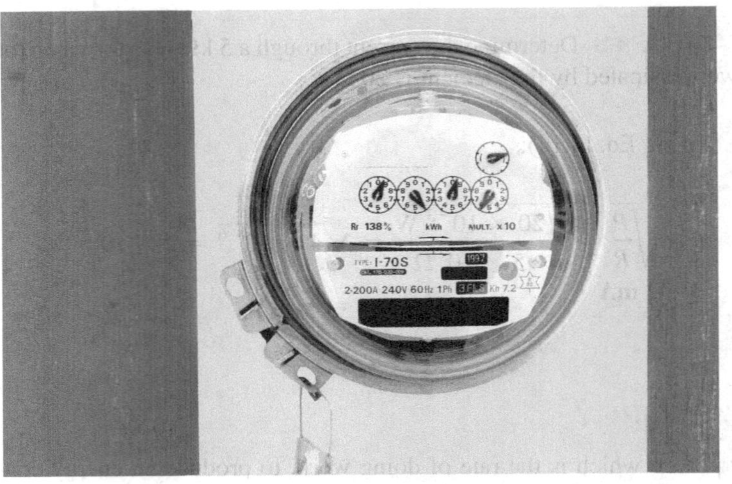

FIG. 4.16
Electric meter.
© Hamiza Bakirci/ Fotolia.com

EXAMPLE 4.10 For the dial positions in Fig. 4.16(a), calculate the electricity bill if the previous reading was 4650 kWh and the average cost in your area is 11¢ per kilowatthour.

Solution:

$$5360 \text{ kWh} - 4650 \text{ kWh} = 710 \text{ kWh used}$$

$$710 \cancel{\text{kWh}}\left(\frac{11¢}{\cancel{\text{kWh}}}\right) = \mathbf{\$78.10}$$

EXAMPLE 4.11 How much energy (in kilowatthours) is required to light a 60 W bulb continuously for 1 year (365 days)?

Solution:

$$W = \frac{Pt}{1000} = \frac{(60 \text{ W})(24 \text{ h/day})(365 \text{ days})}{1000} = \frac{525{,}600 \text{ Wh}}{1000}$$
$$= \mathbf{525.60 \text{ kWh}}$$

EXAMPLE 4.12 How long can a 340 W plasma TV be on before it uses more than 4 kWh of energy?

Solution:

$$W = \frac{Pt}{1000} \Rightarrow t \text{ (hours)} = \frac{(W)(1000)}{P} = \frac{(4 \text{ kWh})(1000)}{340 \text{ W}} = \mathbf{11.76 \text{ h}}$$

EXAMPLE 4.13 What is the cost of using a 5 hp motor for 2 h if the rate is 11¢ per kilowatthour?

Solution:

$$W \text{ (kilowatthours)} = \frac{Pt}{1000} = \frac{(5 \text{ hp} \times 746 \text{ W/hp})(2 \text{ h})}{1000} = 7.46 \text{ kWh}$$
$$\text{Cost} = (7.46 \text{ kWh})(11¢/\text{kWh}) = \mathbf{82.06¢}$$

EXAMPLE 4.14 What is the total cost of using all of the following at 11¢ per kilowatthour?

A 1200 W toaster for 30 min
Six 50 W bulbs for 4 h
A 500 W washing machine for 45 min
A 4300 W electric clothes dryer for 20 min
An 80 W PC for 6 h

Solution:

$$W = \frac{(1200 \text{ W})(\frac{1}{2} \text{ h}) + (6)(50 \text{ W})(4 \text{ h}) + (500 \text{ W})(\frac{3}{4} \text{ h}) + (4300 \text{ W})(\frac{1}{3} \text{ h}) + (80 \text{ W})(6 \text{ h})}{1000}$$

$$= \frac{600 \text{ Wh} + 1200 \text{ Wh} + 375 \text{ Wh} + 1433 \text{ Wh} + 480 \text{ Wh}}{1000} = \frac{4088 \text{ Wh}}{1000}$$

$$W = 4.09 \text{ kWh}$$

$$\text{Cost} = (4.09 \text{ kWh})(11¢/\text{kWh}) \cong \mathbf{45¢}$$

The chart in Fig. 4.17 shows the national average cost per kilowatthour compared to the kilowatthours used per customer. Note that the cost today is just above the level of 1926, but the average customer uses more than 20 times as much electrical energy in a year. Keep in mind that the chart in Fig. 4.17 is the average cost across the nation. Some states have average rates closer to 7¢ per kilowatthour, whereas others approach 20¢ per kilowatthour.

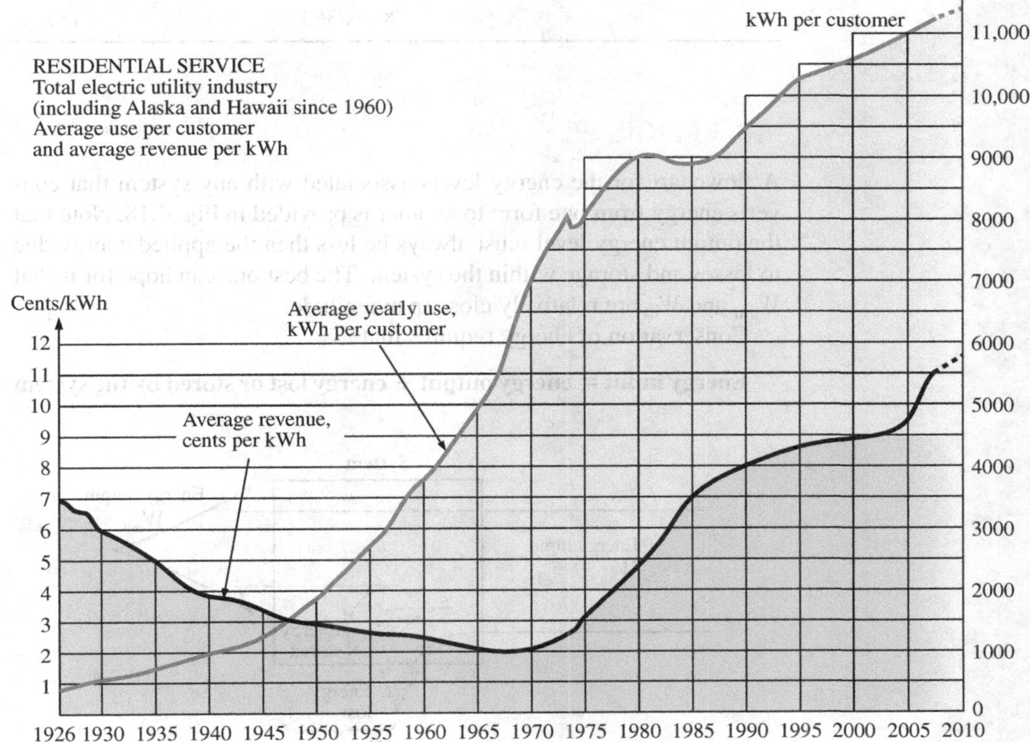

FIG. 4.17

Cost per kWh and average kWh per customer versus time.
(Based on data from Edison Electric Institute.)

Table 4.1 lists some common household appliances with their typical wattage ratings. You might find it interesting to calculate the cost of operating some of these appliances over a period of time, using the chart in Fig. 4.17 to find the cost per kilowatthour.

TABLE 4.1
Typical wattage ratings of some common household items.

Appliance	Wattage Rating	Appliance	Wattage Rating
Air conditioner (room)	1400	Laptop computer:	
		Sleep	<1 (typically 0.3 to 0.5)
Blow dryer	1300		
Cellular phone:		Average use	80
Standby	≅35 mW	Microwave oven	1200
Talk	≅4.3 W	Nintendo Wii	19
Clock	2	Radio	70
Clothes dryer (electric)	4300	Range (self-cleaning)	12,200
Coffee maker	900	Refrigerator (automatic defrost)	1800
Dishwasher	1200	Shaver	15
Fan:		Sun lamp	280
Portable	90	Toaster	1200
Window	200	Trash compactor	400
Heater	1500	TV:	
Heating equipment:		Plasma	340
Furnace fan	320	LCD	220
Oil-burner motor	230	VCR/DVD	25
Iron, dry or steam	1000	Washing machine	500
		Water heater	4500
		Xbox 360	187

4.6 EFFICIENCY

A flowchart for the energy levels associated with any system that converts energy from one form to another is provided in Fig. 4.18. Note that the output energy level must always be less than the applied energy due to losses and storage within the system. The best one can hope for is that W_{out} and W_{in} are relatively close in magnitude.

Conservation of energy requires that

Energy input = energy output + energy lost or stored by the system

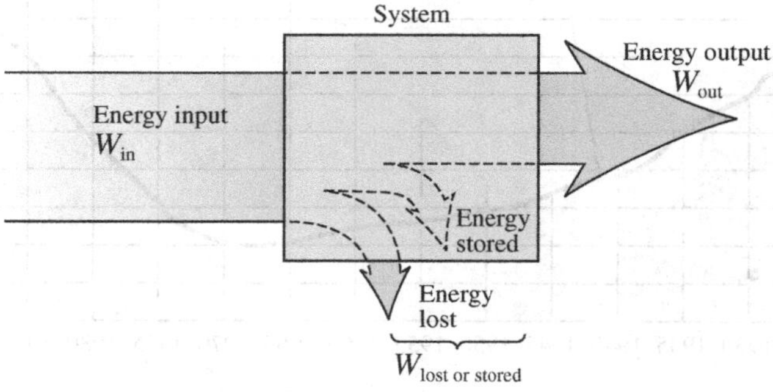

FIG. 4.18
Energy flow through a system.

Dividing both sides of the relationship by t gives

$$\frac{W_{in}}{t} = \frac{W_{out}}{t} + \frac{W_{\text{lost or stored by the system}}}{t}$$

Since $P = W/t$, we have the following:

$$P_i = P_o + P_{\text{lost or stored}} \quad \text{(W)} \tag{4.19}$$

The **efficiency** (η) of the system is then determined by the following equation:

$$\text{Efficiency} = \frac{\text{power output}}{\text{power input}}$$

and

$$\eta = \frac{P_o}{P_i} \quad \text{(decimal number)} \tag{4.20}$$

where η (the lowercase Greek letter *eta*) is a decimal number. Expressed as a percentage,

$$\eta\% = \frac{P_o}{P_i} \times 100\% \quad \text{(persent)} \tag{4.21}$$

In terms of the input and output energy, the efficiency in percent is given by

$$\eta\% = \frac{W_o}{W_i} \times 100\% \quad \text{(persent)} \tag{4.22}$$

The maximum possible efficiency is 100%, which occurs when $P_o = P_i$, or when the power lost or stored in the system is zero. Obviously, the greater the internal losses of the system in generating the necessary output power or energy, the lower is the net efficiency.

EXAMPLE 4.15 A 2 hp motor operates at an efficiency of 75%. What is the power input in watts? If the applied voltage is 220 V, what is the input current?

Solution:

$$\eta\% = \frac{P_o}{P_i} \times 100\%$$

$$0.75 = \frac{(2\text{ hp})(746\text{ W/hp})}{P_i}$$

and

$$P_i = \frac{1492\text{ W}}{0.75} = \mathbf{1989.33\ W}$$

$$P_i = EI \quad \text{or} \quad I = \frac{P_i}{E} = \frac{1989.33\text{ W}}{220\text{ V}} = \mathbf{9.04\ A}$$

EXAMPLE 4.16 What is the output in horsepower of a motor with an efficiency of 80% and an input current of 8 A at 120 V?

Solution:

$$\eta\% = \frac{P_o}{P_i} \times 100\%$$

$$0.80 = \frac{P_o}{(120\text{ V})(8\text{ A})}$$

and $$P_o = (0.80)(120\text{ V})(8\text{ A}) = 768\text{ W}$$

with $$768\text{ W}\left(\frac{1\text{ hp}}{746\text{ W}}\right) = \mathbf{1.03\text{ hp}}$$

EXAMPLE 4.17 If $\eta = 0.85$, determine the output energy level if the applied energy is 50 J.

Solution:

$$\eta = \frac{W_o}{W_i} \Rightarrow W_o = \eta W_i = (0.85)(50\text{ J}) = \mathbf{42.5\text{ J}}$$

The very basic components of a generating (voltage) system are depicted in Fig. 4.19. The source of mechanical power is a structure such as a paddlewheel that is turned by water rushing over the dam. The gear train ensures that the rotating member of the generator is turning at rated speed. The output voltage must then be fed through a transmission system to the load. For each component of the system, an input and output power have been indicated. The efficiency of each system is given by

$$\eta_1 = \frac{P_{o_1}}{P_{i_1}} \qquad \eta_2 = \frac{P_{o_2}}{P_{i_2}} \qquad \eta_3 = \frac{P_{o_3}}{P_{i_3}}$$

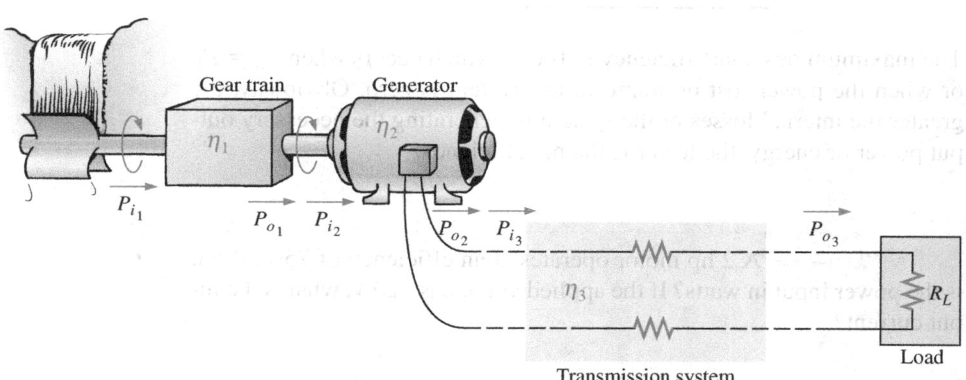

FIG. 4.19
Basic components of a generating system.

If we form the product of these three efficiencies,

$$\eta_1 \cdot \eta_2 \cdot \eta_3 = \frac{\cancel{P_{o_1}}}{P_{i_1}} \cdot \frac{\cancel{P_{o_2}}}{\cancel{P_{i_2}}} \cdot \frac{P_{o_3}}{\cancel{P_{i_3}}} = \frac{P_{o_3}}{P_{i_1}}$$

and substitute the fact that $P_{i_2} = P_{o_1}$ and $P_{i_3} = P_{o_2}$, we find that the quantities indicated above will cancel, resulting in P_{o_3}/P_{i_1}, which is a measure of the efficiency of the entire system.

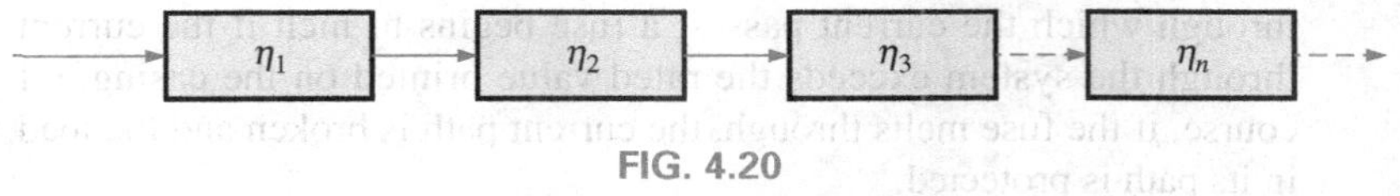

FIG. 4.20
Cascaded system.

In general, for the representative cascaded system in Fig. 4.20,

$$\boxed{\eta_{total} = \eta_1 \cdot \eta_2 \cdot \eta_3 \cdots \eta_n} \quad \textbf{(4.23)}$$

EXAMPLE 4.18 Find the overall efficiency of the system in Fig. 4.19 if $\eta_1 = 90\%$, $\eta_2 = 85\%$, and $\eta_3 = 95\%$.

Solution:

$$\eta_T = \eta_1 \cdot \eta_2 \cdot \eta_3 = (0.90)(0.85)(0.95) = 0.727, \text{ or } \mathbf{72.7\%}$$

EXAMPLE 4.19 If the efficiency η_1 drops to 40%, find the new overall efficiency and compare the result with that obtained in Example 4.18.

Solution:

$$\eta_T = \eta_1 \cdot \eta_2 \cdot \eta_3 = (0.40)(0.85)(0.95) = 0.323, \text{ or } \mathbf{32.3\%}$$

Certainly 32.3% is noticeably less than 72.7%. The total efficiency of a cascaded system is therefore determined primarily by the lowest efficiency (weakest link) and is less than (or equal to if the remaining efficiencies are 100%) the least efficient link of the system.

4.7 CIRCUIT BREAKERS, GFCIs, AND FUSES

The incoming power to any large industrial plant, heavy equipment, simple circuit in the home, or meters used in the laboratory must be limited to ensure that the current through the lines is not above the rated value. Otherwise, the conductors or the electrical or electronic equipment may be damaged, and dangerous side effects such as fire or smoke may result.

To limit the current level, **fuses** or **circuit breakers** are installed where the power enters the installation, such as in the panel in the basement of most homes at the point where the outside feeder lines enter the dwelling. The fuses in Fig. 4.21 have an internal metallic conductor

through which the current passes; a fuse begins to melt if the current through the system exceeds the rated value printed on the casing. Of course, if the fuse melts through, the current path is broken and the load in its path is protected.

In homes built in recent years, fuses have been replaced by circuit breakers such as those appearing in Fig. 4.22. When the current exceeds rated conditions, an electromagnet in the device will have sufficient strength to draw the connecting metallic link in the breaker out of the circuit and open the current path. When conditions have been corrected, the breaker can be reset and used again.

The most recent National Electrical Code requires that outlets in the bathroom and other sensitive areas be of the ground fault circuit interrupt (GFCI) variety; GFCIs (often abbreviated GFI) are designed to trip more quickly than the standard circuit breaker. The commercial unit in Fig. 4.23 trips in 5 ms. It has been determined that 6 mA is the maximum level that most individuals can be exposed to for a short period of time without the risk of serious injury. A current higher than 11 mA can cause involuntary muscle contractions that could prevent a person from letting go of the conductor and possibly cause him or her to enter a state of shock. Higher currents lasting more than a second can cause the heart to go into fibrillation and possibly cause death in a few minutes. The GFCI is able to react as quickly as it does by sensing the difference between the input and output currents to the outlet; the currents should be the same if everything is working properly. An errant path, such as through an individual, establishes a difference in the two current levels and causes the breaker to trip and disconnect the power source.

4.8 APPLICATIONS

Fluorescent versus Incandescent

FIG. 4.24
A 23 W, 380 mA compact fluorescent lamp (CFL).

A hot topic in the discussion of energy conservation is the growing pressure to switch from incandescent bulbs to fluorescent bulbs such as in Fig. 4.24. Countries throughout the world have set goals for the near future, with some mandating a ban of the use of incandescent bulbs by 2012. Japan is currently at an adoption rate of 80%, and the World Heritage sites of Shirakawa-go and Gokayama were completely stripped of incandescent lighting in 2007 in a bold effort to reduce carbon dioxide emissions near the villages. Germany is at a 50% adoption rate, the United Kingdom at 20%, and the United States at about 6%. Australia announced a complete ban on incandescent lighting by 2009, and Canada by 2012.

This enormous shift in usage is primarily due to the higher energy efficiency of fluorescent bulbs and their longer lifespans. For the same number of lumens (a unit of light measurement) the energy dissipated by an incandescent light can range from approximately four to six times greater than that of a fluorescent bulb. The range is a function of the lumens level. The greater the number of lumens available per bulb, the lower is the ratio. **In other words, the energy saved increases with decrease in the wattage rating of the fluorescent bulb.** Table 4.2 compares the wattage ratings of fluorescent bulbs and incandescent bulbs for the same level of lumens generated. For the same period of time the ratio of the energy used requires that we simply divide the wattage ratings at the same lumens level.

TABLE 4.2
Comparison of the lumens generated by incandescent and fluorescent bulbs.

Incandescents	Lumens	Fluorescents
100 W (950 h)	1675	
	1600	23 W (12,000 h)
	1100	15 W (8000 h)
75 W (1500 h)	1040	
	870	13 W (10,000 h)
60 W (1500 h)	830	
	660	11 W (8000 h)
	580	9 W (8000 h)
40 W (1250 h)	495	
	400	7 W (10,000 h)
25 W (2500 h)	250 250	4 W (8000 h)
15 W (3000 h)	115	

As mentioned, the other positive benefit of fluorescent bulbs is the longevity of the bulbs. A 60 W incandescent bulb will have a rated life of 1500 h, whereas a 13 W fluorescent bulb with an equivalent lumens level is rated to last 10,000 h—a life ratio of 6.67. A 25 W incandescent bulb may have a rated lifetime of 2500 h, but a 4 W fluorescent bulb of similar lumens emission has a rated lifetime of 8000 h—a life ratio of only 3.2. It is interesting to note in Table 4.2 that the lifetime of fluorescent bulbs remains quite high at all wattage ratings, whereas the lifetime of incandescent bulbs decreases substantially with increase in wattage level. Finally, we have to consider the cost level of purchase and use. Currently, a 60 W incandescent bulb can be purchased for about 80¢, whereas a similar lumens rated 13 W fluorescent bulb may cost \$2.50—an increase by a factor in excess of 3 : 1. For most people this is an important factor and has without question had an effect on the level of adoption of fluorescent bulbs. However, one must also consider the cost of using the bulbs just described over a period of 1 year. Consider that each is used 5 h/day for 365 days at a cost of 11¢/kWh.

For the incandescent bulb the cost is determined as follows:

$$\text{kWh} = \frac{(5\text{ h})(365\text{ days})(60\text{ W})}{1000} = 109.5\text{ kWh}$$

$$\text{Cost} = (109.5\text{ kWh})(11¢/\text{kWh}) = \mathbf{\$12.05/year}$$

For the fluorescent bulb the cost is determined as follows:

$$\text{kWh} = \frac{(5\text{ h})(365\text{ days})(13\text{ W})}{1000} = 23.73\text{ kWh}$$

$$\text{Cost} = (23.73\text{ kWh})(11¢/\text{kWh}) = \mathbf{\$2.61/year}$$

The cost ratio of fluorescent to incandescent bulbs is about 4.6, which is certainly significant, indicating that the cost of fluorescent lighting is about 22% of that of incandescent lighting. Returning to the initial cost, it is clear that the bulb would pay for itself almost four times over in 1 year.

The other positive factor in using lower-wattage bulbs is the savings in carbon dioxide emissions in the process of producing the necessary electrical power. For the two villages in Japan mentioned earlier, where some 700 bulbs were replaced, the savings will be some 24 tons in 1 year. Consider what that number would be if this policy were adopted all over the world.

As with every innovative approach to saving energy, there are some concerns about its worldwide adoption. They include the mercury that is inherent in the design of the fluorescent bulb. Each bulb contains about 5 mg of mercury, an element that can be harmful to the nervous system and brain development. The effects of breaking a bulb are under investigation, and the concern is such that departments of environmental protection in most countries are setting up standards for the clean-up process. In every approach described it is agreed that a room in which breakage occurs should be cleared out and the windows **opened for ventilation.** Then the materials **should not be vacuumed up** but carefully scooped up and put in a small, sealed container. Some agencies go well beyond these two simple measures in the clean-up, but clearly it is a process that must be handled with care. The other concern is how to dispose of the bulbs when they wear out. At the moment this may not be a big problem because the adoption of fluorescent bulbs has just begun, and the bulbs will last for quite a few years. However, there will become a time when the proper disposal will have to be defined. Fortunately, most developed countries are carefully looking at this problem and setting up facilities designed specifically to dispose of this type of appliance. When disposal does reach a higher level, such as the 350 million per year projected in a country such as Japan, it will represent levels that will have to be addressed efficiently and correctly to remove the mercury levels that will result.

Other concerns relate to the light emitted by fluorescent bulbs as compared to that from incandescent bulbs. In general the light emitted by incandescent lights (more red and less blue components) is a closer match to natural light than that from fluorescent bulbs, which emit a bluish tint. However, by applying the proper phosphor to the inside of the bulb, a **white light** can be established that is more comfortable to the normal eye. Another concern is the fact that currently dimmers—a source of energy conservation in many instances—can only be used with specially designed fluorescent bulbs. However, again research is now going on that will probably remove this problem in the near future. One other concern of some importance is the fact that fluorescent bulbs emit ultraviolet (UV) rays (as in the light used in tanning salons), which are not a component of visible light but are a concern to those with skin problems such as those in individuals with lupus; however, once again, the studies are ongoing. For many years fluorescent lights were relegated to ceiling fixtures, where their distance negated most concerns about UV rays, but now they have been brought closer to the consumer. Recall also that the growing of plants in a dark interior space can only be accomplished using fluorescent bulbs because of the UV radiation. Finally, it turns out that, as with all products, you get what you pay for: Cheaper bulbs seem to fail to meet their lifespan guarantee and emit a poorer light spectrum.

The debate could go on for numerous pages, balancing benefits against disadvantages. For instance, consider that the heat generated by incandescent lamps provides some of the heating in large institutions and therefore more heat will have to be supplied if the switch is made to fluorescents. However, in the summer months the cooler fluorescent bulbs will require less cooling, affording a savings at the same location. In the final analysis, it appears that the decision will come down to individuals (unless it is government mandated) and what they are most comfortable with. Be assured, however, that any strong reaction to a proposed switchover will be well documented and should not be a real concern.

The exponential growth of interest in fluorescents in recent years has been primarily due to the introduction of electronic circuitry that can "fire" or "ignite" the bulb in a way that provides a quicker startup and

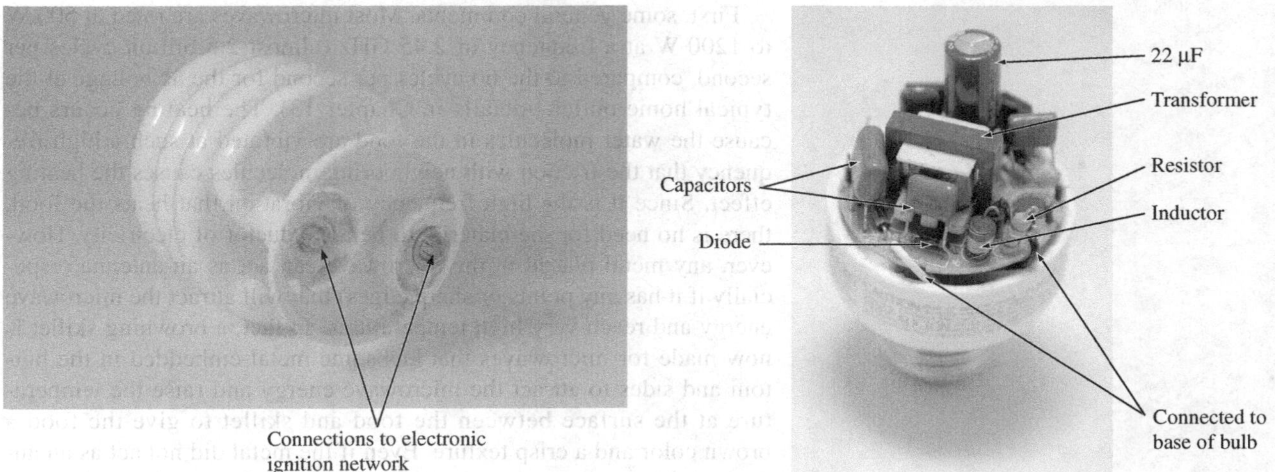

FIGURE 4.25
Internal construction of the CFL of Fig. 4.24.

smaller units. A full description of the older variety of fluorescent bulbs appears in the Application section of Chapter 22, which describes the large size of the ballast transformer and the need for a starter mechanism. A glimpse at the relatively small electronic firing mechanism for a compact fluorescent lamp (CFL) is provided in Fig. 4.25. The bulb section remains completely isolated, with only four leads available to connect to the circuitry, which reduces the possibility of exposure when the bulb is being constructed. The circuitry has been flipped from its position in the base of the bulb. The black and white lead on the edges is connected to the base of the bulb, where it is connected to a 120 V source. Note that the two largest components are the transformer and electrolytic near the center of the printed circuit board (PCB). A number of other elements to be described in the text have been identified.

Microwave Oven

It is probably safe to say that most homes today have a microwave oven (see Fig. 4.26). Most users are not concerned with its operating efficiency. However, it is interesting to learn how the units operate and apply some of the theory presented in this chapter.

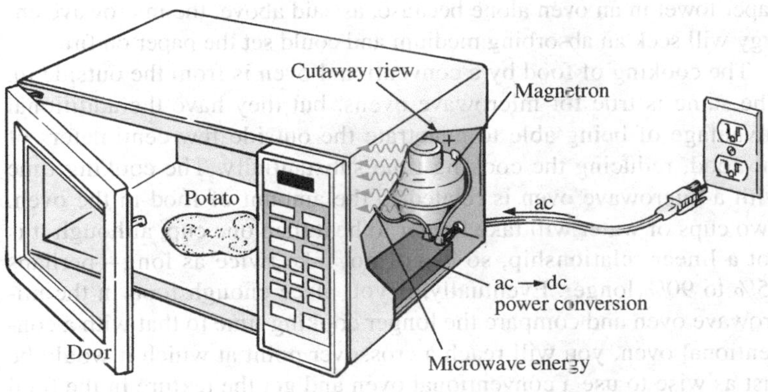

FIG. 4.26
Microwave oven.

First, some general comments. Most microwaves are rated at 500 W to 1200 W at a frequency of 2.45 GHz (almost 2.5 billion cycles per second, compared to the 60 cycles per second for the ac voltage at the typical home outlet—details in Chapter 13). The heating occurs because the water molecules in the food are vibrated at such a high frequency that the friction with neighboring molecules causes the heating effect. Since it is the high frequency of vibration that heats the food, there is no need for the material to be a conductor of electricity. However, any metal placed in the microwave can act as an antenna (especially if it has any points or sharp edges) that will attract the microwave energy and reach very high temperatures. In fact, a browning skillet is now made for microwaves that has some metal embedded in the bottom and sides to attract the microwave energy and raise the temperature at the surface between the food and skillet to give the food a brown color and a crisp texture. Even if the metal did not act as an antenna, it is a good conductor of heat and could get quite hot as it draws heat from the food.

Any container with low moisture content can be used to heat foods in a microwave. Because of this requirement, manufacturers have developed a whole line of microwave cookware that is very low in moisture content. Theoretically, glass and plastic have very little moisture content, but even so, when heated in the oven for a minute or so, they do get warm. It could be the moisture in the air that clings to the surface of each or perhaps the lead used in good crystal. In any case, microwaves should be used only to prepare food. They were not designed to be dryers or evaporators.

The instructions with every microwave specify that the oven should not be turned on when empty. Even though the oven may be empty, microwave energy will be generated and will make every effort to find a channel for absorption. If the oven is empty, the energy might be attracted to the oven itself and could do some damage. To demonstrate that a dry empty glass or plastic container will not attract a significant amount of microwave energy, place two glasses in an oven, one with water and the other empty. After 1 min, you will find the glass with the water quite warm due to the heating effect of the hot water while the other is close to its original temperature. In other words, the water created a heat sink for the majority of the microwave energy, leaving the empty glass as a less attractive path for heat conduction. Dry paper towels and plastic wrap can be used in the oven to cover dishes since they initially have low water molecule content, and paper and plastic are not good conductors of heat. However, it would be very unsafe to place a paper towel in an oven alone because, as said above, the microwave energy will seek an absorbing medium and could set the paper on fire.

The cooking of food by a conventional oven is from the outside in. The same is true for microwave ovens, but they have the additional advantage of being able to penetrate the outside few centimeters of the food, reducing the cooking time substantially. The cooking time with a microwave oven is related to the amount of food in the oven. Two cups of water will take longer to heat than one cup, although it is not a linear relationship, so it will not take twice as long—perhaps 75% to 90% longer. Eventually, if you place enough food in the microwave oven and compare the longer cooking time to that with a conventional oven, you will reach a crossover point at which it would be just as wise to use a conventional oven and get the texture in the food you might prefer.

The basic construction of the microwave oven is depicted in Fig. 4.26. It uses a 120 V ac supply, which is then converted through a high-voltage transformer to one having peak values approaching 5000 V (at substantial current levels)—sufficient warning to leave microwave repair to the local service location. Through the rectifying process briefly described in Chapter 2, a high dc voltage of a few thousand volts is generated that appears across a magnetron. The magnetron, through its very special design (currently the same design as in World War II, when it was invented by the British for use in high-power radar units), generates the required 2.45 GHz signal for the oven. It should be pointed out also that the magnetron has a specific power level of operation that cannot be controlled—once it's on, it's on at a set power level. One may then wonder how the cooking temperature and duration can be controlled. This is accomplished through a controlling network that determines the amount of off and on time during the input cycle of the 120 V supply. Higher temperatures are achieved by setting a high ratio of on tc off time, while low temperatures are set by the reverse action.

One unfortunate characteristic of the magnetron is that in the conversion process, it generates a great deal of heat that does not go toward the heating of the food and that must be absorbed by heat sinks or dispersed by a cooling fan. Typical conversion efficiencies are between 55% and 75%. Considering other losses inherent in any operating system, it is reasonable to assume that most microwaves are between 50% and 60% efficient. However, the conventional oven with its continually operating exhaust fan and heating of the oven, cookware, surrounding air, and so on also has significant losses, even if it is less sensitive to the amount of food to be cooked. All in all, convenience is probably the other factor that weighs the heaviest in this discussion. It also leaves the question of how our time is figured into the efficiency equation.

For specific numbers, let us consider the energy associated with baking a 5-oz potato in a 1200 W microwave oven for 5 min if the conversion efficiency is an average value of 55%. First, it is important to realize that when a unit is rated as 1200 W, that is the rated power drawn from the line during the cooking process. If the microwave is plugged into a 120 V outlet, the current drawn is

$$I = P/V = 1200\text{ W}/120\text{ V} = 10.0\text{ A}$$

which is a significant level of current. Next, we can determine the amount of power dedicated solely to the cooking process by using the efficiency level. That is,

$$P_o = \eta P_i = (0.55)(1200\text{ W}) = 600\text{ W}$$

The energy transferred to the potato over a period of 5 min can then be determined from

$$W = Pt = (660\text{ W})(5\text{ min})(60\text{ s}/1\text{ min}) = 198\text{ kJ}$$

which is about half of the energy (nutritional value) derived from eating a 5-oz potato. The number of kilowatthours drawn by the unit is determined from

$$W = Pt/1000 = (1200\text{ W})(6/60\text{ h})/1000 = 0.1\text{ kWh}$$

At a rate of 10¢/kWh we find that we can cook the potato for 1 penny—relatively speaking, pretty cheap. A typical 1550 W toaster oven would take an hour to heat the same potato, using 1.55 kWh and costing 15.5 cents—a significant increase in cost.

Household Wiring

A number of facets of household wiring can be discussed without examining the manner in which they are physically connected. In the following chapters, additional coverage is provided to ensure that you develop a solid fundamental understanding of the overall household wiring system. At the very least you will establish a background that will permit you to answer questions that you should be able to answer as a student of this field.

The one specification that defines the overall system is the maximum current that can be drawn from the power lines since the voltage is fixed at 120 V or 240 V (sometimes 208 V). For most older homes with a heating system other than electric, a 100 A service is the norm. Today, with all the electronic systems becoming commonplace in the home, many people are opting for the 200 A service even if they do not have electric heat. A 100 A service specifies that the maximum current that can be drawn through the power lines into your home is 100 A. Using the line-to-line rated voltage and the full-service current (and assuming all resistive-type loads), we can determine the maximum power that can be delivered using the basic power equation:

$$P = EI = (240\text{ V})(100\text{ A}) = 24{,}000\text{ W} = 24\text{ kW}$$

This rating reveals that the total rating of all the units turned on in the home cannot exceed 24 kW at any one time. If it did, we could expect the main breaker at the top of the power panel to open. Initially, 24 kW may seem like quite a large rating, but when you consider that a self-cleaning electric oven may draw 12.2 kW, a dryer 4.8 kW, a water heater 4.5 kW, and a dishwasher 1.2 kW, we are already at 22.7 kW (if all the units are operating at peak demand), and we have not turned the lights or TV on yet. Obviously, the use of an electric oven alone may strongly suggest considering a 200 A service. However, seldom are all the burners of a stove used at once, and the oven incorporates a thermostat to control the temperature so that it is not on all the time. The same is true for the water heater and dishwasher, so the chances of all the units in a home demanding full service at the same time is very slim. Certainly, a typical home with electric heat that may draw 16 kW just for heating in cold weather must consider a 200 A service. You must also understand that there is some leeway in maximum ratings for safety purposes. In other words, a system designed for a maximum load of 100 A can accept a slightly higher current for short periods of time without significant damage. For the long term, however, the limit should not be exceeded.

Changing the service to 200 A is not simply a matter of changing the panel in the basement—a new, heavier line must be run from the road to the house. In some areas feeder cables are aluminum because of the reduced cost and weight. In other areas, aluminum is not permitted because of its temperature sensitivity (expansion and contraction), and copper must be used. In any event, when aluminum is used, the contractor must be absolutely sure that the connections at both ends are very secure. The National Electric Code specifies that 100 A service must use a #4 AWG copper conductor or #2 aluminum conductor. For 200 A service, a 2/0 copper wire or a 4/0 aluminum conductor must be used, as shown in Fig. 4.27(a). A 100 A or 200 A service must have two lines and a service neutral as shown in Fig. 4.27(b). Note in Fig. 4.27(b) that the lines are coated and insulated from each other, and the service neutral is spread around the inside of the wire coating. At the terminal point, all

FIG. 4.27
200 A service conductors: (a) 4/0 aluminum and 2/0 copper; (b) three-wire 4/0 aluminum service.

the strands of the service neutral are gathered together and securely attached to the panel. It is fairly obvious that the cables of Fig. 4.27(a) are stranded for added flexibility.

Within the system, the incoming power is broken down into a number of circuits with lower current ratings utilizing 15 A, 20 A, 30 A, and 40 A protective breakers. Since the load on each breaker should not exceed 80% of its rating, in a 15 A breaker the maximum current should be limited to 80% of 15 A, or 12 A, with 16 A for a 20 A breaker, 24 A for a 30 A breaker, and 32 A for a 40 A breaker. The result is that a home with 200 A service can theoretically have a maximum of 12 circuits (200 A/16 A = 12.5) utilizing the 16 A maximum current ratings associated with 20 A breakers. However, if they are aware of the loads on each circuit, electricians can install as many circuits as they feel appropriate. The code further specifies that a #14 wire should not carry a current in excess of 15 A, a #12 in excess of 20 A, and a #10 in excess of 30 A. Thus, #12 wire is now the most common for general home wiring to ensure that it can handle any excursions beyond 15 A on the 20 A breaker (the most common breaker size). The #14 wire is often used in conjunction with the #12 wire in areas where it is known that the current levels are limited. The #10 wire is typically used for high-demand appliances such as dryers and ovens.

The circuits themselves are usually broken down into those that provide lighting, outlets, and so on. Some circuits (such as ovens and dryers) require a higher voltage of 240 V, obtained by using two power lines and the neutral. The higher voltage reduces the current requirement for the same power rating, with the net result that the appliance can usually be smaller. For example, the size of an air conditioner with the same cooling ability is measurably smaller when designed for a 240 V line than when designed for 120 V. Most 240 V lines, however, demand a current level that requires 30 A or 40 A breakers and special outlets to ensure that appliances rated at 120 V are not connected to the same outlet. Check the panel in your home and note the number of circuits—in particular, the rating of each breaker and the number of 240 V lines indicated by breakers requiring two slots of the panel. Determine the total of the current ratings of all the breakers in your panel, and explain, using the above information, why the total exceeds your feed level.

For safety sake, grounding is a very important part of the electrical system in your home. The National Electric Code requires that the neutral wire of a system be grounded to an earth-driven rod, a metallic water piping system of 10 ft or more, or a buried metal plate. That ground is

then passed on through the electrical circuits of the home for further protection. In a later chapter, the details of the connections and grounding methods are discussed.

4.9 COMPUTER ANALYSIS

Now that a complete circuit has been introduced and examined in detail, we can begin the application of computer methods. As mentioned in Chapter 1, two software packages will be introduced to demonstrate the options available with each and the differences that exist. Each has a broad range of support in the educational and industrial communities. The student version of PSpice (OrCAD Release 16.2 from Cadence Design Systems) has received the most attention, followed by Multisim. Each approach has its own characteristics with procedures that must be followed exactly; otherwise, error messages will appear. Do not assume that you can "force" the system to respond the way you would prefer—every step is well defined, and one error on the input side can yield results of a meaningless nature. At times you may believe that the system is in error because you are absolutely sure you followed every step correctly. In such cases, accept the fact that something was entered incorrectly, and review all your work very carefully. All it takes is a comma instead of a period or a decimal point to generate incorrect results.

Be patient with the learning process; keep notes of specific maneuvers that you learn; and don't be afraid to ask for help when you need it. For each approach, there is always the initial concern about how to start and proceed through the first phases of the analysis. However, be assured that with time and exposure you will work through the required maneuvers at a speed you never would have expected. In time you will be absolutely delighted with the results you can obtain using computer methods.

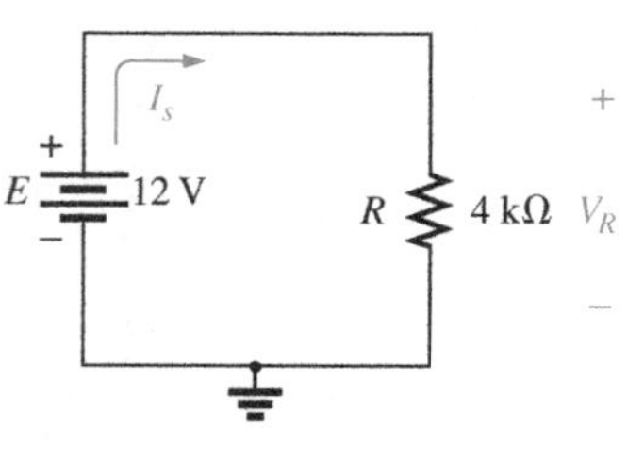

FIG. 4.28
Circuit to be analyzed using PSpice and Multisim.

In this section, Ohm's law is investigated using the software packages PSpice and Multisim to analyze the circuit in Fig. 4.28. Both require that the circuit first be "drawn" on the computer screen and then analyzed (simulated) to obtain the desired results. As mentioned above, the analysis program cannot be changed by the user. The most proficient user is one who can draw the most out of a computer software package.

Although the author feels that there is sufficient material in the text to carry a new student of the material through the programs provided, be aware that this is not a computer text. Rather, it is one whose primary purpose is simply to introduce the different approaches and how they can be applied effectively. Excellent texts and manuals are available that cover the material in a great deal more detail and perhaps at a slower pace. In fact, the quality of the available literature has improved dramatically in recent years.

PSpice

Readers who were familiar with older versions of PSpice will find that the changes in Version 16.2 are primarily in the front end and the simulation process. After executing a few programs, you will find that most of the procedures you learned from older versions will be applicable here also—at least the sequential process has a number of strong similarities.

The installation process for the OrCAD software requires a computer system with DVD capability and the minimum system requirements appearing in Appendix B. After the disk is installed, the question appearing during the license process can be answered by simply replying **port1@host1.** A **Yes** response to the next few questions followed by selecting **Finish** will install the software—a very simple and direct process.

Once OrCAD Version 16.2 has been installed, you must first open a **Folder** in the **C:** drive for storage of the circuit files that result from the analysis. Be aware, however, that

once the folder has been defined, it does not have to be defined for each new project unless you choose to do so. If you are satisfied with one location (folder) for all your projects, this is a one-time operation that does not have to be repeated with each network.

To establish the **Folder,** simply right-click the mouse on **Start** at the bottom left of the screen to obtain a listing that includes **Explore.** Select **Explore** to obtain the **Start Menu** dialog box and then use the sequence **File-New Folder** to open a new folder. Type in **PSpice** (the author's choice) and left-click to install. Then exit (using the **X** at the top right of the screen). The folder **PSpice** will be used for all the projects you plan to work on in this text.

Our first project can now be initiated by double-clicking on the **OrCAD 16.2 Demo** icon on the screen, or you can use the sequence **Start All Programs-CAPTURE CIS DEMO.** The resulting screen has only a few active keys on the top toolbar. The first keypad at the top left is the **Create document** key (or you can use the sequence **File-New Project**). A **New Project** dialog box opens in which you must enter the **Name** of the project. For our purposes we will choose **PSpice 4-1** as shown in Fig. 4.29 and select **Analog or Mixed A/D** (to be used for all the analyses of this text). Note at the bottom of the dialog box that the **Location** appears as **PSpice** as set above. Click **OK,** and another dialog

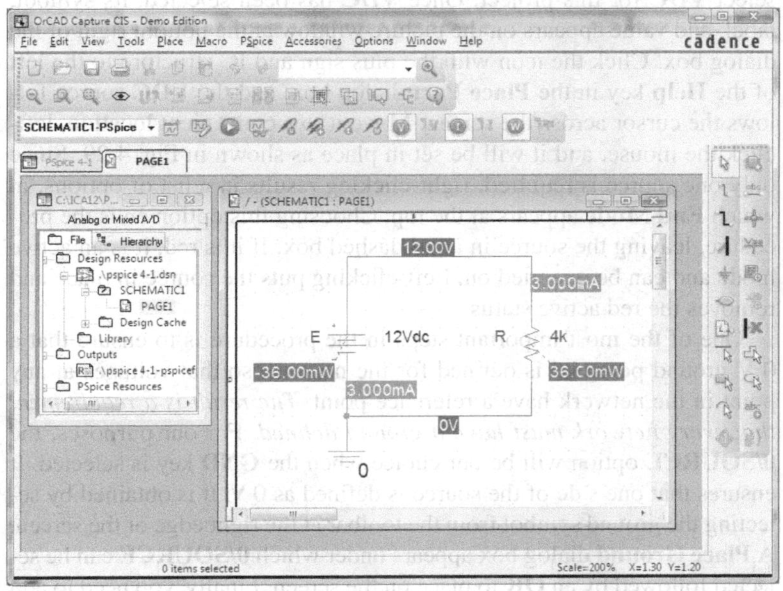

FIG. 4.29

Using PSpice to determine the voltage, current, and power levels for the circuit in Fig. 4.28.

box appears titled **Create PSpice Project.** Select **Create a blank project** (again, for all the analyses to be performed in this text). Click **OK,** and a third toolbar appears at the top of the screen with some of the keys enabled. A **Project Manager Window** appears with **PSpice 4-1** next to an icon and an associated + sign in a small square. Clicking on the + sign will take the listing a step further to **SCHEMATIC1.** Click + again, and **PAGE1** appears; clicking on a − sign reverses the process. Double-clicking on **PAGE1** creates a working window titled **SCHEMATIC1: PAGE1,** revealing that a project can have more than one schematic file and more than one associated page. The width and height of the window can be adjusted by grabbing an edge until you see a double-headed arrow and dragging the border to the desired location. Either window on the screen can be moved by clicking on the top heading to make it dark blue and then dragging it to any location.

Now you are ready to build the simple circuit in Fig. 4.28. Select the **Place a part** key (the key on the right tool bar with a small plus sign and IC structure) to obtain the **Place Part** dialog box. Since this is the first circuit to be constructed, you must ensure that the parts appear in the list of active libraries. Select **Add Library-Browse File,** and select **analog.olb,** and when it appears under the **File name** heading, select **Open.** It will now appear in the **Libraries** listing at the bottom left of the dialog box. Repeat for the **source.olb** and **special.olb** libraries. All three files are required to build the networks appearing in this text. However, it is important to realize that

once the library files have been selected, they will appear in the active listing for each new project without your having to add them each time—a step, such as the Folder step above, that does not have to be repeated with each similar project.

You are now ready to place components on the screen. For the dc voltage source, first select the **Place a part key** and then select **SOURCE** in the library listing. Under **Part List**, a list of available sources appears; select **VDC** for this project. Once **VDC** has been selected, its symbol, label, and value appears on the picture window at the bottom right of the dialog box. Click the icon with the plus sign and IC structure to the left of the **Help** key in the **Place Part** dialog box, and the **VDC** source follows the cursor across the screen. Move it to a convenient location, left-click the mouse, and it will be set in place as shown in Fig. 4.29. Since only one source is required, right-clicking results in a list of options, in which **End Mode** appears at the top. Choosing this option ends the procedure, leaving the source in a red dashed box. If it is red, it is an active mode and can be operated on. Left-clicking puts the source in place and removes the red active status.

One of the most important steps in the procedure is to ensure that a 0 V ground potential is defined for the network so that voltages at any point in the network have a reference point. *The result is a requirement that every network must have a ground defined.* For our purposes, the **0/SOURCE** option will be our choice when the **GND** key is selected. It ensures that one side of the source is defined as 0 V. It is obtained by selecting the ground symbol from the toolbar at the right edge of the screen. A **Place Ground** dialog box appears under which **0/SOURCE** can be selected followed by an **OK** to place on the screen. Finally, you need to add a resistor to the network by selecting the **Place a part** key again and then selecting the **ANALOG** library. Scrolling the options, note that **R** appears and should be selected. Click **OK,** and the resistor appears next to

the cursor on the screen. Move it to the desired location and click it in place. Then right-click and **End Mode,** and the resistor has been entered into the schematic's memory. Unfortunately, the resistor ended up in the horizontal position, and the circuit of Fig. 4.28 has the resistor in the vertical position. No problem: Simply select the resistor again to make it red, and right-click. A listing appears in which **Rotate** is an option. It turns the resistor 90° in the counterclockwise direction.

All the required elements are on the screen, but they need to be connected. To accomplish this, select the **Place a wire** key that looks like a step in the right toolbar. The result is a crosshair with the center that should be placed at the point to be connected. Place the crosshair at the top of the voltage source, and left-click it once to connect it to that point. Then draw a line to the end of the next element, and click again when the crosshair is at the correct point. A red line results with a square at each end to confirm that the connection has been made. Then move the crosshair to the other elements, and build the circuit. Once everything is connected, right-clicking provides the **End Mode** option. Do not forget to connect the source to ground as shown in Fig. 4.29.

Now you have all the elements in place, but their labels and values are wrong. To change any parameter, simply double-click on the parameter (the label or the value) to obtain the **Display Properties** dialog box. Type in the correct label or value, click **OK,** and the quantity is changed on the screen. Before selecting **OK,** be sure to check the **Display Format** to specify what will appear on the screen. The labels and values can be moved by simply clicking on the center of the parameter until it is closely surrounded by the four small squares and then dragging it to the new location. Left-clicking again deposits it in its new location.

Finally, you can initiate the analysis process, called **Simulation,** by selecting the **New simulation profile** key in the bottom toolbar of the heading of the screen that resembles a data page with a varying waveform and yellow star in the top right corner. **A New Simulation** dialog box opens that first asks for the **Name** of the simulation. The New Simulation dialog box can also be obtained by using the sequence **PSpice-New Simulation Profile-Bias Point** is entered for a dc solution, and **none** is left in the **Inherit From** request. Then select **Create,** and a **Simulation Settings** dialog box appears in which **Analysis-Analysis Type: Bias Point** is sequentially selected. Click **OK,** and select the **Run PSpice** key (which looks like a green circular key with an arrowhead) or choose **PSpice-Run** from the menu bar. An output window will appear with the dc voltages of the network: **12 V** and **0 V.** The dc currents and power levels can be displayed as shown in Fig. 4.29 by simply selecting the green circular keys with the **I** and **W** in the bottom tool bar at the top of the screen. Individual values can be removed by simply selecting the value and pressing the **Delete** key or the scissors key in the top menu bar. Resulting values can be moved by simply left-clicking the value and dragging it to the desired location.

Note in Fig. 4.29 that the current is 3 mA (as expected) at each point in the network, and the power delivered by the source and dissipated by the resistor is the same, or 36 mW. There are also 12 V across the resistor as required by the configuration.

There is no question that this procedure seems long for such a simple circuit. However, keep in mind that we needed to introduce many new facets of using PSpice that are not discussed again. By the time you finish analyzing your third or fourth network, the procedure will be routine and easy to do.

Multisim

For comparison purposes, Multisim is also used to analyze the circuit in Fig. 4.28. Although there are differences between PSpice and Multisim, such as in initiating the process, constructing the networks, making the measurements, and setting up the simulation procedure, there are sufficient similarities between the two approaches to make it easier to learn one if you are already familiar with the other. The similarities will be obvious only if you make an attempt to learn both. One of the major differences between the two is the option to use actual instruments in Multisim to make the measurements—a positive trait in preparation for the laboratory experience. However, in Multisim, you may not find the extensive list of options available with PSpice. In general, however, both software packages are well prepared to take us through the types of analyses to be encountered in this text. The installation process for **Multisim** is not as direct as for the **OrCAD** demo version because the software package must be purchased to obtain a serial number. In most cases, the **Multisim** package will be available through the local educational facility.

When the **Multisim** icon is selected from the opening window, a screen appears with the heading **Circuit 1-Multisim.** A menu bar appears across the top of the screen, with seven additional toolbars: **Standard, View, Main, Components, Simulation Switch, Simulation,** and **Instruments.** By selecting **View** from the top menu bar followed by **Toolbars,** you can add or delete toolbars. The heading can be changed to **Multisim 4-1** by selecting **File-Save As** to open the **Save As** dialog box. Enter **Multisim 4-1** as the **File name** to obtain the listing of Fig. 4.30.

For the placement of components, **View-Show Grid** was selected so that a grid would appear on the screen. As you place an element, it will automatically be placed in a relationship specific to the grid structure.

To build the circuit in Fig. 4.28, first take the cursor and place it on the battery symbol in the **Component toolbar.** Left-click, and a

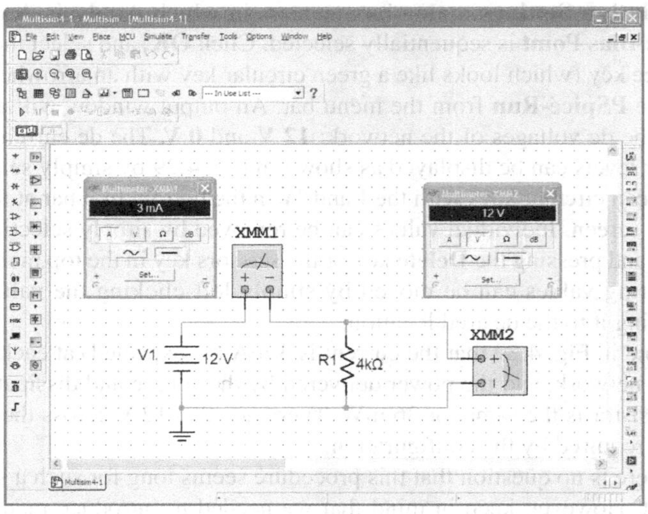

FIG. 4.30
Using Multisim to determine the voltage and current levels for the circuit of Fig. 4.28.

Component dialog box will appear, which provides a list of sources. Under **Component,** select **DC-POWER.** The symbol appears in the adjoining box area. Click **OK.** The battery symbol appears on the screen next to the location of the cursor. Move the cursor to the desired location, and left-click to set the battery symbol in place. The operation is complete. If you want to delete the source, simply left-click on the symbol again to create a dashed rectangle around the source. These rectangles indicate that the source is in the active mode and can be operated on. If you want to delete it, click on the **Delete** key or select the **Scissor** keypad in the **Standard toolbar.** If you want to modify the source, right-click *outside* the rectangle, and you get one list. Right-click *within* the rectangle, and you have a different set of options. At any time, if you want to remove the active state, left-click anywhere on the screen. If you want to move the source, click on the source symbol to create the rectangle, but do not release the mouse. Hold it down and drag the source to the preferred location. When the source is in place, release the mouse. Click again to remove the active state. *From now on, whenever possible, the word "click" means a left-click.* The need for a right click will continue to be spelled out.

For the simple circuit in Fig. 4.30, you need to place a resistor across the source. Select the keypad in the **Components toolbar** that looks like a resistor symbol. A **Select a Component** dialog box opens with a **Family** listing. Selecting **RESISTOR** results in a list of standard values that can be quickly selected for the deposited resistor. However, in this case, you want to use a 4 kΩ resistor, which is not a standard value but can be changed to 4 kΩ by simply changing the value once it has been placed on the screen. Another approach is to add the **Virtual toolbar** (also referred to as the **Basic toolbar**), which provides a list of components for which the value can be set. Selecting the resistor symbol from the **Virtual toolbar** will result in the placement of a resistor with a starting value of 1 kΩ. Once placed on the screen, the value of the resistor can be changed by simply double-clicking on the resistor value to obtain a dialog box that permits the change. The placement of the resistor is exactly the same as that employed for the source above.

In Fig. 4.28, the resistor is in the vertical position, so a rotation must be made. Click on the resistor to obtain the active state, and then right-click within the rectangle. A number of options appear, including **Flip Horizontal, Flip Vertical, 90° Clockwise,** and **90° Counter CW.** To rotate 90° counterclockwise, select that option, and the resistor is automatically rotated 90°.

Finally, you need a ground for all networks. Going back to the **Sources** parts bin, find **GROUND,** which is the fourth option down under **Component.** Select **GROUND** and place it on the screen below the voltage source as shown in Fig. 4.30. Now, before connecting the components together, move the labels and the value of each component to the relative positions shown in Fig. 4.30. Do this by clicking on the label or value to create a small set of squares around the element and then dragging the element to the desired location. Release the mouse, and then click again to set the element in place. To change the label or value, double-click on the label (such as **V1**) to open a **DC_POWER** dialog box. Select **Label** and enter **E** as the **Reference Designation (Ref Des).** Then, before leaving the dialog box, go to **Value** and change the value if necessary. It is very important to realize that you cannot type in the units where the **V** now appears to the right

of the value. The suffix is controlled by the scroll keys at the left of the unit of measure. For practice, try the scroll keys, and you will find that you can go from **pV** to **TV.** For now, leave it as **V.** Click **OK,** and both have been changed on the screen. The same process can be applied to the resistive element to obtain the label and value appearing in Fig. 4.30.

Next, you need to tell the system which results should be generated and how they should be displayed. For this example, we use a multimeter to measure both the current and the voltage of the circuit. The **Multimeter** is the first option in the list of instruments appearing in the toolbar to the right of the screen. When selected, it appears on the screen and can be placed anywhere, using the same procedure defined for the components above. Double-click on the meter symbol, and a **Multimeter** dialog box opens in which the function of the meter must be defined. Since the meter **XMM1** will be used as an ammeter, select the letter **A** and the horizontal line to indicate dc level. There is no need to select **Set** for the default values since they have been chosen for the broad range of applications. The dialog meters can be moved to any location by clicking on their heading bar to make it dark blue and then dragging the meter to the preferred position. For the voltmeter, **V** and the horizontal bar were selected as shown in Fig. 4.30. The voltmeter was turned clockwise 90° using the same procedure as described for the resistor above.

Finally, the elements need to be connected. To do this, bring the cursor to one end of an element, say, the top of the voltage source. A small dot and a crosshair appear at the top end of the element. Click once, follow the path you want, and place the crosshair over the positive terminal of the ammeter. Click again, and the wire appears in place.

At this point, you should be aware that the software package has its preferences about how it wants the elements to be connected. That is, you may try to draw it one way, but the computer generation may be a different path. Eventually, you will learn these preferences and can set up the network to your liking. Now continue making the connections appearing in Fig. 4.30, moving elements or adjusting lines as necessary. Be sure that the small dot appears at any point where you want a connection. Its absence suggests that the connection has not been made and the software program has not accepted the entry.

You are now ready to run the program and view the solution. The analysis can be initiated in a number of ways. One option is to select **Simulate** from the top toolbar, followed by **RUN.** Another is to select the **Simulate** key (the green arrow) in the **Simulation toolbar.** The last option, and the one we use the most, utilizes the **OFF/ON, 0/1 Simulation** switch at the top right of the screen. With this last option, the analysis (called **Simulation**) is initiated by clicking the switch into the 1 position. The analysis is performed, and the current and voltage appear on the meter as shown in Fig. 4.30. Note that both provide the expected results.

One of the most important things to learn about applying Multisim:

Always stop or end the simulation (clicking on 0 or choosing OFF) before making any changes in the network. When the simulation is initiated, it stays in that mode until turned off.

There was obviously a great deal of material to learn in this first exercise using Multisim. Be assured, however, that as we continue with more examples, you will find the procedure quite straightforward and actually enjoyable to apply.

PROBLEMS

SECTION 4.2 Ohm's Law

1. What is the voltage across a 220 Ω resistor if the current through it is 5.6 mA?
2. What is the current through a 6.8 Ω resistor if the voltage drop across it is 24 V?
3. How much resistance is required to limit the current to 1.5 mA if the potential drop across the resistor is 24 V?
4. At starting, what is the current drain on a 12 V car battery if the resistance of the starting motor is 40 mΩ?
5. If the current through a 0.02 MΩ resistor is 3.6 μA, what is the voltage drop across the resistor?
6. If a voltmeter has an internal resistance of 50 kΩ, find the current through the meter when it reads 120 V.
7. If a refrigerator draws 2.2 A at 120 V, what is its resistance?
8. If a clock has an internal resistance of 8 kΩ, find the current through the clock if it is plugged into a 120 V outlet.
9. A washing machine is rated at 4.2 A at 120 V. What is its internal resistance?
10. A CD player draws 125 mA when 4.5 V is applied. What is the internal resistance?
11. The input current to a transistor is 20 μA. If the applied (input) voltage is 24 mV, determine the input resistance of the transistor.
12. The internal resistance of a dc generator is 0.5 Ω. Determine the loss in terminal voltage across this internal resistance if the current is 12 A.

*13. a. If an electric heater draws 9.5 A when connected to a 120 V supply, what is the internal resistance of the heater?
b. Using the basic relationships of Chapter 2, determine how much energy in joules (J) is converted if the heater is used for 2 h during the day.

14. In a TV camera, a current of 2.4 μA passes through a resistor of 3.3 MΩ. What is the voltage drop across the resistor?

SECTION 4.3 Plotting Ohm's Law

15. a. Plot the curve of *I* (vertical axis) versus *V* (horizontal axis) for a 120 Ω resistor. Use a horizontal scale of 0 to 100 V and a vertical scale of 0 to 1 A.
b. Using the graph of part (a), find the current at a voltage of 20 V and 50 V.

16. a. Plot the *I-V* curve for a 5 Ω and a 20 Ω resistor on the same graph. Use a horizontal scale of 0 to 40 V and a vertical scale of 0 to 2 A.
b. Which is the steeper curve? Can you offer any general conclusions based on results?
c. If the horizontal and vertical scales were interchanged, which would be the steeper curve?

17. a. Plot the *I-V* characteristics of a 1 Ω, 100 Ω, and 1000 Ω resistor on the same graph. Use a horizontal axis of 0 to 100 V and a vertical axis of 0 to 100 A.
b. Comment on the steepness of a curve with increasing levels of resistance.

*18. Sketch the internal resistance characteristics of a device that has an internal resistance of 20 Ω from 0 to 10 V, an internal resistance of 4 Ω from 10 V to 15 V, and an internal resistance of 1 Ω for any voltage greater than 15 V. Use a horizontal scale that extends from 0 to 20 V and a vertical scale that permits plotting the current for all values of voltage from 0 to 20 V.

*19. a. Plot the *I-V* characteristics of a 2 kΩ, 1 MΩ, and a 100 Ω resistor on the same graph. Use a horizontal axis of 0 to 20 V and a vertical axis of 0 to 10 mA.
b. Comment on the steepness of the curve with decreasing levels of resistance.
c. Are the curves linear or nonlinear? Why?

SECTION 4.4 Power

20. If 540 J of energy are absorbed by a resistor in 4 min, what is the power delivered to the resistor in watts?
21. The power to a device is 40 joules per second (J/s). How long will it take to deliver 640 J?
22. a. How many joules of energy does a 2 W nightlight dissipate in 8 h?
b. How many kilowatthours does it dissipate?
23. How long must a steady current of 1.4 A exist in a resistor that has 3 V across it to dissipate 12 J of energy?
24. What is the power delivered by a 6 V battery if the current drain is 750 mA?
25. The current through a 4 kΩ resistor is 7.2 mA. What is the power delivered to the resistor?
26. The power consumed by a 2.2 kΩ resistor is 240 mW. What is the current level through the resistor?
27. What is the maximum permissible current in a 120 Ω, 2 W resistor? What is the maximum voltage that can be applied across the resistor?
28. The voltage drop across a transistor network is 22 V. If the total resistance is 16.8 kΩ, what is the current level? What is the power delivered? How much energy is dissipated in 1 h?
29. If the power applied to a system is 324 W, what is the voltage across the line if the current is 2.7 A?
30. A 1 W resistor has a resistance of 4.7 MΩ. What is the maximum current level for the resistor? If the wattage rating is increased to 2 W, will the current rating double?
31. A 2.2 kΩ resistor in a stereo system dissipates 42 mW of power. What is the voltage across the resistor?
32. What are the "hot" resistance level and current rating of a 120 V, 100 W bulb?
33. What are the internal resistance and voltage rating of a 450 W automatic washer that draws 3.75 A?
34. A calculator with an internal 3 V battery draws 0.4 mW when fully functional.
a. What is the current demand from the supply?
b. If the calculator is rated to operate 500 h on the same battery, what is the ampere-hour rating of the battery?
35. A 20 kΩ resistor has a rating of 100 W. What are the maximum current and the maximum voltage that can be applied to the resistor?
36. What is the total horsepower rating of a series of commercial ceiling fans that draw 30 A at 220 V?

SECTION 4.5 Energy

37. A 10 Ω resistor is connected across a 12 V battery.
a. How many joules of energy will it dissipate in 1 min?
b. If the resistor is left connected for 2 min instead of 1 min, will the energy used increase? Will the power dissipation level increase?

38. How much energy in kilowatthours is required to keep a 230 W oil-burner motor running 12 h a week for 5 months? (Use 4 weeks = 1 month.)

39. How long can a 1500 W heater be on before using more than 12 kWh of energy?

40. A 60 W bulb is on for 10 h.
a. What is the energy used in wattseconds?
b. What is the energy dissipated in joules?
c. What is the energy transferred in watthours?
d. How many kilowatthours of energy were dissipated?
e. At 11¢/kWh, what was the total cost?

41. a. In 10 h an electrical system converts 1200 kWh of electrical energy into heat. What is the power level of the system?
b. If the applied voltage is 208 V, what is the current drawn from the supply?
c. If the efficiency of the system is 82%, how much energy is lost or stored in 10 h?

42. At 11¢/kWh, how long can you play a 250 W color television for $1?

43. The electric bill for a family for a month is $74.
a. Assuming 31 days in the month, what is the cost per day?
b. Based on 15-h days, what is the cost per hour?
c. How many kilowatthours are used per hour if the cost is 11¢/kWh?
d. How many 60 W lightbulbs (approximate number) could you have on to use up that much energy per hour?
e. Do you believe the cost of electricity is excessive?

44. How long can you use an Xbox 360 for $1 if it uses 187 W and the cost is 11¢/kWh?

45. The average plasma screen TV draws 339 W of power, whereas the average LCD TV draws 213 W. If each set was used 5 h/day for 365 days, what would be the cost savings for the LCD unit over the year if the cost is 11¢/kWh?

46. The average PC draws 78 W. What is the cost of using the PC for 4 h/day for a month of 31 days if the cost is 11¢/kWh?

***47. a.** If a house is supplied with 120 V, 100 A service, find the maximum power capability.
b. Can the homeowner safely operate the following loads at the same time?
5 hp motor
3000 W clothes dryer
2400 W electric range
1000 W steam iron
c. If all the appliances are used for 2 hours, how much energy is converted in kWh?

***48.** What is the total cost of using the following at 11¢/kWh?
a. 1600 W air conditioner for 6 h
b. 1200 W hair dryer for 15 min
c. 4800 W clothes dryer for 30 min
d. 900 W coffee maker for 10 min
e. 200 W Play Station 3 for 2 h
f. 50 W stereo for 3.5 h

***49.** What is the total cost of using the following at 11¢/kWh?
a. 200 W fan for 4 h
b. Six 60 W bulbs for 6 h
c. 1200 W dryer for 20 min
d. 175 W desktop computer for 3.5 h
e. 250 W color television set for 2 h 10 min
f. 30 W satellite dish for 8 h

SECTION 4.6 Efficiency

50. What is the efficiency of a motor that has an output of 0.5 hp with an input of 410 W?

51. The motor of a power saw is rated 68.5% efficient. If 1.8 hp are required to cut a particular piece of lumber, what is the current drawn from a 120 V supply?

52. What is the efficiency of a dryer motor that delivers 0.8 hp when the input current and voltage are 4 A and 220 V, respectively?

53. A stereo system draws 1.8 A at 120 V. The audio output power is 50 W.
a. How much power is lost in the form of heat in the system?
b. What is the efficiency of the system?

54. If an electric motor having an efficiency of 76% and operating off a 220 V line delivers 3.6 hp, what input current does the motor draw?

55. A motor is rated to deliver 2 hp.
a. If it runs on 110 V and is 90% efficient, how many watts does it draw from the power line?
b. What is the input current?
c. What is the input current if the motor is only 70% efficient?

56. An electric motor used in an elevator system has an efficiency of 90%. If the input voltage is 220 V, what is the input current when the motor is delivering 15 hp?

57. The motor used on a conveyor belt is 85% efficient. If the overall efficiency is 75%, what is the efficiency of the conveyor belt assembly?

58. A 2 hp motor drives a sanding belt. If the efficiency of the motor is 87% and that of the sanding belt is 75% due to slippage, what is the overall efficiency of the system?

59. The overall efficiency of two systems in cascade is 78%. If the efficiency of one is 0.9, what is the efficiency, in percent, of the other?

60. a. What is the total efficiency of three systems in cascade with respective efficiencies of 93%, 87%, and 21%?
b. If the system with the least efficiency (21%) were removed and replaced by one with an efficiency of 80%, what would be the percentage increase in total efficiency?

***61.** If the total input and output power of two systems in cascade are 400 W and 128 W, respectively, what is the efficiency of each system if one has twice the efficiency of the other?

SECTION 4.9 Computer Analysis

62. Using PSpice or Multisim, repeat the analysis of the circuit in Fig. 4.28 with $E = 400$ mV and $R = 0.04$ MΩ.

63. Using PSpice or Multisim, repeat the analysis of the circuit in Fig. 4.28, but reverse the polarity of the battery and use $E = 8$ V and $R = 220$ Ω.

GLOSSARY

Circuit breaker A two-terminal device designed to ensure that current levels do not exceed safe levels. If "tripped," it can be reset with a switch or a reset button.

Diode A semiconductor device whose behavior is much like that of a simple switch; that is, it will pass current ideally in only one direction when operating within specified limits.

Efficiency (η) A ratio of output to input power that provides immediate information about the energy-converting characteristics of a system.

Energy (W) A quantity whose change in state is determined by the product of the rate of conversion (P) and the period involved (t). It is measured in joules (J) or wattseconds (Ws).

Fuse A two-terminal device whose sole purpose is to ensure that current levels in a circuit do not exceed safe levels.

Horsepower (hp) Equivalent to 746 watts in the electrical system.

Kilowatthour meter An instrument for measuring kilowatthours of energy supplied to a residential or commercial user of electricity.

Ohm's law An equation that establishes a relationship among the current, voltage, and resistance of an electrical system.

Power An indication of how much work can be done in a specified amount of time; a *rate* of doing work. It is measured in joules/second (J/s) or watts (W).

Series dc Circuits

Objectives

- **Become familiar with the characteristics of a series circuit and how to solve for the voltage, current, and power to each of the elements.**
- **Develop a clear understanding of Kirchhoff's voltage law and how important it is to the analysis of electric circuits.**
- **Become aware of how an applied voltage will divide among series components and how to properly apply the voltage divider rule.**
- **Understand the use of single- and double-subscript notation to define the voltage levels of a network.**
- **Learn how to use a voltmeter, ammeter, and ohmmeter to measure the important quantities of a network.**

5.1 INTRODUCTION

Two types of current are readily available to the consumer today. One is *direct current* (dc), in which ideally the flow of charge (current) does not change in magnitude (or direction) with time. The other is *sinusoidal alternating current* (ac), in which the flow of charge is continually changing in magnitude (and direction) with time. The next few chapters are an introduction to circuit analysis purely from a dc approach. The methods and concepts are discussed in detail for direct current; when possible, a short discussion suffices to cover any variations we may encounter when we consider ac in the later chapters.

The battery in Fig. 5.1, by virtue of the potential difference between its terminals, has the ability to cause (or "pressure") charge to flow through the simple circuit. The positive terminal attracts the electrons through the wire at the same rate at which electrons are supplied by the negative terminal. As long as the battery is connected in the circuit and maintains its terminal characteristics, the current (dc) through the circuit will not change in magnitude or direction.

If we consider the wire to be an ideal conductor (that is, having no opposition to flow), the potential difference V across the resistor equals the applied voltage of the battery: V (volts) = E (volts).

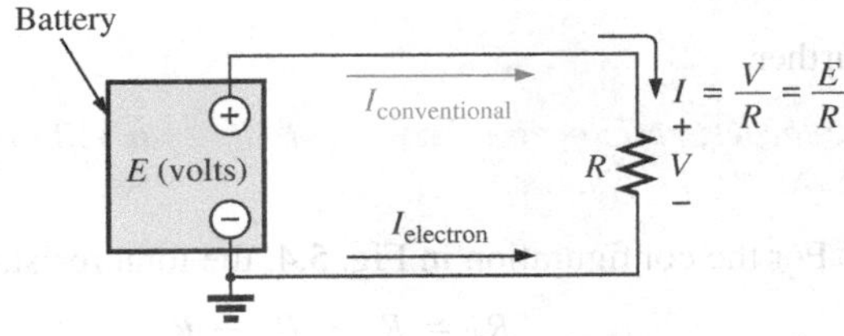

FIG. 5.1
Introducing the basic components of an electric circuit.

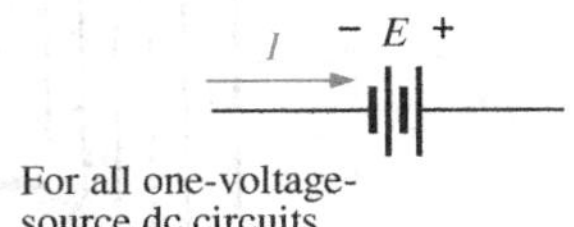

FIG. 5.2
Defining the direction of conventional flow for single-source dc circuits.

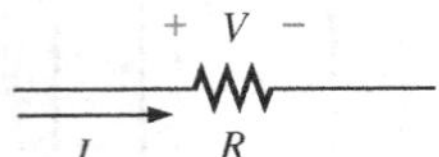

For any combination of voltage sources in the same dc circuit

FIG. 5.3
Defining the polarity resulting from a conventional current I through a resistive element.

The current is limited only by the resistor R. The higher the resistance, the less is the current, and conversely, as determined by Ohm's law.

By convention (as discussed in Chapter 2), the direction of **conventional current flow** ($I_{\text{conventional}}$) as shown in Fig. 5.1 is opposite to that of **electron flow** (I_{electron}). Also, the uniform flow of charge dictates that the direct current I be the same everywhere in the circuit. By following the direction of conventional flow, we notice that there is a rise in potential across the battery ($-$ to $+$) and a drop in potential across the resistor ($+$ to $-$). For single-voltage-source dc circuits, conventional flow always passes from a low potential to a high potential when passing through a voltage source, as shown in Fig. 5.2. However, conventional flow always passes from a high to a low potential when passing through a resistor for any number of voltage sources in the same circuit, as shown in Fig. 5.3.

The circuit in Fig. 5.1 is the simplest possible configuration. This chapter and the following chapters add elements to the system in a very specific manner to introduce a range of concepts that will form a major part of the foundation required to analyze the most complex system. Be aware that the laws, rules, and so on introduced in Chapters 5 and 6 will be used throughout your studies of electrical, electronic, or computer systems. They are not replaced by a more advanced set as you progress to more sophisticated material. It is therefore critical that you understand the concepts thoroughly and are able to apply the various procedures and methods with confidence.

5.2 SERIES RESISTORS

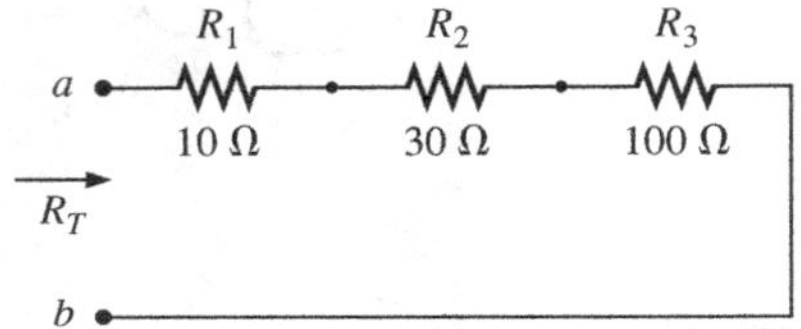

FIG. 5.4
Series connection of resistors.

Before the series connection is described, first recognize that every fixed resistor has only two terminals to connect in a configuration—it is therefore referred to as a **two-terminal device.** In Fig. 5.4, one terminal of resistor R_2 is connected to resistor R_1 on one side, and the remaining terminal is connected to resistor R_3 on the other side, resulting in one, and only one, connection between adjoining resistors. When connected in this manner, the resistors have established a series connection. If three elements were connected to the same point, as shown in Fig. 5.5, there would not be a series connection between resistors R_1 and R_2.

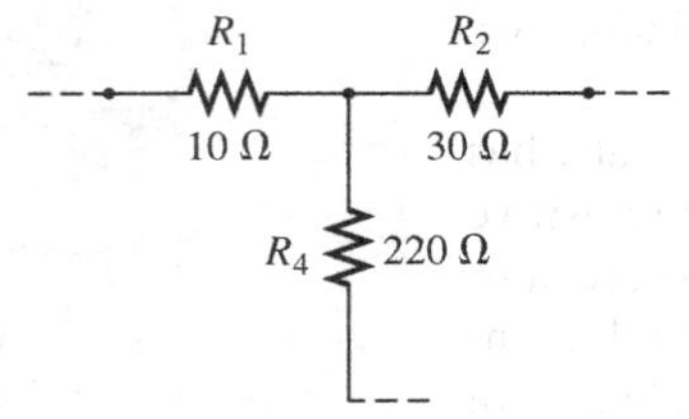

FIG. 5.5
Configuration in which none of the resistors are in series.

For resistors in series,

the total resistance of a series configuration is the sum of the resistance levels.

In equation form for any number (N) of resistors,

$$\boxed{R_T = R_1 + R_2 + R_3 + R_4 + \cdots + R_N} \tag{5.1}$$

A result of Eq. (5.1) is that

the more resistors we add in series, the greater is the resistance, no matter what their value.

Further,

the largest resistor in a series combination will have the most impact on the total resistance.

For the configuration in Fig. 5.4, the total resistance is

$$\begin{aligned} R_T &= R_1 + R_2 + R_3 \\ &= 10\ \Omega + 30\ \Omega + 100\ \Omega \end{aligned}$$

and $$R_T = \mathbf{140\ \Omega}$$

EXAMPLE 5.1 Determine the total resistance of the series connection in Fig. 5.6. Note that all the resistors appearing in this network are standard values.

Solution: Note in Fig. 5.6 that even though resistor R_3 is on the vertical and resistor R_4 returns at the bottom to terminal b, all the resistors are in series since there are only two resistor leads at each connection point.

Applying Eq. (5.1) gives

$$R_T = R_1 + R_2 + R_3 + R_4$$
$$R_T = 20\ \Omega + 220\ \Omega + 1.2\ \text{k}\Omega + 5.6\ \text{k}\Omega$$

and $$R_T = 7040\ \Omega = \mathbf{7.04\ k\Omega}$$

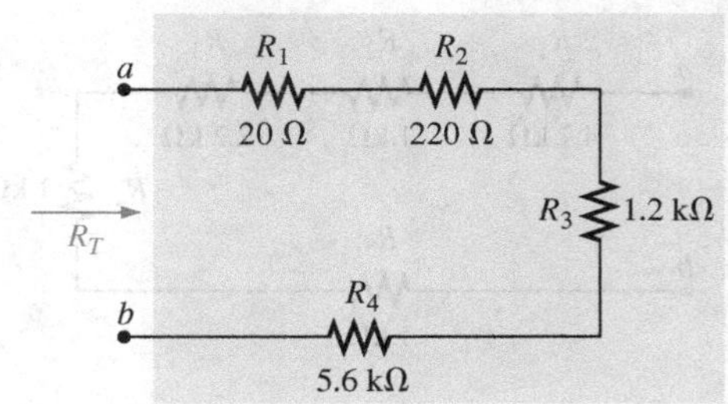

FIG. 5.6
Series connection of resistors for Example 5.1.

For the special case where resistors are the *same value,* Eq. (5.1) can be modified as follows:

$$\boxed{R_T = NR} \qquad (5.2)$$

where N is the number of resistors in series of value R.

EXAMPLE 5.2 Find the total resistance of the series resistors in Fig. 5.7. Again, recognize 3.3 kΩ as a standard value.

Solution: Again, don't be concerned about the change in configuration. Neighboring resistors are connected only at cne point, satisfying the definition of series elements.

Eq. (5.2): $R_T = NR$
$= (4)(3.3\ \text{k}\Omega) = \mathbf{13.2\ k\Omega}$

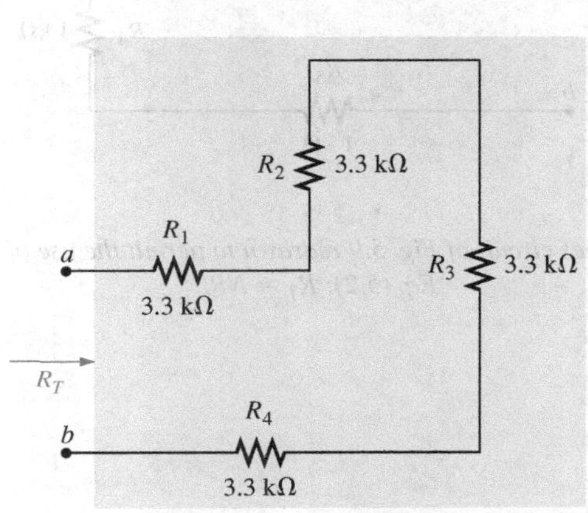

FIG. 5.7
Series connection of four resistors of the same value (Example 5.2).

It is important to realize that since the parameters of Eq. (5.1) can be put in any order,

the total resistance of resistors in series is unaffected by the order in which they are connected.

The result is that the total resistance in Fig. 5.8(a) is the same as in Fig. 5.8(b). Again, note that all the resistors are standard values.

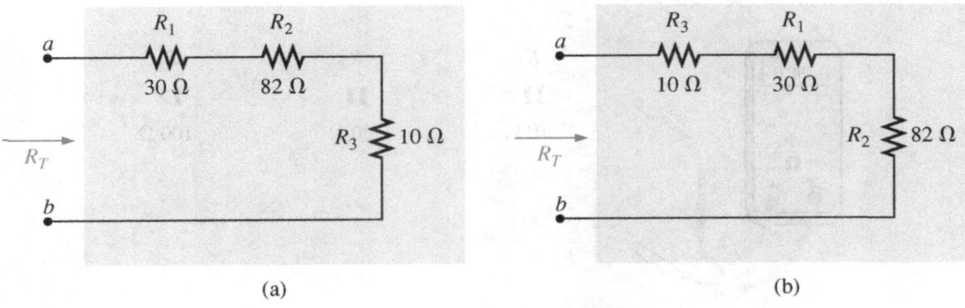

FIG. 5.8
Two series combinations of the same elements with the same total resistance.

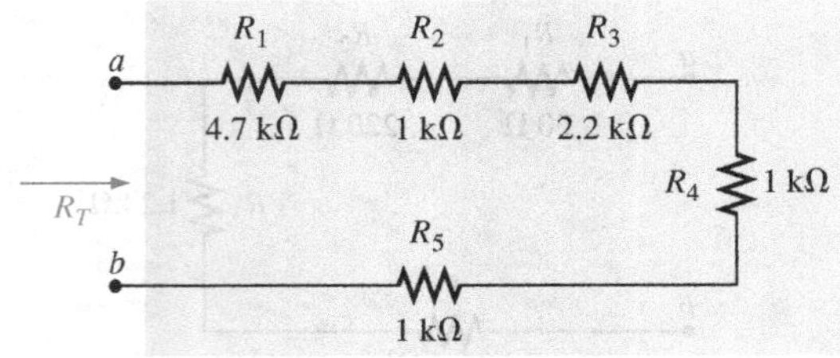

FIG. 5.9
Series combination of resistors for Example 5.3.

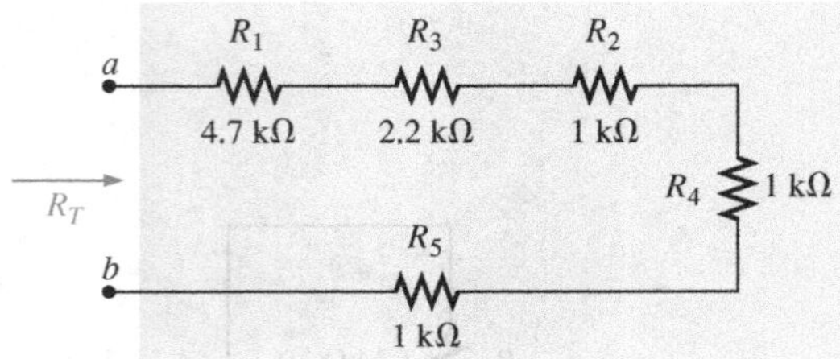

FIG. 5.10
Series circuit of Fig. 5.9 redrawn to permit the use of Eq. (5.2): $R_T = NR$.

EXAMPLE 5.3 Determine the total resistance for the series resistors (standard values) in Fig. 5.9.

Solution: First, the order of the resistors is changed as shown in Fig. 5.10 to permit the use of Eq. (5.2). The total resistance is then

$$\begin{aligned} R_T &= R_1 + R_3 + NR_2 \\ &= 4.7\text{ k}\Omega + 2.2\text{ k}\Omega + (3)(1\text{ k}\Omega) = \mathbf{9.9\text{ k}\Omega} \end{aligned}$$

Analogies

Throughout the text, analogies are used to help explain some of the important fundamental relationships in electrical circuits. An analogy is simply a combination of elements of a different type that are helpful in explaining a particular concept, relationship, or equation.

One analogy that works well for the series combination of elements is connecting different lengths of rope together to make the rope longer. Adjoining pieces of rope are connected at only one point, satisfying the definition of series elements. Connecting a third rope to the common point would mean that the sections of rope are no longer in a series.

Another analogy is connecting hoses together to form a longer hose. Again, there is still only one connection point between adjoining sections, resulting in a series connection.

Instrumentation

The total resistance of any configuration can be measured by simply connecting an ohmmeter across the access terminals as shown in Fig. 5.11 for the circuit in Fig. 5.4. *Since there is no polarity associated with resistance,* either lead can be connected to point *a,* with the other lead connected to point *b.* Choose a scale that will exceed the total resistance of the circuit, and remember when you read the response on the meter, if a kilohm scale was selected, the result will be in kilohms. For Fig. 5.11, the 200 Ω scale of our chosen multimeter was used because the total resistance is 140 Ω. If the 2 kΩ scale of our meter were selected, the digital display would read 0.140, and you must recognize that the result is in kilohms.

In the next section, another method for determining the total resistance of a circuit is introduced using Ohm's law.

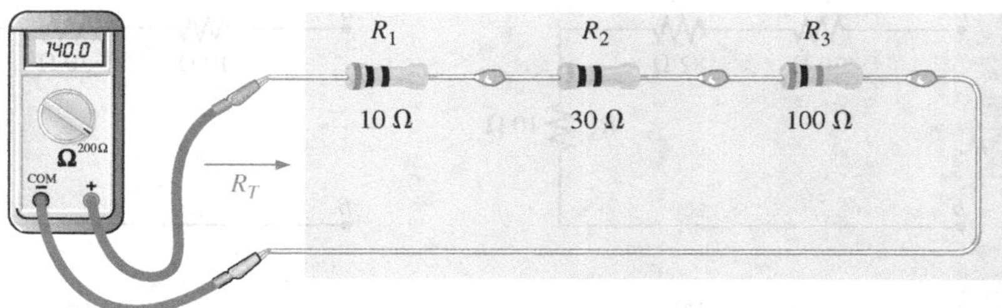

FIG. 5.11
Using an ohmmeter to measure the total resistance of a series circuit.

5.3 SERIES CIRCUITS

If we now take an 8.4 V dc supply and connect it in series with the series resistors in Fig. 5.4, we have the **series circuit** in Fig. 5.12.

A circuit is any combination of elements that will result in a continuous flow of charge, or current, through the configuration.

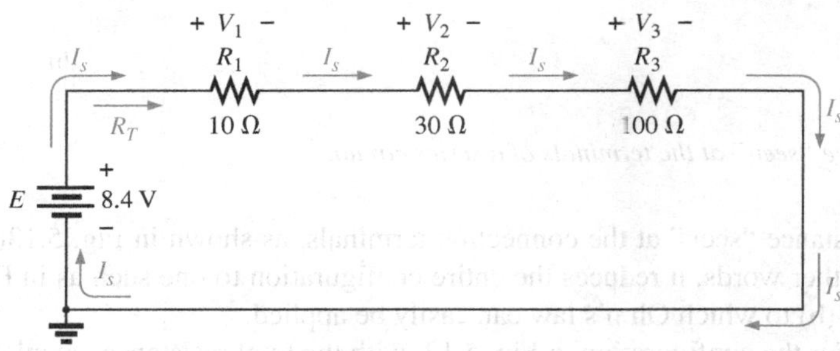

FIG. 5.12
Schematic representation for a dc series circuit.

First, recognize that the *dc supply is also a two-terminal device* with two points to be connected. If we simply ensure that there is only one connection made at each end of the supply to the series combination of resistors, we can be sure that we have established a series circuit.

The manner in which the supply is connected determines the direction of the resulting conventional current. For series dc circuits:

the direction of conventional current in a series dc circuit is such that it leaves the positive terminal of the supply and returns to the negative terminal, as shown in Fig. 5.12.

One of the most important concepts to remember when analyzing series circuits and defining elements that are in series is:

The current is the same at every point in a series circuit.

For the circuit in Fig. 5.12, the above statement dictates that the current is the same through the three resistors and the voltage source. In addition, if you are ever concerned about whether two elements are in series, simply check whether the current is the same through each element.

In any configuration, if two elements are in series, the current must be the same. However, if the current is the same for two adjoining elements, the elements may or may not be in series.

The need for this constraint in the last sentence will be demonstrated in the chapters to follow.

Now that we have a complete circuit and current has been established, the level of current and the voltage across each resistor should be determined. To do this, return to Ohm's law and replace the resistance in the equation by the total resistance of the circuit. That is,

$$I_s = \frac{E}{R_T} \tag{5.3}$$

with the subscript s used to indicate source current.

It is important to realize that when a dc supply is connected, it does not "see" the individual connection of elements but simply the total

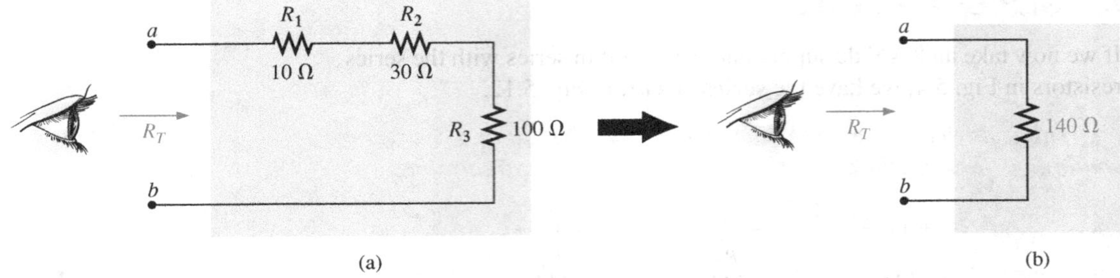

FIG. 5.13
Resistance "seen" at the terminals of a series circuit.

resistance "seen" at the connection terminals, as shown in Fig. 5.13(a). In other words, it reduces the entire configuration to one such as in Fig. 5.13(b) to which Ohm's law can easily be applied.

For the configuration in Fig. 5.12, with the total resistance calculated in the last section, the resulting current is

$$I_s = \frac{E}{R_T} = \frac{8.4\ \text{V}}{140\ \Omega} = 0.06\ \text{A} = \mathbf{60\ mA}$$

Note that the current I_s at every point or corner of the network is the same. Furthermore, note that the current is also indicated on the current display of the power supply.

Now that we have the current level, we can calculate the voltage across each resistor. First recognize that

the polarity of the voltage across a resistor is determined by the direction of the current.

Current entering a resistor creates a drop in voltage with the polarity indicated in Fig. 5.14(a). Reverse the direction of the current, and the polarity will reverse as shown in Fig. 5.14(b). Change the orientation of the resistor, and the same rules apply as shown in Fig. 5.14(c). Applying the above to the circuit in Fig. 5.12 will result in the polarities appearing in that figure.

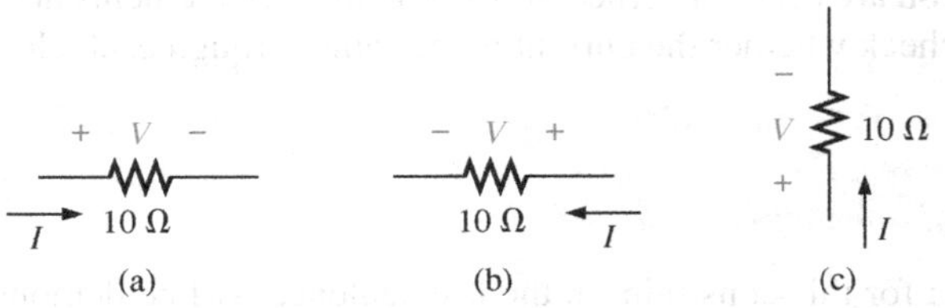

FIG. 5.14
Inserting the polarities across a resistor as determined by the direction of the current.

The magnitude of the voltage drop across each resistor can then be found by applying Ohm's law using only the resistance of each resistor. That is,

$$\boxed{\begin{aligned} V_1 &= I_1R_1 \\ V_2 &= I_2R_2 \\ V_3 &= I_3R_3 \end{aligned}} \qquad \textbf{(5.4)}$$

which for Fig. 5.12 results in

$$V_1 = I_1R_1 = I_sR_1 = (60 \text{ mA})(10 \ \Omega) = \mathbf{0.6 \ V}$$
$$V_2 = I_2R_2 = I_sR_2 = (60 \text{ mA})(30 \ \Omega) = \mathbf{1.8 \ V}$$
$$V_3 = I_3R_3 = I_sR_3 = (60 \text{ mA})(100 \ \Omega) = \mathbf{6.0 \ V}$$

Note that in all the numerical calculations appearing in the text thus far, a unit of measurement has been applied to each calculated quantity. Always remember that a quantity without a unit of measurement is often meaningless.

EXAMPLE 5.4 For the series circuit in Fig. 5.15:

a. Find the total resistance R_T.
b. Calculate the resulting source current I_s.
c. Determine the voltage across each resistor.

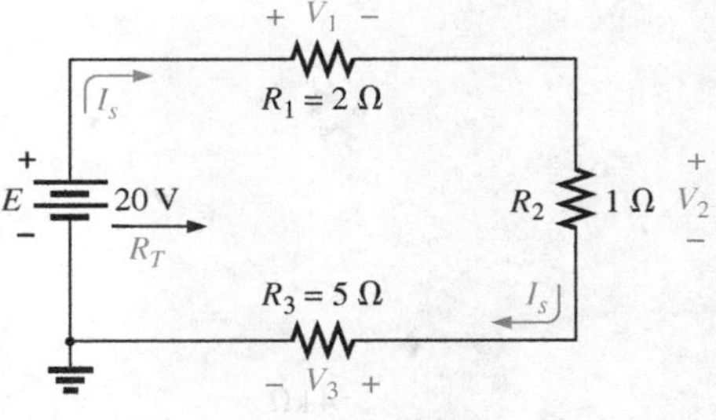

FIG. 5.15
Series circuit to be investigated in Example 5.4.

Solutions:

a. $R_T = R_1 + R_2 + R_3$
$= 2\ \Omega + 1\ \Omega + 5\ \Omega$
$R_T = \mathbf{8\ \Omega}$

b. $I_s = \dfrac{E}{R_T} = \dfrac{20\text{ V}}{8\ \Omega} = \mathbf{2.5\ A}$

c. $V_1 = I_1R_1 = I_sR_1 = (2.5\text{ A})(2\ \Omega) = \mathbf{5\ V}$
$V_2 = I_2R_2 = I_sR_2 = (2.5\text{ A})(1\ \Omega) = \mathbf{2.5\ V}$
$V_3 = I_3R_3 = I_sR_3 = (2.5\text{ A})(5\ \Omega) = \mathbf{12.5\ V}$

EXAMPLE 5.5 For the series circuit in Fig. 5.16:

a. Find the total resistance R_T.
b. Determine the source current I_s and indicate its direction on the circuit.
c. Find the voltage across resistor R_2 and indicate its polarity on the circuit.

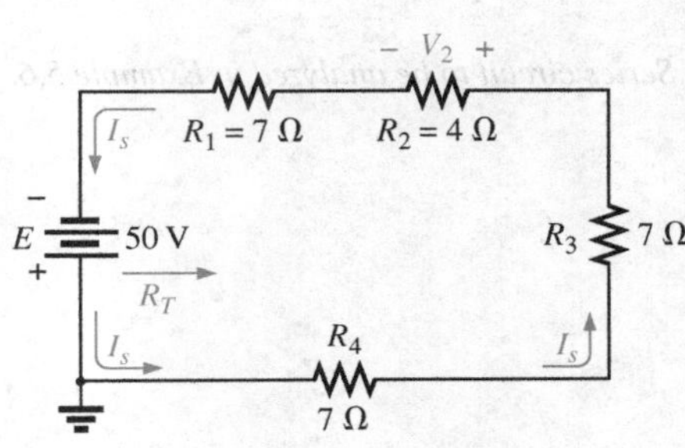

FIG. 5.16
Series circuit to be analyzed in Example 5.5.

Solutions:

a. The elements of the circuit are rearranged as shown in Fig. 5.17.

$$R_T = R_2 + NR$$
$$= 4\ \Omega + (3)(7\ \Omega)$$
$$= 4\ \Omega + 21\ \Omega$$
$$R_T = \mathbf{25\ \Omega}$$

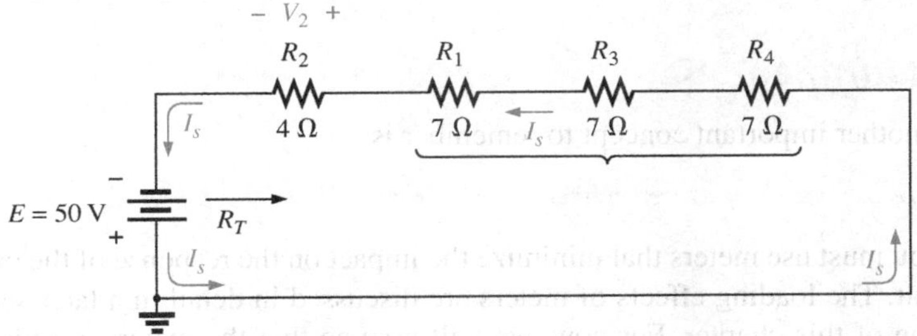

FIG. 5.17
Circuit in Fig. 5.16 redrawn to permit the use of Eq. (5.2).

b. Note that because of the manner in which the dc supply was connected, the current now has a counterclockwise direction as shown in Fig. 5.17:

$$I_s = \frac{E}{R_T} = \frac{50\ \text{V}}{25\ \Omega} = \mathbf{2\ A}$$

c. The direction of the current will define the polarity for V_2 appearing in Fig. 5.17:

$$V_2 = I_2R_2 = I_sR_2 = (2\ \text{A})(4\ \Omega) = \mathbf{8\ V}$$

Examples 5.4 and 5.5 are straightforward, substitution-type problems that are relatively easy to solve with some practice. Example 5.6, however, is another type of problem that requires both a firm grasp of the fundamental laws and equations and an ability to identify which quantity should be determined first. The best preparation for this type of exercise is to work through as many problems of this kind as possible.

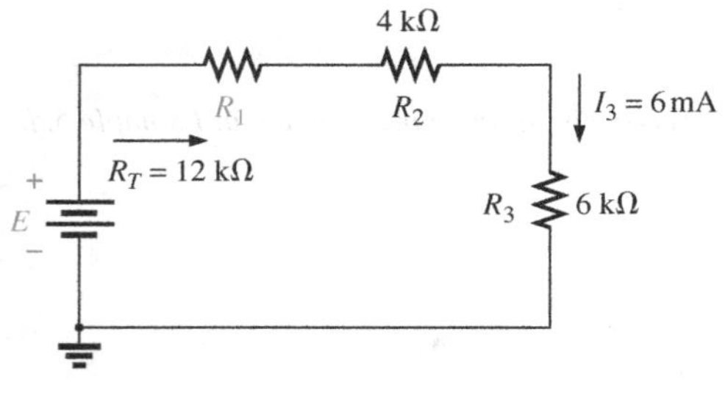

FIG. 5.18
Series circuit to be analyzed in Example 5.6.

EXAMPLE 5.6 Given R_T and I_3, calculate R_1 and E for the circuit in Fig. 5.18.

Solution: Since we are given the total resistance, it seems natural to first write the equation for the total resistance and then insert what we know:

$$R_T = R_1 + R_2 + R_3$$

We find that there is only one unknown, and it can be determined with some simple mathematical manipulations. That is,

$$12\ \text{k}\Omega = R_1 + 4\ \text{k}\Omega + 6\ \text{k}\Omega = R_1 + 10\ \text{k}\Omega$$

and $$12\ \text{k}\Omega - 10\ \text{k}\Omega = R_1$$

so that $$R_1 = \mathbf{2\ k\Omega}$$

The dc voltage can be determined directly from Ohm's law:

$$E = I_sR_T = I_3R_T = (6\ \text{mA})(12\ \text{k}\Omega) = \mathbf{72\ V}$$

Analogies

The analogies used earlier to define the series connection are also excellent for the current of a series circuit. For instance, for the series-connected ropes, the stress on each rope **is the same** as they try to hold the heavy weight. For the water analogy, the flow of water **is the same** through each section of hose as the water is carried to its destination.

Instrumentation

Another important concept to remember is:

The insertion of any meter in a circuit will affect the circuit.

You must use meters that minimize the impact on the response of the circuit. The loading effects of meters are discussed in detail in a later section of this chapter. For now, we will assume that the meters are ideal and do not affect the networks to which they are applied so that we can concentrate on their proper usage.

Furthermore, it is particularly helpful in the laboratory to realize that

the voltages of a circuit can be measured without disturbing (breaking the connections in) the circuit.

In Fig. 5.19, all the voltages of the circuit in Fig. 5.12 are being measured by voltmeters that were connected without disturbing the original configuration. Note that all the voltmeters are placed **across** the resistive elements. In addition, note that the positive (normally red) lead of the voltmeter is connected to the point of higher potential (positive sign), with the negative (normally black) lead of the voltmeter connected to the point of lower potential (negative sign) for V_1 and V_2. The result is a positive reading on the display. If the leads were reversed, the magnitude would remain the same, but a negative sign would appear as shown for V_3.

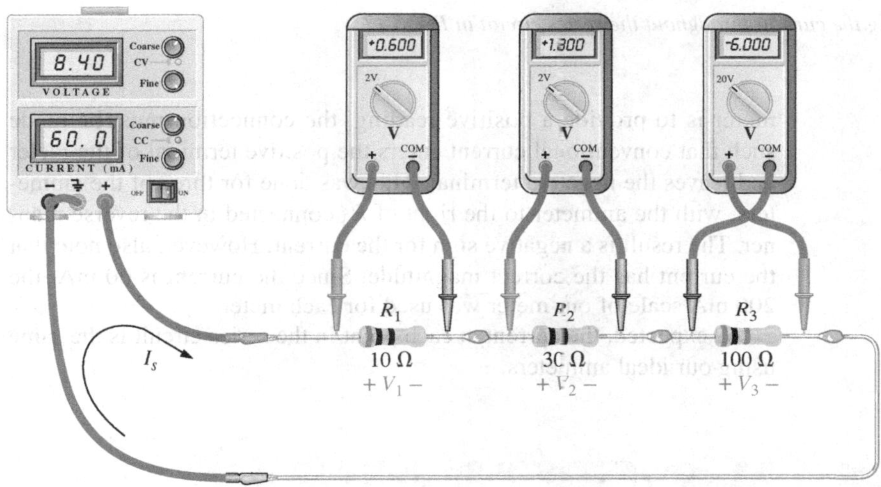

FIG. 5.19
Using voltmeters to measure the voltages across the resistors in Fig. 5.12.

Take special note that the 20 V scale of our meter was used to measure the −6 V level, while the 2 V scale of our meter was used to measure the 0.6 V and 1.8 V levels. The maximum value of the chosen scale must always exceed the maximum value to be measured. In general,

when using a voltmeter, start with a scale that will ensure that the reading is less than the maximum value of the scale. Then work your way down in scales until the reading with the highest level of precision is obtained.

Turning our attention to the current of the circuit, we find that

using an ammeter to measure the current of a circuit requires that the circuit be broken at some point and the meter inserted in series with the branch in which the current is to be determined.

For instance, to measure the current leaving the positive terminal of the supply, the connection to the positive terminal must be removed to create an open circuit between the supply and resistor R_1. The ammeter is then inserted between these two points to form a bridge between the supply and the first resistor, as shown in Fig. 5.20. The ammeter is now in series with the supply and the other elements of the circuit. If each

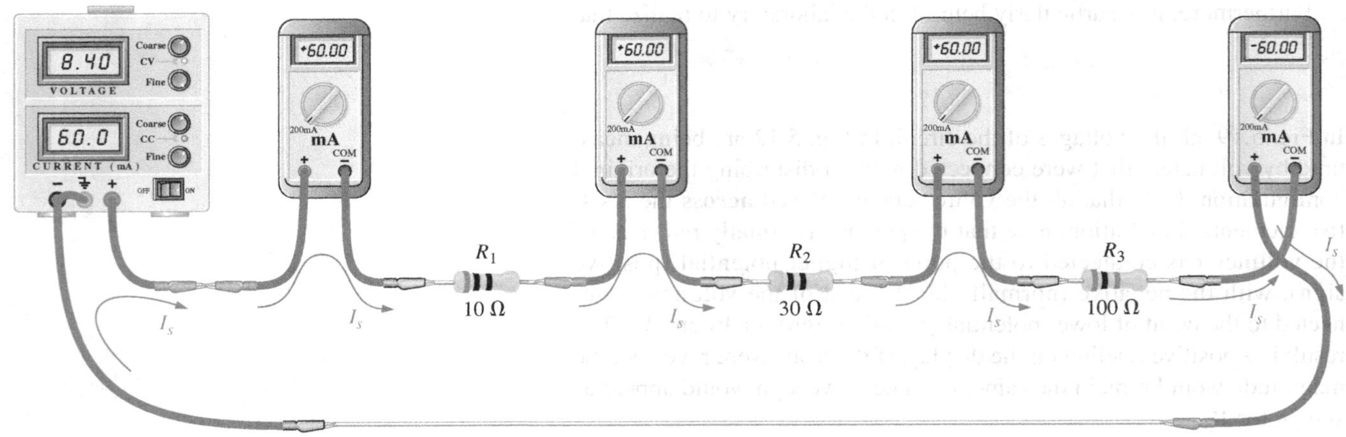

FIG. 5.20
Measuring the current throughout the series circuit in Fig. 5.12.

meter is to provide a positive reading, the connection must be made such that conventional current enters the positive terminal of the meter and leaves the negative terminal. This was done for three of the ammeters, with the ammeter to the right of R_3 connected in the reverse manner. The result is a negative sign for the current. However, also note that the current has the correct magnitude. Since the current is 60 mA, the 200 mA scale of our meter was used for each meter.

As expected, the current at each point in the series circuit is the same using our ideal ammeters.

5.4 POWER DISTRIBUTION IN A SERIES CIRCUIT

In any electrical system, the power applied will equal the power dissipated or absorbed. For any series circuit, such as that in Fig. 5.21,

the power applied by the dc supply must equal that dissipated by the resistive elements.

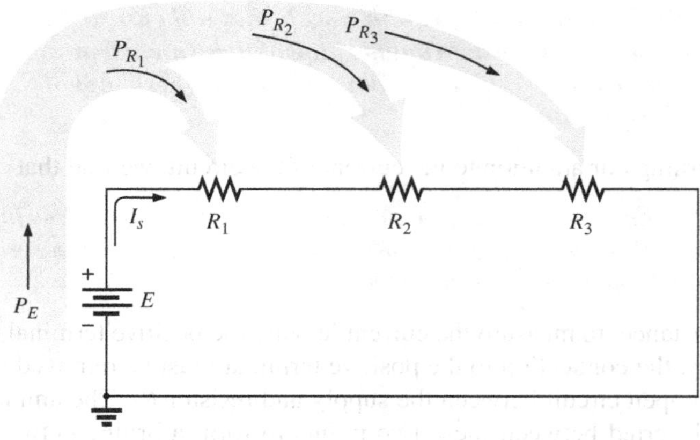

FIG. 5.21
Power distribution in a series circuit.

In equation form,

$$P_E = P_{R_1} + P_{R_2} + P_{R_3} \tag{5.5}$$

The power delivered by the supply can be determined using

$$P_E = EI_s \qquad \text{(watts, W)} \tag{5.6}$$

The power dissipated by the resistive elements can be determined by any of the following forms (shown for resistor R_1 only):

$$P_1 = V_1 I_1 = I_1^2 R_1 = \frac{V_1^2}{R_1} \qquad \text{(watts, W)} \tag{5.7}$$

Since the current is the same through series elements, you will find in the following examples that

in a series configuration, maximum power is delivered to the largest resistor.

EXAMPLE 5.7 For the series circuit in Fig. 5.22 (all standard values):

a. Determine the total resistance R_T.
b. Calculate the current I_s.
c. Determine the voltage across each resistor.
d. Find the power supplied by the battery.
e. Determine the power dissipated by each resistor.
f. Comment on whether the total power supplied equals the total power dissipated.

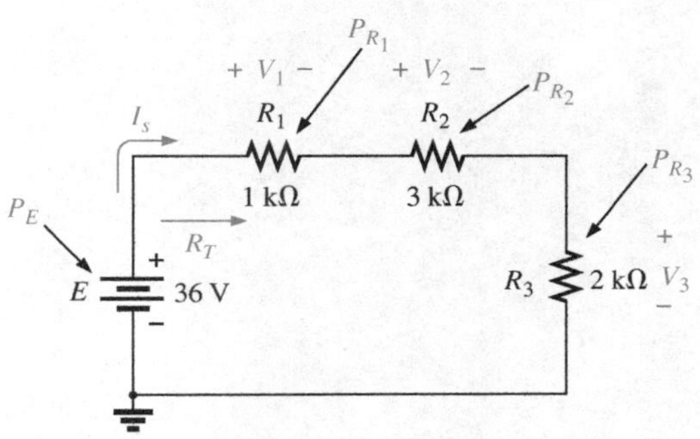

FIG. 5.22
Series circuit to be investigated in Example 5.7.

Solutions:

a. $R_T = R_1 + R_2 + R_3$
$\quad = 1\text{ k}\Omega + 3\text{ k}\Omega + 2\text{ k}\Omega$
$R_T = \mathbf{6\text{ k}\Omega}$

b. $I_s = \dfrac{E}{R_T} = \dfrac{36\text{ V}}{6\text{ k}\Omega} = \mathbf{6\text{ mA}}$

c. $V_1 = I_1R_1 = I_sR_1 = (6\text{ mA})(1\text{ k}\Omega) = \mathbf{6\text{ V}}$
$V_2 = I_2R_2 = I_sR_2 = (6\text{ mA})(3\text{ k}\Omega) = \mathbf{18\text{ V}}$
$V_3 = I_3R_3 = I_sR_3 = (6\text{ mA})(2\text{ k}\Omega) = \mathbf{12\text{ V}}$

d. $P_E = EI_s = (36\text{ V})(6\text{ mA}) = \mathbf{216\text{ mW}}$

e. $P_1 = V_1I_1 = (6\text{ V})(6\text{ mA}) = \mathbf{36\text{ mW}}$
$P_2 = I_2^2R_2 = (6\text{ mA})^2(3\text{ k}\Omega) = \mathbf{108\text{ mW}}$
$P_3 = \dfrac{V_3^2}{R_3} = \dfrac{(12\text{ V})^2}{2\text{ k}\Omega} = \mathbf{72\text{ mW}}$

f. $P_E = P_{R_1} + P_{R_2} + P_{R_3}$
$216\text{ mW} = 36\text{ mW} + 108\text{ mW} + 72\text{ mW} = \mathbf{216\text{ mW}}$ (checks)

5.5 VOLTAGE SOURCES IN SERIES

Voltage sources can be connected in series, as shown in Fig. 5.23, to increase or decrease the total voltage applied to a system. The net voltage is determined by summing the sources with the same polarity and subtracting the total of the sources with the opposite polarity. The net polarity is the polarity of the larger sum.

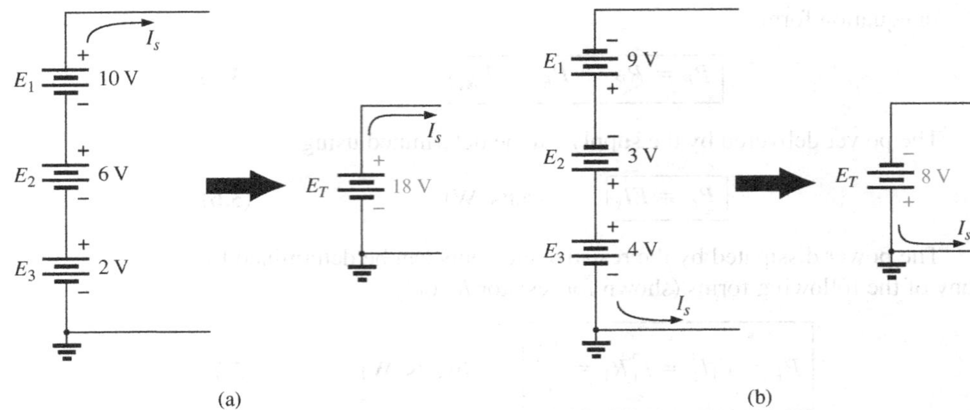

FIG. 5.23
Reducing series dc voltage sources to a single source.

In Fig. 5.23(a), for example, the sources are all "pressuring" current to follow a clockwise path, so the net voltage is

$$E_T = E_1 + E_2 + E_3 = 10 \text{ V} + 6 \text{ V} + 2 \text{ V} = \mathbf{18 \text{ V}}$$

as shown in the figure. In Fig. 5.23(b), however, the 4 V source is "pressuring" current in the clockwise direction while the other two are trying to establish current in the counterclockwise direction. In this case, the applied voltage for a counterclockwise direction is greater than that for the clockwise direction. The result is the counterclockwise direction for the current as shown in Fig. 5.23(b). The net effect can be determined by finding the difference in applied voltage between those supplies "pressuring" current in one direction and the total in the other direction. In this case,

$$E_T = E_1 + E_2 + E_3 = 9 \text{ V} + 3 \text{ V} - 4 \text{ V} = \mathbf{8 \text{ V}}$$

with the polarity shown in the figure.

Instrumentation

The connection of batteries in series to obtain a higher voltage is common in much of today's portable electronic equipment. For example, in Fig. 5.24(a), four 1.5 V AAA batteries have been connected in series to obtain a source voltage of 6 V. Although the voltage has increased, keep in mind that the maximum current for each AAA battery and for the 6 V supply is still the same. However, the power available has increased by a factor of 4 due to the increase in terminal voltage. Note also, as mentioned in Chapter 2, that the negative end of each battery is connected to the spring and the positive end to the solid contact. In addition, note how the connection is made between batteries using the horizontal connecting tabs.

In general, supplies with only two terminals (+ and −) can be connected as shown for the batteries. A problem arises, however, if the supply has an optional or fixed internal ground connection. In Fig. 5.24(b), two laboratory supplies have been connected in series with both grounds connected. The result is a shorting out of the lower source E_1 (which may damage the supply if the protective fuse does not activate quickly enough) because both grounds are at zero potential. In such cases, the supply E_2 must be left ungrounded (floating), as shown in Fig. 5.24(c), to provide the 60 V terminal voltage. If the laboratory supplies have an internal connection from the negative terminal to ground as a protective

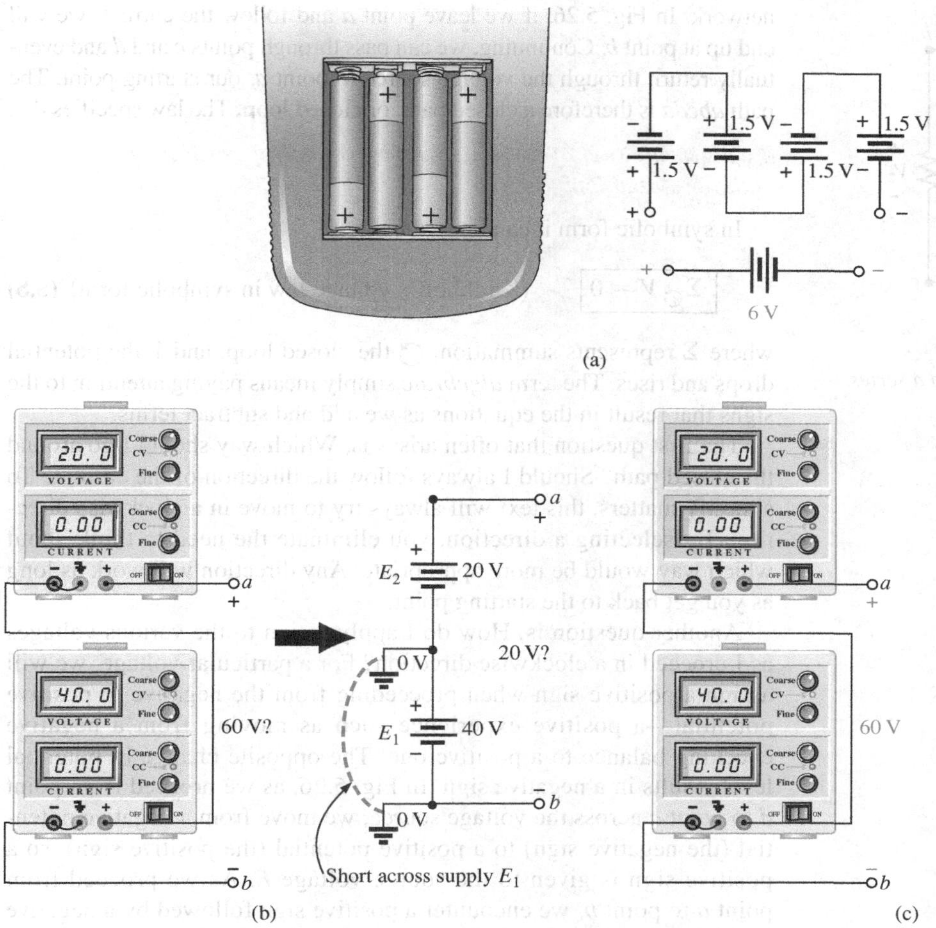

FIG. 5.24
Series connection of dc supplies: (a) four 1.5 V batteries in series to establish a terminal voltage of 6 V; (b) incorrect connections for two series dc supplies; (c) correct connection of two series supplies to establish 60 V at the output terminals.

feature for the users, a series connection of supplies cannot be made. Be aware of this fact, because some educational institutions add an internal ground to the supplies as a protective feature even though the panel still displays the ground connection as an optional feature.

5.6 KIRCHHOFF'S VOLTAGE LAW

The law to be described in this section is one of the most important in this field. It has application not only to dc circuits but also to any type of signal—whether it be ac, digital, and so on. This law is far-reaching and can be very helpful in working out solutions to networks that sometimes leave us lost for a direction of investigation.

The law, called **Kirchhoff's voltage law (KVL),** was developed by Gustav Kirchhoff (Fig. 5.25) in the mid-1800s. It is a cornerstone of the entire field and, in fact, will never be outdated or replaced.

The application of the law requires that we define a closed path of investigation, permitting us to start at one point in the network, travel through the network, and find our way back to the original starting point. The path does not have to be circular, square, or any other defined shape; it must simply provide a way to leave a point and get back to it without leaving the

FIG. 5.25
Gustav Robert Kirchhoff.

German (Königsberg, Berlin)
(1824–87),
Physicist
Professor of Physics, University of Heidelberg

Although a contributor to a number of areas in the physics domain, he is best known for his work in the electrical area with his definition of the relationships between the currents and voltages of a network in 1847. Did extensive research with German chemist Robert Bunsen (developed the *Bunsen burner*), resulting in the discovery of the important elements of *cesium and rubidium.*

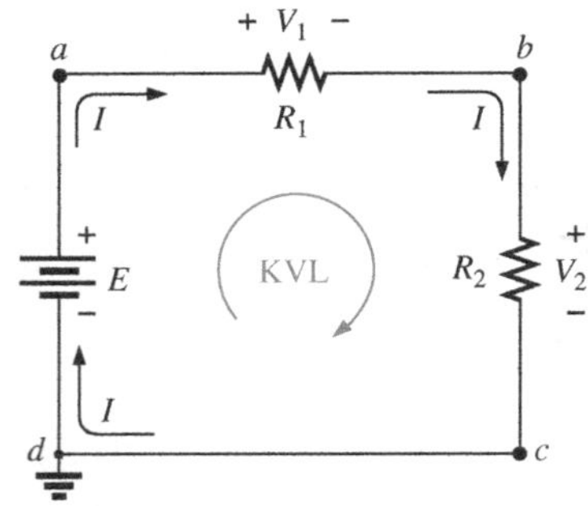

FIG. 5.26
Applying Kirchhoff's voltage law to a series dc circuit.

network. In Fig. 5.26, if we leave point *a* and follow the current, we will end up at point *b.* Continuing, we can pass through points *c* and *d* and eventually return through the voltage source to point *a,* our starting point. The path *abcda* is therefore a closed path, or **closed loop.** The law specifies that

the algebraic sum of the potential rises and drops around a closed path (or closed loop) is zero.

In symbolic form it can be written as

$$\boxed{\Sigma_{\circlearrowright} V = 0} \qquad \text{(Kirchhoff's voltage law in symbolic form)} \quad \textbf{(5.8)}$$

where Σ represents summation, $\circlearrowright$ the closed loop, and *V* the potential drops and rises. The term *algebraic* simply means paying attention to the signs that result in the equations as we add and subtract terms.

The first question that often arises is, Which way should I go around the closed path? Should I always follow the direction of the current? To simplify matters, this text will always try to move in a clockwise direction. By selecting a direction, you eliminate the need to think about which way would be more appropriate. Any direction will work as long as you get back to the starting point.

Another question is, How do I apply a sign to the various voltages as I proceed in a clockwise direction? For a particular voltage, we will assign a positive sign when proceeding from the negative to positive potential—a positive experience such as moving from a negative checking balance to a positive one. The opposite change in potential level results in a negative sign. In Fig. 5.26, as we proceed from point *d* to point *a* across the voltage source, we move from a negative potential (the negative sign) to a positive potential (the positive sign), so a positive sign is given to the source voltage *E.* As we proceed from point *a* to point *b,* we encounter a positive sign followed by a negative sign, so a drop in potential has occurred, and a negative sign is applied. Continuing from *b* to *c,* we encounter another drop in potential, so another negative sign is applied. We then arrive back at the starting point *d,* and the resulting sum is set equal to zero as defined by Eq. (5.8).

Writing out the sequence with the voltages and the signs results in the following:

$$+E - V_1 - V_2 = 0$$

which can be rewritten as $\quad E = V_1 + V_2$

The result is particularly interesting because it tells us that

the applied voltage of a series dc circuit will equal the sum of the voltage drops of the circuit.

Kirchhoff's voltage law can also be written in the following form:

$$\boxed{\Sigma_{\circlearrowright} V_{\text{rises}} = \Sigma_{\circlearrowright} V_{\text{drops}}} \qquad \textbf{(5.9)}$$

revealing that

the sum of the voltage rises around a closed path will always equal the sum of the voltage drops.

To demonstrate that the direction that you take around the loop has no effect on the results, let's take the counterclockwise path and compare results. The resulting sequence appears as

$$-E + V_2 + V_1 = 0$$

yielding the same result of $\quad E = V_1 + V_2$

EXAMPLE 5.8 Use Kirchhoff's voltage law to determine the unknown voltage for the circuit in Fig. 5.27.

Solution: When applying Kirchhoff's voltage law, be sure to concentrate on the polarities of the voltage rise or drop rather than on the type of element. In other words, do not treat a voltage drop across a resistive element differently from a voltage rise (or drop) across a source. If the polarity dictates that a drop has occurred, that is the important fact, not whether it is a resistive element or source.

Application of Kirchhoff's voltage law to the circuit in Fig. 5.27 in the clockwise direction results in

$$+E_1 - V_1 - V_2 - E_2 = 0$$

and $$V_1 = E_1 - V_2 - E_2 = 16\text{ V} - 4.2\text{ V} - 9\text{ V}$$

so $$V_1 = \mathbf{2.8\text{ V}}$$

The result clearly indicates that you do not need to know the values of the resistors or the current to determine the unknown voltage. Sufficient information was carried by the other voltage levels to determine the unknown.

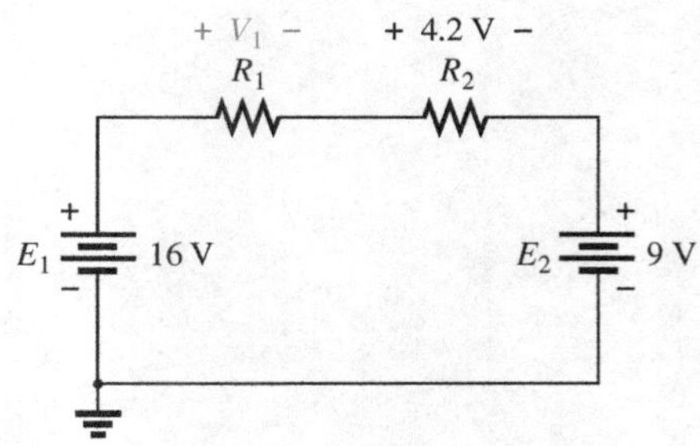

FIG. 5.27
Series circuit to be examined in Example 5.8.

EXAMPLE 5.9 Determine the unknown voltage for the circuit in Fig. 5.28.

Solution: In this case, the unknown voltage is not across a single resistive element but between two arbitrary points in the circuit. Simply apply Kirchhoff's voltage law around a path, including the source or resistor R_3. For the clockwise path, including the source, the resulting equation is the following:

$$+E - V_1 - V_x = 0$$

and $$V_x = E - V_1 = 32\text{ V} - 12\text{ V} = \mathbf{20\text{ V}}$$

For the clockwise path, including resistor R_3, the following results:

$$+V_x - V_2 - V_3 = 0$$

and $$V_x = V_2 + V_3 = 6\text{ V} + 14\text{ V}$$

with $$V_x = \mathbf{20\text{ V}}$$

providing exactly the same solution.

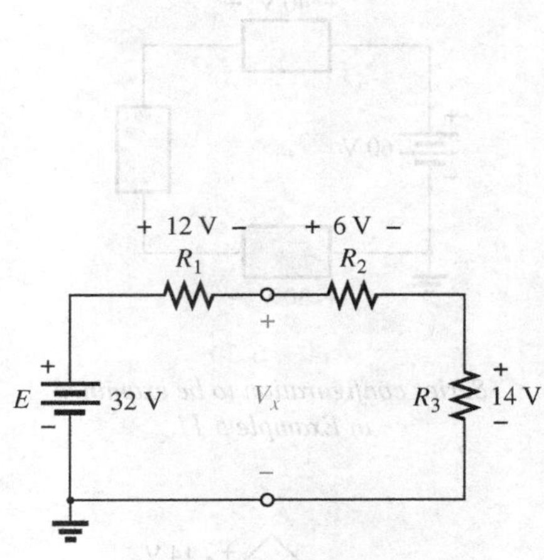

FIG. 5.28
Series dc circuit to be analyzed in Example 5.9.

There is no requirement that the followed path have charge flow or current. In Example 5.10, the current is zero everywhere, but Kirchhoff's voltage law can still be applied to determine the voltage between the points of interest. Also, there will be situations where the actual polarity will not be provided. In such cases, simply assume a polarity. If the answer is negative, the magnitude of the result is correct, but the polarity should be reversed.

EXAMPLE 5.10 Using Kirchhoff's voltage law, determine voltages V_1 and V_2 for the network in Fig. 5.29.

Solution: For path 1, starting at point a in a clockwise direction,

$$+25\text{ V} - V_1 + 15\text{ V} = 0$$

and $$V_1 = \mathbf{40\text{ V}}$$

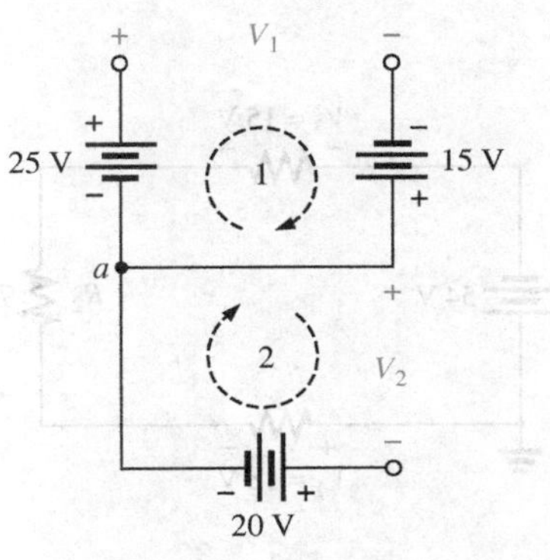

FIG. 5.29
Combination of voltage sources to be examined in Example 5.10.

For path 2, starting at point a in a clockwise direction,

$$-V_2 - 20\text{ V} = 0$$

and $$V_2 = \mathbf{-20\text{ V}}$$

The minus sign in the solution simply indicates that the actual polarities are different from those assumed.

The next example demonstrates that you do not need to know what elements are inside a container when applying Kirchhoff's voltage law. They could all be voltage sources or a mix of sources and resistors. It doesn't matter—simply pay strict attention to the polarities encountered.

Try to find the unknown quantities in the next examples without looking at the solutions. It will help define where you may be having trouble.

Example 5.11 emphasizes the fact that when you are applying Kirchhoff's voltage law, the polarities of the voltage rise or drop are the important parameters, not the type of element involved.

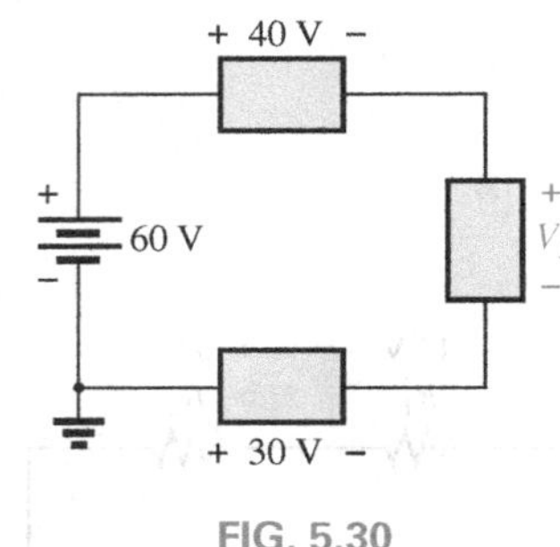

FIG. 5.30
Series configuration to be examined in Example 5.11.

EXAMPLE 5.11 Using Kirchhoff's voltage law, determine the unknown voltage for the circuit in Fig. 5.30.

Solution: Note that in this circuit, there are various polarities across the unknown elements since they can contain any mixture of components. Applying Kirchhoff's voltage law in the clockwise direction results in

$$+60\text{ V} - 40\text{ V} - V_x + 30\text{ V} = 0$$

and $$V_x = 60\text{ V} + 30\text{ V} - 40\text{ V} = 90\text{ V} - 40\text{ V}$$

with $$V_x = \mathbf{50\text{ V}}$$

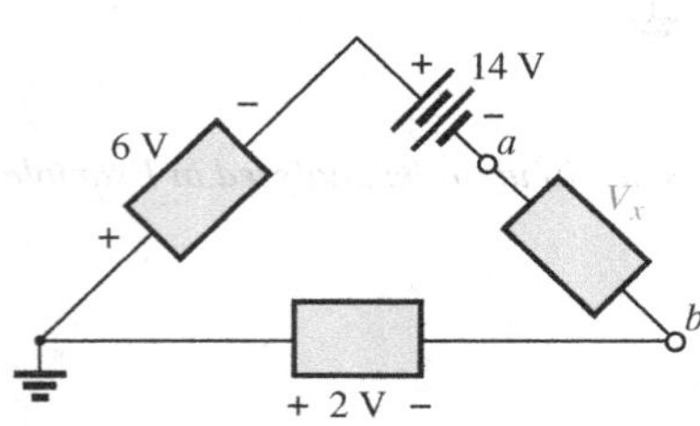

FIG. 5.31
Applying Kirchhoff's voltage law to a circuit in which the polarities have not been provided for one of the voltages (Example 5.12).

EXAMPLE 5.12 Determine the voltage V_x for the circuit in Fig. 5.31. Note that the polarity of V_x was not provided.

Solution: For cases where the polarity is not included, simply make an assumption about the polarity, and apply Kirchhoff's voltage law as before. If the result has a positive sign, the assumed polarity was correct. If the result has a minus sign, the **magnitude is correct,** but the assumed polarity must be reversed. In this case, if we assume point a to be positive and point b to be negative, an application of Kirchhoff's voltage law in the clockwise direction results in

$$-6\text{ V} - 14\text{ V} - V_x + 2\text{ V} = 0$$

and $$V_x = -20\text{ V} + 2\text{ V}$$

so that $$V_x = \mathbf{-18\text{ V}}$$

Since the result is negative, we know that point a should be negative and point b should be positive, but the magnitude of 18 V is correct.

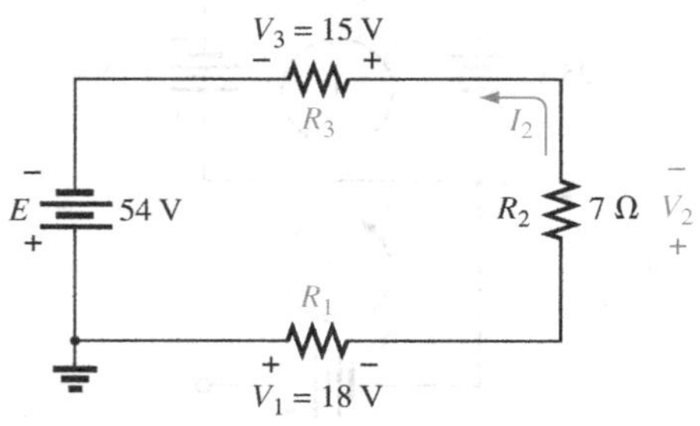

FIG. 5.32
Series configuration to be examined in Example 5.13.

EXAMPLE 5.13 For the series circuit in Fig. 5.32.

a. Determine V_2 using Kirchhoff's voltage law.
b. Determine current I_2.
c. Find R_1 and R_3.

Solutions:

a. Applying Kirchhoff's voltage law in the clockwise direction starting at the negative terminal of the supply results in

$$-E + V_3 + V_2 + V_1 = 0$$

and $\quad E = V_1 + V_2 + V_3$ (as expected)

so that $\quad V_2 = E - V_1 - V_3 = 54\text{ V} - 18\text{ V} - 15\text{ V}$

and $\quad \mathbf{V_2 = 21\text{ V}}$

b. $I_2 = \dfrac{V_2}{R_2} = \dfrac{21\text{ V}}{7\ \Omega}$

$\mathbf{I_2 = 3A}$

c. $R_1 = \dfrac{V_1}{I_1} = \dfrac{18\text{ V}}{3\text{ A}} = \mathbf{6\ \Omega}$

with $R_3 = \dfrac{V_3}{I_3} = \dfrac{15\text{ V}}{3\text{ A}} = \mathbf{5\ \Omega}$

EXAMPLE 5.14 Using Kirchhoff's voltage law and Fig. 5.12, verify Eq. (5.1).

Solution: Applying Kirchhoff's voltage law arcund the closed path:

$$E = V_1 + V_2 + V_3$$

Substituting Ohm's law:

$$I_sR_T = I_1R_1 + I_2R_2 + I_3R_3$$

but $\quad I_s = I_1 = I_2 = I_3$

so that $\quad I_sR_T = I_s(R_1 + R_2 + R_3)$

and $\quad \mathbf{R_T = R_1 + R_2 + R_3}$

which is Eq. (5.1).

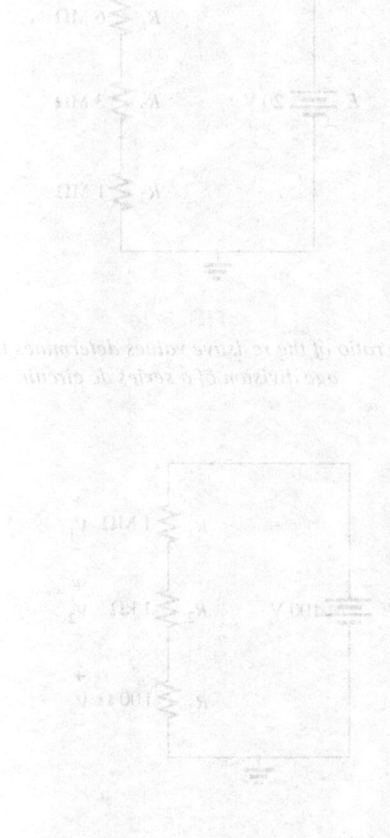

5.7 VOLTAGE DIVISION IN A SERIES CIRCUIT

The previous section demonstrated that the sum of the voltages across the resistors of a series circuit will always equal the applied voltage. It cannot be more or less than that value. The next question is, How will a resistor's value affect the voltage across the resistor? It turns out that ***the voltage across series resistive elements will divide as the magnitude of the resistance levels.***

In other words,

in a series resistive circuit, the larger the resistance, the more of the applied voltage it will capture.

In addition,

the ratio of the voltages across series resistors will be the same as the ratio of their resistance levels.

All of the above statements can best be described by a few simple examples. In Fig. 5.33, all the voltages across the resistive elements are provided. The largest resistor of 6 Ω captures the bulk cf the applied voltage, while the smallest resistor, R_3, has the least. In addition, note that since the resistance level of R_1 is six times that of R_3, the voltage across R_1 is six

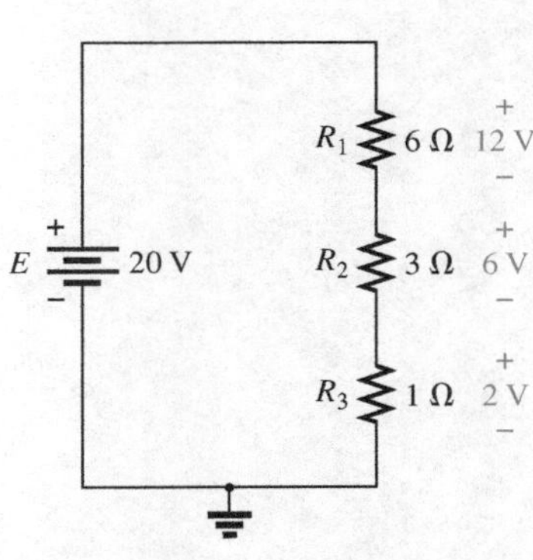

FIG. 5.33
Revealing how the voltage will divide across series resistive elements.

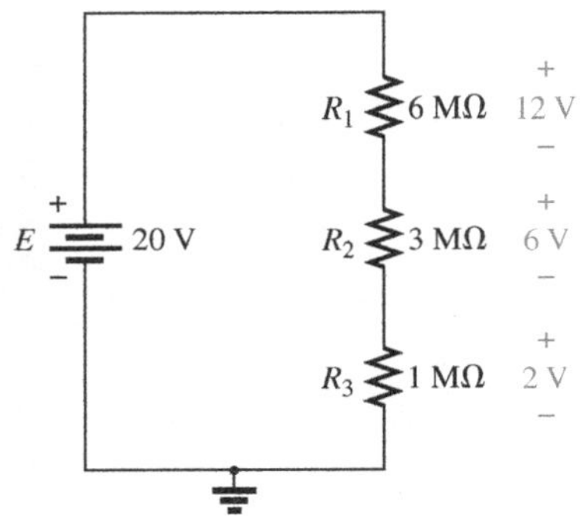

FIG. 5.34
The ratio of the resistive values determines the voltage division of a series dc circuit.

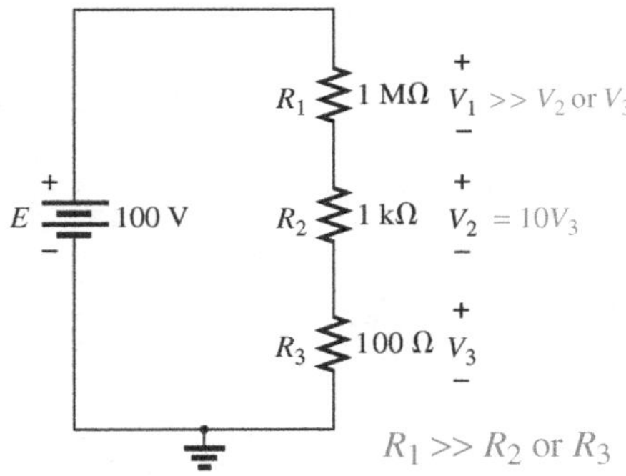

FIG. 5.35
The largest of the series resistive elements will capture the major share of the applied voltage.

times that of R_3. The fact that the resistance level of R_2 is three times that of R_3 results in three times the voltage across R_2. Finally, since R_1 is twice R_2, the voltage across R_1 is twice that of R_2. In general, therefore, the voltage across series resistors will have the same ratio as their resistance levels.

Note that if the resistance levels of all the resistors in Fig. 5.33 are increased by the same amount, as shown in Fig. 5.34, the voltage levels all remain the same. In other words, even though the resistance levels were increased by a factor of 1 million, the voltage ratios remained the same. Clearly, therefore, it is the ratio of resistor values that counts when it comes to voltage division, not the magnitude of the resistors. The current level of the network will be severely affected by this change in resistance level, but the voltage levels remain unaffected.

Based on the above, it should now be clear that when you first encounter a circuit such as that in Fig. 5.35, you will expect that the voltage across the 1 MΩ resistor will be much greater than that across the 1 kΩ or the 100 Ω resistor. In addition, based on a statement above, the voltage across the1 kΩ resistor will be 10 times as great as that across the 100 Ω resistor since the resistance level is 10 times as much. Certainly, you would expect that very little voltage will be left for the 100 Ω resistor. Note that the current was never mentioned in the above analysis. The distribution of the applied voltage is determined solely by the ratio of the resistance levels. Of course, the magnitude of the resistors will determine the resulting current level.

To continue with the above, since 1 MΩ is 1000 times larger than 1 kΩ, voltage V_1 will be 1000 times larger than V_2. In addition, voltage V_2 will be 10 times larger than V_3. Finally, the voltage across the largest resistor of 1 MΩ will be (10)(1000) = 10,000 times larger than V_3.

Now for some details. The total resistance is

$$\begin{aligned} R_T &= R_1 + R_2 + R_3 \\ &= 1\ \text{M}\Omega + 1\ \text{k}\Omega + 100\ \Omega \\ R_T &= \mathbf{1{,}001{,}100\ \Omega} \end{aligned}$$

The current is

$$I_s = \frac{E}{R_T} = \frac{100\ \text{V}}{1{,}001{,}100\ \Omega} \cong 99.89\ \mu\text{A} \qquad \text{(about 100 } \mu\text{A)}$$

with

$$\begin{aligned} V_1 &= I_1R_1 = I_sR_1 = (99.89\ \mu\text{A})(1\ \text{M}\Omega) = \mathbf{99.89\ V} \quad \text{(almost the full 100 V)} \\ V_2 &= I_2R_2 = I_sR_2 = (99.89\ \mu\text{A})(1\ \text{k}\Omega) = \mathbf{99.89\ mV} \quad \text{(about 100 mV)} \\ V_3 &= I_3R_3 = I_sR_3 = (99.89\ \mu\text{A})(100\ \Omega) = \mathbf{9.989\ mV} \quad \text{(about 10 mV)} \end{aligned}$$

As illustrated above, the major part of the applied voltage is across the 1 MΩ resistor. The current is in the microampere range due primarily to the large 1 MΩ resistor. Voltage V_2 is about 0.1 V, compared to almost 100 V for V_1. The voltage across R_3 is only about 10 mV, or 0.010 V.

Before making any detailed, lengthy calculations, you should first examine the resistance levels of the series resistors to develop some idea of how the applied voltage will be divided through the circuit. It will reveal, with a minumum amount of effort, what you should expect when performing the calculations (a checking mechanism). It also allows you to speak intelligently about the response of the circuit without having to resort to any calculations.

Voltage Divider Rule (VDR)

The **voltage divider rule (VDR)** permits the determination of the voltage across a series resistor without first having to determine the current of the circuit. The rule itself can be derived by analyzing the simple series circuit in Fig. 5.36.

First, determine the total resistance as follows:

$$R_T = R_1 + R_2$$

Then $$I_s = I_1 = I_2 = \frac{E}{R_T}$$

Apply Ohm's law to each resistor:

$$V_1 = I_1R_1 = \left(\frac{E}{R_T}\right)R_1 = R_1\frac{E}{R_T}$$

$$V_2 = I_2R_2 = \left(\frac{E}{R_T}\right)R_2 = R_2\frac{E}{R_T}$$

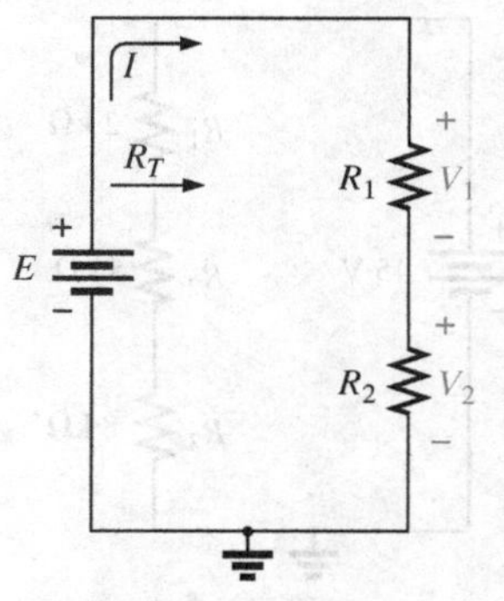

FIG. 5.36
Developing the voltage divider rule.

The resulting format for V_1 and V_2 is

$$V_x = R_x\frac{E}{R_T} \quad \text{(voltage divider rule)} \quad \textbf{(5.10)}$$

where V_x is the voltage across the resistor R_x, E is the impressed voltage across the series elements, and R_T is the total resistance of the series circuit.

The voltage divider rule states that

the voltage across a resistor in a series circuit is equal to the value of that resistor times the total applied voltage divided by the total resistance of the series configuration.

Although Eq. (5.10) was derived using a series circuit of only two elements, it can be used for series circuits with any number of series resistors.

EXAMPLE 5.15 For the series circuit in Fig. 5.37.

a. Without making any calculations, how much larger would you expect the voltage across R_2 to be compared to that across R_1?
b. Find the voltage V_1 using only the voltage divider rule.
c. Using the conclusion of part (a), determine the voltage across R_2.
d. Use the voltage divider rule to determine the voltage across R_2, and compare your answer to your conclusion in part (c).
e. How does the sum of V_1 and V_2 compare to the applied voltage?

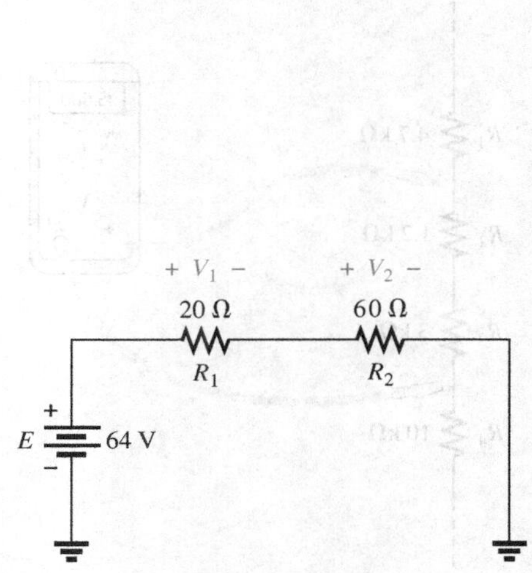

FIG. 5.37
Series circuit to be examined using the voltage divider rule in Example 5.15.

Solutions:

a. Since resistor R_2 is three times R_1, it is expected that $V_2 = 3V_1$.

b. $V_1 = R_1\frac{E}{R_T} = 20\ \Omega\left(\frac{64\ \text{V}}{20\ \Omega + 60\ \Omega}\right) = 20\ \Omega\left(\frac{64\ \text{V}}{80\ \Omega}\right) = \mathbf{16\ V}$

c. $V_2 = 3V_1 = 3(16\ \text{V}) = \mathbf{48\ V}$

d. $V_2 = R_2\frac{E}{R_T} = (60\ \Omega)\left(\frac{64\ \text{V}}{80\ \Omega}\right) = \mathbf{48\ V}$

The results are an exact match.

e. $E = V_1 + V_2$
$64\ \text{V} = 16\ \text{V} + 48\ \text{V} = \mathbf{64\ V}$ (checks)

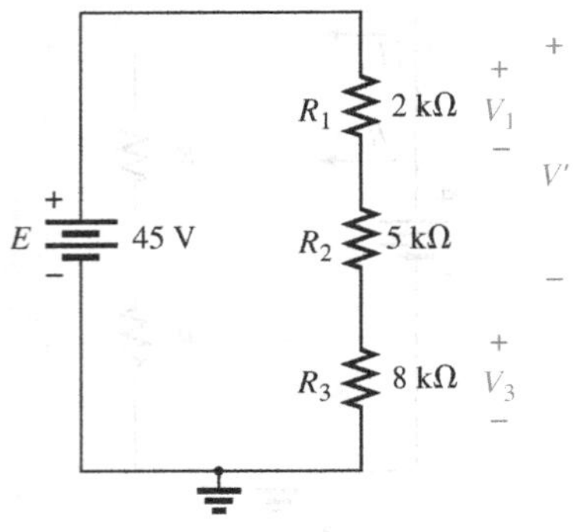

FIG. 5.38
Series circuit to be investigated in Examples 5.16 and 5.17.

EXAMPLE 5.16 Using the voltage divider rule, determine voltages V_1 and V_3 for the series circuit in Fig. 5.38.

Solution:

$$R_T = R_1 + R_2 + R_3$$
$$= 2\text{ k}\Omega + 5\text{ k}\Omega + 8\text{ k}\Omega$$
$$R_T = 15\text{k}\Omega$$
$$V_1 = R_1\frac{E}{R_T} = 2\text{ k}\Omega\left(\frac{45\text{ V}}{15\text{k}\Omega}\right) = \mathbf{6\text{ V}}$$

and
$$V_3 = R_3\frac{E}{R_T} = 8\text{ k}\Omega\left(\frac{45\text{ V}}{15\ \Omega}\right) = \mathbf{24\text{ V}}$$

The voltage divider rule can be extended to the voltage across two or more series elements if the resistance in the numerator of Eq. (5.10) is expanded to include the total resistance of the series resistors across which the voltage is to be found (R'). That is,

$$V' = R'\frac{E}{R_T} \quad \textbf{(5.11)}$$

EXAMPLE 5.17 Determine the voltage (denoted V') across the series combination of resistors R_1 and R_2 in Fig. 5.38.

Solution: Since the voltage desired is across both R_1 and R_2, the sum of R_1 and R_2 will be substituted as R' in Eq. (5.11). The result is

$$R' = R_1 + R_2 = 2\text{ k}\Omega + 5\text{ k}\Omega = 7\text{ k}\Omega$$

and
$$V' = R'\frac{E}{R_T} = 7\text{ k}\Omega\left(\frac{45\text{ V}}{15\text{ k}\Omega}\right) = \mathbf{21\text{ V}}$$

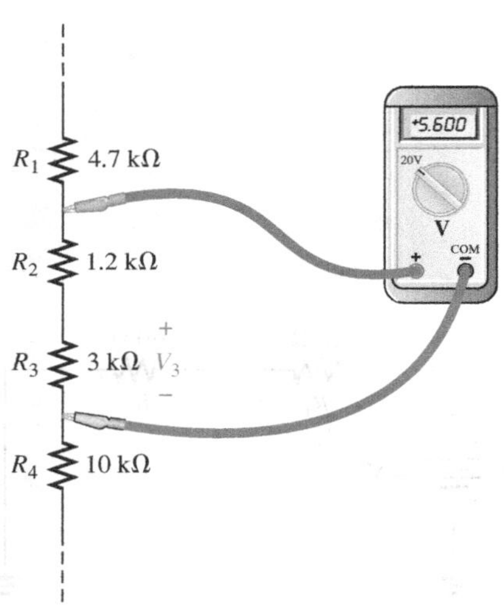

FIG. 5.39
Voltage divider action for Example 5.18.

In the next example you are presented with a problem of the other kind: Given the voltage division, you must determine the required resistor values. In most cases, problems of this kind simply require that you are able to use the basic equations introduced thus far in the text.

EXAMPLE 5.18 Given the voltmeter reading in Fig. 5.39, find voltage V_3.

Solution: Even though the rest of the network is not shown and the current level has not been determined, the voltage divider rule can be applied by using the voltmeter reading as the full voltage across the series combination of resistors. That is,

$$V_3 = R_3\frac{(V_{\text{meter}})}{R_3 + R_2} = \frac{3\text{ k}\Omega(5.6\text{ V})}{3\text{ k}\Omega + 1.2\text{ k}\Omega}$$
$$V_3 = \mathbf{4\text{ V}}$$

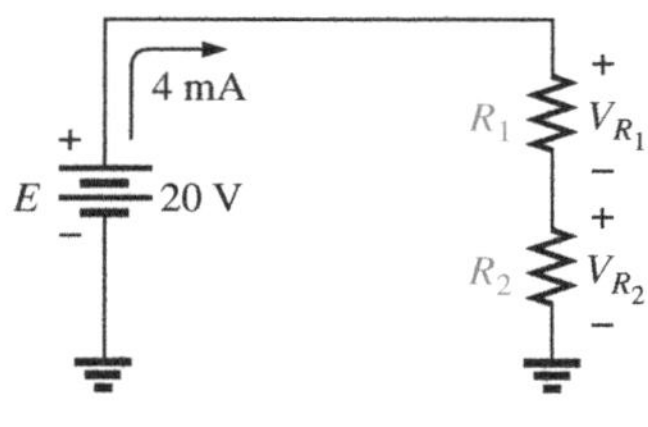

FIG. 5.40
Designing a voltage divider circuit (Example 5.19).

EXAMPLE 5.19 Design the voltage divider circuit in Fig. 5.40 such that the voltage across R_1 will be four times the voltage across R_2; that is, $V_{R_1} = 4V_{R_2}$.

Solution: The total resistance is defined by

$$R_T = R_1 + R_2$$

However, if $V_{R_1} = 4V_{R_2}$

then $R_1 = 4R_2$

so that $R_T = R_1 + R_2 = 4R_2 + R_2 = 5R_2$

Applying Ohm's law, we can determine the total resistance of the circuit:

$$R_T = \frac{E}{I_s} = \frac{20\text{ V}}{4\text{ mA}} = 5\text{ k}\Omega$$

so $R_T = 5R_2 = 5\text{ k}\Omega$

and $R_2 = \frac{5\text{ k}\Omega}{5} = \mathbf{1\text{ k}\Omega}$

Then $R_1 = 4R_2 = 4(1\text{ k}\Omega) = \mathbf{4\text{ k}\Omega}$

5.8 INTERCHANGING SERIES ELEMENTS

The elements of a series circuit can be interchanged without affecting the total resistance, current, or power to each element. For instance, the network in Fig. 5.41 can be redrawn as shown in Fig. 5.42 without affecting I or V_2. The total resistance R_T is 35 Ω in both cases, and I = 70 V/35 Ω = 2 A. The voltage $V_2 = IR_2$ = (2 A)(5 Ω) = 10 V for both configurations.

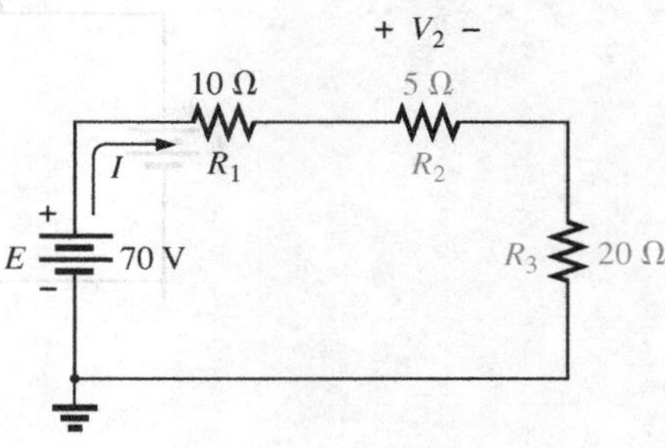

FIG. 5.41
Series dc circuit with elements to be interchanged.

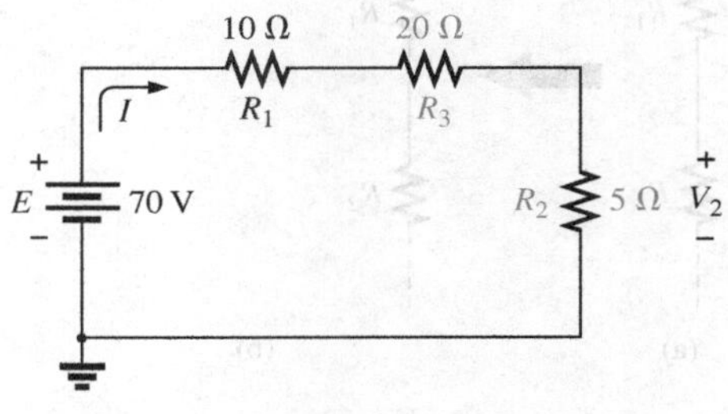

FIG. 5.42
Circuit in Fig. 5.41 with R_2 and R_3 interchanged.

EXAMPLE 5.20 Determine I and the voltage across the 7 Ω resistor for the network in Fig. 5.43.

Solution: The network is redrawn in Fig. 5.44.

$$R_T = (2)(4\ \Omega) + 7\ \Omega = 15\ \Omega$$

$$I = \frac{E}{R_T} = \frac{37.5\text{ V}}{15\ \Omega} = \mathbf{2.5\text{ A}}$$

$$V_{7\Omega} = IR = (2.5\text{ A})(7\ \Omega) = \mathbf{17.5\text{ V}}$$

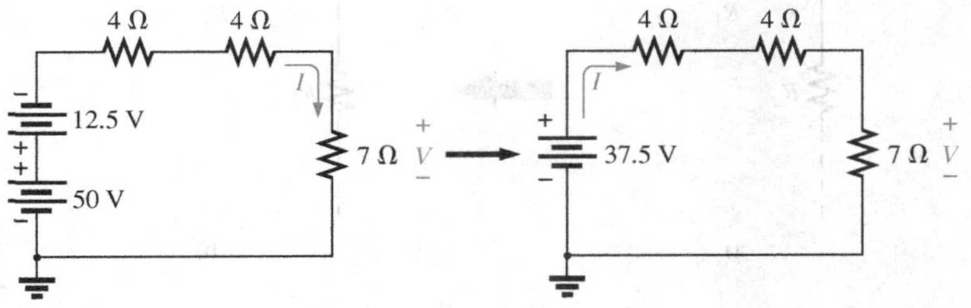

FIG. 5.44
Redrawing the circuit in Fig. 5.43.

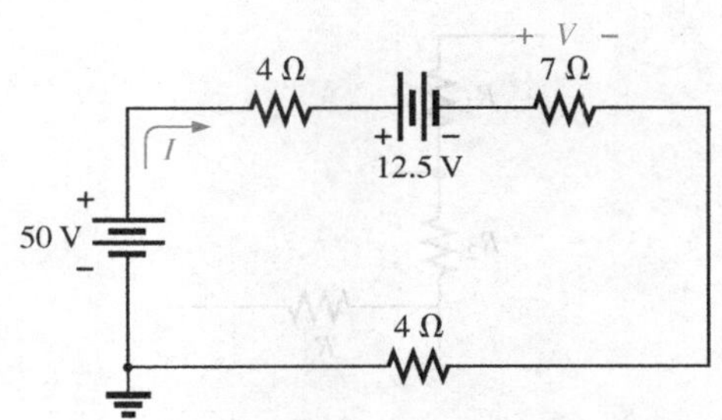

FIG. 5.43
Example 5.20.

5.9 NOTATION

Notation plays an increasingly important role in the analysis to follow. It is important, therefore, that we begin to examine the notation used throughout the industry.

FIG. 5.45
Ground potential.

Voltage Sources and Ground

Except for a few special cases, electrical and electronic systems are grounded for reference and safety purposes. The symbol for the ground connection appears in Fig. 5.45 with its defined potential level—zero volts. A grounded circuit may appear as shown in Fig. 5.46(a), (b), or (c). In any case, it is understood that the negative terminal of the battery and the bottom of the resistor R_2 are at ground potential. Although Fig. 5.46(c) shows no connection between the two grounds, it is recognized that such a connection exists for the continuous flow of charge. If E = 12 V, then point a is 12 V positive with respect to ground potential, and 12 V exist across the series combination of resistors R_1 and R_2. If a voltmeter placed from point b to ground reads 4 V, then the voltage across R_2 is 4 V, with the higher potential at point b.

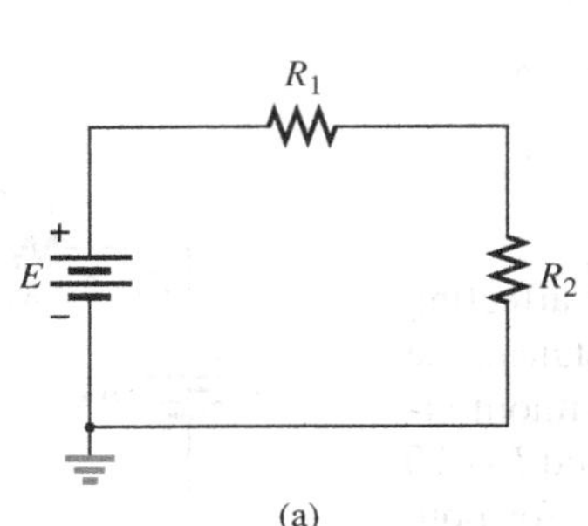

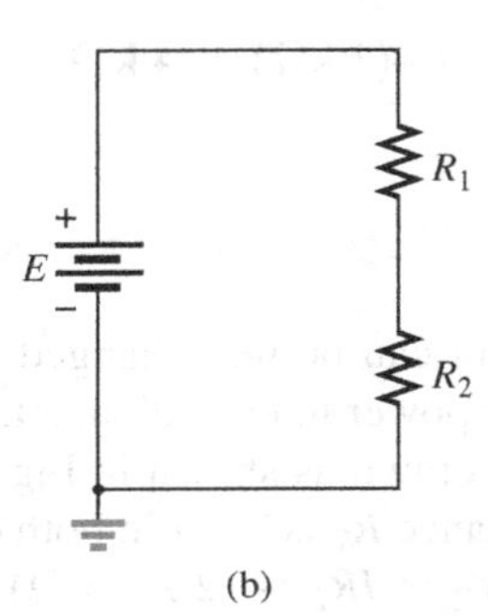

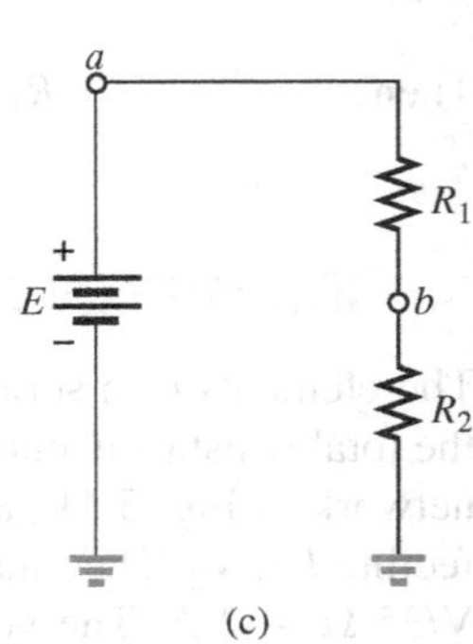

FIG. 5.46
Three ways to sketch the same series dc circuit.

FIG. 5.47
Replacing the special notation for a dc voltage source with the standard symbol.

On large schematics where space is at a premium and clarity is important, voltage sources may be indicated as shown in Figs. 5.47(a) and 5.48(a) rather than as illustrated in Figs. 5.47(b) and 5.48(b). In addition, potential levels may be indicated as in Fig. 5.49, to permit a rapid check of the potential levels at various points in a network with respect to ground to ensure that the system is operating properly.

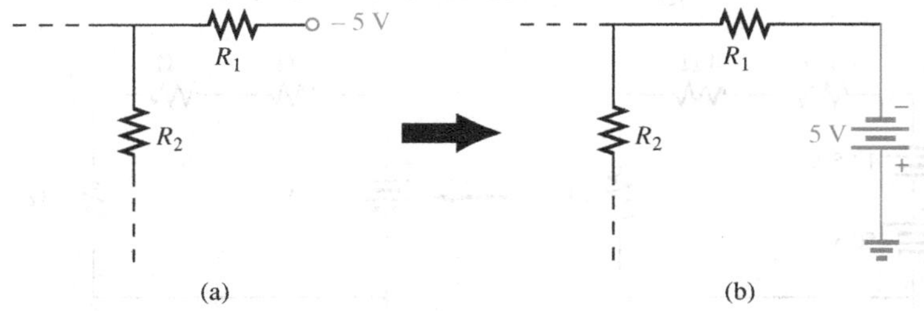

FIG. 5.48
Replacing the notation for a negative dc supply with the standard notation.

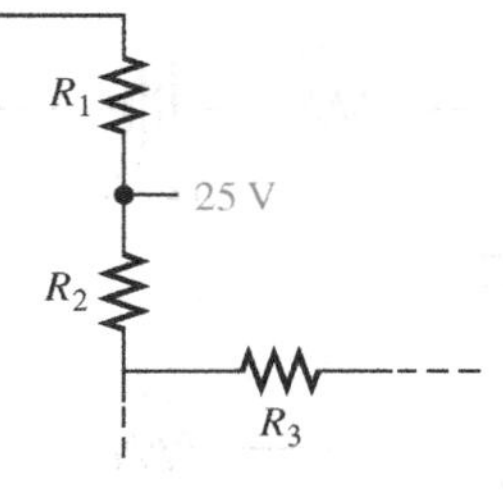

FIG. 5.49
The expected voltage level at a particular point in a network if the system is functioning properly.

Double-Subscript Notation

The fact that voltage is an *across* variable and exists between two points has resulted in a double-subscript notation that defines the first subscript as the higher potential. In Fig. 5.50(a), the two points that define the voltage across the resistor R are denoted by a and b. Since a is the first subscript for V_{ab}, point a must have a higher potential than point b if V_{ab}

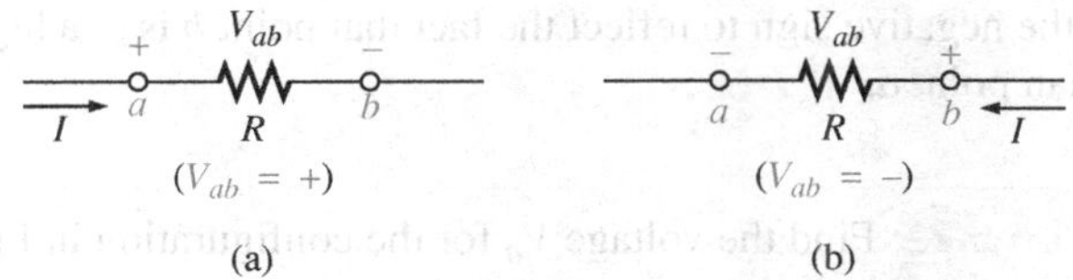

FIG. 5.50
Defining the sign for double-subscript notation.

is to have a positive value. If, in fact, point *b* is at a higher potential than point *a*, V_{ab} will have a negative value, as indicated in Fig. 5.50(b).

In summary:

The double-subscript notation V_{ab} specifies point a as the higher potential. If this is not the case, a negative sign must be associated with the magnitude of V_{ab}.

In other words,

the voltage V_{ab} is the voltage at point a with respect to (w.r.t.) point b.

Single-Subscript Notation

If point *b* of the notation V_{ab} is specified as ground potential (zero volts), then a single-subscript notation can be used that provides the voltage at a point with respect to ground.

In Fig. 5.51, V_a is the voltage from point *a* to ground. In this case, it is obviously 10 V since it is right across the source voltage *E*. The voltage V_b is the voltage from point *b* to ground. Because it is directly across the 4 Ω resistor, $V_b = 4$ V.

In summary:

The single-subscript notation V_a specifies the voltage at point a with respect to ground (zero volts). If the voltage is less than zero volts, a negative sign must be associated with the magnitude of V_a.

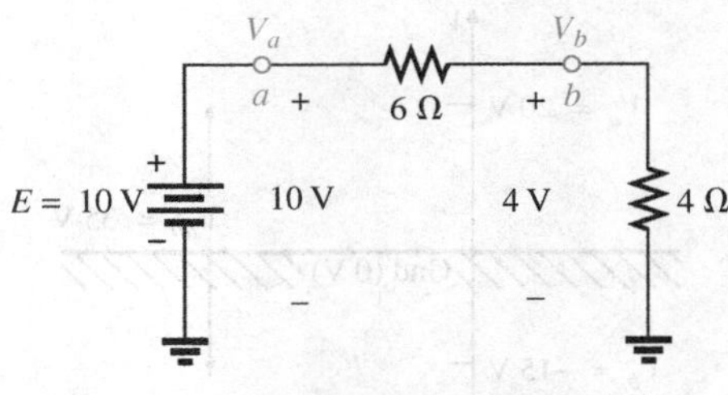

FIG. 5.51
Defining the use of single-subscript notation for voltage levels.

General Comments

A particularly useful relationship can now be established that has extensive applications in the analysis of electronic circuits. For the above notational standards, the following relationship exists:

$$\boxed{V_{ab} = V_a - V_b} \tag{5.12}$$

In other words, if the voltage at points *a* and *b* is known with respect to ground, then the voltage V_{ab} can be determined using Eq. (5.12). In Fig. 5.51, for example,

$$\begin{aligned} V_{ab} &= V_a - V_b = 10\text{ V} - 4\text{ V} \\ &= 6\text{ V} \end{aligned}$$

EXAMPLE 5.21 Find the voltage V_{ab} for the conditions in Fig. 5.52.

Solution: Applying Eq. (5.12) gives

$$\begin{aligned} V_{ab} &= V_a - V_b = 16\text{ V} - 20\text{ V} \\ &= \mathbf{-4\text{ V}} \end{aligned}$$

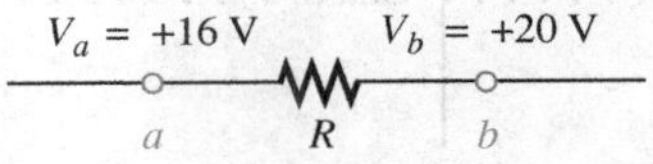

V_a = +16 V V_b = +20 V
a R b

FIG. 5.52
Example 5.21.

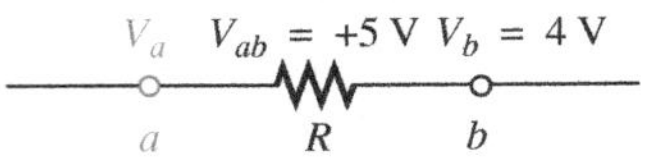

V_a $V_{ab} = +5$ V $V_b = 4$ V

a R b

FIG. 5.53
Example 5.22.

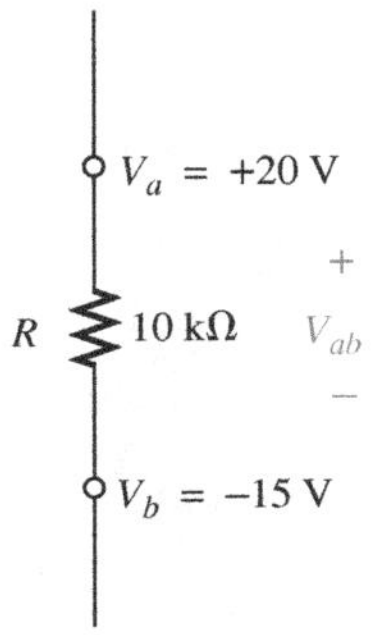

$V_a = +20$ V

R 10 kΩ V_{ab}

$V_b = -15$ V

FIG. 5.54
Example 5.23.

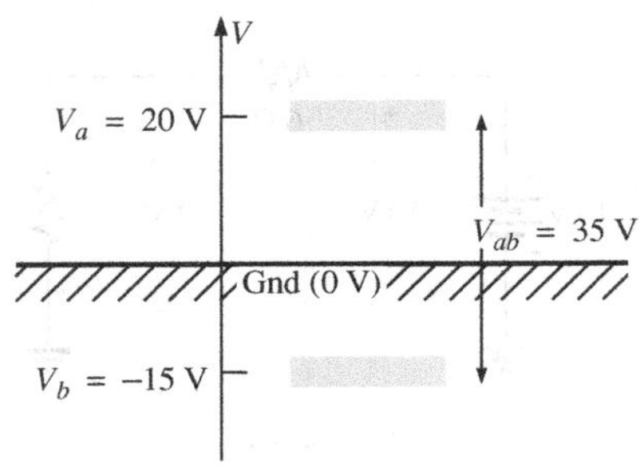

FIG. 5.55
The impact of positive and negative voltages on the total voltage drop.

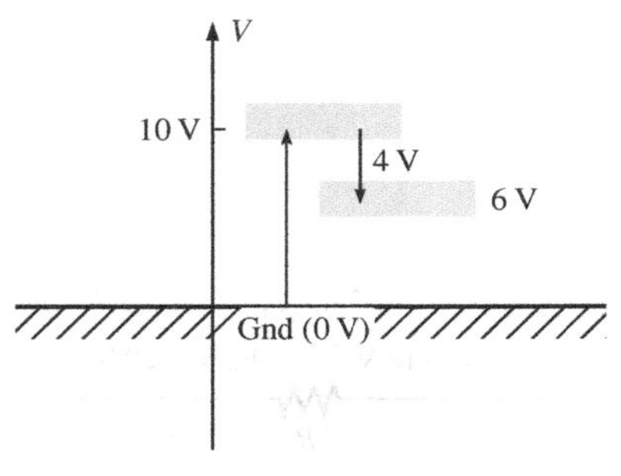

FIG. 5.57
Determining V_b using the defined voltage levels.

Note the negative sign to reflect the fact that point b is at a higher potential than point a.

EXAMPLE 5.22 Find the voltage V_a for the configuration in Fig. 5.53.

Solution: Applying Eq. (5.12) gives

$$V_{ab} = V_a - V_b$$

and

$$V_a = V_{ab} + V_b = 5\text{ V} + 4\text{ V} = \mathbf{9\ V}$$

EXAMPLE 5.23 Find the voltage V_{ab} for the configuration in Fig. 5.54.

Solution: Applying Eq. (5.12) gives

$$V_{ab} = V_a - V_b = 20\text{ V} - (-15\text{ V}) = 20\text{ V} + 15\text{ V} = \mathbf{35\ V}$$

Note in Example 5.23 you must be careful with the signs when applying the equation. The voltage is dropping from a high level of +20 V to a negative voltage of −15 V. As shown in Fig. 5.55, this represents a drop in voltage of 35 V. In some ways it's like going from a positive checking balance of $20 to owing $15; the total expenditure is $35.

EXAMPLE 5.24 Find the voltages V_b, V_c, and V_{ac} for the network in Fig. 5.56.

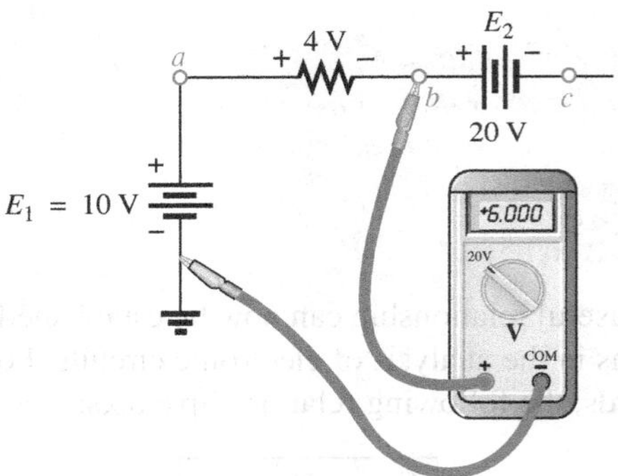

FIG. 5.56
Example 5.24.

Solution: Starting at ground potential (zero volts), we proceed through a rise of 10 V to reach point a and then pass through a drop in potential of 4 V to point b. The result is that the meter reads

$$V_b = +10\text{ V} - 4\text{ V} = \mathbf{6\ V}$$

as clearly demonstrated by Fig. 5.57.

If we then proceed to point c, there is an additional drop of 20 V, resulting in

$$V_c = V_b - 20\text{ V} = 6\text{ V} - 20\text{ V} = \mathbf{-14\ V}$$

as shown in Fig. 5.58:

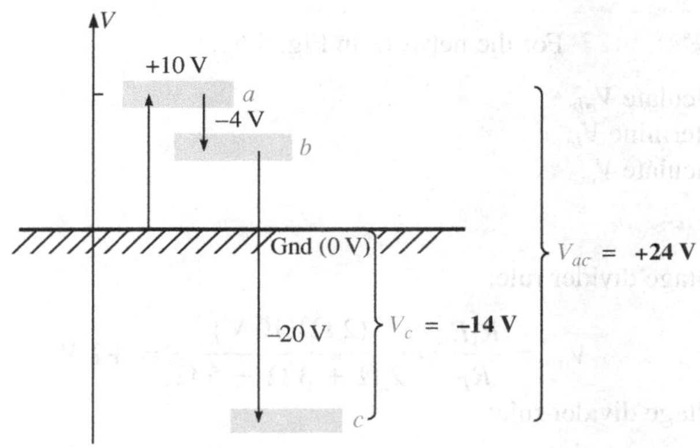

FIG. 5.58
Review of the potential levels for the circuit in Fig. 5.56.

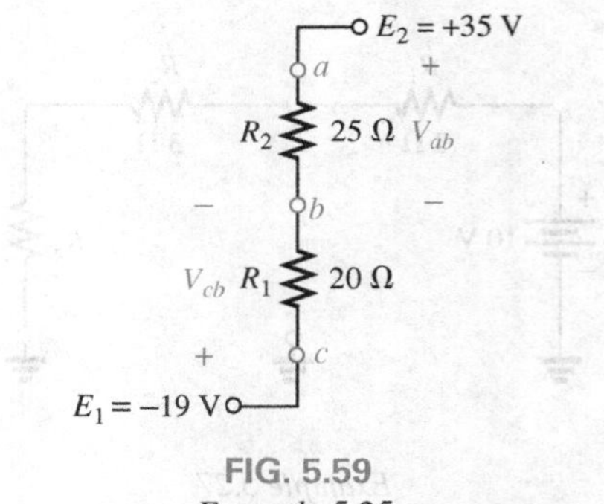

FIG. 5.59
Example 5.25.

The voltage V_{ac} can be obtainted using Eq. (5.12) or by simply referring to Fig. 5.58:

$$V_{ac} = V_a - V_c = 10\text{ V} - (-14\text{ V}) = \mathbf{24\text{ V}}$$

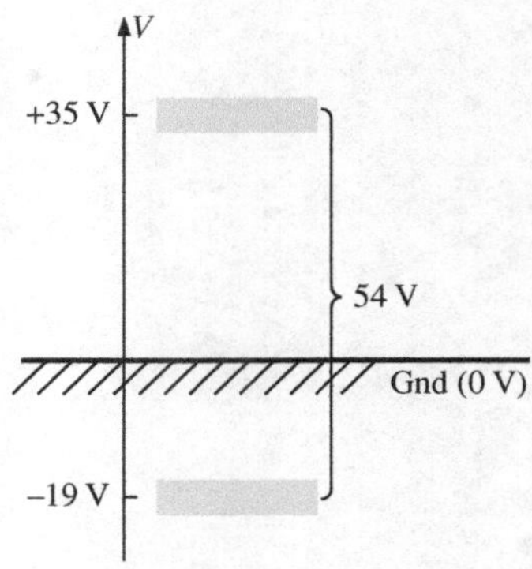

FIG. 5.60
Determining the total voltage drop across the resistive elements in Fig 5.59.

EXAMPLE 5.25 Determine V_{ab}, V_{cb}, and V_c for the network in Fig. 5.59.

Solution: There are two ways to approach this problem. The first is to sketch the diagram in Fig. 5.60 and note that there is a 54 V drop across the series resistors R_1 and R_2. The current can then be determined using Ohm's law and the voltage levels as follows:

$$I = \frac{54\text{ V}}{45\ \Omega} = 1.2\text{ A}$$

$$V_{ab} = IR_2 = (1.2\text{ A})(25\ \Omega) = \mathbf{30\text{ V}}$$

$$V_{cb} = -IR_1 = -(1.2\text{ A})(20\ \Omega) = \mathbf{-24\text{ V}}$$

$$V_c = E_1 = \mathbf{-19\text{ V}}$$

The other approach is to redraw the network as shown in Fig. 5.61 to clearly establish the aiding effect of E_1 and E_2 and then solve the resulting series circuit:

$$I = \frac{E_1 + E_2}{R_T} = \frac{19\text{ V} + 35\text{ V}}{45\ \Omega} = \frac{54\text{ V}}{45\ \Omega} = 1.2\text{ A}$$

and $\quad V_{ab} = \mathbf{30\text{ V}} \quad V_{cb} = \mathbf{-24\text{ V}} \quad V_c = \mathbf{-19\text{ V}}$

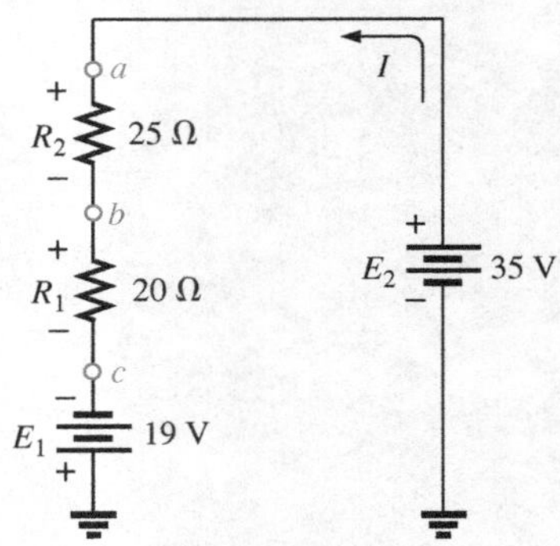

FIG. 5.61
Redrawing the circuit in Fig. 5.59 using standard dc voltage supply symbols.

EXAMPLE 5.26 Using the voltage divider rule, determine the voltages V_1 and V_2 of Fig. 5.62.

Solution: Redrawing the network with the standard battery symbol results in the network in Fig. 5.63. Applying the voltage divider rule gives

$$V_1 = \frac{R_1E}{R_1 + R_2} = \frac{(4\ \Omega)(24\text{ V})}{4\ \Omega + 2\ \Omega} = \mathbf{16\text{ V}}$$

$$V_2 = \frac{R_2E}{R_1 + R_2} = \frac{(2\ \Omega)(24\text{ V})}{4\ \Omega + 2\ \Omega} = \mathbf{8\text{ V}}$$

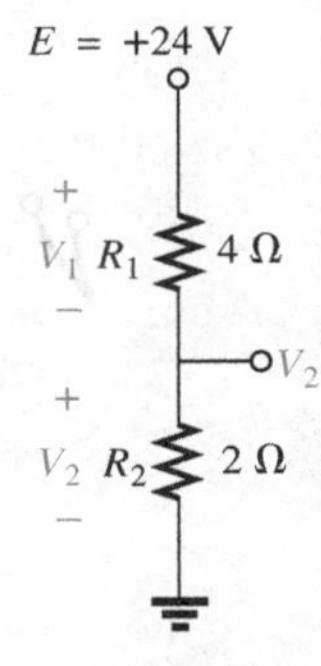

FIG. 5.62
Example 5.26.

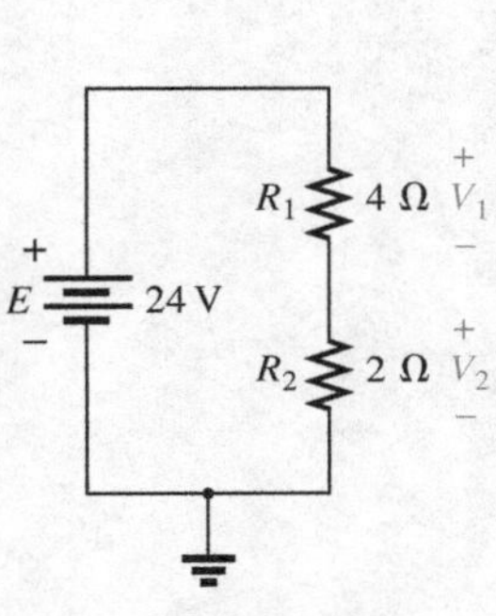

FIG. 5.63
Circuit of Fig. 5.62 redrawn.

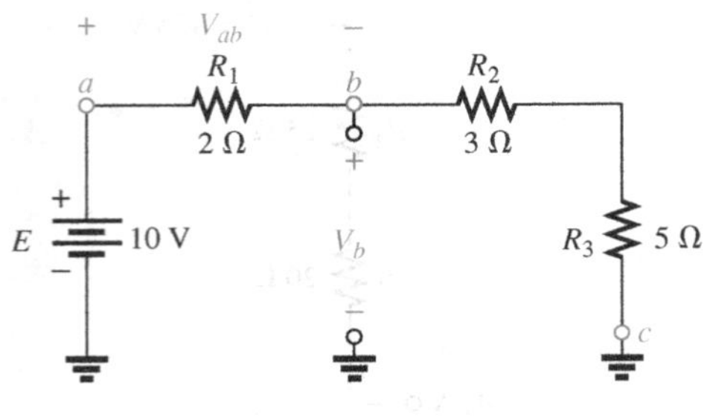

FIG. 5.64
Example 5.27.

EXAMPLE 5.27 For the network in Fig. 5.64.

a. Calculate V_{ab}.
b. Determine V_b.
c. Calculate V_c.

Solutions:

a. Voltage divider rule:

$$V_{ab} = \frac{R_1E}{R_T} = \frac{(2\ \Omega)(10\ \text{V})}{2\ \Omega + 3\ \Omega + 5\ \Omega} = \mathbf{+2\ V}$$

b. Voltage divider rule:

$$V_b = V_{R_2} + V_{R_3} = \frac{(R_2 + R_3)E}{R_T} = \frac{(3\ \Omega + 5\ \Omega)(10\ \text{V})}{10\ \Omega} = \mathbf{8\ V}$$

or $\quad V_b = V_a - V_{ab} = E - V_{ab} = 10\ \text{V} - 2\ \text{V} = \mathbf{8\ V}$

c. V_c = ground potential = **0 V**

5.10 VOLTAGE REGULATION AND THE INTERNAL RESISTANCE OF VOLTAGE SOURCES

When you use a dc supply such as the generator, battery, or supply in Fig. 5.65, you initially assume that it will provide the desired voltage for any resistive load you may hook up to the supply. In other words, if the battery is labeled 1.5 V or the supply is set at 20 V, you assume that they will provide that voltage no matter what load you may apply. Unfortunately, this is not always the case. For instance, if we apply a 1 kΩ resistor to a dc laboratory supply, it is fairly easy to set the voltage across the resistor to 20 V. However, if we remove the 1 kΩ resistor and replace it with a 100 Ω resistor and don't touch the controls on the supply at all, we may find that the voltage has dropped to 19.14 V. Change the load to a 68 Ω resistor, and the terminal voltage drops to 18.72 V. We discover that the load applied affects the terminal voltage of the supply. In fact, this example points out that

a network should always be connected to a supply before the level of supply voltage is set.

The reason the terminal voltage drops with changes in load (current demand) is that

every practical (real-world) supply has an internal resistance in series with the idealized voltage source

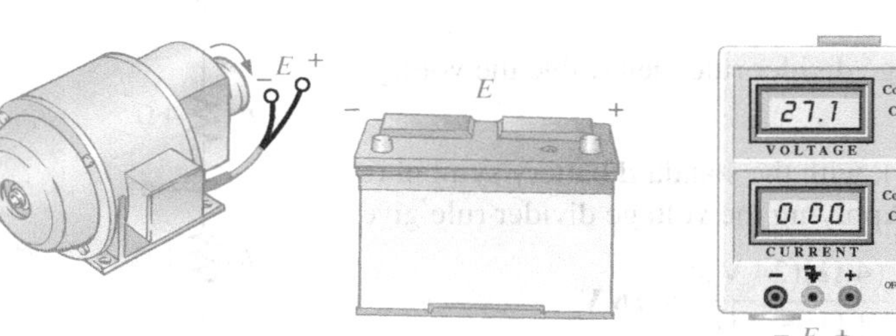

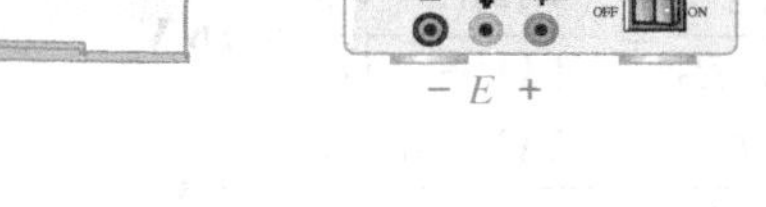

(a)

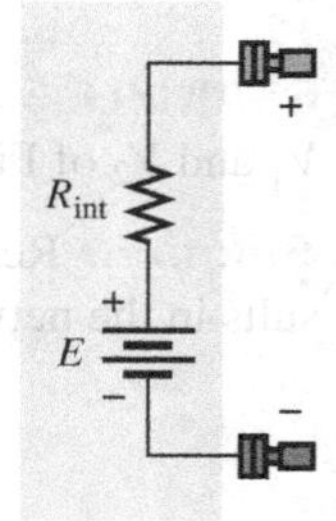

(b)

FIG. 5.65
(a) Sources of dc voltage; (b) equivalent circuit.

as shown in Fig. 5.65(b). The resistance level depends on the type of supply, but it is always present. Every year new supplies come out that are less sensitive to the load applied, but even so, some sensitivity still remains.

The supply in Fig. 5.66 helps explain the action that occurred above as we changed the load resistor. Due to the **internal resistance** of the supply, the ideal internal supply must be set to 20.1 V in Fig. 5.66(a) if 20 V are to appear across the 1 kΩ resistor. The internal resistance will capture 0.1 V of the applied voltage. The current in the circuit is determined by simply looking at the load and using Ohm's law; that is, $I_L = V_L/R_L = 20\text{ V}/1\text{ k}\Omega = 20\text{ mA}$, which is a relatively low current.

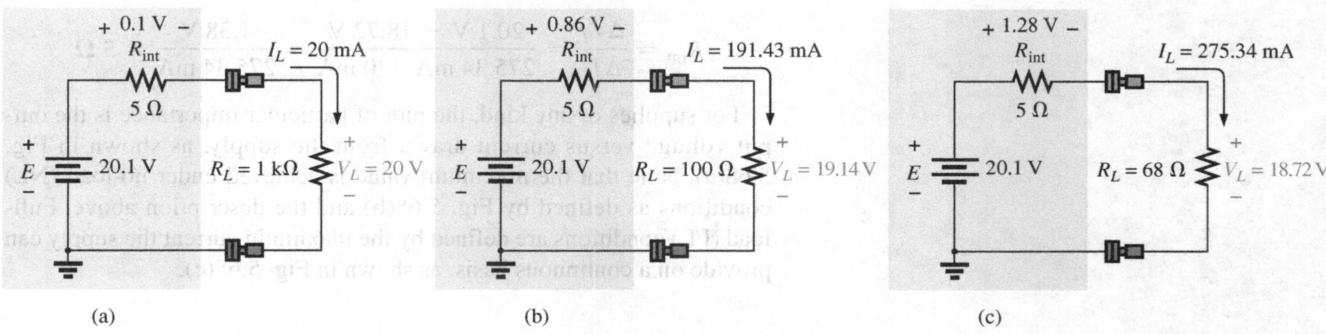

FIG. 5.66
Demonstrating the effect of changing a load on the terminal voltage of a supply.

In Fig. 5.66(b), all the settings of the supply are left untouched, but the 1 kΩ load is replaced by a 100 Ω resistor. The resulting current is now $I_L = E/R_T = 20.1\text{ V}/105\ \Omega = 191.43\text{ mA}$, and the output voltage is $V_L = I_L R = (191.43\text{ mA})(100\ \Omega) = 19.14\text{ V}$, a drop of 0.86 V. In Fig. 5.66(c), a 68 Ω load is applied, and the current increases substantially to 275.34 mA with a terminal voltage of only 18.72 V. This is a drop of 1.28 V from the expected level. Quite obviously, therefore, as the current drawn from the supply increases, the terminal voltage continues to drop.

If we plot the terminal voltage versus current demand from 0 A to 275.34 mA, we obtain the plot in Fig. 5.67. Interestingly enough, it turns out to be a straight line that continues to drop with an increase in current demand. Note, in particular, that the curve begins at a current level of 0 A.

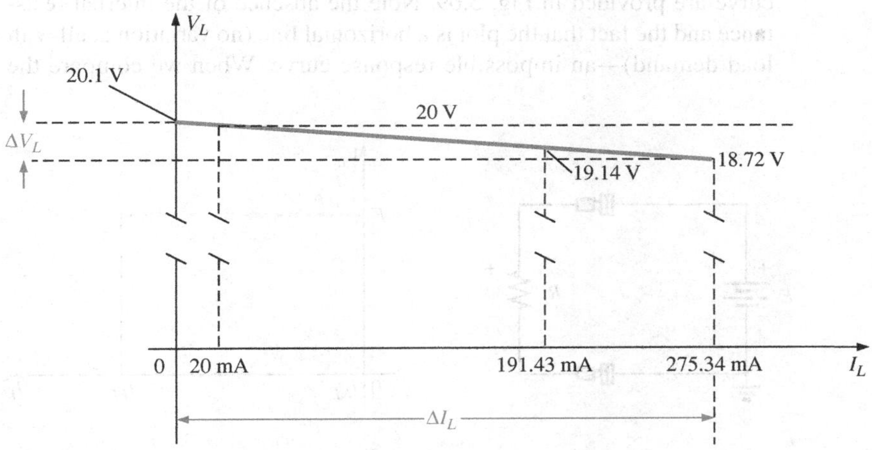

FIG. 5.67
Plotting V_L versus I_L for the supply in Fig. 5.66.

Under no-load conditions, where the output terminals of the supply are not connected to any load, the current will be 0 A due to the absence of a complete circuit. The output voltage will be the internal ideal supply level of 20.1 V.

The slope of the line is defined by the internal resistance of the supply. That is,

$$R_{int} = \frac{\Delta V_L}{\Delta I_L} \quad \text{(ohms, }\Omega\text{)} \tag{5.13}$$

which for the plot in Fig. 5.67 results in

$$R_{int} = \frac{\Delta V_L}{\Delta I_L} = \frac{20.1\text{ V} - 18.72\text{ V}}{275.34\text{ mA} - 0\text{ mA}} = \frac{1.38\text{ V}}{275.34\text{ mA}} = \mathbf{5\ \Omega}$$

For supplies of any kind, the plot of particular importance is the output voltage versus current drawn from the supply, as shown in Fig. 5.68(a). Note that the maximum value is achieved under no-load (NL) conditions as defined by Fig. 5.68(b) and the description above. Full-load (FL) conditions are defined by the maximum current the supply can provide on a continuous basis, as shown in Fig. 5.68(c).

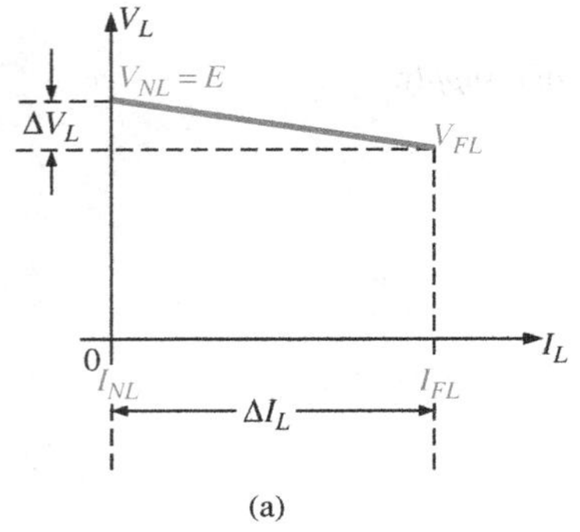

(a)

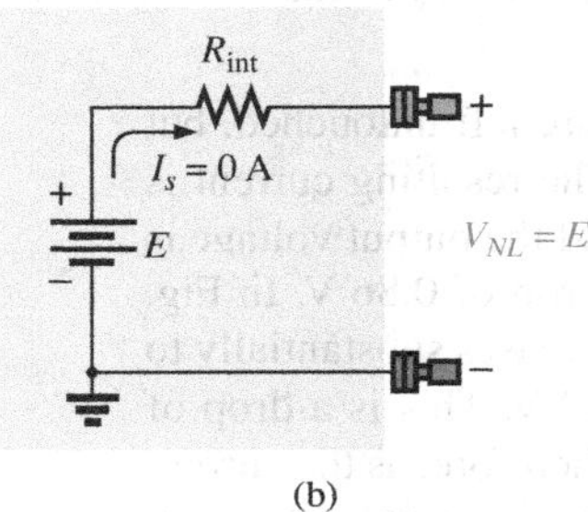

(b)

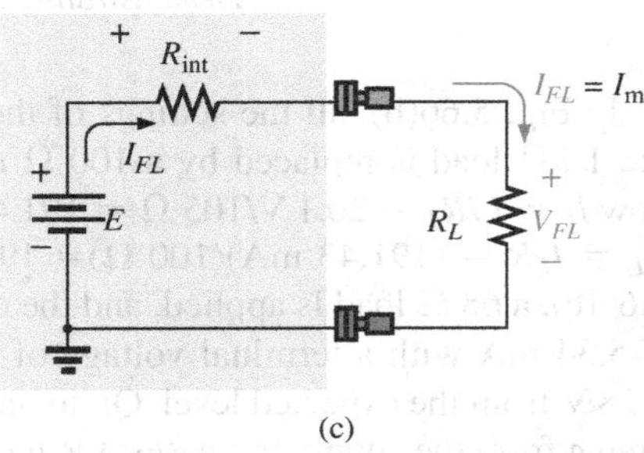

(c)

FIG. 5.68
Defining the properties of importance for a power supply.

As a basis for comparison, an ideal power supply and its response curve are provided in Fig. 5.69. Note the absence of the internal resistance and the fact that the plot is a horizontal line (no variation at all with load demand)—an impossible response curve. When we compare the

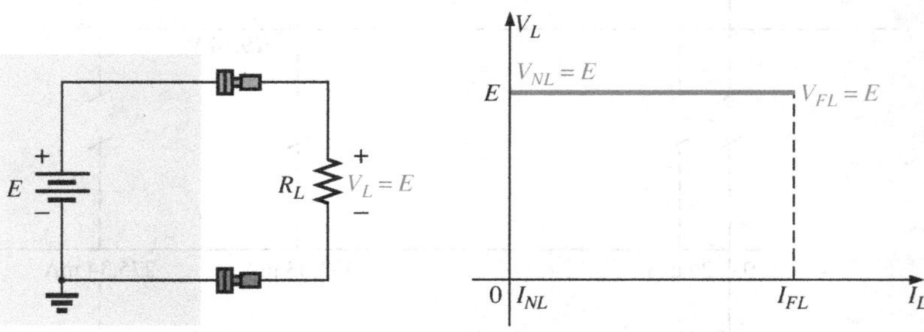

FIG. 5.69
Ideal supply and its terminal characteristics.

curve in Fig. 5.69 with that in Fig. 5.68(a), however, we now realize that the *steeper the slope,* the more sensitive is the supply to the change in load and therefore the *less desirable* it is for many laboratory procedures. In fact,

the larger the internal resistance, the steeper is the drop in voltage with an increase in load demand (current).

To help us anticipate the expected response of a supply, a defining quantity called **voltage regulation** (abbreviated *VR*; often called *load regulation* on specification sheets) was established. The basic equation in terms of the quantities in Fig. 5.68(a) is the following:

$$VR = \frac{V_{NL} - V_{FL}}{V_{FL}} \times 100\% \qquad \textbf{(5.14)}$$

The examples to follow demonstrate that

the smaller the voltage or load regulation of a supply, the less will the terminal voltage change with increasing levels of current demand.

For the supply above with a no-load voltage of 20.1 V and a full-load voltage of 18.72 V, at 275.34 mA the voltage regulation is

$$VR = \frac{V_{NL} - V_{FL}}{V_{FL}} \times 100\% = \frac{20.1\text{ V} - 18.72\text{ V}}{18.72\text{ V}} \times 100\% \cong \mathbf{7.37\%}$$

which is quite high, revealing that we have a very sensitive supply. Most modern commercial supplies have regulation factors less than 1%, with 0.01% being very typical.

EXAMPLE 5.28

a. Given the characteristics in Fig. 5.70, determine the voltage regulation of the supply.
b. Determine the internal resistance of the supply.
c. Sketch the equivalent circuit for the supply.

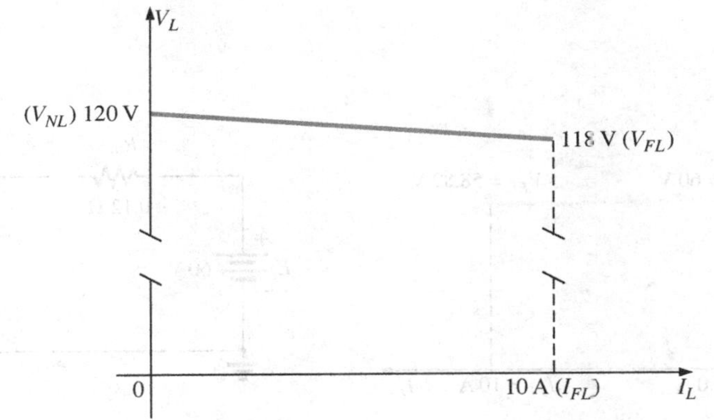

FIG. 5.70
Terminal characteristics for the supply of Example 5.28.

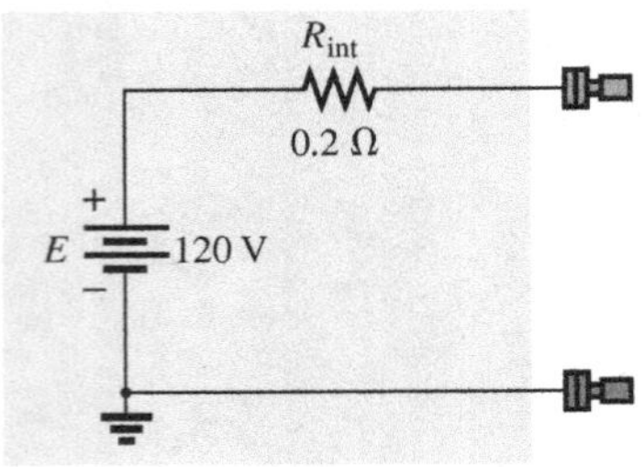

FIG. 5.71
dc supply with the terminal characteristics of Fig. 5.70.

Solutions:

a. $$VR = \frac{V_{NL} - V_{FL}}{V_{FL}} \times 100\%$$
$$= \frac{120\text{ V} - 118\text{ V}}{118\text{ V}} \times 100\% = \frac{2}{118} \times 100\%$$
$$VR \cong \mathbf{1.7\%}$$

b. $$R_{int} = \frac{\Delta V_L}{\Delta I_L} = \frac{120\text{ V} - 118\text{ V}}{10\text{ A} - 0\text{ A}} = \frac{2\text{ V}}{10\text{ A}} = \mathbf{0.2\ \Omega}$$

c. See Fig. 5.71.

EXAMPLE 5.29 Given a 60 V supply with a voltage regulation of 2%:

a. Determine the terminal voltage of the supply under full-load conditions.
b. If the half-load current is 5 A, determine the internal resistance of the supply.
c. Sketch the curve of the terminal voltage versus load demand and the equivalent circuit for the supply.

Solutions:

a. $$VR = \frac{V_{NL} - V_{FL}}{V_{FL}} \times 100\%$$
$$2\% = \frac{60\text{ V} - V_{FL}}{V_{FL}} \times 100\%$$
$$\frac{2\%}{100\%} = \frac{60\text{ V} - V_{FL}}{V_{FL}}$$
$$0.02V_{FL} = 60\text{ V} - V_{FL}$$
$$1.02V_{FL} = 60\text{ V}$$
$$V_{FL} = \frac{60\text{ V}}{1.02} = \mathbf{58.82\ V}$$

b. $I_{FL} = 10\text{ A}$
$$R_{int} = \frac{\Delta V_L}{\Delta I_L} = \frac{60\text{ V} - 58.82\text{ V}}{10\text{ A} - 0\text{ A}} = \frac{1.18\text{ V}}{10\text{ A}} \cong \mathbf{0.12\ \Omega}$$

c. See Fig. 5.72.

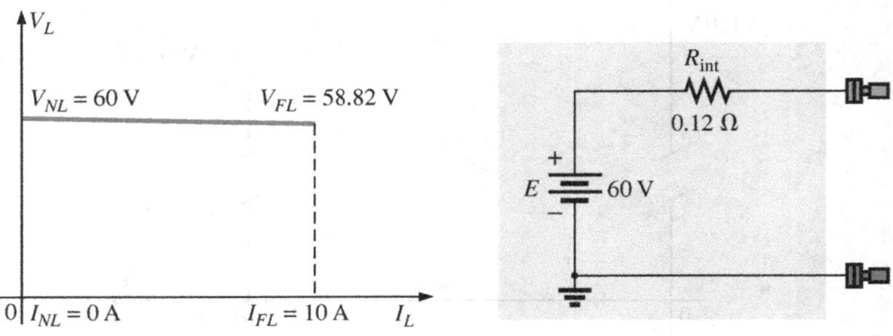

FIG. 5.72
Characteristics and equivalent circuit for the supply of Example 5.29.

5.11 LOADING EFFECTS OF INSTRUMENTS

In the previous section, we learned that power supplies are not the ideal instruments we may have thought they were. The applied load can have an effect on the terminal voltage. Fortunately, since today's supplies have such small load regulation factors, the change in terminal voltage with load can usually be ignored for most applications. If we now turn our attention to the various meters we use in the lab, we again find that they are not totally ideal:

Whenever you apply a meter to a circuit, you change the circuit and the response of the system. Fortunately, however, for most applications, considering the meters to be ideal is a valid approximation as long as certain factors are considered.

For instance,

any ammeter connected in a series circuit will introduce resistance to the series combination that will affect the current and voltages of the configuration.

The resistance between the terminals of an ammeter is determined by the chosen scale of the ammeter. In general,

for ammeters, the higher the maximum value of the current for a particular scale, the smaller will the internal resistance be.

For example, it is not uncommon for the resistance between the terminals of an ammeter to be 250 Ω for a 2 mA scale but only 1.5 Ω for the 2 A scale, as shown in Fig. 5.73(a) and (b). If you are analyzing a circuit in detail, you can include the internal resistance as shown in Fig. 5.73 as a resistor between the two terminals of the meter.

At first reading, such resistance levels at low currents give the impression that ammeters are far from ideal, and that they should be used only to obtain a general idea of the current and should not be expected to provide a true reading. Fortunately, however, when you are reading currents below the 2 mA range, the resistors in series with the ammeter are typically in the kilohm range. For example, in Fig. 5.74(a), for an ideal ammeter, the current

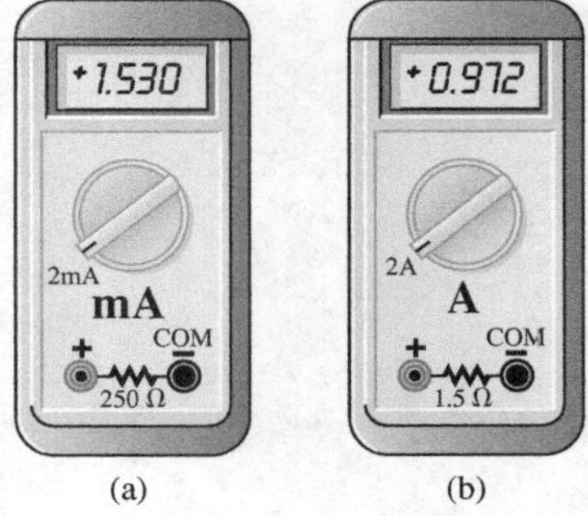

FIG. 5.73
Including the effects of the internal resistance of an ammeter: (a) 2 mA scale; (b) 2 A scale.

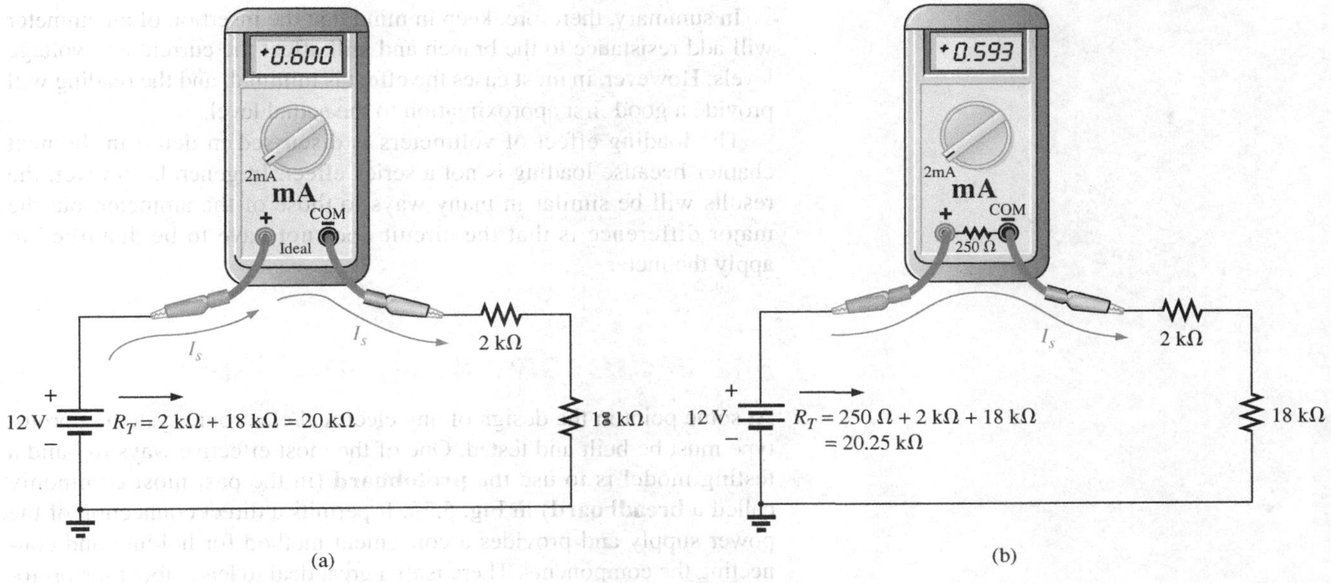

FIG. 5.74
Applying an ammeter set on the 2 mA scale to a circuit with resistors in the kilohm range: (a) ideal; (b) practical.

displayed is 0.6 mA as determined from $I_s = E/R_T = 12\ \text{V}/20\ \text{k}\Omega = 0.6$ mA. If we now insert a meter with an internal resistance of 250 Ω as shown in Fig. 5.74(b), the additional resistance in the circuit will drop the current to 0.593 mA as determined from $I_s = E/R_T = 12\ \text{V}/20.25\ \text{k}\Omega = 0.593$ mA. Now, certainly the current has dropped from the ideal level, but the difference in results is only about 1%—nothing major, and the measurement can be used for most purposes. If the series resistors were in the same range as the 250 Ω resistors, we would have a different problem, and we would have to look at the results very carefully.

Let us go back to Fig. 5.20 and determine the actual current if each meter on the 2 A scale has an internal resistance of 1.5 Ω. The fact that there are four meters will result in an additional resistance of (4)(1.5 Ω) = 6 Ω in the circuit, and the current will be $I_s = E/R_T = 8.4\ \text{V}/146\ \Omega \cong 58$ mA, rather than the 60 mA under ideal conditions. This value is still close enough to be considered a helpful reading. However, keep in mind that if we were measuring the current in the circuit, we would use only one ammeter, and the current would be $I_s = E/R_T = 8.4\ \text{V}/141.5\ \Omega \cong 59$ mA, which can certainly be approximated as 60 mA.

In general, therefore, be aware that this internal resistance must be factored in, but for the reasons just described, most readings can be used as an excellent first approximation to the actual current.

It should be added that because of this *insertion problem* with ammeters, and because of the very important fact that the *circuit must be disturbed* to measure a current, ammeters are not used as much as you might initially expect. Rather than break a circuit to insert a meter, the voltage across a resistor is often measured and the current then calculated using Ohm's law. This eliminates the need to worry about the level of the meter resistance and having to disturb the circuit. Another option is to use the clamp-type ammeters introduced in Chapter 2, removing the concerns about insertion loss and disturbing the circuit. Of course, for many practical applications (such as on power supplies), it is convenient to have an ammeter permanently installed so that the current can quickly be read from the display. In such cases, however, the design is such as to compensate for the insertion loss.

In summary, therefore, keep in mind that the insertion of an ammeter will add resistance to the branch and will affect the current and voltage levels. However, in most cases the effect is minimal, and the reading will provide a good first approximation to the actual level.

The loading effect of voltmeters is discussed in detail in the next chapter because loading is not a series effect. In general, however, the results will be similar in many ways to those of the ammeter, but the major difference is that the circuit does not have to be disturbed to apply the meter.

5.12 PROTOBOARDS (BREADBOARDS)

At some point in the design of any electrical/electronic system, a prototype must be built and tested. One of the most effective ways to build a testing model is to use the **protoboard** (in the past most commonly called a **breadboard**) in Fig. 5.75. It permits a direct connection of the power supply and provides a convenient method for holding and connecting the components. There isn't a great deal to learn about the protoboard, but it is important to point out some of its characteristics, including the way the elements are typically connected.

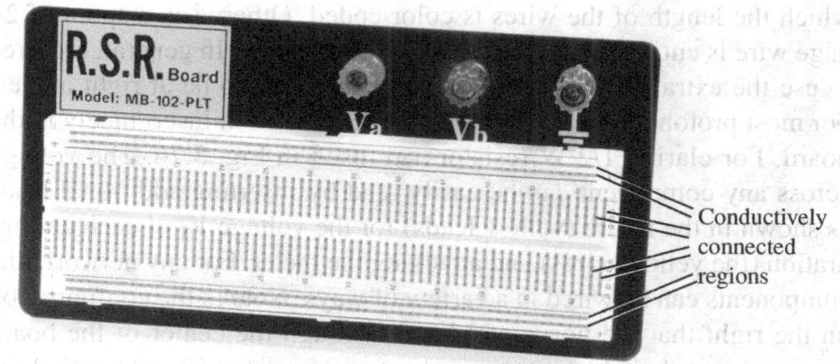

FIG. 5.75
Protoboard with areas of conductivity defined using two different approaches.

The red terminal V_a is connected directly to the positive terminal of the dc power supply, with the black lead V_b connected to the negative terminal and the green terminal used for the ground connection. Under the hole pattern, there are continuous horizontal copper strips under the top and bottom rows, as shown by the copper bands in Fig. 5.75. In the center region, the conductive strips are vertical but do not extend beyond the deep notch running the horizontal length of the board. That's all there is to it, although it will take some practice to make the most effective use of the conductive patterns.

As examples, the network in Fig. 5.12 is connected on the protoboard in the photo in Fig. 5.76 using *two different approaches.* After the dc power supply has been hooked up, a lead is brought down from the positive red terminal to the top conductive strip marked "+." Keep in mind that now the entire strip is connected to the positive terminal of the supply. The negative terminal is connected to the bottom strip marked with a minus sign (−), so that 8.4 V can be read at any point between the top positive strip and the bottom negative strip. A ground connection to the negative terminal of the battery was made at the site of the three terminals. For the user's convenience, kits are available in

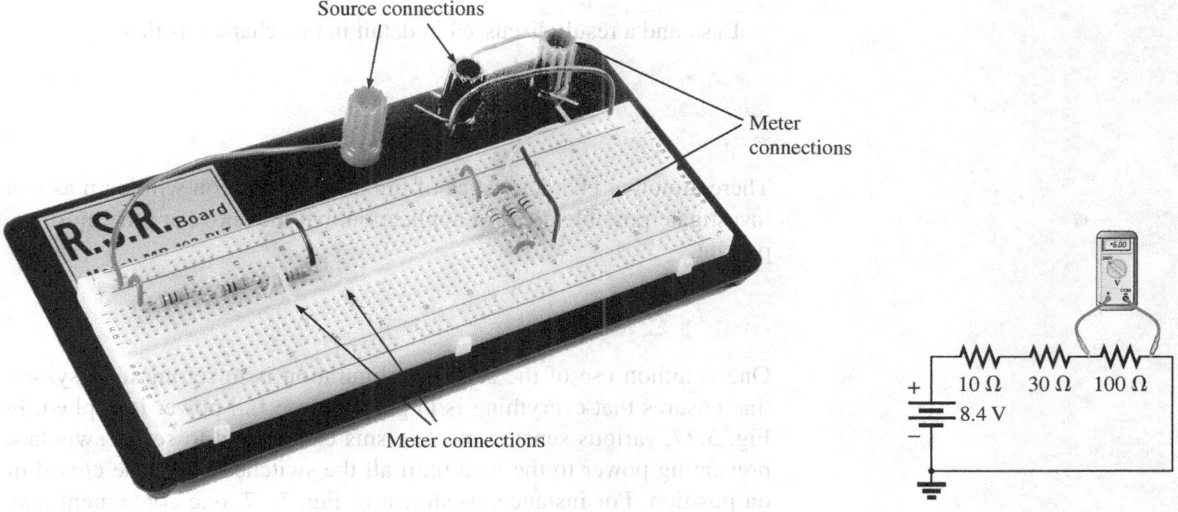

FIG. 5.76
Two setups for the network in Fig. 5.12 on a protoboard with yellow leads added to each configuration to measure voltage V_3 with a voltmeter.

which the length of the wires is color coded. Otherwise, a spool of 24 gage wire is cut to length and the ends are stripped. In general, feel free to use the extra length—everything doesn't have to be at right angles. For most protoboards, 1/4 through 1 W resistors will insert nicely in the board. For clarity, 1/2 W resistors are used in Fig. 5.76. The voltage across any component can be easily read by inserting additional leads as shown in the figure (yellow leads) for the voltage V_3 of each configuration (the yellow wires) and attaching the meter. For any network, the components can be wired in a variety of ways. Note in the configuration on the right that the horizontal break through the center of the board was used to isolate the two terminals of each resistor. Even though there are no set standards, it is important that the arrangement can *easily be understood* by someone else.

Additional setups using the protoboard are in the chapters to follow so that you can become accustomed to the manner in which it is used most effectively. You will probably see the protoboard quite frequently in your laboratory sessions or in an industrial setting.

5.13 APPLICATIONS

Before looking at a few applications, we need to consider a few general characteristics of the series configuration that you should always keep in mind when designing a system. First, and probably the most important, is that

if one element of a series combination of elements should fail, it will disrupt the response of all the series elements. If an open circuit occurs, the current will be zero. If a short circuit results, the voltage will increase across the other elements, and the current will increase in magnitude.

Second, and a thought you should always keep in mind, is that

for the same source voltage, the more elements you place in series, the less is the current and the less is the voltage across all the elements of the series combination.

Last, and a result discussed in detail in this chapter, is that

the current is the same for each element of a series combination, but the voltage across each element is a function of its terminal resistance.

There are other characteristics of importance that you will learn as you investigate possible areas of application, but the above are the most important.

Series Control

One common use of the series configuration is in setting up a system that ensures that everything is in place before full power is applied. In Fig. 5.77, various sensing mechanisms can be tied to series switches, preventing power to the load until all the switches are in the closed or on position. For instance, as shown in Fig. 5.77, one component may test the environment for dangers such as gases, high temperatures, and so on. The next component may be sensitive to the properties of the system to be energized to be sure all components are working. Security is

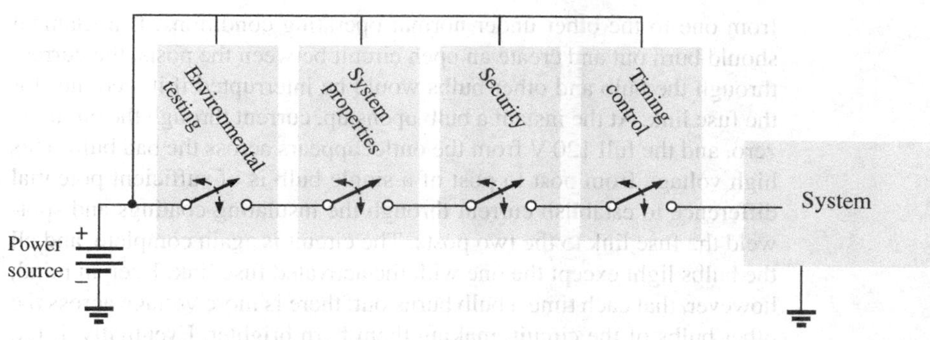

FIG. 5.77
Series control over an operating system.

another factor in the series sequence, and finally a timing mechanism may be present to ensure limited hours of operation or to restrict operating periods. The list is endless, but the fact remains that "all systems must be go" before power reaches the operating system.

Holiday Lights

In recent years, the small blinking holiday lights with 50 to 100 bulbs on a string have become very popular [see Fig. 5.78(a)]. Although holiday lights can be connected in series or parallel (to be described in the next chapter), the smaller blinking light sets are normally connected in series. It is relatively easy to determine if the lights are connected in series. If one wire enters and leaves the bulb casing, they are in series. If two wires enter and leave, they are probably in parallel. *Normally, when bulbs are connected in series, if one burns out (the filament breaks and the circuit opens), all the bulbs go out since the current path has been interrupted.* However, the bulbs in Fig. 5.78(a) are specially designed, as shown in Fig. 5.78(b), to permit current to continue to flow to the other bulbs when the filament burns out. At the base of each bulb, there is a fuse link wrapped around the two posts holding the filament. The fuse link of a soft conducting metal appears to be touching the two vertical posts, but in fact a coating on the posts or fuse link prevents conduction

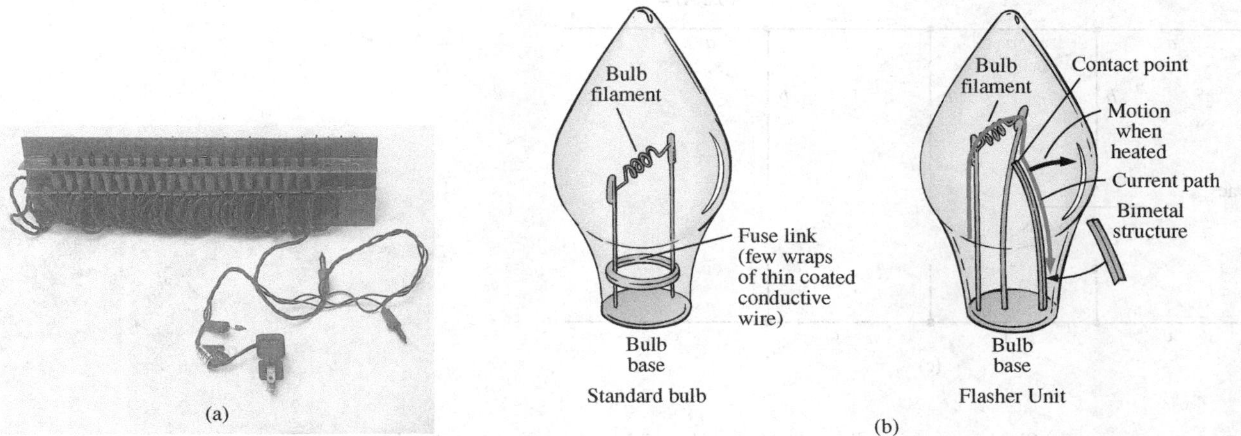

FIG. 5.78
Holiday lights: (a) 50-unit set; (b) bulb construction.

from one to the other under normal operating conditions. If a filament should burn out and create an open circuit between the posts, the current through the bulb and other bulbs would be interrupted if it were not for the fuse link. At the instant a bulb opens up, current through the circuit is zero, and the full 120 V from the outlet appears across the bad bulb. This high voltage from post to post of a single bulb is of sufficient potential difference to establish current through the insulating coatings and spot-weld the fuse link to the two posts. The circuit is again complete, and all the bulbs light except the one with the activated fuse link. Keep in mind, however, that each time a bulb burns out, there is more voltage across the other bulbs of the circuit, making them burn brighter. Eventually, if too many bulbs burn out, the voltage reaches a point where the other bulbs burn out in rapid succession. To prevent this, you must replace burned-out bulbs at the earliest opportunity.

The bulbs in Fig. 5.78(b) are rated 2.5 V at 0.2 A or 200 mA. Since there are 50 bulbs in series, the total voltage across the bulbs will be 50×2.5 V or 125 V, which matches the voltage available at the typical home outlet. Since the bulbs are in series, the current through each bulb will be 200 mA. The power rating of each bulb is therefore $P = VI = (2.5 \text{ V})(0.2 \text{ A}) = 0.5 \text{ W}$ with a total wattage demand of $50 \times 0.5 \text{ W} = 25 \text{ W}$.

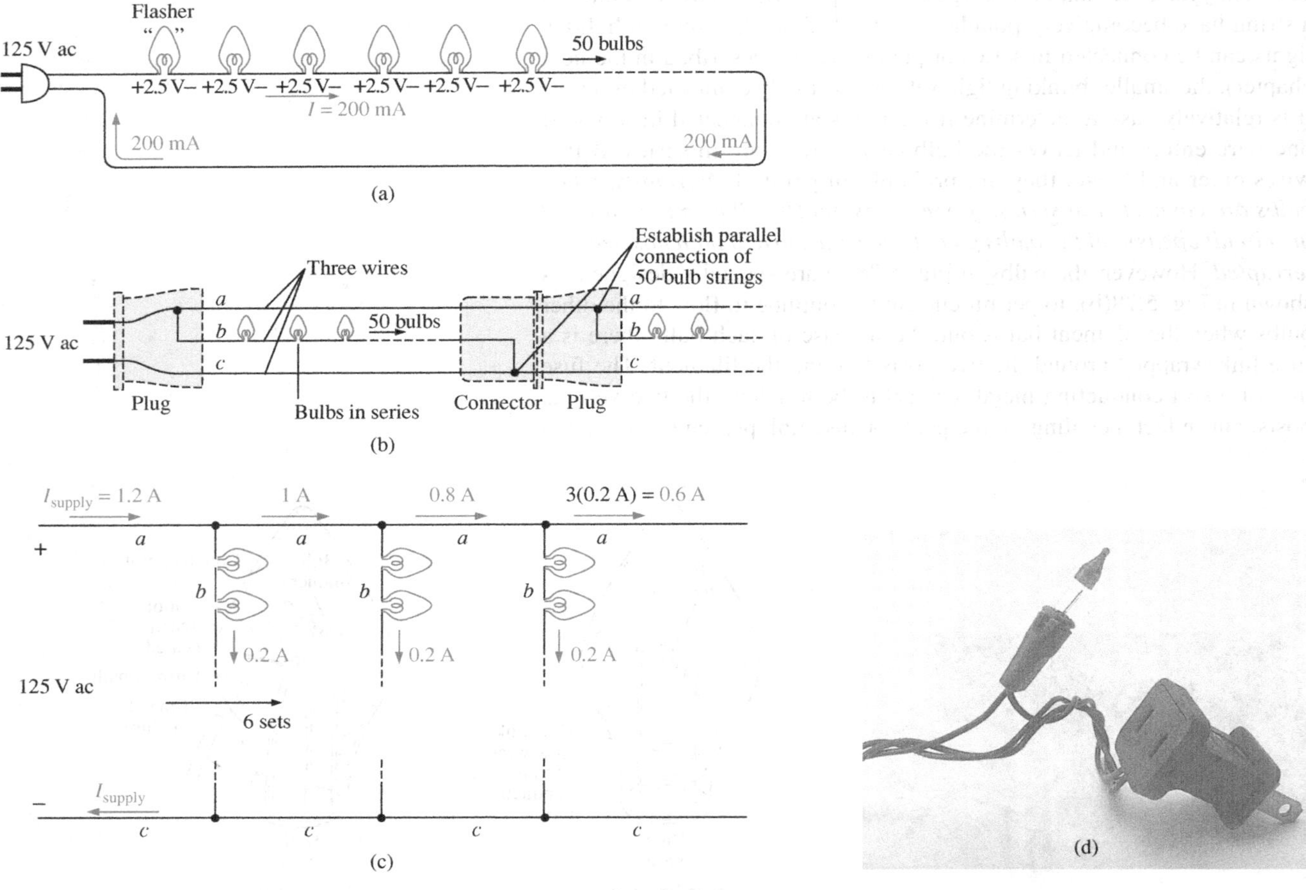

FIG. 5.79
(a) Single-set wiring diagram; (b) special wiring arrangement; (c) redrawn schematic; (d) special plug and flasher unit.

A schematic representation for the set of Fig. 5.78(a) is provided in Fig. 5.79(a). Note that only one flasher unit is required. Since the bulbs are in series, when the flasher unit interrupts the current flow, it turns off all the bulbs. As shown in Fig. 5.78(b), the flasher unit incorporates a bimetal thermal switch that opens when heated by the current to a preset level. As soon as it opens, it begins to cool down and closes again so that current can return to the bulbs. It then heats up again, opens up, and repeats the entire process. The result is an on-and-off action that creates the flashing pattern we are so familiar with. Naturally, in a colder climate (for example, outside in the snow and ice), it initially takes longer to heat up, so the flashing pattern is slow at first, but as the bulbs warm up, the frequency increases.

The manufacturer specifies that no more than six sets should be connected together. How can you connect the sets together, end to end, without reducing the voltage across each bulb and making all the lights dimmer? If you look closely at the wiring, you will find that since the bulbs are connected in series, there is one wire to each bulb with additional wires from plug to plug. Why would they need two additional wires if the bulbs are connected in series? Because when each set is connected together, they are actually in a parallel arrangement (to be discussed in the next chapter). This unique wiring arrangement is shown in Fig. 5.79(b) and redrawn in Fig. 5.79(c). Note that the top line is the hot line to all the connected sets, and the bottom line is the return, neutral, or ground line for all the sets. Inside the plug in Fig. 5.79(d), the hot line and return are connected to each set, with the connections to the metal spades of the plug as shown in Fig. 5.79(b). We will find in the next chapter that the current drawn from the wall outlet for parallel loads is the sum of the current to each branch. The result, as shown in Fig. 5.79(c), is that the current drawn from the supply is $6 \times 200\text{ mA} = 1.2\text{ A}$, and the total wattage for all six sets is the product of the applied voltage and the source current or $(120\text{ V})(1.2\text{ A}) = 144\text{ W}$ with $144\text{ W}/6 = 24\text{ W}$ per set.

Microwave Oven

Series circuits can be very effective in the design of safety equipment. Although we all recognize the usefulness of the microwave oven, it can be quite dangerous if the door is not closed or sealed properly. It is not enough to test the closure at only one point around the door because the door may be bent or distorted from continual use, and leakage can result at some point distant from the test point. One common safety arrangement appears in Fig. 5.80. Note that magnetic switches are located all around the door, with the magnet in the door itself and the magnetic door switch in the main frame. Magnetic switches are simply switches where the magnet draws a magnetic conducting bar between two contacts to complete the circuit—somewhat revealed by the symbol for the device in the circuit diagram in Fig. 5.80. Since the magnetic switches are all in series, they must all be closed to complete the circuit and turn on the power unit. If the door is sufficiently out of shape to prevent a single magnet from getting close enough to the switching mechanism, the circuit will not be complete, and the power cannot be turned on. Within the control unit of the power supply, either the series circuit completes a circuit for operation or a sensing current is established and monitored that controls the system operation.

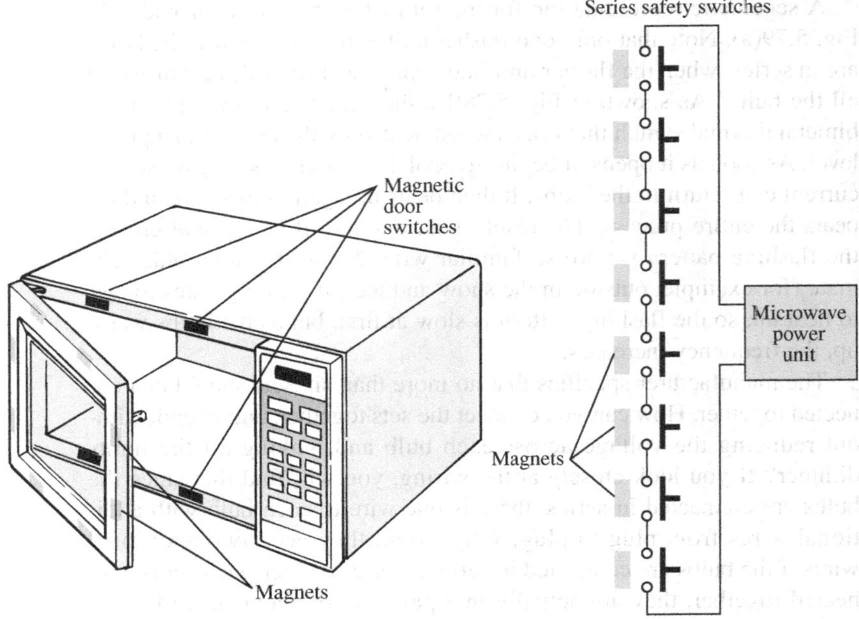

FIG. 5.80
Series safety switches in a microwave oven.

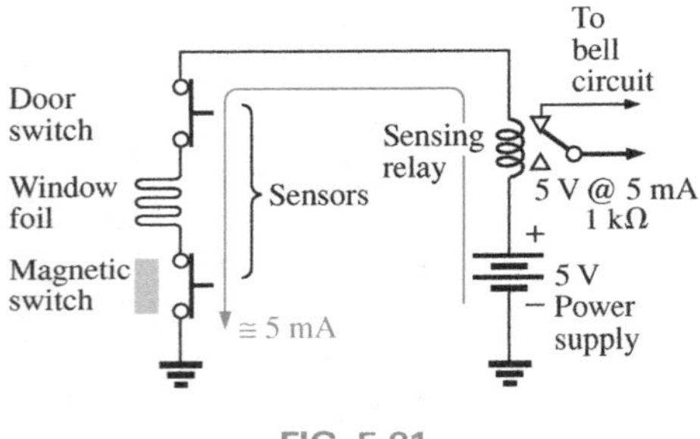

FIG. 5.81
Series alarm circuit.

Series Alarm Circuit

The circuit in Fig. 5.81 is a simple alarm circuit. Note that every element of the design is in a series configuration. The power supply is a 5 V dc supply that can be provided through a design similar to that in Fig. 2.33, a dc battery, or a combination of an ac and a dc supply that ensures that the battery will always be at full charge. If all the sensors are closed, a current of 5 mA results because of the terminal load of the relay of about 1 kΩ. That current energizes the relay and maintains an off position for the alarm. However, if any of the sensors is opened, the current will be interrupted, the relay will let go, and the alarm circuit will be energized. With relatively short wires and a few sensors, the system should work well since the voltage drop across each is minimal. However, since the alarm wire is usually relatively thin, resulting in a measurable resistance level, if the wire to the sensors is too long, a sufficient voltage drop could occur across the line, reducing the voltage across the relay to a point where the alarm fails to operate properly. Thus, wire length is a factor that must be considered if a series configuration is used. Proper sensitivity to the length of the line should remove any concerns about its operation. An improved design is described in Chapter 8.

5.14 COMPUTER ANALYSIS

PSpice

In Section 4.9, the basic procedure for setting up the PSpice folder and running the program were presented. Because of the detail provided in that section, you should review it before proceeding with this example.

Because this is only the second example using PSpice, some detail is provided, but not at the level of Section 4.9.

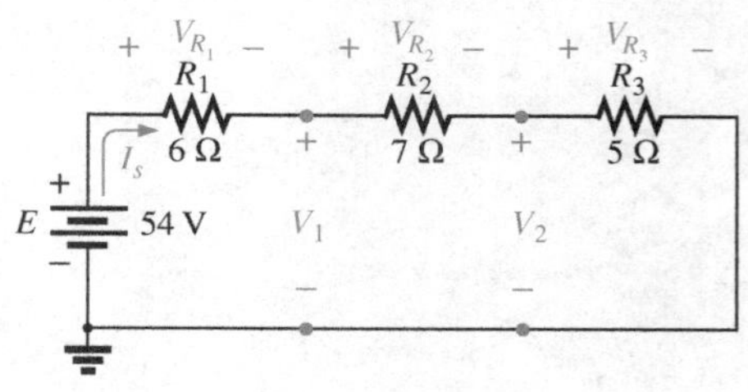

FIG. 5.82
Series dc network to be analyzed using PSpice.

The circuit to be investigated appears in Fig. 5.82. You will use the **PSpice** folder established in Section 4.9. Double-clicking on the **OrCAD 10.0 DEMO/CAPTURE CIS** icon opens the window. A new project is initiated by selecting the **Create document** key at the top left of the screen (it looks like a page with a star in the upper left corner). The result is the **New Project** dialog box in which **PSpice 5–1** is entered as the **Name.** The **Analog or Mixed A/D** is already selected, and **PSpice** appears as the **Location.** Click **OK,** and the **Create PSpice Project** dialog box appears. Select **Create a blank project,** click **OK,** and the working windows appear. Grab the left edge of the **SCHEMATIC1: PAGE1** window to move it to the right so that you can see both screens. Clicking the + sign in the **Project Manager** window allows you to set the sequence down to **PAGE1.** You can change the name of the **SCHEMATIC1** by selecting it and right-clicking. Choose **Rename** from the list. Enter **PSpice 5–1** in the **Rename Schematic** dialog box. In Fig. 5.83 it was left as **SCHEMATIC1.**

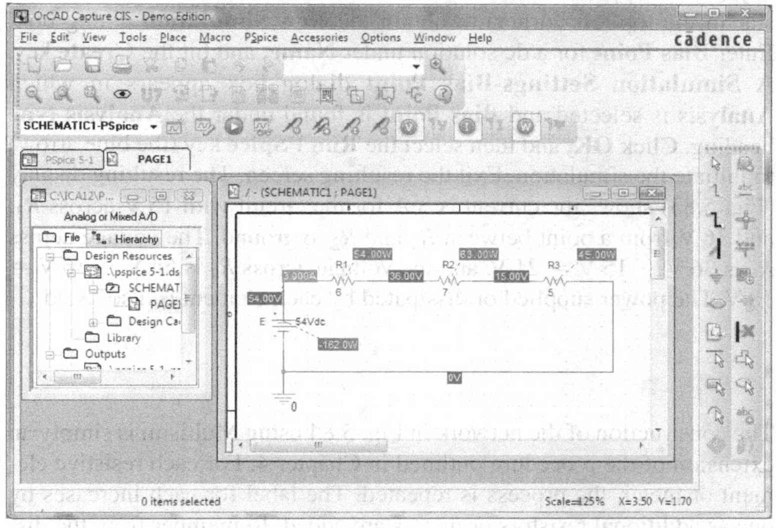

FIG. 5.83
Applying PSpice to a series dc circuit

This next step is important. If the toolbar on the right does not appear, left-click anywhere on the **SCHEMATIC1:PAGE1** screen. To start building the circuit, select the **Place part** key to open the **Place Part** dialog box. Note that the **SOURCE** library is already in place in the **Library** list (from the efforts of Chapter 4). Selecting **SOURCE** results in the list of sources under **Part List,** and **VDC** can be selected. Click **OK,** and the cursor can put it in place with a single left click. Right-click and select **End Mode** to end the process since the network has only one source. One more left click, and the source is in place. Select the **Place a Part** key again, followed by **ANALOG** library to find the resistor **R.** Once the resistor has been selected, click **OK** to place it next to the cursor on the screen. This time, since three resistors need to be placed, there is no need to go to **End Mode** between depositing each. Simply click one in place, then the next, and finally the third. Then right-click to end the process with **End Mode.** Finally, add a **GND** by selecting the ground key from the right toolbar and selecting **0/SOURCE** in the **Place Ground** dialog box. Click **OK,** and place the ground as shown in Fig. 5.83.

Connect the elements by using the **Place a wire** key to obtain the crosshair on the screen. Start at the top of the voltage source with a left click, and draw the wire, left-clicking it at every 90° turn. When a wire is connected from one element to another, move on to the next connection to be made—there is no need to go **End Mode** between connections. Now set the labels and values by double-clicking on each parameter to obtain a **Display Properties** dialog box. Since the dialog box appears with the quantity of interest in a blue background, type in the desired label or value, followed by **OK.** The network is now complete and ready to be analyzed.

Before simulation, select the **V, I,** and **W** in the toolbar at the top of the window to ensure that the voltages, currents, and power are displayed on the screen. To simulate, select the **New Simulation Profile** key (which appears as a data sheet on the second toolbar down with a star in the top left corner) to obtain the **New Simulation** dialog box. Enter **Bias Point** for a dc solution under **Name,** and hit the **Create** key. A **Simulation Settings-Bias Point** dialog box appears in which **Analysis** is selected and **Bias Point** is found under the **Analysis type** heading. Click **OK,** and then select the **Run PSpice** key (the blue arrow) to initiate the simulation. Exit the resulting screen. The resulting display (Fig. 5.83) shows the current is 3 A for the circuit with 15 V across R_3, and 36 V from a point between R_1 and R_2 to ground. The voltage across R_2 is $36\text{ V} - 15\text{ V} = 21\text{ V}$, and the voltage across R_1 is $54\text{ V} - 36\text{ V} = 18\text{ V}$. The power supplied or dissipated by each element is also listed.

Multisim

The construction of the network in Fig. 5.84 using Multisim is simply an extension of the procedure outlined in Chapter 4. For each resistive element or meter, the process is repeated. The label for each increases by one as additional resistors or meters are added. Remember from the discussion of Chapter 4 that you should add the meters before connecting the elements together because the meters take space and must be properly oriented. The current is determined by the **XMM1** ammeter and the voltages by **XMM2** through **XMM5.** Of particular importance, note that

in Multisim the meters are connected in exactly the same way they would be placed in an active circuit in the laboratory. Ammeters are in series with the branch in which the current is to be determined, and voltmeters are connected between the two points of interest (across resistors). In addition, for positive readings, ammeters are connected so that conventional current enters the positive terminal, and voltmeters are connected so that the point of higher potential is connected to the positive terminal.

The meter settings are made by double-clicking on the meter symbol on the schematic. In each case, **V** or **I** had to be chosen, but the horizontal line for dc analysis is the same for each. Again, you can select the **Set** key to see what it controls, but the default values of meter input resistance

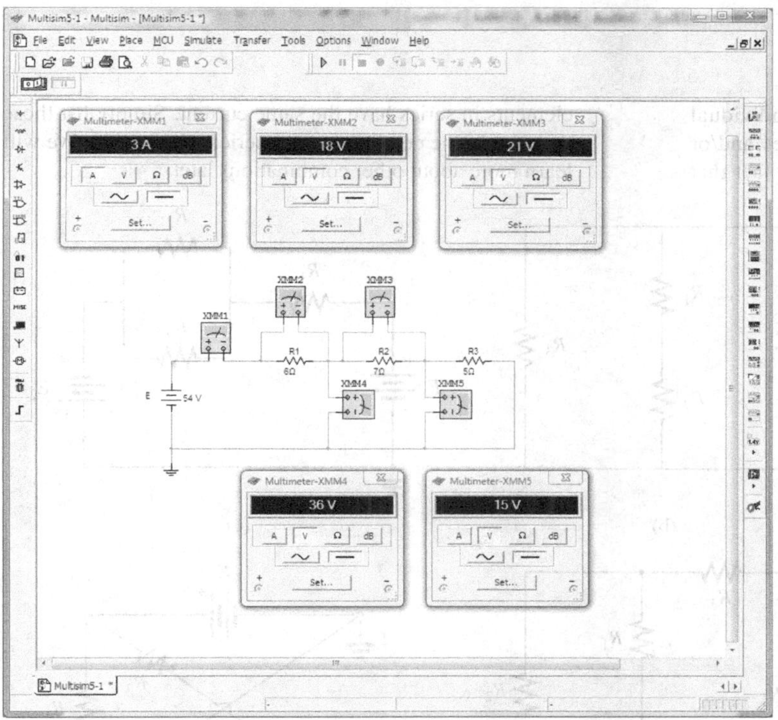

FIG. 5.84
Applying Multisim to a series dc circuit.

levels are fine for all the analyses described in this text. Leave the meters on the screen so that the various voltages and the current level will be displayed after the simulation.

Recall from Chapter 4 that elements can be moved by simply clicking on each schematic symbol and dragging it to the desired location. The same is true for labels and values. Labels and values are set by double-clicking on the label or value and entering your preference. Click **OK,** and they are changed on the schematic. You do not have to first select a special key to connect the elements. Simply bring the cursor to the starting point to generate the small circle and crosshair. Click on the starting point, and follow the desired path to the next connection path. When in the correct location, click again, and the line appears. All connecting lines can make 90° turns. However, you cannot follow a diagonal path from one point to another. To remove any element, label, or line, click on the quantity to obtain the four-square active status, and select the **Delete** key or the scissors key on the top menu bar.

Recall from Chapter 4 that you can initiate simulation through the sequence **Simulate-Run** by selecting the green **Run** key or switching the **Simulate Switch** to the **1** position.

Note from the results that the sum of the voltages measured by XMM2 and XMM4 equals the applied voltage. All the meters are considered ideal, so there is no voltage drop across the XMM1 ammeter. In addition, they do not affect the value of the current measured by XMM1. All the voltmeters have essentially infinite internal resistance, while the ammeters all have zero internal resistance. Of course, the meters can be entered as anything but ideal using the **Set** option. Note also that the sum of the voltages measured by XMM3 and XMM5 equals that measured by XMM4 as required by Kirchhoff's voltage law.

PROBLEMS

SECTION 5.2 Series Resistors

1. For each configuration in Fig. 5.85, find the individual (not combinations of) elements (voltage sources and/or resistors) that are in series. If necessary, use the fact that elements in series have the same current. Simply list those that satisfy the conditions for a series relationship. We will learn more about other combinations later.

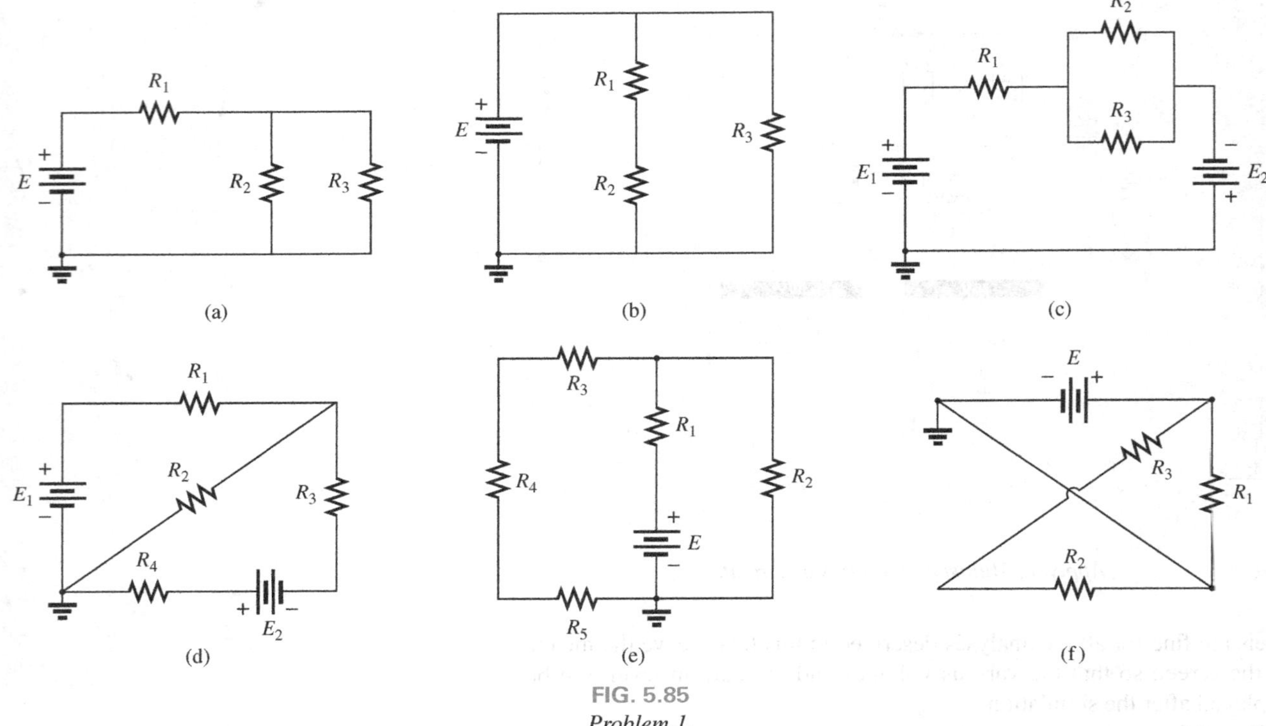

FIG. 5.85
Problem 1.

2. Find the total resistance R_T for each configuration in Fig. 5.86. Note that only standard resistor values were used.

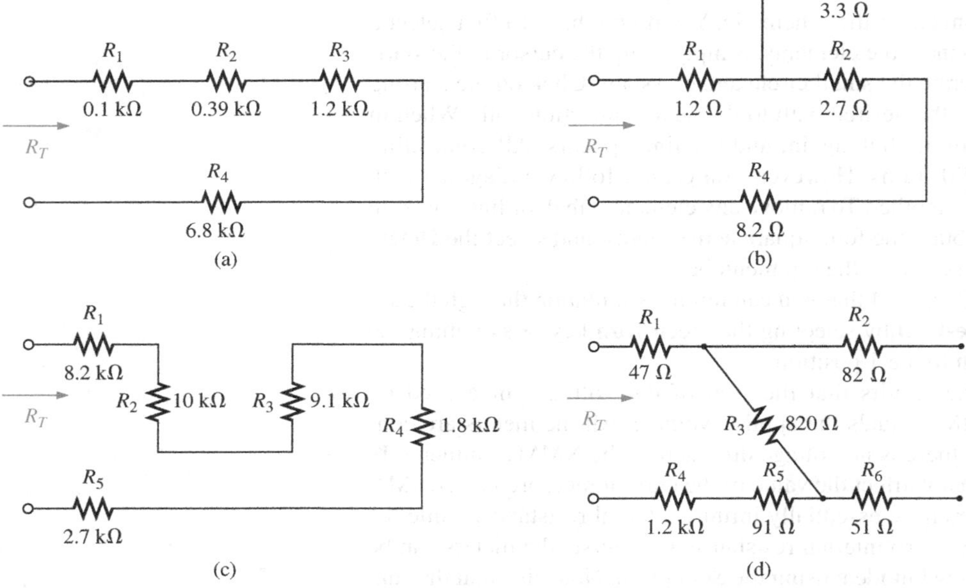

FIG. 5.86
Problem 2.

-S-

3. For each circuit board in Fig. 5.87, find the total resistance between connection tabs 1 and 2.

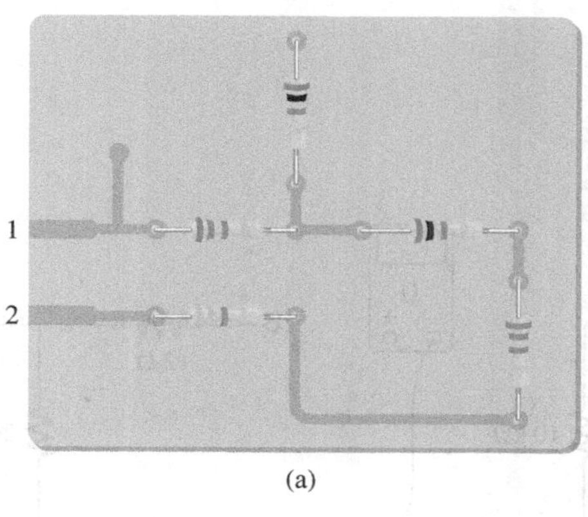

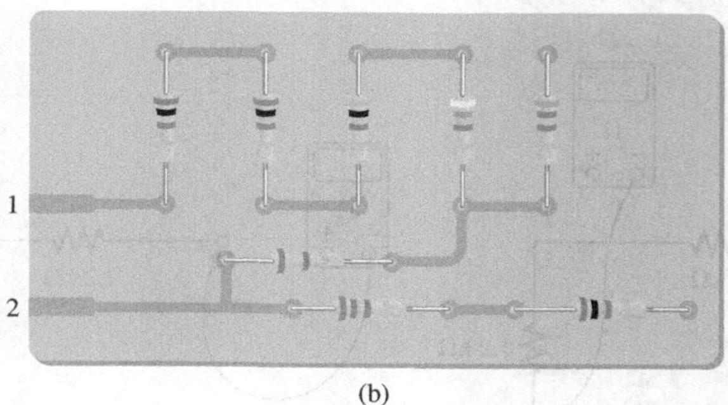

FIG. 5.87
Problem 3.

4. For the circuit in Fig. 5.88, composed of standard values:
   a. Which resistor will have the most impact on the total resistance?
   b. On an approximate basis, which resistors can be ignored when determining the total resistance?
   c. Find the total resistance, and comment on your results for parts (a) and (b).
5. For each configuration in Fig. 5.89, determine the ohmmeter reading.
6. Find the resistance R, given the ohmmeter reading for each configuration of Fig. 5.90.

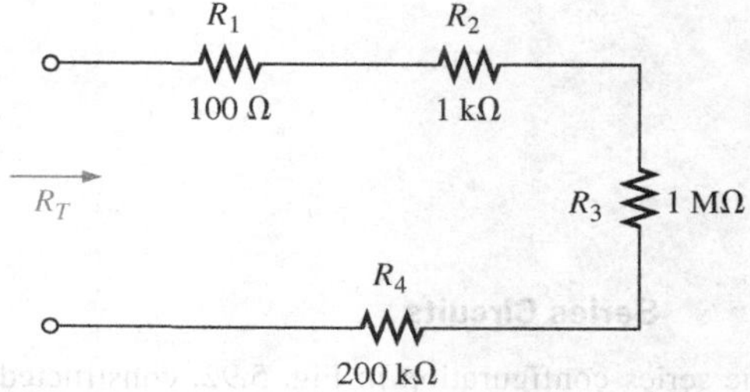

FIG. 5.88
Problem 4.

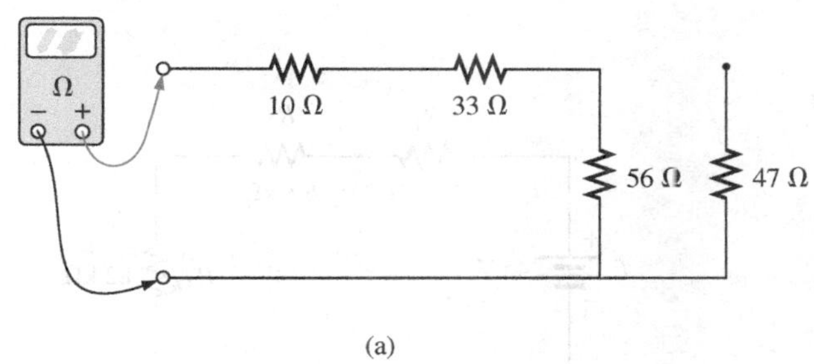

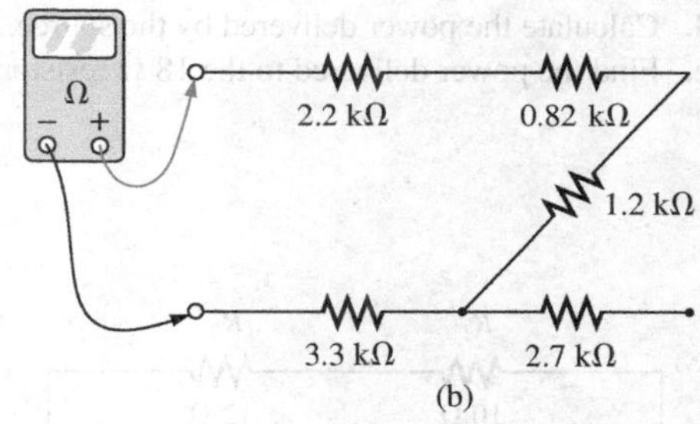

FIG. 5.89
Problem 5.

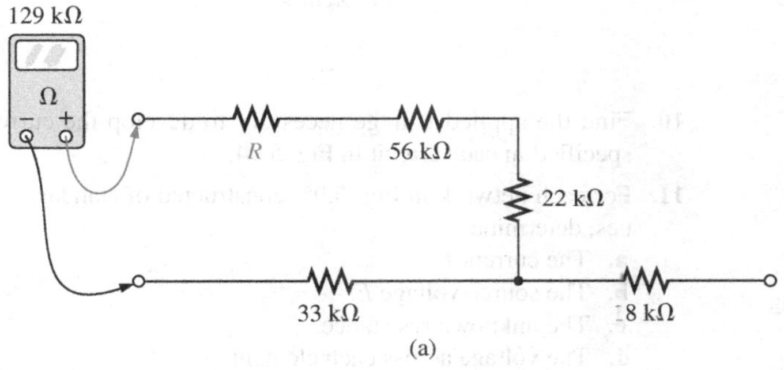

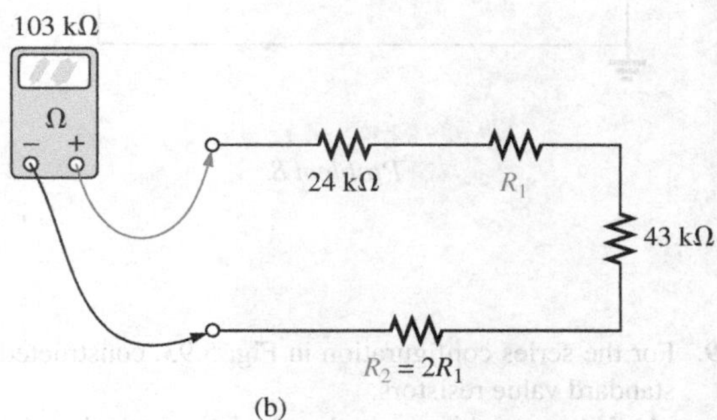

FIG. 5.90
Problem 6.

7. What is the ohmmeter reading for each configuration in Fig. 5.91?

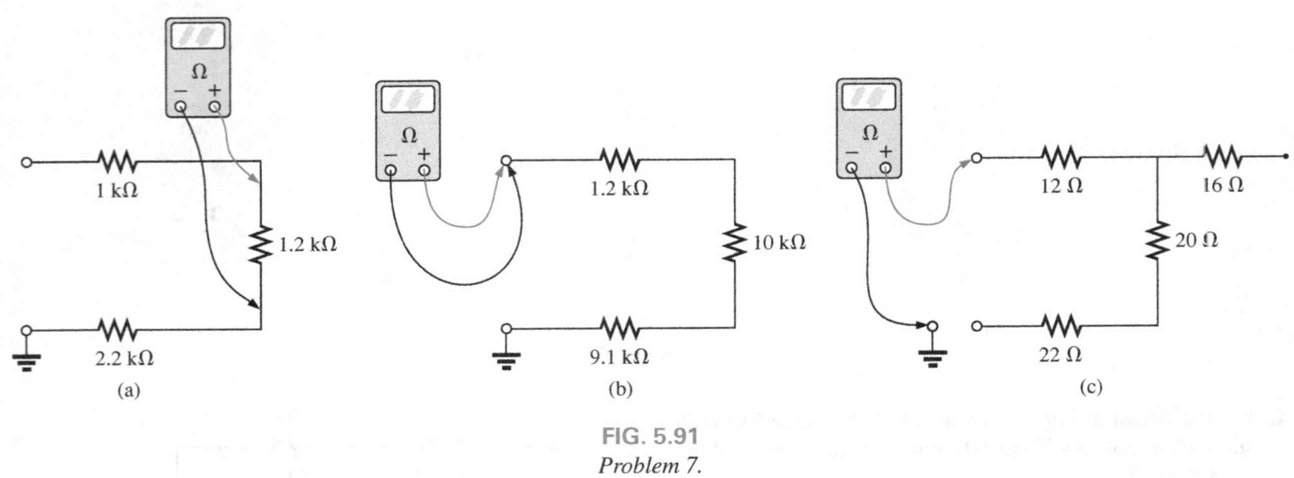

FIG. 5.91
Problem 7.

SECTION 5.3 Series Circuits

8. For the series configuration in Fig. 5.92, constructed of standard values:
   a. Find the total resistance.
   b. Calculate the current.
   c. Find the voltage across each resistive element.
   d. Calculate the power delivered by the source.
   e. Find the power delivered to the 18 Ω resistor.

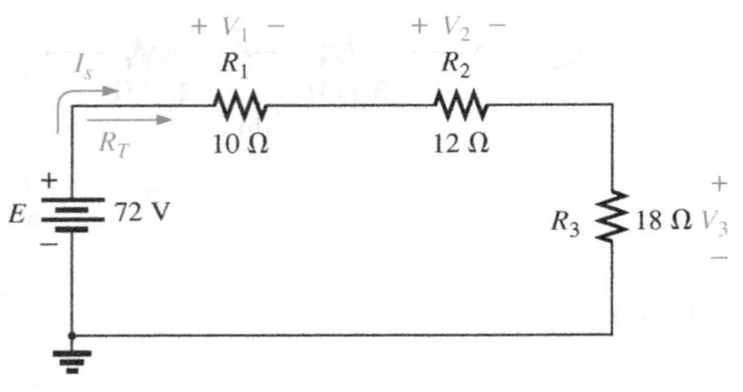

FIG. 5.92
Problem 8.

9. For the series configuration in Fig. 5.93, constructed using standard value resistors:
   a. Without making a single calculation, which resistive element will have the most voltage across it? Which will have the least?
   b. Which resistor will have the most impact on the total resistance and the resulting current? Find the total resistance and the current.
   c. Find the voltage across each element and review your response to part (a).

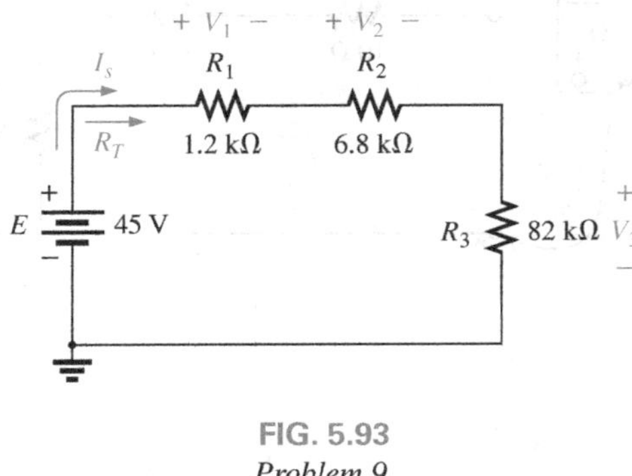

FIG. 5.93
Problem 9.

10. Find the applied voltage necessary to develop the current specified in each circuit in Fig. 5.94.
11. For each network in Fig. 5.95, constructed of standard values, determine:
    a. The current *I*.
    b. The source voltage *E*.
    c. The unknown resistance.
    d. The voltage across each element.
12. For each configuration in Fig. 5.96, what are the readings of the ammeter and the voltmeter?

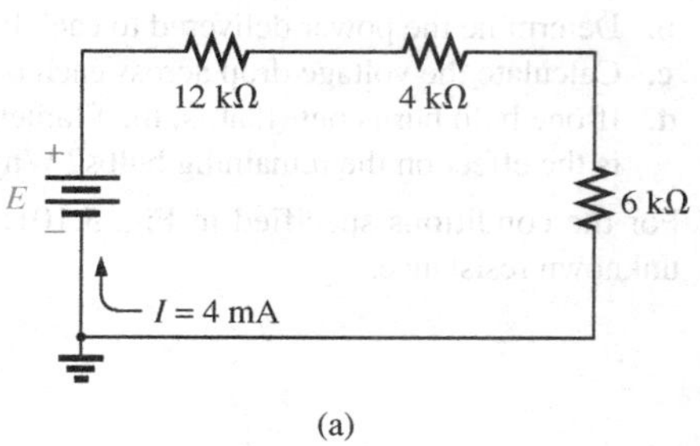

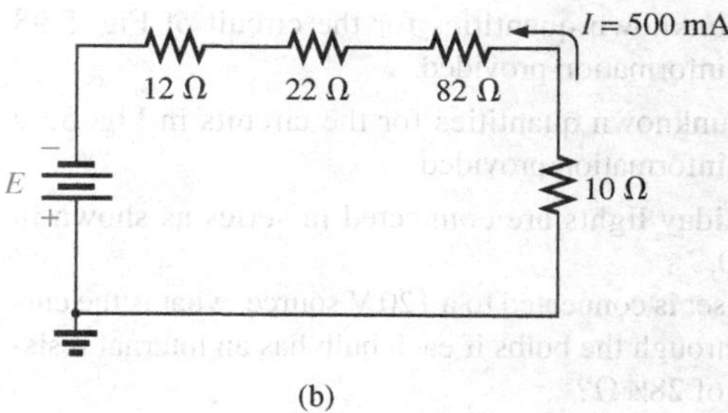

FIG. 5.94
Problem 10.

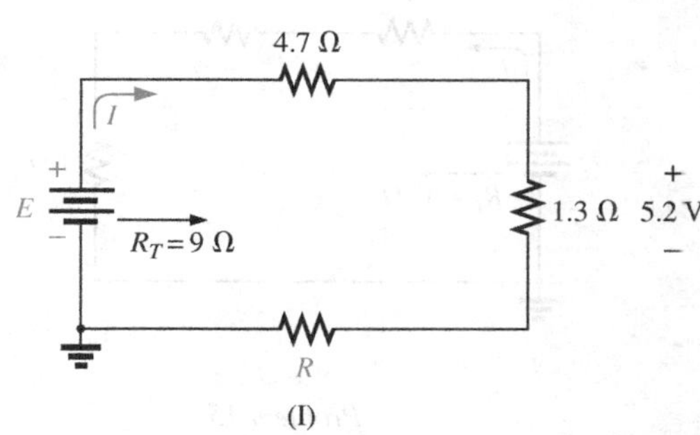

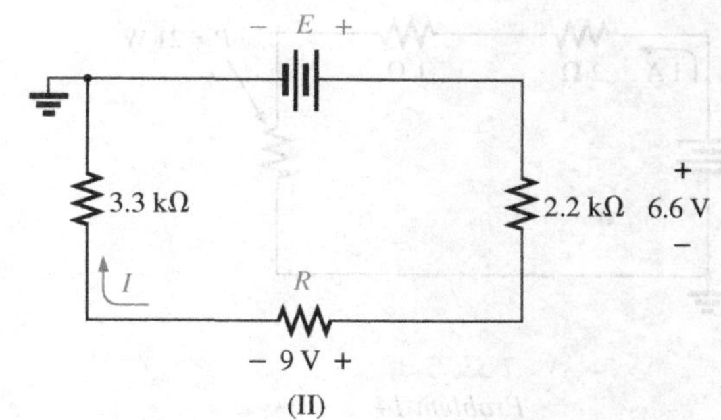

FIG. 5.95
Problem 11.

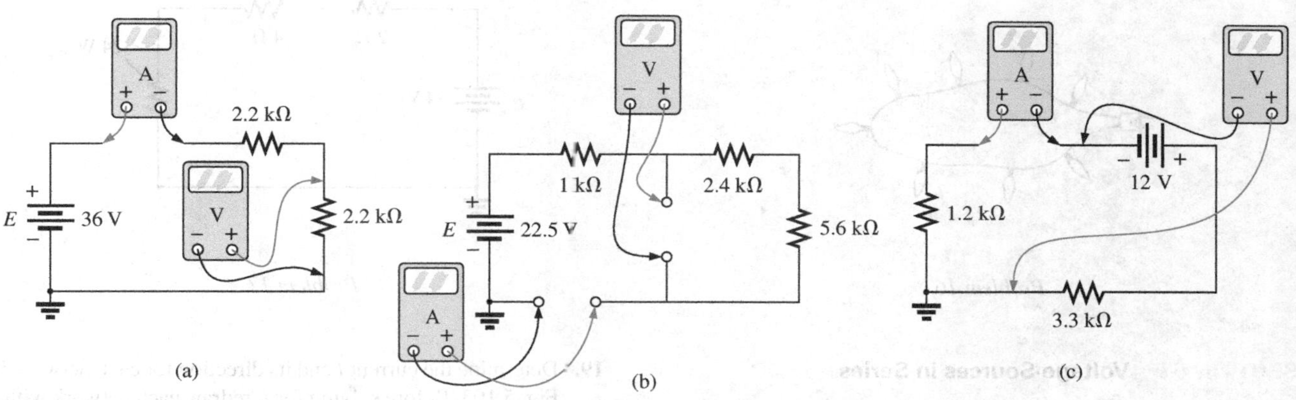

FIG. 5.96
Problem 12.

SECTION 5.4 Power Distribution in a Series Circuit

13. For the circuit in Fig. 5.97, constructed of standard value resistors:

a. Find the total resistance, current, and voltage across each element.

b. Find the power delivered to each resistor.

c. Calculate the total power delivered to all the resistors.

d. Find the power delivered by the source.

e. How does the power delivered by the source compare to that delivered to all the resistors?

f. Which resistor received the most power? Why?

g. What happened to all the power delivered to the resistors?

h. If the resistors are available with wattage ratings of 1/2 W, 1 W, 2 W, and 5 W, what minimum wattage rating can be used for each resistor?

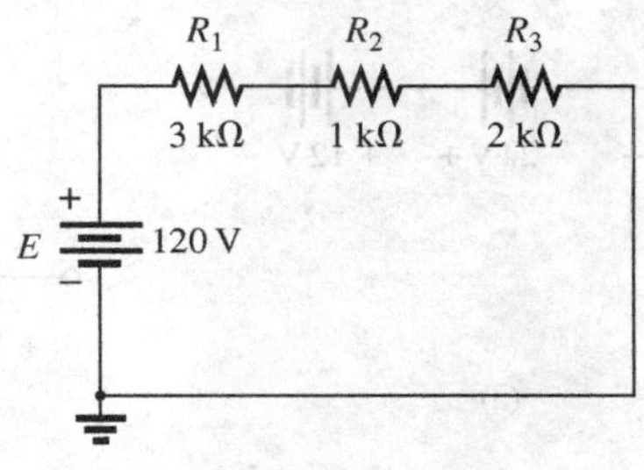

FIG. 5.97
Problem 13.

14. Find the unknown quantities for the circuit of Fig. 5.98 using the information provided.

15. Find the unknown quantities for the circuits in Fig. 5.99 using the information provided.

16. Eight holiday lights are connected in series as shown in Fig. 5.100.
    a. If the set is connected to a 120 V source, what is the current through the bulbs if each bulb has an internal resistance of 28⅛ Ω?
    b. Determine the power delivered to each bulb.
    c. Calculate the voltage drop across each bulb.
    d. If one bulb burns out (that is, the filament opens), what is the effect on the remaining bulbs? Why?

*17. For the conditions specified in Fig. 5.101, determine the unknown resistance.

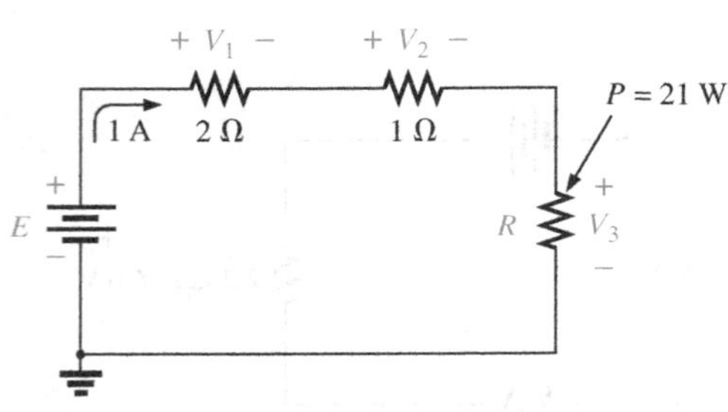

FIG. 5.98
Problem 14.

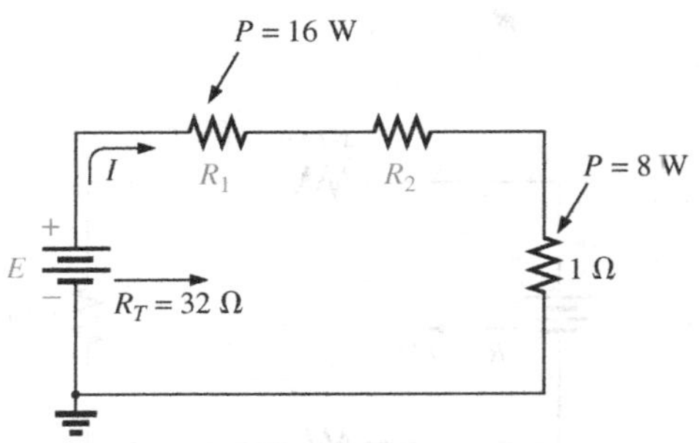

FIG. 5.99
Problem 15.

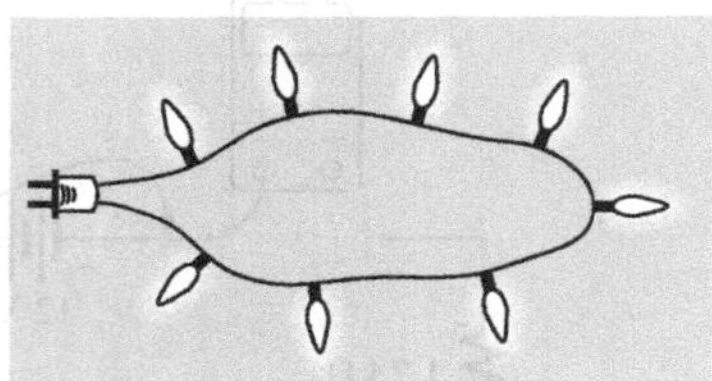

FIG. 5.100
Problem 16.

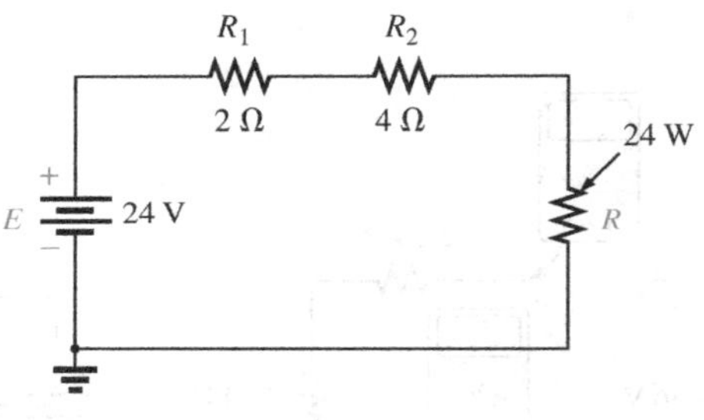

FIG. 5.101
Problem 17.

SECTION 5.5 Voltage Sources in Series

18. Combine the series voltage sources in Fig. 5.102 into a single voltage source between points *a* and *b*.

19. Determine the current *I* and its direction for each network in Fig. 5.103. Before solving for *I*, redraw each network with a single voltage source.

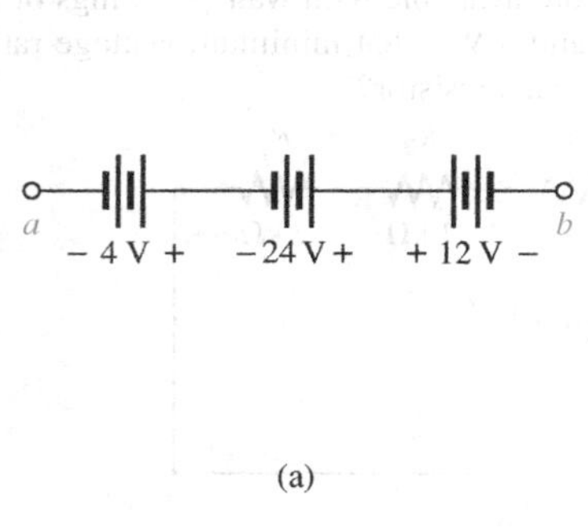

(a)

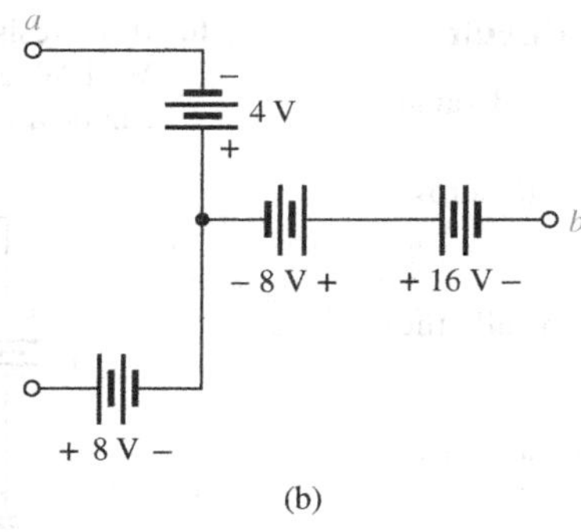

(b)

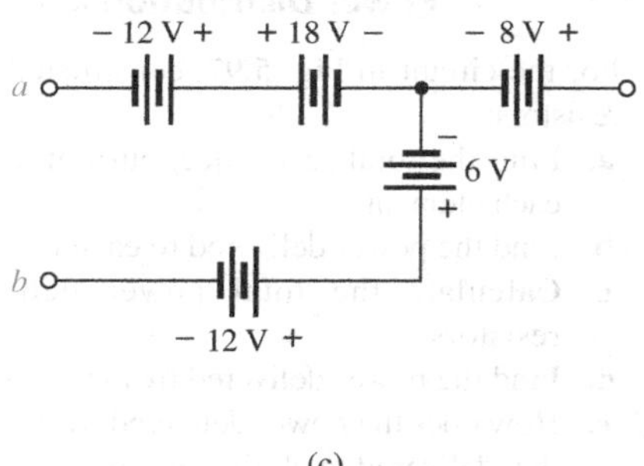

(c)

FIG. 5.102
Problem 18.

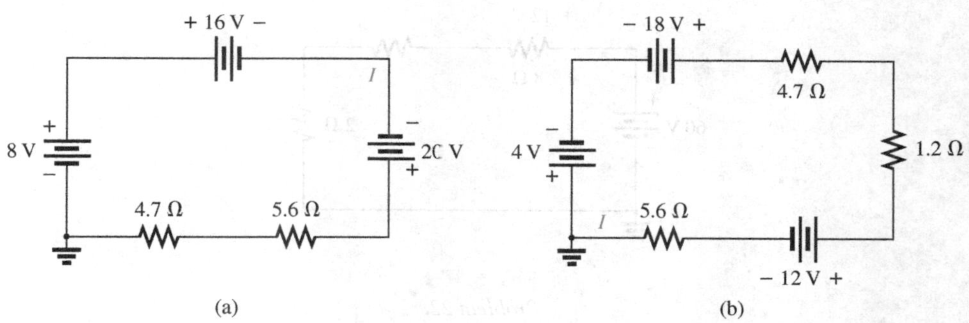

FIG. 5.103
Problem 19.

20. Find the unknown voltage source and resistor for the networks in Fig. 5.104. First combine the series voltage sources into a single source. Indicate the direction of the resulting current.

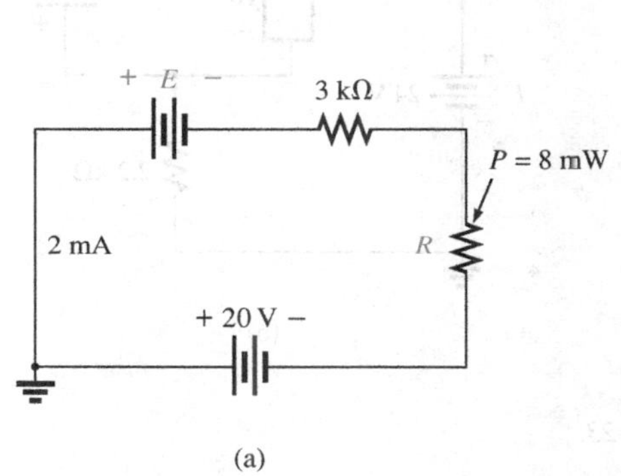

FIG. 5.104
Problem 20.

SECTION 5.6 Kirchhoff's Voltage Law

21. Using Kirchhoff's voltage law, find the unknown voltages for the circuits in Fig. 5.105.

22. **a.** Find the current I for the network of Fig. 5.106.
b. Find the voltage V_2.
c. Find the voltage V_1 using Kirchhoff's voltage law.

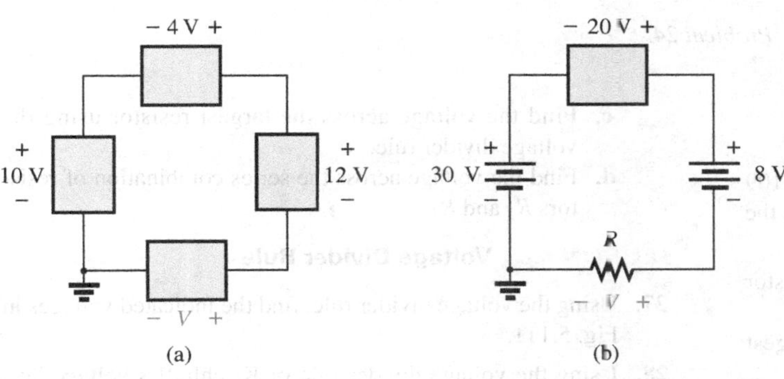

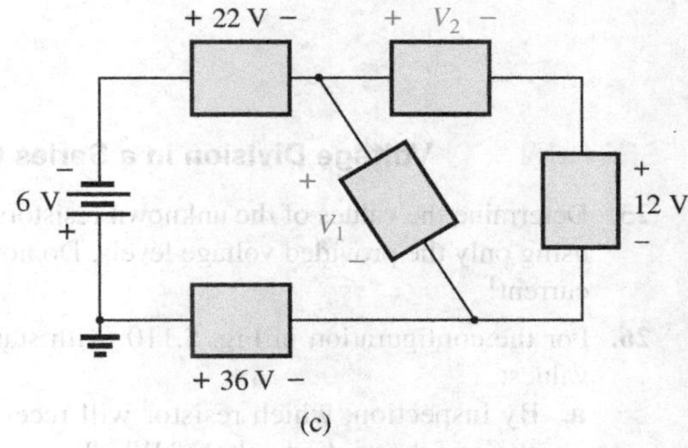

FIG. 5.105
Problem 21.

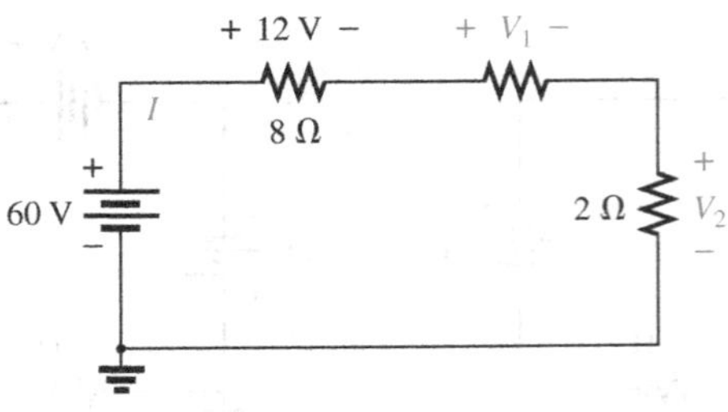

FIG. 5.106
Problem 22.

23. Using Kirchhoff's voltage law, determine the unknown voltages for the series circuits in Fig. 5.107.

24. Using Kirchhoff's voltage law, find the unknown voltages for the configurations in Fig. 5.108.

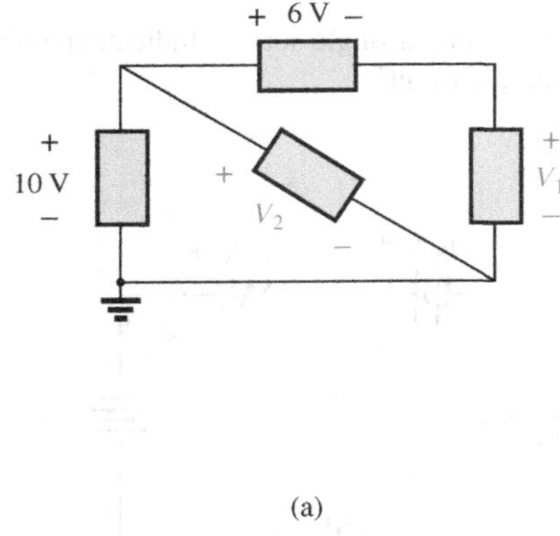

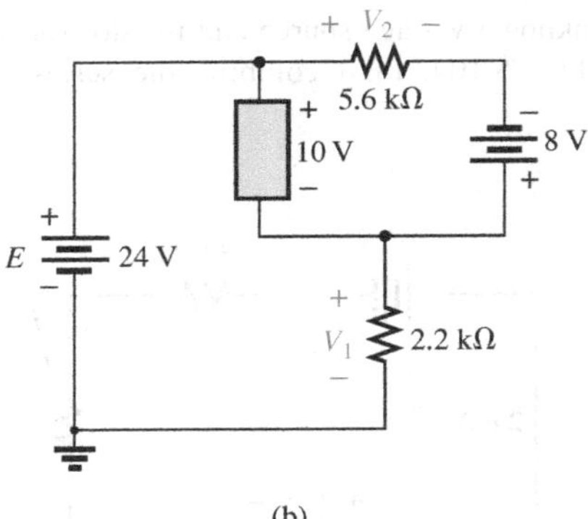

FIG. 5.107
Problem 23.

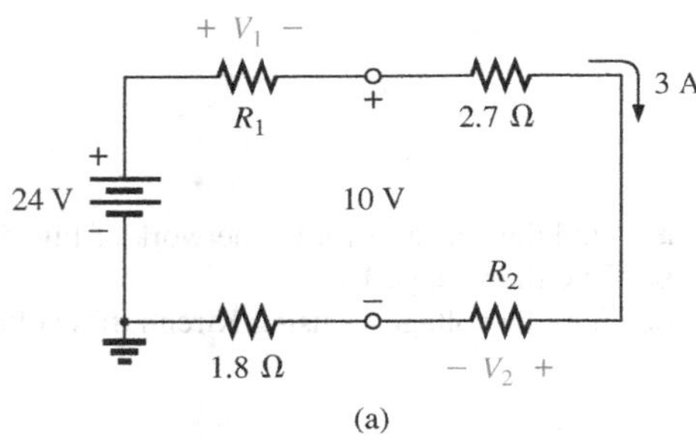

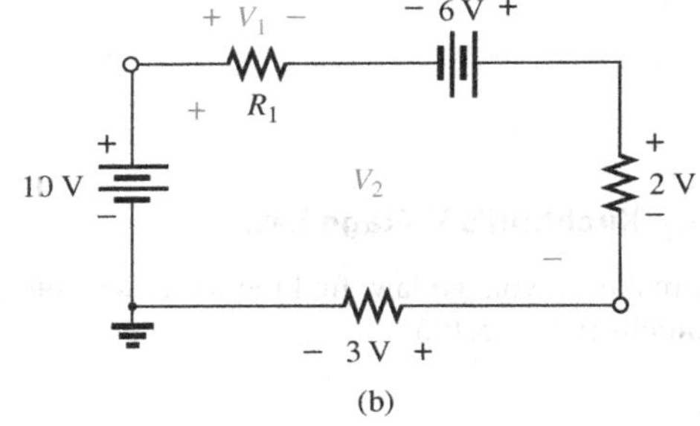

FIG. 5.108
Problem 24.

SECTION 5.7 Voltage Division in a Series Circuit

25. Determine the values of the unknown resistors in Fig. 5.109 using only the provided voltage levels. Do not calculate the current!

26. For the configuration in Fig. 5.110, with standard resistor values:

a. By inspection, which resistor will receive the largest share of the applied voltage? Why?

b. How much larger will voltage V_3 be compared to V_2 and V_1?

c. Find the voltage across the largest resistor using the voltage divider rule.

d. Find the voltage across the series combination of resistors R_2 and R_3.

SECTION 5.8 Voltage Divider Rule

27. Using the voltage divider rule, find the indicated voltages in Fig. 5.111.

28. Using the voltage divider rule or Kirchhoff's voltage law, determine the unknown voltages for the configurations in Fig. 5.112. Do not calculate the current!

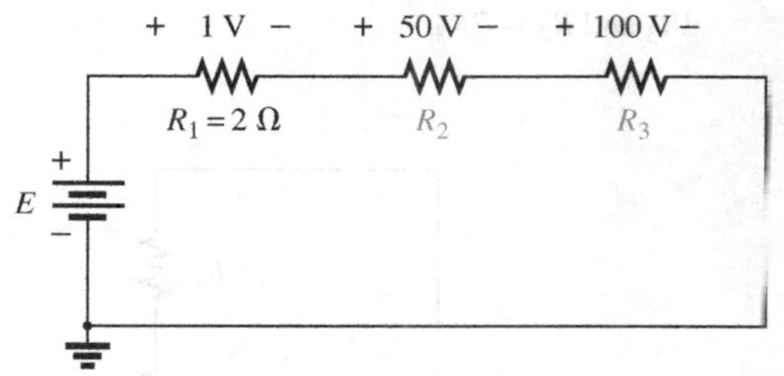

FIG. 5.109
Problem 25.

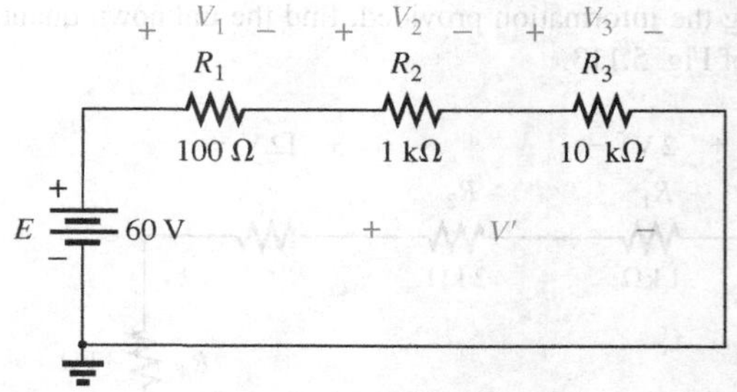

FIG. 5.110
Problem 26.

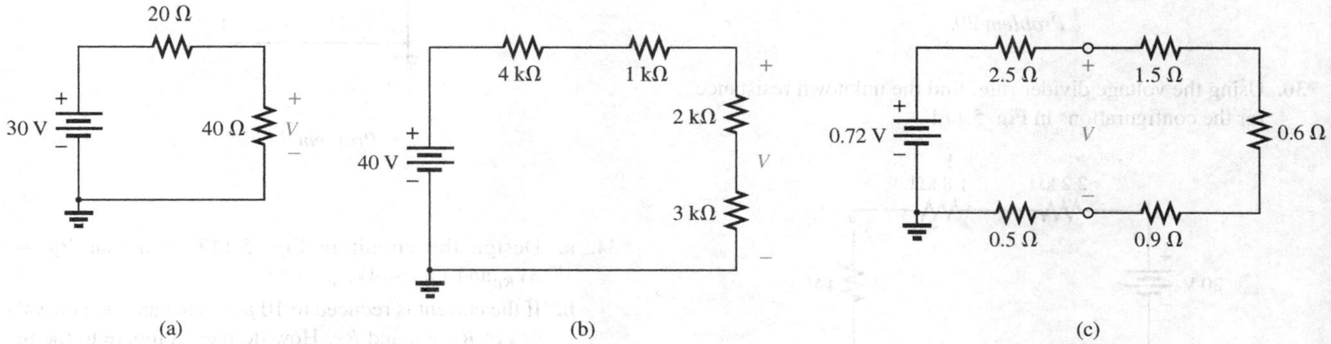

FIG. 5.111
Problem 27.

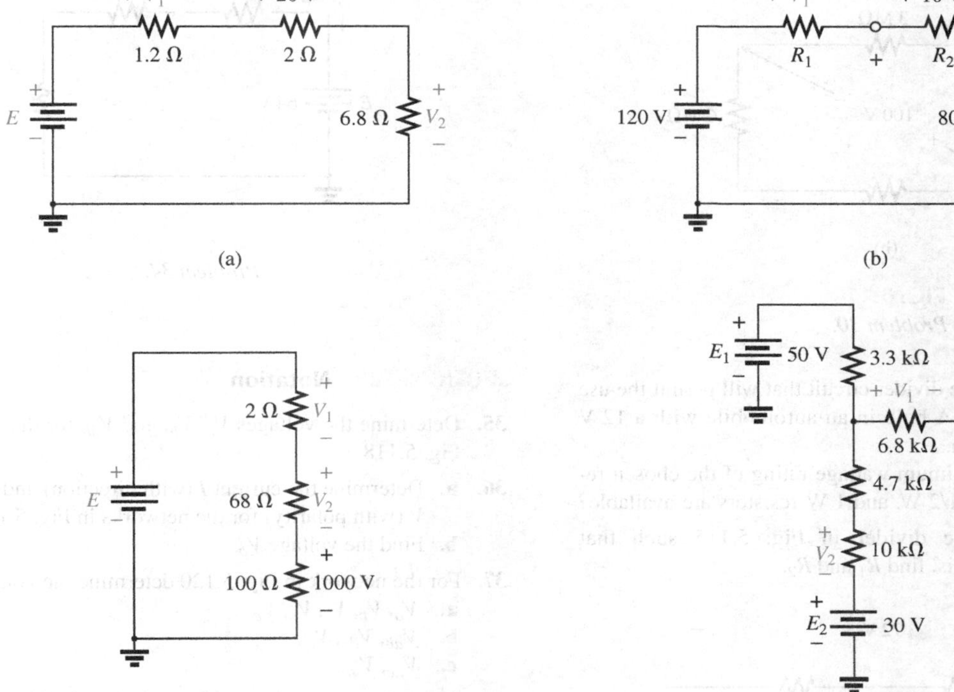

FIG. 5.112
Problem 28.

29. Using the information provided, find the unknown quantities of Fig. 5.113.

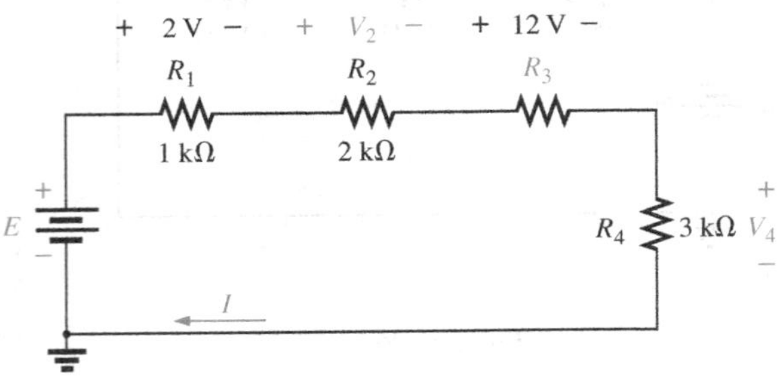

FIG. 5.113
Problem 29.

***30.** Using the voltage divider rule, find the unknown resistance for the configurations in Fig. 5.114.

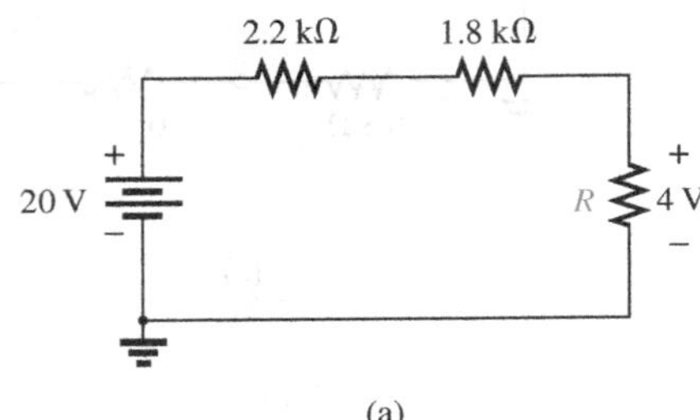

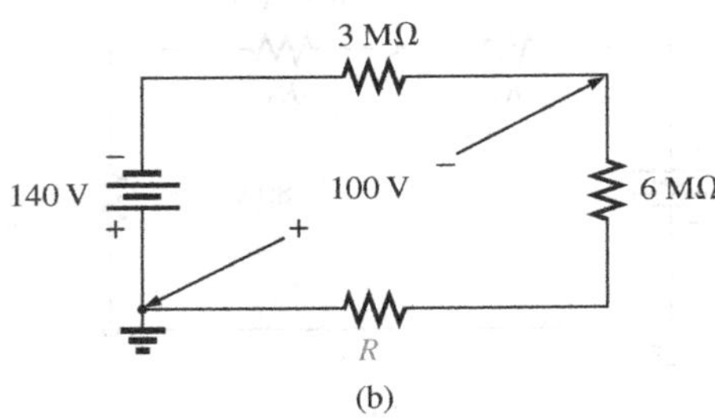

FIG. 5.114
Problem 30.

31. **a.** Design a voltage divider circuit that will permit the use of an 8 V, 50 mA bulb in an automobile with a 12 V electrical system.

b. What is the minimum wattage rating of the chosen resistor if 1/4 W, 1/2 W, and 1 W resistors are available?

***32.** Design the voltage divider in Fig. 5.115 such that $V_{R_1} = 1/5V_{R_1}$. That is, find R_1 and R_2.

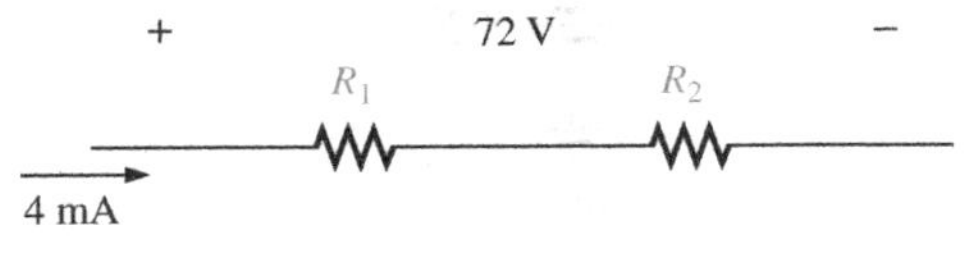

FIG. 5.115
Problem 32.

***33.** Find the voltage across each resistor in Fig. 5.116 if $R_1 = 2R_3$ and $R_2 = 7R_3$.

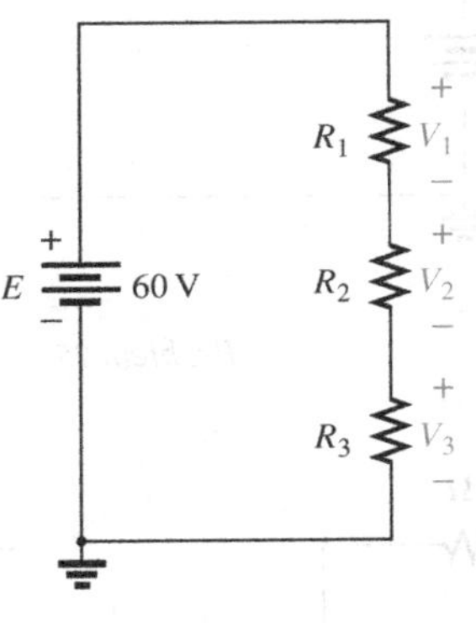

FIG. 5.116
Problem 33.

***34.** **a.** Design the circuit in Fig. 5.117 such that $V_{R_2} = 3V_{R_1}$ and $V_{R_3} = 4V_{R_2}$.

b. If the current is reduced to 10 μA, what are the new values of R_1, R_2, and R_3? How do they compare to the results of part (a)?

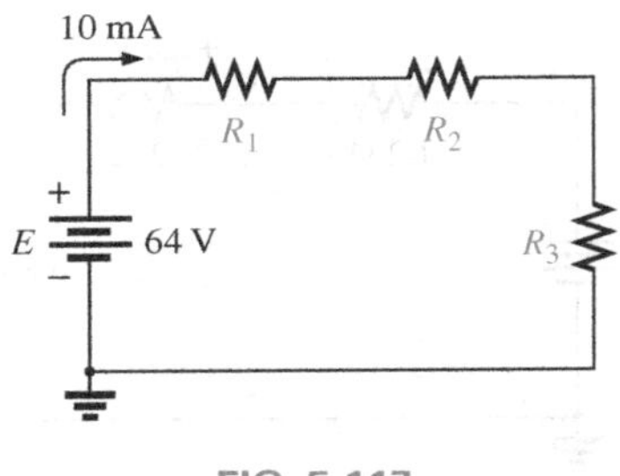

FIG. 5.117
Problem 34.

SECTION 5.10 **Notation**

35. Determine the voltages V_a, V_b, and V_{ab} for the networks in Fig. 5.118.

36. **a.** Determine the current I (with direction) and the voltage V (with polarity) for the networks in Fig. 5.119.

b. Find the voltage V_a.

37. For the network in Fig. 5.120 determine the voltages:

a. V_a, V_b, V_c, V_d, V_e

b. V_{ab}, V_{dc}, V_{cb}

c. V_{ac}, V_{db}

***38.** Given the information appearing in Fig. 5.121, find the level of resistance for R_1 and R_3.

39. Determine the values of R_1, R_2, R_3, and R_4 for the voltage divider of Fig. 5.122 if the source current is 16 mA.

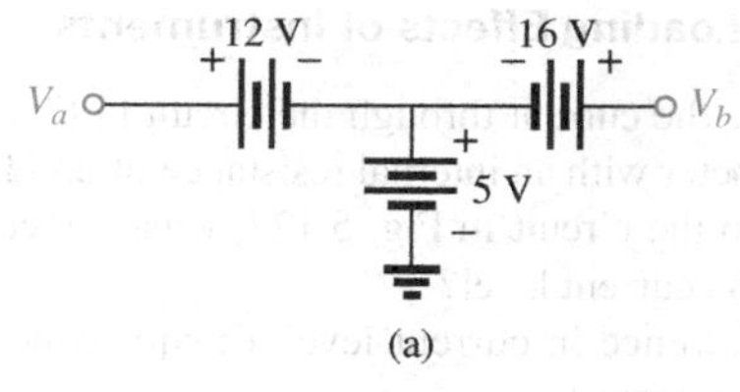

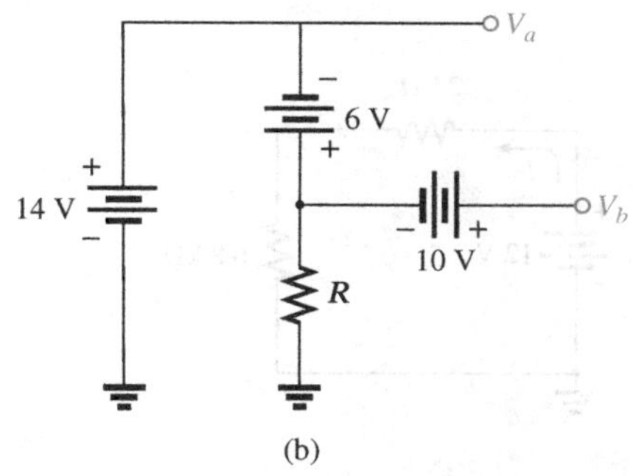

FIG. 5.118
Problem 35.

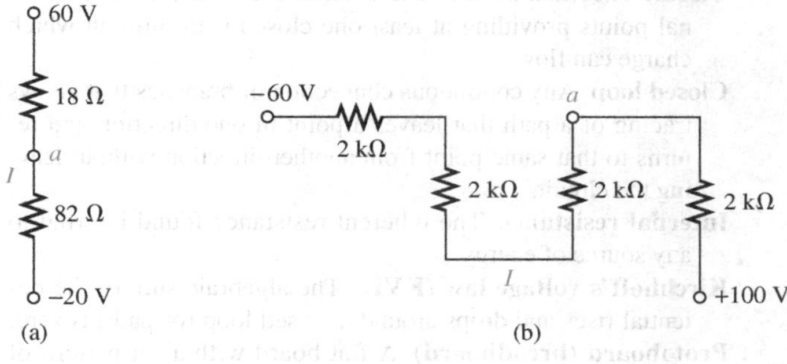

FIG. 5.119
Problem 36.

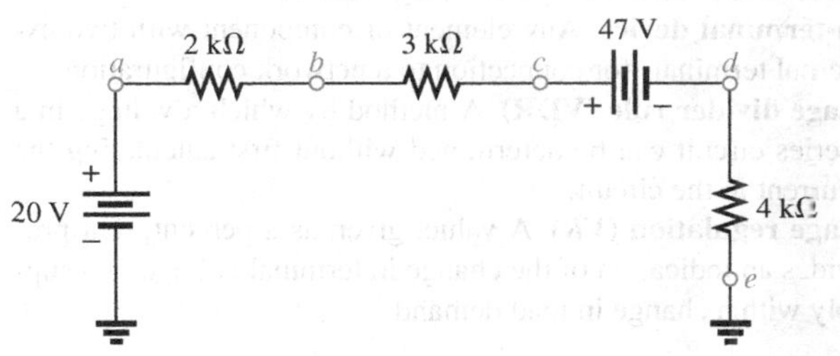

FIG. 5.120
Problem 37.

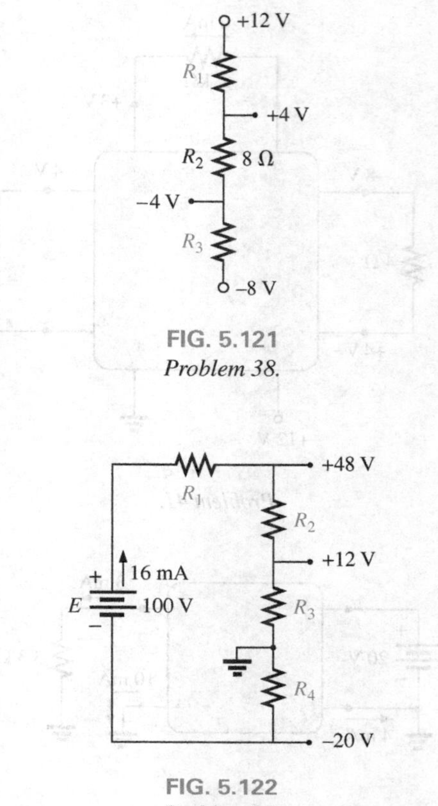

FIG. 5.121
Problem 38.

FIG. 5.122
Problem 39.

40. For the network in Fig. 5.123, determine the voltages:

a. V_a, V_b, V_c, V_d

b. V_{ab}, V_{cb}, V_{cd}

c. V_{ad}, V_{ca}

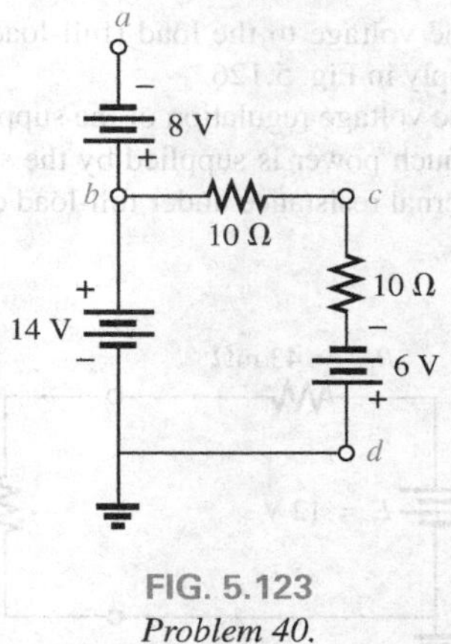

FIG. 5.123
Problem 40.

***41.** For the integrated circuit in Fig. 5.124, determine V_0, V_4, V_7, V_{10}, V_{23}, V_{30}, V_{67}, V_{56}, and I (magnitude and direction.)

***42.** For the integrated circuit in Fig. 5.125, determine V_0, V_{03}, V_2, V_{23}, V_{12}, and I_i.

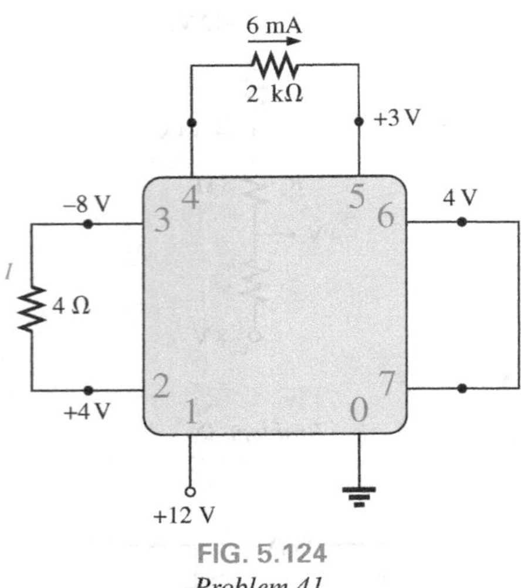

FIG. 5.124
Problem 41.

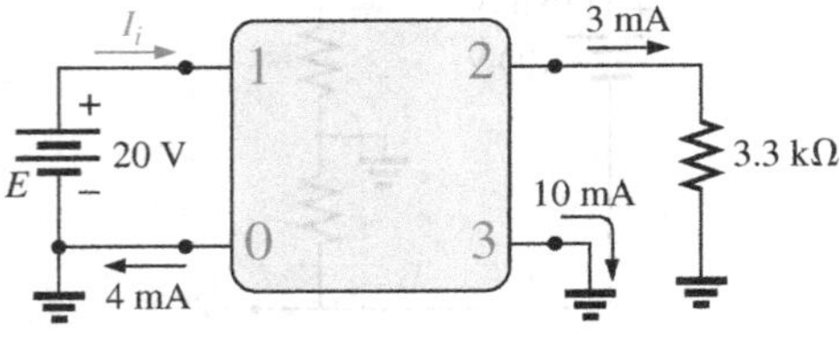

FIG. 5.125
Problem 42.

SECTION 5.9 Voltage Regulation and the Internal Resistance of Voltage Sources

43. a. Find the internal resistance of a battery that has a no-load output of 60 V and that supplies a full-load current of 2 A to a load of 28 Ω.

b. Find the voltage regulation of the supply.

44. a. Find the voltage to the load (full-load conditions) for the supply in Fig. 5.126.

b. Find the voltage regulation of the supply.

c. How much power is supplied by the source and lost to the internal resistance under full-load conditions?

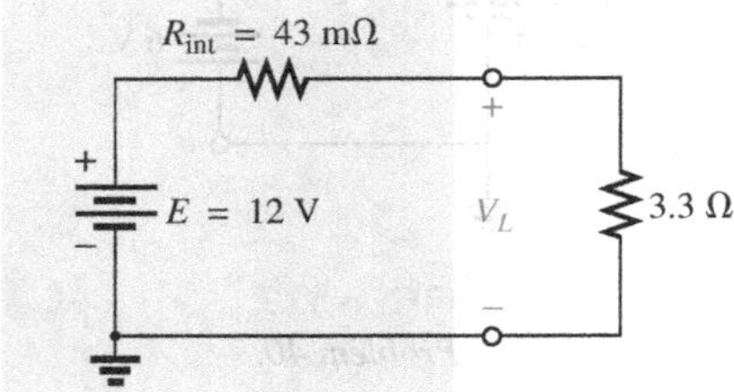

FIG. 5.126
Problem 44.

SECTION 5.10 Loading Effects of Instruments

45. a. Determine the current through the circuit in Fig. 5.127.

b. If an ammeter with an internal resistance of 250 Ω is inserted into the circuit in Fig. 5.127, what effect will it have on the current level?

c. Is the difference in current level a major concern for most applications?

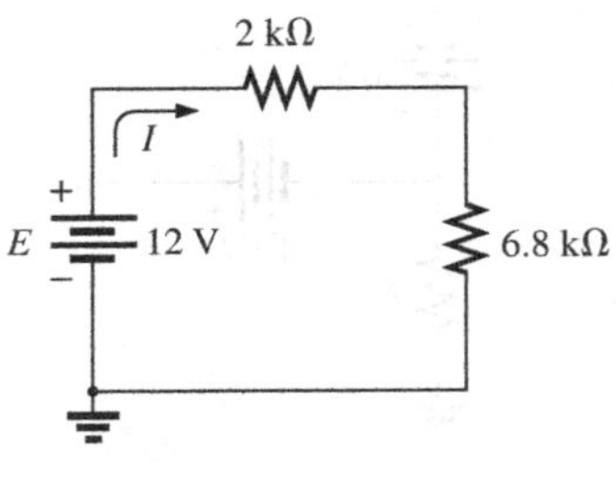

FIG. 5.127
Problem 45.

SECTION 5.11 Computer Analysis

46. Use the computer to verify the results of Example 5.4.

47. Use the computer to verify the results of Example 5.5.

48. Use the computer to verify the results of Example 5.15.

GLOSSARY

Circuit A combination of a number of elements joined at terminal points providing at least one closed path through which charge can flow.

Closed loop Any continuous connection of branches that allows tracing of a path that leaves a point in one direction and returns to that same point from another direction without leaving the circuit.

Internal resistance The inherent resistance found internal to any source of energy.

Kirchhoff's voltage law (KVL) The algebraic sum of the potential rises and drops around a closed loop (or path) is zero.

Protoboard (breadboard) A flat board with a set pattern of conductively connected holes designed to accept 24-gage wire and components with leads of about the same diameter.

Series circuit A circuit configuration in which the elements have only one point in common and each terminal is not connected to a third, current-carrying element.

Two-terminal device Any element or component with two external terminals for connection to a network configuration.

Voltage divider rule (VDR) A method by which a voltage in a series circuit can be determined without first calculating the current in the circuit.

Voltage regulation (*VR*) A value, given as a percent, that provides an indication of the change in terminal voltage of a supply with a change in load demand.

Parallel dc Circuits

6

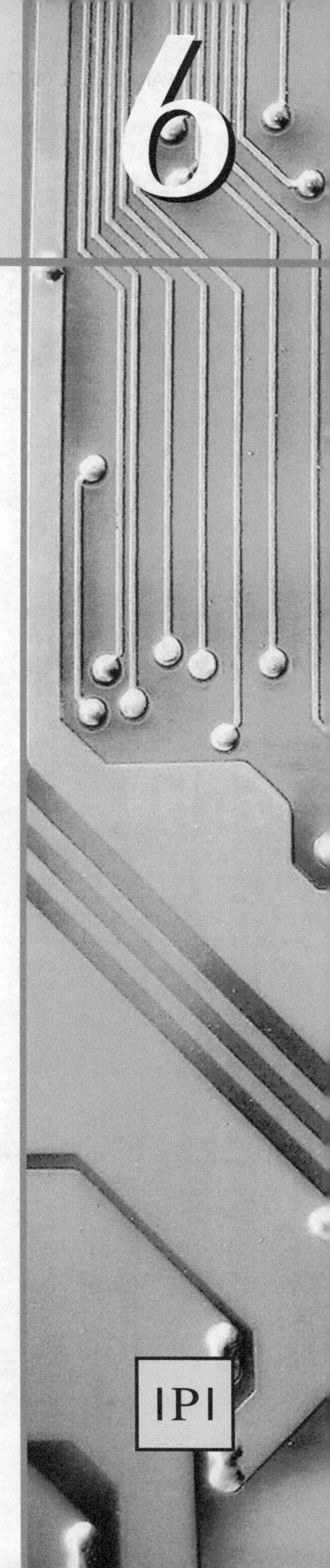

Objectives

- *Become familiar with the characteristics of a parallel network and how to solve for the voltage, current, and power to each element.*
- *Develop a clear understanding of Kirchhoff's current law and its importance to the analysis of electric circuits.*
- *Become aware of how the source current will split between parallel elements and how to properly apply the current divider rule.*
- *Clearly understand the impact of open and short circuits on the behavior of a network.*
- *Learn how to use an ohmmeter, voltmeter, and ammeter to measure the important parameters of a parallel network.*

6.1 INTRODUCTION

Two network configurations, series and parallel, form the framework for some of the most complex network structures. A clear understanding of each will pay enormous dividends as more complex methods and networks are examined. The series connection was discussed in detail in the last chapter. We will now examine the **parallel circuit** and all the methods and laws associated with this important configuration.

6.2 PARALLEL RESISTORS

The term *parallel* is used so often to describe a physical arrangement between two elements that most individuals are aware of its general characteristics.

In general,

two elements, branches, or circuits are in parallel if they have two points in common.

For instance, in Fig. 6.1(a), the two resistors are in parallel because they are connected at points a and b. If both ends were *not* connected as shown, the resistors would not be in parallel. In Fig. 6.1(b), resistors R_1 and R_2 are in parallel because they again have points a and b in common. R_1 is not in parallel with R_3 because they are connected at only one point (b). Further, R_1 and R_3 are not in series because a third connection appears at point b. The same can be said for resistors R_2 and R_3. In Fig. 6.1(c), resistors R_1 and R_2 are in series because they have only one point in common that is not connected elsewhere in the network. Resistors R_1 and R_3 are not in parallel because they have only point a in common. In addition, they are not in series because of the third connection to point a. The same can be said for resistors R_2 and R_3. In a broader context, it can be said that the series combination of resistors R_1 and R_2 is in parallel with resistor R_3 (more will be said about this option in Chapter 7). Furthermore, even though the discussion above was only for resistors, it can be applied to any two-terminal elements such as voltage sources and meters.

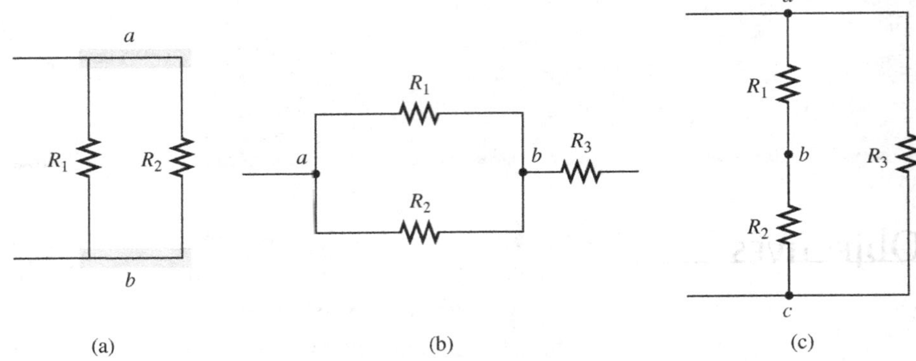

FIG. 6.1
(a) Parallel resistors; (b) R_1 and R_2 are in parallel; (c) R_3 is in parallel with the series combination of R_1 and R_2.

On schematics, the parallel combination can appear in a number of ways, as shown in Fig. 6.2. In each case, the three resistors are in parallel. They all have points *a* and *b* in common.

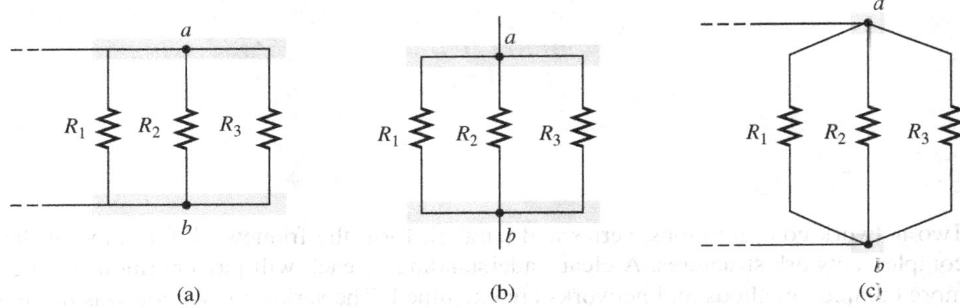

FIG. 6.2
Schematic representations of three parallel resistors.

For resistors in parallel as shown in Fig. 6.3, the total resistance is determined from the following equation:

$$\frac{1}{R_T} = \frac{1}{R_1} + \frac{1}{R_2} + \frac{1}{R_3} + \cdots + \frac{1}{R_N} \qquad \textbf{(6.1)}$$

Since $G = 1/R$, the equation can also be written in terms of conductance levels as follows:

$$G_T = G_1 + G_2 + G_3 + \cdots + G_N \qquad \text{(siemens, S)} \qquad \textbf{(6.2)}$$

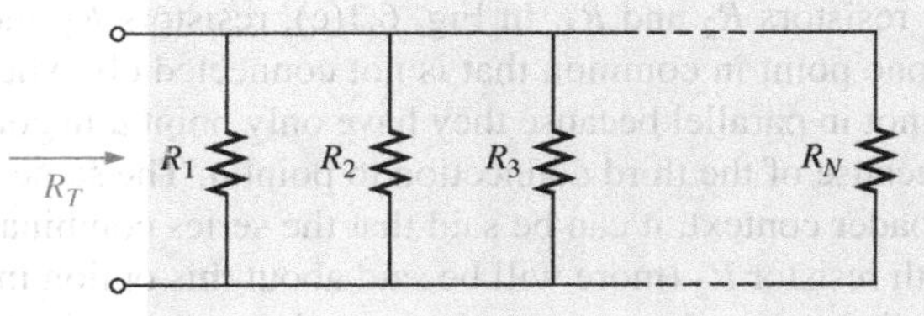

FIG. 6.3
Parallel combination of resistors.

which is an exact match in format with the equation for the total resistance of resistors in series: $R_T = R_1 + R_2 + R_3 + \cdots + R_N$. The result of this duality is that you can go from one equation to the other simply by interchanging R and G.

In general, however, when the total resistance is desired, the following format is applied:

$$R_T = \cfrac{1}{\cfrac{1}{R_1} + \cfrac{1}{R_2} + \cfrac{1}{R_3} + \cdots + \cfrac{1}{R_N}} \qquad \textbf{(6.3)}$$

Quite obviously, Eq. (6.3) is not as "clean" as the equation for the total resistance of series resistors. You must be careful when dealing with all the divisions into 1. The great feature about the equation, however, is that it can be applied to any number of resistors in parallel.

EXAMPLE 6.1

a. Find the total conductance of the parallel network in Fig. 6.4.
b. Find the total resistance of the same network using the results of part (a) and using Eq. (6.3).

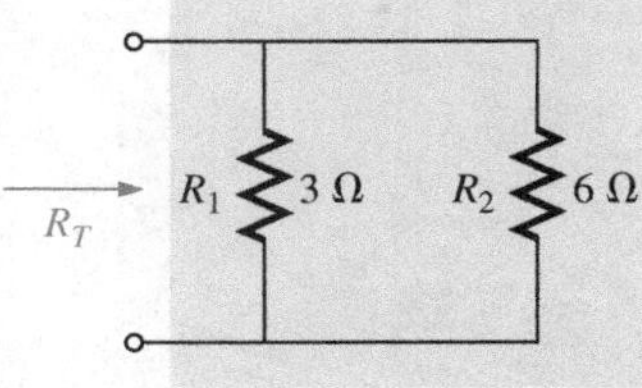

FIG. 6.4
Parallel resistors for Example 6.1.

Solutions:

a. $G_{1\pi} = \dfrac{1}{R_1} = \dfrac{1}{3\ \Omega} = 0.333\ \text{S}, \qquad G_2 = \dfrac{1}{R_2} = \dfrac{1}{6\ \Omega} = 0.167\ \text{S}$

and $G_T = G_1 + G_2 = 0.333\ \text{S} + 0.167\ \text{S} = \mathbf{0.5\ S}$

b. $R_T = \dfrac{1}{G_T} = \dfrac{1}{0.5\ \text{S}} = \mathbf{2\ \Omega}$

Applying Eq. (6.3) gives

$$R_T = \cfrac{1}{\cfrac{1}{R_1} + \cfrac{1}{R_2}} = \cfrac{1}{\cfrac{1}{3\ \Omega} + \cfrac{1}{6\ \Omega}}$$

$$= \frac{1}{0.333\ \text{S} + 0.167\ \text{S}} = \frac{1}{0.5\ \text{S}} = \mathbf{2\ \Omega}$$

EXAMPLE 6.2

a. By inspection, which parallel element in Fig. 6.5 has the least conductance? Determine the total conductance of the network and note whether your conclusion was verified.

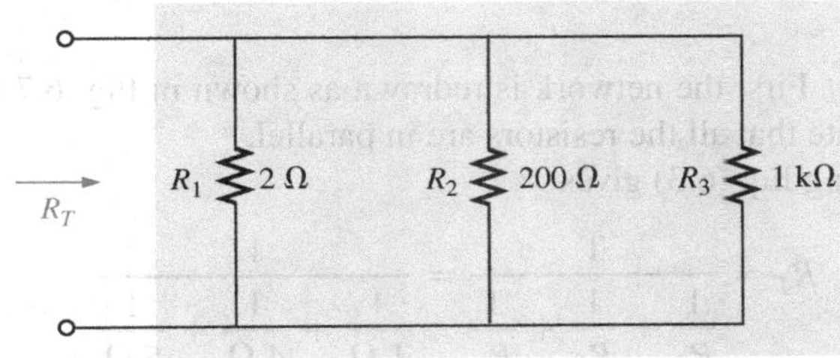

FIG. 6.5
Parallel resistors for Example 6.2.

b. Determine the total resistance from the results of part (a) and by applying Eq. (6.3).

Solutions:

a. Since the 1 kΩ resistor has the largest resistance and therefore the largest opposition to the flow of charge (level of conductivity), it will have the lowest level of conductance:

$$G_1 = \frac{1}{R_1} = \frac{1}{2\,\Omega} = 0.5\text{ S},\ G_2 = \frac{1}{R_2} + \frac{1}{200\,\Omega} = 0.005\text{ S} = 5\text{ mS}$$

$$G_3 = \frac{1}{R_3} = \frac{1}{1\text{ k}\Omega} = \frac{1}{1000\,\Omega} = 0.001\text{ S} = 1\text{ mS}$$

$$G_T = G_1 + G_2 + G_3 = 0.5\text{ S} + 5\text{ mS} + 1\text{ mS} = \mathbf{506\text{ mS}}$$

Note the difference in conductance level between the 2 Ω (500 mS) and the 1 kΩ (1 mS) resistor.

b. $R_T = \dfrac{1}{G_T} = \dfrac{1}{506\text{ mS}} = \mathbf{1.976\,\Omega}$

Applying Eq. (6.3) gives

$$R_T = \frac{1}{\dfrac{1}{R_1} + \dfrac{1}{R_2} + \dfrac{1}{R_3}} = \frac{1}{\dfrac{1}{2\,\Omega} + \dfrac{1}{200\,\Omega} + \dfrac{1}{1\text{ k}\Omega}}$$

$$= \frac{1}{0.5\text{ S} + 0.005\text{ S} + 0.001\text{ S}} = \frac{1}{0.506\text{ S}} = \mathbf{1.98\,\Omega}$$

EXAMPLE 6.3 Find the total resistance of the configuration in Fig. 6.6.

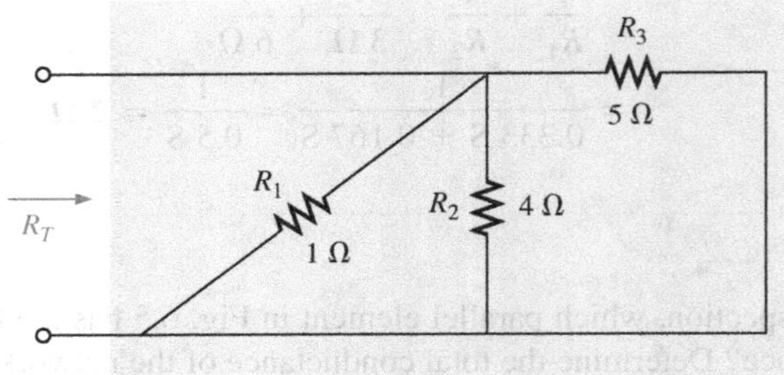

FIG. 6.6
Network to be investigated in Example 6.3.

Solution: First the network is redrawn as shown in Fig. 6.7 to clearly demonstrate that all the resistors are in parallel.

Applying Eq. (6.3) gives

$$R_T = \frac{1}{\dfrac{1}{R_1} + \dfrac{1}{R_2} + \dfrac{1}{R_3}} = \frac{1}{\dfrac{1}{1\,\Omega} + \dfrac{1}{4\,\Omega} + \dfrac{1}{5\,\Omega}}$$

$$= \frac{1}{1\text{ S} + 0.25\text{ S} + 0.2\text{ S}} = \frac{1}{1.45\text{ S}} \cong \mathbf{0.69\,\Omega}$$

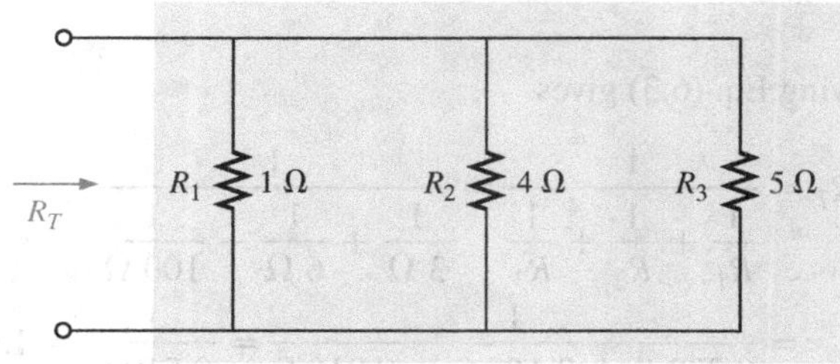

FIG. 6.7
Network in Fig. 6.6 redrawn.

If you review the examples above, you will find that the total resistance is less than the smallest parallel resistor. That is, in Example 6.1, 2 Ω is less than 3 Ω or 6 Ω. In Example 6.2, 1.976 Ω is less than 2 Ω, 100 Ω, or 1 kΩ; and in Example 6.3, 0.69 Ω is less than 1 Ω, 4 Ω, or 5 Ω. In general, therefore,

the total resistance of parallel resistors is always less than the value of the smallest resistor.

This is particularly important when you want a quick estimate of the total resistance of a parallel combination. Simply find the smallest value, and you know that the total resistance will be less than that value. It is also a great check on your calculations. In addition, you will find that

if the smallest resistance of a parallel combination is much smaller than that of the other parallel resistors, the total resistance will be very close to the smallest resistance value.

This fact is obvious in Example 6.2, where the total resistance of 1.976 Ω is very close to the smallest resistance of 2 Ω.

Another interesting characteristic of parallel resistors is demonstrated in Example 6.4.

EXAMPLE 6.4

a. What is the effect of adding another resistor of 100 Ω in parallel with the parallel resistors of Example 6.1 as shown in Fig. 6.8?
b. What is the effect of adding a parallel 1 Ω resistor to the configuration in Fig. 6.8?

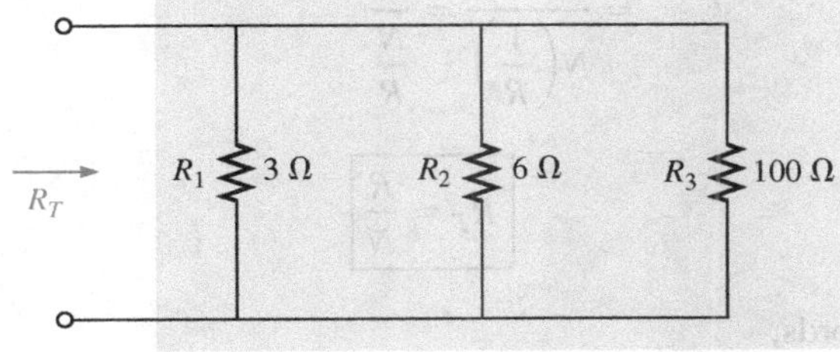

FIG. 6.8
Adding a parallel 100 Ω resistor to the network in Fig. 6.4.

Solutions:

a. Applying Eq. (6.3) gives

$$R_T = \frac{1}{\dfrac{1}{R_1} + \dfrac{1}{R_2} + \dfrac{1}{R_3}} = \frac{1}{\dfrac{1}{3\ \Omega} + \dfrac{1}{6\ \Omega} + \dfrac{1}{100\ \Omega}}$$
$$= \frac{1}{0.333\ \text{S} + 0.167\ \text{S} + 0.010\ \text{S}} = \frac{1}{0.510\ \text{S}} = \mathbf{1.96\ \Omega}$$

The parallel combination of the 3 Ω and 6 Ω resistors resulted in a total resistance of 2 Ω in Example 6.1. The effect of adding a resistor in parallel of 100 Ω had little effect on the total resistance because its resistance level is significantly higher (and conductance level significantly less) than that of the other two resistors. The total change in resistance was less than 2%. However, note that the total resistance dropped with the addition of the 100 Ω resistor.

b. Applying Eq. (6.3) gives

$$R_T = \frac{1}{\dfrac{1}{R_1} + \dfrac{1}{R_2} + \dfrac{1}{R_3} + \dfrac{1}{R_4}} = \frac{1}{\dfrac{1}{3\ \Omega} + \dfrac{1}{6\ \Omega} + \dfrac{1}{100\ \Omega} + \dfrac{1}{1\ \Omega}}$$
$$= \frac{1}{0.333\ \text{S} + 0.167\ \text{S} + 0.010\ \text{S} + 1\ \text{S}} = \frac{1}{0.51\ \text{S}} = \mathbf{0.66\ \Omega}$$

The introduction of the 1 Ω resistor reduced the total resistance from 2 Ω to only 0.66 Ω—a decrease of almost 67%. The fact that the added resistor has a resistance less than that of the other parallel elements and one-third that of the smallest contributed to the significant drop in resistance level.

In part (a) of Example 6.4, the total resistance dropped from 2 Ω to 1.96 Ω. In part (b), it dropped to 0.66 Ω. The results clearly reveal that

the total resistance of parallel resistors will always drop as new resistors are added in parallel, irrespective of their value.

Recall that this is the opposite of what occurs for series resistors, where additional resistors of any value increase the total resistance.

For equal resistors in parallel, the equation for the total resistance becomes significantly easier to apply. For N equal resistors in parallel, Eq. (6.3) becomes

$$R_T = \frac{1}{\dfrac{1}{R} + \dfrac{1}{R} + \dfrac{1}{R} + \cdots + \dfrac{1}{R_N}}$$
$$= \frac{1}{N\left(\dfrac{1}{R}\right)} = \frac{1}{\dfrac{N}{R}}$$

and

$$\boxed{R_T = \frac{R}{N}} \tag{6.4}$$

In other words,

the total resistance of N parallel resistors of equal value is the resistance of one resistor divided by the number (N) of parallel resistors.

EXAMPLE 6.5 Find the total resistance of the parallel resistors in Fig. 6.9.

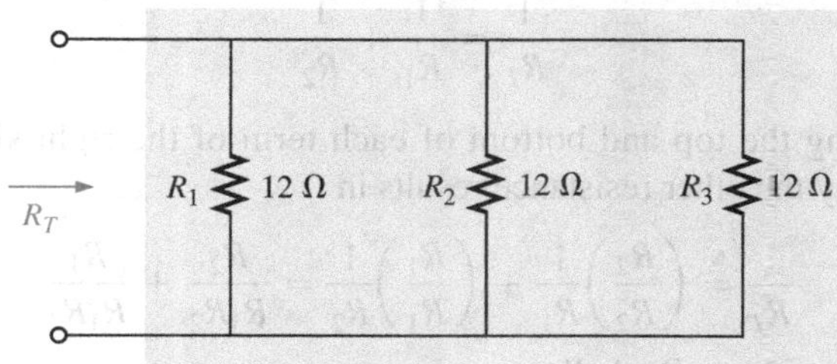

FIG. 6.9
Three equal parallel resistors to be investigated in Example 6.5.

Solution: Applying Eq. (6.4) gives

$$R_T = \frac{R}{N} = \frac{12\ \Omega}{3} = \mathbf{4\ \Omega}$$

EXAMPLE 6.6 Find the total resistance for the configuration in Fig. 6.10.

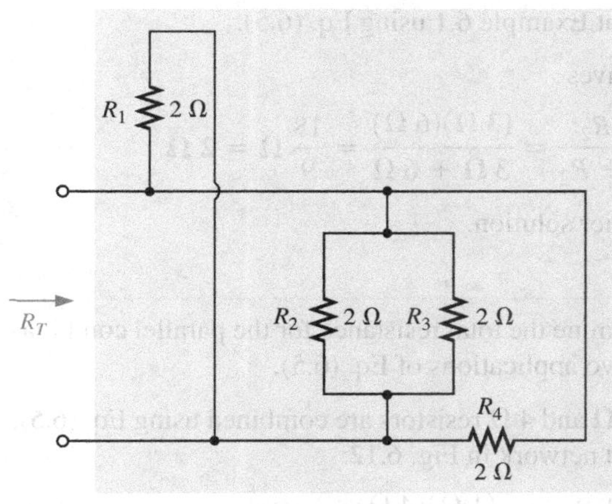

FIG. 6.10
Parallel configuration for Example 6.6.

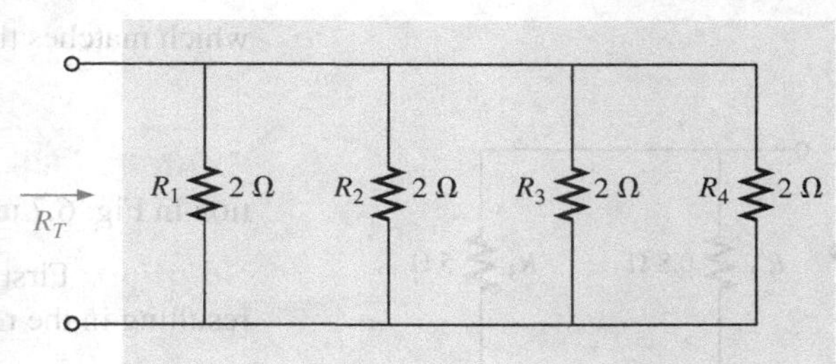

FIG. 6.11
Network in Fig. 6.10 redrawn.

Solution: Redrawing the network results in the parallel network in Fig. 6.11.

Applying Eq. (6.4) gives

$$R_T = \frac{R}{N} = \frac{2\ \Omega}{4} = \mathbf{0.5\ \Omega}$$

Special Case: Two Parallel Resistors

In the vast majority of cases, only two or three parallel resistors will have to be combined. With this in mind, an equation has been derived for two parallel resistors that is easy to apply and removes the need to continually worry about dividing into 1 and possibly misplacing a decimal point. For three parallel resistors, the equation to be derived here can be applied twice, or Eq. (6.3) can be used.

For two parallel resistors, the total resistance is determined by Eq. (6.1):

$$\frac{1}{R_T} = \frac{1}{R_1} + \frac{1}{R_2}$$

Multiplying the top and bottom of each term of the right side of the equation by the other resistance results in

$$\frac{1}{R_T} = \left(\frac{R_2}{R_2}\right)\frac{1}{R_1} + \left(\frac{R_1}{R_1}\right)\frac{1}{R_2} = \frac{R_2}{R_1R_2} + \frac{R_1}{R_1R_2}$$

$$\frac{1}{R_T} = \frac{R_2 + R_1}{R_1R_2}$$

and

$$\boxed{R_T = \frac{R_1R_2}{R_1 + R_2}} \qquad \textbf{(6.5)}$$

In words, the equation states that

the total resistance of two parallel resistors is simply the product of their values divided by their sum.

EXAMPLE 6.7 Repeat Example 6.1 using Eq. (6.5).

Solution: Eq. (6.5) gives

$$R_T = \frac{R_1R_2}{R_1 + R_2} = \frac{(3\ \Omega)(6\ \Omega)}{3\ \Omega + 6\ \Omega} = \frac{18}{9}\ \Omega = \mathbf{2\ \Omega}$$

which matches the earlier solution.

EXAMPLE 6.8 Determine the total resistance for the parallel combination in Fig. 6.7 using two applications of Eq. (6.5).

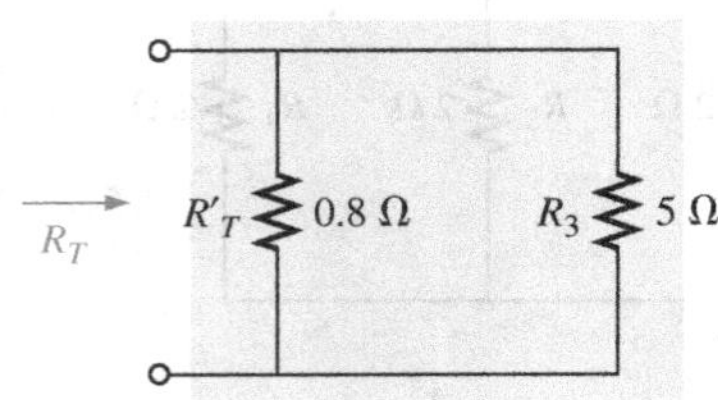

FIG. 6.12
Reduced equivalent in Fig. 6.7.

Solution: First the 1 Ω and 4 Ω resistors are combined using Eq. (6.5), resulting in the reduced network in Fig. 6.12:

$$\text{Eq. (6.4):}\ R'_T = \frac{R_1R_2}{R_1 + R_2} = \frac{(1\ \Omega)(4\ \Omega)}{1\ \Omega + 4\ \Omega} = \frac{4}{5}\ \Omega = 0.8\ \Omega$$

Then Eq. (6.5) is applied again using the equivalent value:

$$R_T = \frac{R'_T R_3}{R'_T + R_3} = \frac{(0.8\ \Omega)(5\ \Omega)}{0.8\ \Omega + 5\ \Omega} = \frac{4}{5.8}\ \Omega = \mathbf{0.69\ \Omega}$$

The result matches that obtained in Example 6.3.

Recall that series elements can be interchanged without affecting the magnitude of the total resistance. In parallel networks,

parallel resistors can be interchanged without affecting the total resistance.

The next example demonstrates this and reveals how redrawing a network can often define which operations or equations should be applied.

EXAMPLE 6.9 Determine the total resistance of the parallel elements in Fig. 6.13.

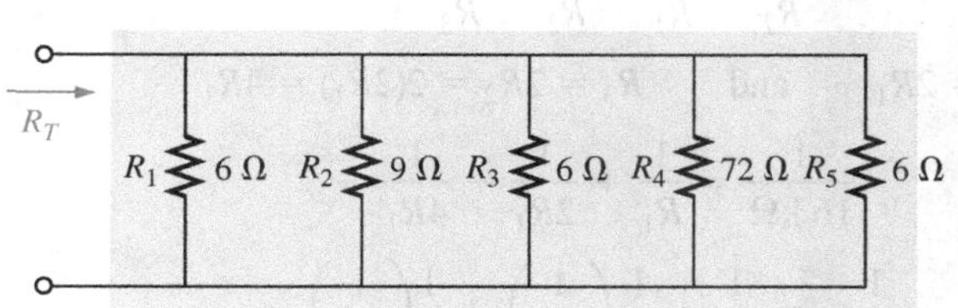

FIG. 6.13
Parallel network for Example 6.9.

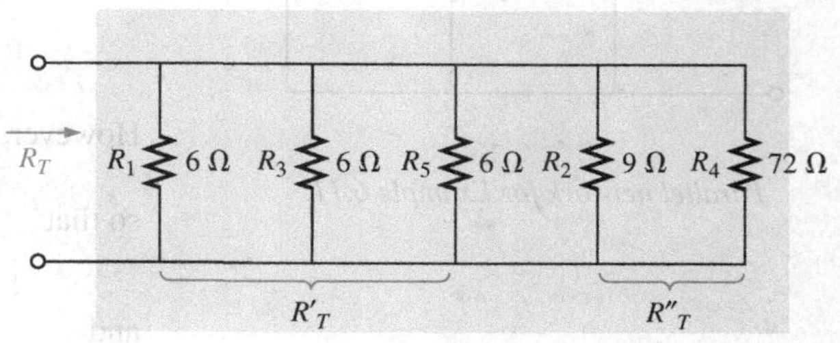

FIG. 6.14
Redrawn network in Fig. 6.13 (Example 6.9).

Solution: The network is redrawn in Fig. 6.14.

Eq. (6.4): $$R'_T = \frac{R}{N} = \frac{6\ \Omega}{3} = 2\ \Omega$$

Eq. (6.5): $$R''_T = \frac{R_2R_4}{R_2 + R_4} = \frac{(9\ \Omega)(72\ \Omega)}{9\ \Omega + 72\ \Omega} = \frac{648}{81}\ \Omega = 8\ \Omega$$

Eq. (6.5): $$R_T = \frac{R'_TR''_T}{R'_T + R''_T} = \frac{(2\ \Omega)(8\ \Omega)}{2\ \Omega + 8\ \Omega} = \frac{16}{10}\ \Omega = \mathbf{1.6\ \Omega}$$

The preceding examples involve direct substitution; that is, once the proper equation has been defined, it is only a matter of plugging in the numbers and performing the required algebraic manipulations. The next two examples have a design orientation, in which specific network parameters are defined and the circuit elements must be determined.

EXAMPLE 6.10 Determine the value of R_2 in Fig. 6.15 to establish a total resistance of 9 kΩ.

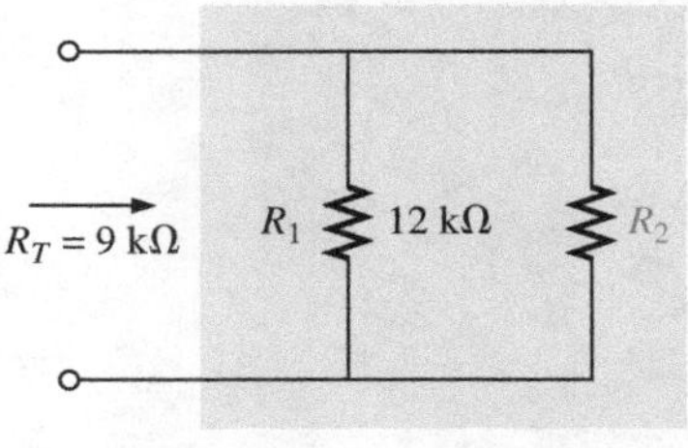

FIG. 6.15
Parallel network for Example 6.10.

Solution:

$$R_T = \frac{R_1R_2}{R_1 + R_2}$$
$$R_T(R_1 + R_2) = R_1R_2$$
$$R_TR_1 + R_TR_2 = R_1R_2$$
$$R_TR_1 = R_1R_2 - R_TR_2$$
$$R_TR_1 = (R_1 - R_T)R_2$$

and $$R_2 = \frac{R_TR_1}{R_1 - R_T}$$

Substituting values gives

$$R_2 = \frac{(9\ \text{k}\Omega)(12\ \text{k}\Omega)}{12\ \text{k}\Omega - 9\ \text{k}\Omega} = \frac{108}{3}\ \text{k}\Omega = \mathbf{36\ k\Omega}$$

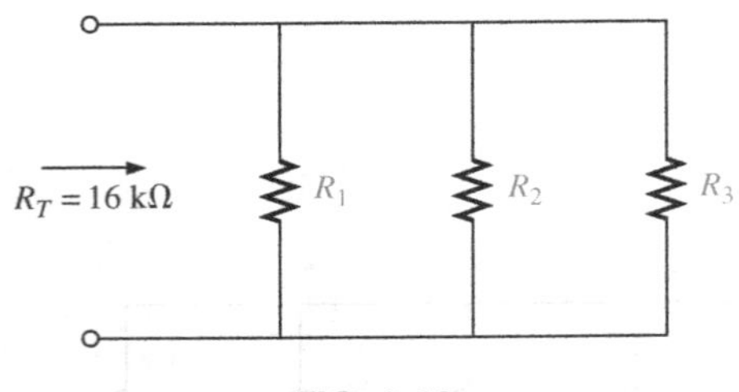

FIG. 6.16
Parallel network for Example 6.11.

EXAMPLE 6.11 Determine the values of R_1, R_2, and R_3 in Fig. 6.16 if $R_2 = 2R_1$, $R_3 = 2R_2$, and the total resistance is 16 kΩ.

Solution: Eq. (6.1) states

$$\frac{1}{R_T} = \frac{1}{R_1} + \frac{1}{R_2} + \frac{1}{R_3}$$

However, $R_2 = 2R_1$ and $R_3 = 2R_2 = 2(2R_1) = 4R_1$

so that
$$\frac{1}{16\ \text{k}\Omega} = \frac{1}{R_1} + \frac{1}{2R_1} + \frac{1}{4R_1}$$

and
$$\frac{1}{16\ \text{k}\Omega} = \frac{1}{R_1} + \frac{1}{2}\left(\frac{1}{R_1}\right) + \frac{1}{4}\left(\frac{1}{R_1}\right)$$

or
$$\frac{1}{16\ \text{k}\Omega} = 1.75\left(\frac{1}{R_1}\right)$$

resulting in $R_1 = 1.75(16\ \text{k}\Omega) = \mathbf{28\ k\Omega}$

so that $R_2 = 2R_1 = 2(28\ \text{k}\Omega) = \mathbf{56\ k\Omega}$

and $R_3 = 2R_2 = 2(56\ \text{k}\Omega) = \mathbf{112\ k\Omega}$

Analogies

Analogies were effectively used to introduce the concept of series elements. They can also be used to help define a *parallel configuration.* On a ladder, the rungs of the ladder form a parallel configuration. When ropes are tied between a grappling hook and a load, they effectively absorb the stress in a parallel configuration. The cables of a suspended roadway form a parallel configuration. There are numerous other analogies that demonstrate how connections between the same two points permit a distribution of stress between the parallel elements.

Instrumentation

As shown in Fig. 6.17, the total resistance of a parallel combination of resistive elements can be found by simply applying an ohmmeter. There is no polarity to resistance, so either lead of the ohmmeter can be connected to either side of the network. Although there are no supplies in Fig. 6.17, always keep in mind that ohmmeters can never be applied to a "live" circuit. It is not enough to set the supply to 0 V or to turn it off. It may still load down (change the network configuration of) the circuit and change the reading. It is best to remove the supply and apply the

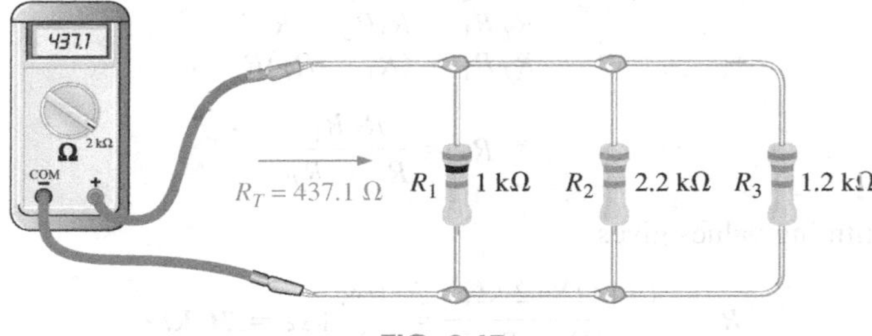

FIG. 6.17
Using an ohmmeter to measure the total resistance of a parallel network.

ohmmeter to the two resulting terminals. Since all the resistors are in the kilohm range, the 20 kΩ scale was chosen first. We then moved down to the 2 kΩ scale for increased precision. Moving down to the 200 Ω scale resulted in an "OL" indication since we were below the measured resistance value.

6.3 PARALLEL CIRCUITS

A **parallel circuit** can now be established by connecting a supply across a set of parallel resistors as shown in Fig. 6.18. The positive terminal of the supply is directly connected to the top of each resistor while the negative terminal is connected to the bottom of each resistor. Therefore, it should be quite clear that the applied voltage is the same across each resistor. In general,

the voltage is always the same across parallel elements.

Therefore, remember that

if two elements are in parallel, the voltage across them must be the same. However, if the voltage across two neighboring elements is the same, the two elements may or may not be in parallel.

The reason for this qualifying comment in the above statement is discussed in detail in Chapter 7.

For the voltages of the circuit in Fig. 6.18, the result is that

$$V_1 = V_2 = E \tag{6.6}$$

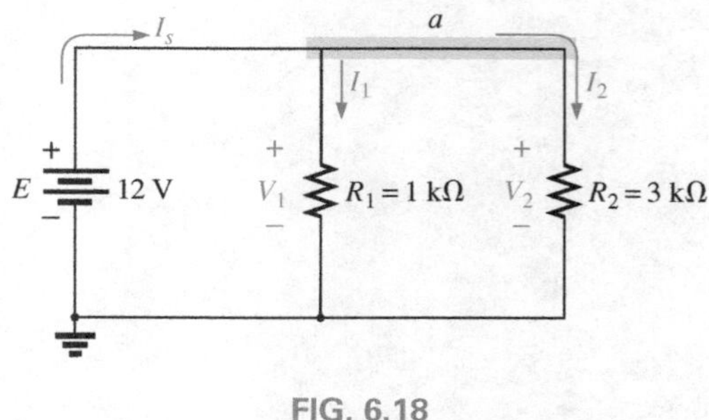

FIG. 6.18
Parallel network.

Once the supply has been connected, a source current is established through the supply that passes through the parallel resistors. The current that results is a direct function of the total resistance of the parallel circuit. The smaller the total resistance, the greater is the current, as occurred for series circuits also.

Recall from series circuits that the source does not "see" the parallel combination of elements. It reacts only to the total resistance of the circuit, as shown in Fig. 6.19. The source current can then be determined using Ohm's law:

$$I_s = \frac{E}{R_T} \tag{6.7}$$

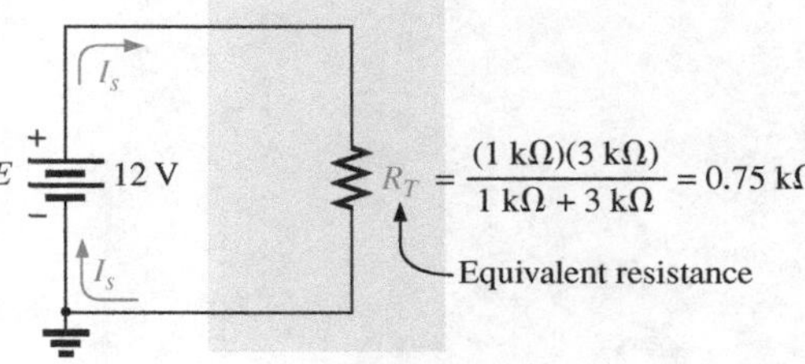

FIG. 6.19
Replacing the parallel resistors in Fig. 6.18 with the equivalent total resistance.

Since the voltage is the same across parallel elements, the current through each resistor can also be determined using Ohm's law. That is,

$$I_1 = \frac{V_1}{R_1} = \frac{E}{R_1} \quad \text{and} \quad I_2 = \frac{V_2}{R_2} = \frac{E}{R_2} \tag{6.8}$$

The direction for the currents is dictated by the polarity of the voltage across the resistors. Recall that for a resistor, current enters the positive side of a potential drop and leaves the negative side. The result, as shown in Fig. 6.18, is that the source current enters point a, and currents I_1 and I_2 leave the same point. An excellent analogy for describing the flow of charge through the network of Fig. 6.18 is the flow of water through the parallel pipes of Fig. 6.20. The larger pipe, with less "resistance" to the

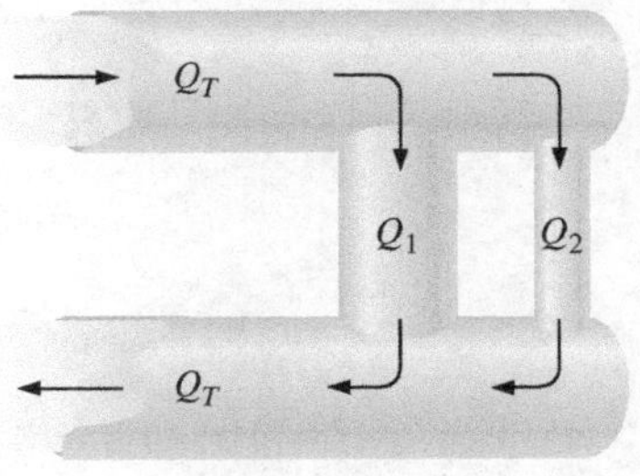

FIG. 6.20
Mechanical analogy for Fig. 6.18.

flow of water, will have a larger flow of water, through it. The thinner pipe, with its increased "resistance" level, will have less water flowing through it. In any case, the total water entering the pipes at the top Q_T must equal that leaving at the bottom, with $Q_T = Q_1 + Q_2$.

The relationship between the source current and the parallel resistor currents can be derived by simply taking the equation for the total resistance in Eq. (6.1):

$$\frac{1}{R_T} = \frac{1}{R_1} + \frac{1}{R_2}$$

Multiplying both sides by the applied voltage gives

$$E\left(\frac{1}{R_T}\right) = E\left(\frac{1}{R_1} + \frac{1}{R_2}\right)$$

resulting in

$$\frac{E}{R_T} = \frac{E}{R_1} + \frac{E}{R_2}$$

Then note that $E/R_1 = I_1$ and $E/R_2 = I_2$ to obtain

$$\boxed{I_s = I_1 + I_2} \qquad \textbf{(6.9)}$$

The result reveals a very important property of parallel circuits:

For single-source parallel networks, the source current (I_s) is always equal to the sum of the individual branch currents.

The duality that exists between series and parallel circuits continues to surface as we proceed through the basic equations for electric circuits. This is fortunate because it provides a way of remembering the characteristics of one using the results of another. For instance, in Fig. 6.21(a), we have a parallel circuit where it is clear that $I_T = I_1 + I_2$. By simply replacing the currents of the equation in Fig. 6.21(a) by a voltage level, as shown in Fig. 6.21(b), we have Kirchhoff's voltage law for a series circuit: $E = V_1 + V_2$. In other words,

for a parallel circuit, the source current equals the sum of the branch currents, while for a series circuit, the applied voltage equals the sum of the voltage drops.

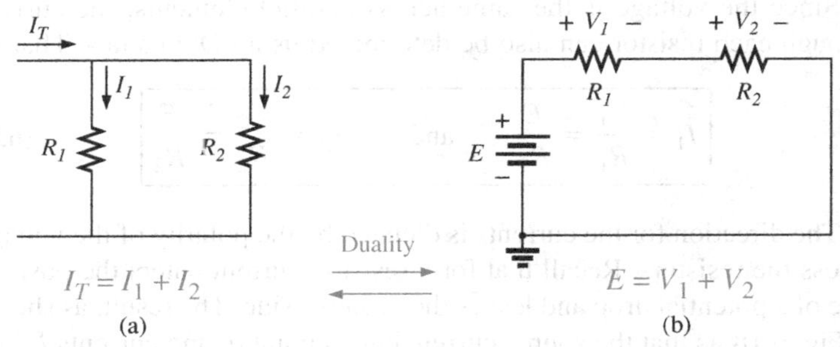

FIG. 6.21
Demonstrating the duality that exists between series and parallel circuits.

EXAMPLE 6.12 For the parallel network in Fig. 6.22:

a. Find the total resistance.
b. Calculate the source current.
c. Determine the current through each parallel branch.
d. Show that Eq. (6.9) is satisfied.

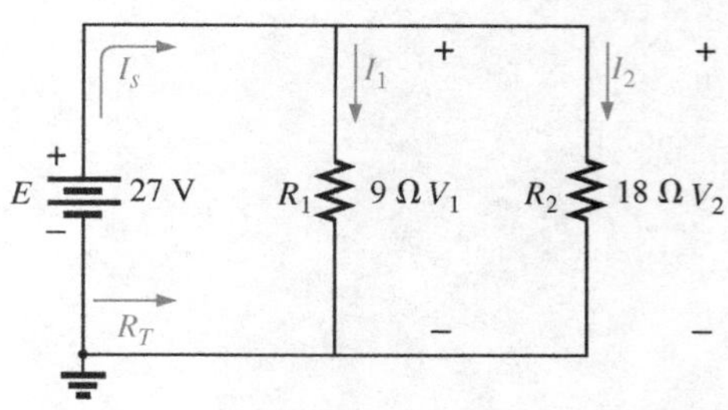

FIG. 6.22
Parallel network for Example 6.12.

Solutions:

a. Using Eq. (6.5) gives

$$R_T = \frac{R_1 R_2}{R_1 + R_2} = \frac{(9\ \Omega)(18\ \Omega)}{9\ \Omega + 18\ \Omega} = \frac{162}{27}\ \Omega = \mathbf{6\ \Omega}$$

b. Applying Ohm's law gives

$$I_s = \frac{E}{R_T} = \frac{27\ \text{V}}{6\ \Omega} = \mathbf{4.5\ A}$$

c. Applying Ohm's law gives

$$I_1 = \frac{V_1}{R_1} = \frac{E}{R_1} = \frac{27\ \text{V}}{9\ \Omega} = \mathbf{3\ A}$$

$$I_2 = \frac{V_2}{R_2} = \frac{E}{R_2} = \frac{27\ \text{V}}{18\ \Omega} = \mathbf{1.5\ A}$$

d. Substituting values from parts (b) and (c) gives

$$I_s = \mathbf{4.5\ A} = I_1 + I_2 = 3\ \text{A} + 1.5\ \text{A} = \mathbf{4.5\ A} \qquad \text{(checks)}$$

EXAMPLE 6.13 For the parallel network in Fig. 6.23.

a. Find the total resistance.
b. Calculate the source current.
c. Determine the current through each branch.

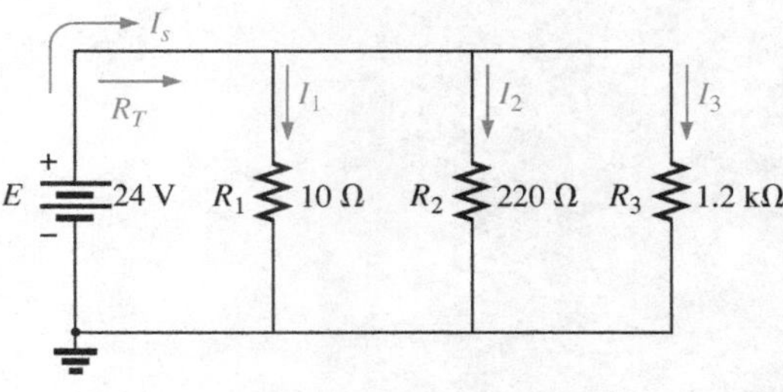

FIG. 6.23
Parallel network for Example 6.13.

Solutions:

a. Applying Eq. (6.3) gives

$$R_T = \frac{1}{\frac{1}{R_1} + \frac{1}{R_2} + \frac{1}{R_3}} = \frac{1}{\frac{1}{10\ \Omega} + \frac{1}{220\ \Omega} + \frac{1}{1.2\ \text{k}\Omega}}$$

$$= \frac{1}{100 \times 10^{-3} + 4.545 \times 10^{-3} + 0.833 \times 10^{-3}} = \frac{1}{105.38 \times 10^{-3}}$$

$$R_T = \mathbf{9.49\ \Omega}$$

Note that the total resistance is less than that of the smallest parallel resistor, and its magnitude is very close to the resistance of the smallest resistor because the other resistors are larger by a factor greater than 10 : 1.

b. Using Ohm's law gives

$$I_s = \frac{E}{R_T} = \frac{24\ \text{V}}{9.49\ \Omega} = \mathbf{2.53\ A}$$

c. Applying Ohm's law gives

$$I_1 = \frac{V_1}{R_1} = \frac{E}{R_1} = \frac{24\text{ V}}{10\ \Omega} = \mathbf{2.4\ A}$$

$$I_2 = \frac{V_2}{R_2} = \frac{E}{R_2} = \frac{24\text{ V}}{220\ \Omega} = \mathbf{0.11\ A}$$

$$I_3 = \frac{V_3}{R_3} = \frac{E}{R_3} = \frac{24\text{ V}}{1.2\text{ k}\Omega} = \mathbf{0.02\ A}$$

A careful examination of the results of Example 6.13 reveals that the larger the parallel resistor, the lower is the branch current. In general, therefore,

for parallel resistors, the greatest current will exist in the branch with the least resistance.

A more powerful statement is that

current always seeks the path of least resistance.

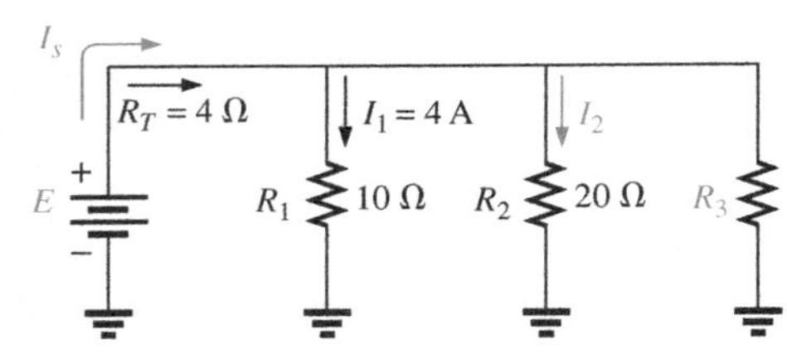

FIG. 6.24
Parallel network for Example 6.14.

EXAMPLE 6.14 Given the information provided in Fig. 6.24:

a. Determine R_3.
b. Find the applied voltage E.
c. Find the source current I_s.
d. Find I_2.

Solutions:

a. Applying Eq. (6.1) gives

$$\frac{1}{R_T} = \frac{1}{R_1} + \frac{1}{R_2} + \frac{1}{R_3}$$

Substituting gives $\displaystyle \frac{1}{4\ \Omega} = \frac{1}{10\ \Omega} + \frac{1}{20\ \Omega} + \frac{1}{R_3}$

so that $\displaystyle 0.25\text{ S} = 0.1\text{ S} + 0.05\text{ S} + \frac{1}{R_3}$

and $\displaystyle 0.25\text{ S} = 0.15\text{ S} + \frac{1}{R_3}$

with $\displaystyle \frac{1}{R_3} = 0.1\text{ S}$

and $\displaystyle R_3 = \frac{1}{0.1\text{ S}} = \mathbf{10\ \Omega}$

b. Using Ohm's law gives

$$E = V_1 = I_1R_1 = (4\text{ A})(10\ \Omega) = \mathbf{40\ V}$$

c. $$I_s = \frac{E}{R_T} = \frac{40\text{ V}}{4\ \Omega} = \mathbf{10\ A}$$

d. Applying Ohm's law gives

$$I_2 = \frac{V_2}{R_2} = \frac{E}{R_2} = \frac{40\text{ V}}{20\ \Omega} = \mathbf{2\ A}$$

Instrumentation

In Fig. 6.25, voltmeters have been connected to verify that the voltage across parallel elements is the same. Note that the positive or red lead of each voltmeter is connected to the high (positive) side of the voltage across each resistor to obtain a positive reading. The 20 V scale was used because the applied voltage exceeded the range of the 2 V scale.

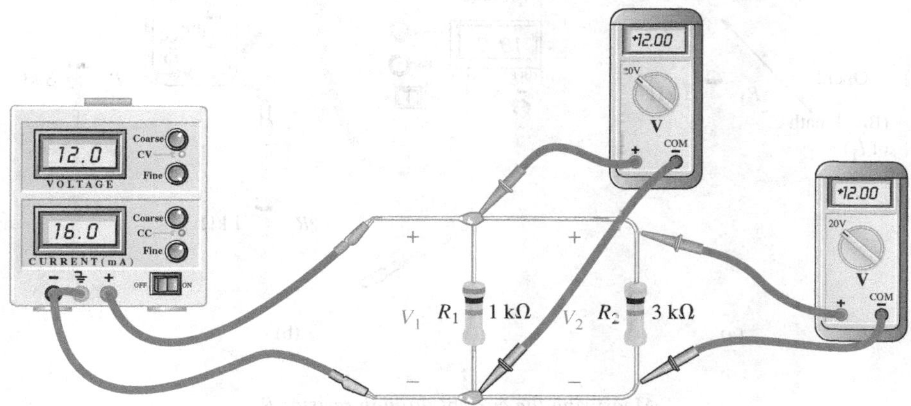

FIG. 6.25
Measuring the voltages of a parallel dc network.

In Fig. 6.26, an ammeter has been hooked up to measure the source current. First, the connection to the supply had to be broken at the positive terminal and the meter inserted as shown. Be sure to use ammeter terminals on your meter for such measurements. The red or positive lead of the meter is connected so that the source current enters that lead and leaves the negative or black lead to ensure a positive reading. The 200 mA scale was used because the source current exceeded the maximum value of the 2 mA scale. For the moment, we assume that the internal resistance of the meter can be ignored. Since the internal resistance of an ammeter on the 200 mA scale is typically only a few ohms, compared to the parallel resistors in the kilohm range, it is an excellent assumption.

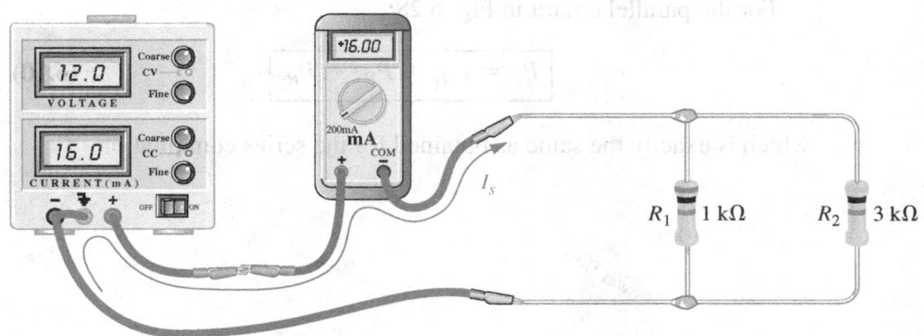

FIG. 6.26
Measuring the source current of a parallel network.

A more difficult measurement is for the current through resistor R_1. This measurement often gives trouble in the laboratory session. First, as shown in Fig. 6.27(a), resistor R_1 must be disconnected from the upper connection point to establish an open circuit. The ammeter is then inserted between the resulting terminals so that the current enters the positive or red terminal, as shown in Fig. 6.27(b). Always remember: When using an ammeter, first establish an open circuit in the branch in which the current is to be measured, and then insert the meter.

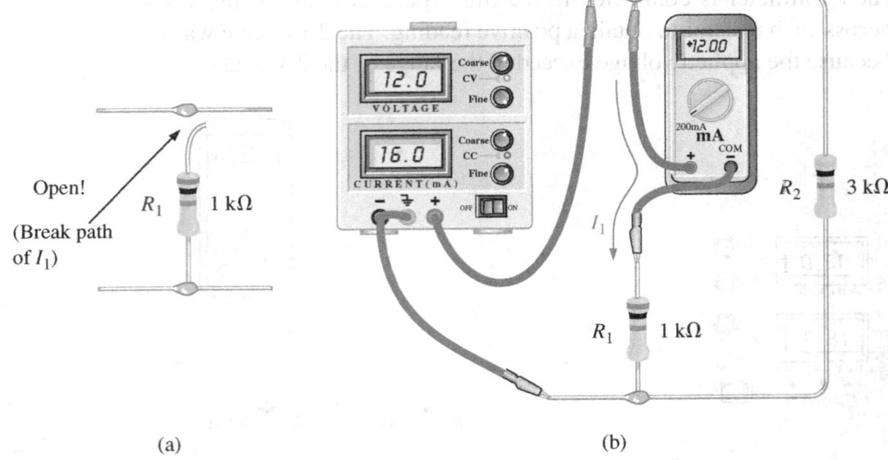

FIG. 6.27
Measuring the current through resistor R_1.

The easiest measurement is for the current through resistor R_2. Break the connection to R_2 above or below the resistor, and insert the ammeter with the current entering the positive or red lead to obtain a positive reading.

6.4 POWER DISTRIBUTION IN A PARALLEL CIRCUIT

Recall from the discussion of series circuits that the power applied to a series resistive circuit equals the power dissipated by the resistive elements. The same is true for parallel resistive networks. In fact,

for any network composed of resistive elements, the power applied by the battery will equal that dissipated by the resistive elements.

For the parallel circuit in Fig. 6.28:

$$P_E = P_{R_1} + P_{R_2} + P_{R_3} \tag{6.10}$$

which is exactly the same as obtained for the series combination.

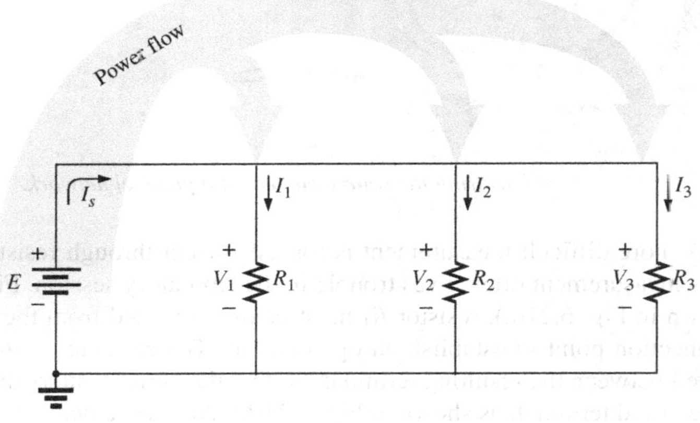

FIG. 6.28
Power flow in a dc parallel network.

The power delivered by the source in the same:

$$P_E = EI_s \quad \text{(watts, W)} \qquad \textbf{(6.11)}$$

as is the equation for the power to each resistor (shown for R_1 only):

$$P_1 = V_1I_1 = I_1^2R_1 = \frac{V_1^2}{R_1} \quad \text{(watts, W)} \qquad \textbf{(6.12)}$$

In the equation $P = V^2/R$, the voltage across each resistor in a parallel circuit will be the same. The only factor that changes is the resistance in the denominator of the equation. The result is that

in a parallel resistive network, the larger the resistor, the less is the power absorbed.

EXAMPLE 6.15 For the parallel network in Fig. 6.29 (all standard values):

a. Determine the total resistance R_T.
b. Find the source current and the current through each resistor.
c. Calculate the power delivered by the source.
d. Determine the power absorbed by each parallel resistor.
e. Verify Eq. (6.10).

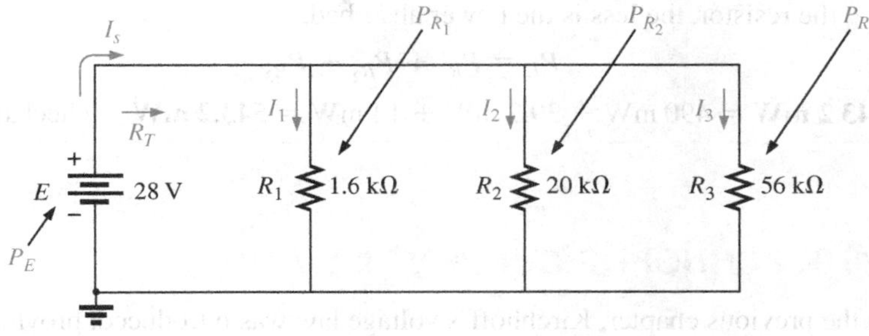

FIG. 6.29
Parallel network for Example 6.15.

Solutions:

a. Without making a single calculation, it should now be apparent from previous examples that the total resistance is less than 1.6 kΩ and very close to this value because of the magnitude of the other resistance levels:

$$R_T = \frac{1}{\frac{1}{R_1} + \frac{1}{R_2} + \frac{1}{R_3}} = \frac{1}{\frac{1}{1.6\text{ k}\Omega} + \frac{1}{20\text{ k}\Omega} + \frac{1}{56\text{ k}\Omega}}$$

$$= \frac{1}{625 \times 10^{-6} + 50 \times 10^{-6} + 17.867 \times 10^{-6}} = \frac{1}{692.867 \times 10^{-6}}$$

and $R_T = \mathbf{1.44\text{ k}\Omega}$

b. Applying Ohm's law gives

$$I_s = \frac{E}{R_T} = \frac{28\text{ V}}{1.44\text{ k}\Omega} = \mathbf{19.44\text{ mA}}$$

Recalling that current always seeks the path of least resistance immediately tells us that the current through the 1.6 kΩ resistor will be the largest and the current through the 56 kΩ resistor the smallest.

Applying Ohm's law again gives

$$I_1 = \frac{V_1}{R_1} = \frac{E}{R_1} = \frac{28\text{ V}}{1.6\text{ k}\Omega} = \mathbf{17.5\text{ mA}}$$

$$I_2 = \frac{V_2}{R_2} = \frac{E}{R_2} = \frac{28\text{ V}}{20\text{ k}\Omega} = \mathbf{1.4\text{ mA}}$$

$$I_3 = \frac{V_3}{R_3} = \frac{E}{R_3} = \frac{28\text{ V}}{56\text{ k}\Omega} = \mathbf{0.5\text{ mA}}$$

c. Applying Eq. (6.11) gives

$$P_E = EI_s = (28\text{ V})(19.4\text{ mA}) = \mathbf{543.2\text{ mW}}$$

d. Applying each form of the power equation gives

$$P_1 = V_1I_1 = EI_1 = (28\text{ V})(17.5\text{ mA}) = \mathbf{490\text{ mW}}$$

$$P_2 = I_2^2R_2 = (1.4\text{ mA})^2(20\text{ k}\Omega) = \mathbf{39.2\text{ mW}}$$

$$P_3 = \frac{V_3^2}{R_3} = \frac{E^2}{R_3} = \frac{(28\text{ V})^2}{56\text{ k}\Omega} = \mathbf{14\text{ mW}}$$

A review of the results clearly substantiates the fact that the larger the resistor, the less is the power absorbed.

e. $$P_E = P_{R_1} + P_{R_2} + P_{R_3}$$

$$\mathbf{543.2\text{ mW}} = 490\text{ mW} + 39.2\text{ mW} + 14\text{ mW} = \mathbf{543.2\text{ mW}} \quad \text{(checks)}$$

6.5 KIRCHHOFF'S CURRENT LAW

In the previous chapter, Kirchhoff's voltage law was introduced, providing a very important relationship among the voltages of a closed path. Kirchhoff is also credited with developing the following equally important relationship between the currents of a network, called **Kirchhoff's current law (KCL):**

The algebraic sum of the currents entering and leaving a junction (or region) of a network is zero.

The law can also be stated in the following way:

The sum of the currents entering a junction (or region) of a network must equal the sum of the currents leaving the same junction (or region).

In equation form, the above statement can be written as follows:

$$\boxed{\Sigma I_i = \Sigma I_o} \qquad \textbf{(6.13)}$$

with I_i representing the current entering, or "in," and I_o representing the current leaving, or "out."

In Fig. 6.30, for example, the shaded area can enclose an entire system or a complex network, or it can simply provide a connection point

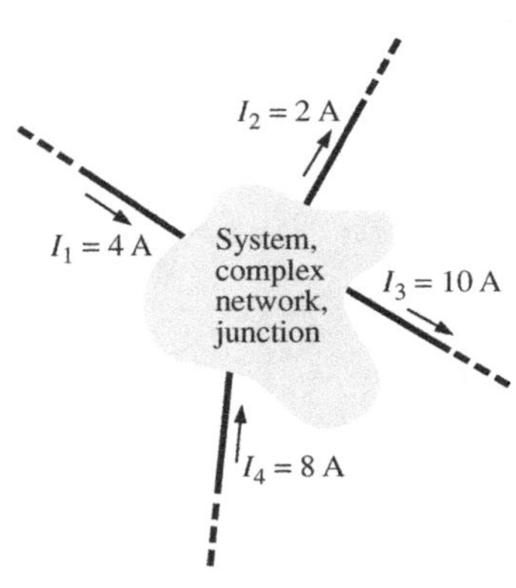

FIG. 6.30
Introducing Kirchhoff's current law.

(junction) for the displayed currents. In each case, the current entering must equal that leaving, as required by Eq. (6.13):

$$\Sigma I_i = \Sigma I_o$$
$$I_1 + I_4 = I_2 + I_3$$
$$4\text{ A} + 8\text{ A} = 2\text{ A} + 10\text{ A}$$
$$\mathbf{12\text{ A} = 12\text{ A}} \quad \text{(checks)}$$

The most common application of the law will be at a junction of two or more current paths, as shown in Fig. 6.31(a). Some students have difficulty initially determining whether a current is entering or leaving a junction. One approach that may help is to use the water analog in Fig. 6.31(b), where the junction in Fig. 6.31(a) is the small bridge across the stream. Simply relate the current of I_1 to the fluid flow of Q_1, the smaller branch current I_2 to the water flow Q_2, and the larger branch current I_3 to the flow Q_3. The water arriving at the bridge must equal the sum of that leaving the bridge, so that $Q_1 = Q_2 + Q_3$. Since the current I_1 is pointing *at* the junction and the fluid flow Q_1 is *toward* the person on the bridge, both quantities are seen as approaching the junction, and can be considered *entering* the junction. The currents I_2 and I_3 are both leaving the junction, just as Q_2 and Q_3 are leaving the fork in the river. The quantities I_2, I_3, Q_2, and Q_3 are therefore all *leaving* the junction.

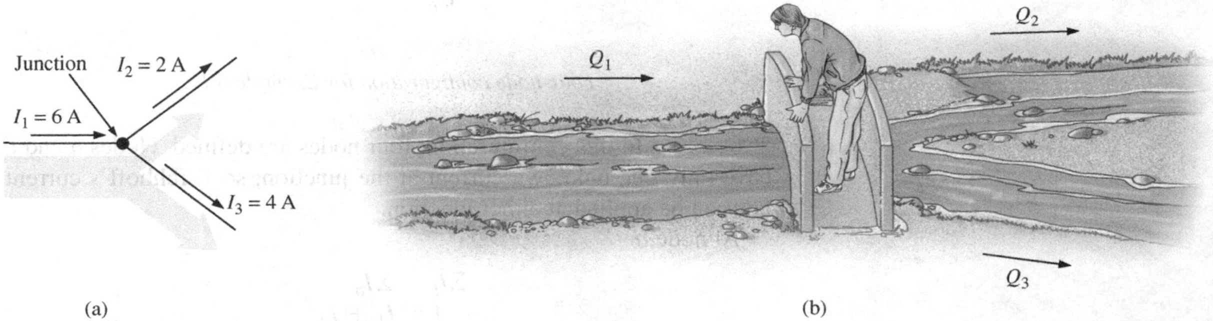

FIG. 6.31
(a) Demonstrating Kirchhoff's current law; (b) the water analogy for the junction in (a).

In the next few examples, unknown currents can be determined by applying Kirchhoff's current law. Remember to place all current levels entering the junction to the left of the equals sign and the sum of all currents leaving the junction to the right of the equals sign.

In technology, the term **node** is commonly used to refer to a junction of two or more branches. Therefore, this term is used frequently in the analyses to follow.

EXAMPLE 6.16 Determine currents I_3 and I_4 in Fig. 6.32 using Kirchhoff's current law.

Solution: There are two junctions or nodes in Fig. 6.32. Node *a* has only one unknown, while node *b* has two unknowns. Since a single equation can be used to solve for only one unknown, we must apply Kirchhoff's current law to node *a* first.

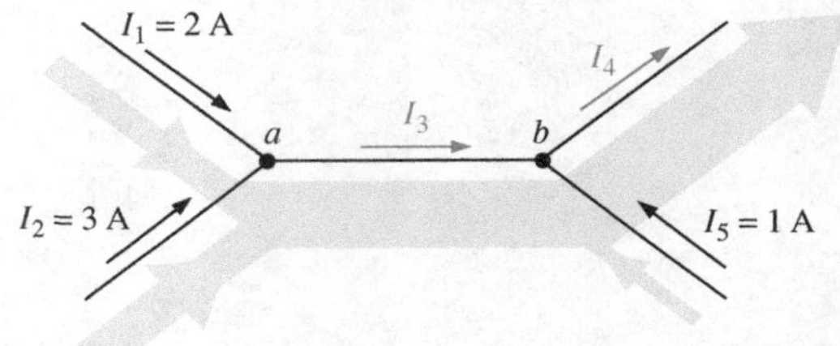

FIG. 6.32
Two-node configuration for Example 6.16.

At node a

$$\Sigma I_i = \Sigma I_o$$
$$I_1 + I_2 = I_3$$
$$2\text{ A} + 3\text{ A} = I_3 = \mathbf{5\ A}$$

At node b, using the result just obtained,

$$\Sigma I_i = \Sigma I_o$$
$$I_3 + I_5 = I_4$$
$$5\text{ A} + 1\text{ A} = I_4 = \mathbf{6\ A}$$

Note that in Fig. 6.32, the width of the blue-shaded regions matches the magnitude of the current in that region.

EXAMPLE 6.17 Determine currents I_1, I_3, I_4, and I_5 for the network in Fig. 6.33.

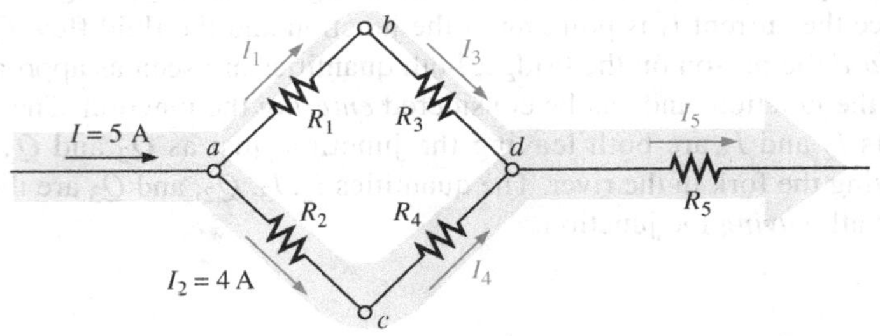

FIG. 6.33
Four-node configuration for Example 6.17.

Solution: In this configuration, four nodes are defined. Nodes a and c have only one unknown current at the junction, so Kirchhoff's current law can be applied at either junction.

At node a

$$\Sigma I_i = \Sigma I_o$$
$$I = I_1 + I_2$$
$$5\text{ A} = I_1 + 4\text{ A}$$

and
$$I_1 = 5\text{ A} - 4\text{ A} = \mathbf{1\ A}$$

At node c

$$\Sigma I_i = \Sigma I_o$$
$$I_2 = I_4$$

and
$$I_4 = I_2 = \mathbf{4\ A}$$

Using the above results at the other junctions results in the following.

At node b

$$\Sigma I_i = \Sigma I_o$$
$$I_1 = I_3$$

and
$$I_3 = I_1 = \mathbf{1\ A}$$

At node d

$$\Sigma I_i = \Sigma I_o$$
$$I_3 + I_4 = I_5$$
$$1\text{ A} + 4\text{ A} = I_5 = \mathbf{5\ A}$$

If we enclose the entire network, we find that the current entering from the far left is $I = 5$ A, while the current leaving from the far right is

$I_5 = 5$ A. The two must be equal since the net current entering any system must equal the net current leaving.

EXAMPLE 6.18 Determine currents I_3 and I_5 in Fig. 6.34 through applications of Kirchhoff's current law.

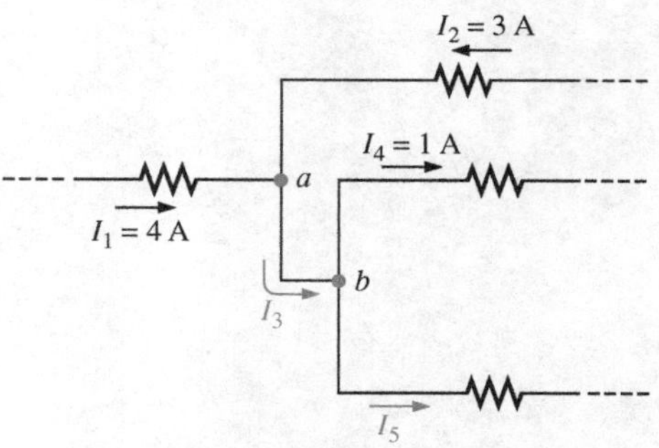

FIG. 6.34
Network for Example 6.18.

Solution: Note first that since node *b* has two unknown quantities (I_3 and I_5), and node *a* has only one, Kirchhoff's current law must first be applied to node *a*. The result is then applied to node *b*.

At node *a*

$$\Sigma I_i = \Sigma I_o$$
$$I_1 + I_2 = I_3$$
$$4\text{ A} + 3\text{ A} = I_3 = \mathbf{7\ A}$$

At node *b*

$$\Sigma I_i = \Sigma I_o$$
$$I_3 = I_4 + I_5$$
$$7\text{ A} = 1\text{ A} + I_5$$

and

$$I_5 = 7\text{ A} - 1\text{ A} = \mathbf{6\ A}$$

EXAMPLE 6.19 For the parallel dc network in Fig. 6.35:

a. Determine the source current I_s.
b. Find the source voltage *E*.
c. Determine R_3.
d. Calculate R_T.

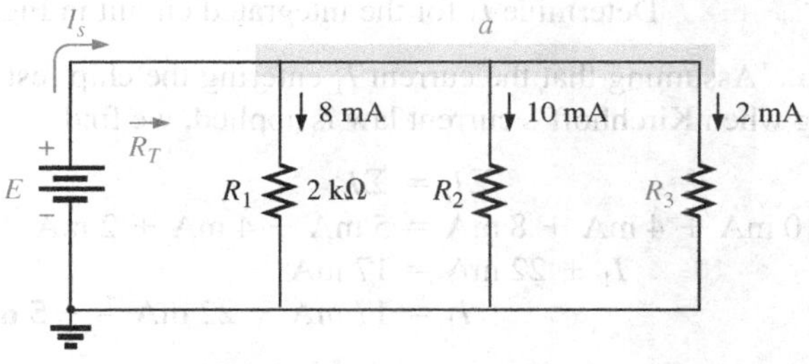

FIG. 6.35
Parallel network for Example 6.19.

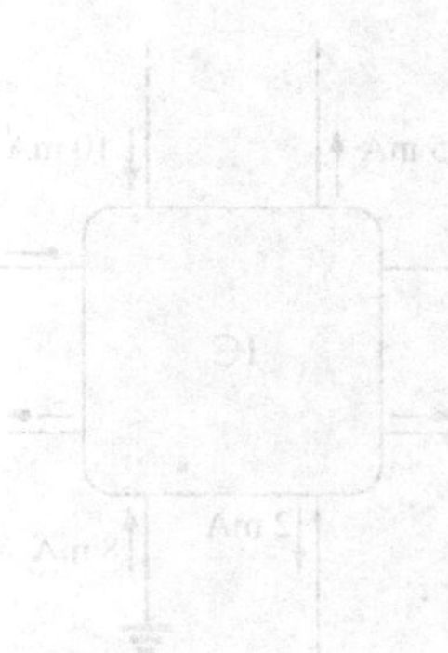

Solutions:

a. First apply Eq. (6.13) at node *a*. Although node *a* in Fig. 6.35 may not initially appear as a single junction, it can be redrawn as shown in Fig. 6.36, where it is clearly a common point for all the branches.

The result is

$$\Sigma I_i = \Sigma I_o$$
$$I_s = I_1 + I_2 + I_3$$

Substituting values: $I_s = 8\text{ mA} + 10\text{ mA} + 2\text{ mA} = \mathbf{20\ mA}$

Note in this solution that you do not need to know the resistor values or the voltage applied. The solution is determined solely by the current levels.

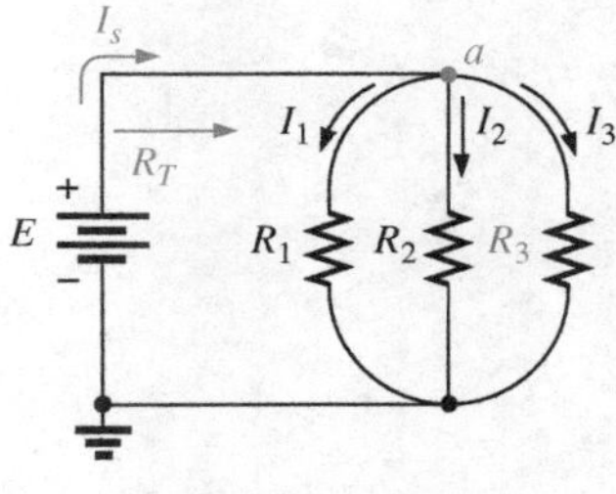

FIG. 6.36
Redrawn network in Fig. 6.35.

b. Applying Ohm's law gives

$$E = V_1 = I_1R_1 = (8 \text{ mA})(2 \text{ k}\Omega) = \mathbf{16 \text{ V}}$$

c. Applying Ohm's law in a different form gives

$$R_3 = \frac{V_3}{I_3} = \frac{E}{I_3} = \frac{16 \text{ V}}{2 \text{ mA}} = \mathbf{8 \text{ k}\Omega}$$

d. Applying Ohm's law again gives

$$R_T = \frac{E}{I_s} = \frac{16 \text{ V}}{20 \text{ mA}} = \mathbf{0.8 \text{ k}\Omega}$$

The application of Kirchhoff's current law is not limited to networks where all the internal connections are known or visible. For instance, all the currents of the integrated circuit in Fig. 6.37 are known except I_1. By treating the entire system (which could contain over a million elements) as a single node, we can apply Kirchhoff's current law as shown in Example 6.20.

Before looking at Example 6.20 in detail, note that the direction of the unknown current I_1 is not provided in Fig. 6.37. On many occasions, this will be true. With so many currents entering or leaving the system, it is difficult to know by inspection which direction should be assigned to I_1. *In such cases, simply make an assumption about the direction and then check out the result. If the result is negative, the wrong direction was assumed. If the result is positive, the correct direction was assumed. In either case, the magnitude of the current will be correct.*

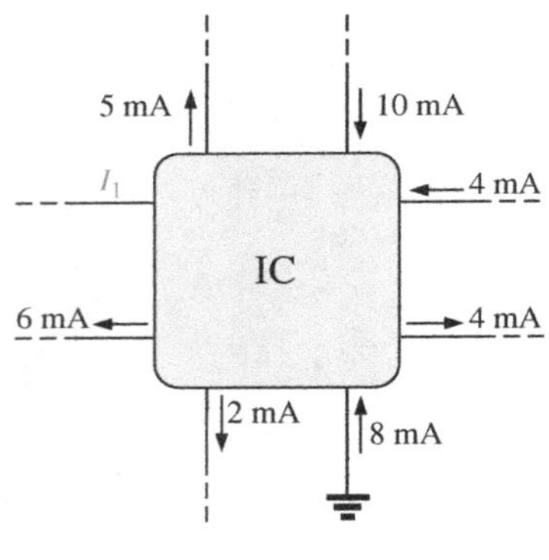

FIG. 6.37
Integrated circuit for Example 6.20.

EXAMPLE 6.20 Determine I_1 for the integrated circuit in Fig. 6.37.

Solution: Assuming that the current I_1 entering the chip results in the following when Kirchhoff's current law is applied, we find

$$\Sigma I_i = \Sigma I_o$$

$$I_1 + 10 \text{ mA} + 4 \text{ mA} + 8 \text{ mA} = 5 \text{ mA} + 4 \text{ mA} + 2 \text{ mA} + 6 \text{ mA}$$

$$I_1 + 22 \text{ mA} = 17 \text{ mA}$$

$$I_1 = 17 \text{ mA} - 22 \text{ mA} = \mathbf{-5 \text{ mA}}$$

We find that the direction for I_1 is *leaving* the IC, although the magnitude of 5 mA is correct.

As we leave this important section, be aware that Kirchhoff's current law will be applied in one form or another throughout the text. *Kirchhoff's laws are unquestionably two of the most important in this field because they are applicable to the most complex configurations in existence today.* They will not be replaced by a more important law or dropped for a more sophisticated approach.

6.6 CURRENT DIVIDER RULE

For series circuits we have the powerful voltage divider rule for finding the voltage across a resistor in a series circuit. We now introduce the equally powerful **current divider rule (CDR)** for finding the current through a resistor in a parallel circuit.

In Section 6.4, it was pointed out that current will always seek the path of least resistance. In Fig. 6.38, for example, the current of 9 A is faced with splitting between the three parallel resistors. Based on the previous sections, it should now be clear without a single calculation that the majority of the current will pass through the smallest resistor of 10 Ω, and the least current will pass through the 1 kΩ resistor. In fact, the current through the 100 Ω resistor will also exceed that through the 1 kΩ resistor. We can take it one step further by recognizing that the resistance of the 100 Ω resistor is 10 times that of the 10 Ω resistor. The result is a current through the 10 Ω resistor that is 10 times that of the 100 Ω resistor. Similarly, the current through the 100 Ω resistor is 10 times that through the 1 kΩ resistor.

In general,

For two parallel elements of equal value, the current will divide equally.

For parallel elements with different values, the smaller the resistance, the greater is the share of input current.

For parallel elements of different values, the current will split with a ratio equal to the inverse of their resistance values.

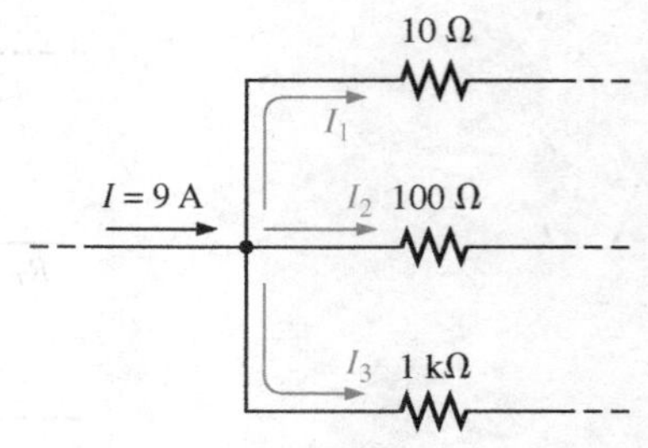

FIG. 6.38
Discussing the manner in which the current will split between three parallel branches of different resistive value.

EXAMPLE 6.21

a. Determine currents I_1 and I_3 for the network in Fig. 6.39.
b. Find the source current I_s.

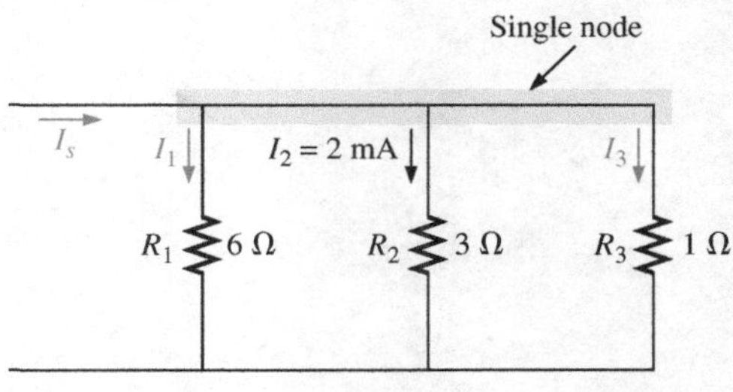

FIG. 6.39
Parallel network for Example 6.21.

Solutions:

a. Since R_1 is twice R_2, the current I_1 must be one-half I_2, and

$$I_1 = \frac{I_2}{2} = \frac{2\text{ mA}}{2} = \mathbf{1\text{ mA}}$$

Since R_2 is three times R_3, the current I_3 must be three times I_2, and

$$I_3 = 3I_2 = 3(2\text{ mA}) = \mathbf{6\text{ mA}}$$

b. Applying Kirchhoff's current law gives

$$\Sigma I_i = \Sigma I_o$$
$$I_s = I_1 + I_2 + I_3$$
$$I_s = 1\text{ mA} + 2\text{ mA} + 6\text{ mA} = \mathbf{9\text{ mA}}$$

Although the above discussions and examples allowed us to determine the relative magnitude of a current based on a known level, they do not provide the magnitude of a current through a branch of a parallel network if only the total entering current is known. The result is a need for the current divider rule, which will be derived using the parallel configuration in Fig. 6.40(a). The current I_T (using the subscript T to indicate the total entering current) splits between the N parallel resistors and then gathers itself together again at the bottom of the configuration. In Fig. 6.40(b), the parallel combination of resistors has been replaced by a single resistor equal to the total resistance of the parallel combination as determined in the previous sections.

The current I_T can then be determined using Ohm's law:

$$I_T = \frac{V}{R_T}$$

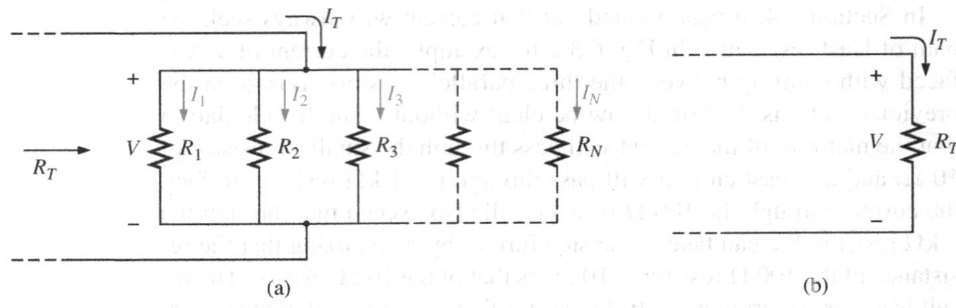

FIG. 6.40
Deriving the current divider rule: (a) parallel network of N parallel resistors; (b) reduced equivalent of part (a).

Since the voltage V is the same across parallel elements, the following is true:

$$V = I_1R_1 = I_2R_2 = I_3R_3 = \cdots = I_xR_x$$

where the product I_xR_x refers to any combination in the series.

Substituting for V in the above equation for I_T, we have

$$I_T = \frac{I_xR_x}{R_T}$$

Solving for I_x, the final result is the **current divider rule:**

$$I_x = \frac{R_T}{R_x}I_T \tag{6.14}$$

which states that

the current through any branch of a parallel resistive network is equal to the total resistance of the parallel network divided by the resistance of the resistor of interest and multiplied by the total current entering the parallel configuration.

Since R_T and I_T are constants, for a particular configuration the larger the value of R_x (in the denominator), the smaller is the value of I_x for that branch, confirming the fact that current always seeks the path of least resistance.

EXAMPLE 6.22 For the parallel network in Fig. 6.41, determine current I_1 using Eq. (6.14).

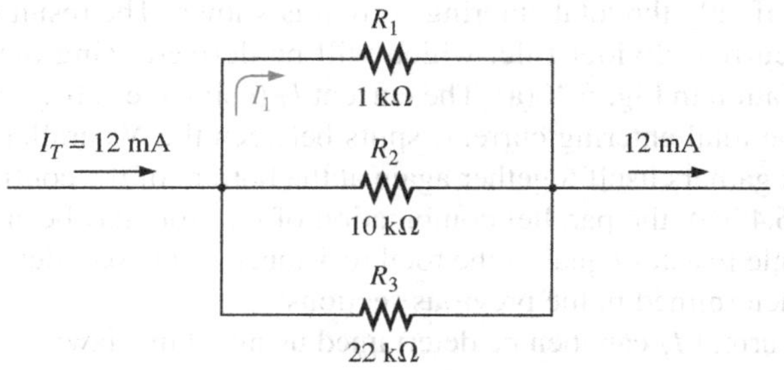

FIG. 6.41
Using the current divider rule to calculate current I_1 in Example 6.22.

Solution: Eq. (6.3):

$$R_T = \frac{1}{\frac{1}{R_1} + \frac{1}{R_2} + \frac{1}{R_3}}$$

$$= \frac{1}{\frac{1}{1\text{ k}\Omega} + \frac{1}{10\text{ k}\Omega} + \frac{1}{22\text{ k}\Omega}}$$

$$= \frac{1}{1 \times 10^{-3} + 100 \times 10^{-6} + 45.46 \times 10^{-6}}$$

$$= \frac{1}{1.145 \times 10^{-3}} = \mathbf{873.01\ \Omega}$$

Eq. (6.14): $I_1 = \frac{R_T}{R_1} I_T$

$$= \frac{(873.01\ \Omega)}{1\text{ k}\Omega}(12\text{ mA}) = (0.873)(12\text{ mA}) = \mathbf{10.48\ mA}$$

and the smallest parallel resistor receives the majority of the current.

Note also that

for a parallel network, the current through the smallest resistor will be very close to the total entering current if the other parallel elements of the configuration are much larger in magnitude.

In Example 6.22, the current through R_1 is very close to the total current because R_1 is 10 times less than the next smallest resistance.

Special Case: Two Parallel Resistors

For the case of two parallel resistors as shown in Fig. 6.42, the total resistance is determined by

$$R_T = \frac{R_1 R_2}{R_1 + R_2}$$

Substituting R_T into Eq. (6.14) for current I_1 results in

$$I_1 = \frac{R_T}{R_1} I_T = \frac{\left(\frac{R_1 R_2}{R_1 + R_2}\right)}{R_1} I_T$$

and

$$\boxed{I_1 = \left(\frac{R_2}{R_1 + R_2}\right) I_T} \qquad \textbf{(6.15a)}$$

Similarly, for I_2,

$$\boxed{I_2 = \left(\frac{R_1}{R_1 + R_2}\right) I_T} \qquad \textbf{(6.15b)}$$

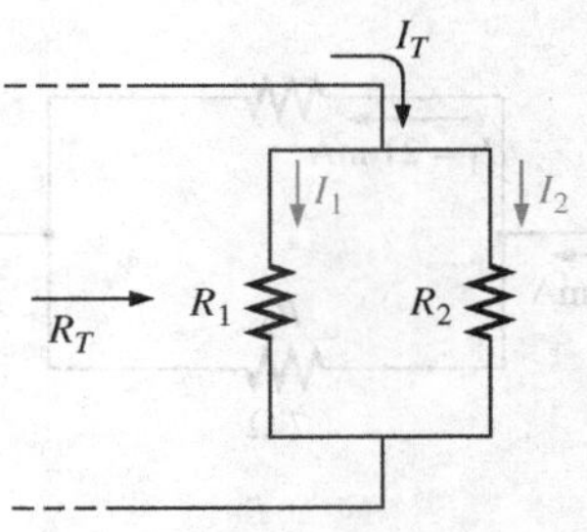

FIG. 6.42
Deriving the current divider rule for the special case of only two parallel resistors.

Eq. (6.15) states that

for two parallel resistors, the current through one is equal to the resistance of the other times the total entering current divided by the sum of the two resistances.

Since the combination of two parallel resistors is probably the most common parallel configuration, the simplicity of the format for Eq. (6.15) suggests that it is worth memorizing. Take particular note, however, that the denominator of the equation is simply the sum, not the total resistance, of the combination.

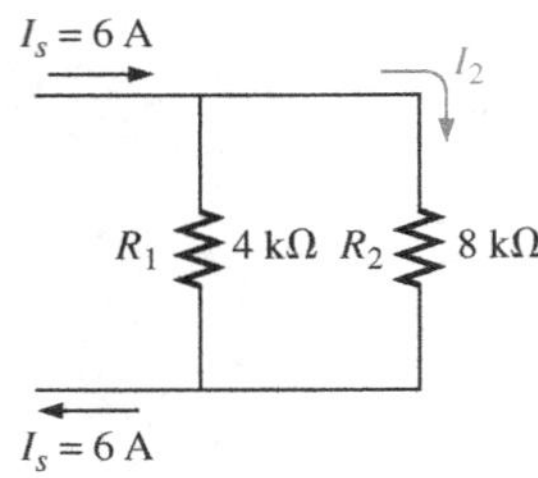

FIG. 6.43
Using the current divider rule to determine current I_2 in Example 6.23.

EXAMPLE 6.23 Determine current I_2 for the network in Fig. 6.43 using the current divider rule.

Solution: Using Eq. (6.15b) gives

$$I_2 = \left(\frac{R_1}{R_1 + R_2}\right)I_T$$
$$= \left(\frac{4\text{ k}\Omega}{4\text{ k}\Omega + 8\text{ k}\Omega}\right)6\text{ A} = (0.333)(6\text{ A}) = \mathbf{2\text{ A}}$$

Using Eq. (6.14) gives

$$I_2 = \frac{R_T}{R_2}I_T$$

with
$$R_T = 4\text{ k}\Omega \parallel 8\text{ k}\Omega = \frac{(4\text{ k}\Omega)(8\text{ k}\Omega)}{4\text{ k}\Omega + 8\text{ k}\Omega} = 2.667\text{ k}\Omega$$

and
$$I_2 = \left(\frac{2.667\text{ k}\Omega}{8\text{ k}\Omega}\right)6\text{ A} = (0.333)(6\text{ A}) = \mathbf{2\text{ A}}$$

matching the above solution.

It would appear that the solution with Eq. 6.15(b) is more direct in Example 6.23. However, keep in mind that Eq. (6.14) is applicable to any parallel configuration, removing the necessity to remember two equations.

Now we present a design-type problem.

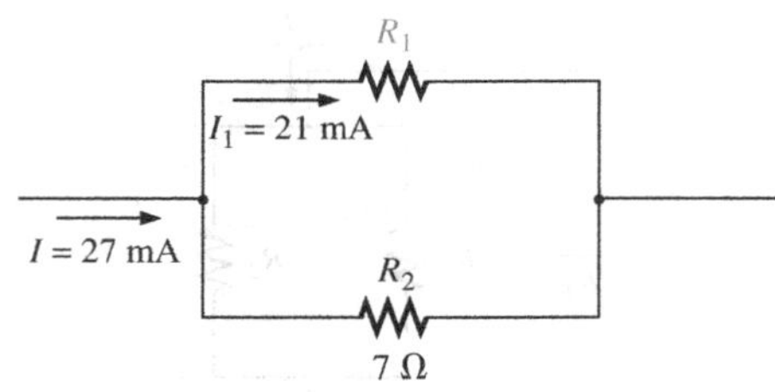

FIG. 6.44
A design-type problem for two parallel resistors (Example 6.24).

EXAMPLE 6.24 Determine resistor R_1 in Fig. 6.44 to implement the division of current shown.

Solution: There are essentially two approaches to this type of problem. One involves the direct substitution of known values into the current divider rule equation followed by a mathematical analysis. The other is the sequential application of the basic laws of electric circuits. First we will use the latter approach.

Applying Kirchhoff's current law gives

$$\Sigma I_i = \Sigma I_o$$
$$I = I_1 + I_2$$
$$27\text{ mA} = 21\text{ mA} + I_2$$

and
$$I_2 = 27\text{ mA} - 21\text{ mA} = 6\text{ mA}$$

The voltage V_2:
$$V_2 = I_2R_2 = (6\text{ mA})(7\ \Omega) = 42\text{ mV}$$

so that
$$V_1 = V_2 = 42\text{ mV}$$

Finally,
$$R_1 = \frac{V_1}{I_1} = \frac{42\text{ mV}}{21\text{ mA}} = \mathbf{2\ \Omega}$$

Now for the other approach using the current divider rule:

$$I_1 = \frac{R_2}{R_1 + R_2}I_T$$
$$21\text{ mA} = \left(\frac{7\ \Omega}{R_1 + 7\ \Omega}\right)27\text{ mA}$$

$$(R_1 + 7\,\Omega)(21\text{ mA}) = (7\,\Omega)(27\text{ mA})$$
$$(21\text{ mA})R_1 + 147\text{ mV} = 189\text{ mV}$$
$$(21\text{ mA})R_1 = 189\text{ mV} - 147\text{ mV} = 42\text{ mV}$$

and
$$R_1 = \frac{42\text{ mV}}{21\text{ mA}} = \mathbf{2\,\Omega}$$

In summary, therefore, remember that current always seeks the path of least resistance, and the ratio of the resistance values is the inverse of the resulting current levels, as shown in Fig 6.45. The thickness of the blue bands in Fig. 6.45 reflects the relative magnitude of the current in each branch.

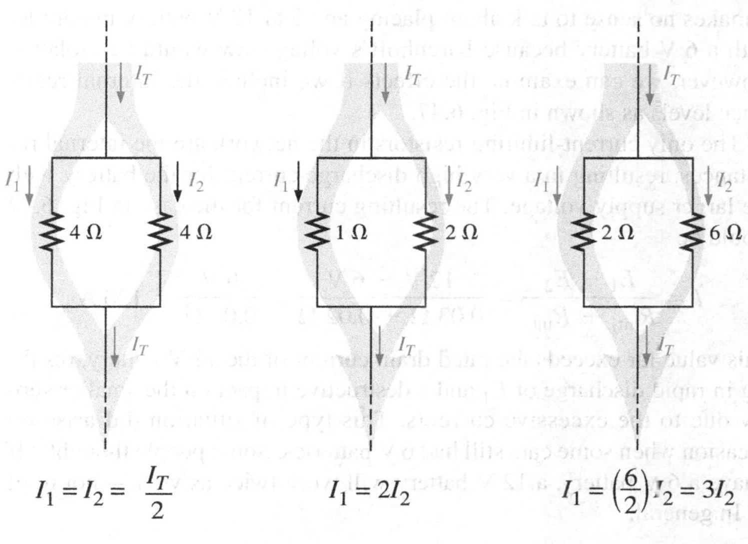

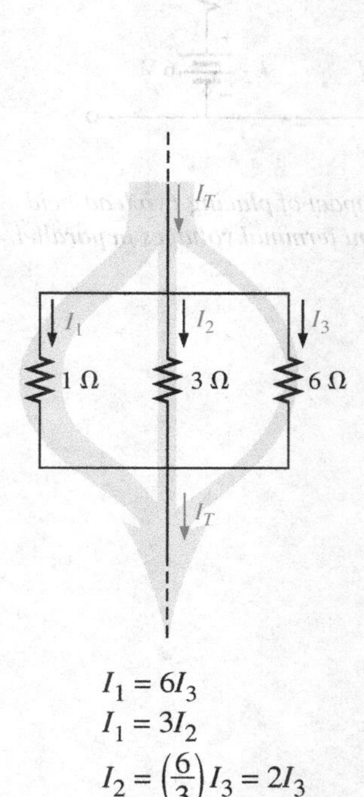

FIG. 6.45
Demonstrating how current divides through equal and unequal parallel resistors.

6.7 VOLTAGE SOURCES IN PARALLEL

Because the voltage is the same across parallel elements,

voltage sources can be placed in parallel only if they have the same voltage.

The primary reason for placing two or more batteries or supplies in parallel is to increase the current rating above that of a single supply. For example, in Fig. 6.46, two ideal batteries of 12 V have been placed in

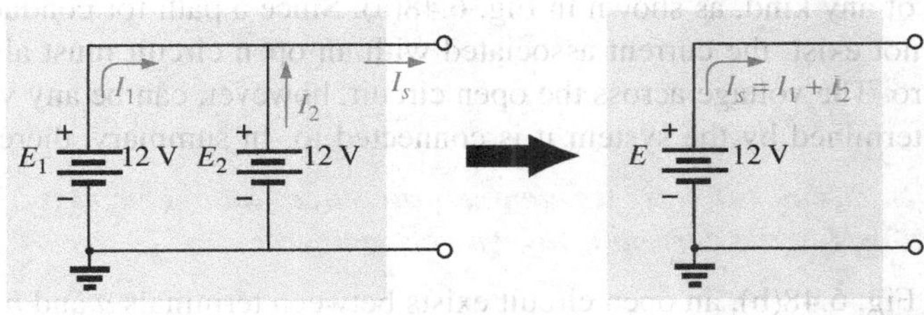

FIG. 6.46
Demonstrating the effect of placing two ideal supplies of the same voltage in parallel.

parallel. The total source current using Kirchhoff's current law is now the sum of the rated currents of each supply. The resulting power available will be twice that of a single supply if the rated supply current of each is the same. That is,

with $I_1 = I_2 = I$

then $P_T = E(I_1 + I_2) = E(I + I) = E(2I) = 2(EI) = 2P_{\text{(one supply)}}$

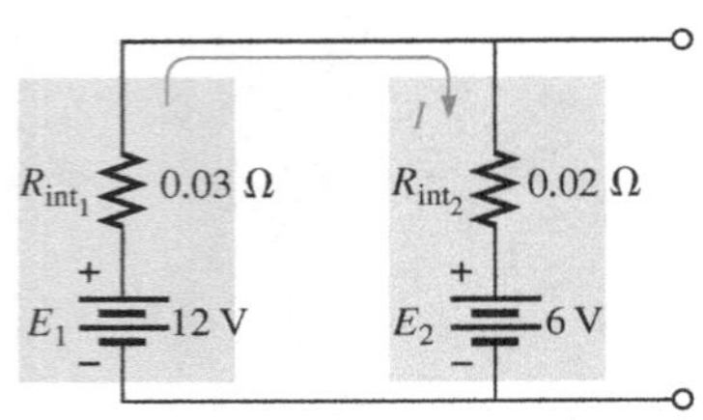

FIG. 6.47
Examining the impact of placing two lead-acid batteries of different terminal voltages in parallel.

If for some reason two batteries of different voltages are placed in parallel, both will become ineffective or damaged because the battery with the larger voltage will rapidly discharge through the battery with the smaller terminal voltage. For example, consider two lead-acid batteries of different terminal voltages placed in parallel as shown in Fig 6.47. It makes no sense to talk about placing an ideal 12 V battery in parallel with a 6 V battery because Kirchhoff's voltage law would be violated. However, we can examine the effects if we include the internal resistance levels as shown in Fig. 6.47.

The only current-limiting resistors in the network are the internal resistances, resulting in a very high discharge current for the battery with the larger supply voltage. The resulting current for the case in Fig. 6.47 would be

$$I = \frac{E_1 - E_2}{R_{\text{int}_1} + R_{\text{int}_2}} = \frac{12\text{ V} - 6\text{ V}}{0.03\ \Omega + 0.02\ \Omega} = \frac{6\text{ V}}{0.05\ \Omega} = 120\text{ A}$$

This value far exceeds the rated drain current of the 12 V battery, resulting in rapid discharge of E_1 and a destructive impact on the smaller supply due to the excessive currents. This type of situation did arise on occasion when some cars still had 6 V batteries. Some people thought, "If I have a 6 V battery, a 12 V battery will work twice as well"—not true!

In general,

it is always recommended that when you are replacing batteries in series or parallel, replace all the batteries.

A fresh battery placed in parallel with an older battery probably has a higher terminal voltage and immediately starts discharging through the older battery. In addition, the available current is less for the older battery, resulting in a higher-than-rated current drain from the newer battery when a load is applied.

6.8 OPEN AND SHORT CIRCUITS

Open circuits and short circuits can often cause more confusion and difficulty in the analysis of a system than standard series or parallel configurations. This will become more obvious in the chapters to follow when we apply some of the methods and theorems.

An **open circuit** is two isolated terminals not connected by an element of any kind, as shown in Fig. 6.48(a). Since a path for conduction does not exist, the current associated with an open circuit must always be zero. The voltage across the open circuit, however, can be any value, as determined by the system it is connected to. In summary, therefore,

an open circuit can have a potential difference (voltage) across its terminals, but the current is always zero amperes.

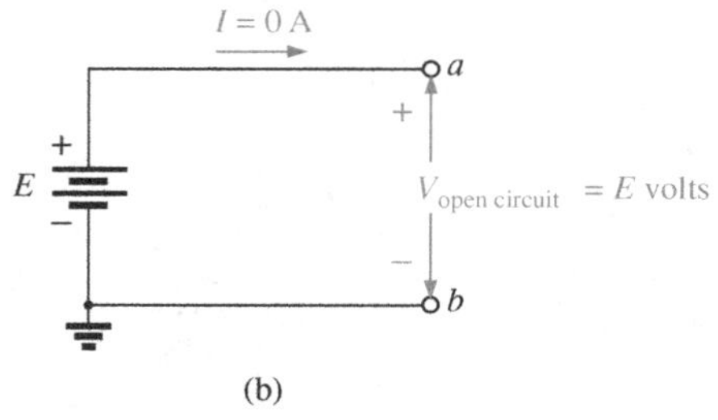

FIG. 6.48
Defining an open circuit.

In Fig. 6.48(b), an open circuit exists between terminals *a* and *b*. The voltage across the open-circuit terminals is the supply voltage, but the current is zero due to the absence of a complete circuit.

Some practical examples of open circuits and their impact are provided in Fig. 6.49. In Fig. 6.49(a), the excessive current demanded by the circuit caused a fuse to fail, creating an open circuit that reduced the current to zero amperes. However, it is important to note that *the full applied voltage is now across the open circuit,* so you must be careful when changing the fuse. If there is a main breaker ahead of the fuse, throw it first to remove the possibility of getting a shock. This situation clearly reveals the benefit of circuit breakers: You can reset the breaker without having to get near the hot wires.

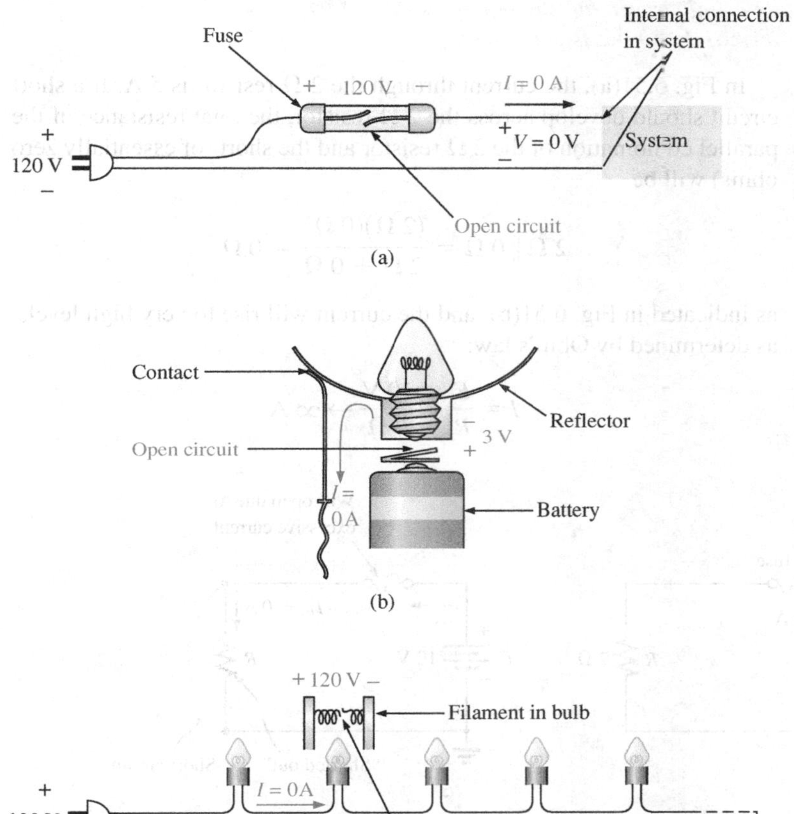

FIG. 6.49
Examples of open circuits.

In Fig. 6.49(b), the pressure plate at the bottom of the bulb cavity in a flashlight was bent when the flashlight was dropped. An open circuit now exists between the contact point of the bulb and the plate connected to the batteries. The current has dropped to zero amperes, but the 3 V provided by the series batteries appears across the open circuit. The situation can be corrected by placing a flat-edge screwdriver under the plate and bending it toward the bulb.

Finally, in Fig. 6.49(c), the filament in a bulb in a series connection has opened due to excessive current or old age, creating an open circuit that knocks out all the bulbs in the series configuration. Again, the current has dropped to zero amperes, but the full 120 V will appear across

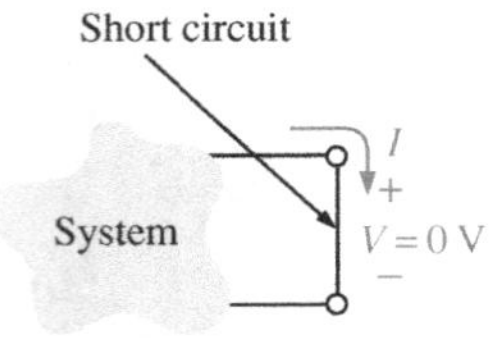

FIG. 6.50
Defining a short circuit.

the contact points of the bad bulb. For situations such as this, *you should remove the plug from the wall before changing the bulb.*

A **short circuit** is a very low resistance, direct connection between two terminals of a network, as shown in Fig. 6.50. The current through the short circuit can be any value, as determined by the system it is connected to, but the voltage across the short circuit is always zero volts because the resistance of the short circuit is assumed to be essentially zero ohms and $V = IR = I(0\ \Omega) = 0\ \text{V}$.

In summary, therefore,

a short circuit can carry a current of a level determined by the external circuit, but the potential difference (voltage) across its terminals is always zero volts.

In Fig. 6.51(a), the current through the 2 Ω resistor is 5 A. If a short circuit should develop across the 2 Ω resistor, the total resistance of the parallel combination of the 2 Ω resistor and the short (of essentially zero ohms) will be

$$2\ \Omega \parallel 0\ \Omega = \frac{(2\ \Omega)(0\ \Omega)}{2\ \Omega + 0\ \Omega} = 0\ \Omega$$

as indicated in Fig. 6.51(b), and the current will rise to very high levels, as determined by Ohm's law:

$$I = \frac{E}{R} = \frac{10\ \text{V}}{0\ \Omega} \rightarrow \infty\ \text{A}$$

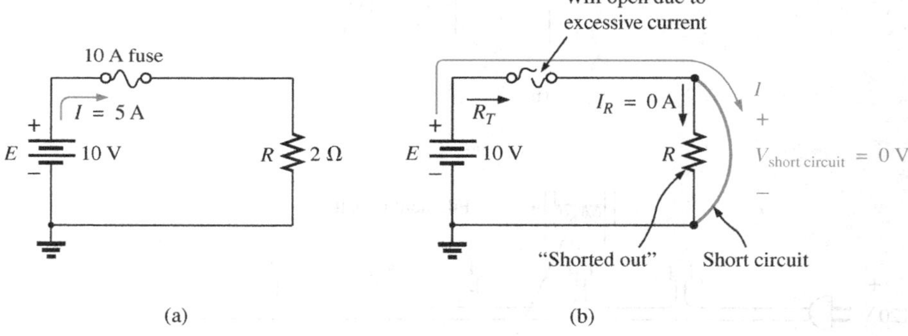

FIG. 6.51
Demonstrating the effect of a short circuit on current levels.

The effect of the 2 Ω resistor has effectively been "shorted out" by the low-resistance connection. The maximum current is now limited only by the circuit breaker or fuse in series with the source.

Some practical examples of short circuits and their impact are provided in Fig. 6.52. In Fig. 6.52(a), a hot (the feed) wire wrapped around a screw became loose and is touching the return connection. A short-circuit connection between the two terminals has been established that could result in a very heavy current and a possible fire hazard. One hopes that the breaker will "pop," and the circuit will be deactivated. Problems such as this are among the reasons aluminum wires (cheaper and lighter than copper) are not permitted in residential or industrial wiring. Aluminum is more sensitive to temperature than copper and will expand and contract due to the heat developed by the current passing

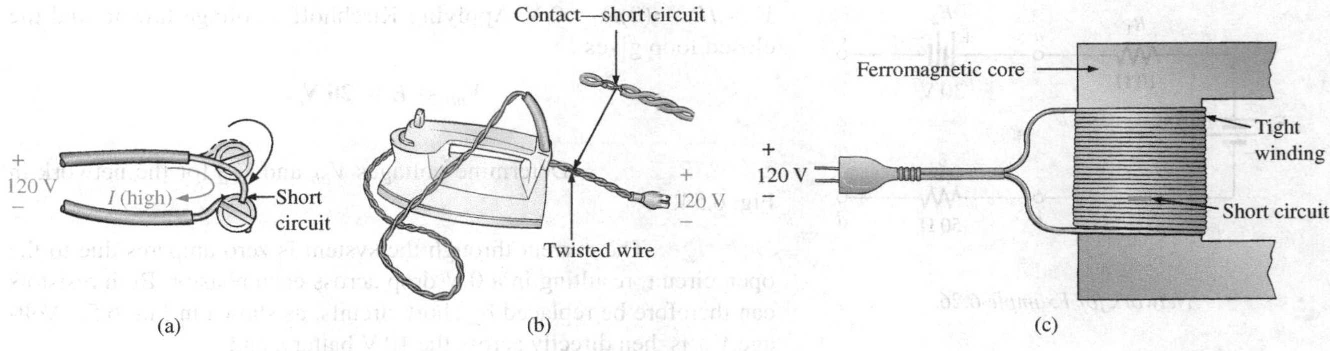

FIG. 6.52
Examples of short circuits.

through the wire. Eventually, this expansion and contraction can loosen the screw, and a wire under some torsional stress from the installation can move and make contact as shown in Fig. 6.52(a). Aluminum is still used in large panels as a bus-bar connection, but it is bolted down.

In Fig. 6.52(b), the wires of an iron have started to twist and crack due to excessive currents or long-term use of the iron. Once the insulation breaks down, the twisting can cause the two wires to touch and establish a short circuit. One can hope that a circuit breaker or fuse will quickly disconnect the circuit. Often, it is not the wire of the iron that causes the problem, but a cheap extension cord with the wrong gage wire. Be aware that you cannot tell the capacity of an extension cord by its outside jacket. It may have a thick orange covering but have a very thin wire inside. Check the gage on the wire the next time you buy an extension cord, and be sure that it is at least #14 gage, with #12 being the better choice for high-current appliances.

Finally, in Fig. 6.52(c), the windings in a transformer or motor for residential or industrial use are illustrated. The windings are wound so tightly together with such a very thin coating of insulation that it is possible with age and use for the insulation to break down and short out the windings. In many cases, shorts can develop, but a short will simply reduce the number of effective windings in the unit. The tool or appliance may still work but with less strength or rotational speed. If you notice such a change in the response, you should check the windings because a short can lead to a dangerous situation. In many cases, the state of the windings can be checked with a simple ohmmeter reading. If a short has occurred, the length of usable wire in the winding has been reduced, and the resistance drops. If you know what the resistance normally is, you can compare and make a judgment.

For the layperson, the terminology *short circuit* or *open circuit* is usually associated with dire situations such as power loss, smoke, or fire. However, in network analysis, both can play an integral role in determining specific parameters of a system. Most often, however, if a short-circuit condition is to be established, it is accomplished with a *jumper*—a lead of negligible resistance to be connected between the points of interest. Establishing an open circuit just requires making sure that the terminals of interest are isolated from each other.

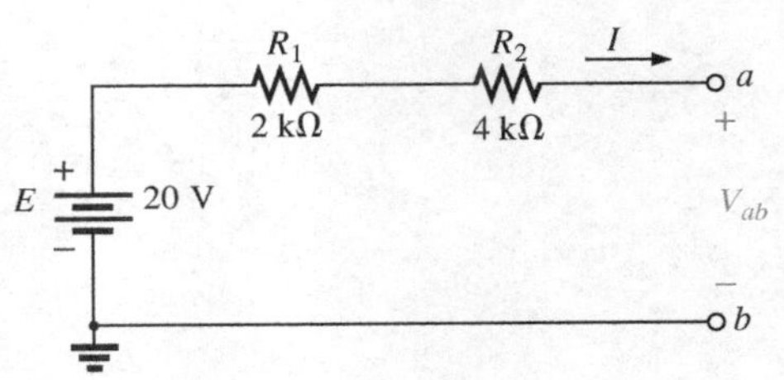

FIG. 6.53
Network for Example 6.25.

EXAMPLE 6.25 Determine voltage V_{ab} for the network in Fig. 6.53.

Solution: The open circuit requires that I be zero amperes. The voltage drop across both resistors is therefore zero volts since

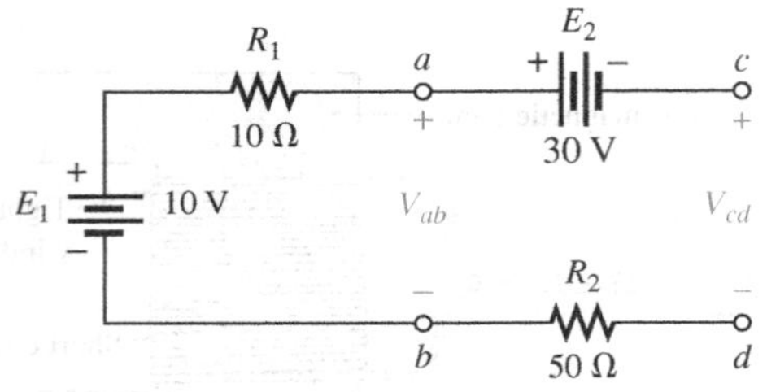

FIG. 6.54
Network for Example 6.26.

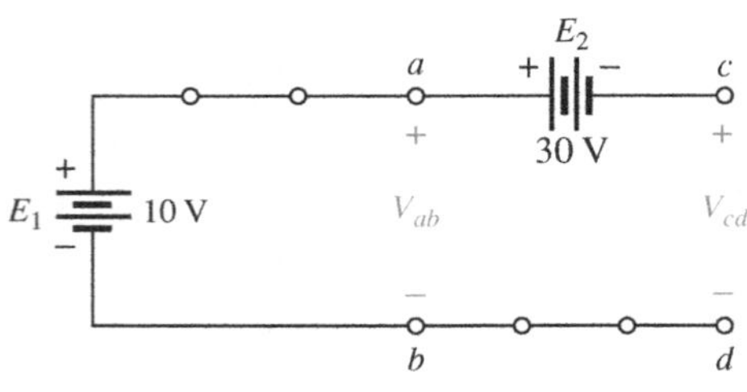

FIG. 6.55
Circuit in Fig. 6.54 redrawn.

$V = IR = (0)R = 0$ V. Applying Kirchhoff's voltage law around the closed loop gives

$$V_{ab} = E = \mathbf{20\ V}$$

EXAMPLE 6.26 Determine voltages V_{ab} and V_{cd} for the network in Fig. 6.54.

Solution: The current through the system is zero amperes due to the open circuit, resulting in a 0 V drop across each resistor. Both resistors can therefore be replaced by short circuits, as shown in Fig. 6.55. Voltage V_{ab} is then directly across the 10 V battery, and

$$V_{ab} = E_1 = \mathbf{10\ V}$$

Voltage V_{cd} requires an application of Kirchhoff's voltage law:

$$+E_1 - E_2 - V_{cd} = 0$$

or
$$V_{cd} = E_1 - E_2 = 10\ \text{V} - 30\ \text{V} = \mathbf{-20\ V}$$

The negative sign in the solution indicates that the actual voltage V_{cd} has the opposite polarity of that appearing in Fig. 6.54.

EXAMPLE 6.27 Determine the unknown voltage and current for each network in Fig. 6.56.

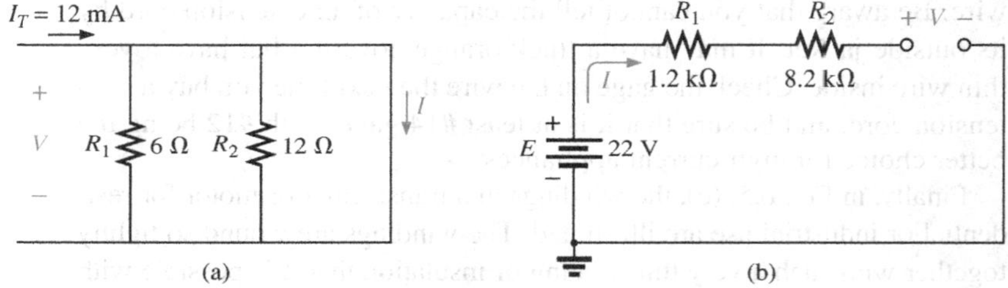

FIG. 6.56
Networks for Example 6.27.

Solution: For the network in Fig. 6.56(a), the current I_T will take the path of least resistance, and since the short-circuit condition at the end of the network is the least-resistance path, all the current will pass through the short circuit. This conclusion can be verified using the current divider rule. The voltage across the network is the same as that across the short circuit and will be zero volts, as shown in Fig. 6.57(a).

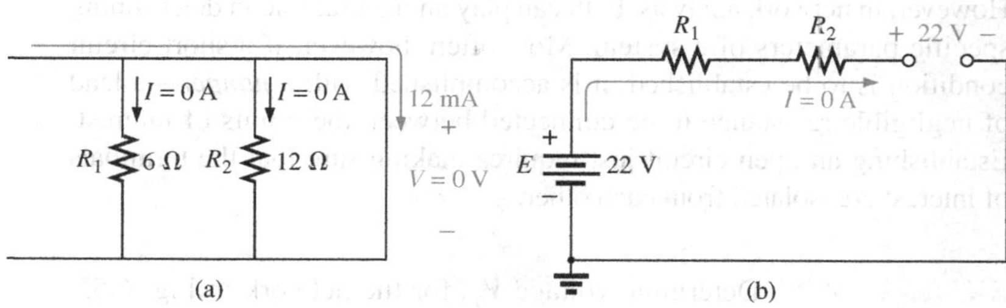

FIG. 6.57
Solutions to Example 6.27.

For the network in Fig. 6.56(b), the open-circuit condition requires that the current be zero amperes. The voltage drops across the resistors must therefore be zero volts, as determined by Ohm's law [$V_R = IR = (0)R = 0$ V], with the resistors acting as a connection from the supply to the open circuit. The result is that the open-circuit voltage is $E = 22$ V, as shown in Fig. 6.57(b).

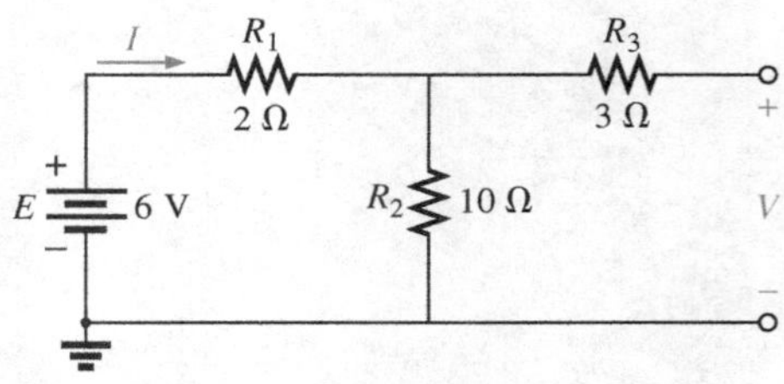

FIG. 6.58
Network for Example 6.28.

EXAMPLE 6.28 Determine V and I for the network in Fig. 6.58 if resistor R_2 is shorted out.

Solution: The redrawn network appears in Fig. 6.59. The current through the 3 Ω resistor is zero due to the open circuit, causing all the current I to pass through the jumper. Since $V_{3\Omega} = IR = (0)R = 0$ V, the voltage V is directly across the short, and

$$V = \mathbf{0\ V}$$

with
$$I = \frac{E}{R_1} = \frac{6\ \text{V}}{2\ \Omega} = \mathbf{3\ A}$$

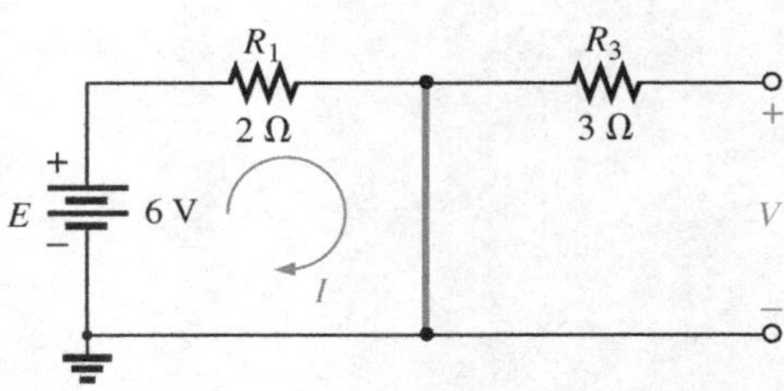

FIG. 6.59
Network in Fig. 6.58 with R_2 replaced by a jumper.

6.9 VOLTMETER LOADING EFFECTS

In previous chapters, we learned that ammeters are not ideal instruments. When you insert an ammeter, you actually introduce an additional resistance in series with the branch in which you are measuring the current. Generally, this is not a serious problem, but it can have a troubling effect on your readings, so it is important to be aware of it.

Voltmeters also have an internal resistance that appears between the two terminals of interest when a measurement is being made. While an ammeter places an additional resistance in series with the branch of interest, a voltmeter places an additional resistance *across* the element, as shown in Fig. 6.60. Since it appears in parallel with the element of interest, *the ideal level for the internal resistance of a voltmeter would be infinite ohms, just as zero ohms would be ideal for an ammeter.* Unfortunately, the internal resistance of any voltmeter is not infinite and changes from one type of meter to another.

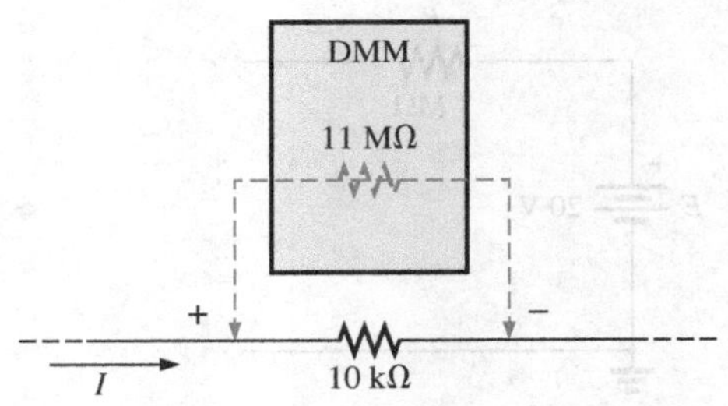

FIG. 6.60
Voltmeter loading.

Most digital meters have a fixed internal resistance level in the megohm range that remains the same *for all its scales.* For example, the meter in Fig. 6.60 has the typical level of 11 MΩ for its internal resistance, no matter which voltage scale is used. When the meter is placed across the 10 kΩ resistor, the total resistance of the combination is

$$R_T = 10\ \text{k}\Omega \parallel 11\ \text{M}\Omega = \frac{(10^4\ \Omega)(11 \times 10^6\ \Omega)}{10^4\ \Omega + (11 \times 10^6)} = 9.99\ \text{k}\Omega$$

and the behavior of the network is not seriously affected. The result, therefore, is that

most digital voltmeters can be used in circuits with resistances up to the high-kilohm range without concern for the effect of the internal resistance on the reading.

However, if the resistances are in the megohm range, you should investigate the effect of the internal resistance.

An analog VOM is a different matter, however, because the internal resistance levels are much lower and the internal resistance levels are a function of the scale used. If a VOM on the 2.5 V scale were placed across the 10 kΩ resistor in Fig. 6.60, the internal resistance might be 50 kΩ, resulting in a combined resistance of

$$R_T = 10\text{ k}\Omega \parallel 50\text{ k}\Omega = \frac{(10^4\ \Omega)(50 \times 10^3\ \Omega)}{10^4\ \Omega + (50 \times 10^3\ \Omega)} = 8.33\text{ k}\Omega$$

and the behavior of the network would be affected because the 10 kΩ resistor would appear as an 8.33 kΩ resistor.

To determine the resistance R_m of any scale of a VOM, simply multiply the **maximum voltage** of the chosen scale by the **ohm/volt (Ω/V) rating** normally appearing at the bottom of the face of the meter. That is,

$$R_m\ (\text{VOM}) = (\text{scale})(\Omega/\text{V rating})$$

For a typical Ω/V rating of 20,000, the 2.5 V scale would have an internal resistance of

$$(2.5\text{ V})(20{,}000\ \Omega/\text{V}) = \mathbf{50\ k\Omega}$$

whereas for the 100 V scale, the internal resistance of the VOM would be

$$(100\text{ V})(20{,}000\ \Omega/\text{V}) = \mathbf{2\ M\Omega}$$

and for the 250 V scale,

$$(250\text{ V})(20{,}000\ \Omega/\text{V}) = \mathbf{5\ M\Omega}$$

EXAMPLE 6.29 For the relatively simple circuit in Fig. 6.61(a):

a. What is the open-circuit voltage V_{ab}?
b. What will a DMM indicate if it has an internal resistance of 11 MΩ? Compare your answer to that of part (a).
c. Repeat part (b) for a VOM with an Ω/V rating of 20,000 on the 100 V scale.

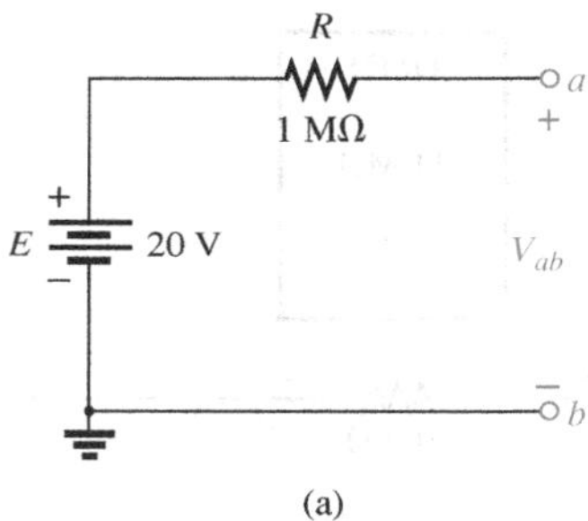

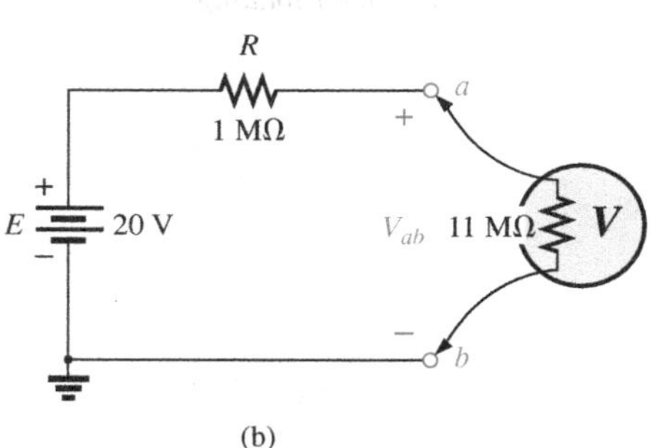

FIG. 6.61
(a) Measuring an open-circuit voltage with a voltmeter; (b) determining the effect of using a digital voltmeter with an internal resistance of 11 MΩ on measuring an open-circuit voltage (Example 6.29).

Solutions:

a. Due to the open circuit, the current is zero, and the voltage drop across the 1 MΩ resistor is zero volts. The result is that the entire source voltage appears between points *a* and *b*, and

$$V_{ab} = \mathbf{20\ V}$$

b. When the meter is connected as shown in Fig. 6.61(b), a complete circuit has been established, and current can pass through the circuit. The voltmeter reading can be determined using the voltage divider rule as follows:

$$V_{ab} = \frac{(11\text{ M}\Omega)(20\text{ V})}{(11\text{ M}\Omega + 1\text{ M}\Omega)} = \mathbf{18.33\ V}$$

and the reading is affected somewhat.

c. For the VOM, the internal resistance of the meter is

$$R_m = (100\text{ V})\ (20{,}000\ \Omega/\text{V}) = 2\text{ M}\Omega$$

and $$V_{ab} = \frac{(2\text{ M}\Omega)(20\text{ V})}{(2\text{ M}\Omega + 1\text{ M}\Omega)} = \mathbf{13.33\ V}$$

which is considerably below the desired level of 20 V.

6.10 SUMMARY TABLE

Now that the series and parallel configurations have been covered in detail, we will review the salient equations and characteristics of each. The equations for the two configurations have a number of similarities. In fact, the equations for one can often be obtained directly from the other by simply applying the **duality** principle. Duality between equations means that the format for an equation can be applied to two different situations by just changing the variable of interest. For instance, the equation for the total resistance of a series circuit is the sum of the resistances. By changing the resistance parameters to conductance parameters, you can obtain the equation for the total conductance of a parallel network—an easy way to remember the two equations. Similarly, by starting with the total conductance equation, you can easily write the total resistance equation for series circuits by replacing the conductance parameters by resistance parameters. Series and parallel networks share two important dual relationships: (1) between resistance of series circuits and conductance of parallel circuits and (2) between the voltage or current of a series circuit and the current or voltage, respectively, of a parallel circuit. Table 6.1 summarizes this duality.

TABLE 6.1
Summary table.

Series and Parallel Circuits		
Series	**Duality**	**Parallel**
$R_T = R_1 + R_2 + R_3 + \cdots + R_N$	$R \rightleftarrows G$	$G_T = G_1 + G_2 + G_3 + \cdots + G_N$
R_T increases (G_T decreases) if additional resistors are added in series	$R \rightleftarrows G$	G_T increases (R_T decreases) if additional resistors are added in parallel
Special case: two elements $R_T = R_1 + R_2$	$R \rightleftarrows G$	$G_T = G_1 + G_2$
I the same through series elements	$I \rightleftarrows V$	V the same across parallel elements
$E = V_1 + V_2 + V_3$	$E, V \rightleftarrows I$	$I_T = I_1 + I_2 + I_3$
Largest V across largest R	$V \rightleftarrows I$ and $R \rightleftarrows G$	Greatest I through largest G (smallest R)
$V_x = \dfrac{R_x E}{R_T}$	$E, V \rightleftarrows I$ and $R \rightleftarrows G$	$I_x = \dfrac{G_x I_T}{G_T}$
$P = EI_T$	$E \rightleftarrows I$ and $I \rightleftarrows E$	$P = I_T E$
$P = I^2R$	$I \rightleftarrows V$ and $R \rightleftarrows G$	$P = V^2G$
$P = V^2/R$	$V \rightleftarrows I$ and $R \rightleftarrows G$	$P = I^2/G$

The format for the total resistance for a series circuit has the same format as the total conductance of a parallel network, as shown in Table 6.1. All that is required to move back and forth between the series and parallel headings is to interchange the letters R and G. For the special case of two elements, the equations have the same format, but the equation applied for the total resistance of the parallel configuration has changed. In the series configuration, the total resistance increases with each added resistor. For parallel networks, the total conductance increases with each additional conductance. The result is that the total conductance of a series circuit drops with added resistive elements, while the total resistance of parallel networks decreases with added elements.

In a series circuit, the current is the same everywhere. In a parallel network, the voltage is the same across each element. The result is a duality between voltage and current for the two configurations. What is true for one in one configuration is true for the other in the other configuration. In a series circuit, the applied voltage divides between the series elements. In a parallel network, the current divides between parallel elements. For series circuits, the largest resistor captures the largest share of the applied voltage. For parallel networks, the branch with the highest conductance captures the greater share of the incoming current. In addition, for series circuits, the applied voltage equals the sum of the voltage drops across the series elements of the circuit, while the source current for parallel branches equals the sum of the currents through all the parallel branches.

The total power delivered to a series or parallel network is determined by the product of the applied voltage and resulting source current. The power delivered to each element is also the same for each configuration. Duality can be applied again, but the equation $P = EI$ results in the same result as $P = IE$. Also, $P = I^2R$ can be replaced by $P = V^2G$ for parallel elements, but essentially each can be used for each configuration. The duality principle can be very helpful in the learning process. Remember this as you progress through the next few chapters. You will find in the later chapters that this duality can also be applied between two important elements—inductors and capacitors.

6.11 TROUBLESHOOTING TECHNIQUES

The art of *troubleshooting* is not limited solely to electrical or electronic systems. In the broad sense,

troubleshooting is a process by which acquired knowledge and experience are used to localize a problem and offer or implement a solution.

There are many reasons why the simplest electrical circuit might not be operating correctly. A connection may be open; the measuring instruments may need calibration; the power supply may not be on or may have been connected incorrectly to the circuit; an element may not be performing correctly due to earlier damage or poor manufacturing; a fuse may have blown; and so on. Unfortunately, a defined sequence of steps does not exist for identifying the wide range of problems that can surface in an electrical system. It is only through experience and a clear understanding of the basic laws of electric circuits that you can become proficient at quickly locating the cause of an erroneous output.

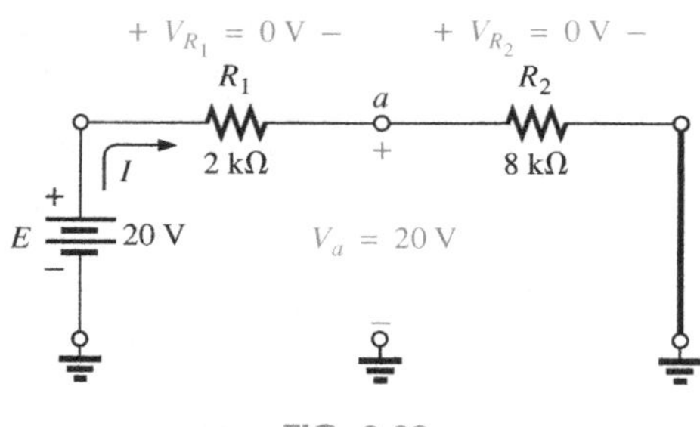

FIG. 6.62
A malfunctioning network.

It should be fairly obvious, however, that the first step in checking a network or identifying a problem area is to have some idea of the expected voltage and current levels. For instance, the circuit in Fig. 6.62 should have a current in the low milliampere range, with the majority of the supply voltage across the 8 kΩ resistor. However, as indicated in Fig. 6.62, $V_{R_1} = V_{R_2} = 0$ V and $V_a = 20$ V. Since $V = IR$, the results immediately suggest that $I = 0$ A and an open circuit exists in the circuit. The fact that $V_a = 20$ V immediately tells us that the connections are true from the ground of the supply to point a. The open circuit must therefore exist between R_1 and R_2 or at the ground connection of R_2. An open circuit at either point results in $I = 0$ A and the readings obtained previously. Keep in mind that, even though $I = 0$ A, R_1 does form a connection between the supply and point *a*. That is, if $I = 0$ A, $V_{R_1} = IR_2 = (0)R_2 = 0$ V, as obtained for a short circuit.

In Fig. 6.62, if $V_{R_1} \cong 20$ V and V_{R_2} is quite small ($\cong 0.08$ V), it first suggests that the circuit is complete, a current does exist, and a problem surrounds the resistor R_2. R_2 is not shorted out since such a condition would result in $V_{R_2} = 0$V. A careful check of the inserted resistor reveals that an 8 Ω resistor was used rather than the 8 kΩ resistor specified—an incorrect reading of the color code. To avoid this, an ohmmeter should be used to check a resistor to validate the color-code reading or to ensure that its value is still in the prescribed range set by the color code.

Occasionally, the problem may be difficult to diagnose. You've checked all the elements, and all the connections appear tight. The supply is on and set at the proper level; the meters appear to be functioning correctly. In situations such as this, experience becomes a key factor. Perhaps you can recall when a recent check of a resistor revealed that the internal connection (not externally visible) was a "make or break" situation or that the resistor was damaged earlier by excessive current levels, so its actual resistance was much lower than called for by the color code. Recheck the supply! Perhaps the terminal voltage was set correctly, but the current control knob was left in the zero or minimum position. Is the ground connection stable? The questions that arise may seem endless. However, as you gain experience, you will be able to localize problems more rapidly. Of course, the more complicated the system, the longer is the list of possibilities, but it is often possible to identify a particular area of the system that is behaving improperly before checking individual elements.

6.12 PROTOBOARDS (BREADBOARDS)

In Section 5.12, the protoboard was introduced with the connections for a simple series circuit. To continue the development, the network in Fig. 6.17 was set up on the board in Fig. 6.63(a) using two different techniques. The possibilities are endless, but these two solutions use a fairly straightforward approach.

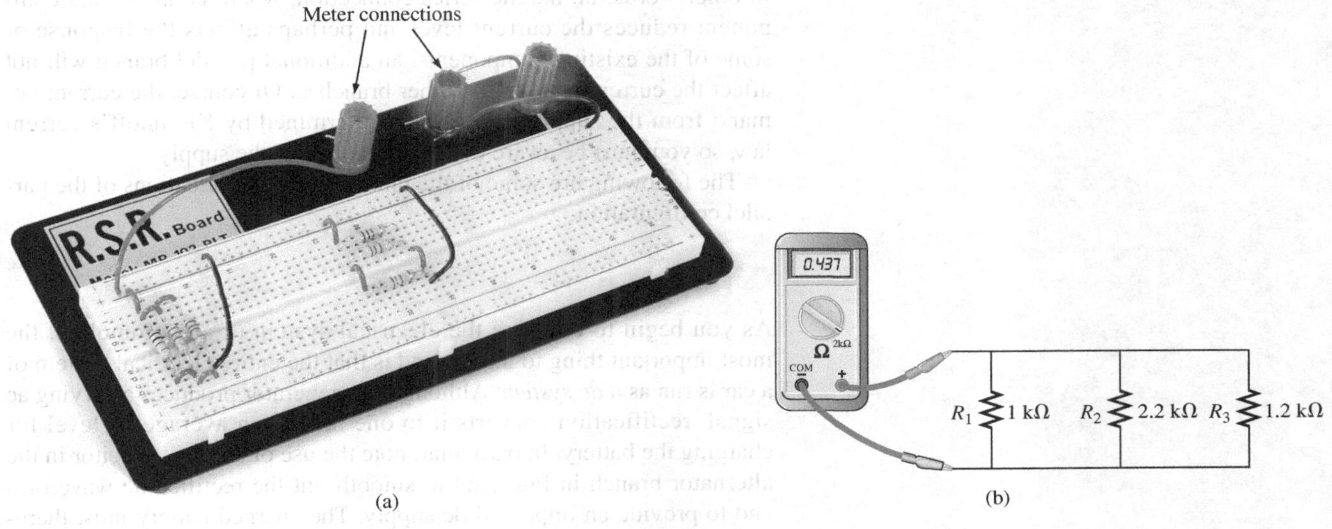

FIG. 6.63
Using a protoboard to set up the circuit in Fig. 6.17.

First, note that the supply lines and ground are established across the length of the board using the horizontal conduction zones at the top and bottom of the board through the connections to the terminals. The network to the left on the board was used to set up the circuit in much the same manner as it appears in the schematic of Fig. 6.63(b). This approach required that the resistors be connected between two vertical conducting strips. If placed perfectly vertical in a single conducting strip, the resistors would have shorted out. Often, setting the network up in a manner that best copies the original can make it easier to check and make measurements. The network to the right in part (a) used the vertical conducting strips to connect the resistors together at each end. Since there wasn't enough room for all three, a connection had to be added from the upper vertical set to the lower set. The resistors are in order R_1, R_2, and R_3 from the top down. For both configurations, the ohmmeter can be connected to the positive lead of the supply terminal and the negative or ground terminal.

Take a moment to review the connections and think of other possibilities. Improvements can often be made, and it can be satisfying to find the most effective setup with the least number of connecting wires.

6.13 APPLICATIONS

One of the most important advantages of the parallel configuration is that

if one branch of the configuration should fail (open circuit), the remaining branches will still have full operating power.

In a home, the parallel connection is used throughout to ensure that if one circuit has a problem and opens the circuit breaker, the remaining circuits still have the full 120 V. The same is true in automobiles, computer systems, industrial plants, and wherever it would be disastrous for one circuit to control the total power distribution.

Another important advantage is that

branches can be added at any time without affecting the behavior of those already in place.

In other words, unlike the series connection, where an additional component reduces the current level and perhaps affects the response of some of the existing components, an additional parallel branch will not affect the current level in the other branches. Of course, the current demand from the supply increases as determined by Kirchhoff's current law, so you must be aware of the limitations of the supply.

The following are some of the most common applications of the parallel configuration.

Car System

As you begin to examine the electrical system of an automobile, the most important thing to understand is that the entire electrical system of a car is run as a *dc system.* Although the generator produces a varying ac signal, rectification converts it to one having an average dc level for charging the battery. In particular, note the use of a filter capacitor in the alternator branch in Fig. 6.64 to smooth out the rectified ac waveform and to provide an improved dc supply. The charged battery must therefore provide the required direct current for the entire electrical system of the car. Thus, the power demand on the battery at any instant is the product

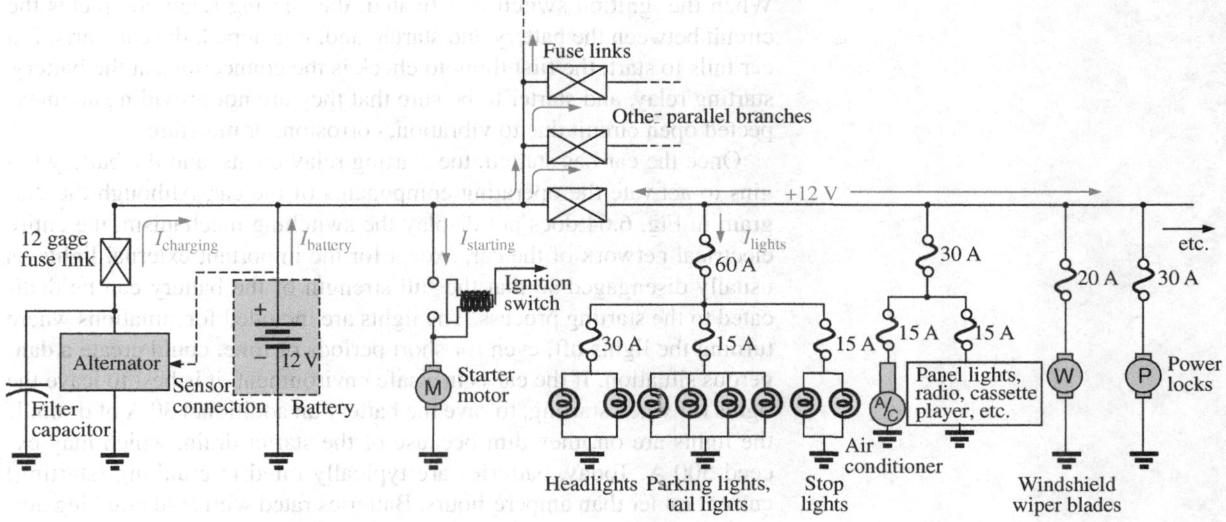

FIG. 6.64
Expanded view of an automobile's electrical system.

of the terminal voltage and the current drain of the total load of every operating system of the car. This certainly places an enormous burden on the battery and its internal chemical reaction and warrants all the battery care we can provide.

Since the electrical system of a car is essentially a parallel system, the total current drain on the battery is the sum of the currents to all the parallel branches of the car connected directly to the battery. In Fig. 6.64, a few branches of the wiring diagram for a car have been sketched to provide some background information on basic wiring, current levels, and fuse configurations. Every automobile has fuse links and fuses, and some also have circuit breakers, to protect the various components of the car and to ensure that a dangerous fire situation does not develop. Except for a few branches that may have series elements, the operating voltage for most components of a car is the terminal voltage of the battery, which we will designate as 12 V even though it will typically vary between 12 V and the charging level of 14.6 V. In other words, each component is connected to the battery at one end and to the ground or chassis of the car at the other end.

Referring to Fig. 6.64, note that the alternator or charging branch of the system is connected directly across the battery to provide the charging current as indicated. Once the car is started, the rotor of the alternator turns, generating an ac varying voltage that then passes through a rectifier network and filter to provide the dc charging voltage for the battery. Charging occurs only when the sensor connected directly to the battery signals that the terminal voltage of the battery is too low. Just to the right of the battery the starter branch was included to demonstrate that there is no fusing action between the battery and starter when the ignition switch is activated. The lack of fusing action is provided because enormous starting currents (hundreds of amperes) flow through the starter to start a car that has not been used for days and/or has been sitting in a cold climate—and high friction occurs between components until the oil starts flowing. The starting level can vary so much that it would be difficult to find the right fuse level, and frequent high currents may damage the fuse link and cause a failure at expected levels of current.

When the ignition switch is activated, the starting relay completes the circuit between the battery and starter, and, it is hoped, the car starts. If a car fails to start, the first thing to check is the connections at the battery, starting relay, and starter to be sure that they are not providing an unexpected open circuit due to vibration, corrosion, or moisture.

Once the car has started, the starting relay opens, and the battery begins to activate the operating components of the car. Although the diagram in Fig. 6.64 does not display the switching mechanism, the entire electrical network of the car, except for the important external lights, is usually disengaged so that the full strength of the battery can be dedicated to the starting process. The lights are included for situations where turning the lights off, even for short periods of time, could create a dangerous situation. If the car is in a safe environment, it is best to leave the lights off when starting, to save the battery an additional 30 A of drain. If the lights are on, they dim because of the starter drain, which may exceed 500 A. Today, batteries are typically rated in cranking (starting) current rather than ampere-hours. Batteries rated with cold cranking ampere ratings between 700 A and 1000 A are typical today.

Separating the alternator from the battery and the battery from the numerous networks of the car are fuse links such as shown in Fig. 6.65. Fuse links are actually wires of a specific gage designed to open at fairly high current levels of 100 A or more. They are included to protect against those situations where there is an unexpected current drawn from the many circuits to which they are connected. That heavy drain can, of course, be from a short circuit in one of the branches, but in such cases the fuse in that branch will probably release. The fuse link is an additional protection for the line if the total current drawn by the parallel-connected branches begins to exceed safe levels. The fuses following the fuse link have the appearance shown in Fig. 6.65(b), where a gap between the legs of the fuse indicates a blown fuse. As shown in Fig. 6.64, the 60 A fuse (often called a *power distribution fuse*) for the lights is a second-tier fuse sensitive to the total drain from the three light circuits. Finally, the third fuse level is for the individual units of a car such as the lights, air conditioner, and power locks. In each case, the fuse rating exceeds the normal load (current level) of the operating component, but the level of each fuse does give some indication of the demand to be expected under normal operating conditions. For instance, headlights typically draw more than 10 A, tail lights more than 5 A, air conditioner about 10 A (when the clutch engages), and power windows 10 A to 20 A, depending on how many are operated at once.

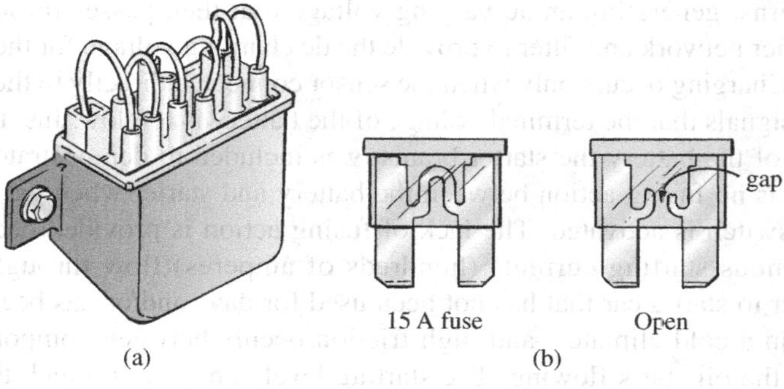

FIG. 6.65
Car fuses: (a) fuse link; (b) plug-in.

Some details for only one section of the total car network are provided in Fig. 6.64. In the same figure, additional parallel paths with their respective fuses have been provided to further reveal the parallel arrangement of all the circuits.

In most vehicles the return path to the battery through the ground connection is through the chassis of the car. That is, there is only one wire to each electrical load, with the other end simply grounded to the chassis. The return to the battery (chassis to negative terminal) is therefore a heavy-gage wire matching that connected to the positive terminal. In some cars constructed of a mixture of materials such as metal, plastic, and rubber, the return path through the metallic chassis may be lost, and two wires must be connected to each electrical load of the car.

House Wiring

In Chapter 4, the basic power levels of importance were discussed for various services to the home. We are now ready to take the next step and examine the actual connection of elements in the home.

First, it is important to realize that except for some very special circumstances, the basic wiring is done in a parallel configuration. Each parallel branch, however, can have a combination of parallel and series elements. Every full branch of the circuit receives the full 120 V or 208 V, with the current determined by the applied load. Figure 6.66(a) provides the detailed wiring of a single circuit having a light bulb and two outlets. Figure 6.66(b) shows the schematic representation. Note that although each load is in parallel with the supply, switches are always connected in series with the load. The power is transmitted to the lamp only when the switch is closed and the full 120 V appears across the bulb. The connection point for the two outlets is in the ceiling box holding the light bulb. Since a switch is not present, both outlets are always "hot" unless the circuit breaker in the main panel is opened. This is important to understand in case you are tempted to change the light fixture by simply turning off the wall switch. True, if you're very careful, you can work with one line at a time (being sure that you don't touch the other line at any time), but it is much safer to throw the circuit breaker on the panel whenever working on a circuit. Note in Fig. 6.66(a) that the *feed* wire (black) into the fixture from the panel is connected to the switch and both outlets at one point. It is not connected directly to the light fixture because the lamp would be on all the time. Power to the light fixture is made available through the switch. The continuous connection to the outlets from the panel ensures that the outlets are "hot" whenever the circuit breaker in the panel is on. Note also how the *return* wire (white) is connected directly to the light switch and outlets to provide a return for each component. There is no need for the white wire to go through the switch since an applied voltage is a two-point connection and the black wire is controlled by the switch.

Proper grounding of the system in total and of the individual loads is one of the most important facets in the installation of any system. There is a tendency at times to be satisfied that the system is working and to pay less attention to proper grounding technique. Always keep in mind that a properly grounded system has a direct path to ground if an undesirable situation should develop. The absence of a direct ground causes the system to determine its own path to ground, and you could be that path if you happened to touch the wrong wire, metal box, metal pipe,

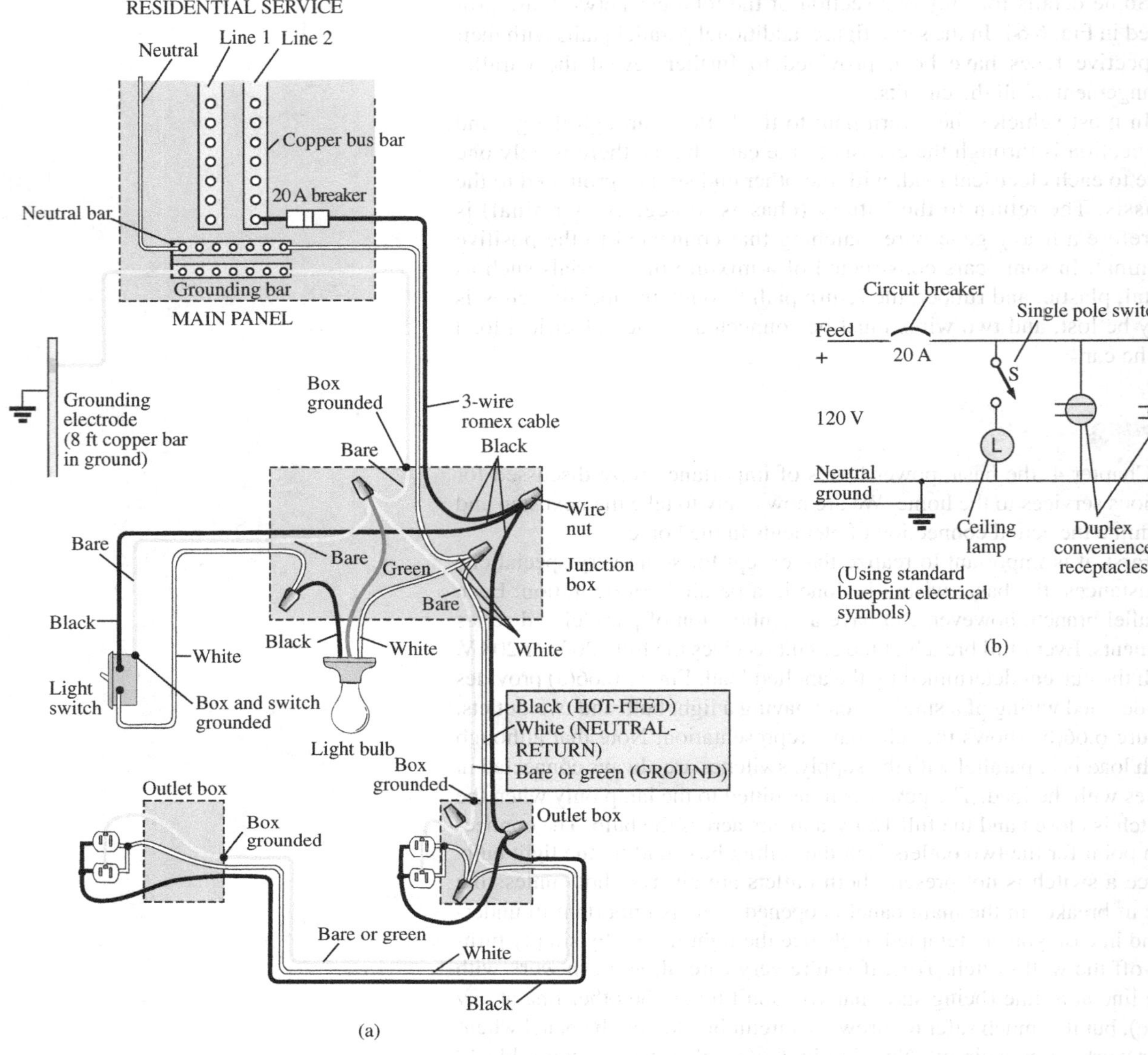

FIG. 6.66
Single phase of house wiring: (a) physical details; (b) schematic representation.

and so on. In Fig. 6.66(a), the connections for the ground wires have been included. For the romex (plastic-coated wire) used in Fig. 6.66(a), the ground wire is provided as a bare copper wire. Note that it is connected to the panel, which in turn is directly connected to the grounded 8 ft copper rod. In addition, note that the ground connection is carried through the entire circuit, including the switch, light fixture, and outlets. It is one continuous connection. If the outlet box, switch box, and housing for the light fixture are made of a conductive material such as metal, the ground will be connected to each. If each is plastic, there is no need for the ground connection. However, the switch, both outlets, and the fixture itself are connected to ground. For the switch and outlets, there is usually a green screw for the ground wire, which is connected to the entire framework of the switch or outlet as shown in Fig. 6.67, including the ground connection of the outlet. For both the switch and the outlet, even the screw or screws used to hold the outside plate in place are grounded since they are screwed into the metal housing of the switch or outlet. When screwed into a metal box, the ground connection can be made by the screws that hold the switch or outlet in the box as shown in Fig. 6.67. *Always pay strict attention to the grounding process whenever installing any electrical equipment.*

On the practical side, whenever hooking up a wire to a screw-type terminal, always wrap the wire around the screw in the clockwise manner so that when you tighten the screw, it grabs the wire and turns it in the same direction. An expanded view of a typical house-wiring arrangement appears in Chapter 15.

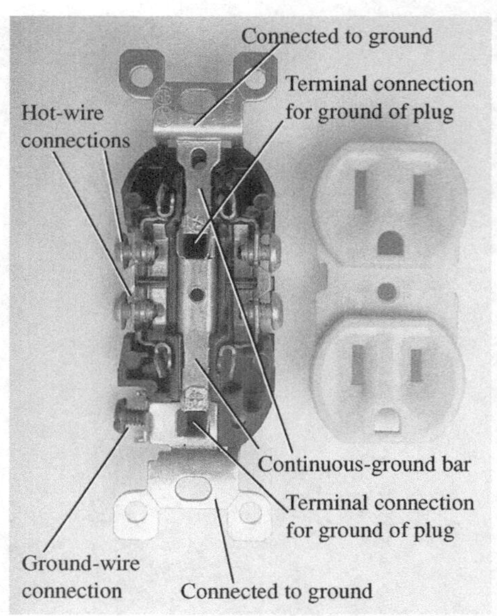

FIG. 6.67
Continuous ground connection in a duplex outlet.

Parallel Computer Bus Connections

The internal construction (hardware) of large mainframe computers and personal computers is set up to accept a variety of adapter cards in the slots appearing in Fig. 6.68(a). The primary board (usually the largest), commonly called the *motherboard,* contains most of the functions

(a)

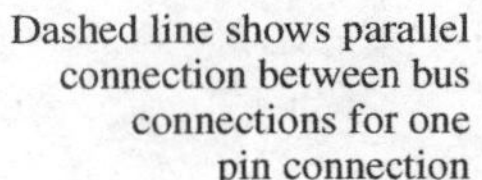

(b)

FIG. 6.68
(a) Motherboard for a desktop computer; (b) the printed circuit board connections for the region indicated in part (a).

required for full computer operation. Adapter cards are normally added to expand the memory, set up a network, add peripheral equipment, and so on. For instance, if you decide to add another hard drive to your computer, you can simply insert the card into the proper channel of Fig. 6.68(a). The bus connectors are connected in parallel with common connections to the power supply, address and data buses, control signals, ground, and so on. For instance, if the bottom connection of each bus connector is a ground connection, that ground connection carries through each bus connector and is immediately connected to any adapter card installed. Each card has a slot connector that fits directly into the bus connector without the need for any soldering or construction. The pins of the adapter card are then designed to provide a path between the motherboard and its components to support the desired function. Note in Fig. 6.68(b), which is a back view of the region identified in Fig. 6.68(a), that if you follow the path of the second pin from the top on the far left, you will see that it is connected to the same pin on the other three bus connectors.

Most small laptop computers today have all the options already installed, thereby bypassing the need for bus connectors. Additional memory and other upgrades are added as direct inserts into the motherboard.

6.14 COMPUTER ANALYSIS

PSpice

Parallel dc Network The computer analysis coverage for parallel dc circuits is very similar to that for series dc circuits. However, in this case the voltage is the same across all the parallel elements, and the current through each branch changes with the resistance value. The parallel network to be analyzed will have a wide range of resistor values to demonstrate the effect on the resulting current. The following is a list of abbreviations for any parameter of a network when using PSpice:

$$\mathbf{f} = 10^{-15}$$
$$\mathbf{p} = 10^{-12}$$
$$\mathbf{n} = 10^{-9}$$
$$\mathbf{u} = 10^{-6}$$
$$\mathbf{m} = 10^{-3}$$
$$\mathbf{k} = 10^{+3}$$
$$\mathbf{MEG} = 10^{+6}$$
$$\mathbf{G} = 10^{+9}$$
$$\mathbf{T} = 10^{+12}$$

In particular, note that **m** (or **M**) is used for "milli" and **MEG** for "megohms." Also, PSpice does not distinguish between upper- and lower-case units, but certain parameters typically use either the upper- or lower-case abbreviation as shown above.

Since the details of setting up a network and going through the simulation process were covered in detail in Sections 4.9 and 5.14 for dc circuits, the coverage here is limited solely to the various steps required. These steps will help you learn how to "draw" a circuit and then run a simulation fairly quickly and easily.

After selecting the **Create document** key (the top left of the screen), the following sequence opens the **Schematic** window: **PSpice 6-1-OK-Create a blank project-OK-PAGE1** (if necessary).

Add the voltage source and resistors as described in detail in earlier sections, but now you need to turn the resistors 90°. You do this by right-clicking before setting a resistor in place. Choose **Rotate** from the list of options, which turns the resistor counterclockwise 90°. It can also be rotated by simultaneously selecting **Ctrl-R.** The resistor can then be placed in position by a left click. An additional benefit of this maneuver is that the remaining resistors to be placed will already be in the vertical position. The values selected for the voltage source and resistors appear in Fig. 6.69.

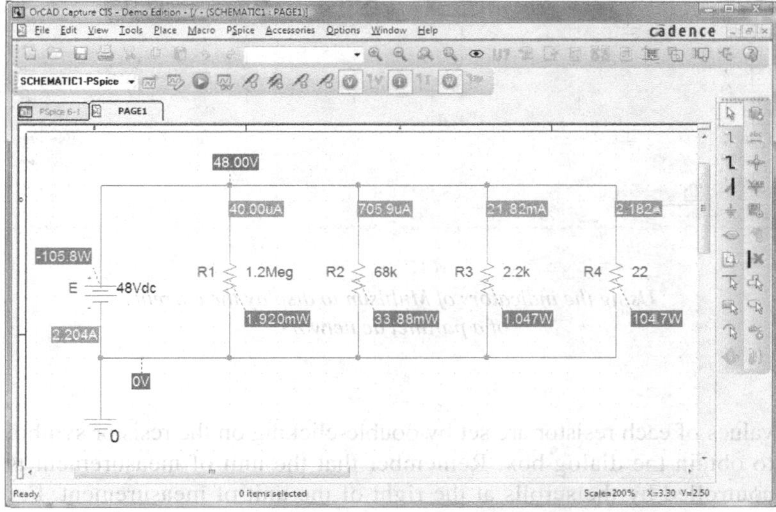

FIG. 6.69

Applying PSpice to a parallel network.

Once the network is complete, you can obtain the simulation and the results through the following sequence: **Select New Simulation Profile** key-**Bias Point-Create-Analysis-Bias Point-OK-Run PSpice** key-**Exit(X).**

The result, shown in Fig. 6.69, reveals that the voltage is the same across all the parallel elements, and the current increases significantly with decrease in resistance. The range in resistor values suggests, by inspection, that the total resistance is just less than the smallest resistance of 22 Ω. Using Ohm's law and the source current of 2.204 A results in a total resistance of $R_T = E/I_s = 48\text{ V}/2.204\text{ A} = 21.78\ \Omega$, confirming the above conclusion.

Multisim

Parallel dc Network For comparison purposes with the PSpice approach, the same parallel network in Fig. 6.69 is now analyzed using Multisim. The source and ground are selected and placed as shown in Fig. 6.70 using the procedure defined in previous chapters. For the resistors, choose the resistor symbol in the **BASIC toolbar** listing. However, you must rotate it 90° to match the configuration of Fig. 6.69. You do this by first clicking on the resistor symbol to place it in the active state. (Be sure that the resulting small black squares surround the symbol, label, and value; otherwise, you may have activated only the label or value.) Then right-click inside the rectangle. Select **90° Clockwise,** and the resistor is turned automatically. Unfortunately, there is no continuum here, so the next resistor has to be turned using the same procedure. The

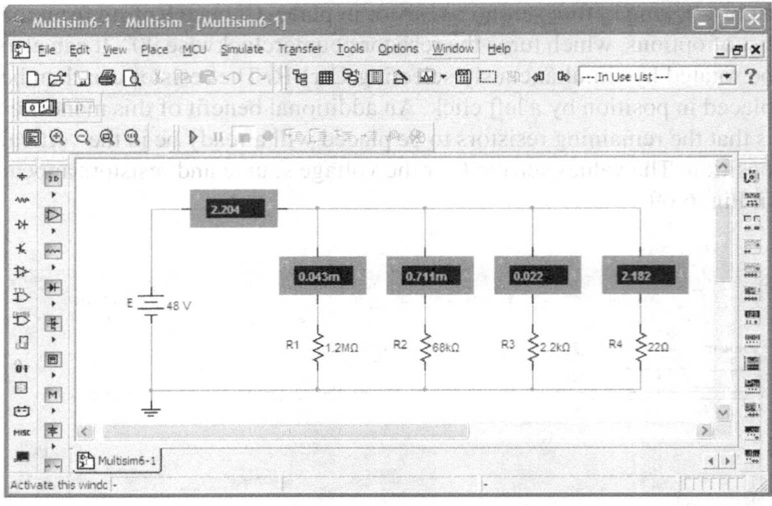

FIG. 6.70
Using the indicators of Multisim to display the currents of a parallel dc network.

values of each resistor are set by double-clicking on the resistor symbol to obtain the dialog box. Remember that the unit of measurement is controlled by the scrolls at the right of the unit of measurement. For Multisim, unlike PSpice, megohm uses capital **M** and milliohm uses lowercase **m.**

This time, rather than use the full meter employed in earlier measurements, let us use the measurement options available in the **Virtual** (also called **BASIC**) **toolbar.** If it is not already available, the toolbar can be obtained through the sequence **View-Toolbars-Virtual**. If the key in the toolbar that looks like a small meter **(Show Measurement Family)** is chosen, it will present four options for the use of an ammeter, four for a voltmeter, and five probes. The four choices for an ammeter simply set the position and the location of the positive and negative connectors. The option **Place Ammeter (Horizontal)** sets the ammeter in the horizontal position as shown in Fig. 6.70 in the top left of the diagram with the plus sign on the left and the minus sign on the right—the same polarity that would result if the current through a resistor in the same position were from left to right. Choosing **Place Ammeter (Vertical)** will result in the ammeters in the vertical sections of the network with the positive connection at the top and negative connection at the bottom, as shown in Fig. 6.70 for the four branches. If you chose **Place Ammeter (Horizontally rotated)** for the source current, it would simply reverse the positions of the positive and negative signs and provide a negative answer for the reading. If **Place Ammeter (Vertically rotated)** was chosen for the vertical branches, the readings would all be correct but with negative signs. Once all the elements are in place and their values set, initiate simulation with the sequence **Simulate-Run.** The results shown in Fig. 6.70 appear.

Note that all the results appear with the meter boxes. All are positive results because the ammeters were all entered with a configuration that would result in conventional current entering the positive current. Also note that, as was true for inserting the meters, the meters are placed in series with the branch in which the current is to be measured.

PROBLEMS

SECTION 6.2 Parallel Resistors

1. For each configuration in Fig. 6.71, find the voltage sources and/or resistors elements (individual elements, not combinations of elements) that are in parallel. Remember that elements in parallel have the same voltage.

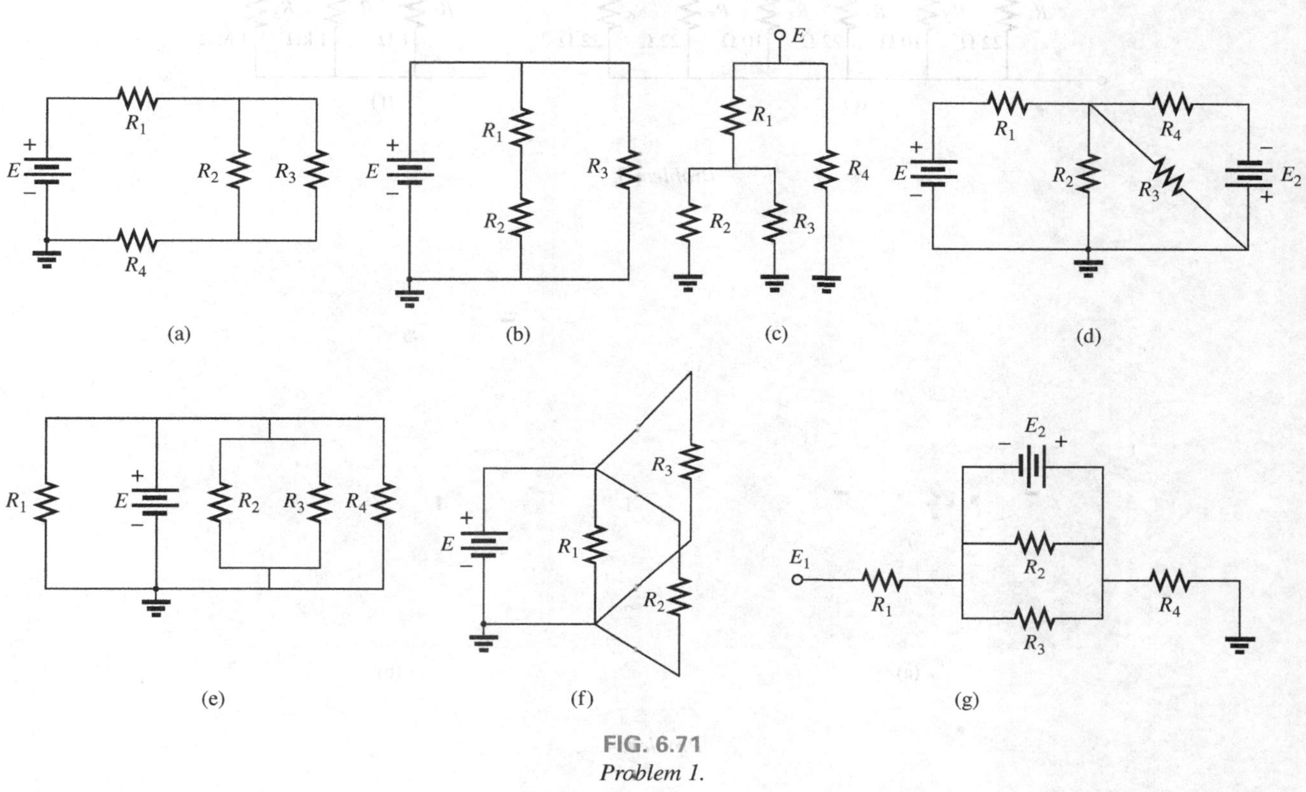

FIG. 6.71
Problem 1.

2. For the network in Fig. 6.72:
   a. Find the elements (voltage sources and/or resistors) that are in parallel.
   b. Find the elements (voltage sources and/or resistors) that are in series.
3. Find the total resistance for each configuration in Fig. 6.73. Note that only standard value resistors were used.
4. For each circuit board in Fig. 6.74, find the total resistance between connection tabs 1 and 2.
5. The total resistance of each of the configurations in Fig. 6.75 is specified. Find the unknown resistance levels.

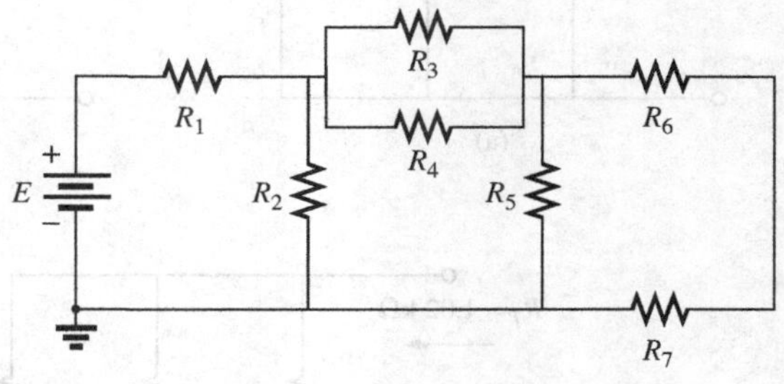

FIG. 6.72
Problem 2.

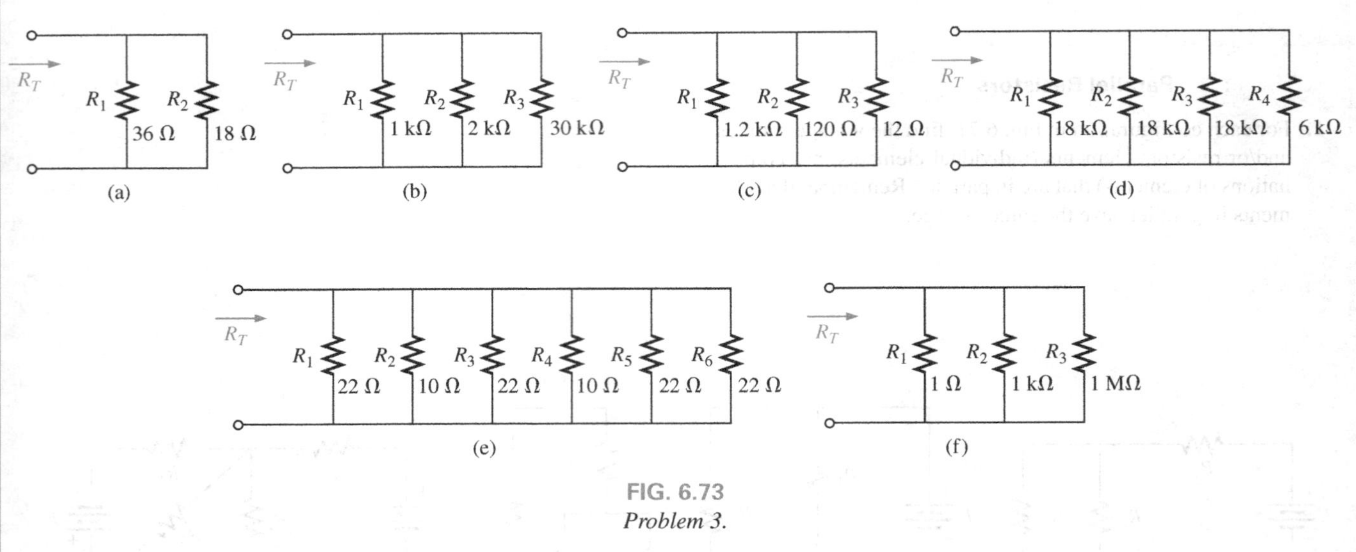

FIG. 6.73
Problem 3.

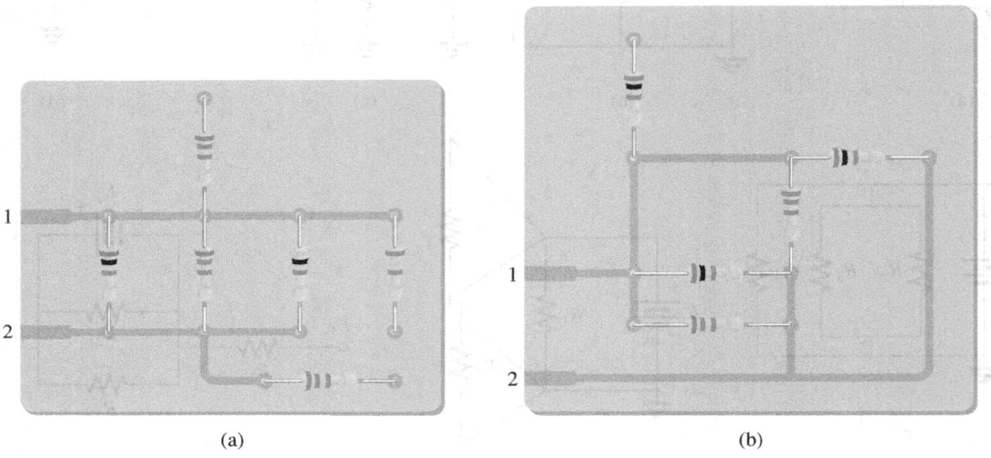

FIG. 6.74
Problem 4.

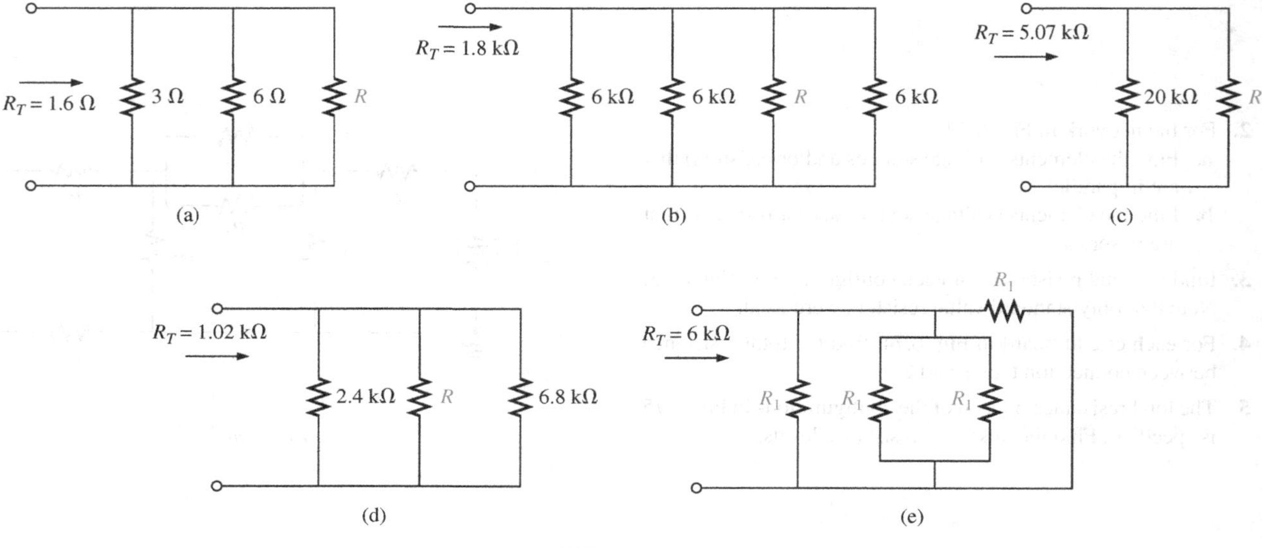

FIG. 6.75
Problem 5.

6. For the parallel network in Fig. 6.76, composed of standard values:
   a. Which resistor has the most impact on the total resistance?
   b. Without making a single calculation, what is an approximate value for the total resistance?
   c. Calculate the total resistance, and comment on your response to part (b).
   d. On an approximate basis, which resistors can be ignored when determining the total resistance?
   e. If we add another parallel resistor of any value to the network, what is the impact on the total resistance?

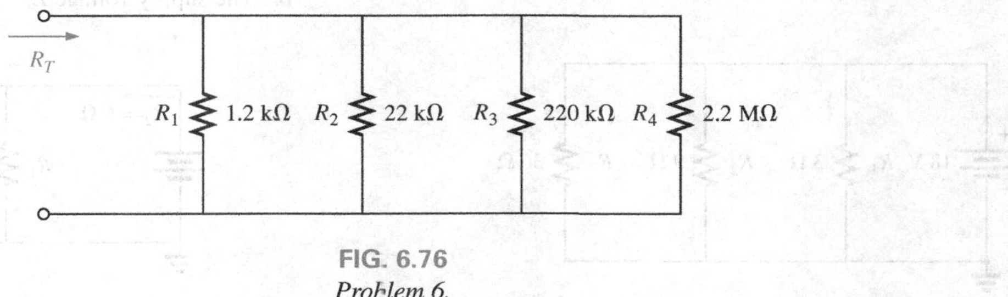

FIG. 6.76
Problem 6.

7. What is the ohmmeter reading for each configuration in Fig. 6.77?

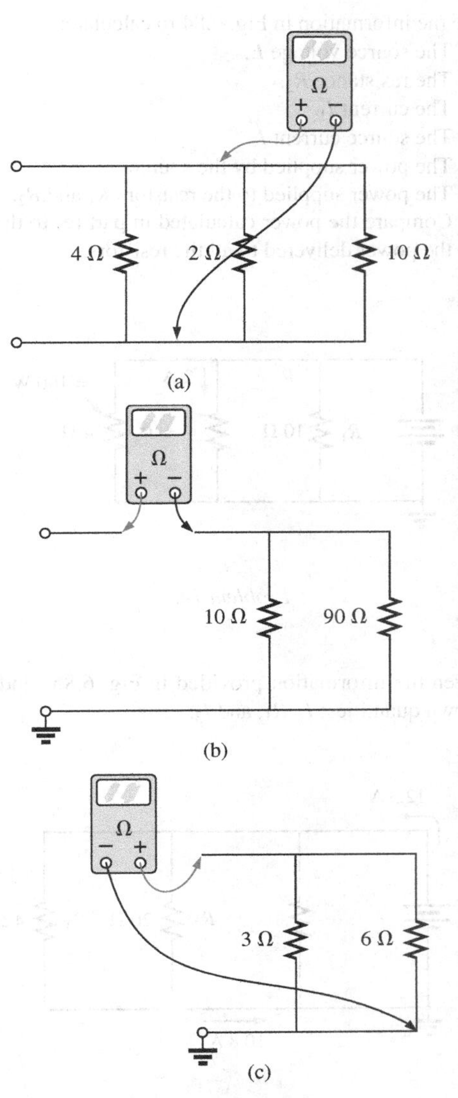

FIG. 6.77
Problem 7.

*8. Determine R_1 for the network in Fig. 6.78.

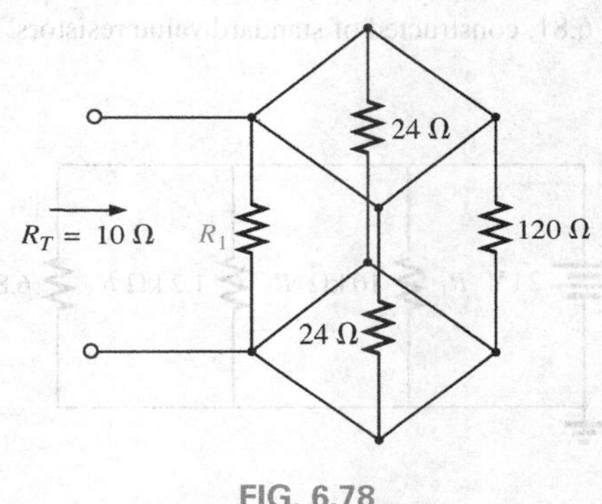

FIG. 6.78
Problem 8.

SECTION 6.3 Parallel Circuits

9. For the parallel network in Fig. 6.79:
   a. Find the total resistance.
   b. What is the voltage across each branch?
   c. Determine the source current and the current through each branch.
   d. Verify that the source current equals the sum of the branch currents.

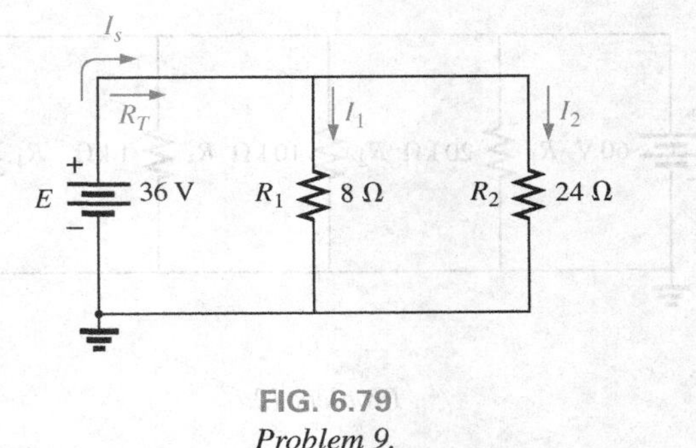

FIG. 6.79
Problem 9.

10. For the network of Fig. 6.80:

a. Find the current through each branch.
b. Find the total resistance.
c. Calculate I_s using the result of part (b).
d. Find the source current using the result of part (a).
e. Compare the results of parts (c) and (d).

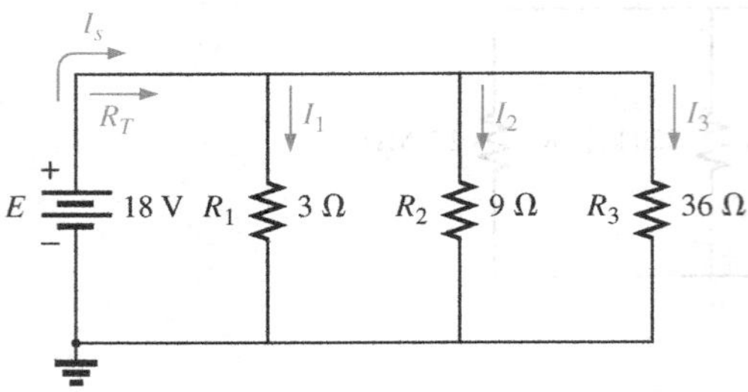

FIG. 6.80
Problem 10.

11. Repeat the analysis of Problem 10 for the network in Fig. 6.81, constructed of standard value resistors.

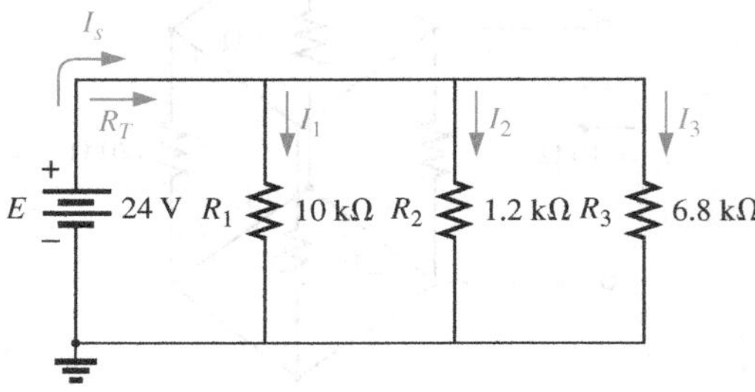

FIG. 6.81
Problem 11.

12. For the parallel network in Fig. 6.82:

a. Without making a single calculation, make a guess on the total resistance.
b. Calculate the total resistance, and compare it to your guess in part (a).
c. Without making a single calculation, which branch will have the most current? Which will have the least?
d. Calculate the current through each branch, and compare your results to the assumptions of part (c).
e. Find the source current and test whether it equals the sum of the branch currents.
f. How does the magnitude of the source current compare to that of the branch currents?

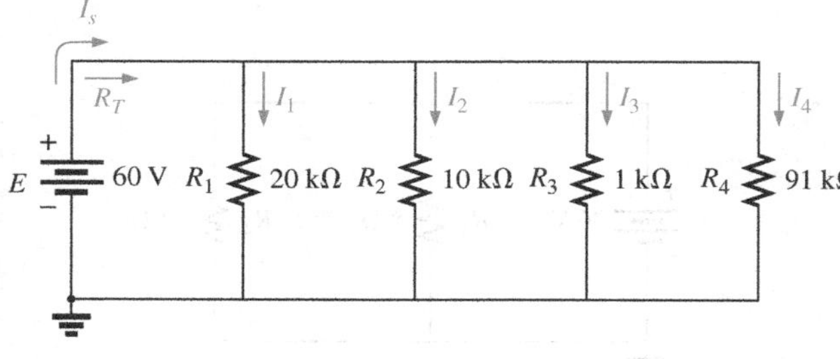

FIG. 6.82
Problem 12.

13. Given the information provided in Fig. 6.83, find:

a. The resistance R_2.
b. The supply voltage E.

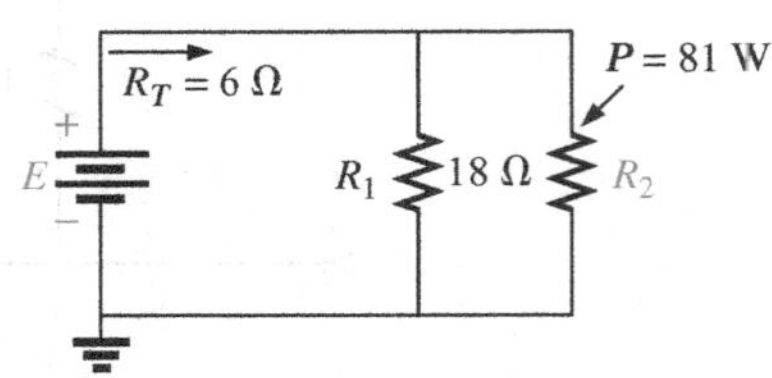

FIG. 6.83
Problem 13.

14. Use the information in Fig. 6.84 to calculate:

a. The source voltage E.
b. The resistance R_2.
c. The current I_1.
d. The source current I_s.
e. The power supplied by the source.
f. The power supplied to the resistors R_1 and R_2.
g. Compare the power calculated in part (e) to the sum of the power delivered to all the resistors.

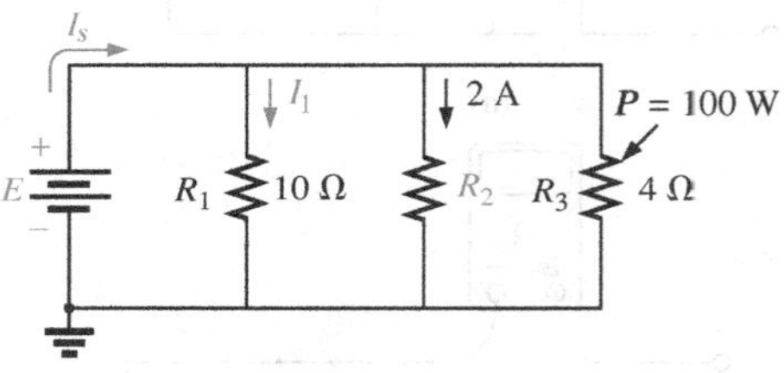

FIG. 6.84
Problem 14.

15. Given the information provided in Fig. 6.85, find the unknown quantities: E, R_1, and I_3.

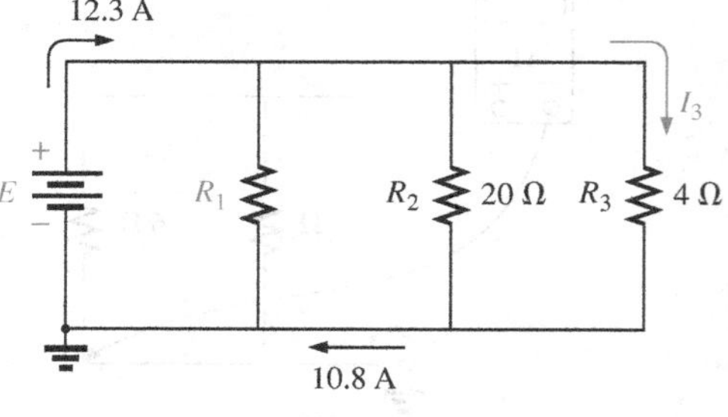

FIG. 6.85
Problem 15.

16. For the network of Fig. 6.86, find:
    a. The voltage V.
    b. The current I_2.
    c. The current I_s.
    d. The power to the 12 kΩ resistor.

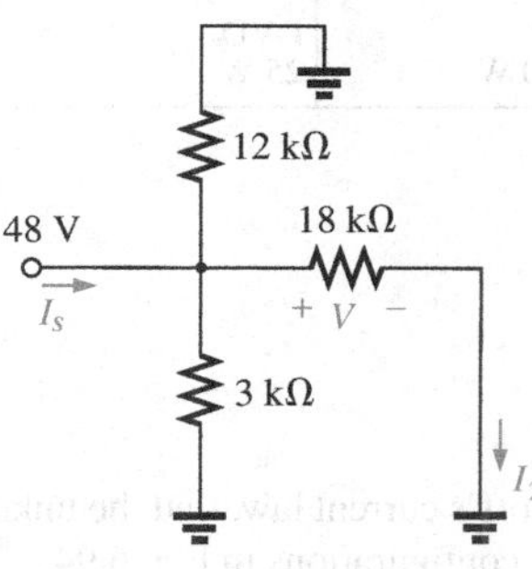

FIG. 6.86
Problem 16.

17. Using the information provided in Fig. 6.87, find:
    a. The resistance R_2.
    b. The resistance R_3.
    c. The current I_s.

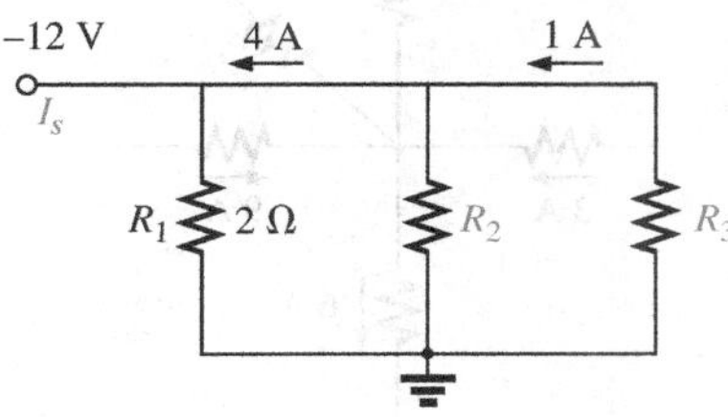

FIG. 6.87
Problem 17.

18. For the network in Fig. 6.81:
    a. Redraw the network and insert ammeters to measure the source current and the current through each branch.
    b. Connect a voltmeter to measure the source voltage and the voltage across resistor R_3. Is there any difference in the connections? Why?

SECTION 6.4 Power Distribution in a Parallel Circuit

19. For the configuration in Fig. 6.88:
    a. Find the total resistance and the current through each branch.
    b. Find the power delivered to each resistor.

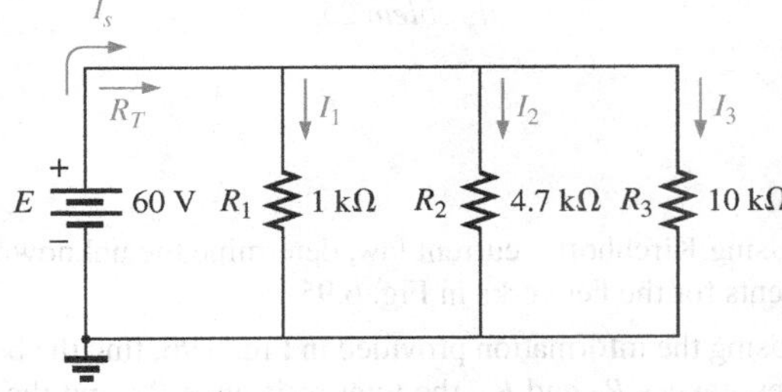

FIG. 6.88
Problem 19.

    c. Calculate the power delivered by the source.
    d. Compare the power delivered by the source to the sum of the powers delivered to the resistors.
    e. Which resistor received the most power? Why?

20. Eight holiday lights are connected in parallel as shown in Fig. 6.89.
    a. If the set is connected to a 120 V source, what is the current through each bulb if each bulb has an internal resistance of 1.8 kΩ?
    b. Determine the total resistance of the network.
    c. Find the current drain from the supply.
    d. What is the power delivered to each bulb?
    e. Using the results of part (d), find the power delivered by the source.
    f. If one bulb burns out (that is, the filament opens up), what is the effect on the remaining bulbs? What is the effect on the source current? Why?

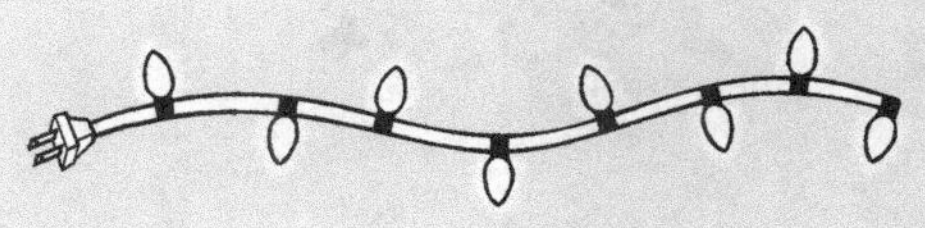

FIG. 6.89
Problem 20.

21. Determine the power delivered by the dc battery in Fig. 6.90.

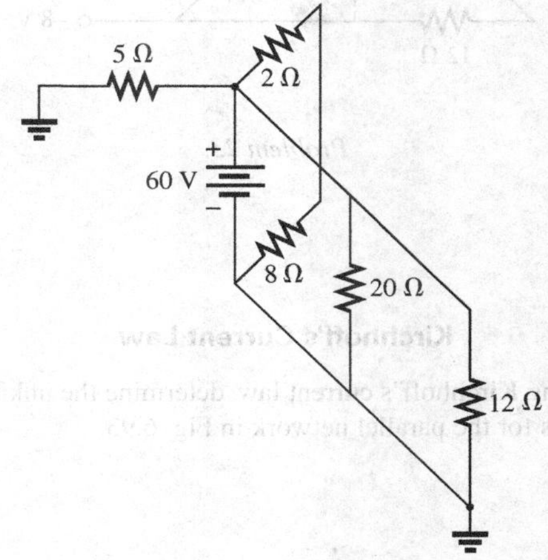

FIG. 6.90
Problem 21.

22. A portion of a residential service to a home is depicted in Fig. 6.91.
    a. Determine the current through each parallel branch of the system.
    b. Calculate the current drawn from the 120 V source. Will the 20 A breaker trip?
    c. What is the total resistance of the network?
    d. Determine the power delivered by the source. How does it compare to the sum of the wattage ratings appearing in Fig. 6.91?

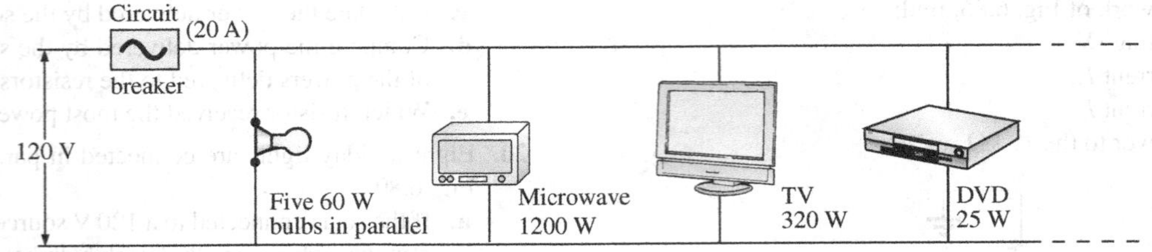

FIG. 6.91
Problem 22.

***23.** For the network in Fig. 6.92:

a. Find the current I_1.
b. Calculate the power dissipated by the 4 Ω resistor.
c. Find the current I_2.

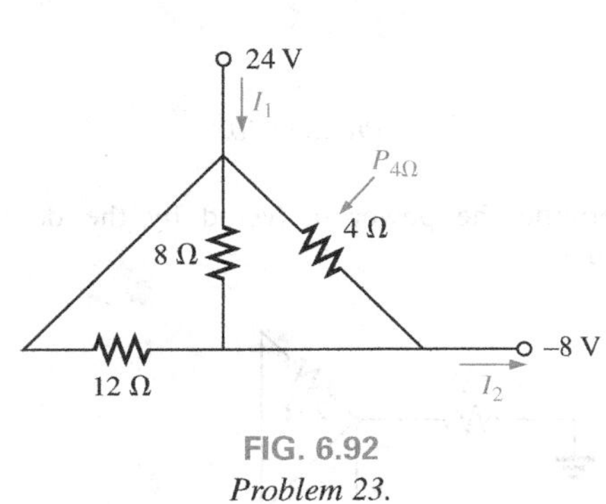

FIG. 6.92
Problem 23.

SECTION 6.5 Kirchhoff's Current Law

24. Using Kirchhoff's current law, determine the unknown currents for the parallel network in Fig. 6.93.

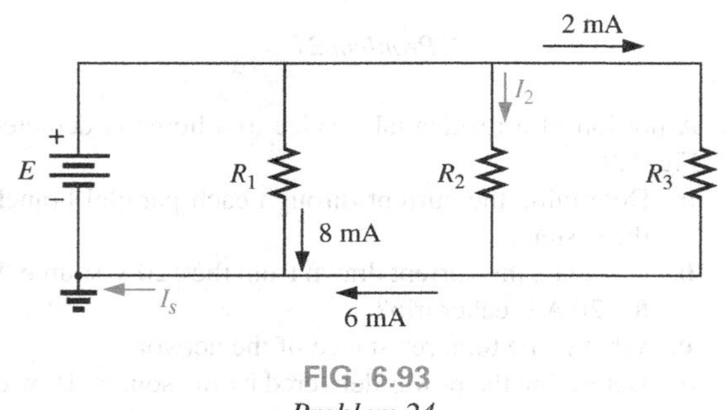

FIG. 6.93
Problem 24.

25. Using Kirchoff's current law, find the unknown currents for the complex configurations in Fig. 6.94.

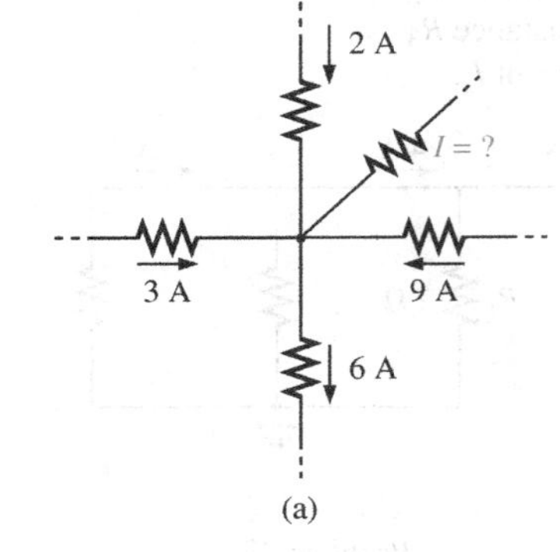

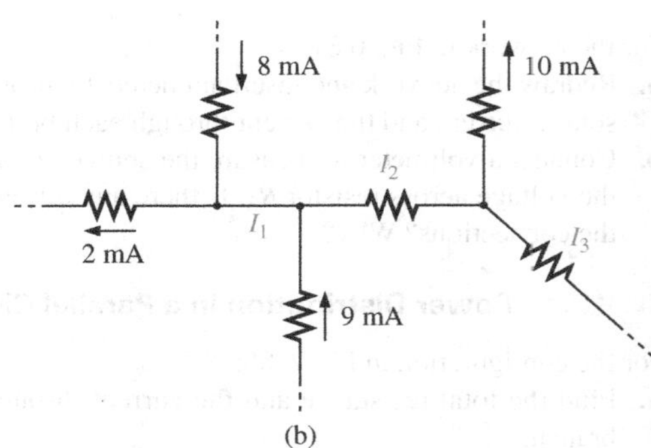

FIG. 6.94
Problem 25.

26. Using Kirchhoff's current law, determine the unknown currents for the networks in Fig. 6.95.

27. Using the information provided in Fig. 6.96, find the branch resistances R_1 and R_3, the total resistance R_T, and the voltage source E.

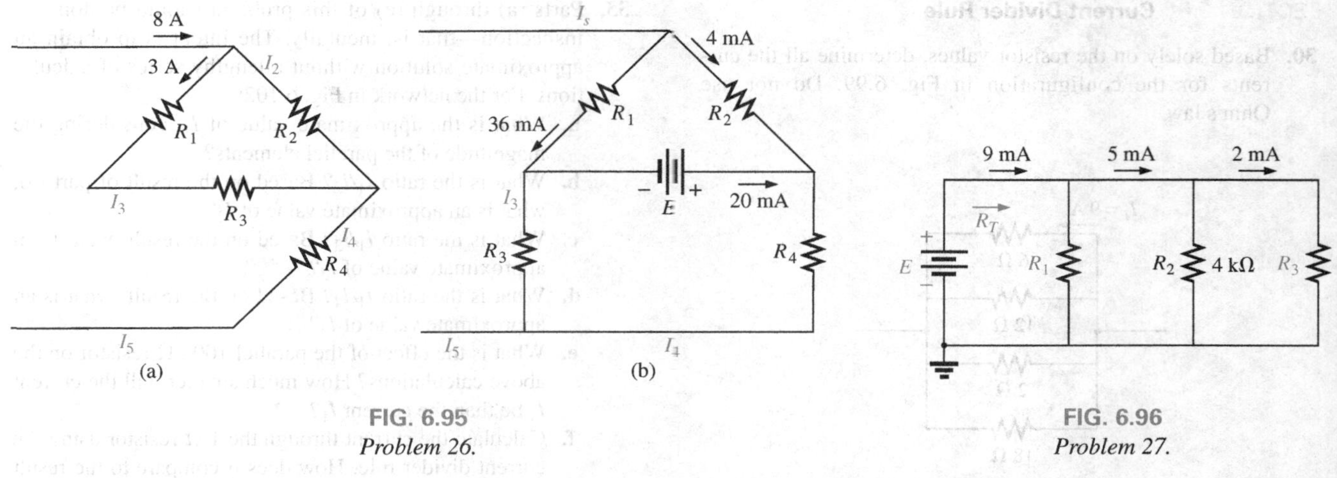

FIG. 6.95
Problem 26.

FIG. 6.96
Problem 27.

28. Find the unknown quantities for the networks in Fig. 6.97 using the information provided.

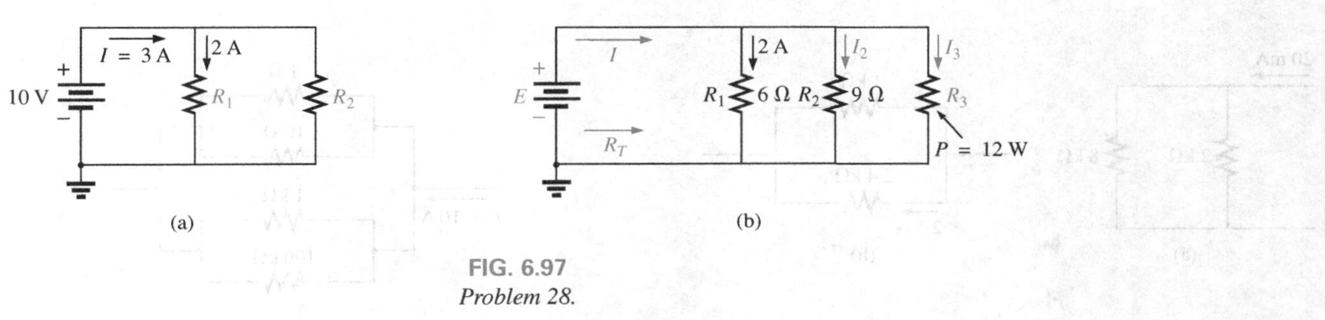

FIG. 6.97
Problem 28.

29. Find the unknown quantities for the networks of Fig. 6.98 using the information provided.

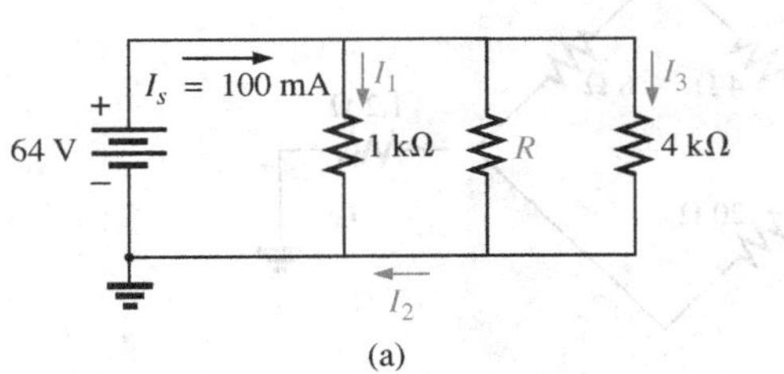

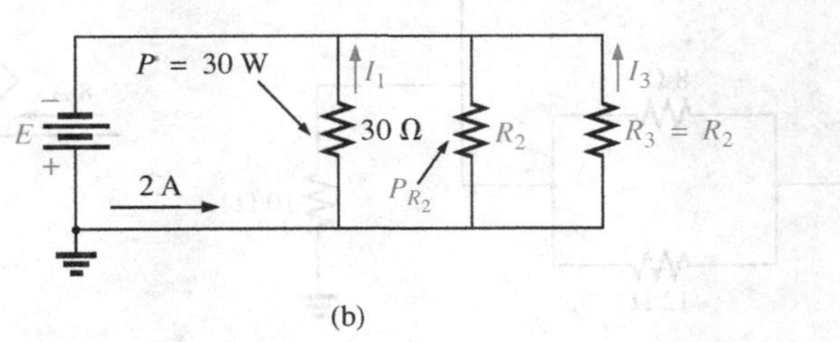

FIG. 6.98
Problem 29.

SECTION 6.6 Current Divider Rule

30. Based solely on the resistor values, determine all the currents for the configuration in Fig. 6.99. Do not use Ohm's law.

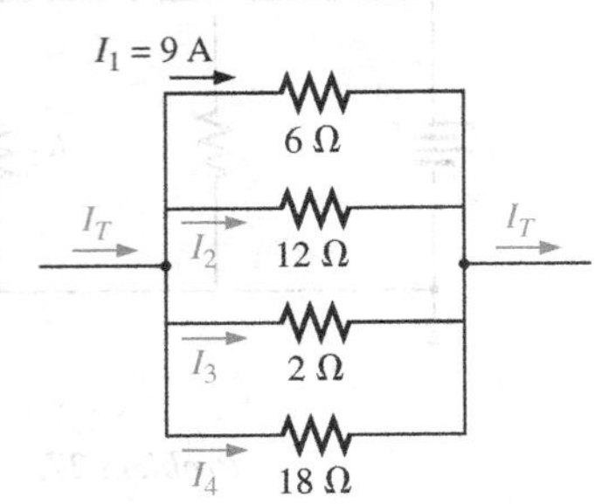

FIG. 6.99
Problem 30.

31. **a.** Determine one of the unknown currents of Fig. 6.100 using the current divider rule.
b. Determine the other current using Kirchhoff's current law.

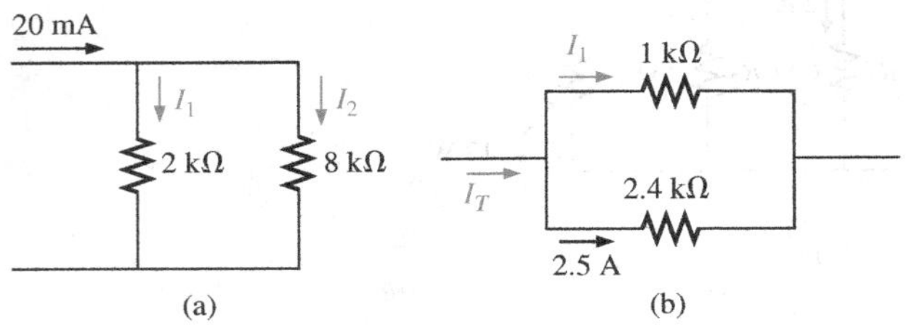

FIG. 6.100
Problem 31.

32. For each network of Fig. 6.101, determine the unknown currents.

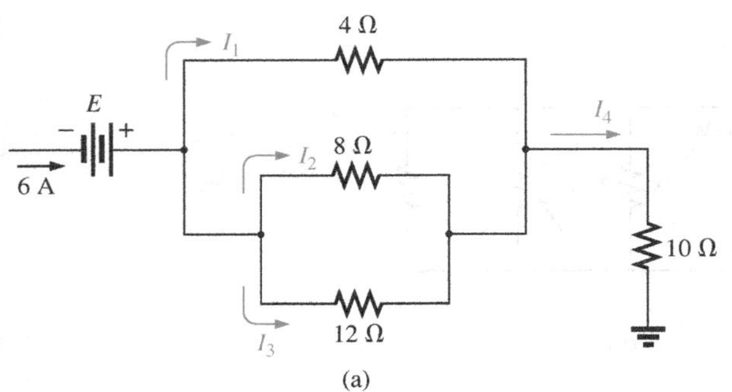

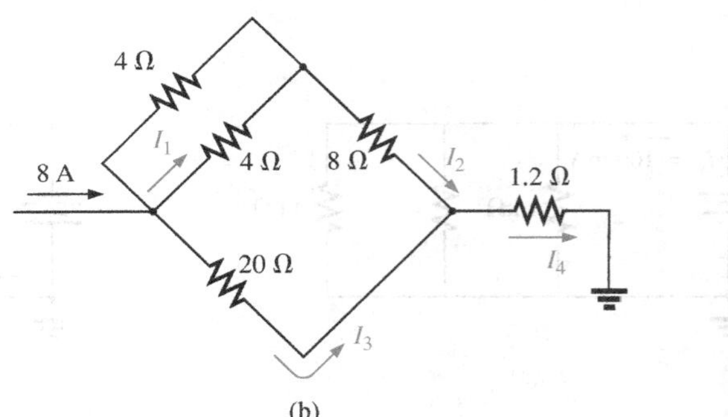

FIG. 6.101
Problem 32.

33. Parts (a) through (e) of this problem should be done by inspection—that is, mentally. The intent is to obtain an approximate solution without a lengthy series of calculations. For the network in Fig. 6.102:

a. What is the approximate value of I_1, considering the magnitude of the parallel elements?
b. What is the ratio I_1/I_2? Based on the result of part (a), what is an approximate value of I_2?
c. What is the ratio I_1/I_3? Based on the result, what is an approximate value of I_3?
d. What is the ratio I_1/I_4? Based on the result, what is an approximate value of I_4?
e. What is the effect of the parallel 100 kΩ resistor on the above calculations? How much smaller will the current I_4 be than the current I_1?
f. Calculate the current through the 1 Ω resistor using the current divider rule. How does it compare to the result of part (a)?
g. Calculate the current through the 10 Ω resistor. How does it compare to the result of part (b)?
h. Calculate the current through the 1 kΩ resistor. How does it compare to the result of part (c)?
i. Calculate the current through the 100 kΩ resistor. How does it compare to the solutions to part (e)?

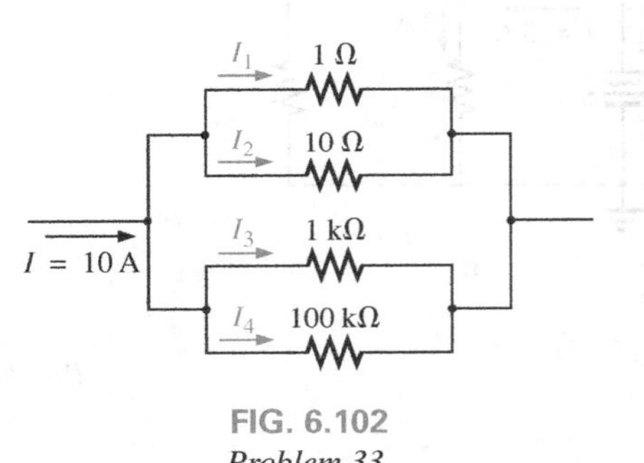

FIG. 6.102
Problem 33.

34. Find the unknown quantities for the networks in Fig. 6.103 using the information provided.

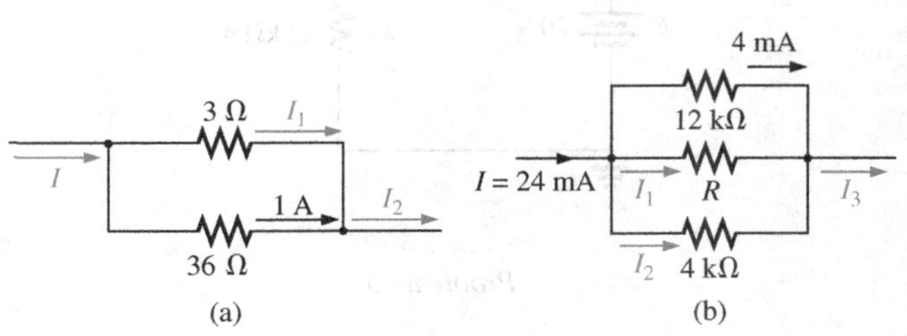

FIG. 6.103
Problem 34.

35. **a.** Find resistance R for the network in Fig. 6.104 that will ensure that $I_1 = 3I_2$.
b. Find I_1 and I_2.

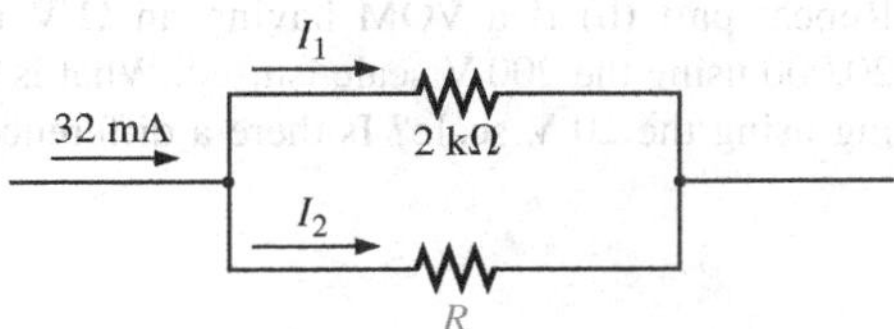

FIG. 6.104
Problem 35.

36. Design the network in Fig. 6.105 such that $I_2 = 2I_1$ and $I_3 = 2I_2$.

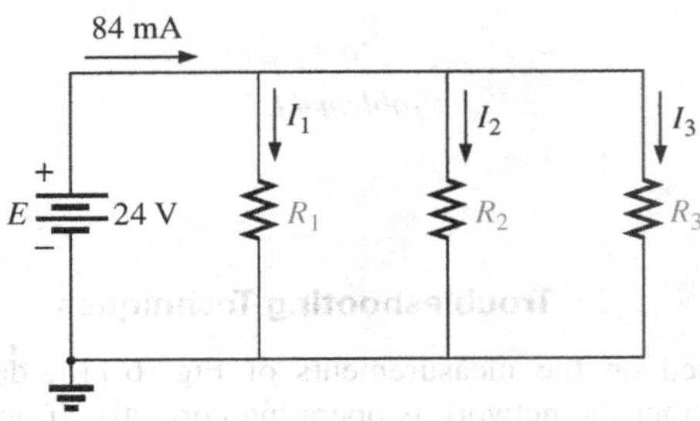

FIG. 6.105
Problem 36.

SECTION 6.7 Voltage Source in Parallel

37. Assuming identical supplies in Fig. 6.106:
a. Find the indicated currents.
b. Find the power delivered by each source.
c. Find the total power delivered by both sources, and compare it to the power delivered to the load R_L.
d. If only source current were available, what would the current drain be to supply the same power to the load? How does the current level compare to the calculated level of part (a)?

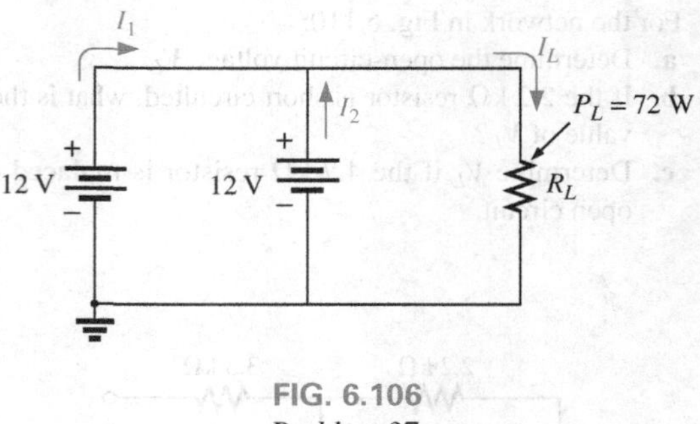

FIG. 6.106
Problem 37.

38. Assuming identical supplies, determine currents I_1, I_2, and I_3 for the configuration in Fig. 6.107.

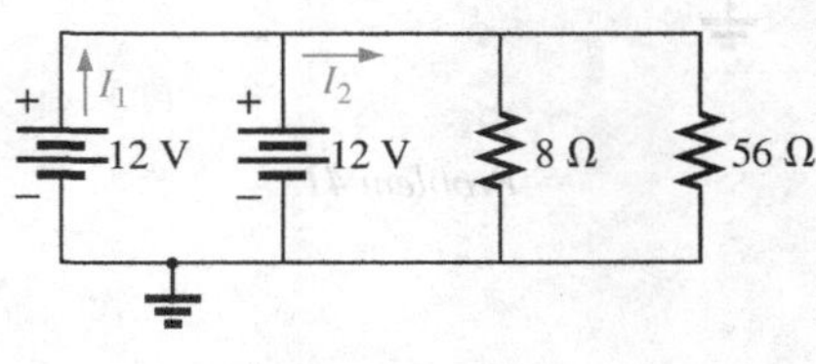

FIG. 6.107
Problem 38.

39. Assuming identical supplies, determine the current I and resistance R for the parallel network in Fig. 6.108.

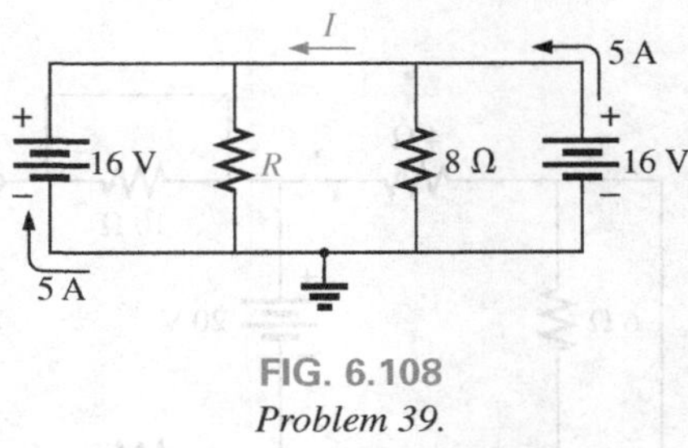

FIG. 6.108
Problem 39.

SECTION 6.8 Open and Short Circuits

40. For the network in Fig. 6.109:
a. Determine I_s and V_L.
b. Determine I_s if R_L is shorted out.
c. Determine V_L if R_L is replaced by an open circuit.

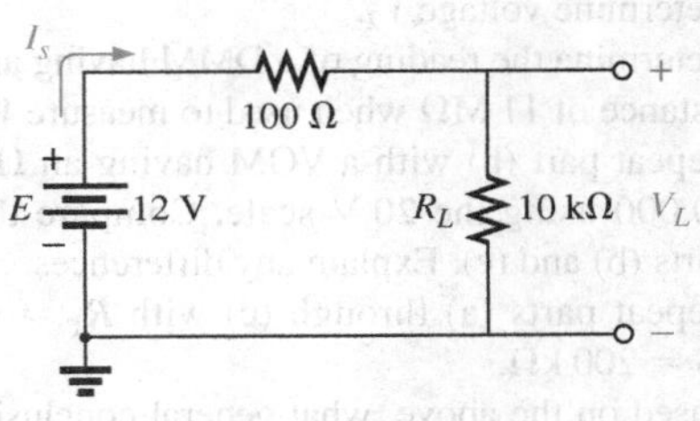

FIG. 6.109
Problem 40.

IPI

41. For the network in Fig. 6.110:
  a. Determine the open-circuit voltage V_L.
  b. If the 2.2 kΩ resistor is short circuited, what is the new value of V_L?
  c. Determine V_L if the 4.7 kΩ resistor is replaced by an open circuit.

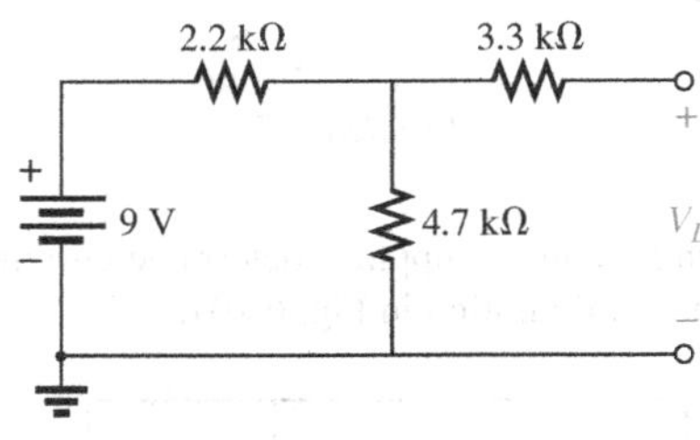

FIG. 6.110
Problem 41.

***42.** For the network in Fig. 6.111, determine:
  a. The short-circuit currents I_1 and I_2.
  b. The voltages V_1 and V_2.
  c. The source current I_s.

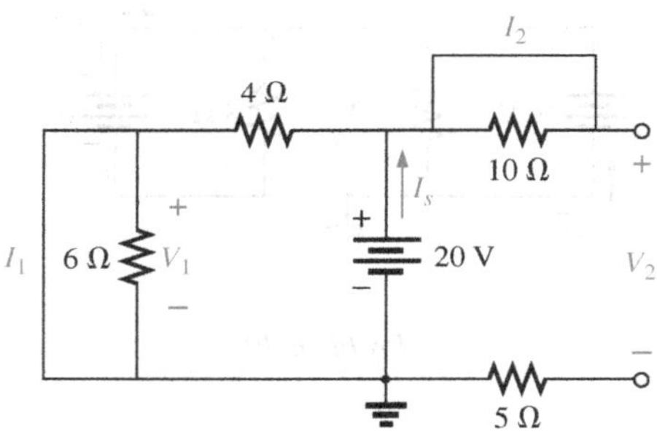

FIG. 6.111
Problem 42.

SECTION 6.9 Voltmeter Loading Effects

43. For the simple series configuration in Fig. 6.112:
  a. Determine voltage V_2.
  b. Determine the reading of a DMM having an internal resistance of 11 MΩ when used to measure V_2.
  c. Repeat part (b) with a VOM having an Ω/V rating of 20,000 using the 20 V scale. Compare the results of parts (b) and (c). Explain any differences.
  d. Repeat parts (a) through (c) with $R_1 = 100$ kΩ and $R_2 = 200$ kΩ.
  e. Based on the above, what general conclusions can you make about the use of a DMM or a VOM in the voltmeter mode?

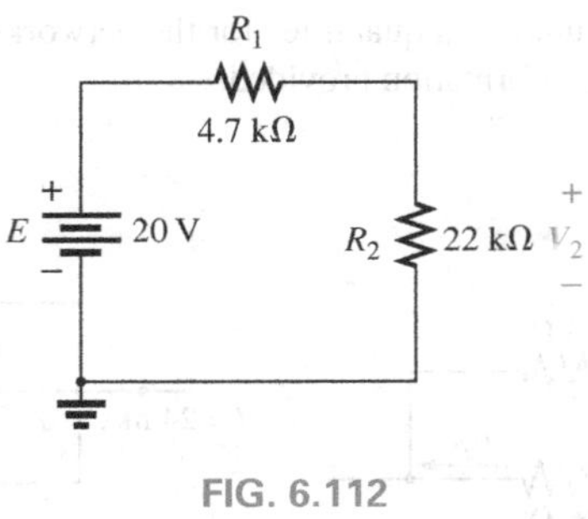

FIG. 6.112
Problem 43.

44. Given the configuration in Fig. 6.113:
  a. What is the voltage between points a and b?
  b. What will the reading of a DMM be when placed across terminals a and b if the internal resistance of the meter is 11 MΩ?
  c. Repeat part (b) if a VOM having an Ω/V rating of 20,000 using the 200 V scale is used. What is the reading using the 20 V scale? Is there a difference? Why?

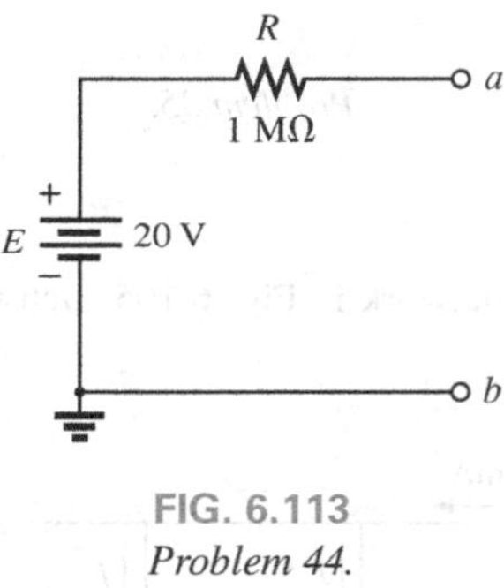

FIG. 6.113
Problem 44.

SECTION 6.10 Troubleshooting Techniques

45. Based on the measurements of Fig. 6.114, determine whether the network is operating correctly. If not, determine why.

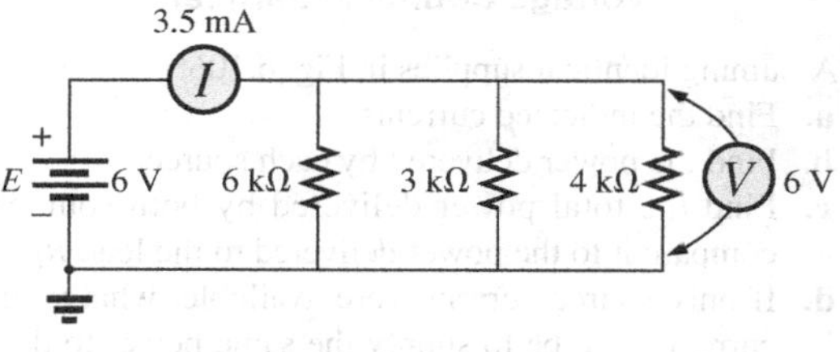

FIG. 6.114
Problem 45.

46. Referring to Fig. 6.115, find the voltage V_{ab} without the meter in place. When the meter is applied to the active network, it reads 8.8 V. If the measured value does not equal the theoretical value, which element or elements may have been connected incorrectly?

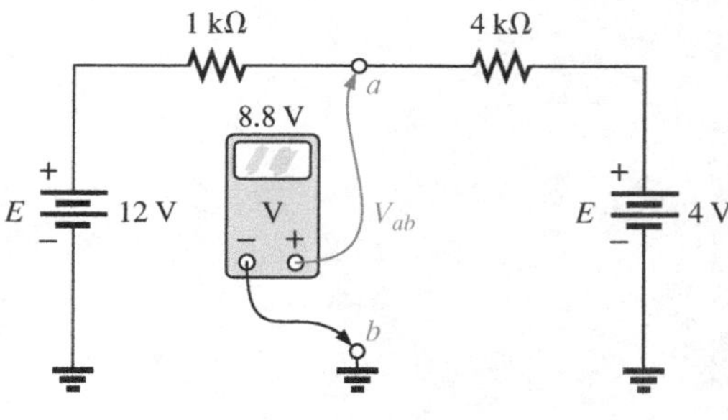

FIG. 6.115
Problem 46.

47. **a.** The voltage V_a for the network in Fig. 6.116 is −1 V. If it suddenly jumped to 20 V, what could have happened to the circuit structure? Localize the problem area.
b. If the voltage V_a is 6 V rather than −1 V, explain what is wrong about the network construction.

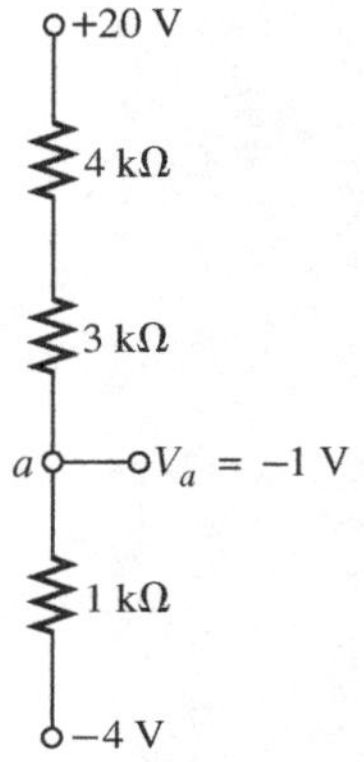

+20 V
4 kΩ
3 kΩ
a
V_a = −1 V
1 kΩ
−4 V

FIG. 6.116
Problem 47.

SECTION 6.14 Computer Analysis

48. Using PSpice or Multisim, verify the results of Example 6.13.

49. Using PSpice or Multisim, determine the solution to Problem 9, and compare your answer to the longhand solution.

50. Using PSpice or Multisim, determine the solution to Problem 11, and compare your answer to the longhand solution.

GLOSSARY

Current divider rule (CDR) A method by which the current through parallel elements can be determined without first finding the voltage across those parallel elements.

Kirchhoff's current law (KCL) The algebraic sum of the currents entering and leaving a node is zero.

Node A junction of two or more branches.

Ohm/volt (Ω/V) rating A rating used to determine both the current sensitivity of the movement and the internal resistance of the meter.

Open circuit The absence of a direct connection between two points in a network.

Parallel circuit A circuit configuration in which the elements have two points in common.

Short circuit A direct connection of low resistive value that can significantly alter the behavior of an element or system.

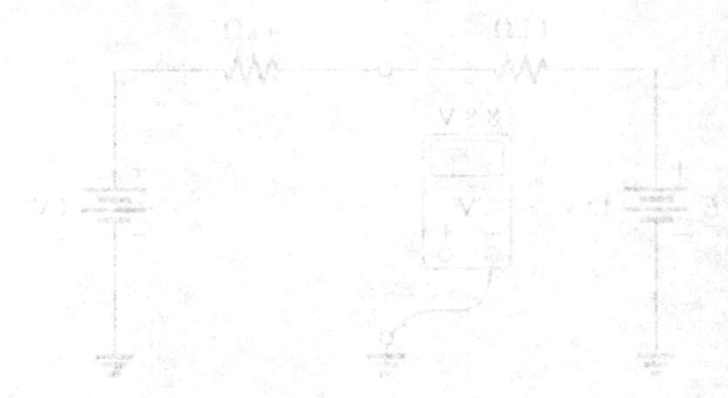

SERIES-PARALLEL CIRCUITS

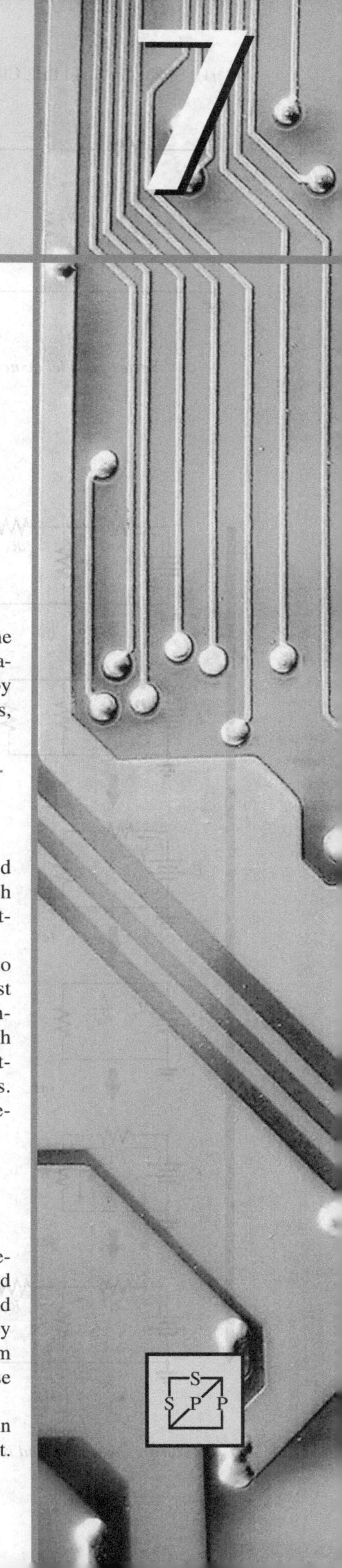

Objectives

- *Learn about the unique characteristics of series-parallel configurations and how to solve for the voltage, current, or power to any individual element or combination of elements.*
- *Become familiar with the voltage divider supply and the conditions needed to use it effectively.*
- *Learn how to use a potentiometer to control the voltage across any given load.*

7.1 INTRODUCTION

Chapters 5 and 6 were dedicated to the fundamentals of series and parallel circuits. In some ways, these chapters may be the most important ones in the text because they form a foundation for all the material to follow. The remaining network configurations cannot be defined by a strict list of conditions because of the variety of configurations that exists. In broad terms, we can look upon the remaining possibilities as either **series-parallel** or **complex**.

A series-parallel configuration is one that is formed by a combination of series and parallel elements.

A complex configuration is one in which none of the elements are in series or parallel.

In this chapter, we examine the series-parallel combination using the basic laws introduced for series and parallel circuits. There are no new laws or rules to learn—simply an approach that permits the analysis of such structures. In the next chapter, we consider complex networks using methods of analysis that allow us to analyze any type of network.

The possibilities for series-parallel configurations are infinite. Therefore, you need to examine each network as a separate entity and define the approach that provides the best path to determining the unknown quantities. In time, you will find similarities between configurations that make it easier to define the best route to a solution, but this occurs only with exposure, practice, and patience. The best preparation for the analysis of series-parallel networks is a firm understanding of the concepts introduced for series and parallel networks. All the rules and laws to be applied in this chapter have already been introduced in the previous two chapters.

7.2 SERIES-PARALLEL NETWORKS

The network in Fig. 7.1 is a series-parallel network. At first, you must be very careful to determine which elements are in series and which are in parallel. For instance, resistors R_1 and R_2 are *not* in series due to resistor R_3 being connected to the common point b between R_1 and R_2. Resistors R_2 and R_4 are *not* in parallel because they are not connected at both ends. They are separated at one end by resistor R_3. The need to be absolutely sure of your definitions from the last two chapters now becomes obvious. In fact, it may be a good idea to refer to those rules as we progress through this chapter.

If we look carefully enough at Fig. 7.1, we do find that the two resistors R_3 and R_4 are in series because they share only point c, and no other element is connected to that point.

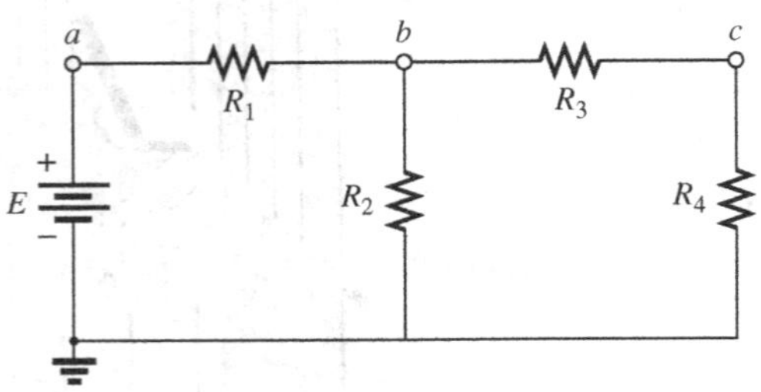

FIG. 7.1
Series-parallel dc network.

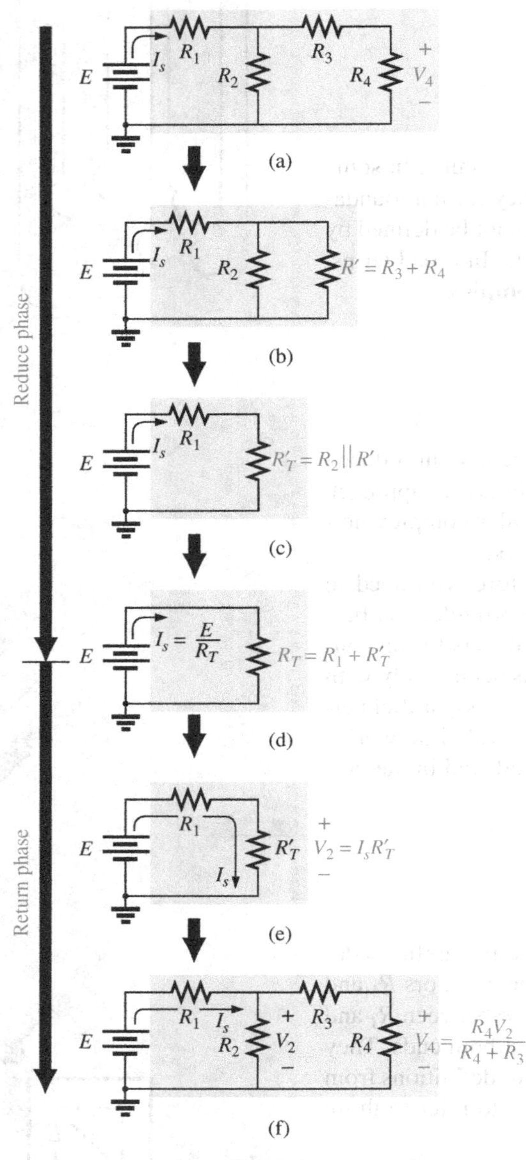

FIG. 7.2
Introducing the reduce and return approach.

Further, the voltage source E and resistor R_1 are in series because they share point a, with no other elements connected to the same point. In the entire configuration, there are no two elements in parallel.

How do we analyze such configurations? The approach is one that requires us to first identify elements that can be combined. Since there are no parallel elements, we must turn to the possibilities with series elements. The voltage source and the series resistor cannot be combined because they are different types of elements. However, resistors R_3 and R_4 can be combined to form a single resistor. The total resistance of the two is their sum as defined by series circuits. The resulting resistance is then in parallel with resistor R_2, and they can be combined using the laws for parallel elements. The process has begun: We are slowly reducing the network to one that will be represented by a single resistor equal to the total resistance "seen" by the source.

The source current can now be determined using Ohm's law, and we can work back through the network to find all the other currents and voltages. The ability to define the first step in the analysis can sometimes be difficult. However, combinations can be made only by using the rules for series or parallel elements, so naturally the first step may simply be to define which elements are in series or parallel. You must then define how to find such things as the total resistance and the source current and proceed with the analysis. In general, the following steps will provide some guidance for the wide variety of possible combinations that you might encounter.

General Approach:

1. ***Take a moment to study the problem "in total" and make a brief mental sketch of the overall approach you plan to use. The result may be time- and energy-saving shortcuts.***
2. ***Examine each region of the network independently before tying them together in series-parallel combinations. This usually simplifies the network and possibly reveals a direct approach toward obtaining one or more desired unknowns. It also eliminates many of the errors that may result due to the lack of a systematic approach.***
3. ***Redraw the network as often as possible with the reduced branches and undisturbed unknown quantities to maintain clarity and provide the reduced networks for the trip back to unknown quantities from the source.***
4. ***When you have a solution, check that it is reasonable by considering the magnitudes of the energy source and the elements in the network. If it does not seem reasonable, either solve the circuit using another approach or review your calculations.***

7.3 REDUCE AND RETURN APPROACH

The network of Fig. 7.1 is redrawn as Fig. 7.2(a). For this discussion, let us assume that voltage V_4 is desired. As described in Section 7.2, first combine the series resistors R_3 and R_4 to form an equivalent resistor R' as shown in Fig. 7.2(b). Resistors R_2 and R' are then in parallel and can be combined to establish an equivalent resistor R'_T as shown in Fig. 7.2(c). Resistors R_1 and R'_T are then in series and can be combined to establish the total resistance of the network as shown in Fig. 7.2(d). The **reduction phase** of the analysis is now complete. The network cannot be put in a simpler form.

We can now proceed with the **return phase** whereby we work our way back to the desired voltage V_4. Due to the resulting series

configuration, the source current is also the current through R_1 and R'_T. The voltage across R'_T (and therefore across R_2) can be determined using Ohm's law as shown in Fig. 7.2(e). Finally, the desired voltage V_4 can be determined by an application of the voltage divider rule as shown in Fig. 7.2(f).

The *reduce and return approach* has now been introduced. This process enables you to reduce the network to its simplest form across the source and then determine the source current. In the return phase, you use the resulting source current to work back to the desired unknown. For most single-source series-parallel networks, the above approach provides a viable option toward the solution. In some cases, shortcuts can be applied that save some time and energy. Now for a few examples.

EXAMPLE 7.1 Find current I_3 for the series-parallel network in Fig. 7.3.

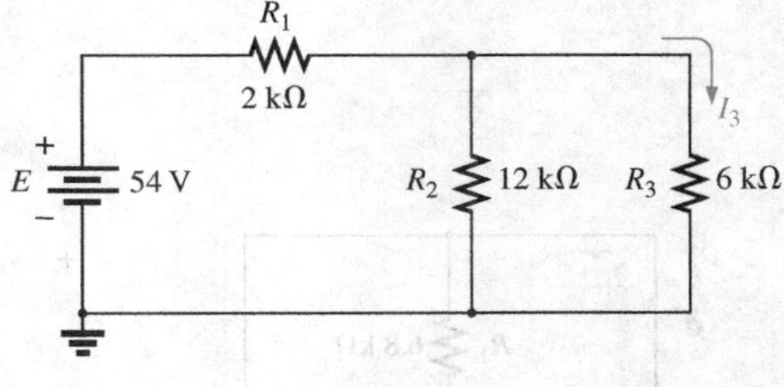

FIG. 7.3
Series-parallel network for Example 7.1.

Solution: Checking for series and parallel elements, we find that resistors R_2 and R_3 are in parallel. Their total resistance is

$$R' = R_2 \| R_3 = \frac{R_2 R_3}{R_2 + R_3} = \frac{(12\text{ k}\Omega)(6\text{ k}\Omega)}{12\text{ k}\Omega + 6\text{ k}\Omega} = 4\text{ k}\Omega$$

Replacing the parallel combination with a single equivalent resistance results in the configuration in Fig. 7.4. Resistors R_1 and R' are then in series, resulting in a total resistance of

$$R_T = R_1 + R' = 2\text{ k}\Omega + 4\text{ k}\Omega = 6\text{ k}\Omega$$

The source current is then determined using Ohm's law:

$$I_s = \frac{E}{R_T} = \frac{54\text{ V}}{6\text{ k}\Omega} = 9\text{ mA}$$

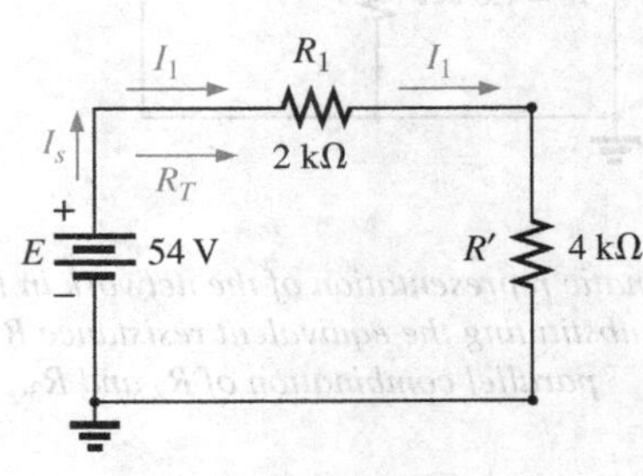

FIG. 7.4
Substituting the parallel equivalent resistance for resistors R_2 and R_3 in Fig. 7.3.

In Fig. 7.4, since R_1 and R' are in series, they have the same current I_s. The result is

$$I_1 = I_s = 9\text{ mA}$$

Returning to Fig. 7.3, we find that I_1 is the total current entering the parallel combination of R_2 and R_3. Applying the current divider rule results in the desired current:

$$I_3 = \left(\frac{R_2}{R_2 + R_3}\right) I_1 = \left(\frac{12\text{ k}\Omega}{12\text{ k}\Omega + 6\text{ k}\Omega}\right) 9\text{ mA} = \mathbf{6\text{ mA}}$$

Note in the solution for Example 7.1 that all of the equations used were introduced in the last two chapters—nothing new was introduced except how to approach the problem and use the equations properly.

EXAMPLE 7.2 For the network in Fig. 7.5:

a. Determine currents I_4 and I_s and voltage V_2.
b. Insert the meters to measure current I_4 and voltage V_2.

Solutions:

a. Checking out the network, we find that there are no two resistors in series, and the only parallel combination is resistors R_2

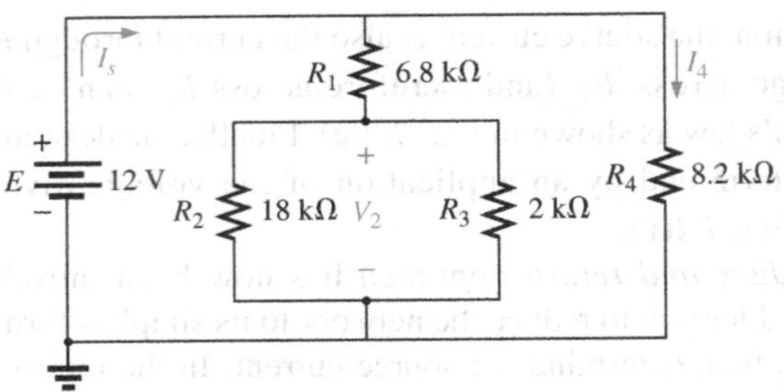

FIG. 7.5
Series-parallel network for Example 7.2.

and R_3. Combining the two parallel resistors results in a total resistance of

$$R' = R_2 \| R_3 = \frac{R_2 R_3}{R_2 + R_3} = \frac{(18\text{ k}\Omega)(2\text{ k}\Omega)}{18\text{ k}\Omega + 2\text{ k}\Omega} = 1.8\text{ k}\Omega$$

Redrawing the network with resistance R' inserted results in the configuration in Fig. 7.6.

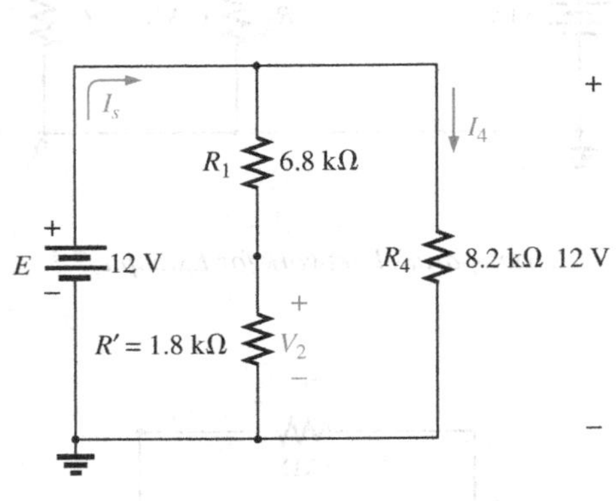

FIG. 7.6
Schematic representation of the network in Fig. 7.5 after substituting the equivalent resistance R' for the parallel combination of R_2 and R_3.

You may now be tempted to combine the series resistors R_1 and R' and redraw the network. However, a careful examination of Fig. 7.6 reveals that since the two resistive branches are in parallel, the voltage is the same across each branch. That is, the voltage across the series combination of R_1 and R' is 12 V and that across resistor R_4 is 12 V. The result is that I_4 can be determined directly using Ohm's law as follows:

$$I_4 = \frac{V_4}{R_4} = \frac{E}{R_4} = \frac{12\text{ V}}{8.2\text{ k}\Omega} = \mathbf{1.46\text{ mA}}$$

In fact, for the same reason, I_4 could have been determined directly from Fig. 7.5. Because the total voltage across the series combination of R_1 and R'_T is 12 V, the voltage divider rule can be applied to determine voltage V_2 as follows:

$$V_2 = \left(\frac{R'}{R' + R_1}\right)E = \left(\frac{1.8\text{ k}\Omega}{1.8\text{ k}\Omega + 6.8\text{ k}\Omega}\right)12\text{ V} = \mathbf{2.51\text{ V}}$$

The current I_s can be found in one of two ways. Find the total resistance and use Ohm's law, or find the current through the other parallel branch and apply Kirchhoff's current law. Since we already have the current I_4, the latter approach will be applied:

$$I_1 = \frac{E}{R_1 + R'} = \frac{12\text{ V}}{6.8\text{ k}\Omega + 1.8\text{ k}\Omega} = 1.40\text{ mA}$$

and $\quad I_s = I_1 + I_4 = 1.40\text{ mA} + 1.46\text{ mA} = \mathbf{2.86\text{ mA}}$

b. The meters have been properly inserted in Fig. 7.7. Note that the voltmeter is across both resistors since the voltage across parallel elements is the same. In addition, note that the ammeter is in series with resistor R_4, forcing the current through the meter to be the same as that through the series resistor. The power supply is displaying the source current.

Clearly, Example 7.2 revealed how a careful study of a network can eliminate unnecessary steps toward the desired solution. It is often worth

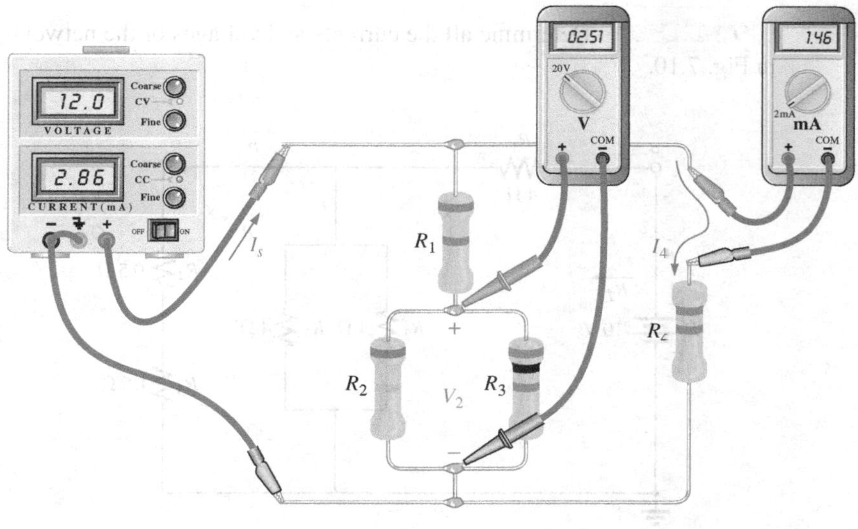

FIG. 7.7
Inserting an ammeter and a voltmeter to measure I_4 and V_2, respectively.

the extra time to sit back and carefully examine a network before trying every equation that seems appropriate.

7.4 BLOCK DIAGRAM APPROACH

In the previous example, we used the reduce and return approach to find the desired unknowns. The direction seemed fairly obvious and the solution relatively easy to understand. However, occasionally the approach is not as obvious, and you may need to look at groups of elements rather than the individual components. Once the grouping of elements reveals the most direct approach, you can examine the impact of the individual components in each group. This grouping of elements is called the *block diagram approach* and is used in the following examples.

In Fig. 7.8, blocks *B* and *C* are in parallel (points *b* and *c* in common), and the voltage source *E* is in series with block *A* (point *a* in common). The parallel combination of *B* and *C* is also in series with *A* and the voltage source *E* due to the common points *b* and *c*, respectively.

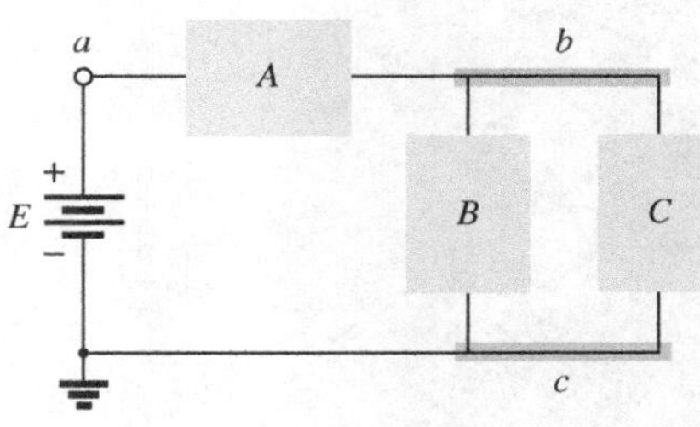

FIG. 7.8
Introducing the block diagram approach.

To ensure that the analysis to follow is as clear and uncluttered as possible, the following notation is used for series and parallel combinations of elements. For series resistors R_1 and R_2, a comma is inserted between their subscript notations, as shown here:

$$R_{1,2} = R_1 + R_2$$

For parallel resistors R_1 and R_2, the parallel symbol is inserted between their subscripted notations, as follows:

$$R_{1\|2} = R_1 \| R_2 = \frac{R_1 R_2}{R_1 + R_2}$$

If each block in Fig. 7.8 were a single resistive element, the network in Fig. 7.9 would result. Note that it is an exact replica of Fig. 7.3 in Example 7.1. Blocks *B* and *C* are in parallel, and their combination is in series with block *A*.

However, as shown in the next example, the same block configuration can result in a totally different network.

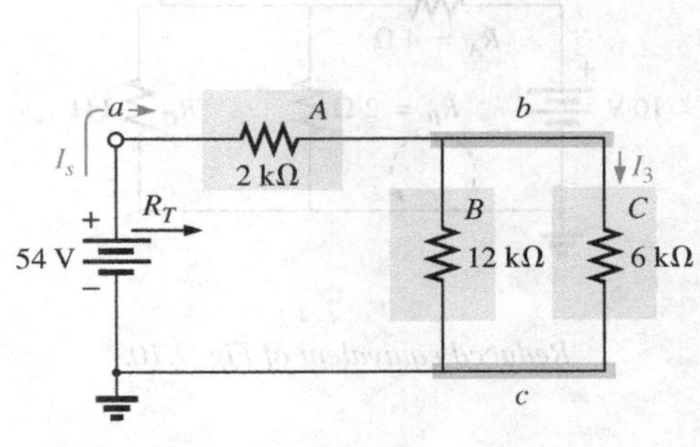

FIG. 7.9
Block diagram format of Fig. 7.3.

EXAMPLE 7.3 Determine all the currents and voltages of the network in Fig. 7.10.

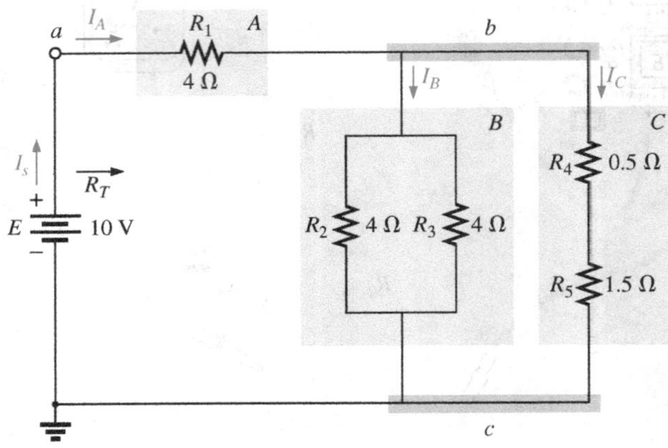

FIG. 7.10
Example 7.3.

Solution: Blocks *A*, *B*, and *C* have the same relative position, but the internal components are different. Note that blocks *B* and *C* are still in parallel, and block *A* is in series with the parallel combination. First, reduce each block into a single element and proceed as described for Example 7.1.

In this case:

$$A\colon R_A = 4\ \Omega$$

$$B\colon R_B = R_2 \parallel R_3 = R_{2\|3} = \frac{R}{N} = \frac{4\ \Omega}{2} = 2\ \Omega$$

$$C\colon R_C = R_4 + R_5 = R_{4,5} = 0.5\ \Omega + 1.5\ \Omega = 2\ \Omega$$

Blocks *B* and *C* are still in parallel, and

$$R_{B\|C} = \frac{R}{N} = \frac{2\ \Omega}{2} = \mathbf{1\ \Omega}$$

with

$$R_T = R_A + R_{B\|C} = 4\ \Omega + 1\ \Omega = \mathbf{5\ \Omega}$$

(Note the similarity between this equation and that obtained for Example 7.1.)

and

$$I_s = \frac{E}{R_T} = \frac{10\ \text{V}}{5\ \Omega} = \mathbf{2\ A}$$

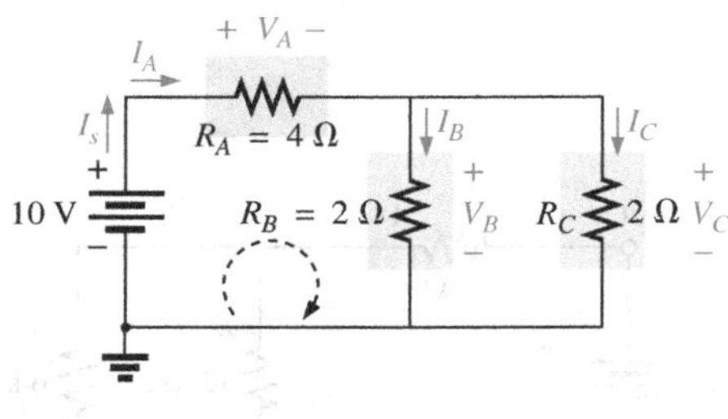

FIG. 7.11
Reduced equivalent of Fig. 7.10.

We can find the currents I_A, I_B, and I_C using the reduction of the network in Fig. 7.10 (recall Step 3) as found in Fig. 7.11. Note that I_A, I_B, and I_C are the same in Figs. 7.10 and Fig. 7.11 and therefore also appear in Fig. 7.11. In other words, the currents I_A, I_B, and I_C in Fig. 7.11 have the same magnitude as the same currents in Fig. 7.10. We have

$$I_A = I_s = \mathbf{2\ A}$$

and

$$I_B = I_C = \frac{I_A}{2} = \frac{I_s}{2} = \frac{2\ \text{A}}{2} = \mathbf{1\ A}$$

Returning to the network in Fig. 7.10, we have

$$I_{R_2} = I_{R_3} = \frac{I_B}{2} = \mathbf{0.5\ A}$$

The voltages V_A, V_B, and V_C from either figure are

$$V_A = I_A R_A = (2\text{ A})(4\ \Omega) = \mathbf{8\ V}$$
$$V_B = I_B R_B = (1\text{ A})(2\ \Omega) = \mathbf{2\ V}$$
$$V_C = V_B = \mathbf{2\ V}$$

Applying Kirchhoff's voltage law for the loop indicated in Fig. 7.11, we obtain

$$\Sigma_{\circlearrowright} V = E - V_A - V_B = 0$$
$$E = E_A + V_B = 8\text{ V} + 2\text{ V}$$

or

$$10\text{ V} = 10\text{ V} \quad \text{(checks)}$$

EXAMPLE 7.4 Another possible variation of Fig. 7.8 appears in Fig. 7.12. Determine all the currents and voltages.

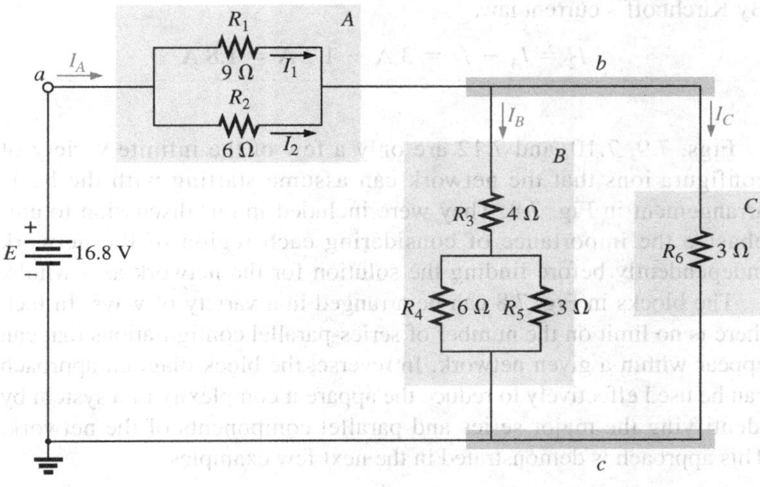

FIG. 7.12
Example 7.4.

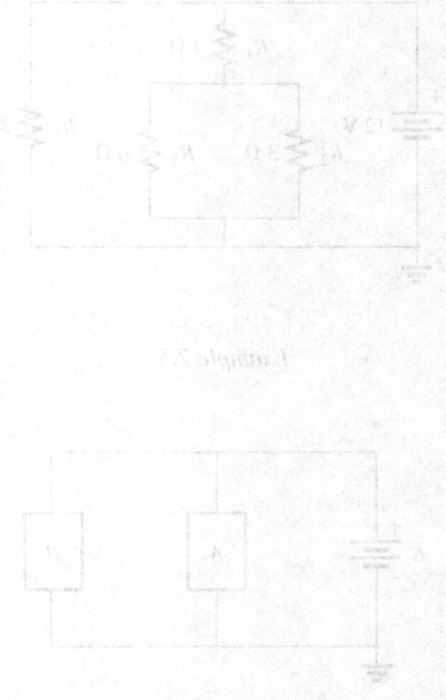

Solution:

$$R_A = R_{1\|2} = \frac{(9\ \Omega)(6\ \Omega)}{9\ \Omega + 6\ \Omega} = \frac{54\ \Omega}{15} = 3.6\ \Omega$$

$$R_B = R_3 + R_{4\|5} = 4\ \Omega + \frac{(6\ \Omega)(3\ \Omega)}{6\ \Omega + 3\ \Omega} = 4\ \Omega + 2\ \Omega = 6\ \Omega$$

$$R_C = 3\ \Omega$$

The network in Fig. 7.12 can then be redrawn in reduced form, as shown in Fig. 7.13. Note the similarities between this circuit and the circuits in Figs. 7.9 and 7.11. We have

$$R_T = R_A + R_{B\|C} = 3.6\ \Omega + \frac{(6\ \Omega)(3\ \Omega)}{6\ \Omega + 3\ \Omega}$$
$$= 3.6\ \Omega + 2\ \Omega = \mathbf{5.6\ \Omega}$$
$$I_s = \frac{E}{R_T} = \frac{16.8\text{ V}}{5.6\ \Omega} = \mathbf{3\ A}$$
$$I_A = I_s = \mathbf{3\ A}$$

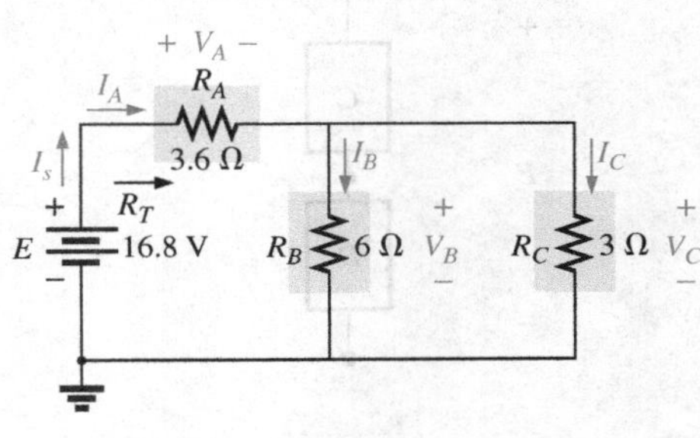

FIG. 7.13
Reduced equivalent of Fig. 7.12.

Applying the current divider rule yields

$$I_B = \frac{R_C I_A}{R_C + R_B} = \frac{(3\ \Omega)(3\ \text{A})}{3\ \Omega + 6\ \Omega} = \frac{9\ \text{A}}{9} = \mathbf{1\ A}$$

By Kirchhoff's current law,

$$I_C = I_A - I_B = 3\ \text{A} - 1\ \text{A} = \mathbf{2\ A}$$

By Ohm's law,

$$V_A = I_A R_A = (3\ \text{A})(3.6\ \Omega) = \mathbf{10.8\ V}$$
$$V_B = I_B R_B = V_C = I_C R_C = (2\ \text{A})(3\ \Omega) = \mathbf{6\ V}$$

Returning to the original network (Fig. 7.12) and applying the current divider rule gives

$$I_1 = \frac{R_2 I_A}{R_2 + R_1} = \frac{(6\ \Omega)(3\ \text{A})}{6\ \Omega + 9\ \Omega} = \frac{18\ \text{A}}{15} = \mathbf{1.2\ A}$$

By Kirchhoff's current law,

$$I_2 = I_A - I_1 = 3\ \text{A} - 1.2\ \text{A} = \mathbf{1.8\ A}$$

Figs. 7.9, 7.10, and 7.12 are only a few of the infinite variety of configurations that the network can assume starting with the basic arrangement in Fig. 7.8. They were included in our discussion to emphasize the importance of considering each region of the network independently before finding the solution for the network as a whole.

The blocks in Fig. 7.8 can be arranged in a variety of ways. In fact, there is no limit on the number of series-parallel configurations that can appear within a given network. In reverse, the block diagram approach can be used effectively to reduce the apparent complexity of a system by identifying the major series and parallel components of the network. This approach is demonstrated in the next few examples.

7.5 DESCRIPTIVE EXAMPLES

EXAMPLE 7.5 Find the current I_4 and the voltage V_2 for the network in Fig. 7.14 using the block diagram approach.

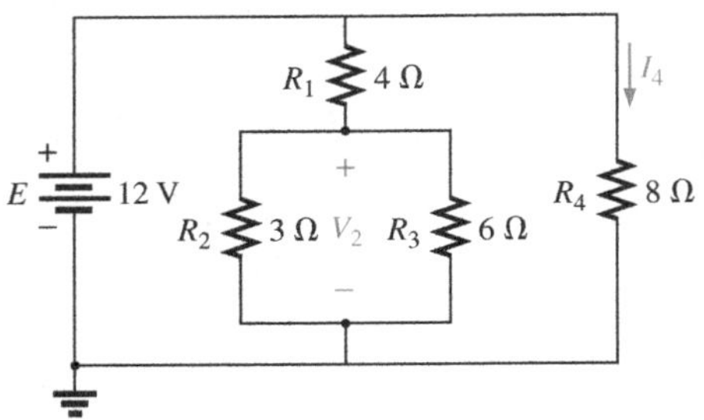

FIG. 7.14
Example 7.5.

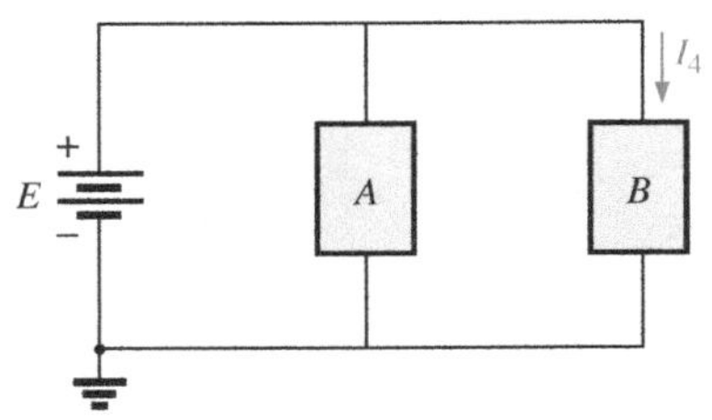

FIG. 7.15
Block diagram of Fig. 7.14.

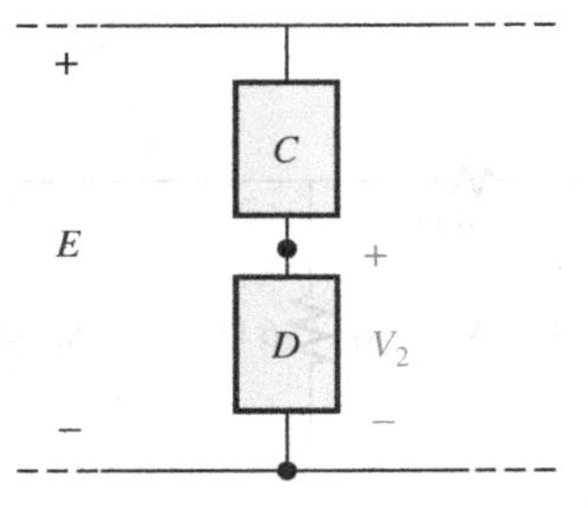

FIG. 7.16
Alternative block diagram for the first parallel branch in Fig. 7.14.

Solution: Note the similarities with the network in Fig. 7.5. In this case, particular unknowns are requested instead of a complete solution. It would, therefore, be a waste of time to find all the currents and voltages of the network. The method used should concentrate on obtaining only the unknowns requested. With the block diagram approach, the network has the basic structure in Fig. 7.15, clearly indicating that the three branches are in parallel and the voltage across A and B is the supply voltage. The current I_4 is now immediately obvious as simply the supply voltage divided by the resultant resistance for B. If desired, block A can be broken down further, as shown in Fig. 7.16, to identify C and D as series elements, with the voltage V_2 capable of being determined using the voltage divider rule once the resistance of C and D is reduced to a single value. This is an example of how making a mental sketch of the approach before applying laws, rules, and so on can help avoid dead ends and frustration.

Applying Ohm's law, we have

$$I_4 = \frac{E}{R_B} = \frac{E}{R_4} = \frac{12\text{ V}}{8\ \Omega} = \mathbf{1.5\ A}$$

Combining the resistors R_2 and R_3 in Fig. 7.14 results in

$$R_D = R_2 \parallel R_3 = 3\ \Omega \parallel 6\ \Omega = \frac{(3\ \Omega)(6\ \Omega)}{3\ \Omega + 6\ \Omega} = \frac{18\ \Omega}{9} = 2\ \Omega$$

and, applying the voltage divider rule, we find

$$V_2 = \frac{R_D E}{R_D + R_C} = \frac{(2\ \Omega)(12\text{ V})}{2\ \Omega + 4\ \Omega} = \frac{24\text{ V}}{6} = \mathbf{4\ V}$$

EXAMPLE 7.6 Find the indicated currents and voltages for the network in Fig. 7.17.

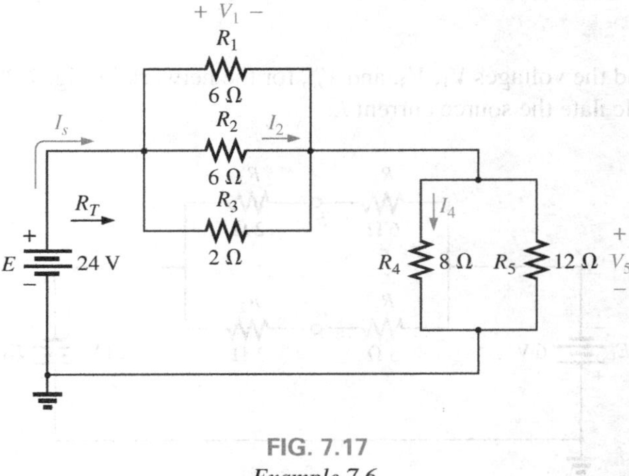

FIG. 7.17
Example 7.6.

Solution: Again, only specific unknowns are requested. When the network is redrawn, be sure to note which unknowns are preserved and which have to be determined using the original configuration. The block diagram of the network may appear as shown in Fig. 7.18, clearly revealing that A and B are in series. Note in this form the number of unknowns that have been preserved. The voltage V_1 is the same across the three parallel branches in Fig. 7.17, and V_5 is the same across R_4 and R_5. The unknown currents I_2 and I_4 are lost since they represent the currents through only one of the parallel branches. However, once V_1 and V_5 are known, you can find the required currents using Ohm's law.

$$R_{1\|2} = \frac{R}{N} = \frac{6\ \Omega}{2} = 3\ \Omega$$

$$R_A = R_{1\|2\|3} = \frac{(3\ \Omega)(2\ \Omega)}{3\ \Omega + 2\ \Omega} = \frac{6\ \Omega}{5} = 1.2\ \Omega$$

$$R_B = R_{4\|5} = \frac{(8\ \Omega)(12\ \Omega)}{8\ \Omega + 12\ \Omega} = \frac{96\ \Omega}{20} = 4.8\ \Omega$$

The reduced form of Fig. 7.17 then appears as shown in Fig. 7.19, and

$$R_T = R_{1\|2\|3} + R_{4\|5} = 1.2\ \Omega + 4.8\ \Omega = \mathbf{6\ \Omega}$$

$$I_s = \frac{E}{R_T} = \frac{24\text{ V}}{6\ \Omega} = \mathbf{4\ A}$$

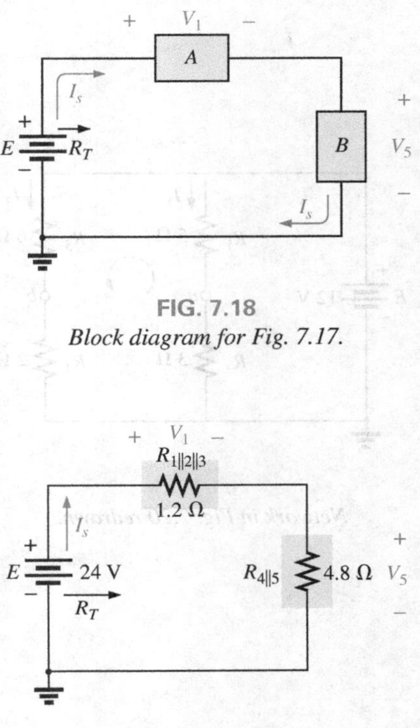

FIG. 7.18
Block diagram for Fig. 7.17.

FIG. 7.19
Reduced form of Fig. 7.17.

with $V_1 = I_s R_{1\|2\|3} = (4 \text{ A})(1.2\ \Omega) = \mathbf{4.8\ V}$

$V_5 = I_s R_{4\|5} = (4 \text{ A})(4.8\ \Omega) = \mathbf{19.2\ V}$

Applying Ohm's law gives

$$I_4 = \frac{V_5}{R_4} = \frac{19.2 \text{ V}}{8\ \Omega} = \mathbf{2.4\ A}$$

$$I_2 = \frac{V_2}{R_2} = \frac{V_1}{R_2} = \frac{4.8 \text{ V}}{6\ \Omega} = \mathbf{0.8\ A}$$

The next example demonstrates that unknown voltages do not have to be across elements but can exist between any two points in a network. In addition, the importance of redrawing the network in a more familiar form is clearly revealed by the analysis to follow.

EXAMPLE 7.7

a. Find the voltages V_1, V_3, and V_{ab} for the network in Fig. 7.20.
b. Calculate the source current I_s.

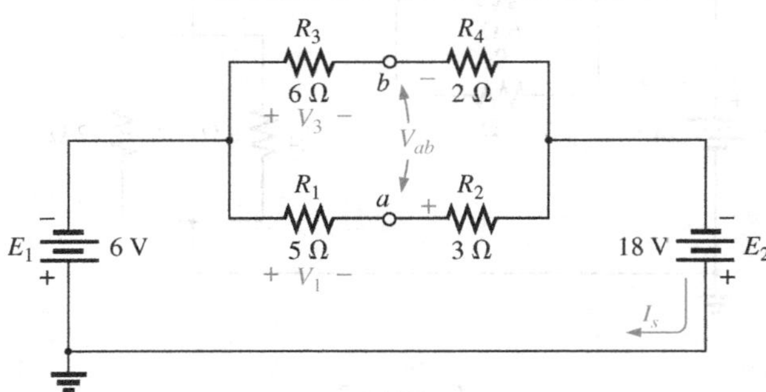

FIG. 7.20
Example 7.7.

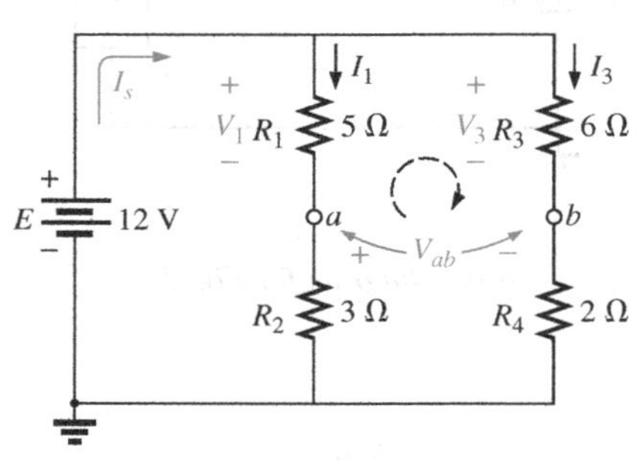

FIG. 7.21
Network in Fig. 7.20 redrawn.

Solutions: This is one of those situations where it may be best to redraw the network before beginning the analysis. Since combining both sources will not affect the unknowns, the network is redrawn as shown in Fig. 7.21, establishing a parallel network with the total source voltage across each parallel branch. The net source voltage is the difference between the two with the polarity of the larger.

a. Note the similarities with Fig. 7.16, permitting the use of the voltage divider rule to determine V_1 and V_3:

$$V_1 = \frac{R_1 E}{R_1 + R_2} = \frac{(5\ \Omega)(12 \text{ V})}{5\ \Omega + 3\ \Omega} = \frac{60 \text{ V}}{8} = \mathbf{7.5\ V}$$

$$V_3 = \frac{R_3 E}{R_3 + R_4} = \frac{(6\ \Omega)(12 \text{ V})}{6\ \Omega + 2\ \Omega} = \frac{72 \text{ V}}{8} = \mathbf{9\ V}$$

The open-circuit voltage V_{ab} is determined by applying Kirchhoff's voltage law around the indicated loop in Fig. 7.21 in the clockwise direction starting at terminal a. We have

$$+V_1 - V_3 + V_{ab} = 0$$

and $V_{ab} = V_3 - V_1 = 9 \text{ V} - 7.5 \text{ V} = \mathbf{1.5\ V}$

b. By Ohm's law,

$$I_1 = \frac{V_1}{R_1} = \frac{7.5\text{ V}}{5\ \Omega} = 1.5\text{ A}$$

$$I_3 = \frac{V_3}{R_3} = \frac{9\text{ V}}{6\ \Omega} = 1.5\text{ A}$$

Applying Kirchhoff's current law gives

$$I_s = I_1 + I_3 = 1.5\text{ A} + 1.5\text{ A} = \mathbf{3\text{ A}}$$

EXAMPLE 7.8 For the network in Fig. 7.22, determine the voltages V_1 and V_2 and the current I.

Solution: It would indeed be difficult to analyze the network in the form in Fig. 7.22 with the symbolic notation for the sources and the reference or ground connection in the upper left corner of the diagram. However, when the network is redrawn as shown in Fig. 7.23, the unknowns and the relationship between branches become significantly clearer. Note the common connection of the grounds and the replacing of the terminal notation by actual supplies.

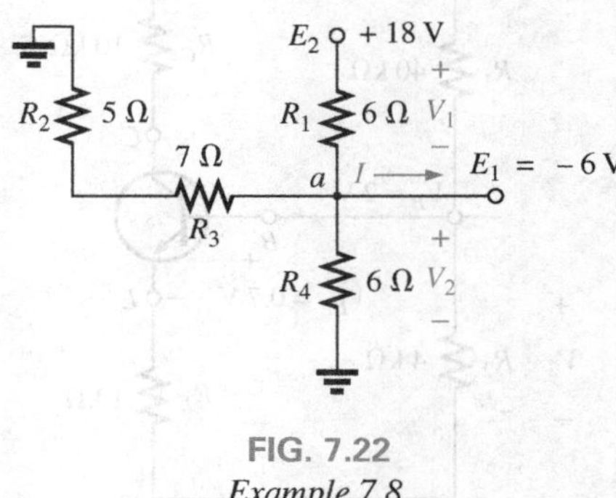

FIG. 7.22
Example 7.8.

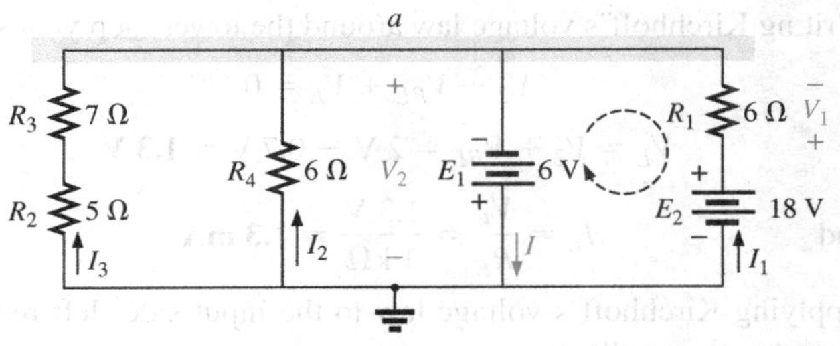

FIG. 7.23
Network in Fig. 7.22 redrawn.

It is now obvious that

$$V_2 = -E_1 = \mathbf{-6\text{ V}}$$

The minus sign simply indicates that the chosen polarity for V_2 in Fig. 7.22 is opposite to that of the actual voltage. Applying Kirchhoff's voltage law to the loop indicated, we obtain

$$-E_1 + V_1 - E_2 = 0$$

and

$$V_1 = E_2 + E_1 = 18\text{ V} + 6\text{ V} = \mathbf{24\text{ V}}$$

Applying Kirchhoff's current law to node a yields

$$\begin{aligned} I &= I_1 + I_2 + I_3 \\ &= \frac{V_1}{R_1} + \frac{E_1}{R_4} + \frac{E_1}{R_2 + R_3} \\ &= \frac{24\text{ V}}{6\ \Omega} + \frac{6\text{ V}}{6\ \Omega} + \frac{6\text{ V}}{12\ \Omega} \\ &= 4\text{ A} + 1\text{ A} + 0.5\text{ A} \\ I &= \mathbf{5.5\text{ A}} \end{aligned}$$

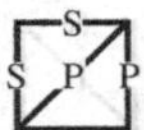

The next example is clear evidence that techniques learned in the current chapters will have far-reaching applications and will not be dropped for improved methods. Even though we have not studied the **transistor** yet, the dc levels of a transistor network can be examined using the basic rules and laws introduced in earlier chapters.

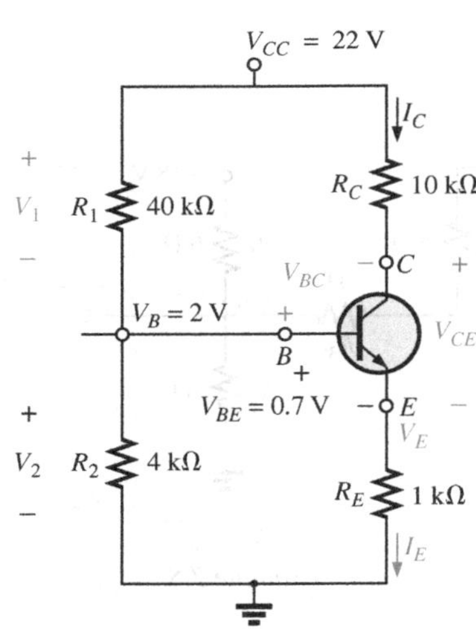

FIG. 7.24
Example 7.9.

EXAMPLE 7.9 For the transistor configuration in Fig. 7.24, in which V_B and V_{BE} have been provided:

a. Determine the voltage V_E and the current I_E.
b. Calculate V_1.
c. Determine V_{BC} using the fact that the approximation $I_C = I_E$ is often applied to transistor networks.
d. Calculate V_{CE} using the information obtained in parts (a) through (c).

Solutions:

a. From Fig. 7.24, we find

$$V_2 = V_B = 2\text{ V}$$

Writing Kirchhoff's voltage law around the lower loop yields

$$V_2 - V_{BE} + V_E = 0$$

or $$V_E = V_2 + V_{BE} = 2\text{ V} - 0.7\text{ V} = \mathbf{1.3\text{ V}}$$

and $$I_E = \frac{V_E}{R_E} = \frac{1.3\text{ V}}{1\text{ k}\Omega} = \mathbf{1.3\text{ mA}}$$

b. Applying Kirchhoff's voltage law to the input side (left region of the network) results in

$$V_2 + V_1 - V_{CC} = 0$$

and $$V_1 = V_{CC} - V_2$$

but $$V_2 = V_B$$

and $$V_1 = V_{CC} - V_2 = 22\text{ V} - 2\text{ V} = \mathbf{20\text{ V}}$$

c. Redrawing the section of the network of immediate interest results in Fig. 7.25, where Kirchhoff's voltage law yields

$$V_C + V_{R_C} - V_{CC} = 0$$

and $$V_C = V_{CC} - V_{R_C} = V_{CC} - I_C R_C$$

but $$I_C = I_E$$

and $$V_C = V_{CC} - I_E R_C = 22\text{ V} - (1.3\text{ mA})(10\text{ k}\Omega) = 9\text{ V}$$

Then $$V_{BC} = V_B - V_C = 2\text{ V} - 9\text{ V} = \mathbf{-7\text{ V}}$$

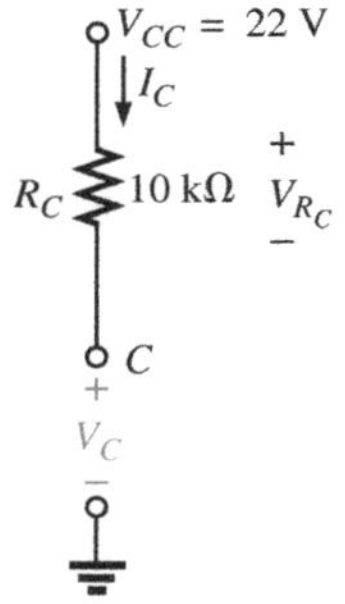

FIG. 7.25
Determining V_C for the network in Fig. 7.24.

d. $$V_{CE} = V_C - V_E = 9\text{ V} - 1.3\text{ V} = \mathbf{7.7\text{ V}}$$

EXAMPLE 7.10 Calculate the indicated currents and voltage in Fig. 7.26.

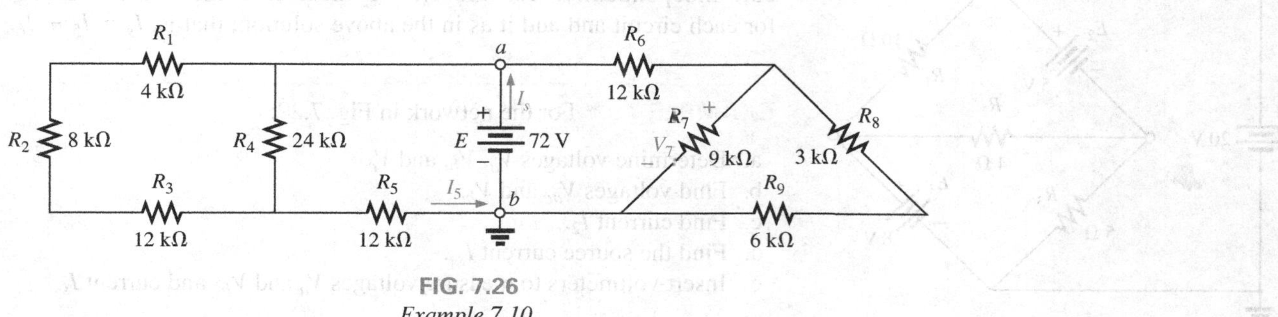

FIG. 7.26
Example 7.10.

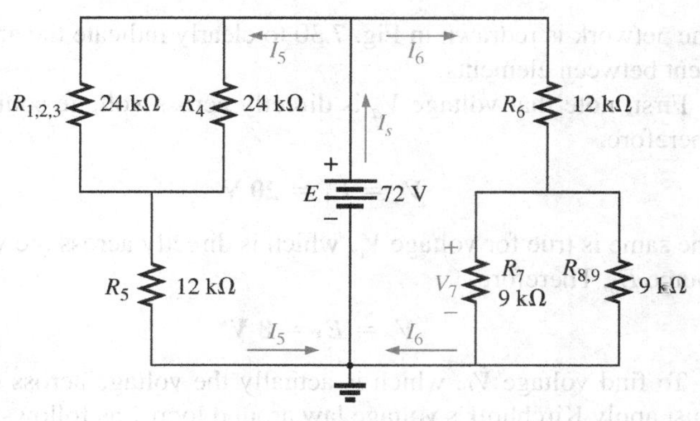

FIG. 7.27
Network in Fig. 7.26 redrawn.

Solution: Redrawing the network after combining series elements yields Fig. 7.27, and

$$I_5 = \frac{E}{R_{(1,2,3)\|4} + R_5} = \frac{72\text{ V}}{12\text{ k}\Omega + 12\text{ k}\Omega} = \frac{72\text{ V}}{24\text{ k}\Omega} = \mathbf{3\text{ mA}}$$

with

$$V_7 = \frac{R_{7\|(8,9)}E}{R_{7\|(8,9)} + R_6} = \frac{(4.5\text{ k}\Omega)(72\text{ V})}{4.5\text{ k}\Omega + 12\text{ k}\Omega} = \frac{324\text{ V}}{16.5} = \mathbf{19.6\text{ V}}$$

$$I_6 = \frac{V_7}{R_{7\|(8,9)}} = \frac{19.6\text{ V}}{4.5\text{ k}\Omega} = \mathbf{4.35\text{ mA}}$$

and

$$I_s = I_5 + I_6 = 3\text{ mA} + 4.35\text{ mA} = \mathbf{7.35\text{ mA}}$$

Since the potential difference between points *a* and *b* in Fig. 7.26 is fixed at *E* volts, the circuit to the right or left is unaffected if the network is reconstructed as shown in Fig. 7.28.

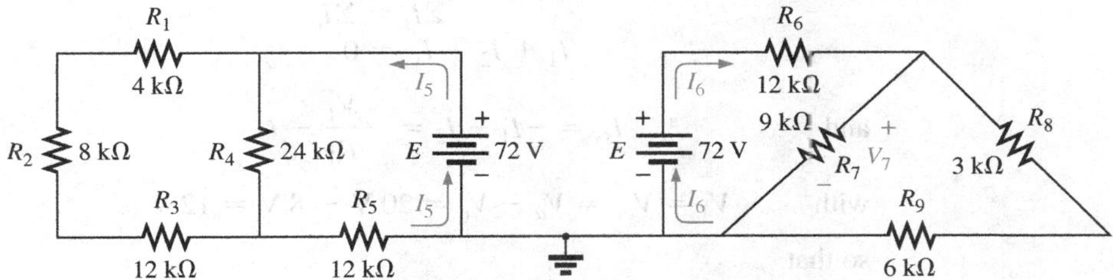

FIG. 7.28
An alternative approach to Example 7.10.

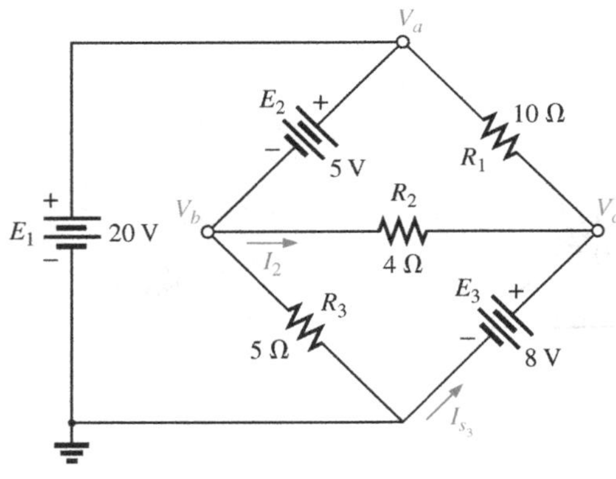

FIG. 7.29
Example 7.11.

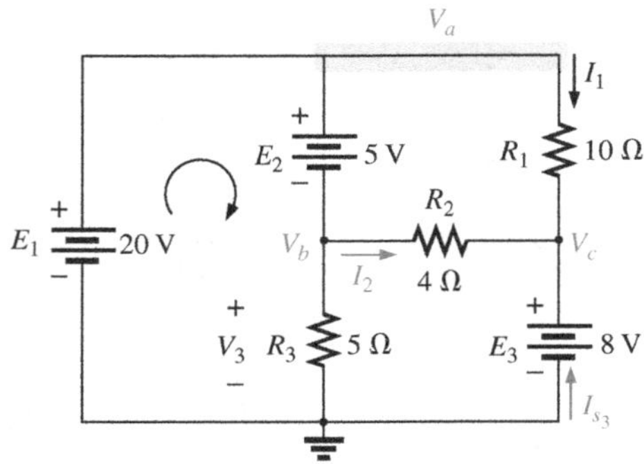

FIG. 7.30
Network in Fig. 7.29 redrawn to better define a path toward the desired unknowns.

We can find each quantity required, except I_s, by analyzing each circuit independently. To find I_s, we must find the source current for each circuit and add it as in the above solution; that is, $I_s = I_5 + I_6$.

EXAMPLE 7.11 For the network in Fig. 7.29:

a. Determine voltages V_a, V_b, and V_c.
b. Find voltages V_{ac} and V_{bc}.
c. Find current I_2.
d. Find the source current I_{s_3}.
e. Insert voltmeters to measure voltages V_a and V_{bc} and current I_{s_3}.

Solutions:

a. The network is redrawn in Fig. 7.30 to clearly indicate the arrangement between elements.

First, note that voltage V_a is directly across voltage source E_1. Therefore,

$$V_a = E_1 = \mathbf{20\ V}$$

The same is true for voltage V_c, which is directly across the voltage source E_3. Therefore,

$$V_c = E_3 = \mathbf{8\ V}$$

To find voltage V_b, which is actually the voltage across R_3, we must apply Kirchhoff's voltage law around loop 1 as follows:

$$+E_1 - E_2 - V_3 = 0$$

and $$V_3 = E_1 - E_2 = 20\ \text{V} - 5\ \text{V} = 15\ \text{V}$$

and $$V_b = V_3 = \mathbf{15\ V}$$

b. Voltage V_{ac}, which is actually the voltage across resistor R_1, can then be determined as follows:

$$V_{ac} = V_a - V_c = 20\ \text{V} - 8\ \text{V} = \mathbf{12\ V}$$

Similarly, voltage V_{bc}, which is actually the voltage across resistor R_2, can then be determined as follows:

$$V_{bc} = V_b - V_c = 15\ \text{V} - 8\ \text{V} = \mathbf{7\ V}$$

c. Current I_2 can be determined using Ohm's law:

$$I_2 = \frac{V_2}{R_2} = \frac{V_{bc}}{R_2} = \frac{7\ \text{V}}{4\ \Omega} = \mathbf{1.75\ A}$$

d. The source current I_{s_3} can be determined using Kirchhoff's current law at node c:

$$\Sigma I_i = \Sigma I_o$$
$$I_1 + I_2 + I_{s_3} = 0$$

and $$I_{s_3} = -I_1 - I_2 = -\frac{V_1}{R_1} - I_2$$

with $$V_1 = V_{ac} = V_a - V_c = 20\ \text{V} - 8\ \text{V} = 12\ \text{V}$$

so that

$$I_{s_3} = -\frac{12\ \text{V}}{10\ \Omega} - 1.75\ \text{A} = -1.2\ \text{A} - 1.75\ \text{A} = \mathbf{-2.95\ A}$$

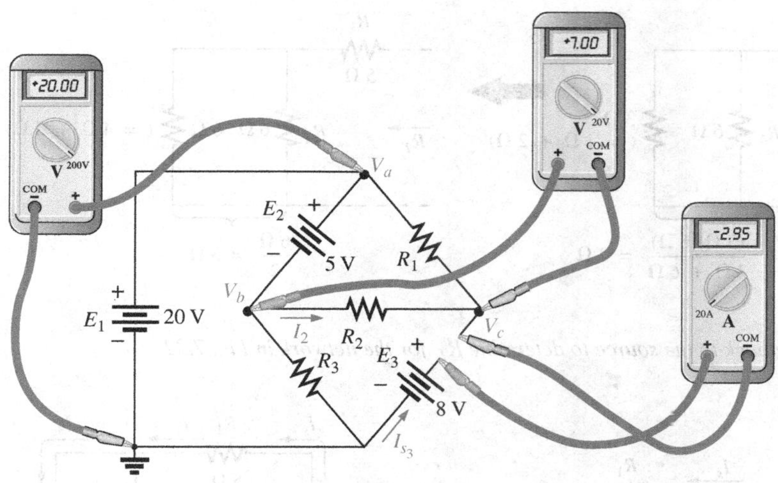

FIG. 7.31
Complex network for Example 7.11.

revealing that current is actually being forced through source E_3 in a direction opposite to that shown in Fig. 7.29.

e. Both voltmeters have a positive reading, as shown in Fig. 7.31, while the ammeter has a negative reading.

7.6 LADDER NETWORKS

A three-section **ladder network** appears in Fig. 7.32. The reason for the terminology is quite obvious for the repetitive structure. Basically two approaches are used to solve networks of this type.

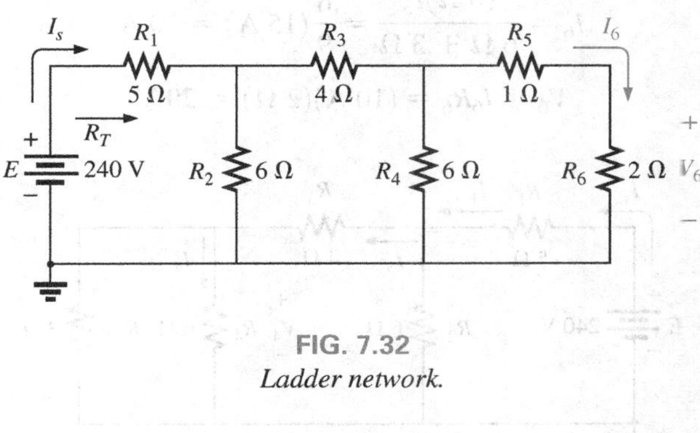

FIG. 7.32
Ladder network.

Method 1

Calculate the total resistance and resulting source current, and then work back through the ladder until the desired current or voltage is obtained. This method is now employed to determine V_6 in Fig. 7.32.

Combining parallel and series elements as shown in Fig. 7.33 results in the reduced network in Fig. 7.34, and

$$R_T = 5\ \Omega + 3\ \Omega = 8\ \Omega$$

$$I_s = \frac{E}{R_T} = \frac{240\ \text{V}}{8\ \Omega} = 30\ \text{A}$$

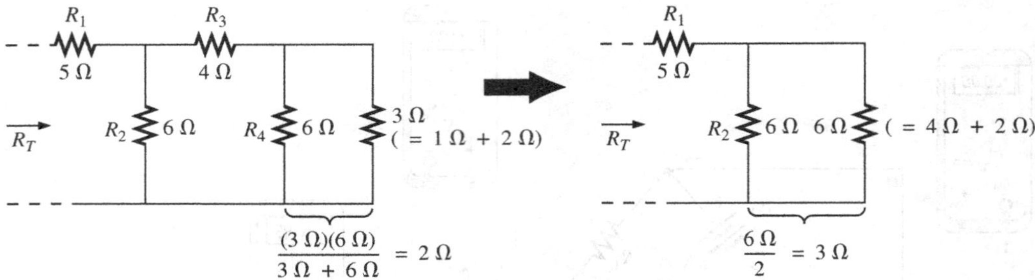

FIG. 7.33
Working back to the source to determine R_T for the network in Fig. 7.32.

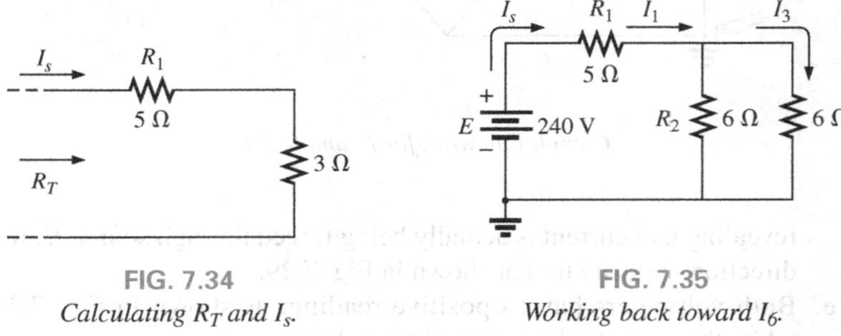

FIG. 7.34
Calculating R_T and I_s.

FIG. 7.35
Working back toward I_6.

Working our way back to I_6 (Fig. 7.35), we find that

$$I_1 = I_s$$

and
$$I_3 = \frac{I_s}{2} = \frac{30\text{ A}}{2} = 15\text{ A}$$

and, finally (Fig. 7.36),

$$I_6 = \frac{(6\,\Omega)I_3}{6\,\Omega + 3\,\Omega} = \frac{6}{9}(15\text{ A}) = 10\text{ A}$$

and
$$V_6 = I_6R_6 = (10\text{ A})(2\,\Omega) = \mathbf{20\text{ V}}$$

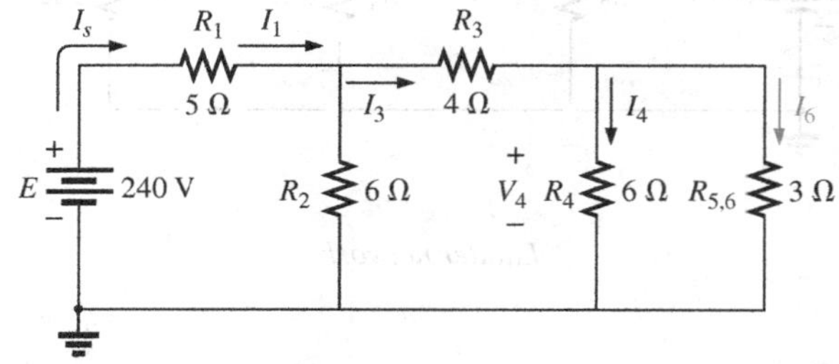

FIG. 7.36
Calculating I_6.

Method 2

Assign a letter symbol to the last branch current and work back through the network to the source, maintaining this assigned current or other current of interest. The desired current can then be found directly. This method can best be described through the analysis of the same network considered in Fig. 7.32, redrawn in Fig. 7.37.

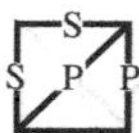

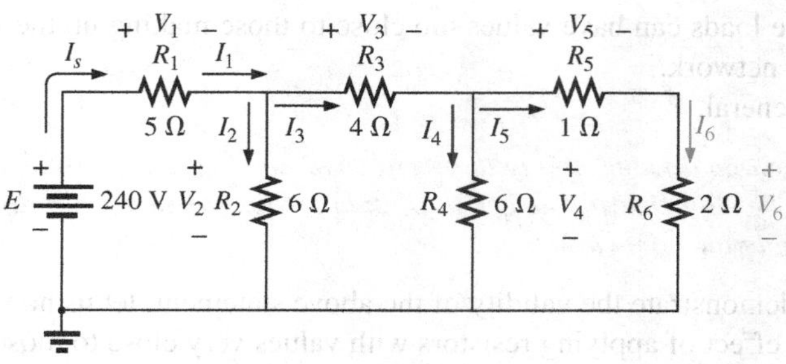

FIG. 7.37
An alternative approach for ladder networks.

The assigned notation for the current through the final branch is I_6:

$$I_6 = \frac{V_4}{R_5 + R_6} = \frac{V_4}{1\ \Omega + 2\ \Omega} = \frac{V_4}{3\ \Omega}$$

or
$$V_4 = (3\ \Omega)I_6$$

so that
$$I_4 = \frac{V_4}{R_4} = \frac{(3\ \Omega)I_6}{6\ \Omega} = 0.5I_6$$

and
$$I_3 = I_4 + I_6 = 0.5I_6 + I_6 = 1.5I_6$$
$$V_3 = I_3R_3 = (1.5I_6)(4\ \Omega) = (6\ \Omega)I_6$$

Also,
$$V_2 = V_3 + V_4 = (6\ \Omega)I_6 + (3\ \Omega)I_6 = (9\ \Omega)I_6$$

so that
$$I_2 = \frac{V_2}{R_2} = \frac{(9\ \Omega)I_6}{6\ \Omega} = 1.5I_6$$

and
$$I_s = I_2 + I_3 = 1.5I_6 + 1.5I_6 = 3I_6$$

with
$$V_1 = I_1R_1 = I_sR_1 = (5\ \Omega)I_s$$

so that
$$E = V_1 + V_2 = (5\ \Omega)I_s + (9\ \Omega)I_6$$
$$= (5\ \Omega)(3I_6) + (9\ \Omega)I_6 = (24\ \Omega)I_6$$

and
$$I_6 = \frac{E}{24\ \Omega} = \frac{240\ \text{V}}{24\ \Omega} = 10\ \text{A}$$

with
$$V_6 = I_6R_6 = (10\ \text{A})(2\ \Omega) = \mathbf{20\ V}$$

as was obtained using method 1.

7.7 VOLTAGE DIVIDER SUPPLY (UNLOADED AND LOADED)

When the term *loaded* is used to describe voltage divider supply, it refers to the application of an element, network, or system to a supply that draws current from the supply. In other words,

the loading down of a system is the process of introducing elements that will draw current from the system. The heavier the current, the greater is the loading effect.

Recall from Section 5.10 that the application of a load can affect the terminal voltage of a supply due to the internal resistance.

No-Load Conditions

Through a voltage divider network such as that in Fig. 7.38, a number of different terminal voltages can be made available from a single supply. Instead of having a single supply of 120 V, we now have terminal voltages of 100 V and 60 V available—a wonderful result for such a simple network. However, there can be disadvantages. One is that the applied

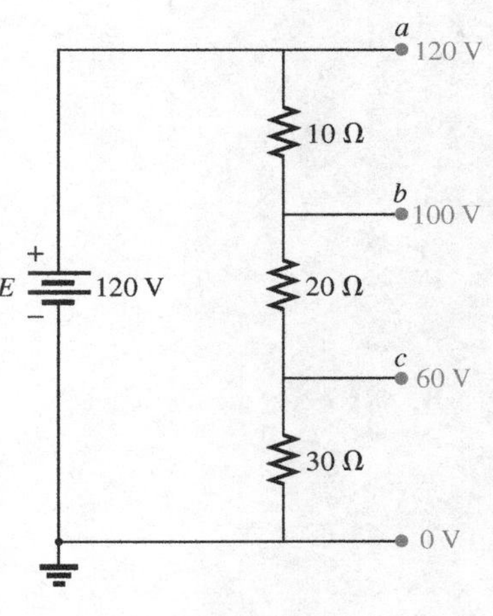

FIG.7.38
Voltage divider supply.

resistive loads can have values too close to those making up the voltage divider network.

In general,

for a voltage divider supply to be effective, the applied resistive loads should be significantly larger than the resistors appearing in the voltage divider network.

To demonstrate the validity of the above statement, let us now examine the effect of applying resistors with values very close to those of the voltage divider network.

Loaded Conditions

In Fig. 7.39, resistors of 20 Ω have been connected to each of the terminal voltages. Note that this value is equal to one of the resistors in the voltage divider network and very close to the other two.

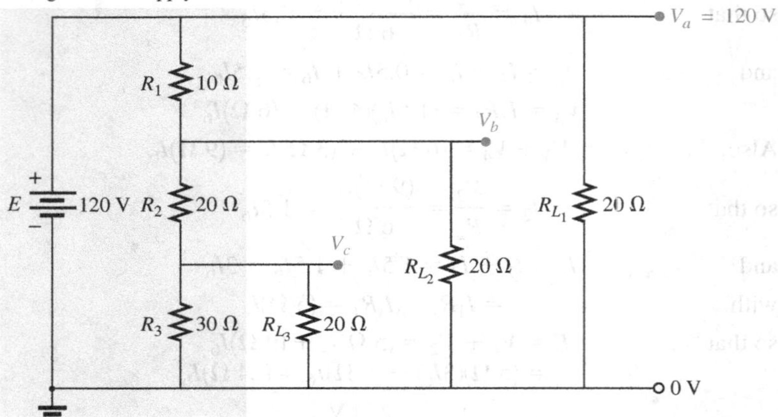

FIG. 7.39
Voltage divider supply with loads equal to the average value of the resistive elements that make up the supply.

Voltage V_a is unaffected by the load R_{L_1} since the load is in parallel with the supply voltage E. The result is $V_a = 120$ V, which is the same as the no-load level. To determine V_b, we must first note that R_3 and R_{L_3} are in parallel and $R'_3 = R_3 \parallel R_{L_3} = 30\ \Omega \parallel 20\ \Omega = 12\ \Omega$. The parallel combination gives

$$\begin{aligned} R'_2 &= (R_2 + R'_3) \parallel R_{L_2} = (20\ \Omega + 12\ \Omega) \parallel 20\ \Omega \\ &= 32\ \Omega \parallel 20\ \Omega = 12.31\ \Omega \end{aligned}$$

Applying the voltage divider rule gives

$$V_b = \frac{(12.31\ \Omega)(120\ \text{V})}{12.31\ \Omega + 10\ \Omega} = \mathbf{66.21\ V}$$

versus 100 V under no-load conditions.

Voltage V_c is

$$V_c = \frac{(12\ \Omega)(66.21\ \text{V})}{12\ \Omega + 20\ \Omega} = \mathbf{24.83\ V}$$

versus 60 V under no-load conditions.

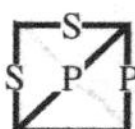

The effect of load resistors close in value to the resistor employed in the voltage divider network is, therefore, to decrease significantly some of the terminal voltages.

If the load resistors are changed to the 1 kΩ level, the terminal voltages will all be relatively close to the no-load values. The analysis is similar to the above, with the following results:

$$V_a = \mathbf{120\ V} \qquad V_b = \mathbf{98.88\ V} \qquad V_c = \mathbf{58.63\ V}$$

If we compare current drains established by the applied loads, we find for the network in Fig. 7.39 that

$$I_{L_2} = \frac{V_{L_2}}{R_{L_2}} = \frac{66.21\ \text{V}}{20\ \Omega} = 3.31\ \text{A}$$

and for the 1 kΩ level,

$$I_{L_2} = \frac{98.88\ \text{V}}{1\ \text{k}\Omega} = 98.88\ \text{mA} < 0.1\ \text{A}$$

As demonstrated above, the greater the current drain, the greater is the change in terminal voltage with the application of the load. This is certainly verified by the fact that I_{L2} is about 33.5 times larger with the 20 Ω loads.

The next example is a design exercise. The voltage and current ratings of each load are provided, along with the terminal ratings of the supply. The required voltage divider resistors must be found.

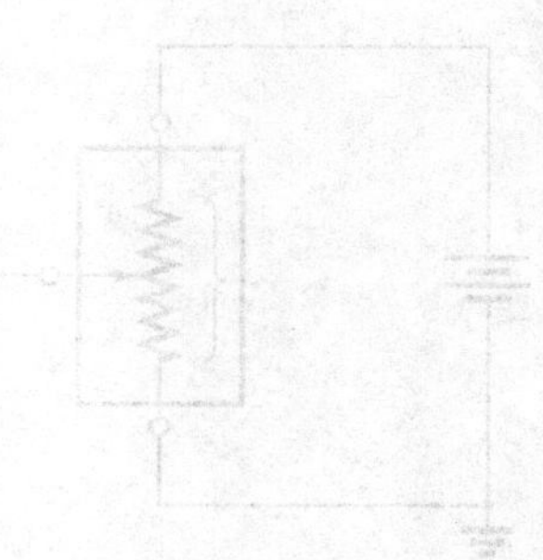

EXAMPLE 7.12 Determine R_1, R_2, and R_3 for the voltage divider supply in Fig. 7.40. Can 2 W resistors be used in the design?

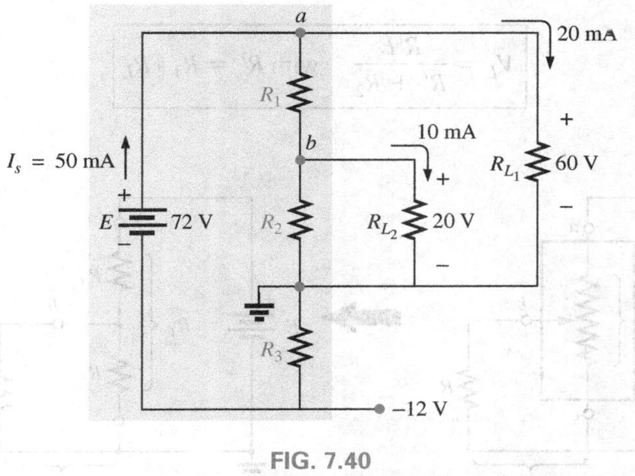

FIG. 7.40
Voltage divider supply for Example 7.12.

Solution: R_3:

$$R_3 = \frac{V_{R_3}}{I_{R_3}} = \frac{V_{R_3}}{I_s} = \frac{12\ \text{V}}{50\ \text{mA}} = \mathbf{240\ \Omega}$$

$$P_{R_3} = (I_{R_3})^2 R_3 = (50\ \text{mA})^2\, 240\ \Omega = 0.6\ \text{W} < 2\ \text{W}$$

R_1: Applying Kirchhoff's current law to node a, we have

$$I_s - I_{R_1} - I_{L_1} = 0$$

and $I_{R_1} = I_s - I_{L_1} = 50\text{ mA} - 20\text{ mA} = 30\text{ mA}$

$$R_1 = \frac{V_{R_1}}{I_{R_1}} = \frac{V_{L_1} - V_{L_2}}{I_{R_1}} = \frac{60\text{ V} - 20\text{ V}}{30\text{ mA}} = \frac{40\text{ V}}{30\text{ mA}} = \mathbf{1.33\ k\Omega}$$

$$P_{R_1} = (I_{R_1})^2 R_1 = (30\text{ mA})^2\, 1.33\text{ k}\Omega = 1.197\text{ W} < 2\text{ W}$$

R_2: Applying Kirchhoff's current law at node b, we have

$$I_{R_1} - I_{R_2} - I_{L_2} = 0$$

and $$I_{R_2} = I_{R_1} - I_{L_2} = 30\text{ mA} - 10\text{ mA} = 20\text{ mA}$$

$$R_2 = \frac{V_{R_2}}{I_{R_2}} = \frac{20\text{ V}}{20\text{ mA}} = \mathbf{1\ k\Omega}$$

$$P_{R_2} = (I_{R_2})^2 R_2 = (20\text{ mA})^2\, 1\text{ k}\Omega = 0.4\text{ W} < 2\text{ W}$$

Since P_{R_1}, P_{R_2}, and P_{R_3} are less than 2 W, 2 W resistors can be used for the design.

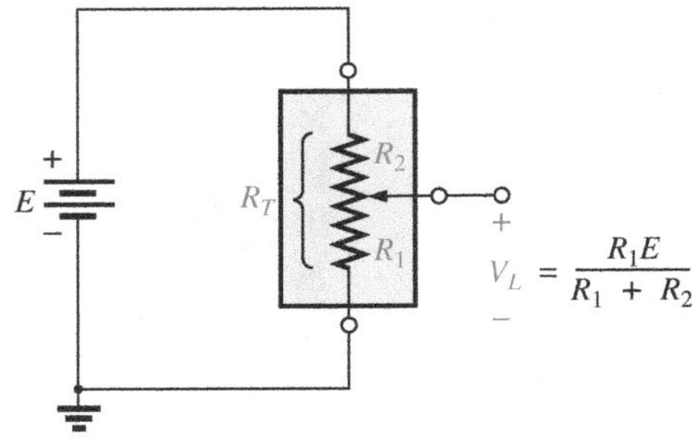

7.8 POTENTIOMETER LOADING

For the unloaded potentiometer in Fig. 7.41, the output voltage is determined by the voltage divider rule, with R_T in the figure representing the total resistance of the potentiometer. Too often it is assumed that the voltage across a load connected to the wiper arm is determined solely by the potentiometer and the effect of the load can be ignored. This is definitely not the case, as is demonstrated here.

When a load is applied as shown in Fig. 7.42, the output voltage V_L is now a function of the magnitude of the load applied since R_1 is not as shown in Fig. 7.41 but is instead the parallel combination of R_1 and R_L.

The output voltage is now

$$V_L = \frac{R'E}{R' + R_2} \quad \text{with } R' = R_1 \| R_L \tag{7.1}$$

FIG. 7.41
Unloaded potentiometer.

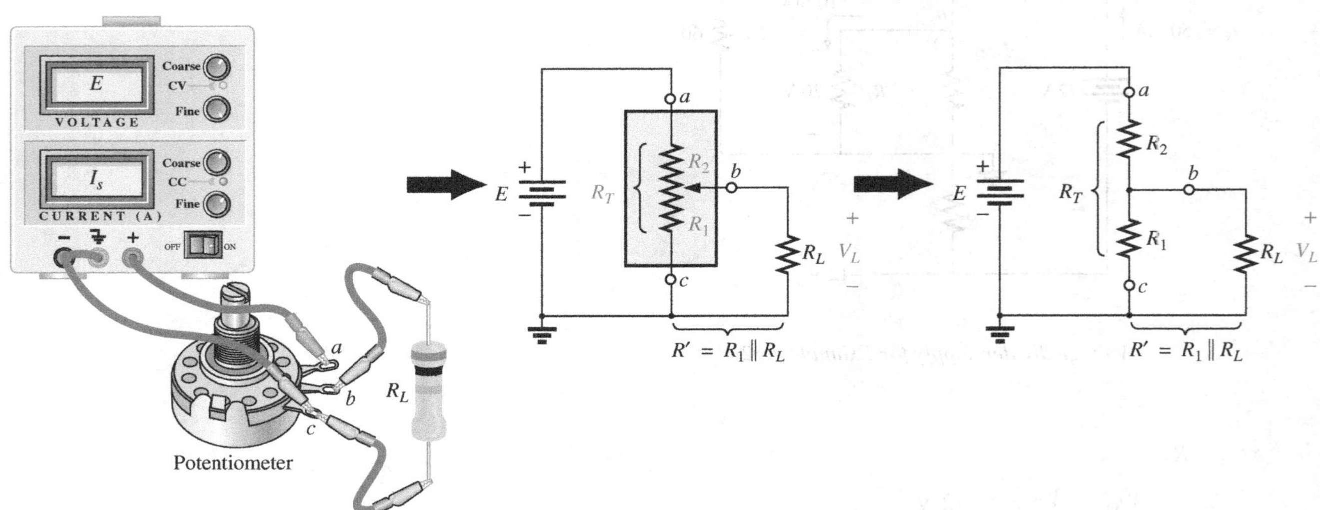

FIG. 7.42
Loaded potentiometer.

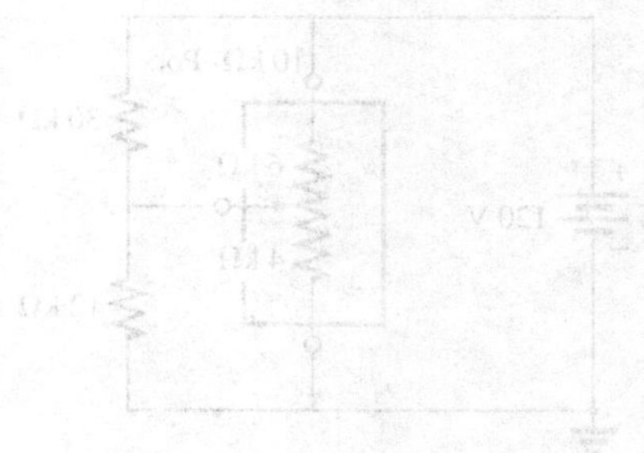

If you want to have good control of the output voltage V_L through the controlling dial, knob, screw, or whatever, you must choose a load or potentiometer that satisfies the following relationship:

$$\boxed{R_L >> R_T} \tag{7.2}$$

In general,

when hooking up a load to a potentiometer, be sure that the load resistance far exceeds the maximum terminal resistance of the potentiometer if good control of the output voltage is desired.

For example, let's disregard Eq. (7.2) and choose a 1 MΩ potentiometer with a 100 Ω load and set the wiper arm to 1/10 the total resistance, as shown in Fig. 7.43. Then

$$R' = 100 \text{ k}\Omega \parallel 100 \ \Omega = 99.9 \ \Omega$$

and
$$V_L = \frac{99.9 \ \Omega(10 \text{ V})}{99.9 \ \Omega + 900 \text{ k}\Omega} \cong 0.001 \text{ V} = 1 \text{ mV}$$

which is extremely small compared to the expected level of 1 V.

In fact, if we move the wiper arm to the midpoint,

$$R' = 500 \text{ k}\Omega \parallel 100 \ \Omega = 99.98 \ \Omega$$

and
$$V_L = \frac{(99.98 \ \Omega)(10 \text{ V})}{99.98 \ \Omega + 500 \text{ k}\Omega} \cong 0.002 \text{ V} = 2 \text{ mV}$$

which is negligible compared to the expected level of 5 V. Even at R_1 = 900 kΩ, V_L is only 0.01 V, or 1/1000 of the available voltage.

Using the reverse situation of R_T = 100 Ω and R_L = 1 MΩ and the wiper arm at the 1/10 position, as in Fig. 7.44, we find

$$R' = 10 \ \Omega \parallel 1 \text{ M}\Omega \cong 10 \ \Omega$$

and
$$V_L = \frac{10 \ \Omega(10 \text{ V})}{10 \ \Omega + 90 \ \Omega} = 1 \text{ V}$$

as desired.

For the lower limit (worst-case design) of $R_L = R_T$ = 100 Ω, as defined by Eq. (7.2) and the halfway position of Fig. 7.42,

$$R' = 50 \ \Omega \parallel 100 \ \Omega = 33.33 \ \Omega$$

and
$$V_L = \frac{33.33 \ \Omega(10 \text{ V})}{33.33 \ \Omega + 50 \ \Omega} \cong 4 \text{ V}$$

It may not be the ideal level of 5 V, but at least 40% of the voltage E has been achieved at the halfway position rather than the 0.02% obtained with R_L = 100 Ω and R_T = 1 MΩ.

In general, therefore, try to establish a situation for potentiometer control in which Eq. (7.2) is satisfied to the highest degree possible.

Someone might suggest that we make R_T as small as possible to bring the percent result as close to the ideal as possible. Keep in mind, however, that the potentiometer has a power rating, and for networks such as Fig. 7.44, $P_{max} \cong E^2/R_T = (10 \text{ V})^2/100 \ \Omega = 1$ W. If R_T is reduced to 10 Ω, $P_{max} = (10 \text{ V})^2/10 \ \Omega = 10$ W, which would require a *much larger* unit.

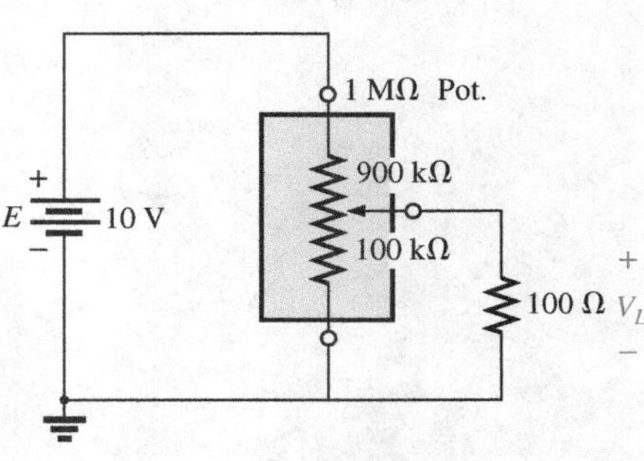

FIG. 7.43
Loaded potentiometer with R_L << R_T.

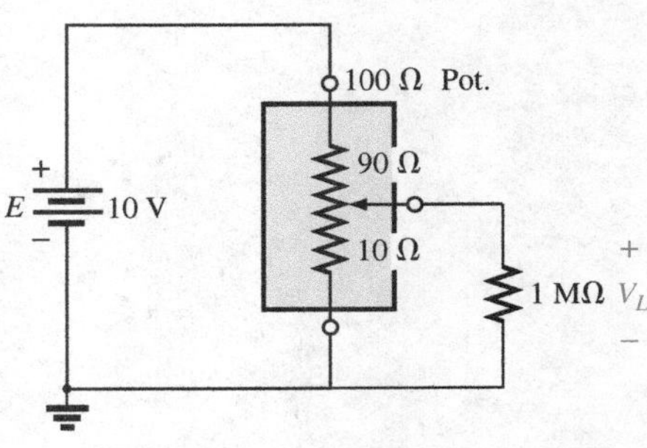

FIG. 7.44
Loaded potentiometer with R_L >> R_T.

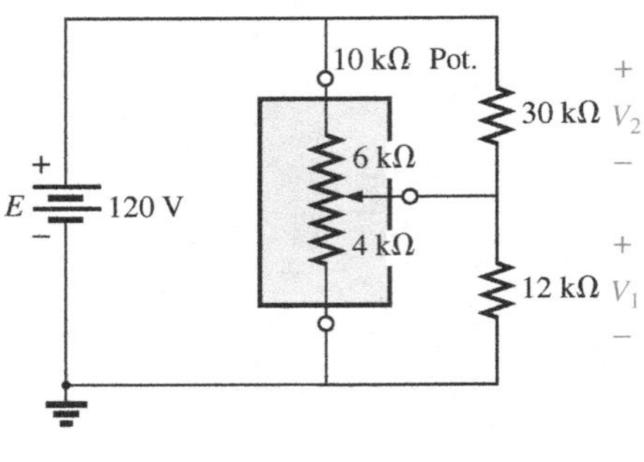

FIG. 7.45
Example 7.13.

EXAMPLE 7.13 Find voltages V_1 and V_2 for the loaded potentiometer of Fig. 7.45.

Solution: Ideal (no load):

$$V_1 = \frac{4\text{ k}\Omega(120\text{ V})}{10\text{ k}\Omega} = \mathbf{48\text{ V}}$$

$$V_2 = \frac{6\text{ k}\Omega(120\text{ V})}{10\text{ k}\Omega} = \mathbf{72\text{ V}}$$

Loaded:

$$R' = 4\text{ k}\Omega \parallel 12\text{ k}\Omega = 3\text{ k}\Omega$$

$$R'' = 6\text{ k}\Omega \parallel 30\text{ k}\Omega = 5\text{ k}\Omega$$

$$V_1 = \frac{3\text{ k}\Omega(120\text{ V})}{8\text{ k}\Omega} = \mathbf{45\text{ V}}$$

$$V_2 = \frac{5\text{ k}\Omega(120\text{ V})}{8\text{ k}\Omega} = \mathbf{75\text{ V}}$$

The ideal and loaded voltage levels are so close that the design can be considered a good one for the applied loads. A slight variation in the position of the wiper arm will establish the ideal voltage levels across the two loads.

7.9 AMMETER, VOLTMETER, AND OHMMETER DESIGN

The designs of this section will use the iron-vane movement of Fig. 7.46 because it is the one that is most frequently used by current instrument manufacturers. It operates using the principle that there is a repulsive force between like magnetic poles. When a current is applied to the coil wrapped around the two vanes, a magnetic field is established within the coil, magnetizing the fixed and moveable vanes. Since both vanes will be magnetized in the same manner, they will have the same polarity, and a force of repulsion will develop between the two vanes. The stronger the applied current, the stronger are the magnetic field and the force of repulsion between

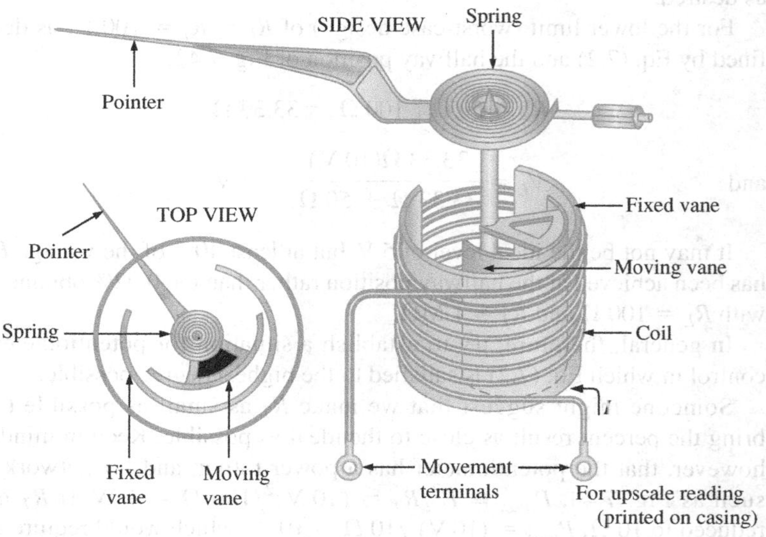

FIG. 7.46
Iron-vane movement.

the vanes. The fixed vane will remain in position, but the moveable vane will rotate and provide a measure of the strength of the applied current.

An iron-vane movement manufactured by the Simpson Company appears in Fig. 7.47(a). Movements of this type are usually rated in terms of current and resistance. The current sensitivity (*CS*) is the current that will result in a full-scale deflection. The resistance (R_m) is the internal resistance of the movement. The graphic symbol for a movement appears in Fig. 7.47(b) with the current sensitivity and internal resistance for the unit of Fig. 7.47(a).

Movements are usually rated by current and resistance. The specifications of a typical movement may be 1 mA, 50 Ω. The 1 mA is the *current sensitivity (CS)* of the movement, which is the current required for a full-scale deflection. It is denoted by the symbol I_{CS}. The 50 Ω represents the internal resistance (R_m) of the movement. A common notation for the movement and its specifications is provided in Fig. 7.48.

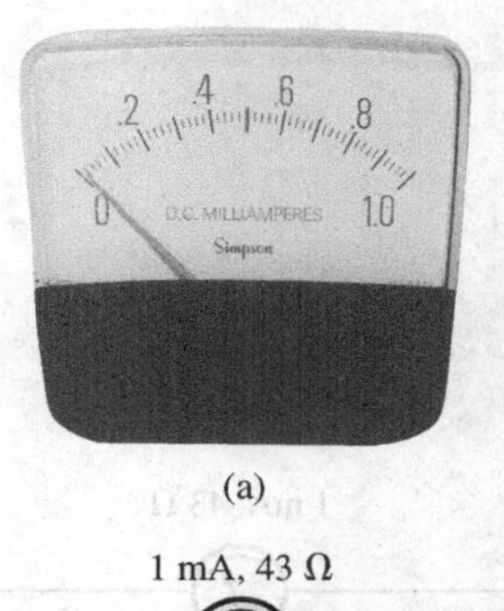

(a)

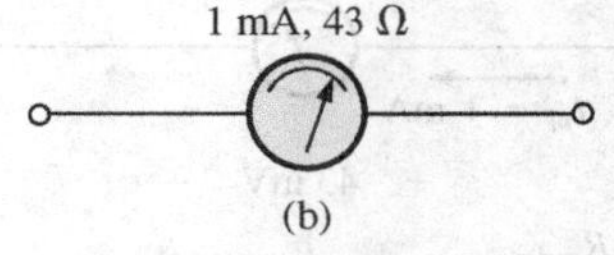

(b)

FIG. 7.47
Iron-vane movement; (a) photo, (b) symbol and ratings.

The Ammeter

The maximum current that the iron-vane movement can read independently is equal to the current sensitivity of the movement. However, higher currents can be measured if additional circuitry is introduced. This additional circuitry, as shown in Fig. 7.48, results in the basic construction of an ammeter.

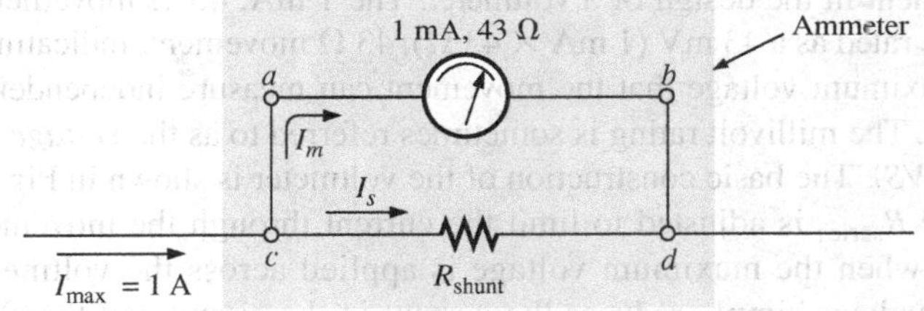

FIG. 7.48
Basic ammeter.

The resistance R_{shunt} is chosen for the ammeter in Fig. 7.49 to allow 1 mA to flow through the movement when a maximum current of 1 A enters the ammeter. If less than 1 A flows through the ammeter, the movement will have less than 1 mA flowing through it and will indicate less than full-scale deflection.

Since the voltage across parallel elements must be the same, the potential drop across *a-b* in Fig. 7.49 must equal that across *c-d*; that is,

$$(1\text{ mA})(43\ \Omega) = R_{shunt}I_s$$

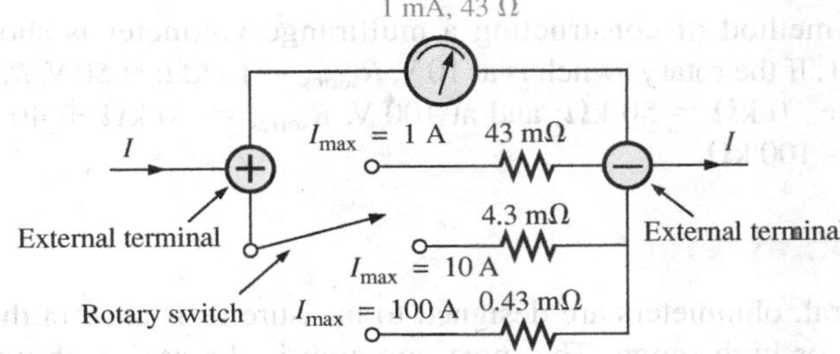

FIG. 7.49
Multirange ammeter.

Also, I_s must equal 1 A − 1 mA = 999 mA if the current is to be limited to 1 mA through the movement (Kirchhoff's current law). Therefore,

$$(1\text{ mA})(43\ \Omega) = R_{\text{shunt}}(999\text{ mA})$$

$$R_{\text{shunt}} = \frac{(1\text{ mA})(43\ \Omega)}{999\text{ mA}}$$

$$\cong 43\text{ m}\Omega\text{ (a standard value)}$$

In general,

$$R_{\text{shunt}} = \frac{R_m I_{CS}}{I_{\max} - I_{CS}} \qquad (7.4)$$

One method of constructing a multirange ammeter is shown in Fig. 7.50, where the rotary switch determines the R_{shunt} to be used for the maximum current indicated on the face of the meter. Most meters use the same scale for various values of maximum current. If you read 375 on the 0–5 mA scale with the switch on the 5 setting, the current is 3.75 mA; on the 50 setting, the current is 37.5 mA; and so on.

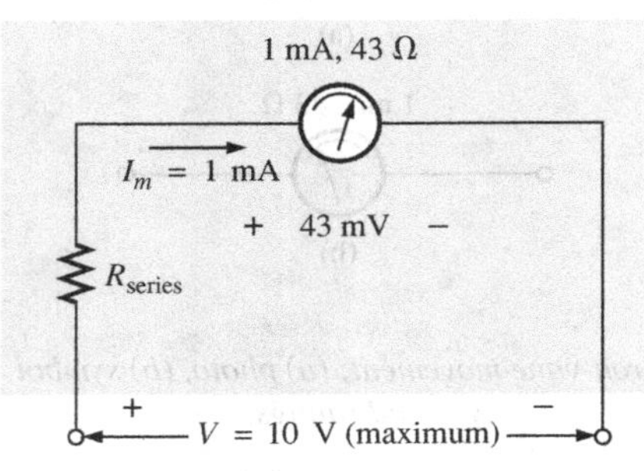

FIG. 7.50
Basic voltmeter.

The Voltmeter

A variation in the additional circuitry permits the use of the iron-vane movement in the design of a voltmeter. The 1 mA, 43 Ω movement can also be rated as a 43 mV (1 mA × 43 Ω), 43 Ω movement, indicating that the maximum voltage that the movement can measure independently is 43 mV. The millivolt rating is sometimes referred to as the *voltage sensitivity (VS).* The basic construction of the voltmeter is shown in Fig. 7.50.

The R_{series} is adjusted to limit the current through the movement to 1 mA when the maximum voltage is applied across the voltmeter. A lower voltage simply reduces the current in the circuit and thereby the deflection of the movement.

Applying Kirchhoff's voltage law around the closed loop of Fig. 7.50, we obtain

$$[10\text{ V} - (1\text{ mA})(R_{\text{series}})] - 43\text{ mV} = 0$$

or

$$R_{\text{series}} = \frac{10\text{ V} - (43\text{ mV})}{1\text{ mA}} = 9957\ \Omega \cong 10\text{ k}\Omega$$

In general,

$$R_{\text{series}} = \frac{V_{\max} - V_{VS}}{I_{CS}} \qquad (7.5)$$

One method of constructing a multirange voltmeter is shown in Fig. 7.51. If the rotary switch is at 10 V, R_{series} = 10 kΩ; at 50 V, R_{series} = 40 kΩ + 10 kΩ = 50 kΩ; and at 100 V, R_{series} = 50 kΩ + 40 kΩ + 10 kΩ = 100 kΩ.

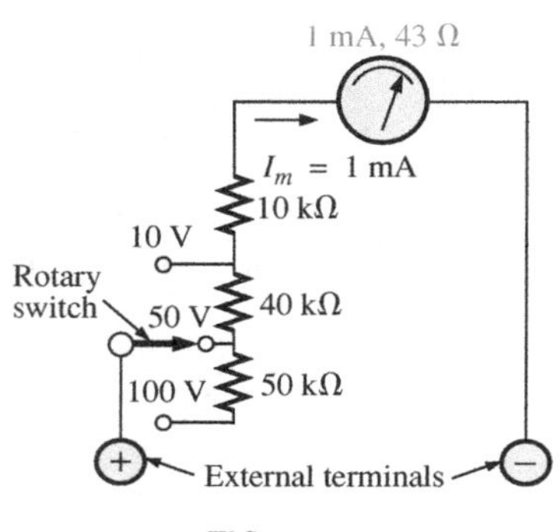

FIG. 7.51
Multirange voltmeter.

The Ohmmeter

In general, ohmmeters are designed to measure resistance in the low, middle, or high range. The most common is the **series ohmmeter,** designed to read resistance levels in the midrange. It uses the series

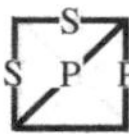

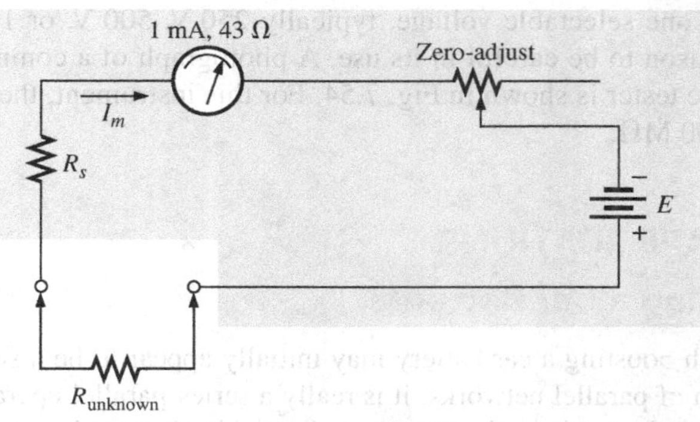

FIG. 7.52
Series ohmmeter.

configuration in Fig. 7.52. The design is quite different from that of the ammeter or voltmeter because it shows a full-scale deflection for zero ohms and no deflection for infinite resistance.

To determine the series resistance R_s, the external terminals are shorted (a direct connection of zero ohms between the two) to simulate zero ohms, and the zero-adjust is set to half its maximum value. The resistance R_s is then adjusted to allow a current equal to the current sensitivity of the movement (1 mA) to flow in the circuit. The zero-adjust is set to half its value so that any variation in the components of the meter that may produce a current more or less than the current sensitivity can be compensated for. The current I_m is

$$I_m\,(\text{full scale}) = I_{CS} = \frac{E}{R_s + R_m + \dfrac{\text{zero-adjust}}{2}} \tag{7.6}$$

and

$$R_s = \frac{E}{I_{CS}} - R_m - \frac{\text{zero-adjust}}{2} \tag{7.7}$$

If an unknown resistance is then placed between the external terminals, the current is reduced, causing a deflection less than full scale. If the terminals are left open, simulating infinite resistance, the pointer does not deflect since the current through the circuit is zero.

An instrument designed to read very low values of resistance and voltage appears in Fig. 7.53. It is capable of reading resistance levels between 10 mΩ (0.01 Ω) and 100 mΩ (0.1 Ω) and voltages between 10 mV and 100 V. Because of its low-range capability, the network design must be a great deal more sophisticated than described above It uses electronic components that eliminate the inaccuracies introduced by lead and contact resistances. It is similar to the above system in the sense that it is completely portable and does require a dc battery to establish measurement conditions. Special leads are used to limit any introduced resistance levels.

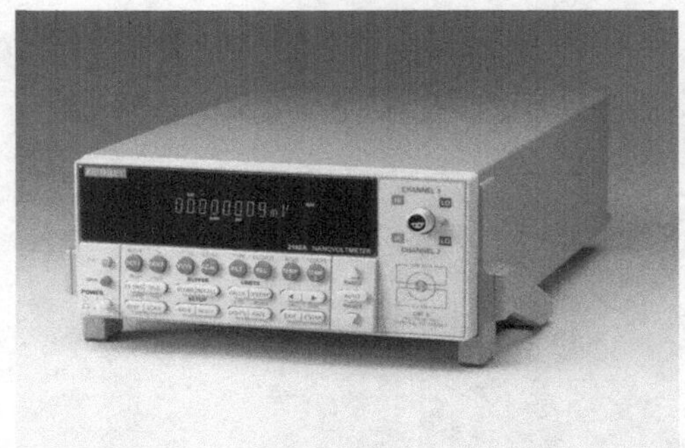

FIG. 7.53
Nanovoltmeter.
(Courtesy of Keithley Instruments.)

The **megohmmeter** (often called a *megger*) is an instrument for measuring very high resistance values. Its primary function is to test the insulation found in power transmission systems, electrical machinery, transformers, and so on. To measure the high-resistance values, a high dc voltage is established by a hand-driven generator. If the shaft is rotated above some set value, the output of the generator is

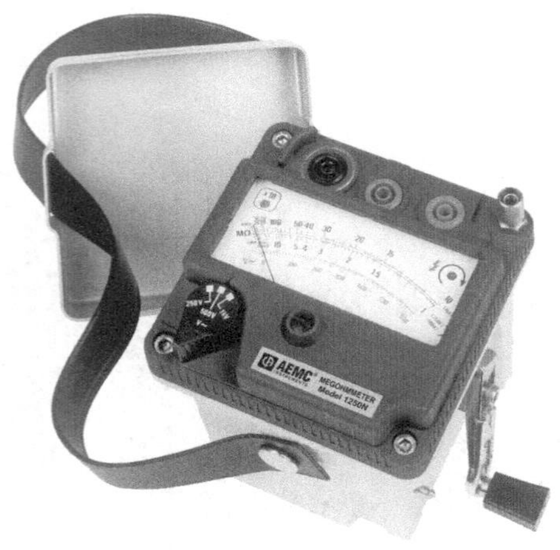

FIG. 7.54
Megohmmeter.
(Courtesy of AEMC® Instruments, Foxborough, MA.)

fixed at one selectable voltage, typically 250 V, 500 V, or 1000 V—good reason to be careful in its use. A photograph of a commercially available tester is shown in Fig. 7.54. For this instrument, the range is 0 to 5000 MΩ.

7.10 APPLICATIONS

Boosting a Car Battery

Although boosting a car battery may initially appear to be a simple application of parallel networks, it is really a series-parallel operation that is worthy of some investigation. As indicated in Chapter 2, every dc supply has some internal resistance. For the typical 12 V lead-acid car battery, the resistance is quite small—in the milliohm range. In most cases, the low internal resistance ensures that most of the voltage (or power) is delivered to the load and not lost on the internal resistance. In Fig. 7.55, battery #2 has discharged because the lights were left on for 3 hours during a movie. Fortunately, a friend who made sure his own lights were off has a fully charged battery #1 and a good set of 16-ft cables with #6 gage stranded wire and well-designed clips. The investment in a good set of cables with sufficient length and heavy wire is a wise one, particularly if you live in a cold climate. Flexibility, as provided by stranded wire, is also a very desirable characteristic under some conditions. Be sure to check the gage of the wire and not just the thickness of the insulating jacket. You get what you pay for, and the copper is the most expensive part of the cables. Too often the label says "heavy-duty," but the gage number of the wire is too high.

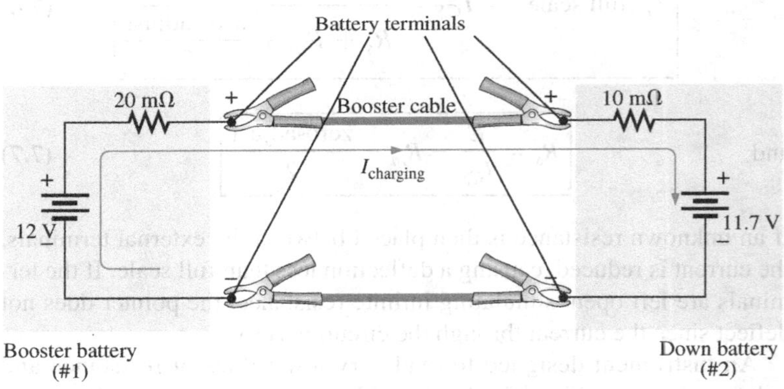

FIG. 7.55
Boosting a car battery.

The proper sequence of events in boosting a car is often a function of to whom you speak or what information you read. For safety's sake, some people recommend that the car with the good battery be turned off when making the connections. This, however, can create an immediate problem if the "dead" battery is in such a bad state that when it is hooked up to the good battery, it immediately drains the good battery to the point that neither car will start. With this in mind, it does make some sense to leave the car running to ensure that the charging process continues until the starting of the disabled car is initiated. *Because accidents do happen, it is strongly recommended that the person making the connections wear the proper type of protective eye equipment. Take sufficient time to be*

sure that you know which are the positive and negative terminals for both cars. If it's not immediately obvious, keep in mind that the negative or ground side is usually connected to the chassis of the car with a relatively short, heavy wire.

When you are sure which are the positive and negative terminals, first connect one of the red wire clamps of the booster cables to the positive terminal of the discharged battery—all the while being sure that the other red clamp *is not touching the battery or car.* Then connect the other end of the red wire to the positive terminal of the fully charged battery. Next, connect one end of the black cable of the booster cables to the negative terminal of the booster battery, and finally connect the other end of the black cable to the engine block of the stalled vehicle (not the negative post of the dead battery) away from the carburetor, fuel lines, or moving parts of the car. Lastly, have someone maintain a constant idle speed in the car with the good battery as you start the car with the bad battery. After the vehicle starts, remove the cables in the *reverse order* starting with the cable connected to the engine block. Always be careful to ensure that clamps don't touch the battery or chassis of the car or get near any moving parts.

Some people feel that the car with the good battery should charge the bad battery for 5 to 10 minutes before starting the disabled car so the disabled car will be essentially using its own battery in the starting process. Keep in mind that the instant the booster cables are connected, the booster car is making a concerted effort to charge both its own battery and the drained battery. At starting, the good battery is asked to supply a heavy current to start the other car. It's a pretty heavy load to put on a single battery. For the situation in Fig. 7.55, the voltage of battery #2 is less than that of battery #1, and the charging current will flow as shown. The resistance in series with the boosting battery is greater because of the long length of the booster cable to the other car. The current is limited only by the series milliohm resistors of the batteries, but the voltage difference is so small that the starting current will be in safe range for the cables involved. The initial charging current will be $I = (12\text{ V} - 11.7\text{ V})/(20\text{ m}\Omega + 10\text{ m}\Omega) = 0.3\text{ V}/30\text{ m}\Omega = 10\text{ A}$. At starting, the current levels will be as shown in Fig. 7.56 for the resistance levels and battery voltages assumed. At starting, an internal resistance for the starting circuit of $0.1\ \Omega = 100\text{ m}\Omega$ is assumed. Note that the battery of the disabled car has now charged up to 11.8 V with an associated increase in its power level. The presence of two batteries requires that the analysis wait for the methods to be introduced in the next chapter.

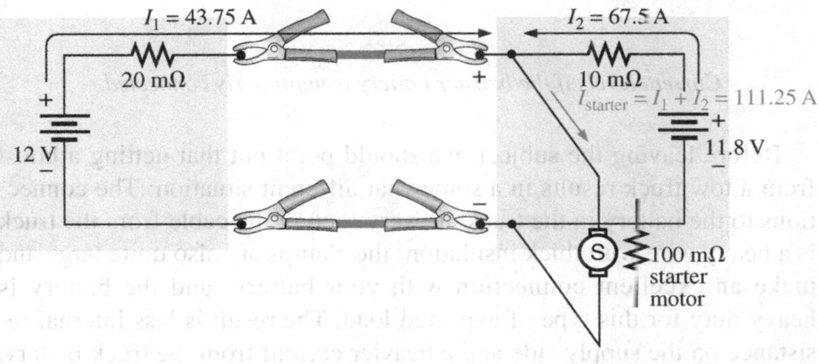

FIG. 7.56
Current levels at starting.

Note also that the current drawn from the starting circuit for the disabled car is over 100 A and that the majority of the starting current is provided by the battery being charged. In essence, therefore, the majority of the starting current is coming from the disabled car. The good battery has provided an initial charge to the bad battery and has provided the additional current necessary to start the car. In total, however, it is the battery of the disabled car that is the primary source of the starting current. For this very reason, the charging action should continue for 5 or 10 minutes before starting the car. If the disabled car is in really bad shape with a voltage level of only 11 V, the resulting levels of current will reverse, with the good battery providing 68.75 A and the bad battery only 37.5 A. Quite obviously, therefore, the worse the condition of the dead battery, the heavier is the drain on the good battery. A point can also be reached where the bad battery is in such bad shape that it cannot accept a good charge or provide its share of the starting current. The result can be continuous cranking of the disabled car without starting and possible damage to the battery of the running car due to the enormous current drain. Once the car is started and the booster cables are removed, the car with the discharged battery will continue to run because the alternator will carry the load (charging the battery and providing the necessary dc voltage) after ignition.

The above discussion was all rather straightforward, but let's investigate what may happen if it is a dark and rainy night, you are rushed, and you hook up the cables incorrectly as shown in Fig. 7.57. The result is two series-aiding batteries and a very low resistance path. The resulting current can then theoretically be extremely high [$I = (12\text{ V} + 11.7\text{ V})/30\text{ m}\Omega = 23.7\text{ V}/30\text{ m}\Omega = 790\text{ A}$], perhaps permanently damaging the electrical system of both cars and, worst of all, causing an explosion that may seriously injure someone. It is therefore very important that you treat the process of boosting a car with great care. Find that flashlight, double-check the connections, and be sure that everyone is clear when you start that car.

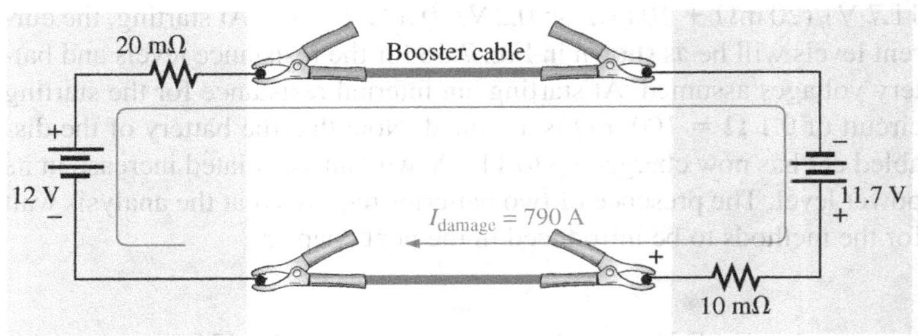

FIG. 7.57
Current levels if the booster battery is improperly connected.

Before leaving the subject, we should point out that getting a boost from a tow truck results in a somewhat different situation: The connections to the battery in the truck are very secure; the cable from the truck is a heavy wire with thick insulation; the clamps are also quite large and make an excellent connection with your battery; and the battery is heavy-duty for this type of expected load. The result is less internal resistance on the supply side and a heavier current from the truck battery. In this case, the truck is really starting the disabled car, which simply reacts to the provided surge of power.

Electronic Circuits

The operation of most electronic systems requires a distribution of dc voltages throughout the design. Although a full explanation of why the dc level is required (since it is an ac signal to be amplified) will have to wait for the introductory courses in electronic circuits, the dc analysis will proceed in much the same manner as described in this chapter. In other words, this chapter and the preceding chapters are sufficient background to perform the dc analysis of the majority of electronic networks you will encounter if given the dc terminal characteristics of the electronic elements. For example, the network in Fig. 7.58 using a transistor will be covered in detail in any introductory electronics course. The dc voltage between the base (B) of the transistor and the emitter (E) is about 0.7 V under normal operating conditions, and the collector (C) is related to the base current by $I_C = \beta I_B = 50 I_B$ (β varies from transistor to transistor). Using these facts will enable us to determine all the dc currents and voltages of the network using the laws introduced in this chapter. In general, therefore, be encouraged that you will use the content of this chapter in numerous applications in the courses to follow.

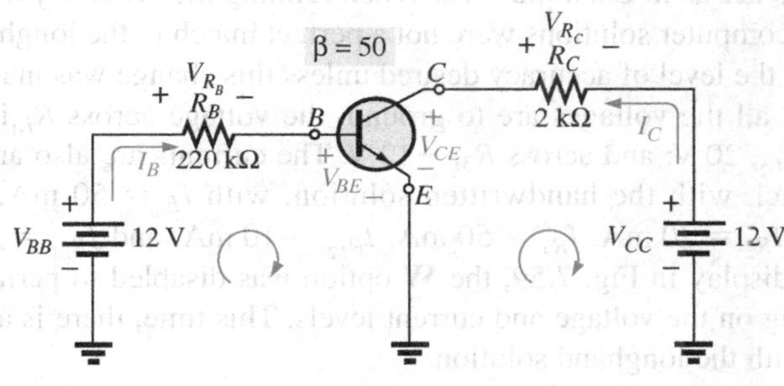

FIG. 7.58
The dc bias levels of a transistor amplifier.

For the network in Fig. 7.58, we begin our analysis by applying Kirchhoff's voltage law to the base circuit (the left loop):

$$+V_{BB} - V_{R_B} - V_{BE} = 0 \quad \text{or} \quad V_{BB} = V_{R_B} + V_{BE}$$

and $$V_{R_B} = V_{BB} - V_{BE} = 12\text{ V} - 0.7\text{ V} = 11.3\text{ V}$$

so that $$V_{R_B} = I_B R_B = 11.3\text{ V}$$

and $$I_B = \frac{V_{R_B}}{R_B} = \frac{11.3\text{ V}}{220\text{ k}\Omega} = \mathbf{51.4\ \mu A}$$

Then $$I_C = \beta I_B = 50 I_B = 50(51.4\ \mu\text{A}) = \mathbf{2.57\ mA}$$

For the output circuit (the right loop)

$$+V_{CE} + V_{R_C} - V_{CC} = 0 \quad \text{or} \quad V_{CC} = V_{R_C} + V_{CE}$$

with $$V_{CE} = V_{CC} - V_{R_C} = V_{CC} - I_C R_C = 12\text{ V} - (2.57\text{ mA})(2\text{ k}\Omega)$$
$$= 12\text{ V} - 5.14\text{ V} = \mathbf{6.86\ V}$$

For a typical dc analysis of a transistor, all the currents and voltages of interest are now known: I_B, V_{BE}, I_C, and V_{CE}. All the remaining voltage, current, and power levels for the other elements of the network can now be found using the basic laws applied in this chapter.

The previous example is typical of the type of exercise you will be asked to perform in your first electronics course. For now you only need to be exposed to the device and to understand the reason for the relationships between the various currents and voltages of the device.

7.11 COMPUTER ANALYSIS

PSpice

Voltage Divider Supply We will now use PSpice to verify the results of Example 7.12. The calculated resistor values will be substituted and the voltage and current levels checked to see if they match the handwritten solution.

As shown in Fig. 7.59, the network is drawn as in earlier chapters using only the tools described thus far—in one way, a practice exercise for everything learned about the **Capture CIS Edition.** Note in this case that rotating the first resistor sets everything up for the remaining resistors. Further, it is a nice advantage that you can place one resistor after another without going to the **End Mode** option. Be especially careful with the placement of the ground, and be sure that **0/SOURCE** is used. Note also that resistor R_1 in Fig. 7.59 was entered as 1.333 kΩ rather than 1.33 kΩ as in Example 7.12. When running the program, we found that the computer solutions were not a perfect match to the longhand solution to the level of accuracy desired unless this change was made.

Since all the voltages are to ground, the voltage across $R_{L_1}$is 60 V; across R_{L_2}, 20 V; and across R_3, −12 V. The currents are also an excellent match with the handwritten solution, with I_E = 50 mA, I_{R_1} = 30 mA, I_{R_2} = 20 mA, I_{R_3} = 50 mA, $I_{R_{L2}}$ = 10 mA, and $I_{R_{L1}}$ = 20 mA. For the display in Fig. 7.59, the **W** option was disabled to permit concentrating on the voltage and current levels. This time, there is an exact match with the longhand solution.

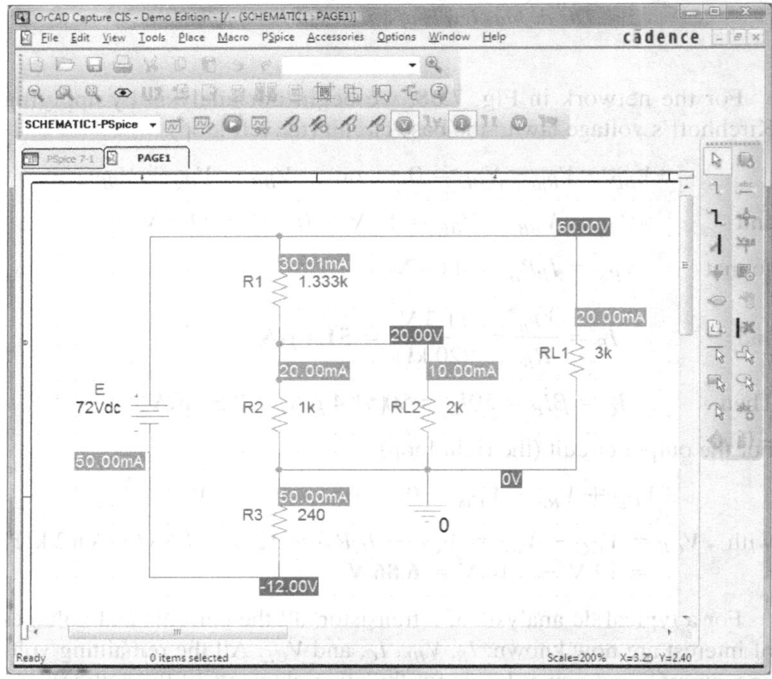

FIG. 7.59
Using PSpice to verify the results of Example 7.12.

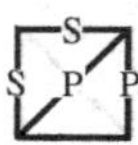

PROBLEMS

SECTION 7.2–7.5 Series-Parallel Networks

1. Which elements (individual elements, not combinations of elements) of the networks in Fig. 7.60 are in series? Which are in parallel? As a check on your assumptions, be sure that the elements in series have the same current and that the elements in parallel have the same voltage. Restrict your decisions to single elements, not combinations of elements.

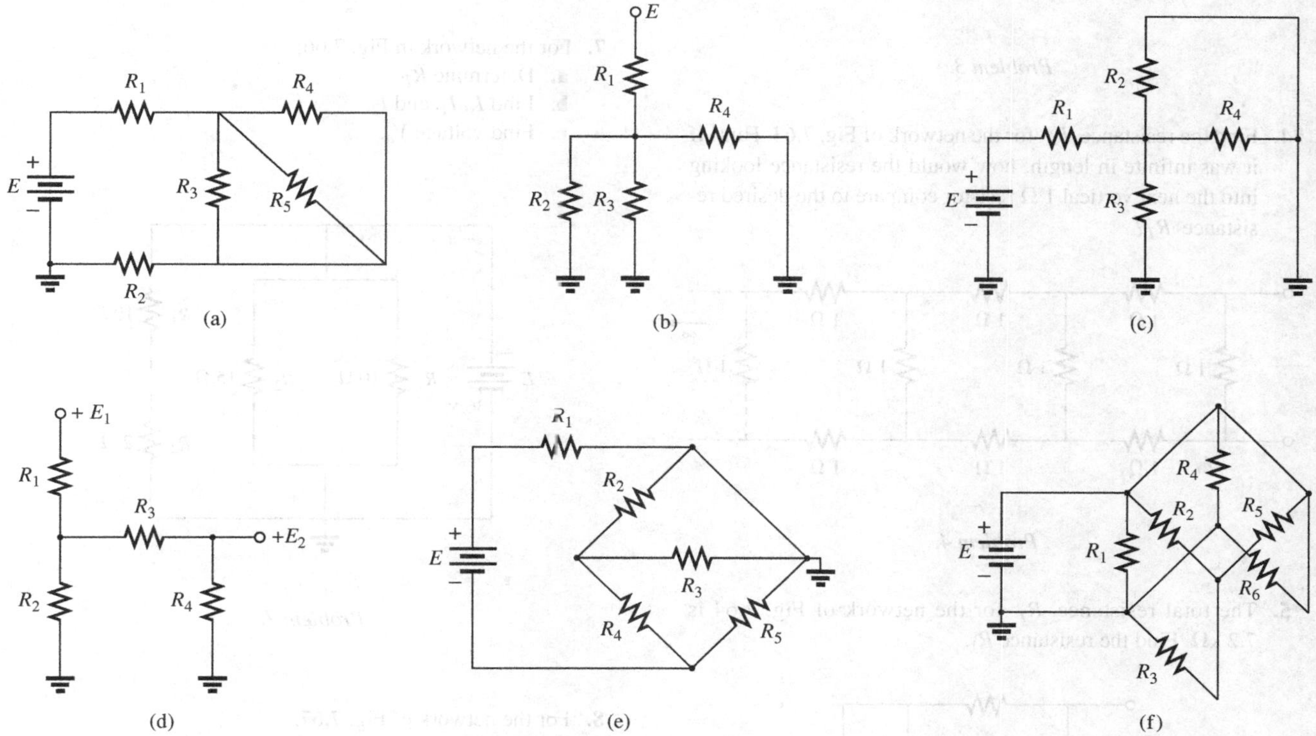

FIG. 7.60
Problem 1.

2. Determine R_T for the networks in Fig. 7.61.

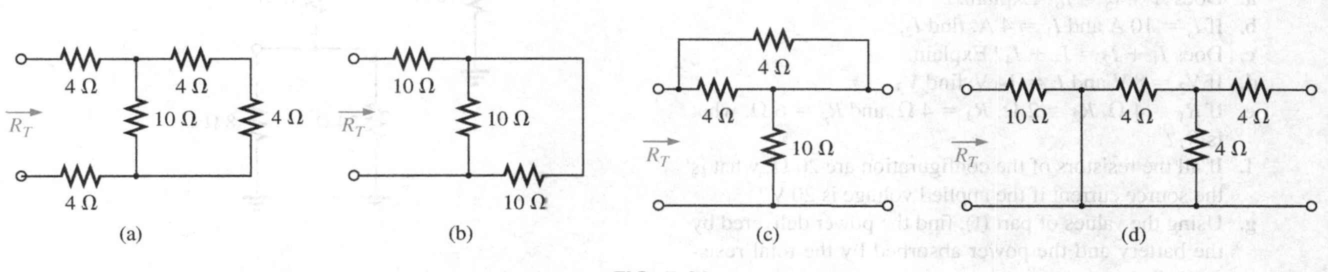

FIG. 7.61
Problem 2.

3. Find the total resistance for the configuration of Fig. 7.62.

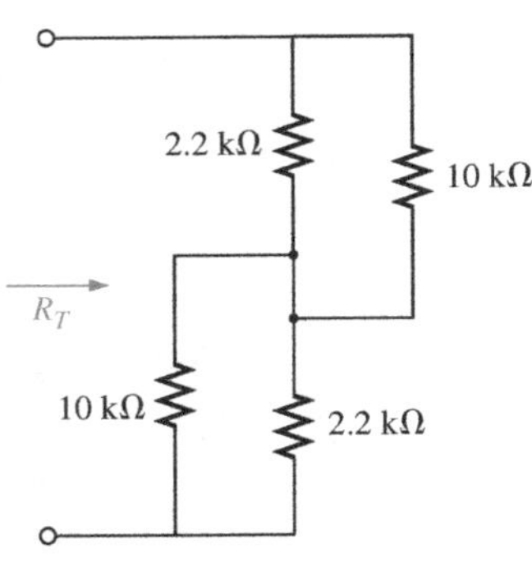

FIG. 7.62
Problem 3.

***4.** Find the resistance R_T for the network of Fig. 7.63. Hint! If it was infinite in length, how would the resistance looking into the next vertical 1 Ω resistor compare to the desired resistance R_T?

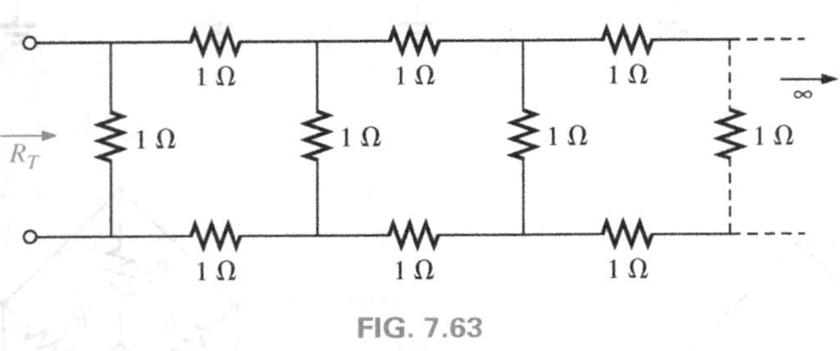

FIG. 7.63
Problem 4.

***5.** The total resistance R_T for the network of Fig. 7.64 is 7.2 kΩ. Find the resistance R_1.

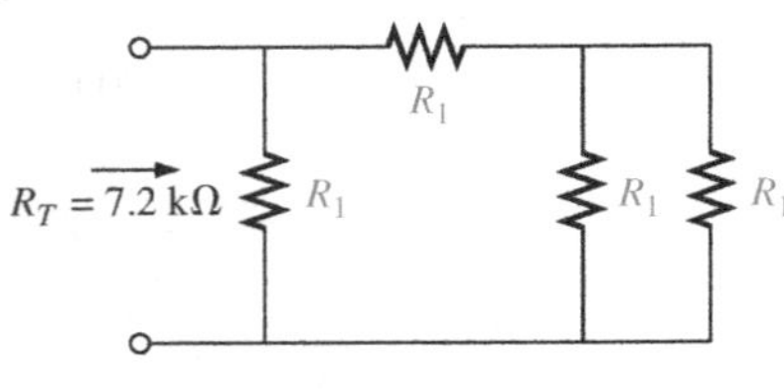

FIG. 7.64
Problem 5.

6. For the network in Fig. 7.65:
  a. Does $I_s = I_5 = I_6$? Explain.
  b. If $I_s = 10$ A and $I_1 = 4$ A, find I_2.
  c. Does $I_1 + I_2 = I_3 + I_4$? Explain.
  d. If $V_2 = 8$ V and $E = 14$ V, find V_3.
  e. If $R_1 = 4\ \Omega$, $R_2 = 2\ \Omega$, $R_3 = 4\ \Omega$, and $R_4 = 6\ \Omega$, what is R_T?
  f. If all the resistors of the configuration are 20 Ω, what is the source current if the applied voltage is 20 V?
  g. Using the values of part (f), find the power delivered by the battery and the power absorbed by the total resistance R_T.

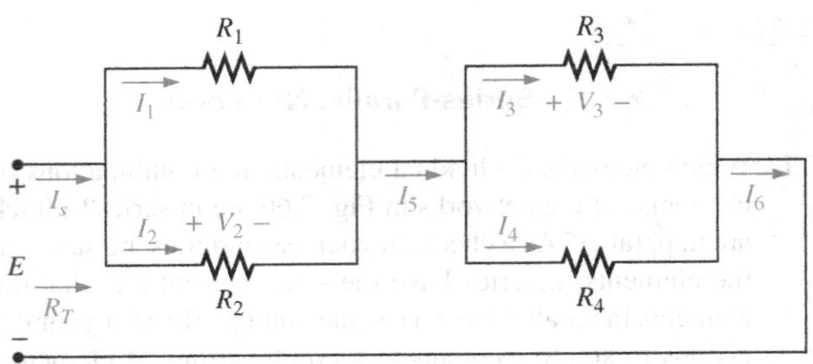

FIG. 7.65
Problem 6.

7. For the network in Fig. 7.66:
  a. Determine R_T.
  b. Find I_s, I_1, and I_2.
  c. Find voltage V_a.

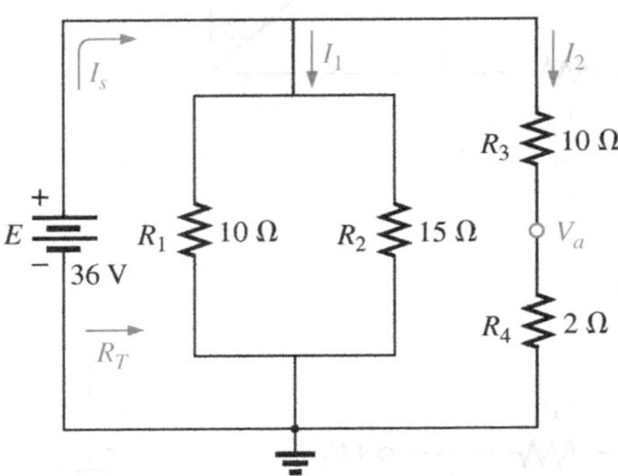

FIG. 7.66
Problem 7.

8. For the network of Fig. 7.67:
  a. Find the voltages V_a and V_b.
  b. Find the currents I_1 and I_s.

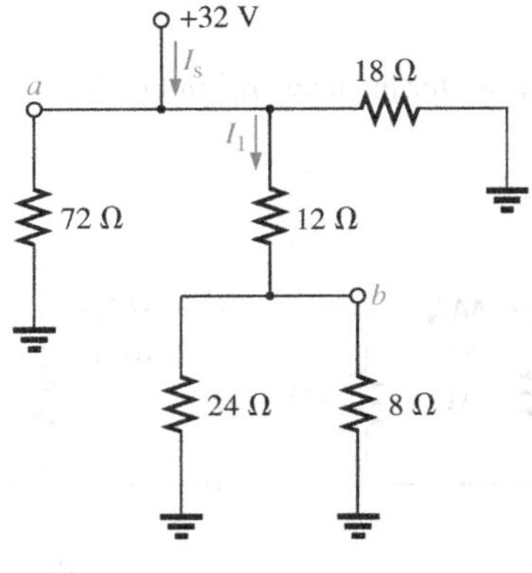

FIG. 7.67
Problem 8.

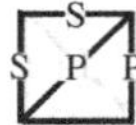

9. For the network of Fig. 7.68:
   a. Find the voltages V_a, V_b, and V_c.
   b. Find the currents I_1 and I_2.

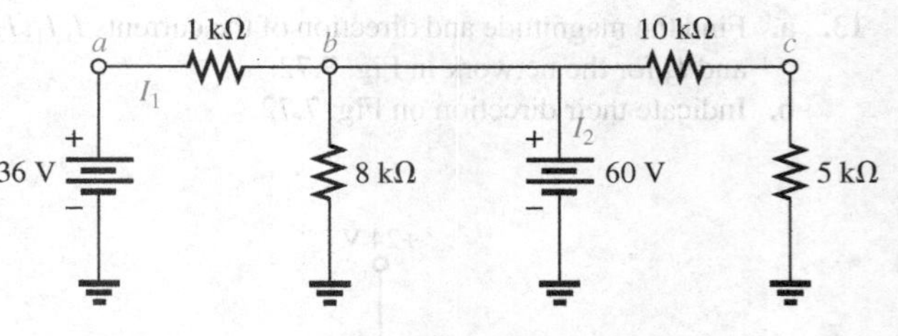

FIG. 7.68
Problem 9.

10. For the circuit board in Fig. 7.69:
    a. Find the total resistance R_T of the configuration.
    b. Find the current drawn from the supply if the applied voltage is 48 V.
    c. Find the reading of the applied voltmeter.

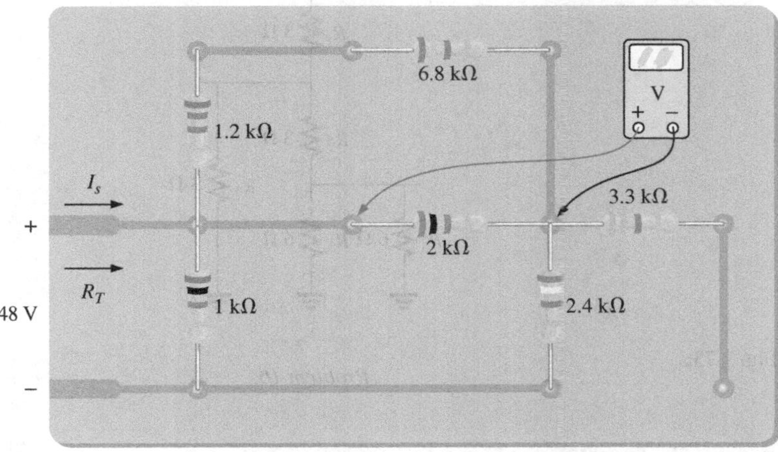

FIG. 7.69
Problem 10.

11. Find the value of each resistor for the network of Fig. 7.70.

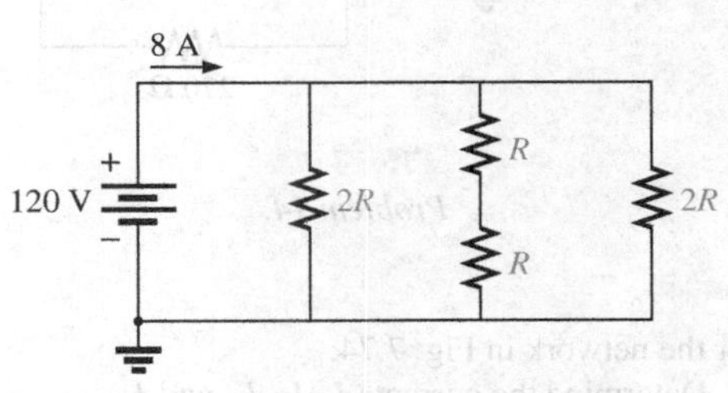

FIG. 7.70
Problem 11.

*12. For the network in Fig. 7.71:
   a. Find currents I_s, I_2, and I_6.
   b. Find voltages V_1 and V_5.
   c. Find the power delivered to the 3 kΩ resistor.

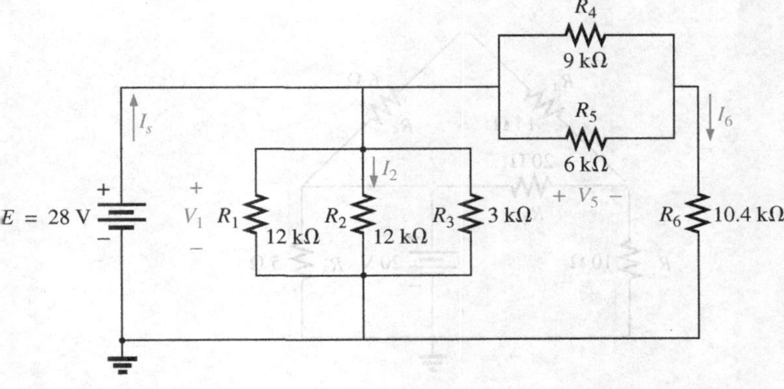

FIG. 7.71
Problem 12.

13. a. Find the magnitude and direction of the currents I, I_1, I_2, and I_3 for the network in Fig. 7.72.
b. Indicate their direction on Fig. 7.72.

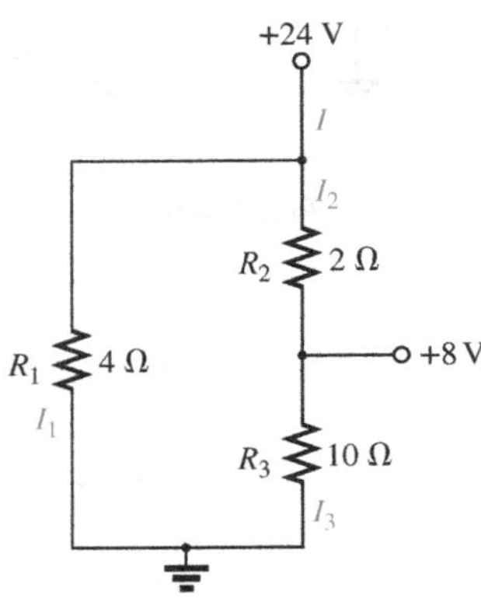

FIG. 7.72
Problem 13.

14. Determine the currents I_1 and I_2 for the network in Fig. 7.73, constructed of standard values.

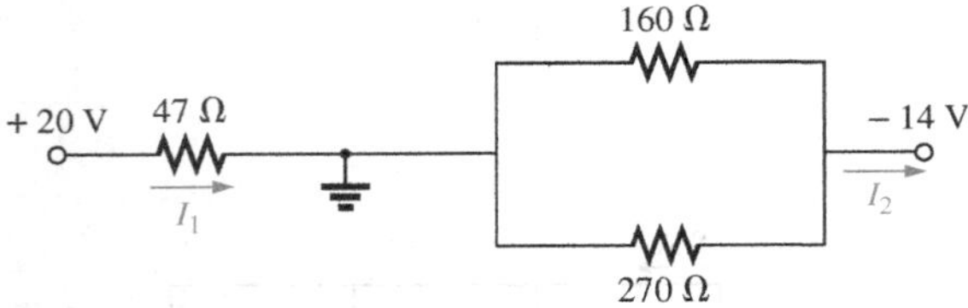

FIG. 7.73
Problem 14.

***15.** For the network in Fig. 7.74:
a. Determine the currents I_s, I_1, I_3, and I_4.
b. Calculate V_a and V_{bc}.

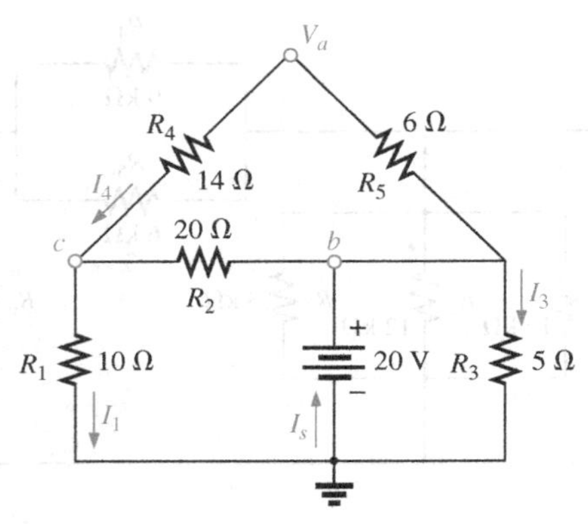

FIG. 7.74
Problem 15.

16. For the network in Fig. 7.75:
a. Determine the current I_1.
b. Calculate the currents I_2 and I_3.
c. Determine the voltage levels V_a and V_b.

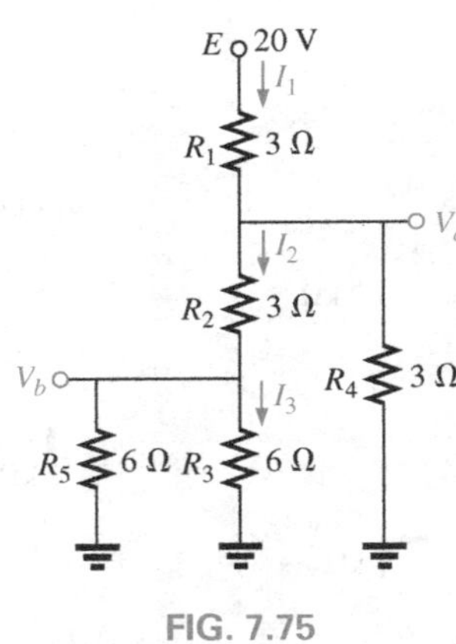

FIG. 7.75
Problem 16.

***17.** Determine the dc levels for the transistor network in Fig. 7.76 using the fact that $V_{BE} = 0.7$ V, $V_E = 2$ V, and $I_C = I_E$. That is:
a. Determine I_E and I_C.
b. Calculate I_B.
c. Determine V_B and V_C.
d. Find V_{CE} and V_{BC}.

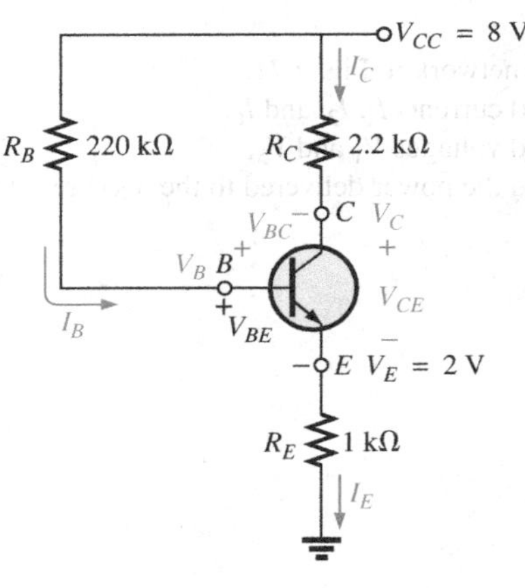

FIG. 7.76
Problem 17.

18. For the network in Fig. 7.77:
  a. Determine the current I.
  b. Find V_1.

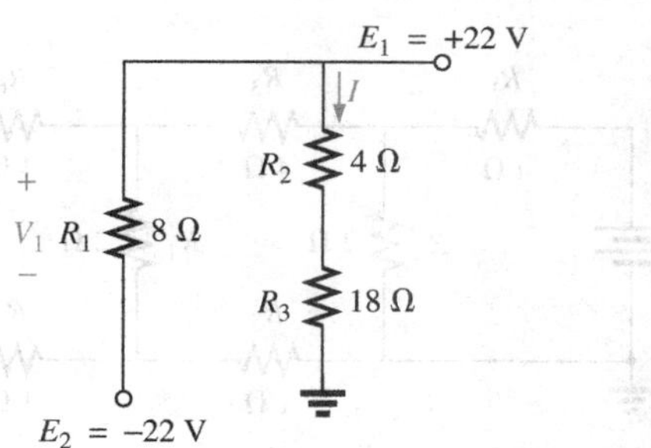

FIG. 7.77
Problem 18.

***19.** For the network in Fig. 7.78:
  a. Determine R_T by combining resistive elements.
  b. Find V_1 and V_4.
  c. Calculate I_3 (with direction).
  d. Determine I_s by finding the current through each element and then applying Kirchhoff's current law. Then calculate R_T from $R_T = E/I_s$, and compare the answer with the solution of part (a).

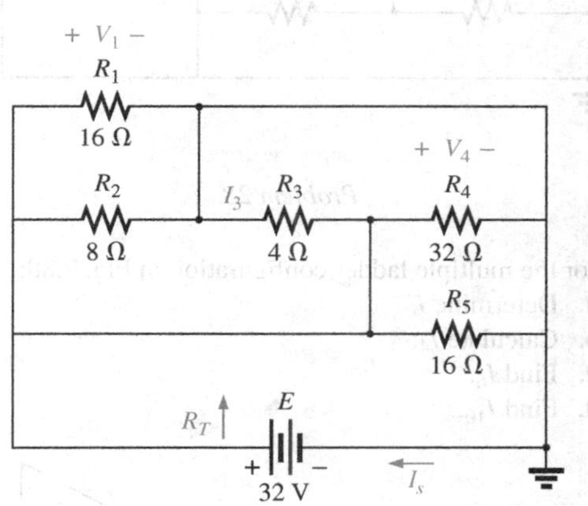

FIG. 7.78
Problem 19.

20. Determine the voltage V_{ab} and the current I for the network of Fig. 7.79. Recall the discussion of short and open circuits in Section 6.8.

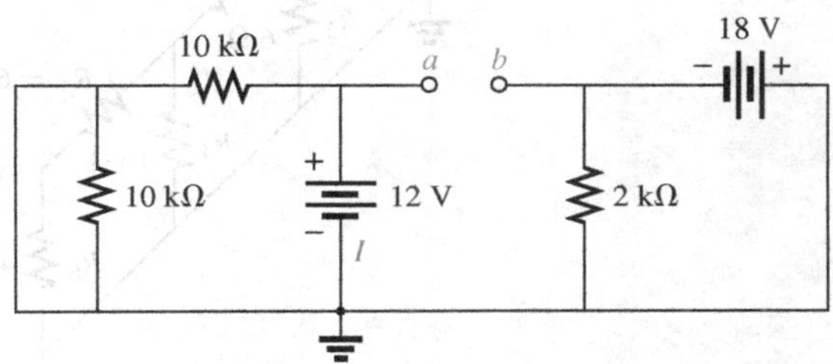

FIG. 7.79
Problem 20.

***21.** For the network of Fig. 7.80:
  a. Determine the voltage V_{ab}.
  b. Calculate the current I.
  c. Find the voltages V_a and V_b.

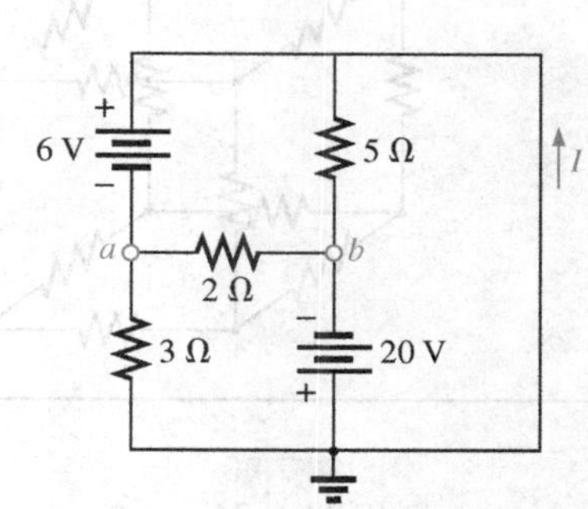

FIG. 7.80
Problem 21.

***22.** For the network in Fig. 7.81:
  a. Determine the current I.
  b. Calculate the open-circuit voltage V.

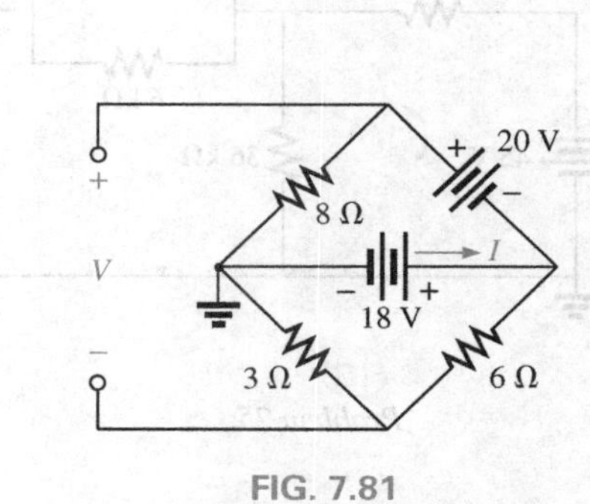

FIG. 7.81
Problem 22.

***23.** For the network in Fig. 7.82, find the resistance R_3 if the current through it is 2 A.

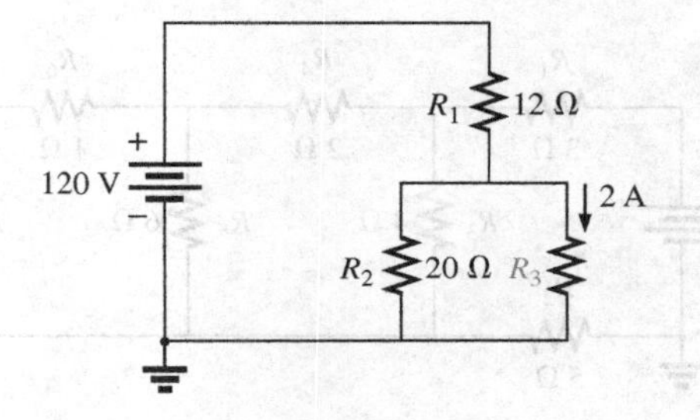

FIG. 7.82
Problem 23.

*24. If all the resistors of the cube in Fig. 7.83 are 10 Ω, what is the total resistance? (*Hint:* Make some basic assumptions about current division through the cube.)

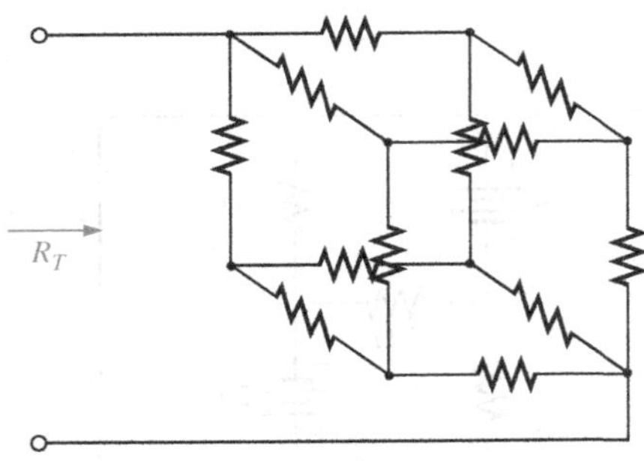

FIG. 7.83
Problem 24.

*25. Given the voltmeter reading $V = 27$ V in Fig. 7.84:

a. Is the network operating properly?

b. If not, what could be the cause of the incorrect reading?

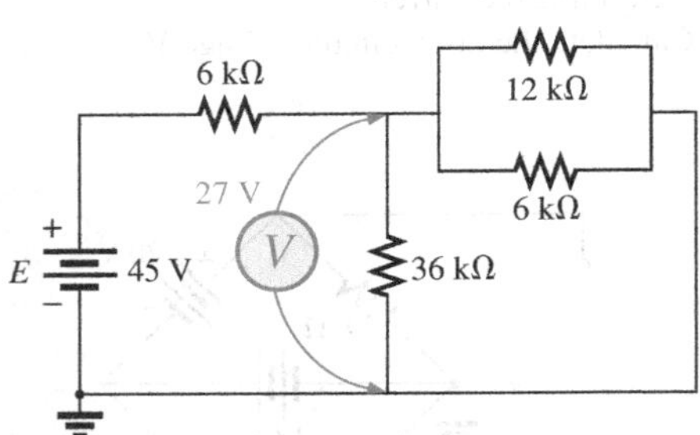

FIG. 7.84
Problem 25.

SECTION 7.6 Ladder Networks

26. For the ladder network in Fig. 7.85:

a. Find the current I.

b. Find the current I_7.

c. Determine the voltages V_3, V_5, and V_7.

d. Calculate the power delivered to R_7, and compare it to the power delivered by the 240 V supply.

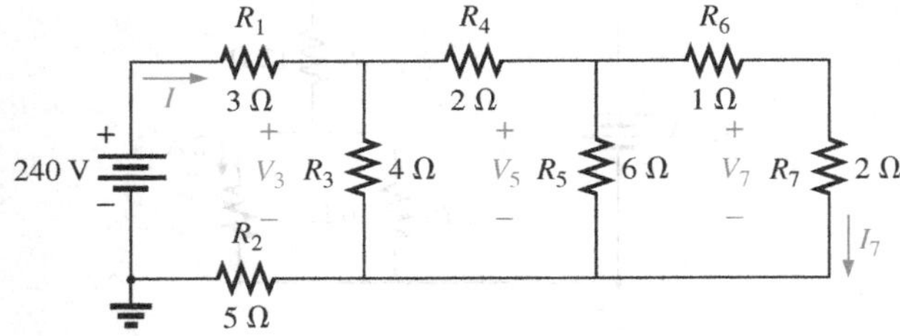

FIG. 7.85
Problem 26.

27. For the ladder network in Fig. 7.86:

a. Determine R_T.

b. Calculate I.

c. Find the power delivered to R_7.

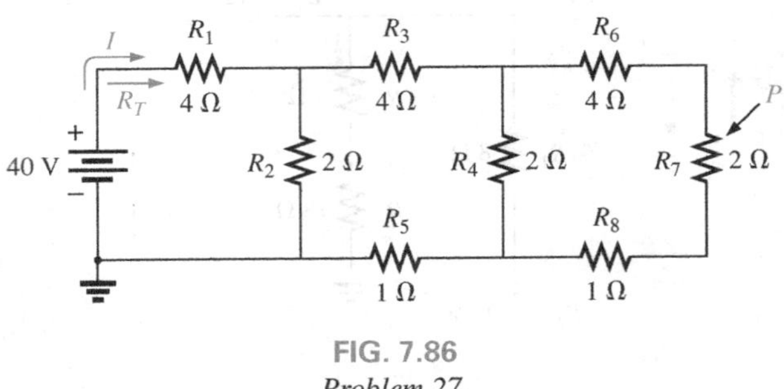

FIG. 7.86
Problem 27.

*28. Determine the power delivered to the 6 Ω load in Fig. 7.87.

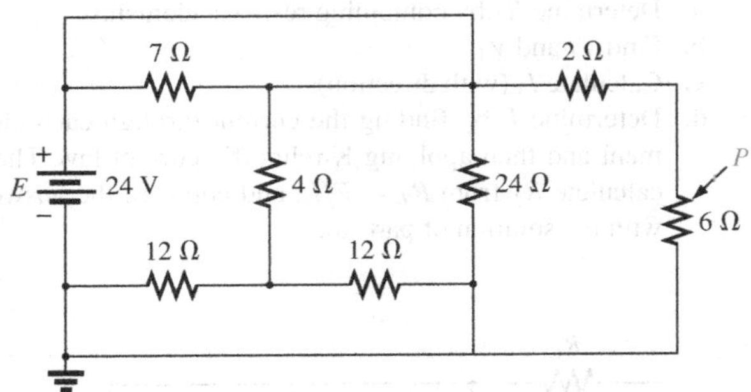

FIG. 7.87
Problem 28.

29. For the multiple ladder configuration in Fig. 7.88:

a. Determine I.

b. Calculate I_4.

c. Find I_6.

d. Find I_{10}.

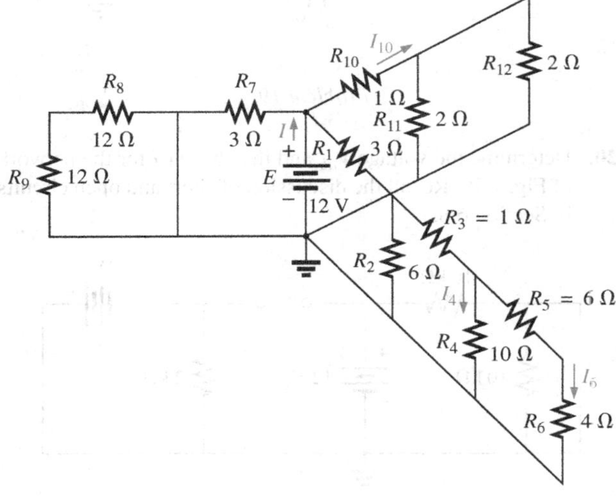

FIG. 7.88
Problem 29.

SECTION 7.7 Voltage Divider Supply (Unloaded and Loaded)

30. Given the voltage divider supply in Fig. 7.89:

a. Determine the supply voltage E.

b. Find the load resistors R_{L_2} and R_{L_3}.

c. Determine the voltage divider resistors R_1, R_2, and R_3.

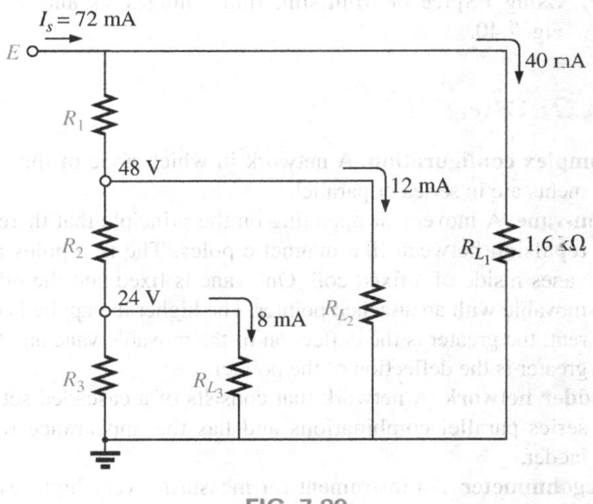

FIG. 7.89
Problem 30.

***31.** Determine the voltage divider supply resistors for the configuration in Fig. 7.90. Also determine the required wattage rating for each resistor, and compare their levels.

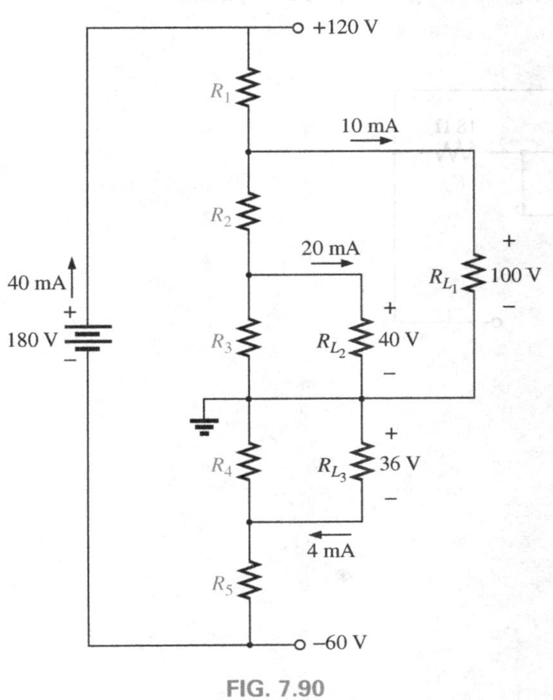

FIG. 7.90
Problem 31.

***32.** A studio lamp requires 40 V at 50 mA to burn brightly. Design a voltage divider arrangement that will work properly off a 120 V source supplying a current of 200 mA. Use resistors as close as possible to standard values, and specify the minimum wattage rating of each.

SECTION 7.8 Potentiometer Loading

***33.** For the system in Fig. 7.91:

a. At first exposure, does the design appear to be a good one?

b. In the absence of the 10 kΩ load, what are the values of R_1 and R_2 to establish 3 V across R_2?

c. Determine the values of R_1 and R_2 to establish $V_{R_L} = 3$ V when the load is applied, and compare them to the results of part (b).

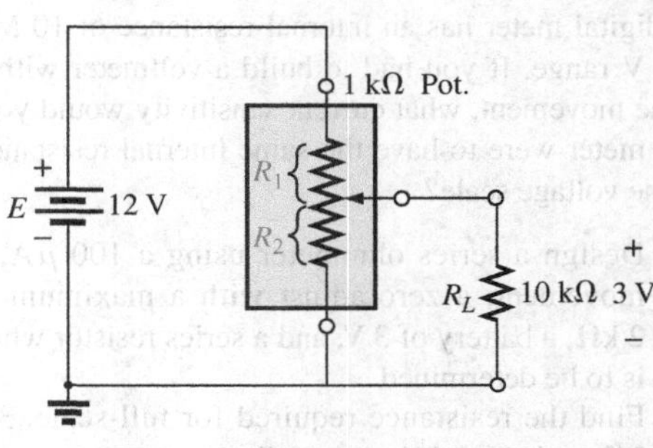

FIG. 7.91
Problem 33.

***34.** For the potentiometer in Fig. 7.92:

a. What are the voltages V_{ab} and V_{bc} with no load applied ($R_{L_1} = R_{L_2} = \infty\ \Omega$)?

b. What are the voltages V_{ab} and V_{bc} with the indicated loads applied?

c. What is the power dissipated by the potentiometer under the loaded conditions in Fig. 7.92?

d. What is the power dissipated by the potentiometer with no loads applied? Compare it to the results of part (c).

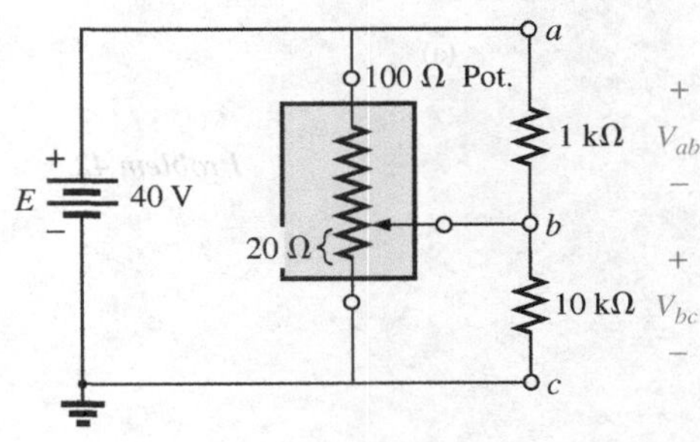

FIG. 7.92
Problem 34.

SECTION 7.9 Ammeter, Voltmeter, and Ohmmeter Design

35. An iron-vane movement is rated 1 mA, 100 Ω.
  a. What is the current sensitivity?
  b. Design a 20 A ammeter using the above movement. Show the circuit and component values.

36. Using a 50 μA, 1000 Ω movement, design a multirange milliammeter having scales of 25 mA, 50 mA, and 100 mA. Show the circuit and component values.

37. An iron-vane movement is rated 50 μA, 1000 Ω.
  a. Design a 15 V dc voltmeter. Show the circuit and component values.
  b. What is the ohm/volt rating of the voltmeter?

38. Using a 1 mA, 1000 Ω movement, design a multirange voltmeter having scales of 5 V, 50 V, and 500 V. Show the circuit and component values.

39. A digital meter has an internal resistance of 10 MΩ on its 0.5 V range. If you had to build a voltmeter with an iron-vane movement, what current sensitivity would you need if the meter were to have the same internal resistance on the same voltage scale?

***40.** **a.** Design a series ohmmeter using a 100 μA, 1000 Ω movement, a zero-adjust with a maximum value of 2 kΩ, a battery of 3 V, and a series resistor whose value is to be determined.
  b. Find the resistance required for full-scale, 3/4-scale, 1/2-scale, and 1/4-scale deflection.
  c. Using the results of part (b), draw the scale to be used with the ohmmeter.

41. Describe the basic construction and operation of the megohmmeter.

***42.** Determine the reading of the ohmmeter for each configuration of Fig. 7.93.

SECTION 7.11 Computer Analysis

43. Using PSpice or Multisim, verify the results of Example 7.2.

44. Using PSpice or Multisim, confirm the solutions of Example 7.5.

45. Using PSpice or Multisim, verify the results of Example 7.10.

46. Using PSpice or Multisim, find voltage V_6 of Fig. 7.32.

47. Using PSpice or Multisim, find voltages V_b and V_c of Fig. 7.40.

GLOSSARY

Complex configuration A network in which none of the elements are in series or parallel.

Iron-vane A movement operating on the principle that there is repulsion between like magnetic poles. The two poles are vanes inside of a fixed coil. One vane is fixed and the other movable with an attached pointer. The higher the applied current, the greater is the deflection of the movable vane and the greater is the deflection of the pointer.

Ladder network A network that consists of a cascaded set of series-parallel combinations and has the appearance of a ladder.

Megohmmeter An instrument for measuring very high resistance levels, such as in the megohm range.

Series ohmmeter A resistance-measuring instrument in which the movement is placed in series with the unknown resistance.

Series-parallel network A network consisting of a combination of both series and parallel branches.

Transistor A three-terminal semiconductor electronic device that can be used for amplification and switching purposes.

Voltage divider supply A series network that can provide a range of voltage levels for an application.

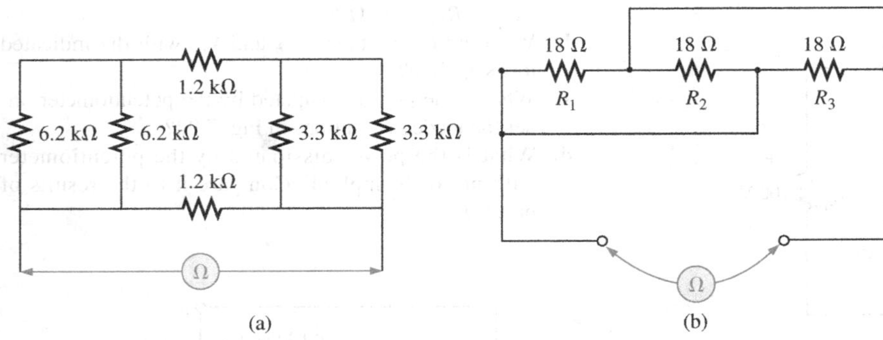

FIG. 7.93
Problem 42.

Methods of Analysis and Selected Topics (dc)

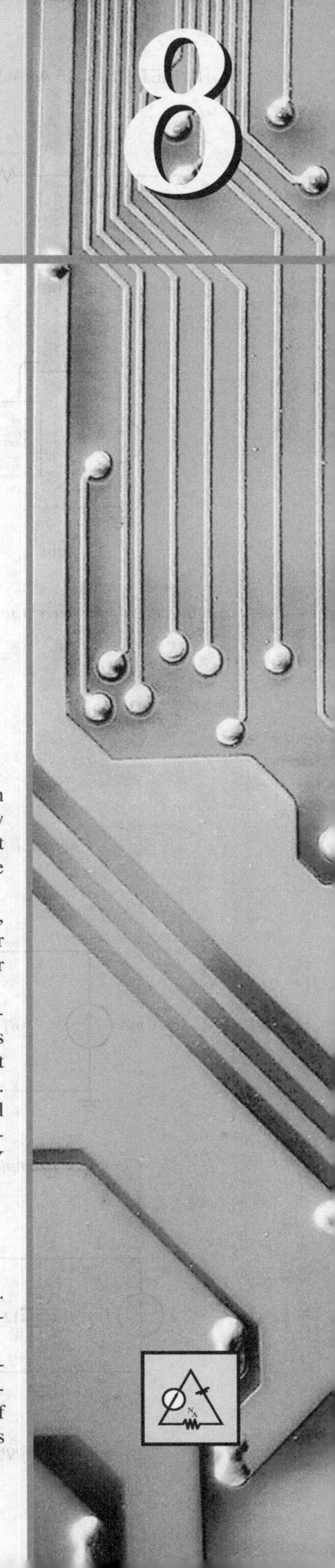

Objectives

- *Become familiar with the terminal characteristics of a current source and how to solve for the voltages and currents of a network using current sources and/or current sources and voltage sources.*
- *Be able to apply branch-current analysis and mesh analysis to find the currents of network with one or more independent paths.*
- *Be able to apply nodal analysis to find all the terminal voltages of any series-parallel network with one or more independent sources.*
- *Become familiar with bridge network configurations and how to perform Δ−Y or Y−Δ conversions.*

8.1 INTRODUCTION

The circuits described in previous chapters had only one source or two or more sources in series or parallel. The step-by-step procedures outlined in those chapters can be applied only if the sources are in series or parallel. There will be an interaction of sources that will not permit the reduction techniques used to find quantities such as the total resistance and the source current.

For such situations, methods of analysis have been developed that allow us to approach, in a systematic manner, networks with any number of sources in any arrangement. To our benefit, the methods to be introduced can also be applied to networks with only *one source* or to networks in which sources are in *series or parallel.*

The methods to be introduced in this chapter include branch-current analysis, mesh analysis, and nodal analysis. Each can be applied to the same network, although usually one is more appropriate than the other. The "best" method cannot be defined by a strict set of rules but can be determined only after developing an understanding of the relative advantages of each.

Before considering the first of the methods, we will examine current sources in detail because they appear throughout the analyses to follow. The chapter concludes with an investigation of a complex network called the *bridge configuration,* followed by the use of Δ-Y and Y-Δ conversions to analyze such configurations.

8.2 CURRENT SOURCES

In previous chapters, the voltage source was the only source appearing in the circuit analysis. This was primarily because voltage sources such as the battery and supply are the most common in our daily lives and in the laboratory environment.

We now turn our attention to a second type of source, called the **current source,** which appears throughout the analyses in this chapter. Although current sources are available as laboratory supplies (introduced in Chapter 2), they appear extensively in the modeling of electronic devices such as the transistor. Their characteristics and their impact on the currents

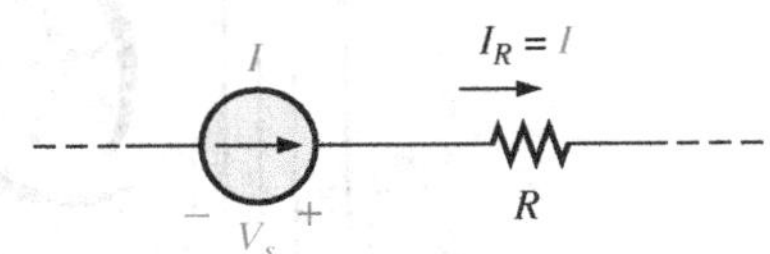

(a)

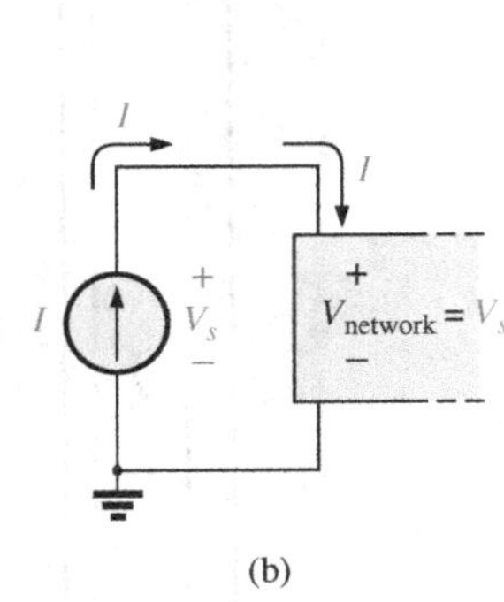

(b)

FIG. 8.1
Introducing the current source symbol.

and voltages of a network must therefore be clearly understood if electronic systems are to be properly investigated.

The current source is often described as the *dual* of the voltage source. Just as a battery provides a fixed voltage to a network, a current source establishes a fixed current in the branch where it is located. Further, the current through a battery is a function of the network to which it is applied, just as the voltage across a current source is a function of the connected network. The term *dual* is applied to any two elements in which the traits of one variable can be interchanged with the traits of another. This is certainly true for the current and voltage of the two types of sources.

The symbol for a current source appears in Fig. 8.1(a). The arrow indicates the direction in which it is supplying current to the branch where it is located. The result is a current equal to the source current through the series resistor. In Fig. 8.1(b), we find that the voltage across a current source is determined by the polarity of the voltage drop caused by the current source. For single-source networks, it always has the polarity of Fig. 8.1(b), but for multisource networks it can have either polarity.

In general, therefore,

a current source determines the direction and magnitude of the current in the branch where it is located.

Furthermore,

the magnitude and the polarity of the voltage across a current source are each a function of the network to which the voltage is applied.

A few examples will demonstrate the similarities between solving for the source current of a voltage source and the terminal voltage of a current source. All the rules and laws developed in the previous chapter still apply, so we just have to remember what we are looking for and properly understand the characteristics of each source.

The simplest possible configuration with a current source appears in Example 8.1.

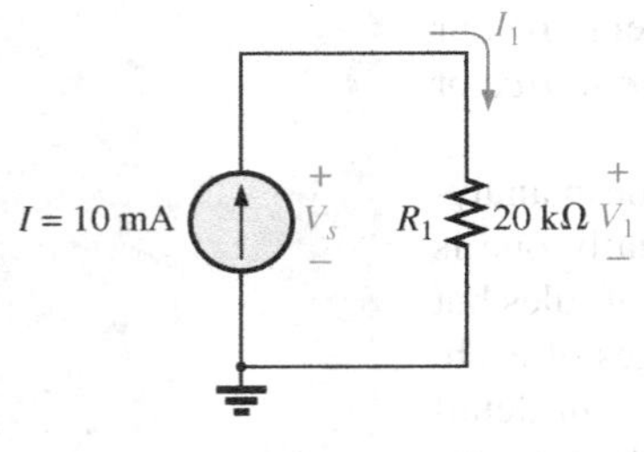

FIG. 8.2
Circuit for Example 8.1.

EXAMPLE 8.1 Find the source voltage, the voltage V_1, and current I_1 for the circuit in Fig. 8.2.

Solution: Since the current source establishes the current in the branch in which it is located, the current I_1 must equal I, and

$$I_1 = I = \mathbf{10\ mA}$$

The voltage across R_1 is then determined by Ohm's law:

$$V_1 = I_1R_1 = (10\text{ mA})(20\text{ k}\Omega) = \mathbf{200\ V}$$

Since resistor R_1 and the current source are in parallel, the voltage across each must be the same, and

$$V_s = V_1 = \mathbf{200\ V}$$

with the polarity shown.

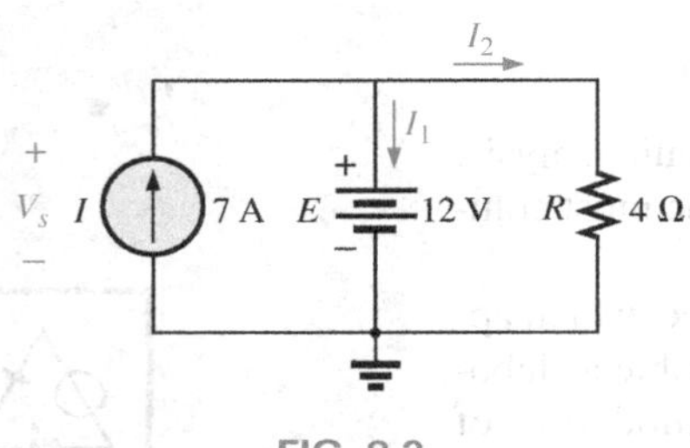

FIG. 8.3
Network for Example 8.2.

EXAMPLE 8.2 Find the voltage V_s and currents I_1 and I_2 for the network in Fig. 8.3.

Solution: This is an interesting problem because it has both a current source and a voltage source. For each source, the dependent (a function

of something else) variable will be determined. That is, for the current source, V_s must be determined, and for the voltage source, I_s must be determined.

Since the current source and voltage source are in parallel,

$$V_s = E = \mathbf{12\ V}$$

Further, since the voltage source and resistor R are in parallel,

$$V_R = E = 12\ \text{V}$$

and $$I_2 = \frac{V_R}{R} = \frac{12\ \text{V}}{4\ \Omega} = \mathbf{3\ A}$$

The current I_1 of the voltage source can then be determined by applying Kirchhoff's current law at the top of the network as follows:

$$\Sigma I_i = \Sigma I_o$$

$$I = I_1 + I_2$$

and $$I_1 = I - I_2 = 7\ \text{A} - 3\ \text{A} = \mathbf{4\ A}$$

EXAMPLE 8.3 Determine the current I_1 and the voltage V_s for the network in Fig. 8.4.

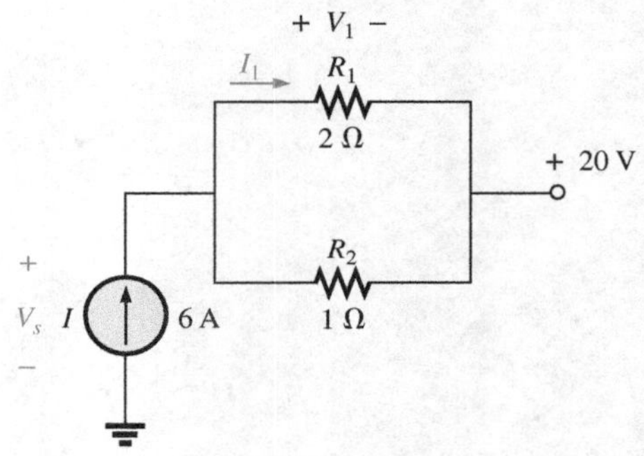

FIG. 8.4
Example 8.3.

Solution: First note that the current in the branch with the current source must be 6 A, no matter what the magnitude of the voltage source to the right. In other words, the currents of the network are defined by I, R_1, and R_2. However, the voltage across the current source is directly affected by the magnitude and polarity of the applied source.

Using the current divider rule gives

$$I_1 = \frac{R_2 I}{R_2 + R_1} = \frac{(1\ \Omega)(6\ \text{A})}{1\ \Omega + 2\ \Omega} = \frac{1}{3}(6\ \text{A}) = \mathbf{2\ A}$$

The voltage V_1 is given by

$$V_1 = I_1 R_1 = (2\ \text{A})(2\ \Omega) = 4\ \text{V}$$

Applying Kirchhoff's voltage rule to determine V_s gives

$$+V_s - V_1 - 20\ \text{V} = 0$$

and $$V_s = V_1 + 20\ \text{V} = 4\ \text{V} + 20\ \text{V} = \mathbf{24\ V}$$

In particular, note the polarity of the voltage V_s as determined by the network.

8.3 SOURCE CONVERSIONS

The current source appearing in the previous section is called an *ideal source* due to the absence of any internal resistance. In reality, all sources—whether they are voltage sources or current sources—have some internal resistance in the relative positions shown in Fig. 8.5. For the voltage source, if $R_s = 0\ \Omega$, or if it is so small compared to any series resistors that it can be ignored, then we have an "ideal" voltage source for all practical purposes. For the current source, since the resistor R_P is in parallel, if $R_P = \infty\ \Omega$, or if it is large enough compared to any parallel resistive elements that it can be ignored, then we have an "ideal" current source.

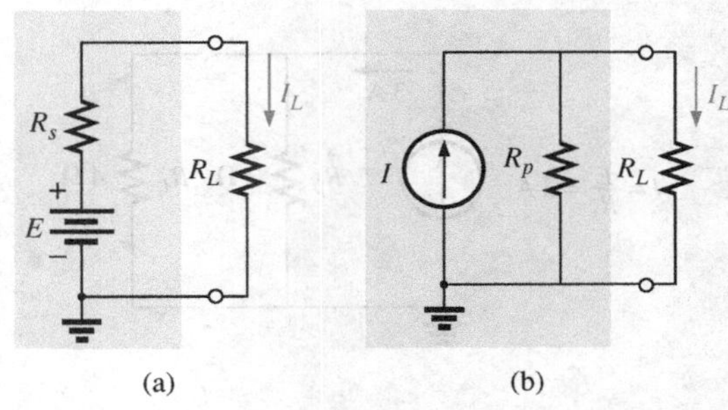

FIG. 8.5
Practical sources: (a) voltage; (b) current.

Unfortunately, however, ideal sources *cannot be converted* from one type to another. That is, a voltage source cannot be converted to a

current source, and vice versa—*the internal resistance must be present.* If the voltage source in Fig. 8.5(a) is to be equivalent to the source in Fig. 8.5(b), any load connected to the sources such as R_L should receive the same current, voltage, and power from each configuration. In other words, if the source were enclosed in a container, the load R_L would not know which source it was connected to.

This type of equivalence is established using the equations appearing in Fig. 8.6. First note that the resistance is the same in each configuration—a nice advantage. For the voltage source equivalent, the voltage is determined by a simple application of Ohm's law to the current source: $E = IR_P$. For the current source equivalent, the current is again determined by applying Ohm's law to the voltage source: $I = E/R_s$. At first glance, it all seems too simple, but Example 8.4 verifies the results.

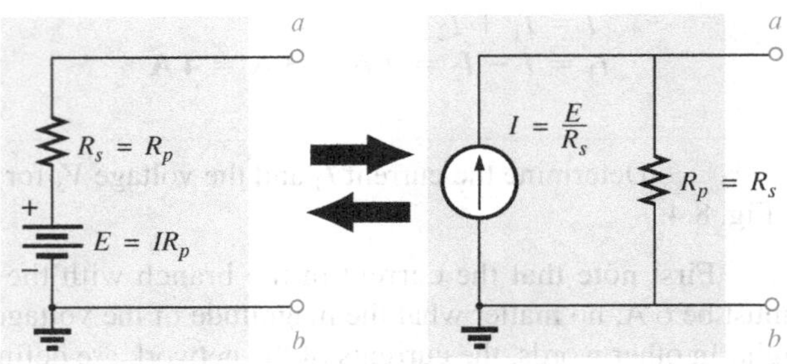

FIG. 8.6
Source conversion.

It is important to realize, however, that

the equivalence between a current source and a voltage source exists only at their external terminals.

The internal characteristics of each are quite different.

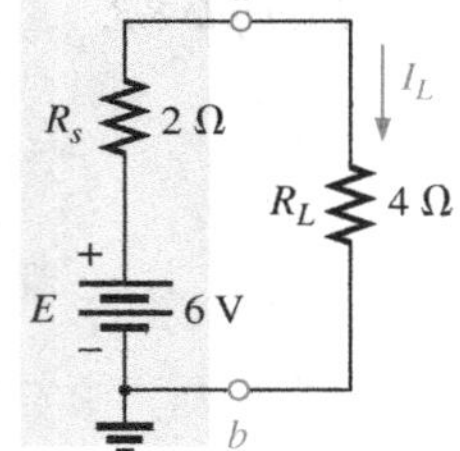

FIG. 8.7
Practical voltage source and load for Example 8.4.

EXAMPLE 8.4 For the circuit in Fig. 8.7:

a. Determine the current I_L.
b. Convert the voltage source to a current source.
c. Using the resulting current source of part (b), calculate the current through the load resistor, and compare your answer to the result of part (a).

Solutions:

a. Applying Ohm's law gives

$$I_L = \frac{E}{R_s + R_L} = \frac{6\text{ V}}{2\ \Omega + 4\ \Omega} = \frac{6\text{ V}}{6\ \Omega} = \mathbf{1\ A}$$

b. Using Ohm's law again gives

$$I = \frac{E}{R_s} = \frac{6\text{ V}}{2\ \Omega} = \mathbf{3\ A}$$

and the equivalent source appears in Fig. 8.8 with the load reapplied.

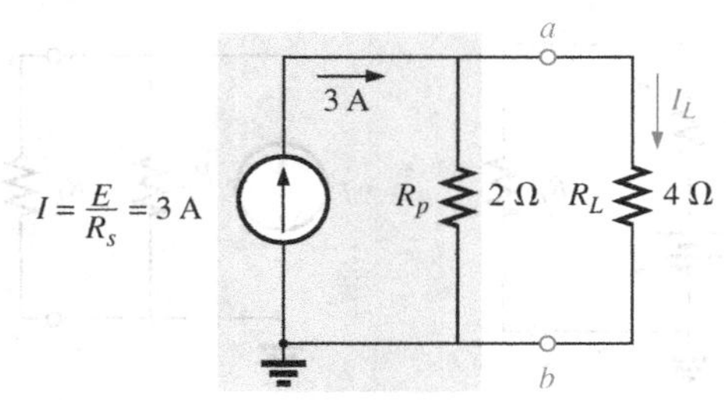

FIG. 8.8
Equivalent current source and load for the voltage source in Fig. 8.7.

c. Using the current divider rule gives

$$I_L = \frac{R_p I}{R_p + R_L} = \frac{(2\ \Omega)(3\ \text{A})}{2\ \Omega + 4\ \Omega} = \frac{1}{3}(3\ \text{A}) = \mathbf{1\ A}$$

We find that the current I_L is the same for the voltage source as it was for the equivalent current source—the sources are therefore equivalent.

As demonstrated in Fig. 8.5 and in Example 8.4, note that

a source and its equivalent will establish current in the same direction through the applied load.

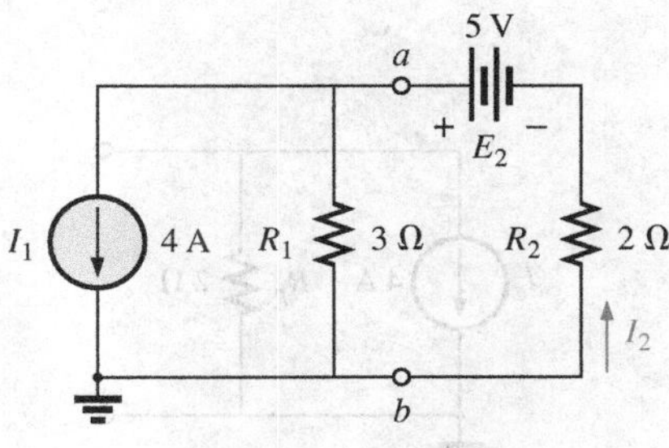

FIG. 8.9
Two-source network for Example 8.5.

In Example 8.4, note that both sources pressure or establish current up through the circuit to establish the same direction for the load current I_L and the same polarity for the voltage V_L.

EXAMPLE 8.5 Determine current I_2 for the network in Fig. 8.9.

Solution: Although it may appear that the network cannot be solved using methods introduced thus far, one source conversion, as shown in Fig. 8.10, results in a simple series circuit. It does not make sense to convert the voltage source to a current source because you would lose the current I_2 in the redrawn network. Note the polarity for the equivalent voltage source as determined by the current source.

For the source conversion

$$E_1 = I_1 R_1 = (4\ \text{A})(3\ \Omega) = 12\ \text{V}$$

and
$$I_2 = \frac{E_1 + E_2}{R_1 + R_2} = \frac{12\ \text{V} + 5\ \text{V}}{3\ \Omega + 2\ \Omega} = \frac{17\ \text{V}}{5\ \Omega} = \mathbf{3.4\ A}$$

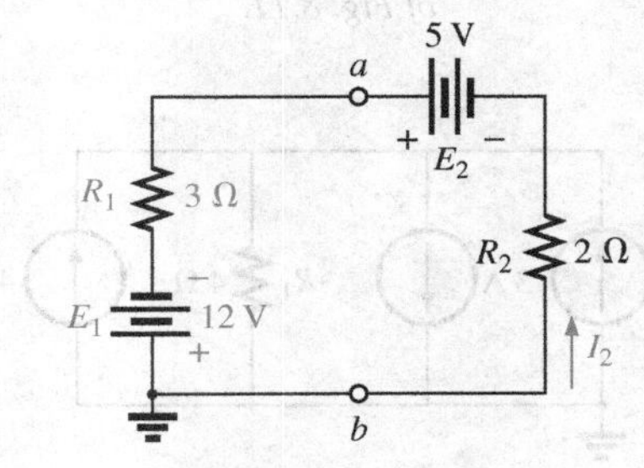

FIG. 8.10
Network in Fig. 8.9 following the conversion of the current source to a voltage source.

8.4 CURRENT SOURCES IN PARALLEL

We found that voltage sources of different terminal voltages cannot be placed in parallel because of a violation of Kirchhoff's voltage law. Similarly,

current sources of different values cannot be placed in series due to a violation of Kirchhoff's current law.

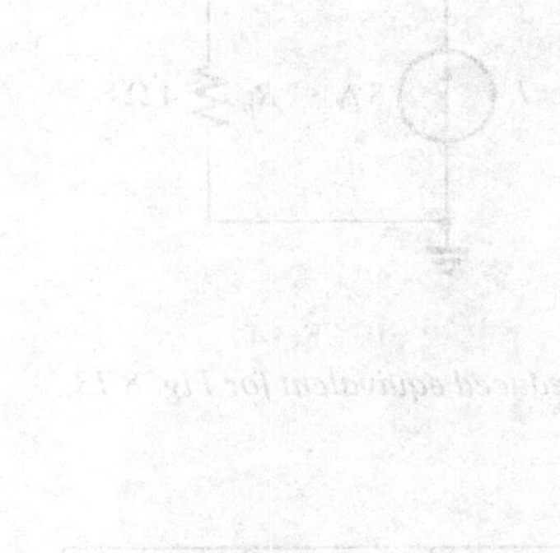

However, current sources can be placed in parallel just as voltage sources can be placed in series. In general,

two or more current sources in parallel can be replaced by a single current source having a magnitude determined by the difference of the sum of the currents in one direction and the sum in the opposite direction. The new parallel internal resistance is the total resistance of the resulting parallel resistive elements.

Consider the following examples.

EXAMPLE 8.6 Reduce the parallel current sources in Fig. 8.11 to a single current source.

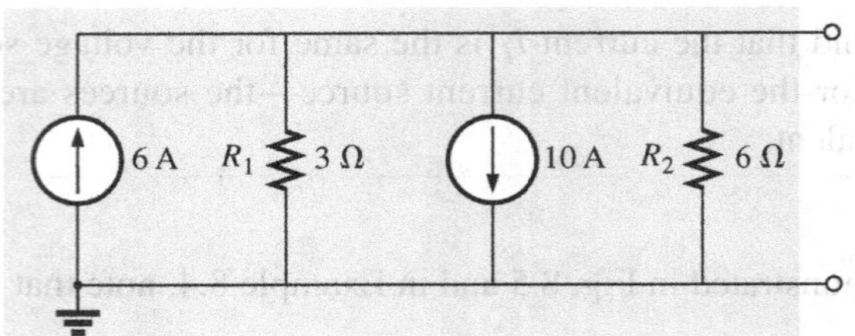

FIG. 8.11
Parallel current sources for Example 8.6.

Solution: The net source current is

$$I = 10\text{ A} - 6\text{ A} = \mathbf{4\text{ A}}$$

with the direction being that of the larger source.

The net internal resistance is the parallel combination of resistances, R_1 and R_2:

$$R_p = 3\ \Omega \parallel \Omega = \mathbf{2\ \Omega}$$

The reduced equivalent appears in Fig. 8.12.

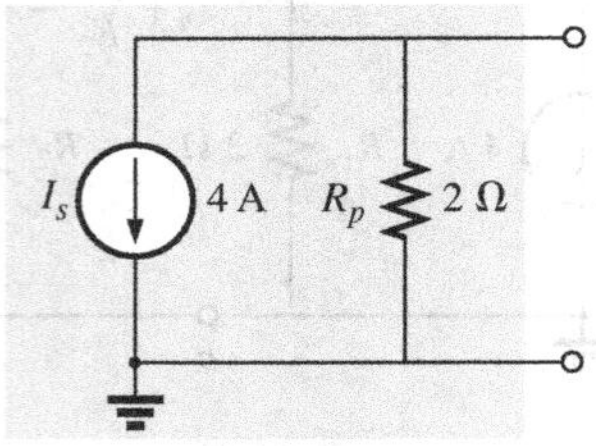

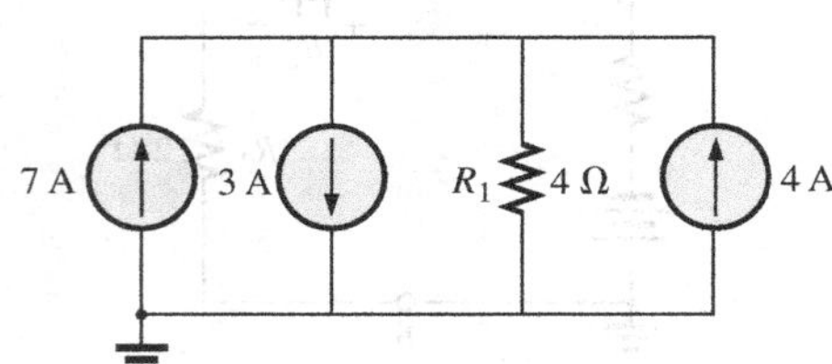

FIG. 8.12
Reduced equivalent for the configuration of Fig. 8.11.

EXAMPLE 8.7 Reduce the parallel current sources in Fig. 8.13 to a single current source.

Solution: The net current is

$$I = 7\text{ A} + 4\text{ A} - 3\text{ A} = \mathbf{8\text{ A}}$$

with the direction shown in Fig. 8.14. The net internal resistance remains the same.

FIG. 8.13
Parallel current sources for Example 8.7.

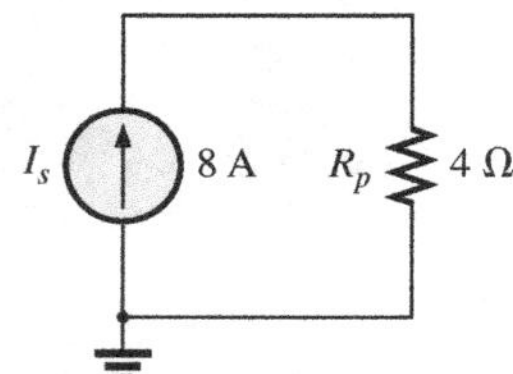

FIG. 8.14
Reduced equivalent for Fig. 8.13.

EXAMPLE 8.8 Reduce the network in Fig. 8.15 to a single current source, and calculate the current through R_L.

Solution: In this example, the voltage source will first be converted to a current source as shown in Fig. 8.16. Combining current sources gives

$$I_s = I_1 + I_2 = 4\text{ A} + 6\text{ A} = \mathbf{10\text{ A}}$$

and

$$R_s = R_1 \| R_2 = 8\ \Omega \| 24\ \Omega = \mathbf{6\ \Omega}$$

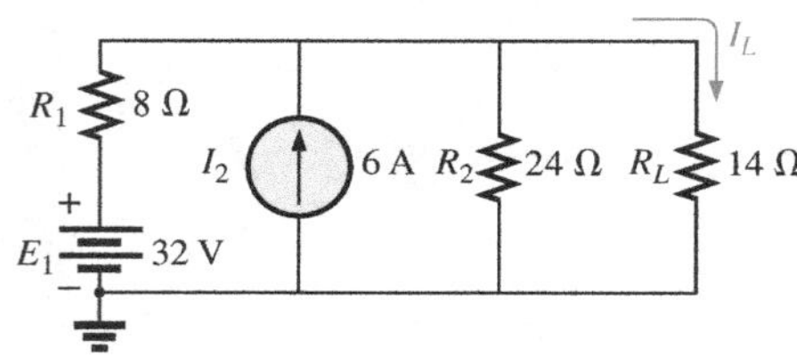

FIG. 8.15
Example 8.8.

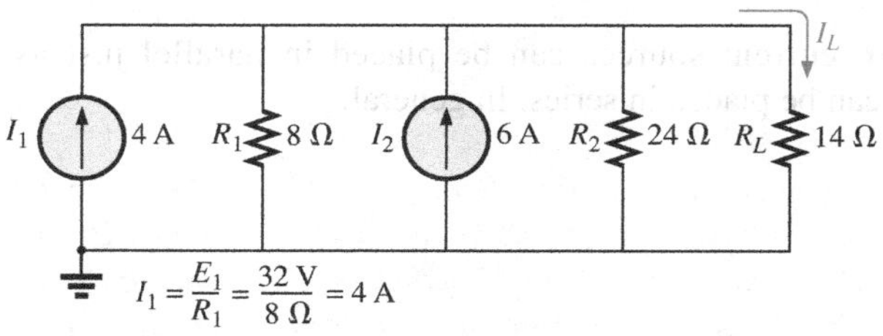

FIG. 8.16
Network in Fig. 8.15 following the conversion of the voltage source to a current source.

Applying the current divider rule to the resulting network in Fig. 8.17 gives

$$I_L = \frac{R_p I_s}{R_p + R_L} = \frac{(6\ \Omega)(10\ \text{A})}{6\ \Omega + 14\ \Omega} = \frac{60\ \text{A}}{20} = \mathbf{3\ A}$$

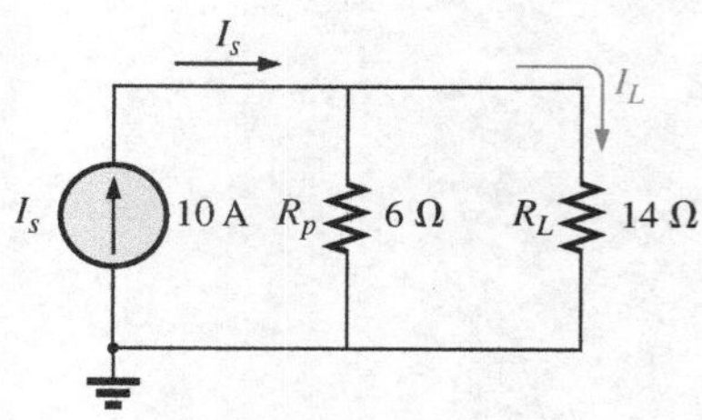

FIG. 8.17
Network in Fig. 8.16 reduced to its simplest form.

8.5 CURRENT SOURCES IN SERIES

The current through any branch of a network can be only single-valued. For the situation indicated at point *a* in Fig. 8.18, we find by application of Kirchhoff's current law that the current leaving that point is greater than that entering—an impossible situation. Therefore,

current sources of different current ratings are not connected in series,

just as voltage sources of different voltage ratings are not connected in parallel.

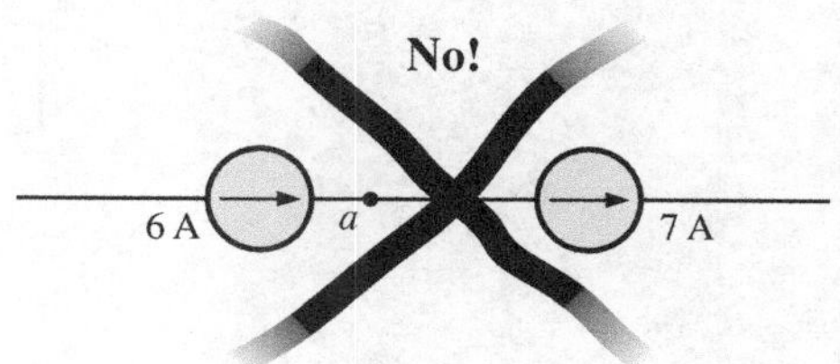

FIG. 8.18
Invalid situation.

8.6 BRANCH-CURRENT ANALYSIS

Before examining the details of the first important method of analysis, let us examine the network in Fig. 8.19 to be sure that you understand the need for these special methods.

Initially, it may appear that we can use the reduce and return approach to work our way back to the source E_1 and calculate the source current I_{s_1}. Unfortunately, however, the series elements R_3 and E_2 cannot be combined because they are different types of elements. A further examination of the network reveals that there are no two like elements that are in series or parallel. No combination of elements can be performed, and it is clear that another approach must be defined.

It must be noted that the network of Fig. 8.19 can be solved if we convert each voltage source to a current source and then combine parallel current sources. However, if a specific quantity of the original network is required, it would require working back using the information determined from the source conversion. Further, there will be complex networks for which source conversions will not permit a solution, so it is important to understand the methods to be described in this chapter.

The first approach to be introduced is called the **branch-current method** because we will define and solve for the currents of each branch of the network. The best way to introduce this method and understand its application is to follow a series of steps, as listed below. Each step is carefully defined in the examples to follow.

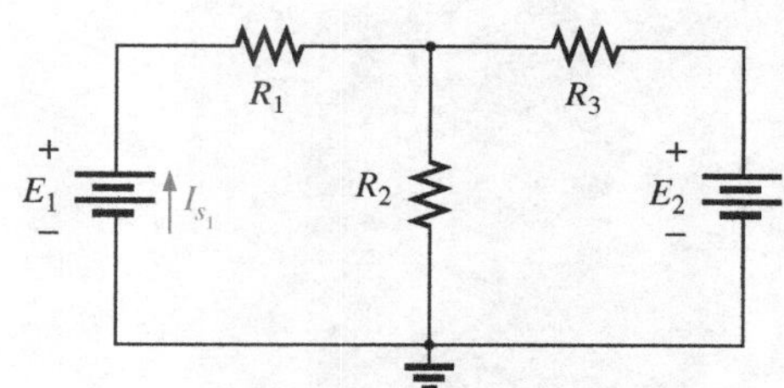

FIG. 8.19
Demonstrating the need for an approach such as branch-current analysis.

Branch-Current Analysis Procedure

1. ***Assign a distinct current of arbitrary direction to each branch of the network.***
2. ***Indicate the polarities for each resistor as determined by the assumed current direction.***
3. ***Apply Kirchhoff's voltage law around each closed, independent loop of the network.***

The best way to determine how many times Kirchhoff's voltage law has to be applied is to determine the number of "windows" in the network. The network in Example 8.9 has a definite similarity to the two-window configuration in Fig. 8.20(a). The result is a need to apply Kirchhoff's voltage law twice. For networks with three windows, as shown in Fig. 8.20(b), three applications of Kirchhoff's voltage law are required, and so on.

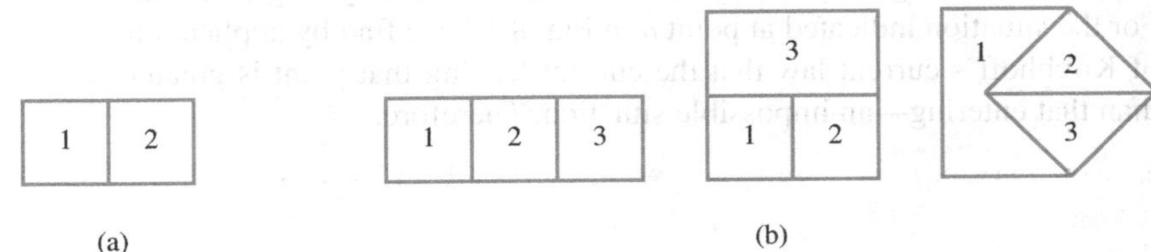

FIG. 8.20
Determining the number of independent closed loops.

4. Apply Kirchhoff's current law at the minimum number of nodes that will include all the branch currents of the network.

The minimum number is one less than the number of independent nodes of the network. For the purposes of this analysis, a **node** is a junction of two or more branches, where a branch is any combination of series elements. Fig. 8.21 defines the number of applications of Kirchhoff's current law for each configuration in Fig. 8.20.

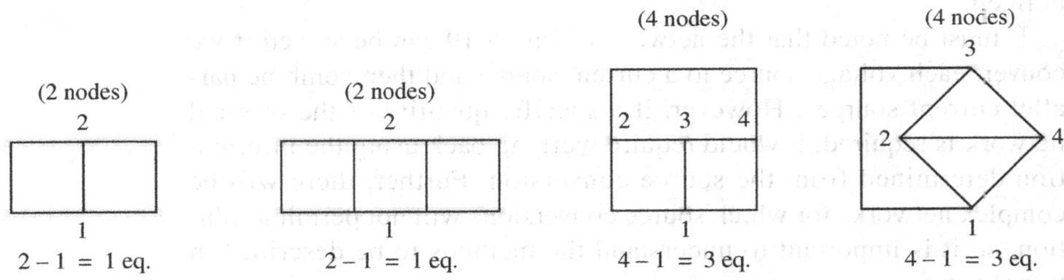

FIG. 8.21
Determining the number of applications of Kirchhoff's current law required.

5. Solve the resulting simultaneous linear equations for assumed branch currents.

It is assumed that the use of the **determinants method** to solve for the currents I_1, I_2, and I_3 is understood and is a part of the student's mathematical background. If not, a detailed explanation of the procedure is provided in Appendix C. Calculators and computer software packages such as MATLAB and Mathcad can find the solutions quickly and accurately.

EXAMPLE 8.9 Apply the branch-current method to the network in Fig. 8.22.

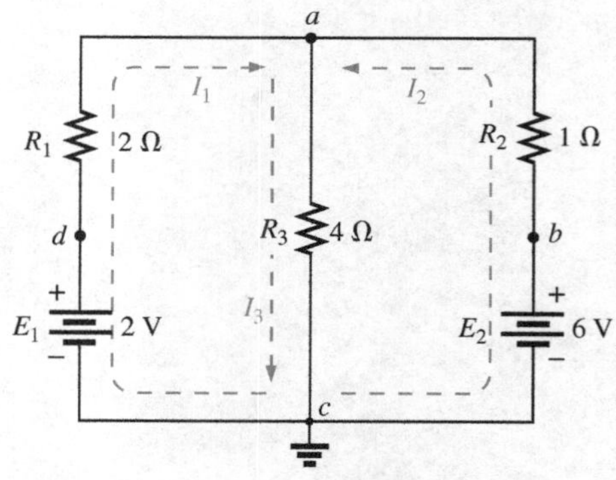

FIG. 8.22
Example 8.9.

Solution 1:

Step 1: Since there are three distinct branches (*cda*, *cba*, *ca*), three currents of arbitrary directions (I_1, I_2, I_3) are chosen, as indicated in Fig. 8.22. The current directions for I_1 and I_2 were chosen to match the "pressure" applied by sources E_1 and E_2, respectively. Since both I_1 and I_2 enter node *a*, I_3 is leaving.

Step 2: Polarities for each resistor are drawn to agree with assumed current directions, as indicated in Fig. 8.23.

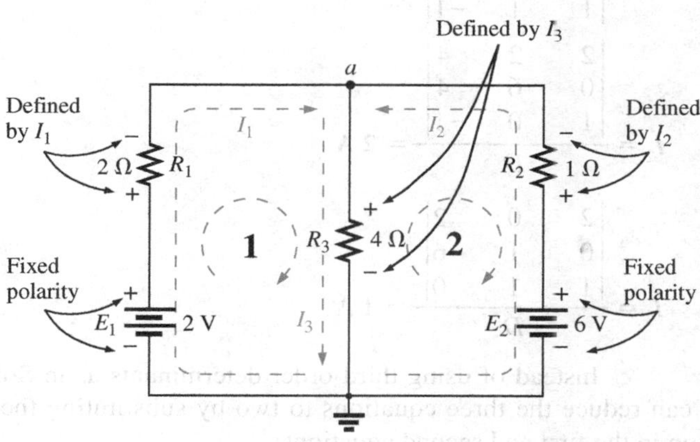

FIG. 8.23
Inserting the polarities across the resistive elements as defined by the chosen branch currents.

Step 3: Kirchhoff's voltage law is applied around each closed loop (1 and 2) in the clockwise direction:

Rise in potential

$$\text{loop 1:}\quad \Sigma_{\circlearrowright} V = +E_1 - V_{R_1} - V_{R_3} = 0$$

Drop in potential

and

Rise in potential

$$\text{loop 2:}\quad \Sigma_{\circlearrowright} V = +V_{R_3} + V_{R_2} - E_2 = 0$$

Drop in potential

$$\text{loop 1:}\quad \Sigma_{\circlearrowright} V = \underbrace{+2\text{ V}}_{\substack{\text{Battery}\\ \text{potential}}} - \underbrace{(2\ \Omega)I_1}_{\substack{\text{Voltage drop}\\ \text{across } 2\ \Omega\\ \text{resistor}}} - \underbrace{(4\ \Omega)I_3}_{\substack{\text{Voltage drop}\\ \text{across } 4\ \Omega\\ \text{resistor}}} = 0$$

$$\text{loop 2:}\quad \Sigma_{\circlearrowright} V = (4\ \Omega)I_3 + (1\ \Omega)I_2 - 6\text{ V} = 0$$

Step 4: Applying Kirchhoff's current law at node *a* (in a two-node network, the law is applied at only one node) gives

$$I_1 + I_2 = I_3$$

Step 5: There are three equations and three unknowns (units removed for clarity):

$$2 - 2I_1 - 4I_3 = 0 \qquad \text{Rewritten: } 2I_1 + 0 + 4I_3 = 2$$
$$4I_3 + 1I_2 - 6 = 0 \qquad 0 + I_2 + 4I_3 = 6$$
$$I_1 + I_2 = I_3 \qquad I_1 + I_2 - I_3 = 0$$

Using third-order determinants (Appendix C), we have

$$I_1 = \frac{\begin{vmatrix} 2 & 0 & 4 \\ 6 & 1 & 4 \\ 0 & 1 & -1 \end{vmatrix}}{\begin{vmatrix} 2 & 0 & 4 \\ 0 & 1 & 4 \\ 1 & 1 & -1 \end{vmatrix}} = \mathbf{-1\ A}$$

A negative sign in front of a branch current indicates only that the actual current is in the direction opposite to that assumed.

$$D = \begin{vmatrix} 2 & 0 & 4 \\ 0 & 1 & 4 \\ 1 & 1 & -1 \end{vmatrix}$$

$$I_2 = \frac{\begin{vmatrix} 2 & 2 & 4 \\ 0 & 6 & 4 \\ 1 & 0 & -1 \end{vmatrix}}{D} = \mathbf{2\ A}$$

$$I_3 = \frac{\begin{vmatrix} 2 & 0 & 2 \\ 0 & 1 & 6 \\ 1 & 1 & 0 \end{vmatrix}}{D} = \mathbf{1\ A}$$

Solution 2: Instead of using third-order determinants as in Solution 1, we can reduce the three equations to two by substituting the third equation in the first and second equations:

$$2 - 2I_1 - 4\overbrace{(I_1 + I_2)}^{I_3} = 0 \qquad 2 - 2I_1 - 4I_1 - 4I_2 = 0$$
$$4\overbrace{(I_1 + I_2)}^{I_3} + I_2 - 6 = 0 \qquad 4I_1 + 4I_2 + I_2 - 6 = 0$$

or

$$-6I_1 - 4I_2 = -2$$
$$+4I_1 + 5I_2 = +6$$

Multiplying through by -1 in the top equation yields

$$6I_1 - 4I_2 = +2$$
$$4I_1 + 5I_2 = +6$$

and using determinants gives

$$I_1 = \frac{\begin{vmatrix} 2 & 4 \\ 6 & 5 \end{vmatrix}}{\begin{vmatrix} 6 & 4 \\ 4 & 5 \end{vmatrix}} = \frac{10 - 24}{30 - 16} = \frac{-14}{14} = \mathbf{-1\ A}$$

TI-89 Solution: The procedure for determining the determinant in Example 8.9 requires some scrolling to obtain the desired math functions, but in time that procedure can be performed quite rapidly. As with any computer or calculator system, it is paramount that you enter all parameters correctly. One error in the sequence negates the entire process. For the TI-89, the entries are shown in Fig. 8.24(a).

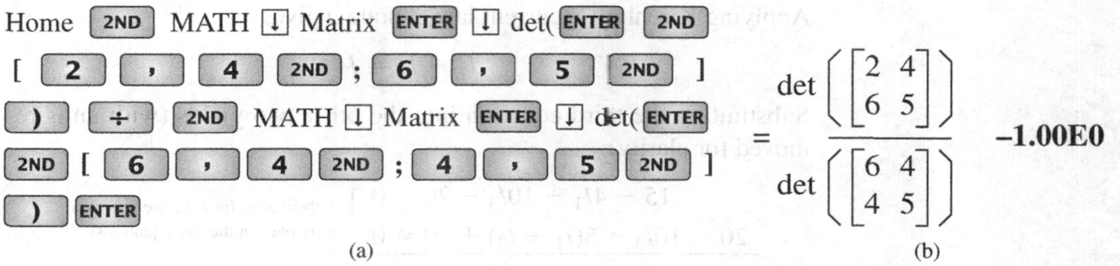

Home [2ND] MATH [↓] Matrix [ENTER] [↓] det([ENTER] [2ND]
[[2] [,] [4] [2ND] ; [6] [,] [5] [2ND]]
[)] [÷] [2ND] MATH [↓] Matrix [ENTER] [↓] det([ENTER]
[2ND] [[6] [,] [4] [2ND] ; [4] [,] [5] [2ND]]
[)] [ENTER]

(a)

$$\frac{\det\left(\begin{bmatrix} 2 & 4 \\ 6 & 5 \end{bmatrix}\right)}{\det\left(\begin{bmatrix} 6 & 4 \\ 4 & 5 \end{bmatrix}\right)} = \quad \mathbf{-1.00E0}$$

(b)

FIG. 8.24
TI-89 solution for the current I_1 of Fig. 8.22.

After you select the last ENTER key, the screen shown in Fig. 8.24(b) appears.

$$I_2 = \frac{\begin{vmatrix} 6 & 2 \\ 4 & 6 \end{vmatrix}}{14} = \frac{36 - 8}{14} = \frac{28}{14} = \mathbf{2\ A}$$

$$I_3 = I_1 + I_2 = -1 + 2 = \mathbf{1\ A}$$

It is now important that the impact of the results obtained be understood. The currents I_1, I_2, and I_3 are the actual currents in the branches in which they were defined. A negative sign in the solution means that the actual current has the opposite direction than initially defined—the magnitude is correct. Once the actual current directions and their magnitudes are inserted in the original network, the various voltages and power levels can be determined. For this example, the actual current directions and their magnitudes have been entered on the original network in Fig. 8.25. Note that the current through the series elements R_1 and E_1 is 1 A; the current through R_3, is 1 A; and the current through the series elements R_2 and E_2 is 2 A. Due to the minus sign in the solution, the direction of I_1 is opposite to that shown in Fig. 8.22. The voltage across any resistor can now be found using Ohm's law, and the power delivered by either source or to any one of the three resistors can be found using the appropriate power equation.

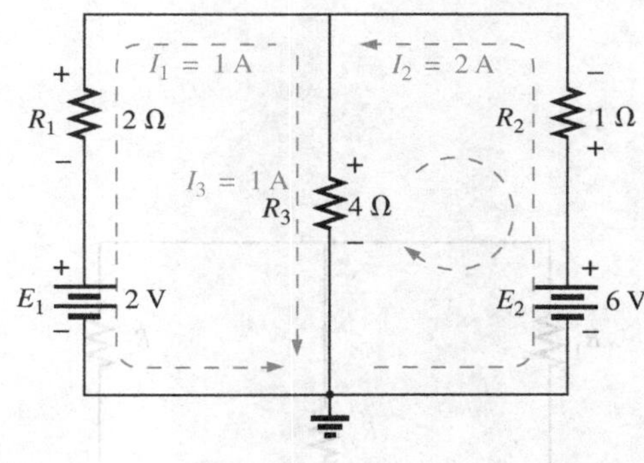

FIG. 8.25
Reviewing the results of the analysis of the network in Fig. 8.22.

Applying Kirchhoff's voltage law around the loop indicated in Fig. 8.25 gives

$$\Sigma_{\circlearrowright} V = +(4\ \Omega)I_3 + (1\ \Omega)I_2 - 6\ \text{V} = 0$$

or $$(4\ \Omega)I_3 + (1\ \Omega)I_2 = 6\ \text{V}$$

and $$(4\ \Omega)(1\ \text{A}) + (1\ \Omega)(2\ \text{A}) = 6\ \text{V}$$

$$4\ \text{V} + 2\ \text{V} = 6\ \text{V}$$

$$6\ \text{V} = 6\ \text{V} \qquad \text{(checks)}$$

EXAMPLE 8.10 Apply branch-current analysis to the network in Fig. 8.26.

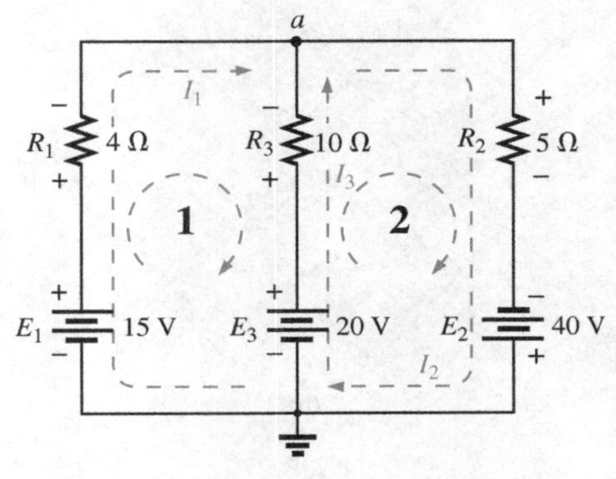

FIG. 8.26
Example 8.10.

Solution: Again, the current directions were chosen to match the "pressure" of each battery. The polarities are then added, and Kirchhoff's voltage law is applied around each closed loop in the clockwise direction. The result is as follows:

$$\text{loop 1:} \quad +15\ \text{V} - (4\ \Omega)I_1 + (10\ \Omega)I_3 - 20\ \text{V} = 0$$

$$\text{loop 2:} \quad +20\ \text{V} - (10\ \Omega)I_3 - (5\ \Omega)I_2 + 40\ \text{V} = 0$$

Applying Kirchhoff's current law at node a gives

$$I_1 + I_3 = I_2$$

Substituting the third equation into the other two yields (with units removed for clarity)

$$\left.\begin{aligned} 15 - 4I_1 + 10I_3 - 20 &= 0 \\ 20 - 10I_3 - 5(I_1 + I_3) + 40 &= 0 \end{aligned}\right\} \text{Substituting for } I_2 \text{ (since it occurs only once in the two equations)}$$

or

$$\begin{aligned} -4I_1 + 10I_3 &= 5 \\ -5I_1 - 15I_3 &= -60 \end{aligned}$$

Multiplying the lower equation by -1, we have

$$\begin{aligned} -4I_1 + 10I_3 &= 5 \\ 5I_1 + 15I_3 &= 60 \end{aligned}$$

$$I_1 = \frac{\begin{vmatrix} 5 & 10 \\ 60 & 15 \end{vmatrix}}{\begin{vmatrix} -4 & 10 \\ 5 & 15 \end{vmatrix}} = \frac{75 - 600}{-60 - 50} = \frac{-525}{-110} = \mathbf{4.77\ A}$$

$$I_3 = \frac{\begin{vmatrix} -4 & 5 \\ 5 & 60 \end{vmatrix}}{-110} = \frac{-240 - 25}{-110} = \frac{-265}{-110} = \mathbf{2.41\ A}$$

$$I_2 = I_1 + I_3 = 4.77\text{ A} + 2.41\text{ A} = \mathbf{7.18\ A}$$

revealing that the assumed directions were the actual directions, with I_2 equal to the sum of I_1 and I_3.

8.7 MESH ANALYSIS (GENERAL APPROACH)

The next method to be described—**mesh analysis**—is actually an extension of the branch-current analysis approach just introduced. By defining a unique array of currents to the network, the information provided by the application of Kirchhoff's current law is already included when we apply Kirchhoff's voltage law. In other words, there is no need to apply step 4 of the branch-current method.

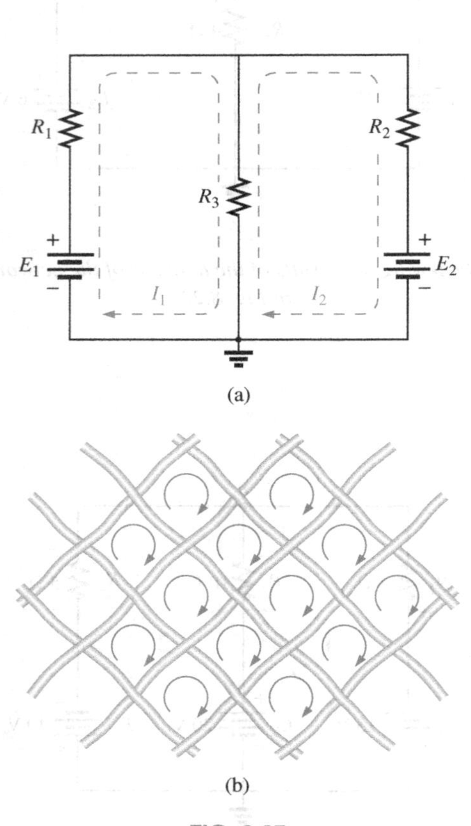

FIG. 8.27
Defining the mesh (loop) current: (a) "two-window" network; (b) wire mesh fence analogy.

The currents to be defined are called **mesh** or **loop currents.** The two terms are used interchangeably. In Fig. 8.27(a), a network with two "windows" has had two mesh currents defined. Note that each forms a closed "loop" around the inside of each window; these loops are similar to the loops defined in the wire mesh fence in Fig. 8.27(b)—hence the use of the term *mesh* for the loop currents. We will find that

the number of mesh currents required to analyze a network will equal the number of "windows" of the configuration.

The defined mesh currents can initially be a little confusing because it appears that two currents have been defined for resistor R_3. There is no problem with E_1 and R_1, which have only current I_1, or with E_2 and R_2, which have only current I_2. However, defining the current through R_3 may seem a little troublesome. Actually, it is quite straightforward. The current through R_3 is simply the difference between I_1 and I_2, with the direction being that of the larger. This is demonstrated in the examples to follow.

Because mesh currents can result in more than one current through an element, branch-current analysis was introduced first. Branch-current

analysis is the straightforward application of the basic laws of electric circuits. Mesh analysis employs a maneuver ("trick," if you prefer) that removes the need to apply Kirchhoff's current law.

Mesh Analysis Procedure

1. Assign a distinct current in the clockwise direction to each independent, closed loop of the network. It is not absolutely necessary to choose the clockwise direction for each loop current. In fact, any direction can be chosen for each loop current with no loss in accuracy, as long as the remaining steps are followed properly. However, by choosing the clockwise direction as a standard, we can develop a shorthand method (Section 8.8) for writing the required equations that will save time and possibly prevent some common errors.

This first step is accomplished most effectively by placing a loop current *within* each "window" of the network, as demonstrated in the previous section, to ensure that they are all independent. A variety of other loop currents can be assigned. In each case, however, be sure that the information carried by any one loop equation is not included in a combination of the other network equations. This is the crux of the terminology *independent*. No matter how you choose your loop currents, the number of loop currents required is always equal to the number of windows of a planar (no-crossovers) network. On occasion, a network may appear to be nonplanar. However, a redrawing of the network may reveal that it is, in fact, planar. This may be true for one or two problems at the end of the chapter.

Before continuing to the next step, let us ensure that the concept of a loop current is clear. For the network in Fig. 8.28, the loop current I_1 is the branch current of the branch containing the 2 Ω resistor and 2 V battery. The current through the 4 Ω resistor is not I_1, however, since there is also a loop current I_2 through it. Since they have opposite directions, $I_{4\Omega}$ equals the difference between the two, $I_1 - I_2$ or $I_2 - I_1$, depending on which you choose to be the defining direction. In other words, *a loop current is a branch current only when it is the only loop current assigned to that branch.*

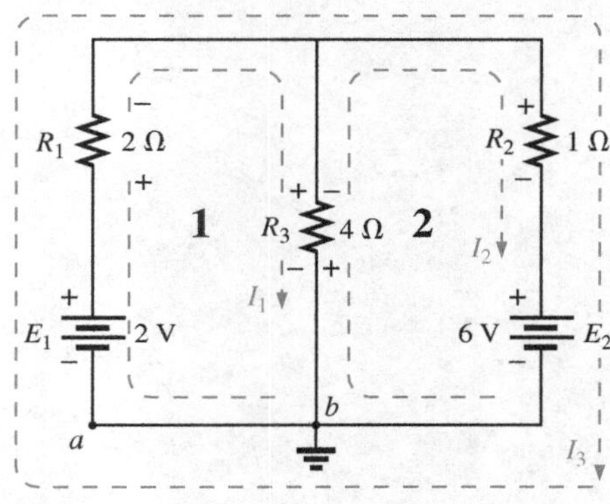

FIG. 8.28
Defining the mesh currents for a "two-window" network.

2. Indicate the polarities within each loop for each resistor as determined by the assumed direction of loop current for that loop. Note the requirement that the polarities be placed within each loop. This requires, as shown in Fig. 8.28, that the 4 Ω resistor have two sets of polarities across it.

3. Apply Kirchhoff's voltage law around each closed loop in the clockwise direction. Again, the clockwise direction was chosen to establish uniformity and prepare us for the method to be introduced in the next section.

a. If a resistor has two or more assumed currents through it, the total current through the resistor is the assumed current of the loop in which Kirchhoff's voltage law is being applied, plus the assumed currents of the other loops passing through in the same direction, minus the assumed currents through in the opposite direction.

b. The polarity of a voltage source is unaffected by the direction of the assigned loop currents.

4. Solve the resulting simultaneous linear equations for the assumed loop currents.

EXAMPLE 8.11 Consider the same basic network as in Example 8.9, now appearing as Fig. 8.28.

Solution:

Step 1: Two loop currents (I_1 and I_2) are assigned in the clockwise direction in the windows of the network. A third loop (I_3) could have been included around the entire network, but the information carried by this loop is already included in the other two.

Step 2: Polarities are drawn within each window to agree with assumed current directions. Note that for this case, the polarities across the 4 Ω resistor are the opposite for each loop current.

Step 3: Kirchhoff's voltage law is applied around each loop in the clockwise direction. Keep in mind as this step is performed that the law is concerned only with the magnitude and polarity of the voltages around the closed loop and not with whether a voltage rise or drop is due to a battery or a resistive element. The voltage across each resistor is determined by $V = IR$. For a resistor with more than one current through it, the current is the loop current of the loop being examined plus or minus the other loop currents as determined by their directions. If clockwise applications of Kirchhoff's voltage law are always chosen, the other loop currents are always subtracted from the loop current of the loop being analyzed.

loop 1: $+E_1 - V_1 - V_3 = 0$ (clockwise starting at point a)

$$+2\text{ V} - (2\ \Omega)I_1 - \overbrace{(4\ \Omega)\underbrace{(I_1 - I_2)}_{\substack{\text{Total current}\\\text{through}\\4\ \Omega\text{ resistor}}}}^{\substack{\text{Voltage drop across}\\4\ \Omega\text{ resistor}}} = 0$$

(Subtracted since I_2 is opposite in direction to I_1.)

loop 2: $-V_3 - V_2 - E_2 = 0$ (clockwise starting at point b)

$$-(4\ \Omega)(I_2 - I_1) - (1\ \Omega)I_2 - 6\text{ V} = 0$$

Step 4: The equations are then rewritten as follows (without units for clarity):

loop 1: $+2 - 2I_1 - 4I_1 + 4I_2 = 0$
loop 2: $-4I_2 + 4I_1 - 1I_2 - 6 = 0$

and

loop 1: $+2 - 6I_1 + 4I_2 = 0$
loop 2: $-5I_2 + 4I_1 - 6 = 0$

or

loop 1: $-6I_1 + 4I_2 = -2$
loop 2: $+4I_1 - 5I_2 = +6$

Applying determinants results in

$$I_1 = \mathbf{-1\ A} \quad \text{and} \quad I_2 = \mathbf{-2\ A}$$

The minus signs indicate that the currents have a direction opposite to that indicated by the assumed loop current.

The actual current through the 2 V source and 2 Ω resistor is therefore 1 A in the other direction, and the current through the 6 V source and 1 Ω resistor is 2 A in the opposite direction indicated on the circuit. The current through the 4 Ω resistor is determined by the following equation from the original network:

loop 1: $I_{4\Omega} = I_1 - I_2 = -1\text{ A} - (-2\text{ A}) = -1\text{ A} + 2\text{ A}$
$= \mathbf{1\ A}$ (in the direction of I_1)

The outer loop (I_3) and *one* inner loop (either I_1 or I_2) would also have produced the correct results. This approach, however, often leads to errors since the loop equations may be more difficult to write. The best method of picking these loop currents is the window approach.

EXAMPLE 8.12 Find the current through each branch of the network in Fig. 8.29.

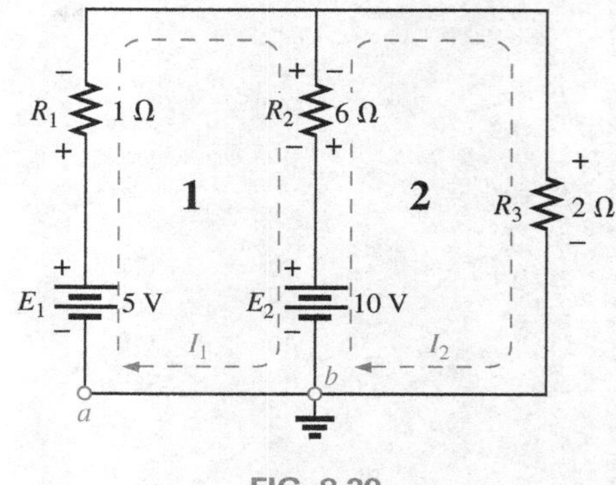

FIG. 8.29
Example 8.12.

Solution:

Steps 1 and 2: These are as indicated in the circuit. Note that the polarities of the 6 Ω resistor are different for each loop current.

Step 3: Kirchhoff's voltage law is applied around each closed loop in the clockwise direction:

loop 1: $+E_1 - V_1 - V_2 - E_2 = 0$ (clockwise starting at point a)

$$+5\text{ V} - (1\ \Omega)I_1 - (6\ \Omega)(I_1 - I_2) - 10\text{ V} = 0$$

↑ I_2 flows through the 6 Ω resistor in the direction opposite to I_1.

loop 2: $E_2 - V_2 - V_3 = 0$ (clockwise starting at point b)

$$+10\text{ V} - (6\ \Omega)(I_2 - I_1) - (2\ \Omega)I_2 = 0$$

The equations are rewritten as

$$\left.\begin{aligned} 5 - I_1 - 6I_1 + 6I_2 - 10 &= 0 \\ 10 - 6I_2 + 6I_1 - 2I_2 &= 0 \end{aligned}\right\} \begin{aligned} -7I_1 + 6I_2 &= 5 \\ +6I_1 - 8I_2 &= -10 \end{aligned}$$

Step 4:

$$I_1 = \frac{\begin{vmatrix} 5 & 6 \\ -10 & -8 \end{vmatrix}}{\begin{vmatrix} -7 & 6 \\ 6 & -8 \end{vmatrix}} = \frac{-40 + 60}{56 - 36} = \frac{20}{20} = \mathbf{1\ A}$$

$$I_2 = \frac{\begin{vmatrix} -7 & 5 \\ 6 & -10 \end{vmatrix}}{20} = \frac{70 - 30}{20} = \frac{40}{20} = \mathbf{2\ A}$$

Since I_1 and I_2 are positive and flow in opposite directions through the 6 Ω resistor and 10 V source, the total current in this branch is equal to the difference of the two currents in the direction of the larger:

$$I_2 > I_1 \quad (2\text{ A} > 1\text{ A})$$

Therefore,

$$I_{R_2} = I_2 - I_1 = 2\text{ A} - 1\text{ A} = \mathbf{1\ A} \quad \text{in the direction of } I_2$$

It is sometimes impractical to draw all the branches of a circuit at right angles to one another. The next example demonstrates how a portion of a network may appear due to various constraints. The method of analysis is no different with this change in configuration.

FIG. 8.30
Example 8.13.

EXAMPLE 8.13 Find the branch currents of the networks in Fig. 8.30.

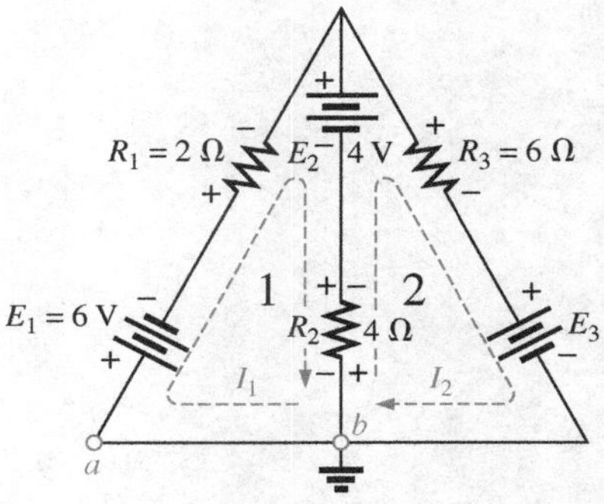

Solution:

Steps 1 and 2: These are as indicated in the circuit.

Step 3: Kirchhoff's voltage law is applied around each closed loop:

loop 1: $-E_1 - I_1R_1 - E_2 - V_2 = 0$ (clockwise from point *a*)
$-6\text{ V} - (2\ \Omega)I_1 - 4\text{ V} - (4\ \Omega)(I_1 - I_2) = 0$

loop 2: $-V_2 + E_2 - V_3 - E_3 = 0$ (clockwise from point *b*)
$-(4\ \Omega)(I_2 - I_1) + 4\text{ V} - (6\ \Omega)(I_2) - 3\text{ V} = 0$

which are rewritten as

$$\left.\begin{aligned} -10 - 4I_1 - 2I_1 + 4I_2 &= 0 \\ +1 + 4I_1 + 4I_2 - 6I_2 &= 0 \end{aligned}\right\} \begin{aligned} -6I_1 + 4I_2 &= +10 \\ +4I_1 - 10I_2 &= -1 \end{aligned}$$

or, by multiplying the top equation by -1, we obtain

$$\begin{aligned} 6I_1 - 4I_2 &= -10 \\ 4I_1 - 10I_2 &= -1 \end{aligned}$$

$$\textit{Step 4: } I_1 = \frac{\begin{vmatrix} -10 & -4 \\ -1 & -10 \end{vmatrix}}{\begin{vmatrix} 6 & -4 \\ 4 & -10 \end{vmatrix}} = \frac{100 - 4}{-60 + 16} = \frac{96}{-44} = \mathbf{-2.18\ A}$$

$$I_2 = \frac{\begin{vmatrix} 6 & -10 \\ 4 & -1 \end{vmatrix}}{-44} = \frac{-6 + 40}{-44} = \frac{34}{-44} = \mathbf{-0.77\ A}$$

The current in the 4 Ω resistor and 4 V source for loop 1 is

$$\begin{aligned} I_1 - I_2 &= -2.18\text{ A} - (-0.77\text{ A}) \\ &= -2.18\text{ A} + 0.77\text{ A} \\ &= \mathbf{-1.41\ A} \end{aligned}$$

revealing that it is 1.41 A in a direction opposite (due to the minus sign) to I_1 in loop 1.

Supermesh Currents

Occasionally, you will find current sources in a network without a parallel resistance. This removes the possibility of converting the source to a voltage source as required by the given procedure. In such cases, you have a choice of two approaches.

The simplest and most direct approach is to place a resistor in parallel with the current source that has a much higher value than the other resistors of the network. For instance, if most of the resistors of the network are in the 1 to 10 Ω range, choosing a resistor of 100 Ω or higher would provide one level of accuracy for the answer. However, choosing a resistor of 1000 Ω or higher would increase the accuracy of the answer. You will never get the exact answer because the network has been modified by this introduced element. However for most applications, the answer will be sufficiently accurate.

The other choice is to use the **supermesh approach** described in the following steps. Although this approach will provide the exact solution, it does require some practice to become proficient in its use. The procedure is as follows.

Start as before, and assign a mesh current to each independent loop, including the current sources, as if they were resistors or voltage sources. Then mentally (redraw the network if necessary) remove the current sources (replace with open-circuit equivalents), and apply

Kirchhoff's voltage law to all the remaining independent paths of the network using the mesh currents just defined. Any resulting path, including two or more mesh currents, is said to be the path of a **supermesh current.** Then relate the chosen mesh currents of the network to the independent current sources of the network, and solve for the mesh currents. The next example clarifies the definition of supermesh current and the procedure.

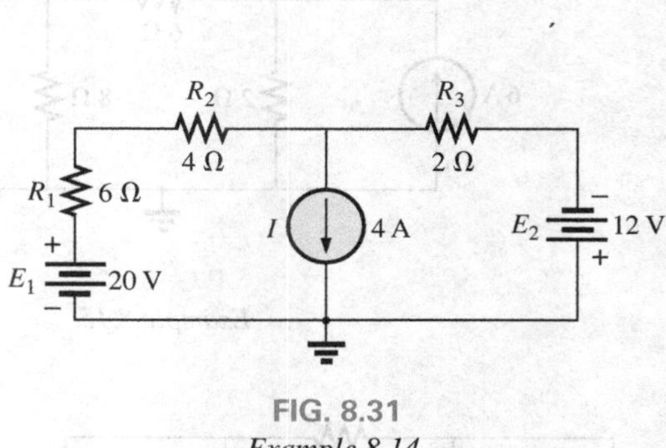

FIG. 8.31
Example 8.14.

EXAMPLE 8.14 Using mesh analysis, determine the currents of the network in Fig. 8.31.

Solution: First, the mesh currents for the network are defined, as shown in Fig. 8.32. Then the current source is mentally removed, as shown in Fig. 8.33, and Kirchhoff's voltage law is applied to the resulting network. The single path now including the effects of two mesh currents is referred to as the path of a *supermesh current.*

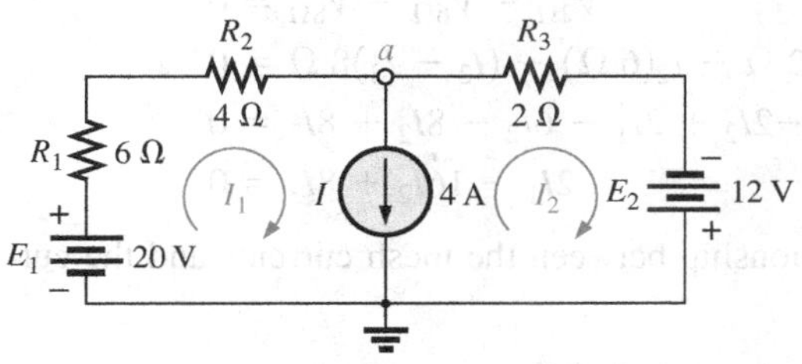

FIG. 8.32
Defining the mesh currents for the network in Fig. 8.31.

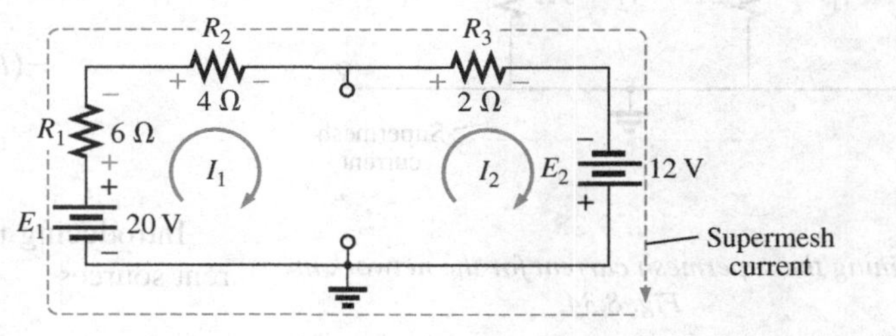

FIG. 8.33
Defining the supermesh current.

Applying Kirchhoff's law gives

$$20\text{ V} - I_1(6\ \Omega) - I_1(4\ \Omega) - I_2(2\ \Omega) + 12\text{ V} = 0$$

or

$$10I_1 + 2I_2 = 32$$

Node a is then used to relate the mesh currents and the current source using Kirchhoff's current law:

$$I_1 = I + I_2$$

The result is two equations and two unknowns:

$$\begin{aligned} 10I_1 + 2I_2 &= 32 \\ I_1 - I_2 &= 4 \end{aligned}$$

Applying determinants gives

$$I_1 = \frac{\begin{vmatrix} 32 & 2 \\ 4 & -1 \end{vmatrix}}{\begin{vmatrix} 10 & 2 \\ 1 & -1 \end{vmatrix}} = \frac{(32)(-1) - (2)(4)}{(10)(-1) - (2)(1)} = \frac{40}{12} = \mathbf{3.33\ A}$$

and

$$I_2 = I_1 - I = 3.33\text{ A} - 4\text{ A} = \mathbf{-0.67\ A}$$

In the above analysis, it may appear that when the current source was removed, $I_1 = I_2$. However, the supermesh approach requires that we stick with the original definition of each mesh current and not alter those definitions when current sources are removed.

EXAMPLE 8.15 Using mesh analysis, determine the currents for the network in Fig. 8.34.

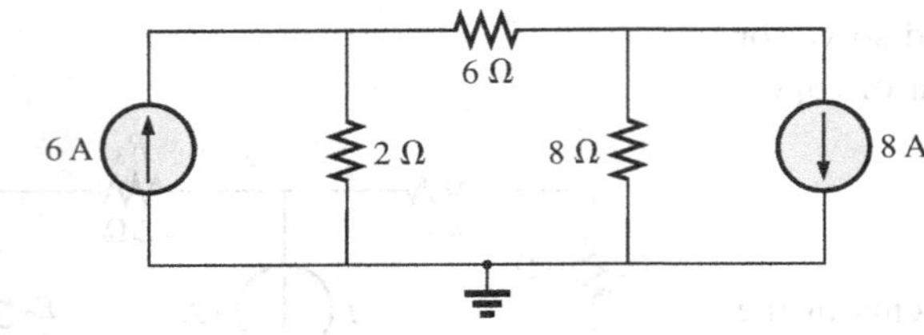

FIG. 8.34
Example 8.15.

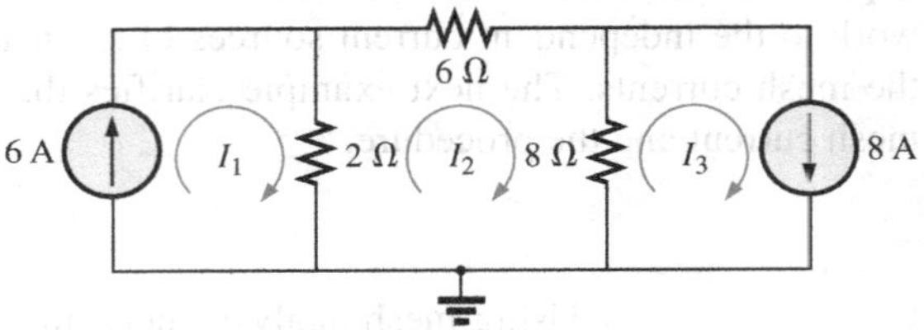

FIG. 8.35
Defining the mesh currents for the network in Fig. 8.34.

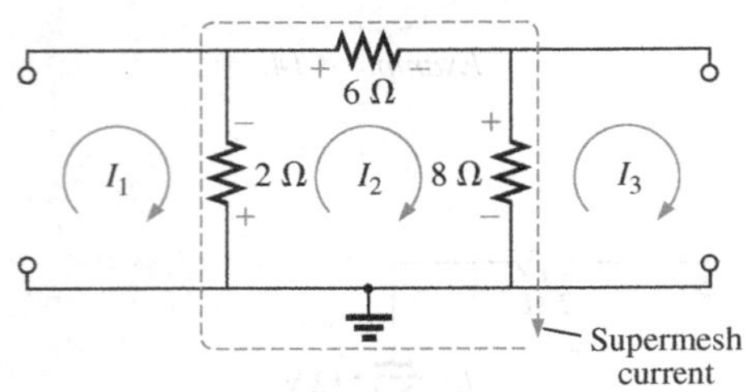

FIG. 8.36
Defining the supermesh current for the network in Fig. 8.34.

Solution: The mesh currents are defined in Fig. 8.35. The current sources are removed, and the single supermesh path is defined in Fig. 8.36.

Applying Kirchhoff's voltage law around the supermesh path gives

$$-V_{2\Omega} - V_{6\Omega} - V_{8\Omega} = 0$$
$$-(I_2 - I_1)2\ \Omega - I_2(6\ \Omega) - (I_2 - I_3)8\ \Omega = 0$$
$$-2I_2 + 2I_1 - 6I_2 - 8I_2 + 8I_3 = 0$$
$$2I_1 - 16I_2 + 8I_3 = 0$$

Introducing the relationship between the mesh currents and the current sources

$$I_1 = 6\text{ A}$$
$$I_3 = 8\text{ A}$$

results in the following solutions:

$$2I_1 - 16I_2 + 8I_3 = 0$$
$$2(6\text{ A}) - 16I_2 + 8(8\text{ A}) = 0$$

and $$I_2 = \frac{76\text{ A}}{16} = \mathbf{4.75\ A}$$

Then $$I_{2\Omega}\downarrow = I_1 - I_2 = 6\text{ A} - 4.75\text{ A} = \mathbf{1.25\ A}$$

and $$I_{8\Omega}\uparrow = I_3 - I_2 = 8\text{ A} - 4.75\text{ A} = \mathbf{3.25\ A}$$

Again, note that you must stick with your original definitions of the various mesh currents when applying Kirchhoff's voltage law around the resulting supermesh paths.

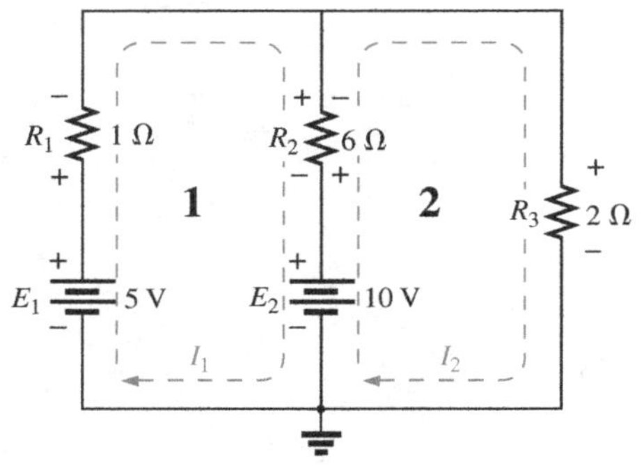

FIG. 8.37
Network in Fig. 8.29 redrawn with assigned loop currents.

8.8 MESH ANALYSIS (FORMAT APPROACH)

Now that the basis for the mesh-analysis approach has been established, we now examine a technique for writing the mesh equations more rapidly and usually with fewer errors. As an aid in introducing the procedure, the network in Example 8.12 (Fig. 8.29) has been redrawn in Fig. 8.37 with the assigned loop currents. (Note that each loop current has a clockwise direction.)

The equations obtained are

$$-7I_1 + 6I_2 = 5$$
$$6I_1 - 8I_2 = -10$$

which can also be written as

$$7I_1 - 6I_2 = -5$$
$$8I_2 - 9I_1 = 10$$

and expanded as

Col. 1		Col. 2	Col. 3
$(1 + 6)I_1$	$-$	$6I_2$	$= (5 - 10)$
$(2 + 6)I_1$	$-$	$6I_2$	$= 10$

Note in the above equations that column 1 is composed of a loop current times the sum of the resistors through which that loop current passes. Column 2 is the product of the resistors common to another loop current times that other loop current. Note that in each equation, this column is subtracted from column 1. Column 3 is the *algebraic* sum of the voltage sources through which the loop current of interest passes. A source is assigned a positive sign if the loop current passes from the negative to the positive terminal, and a negative value is assigned if the polarities are reversed. The comments above are correct only for a standard direction of loop current in each window, the one chosen being the clockwise direction.

The above statements can be extended to develop the following *format approach* to mesh analysis.

Mesh Analysis Procedure

1. *Assign a loop current to each independent, closed loop (as in the previous section) in a clockwise direction.*
2. *The number of required equations is equal to the number of chosen independent, closed loops. Column 1 of each equation is formed by summing the resistance values of those resistors through which the loop current of interest passes and multiplying the result by that loop current.*
3. *We must now consider the mutual terms, which, as noted in the examples above, are always subtracted from the first column. A mutual term is simply any resistive element having an additional loop current passing through it. It is possible to have more than one mutual term if the loop current of interest has an element in common with more than one other loop current. This will be demonstrated in an example to follow. Each term is the product of the mutual resistor and the other loop current passing through the same element.*
4. *The column to the right of the equality sign is the algebraic sum of the voltage sources through which the loop current of interest passes. Positive signs are assigned to those sources of voltage having a polarity such that the loop current passes from the negative to the positive terminal. A negative sign is assigned to those potentials for which the reverse is true.*
5. *Solve the resulting simultaneous equations for the desired loop currents.*

Before considering a few examples, be aware that since the column to the right of the equals sign is the algebraic sum of the voltage sources in that loop, *the format approach can be applied only to networks in which all current sources have been converted to their equivalent voltage source.*

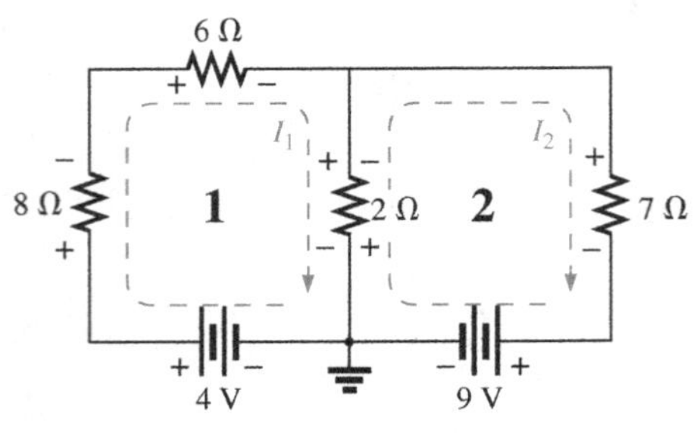

FIG. 8.38
Example 8.16.

EXAMPLE 8.16 Write the mesh equations for the network in Fig. 8.38, and find the current through the 7 Ω resistor.

Solution:

Step 1: As indicated in Fig. 8.38, each assigned loop current has a clockwise direction.

Steps 2 to 4:

$$I_1: \quad (8\ \Omega + 6\ \Omega + 2\ \Omega)I_1 - (2\ \Omega)I_2 = 4\text{ V}$$
$$I_2: \quad (7\ \Omega + 2\ \Omega)I_2 - (2\ \Omega)I_1 = -9\text{ V}$$

and

$$16I_1 - 2I_2 = 4$$
$$9I_2 - 2I_1 = -9$$

which, for determinants, are

$$16I_1 - 2I_2 = 4$$
$$-2I_1 + 9I_2 = -9$$

and

$$I_2 = I_{7\Omega} = \frac{\begin{vmatrix} 16 & 4 \\ -2 & -9 \end{vmatrix}}{\begin{vmatrix} 16 & -2 \\ -2 & 9 \end{vmatrix}} = \frac{-144 + 8}{144 - 4} = \frac{-136}{140}$$
$$= \mathbf{-0.97\ A}$$

EXAMPLE 8.17 Write the mesh equations for the network in Fig. 8.39.

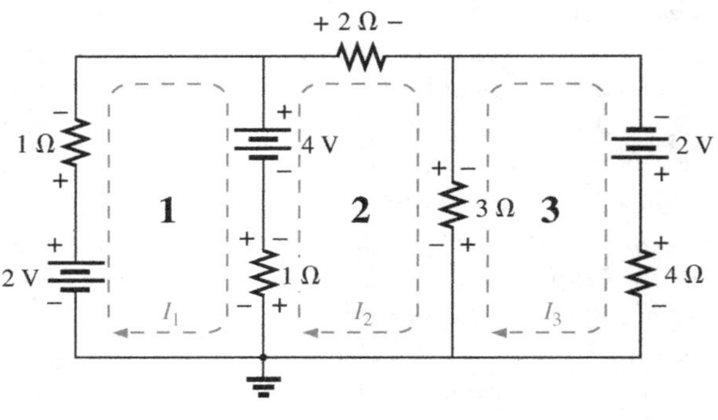

FIG. 8.39
Example 8.17.

Solution: Each window is assigned a loop current in the clockwise direction:

I_1 does not pass through an element mutual with I_3.

$$I_1: \quad (1\ \Omega + 1\ \Omega)I_1 - (1\ \Omega)I_2 + 0 = 2\text{ V} - 4\text{ V}$$
$$I_2: \quad (1\ \Omega + 2\ \Omega + 3\ \Omega)I_2 - (1\ \Omega)I_1 - (3\ \Omega)I_3 = 4\text{ V}$$
$$I_3: \quad (3\ \Omega + 4\ \Omega)I_3 - (3\ \Omega)I_2 + 0 = 2\text{ V}$$

I_3 does not pass through an element mutual with I_1.

Summing terms yields

$$\begin{aligned} 2I_1 - I_2 + 0 &= -2 \\ 6I_2 - I_1 - 3I_3 &= 4 \\ 7I_3 - 3I_2 + 0 &= 2 \end{aligned}$$

which are rewritten for determinants as

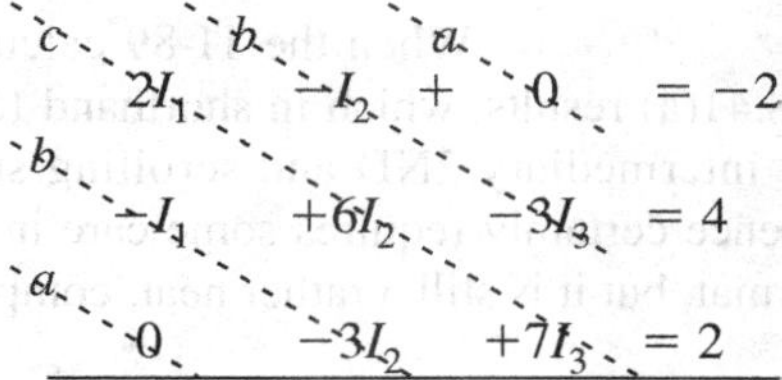

Note that the coefficients of the *a* and *b* diagonals are equal. This *symmetry* about the *c*-axis will always be true for equations written using the format approach. It is a check on whether the equations were obtained correctly.

We now consider a network with only one source of voltage to point out that mesh analysis can be used to advantage in other than multisource networks.

EXAMPLE 8.18 Find the current through the 10 Ω resistor of the network in Fig. 8.40.

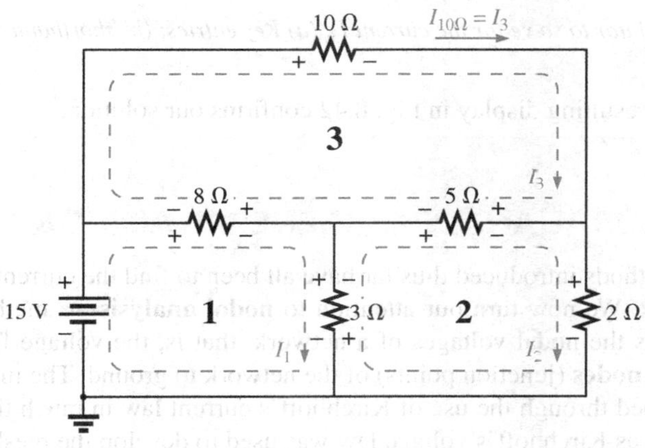

FIG. 8.40
Example 8.18.

Solution:

$$\begin{aligned} I_1&: \quad (8\ \Omega + 3\ \Omega)I_1 - (8\ \Omega)I_3 - (3\ \Omega)I_2 = 15\ \text{V} \\ I_2&: \quad (3\ \Omega + 5\ \Omega + 2\ \Omega)I_2 - (3\ \Omega)I_1 - (5\ \Omega)I_3 = 0 \\ I_3&: \quad (8\ \Omega + 10\ \Omega + 5\ \Omega)I_3 - (8\ \Omega)I_1 - (5\ \Omega)I_2 = 0 \end{aligned}$$

$$\begin{aligned} 11I_1 - 8I_3 - 3I_2 &= 15\ \text{V} \\ 10I_2 - 3I_1 - 5I_3 &= 0 \\ 23I_3 - 8I_1 - 5I_2 &= 0 \end{aligned}$$

or

$$\begin{aligned} 11I_1 - 3I_2 - 8I_3 &= 15\ \text{V} \\ -3I_1 + 10I_2 - 5I_3 &= 0 \\ -8I_1 - 5I_2 + 23I_3 &= 0 \end{aligned}$$

and
$$I_3 = I_{10\Omega} = \frac{\begin{vmatrix} 11 & -3 & 15 \\ -3 & 10 & 0 \\ -8 & -5 & 0 \end{vmatrix}}{\begin{vmatrix} 11 & -3 & -8 \\ -3 & 10 & -5 \\ -8 & -5 & 23 \end{vmatrix}} = \mathbf{1.22\ A}$$

TI-89 Calculator Solution: When the TI-89 calculator is used, the sequence in Fig. 8.41(a) results, which in shorthand form appears as in Fig. 8.41(b). The intermediary 2ND and scrolling steps were not included. This sequence certainly requires some care in entering the data in the required format, but it is still a rather neat, compact format.

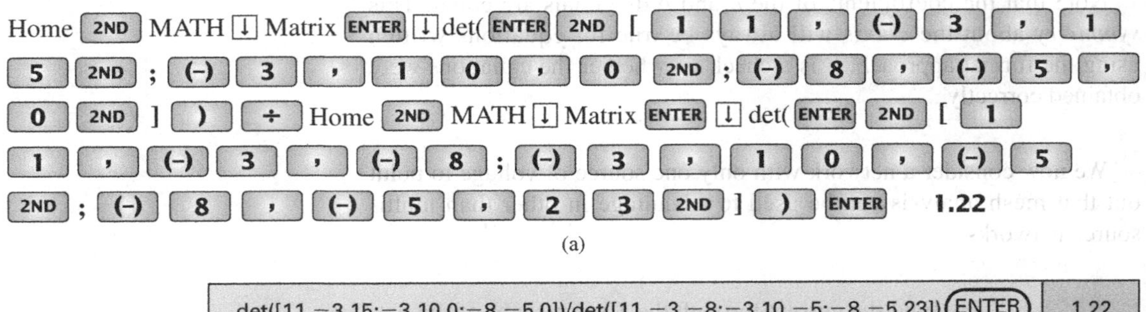

(a)

det([11,−3,15;−3,10,0;−8,−5,0])/det([11,−3,−8;−3,10,−5;−8,−5,23]) ENTER | 1.22

FIG. 8.41
Using the TI-89 calculator to solve for the current I_3. (a) Key entries; (b) shorthand form.

$$\frac{\det\left(\begin{bmatrix} 11 & -3 & 15 \\ -3 & 10 & 0 \\ -8 & -5 & 0 \end{bmatrix}\right)}{\det\left(\begin{bmatrix} 11 & -3 & -8 \\ -3 & 10 & -5 \\ -8 & -5 & 23 \end{bmatrix}\right)} \qquad \mathbf{1.22E0}$$

FIG. 8.42
The resulting display after properly entering the data for the current I_3.

The resulting display in Fig. 8.42 confirms our solution.

8.9 NODAL ANALYSIS (GENERAL APPROACH)

The methods introduced thus far have all been to find the currents of the network. We now turn our attention to **nodal analysis**—a method that provides the nodal voltages of a network, that is, the voltage from the various **nodes** (junction points) of the network to ground. The method is developed through the use of Kirchhoff's current law in much the same manner as Kirchhoff's voltage law was used to develop the mesh analysis approach.

Although it is not a requirement, we make it a policy to make ground our reference node and assign it a potential level of zero volts. All the other voltage levels are then found with respect to this reference level. For a network of N nodes, by assigning one as our reference node, we have $(N - 1)$ nodes for which the voltage must be determined. In other words,

the number of nodes for which the voltage must be determined using nodal analysis is 1 less than the total number of nodes.

The result of the above is $(N - 1)$ nodal voltages that need to be determined, requiring that $(N - 1)$ independent equations be written to find the nodal voltages. In other words,

the number of equations required to solve for all the nodal voltages of a network is 1 less than the total number of independent nodes.

Since each equation is the result of an application of Kirchhoff's current law, Kirchhoff's current law must be applied $(N - 1)$ times for each network.

Nodal analysis, like mesh analysis, can be applied by a series of carefully defined steps. The examples to follow explain each step in detail.

Nodal Analysis Procedure

1. *Determine the number of nodes within the network.*
2. *Pick a reference node, and label each remaining node with a subscripted value of voltage: V_1, V_2, and so on.*
3. *Apply Kirchhoff's current law at each node except the reference. Assume that all unknown currents leave the node for each application of Kirchhoff's current law. In other words, for each node, don't be influenced by the direction that an unknown current for another node may have had. Each node is to be treated as a separate entity, independent of the application of Kirchhoff's current law to the other nodes.*
4. *Solve the resulting equations for the nodal voltages.*

A few examples clarify the procedure defined by step 3. It initially takes some practice writing the equations for Kirchhoff's current law correctly, but in time the advantage of assuming that all the currents leave a node rather than identifying a specific direction for each branch becomes obvious. (The same type of advantage is associated with assuming that all the mesh currents are clockwise when applying mesh analysis.)

As with mesh and branch-current analysis, a number of networks to be encountered in this section can be solved using a simple source conversion. In Example 8.19, for instance, the network of Fig. 8.43 can be easily solved by converting the voltage source to a current source and combining the parallel current sources. However, as noted for mesh and branch-current analysis, this method can also be applied to more complex networks where a source conversion is not possible.

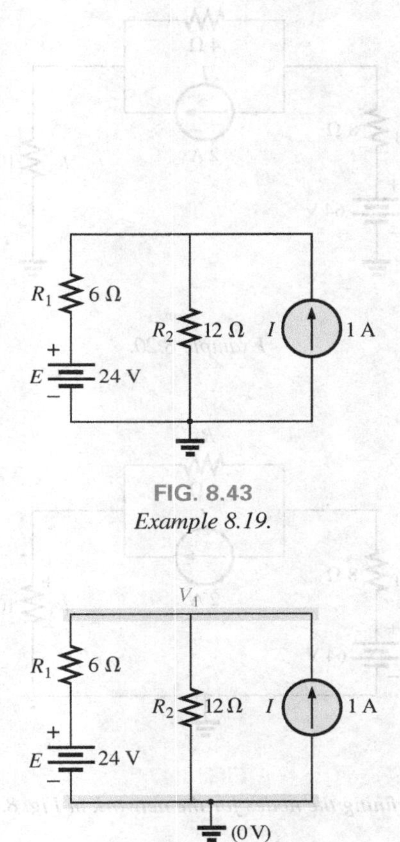

FIG. 8.43
Example 8.19.

FIG. 8.44
Network in Fig. 8.43 with assigned nodes.

EXAMPLE 8.19 Apply nodal analysis to the network in Fig. 8.43.

Solution:

Steps 1 and 2: The network has two nodes, as shown in Fig. 8.44. The lower node is defined as the reference node at ground potential (zero volts), and the other node as V_1, the voltage from node 1 to ground.

Step 3: I_1 and I_2 are defined as leaving the node in Fig 8.45, and Kirchhoff's current law is applied as follows:

$$I = I_2 + I_2$$

The current I_2 is related to the nodal voltage V_1 by Ohm's law:

$$I_2 = \frac{V_{R_2}}{R_2} = \frac{V_1}{R_2}$$

The current I_1 is also determined by Ohm's law as follows:

$$I_1 = \frac{V_{R_1}}{R_1}$$

with
$$V_{R_1} = V_1 - E$$

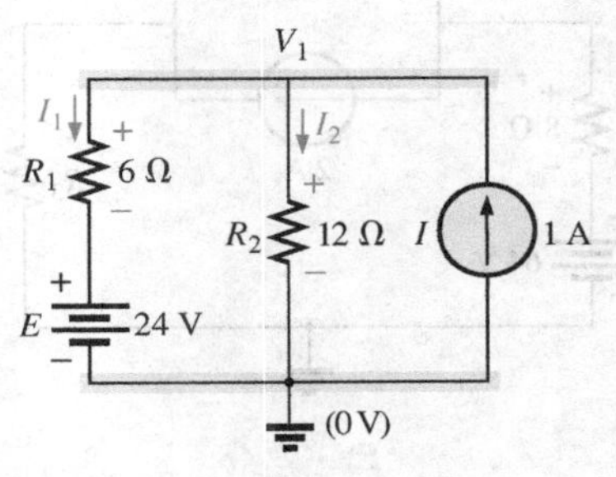

FIG. 8.45
Applying Kirchhoff's current law to the node V_1.

Substituting into the Kirchhoff's current law equation

$$I = \frac{V_1 - E}{R_1} + \frac{V_1}{R_2}$$

and rearranging, we have

$$I = \frac{V_1}{R_1} - \frac{E}{R_1} + \frac{V_1}{R_2} = V_1\left(\frac{1}{R_1} + \frac{1}{R_2}\right) - \frac{E}{R_1}$$

or

$$V_1\left(\frac{1}{R_1} + \frac{1}{R_2}\right) = \frac{E}{R_1} + 1$$

Substituting numerical values, we obtain

$$V_1\left(\frac{1}{6\ \Omega} + \frac{1}{12\ \Omega}\right) = \frac{24\ \text{V}}{6\ \Omega} + 1\ \text{A} = 4\ \text{A} + 1\ \text{A}$$

$$V_1\left(\frac{1}{4\ \Omega}\right) = 5\ \text{A}$$

$$V_1 = \mathbf{20\ V}$$

The currents I_1 and I_2 can then be determined by using the preceding equations:

$$I_1 = \frac{V_1 - E}{R_1} = \frac{20\ \text{V} - 24\ \text{V}}{6\ \Omega} = \frac{-4\ \text{V}}{6\ \Omega}$$

$$= \mathbf{-0.67\ A}$$

The minus sign indicates that the current I_1 has a direction opposite to that appearing in Fig. 8.45. In addition,

$$I_2 = \frac{V_1}{R_2} = \frac{20\ \text{V}}{12\ \Omega} = \mathbf{1.67\ A}$$

EXAMPLE 8.20 Apply nodal analysis to the network in Fig. 8.46.

Solution:

Steps 1 and 2: The network has three nodes, as defined in Fig. 8.47, with the bottom node again defined as the reference node (at ground potential, or zero volts), and the other nodes as V_1 and V_2.

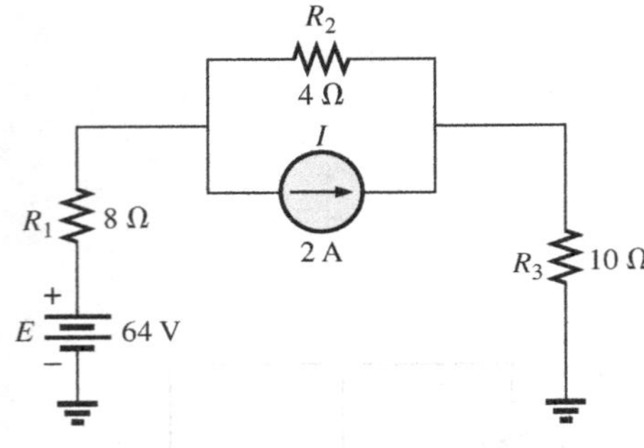

FIG. 8.46
Example 8.20.

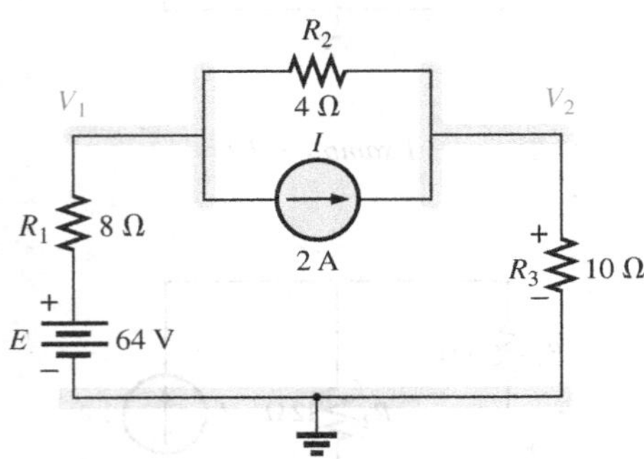

FIG. 8.47
Defining the nodes for the network in Fig. 8.46.

Step 2: For node V_1, the currents are defined as shown in Fig. 8.48, and Kirchhoff's current law is applied:

$$0 = I_1 + I_2 + I$$

with

$$I_1 = \frac{V_1 - E}{R_1}$$

and

$$I_2 = \frac{V_{R_2}}{R_2} = \frac{V_1 - V_2}{R_2}$$

so that

$$\frac{V_1 - E}{R_1} + \frac{V_1 - V_2}{R_2} + I = 0$$

or

$$\frac{V_1}{R_1} - \frac{E}{R_1} + \frac{V_1}{R_2} - \frac{V_2}{R_2} + I = 0$$

and

$$V_1\left(\frac{1}{R_1} + \frac{1}{R_2}\right) - V_2\left(\frac{1}{R_2}\right) = -I + \frac{E}{R_1}$$

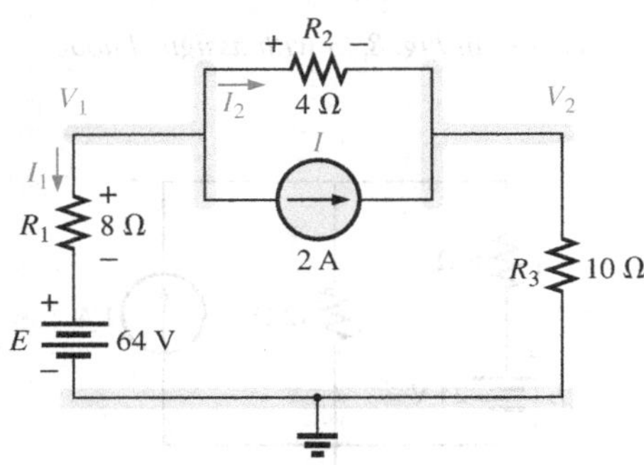

FIG. 8.48
Applying Kirchhoff's current law to node V_1.

Substituting values gives

$$V_1\left(\frac{1}{8\ \Omega}+\frac{1}{4\ \Omega}\right)-V_2\left(\frac{1}{4\ \Omega}\right)=-2\text{ A}+\frac{64\text{ V}}{8\ \Omega}=6\text{ A}$$

For node V_2, the currents are defined as shown in Fig. 8.49, and Kirchhoff's current law is applied:

$$I=I_2+I_3$$

with

$$I=\frac{V_2-V_1}{R_2}+\frac{V_2}{R_3}$$

or

$$I=\frac{V_2}{R_2}-\frac{V_1}{R_2}+\frac{V_2}{R_3}$$

and

$$V_2\left(\frac{1}{R_2}+\frac{1}{R_3}\right)-V_1\left(\frac{1}{R_2}\right)=I$$

Substituting values gives

$$V_2\left(\frac{1}{4\ \Omega}+\frac{1}{10\ \Omega}\right)-V_1\left(\frac{1}{4\ \Omega}\right)=2\text{ A}$$

Step 4: The result is two equations and two unknowns:

$$V_1\left(\frac{1}{8\ \Omega}+\frac{1}{4\ \Omega}\right)-V_2\left(\frac{1}{4\ \Omega}\right)=6\text{ A}$$
$$-V_1\left(\frac{1}{4\ \Omega}\right)+V_2\left(\frac{1}{4\ \Omega}+\frac{1}{10\ \Omega}\right)=2\text{ A}$$

which become

$$0.375V_1-0.25V_2=6$$
$$-0.25V_1+0.35V_2=2$$

Using determinants, we obtain

$$V_1=\mathbf{37.82\ V}$$
$$V_2=\mathbf{32.73\ V}$$

Since E is greater than V_1, the current I_1 flows from ground to V_1 and is equal to

$$I_{R_1}=\frac{E-V_1}{R_1}=\frac{64\text{ V}-37.82\text{ V}}{8\ \Omega}=\mathbf{3.27\ A}$$

The positive value for V_2 results in a current I_{R_3} from node V_2 to ground equal to

$$I_{R_3}=\frac{V_{R_3}}{R_3}=\frac{V_2}{R_3}=\frac{32.73\text{ V}}{10\ \Omega}=\mathbf{3.27\ A}$$

Since V_1 is greater than V_2, the current I_{R_2} flows from V_1 to V_2 and is equal to

$$I_{R_2}=\frac{V_1-V_2}{R_2}=\frac{37.82\text{ V}-32.73\text{ V}}{4\ \Omega}=\mathbf{1.27\ A}$$

The results of $V_1 = 37.82$ V and $V_2 = 32.73$ V confirm the theoretical solution.

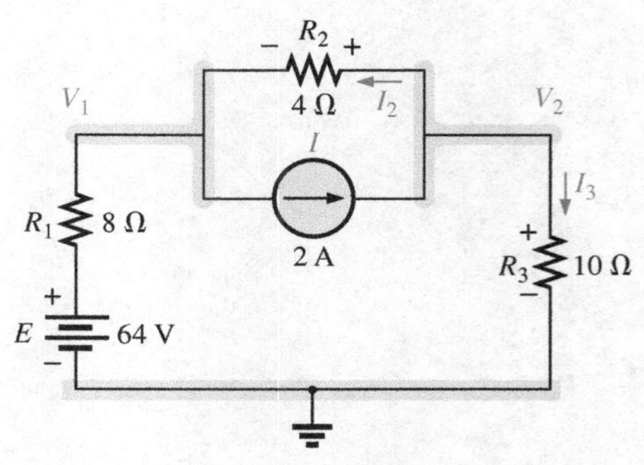

FIG. 8.49
Applying Kirchhoff's current law to node V_2.

EXAMPLE 8.21 Determine the nodal voltages for the network in Fig. 8.50.

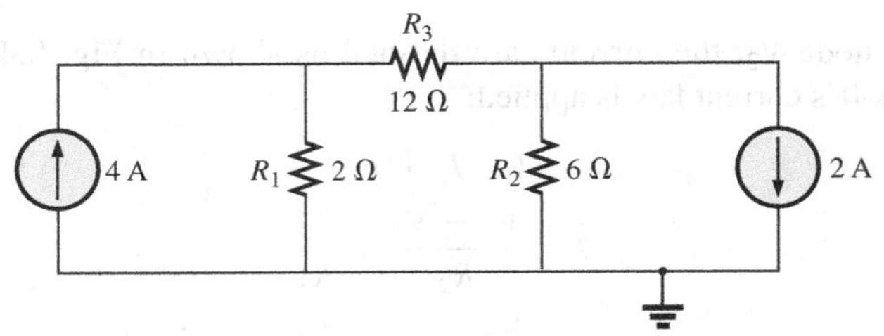

FIG. 8.50
Example 8.21.

Solution:

Steps 1 and 2: As indicated in Fig. 8.51:

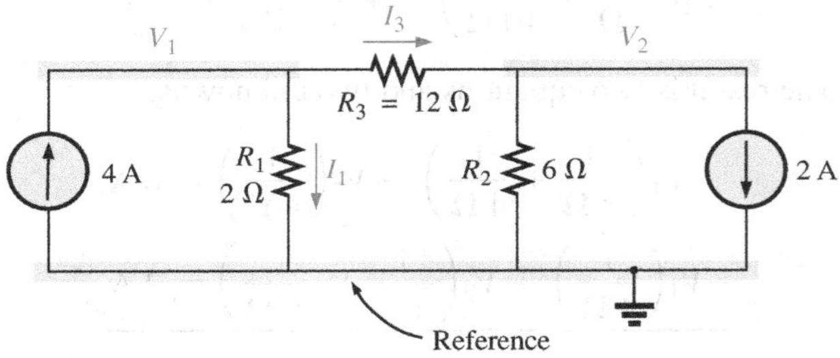

FIG. 8.51
Defining the nodes and applying Kirchhoff's current law to the node V_1.

Step 3: Included in Fig. 8.51 for the node V_1. Applying Kirchhoff's current law gives

$$4 \text{ A} = I_1 + I_3$$

and
$$4 \text{ A} = \frac{V_1}{R_1} + \frac{V_1 - V_2}{R_3} = \frac{V_1}{2\,\Omega} + \frac{V_1 - V_2}{12\,\Omega}$$

Expanding and rearranging gives

$$V_1\left(\frac{1}{2\,\Omega} + \frac{1}{12\,\Omega}\right) - V_2\left(\frac{1}{12\,\Omega}\right) = 4 \text{ A}$$

For node V_2, the currents are defined as in Fig. 8.52.

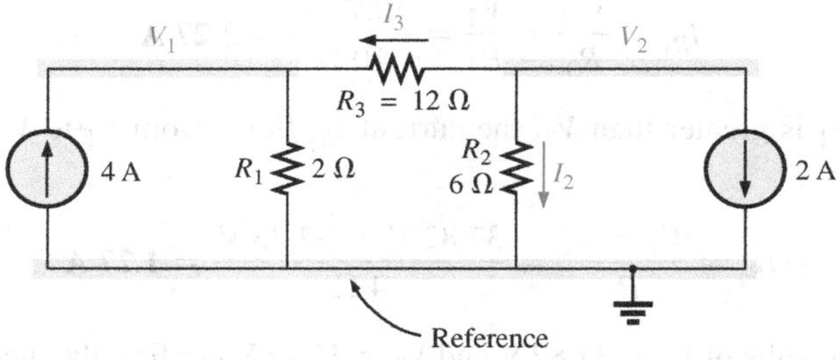

FIG. 8.52
Applying Kirchhoff's current law to the node V_2.

Applying Kirchhoff's current law gives

$$0 = I_3 + I_2 + 2\text{ A}$$

and $\dfrac{V_2 - V_1}{R_3} + \dfrac{V_2}{R_2} + 2\text{ A} = 0 \longrightarrow \dfrac{V_2 - V_1}{12\ \Omega} + \dfrac{V_2}{2\ \Omega} + 2\text{ A} = 0$

Expanding and rearranging gives

$$V_2\left(\frac{1}{12\ \Omega} + \frac{1}{6\ \Omega}\right) - V_1\left(\frac{1}{12\ \Omega}\right) = -2\text{ A}$$

resulting in the following two equations and two unknowns:

$$\left.\begin{aligned} V_1\left(\frac{1}{2\ \Omega} + \frac{1}{12\ \Omega}\right) - V_2\left(\frac{1}{12\ \Omega}\right) &= +4\text{ A} \\ V_2\left(\frac{1}{12\ \Omega} + \frac{1}{6\ \Omega}\right) - V_1\left(\frac{1}{12\ \Omega}\right) &= -2\text{ A} \end{aligned}\right\} \quad \textbf{(8.1)}$$

producing

$$\left.\begin{aligned} \frac{7}{12}V_1 - \frac{1}{12}V_2 &= +4 \\ -\frac{1}{12}V_1 + \frac{3}{12}V_2 &= -2 \end{aligned}\right\} \quad \begin{aligned} 7V_1 - \ \ V_2 &= 48 \\ -1V_1 + 3V_2 &= -24 \end{aligned}$$

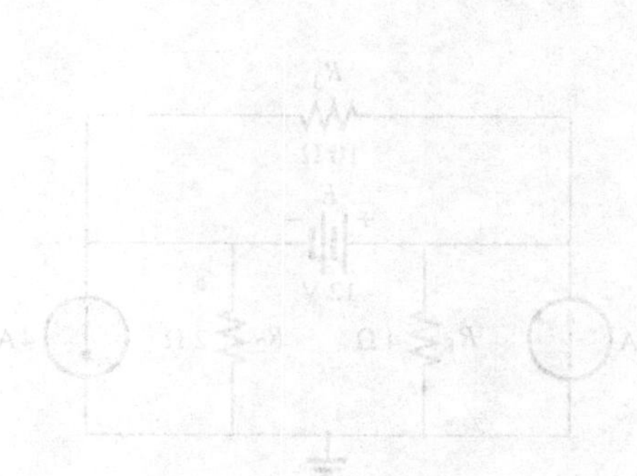

and

$$V_1 = \frac{\begin{vmatrix} 48 & -1 \\ -24 & 3 \end{vmatrix}}{\begin{vmatrix} 7 & -1 \\ -1 & 3 \end{vmatrix}} = \frac{120}{20} = +\mathbf{6\ V}$$

$$V_2 = \frac{\begin{vmatrix} 7 & 48 \\ -1 & -24 \end{vmatrix}}{20} = \frac{-120}{20} = -\mathbf{6\ V}$$

Since V_1 is greater than V_2, the current through R_3 passes from V_1 to V_2. Its value is

$$I_{R_3} = \frac{V_1 - V_2}{R_3} = \frac{6\text{ V} - (-6\text{ V})}{12\ \Omega} = \frac{12\text{ V}}{12\ \Omega} = \mathbf{1\ A}$$

The fact that V_1 is positive results in a current I_{R_1} from V_1 to ground equal to

$$I_{R_1} = \frac{V_{R_1}}{R_1} = \frac{V_1}{R_1} = \frac{6\text{ V}}{2\ \Omega} = \mathbf{3\ A}$$

Finally, since V_2 is negative, the current I_{R_2} flows from ground to V_2 and is equal to

$$I_{R_2} = \frac{V_{R_2}}{R_2} = \frac{V_2}{R_2} = \frac{6\text{ V}}{6\ \Omega} = \mathbf{1\ A}$$

Supernode

Occasionally, you may encounter voltage sources in a network that do not have a series internal resistance that would permit a conversion to a current source. In such cases, you have two options.

The simplest and most direct approach is to *place a resistor in series with the source of a very small value compared to the other resistive*

elements of the network. For instance, if most of the resistors are 10 Ω or larger, placing a 1 Ω resistor in series with a voltage source provides one level of accuracy for your answer. However, choosing a resistor of 0.1 Ω or less increases the accuracy of your answer. You will never get an exact answer because the network has been modified by the introduced element. However, for most applications, the accuracy will be sufficiently high.

The other approach is to use the **supernode approach** described below. This approach provides an exact solution but requires some practice to become proficient.

Start as usual and assign a nodal voltage to each independent node of the network, including each independent voltage source as if it were a resistor or current source. Then mentally replace the independent voltage sources with short-circuit equivalents, and apply Kirchhoff's current law to the defined nodes of the network. Any node including the effect of elements tied only to *other* nodes is referred to as a *supernode* (since it has an additional number of terms). Finally, relate the defined nodes to the independent voltage sources of the network, and solve for the nodal voltages. The next example clarifies the definition of *supernode.*

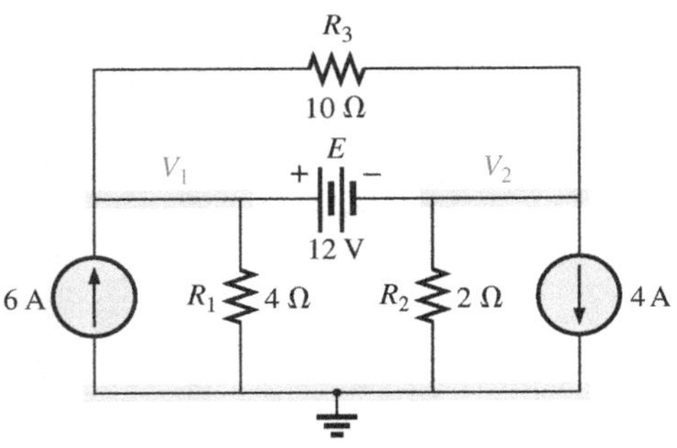

FIG. 8.53
Example 8.22.

EXAMPLE 8.22 Determine the nodal voltages V_1 and V_2 in Fig. 8.53 using the concept of a supernode.

Solution: Replacing the independent voltage source of 12 V with a short-circuit equivalent results in the network in Fig. 8.54. Even though the mental application of a short-circuit equivalent is discussed above, it would be wise in the early stage of development to redraw the network as shown in Fig. 8.54. The result is a single supernode for which Kirchhoff's current law must be applied. Be sure to leave the other defined nodes in place, and use them to define the currents from that region of the network. In particular, note that the current I_3 leaves the supernode at V_1 and then enters the same supernode at V_2. It must therefore appear twice when applying Kirchhoff's current law, as shown below:

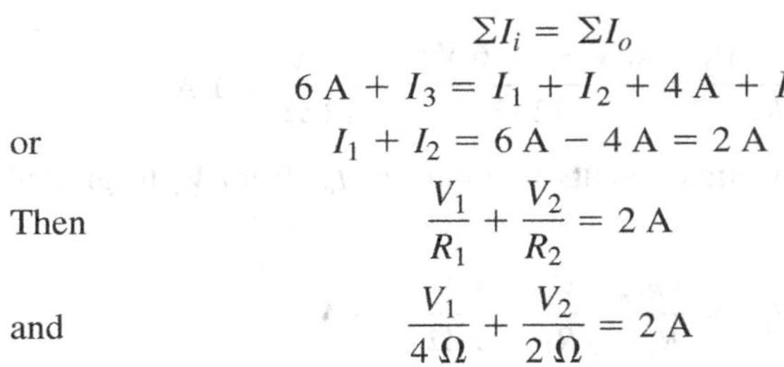

$$\Sigma I_i = \Sigma I_o$$

$$6\text{ A} + I_3 = I_1 + I_2 + 4\text{ A} + I_3$$

or

$$I_1 + I_2 = 6\text{ A} - 4\text{ A} = 2\text{ A}$$

Then

$$\frac{V_1}{R_1} + \frac{V_2}{R_2} = 2\text{ A}$$

and

$$\frac{V_1}{4\,\Omega} + \frac{V_2}{2\,\Omega} = 2\text{ A}$$

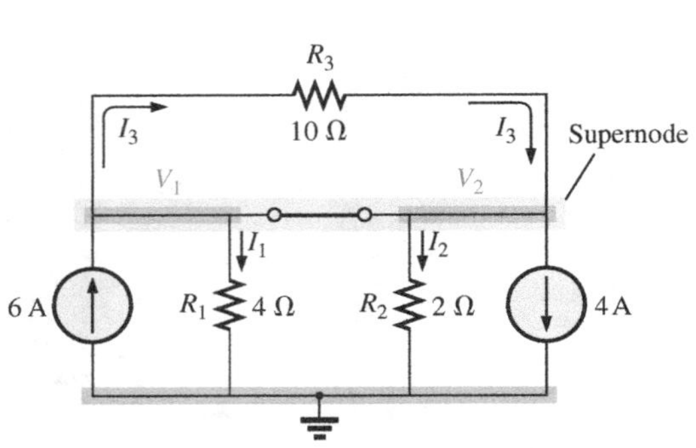

FIG. 8.54
Defining the supernode for the network in Fig. 8.53.

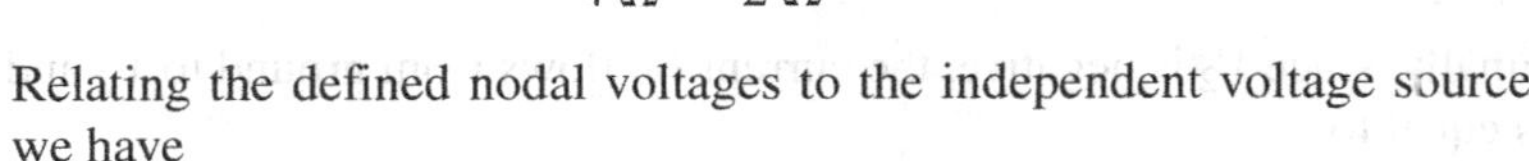

Relating the defined nodal voltages to the independent voltage source, we have

$$V_1 - V_2 = E = 12\text{ V}$$

which results in two equations and two unknowns:

$$\begin{aligned} 0.25V_1 + 0.5V_2 &= 2 \\ V_1 - 1V_2 &= 12 \end{aligned}$$

Substituting gives

$$V_1 = V_2 + 12$$

$$0.25(V_2 + 12) + 0.5V_2 = 2$$

and

$$0.75V_2 = 2 - 3 = -1$$

so that $$V_2 = \frac{-1}{0.75} = \mathbf{-1.33\ V}$$

and $$V_1 = V_2 + 12\text{ V} = -1.33\text{ V} + 12\text{ V} = \mathbf{-10.67\ V}$$

The current of the network can then be determined as follows:

$$I_1\downarrow = \frac{V}{R_1} = \frac{10.67\text{ V}}{4\ \Omega} = \mathbf{2.67\ A}$$

$$I_2\uparrow = \frac{V_2}{R_2} = \frac{1.33\text{ V}}{2\ \Omega} = \mathbf{0.67\ A}$$

$$\underset{\rightarrow}{I_3} = \frac{V_1 - V_2}{10\ \Omega} = \frac{10.67\text{ V} - (-1.33\text{ V})}{10\ \Omega} = \frac{12\ \Omega}{10\ \Omega} = \mathbf{1.2\ A}$$

A careful examination of the network at the beginning of the analysis would have revealed that the voltage across the resistor R_3 must be 12 V and I_3 must be equal to 1.2 A.

8.10 NODAL ANALYSIS (FORMAT APPROACH)

A close examination of Eq. (8.1) appearing in Example 8.21 reveals that the subscripted voltage at the node in which Kirchhoff's current law is applied is multiplied by the sum of the conductances attached to that node. Note also that the other nodal voltages within the same equation are multiplied by the negative of the conductance between the two nodes. The current sources are represented to the right of the equals sign with a positive sign if they supply current to the node and with a negative sign if they draw current from the node.

These conclusions can be expanded to include networks with any number of nodes. This allows us to write nodal equations rapidly and in a form that is convenient for the use of determinants. A major requirement, however, is that *all voltage sources must first be converted to current sources before the procedure is applied.* Note the parallelism between the following four steps of application and those required for mesh analysis in Section 8.8.

Nodal Analysis Procedure

1. *Choose a reference node, and assign a subscripted voltage label to the (N − 1) remaining nodes of the network.*
2. *The number of equations required for a complete solution is equal to the number of subscripted voltages (N − 1). Column 1 of each equation is formed by summing the conductances tied to the node of interest and multiplying the result by that subscripted nodal voltage.*
3. *We must now consider the mutual terms, which, as noted in the preceding example, are always subtracted from the first column. It is possible to have more than one mutual term if the nodal voltage of current interest has an element in common with more than one other nodal voltage. This is demonstrated in an example to follow. Each mutual term is the product of the mutual conductance and the other nodal voltage, tied to that conductance.*
4. *The column to the right of the equality sign is the algebraic sum of the current sources tied to the node of interest. A current source is*

assigned a positive sign if it supplies current to a node and a negative sign if it draws current from the node.

5. *Solve the resulting simultaneous equations for the desired voltages.*

Let us now consider a few examples.

EXAMPLE 8.23 Write the nodal equations for the network in Fig. 8.55.

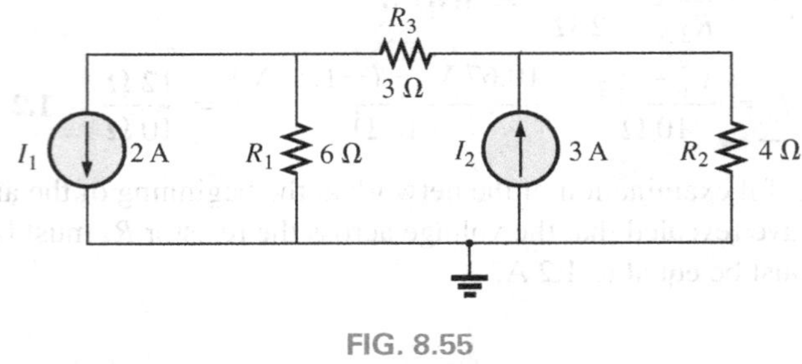

FIG. 8.55
Example 8.23.

Solution:

Step 1: Redraw the figure with assigned subscripted voltages in Fig. 8.56.

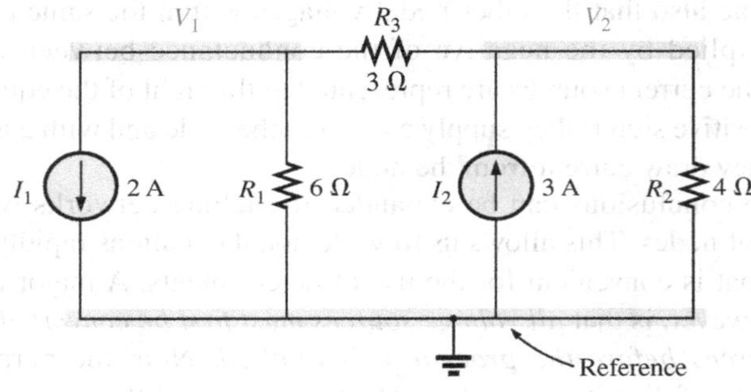

FIG. 8.56
Defining the nodes for the network in Fig. 8.55.

Steps 2 to 4:

Drawing current from node 1

$$V_1: \quad \underbrace{\left(\frac{1}{6\,\Omega} + \frac{1}{3\,\Omega}\right)}_{\substack{\text{Sum of}\\\text{conductances}\\\text{connected}\\\text{to node 1}}} V_1 - \underbrace{\left(\frac{1}{3\,\Omega}\right)}_{\substack{\text{Mutual}\\\text{conductance}}} V_2 = -2\text{ A}$$

Supplying current to node 2

$$V_2: \quad \underbrace{\left(\frac{1}{4\,\Omega} + \frac{1}{3\,\Omega}\right)}_{\substack{\text{Sum of}\\\text{conductances}\\\text{connected}\\\text{to node 2}}} V_2 - \underbrace{\left(\frac{1}{3\,\Omega}\right)}_{\substack{\text{Mutual}\\\text{conductance}}} V_1 = +3\text{ A}$$

and
$$\frac{1}{2}V_1 - \frac{1}{3}V_2 = -2$$
$$-\frac{1}{3}V_1 + \frac{7}{12}V_2 = 3$$

EXAMPLE 8.24 Find the voltage across the 3 Ω resistor in Fig. 8.57 by nodal analysis.

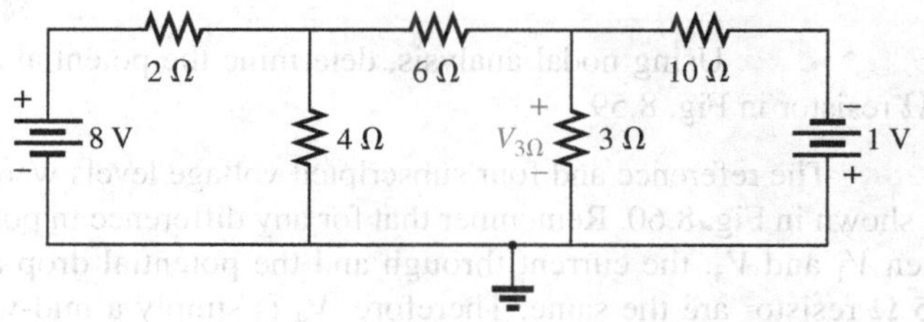

FIG. 8.57
Example 8.24.

Solution: Converting sources and choosing nodes (Fig. 8.58), we have

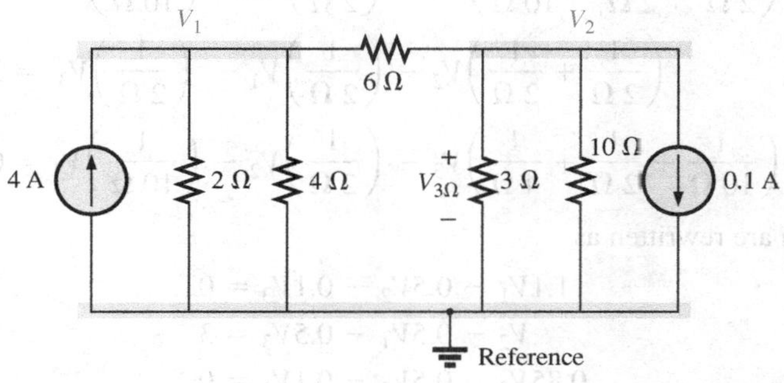

FIG. 8.58
Defining the nodes for the network in Fig. 8.57.

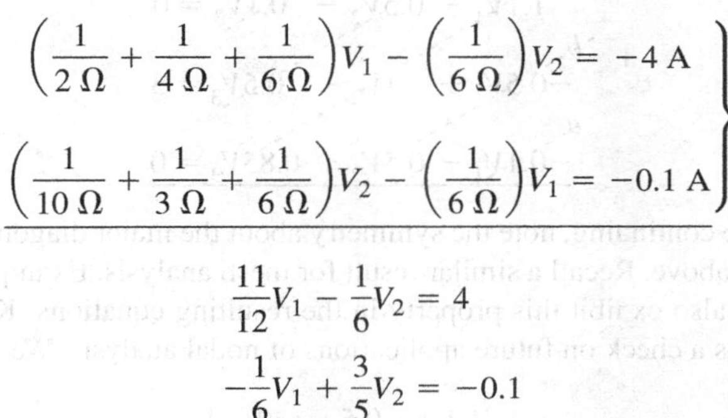

$$\left(\frac{1}{2\ \Omega} + \frac{1}{4\ \Omega} + \frac{1}{6\ \Omega}\right)V_1 - \left(\frac{1}{6\ \Omega}\right)V_2 = -4\ \text{A}$$
$$\left(\frac{1}{10\ \Omega} + \frac{1}{3\ \Omega} + \frac{1}{6\ \Omega}\right)V_2 - \left(\frac{1}{6\ \Omega}\right)V_1 = -0.1\ \text{A}$$

$$\frac{11}{12}V_1 - \frac{1}{6}V_2 = 4$$
$$-\frac{1}{6}V_1 + \frac{3}{5}V_2 = -0.1$$

resulting in

$$11V_1 - 2V_2 = +48$$
$$-5V_1 + 18V_2 = -3$$

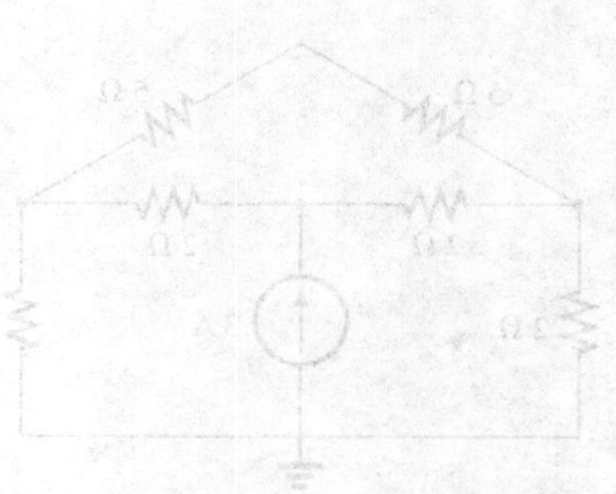

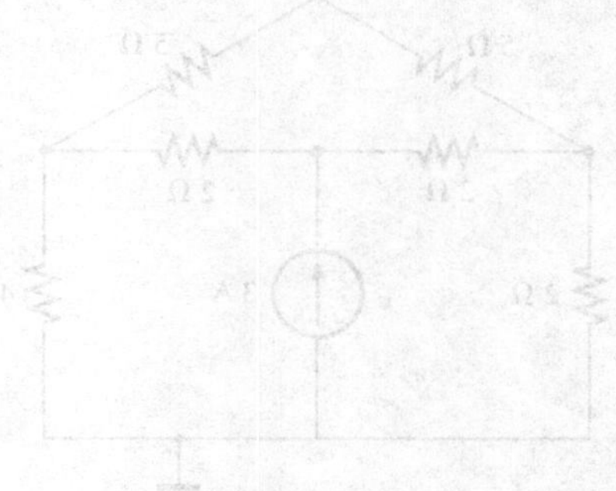

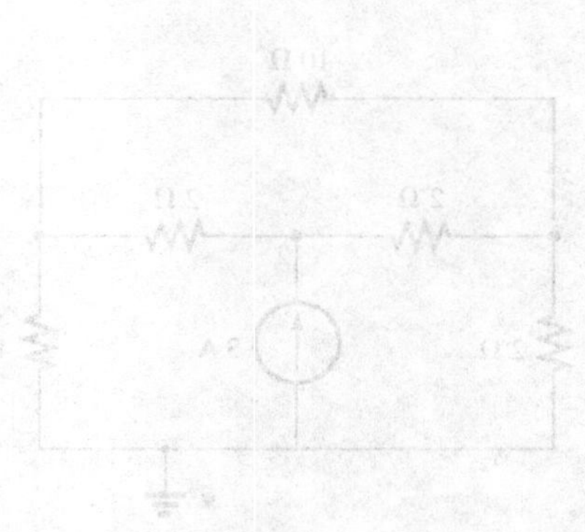

and

$$V_2 = V_{3\Omega} = \frac{\begin{vmatrix} 11 & 48 \\ -5 & -3 \end{vmatrix}}{\begin{vmatrix} 11 & -2 \\ -5 & 18 \end{vmatrix}} = \frac{-33 + 240}{198 - 10} = \frac{207}{188} = \mathbf{1.10\ V}$$

As demonstrated for mesh analysis, nodal analysis can also be a very useful technique for solving networks with only one source.

EXAMPLE 8.25 Using nodal analysis, determine the potential across the 4 Ω resistor in Fig. 8.59.

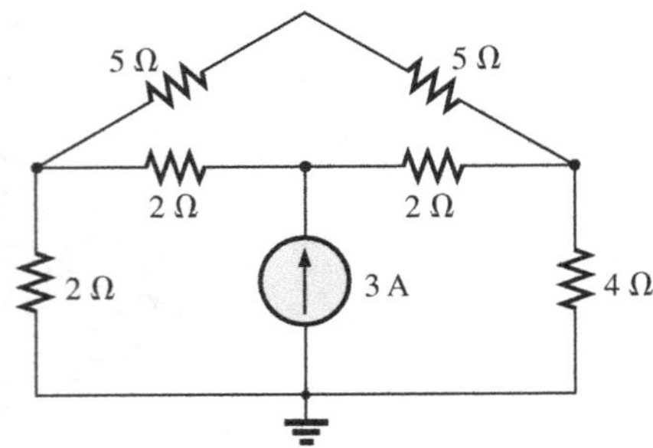

FIG. 8.59
Example 8.25.

Solution: The reference and four subscripted voltage levels were chosen as shown in Fig. 8.60. Remember that for any difference in potential between V_1 and V_3, the current through and the potential drop across each 5 Ω resistor are the same. Therefore, V_4 is simply a mid-voltage level between V_1 and V_3 and is known if V_1 and V_3 are available. We will therefore not include it in a nodal voltage and will redraw the network as shown in Fig. 8.61. Understand, however, that V_4 can be included if desired, although four nodal voltages will result rather than three as in the solution of this problem. We have

$$V_1\text{:} \quad \left(\frac{1}{2\ \Omega} + \frac{1}{2\ \Omega} + \frac{1}{10\ \Omega}\right)V_1 - \left(\frac{1}{2\ \Omega}\right)V_2 - \left(\frac{1}{10\ \Omega}\right)V_3 = 0$$

$$V_2\text{:} \quad \left(\frac{1}{2\ \Omega} + \frac{1}{2\ \Omega}\right)V_2 - \left(\frac{1}{2\ \Omega}\right)V_1 - \left(\frac{1}{2\ \Omega}\right)V_3 = 3\text{ A}$$

$$V_3\text{:} \quad \left(\frac{1}{10\ \Omega} + \frac{1}{2\ \Omega} + \frac{1}{4\ \Omega}\right)V_3 - \left(\frac{1}{2\ \Omega}\right)V_2 - \left(\frac{1}{10\ \Omega}\right)V_1 = 0$$

which are rewritten as

$$\begin{aligned} 1.1V_1 - 0.5V_2 - 0.1V_3 &= 0 \\ V_2 - 0.5V_1 - 0.5V_3 &= 3 \\ 0.85V_3 - 0.5V_2 - 0.1V_1 &= 0 \end{aligned}$$

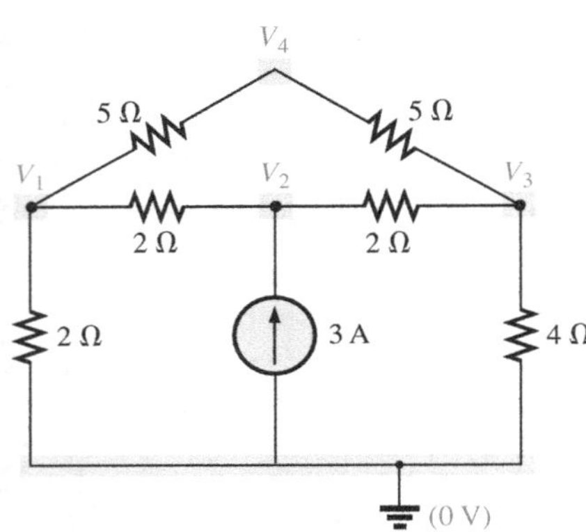

FIG. 8.60
Defining the nodes for the network in Fig. 8.59.

For determinants, we have

$$\begin{aligned} 1.1V_1 - 0.5V_2 - 0.1V_3 &= 0 \\ -0.5V_1 + 1V_2 - 0.5V_3 &= 3 \\ -0.1V_1 - 0.5V_2 + 0.85V_3 &= 0 \end{aligned}$$

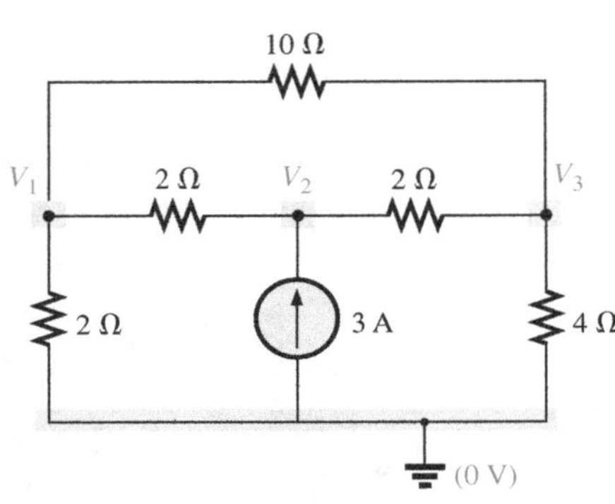

FIG. 8.61
Reducing the number of nodes for the network in Fig. 8.59 by combining the two 5 Ω resistors.

Before continuing, note the symmetry about the major diagonal in the equation above. Recall a similar result for mesh analysis. Examples 8.23 and 8.24 also exhibit this property in the resulting equations. Keep this in mind as a check on future applications of nodal analysis. We have

$$V_3 = V_{4\Omega} = \frac{\begin{vmatrix} 1.1 & -0.5 & 0 \\ -0.5 & +1 & 3 \\ -0.1 & -0.5 & 0 \end{vmatrix}}{\begin{vmatrix} 1.1 & -0.5 & -0.1 \\ -0.5 & +1 & -0.5 \\ -0.1 & -0.5 & +0.85 \end{vmatrix}} = \mathbf{4.65\ V}$$

The next example has only one source applied to a ladder network.

EXAMPLE 8.26 Write the nodal equations and find the voltage across the 2 Ω resistor for the network in Fig. 8.62.

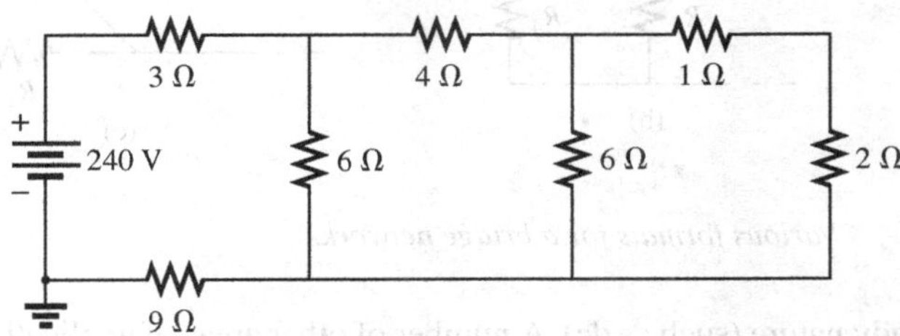

FIG. 8.62
Example 8.26.

Solution: The nodal voltages are chosen as shown in Fig. 8.63. We have

$$V_1\text{:}\quad \left(\frac{1}{12\ \Omega}+\frac{1}{6\ \Omega}+\frac{1}{4\ \Omega}\right)V_1-\left(\frac{1}{4\ \Omega}\right)V_2+0=20\text{ A}$$

$$V_2\text{:}\quad \left(\frac{1}{4\ \Omega}+\frac{1}{6\ \Omega}+\frac{1}{1\ \Omega}\right)V_2-\left(\frac{1}{4\ \Omega}\right)V_1-\left(\frac{1}{1\ \Omega}\right)V_3=0$$

$$V_3\text{:}\quad \left(\frac{1}{1\ \Omega}+\frac{1}{2\ \Omega}\right)V_3-\left(\frac{1}{1\ \Omega}\right)V_2+0=0$$

and

$$\begin{aligned}0.5V_1-0.25V_2+0&=20\\-0.25V_1+\frac{17}{12}V_2-1V_3&=0\\0-1V_2+1.5V_3&=0\end{aligned}$$

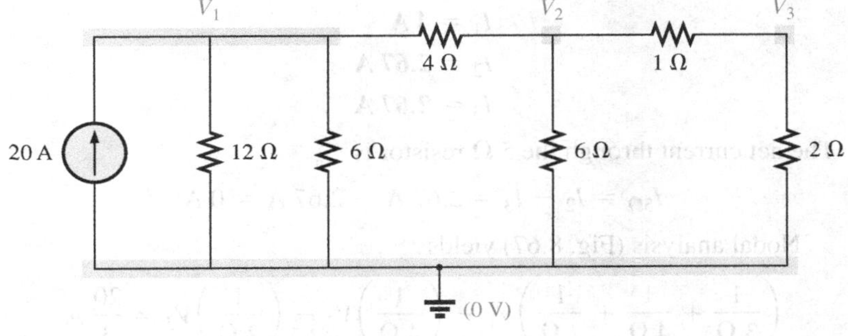

FIG. 8.63
Converting the voltage source to a current source and defining the nodes for the network in Fig. 8.62.

Note the symmetry present about the major axis. Application of determinants reveals that

$$V_3 = V_{2\Omega} = \mathbf{10.67\ V}$$

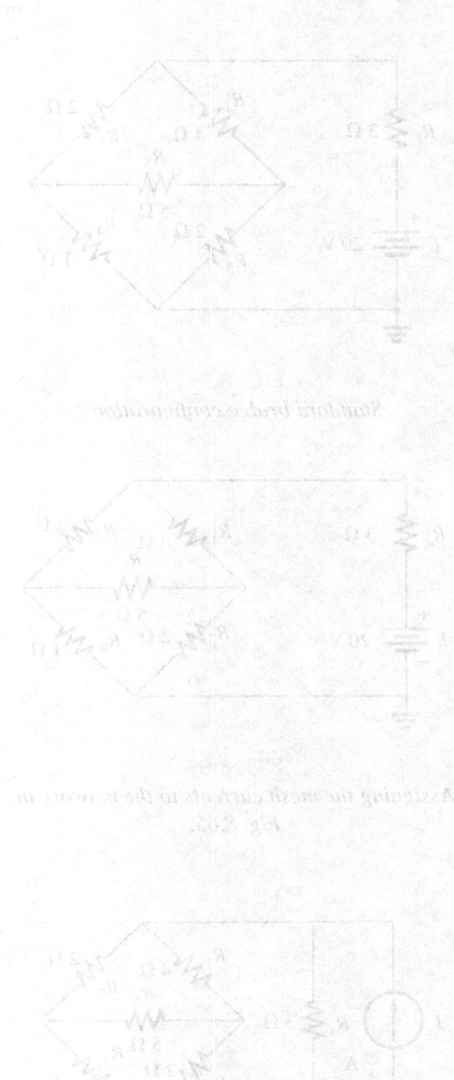

8.11 BRIDGE NETWORKS

This section introduces the **bridge network,** a configuration that has a multitude of applications. In the following chapters, this type of network is used in both dc and ac meters. Electronics courses introduce these in the discussion of rectifying circuits used in converting a varying signal to one

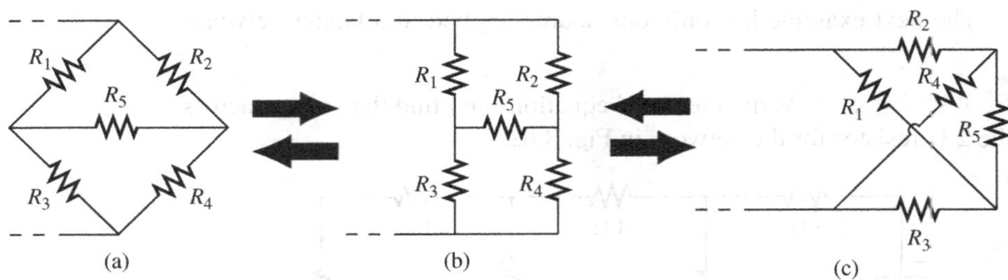

FIG. 8.64
Various formats for a bridge network.

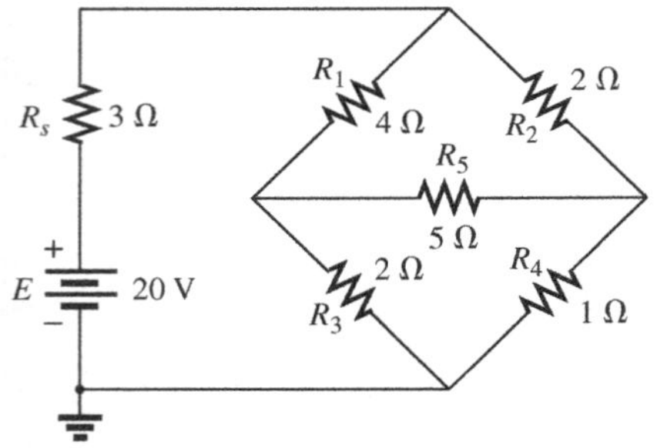

FIG. 8.65
Standard bridge configuration.

of a steady nature (such as dc). A number of other areas of application also require some knowledge of ac networks; these areas are discussed later.

The bridge network may appear in one of the three forms as indicated in Fig. 8.64. The network in Fig. 8.64(c) is also called a *symmetrical lattice network* if $R_2 = R_3$ and $R_1 = R_4$. Fig. 8.64(c) is an excellent example of how a planar network can be made to appear nonplanar. For the purposes of investigation, let us examine the network in Fig. 8.65 using mesh and nodal analysis.

Mesh analysis (Fig. 8.66) yields

$$(3\,\Omega + 4\,\Omega + 2\,\Omega)I_1 - (4\,\Omega)I_2 - (2\,\Omega)I_3 = 20\text{ V}$$
$$(4\,\Omega + 5\,\Omega + 2\,\Omega)I_2 - (4\,\Omega)I_1 - (5\,\Omega)I_3 = 0$$
$$(2\,\Omega + 5\,\Omega + 1\,\Omega)I_3 - (2\,\Omega)I_1 - (5\,\Omega)I_2 = 0$$

and

$$9I_1 - 4I_2 - 2I_3 = 20$$
$$-4I_1 + 11I_2 - 5I_3 = 0$$
$$-2I_1 - 5I_2 + 8I_3 = 0$$

with the result that

$$I_1 = \mathbf{4\text{ A}}$$
$$I_2 = \mathbf{2.67\text{ A}}$$
$$I_3 = \mathbf{2.67\text{ A}}$$

The net current through the 5 Ω resistor is

$$I_{5\Omega} = I_2 - I_3 = 2.67\text{ A} - 2.67\text{ A} = 0\text{ A}$$

Nodal analysis (Fig. 8.67) yields

$$\left(\frac{1}{3\,\Omega} + \frac{1}{4\,\Omega} + \frac{1}{2\,\Omega}\right)V_1 - \left(\frac{1}{4\,\Omega}\right)V_2 - \left(\frac{1}{2\,\Omega}\right)V_3 = \frac{20}{3}\text{ A}$$
$$\left(\frac{1}{4\,\Omega} + \frac{1}{2\,\Omega} + \frac{1}{5\,\Omega}\right)V_2 - \left(\frac{1}{4\,\Omega}\right)V_1 - \left(\frac{1}{5\,\Omega}\right)V_3 = 0$$
$$\left(\frac{1}{5\,\Omega} + \frac{1}{2\,\Omega} + \frac{1}{1\,\Omega}\right)V_3 - \left(\frac{1}{2\,\Omega}\right)V_1 - \left(\frac{1}{5\,\Omega}\right)V_2 = 0$$

and

$$\left(\frac{1}{3\,\Omega} + \frac{1}{4\,\Omega} + \frac{1}{2\,\Omega}\right)V_1 - \left(\frac{1}{4\,\Omega}\right)V_2 - \left(\frac{1}{2\,\Omega}\right)V_3 = 6.67\text{ A}$$
$$-\left(\frac{1}{4\,\Omega}\right)V_1 + \left(\frac{1}{4\,\Omega} + \frac{1}{2\,\Omega} + \frac{1}{5\,\Omega}\right)V_2 - \left(\frac{1}{5\,\Omega}\right)V_3 = 0$$
$$-\left(\frac{1}{2\,\Omega}\right)V_1 - \left(\frac{1}{5\,\Omega}\right)V_2 + \left(\frac{1}{5\,\Omega} + \frac{1}{2\,\Omega} + \frac{1}{1\,\Omega}\right)V_3 = 0$$

Note the symmetry of the solution.

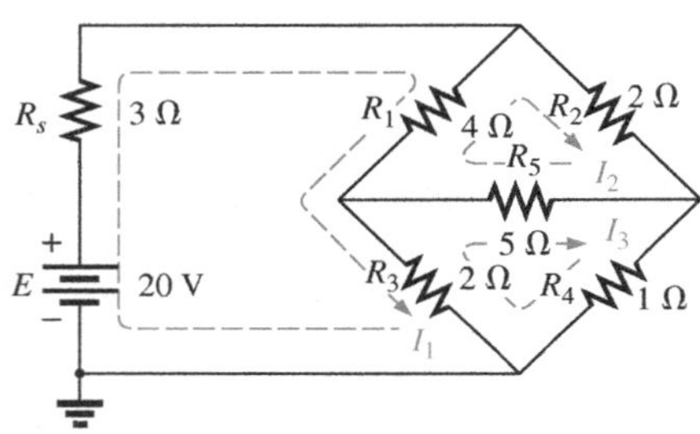

FIG. 8.66
Assigning the mesh currents to the network in Fig. 8.65.

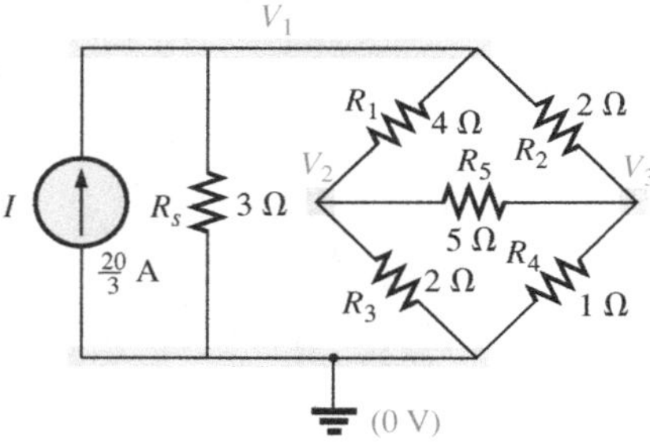

FIG. 8.67
Defining the nodal voltages for the network in Fig. 8.65.

TI-89 Calculator Solution

With the TI-89 calculator, the top part of the determinant is determined by the sequence in Fig. 8.68 (take note of the calculations within parentheses):

det([6.67,−1/4,−1/2;0,(1/4+1/2+1/5),−1/5;0,−1/5,(1/5+1/2+1/1)]) (ENTER)	10.51E0

FIG. 8.68
TI-89 solution for the numerator of the solution for V_1.

with the bottom of the determinant determined by the sequence in Fig. 8.69.

det([(1/3+1/4+1/2),−1/4,−1/2;−1/4,(1/4+1/2+1/5),−1/5;−1/2,−1/5,(1/5+1/2+1/1)]) (ENTER)	1.31E0

FIG. 8.69
TI-89 solution for the denominator of the equation for V_1.

Finally, the simple division in Fig. 8.70 provides the desired result.

10.51/1.31 (ENTER)	8.02

FIG. 8.70
TI-89 solution for V_1.

and $\qquad V_1 = \mathbf{8.02\ V}$

Similarly, $\qquad V_2 = \mathbf{2.67\ V} \quad$ and $\quad V_3 = \mathbf{2.67\ V}$

and the voltage across the 5 Ω resistor is

$$V_{5\Omega} = V_2 - V_3 = 2.67\ \text{A} - 2.67\ \text{A} = \mathbf{0\ V}$$

Since $V_{5\Omega} = 0$ V, we can insert a short in place of the bridge arm without affecting the network behavior. (Certainly $V = IR = I \cdot (0) = 0$ V.) In Fig. 8.71, a short circuit has replaced the resistor R_5, and the voltage across R_4 is to be determined. The network is redrawn in Fig. 8.72, and

$$V_{1\Omega} = \frac{(2\ \Omega \parallel 1\ \Omega)20\ \text{V}}{(2\ \Omega \parallel 1\ \Omega) + (4\ \Omega \parallel 2\ \Omega) + 3\ \Omega} \quad \text{(voltage divider rule)}$$

$$= \frac{\frac{2}{3}(20\ \text{V})}{\frac{2}{3} + \frac{8}{6} + 3} = \frac{\frac{2}{3}(20\ \text{V})}{\frac{2}{3} + \frac{4}{3} + \frac{9}{3}}$$

$$= \frac{2(20\ \text{V})}{2 + 4 + 9} = \frac{40\ \text{V}}{15} = \mathbf{2.67\ V}$$

as obtained earlier.

We found through mesh analysis that $I_{5\Omega} = 0$ A, which has as its equivalent an open circuit as shown in Fig. 8.73(a). (Certainly $I = V/R =$

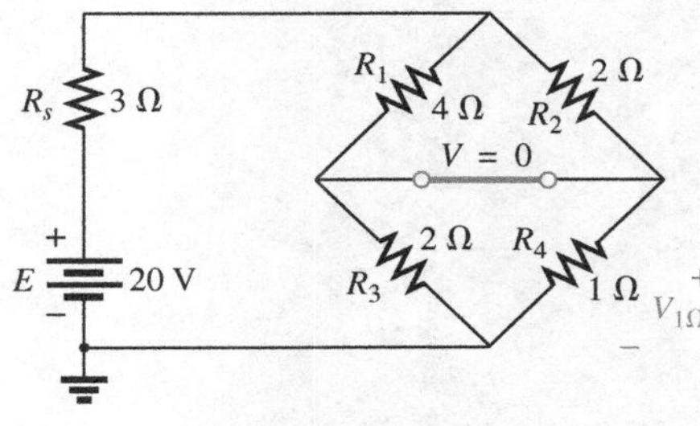

FIG. 8.71
Substituting the short-circuit equivalent for the balance arm of a balanced bridge.

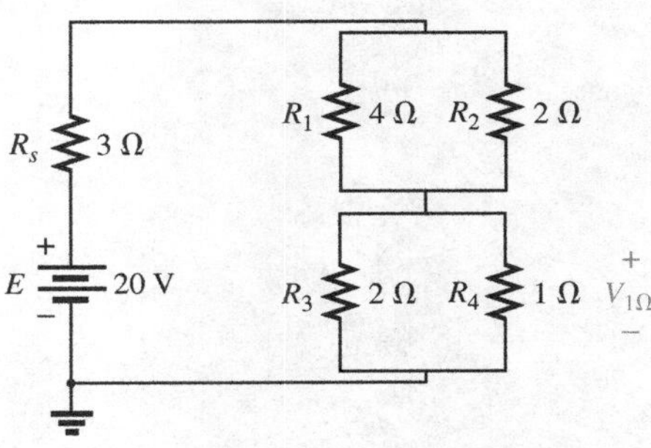

FIG. 8.72
Redrawing the network in Fig. 8.71.

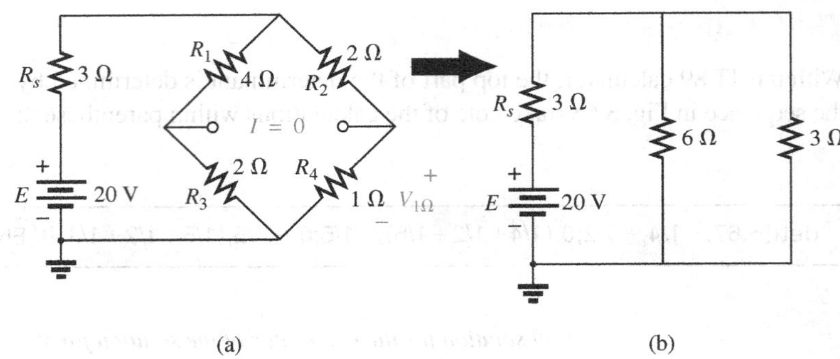

FIG. 8.73
Substituting the open-circuit equivalent for the balance arm of a balanced bridge.

$0/(\infty\ \Omega) = 0$ A.) The voltage across the resistor R_4 is again determined and compared with the result above.

The network is redrawn after combining series elements as shown in Fig. 8.73(b), and

$$V_{3\Omega} = \frac{(6\ \Omega \parallel 3\ \Omega)(20\ \text{V})}{6\ \Omega \parallel 3\ \Omega + 3\ \Omega} = \frac{2\ \Omega(20\ \text{V})}{2\ \Omega + 3\ \Omega} = 8\ \text{V}$$

and
$$V_{1\Omega} = \frac{1\ \Omega(8\ \text{V})}{1\ \Omega + 2\ \Omega} = \frac{8\ \text{V}}{3} = \mathbf{2.67\ V}$$

as above.

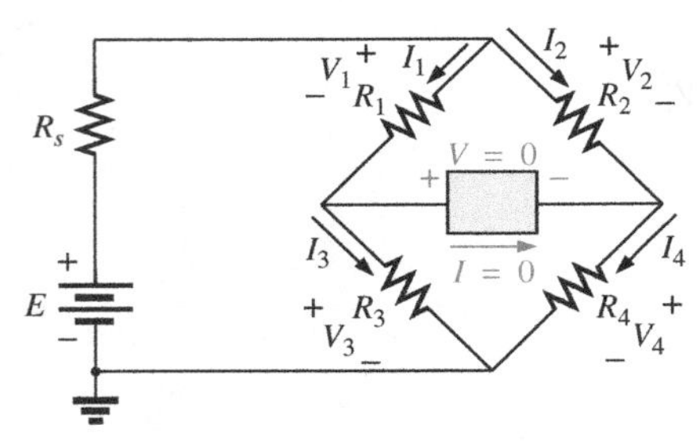

FIG. 8.74
Establishing the balance criteria for a bridge network.

The condition $V_{5\Omega} = 0$ V or $I_{5\Omega} = 0$ A exists only for a particular relationship between the resistors of the network. Let us now derive this relationship using the network in Fig. 8.74, in which it is indicated that $I = 0$ A and $V = 0$ V. Note that resistor R_s of the network in Fig. 8.73 does not appear in the following analysis.

The bridge network is said to be *balanced* when the condition of $I = 0$ A or $V = 0$ V exists.

If $V = 0$ V (short circuit between *a* and *b*), then

$$V_1 = V_2$$

and
$$I_1R_1 = I_2R_2$$

or
$$I_1 = \frac{I_2R_2}{R_1}$$

In addition, when $V = 0$ V,

$$V_3 = V_4$$

and
$$I_3R_3 = I_4R_4$$

If we set $I = 0$ A, then $I_3 = I_1$ and $I_4 = I_2$, with the result that the above equation becomes

$$I_1R_3 = I_2R_4$$

Substituting for I_1 from above yields

$$\left(\frac{I_2R_2}{R_1}\right)R_3 = I_2R_4$$

or, rearranging, we have

$$\boxed{\frac{R_1}{R_3} = \frac{R_2}{R_4}} \qquad \textbf{(8.2)}$$

This conclusion states that if the ratio of R_1 to R_3 is equal to that of R_2 to R_4, the bridge is balanced, and $I = 0$ A or $V = 0$ V. A method of memorizing this form is indicated in Fig. 8.75.

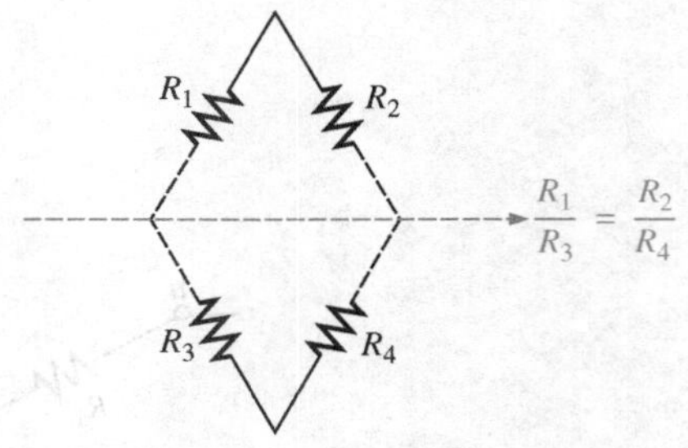

FIG. 8.75
A visual approach to remembering the balance condition.

For the example above, $R_1 = 4\ \Omega$, $R_2 = 2\ \Omega$, $R_3 = 2\ \Omega$, $R_4 = 1\ \Omega$, and

$$\frac{R_1}{R_3} = \frac{R_2}{R_4} \rightarrow \frac{4\ \Omega}{2\ \Omega} = \frac{2\ \Omega}{1\ \Omega} = 2$$

The emphasis in this section has been on the balanced situation. Understand that if the ratio is not satisfied, there will be a potential drop across the balance arm and a current through it. The methods just described (mesh and nodal analysis) will yield any and all potentials or currents desired, just as they did for the balanced situation.

8.12 Y-Δ (T-π) AND Δ-Y (π-T) CONVERSIONS

Circuit configurations are often encountered in which the resistors do not appear to be in series or parallel. Under these conditions, it may be necessary to convert the circuit from one form to another to solve for any unknown quantities if mesh or nodal analysis is not applied. Two circuit configurations that often account for these difficulties are the **wye (Y)** and **delta (Δ) configurations** depicted in Fig. 8.76(a). They are also referred to as the **tee (T)** and **pi (π),** respectively, as indicated in Fig. 8.76(b). Note that the pi is actually an inverted delta.

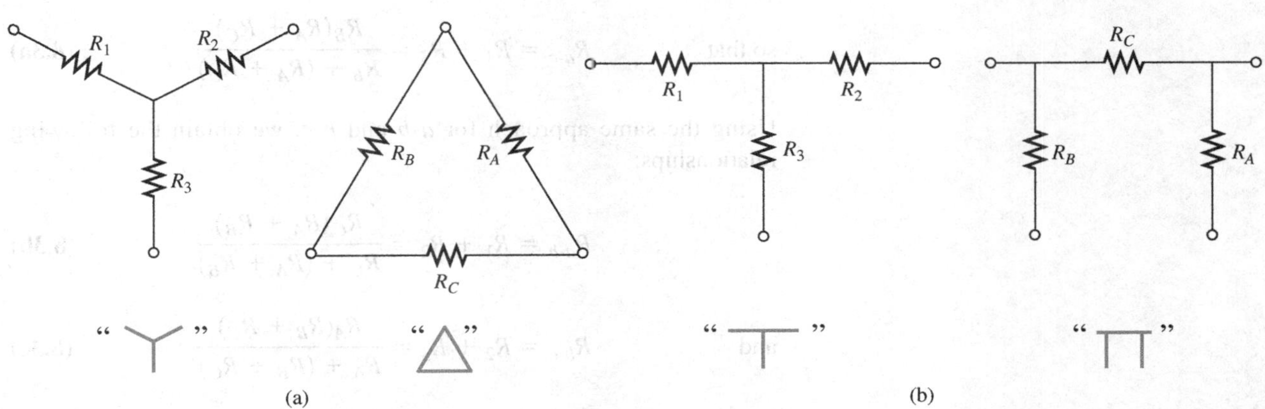

FIG. 8.76
The Y (T) and Δ (π) configurations.

The purpose of this section is to develop the equations for converting from Δ to Y, or vice versa. This type of conversion normally leads to a network that can be solved using techniques such as those described in Chapter 7. In other words, in Fig. 8.77, with terminals *a*, *b*, and *c* held fast, if the wye (Y) configuration were desired *instead of* the inverted delta (Δ) configuration, all that would be necessary is a direct application of the equations to be derived. The phrase *instead of* is emphasized to ensure that it is understood that only one of these configurations is to appear at one time between the indicated terminals.

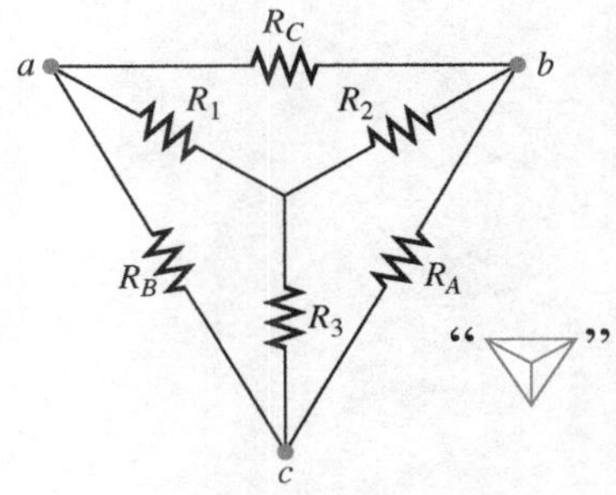

FIG. 8.77
Introducing the concept of Δ-Y or Y-Δ conversions.

It is our purpose (referring to Fig. 8.77) to find some expression for R_1, R_2, and R_3 in terms of R_A, R_B, and R_C, and vice versa, that will ensure that the resistance between any two terminals of the Y configuration will

be the same with the Δ configuration inserted in place of the Y configuration (and vice versa). If the two circuits are to be equivalent, the total resistance between any two terminals must be the same. Consider terminals *a-c* in the Δ-Y configurations in Fig. 8.78.

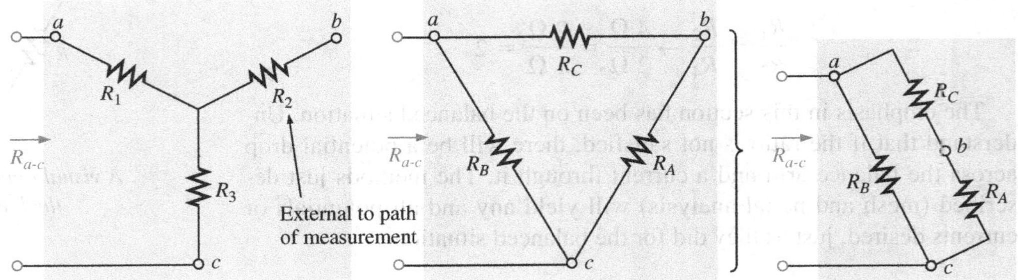

FIG. 8.78
Finding the resistance $R_{a\text{-}c}$ for the Y and Δ configurations.

Let us first assume that we want to convert the Δ (R_A, R_B, R_C) to the Y (R_1, R_2, R_3). This requires that we have a relationship for R_1, R_2, and R_3 in terms of R_A, R_B, and R_C. If the resistance is to be the same between terminals *a-c* for both the Δ and the Y, the following must be true:

$$R_{a\text{-}c}(\text{Y}) = R_{a\text{-}c}(\Delta)$$

so that

$$R_{a\text{-}c} = R_1 + R_3 = \frac{R_B(R_A + R_C)}{R_B + (R_A + R_C)} \tag{8.3a}$$

Using the same approach for *a-b* and *b-c*, we obtain the following relationships:

$$R_{a\text{-}b} = R_1 + R_2 = \frac{R_C(R_A + R_B)}{R_C + (R_A + R_B)} \tag{8.3b}$$

and

$$R_{b\text{-}c} = R_2 + R_3 = \frac{R_A(R_B + R_C)}{R_A + (R_B + R_C)} \tag{8.3c}$$

Subtracting Eq. (8.3a) from Eq. (8.3b), we have

$$(R_1 + R_2) - (R_1 + R_3) = \left(\frac{R_C R_B + R_C R_A}{R_A + R_B + R_C}\right) - \left(\frac{R_B R_A + R_B R_A}{R_A + R_B + R_C}\right)$$

so that

$$R_2 - R_3 = \frac{R_A R_C - R_B R_A}{R_A + R_B + R_C} \tag{8.4}$$

Subtracting Eq. (8.4) from Eq. (8.3c) yields

$$(R_2 + R_3) - (R_2 - R_3) = \left(\frac{R_A R_B + R_A R_C}{R_A + R_B + R_C}\right) - \left(\frac{R_A R_C - R_B R_A}{R_A + R_B + R_C}\right)$$

so that

$$2R_3 = \frac{2R_B R_A}{R_A + R_B + R_C}$$

resulting in the following expression for R_3 in terms of R_A, R_B, and R_C:

$$R_3 = \frac{R_A R_B}{R_A + R_B + R_C} \quad \textbf{(8.5a)}$$

Following the same procedure for R_1 and R_2, we have

$$R_1 = \frac{R_B R_C}{R_A + R_B + R_C} \quad \textbf{(8.5b)}$$

and

$$R_2 = \frac{R_A R_C}{R_A + R_B + R_C} \quad \textbf{(8.5c)}$$

Note that each resistor of the Y is equal to the product of the resistors in the two closest branches of the Δ divided by the sum of the resistors in the Δ.

To obtain the relationships necessary to convert from a Y to a Δ, first divide Eq. (8.5a) by Eq. (8.5b):

$$\frac{R_3}{R_1} = \frac{(R_A R_B)/(R_A + R_B + R_C)}{(R_B R_C)/(R_A + R_B + R_C)} = \frac{R_A}{R_C}$$

or

$$R_A = \frac{R_C R_3}{R_1}$$

Then divide Eq. (8.5a) by Eq. (8.5c):

$$\frac{R_3}{R_2} = \frac{(R_A R_B)/(R_A + R_B + R_C)}{(R_A R_C)/(R_A + R_B + R_C)} = \frac{R_B}{R_C}$$

or

$$R_B = \frac{R_3 R_C}{R_2}$$

Substituting for R_A and R_B in Eq. (8.5c) yields

$$R_2 = \frac{(R_C R_3/R_1)R_C}{(R_3 R_C/R_2) + (R_C R_3/R_1) + R_C}$$

$$= \frac{(R_3/R_1)R_C}{(R_3/R_2) + (R_3/R_1) + 1}$$

Placing these over a common denominator, we obtain

$$R_2 = \frac{(R_3 R_C/R_1)}{(R_1 R_2 + R_1 R_3 + R_2 R_3)/(R_1 R_2)}$$

$$= \frac{R_2 R_3 R_C}{R_1 R_2 + R_1 R_3 + R_2 R_3}$$

and

$$R_C = \frac{R_1 R_2 + R_1 R_3 + R_2 R_3}{R_3} \quad \textbf{(8.6a)}$$

We follow the same procedure for R_B and R_A:

$$R_A = \frac{R_1 R_2 + R_1 R_3 + R_2 R_3}{R_1} \quad \textbf{(8.6b)}$$

and

$$R_B = \frac{R_1R_2 + R_1R_3 + R_2R_3}{R_2} \tag{8.6c}$$

Note that the value of each resistor of the Δ is equal to the sum of the possible product combinations of the resistances of the Y divided by the resistance of the Y farthest from the resistor to be determined.

Let us consider what would occur if all the values of a Δ or Y were the same. If $R_A = R_B = R_C$, Eq. (8.5a) would become (using R_A only) the following:

$$R_3 = \frac{R_AR_B}{R_A + R_B + R_C} = \frac{R_AR_A}{R_A + R_A + R_A} = \frac{R_A^2}{3R_A} = \frac{R_A}{3}$$

and, following the same procedure,

$$R_1 = \frac{R_A}{3} \quad R_2 = \frac{R_A}{3}$$

In general, therefore,

$$R_Y = \frac{R_\Delta}{3} \tag{8.7}$$

or

$$R_\Delta = 3R_Y \tag{8.8}$$

which indicates that *for a Y of three equal resistors, the value of each resistor of the Δ is equal to three times the value of any resistor of the Y.* If only two elements of a Y or a Δ are the same, the corresponding Δ or Y of each will also have two equal elements. The converting of equations is left as an exercise for you.

The Y and the Δ often appear as shown in Fig. 8.79. They are then referred to as a **tee (T)** and a **pi (π)** network, respectively. The equations used to convert from one form to the other are exactly the same as those developed for the Y and Δ transformation.

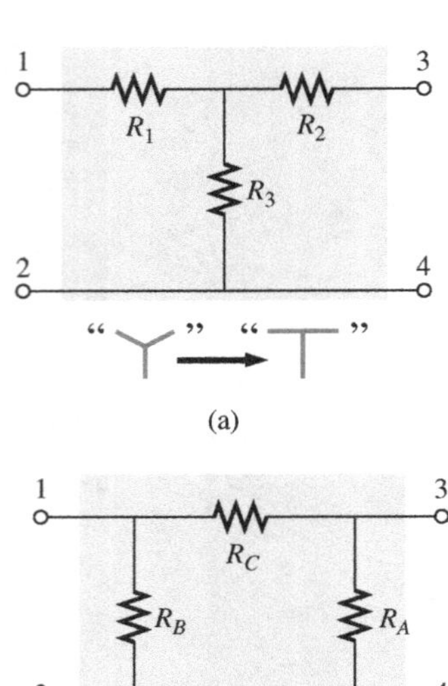

FIG. 8.79
The relationship between the Y and T configurations and the Δ and π configurations.

EXAMPLE 8.27 Convert the Δ in Fig. 8.80 to a Y.

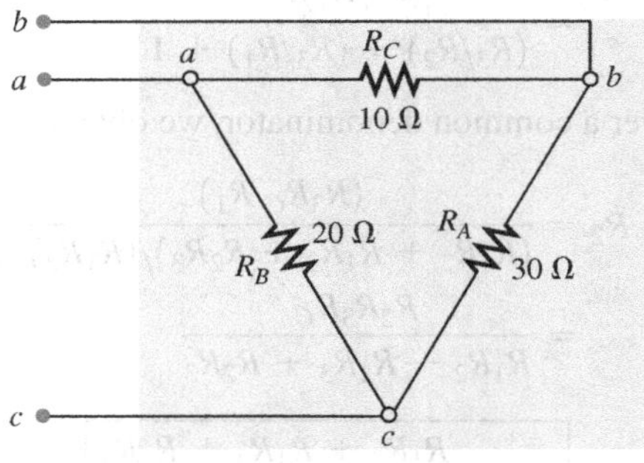

FIG. 8.80
Example 8.27.

Solution:

$$R_1 = \frac{R_BR_C}{R_A + R_B + R_C} = \frac{(20\ \Omega)(10\ \Omega)}{30\ \Omega + 20\ \Omega + 10\ \Omega} = \frac{200\ \Omega}{60} = 3\tfrac{1}{3}\ \Omega$$

$$R_2 = \frac{R_A R_C}{R_A + R_B + R_C} = \frac{(30\ \Omega)(10\ \Omega)}{60\ \Omega} = \frac{300\ \Omega}{60} = \mathbf{5\ \Omega}$$

$$R_3 = \frac{R_A R_B}{R_A + R_B + R_C} = \frac{(20\ \Omega)(30\ \Omega)}{60\ \Omega} = \frac{600\ \Omega}{60} = \mathbf{10\ \Omega}$$

The equivalent network is shown in Fig. 8.81.

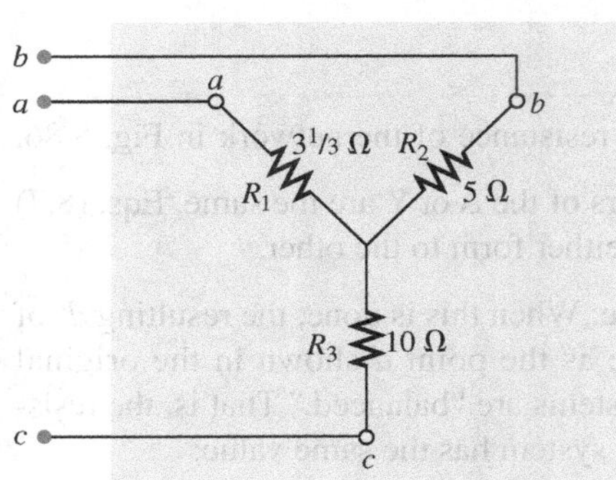

FIG. 8.81
The Y equivalent for the Δ in Fig. 8.80.

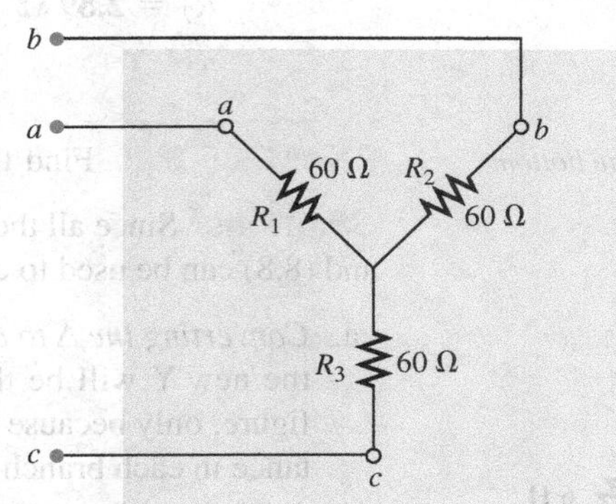

FIG. 8.82
Example 8.28.

EXAMPLE 8.28 Convert the Y in Fig. 8.82 to a Δ.

Solution:

$$R_A = \frac{R_1R_2 + R_1R_3 + R_2R_3}{R_1}$$

$$= \frac{(60\ \Omega)(60\ \Omega) + (60\ \Omega)(60\ \Omega) + (60\ \Omega)(60\ \Omega)}{60\ \Omega}$$

$$= \frac{3600\ \Omega + 3600\ \Omega + 3600\ \Omega}{60} = \frac{10{,}800\ \Omega}{60}$$

$$R_A = \mathbf{180\ \Omega}$$

However, the three resistors for the Y are equal, permitting the use of Eq. (8.8) and yielding

$$R_\Delta = 3R_Y = 3(60\ \Omega) = 180\ \Omega$$

and

$$R_B = R_C = \mathbf{180\ \Omega}$$

The equivalent network is shown in Fig. 8.83.

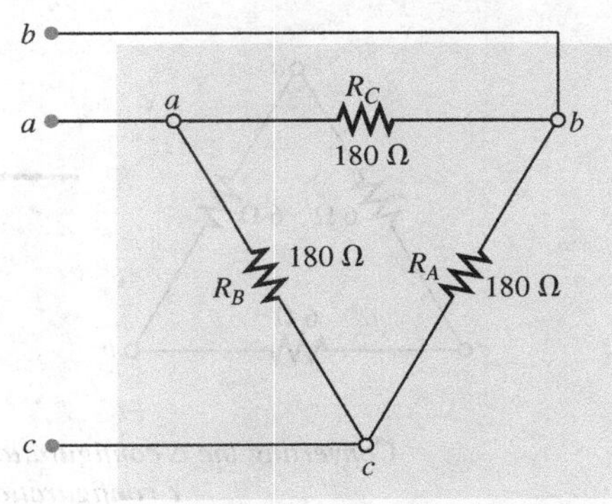

FIG. 8.83
The Δ equivalent for the Y in Fig. 8.82.

EXAMPLE 8.29 Find the total resistance of the network in Fig. 8.84, where $R_A = 3\ \Omega$, $R_B = 3\ \Omega$, and $R_C = 6\ \Omega$.

Solution:

Two resistors of the Δ were equal; therefore, two resistors of the Y will be equal.

$$R_1 = \frac{R_B R_C}{R_A + R_B + R_C} = \frac{(3\ \Omega)(6\ \Omega)}{3\ \Omega + 3\ \Omega + 6\ \Omega} = \frac{18\ \Omega}{12} = \mathbf{1.5\ \Omega}$$

$$R_2 = \frac{R_A R_C}{R_A + R_B + R_C} = \frac{(3\ \Omega)(6\ \Omega)}{12\ \Omega} = \frac{18\ \Omega}{12} = \mathbf{1.5\ \Omega}$$

$$R_3 = \frac{R_A R_B}{R_A + R_B + R_C} = \frac{(3\ \Omega)(3\ \Omega)}{12\ \Omega} = \frac{9\ \Omega}{12} = \mathbf{0.75\ \Omega}$$

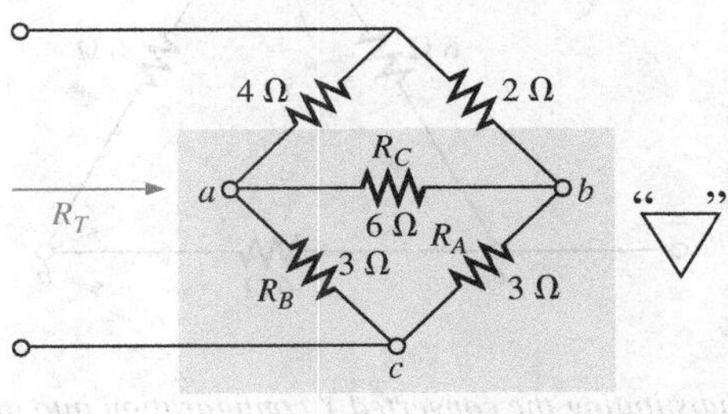

FIG. 8.84
Example 8.29.

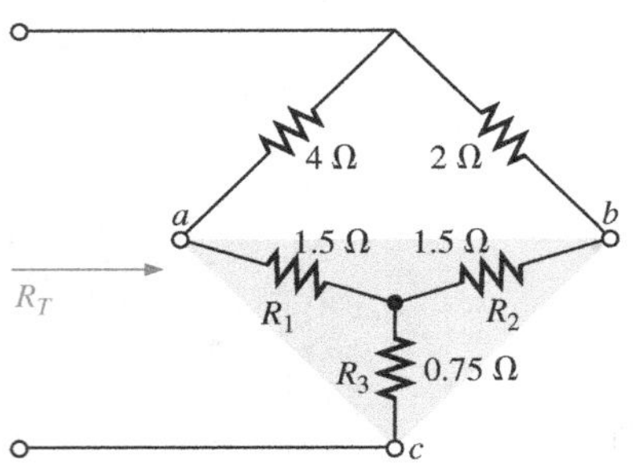

FIG. 8.85
Substituting the Y equivalent for the bottom Δ in Fig. 8.84.

Replacing the Δ by the Y, as shown in Fig. 8.85, yields

$$\begin{aligned} R_T &= 0.75\ \Omega + \frac{(4\ \Omega + 1.5\ \Omega)(2\ \Omega + 1.5\ \Omega)}{(4\ \Omega + 1.5\ \Omega) + (2\ \Omega + 1.5\ \Omega)} \\ &= 0.75\ \Omega + \frac{(5.5\ \Omega)(3.5\ \Omega)}{5.5\ \Omega + 3.5\ \Omega} \\ &= 0.75\ \Omega + 2.139\ \Omega \\ R_T &= \mathbf{2.89\ \Omega} \end{aligned}$$

EXAMPLE 8.30 Find the total resistance of the network in Fig. 8.86.

Solutions: Since all the resistors of the Δ or Y are the same, Eqs. (8.7) and (8.8) can be used to convert either form to the other.

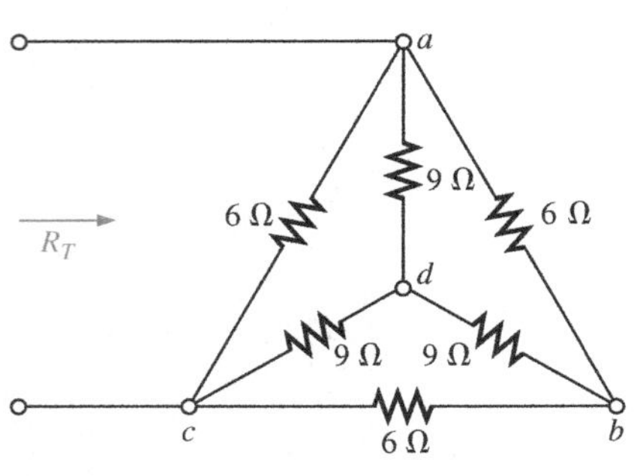

FIG. 8.86
Example 8.30.

a. *Converting the Δ to a Y:* Note: When this is done, the resulting d' of the new Y will be the same as the point d shown in the original figure, only because both systems are "balanced." That is, the resistance in each branch of each system has the same value:

$$R_Y = \frac{R_\Delta}{3} = \frac{6\ \Omega}{3} = 2\ \Omega \qquad \text{(Fig. 8.87)}$$

The network then appears as shown in Fig. 8.88. We have

$$R_T = 2\left[\frac{(2\ \Omega)(9\ \Omega)}{2\ \Omega + 9\ \Omega}\right] = \mathbf{3.27\ \Omega}$$

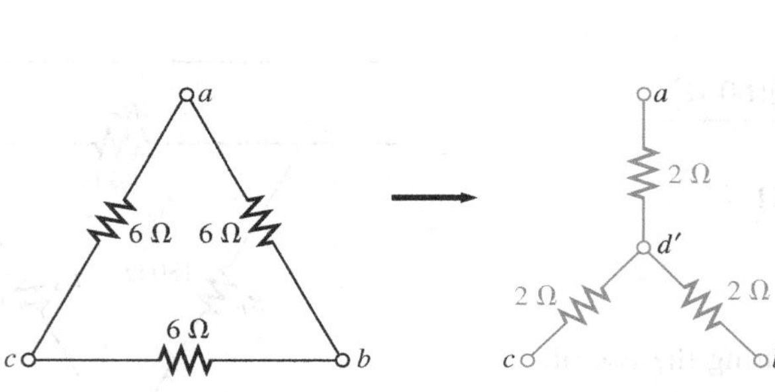

FIG. 8.87
Converting the Δ configuration of Fig. 8.86 to a Y configuration.

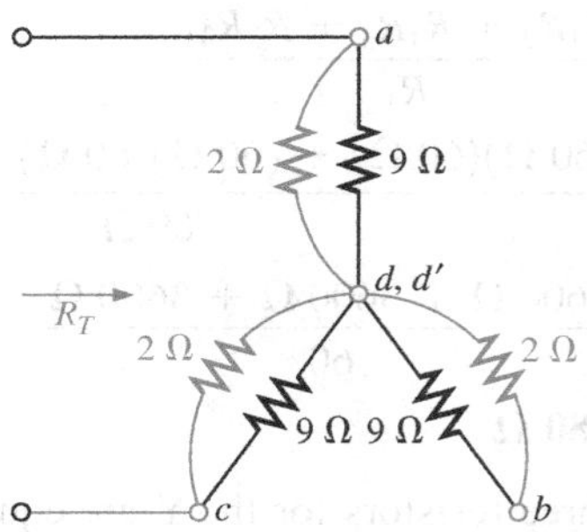

FIG. 8.88
Substituting the Y configuration for the converted Δ into the network in Fig. 8.86.

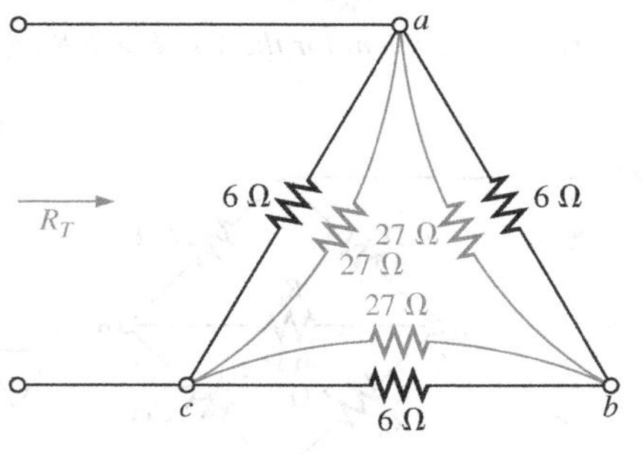

FIG. 8.89
Substituting the converted Y configuration into the network in Fig. 8.86.

b. *Converting the Y to a Δ:*

$$R_\Delta = 3R_Y = (3)(9\ \Omega) = 27\ \Omega \qquad \text{(Fig. 8.89)}$$

$$R'_T = \frac{(6\ \Omega)(27\ \Omega)}{6\ \Omega + 27\ \Omega} = \frac{162\ \Omega}{33} = 4.91\ \Omega$$

$$R_T = \frac{R'_T(R'_T + R'_T)}{R'_T + (R'_T + R'_T)} = \frac{R'_T 2R'_T}{3R'_T} = \frac{2R'_T}{3}$$

$$= \frac{2(4.91\ \Omega)}{3} = \mathbf{3.27\ \Omega}$$

which checks with the previous solution.

8.13 APPLICATIONS

This section discusses the constant-current characteristic in the design of security systems, the bridge circuit in a common residential smoke detector, and the nodal voltages of a digital logic probe.

Constant-Current Alarm Systems

The basic components of an alarm system using a constant-current supply are provided in Fig. 8.90. This design is improved over that provided in Chapter 5 in the sense that it is less sensitive to changes in resistance in the circuit due to heating, humidity, changes in the length of the line to the sensors, and so on. The 1.5 kΩ rheostat (total resistance between points *a* and *b*) is adjusted to ensure a current of 5 mA through the single-series security circuit. The adjustable rheostat is necessary to compensate for variations in the total resistance of the circuit introduced by the resistance of the wire, sensors, sensing relay, and milliammeter. The milliammeter is included to set the rheostat and ensure a current of 5 mA.

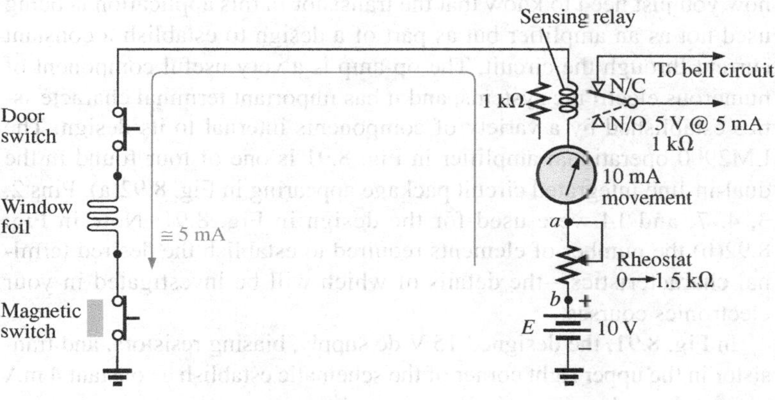

FIG. 8.90
Constant-current alarm system.

If any of the sensors opens the current through the entire circuit drops to zero, the coil of the relay releases the plunger, and contact is made with the N/C position of the relay. This action completes the circuit for the bell circuit, and the alarm sounds. For the future, keep in mind that switch positions for a relay are always shown with no power to the network, resulting in the N/C position in Fig. 8.90. When power is applied, the switch will have the position indicated by the dashed line. That is, various factors, such as a change in resistance of any of the elements due to heating, humidity, and so on, cause the applied voltage to redistribute itself and create a sensitive situation. With an adjusted 5 mA, the loading can change, but the current will always be 5 mA and the chance of tripping reduced. Note that the relay is rated as 5 V at 5 mA, indicating that in the on state the voltage across the relay is 5 V and the current through the relay is 5 mA. Its internal resistance is therefore 5 V/5 mA = 1 kΩ in this state.

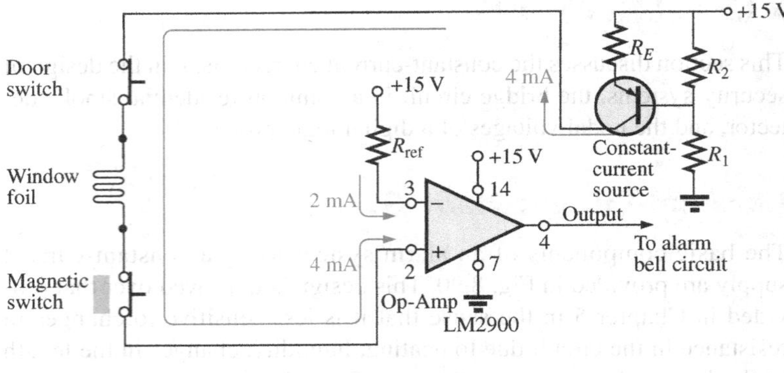

FIG. 8.91
Constant-current alarm system with electronic components.

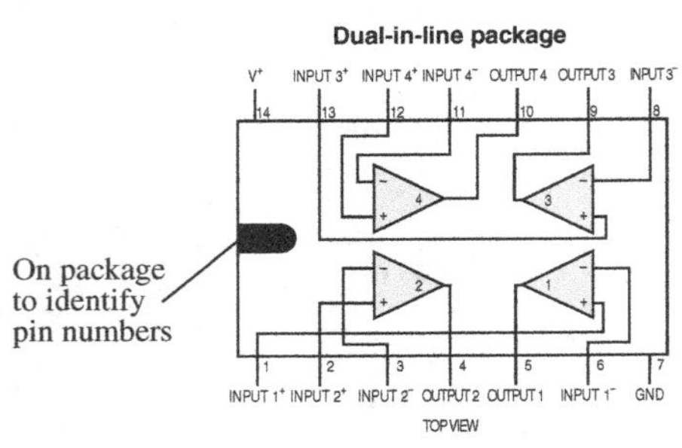

(a)

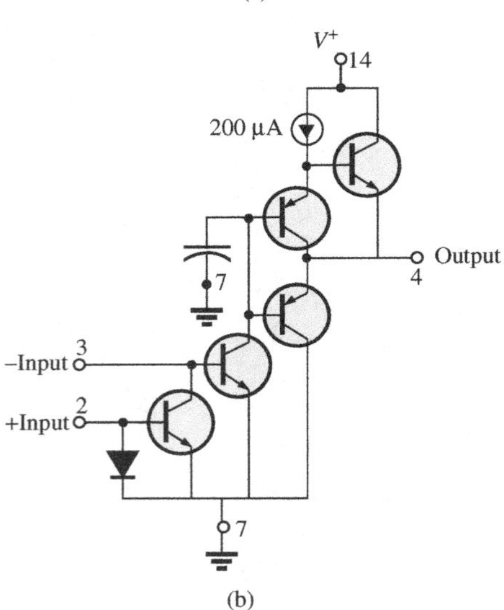

(b)

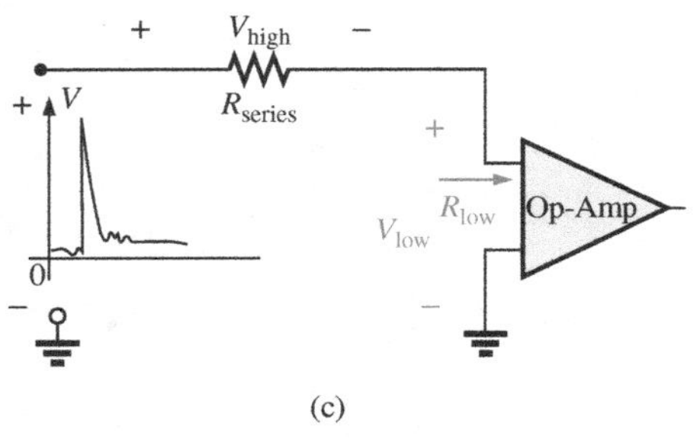

(c)

FIG. 8.92
LM2900 operational amplifier: (a) dual-in-line package (DIP); (b) components; (c) impact of low-input impedance.

A more advanced alarm system using a constant current is illustrated in Fig. 8.91. In this case, an electronic system using a single transistor, biasing resistors, and a dc battery are establishing a current of 4 mA through the series sensor circuit connected to the positive side of an operational amplifier (op-amp). Transistors and op-amp devices may be new to you (these are discussed in detail in electronics courses), but for now you just need to know that the transistor in this application is being used not as an amplifier but as part of a design to establish a constant current through the circuit. The op-amp is a very useful component of numerous electronic systems, and it has important terminal characteristics established by a variety of components internal to its design. The LM2900 operational amplifier in Fig. 8.91 is one of four found in the dual-in-line integrated circuit package appearing in Fig. 8.92(a). Pins 2, 3, 4, 7, and 14 were used for the design in Fig. 8.91. Note in Fig. 8.92(b) the number of elements required to establish the desired terminal characteristics—the details of which will be investigated in your electronics courses.

In Fig. 8.91, the designed 15 V dc supply, biasing resistors, and transistor in the upper right corner of the schematic establish a constant 4 mA current through the circuit. It is referred to as a *constant-current source* because the current remains fairly constant at 4 mA even though there may be moderate variations in the total resistance of the series sensor circuit connected to the transistor. Following the 4 mA through the circuit, we find that it enters terminal 2 (positive side of the input) of the op-amp. A second current of 2 mA, called the *reference current,* is established by the 15 V source and resistance R and enters terminal 3 (negative side of the input) of the op-amp. The reference current of 2 mA is necessary to establish a current for the 4 mA current of the network to be compared against. As long as the 4 mA current exists, the operational amplifier provides a "high" output voltage that exceeds 13.5 V, with a typical level of 14.2 V (according to the specification sheet for the op-amp). However, if the sensor current drops from 4 mA to a level below the reference level of 2 mA, the op-amp responds with a "low" output voltage that is typically about 0.1 V. The output of the operational amplifier then signals the alarm circuit about the disturbance. Note from the above that it is not necessary for the sensor current to drop to 0 mA to signal the alarm circuit—just a variation around the reference level that appears unusual.

One very important characteristic of this particular op-amp is that the input impedance to the op-amp is relatively low. This feature is

important because you don't want alarm circuits reacting to every voltage spike or turbulence that comes down the line because of external switching action or outside forces such as lightning. In Fig. 8.92(c), for instance, if a high voltage should appear at the input to the series configuration, most of the voltage would be absorbed by the series resistance of the sensor circuit rather than travel across the input terminals of the operational amplifier—thus preventing a false output and an activation of the alarm.

Wheatstone Bridge Smoke Detector

The Wheatstone bridge is a popular network configuration whenever detection of small changes in a quantity is required. In Fig. 8.93(a), the dc

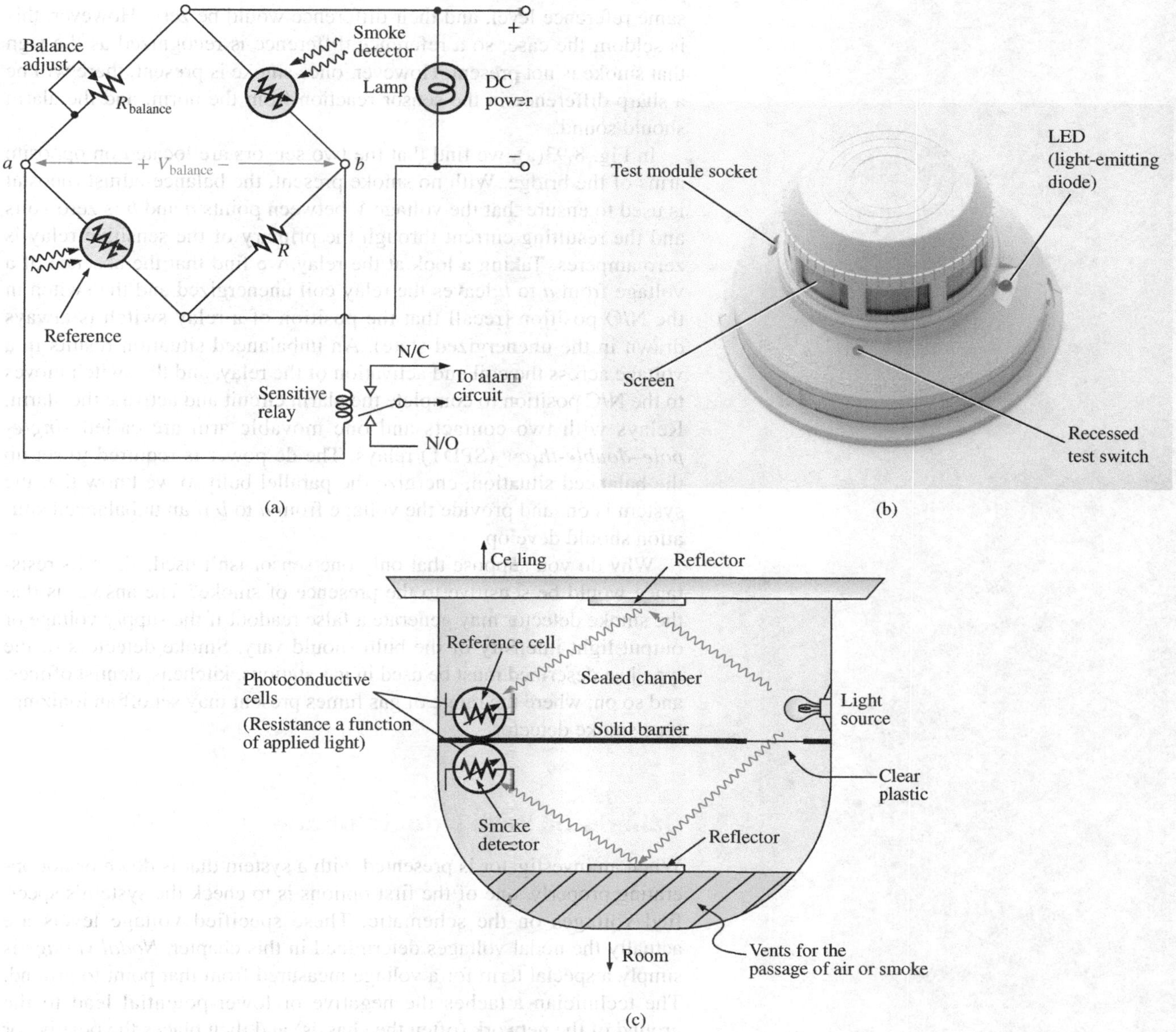

(a)

(b)

(c)

FIG. 8.93
Wheatstone bridge smoke detector: (a) dc bridge configuration; (b) outside appearance; (c) internal construction.

bridge configuration uses a photoelectric device to detect the presence of smoke and to sound the alarm. A photograph of a photoelectric smoke detector appears in Fig. 8.93(b), and the internal construction of the unit is shown in Fig. 8.93(c). First, note that air vents are provided to permit the smoke to enter the chamber below the clear plastic. The clear plastic prevents the smoke from entering the upper chamber but permits the light from the bulb in the upper chamber to bounce off the lower reflector to the semiconductor light sensor (a cadmium photocell) at the left side of the chamber. The clear plastic separation ensures that the light hitting the light sensor in the upper chamber is not affected by the entering smoke. It establishes a reference level to compare against the chamber with the entering smoke. If no smoke is present, the difference in response between the sensor cells will be registered as the normal situation. Of course, if both cells were exactly identical, and if the clear plastic did not cut down on the light, both sensors would establish the same reference level, and their difference would be zero. However, this is seldom the case, so a reference difference is recognized as the sign that smoke is not present. However, once smoke is present, there will be a sharp difference in the sensor reaction from the norm, and the alarm should sound.

In Fig. 8.93(a), we find that the two sensors are located on opposite arms of the bridge. With no smoke present, the balance-adjust rheostat is used to ensure that the voltage V between points a and b is zero volts and the resulting current through the primary of the sensitive relay is zero amperes. Taking a look at the relay, we find that the absence of a voltage from a to b leaves the relay coil unenergized and the switch in the N/O position (recall that the position of a relay switch is always drawn in the unenergized state). An unbalanced situation results in a voltage across the coil and activation of the relay, and the switch moves to the N/C position to complete the alarm circuit and activate the alarm. Relays with two contacts and one movable arm are called *single-pole–double-throw* (SPDT) relays. The dc power is required to set up the balanced situation, energize the parallel bulb so we know that the system is on, and provide the voltage from a to b if an unbalanced situation should develop.

Why do you suppose that only one sensor isn't used, since its resistance would be sensitive to the presence of smoke? The answer is that the smoke detector may generate a false readout if the supply voltage or output light intensity of the bulb should vary. Smoke detectors of the type just described must be used in gas stations, kitchens, dentist offices, and so on, where the range of gas fumes present may set off an ionizing-type smoke detector.

Schematic with Nodal Voltages

When an investigator is presented with a system that is down or not operating properly, one of the first options is to check the system's specified voltages on the schematic. These specified voltage levels are actually the nodal voltages determined in this chapter. *Nodal voltage* is simply a special term for a voltage measured from that point to ground. The technician attaches the negative or lower-potential lead to the ground of the network (often the chassis) and then places the positive or higher-potential lead on the specified points of the network to check the nodal voltages. If they match, it means that section of the system is op-

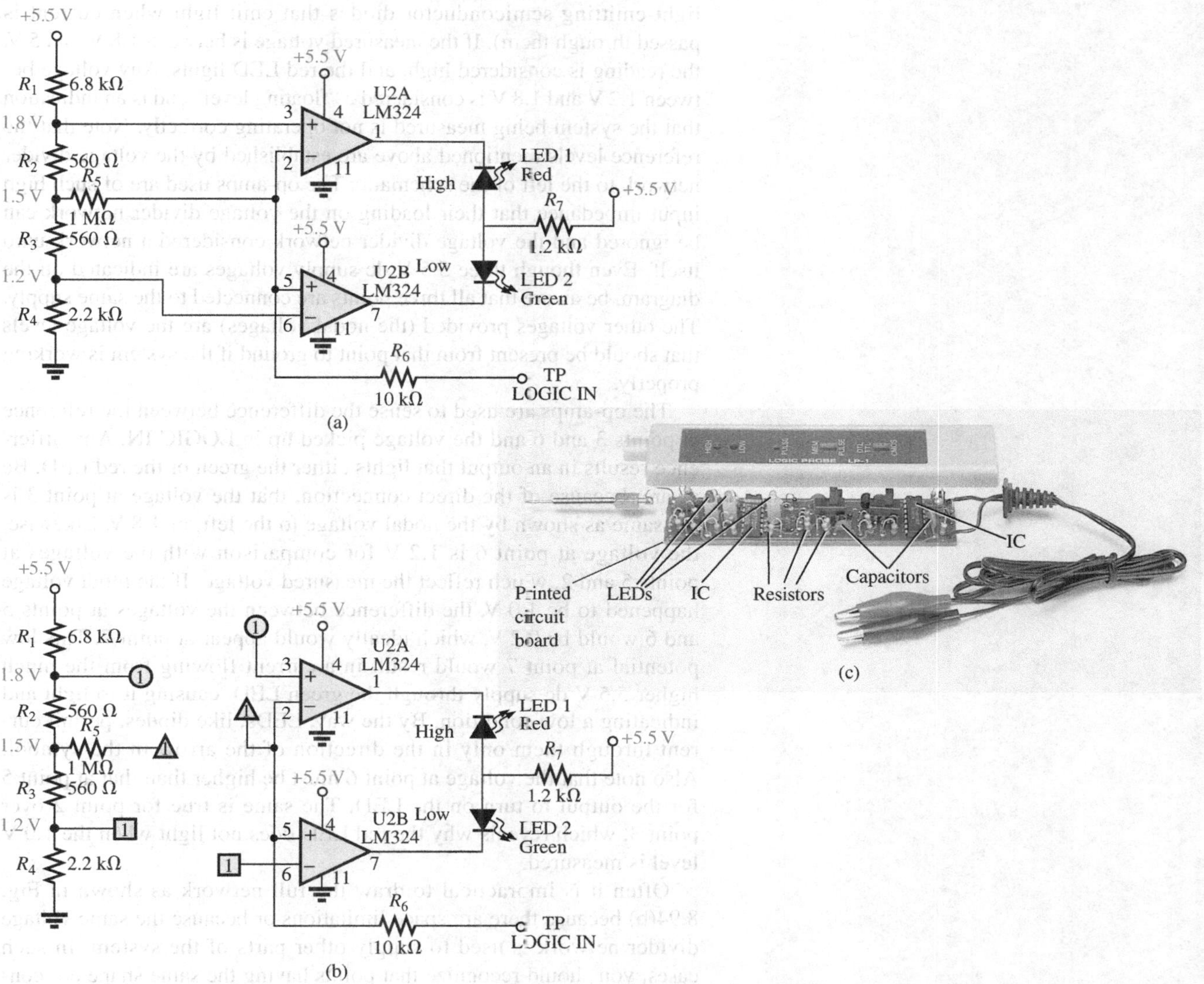

FIG. 8.94

Logic probe: (a) schematic with nodal voltages; (b) network with global connections; (c) photograph of commercially available unit.

erating properly. If one or more fail to match the given values, the problem area can usually be identified. Be aware that a reading of −15.87 V is significantly different from an expected reading of +16 V if the leads have been properly attached. Although the actual numbers seem close, the difference is actually more than 30 V. You must expect some deviation from the given value as shown, but always be very sensitive to the resulting sign of the reading.

The schematic in Fig. 8.94(a) includes the nodal voltages for a logic probe used to measure the input and output states of integrated circuit logic chips. In other words, the probe determines whether the measured voltage is one of two states: high or low (often referred to as "on" or "off" or 1 or 0). If the LOGIC IN terminal of the probe is placed on a chip at a location where the voltage is between 0 and 1.2 V, the voltage is considered to be a low level, and the green LED lights (LEDs are

light-emitting semiconductor diodes that emit light when current is passed through them). If the measured voltage is between 1.8 V and 5 V, the reading is considered high, and the red LED lights. Any voltage between 1.2 V and 1.8 V is considered a "floating level" and is an indication that the system being measured is not operating correctly. Note that the reference levels mentioned above are established by the voltage divider network to the left of the schematic. The op-amps used are of such high input impedance that their loading on the voltage divider network can be ignored and the voltage divider network considered a network unto itself. Even though three 5.5 V dc supply voltages are indicated on the diagram, be aware that all three points are connected to the same supply. The other voltages provided (the nodal voltages) are the voltage levels that should be present from that point to ground if the system is working properly.

The op-amps are used to sense the difference between the reference at points 3 and 6 and the voltage picked up in LOGIC IN. Any difference results in an output that lights either the green or the red LED. Be aware, because of the direct connection, that the voltage at point 3 is the same as shown by the nodal voltage to the left, or 1.8 V. Likewise, the voltage at point 6 is 1.2 V for comparison with the voltages at points 5 and 2, which reflect the measured voltage. If the input voltage happened to be 1.0 V, the difference between the voltages at points 5 and 6 would be 0.2 V, which ideally would appear at point 7. This low potential at point 7 would result in a current flowing from the much higher 5.5 V dc supply through the green LED, causing it to light and indicating a low condition. By the way, LEDs, like diodes, permit current through them only in the direction of the arrow in the symbol. Also note that the voltage at point 6 must be higher than that at point 5 for the output to turn on the LED. The same is true for point 2 over point 3, which reveals why the red LED does not light when the 1.0 V level is measured.

Often it is impractical to draw the full network as shown in Fig. 8.94(b) because there are space limitations or because the same voltage divider network is used to supply other parts of the system. In such cases, you should recognize that points having the same shape are connected, and the number in the figure reveals how many connections are made to that point.

A photograph of the outside and inside of a commercially available logic probe is shown in Fig. 8.94(c). Note the increased complexity of system because of the variety of functions that the probe can perform.

8.14 COMPUTER ANALYSIS

PSpice

We will now analyze the bridge network in Fig. 8.67 using PSpice to ensure that it is in the balanced state. The only component that has not been introduced in earlier chapters is the dc current source. To obtain it, first select the **Place a part** key and then the **SOURCE** library. Scrolling the **Part List** results in the option **IDC.** A left click of **IDC** followed by **OK** results in a dc current source whose direction is toward the bottom of the screen. One left click (to make it red, or active) followed by a right click results in a listing having a **Mirror Vertically** option. Selecting that option flips the source and gives it the direction in Fig. 8.67.

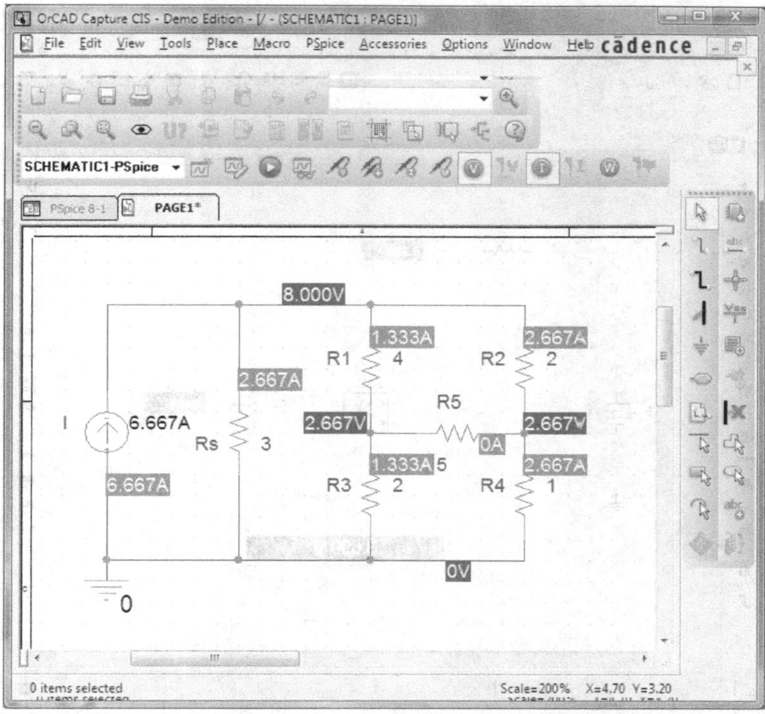

FIG. 8.95
Applying PSpice to the bridge network in Fig. 8.67.

The remaining parts of the PSpice analysis are pretty straightforward, with the results in Fig. 8.95 matching those obtained in the analysis of Fig. 8.67. The voltage across the current source is 8 V positive to ground, and the voltage at either end of the bridge arm is 2.667 V. The voltage across R_5 is obviously 0 V for the level of accuracy displayed, and the current is such a small magnitude compared to the other current levels of the network that it can essentially be considered 0 A. Note also for the balanced bridge that the current through R_1 equals that of R_3, and the current through R_2 equals that of R_4.

Multisim

We will now use Multisim to verify the results in Example 8.18. All the elements of creating the schematic in Fig. 8.96 have been presented in earlier chapters; they are not repeated here to demonstrate how little documentation is now necessary to carry you through a fairly complex network.

For the analysis, both the standard **Multimeter** and meters from the **Show Measurement Family** of the **BASIC toolbar** listing were employed. For the current through the resistor R_5, the **Place Ammeter (Horizontal)** was used, while for the voltage across the resistor R_4, the **Place Voltmeter (Vertical)** was used. The **Multimeter** is reading the voltage across the resistor R_2. In actuality, the ammeter is reading the loop current for the top window, and the voltmeters are showing the nodal voltages of the network.

After simulation, the results displayed are an exact match with those in Example 8.18.

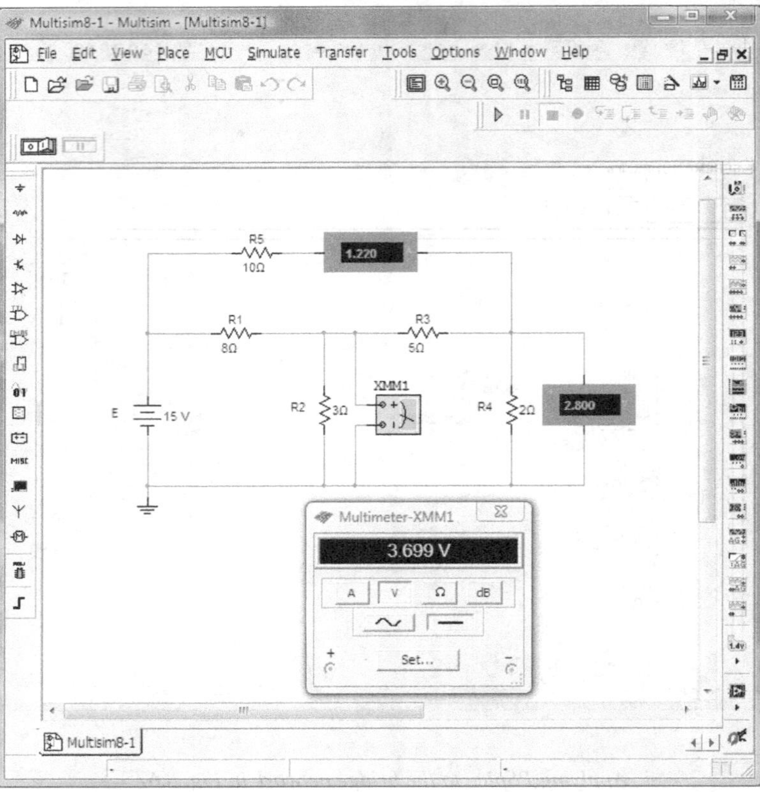

FIG. 8.96
Using Multisim to verify the results in Example 8.18.

PROBLEMS

SECTION 8.2 Current Sources

1. For the network of Fig. 8.97:
   a. Find the currents I_1 and I_2.
   b. Determine the voltage V_s.

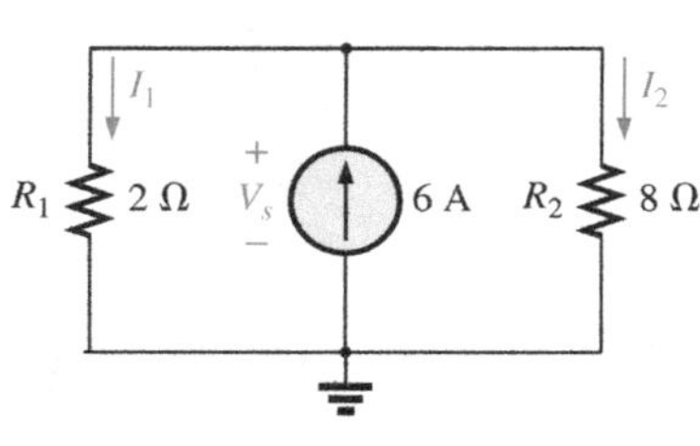

FIG. 8.97
Problem 1.

2. For the network of Fig. 8.98:
   a. Determine the currents I_1 and I_2.
   b. Calculate the voltages V_2 and V_s.

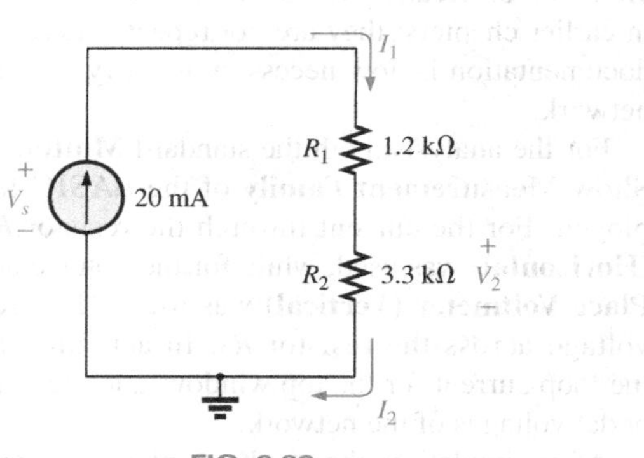

FIG. 8.98
Problem 2.

3. Find voltage V_s (with polarity) across the ideal current source in Fig. 8.99.

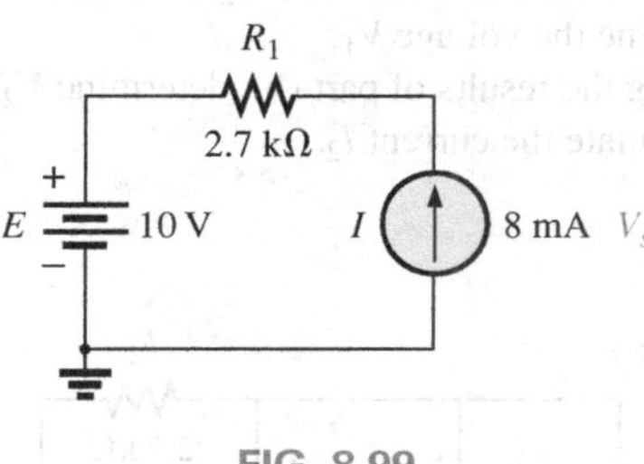

FIG. 8.99
Problem 3.

4. For the network in Fig. 8.100:
   a. Find voltage V_s.
   b. Calculate current I_2.
   c. Find the source current I_s.

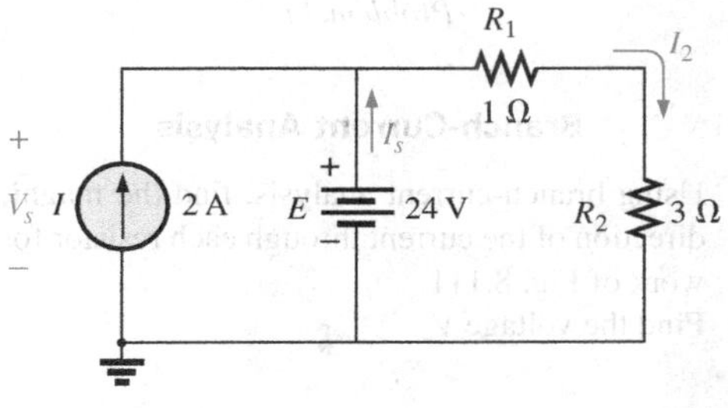

FIG. 8.100
Problem 4.

5. Find the voltage V_3 and the current I_2 for the network in Fig. 8.101.

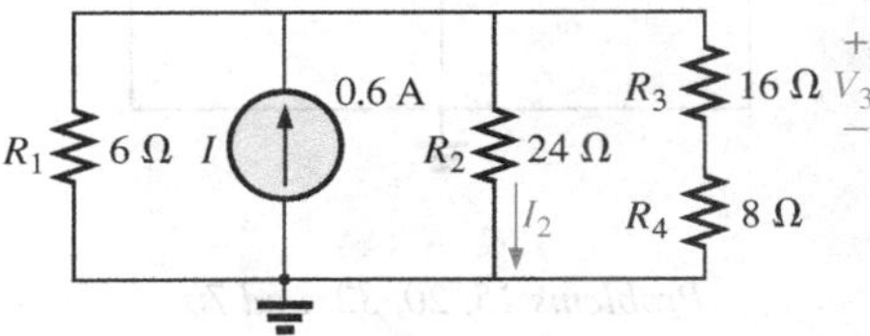

FIG. 8.101
Problem 5.

6. For the network in Fig. 8.102:
   a. Find the currents I_1 and I_s.
   b. Find the voltages V_s and V_3.

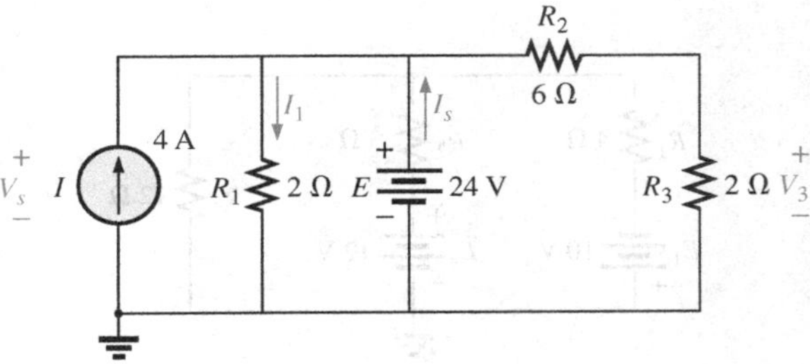

FIG. 8.102
Problem 6.

SECTION 8.3 Source Conversions

7. Convert the voltage sources in Fig. 8.103 to current sources.

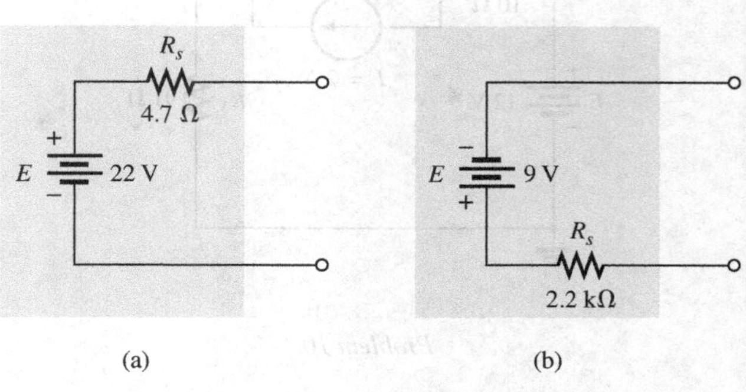

FIG. 8.103
Problem 7.

8. Convert the current sources in Fig. 8.104 to voltage sources.

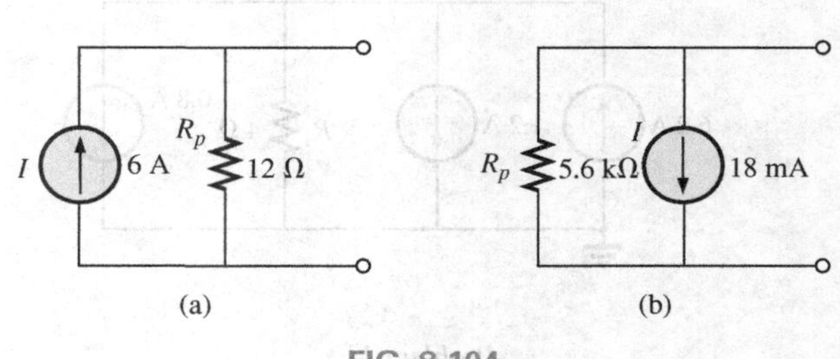

FIG. 8.104
Problem 8.

9. For the network in Fig. 8.105:
   a. Find the current through the 10 Ω resistor. Nothing that the resistance R_L is significantly less than R_p, what was the impact on the current through R_L?
   b. Convert the current source to a voltage source, and recalculate the current through the 10 Ω resistor.
   Did you obtain the same result?

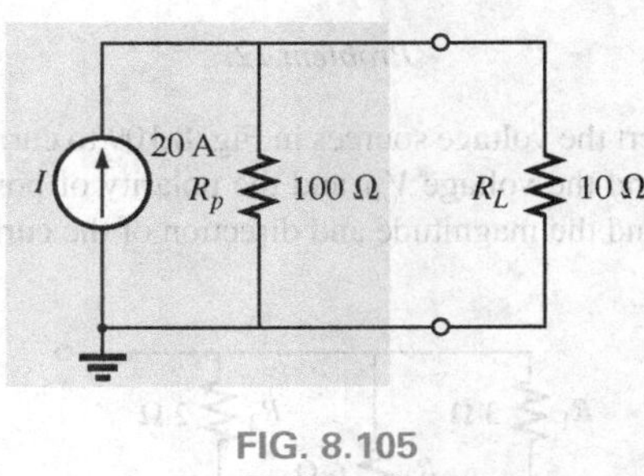

FIG. 8.105
Problem 9.

10. For the configuration of Fig. 8.106:
    a. Convert the current source to a voltage source.
    b. Combine the two series voltage sources into one source.
    c. Calculate the current through the 91 Ω resistor.

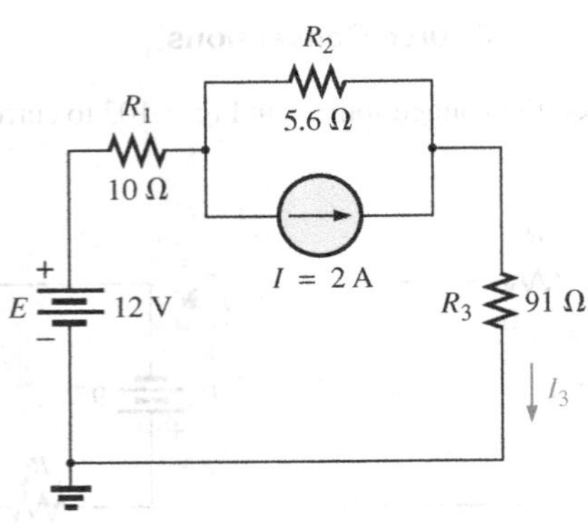

FIG. 8.106
Problem 10.

SECTION 8.4 Current Sources in Parallel

11. For the network in Fig. 8.107:

a. Replace all the current sources by a single current source.

b. Find the source voltage V_s.

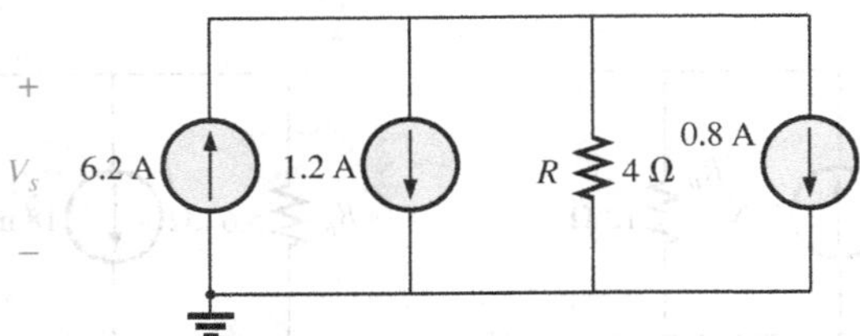

FIG. 8.107
Problem 11.

12. Find the voltage V_s and the current I_1 for the network in Fig. 8.108.

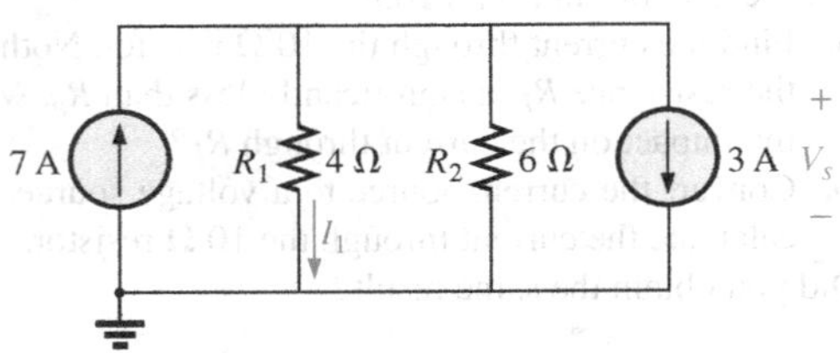

FIG. 8.108
Problem 12.

13. Convert the voltage sources in Fig. 8.109 to current sources.

a. Find the voltage V_{ab} and the polarity of points a and b.

b. Find the magnitude and direction of the current I_3.

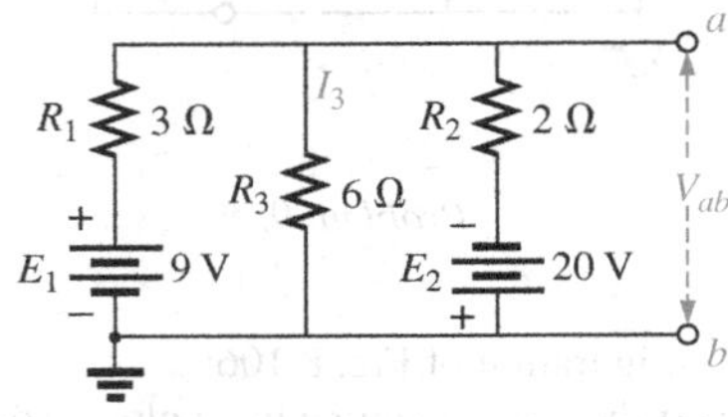

FIG. 8.109
Problems 13 and 37.

14. For the network in Fig. 8.110:

a. Convert the voltage source to a current source.

b. Reduce the network to a single current source, and determine the voltage V_1.

c. Using the results of part (b), determine V_2.

d. Calculate the current I_2.

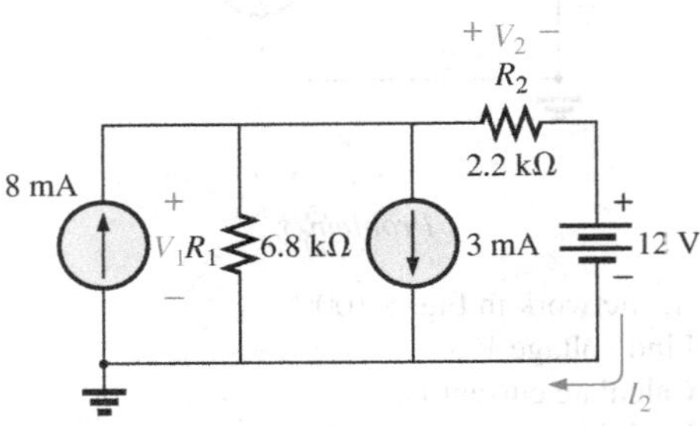

FIG. 8.110
Problem 14.

SECTION 8.6 Branch-Current Analysis

15. a. Using branch-current analysis, find the magnitude and direction of the current through each resistor for the network of Fig. 8.111.

b. Find the voltage V_a.

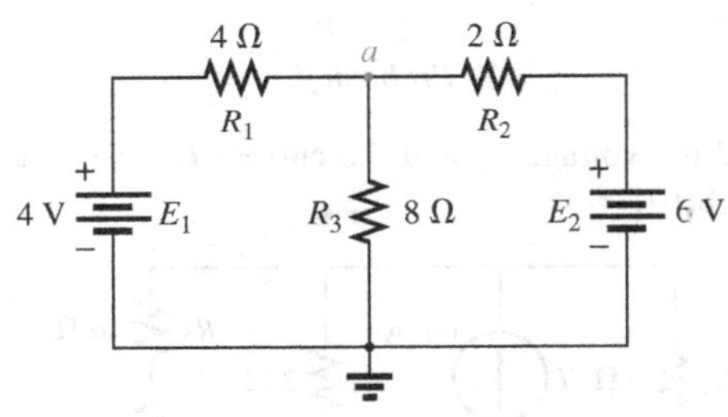

FIG. 8.111
Problems 15, 20, 32, and 70.

16. For the network of Fig. 8.112:

a. Determine the current through the 12 Ω resistor using branch-current analysis.

b. Convert the two voltage sources to current sources, and then determine the current through the 12 Ω resistor.

c. Compare the results of parts (a) and (b).

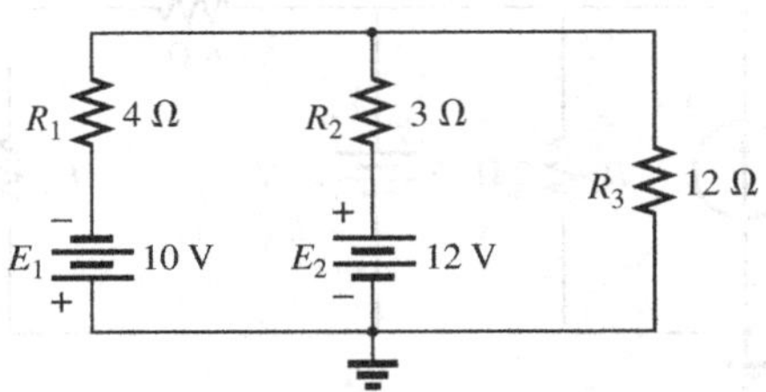

FIG. 8.112
Problems 16, 21, and 33.

*17. Using branch-current analysis, find the current through each resistor for the network of Fig. 8.113. The resistors are all standard values.

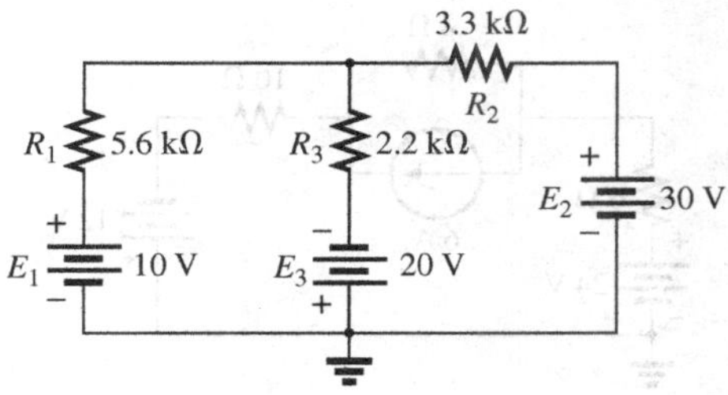

FIG. 8.113
Problems 17, 22, and 34.

*18. a. Using branch-current analysis, find the current through the 9.1 kΩ resistor in Fig. 8.114. Note that all the resistors are standard values.
b. Using the results of part (a), determine the voltage V_a.

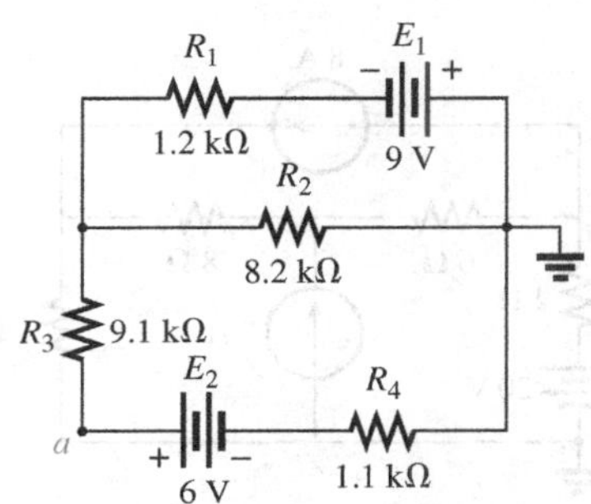

FIG. 8.114
Problems 18 and 23.

*19. For the network in Fig. 8.115:
a. Write the equations necessary to solve for the branch currents.
b. By substitution of Kirchhoff's current law, reduce the set to three equations.
c. Rewrite the equations in a format that can be solved using third-order determinants.
d. Solve for the branch current through the resistor R_3.

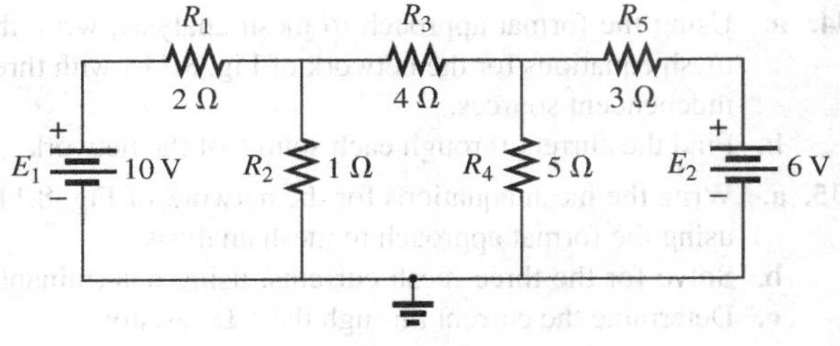

FIG. 8.115
Problems 19, 24, and 35.

SECTION 8.7 Mesh Analysis (General Approach)

20. a. Using the general approach to mesh analysis, determine the current through each resistor of Fig. 8.111.
b. Using the results of part (a), find the voltage V_a.

21. a. Using the general approach to mesh analysis, determine the current through each voltage source in Fig. 8.112.
b. Using the results of part (a), find the power delivered by the source E_2 and to the resistor R_3.

22. a. Using the general approach to mesh analysis, determine the current through each resistor of Fig. 8.113.
b. Using the results of part (a), determine the voltage across the 3.3 kΩ resistor.

23. a. Using the general approach to mesh analysis, determine the current through each resistor of Fig. 8.114.
b. Using the results of part (a), find the voltage V_a.

*24. a. Determine the mesh currents for the network of Fig. 8.115 using the general approach.
b. Through the proper use of Kirchhoff's current law, reduce the resulting set of equations to three.
c. Use determinants to find the three mesh currents.
d. Determine the current through each source, using the results of part (c).

*25. a. Write the mesh equations for the network of Fig. 8.116 using the general approach.
b. Using determinants, calculate the mesh currents.
c. Using the results of part (b), find the current through each source.

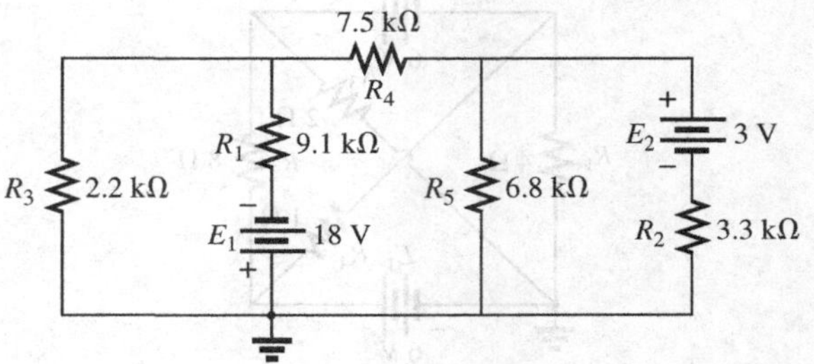

FIG. 8.116
Problems 25 and 36.

*26. a. Write the mesh equations for the network of Fig. 8.117 using the general approach.
b. Using determinants, calculate the mesh currents.
c. Using the results of part (b), calculate the current through the resistor R_5.

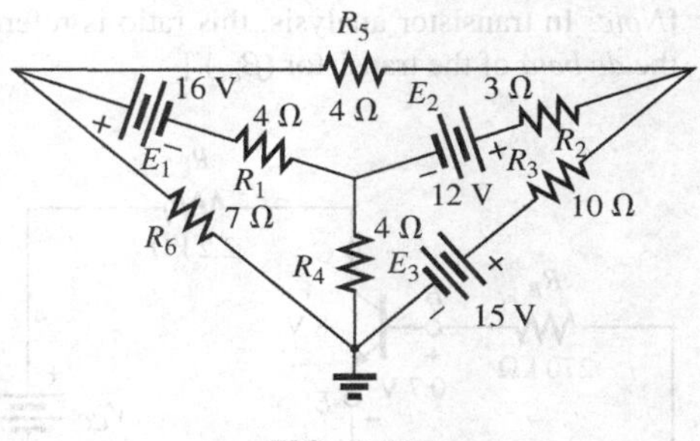

FIG. 8.117
Problem 26.

*27. a. Write the mesh currents for the network of Fig. 8.118 using the general approach.
b. Using determinants, calculate the mesh currents.
c. Using the results of part (b), find the power delivered by the 6 V source.

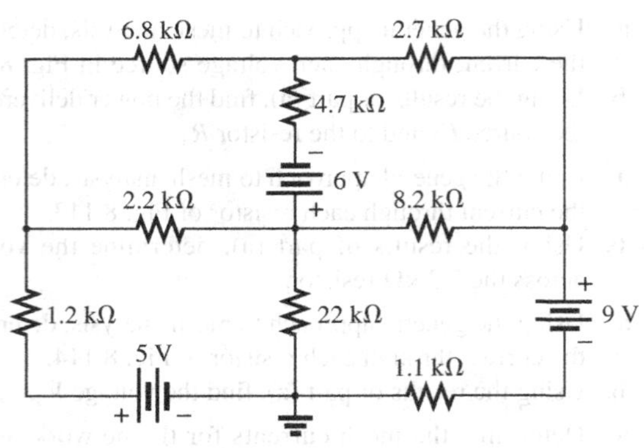

FIG. 8.118
Problems 27, 38, and 71.

***28. a.** Redraw the network of Fig. 8.119 in a manner that will remove the crossover.

b. Write the mesh equations for the network using the general approach.

c. Calculate the mesh currents for the network.

d. Find the total power delivered by the two sources.

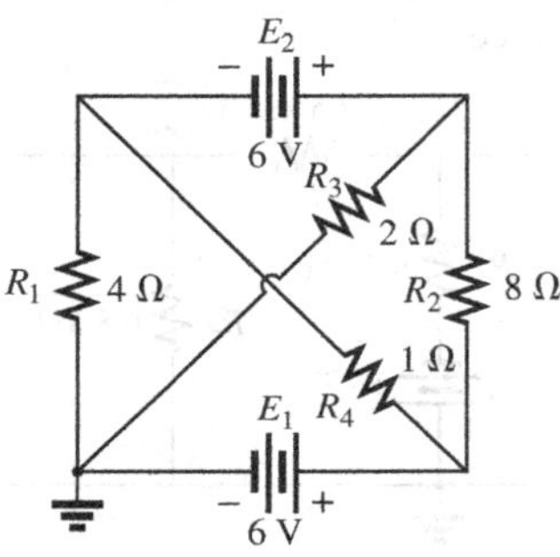

FIG. 8.119
Problem 28.

***29.** For the transistor configuration in Fig. 8.120:

a. Solve for the currents I_B, I_C, and I_E, using the fact that $V_{BE} = 0.7$ V and $V_{CE} = 8$ V.

b. Find the voltages V_B, V_C, and V_E with respect to ground.

c. What is the ratio of output current I_C to input current I_B? [*Note:* In transistor analysis, this ratio is referred to as the *dc beta* of the transistor (β_{dc}).]

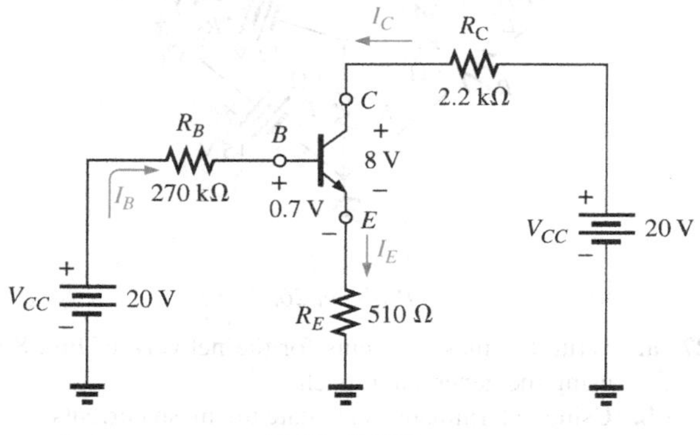

FIG. 8.120
Problem 29.

***30.** Using the supermesh approach, find the current through each element of the network of Fig. 8.121.

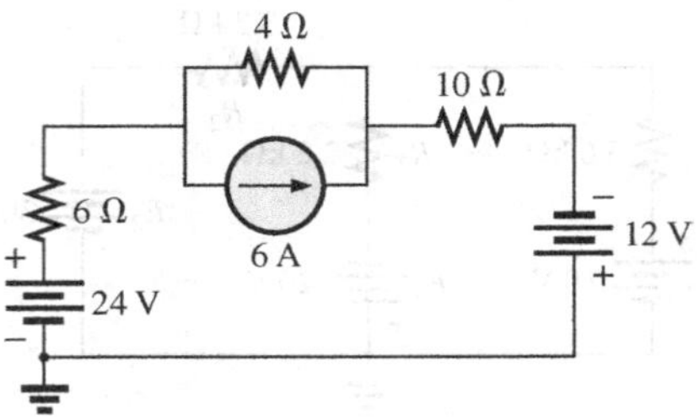

FIG. 8.121
Problem 30.

***31.** Using the supermesh approach, find the current through each element of the network of Fig. 8.122.

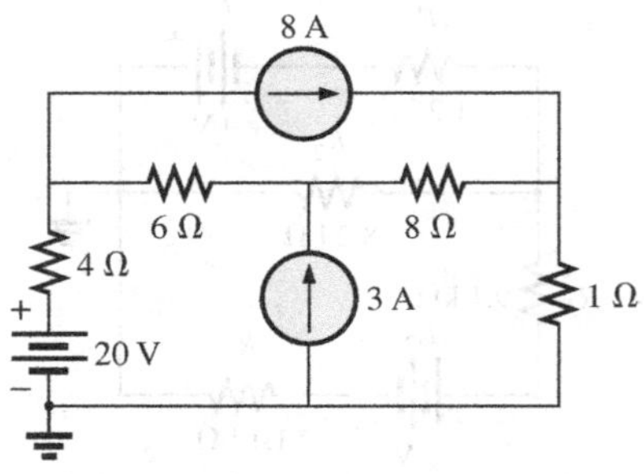

FIG. 8.122
Problem 31.

SECTION 8.8 **Mesh Analysis (Format Approach)**

32. a. Using the format approach to mesh analysis, write the mesh equations for the network of Fig. 8.111.

b. Solve for the current through the 8 Ω resistor.

33. a. Using the format approach to mesh analysis, write the mesh equations for the network of Fig. 8.112.

b. Solve for the current through the 3 Ω resistor.

34. a. Using the format approach to mesh analysis, write the mesh equations for the network of Fig. 8.113 with three independent sources.

b. Find the current through each source of the network.

***35. a.** Write the mesh equations for the network of Fig. 8.115 using the format approach to mesh analysis.

b. Solve for the three mesh currents, using determinants.

c. Determine the current through the 1 Ω resistor.

***36. a.** Write the mesh equations for the network of Fig. 8.116 using the format approach to mesh analysis.

b. Solve for the three mesh currents, using determinants.

c. Find the current through each source of the network.

37. a. Write the mesh equations for the network of Fig. 8.109 using the format approach.

b. Find the voltage V_{ab} using the result of part (a).

*38. a. Write the mesh equations for the network of Fig. 8.18 using the format approach to mesh analysis.
b. Solve for the four mesh currents using determinants.
c. Find the voltage at the common connection at the center of the diagram.

*39. a. Write the mesh equations for the network of Fig. 8.123 using the format approach to mesh analysis.
b. Use determinants to determine the mesh currents.
c. Find the voltages V_a and V_b.
d. Determine the voltage V_{ab}.

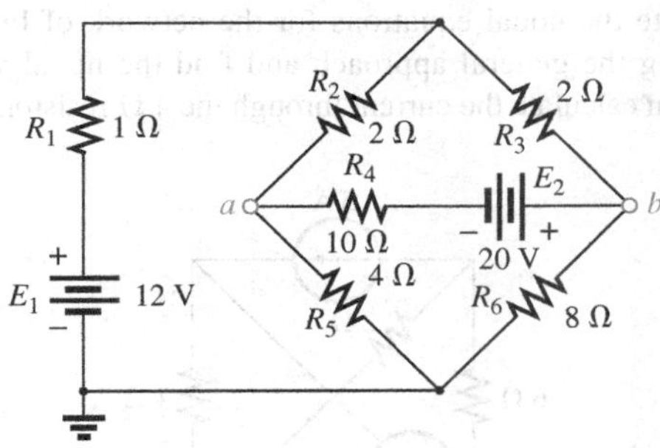

FIG. 8.123
Problems 39 and 56.

SECTION 8.9 Nodal Analysis (General Approach)

40. a. Write the nodal equations using the general approach for the network of Fig. 8.124.
b. Find the nodal voltages using determinants.
c. Use the results of part (b) to find the voltage across the 8 Ω resistor.
d. Use the results of part (b) to find the current through the 2 Ω and 4 Ω resistors.

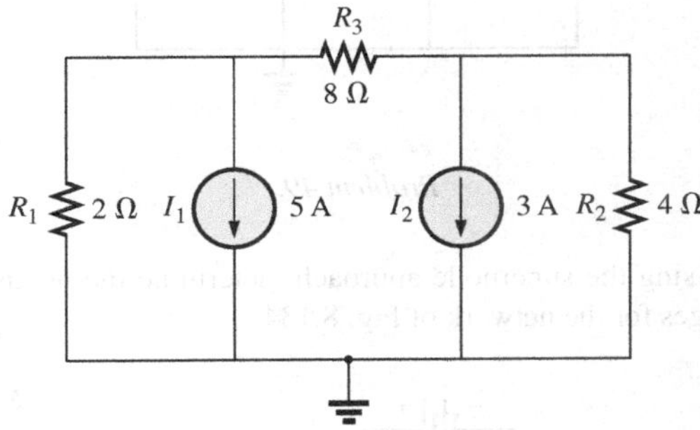

FIG. 8.124
Problems 40 and 51.

41. a. Write the nodal equations using the general approach for the network of Fig. 8.125.
b. Find the nodal voltages using determinants.
c. Using the results of part (a), calculate the current through the 20 Ω resistor.

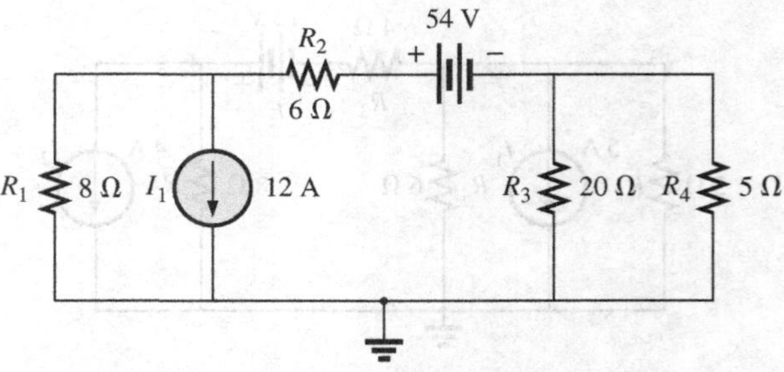

FIG. 8.125
Problems 41 and 52.

42. a. Write the nodal equations using the general approach for the network of Fig. 8.126.
b. Find the nodal voltages using determinants.
c. What is the total power supplied by the current sources?

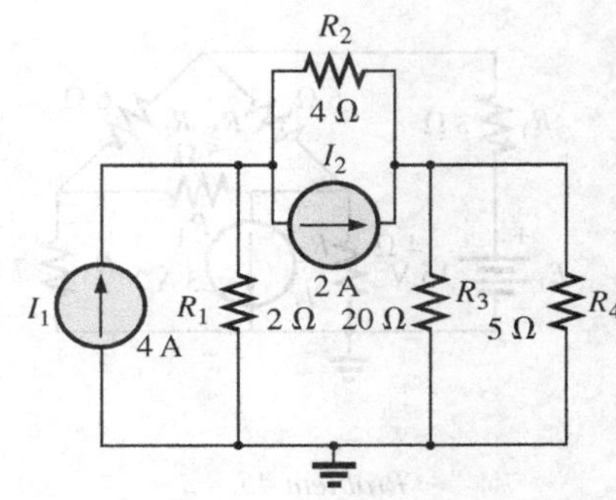

FIG. 8.126
Problem 42.

*43. a. Write the nodal equations for the network of Fig. 8.127 using the general approach.
b. Using determinants, solve for the nodal voltages.
c. Determine the magnitude and polarity of the voltage across each resistor.

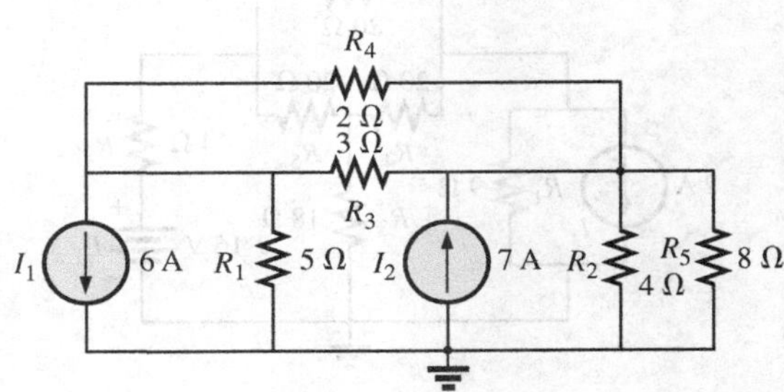

FIG. 8.127
Problem 43.

*44. a. Write the nodal equations for the network of Fig. 8.128 using the general approach.
b. Solve for the nodal voltages using determinants.
c. Find the current through the 6 Ω resistor.

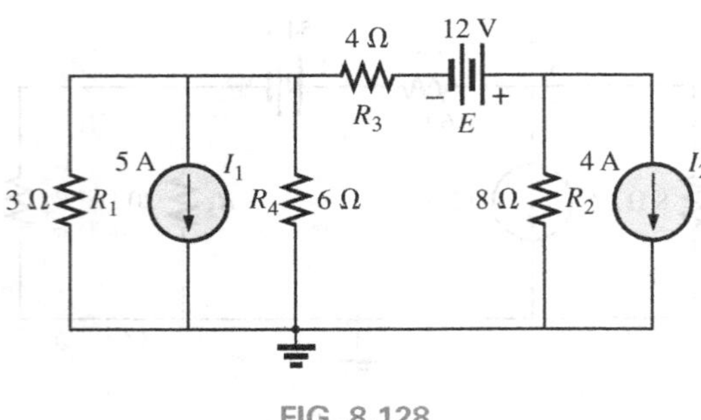

FIG. 8.128
Problem 44.

***45. a.** Write the nodal equations for the network of Fig. 8.129 using the general approach.
b. Solve for the nodal voltages using determinants.
c. Find the voltage across the 5 Ω resistor.

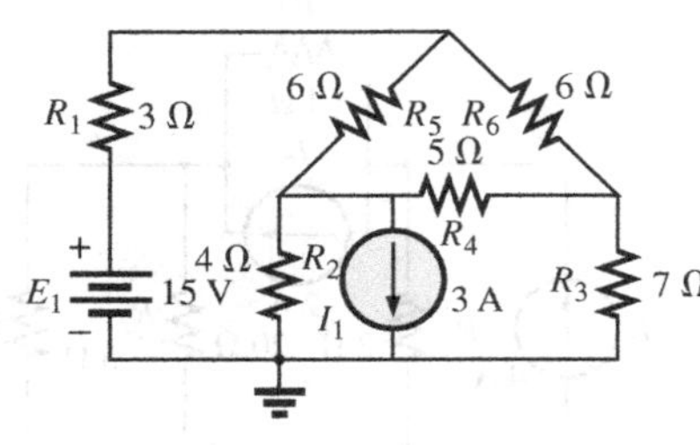

FIG. 8.129
Problem 45.

***46. a.** Write the nodal equations for the network of Fig. 8.130 using the general approach.
b. Solve for the nodal voltages using determinants.
c. Find the voltage across the resistor R_6.

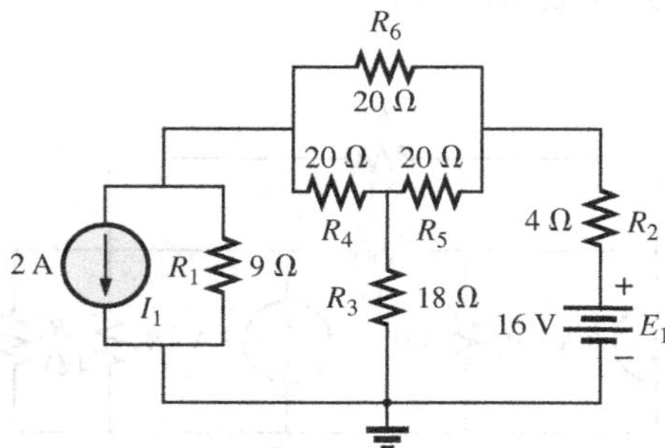

FIG. 8.130
Problems 46 and 53.

***47. a.** Write the nodal equations for the network of Fig. 8.131 using the general approach.
b. Find the nodal voltages using determinants.
c. Determine the current through the 9 Ω resistor.

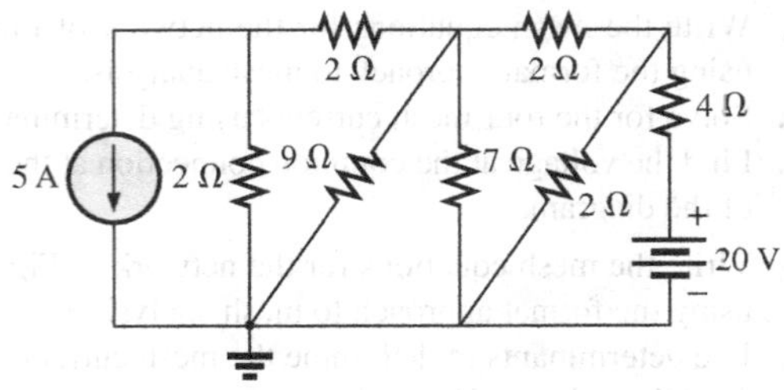

FIG. 8.131
Problems 47, 54, and 72.

***48.** Write the nodal equations for the network of Fig. 8.132 using the general approach and find the nodal voltages. Then calculate the current through the 4 Ω resistor.

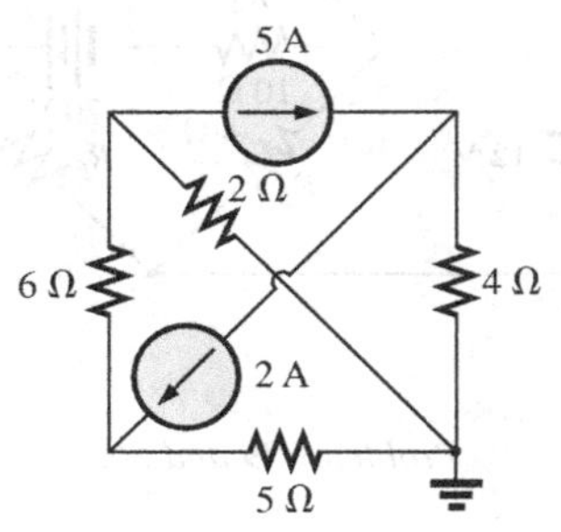

FIG. 8.132
Problems 48 and 55.

***49.** Using the supernode approach, determine the nodal voltages for the network of Fig. 8.133.

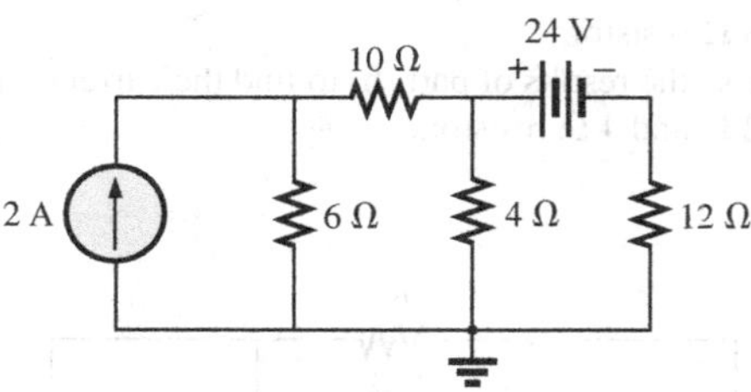

FIG. 8.133
Problem 49.

***50.** Using the supernode approach, determine the nodal voltages for the network of Fig. 8.134.

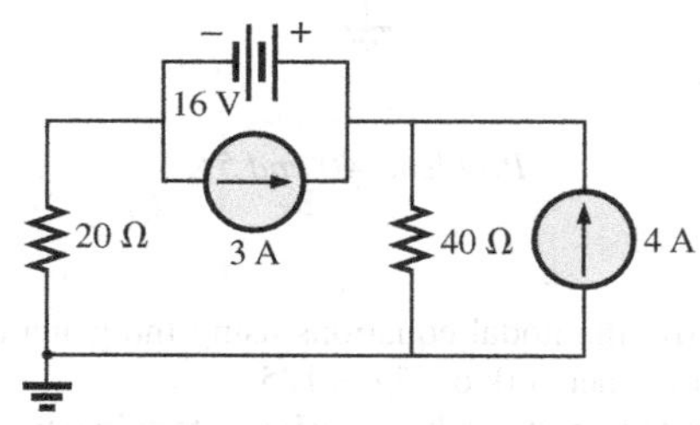

FIG. 8.134
Problem 50.

SECTION 8.10 Nodal Analysis (Format Approach)

51. **a.** Determine the nodal voltages of Fig. 8.124 using the format approach to nodal analysis.
b. Then find the voltage across each current source.

52. **a.** Convert the voltage source of Fig 8.125 to a current source, and then find the nodal voltages using the format approach to nodal analysis.
b. Use the results of part (a) to find the voltage across the 6 Ω resistor of Fig. 8.125.

***53.** **a.** Convert the voltage source of Fig. 8.130 to a current source, and then apply the format approach to nodal analysis to find the nodal voltages.
b. Use the results of part (a) to find the current through the 4 Ω resistor.

***54.** **a.** Convert the voltage source of Fig. 8.131 to a current source, and then apply the format approach to nodal analysis to find the nodal voltages.
b. Use the results of part (a) to find the current through the 9 Ω resistor.

***55.** **a.** Using the format approach, find the nodal voltages of Fig. 8.132 using nodal analysis.
b. Using the results of part (a), find the current through the 2 Ω resistor.

***56.** **a.** Convert the voltage sources of Fig. 8.123 to current sources, and then find the nodal voltages of the resulting network using the format approach to nodal analysis.
b. Using the results of part (a), find the voltage between points *a* and *b*.

SECTION 8.11 Bridge Networks

57. For the bridge network in Fig. 8.135:
a. Write the mesh equations using the format approach.
b. Determine the current through R_5.
c. Is the bridge balanced?
d. Is Eq. (8.2) satisfied?

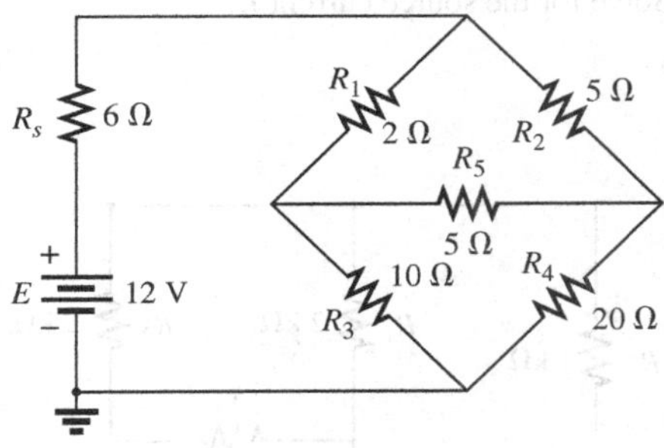

FIG. 8.135
Problems 57 and 58.

58. For the network in Fig. 8.135:
a. Write the nodal equations using the format approach.
b. Determine the voltage across R_5.
c. Is the bridge balanced?
d. Is Eq. (8.2) satisfied?

59. For the bridge in Fig. 8.136:
a. Write the mesh equations using the format approach.
b. Determine the current through R_5.
c. Is the bridge balanced?
d. Is Eq. (8.2) satisfied?

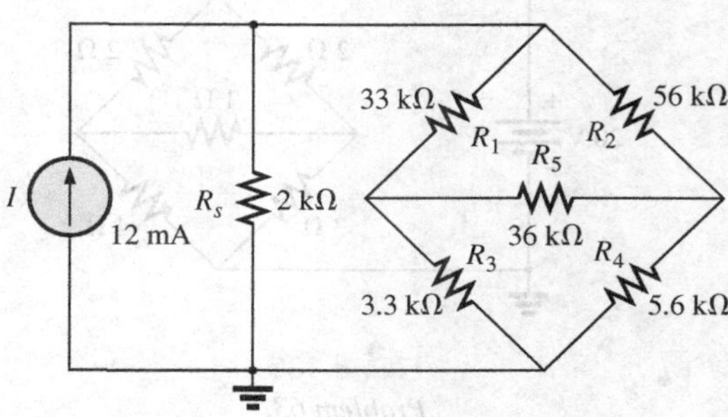

FIG. 8.136
Problems 59 and 60.

60. For the bridge network in Fig. 8.136:
a. Write the nodal equations using the format approach.
b. Determine the current across R_5.
c. Is the bridge balanced?
d. Is Eq. (8.2) satisfied?

***61.** Determine the current through the source resistor R_s in Fig. 8.137 using either mesh or nodal analysis. Explain why you chose one method over the other.

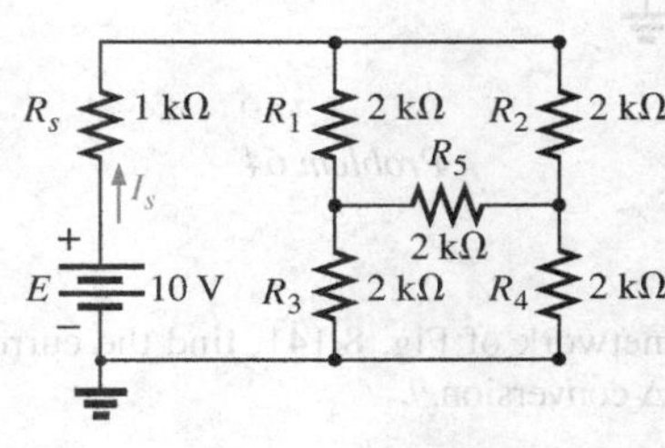

FIG. 8.137
Problem 61.

***62.** Repeat Problem 61 for the network of Fig. 8.138.

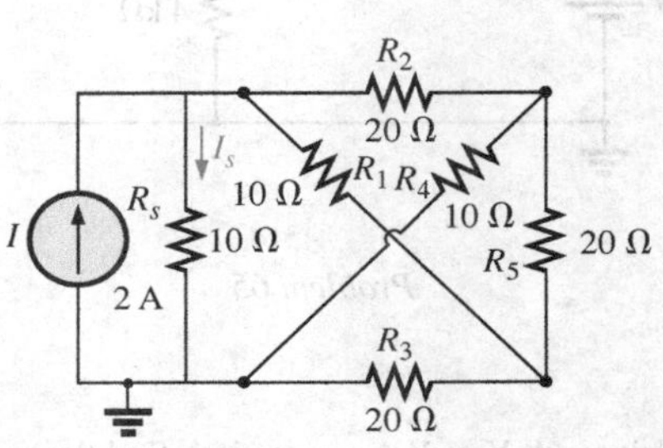

FIG. 8.138
Problem 62.

SECTION 8.12 Y-Δ (T-π) and Δ-Y (π-T) Conversions

63. Using a Δ-Y or Y-Δ conversion, find the current I for the network of Fig. 8.139.

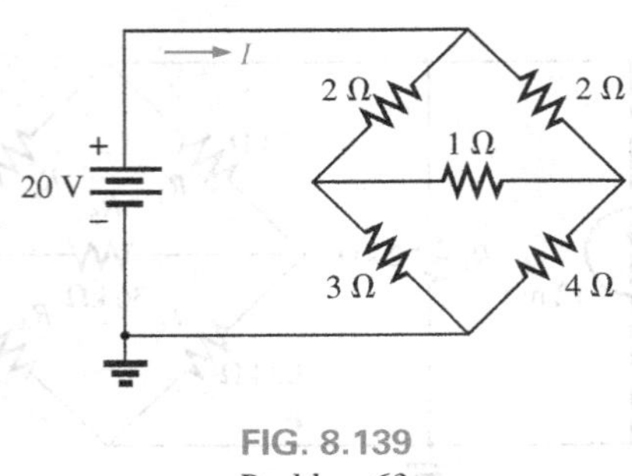

FIG. 8.139
Problem 63.

64. Convert the Δ of 6.8 kΩ resistors in Fig. 8.140 to a T configuration and find the current I.

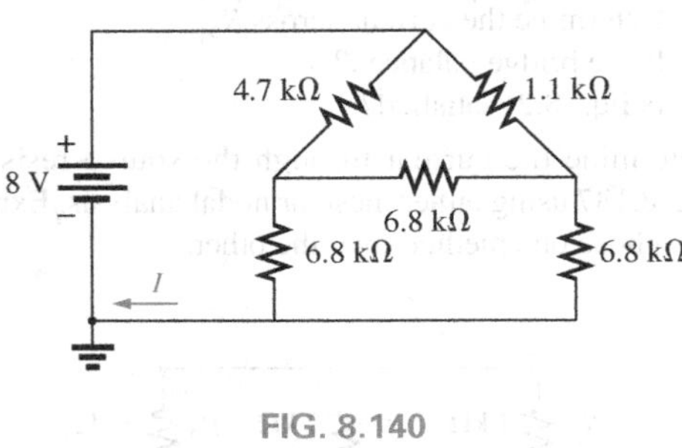

FIG. 8.140
Problem 64.

65. For the network of Fig. 8.141, find the current I without using Y-Δ conversion.

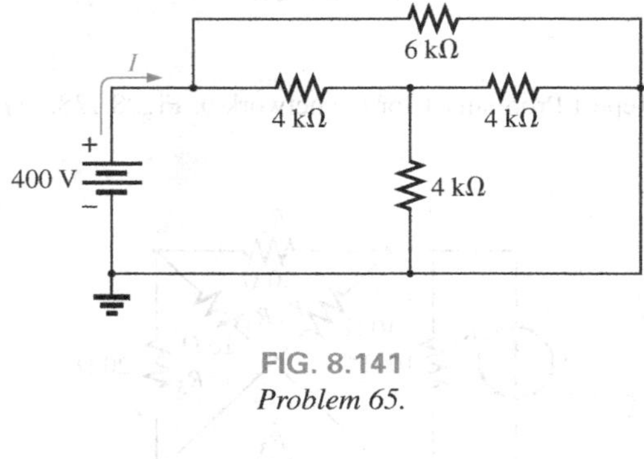

FIG. 8.141
Problem 65.

66. a. Using a Δ-Y or Y-Δ conversion, find the current I in the network of Fig. 8.142.

b. What other method could be used to find the current I?

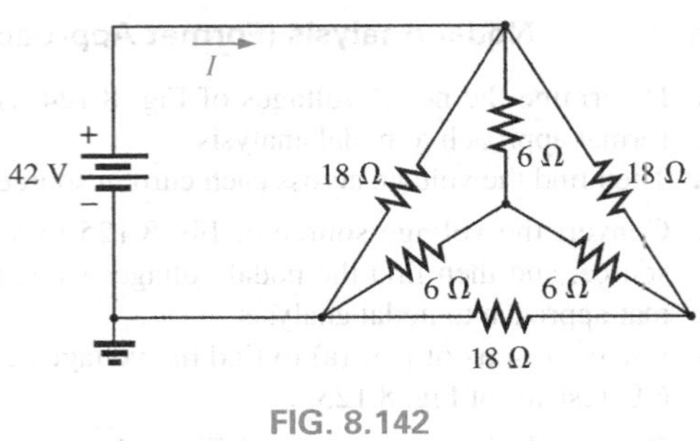

FIG. 8.142
Problem 66.

67. The network of Fig. 8.143 is very similar to the two-source networks solved using mesh or nodal analysis. We will now use a Y-Δ conversion to solve the same network. Find the source current I_{s_1} using a Y-Δ conversion.

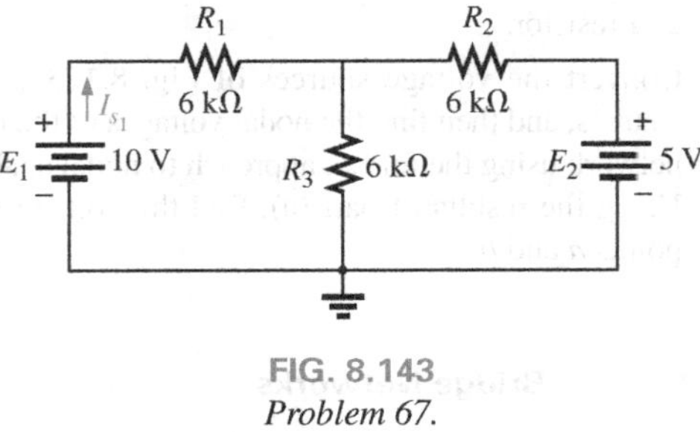

FIG. 8.143
Problem 67.

68. a. Replace the π configuration in Fig. 8.144 (composed of 3 kΩ resistors) with a T configuration.

b. Solve for the source current I_s.

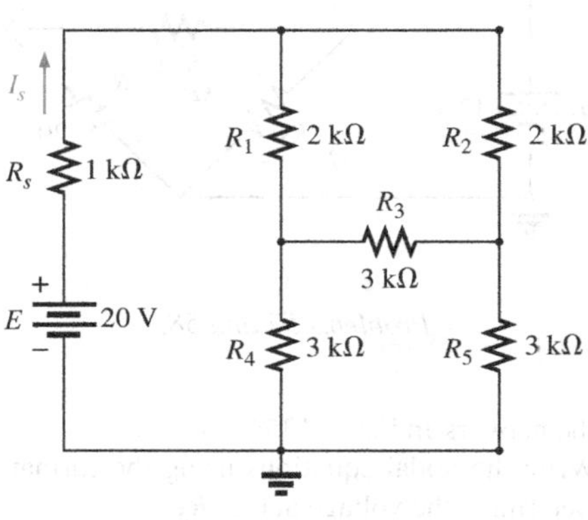

FIG. 8.144
Problem 68.

***69.** Using Y-Δ or Δ-Y conversion, determine the total resistance of the network in Fig. 8.145.

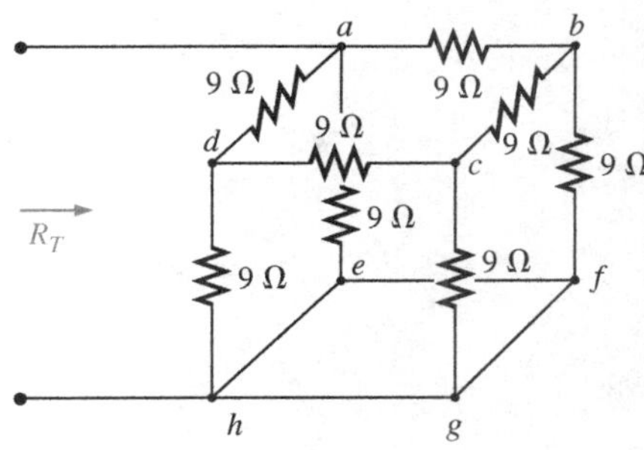

FIG. 8.145
Problem 69.

SECTION 8.14 Computer Analysis

PSpice or Multisim

70. Using schematics, find the current through each element in Fig. 8.111.

***71.** Using schematics, find the mesh currents for the network in Fig. 8.118.

***72.** Using schematics, determine the nodal voltages for the network in Fig. 8.131.

GLOSSARY

Branch-current method A technique for determining the branch currents of a multiloop network.

Bridge network A network configuration typically having a diamond appearance in which no two elements are in series or parallel.

Current sources Sources that supply a fixed current to a network and have a terminal voltage dependent on the network to which they are applied.

Delta (Δ), pi (π) configuration A network structure that consists of three branches and has the appearance of the Greek letter delta (Δ) or pi (π).

Determinants method A mathematical technique for finding the unknown variables of two or more simultaneous linear equations.

Mesh analysis A technique for determining the mesh (loop) currents of a network that results in a reduced set of equations compared to the branch-current method.

Mesh (loop) current A labeled current assigned to each distinct closed loop of a network that can, individually or in combination with other mesh currents, define all of the branch currents of a network.

Nodal analysis A technique for determining the nodal voltages of a network.

Node A junction of two or more branches in a network.

Supermesh current A current defined in a network with ideal current sources that permits the use of mesh analysis.

Supernode A node defined in a network with ideal voltage sources that permits the use of nodal analysis.

Wye (Y), tee (T) configuration A network structure that consists of three branches and has the appearance of the capital letter Y or T.

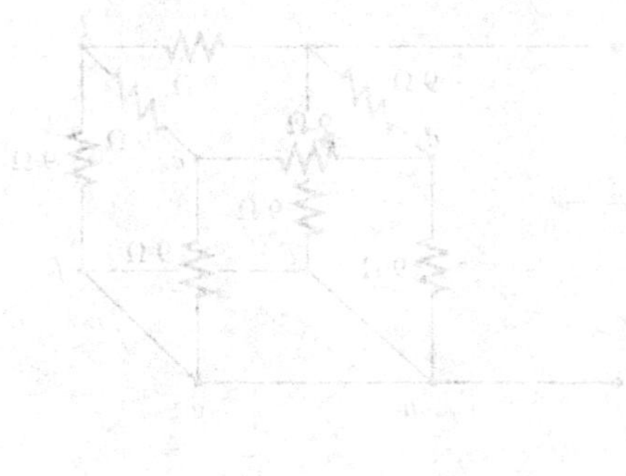

Network Theorems

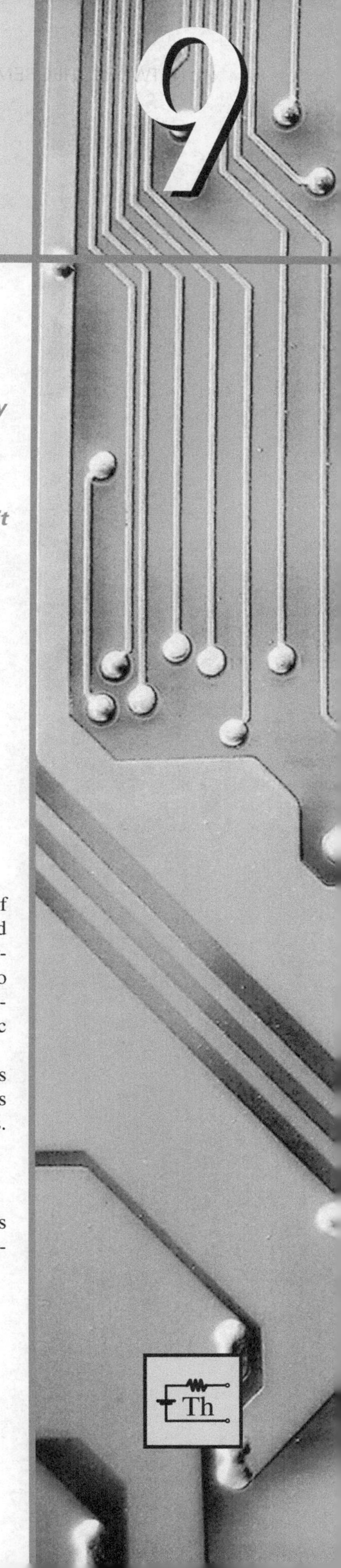

Objectives

- *Become familiar with the superposition theorem and its unique ability to separate the impact of each source on the quantity of interest.*
- *Be able to apply Thévenin's theorem to reduce any two-terminal, series-parallel network with any number of sources to a single voltage source and series resistor.*
- *Become familiar with Norton's theorem and how it can be used to reduce any two-terminal, series-parallel network with any number of sources to a single current source and a parallel resistor.*
- *Understand how to apply the maximum power transfer theorem to determine the maximum power to a load and to choose a load that will receive maximum power.*
- *Become aware of the reduction powers of Millman's theorem and the powerful implications of the substitution and reciprocity theorems.*

9.1 INTRODUCTION

This chapter introduces a number of theorems that have application throughout the field of electricity and electronics. Not only can they be used to solve networks such as encountered in the previous chapter, but they also provide an opportunity to determine the impact of a particular source or element on the response of the entire system. In most cases, the network to be analyzed and the mathematics required to find the solution are simplified. All of the theorems appear again in the analysis of ac networks. In fact, the application of each theorem to ac networks is very similar in content to that found in this chapter.

The first theorem to be introduced is the superposition theorem, followed by Thévenin's theorem, Norton's theorem, and the maximum power transfer theorem. The chapter concludes with a brief introduction to Millman's theorem and the substitution and reciprocity theorems.

9.2 SUPERPOSITION THEOREM

The **superposition theorem** is unquestionably one of the most powerful in this field. It has such widespread application that people often apply it without recognizing that their maneuvers are valid only because of this theorem.

In general, the theorem can be used to do the following:

- *Analyze networks such as introduced in the last chapter that have two or more sources that are not in series or parallel.*
- *Reveal the effect of each source on a particular quantity of interest.*
- *For sources of different types (such as dc and ac, which affect the parameters of the network in a different manner) and apply a separate analysis for each type, with the total result simply the algebraic sum of the results.*

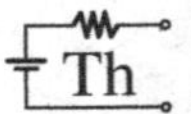

The first two areas of application are described in detail in this section. The last are covered in the discussion of the superposition theorem in the ac portion of the text.

The superposition theorem states the following:

The current through, or voltage across, any element of a network is equal to the algebraic sum of the currents or voltages produced independently by each source.

In other words, this theorem allows us to find a solution for a current or voltage using *only one source at a time*. Once we have the solution for each source, we can combine the results to obtain the total solution. The term *algebraic* appears in the above theorem statement because the currents resulting from the sources of the network can have different directions, just as the resulting voltages can have opposite polarities.

If we are to consider the effects of each source, the other sources obviously must be removed. Setting a voltage source to zero volts is like placing a short circuit across its terminals. Therefore,

when removing a voltage source from a network schematic, replace it with a direct connection (short circuit) of zero ohms. Any internal resistance associated with the source must remain in the network.

Setting a current source to zero amperes is like replacing it with an open circuit. Therefore,

when removing a current source from a network schematic, replace it by an open circuit of infinite ohms. Any internal resistance associated with the source must remain in the network.

The above statements are illustrated in Fig. 9.1.

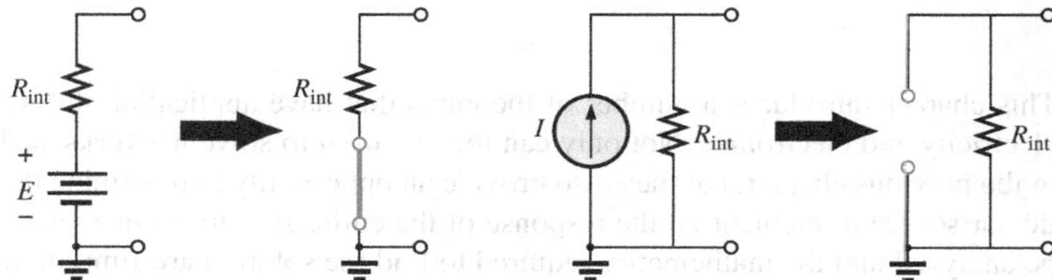

FIG. 9.1
Removing a voltage source and a current source to permit the application of the superposition theorem.

Since the effect of each source will be determined independently, the number of networks to be analyzed will equal the number of sources.

If a particular current of a network is to be determined, the contribution to that current must be determined for *each source*. When the effect of each source has been determined, those currents in the same direction are added, and those having the opposite direction are subtracted; the algebraic sum is being determined. The total result is the direction of the larger sum and the magnitude of the difference.

Similarly, if a particular voltage of a network is to be determined, the contribution to that voltage must be determined for each source. When the effect of each source has been determined, those voltages with the same polarity are added, and those with the opposite polarity are subtracted; the algebraic sum is being determined. The total result has the polarity of the larger sum and the magnitude of the difference.

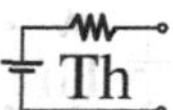

Superposition cannot be applied to power effects because the power is related to the square of the voltage across a resistor or the current through a resistor. The squared term results in a nonlinear (a curve, not a straight line) relationship between the power and the determining current or voltage. For example, doubling the current through a resistor does not double the power to the resistor (as defined by a linear relationship) but, in fact, increases it by a factor of 4 (due to the squared term). Tripling the current increases the power level by a factor of 9. Example 9.1 demonstrates the differences between a linear and a nonlinear relationship.

A few examples clarify how sources are removed and total solutions obtained.

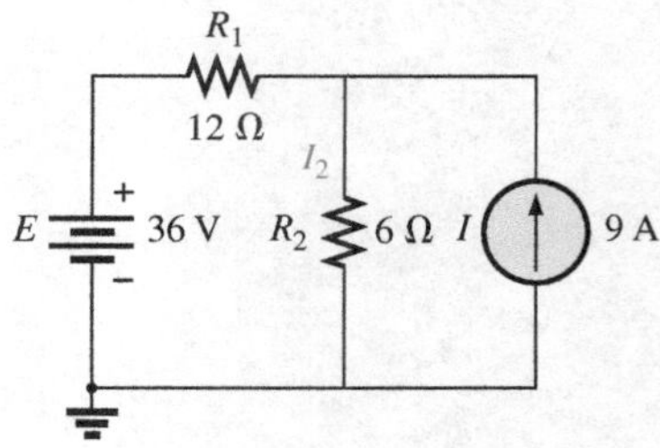

FIG. 9.2
Network to be analyzed in Example 9.1 using the superposition theorem.

EXAMPLE 9.1

a. Using the superposition theorem, determine the current through resistor R_2 for the network in Fig. 9.2.
b. Demonstrate that the superposition theorem is not applicable to power levels.

Solutions:

a. In order to determine the effect of the 36 V voltage source, the current source must be replaced by an open-circuit equivalent as shown in Fig. 9.3. The result is a simple series circuit with a current equal to

$$I'_2 = \frac{E}{R_T} = \frac{E}{R_1 + R_2} = \frac{36\text{ V}}{12\ \Omega + 6\ \Omega} = \frac{36\text{ V}}{18\ \Omega} = 2\text{ A}$$

Examining the effect of the 9 A current source requires replacing the 36 V voltage source by a short-circuit equivalent as shown in Fig. 9.4. The result is a parallel combination of resistors R_1 and R_2. Applying the current divider rule results in

$$I''_2 = \frac{R_1(I)}{R_1 + R_2} = \frac{(12\ \Omega)(9\text{ A})}{12\ \Omega + 6\ \Omega} = 6\text{ A}$$

Since the contribution to current I_2 has the same direction for each source, as shown in Fig. 9.5, the total solution for current I_2 is the sum of the currents established by the two sources. That is,

$$I_2 = I'_2 + I''_2 = 2\text{ A} + 6\text{ A} = \mathbf{8\text{ A}}$$

b. Using Fig. 9.3 and the results obtained, we find the power delivered to the 6 Ω resistor

$$P_1 = (I'_2)^2(R_2) = (2\text{ A})^2(6\ \Omega) = \mathbf{24\text{ W}}$$

Using Fig. 9.4 and the results obtained, we find the power delivered to the 6 Ω resistor

$$P_2 = (I''_2)^2(R_2) = (6\text{ A})^2(6\ \Omega) = \mathbf{216\text{ W}}$$

Using the total results of Fig. 9.5, we obtain the power delivered to the 6 Ω resistor

$$P_T = I_2^2R_2 = (8\text{ A})^2(6\ \Omega) = \mathbf{384\text{ W}}$$

It is now quite clear that the power delivered to the 6 Ω resistor using the total current of 8 A is not equal to the sum of the power levels due to each source independently. That is,

$$P_1 + P_2 = 24\text{ W} + 216\text{ W} = 240\text{ W} \neq P_T = 348\text{ W}$$

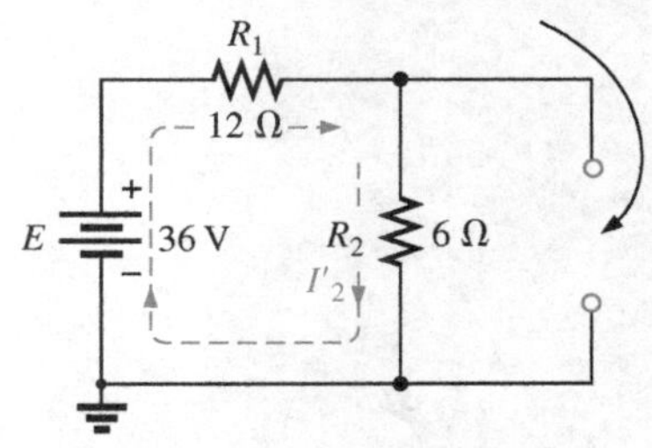

FIG. 9.3
Replacing the 9 A current source in Fig. 9.2 by an open circuit to determine the effect of the 36 V voltage source on current I_2.

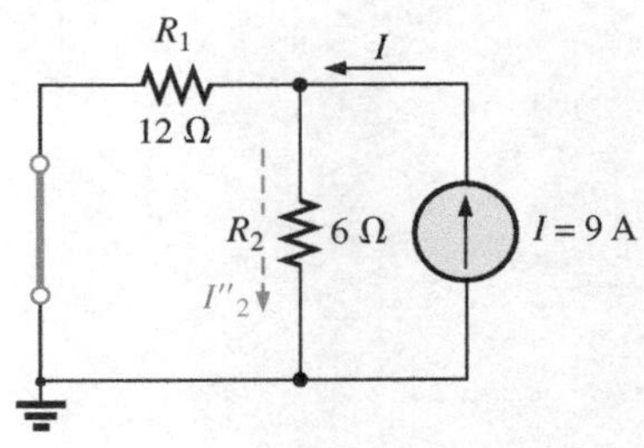

FIG. 9.4
Replacing the 36 V voltage source by a short-circuit equivalent to determine the effect of the 9 A current source on current I_2.

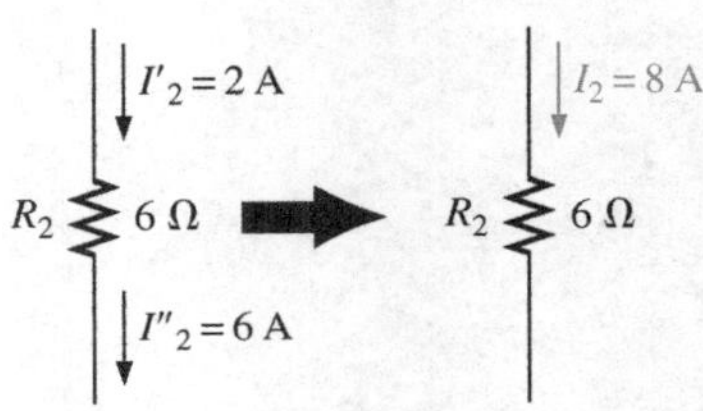

FIG. 9.5
Using the results of Figs. 9.3 and 9.4 to determine current I_2 for the network in Fig. 9.2.

To expand on the above conclusion and further demonstrate what is meant by a *nonlinear relationship,* the power to the 6 Ω resistor versus current through the 6 Ω resistor is plotted in Fig. 9.6. Note that the curve is not a straight line but one whose rise gets steeper with increase in current level.

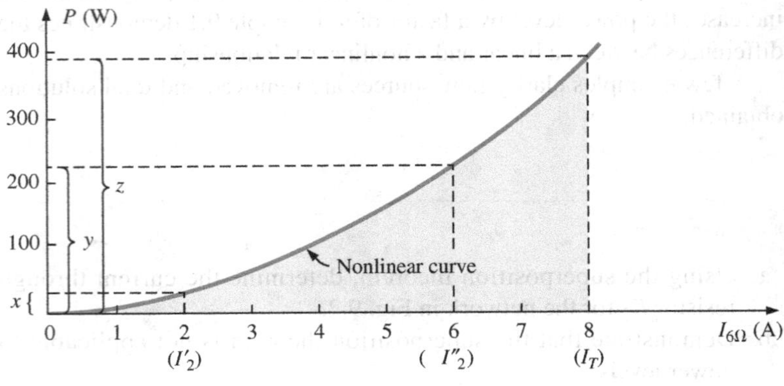

FIG. 9.6
Plotting power delivered to the 6 Ω resistor versus current through the resistor.

Recall from Fig. 9.4 that the power level was 24 W for a current of 2 A developed by the 36 V voltage source, shown in Fig. 9.6. From Fig. 9.5, we found that the current level was 6 A for a power level of 216 W, shown in Fig. 9.6. Using the total current of 8 A, we find that the power level in 384 W, shown in Fig. 9.6. Quite clearly, the sum of power levels due to the 2 A and 6 A current levels does not equal that due to the 8 A level. That is,

$$x + y \neq z$$

Now, the relationship between the voltage across a resistor and the current through a resistor is a linear (straight line) one as shown in Fig. 9.7, with

$$c = a + b$$

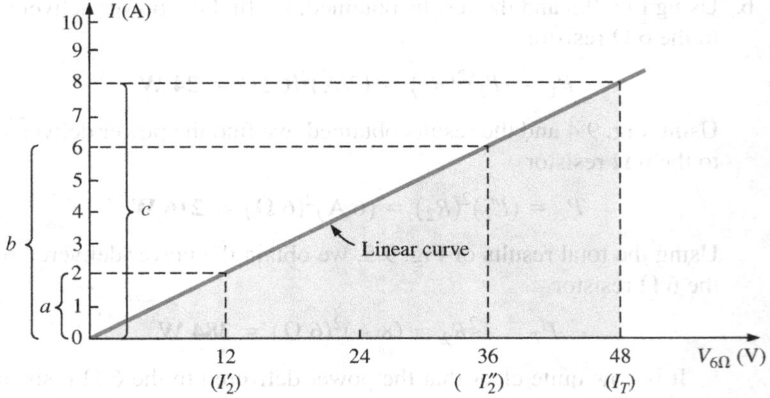

FIG. 9.7
Plotting I versus V for the 6 Ω resistor.

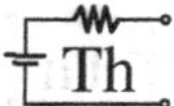

EXAMPLE 9.2 Using the superposition theorem, determine the current through the 12 Ω resistor in Fig. 9.8. Note that this is a two-source network of the type examined in the previous chapter when we applied branch-current analysis and mesh analysis.

Solution: Considering the effects of the 54 V source requires replacing the 48 V source by a short-circuit equivalent as shown in Fig. 9.9. The result is that the 12 Ω and 4 Ω resistors are in parallel.

The total resistance seen by the source is therefore

$$R_T = R_1 + R_2 \parallel R_3 = 24\ \Omega + 12\ \Omega \parallel 4\ \Omega = 24\ \Omega + 3\ \Omega = 27\ \Omega$$

and the source current is

$$I_s = \frac{E_1}{R_T} = \frac{54\ \text{V}}{27\ \Omega} = 2\ \text{A}$$

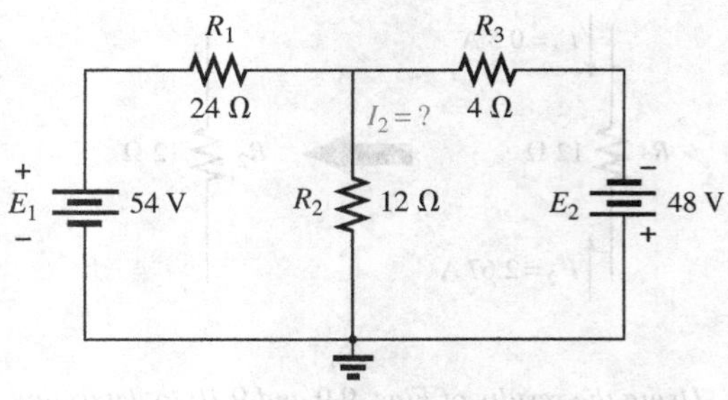

FIG. 9.8
Using the superposition theorem to determine the current through the 12 Ω resistor (Example 9.2).

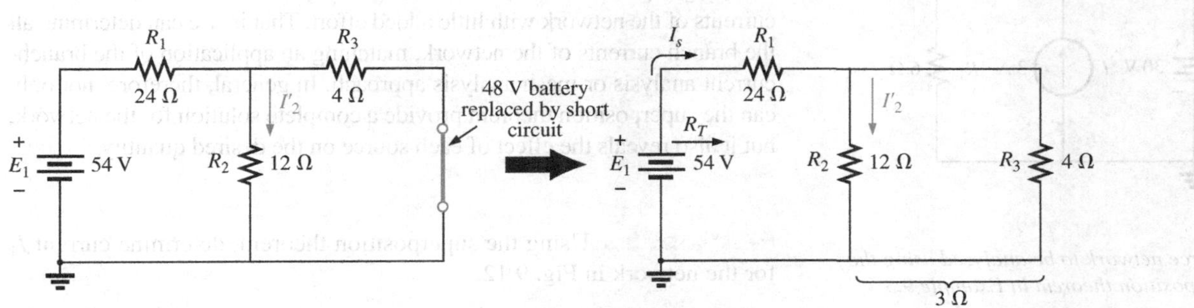

FIG. 9.9
Using the superposition theorem to determine the effect of the 54 V voltage source on current I_2 in Fig. 9.8.

Using the current divider rule results in the contribution to I_2 due to the 54 V source:

$$I'_2 = \frac{R_3 I_s}{R_3 + R_2} = \frac{(4\ \Omega)(2\ \text{A})}{4\ \Omega + 12\ \Omega} = 0.5\ \text{A}$$

If we now replace the 54 V source by a short-circuit equivalent, the network in Fig. 9.10 results. The result is a parallel connection for the 12 Ω and 24 Ω resistors.

Therefore, the total resistance seen by the 48 V source is

$$R_T = R_3 + R_2 \parallel R_1 = 4\ \Omega + 12\ \Omega \parallel 24\ \Omega = 4\ \Omega + 8\ \Omega = 12\ \Omega$$

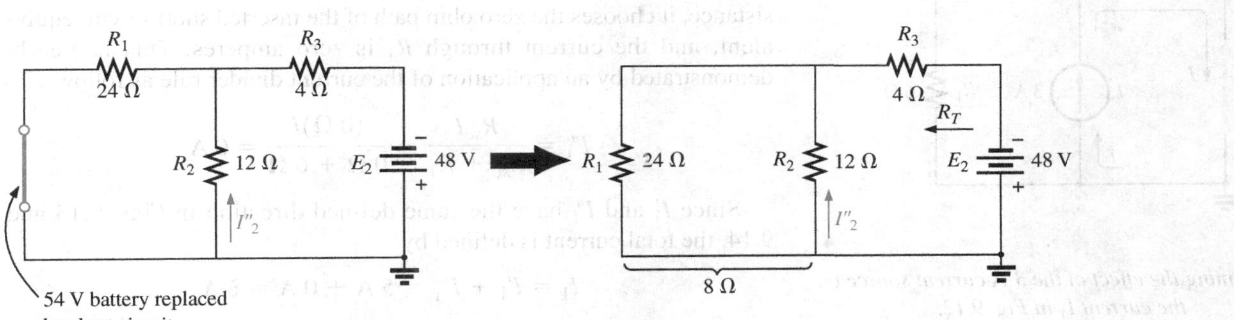

FIG. 9.10
Using the superposition theorem to determine the effect of the 48 V voltage source on current I_2 in Fig. 9.8.

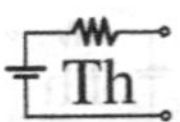

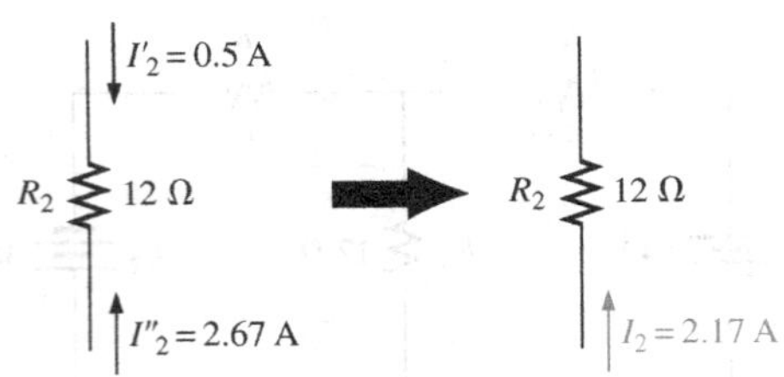

FIG. 9.11
Using the results of Figs. 9.9 and 9.10 to determine current I_2 for the network in Fig. 9.8.

and the source current is

$$I_s = \frac{E_2}{R_T} = \frac{48\text{ V}}{12\ \Omega} = 4\text{ A}$$

Applying the current divider rule results in

$$I''_2 = \frac{R_1(I_s)}{R_1 + R_2} = \frac{(24\ \Omega)(4\text{ A})}{24\ \Omega + 12\ \Omega} = 2.67\text{ A}$$

It is now important to realize that current I_2 due to each source has a different direction, as shown in Fig. 9.11. The net current therefore is the difference of the two and in the direction of the larger as follows:

$$I_2 = I''_2 - I'_2 = 2.67\text{ A} - 0.5\text{ A} = \mathbf{2.17\text{ A}}$$

Using Figs. 9.9 and 9.10 in Example 9.2, we can determine the other currents of the network with little added effort. That is, we can determine all the branch currents of the network, matching an application of the branch-current analysis or mesh analysis approach. In general, therefore, not only can the superposition theorem provide a complete solution for the network, but it also reveals the effect of each source on the desired quantity.

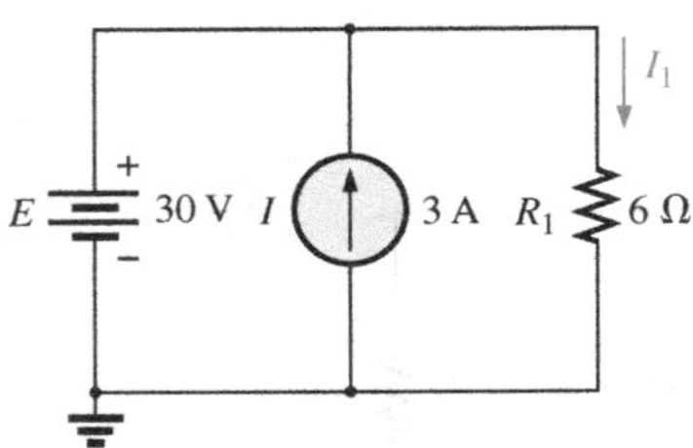

FIG. 9.12
Two-source network to be analyzed using the superposition theorem in Example 9.3.

EXAMPLE 9.3 Using the superposition theorem, determine current I_1 for the network in Fig. 9.12.

Solution: Since two sources are present, there are two networks to be analyzed. First let us determine the effects of the voltage source by setting the current source to zero amperes as shown in Fig. 9.13. Note that the resulting current is defined as I'_1 because it is the current through resistor R_1 due to the voltage source only.

Due to the open circuit, resistor R_1 is in series (and, in fact, in parallel) with the voltage source E. The voltage across the resistor is the applied voltage, and current I'_1 is determined by

$$I'_1 = \frac{V_1}{R_1} = \frac{E}{R_1} = \frac{30\text{ V}}{6\ \Omega} = 5\text{ A}$$

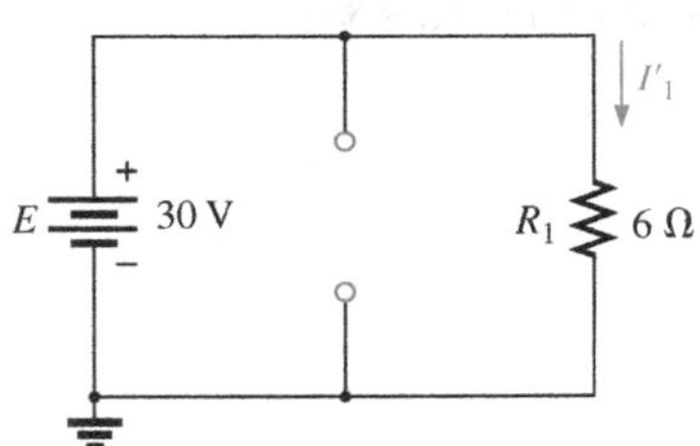

FIG. 9.13
Determining the effect of the 30 V supply on the current I_1 in Fig. 9.12.

Now for the contribution due to the current source. Setting the voltage source to zero volts results in the network in Fig. 9.14, which presents us with an interesting situation. The current source has been replaced with a short-circuit equivalent that is directly across the current source and resistor R_1. Since the source current takes the path of least resistance, it chooses the zero ohm path of the inserted short-circuit equivalent, and the current through R_1 is zero amperes. This is clearly demonstrated by an application of the current divider rule as follows:

$$I''_1 = \frac{R_{sc}I}{R_{sc} + R_1} = \frac{(0\ \Omega)I}{0\ \Omega + 6\ \Omega} = 0\text{ A}$$

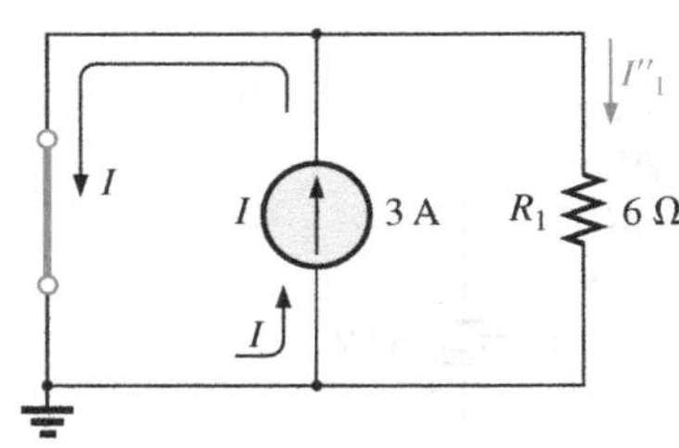

FIG. 9.14
Determining the effect of the 3 A current source on the current I_1 in Fig. 9.12.

Since I'_1 and I''_1 have the same defined direction in Figs. 9.13 and 9.14, the total current is defined by

$$I_1 = I'_1 + I''_1 = 5\text{ A} + 0\text{ A} = \mathbf{5\text{ A}}$$

Although this has been an excellent introduction to the application of the superposition theorem, it should be immediately clear in Fig. 9.12 that the voltage source is in parallel with the current source and load

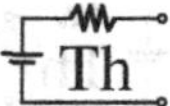

resistor R_1, so the voltage across each must be 30 V. The result is that I_1 must be determined solely by

$$I_1 = \frac{V_1}{R_1} = \frac{E}{R_1} = \frac{30\text{ V}}{6\ \Omega} = \mathbf{5\ A}$$

EXAMPLE 9.4 Using the principle of superposition, find the current I_2 through the 12 kΩ resistor in Fig. 9.15.

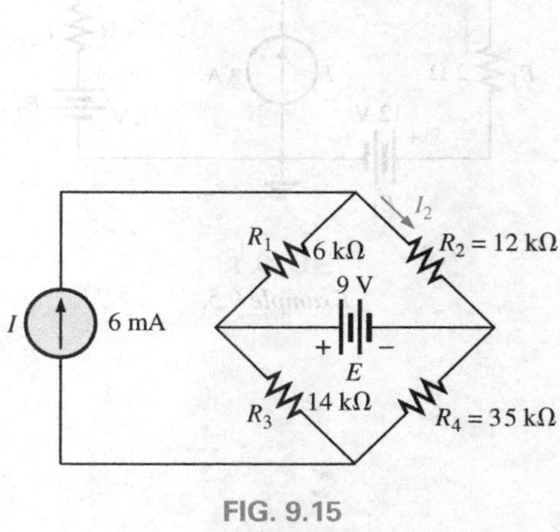

FIG. 9.15
Example 9.4.

Solution: Consider the effect of the 6 mA current source (Fig. 9.16).

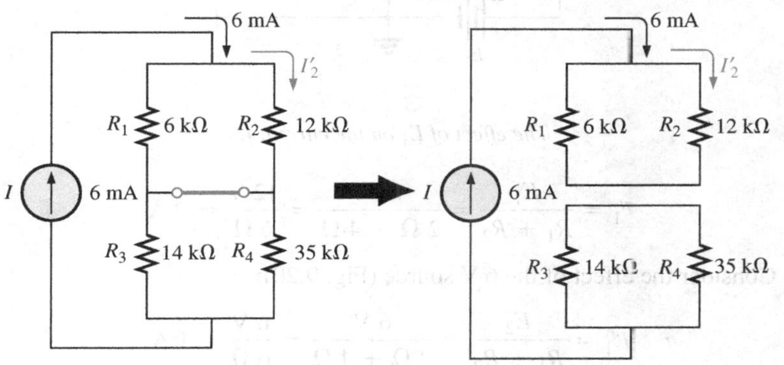

FIG. 9.16
The effect of the current source I on the current I_2.

The current divider rule gives

$$I'_2 = \frac{R_1 I}{R_1 + R_2} = \frac{(6\text{ k}\Omega)(6\text{ mA})}{6\text{ k}\Omega + 12\text{ k}\Omega} = 2\text{ mA}$$

Considering the effect of the 9 V voltage source (Fig. 9.17) gives

$$I''_2 = \frac{E}{R_1 + R_2} = \frac{9\text{ V}}{6\text{ k}\Omega + 12\text{ k}\Omega} = 0.5\text{ mA}$$

Since I'_2 and I''_2 have the same direction through R_2, the desired current is the sum of the two:

$$\begin{aligned} I_2 &= I'_2 + I''_2 \\ &= 2\text{ mA} + 0.5\text{ mA} \\ &= \mathbf{2.5\ mA} \end{aligned}$$

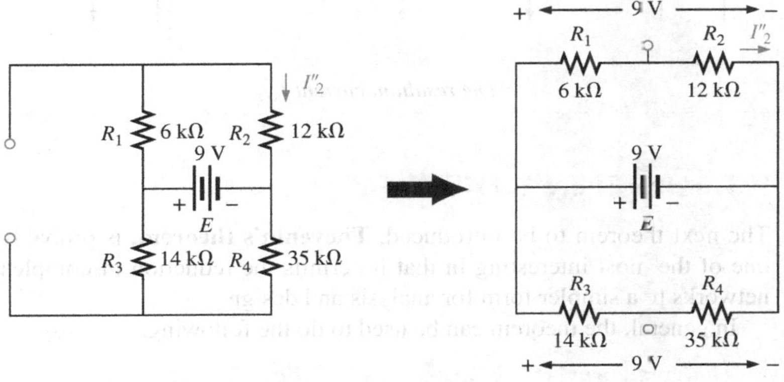

FIG. 9.17
The effect of the voltage source E on the current I_2.

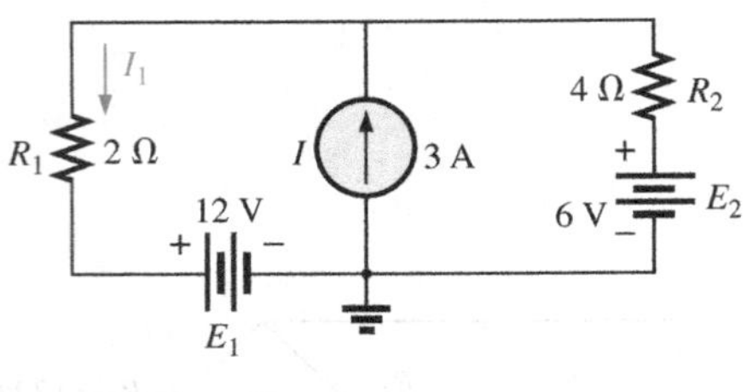

FIG. 9.18
Example 9.5.

EXAMPLE 9.5 Find the current through the 2 Ω resistor of the network in Fig. 9.18. The presence of three sources results in three different networks to be analyzed.

Solution: Consider the effect of the 12 V source (Fig. 9.19):

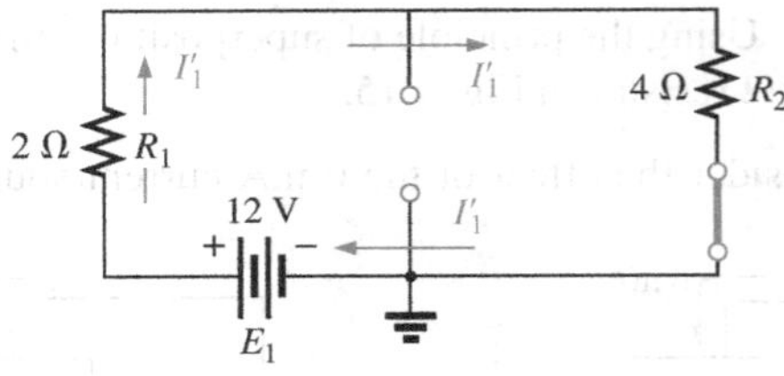

FIG. 9.19
The effect of E_1 on the current I.

$$I'_1 = \frac{E_1}{R_1 + R_2} = \frac{12\text{ V}}{2\ \Omega + 4\ \Omega} = \frac{12\text{ V}}{6\ \Omega} = 2\text{ A}$$

Consider the effect of the 6 V source (Fig. 9.20):

$$I''_1 = \frac{E_2}{R_1 + R_2} = \frac{6\text{ V}}{2\ \Omega + 4\ \Omega} = \frac{6\text{ V}}{6\ \Omega} = 1\text{ A}$$

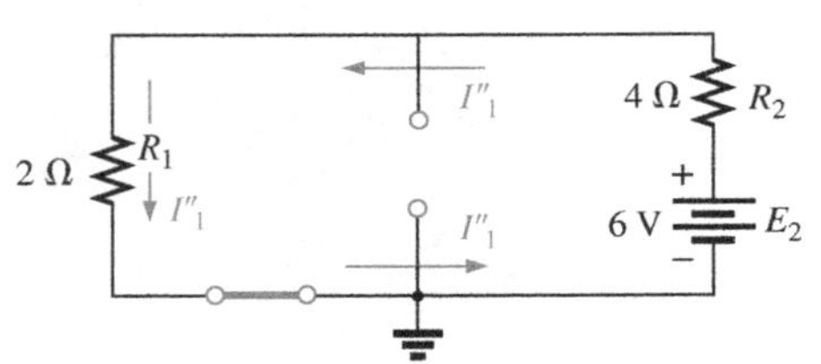

FIG. 9.20
The effect of E_2 on the current I_1.

Consider the effect of the 3 A source (Fig. 9.21): Applying the current divider rule gives

$$I'''_1 = \frac{R_2 I}{R_1 + R_2} = \frac{(4\ \Omega)(3\text{ A})}{2\ \Omega + 4\ \Omega} = \frac{12\text{ A}}{6} = 2\text{ A}$$

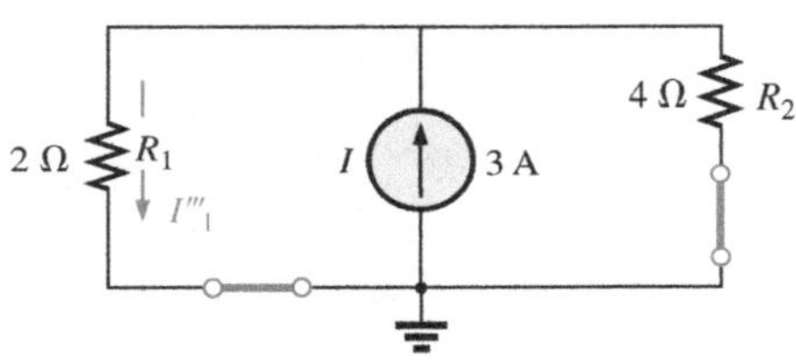

FIG. 9.21
The effect of I on the current I_1.

The total current through the 2 Ω resistor appears in Fig. 9.22, and

Same direction as I_1 in Fig. 9.18 — Opposite direction to I_1 in Fig. 9.18

$$\begin{aligned} I_1 &= I''_1 + I'''_1 - I'_1 \\ &= 1\text{ A} + 2\text{ A} - 2\text{ A} = \mathbf{1\ A} \end{aligned}$$

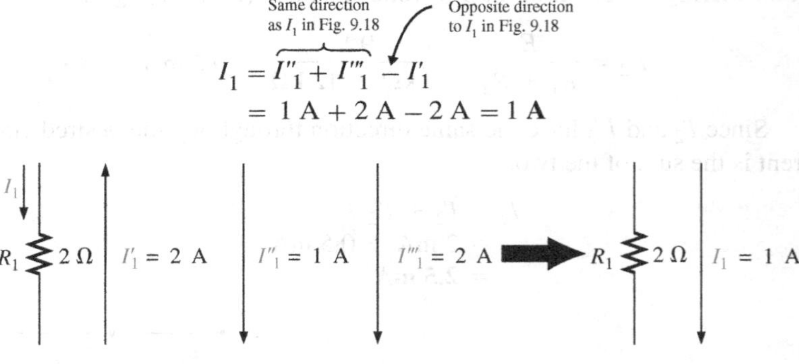

FIG. 9.22
The resultant current I_1.

9.3 THÉVENIN'S THEOREM

The next theorem to be introduced, **Thévenin's theorem,** is probably one of the most interesting in that it permits the reduction of complex networks to a simpler form for analysis and design.

In general, the theorem can be used to do the following:

- ***Analyze networks with sources that are not in series or parallel.***
- ***Reduce the number of components required to establish the same characteristics at the output terminals.***

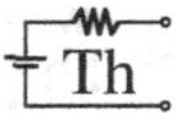

- *Investigate the effect of changing a particular component on the behavior of a network without having to analyze the entire network after each change.*

All three areas of application are demonstrated in the examples to follow.

Thévenin's theorem states the following:

Any two-terminal dc network can be replaced by an equivalent circuit consisting solely of a voltage source and a series resistor as shown in Fig. 9.23.

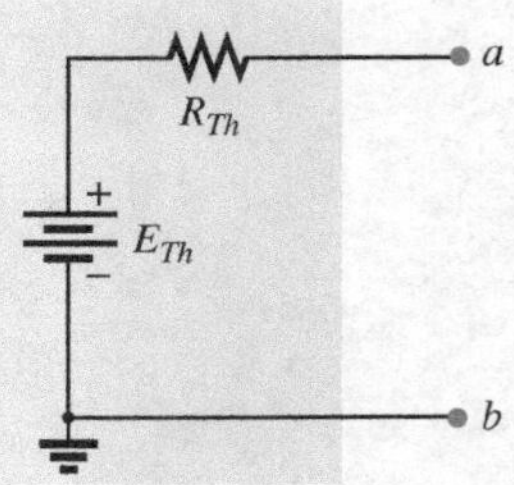

FIG. 9.23
Thévenin equivalent circuit.

The theorem was developed by Commandant Leon-Charles Thévenin in 1883 as described in Fig. 9.24.

To demonstrate the power of the theorem, consider the fairly complex network of Fig. 9.25(a) with its two sources and series-parallel connections. The theorem states that the entire network inside the blue shaded area can be replaced by one voltage source and one resistor as shown in Fig. 9.25(b). If the replacement is done properly, the voltage across, and the current through, the resistor R_L will be the same for each network. The value of R_L can be changed to any value, and the voltage, current, or power to the load resistor is the same for each configuration. Now, this is a very powerful statement—one that is verified in the examples to follow.

The question then is, How can you determine the proper value of Thévenin voltage and resistance? In general, finding the Thévenin *resistance* value is quite straightforward. Finding the Thévenin *voltage* can be more of a challenge and, in fact, may require using the superposition theorem or one of the methods described in Chapter 8.

Fortunately, there are a series of steps that will lead to the proper value of each parameter. Although a few of the steps may seem trivial at first, they can become quite important when the network becomes complex.

FIG. 9.24
Leon-Charles Thévenin.
Courtesy of Ecole Polytechnique Library (Palaiseau, France)

French (Meaux, Paris)
(1857–1927)
Telegraph Engineer, Commandant and Educator
École Polytechnique and École Supérieure de Télégraphie

Although active in the study and design of telegraphic systems (including underground transmission), cylindrical condensers (capacitors), and electromagnetism, he is best known for a theorem first presented in the French *Journal of Physics—Theory and Applications* in 1883. It appeared under the heading of "Sur un nouveau théorème d'électricité dynamique" ("On a new theorem of dynamic electricity") and was originally referred to as the *equivalent generator theorem.* There is some evidence that a similar theorem was introduced by Hermann von Helmholtz in 1853. However, Professor Helmholtz applied the theorem to animal physiology and not to communication or generator systems, and therefore he has not received the credit in this field that he might deserve. In the early 1920s AT&T did some pioneering work using the equivalent circuit and may have initiated the reference to the theorem as simply Thévenin's theorem. In fact, Edward L. Norton, an engineer at AT&T at the time, introduced a current source equivalent of the Thévenin equivalent currently referred to as the Norton equivalent circuit. As an aside, Commandant Thévenin was an avid skier and in fact was commissioner of an international ski competition in Chamonix, France, in 1912.

Thévenin's Theorem Procedure

Preliminary:

1. *Remove that portion of the network where the Thévenin equivalent circuit is found. In Fig. 9.25(a), this requires that the load resistor R_L be temporarily removed from the network.*
2. *Mark the terminals of the remaining two-terminal network. (The importance of this step will become obvious as we progress through some complex networks.)*

R_{Th}:

3. *Calculate R_{Th} by first setting all sources to zero (voltage sources are replaced by short circuits and current sources by open circuits) and then finding the resultant resistance between the two marked terminals. (If the internal resistance of the voltage and/or current sources is included in the original network, it must remain when the sources are set to zero.)*

E_{Th}:

4. *Calculate E_{Th} by first returning all sources to their original position and finding the open-circuit voltage between the marked terminals. (This step is invariably the one that causes most confusion and errors. In all cases, keep in mind that it is the open-circuit potential between the two terminals marked in step 2.)*

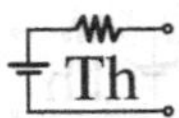

Conclusion:

5. Draw the Thévenin equivalent circuit with the portion of the circuit previously removed replaced between the terminals of the equivalent circuit. This step is indicated by the placement of the resistor R_L between the terminals of the Thévenin equivalent circuit as shown in Fig. 9.25(b).

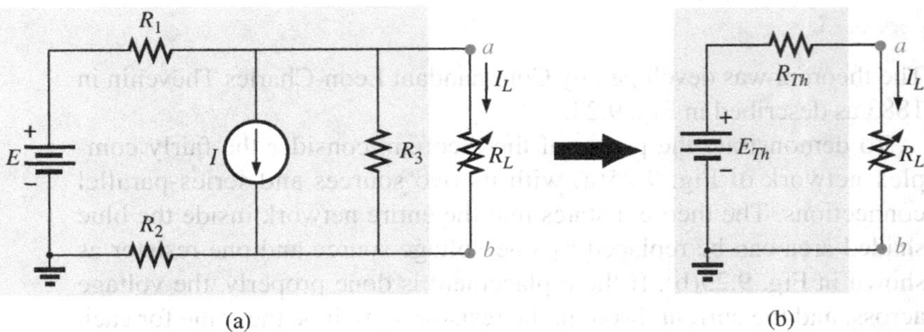

FIG. 9.25
Substituting the Thévenin equivalent circuit for a complex network.

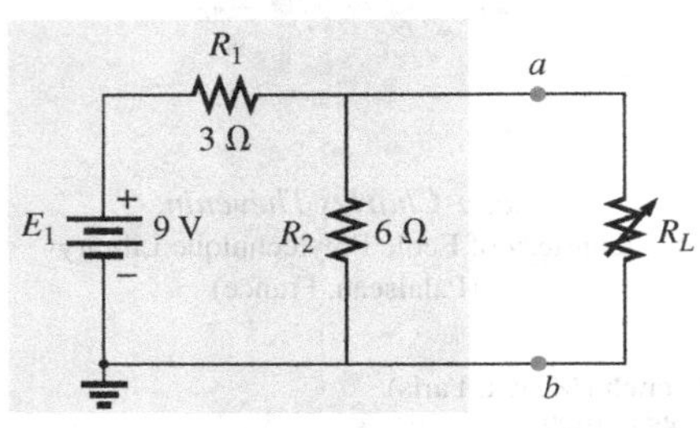

FIG. 9.26
Example 9.6.

EXAMPLE 9.6 Find the Thévenin equivalent circuit for the network in the shaded area of the network in Fig. 9.26. Then find the current through R_L for values of 2 Ω, 10 Ω, and 100 Ω.

Solution:

Steps 1 and 2: These produce the network in Fig. 9.27. Note that the load resistor R_L has been removed and the two "holding" terminals have been defined as *a* and *b*.

Step 3: Replacing the voltage source E_1 with a short-circuit equivalent yields the network in Fig. 9.28(a), where

$$R_{Th} = R_1 \| R_2 = \frac{(3\ \Omega)(6\ \Omega)}{3\ \Omega + 6\ \Omega} = \mathbf{2\ \Omega}$$

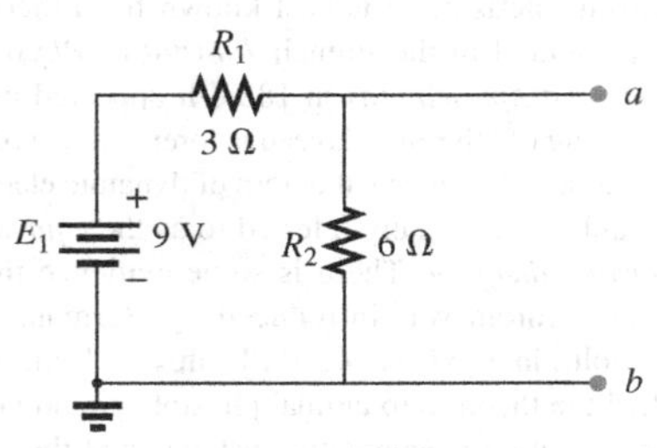

FIG. 9.27
Identifying the terminals of particular importance when applying Thévenin's theorem.

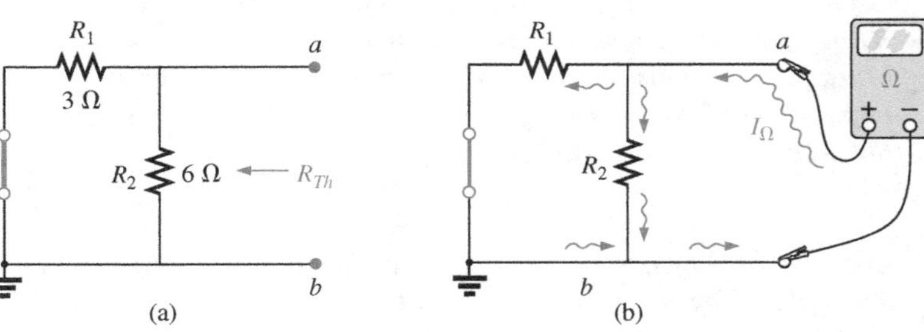

FIG. 9.28
Determining R_{Th} for the network in Fig. 9.27.

The importance of the two marked terminals now begins to surface. They are the two terminals across which the Thévenin resistance is measured. It is no longer the total resistance as seen by the source, as determined in the majority of problems of Chapter 7. If some difficulty develops when determining R_{Th} with regard to whether the resistive elements are in series or parallel, consider recalling that the ohmmeter sends out a trickle current into a resistive combination and senses the

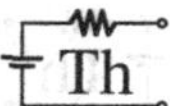

level of the resulting voltage to establish the measured resistance level. In Fig. 9.28(b), the trickle current of the ohmmeter approaches the network through terminal a, and when it reaches the junction of R_1 and R_2, it splits as shown. The fact that the trickle current splits and then recombines at the lower node reveals that the resistors are in parallel as far as the ohmmeter reading is concerned. In essence, the path of the sensing current of the ohmmeter has revealed how the resistors are connected to the two terminals of interest and how the Thévenin resistance should be determined. Remember this as you work through the various examples in this section.

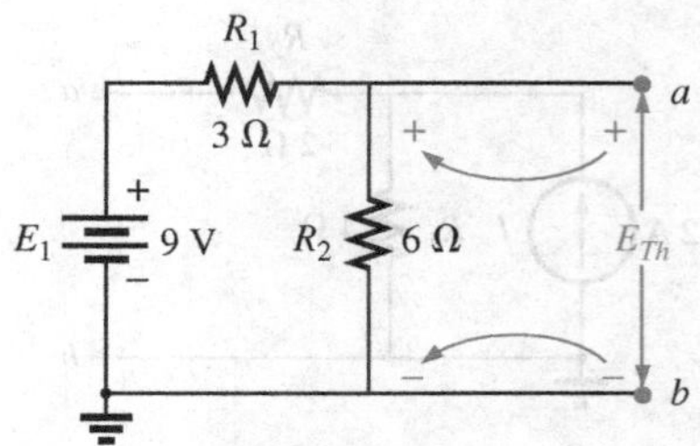

FIG. 9.29
Determining E_{Th} for the network in Fig. 9.27.

Step 4: Replace the voltage source (Fig. 9.29). For this case, the open-circuit voltage E_{Th} is the same as the voltage drop across the 6 Ω resistor. Applying the voltage divider rule gives

$$E_{Th} = \frac{R_2E_1}{R_2 + R_1} = \frac{(6\,\Omega)(9\text{ V})}{6\,\Omega + 3\,\Omega} = \frac{54\text{ V}}{9} = \mathbf{6\text{ V}}$$

It is particularly important to recognize that E_{Th} is the open-circuit potential between points a and b. Remember that an open circuit can have any voltage across it, but the current must be zero. In fact, the current through any element in series with the open circuit must be zero also. The use of a voltmeter to measure E_{Th} appears in Fig. 9.30. Note that it is placed directly across the resistor R_2 since E_{Th} and V_{R_2} are in parallel.

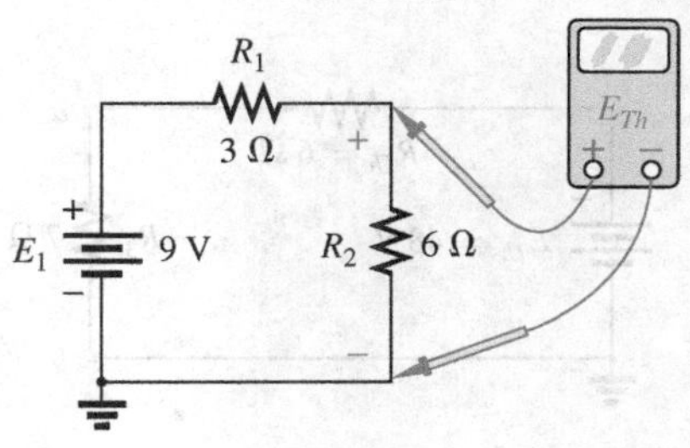

FIG. 9.30
Measuring E_{Th} for the network in Fig. 9.27.

Step 5: (Fig. 9.31):

$$I_L = \frac{E_{Th}}{R_{Th} + R_L}$$

$$R_L = 2\,\Omega: \qquad I_L = \frac{6\text{ V}}{2\,\Omega + 2\,\Omega} = \mathbf{1.5\text{ A}}$$

$$R_L = 10\,\Omega: \qquad I_L = \frac{6\text{ V}}{2\,\Omega + 10\,\Omega} = \mathbf{0.5\text{ A}}$$

$$R_L = 100\,\Omega: \qquad I_L = \frac{6\text{ V}}{2\,\Omega + 100\,\Omega} = \mathbf{0.06\text{ A}}$$

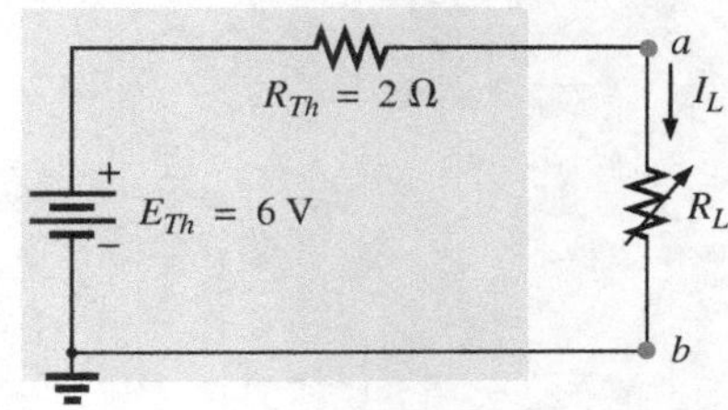

FIG. 9.31
Substituting the Thévenin equivalent circuit for the network external to R_L in Fig. 9.26.

If Thévenin's theorem were unavailable, each change in R_L would require that the entire network in Fig. 9.26 be reexamined to find the new value of R_L.

EXAMPLE 9.7 Find the Thévenin equivalent circuit for the network in the shaded area of the network in Fig. 9.32.

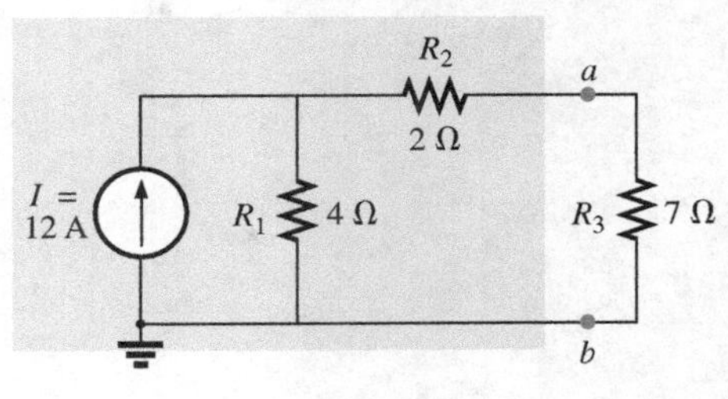

FIG. 9.32
Example 9.7.

Solution:

Steps 1 and 2: See Fig. 9.33.

Step 3: See Fig. 9.34. The current source has been replaced with an open-circuit equivalent and the resistance determined between terminals a and b.

In this case, an ohmmeter connected between terminals a and b sends out a sensing current that flows directly through R_1 and R_2 (at the same level). The result is that R_1 and R_2 are in series and the Thévenin resistance is the sum of the two,

$$R_{Th} = R_1 + R_2 = 4\,\Omega + 2\,\Omega = \mathbf{6\,\Omega}$$

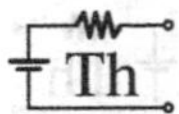

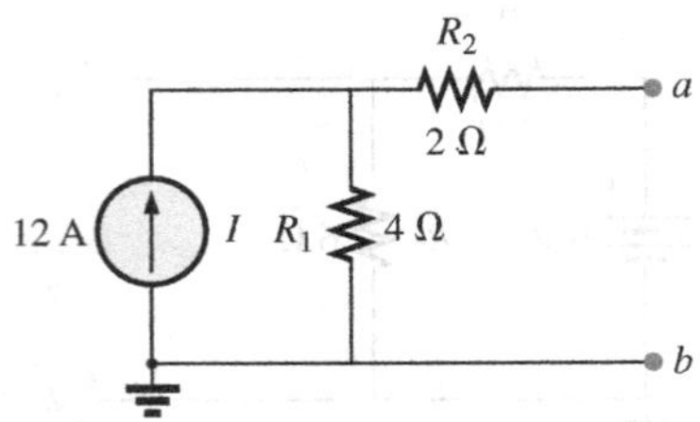

FIG. 9.33
Establishing the terminals of particular interest for the network in Fig. 9.32.

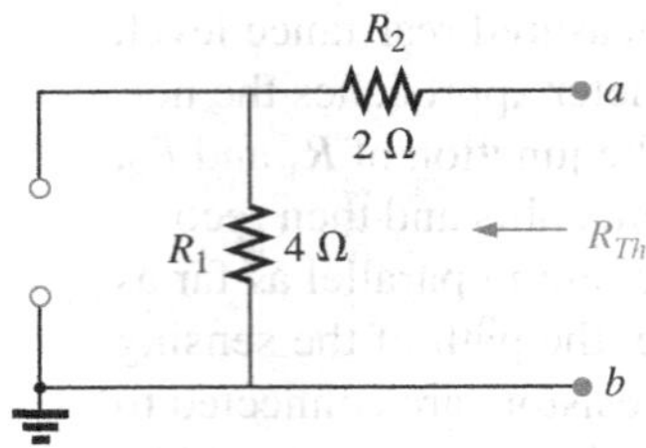

FIG. 9.34
Determining R_{Th} for the network in Fig. 9.33.

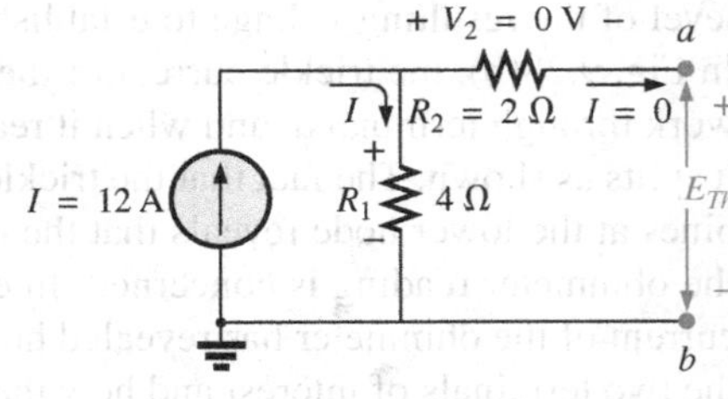

FIG. 9.35
Determining E_{Th} for the network in Fig. 9.33.

Step 4: See Fig. 9.35. In this case, since an open circuit exists between the two marked terminals, the current is zero between these terminals and through the 2 Ω resistor. The voltage drop across R_2 is, therefore,

$$V_2 = I_2R_2 = (0)R_2 = 0 \text{ V}$$

and $$E_{Th} = V_1 = I_1R_1 = IR_1 = (12 \text{ A})(4\Omega) = \mathbf{48 \text{ V}}$$

Step 5: See Fig. 9.36.

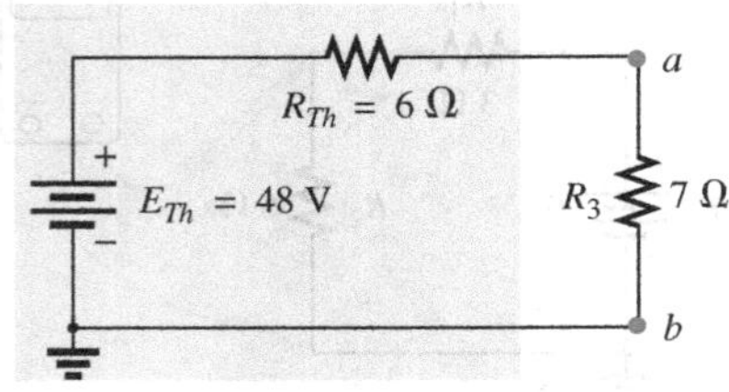

FIG. 9.36
Substituting the Thévenin equivalent circuit in the network external to the resistor R_3 in Fig. 9.32.

EXAMPLE 9.8 Find the Thévenin equivalent circuit for the network in the shaded area of the network in Fig. 9.37. Note in this example that there is no need for the section of the network to be preserved to be at the "end" of the configuration.

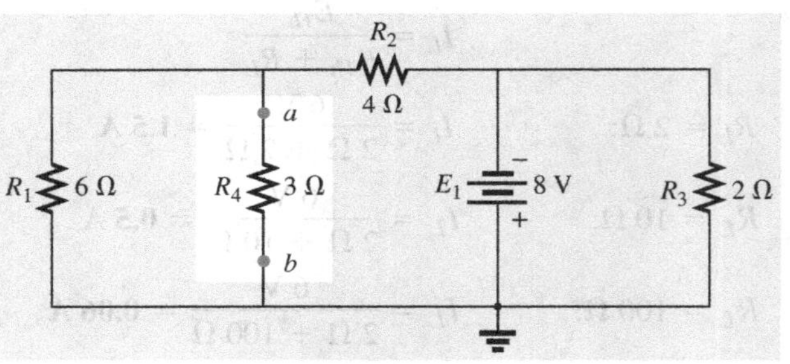

FIG. 9.37
Example 9.8.

Solution:

Steps 1 and 2: See Fig. 9.38.

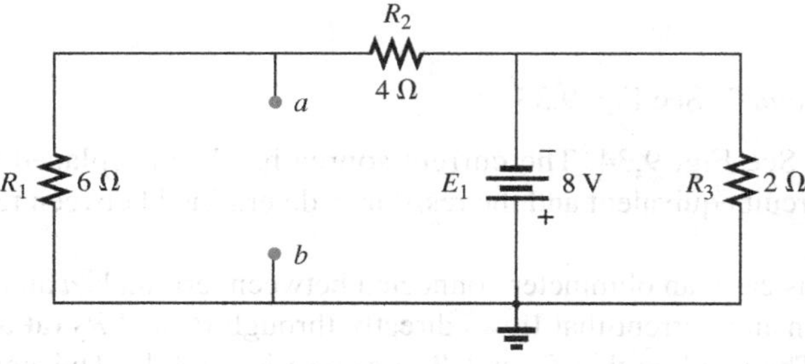

FIG. 9.38
Identifying the terminals of particular interest for the network in Fig. 9.37.

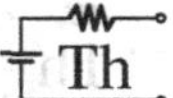

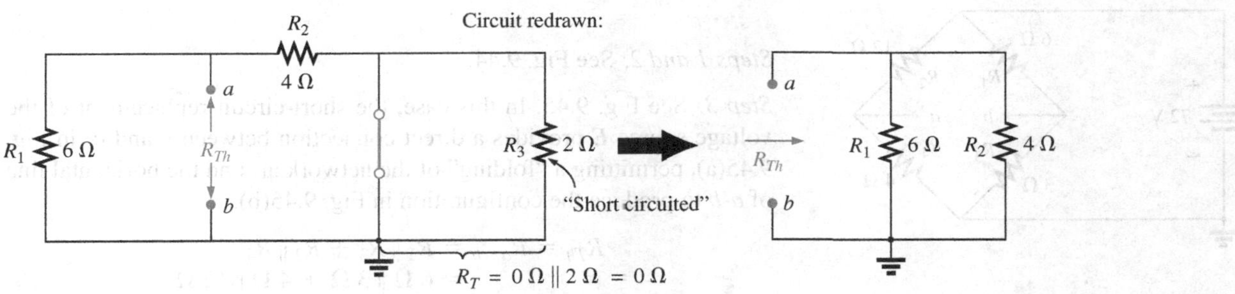

FIG. 9.39
Determining R_{Th} for the network in Fig. 9.38.

Step 3: See Fig. 9.39. Steps 1 and 2 are relatively easy to apply, but now we must be careful to "hold" onto the terminals a and b as the Thévenin resistance and voltage are determined. In Fig. 9.39, all the remaining elements turn out to be in parallel, and the network can be redrawn as shown. We have

$$R_{Th} = R_1 \parallel R_2 = \frac{(6\,\Omega)(4\,\Omega)}{6\,\Omega + 4\,\Omega} = \frac{24\,\Omega}{10} = \mathbf{2.4\,\Omega}$$

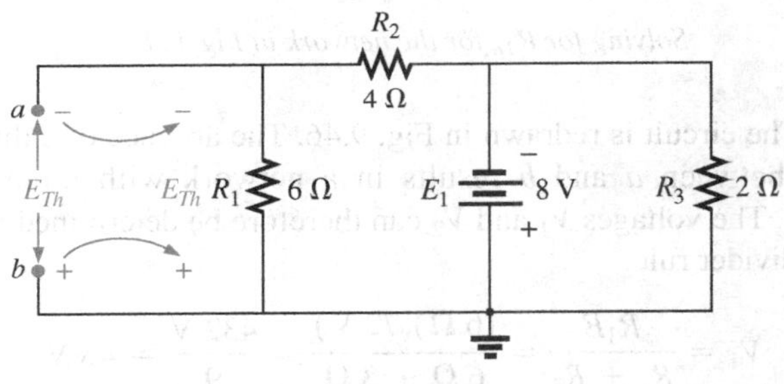

FIG. 9.40
Determining E_{Th} for the network in Fig. 9.38.

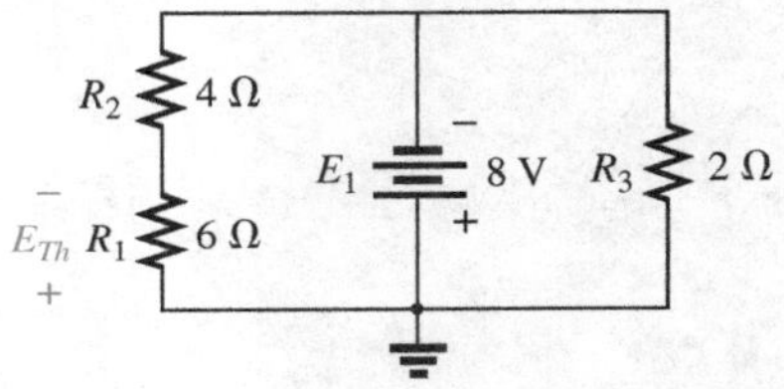

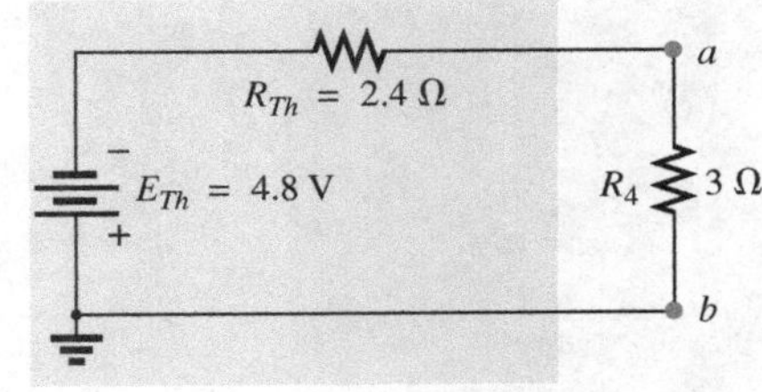

FIG. 9.41
Network of Fig. 9.40 redrawn.

Step 4: See Fig. 9.40. In this case, the network can be redrawn as shown in Fig. 9.41. Since the voltage is the same across parallel elements, the voltage across the series resistors R_1 and R_2 is E_1, or 8 V. Applying the voltage divider rule gives

$$E_{Th} = \frac{R_1E_1}{R_1 + R_2} = \frac{(6\,\Omega)(8\,\text{V})}{6\,\Omega + 4\,\Omega} = \frac{48\,\text{V}}{10} = \mathbf{4.8\,V}$$

Step 5: See Fig. 9.42.

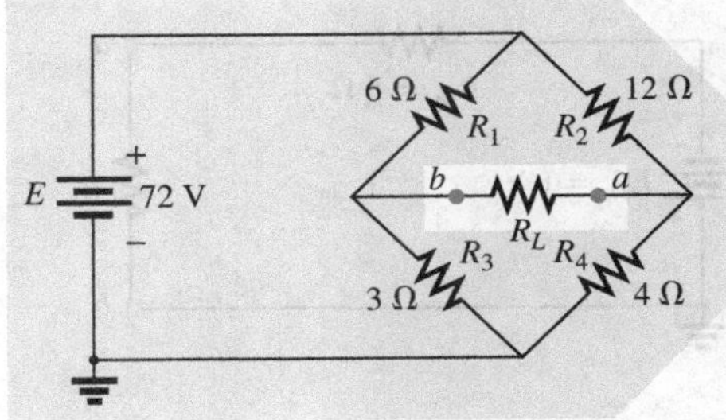

FIG. 9.42
Substituting the Thévenin equivalent circuit for the network external to the resistor R_4 in Fig. 9.37.

The importance of marking the terminals should be obvious from Example 9.8. Note that there is no requirement that the Thévenin voltage have the same polarity as the equivalent circuit originally introduced.

EXAMPLE 9.9 Find the Thévenin equivalent circuit for the network in the shaded area of the bridge network in Fig. 9.43.

FIG. 9.43
Example 9.9.

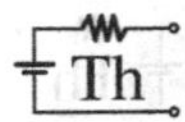

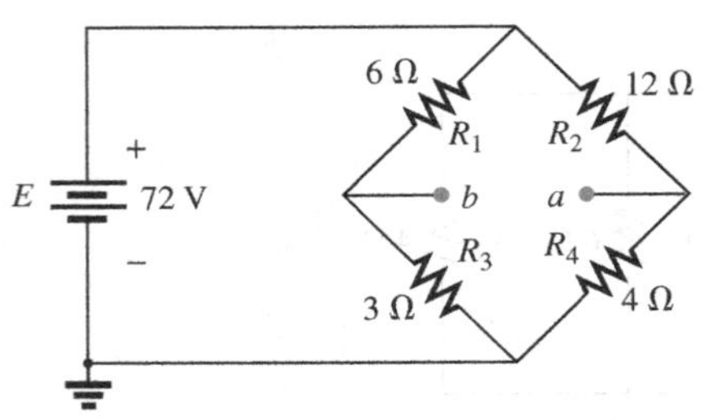

FIG. 9.44
Identifying the terminals of particular interest for the network in Fig. 9.43.

Solution:

Steps 1 and 2: See Fig. 9.44.

Step 3: See Fig. 9.45. In this case, the short-circuit replacement of the voltage source E provides a direct connection between c and c' in Fig. 9.45(a), permitting a "folding" of the network around the horizontal line of a-b to produce the configuration in Fig. 9.45(b).

$$\begin{aligned} R_{Th} = R_{a-b} &= R_1 \parallel R_3 + R_2 \parallel R_4 \\ &= 6\,\Omega \parallel 3\,\Omega + 4\,\Omega \parallel 12\,\Omega \\ &= 2\,\Omega + 3\,\Omega = \mathbf{5\,\Omega} \end{aligned}$$

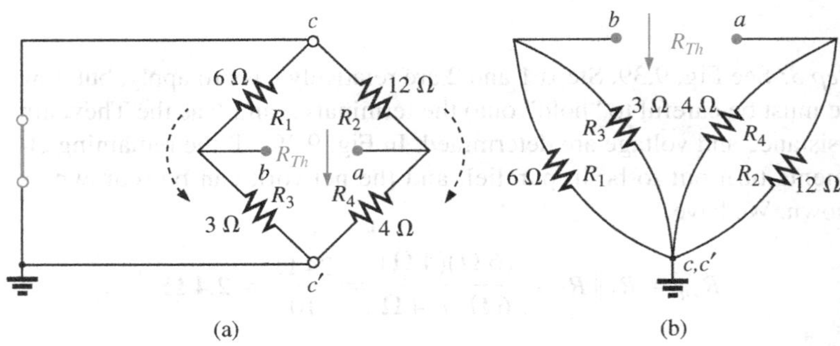

FIG. 9.45
Solving for R_{Th} for the network in Fig. 9.44.

Step 4: The circuit is redrawn in Fig. 9.46. The absence of a direct connection between a and b results in a network with three parallel branches. The voltages V_1 and V_2 can therefore be determined using the voltage divider rule:

$$V_1 = \frac{R_1 E}{R_1 + R_3} = \frac{(6\,\Omega)(72\text{ V})}{6\,\Omega + 3\,\Omega} = \frac{432\text{ V}}{9} = 48\text{ V}$$

$$V_2 = \frac{R_2 E}{R_2 + R_4} = \frac{(12\,\Omega)(72\text{ V})}{12\,\Omega + 4\,\Omega} = \frac{864\text{ V}}{16} = 54\text{ V}$$

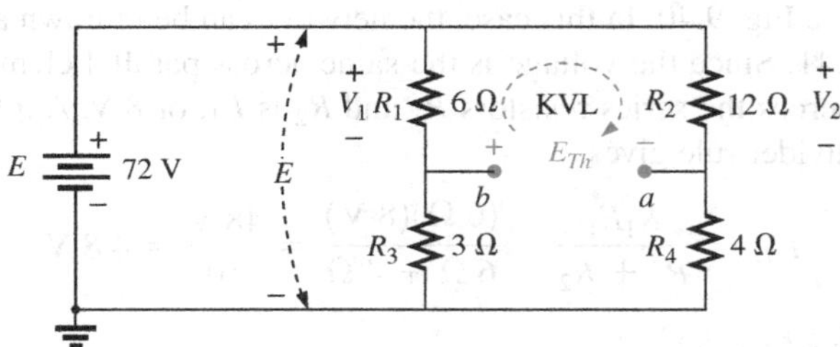

FIG. 9.46
Determining E_{Th} for the network in Fig. 9.44.

Assuming the polarity shown for E_{Th} and applying Kirchhoff's voltage law to the top loop in the clockwise direction results in

$$\Sigma_C V = +E_{Th} + V_1 - V_2 = 0$$

and
$$E_{Th} = V_2 - V_1 = 54\text{ V} - 48\text{ V} = \mathbf{6\text{ V}}$$

Step 5: See Fig. 9.47.

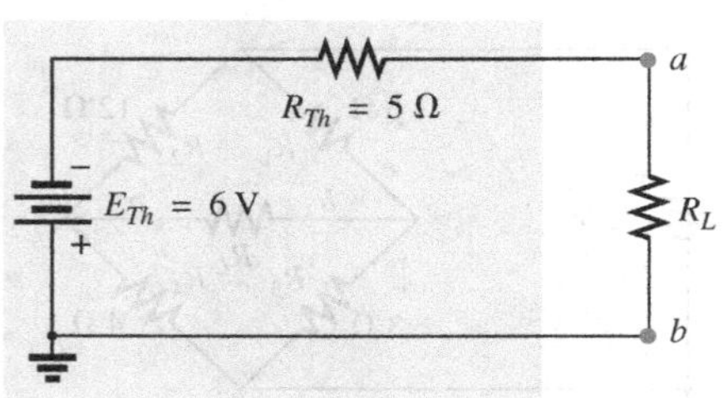

FIG. 9.47
Substituting the Thévenin equivalent circuit for the network external to the resistor R_L in Fig. 9.43.

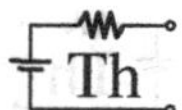

Thévenin's theorem is not restricted to a single passive element, as shown in the preceding examples, but can be applied across sources, whole branches, portions of networks, or any circuit configuration, as shown in the following example. It is also possible that you may have to use one of the methods previously described, such as mesh analysis or superposition, to find the Thévenin equivalent circuit.

EXAMPLE 9.10 (Two sources) Find the Thévenin circuit for the network within the shaded area of Fig. 9.48.

Solution:

Steps 1 and 2: See Fig. 9.49. The network is redrawn.

Step 3: See Fig. 9.50.

$$\begin{aligned} R_{Th} &= R_4 + R_1 \| R_2 \| R_3 \\ &= 1.4\ \text{k}\Omega + 0.8\ \text{k}\Omega \| 4\ \text{k}\Omega \| 6\ \text{k}\Omega \\ &= 1.4\ \text{k}\Omega + 0.8\ \text{k}\Omega \| 2.4\ \text{k}\Omega \\ &= 1.4\ \text{k}\Omega + 0.6\ \text{k}\Omega \\ &= \mathbf{2\ k\Omega} \end{aligned}$$

Step 4: Applying superposition, we will consider the effects of the voltage source E_1 first. Note Fig. 9.51. The open circuit requires that $V_4 = I_4R_4 = (0)R_4 = 0$ V, and

$$E'_{Th} = V_3$$
$$R'_T = R_2 \| R_3 = 4\ \text{k}\Omega \| 6\ \text{k}\Omega = 2.4\ \text{k}\Omega$$

Applying the voltage divider rule gives

$$V_3 = \frac{R'_T E_1}{R'_T} = \frac{(2.4\ \text{k}\Omega)(6\ \text{V})}{2.4\ \text{k}\Omega + 0.8\ \text{k}\Omega} = \frac{14.4\ \text{V}}{3.2} = 4.5\ \text{V}$$
$$E'_{Th} = V_3 = 4.5\ \text{V}$$

For the source E_2, the network in Fig. 9.52 results. Again, $V_4 = I_4R_4 = (0)R_4 = 0$ V, and

$$E''_{Th} = V_3$$
$$R'_T = R_1 \| R_3 = 0.8\ \text{k}\Omega \| 6\ \text{k}\Omega = 0.706\ \text{k}\Omega$$

and $$V_3 = \frac{R'_T E_2}{R'_T + R_2} = \frac{(0.706\ \text{k}\Omega)(10\ \text{V})}{0.706\ \text{k}\Omega + 4\ \text{k}\Omega} = \frac{7.06\ \text{V}}{4.706} = 1.5\ \text{V}$$
$$E''_{Th} = V_3 = 1.5\ \text{V}$$

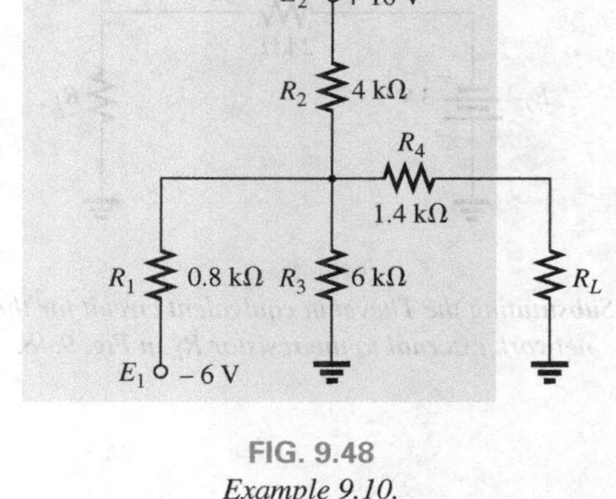

FIG. 9.48
Example 9.10.

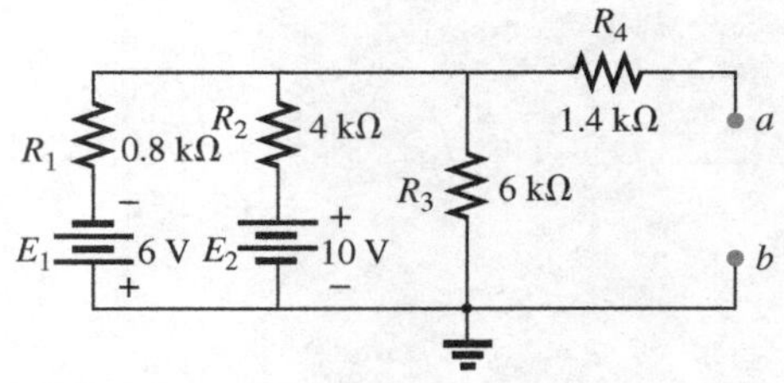

FIG. 9.49
Identifying the terminals of particular interest for the network in Fig. 9.48.

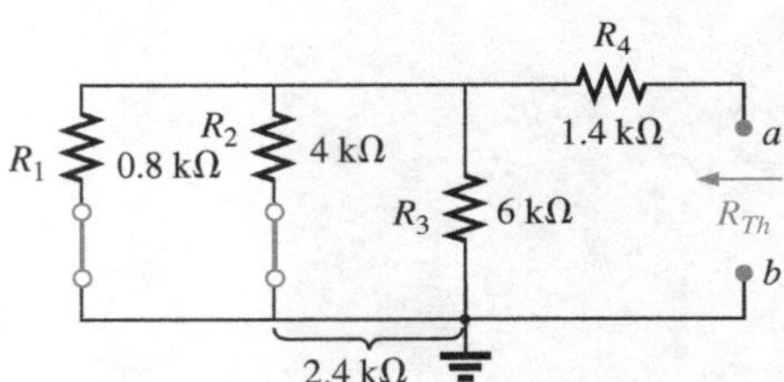

FIG. 9.50
Determining R_{Th} for the network in Fig. 9.49.

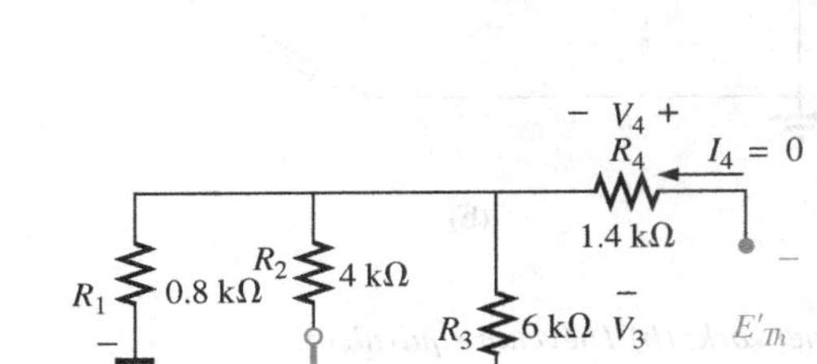

FIG. 9.51
Determining the contribution to E_{Th} from the source E_1 for the network in Fig. 9.49.

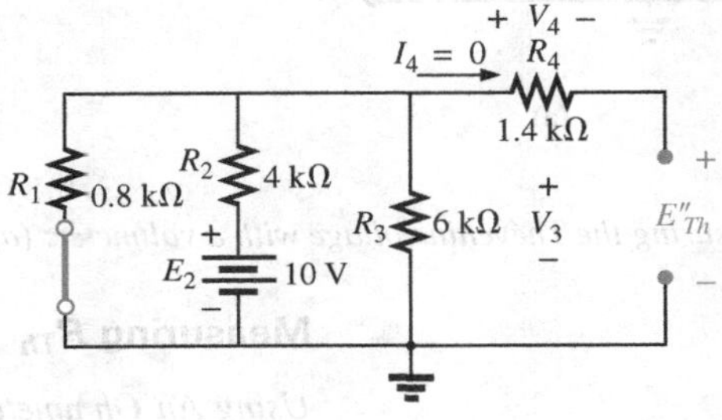

FIG. 9.52
Determining the contribution to E_{Th} from the source E_2 for the network in Fig. 9.49.

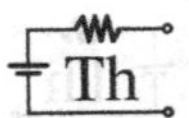

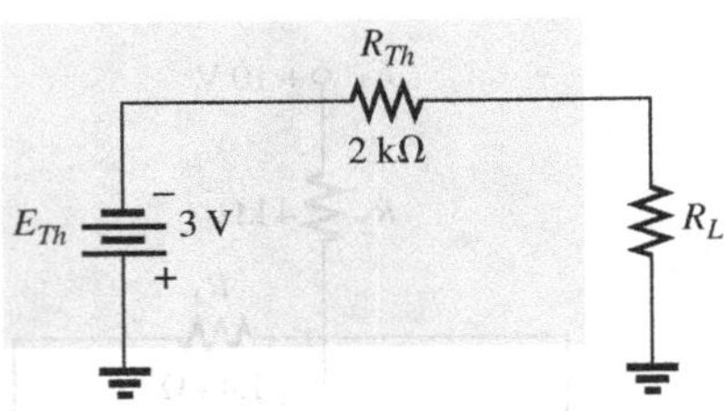

FIG. 9.53
Substituting the Thévenin equivalent circuit for the network external to the resistor R_L in Fig. 9.48.

Since E'_{Th} and E''_{Th} have opposite polarities,

$$\begin{aligned} E_{Th} &= E'_{Th} - E''_{Th} \\ &= 4.5\text{ V} - 1.5\text{ V} \\ &= \mathbf{3\text{ V}} \qquad \text{(polarity of } E'_{Th}\text{)} \end{aligned}$$

Step 5: See Fig. 9.53.

Experimental Procedures

Now that the analytical procedure has been described in detail and a sense for the Thévenin impedance and voltage established, it is time to investigate how both quantities can be determined using an experimental procedure.

Even though the Thévenin resistance is usually the easiest to determine analytically, the Thévenin voltage is often the easiest to determine experimentally, and therefore it will be examined first.

Measuring E_{Th} The network of Fig. 9.54(a) has the equivalent Thévenin circuit appearing in Fig. 9.54(b). The open-circuit Thévenin voltage can be determined by simply placing a voltmeter on the output terminals in Fig. 9.54(a) as shown. This is due to the fact that the open circuit in Fig. 9.54(b) dictates that the current through and the voltage across the Thévenin resistance must be zero. The result for Fig. 9.54(b) is that

$$V_{oc} = E_{Th} = \mathbf{4.5\text{ V}}$$

In general, therefore,

the Thévenin voltage is determined by connecting a voltmeter to the output terminals of the network. Be sure the internal resistance of the voltmeter is significantly more than the expected level of R_{Th}.

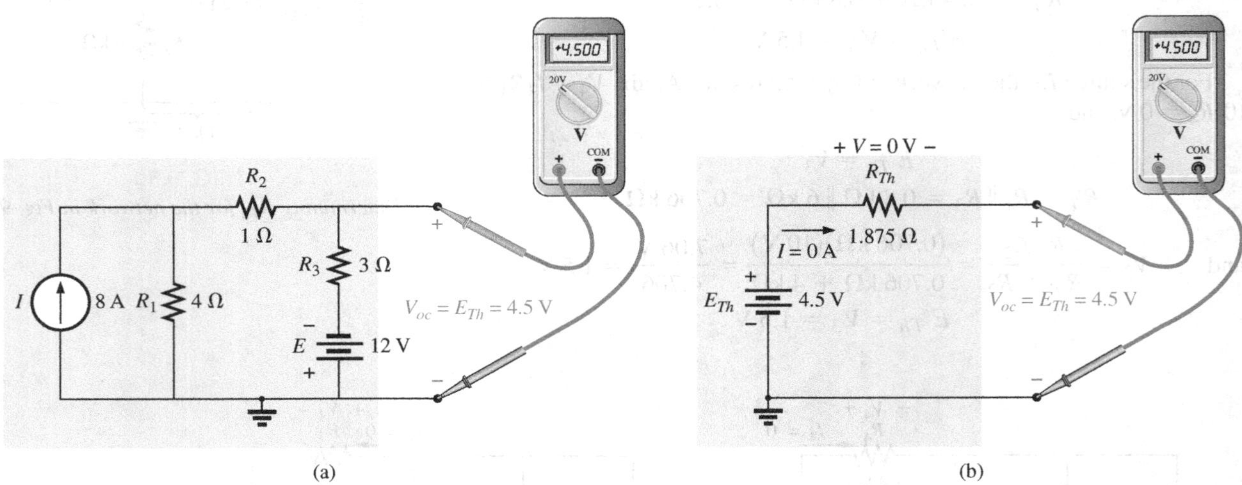

FIG. 9.54
Measuring the Thévenin voltage with a voltmeter: (a) actual network; (b) Thévenin equivalent.

Measuring R_{Th}

Using An Ohmmeter In Fig. 9.55, the sources in Fig. 9.54(a) have been set to zero, and an ohmmeter has been applied to measure the Thévenin resistance. In Fig. 9.54(b), it is clear that if the Thévenin voltage is set to zero volts, the ohmmeter will read the Thévenin resistance directly.

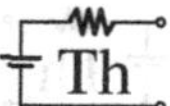

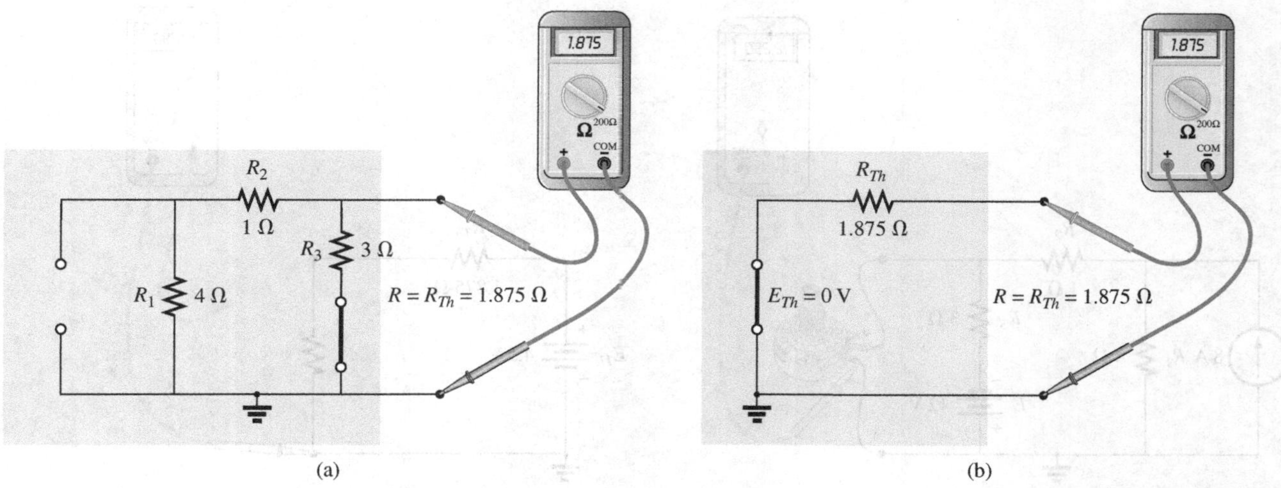

FIG. 9.55
Measuring R_{Th} with an ohmmeter: (a) actual network; (b) Thévenin equivalent.

In general, therefore,

the Thévenin resistance can be measured by setting all the sources to zero and measuring the resistance at the output terminals.

It is important to remember, however, that ohmmeters cannot be used on live circuits, and you cannot set a voltage source by putting a short circuit across it—it causes instant damage. The source must either be set to zero or removed entirely and then replaced by a direct connection. For the current source, the open-circuit condition must be clearly established; otherwise, the measured resistance will be incorrect. For most situations, it is usually best to remove the sources and replace them by the appropriate equivalent.

Using a Potentiometer If we use a potentiometer to measure the Thévenin resistance, the sources can be left as is. For this reason alone, this approach is one of the more popular. In Fig. 9.56(a), a potentiometer has been connected across the output terminals of the network to establish the condition appearing in Fig. 9.56(b) for the Thévenin equivalent. If the resistance of the potentiometer is now adjusted so that the voltage across the potentiometer is one-half the measured Thévenin voltage, the Thévenin resistance must match that of the potentiometer. Recall that for a series circuit, the applied voltage will divide equally across two equal series resistors.

If the potentiometer is then disconnected and the resistance measured with an ohmmeter as shown in Fig. 9.56(c), the ohmmeter displays the Thévenin resistance of the network. In general, therefore,

the Thévenin resistance can be measured by applying a potentiometer to the output terminals and varying the resistance until the output voltage is one-half the measured Thévenin voltage. The resistance of the potentiometer is the Thévenin resistance for the network.

Using the Short-Circuit Current The Thévenin resistance can also be determined by placing a short circuit across the output terminals and finding the current through the short circuit. Since ammeters ideally have zero internal ohms between their terminals, hooking up an ammeter as shown in Fig. 9.57(a) has the effect of both hooking up a short circuit across the

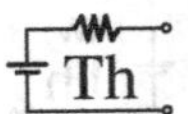

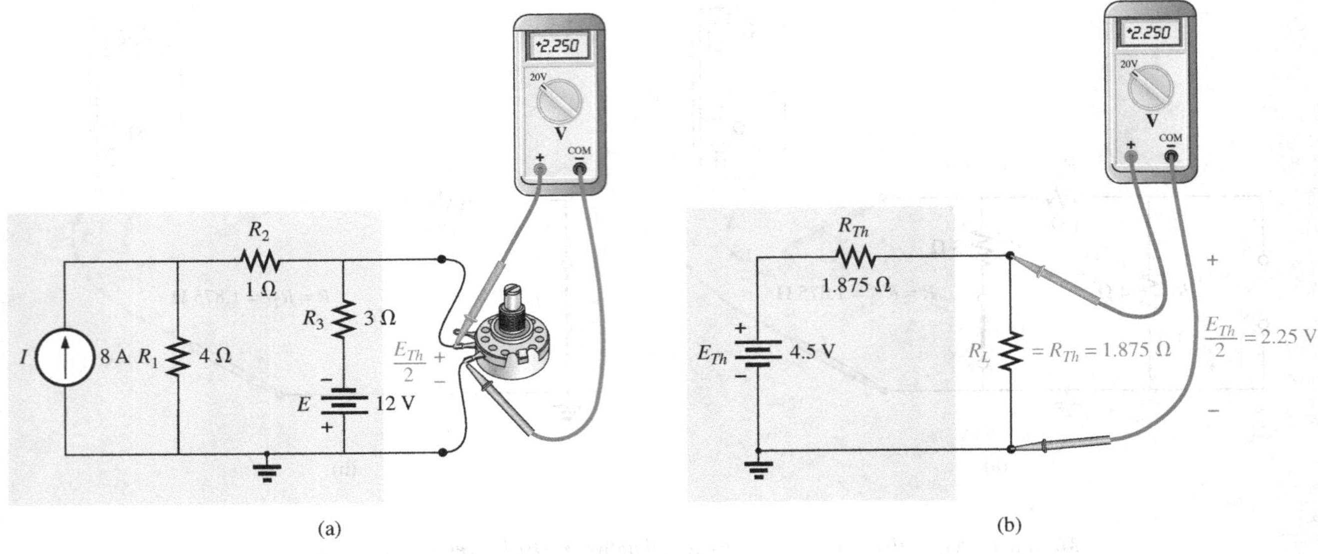

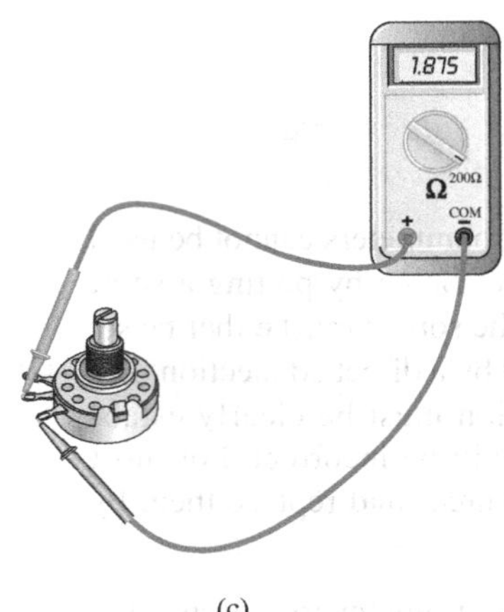

FIG. 9.56
Using a potentiometer to determine R_{Th}: (a) actual network; (b) Thévenin equivalent; (c) measuring R_{Th}.

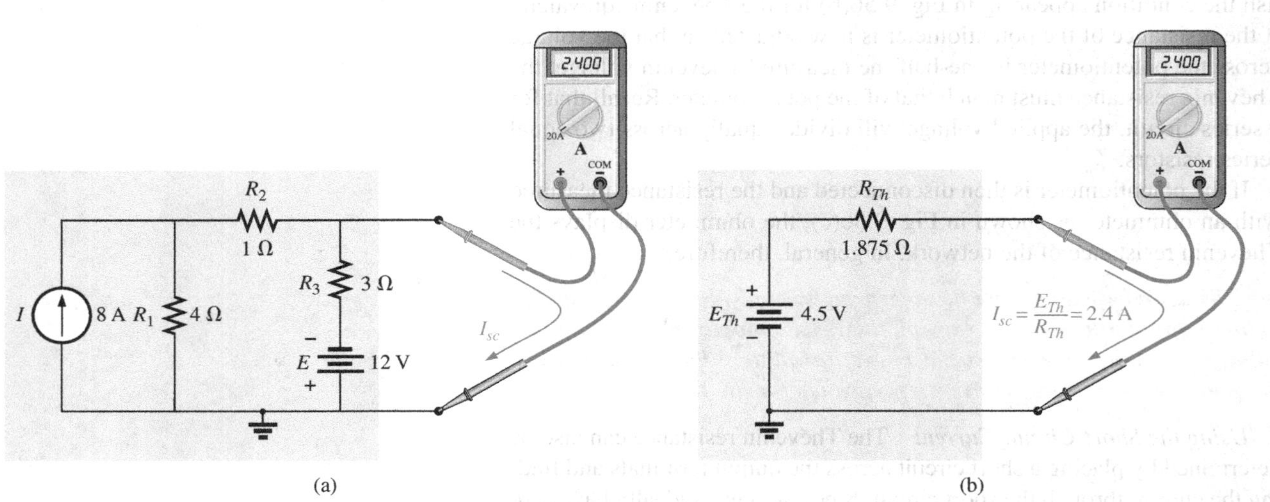

FIG. 9.57
Determining R_{Th} using the short-circuit current: (a) actual network; (b) Thévenin equivalent.

terminals and measuring the resulting current. The same ammeter was connected across the Thévenin equivalent circuit in Fig. 9.57(b).

On a practical level, it is assumed, of course, that the internal resistance of the ammeter is approximately zero ohms in comparison to the other resistors of the network. It is also important to be sure that the resulting current does not exceed the maximum current for the chosen ammeter scale.

In Fig. 9.57(b), since the short-circuit current is

$$I_{sc} = \frac{E_{Th}}{R_{Th}}$$

the Thévenin resistance can be determined by

$$R_{Th} = \frac{E_{Th}}{I_{sc}}$$

In general, therefore,

the Thévenin resistance can be determined by hooking up an ammeter across the output terminals to measure the short-circuit current and then using the open-circuit voltage to calculate the Thévenin resistance in the following manner:

$$\boxed{R_{Th} = \frac{V_{oc}}{I_{sc}}} \qquad \textbf{(9.1)}$$

As a result, we have three ways to measure the Thévenin resistance of a configuration. Because of the concern about setting the sources to zero in the first procedure and the concern about current levels in the last, the second method is often chosen.

FIG. 9.58
Edward L. Norton.
Reprinted with permission of Alcatel-Lucent USA Inc.

American (Rockland, Maine; Summit, New Jersey)
1898–1983
Electrical Engineer, Scientist, Inventor
Department Head: Bell Laboratories
Fellow: Acoustical Society and Institute of Radio Engineers

Although interested primarily in communications circuit theory and the transmission of data at high speeds over telephone lines, Edward L. Norton is best remembered for development of the dual of Thévenin equivalent circuit, currently referred to as *Norton's equivalent circuit.* In fact, Norton and his associates at AT&T in the early 1920s are recognized as being among the first to perform work applying Thévenin's equivalent circuit and referring to this concept simply as Thévenin's theorem. In 1926, he proposed the equivalent circuit using a current source and parallel resistor to assist in the design of recording instrumentation that was primarily current driven. He began his telephone career in 1922 with the Western Electric Company's Engineering Department, which later became Bell Laboratories. His areas of active research included network theory, acoustical systems, electromagnetic apparatus, and data transmission. A graduate of MIT and Columbia University, he held nineteen patents on his work.

9.4 NORTON'S THEOREM

In Section 8.3, we learned that every voltage source with a series internal resistance has a current source equivalent. The current source equivalent can be determined by **Norton's theorem** (Fig. 9.58). It can also be found through the conversions of Section 8.3.

The theorem states the following:

Any two-terminal linear bilateral dc network can be replaced by an equivalent circuit consisting of a current source and a parallel resistor, as shown in Fig. 9.59.

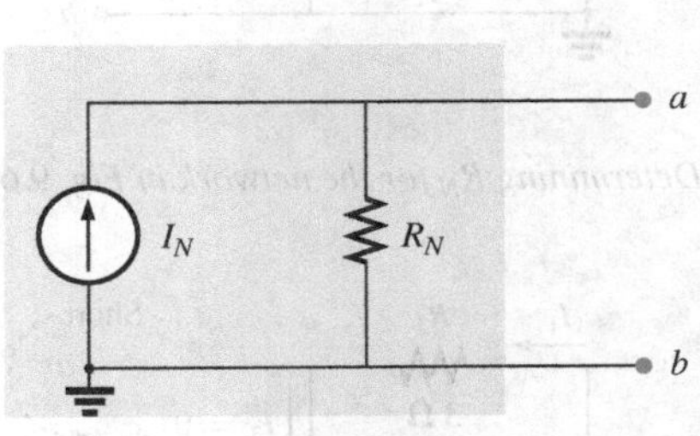

FIG. 9.59
Norton equivalent circuit.

The discussion of Thévenin's theorem with respect to the equivalent circuit can also be applied to the Norton equivalent circuit. The steps leading to the proper values of I_N and R_N are now listed.

Norton's Theorem Procedure

Preliminary:

1. Remove that portion of the network across which the Norton equivalent circuit is found.
2. Mark the terminals of the remaining two-terminal network.

R_N:

3. Calculate R_N by first setting all sources to zero (voltage sources are replaced with short circuits and current sources with open circuits) and then finding the resultant resistance between the two

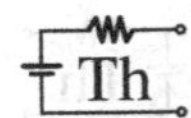

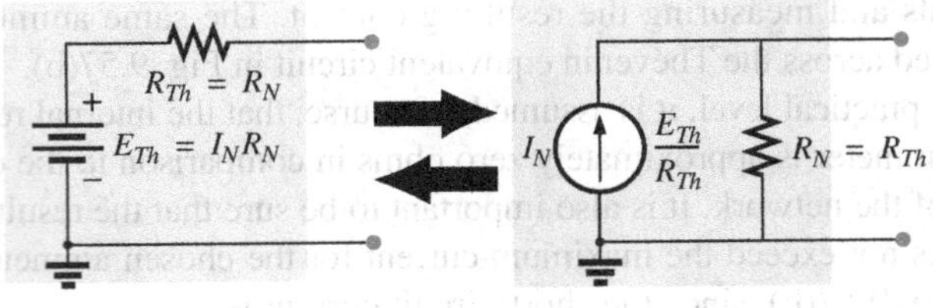

FIG. 9.60
Converting between Thévenin and Norton equivalent circuits.

marked terminals. (If the internal resistance of the voltage and/or current sources is included in the original network, it must remain when the sources are set to zero.) Since $R_N = R_{Th}$, the procedure and value obtained using the approach described for Thévenin's theorem will determine the proper value of R_N.

I_N:

4. Calculate I_N by first returning all sources to their original position and then finding the short-circuit current between the marked terminals. It is the same current that would be measured by an ammeter placed between the marked terminals.

Conclusion:

5. Draw the Norton equivalent circuit with the portion of the circuit previously removed replaced between the terminals of the equivalent circuit.

The Norton and Thévenin equivalent circuits can also be found from each other by using the source transformation discussed earlier in this chapter and reproduced in Fig. 9.60.

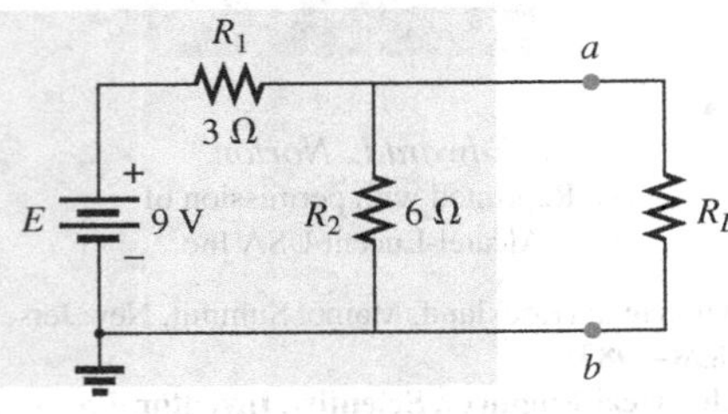

FIG. 9.61
Example 9.11.

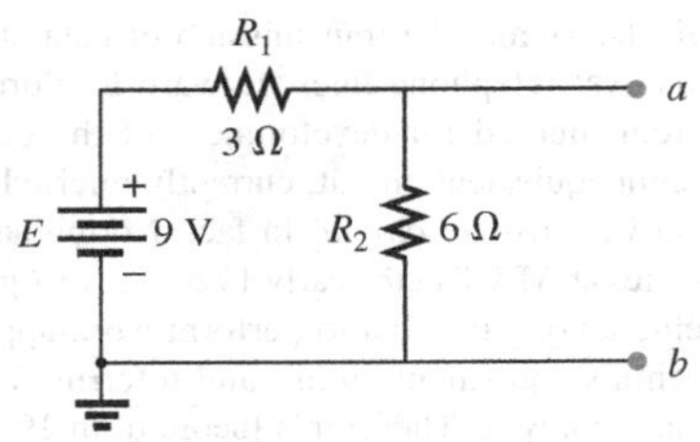

FIG. 9.62
Identifying the terminals of particular interest for the network in Fig. 9.61.

EXAMPLE 9.11 Find the Norton equivalent circuit for the network in the shaded area in Fig. 9.61.

Solution:

Steps 1 and 2: See Fig. 9.62.

Step 3: See Fig. 9.63, and

$$R_N = R_1 \| R_2 = 3\ \Omega \| 6\ \Omega = \frac{(3\ \Omega)(6\ \Omega)}{3\ \Omega + 6\ \Omega} = \frac{18\ \Omega}{9} = \mathbf{2\ \Omega}$$

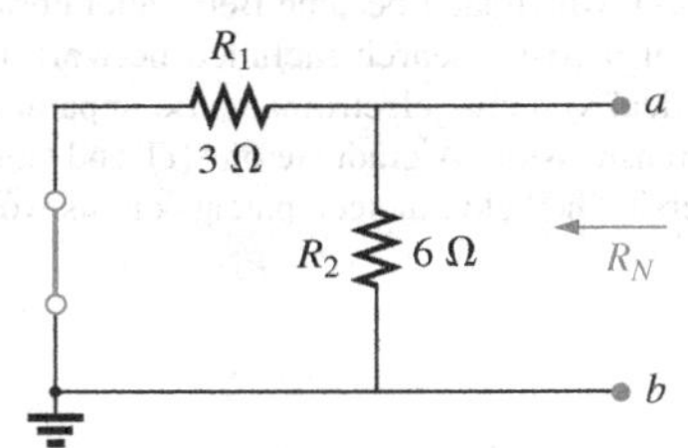

FIG. 9.63
Determining R_N for the network in Fig. 9.62.

Step 4: See Fig. 9.64, which clearly indicates that the short-circuit connection between terminals a and b is in parallel with R_2 and eliminates its effect. I_N is therefore the same as through R_1, and the full battery voltage appears across R_1 since

$$V_2 = I_2 R_2 = (0)6\ \Omega = 0\ \text{V}$$

Therefore,

$$I_N = \frac{E}{R_1} = \frac{9\ \text{V}}{3\ \Omega} = \mathbf{3\ A}$$

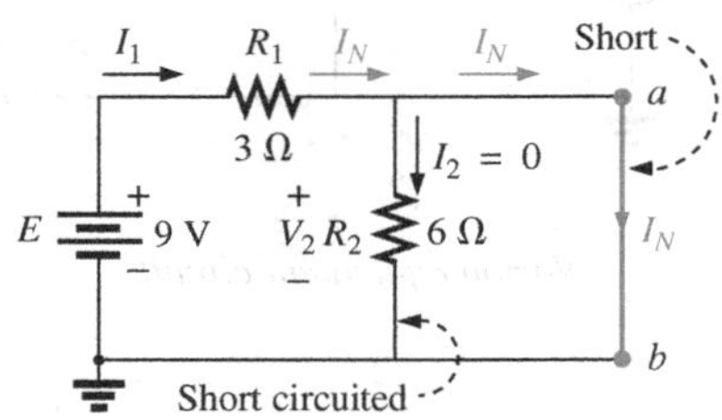

FIG. 9.64
Determining I_N for the network in Fig. 9.62.

Step 5: See Fig. 9.65. This circuit is the same as the first one considered in the development of Thévenin's theorem. A simple conversion indicates that the Thévenin circuits are, in fact, the same (Fig. 9.66).

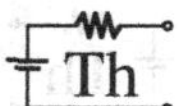

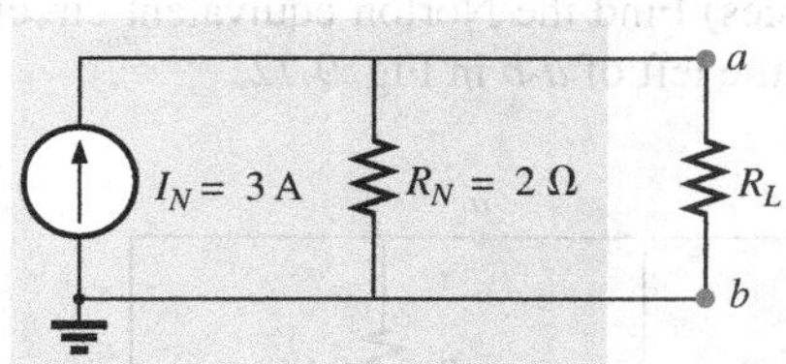

FIG. 9.65
Substituting the Norton equivalent circuit for the network external to the resistor R_L in Fig. 9.61.

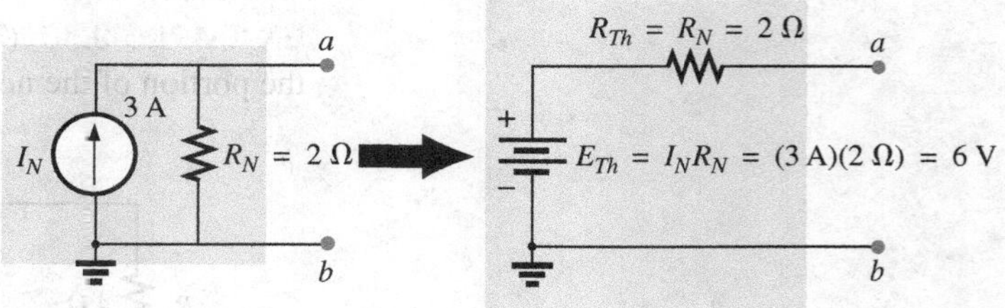

FIG. 9.66
Converting the Norton equivalent circuit in Fig. 9.65 to a Thévenin equivalent circuit.

EXAMPLE 9.12 Find the Norton equivalent circuit for the network external to the 9 Ω resistor in Fig. 9.67.

Solution:

Steps 1 and 2: See Fig. 9.68.

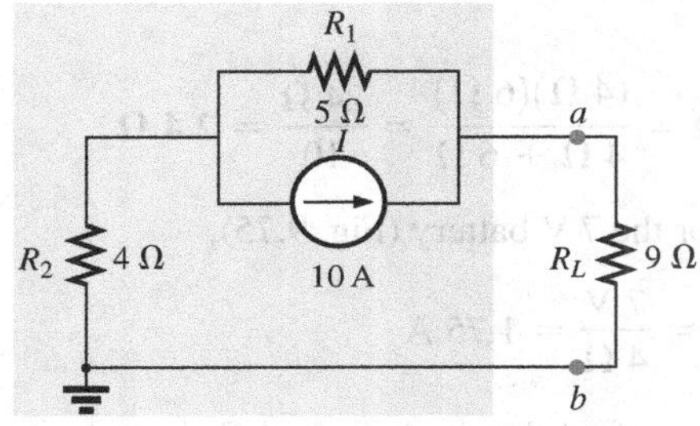

FIG. 9.67
Example 9.12.

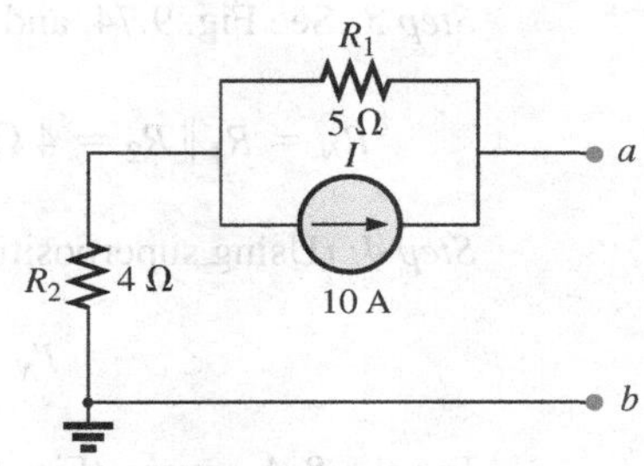

FIG. 9.68
Identifying the terminals of particular interest for the network in Fig. 9.67.

Step 3: See Fig. 9.69, and

$$R_N = R_1 + R_2 = 5\ \Omega + 4\ \Omega = \mathbf{9\ \Omega}$$

Step 4: As shown in Fig. 9.70, the Norton current is the same as the current through the 4 Ω resistor. Applying the current divider rule gives

$$I_N = \frac{R_1 I}{R_1 + R_2} = \frac{(5\ \Omega)(10\ \text{A})}{5\ \Omega + 4\ \Omega} = \frac{50\ \text{A}}{9} = \mathbf{5.56\ A}$$

Step 5: See Fig. 9.71.

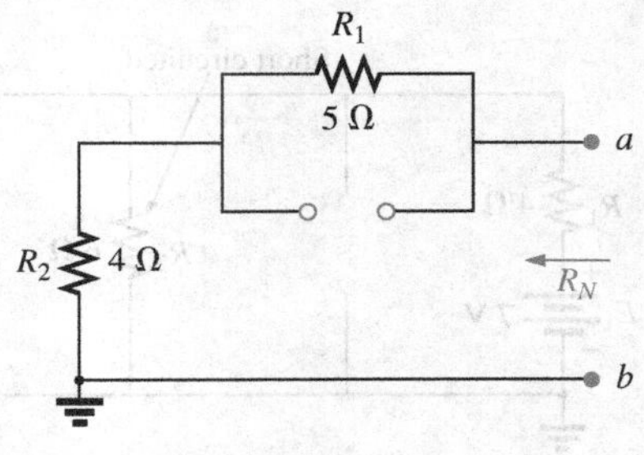

FIG. 9.69
Determining R_N for the network in Fig. 9.68.

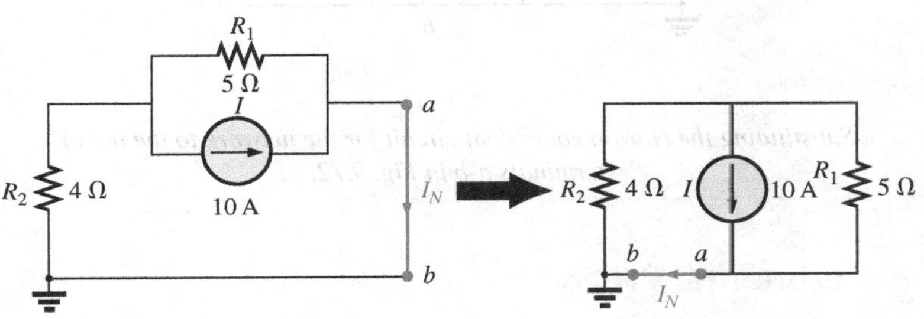

FIG. 9.70
Determining I_N for the network in Fig. 9.68.

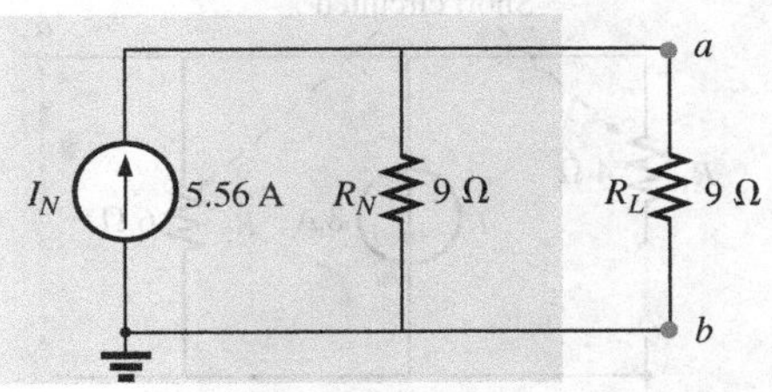

FIG. 9.71
Substituting the Norton equivalent circuit for the network external to the resistor R_L in Fig. 9.67.

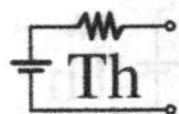

EXAMPLE 9.13 (Two sources) Find the Norton equivalent circuit for the portion of the network to the left of *a-b* in Fig. 9.72.

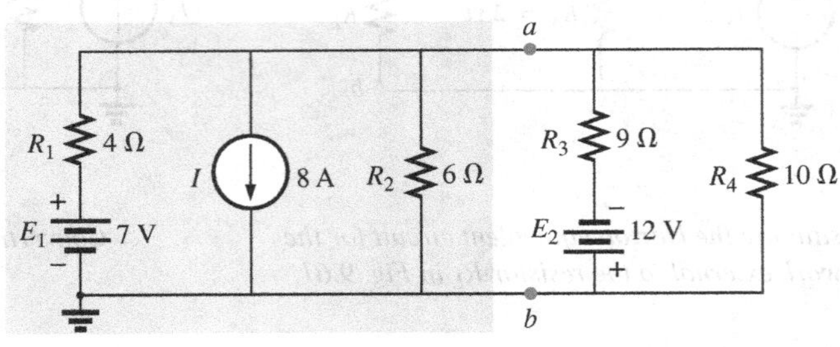

FIG. 9.72
Example 9.13.

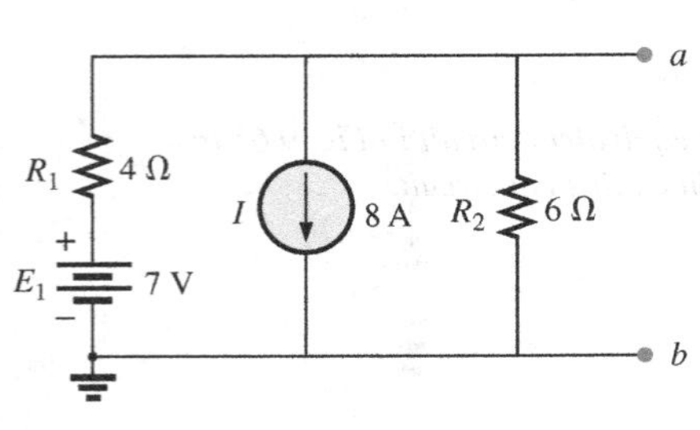

FIG. 9.73
Identifying the terminals of particular interest for the network in Fig. 9.72.

Solution:

Steps 1 and 2: See Fig. 9.73.

Step 3: See Fig. 9.74, and

$$R_N = R_1 \| R_2 = 4\ \Omega \| 6\ \Omega = \frac{(4\ \Omega)(6\ \Omega)}{4\ \Omega + 6\ \Omega} = \frac{24\ \Omega}{10} = \mathbf{2.4\ \Omega}$$

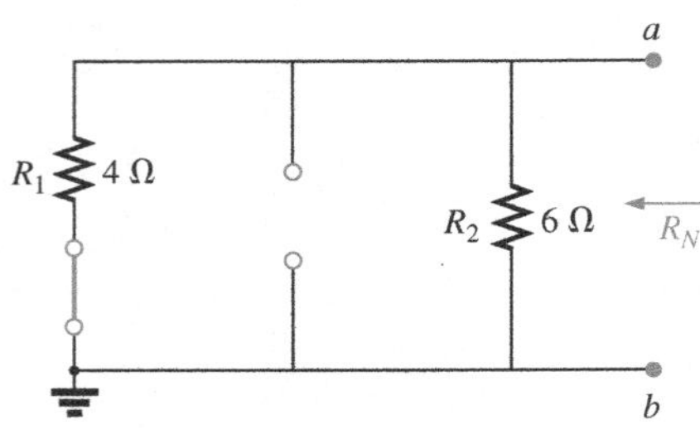

FIG. 9.74
Determining R_N for the network in Fig. 9.73.

Step 4: (Using superposition) For the 7 V battery (Fig. 9.75),

$$I'_N = \frac{E_1}{R_1} = \frac{7\ \text{V}}{4\ \Omega} = 1.75\ \text{A}$$

For the 8 A source (Fig. 9.76), we find that both R_1 and R_2 have been "short circuited" by the direct connection between *a* and *b*, and

$$I''_N = I = 8\ \text{A}$$

The result is

$$I_N = I''_N - I'_N = 8\ \text{A} - 1.75\ \text{A} = \mathbf{6.25\ A}$$

Step 5: See Fig. 9.77.

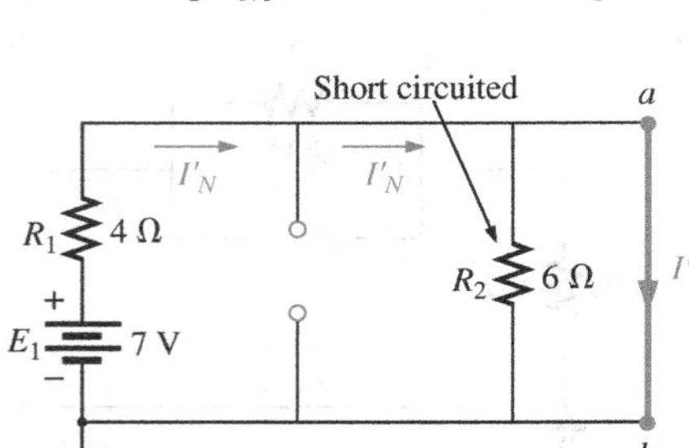

FIG. 9.75
Determining the contribution to I_N from the voltage source E_1.

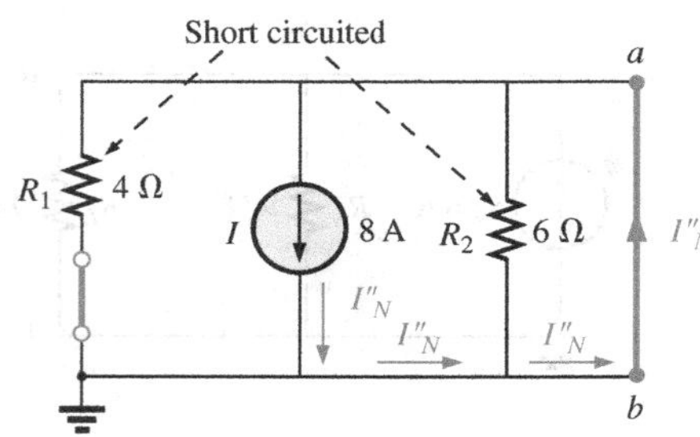

FIG. 9.76
Determining the contribution to I_N from the current source I.

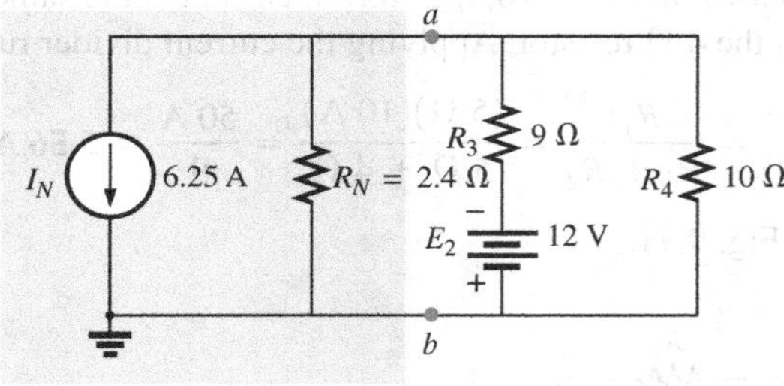

FIG. 9.77
Substituting the Norton equivalent circuit for the network to the left of terminals a-b in Fig. 9.72.

Experimental Procedure

The Norton current is measured in the same way as described for the short-circuit current (I_{sc}) for the Thévenin network. Since the Norton and Thévenin resistances are the same, the same procedures can be followed as described for the Thévenin network.

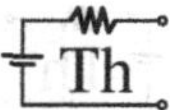

9.5 MAXIMUM POWER TRANSFER THEOREM

When designing a circuit, it is often important to be able to answer one of the following questions:

What load should be applied to a system to ensure that the load is receiving maximum power from the system?

Conversely:

For a particular load, what conditions should be imposed on the source to ensure that it will deliver the maximum power available?

Even if a load cannot be set at the value that would result in maximum power transfer, it is often helpful to have some idea of the value that will draw maximum power so that you can compare it to the load at hand. For instance, if a design calls for a load of 100 Ω, to ensure that the load receives maximum power, using a resistor of 1 Ω or 1 kΩ results in a power transfer that is much less than the maximum possible. However, using a load of 82 Ω or 120 Ω probably results in a fairly good level of power transfer.

Fortunately, the process of finding the load that will receive maximum power from a particular system is quite straightforward due to the **maximum power transfer theorem,** which states the following:

A load will receive maximum power from a network when its resistance is exactly equal to the Thévenin resistance of the network applied to the load. That is,

$$\boxed{R_L = R_{Th}} \tag{9.2}$$

In other words, for the Thévenin equivalent circuit in Fig. 9.78, when the load is set equal to the Thévenin resistance, the load will receive maximum power from the network.

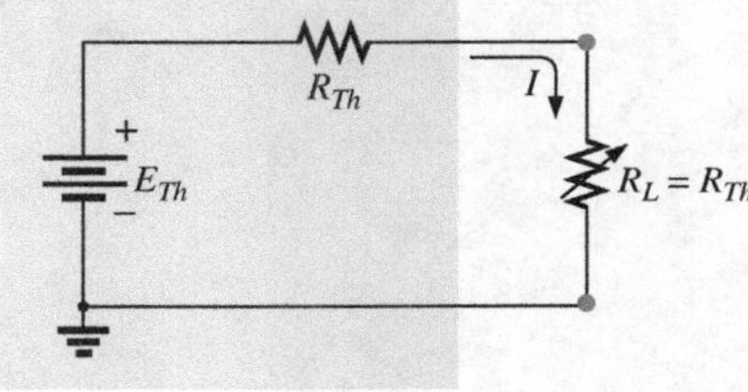

FIG. 9.78
Defining the conditions for maximum power to a load using the Thévenin equivalent circuit.

Using Fig. 9.78 with $R_L = R_{Th}$, we can determine the maximum power delivered to the load by first finding the current:

$$I_L = \frac{E_{Th}}{R_{Th} + R_L} = \frac{E_{Th}}{R_{Th} + R_{Th}} = \frac{E_{Th}}{2R_{Th}}$$

Then we substitute into the power equation:

$$P_L = I_L^2 R_L = \left(\frac{E_{Th}}{2R_{Th}}\right)^2 (R_{Th}) = \frac{E_{Th}^2 R_{Th}}{4R_{Th}^2}$$

and

$$\boxed{P_{L_{max}} = \frac{E_{Th}^2}{4R_{Th}}} \tag{9.3}$$

To demonstrate that maximum power is indeed transferred to the load under the conditions defined above, consider the Thévenin equivalent circuit in Fig. 9.79.

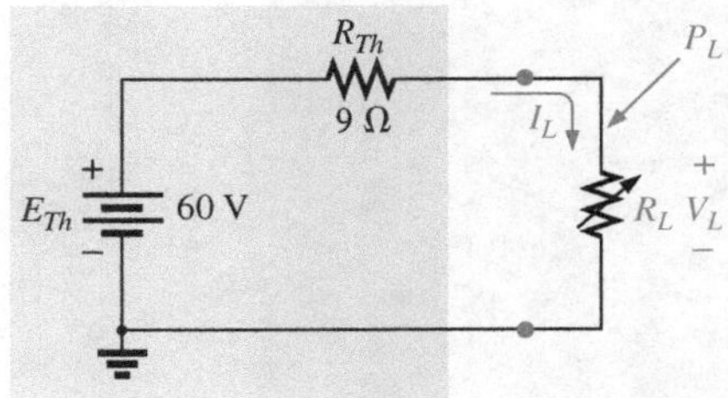

FIG. 9.79
Thévenin equivalent network to be used to validate the maximum power transfer theorem.

Before getting into detail, however, if you were to guess what value of R_L would result in maximum power transfer to R_L, you might think that the smaller the value of R_L, the better it is because the current reaches a maximum when it is squared in the power equation. The problem is, however, that in the equation $P_L = I_L^2 R_L$, the load resistance is a multiplier. As it gets smaller, it forms a smaller product. Then again, you might suggest larger values of R_L because the output voltage increases, and power

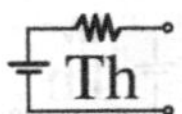

is determined by $P_L = V_L^2/R_L$. This time, however, the load resistance is in the denominator of the equation and causes the resulting power to decrease. A balance must obviously be made between the load resistance and the resulting current or voltage. The following discussion shows that

maximum power transfer occurs when the load voltage and current are one-half their maximum possible values.

For the circuit in Fig. 9.79, the current through the load is determined by

$$I_L = \frac{E_{Th}}{R_{Th} + R_L} = \frac{60\text{ V}}{9\ \Omega + R_L}$$

The voltage is determined by

$$V_L = \frac{R_L E_{Th}}{R_L + R_{Th}} = \frac{R_L(60\text{ V})}{R_L + R_{Th}}$$

and the power by

$$P_L = I_L^2 R_L = \left(\frac{60\text{ V}}{9\ \Omega + R_L}\right)^2 (R_L) = \frac{3600 R_L}{(9\ \Omega + R_L)^2}$$

If we tabulate the three quantities versus a range of values for R_L from 0.1 Ω to 30 Ω, we obtain the results appearing in Table 9.1. Note in particular that when R_L is equal to the Thévenin resistance of 9 Ω, the power

TABLE 9.1

R_L (Ω)	P_L (W)		I_L (A)		V_L (V)	
0.1	4.35		6.60		0.66	
0.2	8.51		6.52		1.30	
0.5	19.94		6.32		3.16	
1	36.00		6.00		6.00	
2	59.50		5.46		10.91	
3	75.00		5.00		15.00	
4	85.21		4.62		18.46	
5	91.84		4.29		21.43	
6	96.00		4.00		24.00	
7	98.44	Increase	3.75	Decrease	26.25	Increase
8	99.65		3.53		28.23	
9 (R_{Th})	100.00 (Maximum)		3.33 ($I_{max}/2$)		30.00 ($E_{Th}/2$)	
10	99.72		3.16		31.58	
11	99.00		3.00		33.00	
12	97.96		2.86		34.29	
13	96.69		2.73		35.46	
14	95.27		2.61		36.52	
15	93.75		2.50		37.50	
16	92.16		2.40		38.40	
17	90.53		2.31		39.23	
18	88.89		2.22		40.00	
19	87.24		2.14		40.71	
20	85.61		2.07		41.38	
25	77.86		1.77		44.12	
30	71.00		1.54		46.15	
40	59.98		1.22		48.98	
100	30.30		0.55		55.05	
500	6.95	Decrease	0.12	Decrease	58.94	Increase
1000	3.54		0.06		59.47	

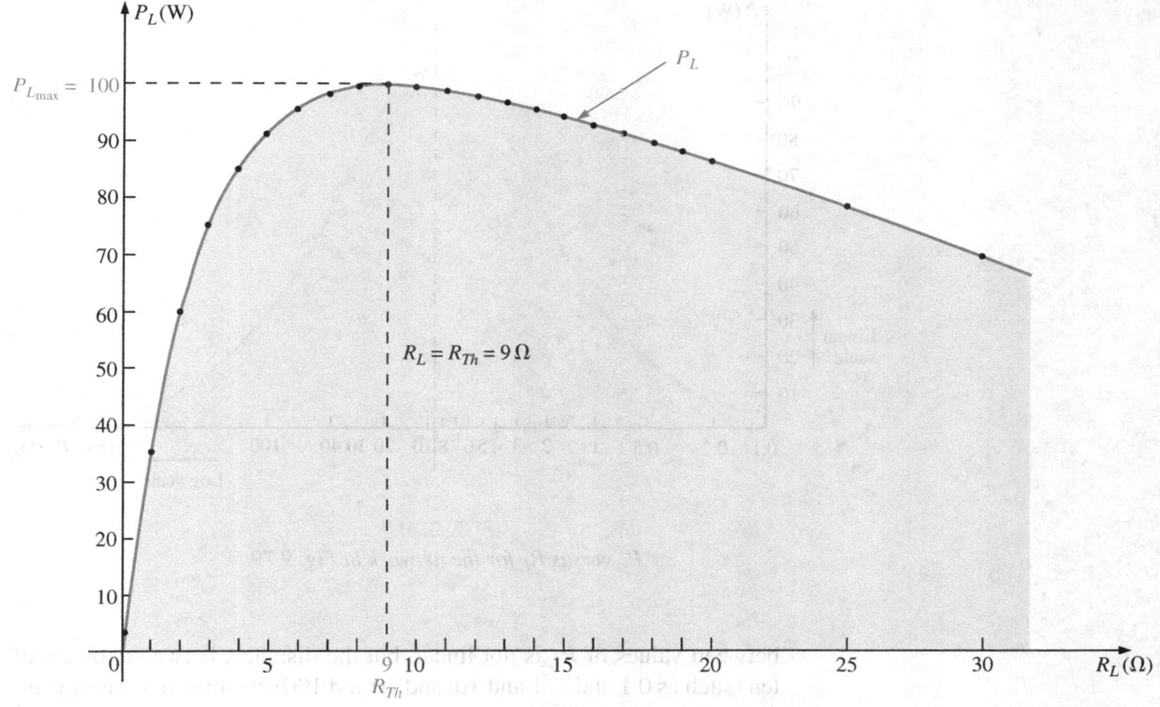

FIG. 9.80
P_L versus R_L for the network in Fig. 9.79.

has a maximum value of 100 W, the current is 3.33 A, or one-half its maximum value of 6.60 A (as would result with a short circuit across the output terminals), and the voltage across the load is 30 V, or one-half its maximum value of 60 V (as would result with an open circuit across its output terminals). As you can see, there is no question that maximum power is transferred to the load when the load equals the Thévenin value.

The power to the load versus the range of resistor values is provided in Fig. 9.80. Note in particular that for values of load resistance less than the Thévenin value, the change is dramatic as it approaches the peak value. However, for values greater than the Thévenin value, the drop is a great deal more gradual. This is important because it tells us the following:

If the load applied is less than the Thévenin resistance, the power to the load will drop off rapidly as it gets smaller. However, if the applied load is greater than the Thévenin resistance, the power to the load will not drop off as rapidly as it increases.

For instance, the power to the load is at least 90 W for the range of about 4.5 Ω to 9 Ω below the peak value, but it is at least the same level for a range of about 9 Ω to 18 Ω above the peak value. The range below the peak is 4.5 Ω, while the range above the peak is almost twice as much at 9 Ω. As mentioned above, if maximum transfer conditions cannot be established, at least we now know from Fig. 9.80 that any resistance relatively close to the Thévenin value results in a strong transfer of power. More distant values such as 1 Ω or 100 Ω result in much lower levels.

It is particularly interesting to plot the power to the load versus load resistance using a log scale, as shown in Fig. 9.81. Logarithms will be discussed in detail in Chapter 21, but for now notice that the spacing

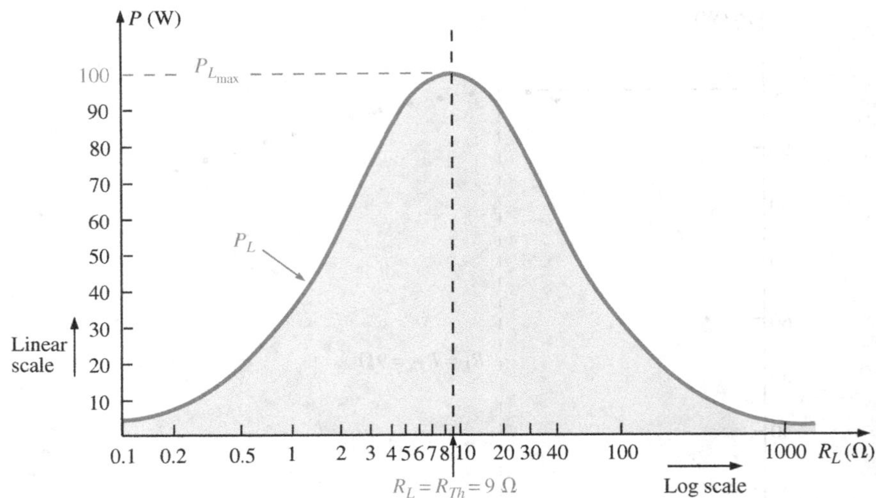

FIG. 9.81
P_L versus R_L for the network in Fig. 9.79.

between values of R_L is not linear, but the distance, between powers of ten (such as 0.1 and 1, 1 and 10, and 10 and 100) are all equal. The advantage of the log scale is that a wide resistance range can be plotted on a relatively small graph.

Note in Fig. 9.81 that a smooth, bell-shaped curve results that is symmetrical about the Thévenin resistance of 9 Ω. At 0.1 Ω, the power has dropped to about the same level as that at 1000 Ω, and at 1 Ω and 100 Ω, the power has dropped to the neighborhood of 30 W.

Although all of the above discussion centers on the power to the load, it is important to remember the following:

The total power delivered by a supply such as E_{Th} is absorbed by both the Thévenin equivalent resistance and the load resistance. Any power delivered by the source that does not get to the load is lost to the Thévenin resistance.

Under maximum power conditions, only half the power delivered by the source gets to the load. Now, that sounds disastrous, but remember that we are starting out with a fixed Thévenin voltage and resistance, and the above simply tells us that we must make the two resistance levels equal if we want maximum power to the load. On an efficiency basis, we are working at only a 50% level, but we are content because *we are getting maximum power out of our system.*

The dc operating efficiency is defined as the ratio of the power delivered to the load (P_L) to the power delivered by the source (P_s). That is,

$$\eta\% = \frac{P_L}{P_s} \times 100\% \tag{9.4}$$

For the situation where $R_L = R_{Th}$,

$$\eta\% = \frac{I_L^2 R_L}{I_L^2 R_T} \times 100\% = \frac{R_L}{R_T} \times 100\% = \frac{R_{Th}}{R_{Th} + R_{Th}} \times 100\%$$

$$= \frac{R_{Th}}{2R_{Th}} \times 100\% = \frac{1}{2} \times 100\% = \mathbf{50\%}$$

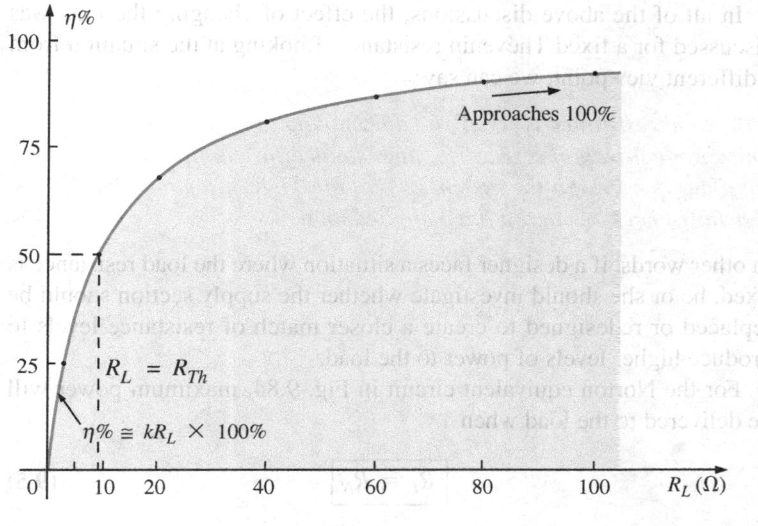

FIG. 9.82
Efficiency of operation versus increasing values of R_L.

For the circuit in Fig. 9.79, if we plot the efficiency of operation versus load resistance, we obtain the plot in Fig. 9.82, which clearly shows that the efficiency continues to rise to a 100% level as R_L gets larger. Note in particular that the efficiency is 50% when $R_L = R_{Th}$.

To ensure that you completely understand the effect of the maximum power transfer theorem and the efficiency criteria, consider the circuit in Fig. 9.83, where the load resistance is set at 100 Ω and the power to the Thévenin resistance and to the load are calculated as follows:

$$I_L = \frac{E_{Th}}{R_{Th} + R_L} = \frac{60\text{ V}}{9\ \Omega + 100\ \Omega} = \frac{60\text{ V}}{109\ \Omega} = 550.5\text{ mA}$$

with $\quad P_{R_{Th}} = I_L^2 R_{Th} = (550.5\text{ mA})^2(9\ \Omega) \cong \mathbf{2.73\ W}$

and $\quad P_L = I_L^2 R_L = (550.5\text{ mA})^2(100\ \Omega) \cong \mathbf{30.3\ W}$

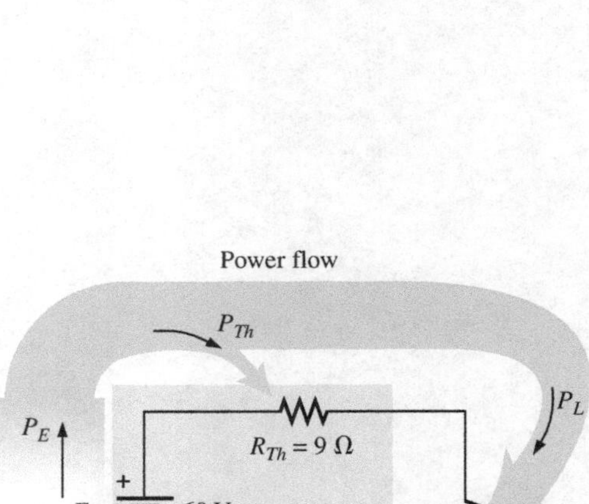

FIG. 9.83
Examining a circuit with high efficiency but a relatively low level of power to the load.

The results clearly show that most of the power supplied by the battery is getting to the load—a desirable attribute on an efficiency basis. However, the power getting to the load is only 30.3 W compared to the 100 W obtained under maximum power conditions. In general, therefore, the following guidelines apply:

If efficiency is the overriding factor, then the load should be much larger than the internal resistance of the supply. If maximum power transfer is desired and efficiency less of a concern, then the conditions dictated by the maximum power transfer theorem should be applied.

A relatively low efficiency of 50% can be tolerated in situations where power levels are relatively low, such as in a wide variety of electronic systems, where maximum power transfer for the given system is usually more important. However, when large power levels are involved, such as at generating plants, efficiencies of 50% cannot be tolerated. In fact, a great deal of expense and research is dedicated to raising power generating and transmission efficiencies a few percentage points. Raising an efficiency level of a 10 MkW power plant from 94% to 95% (a 1% increase) can save 0.1 MkW, or 100 million watts, of power—an enormous saving.

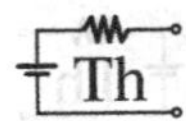

In all of the above discussions, the effect of changing the load was discussed for a fixed Thévenin resistance. Looking at the situation from a different viewpoint, we can say

if the load resistance is fixed and does not match the applied Thévenin equivalent resistance, then some effort should be made (if possible) to redesign the system so that the Thévenin equivalent resistance is closer to the fixed applied load.

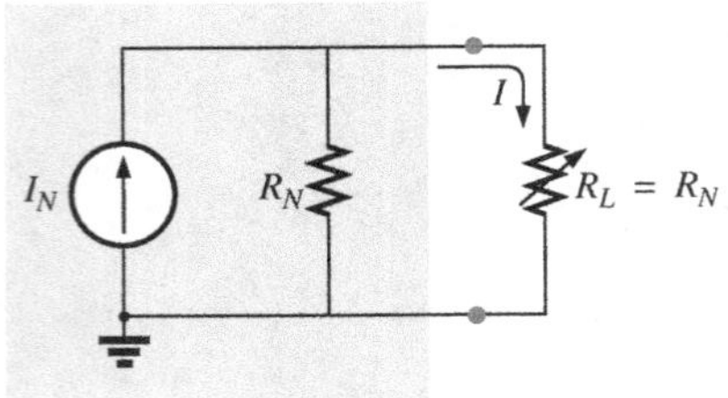

FIG. 9.84
Defining the conditions for maximum power to a load using the Norton equivalent circuit.

In other words, if a designer faces a situation where the load resistance is fixed, he or she should investigate whether the supply section should be replaced or redesigned to create a closer match of resistance levels to produce higher levels of power to the load.

For the Norton equivalent circuit in Fig. 9.84, maximum power will be delivered to the load when

$$\boxed{R_L = R_N} \qquad \textbf{(9.5)}$$

This result [Eq. (9.5)] will be used to its fullest advantage in the analysis of transistor networks, where the most frequently applied transistor circuit model uses a current source rather than a voltage source.

For the Norton circuit in Fig. 9.84,

$$\boxed{P_{L_{max}} = \frac{I_N^2 R_N}{4} \quad (\text{W})} \qquad \textbf{(9.6)}$$

EXAMPLE 9.14 A dc generator, battery, and laboratory supply are connected to resistive load R_L in Fig. 9.85.

a. For each, determine the value of R_L for maximum power transfer to R_L.
b. Under maximum power conditions, what are the current level and the power to the load for each configuration?
c. What is the efficiency of operation for each supply in part (b)?
d. If a load of 1 kΩ were applied to the laboratory supply, what would the power delivered to the load be? Compare your answer to the level of part (b). What is the level of efficiency?
e. For each supply, determine the value of R_L for 75% efficiency.

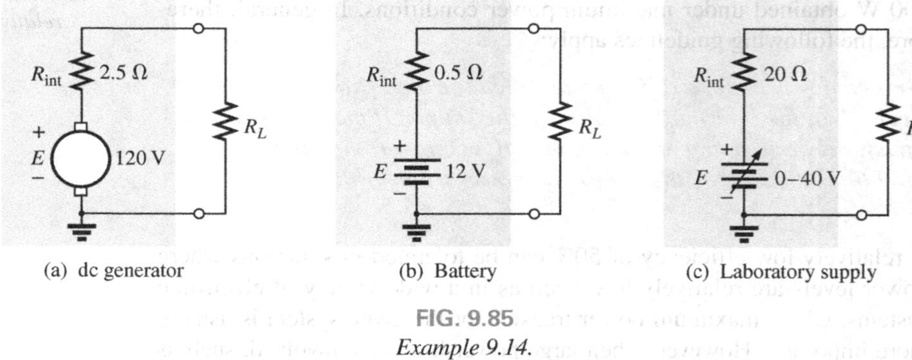

FIG. 9.85
Example 9.14.

Solutions:

a. For the dc generator,

$$R_L = R_{Th} = R_{int} = \mathbf{2.5\ \Omega}$$

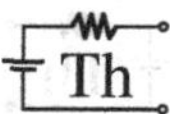

For the 12 V car battery,

$$R_L = R_{Th} = R_{int} = \mathbf{0.05\ \Omega}$$

For the dc laboratory supply,

$$R_L = R_{Th} = R_{int} = \mathbf{20\ \Omega}$$

b. For the dc generator,

$$P_{L_{max}} = \frac{E_{Th}^2}{4R_{Th}} = \frac{E^2}{4R_{int}} = \frac{(120\text{ V})^2}{4(2.5\ \Omega)} = \mathbf{1.44\ kW}$$

For the 12 V car battery,

$$P_{L_{max}} = \frac{E_{Th}^2}{4R_{Th}} = \frac{E^2}{4R_{int}} = \frac{(12\text{ V})^2}{4(0.05\ \Omega)} = \mathbf{720\ W}$$

For the dc laboratory supply,

$$P_{L_{max}} = \frac{E_{Th}^2}{4R_{Th}} = \frac{E^2}{4R_{int}} = \frac{(40\text{ V})^2}{4(20\ \Omega)} = \mathbf{20\ W}$$

c. They are all operating under a 50% efficiency level because $R_L = R_{Th}$.

d. The power to the load is determined as follows:

$$I_L = \frac{E}{R_{int} + R_L} = \frac{40\text{ V}}{20\ \Omega + 1000\ \Omega} = \frac{40\text{ V}}{1020\ \Omega} = 39.22\text{ mA}$$

and $\quad P_L = I_L^2 R_L = (39.22\text{ mA})^2(1000\ \Omega) = \mathbf{1.54\ W}$

The power level is significantly less than the 20 W achieved in part (b). The efficiency level is

$$\eta\% = \frac{P_L}{P_s} \times 100\% = \frac{1.54\text{ W}}{EI_s} \times 100\% = \frac{1.54\text{ W}}{(40\text{ V})(39.22\text{ mA})} \times 100\%$$

$$= \frac{1.54\text{ W}}{1.57\text{ W}} \times 100\% = \mathbf{98.09\%}$$

which is markedly higher than achieved under maximum power conditions—albeit at the expense of the power level.

e. For the dc generator,

$$\eta = \frac{P_o}{P_s} = \frac{R_L}{R_{Th} + R_L} \quad (\eta \text{ in decimal form})$$

and $\quad \eta = \dfrac{R_L}{R_{Th} + R_L}$

$$\eta(R_{Th} + R_L) = R_L$$

$$\eta R_{Th} + \eta R_L = R_L$$

$$R_L(1 - \eta) = \eta R_{Th}$$

and

$$\boxed{R_L = \frac{\eta R_{Th}}{1 - \eta}} \qquad (9.7)$$

$$R_L = \frac{0.75(2.5\ \Omega)}{1 - 0.75} = \mathbf{7.5\ \Omega}$$

For the battery,

$$R_L = \frac{0.75(0.05\ \Omega)}{1 - 0.75} = \mathbf{0.15\ \Omega}$$

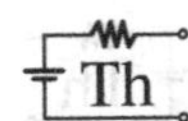

For the laboratory supply,

$$R_L = \frac{0.75(20\ \Omega)}{1 - 0.75} = \mathbf{60\ \Omega}$$

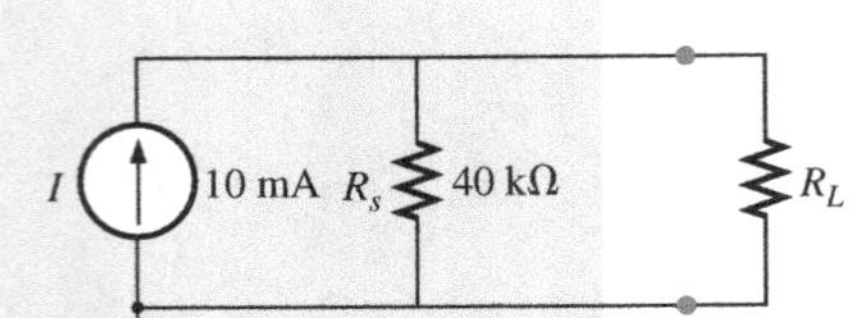

FIG. 9.86
Example 9.15.

EXAMPLE 9.15 The analysis of a transistor network resulted in the reduced equivalent in Fig. 9.86.

a. Find the load resistance that will result in maximum power transfer to the load, and find the maximum power delivered.
b. If the load were changed to 68 kΩ, would you expect a fairly high level of power transfer to the load based on the results of part (a)? What would the new power level be? Is your initial assumption verified?
c. If the load were changed to 8.2 kΩ, would you expect a fairly high level of power transfer to the load based on the results of part (a)? What would the new power level be? Is your initial assumption verified?

Solutions:

a. Replacing the current source by an open-circuit equivalent results in

$$R_{Th} = R_s = 40\ \text{k}\Omega$$

Restoring the current source and finding the open-circuit voltage at the output terminals results in

$$E_{Th} = V_{oc} = IR_s = (10\ \text{mA})(40\ \text{k}\Omega) = 400\ \text{V}$$

For maximum power transfer to the load,

$$R_L = R_{Th} = \mathbf{40\ k\Omega}$$

with a maximum power level of

$$P_{L_{\text{max}}} = \frac{E_{Th}^2}{4R_{Th}} = \frac{(400\ \text{V})^2}{4(40\ \text{k}\Omega)} = \mathbf{1\ W}$$

b. Yes, because the 68 kΩ load is greater (note Fig. 9.80) than the 40 kΩ load, but relatively close in magnitude.

$$I_L = \frac{E_{Th}}{R_{Th} + R_L} = \frac{400\ \text{V}}{40\ \text{k}\Omega + 68\ \text{k}\Omega} = \frac{400}{108\ \text{k}\Omega} \cong 3.7\ \text{mA}$$

$$P_L = I_L^2 R_L = (3.7\ \text{mA})^2(68\ \text{k}\Omega \cong \mathbf{0.93\ W}$$

Yes, the power level of 0.93 W compared to the 1 W level of part (a) verifies the assumption.

c. No, 8.2 kΩ is quite a bit less (note Fig. 9.80) than the 40 kΩ value.

$$I_L = \frac{E_{Th}}{R_{Th} + R_L} = \frac{400\ \text{V}}{40\ \text{k}\Omega + 8.2\ \text{k}\Omega} = \frac{400\ \text{V}}{48.2\ \text{k}\Omega} \cong 8.3\ \text{mA}$$

$$P_L = I_L^2 R_L = (8.3\ \text{mA})^2(8.2\ \text{k}\Omega) \cong \mathbf{0.57\ W}$$

Yes, the power level of 0.57 W compared to the 1 W level of part (a) verifies the assumption.

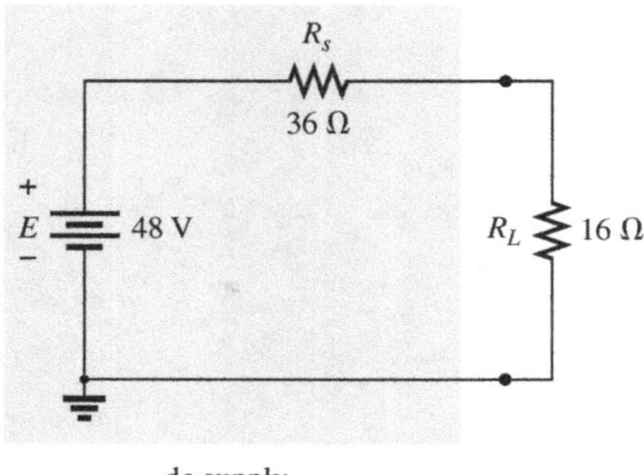

dc supply

FIG. 9.87
dc supply with a fixed 16 Ω load (Example 9.16).

EXAMPLE 9.16 In Fig. 9.87, a fixed load of 16 Ω is applied to a 48 V supply with an internal resistance of 36 Ω.

a. For the conditions in Fig. 9.87, what is the power delivered to the load and lost to the internal resistance of the supply?
b. If the designer has some control over the internal resistance level of the supply, what value should he or she make it for maximum power to the load? What is the maximum power to the load? How does it compare to the level obtained in part (a)?
c. Without making a single calculation, find the value that would result in more power to the load if the designer could change the internal resistance to 22 Ω or 8.2 Ω. Verify your conclusion by calculating the power to the load for each value.

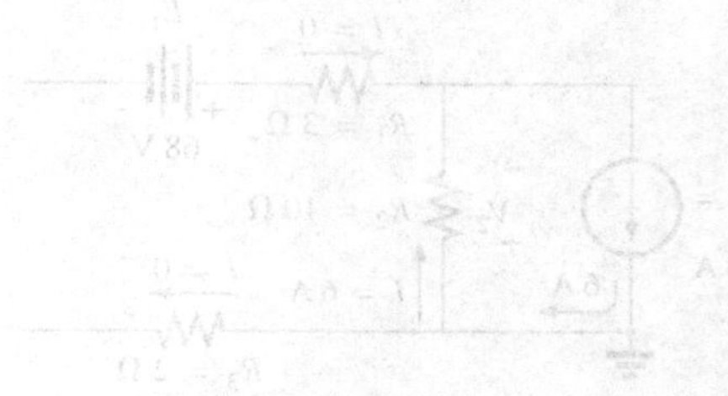

Solutions:

a. $$I_L = \frac{E}{R_s + R_L} = \frac{48\text{ V}}{36\ \Omega + 16\ \Omega} = \frac{48\text{ V}}{52\ \Omega} = 923.1\text{ mA}$$

$$P_{R_s} = I_L^2 R_s = (923.1\text{ mA})^2(36\ \Omega) = \mathbf{30.68\text{ W}}$$

$$P_L = I_L^2 R_L = (923.1\text{ mA})^2(16\ \Omega) = \mathbf{13.63\text{ W}}$$

b. Be careful here. The quick response is to make the source resistance R_s equal to the load resistance to satisfy the criteria of the maximum power transfer theorem. However, this is a totally different type of problem from what was examined earlier in this section. If the load is fixed, the smaller the source resistance R_s, the more applied voltage will reach the load and the less will be lost in the internal series resistor. In fact, the source resistance should be made as small as possible. If zero ohms were possible for R_s, the voltage across the load would be the full supply voltage, and the power delivered to the load would equal

$$P_L = \frac{V_L^2}{R_L} = \frac{(48\text{ V})^2}{16\ \Omega} = \mathbf{144\text{ W}}$$

which is more than 10 times the value with a source resistance of 36 Ω.

c. Again, forget the impact in Fig. 9.80: The smaller the source resistance, the greater is the power to the fixed 16 Ω load. Therefore, the 8.2 Ω resistance level results in a higher power transfer to the load than the 22 Ω resistor.

For $R_s = 8.2\ \Omega$

$$I_L = \frac{E}{R_s + R_L} = \frac{48\text{ V}}{8.2\ \Omega + 16\ \Omega} = \frac{48\text{ V}}{24.2\ \Omega} = 1.983\text{ A}$$

and $$P_L = I_L^2 R_L = (1.983\text{ A})^2(16\ \Omega) \cong \mathbf{62.92\text{ W}}$$

For $R_s = 22\ \Omega$

$$I_L = \frac{E}{R_s + R_L} = \frac{48\text{ V}}{22\ \Omega + 16\ \Omega} = \frac{48\text{ V}}{38\ \Omega} = 1.263\text{ A}$$

and $$P_L = I_L^2 R_L = (1.263\text{ A})^2(16\ \Omega) \cong \mathbf{25.52\text{ W}}$$

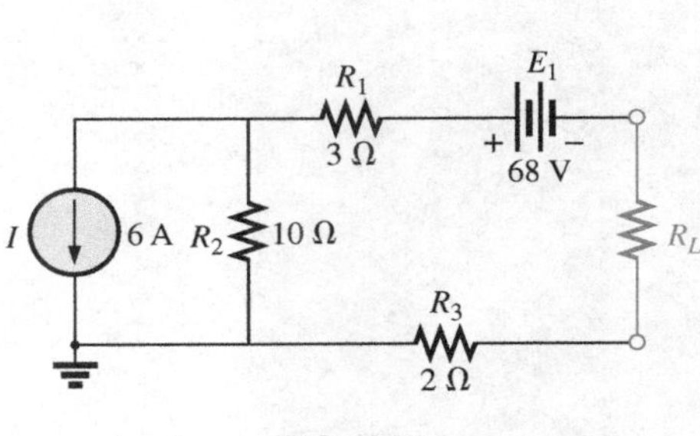

FIG. 9.88
Example 9.17.

EXAMPLE 9.17 Given the network in Fig. 9.88, find the value of R_L for maximum power to the load, and find the maximum power to the load.

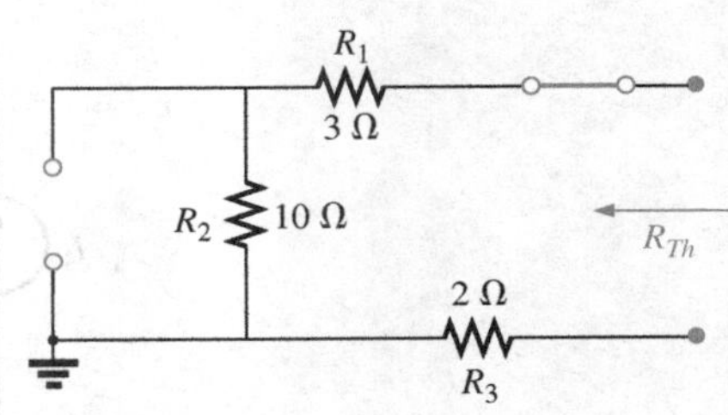

FIG. 9.89
Determining R_{Th} for the network external to resistor R_L in Fig. 9.88.

Solution: The Thévenin resistance is determined from Fig. 9.89:

$$R_{Th} = R_1 + R_2 + R_3 = 3\ \Omega + 10\ \Omega + 2\ \Omega = 15\ \Omega$$

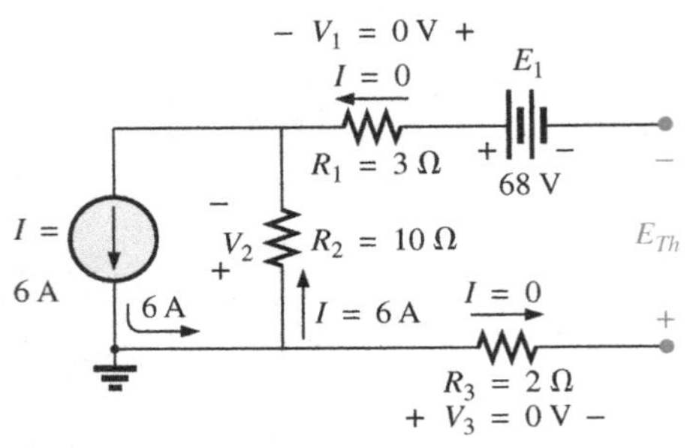

FIG. 9.90
Determining E_{Th} for the network external to resistor R_L in Fig. 9.88.

so that $$R_L = R_{Th} = \mathbf{15\ \Omega}$$

The Thévenin voltage is determined using Fig. 9.90, where

$$V_1 = V_3 = 0\text{ V} \quad \text{and} \quad V_2 = I_2R_2 = IR_2 = (6\text{ A})(10\ \Omega) = 60\text{ V}$$

Applying Kirchhoff's voltage law gives

$$-V_2 - E + E_{Th} = 0$$

and $$E_{Th} = V_2 + E = 60\text{ V} + 68\text{ V} = \mathbf{128\ V}$$

with the maximum power equal to

$$P_{L_{max}} = \frac{E_{Th}^2}{4R_{Th}} = \frac{(128\text{ V})^2}{4(15\text{ k}\Omega)} = \mathbf{273.07\ W}$$

9.6 MILLMAN'S THEOREM

Through the application of **Millman's theorem,** any number of parallel voltage sources can be reduced to one. In Fig. 9.91, for example, the three voltage sources can be reduced to one. This permits finding the current through or voltage across R_L without having to apply a method such as mesh analysis, nodal analysis, superposition, and so on. The theorem can best be described by applying it to the network in Fig. 9.91. Basically, three steps are included in its application.

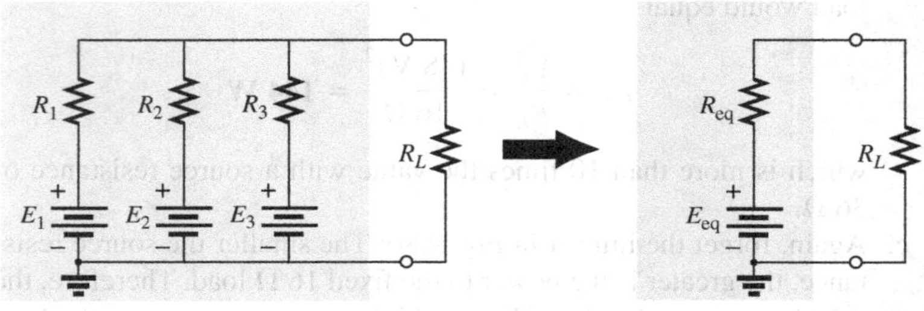

FIG. 9.91
Demonstrating the effect of applying Millman's theorem.

Step 1: Convert all voltage sources to current sources as outlined in Section 8.3. This is performed in Fig. 9.92 for the network in Fig. 9.91.

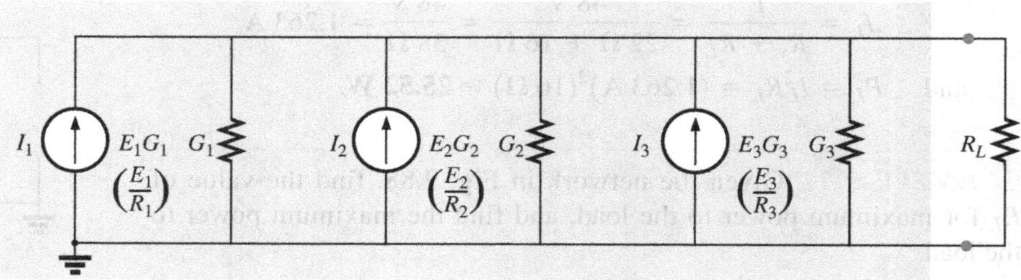

FIG. 9.92
Converting all the sources in Fig. 9.91 to current sources.

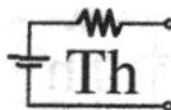

Step 2: Combine parallel current sources as described in Section 8.4. The resulting network is shown in Fig. 9.93, where

$$I_T = I_1 + I_2 + I_3 \quad \text{and} \quad G_T = G_1 + G_2 + G_3$$

Step 3: Convert the resulting current source to a voltage source, and the desired single-source network is obtained, as shown in Fig. 9.94.

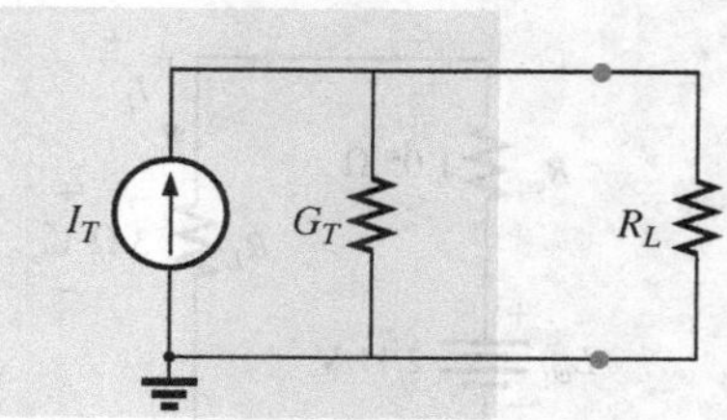

FIG. 9.93
Reducing all the current sources in Fig. 9.92 to a single current source.

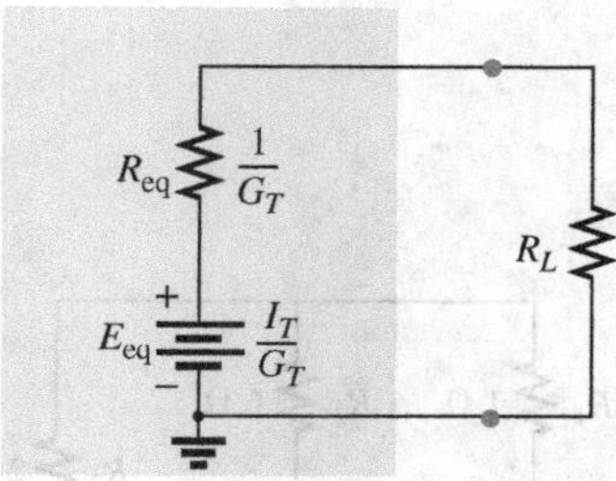

FIG. 9.94
Converting the current source in Fig. 9.93 to a voltage source.

In general, Millman's theorem states that for any number of parallel voltage sources,

$$E_{eq} = \frac{I_T}{G_T} = \frac{\pm I_1 \pm I_2 \pm I_3 \pm \cdots \pm I_N}{G_1 + G_2 + G_3 + \cdots + G_N}$$

or

$$E_{eq} = \frac{\pm E_1G_1 \pm E_2G_2 \pm E_3G_3 \pm \cdots \pm E_NG_N}{G_1 + G_2 + G_3 + \cdots + G_N} \tag{9.8}$$

The plus-and-minus signs appear in Eq. (9.8) to include those cases where the sources may not be supplying energy in the same direction. (Note Example 9.18.)

The equivalent resistance is

$$R_{eq} = \frac{1}{G_T} = \frac{1}{G_1 + G_2 + G_3 + \cdots + G_N} \tag{9.9}$$

In terms of the resistance values,

$$E_{eq} = \frac{\pm \dfrac{E_1}{R_1} \pm \dfrac{E_2}{R_2} \pm \dfrac{E_3}{R_3} \pm \cdots \pm \dfrac{E_N}{R_N}}{\dfrac{1}{R_1} + \dfrac{1}{R_2} + \dfrac{1}{R_3} + \cdots + \dfrac{1}{R_N}} \tag{9.10}$$

and

$$R_{eq} = \frac{1}{\dfrac{1}{R_1} + \dfrac{1}{R_2} + \dfrac{1}{R_3} + \cdots + \dfrac{1}{R_N}} \tag{9.11}$$

Because of the relatively few direct steps required, you may find it easier to apply each step rather than memorizing and employing Eqs. (9.8) through (9.11).

EXAMPLE 9.18 Using Millman's theorem, find the current through and voltage across the resistor R_L in Fig. 9.95.

Solution: By Eq. (9.10),

$$E_{eq} = \frac{+\dfrac{E_1}{R_1} - \dfrac{E_2}{R_2} + \dfrac{E_3}{R_3}}{\dfrac{1}{R_1} + \dfrac{1}{R_2} + \dfrac{1}{R_3}}$$

The minus sign is used for E_2/R_2 because that supply has the opposite polarity of the other two. The chosen reference direction is therefore that of E_1 and E_3. The total conductance is unaffected by the direction, and

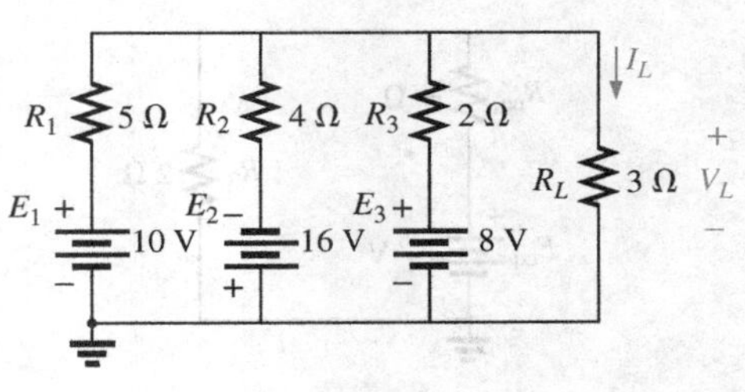

FIG. 9.95
Example 9.18.

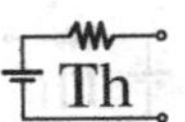

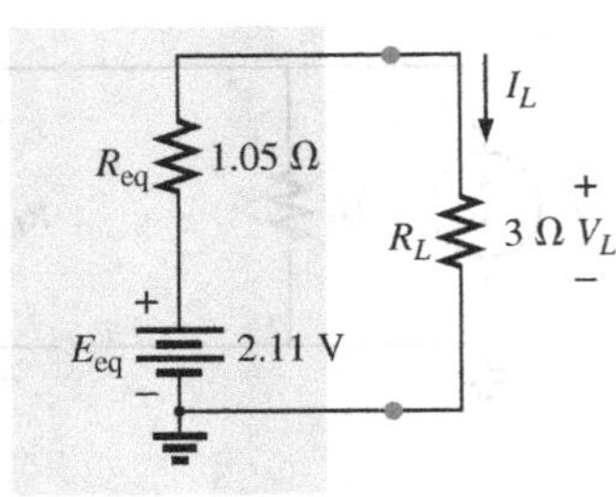

FIG. 9.96
The result of applying Millman's theorem to the network in Fig. 9.95.

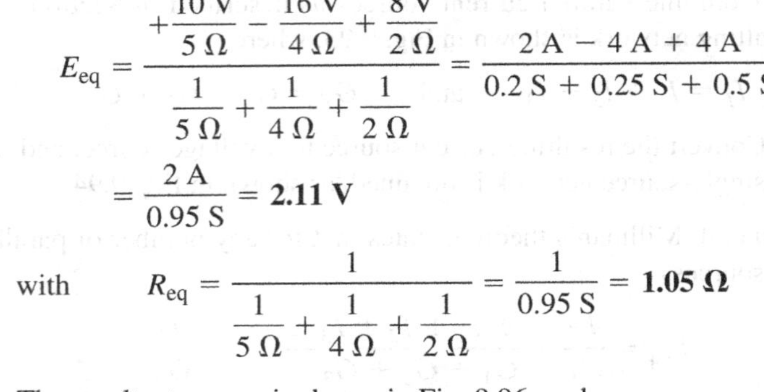

$$E_{eq} = \frac{+\dfrac{10\text{ V}}{5\ \Omega} - \dfrac{16\text{ V}}{4\ \Omega} + \dfrac{8\text{ V}}{2\ \Omega}}{\dfrac{1}{5\ \Omega} + \dfrac{1}{4\ \Omega} + \dfrac{1}{2\ \Omega}} = \frac{2\text{ A} - 4\text{ A} + 4\text{ A}}{0.2\text{ S} + 0.25\text{ S} + 0.5\text{ S}}$$

$$= \frac{2\text{ A}}{0.95\text{ S}} = \mathbf{2.11\ V}$$

with
$$R_{eq} = \frac{1}{\dfrac{1}{5\ \Omega} + \dfrac{1}{4\ \Omega} + \dfrac{1}{2\ \Omega}} = \frac{1}{0.95\text{ S}} = \mathbf{1.05\ \Omega}$$

The resultant source is shown in Fig. 9.96, and

$$I_L = \frac{2.11\text{ V}}{1.05\ \Omega + 3\ \Omega} = \frac{2.11\text{ V}}{4.05\ \Omega} = \mathbf{0.52\ A}$$

with
$$V_L = I_L R_L = (0.52\text{ A})(3\ \Omega) = \mathbf{1.56\ V}$$

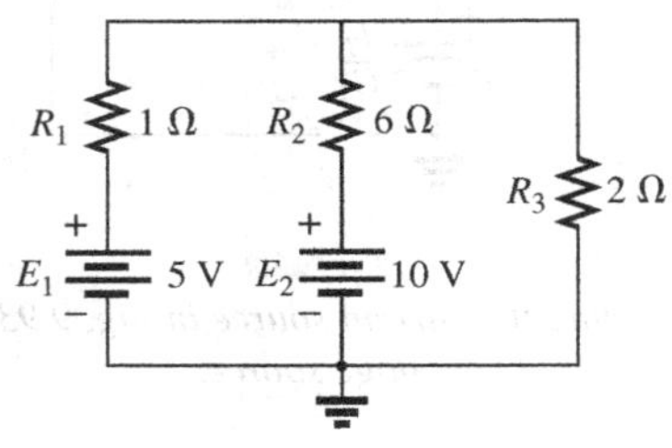

FIG. 9.97
Example 9.19.

EXAMPLE 9.19 Let us now consider the type of problem encountered in the introduction to mesh and nodal analysis in Chapter 8. Mesh analysis was applied to the network of Fig. 9.97 (Example 8.12). Let us now use Millman's theorem to find the current through the 2 Ω resistor and compare the results.

Solutions:

a. Let us first apply each step and, in the (b) solution, Eq. (9.10). Converting sources yields Fig. 9.98. Combining sources and parallel conductance branches (Fig. 9.99) yields

$$I_T = I_1 + I_2 = 5\text{ A} + \frac{5}{3}\text{ A} = \frac{15}{3}\text{ A} + \frac{5}{3}\text{ A} = \frac{20}{3}\text{ A}$$

$$G_T = G_1 + G_2 = 1\text{ S} + \frac{1}{6}\text{ S} = \frac{6}{6}\text{ S} + \frac{1}{6}\text{ S} = \frac{7}{6}\text{ S}$$

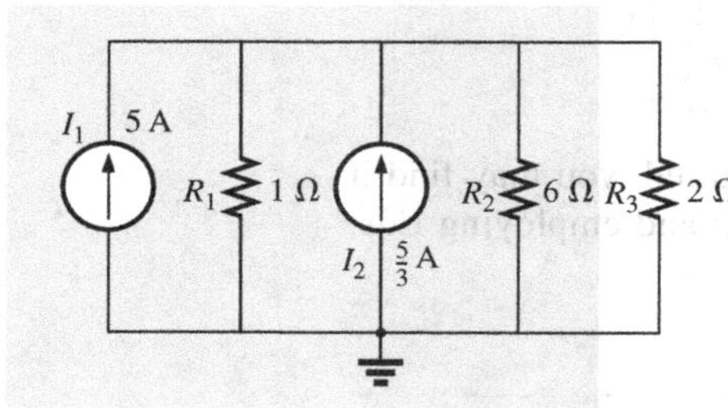

FIG. 9.98
Converting the sources in Fig. 9.97 to current sources.

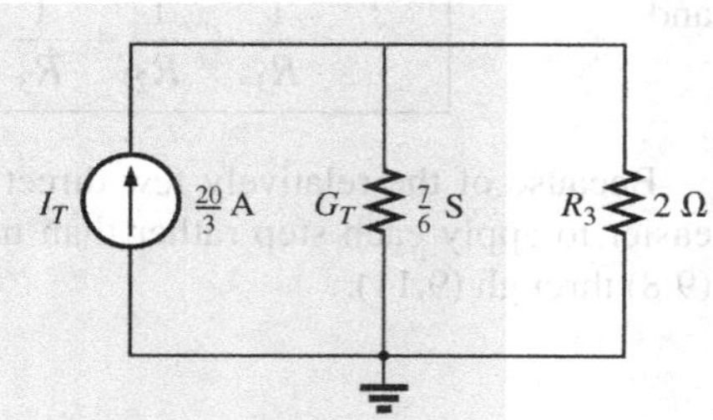

FIG. 9.99
Reducing the current sources in Fig. 9.98 to a single source.

Converting the current source to a voltage source (Fig. 9.100), we obtain

$$E_{eq} = \frac{I_T}{G_T} = \frac{\dfrac{20}{3}\text{ A}}{\dfrac{7}{6}\text{ S}} = \frac{(6)(20)}{(3)(7)}\text{ V} = \mathbf{\frac{40}{7}\ V}$$

and
$$R_{eq} = \frac{1}{G_T} = \frac{1}{\dfrac{7}{6}\text{ S}} = \mathbf{\frac{6}{7}\ \Omega}$$

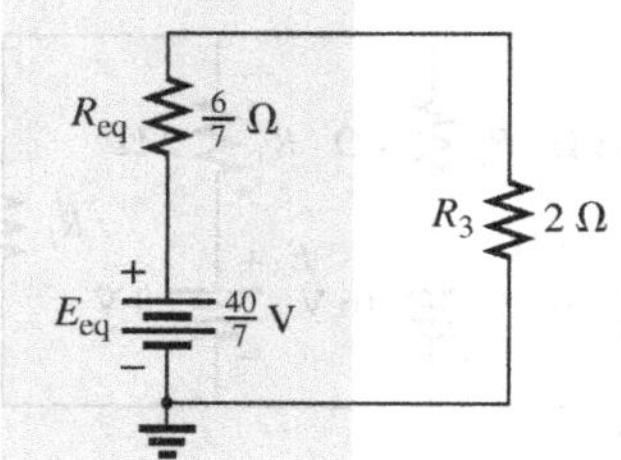

FIG. 9.100
Converting the current source in Fig. 9.99 to a voltage source.

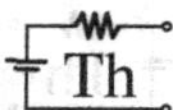

so that

$$I_{2\Omega} = \frac{E_{eq}}{R_{eq} + R_3} = \frac{\frac{40}{7}\text{ V}}{\frac{6}{7}\,\Omega + 2\,\Omega} = \frac{\frac{40}{7}\text{ V}}{\frac{6}{7}\,\Omega + \frac{14}{7}\,\Omega} = \frac{40\text{ V}}{20\,\Omega} = \mathbf{2\ A}$$

which agrees with the result obtained in Example 8.18.

b. Let us now simply apply the proper equation, Eq. (9.10):

$$E_{eq} = \frac{+\frac{5\text{ V}}{1\,\Omega} + \frac{10\text{ V}}{6\,\Omega}}{\frac{1}{1\,\Omega} + \frac{1}{6\,\Omega}} = \frac{\frac{30\text{ V}}{6\,\Omega} + \frac{10\text{ V}}{6\,\Omega}}{\frac{6}{6\,\Omega} + \frac{1}{6\,\Omega}} = \mathbf{\frac{40}{7}\ V}$$

and

$$R_{eq} = \frac{1}{\frac{1}{1\,\Omega} + \frac{1}{6\,\Omega}} = \frac{1}{\frac{6}{6\,\Omega} + \frac{1}{6\,\Omega}} = \frac{1}{\frac{7}{6}\text{ S}} = \mathbf{\frac{6}{7}\ \Omega}$$

which are the same values obtained above.

The dual of Millman's theorem (Fig. 9.91) appears in Fig. 9.101. It can be shown that I_{eq} and R_{eq}, as in Fig. 9.101, are given by

$$I_{eq} = \frac{\pm I_1R_1 \pm I_2R_2 \pm I_3R_3}{R_1 + R_2 + R_3} \tag{9.12}$$

and

$$R_{eq} = R_1 + R_2 + R_3 \tag{9.13}$$

The derivation appears as a problem at the end of the chapter.

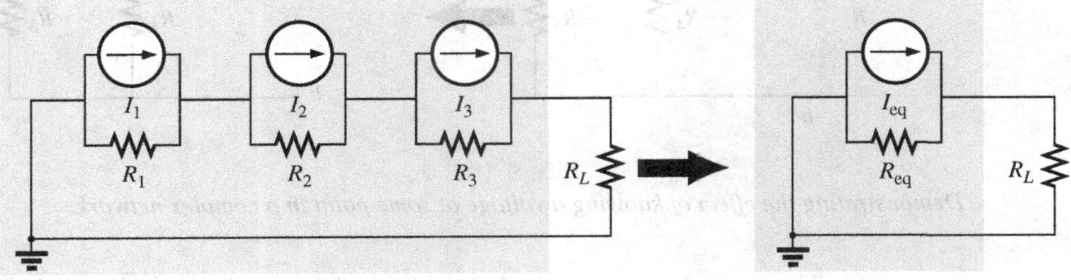

FIG. 9.101
The dual effect of Millman's theorem.

9.7 SUBSTITUTION THEOREM

The **substitution theorem** states the following:

If the voltage across and the current through any branch of a dc bilateral network are known, this branch can be replaced by any combination of elements that will maintain the same voltage across and current through the chosen branch.

More simply, the theorem states that for branch equivalence, the terminal voltage and current must be the same. Consider the circuit in Fig. 9.102, in which the voltage across and current through the branch

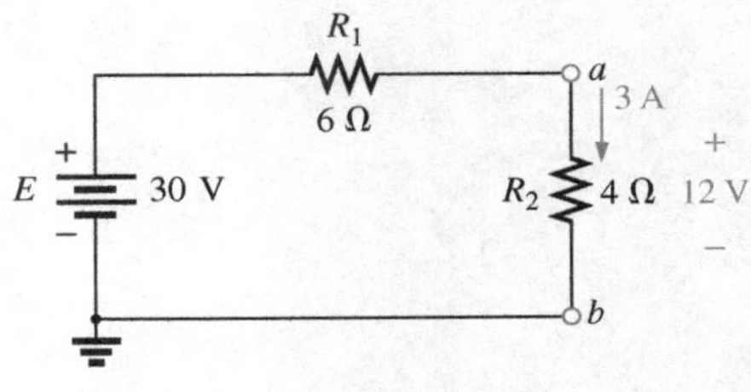

FIG. 9.102
Demonstrating the effect of the substitution theorem.

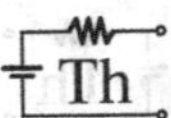

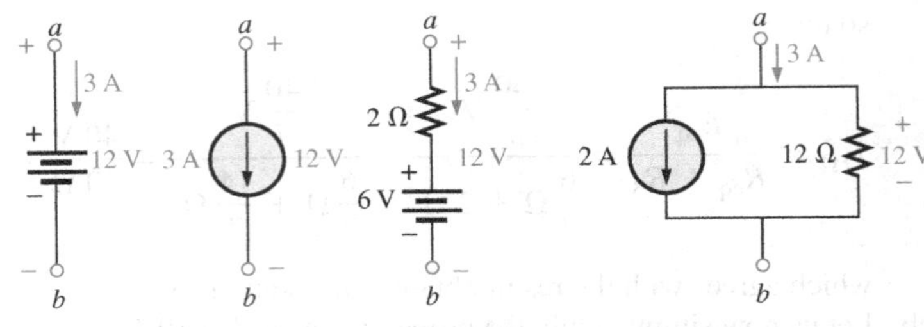

FIG. 9.103
Equivalent branches for the branch a-b in Fig. 9.102.

a-b are determined. Through the use of the substitution theorem, a number of equivalent *a-a′* branches are shown in Fig. 9.103.

Note that for each equivalent, the terminal voltage and current are the same. Also consider that the response of the remainder of the circuit in Fig. 9.102 is unchanged by substituting any one of the equivalent branches. As demonstrated by the single-source equivalents in Fig. 9.103, *a known potential difference and current in a network can be replaced by an ideal voltage source and current source, respectively.*

Understand that this theorem cannot be used to *solve* networks with two or more sources that are not in series or parallel. For it to be applied, a potential difference or current value must be known or found using one of the techniques discussed earlier. One application of the theorem is shown in Fig. 9.104. Note that in the figure the known potential difference V was replaced by a voltage source, permitting the isolation of the portion of the network including R_3, R_4, and R_5. Recall that this was basically the approach used in the analysis of the ladder network as we worked our way back toward the terminal resistance R_5.

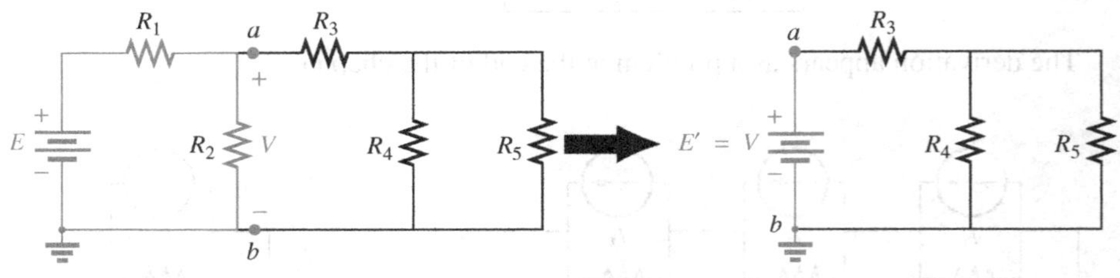

FIG. 9.104
Demonstrating the effect of knowing a voltage at some point in a complex network.

The current source equivalence of the above is shown in Fig. 9.105, where a known current is replaced by an ideal current source, permitting the isolation of R_4 and R_5.

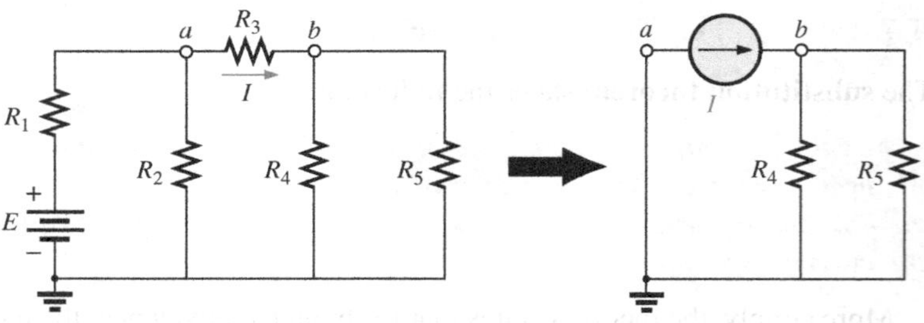

FIG. 9.105
Demonstrating the effect of knowing a current at some point in a complex network.

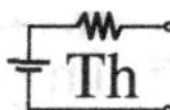

Recall from the discussion of bridge networks that $V = 0$ and $I = 0$ were replaced by a short circuit and an open circuit, respectively. This substitution is a very specific application of the substitution theorem.

9.8 RECIPROCITY THEOREM

The **reciprocity theorem** is applicable only to single-source networks. It is, therefore, not a theorem used in the analysis of multisource networks described thus far. The theorem states the following:

The current I in any branch of a network due to a single voltage source E anywhere else in the network will equal the current through the branch in which the source was originally located if the source is placed in the branch in which the current I was originally measured.

In other words, the location of the voltage source and the resulting current may be interchanged without a change in current. The theorem requires that the polarity of the voltage source have the same correspondence with the direction of the branch current in each position.

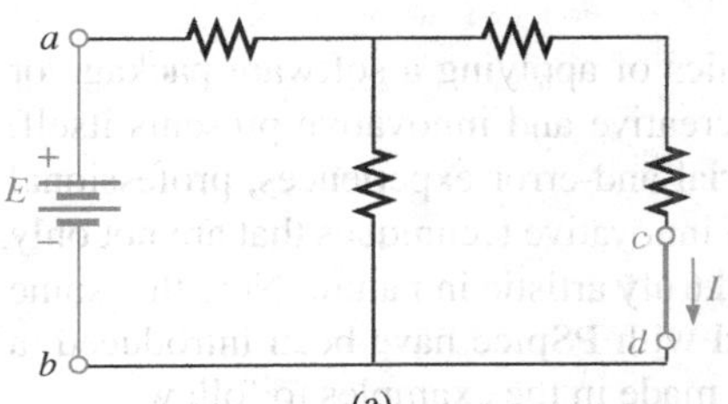

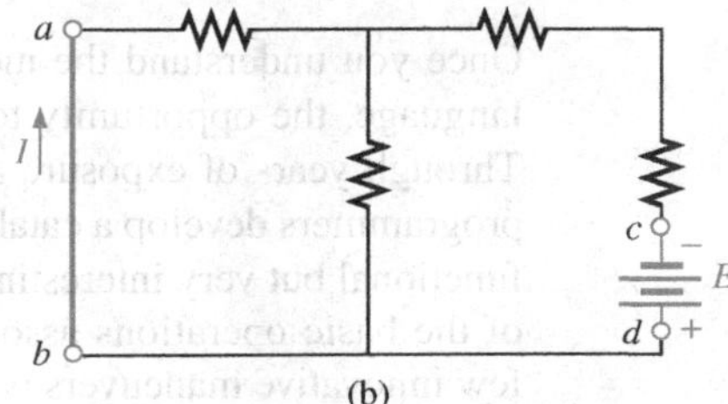

FIG. 9.106
Demonstrating the impact of the reciprocity theorem.

In the representative network in Fig. 9.106(a), the current I due to the voltage source E was determined. If the position of each is interchanged as shown in Fig. 9.106(b), the current I will be the same value as indicated. To demonstrate the validity of this statement and the theorem, consider the network in Fig. 9.107, in which values for the elements of Fig. 9.106(a) have been assigned.

The total resistance is

$$R_T = R_1 + R_2 \parallel (R_3 + R_4) = 12\ \Omega + 6\ \Omega \parallel (2\ \Omega + 4\ \Omega)$$
$$= 12\ \Omega + 6\ \Omega \parallel 6\ \Omega = 12\ \Omega + 3\ \Omega = 15\ \Omega$$

and $$I_s = \frac{E}{R_T} = \frac{45\text{ V}}{15\ \Omega} = 3\text{ A}$$

with $$I = \frac{3\text{ A}}{2} = \mathbf{1.5\text{ A}}$$

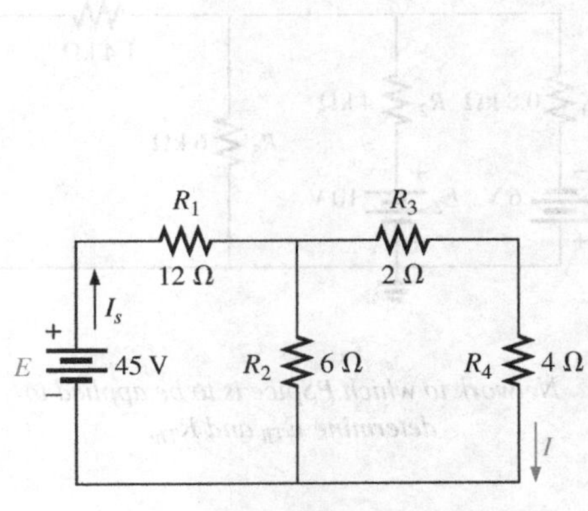

FIG. 9.107
Finding the current I due to a source E.

For the network in Fig. 9.108, which corresponds to that in Fig. 9.106(b), we find

$$R_T = R_4 + R_3 + R_1 \parallel R_2$$
$$= 4\ \Omega + 2\ \Omega + 12\ \Omega \parallel 6\ \Omega = 10\ \Omega$$

and $$I_s = \frac{E}{R_T} = \frac{45\text{ V}}{10\ \Omega} = 4.5\text{ A}$$

so that $$I = \frac{(6\ \Omega)(4.5\text{ A})}{12\ \Omega + 6\ \Omega} = \frac{4.5\text{ A}}{3} = \mathbf{1.5\text{ A}}$$

which agrees with the above.

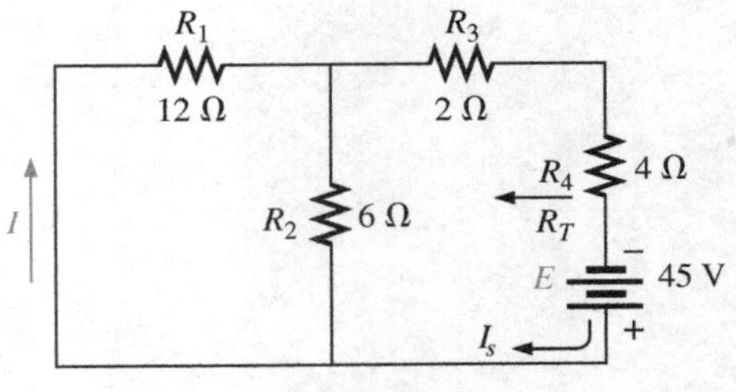

FIG. 9.108
Interchanging the location of E and I of Fig. 9.107 to demonstrate the validity of the reciprocity theorem.

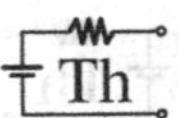

The uniqueness and power of this theorem can best be demonstrated by considering a complex, single-source network such as the one shown in Fig. 9.109.

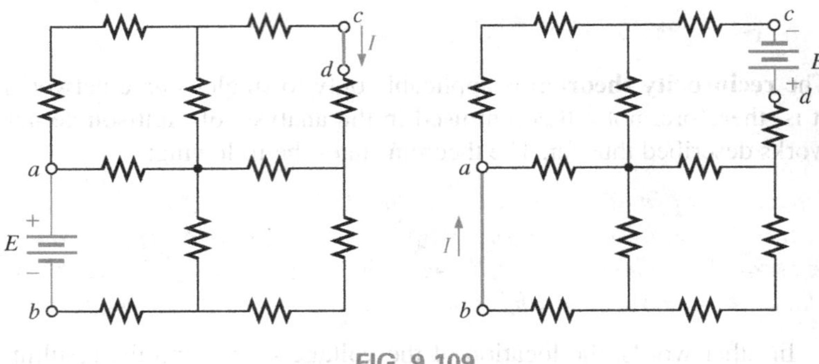

FIG. 9.109
Demonstrating the power and uniqueness of the reciprocity theorem.

9.9 COMPUTER ANALYSIS

Once you understand the mechanics of applying a software package or language, the opportunity to be creative and innovative presents itself. Through years of exposure and trial-and-error experiences, professional programmers develop a catalog of innovative techniques that are not only functional but very interesting and truly artistic in nature. Now that some of the basic operations associated with PSpice have been introduced, a few innovative maneuvers will be made in the examples to follow.

PSpice

Thévenin's Theorem The application of Thévenin's theorem requires an interesting maneuver to determine the Thévenin resistance. It is a maneuver, however, that has application beyond Thévenin's theorem whenever a resistance level is required. The network to be analyzed appears in Fig. 9.110 and is the same one analyzed in Example 9.10 (Fig. 9.48).

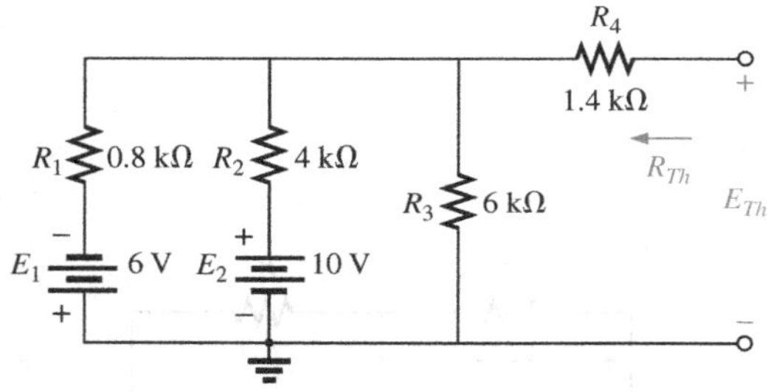

FIG. 9.110
Network to which PSpice is to be applied to determine E_{Th} and R_{Th}.

Since PSpice is not set up to measure resistance levels directly, a 1 A current source can be applied as shown in Fig. 9.111, and Ohm's law can be used to determine the magnitude of the Thévenin resistance in the following manner:

$$|R_{Th}| = \left|\frac{V_s}{I_s}\right| = \left|\frac{V_s}{1\text{ A}}\right| = |V_s| \tag{9.14}$$

In Eq. (9.14), since $I_s = 1$ A, the magnitude of R_{Th} in ohms is the same as the magnitude of the voltage V_s (in volts) across the current source. The result is that when the voltage across the current source is displayed, it can be read as ohms rather than volts.

When PSpice is applied, the network appears as shown in Fig. 9.111. Flip the voltage source E_1 and the current source by right-clicking on the source and choosing the **Mirror Vertically** option. Set both voltage sources to zero through the **Display Properties** dialog box obtained by double-clicking on the source symbol. The result of the **Bias Point** simulation is 2 kV across the current source. The Thévenin resistance is therefore 2 kΩ between the two terminals of the network to the left of the current source (to match the results of Example 9.10). In total, by

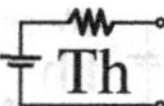

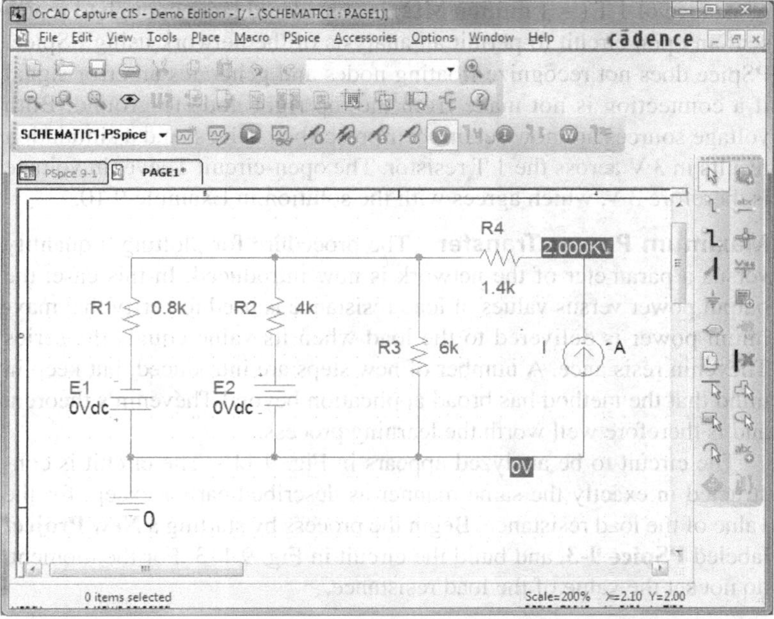

FIG. 9.111
Using PSpice to determine the Thévenin resistance of a network through the application of a 1 A current source.

setting the voltage source to 0 V, we have dictated that the voltage is the same at both ends of the voltage source, replicating the effect of a short-circuit connection between the two points.

For the open-circuit Thévenin voltage between the terminals of interest, the network must be constructed as shown in Fig. 9.112. The

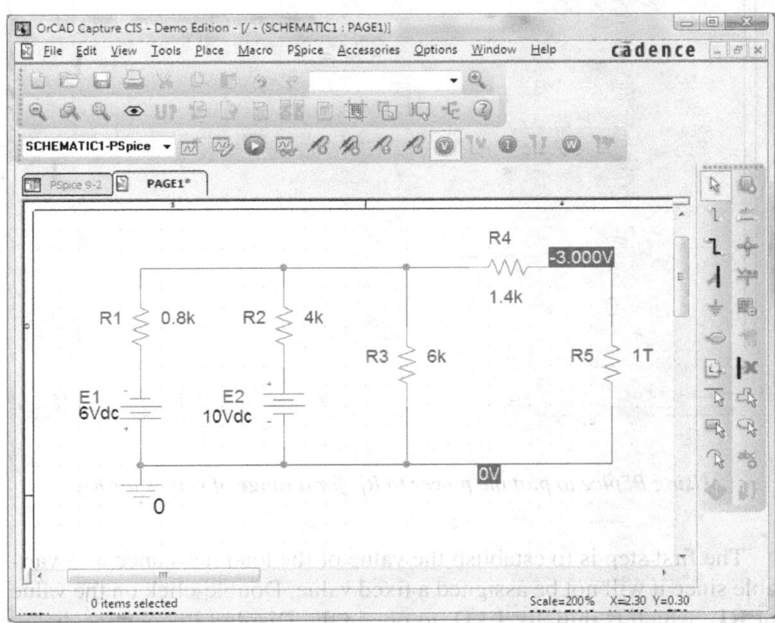

FIG. 9.112
Using PSpice to determine the Thévenin voltage for a network using a very large resistance value to represent the open-circuit condition between the terminals of interest.

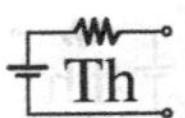

resistance of 1 T (= 1 million MΩ) is considered large enough to represent an open circuit to permit an analysis of the network using PSpice. PSpice does not recognize floating nodes and generates an error signal if a connection is not made from the top right node to ground. Both voltage sources are now set on their prescribed values, and a simulation results in 3 V across the 1 T resistor. The open-circuit Thévenin voltage is therefore 3 V, which agrees with the solution in Example 9.10.

Maximum Power Transfer The procedure for plotting a quantity versus a parameter of the network is now introduced. In this case, the output power versus values of load resistance is used to verify that maximum power is delivered to the load when its value equals the series Thévenin resistance. A number of new steps are introduced, but keep in mind that the method has broad application beyond Thévenin's theorem and is therefore well worth the learning process.

The circuit to be analyzed appears in Fig. 9.113. The circuit is constructed in exactly the same manner as described earlier except for the value of the load resistance. Begin the process by starting a **New Project** labeled **PSpice 9-3,** and build the circuit in Fig. 9.113. For the moment, do not set the value of the load resistance.

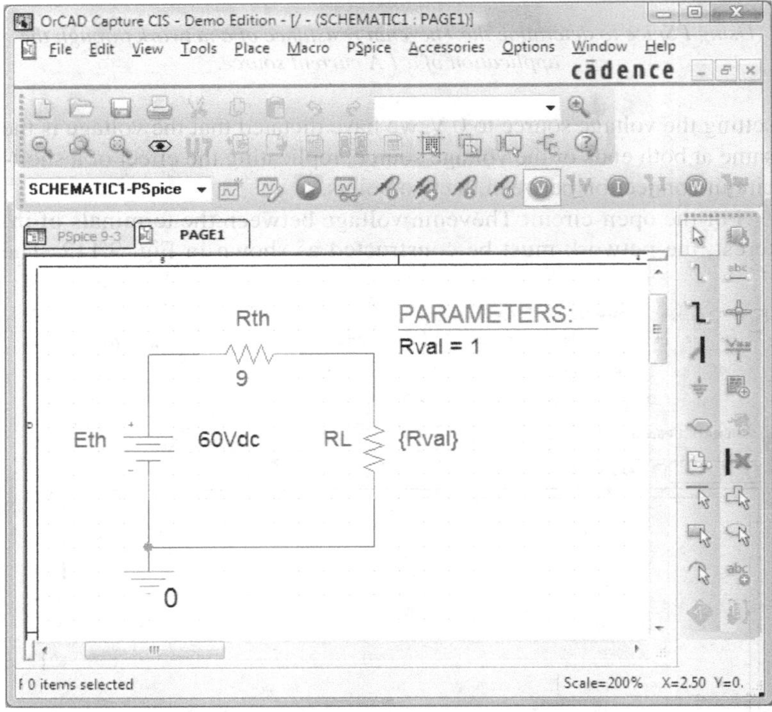

FIG. 9.113
Using PSpice to plot the power to R_L for a range of values for R_L.

The first step is to establish the value of the load resistance as a variable since it will not be assigned a fixed value. Double-click on the value of **RL,** which is initially 1 kΩ, to obtain the **Display Properties** dialog box. For **Value,** type in **{Rval}** and click in place. The brackets (*not* parentheses) are required, but the variable does not have to be called **Rval**—it is the choice of the user. Next select the **Place part** key to obtain the **Place Part** dialog box. If you are not already in the **Libraries**

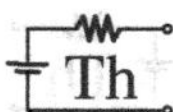

list, choose **Add Library** and add **SPECIAL** to the list. Select the **SPECIAL** library and scroll the **Part List** until **PARAM** appears. Select it; then click **OK** to obtain a rectangular box next to the cursor on the screen. Select a spot near **Rval,** and deposit the rectangle. The result is **PARAMETERS:** as shown in Fig. 9.113.

Next double-click on **PARAMETERS:** to obtain a **Property Editor** dialog box, which should have **SCHEMATIC1:PAGE1** in the second column from the left. Now select the **New Column** option from the top list of choices to obtain the **Add New Column** dialog box. Under **Name,** enter **Rval** and under **Value,** enter **1** followed by an **OK** to leave the dialog box. The result is a return to the **Property Editor** dialog box but with **Rval** and its value (below **Rval**) added to the horizontal list. Now select **Rval/1** by clicking on **Rval** to surround **Rval** by a dashed line and add a black background around the **1.** Choose **Display** to produce the **Display Properties** dialog box, and select **Name and Value** followed by **OK.** Then exit the **Property Editor** dialog box **(X)** to display the screen in Fig. 9.113. Note that now the first value (1 Ω) of **Rval** is displayed.

We are now ready to set up the simulation process. Under **PSpice,** select the **New Simulation Profile** key to open the **New Simulation** dialog box. Enter **DC Sweep** under **Name** followed by **Create.** The **Simulation Settings-DC Sweep** dialog box appears. After selecting **Analysis,** select **DC Sweep** under the **Analysis type** heading. Then leave the **Primary Sweep** under the **Options** heading, and select **Global parameter** under the **Sweep variable.** The **Parameter name** should then be entered as **Rval.** For the **Sweep type,** the **Start value** should be 1 Ω; but if we use 1 Ω, the curve to be generated will start at 1 Ω, leaving a blank from 0 to 1 Ω. The curve will look incomplete. To solve this problem, select 0.001 Ω as the **Start value** (very close to 0 Ω) with an **Increment** of 1 Ω. Enter the **End value** as 30.001 Ω to ensure a calculation at $R_L = 30\ \Omega$. If we used 30 Ω as the end value, the last calculation would be at 29.001 Ω since 29.001 Ω + 1 Ω = 30.001 Ω, which is beyond the range of 30 Ω. The values of **RL** will therefore be 0.001 Ω, 1.001 Ω, 2.001 Ω, . . . 29.001 Ω, 30.001 Ω, and so on, although the plot will look as if the values were 0 Ω, 1 Ω, 2 Ω, 29 Ω, 30 Ω, and so on. Click **OK,** and select **Run** under **PSpice** to obtain the display in Fig. 9.114.

Note that there are no plots on the graph, and that the graph extends to 32 Ω rather than 30 Ω as desired. It did not respond with a plot of power versus **RL** because we have not defined the plot of interest for the computer. To do this, select the **Add Trace** key (the key that has a red curve peaking in the middle of the plot) or **Trace-Add Trace** from the top menu bar. Either choice results in the **Add Traces** dialog box. The most important region of this dialog box is the **Trace Expression** listing at the bottom. The desired trace can be typed in directly, or the quantities of interest can be chosen from the list of **Simulation Output Variables** and deposited in the **Trace Expression** listing. To find the power to **RL** for the chosen range of values for **RL,** select **W(RL)** in the listing; it then appears as the **Trace Expression.** Click **OK,** and the plot in Fig. 9.115 appears. Originally, the plot extended from 0 Ω to 35 Ω. We reduced the range to 0 Ω to 30 Ω by selecting **Plot-Axis Settings-X Axis-User Defined 0 to 30-OK.**

Select the **Toggle cursor** key (which has an arrow set in a blue background), and seven options will open to the right of the key that include **Cursor Peak, Cursor Trough, Cursor Slope, Cursor Min, Cursor Max, Cursor Point,** and **Cursor Search.** Select **Cursor Max,** and the

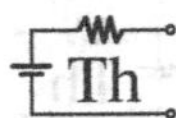

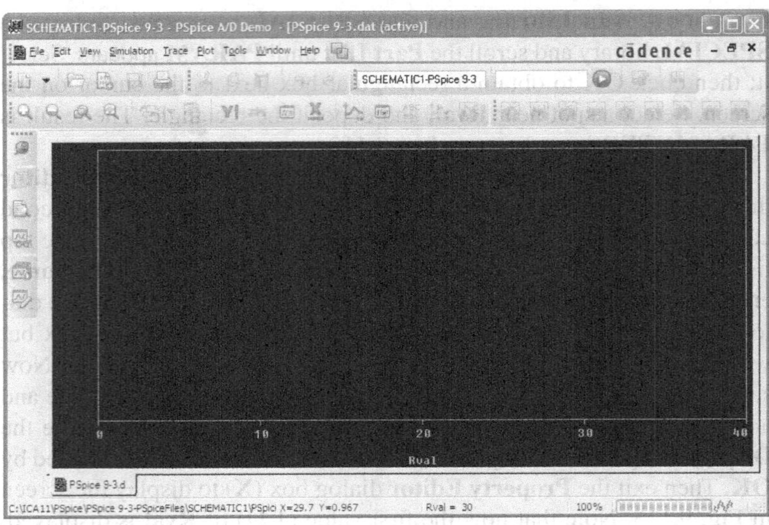

FIG. 9.114
Plot resulting from the dc sweep of R_L for the network in Fig. 9.113 before defining the parameters to be displayed.

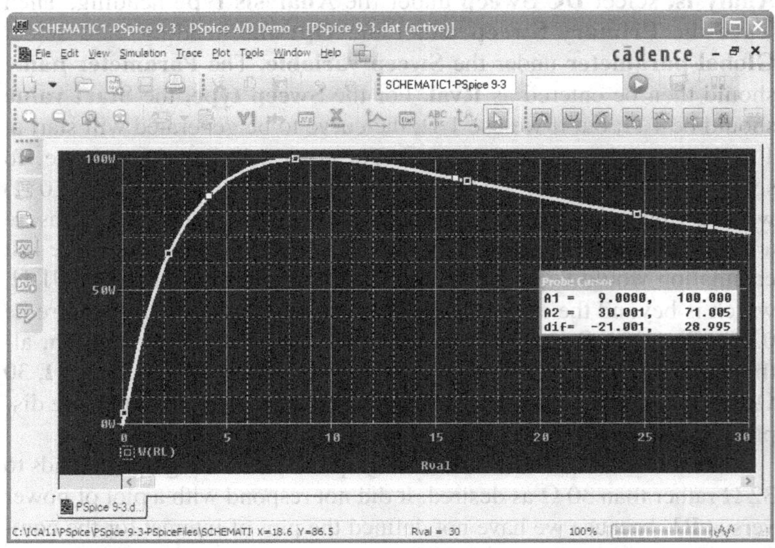

FIG. 9.115
A plot of the power delivered to R_L in Fig. 9.113 for a range of values for R_L extending from 0 Ω to 30 Ω.

Probe Cursor dialog box at the bottom right of the screen will reveal where the peak occurred and the power level at that point. Note that **A1** is 9.001 to reflect a load of 9 Ω, which equals the Thévenin resistance. The maximum power at this point is 100 W, as also indicated to the right of the resistance value. The **Probe Cursor** box can be moved to any position on the screen simply by selecting it and dragging it to the desired position. A second cursor can be generated by right-clicking the mouse on the **Cursor Point** option and moving it to a resistance of 30 Ω. The result is **A2** = 30 Ω, with a power level of 71.005 W, as shown on the plot. Notice also that the plot generated appears as a listing at the bottom left of the screen as **W(RL)**.

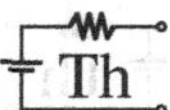

Multisim

Superposition Let us now apply superposition to the network in Fig. 9.116, which appeared earlier as Fig. 9.2 in Example 9.1, to permit a comparison of resulting solutions. The current through R_2 is to be determined. With the use of methods described in earlier chapters for the application of Multisim, the network in Fig. 9.117 results, which allows us to determine the effect of the 36 V voltage source. Note in Fig. 9.117 that both the voltage source and current source are present even though we are finding the contribution due solely to the voltage source. Obtain the voltage source by selecting the **Place Source** option at the top of the left toolbar to open the **Select a Component** dialog box. Then select **POWER_SOURCES** followed by **DC_POWER** as described in earlier chapters. You can also obtain the current source from the same dialog box by selecting **SIGNAL_CURRENT** under **Family** followed by **DC_CURRENT** under **Component.** The current source can be flipped vertically by right-clicking the source and selecting **Flip Vertical.** Set the current source to zero by left-clicking the source twice to obtain the **DC_CURRENT** dialog box. After choosing **Value,** set **Current(I)** to 0 A.

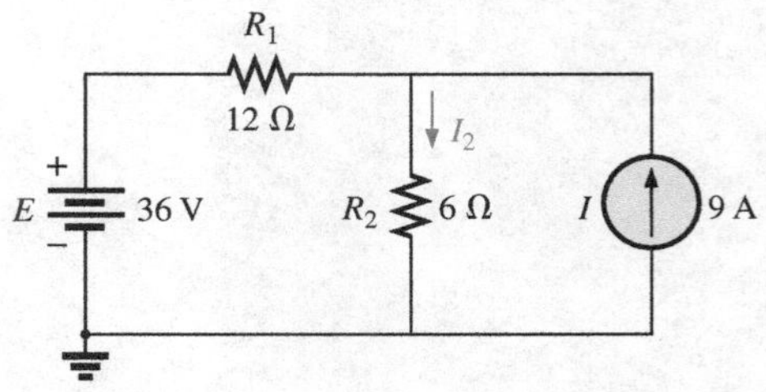

FIG. 9.116
Applying Multisim to determine the current I_2 using superposition.

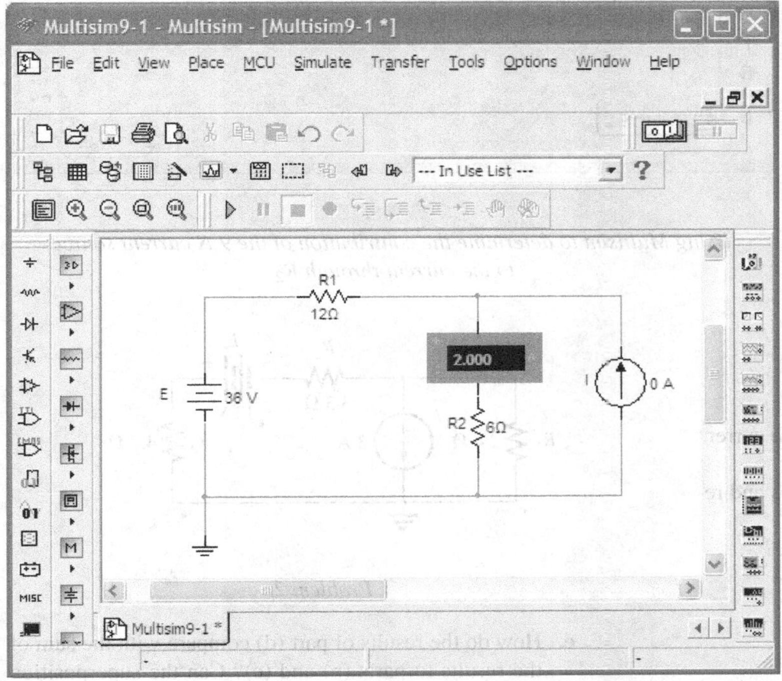

FIG. 9.117
Using Multisim to determine the contribution of the 36 V voltage source to the current through R_2.

Following simulation, the results appear as in Fig. 9.117. The current through the 6 Ω resistor is 2 A due solely to the 36 V voltage source. The positive value for the 2 A reading reveals that the current due to the 36 V source is down through resistor R_2.

For the effects of the current source, the voltage source is set to 0 V as shown in Fig. 9.118. The resulting current is then 6 A through R_2, with the same direction as the contribution due to the voltage source.

The resulting current for the resistor R_2 is the sum of the two currents: $I_T = 2\text{ A} + 6\text{ A} = 8\text{ A}$, as determined in Example 9.1.

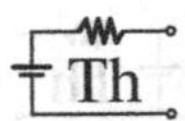

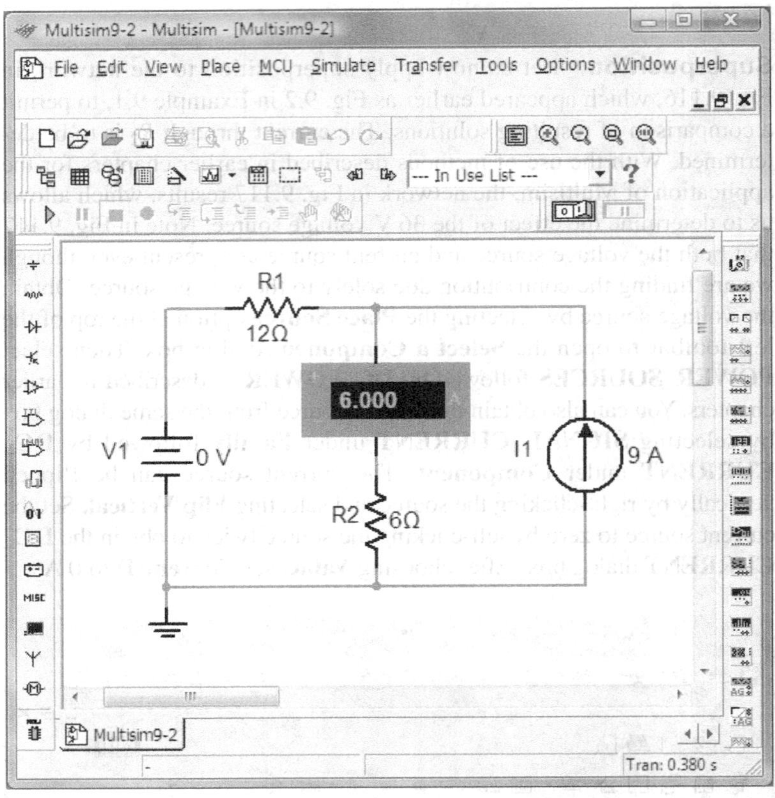

FIG. 9.118
Using Multisim to determine the contribution of the 9 A current source to the current through R_2.

PROBLEMS

SECTION 9.2 Superposition Theorem

1. **a.** Using the superposition theorem, determine the current through the 12 Ω resistor of Fig. 9.119.
   b. Convert both voltage sources to current sources and recalculate the current to the 12 Ω resistor.
   c. How do the results of parts (a) and (b) compare?

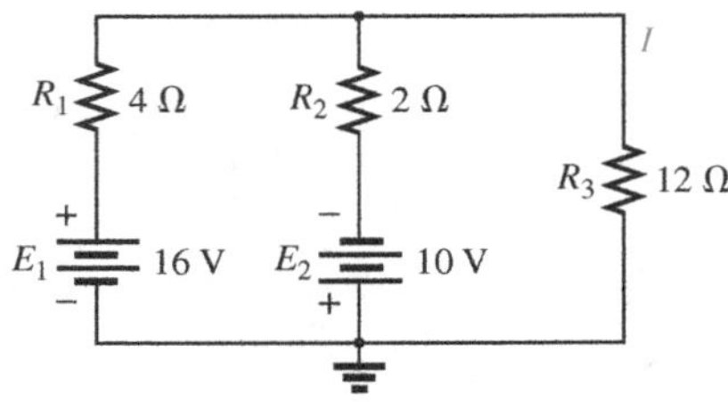

FIG. 9.119
Problem 1.

2. **a.** Using the superposition theorem, determine the voltage across the 4.7 Ω resistor of Fig. 9.120.
   b. Find the power delivered to the 4.7 Ω resistor due solely to the current source.
   c. Find the power delivered to the 4.7 Ω resistor due solely to the voltage source.
   d. Find the power delivered to the 4.7 Ω resistor using the voltage found in part (a).

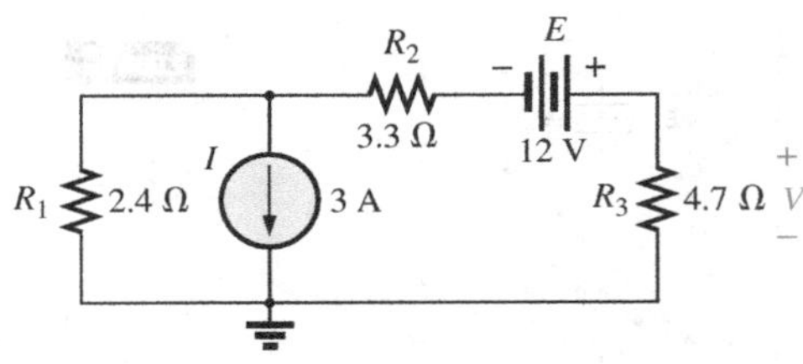

FIG. 9.120
Problem 2.

   e. How do the results of part (d) compare with the sum of the results to parts (b) and (c)? Can the superposition theorem be applied to power levels?

3. Using the superposition theorem, determine the current through the 56 Ω resistor of Fig. 9.121.

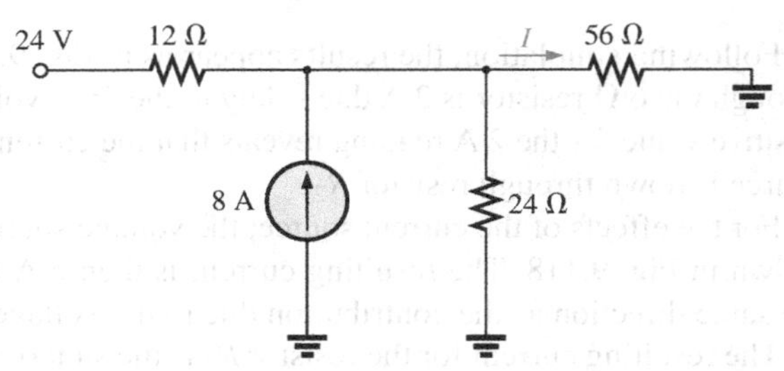

FIG. 9.121
Problem 3.

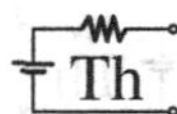

4. Using superposition, find the current I through the 24 V source in Fig. 9.122.

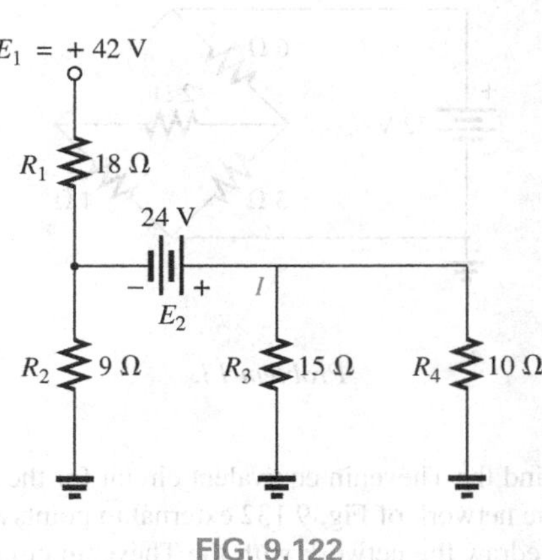

FIG. 9.122
Problem 4.

5. Using superposition, find the voltage V_2 for the network in Fig. 9.123.

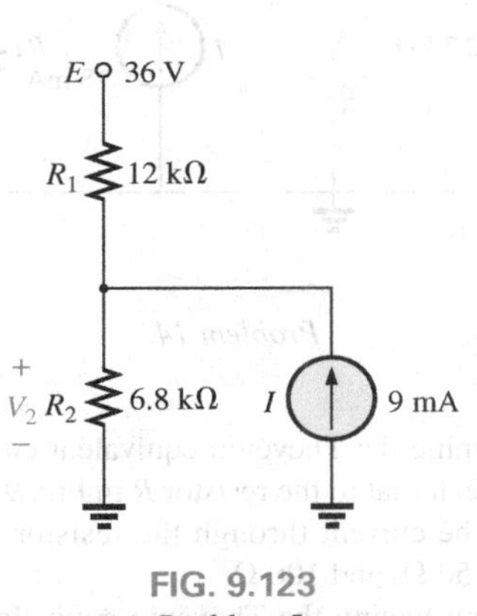

FIG. 9.123
Problem 5.

*6. Using superposition, find the current through R_1 for the network in Fig. 9.124.

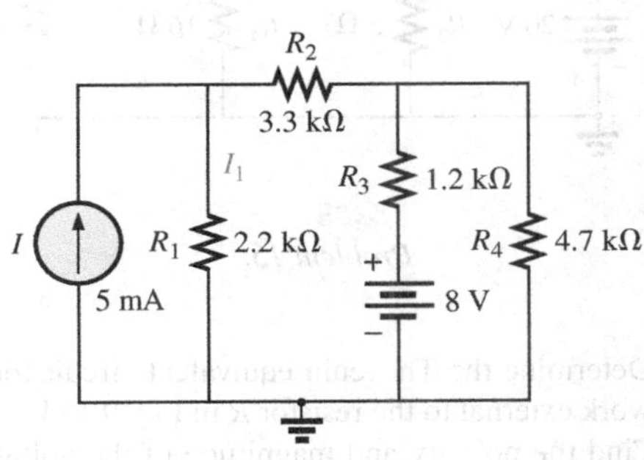

FIG. 9.124
Problem 6.

*7. Using superposition, find the voltage across the 6 A source in Fig. 9.125.

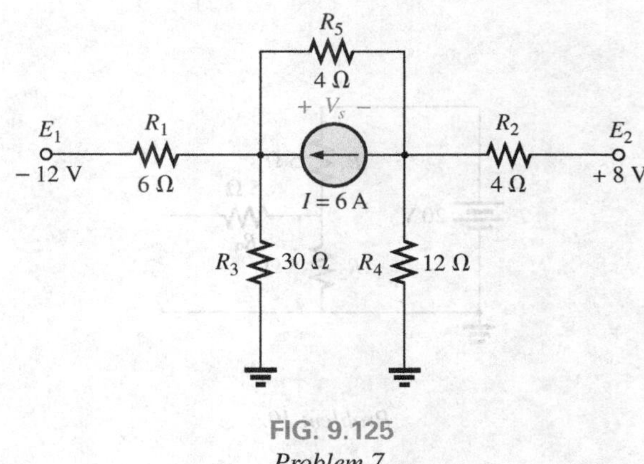

FIG. 9.125
Problem 7.

SECTION 9.3 Thévenin's Theorem

8. **a.** Find the Thévenin equivalent circuit for the network external to the resistor R in Fig. 9.126.
   b. Find the current through R when R is 2 Ω, 30 Ω, and 100 Ω.

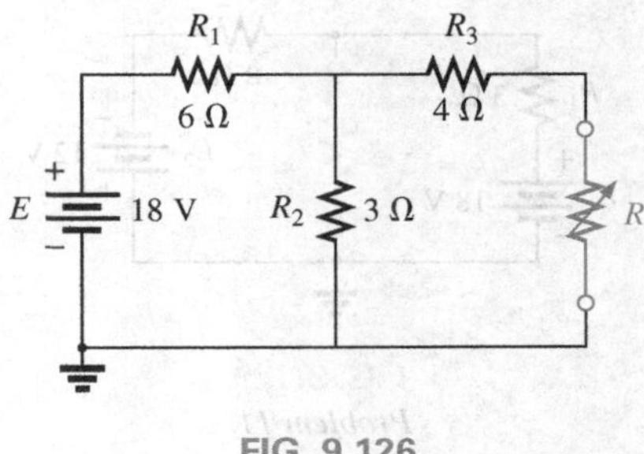

FIG. 9.126
Problem 8.

9. **a.** Find the Thévenin equivalent circuit for the network external to the resistor R for the network in Fig. 9.127.
   b. Find the power delivered to R when R is 2 kΩ and 100 kΩ.

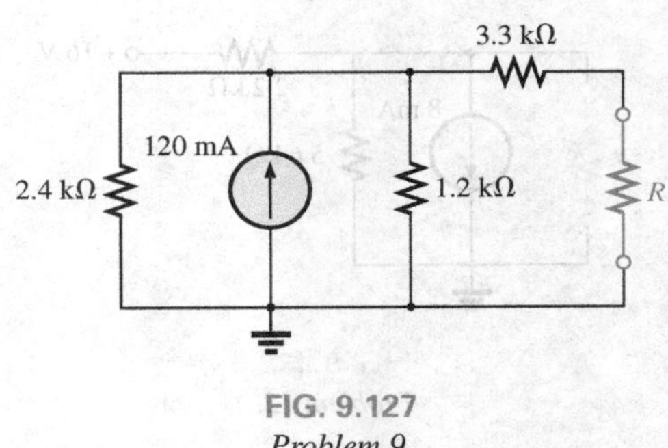

FIG. 9.127
Problem 9.

10. a. Find the Thévenin equivalent circuit for the network external to the resistor R for the network in Fig. 9.128.
    b. Find the power delivered to R when R is 2 Ω and 100 Ω.

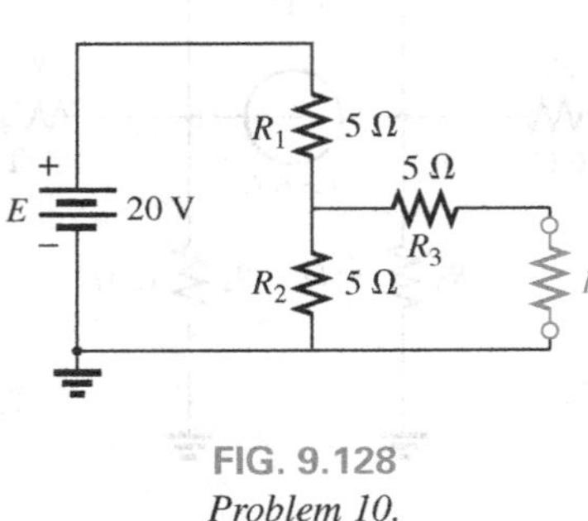

FIG. 9.128
Problem 10.

11. Find the Thévenin equivalent circuit for the network external to the resistor R for the network in Fig. 9.129.

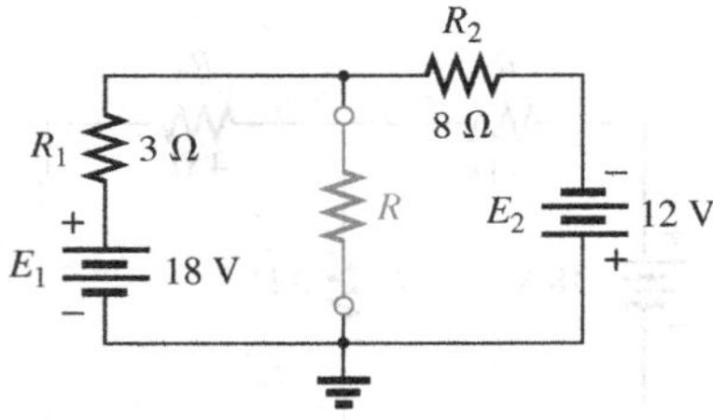

FIG. 9.129
Problem 11.

12. Find the Thévenin equivalent circuit for the network external to the resistor R for the network in Fig. 9.130.

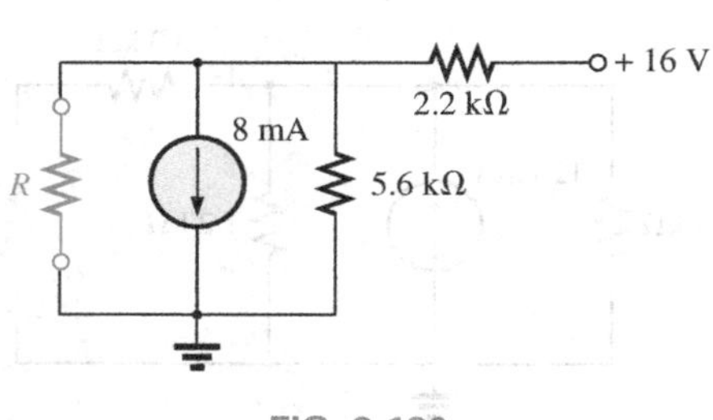

FIG. 9.130
Problem 12.

*13. Find the Thévenin equivalent circuit for the network external to the resistor R in Fig. 9.131.

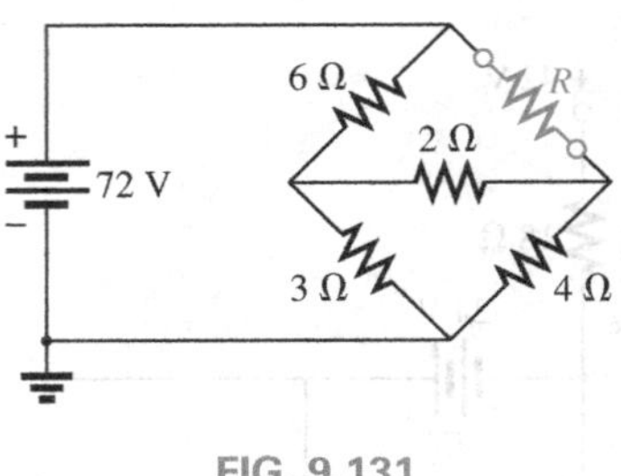

FIG. 9.131
Problem 13.

14. a. Find the Thévenin equivalent circuit for the portions of the network of Fig. 9.132 external to points a and b.
    b. Redraw the network with the Thévenin circuit in place and find the current through the 1.2 kΩ resistor.

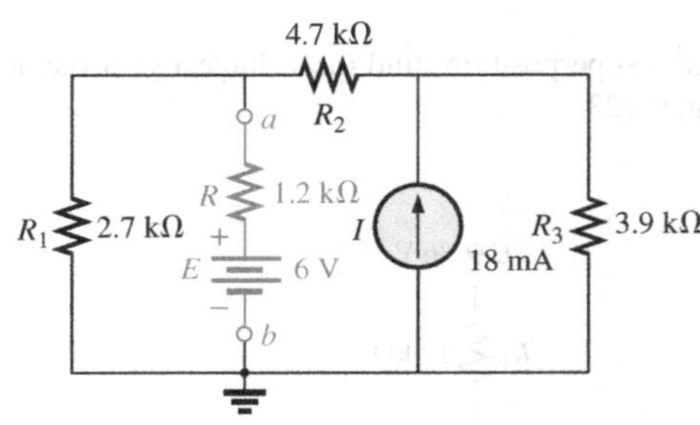

FIG. 9.132
Problem 14.

*15. a. Determine the Thevénin equivalent circuit for the network external to the resistor R in Fig. 9.133.
    b. Find the current through the resistor R if its value is 20 Ω, 50 Ω, and 100 Ω.
    c. Without having the Thévenin equivalent circuit, what would you have to do to find the current through the resistor R for all the values of part (b)?

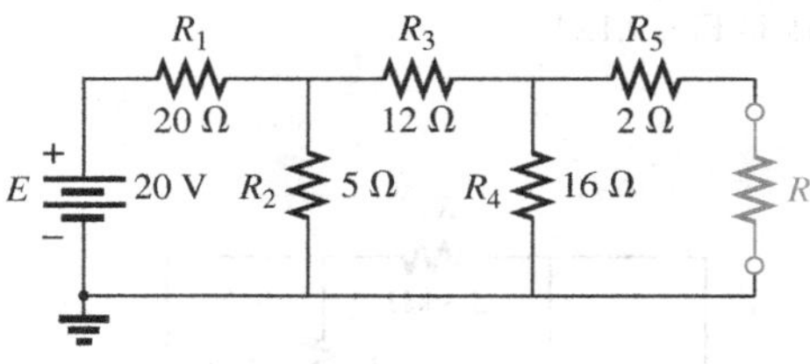

FIG. 9.133
Problem 15.

*16. a. Determine the Thévenin equivalent circuit for the network external to the resistor R in Fig. 9.134.
    b. Find the polarity and magnitude of the voltage across the resistor R if its value is 1.2 kΩ.

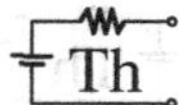

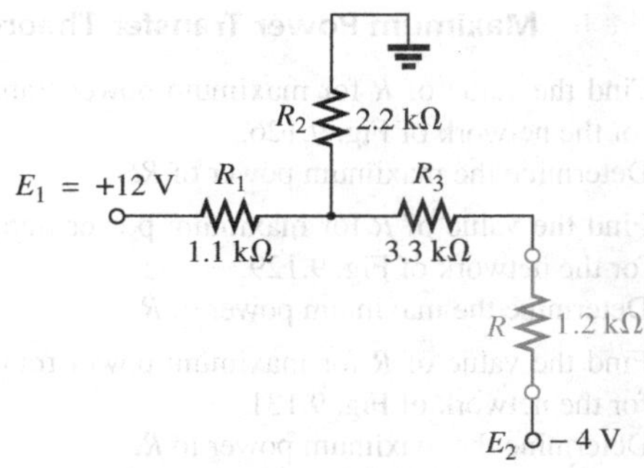

FIG. 9.134
Problem 16.

***17.** For the network in Fig. 9.135, find the Thévenin equivalent circuit for the network external to the load resistor R_L.

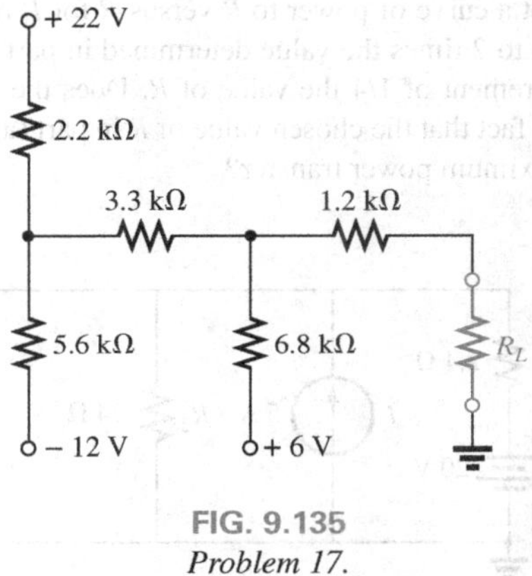

FIG. 9.135
Problem 17.

***18.** For the transistor network in Fig. 9.136:

a. Find the Thévenin equivalent circuit for that portion of the network to the left of the base (*B*) terminal.

b. Using the fact that $I_C = I_E$ and $V_{CE} = 8$ V, determine the magnitude of I_E.

c. Using the results of parts (a) and (b), calculate the base current I_B if $V_{BE} = 0.7$ V.

d. What is the voltage V_C?

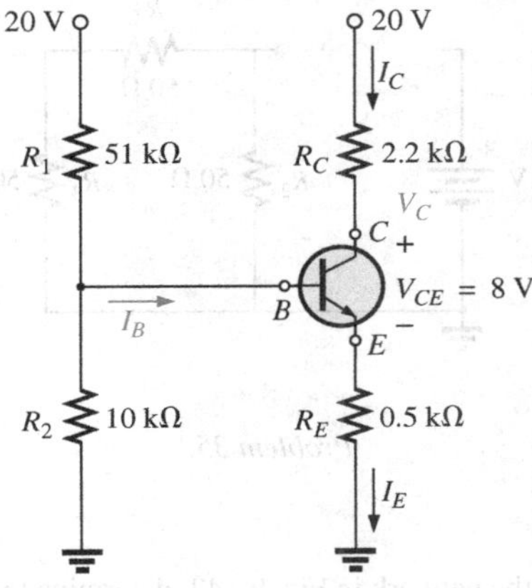

FIG. 9.136
Problem 18.

19. For each vertical set of measurements appearing in Fig. 9.137, determine the Thévenin equivalent circuit.

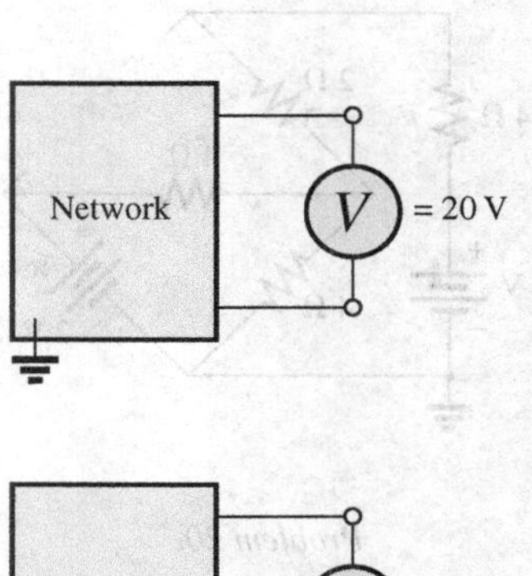

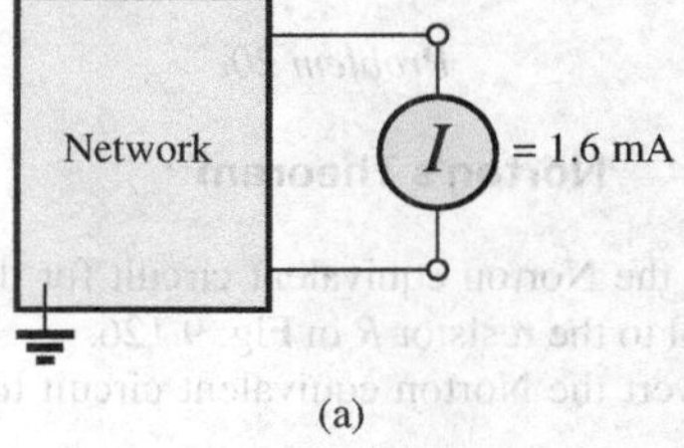

(a)

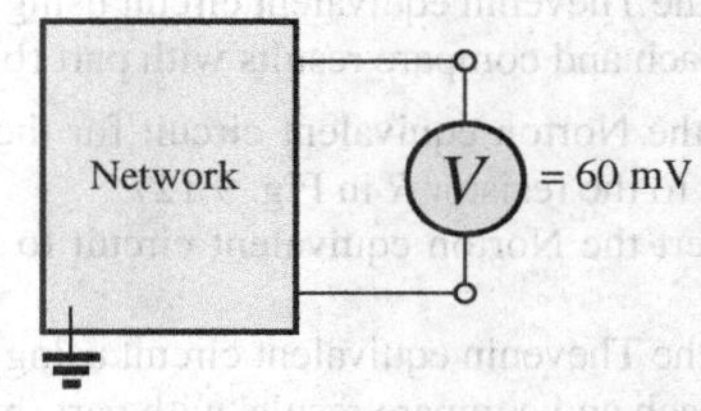

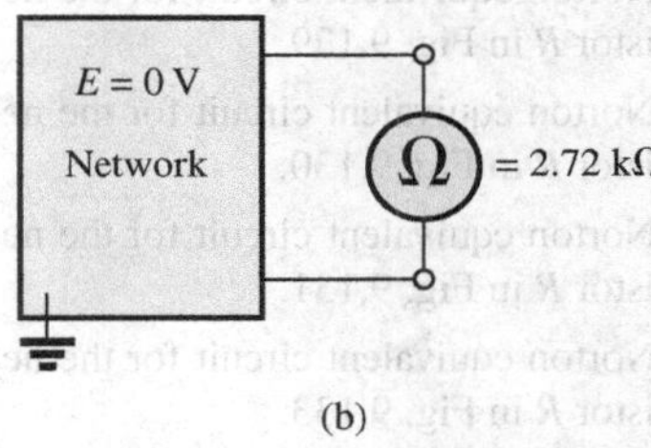

(b)

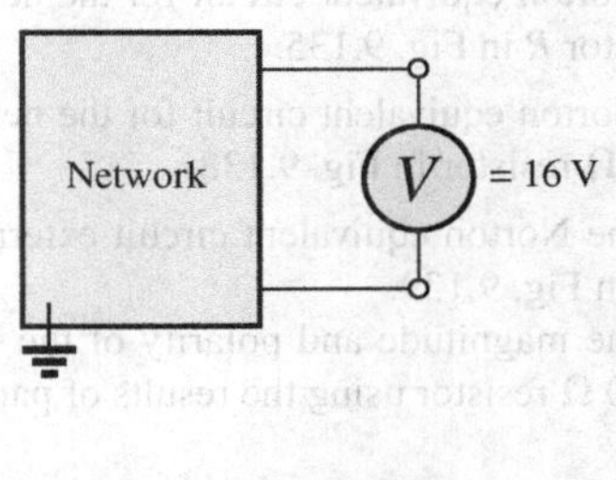

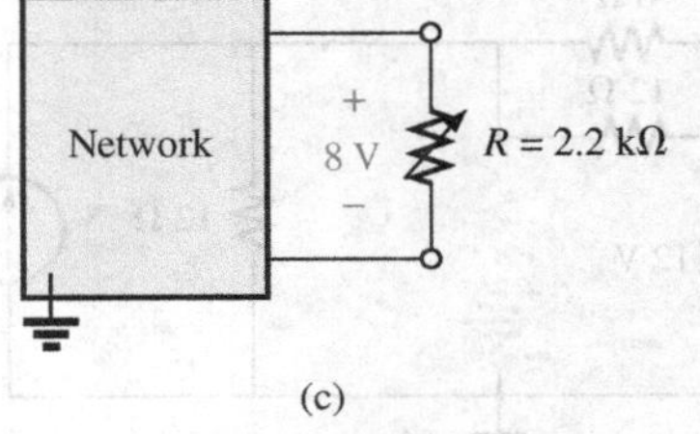

(c)

FIG. 9.137
Problem 19.

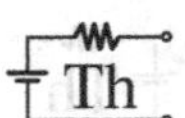

***20.** For the network of Fig. 9.138, find the Thévenin equivalent circuit for the network external to the 300 Ω resistor.

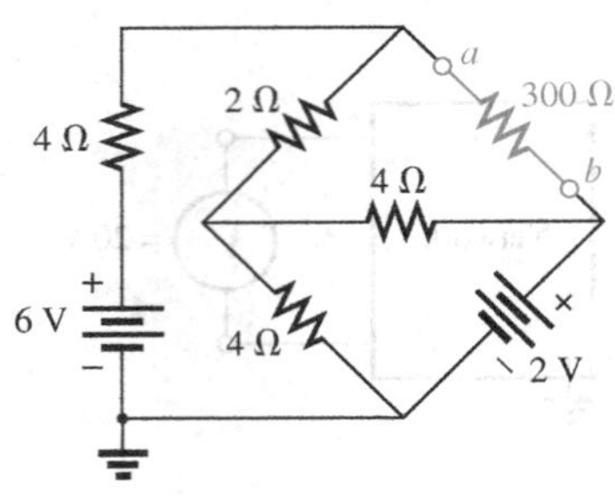

FIG. 9.138
Problem 20.

SECTION 9.4 Norton's Theorem

21. a. Find the Norton equivalent circuit for the network external to the resistor R in Fig. 9.126.
b. Convert the Norton equivalent circuit to the Thévenin form.
c. Find the Thévenin equivalent circuit using the Thévenin approach and compare results with part (b).

22. a. Find the Norton equivalent circuit for the network external to the resistor R in Fig. 9.127.
b. Convert the Norton equivalent circuit to the Thévenin form.
c. Find the Thévenin equivalent circuit using the Thévenin approach and compare results with part (b).

23. Find the Norton equivalent circuit for the network external to the resistor R in Fig. 9.129.

24. Find the Norton equivalent circuit for the network external to the resistor R in Fig. 9.130.

***25.** Find the Norton equivalent circuit for the network external to the resistor R in Fig. 9.131.

***26.** Find the Norton equivalent circuit for the network external to the resistor R in Fig. 9.133.

***27.** Find the Norton equivalent circuit for the network external to the resistor R in Fig. 9.135.

***28.** Find the Norton equivalent circuit for the network external to the 300 Ω resistor in Fig. 9.138.

***29. a.** Find the Norton equivalent circuit external to points a and b in Fig. 9.139.
b. Find the magnitude and polarity of the voltage across the 100 Ω resistor using the results of part (a).

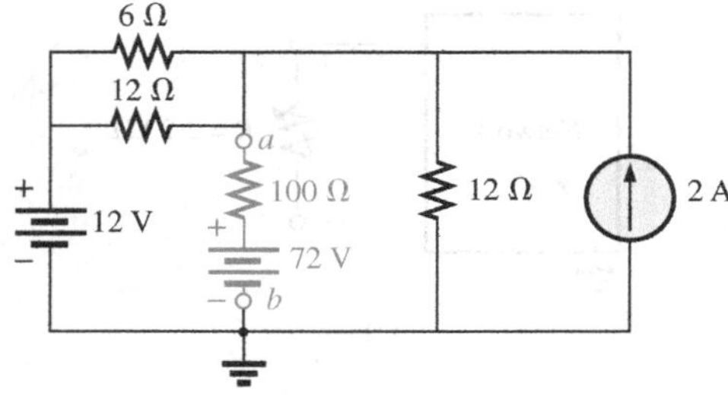

FIG. 9.139
Problem 29.

SECTION 9.5 Maximum Power Transfer Theorem

30. a. Find the value of R for maximum power transfer to R for the network of Fig. 9.126.
b. Determine the maximum power of R.

31. a. Find the value of R for maximum power transfer to R for the network of Fig. 9.129.
b. Determine the maximum power of R.

32. a. Find the value of R for maximum power transfer to R for the network of Fig. 9.131.
b. Determine the maximum power to R.

***33. a.** Find the value of R_L in Fig. 9.135 for maximum power transfer to R_L.
b. Find the maximum power to R_L.

34. a. For the network of Fig. 9.140, determine the value of R for maximum power to R.
b. Determine the maximum power to R.
c. Plot a curve of power to R versus R for R ranging from 1/4 to 2 times the value determined in part (a) using an increment of 1/4 the value of R. Does the curve verify the fact that the chosen value of R in part (a) will ensure maximum power transfer?

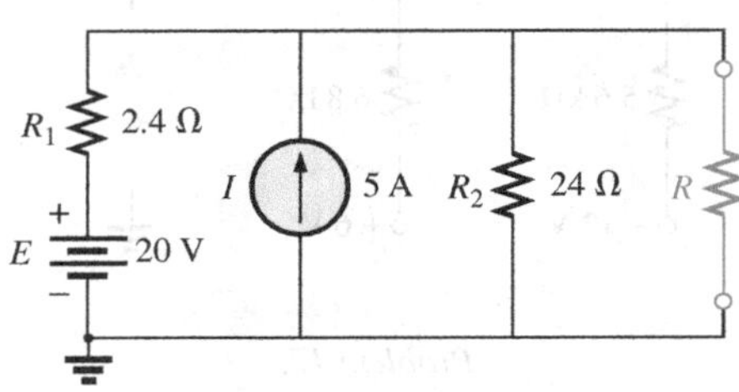

FIG. 9.140
Problem 34.

***35.** Find the resistance R_1 in Fig. 9.141 such that the resistor R_4 will receive maximum power. Think!

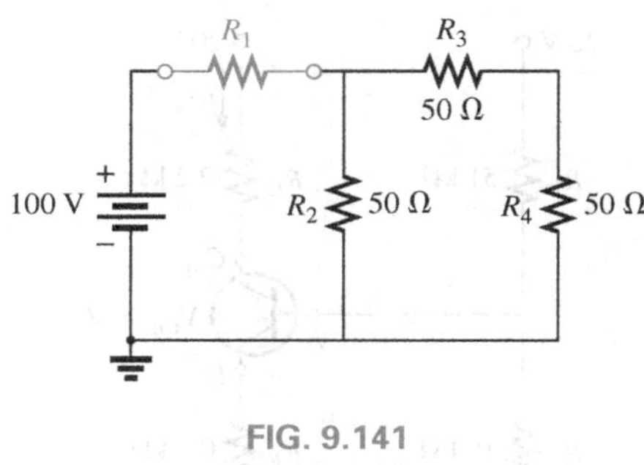

FIG. 9.141
Problem 35.

***36. a.** For the network in Fig. 9.142, determine the value of R_2 for maximum power to R_4.
b. Is there a general statement that can be made about situations such as those presented here and in Problem 35?

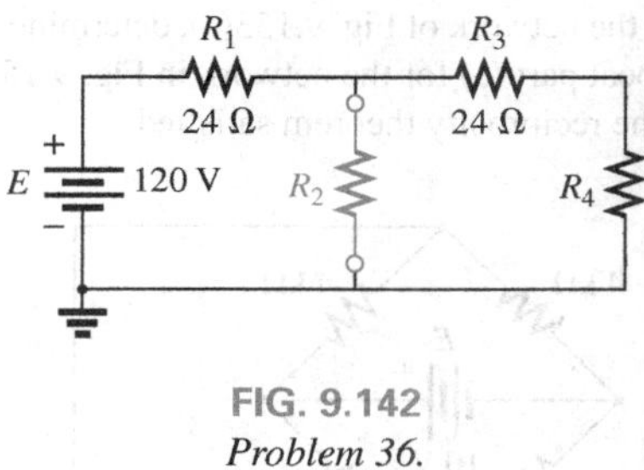

FIG. 9.142
Problem 36.

*37. For the network in Fig. 9.143, determine the level of R that will ensure maximum power to the 100 Ω resistor. Find the maximum power to R_L.

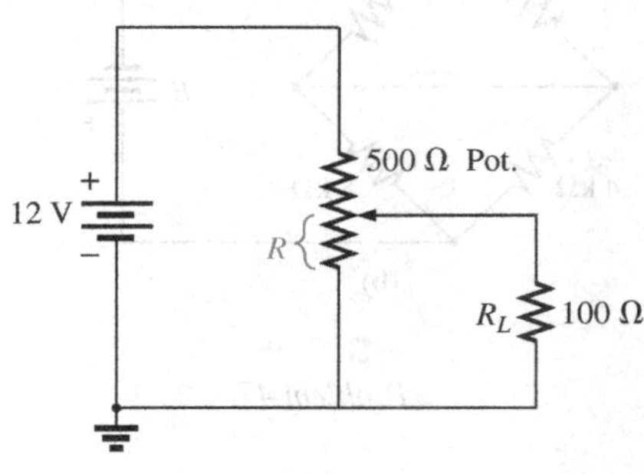

FIG. 9.143
Problem 37.

SECTION 9.6 Millman's Theorem

38. Using Millman's theorem, find the current through and voltage across the resistor R_L in Fig. 9.144.

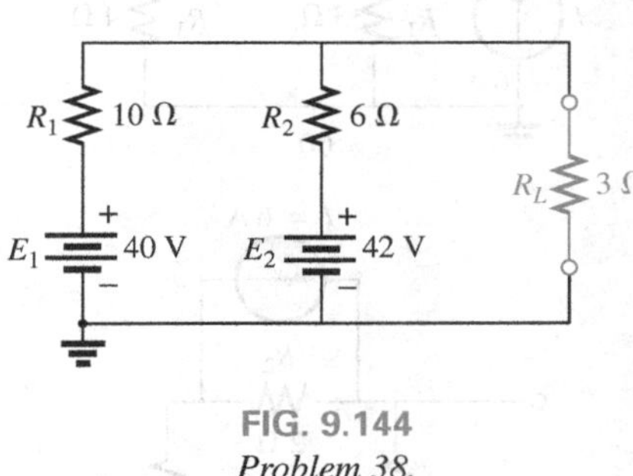

FIG. 9.144
Problem 38.

39. Repeat Problem 38 for the network in Fig. 9.145.

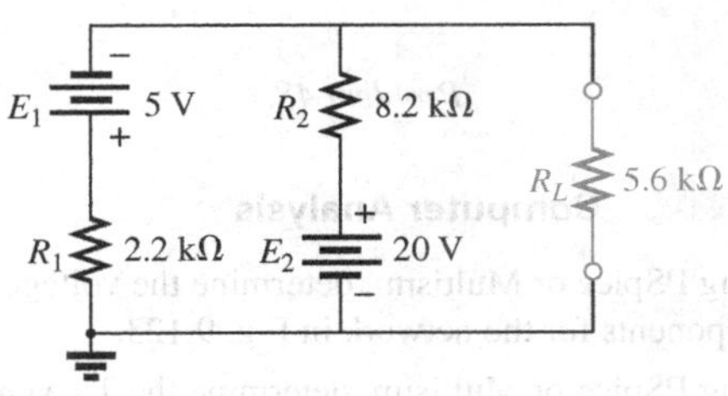

FIG. 9.145
Problem 39.

40. Using Millman's theorem, find the current through and voltage across the resistor R_L in Fig. 9.146.

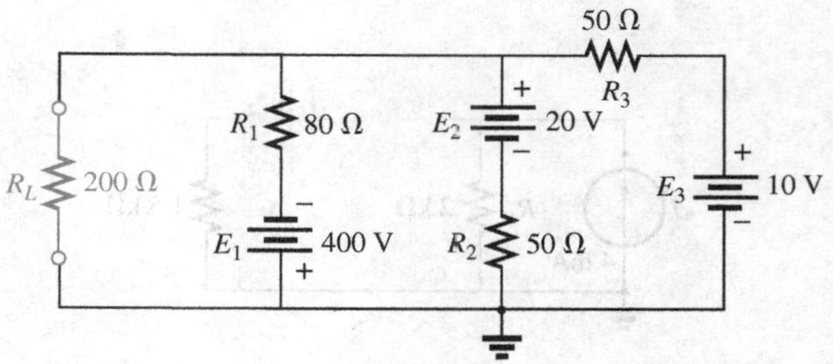

FIG. 9.146
Problem 40.

41. Using the dual of Millman's theorem, find the current through and voltage across the resistor R_L in Fig. 9.147.

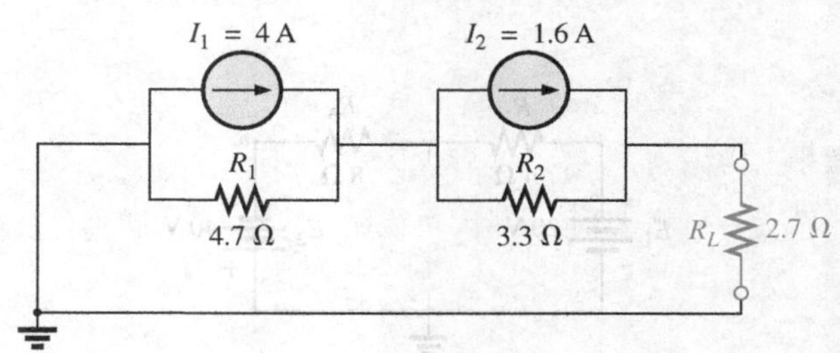

FIG. 9.147
Problem 41.

42. Using the dual of Millman's theorem, find the current through and voltage across the resistor R_L in Fig. 9.148.

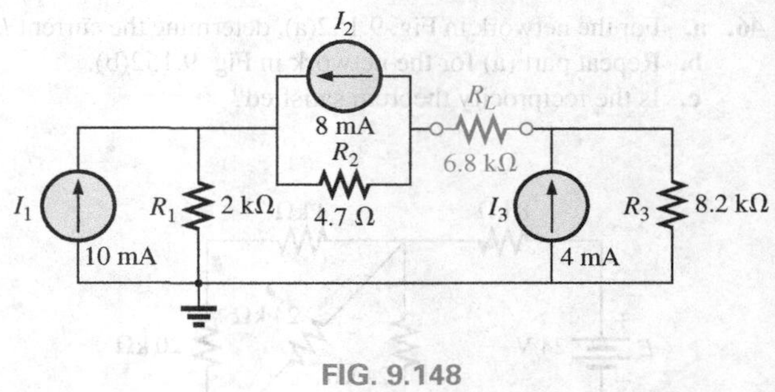

FIG. 9.148
Problem 42.

SECTION 9.7 Substitution Theorem

43. Using the substitution theorem, draw three equivalent branches for the branch a-b of the network in Fig. 9.149.

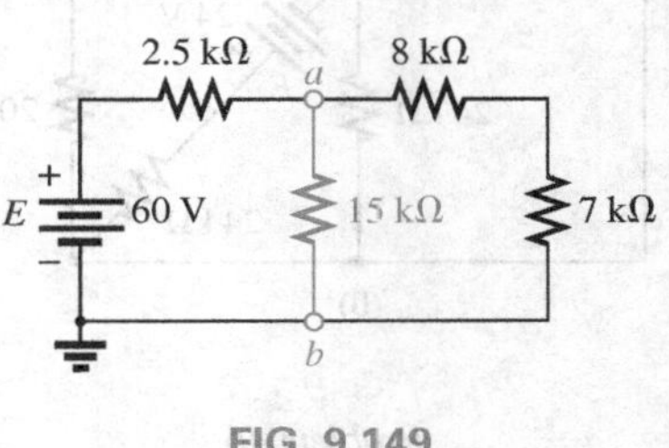

FIG. 9.149
Problem 43.

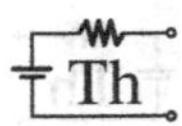

44. Using the substitution theorem, draw three equivalent branches for the branch *a-b* of the network in Fig. 9.150.

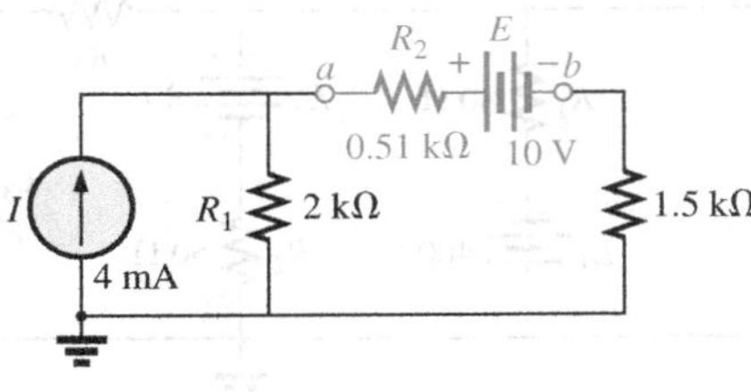

FIG. 9.150
Problem 44.

***45.** Using the substitution theorem, draw three equivalent branches for the branch *a-b* of the network of Fig. 9.151.

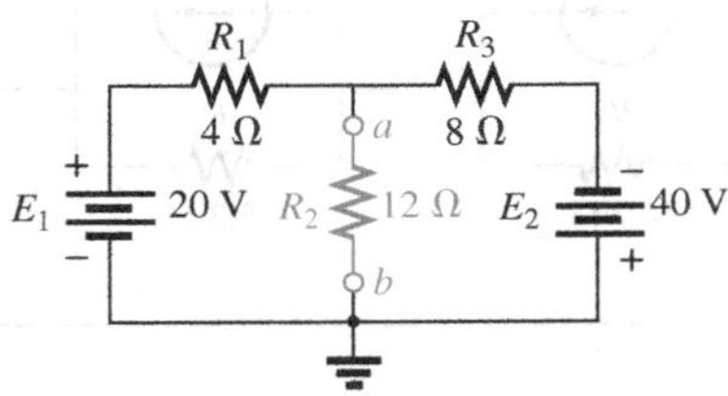

FIG. 9.151
Problem 45.

SECTION 9.8 Reciprocity Theorem

46. a. For the network in Fig. 9.152(a), determine the current *I*.
b. Repeat part (a) for the network in Fig. 9.152(b).
c. Is the reciprocity theorem satisfied?

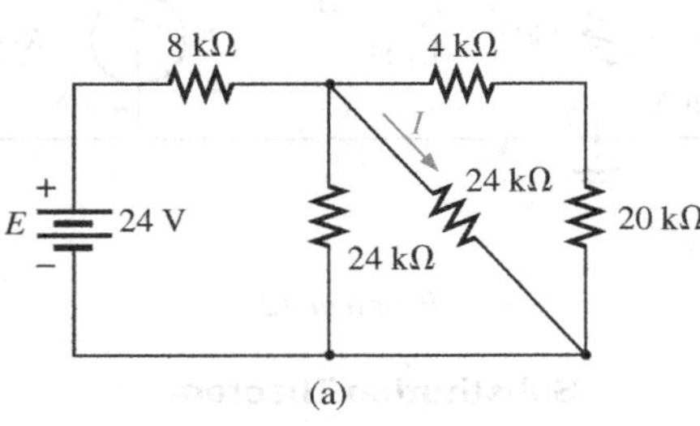

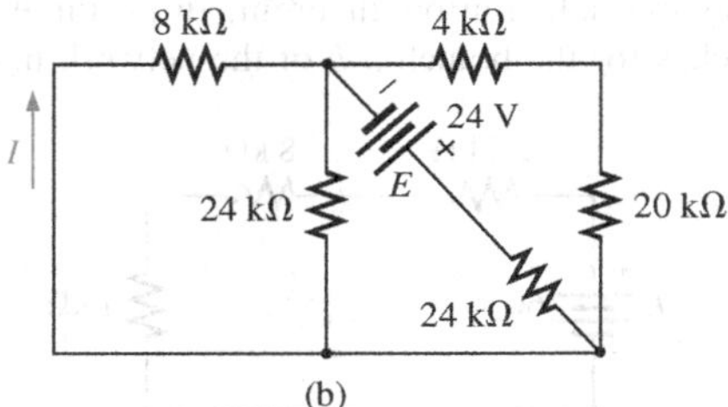

FIG. 9.152
Problem 46.

47. a. For the network of Fig. 9.153(a), determine the current *I*.
b. Repeat part (a) for the network in Fig. 9.153(b).
c. Is the reciprocity theorem satisfied?

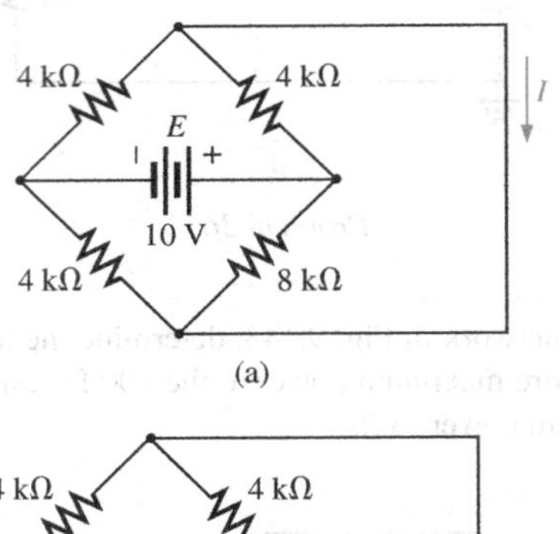

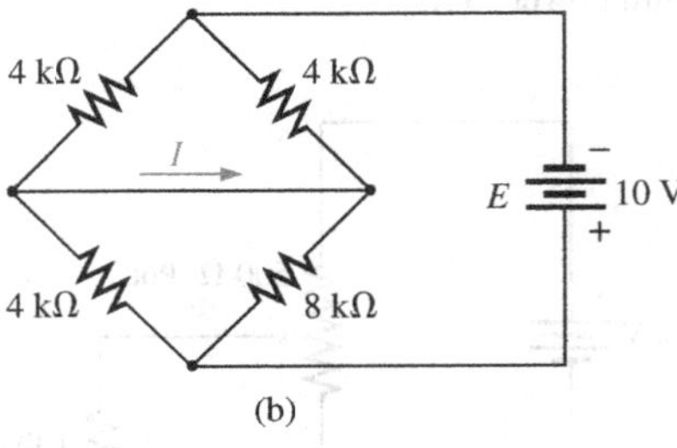

FIG. 9.153
Problem 47.

48. a. Determine the voltage *V* for the network in Fig. 9.154(a).
b. Repeat part (a) for the network in Fig. 9.154(b).
c. Is the dual of the reciprocity theorem satisfied?

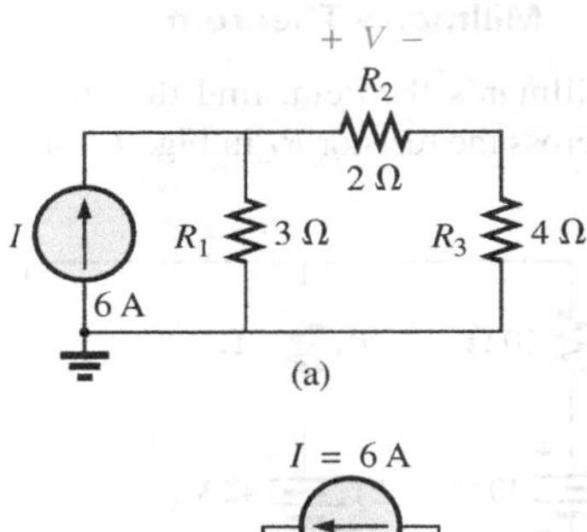

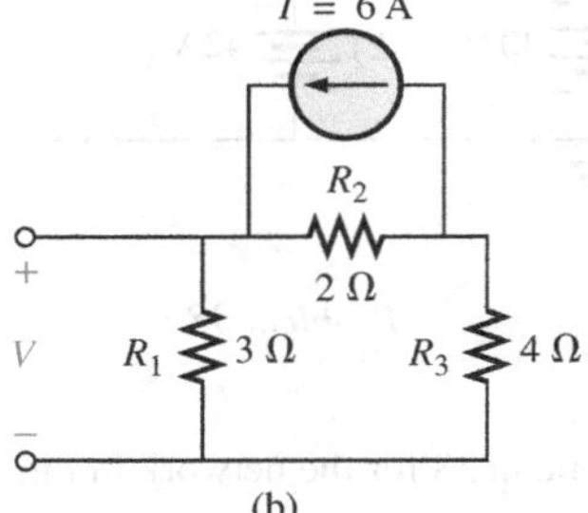

FIG. 9.154
Problem 48.

SECTION 9.9 Computer Analysis

49. Using PSpice or Multisim, determine the voltage V_2 and its components for the network in Fig. 9.123.

50. Using PSpice or Multisim, determine the Thévenin equivalent circuit for the network in Fig. 9.131.

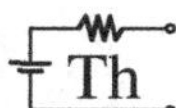

*51. **a.** Using PSpice, plot the power delivered to the resistor R in Fig. 9.128 for R having values from 1 Ω to 10 Ω.

b. From the plot, determine the value of R resulting in the maximum power to R and the maximum power to R.

c. Compare the results of part (a) to the numerical solution.

d. Plot V_R and I_R versus R, and find the value of each under maximum power conditions.

*52. Change the 300 Ω resistor in Fig. 9.138 to a variable resistor, and using Pspice, plot the power delivered to the resistor versus values of the resistor. Determine the range of resistance by trial and error rather than first performing a longhand calculation. Determine the Norton equivalent circuit from the results. The Norton current can be determined from the maximum power level.

GLOSSARY

Maximum power transfer theorem A theorem used to determine the load resistance necessary to ensure maximum power transfer to the load.

Millman's theorem A method using source conversions that will permit the determination of unknown variables in a multiloop network.

Norton's theorem A theorem that permits the reduction of any two-terminal linear dc network to one having a single current source and parallel resistor.

Reciprocity theorem A theorem that states that for single-source networks, the current in any branch of a network due to a single voltage source in the network will equal the current through the branch in which the source was originally located if the source is placed in the branch in which the current was originally measured.

Substitution theorem A theorem that states that if the voltage across and current through any branch of a dc bilateral network are known, the branch can be replaced by any combination of elements that will maintain the same voltage across and current through the chosen branch.

Superposition theorem A network theorem that permits considering the effects of each source independently. The resulting current and/or voltage is the algebraic sum of the currents and/or voltages developed by each source independently.

Thévenin's theorem A theorem that permits the reduction of any two-terminal, linear dc network to one having a single voltage source and series resistor.

Capacitors

Objectives

- *Become familiar with the basic construction of a capacitor and the factors that affect its ability to store charge on its plates.*
- *Be able to determine the transient (time-varying) response of a capacitive network and plot the resulting voltages and currents.*
- *Understand the impact of combining capacitors in series or parallel and how to read the nameplate data.*
- *Develop some familiarity with the use of computer methods to analyze networks with capacitive elements.*

10.1 INTRODUCTION

Thus far, the resistor has been the only network component appearing in our analyses. In this chapter, we introduce the **capacitor,** which has a significant impact on the types of networks that you will be able to design and analyze. Like the resistor, it is a two-terminal device, but its characteristics are totally different from those of a resistor. In fact, *the capacitor displays its true characteristics only when a change in the voltage or current is made in the network.* All the power delivered to a resistor is dissipated in the form of heat. An ideal capacitor, however, stores the energy delivered to it in a form that can be returned to the system.

Although the basic construction of capacitors is actually quite simple, it is a component that opens the door to all types of practical applications, extending from touch pads to sophisticated control systems. A few applications are introduced and discussed in detail later in this chapter.

10.2 THE ELECTRIC FIELD

Recall from Chapter 2 that a force of attraction or repulsion exists between two charged bodies. We now examine this phenomenon in greater detail by considering the electric field that exists in the region around any charged body. This electric field is represented by **electric flux lines,** which are drawn to indicate the strength of the electric field at any point around the charged body. The denser the lines of flux, the stronger is the electric field. In Fig. 10.1, for example, the electric field strength is stronger in region a than region b because the flux lines are denser in region a than in b. That is, the same number of flux lines pass through each region, but the area A_1 is much smaller than area A_2. The symbol for electric flux is the Greek letter ψ (*psi*). The flux per unit area (flux density) is represented by the capital letter D and is determined by

$$D = \frac{\psi}{A} \qquad \text{(flux/unit area)} \qquad (10.1)$$

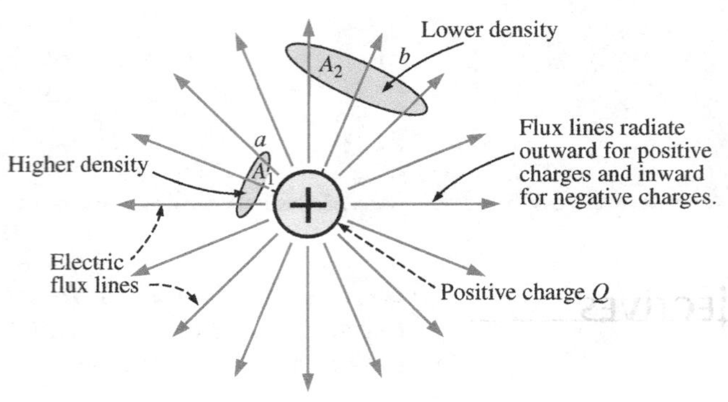

FIG. 10.1
Flux distribution from an isolated positive charge.

The larger the charge Q in coulombs, the greater is the number of flux lines extending or terminating per unit area, independent of the surrounding medium. Twice the charge produces twice the flux per unit area. The two can therefore be equated:

$$\boxed{\psi \equiv Q} \qquad \text{(coulombs, C)} \tag{10.2}$$

By definition, the **electric field strength** (designated by the capital script letter $\mathscr{E}$) at a point is the force acting on a unit positive charge at that point; that is,

$$\boxed{\mathscr{E} = \frac{F}{Q}} \qquad \text{(newtons/coulomb, N/C)} \tag{10.3}$$

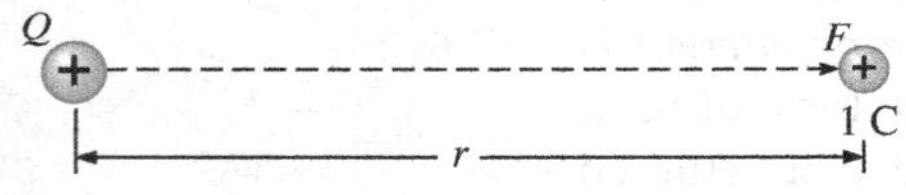

FIG. 10.2
Determining the force on a unit charge r meters from a charge Q of similar polarity.

In Fig. 10.2, the force exerted on a unit (1 coulomb) positive charge by a charge Q, r meters away, can be determined using **Coulomb's law** (Eq. 2.1) as follows:

$$F = k\frac{Q_1Q_2}{r^2} = k\frac{Q(1\ \text{C})}{r^2} = \frac{kQ}{r^2}\ (k = 9 \times 10^9\ \text{N}\cdot\text{m}^2/\text{C}^2)$$

Substituting the result into Eq. 10.3 for a unit positive charge results in

$$\mathscr{E} = \frac{F}{Q} = \frac{kQ/r^2}{1/\text{C}}$$

and

$$\boxed{\mathscr{E} = \frac{kQ}{r^2}} \qquad \text{(N/C)} \tag{10.4}$$

The result clearly reveals that the electric field strength is directly related to the size of the charge Q. The greater the charge Q, the greater is the electric field intensity on a unit charge at any point in the neighborhood. However, the distance is a squared term in the denominator. The result is that the greater the distance from the charge Q, the less is the electric field strength, and dramatically so because of the squared term. In Fig. 10.1, the electric field strength at region A_2 is therefore significantly less than at region A_1.

For two charges of similar and opposite polarities, the flux distribution appears as shown in Fig. 10.3. In general,

electric flux lines always extend from a positively charged body to a negatively charged body, always extend or terminate perpendicular to the charged surfaces, and never intersect.

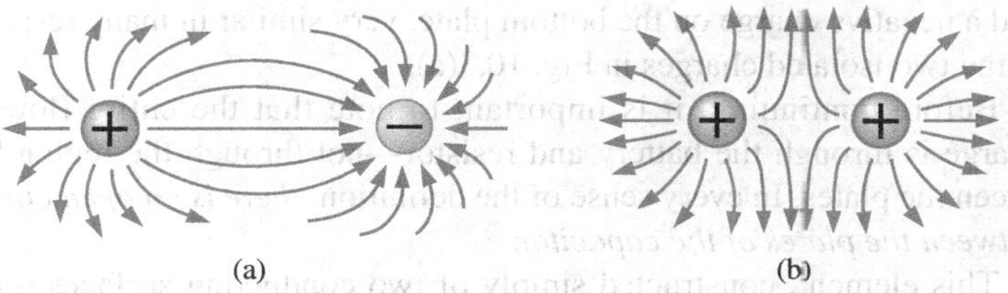

FIG. 10.3
Electric flux distributions: (a) opposite charges; (b) like charges.

Note in Fig. 10.3(a) that the electric flux lines establish the most direct pattern possible from the positive to negative charge. They are evenly distributed and have the shortest distance on the horizontal between the two charges. This pattern is a direct result of the fact that electric flux lines strive to establish the shortest path from one charged body to another. The result is a natural pressure to be as close as possible. If two bodies of the same polarity are in the same vicinity, as shown in Fig. 10.3(b), the result is the direct opposite. The flux lines tend to establish a buffer action between the two with a repulsive action that grows as the two charges are brought closer to one another.

10.3 CAPACITANCE

Thus far, we have examined only isolated positive and negative spherical charges, but the description can be extended to charged surfaces of any shape and size. In Fig. 10.4, for example, two parallel plates of a material such as aluminum (the most commonly used metal in the construction of capacitors) have been connected through a switch and a resistor to a battery. If the parallel plates are initially uncharged and the switch is left open, no net positive or negative charge exists on either plate. The instant the switch is closed, however, electrons are drawn from the upper plate through the resistor to the positive terminal of the battery. There will be a surge of current at first, limited in magnitude by the resistance present. The level of flow then declines, as will be demonstrated in the sections to follow. This action creates a net positive charge on the top plate. Electrons are being repelled by the negative terminal through the lower conductor to the bottom plate at the same rate they are being drawn to the positive terminal. This transfer of electrons continues until the potential difference across the parallel plates is exactly equal to the battery voltage. The final result is a net positive charge on the top plate

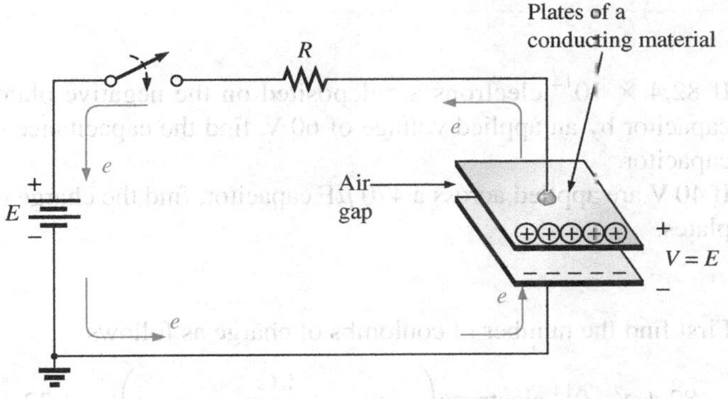

FIG. 10.4
Fundamental charging circuit.

FIG. 10.5
Michael Faraday.

English (London)
(1791–1867)
Chemist and Electrical Experimenter
Honorary Doctorate, Oxford University, 1832

An experimenter with no formal education, he began his research career at the Royal Institute in London as a laboratory assistant. Intrigued by the interaction between electrical and magnetic effects, he discovered *electromagnetic induction,* demonstrating that electrical effects can be generated from a magnetic field (the birth of the generator as we know it today). He also discovered *self-induced currents* and introduced the concept of *lines and fields of magnetic force.* Having received over one hundred academic and scientific honors, he became a Fellow of the Royal Society in 1824 at the young age of 32.

and a negative charge on the bottom plate, very similar in many respects to the two isolated charges in Fig. 10.3(a).

Before continuing, it is important to note that the entire flow of charge is through the battery and resistor—not through the region between the plates. In every sense of the definition, *there is an open circuit between the plates of the capacitor.*

This element, constructed simply of two conducting surfaces separated by the air gap, is called a **capacitor.**

Capacitance is a measure of a capacitor's ability to store charge on its plates—in other words, its storage capacity.

In addition,

the higher the capacitance of a capacitor, the greater is the amount of charge stored on the plates for the same applied voltage.

The unit of measure applied to capacitors is the farad (F), named after an English scientist, Michael Faraday, who did extensive research in the field (Fig. 10.5). In particular,

a capacitor has a capacitance of 1 F if 1 C of charge (6.242×10^{18} electrons) is deposited on the plates by a potential difference of 1 V across its plates.

The farad, however, is generally too large a measure of capacitance for most practical applications, so the microfarad (10^{-6}) or picofarad (10^{-12}) are more commonly encountered.

The relationship connecting the applied voltage, the charge on the plates, and the capacitance level is defined by the following equation:

$$\boxed{C = \frac{Q}{V}} \qquad \begin{aligned} C &= \text{farads (F)} \\ Q &= \text{coulombs (C)} \\ V &= \text{volts (V)} \end{aligned} \qquad \textbf{(10.5)}$$

Eq. (10.5) reveals that for the same voltage (V), the greater the charge (Q) on the plates (in the numerator of the equation), the higher is the capacitance level (C).

If we write the equation in the form

$$\boxed{Q = CV} \qquad \text{(coulombs, C)} \qquad \textbf{(10.6)}$$

it becomes obvious through the product relationship that the higher the capacitance (C) or applied voltage (V), the greater is the charge on the plates.

EXAMPLE 10.1

a. If 82.4×10^{14} electrons are deposited on the negative plate of a capacitor by an applied voltage of 60 V, find the capacitance of the capacitor.
b. If 40 V are applied across a 470 μF capacitor, find the charge on the plates.

Solutions:

a. First find the number of coulombs of charge as follows:

$$82.4 \times 10^{14} \cancel{\text{electrons}} \left(\frac{1\text{ C}}{6.242 \times 10^{18} \cancel{\text{electrons}}} \right) = 1.32 \text{ mC}$$

and then

$$C = \frac{Q}{V} = \frac{1.32\text{ mC}}{60\text{ V}} = \mathbf{22\ \mu F} \quad \text{(a standard value)}$$

b. Applying Eq. (10.6) gives

$$Q = CV = (470\ \mu\text{F})(40\text{ V}) = \mathbf{18.8\ mC}$$

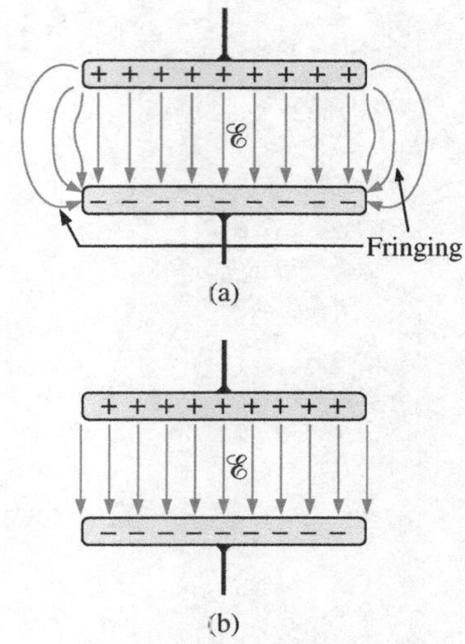

FIG. 10.6

Electric flux distribution between the plates of a capacitor: (a) including fringing; (b) ideal.

A cross-sectional view of the parallel plates in Fig. 10.4 is provided in Fig. 10.6(a). Note the **fringing** that occurs at the edges as the flux lines originating from the points farthest away from the negative plate strive to complete the connection. This fringing, which has the effect of reducing the net capacitance somewhat, can be ignored for most applications. Ideally, and the way we will assume the distribution to be in this text, the electric flux distribution appears as shown in Fig. 10.6(b), where all the flux lines are equally distributed and "fringing" does not occur.

The **electric field strength** between the plates is determined by the voltage across the plates and the distance between the plates as follows:

$$\boxed{\mathscr{E} = \frac{V}{d}} \qquad \begin{aligned} \mathscr{E} &= \text{volts/m (V/m)} \\ V &= \text{volts (V)} \\ d &= \text{meters (m)} \end{aligned} \qquad \mathbf{(10.7)}$$

Note that the distance between the plates is measured in meters, not centimeters or inches.

The equation for the electric field strength is determined by two factors only: *the applied voltage and the distance between the plates.* The charge on the plates does not appear in the equation, nor does the size of the capacitor or the plate material.

Many values of capacitance can be obtained for the same set of parallel plates by the addition of certain insulating materials between the plates. In Fig. 10.7, an insulating material has been placed between a set of parallel plates having a potential difference of V volts across them.

Since the material is an insulator, the electrons within the insulator are unable to leave the parent atom and travel to the positive plate. The positive components (protons) and negative components (electrons) of each atom do shift, however [as shown in Fig. 10.7(a)], to form *dipoles.* When the dipoles align themselves as shown in Fig. 10.7(a), the material is *polarized.* A close examination within this polarized material reveals that the positive and negative components of adjoining dipoles

are neutralizing the effects of each other [note the oval area in Fig. 10.7(a)]. The layer of positive charge on one surface and the negative charge on the other are not neutralized, however, resulting in the establishment of an electric field within the insulator [$\mathscr{E}_{\text{dielectric}}$; Fig. 10.7(b)].

In Fig. 10.8(a), two plates are separated by an air gap and have layers of charge on the plates as established by the applied voltage and the distance between the plates. The electric field strength is $\mathscr{E}_1$ as defined by Eq. (10.7). In Fig. 10.8(b), a slice of mica is introduced, which, through an alignment of cells within the dielectric, establishes an electric field $\mathscr{E}_2$ that will oppose electric field $\mathscr{E}_1$. The effect is to try to reduce the electric field strength between the plates. However, Eq. (10.7) states that the electric field strength *must be* the value established by the applied voltage and the distance between the plates. This condition is maintained by placing more charge on the plates, thereby increasing the electric field strength between the plates to a level that cancels out the opposing electric field introduced by the mica sheet. The net result is an increase in charge on the plates and an increase in the capacitance level as established by Eq. (10.5).

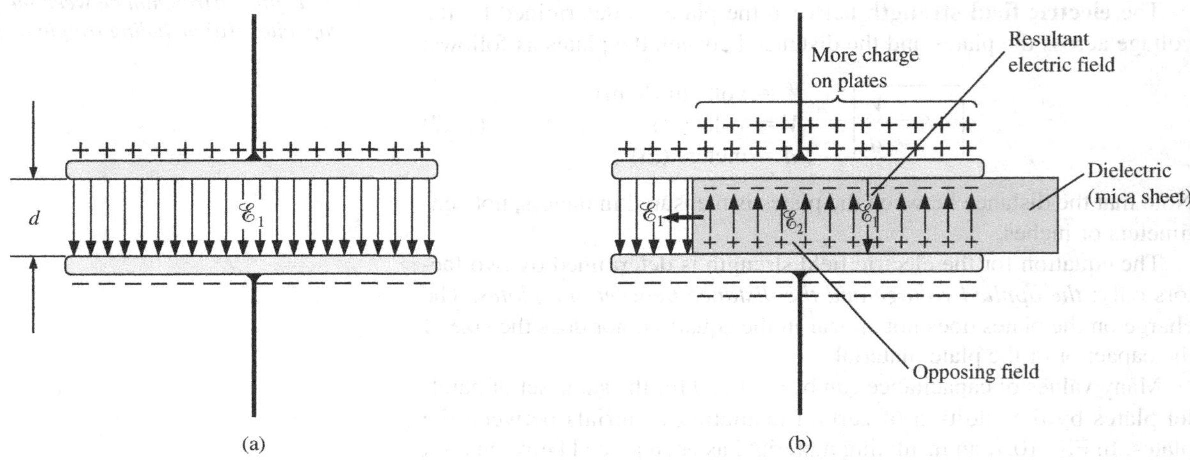

FIG. 10.8
Demonstrating the effect of inserting a dielectric between the plates of a capacitor: (a) air capacitor; (b) dielectric being inserted.

Different materials placed between the plates establish different amounts of additional charge on the plates. All, however, must be insulators and must have the ability to set up an electric field within the structure. A list of common materials appears in Table 10.1 using air as the reference level of 1.* All of these materials are referred to as **dielectrics,** the "di" for *opposing,* and the "electric" from *electric field.* The symbol ϵ_r in Table 10.1 is called the **relative permittivity** (or **dielectric constant**). The term **permittivity** is applied as a measure of how easily a material "permits" the establishment of an electric field in the material. The relative permittivity compares the permittivity of a material to that of air. For instance, Table 10.1 reveals that mica, with a relative permittivity of 5, "permits" the establishment of an opposing electric field in the material five times better than in air. Note the ceramic material at the bottom of the chart with a relative permittivity of 7500—a relative permittivity that makes it a very special dielectric in the manufacture of capacitors.

*Although there is a difference in dielectric characteristics between air and a vacuum, the difference is so small that air is commonly used as the reference level.

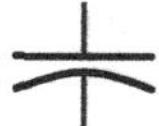

TABLE 10.1
Relative permittivity (dielectric constant) ϵ_r of various dielectrics.

Dielectric	ϵ_r (Average Values)
Vacuum	1.0
Air	1.0006
Teflon®	2.0
Paper, paraffined	2.5
Rubber	3.0
Polystyrene	3.0
Oil	4.0
Mica	5.0
Porcelain	6.0
Bakelite®	7.0
Aluminum oxide	7
Glass	7.5
Tantalum oxide	30
Ceramics	20–7500
Barium-strontium titanite (ceramic)	7500.0

Defining ϵ_o as the permittivity of air, we define the relative permittivity of a material with a permittivity ϵ by

$$\epsilon_r = \frac{\epsilon}{\epsilon_o} \quad \text{(dimensionless)} \qquad \textbf{(10.8)}$$

Note that ϵ_r, which (as mentioned previously) is often called the **dielectric constant,** is a dimensionless quantity because it is a ratio of similar quantities. However, permittivity does have the units of farads/meter (F/m) and is 8.85×10^{-12} F/m for air. Although the relative permittivity for the air we breathe is listed as 1.006, a value of 1 is normally used for the relative permittivity of air.

For every dielectric there is a potential that, if applied across the dielectric, will break down the bonds within it and cause current to flow through it. The voltage required per unit length is an indication of its **dielectric strength** and is called the **breakdown voltage.** When breakdown occurs, the capacitor has characteristics very similar to those of a conductor. A typical example of dielectric breakdown is lightning, which occurs when the potential between the clouds and the earth is so high that charge can pass from one to the other through the atmosphere (the dielectric). The average dielectric strengths for various dielectrics are tabulated in volts/mil in Table 10.2 (1 mil = 1/1000 inch).

One of the important parameters of a capacitor is the **maximum working voltage.** It defines the maximum voltage that can be placed across the capacitor on a continuous basis without damaging it or changing its characteristics. For most capacitors, it is the dielectric strength that defines the maximum working voltage.

TABLE 10.2
Dielectric strength of some dielectric materials.

Dielectric	Dielectric Strength (Average Value) in Volts/Mil
Air	75
Barium-strontium titanite (ceramic)	75
Ceramics	75–1000
Porcelain	200
Oil	400
Bakelite®	400
Rubber	700
Paper paraffined	1300
Teflon®	1500
Glass	3000
Mica	5000

10.4 CAPACITORS

Capacitor Construction

We are now aware of the basic components of a capacitor: conductive plates, separation, and dielectric. However, the question remains, How do all these factors interact to determine the capacitance of a capacitor?

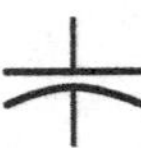

Larger plates permit an increased area for the storage of charge, so the area of the plates should be in the numerator of the defining equation. *The smaller the distance between the plates,* the larger is the capacitance, so this factor should appear in the numerator of the equation. Finally, since *higher levels of permittivity* result in higher levels of capacitance, the factor ϵ should appear in the numerator of the defining equation.

The result is the following general equation for capacitance:

$$C = \epsilon\frac{A}{d} \qquad \begin{aligned} C &= \text{farads (F)} \\ \epsilon &= \text{permittivity (F/m)} \\ A &= \text{m}^2 \\ d &= \text{m} \end{aligned} \tag{10.9}$$

If we substitute Eq. (10.8) for the permittivity of the material, we obtain the following equation for the capacitance:

$$C = \epsilon_o\epsilon_r\frac{A}{d} \qquad \text{(farads, F)} \tag{10.10}$$

or if we substitute the known value for the permittivity of air, we obtain the following useful equation:

$$C = 8.85 \times 10^{-12}\,\epsilon_r\frac{A}{d} \qquad \text{(farads, F)} \tag{10.11}$$

It is important to note in Eq. (10.11) that the area of the plates (actually the area of only one plate) is in meters squared (m^2); the distance between the plates is measured in meters; and the numerical value of ϵ_r is simply taken from Table 10.1.

You should also be aware that most capacitors are in the μF, nF, or pF range, not the 1 F or greater range. A 1 F capacitor can be as large as a typical flashlight, requiring that the housing for the system be quite large. Most capacitors in electronic systems are the size of a thumbnail or smaller.

If we form the ratio of the equation for the capacitance of a capacitor with a specific dielectric to that of the same capacitor with air as the dielectric, the following results:

$$\frac{C = \epsilon\dfrac{A}{d}}{C_o = \epsilon_o\dfrac{A}{d}} \Rightarrow \frac{C}{C_o} = \frac{\epsilon}{\epsilon_o} = \epsilon_r$$

and

$$C = \epsilon_r C_o \tag{10.12}$$

The result is that

the capacitance of a capacitor with a dielectric having a relative permittivity of ϵ_r is ϵ_r times the capacitance using air as the dielectric.

The next few examples review the concepts and equations just presented.

EXAMPLE 10.2 In Fig. 10.9, if each air capacitor in the left column is changed to the type appearing in the right column, find the new capacitance level. For each change, the other factors remain the same.

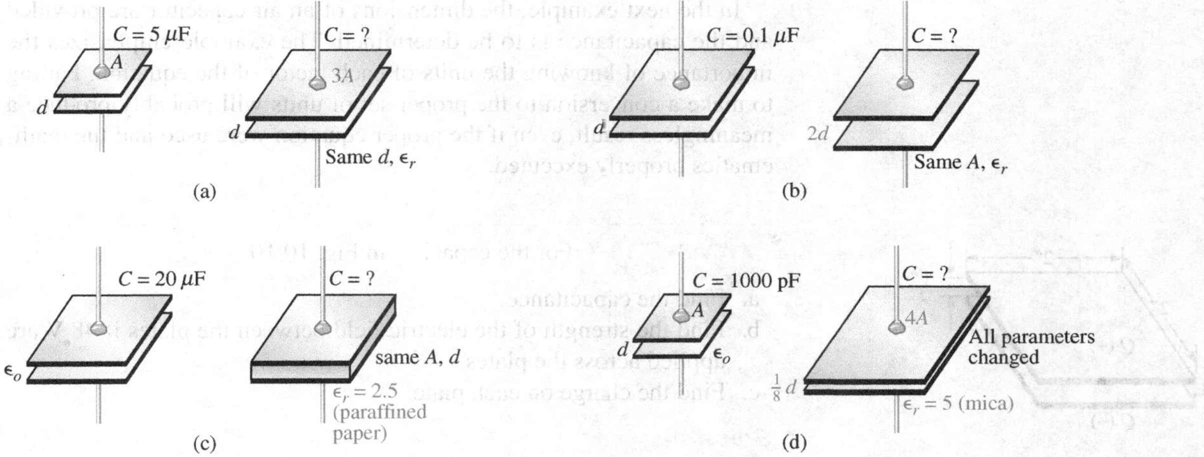

FIG. 10.9
Example 10.2.

Solutions:

a. In Fig. 10.9(a), the area has increased by a factor of three, providing more space for the storage of charge on each plate. Since the area appears in the numerator of the capacitance equation, the capacitance increases by a factor of three. That is,

$$C = 3(C_o) = 3(5\ \mu\text{F}) = \mathbf{15\ \mu F}$$

b. In Fig. 10.9(b), the area stayed the same, but the distance between the plates was increased by a factor of two. Increasing the distance reduces the capacitance level, so the resulting capacitance is one-half of what it was before. That is,

$$C = \frac{1}{2}(0.1\ \mu\text{F}) = \mathbf{0.05\ \mu F}$$

c. In Fig. 10.9(c), the area and the distance between the plates were maintained, but a dielectric of paraffined (waxed) paper was added between the plates. Since the permittivity appears in the numerator of the capacitance equation, the capacitance increases by a factor determined by the relative permittivity. That is,

$$C = \epsilon_r C_o = 2.5(20\ \mu\text{F}) = \mathbf{50\ \mu F}$$

d. In Fig. 10.9(d), a multitude of changes are happening at the same time. However, solving the problem is simply a matter of determining whether the change increases or decreases the capacitance and then placing the multiplying factor in the numerator or denominator of the equation. The increase in area by a factor of four produces a multiplier of four in the numerator, as shown in the equation below. Reducing the distance by a factor of 1/8 will increase the capacitance by its inverse, or a factor of eight. Inserting the mica dielectric increases the capacitance by a factor of five. The result is

$$C = (5)\frac{4}{(1/8)}(C_o) = 160(1000\ \text{pF}) = \mathbf{0.16\ \mu F}$$

In the next example, the dimensions of an air capacitor are provided and the capacitance is to be determined. The example emphasizes the importance of knowing the units of each factor of the equation. Failing to make a conversion to the proper set of units will probably produce a meaningless result, even if the proper equation were used and the mathematics properly executed.

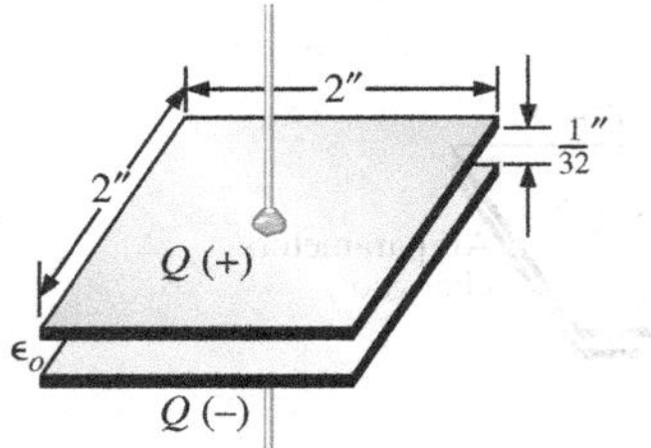

FIG. 10.10
Air capacitor for Example 10.3.

EXAMPLE 10.3 For the capacitor in Fig. 10.10:

a. Find the capacitance.
b. Find the strength of the electric field between the plates if 48 V are applied across the plates.
c. Find the charge on each plate.

Solutions:

a. First, the area and the distance between the plates must be converted to the SI system as required by Eq. (10.11):

$$d = \frac{1}{32}\,\text{in.}\left(\frac{1\text{ m}}{39.37\text{ in.}}\right) = 0.794\text{ mm}$$

and $A = (2\text{ in.})(2\text{ in.})\left(\frac{1\text{ m}}{39.37\text{ in.}}\right)\left(\frac{1\text{ m}}{39.37\text{ in.}}\right) = 2.581 \times 10^{-3}\text{ m}^2$

Eq. (10.11):

$$C = 8.85 \times 10^{-12}\,\epsilon_r \frac{A}{d} = 8.85 \times 10^{-12}(1)\frac{(2.581 \times 10^{-3}\text{m}^2)}{0.794\text{ mm}} = \mathbf{28.8\ pF}$$

b. The electric field between the plates is determined by Eq. (10.7):

$$\mathscr{E} = \frac{V}{d} = \frac{48\text{ V}}{0.794\text{ mm}} = \mathbf{60.5\ kV/m}$$

c. The charge on the plates is determined by Eq. (10.6):

$$Q = CV = (28.8\text{ pF})(48\text{ V}) = \mathbf{1.38\ nC}$$

In the next example, we will insert a ceramic dielectric between the plates of the air capacitor in Fig. 10.10 and see how it affects the capacitance level, electric field, and charge on the plates.

EXAMPLE 10.4

a. Insert a ceramic dielectric with an ϵ_r of 250 between the plates of the capacitor in Fig. 10.10. Then determine the new level of capacitance. Compare your results to the solution in Example 10.3.
b. Find the resulting electric field strength between the plates, and compare your answer to the result in Example 10.3.
c. Determine the charge on each of the plates, and compare your answer to the result in Example 10.3.

Solutions:

a. From Eq. (10.12), the new capacitance level is

$$C = \epsilon_r C_o = (250)(28.8\text{ pF}) = \mathbf{7200\ pF = 7.2\ nF = 0.0072\ \mu F}$$

which is *significantly higher* than the level in Example 10.3.

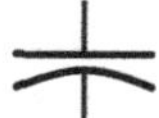

b. $\mathscr{E} = \dfrac{V}{d} = \dfrac{48\text{ V}}{0.794\text{ mm}} = \mathbf{60.5\ kV/m}$

Since the applied voltage and the distance between the plates did not change, *the electric field between the plates remains the same.*

c. $Q = CV = (7200\text{ pF})(48\text{ V}) = \mathbf{345.6\ nC} = \mathbf{0.35\ \mu C}$

We now know that the insertion of a dielectric between the plates increases the amount of charge stored on the plates. In Example 10.4, since the relative permittivity increased by a factor of 250, the charge on the plates *increased by the same amount.*

EXAMPLE 10.5 Find the maximum voltage that can be applied across the capacitor in Example 10.4 if the dielectric strength is 80 V/mil.

Solution:

$$d = \frac{1}{32}\text{ in.}\left(\frac{1000\text{ mils}}{1\text{ in.}}\right) = 31.25\text{ mils}$$

and
$$V_{\text{max}} = 31.25\text{ mils}\left(\frac{80\text{ V}}{\text{mil}}\right) = \mathbf{2.5\ kV}$$

although the provided working voltage may be only 2 kV to provide a margin of safety.

Types of Capacitors

Capacitors, like resistors, can be listed under two general headings: **fixed** and **variable.** The symbol for the fixed capacitor appears in Fig. 10.11(a). Note that the curved side is normally connected to ground or to the point of lower dc potential. The symbol for variable capacitors appears in Fig. 10.11(b).

FIG. 10.11
Symbols for the capacitor: (a) fixed; (b) variable.

Fixed Capacitors Fixed-type capacitors come in all shapes and sizes. However,

in general, for the same type of construction and dielectric, the larger the required capacitance, the larger is the physical size of the capacitor.

In Fig. 10.12(a), the 10,000 μF electrolytic capacitor is significantly larger than the 1 μF capacitor. However, it is certainly not 10,000 times larger. For the polyester-film type of Fig. 10.12(b), the 2.2 μF capacitor is significantly larger than the 0.01 μF capacitor, but again it is not 220 times larger. The 22 μF tantalum capacitor of Fig. 10.12(c) is about 6 times larger than the 1.5 μF capacitor, even though the capacitance level is about 15 times higher. It is particularly interesting to note that due to the difference in dielectric and construction, the 22 μF tantalum capacitor is significantly smaller than the 2.2 μF polyester-film capacitor and much smaller than 1/5 the size of the 100 μF electrolytic capacitor. The relatively large 10,000 μF electrolytic capacitor is normally used for high-power applications, such as in power supplies and high-output speaker systems. All the others may appear in any commercial electronic system.

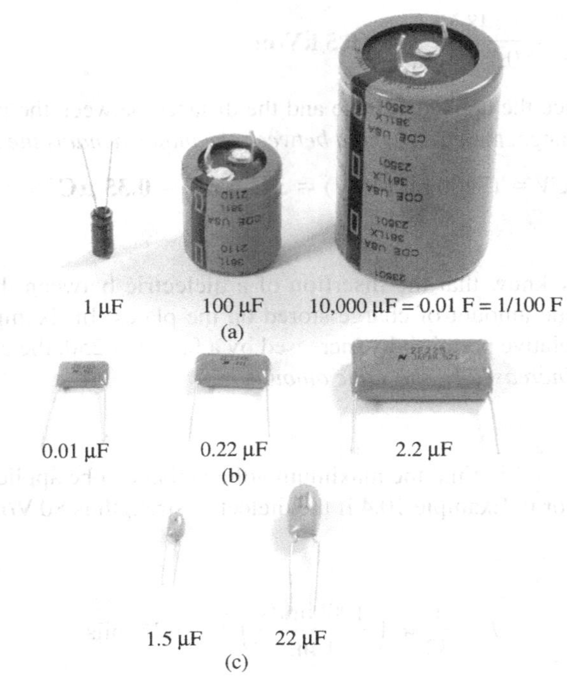

FIG. 10.12
Demonstrating that, in general, for each type of construction, the size of a capacitor increases with the capacitance value: (a) electrolytic; (b) polyester-film; (c) tantalum.

The increase in size is due primarily to the effect of area and thickness of the dielectric on the capacitance level. There are a number of ways to increase the area without making the capacitor too large. One is to lay out the plates and the dielectric in long, narrow strips and then roll them all together, as shown in Fig. 10.13(a). The dielectric (remember that it has the characteristics of an insulator) between the conducting strips ensures the strips never touch. Of course, the dielectric must be the type that can be rolled without breaking up. Depending on how the materials are wrapped, the capacitor can be either a cylindrical or a rectangular, box-type shape.

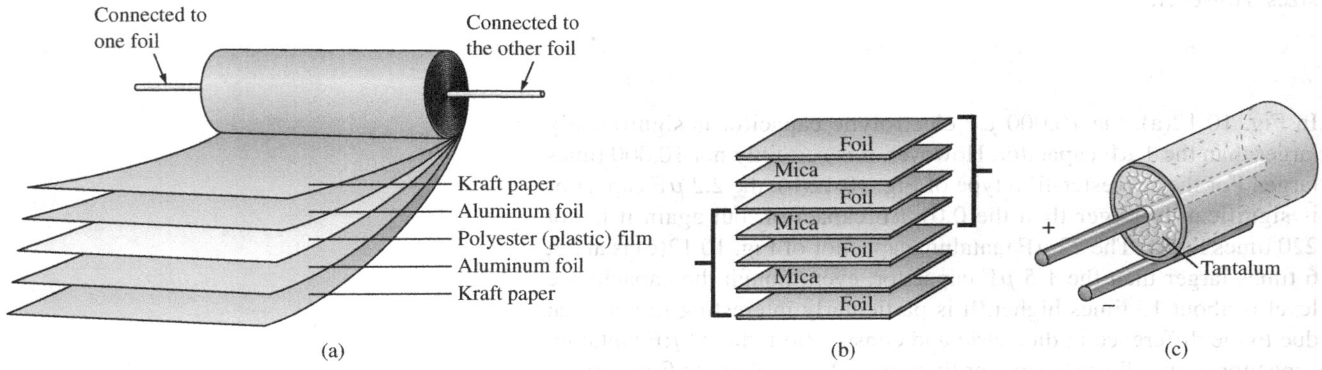

FIG. 10.13
Three ways to increase the area of a capacitor: (a) rolling; (b) stacking; (c) insertion.

A second popular method is to stack the plates and the dielectrics, as shown in Fig. 10.14(b). The area is now a multiple of the number of dielectric layers. This construction is very popular for smaller capacitors.

A third method is to use the dielectric to establish the body shape [a cylinder in Fig. 10.13(c)]. Then simply insert a rod for the positive plate, and coat the surface of the cylinder to form the negative plate, as shown in Fig. 10.13(c). Although the resulting "plates" are not the same in construction or surface area, the effect is to provide a large surface area for storage (the density of electric field lines will be different on the two "plates"), although the resulting distance factor may be larger than desired. Using a dielectric with a high ϵ_r, however, compensates for the increased distance between the plates.

There are other variations of the above to increase the area factor, but the three depicted in Fig. 10.13 are the most popular.

The next controllable factor is the distance between the plates. This factor, however, is very sensitive to how thin the dielectric can be made, with natural concerns because the working voltage (the breakdown voltage) drops as the gap decreases. Some of the thinnest dielectrics are just oxide coatings on one of the conducting surfaces (plates). A very thin polyester material, such as Mylar®, Teflon®, or even paper with a paraffin coating, provides a thin sheet of material than can easily be wrapped for increased areas. Materials such as mica and some ceramic materials can be made only so thin before crumbling or breaking down under stress.

The last factor is the dielectric, for which there is a wide range of possibilities. However, the following factors greatly influence which dielectric is used:

The level of capacitance desired
The resulting size
The possibilities for rolling, stacking, and so on
Temperature sensitivity
Working voltage

The range of relative permittivities is enormous, as shown in Table 10.2, but all the factors listed above must be considered in the construction process.

In general, the most common fixed capacitors are the electrolytic, film, polyester, foil, ceramic, mica, dipped, and oil types.

The **electrolytic capacitors** in Fig. 10.14 are usually easy to identify by their shape and the fact that they usually have a polarity marking on the body (although special-application electrolytics are available that are not polarized). Few capacitors have a polarity marking, but those that do must be connected with the negative terminal connected to ground or to the point of lower potential. The markings often used to denote the positive terminal or plate include +, □, and Δ. In general, electrolytic

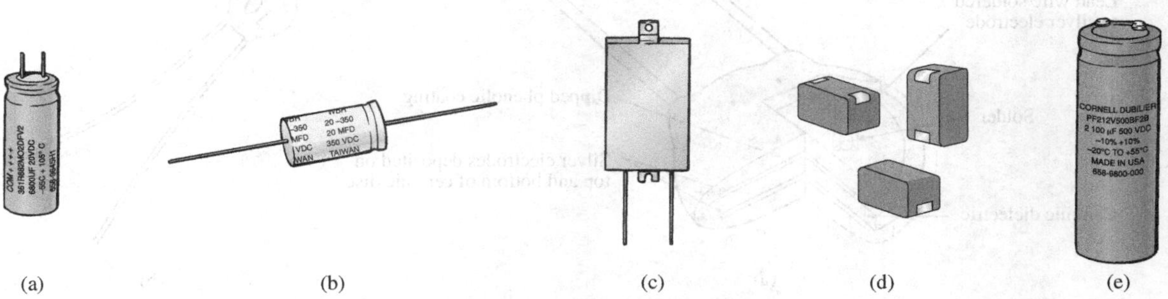

FIG. 10.14
Various types of electrolytic capacitors: (a) miniature radial leads; (b) axial leads; (c) flatpack; (d) surface-mount; (e) screw-in terminals.

capacitors offer some of the highest capacitance values available, although their working voltage levels are limited. Typical values range from 0.1 μF to 15,000 μF, with working voltages from 5 V to 450 V. The basic construction uses the rolling process in Fig. 10.13(a) in which a roll of aluminum foil is coated on one side with aluminum oxide—the aluminum being the positive plate and the oxide the dielectric. A layer of paper or gauze saturated with an electrolyte (a solution or paste that forms the conducting medium between the electrodes of the capacitor) is placed over the aluminum oxide coating of the positive plate. Another layer of aluminum without the oxide coating is then placed over this layer to assume the role of the negative plate. In most cases, the negative plate is connected directly to the aluminum container, which then serves as the negative terminal for external connections. Because of the size of the roll of aluminum foil, the overall size of the electrolytic capacitor is greater than most.

Film, polyester, foil, polypropylene, or **Teflon® capacitors** use a rolling or stacking process to increase the surface area, as shown in Fig. 10.15. The resulting shape can be either round or rectangular, with radial or axial leads. The typical range for such capacitors is 100 pF to 10 μF, with units available up to 100 μF. The name of the unit defines the type of dielectric employed. Working voltages can extend from a few volts to 2000 V, depending on the type of unit.

FIG. 10.15
(a) Film/foil polyester radial lead; (b) metalized polyester-film axial lead; (c) surface-mount polyester-film; (d) polypropylene-film, radial lead.

Ceramic capacitors (often called **disc capacitors**) use a ceramic dielectric, as shown in Fig. 10.16(a), to utilize the excellent ϵ_r values and high working voltages associated with a number of ceramic materials.

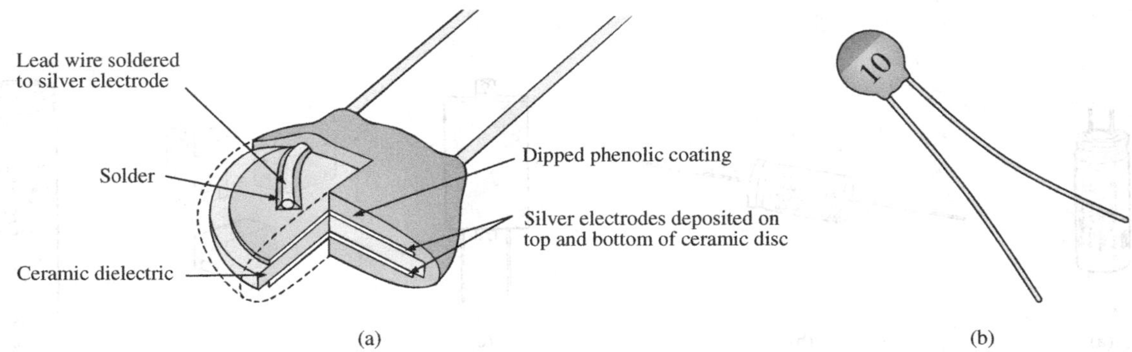

FIG. 10.16
Ceramic (disc) capacitor: (a) construction; (b) appearance.

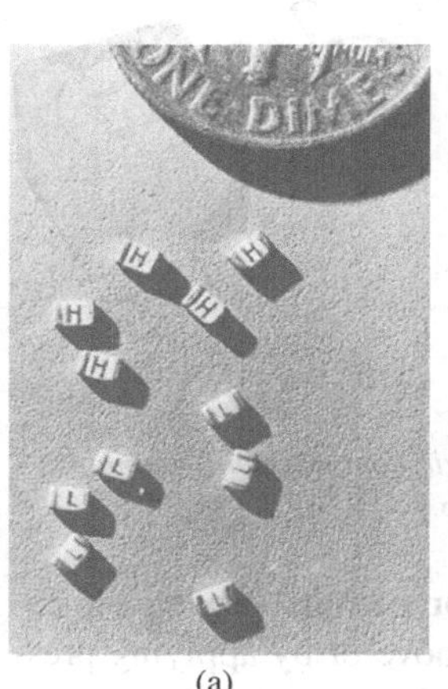

(a)

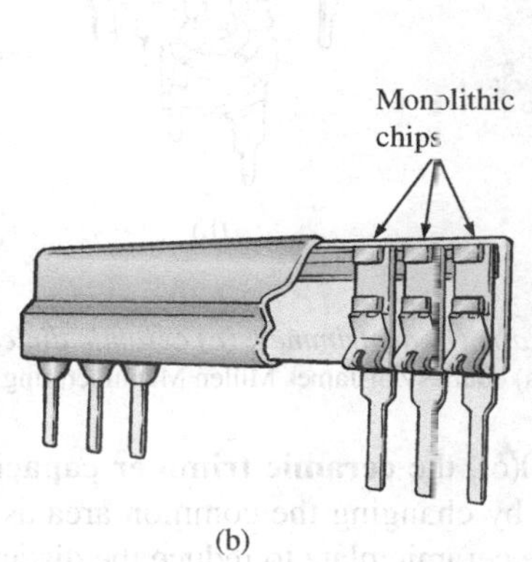

(b)

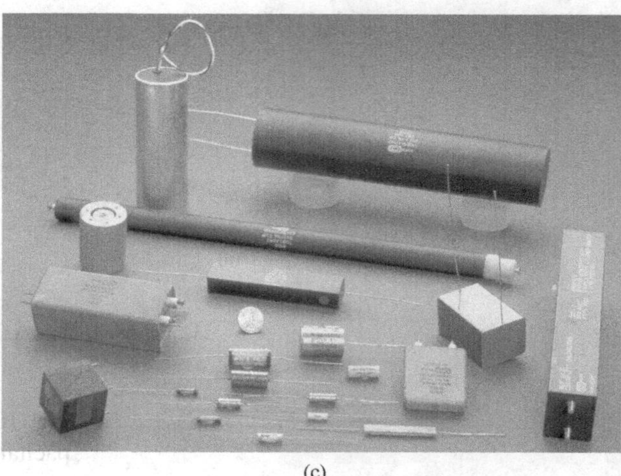

(c)

FIG. 10.17

Mica capacitors: (a) and (b) surface-mount monolithic chips; (c) high-voltage/temperature mica paper capacitors.
[(a) and (b) courtesy of Vishay Intertechnology, Inc.; (c) courtesy of Custom Electronics, Inc.]

Stacking can also be applied to increase the surface area. An example of the disc variety appears in Fig. 10.16(b). Ceramic capacitors typically range in value from 10 pF to 0.047 μF, with high working voltages that can reach as high as 10 kV.

Mica capacitors use a mica dielectric that can be monolithic (single chip) or stacked. The relatively small size of monolithic mica chip capacitors is demonstrated in Fig. 10.17(a), with their placement shown in Fig. 10.17(b). A variety of high-voltage mica paper capacitors are displayed in Fig. 10.17(c). Mica capacitors typically range in value from 2 pF to several microfarads, with working voltages up to 20 kV.

Dipped capacitors are made by dipping the dielectric (tantalum or mica) into a conductor in a molten state to form a thin, conductive sheet on the dielectric. Due to the presence of an electrolyte in the manufacturing process, dipped tantalum capacitors require a polarity marking to ensure that the positive plate is always at a higher potential than the negative plate, as shown in Fig. 10.18(a). A series of small positive signs is typically applied to the casing near the positive lead. A group of nonpolarized, mica dipped capacitors are shown in Fig. 10.18(b). They typically range in value from 0.1 μF to 680 μF, but with lower working voltages ranging from 6 V to 50 V.

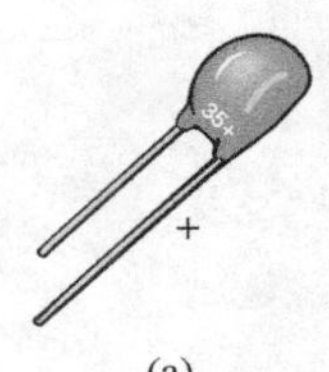

(a)

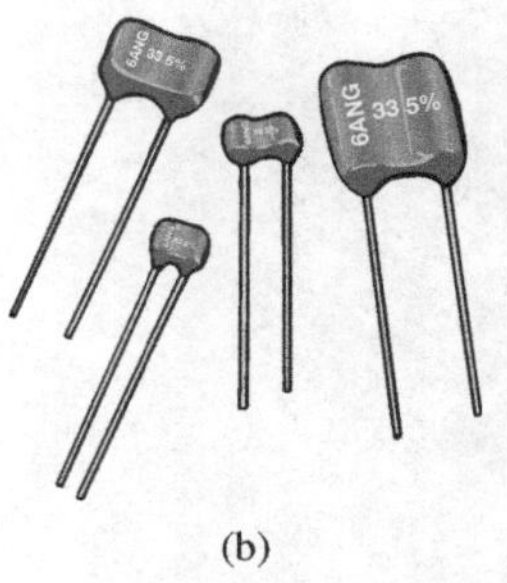

(b)

FIG. 10.18

Dipped capacitors: (a) polarized tantalum; (b) nonpolarized mica.

Most **oil capacitors** such as appearing in Fig. 10.19 are used for industrial applications such as welding, high-voltage power supplies, surge protection, and power-factor correction (Chapter 19). They can provide capacitance levels extending from 0.001 μF all the way up to 10,000 μF, with working voltages up to 150 kV. Internally, there are a number of parallel plates sitting in a bath of oil or oil-impregnated material (the dielectric).

FIG. 10.19

Oil-filled, metallic oval case snubber capacitor (the snubber removes unwanted voltage spikes).

Variable Capacitors All the parameters in Eq (10.11) can be changed to some degree to create a **variable capacitor.** For example, in Fig. 10.20(a), the capacitance of the variable air capacitor is changed by turning the shaft at the end of the unit. By turning the shaft, you control the amount of common area between the plates: The less common area there is, the lower is the capacitance. In Fig. 10 20(b), we have a much smaller **air trimmer capacitor.** It works under the same principle, but the rotating blades are totally hidden inside the structure. In

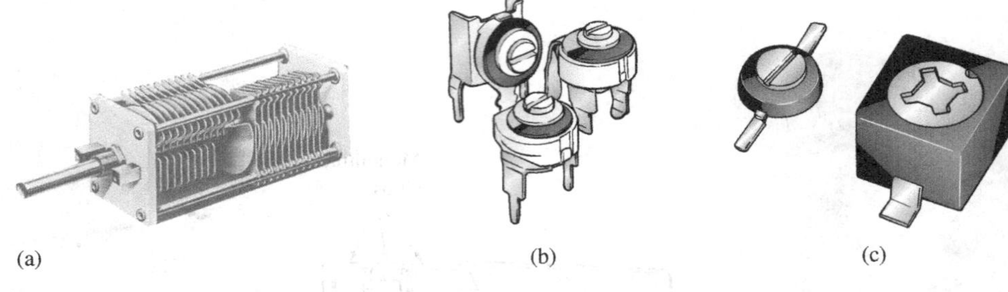

FIG. 10.20
Variable capacitors: (a) air; (b) air trimmer; (c) ceramic dielectric compression trimmer.
[(a) courtesy of James Millen Manufacturing Co.]

Fig. 10.20(c), the **ceramic trimmer capacitor** permits varying the capacitance by changing the common area as above or by applying pressure to the ceramic plate to reduce the distance between the plates.

Leakage Current and ESR

Although we would like to think of capacitors as ideal elements, unfortunately, this is not the case. There is a dc resistance appearing as R_s in the equivalent model of Fig. 10.21 due to the resistance introduced by the contacts, the leads, or the plate or foil materials. In addition, up to this point, we have assumed that the insulating characteristics of dielectrics prevent any flow of charge between the plates unless the breakdown voltage is exceeded. In reality, however, dielectrics are not perfect insulators, and they do carry a few free electrons in their atomic structure.

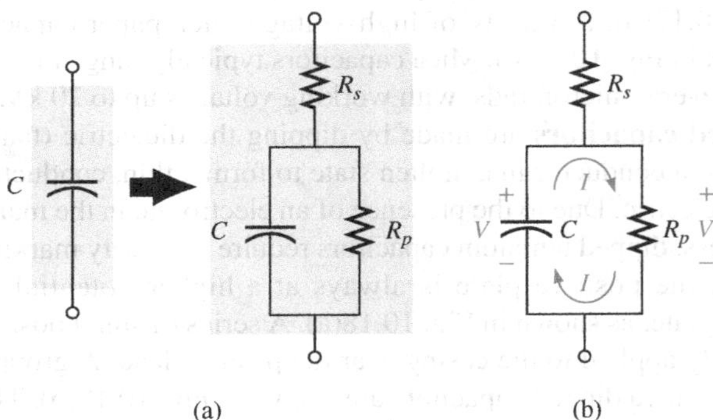

FIG. 10.21
Leakage current: (a) including the leakage resistance in the equivalent model for a capacitor; (b) internal discharge of a capacitor due to the leakage current.

When a voltage is applied across a capacitor, a **leakage current** is established between the plates. This current is usually so small that it can be ignored for the application under investigation. The availability of free electrons to support current flow is represented by a large parallel resistor R_p in the equivalent circuit for a capacitor as shown in Fig. 10.21(a). If we apply 10 V across a capacitor with an internal resistance of 1000 MΩ, the current will be 0.01 μA—a level that can be ignored for most applications.

The real problem associated with leakage currents is not evident until you ask the capacitors to sit in a charged state for long periods of time. As shown in Fig. 10.21(b), the voltage ($V = Q/C$) across a charged capacitor also appears across the parallel leakage resistance and establishes a discharge current through the resistor. In time, the capacitor is totally

discharged. Capacitors such as the electrolytic that have high leakage currents (a leakage resistance of 0.5 MΩ is typical) usually have a limited shelf life due to this internal discharge characteristic. Ceramic, tantalum, and mica capacitors typically have unlimited shelf life due to leakage resistances in excess of 1000 MΩ. Thin-film capacitors have lower levels of leakage resistances that result in some concern about shelf life.

There is another quantity of importance when defining the complete capacitive equivalent: the **equivalent series resistance (ESR).** It is a quantity of such importance to the design of switching and linear power supplies that it holds equal weight with the actual capacitance level. It is a frequency-sensitive characteristic that will be examined in Chapter 14 after the concept of frequency response has been introduced in detail. As the name implied, it is included in the equivalent model for the capacitor as a series resistor that includes all the dissipative factors in an actual capacitor that go beyond just the dc resistance.

Temperature Effects: ppm

Every capacitor is temperature sensitive, with the nameplate capacitance level specified at room temperature. Depending on the type of dielectric, increasing or decreasing temperatures can cause either a drop or a rise in capacitance. If temperature is a concern for a particular application, the manufacturer will provide a temperature plot, such as shown in Fig. 10.22, or a **ppm/°C** (parts per million per degree Celsius) rating for the capacitor. Note in Fig. 10.20 the 0% variation from the nominal (nameplate) value at 25°C (room temperature). At 0°C (freezing), it has dropped 20%, while at 100°C (the boiling point of water), it has dropped 70%—a factor to consider for some applications.

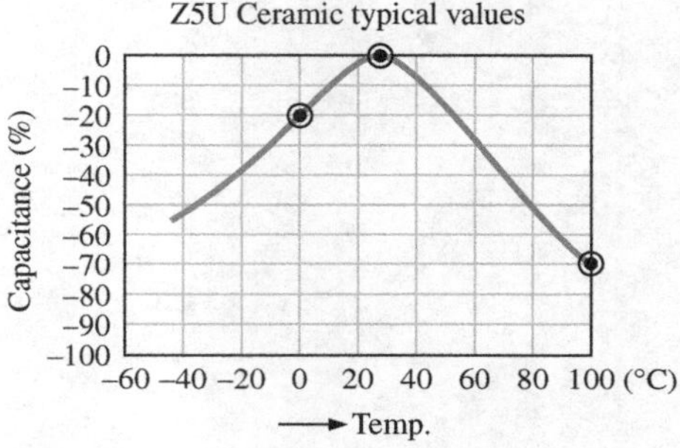

FIG. 10.22
Variation of capacitor value with temperature.

As an example of using the ppm level, consider a 100 μF capacitor with a **temperature coefficient** or **ppm** of −150 ppm/°C. It is important to note the negative sign in front of the ppm value because it reveals that the capacitance will drop with increase in temperature. It takes a moment to fully appreciate a term such as *parts per million.* In equation form, 150 parts per million can be written as

$$\frac{150}{1{,}000{,}000} \times$$

If we then multiply this term by the capacitor value, we can obtain the change in capacitance for each 1°C change in temperature. That is,

$$-\frac{150}{1{,}000{,}000}(100\ \mu\text{F})/°\text{C} = \mathbf{-0.015\ \mu F/°C} = \mathbf{-15{,}000\ pF/°C}$$

If the temperature should rise by 25°C, the capacitance would decrease by

$$-\frac{15{,}000\ \text{pF}}{°\cancel{\text{C}}}(25\ °\cancel{\text{C}}) = \mathbf{-0.38\ \mu F}$$

changing the capacitance level to

$$100\ \mu\text{F} - 0.38\ \mu\text{F} = \mathbf{99.62\ \mu F}$$

Capacitor Labeling

Due to the small size of some capacitors, various marking schemes have been adopted to provide the capacitance level, tolerance, and, if possible, working voltage. In general, however, as pointed out above, *the size of*

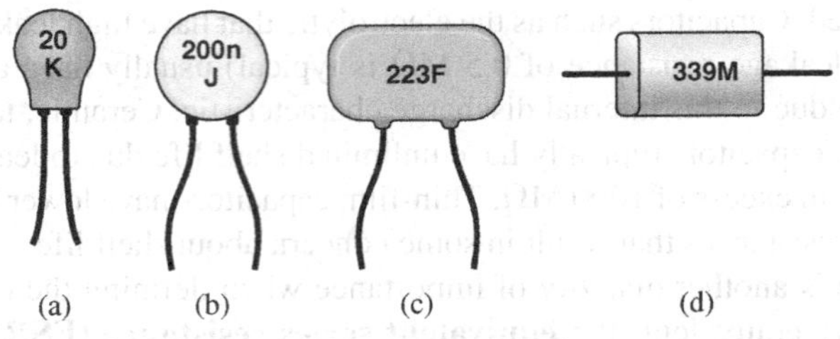

FIG. 10.23
Various marking schemes for small capacitors.

the capacitor is the first indicator of its value. In fact, most marking schemes do not indicate whether it is in μF or pF. It is assumed that you can make that judgment purely from the size. The smaller units are typically in pF and the larger units in μF. Unless indicated by an **n** or **N,** most units are not provided in nF. On larger μF units, the value can often be printed on the jacket with the tolerance and working voltage. However, smaller units need to use some form of abbreviation as shown in Fig. 10.23. For very small units such as those in Fig. 10.23(a) with only two numbers, the value is recognized immediately as being in pF with the **K** an indicator of a $\pm 10\%$ tolerance level. Too often the K is read as a multiplier of 10^3, and the capacitance is read as 20,000 pF or 20 nF rather than the actual 20 pF.

For the unit in Fig. 10.23(b), there was room for a lowercase **n** to represent a multiplier of 10^{-9}, resulting in a value of 200 nF. To avoid unnecessary confusion, the letters used for tolerance do not include **N, U,** or **P,** so the presence of any of these letters in upper- or lowercase normally refers to the multiplier level. The **J** appearing on the unit in Fig. 10.23(b) represents a $\pm 5\%$ tolerance level. For the capacitor in Fig. 10.23(c), the first two numbers are the numerical value of the capacitor, while the third number is the power of the multiplier (or number of zeros to be added to the first two numbers). The question then remains whether the units are μF or pF. With the 223 representing a number of 22,000, the units are certainly not μF because the unit is too small for such a large capacitance. It is a 22,000 pF = 22 nF capacitor. The **F** represents a $\pm 1\%$ tolerance level. Multipliers of 0.01 use an 8 for the third digit, while multipliers of 0.1 use a 9. The capacitor in Fig. 10.23(d) is a $33 \times 0.1 = 3.3\ \mu$F capacitor with a tolerance of $\pm 20\%$ as defined by the capital letter **M.** The capacitance is not 3.3 pF because the unit is too large; again, the factor of size is very helpful in making a judgment about the capacitance level. It should also be noted that **MFD** is sometimes used to signify microfarads.

Measurement and Testing of Capacitors

The capacitance of a capacitor can be read directly using a meter such as the Universal LCR Meter in Fig. 10.24. If you set the meter on **C** for *capacitance,* it will automatically choose the most appropriate unit of measurement for the element, that is, F, μF, nF, or pF. Note the polarity markings on the meter for capacitors that have a specified polarity.

FIG. 10.24
Digital reading capacitance meter.
(B&K Precision Corp.)

The best check is to use a meter such as the one in Fig. 10.24. However, if it is unavailable, an ohmmeter can be used to determine whether the dielectric is still in good working order or whether it has deteriorated due to age or use (especially for paper and electrolytics). As the dielectric breaks down, the insulating qualities of the material decrease to the point where the resistance between the plates drops to a relatively low

level. To use an ohmmeter, be sure that the capacitor is fully discharged by placing a lead directly across its terminals. Then hook up the meter (paying attention to the polarities if the unit is polarized) as shown in Fig. 10.25, and note whether the resistance has dropped to a relatively low value (0 to a few kilohms). If so, the capacitor should be discarded. You may find that the reading changes when the meter is first connected. This change is due to the charging of the capacitor by the internal supply of the ohmmeter. In time the capacitor becomes stable, and the correct reading can be observed. Typically, it should pin at the highest level on the megohm scales or indicate OL on a digital meter.

The above ohmmeter test is not all-inclusive because some capacitors exhibit the breakdown characteristics only when a large voltage is applied. The test, however, does help isolate capacitors in which the dielectric has deteriorated.

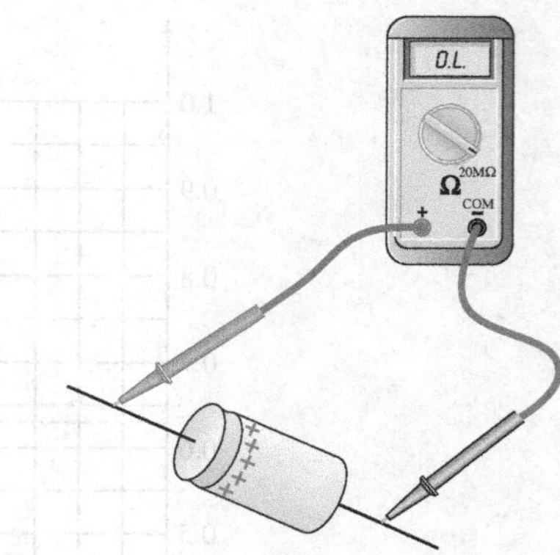

FIG. 10.25
Checking the dielectric of an electrolytic capacitor.

Standard Capacitor Values

The most common capacitors use the same numerical multipliers encountered for resistors.

The vast majority are available with 5%, 10%, or 20% tolerances. There are capacitors available, however, with tolerances of 1%, 2%, or 3%, if you are willing to pay the price. Typical values include 0.1 μF, 0.15 μF, 0.22 μF, 0.33 μF, 0.47 μF, 0.68 μF; and 1 μF, 1.5 μF, 2.2 μF, 3.3 μF, 4.7 μF, 6.8 μF; and 10 pF, 22 pF, 33 pF, 100 pF; and so on.

10.5 TRANSIENTS IN CAPACITIVE NETWORKS: THE CHARGING PHASE

The placement of charge on the plates of a capacitor does not occur instantaneously. Instead, it occurs over a period of time determined by the components of the network. The charging phase—the phase during which charge is deposited on the plates—can be described by reviewing the response of the simple series circuit in Fig. 10.4. The circuit has been redrawn in Fig. 10.26 with the symbol for a fixed capacitor.

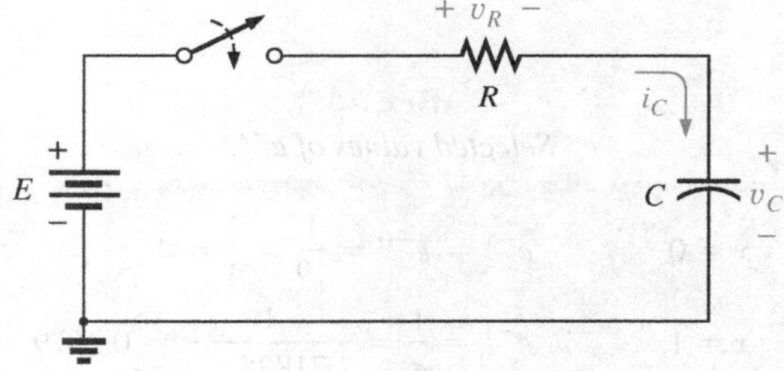

FIG. 10.26
Basic R-C charging network.

Recall that the instant the switch is closed, electrons are drawn from the top plate and deposited on the bottom plate by the battery, resulting in a net positive charge on the top plate and a negative charge on the bottom plate. The transfer of electrons is very rapid at first, slowing down as the potential across the plates approaches the applied voltage of the battery. Eventually, when the voltage across the capacitor equals the applied voltage, the transfer of electrons ceases, and the plates have a net charge determined by $Q = CV_C = CE$. This period of time during which charge is being deposited on the plates is called the **transient period**—a period of time where the voltage or current changes from one steady-state level to another.

Since the voltage across the plates is directly related to the charge on the plates by $V = Q/C$, a plot of the voltage across the capacitor will have the same shape as a plot of the charge on the plates over time. As shown in Fig. 10.27, the voltage across the capacitor is zero volts when the switch is closed ($t = 0$ s). It then builds up very quickly at first since charge is being deposited at a very high rate of speed. As time passes, the charge is deposited at a slower rate, and the change in voltage drops off. The voltage continues to grow, but at a much slower rate. Eventually, as the voltage across the plates approaches the applied voltage, the charging rate is very slow, until finally the voltage across the plates is equal to the applied voltage—the transient phase has passed.

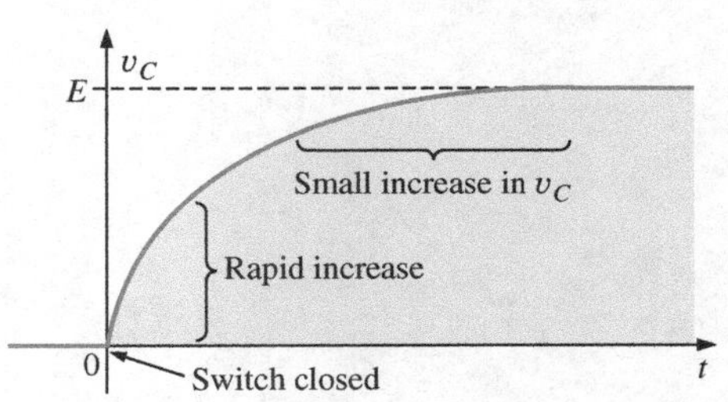

FIG. 10.27
v_C during the charging phase.

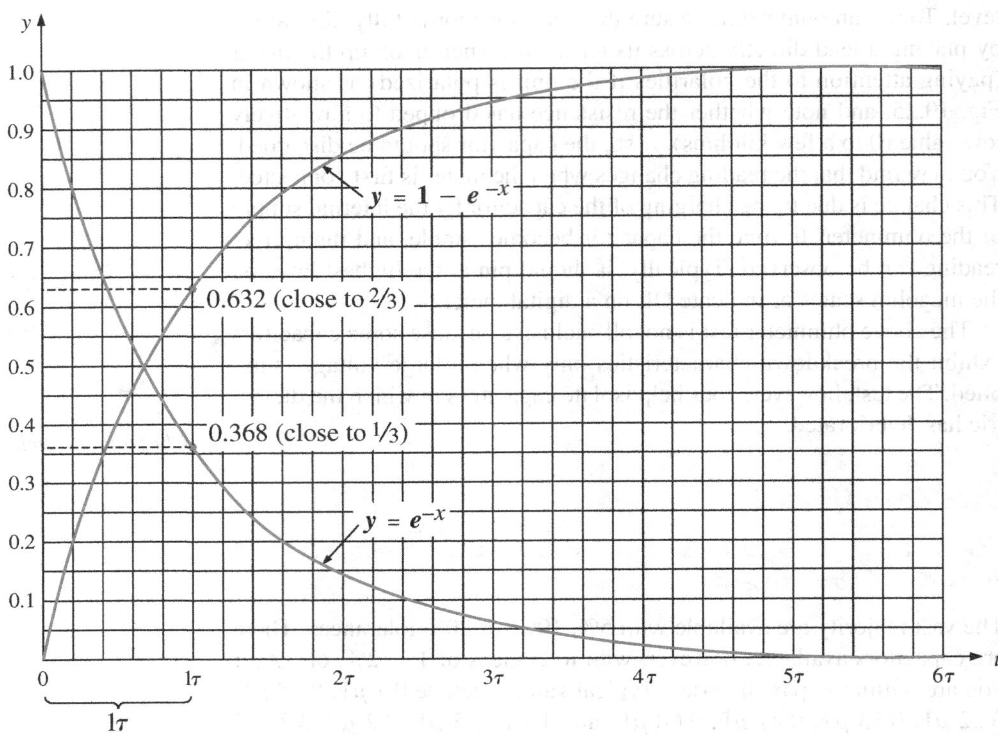

FIG. 10.28
Universal time constant chart.

TABLE 10.3
Selected values of e^{-x}.

$x = 0$	$e^{-x} = e^{-0} = \frac{1}{e^0} = \frac{1}{1} = 1$
$x = 1$	$e^{-1} = \frac{1}{e} = \frac{1}{2.71828\ldots} = 0.3679$
$x = 2$	$e^{-2} = \frac{1}{e^2} = 0.1353$
$x = 5$	$e^{-5} = \frac{1}{e^5} = 0.00674$
$x = 10$	$e^{-10} = \frac{1}{e^{10}} = 0.0000454$
$x = 100$	$e^{-100} = \frac{1}{e^{100}} = 3.72 \times 10^{-44}$

Fortunately, the waveform in Fig. 10.27 from beginning to end can be described using the mathematical function e^{-x}. It is an exponential function that decreases with time, as shown in Fig. 10.28. If we substitute zero for x, we obtain e^{-0}, which by definition is 1, as shown in Table 10.3 and on the plot in Fig. 10.28. Table 10.3 reveals that as x increases, the function e^{-x} decreases in magnitude until it is very close to zero after $x = 5$. As noted in Table 10.3, the exponential factor $e^1 = e = 2.71828$.

A plot of $1 - e^{-x}$ is also provided in Fig. 10.28 since it is a component of the voltage v_C in Fig. 10.27. When e^{-x} is 1, $1 - e^{-x}$ is zero, as shown in Fig. 10.28, and when e^{-x} decreases in magnitude, $1 - e^{-x}$ approaches 1, as shown in the same figure.

You may wonder how this function can help us if it decreases with time and the curve for the voltage across the capacitor increases with time. We simply place the exponential in the proper mathematical form as follows:

$$\boxed{v_C = E(1 - e^{-t/\tau})}_{\text{charging}} \qquad \text{(volts, V)} \qquad \textbf{(10.13)}$$

First note in Eq. (10.13) that the voltage v_C is written in *lowercase (not capital) italic* to point out that it is a function that will change with time—it is not a constant. The exponent of the exponential function is no longer just x, but now is time (t) divided by a constant τ, the Greek letter *tau*. The quantity τ is defined by

$$\boxed{\tau = RC} \qquad \text{(time, s)} \qquad \textbf{(10.14)}$$

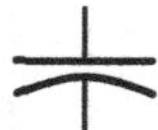

The factor τ, called the **time constant** of the network, has the units of time, as shown below using some of the basic equations introduced earlier in this text:

$$\tau = RC = \left(\frac{V}{I}\right)\left(\frac{Q}{V}\right) = \left(\frac{\cancel{V}}{\cancel{Q}/t}\right)\left(\frac{\cancel{Q}}{\cancel{V}}\right) = t \text{ (seconds)}$$

A plot of Eq. (10.13) results in the curve in Fig. 10.29, whose shape is an exact match with that in Fig. 10.27.

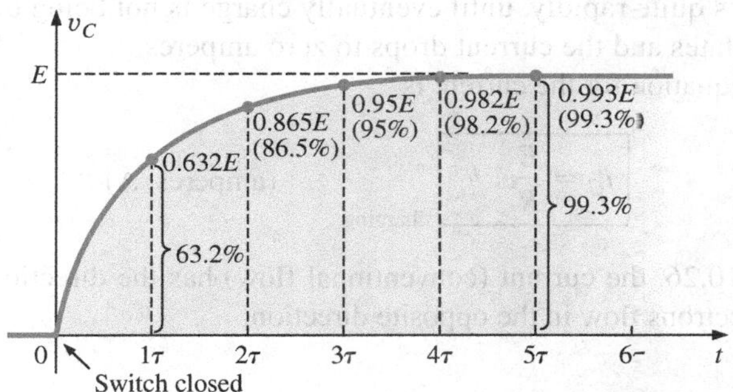

FIG. 10.29
Plotting the equation $v_C = E(1 - e^{-t/\tau})$ versus time (t).

In Eq. (10.13), if we substitute $t = 0$ s, we find that

$$e^{-t/\tau} = e^{-0/\tau} = e^{-0} = \frac{1}{e^0} = \frac{1}{1} = 1$$

and $$v_C = E(1 - e^{-t/\tau}) = E(1 - 1) = \mathbf{0\ V}$$

as appearing in the plot in Fig. 10.29.

It is important to realize at this point that the plot in Fig. 10.29 is not against simply time but against τ, the time constant of the network. If we want to know the voltage across the plates after one time constant, we simply plug $t = 1\tau$ into Eq. (10.13). The result is

$$e^{-t/\tau} = e^{-\tau/\tau} = e^{-1} \cong 0.368$$

and $$v_C = E(1 - e^{-t/\tau}) = E(1 - 0.368) = \mathbf{0.632E}$$

as shown in Fig. 10.29.

At $t = 2\tau$

$$e^{-t/\tau} = e^{-2\tau/\tau} = e^{-2} \cong 0.135$$

and $$v_C = E(1 - e^{-t/\tau}) = E(1 - 0.135) \cong \mathbf{0.865E}$$

as shown in Fig. 10.29.

As the number of time constants increases, the voltage across the capacitor does indeed approach the applied voltage.

At $t = 5\tau$

$$e^{-t/\tau} = e^{-5\tau/\tau} = e^{-5} \cong 0.007$$

and $$v_C = E(1 - e^{-t/\tau}) = E(1 - 0.007) = \mathbf{0.993E} \cong E$$

In fact, we can conclude from the results just obtained that

the voltage across a capacitor in a dc network is essentially equal to the applied voltage after five time constants of the charging phase have passed.

Or, in more general terms,

the transient or charging phase of a capacitor has essentially ended after five time constants.

It is indeed fortunate that the same exponential function can be used to plot the current of the capacitor versus time. When the switch is first closed, the flow of charge or current jumps very quickly to a value limited by the applied voltage and the circuit resistance, as shown in Fig. 10.30. The rate of deposit, and hence the current, then decreases quite rapidly, until eventually charge is not being deposited on the plates and the current drops to zero amperes.

The equation for the current is

$$\boxed{i_C = \frac{E}{R}e^{-t/\tau}}_{\text{charging}} \qquad \text{(amperes, A)} \qquad \textbf{(10.15)}$$

In Fig. 10.26, the current (conventional flow) has the direction shown since electrons flow in the opposite direction.

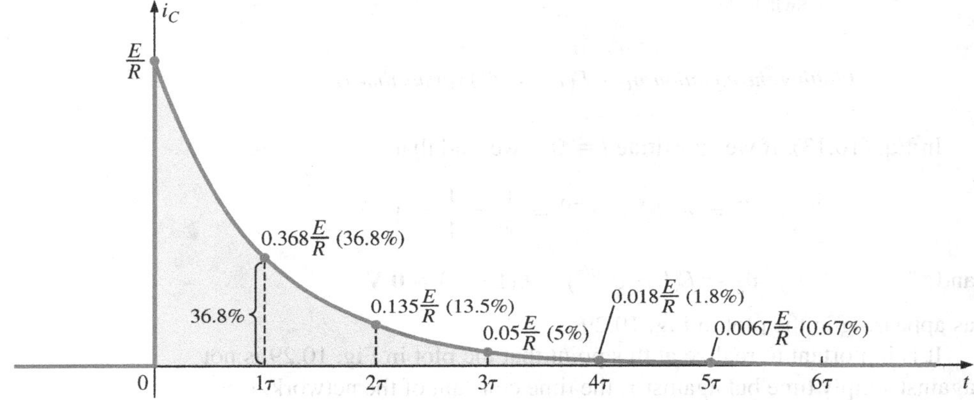

FIG. 10.30

Plotting the equation $i_C = \frac{E}{R}e^{-t/\tau}$ versus time (t).

At $t = 0$ s

$$e^{-t/\tau} = e^{-0} = 1$$

and
$$i_C = \frac{E}{R}e^{-t/\tau} = \frac{E}{R}(1) = \frac{\boldsymbol{E}}{\boldsymbol{R}}$$

At $t = 1\tau$

$$e^{-t/\tau} = e^{-\tau/\tau} = e^{-1} \cong 0.368$$

and
$$i_C = \frac{E}{R}e^{-t/\tau} = \frac{E}{R}(0.368) = \mathbf{0.368}\frac{\boldsymbol{E}}{\boldsymbol{R}}$$

In general, Fig. 10.30 clearly reveals that

the current of a capacitive dc network is essentially zero amperes after five time constants of the charging phase have passed.

It is also important to recognize that

during the charging phase, the major change in voltage and current occurs during the first time constant.

The voltage across the capacitor reaches about 63.2% (about 2/3) of its final value, whereas the current drops to 36.8% (about 1/3) of its peak value. During the next time constant, the voltage increases only about 23.3%, whereas the current drops to 13.5%. The first time constant is therefore a very dramatic time for the changing parameters. Between the fourth and fifth time constants, the voltage increases only about 1.2%, whereas the current drops to less than 1% of its peak value.

Returning to Figs. 10.29 and 10.30, note that when the voltage across the capacitor reaches the applied voltage E, the current drops to zero amperes, as reviewed in Fig. 10.31. These conditions match those of an open circuit, permitting the following conclusion:

A capacitor can be replaced by an open-circuit equivalent once the charging phase in a dc network has passed.

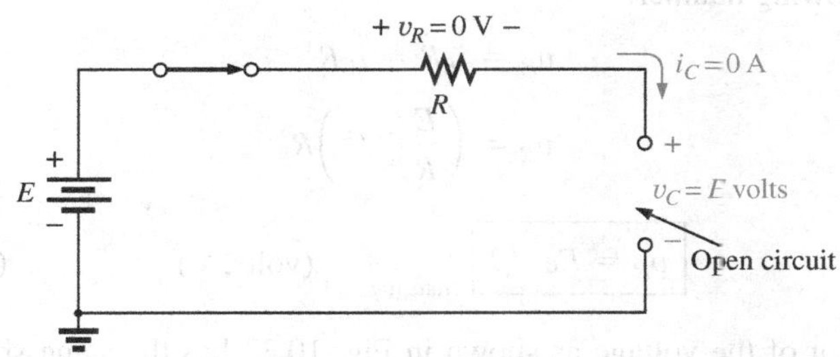

FIG. 10.31

Demonstrating that a capacitor has the characteristics of an open circuit after the charging phase has passed.

This conclusion will be particularly useful when analyzing dc networks that have been on for a long period of time or have passed the transient phase that normally occurs when a system is first turned on.

A similar conclusion can be reached if we consider the instant the switch is closed in the circuit in Fig. 10.26. Referring to Figs. 10.29 and 10.30 again, we find that the current is a peak value at $t = 0$ s, whereas the voltage across the capacitor is 0 V, as shown in the equivalent circuit in Fig. 10.32. The result is that

a capacitor has the characteristics of a short-circuit equivalent at the instant the switch is closed in an uncharged series R-C circuit.

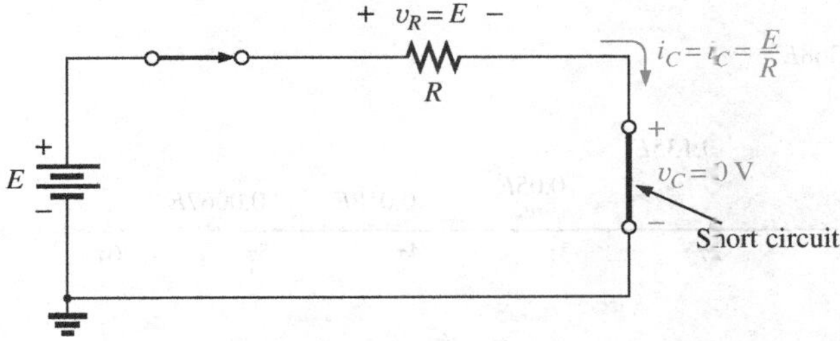

FIG. 10.32

Revealing the short-circuit equivalent for the capacitor that occurs when the switch is first closed.

In Eq. (10.13), the time constant τ will always have some value because some resistance is always present in a capacitive network. In some cases, the value of τ may be very small, but five times that value of τ, no

matter how small, must therefore always exist; it cannot be zero. The result is the following very important conclusion:

The voltage across a capacitor cannot change instantaneously.

In fact, we can take this statement a step further by saying that the capacitance of a network is a measure of how much it will oppose a change in voltage in a network. The larger the capacitance, the larger is the time constant, and the longer it will take the voltage across the capacitor to reach the applied value. This can prove very helpful when lightning arresters and surge suppressors are designed to protect equipment from unexpected high surges in voltage.

Since the resistor and the capacitor in Fig. 10.26 are in series, the current through the resistor is the same as that associated with the capacitor. The voltage across the resistor can be determined by using Ohm's law in the following manner:

$$\upsilon_R = i_R R = i_C R$$

so that

$$\upsilon_R = \left(\frac{E}{R} e^{-t/\tau}\right) R$$

and

$$\boxed{\upsilon_R = Ee^{-t/\tau}}_{\text{charging}} \qquad \text{(volts, V)} \qquad \textbf{(10.16)}$$

A plot of the voltage as shown in Fig. 10.33 has the same shape as that for the current because they are related by the constant R. Note, however, that the voltage across the resistor starts at a level of E volts because the voltage across the capacitor is zero volts and Kirchhoff's voltage law must always be satisfied. When the capacitor has reached the applied voltage, the voltage across the resistor must drop to zero volts for the same reason. Always remember that

Kirchhoff's voltage law is applicable at any instant of time for any type of voltage in any type of network.

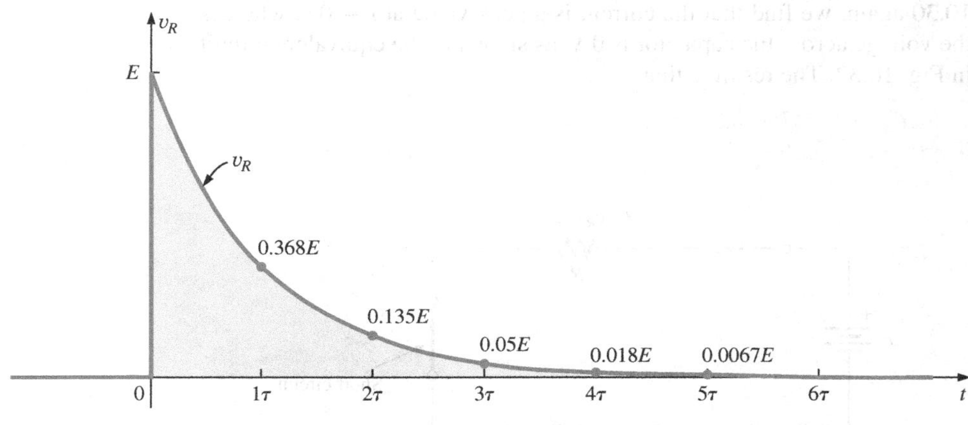

FIG. 10.33
Plotting the equation $\upsilon_R = Ee^{-t/\tau}$ versus time (t).

Using the Calculator to Solve Exponential Functions

Before looking at an example, we will first discuss the use of the TI-89 calculator with exponential functions. The process is actually quite simple for a number such as $e^{-1.2}$. Just select the 2nd function (diamond) key, followed by the function e^x. Then insert the (−) sign from the numerical

♦ e^x (–) 1 . 2) ENTER 301.2E-3

FIG. 10.34
Calculator key strokes to determine $e^{-1.2}$.

keyboard (not the mathematical functions), and insert the number 1.2 followed by ENTER to obtain the result of 0.301, as shown in Fig. 10.34. The use of the computer software program Mathcad is demonstrated in a later example.

EXAMPLE 10.6 For the circuit in Fig. 10.35:

a. Find the mathematical expression for the transient behavior of v_C, i_C, and v_R if the switch is closed at $t = 0$ s.
b. Plot the waveform of v_C versus the time constant of the network.
c. Plot the waveform of v_C versus time.
d. Plot the waveforms of i_C and v_R versus the time constant of the network.
e. What is the value of v_C at $t = 20$ ms?
f. On a practical basis, how much time must pass before we can assume that the charging phase has passed?
g. When the charging phase has passed, how much charge is sitting on the plates?
h. If the capacitor has a leakage resistance of 10,000 MΩ, what is the initial leakage current? Once the capacitor is separated from the circuit, how long will it take to totally discharge, assuming a linear (unchanging) discharge rate?

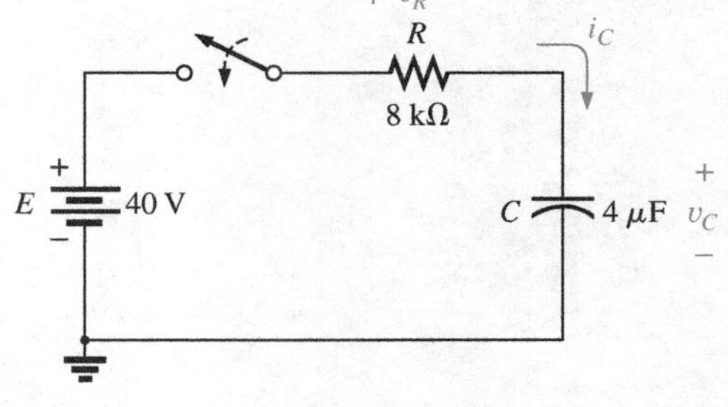

FIG. 10.35
Transient network for Example 10.6.

Solutions:

a. The time constant of the network is

$$\tau = RC = (8\text{ k}\Omega)(4\ \mu\text{F}) = 32\text{ ms}$$

resulting in the following mathematical equations:

$$v_C = E(1 - e^{-t/\tau}) = \mathbf{40\ V(1 - e^{-t/32ms})}$$

$$i_C = \frac{E}{R}e^{-t/\tau} = \frac{40\text{ V}}{8\text{ k}\Omega}e^{-t/32\text{ms}} = \mathbf{5\ mAe^{-t/32ms}}$$

$$v_R = Ee^{-t/\tau} = \mathbf{40\ Ve^{-t/32ms}}$$

b. The resulting plot appears in Fig. 10.36.
c. The horizontal scale will now be against time rather than time constants, as shown in Fig. 10.37. The plot points in Fig. 10.37 were taken from Fig. 10.36.

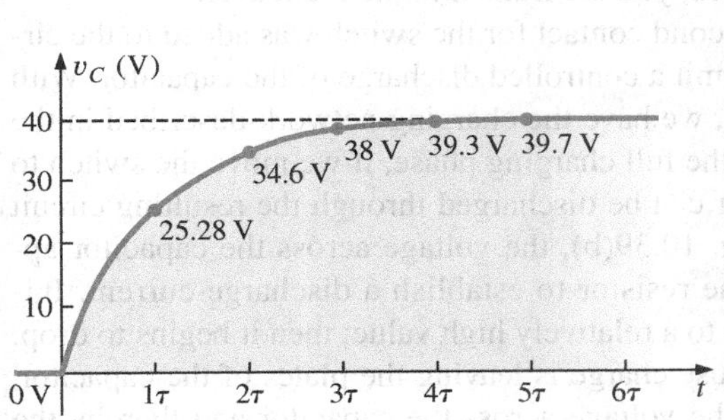

FIG. 10.36
v_C versus time for the charging network in Fig. 10.35.

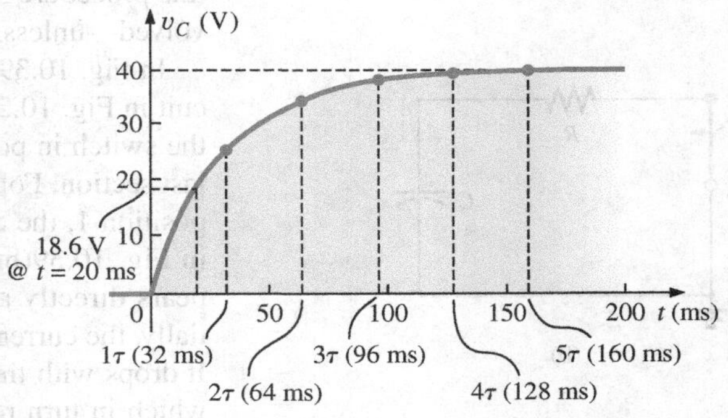

FIG. 10.37
Plotting the waveform in Fig. 10.36 versus time (t).

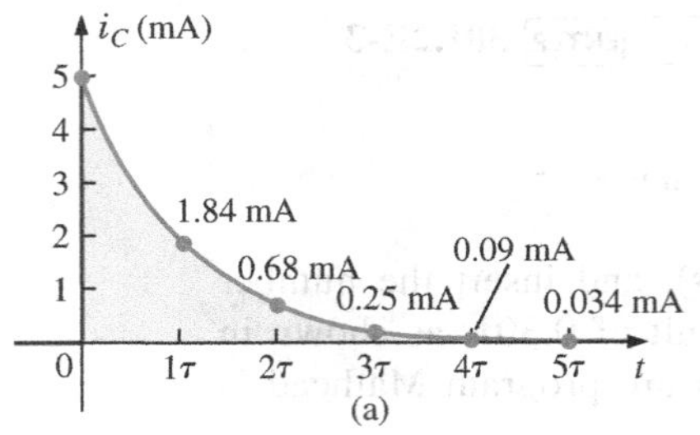

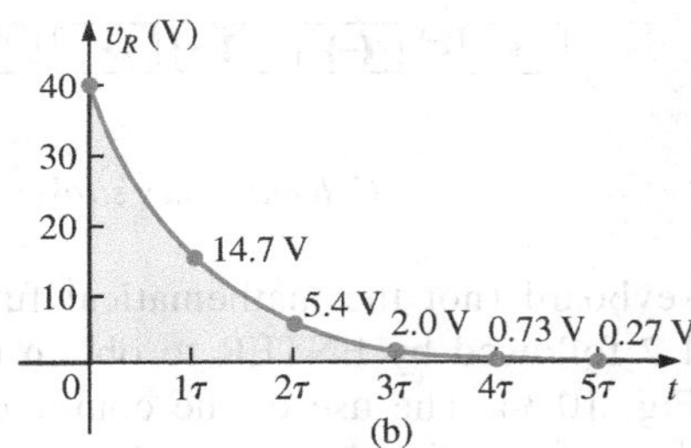

FIG. 10.38
i_C and v_R for the charging network in Fig. 10.36.

d. Both plots appear in Fig. 10.38.

e. Substituting the time $t = 20$ ms results in the following for the exponential part of the equation:

$$e^{t/\tau} = e^{-20\text{ ms}/32\text{ ms}} = e^{-0.625} = 0.535 \qquad \text{(using a calculator)}$$

so that $v_C = 40\text{ V}(1 - e^{\tau/32\text{ ms}}) = 40\text{ V }(1 - 0.535)$

$$= (40\text{ V})(0.465) = \mathbf{18.6\text{ V}} \qquad \text{(as verified by Fig. 10.37)}$$

f. Assuming a full charge in five time constants results in

$$5\tau = 5(32\text{ ms}) = \mathbf{160\text{ ms} = 0.16\text{ s}}$$

g. Using Eq. (10.6) gives

$$Q = CV = (4\ \mu\text{F})(40\text{ V}) = \mathbf{160\ \mu C}$$

h. Using Ohm's law gives

$$I_{\text{leakage}} = \frac{40\text{ V}}{10{,}000\text{ M}\Omega} = 4\text{ nA}$$

Finally, the basic equation $I = Q/t$ results in

$$t = \frac{Q}{I} = \frac{160\ \mu C}{4\text{ nA}} = (40{,}000\ \cancel{s})\left(\frac{1\ \cancel{\text{min}}}{60\ \cancel{s}}\right)\left(\frac{1\text{ h}}{60\ \cancel{\text{min}}}\right) = \mathbf{11.11\text{ h}}$$

10.6 TRANSIENTS IN CAPACITIVE NETWORKS: THE DISCHARGING PHASE

We now investigate how to discharge a capacitor while exerting some control on how long the discharge time will be. You can, of course, place a lead directly across a capacitor to discharge it very quickly—and possibly cause a visible spark. For larger capacitors such those in TV sets, this procedure should not be attempted because of the high voltages involved—unless, of course, you are trained in the maneuver.

In Fig. 10.39(a), a second contact for the switch was added to the circuit in Fig. 10.26 to permit a controlled discharge of the capacitor. With the switch in position 1, we have the charging network described in the last section. Following the full charging phase, if we move the switch to position 1, the capacitor can be discharged through the resulting circuit in Fig. 10.39(b). In Fig. 10.39(b), the voltage across the capacitor appears directly across the resistor to establish a discharge current. Initially, the current jumps to a relatively high value; then it begins to drop. It drops with time because charge is leaving the plates of the capacitor, which in turn reduces the voltage across the capacitor and thereby the voltage across the resistor and the resulting current.

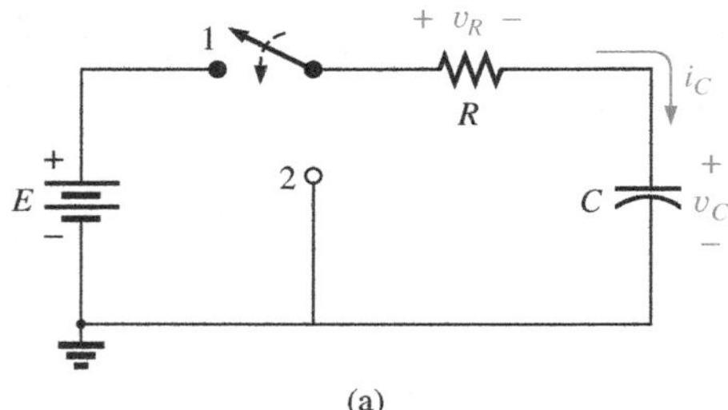

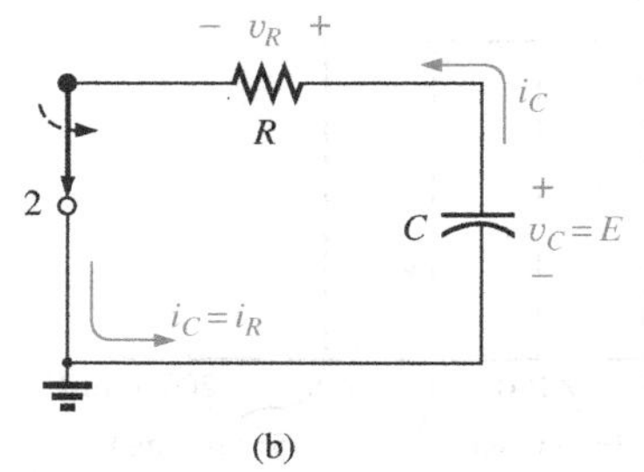

FIG. 10.39
(a) Charging network; (b) discharging configuration.

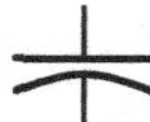

Before looking at the wave shapes for each quantity of interest, note that current i_C has now reversed direction as shown in Fig. 10.39(b). As shown in parts (a) and (b) in Fig. 10.39, the voltage across the capacitor does not reverse polarity, but the current reverses direction. We will show the reversals on the resulting plots by sketching the waveforms in the negative regions of the graph. In all the waveforms, note that all the mathematical expressions use the same e^{-x} factor appearing during the charging phase.

For the voltage across the capacitor that is decreasing with time, the mathematical expression is

$$\boxed{v_C = Ee^{-t/\tau}}_{\text{discharging}} \tag{10.17}$$

For this circuit, the time constant τ is defined by the same equation as used for the charging phase. That is,

$$\boxed{\tau = RC}_{\text{discharging}} \tag{10.18}$$

Since the current decreases with time, it will have a similar format:

$$\boxed{i_C = \frac{E}{R}e^{-t/\tau}}_{\text{discharging}} \tag{10.19}$$

For the configuration in Fig. 10.39(b), since $v_R = v_C$ (in parallel), the equation for the voltage v_R has the same format:

$$\boxed{v_R = Ee^{-t/\tau}}_{\text{discharging}} \tag{10.20}$$

The complete discharge will occur, for all practical purposes, in five time constants. If the switch is moved between terminals 1 and 2 every five time constants, the wave shapes in Fig. 10.40 will result for v_C, i_C,

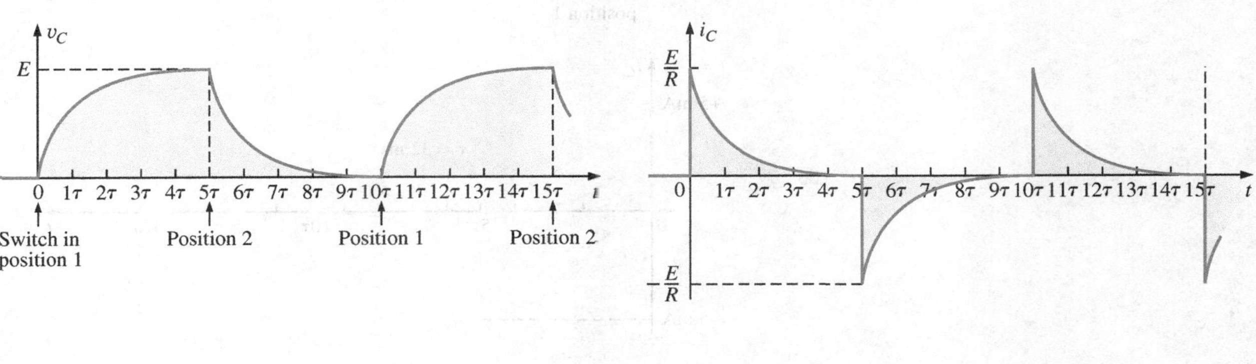

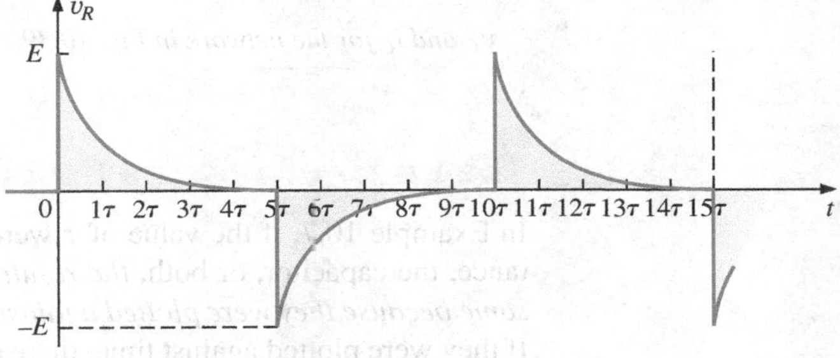

FIG. 10.40
v_C, i_C, and v_R for 5τ switching between contacts in Fig. 10.39(a).

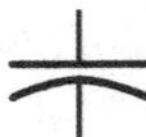

and v_R. For each curve, the current directions and voltage polarities are as defined by the configurations in Fig. 10.39. Note, as pointed out above, that the current reverses direction during the discharge phase.

The discharge rate does not have to equal the charging rate if a different switching arrangement is used. In fact, Example 10.8 will demonstrate how to change the discharge rate.

EXAMPLE 10.7 Using the values in Example 10.6, plot the waveforms for v_C and i_C resulting from switching between contacts 1 and 2 in Fig. 10.39 every five time constants.

Solution: The time constant is the same for the charging and discharging phases. That is,

$$\tau = RC = (8\ \text{k}\Omega)(4\ \mu\text{F}) = 32\ \text{ms}$$

For the discharge phase, the equations are

$$v_C = Ee^{-t/\tau} = \mathbf{40\ Ve^{-t/32\ ms}}$$

$$i_C = -\frac{E}{R}e^{-t/\tau} = \frac{40\ \text{V}}{8\ \text{k}\Omega}e^{-t/32\ \text{ms}} = \mathbf{-5\ mAe^{-t/32\ ms}}$$

$$v_R = v_C = \mathbf{40\ Ve^{-t/32\ ms}}$$

A continuous plot for the charging and discharging phases appears in Fig. 10.41.

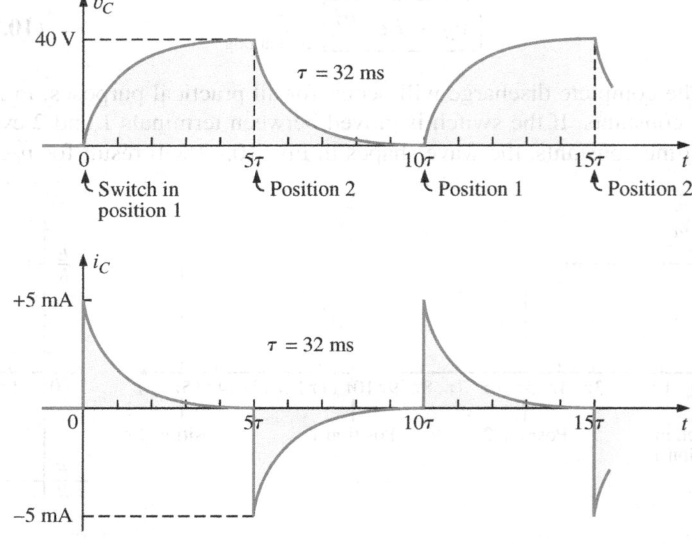

FIG. 10.41
v_C and i_C for the network in Fig. 10.39(a) with the values in Example 10.6.

The Effect of τ on the Response

In Example 10.7, if the value of τ were changed by changing the resistance, the capacitor, or both, *the resulting waveforms would appear the same because they were plotted against the time constant of the network.* If they were plotted against time, there could be a dramatic change in the appearance of the resulting plots. In fact, on an oscilloscope, an instrument designed to display such waveforms, the plots are against time, and the change will be immediately apparent. In Fig. 10.42(a), the waveforms in

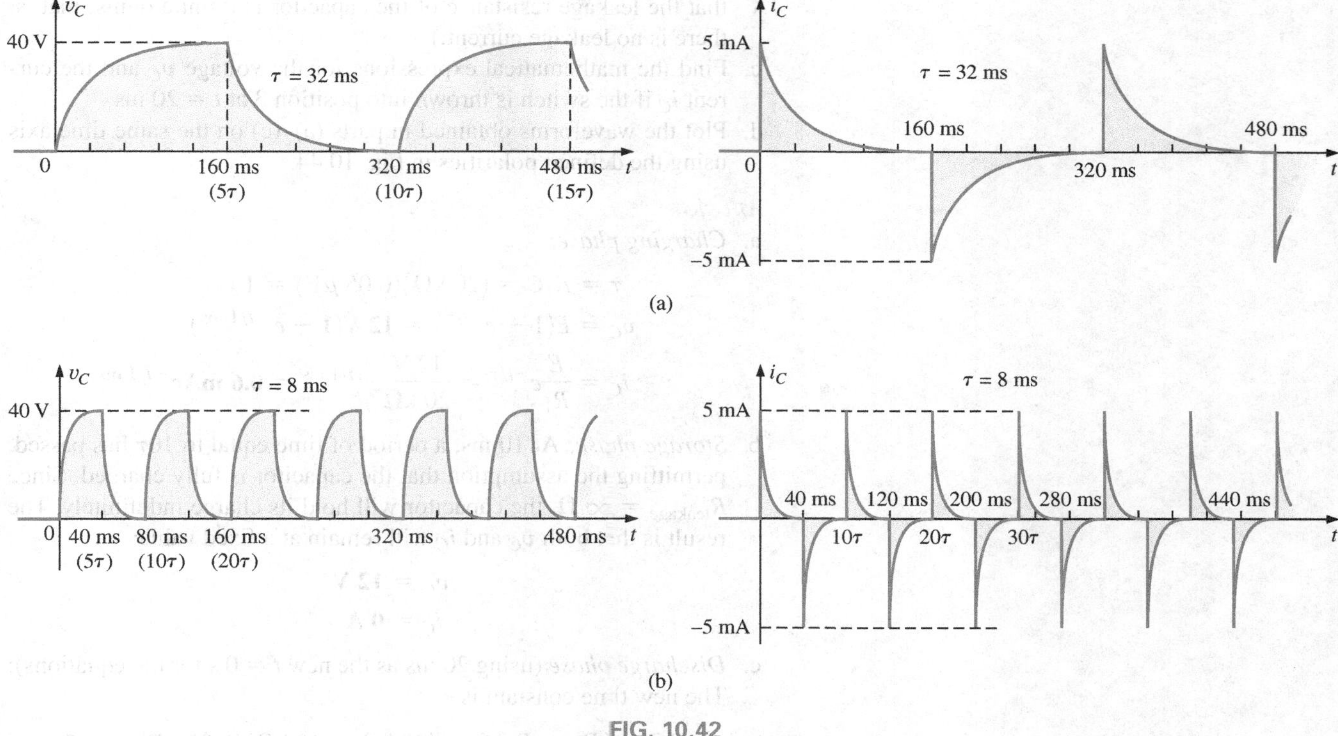

FIG. 10.42
Plotting v_C and i_C versus time in ms: (a) τ = 32 ms; (b) τ = 8 ms.

Fig. 10.41 for v_C and i_C were plotted against time. In Fig. 10.42(b), the capacitance was decreased to 1 μF, which reduces the time constant to 8 ms. Note the dramatic effect on the appearance of the waveform.

For a fixed-resistance network, the effect of increasing the capacitance is clearly demonstrated in Fig. 10.43. The larger the capacitance, and hence the time constant, the longer it takes the capacitor to charge up—there is more charge to be stored. The same effect can be created by holding the capacitance constant and increasing the resistance, but now the longer time is due to the lower currents that are a result of the higher resistance.

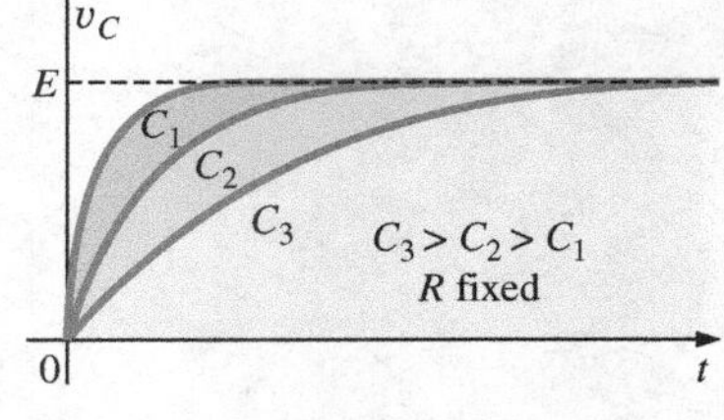

FIG. 10.43
Effect of increasing values of C (with R constant) on the charging curve for v_C.

EXAMPLE 10.8 For the circuit in Fig. 10.44:

a. Find the mathematical expressions for the transient behavior of the voltage v_C and the current i_C if the capacitor was initially uncharged and the switch is thrown into position 1 at $t = 0$ s.
b. Find the mathematical expressions for the voltage v_C and the current i_C if the switch is moved to position 2 at $t = 10$ ms. (Assume

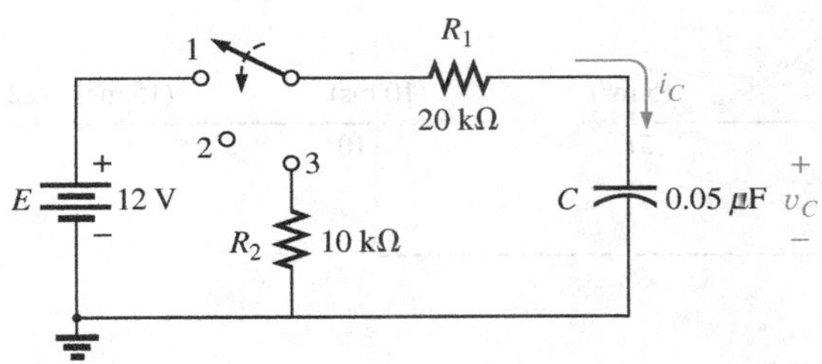

FIG. 10.44
Network to be analyzed in Example 10.8.

that the leakage resistance of the capacitor is infinite ohms; that is, there is no leakage current.)

c. Find the mathematical expressions for the voltage v_C and the current i_C if the switch is thrown into position 3 at $t = 20$ ms.

d. Plot the waveforms obtained in parts (a)–(c) on the same time axis using the defined polarities in Fig. 10.44.

Solutions:

a. *Charging phase:*

$$\tau = R_1C = (20\text{ k}\Omega)(0.05\ \mu\text{F}) = 1\text{ ms}$$

$$v_C = E(1 - e^{-t/\tau}) = \mathbf{12\ V(1 - e^{-t/1\,ms})}$$

$$i_C = \frac{E}{R_1}e^{-t/\tau} = \frac{12\text{ V}}{20\text{ k}\Omega}e^{-t/1\text{ ms}} = \mathbf{0.6\ mAe^{-t/1\,ms}}$$

b. *Storage phase:* At 10 ms, a period of time equal to 10τ has passed, permitting the assumption that the capacitor is fully charged. Since $R_{\text{leakage}} = \infty\ \Omega$, the capacitor will hold its charge indefinitely. The result is that both v_C and i_C will remain at a fixed value:

$$v_C = \mathbf{12\ V}$$

$$i_C = \mathbf{0\ A}$$

c. *Discharge phase* (using 20 ms as the new $t = 0$ s for the equations): The new time constant is

$$\tau' = RC = (R_1 + R_2)C = (20\text{ k}\Omega + 10\text{ k}\Omega)(0.05\ \mu\text{F}) = 1.5\text{ ms}$$

$$v_C = Ee^{-t/\tau'} = \mathbf{12\ Ve^{-t/1.5\,ms}}$$

$$i_C = -\frac{E}{R}e^{-t/\tau'} = -\frac{E}{R_1 + R_2}e^{-t/\tau'}$$

$$= -\frac{12\text{ V}}{20\text{ k}\Omega + 10\text{ k}\Omega}e^{-t/1.5\text{ ms}} = \mathbf{-0.4\ mAe^{-t/1.5\,ms}}$$

d. See Fig. 10.45.

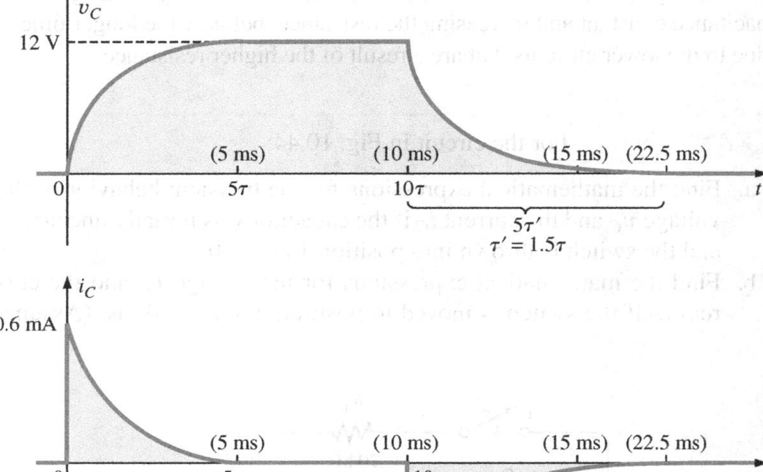

FIG. 10.45
v_C and i_C for the network in Fig. 10.44.

EXAMPLE 10.9 For the network in Fig. 10.46:

a. Find the mathematical expression for the transient behavior of the voltage across the capacitor if the switch is thrown into position 1 at $t = 0$ s.
b. Find the mathematical expression for the transient behavior of the voltage across the capacitor if the switch is moved to position 2 at $t = 1\tau$.
c. Plot the resulting waveform for the voltage v_C as determined by parts (a) and (b).
d. Repeat parts (a)–(c) for the current i_C.

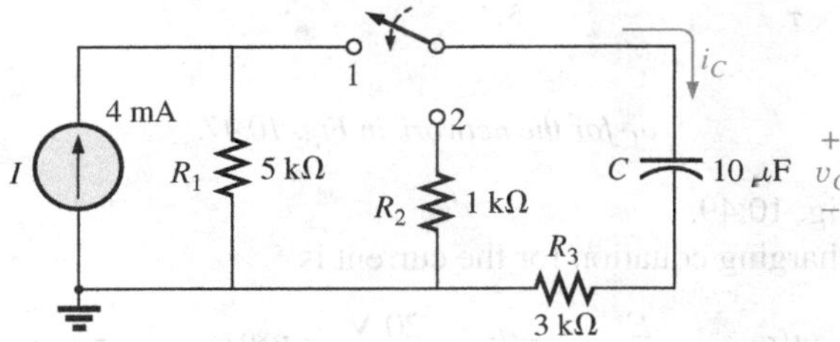

FIG. 10.46
Network to be analyzed in Example 10.9.

Solutions:

a. Converting the current source to a voltage source results in the configuration in Fig. 10.47 for the charging phase.

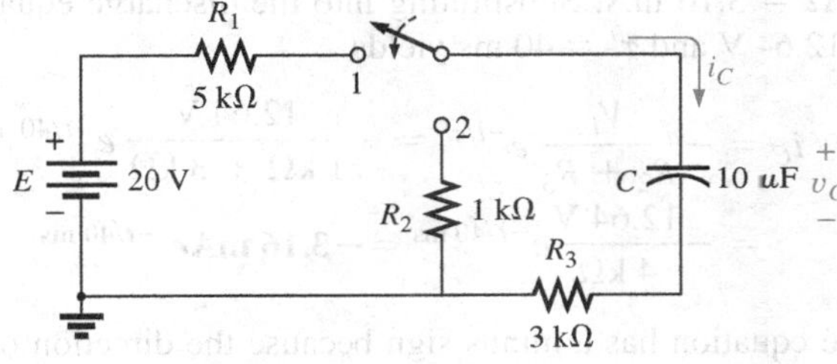

FIG. 10.47
The charging phase for the network in Fig. 10.46.

For the source conversion

$$E = IR = (4\text{ mA})(5\text{ k}\Omega) = 20\text{ V}$$

and
$$R_s = R_p = 5\text{ k}\Omega$$

$$\tau = RC = (R_1 + R_3)C = (5\text{ k}\Omega + 3\text{ k}\Omega)(10\ \mu\text{F}) = 80\text{ ms}$$
$$v_C = E(1 - e^{-t/\tau}) = \mathbf{20\ V\,(1 - e^{-t/80\ ms})}$$

b. With the switch in position 2, the network appears as shown in Fig. 10.48. The voltage at 1τ can be found by using the fact that the voltage is 63.2% of its final value of 20 V, so that 0.632(20 V) = 12.64 V. Alternatively, you can substitute into the derived equation as follows:

$$e^{-t/\tau} = e^{-\tau/\tau} = e^{-1} = 0.368$$

and
$$v_C = 20\text{ V}(1 - e^{-t/80\text{ ms}}) = 20\text{ V}(1 - 0.368)$$
$$= (20\text{ V})(0.632) = 12.64\text{ V}$$

Using this voltage as the starting point and substituting into the discharge equation results in

$$\tau' = RC = (R_2 + R_3)C = (1\text{ k}\Omega + 3\text{ k}\Omega)(10\ \mu\text{F}) = 40\text{ ms}$$
$$v_C = Ee^{-t/\tau'} = \mathbf{12.64\ Ve^{-t/40\ ms}}$$

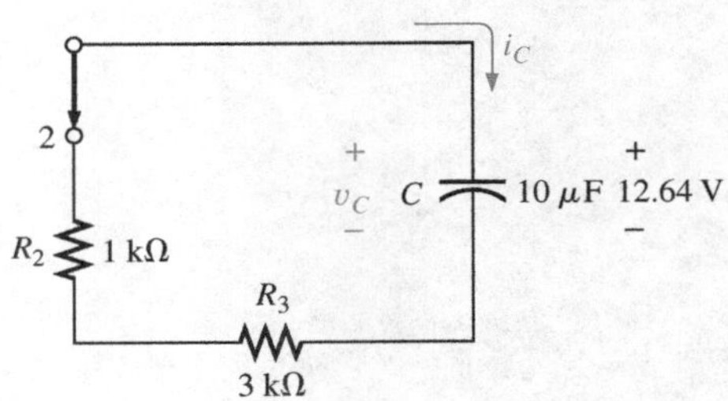

FIG. 10.48
Network in Fig. 10.47 when the switch is moved to position 2 at $t = 1\tau_1$.

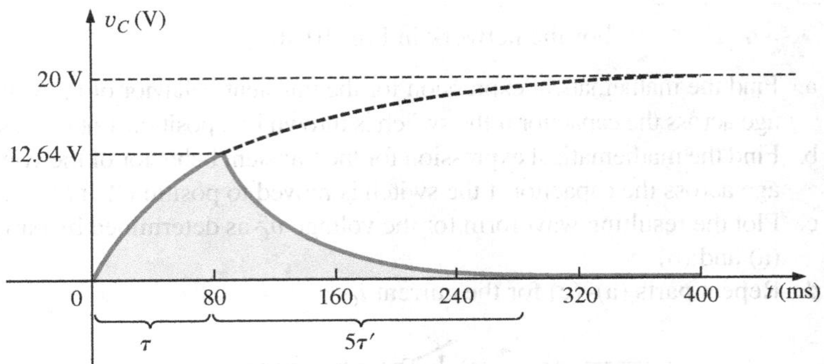

FIG. 10.49
v_C for the network in Fig. 10.47.

c. See Fig. 10.49.
d. The charging equation for the current is

$$i_C = \frac{E}{R}e^{-t/\tau} = \frac{E}{R_1 + R_3}e^{-t/\tau} = \frac{20\text{ V}}{8\text{ k}\Omega}e^{-t/80\text{ ms}} = \mathbf{2.5\text{ mA}e^{-t/80\text{ ms}}}$$

which, at $t = 80$ ms, results in

$$i_C = 2.5\text{ mA}e^{-80\text{ ms}/80\text{ ms}} = 2.5\text{ mA}e^{-1} = (2.5\text{ mA})(0.368) = 0.92\text{ mA}$$

When the switch is moved to position 2, the 12.64 V across the capacitor appears across the resistor to establish a current of 12.64 V/4 kΩ = 3.16 mA. Substituting into the discharge equation with $V_i = 12.64$ V and $\tau' = 40$ ms yields

$$i_C = -\frac{V_i}{R_2 + R_3}e^{-t/\tau'} = -\frac{12.64\text{ V}}{1\text{ k}\Omega + 3\text{ k}\Omega}e^{-t/40\text{ ms}}$$
$$= -\frac{12.64\text{ V}}{4\text{ k}\Omega}e^{-t/40\text{ ms}} = \mathbf{-3.16\text{ mA}e^{-t/40\text{ ms}}}$$

The equation has a minus sign because the direction of the discharge current is opposite to that defined for the current in Fig. 10.48. The resulting plot appears in Fig. 10.50.

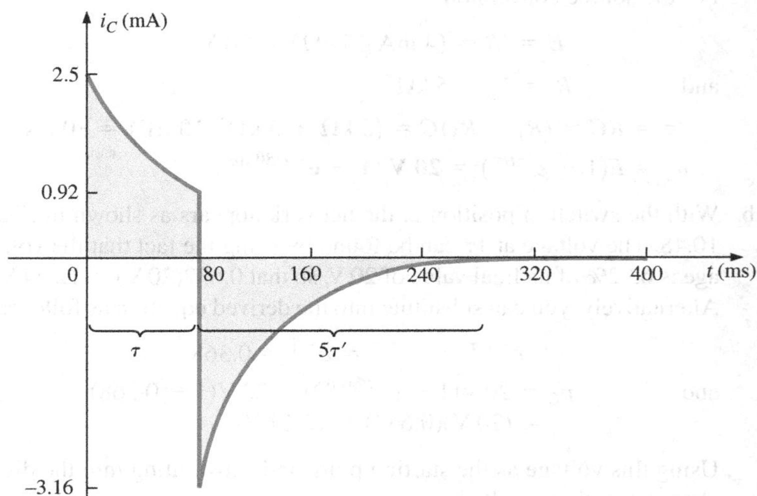

FIG. 10.50
i_c for the network in Fig. 10.47.

10.7 INITIAL CONDITIONS

In all the examples in the previous sections, the capacitor was uncharged before the switch was thrown. We now examine the effect of a charge, and therefore a voltage ($V = Q/C$), on the plates at the instant the switching action takes place. The voltage across the capacitor at this instant is called the **initial value,** as shown for the general waveform in Fig. 10.51.

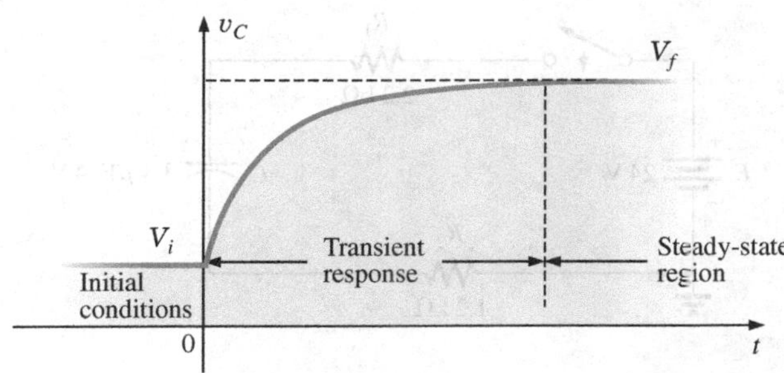

FIG. 10.51
Defining the regions associated with a transient response.

Once the switch is thrown, the transient phase commences until a leveling off occurs after five time constants. This region of relatively fixed value that follows the transient response is called the **steady-state region,** and the resulting value is called the **steady-state** or **final value.** The steady-state value is found by substituting the open-circuit equivalent for the capacitor and finding the voltage across the plates. Using the transient equation developed in the previous section, we can write an equation for the voltage v_C for the entire time interval in Fig. 10.51. That is, for the transient period, the voltage rises from V_i (previously 0 V) to a final value of V_f. Therefore,

$$v_C = E(1 - e^{-t/\tau}) = (V_f - V_i)(1 - e^{-t/\tau})$$

Adding the starting value of V_i to the equation results in

$$v_C = V_i = (V_f - V_i)(1 - e^{-t/\tau})$$

However, by multiplying through and rearranging terms, we obtain

$$\begin{aligned} v_C &= V_i + V_f - V_f e^{-t/\tau} V_i + V_i e^{-t/\tau} \\ &= V_f V_f e^{-t/\tau} + V_i e^{-t/\tau} \end{aligned}$$

We find

$$\boxed{v_C = V_f + (V_i - V_f)e^{-t/\tau}} \qquad \textbf{(10.21)}$$

Now that the equation has been developed, it is important to recognize that

Eq. (10.21) is a universal equation for the transient response of a capacitor.

That is, it can be used whether or not the capacitor has an initial value. If the initial value is 0 V as it was in all the previous examples, simply set V_i equal to zero in the equation, and the desired equation results. The final value is the voltage across the capacitor when the open-circuit equivalent is substituted.

EXAMPLE 10.10 The capacitor in Fig. 10.52 has an initial voltage of 4 V.

a. Find the mathematical expression for the voltage across the capacitor once the switch is closed.
b. Find the mathematical expression for the current during the transient period.
c. Sketch the waveform for each from initial value to final value.

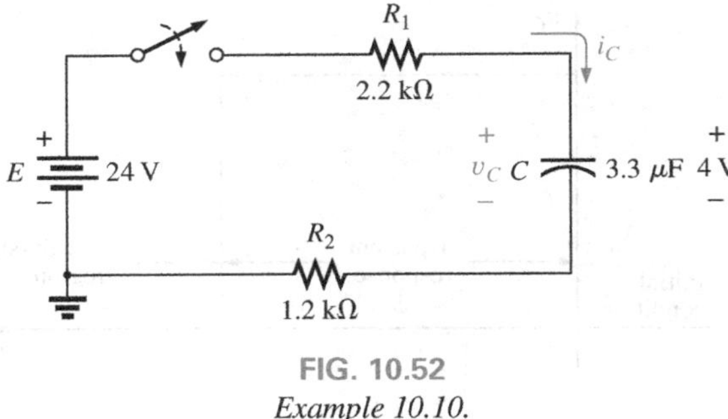

FIG. 10.52
Example 10.10.

Solutions:

a. Substituting the open-circuit equivalent for the capacitor results in a final or steady-state voltage v_C of 24 V.

The time constant is determined by

$$\tau = (R_1 + R_2)C$$
$$= (2.2\text{ k}\Omega + 1.2\text{ k}\Omega)(3.3\ \mu\text{F}) = 11.22\text{ ms}$$

with $5\tau = 56.1\text{ ms}$

Applying Eq. (10.21) gives

$$v_C = V_f + (V_i - V_f)e^{-t/\tau} = 24\text{ V} + (4\text{ V} - 24\text{ V})e^{-t/11.22\text{ ms}}$$

and $$v_C = \mathbf{24\ V - 20\ Ve^{-t/11.22\ ms}}$$

b. Since the voltage across the capacitor is constant at 4 V prior to the closing of the switch, the current (whose level is sensitive only to changes in voltage across the capacitor) must have an initial value of 0 mA. At the instant the switch is closed, the voltage across the capacitor cannot change instantaneously, so the voltage across the resistive elements at this instant is the applied voltage less the initial voltage across the capacitor. The resulting peak current is

$$I_m = \frac{E - V_C}{R_1 + R_2} = \frac{24\text{ V} - 4\text{ V}}{2.2\text{ k}\Omega + 1.2\text{ k}\Omega} = \frac{20\text{ V}}{3.4\text{ k}\Omega} = 5.88\text{ mA}$$

The current then decays (with the same time constant as the voltage v_C) to zero because the capacitor is approaching its open-circuit equivalence.

The equation for i_C is therefore

$$i_C = \mathbf{5.88\ mAe^{-t/11.22\ ms}}$$

c. See Fig. 10.53. The initial and final values of the voltage were drawn first, and then the transient response was included between these levels. For the current, the waveform begins and ends at zero, with the peak value having a sign sensitive to the defined direction of i_C in Fig. 10.52.

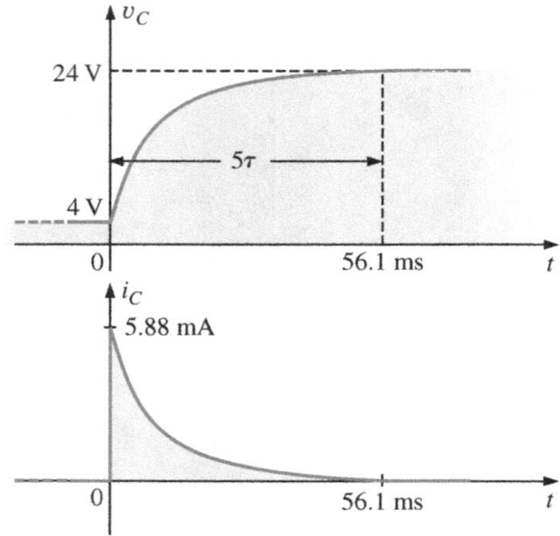

FIG. 10.53
v_C and i_C for the network in Fig. 10.52.

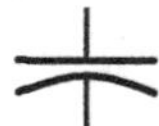

Let us now test the validity of the equation for v_C by substituting $t = 0$ s to reflect the instant the switch is closed. We have

$$e^{-t/\tau} = e^{-0} = 1$$

and $\quad v_C = 24\text{ V} - 20\text{ V}e^{t/\tau} = 24\text{ V} - 20\text{ V} = 4\text{ V}$

When $t > 5\tau$,

$$e^{-t/\tau} \cong 0$$

and $\quad v_C = 24\text{ V} - 20\text{ V}e^{t/\tau} = 24\text{ V} - 0\text{ V} = 24\text{ V}$

Eq. (10.21) can also be applied to the discharge phase by applying the correct levels of V_i and V_f.

For the discharge pattern in Fig. 10.54, $V_f = 0$ V, and Eq. (10.21) becomes

$$v_C = V_f + (V_i - V_f)e^{-t/\tau} = 0\text{ V} + (V_i - 0\text{ V})e^{-t/\tau}$$

and $\quad \boxed{v_C = V_i e^{-t/\tau}}_{\text{discharging}} \qquad$ **(10.22)**

Substituting $V_i = E$ volts results in Eq. (10.17).

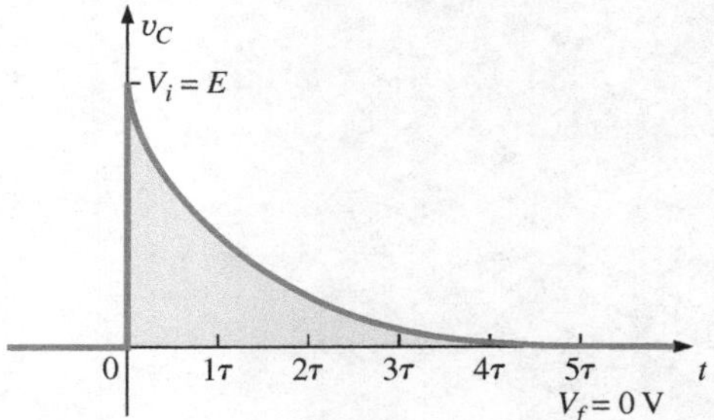

FIG. 10.54
Defining the parameters in Eq. (10.21) for the discharge phase.

10.8 INSTANTANEOUS VALUES

Occasionally, you may need to determine the voltage or current at a particular instant of time that is not an integral multiple of τ, as in the previous sections. For example, if

$$v_C = 20\text{ V}(1 - e^{(-t/2\text{ ms})})$$

the voltage v_C may be required at $t = 5$ ms, which does not correspond to a particular value of τ. Fig. 10.28 reveals that $(1 - e^{t/\tau})$ is approximately 0.93 at $t = 5\text{ ms} = 2.5\tau$, resulting in $v_C - 20(0.93) - 18.6$ V. Additional accuracy can be obtained by substituting $v = 5$ ms into the equation and solving for v_C using a calculator or table to determine $e^{-2.5}$. Thus,

$$v_C = 20\text{ V}(1 - e^{-5\text{ ms}/2\text{ ms}}) = (20\text{ V})(1 - e^{-2.5}) = (20\text{ V})(1 - 0.082)$$
$$= (20\text{ V})(0.918) = \mathbf{18.36\text{ V}}$$

The results are close, but accuracy beyond the tenths place is suspect using Fig. 10.29. The above procedure can also be applied to any other equation introduced in this chapter for currents or other voltages.

Occasionally, you may need to determine the time required to reach a particular voltage or current. The procedure is complicated somewhat by the use of natural logs ($\log_e$, or ln), but today's calculators are equipped to handle the operation with ease.

For example, solving for t in the equation

$$v_C = V_f + (V_i - V_f)e^{-t/\tau}$$

results in

$$\boxed{t = \tau(\log_e)\frac{(V_i - V_f)}{(v_C - V_f)}} \qquad \textbf{(10.23)}$$

For example, suppose that

$$v_C = 20\text{ V}(1 - e^{-t/2\text{ ms}})$$

and the time t to reach 10 V is desired. Since $V_i = 0$ V, and $V_f = 20$ V, we have

$$\begin{aligned} t &= \tau(\log_e)\frac{(V_i - V_f)}{(v_C - V_f)} = (2\text{ ms})(\log_e)\frac{(0\text{ V} - 20\text{ V})}{(10\text{ V} - 20\text{ V})} \\ &= (2\text{ ms})\left[\log_e\left(\frac{-20\text{ V}}{-10\text{ V}}\right)\right] = (2\text{ ms})(\log_e 2) = (2\text{ ms})(0.693) \\ &= \mathbf{1.386\text{ ms}} \end{aligned}$$

The TI-89 calculator key strokes appear in Fig. 10.55.

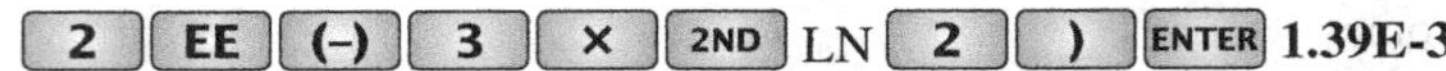

FIG. 10.55
Key strokes to determine (2 ms)(log$_e$2) using the TI-89 calculator.

For the discharge equation,

$$v_C = Ee^{-t/\tau} = V_i\,(e^{-t/\tau}) \qquad \text{with } V_f = 0\text{ V}$$

Using Eq. (10.23) gives

$$t = \tau(\log_e)\frac{(V_i - V_f)}{(v_C - V_f)} = \tau(\log_e)\frac{(V_i - 0\text{ V})}{(v_C - 0\text{ V})}$$

and

$$\boxed{t = \tau \log_e \frac{V_i}{v_C}} \qquad \textbf{(10.24)}$$

For the current equation,

$$i_C = \frac{E}{R}e^{-t/\tau} \qquad I_i = \frac{E}{R} \qquad I_f = 0\text{ A}$$

and

$$\boxed{t = \log_e \frac{I_i}{i_C}} \qquad \textbf{(10.25)}$$

10.9 THÉVENIN EQUIVALENT: $\tau = R_{Th}C$

You may encounter instances in which the network does not have the simple series form in Fig. 10.26. You then need to find the Thévenin equivalent circuit for the network external to the capacitive element. E_{Th} will be the source voltage E in Eqs. (10.13) through (10.25), and R_{Th} will be the resistance R. The time constant is then $\tau = R_{Th}C$.

EXAMPLE 10.11 For the network in Fig. 10.56:

a. Find the mathematical expression for the transient behavior of the voltage v_C and the current i_C following the closing of the switch (position 1 at $t = 0$ s).

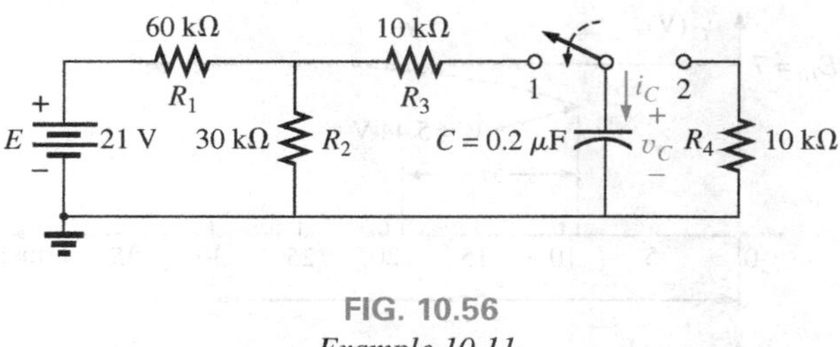

FIG. 10.56
Example 10.11.

b. Find the mathematical expression for the voltage v_C and the current i_C as a function of time if the switch is thrown into position 2 at $t = 9$ ms.
c. Draw the resultant waveforms of parts (a) and (b) on the same time axis.

Solutions:

a. Applying Thévenin's theorem to the 0.2 μF capacitor, we obtain Fig. 10.57. We have

$$R_{Th} = R_1 \parallel R_2 + R_3 = \frac{(60\text{ k}\Omega)(30\text{ k}\Omega)}{90\text{ k}\Omega} + 10\text{ k}\Omega$$
$$= 20\text{ k}\Omega + 10\text{ k}\Omega = 30\text{ k}\Omega$$
$$E_{Th} = \frac{R_2 E}{R_2 + R_1} = \frac{(30\text{ k}\Omega)(21\text{ V})}{30\text{ k}\Omega + 60\text{ k}\Omega} = \frac{1}{3}(21\text{ V}) = 7\text{ V}$$

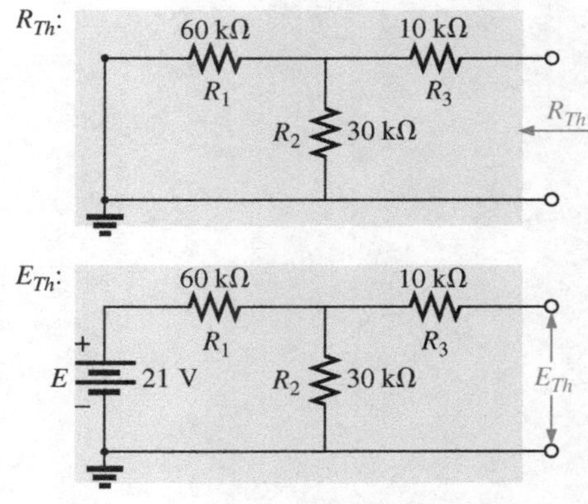

FIG. 10.57
Applying Thévenin's theorem to the network in Fig. 10.56.

The resultant Thévenin equivalent circuit with the capacitor replaced is shown in Fig. 10.58.
Using Eq. (10.21) with $V_f = E_{Th}$ and $V_i = 0$ V, we find that

$$v_C = V_f + (V_i - V_f)e^{-t/\tau}$$

becomes
$$v_C = E_{Th} + (0\text{ V} - E_{Th})e^{-t/\tau}$$

or
$$v_C = E_{Th}(1 - e^{-t/\tau})$$

with
$$\tau = RC = (30\text{ k}\Omega)(0.2\ \mu\text{F}) = 6\text{ ms}$$

Therefore,
$$\mathbf{v_C = 7\ V(1 - e^{-t/6\ ms})}$$

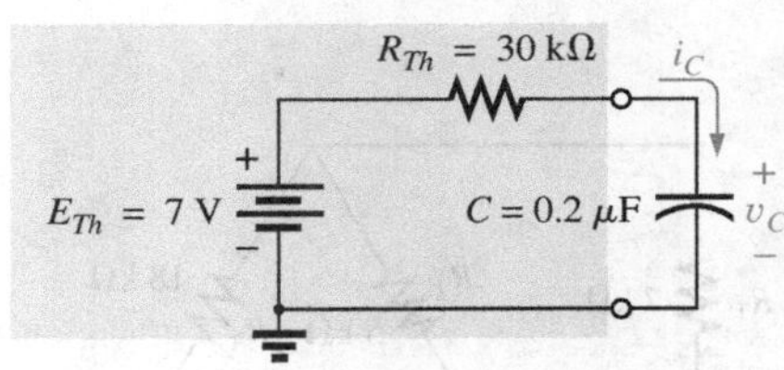

FIG. 10.58
Substituting the Thévenin equivalent for the network in Fig. 10.56.

For the current i_C:

$$i_C = \frac{E_{Th}}{R}e^{-t/RC} = \frac{7\text{ V}}{30\text{ k}\Omega}e^{-t/6\text{ ms}}$$
$$= \mathbf{0.23\ mAe^{-t/6\ ms}}$$

b. At $t = 9$ ms,

$$v_C = E_{Th}(1 - e^{-t/\tau}) = 7\text{ V}(1 - e^{-(9\text{ ms}/6\text{ ms})})$$
$$= (7\text{ V})(1 - e^{-1.5}) = (7\text{ V})(1 - 0.223)$$
$$= (7\text{ V})(0.777) = 5.44\text{ V}$$

and
$$i_C = \frac{E_{Th}}{R}e^{-t/\tau} = 0.23\text{ mA}e^{-1.5}$$
$$= (0.23 \times 10^{-3})(0.233) = 0.052 \times 10^{-3} = 0.05\text{ mA}$$

Using Eq. (10.21) with $V_f = 0$ V and $V_i = 5.44$ V, we find that

$$v_C = V_f + (V_i - V_f)e^{-t/\tau'}$$

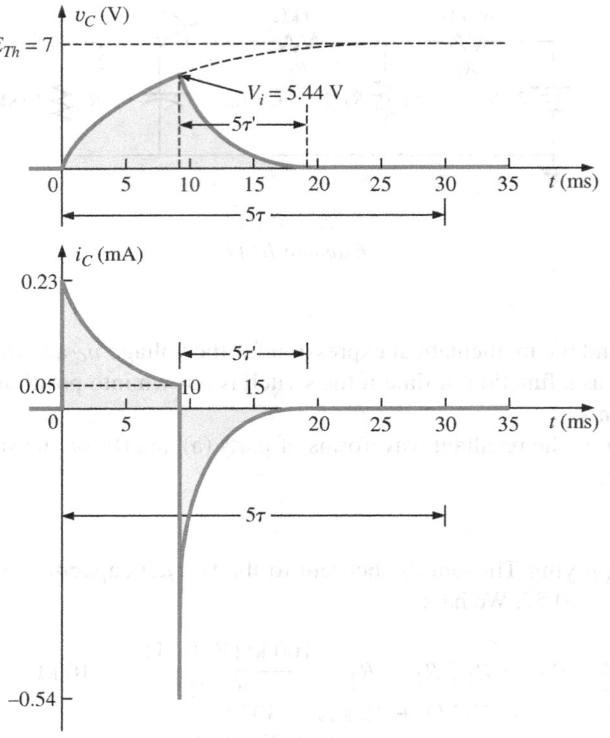

FIG. 10.59
The resulting waveforms for the network in Fig. 10.56.

becomes
$$v_C = 0\text{ V} + (5.44\text{ V} - 0\text{ V})e^{-t/\tau'}$$
$$= 5.44\text{ V}e^{-t/\tau'}$$

with
$$\tau' = R_4C = (10\text{ k}\Omega)(0.2\ \mu\text{F}) = 2\text{ ms}$$

and
$$v_C = \mathbf{5.44\ Ve^{-t/2\ ms}}$$

By Eq. (10.19),

$$I_i = \frac{5.44\text{ V}}{10\text{ k}\Omega} = 0.54\text{ mA}$$

and
$$i_C = I_i e^{-t/\tau} - \mathbf{0.54\ mAe^{-t/2\ ms}}$$

c. See Fig. 10.59.

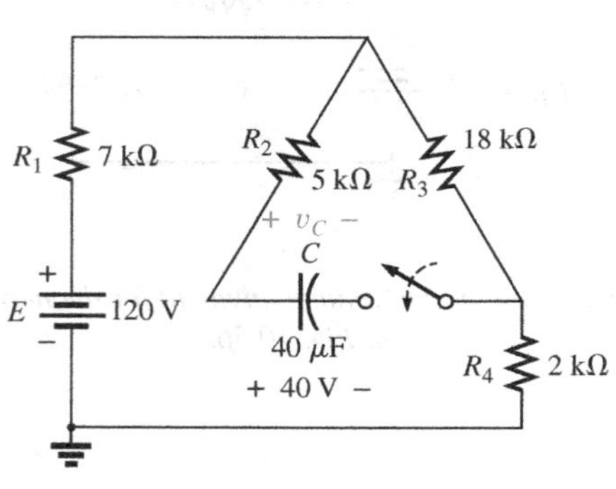

FIG. 10.60
Example 10.12.

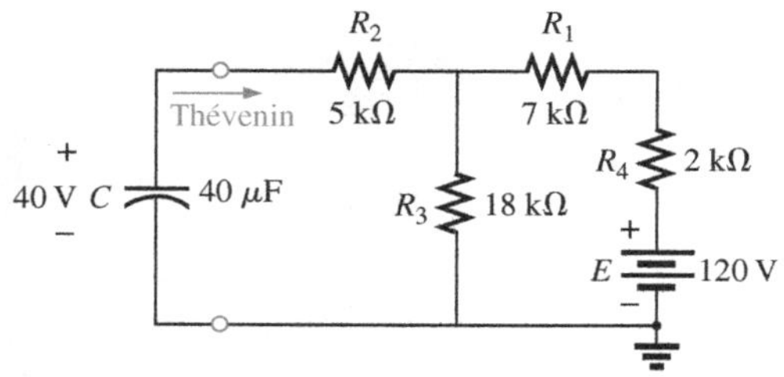

FIG. 10.61
Network in Fig. 10.60 redrawn.

EXAMPLE 10.12 The capacitor in Fig. 10.60 is initially charged to 40 V. Find the mathematical expression for v_C after the closing of the switch. Plot the waveform for v_C.

Solution: The network is redrawn in Fig. 10.61.

E_{Th}:

$$E_{Th} = \frac{R_3E}{R_3 + R_1 + R_4} = \frac{(18\text{ k}\Omega)(120\text{ V})}{18\text{ k}\Omega + 7\text{ k}\Omega + 2\text{ k}\Omega} = 80\text{ V}$$

R_{Th}:

$$R_{Th} = 5\text{ k}\Omega + (18\text{ k}\Omega) \parallel (7\text{ k}\Omega + 2\text{ k}\Omega)$$
$$= 5\text{ k}\Omega + 6\text{ k}\Omega = 11\text{ k}\Omega$$

Therefore, $V_i = 40\text{ V}$ and $V_f = 80\text{ V}$

and $\tau = R_{Th}C = (11\text{ k}\Omega)(40\ \mu\text{F}) = 0.44\text{ s}$

Eq. (10.21):
$$v_C = V_f + (V_i - V_f)e^{-t/\tau}$$
$$= 80\text{ V} + (40\text{ V} - 80\text{ V})e^{-t/0.44\text{ s}}$$

and
$$\mathbf{v_C = 80\text{ V} - 40\text{ V}e^{-t/0.44\text{ s}}}$$

The waveform appears as in Fig. 10.62.

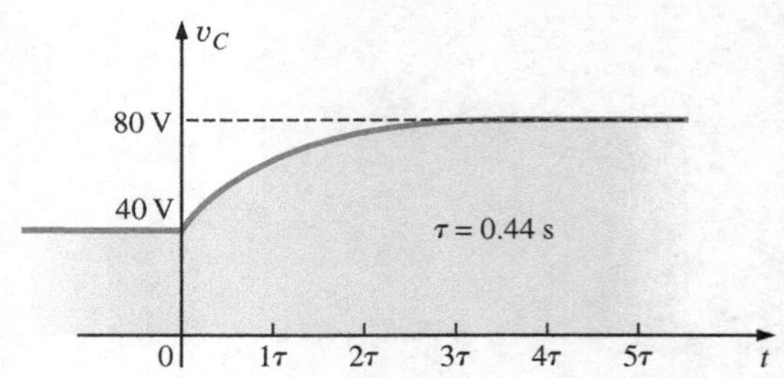

FIG. 10.62
v_C for the network in Fig. 10.60.

EXAMPLE 10.13 For the network in Fig. 10.63, find the mathematical expression for the voltage v_C after the closing of the switch (at $t = 0$).

Solution:

$$R_{Th} = R_1 + R_2 = 6\ \Omega + 10\ \Omega = 16\ \Omega$$
$$E_{Th} = V_1 + V_2 = IR_1 + 0$$
$$= (20 \times 10^{-3}\text{ A})(6\ \Omega) = 120 \times 10^{-3}\text{ V} = 0.12\text{ V}$$

and $\tau = R_{Th}C = (16\ \Omega)(500 \times 10^{-6}\text{ F}) = 8\text{ ms}$

so that $\mathbf{v_C = 0.12\text{ V}(1 - e^{-t/8\text{ ms}})}$

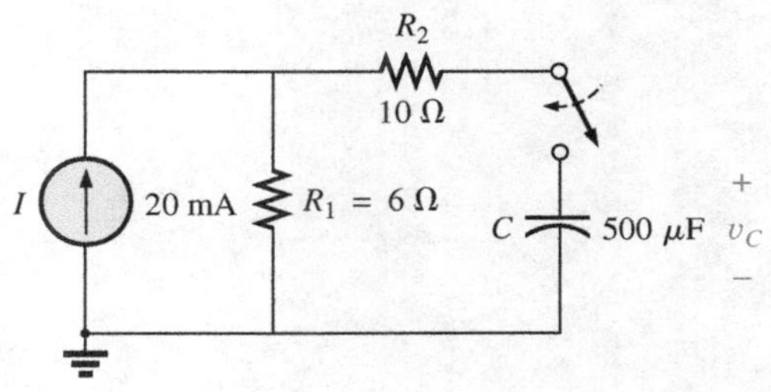

FIG. 10.63
Example 10.13.

10.10 THE CURRENT i_C

There is a very special relationship between the current of a capacitor and the voltage across it. For the resistor, it is defined by Ohm's law: $i_R = v_R/R$. The current through and the voltage across the resistor are related by a constant R—a very simple direct linear relationship. For the capacitor, it is the more complex relationship defined by

$$\boxed{i_C = C\frac{dv_C}{dt}} \qquad \textbf{(10.26)}$$

The factor C reveals that the higher the capacitance, the greater is the resulting current. Intuitively, this relationship makes sense because higher capacitance levels result in increased levels of stored charge, providing a source for increased current levels. The second term, dv_C/dt, is sensitive to the *rate of change* of v_C with time. The function dv_C/dt is called the **derivative** (calculus) of the voltage v_C with respect to time t. The faster the voltage v_C changes with time, the larger will be the factor dv_C/dt and the larger will be the resulting current i_C. That is why the current jumps to its maximum of E/R in a charging circuit the instant the switch is closed. At that instant, if you look at the charging curve for v_C, the voltage is *changing* at its greatest rate. As it approaches its final value, the rate of change decreases, and, as confirmed by Eq. (10.26), the level of current decreases.

Take special note of the following:

The capacitive current is directly related to the rate of change of the voltage across the capacitor, not the levels of voltage involved.

For example, the current of a capacitor will be greater when the voltage changes from 1 V to 10 V in 1 ms than when it changes from 10 V to 100 V in 1 s; in fact, it will be 100 times more.

If the voltage fails to change over time, then

$$\frac{dv_C}{dt} = 0$$

and
$$i_C = C\frac{dv_C}{dt} = C(0) = 0\text{ A}$$

In an effort to develop a clearer understanding of Eq. (10.26), let us calculate the **average current** associated with a capacitor for various voltages impressed across the capacitor. The average current is defined by the equation

$$\boxed{i_{C_{av}} = C\frac{\Delta v_C}{\Delta t}} \qquad \textbf{(10.27)}$$

where Δ indicates a finite (measurable) change in voltage or time. The instantaneous current can be derived from Eq. (10.27) by letting Δt become vanishingly small; that is,

$$i_{C_{inst}} = \lim_{\Delta t \to 0} C\frac{\Delta v_C}{\Delta t} = C\frac{dv_C}{dt}$$

In the following example, the change in voltage Δv_C will be considered for each slope of the voltage waveform. If the voltage increases with time, the average current is the change in voltage divided by the change in time, with a positive sign. If the voltage decreases with time, the average current is again the change in voltage divided by the change in time, but with a negative sign.

EXAMPLE 10.14 Find the waveform for the average current if the voltage across a 2 μF capacitor is as shown in Fig. 10.64.

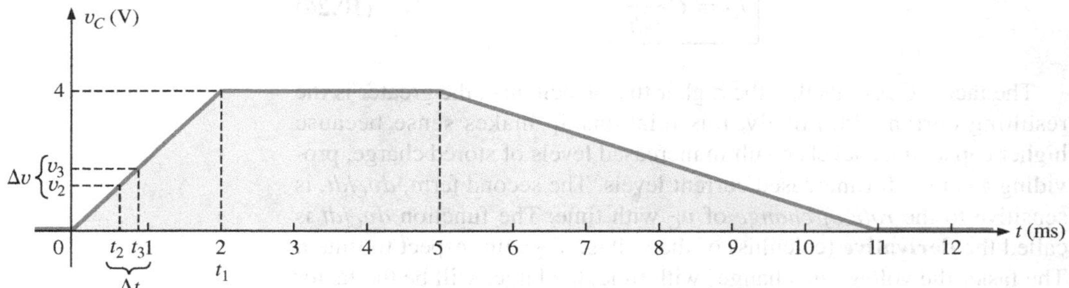

FIG. 10.64
v_C for Example 10.14.

Solutions:

a. From 0 ms to 2 ms, the voltage increases linearly from 0 V to 4 V; the change in voltage $\Delta v = 4\text{ V} - 0 = 4\text{ V}$ (with a positive sign since the voltage increases with time). The change in time $\Delta t = 2\text{ ms} - 0 = 2\text{ ms}$, and

$$i_{C_{av}} = C\frac{\Delta v_C}{\Delta t} = (2 \times 10^{-6}\text{ F})\left(\frac{4\text{ V}}{2 \times 10^{-3}\text{ s}}\right)$$
$$= 4 \times 10^{-3}\text{ A} = \mathbf{4\text{ mA}}$$

b. From 2 ms to 5 ms, the voltage remains constant at 4 V; the change in voltage $\Delta v = 0$. The change in time $\Delta t = 3$ ms, and

$$i_{C_{av}} = C\frac{\Delta v_C}{\Delta t} = C\frac{0}{\Delta t} = \mathbf{0\ mA}$$

c. From 5 ms to 11 ms, the voltage decreases from 4 V to 0 V. The change in voltage Δv is, therefore, 4 V − 0 = 4 V (with a negative sign since the voltage is decreasing with time). The change in time $\Delta t = 11$ ms − 5 ms = 6 ms, and

$$\begin{aligned} i_{C_{av}} &= C\frac{\Delta v_C}{\Delta t} = -(2 \times 10^{-6}\ \text{F})\left(\frac{4\ \text{V}}{6 \times 10^{-3}\ \text{s}}\right) \\ &= -1.33 \times 10^{-3}\ \text{A} = \mathbf{-1.33\ mA} \end{aligned}$$

d. From 11 ms on, the voltage remains constant at 0 and $\Delta v = 0$, so $i_{C_{av}} = 0$ mA. The waveform for the average current for the impressed voltage is as shown in Fig. 10.65.

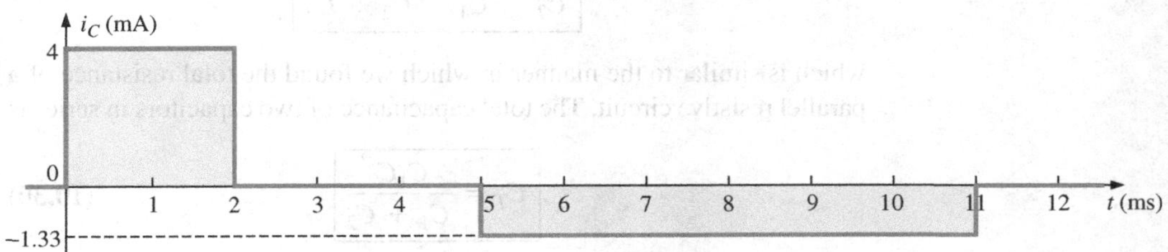

FIG. 10.65
The resulting current i_C for the applied voltage in Fig. 10.64.

Note in Example 10.14 that, in general, the steeper the slope, the greater is the current, and when the voltage fails to change, the current is zero. In addition, the average value is the same as the instantaneous value at any point along the slope over which the average value was found. For example, if the interval Δt is reduced from $0 \rightarrow t_1$ to $t_2 - t_3$, as noted in Fig. 10.64, $\Delta v/\Delta t$ is still the same. In fact, no matter how small the interval Δt, the slope will be the same, and therefore the current $i_{C_{av}}$ will be the same. If we consider the limit as $\Delta t \rightarrow 0$, the slope will still remain the same, and therefore $i_{C_{av}} = i_{C_{inst}}$ at any instant of time between 0 and t_1. The same can be said about any portion of the voltage waveform that has a constant slope.

An important point to be gained from this discussion is that it is not the magnitude of the voltage across a capacitor that determines the current but rather how quickly the voltage *changes* across the capacitor. An applied steady dc voltage of 10,000 V would (ideally) not create any flow of charge (current), but a change in voltage of 1 V in a very brief period of time could create a significant current.

The method described above is only for waveforms with straight-line (linear) segments. For nonlinear (curved) waveforms, a method of calculus (differentiation) must be used.

10.11 CAPACITORS IN SERIES AND IN PARALLEL

Capacitors, like resistors, can be placed in series and in parallel. Increasing levels of capacitance can be obtained by placing capacitors in parallel, while decreasing levels can be obtained by placing capacitors in series.

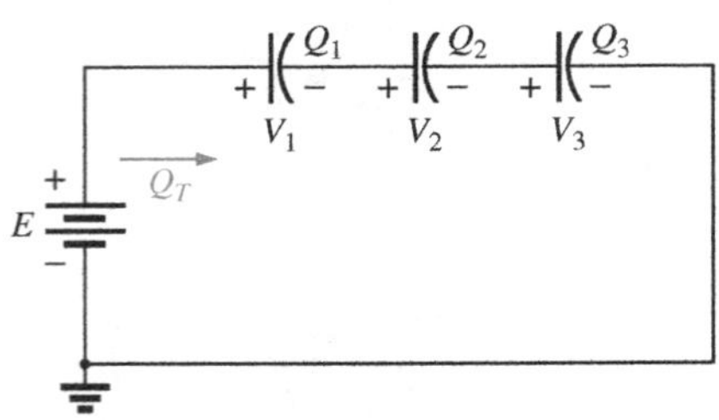

FIG. 10.66
Series capacitors.

For capacitors in series, the charge is the same on each capacitor (Fig. 10.66):

$$Q_T = Q_1 = Q_2 = Q_3 \tag{10.28}$$

Applying Kirchhoff's voltage law around the closed loop gives

$$E = V_1 + V_2 + V_3$$

However, $$V = \frac{Q}{C}$$

so that $$\frac{Q_T}{C_T} = \frac{Q_1}{C_1} + \frac{Q_2}{C_2} + \frac{Q_3}{C_3}$$

Using Eq. (10.28) and dividing both sides by Q yields

$$\frac{1}{C_T} = \frac{1}{C_1} + \frac{1}{C_2} + \frac{1}{C_3} \tag{10.29}$$

which is similar to the manner in which we found the total resistance of a parallel resistive circuit. The total capacitance of two capacitors in series is

$$C_T = \frac{C_1 C_2}{C_1 + C_2} \tag{10.30}$$

The voltage across each capacitor in Fig. 10.66 can be found by first recognizing that

$$Q_T = Q_1$$

or $$C_T E = C_1 V_1$$

Solving for V_1 gives $$V_1 = \frac{C_T E}{C_1}$$

and substituting for C_T gives

$$V_1 = \left(\frac{1/C_1}{1/C_1 + 1/C_2 + 1/C_3}\right)E \tag{10.31}$$

A similar equation results for each capacitor of the network.

For capacitors in parallel, as shown in Fig. 10.67, the voltage is the same across each capacitor, and the total charge is the sum of that on each capacitor:

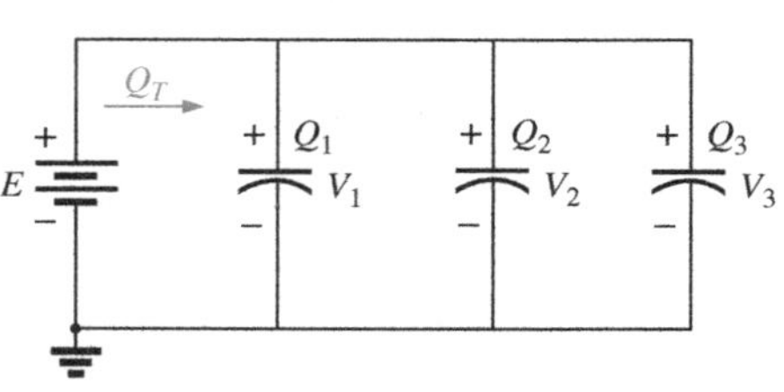

FIG. 10.67
Parallel capacitors.

$$Q_T = Q_1 + Q_2 + Q_3 \tag{10.32}$$

However, $$Q = CV$$

Therefore, $$C_T E = C_1 V_1 = C_2 V_2 = C_3 V_3$$

but $$E = V_1 = V_2 = V_3$$

Thus, $$C_T = C_1 + C_2 + C_3 \tag{10.33}$$

which is similar to the manner in which the total resistance of a series circuit is found.

EXAMPLE 10.15 For the circuit in Fig. 10.68:

a. Find the total capacitance.
b. Determine the charge on each plate.
c. Find the voltage across each capacitor.

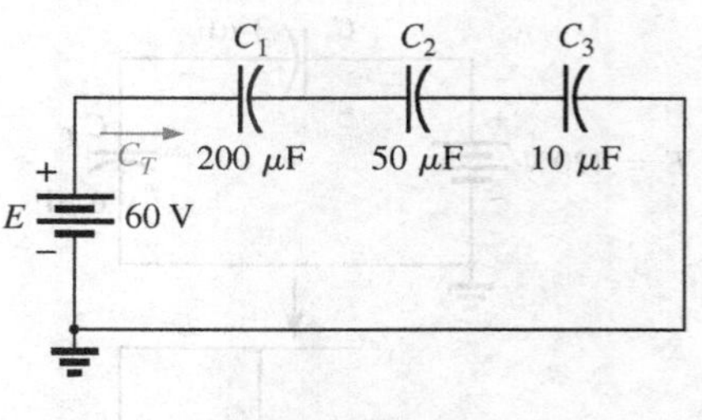

FIG. 10.68
Example 10.15.

Solutions:

a.
$$\frac{1}{C_T} = \frac{1}{C_1} + \frac{1}{C_2} + \frac{1}{C_3}$$
$$= \frac{1}{200 \times 10^{-6}\,\text{F}} + \frac{1}{50 \times 10^{-6}\,\text{F}} + \frac{1}{10 \times 10^{-6}\,\text{F}}$$
$$= 0.005 \times 10^{6} + 0.02 \times 10^{6} + 0.1 \times 10^{6}$$
$$= 0.125 \times 10^{6}$$

and
$$C_T = \frac{1}{0.125 \times 10^{6}} = \mathbf{8\,\mu F}$$

b.
$$Q_T = Q_1 = Q_2 = Q_3$$
$$= C_T E = (8 \times 10^{-6}\,\text{F})(60\,\text{V}) = \mathbf{480\,\mu C}$$

c.
$$V_1 = \frac{Q_1}{C_1} = \frac{480 \times 10^{-6}\,\text{C}}{200 \times 10^{-6}\,\text{F}} = \mathbf{2.4\,V}$$
$$V_2 = \frac{Q_2}{C_2} = \frac{480 \times 10^{-6}\,\text{C}}{50 \times 10^{-6}\,\text{F}} = \mathbf{9.6\,V}$$
$$V_3 = \frac{Q_3}{C_3} = \frac{480 \times 10^{-6}\,\text{C}}{10 \times 10^{-6}\,\text{F}} = \mathbf{48.0\,V}$$

and $E = V_1 + V_2 + V_3 = 2.4\,\text{V} + 9.6\,\text{V} + 48\,\text{V} = \mathbf{60\,V}$ (checks)

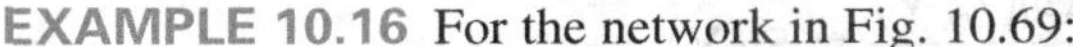

EXAMPLE 10.16 For the network in Fig. 10.69:

a. Find the total capacitance.
b. Determine the charge on each plate.
c. Find the total charge.

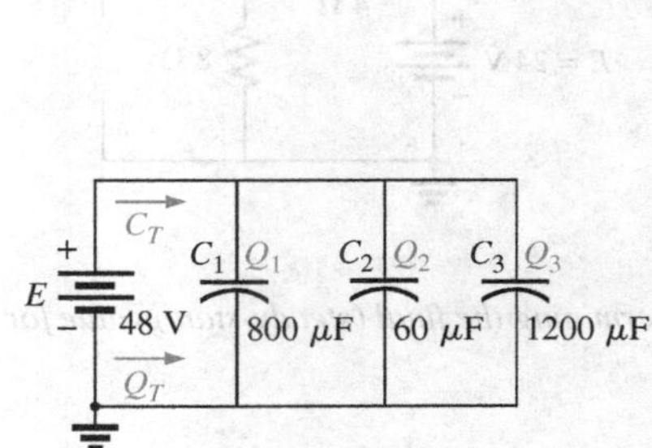

FIG. 10.69
Example 10.16.

Solutions:

a. $C_T = C_1 + C_2 + C_3 = 800\,\mu\text{F} + 60\,\mu\text{F} + 1200\,\mu\text{F} = \mathbf{2060\,\mu F}$

b. $Q_1 = C_1E = (800 \times 10^{-6}\,\text{F})(48\,\text{V}) = \mathbf{38.4\,mC}$

$Q_2 = C_2E = (60 \times 10^{-6}\,\text{F})(48\,\text{V}) = \mathbf{2.88\,mC}$

$Q_3 = C_3E = (1200 \times 10^{-6}\,\text{F})(48\,\text{V}) = \mathbf{57.6\,mC}$

c. $Q_T = Q_1 + Q_2 + Q_3 = 38.4\,\text{mC} + 2.88\,\text{mC} + 57.6\,\text{mC} = \mathbf{98.88\,mC}$

EXAMPLE 10.17 Find the voltage across and the charge on each capacitor for the network in Fig. 10.70.

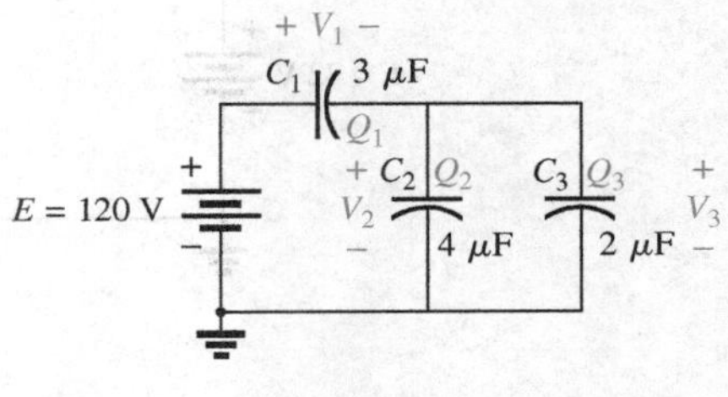

FIG. 10.70
Example 10.17.

Solution:

$$C'_T = C_2 + C_3 = 4\,\mu\text{F} + 2\,\mu\text{F} = 6\,\mu\text{F}$$
$$C_T = \frac{C_1 C'_T}{C_1 + C'_T} = \frac{(3\,\mu\text{F})(6\,\mu\text{F})}{3\,\mu\text{F} + 6\,\mu\text{F}} = 2\,\mu\text{F}$$
$$Q_T = C_T E = (2 \times 10^{-6}\,\text{F})(120\,\text{V}) = \mathbf{240\,\mu C}$$

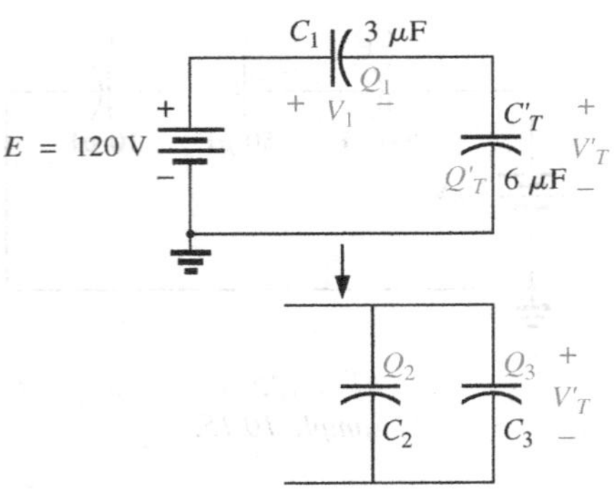

FIG. 10.71
Reduced equivalent for the network in Fig. 10.70.

An equivalent circuit (Fig. 10.71) has

$$Q_T = Q_1 = Q'_T$$

and, therefore,
$$Q_1 = \mathbf{240\ \mu C}$$

and
$$V_1 = \frac{Q_1}{C_1} = \frac{240 \times 10^{-6}\ \text{C}}{3 \times 10^{-6}\ \text{F}} = \mathbf{80\ V}$$

$$Q'_T = 240\ \mu\text{C}$$

Therefore,
$$V'_T = \frac{Q'_T}{C'_T} = \frac{240 \times 10^{-6}\ \text{C}}{6 \times 10^{-6}\ \text{F}} = \mathbf{40\ V}$$

and
$$Q_2 = C_2V'_T = (4 \times 10^{-6}\ \text{F})(40\ \text{V}) = \mathbf{160\ \mu C}$$
$$Q_3 = C_3V'_T = (2 \times 10^{-6}\ \text{F})(40\ \text{V}) = \mathbf{80\ \mu C}$$

EXAMPLE 10.18 Find the voltage across and the charge on capacitor C_1 in Fig. 10.72 after it has charged up to its final value.

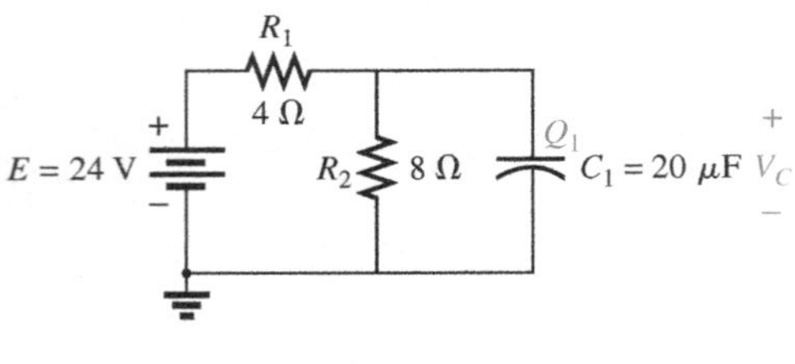

FIG. 10.72
Example 10.18.

Solution: As previously discussed, the capacitor is effectively an open circuit for dc after charging up to its final value (Fig. 10.73).

Therefore,

$$V_C = \frac{(8\ \Omega)(24\ \text{V})}{4\ \Omega + 8\ \Omega} = \mathbf{16\ V}$$
$$Q_1 = C_1V_C = (20 \times 10^{-6}\ \text{F})(16\ \text{V}) = \mathbf{320\ \mu C}$$

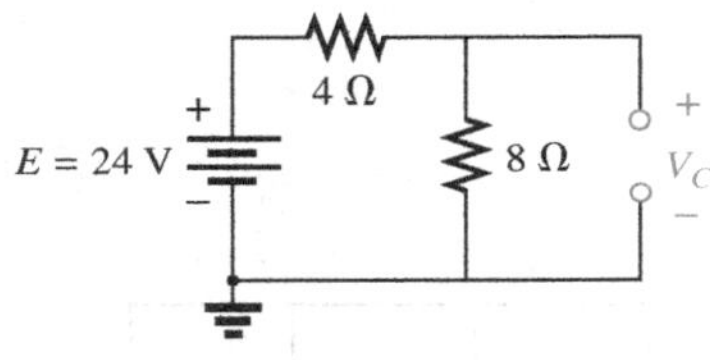

FIG. 10.73
Determining the final (steady-state) value for v_C.

EXAMPLE 10.19 Find the voltage across and the charge on each capacitor of the network in Fig. 10.74(a) after each has charged up to its final value.

Solution: See Fig. 10.74(b). We have

$$V_{C_2} = \frac{(7\ \Omega)(72\ \text{V})}{7\ \Omega + 2\ \Omega} = \mathbf{56\ V}$$

$$V_{C_1} = \frac{(2\ \Omega)(72\ \text{V})}{2\ \Omega + 7\ \Omega} = \mathbf{16\ V}$$

$$Q_1 = C_1V_{C_1} = (2 \times 10^{-6}\ \text{F})(16\ \text{V}) = \mathbf{32\ \mu C}$$
$$Q_2 = C_2V_{C_2} = (3 \times 10^{-6}\ \text{F})(56\ \text{V}) = \mathbf{168\ \mu C}$$

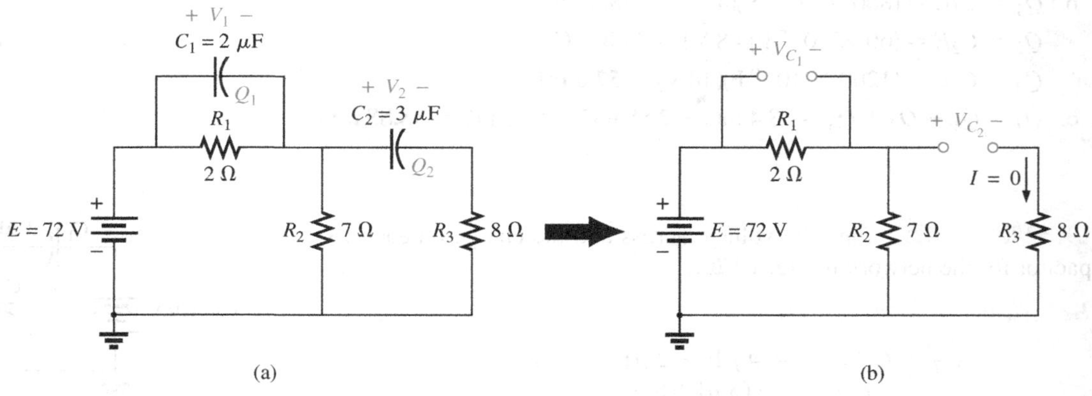

FIG. 10.74
Example 10.19.

10.12 ENERGY STORED BY A CAPACITOR

An ideal capacitor does not dissipate any of the energy supplied to it. It stores the energy in the form of an electric field between the conducting surfaces. A plot of the voltage, current, and power to a capacitor during the charging phase is shown in Fig. 10.75. The power curve can be obtained by finding the product of the voltage and current at selected instants of time and connecting the points obtained. *The energy stored is represented by the shaded area under the power curve.* Using calculus, we can determine the area under the curve:

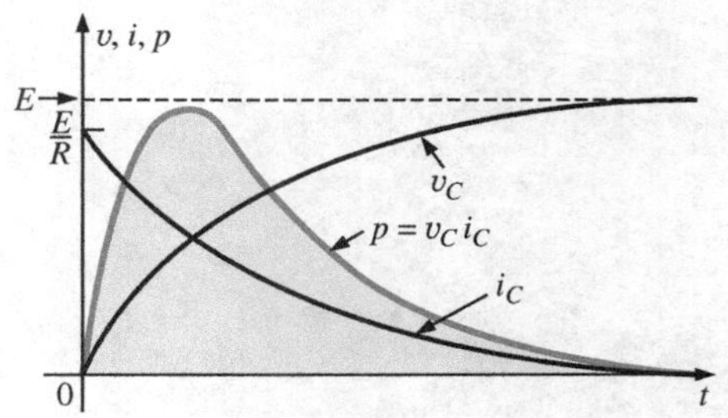

FIG. 10.75
Plotting the power to a capacitive element during the transient phase.

$$W_C = \frac{1}{2}CE^2$$

In general,

$$W_C = \frac{1}{2}CV^2 \quad \text{(J)} \qquad \textbf{(10.34)}$$

where V is the steady-state voltage across the capacitor. In terms of Q and C,

$$W_C = \frac{1}{2}C\left(\frac{Q}{C}\right)^2$$

or

$$W_C = \frac{Q^2}{2C} \quad \text{(J)} \qquad \textbf{(10.35)}$$

EXAMPLE 10.20 For the network in Fig. 10.74(a), determine the energy stored by each capacitor.

Solution: For C_1:

$$W_C = \frac{1}{2}CV^2$$
$$= \frac{1}{2}(2 \times 10^{-6}\,\text{F})(16\,\text{V})^2 = (1 \times 10^{-6})(256) = \mathbf{256\,\mu J}$$

For C_2:

$$W_C = \frac{1}{2}CV^2$$
$$= \frac{1}{2}(3 \times 10^{-6}\,\text{F})(56\,\text{V})^2 = (1.5 \times 10^{-6})(3136) = \mathbf{4704\,\mu J}$$

Due to the squared term, the energy stored increases rapidly with increasing voltages.

10.13 STRAY CAPACITANCES

In addition to the capacitors discussed so far in this chapter, there are **stray capacitances** that exist not through design but simply because two conducting surfaces are relatively close to each other. Two conducting wires in the same network have a capacitive effect between them, as shown in Fig. 10.76(a). In electronic circuits, capacitance levels exist between conducting surfaces of the transistor, as shown in Fig. 10.76(b). In Chapter 11, we will discuss another element, called the *inductor,* which

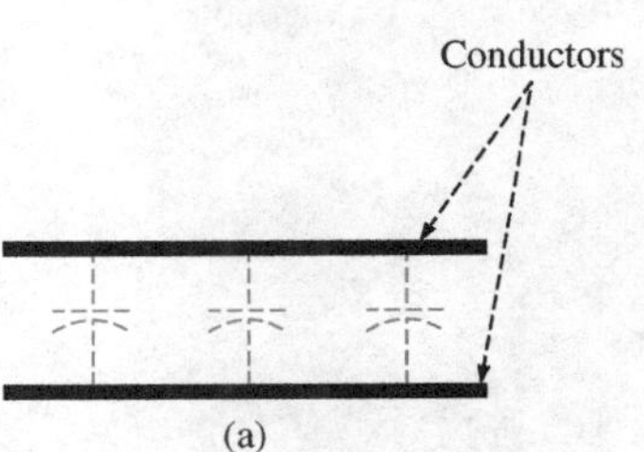

(a)

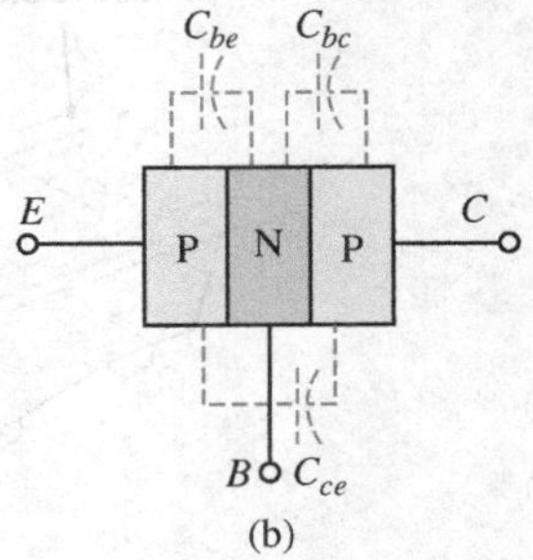

(b)

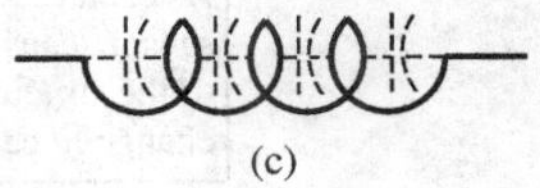

(c)

FIG. 10.76
Examples of stray capacitance.

has capacitive effects between the windings [Fig. 10.76(c)]. Stray capacitances can often lead to serious errors in system design if they are not considered carefully.

10.14 APPLICATIONS

This section includes a description of the operation of touch pads and one of the less expensive, throwaway cameras that have become so popular, as well as a discussion of the use of capacitors in the line conditioners (surge protectors) that are used in many homes and throughout the business world. Additional examples of the use of capacitors appear in Chapter 11.

Touch Pad

FIG. 10.77
Laptop touch pad.

The touch pad on the computer of Fig. 10.77 is used to control the position of the pointer on the computer screen by providing a link between the position of a finger on the pad to a position on the screen. There are two general approaches to providing this linkage: **capacitance sensing** and **conductance sensing.** Capacitance sensing depends on the charge carried by the human body, while conductance sensing only requires that pressure be applied to a particular position on the pad. In other words, the wearing of gloves or using a pencil will not work with capacitance sensing but is effective with conductance sensing.

There are two methods commonly employed for capacitance testing. One is referred to as the **matrix approach,** and the other is called the **capacitive shunt approach.** The matrix approach requires two sets of parallel conductors separated by a dielectric and perpendicular to each other as shown in Fig. 10.78. Two sets of perpendicular wires are required to permit the determination of the location of the point on the two-dimensional plane—one for the horizontal displacement and the other

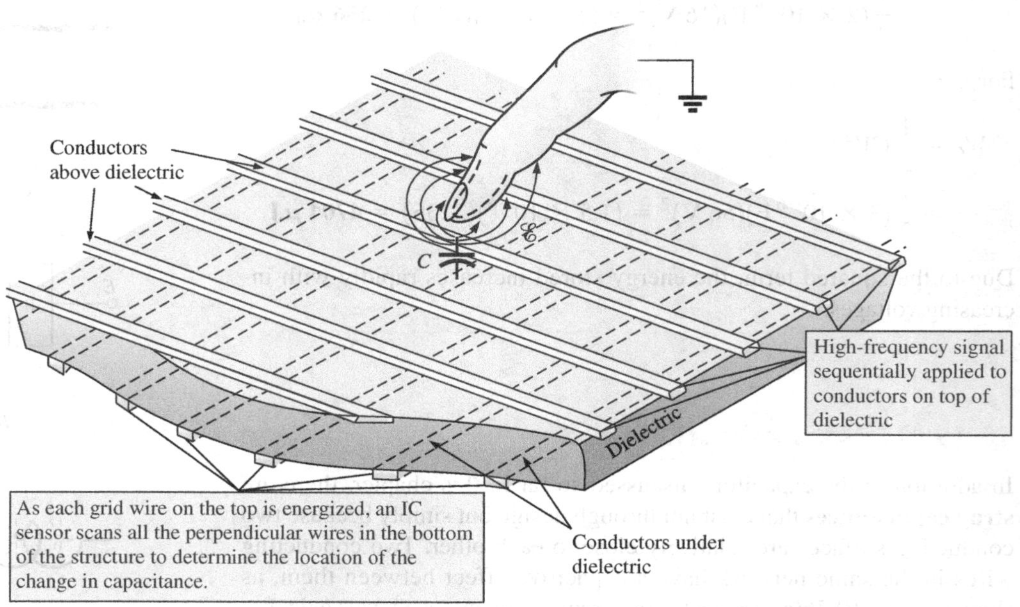

FIG. 10.78
Matrix approach to capacitive sensing in a touch pad.

for the vertical displacement. The result when looking down at the pad is a two-dimensional grid with intersecting points or nodes. Its operation requires the application of a high-frequency signal that will permit the monitoring of the capacitance between each set of wires at each intersection as shown in Fig. 10.78 using ICs connected to each set of wires. When a finger approaches a particular intersection the charge on the finger will change the field distribution at that point by drawing some of the field lines away from the intersection. Some like to think of the finger as applying a **virtual ground** to the point as shown in the figure. Recall from the discussion in Section 10.3 that any change in electric field strength for a fixed capacitor (such as the insertion of a dielectric between the plates of a capacitor) will change the charge on the plates and the level of capacitance determined by $C = Q/V$. The change in capacitance at the intersection will be noted by the ICs. That change in capacitance can then be translated by a **capacitance to digital converter** (CDC) and used to define the location on the screen. Recent experiments have found that this type of sensing is most effective with a soft, delicate touch on the pad rather than hard, firm pressure.

The capacitive shunt approach takes a totally different approach. Rather than establish a grid, a sensor is used to detect changes in capacitive levels. The basic construction for an analog device appears in Fig. 10.79. The sensor has a transmitter and a receiver, both of which are formed on separate printed circuit board (PCB) platforms with a plastic cover over the transmitter to avoid actual contact with the finger. When the excitation signal of 250 kHz is applied to the transmitter platform, an electric field is established between the transmitter and receiver, with a strong fringing effect on the surface of the sensor. If a finger with its negative charge is brought close to the transmitter surface, it will distort the fringing effect by attracting some of the electric field as shown in the figure. The resulting change in total field strength will affect the charge level on the plates of the sensor and therefore the capacitance between the transmitter and receiver. This will be detected by the sensor and provide either the horizontal or vertical position of the contact. The resultant change in capacitance is only in the order of femtofarads, as compared to the picofarads for the sensor, but is still sufficient to be detected by the sensor. The change in capacitance is picked up by a 16-bit Σ-Δ capacitor to digital convertor (CDC) and the

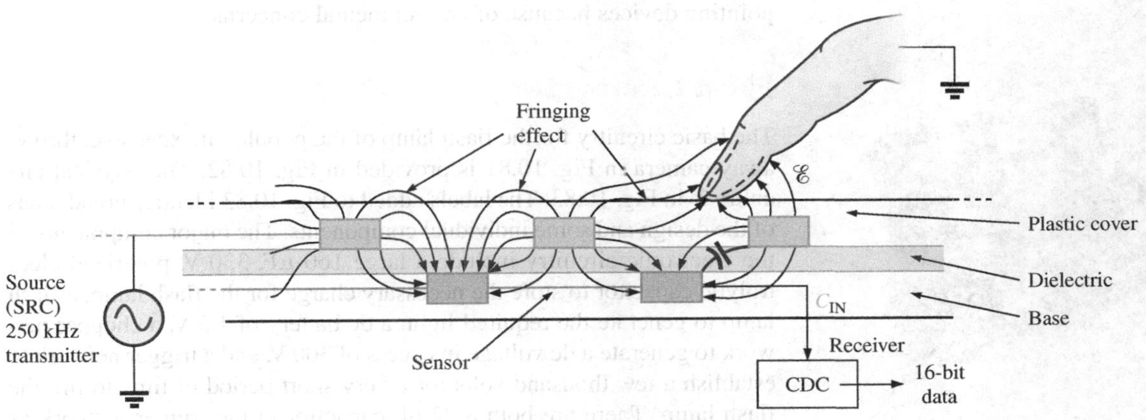

FIG. 10.79
Capacitive shunt approach.

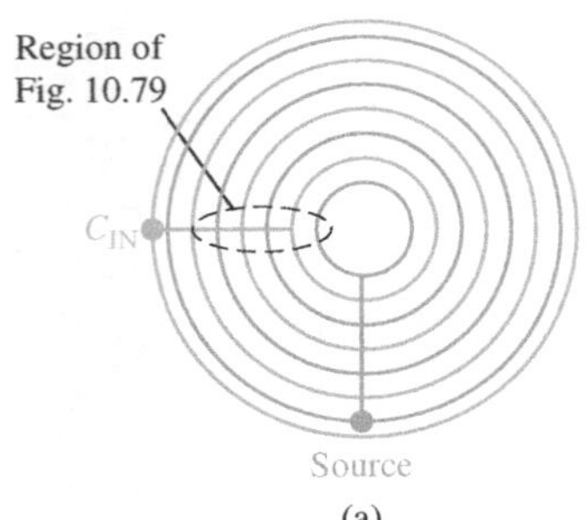

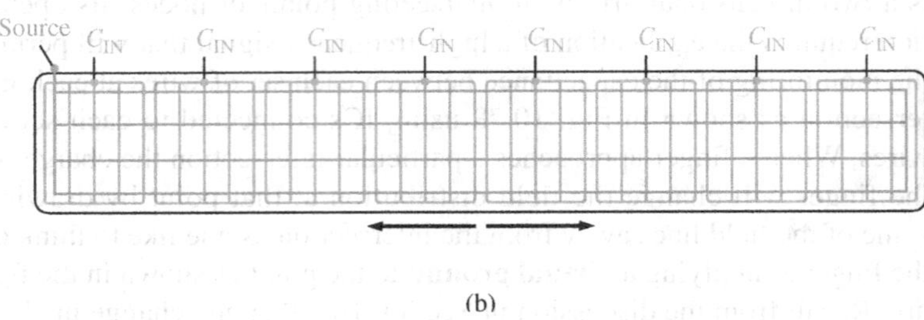

FIG. 10.80
Capacitive shunt sensors: (a) bottom, (b) slice.

results fed into the contoller for the system to which the sensor is connected. The term *shunt* comes from the fact that some of the electric field is "shunted" away from the sensor. The sensors themselves can be made of many different shapes and sizes. For applications such as the circular button for an elevator, the circular pattern of Fig. 10.80(a) may be applied, while for a slide control, it may appear as shown in Fig. 10.80(b). In each case the excitation is applied to the red lines and regions and the capacitance level measured by the C_{IN} blue lines and regions. In other words, a field is established between the red and blue lines throughout the pattern, and touching the pads in any area will reveal a change in capacitance. For a computer touch pad the number of C_{IN} inputs required is one per row and one per column to provide the location in a two-dimensional space.

The last method to be described is the conductance-sensing approach. Basically, it employs two thin metallic conducting surfaces separated by a very thin space. The top surface is usually flexible, while the bottom is fixed and coated with a layer of small conductive nipples. When the top surface is touched, it drops down and touches a nipple, causing the conductance between the two surfaces to increase dramatically in that one location. This change in conductance is then picked up by the ICs on each side of the grid and the location determined for use in setting the position on the screen of the computer. This type of mouse pad permits the use of a pen, pencil, or other nonconductive instrument to set the location on the screen, which is useful in situations in which one may have to wear gloves continually or need to use nonconductive pointing devices because of environmental concerns.

Flash Lamp

FIG. 10.81
Flash camera: general appearance.

The basic circuitry for the flash lamp of the popular, inexpensive, throwaway camera in Fig. 10.81 is provided in Fig. 10.82. The physical circuitry is in Fig. 10.83. The labels added to Fig. 10.83 identify broad areas of the design and some individual components. The major components of the electronic circuitry include a large 160 μF, 330 V, polarized electrolytic capacitor to store the necessary charge for the flash lamp, a flash lamp to generate the required light, a dc battery of 1.5 V, a chopper network to generate a dc voltage in excess of 300 V, and a trigger network to establish a few thousand volts for a very short period of time to fire the flash lamp. There are both a 22 nF capacitor in the trigger network as shown in Figs. 10.82 and 10.83 and a third capacitor of 470 pF in the high-frequency oscillator of the chopper network. In particular, note that

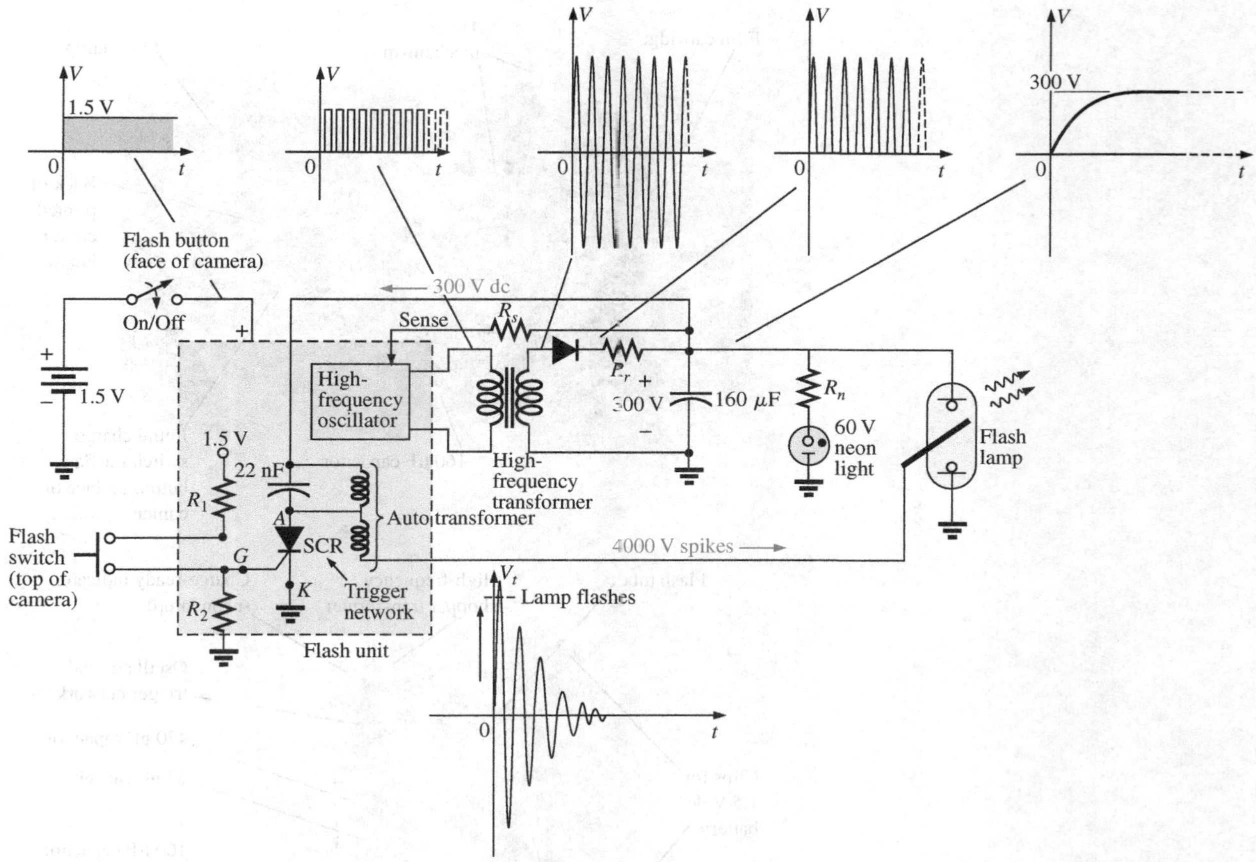

FIG. 10.82
Flash camera: basic circuitry.

the size of each capacitor is directly related to its capacitance level. It should certainly be of some interest that a single source of energy of only 1.5 V dc can be converted to one of a few thousand volts (albeit for a very short period of time) to fire the flash lamp. In fact, that single, small battery has sufficient power for the entire run of film through the camera. Always keep in mind that energy is related to power and time by $W = Pt = (VI)t$. That is, a high level of voltage can be generated for a defined energy level as long as the factors I and t are sufficiently small.

When you first use the camera, you are directed to press the flash button on the face of the camera and wait for the flash-ready light to come on. As soon as the flash button is depressed, the full 1.5 V of the dc battery are applied to an electronic network (a variety of networks can perform the same function) that generates an oscillating waveform of very high frequency (with a high repetitive rate) as shown in Fig. 10.83. The high-frequency transformer then significantly increases the magnitude of the generated voltage and passes it on to a half-wave rectification system (introduced in earlier chapters), resulting in a dc voltage of about 300 V across the 160 μF capacitor to charge the capacitor (as determined by $Q = CV$). Once the 300 V level is reached, the lead marked "sense" in Fig. 10.82 feeds the information back to the oscillator and turns it off until the output dc voltage drops to a low threshold level. When the capacitor is fully charged, the neon light in parallel with the capacitor turns on (labeled "flash-ready lamp" on the camera) to let you know that the camera is ready to use.

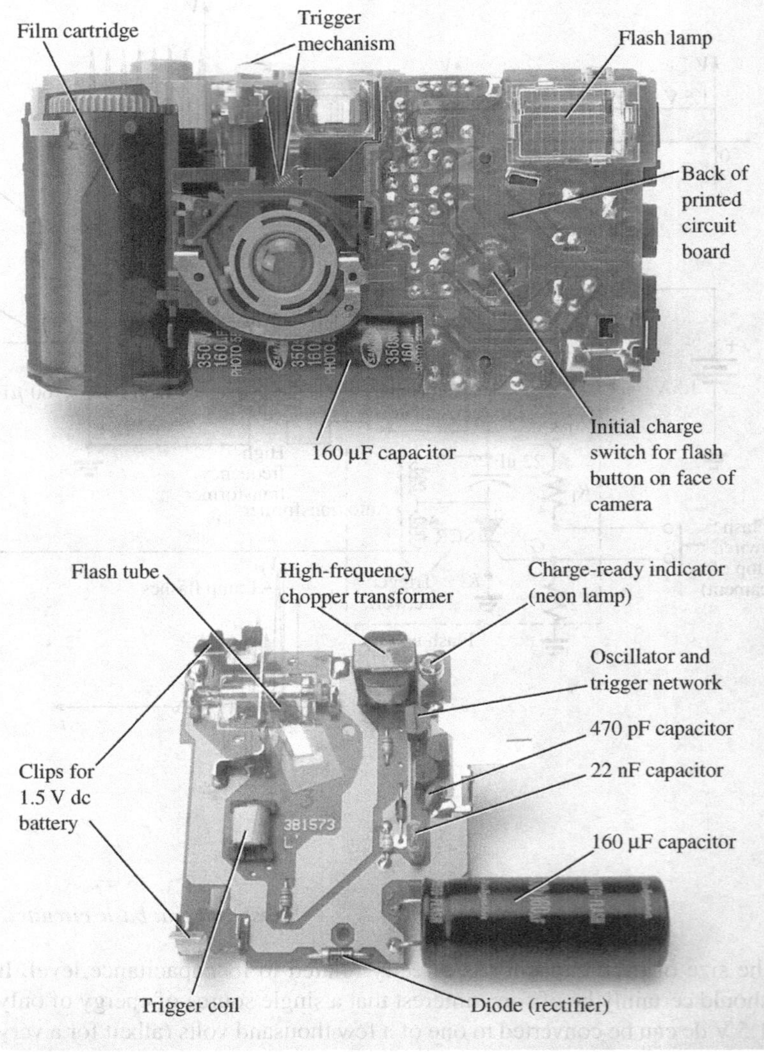

FIG. 10.83
Flash camera: internal construction.

The entire network from the 1.5 V dc level to the final 300 V level is called a *dc-dc converter.* The terminology *chopper network* comes from the fact that the applied dc voltage of 1.5 V was chopped up into one that changes level at a very high frequency so that the transformer can perform its function.

Even though the camera may use a 60 V neon light, the neon light and series resistor R_n must have a full 300 V across the branch before the neon light turns on. Neon lights are simply bulbs with a neon gas that support conduction when the voltage across the terminals reaches a sufficiently high level. There is no filament or hot wire as in a light bulb, but simply conduction through the gaseous medium. For new cameras, the first charging sequence may take 12 s to 15 s. Succeeding charging cycles may only take some 7 s or 8 s because the capacitor still has some residual charge on its plates. If the flash unit is not used, the neon light begins to drain the 300 V dc supply with a drain current in microamperes. As the terminal voltage drops, the neon light eventually turns off. For the unit in Fig. 10.81, it takes about 15 min before the light turns off.

Once off, the neon light no longer drains the capacitor, and the terminal voltage of the capacitor remains fairly constant. Eventually, however, the capacitor discharges due to its own leakage current, and the terminal voltage drops to very low levels. The discharge process is very rapid when the flash unit is used, causing the terminal voltage to drop very quickly ($V = Q/C$) and, through the feedback-sense connection signal, causing the oscillator to start up again and recharge the capacitor. You may have noticed when using a camera of this type that once the camera has its initial charge, you do not need to press the charge button between pictures—it is done automatically. However, if the camera sits for a long period of time, you must depress the charge button, but the charge time is only 3 s or 4 s due to the residual charge on the plates of the capacitor.

The 300 V across the capacitor are insufficient to fire the flash lamp. Additional circuitry, called the *trigger network,* must be incorporated to generate the few thousand volts necessary to fire the flash lamp. The resulting high voltage is one reason that there is a CAUTION note on each camera regarding the high internal voltages generated and the possibility of electrical shock if the camera is opened.

The thousands of volts required to fire the flash lamp require a discussion that introduces elements and concepts beyond the current level of the text. This description is simply a first exposure to some of the interesting possibilities available from the right mix of elements. When the flash switch at the bottom left of Fig. 10.82 is closed, it establishes a connection between the resistors R_1 and R_2. Through a voltage divider action, a dc voltage appears at the gate (G) terminal of the SCR (silicon-controlled rectifier—a device whose state is controlled by the voltage at the gate terminal). This dc voltage turns the SCR "on" and establishes a very low resistance path (like a short circuit) between its anode (A) and cathode (K) terminals. At this point the trigger capacitor, which is connected directly to the 300 V sitting across the capacitor, rapidly charges to 300 V because it now has a direct, low-resistance path to ground through the SCR. Once it reaches 300 V, the charging current in this part of the network drops to 0 A, and the SCR opens up again since it is a device that needs a steady current in the anode circuit to stay on. The capacitor then sits across the parallel coil (with no connection to ground through the SCR) with its full 300 V and begins to quickly discharge through the coil because the only resistance in the circuit affecting the time constant is the resistance of the parallel coil. As a result, a rapidly changing current through the coil generates a high voltage across the coil for reasons to be introduced in Chapter 11.

When the capacitor decays to zero volts, the current through the coil will be zero amperes, but a strong magnetic field has been established around the coil. This strong magnetic field then quickly collapses, establishing a current in the parallel network that recharges the capacitor again. This continual exchange between the two storage elements continues for a period of time, depending on the resistance in the circuit. The more the resistance, the shorter is the "ringing" of the voltage at the output. This action of the energy "flying back" to the other element is the basis for the "flyback" effect that is frequently used to generate high dc voltages such as needed in TVs. In Fig. 10.82, you will find that the trigger coil is connected directly to a second coil to form an autotransformer (a transformer with one end connected). Through transformer action, the high voltage generated across the trigger coil increases further, resulting in the 4000 V necessary to fire the flash lamp. Note in Fig. 10.83 that the 4000 V are applied to a grid that actually lies on the surface of the glass tube of the flash

lamp (not internally connected or in contact with the gases). When the trigger voltage is applied, it excites the gases in the lamp, causing a very high current to develop in the bulb for a very short period of time and producing the desired bright light. The current in the lamp is supported by the charge on the 160 μF capacitor, which is dissipated very quickly. The capacitor voltage drops very quickly, the photo lamp shuts down, and the charging process begins again. If the entire process didn't occur as quickly as it does, the lamp would burn out after a single use.

Surge Protector (Line Conditioner)

In recent years we have all become familiar with the surge protector as a safety measure for our computers, TVs, DVD players, and other sensitive instrumentation. In addition to protecting equipment from unexpected surges in voltage and current, most quality units also filter out (remove) electromagnetic interference (EMI) and radio-frequency interference (RFI). EMI encompasses any unwanted disturbances down the power line established by any combination of electromagnetic effects such as those generated by motors on the line, power equipment in the area emitting signals picked up by the power line acting as an antenna, and so on. RFI includes all signals in the air in the audio range and beyond that may also be picked up by power lines inside or outside the house.

The unit in Fig. 10.84 has all the design features expected in a good line conditioner. Figure 10.84 reveals that it can handle the power drawn by six outlets and that it is set up for FAX/MODEM protection. Also note that it has both LED (light-emitting diode) displays, which reveal whether there is fault on the line or whether the line is OK, and an external circuit breaker to reset the system. In addition, when the surge protector is on, a red light is visible at the power switch.

FIG. 10.84
Surge protector: general appearance.

The schematic in Fig. 10.85 does not include all the details of the design, but it does include the major components that appear in most good line conditioners. First note in the photograph in Fig. 10.86 that the outlets are all connected in parallel, with a ground bar used to establish a ground connection for each outlet. The circuit board had to be flipped over to show the components, so it will take some adjustment to relate the position of the elements on the board to the casing. The *feed line* or *hot lead wire* (black in the actual unit) is connected directly from the line to the circuit breaker. The other end of the circuit breaker is connected to the other side of the circuit board. All the large discs that you see are 2 nF capacitors [not all have been included in Fig 10.86 for clarity]. There are quite a few capacitors to handle all the possibilities. For instance, there are capacitors from line to return (black wire to white wire), from line to ground (black to green), and from return to ground (white to ground). Each has two functions. The first and most obvious function is to prevent any spikes in voltage that may come down the line because of external effects such as lightning from reaching the equipment plugged into the unit. Recall from this chapter that the voltage across capacitors cannot change instantaneously and, in fact, acts to squelch any rapid change in voltage across its terminals. The capacitor, therefore, prevents the line to neutral voltage from changing too quickly, and any spike that tries to come down the line has to find another point in the feed circuit to fall across. In this way, the appliances plugged into the surge protector are well protected.

The second function requires some knowledge of the reaction of capacitors to different frequencies and is discussed in more detail in later

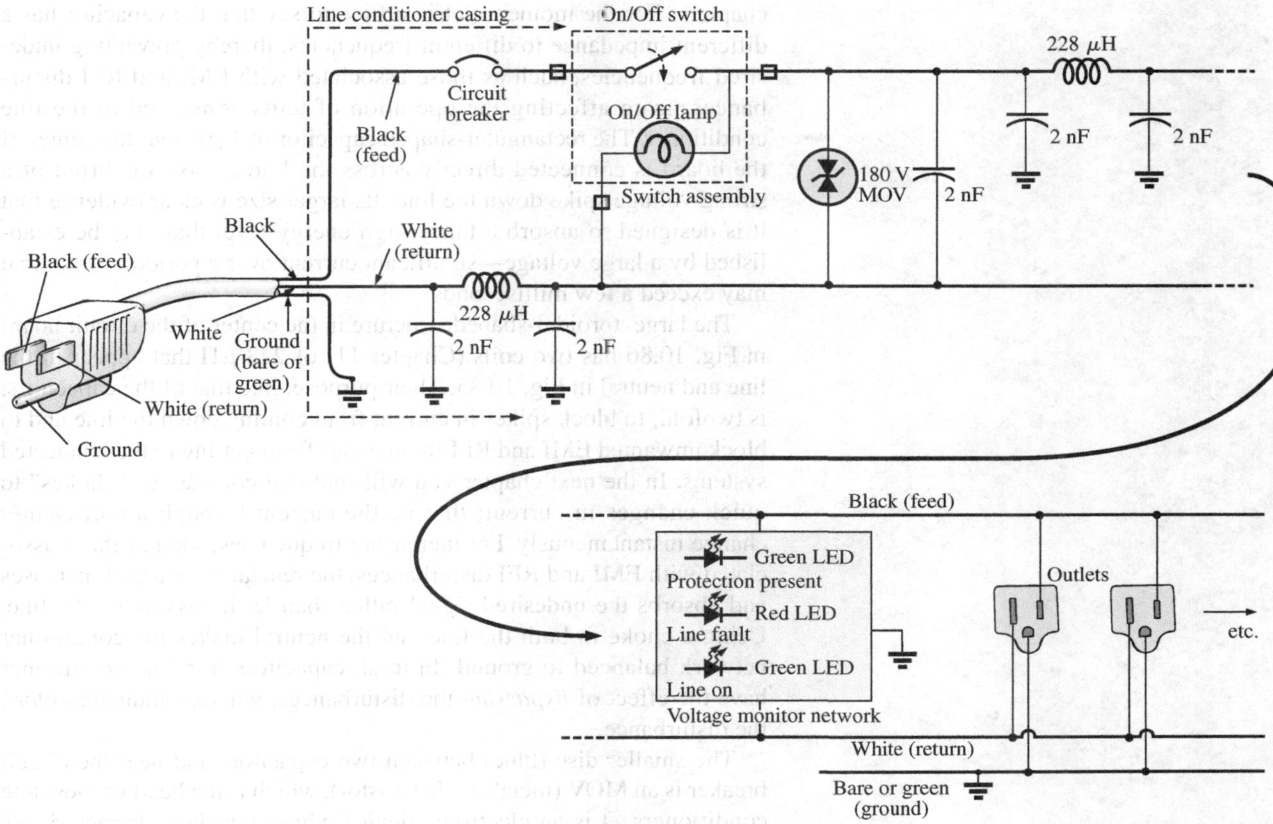

FIG. 10.85
Electrical schematic.

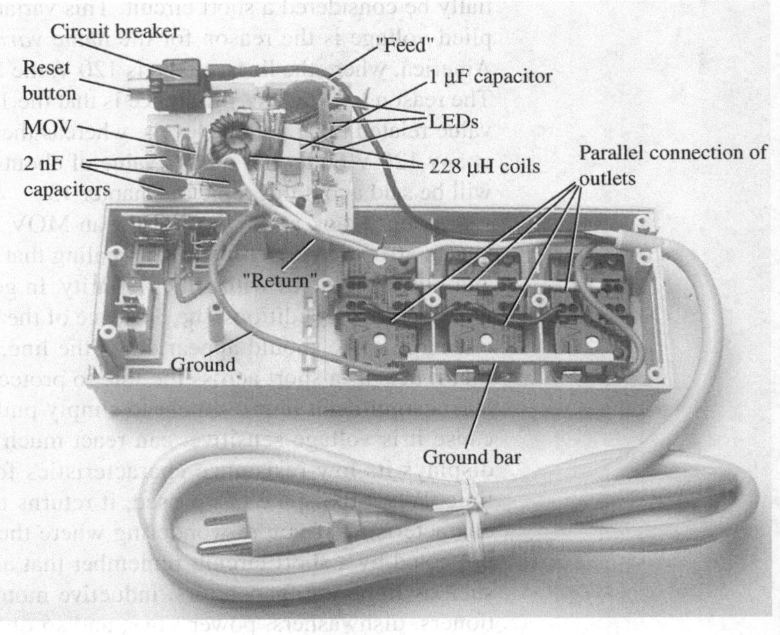

FIG. 10.86
Internal construction of surge protector.

chapters. For the moment, let it suffice to say that the capacitor has a different impedance to different frequencies, thereby preventing undesired frequencies, such as those associated with EMI and RFI disturbances, from affecting the operation of units connected to the line conditioner. The rectangular-shaped capacitor of 1 μF near the center of the board is connected directly across the line to take the brunt of a strong voltage spike down the line. Its larger size is clear evidence that it is designed to absorb a fairly high energy level that may be established by a large voltage—significant current over a period of time that may exceed a few milliseconds.

The large, toroidal-shaped structure in the center of the circuit board in Fig. 10.86 has two coils (Chapter 11) of 228 μH that appear in the line and neutral in Fig. 10.85. Their purpose, like that of the capacitors, is twofold: to block spikes in current from coming down the line and to block unwanted EMI and RFI frequencies from getting to the connected systems. In the next chapter you will find that coils act as "chokes" to quick changes in current; that is, the current through a coil cannot change instantaneously. For increasing frequencies, such as those associated with EMI and RFI disturbances, the reactance of a coil increases and absorbs the undesired signal rather than let it pass down the line. Using a choke in both the line and the neutral makes the conditioner network balanced to ground. In total, capacitors in a line conditioner have the effect of *bypassing* the disturbances, whereas inductors *block* the disturbance.

The smaller disc (blue) between two capacitors and near the circuit breaker is an MOV (metal-oxide varistor), which is the heart of most line conditioners. It is an electronic device whose terminal characteristics change with the voltage applied across its terminals. For the normal range of voltages down the line, its terminal resistance is sufficiently large to be considered an open circuit, and its presence can be ignored. However, if the voltage is too large, its terminal characteristics change from a very large resistance to a very small resistance that can essentially be considered a short circuit. This variation in resistance with applied voltage is the reason for the name *varistor.* For MOVs in North America, where the line voltage is 120 V, the MOVs are 180 V or more. The reason for the 60 V difference is that the 120 V rating is an effective value related to dc voltage levels, whereas the waveform for the voltage at any 120 V outlet has a peak value of about 170 V. A great deal more will be said about this topic in Chapter 13.

Taking a look at the symbol for an MOV in Fig. 10.86, note that it has an arrow in each direction, revealing that the MOV is bidirectional and blocks voltages with either polarity. In general, therefore, for normal operating conditions, the presence of the MOV can be ignored, but if a large spike should appear down the line, exceeding the MOV rating, it acts as a short across the line to protect the connected circuitry. It is a significant improvement to simply putting a fuse in the line because it is voltage sensitive, can react much quicker than a fuse, and displays its low-resistance characteristics for only a short period of time. When the spike has passed, it returns to its normal open-circuit characteristic. If you're wondering where the spike goes if the load is protected by a short circuit, remember that all sources of disturbance, such as lightning, generators, inductive motors (such as in air conditioners, dishwashers, power saws, and so on), have their own "source resistance," and there is always some resistance down the line to absorb the disturbance.

Most line conditioners, as part of their advertising, mention their energy absorption level. The rating of the unit in Fig. 10.84 is 1200 J, which is actually higher than most. Remembering that $W = Pt = EIt$ from the earlier discussion of cameras, we now realize that if a 5000 V spike occurred, we would be left with the product $It = W/E =$ 1200 J/5000 V = 240 mAs. Assuming a linear relationship between all quantities, the rated energy level reveals that a current of 100 A could be sustained for $t =$ 240 mAs/100 A = 2.4 μs, a current of 1000 A for 240 μs, and a current of 10,000 A for 24 μs. Obviously, the higher the power product of E and I, the less is the time element.

The technical specifications of the unit in Fig. 10.84 include an instantaneous response time in the order of picoseconds, with a phone line protection of 5 ns. The unit is rated to dissipate surges up to 6000 V and current spikes up to 96,000 A. It has a very high noise suppression ratio (80 dB; see Chapter 21) at frequencies from 50 kHz to 1000 MHz, and (a credit to the company) it has a lifetime warranty.

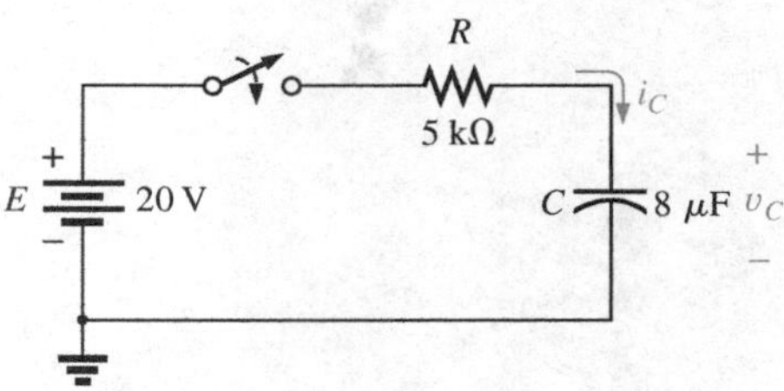

FIG. 10.87
Circuit to be analyzed using PSpice.

10.15 COMPUTER ANALYSIS

PSpice

Transient *RC* Response We now use PSpice to investigate the transient response for the voltage across the capacitor in Fig. 10.87. In all the examples in the text involving a transient response, a switch appeared in series with the source as shown in Fig. 10.88(a). When applying PSpice, we establish this instantaneous change in voltage level by applying a pulse waveform as shown in Fig. 10.88(b) with a pulse width *(PW)* longer than the period (5τ) of interest for the network.

To obtain a pulse source, start with the sequence **Place part key-Libraries-SOURCE-VPULSE-OK.** Once in place, set the label and all the parameters by double-clicking on each to obtain the **Display Properties** dialog box. As you scroll down the list of attributes, you will see the following parameters defined by Fig. 10.89:

V1 is the initial value.
V2 is the pulse level.
TD is the delay time.
TR is the rise time.
TF is the fall time.
PW is the pulse width at the V_2 level.
PER is the period of the waveform.

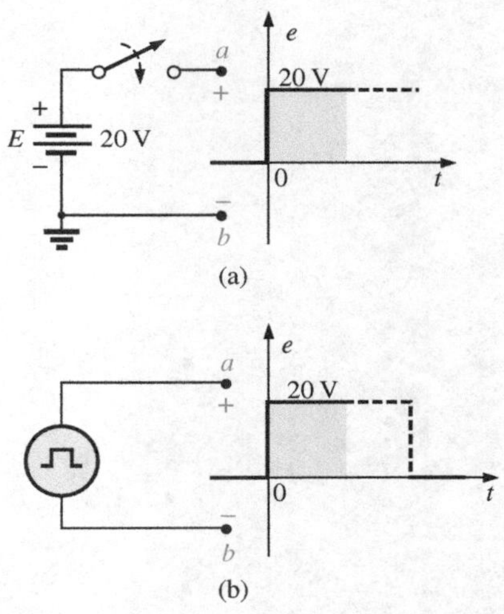

FIG. 10.88
Establishing a switching dc voltage level: (a) series dc voltage-switch combination; (b) PSpice pulse option.

All the parameters have been set as shown on the schematic in Fig. 10.90 for the network in Fig. 10.87. Since a rise and fall time of 0 s is unrealistic from a practical standpoint, 0.1 ms was chosen for each in this example. Further, since $\tau = RC = (5\text{ k}\Omega) \times (8\ \mu\text{F}) = 20$ ms and $5\tau =$ 200 ms, a pulse width of 500 ms was selected. The period was simply chosen as twice the pulse width.

Now for the simulation process. First select the **New Simulation Profile** key to obtain the **New Simulation** dialog box in which **PSpice 10-1** is inserted for the **Name** and **Create** is chosen to leave the dialog box. The **Simulation Settings-PSpice 10-1** dialog box results, and under **Analysis,** choose the **Time Domain (Transient)** option under **Analysis type.** Set the **Run to time** at 200 ms so that only the first five time constants will be plotted. Set the **Start saving data after** option at 0 s to ensure that the data are collected immediately. The **Maximum**

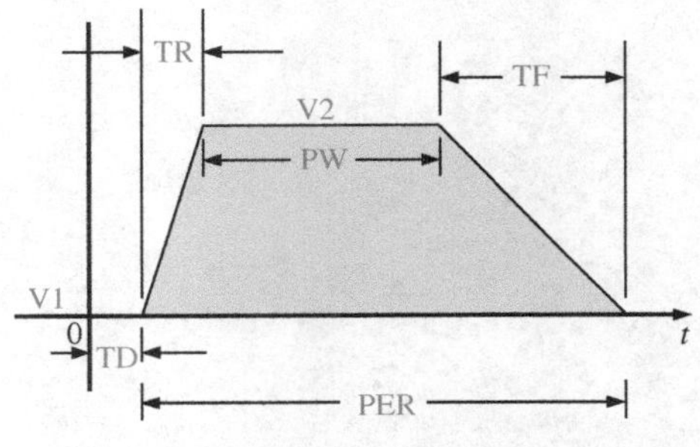

FIG. 10.89
The defining parameters of PSpice VPulse.

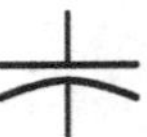

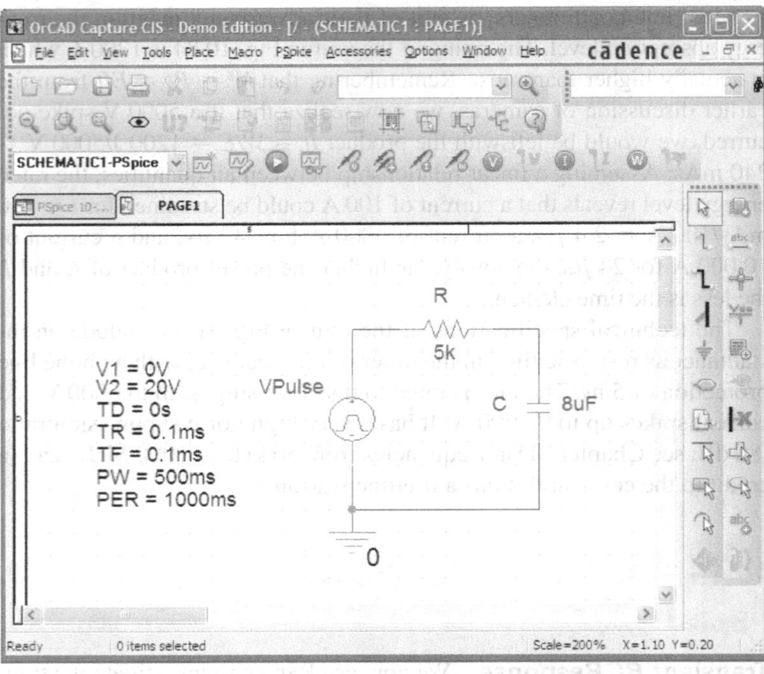

FIG. 10.90
Using PSpice to investigate the transient response of the series R-C circuit in Fig. 10.87.

step size is 1 ms to provide sufficient data points for a good plot. Click **OK,** and you are ready to select the **Run PSpice** key. The result is a graph without a plot (since it has not been defined yet) and an x-axis that extends from 0 s to 200 ms as defined above. If the graph fails to appear, check the **Probe Window** in the **Simulation Settings** to ensure that the **Display Probe Window** (with the **after simulation has completed** option selected) is checked, and **Run-PSpice** again. If problems continue and warning messages do not appear, close the screen by selecting the **X** in the top right corner and respond with a **No** to the request to **Save Files in Project.** The graphs should then appear. Finally, if all else seems to fail, try selecting **View Simulation Results** before the **PSpice-Run** sequence. The response will be a **PSpice** dialog box, indicating that the simulation has not been applied and the data are not available. Respond with a **Yes** to perform the simulation, and the graph should appear. To obtain a plot of the voltage across the capacitor versus time, apply the following sequence: **Add Trace** key-**Add Traces** dialog box-**V1(C)-OK.** The plot in Fig. 10.91 results. The color and thickness of the plot and the axis can be changed by placing the cursor on the plot line and right-clicking. Select **Properties** from the list that appears. A **Trace Properties** dialog box appears in which you can change the color and thickness of the line. Since the plot is against a black background, a better printout occurred when yellow was selected and the line was made thicker as shown in Fig. 10.91. For comparison, plot the applied pulse signal also. This is accomplished by going back to **Trace** and selecting **Add Trace** followed by **V(Vpulse:+)** and **OK.** Now both waveforms appear on the same screen as shown in Fig. 10.84. In this case, the plot has a reddish tint so it can be distinguished from the axis and the other plot. Note that it follows the left axis to the top and travels across the screen at 20 V.

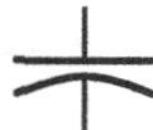

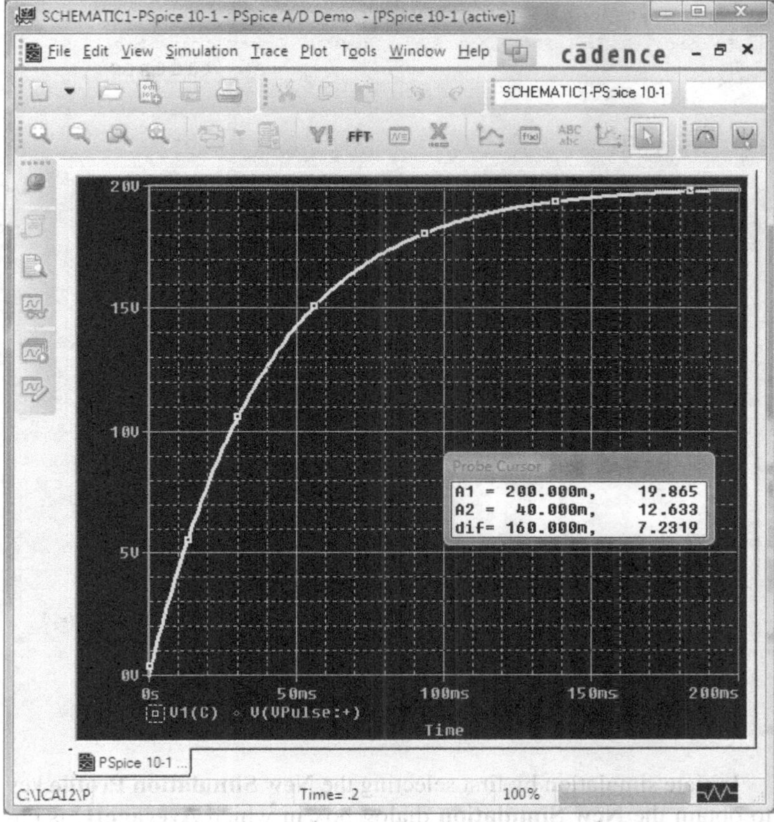

FIG. 10.91

Transient response for the voltage across the capacitor in Fig. 10.87 when VPulse is applied.

If you want the magnitude of either plot at any instant, simply select the **Toggle cursor** key. Then click on **V1(C)** at the bottom left of the screen. A box appears around **V1(C)** that reveals the spacing between the dots of the cursor on the screen. This is important when more than one cursor is used. By moving the cursor to 200 ms, you find that the magnitude **(A1)** is 19.865 V (in the **Probe Cursor** dialog box), clearly showing how close it is to the final value of 20 V. A second cursor can be placed on the screen with a right click and then a click on the same **V1(C)** on the bottom of the screen. The box around **V1(C)** cannot show two boxes, but the spacing and the width of the lines of the box have definitely changed. There is no box around the **Pulse** symbol since it was not selected—although it could have been selected by either cursor. If you now move the second cursor to one time constant of 40 ms, you find that the voltage is 12.659 V as shown in the **Probe Cursor** dialog box. This confirms that the voltage should be 63.2% of its final value of 20 V in one time constant ($0.632 \times 20\text{ V} = 12.4\text{ V}$). Two separate plots could have been obtained by going to **Plot-Add Plot to Window** and then using the trace sequence again.

Average Capacitive Current As an exercise in using the pulse source and to verify our analysis of the average current for a purely capacitive network, the description to follow verifies the results of Example 10.14. For the pulse waveform in Fig. 10.64, the parameters of the pulse supply appear in Fig. 10.92. Note that the rise time is now 2 ms, starting at 0 s, and the fall time is 6 ms. The period was set at 15 ms to permit monitoring the current after the pulse had passed.

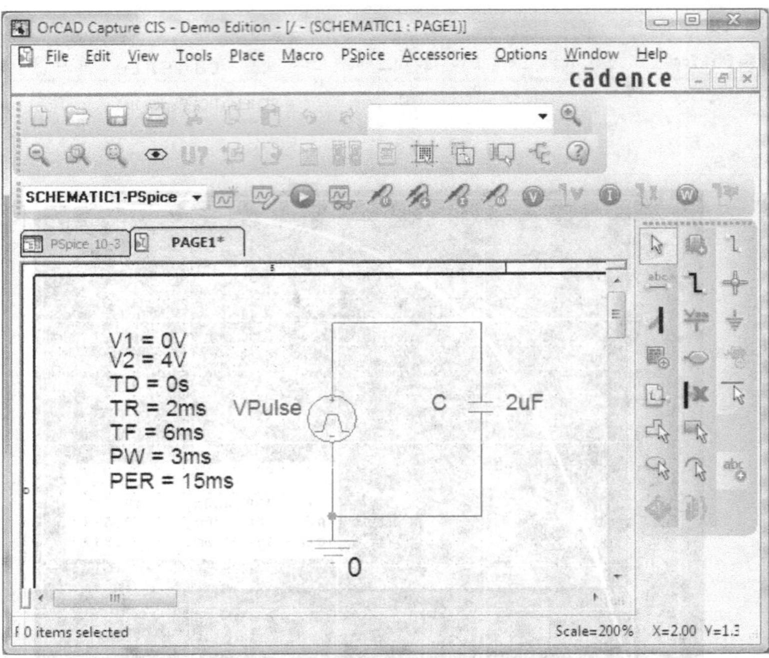

FIG. 10.92
Using PSpice to verify the results in Example 10.14.

Initiate simulation by first selecting the **New Simulation Profile** key to obtain the **New Simulation** dialog box in which **AverageIC** is entered as the **Name.** Choose **Create** to obtain the **Simulation Settings-AverageIC** dialog box. Select **Analysis,** and choose **Time Domain (Transient)** under the **Analysis type** options. Set the **Run to time** to 15 ms to encompass the period of interest, and set the **Start saving data after** at 0 s to ensure data points starting at $t = 0$ s. Select the **Maximum step size** from 15 ms/1000 = 15 μs to ensure 1000 data points for the plot. Click **OK,** and select the **Run PSpice** key. A window appears with a horizontal scale that extends from 0 to 15 ms as defined above. Then select the **Add Trace** key, and choose **I(C)** to appear in the **Trace Expression** below. Click **OK,** and the plot of **I(C)** appears in the bottom of Fig. 10.93. This time it would be nice to see the pulse waveform in the same window but as a separate plot. Therefore, continue with **Plot-Add Plot to Window-Trace-Add Trace-V(Vpulse:+)-OK,** and both plots appear as shown in Fig. 10.93.

Now use the cursors to measure the resulting average current levels. First, select the **I(C)** plot to move the **SEL>>** notation to the lower plot. The **SEL>>** defines which plot for multiplot screens is active. Then select the **Toggle cursor** key, and left-click on the **I(C)** plot to establish the crosshairs of the cursor. Set the value at 1 ms, and the magnitude **A1** is displayed as 4 mA. Right-click on the same plot, and a second cursor results that can be placed at 6 ms to get a response of −1.33 mA **(A2)** as expected from Example 10.14. The plot for **I(C)** was set in the yellow color with a wider line by right-clicking on the curve and choosing **Properties.** You will find after using the **DEMO** version for a while that it informs you that there is a limit of nine files that can be saved under the **File** listing. The result is that any further use of the **DEMO** version requires opening one of the nine files and deleting the contents if you want to run another program. That is, clear the screen and enter the new network.

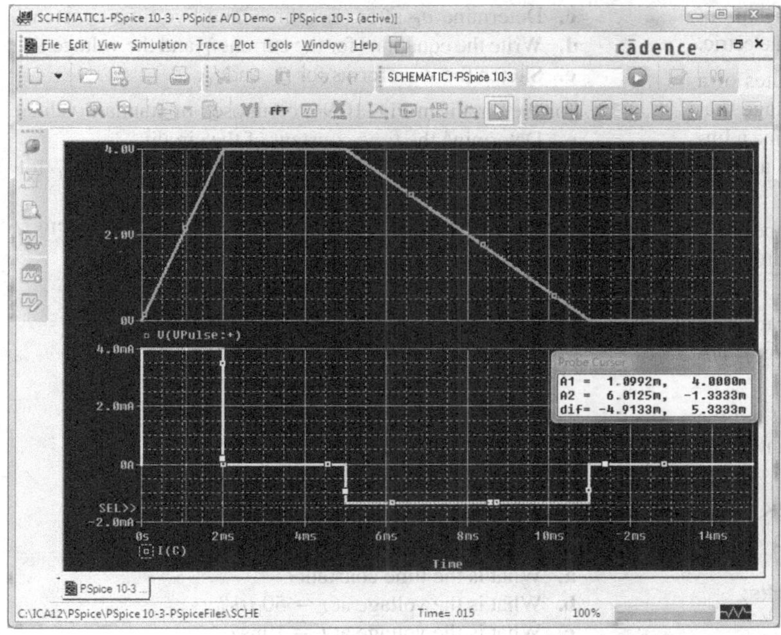

FIG. 10.93
The applied pulse and resulting current for the 2 μF capacitor in Fig. 10.92.

PROBLEMS

SECTION 10.2 The Electric Field

1. a. Find the electric field strength at a point 1 m from a charge of 4 μC.
   b. Find the electric field strength at a point 1 mm [1/1000 the distance of part (a)] from the same charge as part (a) and compare results.
2. The electric field strength is 72 newtons/coulomb (N/C) at a point r meters from a charge of 2 μC. Find the distance r.

SECTIONS 10.3 AND 10.4 Capacitance and Capacitors

3. Find the capacitance of a parallel plate capacitor if 1200 μC of charge are deposited on its plates when 24 V are applied across the plates.
4. How much charge is deposited on the plates of a 0.15 μF capacitor if 45 V are applied across the capacitor?
5. a. Find the electric field strength between the plates of a parallel plate capacitor if 500 mV are applied across the plates and the plates are 1 inch apart.
   b. Repeat part (a) if the distance between the plates is 1/100 inch.
   c. Compare the results of parts (a) and (b). Is the difference in field strength significant?
6. A 6.8 μF parallel plate capacitor has 160 μC of charge on its plates. If the plates are 5 mm apart, find the electric field strength between the plates.
7. Find the capacitance of a parallel plate capacitor if the area of each plate is 0.1 m^2 and the distance between the plates is 0.1 inch. The dielectric is air.
8. Repeat Problem 7 if the dielectric is paraffin-coated paper.
9. Find the distance in mils between the plates of a 2 μF capacitor if the area of each plate is 0.15 m^2 and the dielectric is transformer oil.
10. The capacitance of a capacitor with a dielectric of air is 1360 pF. When a dielectric is inserted between the plates, the capacitance increases to 6.8 nF. Of what material is the dielectric made?
11. The plates of a parallel plate capacitor with a dielectric of Bakelite are 0.2 mm apart and have an area of 0.08 m^2, and 200 V are applied across the plates.
    a. Determine the capacitance.
    b. Find the electric field intensity between the plates.
    c. Find the charge on each plate.
12. A parallel plate air capacitor has a capacitance of 4.7 μF. Find the new capacitance if:
    a. The distance between the plates is doubled (everything else remains the same).
    b. The area of the plates is doubled (everything else remains the same as for the 4.7 μF level).
    c. A dielectric with a relative permittivity of 20 is inserted between the plates (everything else remains the same as for the 4.7 μF level).
    d. A dielectric is inserted with a relative permittivity of 4, and the area is reduced to 1/3 and the distance to 1/4 of their original dimensions.

*13. Find the maximum voltage that can be applied across a parallel plate capacitor of 6800 pF if the area of one plate is 0.02 m^2

and the dielectric is mica. Assume a linear relationship between the dielectric strength and the thickness of the dielectric.

***14.** Find the distance in micrometers between the plates of a parallel plate mica capacitor if the maximum voltage that can be applied across the capacitor is 1200 V. Assume a linear relationship between the breakdown strength and the thickness of the dielectric.

15. A 22 μF capacitor has -200 ppm/°C at room temperature of 20°C. What is the capacitance if the temperature increases to 100°C, the boiling point of water?

16. What is the capacitance of a small teardrop capacitor labeled 40 J? What is the range of expected values as established by the tolerance?

17. A large, flat, mica capacitor is labeled 471F. What are the capacitance and the expected range of values guaranteed by the manufacturer?

18. A small, flat, disc ceramic capacitor is labeled 182K. What are the capacitance level and the expected range of values?

SECTION 10.5 Transients in Capacitive Networks: The Charging Phase

19. For the circuit in Fig. 10.94, composed of standard values:
   a. Determine the time constant of the circuit.
   b. Write the mathematical equation for the voltage v_C following the closing of the switch.
   c. Determine the voltage v_C after one, three, and five time constants.
   d. Write the equations for the current i_C and the voltage v_R.
   e. Sketch the waveforms for v_C and i_C.

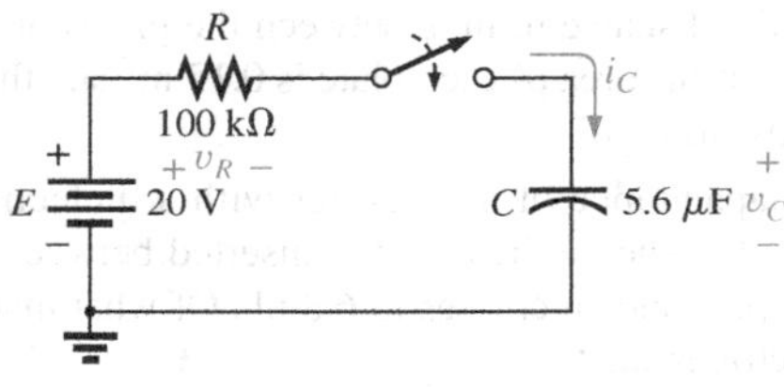

FIG. 10.94
Problems 19 and 20.

20. Repeat Problem 19 for $R = 1$ MΩ, and compare the results.

21. For the circuit in Fig. 10.95, composed of standard values:
   a. Determine the time constant of the circuit.
   b. Write the mathematical equation for the voltage v_C following the closing of the switch.
   c. Determine v_C after one, three, and five time constants.
   d. Write the equations for the current i_C and the voltage v_{R_2}.
   e. Sketch the waveforms for v_C and i_C.

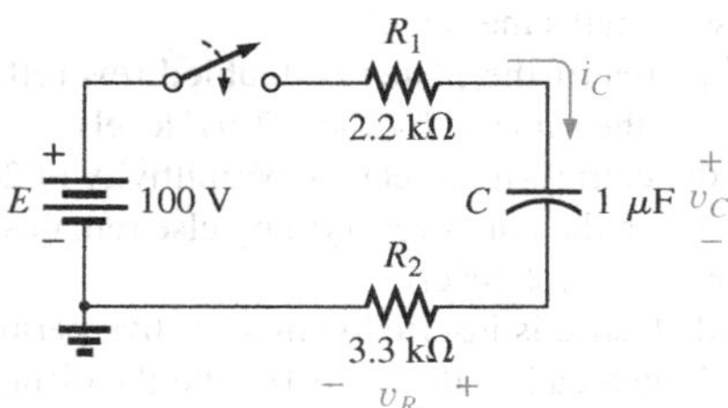

FIG. 10.95
Problem 21.

***22.** For the circuit in Fig. 10.96, composed of standard values:
   a. Determine the time constant of the circuit.
   b. Write the mathematical equation for the voltage v_C following the closing of the switch.
   c. Write the mathematical expression for the current i_C following the closing of the switch.
   d. Sketch the waveforms of v_C and i_C.

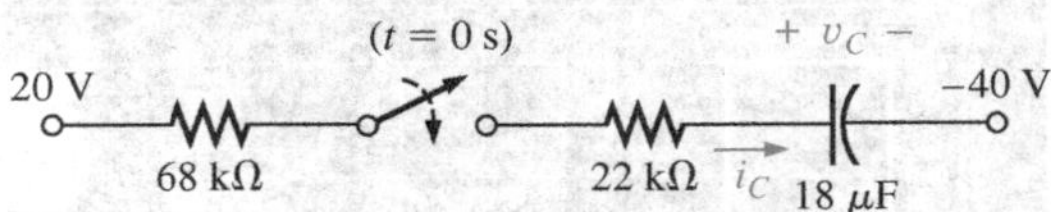

FIG. 10.96
Problem 22.

23. Given the voltage $v_C = 12\text{ V}(1 - e^{-t/100\,\mu s})$:
   a. What is the time constant?
   b. What is the voltage at $t = 50\ \mu$s?
   c. What is the voltage at $t = 1$ ms?

24. The voltage across a 10 μF capacitor in a series R-C circuit is $v_C = 40\text{ mV}(1 - e^{-t/20\text{ ms}})$.
   a. On a practical basis, how much time must pass before the charging phase has passed?
   b. What is the resistance of the circuit?
   c. What is the voltage at $t = 20$ ms?
   d. What is the voltage at 10 time constants?
   e. Under steady-state conditions, how much charge is on the plates?
   f. If the leakage resistance is 1000 MΩ, how long will it take (in hours) for the capacitor to discharge if we assume that the discharge rate is constant throughout the discharge period?

SECTION 10.6 Transients in Capacitive Networks: The Discharging Phase

25. For the R-C circuit in Fig. 10.97, composed of standard values:
   a. Determine the time constant of the circuit when the switch is thrown into position 1.
   b. Find the mathematical expression for the voltage across the capacitor and the current after the switch is thrown into position 1.

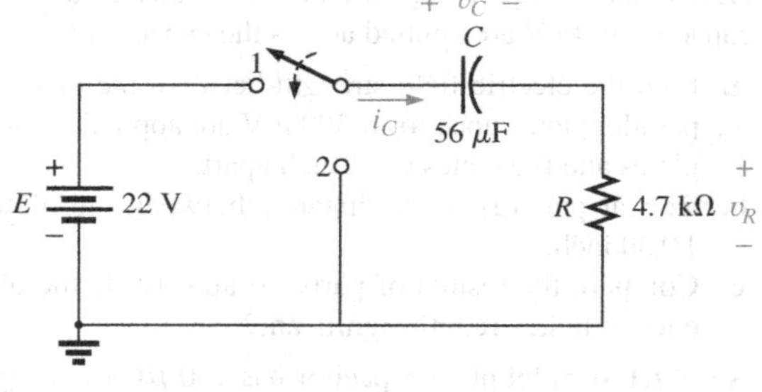

FIG. 10.97
Problem 25.

c. Determine the magnitude of the voltage v_C and the current i_C the instant the switch is thrown into position 2 at $t = 1$ s.
d. Determine the mathematical expression for the voltage v_C and the current i_C for the discharge phase.
e. Plot the waveforms of v_C and i_C for a period of time extending from 0 to 2 s from when the switch was thrown into position 1.

26. For the network in Fig. 10.98, composed of standard values:
a. Write the mathematical expressions for the voltages v_C, and v_{R_1} and the current i_C after the switch is thrown into position 1.
b. Find the values of v_C, v_{R_1}, and i_C when the switch is moved to position 2 at $t = 100$ ms.
c. Write the mathematical expressions for the voltages v_C and v_{R_2} and the current i_C if the switch is moved to position 3 at $t = 200$ ms.
d. Plot the waveforms of v_C, v_{R_2}, and i_C for the time period extending from 0 to 300 ms.

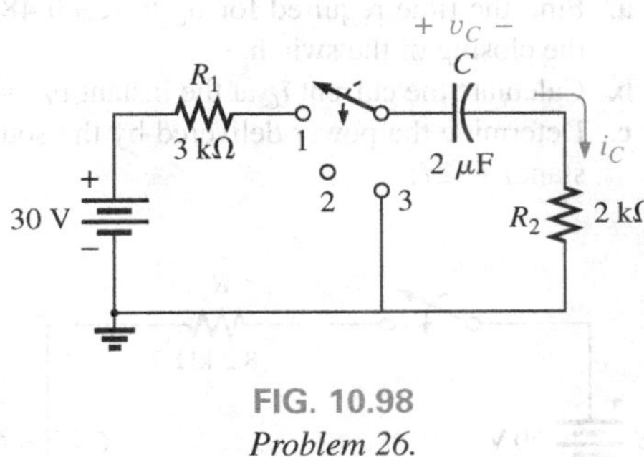

FIG. 10.98
Problem 26.

***27.** For the network in Fig. 10.99, composed of standard values:
a. Find the mathematical expressions for the voltage v_C and the current i_C when the switch is thrown into position 1.
b. Find the mathematical expressions for the voltage v_C and the current i_C if the switch is thrown into position 2 at a time equal to five time constants of the charging circuit.
c. Plot the waveforms of v_C and i_C for a period of time extending from 0 to 30 μs.
d. Plot the waveform of v_R for the same period as in part (a).

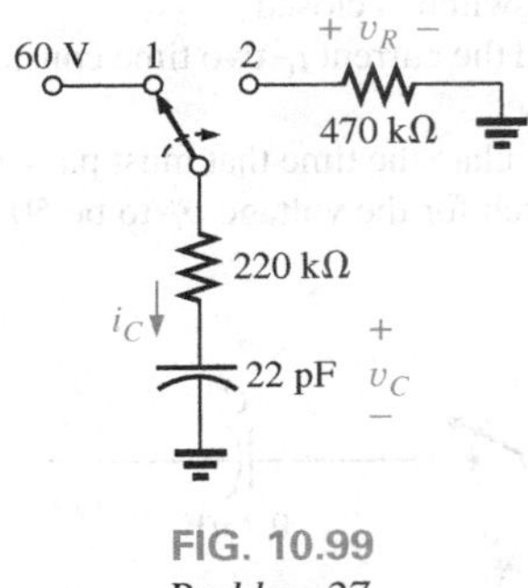

FIG. 10.99
Problem 27.

28. The 1000 μF capacitor in Fig. 10.100 is charged to 12 V in an automobile. To discharge the capacitor before further use, a wire with a resistance of 2 mΩ is placed across the capacitor.

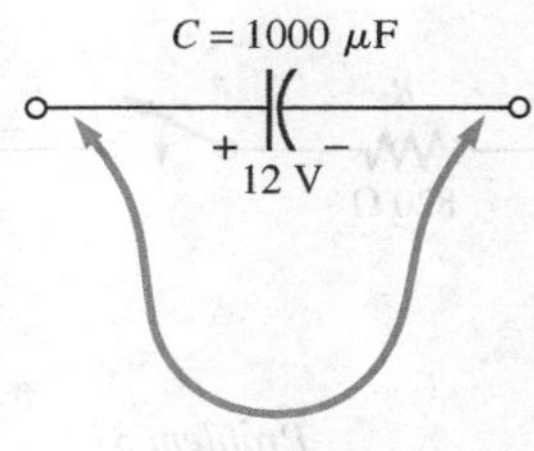

FIG. 10.100
Problem 28.

a. How long will it take to discharge the capacitor?
b. What is the peak value of the current?
c. Based on the answer to part (b), is a spark expected when contact is made with both ends of the capacitor?

SECTION 10.7 Initial Conditions

29. The capacitor in Fig. 10.101 is initially charged to 6 V with the polarity shown.
a. Write the expression for the voltage v_C after the switch is closed.
b. Write the expression for the current i_C after the switch is closed.
c. Plot the results of parts (a) and (b).

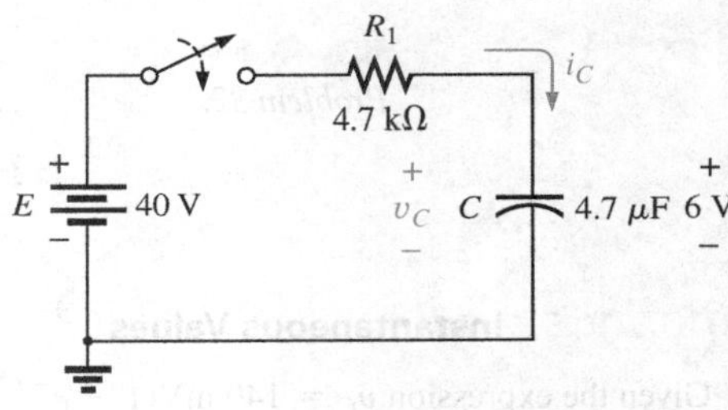

FIG. 10.101
Problem 29.

30. The capacitor in Fig. 10.102 is initially charged to 40 V before the switch is closed. Write the expressions for the voltages v_C and v_R and the current i_C following the closing of the switch. Plot the resulting waveforms.

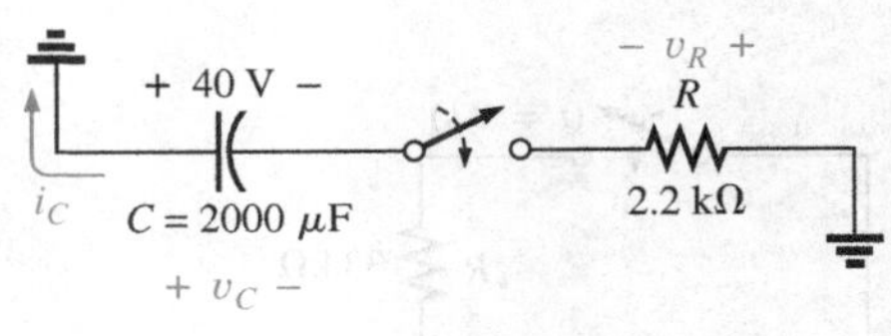

FIG. 10.102
Problem 30.

***31.** The capacitor in Fig. 10.103 is initially charged to 10 V with the polarity shown. Write the expressions for the voltage v_C and the current i_C following the closing of the switch. Plot the resulting waveforms.

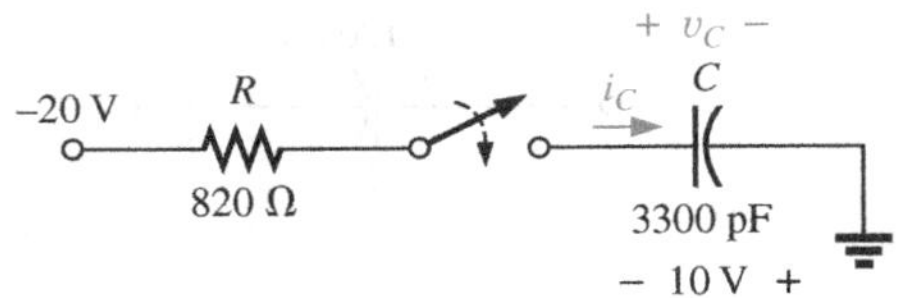

FIG. 10.103
Problem 31.

***32.** The capacitor in Fig. 10.104 is initially charged to 8 V with the polarity shown.
- **a.** Find the mathematical expressions for the voltage v_C and the current i_C when the switch is closed.
- **b.** Sketch the waveforms of v_C and i_C.

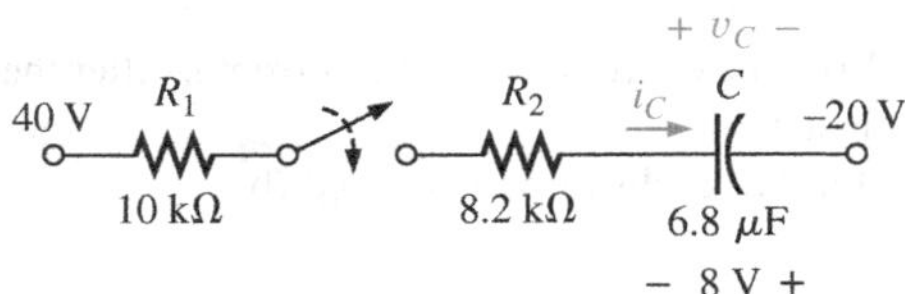

FIG. 10.104
Problem 32.

SECTION 10.8 Instantaneous Values

33. Given the expression $v_C = 140\text{ mV}(1 - e^{-t/2\text{ ms}})$
- **a.** Determine v_C at $t = 1$ ms.
- **b.** Determine v_C at $t = 20$ ms.
- **c.** Find the time t for v_C to reach 100 mV.
- **d.** Find the time t for v_C to reach 138 mV.

34. For the automobile circuit of Fig. 10.105, V_L must be 8 V before the system is activated. If the switch is closed at $t = 0$ s, how long will it take for the system to be activated?

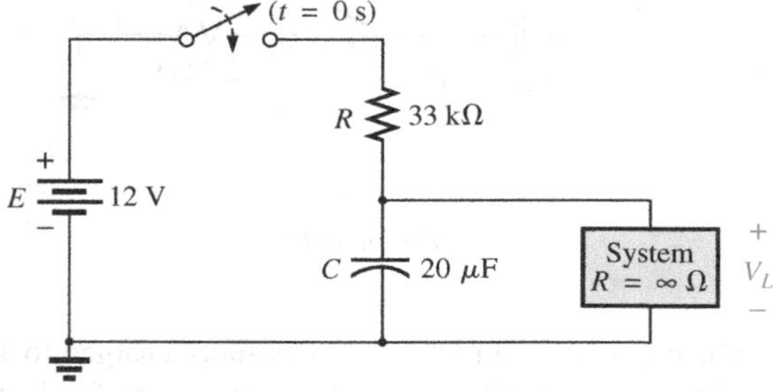

FIG. 10.105
Problem 34.

***35.** Design the network in Fig. 10.106 such that the system turns on 10 s after the switch is closed.

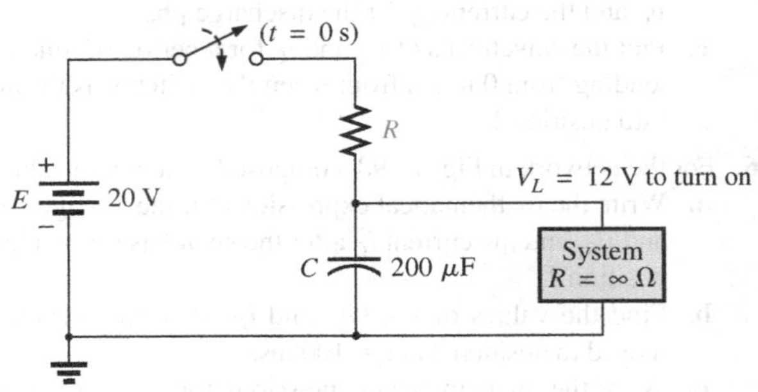

FIG. 10.106
Problem 35.

36. For the circuit in Fig. 10.107:
- **a.** Find the time required for v_C to reach 48 V following the closing of the switch.
- **b.** Calculate the current i_C at the instant $v_C = 48$ V.
- **c.** Determine the power delivered by the source at the instant $t = 2\tau$.

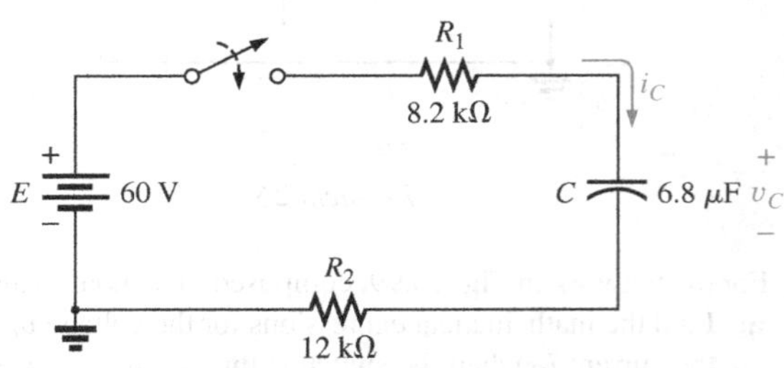

FIG. 10.107
Problem 36.

37. For the system in Fig. 10.108, using a DMM with a 10 MΩ internal resistance in the voltmeter mode:
- **a.** Determine the voltmeter reading one time constant after the switch is closed.
- **b.** Find the current i_C two time constants after the switch is closed.
- **c.** Calculate the time that must pass after the closing of the switch for the voltage v_C to be 50 V.

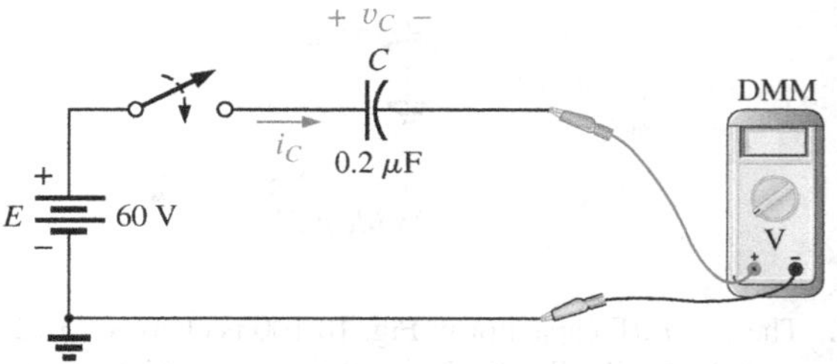

FIG. 10.108
Problem 37.

SECTION 10.9 Thévenin Equivalent: $\tau = R_{Th}C$

38. For the circuit in Fig. 10.109:

a. Find the mathematical expressions for the transient behavior of the voltage v_C and the current i_C following the closing of the switch.

b. Sketch the waveforms of v_C and i_C.

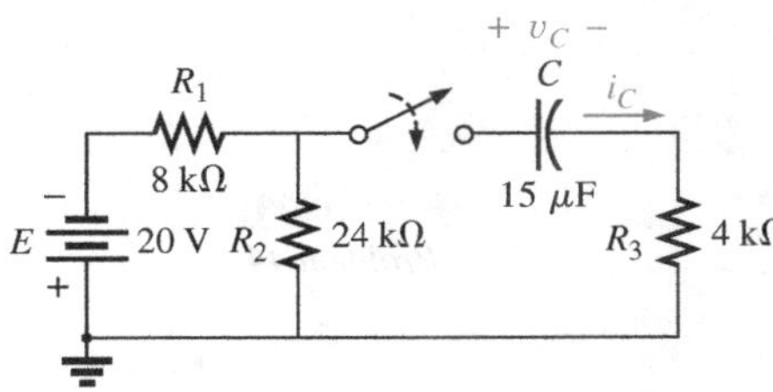

FIG. 10.109
Problem 38.

39. The capacitor in Fig. 10.110 is initially charged to 10 V with the polarity shown.

a. Write the mathematical expressions for the voltage v_C and the current i_C when the switch is closed.

b. Sketch the waveforms of v_C and i_C.

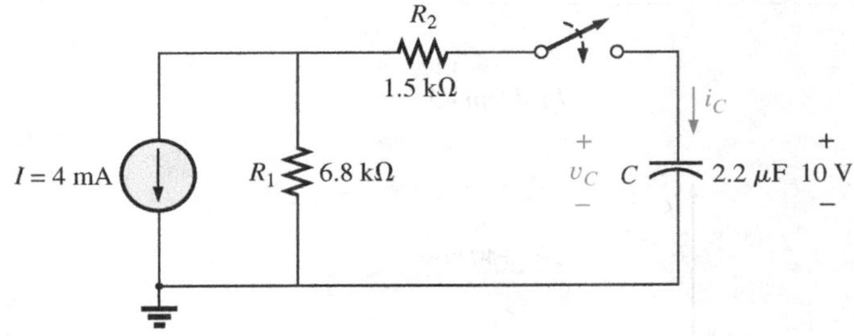

FIG. 10.110
Problem 39.

40. The capacitor in Fig. 10.111 is initially charged to 12 V with the polarity shown.

a. Write the mathematical expressions for the voltage v_C and the current i_C when the switch is closed.

b. Sketch the waveforms of v_C and i_C.

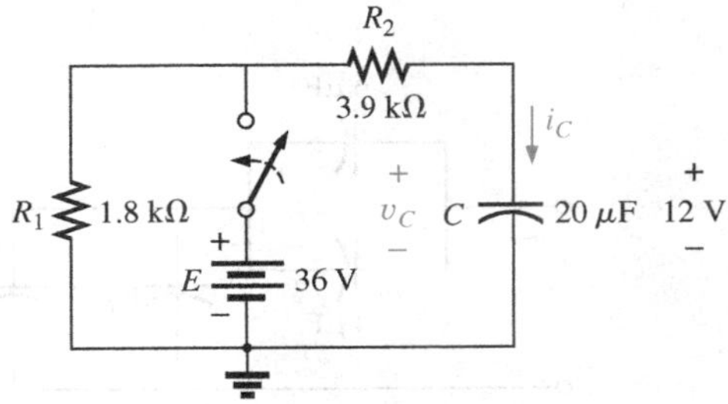

FIG. 10.111
Problem 40.

41. For the circuit in Fig. 10.112:

a. Find the mathematical expressions for the transient behavior of the voltage v_C and the current i_C following the closing of the switch.

b. Sketch the waveforms of v_C and i_C.

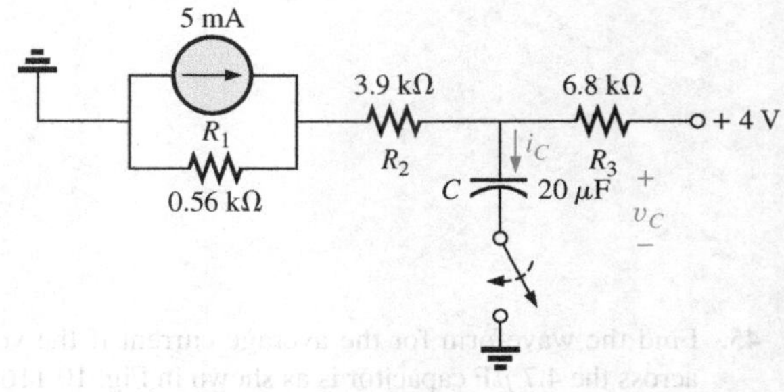

FIG. 10.112
Problem 41.

***42.** The capacitor in Fig. 10.113 is initially charged to 8 V with the polarity shown.

a. Write the mathematical expressions for the voltage v_C and the current i_C when the switch is closed.

b. Sketch the waveforms of v_C and i_C.

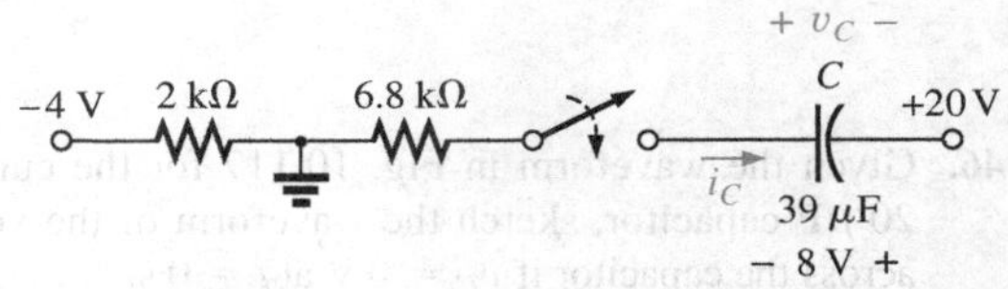

FIG. 10.113
Problem 42.

43. For the system in Fig. 10.114, using a DMM with a 10 MΩ internal resistance in the voltmeter mode:

a. Determine the voltmeter reading four time constants after the switch is closed.

b. Find the time that must pass before i_C drops to 3 μA.

c. Find the time that must pass after the closing of the switch for the voltage across the meter to reach 10 V.

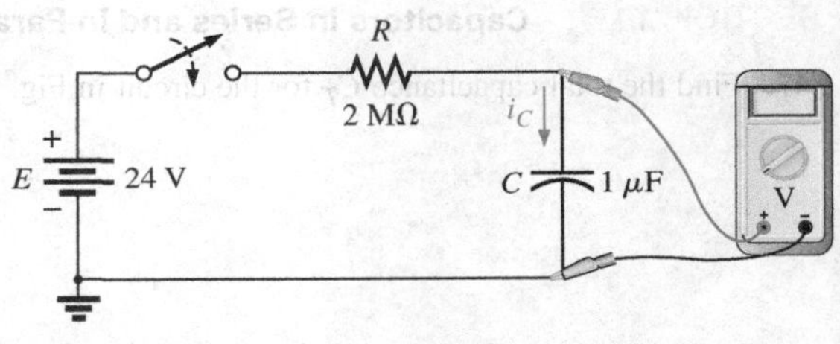

FIG. 10.114
Problem 43.

SECTION 10.10 The Current i_C

44. Find the waveform for the average current if the voltage across the 2 μF capacitor is as shown in Fig. 10.115.

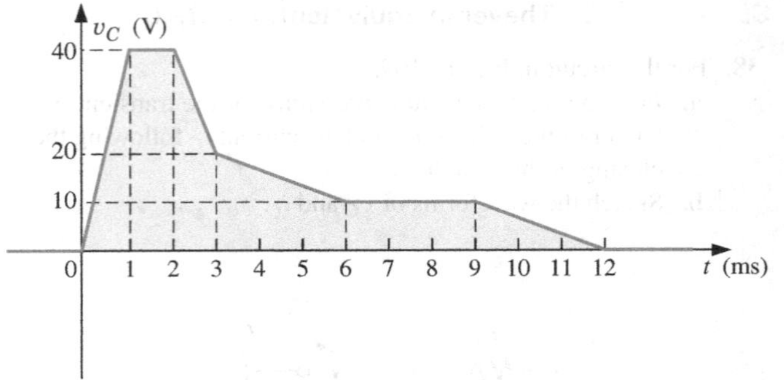

FIG. 10.115
Problem 44.

45. Find the waveform for the average current if the voltage across the 4.7 μF capacitor is as shown in Fig. 10.116.

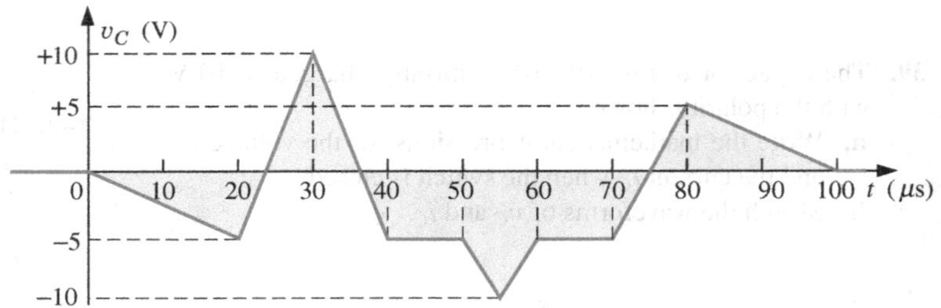

FIG. 10.116
Problem 45.

46. Given the waveform in Fig. 10.117 for the current of a 20 μF capacitor, sketch the waveform of the voltage v_C across the capacitor if $v_C = 0$ V at $t = 0$ s.

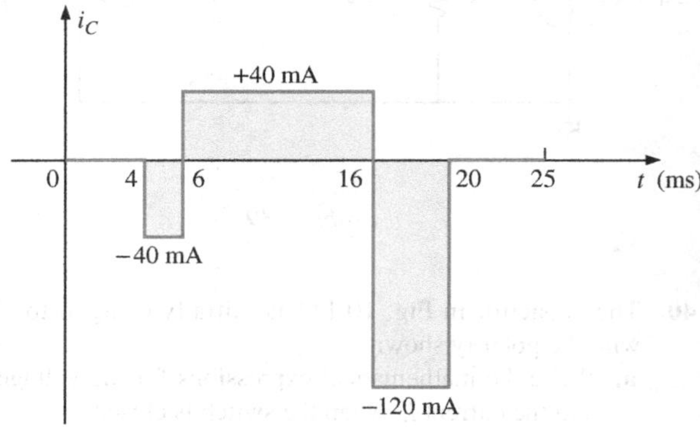

FIG. 10.117
Problem 46.

SECTION 10.11 Capacitors in Series and in Parallel

47. Find the total capacitance C_T for the circuit in Fig. 10.118.

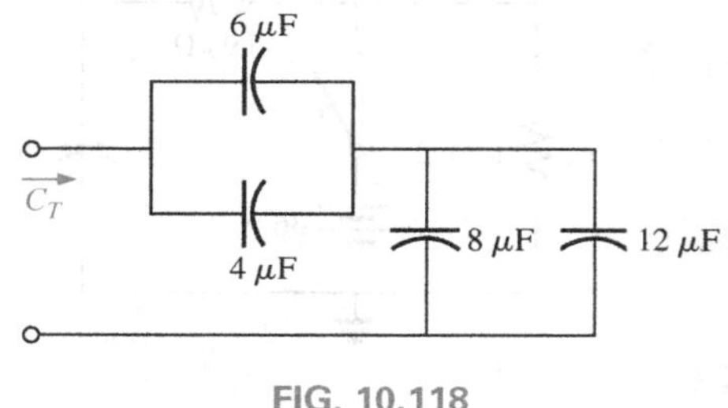

FIG. 10.118
Problem 47.

48. Find the total capacitance C_T for the circuit in Fig. 10.119.

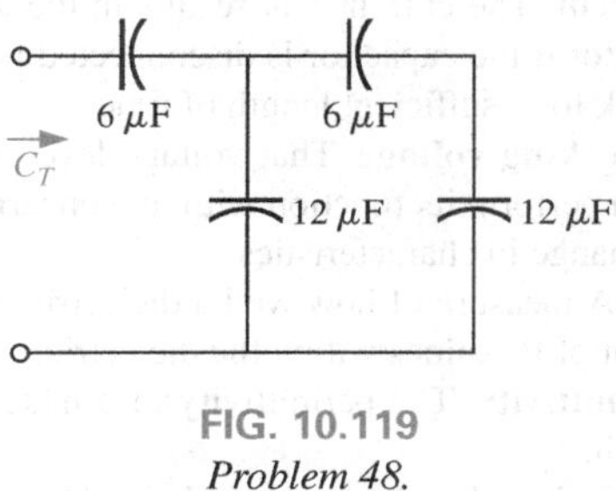

FIG. 10.119
Problem 48.

49. Find the voltage across and the charge on each capacitor for the circuit in Fig. 10.120.

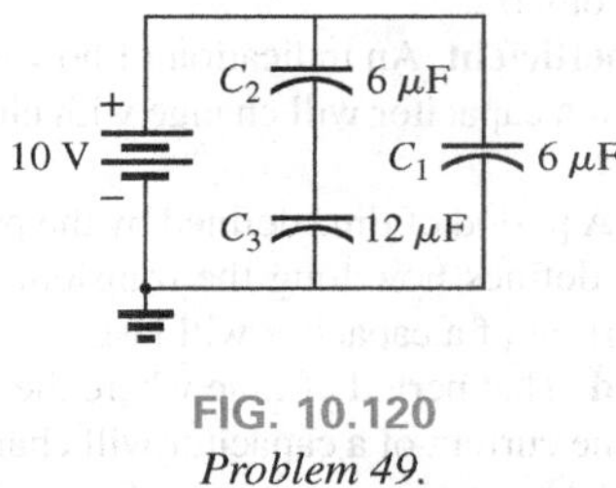

FIG. 10.120
Problem 49.

50. Find the voltage across and the charge on each capacitor for the circuit in Fig. 10.121.

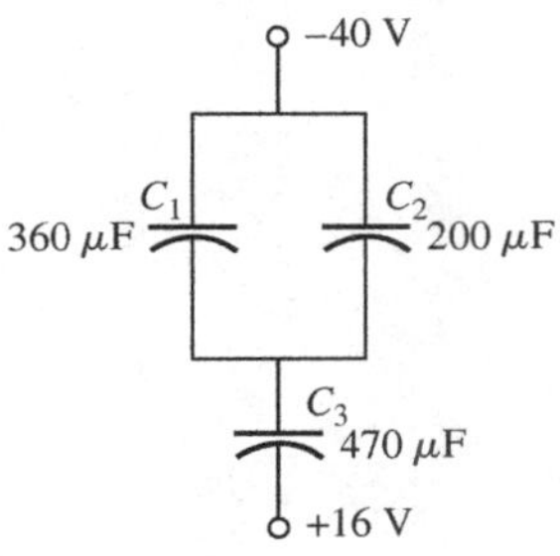

FIG. 10.121
Problem 50.

51. For the configuration in Fig. 10.122, determine the voltage across each capacitor and the charge on each capacitor under steady-state conditions.

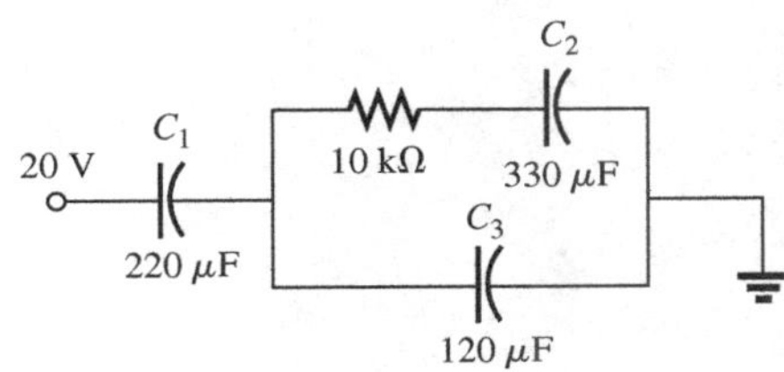

FIG. 10.122
Problem 51.

52. For the configuration in Fig. 10.123, determine the voltage across each capacitor and the charge on each capacitor.

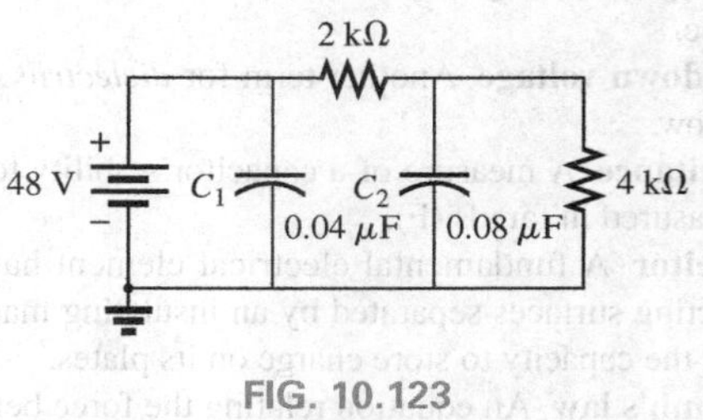

FIG. 10.123
Problem 52.

SECTION 10.12 Energy Stored by a Capacitor

53. Find the energy stored by a 120 pF capacitor with 12 V across its plates.

54. If the energy stored by a 6 μF capacitor is 1200 J, find the charge Q on each plate of the capacitor.

***55.** For the network in Fig. 10.124, determine the energy stored by each capacitor under steady-state conditions.

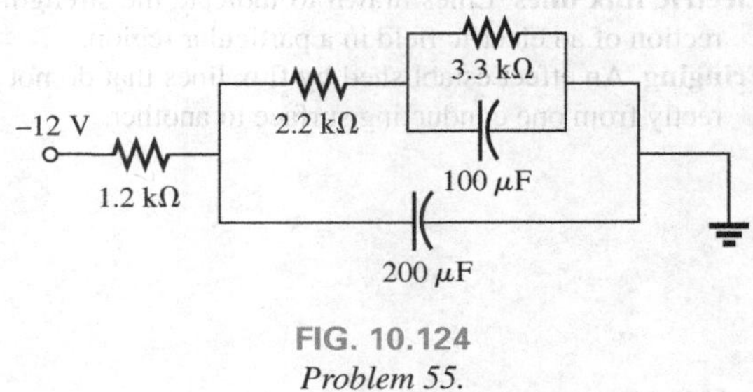

FIG. 10.124
Problem 55.

***56.** An electronic flashgun has a 1000 μF capacitor that is charged to 100 V.

a. How much energy is stored by the capacitor?

b. What is the charge on the capacitor?

c. When the photographer takes a picture, the flash fires for 1/2000 s. What is the average current through the flashtube?

d. Find the power delivered to the flashtube.

e. After a picture is taken, the capacitor has to be recharged by a power supply that delivers a maximum current of 10 mA. How long will it take to charge the capacitor?

SECTION 10.15 Computer Analysis

57. Using PSpice or Multisim, verify the results in Example 10.6.

58. Using the initial condition operator, verify the results in Example 10.8 for the charging phase using PSpice or Multisim.

59. Using PSpice or Multisim, verify the results for v_C during the charging phase in Example 10.11.

60. Using PSpice or Multisim, verify the results in Problem 42.

GLOSSARY

Average current The current defined by a linear (straight line) change in voltage across a capacitor for a specific period of time.

Breakdown voltage Another term for *dielectric strength,* listed below.

Capacitance A measure of a capacitor's ability to store charge; measured in farads (F).

Capacitor A fundamental electrical element having two conducting surfaces separated by an insulating material and having the capacity to store charge on its plates.

Coulomb's law An equation relating the force between two like or unlike charges.

Derivative The instantaneous change in a quantity at a particular instant in time.

Dielectic The insulating material between the plates of a capacitor that can have a pronounced effect on the charge stored on the plates of a capacitor.

Dielectric constant Another term for *relative permittivity,* listed below.

Dielectric strength An indication of the voltage required for unit length to establish conduction in a dielectric.

Electric field strength The force acting on a unit positive charge in the region of interest.

Electric flux lines Lines drawn to indicate the strength and direction of an electric field in a particular region.

Fringing An effect established by flux lines that do not pass directly from one conducting surface to another.

Initial value The steady-state voltage across a capacitor before a transient period begins.

Leakage current The current that results in the total discharge of a capacitor if the capacitor is disconnected from the charging network for a sufficient length of time.

Maximum working voltage That voltage level at which a capacitor can perform its function without concern about breakdown or change in characteristics.

Permittivity A measure of how well a dielectric *permits* the establishment of flux lines within the dielectric.

Relative permittivity The permittivity of a material compared to that of air.

Steady-state region A period of time defined by the fact that the voltage across a capacitor has reached a level that, for all practical purposes, remains constant.

Stray capacitance Capacitances that exist not through design but simply because two conducting surfaces are relatively close to each other.

Temperature coefficient An indication of how much the capacitance value of a capacitor will change with change in temperature.

Time constant A period of time defined by the parameters of the network that defines how long the transient behavior of the voltage or current of a capacitor will last.

Transient period That period of time where the voltage across a capacitor or the current of a capacitor will change in value at a rate determined by the time constant of the network.

Inductors

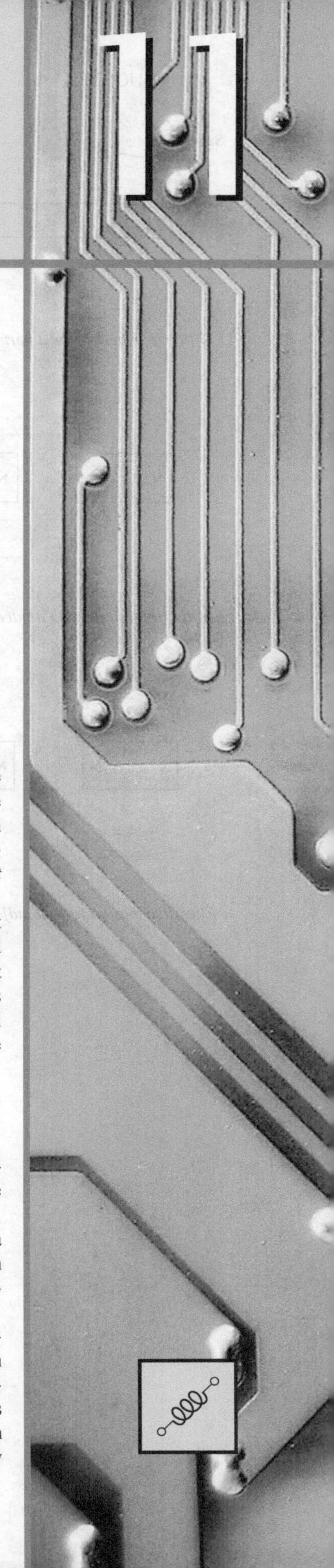

Objectives

- *Become familiar with the basic construction of an inductor, the factors that affect the strength of the magnetic field established by the element, and how to read the nameplate data.*
- *Be able to determine the transient (time-varying) response of an inductive network and plot the resulting voltages and currents.*
- *Understand the impact of combining inductors in series or parallel.*
- *Develop some familiarity with the use of PSpice or Multisim to analyze networks with inductive elements.*

11.1 INTRODUCTION

Three basic components appear in the majority of electrical/electronic systems in use today. They include the *resistor* and the *capacitor,* which have already been introduced, and the **inductor,** to be examined in detail in this chapter. In many ways, the inductor is the dual of the capacitor; that is, the voltage of one is applicable to the current of the other, and vice versa. In fact, some sections in this chapter parallel those in Chapter 10 on the capacitor. Like the capacitor, *the inductor exhibits its true characteristics only when a change in voltage or current is made in the network.*

Recall from Chapter 10 that a capacitor can be replaced by an open-circuit equivalent under steady-state conditions. You will see in this chapter that an inductor can be replaced by a short-circuit equivalent under steady-state conditions. Finally, you will learn that while resistors dissipate the power delivered to them in the form of heat, ideal capacitors store the energy delivered to them in the form of an electric field. Inductors, in the ideal sense, are like capacitors in that they also store the energy delivered to them—but in the form of a magnetic field.

11.2 MAGNETIC FIELD

Magnetism plays an integral part in almost every electrical device used today in industry, research, or the home. Generators, motors, transformers, circuit breakers, televisions, computers, tape recorders, and telephones all employ magnetic effects to perform a variety of important tasks.

The compass, used by Chinese sailors as early as the second century A.D., relies on a **permanent magnet** for indicating direction. A permanent magnet is made of a material, such as steel or iron, that remains magnetized for long periods of time without the need for an external source of energy.

In 1820, the Danish physicist Hans Christian Oersted discovered that the needle of a compass deflects if brought near a current-carrying conductor. This was the first demonstration that electricity and magnetism were related. In the same year, the French physicist André-Marie Ampère performed experiments in this area and developed what is presently known as **Ampère's circuital law.** In subsequent years, others, such as Michael Faraday, Karl Friedrich Gauss, and James Clerk Maxwell, continued to experiment in this area and developed many

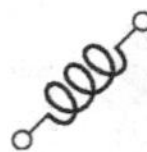

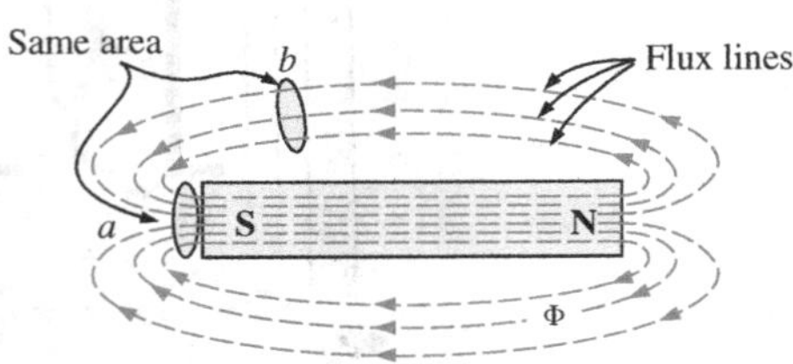

FIG. 11.1
Flux distribution for a permanent magnet.

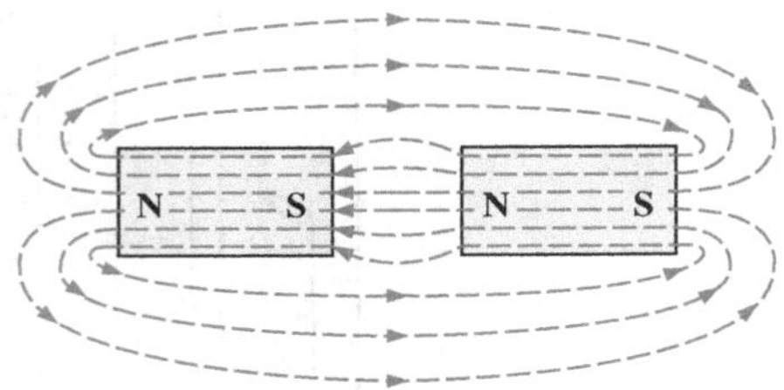

FIG. 11.2
Flux distribution for two adjacent, opposite poles.

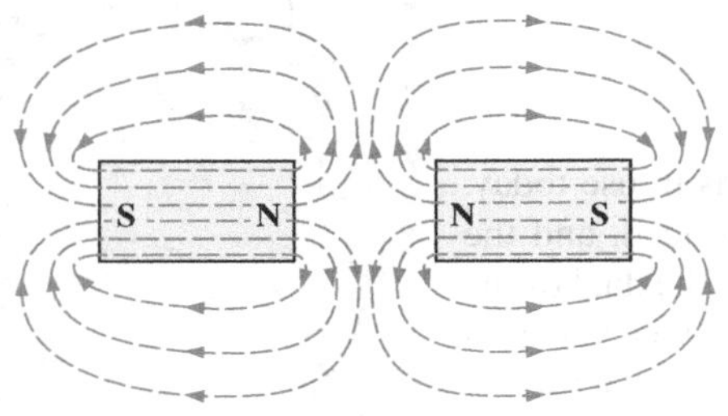

FIG. 11.3
Flux distribution for two adjacent, like poles.

of the basic concepts of **electromagnetism**—magnetic effects induced by the flow of charge, or current.

A magnetic field exists in the region surrounding a permanent magnet, which can be represented by **magnetic flux lines** similar to electric flux lines. Magnetic flux lines, however, do not have origins or terminating points as do electric flux lines but exist in *continuous loops,* as shown in Fig. 11.1.

The magnetic flux lines radiate from the north pole to the south pole, returning to the north pole through the metallic bar. Note the equal spacing between the flux lines within the core and the symmetric distribution outside the magnetic material. These are additional properties of magnetic flux lines in homogeneous materials (that is, materials having uniform structure or composition throughout). It is also important to realize that the continuous magnetic flux line will strive to occupy as small an area as possible. This results in magnetic flux lines of minimum length between the unlike poles, as shown in Fig. 11.2. The strength of a magnetic field in a particular region is directly related to the density of flux lines in that region. In Fig. 11.1, for example, the magnetic field strength at point *a* is twice that at point *b* since twice as many magnetic flux lines are associated with the perpendicular plane at point *a* than at point *b.* Recall from childhood experiments that the strength of permanent magnets is always stronger near the poles.

If unlike poles of two permanent magnets are brought together, the magnets attract, and the flux distribution is as shown in Fig. 11.2. If like poles are brought together, the magnets repel, and the flux distribution is as shown in Fig. 11.3.

If a nonmagnetic material, such as glass or copper, is placed in the flux paths surrounding a permanent magnet, an almost unnoticeable change occurs in the flux distribution (Fig. 11.4). However, if a magnetic material, such as soft iron, is placed in the flux path, the flux lines pass through the soft iron rather than the surrounding air because flux lines pass with greater ease through magnetic materials than through air. This principle is used in shielding sensitive electrical elements and instruments that can be affected by stray magnetic fields (Fig. 11.5).

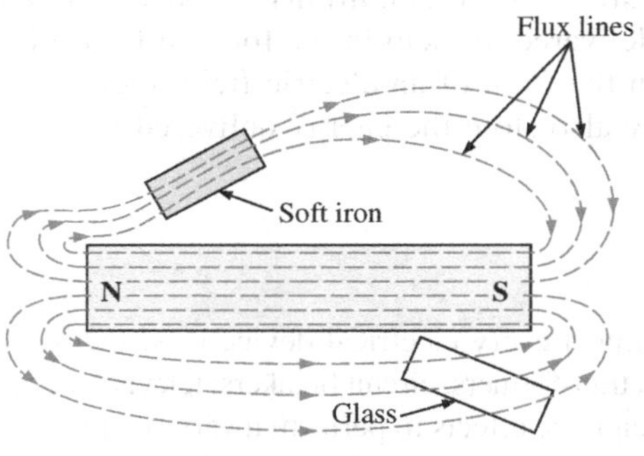

FIG. 11.4
Effect of a ferromagnetic sample on the flux distribution of a permanent magnet.

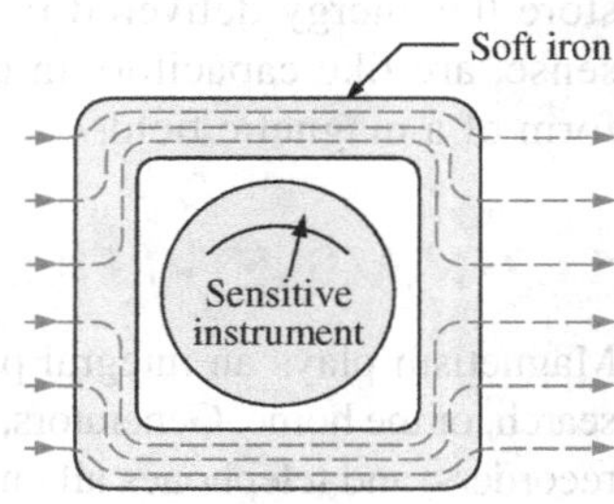

FIG. 11.5
Effect of a magnetic shield on the flux distribution.

A magnetic field (represented by concentric magnetic flux lines, as in Fig. 11.6) is present around every wire that carries an electric current. The direction of the magnetic flux lines can be found simply by placing the thumb of the *right* hand in the direction of *conventional* current flow and noting the direction of the fingers. (This method is commonly called the *right-hand rule.*) If the conductor is wound in a

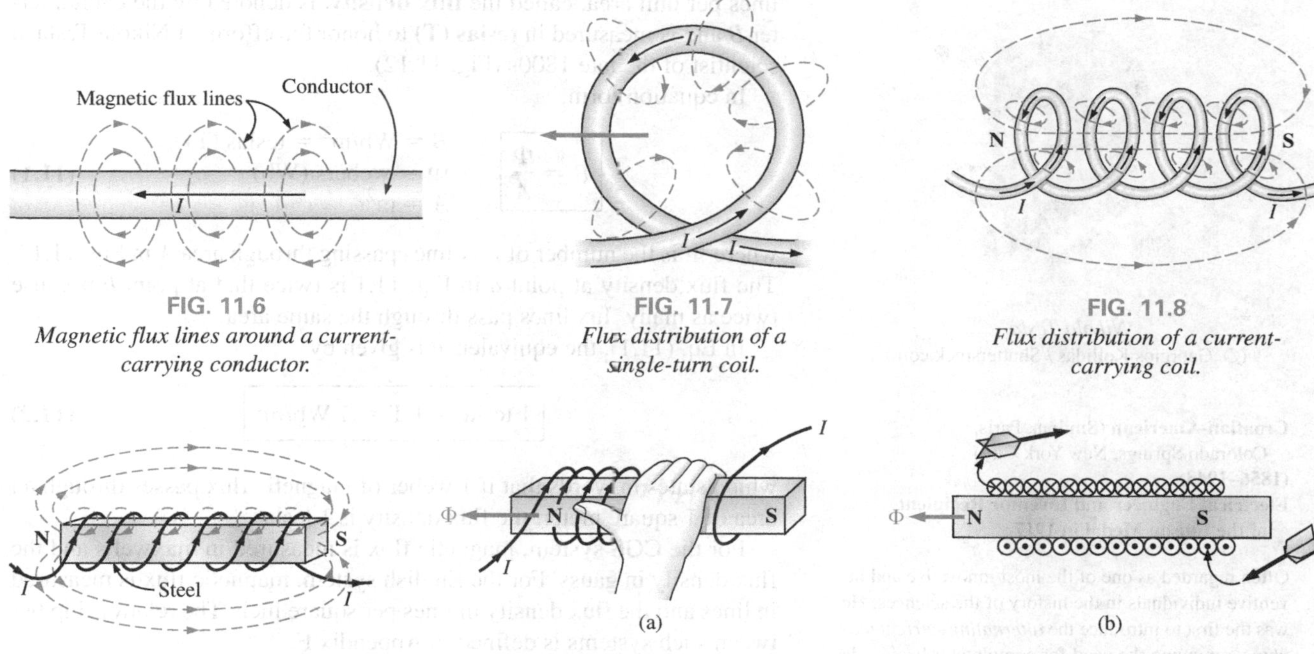

FIG. 11.6
Magnetic flux lines around a current-carrying conductor.

FIG. 11.7
Flux distribution of a single-turn coil.

FIG. 11.8
Flux distribution of a current-carrying coil.

FIG. 11.9
Electromagnet.

FIG. 11.10
Determining the direction of flux for an electromagnet: (a) method; (b) notation.

single-turn coil (Fig. 11.7), the resulting flux flows in a common direction through the center of the coil. A coil of more than one turn produces a magnetic field that exists in a continuous path through and around the coil (Fig. 11.8).

The flux distribution of the coil is quite similar to that of the permanent magnet. The flux lines leaving the coil from the left and entering to the right simulate a north and a south pole, respectively. The principal difference between the two flux distributions is that the flux lines are more concentrated for the permanent magnet than for the coil. Also, since *the strength of a magnetic field is determined by the density of the flux lines,* the coil has a weaker field strength. The field strength of the coil can be effectively increased by placing certain materials, such as iron, steel, or cobalt, within the coil to increase the flux density within the coil. By increasing the field strength with the addition of the core, we have devised an *electromagnet* (Fig. 11.9) that not only has all the properties of a permanent magnet but also has a field strength that can be varied by changing one of the component values (current, turns, and so on). Of course, current must pass through the coil of the electromagnet for magnetic flux to be developed, whereas there is no need for the coil or current in the permanent magnet. The direction of flux lines can be determined for the electromagnet (or in any core with a wrapping of turns) by placing the fingers of your right hand in the direction of current flow around the core. Your thumb then points in the direction of the north pole of the induced magnetic flux, as demonstrated in Fig. 11.10(a). A cross section of the same electromagnet is in Fig. 11.10(b) to introduce the convention for directions perpendicular to the page. The cross and the dot refer to the tail and the head of the arrow, respectively.

In the SI system of units, magnetic flux is measured in **webers (Wb)** as derived from the surname of Wilhelm Eduard Weber (Fig. 11.11). The applied symbol is the capital Greek letter *phi,* ϕ. The number of flux

FIG. 11.11
Wilhelm Eduard Weber.

German (Wittenberg, Göttingen)
(1804–91)
Physicist
Professor of Physics, University of Göttingen

An important contributor to the establishment of a system of *absolute units* for the electrical sciences, which was beginning to become a very active area of research and development. Established a definition of electric current in an electromagnetic system based on the magnetic field produced by the current. He was politically active and, in fact, was dismissed from the faculty of the University of Göttingen for protesting the suppression of the constitution by the King of Hanover in 1837. However, he found other faculty positions and eventually returned to Göttingen as director of the astronomical observatory. He received honors from England, France, and Germany, including the Copley Medal of the Royal Society of London.

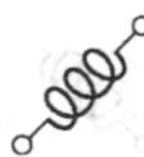

FIG. 11.12
Nikola Tesla.
(© Georgios Kollidas / Shutterstock.com)

Croatian-American (Smiljan, Paris, Colorado Springs, New York City)
(1856–1943)
Electrical Engineer and Inventor Recipient of the Edison Medal in 1917

Often regarded as one of the most innovative and inventive individuals in the history of the sciences. He was the first to introduce the *alternating-current machine,* removing the need for commutator bars of dc machines. After emigrating to the United States in 1884, he sold a number of his patents on *ac machines, transformers,* and *induction coils* (including the Tesla coil as we know it today) to the Westinghouse Electric Company. Some say that his most important discovery was made at his laboratory in Colorado Springs, where in 1900 he discovered *terrestrial stationary waves.* The range of his discoveries and inventions is too extensive to list here but extends from lighting systems to *polyphase power systems* to a *wireless world broadcasting system.*

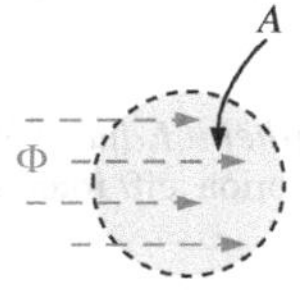

FIG. 11.13
Defining the flux density B.

lines per unit area, called the **flux density,** is denoted by the capital letter B and is measured in **teslas (T)** to honor the efforts of Nikola Tesla, a scientist of the late 1800s (Fig. 11.12).

In equation form,

$$\boxed{B = \frac{\Phi}{A}} \qquad \begin{aligned} B &= \text{Wb/m}^2 = \text{teslas (T)} \\ \Phi &= \text{webers (Wb)} \\ A &= \text{m}^2 \end{aligned} \tag{11.1}$$

where Φ is the number of flux lines passing through area A in Fig. 11.13. The flux density at point a in Fig. 11.1 is twice that at point b because twice as many flux lines pass through the same area.

In Eq. (11.1), the equivalence is given by

$$\boxed{1 \text{ tesla} = 1 \text{ T} = 1 \text{ Wb/m}^2} \tag{11.2}$$

which states in words that if 1 weber of magnetic flux passes through an area of 1 square meter, the flux density is 1 tesla.

For the CGS system, magnetic flux is measured in maxwells and the flux density in gauss. For the English system, magnetic flux is measured in lines and the flux density in lines per square inch. The relationship between such systems is defined in Appendix E.

The flux density of an electromagnet is directly related to the number of turns of, and current through, the coil. The product of the two, called the **magnetomotive force,** is measured in **ampere-turns (At)** as defined by

$$\boxed{\mathcal{F} = NI} \qquad \text{(ampere-turns, At)} \tag{11.3}$$

In other words, if you increase the number of turns around a core and/or increase the current through the coil, the magnetic field strength also increases. In many ways, the magnetomotive force for magnetic circuits is similar to the applied voltage in an electric circuit. Increasing either one results in an increase in the desired effect: magnetic flux for magnetic circuits and current for electric circuits.

For the CGS system, the magnetomotive force is measured in gilberts, while for the English system, it is measured in ampere-turns.

Another factor that affects the magnetic field strength is the type of core used. Materials in which magnetic flux lines can readily be set up are said to be **magnetic** and to have a high **permeability.** Again, note the similarity with the word "permit" used to describe permittivity for the dielectrics of capacitors. Similarly, the permeability (represented by the Greek letter *mu,* μ) of a material is a measure of the ease with which magnetic flux lines can be established in the material.

Just as there is a specific value for the permittivity of air, there is a specific number associated with the permeability of air:

$$\boxed{\mu_o = 4\pi \times 10^{-7} \text{ Wb/A}\cdot\text{m}} \tag{11.4}$$

Practically speaking, the permeability of all nonmagnetic materials, such as copper, aluminum, wood, glass, and air, is the same as that for free space. Materials that have permeabilities slightly less than that of free space are said to be **diamagnetic,** and those with permeabilities slightly greater than that of free space are said to be **paramagnetic.** Magnetic materials, such as iron, nickel, steel, cobalt, and alloys of these metals, have permeabilities hundreds and even thousands of times that of free space. Materials with these very high permeabilities are referred to as **ferromagnetic.**

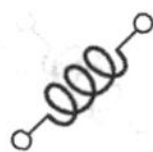

The ratio of the permeability of a material to that of free space is called its **relative permeability;** that is,

$$\mu_r = \frac{\mu}{\mu_o} \qquad \textbf{(11.5)}$$

In general, for ferromagnetic materials, $\mu_r \geq 100$, and for nonmagnetic materials, $\mu_r = 1$.

A table of values for μ to match the provided table for permittivity levels of specific dielectrics would be helpful. Unfortunately, such a table cannot be provided because *relative permeability is a function of the operating conditions.* If you change the magnetomotive force applied, the level of μ can vary between extreme limits. At one level of magnetomotive force, the permeability of a material can be 10 times that at another level.

An instrument designed to measure flux density in milligauss (CGS system) appears in Fig. 11.14. The meter has two sensitivities, 0.5 to 100 milligauss at 60 Hz and 0.2 to 3 milligauss at 60 Hz. It can be used to measure the electric field strength discussed in Chapter 10 on switching to the ELECTRIC setting. The top scale will then provide a reading in kilovolts/meter. (As an aside, the meter of Fig. 11.14 has appeared in TV programs as a device for detecting a "paranormal" response.) Appendix E reveals that 1 T = 10^4 gauss. The magnitude of the reading of 20 milligauss would be equivalent to

$$20 \text{ milligauss}\left(\frac{1 \text{ T}}{10^4 \text{ gauss}}\right) = 2\ \mu\text{T}$$

Although our emphasis in this chapter is to introduce the parameters that affect the nameplate data of an inductor, the use of magnetics has widespread application in the electrical/electronics industry, as shown by a few areas of application in Fig. 11.15.

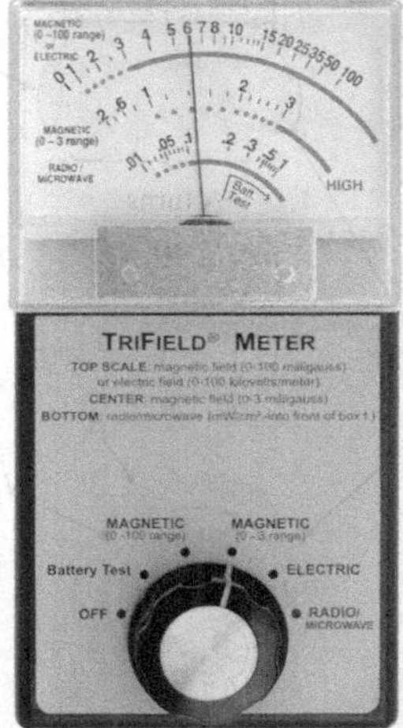

FIG. 11.14
Milligaussmeter.
(Courtesy of AlphaLab. Inc.)

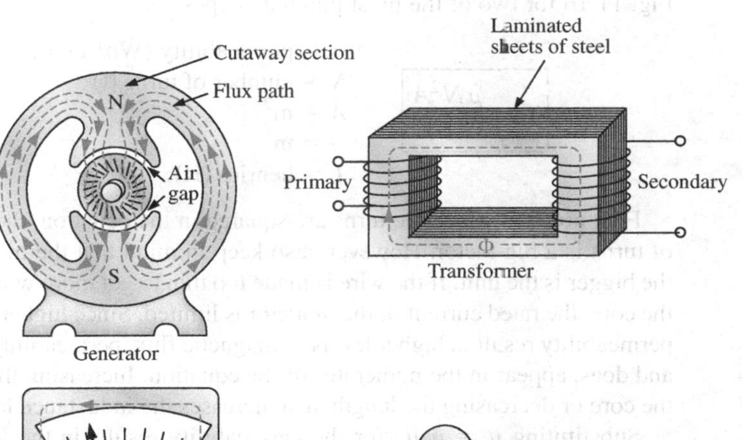

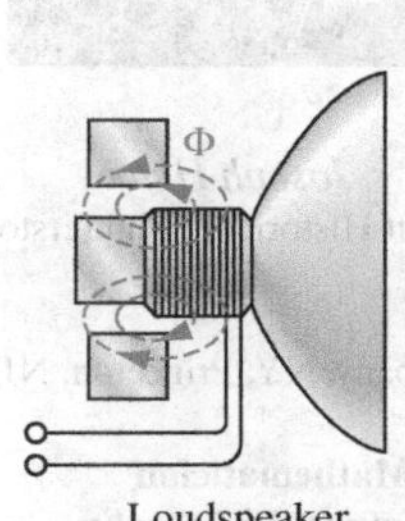

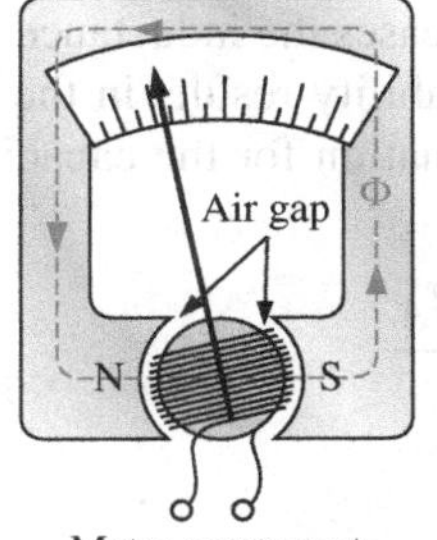

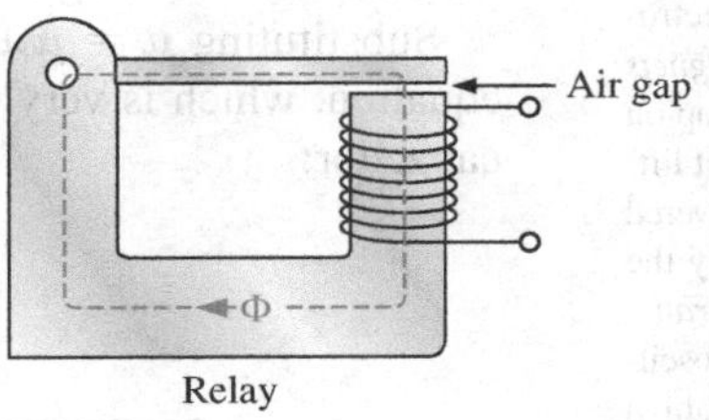

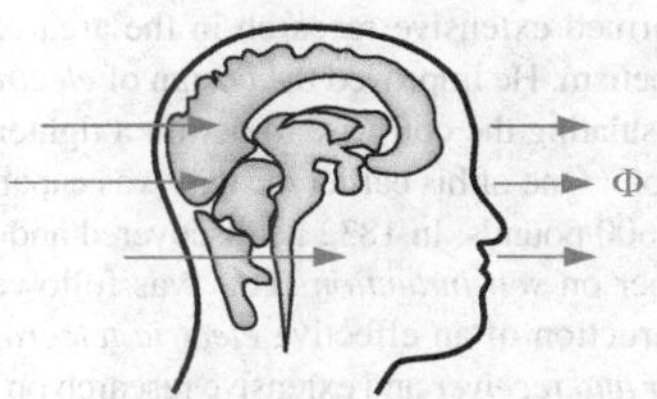

FIG. 11.15
Some areas of application of magnetic effects.

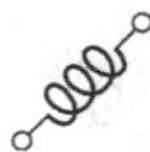

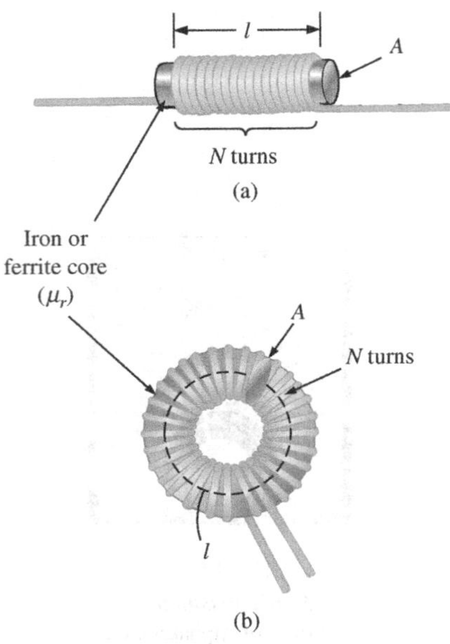

FIG. 11.16
Defining the parameters for Eq. (11.6).

11.3 INDUCTANCE

In the previous section, we learned that sending a current through a coil of wire, with or without a core, establishes a magnetic field through and surrounding the unit. This component, of rather simple construction (see Fig. 11.16), is called an **inductor** (often referred to as a **coil**). Its **inductance** level determines the strength of the magnetic field around the coil due to an applied current. The higher the inductance level, the greater is the strength of the magnetic field. In total, therefore,

inductors are designed to set up a strong magnetic field linking the unit, whereas capacitors are designed to set up a strong electric field between the plates.

Inductance is measured in **henries (H),** after the American physicist Joseph Henry (Fig. 11.17). However, just as the farad is too large a unit for most applications, most inductors are of the millihenry (mH) or microhenry (μH) range.

In Chapter 10, 1 farad was defined as a capacitance level that would result in 1 coulomb of charge on the plates due to the application of 1 volt across the plates. For inductors,

1 henry is the inductance level that will establish a voltage of 1 volt across the coil due to a change in current of 1 A/s through the coil.

FIG. 11.17
Joseph Henry.
(© Everett Historical / Shutterstock.com)

American (Albany, NY; Princeton, NJ)
(1797–1878)
Physicist and Mathematician
Professor of Natural Philosophy,
Princeton University

In the early 1800s the title Professor of Natural Philosophy was applied to educators in the sciences. As a student and teacher at the Albany Academy, Henry performed extensive research in the area of electromagnetism. He improved the design of *electromagnets* by insulating the coil wire to permit a tighter wrap on the core. One of his earlier designs was capable of lifting 3600 pounds. In 1832 he discovered and delivered a paper on *self-induction*. This was followed by the construction of an effective *electric telegraph transmitter and receiver* and extensive research on the oscillatory nature of lightning and discharges from a *Leyden jar*. In 1845 he was appointed the first Secretary of the Smithsonian.

Inductor Construction

In Chapter 10, we found that capacitance is sensitive to the area of the plates, the distance between the plates, and the dielectic employed. The level of inductance has similar construction sensitivities in that it is dependent on the area within the coil, the length of the unit, and the permeability of the core material. It is also sensitive to the number of turns of wire in the coil as dictated by the following equation and defined in Fig. 11.16 for two of the most popular shapes:

$$\boxed{L = \frac{\mu N^2 A}{l}} \qquad \begin{aligned} \mu &= \text{permeability (Wb/A}\cdot\text{m)} \\ N &= \text{number of turns (t)} \\ A &= \text{m}^2 \\ l &= \text{m} \\ L &= \text{henries (H)} \end{aligned} \tag{11.6}$$

First note that since the turns are squared in the equation, the number of turns is a big factor. However, also keep in mind that the more turns, the bigger is the unit. If the wire is made too thin to get more windings on the core, the rated current of the inductor is limited. Since higher levels of permeability result in higher levels of magnetic flux, permeability should, and does, appear in the numerator of the equation. Increasing the area of the core or decreasing the length also increases the inductance level.

Substituting $\mu = \mu_r\mu_o$ for the permeability results in the following equation, which is very similar to the equation for the capacitance of a capacitor:

$$L = \frac{\mu_r \mu_o N^2 A}{l}$$

or

$$\boxed{L = 4\pi \times 10^{-7}\frac{\mu_r N^2 A}{l}} \qquad \text{(henries, H)} \tag{11.7}$$

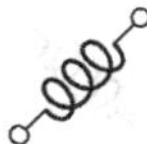

If we break out the relative permeability as

$$L = \mu_r\left(\frac{\mu_o N^2 A}{l}\right)$$

we obtain the following useful equation:

$$\boxed{L = \mu_r L_o} \qquad \textbf{(11.8)}$$

which is very similar to the equation $C = \epsilon_r C_o$. Eq. (11.8) states the following:

The inductance of an inductor with a ferromagnetic core is μ_r times the inductance obtained with an air core.

Although Eq. (11.6) is approximate at best, the equations for the inductance of a wide variety of coils can be found in reference handbooks. Most of the equations are mathematically more complex than Eq. (11.6), but the impact of each factor is the same in each equation.

EXAMPLE 11.1 For the air-core coil in Fig. 11.18:

a. Find the inductance.
b. Find the inductance if a metallic core with $\mu_r = 2000$ is inserted in the coil.

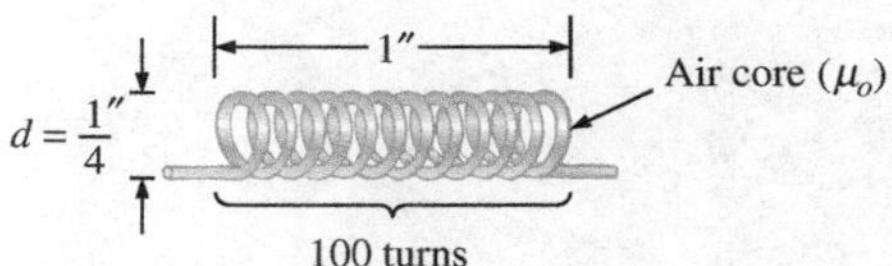

FIG. 11.18
Air-core coil for Example 11.1.

Solutions:

a. $d = \frac{1}{4}\text{ in.}\left(\frac{1\text{ m}}{39.37\text{ in.}}\right) = 6.35\text{ mm}$

$$A = \frac{\pi d^2}{4} = \frac{\pi(6.35\text{ mm})^2}{4} = 31.7\ \mu\text{m}^2$$

$$l = 1\text{ in.}\left(\frac{1\text{ m}}{39.37\text{ in.}}\right) = 25.4\text{ mm}$$

$$L = 4\pi \times 10^{-7}\frac{\mu_r N^2 A}{l}$$

$$= 4\pi \times 10^{-7}\frac{(1)(100\text{ t})^2(31.7\ \mu\text{m}^2)}{25.4\text{ mm}} = \mathbf{15.68\ \mu H}$$

b. Eq. (11.8): $L = \mu_r L_o = (2000)(15.68\ \mu\text{H}) = \mathbf{31.36\ mH}$

EXAMPLE 11.2 In Fig. 11.19, if each inductor in the left column is changed to the type appearing in the right column, find the new inductance level. For each change, assume that the other factors remain the same.

Solutions:

a. The only change was the number of turns, but it is a squared factor, resulting in

$$L = (2)^2 L_o = (4)(20\ \mu\text{H}) = \mathbf{80\ \mu H}$$

b. In this case, the area is three times the original size, and the number of turns is 1/2. Since the area is in the numerator, it increases the

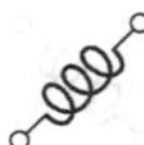

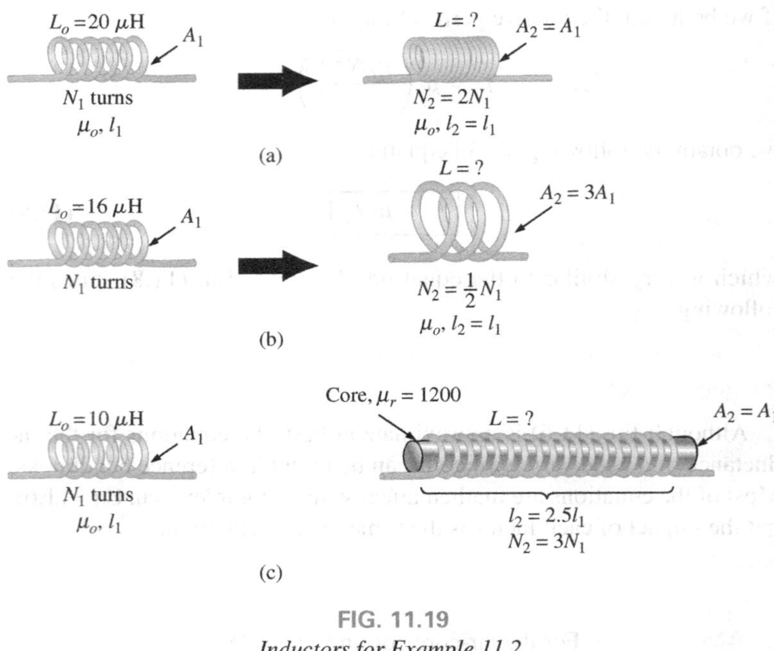

FIG. 11.19
Inductors for Example 11.2.

inductance by a factor of three. The drop in the number of turns reduces the inductance by a factor of $(1/2)^2 = 1/4$. Therefore,

$$L = (3)\left(\frac{1}{4}\right)L_o = \frac{3}{4}(16\ \mu\text{H}) = \mathbf{12\ \mu H}$$

c. Both μ and the number of turns have increased, although the increase in the number of turns is squared. The increased length reduces the inductance. Therefore,

$$L = \frac{(3)^2(1200)}{2.5}L_o = (4.32 \times 10^3)(10\ \mu\text{H}) = \mathbf{43.2\ mH}$$

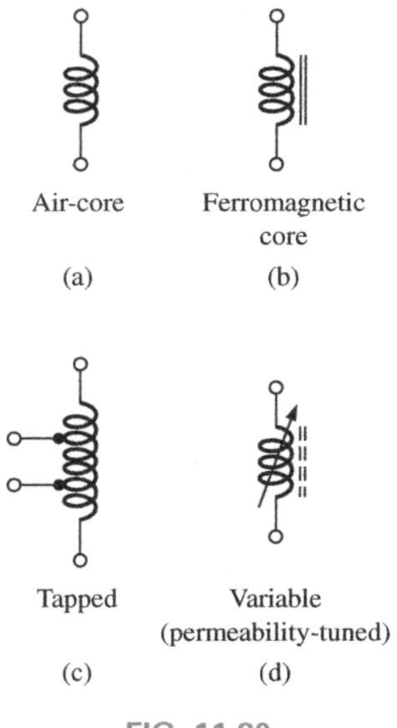

FIG. 11.20
Inductor (coil) symbols.

Types of Inductors

Inductors, like capacitors and resistors, can be categorized under the general headings **fixed** or **variable.** The symbol for a fixed air-core inductor is provided in Fig. 11.20(a), for an inductor with a ferromagnetic core in Fig. 11.20(b), for a tapped coil in Fig. 11.20(c), and for a variable inductor in Fig. 11.20(d).

Fixed Fixed-type inductors come in all shapes and sizes. However, *in general, the size of an inductor is determined primarily by the type of construction, the core used, and the current rating.*

In Fig. 11.21(a), the 10 μH and 1 mH coils are about the same size because a thinner wire was used for the 1 mH coil to permit more turns in the same space. The result, however, is a drop in rated current from 10 A to only 1.3 A. If the wire of the 10 μH coil had been used to make the 1 mH coil, the resulting coil would have been many times the size of the 10 μH

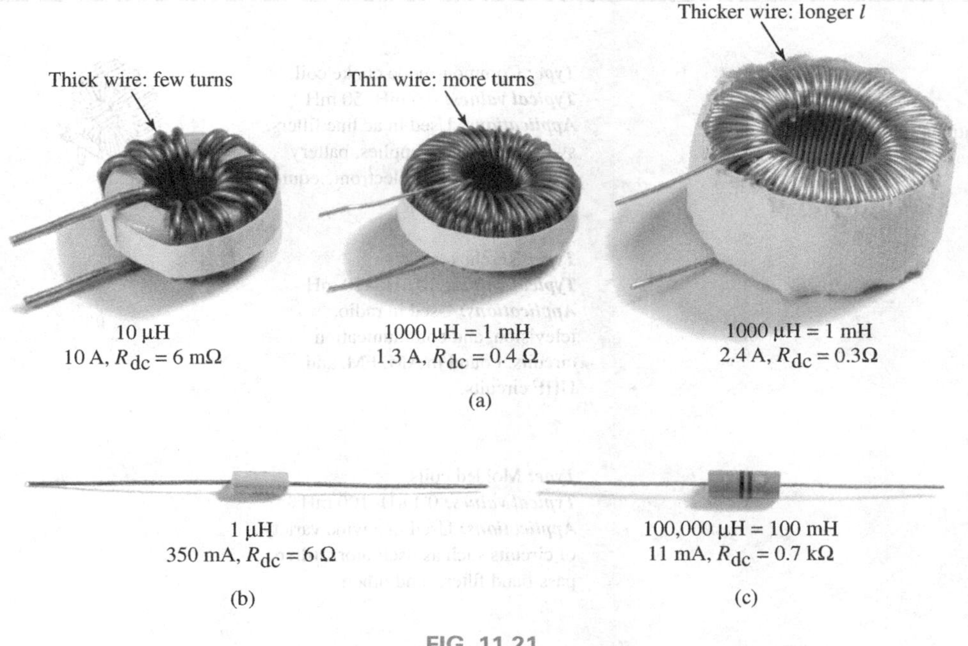

FIG. 11.21
Relative sizes of different types of inductors: (a) toroid, high-current; (b) phenolic (resin or plastic core); (c) ferrite core.

coil. The impact of the wire thickness is clearly revealed by the 1 mH coil at the far right in Fig. 11.21(a), where a thicker wire was used to raise the rated current level from 1.3 A to 2.4 A. Even though the inductance level is the same, the size of the toroid is four to five greater.

The phenolic inductor (using a nonferromagnetic core of resin or plastic) in Fig. 11.21(b) is quite small for its level of inductance. We must assume that it has a high number of turns of very thin wire. Note, however, that the use of a very thin wire has resulted in a relatively low current rating of only 350 mA (0.35 A). The use of a ferrite (ferromagnetic) core in the inductor in Fig. 11.21(c) has resulted in an amazingly high level of inductance for its size. However, the wire is so thin that the current rating is only 11 mA = 0.011 A. Note that for all the inductors, the dc resistance of the inductor increases with a decrease in the thickness of the wire. The 10 μH toroid has a dc resistance of only 6 mΩ, whereas the dc resistance of the 100 mH ferrite inductor is 700 Ω—a price to be paid for the smaller size and high inductance level.

Different types of fixed inductive elements are displayed in Fig. 11.22, including their typical range of values and common areas of application. Based on the earlier discussion of inductor construction, it is fairly easy to identify an inductive element. The shape of a molded film resistor is similar to that of an inductor. However, careful examination of the typical shapes of each reveals some differences, such as the ridges at each end of a resistor that do not appear on most inductors.

Variable A number of variable inductors are depicted in Fig. 11.23. In each case, the inductance is changed by turning the slot at the end of the core to move it in and out of the unit. The farther in the core is, the more the ferromagnetic material is part of the magnetic circuit, and the higher is the magnetic field strength and the inductance level.

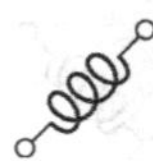

Type: Air-core inductors (1–32 turns)
Typical values: 2.5 nH–1 μH
Applications: High-frequency applications

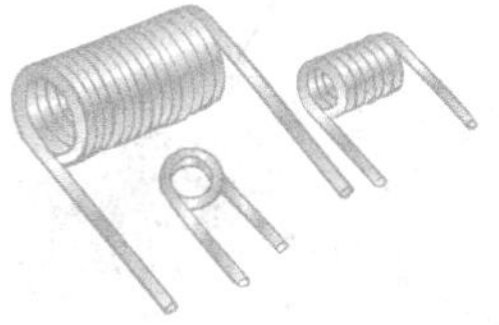

Type: Toroid coil
Typical values: 10 μH–30 mH
Applications: Used as a choke in ac power line circuits to filter transient and reduce EMI interference. This coil is found in many electronic appliances.

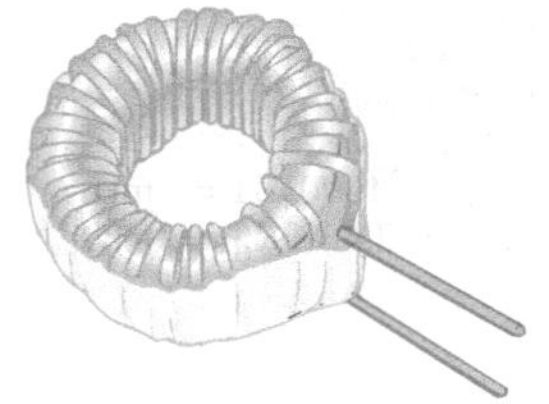

Type: Hash choke coil
Typical values: 3 μH–1 mH
Applications: Used in ac supply lines that deliver high currents.

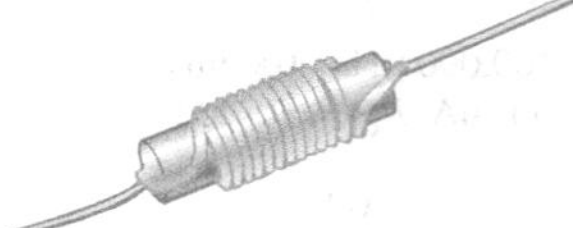

Type: Delay line coil
Typical values: 10 μH–50 μH
Applications: Used in color televisions to correct for timing differences between the color signal and the black-and-white signal.

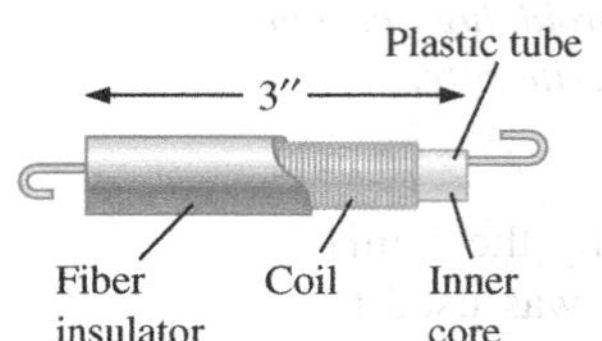

Type: Common-mode choke coil
Typical values: 0.6 mH–50 mH
Applications: Used in ac line filters, switching power supplies, battery chargers, and other electronic equipment.

Type: RF chokes
Typical values: 10 μH–470 mH
Applications: Used in radio, television, and communication circuits. Found in AM, FM, and UHF circuits.

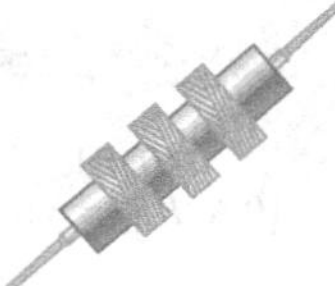

Type: Molded coils
Typical values: 0.1 μH–100 mH
Applications: Used in a wide variety of circuits such as oscillators, filters, pass-band filters, and others.

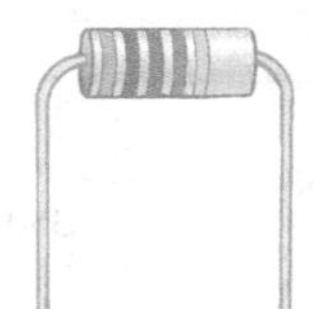

Type: Surface-mount inductors
Typical values: 0.01 μH–250 μH
Applications: Found in many electronic circuits that require miniature components on multilayered PCBs (printed circuit boards).

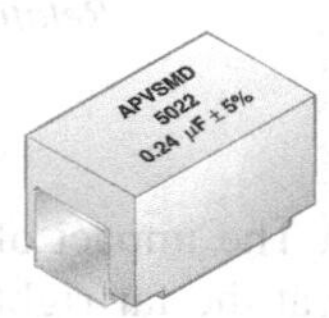

FIG. 11.22
Typical areas of application for inductive elements.

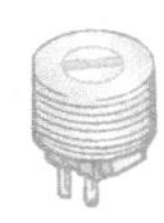

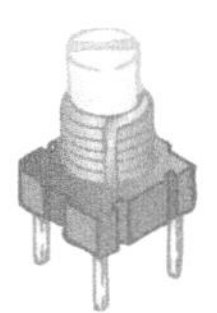

FIG. 11.23
Variable inductors with a typical range of values from 1 μH to 100 μH; commonly used in oscillators and various RF circuits such as CB transceivers, televisions, and radios.

Practical Equivalent Inductors

Inductors, like capacitors, are not ideal. Associated with every inductor is a resistance determined by the resistance of the turns of wire (the thinner the wire, the greater is the resistance for the same material) and by the core losses (radiation and skin effect, eddy current and hysteresis losses—all discussed in more advanced texts). There is also some stray capacitance due to the capacitance between the current-carrying turns of wire of the coil. Recall that capacitance appears whenever there are two conducting surfaces separated by an insulator, such as air, and when those wrappings are fairly tight and are parallel. Both elements are included in the equivalent circuit in Fig. 11.24. For most applications in this text, the capacitance can be ignored, resulting in the equivalent model in Fig. 11.25. The resistance R_l plays an important part in some areas (such as resonance, discussed in Chapter 20)

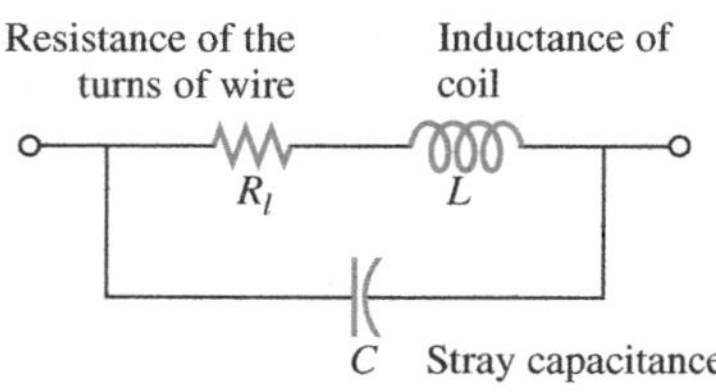

FIG. 11.24
Complete equivalent model for an inductor.

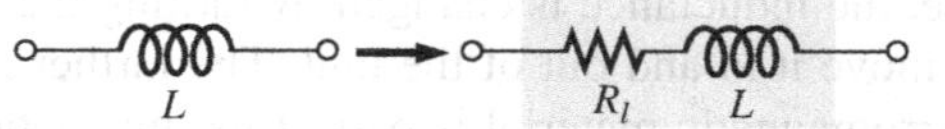

FIG. 11.25
Practical equivalent model for an inductor.

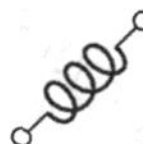

because the resistance can extend from a few ohms to a few hundred ohms, depending on the construction. For this chapter, the inductor is considered an ideal element, and the series resistance is dropped from Fig. 11.25.

Inductor Labeling

Because some inductors are larger in size, their nameplate value can often be printed on the body of the element. However, on smaller units, there may not be enough room to print the actual value, so an abbreviation is used that is fairly easy to understand. First, realize that the *microhenry* (μH) is the fundamental unit of measurement for this marking. Most manuals list the inductance value in μH even if the value must be reported as 470,000 μH rather than as 470 mH. If the label reads 223K, the third number (3) is the power to be applied to the first two. The K is not from *kilo,* representing a power of three, but is used to denote a tolerance of $\pm 10\%$ as described for capacitors. The resulting number of 22,000 is, therefore, in μH, so the 223K unit is a 22,000 μH or 22 mH inductor. The letters J and M indicate a tolerance of $\pm 5\%$ and $\pm 20\%$, respectively.

For molded inductors, a color-coding system very similar to that used for resistors is used. The major difference is that *the resulting value is always in μH,* and a wide band at the beginning of the labeling is an MIL ("meets military standards") indicator. Always read the colors in sequence, starting with the band closest to one end as shown in Fig. 11.26.

The standard values for inductors employ the same numerical values and multipliers used with resistors and capacitors. In general, therefore,

Color Code Table

Color[1]	Significant Figure	Multiplier[2]	Inductance Tolerance (%)
Black	0	1	
Brown	1	10	
Red	2	100	
Orange	3	1000	
Yellow	4		
Green	5		
Blue	6		
Violet	7		
Gray	8		
White	9		
None[2]			±20
Silver			±10
Gold	Decimal point		±5

[1] Indicates body color.
[2] The multiplier is the factor by which the two significant figures are multiplied to yield the nominal inductance value.

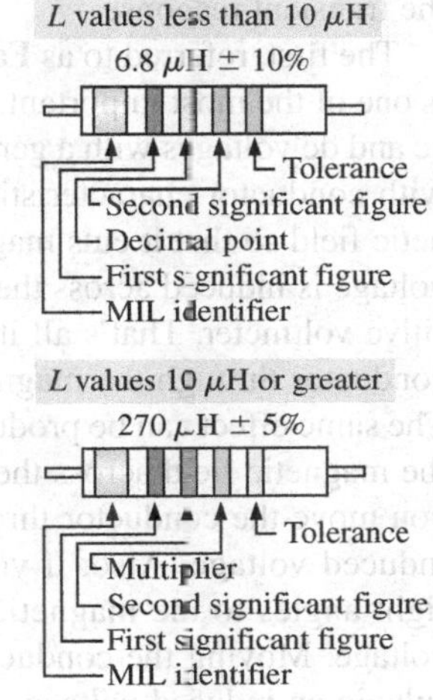

Cylindrical molded choke coils are marked with five colored bands. A wide silver band, located at one end of the coil, identifies military radio-frequency coils. The next three bands indicate the inductance in microhenries, and the fourth band is the tolerance.

Color coding is in accordance with the color code table, shown on the left. If the first or second band is gold, it represents the decimal point for inductance values less than 10. Then the following two bands are significant figures. For inductance values of 10 or more, the first two bands represent significant figures, and the third is the multiplier.

FIG. 11.26
Molded inductor color coding.

expect to find inductors with the following multipliers: 1 μH, 1.5 μH, 2.2 μH, 3.3 μH, 4.7 μH, 6.8 μH, 10 μH, and so on.

Measurement and Testing of Inductors

The inductance of an inductor can be read directly using a meter such as the Universal LCR Meter (Fig. 11.27), also discussed in Chapter 10 on capacitors. Set the meter to L for inductance, and the meter automatically chooses the most appropriate unit of measurement for the element, that is, H, mH, μH, or pH.

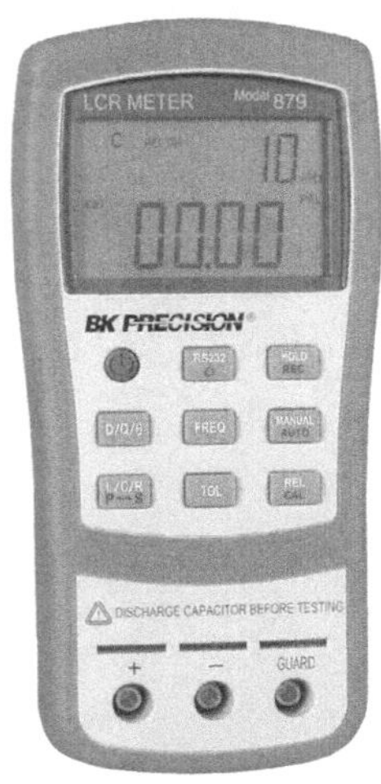

FIG. 11.27
Digital reading inductance meter.
(B&K Precision Corp.)

An inductance meter is the best choice, but an ohmmeter can also be used to check whether a short has developed between the windings or whether an open circuit has developed. The open-circuit possibility is easy to check because a reading of infinite ohms or very high resistance results. The short-circuit condition is harder to check because the resistance of many good inductors is relatively small, and the shorting of a few windings may not adversely affect the total resistance. Of course, if you are aware of the typical resistance of the coil, you can compare it to the measured value. A short between the windings and the core can be checked by simply placing one lead of the ohmmeter on one wire (perhaps a terminal) and the other on the core itself. A reading of zero ohms reveals a short between the two that may be due to a breakdown in the insulation jacket around the wire resulting from excessive currents, environmental conditions, or simply old age and cracking.

11.4 INDUCED VOLTAGE v_L

Before analyzing the response of inductive elements to an applied dc voltage, we must introduce a number of laws and equations that affect the transient response.

The first, referred to as **Faraday's law of electromagnetic induction,** is one of the most important in this field because it enables us to establish ac and dc voltages with a generator. If we move a conductor (any material with conductor characteristics as defined in Chapter 2) through a magnetic field so that it cuts magnetic lines of flux as shown in Fig. 11.28, a voltage is induced across the conductor that can be measured with a sensitive voltmeter. That's all it takes, and, in fact, the faster you move the conductor through the magnetic flux, the greater is the induced voltage. The same effect can be produced if you hold the conductor still and move the magnetic field across the conductor. Note that the direction in which you move the conductor through the field determines the polarity of the induced voltage. Also, if you move the conductor through the field at right angles to the magnetic flux, you generate the maximum induced voltage. Moving the conductor parallel with the magnetic flux lines results in an induced voltage of zero volts since magnetic lines of flux are not crossed.

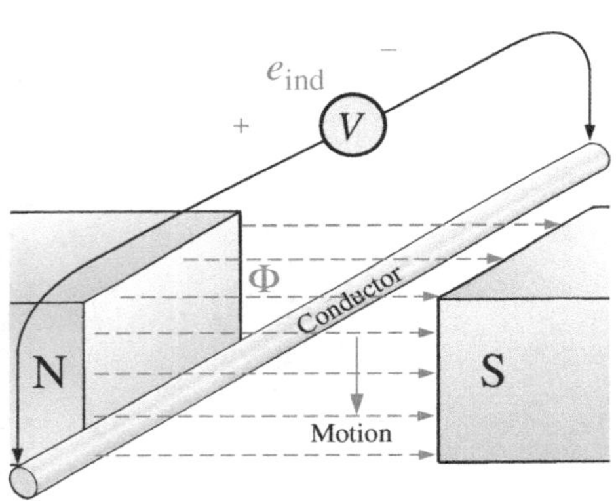

FIG. 11.28
Generating an induced voltage by moving a conductor through a magnetic field.

If we now go a step further and move a coil of N turns through the magnetic field as shown in Fig. 11.29, a voltage will be induced across the coil as determined by **Faraday's law:**

$$e = N\frac{d\phi}{dt} \quad \text{(volts, V)} \qquad \textbf{(11.9)}$$

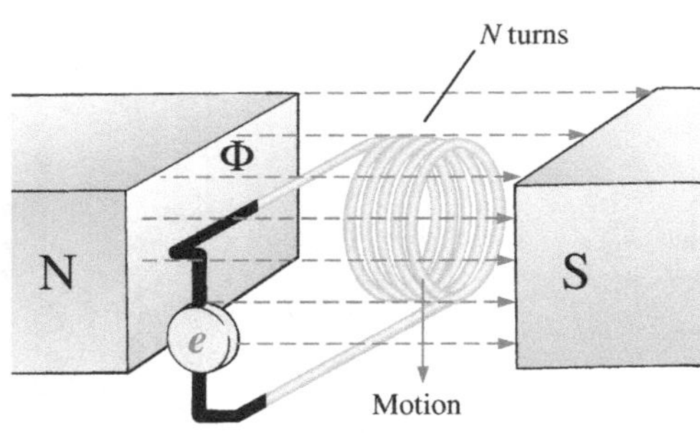

FIG. 11.29
Demonstrating Faraday's law.

The greater the number of turns or the faster the coil is moved through the magnetic flux pattern, the greater is the induced voltage. The term

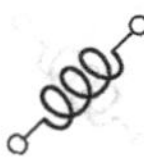

$d\phi/dt$ is the differential change in magnetic flux through the coil at a particular instant in time. If the magnetic flux passing through a coil remains constant—no matter how strong the magnetic field—the term will be zero, and the induced voltage zero volts. It doesn't matter whether the changing flux is due to moving the magnetic field or moving the coil in the vicinity of a magnetic field: The only requirement is that the flux linking (passing through) the coil changes with time. Before the coil passes through the magnetic poles, the induced voltage is zero because there are no magnetic flux lines passing through the coil. As the coil enters the flux pattern, the number of flux lines cut per instant of time increases until it peaks at the center of the poles. The induced voltage then decreases with time as it leaves the magnetic field.

This important phenomenon can now be applied to the inductor in Fig. 11.30, which is simply an extended version of the coil in Fig. 11.29. In Section 11.2, we found that the magnetic flux linking the coil of N turns with a current I has the distribution shown in Fig. 11.30. If the current through the coil increases in magnitude, the flux linking the coil also increases. We just learned through Faraday's law, however, that a coil in the vicinity of a changing magnetic flux will have a voltage induced across it. The result is that a voltage is induced across the coil in Fig. 11.30 due to the *change in current through the coil.*

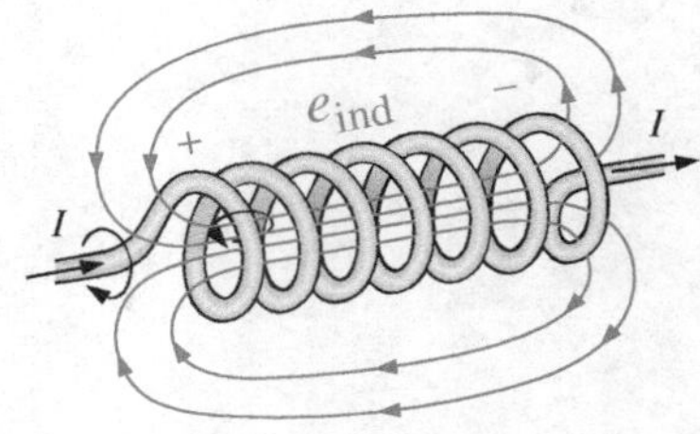

FIG. 11.30
Demonstrating the effect of Lenz's law.

It is very important to note in Fig. 11.30 that the polarity of the induced voltage across the coil is such that it opposes the increasing level of current in the coil. In other words, the changing current through the coil induces a voltage across the coil that is opposing the applied voltage that establishes the increase in current in the first place. The quicker the change in current through the coil, the greater is the opposing induced voltage to squelch the attempt of the current to increase in magnitude. The "choking" action of the coil is the reason inductors or coils are often referred to as **chokes.** This effect is a result of an important law referred to as **Lenz's law,** which states that

an induced effect is always such as to oppose the cause that produced it.

The inductance of a coil is also a measure of the change in flux linking the coil due to a change in current through the coil. That is,

$$L = N\frac{d\phi}{di_L} \qquad \text{(henries, H)} \qquad \mathbf{(11.10)}$$

The equation reveals that the greater the number of turns or the greater the change in flux linking the coil due to a particular change in current, the greater is the level of inductance. In other words, coils with smaller levels of inductance generate smaller changes in flux linking the coil for the same change in current through the coil. If the inductance level is very small, there will be almost no change in flux linking the coil, and the induced voltage across the coil will be very small. In fact, if we now write Eq. (11.9) in the form

$$e = N\frac{d\phi}{dt} = \left(N\frac{d\phi}{di_L}\right)\left(\frac{di_L}{dt}\right)$$

and substitute Eq. (11.10), we obtain

$$e_L = L\frac{di_L}{dt} \qquad \text{(volts, V)} \qquad \mathbf{(11.11)}$$

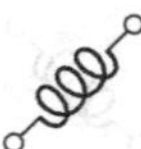

which relates the voltage across a coil to the number of turns of the coil and the change in current through the coil.

When induced effects are used in the generation of voltages such as those from dc or ac generators, the symbol e is applied to the induced voltage. However, in network analysis, the voltage induced across an inductor will always have a polarity that opposes the applied voltage (like the voltage across a resistor). Therefore, the following notation is used for the induced voltage across an inductor:

$$\boxed{v_L = L\frac{di_L}{dt}} \qquad \text{(volts, V)} \qquad \textbf{(11.12)}$$

The equation clearly states that

the larger the inductance and/or the more rapid the change in current through a coil, the larger will be the induced voltage across the coil.

If the current through the coil fails to change with time, the induced voltage across the coil will be zero. We will find in the next section that for dc applications, when the transient phase has passed, $di_L/dt = 0$, and the induced voltage across the coil is

$$v_L = L\frac{di_L}{dt} = L(0) = 0\text{ V}$$

The duality that exists between inductive and capacitive elements is now abundantly clear. Simply interchange the voltages and currents of Eq. (11.12), and interchange the inductance and capacitance. The following equation for the current of a capacitor results:

$$v_L = L\frac{di_L}{dt}$$

$$\downarrow \qquad \downarrow$$

$$i_C = C\frac{dv_C}{dt}$$

We are now at a point where we have all the background relationships necessary to investigate the transient behavior of inductive elements.

11.5 *R-L* TRANSIENTS: THE STORAGE PHASE

A great number of similarities exist between the analyses of inductive and capacitive networks. That is, what is true for the voltage of a capacitor is also true for the current of an inductor, and what is true for the current of a capacitor can be matched in many ways by the voltage of an inductor. The storage waveforms have the same shape, and time constants are defined for each configuration. Because these concepts are so similar (refer to Section 10.5 on the charging of a capacitor), you have an opportunity to reinforce concepts introduced earlier and still learn more about the behavior of inductive elements.

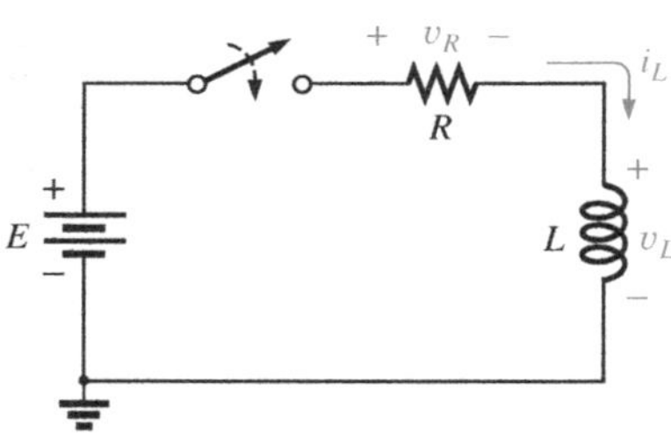

FIG. 11.31
Basic R-L transient network.

The circuit in Fig. 11.31 is used to describe the storage phase. Note that it is the same circuit used to describe the charging phase of capacitors, with a simple replacement of the capacitor by an ideal inductor. Throughout the analysis, it is important to remember that energy is stored in the form of an electric field between the plates of a capacitor. For inductors, on the other hand, energy is stored in the form of a magnetic field linking the coil.

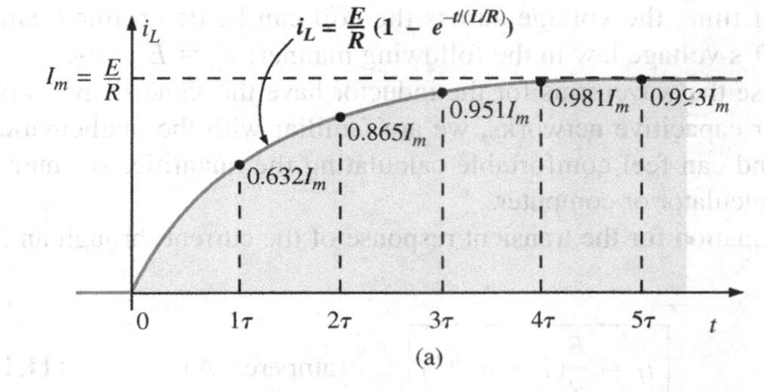

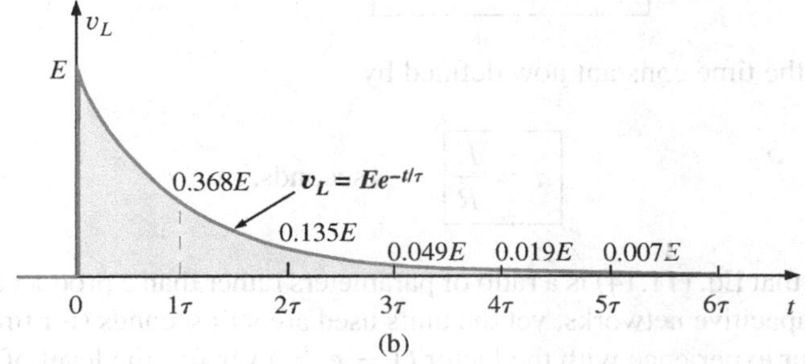

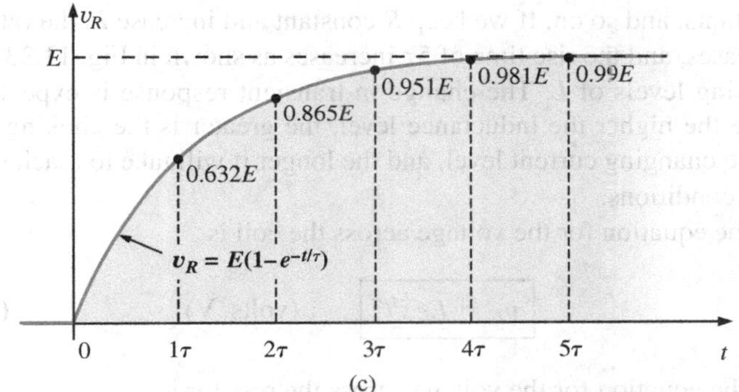

FIG. 11.32
i_L, v_L, and v_R for the circuit in Fig. 11.31 following the closing of the switch.

At the instant the switch is closed, the choking action of the coil prevents an instantaneous change in current through the coil, resulting in $i_L = 0$ A, as shown in Fig. 11.32(a). The absence of a current through the coil and circuit at the instant the switch is closed results in zero volts across the resistor as determined by $v_R = i_R R = i_L R = (0 \text{ A})R = 0$ V, as shown in Fig. 11.32(c). Applying Kirchhoff's voltage law around the closed loop results in E volts across the coil at the instant the switch is closed, as shown in Fig. 11.32(b).

Initially, the current increases very rapidly, as shown in Fig. 11.32(a) and then at a much slower rate as it approaches its steady-state value determined by the parameters of the network (E/R). The voltage across the resistor rises at the same rate because $v_R = i_R R = i_L R$. Since the voltage across the coil is sensitive to the rate of change of current through the coil, the voltage will be at or near its maximum value early in the storage phase. Finally, when the current reaches its steady-state value of E/R amperes, the current through the coil ceases to change, and the voltage across the coil drops to zero volts. At any

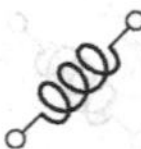

instant of time, the voltage across the coil can be determined using Kirchhoff's voltage law in the following manner: $v_L = E - v_R$.

Because the waveforms for the inductor have the same shape as obtained for capacitive networks, we are familiar with the mathematical format and can feel comfortable calculating the quantities of interest using a calculator or computer.

The equation for the transient response of the current through an inductor is

$$i_L = \frac{E}{R}(1 - e^{-t/\tau}) \quad \text{(amperes, A)} \tag{11.13}$$

with the time constant now defined by

$$\tau = \frac{L}{R} \quad \text{(seconds, s)} \tag{11.14}$$

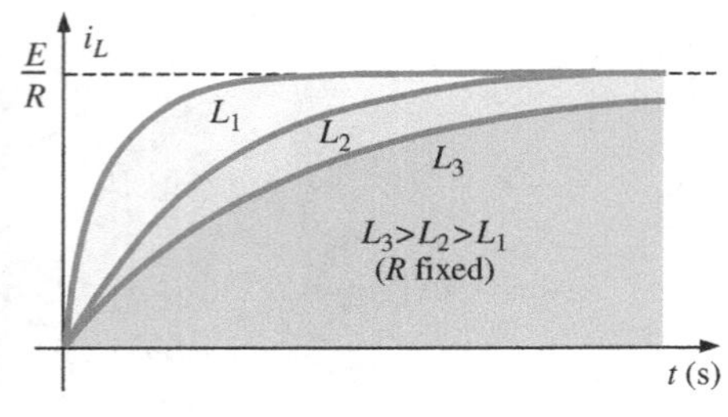

FIG. 11.33
Effect of L on the shape of the i_L storage waveform.

Note that Eq. (11.14) is a ratio of parameters rather than a product as used for capacitive networks, yet the units used are still seconds (for time).

Our experience with the factor $(1 - e^{-t/\tau})$ verifies the level of 63.2% for the inductor current after one time constant, 86.5% after two time constants, and so on. If we keep R constant and increase L, the ratio L/R increases, and the rise time of 5τ increases as shown in Fig. 11.33 for increasing levels of L. The change in transient response is expected because the higher the inductance level, the greater is the choking action on the changing current level, and the longer it will take to reach steady-state conditions.

The equation for the voltage across the coil is

$$v_L = Ee^{-t/\tau} \quad \text{(volts, V)} \tag{11.15}$$

and the equation for the voltage across the resistor is

$$v_R = E(1 - e^{-t/\tau}) \quad \text{(volts, V)} \tag{11.16}$$

As mentioned earlier, the shape of the response curve for the voltage across the resistor must match that of the current i_L since $v_R = i_R R = i_L R$.

Since the waveforms are similar to those obtained for capacitive networks, we will assume that

the storage phase has passed and steady-state conditions have been established once a period of time equal to five time constants has occurred.

In addition, since $\tau = L/R$ will always have some numerical value, even though it may be very small at times, the transient period of 5τ will always have some numerical value. Therefore,

the current cannot change instantaneously in an inductive network.

If we examine the conditions that exist at the instant the switch is closed, we find that the voltage across the coil is E volts, although the current is zero amperes as shown in Fig. 11.34. In essence, therefore,

the inductor takes on the characteristics of an open circuit at the instant the switch is closed.

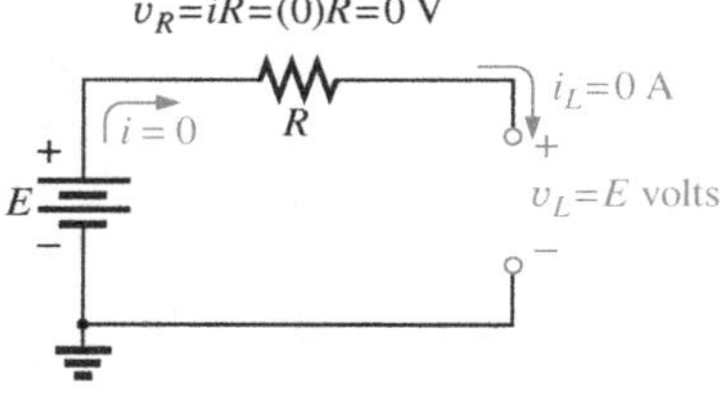

FIG. 11.34
Circuit in Figure 11.31 the instant the switch is closed.

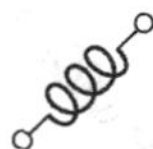

However, if we consider the conditions that exist when steady-state conditions have been established, we find that the voltage across the coil is zero volts and the current is a maximum value of E/R amperes, as shown in Fig. 11.35. In essence, therefore,

the inductor takes on the characteristics of a short circuit when steady-state conditions have been established.

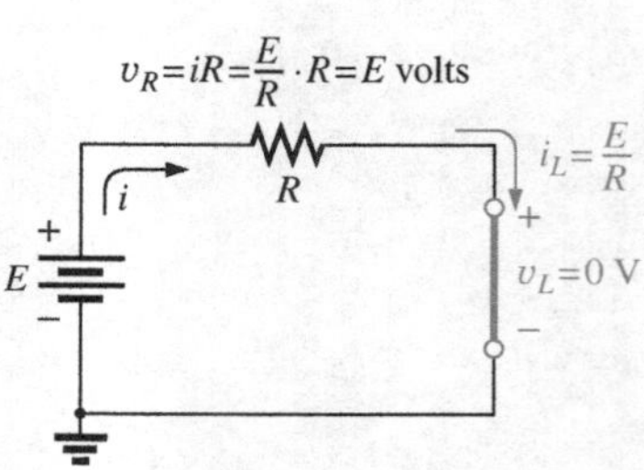

FIG. 11.35
Circuit in Fig. 11.31 under steady-state conditions.

EXAMPLE 11.3 Find the mathematical expressions for the transient behavior of i_L and v_L for the circuit in Fig. 11.36 if the switch is closed at $t = 0$ s. Sketch the resulting curves.

Solution: First, we determine the time constant:

$$\tau = \frac{L}{R_1} = \frac{4\text{ H}}{2\text{ k}\Omega} = 2\text{ ms}$$

Then the maximum or steady-state current is

$$I_m = \frac{E}{R_1} = \frac{50\text{ V}}{2\text{ k}\Omega} = 25 \times 10^{-3}\text{A} = 25\text{ mA}$$

Substituting into Eq. (11.13) gives

$$\mathbf{i_L = 25\text{ mA}\,(1 - e^{-t/2\text{ ms}})}$$

Using Eq. (11.15) gives

$$\mathbf{v_L = 50\text{ V}e^{-t/2\text{ ms}}}$$

The resulting waveforms appear in Fig. 11.37.

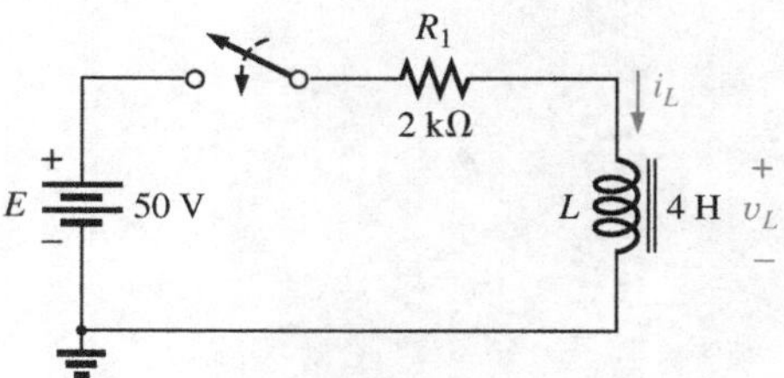

FIG. 11.36
Series R-L circuit for Example 11.3.

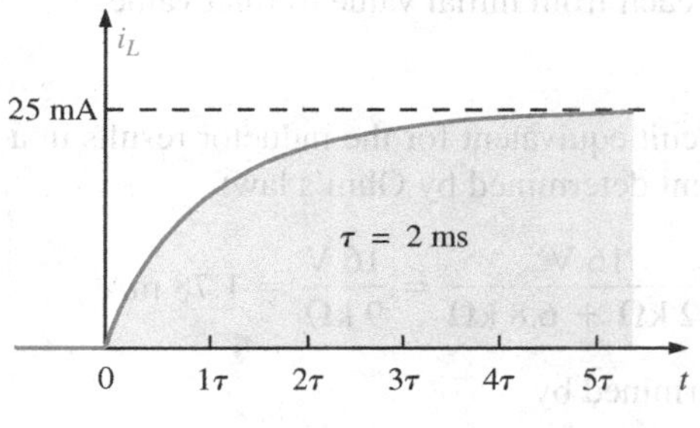

FIG. 11.37
i_L and v_L for the network in Fig. 11.36.

11.6 INITIAL CONDITIONS

This section parallels Section 10.7 on the effect of initial values on the transient phase. Since the current through a coil cannot change instantaneously, the current through a coil begins the transient phase at the initial value established by the network (note Fig. 11.38) before the switch was closed. It then passes through the transient phase until it reaches the steady-state (or final) level after about five time constants. The steady-state level of the inductor current can be found by substituting its short-circuit equivalent (or R_l for the practical equivalent) and finding the resulting current through the element.

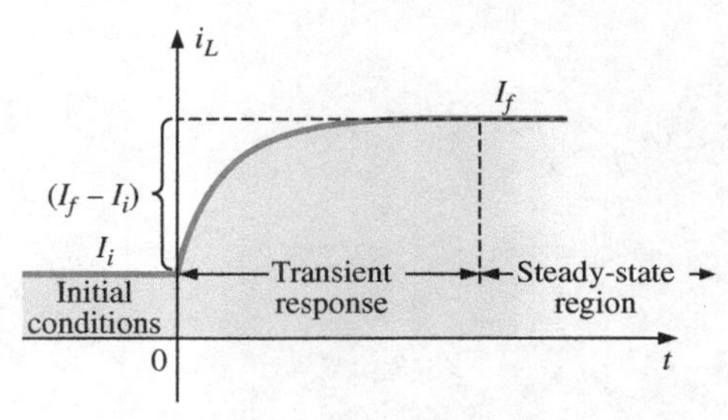

FIG. 11.38
Defining the three phases of a transient waveform.

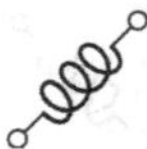

Using the transient equation developed in the previous section, we can write an equation for the current i_L for the entire time interval in Fig. 11.38; that is,

$$i_L = I_i + (I_f - I_i)(1 - e^{-t/\tau})$$

with $(I_f - I_i)$ representing the total change during the transient phase. However, by multiplying through and rearranging terms as

$$\begin{aligned} i_L &= I_i + I_f - I_f e^{-t/\tau} - I_i + I_i e^{-t/\tau} \\ &= I_f - I_f e^{-t/\tau} + I_i e^{-t/\tau} \end{aligned}$$

we find

$$\boxed{i_L = I_f + (I_i - I_f)e^{-t/\tau}} \tag{11.17}$$

If you are required to draw the waveform for the current i_L from initial value to final value, start by drawing a line at the initial value and steady-state levels, and then add the transient response (sensitive to the time constant) between the two levels. The following example will clarify the procedure.

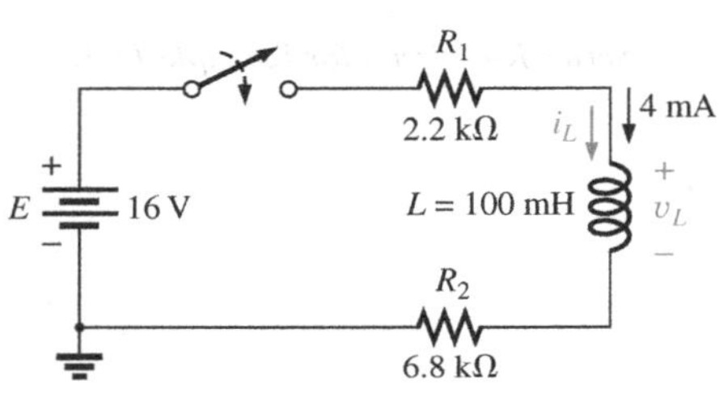

FIG. 11.39
Example 11.4.

EXAMPLE 11.4 The inductor in Fig. 11.39 has an initial current level of 4 mA in the direction shown. (Specific methods to establish the initial current are presented in the sections and problems to follow.)

a. Find the mathematical expression for the current through the coil once the switch is closed.
b. Find the mathematical expression for the voltage across the coil during the same transient period.
c. Sketch the waveform for each from initial value to final value.

Solutions:

a. Substituting the short-circuit equivalent for the inductor results in a final or steady-state current determined by Ohm's law:

$$I_f = \frac{E}{R_1 + R_2} = \frac{16\text{ V}}{2.2\text{ k}\Omega + 6.8\text{ k}\Omega} = \frac{16\text{ V}}{9\text{ k}\Omega} = 1.78\text{ mA}$$

The time constant is determined by

$$\tau = \frac{L}{R_T} = \frac{100\text{ mH}}{2.2\text{ k}\Omega + 6.8\text{ k}\Omega} = \frac{100\text{ mH}}{9\text{ k}\Omega} = 11.11 \mu\text{s}$$

Applying Eq. (11.17) gives

$$\begin{aligned} i_L &= I_f + (I_i - I_f)e^{-t/\tau} = 1.78\text{ mA} + (4\text{ mA} - 1.78\text{ mA})e^{-t/11.11\,\mu\text{s}} \\ &= \mathbf{1.78\text{ mA} + 2.22\text{ mA}e^{-t/11.11\,\mu s}} \end{aligned}$$

b. Since the current through the inductor is constant at 4 mA prior to the closing of the switch, the voltage (whose level is sensitive only to changes in current through the coil) must have an initial value of 0 V. At the instant the switch is closed, the current through the coil cannot change instantaneously, so the current through the resistive elements is 4 mA. The resulting peak voltage at $t = 0$ s can then be found using Kirchhoff's voltage law as follows:

$$\begin{aligned} V_m &= E - V_{R_1} - V_{R_2} = 16\text{ V} - (4\text{ mA})(2.2\text{ k}\Omega) - (4\text{ mA})(6.8\text{ k}\Omega) \\ &= 16\text{ V} - 8.8\text{ V} - 27.2\text{ V} = 16\text{ V} - 36\text{ V} = -20\text{ V} \end{aligned}$$

Note the minus sign to indicate that the polarity of the voltage v_L is opposite to the defined polarity of Fig. 11.39.

The voltage then decays (with the same time constant as the current i_L) to zero because the inductor is approaching its short-circuit equivalence.

The equation for v_L is therefore

$$v_L = -20\ \text{V}e^{-t/11.11\ \mu s}$$

c. See Fig. 11.40. The initial and final values of the current were drawn first, and then the transient response was included between these levels. For the voltage, the waveform begins and ends at zero, with the peak value having a sign sensitive to the defined polarity of v_L in Fig. 11.39.

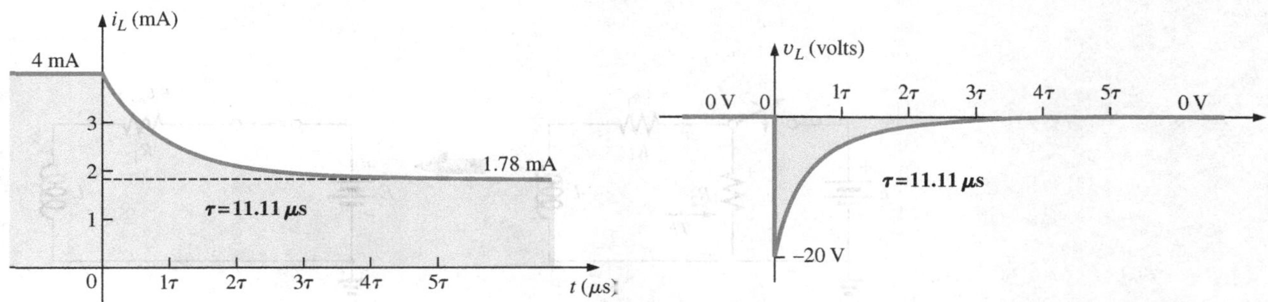

FIG. 11.40
i_L and v_L for the network in Fig. 11.39.

Let us now test the validity of the equation for i_L by substituting $t = 0$ s to reflect the instant the switch is closed. We have

$$e^{-t/\tau} = e^{-0} = 1$$

and $i_L = 1.78\ \text{mA} + 2.22\ \text{mA}e^{-t/\tau} = 1.78\ \text{mA} + 2.22\ \text{mA} = 4\ \text{mA}$

When $t > 5\tau$, $\quad e^{-t/\tau} \cong 0$

and $\quad i_L = 1.78\ \text{mA} + 2.22\ \text{mA}e^{-t/\tau} = 1.78\ \text{mA}$

11.7 *R-L* TRANSIENTS: THE RELEASE PHASE

In the analysis of *R-C* circuits, we found that the capacitor could hold its charge and store energy in the form of an electric field for a period of time determined by the leakage factors. In *R-L* circuits, the energy is stored in the form of a magnetic field established by the current through the coil. Unlike the capacitor, however, an isolated inductor cannot continue to store energy because the absence of a closed path causes the current to drop to zero, releasing the energy stored in the form of a magnetic field. If the series *R-L* circuit in Fig. 11.41 reaches steady-state conditions and the switch is quickly opened, a spark will occur across the contacts due to the rapid change in current from a maximum of E/R to zero amperes. The change in current di/dt of the equation $v_L = L(di/dt)$ establishes a high voltage v_L across the coil that, in conjunction with the applied voltage E, appears across the points of the switch. This is the same mechanism used in the ignition system of a car to ignite the fuel in the cylinder. Some 25,000 V are generated by the rapid decrease in ignition

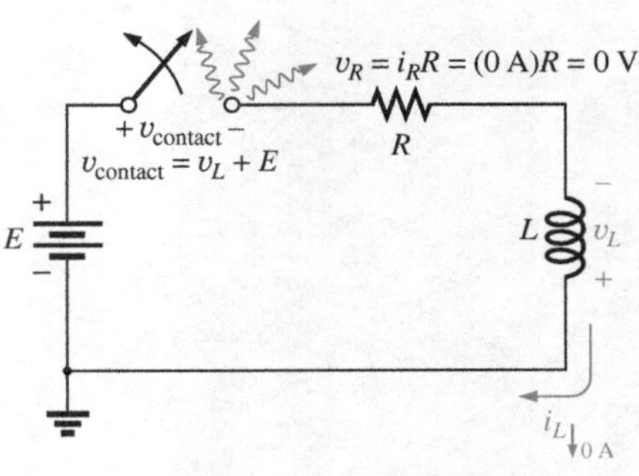

FIG. 11.41
Demonstrating the effect of opening a switch in series with an inductor with a steady-state current.

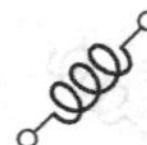

coil current that occurs when the switch in the system is opened. (In older systems, the "points" in the distributor served as the switch.) This inductive reaction is significant when you consider that the only independent source in a car is a 12 V battery.

If opening the switch to move it to another position causes such a rapid discharge in stored energy, how can the decay phase of an *R-L* circuit be analyzed in much the same manner as for the *R-C* circuit? The solution is to use a network like that in Fig. 11.42(a). When the switch is closed, the voltage across resistor R_2 is E volts, and the *R-L* branch responds in the same manner as described above, with the same waveforms and levels. A Thévenin network of E in parallel with R_2 results in the source as shown in Fig. 11.42(b) since R_2 will be shorted out by the short-circuit replacement of the voltage source E when the Thévenin resistance is determined.

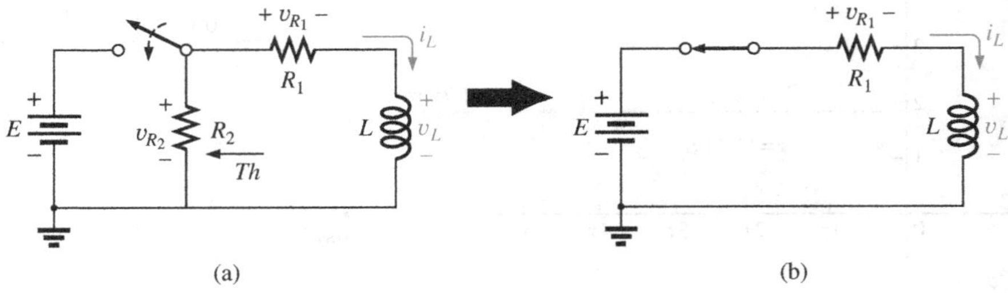

FIG. 11.42
Initiating the storage phase for an inductor by closing the switch.

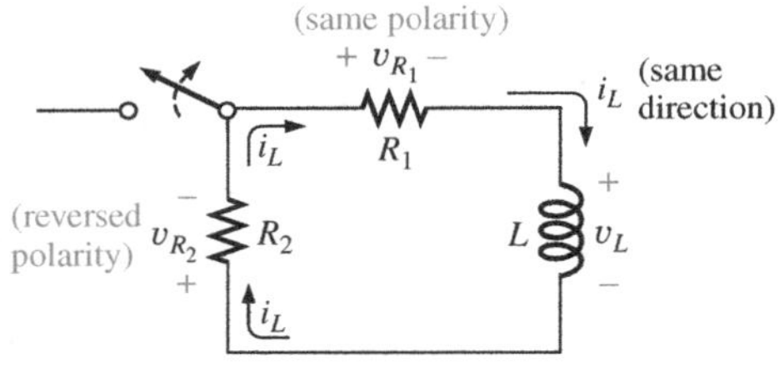

FIG. 11.43
Network in Fig. 11.42 the instant the switch is opened.

After the storage phase has passed and steady-state conditions are established, the switch can be opened without the sparking effect or rapid discharge due to resistor R_2, which provides a complete path for the current i_L. In fact, for clarity the discharge path is isolated in Fig. 11.43. The voltage v_L across the inductor reverses polarity and has a magnitude determined by

$$\boxed{v_L = -(v_{R_1} + v_{R_2})} \quad \textbf{(11.18)}$$

Recall that the voltage across an inductor can change instantaneously but the current cannot. The result is that the current i_L must maintain the same direction and magnitude, as shown in Fig. 11.43. Therefore, the instant after the switch is opened, i_L is still $I_m = E/R_1$, and

$$\begin{aligned} v_L &= -(v_{R_1} + v_{R_2}) = -(i_1R_1 + i_2R_2) \\ &= -i_L(R_1 + R_2) = -\frac{E}{R_1}(R_1 + R_2) = -\left(\frac{R_1}{R_1} + \frac{R_2}{R_1}\right)E \end{aligned}$$

and

$$\boxed{v_L = -\left(1 + \frac{R_2}{R_1}\right)E} \quad \text{switch opened} \quad \textbf{(11.19)}$$

which is bigger than E volts by the ratio R_2/R_1. In other words, when the switch is opened, the voltage across the inductor reverses polarity and drops instantaneously from E to $-[1 + (R_2/R_1)]E$ volts.

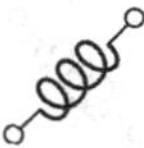

As an inductor releases its stored energy, the voltage across the coil decays to zero in the following manner:

$$\boxed{v_L = -V_i e^{-t/\tau'}} \tag{11.20}$$

with $$V_i = \left(1 + \frac{R_2}{R_1}\right)E$$

and $$\tau' = \frac{L}{R_T} = \frac{L}{R_1 + R_2}$$

The current decays from a maximum of $I_m = E/R_1$ to zero.

Using Eq. (11.17) gives

$$I_i = \frac{E}{R_1} \qquad \text{and} \qquad I_f = 0 \text{ A}$$

so that $$i_L = I_f + (I_i - I_f)e^{-t/\tau'} = 0 \text{ A} + \left(\frac{E}{R_1} - 0 \text{ A}\right)e^{-t/\tau'}$$

and $$\boxed{i_L = \frac{E}{R_1} e^{-t/\tau'}} \tag{11.21}$$

with $$\tau' = \frac{L}{R_1 + R_2}$$

The mathematical expression for the voltage across either resistor can then be determined using Ohm's law:

$$v_{R_1} = i_{R_1}R_1 = i_L R_1 = \frac{E}{R_1} R_1 e^{-t/\tau'}$$

and $$\boxed{v_{R_1} = Ee^{-t/\tau'}} \tag{11.22}$$

The voltage v_{R_1} has the same polarity as during the storage phase since the current i_L has the same direction. The voltage v_{R_2} is expressed as follows using the defined polarity of Fig. 11.42:

$$v_{R_2} = -i_{R_2}R_2 = -i_L R_2 = -\frac{E}{R_1} R_2 e^{-t/\tau'}$$

and $$\boxed{v_{R_2} = -\frac{R_2}{R_1} Ee^{-t/\tau'}} \tag{11.23}$$

EXAMPLE 11.5 Resistor R_2 was added to the network in Fig. 11.36 as shown in Fig. 11.44.

a. Find the mathematical expressions for i_L, v_L, v_{R_1}, and v_{R_2} for five time constants of the storage phase.
b. Find the mathematical expressions for i_L, v_L, v_{R_1}, and v_{R_2} if the switch is opened after five time constants of the storage phase.
c. Sketch the waveforms for each voltage and current for both phases covered by this example. Use the defined polarities in Fig. 11.43.

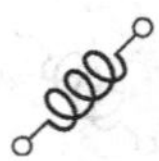

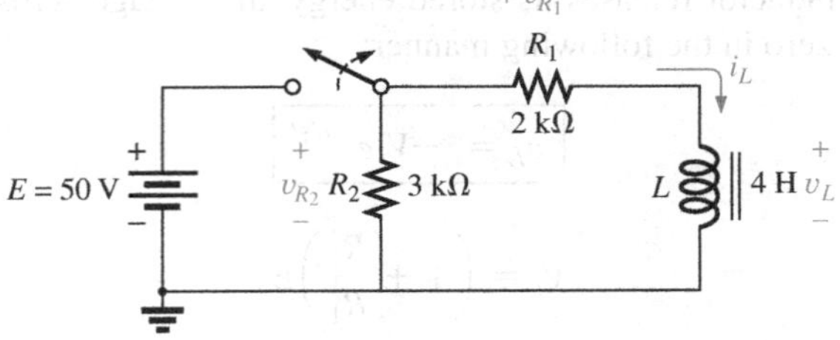

FIG. 11.44
Defined polarities for v_{R_1}, v_{R_2}, v_L, and current direction for i_L for Example 11.5.

Solutions:

a. From Example 11.3:

$$i_L = \mathbf{25\ mA\ (1 - e^{-t/2\ ms})}$$
$$v_L = \mathbf{50\ Ve^{-t/2\ ms}}$$
$$v_{R_1} = i_{R_1}R_1 = i_L R_1$$
$$= \left[\frac{E}{R_1}(1 - e^{-t/\tau})\right]R_1$$
$$= E(1 - e^{-t/\tau})$$

and
$$v_{R_1} = \mathbf{50\ V(1 - e^{-t/2\ ms})}$$
$$v_{R_2} = E = \mathbf{50\ V}$$

b. $$\tau' = \frac{L}{R_1 + R_2} = \frac{4\text{ H}}{2\text{ k}\Omega + 3\text{ k}\Omega} = \frac{4\text{ H}}{5 \times 10^3\ \Omega}$$
$$= 0.8 \times 10^{-3}\text{ s} = 0.8\text{ ms}$$

By Eqs. (11.19) and (11.20):

$$V_i = \left(1 + \frac{R_2}{R_1}\right)E = \left(1 + \frac{3\text{ k}\Omega}{2\text{ k}\Omega}\right)(50\text{ V}) = 125\text{ V}$$

and
$$v_L = -V_i e^{-t/\tau'} = \mathbf{-\ 125\ Ve^{-t/-0.8\ ms}}$$

By Eq. (11.21):

$$I_m = \frac{E}{R_1} = \frac{50\text{ V}}{2\text{ k}\Omega} = 25\text{ mA}$$

and
$$i_L = I_m e^{-t/\tau'} = \mathbf{25\ mAe^{-t/0.8\ ms}}$$

By Eq. (11.22):

$$v_{R_1} = Ee^{-t/\tau'} = \mathbf{50\ Ve^{-t/0.8\ ms}}$$

By Eq. (11.23):

$$v_{R_2} = -\frac{R_2}{R_1}Ee^{-t/\tau'} = -\frac{3\text{ k}\Omega}{2\text{ k}\Omega}(50\text{ V})e^{-t/\tau'} = \mathbf{-\ 75\ Ve^{-t/0.8\ ms}}$$

c. See Fig. 11.45.

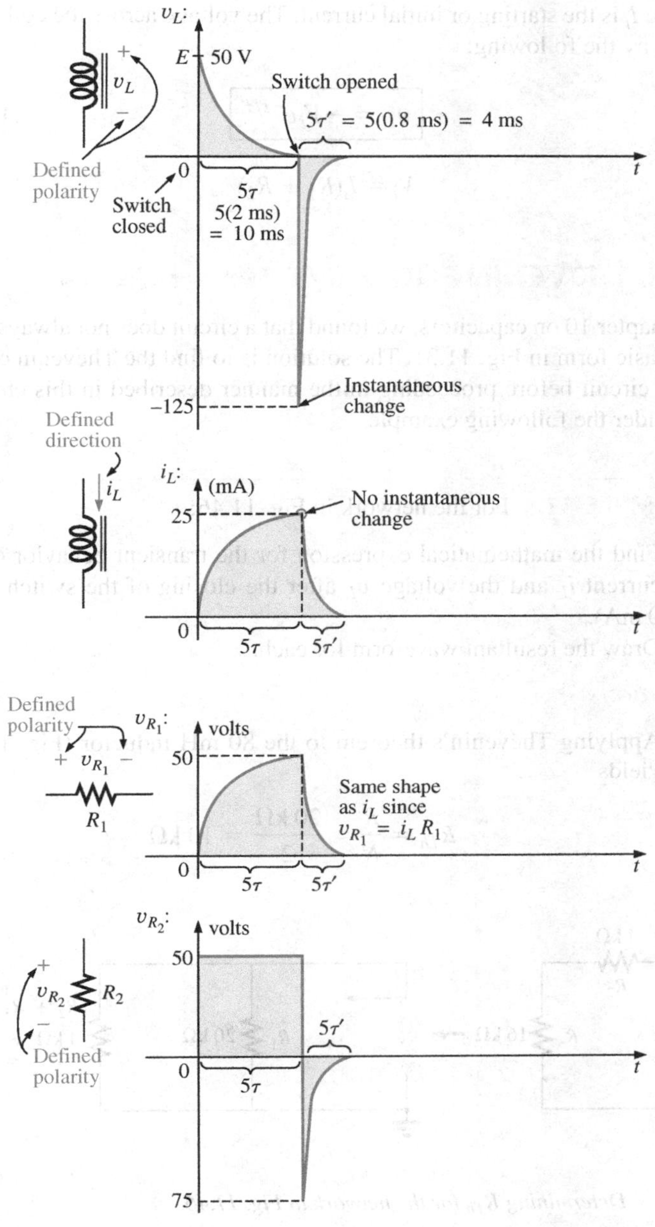

FIG. 11.45

The various voltages and the current for the network in Fig. 11.44.

In the preceding analysis, it was assumed that steady-state conditions were established during the charging phase and $I_m = E/R_1$, with $v_L = 0$ V. However, if the switch in Fig. 11.42 is opened before i_L reaches its maximum value, the equation for the decaying current of Fig. 11.42 must change to

$$i_L = I_i e^{-t/\tau'} \tag{11.24}$$

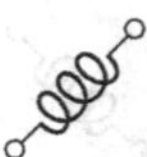

where I_i is the starting or initial current. The voltage across the coil is defined by the following:

$$v_L = -V_i e^{-t/\tau'} \tag{11.25}$$

with $$V_i = I_i(R_1 + R_2)$$

11.8 THÉVENIN EQUIVALENT: $\tau = L/R_{Th}$

In Chapter 10 on capacitors, we found that a circuit does not always have the basic form in Fig. 11.31. The solution is to find the Thévenin equivalent circuit before proceeding in the manner described in this chapter. Consider the following example.

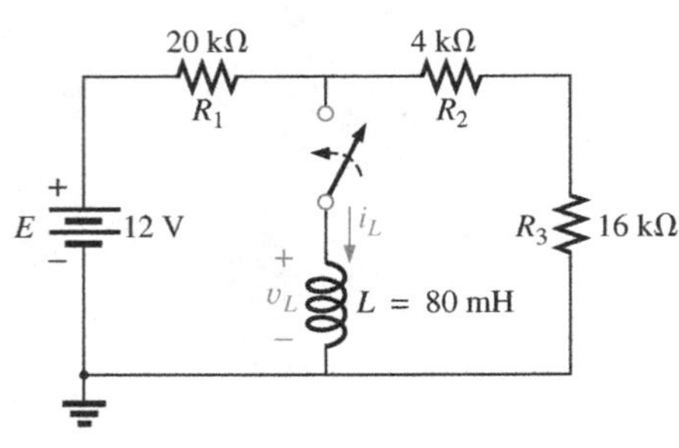

FIG. 11.46
Example 11.6.

EXAMPLE 11.6 For the network in Fig. 11.46:

a. Find the mathematical expression for the transient behavior of the current i_L and the voltage v_L after the closing of the switch (I_i = 0 mA).
b. Draw the resultant waveform for each.

Solutions:

a. Applying Thévenin's theorem to the 80 mH inductor (Fig. 11.47) yields

$$R_{Th} = \frac{R}{N} = \frac{20\text{ k}\Omega}{2} = 10\text{ k}\Omega$$

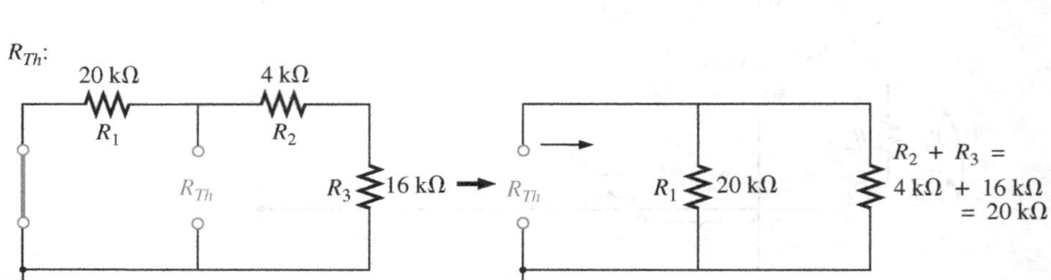

FIG. 11.47
Determining R_{Th} for the network in Fig. 11.46.

Applying the voltage divider rule (Fig. 11.48), we obtain

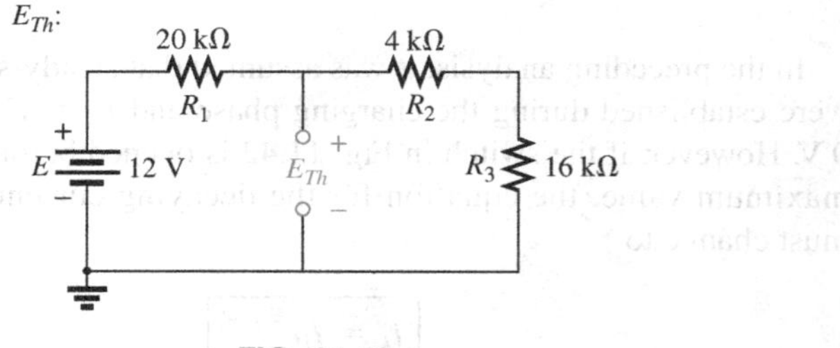

FIG. 11.48
Determining E_{Th} for the network in Fig. 11.46.

$$E_{Th} = \frac{(R_2 + R_3)E}{R_1 + R_2 + R_3}$$
$$= \frac{(4\text{ k}\Omega + 16\text{ k}\Omega)(12\text{ V})}{20\text{ k}\Omega + 4\text{ k}\Omega + 16\text{ k}\Omega} = \frac{(20\text{ k}\Omega)(12\text{ V})}{40\text{ k}\Omega} = 6\text{ V}$$

The Thévenin equivalent circuit is shown in Fig. 11.49. Using Eq. (11.13) gives

$$i_L = \frac{E_{Th}}{R}(1 - e^{-t/\tau})$$

$$\tau = \frac{L}{R_{Th}} = \frac{80 \times 10^{-3}\text{ H}}{10 \times 10^3\ \Omega} = 8 \times 10^{-6}\text{ s} = 8\ \mu\text{s}$$

$$I_m = \frac{E_{Th}}{R_{Th}} = \frac{6\text{ V}}{10 \times 10^3\ \Omega} = 0.6 \times 10^{-3}\text{ A} = 0.6\text{ mA}$$

and $\quad i_L = \mathbf{0.6\ mA\ (1 - e^{-t/8\ \mu s})}$

Using Eq. (11.15) gives

$$v_L = E_{Th}e^{-t/\tau}$$

so that $\quad v_L = \mathbf{6\ Ve^{-t/8\ \mu s}}$

b. See Fig. 11.50.

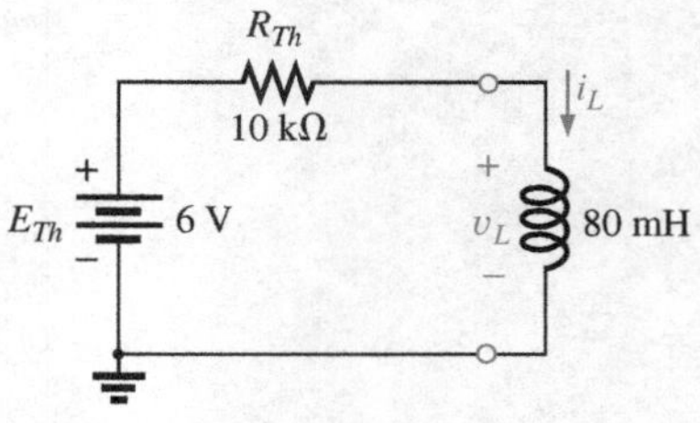

FIG. 11.49
The resulting Thévenin equivalent circuit for the network in Fig. 11.46.

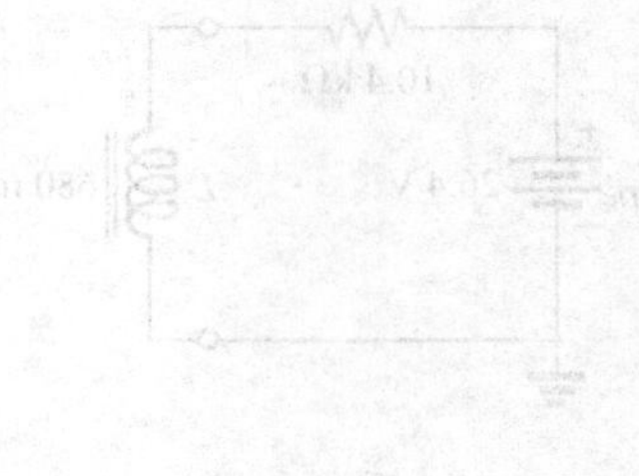

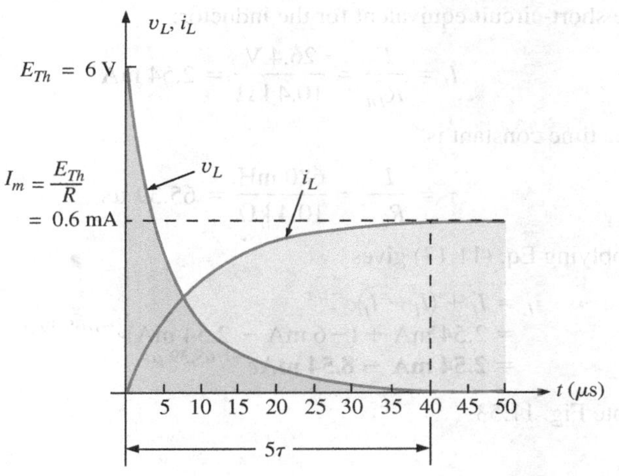

FIG. 11.50
The resulting waveforms for i_L and v_L for the network in Fig. 11.46.

EXAMPLE 11.7 Switch S_1 in Fig. 11.51 has been closed for a long time. At $t = 0$ s, S_1 is opened at the same instant that S_2 is closed to avoid an interruption in current through the coil.

a. Find the initial current through the coil. Pay particular attention to its direction.
b. Find the mathematical expression for the current i_L following the closing of switch S_2.
c. Sketch the waveform for i_L.

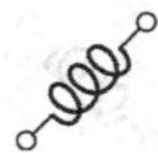

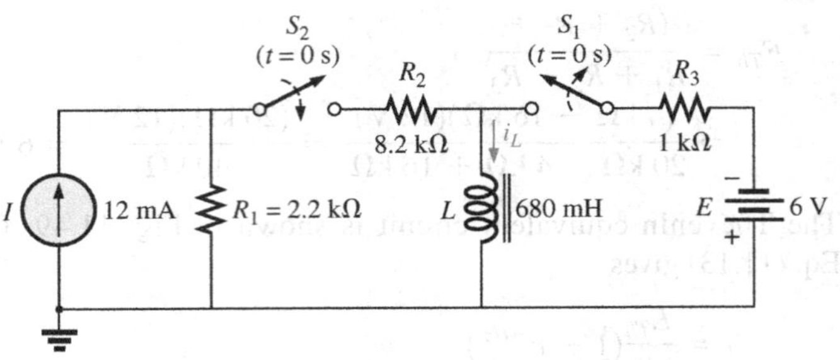

FIG. 11.51
Example 11.7.

Solutions:

a. Using Ohm's law, we find the initial current through the coil:

$$I_i = -\frac{E}{R_3} = -\frac{6\text{ V}}{1\text{ k}\Omega} = -6\text{ mA}$$

b. Applying Thévenin's theorem gives

$$R_{Th} = R_1 + R_2 = 2.2\text{ k}\Omega + 8.2\text{ k}\Omega = 10.4\text{ k}\Omega$$
$$E_{Th} = IR_1 = (12\text{ mA})(2.2\text{ k}\Omega) = 26.4\text{ V}$$

The Thévenin equivalent network appears in Fig. 11.52.

The steady-state current can then be determined by substituting the short-circuit equivalent for the inductor:

$$I_f = \frac{E}{R_{Th}} = \frac{26.4\text{ V}}{10.4\text{ k}\Omega} = 2.54\text{ mA}$$

The time constant is

$$\tau = \frac{L}{R_{Th}} = \frac{680\text{ mH}}{10.4\text{ k}\Omega} = 65.39\ \mu\text{s}$$

Applying Eq. (11.17) gives

$$\begin{aligned} i_L &= I_f + (I_i - I_f)e^{-t/\tau} \\ &= 2.54\text{ mA} + (-6\text{ mA} - 2.54\text{ mA})e^{-t/65.39\ \mu\text{s}} \\ &= \mathbf{2.54\ mA - 8.54\ mA}e^{\mathbf{-t/65.39\ \mu s}} \end{aligned}$$

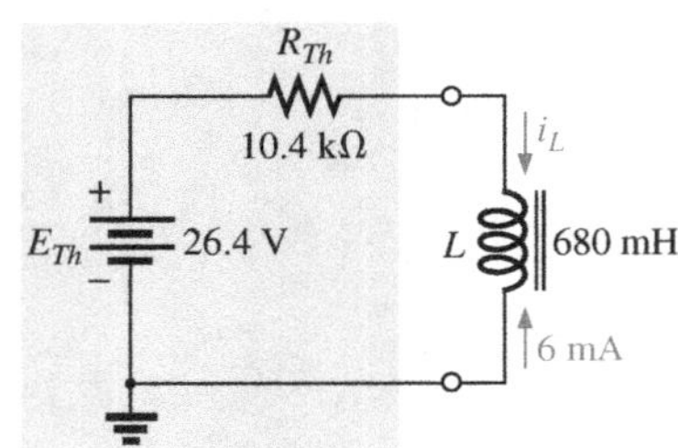

FIG. 11.52
Thévenin equivalent circuit for the network in Fig. 11.51 for t ≥ 0 s.

c. Note Fig. 11.53.

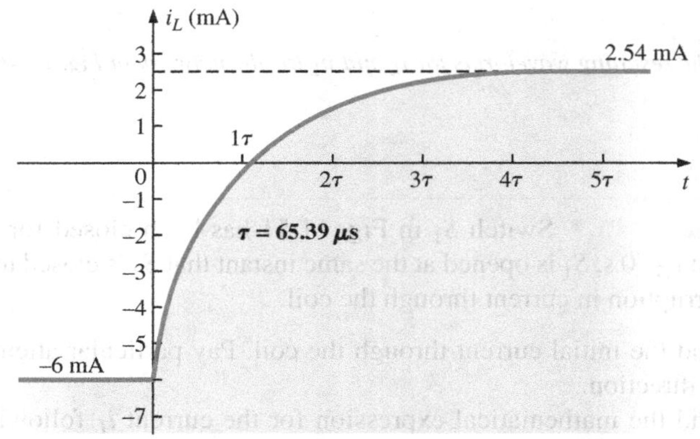

FIG. 11.53
The current i_L for the network in Fig. 11.51.

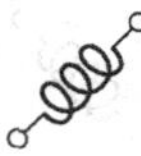

11.9 INSTANTANEOUS VALUES

The development presented in Section 10.8 for capacitive networks can also be applied to *R-L* networks to determine instantaneous voltages, currents, and time. The instantaneous values of any voltage or current can be determined by simply inserting *t* into the equation and using a calculator or table to determine the magnitude of the exponential term.

The similarity between the equations

$$v_C = V_f + (V_i + V_f)e^{-t/\tau}$$

and

$$i_L = I_f + (I_i - I_f)e^{-t/\tau}$$

results in a derivation of the following for *t* that is identical to that used to obtain Eq. (10.23):

$$t = \tau \log_e \frac{(I_i - I_f)}{(i_L - I_f)} \qquad \text{(seconds, s)} \qquad \textbf{(11.26)}$$

For the other form, the equation $v_C = Ee^{-t/\tau}$ is a close match with $v_L = Ee^{-t/\tau} = V_i e^{-t/\tau}$, permitting a derivation similar to that employed for Eq. (10.23):

$$t = \tau \log_e \frac{V_i}{v_L} \qquad \text{(seconds, s)} \qquad \textbf{(11.27)}$$

For the voltage v_R, $V_i = 0$ V and $V_f = EV$ since $v_R = E(1 - e^{-t/\tau})$. Solving for *t* yields

$$t = \tau \log_e \left(\frac{E}{E - v_R} \right)$$

or

$$t = \tau \log_e \left(\frac{V_f}{V_f - v_R} \right) \qquad \text{(seconds, s)} \qquad \textbf{(11.28)}$$

11.10 AVERAGE INDUCED VOLTAGE: $v_{L_{av}}$

In an effort to develop some feeling for the impact of the derivative in an equation, the average value was defined for capacitors in Section 10.10, and a number of plots for the current were developed for an applied voltage. For inductors, a similar relationship exists between the induced voltage across a coil and the current through the coil. For inductors, the average induced voltage is defined by

$$v_{L_{av}} = L \frac{\Delta i_L}{\Delta t} \qquad \text{(volts, V)} \qquad \textbf{(11.29)}$$

where Δ indicates a finite (measurable) change in current or time. Eq. (11.12) for the instantaneous voltage across a coil can be derived from Eq. (11.29) by letting V_L become vanishingly small. That is,

$$v_{L_{inst}} = \lim_{\Delta t \to 0} L \frac{\Delta i_L}{\Delta t} = L \frac{di_L}{dt}$$

In the following example, the change in current Δi_L is considered for each slope of the current waveform. *If the current increases with time, the average current is the change in current divided by the change in*

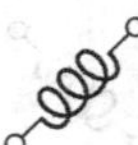

time, with a positive sign. If the current decreases with time, a negative sign is applied. Note in the example that the faster the current changes with time, the greater is the induced voltage across the coil. When making the necessary calculations, do not forget to multiply by the inductance of the coil. Larger inductances result in increased levels of induced voltage for the same change in current through the coil.

EXAMPLE 11.8 Find the waveform for the average voltage across the coil if the current through a 4 mH coil is as shown in Fig. 11.54.

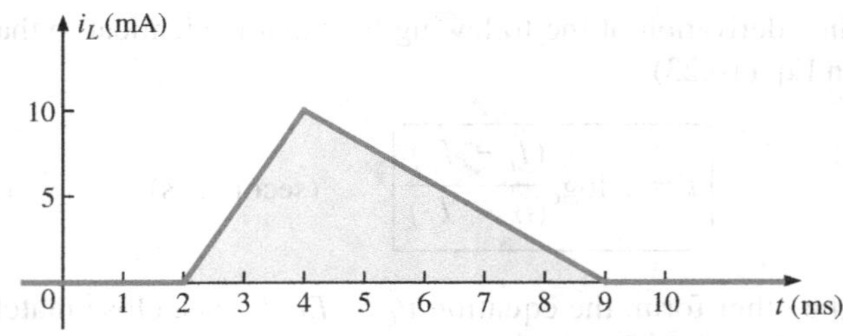

FIG. 11.54
Current i_L to be applied to a 4 mH coil in Example 11.8.

Solutions:

a. *0 to 2 ms:* Since there is no change in current through the coil, there is no voltage induced across the coil. That is,

$$v_L = L\frac{\Delta i}{\Delta t} = L\frac{0}{\Delta t} = \mathbf{0\ V}$$

b. *2 ms to 4 ms:*

$$v_L = L\frac{\Delta i}{\Delta t} = (4 \times 10^{-3}\,\text{H})\left(\frac{10 \times 10^{-3}\,\text{A}}{2 \times 10^{-3}\,\text{s}}\right) = 20 \times 10^{-3}\,\text{V} = \mathbf{20\ mV}$$

c. *4 ms to 9 ms:*

$$v_L = L\frac{\Delta i}{\Delta t} = (-4 \times 10^{-3}\,\text{H})\left(\frac{10 \times 10^{-3}\,\text{A}}{5 \times 10^{-3}\,\text{s}}\right) = -8 \times 10^{-3}\,\text{V} = \mathbf{-8\ mV}$$

d. *9 ms to ∞:*

$$v_L = L\frac{\Delta i}{\Delta t} = L\frac{0}{\Delta t} = \mathbf{0\ V}$$

The waveform for the average voltage across the coil is shown in Fig. 11.55. Note from the curve that

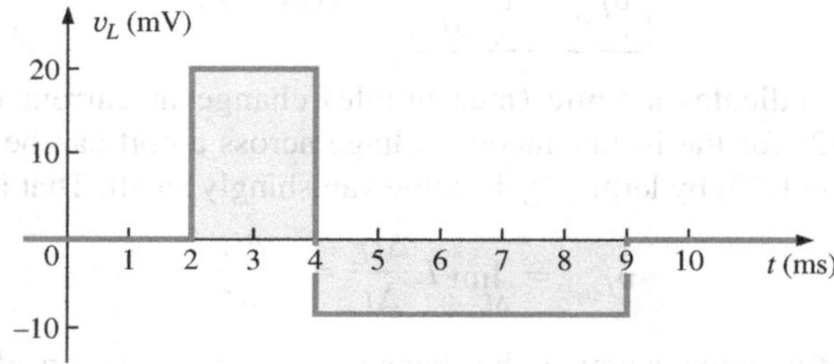

FIG. 11.55
Voltage across a 4 mH coil due to the current in Fig. 11.54.

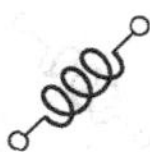

the voltage across the coil is not determined solely by the magnitude of the change in current through the coil (Δi), but by the rate of change of current through the coil ($\Delta i/\Delta t$).

A similar statement was made for the current of a capacitor due to change in voltage across the capacitor.

A careful examination of Fig. 11.55 also reveals that the area under the positive pulse from 2 ms to 4 ms equals the area under the negative pulse from 4 ms to 9 ms. In Section 11.13, we will find that the area under the curves represents the energy stored or released by the inductor. From 2 ms to 4 ms, the inductor is storing energy, whereas from 4 ms to 9 ms, the inductor is releasing the energy stored. For the full period from 0 ms to 10 ms, energy has been stored and released; there has been no dissipation as experienced for the resistive elements. Over a full cycle, both the ideal capacitor and inductor do not consume energy but store and release it in their respective forms.

11.11 INDUCTORS IN SERIES AND IN PARALLEL

Inductors, like resistors and capacitors, can be placed in series or in parallel. Increasing levels of inductance can be obtained by placing inductors in series, while decreasing levels can be obtained by placing inductors in parallel.

For inductors in series, the total inductance is found in the same manner as the total resistance of resistors in series (Fig. 11.56):

$$L_T = L_1 + L_2 + L_3 + \cdots + L_N \tag{11.30}$$

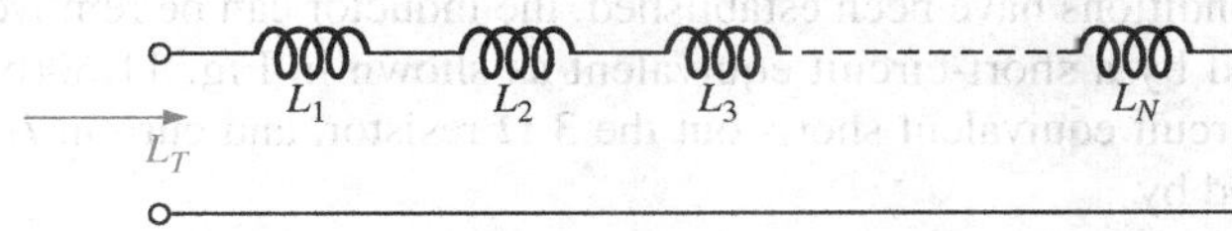

FIG. 11.56
Inductors in series.

For inductors in parallel, the total inductance is found in the same manner as the total resistance of resistors in parallel (Fig 11.57):

$$\frac{1}{L_T} = \frac{1}{L_1} + \frac{1}{L_2} + \frac{1}{L_3} + \cdots + \frac{1}{L_N} \tag{11.31}$$

For two inductors in parallel,

$$L_T = \frac{L_1 L_2}{L_1 + L_2} \tag{11.32}$$

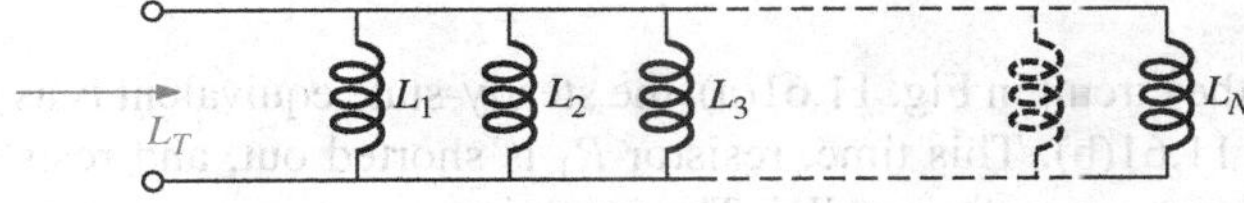

FIG. 11.57
Inductors in parallel.

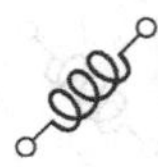

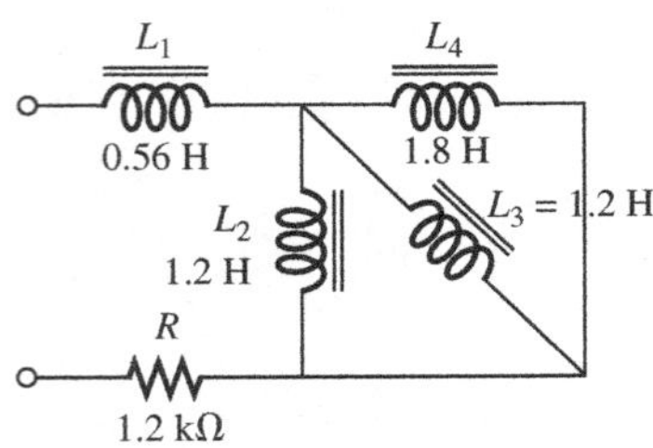

FIG. 11.58
Example 11.9.

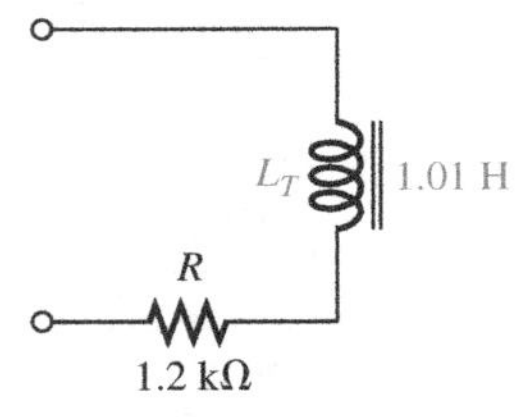

FIG. 11.59
Terminal equivalent of the network in Fig. 11.58.

EXAMPLE 11.9 Reduce the network in Fig. 11.58 to its simplest form.

Solution: Inductors L_2 and L_3 are equal in value and they are in parallel, resulting in an equivalent parallel value of

$$L'_T = \frac{L}{N} = \frac{1.2\text{ H}}{2} = 0.6\text{ H}$$

The resulting 0.6 H is then in parallel with the 1.8 H inductor, and

$$L''_T = \frac{(L'_T)(L_4)}{L'_T + L_4} = \frac{(0.6\text{ H})(1.8\text{ H})}{0.6\text{ H} + 1.8\text{ H}} = 0.45\text{ H}$$

Inductor L_1 is then in series with the equivalent parallel value, and

$$L_T = L_1 + L''_T = 0.56\text{ H} + 0.45\text{ H} = \mathbf{1.01\text{ H}}$$

The reduced equivalent network appears in Fig. 11.59.

11.12 STEADY-STATE CONDITIONS

We found in Section 11.5 that, for all practical purposes, an ideal (ignoring internal resistance and stray capacitances) inductor can be replaced by a short-circuit equivalent once steady-state conditions have been established. Recall that the term *steady state* implies that the voltage and current levels have reached their final resting value and will no longer change unless a change is made in the applied voltage or circuit configuration. For all practical purposes, our assumption is that steady-state conditions have been established after five time constants of the storage or release phase have passed.

For the circuit in Fig. 11.60(a), for example, if we assume that steady-state conditions have been established, the inductor can be removed and replaced by a short-circuit equivalent as shown in Fig. 11.60(b). The short-circuit equivalent shorts out the 3 Ω resistor, and current I_1 is determined by

$$I_1 = \frac{E}{R_1} = \frac{10\text{ V}}{2\ \Omega} = \mathbf{5\text{ A}}$$

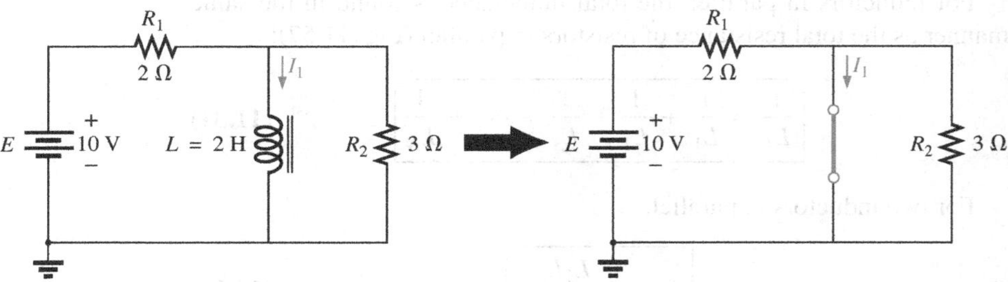

FIG. 11.60
Substituting the short-circuit equivalent for the inductor for t > 5τ.

For the circuit in Fig. 11.61(a), the steady-state equivalent is as shown in Fig. 11.61(b). This time, resistor R_1 is shorted out, and resistors R_2 and R_3 now appear in parallel. The result is

$$I = \frac{E}{R_2 \| R_3} = \frac{21\text{ V}}{2\ \Omega} = \mathbf{10.5\text{ A}}$$

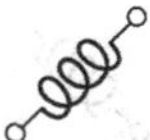

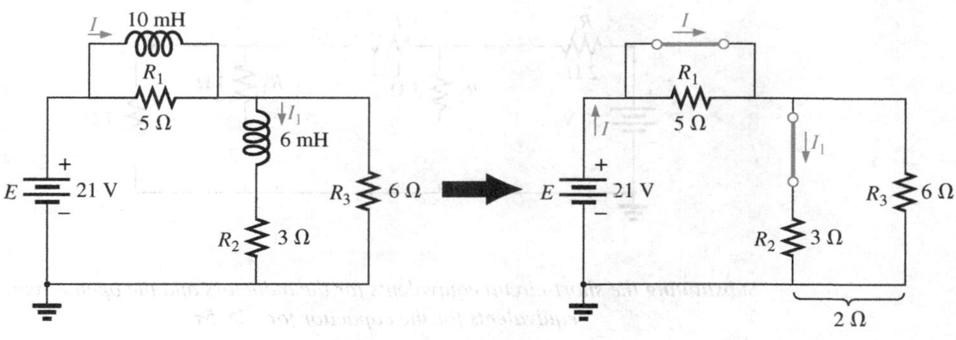

FIG. 11.61
Establishing the equivalent network for t > 5τ.

Applying the current divider rule yields

$$I_1 = \frac{R_3 I}{R_3 + R_2} = \frac{(6\ \Omega)(10.5\ \text{A})}{6\ \Omega + 3\ \Omega} = \frac{63}{9}\text{A} = \mathbf{7\ A}$$

In the examples to follow, it is assumed that steady-state conditions have been established.

EXAMPLE 11.10 Find the current I_L and the voltage V_C for the network in Fig. 11.62.

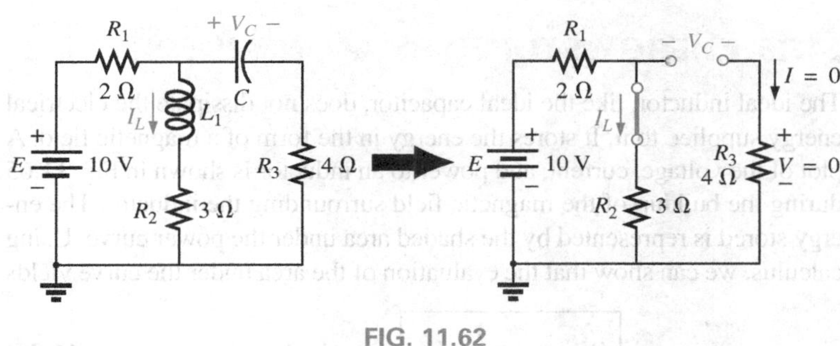

FIG. 11.62
Example 11.10.

Solution:

$$I_L = \frac{E}{R_1 + R_2} = \frac{10\ \text{V}}{5\ \Omega} = \mathbf{2\ A}$$

$$V_C = \frac{R_2 E}{R_2 + R_1} = \frac{(3\ \Omega)(10\ \text{V})}{3\ \Omega + 2\ \Omega} = \mathbf{6\ V}$$

EXAMPLE 11.11 Find currents I_1 and I_2 and voltages V_1 and V_2 for the network in Fig. 11.63.

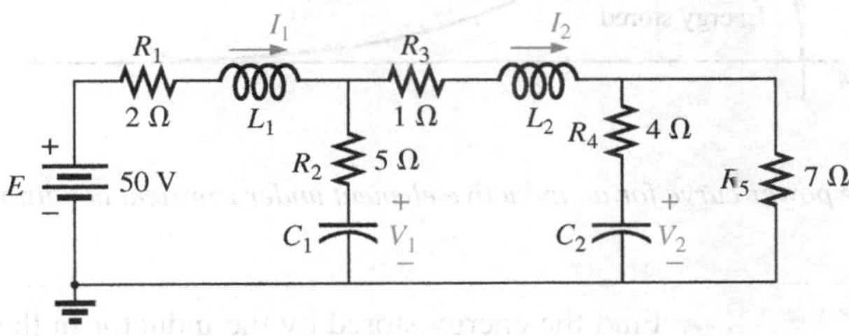

FIG. 11.63
Example 11.11.

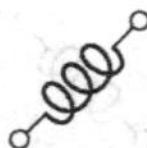

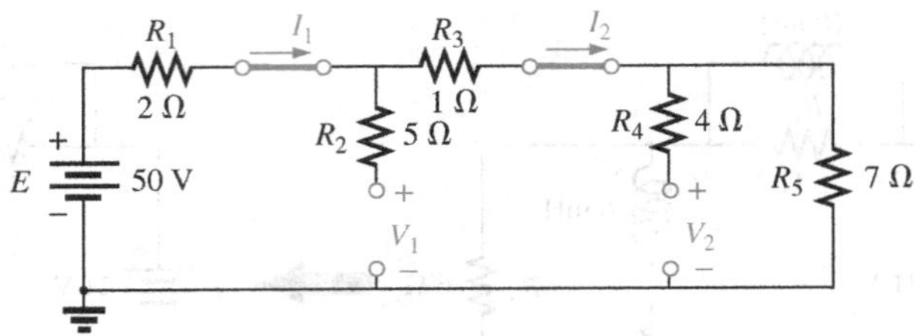

FIG. 11.64
Substituting the short-circuit equivalents for the inductors and the open-circuit equivalents for the capacitor for $t > 5\tau$.

Solution: Note Fig. 11.64.

$$I_1 = I_2$$
$$= \frac{E}{R_1 + R_3 + R_5} = \frac{50\ \text{V}}{2\ \Omega + 1\ \Omega + 7\ \Omega} = \frac{50\ \text{V}}{10\ \Omega} = \mathbf{5\ A}$$
$$V_2 = I_2R_5 = (5\ \text{A})(7\ \Omega) = \mathbf{35\ V}$$

Applying the voltage divider rule yields

$$V_1 = \frac{(R_3 + R_5)E}{R_1 + R_3 + R_5} = \frac{(1\ \Omega + 7\ \Omega)(50\ \text{V})}{2\ \Omega + 1\ \Omega + 7\ \Omega} = \frac{(8\ \Omega)(50\ \text{V})}{10\ \Omega} = \mathbf{40\ V}$$

11.13 ENERGY STORED BY AN INDUCTOR

The ideal inductor, like the ideal capacitor, does not dissipate the electrical energy supplied to it. It stores the energy in the form of a magnetic field. A plot of the voltage, current, and power to an inductor is shown in Fig. 11.65 during the buildup of the magnetic field surrounding the inductor. The energy stored is represented by the shaded area under the power curve. Using calculus, we can show that the evaluation of the area under the curve yields

$$\boxed{W_{\text{stored}} = \frac{1}{2}LI_m^2} \qquad \text{(joules, J)} \qquad \textbf{(11.33)}$$

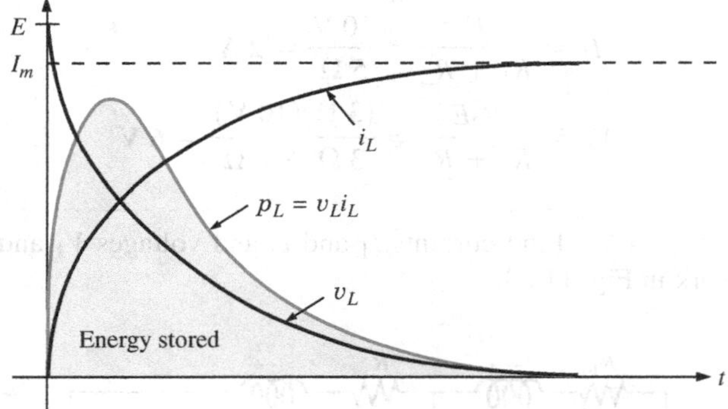

FIG. 11.65
The power curve for an inductive element under transient conditions.

EXAMPLE 11.12 Find the energy stored by the inductor in the circuit in Fig. 11.66 when the current through it has reached its final value.

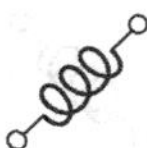

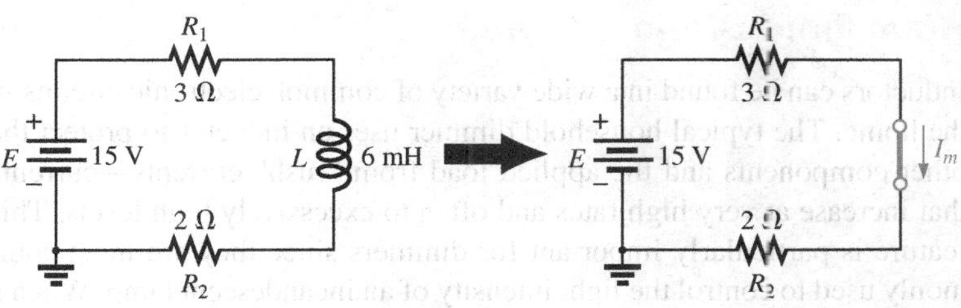

FIG. 11.66
Example 11.12.

Solution:

$$I_m = \frac{E}{R_1 + R_2} = \frac{15\text{ V}}{3\ \Omega + 2\ \Omega} = \frac{15\text{ V}}{5\ \Omega} = 3\text{ A}$$

$$W_{\text{stored}} = \frac{1}{2}LI_m^2 = \frac{1}{2}(6 \times 10^{-3}\text{ H})(3\text{ A})^2 = \frac{54}{2} \times 10^{-3}\text{ J} = \mathbf{27\text{ mJ}}$$

11.14 APPLICATIONS

Camera Flash Lamp

The inductor played an important role in the camera flash lamp circuitry described in the applications section of Chapter 10 on capacitors. For the camera, the inductor was the important component that resulted in the high spike voltage across the trigger coil, which was then magnified by the autotransformer action of the secondary to generate the 4000 V necessary to ignite the flash lamp. Recall that the capacitor in parallel with the trigger coil charged up to 300 V using the low-resistance path provided by the SCR (silicon-controlled rectifier). However, once the capacitor was fully charged, the short-circuit path to ground provided by the SCR was removed, and the capacitor immediately started to discharge through the trigger coil. Since the only resistance in the time constant for the inductive network is the relatively low resistance of the coil itself, the current through the coil grew at a very rapid rate. A significant voltage was then developed across the coil as defined by Eq. (11.12): $v_L = L\,(di_L/dt)$. This voltage was in turn increased by transformer action to the secondary coil of the autotransformer, and the flash lamp was ignited. That high voltage generated across the trigger coil also appears directly across the capacitor of the trigger network. The result is that it begins to charge up again until the generated voltage across the coil drops to zero volts. However, when it does drop, the capacitor again discharges through the coil, establishes another charging current through the coil, and again develops a voltage across the coil. The high-frequency exchange of energy between the coil and capacitor is called *flyback* because of the "flying back" of energy from one storage element to the other. It begins to decay with time because of the resistive elements in the loop. The more resistance, the more quickly it dies out. If the capacitor-inductor pairing is isolated and "tickled" along the way with the application of a dc voltage, the high frequency-generated voltage across the coil can be maintained and put to good use. In fact, it is this flyback effect that is used to generate a steady dc voltage (using rectification to convert the oscillating waveform to one of a steady dc nature) that is commonly used in TVs.

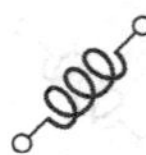

Household Dimmer Switch

Inductors can be found in a wide variety of common electronic circuits in the home. The typical household dimmer uses an inductor to protect the other components and the applied load from "rush" currents—currents that increase at very high rates and often to excessively high levels. This feature is particularly important for dimmers since they are most commonly used to control the light intensity of an incandescent lamp. When a lamp is turned on, the resistance is typically very low, and relatively high currents may flow for short periods of time until the filament of the bulb heats up. The inductor is also effective in blocking high-frequency noise (RFI) generated by the switching action of the triac in the dimmer. A capacitor is also normally included from line to neutral to prevent any voltage spikes from affecting the operation of the dimmer and the applied load (lamp, etc.) and to assist with the suppression of RFI disturbances.

A photograph of one of the most common dimmers is provided in Fig. 11.67(a), with an internal view shown in Fig. 11.67(b). The basic

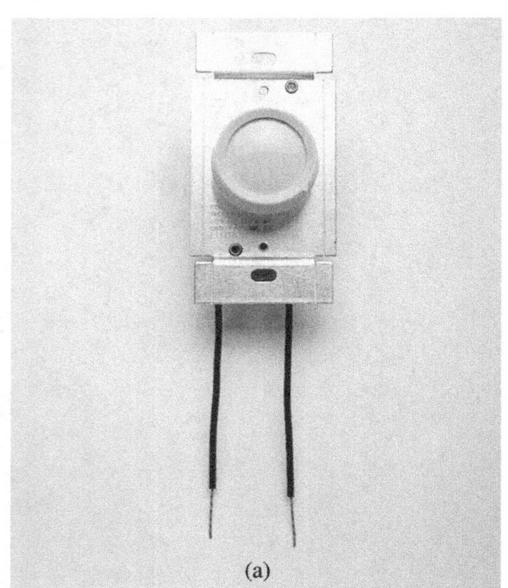

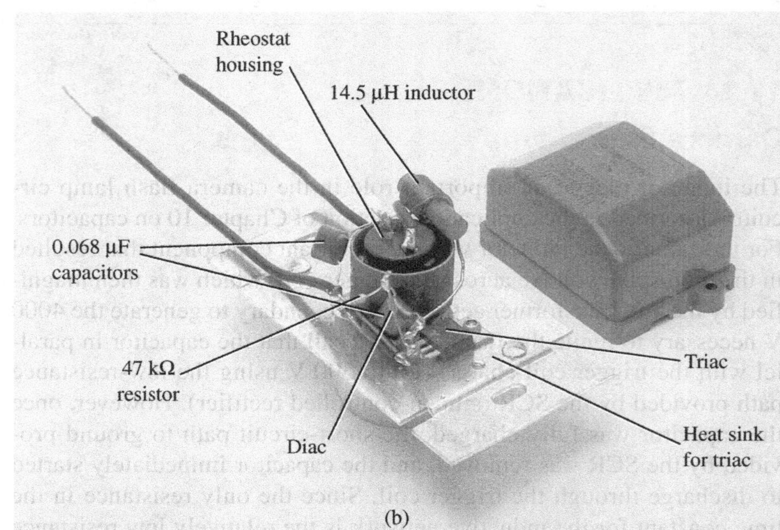

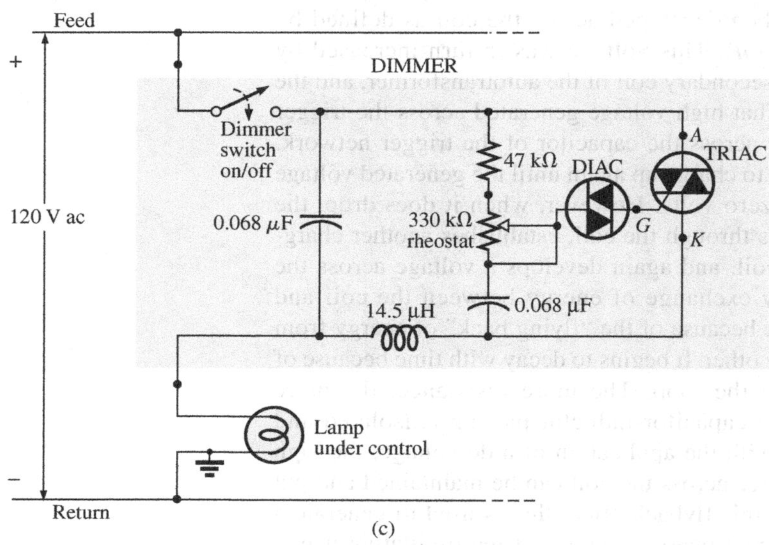

FIG.11.67
Dimmer control: (a) external appearance; (b) internal construction; (c) schematic.

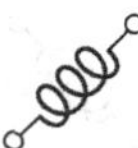

components of most commercially available dimmers appear in the schematic in Fig. 11.67(c). In this design, a 14.5 μH inductor is used in the choking capacity described above, with a 0.068 μF capacitor for the "bypass" operation. Note the size of the inductor with its heavy wire and large ferromagnetic core and the relatively large size of the two 0.068 μF capacitors. Both suggest that they are designed to absorb high-energy disturbances.

The general operation of a dimmer is shown in Fig. 11.68. The controlling network is in series with the lamp and essentially acts as an impedance (like resistance—to be introduced in Chapter 15) that can vary between very low and very high levels. Very low impedance levels resemble a short circuit, so that the majority of the applied voltage appears across the lamp [Fig. 11.68 (a)], and very high impedances approach an open circuit where very little voltage appears across the lamp [Fig. 11.68 (b)]. Intermediate levels of impedance control the terminal voltage of the bulb accordingly. For instance, if the controlling network has a very high impedance (open-circuit equivalent) through half the cycle, as shown in Fig. 11.68(c), the brightness of the bulb will be less than full voltage but not 50% due to the nonlinear relationship between the brightness of a bulb and the applied voltage. A lagging effect is also present in the actual operation of the dimmer, which we will learn about when leading and lagging networks are examined in the ac chapters.

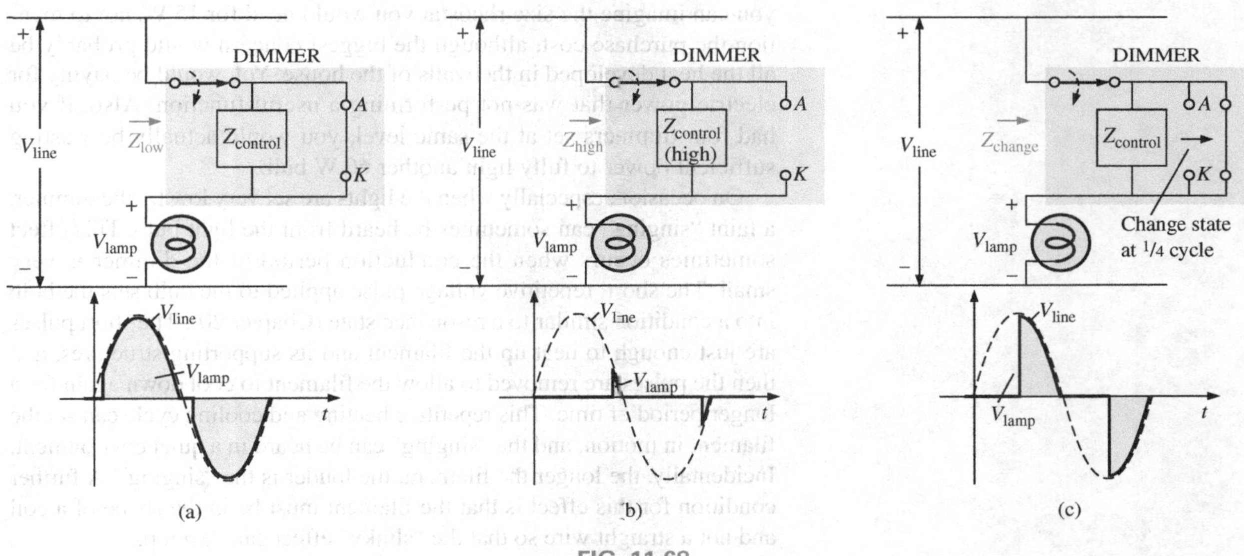

FIG. 11.68

Basic operation of the dimmer in Fig. 11.67: (a) full voltage to the lamp; (b) approaching the cutoff point for the bulb; (c) reduced illumination of the lamp.

The controlling knob, slide, or whatever other method is used on the face of the switch to control the light intensity is connected directly to the rheostat in the branch parallel to the triac. Its setting determines when the voltage across the capacitor reaches a sufficiently high level to turn on the diac (a bidirectional diode) and establish a voltage at the gate (G) of the triac to turn it on. When it does, it establishes a very low resistance path from the anode (A) to the cathode (K), and the applied voltage appears directly across the lamp. When the SCR is off, its terminal resistance between anode and cathode is very high and can be approximated by an open circuit. During this period, the applied voltage does not reach

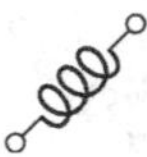

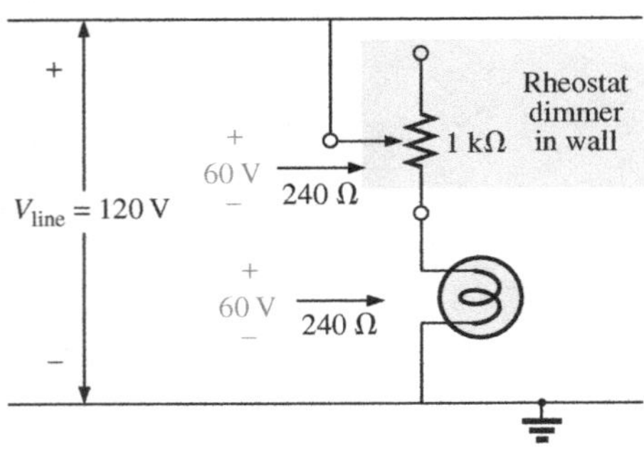

FIG. 11.69
Direct rheostat control of the brightness of a 60 W bulb.

the load (lamp). At this time, the impedance of the parallel branch containing the rheostat, fixed resistor, and capacitor is sufficiently high compared to the load that it can also be ignored, completing the open-circuit equivalent in series with the load. Note the placement of the elements in the photograph in Fig. 11.67(b) and that the metal plate to which the triac is connected is actually a heat sink for the device. The on/off switch is in the same housing as the rheostat. The total design is certainly well planned to maintain a relatively small size for the dimmer.

Since the effort here is to control the amount of power getting to the load, the question is often asked, Why don't we just use a rheostat in series with the lamp? The question is best answered by examining Fig. 11.69, which shows a rather simple network with a rheostat in series with the lamp. At full wattage, a 60 W bulb on a 120 V line theoretically has an internal resistance of $R = V^2/P$ (from the equation $P = V^2/R$) $= (120\text{ V})^2/60\text{ W} = 240\ \Omega$. Although the resistance is sensitive to the applied voltage, we will assume this level for the following calculations.

If we consider the case where the rheostat is set for the same level as the bulb, as shown in Fig. 11.69, there will be 60 V across the rheostat and the bulb. The power to each element is then $P = V^2/R = (60\text{ V})^2/240\ \Omega = 15\text{ W}$. The bulb is certainly quite dim, but the rheostat inside the dimmer switch is dissipating 15 W of power on a continuous basis. When you consider the size of a 2 W potentiometer in your laboratory, you can imagine the size rheostat you would need for 15 W, not to mention the purchase cost, although the biggest concern would probably be all the heat developed in the walls of the house. You would be paying for electric power that was not performing a useful function. Also, if you had four dimmers set at the same level, you would actually be wasting sufficient power to fully light another 60 W bulb.

On occasion, especially when the lights are set very low by the dimmer, a faint "singing" can sometimes be heard from the light bulb. This effect sometimes occurs when the conduction period of the dimmer is very small. The short, repetitive voltage pulse applied to the bulb sets the bulb into a condition similar to a resonance state (Chapter 20). The short pulses are just enough to heat up the filament and its supporting structures, and then the pulses are removed to allow the filament to cool down again for a longer period of time. This repetitive heating and cooling cycle can set the filament in motion, and the "singing" can be heard in a quiet environment. Incidentally, the longer the filament, the louder is the "singing." A further condition for this effect is that the filament must be in the shape of a coil and not a straight wire so that the "slinky" effect can develop.

11.15 COMPUTER ANALYSIS

PSpice

Transient *RL* Response The computer analysis begins with a transient analysis of the network of parallel inductive elements in Fig. 11.70. The inductors are picked up from the **ANALOG** library in the **Place Part** dialog box. As noted in Fig. 11.70, the inductor is displayed with a dot at one end of the coil. The dot is defined by a convention that is used when two or more coils have a mutual inductance, a topic that will be discussed in detail in Chapter 22. In this example, there are no assumed mutual effects, so the dots have no effect on this investigation. However, for this software, the dot is always placed closed to terminal 1 of the

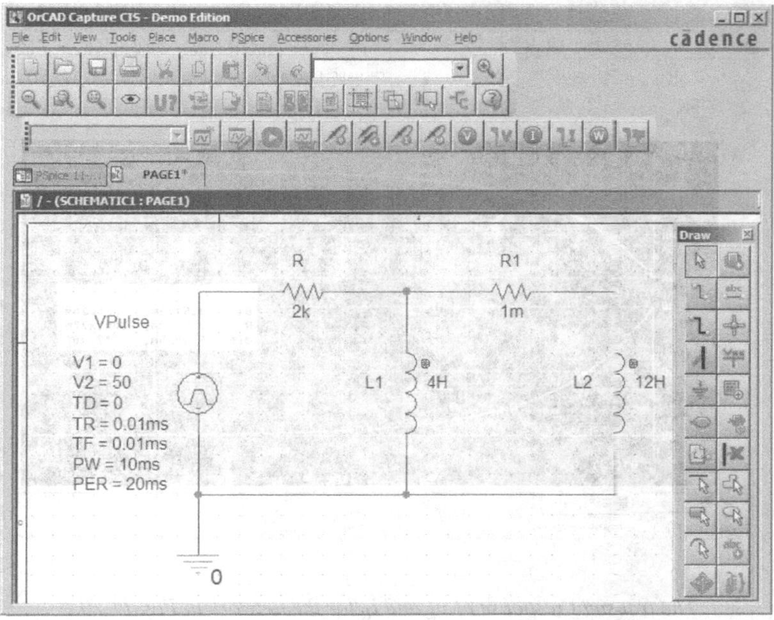

FIG. 11.70
Using PSpice to obtain the transient response of a parallel inductive network due to an applied pulse of 50 V.

inductor. If you bring the pointer controlled by the mouse close to the end of the coil **L1** with the dot, the following will result: **[/L1/1Number:1]**. The number is important because it will define which plot we want to see in the probe response later. When the inductors are placed on the screen, they have to be rotated 270°, which can be accomplished with the **Rotate-Mirror Vertically** sequence.

Also note in Fig. 11.70 the need for a series resistor R_1 within the parallel loop of inductors. In PSpice, inductors must have a series resistor to reflect real-world conditions. The chosen value of 1 mΩ is so small, however, that it will not affect the response of the system. For **VPulse** (obtained from the SOURCE Library), the rise and fall times were selected as 0.01 ms, and the pulse width was chosen as 10 ms because the time constant of the network is $\tau = L_T/R = (4\text{ H} \| 12\text{ H})/2\text{k}\Omega = 1.5$ ms and $5\tau = 7.5$ ms.

The simulation is the same as applied when obtaining the transient response of capacitive networks. In condensed form, the sequence to obtain a plot of the voltage across the coils versus time is as follows: **New SimulationProfile** key-**PSpice 11-1-Create-TimeDomain(Transient)-Run to time:**10ms-**Start saving data after:**0s and **Maximum step size:**5μs-**OK-Run PSpice** key-**Add Trace** key-**V(L2)-OK.** The resulting trace appears in the bottom of Fig. 11.71. A maximum step size of 5 μs was chosen to ensure that it was less than the rise or fall times of 10 μs. Note that the voltage across the coil jumps to the 50 V level almost immediately; then it decays to 0 V in about 8 ms. A plot of the total current through the parallel coils can be obtained through **Plot-Plot to Window-Add Trace** key-**I(R)-OK,** resulting in the trace appearing at the top of Fig. 11.71. When the trace first appeared, the vertical scale extended from 0 A to 30 mA even though the maximum value of i_R was 25 mA. To bring the maximum value to the top of the graph, **Plot** was selected followed by **Axis Settings-Y Axis-User Defined-0A to 25mA-OK.**

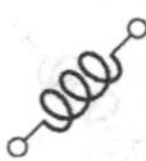

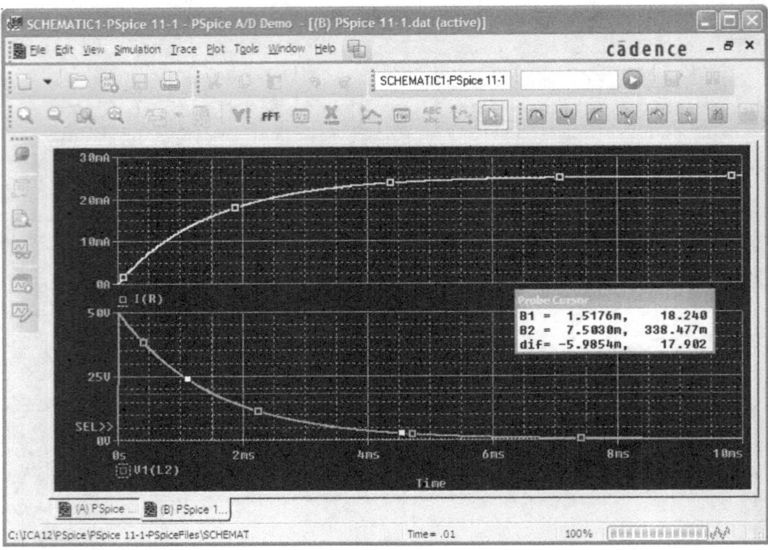

FIG. 11.71
The transient response of v_L and i_R for the network in Fig. 11.70.

For values, the voltage plot was selected, **SEL>>,** followed by the **Toggle cursor** key and a click on the screen to establish the crosshairs. The left-click cursor was set on one time constant of 1.5 ms to reveal a value of 18.24 V for **A1** (about 36.5% of the maximum as defined by the exponential waveform). The right-click cursor was set at 7.5 ms or five time constants, resulting in a relatively low 0.338 V for **A2.**

Transient Response with Initial Conditions The next application verifies the results of Example 11.4, which has an initial condition associated with the inductive element. **VPulse** is again employed with the parameters appearing in Fig. 11.72. Since $\tau = L/R = 100\text{ mH}/(2.2\text{ k}\Omega + 6.8\text{ k}\Omega) = 100\text{ mH}/9\text{ k}\Omega = 11.11\ \mu\text{s}$ and $5\tau = 55.55\ \mu\text{s}$, the pulse width (**PW**) was set to 100 μs. The rise and fall times were set at $100\ \mu\text{s}/1000 = 0.1\ \mu\text{s}$.

Setting the initial conditions for the inductor requires a procedure that has not been described as yet. First double-click on the inductor symbol to obtain the **Property Editor** dialog box. Then select **Parts** at the bottom of the dialog box, and select **New Column** to obtain the **Add New Column** dialog box. Under **Name,** enter **IC** (an abbreviation for "initial condition"—not "capacitive current") followed by the initial condition of 4 mA under **Value;** then click **OK.** The **Property Editor** dialog box appears again, but now the initial condition appears as a **New Column** in the horizontal listing dedicated to the inductive element. Now select **Display** to obtain the **Display Properties** dialog box, and under **Display Format** choose **Name and Value** so that both **IC** and **4 mA** appear. Click **OK** to return to the **Property Editor** dialog box. Finally, click on **Apply** and exit the dialog box (**X**). The result is the display in Fig. 11.72 for the inductive element.

Now for the simulation. First select the **New Simulation Profile** key, insert the name **PSpice 11-3,** and follow up with **Create.** Then in the **Simulation Settings** dialog box, select **Time Domain(Transient)** for the **Analysis type** and **General Settings** for the **Options.** The **Run to time** should be 200 μs so that you can see the full effect of the pulse source on the transient response. The **Start saving data after** should remain at 0 s,

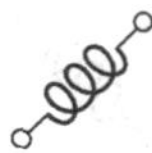

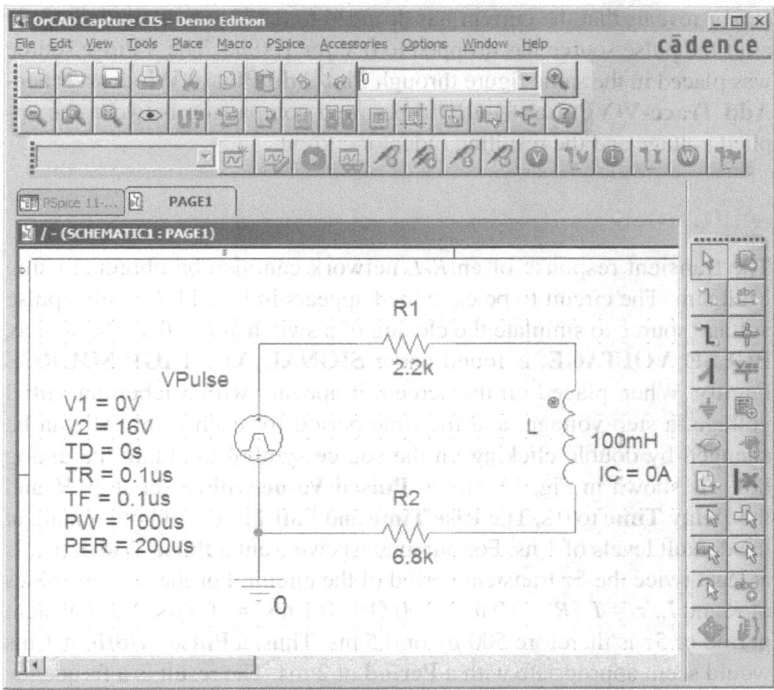

FIG. 11.72
Using PSpice to determine the transient response for a circuit in which the inductive element has an initial condition.

and the **Maximum step size** should be 200 μs/1000 = 200 ns. Click **OK** and then select the **Run PSpice** key. The result is a screen with an x-axis extending from 0 to 200 μs. Selecting **Trace** to get to the **Add Traces** dialog box and then selecting **I(L)** followed by **OK** results in the display in Fig. 11.73. The plot for **I(L)** clearly starts at the initial value of 4 mA and then decays to 1.78 mA as defined by the left-click cursor. The right-click

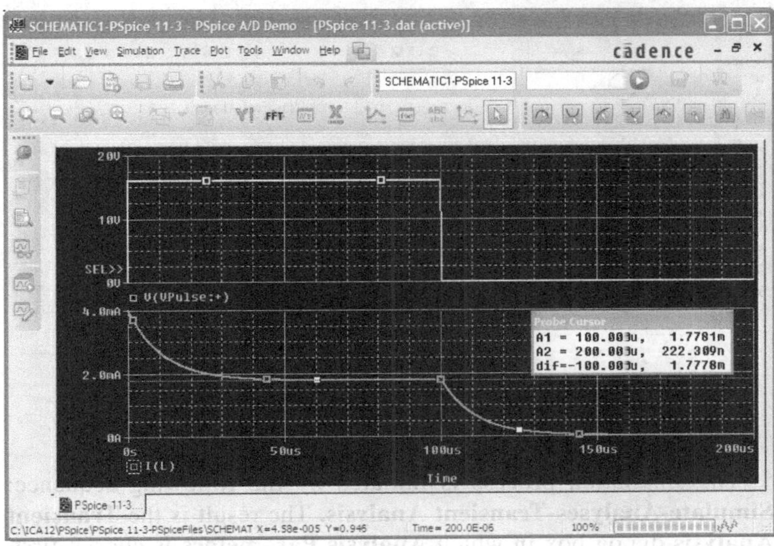

FIG. 11.73
A plot of the applied pulse and resulting current for the circuit in Fig. 11.72.

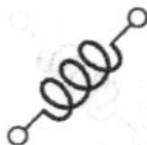

cursor reveals that the current has dropped to 0.222 μA (essentially 0 A) after the pulse source has dropped to 0 V for 100 μs. The **VPulse** source was placed in the same figure through **Plot-Add Plot to Window-Trace-Add Trace-V(VPulse:+)-OK** to permit a comparison between the applied voltage and the resulting inductor current.

Multisim

The transient response of an *R-L* network can also be obtained using Multisim. The circuit to be examined appears in Fig. 11.74 with a pulse voltage source to simulate the closing of a switch at $t = 0$ s. The source, **PULSE_VOLTAGE,** is found under **SIGNAL_VOLTAGE SOURCE** Family. When placed on the screen, it appears with a label, an initial voltage, a step voltage, and the time period for each level. All can be changed by double-clicking on the source symbol to obtain the dialog box. As shown in Fig. 11.74, the **Pulsed Value** will be set at 20 V, and the **Delay Time** to 0 s. The **Rise Time** and **Fall Time** will both remain at the default levels of 1 ns. For our analysis we want a **Pulse Width** that is at least twice the 5τ transient period of the circuit. For the chosen values of R and L, $\tau = L/R = 10\text{ mH}/100\ \Omega = 0.1\text{ ms} = 100\ \mu\text{s}$. The transient period of 5τ is therefore 500 μs or 0.5 ms. Thus, a **Pulse Width** of 1 ms would seem appropriate with a **Period** of 2 ms. The result is a frequency of $f = I/T = 1/2\text{ ms} = 500$ Hz. When the value of the inductor is set at 10 mH using a procedure defined in earlier chapters, an initial value for the current of the inductor can also be set under the heading of **Additional SPICE Simulation Parameters.** In this case, since it is not part of our analysis, it was set at 0 A, as shown in Fig. 11.74. When all have been set and selected, the parameters of the pulse source appear as shown in Fig. 11.74. Next the resistor, inductor, and ground are placed on the screen to complete the circuit.

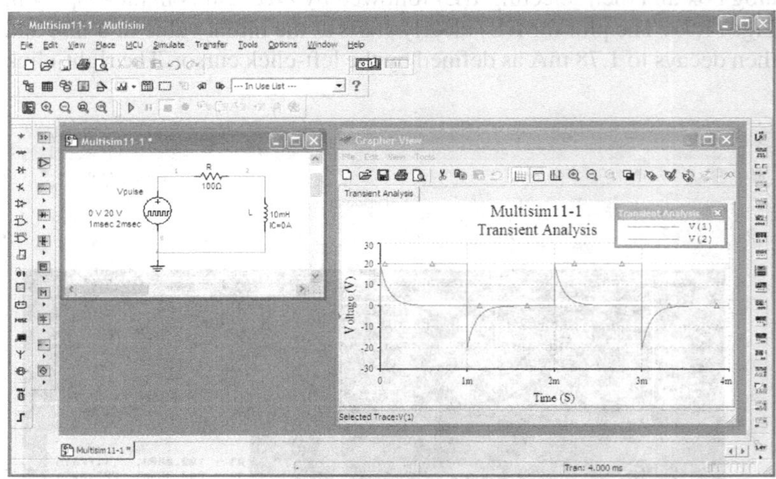

FIG. 11.74
Using Multisim to obtain the transient response for an inductive circuit.

The simulation process is initiated by the following sequence: **Simulate-Analyses-Transient Analysis.** The result is the **Transient Analysis** dialog box in which **Analysis Parameters** is chosen first. Under **Parameters,** use 0 s as the **Start time** and 4 ms (4E-3) as the **End time** so that we get two full cycles of the applied voltage. After enabling

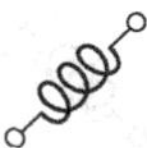

the **Maximum time step settings(TMAX)**, select the **Minimum number of time points** and set at 1000 to get a reasonably good plot during the rapidly changing transient period. Next, select the **Output variables** section and tell the program which voltage and current levels you are interested in. On the left side of the dialog box is a list of **Variables** that have been defined for the circuit. On the right is a list of **Selected variables for analysis.** In between you see **Add** or **Remove.** To move a variable from the left to the right column, select it in the left column and choose **Add.** It then appears in the right column. To plot both the applied voltage and the voltage across the coil, move **V(1)** and **V(2)** to the right column. Then select **Simulate.** A window titled **Grapher View** appears with the selected plots as shown in Fig. 11.74. Click on the **Show/Hide Grid** key (a red grid on a black axis), and the grid lines appear. Selecting the **Show/Hide Legend** key on the immediate right results in the small **Transient Analysis** dialog box that identifies the color that goes with each nodal voltage. In this case, red is the color of the applied voltage, and blue is the color of the voltage across the coil.

The source voltage appears as expected with its transition to 20 V, 50% duty cycle, and the period of 2 ms. The voltage across the coil jumped immediately to the 20 V level and then began its decay to 0 V in about 0.5 ms as predicted. When the source voltage dropped to zero, the voltage across the coil reversed polarity to maintain the same direction of current in the inductive circuit. Remember that for a coil, the voltage can change instantaneously, but the inductor "chokes" any instantaneous change in current. By reversing its polarity, the voltage across the coil ensures the same polarity of voltage across the resistor and therefore the same direction of current through the coil and circuit.

PROBLEMS

SECTION 11.2 Magnetic Field

1. For the electromagnet in Fig. 11.75:
   a. Find the flux density in Wb/m^2.
   b. What is the flux density in teslas?
   c. What is the applied magnetomotive force?
   d. What would the reading of the meter in Fig. 11.14 read in gauss?

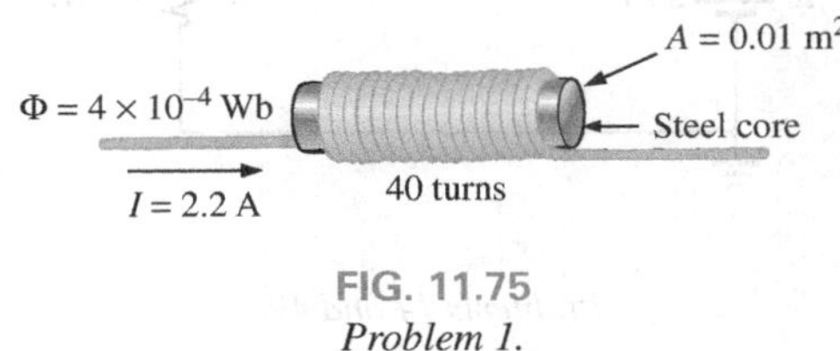

FIG. 11.75
Problem 1.

SECTION 11.3 Inductance

2. For the inductor in Fig. 11.76, find the inductance L in henries.

3. a. Repeat Problem 2 with a ferromagnetic core with $\mu_r = 500$.
   b. How is the new inductance related to the old one? How does it relate to the value of μ_r?

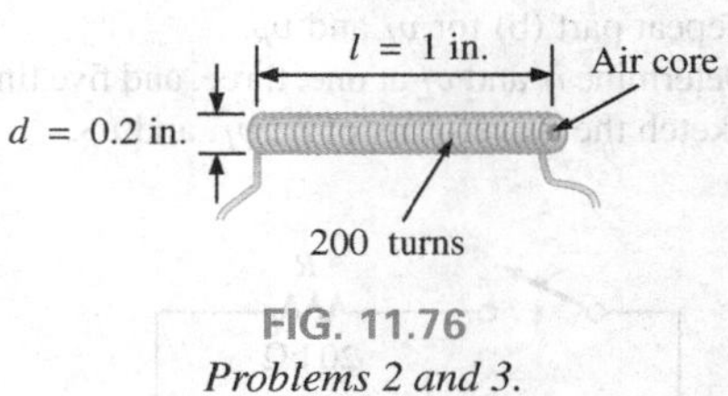

FIG. 11.76
Problems 2 and 3.

4. For the inductor in Fig. 11.77, find the approximate inductance L in henries.

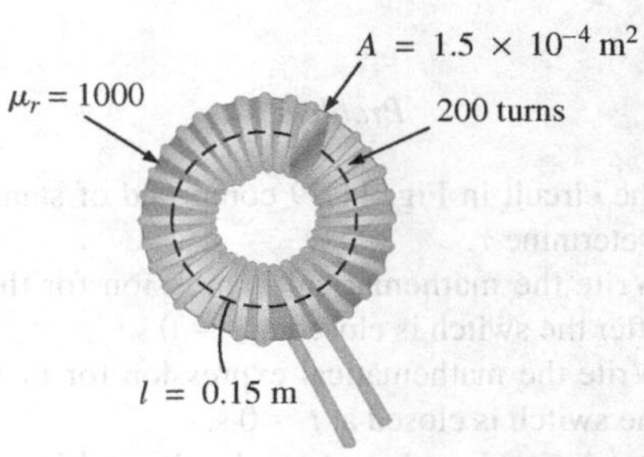

FIG. 11.77
Problem 4.

5. An air-core inductor has a total inductance of 4.7 mH.
   a. What is the inductance if the only change is to increase the number of turns by a factor of three?
   b. What is the inductance if the only change is to increase the length by a factor of three?
   c. What is the inductance if the area is doubled, the length cut in half, and the number of turns doubled?
   d. What is the inductance if the area, length, and number of turns are cut in half and a ferromagnetic core with a μ_r of 1500 is inserted?
6. What are the inductance and the range of expected values for an inductor with the following label?
   a. 392K
   b. blue gray black J
   c. 47K
   d. brown green red K

SECTION 11.4 Indced Voltage v_L

7. If the flux linking a coil of 50 turns changes at a rate of 120 mW/s, what is the induced voltage across the coil?
8. Determine the rate of change of flux linking a coil if 20 V are induced across a coil of 200 turns.
9. How many turns does a coil have if 42 mV are induced across the coil by a change in flux of 3 mW/s?
10. Find the voltage induced across a coil of 22 mH if the rate of change of current through the coil is:
    a. 1 A/s
    b. 1 mA/ms
    c. 2 mA/10 μs

SECTION 11.5 *R-L* Transients: The Storage Phase

11. For the circuit of Fig. 11.78 composed of standard values:
    a. Determine the time constant.
    b. Write the mathematical expression for the current i_L after the switch is closed.
    c. Repeat part (b) for v_L and v_R.
    d. Determine i_L and v_L at one, three, and five time constants.
    e. Sketch the waveforms of i_L, v_L, and v_R.

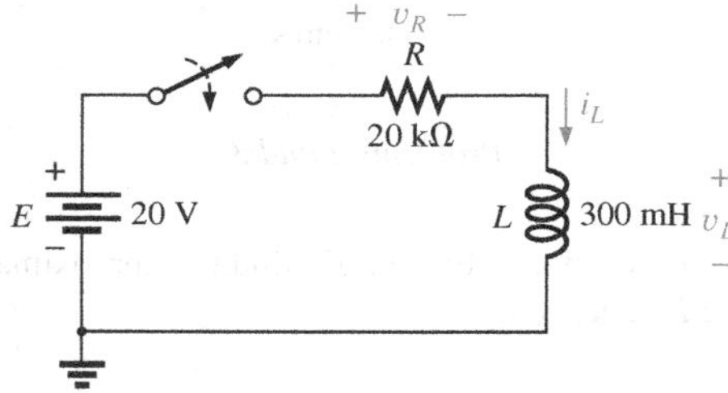

FIG. 11.78
Problem 11.

12. For the circuit in Fig. 11.79 composed of standard values:
    a. Determine τ.
    b. Write the mathematical expression for the current i_L after the switch is closed at $t = 0$ s.
    c. Write the mathematical expression for v_L and v_R after the switch is closed at $t = 0$ s.
    d. Determine i_L and v_L at $t = 1\tau$, 3τ, and 5τ.
    e. Sketch the waveforms of i_L, v_L, and v_R for the storage phase.

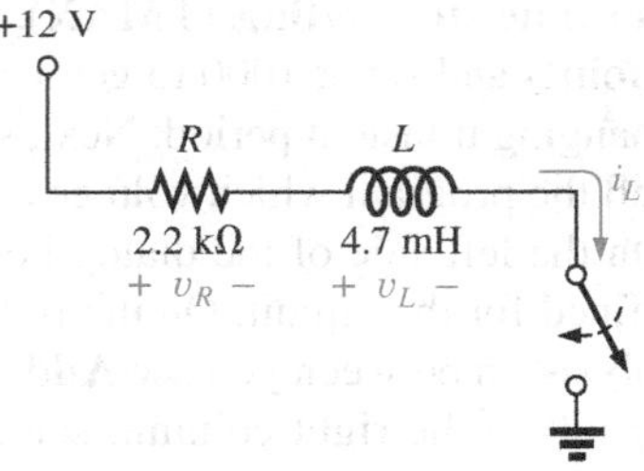

FIG. 11.79
Problem 12.

13. Given a supply of 18 V, use standard values to design a circuit to have the response of Fig. 11.80.

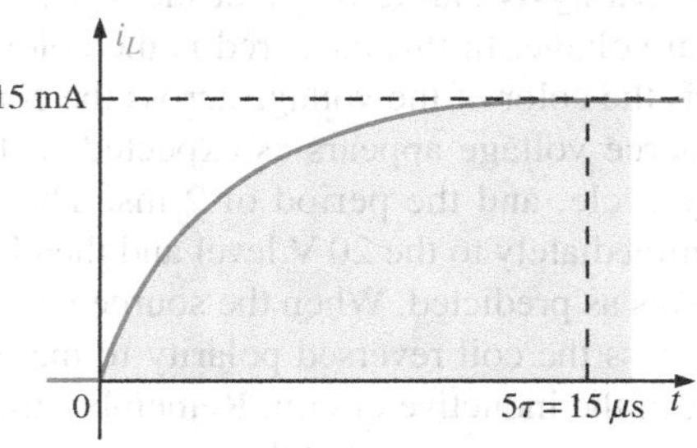

FIG. 11.80
Problem 13.

SECTION 11.6 Initial Conditions

14. For the circuit in Fig. 11.81:
    a. Write the mathematical expressions for the current i_L and the voltage v_L following the closing of the switch. Note the magnitude and the direction of the initial current.
    b. Sketch the waveform of i_L and v_L for the entire period from initial value to steady-state level.

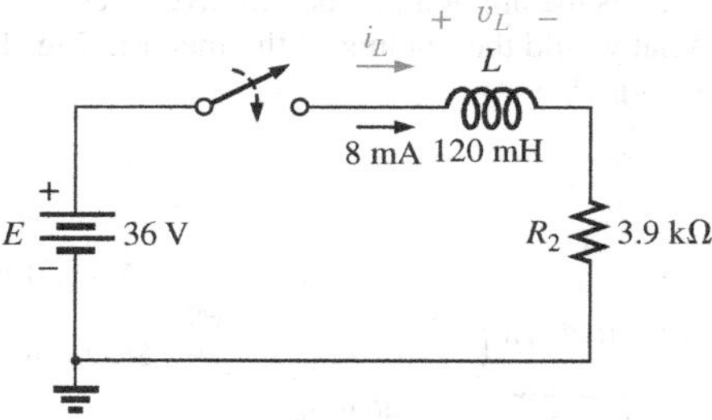

FIG. 11.81
Problems 14 and 49.

15. In this problem, the effect of reversing the initial current is investigated. The circuit in Fig. 11.82 is the same as that appearing in Fig. 11.81, with the only change being the direction of the initial current.
    a. Write the mathematical expressions for the current i_L and the voltage v_L following the closing of the switch. Take careful note of the defined polarity for v_L and the direction for i_L.

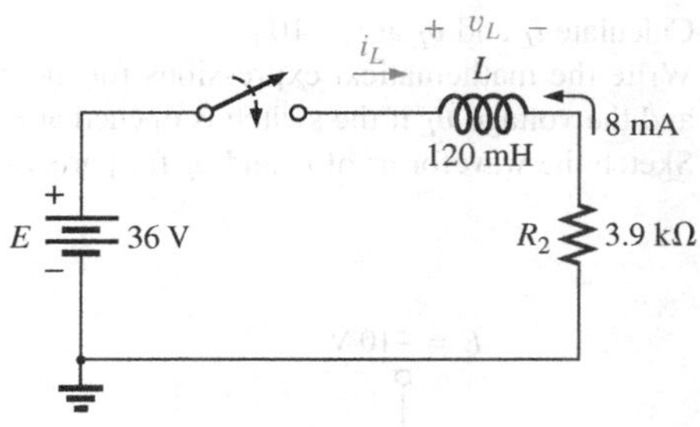

FIG. 11.82
Problem 15.

b. Sketch the waveform of i_L and v_L for the entire period from initial value to steady-state level.
c. Compare the results with those of Problem 14.

16. For the network in Fig. 11.83:
a. Write the mathematical expressions for the current i_L and the voltage v_L following the closing of the switch. Note the magnitude and the direction of the initial current.
b. Sketch the waveform of i_L and v_L for the entire period from initial value to steady-state level.

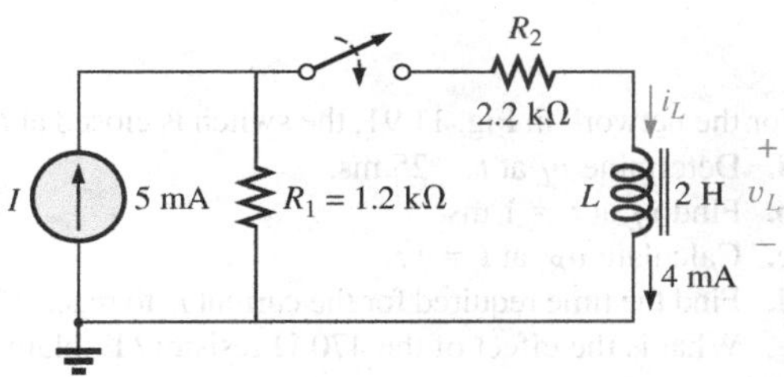

FIG. 11.83
Problem 16.

***17.** For the network in Fig. 11.84:
a. Write the mathematical expressions for the current i_L and the voltage v_L following the closing of the switch. Note the magnitude and direction of the initial current.
b. Sketch the waveform of i_L and v_L for the entire period from initial value to steady-state level.

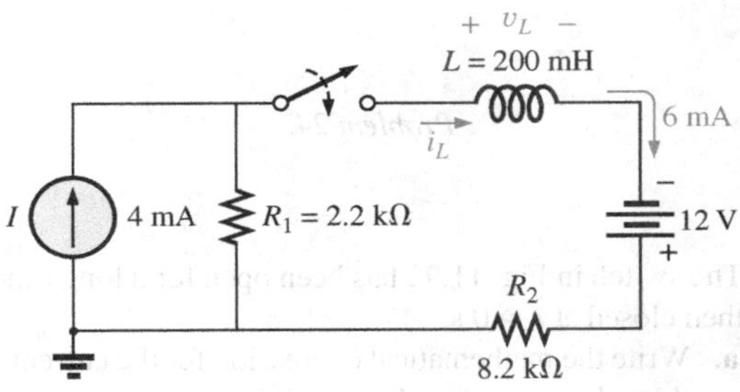

FIG. 11.84
Problem 17.

SECTION 11.7 *R-L* Transients: The Release Phase

18. For the network in Fig. 11.85:
a. Determine the mathematical expressions for the current i_L and the voltage v_L when the switch is closed.
b. Repeat part (a) if the switch is opened after a period of five time constants has passed.
c. Sketch the waveforms of parts (a) and (b) on the same set of axes.

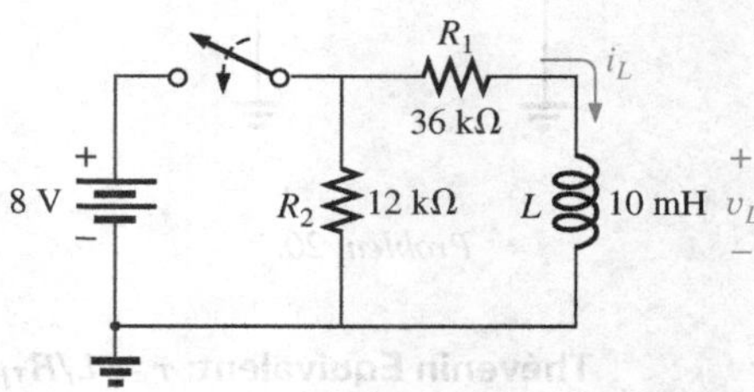

FIG. 11.85
Problem 18.

***19.** For the network in Fig. 11.86:
a. Determine the mathematical expressions for the current i_L and the voltage v_L following the closing of the switch.
b. Repeat part (a) if the switch is opened at $t = 1\ \mu s$.
c. Sketch the waveforms of parts (a) and (b) on the same set of axes.

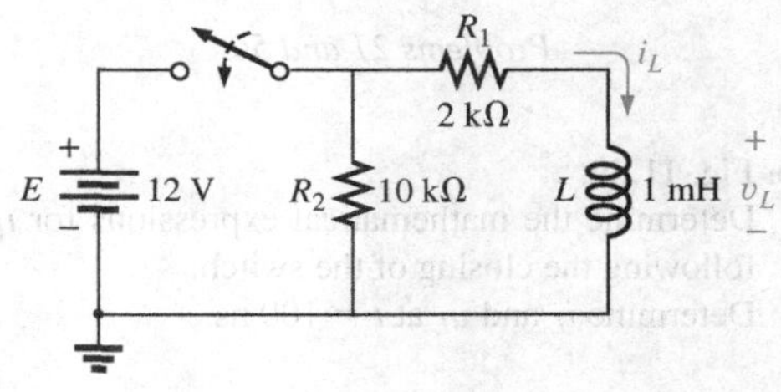

FIG. 11.86
Problem 19.

***20.** For the network in Fig. 11.87:
a. Write the mathematical expression for the current i_L and the voltage v_L following the closing of the switch.
b. Determine the mathematical expressions for i_L and v_L if the switch is opened after a period of five time constants has passed.
c. Sketch the waveforms of i_L and v_L for the time periods defined by parts (a) and (b).
d. Sketch the waveform for the voltage across R_2 for the same period of time encompassed by i_L and v_L. Take careful note of the defined polarities and directions in Fig. 11.87.

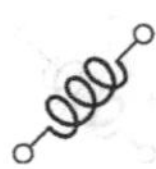

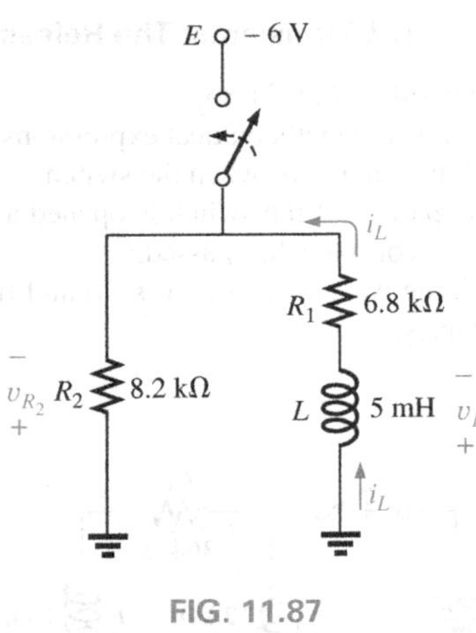

FIG. 11.87
Problem 20.

SECTION 11.8 Thévenin Equivalent: $\tau = L/R_{Th}$

21. For Fig. 11.88:

a. Determine the mathematical expressions for i_L and v_L following the closing of the switch.

b. Determine i_L and v_L after one time constant.

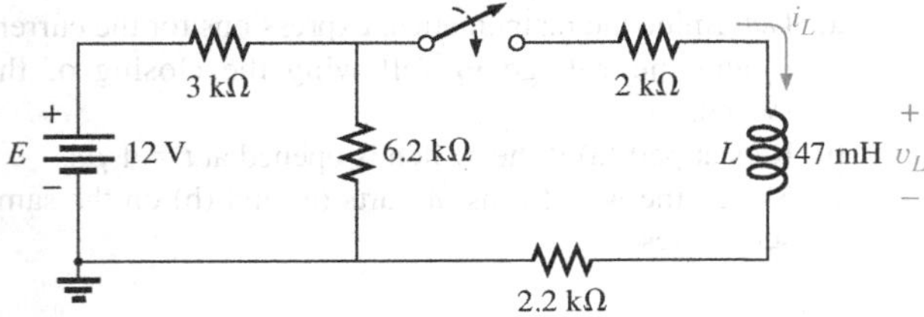

FIG. 11.88
Problems 21 and 50.

22. For Fig. 11.89:

a. Determine the mathematical expressions for i_L and v_L following the closing of the switch.

b. Determine i_L and v_L at $t = 100$ ns.

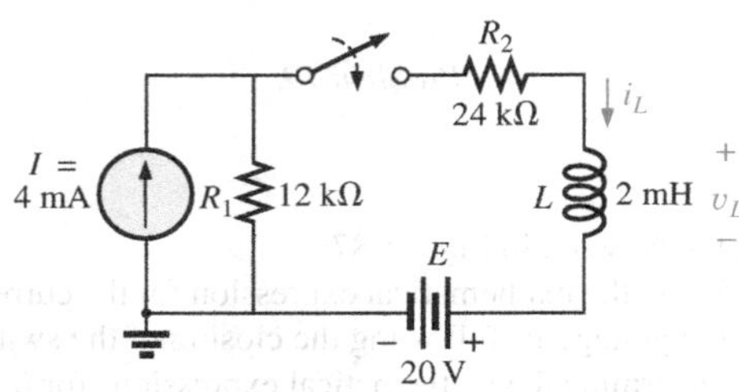

FIG. 11.89
Problem 22.

***23.** For Fig. 11.90:

a. Determine the mathematical expressions for i_L and v_L following the closing of the switch. Note the defined direction for i_L and polarity for v_L.

b. Calculate i_L and v_L at $t = 10\ \mu s$.

c. Write the mathematical expressions for the current i_L and the voltage v_L if the switch is opened at $t = 10\ \mu s$.

d. Sketch the waveforms of i_L and v_L for parts (a) and (c).

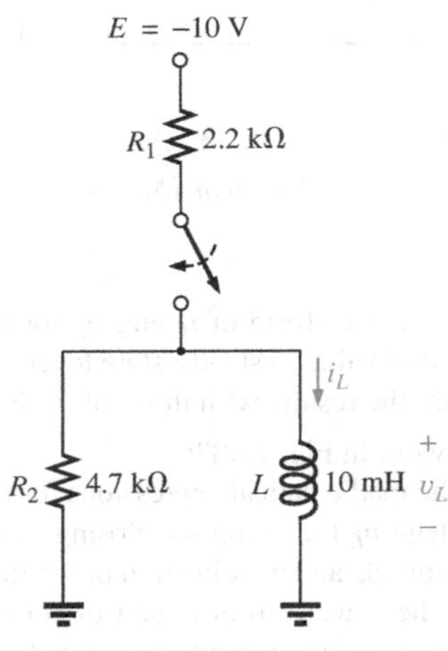

FIG. 11.90
Problem 23.

***24.** For the network in Fig. 11.91, the switch is closed at $t = 0$ s.

a. Determine v_L at $t = 25$ ms.

b. Find v_L at $t = 1$ ms.

c. Calculate v_{R_1} at $t = 1\tau$.

d. Find the time required for the current i_L to reach 100 mA.

e. What is the effect of the 470 Ω resistor? Explain.

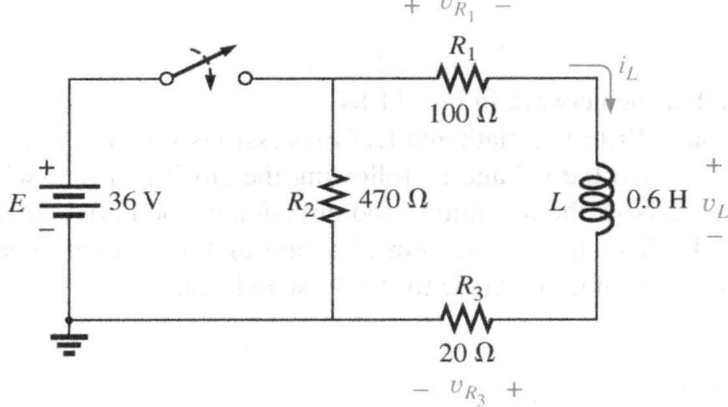

FIG. 11.91
Problem 24.

***25.** The switch in Fig. 11.92 has been open for a long time. It is then closed at $t = 0$ s.

a. Write the mathematical expression for the current i_L and the voltage v_L after the switch is closed.

b. Sketch the waveform of i_L and v_L from the initial value to the steady-state level.

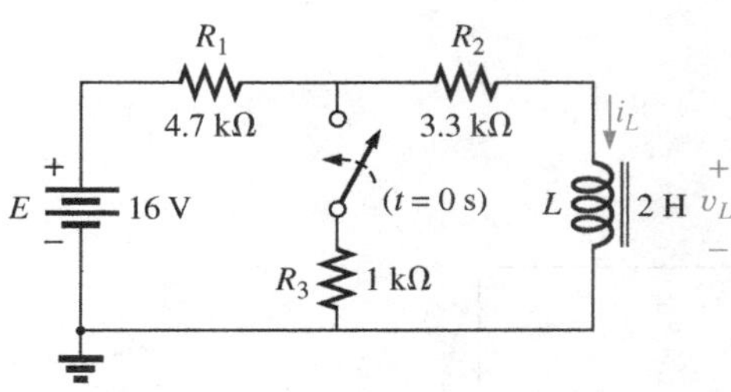

FIG. 11.92
Problem 25.

***26. a.** Determine the mathematical expressions for i_L and v_L following the closing of the switch in Fig. 11.93. The steady-state values of i_L and v_L are established before the switch is closed.

b. Determine i_L and v_L after two time constants of the storage phase.

c. Write the mathematical expressions for the current i_L and the voltage v_L if the switch is opened at the instant defined by part (b).

d. Sketch the waveforms of i_L and v_L for parts (a) and (c).

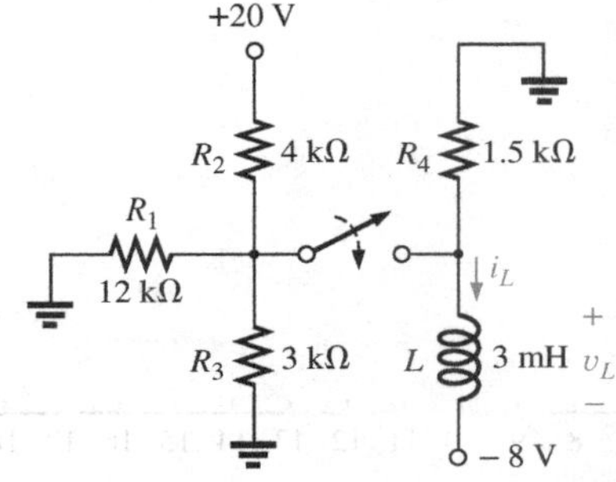

FIG. 11.93
Problem 26.

***27.** The switch for the network in Fig. 11.94 has been closed for about 1 h. It is then opened at the time defined as $t = 0$ s.

a. Determine the time required for the current i_L to drop to 10 μA.

b. Find the voltage v_L at $t = 10\ \mu$s.

c. Calculate v_L at $t = 5\tau$.

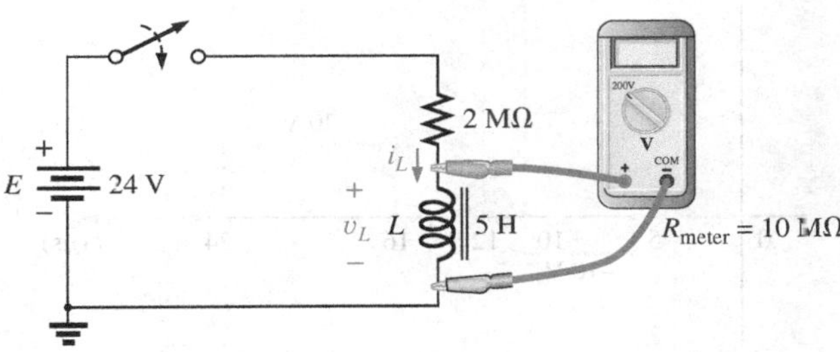

FIG. 11.94
Problem 27.

***28.** The switch in Fig. 11.95 has been closed for a long time. It is then opened at $t = 0$ s.

a. Write the mathematical expression for the current i_L and the voltage v_L after the switch is opened.

b. Sketch the waveform of i_L and v_L from initial value to the steady-state level.

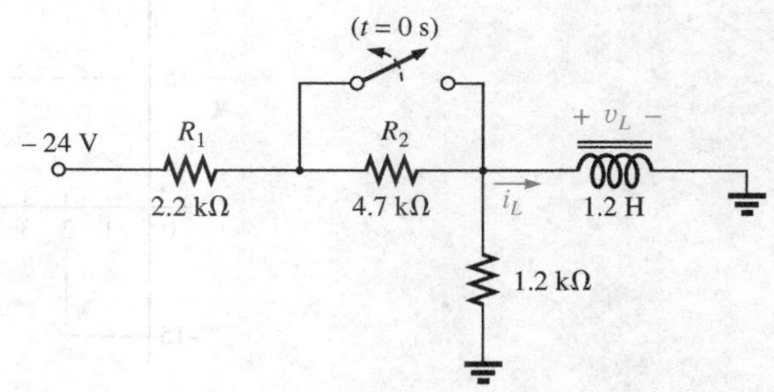

FIG. 11.95
Problem 28.

SECTION 11.9 Instantaneous Values

29. Given $i_L = 100\ \text{mA}\ (1 - e^{-t/20\ \text{ms}})$:

a. Determine i_L at $t = 1$ ms.

b. Determine i_L at $t = 100$ ms.

c. Find the time t when i_L will equal 50 mA.

d. Find the time t when i_L will equal 99 mA.

30. a. If the measured current for an inductor during the storage phase is 126.4 μA at after a period of one time constant has passed, what is the maximum level of current to be achieved?

b. When the current of part (a) reaches 160 μA, 64.4 μs have passed. Find the time constant of the network.

c. If the circuit's resistance is 500 Ω, what is the value of the series inductor to establish the current of part (a)? Is the resulting inductance a standard value?

d. What is the required supply voltage?

31. The network in Fig. 11.96 employs a DMM with an internal resistance of 10 MΩ in the voltmeter mode. The switch is closed at $t = 0$ s.

a. Find the voltage across the coil the instant after the switch is closed.

b. What is the final value of the current i_L?

c. How much time must pass before i_L reaches 10 μA?

d. What is the voltmeter reading at $t = 12\ \mu$s?

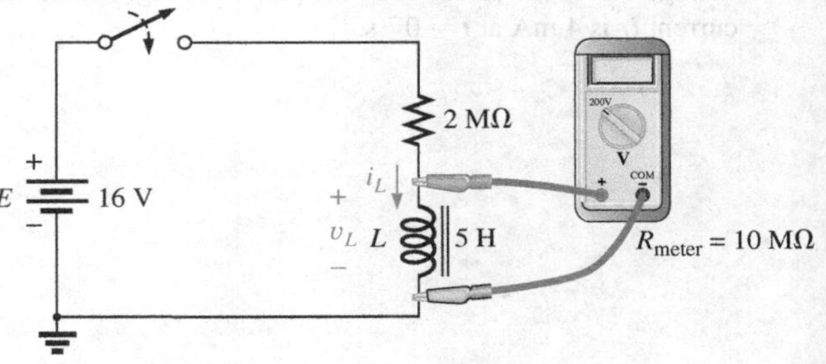

FIG. 11.96
Problem 31.

SECTION 11.10 Average Induced Voltage: $v_{L_{av}}$

32. Find the waveform for the voltage induced across a 200 mH coil if the current through the coil is as shown in Fig. 11.97.

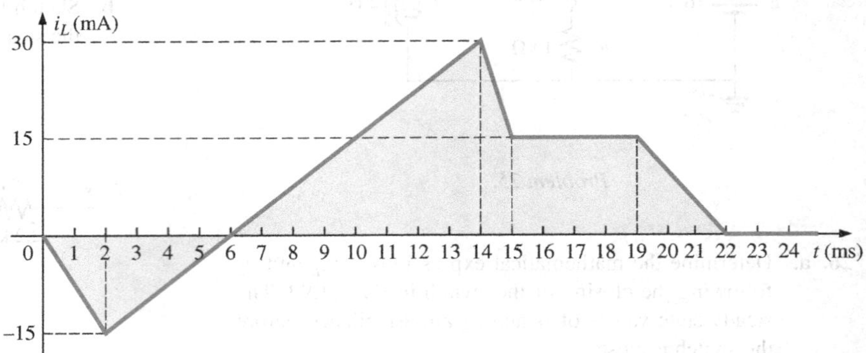

FIG. 11.97
Problem 32.

33. Find the waveform for the voltage induced across a 5 mH coil if the current through the coil is as shown in Fig. 11.98.

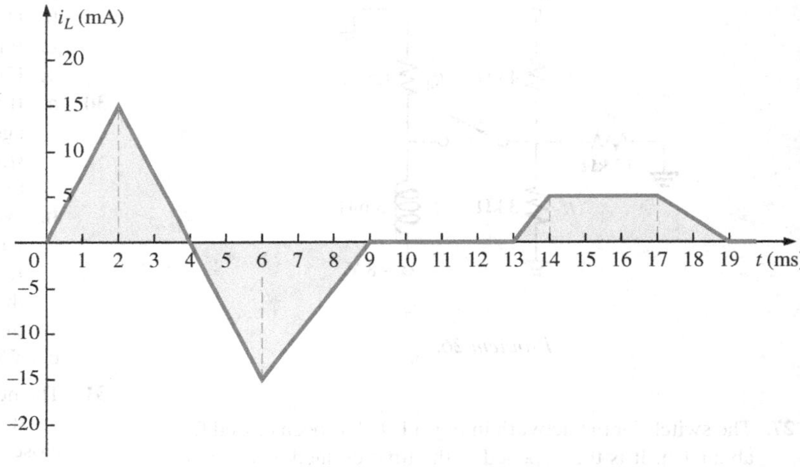

FIG. 11.98
Problem 33.

***34.** Find the waveform for the current of a 10 mH coil if the voltage across the coil follows the pattern in Fig. 11.99. The current i_L is 4 mA at $t = 0^-$ s.

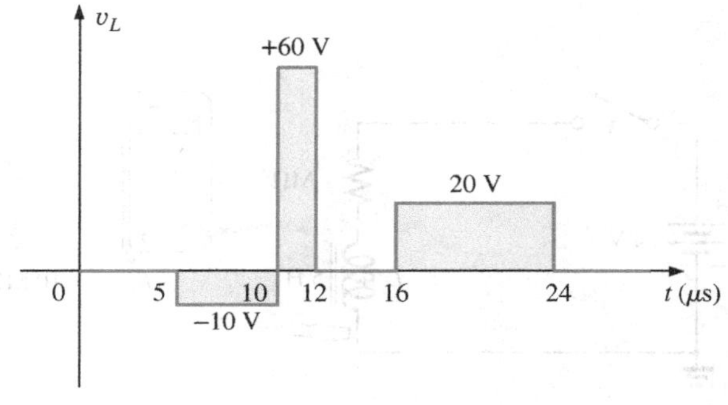

FIG. 11.99
Problem 34.

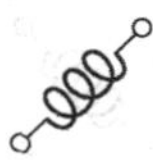

SECTION 11.11 Inductors in Series and in Parallel

35. Find the total inductance of the circuit of Fig. 11.100.

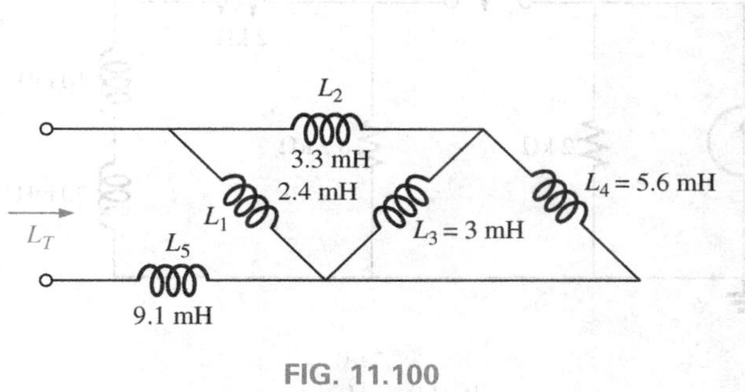

FIG. 11.100
Problem 35.

36. Find the total inductance for the network of Fig. 11.101.

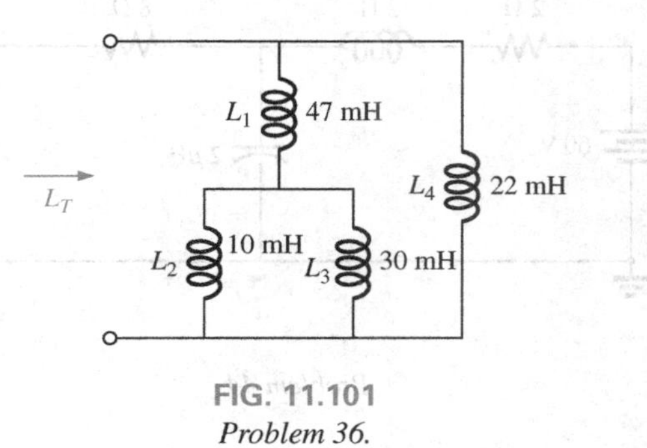

FIG. 11.101
Problem 36.

37. Reduce the network in Fig. 11.102 to the fewest number of components.

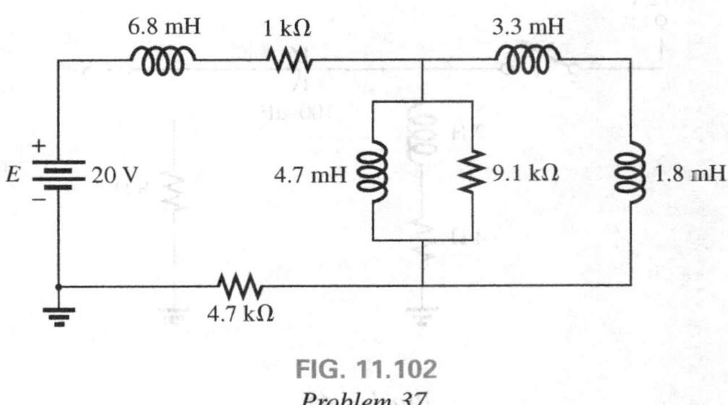

FIG. 11.102
Problem 37.

38. Reduce the network in Fig. 11.103 to the fewest elements.

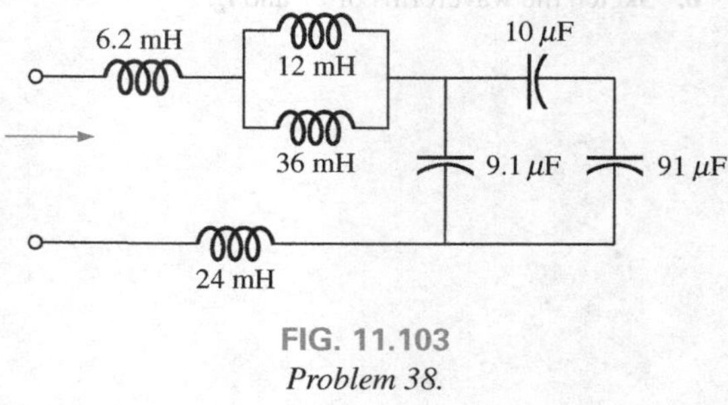

FIG. 11.103
Problem 38.

39. Reduce the network of Fig. 11.104 to the fewest elements.

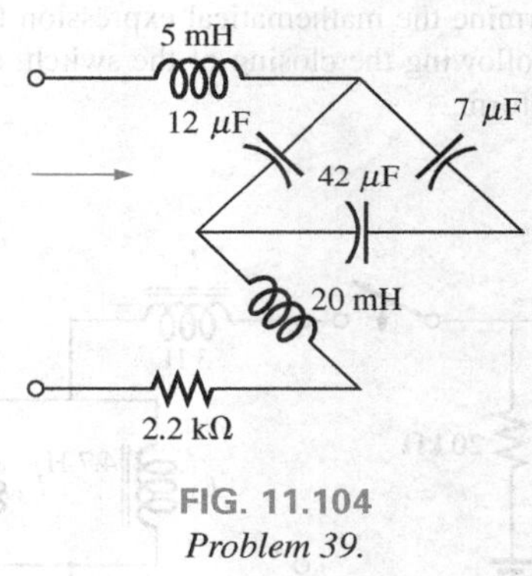

FIG. 11.104
Problem 39.

***40.** For the network in Fig. 11.105:

a. Write the mathematical expressions for the voltages v_L and v_R and the current i_L if the switch is closed at $t = 0$ s.

b. Sketch the waveforms of v_L, v_R, and i_L.

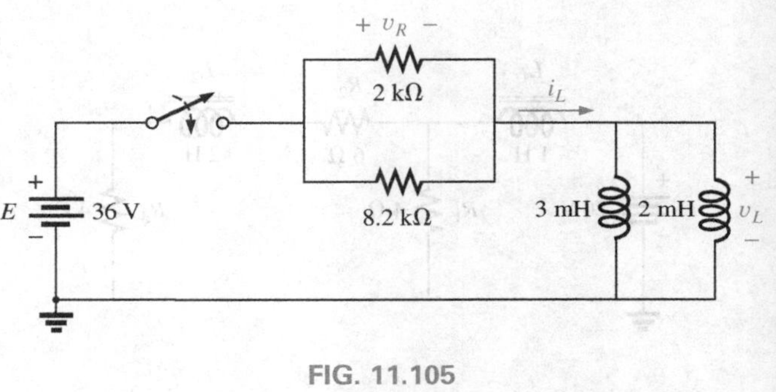

FIG. 11.105
Problem 40.

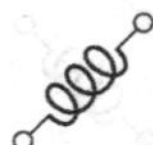

***41.** For the network in Fig. 11.106:

a. Write the mathematical expressions for the voltage v_L and the current i_L if the switch is closed at $t = 0$ s. Take special note of the required v_L.

b. Sketch the waveforms of v_L and i_L.

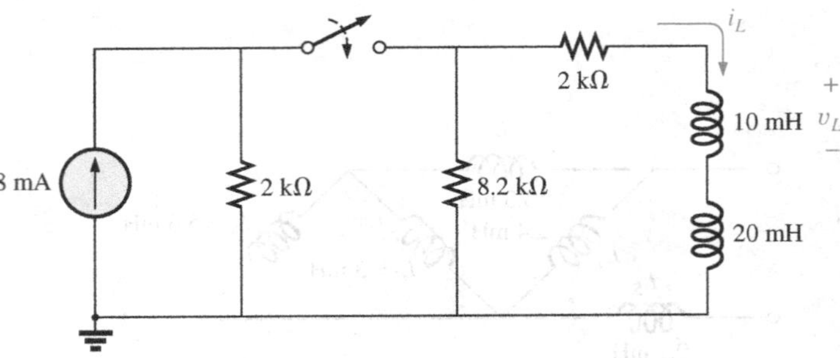

FIG. 11.106
Problem 41.

***42.** For the network in Fig. 11.107:

a. Find the mathematical expressions for the voltage v_L and the current i_L following the closing of the switch.

b. Sketch the waveforms of v_L and i_L obtained in part (a).

c. Determine the mathematical expression for the voltage v_{L_3} following the closing of the switch, and sketch the waveform.

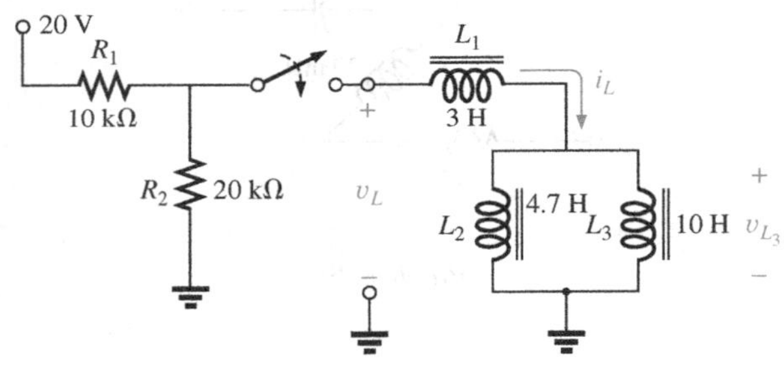

FIG. 11.107
Problem 42.

SECTION 11.12 Steady-State Conditions

43. Find the steady-state currents I_1 and I_2 for the network in Fig. 11.108.

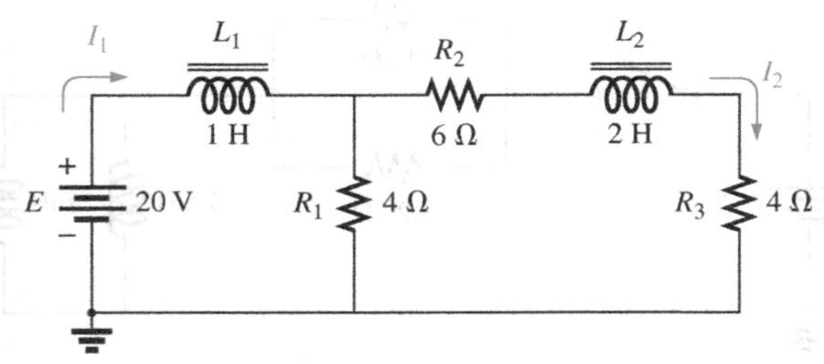

FIG. 11.108
Problem 43.

44. Find the steady-state currents and voltages for the network in Fig. 11.109.

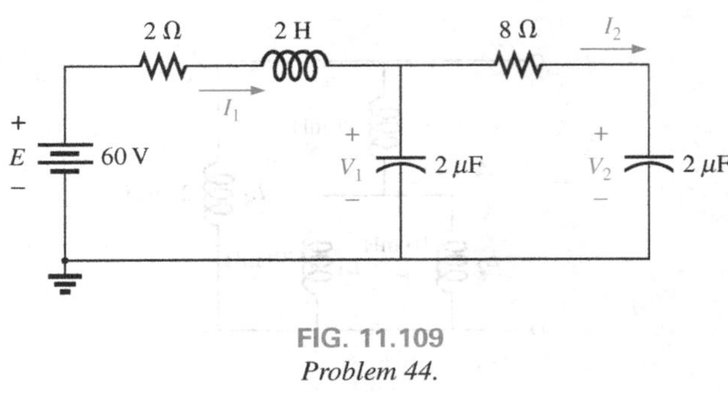

FIG. 11.109
Problem 44.

45. Find the steady-state currents and voltages for the network in Fig. 11.110 after the switch is closed.

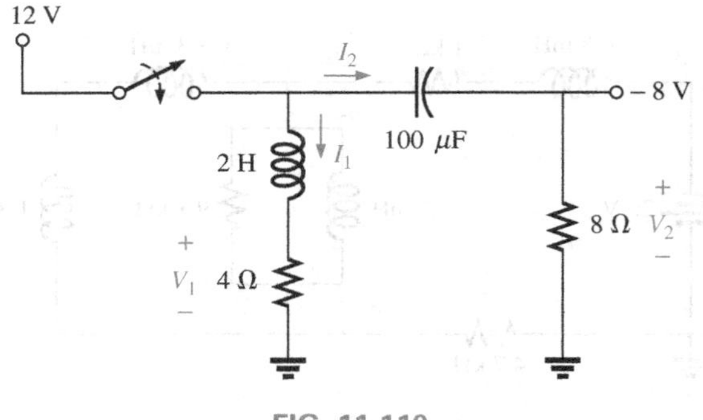

FIG. 11.110
Problem 45.

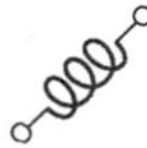

46. Find the indicated steady-state currents and voltages for the network in Fig. 11.111.

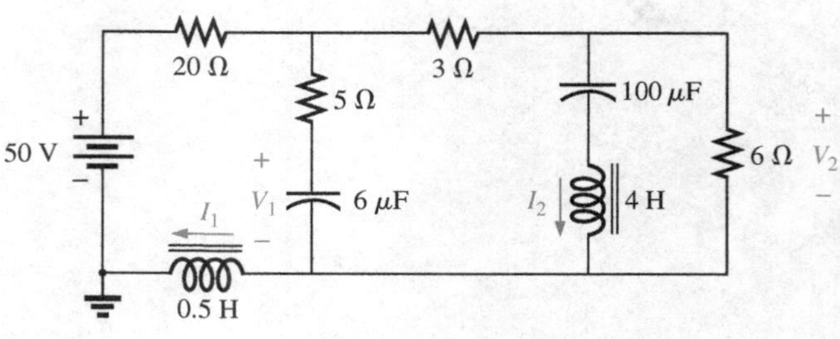

FIG. 11.111
Problem 46.

SECTION 11.15 Computer Analysis

47. Using PSpice or Multisim, verify the results of Example 11.3.
48. Using PSpice or Multisim, verify the results of Example 11.4.
49. Using PSpice or Multisim, find the solution to Problem 14.
50. Using PSpice or Multisim, find the solution to Problem 21.
51. Using PSpice or Multisim, verify the results of Example 11.8.

GLOSSARY

Ampère's circuital law A law establishing the fact that the algebraic sum of the rises and drops of the magnetomotive force (mmf) around a closed loop of a magnetic circuit is equal to zero.

Choke A term often applied to an inductor, due to the ability of an inductor to resist a change in current through it.

Diamagnetic materials Materials that have permeabilities slightly less than that of free space.

Electromagnetism Magnetic effects introduced by the flow of charge, or current.

Faraday's law A law stating the relationship between the voltage induced across a coil and the number of turns in the coil and the rate at which the flux linking the coil is changing.

Ferromagnetic materials Materials having permeabilities hundreds and thousands of times greater than that of free space.

Flux density (B) A measure of the flux per unit area perpendicular to a magnetic flux path. It is measured in teslas (T) or webers per square meter (Wb/m^2).

Inductance (L) A measure of the ability of a coil to oppose any change in current through the coil and to store energy in the form of a magnetic field in the region surrounding the coil.

Inductor (coil) A fundamental element of electrical systems constructed of numerous turns of wire around a ferromagnetic core or an air core.

Lenz's law A law stating that an induced effect is always such as to oppose the cause that produced it.

Magnetic flux lines Lines of a continuous nature that reveal the strength and direction of a magnetic field.

Magnetomotive force (mmf) ($\mathscr{F}$) The "pressure" required to establish magnetic flux in a ferromagnetic material. It is measured in ampere-turns (At).

Paramagnetic materials Materials that have permeabilities slightly greater than that of free space.

Permanent magnet A material such as steel or iron that will remain magnetized for long periods of time without the aid of external means.

Permeability (μ) A measure of the ease with which magnetic flux can be established in a material. It is measured in $Wb/A \cdot m$.

Relative permeability (μ_r) The ratio of the permeability of a material to that of free space.

Magnetic Circuits

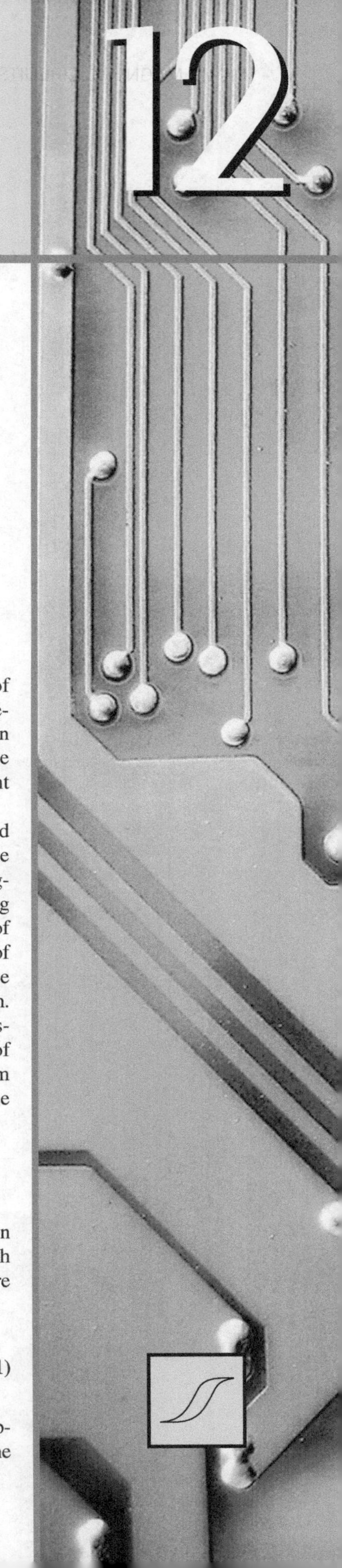

Objectives

- *Become aware of the similarities between the analysis of magnetic circuits and electric circuits.*
- *Develop a clear understanding of the important parameters of a magnetic circuit and how to find each quantity for a variety of magnetic circuit configurations.*
- *Begin to appreciate why a clear understanding of magnetic circuit parameters is an important component in the design of electrical/electronic systems.*

12.1 INTRODUCTION

Magnetic and electromagnetic effects play an important role in the design of a wide variety of electrical/electronic systems in use today. Motors, generators, transformers, loudspeakers, relays, medical equipment and movements of all kinds depend on magnetic effects to function properly. The response and characteristics of each have an impact on the current and voltage levels of the system, the efficiency of the design, the resulting size, and many other important considerations.

Fortunately, there is a great deal of similarity between the analyses of electric circuits and magnetic circuits. The magnetic flux of magnetic circuits has properties very similar to the current of electric circuits. As shown in Fig. 11.15, it has a direction and a closed path. The magnitude of the established flux is a direct function of the applied **magnetomotive force,** resulting in a duality with electric circuits, where the resulting current is a function of the magnitude of the applied voltage. The flux established is also inversely related to the structural opposition of the magnetic path in the same way the current in a network is inversely related to the resistance of the network. All of these similarities are used throughout the analysis to clarify the approach.

One of the difficulties associated with studying magnetic circuits is that three different systems of units are commonly used in the industry. The manufacturer, application, and type of component all have an impact on which system is used. To the extent practical, the SI system is applied throughout the chapter. References to the CGS and English systems require the use of Appendix E.

12.2 MAGNETIC FIELD

The magnetic field distribution around a permanent magnet or **electromagnet** was covered in detail in Chapter 11. Recall that flux lines strive to be as short as possible and take the path with the highest permeability. The **flux density** is defined as follows [Eq. (11.1) repeated here for convenience]:

$$\boxed{B = \frac{\Phi}{A}} \qquad \begin{aligned} B &= \text{Wb/m}^2 = \text{teslas (T)} \\ \Phi &= \text{webers (Wb)} \\ A &= \text{m}^2 \end{aligned} \qquad \textbf{(12.1)}$$

The "pressure" on the system to establish magnetic lines of force is determined by the applied magnetomotive force, which is directly related to the number of turns and current of the

magnetizing coil as appearing in the following equation [Eq. (11.3) repeated here for convenience]:

$$\boxed{\mathscr{F} = NI} \qquad \begin{aligned} \mathscr{F} &= \text{ampere-turns (At)} \\ N &= \text{turns (t)} \\ I &= \text{amperes (A)} \end{aligned} \tag{12.2}$$

The level of magnetic flux established in a ferromagnetic core is a direction function of the permeability of the material. **Ferromagnetic materials** have a very high level of **permeability,** while nonmagnetic materials such as air and wood have very low levels. The ratio of the permeability of the material to that of air is called the **relative permeability** and is defined by the following equation [Eq. (11.5) repeated here for convenience]:

$$\boxed{\mu_r = \frac{\mu}{\mu_o}} \qquad \mu_o = 4\pi \times 10^{-7}\ \text{Wb/A} \cdot \text{m} \tag{12.3}$$

As mentioned in Chapter 11, the values of μ_r are not provided in a table format because the value is determined by the other quantities of the magnetic circuit. Change the magnetomotive force, and the relative permeability changes.

12.3 RELUCTANCE

The resistance of a material to the flow of charge (current) is determined for electric circuits by the equation

$$R = \rho \frac{l}{A} \qquad (\text{ohms}, \Omega)$$

The **reluctance** of a material to the setting up of magnetic flux lines in the material is determined by the following equation:

$$\boxed{\mathscr{R} = \frac{l}{\mu A}} \qquad (\text{rels, or At/Wb}) \tag{12.4}$$

where $\mathscr{R}$ is the reluctance, l is the length of the magnetic path, and A is the cross-sectional area. The t in the units At/Wb is the number of turns of the applied winding. More is said about ampere-turns (At) in the next section. Note that the resistance and reluctance are inversely proportional to the area, indicating that an increase in area results in a reduction in each and an *increase* in the desired result: current and flux. For an increase in length, the opposite is true, and the desired effect is reduced. The reluctance, however, is inversely proportional to the permeability, while the resistance is directly proportional to the resistivity. The larger the μ or the smaller the ρ, the smaller are the reluctance and resistance, respectively. Obviously, therefore, materials with high permeability, such as the ferromagnetics, have very small reluctances and result in an increased measure of flux through the core. There is no widely accepted unit for reluctance, although the *rel* and the At/Wb are usually applied.

12.4 OHM'S LAW FOR MAGNETIC CIRCUITS

Recall the equation

$$\text{Effect} = \frac{\text{cause}}{\text{opposition}}$$

appearing in Chapter 4 to introduce Ohm's law for electric circuits. For magnetic circuits, the effect desired is the flux Φ. The cause is the **magnetomotive force (mmf)** $\mathscr{F}$, which is the external force (or "pressure") required to set up the **magnetic flux lines** within the magnetic material. The opposition to the setting up of the flux Φ is the reluctance $\mathscr{R}$.

Substituting, we have

$$\Phi = \frac{\mathscr{F}}{\mathscr{R}} \quad (12.5)$$

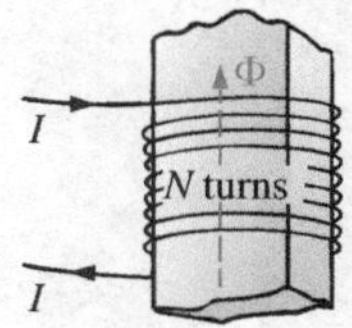

FIG. 12.1
Defining the components of a magnetomotive force.

Since $\mathscr{F} = NI$, Eq. (12.5) clearly reveals that an increase in the number of turns or the current through the wire in Fig. 12.1 results in an increased "pressure" on the system to establish the flux lines through the core.

Although there is a great deal of similarity between electric and magnetic circuits, you must understand that the flux Φ is not a "flow" variable such as current in an electric circuit. Magnetic flux is established in the core through the alteration of the atomic structure of the core due to external pressure and is not a measure of the flow of some charged particles through the core.

12.5 MAGNETIZING FORCE

The magnetomotive force per unit length is called the **magnetizing force** (H). In equation form,

$$H = \frac{\mathscr{F}}{l} \quad \text{(At/m)} \quad (12.6)$$

Substituting for the magnetomotive force results in

$$H = \frac{NI}{l} \quad \text{(At/m)} \quad (12.7)$$

For the magnetic circuit in Fig. 12.2, if $NI = 40$ At and $l = 0.2$ m, then

$$H = \frac{NI}{l} = \frac{40 \text{ At}}{0.2 \text{ m}} = 200 \text{ At/m}$$

In words, the result indicates that there are 200 At of "pressure" per meter to establish flux in the core.

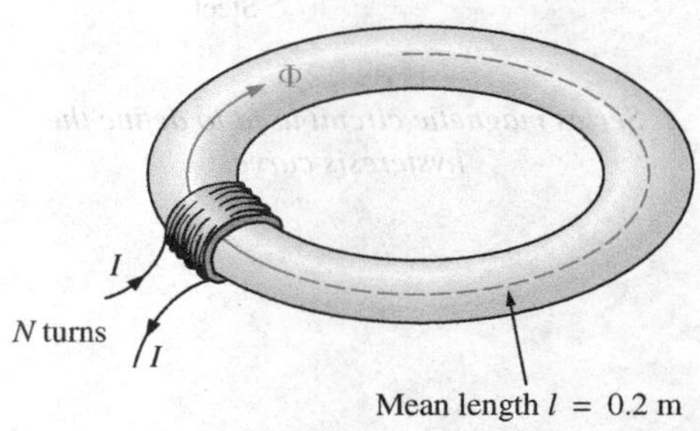

FIG. 12.2
Defining the magnetizing force of a magnetic circuit.

Note in Fig. 12.2 that the direction of the flux Φ can be determined by placing the fingers of your right hand in the direction of current around the core and noting the direction of the thumb. It is interesting to realize that *the magnetizing force is independent of the type of core material*—it is determined solely by the number of turns, the current, and the length of the core.

The applied magnetizing force has a pronounced effect on the resulting permeability of a magnetic material. As the magnetizing force increases, the permeability rises to a maximum and then drops to a minimum, as shown in Fig. 12.3 for three commonly employed magnetic materials.

The flux density and the magnetizing force are related by the following equation:

$$B = \mu H \quad (12.8)$$

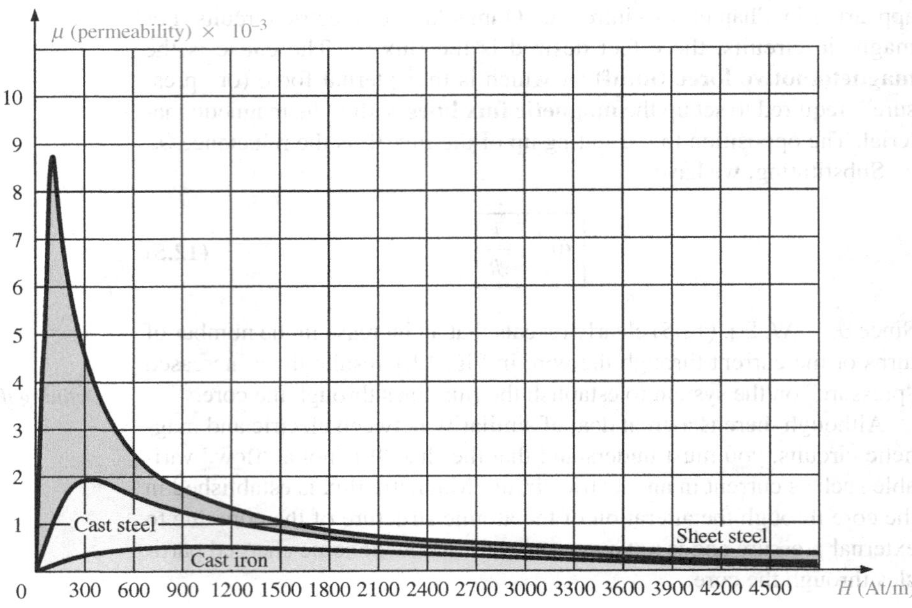

FIG.12.3
Variation of μ with the magnetizing force.

This equation indicates that for a particular magnetizing force, the greater the permeability, the greater is the induced flux density.

Since henries (H) and the magnetizing force (H) use the same capital letter, it must be pointed out that all units of measurement in the text, such as henries, use roman letters, such as H, whereas variables such as the magnetizing force use italic letters, such as H.

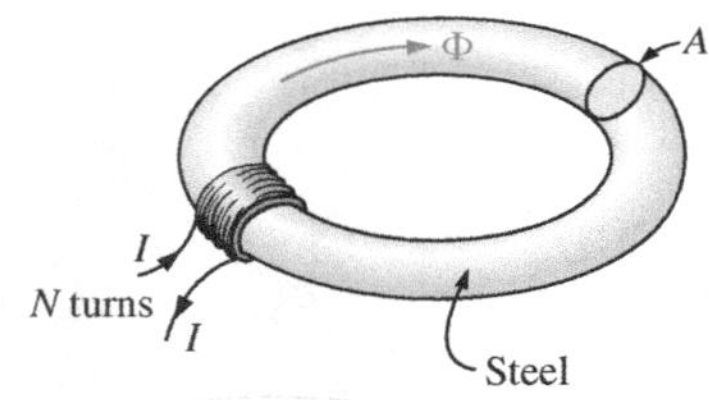

FIG.12.4
Series magnetic circuit used to define the hysteresis curve.

12.6 HYSTERESIS

A curve of the flux density B versus the magnetizing force H of a material is of particular importance to the engineer. Curves of this type can usually be found in manuals, descriptive pamphlets, and brochures published by manufacturers of magnetic materials. A typical B-H curve for a ferromagnetic material such as steel can be derived using the setup in Fig. 12.4.

The core is initially unmagnetized, and the current $I = 0$. If the current I is increased to some value above zero, the magnetizing force H increases to a value determined by

$$H\uparrow = \frac{NI\uparrow}{l}$$

The flux ϕ and the flux density B ($B = \phi/A$) also increase with the current I (or H). If the material has no residual magnetism, and the magnetizing force H is increased from zero to some value H_a, the B-H curve follows the path shown in Fig. 12.5 between o and a. If the magnetizing force H is increased until saturation (H_s) occurs, the curve continues as shown in the figure to point b. When saturation occurs, the flux density has, *for all practical purposes,* reached its maximum value. Any further increase in current through the coil increasing $H = NI/l$ results in a very small increase in flux density B.

If the magnetizing force is reduced to zero by letting I decrease to zero, the curve follows the path of the curve between b and c. The flux

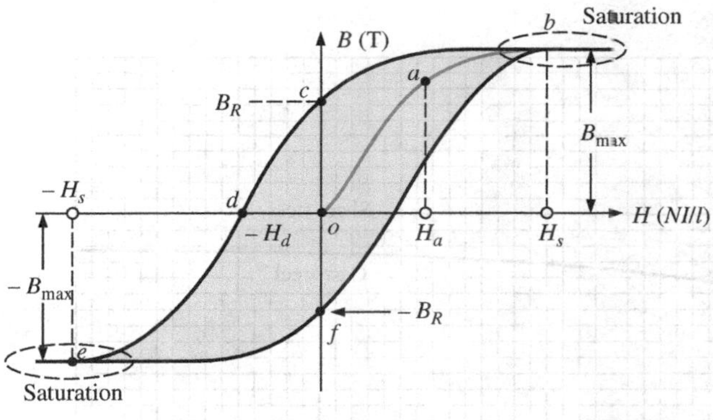

FIG.12.5
Hysteresis curve.

density B_R, which remains when the magnetizing force is zero, is called the *residual flux density.* It is this residual flux density that makes it possible to create permanent magnets. If the coil is now removed from the core in Fig. 12.4, the core will still have the magnetic properties determined by the residual flux density, a measure of its "retentivity." If the current I is reversed, developing a magnetizing force, $-H$, the flux density B decreases with an increase in I. Eventually, the flux density will be zero when $-H_d$ (the portion of curve from c to d) is reached. The magnetizing force $-H_d$ required to "coerce" the flux density to reduce its level to zero is called the *coercive force,* a measure of the coercivity of the magnetic sample. As the force $-H$ is increased until saturation again occurs and is then reversed and brought back to zero, the path *def* results. If the magnetizing force is increased in the positive direction ($+H$), the curve traces the path shown from f to b. The entire curve represented by *bcdefb* is called the **hysteresis** curve for the ferromagnetic material, from the Greek *hysterein,* meaning "to lag behind." The flux density B *lagged* behind the magnetizing force H during the entire plotting of the curve. When H was zero at c, B was not zero but had only begun to decline. Long after H had passed through zero and had become equal to $-H_d$ did the flux density B finally become equal to zero.

If the entire cycle is repeated, the curve obtained for the same core will be determined by the maximum H applied. Three hysteresis loops for the same material for maximum values of H less than the saturation value are shown in Fig. 12.6. In addition, the saturation curve is repeated for comparison purposes.

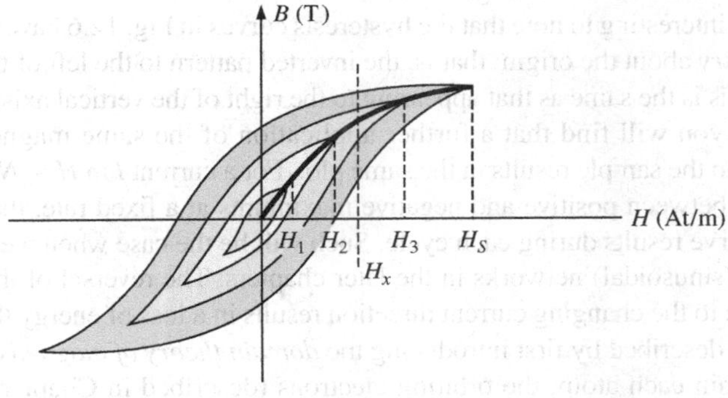

FIG.12.6
Defining the normal magnetization curve.

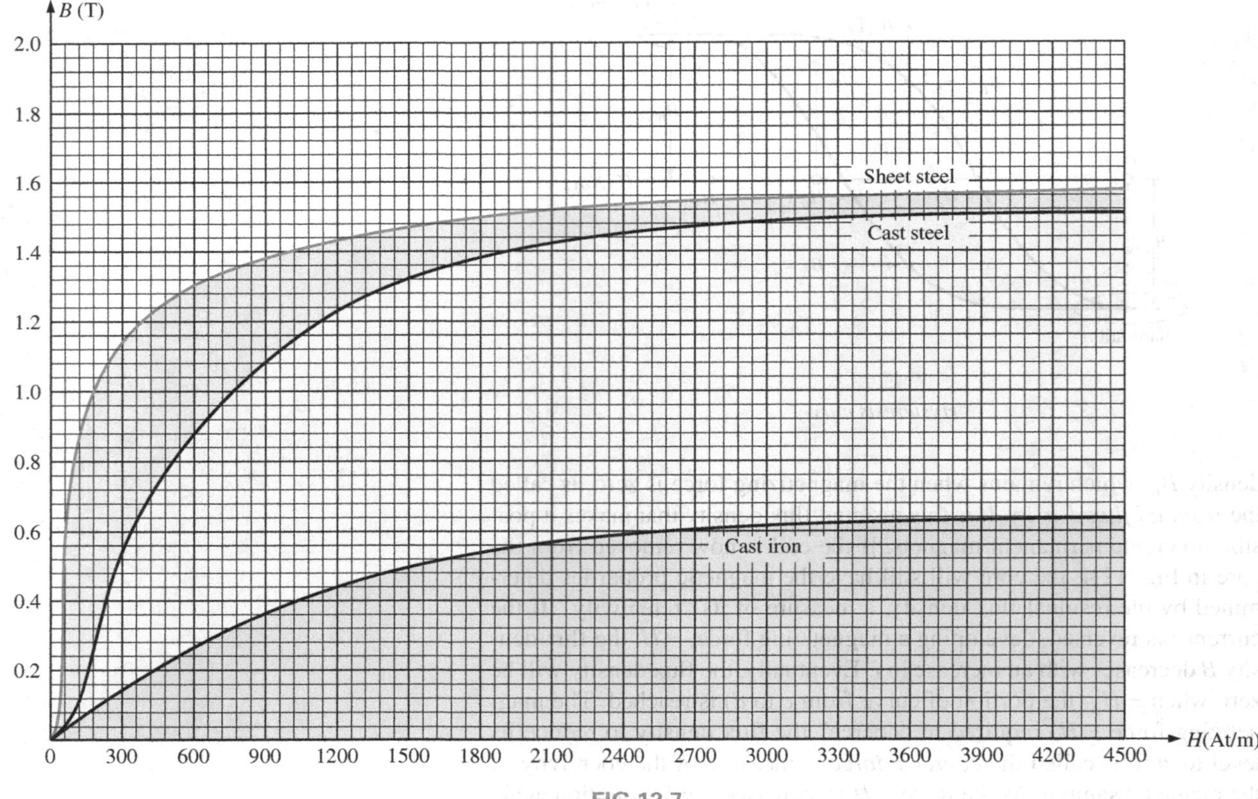

FIG. 12.7
Normal magnetization curve for three ferromagnetic materials.

Note from the various curves that for a particular value of H, say, H_x, the value of B can vary widely, as determined by the history of the core. In an effort to assign a particular value of B to each value of H, we compromise by connecting the tips of the hysteresis loops. The resulting curve, shown by the heavy, solid line in Fig. 12.6 and for various materials in Fig. 12.7, is called the *normal magnetization curve.* An expanded view of one region appears in Fig. 12.8.

A comparison of Figs. 12.3 and 12.7 shows that for the same value of H, the value of B is higher in Fig. 12.7 for the materials with the higher μ in Fig. 12.3. This is particularly obvious for low values of H. This correspondence between the two figures must exist since $B = \mu H$. In fact, if in Fig. 12.7 we find μ for each value of H using the equation $\mu = B/H$, we obtain the curves in Fig. 12.3.

It is interesting to note that the hysteresis curves in Fig. 12.6 have a *point symmetry* about the origin; that is, the inverted pattern to the left of the vertical axis is the same as that appearing to the right of the vertical axis. In addition, you will find that a further application of the same magnetizing forces to the sample results in the same plot. For a current I in $H = NI/l$ that moves between positive and negative maximums at a fixed rate, the same B-H curve results during each cycle. Such will be the case when we examine ac (sinusoidal) networks in the later chapters. The reversal of the field (ϕ) due to the changing current direction results in a loss of energy that can best be described by first introducing the *domain theory of magnetism.*

Within each atom, the orbiting electrons (described in Chapter 2) are also spinning as they revolve around the nucleus. The atom, due to its spinning electrons, has a magnetic field associated with it. In nonmagnetic

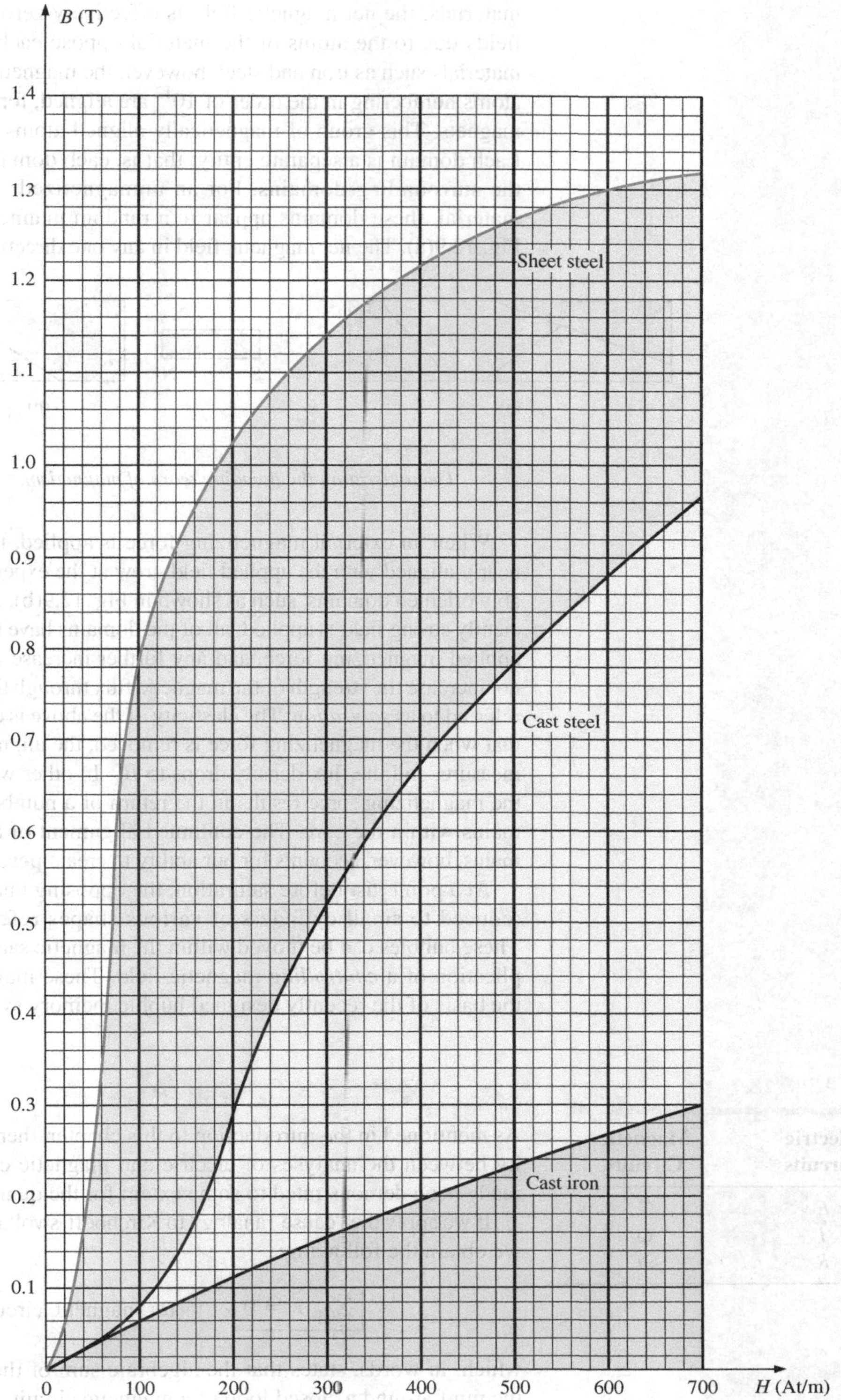

FIG.12.8

Expanded view of Fig. 12.7 for the low magnetizing force region.

materials, the net magnetic field is effectively zero since the magnetic fields due to the atoms of the material oppose each other. In magnetic materials such as iron and steel, however, the magnetic fields of groups of atoms numbering in the order of 10^{12} are aligned, forming very small bar magnets. This group of magnetically aligned atoms is called a **domain.** Each domain is a separate entity; that is, each domain is independent of the surrounding domains. For an unmagnetized sample of magnetic material, these domains appear in a random manner, such as shown in Fig. 12.9(a). The net magnetic field in any one direction is zero.

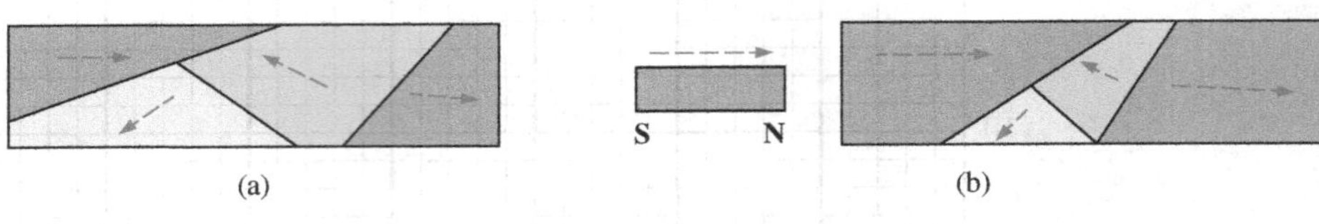

FIG.12.9
Demonstrating the domain theory of magnetism.

When an external magnetizing force is applied, the domains that are nearly aligned with the applied field grow at the expense of the less favorably oriented domains, such as shown in Fig. 12.9(b). Eventually, if a sufficiently strong field is applied, all of the domains have the orientation of the applied magnetizing force, and any further increase in external field will not increase the strength of the magnetic flux through the core—a condition referred to as *saturation.* The elasticity of the above is evidenced by the fact that when the magnetizing force is removed, the alignment is lost to some measure, and the flux density drops to B_R. In other words, the removal of the magnetizing force results in the return of a number of misaligned domains within the core. The continued alignment of a number of the domains, however, accounts for our ability to create **permanent magnets.**

At a point just before saturation, the opposing unaligned domains are reduced to small cylinders of various shapes referred to as *bubbles.* These bubbles can be moved within the magnetic sample through the application of a *controlling* magnetic field. These magnetic bubbles form the basis of the recently designed bubble memory system for computers.

12.7 AMPÈRE'S CIRCUITAL LAW

TABLE 12.1

	Electric Circuits	Magnetic Circuits
Cause	E	$\mathscr{F}$
Effect	I	Φ
Opposition	R	$\mathscr{R}$

As mentioned in the introduction to this chapter, there is a broad similarity between the analyses of electric and magnetic circuits. This has already been demonstrated to some extent for the quantities in Table 12.1.

If we apply the "cause" analogy to Kirchhoff's voltage law ($\Sigma_{\circlearrowright} V = 0$), we obtain the following:

$$\boxed{\Sigma_{\circlearrowright}\, \mathscr{F} = 0} \qquad \text{(for magnetic circuits)} \qquad \textbf{(12.9)}$$

which, in words, states that the algebraic sum of the rises and drops of the mmf around a closed loop of a magnetic circuit is equal to zero; that is, the sum of the rises in mmf equals the sum of the drops in mmf around a closed loop.

Eq. (12.9) is referred to as **Ampère's circuital law.** When it is applied to magnetic circuits, sources of mmf are expressed by the equation

$$\boxed{\mathscr{F} = NI} \qquad \text{(At)} \qquad \textbf{(12.10)}$$

The equation for the mmf drop across a portion of a magnetic circuit can be found by applying the relationships listed in Table 12.1; that is, for electric circuits,

$$V = IR$$

resulting in the following for magnetic circuits:

$$\boxed{\mathscr{F} = \Phi\mathscr{R}} \quad \text{(At)} \qquad \textbf{(12.11)}$$

where ϕ is the flux passing through a section of the magnetic circuit and $\mathscr{R}$ is the reluctance of that section. The reluctance, however, is seldom calculated in the analysis of magnetic circuits. A more practical equation for the mmf drop is

$$\boxed{\mathscr{F} = Hl} \quad \text{(At)} \qquad \textbf{(12.12)}$$

as derived from Eq. (12.6), where H is the magnetizing force on a section of a magnetic circuit and l is the length of the section.

As an example of Eq. (12.9), consider the magnetic circuit appearing in Fig. 12.10 constructed of three different ferromagnetic materials.

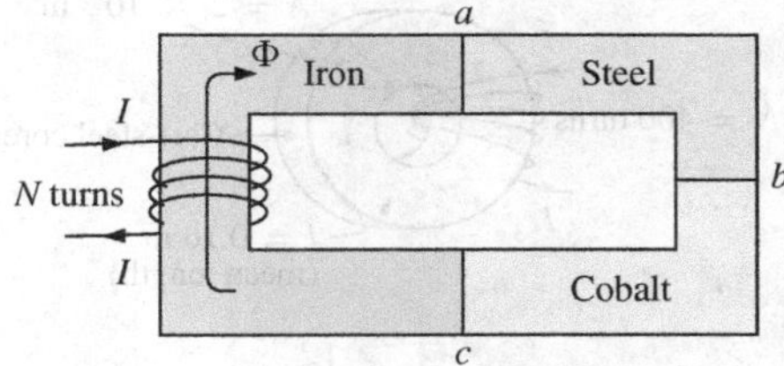

FIG. 12.10
Series magnetic circuit of three different materials.

Applying Ampère's circuital law, we have

$$\Sigma_{\mathcal{C}}\, \mathscr{F} = 0$$

$$\underbrace{+NI}_{\text{Rise}} - \underbrace{H_{ab}l_{ab}}_{\text{Drop}} - \underbrace{H_{bc}l_{bc}}_{\text{Drop}} - \underbrace{H_{ca}l_{ca}}_{\text{Drop}} = 0$$

$$\underbrace{NI}_{\substack{\text{Impressed}\\ \text{mmf}}} = \underbrace{H_{ab}l_{ab} + H_{bc}l_{bc} + H_{ca}l_{ca}}_{\text{mmf drops}}$$

All the terms of the equation are known except the magnetizing force for each portion of the magnetic circuit, which can be found by using the *B-H* curve if the flux density *B* is known.

12.8 FLUX Φ

If we continue to apply the relationships described in the previous section to Kirchhoff's current law, we find that the sum of the fluxes entering a junction is equal to the sum of the fluxes leaving a junction; that is, for the circuit in Fig. 12.11,

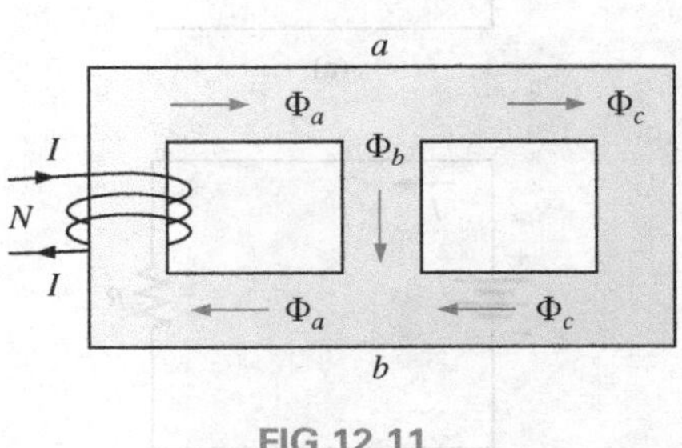

FIG. 12.11
Flux distribution of a series-parallel magnetic network.

$$\Phi_a = \Phi_b + \Phi_c \qquad \text{(at junction } a)$$

or

$$\Phi_b + \Phi_c = \Phi_a \qquad \text{(at junction } b)$$

which are equivalent.

12.9 SERIES MAGNETIC CIRCUITS: DETERMINING *NI*

We are now in a position to solve a few magnetic circuit problems, which are basically of two types. *In one type, Φ is given, and the impressed mmf NI must be computed.* This is the type of problem encountered in the design of motors, generators, and transformers. *In the other type, NI is given, and the flux Φ of the magnetic circuit must be found.* This type of problem is encountered primarily in the design of magnetic amplifiers and is more difficult since the approach is "hit or miss."

As indicated in earlier discussions, the value of μ varies from point to point along the magnetization curve. This eliminates the possibility of finding the reluctance of each "branch" or the "total reluctance" of a network, as was done for electric circuits, where ρ had a fixed value for any applied current or voltage. If the total reluctance can be determined, Φ can then be determined using the Ohm's law analogy for magnetic circuits.

For magnetic circuits, the level of B or H is determined from the other using the B-H curve, and μ is seldom calculated unless asked for.

An approach frequently used in the analysis of magnetic circuits is the *table* method. Before a problem is analyzed in detail, a table is prepared listing in the far left column the various sections of the magnetic circuit (see Table 12.2). The columns on the right are reserved for the quantities to be found for each section. In this way, when you are solving a problem, you can keep track of what the next step should be and what is required to complete the problem. After a few examples, the usefulness of this method should become clear.

This section considers only *series* magnetic circuits in which the flux ϕ is the same throughout. In each example, the magnitude of the magnetomotive force is to be determined.

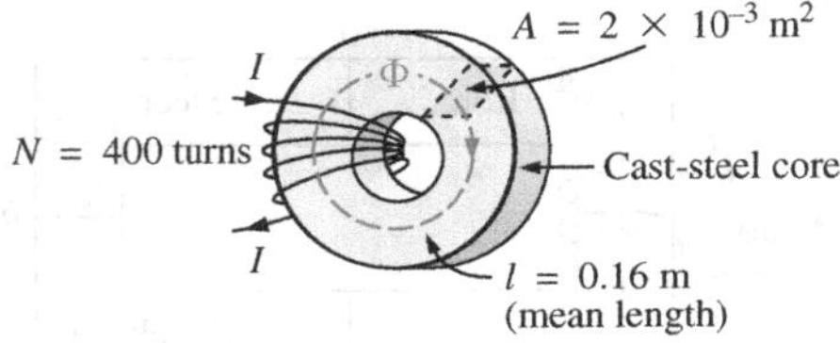

FIG. 12.12
Example 12.1.

EXAMPLE 12.1 For the series magnetic circuit in Fig. 12.12:

a. Find the value of I required to develop a magnetic flux of $\Phi = 4 \times 10^{-4}$ Wb.
b. Determine μ and μ_r for the material under these conditions.

Solutions: The magnetic circuit can be represented by the system shown in Fig. 12.13(a). The electric circuit analogy is shown in Fig. 12.13(b). Analogies of this type can be very helpful in the solution of magnetic circuits. Table 12.2 is for part (a) of this problem. The table is fairly trivial for this example, but it does define the quantities to be found.

a. The flux density B is

$$B = \frac{\Phi}{A} = \frac{4 \times 10^{-4}\ \text{Wb}}{2 \times 10^{-3}\ \text{m}^2} = 2 \times 10^{-1}\ \text{T} = 0.2\ \text{T}$$

Using the B-H curves in Fig. 12.8, we can determine the magnetizing force H:

$$H\,(\text{cast steel}) = 170\ \text{At/m}$$

Applying Ampère's circuital law yields

$$NI = Hl$$

and $$I = \frac{Hl}{N} = \frac{(170\ \text{At/m})(0.16\ \text{m})}{400\ \text{t}} = \mathbf{68\ mA}$$

(Recall that t represents turns.)

b. The permeability of the material can be found using Eq. (12.8):

$$\mu = \frac{B}{H} = \frac{0.2\ \text{T}}{170\ \text{At/m}} = \mathbf{1.176 \times 10^{-3}\ Wb/A \cdot m}$$

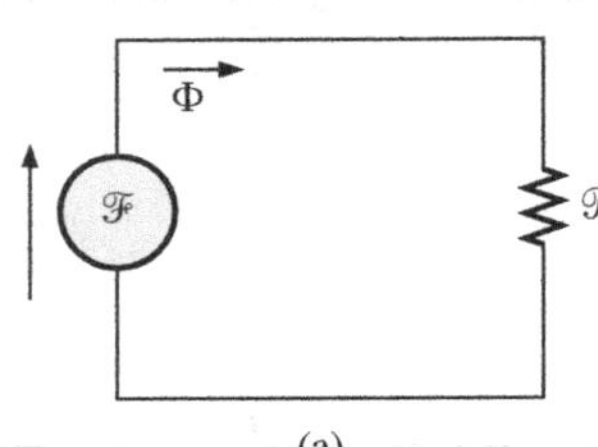

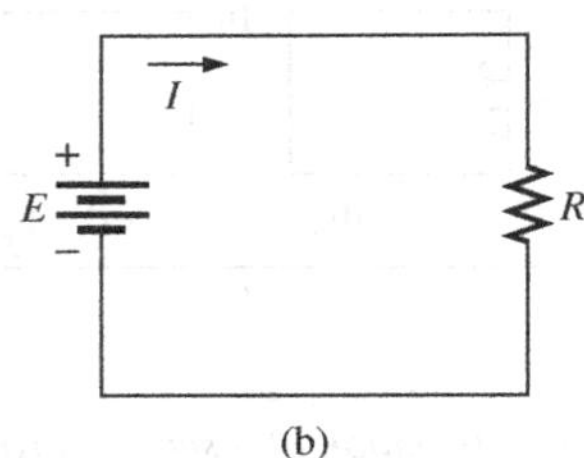

FIG. 12.13
(a) Magnetic circuit equivalent and (b) electric circuit analogy.

TABLE 12.2

Section	Φ (Wb)	A (m²)	B (T)	H (At/m)	l (m)	Hl (At)
One continuous section	4×10^{-4}	2×10^{-3}			0.16	

and the relative permeability is

$$\mu_r = \frac{\mu}{\mu_o} = \frac{1.176 \times 10^{-3}}{4\pi \times 10^{-7}} = \mathbf{935.83}$$

EXAMPLE 12.2 The electromagnet in Fig. 12.14 has picked up a section of cast iron. Determine the current I required to establish the indicated flux in the core.

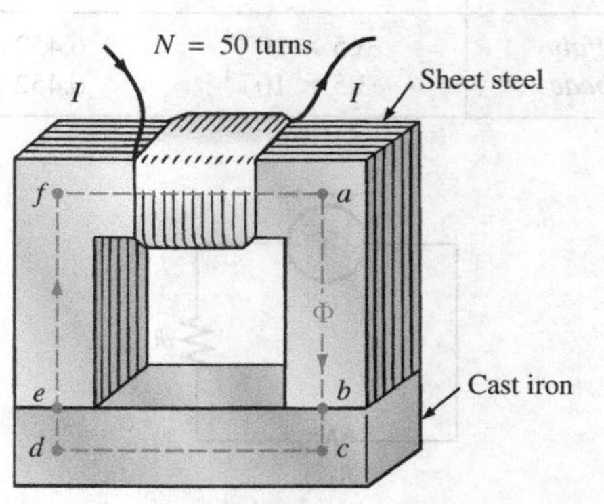

$l_{ab} = l_{cd} = l_{ef} = l_{fa} = 4$ in.
$l_{bc} = l_{de} = 0.5$ in.
Area (throughout) = 1 in.^{2}
$\Phi = 3.5 \times 10^{-4}$ Wb

FIG. 12.12
Electromagnet for Example 12.2.

Solution: To be able to use Figs. 12.7 and 12.8, we must first convert to the metric system. However, since the area is the same throughout, we can determine the length for each material rather than work with the individual sections:

$$l_{efab} = 4 \text{ in.} + 4 \text{ in.} + 4 \text{ in.} = 12 \text{ in.}$$
$$l_{bcde} = 0.5 \text{ in.} + 4 \text{ in.} + 0.5 \text{ in.} = 5 \text{ in.}$$
$$12 \text{ in.}\left(\frac{1 \text{ m}}{39.37 \text{ in.}}\right) = 304.8 \times 10^{-3} \text{ m}$$
$$5 \text{ in.}\left(\frac{1 \text{ m}}{39.37 \text{ in.}}\right) = 127 \times 10^{-3} \text{ m}$$
$$1 \text{ in.}^2\left(\frac{1 \text{ m}}{39.37 \text{ in.}}\right)\left(\frac{1 \text{ m}}{39.37 \text{ in.}}\right) = 6.452 \times 10^{-4} \text{ m}^2$$

The information available from the *efab* and *bcde* specifications of the problem has been inserted in Table 12.3. When the problem has been completed, each space will contain some information. Sufficient data to complete the problem can be found if we fill in each column from left to right. As the various quantities are calculated, they will be placed in a similar table found at the end of the example.

TABLE 12.3

Section	Φ (Wb)	*A* (m²)	*B* (T)	*H* (At/m)	*l* (m)	*Hl* (At)
efab	3.5×10^{-4}	6.452×10^{-4}			304.8×10^{-3}	
bcde	3.5×10^{-4}	6.452×10^{-4}			127×10^{-3}	

The flux density for each section is

$$B = \frac{\Phi}{A} = \frac{3.5 \times 10^{-4} \text{ Wb}}{6.452 \times 10^{-4} \text{ m}^2} = 0.542 \text{ T}$$

and the magnetizing force is

$$H \text{ (sheet steel, Fig. 12.8)} \cong 70 \text{ At/m}$$
$$H \text{ (cast iron, Fig. 12.7)} \cong 1600 \text{ At/m}$$

Note the extreme difference in magnetizing force for each material for the required flux density. In fact, when we apply Ampère's circuital law, we find that the sheet steel section can be ignored with a minimal error in the solution.

Determining *Hl* for each section yields

$$H_{efab}l_{efab} = (70 \text{ At/m})(304.8 \times 10^{-3} \text{ m}) = 21.34 \text{ At}$$
$$H_{bcde}l_{bcde} = (1600 \text{ At/m})(127 \times 10^{-3} \text{ m}) = 203.2 \text{ At}$$

Inserting the above data in Table 12.3 results in Table 12.4.

TABLE 12.4

Section	Φ (Wb)	A (m^2)	B (T)	H (At/m)	l (m)	Hl (At)
efab	3.5×10^{-4}	6.452×10^{-4}	0.542	70	304.8×10^{-3}	21.34
bcde	3.5×10^{-4}	6.452×10^{-4}	0.542	1600	127×10^{-3}	203.2

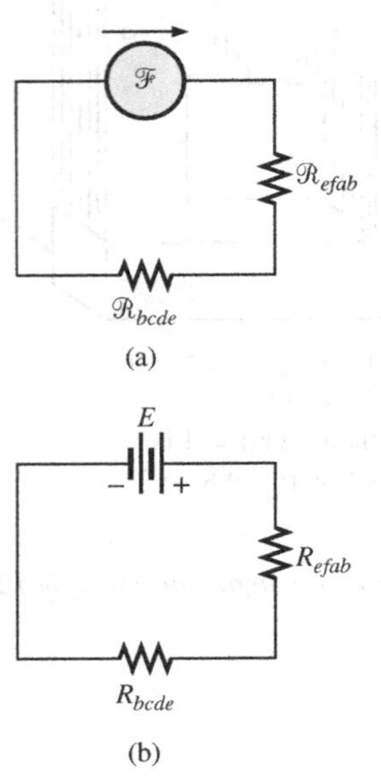

FIG. 12.15
(a) Magnetic circuit equivalent and (b) electric circuit analogy for the electromagnet in Fig. 12.14.

The magnetic circuit equivalent and the electric circuit analogy for the system in Fig. 12.14 appear in Fig. 12.15.

Applying Ampère's circuital law, we obtain

$$NI = H_{efab}l_{efab} + H_{bcde}l_{bcde}$$
$$= 21.34 \text{ At} + 203.2 \text{ At} = 224.54 \text{ At}$$

and

$$(50 \text{ t})I = 224.54 \text{ At}$$

so that

$$I = \frac{224.54 \text{ At}}{50 \text{ t}} = \mathbf{4.49\ A}$$

EXAMPLE 12.3 Determine the secondary current I_2 for the transformer in Fig. 12.16 if the resultant clockwise flux in the core is 1.5×10^{-5} Wb.

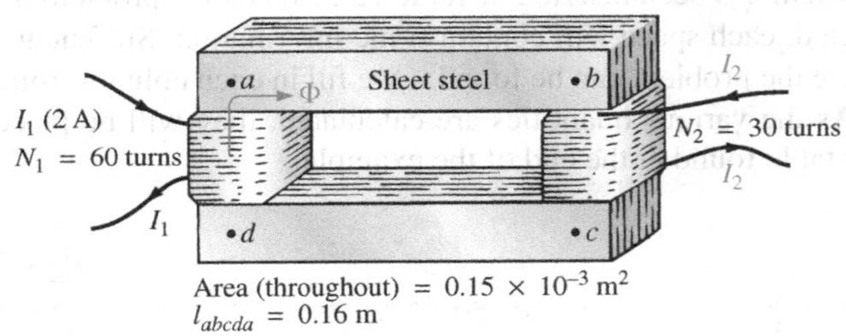

FIG. 12.16
Transformer for Example 12.3.

Solution: This is the first example with two magnetizing forces to consider. In the analogies in Fig. 12.17, note that the resulting flux of each is opposing, just as the two sources of voltage are opposing in the electric circuit analogy.

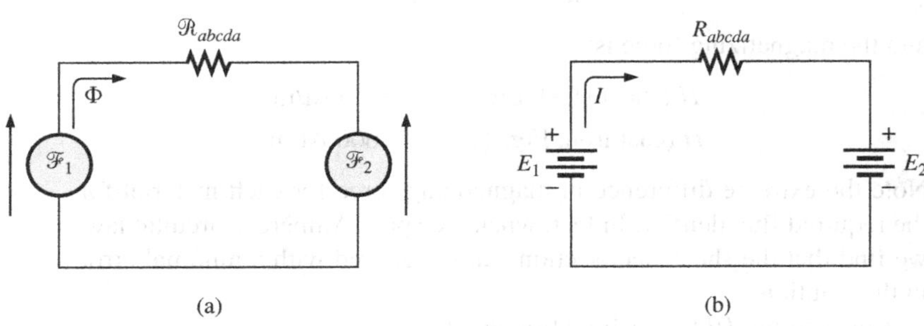

FIG. 12.17
(a) Magnetic circuit equivalent and (b) electric circuit analogy for the transformer in Fig. 12.16.

The *abcda* structural data appear in Table 12.5.

TABLE 12.5

Section	Φ (Wb)	A (m^2)	B (T)	H (At/m)	l (m)	Hl (At)
abcda	1.5×10^{-5}	0.15×10^{-3}			0.16	

The flux density throughout is

$$B = \frac{\phi}{A} = \frac{1.5 \times 10^{-5}\text{ Wb}}{0.15 \times 10^{-3}\text{ m}^2} = 10 \times 10^{-2}\text{ T} = 0.10\text{ T}$$

and

$$H\,(\text{from Fig. 12.8}) \cong \frac{1}{5}(100\text{ At/m}) = 20\text{ At/m}$$

Applying Ampère's circuital law, we obtain

$$N_1I_1 - N_2I_2 = H_{abcda}l_{abcda}$$
$$(60\text{ t})(2\text{ A}) - (30\text{ t})(I_2) = (20\text{ At/m})(0.16\text{m})$$
$$120\text{ At} - (30\text{ t})I_2 = 3.2\text{ At}$$

and

$$(30\text{ t})I_2 = 120\text{ At} - 3.2\text{ At}$$

or

$$I_2 = \frac{116.8\text{ At}}{30\text{ t}} = \mathbf{3.89\text{ A}}$$

For the analysis of most transformer systems, the equation $N_1I_1 = N_2I_2$ is used. This results in 4 A versus 3.89 A above. This difference is normally ignored, however, and the equation $N_1I_1 = N_2I_2$ considered exact.

Because of the nonlinearity of the *B-H* curve, *it is not possible to apply superposition to magnetic circuits;* that is, in Example 12.3, we cannot consider the effects of each source independently and then find the total effects by using superposition.

12.10 AIR GAPS

Before continuing with the illustrative examples, let us consider the effects that an air gap has on a magnetic circuit. Note the presence of air gaps in the magnetic circuits of the motor and meter in Fig. 11.15. The spreading of the flux lines outside the common area of the core for the air gap in Fig. 12.18(a) is known as *fringing.* For our purposes, we shall ignore this effect and assume the flux distribution to be as in Fig. 12.18(b).

The flux density of the air gap in Fig. 12.18(b) is given by

$$B_g = \frac{\Phi_g}{A_g} \tag{12.13}$$

where, for our purposes,

$$\Phi_g = \Phi_{core}$$

and

$$A_g = A_{core}$$

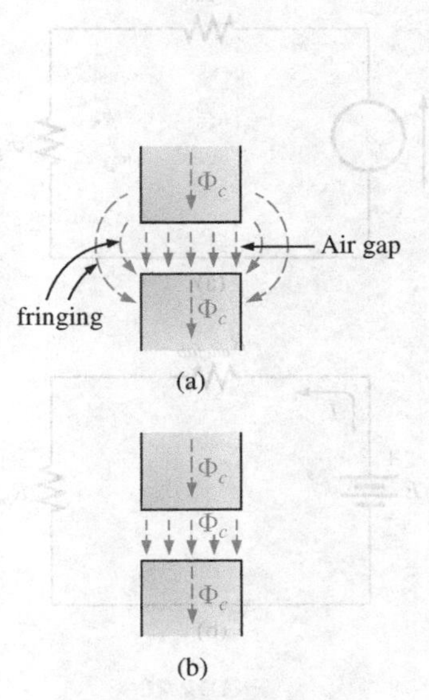

FIG. 12.18
Air gaps: (a) with fringing; (b) ideal.

For most practical applications, the permeability of air is taken to be equal to that of free space. The magnetizing force of the air gap is then determined by

$$H_g = \frac{B_g}{\mu_o} \quad (12.14)$$

and the mmf drop across the air gap is equal to $H_g L_g$. An equation for H_g is as follows:

$$H_g = \frac{B_g}{\mu_o} = \frac{B_g}{4\pi \times 10^{-7}}$$

and

$$H_g = (7.96 \times 10^5)B_g \quad \text{(At/m)} \quad (12.15)$$

EXAMPLE 12.4 Find the value of I required to establish a magnetic flux of $\phi = 0.75 \times 10^{-4}$ Wb in the series magnetic circuit in Fig. 12.19.

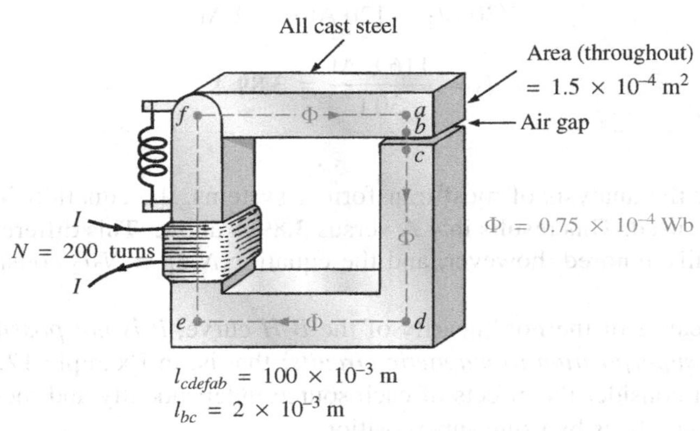

FIG.12.19
Relay for Example 12.4.

Solution: An equivalent magnetic circuit and its electric circuit analogy are shown in Fig. 12.20.

The flux density for each section is

$$B = \frac{\Phi}{A} = \frac{0.75 \times 10^{-4}\,\text{Wb}}{1.5 \times 10^{-4}\,\text{m}^2} = 0.5\ \text{T}$$

From the *B-H* curves in Fig. 12.8,

$$H\,\text{(cast steel)} \cong 280\ \text{At/m}$$

Applying Eq. (12.15),

$$H_g = (7.96 \times 10^5)B_g = (7.96 \times 10^5)(0.5\ \text{T}) = 3.98 \times 10^5\ \text{At/m}$$

The mmf drops are

$$H_{core}l_{core} = (280\ \text{At/m})(100 \times 10^{-3}\,\text{m}) = 28\ \text{At}$$
$$H_g l_g = (3.98 \times 10^5\ \text{At/m})(2 \times 10^{-3}\,\text{m}) = 796\ \text{At}$$

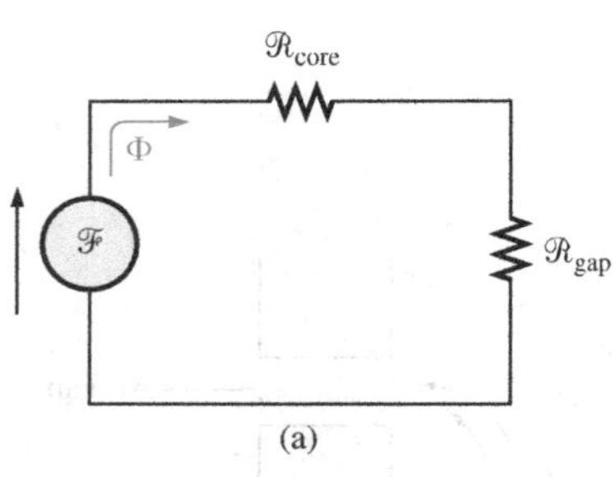

(a)

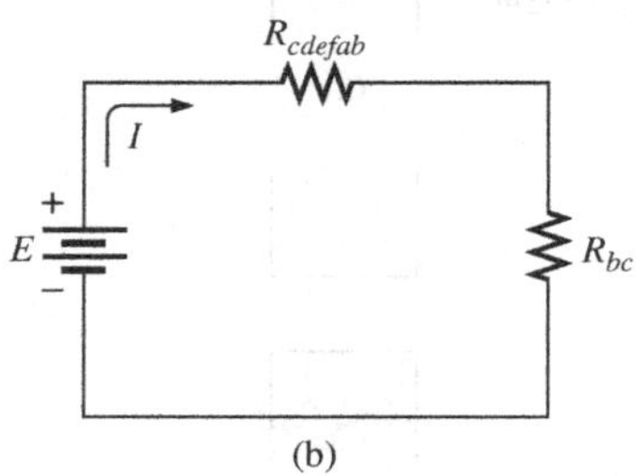

(b)

FIG.12.20
(a) Magnetic circuit equivalent and (b) electric circuit analogy for the relay in Fig. 12.19.

Applying Ampère's circuital law, we obtain

$$NI = H_{core}l_{core} + H_g l_g$$
$$= 28\text{ At} + 796\text{ At}$$
$$(200\text{ t})I = 824\text{ At}$$
$$I = \mathbf{4.12\ A}$$

Note from the above that the air gap requires the biggest share (by far) of the impressed NI because air is nonmagnetic.

12.11 SERIES-PARALLEL MAGNETIC CIRCUITS

As one might expect, the close analogies between electric and magnetic circuits eventually lead to series-parallel magnetic circuits similar in many respects to those encountered in Chapter 7. In fact, the electric circuit analogy will prove helpful in defining the procedure to follow toward a solution.

EXAMPLE 12.5 Determine the current I required to establish a flux of 1.5×10^{-4} Wb in the section of the core indicated in Fig. 12.21.

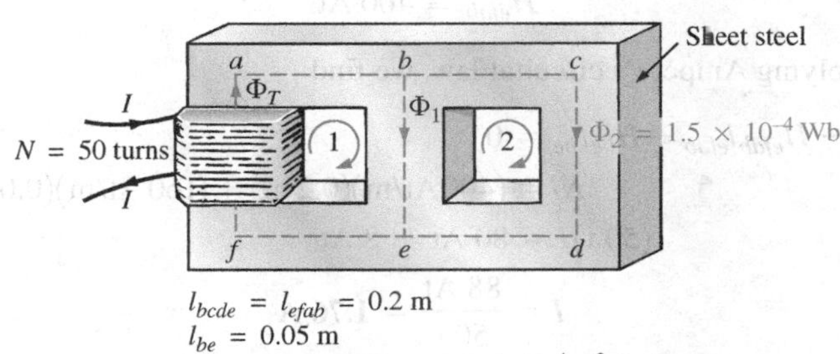

FIG. 12.21
Example 12.5.

Solution: The equivalent magnetic circuit and the electric circuit analogy appear in Fig. 12.22. We have

$$B_2 = \frac{\Phi_2}{A} = \frac{1.5 \times 10^{-4}\text{ Wb}}{6 \times 10^{-4}\text{ m}^2} = 0.25\text{ T}$$

From Fig. 12.8,

$$H_{bcde} \cong 40\text{ At/m}$$

Applying Ampère's circuital law around loop 2 in Figs. 12.21 and 12.22,

$$\Sigma_C\, \mathcal{F} = 0$$
$$H_{be}l_{be} - H_{bcde}l_{bcde} = 0$$
$$H_{be}(0.05\text{ m}) - (40\text{ At/m})(0.2\text{m}) = 0$$
$$H_{be} = \frac{8\text{ At}}{0.05\text{ m}} = 160\text{ At/m}$$

From Fig. 12.8,

$$B_1 \cong 0.97\text{ T}$$

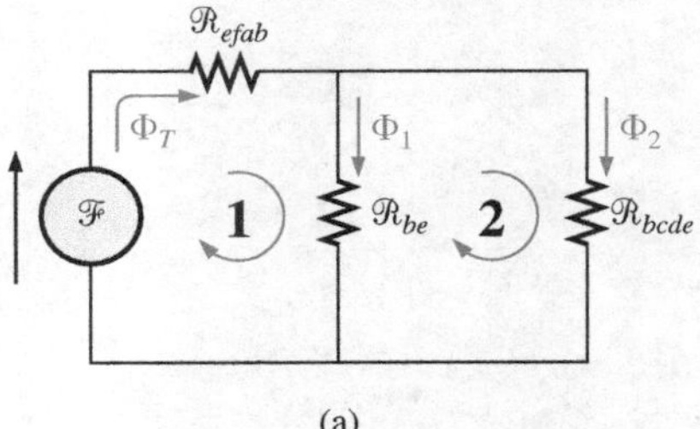

(a)

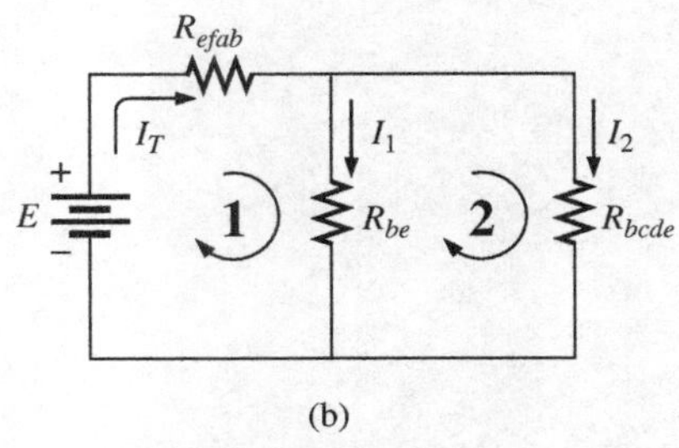

(b)

FIG. 12.22
(a) Magnetic circuit equivalent and (b) electric circuit analogy for the series-parallel system in Fig. 12.21.

and

$$\Phi_1 = B_1 A = (0.97\text{ T})(6 \times 10^{-4}\text{ m}^2) = 5.82 \times 10^{-4}\text{ Wb}$$

The results for *bcde*, *be*, and *efab* are entered in Table 12.6.

TABLE 12.6

Section	Φ (Wb)	A (m^2)	B (T)	H (At/m)	l (m)	Hl (At)
bcde	1.5×10^{-4}	6×10^{-4}	0.25	40	0.2	8
be	5.82×10^{-4}	6×10^{-4}	0.97	160	0.05	8
efab		6×10^{-4}			0.2	

Table 12.6 reveals that we must now turn our attention to section *efab:*

$$\Phi_T = \Phi_1 + \Phi_2 = 5.82 \times 10^{-4}\text{ Wb} + 1.5 \times 10^{-4}\text{ Wb} = 7.32 \times 10^{-4}\text{ Wb}$$

$$B = \frac{\Phi_T}{A} = \frac{7.32 \times 10^{-4}\text{ Wb}}{6 \times 10^{-4}\text{ m}^2} = 1.22\text{ T}$$

From Fig. 12.7,

$$H_{efab} \cong 400\text{ At}$$

Applying Ampère's circuital law, we find

$$+NI - H_{efab}l_{efab} - H_{be}l_{be} = 0$$

$$NI = (400\text{ At/m})(0.2\text{ m}) + (160\text{ At/m})(0.05\text{ m})$$

$$(50\text{ t})I = 80\text{ At} + 8\text{ At}$$

$$I = \frac{88\text{ At}}{50\text{ t}} = \mathbf{1.76\text{ A}}$$

To demonstrate that μ is sensitive to the magnetizing force H, the permeability of each section is determined as follows. For section *bcde*,

$$\mu = \frac{B}{H} = \frac{0.25\text{ T}}{40\text{ At/m}} = 6.25 \times 10^{-3}$$

and $$\mu_r = \frac{\mu}{\mu_o} = \frac{6.25 \times 10^{-3}}{12.57 \times 10^{-7}} = \mathbf{4972.2}$$

For section *be*,

$$\mu = \frac{B}{H} = \frac{0.97\text{ T}}{160\text{ At/m}} = 6.06 \times 10^{-3}$$

and $$\mu_r = \frac{\mu}{\mu_o} = \frac{6.06 \times 10^{-3}}{12.57 \times 10^{-7}} = \mathbf{4821}$$

For section *efab*,

$$\mu = \frac{B}{H} = \frac{1.22\text{ T}}{400\text{ At/m}} = 3.05 \times 10^{-3}$$

and $$\mu_r = \frac{\mu}{\mu_o} = \frac{3.05 \times 10^{-3}}{12.57 \times 10^{-7}} = \mathbf{2426.41}$$

12.12 DETERMINING Φ

The examples of this section are of the second type, where NI is given and the flux Φ must be found. This is a relatively straightforward problem if only one magnetic section is involved. Then

$$H = \frac{NI}{l} \quad H \rightarrow B \qquad (B\text{-}H \text{ curve})$$

and

$$\Phi = BA$$

For magnetic circuits with more than one section, there is no set order of steps that lead to an exact solution for every problem on the first attempt. In general, however, we proceed as follows. We must find the impressed mmf for a *calculated guess* of the flux Φ and then compare this with the specified value of mmf. We can then make adjustments to our guess to bring it closer to the actual value. For most applications, a value within ±5% of the actual Φ or specified NI is acceptable.

We can make a reasonable guess at the value of Φ if we realize that the maximum mmf drop appears across the material with the smallest permeability if the length and area of each material are the same. As shown in Example 12.4, if there is an air gap in the magnetic circuit, there will be a considerable drop in mmf across the gap. As a starting point for problems of this type, therefore, we shall assume that the total mmf (NI) is across the section with the lowest μ or greatest $\mathcal{R}$ (if the other physical dimensions are relatively similar). This assumption gives a value of Φ that will produce a calculated NI greater than the specified value. Then, after considering the results of our original assumption very carefully, we shall *cut* Φ and NI by introducing the effects (reluctance) of the other portions of the magnetic circuit and *try* the new solution. For obvious reasons, this approach is frequently called the *cut and try* method.

EXAMPLE 12.6 Calculate the magnetic flux Φ for the magnetic circuit in Fig. 12.23.

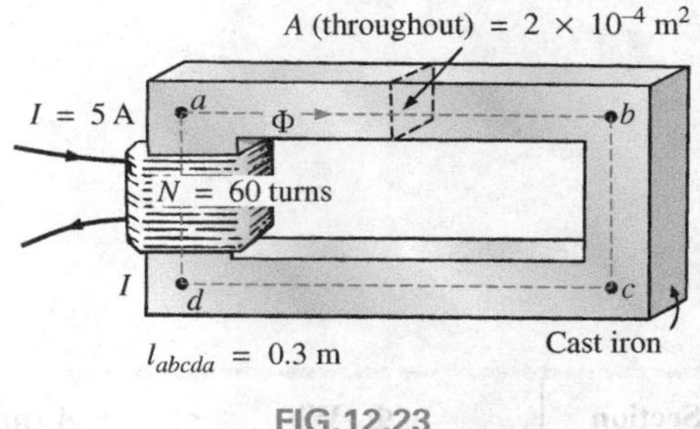

FIG.12.23
Example 12.6.

Solution: By Ampère's circuital law,

$$NI = H_{abcda}l_{abcda}$$

or

$$H_{abcda} = \frac{NI}{l_{abcda}} = \frac{(60\text{ t})(5\text{ A})}{0.3\text{ m}} = \frac{300\text{ At}}{0.3\text{ m}} = 1000\text{ At/m}$$

and

$$B_{abcda}\text{ (from Fig. 12.7)} \cong 0.39\text{ T}$$

Since $B = \Phi/A$, we have

$$\Phi = BA = (0.39\text{ T})(2 \times 10^{-4}\text{ m}^2) = \mathbf{0.78 \times 10^{-4}\ Wb}$$

EXAMPLE 12.7 Find the magnetic flux Φ for the series magnetic circuit in Fig. 12.24 for the specified impressed mmf.

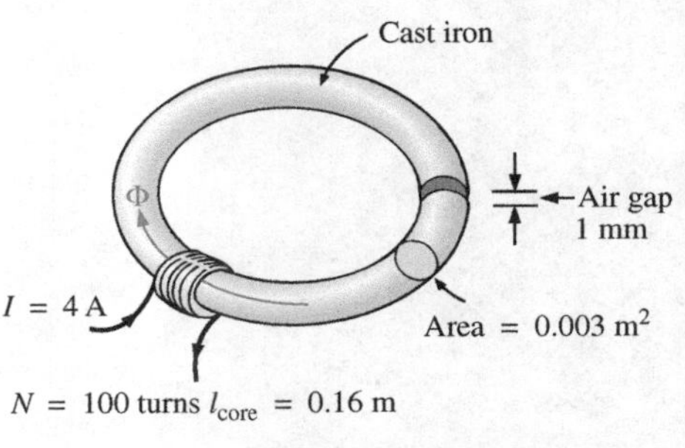

FIG.12.24
Example 12.7.

Solution: Assuming that the total impressed mmf NI is across the air gap, we obtain

$$NI = H_g l_g$$

or

$$H_g = \frac{NI}{l_g} = \frac{400\text{ At}}{0.001\text{ m}} = 4 \times 10^5\text{ At/m}$$

and $$B_g = \mu_o H_g = (4\pi \times 10^{-7})(4 \times 10^5 \text{ At/m}) = 0.503 \text{ T}$$

The flux is given by

$$\Phi_g = \Phi_{core} = B_g A = (0.503 \text{ T})(0.003 \text{ m}^2)$$
$$\Phi_{core} = 1.51 \times 10^{-3} \text{ Wb}$$

Using this value of Φ, we can find NI. The core and gap data are inserted in Table 12.7.

TABLE 12.7

Section	Φ (Wb)	A (m²)	B (T)	H (At/m)	l (m)	Hl (At)
Core	1.51×10^{-3}	0.003	0.503	1500 (*B-H* curve)	0.16	
Gap	1.51×10^{-3}	0.003	0.503	4×10^5	0.001	400

$$H_{core} l_{core} = (1500 \text{ At/m})(0.16 \text{ m}) = 240 \text{ At}$$

Applying Ampère's circuital law results in

$$NI = H_{core} l_{core} + H_g l_g = 240 \text{ At} + 400 \text{ At}$$
$$400 \text{ At} \neq 640 \text{ At}$$

Since we neglected the reluctance of all the magnetic paths but the air gap, the calculated value is greater than the specified value. We must therefore reduce this value by including the effect of these reluctances. Since approximately (640 At − 400 At)/640 At = 240 At/640 At ≅ 37.5% of our calculated value is above the desired value, let us reduce Φ by 30% and see how close we come to the impressed mmf of 400 At:

$$\Phi = (1 - 0.3)(1.51 \times 10^{-3} \text{ Wb}) = 1.057 \times 10^{-3} \text{ Wb}$$

See Table 12.8. We have

TABLE 12.8

Section	Φ (Wb)	A (m²)	B (T)	H (At/m)	l (m)	Hl (At)
Core	1.057×10^{-3}	0.003			0.16	
Gap	1.057×10^{-3}	0.003			0.001	

$$B = \frac{\Phi}{A} = \frac{1.057 \times 10^{-3} \text{ Wb}}{0.003 \text{ m}^3} \cong 0.352 \text{ T}$$

$$H_g l_g = (7.96 \times 10^5) B_g l_g = (7.96 \times 10^5)(0.352 \text{ T})(0.001 \text{ m}) \cong 280.19 \text{ At}$$

From the *B-H* curves,

$$H_{core} \cong 850 \text{ At/m}$$
$$H_{core} l_{core} = (850 \text{ At/m})(0.16 \text{ m}) = 136 \text{ At}$$

Applying Ampère's circuital law yields

$$NI = H_{core}l_{core} + H_g l_g$$
$$= 136\text{ At} + 280.19\text{ At}$$
$$400\text{ At} = \mathbf{416.19\text{ At}} \quad \text{(but within } \pm 5\% \text{ and therefore acceptable)}$$

The solution is, therefore,

$$\Phi \cong \mathbf{1.057 \times 10^{-3}\text{ Wb}}$$

12.13 APPLICATIONS

Speakers and Microphones

Electromagnetic effects are the moving force in the design of speakers such as the one shown in Fig. 12.25. The shape of the pulsating waveform of the input current is determined by the sound to be reproduced by the speaker at a high audio level. As the current peaks and returns to the valleys of the sound pattern, the strength of the electromagnet varies in exactly the same manner. This causes the cone of the speaker to vibrate at a frequency directly proportional to the pulsating input. The higher the pitch of the sound pattern, the higher is the oscillating frequency between the peaks and valleys and the higher is the frequency of vibration of the cone.

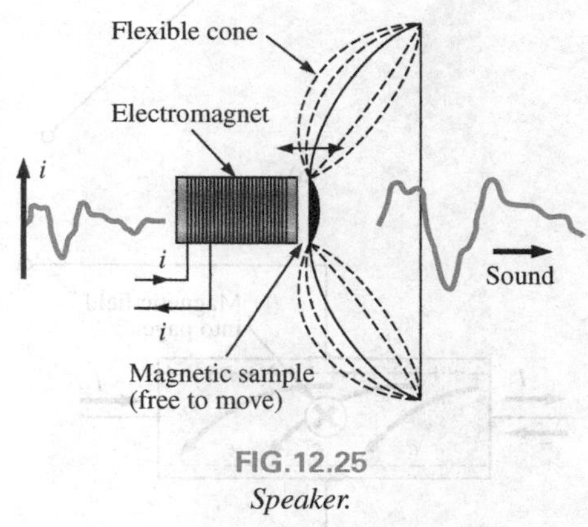

FIG.12.25
Speaker.

A second design used more frequently in more expensive speaker systems appears in Fig. 12.26. In this case, the permanent magnet is fixed, and the input is applied to a movable core within the magnet, as shown in the figure. High peaking currents at the input produce a strong flux pattern in the voice coil, causing it to be drawn well into the flux pattern of the permanent magnet. As occurred for the speaker in Fig. 12.25, the core then vibrates at a rate determined by the input and provides the audible sound.

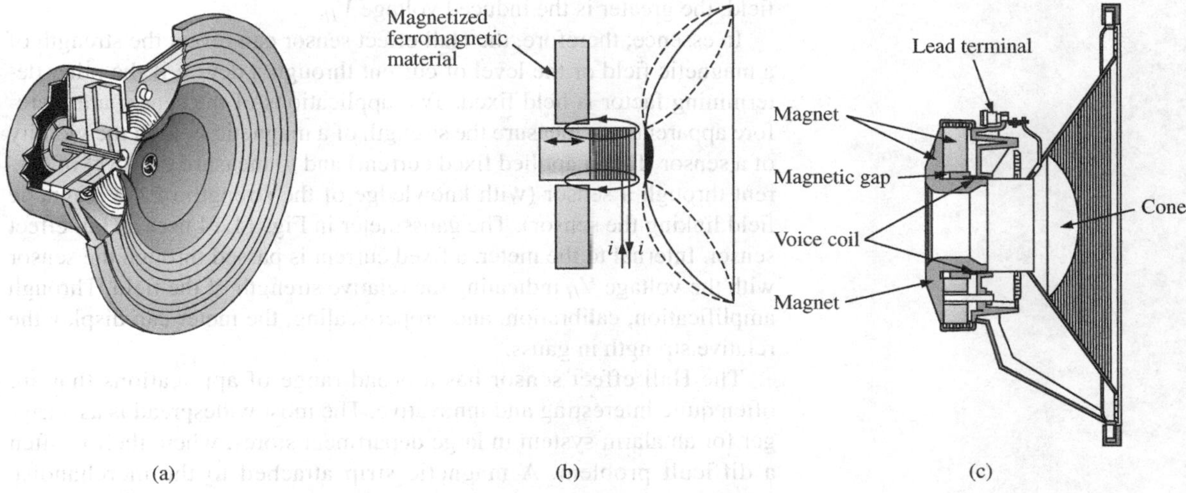

FIG.12.26
Coaxial high-fidelity loudspeaker: (a) construction: (b) basic operation; (c) cross section of actual unit.

Microphones also employ electromagnetic effects. The incoming sound causes the core and attached moving coil to move within the magnetic field of the permanent magnet. Through Faraday's law

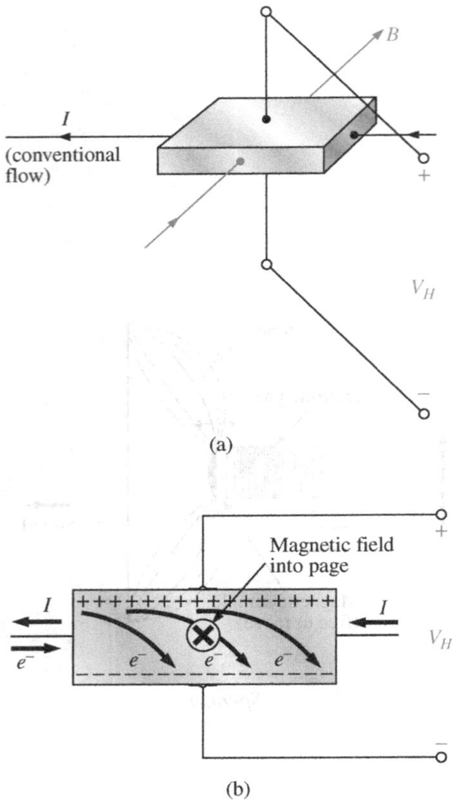

FIG.12.27
Hall effect sensor: (a) orientation of controlling parameters; (b) effect on electron flow.

($e = N\,d\phi/dt$), a voltage is induced across the movable coil proportional to the speed with which it is moving through the magnetic field. The resulting induced voltage pattern can then be amplified and reproduced at a much higher audio level through the use of speakers, as described earlier. Microphones of this type are the most frequently employed, although other types that use capacitive, carbon granular, and piezoelectric* effects are available. This particular design is commercially referred to as a *dynamic* microphone.

Hall Effect Sensor

The Hall effect sensor is a semiconductor device that generates an output voltage when exposed to a magnetic field. The basic construction consists of a slab of semiconductor material through which a current is passed, as shown in Fig. 12.27(a). If a magnetic field is applied, as shown in the figure, perpendicular to the direction of the current, a voltage V_H is generated between the two terminals, as indicated in Fig. 12.27(a). The difference in potential is due to the separation of charge established by the Lorentz force first studied by Professor Hendrick Lorentz in the late 1800s. He found that electrons in a magnetic field are subjected to a force proportional to the velocity of the electrons through the field and the strength of the magnetic field. The direction of the force is determined by the left-hand rule. Simply place the index finger of your left hand in the direction of the magnetic field, with the second finger at right angles to the index finger in the direction of conventional current through the semiconductor material, as shown in Fig. 12.27(b). The thumb, if placed at right angles to the index finger, will indicate the direction of the force on the electrons. In Fig. 12.27(b), the force causes the electrons to accumulate in the bottom region of the semiconductor (connected to the negative terminal of the voltage V_H), leaving a net positive charge in the upper region of the material (connected to the positive terminal of V_H). The stronger the current or strength of the magnetic field, the greater is the induced voltage V_H.

In essence, therefore, the Hall effect sensor can reveal the strength of a magnetic field or the level of current through a device if the other determining factor is held fixed. Two applications of the sensor are therefore apparent—to measure the strength of a magnetic field in the vicinity of a sensor (for an applied fixed current) and to measure the level of current through a sensor (with knowledge of the strength of the magnetic field linking the sensor). The gaussmeter in Fig. 11.14 uses a Hall effect sensor. Internal to the meter, a fixed current is passed through the sensor with the voltage V_H indicating the relative strength of the field. Through amplification, calibration, and proper scaling, the meter can display the relative strength in gauss.

The Hall effect sensor has a broad range of applications that are often quite interesting and innovative. The most widespread is as a trigger for an alarm system in large department stores, where theft is often a difficult problem. A magnetic strip attached to the merchandise sounds an alarm when a customer passes through the exit gates without paying for the product. The sensor, control current, and monitoring system are housed in the exit fence and react to the presence of the magnetic

*Piezoelectricity is the generation of a small voltage by exerting pressure across certain crystals.

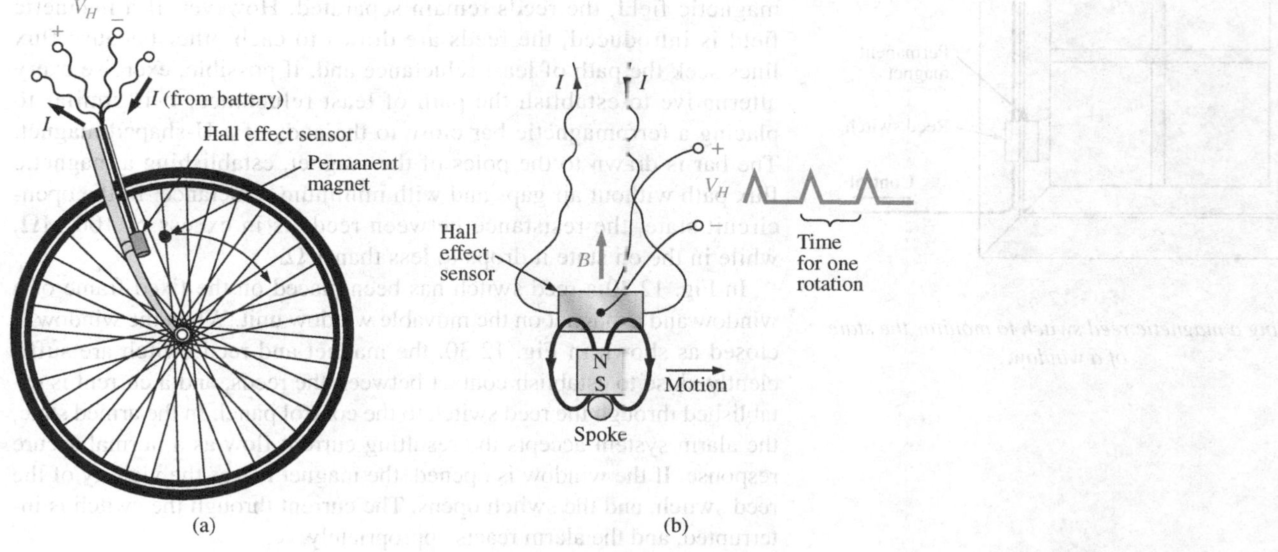

FIG.12.28
Obtaining a speed indication for a bicycle using a Hall effect sensor: (a) mounting the components; (b) Hall effect response.

field as the product leaves the store. When the product is paid for, the cashier removes the strip or demagnetizes the strip by applying a magnetizing force that reduces the residual magnetism in the strip to essentially zero.

The Hall effect sensor is also used to indicate the speed of a bicycle on a digital display conveniently mounted on the handlebars. As shown in Fig. 12.28(a), the sensor is mounted on the frame of the bike, and a small permanent magnet is mounted on a spoke of the front wheel. The magnet must be carefully mounted to be sure that it passes over the proper region of the sensor. When the magnet passes over the sensor, the flux pattern in Fig. 12.28(b) results, and a voltage with a sharp peak is developed by the sensor. For a bicycle with a 26-in.-diameter wheel, the circumference will be about 82 in. Over 1 mi, the number of rotations is

$$5280\,\cancel{\text{ft}}\left(\frac{12\,\cancel{\text{in.}}}{1\,\cancel{\text{ft}}}\right)\left(\frac{1\text{ rotation}}{82\,\cancel{\text{in.}}}\right) \cong 773\text{ rotations}$$

If the bicycle is traveling at 20 mph, an output pulse occurs at a rate of 4.29 per second. It is interesting to note that at a speed of 20 mph, the wheel is rotating at more than 4 revolutions per second, and the total number of rotations over 20 mi is 15,460.

Magnetic Reed Switch

One of the most frequently employed switches in alarm systems is the *magnetic reed switch* shown in Fig. 12.29. As shown by the figure, there are two components of the reed switch—a permanent magnet embedded in one unit that is normally connected to the movable element (door, window, and so on) and a reed switch in the other unit that is connected to the electrical control circuit. The reed switch is constructed of two iron-alloy (ferromagnetic) reeds in a hermetically sealed capsule. The cantilevered ends of the two reeds do not touch but are in very close proximity to one another. In the absence of a

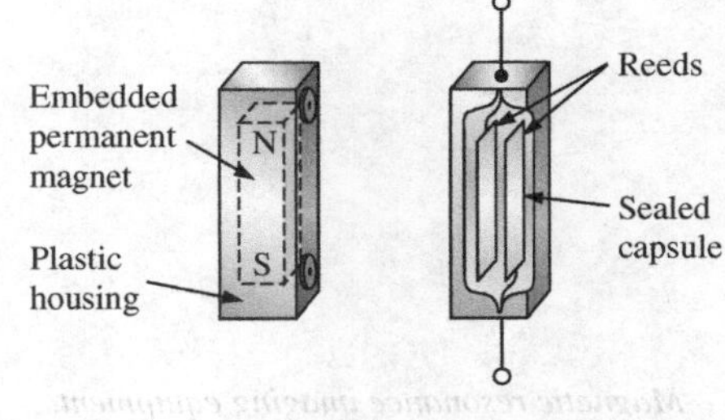

FIG.12.29
Magnetic reed switch.

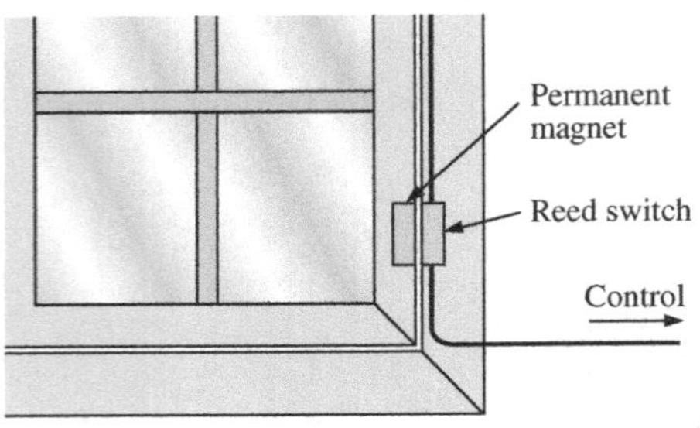

FIG. 12.30
Using a magnetic reed switch to monitor the state of a window.

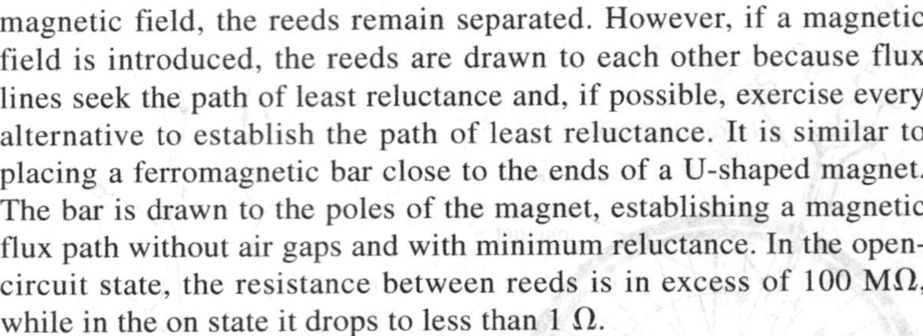

magnetic field, the reeds remain separated. However, if a magnetic field is introduced, the reeds are drawn to each other because flux lines seek the path of least reluctance and, if possible, exercise every alternative to establish the path of least reluctance. It is similar to placing a ferromagnetic bar close to the ends of a U-shaped magnet. The bar is drawn to the poles of the magnet, establishing a magnetic flux path without air gaps and with minimum reluctance. In the open-circuit state, the resistance between reeds is in excess of 100 MΩ, while in the on state it drops to less than 1 Ω.

In Fig. 12.30 a reed switch has been placed on the fixed frame of a window and a magnet on the movable window unit. When the window is closed as shown in Fig. 12.30, the magnet and reed switch are sufficiently close to establish contact between the reeds, and a current is established through the reed switch to the control panel. In the armed state, the alarm system accepts the resulting current flow as a normal secure response. If the window is opened, the magnet leaves the vicinity of the reed switch, and the switch opens. The current through the switch is interrupted, and the alarm reacts appropriately.

One of the distinct advantages of the magnetic reed switch is that the proper operation of any switch can be checked with a portable magnetic element. Simply bring the magnet to the switch and note the output response. There is no need to continually open and close windows and doors. In addition, the reed switch is hermetically enclosed so that oxidation and foreign objects cannot damage it, and the result is a unit that can last indefinitely. Magnetic reed switches are also available in other shapes and sizes, allowing them to be concealed from obvious view. One is a circular variety that can be set into the edge of a door and door jam, resulting in only two small visible disks when the door is open.

Magnetic Resonance Imaging

Magnetic resonance imaging (MRI) provides quality cross-sectional images of the body for medical diagnosis and treatment. MRI does not expose the patient to potentially hazardous X-rays or injected contrast materials such as those used to obtain computerized axial tomography (CAT) scans.

The three major components of an MRI system are a strong magnet, a table for transporting the patient into the circular hole in the magnet, and a control center, as shown in Fig. 12.31. The image is obtained by placing the patient in the tube to a precise depth depending on the cross section to be obtained and applying a strong magnetic field that causes the nuclei of certain atoms in the body to line up. Radio waves of different frequencies are then applied to the patient in the region of interest, and if the frequency of the wave matches the natural frequency of the atom, the nuclei is set into a state of resonance and absorbs energy from the applied signal. When the signal is removed, the nuclei release the acquired energy in the form of weak but detectable signals. The strength and duration of the energy emission vary from one tissue of the body to another. The weak signals are then amplified, digitized, and translated to provide a cross-sectional image such as the one shown in Fig. 12.32. For some patients the claustrophobic feeling they experience while in the circular tube is

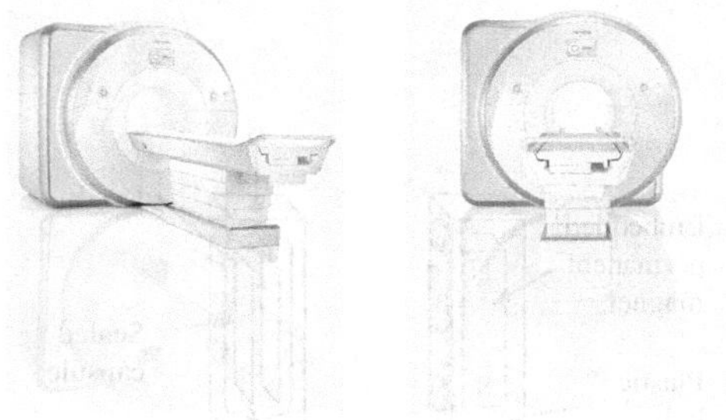

FIG. 12.31
Magnetic resonance imaging equipment.
(© Siemens Inc.)

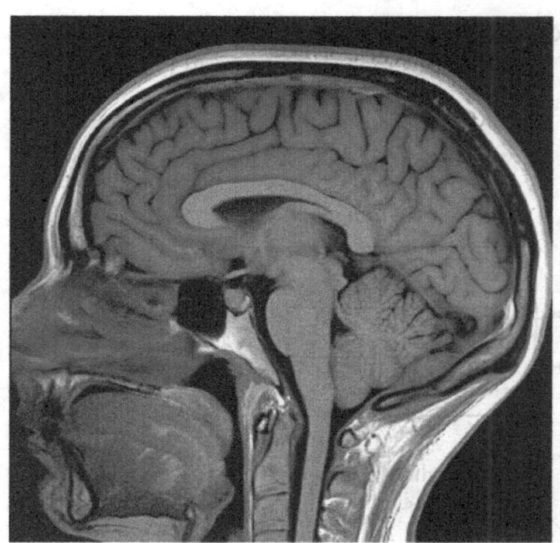

FIG. 12.32
Magnetic resonance image.
(© Siemens Inc.)

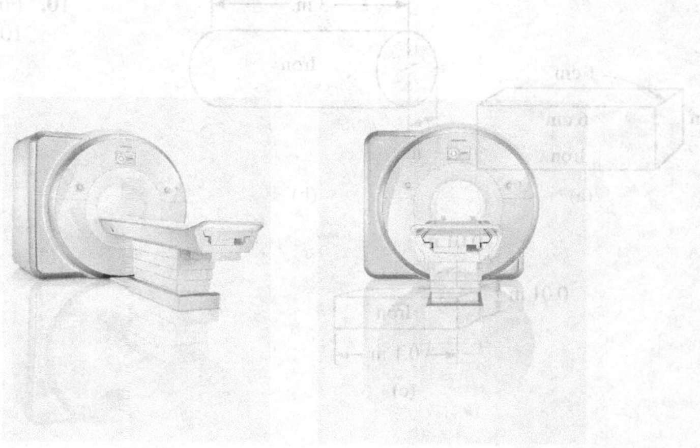

FIG. 12.33
Magnetic resonance imaging equipment (open variety).
(© Siemens Inc.)

difficult to contend with. A more open unit has been developed, as shown in Fig. 12.33, that has removed most of this discomfort.

Patients who have metallic implants or pacemakers or those who have worked in industrial environments where minute ferromagnetic particles may have become lodged in open, sensitive areas such as the eyes, nose, and so on, may have to use a CAT scan system because it does not employ magnetic effects. The attending physician is well trained in such areas of concern and will remove any unfounded fears or suggest alternative methods.

PROBLEMS

SECTION 12.2 Magnetic Field

1. Using Appendix E, fill in the blanks in the following table. Indicate the units for each quantity.

	Φ	B
SI	5×10^{-4} Wb	8×10^{-4} T
CGS	______	______
English	______	______

2. Repeat Problem 1 for the following table if area = 2 in.^{2}:

	Φ	B
SI		______
CGS	60,000 maxwells	______
English	______	______

3. For the electromagnet in Fig. 12.34:
   a. Find the flux density in the core.
   b. Sketch the magnetic flux lines and indicate their direction.
   c. Indicate the north and south poles of the magnet.

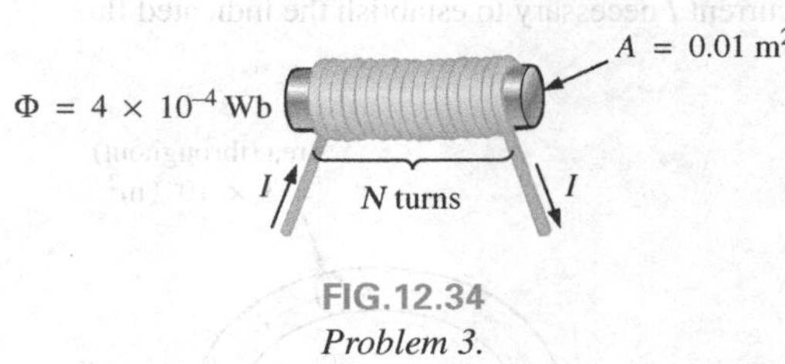

FIG. 12.34
Problem 3.

SECTION 12.3 Reluctance

4. Which section of Fig. 12.35—(a), (b), or (c)—has the largest reluctance to the setting up of flux lines through its longest dimension?

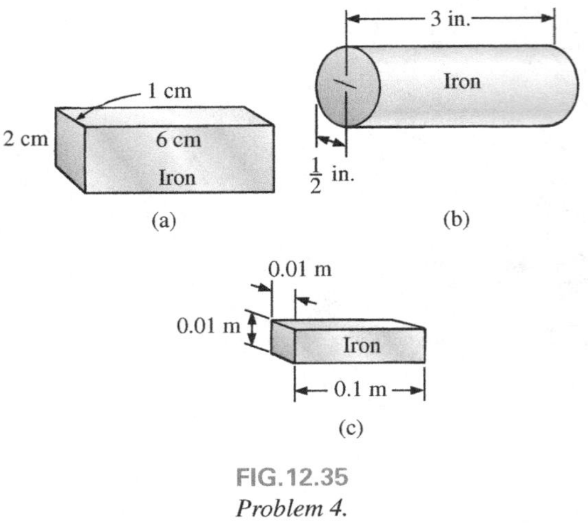

FIG. 12.35
Problem 4.

SECTION 12.4 Ohm's Law for Magnetic Circuits

5. Find the reluctance of a magnetic circuit if a magnetic flux $\Phi = 4.2 \times 10^{-4}$ Wb is established by an impressed mmf of 400 At.

6. Repeat Problem 5 for $\Phi = 72{,}000$ maxwells and an impressed mmf of 120 gilberts.

SECTION 12.5 Magnetizing Force

7. Find the magnetizing force H for Problem 5 in SI units if the magnetic circuit is 6 in. long.

8. If a magnetizing force H of 600 At/m is applied to a magnetic circuit, a flux density B of 1200×10^{-4} Wb/m² is established. Find the permeability μ of a material that will produce twice the original flux density for the same magnetizing force.

SECTIONS 12.6–12.9 Hysteresis through Series Magnetic Circuits

9. For the series magnetic circuit in Fig. 12.36, determine the current I necessary to establish the indicated flux.

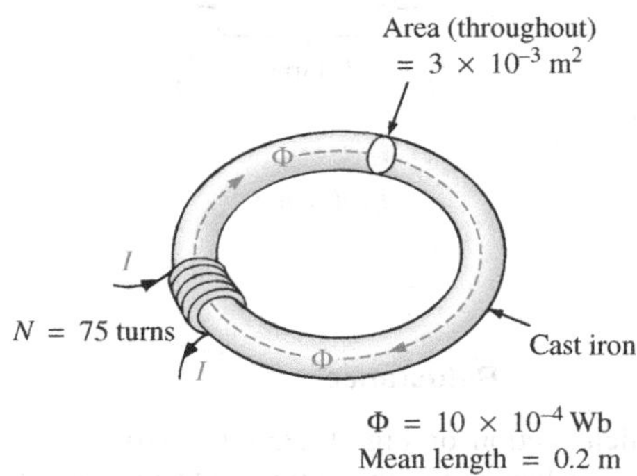

FIG. 12.36
Problem 9.

10. Find the current necessary to establish a flux of $\Phi = 3 \times 10^{-4}$ Wb in the series magnetic circuit in Fig. 12.37.

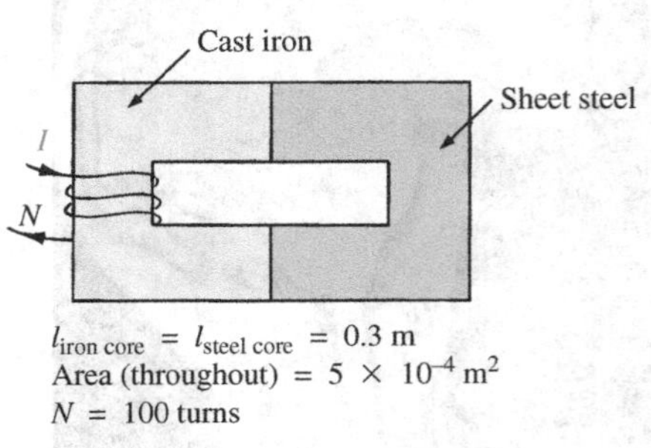

FIG. 12.37
Problem 10.

11. a. Find the number of turns N_1 required to establish a flux $\Phi = 12 \times 10^{-4}$ Wb in the magnetic circuit in Fig. 12.38.
    b. Find the permeability μ of the material.

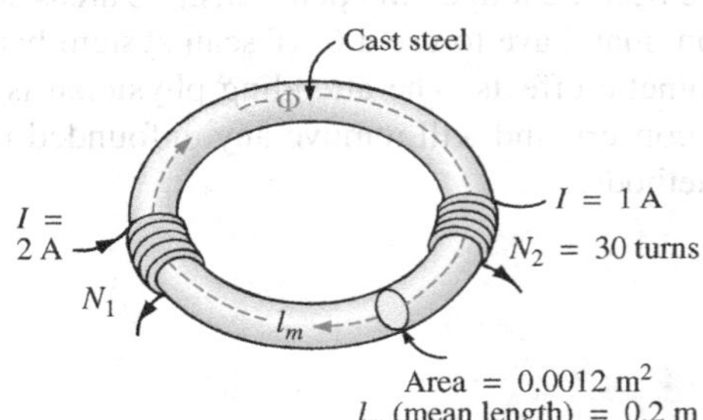

FIG. 12.38
Problem 11.

12. a. Find the mmf (NI) required to establish a flux $\Phi = 80{,}000$ lines in the magnetic circuit in Fig. 12.39.
    b. Find the permeability of each material.

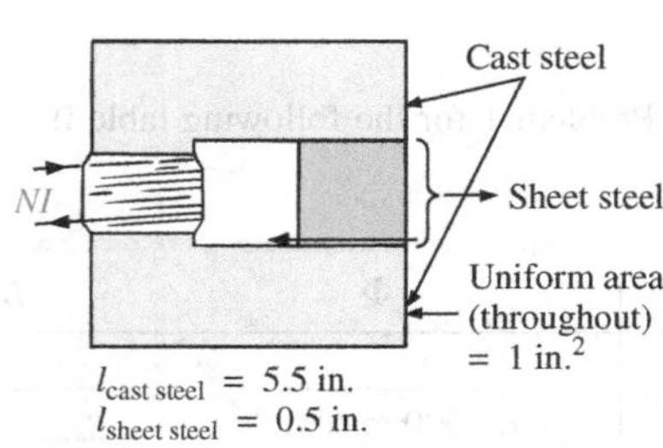

FIG. 12.39
Problem 12.

***13.** For the series magnetic circuit in Fig. 12.40 with two impressed sources of magnetic "pressure," determine the current I. Each applied mmf establishes a flux pattern in the clockwise direction.

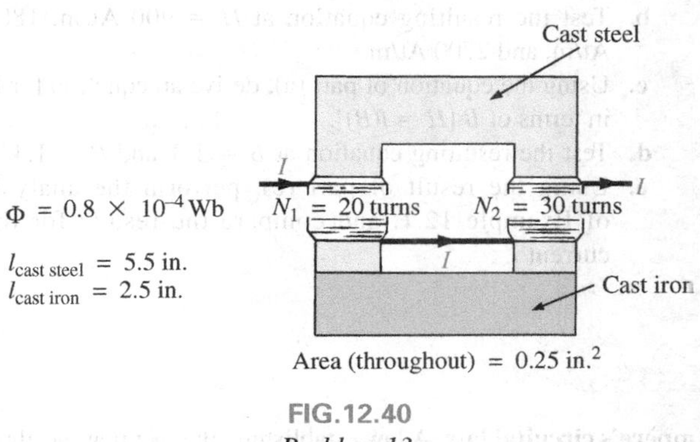

FIG. 12.40
Problem 13.

SECTION 12.10 Air Gaps

14. a. Find the current I required to establish a flux $\Phi = 2.4 \times 10^{-4}$ Wb in the magnetic circuit in Fig. 12.41.

b. Compare the mmf drop across the air gap to that across the rest of the magnetic circuit. Discuss your results using the value of μ for each material.

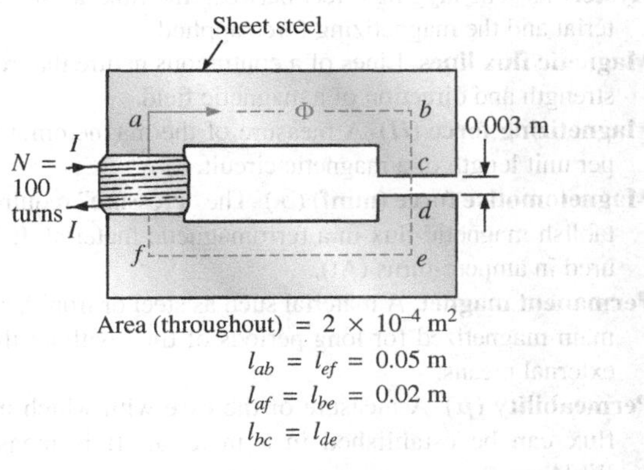

FIG. 12.41
Problem 14.

***15.** The force carried by the plunger of the door chime in Fig. 12.42 is determined by

$$f = \frac{1}{2}NI\frac{d\phi}{dx} \qquad \text{(newtons)}$$

where $d\phi/dx$ is the rate of change of flux linking the coil as the core is drawn into the coil. The greatest rate of change of flux occurs when the core is ¹/₄ to ³/₄ the way through. In this region, if Φ changes from 0.5×10^{-4} Wb to 8×10^{-4} Wb, what is the force carried by the plunger?

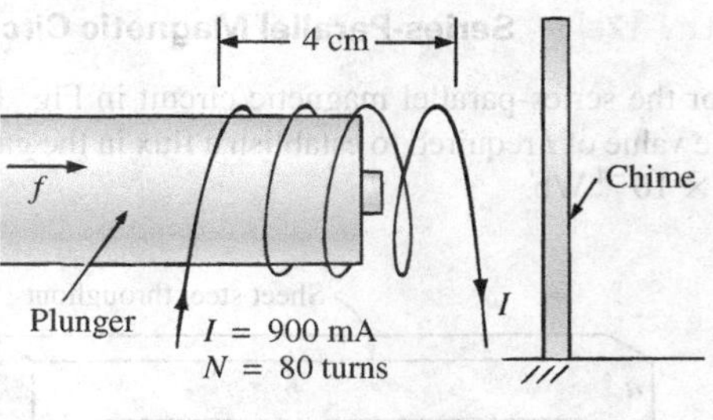

FIG. 12.42
Door chime for Problem 15.

16. Determine the current I_1 required to establish a flux of $\Phi = 2 \times 10^{-4}$ Wb in the magnetic circuit in Fig. 12.43.

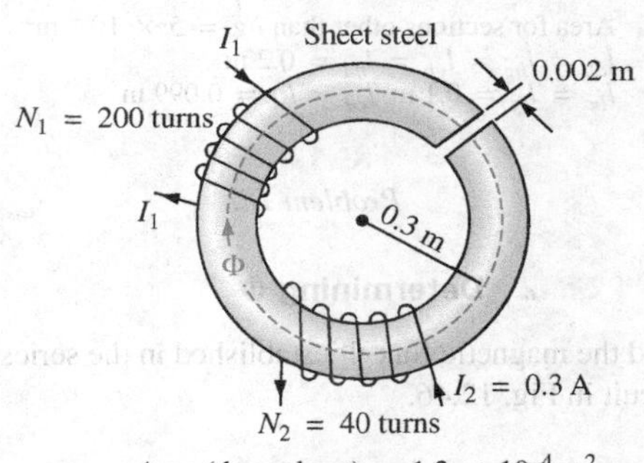

FIG. 12.43
Problem 16.

***17. a.** A flux of 0.2×10^{-4} Wb will establish sufficient attractive force for the armature of the relay in Fig. 12.44 to close the contacts. Determine the required current to establish this flux level if we assume that the total mmf drop is across the air gap.

b. The force exerted on the armature is determined by the equation

$$F(\text{newtons}) = \frac{1}{2} \cdot \frac{B_g^2 A}{\mu_o}$$

where B_g is the flux density within the air gap and A is the common area of the air gap. Find the force in newtons exerted when the flux Φ specified in part (a) is established.

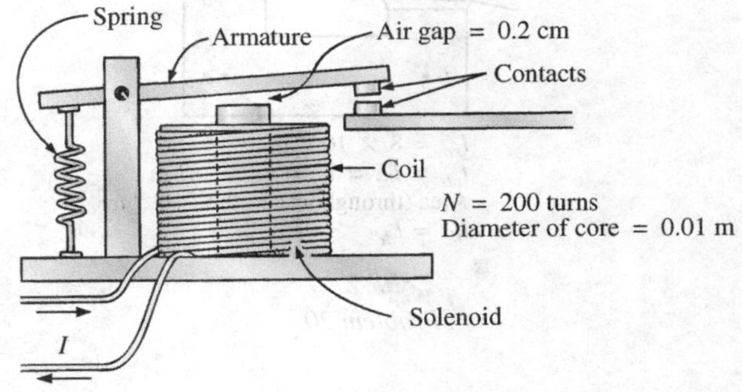

FIG. 12.44
Relay for Problem 17.

SECTION 12.11 Series-Parallel Magnetic Circuits

***18.** For the series-parallel magnetic circuit in Fig. 12.45, find the value of I required to establish a flux in the gap of $\Phi_g = 2 \times 10^{-4}$ Wb.

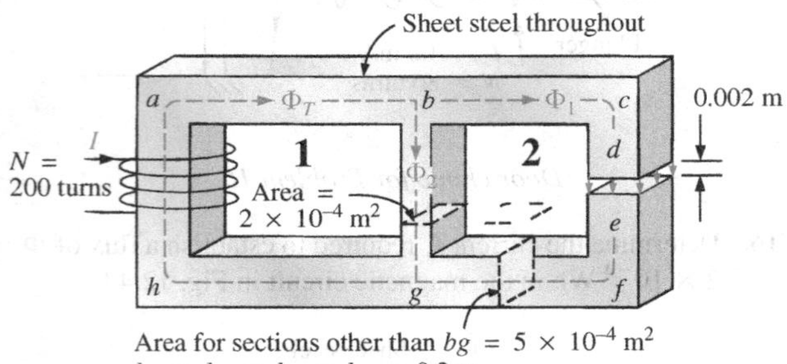

FIG. 12.45
Problem 18.

SECTION 12.12 Determining Φ

19. Find the magnetic flux Φ established in the series magnetic circuit in Fig. 12.46.

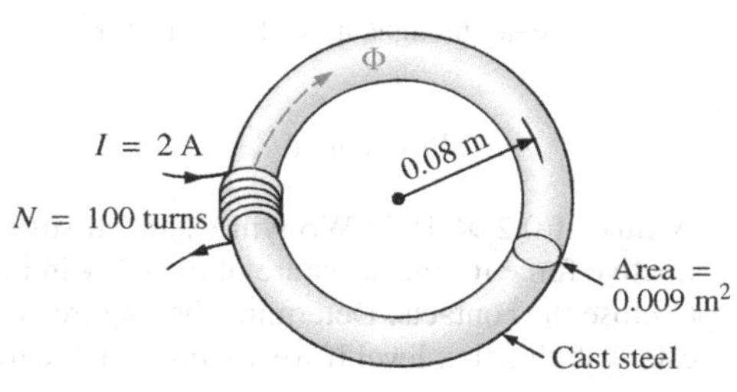

FIG. 12.46
Problem 19.

***20.** Determine the magnetic flux Φ established in the series magnetic circuit in Fig. 12.47.

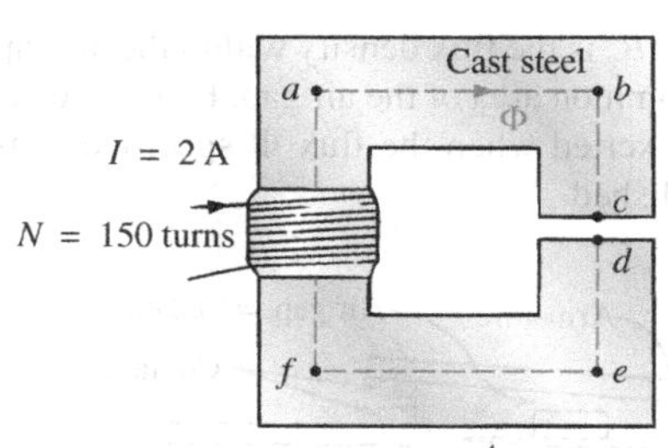

FIG. 12.47
Problem 20.

***21.** Note how closely the B-H curve of cast steel in Fig. 12.7 matches the curve for the voltage across a capacitor as it charges from zero volts to its final value.

a. Using the equation for the charging voltage as a guide, write an equation for B as a function of H [$B = f(H)$] for cast steel.

b. Test the resulting equation at H = 900 At/m, 1800 At/m, and 2700 At/m.

c. Using the equation of part (a), derive an equation for H in terms of B [$H = f(B)$].

d. Test the resulting equation at B = 1 T and B = 1.4 T.

e. Using the result of part (c), perform the analysis of Example 12.1, and compare the results for the current I.

GLOSSARY

Ampère's circuital law A law establishing the fact that the algebraic sum of the rises and drops of the mmf around a closed loop of a magnetic circuit is equal to zero.

Domain A group of magnetically aligned atoms.

Electromagnetism Magnetic effects introduced by the flow of charge or current.

Ferromagnetic materials Materials having permeabilities hundreds and thousands of times greater than that of free space.

Flux density (*B*) A measure of the flux per unit area perpendicular to a magnetic flux path. It is measured in teslas (T) or webers per square meter (Wb/m^2).

Hysteresis The lagging effect between the flux density of a material and the magnetizing force applied.

Magnetic flux lines Lines of a continuous nature that reveal the strength and direction of a magnetic field.

Magnetizing force (*H*) A measure of the magnetomotive force per unit length of a magnetic circuit.

Magnetomotive force (mmf) ($\mathscr{F}$) The "pressure" required to establish magnetic flux in a ferromagnetic material. It is measured in ampere-turns (At).

Permanent magnet A material such as steel or iron that will remain magnetized for long periods of time without the aid of external means.

Permeability (μ) A measure of the ease with which magnetic flux can be established in a material. It is measured in Wb/Am.

Relative permeability (μ_r) The ratio of the permeability of a material to that of free space.

Reluctance ($\mathscr{R}$) A quantity determined by the physical characteristics of a material that will provide an indication of the "reluctance" of that material to the setting up of magnetic flux lines in the material. It is measured in rels or At/Wb.

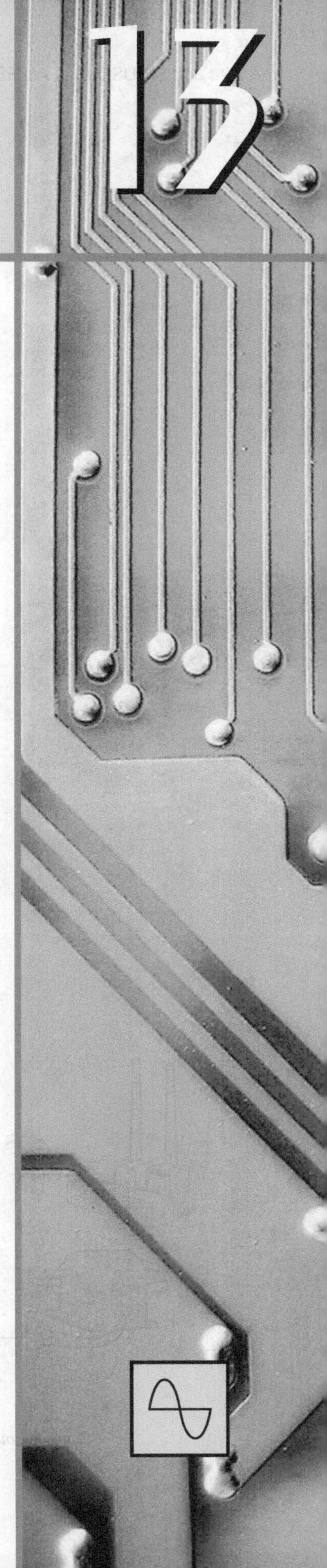

Sinusoidal Alternating Waveforms

13

Objectives

- *Become familiar with the characteristics of a sinusoidal waveform, including its general format, average value, and effective value.*
- *Be able to determine the phase relationship between two sinusoidal waveforms of the same frequency.*
- *Understand how to calculate the average and effective values of any waveform.*
- *Become familiar with the use of instruments designed to measure ac quantities.*

13.1 INTRODUCTION

The analysis thus far has been limited to dc networks—networks in which the currents or voltages are fixed in magnitude except for transient effects. We now turn our attention to the analysis of networks in which the magnitude of the source varies in a set manner. Of particular interest is the time-varying voltage that is commercially available in large quantities and is commonly called the *ac voltage.* (The letters *ac* are an abbreviation for *alternating current.*) To be absolutely rigorous, the terminology *ac voltage* or *ac current* is not sufficient to describe the type of signal we will be analyzing. Each waveform in Fig. 13.1 is an **alternating waveform** available from commercial supplies. The term *alternating* indicates only that the waveform alternates between two prescribed levels in a set time sequence. To be absolutely correct, the term *sinusoidal, square-wave,* or *triangular* must also be applied.

The pattern of particular interest is the **sinusoidal ac voltage** in Fig. 13.1. Since this type of signal is encountered in the vast majority of instances, the abbreviated phrases *ac voltage* and *ac current* are commonly applied without confusion. For the other patterns in Fig. 13.1, the descriptive term is always present, but frequently the *ac* abbreviation is dropped, resulting in the designation *square-wave* or *triangular* waveforms.

One of the important reasons for concentrating on the sinusoidal ac voltage is that it is the voltage generated by utilities throughout the world. Other reasons include its application throughout electrical, electronic, communication, and industrial systems. In addition, the chapters to follow will reveal that the waveform itself has a number of characteristics that result in a unique response when it is applied to basic electrical elements. The wide range of theorems and methods introduced for dc networks will also be applied to sinusoidal ac systems. Although the application of sinusoidal signals raises the required math level, once the

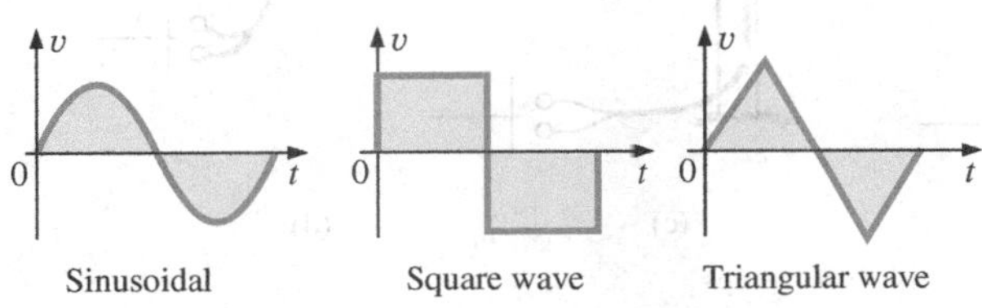

FIG. 13.1
Alternating waveforms.

notation given in Chapter 14 is understood, most of the concepts introduced in the dc chapters can be applied to ac networks with a minimum of added difficulty.

13.2 SINUSOIDAL ac VOLTAGE CHARACTERISTICS AND DEFINITIONS

Generation

Sinusoidal ac voltages are available from a variety of sources. The most common source is the typical home outlet, which provides an ac voltage that originates at a power plant. Most power plants are fueled by water power, oil, gas, or nuclear fusion. In each case, an **ac generator** (also called an *alternator*), as shown in Fig. 13.2(a), is the primary component in the energy-conversion process. The power to the shaft developed by one of the energy sources listed turns a *rotor* (constructed of alternating magnetic poles) inside a set of windings housed in the *stator* (the stationary part of the dynamo) and induces a voltage across the windings of the stator, as defined by Faraday's law:

$$e = N\frac{d\phi}{dt}$$

Through proper design of the generator, a sinusoidal ac voltage is developed that can be transformed to higher levels for distribution through the power lines to the consumer. For isolated locations where power lines have not been installed, portable ac generators [Fig. 13.2 (b)] are available that run on gasoline. As in the larger power plants, however, an ac generator is an integral part of the design.

In an effort to conserve our natural resources and reduce pollution, wind power, solar energy, and fuel cells are receiving increasing interest from various districts of the world that have such energy sources available in level and duration that make the conversion process viable. The turning propellers of the wind-power station [Fig. 13.2 (c)] are connected directly to the shaft of an ac generator to provide the ac voltage described above. Through light energy absorbed in the form of *photons,*

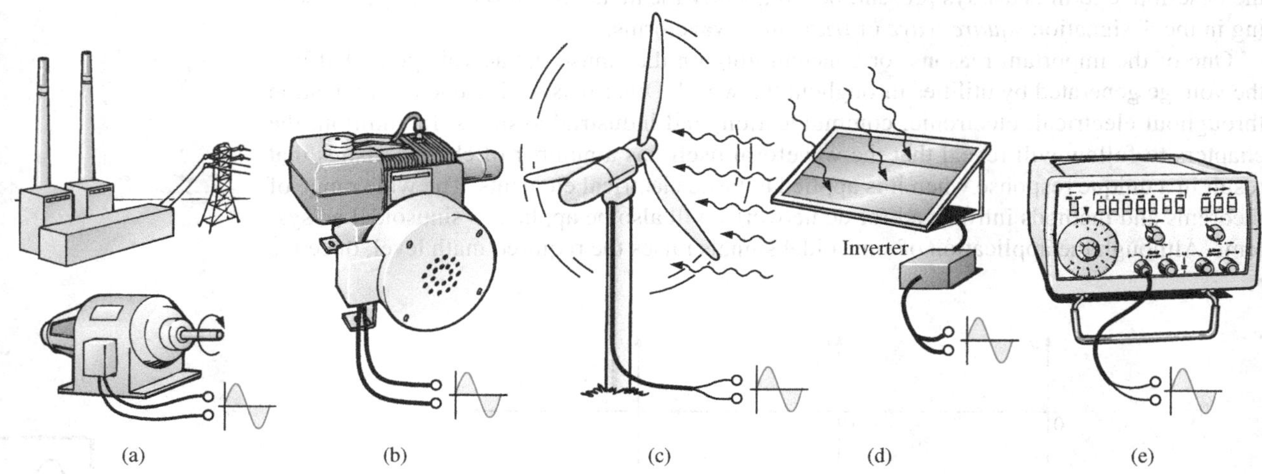

FIG. 13.2
Various sources of ac power: (a) generating plant; (b) portable ac generator; (c) wind-power station; (d) solar panel; (e) function generator.

solar cells [Fig. 13.2 (d)] can generate dc voltages. Through an electronic package called an *inverter*, the dc voltage can be converted to one of a sinusoidal nature. Boats, recreational vehicles (RVs), and so on, make frequent use of the inversion process in isolated areas.

Sinusoidal ac voltages with characteristics that can be controlled by the user are available from **function generators,** such as the one in Fig. 13.2(e). By setting the various switches and controlling the position of the knobs on the face of the instrument, you can make available sinusoidal voltages of different peak values and different repetition rates. The function generator plays an integral role in the investigation of the variety of theorems, methods of analysis, and topics to be introduced in the chapters that follow.

Definitions

The sinusoidal waveform in Fig. 13.3 with its additional notation will now be used as a model in defining a few basic terms. These terms, however, can be applied to any alternating waveform. It is important to remember, as you proceed through the various definitions, that the vertical scaling is in volts or amperes and the horizontal scaling is in units of time.

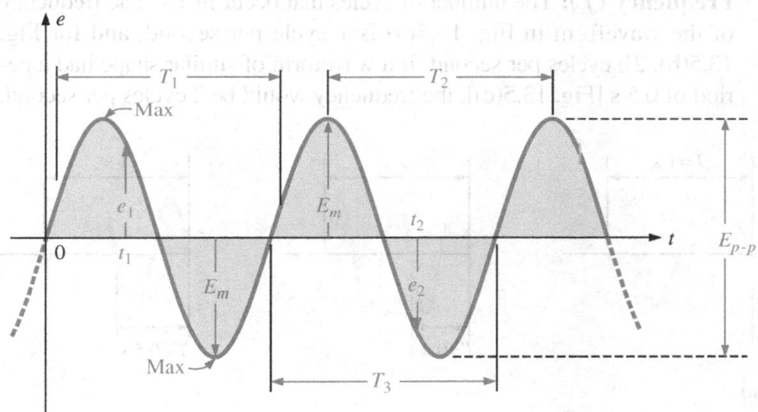

FIG. 13.3
Important parameters for a sinusoidal voltage.

Waveform: The path traced by a quantity, such as the voltage in Fig. 13.3, plotted as a function of some variable, such as time (as above), position, degrees, radians, temperature, and so on.

Instantaneous value: The magnitude of a waveform at any instant of time; denoted by lowercase letters (e_1, e_2 in Fig. 13.3).

Peak amplitude: The maximum value of a waveform as measured from its *average*, or *mean*, value, denoted by uppercase letters [such as E_m (Fig. 13.3) for sources of voltage and V_m for the voltage drop across a load]. For the waveform in Fig. 13.3, the average value is zero volts, and E_m is as defined by the figure.

Peak value: The maximum instantaneous value of a function as measured from the zero volt level. For the waveform in Fig. 13.3, the peak amplitude and peak value are the same since the average value of the function is zero volts.

Peak-to-peak value: Denoted by E_{p-p} or V_{p-p} (as shown in Fig. 13.3), the full voltage between positive and negative peaks of the waveform, that is, the sum of the magnitude of the positive and negative peaks.

Periodic waveform: A waveform that continually repeats itself after the same time interval. The waveform in Fig. 13.3 is a periodic waveform.
Period (T): The time of a periodic waveform.
Cycle: The portion of a waveform contained in one period of time. The cycles within T_1, T_2, and T_3 in Fig. 13.3 may appear different in Fig. 13.4, but they are all bounded by one period of time and therefore satisfy the definition of a cycle.

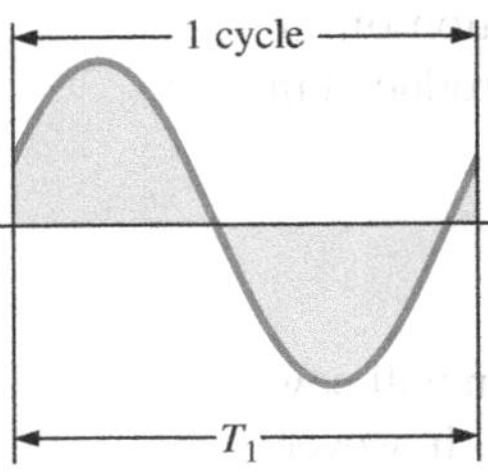

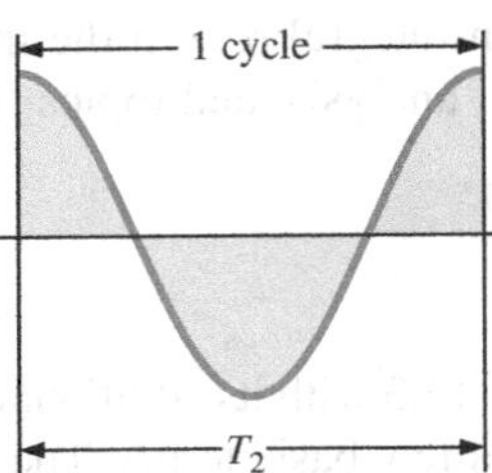

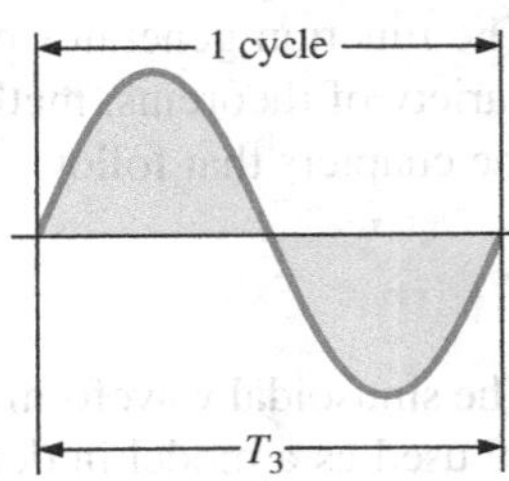

FIG. 13.4
Defining the cycle and period of a sinusoidal waveform.

Frequency (f): The number of cycles that occur in 1 s. The frequency of the waveform in Fig. 13.5(a) is 1 cycle per second, and for Fig. 13.5(b), 2½ cycles per second. If a waveform of similar shape had a period of 0.5 s [Fig. 13.5(c)], the frequency would be 2 cycles per second.

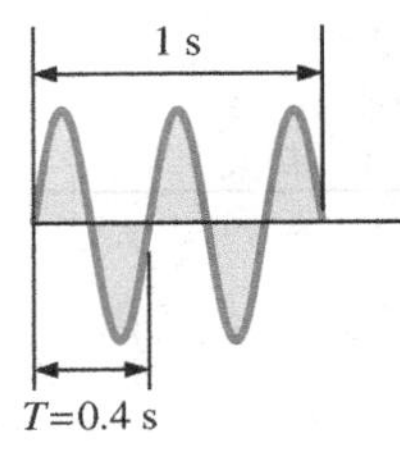

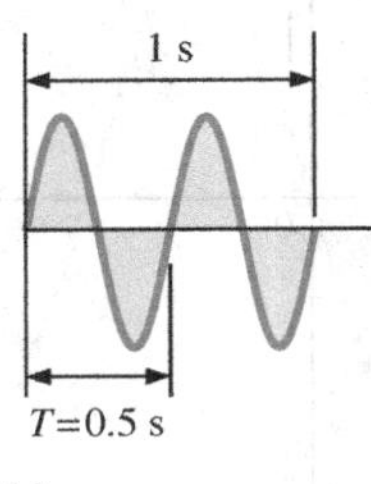

FIG. 13.5
Demonstrating the effect of a changing frequency on the period of a sinusoidal waveform.

The unit of measure for frequency is the *hertz* (Hz), where

$$\boxed{1 \text{ hertz (Hz)} = 1 \text{ cycle per second (cps)}} \qquad (13.1)$$

The unit hertz is derived from the surname of Heinrich Rudolph Hertz (Fig. 13.6), who did original research in the area of alternating currents and voltages and their effect on the basic R, L, and C elements. The frequency standard for North America is 60 Hz, whereas for Europe it is predominantly 50 Hz.

As with all standards, any variation from the norm will cause difficulties. In 1993, Berlin, Germany, received all its power from plants generating ac voltages whose output frequency was varying between 50.03 Hz and 51 Hz. The result was that clocks were gaining as much as 4 minutes a day. Alarms went off too soon, VCRs clicked off before the end of the program, and so on, requiring that clocks be continually reset. In 1994, however, when power was linked with the rest of Europe, the precise standard of 50 Hz was reestablished and everyone was on time again.

FIG. 13.6
Heinrich Rudolph Hertz.
(© Hulton Archive / Stringer)

German (Hamburg, Berlin, Karlsruhe)
(1857–94)
Physicist
Professor of Physics, Karlsruhe Polytechnic and University of Bonn

Spurred on by the earlier predictions of the English physicist James Clerk Maxwell, Heinrich Hertz produced *electromagnetic waves* in his laboratory at the Karlsruhe Polytechnic while in his early 30s. The rudimentary *transmitter* and *receiver* were in essence the first to broadcast and receive radio waves. He was able to measure the *wavelength* of the electromagnetic waves and confirmed that the *velocity of propagation* is in the same order of magnitude as that of light. In addition, he demonstrated that the *reflective* and *refractive* properties of electromagnetic waves are the same as those for heat and light waves. It was indeed unfortunate that such an ingenious, industrious individual should pass away at the very early age of 37 due to a bone disease.

EXAMPLE 13.1 For the sinusoidal waveform in Fig. 13.7.

a. What is the peak value?
b. What is the instantaneous value at 0.3 s and 0.6 s?
c. What is the peak-to-peak value of the waveform?
d. What is the period of the waveform?
e. How many cycles are shown?
f. What is the frequency of the waveform?

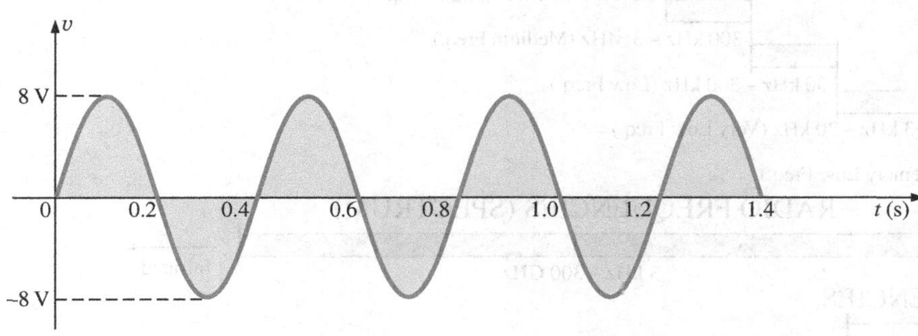

FIG. 13.7
Example 13.1.

Solutions:

a. **8 V.**
b. At 0.3 s, **−8 V;** at 0.6 s, **0 V.**
c. **16 V.**
d. **0.4 s.**
e. **3.5 cycles.**
f. **2.5 cps,** or **2.5 Hz.**

13.3 FREQUENCY SPECTRUM

Using a log scale (described in detail in Chapter 20), we can scale a frequency spectrum from 1 Hz to 1000 GHz on the same axis, as shown in Fig. 13.8. A number of terms in the various portions of the spectrum are probably familiar to you from everyday experiences. Note that the audio range (human ear) extends from only 15 Hz to 20 kHz, but the transmission of radio signals can occur between 3 kHz and 300 GHz. The uniform process of defining the intervals of the radio-frequency spectrum from VLF to EHF is quite evident from the length of the bars in the figure (although keep in mind that it is a log scale, so the frequencies encompassed within each segment are quite different). Other frequencies of particular interest (TV, CB, microwave, and so on) are also included for reference purposes. Although it is numerically easy to talk about frequencies in the megahertz and gigahertz range, keep in mind that a frequency of 100 MHz, for instance, represents a sinusoidal waveform that passes through 100,000,000 cycles in only 1 s—an incredible number when we compare it to the 60 Hz of our conventional power sources. The Intel® Core 2 Extreme processor can run at speeds of 3 GHz. Imagine a product able to handle 3 billion instructions per second—an incredible achievement.

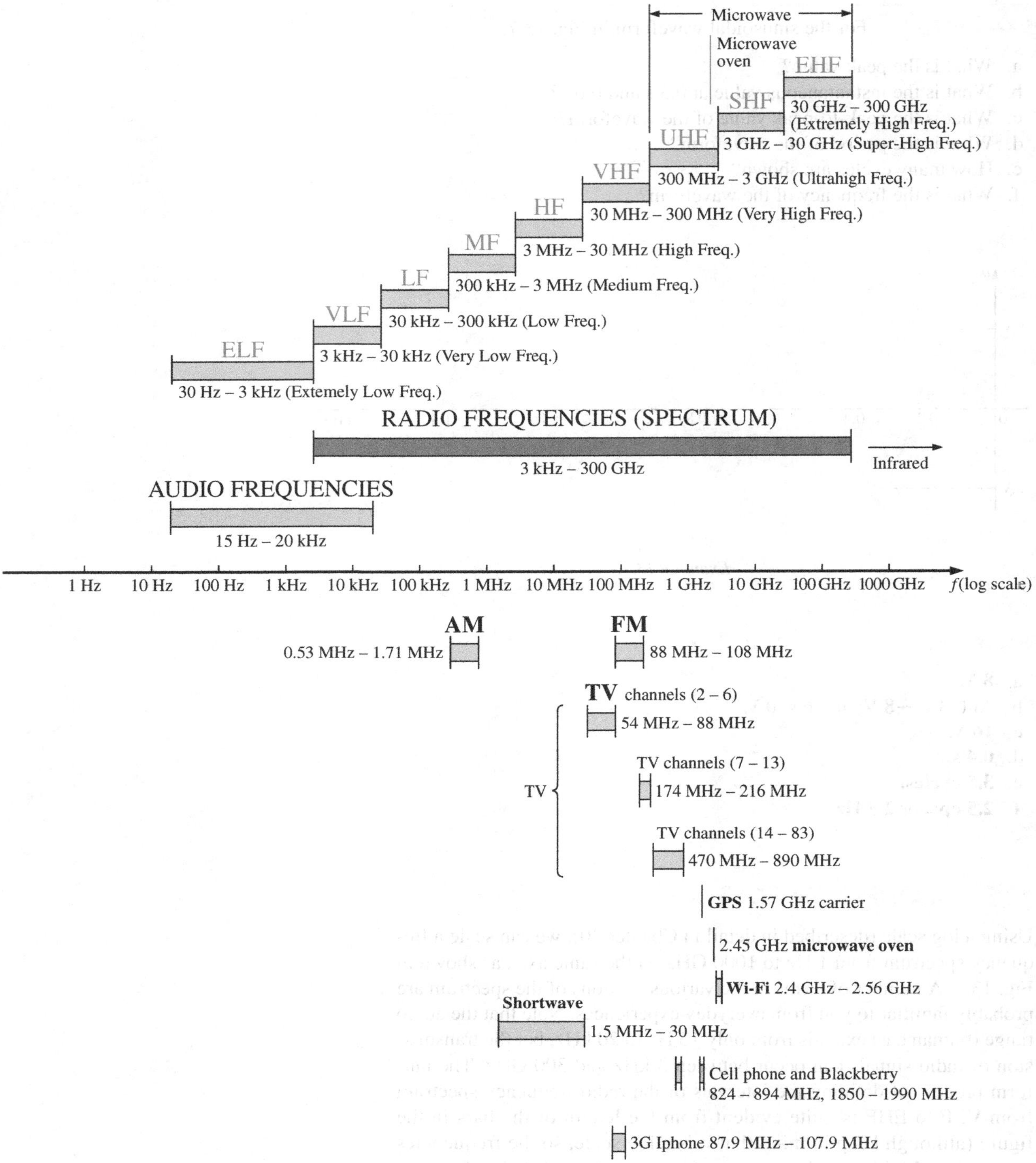

Microwave
Microwave oven
EHF
30 GHz – 300 GHz (Extremely High Freq.)
SHF
3 GHz – 30 GHz (Super-High Freq.)
UHF
300 MHz – 3 GHz (Ultrahigh Freq.)
VHF
30 MHz – 300 MHz (Very High Freq.)
HF
3 MHz – 30 MHz (High Freq.)
MF
300 kHz – 3 MHz (Medium Freq.)
LF
30 kHz – 300 kHz (Low Freq.)
VLF
3 kHz – 30 kHz (Very Low Freq.)
ELF
30 Hz – 3 kHz (Extemely Low Freq.)

RADIO FREQUENCIES (SPECTRUM)
3 kHz – 300 GHz
Infrared

AUDIO FREQUENCIES
15 Hz – 20 kHz

1 Hz | 10 Hz | 100 Hz | 1 kHz | 10 kHz | 100 kHz | 1 MHz | 10 MHz | 100 MHz | 1 GHz | 10 GHz | 100 GHz | 1000 GHz | f (log scale)

AM
0.53 MHz – 1.71 MHz

FM
88 MHz – 108 MHz

TV
TV channels (2 – 6)
54 MHz – 88 MHz
TV channels (7 – 13)
174 MHz – 216 MHz
TV channels (14 – 83)
470 MHz – 890 MHz

GPS 1.57 GHz carrier

2.45 GHz **microwave oven**

Wi-Fi 2.4 GHz – 2.56 GHz

Shortwave
1.5 MHz – 30 MHz

Cell phone and Blackberry
824 – 894 MHz, 1850 – 1990 MHz

3G Iphone 87.9 MHz – 107.9 MHz

FIG. 13.8
Areas of application for specific frequency bands.

Since the frequency is inversely related to the period—that is, as one increases, the other decreases by an equal amount—the two can be related by the following equation:

$$f = \frac{1}{T} \qquad \begin{aligned} f &= \text{Hz} \\ T &= \text{seconds (s)} \end{aligned} \tag{13.2}$$

or

$$T = \frac{1}{f} \tag{13.3}$$

EXAMPLE 13.2 Find the period of periodic waveform with a frequency of

a. 60 Hz.
b. 1000 Hz.

Solutions:

a. $T = \dfrac{1}{f} = \dfrac{1}{60\text{ Hz}} \cong 0.01667\text{ s or } \mathbf{16.67\ ms}$

(a recurring value since 60 Hz is so prevalent)

b. $T = \dfrac{1}{f} = \dfrac{1}{1000\text{ Hz}} = 10^{-3}\text{ s} = \mathbf{1\ ms}$

EXAMPLE 13.3 Determine the frequency of the waveform in Fig. 13.9.

Solution: From the figure, $T = (25\text{ ms} - 5\text{ ms})$ or $(35\text{ ms} - 15\text{ ms}) = 20\text{ ms}$, and

$$f = \frac{1}{T} = \frac{1}{20 \times 10^{-3}\text{ s}} = \mathbf{50\ Hz}$$

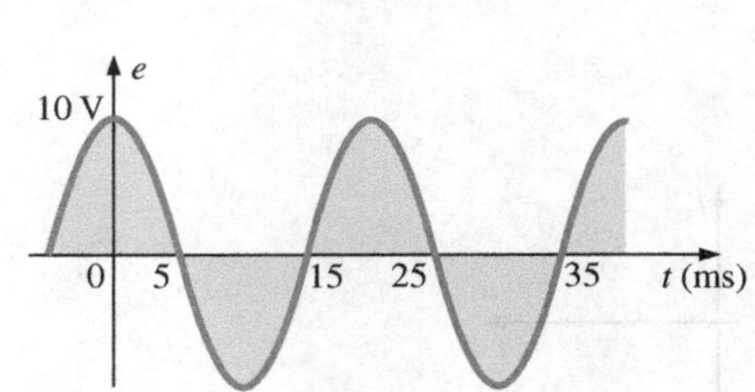

FIG. 13.9
Example 13.3.

In Fig. 13.10, the seismogram resulting from a seismometer near an earthquake is displayed. Prior to the disturbance, the waveform has a relatively steady level, but as the event is about to occur, the frequency begins

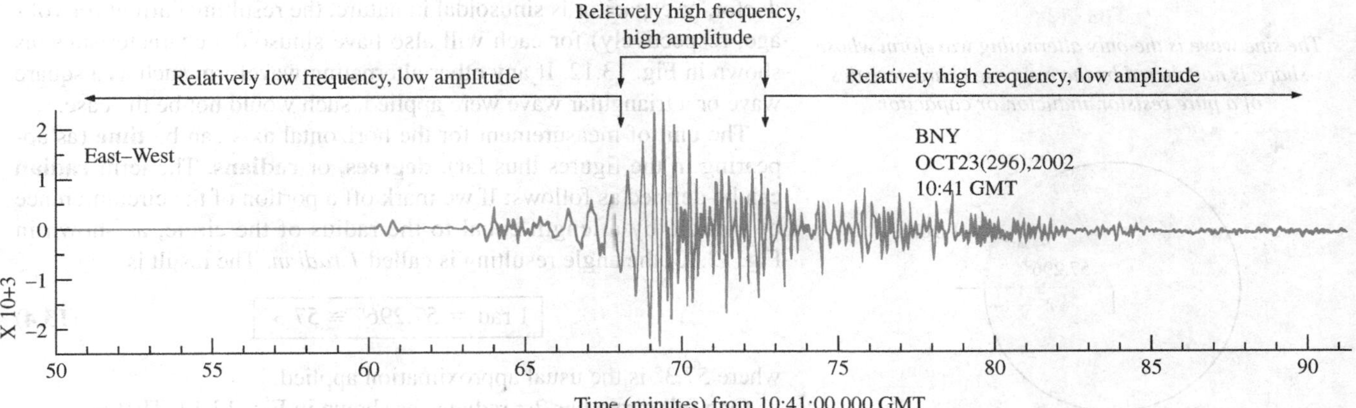

FIG. 13.10
Seismogram from station BNY (Binghamton University) in New York due to magnitude 6.7 earthquake in Central Alaska that occurred at 63.62°N, 148.04°W, with a depth of 10 km, on Wednesday, October 23, 2002.

to increase along with the amplitude. Finally, the earthquake occurs, and the frequency and the amplitude increase dramatically. In other words, the relative frequencies can be determined simply by looking at the tightness of the waveform and the associated period. The change in amplitude is immediately obvious from the resulting waveform. The fact that the earthquake lasts for only a few minutes is clear from the horizontal scale.

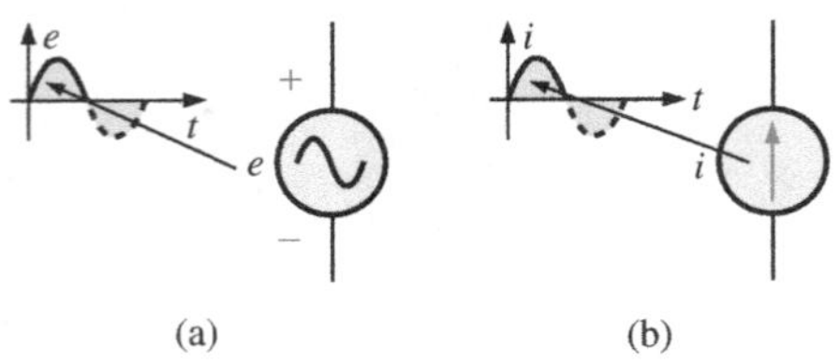

FIG. 13.11
(a) Sinusoidal ac voltage sources; (b) sinusoidal current sources.

Defined Polarities and Direction

You may be wondering how a polarity for a voltage or a direction for a current can be established if the waveform moves back and forth from the positive to the negative region. For a period of time, a voltage has one polarity, while for the next equal period it reverses. To take care of this problem, a positive sign is applied if the voltage is above the axis, as shown in Fig. 13.11(a). For a current source, the direction in the symbol corresponds with the positive region of the waveform, as shown in Fig. 13.11(b).

For any quantity that will not change with time, an uppercase letter such as V or I is used. For expressions that are time dependent or that represent a particular instant of time, a lowercase letter such as e or i is used.

The need for defining polarities and current direction becomes quite obvious when we consider multisource ac networks. Note in the last sentence the absence of the term *sinusoidal* before the phrase *ac networks*. This phrase will be used to an increasing degree as we progress; *sinusoidal* is to be understood unless otherwise indicated.

13.4 THE SINUSOIDAL WAVEFORM

The terms defined in the previous section can be applied to any type of periodic waveform, whether smooth or discontinuous. The sinusoidal waveform is of particular importance, however, since it lends itself readily to the mathematics and the physical phenomena associated with electric circuits. Consider the power of the following statement:

The sinusoidal waveform is the only alternating waveform whose shape is unaffected by the response characteristics of R, L, and C elements.

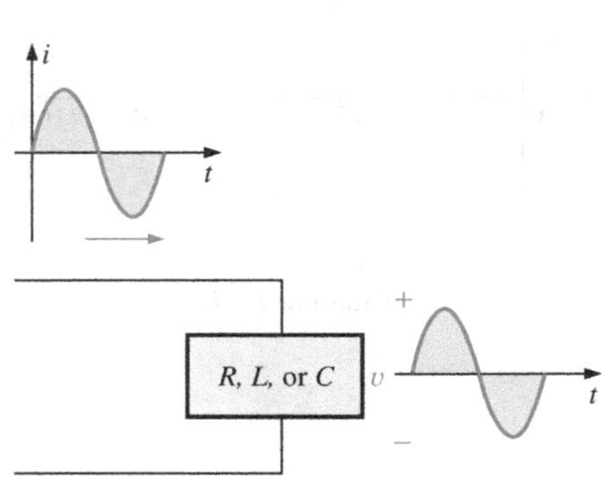

FIG. 13.12
The sine wave is the only alternating waveform whose shape is not altered by the response characteristics of a pure resistor, inductor, or capacitor.

In other words, if the voltage across (or current through) a resistor, inductor, or capacitor is sinusoidal in nature, the resulting current (or voltage, respectively) for each will also have sinusoidal characteristics, as shown in Fig. 13.12. If any other alternating waveform such as a square wave or a triangular wave were applied, such would not be the case.

The unit of measurement for the horizontal axis can be **time** (as appearing in the figures thus far), **degrees,** or **radians.** The term **radian** can be defined as follows: If we mark off a portion of the circumference of a circle by a length equal to the radius of the circle, as shown in Fig. 13.13, the angle resulting is called *1 radian.* The result is

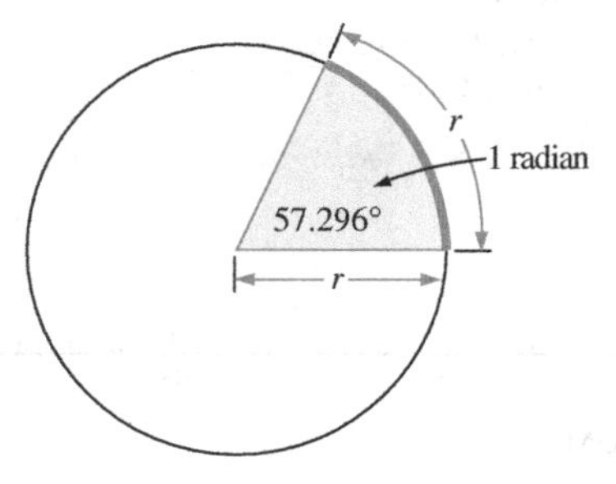

FIG. 13.13
Defining the radian.

$$\boxed{1 \text{ rad} = 57.296^\circ \cong 57.3^\circ} \tag{13.4}$$

where 57.3° is the usual approximation applied.

One full circle has 2π radians, as shown in Fig. 13.14. That is,

$$\boxed{2\pi \text{ rad} = 360^\circ} \tag{13.5}$$

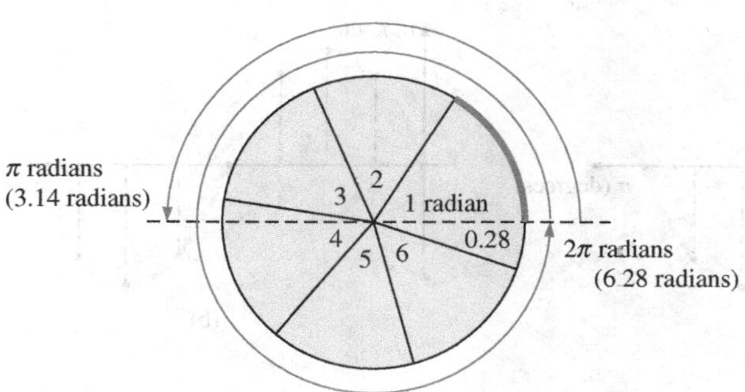

FIG. 13.14
There are 2π radians in one full circle of 360°.

so that $$2\pi = 2(3.142) = 6.28$$
and $$2\pi(57.3^\circ) = 6.28(57.3^\circ) = 359.84^\circ \cong 360^\circ$$

A number of electrical formulas contain a multiplier of π. For this reason, it is sometimes preferable to measure angles in radians rather than in degrees.

The quantity π is the ratio of the circumference of a circle to its diameter.

π has been determined to an extended number of places, primarily in an attempt to see if a repetitive sequence of numbers appears. It does not. A sampling of the effort appears below:

$$\pi = 3.14159\ 26535\ 89793\ 23846\ 26433\ldots$$

Although the approximation $\pi \cong 3.14$ is often applied, all the calculations in the text use the π function as provided on all scientific calculators.

For 180° and 360°, the two units of measurement are related as shown in Fig. 13.14. The conversions equations between the two are the following:

$$\text{Radians} = \left(\frac{\pi}{180^\circ}\right) \times (\text{degrees}) \qquad \textbf{(13.6)}$$

$$\text{Degrees} = \left(\frac{180^\circ}{\pi}\right) \times (\text{radians}) \qquad \textbf{(13.7)}$$

Applying these equations, we find

$$\mathbf{90^\circ}\text{: Radians} = \frac{\pi}{180^\circ}(90^\circ) = \frac{\boldsymbol{\pi}}{\mathbf{2}}\ \textbf{rad}$$

$$\mathbf{30^\circ}\text{: Radians} = \frac{\pi}{180^\circ}(30^\circ) = \frac{\boldsymbol{\pi}}{\mathbf{6}}\ \textbf{rad}$$

$$\frac{\boldsymbol{\pi}}{\mathbf{3}}\ \textbf{rad: }\text{Degrees} = \frac{180^\circ}{\pi}\left(\frac{\pi}{3}\right) = \mathbf{60^\circ}$$

$$\frac{\mathbf{3}\boldsymbol{\pi}}{\mathbf{2}}\ \textbf{rad: }\text{Degrees} = \frac{180^\circ}{\pi}\left(\frac{3\pi}{2}\right) = \mathbf{270^\circ}$$

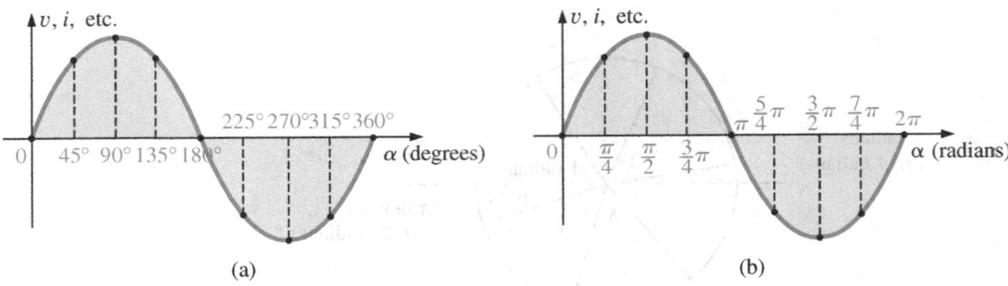

FIG. 13.15
Plotting a sine wave versus (a) degrees and (b) radians.

For comparison purposes, two sinusoidal voltages are plotted in Fig. 13.15 using degrees and radians as the units of measurement for the horizontal axis.

It is of particular interest that the sinusoidal waveform can be derived from the length of the *vertical projection* of a radius vector rotating in a uniform circular motion about a fixed point. Starting as shown in Fig. 13.16(a) and plotting the amplitude (above and below zero) on the coordinates drawn to the right [Figs. 13.16(b) through (i)], we will trace a complete sinusoidal waveform after the radius vector has completed a 360° rotation about the center.

The velocity with which the radius vector rotates about the center, called the **angular velocity,** can be determined from the following equation:

$$\text{Angular velocity} = \frac{\text{distance (degrees or radians)}}{\text{time (seconds)}} \tag{13.8}$$

Substituting into Eq. (13.8) and assigning the lowercase Greek letter *omega* (ω) to the angular velocity, we have

$$\omega = \frac{\alpha}{t} \tag{13.9}$$

and

$$\alpha = \omega t \tag{13.10}$$

Since ω is typically provided in radians per second, the angle α obtained using Eq. (13.10) is usually in radians. If α is required in degrees, Eq. (13.7) must be applied. The importance of remembering the above will become obvious in the examples to follow.

In Fig. 13.16, the time required to complete one revolution is equal to the period (T) of the sinusoidal waveform in Fig. 13.16(i). The radians subtended in this time interval are 2π. Substituting, we have

$$\omega = \frac{2\pi}{T} \quad \text{(rad/s)} \tag{13.11}$$

In words, this equation states that the smaller the period of the sinusoidal waveform of Fig. 13.16(i), or the smaller the time interval before one complete cycle is generated, the greater must be the angular velocity of the rotating radius vector. Certainly this statement agrees with what we have learned thus far. We can now go one step further and apply the

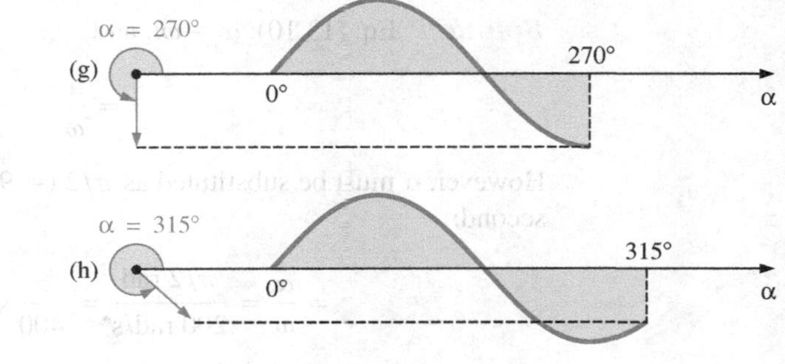

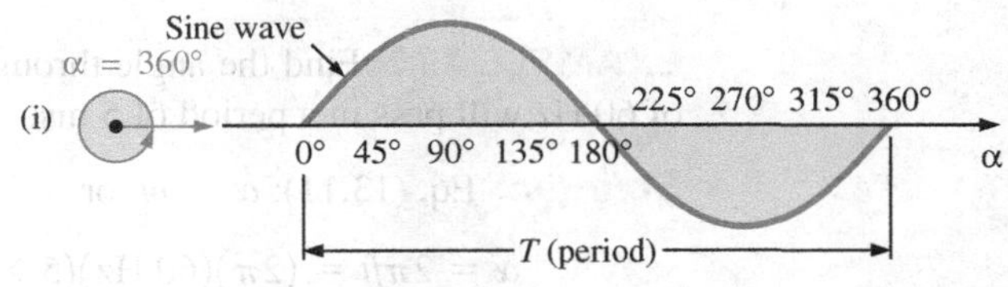

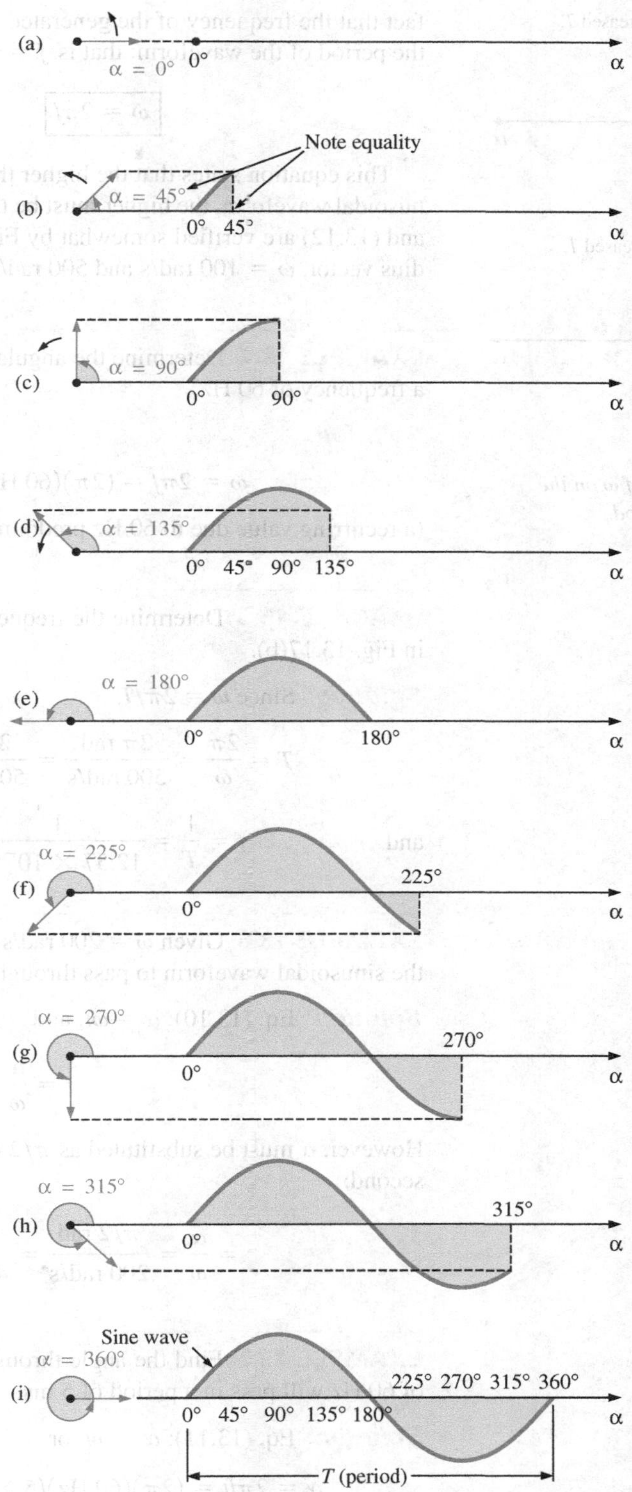

FIG. 13.16
Generating a sinusoidal waveform through the vertical projection of a rotating vector.

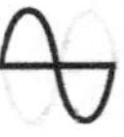

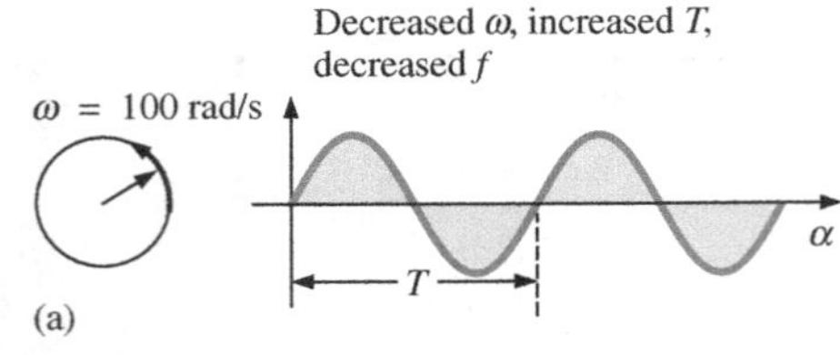

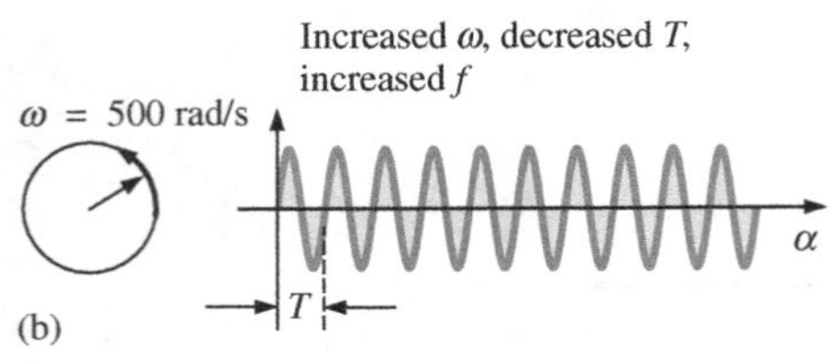

FIG. 13.17
Demonstrating the effect of ω on the frequency and period.

fact that the frequency of the generated waveform is inversely related to the period of the waveform; that is, $f = 1/T$. Thus,

$$\boxed{\omega = 2\pi f} \qquad \text{(rad/s)} \qquad \textbf{(13.12)}$$

This equation states that the higher the frequency of the generated sinusoidal waveform, the higher must be the angular velocity. Eqs. (13.11) and (13.12) are verified somewhat by Fig. 13.17, where for the same radius vector, $\omega = 100$ rad/s and 500 rad/s.

EXAMPLE 13.4 Determine the angular velocity of a sine wave having a frequency of 60 Hz.

Solution:

$$\omega = 2\pi f = (2\pi)(60 \text{ Hz}) \cong \mathbf{377\ rad/s}$$

(a recurring value due to 60 Hz predominance).

EXAMPLE 13.5 Determine the frequency and period of the sine wave in Fig. 13.17(b).

Solution: Since $\omega = 2\pi/T$,

$$T = \frac{2\pi}{\omega} = \frac{2\pi \text{ rad}}{500 \text{ rad/s}} = \frac{2\pi \text{ rad}}{500 \text{ rad/s}} = \mathbf{12.57\ ms}$$

and

$$f = \frac{1}{T} = \frac{1}{12.57 \times 10^{-3} \text{ s}} = \mathbf{79.58\ Hz}$$

EXAMPLE 13.6 Given $\omega = 200$ rad/s, determine how long it will take the sinusoidal waveform to pass through an angle of 90°.

Solution: Eq. (13.10): $\alpha = \omega t$, and

$$t = \frac{\alpha}{\omega}$$

However, α must be substituted as $\pi/2$ (= 90°) since ω is in radians per second:

$$t = \frac{\alpha}{\omega} = \frac{\pi/2 \text{ rad}}{200 \text{ rad/s}} = \frac{\pi}{400} \text{ s} = \mathbf{7.85\ ms}$$

EXAMPLE 13.7 Find the angle through which a sinusoidal waveform of 60 Hz will pass in a period of 5 ms.

Solution: Eq. (13.11): $\alpha = \omega t$, or

$$\alpha = 2\pi f t = (2\pi)(60 \text{ Hz})(5 \times 10^{-3} \text{ s}) = \mathbf{1.885\ rad}$$

If not careful, you might be tempted to interpret the answer as 1.885°. However,

$$\alpha(°) = \frac{180°}{\pi \cancel{\text{rad}}}(1.885 \cancel{\text{rad}}) = \mathbf{108°}$$

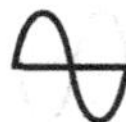

13.5 GENERAL FORMAT FOR THE SINUSOIDAL VOLTAGE OR CURRENT

The basic mathematical format for the sinusoidal waveform is

$$\boxed{A_m \sin \alpha} \qquad \textbf{(13.13)}$$

where A_m is the peak value of the waveform and α is the unit of measure for the horizontal axis, as shown in Fig. 13.18.

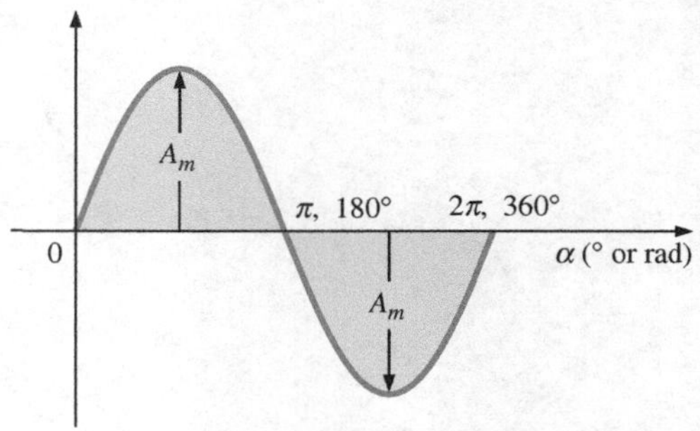

FIG. 13.18
Basic sinusoidal function.

The equation $\alpha = \omega t$ states that the angle α through which the rotating vector in Fig. 13.16 will pass is determined by the angular velocity of the rotating vector and the length of time the vector rotates. For example, for a particular angular velocity (fixed ω), the longer the radius vector is permitted to rotate (that is, the greater the value of t), the greater is the number of degrees or radians through which the vector will pass. Relating this statement to the sinusoidal waveform, we have that, for a particular angular velocity, the longer the time, the greater is the number of cycles shown. For a fixed time interval, the greater is the angular velocity, the greater is the number of cycles generated.

Due to Eq. (13.10), the general format of a sine wave can also be written

$$\boxed{A_m \sin \omega t} \qquad \textbf{(13.14)}$$

with ωt as the horizontal unit of measure.

For electrical quantities such as current and voltage, the general format is

$$i = I_m \sin \omega t = I_m \sin \alpha$$
$$e = E_m \sin \omega t = E_m \sin \alpha$$

where the capital letters with the subscript m represent the amplitude, and the lowercase letters i and e represent the instantaneous value of current and voltage, respectively, at any time t. This format is particularly important because it presents the sinusoidal voltage or current as a function of time, which is the horizontal scale for the oscilloscope. Recall that the horizontal sensitivity of a scope is in time per division, not degrees per centimeter.

EXAMPLE 13.8 Given $e = 5 \sin \alpha$, determine e at $\alpha = 40°$ and $\alpha = 0.8\pi$.

Solution: For $\alpha = 40°$,

$$e = 5 \sin 40° = 5(0.6428) = \mathbf{3.21\ V}$$

For $\alpha = 0.8\pi$,

$$\alpha(°) = \frac{180°}{\pi}(0.8\ \pi) = 144°$$

and
$$e = 5 \sin 144° = 5(0.5878) = \mathbf{2.94\ V}$$

The angle at which a particular voltage level is attained can be determined by rearranging the equation

$$e = E_m \sin \alpha$$

in the following manner:

$$\sin\alpha = \frac{e}{E_m}$$

which can be written

$$\boxed{\alpha = \sin^{-1}\frac{e}{E_m}} \qquad \textbf{(13.15)}$$

Similarly, for a particular current level,

$$\boxed{\alpha = \sin^{-1}\frac{i}{I_m}} \qquad \textbf{(13.16)}$$

EXAMPLE 13.9

a. Determine the angle at which the magnitude of the sinusoidal function $v = 10 \sin 377t$ is 4 V.
b. Determine the time at which the magnitude is attained.

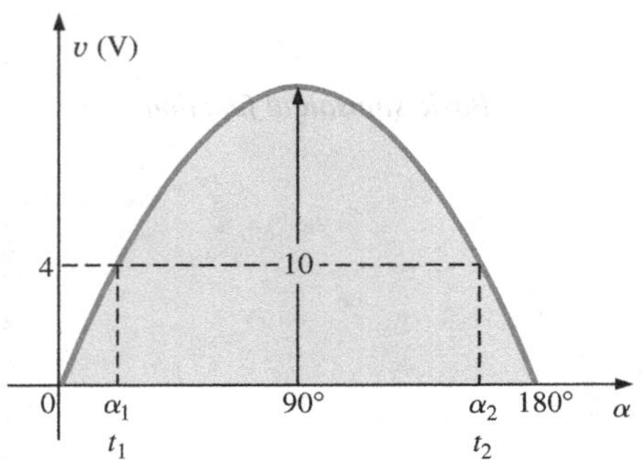

FIG. 13.19
Example 13.9.

Solutions:

a. Eq. (13.15):

$$\alpha_1 = \sin^{-1}\frac{v}{E_m} = \sin^{-1}\frac{4\text{ V}}{10\text{ V}} = \sin^{-1}0.4 = \mathbf{23.58^\circ}$$

However, Fig. 13.19 reveals that the magnitude of 4 V (positive) will be attained at two points between 0° and 180°. The second intersection is determined by

$$\alpha_2 = 180^\circ - 23.578^\circ = \mathbf{156.42^\circ}$$

In general, therefore, keep in mind that Eqs. (13.15) and (13.16) will provide an angle with a magnitude between 0° and 90°.

b. Eq. (13.10): $\alpha = \omega t$, and so $t = \alpha/\omega$. However, α must be in radians. Thus,

$$\alpha(\text{rad}) = \frac{\pi}{180^\circ}(23.578^\circ) = 0.412 \text{ rad}$$

and
$$t_1 = \frac{\alpha}{\omega} = \frac{0.412 \text{ rad}}{377 \text{ rad/s}} = \mathbf{1.09 \text{ ms}}$$

For the second intersection,

$$\alpha(\text{rad}) = \frac{\pi}{180^\circ}(156.422^\circ) = 2.73 \text{ rad}$$

$$t_2 = \frac{\alpha}{\omega} = \frac{2.73 \text{ rad}}{377 \text{ rad/s}} = \mathbf{7.24 \text{ ms}}$$

Calculator Operations

Both sin and $\sin^{-1}$ are available on all scientific calculators. You can also use them to work with the angle in degrees or radians without having to convert from one form to the other. That is, if the angle is in radians and the mode setting is for radians, you can enter the radian measure directly.

To set the DEGREE mode, proceed as outlined in Fig. 13.20(a) using the TI-89 calculator. The magnitude of the voltage e at 40° can then be found using the sequence in Fig. 13.20(b).

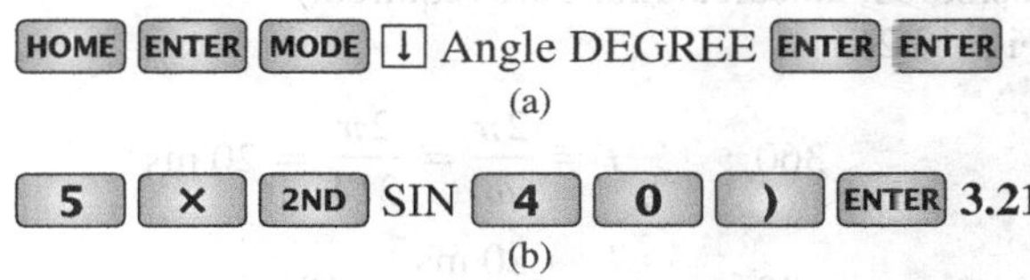

FIG. 13.20
(a) Setting the DEGREE mode; (b) evaluating 5 sin 40°.

After establishing the RADIAN mode, the sequence in Fig. 13.21 determines the voltage at 0.8π.

5 × 2ND SIN 0 • 8 2ND π) ENTER **2.94**

FIG. 13.21
Finding $e = 5 \sin 0.8\pi$ using the calculator in the RADIAN mode.

Finally, the angle in degrees for α_1 in part (a) of Example 13.9 can be determined by the sequence in Fig. 13.22 with the mode set in degrees, whereas the angle in radians for part (a) of Example 13.9 can be determined by the sequence in Fig. 13.23 with the mode set in radians.

♦ SIN^{-1} 4 ÷ 1 0) ENTER **23.60**

FIG. 13.22
Finding $\alpha_1 = \sin^{-1}(4/10)$ using the calculator in the DEGREE mode.

♦ SIN^{-1} 4 ÷ 1 0) ENTER **0.41**

FIG. 13.23
Finding $\alpha_1 = \sin^{-1}(4/10)$ using the calculator in the RADIAN mode.

The sinusoidal waveform can also be plotted against *time* on the horizontal axis. The time period for each interval can be determined from $t = \alpha/\omega$, but the most direct route is simply to find the period T from $T = 1/f$ and break it up into the required intervals. This latter technique is demonstrated in Example 13.10.

Before reviewing the example, take special note of the relative simplicity of the mathematical equation that can represent a sinusoidal waveform. Any alternating waveform whose characteristics differ from those of the sine wave cannot be represented by a single term, but may require two, four, six, or perhaps an infinite number of terms to be represented accurately.

EXAMPLE 13.10 Sketch $e = 10 \sin 314t$ with the abscissa

a. angle (α) in degrees.
b. angle (α) in radians.
c. time (t) in seconds.

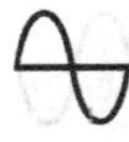

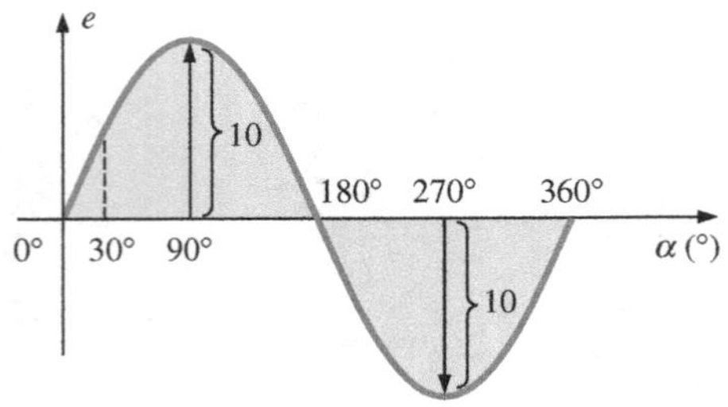

FIG. 13.24
Example 13.10, horizontal axis in degrees.

Solutions:

a. See Fig. 13.24. (Note that no calculations are required.)
b. See Fig. 13.25. (Once the relationship between degrees and radians is understood, no calculations are required.)
c. See Fig. 13.26.

$$360°: \quad T = \frac{2\pi}{\omega} = \frac{2\pi}{314} = 20 \text{ ms}$$

$$180°: \quad \frac{T}{2} = \frac{20 \text{ ms}}{2} = 10 \text{ ms}$$

$$90°: \quad \frac{T}{4} = \frac{20 \text{ ms}}{4} = 5 \text{ ms}$$

$$30°: \quad \frac{T}{12} = \frac{20 \text{ ms}}{12} = 1.67 \text{ ms}$$

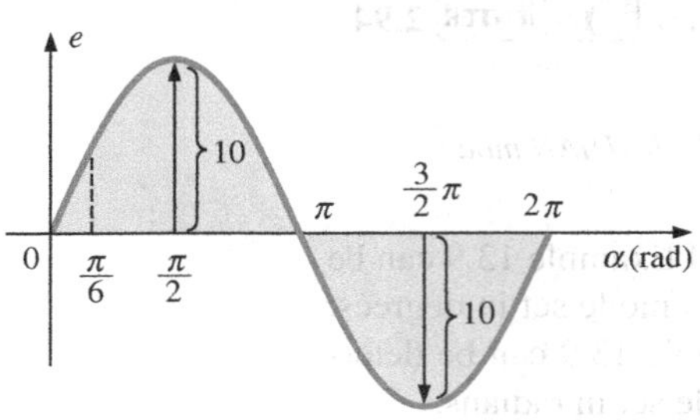

FIG. 13.25
Example 13.10, horizontal axis in radians.

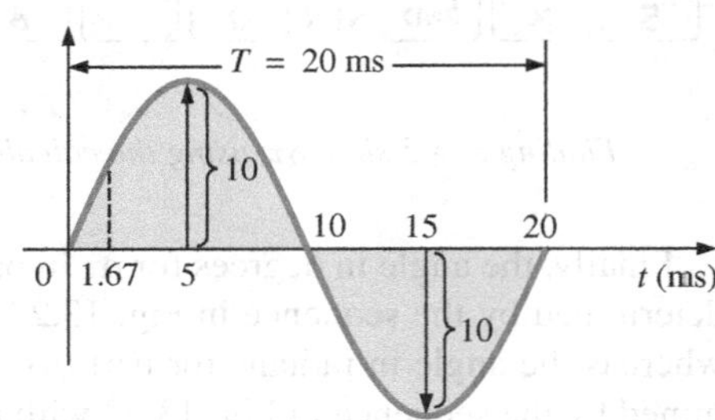

FIG. 13.26
Example 13.10, horizontal axis in milliseconds.

EXAMPLE 13.11 Given $i = 6 \times 10^{-3} \sin 1000t$, determine i at $t = 2$ ms.

Solution:

$$\alpha = \omega t = 1000t = (1000 \text{ rad/s})(2 \times 10^{-3} \text{ s}) = 2 \text{ rad}$$

$$\alpha(°) = \frac{180°}{\pi \text{ rad}}(2 \text{ rad}) = 114.59°$$

$$i = (6 \times 10^{-3})(\sin 114.59°) = (6 \text{ mA})(0.9093) = \mathbf{5.46 \text{ mA}}$$

13.6 PHASE RELATIONS

Thus far, we have considered only sine waves that have maxima at $\pi/2$ and $3\pi/2$, with a zero value at 0, π, and 2π, as shown in Fig. 13.25. If the waveform is shifted to the right or left of 0°, the expression becomes

$$\boxed{A_m \sin(\omega t \pm \theta)} \qquad \textbf{(13.17)}$$

where θ is the angle in degrees or radians that the waveform has been shifted.

If the waveform passes through the horizontal axis with a *positive-going* (increasing with time) slope *before* 0°, as shown in Fig. 13.27, the expression is

$$\boxed{A_m \sin(\omega t + \theta)} \qquad \textbf{(13.18)}$$

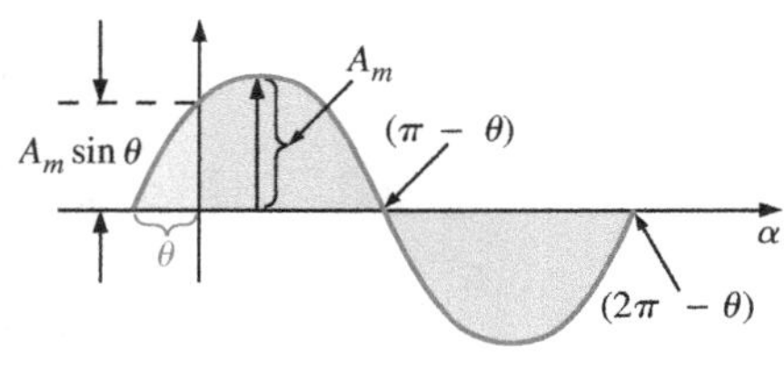

FIG. 13.27
Defining the phase shift for a sinusoidal function that crosses the horizontal axis with a positive slope before 0°.

At $\omega t = \alpha = 0°$, the magnitude is determined by $A_m \sin \theta$. If the waveform passes through the horizontal axis with a positive-going slope *after* 0°, as shown in Fig. 13.28, the expression is

$$\boxed{A_m \sin(\omega t - \theta)} \tag{13.19}$$

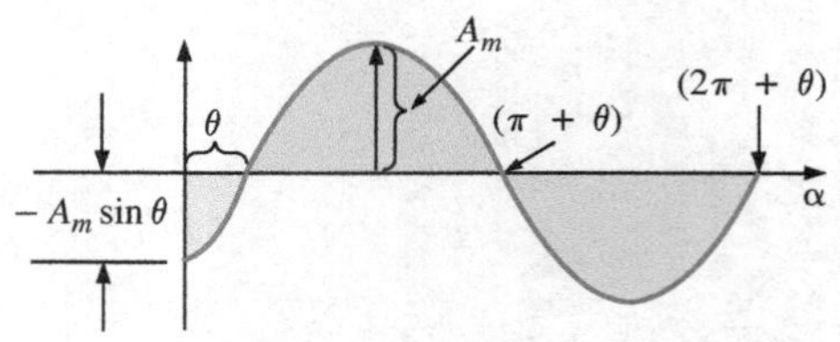

FIG. 13.28
Defining the phase shift for a sinusoidal function that crosses the horizontal axis with a positive slope after 0°.

Finally, at $\omega t = \alpha = 0°$, the magnitude is $A_m \sin(-\theta)$, which, by a trigonometric identity, is $-A_m \sin \theta$.

If the waveform crosses the horizontal axis with a positive-going slope 90° ($\pi/2$) sooner, as shown in Fig. 13.29, it is called a *cosine wave;* that is,

$$\boxed{\sin(\omega t + 90°) = \sin\left(\omega t + \frac{\pi}{2}\right) = \cos \omega t} \tag{13.20}$$

or

$$\boxed{\sin \omega t = \cos(\omega t - 90°) = \cos\left(\omega t - \frac{\pi}{2}\right)} \tag{13.21}$$

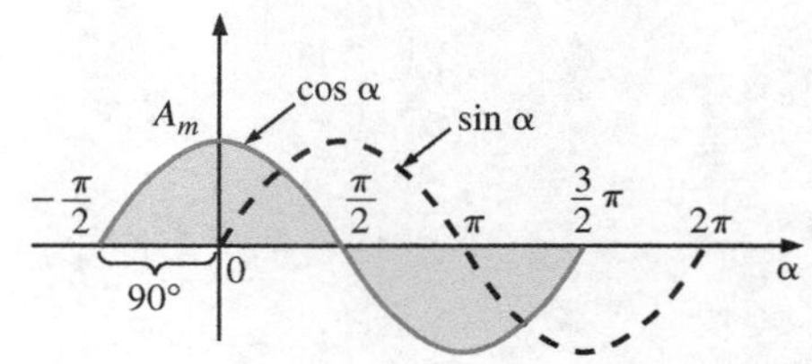

FIG. 13.29
Phase relationship between a sine wave and a cosine wave.

The terms **leading** and **lagging** are used to indicate the relationship between two sinusoidal waveforms of the *same frequency* plotted on the same set of axes. In Fig. 13.29, the cosine curve is said to *lead* the sine curve by 90°, and the sine curve is said to *lag* the cosine curve by 90°. The 90° is referred to as the phase angle between the two waveforms. In language commonly applied, the waveforms are *out of phase* by 90°. Note that the phase angle between the two waveforms is measured between those two points on the horizontal axis through which each passes with the *same slope.* If both waveforms cross the axis at the same point with the same slope, they are *in phase.*

The geometric relationship between various forms of the sine and cosine functions can be derived from Fig. 13.30. For instance, starting at the $+\sin \alpha$ position, we find that $+\cos \alpha$ is an additional 90° in the counterclockwise direction. Therefore, $\cos \alpha = \sin(\alpha + 90°)$. For $-\sin \alpha$ we must travel 180° in the counterclockwise (or clockwise) direction so that $-\sin \alpha = \sin(\alpha \pm 180°)$, and so on, as listed below:

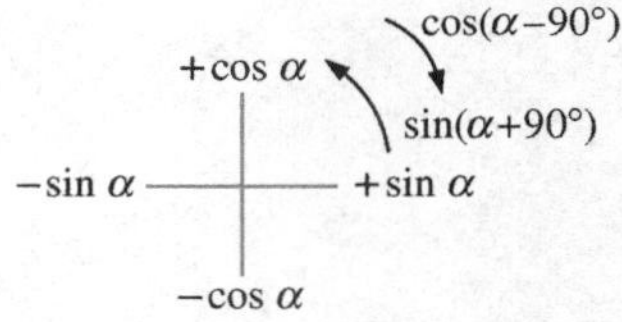

FIG. 13.30
Graphic tool for finding the relationship between specific sine and cosine functions.

$$\begin{aligned} \cos\alpha &= \sin(\alpha + 90°) \\ \sin\alpha &= \cos(\alpha - 90°) \\ -\sin\alpha &= \sin(\alpha \pm 180°) \\ -\cos\alpha &= \sin(\alpha + 270°) = \sin(\alpha - 90°) \\ &\text{etc.} \end{aligned} \tag{13.22}$$

In addition, note that

$$\begin{aligned} \sin(-\alpha) &= -\sin \alpha \\ \cos(-\alpha) &= \cos \alpha \end{aligned} \tag{13.23}$$

If a sinusoidal expression appears as

$$e = -E_m \sin \omega t$$

the negative sign is associated with the sine portion of the expression, not the peak value E_m. In other words, the expression, if not for convenience, would be written

$$e = E_m(-\sin \omega t)$$

Since $$-\sin\omega t = \sin(\omega t \pm 180°)$$

the expression can also be written

$$e = E_m \sin(\omega t \pm 180°)$$

revealing that a negative sign can be replaced by a 180° change in phase angle (+ or −); that is,

$$e = -E_m \sin\omega t = E_m \sin(\omega t + 180°) = E_m \sin(\omega t - 180°)$$

A plot of each will clearly show their equivalence. There are, therefore, two correct mathematical representations for the functions.

The **phase relationship** between two waveforms indicates which one leads or lags the other and by how many degrees or radians.

EXAMPLE 13.12 What is the phase relationship between the sinusoidal waveforms of each of the following sets?

a. $v = 10 \sin(\omega t + 30°)$
   $i = 5 \sin(\omega t + 70°)$
b. $i = 15 \sin(\omega t + 60°)$
   $v = 10 \sin(\omega t - 20°)$
c. $i = 2 \cos(\omega t + 10°)$
   $v = 3 \sin(\omega t - 10°)$
d. $i = -\sin(\omega t + 30°)$
   $v = 2 \sin(\omega t + 10°)$
e. $i = -2 \cos(\omega t - 60°)$
   $v = 3 \sin(\omega t - 150°)$

Solutions:

a. See Fig. 13.31.
   ***i* leads *v* by 40°, or *v* lags *i* by 40°.**

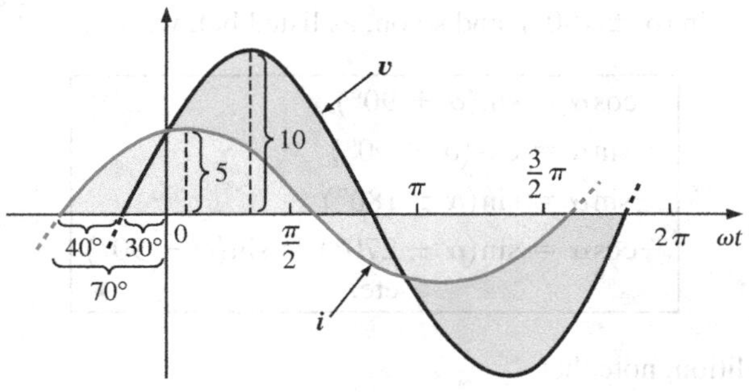

FIG. 13.31
Example 13.12(a): i leads v by 40°.

b. See Fig. 13.32.
   ***i* leads *v* by 80°, or *v* lags *i* by 80°.**

c. See Fig. 13.33.

$$i = 2\cos(\omega t + 10°) = 2\sin(\omega t + 10° + 90°)$$
$$= 2\sin(\omega t + 100°)$$

***i* leads *v* by 110°, or *v* lags *i* by 110°.**

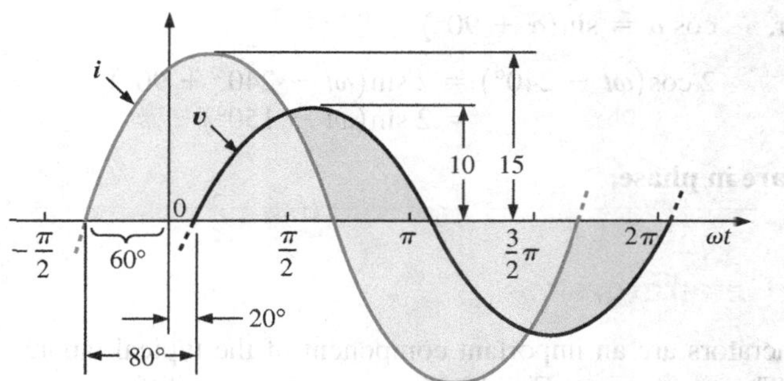

FIG. 13.32

Example 13.12(b): i leads v by 80°.

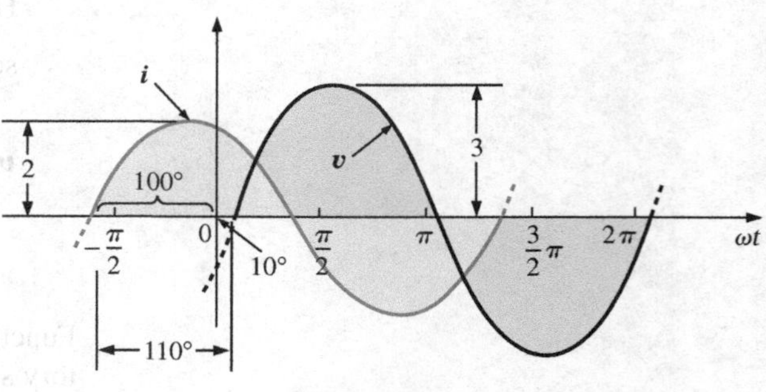

FIG. 13.33

Example 13.12(c): i leads v by 110°.

d. See Fig. 13.34.

$$-\sin(\omega t + 30°) = \sin(\omega t + 30° - 180°) \quad \text{(Note)}$$
$$= \sin(\omega t - 150°)$$

***v* leads *i* by 160°, or *i* lags *v* by 160°.**

Or using

$$-\sin(\omega t + 30°) = \sin(\omega t + 30° + 180°) \quad \text{(Note)}$$
$$= \sin(\omega t + 210°)$$

***i* leads *v* by 200°, or *v* lags *i* by 200°.**

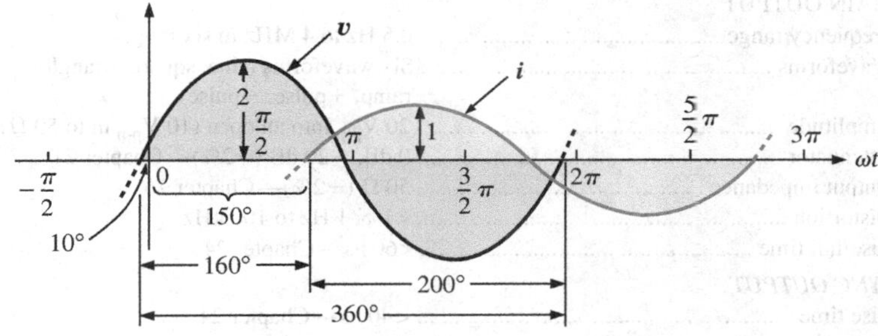

FIG. 13.34

Example 13.12(d): v leads i by 160°.

e. See Fig. 13.35.

$$i = -2\cos(\omega t - 60°) = 2\cos(\omega t - 60° - 180°) \quad \text{(By choice)}$$
$$= 2\cos(\omega t - 240°)$$

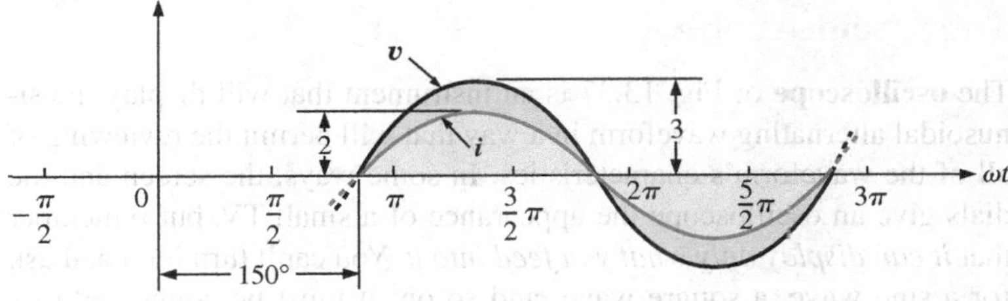

FIG. 13.35

Example 13.12(e): v and i are in phase.

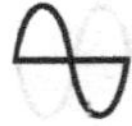

However, $\cos \alpha = \sin(\alpha + 90°)$

so that $2 \cos(\omega t - 240°) = 2 \sin(\omega t - 240° + 90°)$
$= 2 \sin(\omega t - 150°)$

***v* and *i* are in phase.**

Function Generators

Function generators are an important component of the typical laboratory setting. The generator of Fig. 13.36 can generate six different outputs; sine, triangular, and square wave, ramp, +pulse, and −pulse, with frequencies extending from 0.5 Hz to 4 MHz. However, as shown in the output listing, it has a maximum amplitude of 20 $V_{p\text{-}p}$. A number of other characteristics are included to demonstrate how the text will cover each in some detail.

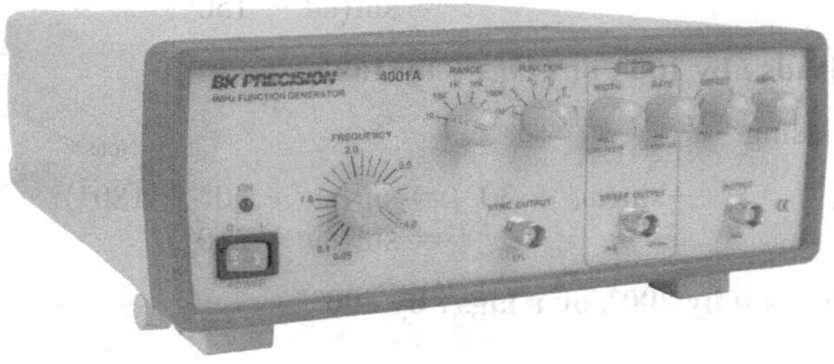

MAIN OUTPUT	
Frequency range	0.5 Hz to 4 MHz in six ranges
Waveforms	Six waveforms (sine, square, triangle, ramp, +pulse, −pulse)
Amplitude	20 $V_{p\text{-}p}$ into an open (10 $V_{p\text{-}p}$ in to 50 Ω)
Attenuator	0 dB, −20 dB (+2%)—Chapter 21
Output impedance	50 Ω (+2%)—Chapter 26
Distortion	<1%, 1 Hz to 100 kHz
Rise/fall time	<60 ns—Chapter 24
SYNC OUTPUT	
Rise time	<40 ns—Chapter 24
Waveforms	Square, pulse—Chapter 24
SWEEP	
Mode	Linear/log sweep—Chapter 21
Rate	From 10 ms to 5 s continuously variable
Sweep output	10 $V_{p\text{-}p}$ (open)
Output impedance	1 kΩ +2%—Chapter 26

FIG. 13.36
Function generator.
(B&K Precision Corp.)

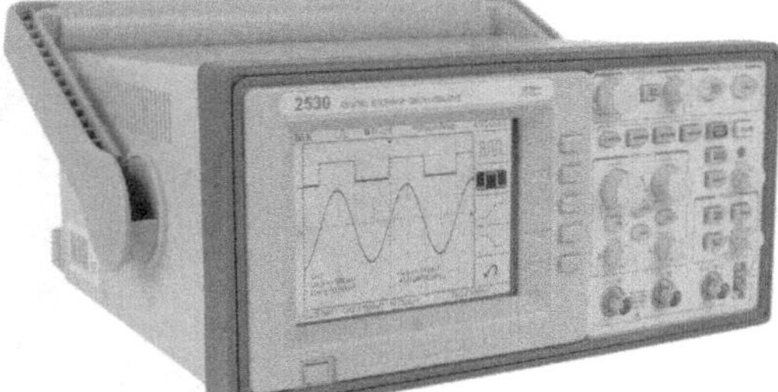

FIG. 13.37
Digital storage oscilloscope.
(B&K Precision Corp.)

The Oscilloscope

The **oscilloscope** of Fig. 13.37 is an instrument that will display the sinusoidal alternating waveform in a way that will permit the reviewing of all of the waveform's characteristics. In some ways, the screen and the dials give an oscilloscope the appearance of a small TV, but remember that *it can display only what you feed into it.* You can't turn it on and ask for a sine wave, a square wave, and so on; it must be connected to a source or an active circuit to pick up the desired waveform.

The screen has a standard appearance, with 10 horizontal divisions and 8 vertical divisions. The distance between divisions is 1 cm on the vertical

and horizontal scales, providing you with an excellent opportunity to become aware of the length of 1 cm. *The vertical scale is set to display voltage levels, whereas the horizontal scale is always in units of time.* The vertical sensitivity control sets the voltage level for each division, whereas the horizontal sensitivity control sets the time associated with each division. In other words, if the vertical sensitivity is set at 1 V/div., each division displays a 1 V swing, so that a total vertical swing of 8 divisions represents 8 V peak-to-peak. If the horizontal control is set on 10 μs/div., 4 divisions equal a time period of 40 μs. Remember, the oscilloscope display presents a sinusoidal voltage versus time, not degrees or radians. Further, the vertical scale is always a voltage sensitivity, never units of amperes.

The oscilloscope of Fig. 13.37 is a digital storage scope, where *storage* indicates that it can store waveform in digital form. The digital storage scope (DSO) is the standard for most laboratories today. At the input to the scope, an analog-to-digital converter (ADC) will convert the analog signal into digital at the rate of 250 MSa/s, or 250,000,000 samples per second—an enormous number—capable of picking up any distortion in the waveform.

EXAMPLE 13.13 Find the period, frequency, and peak value of the sinusoidal waveform appearing on the screen of the oscilloscope in Fig. 13.38. Note the sensitivities provided in the figure.

Solution: One cycle spans 4 divisions. Therefore, the period is

$$T = 4\ \text{div.}\left(\frac{50\ \mu\text{s}}{\text{div.}}\right) = \mathbf{200\ \mu s}$$

and the frequency is

$$f = \frac{1}{T} = \frac{1}{200 \times 10^{-6}\ \text{s}} = \mathbf{5\ kHz}$$

The vertical height above the horizontal axis encompasses 2 divisions. Therefore,

$$V_m = 2\ \text{div.}\left(\frac{0.1\ \text{V}}{\text{div.}}\right) = \mathbf{0.2\ V}$$

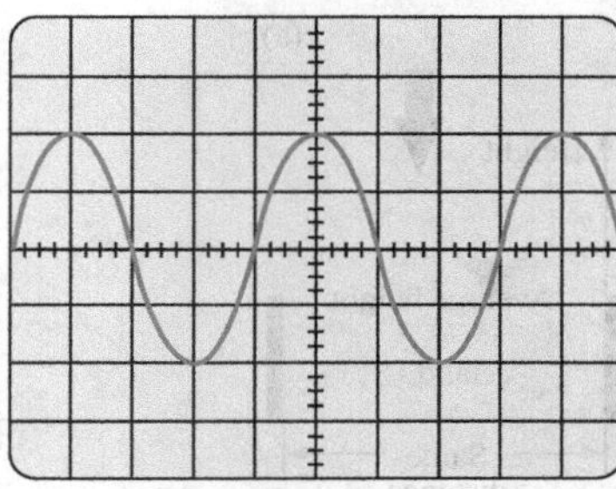

Vertical sensitivity=0.1 V/div.
Horizontal sensitivity=50 μs/div.

FIG. 13.38
Example 13.13.

An oscilloscope can also be used to make phase measurements between two sinusoidal waveforms. Virtually all laboratory oscilloscopes today have the dual-trace option, that is, the ability to show two waveforms at the same time. It is important to remember, however, that both waveforms will and must have the same frequency. The hookup procedure for using an oscilloscope to measure phase angles is covered in detail in Section 15.13. However, the equation for determining the phase angle can be introduced using Fig. 13.39.

First, note that each sinusoidal function *has the same frequency,* permitting the use of either waveform to determine the period. For the waveform chosen in Fig. 13.39, the period encompasses 5 divisions at 0.2 ms/div. The phase shift between the waveforms (irrespective of which is leading or lagging) is 2 divisions. Since the full period represents a cycle of 360°, the following ratio [from which Eq. (13.24) can be derived] can be formed:

$$\frac{360°}{T\ (\text{no. of div.})} = \frac{\theta}{\text{phase shift (no. of div.)}}$$

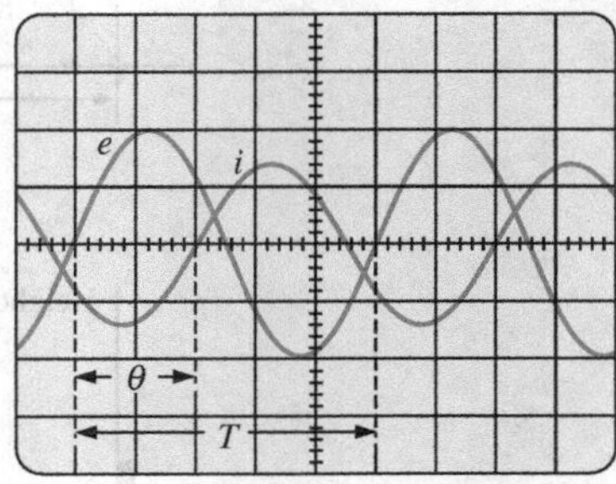

Vertical sensitivity = 2 V/div.
Horizontal sensitivity = 0.2 ms/div.

FIG. 13.39
Finding the phase angle between waveforms using a dual-trace oscilloscope.

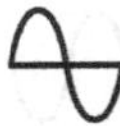

and
$$\theta = \frac{\text{phase shift (no. of div.)}}{T\text{ (no. of div.)}} \times 360° \qquad \textbf{(13.24)}$$

Substituting into Eq. (13.24) results in

$$\theta = \frac{(2\text{ div.})}{(5\text{ div.})} \times 360° = \mathbf{144°}$$

and e leads i by 144°.

13.7 AVERAGE VALUE

Even though the concept of the **average value** is an important one in most technical fields, its true meaning is often misunderstood. In Fig. 13.40(a), for example, the average height of the sand may be required to determine the volume of sand available. The average height of the sand is that height obtained if the distance from one end to the other is maintained while the sand is leveled off, as shown in Fig. 13.40(b). The area under the mound in Fig. 13.40(a) then equals the area under the rectangular shape in Fig. 13.40(b) as determined by $A = b \times h$. Of course, the depth (into the page) of the sand must be the same for Fig. 13.40(a) and (b) for the preceding conclusions to have any meaning.

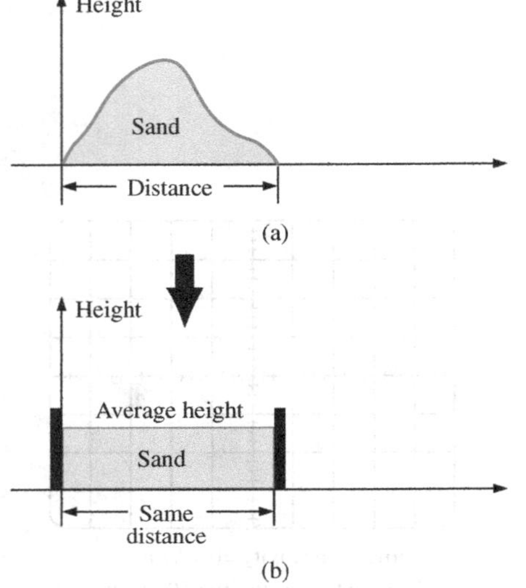

FIG. 13.40
Defining average value.

In Fig. 13.40, the distance was measured from one end to the other. In Fig. 13.41(a), the distance extends beyond the end of the original pile in Fig. 13.40. The situation could be one where a landscaper wants to know the average height of the sand if it is spread out over a distance such as defined in Fig. 13.41(a). The result of an increased distance is shown in Fig. 13.41(b). The average height has decreased compared to Fig. 13.40. Quite obviously, therefore, the longer the distance, the lower is the average value.

If the distance parameter includes a depression, as shown in Fig. 13.42(a), some of the sand will be used to fill the depression, resulting in

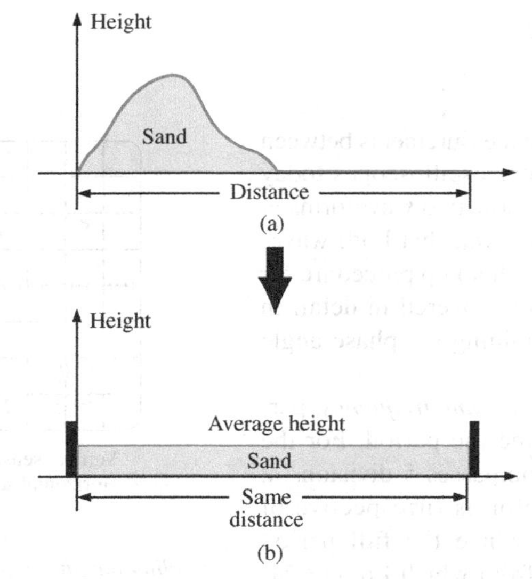

FIG. 13.41
Effect of distance (length) on average value.

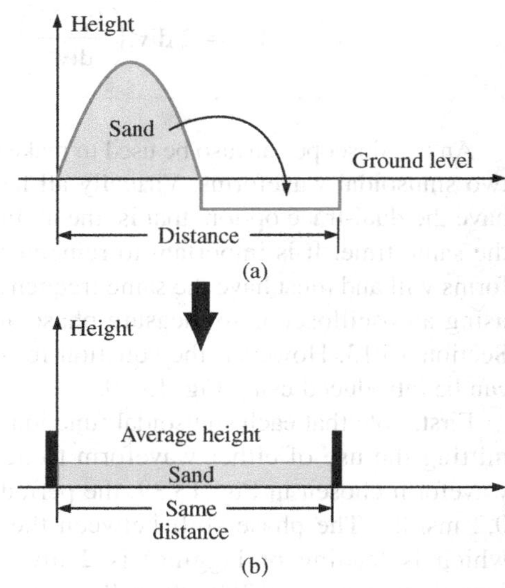

FIG. 13.42
Effect of depressions (negative excursions) on average value.

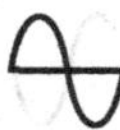

an even lower average value for the landscaper, as shown in Fig. 13.42(b). For a sinusoidal waveform, the depression would have the same shape as the mound of sand (over one full cycle), resulting in an average value at ground level (or zero volts for a sinusoidal voltage over one full period).

After traveling a considerable distance by car, some drivers like to calculate their average speed for the entire trip. This is usually done by dividing the miles traveled by the hours required to drive that distance. For example, if a person traveled 225 mi in 5 h, the average speed was 225 mi/5 h, or 45 mi/h. This same distance may have been traveled at various speeds for various intervals of time, as shown in Fig. 13.43.

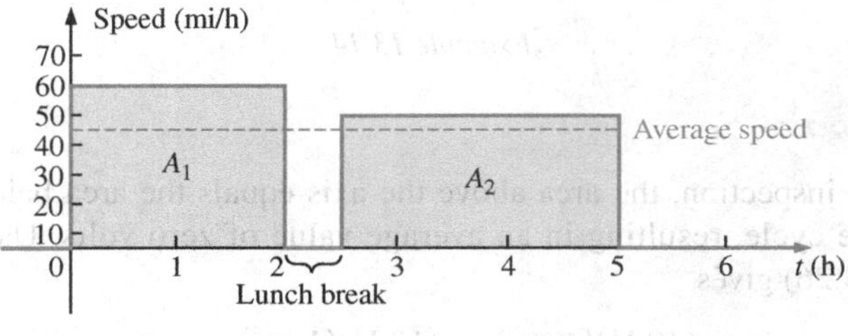

FIG. 13.43
Plotting speed versus time for an automobile excursion.

By finding the total area under the curve for the 5 h and then dividing the area by 5 h (the total time for the trip), we obtain the same result of 45 mi/h; that is,

$$\text{Average speed} = \frac{\text{area under curve}}{\text{length of curve}} \qquad \textbf{(13.25)}$$

$$\begin{aligned}\text{Average speed} &= \frac{A_1 + A_2}{5\text{ h}} = \frac{(60\text{ mi/h})(2\text{ h}) + (50\text{ mi/h})(2.5\text{ h})}{5\text{ h}}\\ &= \frac{225}{5}\text{ mi/h} = \mathbf{45\ mi/h}\end{aligned}$$

Eq. (13.25) can be extended to include any variable quantity, such as current or voltage, if we let G denote the average value, as follows:

$$G\text{ (average value)} = \frac{\text{algebraic sum of areas}}{\text{length of curve}} \qquad \textbf{(13.26)}$$

The *algebraic* sum of the areas must be determined since some area contributions are from below the horizontal axis. Areas above the axis are assigned a positive sign and those below it a negative sign. A positive average value is then above the axis, and a negative value is below it.

The average value of *any* current or voltage is the value indicated on a dc meter. In other words, over a complete cycle, the average value is the equivalent dc value. In the analysis of electronic circuits to be considered in a later course, both dc and ac sources of voltage will be applied to the same network. You will then need to know or determine the dc (or average value) and ac components of the voltage or current in various parts of the system.

EXAMPLE 13.14 Determine the average value of the waveforms in Fig. 13.44.

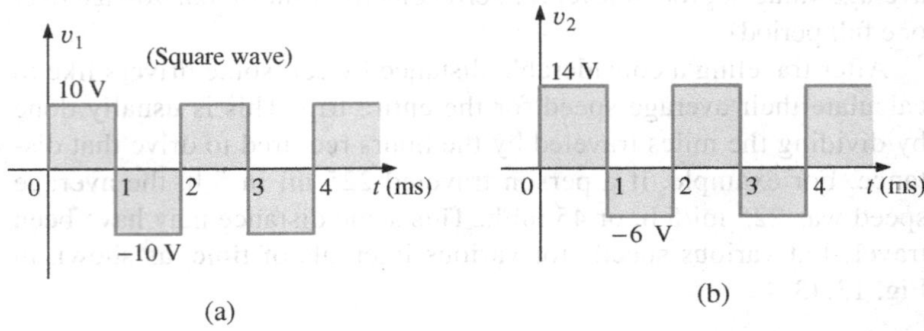

FIG. 13.44
Example 13.14.

Solutions:

a. By inspection, the area above the axis equals the area below over one cycle, resulting in an average value of zero volts. Using Eq. (13.26) gives

$$G = \frac{(10\text{ V})(1\text{ ms}) - (10\text{ V})(1\text{ ms})}{2\text{ ms}} = \frac{0}{2\text{ ms}} = \mathbf{0\text{ V}}$$

b. Using Eq. (13.26) gives

$$G = \frac{(14\text{ V})(1\text{ ms}) - (6\text{ V})(1\text{ ms})}{2\text{ ms}} = \frac{14\text{ V} - 6\text{ V}}{2} = \frac{8\text{ V}}{2} = \mathbf{4\text{ V}}$$

as shown in Fig. 13.45.

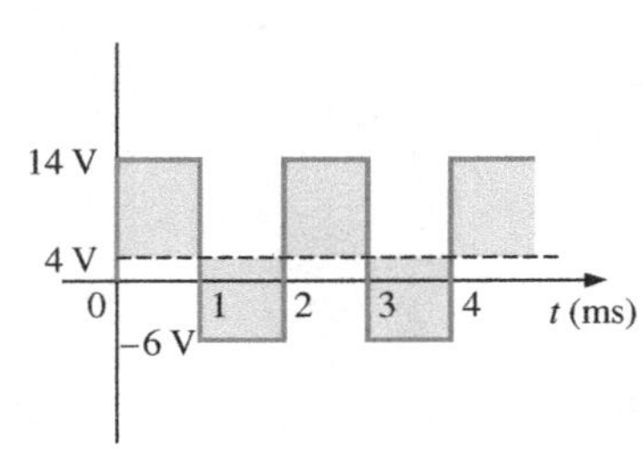

FIG. 13.45
Defining the average value for the waveform in Fig. 13.44(b).

In reality, the waveform in Fig. 13.44(b) is simply the square wave in Fig. 13.44(b) with a dc shift of 4 V; that is,

$$v_2 = v_1 + 4\text{ V}$$

EXAMPLE 13.15 Find the average values of the following waveforms over one full cycle:

a. Fig. 13.46.
b. Fig. 13.47.

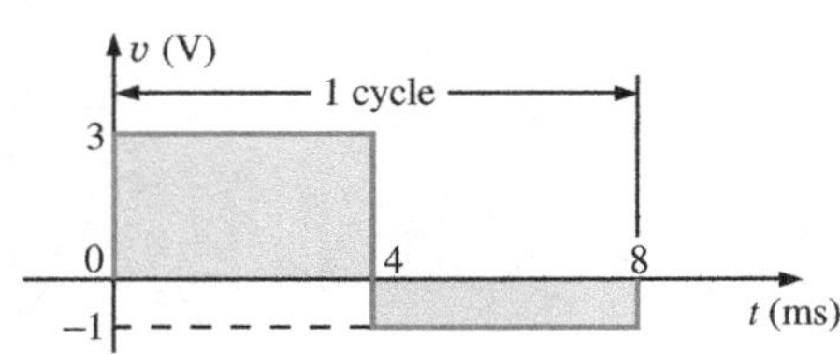

FIG. 13.46
Example 13.15(a).

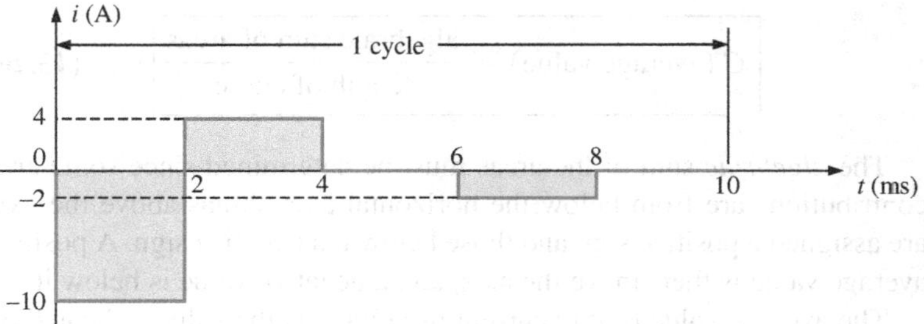

FIG. 13.47
Example 13.15(b).

Solutions:

a. $G = \dfrac{+(3\text{ V})(4\text{ ms}) - (1\text{ V})(4\text{ ms})}{8\text{ ms}} = \dfrac{12\text{ V} - 4\text{ V}}{8} = \mathbf{1\text{ V}}$

Note Fig. 13.48.

b. $$G = \frac{-(10\text{ V})(2\text{ ms}) - (4\text{ V})(2\text{ ms}) - (2\text{ V})(2\text{ ms})}{10\text{ ms}}$$
$$= \frac{-20\text{ V} + 8\text{ V} - 4\text{ V}}{10} = -\frac{16\text{ V}}{10} = \mathbf{-1.6\text{ V}}$$

Note Fig. 13.49.

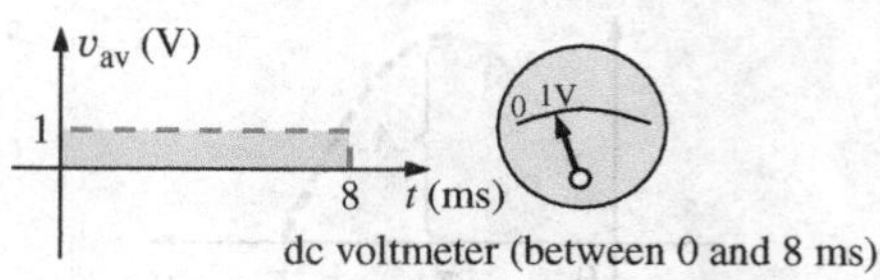

FIG. 13.48
The response of a dc meter to the waveform in Fig. 13.46.

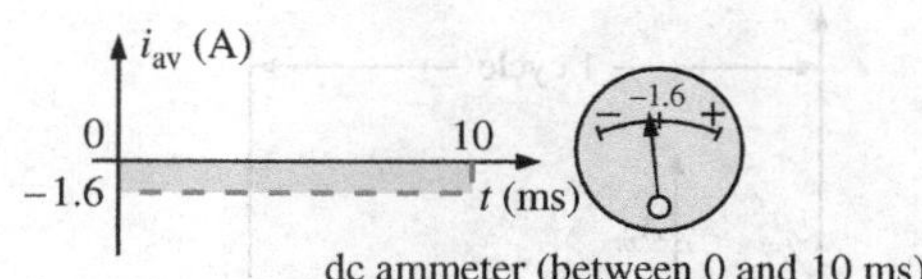

FIG. 13.49
The response of a dc meter to the waveform in Fig. 13.47.

We found the areas under the curves in Example 13.15 by using a simple geometric formula. If we encounter a sine wave or any other unusual shape, however, we must find the area by some other means. We can obtain a good approximation of the area by attempting to reproduce the original wave shape using a number of small rectangles or other familiar shapes, the area of which we already know through simple geometric formulas. For example,

the area of the positive (or negative) pulse of a sine wave is $2A_m$.

Approximating this waveform by two triangles (Fig. 13.50), we obtain (using *area* = *1/2 base* × *height* for the area of a triangle) a rough idea of the actual area:

$$\text{Area shaded} = 2\left(\frac{1}{2}bh\right) = 2\left[\left(\frac{1}{2}\right)\overbrace{\left(\frac{\pi}{2}\right)}^{b}\overbrace{(A_m)}^{h}\right] = \frac{\pi}{2}A_m \cong 1.58A_m$$

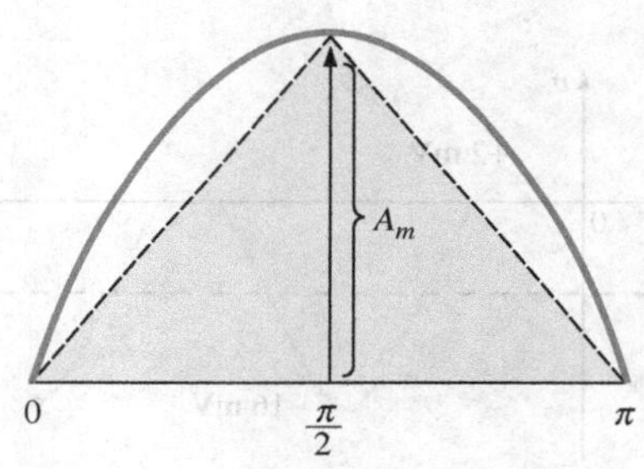

FIG. 13.50
Approximating the shape of the positive pulse of a sinusoidal waveform with two right triangles.

A closer approximation may be a rectangle with two similar triangles (Fig. 13.51):

$$\text{Area} = A_m\frac{\pi}{3} + 2\left(\frac{1}{2}bh\right) = A_m\frac{\pi}{3} + \frac{\pi}{3}A_m = \frac{2}{3}\pi A_m = 2.094A_m$$

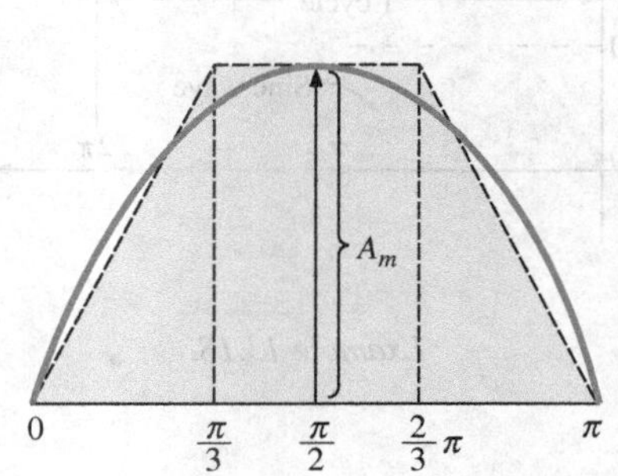

FIG. 13.51
A better approximation for the shape of the positive pulse of a sinusoidal waveform.

which is certainly close to the actual area. If an infinite number of forms is used, an exact answer of $2A_m$ can be obtained. For irregular waveforms, this method can be especially useful if data such as the average value are desired.

The procedure of calculus that gives the exact solution $2A_m$ is known as *integration.* Integration is presented here only to make the method recognizable to you; it is not necessary to be proficient in its use to continue with this text. It is a useful mathematical tool, however, and should be learned. Finding the area under the positive pulse of a sine wave using integration, we have

$$\text{Area} = \int_0^{\pi} A_m \sin\alpha\, d\alpha$$

where $\int$ is the sign of integration, 0 and π are the limits of integration, $A_m \sin\alpha$ is the function to be integrated, and $d\alpha$ indicates that we are integrating with respect to α.

Integrating, we obtain

$$\begin{aligned}\text{Area} &= A_m[-\cos\alpha]_0^{\pi} \\ &= -A_m(\cos\pi - \cos 0°) \\ &= -A_m[-1 - (+1)] = -A_m(-2)\end{aligned}$$

$$\boxed{\text{Area} = 2A_m} \qquad (13.27)$$

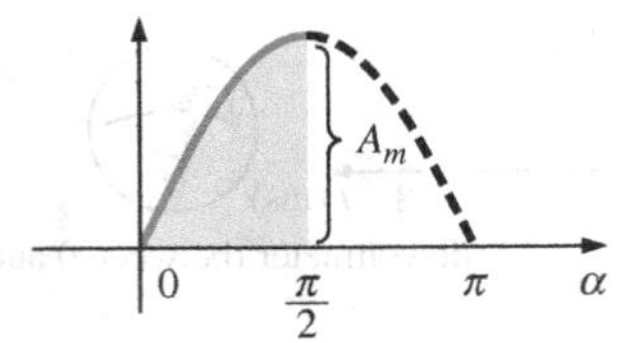

FIG. 13.52
Finding the average value of one-half the positive pulse of a sinusoidal waveform.

Since we know the area under the positive (or negative) pulse, we can easily determine the average value of the positive (or negative) region of a sine wave pulse by applying Eq. (13.26):

$$G = \frac{2A_m}{\pi}$$

and

$$\boxed{G = \frac{2A_m}{\pi} = 0.637A_m} \qquad \textbf{(13.28)}$$

For the waveform in Fig. 13.52,

$$G = \frac{(2A_m/2)}{\pi/2} = \frac{2A_m}{\pi} \qquad \text{(The average is the same as for a full pulse.)}$$

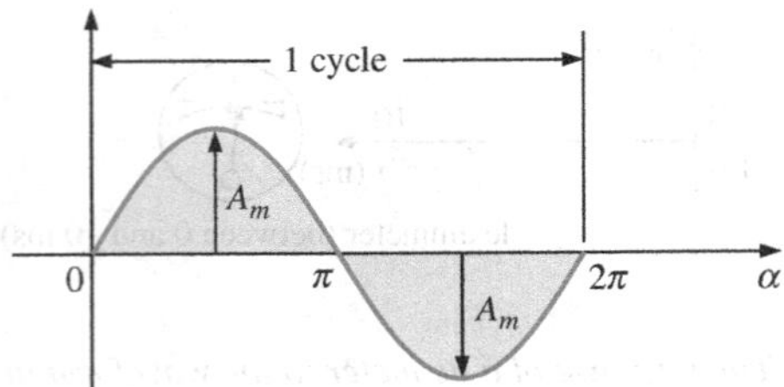

FIG. 13.53
Example 13.16.

EXAMPLE 13.16 Determine the average value of the sinusoidal waveform in Fig. 13.53.

Solution: By inspection it is fairly obvious that

the average value of a pure sinusoidal waveform over one full cycle is zero.

Eq. (13.26):

$$G = \frac{+2A_m - 2A_m}{2\pi} = \mathbf{0\ V}$$

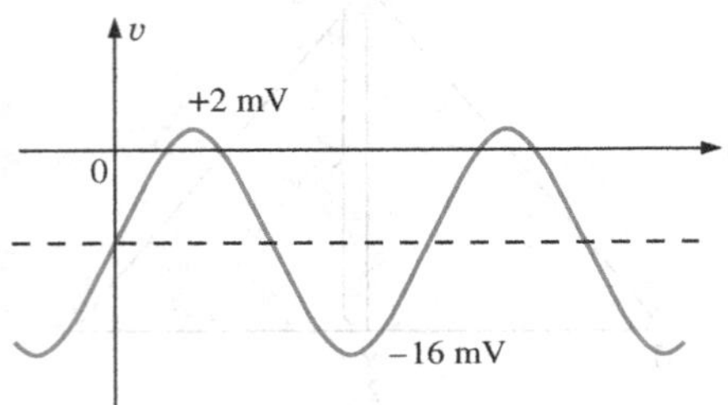

FIG. 13.54
Example 13.17.

EXAMPLE 13.17 Determine the average value of the waveform in Fig. 13.54.

Solution: The peak-to-peak value of the sinusoidal function is 16 mV + 2 mV = 18 mV. The peak amplitude of the sinusoidal waveform is, therefore, 18 mV/2 = 9 mV. Counting down 9 mV from 2 mV (or 9 mV up from −16 mV) results in an average or dc level of **−7 mV,** as noted by the dashed line in Fig. 13.54.

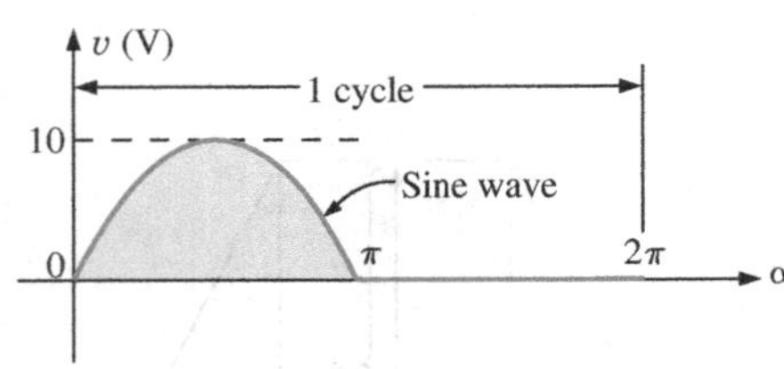

FIG. 13.55
Example 13.18.

EXAMPLE 13.18 Determine the average value of the waveform in Fig. 13.55.

Solution:

$$G = \frac{2A_m + 0}{2\pi} = \frac{2(10\text{ V})}{2\pi} \cong \mathbf{3.18\ V}$$

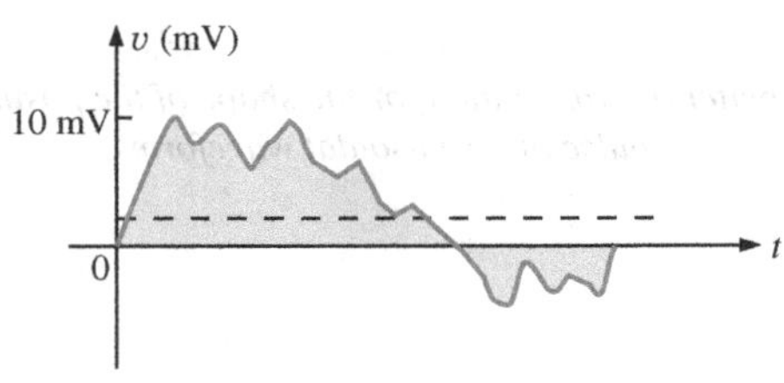

FIG. 13.56
Example 13.19.

EXAMPLE 13.19 For the waveform in Fig. 13.56, determine whether the average value is positive or negative, and determine its approximate value.

Solution: From the appearance of the waveform, the average value is positive and in the vicinity of 2 mV. Occasionally, judgments of this type will have to be made.

Instrumentation

The dc level or average value of any waveform can be found using a digital multimeter (DMM) or an **oscilloscope.** For purely dc circuits, set the

DMM on dc, and read the voltage or current levels. Oscilloscopes are limited to voltage levels using the sequence of steps listed below:

1. First choose GND from the DC-GND-AC option list associated with each vertical channel. The GND option blocks any signal to which the oscilloscope probe may be connected from entering the oscilloscope and responds with just a horizontal line. Set the resulting line in the middle of the vertical axis on the horizontal axis, as shown in Fig. 13.57(a).
2. Apply the oscilloscope probe to the voltage to be measured (if not already connected), and switch to the DC option. If a dc voltage is present, the horizontal line shifts up or down, as demonstrated in Fig. 13.57(b). Multiplying the shift by the vertical sensitivity results in the dc voltage. An upward shift is a positive voltage (higher potential at the red or positive lead of the oscilloscope), while a downward shift is a negative voltage (lower potential at the red or positive lead of the oscilloscope).

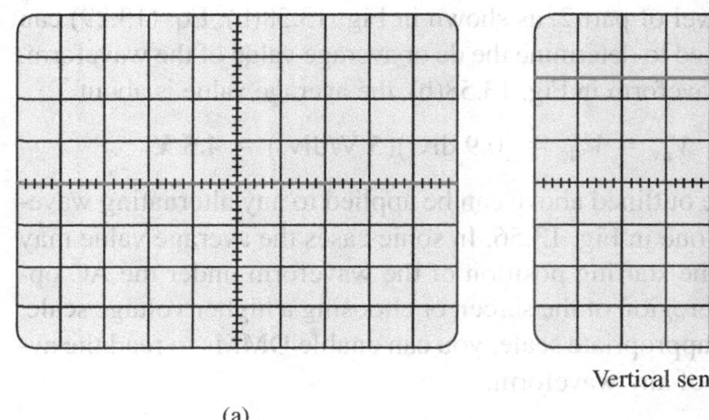

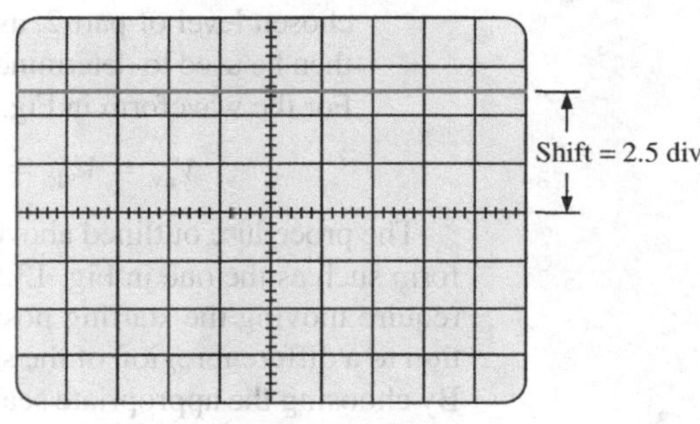

Vertical sensitivity = 50 mV/div.

(a) (b)

FIG. 13.57
Using the oscilloscope to measure dc voltages; (a) setting the GND condition; (b) the vertical shift resulting from a dc voltage when shifted to the DC option.

In general,

$$V_{dc} = (\text{vertical shift in div.}) \times (\text{vertical sensitivity in V/div.}) \quad \textbf{(13.29)}$$

For the waveform in Fig. 13.57(b),

$$V_{dc} = (2.5 \text{ div.})(50 \text{ mV/div.}) = \mathbf{125 \text{ mV}}$$

The oscilloscope can also be used to measure the dc or average level of any waveform using the following sequence:

1. Using the GND option, reset the horizontal line to the middle of the screen.
2. Switch to AC (all dc components of the signal to which the probe is connected will be blocked from entering the oscilloscope—only the alternating, or changing, components are displayed). Note the location of some definitive point on the waveform, such as the bottom of the half-wave rectified waveform of Fig. 13.58(a); that is, note its position on the vertical scale. For the future, *whenever you use the AC option, keep in mind that the computer will distribute the waveform above and below the horizontal axis such that the average value is zero;* that is, the area above the axis will equal the area below.

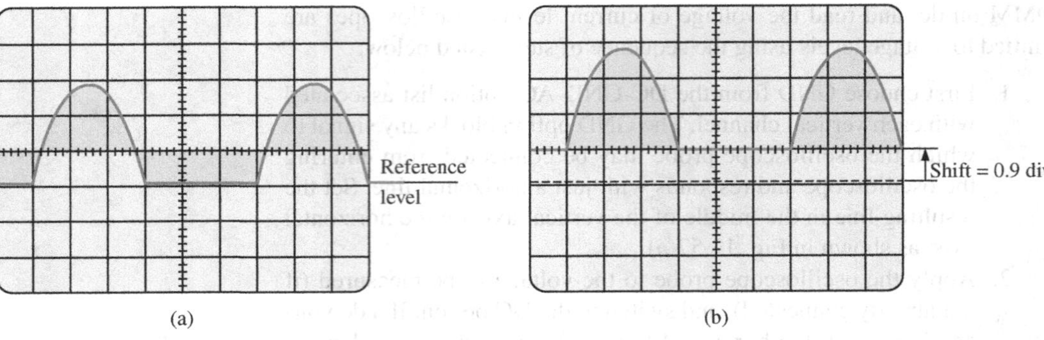

FIG. 13.58
Determining the average value of a nonsinusoidal waveform using the oscilloscope: (a) vertical channel on the ac mode; (b) vertical channel on the dc mode.

3. Then switch to DC (to permit both the dc and the ac components of the waveform to enter the oscilloscope), and note the shift in the chosen level of part 2, as shown in Fig. 13.58(b). Eq. (13.29) can then be used to determine the dc or average value of the waveform. For the waveform in Fig. 13.58(b), the average value is about

$$V_{av} = V_{dc} = (0.9 \text{ div.})(5 \text{ V/div.}) = \mathbf{4.5 \text{ V}}$$

The procedure outlined above can be applied to any alternating waveform such as the one in Fig. 13.56. In some cases the average value may require moving the starting position of the waveform under the AC option to a different region of the screen or choosing a higher voltage scale. By choosing the appropriate scale, you can enable DMMs to read the average or dc level of any waveform.

13.8 EFFECTIVE (rms) VALUES

This section begins to relate dc and ac quantities with respect to the power delivered to a load. It will help us determine the amplitude of a sinusoidal ac current required to deliver the same power as a particular dc current. The question frequently arises, How is it possible for a sinusoidal ac quantity to deliver a net power if, over a full cycle, the net current in any one direction is zero (average value $= 0$)? It would almost appear that the power delivered during the positive portion of the sinusoidal waveform is withdrawn during the negative portion, and since the two are equal in magnitude, the net power delivered is zero. However, understand that regardless of *direction,* current of any magnitude through a resistor delivers power *to that resistor.* In other words, during the positive or negative portions of a sinusoidal ac current, power is being delivered at *each instant of time* to the resistor. The power delivered at each instant, of course, varies with the magnitude of the sinusoidal ac current, but there will be a net flow during either the positive or the negative pulses with a net flow over the full cycle. The net power flow equals twice that delivered by either the positive or the negative regions of sinusoidal quantity.

A fixed relationship between ac and dc voltages and currents can be derived from the experimental setup shown in Fig. 13.59. A resistor in a water bath is connected by switches to a dc and an ac supply. If switch 1 is closed, a dc current I, determined by the resistance R and battery voltage E, is established through the resistor R. The temperature reached by

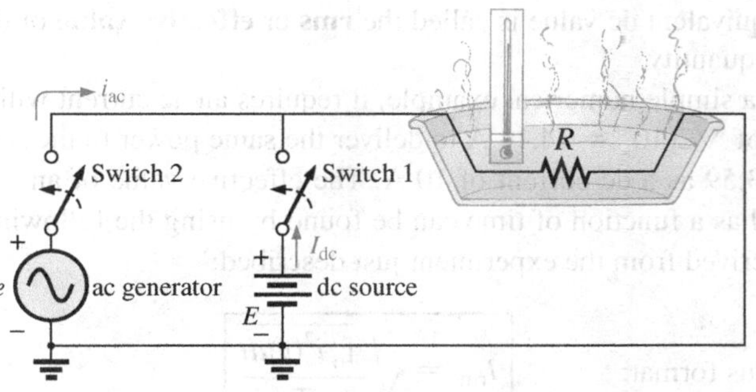

FIG. 13.59
An experimental setup to establish a relationship between dc and ac quantities.

the water is determined by the dc power dissipated in the form of heat by the resistor.

If switch 2 is closed and switch 1 left open, the ac current through the resistor has a peak value of I_m. The temperature reached by the water is now determined by the ac power dissipated in the form of heat by the resistor. The ac input is varied until the temperature is the same as that reached with the dc input. When this is accomplished, the average electrical power delivered to the resistor R by the ac source is the same as that delivered by the dc source.

The power delivered by the ac supply at any instant of time is

$$P_{ac} = (i_{ac})^2R = (I_m \sin \omega t)^2R = (I_m^2 \sin^2 \omega t)R$$

However,

$$\sin^2 \omega t = \frac{1}{2}(1 - \cos 2\omega t) \quad \text{(trigonometric identity)}$$

Therefore,

$$P_{ac} = I_m^2\left[\frac{1}{2}(1 - \cos 2\omega t)\right]R$$

and

$$P_{ac} = \frac{I_m^2 R}{2} - \frac{I_m^2 R}{2}\cos 2\omega t \qquad \textbf{(13.30)}$$

The *average power* delivered by the ac source is just the first term, since the average value of a cosine wave is zero even though the wave may have twice the frequency of the original input current waveform. Equating the average power delivered by the ac generator to that delivered by the dc source,

$$P_{av(ac)} = P_{dc}$$

$$\frac{I_m^2 R}{2} = I_{dc}^2 R$$

and

$$I_{dc} = \frac{I_m}{\sqrt{2}} = 0.707 I_m$$

which, in words, states that

the equivalent dc value of a sinusoidal current or voltage is $1/\sqrt{2}$ or 0.707 of its peak value.

The equivalent dc value is called the **rms** or **effective value** of the sinusoidal quantity.

As a simple numerical example, it requires an ac current with a peak value of $\sqrt{2}(10) = 14.14$ A to deliver the same power to the resistor in Fig. 13.59 as a dc current of 10 A. The effective value of any quantity plotted as a function of time can be found by using the following equation derived from the experiment just described:

Calculus format:

$$I_{\text{rms}} = \sqrt{\frac{\int_0^T i^2(t)\,dt}{T}} \tag{13.31}$$

which means

$$I_{\text{rms}} = \sqrt{\frac{\text{area}(i^2(t))}{T}} \tag{13.32}$$

In words, Eqs. (13.31) and (13.32) state that to find the rms value, the function $i(t)$ must first be squared. After $i(t)$ is squared, the area under the curve is found by integration. It is then divided by T, the length of the cycle or the period of the waveform, to obtain the average or *mean* value of the squared waveform. The final step is to take the *square root* of the mean value. This procedure is the source for the other designation for the effective value, the **root-mean-square (rms) value.** In fact, since *rms* is the most commonly used term in the educational and industrial communities, it is used throughout this text.

The relationship between the peak value and the rms value is the same for voltages, resulting in the following set of relationships for the examples and text material to follow:

$$\begin{aligned} I_{\text{rms}} &= \frac{1}{\sqrt{2}}I_m = 0.707I_m \\ E_{\text{rms}} &= \frac{1}{\sqrt{2}}E_m = 0.707E_m \end{aligned} \tag{13.33}$$

Similarly,

$$\begin{aligned} I_m &= \sqrt{2}I_{\text{rms}} = 1.414I_{\text{rms}} \\ E_m &= \sqrt{2}E_{\text{rms}} = 1.414E_{\text{rms}} \end{aligned} \tag{13.34}$$

EXAMPLE 13.20 Find the rms values of the sinusoidal waveform in each part in Fig. 13.60.

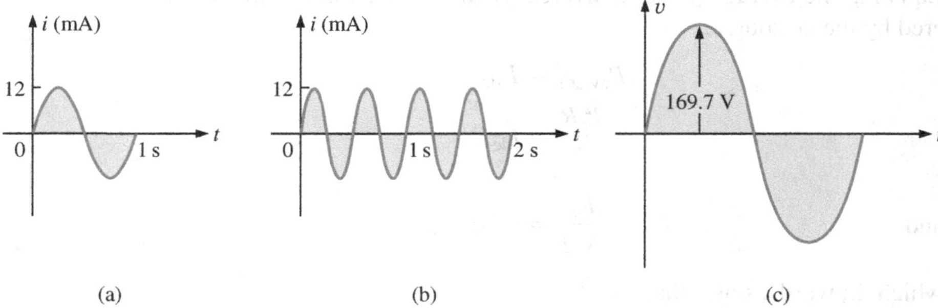

FIG. 13.60
Example 13.20.

Solution: For part (a), $I_{rms} = 0.707(12 \times 10^{-3}\ \text{A}) =$ **8.48 mA.** For part (b), again $I_{rms} =$ **8.48 mA.** Note that frequency did not change the effective value in (b) compared to (a). For part (c), $V_{rms} = 0.707(169.73\ \text{V}) \cong$ **120 V,** the same as available from a home outlet.

EXAMPLE 13.21 The 120 V dc source in Fig. 13.61(a) delivers 3.6 W to the load. Determine the peak value of the applied voltage (E_m) and the current (I_m) if the ac source [Fig. 13.61(b)] is to deliver the same power to the load.

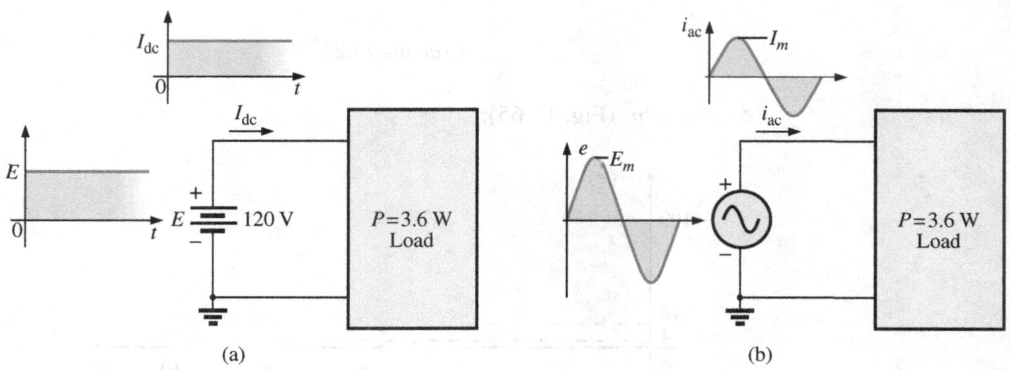

FIG. 13.61
Example 13.21.

Solution:

$$P_{dc} = V_{dc}I_{dc}$$

and

$$I_{dc} = \frac{P_{dc}}{V_{dc}} = \frac{3.6\ \text{W}}{120\ \text{V}} = 30\ \text{mA}$$

$$I_m = \sqrt{2}I_{dc} = (1.414)(30\ \text{mA}) = \mathbf{42.42\ mA}$$

$$E_m = \sqrt{2}E_{dc} = (1.414)(120\ \text{V}) = \mathbf{169.68\ V}$$

EXAMPLE 13.22 Find the rms value of the waveform in Fig. 13.62.

Solution: v^2 (Fig. 13.63):

$$V_{rms} = \sqrt{\frac{(9)(4) + (1)(4)}{8}} = \sqrt{\frac{40}{8}} = \mathbf{2.24\ V}$$

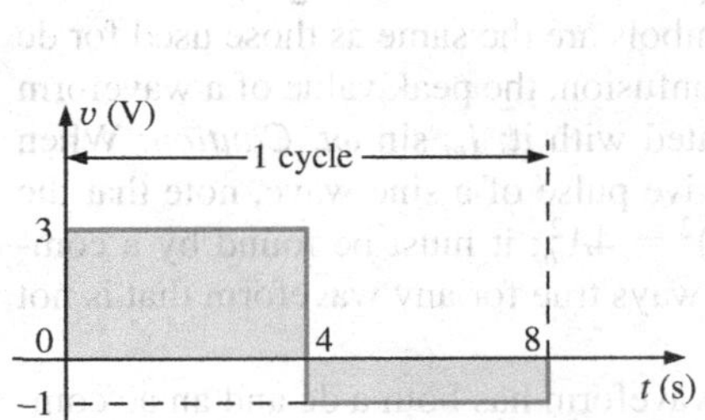

FIG. 13.62
Example 13.22.

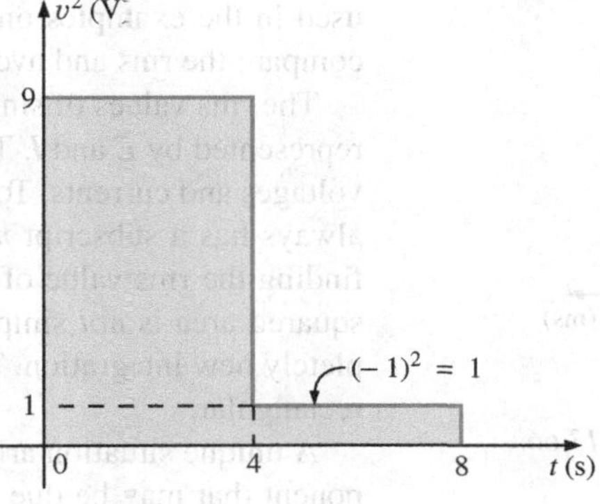

FIG. 13.63
The squared waveform of Fig. 13.62.

EXAMPLE 13.23 Calculate the rms value of the voltage in Fig. 13.64.

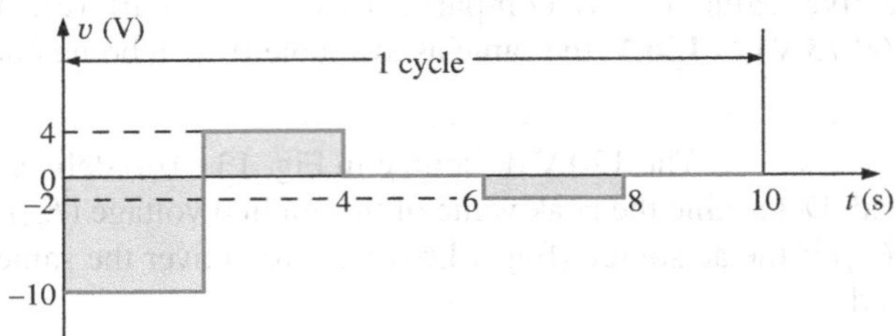

FIG. 13.64
Example 13.23.

Solution: v^2 (Fig. 13.65):

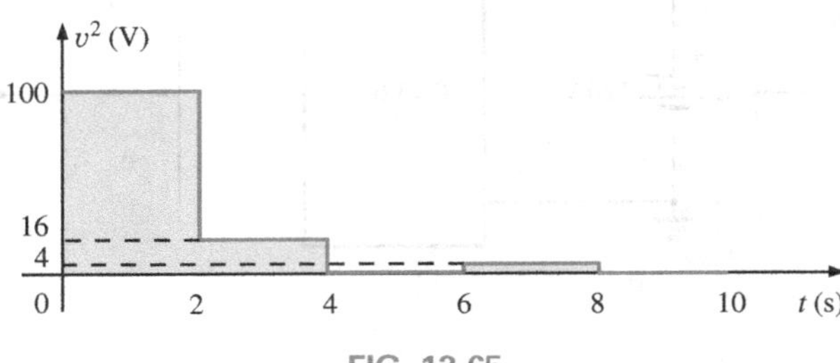

FIG. 13.65
The squared waveform of Fig. 13.64.

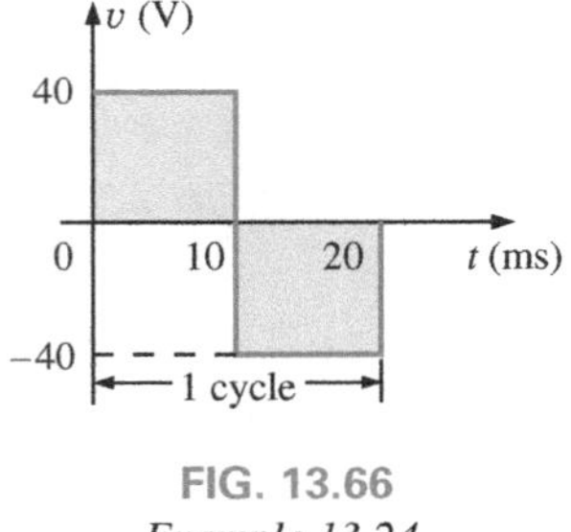

FIG. 13.66
Example 13.24.

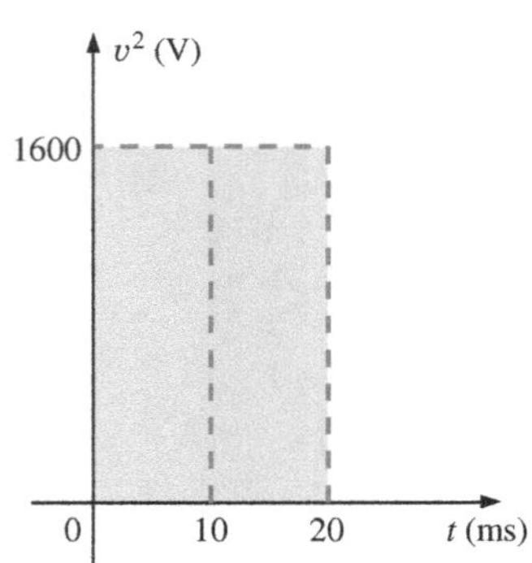

FIG. 13.67
The squared waveform of Fig. 13.66.

EXAMPLE 13.24 Determine the average and rms values of the square wave in Fig. 13.66.

Solution: By inspection, the average value is zero.

v^2 (Fig. 13.67):

$$V_{\text{rms}} = \sqrt{\frac{(1600)(10 \times 10^{-3}) + (1600)(10 \times 10^{-3})}{20 \times 10^{-3}}}$$

$$= \sqrt{\frac{(32{,}000 \times 10^{-3})}{20 \times 10^{-3}}} = \sqrt{1600} = \mathbf{40\ V}$$

(the maximum value of the waveform in Fig. 13.66).

The waveforms appearing in these examples are the same as those used in the examples on the average value. It may prove interesting to compare the rms and average values of these waveforms.

The rms values of sinusoidal quantities such as voltage or current are represented by E and I. These symbols are the same as those used for dc voltages and currents. To avoid confusion, the peak value of a waveform always has a subscript m associated with it: $I_m \sin \omega t$. *Caution:* When finding the rms value of the positive pulse of a sine wave, note that the squared area is *not* simply $(2A_m)^2 = 4A_m^2$; it must be found by a completely new integration. This is always true for any waveform that is not rectangular.

A unique situation arises if a waveform has both a dc and an ac component that may be due to a source, such as the one in Fig. 13.68. The combination appears frequently in the analysis of electronic networks where both dc and ac levels are present in the same system.

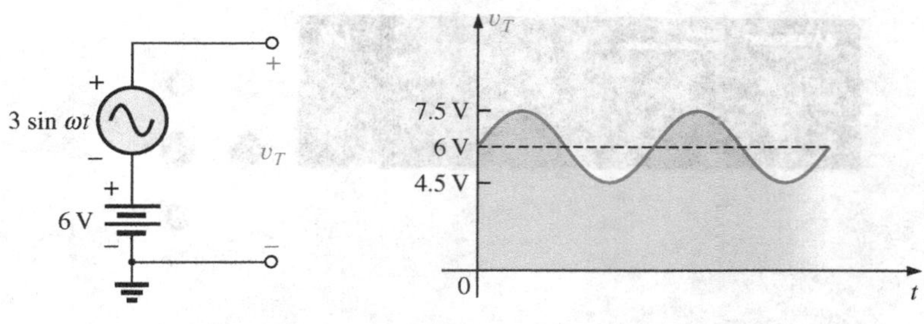

FIG. 13.68
Generation and display of a waveform having a dc and an ac component.

The question arises, What is the rms value of the voltage v_T? You may be tempted to assume that it is the sum of the rms values of each component of the waveform; that is, $V_{T_{rms}} = 0.7071(1.5\text{ V}) + 6\text{ V} = 1.06\text{ V} + 6\text{ V} = 7.06\text{ V}$. However, the rms value is actually determined by

$$\boxed{V_{rms} = \sqrt{V_{dc}^2 + V_{ac(rms)}^2}} \qquad \textbf{(13.35)}$$

which for the waveform in Fig. 13.68 is

$$V_{rms} = \sqrt{(6\text{ V})^2 + (1.06\text{ V})^2} = \sqrt{37.124}\text{ V} \cong \mathbf{6.1\text{ V}}$$

This result is noticeably less than the solution of 7.06 V.

True rms Meters

Throughout this section, the rms value of a variety of waveforms was determined to help ensure that the concept is correctly understood. However, to use a meter to measure the rms value of the same waveforms would require a specially designed meter. Too often, the face of a meter will read **True rms Multimeter** or such. However, in most cases the meter is only designed to read the rms value of periodic signals with no dc level and have a symmetry about the zero axis. Most multimeters are ac coupled (the dc component of the signal is blocked by a capacitor at the input terminals), so only the ac portion is measured. For such cases one may be able to first determine the rms value of the ac portion of the waveform and then use the dc section of the meter to measure the dc level. Then Eq. (13.35) can be used to determine the correct rms value.

The problem, however, is that many waveforms are not symmetric about the zero axis—How is an rms reading obtained? In general, the rms value of any waveform is a measure of the "heating" potential of the applied waveform, as discussed earlier in this section. A direct result is the development of meters that use a thermal converter calibrated to display the proper rms value. A drawback of this approach, however, is that the meter will draw power from the circuit during the heating process, and the results have a low precision standard. A better approach that is commonly used uses an analog-to-digital converter (ADC) mentioned earlier to digitize the signal, so that the rms value then can be determined to a high degree of accuracy. One such meter appears in Fig. 13.69, which samples the input signal at 1.4 MHz, or 1,400,000 samples per second—certainly sufficient for a wide variety of signals. This meter will run the sampling rate at all times, even when making dc measurements, so both the dc and ac content of a waveform can be displayed at the same time.

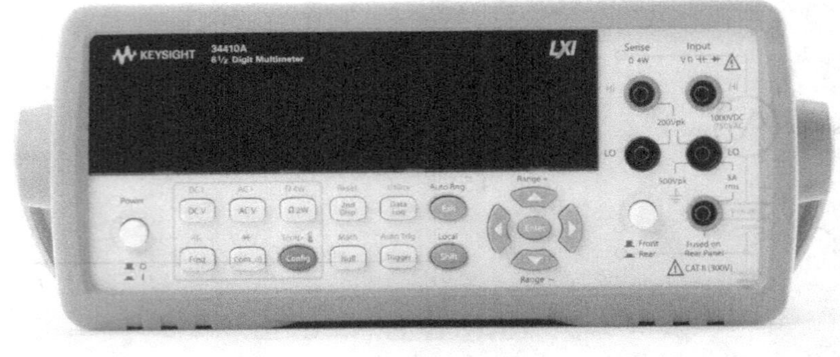

FIG. 13.69
True rms multimeter.

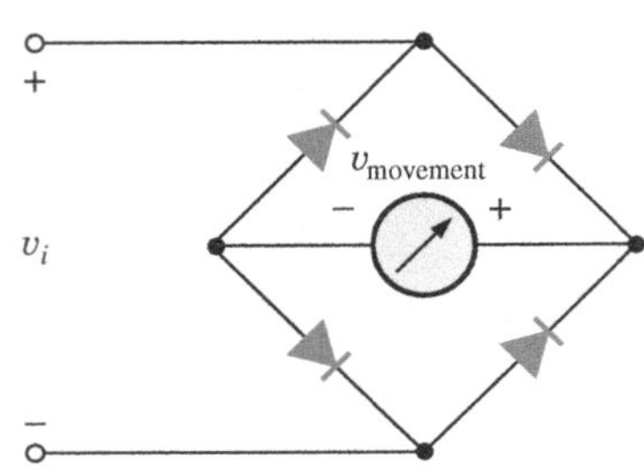

FIG. 13.70
Full-wave bridge rectifier.

13.9 ac METERS AND INSTRUMENTS

Iron-Vane or d'Arsonval Movement

If an average reading movement such as the iron-vane movement used in the **VOM** of Fig. 2.29 is used to measure an ac current or voltage, the level indicated by the movement must be multiplied by a **calibration factor.** In other words, if the movement of any voltmeter or ammeter is reading the average value, that level must be multiplied by a specific constant, or calibration factor, to indicate the rms level. For ac waveforms, the signal must first be converted to one having an average value over the time period. Recall that it is zero over a full period for a sinusoidal waveform. This is usually accomplished for sinusoidal waveforms using a bridge rectifier such as in Fig. 13.70. The conversion process, involving four diodes in a bridge configuration, is well documented in most electronic texts.

Fundamentally, conduction is permitted through the diodes in such a manner as to convert the sinusoidal input of Fig. 13.71(a) to one having the appearance of Fig. 13.71(b). The negative portion of the input has been effectively "flipped over" by the bridge configuration. The resulting waveform in Fig. 13.71(b) is called a *full-wave rectified waveform.*

The zero average value in Fig. 13.71(a) has been replaced by a pattern having an average value determined by

$$G = \frac{2V_m + 2V_m}{2\pi} = \frac{4V_m}{2\pi} = \frac{2V_m}{\pi} = 0.637V_m$$

The movement of the pointer is therefore directly related to the peak value of the signal by the factor 0.637.

Forming the ratio between the rms and dc levels results in

$$\frac{V_{\text{rms}}}{V_{\text{dc}}} = \frac{0.707V_m}{0.637V_m} \cong 1.11$$

revealing that the scale indication is 1.11 times the dc level measured by the movement; that is,

$$\boxed{\text{Meter indication} = 1.11 \text{ (dc or average value)}} \quad \text{full-wave} \quad \textbf{(13.36)}$$

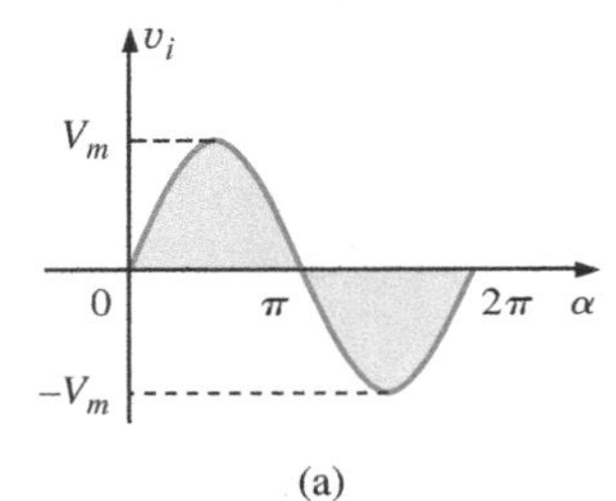

(a)

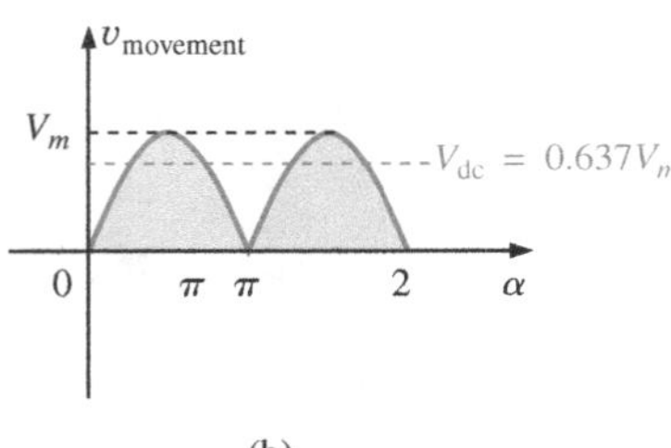

(b)

FIG. 13.71
(a) Sinusoidal input; (b) full-wave rectified signal.

Some ac meters use a half-wave rectifier arrangement that results in the waveform in Fig. 13.72, which has half the average value in Fig. 13.71(b) over one full cycle. The result is

$$\boxed{\text{Meter indication} = 2.22 \text{ (dc or average value)}} \quad \text{half-wave} \quad \textbf{(13.37)}$$

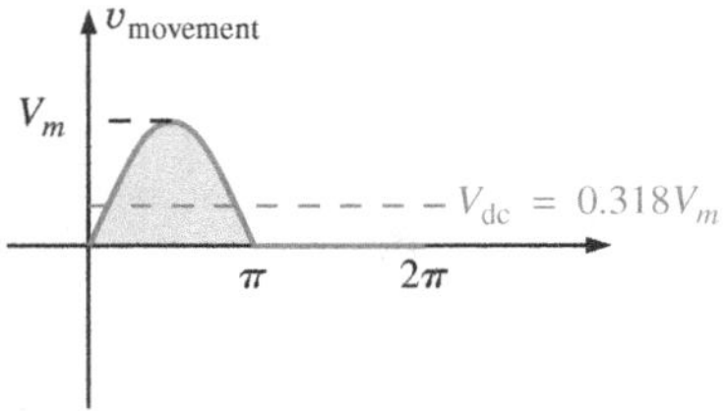

FIG. 13.72
Half-wave rectified signal.

Electrodynamometer Movement

The electrodynamometer movement is a movement that has the distinct advantage of being able to read the turn rms value of any current, voltage, or power measurement without additional circuitry. The basic construction appears in Fig. 13.73, which shows two fixed coils and a rotating coil. The two fixed coils establish a field similar to that established by the permanent magnet in an iron-vane movement. However, in this case, the same current that establishes the field in the fixed coils will also establish the field in the movable coil. The result is opposing polarities between the rotating and fixed coils that will establish a torque on the movable coil and cause it to rotate and provide a reading using the attached pointer. Removing the excitation force will allow the attached spring to bring the pointer back to the rest position. Although the electrodynamometer movement would be very effective in reading the rms value of any voltage or current, it is used almost exclusively in dc/ac wattmeters for any shape of input. It can also be used for phase shift measurements, harmonic analysis, and frequency measurements, although improving digital electronic technology is the new direction for these areas of application.

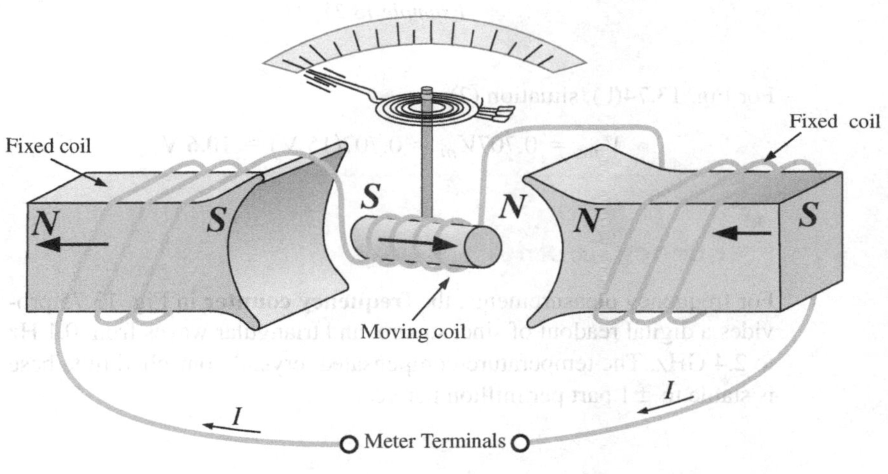

FIG. 13.73
Electrodynamometer movement.

EXAMPLE 13.25 Determine the reading of each meter for each situation in Fig. 13.74(a) and (b).

Solution: For Fig. 13.74(a), situation (1): By Eq. (13.36),

$$\text{Meter indication} = 1.11(20\text{ V}) = \mathbf{22.2\text{ V}}$$

For Fig. 13.74(a), situation (2):

$$V_{\text{rms}} = 0.707V_m = (0.707)(20\text{ V}) = \mathbf{14.14\text{ V}}$$

For Fig. 13.74(b), situation (1):

$$V_{\text{rms}} = V_{\text{dc}} = \mathbf{25\text{ V}}$$

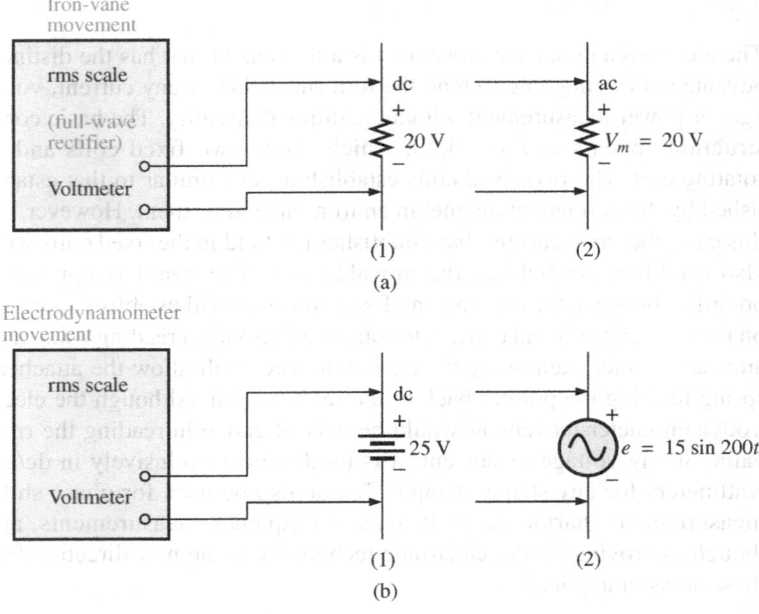

FIG. 13.74
Example 13.25.

For Fig. 13.74(b), situation (2):

$$V_{\text{rms}} = 0.707V_m = 0.707(15\text{ V}) \cong \mathbf{10.6\text{ V}}$$

Frequency Counter

For frequency measurements, the **frequency counter** in Fig. 13.75 provides a digital readout of sine, square, and triangular waves from 0.1 Hz to 2.4 GHz. The temperature-compensated, crystal-controlled time base is stable to ± 1 part per million per year.

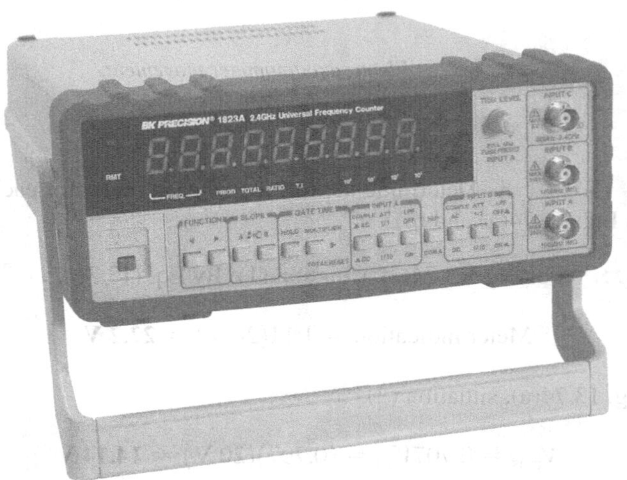

FIG. 13.75
Frequency counter, 2.4 GHz multifunctional instrument.
(B&K Precision Corp.)

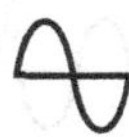

Clamp-on Meters

The AEMC® **Clamp Meter** in Fig. 13.76 is an instrument that can measure alternating current in the ampere range without having to open the circuit. The loop is opened by squeezing the "trigger"; then it is placed around the current-carrying conductor. Through transformer action, the level of current in rms units appears on the appropriate scale. The Model 501 is auto-ranging (that is, each scale changes automatically) and can measure dc or ac currents up to 400 mA. Through the use of additional leads, it can also be used as a voltmeter (up to 400 V, dc or ac) and an ohmmeter (from zero to 400 Ω).

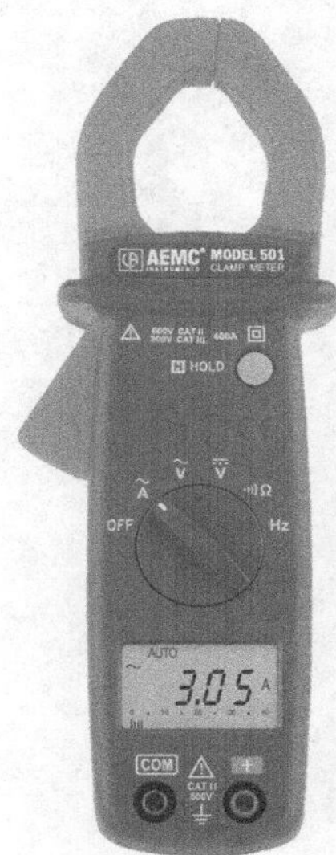

FIG. 13.76
Clamp-on ammeter and voltmeter.
Photo courtesy of AEMC® Instruments, a leader in electrical test and measurement instruments.

Impedance Measurements

Before we leave the subject of ac meters and instrumentation, you should understand that

an ohmmeter cannot be used to measure the ac reactance or impedance of an element or system even though reactance and impedance are measured in ohms.

Recall that ohmmeters cannot be used on energized networks—the power must be shut off or disconnected. For an inductor if the ac power is removed, the reactance of the coil is simply the dc resistance of the windings because the applicable frequency will be 0 Hz. For a capacitor, if the ac power is removed, the reactance of the capacitor is simply the leakage resistance of the capacitor. In general, therefore, always keep in mind that *ohmmeters can read only the dc resistance of an element or network, and only after the applied power has been removed.*

13.10 APPLICATIONS

(120 V at 60 Hz) versus (220 V at 50 Hz)

In North and South America, the most common available ac supply is 120 V at 60 Hz; in Western and Central Europe, Africa; Asia, and Australia, 220 V at 50 Hz is the most common. Japan is unique in that the eastern part of the country uses 100 V at 50 Hz, whereas most of the western part uses 100 V at 60 Hz. It is 220 V at 50 Hz. The choices of rms value and frequency were obviously made carefully because they have such an important impact on the design and operation of so many systems.

The fact that the frequency difference is only 10 Hz reveals that there was agreement on the general frequency range that should be used for power generation and distribution. History suggests that the question of frequency selection originally focused on the frequency that would not exhibit *flicker in the incandescent lamps* available in those days. Technically, however, there really wouldn't be a noticeable difference between 50 and 60 cycles per second based on this criterion. Another important factor in the early design stages was the effect of frequency on the size of transformers, which play a major role in power generation and distribution. Working through the fundamental equations for transformer design, you will find that *the size of a transformer is inversely proportional to frequency.* The result is that transformers operating at 50 Hz must be larger (on a purely mathematical basis about 17% larger) than those operating at 60 Hz. You will therefore find that transformers designed for the international market, where they can operate on 50 Hz or 60 Hz, are designed around the 50 Hz frequency. On the other side of the coin, however,

higher frequencies result in increased concerns about arcing, increased losses in the transformer core due to eddy current and hysteresis losses, and skin effect phenomena. Somewhere in the discussion we may wonder about the fact that 60 Hz is an exact multiple of 60 seconds in a minute and 60 minutes in an hour. On the other side of the coin, however, a 60 Hz signal has a period of 16.67 ms (an awkward number), but the period of a 50 Hz signal is exactly 20 ms. Since accurate timing is such a critical part of our technological design, was this a significant motive in the final choice? There is also the question about whether the 50 Hz is a result of the close affinity of this value to the metric system. Keep in mind that powers of ten are all-powerful in the metric system, with 100 cm in a meter, 100°C the boiling point of water, and so on. Note that 50 Hz is exactly half of this special number. All in all, it would seem that both sides have an argument that is worth defending. However, in the final analysis, we must also wonder whether the difference is simply political in nature.

The difference in voltage between the Americas and Europe is a different matter entirely, in the sense that the difference is close to 100%. Again, however, there are valid arguments for both sides. There is no question that larger voltages such as 220 V *raise safety issues* beyond those raised by voltages of 120 V. However, when higher voltages are supplied, there is less current in the wire for the same power demand, permitting the use of smaller conductors—a real money saver. In addition, motors and some appliances *can be smaller in size.* Higher voltages, however, also bring back the concern about arcing effects, insulation requirements, and, due to real safety concerns, higher installation costs. In general, however, international travelers are prepared for most situations if they have a transformer that can convert from their home level to that of the country they plan to visit. Most equipment (not clocks, of course) can run quite well on 50 Hz or 60 Hz for most travel periods. For any unit not operating at its design frequency, it simply has to "work a little harder" to perform the given task. The major problem for the traveler is not the transformer itself but the wide variety of plugs used from one country to another. Each country has its own design for the "female" plug in the wall. For a three-week tour, this could mean as many as 6 to 10 different plugs of the type shown in Fig. 13.77. For a 120 V, 60 Hz supply, the plug is quite standard in appearance with its two spade leads (and possible ground connection).

FIG. 13.77
Variety of plugs for a 220 V, 50 Hz connection.

In any event, both the 120 V at 60 Hz and the 220 V at 50 Hz are obviously meeting the needs of the consumer. It is a debate that could go on at length without an ultimate victor.

Safety Concerns (High Voltages and dc versus ac)

Be aware that any "live" network should be treated with a calculated level of respect. Electricity in its various forms is not to be feared but used with some awareness of its potentially dangerous side effects. It is common knowledge that electricity and water do not mix (never use extension cords or plug in TVs or radios in the bathroom) because a full 120 V in a layer of water of any height (from a shallow puddle to a full bath) can be *lethal.* However, other effects of dc and ac voltages are less known. In general, as the voltage and current increase, your concern about safety should increase exponentially. For instance, under dry conditions, most human beings can survive a 120 V ac shock such as obtained when changing a light bulb, turning on a switch, and so on. Most electricians have experienced such a jolt many times in their careers. However, ask an electrician to relate how it feels to hit 220 V, and the response (if he or she has been unfortunate to have had such an experience) will be totally different. How often have you

heard of a back-hoe operator hitting a 220 V line and having a fatal heart attack? Remember, the operator is sitting in a metal container on a damp ground, which provides an excellent path for the resulting current to flow from the line to ground. If only for a short period of time, with the best environment (rubber-sole shoes, and so on), in a situation where you can quickly escape the situation, most human beings can also survive a 220 V shock. However, as mentioned above, it is one you will not quickly forget. For voltages beyond 220 V rms, the chances of survival go down exponentially with increase in voltage. It takes only about 10 mA of steady current through the heart to put it in defibrillation. In general, therefore, always be sure that the power is disconnected when working on the repair of electrical equipment. Don't assume that throwing a wall switch will disconnect the power. Throw the main circuit breaker and test the lines with a voltmeter before working on the system. Since voltage is a two-point phenomenon, be sure to work with only one line at at time—accidents happen!

You should also be aware that the reaction to dc voltages is quite different from that to ac voltages. You have probably seen in movies or comic strips that people are often unable to let go of a *hot* wire. This is evidence of the most important difference between the two types of voltages. As mentioned above, if you happen to touch a "hot' 120 V ac line, you will probably get a good sting, but *you can let go.* If it happens to be a "hot" 120 V dc line, you will probably not be able to let go, and you could die. Time plays an important role when this happens, because the longer you are subjected to the dc voltage, the more the resistance in the body decreases, until a fatal current can be established. The reason that we can let go of an ac line is best demonstrated by carefully examining the 120 V rms, 60 Hz voltage in Fig. 13.78. Since the voltage is oscillating, there is a period when the voltage is near zero or less than, say, 20 V, and is reversing in direction. Although this time interval is very short, it appears every 8.3 ms and provides a window for you to *let go.*

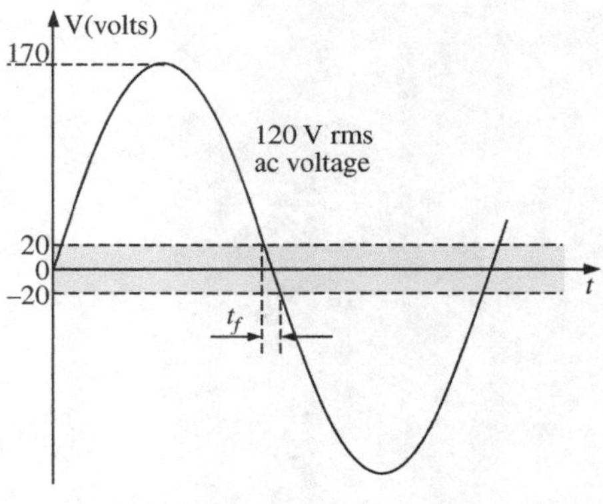

FIG. 13.78
Interval of time when sinusoidal voltage is near zero volts.

Now that we are aware of the additional dangers of dc voltages, it is important to mention that under the wrong conditions, dc voltages as low as 12 V, such as from a car battery, can be quite dangerous. If you happen to be working on a car under wet conditions, or if you are sweating badly for some reason or, worse yet, wearing a wedding ring that may have moisture and body salt underneath, touching the positive terminal may initiate the process whereby the body resistance begins to drop, and serious injury could take place. It is one of the reasons you seldom see a professional electrician wearing any rings or jewelry—it is just not worth the risk.

Before leaving this topic of safety concerns, you should also be aware of the dangers of high-frequency supplies. We are all aware of what 2.45 GHz at 120 V can do to a meat product in a microwave oven, and it is therefore very important that the seal around the oven be as tight as possible. However, don't ever assume that anything is absolutely perfect in design—so don't make it a habit to view the cooking process in the microwave 6 in. from the door on a continuing basis. Find something else to do, and check the food only when the cooking process is complete. If you ever visit the Empire State Building, you will notice that you are unable to get close to the antenna on the dome due to the high-frequency signals being emitted with a great deal of power. Also note the large KEEP OUT signs near radio transmission towers for local radio stations. Standing within 10 ft of an AM transmitter working at 540 kHz would bring on disaster. Simply holding (do not try!) a fluorescent bulb near the tower could make it light up due to the excitation of the molecules inside the bulb.

In total, therefore, treat any situation with high ac voltages or currents, high-energy dc levels, and high frequencies with added care.

13.11 COMPUTER ANALYSIS

PSpice

OrCAD Capture offers a variety of ac voltage and current sources. However, for the purposes of this text, the voltage source **VSIN** and the current source **ISIN** are the most appropriate because they have a list of attributes that cover current areas of interest. Under the library **SOURCE,** a number of others are listed, but they don't have the full range of the above, or they are dedicated to only one type of analysis. On occasion, **ISRC** is used because it has an arrow symbol like that appearing in the text, and it can be used for dc, ac, and some transient analyses. The symbol for **ISIN** is a sine wave that utilizes the plus-and-minus sign (±) to indicate direction. The sources **VAC, IAC, VSRC,** and **ISRC** are fine if the magnitude and the phase of a specific quantity are desired or if a transient plot against frequency is desired. However, they will not provide a transient response against time even if the frequency and the transient information are provided for the simulation.

For all of the sinusoidal sources, the magnitude (**VAMPL**) is the peak value of the waveform, not the rms value. This becomes clear when a plot of a quantity is desired and the magnitude calculated by PSpice is the peak value of the transient response. However, for a purely steady-state ac response, the magnitude provided can be the rms value and the output read as the rms value. Only when a plot is desired will it be clear that PSpice is accepting every ac magnitude as the peak value of the waveform. Of course, the phase angle is the same whether the magnitude is the peak or the rms value.

Before examining the mechanics of getting the various sources, remember that

Transient Analysis provides an ac or a dc output versus time, while AC Sweep is used to obtain a plot versus frequency.

To obtain any of the sources listed above, apply the following sequence: **Place part** key-**Place Part** dialog box-**Source**-(enter type of source). Once you select the source, the ac source **VSIN** appears on the schematic with **OFF, VAMPL,** and **FREQ**. Always specify **VOFF** as 0 V (unless a specific value is part of the analysis), and provide a value for the amplitude and frequency. Enter the remaining quantities of **PHASE, AC, DC, DF,** and **TD** by double-clicking on the source symbol to obtain the **Property Editor**, although **PHASE, DF** (damping factor), and **TD** (time delay) do have a default of 0 s. To add a phase angle, click on **PHASE**, enter the phase angle in the box below, and then select **Apply**. If you want to display a factor such as a phase angle of 60°, click on **PHASE** followed by **Display** to obtain the **Display Properties** dialog box. Then choose **Name and Value** followed by **OK** and **Apply**, and leave the **Properties Editor** dialog box (**X**) to see **PHASE=60** next to the **VSIN** source. The next chapter includes the use of the ac source in a simple circuit.

Multisim

For Multisim, the ac voltage source is available from three sources—the **Place Source** key pad in the **Components** toolbar, the **Show Power Source Family** in the **Virtual** or **BASIC** toolbar, and the **Function Generator.** The major difference among the options is that the phase angle cannot be set using the **Function Generator.**

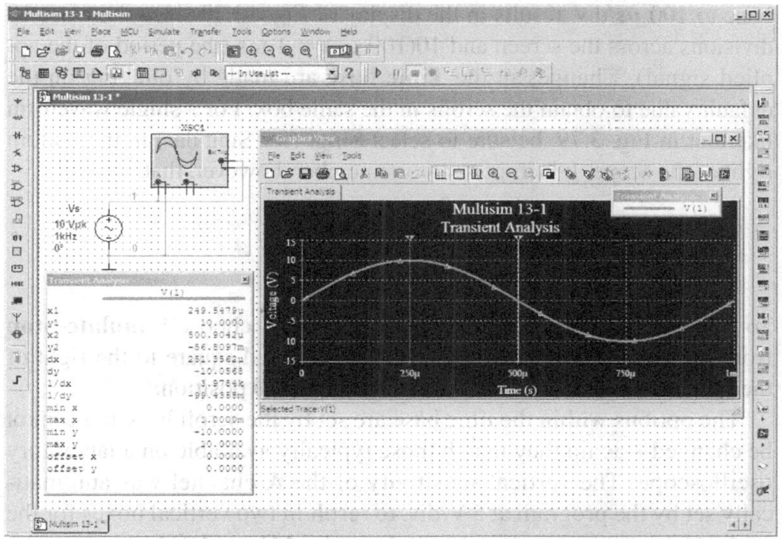

FIG. 13.79
Using the oscilloscope to display the sinusoidal ac voltage source available in the Multisim Sources tool bin.

Using the **Place Source** option, select **SIGNAL_VOLTAGE_SOURCES** group under the **Family** heading, followed by **AC_VOLTAGE.** When selected and placed, it displays the default values for the amplitude, frequency, and phase. All the parameters of the source can be changed by double-clicking on the source symbol to obtain the dialog box. The listing clearly indicates that the set voltage is the peak value. Note that the unit of measurement is controlled by the scrolls to the right of the default label and cannot be set by typing in the desired unit of measurement. The label can be changed by switching the **Label** heading and inserting the desired label. After all the changes have been made in the dialog box, click **OK,** and all the changes appear next to the ac voltage source symbol. In Fig. 13.79, the label was changed to **Vs** and the amplitude to 10 V, while the freqency and phase angle were left with their default values. It is particularly important to realize that

for any frequency analysis (that is, where the frequency will change), the AC Magnitude of the ac source must be set under Analysis Setup in the SIGNAL_VOLTAGE_SOURCES dialog box. Failure to do so will create results linked to the default values rather than the value set under the Value heading.

To view the sinusoidal voltage set in Fig. 13.79, select an oscilloscope from the **Instrument** toolbar at the right of the screen. When hooking up the oscilloscope, do not worry about overlapping wires. Connections are shown by small, solid dots. It is the fourth option down and has the appearance shown in Fig. 13.79 when selected. Note that it is a dual-channel oscilloscope with an **A** channel and a **B** channel. It has a ground **(G)** connection and a trigger **(T)** connection. The connections for viewing the ac voltage source on the **A** channel are provided in Fig. 13.79. Note that the trigger control is also connected to the **A** channel for sync control. The screen appearing in Fig. 13.79 can be displayed by double-clicking on the oscilloscope symbol on the screen. It has all the major controls of a typical laboratory oscilloscope. When you select **Simulate-Run** or select **1** on the **Simulate Switch,** the ac voltage appears on the screen. Changing the **Time**

base to 100 μs/div. results in the display of Fig. 13.79 since there are 10 divisions across the screen and 10(100 μs) = 1 ms (the period of the applied signal). Changes in the **Time base** are made by clicking on the default value to obtain the scrolls in the same box. For a single waveform like that in Fig. 3.79, be sure to select **Sing.** (for Singular) in the bottom right of the scope. It is important to remember, however, that

changes in the oscilloscope setting or any network should not be made until the simulation is ended by disabling the Simulate-Run option or placing the Simulate switch in the 0 mode.

To stop the simulation, there are three options: choose **Simulate-Stop** from the top toolbar on the screen; select the red square to the right of the green arrow; or click the switch back to the **0** position.

The options within the time base are set by the scroll bars and cannot be changed—again they match those typically available on a laboratory oscilloscope. The vertical sensitivity of the **A** channel was automatically set by the program at 5 V/div. to result in two vertical boxes for the peak value as shown in Fig. 13.79. Note the **AC** and **DC** keypads below Channel **A**. Since there is no dc component in the applied signal, either one results in the same display. The **Trigger** control is set on the positive transition at a level of 0 V. The **T1** and **T2** refer to the cursor positions on the horizontal time axis. By clicking on the small green triangle at the top of the green line at the far left edge of the screen and dragging the triangle, you can move the vertical green line to any position along the axis. In Fig. 13.79, it was moved to the peak value of the waveform at one-quarter of the total period or 0.25 ms = 250 μs. Note the value of **T1** (250 μs) and the corresponding value of **VA1** (9.995 V $\cong$ 10.0 V). By selecting the other cursor with a yellow triangle at the top to one-half the total period or 0.5 ms = 500 μs, we find that the value at **T2** (500 μs) is 0.008 pV (**VA2**), which is essentially 0 V for a waveform with a peak value of 10 V. The accuracy is controlled by the number of data points called for in the simulation setup. The more data points, the higher is the likelihood of a higher degree of accuracy for the desired quantity. However, an increased number of data points also extends the running time of the simulation. The third line provides the difference between **T2** and **T1** as 250 μs and difference between their magnitudes (**VA2–VA1**) as −9.995 V, with the negative sign appearing because **VA1** is greater than **VA2**.

As mentioned above, you can also obtain an ac voltage from the **Function Generator** appearing as the second option down on the **Instrument** toolbar. Its symbol appears in Fig. 13.80 with positive, negative, and ground connections. Double-click on the generator graphic symbol, and the **Function Generator** dialog box appears in which selections can be made. For this example, the sinusoidal waveform is chosen. To set the frequency, click on the unit of measurement to produce a list of options. For this case, kHz was chosen and the **1** left as is. The **Amplitude** (peak value) is set as V_p = 10 V and the **Offset** at 0 V. Note that there is no option to set the phase angle as was possible for the source above. Double-clicking on the oscilloscope generates the **Oscilloscope-XSCI** dialog box in which a **Timebase** of 100 μs/div. can be set again with a vertical sensitivity of 5 V/div. Select **1** on the **Simulate** switch, and the waveform of Fig. 13.80 appears. Choosing **Sing.** under **Trigger** results in a fixed display. Set the **Simulate** switch on **0** to end the simulation. Placing the cursors in the same position shows that the waveforms for Figs. 13.79 and 13.80 are the same.

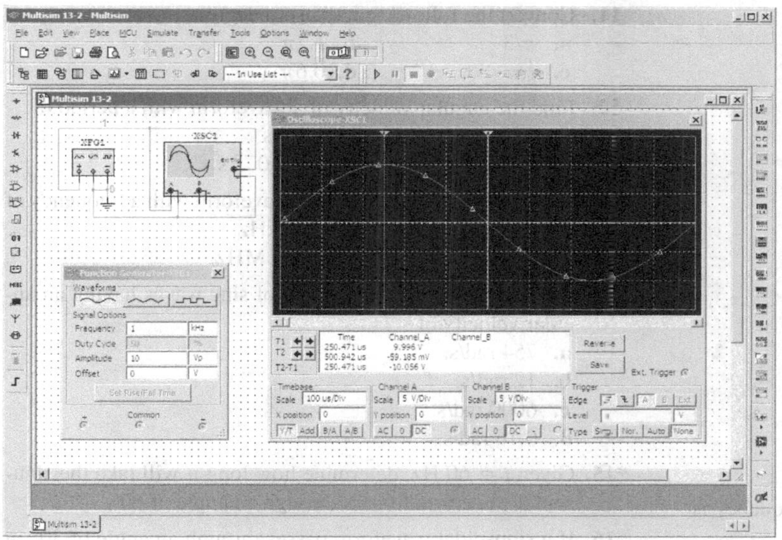

FIG. 13.80
Using the function generator to place a sinusoidal ac voltage waveform on the screen of the oscilloscope.

For most of the Multisim analyses to appear in this text, the **AC_VOLTAGE** under **Place Source** will be employed. However, with such a limited introduction to Multisim, it seemed appropriate to introduce the use of the **Function Generator** because of its close linkage to the laboratory experience.

PROBLEMS

SECTION 13.2 Sinusoidal ac Voltage Characteristics and Definitions

1. For the sinusoidal waveform in Fig. 13.81:
   - **a.** What is the peak value?
   - **b.** What is the instantaneous value at 15 ms and at 20 ms?
   - **c.** What is the peak-to-peak value of the waveform?
   - **d.** What is the period of the waveform?
   - **e.** How many cycles are shown?

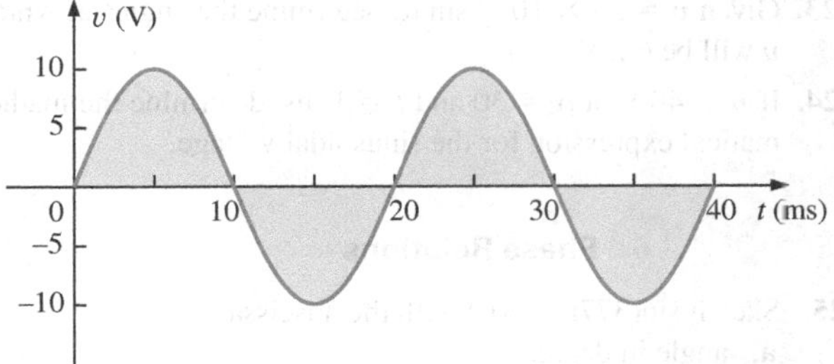

FIG. 13.81
Problem 1.

2. For the sinusoidal signal in Fig. 13.82:
   - **a.** What is the peak value?
   - **b.** What is the instantaneous value at 1 μs and at 7 μs.
   - **c.** What is the peak-to-peak value of the waveform?
   - **d.** What is the period of the waveform?
   - **e.** How many cycles are shown?

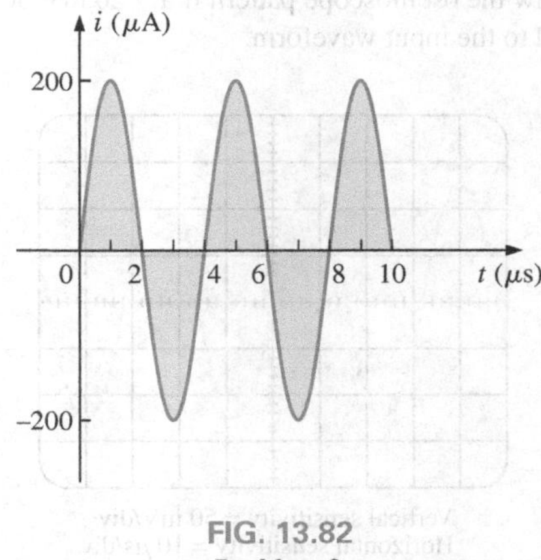

FIG. 13.82
Problem 2.

3. For the periodic square-wave waveform in Fig. 13.83:
   - **a.** What is the peak value?
   - **b.** What is the instantaneous value at 1.5 ms and at 5.1 ms?
   - **c.** What is the peak-to-peak value of the waveform?
   - **d.** What is the period of the waveform?
   - **e.** How many cycles are shown?

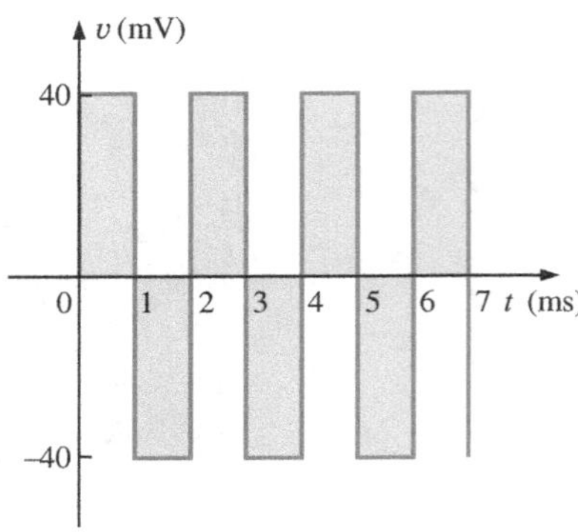

FIG. 13.83
Problem 3.

SECTION 13.3 Frequency Spectrum

4. Find the period of a periodic waveform whose frequency is
   - a. 200 Hz.
   - b. 40 MHz.
   - c. 20 kHz.
   - d. 1 Hz.
5. Find the frequency of a repeating waveform whose period is
   - a. 1 s.
   - b. $\frac{1}{16}$ s.
   - c. 40 ms.
   - d. 25 μs.
6. If a periodic waveform has a frequency of 1 kHz, how long (in seconds) will it take to complete five cycles?
7. Find the period of a sinusoidal waveform that completes 80 cycles in 24 ms.
8. What is the frequency of a periodic waveform that completes 42 cycles in 6 s?
9. For the oscilloscope pattern of Fig. 13.84:
   - a. Determine the peak amplitude.
   - b. Find the period.
   - c. Calculate the frequency.

   Redraw the oscilloscope pattern if a +20 mV dc level were added to the input waveform.

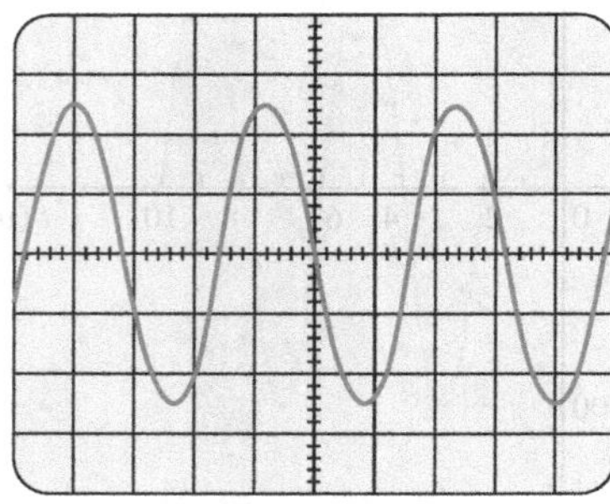

Vertical sensitivity = 50 mV/div.
Horizontal sensitivity = 10 μs/div.

FIG. 13.84
Problem 9.

SECTION 13.4 The Sinusoidal Waveform

10. Convert the following degrees to radians:
    - a. 40°
    - b. 60°
    - c. 135°
    - d. 170°
11. Convert the following radians to degrees:
    - a. $\pi/3$
    - b. 1.2π
    - c. $\frac{1}{10}\pi$
    - d. 0.6π
12. Find the angular velocity of a waveform with a period of
    - a. 1.8 s.
    - b. 0.3 ms.
    - c. 8 μs.
    - d. 4×10^{-6} s.
13. Find the angular velocity of a waveform with a frequency of
    - a. 100 Hz.
    - b. 0.25 kHz.
    - c. 2 kHz.
    - d. 0.004 MHz.
14. Find the frequency and period of sine waves having an angular velocity of
    - a. 754 rad/s.
    - b. 12 rad/s.
    - c. 6000 rad/s.
    - d. 0.16 rad/s.

*15. Given $f = 60$ Hz, determine how long it will take the sinusoidal waveform to pass through an angle of 60°.

*16. If a sinusoidal waveform passes through an angle of 30° in 5 ms, determine the angular velocity of the waveform.

SECTION 13.5 General Format for the Sinusoidal Voltage or Current

17. Find the amplitude and frequency of the following waves:
    - a. $20 \sin 377t$
    - b. $12 \sin 2\pi\, 120t$
    - c. $10^6 \sin 10{,}000t$
    - d. $-8 \sin 10{,}058t$
18. Sketch $6 \sin 754t$ with the abscissa
    - a. angle in degrees.
    - b. angle in radians.
    - c. time in seconds.

*19. Sketch $-8 \sin 2\pi\, 80t$ with the abscissa
   - a. angle in degrees.
   - b. angle in radians.
   - c. time in seconds.

20. If $e = 300 \sin 157t$, how long (in seconds) does it take this waveform to complete 1/2 cycle?
21. Given $i = 0.5 \sin \alpha$, determine i at $\alpha = 72°$.
22. Given $v = 20 \sin \alpha$, determine v at $\alpha = 1.2\pi$.

*23. Given $v = 30 \times 10^{-3} \sin \alpha$, determine the angles at which v will be 6 mV.

*24. If $v = 40$ V at $\alpha = 30$ and $t = 1$ ms, determine the mathematical expression for the sinusoidal voltage.

SECTION 13.6 Phase Relations

25. Sketch $\sin(377t + 60°)$ with the abscissa
    - a. angle in degrees.
    - b. angle in radians.
    - c. time in seconds.
26. Sketch the following waveforms:
    - a. $50 \sin(\omega t + 0°)$
    - b. $5 \sin(\omega t + 120°)$
    - c. $2 \cos(\omega t + 10°)$
    - d. $-20 \sin(\omega t + 10°)$

27. Write the analytical expression for the waveforms of Fig. 13.85 with the phase angle in degrees.

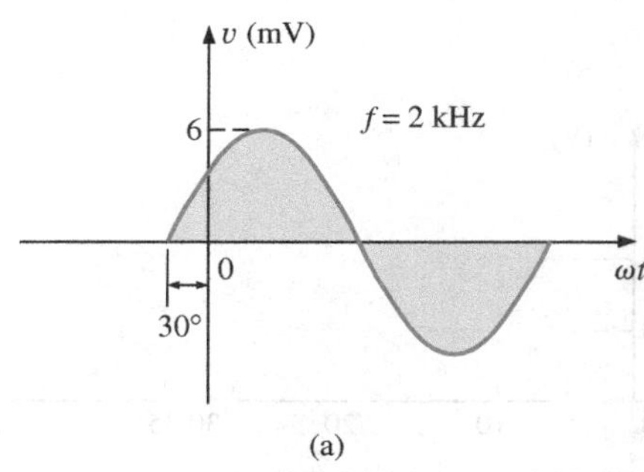

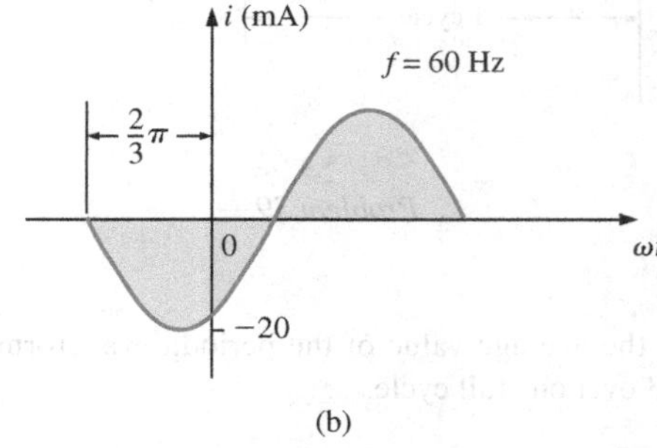

FIG. 13.85
Problem 27.

28. Write the analytical expression for the waveform of Fig. 13.86 with the phase angle in degrees.

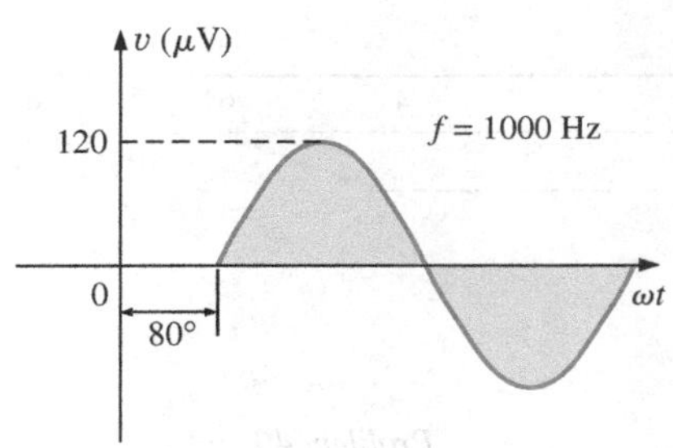

FIG. 13.86
Problem 28.

29. Write the analytical expression for the waveform of Fig. 13.87 with the phase angle in degrees.

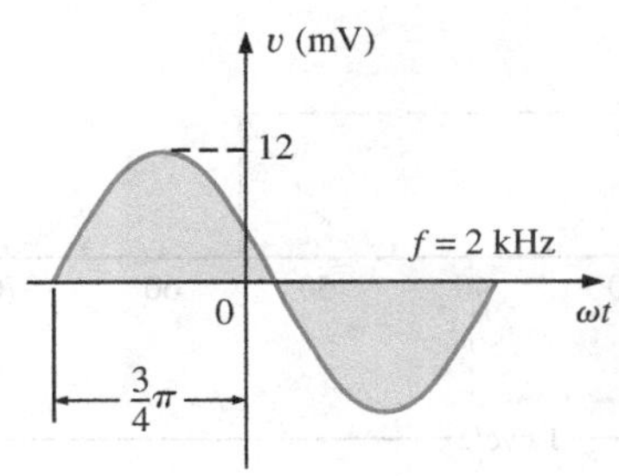

FIG. 13.87
Problem 29.

30. Write the analytical expression for the waveform of Fig. 13.88 with the phase angle in radians.

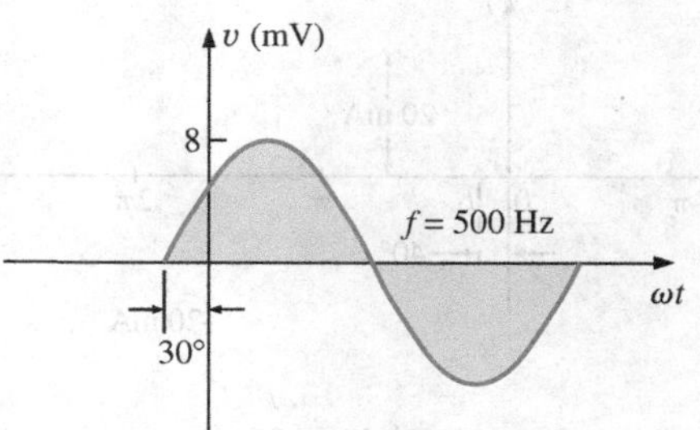

FIG. 13.88
Problem 30.

31. Find the phase relationship between the following waveforms:

$$v = 25 \sin(\omega t + 80°)$$
$$i = 4 \sin(\omega t - 10°)$$

32. Find the phase relationship between the following waveforms:

$$v = 0.2 \sin(\omega t - 60°)$$
$$i = 0.1 \sin(\omega t - 20°)$$

***33.** Find the phase relationship between the following waveforms:

$$v = 2 \cos(\omega t - 30°)$$
$$i = 5 \sin(\omega t + 60°)$$

***34.** Find the phase relationship between the following waveforms:

$$v = -4 \cos(\omega t + 90°)$$
$$i = -2 \sin(\omega t + 10°)$$

***35.** The sinusoidal voltage $v = 160 \sin(2\pi 1000t + 60°)$ is plotted in Fig. 13.89. Determine the time t_1 when the waveform crosses the axis.

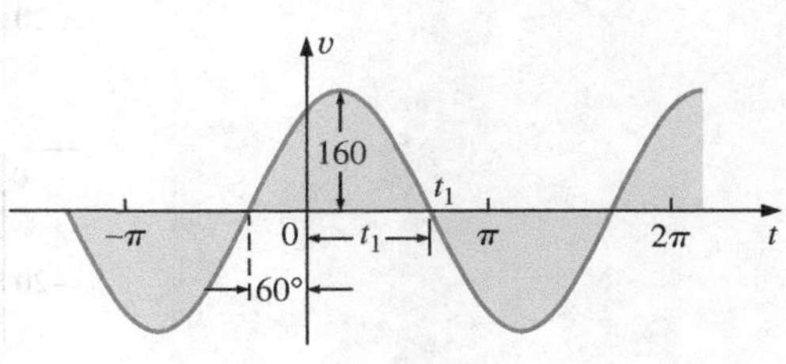

FIG. 13.89
Problem 35.

***36.** The sinusoidal current $i = 20 \times 10^{-3} \sin(50{,}000t - 40°)$ is plotted in Fig. 13.90. Determine the time t_1 when the waveform crosses the axis.

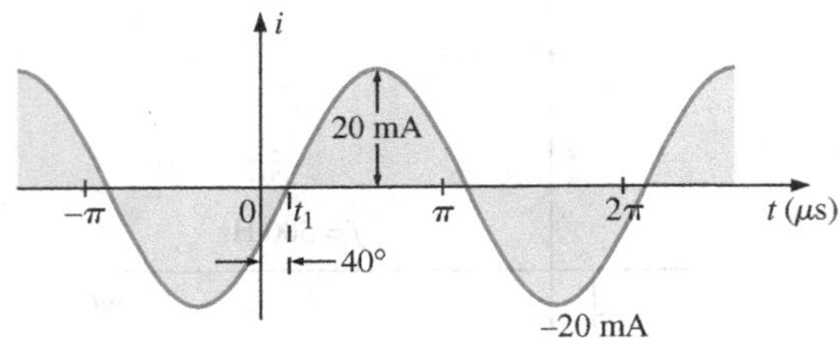

FIG. 13.90
Problem 36.

37. For the waveform of Fig. 13.89, find the time when the waveform has its peak value.

38. For the oscilloscope display in Fig. 13.91:

a. Determine the period of the waveform.
b. Determine the frequency of each waveform.
c. Find the rms value of each waveform.
d. Determine the phase shift between the two waveforms and determine which leads and which lags.

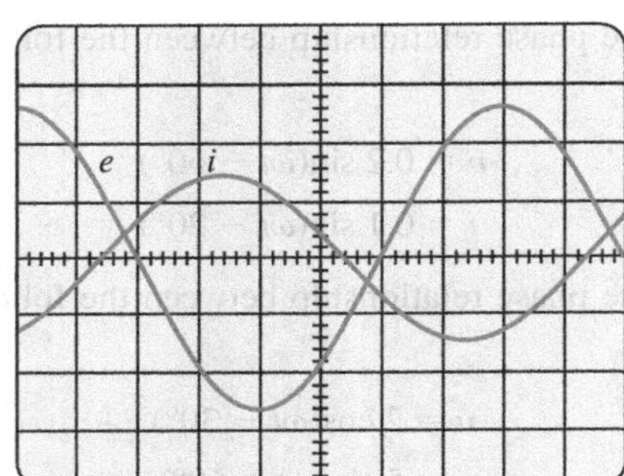

FIG. 13.91
Problem 38.

SECTION 13.7 Average Value

39. Find the average value of the periodic waveform in Fig. 13.92.

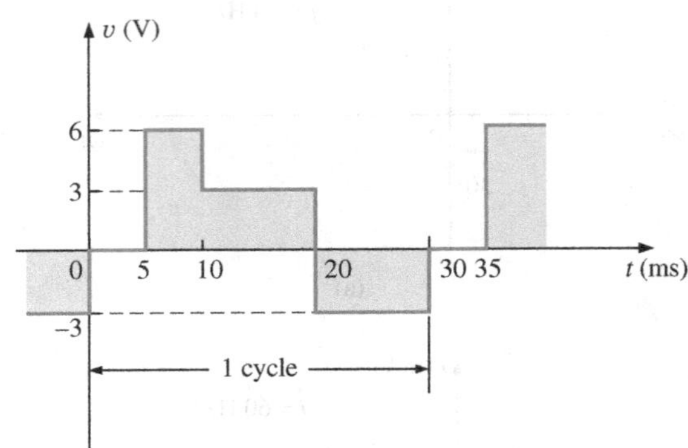

FIG. 13.92
Problem 39.

40. Find the average value of the periodic waveform in Fig. 13.93 over one full cycle.

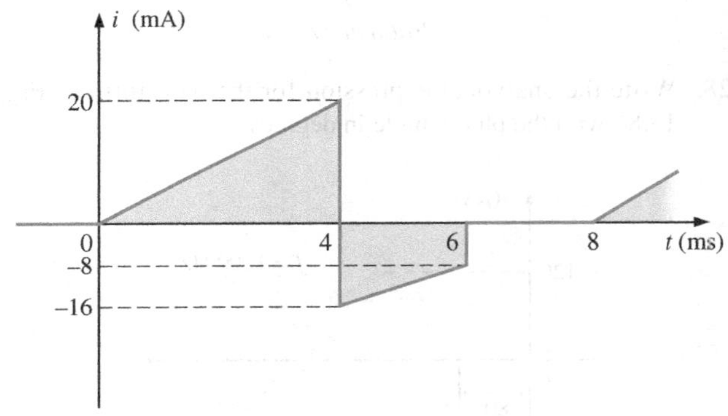

FIG. 13.93
Problem 40.

41. Find the average value of the periodic waveform of Fig. 13.94 over one full cycle.

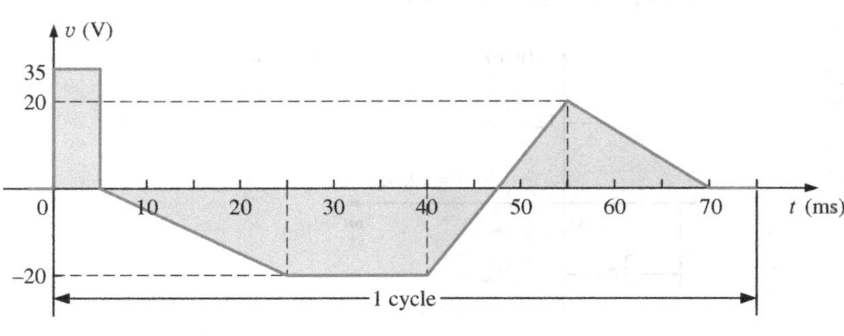

FIG. 13.94
Problem 41.

42. Find the average value of the periodic waveform of Fig. 13.95 over one full cycle.

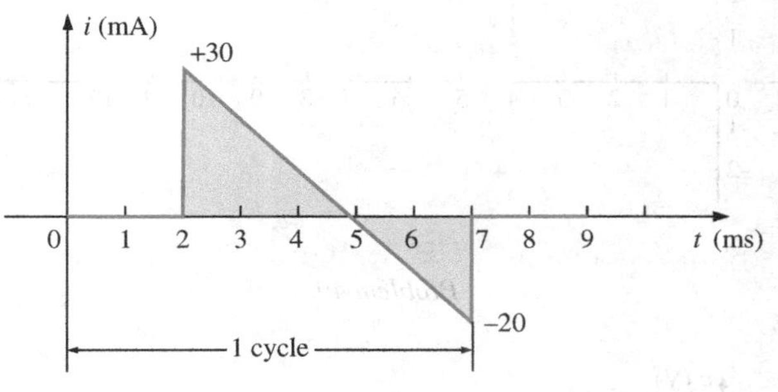

FIG. 13.95
Problem 42.

43. Find the average value of the periodic function of Fig. 13.96:

a. By inspection.
b. Through calculations.
c. Compare the results of parts (a) and (b).

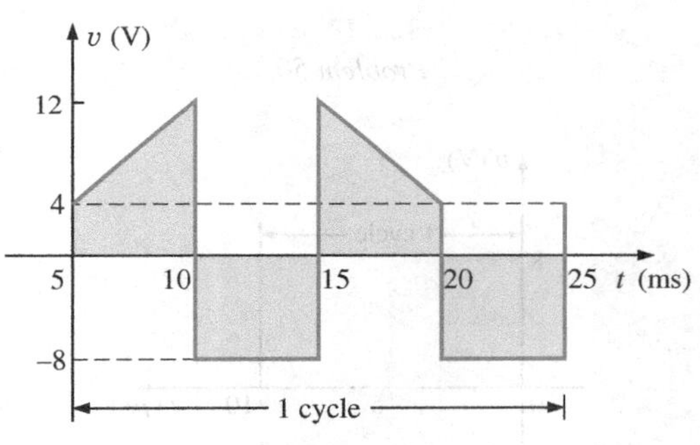

FIG. 13.96
Problem 43.

44. Find the average value of the periodic waveform in Fig. 13.97.

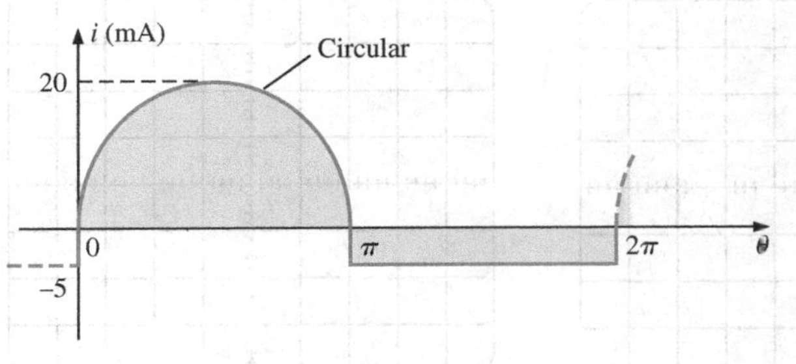

FIG. 13.97
Problem 44.

45. For the waveform in Fig. 13.98:

a. Determine the period.
b. Find the frequency.
c. Determine the average value.
d. Sketch the resulting oscilloscope display if the vertical channel is switched from dc to ac.

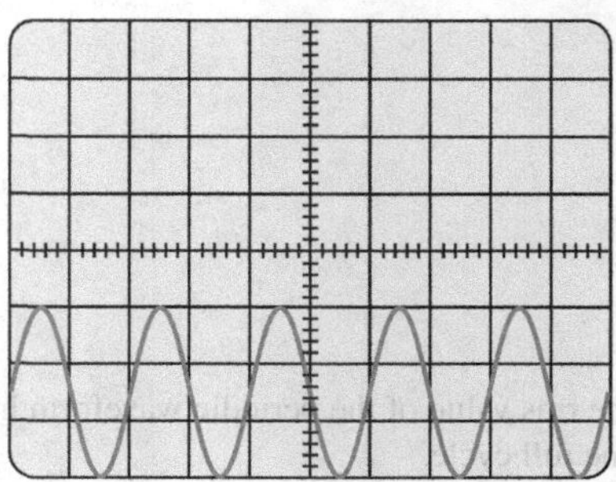

Vertical sensitivity = 10 mV/div.
Horizontal sensitivity = 0.2 ms/div.

FIG. 13.98
Problem 45.

***46.** For the waveform in Fig. 13.99:

a. Determine the period.
b. Find the frequency.
c. Determine the average value.
d. Sketch the resulting oscilloscope display if the vertical channel is switched from dc to ac.

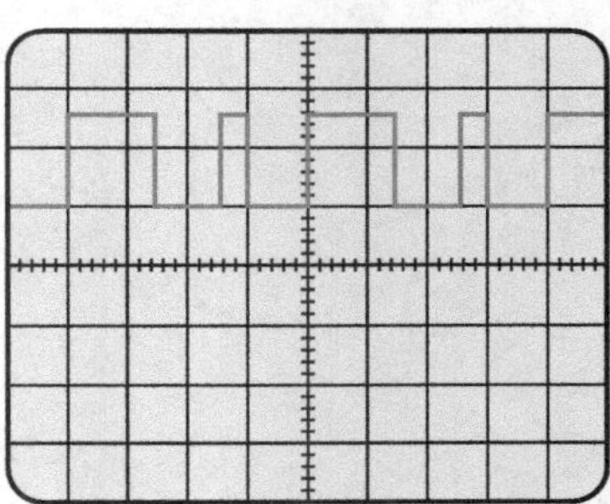

Vertical sensitivity = 10 mV/div.
Horizontal sensitivity = 10 μs/div.

FIG. 13.99
Problem 46.

SECTION 13.8 Effective (rms) Values

47. Find the rms values of the following sinusoidal waveforms:

a. $v = 120 \sin(377t + 60°)$
b. $i = 6 \times 10^{-3} \sin(2\pi\ 1000t)$
c. $v = 8 \times 10^{-6} \sin(2\pi\ 5000t + 30°)$

48. Write the sinusoidal expressions for voltages and currents having the following rms values at a frequency of 60 Hz with zero phase shift:

a. 4.8 V
b. 50 mA
c. 2 kV

49. Find the rms value of the periodic waveform in Fig. 13.100 over one full cycle.

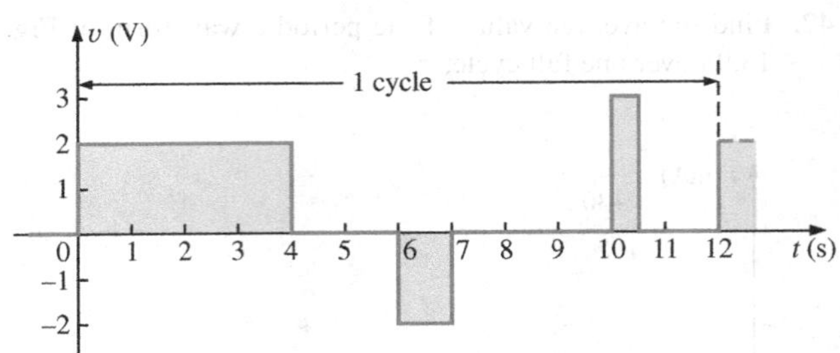

FIG. 13.100
Problem 49.

50. Find the rms value of the periodic waveform in Fig. 13.101 over one full cycle.

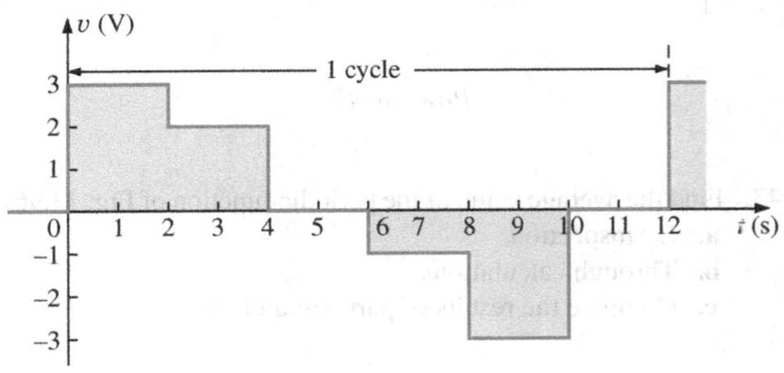

FIG. 13.101
Problem 50.

51. What are the average and rms values of the square wave in Fig. 13.102?

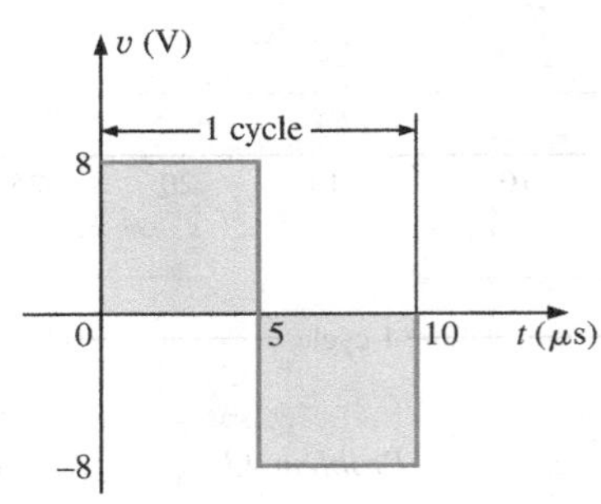

FIG. 13.102
Problem 51.

***52.** For each waveform in Fig. 13.103, determine the period, frequency, average value, and rms value.

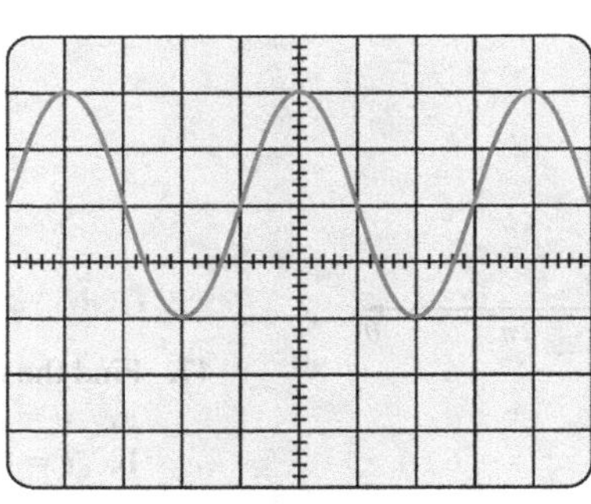

Vertical sensitivity = 20 mV/div.
Horizontal sensitivity = 10 μs/div.

(a)

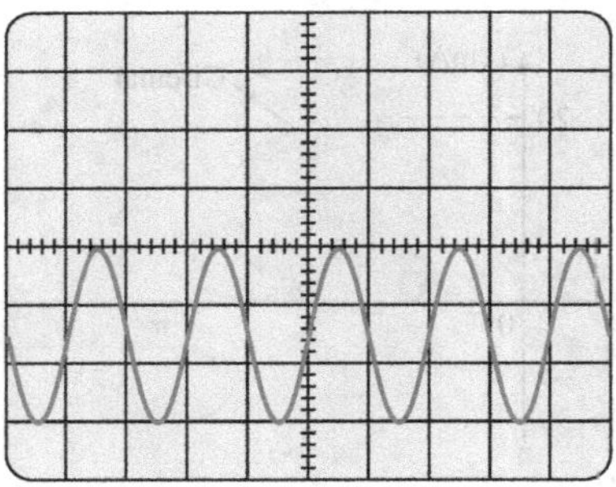

Vertical sensitivity = 0.2 V/div.
Horizontal sensitivity = 50 μs/div.

(b)

FIG. 13.103
Problem 52.

***53.** For the waveform of Fig. 13.104:

a. Carefully sketch the squared waveform. Note that the equation for the sloping line must first be determined.

b. Using some basic area equations and the approximate approach, find the approximate area under the squared curve.

c. Determine the rms value of the original waveform.

d. Find the average value of the original waveform.

e. How does the average value of the waveform compare to the rms value?

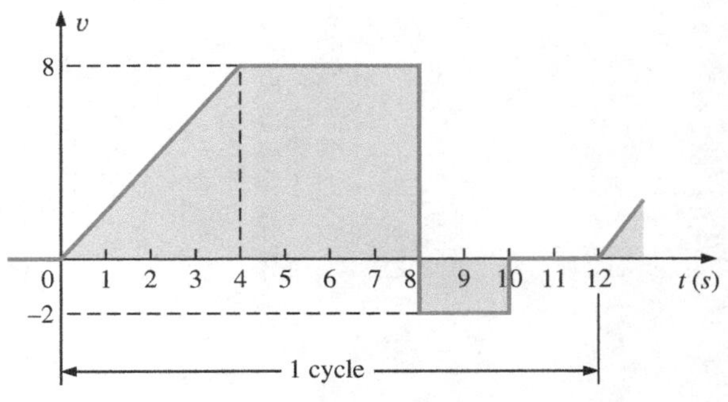

FIG. 13.104
Problem 53.

SECTION 13.9 ac Meters and Instruments

54. Determine the reading of the meter for each situation in Fig. 13.105.

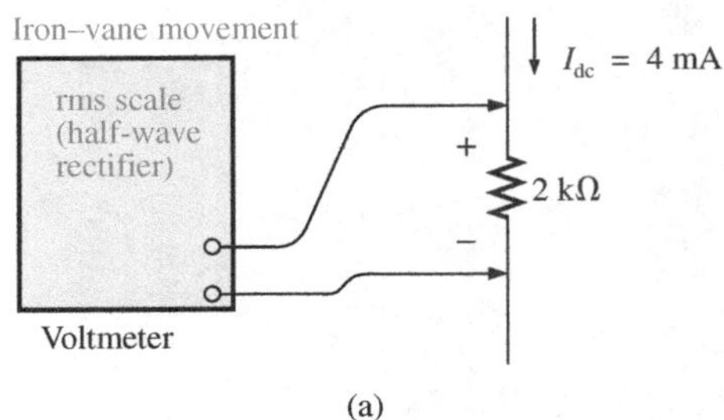

(a)

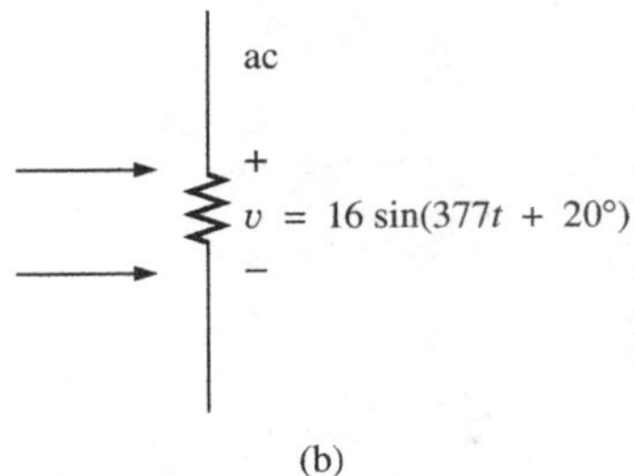

(b)

FIG. 13.105
Problem 54.

GLOSSARY

Alternating waveform A waveform that oscillates above and below a defined reference level.

Angular velocity The velocity with which a radius vector projecting a sinusoidal function rotates about its center.

Average value The level of a waveform defined by the condition that the area enclosed by the curve above this level is exactly equal to the area enclosed by the curve below this level.

Calibration factor A multiplying factor used to convert from one meter indication to another.

Clamp Meter® A clamp-type instrument that will permit noninvasive current measurements and that can be used as a conventional voltmeter or ohmmeter.

Cycle A portion of a waveform contained in one period of time.

Effective value The equivalent dc value of any alternating voltage or current.

Electrodynamometer meters Instruments that can measure both ac and dc quantities without a change in internal circuitry.

Frequency (*f*) The number of cycles of a periodic waveform that occur in 1 s.

Frequency counter An instrument that will provide a digital display of the frequency or period of a periodic time-varying signal.

Instantaneous value The magnitude of a waveform at any instant of time, denoted by lowercase letters.

Lagging waveform A waveform that crosses the time axis at a point in time later than another waveform of the same frequency.

Leading waveform A waveform that crosses the time axis at a point in time ahead of another waveform of the same frequency.

Oscilloscope An instrument that will display, through the use of a cathode-ray tube, the characteristics of a time-varying signal.

Peak amplitude The maximum value of a waveform as measured from its average, or mean, value, denoted by uppercase letters.

Peak-to-peak value The magnitude of the total swing of a signal from positive to negative peaks. The sum of the absolute values of the positive and negative peak values.

Peak value The maximum value of a waveform, denoted by uppercase letters.

Period (*T*) The time interval necessary for one cycle of a periodic waveform.

Periodic waveform A waveform that continually repeats itself after a defined time interval.

Phase relationship An indication of which of two waveforms leads or lags the other, and by how many degrees or radians.

Radian (rad) A unit of measure used to define a particular segment of a circle. One radian is approximately equal to 57.3°; 2π rad are equal to 360°.

Root-mean-square (rms) value The root-mean-square or effective value of a waveform.

Sinusoidal ac waveform An alternating waveform of unique characteristics that oscillates with equal amplitude above and below a given axis.

VOM A multimeter with the capability to measure resistance and both ac and dc levels of current and voltage.

Waveform The path traced by a quantity, plotted as a function of some variable such as position, time, degrees, temperature, and so on.

GLOSSARY

Alternating waveform A waveform that oscillates above and below a defined reference level.

Angular velocity The velocity with which a radius vector projecting a sinusoidal function rotates about its center.

Average value The level of a waveform defined by the condition that the area enclosed by the curve above this level is exactly equal to the area enclosed by the curve below this level.

Calibration factor A multiplying factor used to convert from one meter indication to another.

Clamp Meter® A clamp-type instrument that will permit noninvasive current measurements and that can be used as a conventional voltmeter or ohmmeter.

Cycle A portion of a waveform contained in one period of time.

Effective value The equivalent dc value of any alternating voltage or current.

Electrodynamometer meters Instruments that can measure both ac and dc quantities without a change in internal circuitry.

Frequency (f) The number of cycles of a periodic waveform that occur in 1 s.

Frequency counter An instrument that will provide a digital display of the frequency or period of a periodic time-varying signal.

Instantaneous value The magnitude of a waveform at any instant of time; denoted by lowercase letters.

Lagging waveform A waveform that crosses the time axis at a point in time later than another waveform of the same frequency.

Leading waveform A waveform that crosses the time axis at a point in time ahead of another waveform of the same frequency.

Oscilloscope An instrument that will display, through the use of a cathode-ray tube, the characteristics of a time-varying signal.

Peak amplitude The maximum value of a waveform as measured from its average, or mean, value, denoted by uppercase letters.

Peak-to-peak value The magnitude of the total swing of a signal from positive to negative peaks. The sum of the absolute values of the positive and negative peak values.

Peak value The maximum value of a waveform, denoted by uppercase letters.

Period (T) The time interval necessary for one cycle of a periodic waveform.

Periodic waveform A waveform that continually repeats itself after a defined time interval.

Phase relationship An indication of which of two waveforms leads or lags the other, and by how many degrees or radians.

Radian (rad) A unit of measure used to define a particular segment of a circle. One radian is approximately equal to 57.3°; 2π rad are equal to 360°.

Root-mean-square (rms) value The root-mean-square or effective value of a waveform.

Sinusoidal ac waveform An alternating waveform of unique characteristics that oscillates with equal amplitude above and below a given axis.

VOM A multimeter with the capability to measure resistance and both ac and dc levels of current and voltage.

Waveform The path traced by a quantity, plotted as a function of some variable such as position, time, degrees, temperature, and so on.

53. For the waveform of Fig. 13.104:
   a. Carefully sketch the squared waveform. Note that the equation for the sloping line must first be determined.
   b. Using some basic area equations and the approximate approach, find the approximate area under the squared curve.
   c. Determine the rms value of the original waveform.
   d. Find the average value of the original waveform.
   e. How does the average value of the waveform compare to the rms value?

Problem 53.

SECTION 13.9 ac Meters and Instruments

54. Determine the reading of the meter for each situation in Fig. 13.105.

Problem 54.

The Basic Elements and Phasors

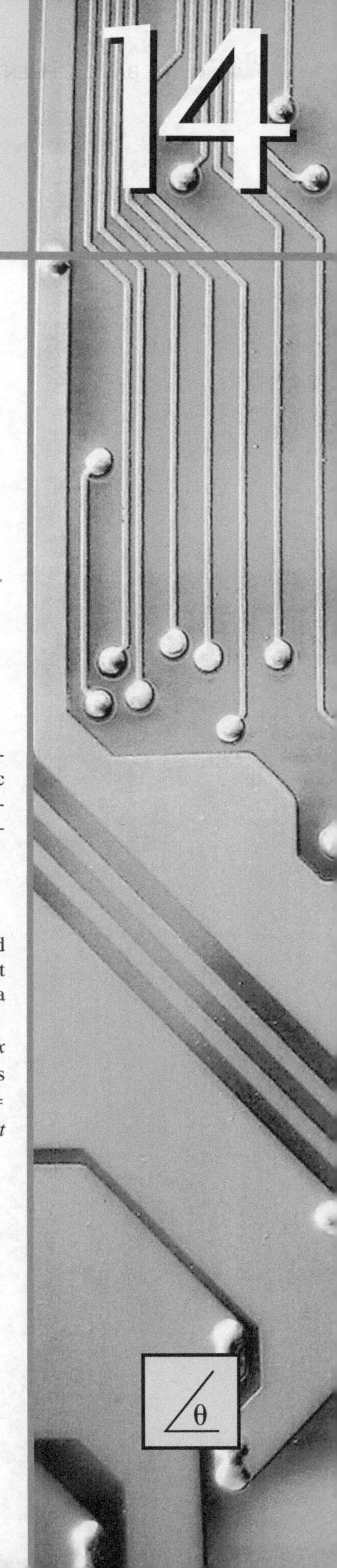

Objectives

- *Become familiar with the response of a resistor, an inductor, and a capacitor to the application of a sinusoidal voltage or current.*
- *Learn how to apply the phasor format to add and subtract sinusoidal waveforms.*
- *Understand how to calculate the real power to resistive elements and the reactive power to inductive and capacitive elements.*
- *Become aware of the differences between the frequency response of ideal and practical elements.*
- *Become proficient in the use of a calculator to work with complex numbers.*

14.1 INTRODUCTION

The response of the basic *R*, *L*, and *C* elements to a sinusoidal voltage and current are examined in this chapter, with special note of how frequency affects the "opposing" characteristic of each element. Phasor notation is then introduced to establish a method of analysis that permits a direct correspondence with a number of the methods, theorems, and concepts introduced in the dc chapters.

14.2 DERIVATIVE

To understand the response of the basic *R*, *L*, and *C* elements to a sinusoidal signal, you need to examine the concept of the **derivative** in some detail. You do not have to become proficient in the mathematical technique but simply understand the impact of a relationship defined by a derivative.

Recall from Section 10.10 that the derivative dx/dt is defined as the rate of change of *x* with respect to time. If *x* fails to change at a particular instant, $dx = 0$, and the derivative is zero. For the sinusoidal waveform, dx/dt is zero only at the positive and negative peaks ($\omega t = \pi/2$ and $\frac{3}{2}\pi$ in Fig. 14.1), since *x* fails to change at these instants of time. The derivative dx/dt is actually the slope of the graph at any instant of time.

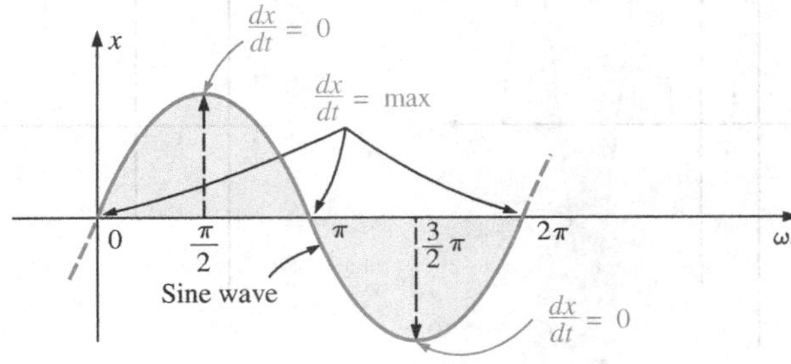

FIG. 14.1
Defining those points in a sinusoidal waveform that have maximum and minimum derivatives.

A close examination of the sinusoidal waveform will also indicate that the greatest change in x occurs at the instants $\omega t = 0$, π, and 2π. The derivative is therefore a maximum at these points. At 0 and 2π, x increases at its greatest rate, and the derivative is given a positive sign since x increases with time. At π, dx/dt decreases at the same rate as it increases at 0 and 2π, but the derivative is given a negative sign since x decreases with time. Since the rate of change at 0, π, and 2π is the same, the magnitude of the derivative at these points is the same also. For various values of ωt between these maxima and minima, the derivative will exist and have values from the minimum to the maximum inclusive. A plot of the derivative in Fig. 14.2 shows that

the derivative of a sine wave is a cosine wave.

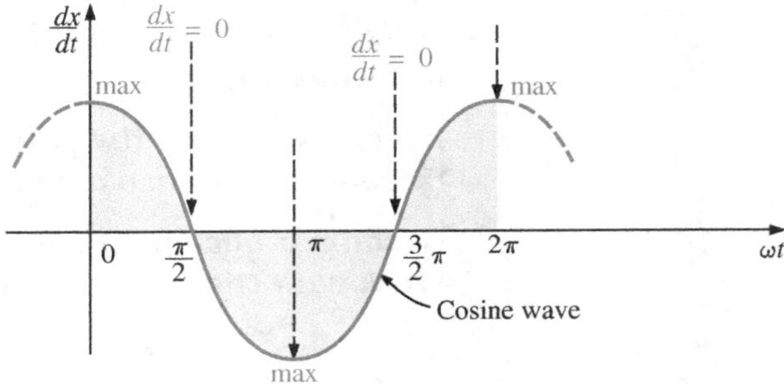

FIG. 14.2
Derivative of the sine wave of Fig. 14.1.

The peak value of the cosine wave is directly related to the frequency of the original waveform. The higher the frequency, the steeper is the slope at the horizontal axis and the greater is the value of dx/dt, as shown in Fig. 14.3 for two different frequencies.

Note in Fig. 14.3 that even though both waveforms (x_1 and x_2) have the same peak value, the sinusoidal function with the higher

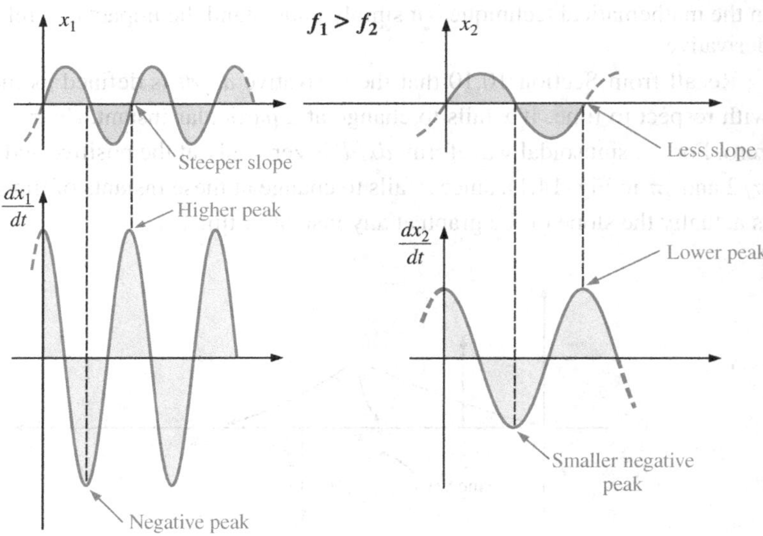

FIG. 14.3
Effect of frequency on the peak value of the derivative.

frequency produces the larger peak value for the derivative. In addition, note that

the derivative of a sine wave has the same period and frequency as the original sinusoidal waveform.

For the sinusoidal voltage

$$e(t) = E_m \sin(\omega t \pm \theta)$$

the derivative can be found directly by differentiation (calculus) to produce the following:

$$\frac{d}{dt}e(t) = \omega E_m \cos(\omega t \pm \theta) = 2\pi f E_m \cos(\omega t \pm \theta) \tag{14.1}$$

The mechanics of the differentiation process are not discussed or investigated here, nor are they required to continue with the text. Note, however, that the peak value of the derivative, $2\pi f E_m$, is a function of the frequency of $e(t)$, and the derivative of a sine wave is a cosine wave.

14.3 RESPONSE OF BASIC R, L, AND C ELEMENTS TO A SINUSOIDAL VOLTAGE OR CURRENT

Now that we are familiar with the characteristics of the derivative of a sinusoidal function, we can investigate the response of the basic elements R, L, and C to a sinusoidal voltage or current.

Resistor

For power-line frequencies and frequencies up to a few hundred kilohertz, resistance is, for all practical purposes, unaffected by the frequency of the applied sinusoidal voltage or current. For this frequency region, the resistor R in Fig. 14.4 can be treated as a constant, and Ohm's law can be applied as follows. For $v = V_m \sin \omega t$,

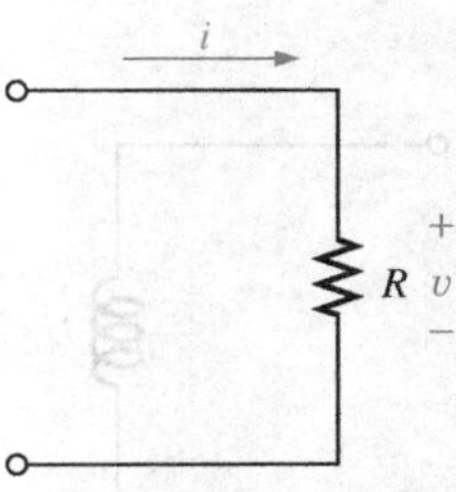

FIG. 14.4
Determining the sinusoidal response for a resistive element.

$$i = \frac{v}{R} = \frac{V_m \sin \omega t}{R} = \frac{V_m}{R} \sin \omega t = I_m \sin \omega t$$

where

$$I_m = \frac{V_m}{R} \tag{14.2}$$

In addition, for a given i,

$$v = iR = (I_m \sin \omega t)R = I_m R \sin \omega t = V_m \sin \omega t$$

where

$$V_m = I_m R \tag{14.3}$$

A plot of v and i in Fig. 14.5 reveals that

for a purely resistive element, the voltage across and the current through the element are in phase, with their peak values related by Ohm's law.

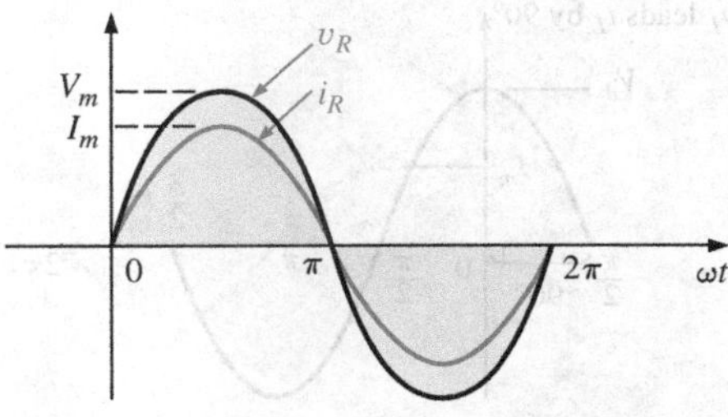

FIG. 14.5
The voltage and current of a resistive element are in phase.

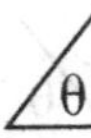

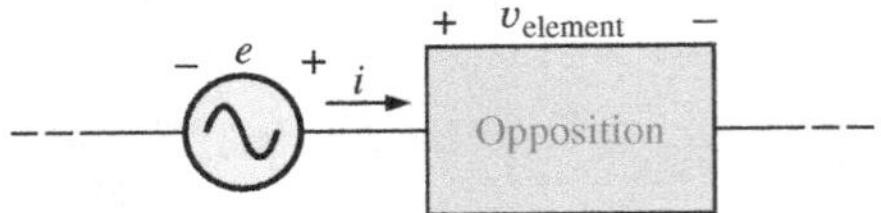

FIG. 14.6
Defining the opposition of an element to the flow of charge through the element.

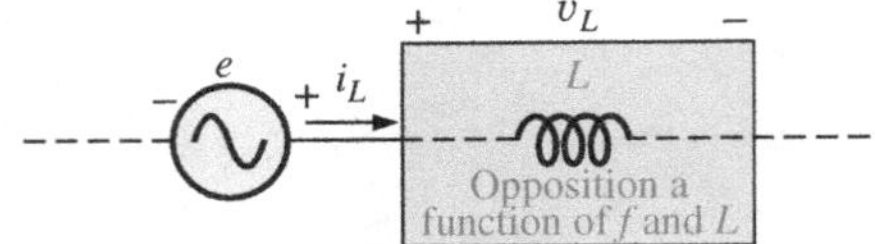

FIG. 14.7
Defining the parameters that determine the opposition of an inductive element to the flow of charge.

Inductor

For the series configuration in Fig. 14.6, the voltage $v_{element}$ of the boxed-in element opposes the source e and thereby reduces the magnitude of the current i. The magnitude of the voltage across the element is determined by the opposition of the element to the flow of charge, or current i. For a resistive element, we have found that the opposition is its resistance and that $v_{element}$ and i are determined by $v_{element} = iR$.

We found in Chapter 11 that the voltage across an inductor is directly related to the rate of change of current through the coil. Consequently, the higher the frequency, the greater is the rate of change of current through the coil, and the greater is the magnitude of the voltage. In addition, we found in the same chapter that the inductance of a coil determines the rate of change of the flux linking a coil for a particular change in current through the coil. The higher the inductance, the greater is the rate of change of the flux linkages, and the greater is the resulting voltage across the coil.

The inductive voltage, therefore, is directly related to the frequency (or, more specifically, the angular velocity of the sinusoidal ac current through the coil) and the inductance of the coil. For increasing values of f and L in Fig. 14.7, the magnitude of v_L increases as described above.

Using the similarities between Figs. 14.6 and 14.7, we find that increasing levels of v_L are directly related to increasing levels of opposition in Fig. 14.6. Since v_L increases with both ω $(= 2\pi f)$ and L, the opposition of an inductive element is as defined in Fig. 14.7.

We will now verify some of the preceding conclusions using a more mathematical approach and then define a few important quantities to be used in the sections and chapters to follow.

For the inductor in Fig. 14.8, we recall from Chapter 11 that

$$v_L = L\frac{di_L}{dt}$$

and, applying differentiation,

$$\frac{di_L}{dt} = \frac{d}{dt}(I_m \sin \omega t) = \omega I_m \cos \omega t$$

Therefore, $$v_L = L\frac{di_L}{dt} = L(\omega I_m \cos \omega t) = \omega L I_m \cos \omega t$$

or $$v_L = V_m \sin(\omega t + 90°)$$

where $$V_m = \omega L I_m$$

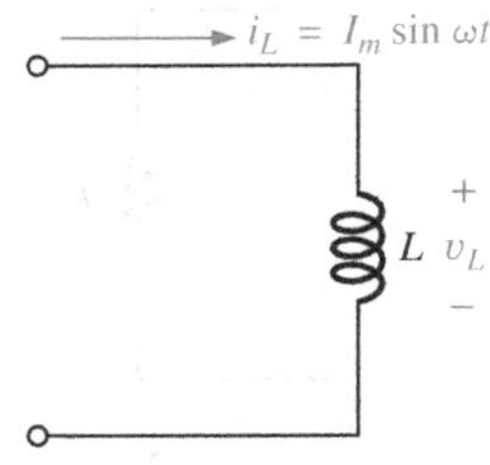

FIG. 14.8
Investigating the sinusoidal response of an inductive element.

Note that the peak value of v_L is directly related to $\omega(= 2\pi f)$ and L as predicted in the discussion above.

A plot of v_L and i_L in Fig. 14.9 reveals that

for an inductor, v_L leads i_L by 90°, or i_L lags v_L by 90°.

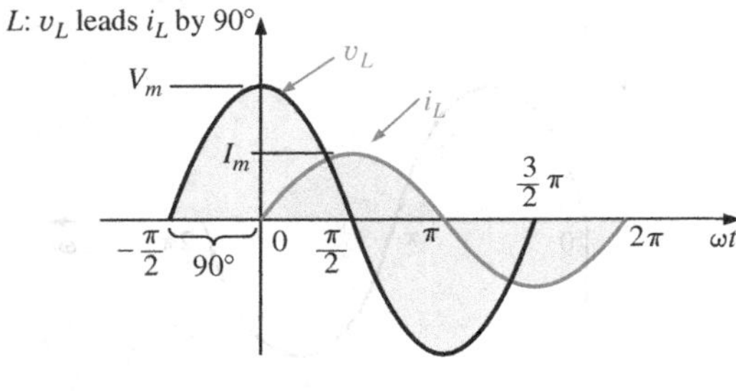

FIG. 14.9
For a pure inductor, the voltage across the coil leads the current through the coil by 90°.

If a phase angle is included in the sinusoidal expression for i_L, such as

$$i_L = I_m \sin(\omega t \pm \theta)$$

then $$v_L = \omega L I_m \sin(\omega t \pm \theta + 90°)$$

The opposition established by an inductor in a sinusoidal ac network can now be found by applying Eq. (4.1):

$$\text{Effect} = \frac{\text{cause}}{\text{opposition}}$$

which, for our purposes, can be written

$$\text{Opposition} = \frac{\text{cause}}{\text{effect}}$$

Substituting values, we have

$$\text{Opposition} = \frac{V_m}{I_m} = \frac{\omega L I_m}{I_m} = \omega L$$

revealing that the opposition established by an inductor in an ac sinusoidal network is directly related to the product of the angular velocity ($\omega = 2\pi f$) and the inductance, verifying our earlier conclusions.

The quantity ωL, called the **reactance** (from the word *reaction*) of an inductor, is symbolically represented by X_L and is measured in ohms; that is,

$$\boxed{X_L = \omega L} \qquad (\text{ohms}, \Omega) \qquad \textbf{(14.4)}$$

In an Ohm's law format, its magnitude can be determined from

$$\boxed{X_L = \frac{V_m}{I_m}} \qquad (\text{ohms}, \Omega) \qquad \textbf{(14.5)}$$

Inductive reactance is the opposition to the flow of current, which results in the continual interchange of energy between the source and the magnetic field of the inductor. In other words, inductive reactance, unlike resistance (which dissipates energy in the form of heat), does not dissipate electrical energy (ignoring the effects of the internal resistance of the inductor.)

Capacitor

Let us now return to the series configuration in Fig. 14 6 and insert the capacitor as the element of interest. For the capacitor, however, we will determine i for a particular voltage across the element. When this approach reaches its conclusion, we will know the relationship between the voltage and current and can determine the opposing voltage (v_{element}) for any sinusoidal current i.

Our investigation of the inductor revealed that the inductive voltage across a coil opposes the instantaneous change in current through the coil. For capacitive networks, the voltage across the capacitor is limited by the rate at which charge can be deposited on, or released by, the plates of the capacitor during the charging and discharging phases, respectively. In other words, an instantaneous change in voltage across a capacitor is opposed by the fact that there is an element of time required to deposit charge on (or release charge from) the plates of a capacitor, and $V = Q/C$.

Since capacitance is a measure of the rate at which a capacitor will store charge on its plates,

for a particular change in voltage across the capacitor, the greater the value of capacitance, the greater is the resulting capacitive current.

In addition, the fundamental equation relating the voltage across a capacitor to the current of a capacitor [$i = C(dv/dt)$] indicates that

for a particular capacitance, the greater the rate of change of voltage across the capacitor, the greater is the capacitive current.

Certainly, an increase in frequency corresponds to an increase in the rate of change of voltage across the capacitor and to an increase in the current of the capacitor.

The current of a capacitor is therefore directly related to the frequency (or, again more specifically, the angular velocity) and the capacitance of the capacitor. An increase in either quantity results in an increase in the current of the capacitor. For the basic configuration in Fig. 14.10, however, we are interested in determining the opposition of the capacitor as related to the resistance of a resistor and ωL for the inductor. Since an increase in current corresponds to a decrease in opposition, and i_C is proportional to ω and C, the opposition of a capacitor is inversely related to ω $(= 2\pi f)$ and C.

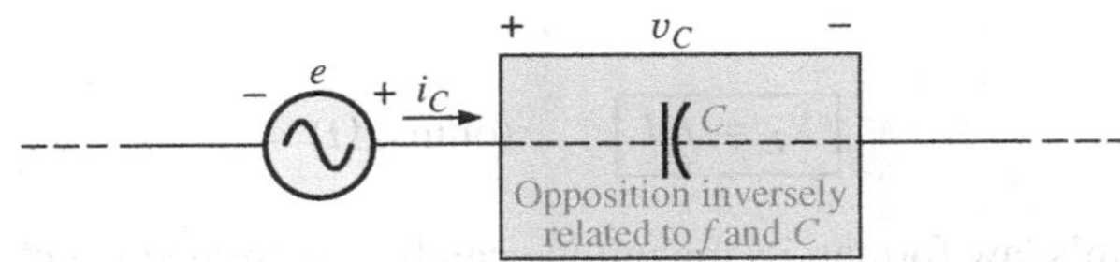

FIG. 14.10
Defining the parameters that determine the opposition of a capacitive element to the flow of charge.

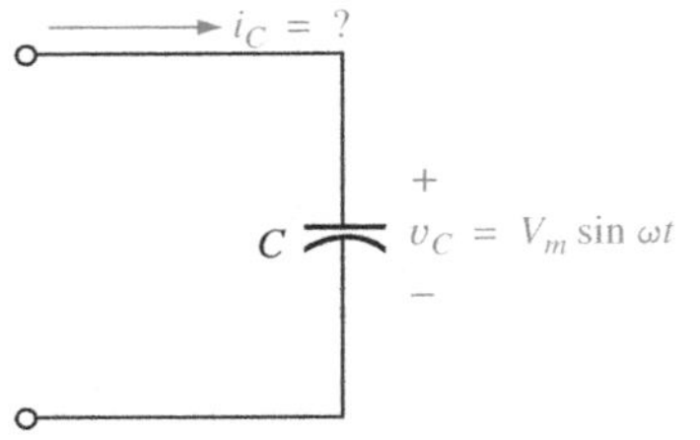

FIG. 14.11
Investigating the sinusoidal response of a capacitive element.

We will now verify, as we did for the inductor, some of the above conclusions using a more mathematical approach.

For the capacitor of Fig. 14.11, we recall from Chapter 11 that

$$i_C = C\frac{dv_C}{dt}$$

and, applying differentiation, we obtain

$$\frac{dv_C}{dt} = \frac{d}{dt}(V_m \sin \omega t) = \omega V_m \cos \omega t$$

Therefore,

$$i_C = C\frac{dv_C}{dt} = C(\omega V_m \cos \omega t) = \omega C V_m \cos \omega t$$

or $$i_C = I_m \sin(\omega t + 90°)$$

where $$I_m = \omega C V_m$$

Note that the peak value of i_C is directly related to $\omega(= 2\pi f)$ and C, as predicted in the discussion above.

A plot of v_C and i_C in Fig. 14.12 reveals that

*for a capacitor, i_C leads v_C by 90°, or v_C lags i_C by 90°.**

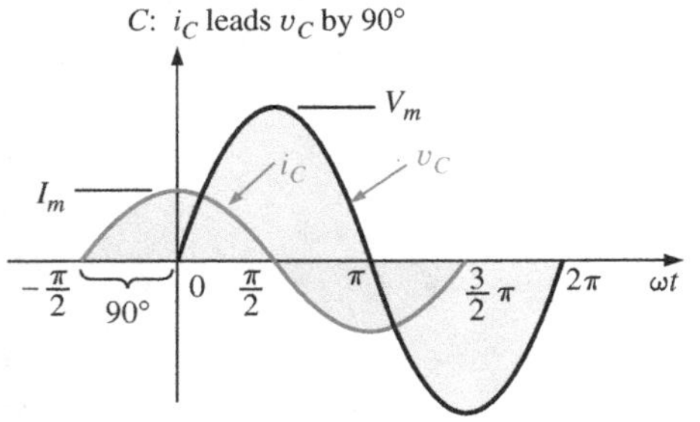

If a phase angle is included in the sinusoidal expression for v_C, such as

$$v_C = V_m \sin(\omega t \pm \theta)$$

then $$i_C = \omega C V_m \sin(\omega t \pm \theta + 90°)$$

FIG. 14.12
The current of a purely capacitive element leads the voltage across the element by 90°.

*A mnemonic phrase sometimes used to remember the phase relationship between the voltage and current of a coil and capacitor is "*ELI* the *ICE* man." Note that the *L* (inductor) has the *E* before the *I* (*e* leads *i* by 90°), and the *C* (capacitor) has the *I* before the *E* (*i* leads *e* by 90°).

Applying

$$\text{Opposition} = \frac{\text{cause}}{\text{effect}}$$

and substituting values, we obtain

$$\text{Opposition} = \frac{V_m}{I_m} = \frac{V_m}{\omega C V_m} = \frac{1}{\omega C}$$

which agrees with the results obtained above.

The quantity $1/\omega C$, called the **reactance** of a capacitor, is symbolically represented by X_C and is measured in ohms; that is,

$$X_C = \frac{1}{\omega C} \quad \text{(ohms, } \Omega\text{)} \tag{14.6}$$

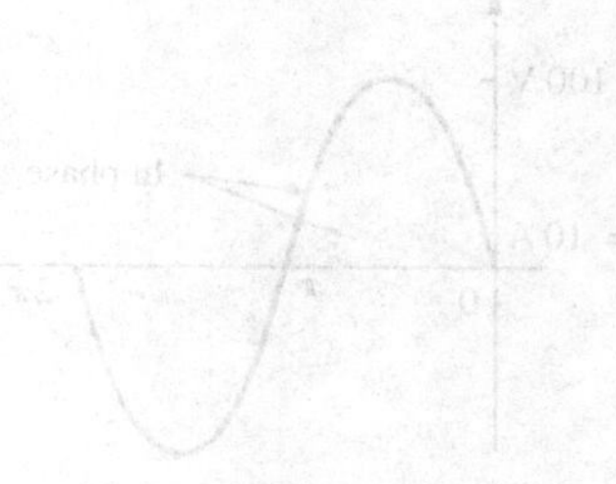

In an Ohm's law format, its magnitude can be determined from

$$X_C = \frac{V_m}{I_m} \quad \text{(ohms, } \Omega\text{)} \tag{14.7}$$

Capacitive reactance is the opposition to the flow of charge, which results in the continual interchange of energy between the source and the electric field of the capacitor. Like the inductor, the capacitor does *not* dissipate energy in any form (ignoring the effects of the leakage resistance).

In the circuits just considered, the current was given in the inductive circuit and the voltage in the capacitive circuit. This was done to avoid the use of integration in finding the unknown quantities. In the inductive circuit,

$$v_L = L\frac{di_L}{dt}$$

but

$$i_L = \frac{1}{L}\int v_L\, dt \tag{14.8}$$

In the capacitive circuit,

$$i_C = C\frac{dv_C}{dt}$$

but

$$v_C = \frac{1}{C}\int i_C\, dt \tag{14.9}$$

Soon, we shall consider a method of analyzing ac circuits that will permit us to solve for an unknown quantity with sinusoidal input without having to use direct integration or differentiation.

It is possible to determine whether a network with one or more elements is predominantly capacitive or inductive by noting the phase relationship between the input voltage and current.

If the source current leads the applied voltage, the network is predominantly capacitive, and if the applied voltage leads the source current, it is predominantly inductive.

Since we now have an equation for the reactance of an inductor or capacitor, we do not need to use derivatives or integration in the examples to be considered. Simply applying Ohm's law, $I_m = E_m/X_L$ (or X_C), and keeping in mind the phase relationship between the voltage and current for each element will be sufficient to complete the examples.

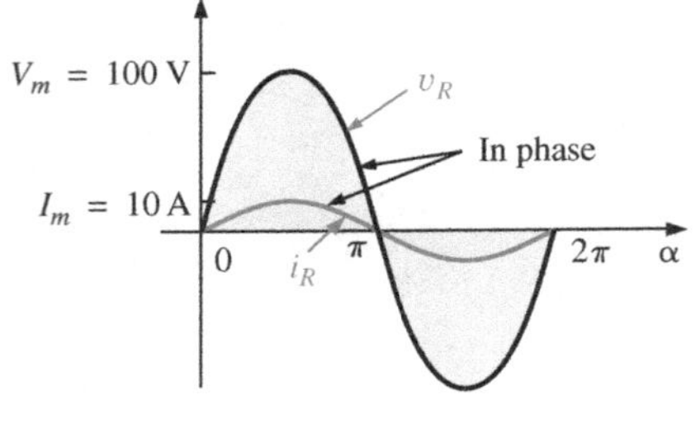

FIG. 14.13
Example 14.1(a).

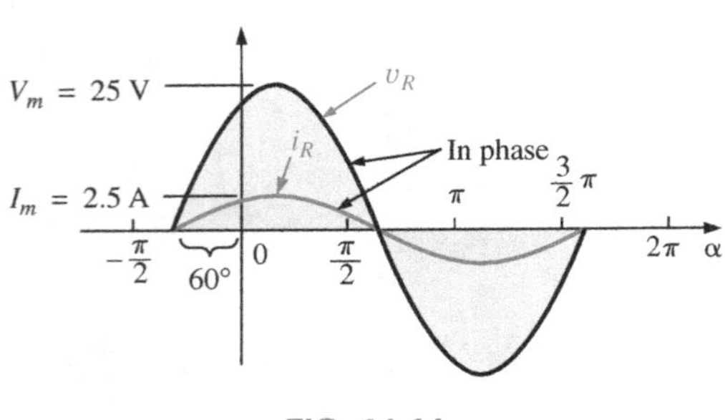

FIG. 14.14
Example 14.1(b).

EXAMPLE 14.1 The voltage across a resistor is indicated. Find the sinusoidal expression for the current if the resistor is 10 Ω. Sketch the curves for v and i.

a. $v = 100 \sin 377t$
b. $v = 25 \sin(377t + 60°)$

Solutions:

a. Eq. (14.2): $I_m = \dfrac{V_m}{R} = \dfrac{100\text{ V}}{10\ \Omega} = 10\text{ A}$

(v and i are in phase), resulting in

$$\mathbf{i = 10 \sin 377t}$$

The curves are sketched in Fig. 14.13.

b. Eq. (14.2): $I_m = \dfrac{V_m}{R} = \dfrac{25\text{ V}}{10\ \Omega} = 2.5\text{ A}$

(v and i are in phase), resulting in

$$\mathbf{i = 2.5 \sin(377t + 60°)}$$

The curves are sketched in Fig. 14.14.

EXAMPLE 14.2 The current through a 5 Ω resistor is given. Find the sinusoidal expression for the voltage across the resistor for $i = 40 \sin(377t + 30°)$.

Solution: Eq. (14.3): $V_m = I_mR = (40\text{ A})(5\ \Omega) = 200\text{ V}$

(v and i are in phase), resulting in

$$\mathbf{v = 200 \sin(377t + 30°)}$$

EXAMPLE 14.3 The current through a 0.1 H coil is provided. Find the sinusoidal expression for the voltage across the coil. Sketch the v and i curves.

a. $i = 10 \sin 377t$
b. $i = 7 \sin(377t - 70°)$

Solutions:

a. Eq. (14.4): $X_L = \omega L = (377\text{ rad/s})(0.1\text{ H}) = 37.7\ \Omega$

Eq. (14.5): $V_m = I_mX_L = (10\text{ A})(37.7\ \Omega) = 377\text{ V}$

and we know that for a coil v leads i by 90°. Therefore,

$$\mathbf{v = 377 \sin(377t + 90°)}$$

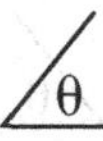

The curves are sketched in Fig. 14.15.

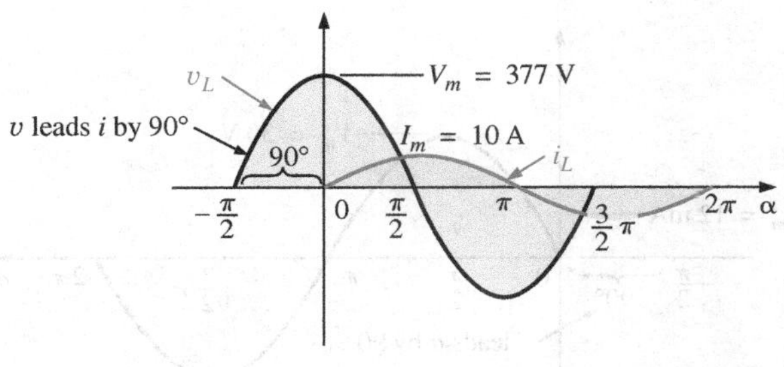

FIG. 14.15
Example 14.3(a).

b. X_L remains at 37.7 Ω.

$$V_m = I_m X_L = (7\text{ A})(37.7\ \Omega) = 263.9\text{ V}$$

and we know that for a coil v leads i by 90°. Therefore,

$$v = 263.9 \sin(377t - 70° + 90°)$$

and

$$\mathbf{v = 263.9 \sin(377t + 20°)}$$

The curves are sketched in Fig. 14.16.

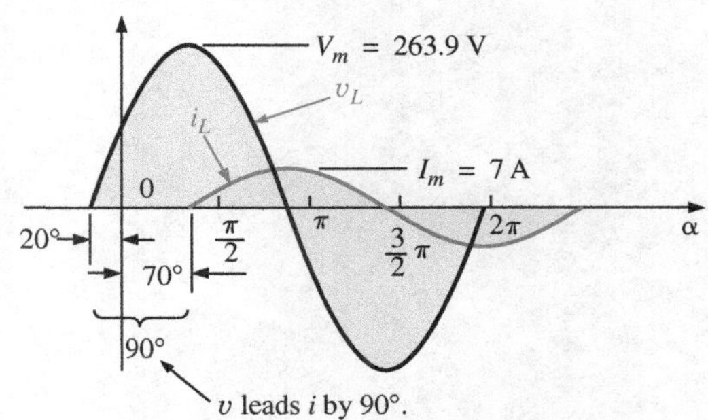

FIG. 14.16
Example 14.3(b).

EXAMPLE 14.4 The voltage across a 0.5 H coil is provided below. What is the sinusoidal expression for the current?

$$v = 100 \sin 20t$$

Solution:

$$X_L = \omega L = (20\text{ rad/s})(0.5\text{ H}) = 10\ \Omega$$

$$I_m = \frac{V_m}{X_L} = \frac{100\text{ V}}{10\ \Omega} = 10\text{ A}$$

and we know the i lags v by 90°. Therefore,

$$\mathbf{i = 10 \sin(20t - 90°)}$$

EXAMPLE 14.5 The voltage across a 1 μF capacitor is provided below. What is the sinusoidal expression for the current? Sketch the v and i curves.

$$v = 30 \sin 400t$$

Solution:

$$\text{Eq. (14.6): } X_C = \frac{1}{\omega C} = \frac{1}{(400\text{ rad/s})(1 \times 10^{-6}\text{ F})} = \frac{10^6\ \Omega}{400} = 2500\ \Omega$$

$$\text{Eq. (14.7): } I_m = \frac{V_m}{X_C} = \frac{30\text{ V}}{2500\ \Omega} = 0.0120\text{ A} = 12\text{ mA}$$

and we know that for a capacitor i leads v by 90°. Therefore,

$$\mathbf{i = 12 \times 10^{-3} \sin(400t + 90°)}$$

The curves are sketched in Fig. 14.17.

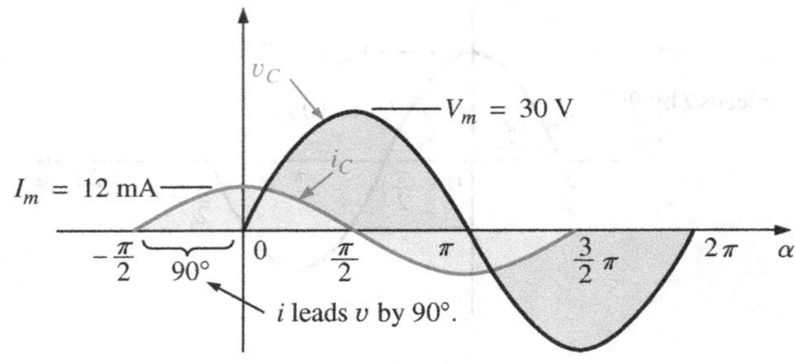

FIG. 14.17
Example 14.5.

EXAMPLE 14.6 The current through a 100 μF capacitor is given. Find the sinusoidal expression for the voltage across the capacitor.

$$i = 40 \sin(500t + 60°)$$

Solution:

$$X_C = \frac{1}{\omega C} = \frac{1}{(500 \text{ rad/s})(100 \times 10^{-6} \text{ F})} = \frac{10^6\ \Omega}{5 \times 10^4} = \frac{10^2\ \Omega}{5} = 20\ \Omega$$

$$V_M = I_M X_C = (40 \text{ A})(20\ \Omega) = 800 \text{ V}$$

and we know that for a capacitor, v lags i by 90°. Therefore,

$$v = 800 \sin(500t + 60° - 90°)$$

and

$$v = \mathbf{800 \sin(500t - 30°)}$$

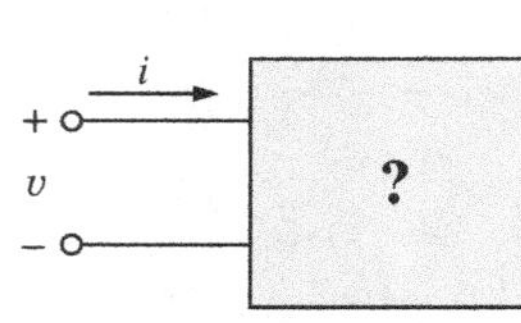

FIG. 14.18
Example 14.7.

EXAMPLE 14.7 For the following pairs of voltages and currents, determine whether the element involved is a capacitor, an inductor, or a resistor. Determine the value of C, L, or R if sufficient data are provided (Fig. 14.18):

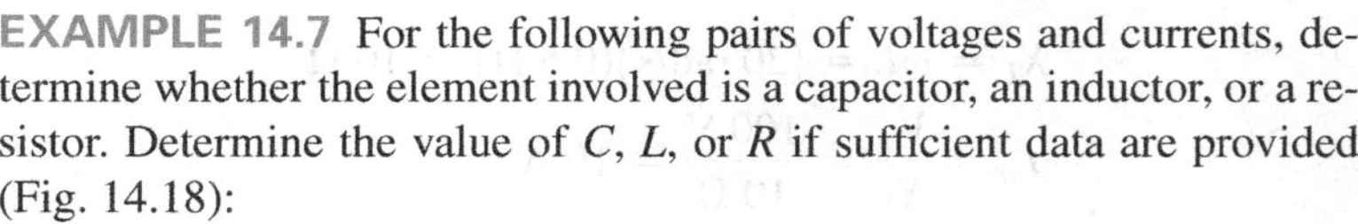

a. $v = 100 \sin(\omega t + 40°)$
   $i = 20 \sin(\omega t + 40°)$
b. $v = 1000 \sin(377t + 10°)$
   $i = 5 \sin(377t - 80°)$
c. $v = 500 \sin(157t + 30°)$
   $i = 1 \sin(157t + 120°)$
d. $v = 50 \cos(\omega t + 20°)$
   $i = 5 \sin(\omega t + 110°)$

Solutions:

a. Since v and i are *in phase*, the element is a *resistor*, and

$$R = \frac{V_m}{I_m} = \frac{100 \text{ V}}{20 \text{ A}} = \mathbf{5\ \Omega}$$

b. Since v *leads* i by 90°, the element is an *inductor*, and

$$X_L = \frac{V_m}{I_m} = \frac{1000 \text{ V}}{5 \text{ A}} = 200\ \Omega$$

so that $X_L = \omega L = 200\ \Omega$ or

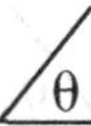

$$L = \frac{200\ \Omega}{\omega} = \frac{200\ \Omega}{377\ \text{rad/s}} = \mathbf{0.53\ H}$$

c. Since i *leads* v by 90°, the element is a *capacitor,* and

$$X_C = \frac{V_m}{I_m} = \frac{500\ \text{V}}{1\ \text{A}} = 500\ \Omega$$

so that $X_C = \dfrac{1}{\omega C} = 500\ \Omega$ or

$$C = \frac{1}{\omega 500\ \Omega} = \frac{1}{(157\ \text{rad/s})(500\ \Omega)} = \mathbf{12.74\ F}$$

d. $v = 50 \cos(\omega t + 20°) = 50 \sin(\omega t + 20° + 90°)$
$= 50 \sin(\omega t + 110°)$

Since v and i are *in phase,* the element is a *resistor,* and

$$R = \frac{V_m}{I_m} = \frac{50\ \text{V}}{5\ \text{A}} = \mathbf{10\ \Omega}$$

14.4 FREQUENCY RESPONSE OF THE BASIC ELEMENTS

Thus far, each description has been for a set frequency, resulting in a fixed level of impedance for each of the basic elements. We must now investigate how a change in frequency affects the impedance level of the basic elements. It is an important consideration because most signals other than those provided by a power plant contain a variety of frequency levels. The last section made it quite clear that the reactance of an inductor or a capacitor is sensitive to the applied frequency. However, the question is, How will these reactance levels change if we steadily increase the frequency from a very low level to a much higher level?

Although we would like to think of every element as ideal, it is important to realize that every commercial element available today *will not respond in an ideal fashion for the full range of possible frequencies.* That is, each element is such that for a particular range of frequencies, it performs in an essentially ideal manner. However, there is always a range of frequencies in which the performance varies from the ideal. Fortunately, the designer is aware of these limitations and will take them into account in the design.

The discussion begins with a look at the response of the *ideal elements*—a response that will be assumed for the remaining chapters of this text and one that can be assumed for any initial investigation of a network. This discussion is followed by a look at the factors that cause an element to deviate from an ideal response as frequency levels become too low or high.

Ideal Response

Resistor *R* For an ideal resistor, you can assume that *frequency will have absolutely no effect on the impedance level,* as shown by the response in Fig. 14.19. Note that at 5 kHz or 20 kHz, the resistance of the resistor remains at 22 Ω; there is no change whatsoever. For the rest of the analyses in this text, the resistance level remains as the nameplate value, no matter what frequency is applied.

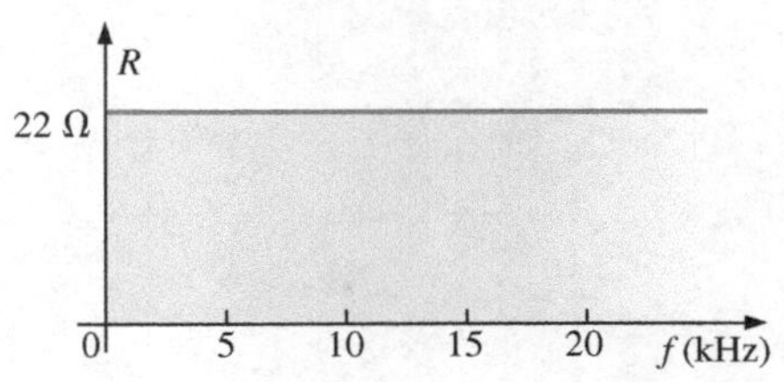

FIG. 14.19
R versus f for the range of interest.

Inductor *L* For the ideal inductor, the equation for the reactance can be written as follows to isolate the frequency term in the equation. The result is a constant times the frequency variable that changes as we move down the horizontal axis of a plot:

$$X_L = \omega L = 2\pi f L = (2\pi L)f = kf \qquad \text{with } k = 2\pi L$$

The resulting equation can be compared directly with the equation for a straight line:

$$y = mx + b = kf + 0 = kf$$

where $b = 0$ and the slope is k or $2\pi L$. X_L is the y variable, and f is the x variable, as shown in Fig. 14.20. Since the inductance determines the slope of the curve, the higher the inductance, the steeper is the straight-line plot, as shown in Fig. 14.20 for two levels of inductance.

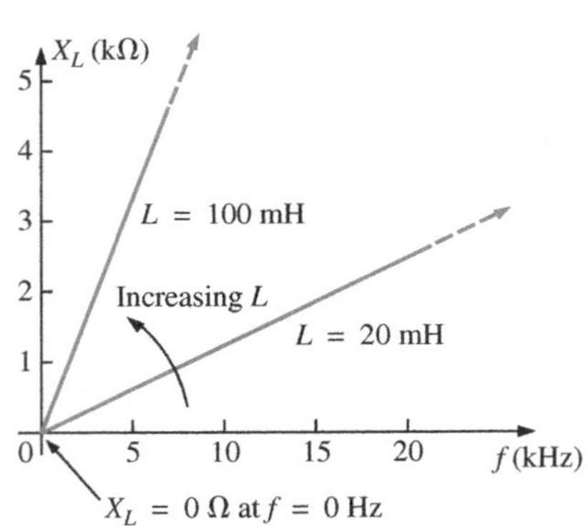

FIG. 14.20
X_L versus frequency.

In particular, note that at $f = 0$ Hz, the reactance of each plot is zero ohms, as determined by substituting $f = 0$ Hz into the basic equation for the reactance of an inductor:

$$X_L = 2\pi f L = 2\pi(0 \text{ Hz})L = 0\ \Omega$$

Since a reactance of zero ohms corresponds with the characteristics of a short circuit, we can conclude that

at a frequency of 0 Hz, an inductor takes on the characteristics of a short circuit, as shown in Fig. 14.21.

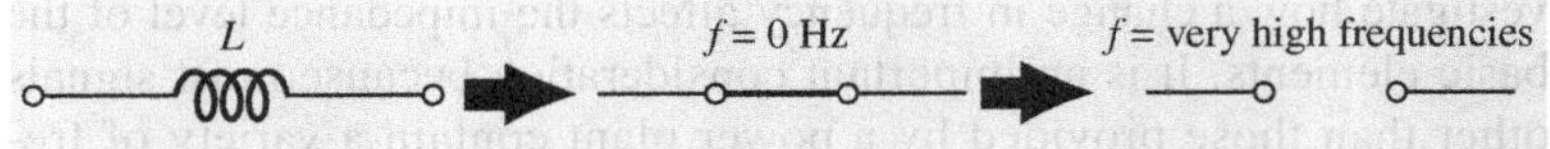

FIG. 14.21
Effect of low and high frequencies on the circuit model of an inductor.

As shown in Fig. 14.21, as the frequency increases, the reactance increases, until it reaches an extremely high level at very high frequencies. The result is that

at very high frequencies, the characteristics of an inductor approach those of an open circuit, as shown in Fig. 14.21.

The inductor, therefore, is capable of handling impedance levels that cover the entire range, from zero ohms to infinite ohms, changing at a *steady rate* determined by the inductance level. The higher the inductance, the faster it approaches the open-circuit equivalent.

Capacitor *C* For the capacitor, the equation for the reactance

$$X_C = \frac{1}{2\pi f C}$$

can be written as

$$X_C f = \frac{1}{2\pi C} = k \qquad \text{(a constant)}$$

which matches the basic format for a hyberbola:

$$yx = k$$

where X_C is the y variable, f the x variable, and k a constant equal to $1/(2\pi C)$.

Hyberbolas have the shape appearing in Fig. 14.22 for two levels of capacitance. Note that the higher the capacitance, the closer the curve approaches the vertical and horizontal axes at low and high frequencies.

At or near 0 Hz, the reactance of any capacitor is extremely high, as determined by the basic equation for capacitance:

$$X_C = \frac{1}{2\pi fC} = \frac{1}{2\pi(0\text{ Hz})C} \Rightarrow \infty\ \Omega$$

The result is that

at or near 0 Hz, the characteristics of a capacitor approach those of an open circuit, as shown in Fig. 14.23.

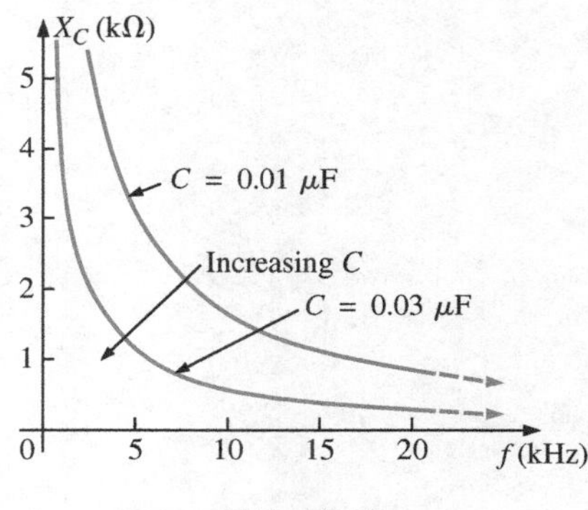

FIG. 14.22
X_C versus frequency.

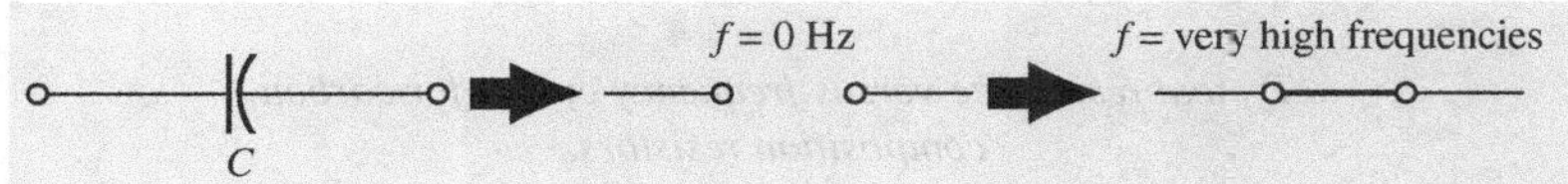

FIG. 14.23
Effect of low and high frequencies on the circuit model of a capacitor.

As the frequency increases, the reactance approaches a value of zero ohms. The result is that

at very high frequencies, a capacitor takes on the characteristics of a short circuit, as shown in Fig. 14.23.

It is important to note in Fig. 14.22 that the reactance drops very rapidly as the frequency increases. It is not a gradual drop as encountered for the rise in inductive reactance. In addition, the reactance sits at a fairly low level for a broad range of frequencies. In general, therefore, recognize that for capacitive elements, the change in reactance level can be dramatic with a relatively small change in frequency level.

Finally, recognize the following:

As frequency increases, the reactance of an inductive element increases, while that of a capacitor decreases, with one approaching an open-circuit equivalent as the other approaches a short-circuit equivalent.

Practical Response

Resistor *R* In the manufacturing process, every resistive element inherits some stray capacitance levels and lead inductances. For most applications, the levels are so low that their effects can be ignored. However, as the frequency extends beyond a few megahertz, it may be necessary to be aware of their effects. For instance, a number of carbon composition resistors have the frequency response appearing in Fig. 14.24. The 100 Ω resistor is essentially stable up to about 300 MHz, whereas the 100 kΩ resistor starts to drop off at about 15 MHz. In general, therefore, this type of carbon composition resistor has the ideal characteristics of Fig. 14.19 for frequencies up to about 15 MHz. For frequencies of 100 Hz, 1 kHz, 150 kHz, and so on, the resistor can be considered ideal.

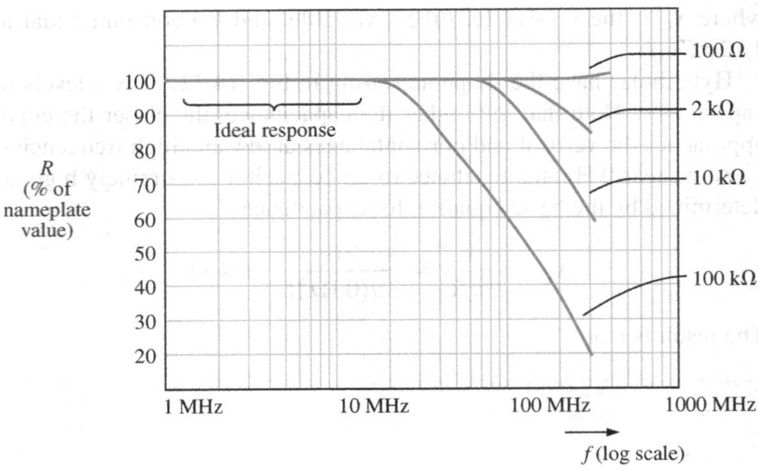

FIG. 14.24
Typical resistance-versus-frequency curves for carbon composition resistors.

The horizontal scale of Fig. 14.24 is a log scale that starts at 1 MHz rather than zero as applied to the vertical scale. Logarithms are discussed in detail in Chapter 21, which describes why the scale cannot start at zero and the fact that the major intervals are separated by powers of 10. For now, simply note that log scales permit the display of a range of frequencies not possible with a linear scale such as was used for the vertical scale of Fig. 14.24. Imagine trying to draw a linear scale from 1 MHz to 1000 MHz using a linear scale. It would be an impossible task unless the horizontal length of the plot was enormous. As indicated above, a great deal more will be said about log scales in Chapter 21.

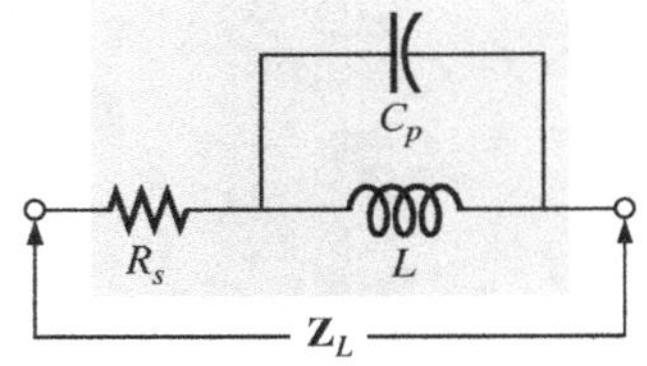

FIG. 14.25
Practical equivalent for an inductor.

Inductor *L* In reality, inductance can be affected by frequency, temperature, and current. A true equivalent for an inductor appears in Fig. 14.25. The series resistance R_s represents the copper losses (resistance of the many turns of thin copper wire); the eddy current losses (losses due to small circular currents in the core when an ac voltage is applied); and the hysteresis losses (losses due to core losses created by the rapidly reversing field in the core). The capacitance C_p is the stray capacitance that exists between the windings of the inductor.

For most inductors, the construction is usually such that the larger the inductance, the lower is the frequency at which the parasitic elements become important. That is, for inductors in the millihenry range (which is very typical), frequencies approaching 100 kHz can have an effect on the ideal characteristics of the element. For inductors in the microhenry range, a frequency of 1 MHz may introduce negative effects. This is not to suggest that the inductors lose their effect at these frequencies but rather that they can no longer be considered ideal (purely inductive elements).

Fig. 14.26 is a plot of the magnitude of the impedance Z_L of Fig. 14.25 versus frequency. Note that up to about 2 MHz, the impedance increases almost linearly with frequency, clearly suggesting that the 100 μH inductor is essentially ideal. However, above 2 MHz, all the factors contributing to R_s start to increase, while the reactance due to the capacitive element C_p is more pronounced. The dropping level of capacitive reactance begins to have a shorting effect across the windings of the

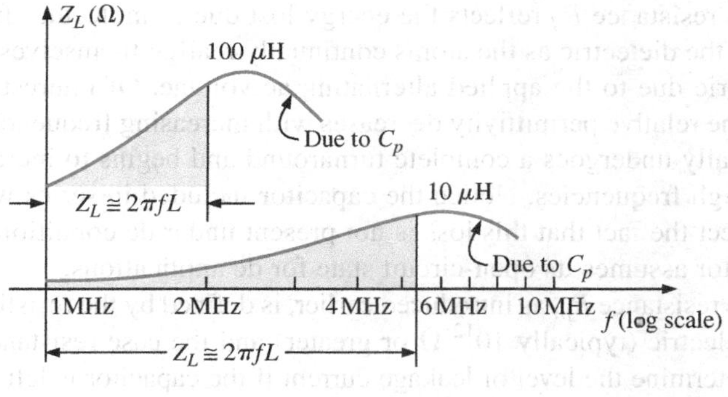

FIG. 14.26
Z_L versus frequency for the practical inductor equivalent of Fig. 14.25.

inductor and reduces the overall inductive effect. Eventually, if the frequency continues to increase, the capacitive effects overcome the inductive effects, and the element actually begins to behave in a capacitive fashion. Note the similarities of this region with the curves in Fig. 14.22. Also, note that decreasing levels of inductance (available with fewer turns and therefore lower levels of C_p) do not demonstrate the degrading effect until higher frequencies are applied.

In general, therefore, the frequency of application for a coil becomes important at increasing frequencies. Inductors lose their ideal characteristics and, in fact, begin to act as capacitive elements with increasing losses at very high frequencies.

Capacitor *C* The capacitor, like the inductor, is not ideal for the full frequency range. In fact, a transition point exists where the characteristics of a capacitor actually take on those of an inductor. The equivalent model for an inductor appearing in Fig. 14.27(a) is an expanded version of that appearing in Fig. 10.21. An inductor L_s was added to reflect the inductance present due to the capacitor leads and any inductance introduced by the design of the capacitor. The inductance of the leads is typically about 0.05 μH per centimeter, which is about 0.2 μH for a capacitor with 2 cm leads at each end—a level of inductance that can be important at very high frequencies.

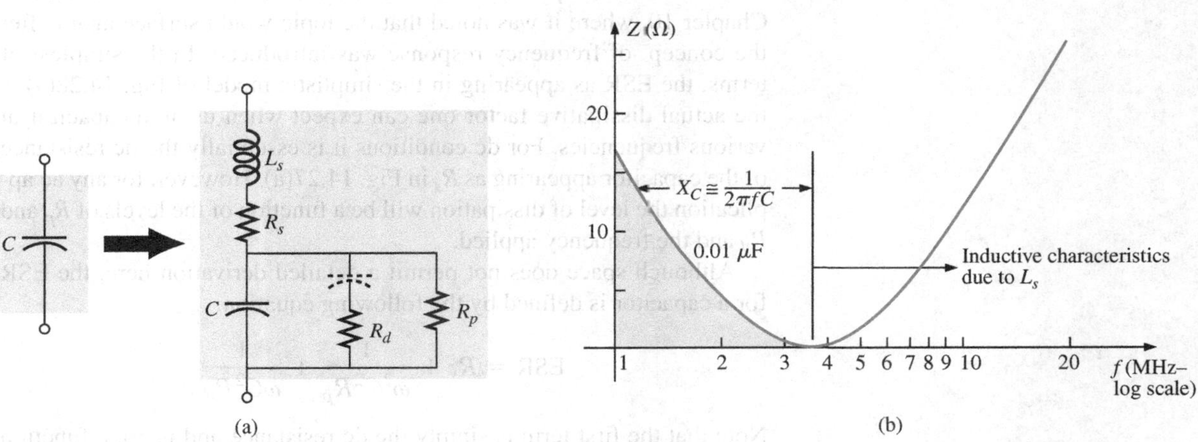

FIG. 14.27
Practical equivalent for a capacitor; (a) network; (b) response.

The resistance R_d reflects the energy lost due to molecular friction within the dielectric as the atoms continually realign themselves in the dielectric due to the applied alternating ac voltage. Of interest, however, the relative permittivity decreases with increasing frequencies but eventually undergoes a complete turnaround and begins to increase at very high frequencies. Notice the capacitor included in series with R_d to reflect the fact that this loss is not present under dc conditions. The capacitor assumes its open-circuit state for dc applications.

The resistance R_p, as introduced earlier, is defined by the resistivity of the dielectric (typically 10^{12} Ω or greater) and the case resistance and will determine the level of leakage current if the capacitor is left to discharge. Depending on the capacitor, the discharge time can extend from a few seconds for some electrolytics to hours (paper) or days (polystyrene), revealing that electrolytics typically have much lower levels of R_p than most other capacitors.

The effect of all the elements on the actual response of a 0.01 μF metallized film capacitor with 2 cm leads is provided in Fig. 14.27(b), where the response is almost ideal for the low and mid-frequency range but then at about 3.7 MHz begins to show an inductive response due to L_s.

In general, therefore, the frequency of application is important for capacitive elements because when the frequency increases to a certain level, the element takes on inductive characteristics. Also, the frequency of application defines the type of capacitor (or inductor) that is applied: Electrolytics are limited to frequencies to perhaps 10 kHz, while ceramic or mica can handle frequencies higher than 10 MHz.

The expected temperature range of operation can have an important impact on the type of capacitor chosen for a particular application. Electrolytics, tantalum, and some high-*k* ceramic capacitors are very sensitive to colder temperatures. In fact, most electrolytics lose 20% of their room-temperature capacitance at 0°C (freezing). Higher temperatures (up to 100°C or 212°F) seem to have less impact in general than colder temperatures, but high-*k* ceramics can lose up to 30% of their capacitance level at 100°C compared to room temperature. With experience, you will learn the type of capacitor to use for each application and only be concerned when you encounter very high frequencies, extreme temperatures, or very high currents or voltages.

ESR The term *equivalent series resistance* (ESR) was introduced in Chapter 10, where it was noted that the topic would surface again after the concept of frequency response was introduced. In the simplest of terms, the ESR as appearing in the simplistic model of Fig. 14.28(a) is the actual dissipative factor one can expect when using a capacitor at various frequencies. For dc conditions it is essentially the dc resistance of the capacitor appearing as R_s in Fig. 14.27(a). However, for any ac application the level of dissipation will be a function of the levels of R_p and R_d and the frequency applied.

Although space does not permit a detailed derivation here, the ESR for a capacitor is defined by the following equation:

$$\text{ESR} = R_s + \frac{1}{\omega^2 C^2 R_p} + \frac{1}{\omega C^2 R_d}$$

Note that the first term is simply the dc resistance and is not a function of frequency. However, the next two terms are a function of frequency in

the denominator, revealing that they will increase very quickly as the frequency drops. The result is the valid concern about levels of ESR at low frequencies. At high frequencies, the second two terms will die off quickly, leaving only the dc resistance. In general, therefore, keep in mind that

the level of ESR or equivalent series resistance is frequency sensitive and considerably greater at low frequencies than just the dc resistance. At very high frequencies, it approaches the dc level.

It is such as important factor in some designs that instruments have been developed primarily to measure this quantity. One such instrument appears in Fig. 14.28(b).

There are some general rules about the level of ESR associated with various capacitors. For all applications, the lower the ESR, the better. Electrolytic capacitors typically have much higher levels of ESR than film, ceramic, or paper capacitors. A standard electrolytic 22 μF capacitor may have an ESR between 5 and 30 Ω, while a standard ceramic may have only 10 to 100 mΩ, a significant difference. Electrolytics, however, because of their other characteristics, are still very popular in power supply design—it is simply a matter of balancing the ESR level with other important factors.

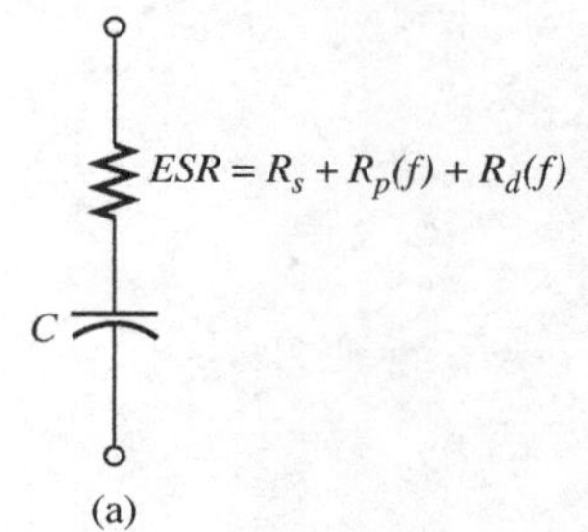

(a)

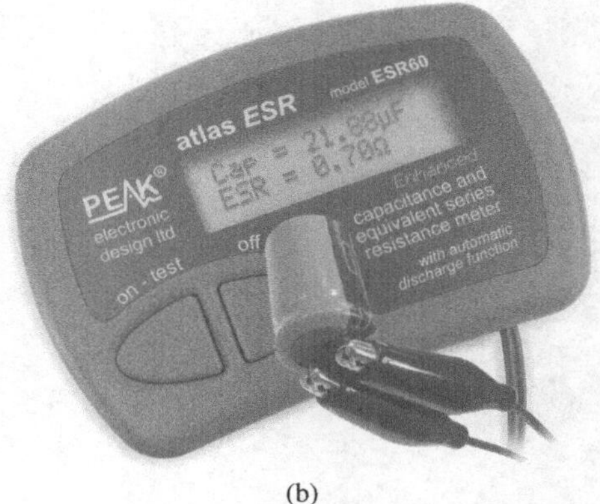

(b)

FIG. 14.28
ESR. (a) Impact on equivalent model; (b) Measuring instrument.
[Part (b) courtesy of Peak Electronic Design Limited.]

EXAMPLE 14.8 At what frequency will the reactance of a 200 mH inductor match the resistance level of a 5 kΩ resistor?

Solution: The resistance remains constant at 5 kΩ for the frequency range of the inductor. Therefore,

$$R = 5000\ \Omega = X_L = 2\pi fL = 2\pi Lf$$
$$= 2\pi(200 \times 10^{-3}\ \text{H})f = 1.257f$$

and
$$f = \frac{5000\ \text{Hz}}{1.257} \cong \mathbf{3.98\ kHz}$$

EXAMPLE 14.9 At what frequency will an inductor of 5 mH have the same reactance as a capacitor of 0.1 μF?

Solution:

$$X_L = X_C$$
$$2\pi fL = \frac{1}{2\pi fC}$$
$$f^2 = \frac{1}{4\pi^2 LC}$$

and

$$f = \frac{1}{2\pi\sqrt{LC}} = \frac{1}{2\pi\sqrt{(5 \times 10^{-3}\ \text{H})(0.1 \times 10^{-6}\ \text{F})}}$$
$$= \frac{1}{2\pi\sqrt{5 \times 10^{-10}}} = \frac{1}{(2\pi)(2.236 \times 10^{-5})} = \frac{10^5\ \text{Hz}}{14.05} \cong \mathbf{7.12\ kHz}$$

14.5 AVERAGE POWER AND POWER FACTOR

A common question is, How can a sinusoidal voltage or current deliver power to a load if it seems to be delivering power during one part of its cycle and taking it back during the negative part of the sinusoidal cycle? The equal oscillations above and below the axis seem to suggest that over one full cycle there is no net transfer of power or energy. However, as mentioned in the last chapter, there is a net transfer of power over one full cycle because power is delivered to the load *at each instant* of the applied voltage or current (except when either is crossing the axis) no matter what the direction is of the current or polarity of the voltage.

To demonstrate this, consider the relatively simple configuration in Fig. 14.29, where an 8 V peak sinusoidal voltage is applied across a 2 Ω resistor. When the voltage is at its positive peak, the power delivered at that instant is 32 W, as shown in the figure. At the midpoint of 4 V, the instantaneous power delivered drops to 8 W; when the voltage crosses the axis, it drops to 0 W. Note, however, that when the applied voltage is at its negative peak, the current may reverse, but at that instant, 32 W is still being delivered to the resistor.

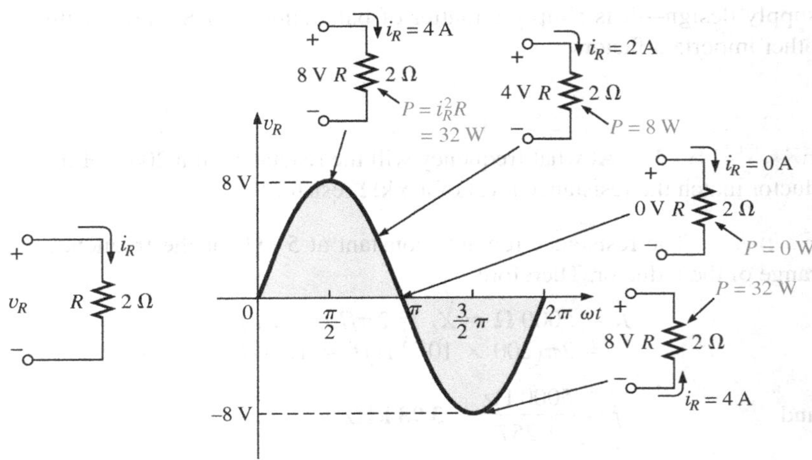

FIG. 14.29
Demonstrating that power is delivered at every instant of a sinusoidal voltage waveform (except $v_R = 0$ V).

In total, therefore,

even though the current through and the voltage across reverse direction and polarity, respectively, power is delivered to the resistive load at each instant of time.

If we plot the power delivered over a full cycle, we obtain the curve in Fig. 14.30. Note that the applied voltage and resulting current are in phase and have twice the frequency of the power curve. For one full cycle of the applied voltage having a period T, the power level peaks for each pulse of the sinusoidal waveform.

The fact that the power curve is always above the horizontal axis reveals that power is being delivered to the load at each instant of time of the applied sinusoidal voltage.

Any portion of the power curve below the axis reveals that power is being returned to the source. The average value of the power curve

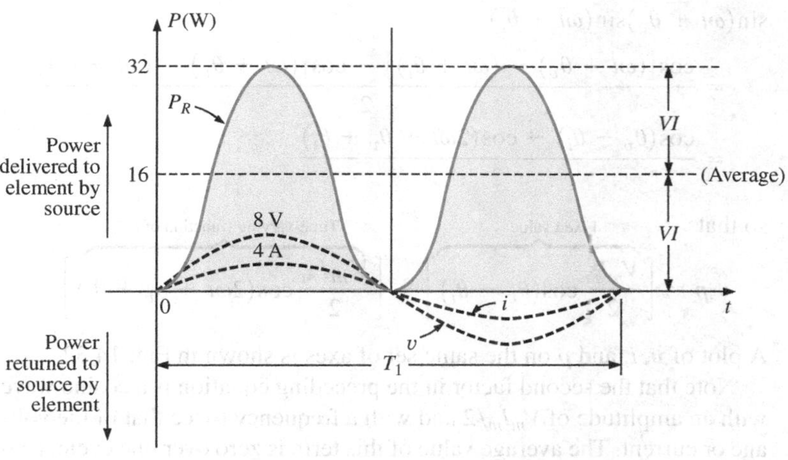

FIG. 14.30
Power versus time for a purely resistive load.

occurs at a level equal to $V_m I_m/2$, as shown in Fig. 14.30. This power level is called the **average** or **real power** level. It establishes a particular level of power transfer for the full cycle, so that we do not have to determine the level of power to apply to a quantity that varies in a sinusoidal nature.

If we substitute the equation for the peak value in terms of the rms value as

$$P_{av} = \frac{V_m I_m}{2} = \frac{(\sqrt{2}\, V_{rms})(\sqrt{2}\, I_{rms})}{2} = \frac{2\, V_{rms}\, I_{rms}}{2}$$

we find that the average or real power delivered to a resistor takes on the following very convenient form:

$$\boxed{P_{av} = V_{rms}\, I_{rms}} \qquad (14.10)$$

Note that the power equation is exactly the same when applied to dc networks as long as we work with rms values.

The above analysis was for a purely resistive load. If the sinusoidal voltage is applied to a network with a combination of *R*, *L*, and *C* components, the instantaneous equation for the power levels is more complex. However, if we are careful in developing the general equation and examine the results, we find some general conclusions that will be very helpful in the analysis to follow.

In Fig. 14.31, a voltage with an initial phase angle is applied to a network with any combination of elements that results in a current with the indicated phase angle.

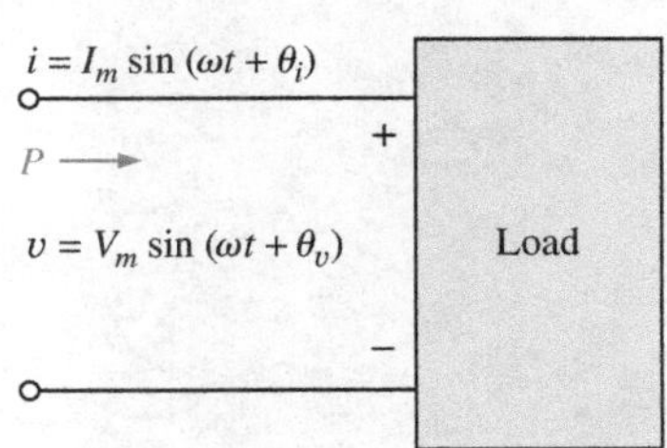

FIG. 14.31
Determining the power delivered in a sinusoidal ac network.

The power delivered at each instant of time is then defined by

$$\begin{aligned} p = vi &= V_m \sin(\omega t + \theta_v) I_m \sin(\omega t + \theta_i) \\ &= V_m I_m \sin(\omega t + \theta_v)\sin(\omega t + \theta_i) \end{aligned}$$

Using the trigonometric identity

$$\sin A \sin B = \frac{\cos(A - B) - \cos(A + B)}{2}$$

we see that the function $\sin(\omega t + \theta_v)\,\sin(\omega t + \theta_i)$ becomes

$$\sin(\omega t + \theta_v)\sin(\omega t + \theta_i)$$
$$= \frac{\cos[(\omega t + \theta_v) - (\omega t + \theta_i)] - \cos[(\omega t + \theta_v) + (\omega t + \theta_i)]}{2}$$
$$= \frac{\cos(\theta_v - \theta_i) - \cos(2\omega t + \theta_v + \theta_i)}{2}$$

so that

$$p = \overbrace{\left[\frac{V_m I_m}{2}\cos(\theta_v - \theta_i)\right]}^{\text{Fixed value}} - \overbrace{\left[\frac{V_m I_m}{2}\cos(2\omega t + \theta_v + \theta_i)\right]}^{\text{Time-varying (function of } t)}$$

A plot of v, i, and p on the same set of axes is shown in Fig. 14.32.

Note that the second factor in the preceding equation is a cosine wave with an amplitude of $V_m I_m/2$ and with a frequency twice that of the voltage or current. The average value of this term is zero over one cycle, producing no net transfer of energy in any one direction.

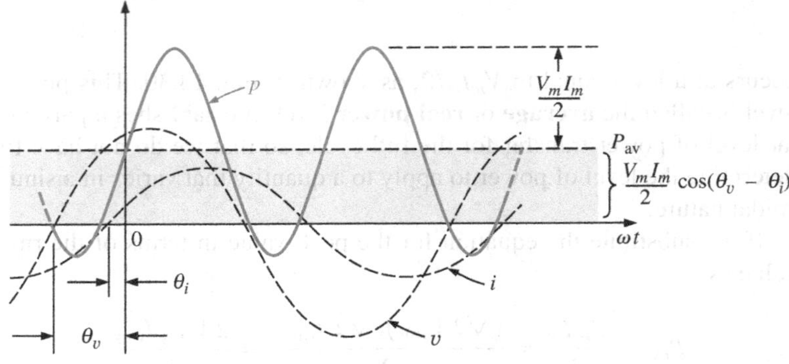

FIG. 14.32
Defining the average power for a sinusoidal ac network.

The first term in the preceding equation, however, has a constant magnitude (no time dependence) and therefore provides some net transfer of energy. This term is referred to as the **average power** or **real power** as introduced earlier. The angle $(\theta_v - \theta_i)$ is the phase angle between v and i. Since $\cos(-\alpha) = \cos\alpha$,

the magnitude of average power delivered is independent of whether v leads i or i leads v.

Defining θ as equal to $|\theta_v - \theta_i|$, where $|\;|$ indicates that only the magnitude is important and the sign is immaterial, we have

$$P = \frac{V_m I_m}{2}\cos\theta \qquad \text{(watts, W)} \qquad \textbf{(14.11)}$$

where P is the average power in watts. This equation can also be written

$$P = \left(\frac{V_m}{\sqrt{2}}\right)\left(\frac{I_m}{\sqrt{2}}\right)\cos\theta$$

or, since $\quad V_{\text{eff}} = \frac{V_m}{\sqrt{2}} \quad$ and $\quad I_{\text{eff}} = \frac{I_m}{\sqrt{2}}$

Eq. (14.11) becomes

$$P = V_{\text{rms}} I_{\text{rms}} \cos\theta \qquad \textbf{(14.12)}$$

Let us now apply Eqs. (14.11) and (14.12) to the basic *R*, *L*, and *C* elements.

Resistor

In a purely resistive circuit, since v and i are in phase, $|\theta_v - \theta_i| = \theta = 0°$, and $\cos\theta = \cos 0° = 1$, so that

$$P = \frac{V_m I_m}{2} = V_{\text{rms}} I_{\text{rms}} \quad \text{(W)} \qquad \textbf{(14.13)}$$

or, since

$$I_{\text{rms}} = \frac{V_{\text{rms}}}{R}$$

then

$$P = \frac{V_{\text{rms}}^2}{R} = I_{\text{rms}}^2 R \quad \text{(W)} \qquad \textbf{(14.14)}$$

Inductor

In a purely inductive circuit, since v leads i by 90°, $|\theta_v - \theta_i| = \theta = |-90°| = 90°$. Therefore,

$$P = \frac{V_m I_m}{2}\cos 90° = \frac{V_m I_m}{2}(0) = \mathbf{0\ W}$$

The average power or power dissipated by the ideal inductor (no associated resistance) is zero watts.

Capacitor

In a purely capacitive circuit, since i leads v by 90°, $|\theta_v - \theta_i| = \theta = |-90°| = 90°$. Therefore,

$$P = \frac{V_m I_m}{2}\cos(90°) = \frac{V_m I_m}{2}(0) = \mathbf{0\ W}$$

The average power or power dissipated by the ideal capacitor (no associated resistance) is zero watts.

EXAMPLE 14.10 Find the average power dissipated in a network whose input current and voltage are the following:

$$i = 5\sin(\omega t + 40°)$$
$$v = 10\sin(\omega t + 40°)$$

Solution: Since v and i are in phase, the circuit appears to be purely resistive at the input terminals. Therefore,

$$P = \frac{V_m I_m}{2} = \frac{(10\text{ V})(5\text{ A})}{2} = \mathbf{25\ W}$$

or

$$R = \frac{V_m}{I_m} = \frac{10\text{ V}}{5\text{ A}} = 2\ \Omega$$

and

$$P = \frac{V_{\text{rms}}^2}{R} = \frac{[(0.707)(10\text{ V})]^2}{2} = \mathbf{25\ W}$$

or

$$P = I_{\text{rms}}^2 R = [(0.707)(5\text{ A})]^2(2) = \mathbf{25\ W}$$

For the following example, the circuit consists of a combination of resistances and reactances producing phase angles between the input current and voltage different from 0° or 90°.

EXAMPLE 14.11 Determine the average power delivered to networks having the following input voltage and current:

a. $v = 100 \sin(\omega t + 40°)$
$i = 20 \sin(\omega t + 70°)$

b. $v = 150 \sin(\omega t - 70°)$
$i = 3 \sin(\omega t - 50°)$

Solutions:

a. $V_m = 100, \quad \theta_v = 40°$
$I_m = 20\text{ A}, \quad \theta_i = 70°$
$\theta = |\theta_v - \theta_i| = |40° - 70°| = |-30°| = 30°$

and

$$P = \frac{V_m I_m}{2}\cos\theta = \frac{(100\text{ V})(20\text{ A})}{2}\cos(30°) = (1000\text{ W})(0.866)$$
$$= \mathbf{866\text{ W}}$$

b. $V_m = 150\text{ V}, \quad \theta_v = -70°$
$I_m = 3\text{ A}, \quad \theta_i = -50°$
$\theta = |\theta_v - \theta_i| = |-70° - (-50°)|$
$= |-70° + 50°| = |-20°| = 20°$

and

$$P = \frac{V_m I_m}{2}\cos\theta = \frac{(150\text{ V})(3\text{ A})}{2}\cos(20°) = (225\text{ W})(0.9397)$$
$$= \mathbf{211.43\text{ W}}$$

Power Factor

In the equation $P = (V_m I_m/2)\cos\theta$, the factor that has significant control over the delivered power level is the $\cos\theta$. No matter how large the voltage or current, if $\cos\theta = 0$, the power is zero; if $\cos\theta = 1$, the power delivered is a maximum. Since it has such control, the expression was given the name **power factor** and is defined by

$$\boxed{\text{Power factor} = F_p = \cos\theta} \qquad \textbf{(14.15)}$$

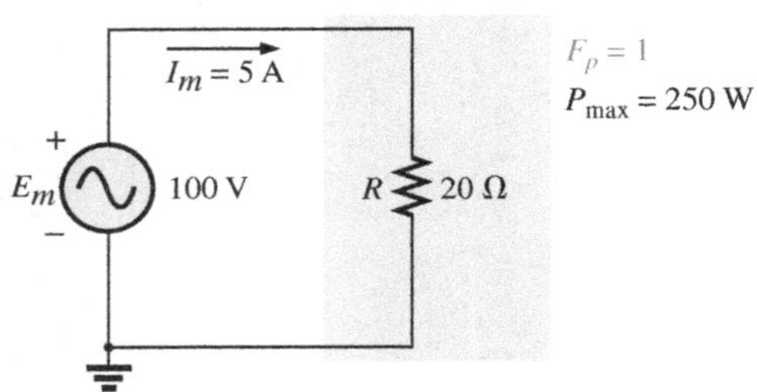

FIG. 14.33
Purely resistive load with $F_p = 1$.

For a purely resistive load such as the one shown in Fig. 14.33, the phase angle between v and i is 0° and $F_p = \cos\theta = \cos 0° = 1$. The power delivered is a maximum of $(V_m I_m/2)\cos\theta = ((100\text{ V})(5\text{ A})/2)(1) = 250\text{ W}$.

For a purely reactive load (inductive or capactitive) such as the one shown in Fig. 14.34, the phase angle between v and i is 90° and $F_p = \cos\theta = \cos 90° = 0$. The power delivered is then the minimum value of zero watts, *even though the current has the same peak value* as that encountered in Fig. 14.33.

For situations where the load is a combination of resistive and reactive elements, the power factor varies between 0 and 1. The more resistive the total impedance, the closer is the power factor to 1; the more reactive the total impedance, the closer is the power factor to 0.

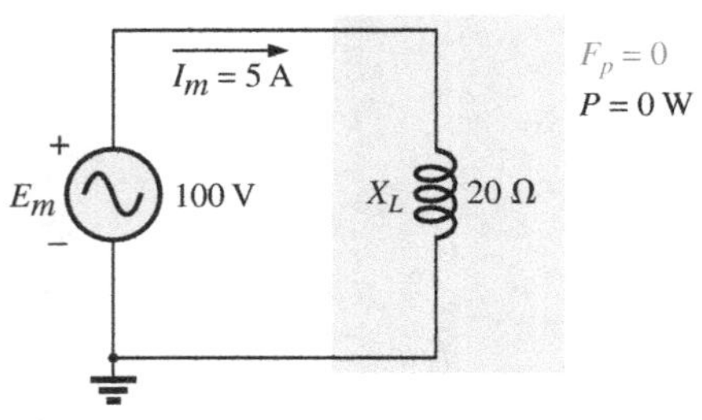

FIG. 14.34
Purely inductive load with $F_p = 0$.

In terms of the average power and the terminal voltage and current,

$$F_p = \cos\theta = \frac{P}{V_{\text{rms}} I_{\text{rms}}} \tag{14.16}$$

The terms *leading* and *lagging* are often written in conjunction with the power factor. *They are defined by the current through the load.* If the current leads the voltage across a load, the load has a **leading power factor.** If the current lags the voltage across the load, the load has a **lagging power factor.** In other words,

capacitive networks have leading power factors, and inductive networks have lagging power factors.

The importance of the power factor to power distribution systems is examined in Chapter 19. In fact, one section is devoted to power-factor correction.

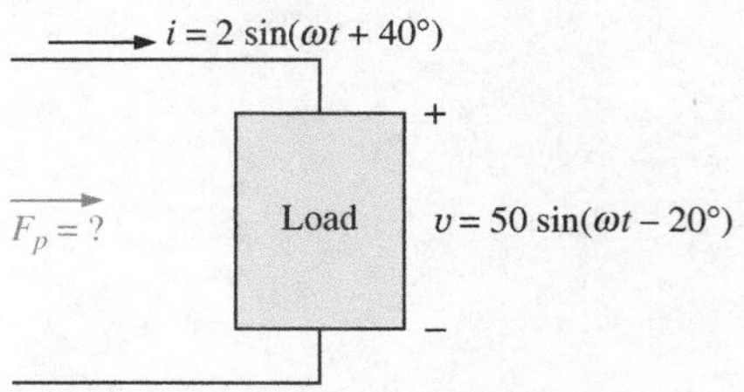

FIG. 14.35
Example 14.12(a).

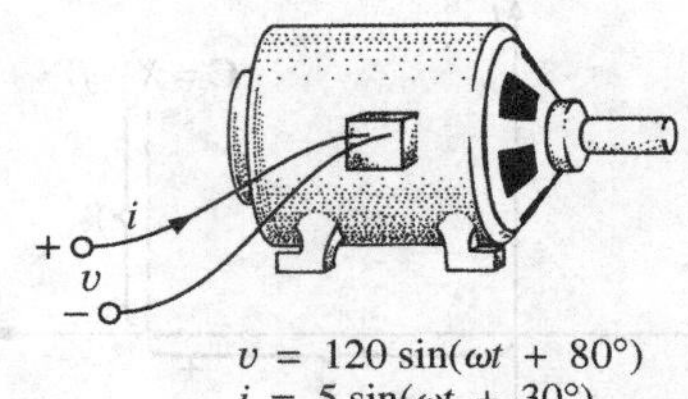

FIG. 14.36
Example 14.12(b).

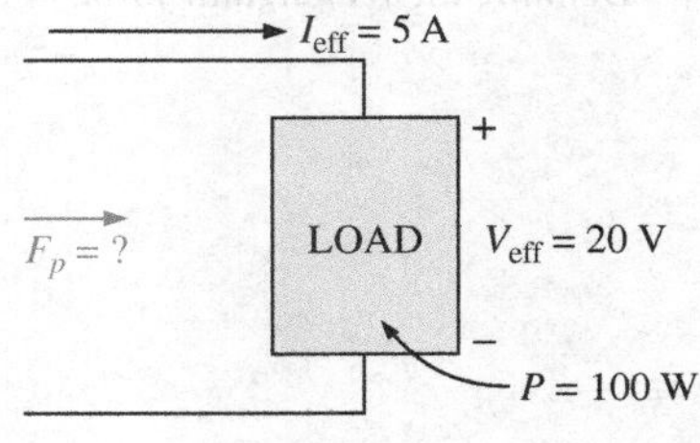

FIG. 14.37
Example 14.12(c).

EXAMPLE 14.12 Determine the power factors of the following loads, and indicate whether they are leading or lagging:

a. Fig. 14.35
b. Fig. 14.36
c. Fig. 14.37

Solutions:

a. $F_p = \cos\theta = \cos|40° - (-20°)| = \cos 60° = \mathbf{0.5\ leading}$
b. $F_p = \cos\theta\,|80° - 30°| = \cos 50° = \mathbf{0.64\ lagging}$
c. $F_p = \cos\theta = \dfrac{P}{V_{\text{eff}} I_{\text{eff}}} = \dfrac{100\text{ W}}{(20\text{ V})(5\text{ A})} = \dfrac{100\text{ W}}{100\text{ W}} = \mathbf{1}$

The load is resistive, and F_p is neither leading nor lagging.

14.6 COMPLEX NUMBERS

In our analysis of dc networks, we found it necessary to determine the algebraic sum of voltages and currents. Since the same will also be true for ac networks, the question arises, How do we determine the algebraic sum of two or more voltages (or currents) that are varying sinusoidally? Although one solution would be to find the algebraic sum on a point-to-point basis (as shown in Section 14.12), this would be a long and tedious process in which accuracy would be directly related to the scale used.

It is the purpose of this chapter to introduce a system of **complex numbers** that, when related to the sinusoidal ac waveform, results in a technique for finding the algebraic sum of sinusoidal waveforms that is quick, direct, and accurate. In the following chapters, the technique is extended to permit the analysis of sinusoidal ac networks in a manner very similar to that applied to dc networks. The methods and theorems as described for dc networks can then be applied to sinusoidal ac networks with little difficulty.

A **complex number** represents a point in a two-dimensional plane located with reference to two distinct axes. This point can also determine a radius vector drawn from the origin to the point. The horizontal axis is called the *real* axis, while the vertical axis is called the *imaginary* axis. Both are labeled in Fig. 14.38. Every number from zero to $\pm\infty$ can be represented by some point along the real axis. Prior to the development of

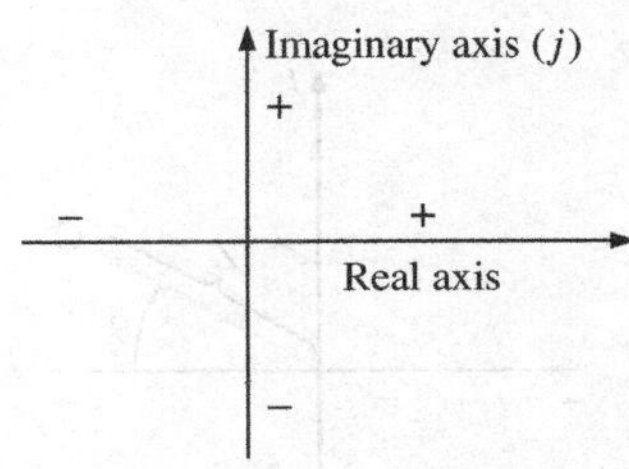

FIG. 14.38
Defining the real and imaginary axes of a complex plane.

this system of complex numbers, it was believed that any number not on the real axis did not exist—hence the term *imaginary* for the vertical axis.

In the complex plane, the horizontal or real axis represents all positive numbers to the right of the imaginary axis and all negative numbers to the left of the imaginary axis. All positive imaginary numbers are represented above the real axis, and all negative imaginary numbers, below the real axis. The symbol j (or sometimes i) is used to denote the imaginary component.

Two forms are used to represent a complex number: **rectangular** and **polar.** Each can represent a point in the plane or a radius vector drawn from the origin to that point.

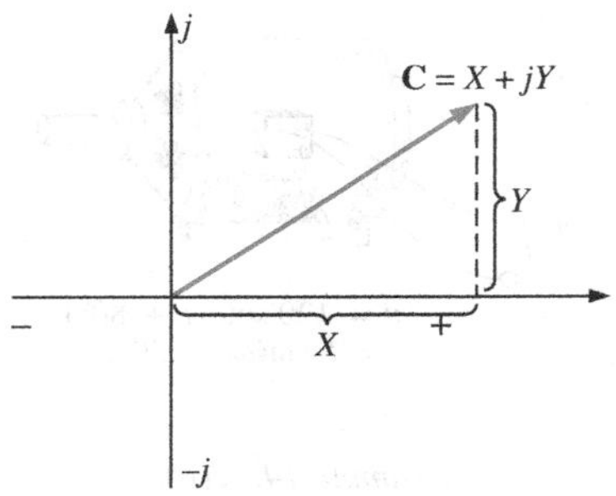

FIG. 14.39
Defining the rectangular form.

14.7 RECTANGULAR FORM

The format for the **rectangular form** is

$$\boxed{\mathbf{C} = X + jY} \tag{14.17}$$

as shown in Fig. 14.39. The letter **C** was chosen from the word "complex." The **boldface** notation is for any number with magnitude and direction. The *italic* is for magnitude only.

EXAMPLE 14.13 Sketch the following complex numbers in the complex plane:

a. $\mathbf{C} = 3 + j4$ b. $\mathbf{C} = 0 - j6$ c. $\mathbf{C} = -10 - j20$

Solutions:

a. See Fig. 14.40. b. See Fig. 14.41. c. See Fig. 14.42.

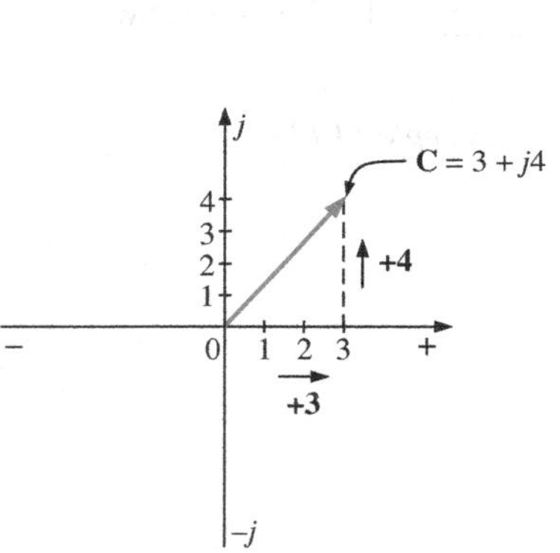

FIG. 14.40
Example 14.13(a).

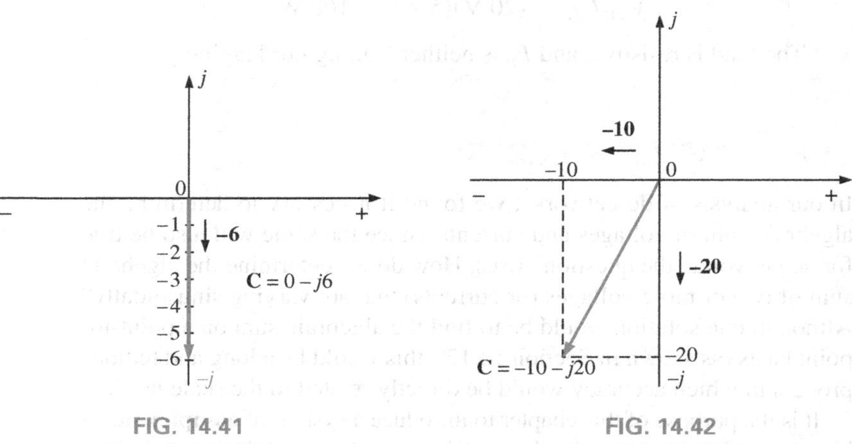

FIG. 14.41
Example 14.13(b).

FIG. 14.42
Example 14.13(c).

14.8 POLAR FORM

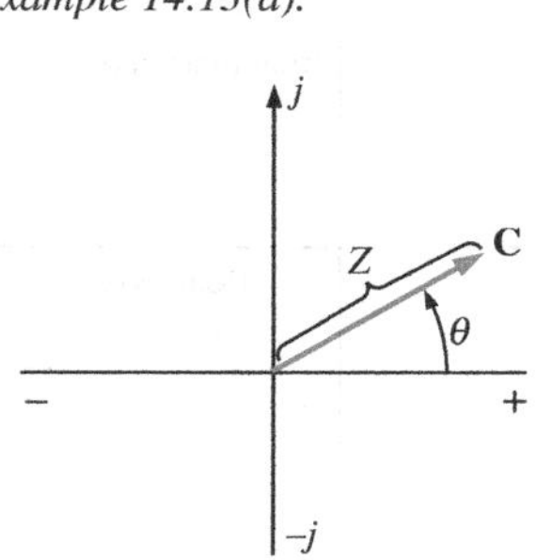

FIG. 14.43
Defining the polar form.

The format for the **polar form** is

$$\boxed{\mathbf{C} = Z \angle \theta} \tag{14.18}$$

with the letter Z chosen from the sequence X, Y, Z.

Z indicates magnitude only, and θ is *always measured counterclockwise (CCW) from the positive real axis,* as shown in Fig. 14.43. Angles measured in the clockwise direction from the positive real axis must have a negative sign associated with them.

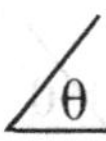

A negative sign in front of the polar form has the effect shown in Fig. 14.44. Note that it results in a complex number directly opposite the complex number with a positive sign.

$$-\mathbf{C} = -Z\angle\theta = Z\angle\theta \pm 180° \tag{14.19}$$

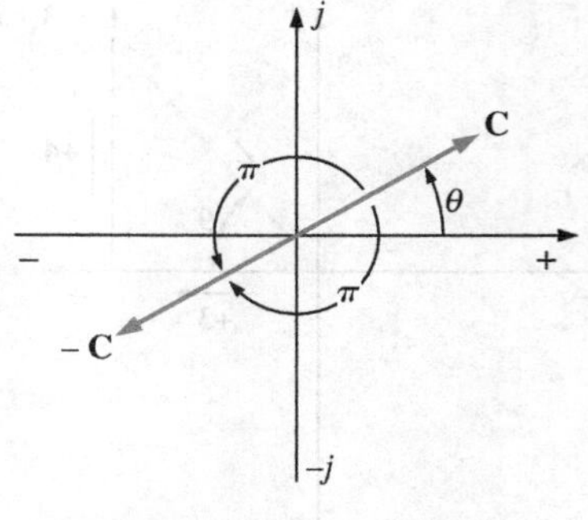

FIG. 14.44
Demonstrating the effect of a negative sign on the polar form.

EXAMPLE 14.14 Sketch the following complex numbers in the complex plane:

a. $\mathbf{C} = 5\angle 30°$
b. $\mathbf{C} = 7\angle -120°$
c. $\mathbf{C} = -4.2\angle 60°$

Solutions:

a. See Fig. 14.45.
b. See Fig. 14.46.
c. See Fig. 14.47.

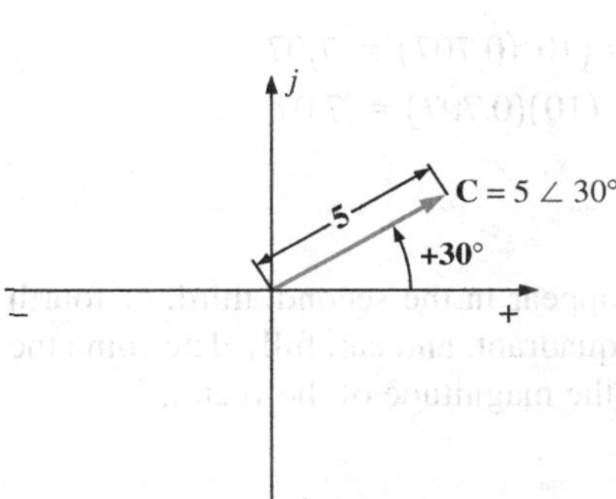

FIG. 14.45
Example 14.14(a).

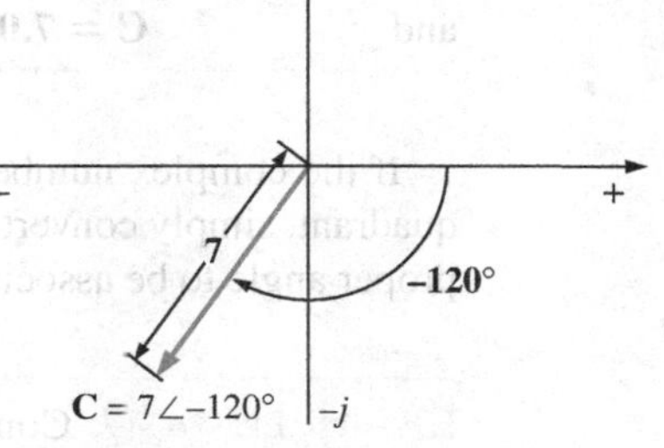

FIG. 14.46
Example 14.14(b).

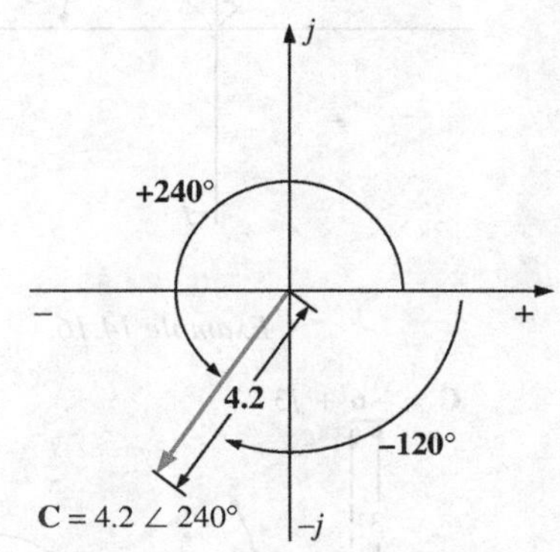

FIG. 14.47
Example 14.14(c).

14.9 CONVERSION BETWEEN FORMS

The two forms are related by the following equations, as illustrated in Fig. 14.48.

Rectangular to Polar

$$Z = \sqrt{X^2 + Y^2} \tag{14.20}$$

$$\theta = \tan^{-1}\frac{Y}{X} \tag{14.21}$$

Polar to Rectangular

$$X = Z\cos\theta \tag{14.22}$$

$$Y = Z\sin\theta \tag{14.23}$$

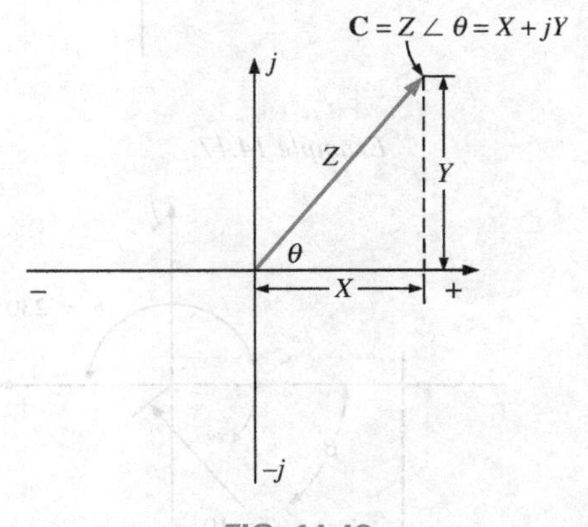

FIG. 14.48
Conversion between forms.

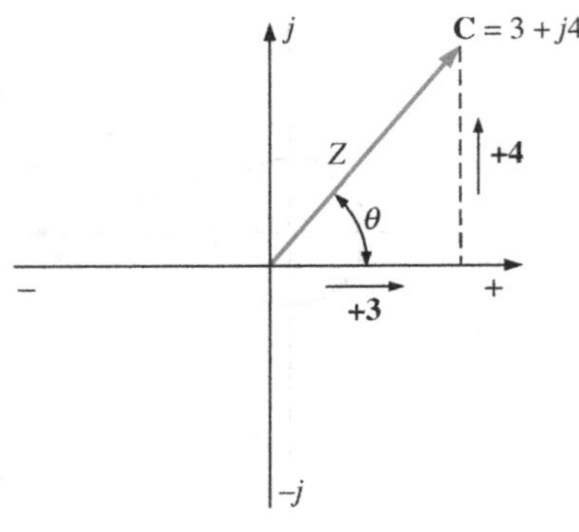

FIG. 14.49
Example 14.15.

EXAMPLE 14.15 Convert the following from rectangular to polar form:

$$\mathbf{C} = 3 + j4 \qquad \text{(Fig. 14.49)}$$

Solution:

$$Z = \sqrt{(3)^2 + (4)^2} = \sqrt{25} = 5$$

$$\theta = \tan^{-1}\left(\frac{4}{3}\right) = 53.13^\circ$$

and $\mathbf{C} = \mathbf{5\angle 53.13^\circ}$

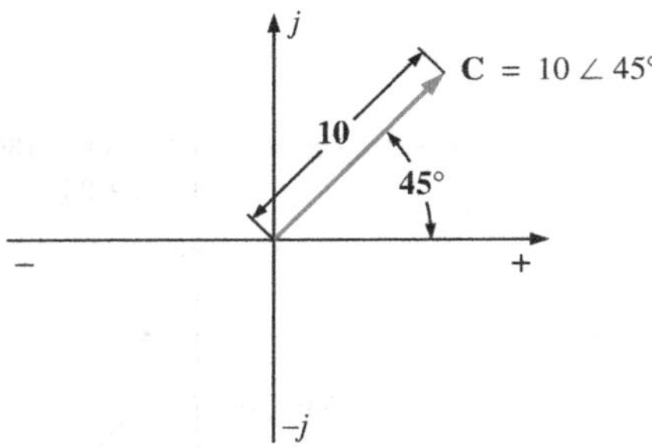

FIG. 14.50
Example 14.16.

EXAMPLE 14.16 Convert the following from polar to rectangular form:

$$\mathbf{C} = 10\angle 45^\circ \qquad \text{(Fig. 14.50)}$$

Solution:

$$X = 10\cos 45^\circ = (10)(0.707) = 7.07$$

$$Y = 10\sin 45^\circ = (10)(0.707) = 7.07$$

and $\mathbf{C} = \mathbf{7.07 + j7.07}$

If the complex number should appear in the second, third, or fourth quadrant, simply convert it in that quadrant, and carefully determine the proper angle to be associated with the magnitude of the vector.

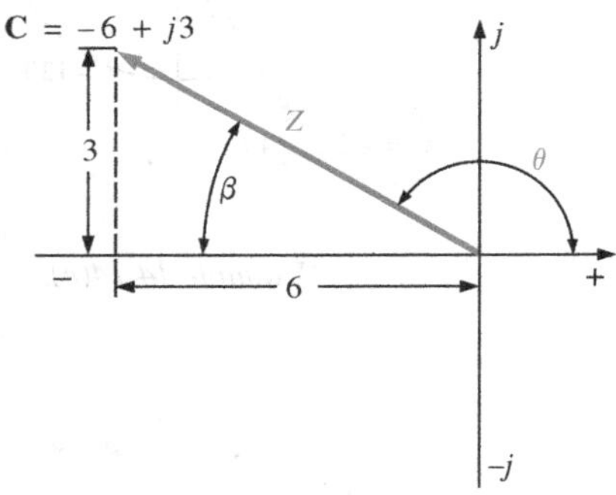

FIG. 14.51
Example 14.17.

EXAMPLE 14.17 Convert the following from rectangular to polar form:

$$\mathbf{C} = -6 + j3 \qquad \text{(Fig. 14.51)}$$

Solution:

$$Z = \sqrt{(6)^2 + (3)^2} = \sqrt{45} = 6.71$$

$$\beta = \tan^{-1}\left(\frac{3}{6}\right) = 26.57^\circ$$

$$\theta = 180^\circ - 26.57^\circ = 153.43^\circ$$

and $\mathbf{C} = \mathbf{6.71\angle 153.43^\circ}$

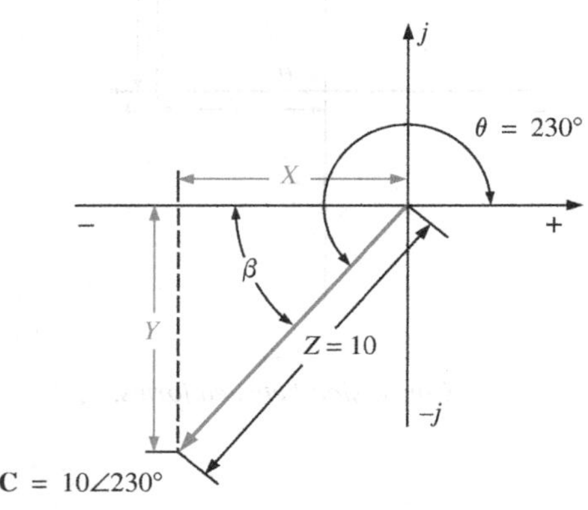

FIG. 14.52
Example 14.18.

EXAMPLE 14.18 Convert the following from polar to rectangular form:

$$\mathbf{C} = 10\angle 230^\circ \qquad \text{(Fig. 14.52)}$$

Solution:

$$X = Z\cos\beta = 10\cos(230^\circ - 180^\circ) = 10\cos 50^\circ$$
$$= (10)(0.6428) = 6.428$$

$$Y = Z\sin\beta = 10\sin 50^\circ = (10)(0.7660) = 7.660$$

and $\mathbf{C} = \mathbf{-6.43 - j\,7.66}$

14.10 MATHEMATICAL OPERATIONS WITH COMPLEX NUMBERS

Complex numbers lend themselves readily to the basic mathematical operations of addition, subtraction, multiplication, and division. A few basic rules and definitions must be understood before considering these operations.

Let us first examine the symbol j associated with imaginary numbers. By definition,

$$\boxed{j = \sqrt{-1}} \tag{14.24}$$

Thus,

$$\boxed{j^2 = -1} \tag{14.25}$$

and $\quad j^3 = j^2 j = -1j = -j$

with $\quad j^4 = j^2 j^2 = (-1)(-1) = +1$

$$j^5 = j$$

and so on. Further,

$$\frac{1}{j} = (1)\left(\frac{1}{j}\right) = \left(\frac{j}{j}\right)\left(\frac{1}{j}\right) = \frac{j}{j^2} = \frac{j}{-1}$$

and

$$\boxed{\frac{1}{j} = -j} \tag{14.26}$$

Complex Conjugate

The **conjugate** or **complex conjugate** of a complex number can be found by simply changing the sign of the imaginary part in the rectangular form or by using the negative of the angle of the polar form. For example, the conjugate of

$$\mathbf{C} = 2 + j3$$

is

$$2 - j3$$

as shown in Fig. 14.53. The conjugate of

$$\mathbf{C} = 2 \angle 30°$$

is

$$2 \angle -30°$$

as shown in Fig. 14.54.

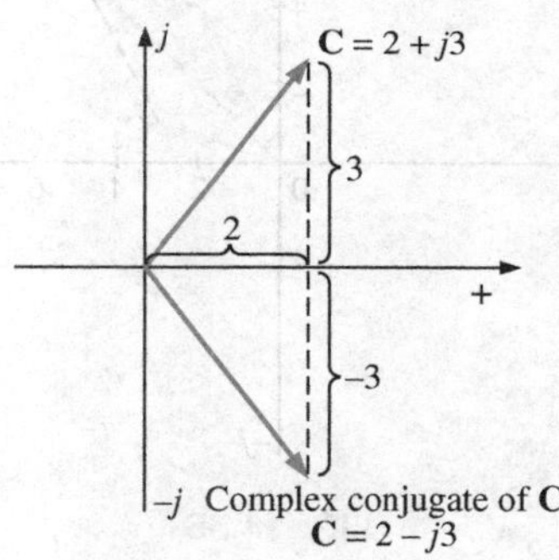

FIG. 14.53
Defining the complex conjugate of a complex number in rectangular form.

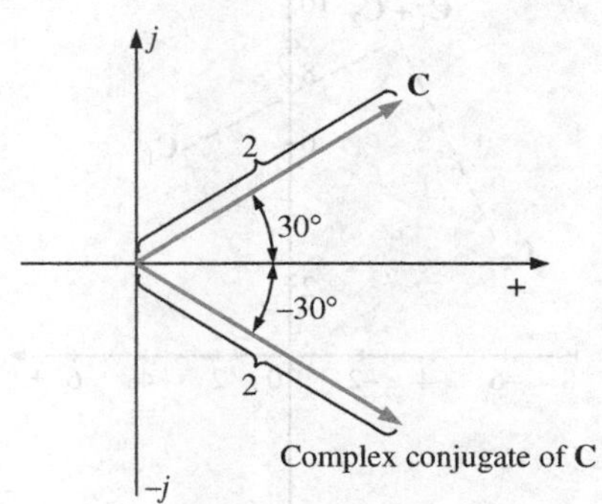

FIG. 14.54
Defining the complex conjugate of a complex number in polar form.

Reciprocal

The **reciprocal** of a complex number is 1 divided by the complex number. For example, the reciprocal of

$$\mathbf{C} = X + jY$$

is

$$\frac{1}{X + jY}$$

and that of $Z \angle \theta$ is

$$\frac{1}{Z \angle \theta}$$

We are now prepared to consider the four basic operations of *addition, subtraction, multiplication,* and *division* with complex numbers.

Addition

To add two or more complex numbers, add the real and imaginary parts separately. For example, if

$$\mathbf{C}_1 = \pm X_1 \pm jY_1 \qquad \text{and} \qquad \mathbf{C}_2 = \pm X_2 \pm jY_2$$

then

$$\boxed{\mathbf{C}_1 + \mathbf{C}_2 = (\pm X_1 \pm X_2) + j(\pm Y_1 \pm Y_2)} \qquad \textbf{(14.27)}$$

There is really no need to memorize the equation. Simply set one above the other and consider the real and imaginary parts separately, as shown in Example 14.19.

EXAMPLE 14.19

a. Add $\mathbf{C}_1 = 2 + j\,4$ and $\mathbf{C}_2 = 3 + j\,1$.
b. Add $\mathbf{C}_1 = 3 + j\,6$ and $\mathbf{C}_2 = -6 + j\,3$.

Solutions:

a. By Eq. (14.27),

$$\mathbf{C}_1 + \mathbf{C}_2 = (2 + 3) + j(4 + 1) = \mathbf{5 + j\,5}$$

Note Fig. 14.55. An alternative method is

$$\begin{array}{c} 2 + j\,4 \\ \underline{3 + j\,1} \\ \downarrow \quad \downarrow \\ \mathbf{5 + j\,5} \end{array}$$

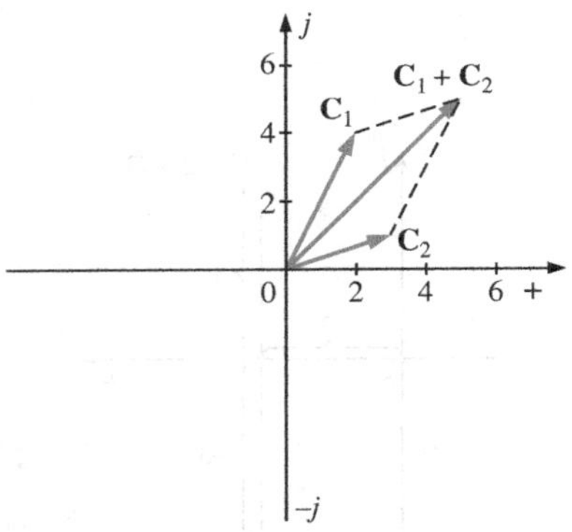

FIG. 14.55
Example 14.19(a).

b. By Eq. (14.27),

$$\mathbf{C}_1 + \mathbf{C}_2 = (3 - 6) + j(6 + 3) = \mathbf{-3 + j\,9}$$

Note Fig. 14.56. An alternative method is

$$\begin{array}{c} 3 + j\,6 \\ \underline{-6 + j\,3} \\ \downarrow \quad \downarrow \\ \mathbf{-3 + j\,9} \end{array}$$

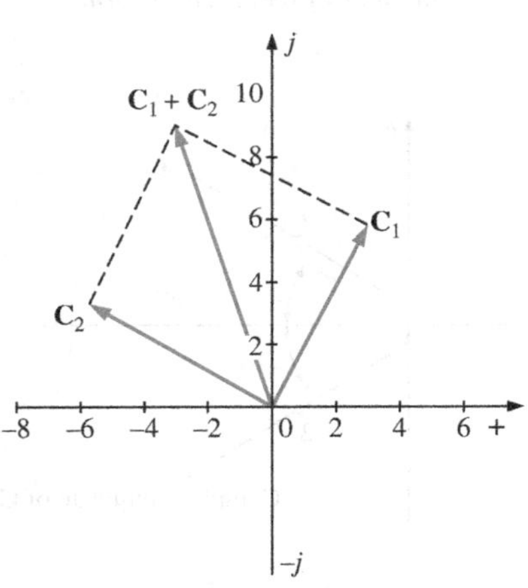

FIG. 14.56
Example 14.19(b).

Subtraction

In subtraction, the real and imaginary parts are again considered separately. For example, if

$$\mathbf{C}_1 = \pm X_1 \pm jY_1 \qquad \text{and} \qquad \mathbf{C}_2 = \pm X_2 \pm jY_2$$

then

$$\boxed{\mathbf{C}_1 - \mathbf{C}_2 = [\pm X_1 - (\pm X_2)] + j[\pm Y_1 - (\pm Y_2)]} \qquad \textbf{(14.28)}$$

Again, there is no need to memorize the equation if the alternative method of Example 14.20 is used.

EXAMPLE 14.20

a. Subtract $\mathbf{C}_2 = 1 + j\,4$ from $\mathbf{C}_1 = 4 + j\,6$.
b. Subtract $\mathbf{C}_2 = -2 + j\,5$ from $\mathbf{C}_1 = +3 + j\,3$.

Solutions:

a. By Eq. (14.28),

$$\mathbf{C}_1 - \mathbf{C}_2 = (4 - 1) + j(6 - 4) = \mathbf{3 + j\,2}$$

Note Fig. 14.57. An alternative method is

$$\begin{array}{r} 4 + j\,6 \\ \underline{-(1 + j\,4)} \\ \downarrow \quad \downarrow \\ \mathbf{3 + j\,2} \end{array}$$

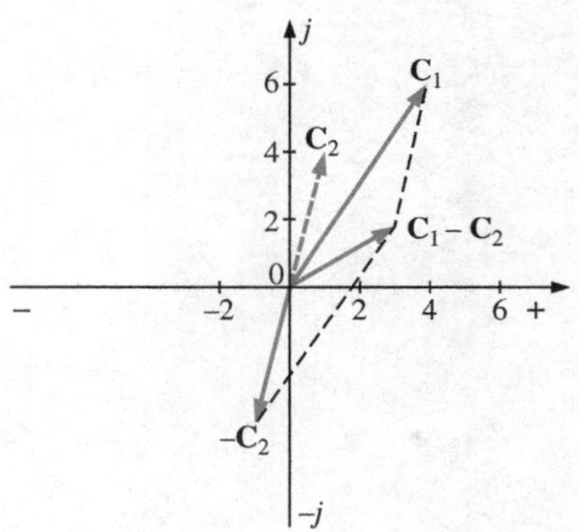

FIG. 14.57
Example 14.20(a).

b. By Eq. (14.28),

$$\mathbf{C}_1 - \mathbf{C}_2 = [3 - (-2)] + j(3 - 5) = \mathbf{5 - j\,2}$$

Note Fig. 14.58. An alternative method is

$$\begin{array}{r} 3 + j\,3 \\ \underline{-(-2 + j\,5)} \\ \downarrow \quad \downarrow \\ \mathbf{5 - j\,2} \end{array}$$

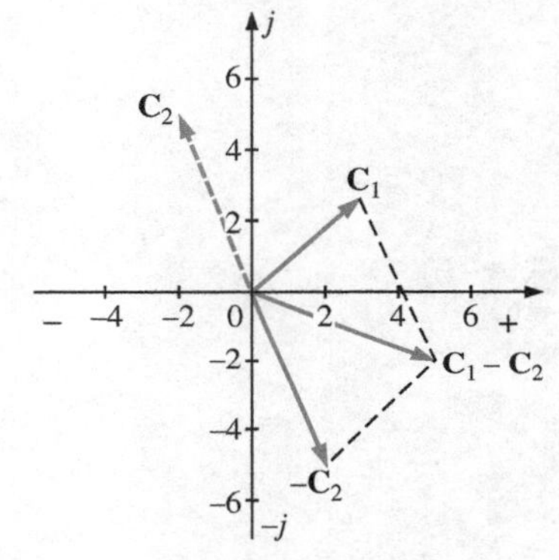

FIG. 14.58
Example 14.20(b).

Addition or subtraction cannot be performed in polar form unless the complex numbers have the same angle θ or unless they differ only by multiples of 180°.

EXAMPLE 14.21

a. $2\,\angle 45° + 3\,\angle 45° = \mathbf{5\,\angle 45°}$. Note Fig. 14.59.
b. $2\,\angle 0° - 4\,\angle 180° = \mathbf{6\,\angle 0°}$. Note Fig. 14.60.

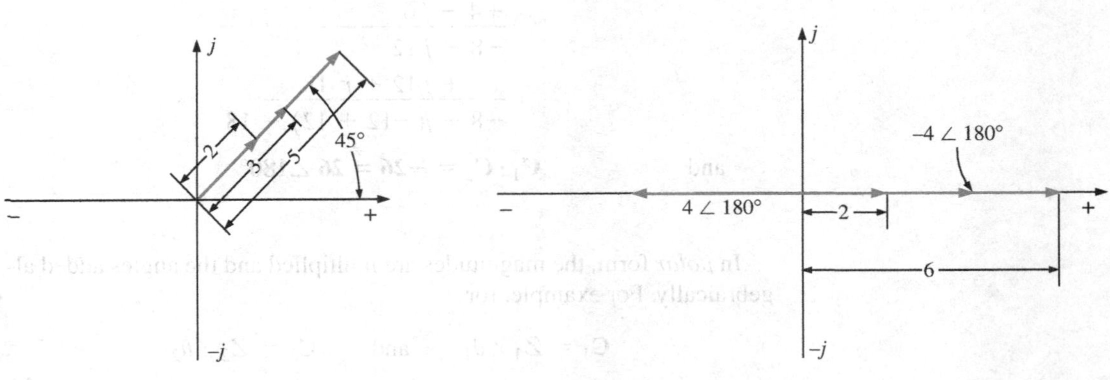

FIG. 14.59
Example 14.21(a).

FIG. 14.60
Example 14.21(b).

Multiplication

To multiply two complex numbers in *rectangular* form, multiply the real and imaginary parts of one in turn by the real and imaginary parts of the other. For example, if

$$\mathbf{C}_1 = X_1 + jY_1 \quad \text{and} \quad \mathbf{C}_2 = X_2 + jY_2$$

then $\mathbf{C}_1 \cdot \mathbf{C}_2$:

$$\begin{array}{r} X_1 + jY_1 \\ \underline{X_2 + jY_2} \\ X_1X_2 + jY_1X_2 \qquad\qquad \\ \underline{+\, jX_1Y_2 + j^2Y_1Y_2} \\ X_1X_2 + j(Y_1X_2 + X_1Y_2) + Y_1Y_2(-1) \end{array}$$

and

$$\boxed{\mathbf{C}_1 \cdot \mathbf{C}_2 = (X_1X_2 - Y_1Y_2) + j(Y_1X_2 + X_1Y_2)} \tag{14.29}$$

In Example 14.22(b), we obtain a solution without resorting to memorizing Eq. (14.29). Simply carry along the j factor when multiplying each part of one vector with the real and imaginary parts of the other.

EXAMPLE 14.22

a. Find $\mathbf{C}_1 \cdot \mathbf{C}_2$ if

$$\mathbf{C}_1 = 2 + j\,3 \quad \text{and} \quad \mathbf{C}_2 = 5 + j\,10$$

b. Find $\mathbf{C}_1 \cdot \mathbf{C}_2$ if

$$\mathbf{C}_1 = -2 - j\,3 \quad \text{and} \quad \mathbf{C}_2 = +4 - j\,6$$

Solutions:

a. Using the format above, we have

$$\begin{aligned} \mathbf{C}_1 \cdot \mathbf{C}_2 &= [(2)(5) - (3)(10)] + j[(3)(5) + (2)(10)] \\ &= \mathbf{-20 + j\,35} \end{aligned}$$

b. Without using the format, we obtain

$$\begin{array}{r} -2 - j\,3 \qquad \\ \underline{+4 - j\,6 \qquad} \\ -8 - j\,12 \qquad\qquad \\ \underline{+\, j\,12 + j^2\,18} \\ -8 + j(-12 + 12) - 18 \end{array}$$

and

$$\mathbf{C}_1 \cdot \mathbf{C}_2 = \mathbf{-26 = 26 \angle 180^\circ}$$

In *polar* form, the magnitudes are multiplied and the angles added algebraically. For example, for

$$\mathbf{C}_1 = Z_1 \angle \theta_1 \quad \text{and} \quad \mathbf{C}_2 = Z_2 \angle \theta_2$$

we write

$$\boxed{\mathbf{C}_1 \cdot \mathbf{C}_2 = Z_1Z_2 \angle \theta_1 + \theta_2} \tag{14.30}$$

EXAMPLE 14.23

a. Find $\mathbf{C}_1 \cdot \mathbf{C}_2$ if

$$\mathbf{C}_1 = 5 \angle 20^\circ \quad \text{and} \quad \mathbf{C}_2 = 10 \angle 30^\circ$$

b. Find $\mathbf{C}_1 \cdot \mathbf{C}_2$ if

$$\mathbf{C}_1 = 2 \angle -40^\circ \quad \text{and} \quad \mathbf{C}_2 = 7 \angle +120^\circ$$

Solutions:

a. $\mathbf{C}_1 \cdot \mathbf{C}_2 = (5 \angle 20^\circ)(10 \angle 30^\circ) = (5)(10) \underline{/20^\circ + 30^\circ} = \mathbf{50 \angle 50^\circ}$

b. $\mathbf{C}_1 \cdot \mathbf{C}_2 = (2 \angle -40^\circ)(7 \angle +120^\circ) = (2)(7)\underline{/-40^\circ + 120^\circ}$
$= \mathbf{14 \angle +80^\circ}$

To multiply a complex number in rectangular form by a real number requires that both the real part and the imaginary part be multiplied by the real number. For example,

$$(10)(2 + j\,3) = \mathbf{20 + j\,30}$$

and
$$50 \angle 0^\circ (0 + j\,6) = j\,300 = \mathbf{300 \angle 90^\circ}$$

Division

To divide two complex numbers in *rectangular* form, multiply the numerator and denominator by the conjugate of the denominator and the resulting real and imaginary parts collected. That is, if

$$\mathbf{C}_1 = X_1 + jY_1 \quad \text{and} \quad \mathbf{C}_2 = X_2 + jY_2$$

then
$$\frac{\mathbf{C}_1}{\mathbf{C}_2} = \frac{(X_1 + jY_1)(X_2 - jY_2)}{(X_2 + jY_2)(X_2 - jY_2)} = \frac{(X_1X_2 + Y_1Y_2) + j(X_2Y_1 - X_1Y_2)}{X_2^2 + Y_2^2}$$

and
$$\boxed{\frac{\mathbf{C}_1}{\mathbf{C}_2} = \frac{X_1X_2 + Y_1Y_2}{X_2^2 + Y_2^2} + j\frac{X_2Y_1 - X_1Y_2}{X_2^2 + Y_2^2}} \qquad \textbf{(14.31)}$$

The equation does not have to be memorized if the steps above used to obtain it are employed. That is, first multiply the numerator by the complex conjugate of the denominator and separate the real and imaginary terms. Then divide each term by the sum of each term of the denominator squared.

EXAMPLE 14.24

a. Find $\mathbf{C}_1/\mathbf{C}_2$ if $\mathbf{C}_1 = 1 + j\,4$ and $\mathbf{C}_2 = 4 + j\,5$.
b. Find $\mathbf{C}_1/\mathbf{C}_2$ if $\mathbf{C}_1 = -4 - j\,8$ and $\mathbf{C}_2 = +6 - j\,1$.

Solutions:

a. By Eq. (14.31),

$$\frac{\mathbf{C}_1}{\mathbf{C}_2} = \frac{(1)(4) + (4)(5)}{4^2 + 5^2} + j\frac{(4)(4) - (1)(5)}{4^2 + 5^2}$$
$$= \frac{24}{41} + \frac{j\,11}{41} \cong \mathbf{0.59 + j\,0.27}$$

b. Using an alternative method, we obtain

$$\begin{array}{r} -4 - j\,8 \\ +6 + j\,1 \\ \hline -24 - j\,48 \\ -j\,4 - j^2\,8 \\ \hline -24 - j\,52 + 8 = -16 - j\,52 \end{array}$$

$$\begin{array}{r} +6 - j\,1 \\ +6 + j\,1 \\ \hline 36 + j\,6 \\ -j\,6 - j^2\,1 \\ \hline 36 + 0 + 1 = 37 \end{array}$$

and $$\frac{\mathbf{C}_1}{\mathbf{C}_2} = \frac{-16}{37} - \frac{j\,52}{37} = \mathbf{-0.43 - j\,1.41}$$

To divide a complex number in rectangular form by a real number, both the real part and the imaginary part must be divided by the real number. For example,

$$\frac{8 + j\,10}{2} = \mathbf{4 + j\,5}$$

and $$\frac{6.8 - j\,0}{2} = 3.4 - j\,0 = \mathbf{3.4 \angle 0^\circ}$$

In *polar* form, division is accomplished by dividing the magnitude of the numerator by the magnitude of the denominator and subtracting the angle of the denominator from that of the numerator. That is, for

$$\mathbf{C}_1 = Z_1 \angle \theta_1 \qquad \text{and} \qquad \mathbf{C}_2 = Z_2 \angle \theta_2$$

we write

$$\boxed{\frac{\mathbf{C}_1}{\mathbf{C}_2} = \frac{Z_1}{Z_2} \underline{/\theta_1 - \theta_2}} \qquad \textbf{(14.32)}$$

EXAMPLE 14.25

a. Find $\mathbf{C}_1/\mathbf{C}_2$ if $\mathbf{C}_1 = 15 \angle 10^\circ$ and $\mathbf{C}_2 = 2 \angle 7^\circ$.
b. Find $\mathbf{C}_1/\mathbf{C}_2$ if $\mathbf{C}_1 = 8 \angle 120^\circ$ and $\mathbf{C}_2 = 16 \angle -50^\circ$.

Solutions:

a. $$\frac{\mathbf{C}_1}{\mathbf{C}_2} = \frac{15 \angle 10^\circ}{2 \angle 7^\circ} = \frac{15}{2} \underline{/10^\circ - 7^\circ} = \mathbf{7.5 \angle 3^\circ}$$

b. $$\frac{\mathbf{C}_1}{\mathbf{C}_2} = \frac{8 \angle 120^\circ}{16 \angle -50^\circ} = \frac{8}{16} \underline{/120^\circ - (-50^\circ)} = \mathbf{0.5 \angle 170^\circ}$$

We obtain the *reciprocal* in the rectangular form by multiplying the numerator and denominator by the complex conjugate of the denominator:

$$\frac{1}{X + jY} = \left(\frac{1}{X + jY}\right)\left(\frac{X - jY}{X - jY}\right) = \frac{X - jY}{X^2 + Y^2}$$

and
$$\frac{1}{X + jY} = \frac{X}{X^2 + Y^2} - j\frac{Y}{X^2 + Y^2} \qquad \textbf{(14.33)}$$

In polar form, the reciprocal is

$$\frac{1}{Z \angle \theta} = \frac{1}{Z}\angle -\theta \qquad \textbf{(14.34)}$$

A concluding example using the four basic operations follows.

EXAMPLE 14.26 Perform the following operations, leaving the answer in polar or rectangular form:

a.
$$\frac{(2 + j\,3) + (4 + j\,6)}{(7 + j\,7) - (3 - j\,3)} = \frac{(2 + 4) + j(3 + 6)}{(7 - 3) + j(7 + 3)}$$
$$= \frac{(6 + j\,9)(4 - j\,10)}{(4 + j\,10)(4 - j\,10)}$$
$$= \frac{[(6)(4) + (9)(10)] + j[(4)(9) - (6)(10)]}{4^2 + 10^2}$$
$$= \frac{114 - j\,24}{116} = \mathbf{0.98 - j\,0.21}$$

b.
$$\frac{(50 \angle 30^\circ)(5 + j\,5)}{10 \angle -20^\circ} = \frac{(50 \angle 30^\circ)(7.07 \angle 45^\circ)}{10 \angle -20^\circ} = \frac{353.5 \angle 75^\circ}{10 \angle -20^\circ}$$
$$= 35.35 \underline{/75^\circ - (-20^\circ)} = \mathbf{35.35 \angle 95^\circ}$$

c.
$$\frac{(2 \angle 20^\circ)^2(3 + j\,4)}{8 - j\,6} = \frac{(2 \angle 20^\circ)(2 \angle 20^\circ)(5 \angle 53.13^\circ)}{10 \angle -36.87^\circ}$$
$$= \frac{(4 \angle 40^\circ)(5 \angle 53.13^\circ)}{10 \angle -36.87^\circ} = \frac{20 \angle 93.13^\circ}{10 \angle -36.87^\circ}$$
$$= 2 \underline{/93.13^\circ - (-36.87^\circ)} = \mathbf{2.0 \angle 130^\circ}$$

d.
$$3 \angle 27^\circ - 6 \angle -40^\circ = (2.673 + j\,1.362) - (4.596 - j\,3.857)$$
$$= (2.673 - 4.596) + j(1.362 + 3.857)$$
$$= \mathbf{-1.92 + j\,5.22}$$

14.11 CALCULATOR METHODS WITH COMPLEX NUMBERS

The process of converting from one form to another or working through lengthy operations with complex numbers can be time-consuming and often frustrating if one lost minus sign or decimal point invalidates the solution. Fortunately, technologists of today have calculators and computer methods that make the process measurably easier with higher degrees of reliability and accuracy.

Calculators

The TI-89 calculator in Fig. 14.61 is only one of numerous calculators that can convert from one form to another and perform lengthy

FIG. 14.61
TI-89 scientific calculator.
(Courtesy of Texas Instruments, Inc.)

calculations with complex numbers in a concise, neat form. Not all of the details of using a specific calculator are included here because each has its own format and sequence of steps. However, the basic operations with the TI-89 are included primarily to demonstrate the ease with which the conversions can be made and the format for more complex operations.

There are different routes to perform the conversions and operations below, but these instructions give you one approach that is fairly direct and straightforward. Since most operations are in the DEGREE rather than RADIAN mode, the sequence in Fig. 14.62 shows how to set the DEGREE mode for the operations to follow. A similar sequence sets the RADIAN mode if required.

MODE ↓ Angle → ↓ DEGREE ENTER ENTER

FIG. 14.62
Setting the DEGREE mode on the TI-89 calculator.

Rectangular to Polar Conversion The sequence in Fig. 14.63 provides a detailed listing of the steps needed to convert from rectangular to polar form. In the examples to follow, the scrolling steps are not listed to simplify the sequence.

In the sequence in Fig. 14.63, an up scroll is chosen after Matrix because that is a more direct path to Vector ops. A down scroll generates the same result, but it requires going through the whole listing. The sequence seems quite long for such a simple conversion, but with practice you will be able to perform the scrolling steps quite rapidly. Always be sure the input data are entered correctly, such as including the *i* after the *y* component. Any incorrect entry will result in an error listing.

(3 + 5 2ND *i*) 2ND MATH ↓ Matrix →
↑ Vector ops → ↓ ▶ Polar ENTER ENTER **5.83E0 ∠ 59.0E0**

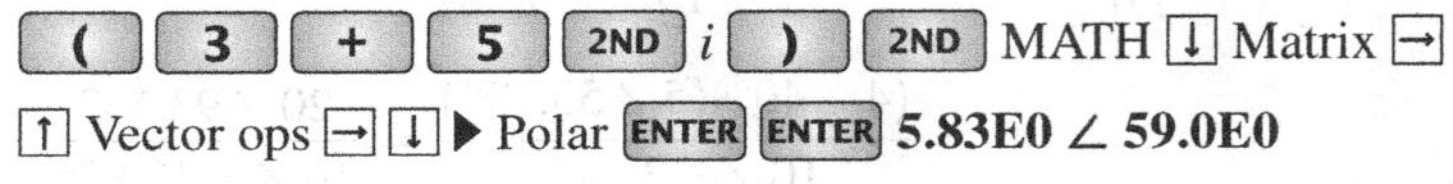

FIG. 14.63
Converting $3 + j\,5$ to the polar form using the TI-89 calculator.

Polar to Rectangular Conversion The sequence in Fig. 14.64 is a detailed listing of the steps needed to convert from polar to rectangular form. Note in the format that the brackets must surround the polar form. Also, the degree sign must be included with the angle to perform the calculation. The answer is displayed in the engineering notation selected.

(5 2ND ∠ 5 3 • 1 2ND MATH ↓
Angle → ° ENTER) 2ND MATH ↓ Matrix → ↑ Vector ops → ↓
Rect ENTER ENTER **3.00E0+4.00E0i**

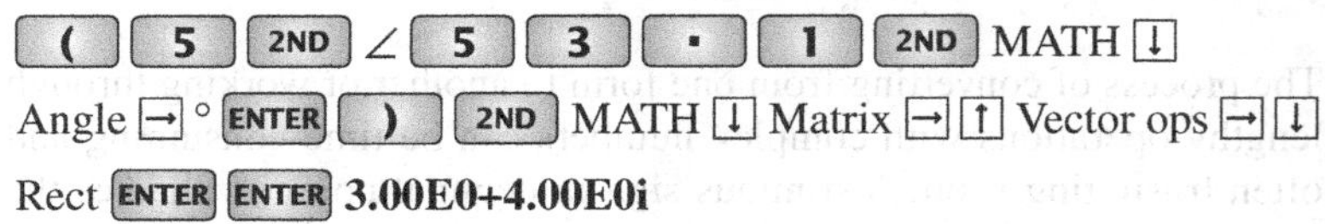

FIG. 14.64
Converting $5\angle 53.1°$ to the rectangular form using the TI-89 calculator.

Mathematical Operations Mathematical operations are performed in the natural order of operations, but you must remember to select the format for the solution. For instance, if the sequence in Fig. 14.65 did not include the polar designation, the answer would be in rectangular

(1 0 ∠ 5 0 °) × (2 ∠ 2 0 °)

▶ Polar ENTER **20.00E0 ∠ 70.00E0**

FIG. 14.65
Performing the operation (10 ∠50°)(2 ∠20°).

form even though both quantities in the calculation are in polar form. In the rest of the examples, the scrolling required to obtain mathematical functions is not included to minimize the length of the sequence.

For the product of mixed complex numbers, the sequence of Fig. 14.66 results. Again, the polar form was selected for the solution.

(5 ∠ 5 3 . 1 °) × (2 + 2 *i*)

▶ Polar ENTER ENTER **14.14E0 ∠ 98.10E0**

FIG. 14.66
Performing the operation (5 ∠53.1°)(2 + j 2).

Finally, Example 14.26(c) is entered as shown by the sequence in Fig. 14.67. Note that the results exactly match those obtained earlier.

(2 ∠ 2 0 °) ^ 2 × (3 + 4 *i*)

÷ (8 − 6 *i*) ▶ Polar ENTER ENTER **2.00E0 ∠ 130.0E0**

FIG. 14.67
Verifying the results of Example 14.26(c).

14.12 PHASORS

As noted earlier in this chapter, the addition of sinusoidal voltages and currents is frequently required in the analysis of ac circuits. One lengthy but valid method of performing this operation is to place both sinusoidal waveforms on the same set of axes and add algebraically the magnitudes of each at every point along the abscissa, as shown for $c = a + b$ in Fig. 14.68. This, however, can be a long and tedious process

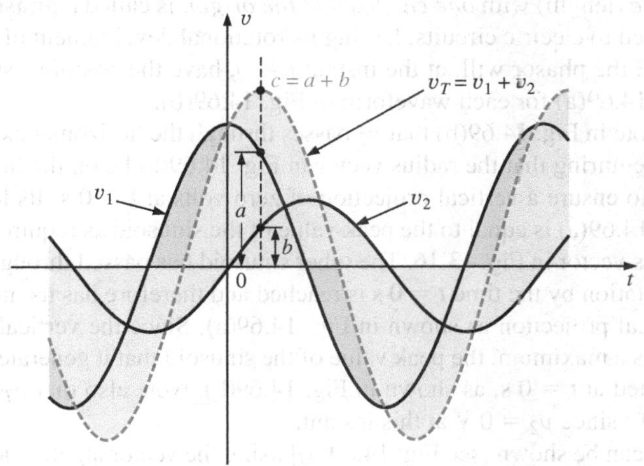

FIG. 14.68
Adding two sinusoidal waveforms on a point-by-point basis.

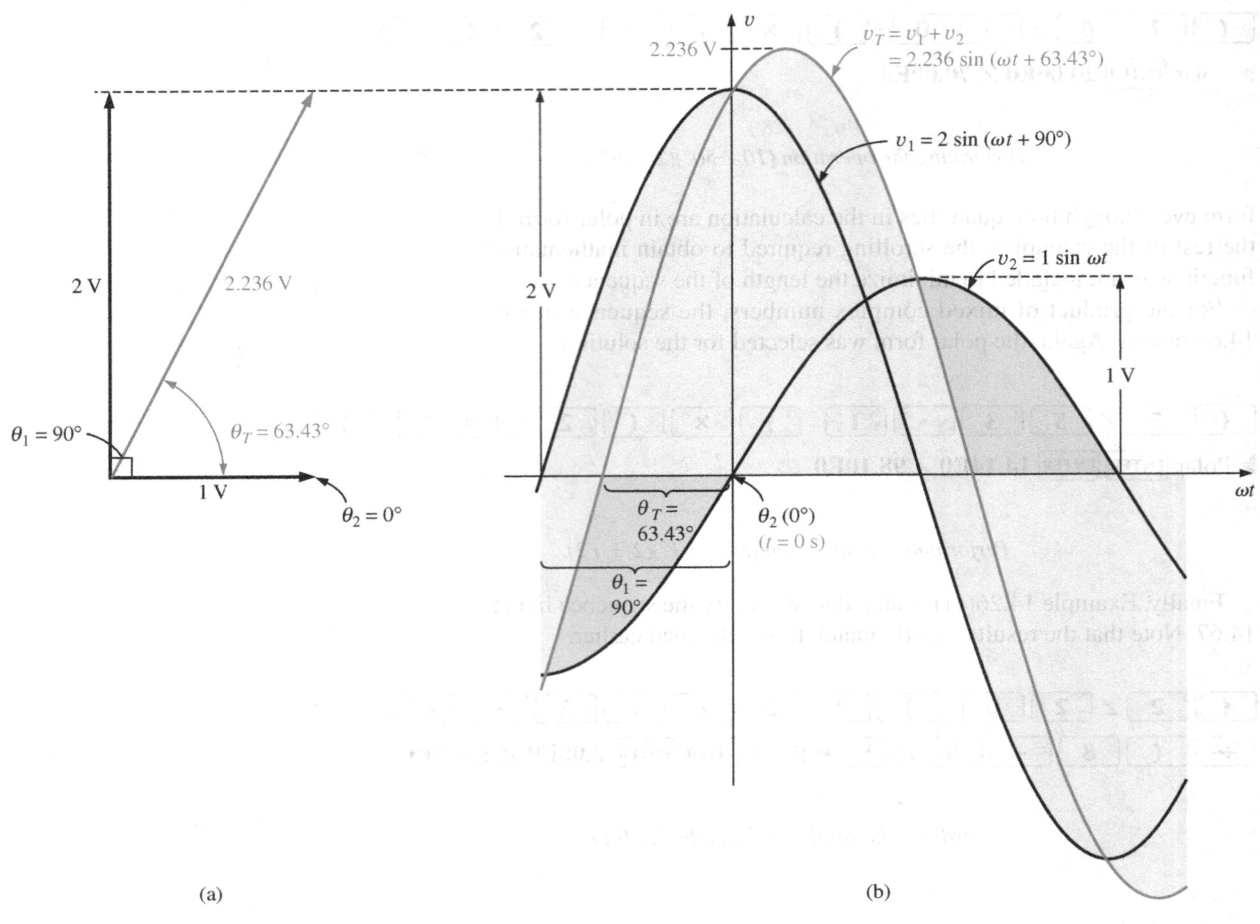

FIG. 14.69

(a) The phasor representation of the sinusoidal waveforms of part (b); (b) finding the sum of two sinusoidal waveforms of v_1 and v_2.

with limited accuracy. A shorter method uses the rotating radius vector first appearing in Fig. 13.16. This *radius vector,* having a *constant magnitude* (length) with *one end fixed at the origin,* is called a **phasor** when applied to electric circuits. During its rotational development of the sine wave, the phasor will, at the instant $t = 0$, have the positions shown in Fig. 14.69(a) for each waveform in Fig. 14.69(b).

Note in Fig. 14.69(b) that v_2 passes through the horizontal axis at $t =$ 0 s, requiring that the radius vector in Fig. 14.69(a) be on the horizontal axis to ensure a vertical projection of zero volts at $t = 0$ s. Its length in Fig. 14.69(a) is equal to the peak value of the sinusoid as required by the radius vector in Fig. 13.16. The other sinusoid has passed through 90° of its rotation by the time $t = 0$ s is reached and therefore has its maximum vertical projection as shown in Fig. 14.69(a). Since the vertical projection is a maximum, the peak value of the sinusoid that it generates is also attained at $t = 0$ s, as shown in Fig. 14.69(b). Note also that $v_T = v_1$ at $t = 0$ s since $v_2 = 0$ V at this instant.

It can be shown [see Fig. 14.69(a)] using the vector algebra described in Section 14.10 that

$$1 \text{ V} \angle 0° + 2 \text{ V} \angle 90° = 2.236 \text{ V} \angle 63.43°$$

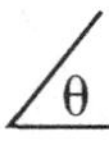

In other words, if we convert v_1 and v_2 to the phasor form using

$$v = V_m \sin(\omega t \pm \theta) \Rightarrow V_m \angle \pm\theta$$

and add them using complex number algebra, we can find the phasor form for v_T with very little difficulty. It can then be converted to the time domain and plotted on the same set of axes, as shown in Fig. 14.69(b). Fig. 14.69(a), showing the magnitudes and relative positions of the various phasors, is called a **phasor diagram.** It is actually a "snapshot" of the rotating radius vectors at $t = 0$ s.

Therefore, if the addition of two sinusoids is required, you should first convert them to the phasor domain and find the sum using complex algebra. You can then convert the result to the time domain.

The case of two sinusoidal functions having phase angles different from 0° and 90° appears in Fig. 14.70. Note again that the vertical height of the functions in Fig. 14.70(b) at $t = 0$ s is determined by the rotational positions of the radius vectors in Fig. 14.70(a).

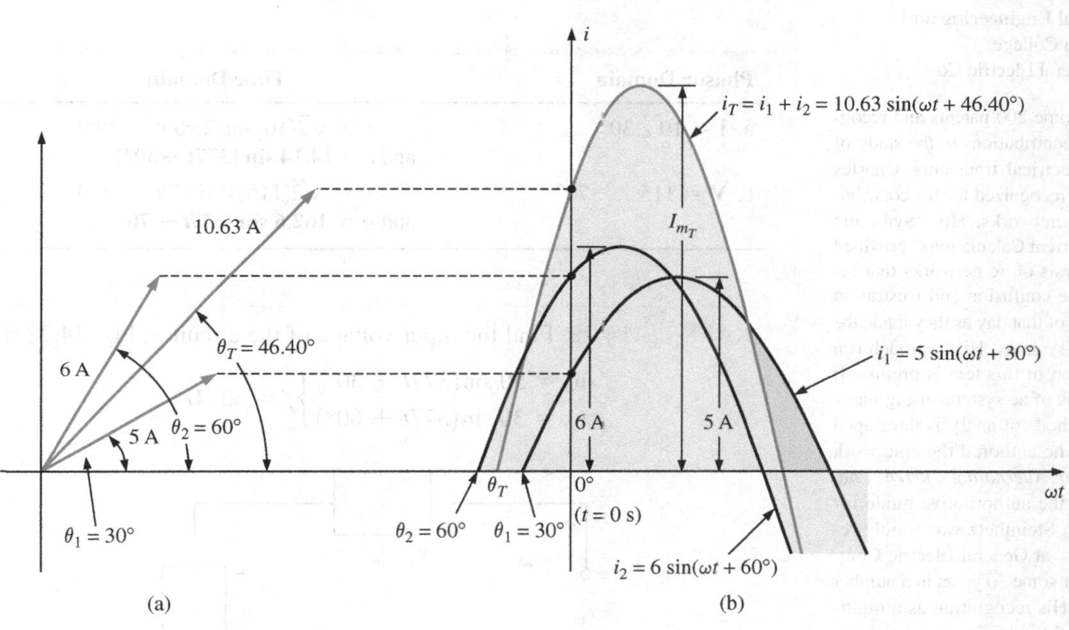

FIG. 14.70
Adding two sinusoidal currents with phase angles other than 90°.

Since the rms, rather than the peak, values are used almost exclusively in the analysis of ac circuits, the phasor will now be redefined for the purposes of practicality and uniformity as having a magnitude equal to the *rms value* of the sine wave it represents. The angle associated with the phasor will remain as previously described—the phase angle.

In general, for all of the analyses to follow, the phasor form of a sinusoidal voltage or current will be

$$\mathbf{V} = V\angle\theta \qquad \text{and} \qquad \mathbf{I} = I\angle\theta$$

where V and I are rms values and θ is the phase angle. It should be pointed out that in phasor notation, the sine wave is always the reference, and the frequency is not represented.

Phasor algebra for sinusoidal quantities is applicable only for waveforms having the same frequency.

FIG. 14.71
Charles Proteus Steinmetz.
Courtesy of the Hall of History Foundation, Schenectady, New York

German-American (Breslau, Germany; Yonkers and Schenectady, NY, USA)
(1865–1923)
Mathematician, Scientist, Engineer, Inventor, Professor of Electrical Engineering and Electrophysics, Union College
Department Head, General Electric Co.

Although the holder of some 200 patents and recognized worldwide for his contributions to the study of hysteresis losses and electrical transients, Charles Proteus Steinmetz is best recognized for his contribution to the study of ac networks. His "Symbolic Method of Alternating-current Calculations" provided an approach to the analysis of ac networks that removed a great deal of the confusion and frustration experienced by engineers of that day as they made the transition from dc to ac systems. His approach (on which the phasor notation of this text is premised) permitted a direct analysis of ac systems using many of the theorems and methods of analysis developed for dc systems. In 1897 he authored the epic work *Theory and Calculation of Alternating Current Phenomena,* which became the authoritative guide for practicing engineers. Dr. Steinmetz was fondly referred to as "The Doctor" at General Electric Company where he worked for some 30 years in a number of important capacities. His recognition as a multi-gifted genius is supported by the fact that he maintained active friendships with such individuals as Albert Einstein, Guglielmo Marconi, and Thomas A. Edison, to name just a few. He was President of the American Institute of Electrical Engineers (AIEE) and the National Association of Corporation Schools and actively supported his local community (Schenectady) as president of the Board of Education and the Commission on Parks and City Planning.

The use of phasor notation in the analysis of ac networks was first introduced by Charles Proteus Steinmetz in 1897 (Fig. 14.71).

EXAMPLE 14.27 Convert the following from the time to the phasor domain:

Time Domain	Phasor Domain
a. $\sqrt{2}(50)\sin\omega t$	$\mathbf{50\ \angle 0^\circ}$
b. $69.6\sin(\omega t + 72^\circ)$	$(0.707)(69.6)\ \angle 72^\circ = \mathbf{49.21\ \angle 72^\circ}$
c. $45\cos\omega t$	$(0.707)(45)\ \angle 90^\circ = \mathbf{31.82\ \angle 90^\circ}$

EXAMPLE 14.28 Write the sinusoidal expression for the following phasors if the frequency is 60 Hz:

Phasor Domain	Time Domain
a. $\mathbf{I} = 10\ \angle 30^\circ$	$i = \sqrt{2}(10)\sin(2\pi 60t + 30^\circ)$ and $i = \mathbf{14.14\ sin\ (377t + 30^\circ)}$
b. $\mathbf{V} = 115\ \angle -70^\circ$	$v = \sqrt{2}(115)\sin(377t - 70^\circ)$ and $v = \mathbf{162.6\ sin\ (377t - 70^\circ)}$

EXAMPLE 14.29 Find the input voltage of the circuit in Fig. 14.72 if

$$\left.\begin{aligned} v_a &= 50\sin(377t + 30^\circ) \\ v_b &= 30\sin(377t + 60^\circ) \end{aligned}\right\} f = 60\text{ Hz}$$

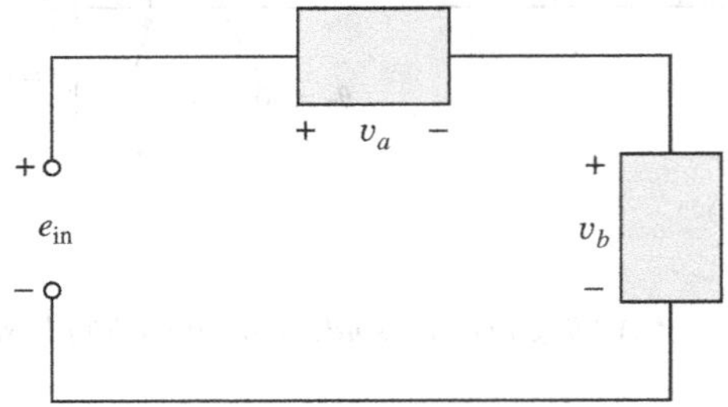

FIG. 14.72
Example 14.29.

Solution: Applying Kirchhoff's voltage law, we have

$$e_{in} = v_a + v_b$$

Converting from the time to the phasor domain yields

$$v_a = 50\sin(377t + 30^\circ) \Rightarrow \mathbf{V}_a = 35.35\text{ V}\ \angle 30^\circ$$
$$v_b = 30\sin(377t + 60^\circ) \Rightarrow \mathbf{V}_b = 21.21\text{ V}\ \angle 60^\circ$$

Converting from polar to rectangular form for addition yields

$$\mathbf{V}_a = 35.35\text{ V}\ \angle 30^\circ = 30.61\text{ V} + j\,17.68\text{ V}$$
$$\mathbf{V}_b = 21.21\text{ V}\ \angle 60^\circ = 10.61\text{ V} + j\,18.37\text{ V}$$

Then

$$\mathbf{E}_{\text{in}} = \mathbf{V}_a + \mathbf{V}_b = (30.61 \text{ V} + j\,17.68 \text{ V}) + (10.61 \text{ V} + j\,18.37 \text{ V})$$
$$= 41.22 \text{ V} + j\,36.05 \text{ V}$$

Converting from rectangular to polar form, we have

$$\mathbf{E}_{\text{in}} = 41.22 \text{ V} + j\,36.05 \text{ V} = 54.76 \text{ V} \angle 41.17^\circ$$

Converting from the phasor to the time domain, we obtain

$$\mathbf{E}_{\text{in}} = 54.76 \text{ V} \angle 41.17^\circ \Rightarrow e_{\text{in}} = \sqrt{2}(54.76)\sin(377t + 41.17^\circ)$$

and $$e_{\text{in}} = \mathbf{77.43 \sin(377}\boldsymbol{t}\mathbf{ + 41.17^\circ)}$$

A plot of the three waveforms is shown in Fig. 14.73. Note that at each instant of time, the sum of the two waveforms does in fact add up to e_{in}. At $t = 0$ ($\omega t = 0$), e_{in} is the sum of the two positive values, while at a value of ωt, almost midway between $\pi/2$ and π, the sum of the positive value of v_a and the negative value of v_b results in $e_{\text{in}} = 0$.

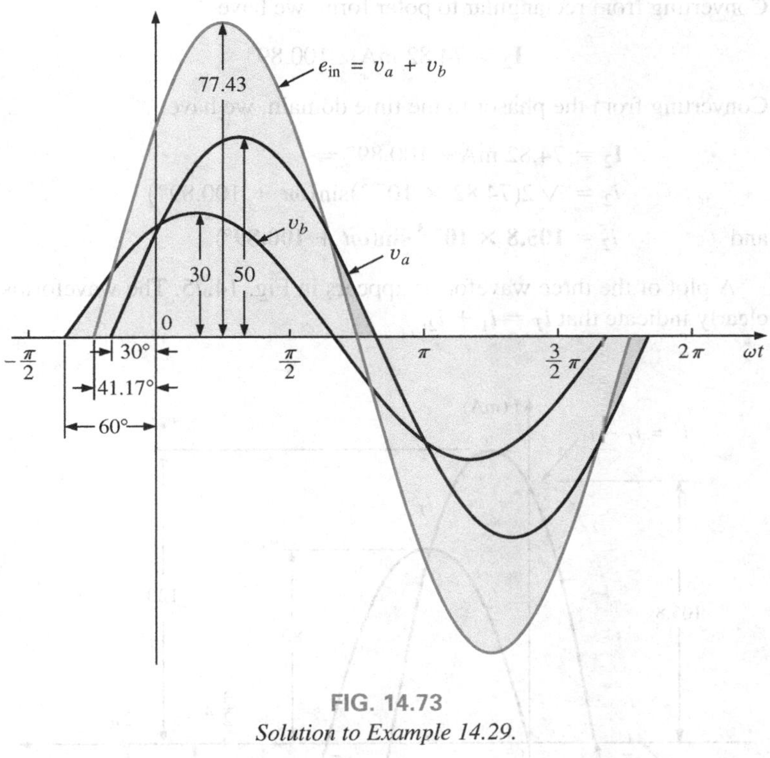

FIG. 14.73
Solution to Example 14.29.

EXAMPLE 14.30 Determine the current i_2 for the network in Fig. 14.74.

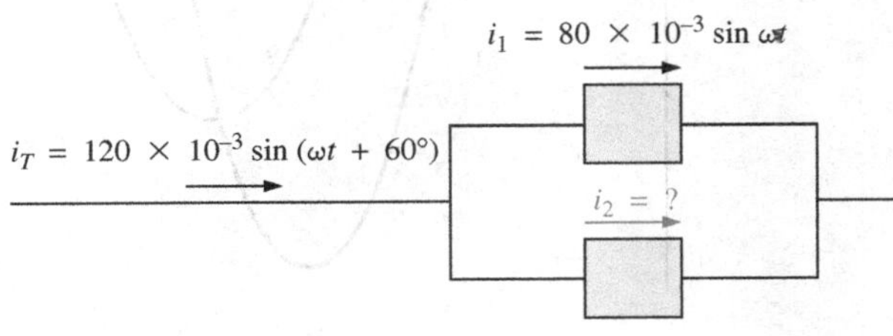

FIG. 14.74
Example 14.30.

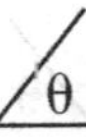

Solution: Applying Kirchhoff's current law, we obtain

$$i_T = i_1 + i_2 \quad \text{or} \quad i_2 = i_T - i_1$$

Converting from the time to the phasor domain yields

$$i_T = 120 \times 10^{-3} \sin(\omega t + 60°) \Rightarrow 84.84 \text{ mA} \angle 60°$$
$$i_1 = 80 \times 10^{-3} \sin \omega t \Rightarrow 56.56 \text{ mA} \angle 0°$$

Converting from polar to rectangular form for subtraction yields

$$\mathbf{I}_T = 84.84 \text{ mA} \angle 60° = 42.42 \text{ mA} + j\,73.47 \text{ mA}$$
$$\mathbf{I}_1 = 56.56 \text{ mA} \angle 0° = 56.56 \text{ mA} + j\,0$$

Then

$$\mathbf{I}_2 = \mathbf{I}_T - \mathbf{I}_1$$
$$= (42.42 \text{ mA} + j\,73.47 \text{ mA}) - (56.56 \text{ mA} + j\,0)$$

and $\quad \mathbf{I}_2 = -14.14 \text{ mA} + j\,73.47 \text{ mA}$

Converting from rectangular to polar form, we have

$$\mathbf{I}_2 = 74.82 \text{ mA} \angle 100.89°$$

Converting from the phasor to the time domain, we have

$$\mathbf{I}_2 = 74.82 \text{ mA} \angle 100.89° \Rightarrow$$
$$i_2 = \sqrt{2}(74.82 \times 10^{-3}) \sin(\omega t + 100.89°)$$

and $\quad \boldsymbol{i_2 = 105.8 \times 10^{-3} \sin(\omega t + 100.89°)}$

A plot of the three waveforms appears in Fig. 14.75. The waveforms clearly indicate that $i_T = i_1 + i_2$.

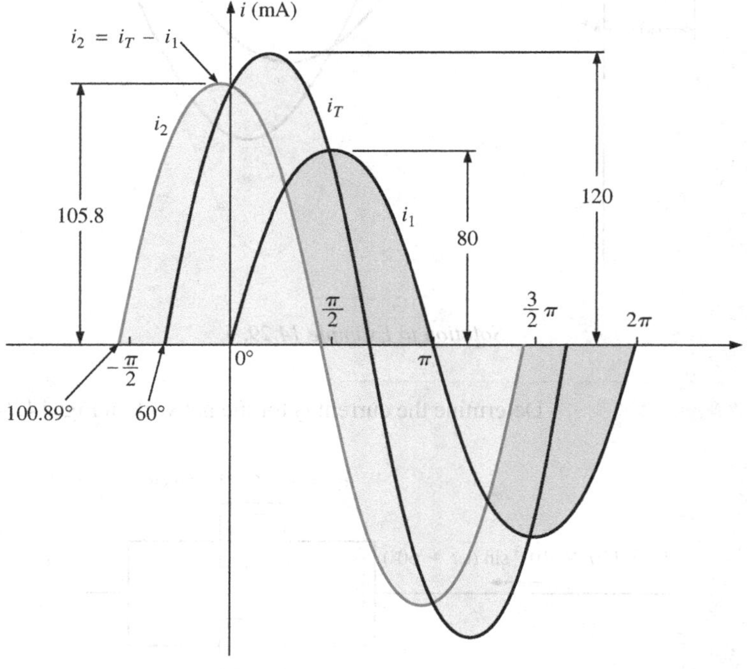

FIG. 14.75
Solution to Example 14.30.

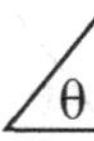

14.13 COMPUTER ANALYSIS

PSpice

Capacitors and the ac Response The simplest of ac capacitive circuits is now analyzed to introduce the process of setting up an ac source and running an ac transient simulation. The ac source in Fig. 14.76 is obtained through **Place part** key-**SOURCE-VSIN-OK.** Change the name or value of any parameter by double-clicking on the parameter on the display or by double-clicking on the source symbol to get the **Property Editor** dialog box. Within the dialog box, set the values appearing in Fig. 14.76. This is done by scrolling across the dialog box and selecting the desired quantity from the top listing. When selected, a black box will appear under the quantity of interest. Click on the black box, and it will turn white. Enter the value followed by **Display-Name and Value** if you want the quantity and its value to appear on the screen. If you do not want it to appear on the screen, do not use the **Display** option. Once each quantity is set, the most important step of all must be applied, which is to select the **Apply** key. If you forget to apply the changes, none will be used in the analysis.

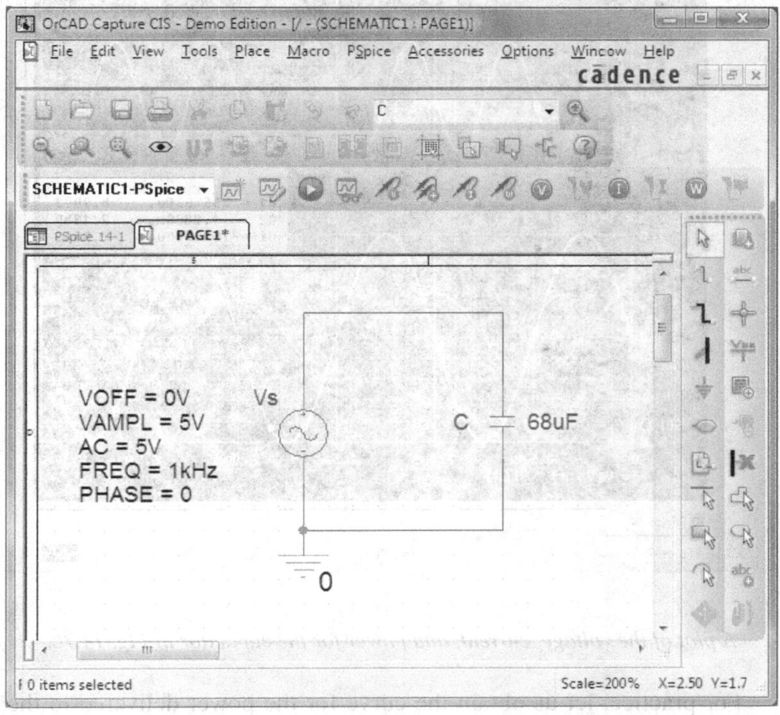

FIG. 14.76

Using PSpice to analyze the response of a capacitor to a sinusoidal ac signal.

The simulation process is initiated by selecting the **New Simulation Profile.** Under **New Simulation,** enter **PSpice 14-1** for the **Name** followed by **Create.** In the **Simulation Settings** dialog box, select **Analysis** and choose **Time Domain(Transient)** under **Analysis type.** Set the **Run to time** at 3 ms to permit a display of three cycles of the sinusoidal waveforms ($T = 1/f = 1/1000$ Hz $= 1$ ms). Leave the **Start saving data after** at 0 s, and set the **Maximum step size** at 3 ms/1000 = 3 μs. Clicking **OK** and then selecting the **Run PSpice** icon results in a plot having a horizontal axis that extends from 0 to 3 ms.

Now you must tell the computer which waveforms you are interested in. First, take a look at the applied ac source by selecting **Trace-Add Trace-V(Vs:+)** followed by **OK.** The result is the sweeping ac voltage in the bottom region of the screen in Fig. 14.77. Note that it has a peak value of 5 V, and three cycles appear in the 3 ms time frame. The current for the capacitor can be added by selecting **Trace-Add Trace** and choosing **I(C)** followed by **OK.** The resulting waveform for **I(C)** appears at a 90° phase shift from the applied voltage, with the current leading the voltage (the current has already peaked as the voltage crosses the 0 V axis). Since the peak value of each plot is in the same magnitude range, the 5 appearing on the vertical scale can be used for both. A theoretical analysis results in $X_C = 2.34\Omega$, and the peak value of $I_C = E/X_C = 5\text{ V}/2.34 = 2.136\text{ A}$, as shown in Fig. 14.77.

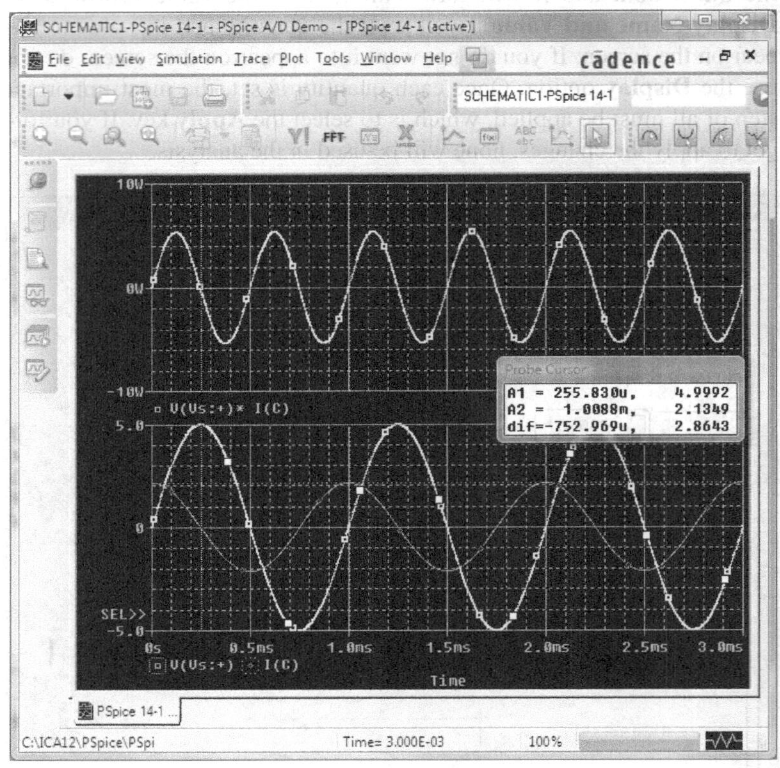

FIG. 14.77
A plot of the voltage, current, and power for the capacitor in Fig. 14.76.

For practice, let us obtain the curve for the power delivered to the capacitor over the same time period. First select **Plot-Add Plot to Window-Trace-Add Trace** to obtain the **Add Traces** dialog box. Then choose **V(Vs:+)**, follow it with a * for multiplication obtained from the **Function** listing on the right side of the **Add Traces** dialog box, and finish by selecting **I(C).** The result is the expression **V(Vs:+)*I(C)** of the power format: $p = vi$. Click **OK,** and the power plot at the top of Fig. 14.77 appears. Note that over the full three cycles, the area above the axis equals the area below—there is no net transfer of power over the 3 ms period. Note also that the power curve is sinusoidal (which is quite interesting) with a frequency twice that of the applied signal. Using the cursor control, we can determine that the maximum power (peak value of the sinusoidal waveform) is 5.34 W.

The cursors, in fact, have been added to the lower curves to show the peak value of the applied sinusoid and the resulting current.

After selecting the **Toggle cursor** icon, left-click to surround the symbol to the left of **V(Vs:+)** at the bottom of the plot with a dashed line to show that the cursor is providing the levels of that quantity. When placed at ¼ of the total period of 250 μs **(A1),** the peak value is exactly 5 V as shown in the **Probe Cursor** dialog box. Placing the cursor over the symbol next to **I(C)** at the bottom of the plot and right-clicking assigns the right cursor to the current. Placing it at exactly 1 ms **(A2)** results in a peak value of 2.136 A to match the solution above. To further distinguish between the voltage and current waveforms, the color and the width of the lines of the traces were changed. Place the cursor right on the plot line and right-click. The **Properties** option appears. When **Properties** is selected, a **Trace Properties** dialog box appears in which the yellow color can be selected and the width widened to improve the visibility on the black background. Note that yellow was chosen for **Vs** and green for **I(C).** Note also that the axis and the grid have been changed to a more visible color using the same procedure.

Multisim

Since PSpice reviewed the response of a capacitive element to an ac voltage, Multisim repeats the analysis for an inductive element. The ac voltage source was derived from the **Place Source** parts bin as described in Chapter 13 with the values appearing in Fig. 14.78 set in the **AC-Voltage** dialog box.

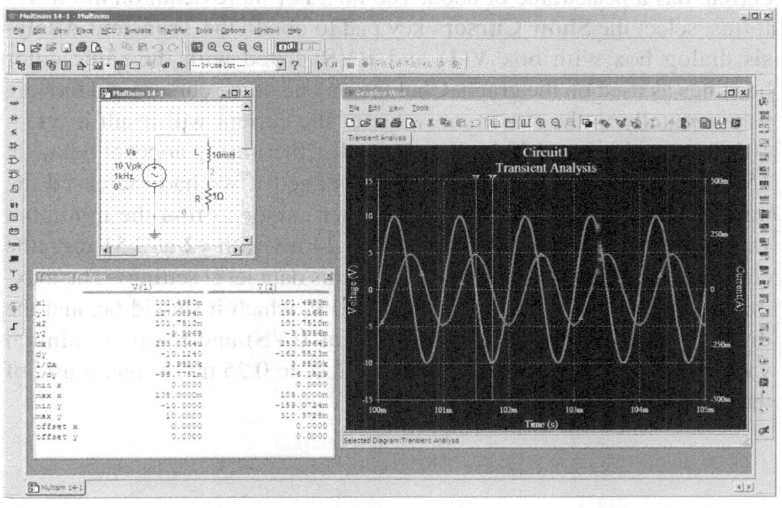

FIG. 14.78

Using Multisim to review the response of an inductive element to a sinusoidal ac signal.

Once the circuit has been constructed, the sequence **Simulate-Analyses-Transient Analysis** results in a **Transient Analysis** dialog box in which the **Start time** is set at 0 s and the **End time** at 105 ms. The 105 ms was set as the **End time** to give the network 100 ms to settle down in its steady-state mode and 5 ms for five cycles in the output display. The **Minimum number of time points** was set at 10,000 to ensure a good display for the rapidly changing waveforms.

Next the **Output** heading was chosen within the dialog box, and the source voltage **V(1)** and source current **I(VS)** were moved from the **Variables in Circuit** to **Selected variables for analysis** using the **Add** option. Choosing **Simulate** results in a waveform that extends from 0 s to 105 ms. Even though we plan to save only the response that occurs after 100 ms, the computer is unaware of our interest, and it plots the response for the entire period. This is corrected by selecting the **Properties** keypad in the toolbar at the top of the graph (it looks like a tag and pencil) to obtain the **Graph Properties** dialog box. Selecting **Bottom Axis** permits setting the **Range** from a **Minimum of 0.100 s = 100 ms** to a **Maximum of 0.105 s = 105 ms.** Click **OK,** and the time period of Fig. 14.78 is displayed. The grid structure is added by selecting the **Show Grid** keypad, and the color associated with each curve is displayed if we choose the **Show Legend** key next to it.

It is clear from the plot that the scale for the source current has to be improved for us to be able to clearly read its peak and negative values. This is done by first clicking on the **I(VS)** curve to set the **Selected Trace** at the bottom of the graph as **I(VS).** A right click, and one can choose the **Trace Properties** option to obtain the **Graph Properties** dialog box. Under **Y-Vertical Axis,** select **Right Axis** to establish the right axis as the scale to be used for the source current. Then select **Right Axis** and insert the **Label: Current(A),** select **Enabled** under the **Axis** heading, and finally choose **Pen Size** as 1. The **Scale** is **Linear** and of range −500E-3 to 500E-3 (−500 mA to 500 mA), with **Total Ticks** of 8 and **Minor Ticks** of 2. The result is the plot of Fig. 14.78. The right axis can now be improved by selecting **Graph Properties** again, followed by **Left Axis,** whereby the **Current(A)** can be deleted. We can now see that the source current has a peak value of about 160 mA. For more detail on the waveforms, select the **Show Cursors** key pad to obtain the **Transient Analysis** dialog box with box **V(1)** and **I(VS)** listed with the same color headings as used on the graph. Clicking on one of the cursors and moving it horizontally to the maximum value of the current will result in **x1** = 101.0 ms with **y1** at 158.91 mA. Actually, the **max y** appears below at 159.07 mA, which could have been obtained if we had increased the number of data points. Moving the other cursor to find the minimum value of current will result in **x2** = 101.24 ms with **y2** at 2.51 mA (the closest to the value of 1 obtainable with this data level setting). The maximum value of **V(1)** appears below as 10 V, which it should be, and the distance between the maximum value of **I(VS)** and the its minimum value is **dx** = 242.91 μs which is very close to 0.25 ms, or one fourth of the period of the applied signal.

PROBLEMS

SECTION 14.2 Derivative

1. Plot the following waveform versus time showing one clear, complete cycle. Then determine the derivative of the waveform using Eq. (14.1), and sketch one complete cycle of the derivative directly under the original waveform. Compare the magnitude of the derivative at various points versus the slope of the original sinusoidal function.

$$v = 1 \sin 6.28t$$

2. Repeat Problem 1 for the following sinusoidal function, and compare results. In particular, determine the frequency of the waveforms of Problems 1 and 2, and compare the magnitude of the derivative.

$$v = 1 \sin 314.2t$$

3. What is the derivative of each of the following sinusoidal expressions?
   a. $10 \sin 377t$ **b.** $0.6 \sin(200t + 20°)$
   c. $\sqrt{2}\, 20 \sin(157t - 20°)$ **d.** $-200 \sin(t + 180°)$

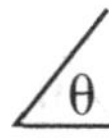

SECTION 14.3 Response of Basic *R*, *L*, and *C* Elements to a Sinusoidal Voltage or Current

4. The voltage across a 3 Ω resistor is as indicated. Find the sinusoidal expression for the current. In addition, sketch the v and i sinusoidal waveforms on the same axis.
  a. $150 \sin 200t$
  b. $30 \sin(377t + 20°)$
  c. $6 \cos(\omega t + 10°)$
  d. $-12 \sin(\omega t + 40°)$

5. The current through a 7 kΩ resistor is as indicated. Find the sinusoidal expression for the voltage. In addition, sketch the v and i sinusoidal waveforms on the same axis.
  a. $0.1 \sin 1000t$
  b. $2 \times 10^{-3} \sin(400t - 120°)$

6. Determine the inductive reactance (in ohms) of a 2 mH coil for
  a. dc
  and for the following frequencies:
  b. 60 Hz
  c. 4 kHz
  d. 1.2 MHz

7. Determine the closest standard value inductance that has a reactance of
  a. 2 kΩ at $f = 14.47$ kHz.
  b. 40 kΩ at $f = 5.3$ kHz.

8. Determine the frequency at which a 1 mH inductance has the following inductive reactances:
  a. 10 Ω
  b. 4 kΩ
  c. 12 kΩ

9. The current through a 20 Ω inductive reactance is given. What is the sinusoidal expression for the voltage? Sketch the v and i sinusoidal waveforms on the same axis.
  a. $i = 5 \sin \omega t$
  b. $i = 40 \times 10^{-3} \sin(\omega t + 60°)$
  c. $i = -6 \sin(\omega t - 30°)$

10. The current through a 0.1 H coil is given. What is the sinusoidal expression for the voltage?
  a. $10 \sin 100t$
  b. $5 \times 10^{-6} \sin(400t + 20°)$

11. The voltage across a 50 Ω inductive reactance is given. What is the sinusoidal expression for the current? Sketch the v and i sinusoidal waveforms on the same set of axes.
  a. $120 \sin \omega t$
  b. $30 \sin(\omega t + 20°)$

12. The voltage across a 0.2 H coil is given. What is the sinusoidal expression for the current?
  a. $1.5 \sin 60t$
  b. $16 \times 10^{-3} \sin(10t + 2°)$

13. Determine the capacitive reactance (in ohms) of a 5 μF capacitor for
  a. dc
  and for the following frequencies:
  b. 60 Hz
  c. 2 kHz
  d. 2 MHz

14. Determine the closest standard value capacitance that has a reactance of
  a. 60 Ω at $f = 265$ Hz.
  b. 1.2 kΩ at 34 kHz.

15. Determine the frequency at which a 3.9 μF capacitor has the following capacitive reactances:
  a. 10 Ω **b.** 1.2 kΩ
  c. 0.1 Ω **d.** 2000 Ω

16. The voltage across a 2.5 Ω capacitive reactance is given. What is the sinusoidal expression for the current? Sketch the v and i sinusoidal waveforms on the same set of axes.
  a. $120 \sin \omega t$
  b. $4 \times 10^{-3} \sin(\omega t + 40°)$

17. The voltage across a 1 μF capacitor is given. What is the sinusoidal expression for the current?
  a. $30 \sin 200t$
  b. $60 \times 10^{-3} \sin 377t$

18. The current through a 10 Ω capacitive reactance is given. Write the sinusoidal expression for the voltage. Sketch the v and i sinusoidal waveforms on the same set of axes.
  a. $i = 50 \times 10^{-3} \sin \omega t$
  b. $i = 2 \times 10^{-6} \sin(\omega t + 60°)$

19. The current through a 0.56 μF capacitor is given. What is the sinusoidal expression for the voltage?
  a. $0.20 \sin 300t$
  b. $8 \times 10^{-3} \sin(377t - 30°)$

***20.** For the following pairs of voltages and currents, indicate whether the element involved is a capacitor, an inductor, or a resistor, and find the value of *C*, *L*, or *R* if sufficient data are given:
  a. $v = 550 \sin(377t + 50°)$
  $i = 11 \sin(377t - 40°)$
  b. $v = 36 \sin(754t - 80°)$
  $i = 4 \sin(754t - 170°)$
  c. $v = 10.5 \sin(\omega t - 13°)$
  $i = 1.5 \sin(\omega t - 13°)$

***21.** Repeat Problem 20 for the following pairs of voltages and currents:
  a. $v = 2000 \sin \omega t$
  $i = 5 \cos \omega t$
  b. $v = 80 \sin(157t + 150°)$
  $i = 2 \sin(157t + 60°)$
  c. $v = 35 \sin(\omega t - 20°)$
  $i = 7 \cos(\omega t - 110°)$

SECTION 14.4 Frequency Response of the Basic Elements

22. Plot X_L versus frequency for a 3 mH coil using a frequency range of zero to 100 kHz on a linear scale.

23. Plot X_C versus frequency for a 1 μF capacitor using a frequency range of zero to 10 kHz on a linear scale.

24. At what frequency will the reactance of a 1 μF capacitor equal the resistance of a 2 kΩ resistor?

25. The reactance of a coil equals the resistance of a 10 kΩ resistor at a frequency of 5 kHz. Determine the inductance of the coil.

26. Determine the frequency at which a 1 μF capacitor and a 10 mH inductor will have the same reactance.

27. Determine the capacitance required to establish a capacitive reactance that will match that of a 2 mH coil at a frequency of 50 kHz.

SECTION 14.5 Average Power and Power Factor

*28. Find the average power loss and power factor for each of the circuits whose input current and voltage are as follows:
  a. $v = 60 \sin(\omega t + 30°)$
     $i = 15 \sin(\omega t + 60°)$
  b. $v = -50 \sin(\omega t - 20°)$
     $i = -2 \sin(\omega t - 20°)$
  c. $v = 50 \sin(\omega t + 80°)$
     $i = 3 \cos(\omega t - 20°)$
  d. $v = 75 \sin(\omega t - 5°)$
     $i = 0.08 \sin(\omega t + 35°)$

29. If the current through and voltage across an element are $i = 8 \sin(\omega t + 40°)$ and $v = 48 \sin(\omega t + 40°)$, respectively, compute the power by I^2R, $(V_mI_m/2) \cos \theta$, and $VI \cos \theta$, and compare answers.

30. A circuit dissipates 100 W (average power) at 150 V (effective input voltage) and 2 A (effective input current). What is the power factor? Repeat if the power is 0 W; 300 W.

*31. The power factor of a circuit is 0.5 lagging. The power delivered in watts is 500. If the input voltage is $50 \sin(\omega t + 10°)$, find the sinusoidal expression for the input current.

32. In Fig. 14.79, $e = 34 \sin(2\pi 60t + 20°)$.
  a. What is the sinusoidal expression for the current?
  b. Find the power loss in the circuit.
  c. How long (in seconds) does it take the current to complete six cycles?

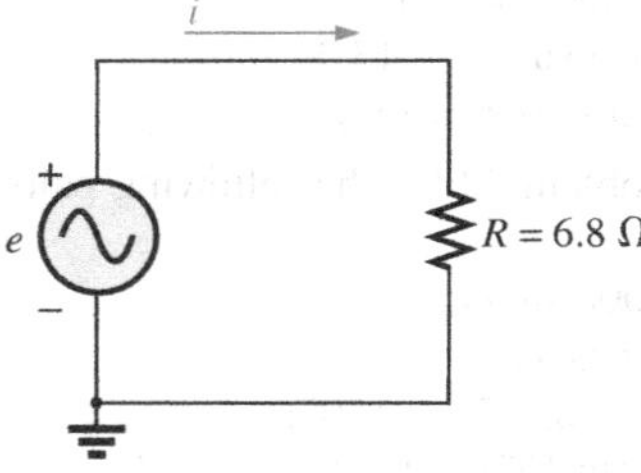

FIG. 14.79
Problem 32.

33. In Fig. 14.80, $e = 128 \sin(1000t + 60°)$.
  a. Find the sinusoidal expression for i.
  b. Find the value of the inductance L. What is its probable standard value in mH?
  c. Find the average power loss by the inductor.

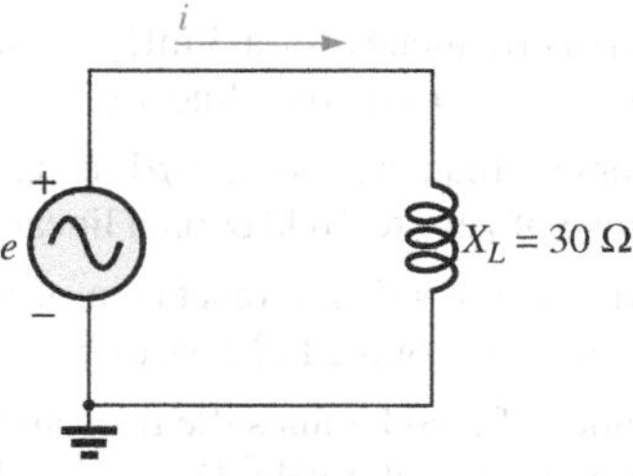

FIG. 14.80
Problem 33.

34. In Fig. 14.81, $i = 30 \times 10^{-3} \sin(2\pi 500t - 20°)$.
  a. Find the sinusoidal expression for e.
  b. Find the value of the capacitance C in microfarads. What is its probable standard value in μF?
  c. Find the average power loss in the capacitor.

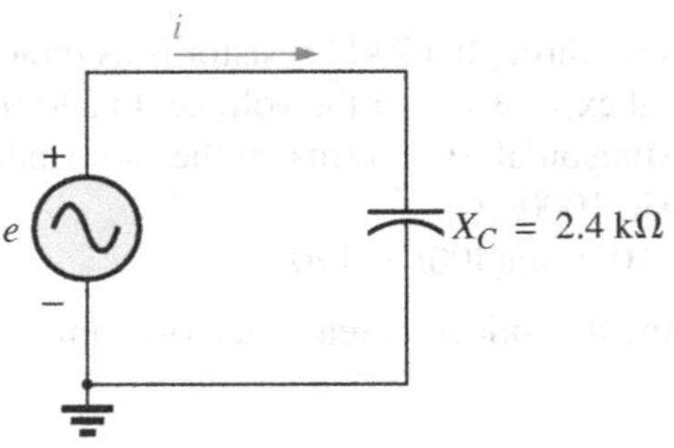

FIG. 14.81
Problem 34.

*35. For the network in Fig. 14.82 and the applied signal:
  a. Determine the sinusoidal expressions for i_1 and i_2.
  b. Find the sinusoidal expression for i_s by combining the two parallel capacitors.

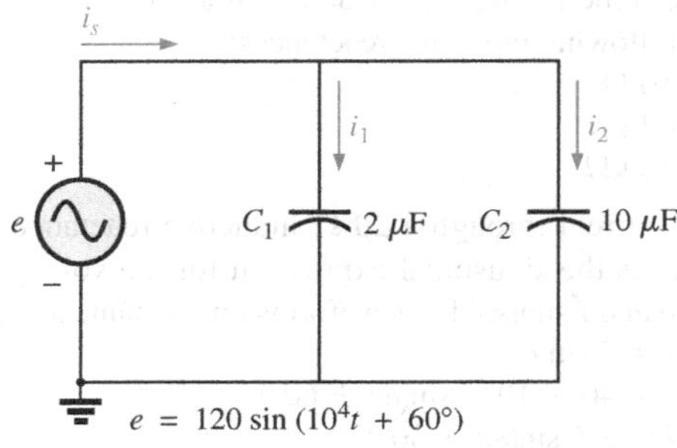

FIG. 14.82
Problem 35.

*36. For the network in Fig. 14.83 and the applied source:
  a. Determine the sinusoidal expression for the source voltage v_s.
  b. Find the sinusoidal expression for the currents i_1 and i_2.

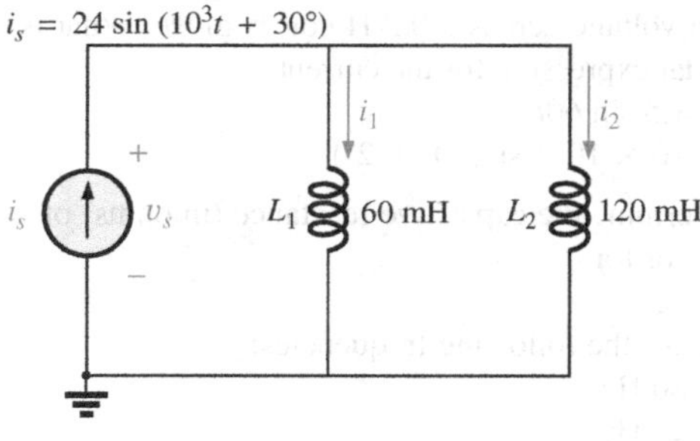

FIG. 14.83
Problem 36.

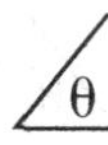

SECTION 14.9 Conversion between Forms

37. Convert the following from rectangular to polar form:
a. $4 + j3$ **b.** $2 + j2$
c. $4 + j12$ **d.** $1000 + j50$
e. $-1000 + j4000$ **f.** $-0.4 + j0.8$

***38.** Convert the following from rectangular to polar form:
a. $-8 - j16$
b. $+8 - j4$
c. $0.02 - j0.003$
d. $-6 \times 10^{-3} - j6 \times 10^{-3}$
e. $200 + j0.02$
f. $-1000 + j20$

39. Convert the following from polar to rectangular form:
a. $6 \angle 40°$ **b.** $12 \angle 120°$
c. $2000 \angle -90°$ **d.** $0.0064 \angle +200°$
e. $48 \angle 2°$ **f.** $5 \times 10^{-4} \angle -20°$

40. Convert the following from polar to rectangular form:
a. $42 \angle 0.15°$
b. $2002 \angle -60°$
c. $0.006 \angle -120°$
d. $8 \times 10^{-3} \angle -220°$
e. $15 \angle +180°$
f. $1.2 \angle -89.9°$

SECTION 14.10 Mathematical Operations with Complex Numbers

41. Perform the following additions in rectangular form:
a. $(4.2 + j6.8) + (7.6 + j0.2)$
b. $(142 + j7) + (9.8 + j42) + (0.1 + j0.9)$
c. $(4 \times 10^{-6} + j76) + (7.2 \times 10^{-7} - j5)$

42. Perform the following subtractions in rectangular form:
a. $(9.8 + j6.2) - (4.6 + j4.6)$
b. $(167 + j243) - (-42.3 - j68)$
c. $(-36.0 + j78) - (-4 - j6) + (10.8 - j72)$

43. Perform the following operations with polar numbers, and leave the answer in polar form:
a. $6 \angle 20° + 8 \angle 80°$
b. $42 \angle 45° + 62 \angle 60° - 70 \angle 120°$
c. $20 \angle -120° - 10 \angle -150° + 8 \angle -210° + 8 \angle +240°$

44. Perform the following multiplications in rectangular form:
a. $(2 + j3)(6 + j8)$
b. $(7.8 + j1)(4 + j2)(7 + j6)$
c. $(400 - j200)(-0.01 - j0.5)(-1 + j3)$

45. Perform the following multiplications in polar form:
a. $(2 \angle 60°)(4 \angle -40°)$
b. $(6.9 \angle 8°)(7.2 \angle -72°)$
c. $(0.002 \angle 120°)(0.5 \angle 200°)(40 \angle +80°)$

46. Perform the following divisions in polar form:
a. $(42 \angle 10°)/(7 \angle 60°)$
b. $(0.006 \angle 120°)/(30 \angle +60°)$
c. $(4360 \angle -20°)/(40 \angle -210°)$

47. Perform the following divisions, and leave the answer in rectangular form:
a. $(8 + j8)/(2 + j2)$
b. $(8 + j42)/(-6 - j4)$
c. $(-4.5 - j6)/(0.1 - j0.8)$

***48.** Perform the following operations, and express your answer in rectangular form:
a. $\dfrac{(4 + j3) + (6 - j8)}{(3 + j3) - (2 + j3)}$
b. $\dfrac{8 \angle 60°}{(2 \angle 0°) + (100 + j400)}$
c. $\dfrac{(6 \angle 20°)(120 \angle -40°)(3 + j8)}{2 \angle -30°}$

***49.** Perform the following operations, and express your answer in polar form:
a. $\dfrac{(0.4 \angle 60°)^2(300 \angle 40°)}{3 + j9}$
b. $\left(\dfrac{1}{(0.02 \angle 10°)^2}\right)\left(\dfrac{2}{j}\right)^3\left(\dfrac{1}{6^2 - j\sqrt{900}}\right)$

***50.** **a.** Determine a solution for x and y if
$$(x + j4) + (3x + jy) - j7 = 16 \angle 0°$$
b. Determine x if
$$(10 \angle 20°)(x \angle -60°) = 30.64 - j25.72$$

***51.** **a.** Determine a solution for x and y if
$$(5x + j10)(2 - jy) = 90 - j70$$
b. Determine θ if
$$\frac{80 \angle 0°}{20 \angle \theta} = 3.464 - j2$$

SECTION 14.12 Phasors

52. Express the following in phasor form:
a. $\sqrt{2}(160)\sin(\omega t + 30°)$
b. $\sqrt{2}(25 \times 10^{-3})\sin(157t - 40°)$
c. $100 \sin(\omega t - 90°)$

***53.** Express the following in phasor form:
a. $20 \sin(377t - 180°)$
b. $6 \times 10^{-6} \cos \omega t$
c. $3.6 \times 10^{-6} \cos(754t - 20°)$

54. Express the following phasor currents and voltages as sine waves if the frequency is 60 Hz:
a. $\mathbf{I} = 40\text{ A} \angle 20°$
b. $\mathbf{V} = 120\text{ V} \angle 10°$
c. $\mathbf{I} = 8 \times 10^{-3}\text{ A} \angle -110°$
d. $\mathbf{V} = \dfrac{6000}{\sqrt{2}}\text{ V} \angle -180°$

55. For the system in Fig. 14.84, find the sinusoidal expression for the unknown voltage v_a if
$$e_{in} = 60 \sin(377t + 45°)$$
$$v_b = 20 \sin(377t - 45°)$$

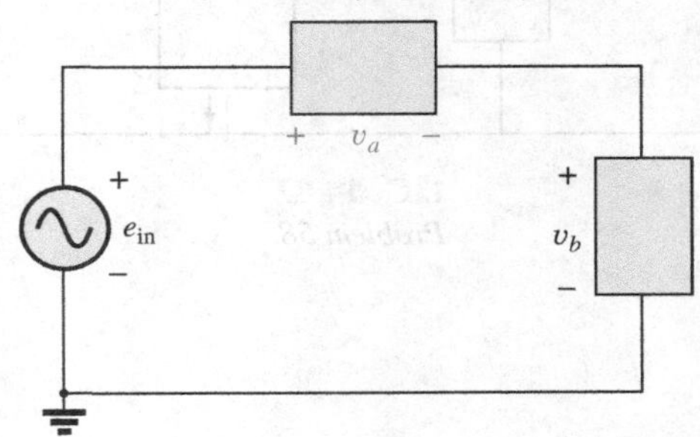

FIG. 14.84
Problem 55.

56. For the system in Fig. 14.85, find the sinusoidal expression for the unknown current i_1 if

$$i_s = 20 \times 10^{-6} \sin(\omega t + 60°)$$
$$i_2 = 6 \times 10^{-6} \sin(\omega t - 30°)$$

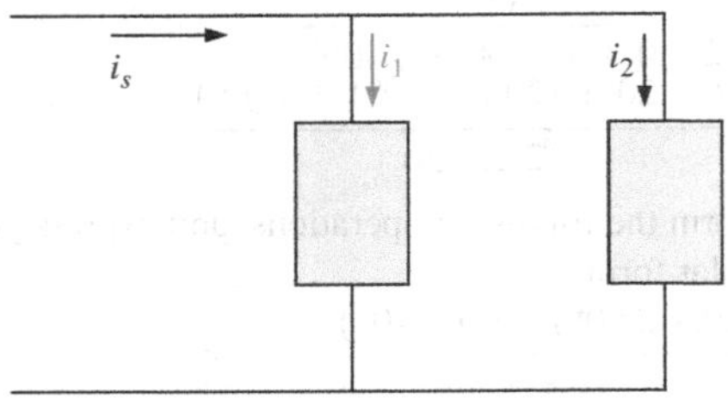

FIG. 14.85
Problem 56.

57. Find the sinusoidal expression for the voltage v_a for the system in Fig. 14.86 if

$$e_{in} = 120 \sin(\omega t + 30°)$$
$$v_b = 30 \sin(\omega t + 60°)$$
$$v_c = 40 \sin(\omega t + 120°)$$

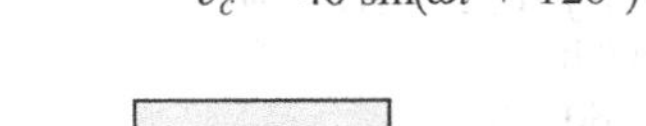

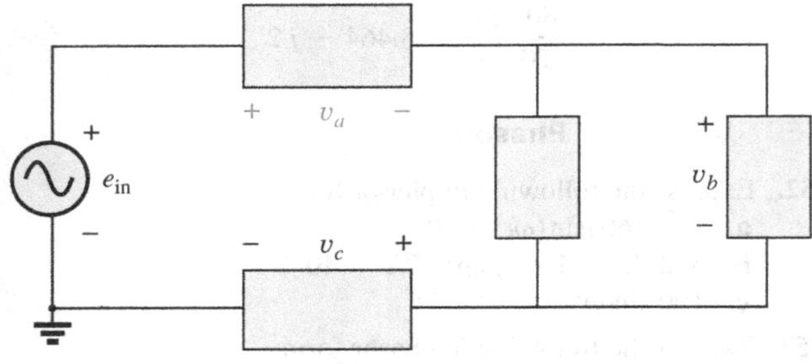

FIG. 14.86
Problem 57.

***58.** Find the sinusoidal expression for the current i_1 for the system in Fig. 14.87 if

$$i_s = 18 \times 10^{-3} \sin(377t + 180°)$$
$$i_2 = 8 \times 10^{-3} \sin(377t - 180°)$$
$$i_3 = 2i_2$$

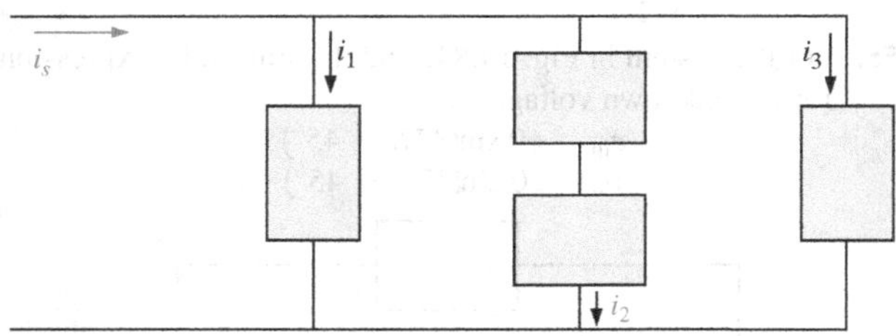

FIG. 14.87
Problem 58.

SECTION 14.13 Computer Analysis

PSpice or Multisim

59. Plot i_c and v_c versus time for the network in Fig. 14.76 for two cycles if the frequency is 0.2 kHz.

60. Plot the magnitude and phase angle of the current i_C versus frequency (100 Hz to 100 kHz) for the network in Fig. 14.76.

***61.** Plot the total impedance of the configuration in Fig. 14.27(a) versus frequency (100 kHz to 100 MHz) for the following parameter values: $C = 0.1\ \mu F$, $L_s = 0.2\ \mu H$, $R_s = 2\ M\Omega$, and $R_p = 100\ M\Omega$. For what frequency range is the capacitor "capacitive"?

GLOSSARY

Average or real power The power delivered to and dissipated by the load over a full cycle.

Complex conjugate A complex number defined by simply changing the sign of an imaginary component of a complex number in the rectangular form.

Complex number A number that represents a point in a two-dimensional plane located with reference to two distinct axes. It defines a vector drawn from the origin to that point.

Derivative The instantaneous rate of change of a function with respect to time or another variable.

Leading and lagging power factors An indication of whether a network is primarily capacitive or inductive in nature. Leading power factors are associated with capacitive networks and lagging power factors with inductive networks.

Phasor A radius vector that has a constant magnitude at a fixed angle from the positive real axis and that represents a sinusoidal voltage or current in the vector domain.

Phasor diagram A "snapshot" of the phasors that represent a number of sinusoidal waveforms at $t = 0$.

Polar form A method of defining a point in a complex plane that includes a single magnitude to represent the distance from the origin and an angle to reflect the counterclockwise distance from the positive real axis.

Power factor (F_p) An indication of how reactive or resistive an electrical system is. The higher the power factor, the greater is the resistive component.

Reactance The opposition of an inductor or a capacitor to the flow of charge that results in the continual exchange of energy between the circuit and magnetic field of an inductor or the electric field of a capacitor.

Reciprocal A format defined by 1 divided by the complex number.

Rectangular form A method of defining a point in a complex plane that includes the magnitude of the real component and the magnitude of the imaginary component, the latter component being defined by an associated letter *j*.

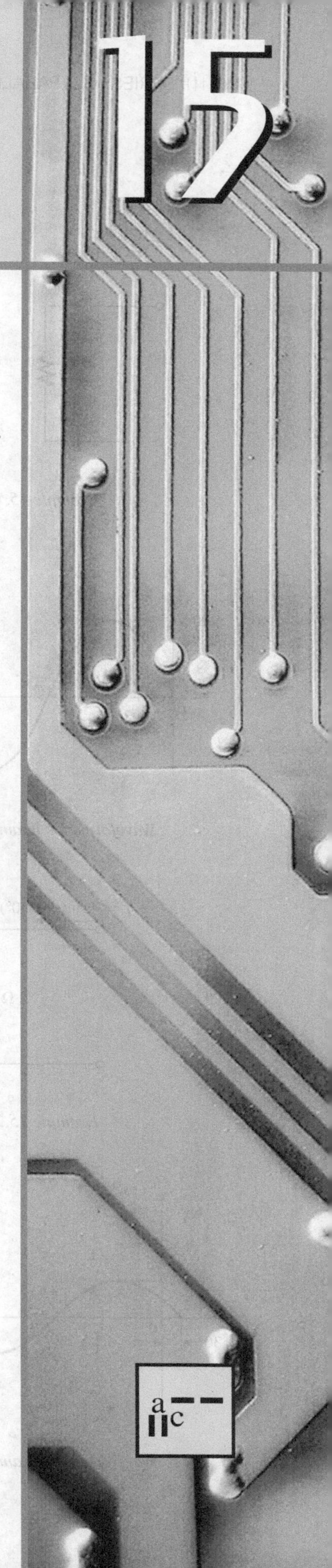

Series and Parallel ac Circuits

Objectives

- **Become familiar with the characteristics of series and parallel ac networks and be able to find current, voltage, and power levels for each element.**
- **Be able to find the total impedance of any series or parallel ac network and sketch the impedance and admittance diagram of each.**
- **Develop confidence in applying Kirchhoff's current and voltage laws to any series or parallel configuration.**
- **Be able to apply the voltage divider rule or current divider rule to any ac network.**
- **Become adept at finding the frequency response of a series or parallel combination of elements.**

15.1 INTRODUCTION

In this chapter, phasor algebra is used to develop a quick, direct method for solving both series and parallel ac circuits. The close relationship that exists between this method for solving for unknown quantities and the approach used for dc circuits will become apparent after a few simple examples are considered. Once this association is established, many of the rules (current divider rule, voltage divider rule, and so on) for dc circuits can be readily applied to ac circuits.

SERIES ac CIRCUITS

15.2 IMPEDANCE AND THE PHASOR DIAGRAM

Resistive Elements

In Chapter 14, we found, for the purely resistive circuit in Fig. 15.1, that v and i were in phase, and the magnitude

$$I_m = \frac{V_m}{R} \qquad \text{or} \qquad V_m = I_m R$$

$i = I_m \sin \omega t$

R $v = V_m \sin \omega t$

FIG. 15.1
Resistive ac circuit.

In phasor form,

$$v = V_m \sin \omega t \Rightarrow \mathbf{V} = V \angle 0^\circ$$

where $V = 0.707V_m$.

Applying Ohm's law and using phasor algebra, we have

$$\mathbf{I} = \frac{V \angle 0^\circ}{R \angle \theta_R} = \frac{V}{R} \underline{/0^\circ - \theta_R}$$

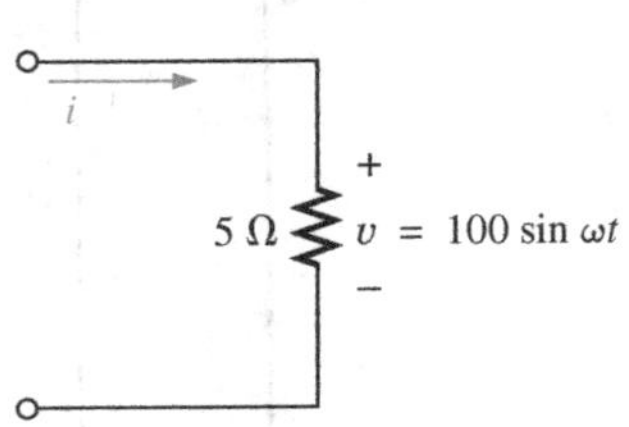

FIG. 15.2
Example 15.1.

Since i and v are in phase, the angle associated with i also must be 0°. To satisfy this condition, θ_R must equal 0°. Substituting $\theta_R = 0^\circ$, we find

$$\mathbf{I} = \frac{V \angle 0^\circ}{R \angle 0^\circ} = \frac{V}{R} \underline{/0^\circ - 0^\circ} = \frac{V}{R} \angle 0^\circ$$

so that in the time domain,

$$i = \sqrt{2}\left(\frac{V}{R}\right) \sin \omega t$$

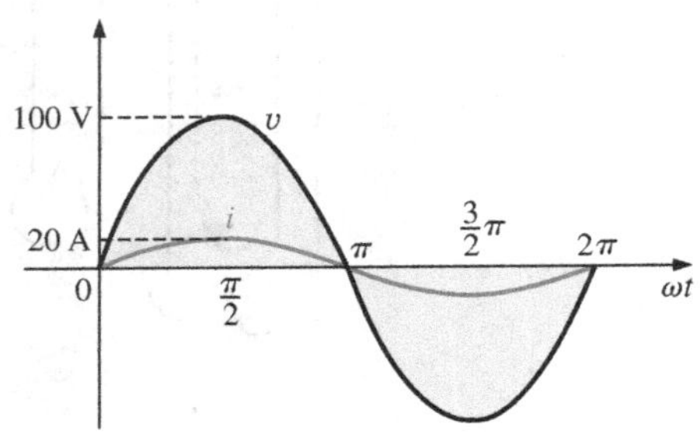

FIG. 15.3
Waveforms for Example 15.1.

We use the fact that $\theta_R = 0^\circ$ in the following polar format to ensure the proper phase relationship between the voltage and current of a resistor:

$$\boxed{\mathbf{Z}_R = R \angle 0^\circ} \qquad \textbf{(15.1)}$$

The boldface roman quantity $\mathbf{Z}_R$, having both magnitude and an associated angle, is referred to as the *impedance* of a resistive element. It is measured in ohms and is a measure of how much the element will "impede" the flow of charge through the network. The above format will prove to be a useful "tool" when the networks become more complex and phase relationships become less obvious. It is important to realize, however, that $\mathbf{Z}_R$ is *not a phasor,* even though the format $R \angle 0^\circ$ is very similar to the phasor notation for sinusoidal currents and voltages. The term *phasor* is reserved for quantities that vary with time, and R and its associated angle of 0° are fixed, nonvarying quantities.

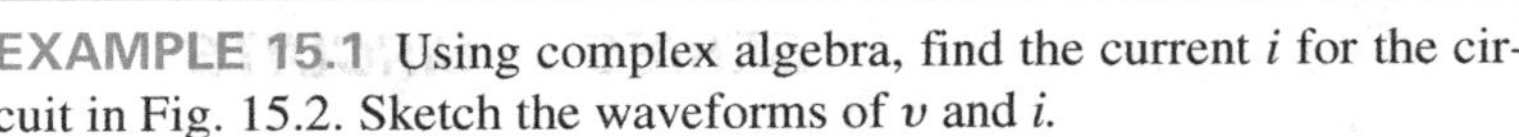

EXAMPLE 15.1 Using complex algebra, find the current i for the circuit in Fig. 15.2. Sketch the waveforms of v and i.

Solution: Note Fig. 15.3:

$$v = 100 \sin \omega t \Rightarrow \text{phasor form } \mathbf{V} = 70.71 \text{ V} \angle 0^\circ$$

$$\mathbf{I} = \frac{\mathbf{V}}{\mathbf{Z}_R} = \frac{V \angle \theta}{R \angle 0^\circ} = \frac{70.71 \text{ V} \angle 0^\circ}{5 \ \Omega \angle 0^\circ} = 14.14 \text{ A} \angle 0^\circ$$

and $$i = \sqrt{2}(14.14) \sin \omega t = \mathbf{20 \sin \omega t}$$

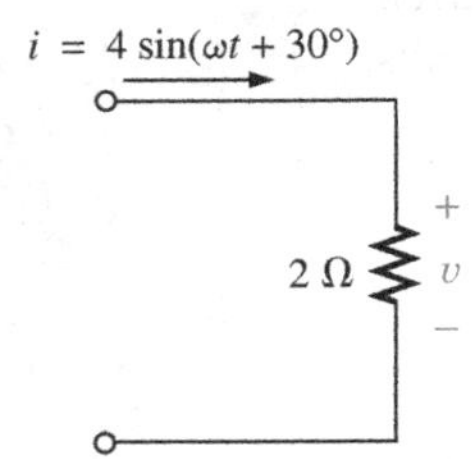

FIG. 15.4
Example 15.2.

EXAMPLE 15.2 Using complex algebra, find the voltage v for the circuit in Fig. 15.4. Sketch the waveforms of v and i.

Solution: Note Fig. 15.5:

$$i = 4 \sin(\omega t + 30^\circ) \Rightarrow \text{phasor form } \mathbf{I} = 2.828 \text{ A} \angle 30^\circ$$

$$\mathbf{V} = \mathbf{I}\mathbf{Z}_R = (I \angle \theta)(R \angle 0^\circ) = (2.828 \text{ A} \angle 30^\circ)(2 \ \Omega \angle 0^\circ)$$
$$= 5.656 \text{ V} \angle 30^\circ$$

and $$v = \sqrt{2}(5.656) \sin(\omega t + 30^\circ) = \mathbf{8.0 \sin(\omega t + 30^\circ)}$$

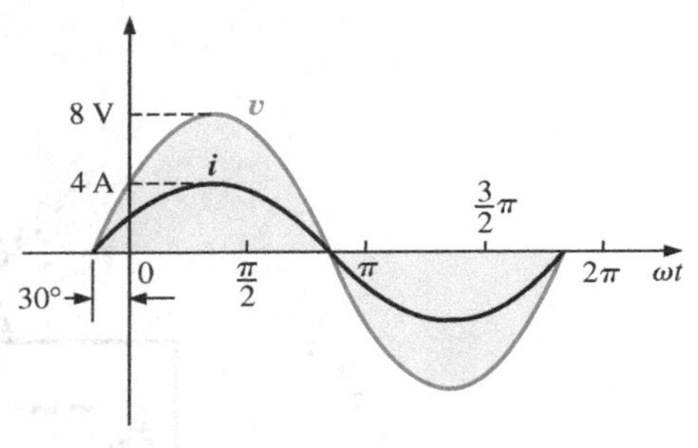

FIG. 15.5
Waveforms for Example 15.2.

It is often helpful in the analysis of networks to have a **phasor diagram,** which shows at a glance the *magnitudes* and *phase relations* among the various quantities within the network. For example, the phasor diagrams of the circuits considered in the two preceding examples would be as shown in Fig. 15.6. In both cases, it is immediately obvious that v and i are in phase since they both have the same phase angle.

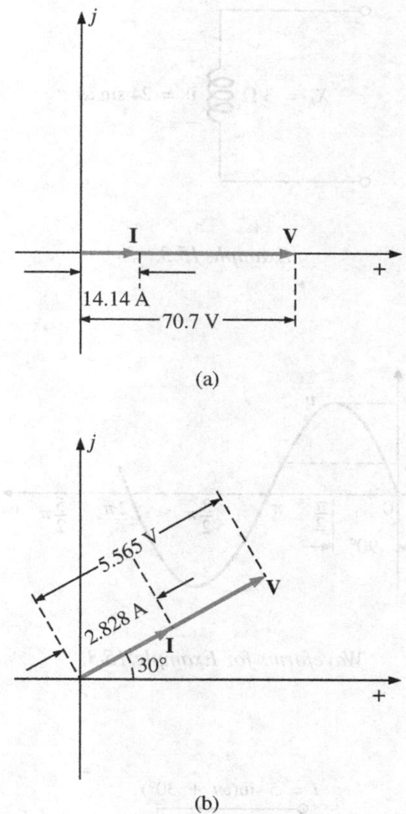

FIG. 15.6
Phasor diagrams for Examples 15.1 and 15.2.

Inductive Reactance

We learned in Chapter 13 that for the pure inductor in Fig. 15.7, the voltage leads the current by 90° and that the reactance of the coil X_L is determined by ωL. We have

$$v = V_m \sin \omega t \Rightarrow \text{phasor form } \mathbf{V} = V \angle 0°$$

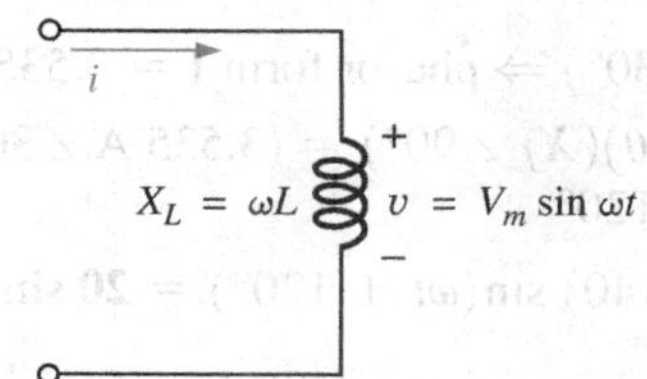

FIG. 15.7
Inductive ac circuit.

By Ohm's law,

$$\mathbf{I} = \frac{V \angle 0°}{X_L \angle \theta_L} = \frac{V}{X_L} \underline{/0° - \theta_L}$$

Since v leads i by 90°, i must have an angle of $-90°$ associated with it. To satisfy this condition, θ_L must equal $+90°$. Substituting $\theta_L = 90°$, we obtain

$$\mathbf{I} = \frac{V \angle 0°}{X_L \angle 90°} = \frac{V}{X_L} \underline{/0° - 90°} = \frac{V}{X_L} \angle -90°$$

so that in the time domain,

$$i = \sqrt{2}\left(\frac{V}{X_L}\right)\sin(\omega t - 90°)$$

We use the fact that $\theta_L = 90°$ in the following polar format for inductive reactance to ensure the proper phase relationship between the voltage and current of an inductor:

$$\boxed{\mathbf{Z}_L = X_L \angle 90°} \qquad (15.2)$$

The boldface roman quantity $\mathbf{Z}_L$, having both magnitude and an associated angle, is referred to as the *impedance* of an inductive element. It is measured in ohms and is a measure of how much the inductive element "controls or impedes" the level of current through the network (always keep in mind that inductive elements are storage devices and do not dissipate like resistors). The above format, like that defined for the resistive element, will prove to be a useful tool in the analysis of ac networks. Again, be aware that $\mathbf{Z}_L$ is not a phasor quantity, for the same reasons indicated for a resistive element.

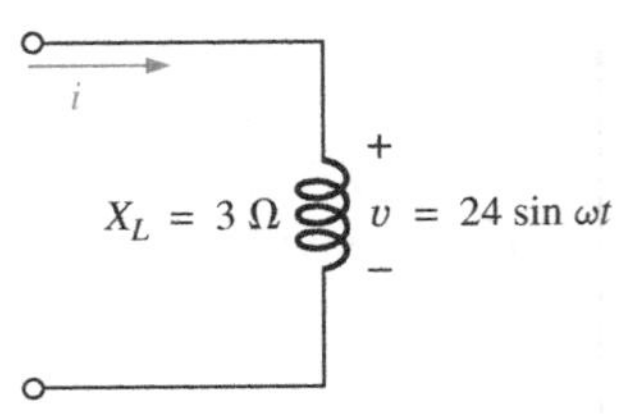

FIG. 15.8
Example 15.3.

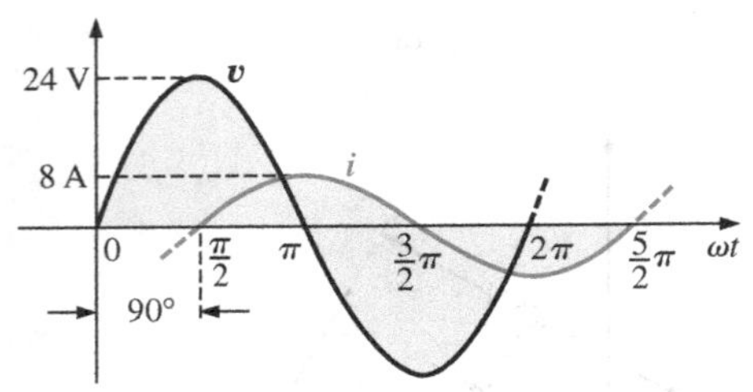

FIG. 15.9
Waveforms for Example 15.3.

EXAMPLE 15.3 Using complex algebra, find the current i for the circuit in Fig. 15.8. Sketch the v and i curves.

Solution: Note Fig. 15.9:

$$v = 24 \sin \omega t \Rightarrow \text{phasor form } \mathbf{V} = 16.968 \text{ V} \angle 0^\circ$$

$$\mathbf{I} = \frac{\mathbf{V}}{\mathbf{Z}_L} = \frac{V \angle \theta}{X_L \angle 90^\circ} = \frac{16.968 \text{ V} \angle 0^\circ}{3\ \Omega \angle 90^\circ} = 5.656 \text{ A} \angle -90^\circ$$

and $\quad i = \sqrt{2}(5.656) \sin(\omega t - 90^\circ) = \mathbf{8.0 \sin(\omega t - 90^\circ)}$

EXAMPLE 15.4 Using complex algebra, find the voltage v for the circuit in Fig. 15.10. Sketch the v and i curves.

Solution: Note Fig. 15.11:

$$i = 5 \sin(\omega t + 30^\circ) \Rightarrow \text{phasor form } \mathbf{I} = 3.535 \text{ A} \angle 30^\circ$$

$$\mathbf{V} = \mathbf{I}\mathbf{Z}_L = (I \angle \theta)(X_L \angle 90^\circ) = (3.535 \text{ A} \angle 30^\circ)(4\ \Omega \angle +90^\circ)$$
$$= 14.140 \text{ V} \angle 120^\circ$$

and $\quad v = \sqrt{2}(14.140) \sin(\omega t + 120^\circ) = \mathbf{20 \sin(\omega t + 120^\circ)}$

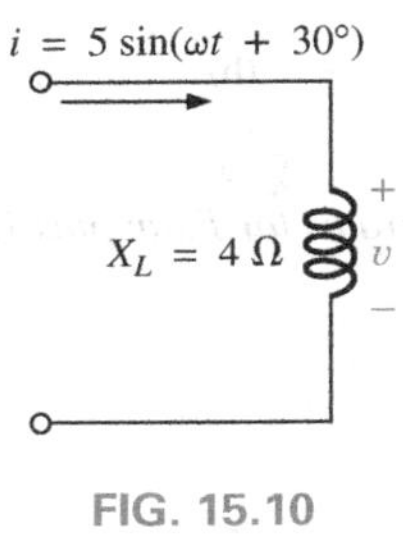

FIG. 15.10
Example 15.4.

The phasor diagrams for the two circuits of the two preceding examples are shown in Fig. 15.12. Both indicate quite clearly that the voltage leads the current by 90°.

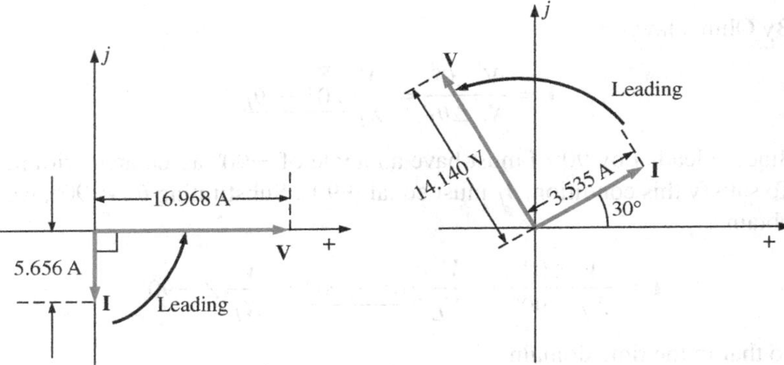

FIG. 15.12
Phasor diagrams for Examples 15.3 and 15.4.

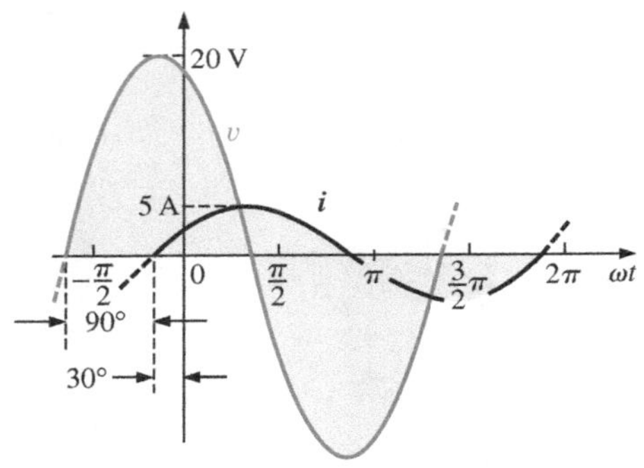

FIG. 15.11
Waveforms for Example 15.4.

Capacitive Reactance

We learned in Chapter 13 that for the pure capacitor in Fig. 15.13, the current leads the voltage by 90° and that the reactance of the capacitor X_C is determined by $1/\omega C$. We have

$$v = V_m \sin \omega t \Rightarrow \text{phasor form } \mathbf{V} = V \angle 0^\circ$$

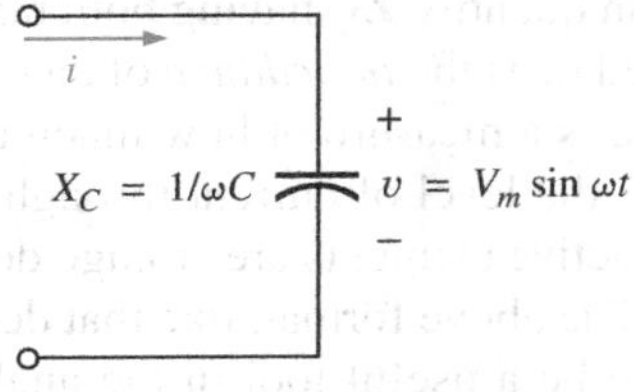

FIG. 15.13
Capacitive ac circuit.

Applying Ohm's law and using phasor algebra, we find

$$\mathbf{I} = \frac{V \angle 0^\circ}{X_C \angle \theta_C} = \frac{V}{X_C} \underline{/0^\circ - \theta_C}$$

Since i leads v by 90°, i must have an angle of +90° associated with it. To satisfy this condition, θ_C must equal −90°. Substituting $\theta_C = -90^\circ$ yields

$$\mathbf{I} = \frac{V \angle 0^\circ}{X_C \angle -90^\circ} = \frac{V}{X_C} \underline{/0^\circ - (-90^\circ)} = \frac{V}{X_C} \angle 90^\circ$$

so, in the time domain,

$$i = \sqrt{2}\left(\frac{V}{X_C}\right) \sin(\omega t + 90^\circ)$$

We use the fact that $\theta_C = -90^\circ$ in the following polar format for capacitive reactance to ensure the proper phase relationship between the voltage and current of a capacitor:

$$\boxed{\mathbf{Z}_C = X_C \angle -90^\circ} \qquad (15.3)$$

The boldface roman quantity $\mathbf{Z}_C$, having both magnitude and an associated angle, is referred to as the *impedance* of a capacitive element. It is measured in ohms and is a measure of how much the capacitive element "controls or impedes" the level of current through the network (always keep in mind that capacitive elements are storage devices and do not dissipate like resistors). The above format, like that defined for the resistive element, will prove a very useful tool in the analysis of ac networks. Again, be aware that $\mathbf{Z}_C$ is not a phasor quantity, for the same reasons indicated for a resistive element.

EXAMPLE 15.5 Using complex algebra, find the current i for the circuit in Fig. 15.14. Sketch the v and i curves.

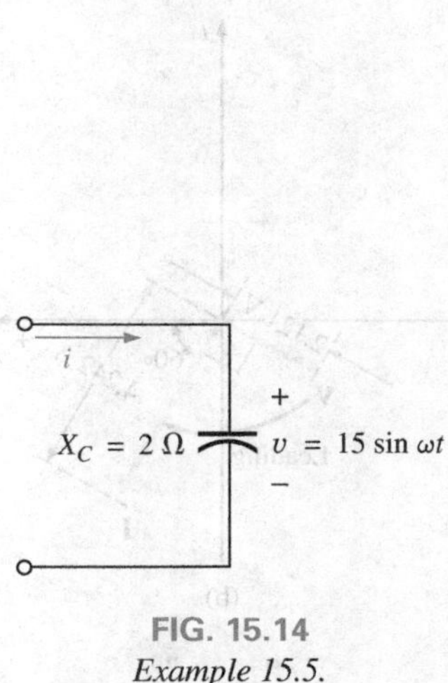

FIG. 15.14
Example 15.5.

Solution: Note Fig. 15.15:

$$v = 15 \sin \omega t \Rightarrow \text{phasor notation } \mathbf{V} = 10.605\text{ V} \angle 0^\circ$$

$$\mathbf{I} = \frac{\mathbf{V}}{\mathbf{Z}_C} = \frac{V \angle \theta}{X_C \angle -90^\circ} = \frac{10.605\text{ V} \angle 0^\circ}{2\ \Omega \angle -90^\circ} = 5.303\text{ A} \angle 90^\circ$$

and $\quad i = \sqrt{2}(5.303)\sin(\omega t + 90^\circ) = \mathbf{7.5 \sin(\omega t + 90^\circ)}$

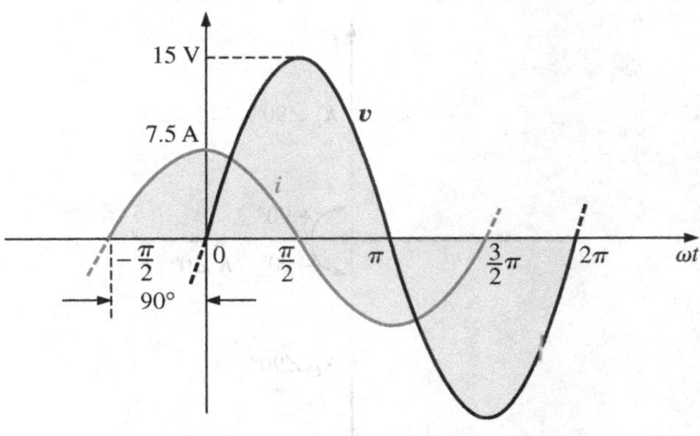

FIG. 15.15
Waveforms for Example 15.5.

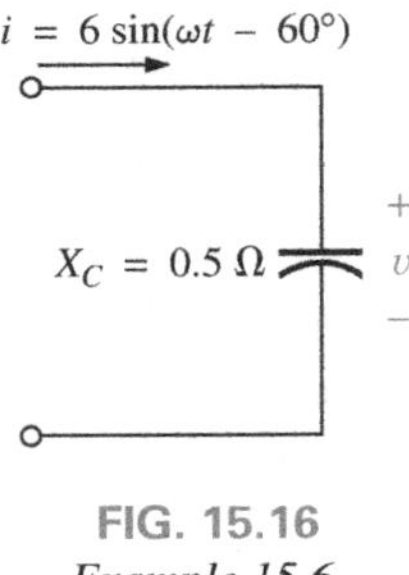

FIG. 15.16
Example 15.6.

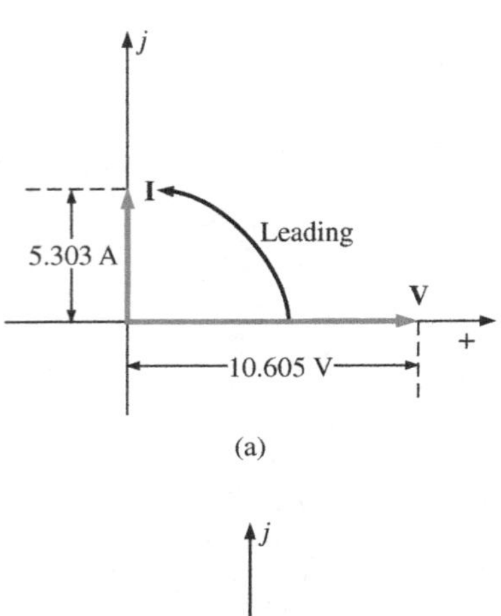

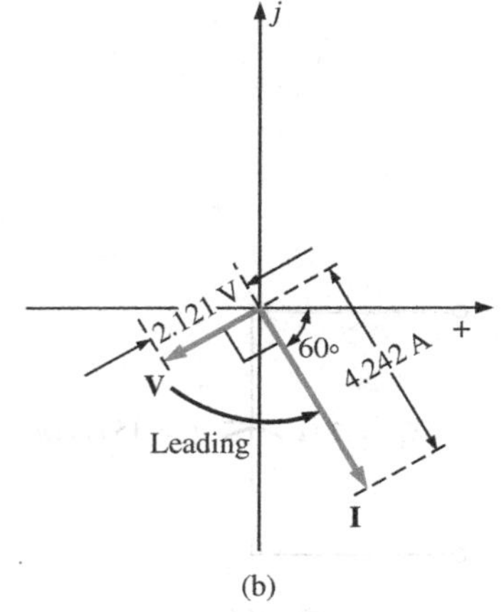

FIG. 15.18
Phasor diagrams for Examples 15.5 and 15.6.

EXAMPLE 15.6 Using complex algebra, find the voltage v for the circuit in Fig. 15.16. Sketch the v and i curves.

Solution: Note Fig. 15.17:

$$i = 6 \sin(\omega t - 60°) \Rightarrow \text{phasor notation } \mathbf{I} = 4.242 \text{ A} \angle -60°$$
$$\mathbf{V} = \mathbf{IZ}_C = (I\angle\theta)(X_C\angle -90°) = (4.242 \text{ A} \angle -60°)(0.5\ \Omega\angle -90°)$$
$$= 2.121 \text{ V} \angle -150°$$

and

$$v = \sqrt{2}(2.121)\sin(\omega t - 150°) = \mathbf{3.0 \sin(\boldsymbol{\omega} t - 150°)}$$

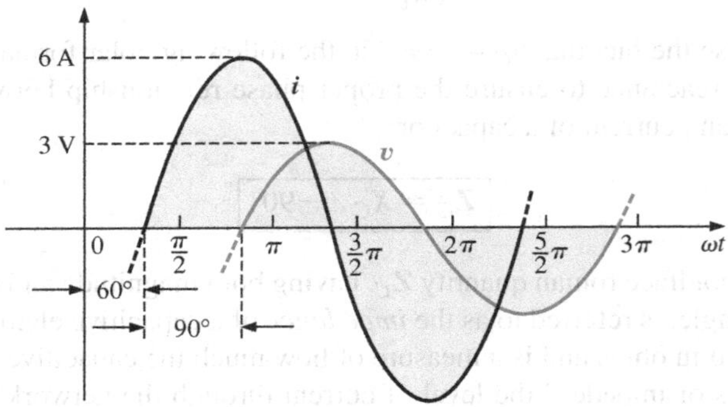

FIG. 15.17
Waveforms for Example 15.6.

The phasor diagrams for the two circuits of the two preceding examples are shown in Fig. 15.18. Both indicate quite clearly that the current i leads the voltage v by 90°.

Impedance Diagram

Now that an angle is associated with resistance, inductive reactance, and capacitive reactance, each can be placed on a complex plane diagram, as shown in Fig. 15.19. For any network, the resistance will *always* appear on the positive real axis, the inductive reactance on the positive imaginary axis, and the capacitive reactance on the negative imaginary axis.

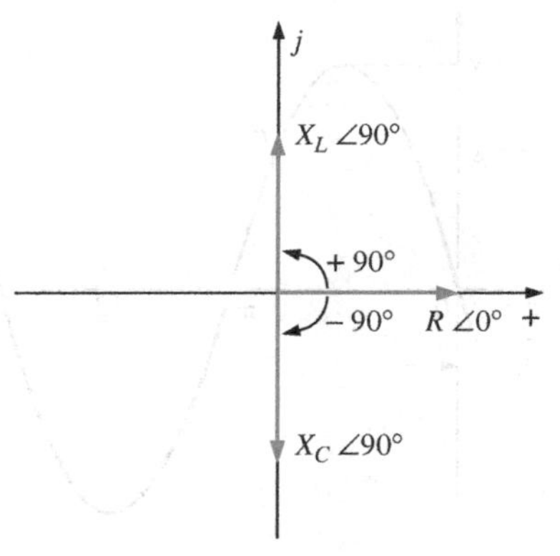

FIG. 15.19
Impedance diagram.

The result is an **impedance diagram** that can reflect the individual and total impedance levels of an ac network.

We will find in the rest of this text that networks combining different types of elements will have total impedances that extend from $-90°$ to $+90°$. If the total impedance has an angle of 0°, it is said to be resistive in nature. If it is closer to 90°, it is inductive in nature. If it is closer to $-90°$, it is capacitive in nature.

Of course, for single-element networks, the angle associated with the impedance will be the same as that of the resistive or reactive element, as revealed by Eqs. (15.1) through (15.3). It is important to remember that impedance, like resistance or reactance, is not a phasor quantity representing a time-varying function with a particular phase shift. It is simply an operating tool that is extremely useful in determining the magnitude and angle of quantities in a sinusoidal ac network.

Once the total impedance of a network is determined, its magnitude will define the resulting current level (through Ohm's law), whereas its angle will reveal whether the network is primarily inductive or capacitive or simply resistive.

For any configuration (series, parallel, series-parallel, and so on), the angle associated with the total impedance is the angle by which the applied voltage leads the source current. For inductive networks, θ_T will be positive, whereas for capacitive networks, θ_T will be negative.

15.3 SERIES CONFIGURATION

The overall properties of series ac circuits (Fig. 15.20) are the same as those for dc circuits. For instance, the total impedance of a system is the sum of the individual impedances:

$$\mathbf{Z}_T = \mathbf{Z}_1 + \mathbf{Z}_2 + \mathbf{Z}_3 + \cdots + \mathbf{Z}_N \qquad (15.4)$$

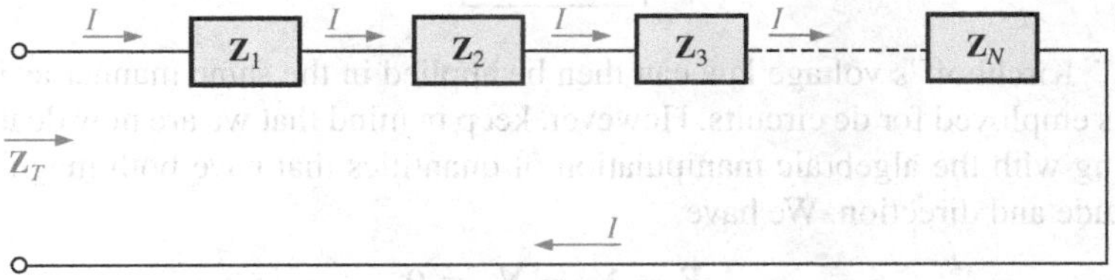

FIG. 15.20
Series impedances.

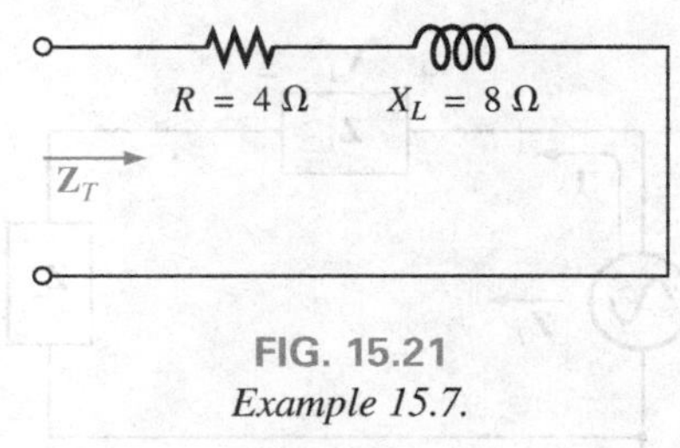

FIG. 15.21
Example 15.7.

EXAMPLE 15.7 Draw the impedance diagram for the circuit in Fig. 15.21, and find the total impedance.

Solution: As indicated by Fig. 15.22, the input impedance can be found graphically from the impedance diagram by properly scaling the real and imaginary axes and finding the length of the resultant vector Z_T and angle θ_T. Or, by using vector algebra, we obtain

$$\begin{aligned}\mathbf{Z}_T &= \mathbf{Z}_1 + \mathbf{Z}_2\\ &= R\angle 0° + X_L\angle 90°\\ &= R + jX_L = 4\ \Omega + j\,8\ \Omega\\ \mathbf{Z}_T &= \mathbf{8.94\ \Omega\ \angle 63.43°}\end{aligned}$$

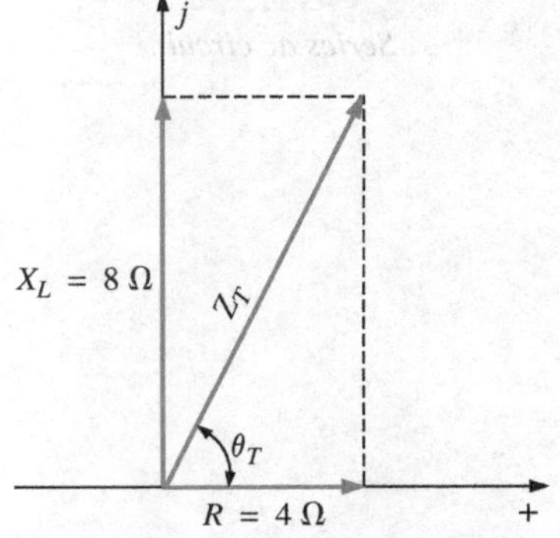

FIG. 15.22
Impedance diagram for Example 15.7.

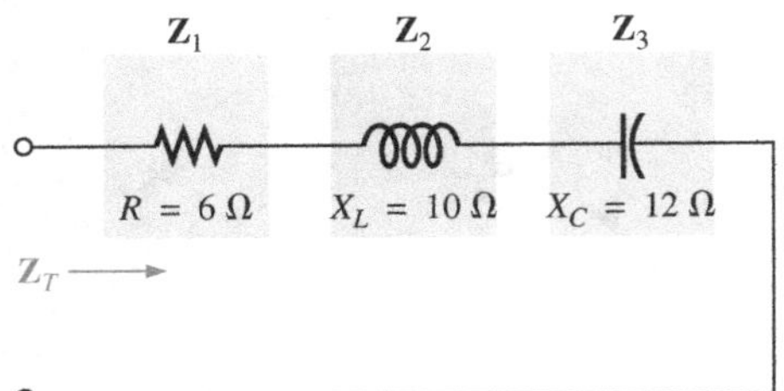

FIG. 15.23
Example 15.8

EXAMPLE 15.8 Determine the input impedance to the series network in Fig. 15.23. Draw the impedance diagram.

Solution:

$$\mathbf{Z}_T = \mathbf{Z}_1 + \mathbf{Z}_2 + \mathbf{Z}_3$$
$$= R \angle 0^\circ + X_L \angle 90^\circ + X_C \angle -90^\circ$$
$$= R + jX_L - jX_C$$
$$= R + j(X_L - X_C) = 6\ \Omega + j(10\ \Omega - 12\ \Omega) = 6\ \Omega - j\,2\ \Omega$$
$$\mathbf{Z}_T = \mathbf{6.32\ \Omega \angle -18.43^\circ}$$

The impedance diagram appears in Fig. 15.24. Note that in this example, series inductive and capacitive reactances are in direct opposition. For the circuit in Fig. 15.23, if the inductive reactance were equal to the capacitive reactance, the input impedance would be purely resistive. We will have more to say about this particular condition in a later chapter.

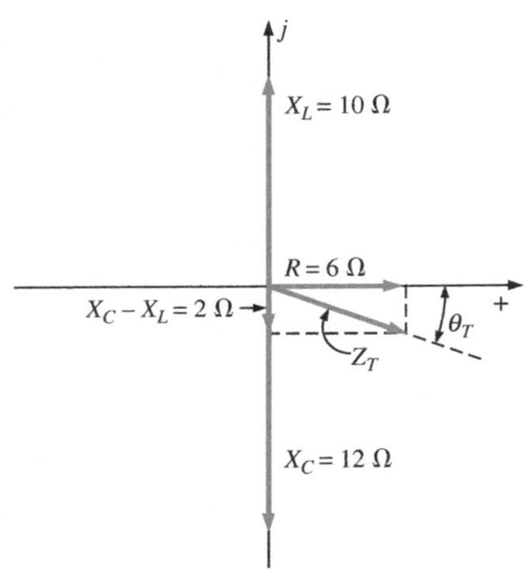

FIG. 15.24
Impedance diagram for Example 15.8.

For the representative **series ac configuration** in Fig. 15.25 having two impedances, *the current is the same through each element* (as it was for the series dc circuits) and is determined by Ohm's law:

$$\mathbf{Z}_T = \mathbf{Z}_1 + \mathbf{Z}_2$$

and

$$\mathbf{I} = \frac{\mathbf{E}}{\mathbf{Z}_T} \tag{15.5}$$

The voltage across each element can then be found by another application of Ohm's law:

$$\mathbf{V}_1 = \mathbf{I}\mathbf{Z}_1 \tag{15.6a}$$

$$\mathbf{V}_2 = \mathbf{I}\mathbf{Z}_2 \tag{15.6b}$$

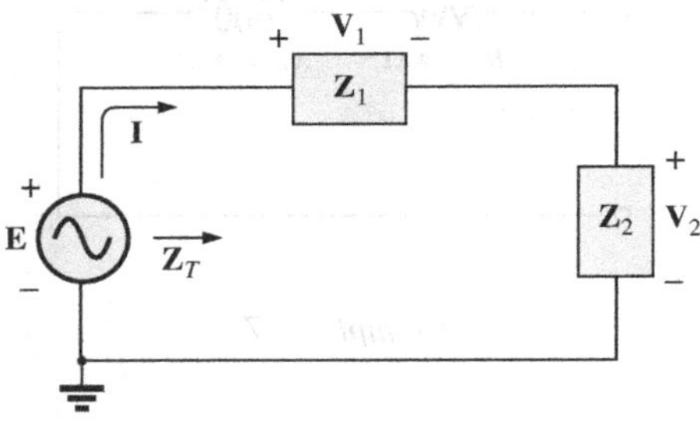

FIG. 15.25
Series ac circuit.

Kirchhoff's voltage law can then be applied in the same manner as it is employed for dc circuits. However, keep in mind that we are now dealing with the algebraic manipulation of quantities that have both magnitude and direction. We have

$$\mathbf{E} - \mathbf{V}_1 - \mathbf{V}_2 = 0$$

or

$$\mathbf{E} = \mathbf{V}_1 + \mathbf{V}_2 \tag{15.7}$$

The power to the circuit can be determined by

$$P = EI \cos \theta_T \tag{15.8}$$

where θ_T is the phase angle between **E** and **I.**

Now that a general approach has been introduced, the simplest of series configurations will be investigated in detail to further emphasize the similarities in the analysis of dc circuits. In many of the circuits to be considered, $3 + j\,4 = 5 \angle 53.13^\circ$ and $4 + j\,3 = 5 \angle 36.87^\circ$ are used quite frequently to ensure that the approach is as clear as possible and not lost in mathematical complexity. Of course, the problems at the end of the chapter will provide plenty of experience with random values.

R-L

Refer to Fig. 15.26.

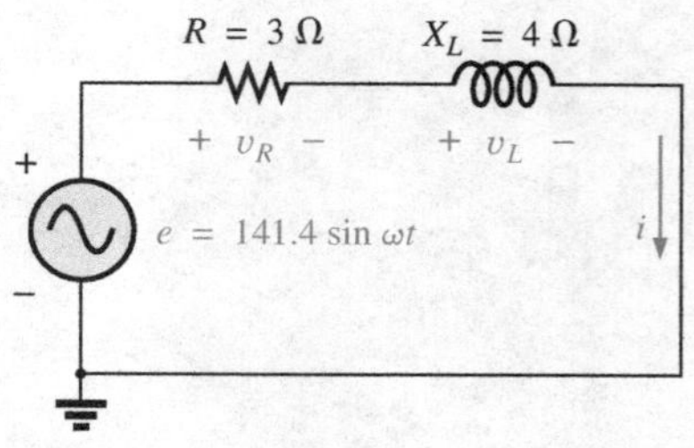

FIG. 15.26
Series R-L circuit.

Phasor Notation

$$e = 141.4 \sin \omega t \Rightarrow \mathbf{E} = 100\ \text{V} \angle 0°$$

Note Fig. 15.27.

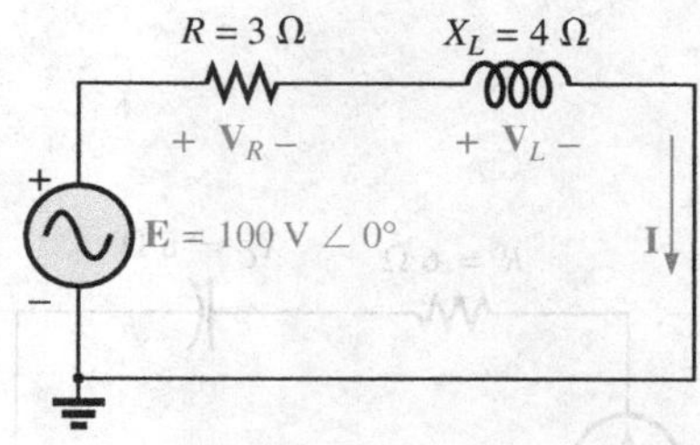

FIG. 15.27
Applying phasor notation to the network in Fig. 15.26.

$\mathbf{Z}_T$

$$\mathbf{Z}_T = \mathbf{Z}_1 + \mathbf{Z}_2 = 3\ \Omega \angle 0° + 4\ \Omega \angle 90° = 3\ \Omega + j\,4\ \Omega$$

and
$$\mathbf{Z}_T = \mathbf{5\ \Omega \angle 53.13°}$$

Impedance diagram: See Fig. 15.28.

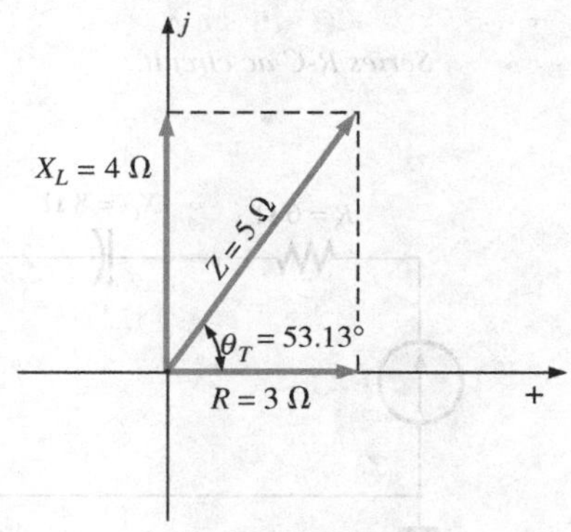

FIG. 15.28
Impedance diagram for the series R-L circuit in Fig. 15.26.

I

$$\mathbf{I} = \frac{\mathbf{E}}{\mathbf{Z}_T} = \frac{100\ \text{V} \angle 0°}{5\ \Omega \angle 53.13°} = \mathbf{20\ A \angle -53.13°}$$

$\mathbf{V}_R$ and $\mathbf{V}_L$

Ohm's law:

$$\mathbf{V}_R = \mathbf{I}\mathbf{Z}_R = (20\ \text{A} \angle -53.13°)(3\ \Omega \angle 0°) = \mathbf{60\ V \angle -53.13°}$$
$$\mathbf{V}_L = \mathbf{I}\mathbf{Z}_L = (20\ \text{A} \angle -53.13°)(4\ \Omega \angle 90°) = \mathbf{80\ V \angle 36.87°}$$

Kirchhoff's voltage law:

$$\Sigma_{\circlearrowleft} \mathbf{V} = \mathbf{E} - \mathbf{V}_R - \mathbf{V}_L = 0$$

or
$$\mathbf{E} = \mathbf{V}_R + \mathbf{V}_L$$

In rectangular form,

$$\mathbf{V}_R = 60\ \text{V} \angle -53.13° = 36\ \text{V} - j\,48\ \text{V}$$
$$\mathbf{V}_L = 80\ \text{V} \angle +36.87° = 64\ \text{V} + j\,48\ \text{V}$$

and

$$\mathbf{E} = \mathbf{V}_R + \mathbf{V}_L = (36\ \text{V} - j\,48\ \text{V}) + (64\ \text{V} + j\,48\ \text{V}) = 100\ \text{V} + j\,0 = 100\ \text{V} \angle 0°$$

as applied.

Phasor diagram: Note that for the phasor diagram in Fig. 15.29, **I** is in phase with the voltage across the resistor and lags the voltage across the inductor by 90°.

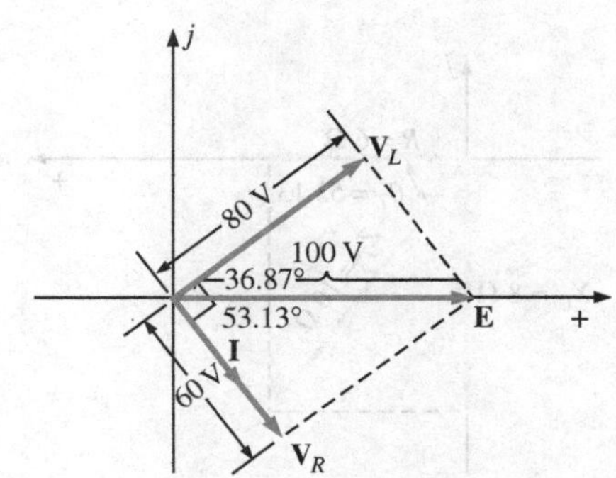

FIG. 15.29
Phasor diagram for the series R-L circuit in Fig. 15.26.

Power: The total power in watts delivered to the circuit is

$$P_T = EI \cos \theta_T = (100\ \text{V})(20\ \text{A}) \cos 53.13° = (2000\ \text{W})(0.6) = \mathbf{1200\ W}$$

where E and I are effective values and θ_T is the phase angle between E and I, or

$$P_T = I^2R = (20\ \text{A})^2(3\ \Omega) = (400)(3) = \mathbf{1200\ W}$$

where I is the effective value, or, finally,

$$\begin{aligned} P_T = P_R + P_L &= V_R I \cos\theta_R + V_L I \cos\theta_L \\ &= (60\text{ V})(20\text{ A})\cos 0^\circ + (80\text{ V})(20\text{ A})\cos 90^\circ \\ &= 1200\text{ W} + 0 \\ &= \mathbf{1200\ W} \end{aligned}$$

where θ_R is the phase angle between $\mathbf{V}_R$ and $\mathbf{I}$, and θ_L is the phase angle between $\mathbf{V}_L$ and $\mathbf{I}$.

Power factor: The power factor F_p of the circuit is $\cos 53.13^\circ = \mathbf{0.6}$ **lagging,** where 53.13° is the phase angle between **E** and **I.**

If we write the basic power equation $P = EI\cos\theta$ as

$$\cos\theta = \frac{P}{EI}$$

where E and I are the input quantities and P is the power delivered to the network, and then perform the following substitutions from the basic series ac circuit as

$$\cos\theta = \frac{P}{EI} = \frac{I^2R}{EI} = \frac{IR}{E} = \frac{R}{E/I} = \frac{R}{Z_T}$$

we find

$$\boxed{F_p = \cos\theta_T = \frac{R}{Z_T}} \tag{15.9}$$

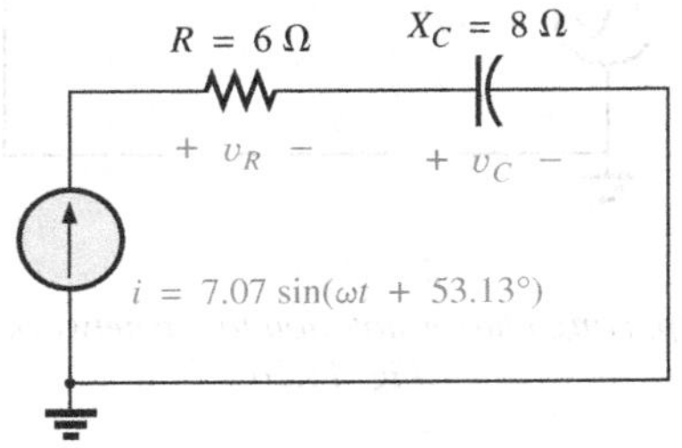

FIG. 15.30
Series R-C ac circuit.

Reference to Fig. 15.28 also indicates that θ is the impedance angle θ_T as written in Eq. (15.9), further supporting the fact that the impedance angle θ_T is also the phase angle between the input voltage and current for a series ac circuit. To determine the power factor, it is necessary only to form the ratio of the total resistance to the magnitude of the input impedance. For the case at hand,

$$F_p = \cos\theta = \frac{R}{Z_T} = \frac{3\ \Omega}{5\ \Omega} = \mathbf{0.6\ lagging}$$

as found above.

R-C

Refer to Fig. 15.30.

Phasor Notation

$$i = 7.07\sin(\omega t + 53.13^\circ) \Rightarrow \mathbf{I} = 5\text{ A}\ \angle 53.13^\circ$$

Note Fig. 15.31.

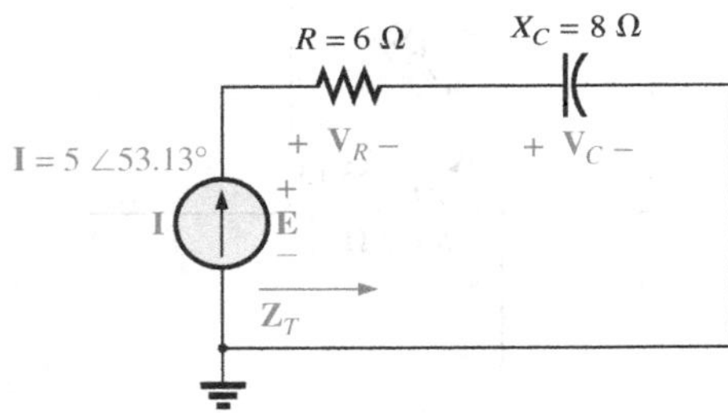

FIG. 15.31
Applying phasor notation to the circuit in Fig. 15.30.

Z_T

$$\mathbf{Z}_T = \mathbf{Z}_1 + \mathbf{Z}_2 = 6\ \Omega\ \angle 0^\circ + 8\ \Omega\ \angle -90^\circ = 6\ \Omega - j\,8\ \Omega$$

and

$$\mathbf{Z}_T = \mathbf{10\ \Omega\ \angle -53.13^\circ}$$

Impedance diagram: As shown in Fig. 15.32.

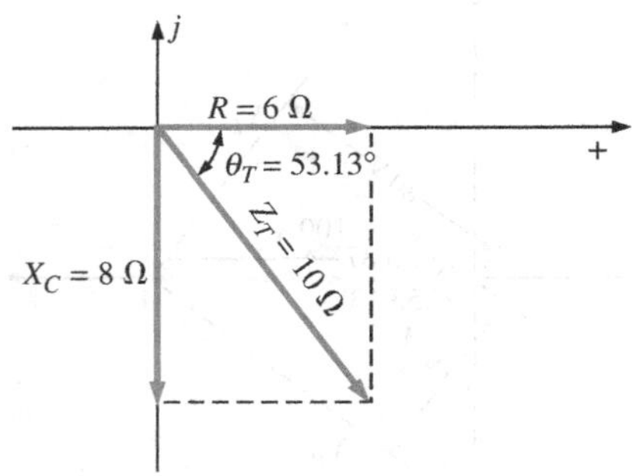

FIG. 15.32
Impedance diagram for the series R-C circuit in Fig. 15.30.

E

$$\mathbf{E} = \mathbf{I}\mathbf{Z}_T = (5\text{ A}\ \angle 53.13^\circ)(10\ \Omega\ \angle -53.13^\circ) = \mathbf{50\ V\ \angle 0^\circ}$$

V_R and V_C

$$\mathbf{V}_R = \mathbf{IZ}_R = (I\angle\theta)(R\angle 0^\circ) = (5\text{ A}\angle 53.13^\circ)(6\ \Omega\angle 0^\circ)$$
$$= \mathbf{30\ V\angle 53.13^\circ}$$
$$\mathbf{V}_C = \mathbf{IZ}_C = (I\angle\theta)(X_C\angle -90^\circ) = (5\text{ A}\angle 53.13^\circ)(8\ \Omega\angle -90^\circ)$$
$$= \mathbf{40\ V\angle -36.87^\circ}$$

Kirchhoff's voltage law:

$$\Sigma_{\circlearrowleft}\mathbf{V} = \mathbf{E} - \mathbf{V}_R - \mathbf{V}_C = 0$$

or

$$\mathbf{E} = \mathbf{V}_R + \mathbf{V}_C$$

which can be verified by vector algebra as demonstrated for the *R-L* circuit.

Phasor diagram: Note on the phasor diagram in Fig. 15.33 that the current **I** is in phase with the voltage across the resistor and leads the voltage across the capacitor by 90°.

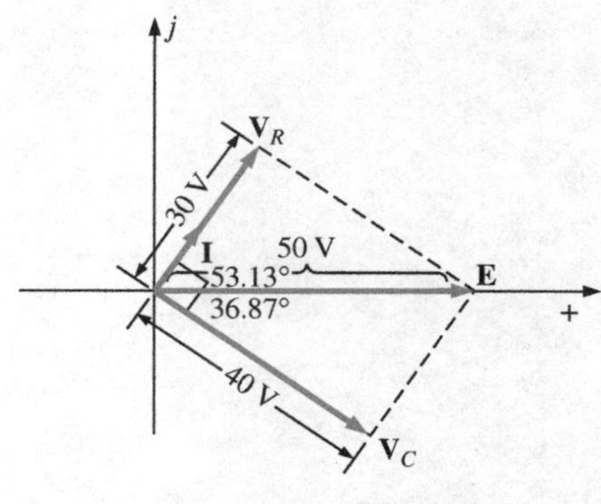

FIG. 15.33
Phasor diagram for the series R-C circuit in Fig. 15.30.

Time domain: In the time domain,

$$e = \sqrt{2}(50)\sin\omega t = \mathbf{70.70\sin\omega t}$$
$$v_R = \sqrt{2}(30)\sin(\omega t + 53.13^\circ) = \mathbf{42.42\sin(\omega t + 53.13^\circ)}$$
$$v_C = \sqrt{2}(40)\sin(\omega t - 36.87^\circ) = \mathbf{56.56\sin(\omega t - 36.87^\circ)}$$

A plot of all of the voltages and the current of the circuit appears in Fig. 15.34. Note again that i and v_R are in phase and that v_C lags i by 90°.

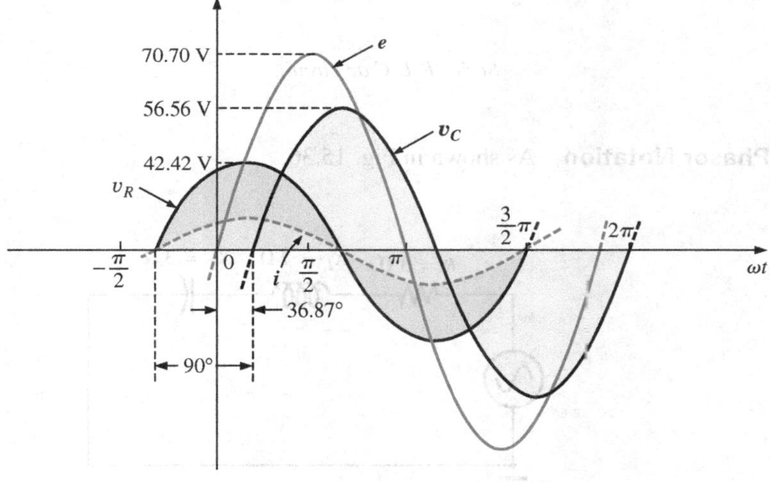

FIG. 15.34
Waveforms for the series R-C circuit in Fig. 15.30.

Power: The total power in watts delivered to the circuit is

$$P_T = EI\cos\theta_T = (50\text{ V})(5\text{ A})\cos 53.13^\circ$$
$$= (250)(0.6) = \mathbf{150\ W}$$

or

$$P_T = I^2R = (5\text{A})^2(6\ \Omega) = (25)(6)$$
$$= \mathbf{150\ W}$$

or, finally,

$$P_T = P_R + P_C = V_RI\cos\theta_R + V_CI\cos\theta_C$$
$$= (30\text{ V})(5\text{ A})\cos 0^\circ + (40\text{ V})(5\text{ A})\cos 90^\circ$$
$$= 150\text{ W} + 0$$
$$= \mathbf{150\ W}$$

Power factor: The power factor of the circuit is

$$F_p = \cos\theta = \cos 53.13° = \mathbf{0.6\ leading}$$

Using Eq. (15.9), we obtain

$$F_p = \cos\theta = \frac{R}{Z_T} = \frac{6\ \Omega}{10\ \Omega} = \mathbf{0.6\ leading}$$

as determined above.

R-L-C

Refer to Fig. 15.35.

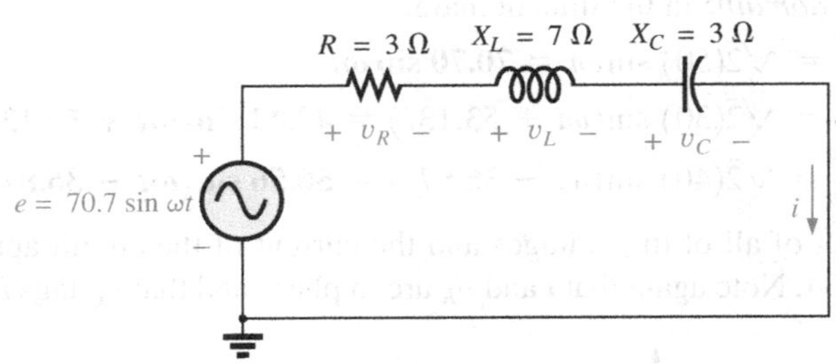

FIG. 15.35
Series R-L-C ac circuit.

Phasor Notation As shown in Fig. 15.36.

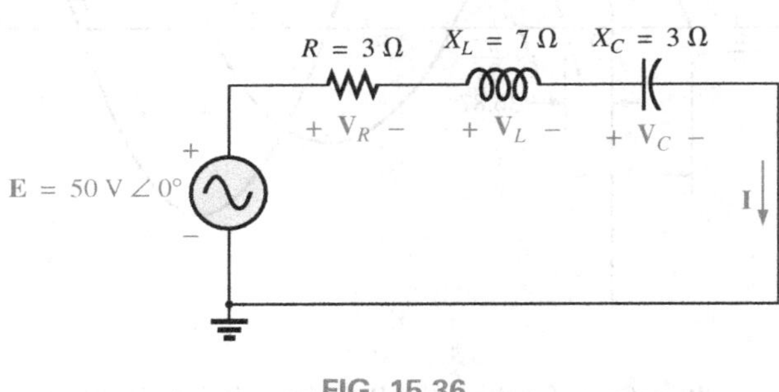

FIG. 15.36
Applying phasor notation to the circuit in Fig. 15.35.

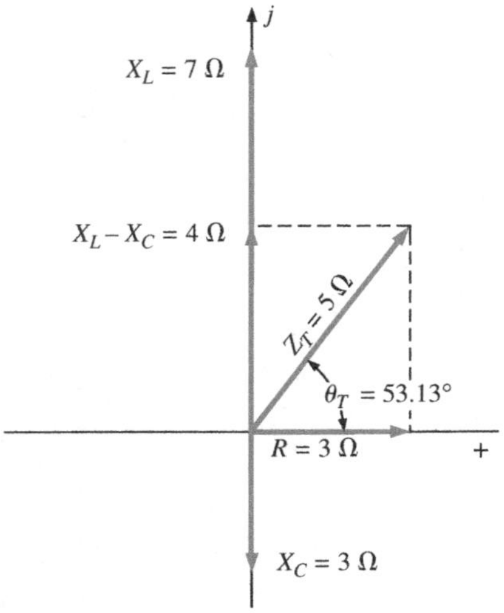

FIG. 15.37
Impedance diagram for the series R-L-C circuit in Fig. 15.35.

Z_T

$$\mathbf{Z}_T = \mathbf{Z}_1 + \mathbf{Z}_2 + \mathbf{Z}_3 = R\angle 0° + X_L\angle 90° + X_C\angle -90°$$
$$= 3\ \Omega + j\,7\ \Omega - j\,3\ \Omega = 3\ \Omega + j\,4\ \Omega$$

and

$$\mathbf{Z}_T = \mathbf{5\ \Omega\ \angle 53.13°}$$

Impedance diagram: As shown in Fig. 15.37.

I

$$\mathbf{I} = \frac{\mathbf{E}}{\mathbf{Z}_T} = \frac{50\ \text{V}\ \angle 0°}{5\ \Omega\ \angle 53.13°} = \mathbf{10\ A\ \angle -53.13°}$$

V_R, V_L, and V_C

$$\begin{aligned}\mathbf{V}_R &= \mathbf{I}\mathbf{Z}_R = (I\angle\theta)(R\angle 0^\circ) = (10\text{ A}\angle -53.13^\circ)(3\ \Omega\angle 0^\circ)\\ &= \mathbf{30\ V\angle -53.13^\circ}\end{aligned}$$

$$\begin{aligned}\mathbf{V}_L &= \mathbf{I}\mathbf{Z}_L = (I\angle\theta)(X_L\angle 90^\circ) = (10\text{ A}\angle -53.13^\circ)(7\ \Omega\angle 90^\circ)\\ &= \mathbf{70\ V\angle 36.87^\circ}\end{aligned}$$

$$\begin{aligned}\mathbf{V}_C &= \mathbf{I}\mathbf{Z}_C = (I\angle\theta)(X_C\angle -90^\circ) = (10\text{ A}\angle -53.13^\circ)(3\ \Omega\angle -90^\circ)\\ &= \mathbf{30\ V\angle -143.13^\circ}\end{aligned}$$

Kirchhoff's voltage law:

$$\Sigma_{\circlearrowright}\mathbf{V} = \mathbf{E} - \mathbf{V}_R - \mathbf{V}_L - \mathbf{V}_C = 0$$

or

$$\mathbf{E} = \mathbf{V}_R + \mathbf{V}_L + \mathbf{V}_C$$

which can also be verified through vector algebra.

Phasor diagram: The phasor diagram in Fig. 15.38 indicates that the current **I** is in phase with the voltage across the resistor, lags the voltage across the inductor by 90°, and leads the voltage across the capacitor by 90°.

Time domain:

$$\begin{aligned}i &= \sqrt{2}(10)\sin(\omega t - 53.13^\circ) = \mathbf{14.14\sin(\omega t - 53.13^\circ)}\\ v_R &= \sqrt{2}(30)\sin(\omega t - 53.13^\circ) = \mathbf{42.42\sin(\omega t - 53.13^\circ)}\\ v_L &= \sqrt{2}(70)\sin(\omega t + 36.87^\circ) = \mathbf{98.98\sin(\omega t + 36.87^\circ)}\\ v_C &= \sqrt{2}(30)\sin(\omega t - 143.13^\circ) = \mathbf{42.42\sin(\omega t - 143.13^\circ)}\end{aligned}$$

A plot of all the voltages and the current of the circuit appears in Fig. 15.39.

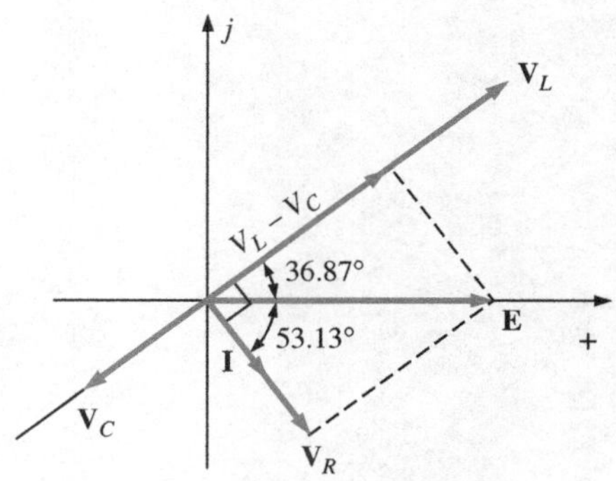

FIG. 15.38
Phasor diagram for the series R-L-C circuit in Fig. 15.35.

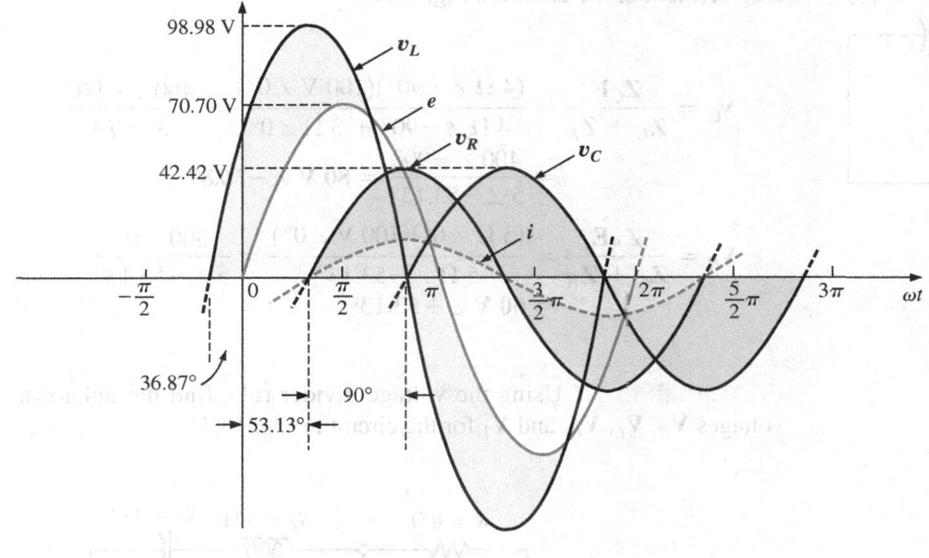

FIG. 15.39
Waveforms for the series R-L-C circuit in Fig. 15.35.

Power: The total power in watts delivered to the circuit is

$$P_T = EI\cos\theta_T = (50\text{ V})(10\text{ A})\cos 53.13^\circ = (500)(0.6) = \mathbf{300\ W}$$

or

$$P_T = I^2R = (10\text{ A})^2(3\ \Omega) = (100)(3) = \mathbf{300\ W}$$

or

$$\begin{aligned} P_T &= P_R + P_L + P_C \\ &= V_R I \cos\theta_R + V_L I \cos\theta_L + V_C I \cos\theta_C \\ &= (30\text{ V})(10\text{ A})\cos 0^\circ + (70\text{ V})(10\text{ A})\cos 90^\circ + (30\text{ V})(10\text{ A})\cos 90^\circ \\ &= (30\text{ V})(10\text{ A}) + 0 + 0 = \mathbf{300\text{ W}} \end{aligned}$$

Power factor: The power factor of the circuit is

$$F_p = \cos\theta_T = \cos 53.13^\circ = \mathbf{0.6\ lagging}$$

Using Eq. (15.9), we obtain

$$F_p = \cos\theta = \frac{R}{Z_T} = \frac{3\ \Omega}{5\ \Omega} = \mathbf{0.6\ lagging}$$

15.4 VOLTAGE DIVIDER RULE

The basic format for the **voltage divider rule** in ac circuits is exactly the same as that for dc circuits:

$$\mathbf{V}_x = \frac{\mathbf{Z}_x\mathbf{E}}{\mathbf{Z}_T} \tag{15.10}$$

where $\mathbf{V}_x$ is the voltage across one or more elements in a series that have total impedance $\mathbf{Z}_x$, $\mathbf{E}$ is the total voltage appearing across the series circuit, and $\mathbf{Z}_T$ is the total impedance of the series circuit.

EXAMPLE 15.9 Using the voltage divider rule, find the voltage across each element of the circuit in Fig. 15.40.

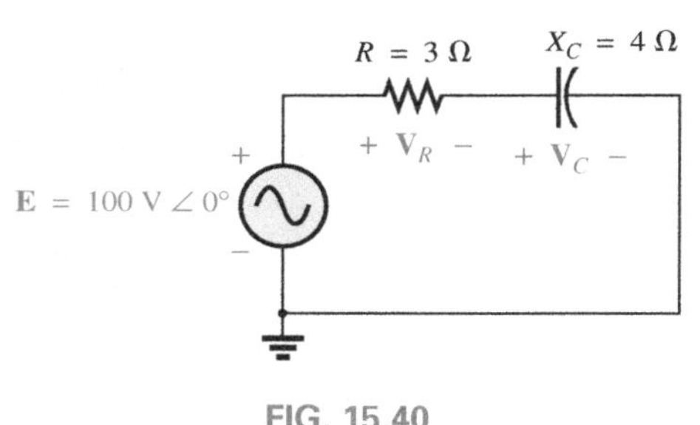

FIG. 15.40
Example 15.9.

Solution:

$$\mathbf{V}_C = \frac{\mathbf{Z}_C\mathbf{E}}{\mathbf{Z}_C + \mathbf{Z}_R} = \frac{(4\ \Omega\ \angle -90^\circ)(100\text{ V}\ \angle 0^\circ)}{4\ \Omega\ \angle -90^\circ + 3\ \Omega\ \angle 0^\circ} = \frac{400\ \angle -90^\circ}{3 - j4}$$

$$= \frac{400\ \angle -90^\circ}{5\ \angle -53.13^\circ} = \mathbf{80\text{ V}\ \angle -36.87^\circ}$$

$$\mathbf{V}_R = \frac{\mathbf{Z}_R\mathbf{E}}{\mathbf{Z}_C + \mathbf{Z}_R} = \frac{(3\ \Omega\ \angle 0^\circ)(100\text{ V}\ \angle 0^\circ)}{5\ \Omega\ \angle -53.13^\circ} = \frac{300\ \angle 0^\circ}{5\ \angle -53.13^\circ}$$

$$= \mathbf{60\text{ V}\ \angle +53.13^\circ}$$

EXAMPLE 15.10 Using the voltage divider rule, find the unknown voltages $\mathbf{V}_R$, $\mathbf{V}_L$, $\mathbf{V}_C$, and $\mathbf{V}_1$ for the circuit in Fig. 15.41.

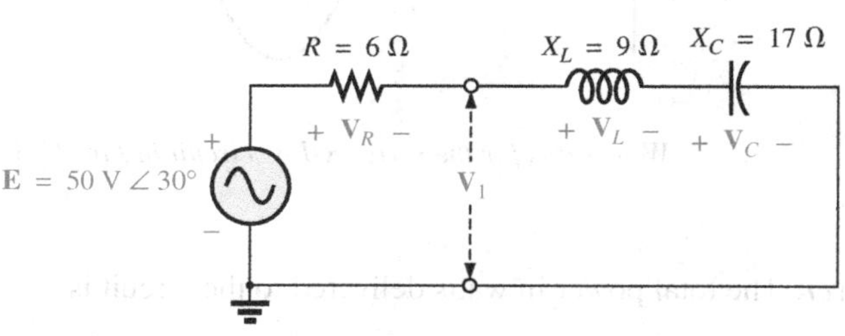

FIG. 15.41
Example 15.10.

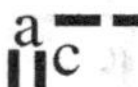

Solution:

$$\mathbf{V}_R = \frac{\mathbf{Z}_R\mathbf{E}}{\mathbf{Z}_R + \mathbf{Z}_L + \mathbf{Z}_C} = \frac{(6\ \Omega\ \angle 0^\circ)(50\ \text{V}\ \angle 30^\circ)}{6\ \Omega\ \angle 0^\circ + 9\ \Omega\ \angle 90^\circ + 17\ \Omega\ \angle -90^\circ}$$
$$= \frac{300\ \angle 30^\circ}{6 + j9 - j17} = \frac{300\ \angle 30^\circ}{6 - j8}$$
$$= \frac{300\ \angle 30^\circ}{10\ \angle -53.13^\circ} = \mathbf{30\ V\ \angle 83.13^\circ}$$

Calculator The above calculation provides an excellent opportunity to demonstrate the power of today's calculators. For the TI-89 calculator, the sequence of steps to calculate $\mathbf{V}_R$ are shown in Fig. 15.42.

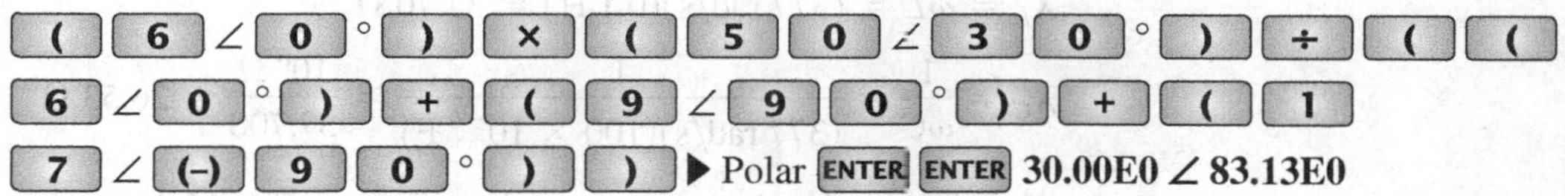

FIG. 15.42
Using the TI-89 calculator to determine V_R in Example 15.10.

$$\mathbf{V}_L = \frac{\mathbf{Z}_L\mathbf{E}}{\mathbf{Z}_T} = \frac{(9\ \Omega\ \angle 90^\circ)(50\ \text{V}\ \angle 30^\circ)}{10\ \Omega\ \angle -53.13^\circ} = \frac{450\ \text{V}\ \angle 120^\circ}{10\ \angle -53.13^\circ}$$
$$= \mathbf{45\ V\ \angle 173.13^\circ}$$

$$\mathbf{V}_C = \frac{\mathbf{Z}_C\mathbf{E}}{\mathbf{Z}_T} = \frac{(17\ \Omega\ \angle -90^\circ)(50\ \text{V}\ \angle 30^\circ)}{10\ \Omega\ \angle -53.13^\circ} = \frac{850\ \text{V}\ \angle -60^\circ}{10\ \angle -53^\circ}$$
$$= \mathbf{85\ V\ \angle -6.87^\circ}$$

$$\mathbf{V}_1 = \frac{(\mathbf{Z}_L + \mathbf{Z}_C)\mathbf{E}}{\mathbf{Z}_T} = \frac{(9\ \Omega\ \angle 90^\circ + 17\ \Omega\ \angle -90^\circ)(50\ \text{V}\ \angle 30^\circ)}{10\ \Omega\ \angle -53.13^\circ}$$
$$= \frac{(8\ \angle -90^\circ)(50\ \angle 30^\circ)}{10\ \angle -53.13^\circ}$$
$$= \frac{400\ \angle -60^\circ}{10\ \angle -53.13^\circ} = \mathbf{40\ V\ \angle -6.87^\circ}$$

EXAMPLE 15.11 For the circuit in Fig. 15.43:

a. Calculate **I**, $\mathbf{V}_R$, $\mathbf{V}_L$, and $\mathbf{V}_C$ in phasor form.
b. Calculate the total power factor.
c. Calculate the average power delivered to the circuit.

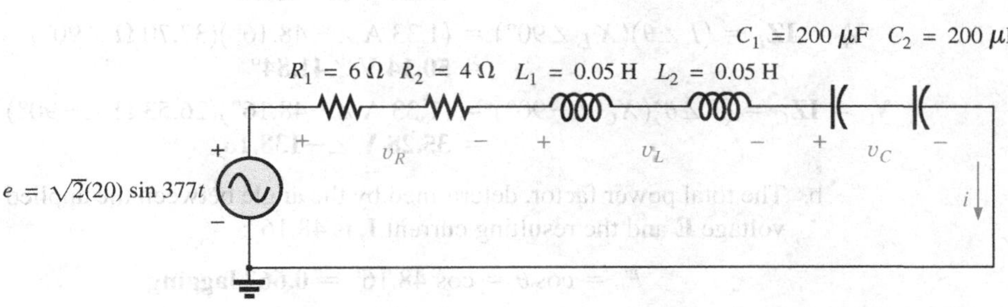

FIG. 15.43
Example 15.11.

d. Draw the phasor diagram.
e. Obtain the phasor sum of $\mathbf{V}_R$, $\mathbf{V}_L$, and $\mathbf{V}_C$, and show that it equals the input voltage **E**.
f. Find $\mathbf{V}_R$ and $\mathbf{V}_C$ using the voltage divider rule.

Solutions:

a. Combining common elements and finding the reactance of the inductor and capacitor, we obtain

$$R_T = 6\ \Omega + 4\ \Omega = 10\ \Omega$$
$$L_T = 0.05\ \text{H} + 0.05\ \text{H} = 0.1\ \text{H}$$
$$C_T = \frac{200\ \mu\text{F}}{2} = 100\ \mu\text{F}$$

$$X_L = \omega L = (377\ \text{rad/s})(0.1\ \text{H}) = 37.70\ \Omega$$

$$X_C = \frac{1}{\omega C} = \frac{1}{(377\ \text{rad/s})(100 \times 10^{-6}\ \text{F})} = \frac{10^6\ \Omega}{37{,}700} = 26.53\ \Omega$$

Redrawing the circuit using phasor notation results in Fig. 15.44.

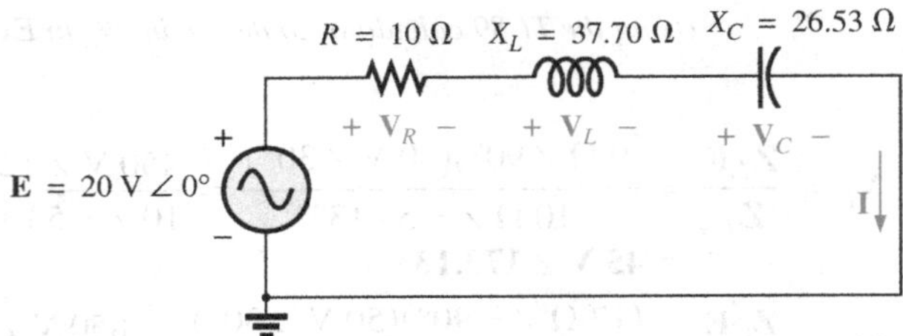

FIG. 15.44
Applying phasor notation to the circuit in Fig. 15.43.

For the circuit in Fig. 15.44,

$$\begin{aligned}\mathbf{Z}_T &= R\ \angle 0° + X_L\ \angle 90° + X_C\ \angle -90°\\ &= 10\ \Omega + j\,37.70\ \Omega - j\,26.53\ \Omega\\ &= 10\ \Omega + j\,11.17\ \Omega = \mathbf{15\ \Omega\ \angle 48.16°}\end{aligned}$$

The current **I** is

$$\mathbf{I} = \frac{\mathbf{E}}{\mathbf{Z}_T} = \frac{20\ \text{V}\ \angle 0°}{15\ \Omega\ \angle 48.16°} = \mathbf{1.33\ A\ \angle -48.16°}$$

The voltage across the resistor, inductor, and capacitor can be found using Ohm's law:

$$\begin{aligned}\mathbf{V}_R = \mathbf{I}\mathbf{Z}_R = (I\ \angle\theta)(R\ \angle 0°) &= (1.33\ \text{A}\ \angle -48.16°)(10\ \Omega\ \angle 0°)\\ &= \mathbf{13.30\ V\ \angle -48.16°}\end{aligned}$$

$$\begin{aligned}\mathbf{V}_L = \mathbf{I}\mathbf{Z}_L = (I\ \angle\theta)(X_L\ \angle 90°) &= (1.33\ \text{A}\ \angle -48.16°)(37.70\ \Omega\ \angle 90°)\\ &= \mathbf{50.14\ V\ \angle 41.84°}\end{aligned}$$

$$\begin{aligned}\mathbf{V}_C = \mathbf{I}\mathbf{Z}_C = (I\ \angle\theta)(X_C\ \angle -90°) &= (1.33\ \text{A}\ \angle -48.16°)(26.53\ \Omega\ \angle -90°)\\ &= \mathbf{35.28\ V\ \angle -138.16°}\end{aligned}$$

b. The total power factor, determined by the angle between the applied voltage **E** and the resulting current **I**, is 48.16°:

$$F_p = \cos\theta = \cos 48.16° = \mathbf{0.667\ lagging}$$

or

$$F_p = \cos\theta = \frac{R}{Z_T} = \frac{10\ \Omega}{15\ \Omega} = \mathbf{0.667\ lagging}$$

c. The total power in watts delivered to the circuit is

$$P_T = EI\cos\theta = (20\text{ V})(1.33\text{ A})(0.667) = \mathbf{17.74\ W}$$

d. The phasor diagram appears in Fig. 15.45.

e. The phasor sum of $\mathbf{V}_R$, $\mathbf{V}_L$, and $\mathbf{V}_C$ is

$$\mathbf{E} = \mathbf{V}_R + \mathbf{V}_L + \mathbf{V}_C$$
$$= 13.30\text{ V}\angle -48.16° + 50.14\text{ V}\angle 41.84° + 35.28\text{ V}\angle -138.16°$$
$$\mathbf{E} = 13.30\text{ V}\angle -48.16° + 14.86\text{ V}\angle 41.84°$$

Therefore,

$$E = \sqrt{(13.30\text{ V})^2 + (14.86\text{ V})^2} = \mathbf{20\ V}$$

and $\theta_E = \mathbf{0°}$ (from phasor diagram)

and $\mathbf{E} = 20\angle 0°$

f. $$\mathbf{V}_R = \frac{\mathbf{Z}_R\mathbf{E}}{\mathbf{Z}_T} = \frac{(10\ \Omega\angle 0°)(20\text{ V}\angle 0°)}{15\ \Omega\angle 48.16°} = \frac{200\text{ V}\angle 0°}{15\angle 48.16°}$$
$$= \mathbf{13.3\ V\angle -48.16°}$$

$$\mathbf{V}_C = \frac{\mathbf{Z}_C\mathbf{E}}{\mathbf{Z}_T} = \frac{(26.5\ \Omega\angle -90°)(20\text{ V}\angle 0°)}{15\ \Omega\angle 48.16°} = \frac{530.6\text{ V}\angle -90°}{15\angle 48.16°}$$
$$= \mathbf{35.37\ V\angle -138.16°}$$

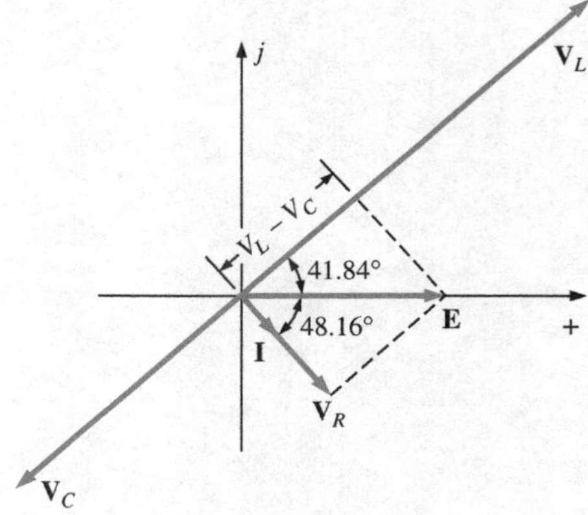

FIG. 15.45
Phasor diagram for the circuit in Fig. 15.43.

15.5 FREQUENCY RESPONSE FOR SERIES ac CIRCUITS

Thus far, the analysis has been for a fixed frequency, resulting in a fixed value for the reactance of an inductor or a capacitor. We now examine how the response of a series circuit changes as the frequency changes. We assume ideal elements throughout the discussion, so that the response of each element will be as shown in Fig. 15.46. Each response in Fig. 15.46 was discussed in detail in Chapter 14.

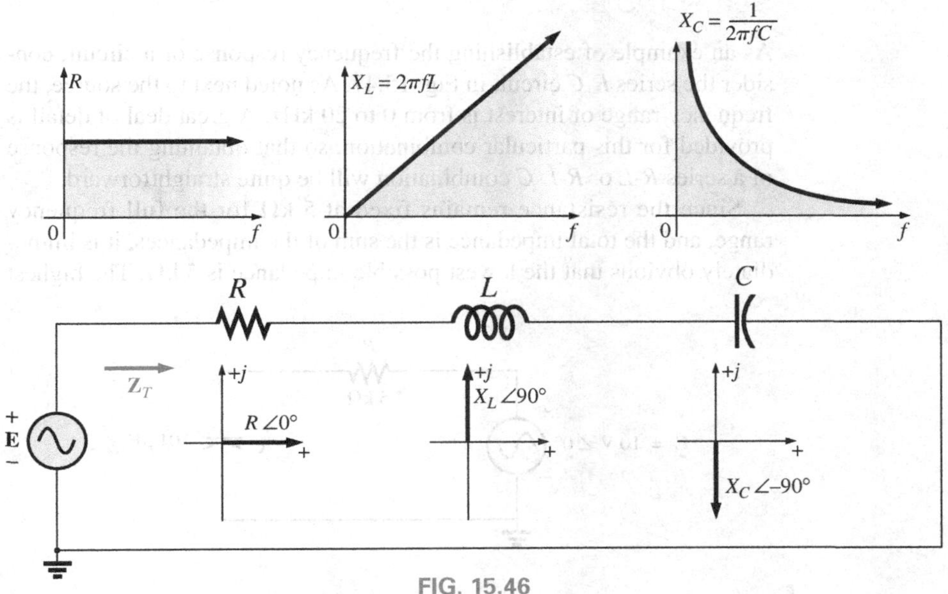

FIG. 15.46
Reviewing the frequency response of the basic elements.

When considering elements in series, remember that the total impedance is the sum of the individual elements and that the reactance of an inductor is in direct opposition to that of a capacitor. For Fig. 15.46, we are first aware that the resistance will remain fixed for the full range of frequencies: It will always be there, but, more important, its magnitude will not change. The inductor, however, will provide increasing levels of impedance as the frequency increases, while the capacitor will provide lower levels of impedance.

We are also aware from Chapter 14 that the inductor has a short-circuit equivalence at $f = 0$ Hz or very low frequencies, while the capacitor is nearly an open circuit for the same frequency range. For very high frequencies, the capacitor approaches the short-circuit equivalence, and the inductor approaches the open-circuit equivalence.

In general, therefore, if we encounter a series *R-L-C* circuit at very low frequencies, we can assume that the capacitor, with its very large impedance, will be the predominant factor. If the circuit is just an *R-L* series circuit, the impedance may be determined primarily by the resistive element since the reactance of the inductor is so small. As the frequency increases, the reactance of the coil increases to the point where it totally outshadows the impedance of the resistor. For an *R-L-C* combination, as the frequency increases, the reactance of the capacitor begins to approach a short-circuit equivalence, and the total impedance will be determined primarily by the inductive element. At very high frequencies, for an *R-C* series circuit, the total impedance eventually approaches that of the resistor since the impedance of the capacitor is dropping off so quickly.

In total, therefore,

when encountering a series ac circuit of any combination of elements, always use the idealized response of each element to establish some feeling for how the circuit will respond as the frequency changes.

Once you have a logical, overall sense for what the response will be, you can concentrate on working out the details.

Series *R-C* ac Circuit

As an example of establishing the frequency response of a circuit, consider the series *R-C* circuit in Fig. 15.47. As noted next to the source, the frequency range of interest is from 0 to 20 kHz. A great deal of detail is provided for this particular combination, so that obtaining the response of a series *R-L* or *R-L-C* combination will be quite straightforward.

Since the resistance remains fixed at 5 kΩ for the full frequency range, and the total impedance is the sum of the impedances, it is immediately obvious that the lowest possible impedance is 5 kΩ. The highest

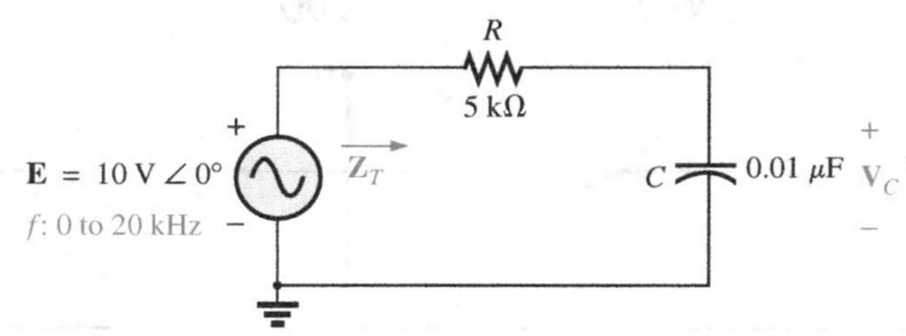

FIG. 15.47
Determining the frequency response of a series R-C circuit.

impedance, however, is dependent on the capacitive element since its impedance at very low frequencies is extremely high. At very low frequencies we can conclude, without a single calculation, that the impedance is determined primarily by the impedance of the capacitor. At the highest frequencies, we can assume that the reactance of the capacitor has dropped to such low levels that the impedance of the combination will approach that of the resistance.

The frequency at which the reactance of the capacitor drops to that of the resistor can be determined by setting the reactance of the capacitor equal to that of the resistor as follows:

$$X_C = \frac{1}{2\pi f_1 C} = R$$

Solving for the frequency yields

$$\boxed{f_1 = \frac{1}{2\pi RC}} \qquad \textbf{(15.11)}$$

This significant point appears in the frequency plots in Fig. 15.48. Substituting values, we find that it occurs at

$$f_1 = \frac{1}{2\pi RC} = \frac{1}{2\pi(5\text{ k}\Omega)(0.01\ \mu\text{F})} \cong 3.18\text{ kHz}$$

We now know that for frequencies greater than f_1, $R > X_C$ and that for frequencies less than f_1, $X_C > R$, as shown in Fig. 15.48.

Now for the details. The total impedance is determined by the following equation:

$$\mathbf{Z}_T = R - jX_C$$

and

$$\boxed{\mathbf{Z}_T = Z_T \angle \theta_T = \sqrt{R^2 + X_C^2} \angle -\tan^{-1}\frac{X_C}{R}} \qquad \textbf{(15.12)}$$

The magnitude and angle of the total impedance can now be found at any frequency of interest by simply substituting into Eq. (15.12). The presence of the capacitor suggests that we start from a low frequency (100 Hz) and then open the spacing until we reach the upper limit of interest (20 kHz).

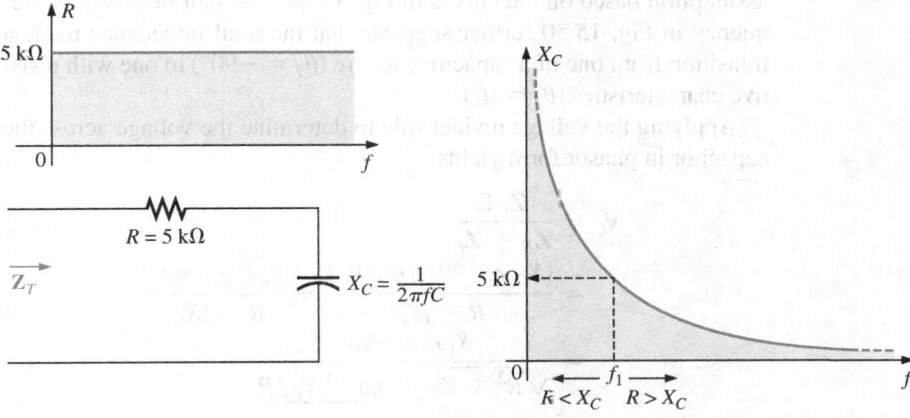

FIG. 15.48
The frequency response for the individual elements of a series R-C circuit.

***f* = 100 Hz**

$$X_C = \frac{1}{2\pi fC} = \frac{1}{2\pi(100\text{ Hz})(0.01\ \mu\text{F})} = 159.16\text{ k}\Omega$$

and $$Z_T = \sqrt{R^2 + X_C^2} = \sqrt{(5\text{ k}\Omega)^2 + (159.16\text{ k}\Omega)^2} = 159.24\text{ k}\Omega$$

with $$\theta_T = -\tan^{-1}\frac{X_C}{R} = -\tan^{-1}\frac{159.16\text{ k}\Omega}{5\text{ k}\Omega} = -\tan^{-1} 31.83$$
$$= -88.2^\circ$$

and $$\mathbf{Z}_T = \mathbf{159.24\ k\Omega \angle -88.2^\circ}$$

which compares very closely with $\mathbf{Z}_C = 159.16\text{ k}\Omega \angle -90^\circ$ if the circuit were purely capacitive ($R = 0\ \Omega$). Our assumption that the circuit is primarily capacitive at low frequencies is therefore confirmed.

***f* = 1 kHz**

$$X_C = \frac{1}{2\pi fC} = \frac{1}{2\pi(1\text{ kHz})(0.01\ \mu\text{F})} = 15.92\text{ k}\Omega$$

and $$Z_T = \sqrt{R^2 + X_C^2} = \sqrt{(5\text{ k}\Omega)^2 + (15.92\text{ k}\Omega)^2} = 16.69\text{ k}\Omega$$

with $$\theta_T = -\tan^{-1}\frac{X_C}{R} = -\tan^{-1}\frac{15.92\text{ k}\Omega}{5\text{ k}\Omega}$$
$$= -\tan^{-1} 3.18 = -72.54^\circ$$

and $$\mathbf{Z}_T = \mathbf{16.69\ k\Omega \angle -72.54^\circ}$$

A noticeable drop in the magnitude has occurred, and the impedance angle has dropped almost 17° from the purely capacitive level.

Continuing, we obtain

$$f = 5\text{ kHz:}\quad \mathbf{Z}_T = \mathbf{5.93\ k\Omega \angle -32.48^\circ}$$
$$f = 10\text{ kHz:}\quad \mathbf{Z}_T = \mathbf{5.25\ k\Omega \angle -17.66^\circ}$$
$$f = 15\text{ kHz:}\quad \mathbf{Z}_T = \mathbf{5.11\ k\Omega \angle -11.98^\circ}$$
$$f = 20\text{ kHz:}\quad \mathbf{Z}_T = \mathbf{5.06\ k\Omega \angle -9.04^\circ}$$

Note how close the magnitude of Z_T at $f = 20$ kHz is to the resistance level of 5 kΩ. In addition, note how the phase angle is approaching that associated with a pure resistive network (0°).

A plot of Z_T versus frequency in Fig. 15.49 completely supports our assumption based on the curves in Fig. 15.48. The plot of θ_T versus frequency in Fig. 15.50 further suggests that the total impedance made a transition from one of a capacitive nature ($\theta_T = -90^\circ$) to one with resistive characteristics ($\theta_T = 0^\circ$).

Applying the voltage divider rule to determine the voltage across the capacitor in phasor form yields

$$\mathbf{V}_C = \frac{\mathbf{Z}_C\mathbf{E}}{\mathbf{Z}_R + \mathbf{Z}_C}$$
$$= \frac{(X_C \angle -90^\circ)(E \angle 0^\circ)}{R - jX_C} = \frac{X_C E \angle -90^\circ}{R - jX_C}$$
$$= \frac{X_C E \angle -90^\circ}{\sqrt{R^2 + X_C^2}\ \angle -\tan^{-1} X_C/R}$$

or $$\mathbf{V}_C = V_C \angle \theta_C = \frac{X_C E}{\sqrt{R^2 + X_C^2}} \angle -90^\circ + \tan^{-1}(X_C/R)$$

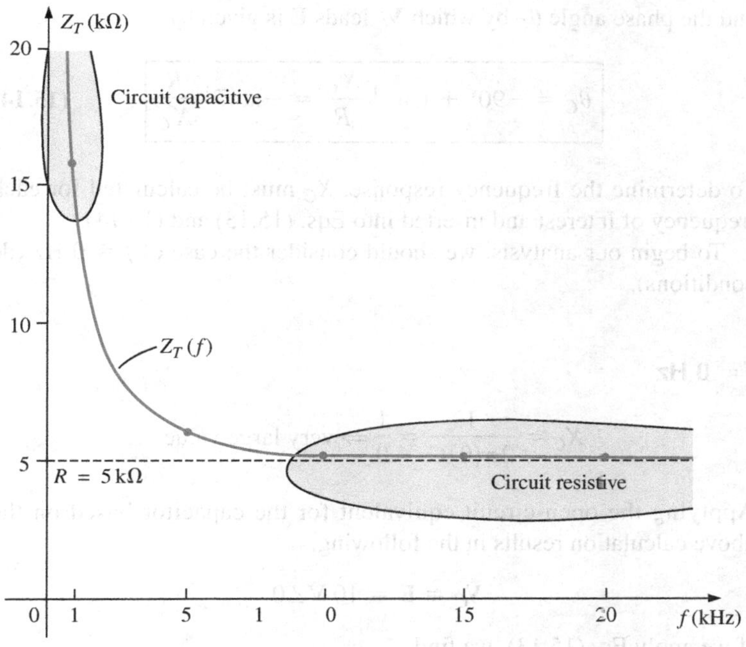

FIG. 15.49
The magnitude of the input impedance versus frequency for the circuit in Fig. 15.47.

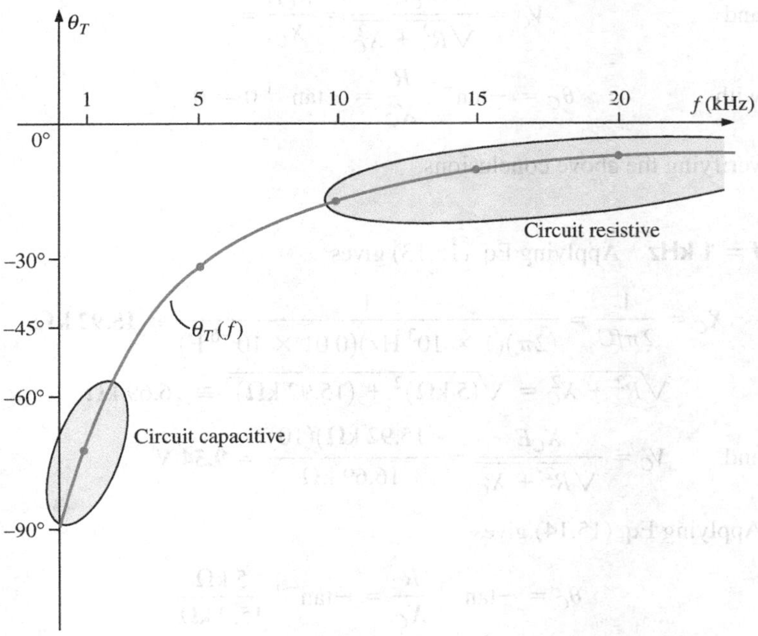

FIG. 15.50
The phase angle of the input impedance versus frequency for the circuit in Fig. 15.47.

The magnitude of $\mathbf{V}_C$ is therefore determined by

$$V_C = \frac{X_C E}{\sqrt{R^2 + X_C^2}} \qquad (15.13)$$

and the phase angle θ_C by which $\mathbf{V}_C$ leads $\mathbf{E}$ is given by

$$\theta_C = -90° + \tan^{-1}\frac{X_C}{R} = -\tan^{-1}\frac{R}{X_C} \quad \textbf{(15.14)}$$

To determine the frequency response, X_C must be calculated for each frequency of interest and inserted into Eqs. (15.13) and (15.14).

To begin our analysis, we should consider the case of $f = 0$ Hz (dc conditions).

***f* = 0 Hz**

$$X_C = \frac{1}{2\pi(0)C} = \frac{1}{0} \Rightarrow \text{very large value}$$

Applying the open-circuit equivalent for the capacitor based on the above calculation results in the following:

$$\mathbf{V}_C = \mathbf{E} = 10\text{ V}\angle 0°$$

If we apply Eq. (15.13), we find

$$X_C^2 \gg R^2$$

and $$\sqrt{R^2 + X_C^2} \cong \sqrt{X_C^2} = X_C$$

and $$V_C = \frac{X_C E}{\sqrt{R^2 + X_C^2}} = \frac{X_C E}{X_C} = E$$

with $$\theta_C = -\tan^{-1}\frac{R}{X_C} = -\tan^{-1} 0 = 0°$$

verifying the above conclusions.

***f* = 1 kHz** Applying Eq. (15.13) gives

$$X_C = \frac{1}{2\pi fC} = \frac{1}{(2\pi)(1 \times 10^3\text{ Hz})(0.01 \times 10^{-6}\text{ F})} \cong \mathbf{15.92\ k\Omega}$$

$$\sqrt{R^2 + X_C^2} = \sqrt{(5\text{ k}\Omega)^2 + (15.92\text{ k}\Omega)^2} \cong 16.69\text{ k}\Omega$$

and $$V_C = \frac{X_C E}{\sqrt{R^2 + X_C^2}} = \frac{(15.92\text{ k}\Omega)(10)}{16.69\text{ k}\Omega} = \mathbf{9.54\ V}$$

Applying Eq. (15.14) gives

$$\theta_C = -\tan^{-1}\frac{R}{X_C} = -\tan^{-1}\frac{5\text{ k}\Omega}{15.9\text{ k}\Omega}$$

$$= -\tan^{-1}0.314 = \mathbf{-17.46°}$$

and $$\mathbf{V}_C = \mathbf{9.83\ V\ \angle -17.46°}$$

As expected, the high reactance of the capacitor at low frequencies has resulted in the major part of the applied voltage appearing across the capacitor.

If we plot the phasor diagrams for $f = 0$ Hz and $f = 1$ kHz, as shown in Fig. 15.51, we find that $\mathbf{V}_C$ is beginning a clockwise rotation with an increase in frequency that will increase the angle θ_C and decrease the phase

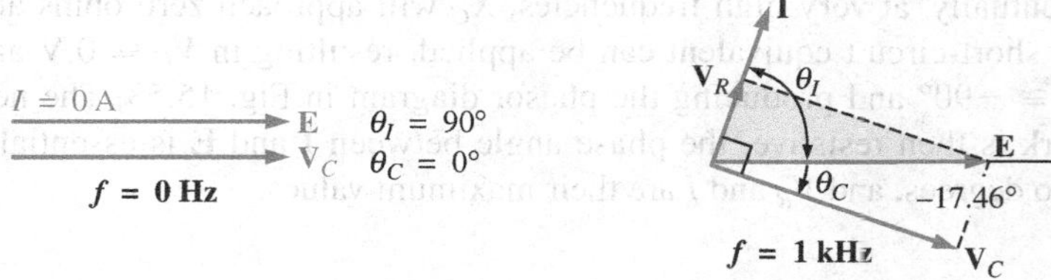

FIG. 15.51
The phasor diagram for the circuit in Fig. 15.47 for f = 0 Hz and 1 kHz.

angle between **I** and **E.** Recall that for a purely capacitive network, **I** leads **E** by 90°. As the frequency increases, therefore, the capacitive reactance is decreasing, and eventually $R >> X_C$ with $\theta_C = -90°$, and the angle between **I** and **E** will approach 0°. Keep in mind as we proceed through the other frequencies that θ_C is the phase angle between $\mathbf{V}_C$ and **E** and that the magnitude of the angle by which **I** leads **E** is determined by

$$|\theta_I| = 90° - |\theta_C| \tag{15.15}$$

f = 5 kHz Applying Eq. (15.13) gives

$$X_C = \frac{1}{2\pi fC} = \frac{1}{(2\pi)(5 \times 10^3\ \text{Hz})(0.01 \times 10^{-6}\ \text{F})} \cong \mathbf{3.18\ k\Omega}$$

Note the dramatic drop in X_C from 1 kHz to 5 kHz. In fact, X_C is now less than the resistance R of the network, and the phase angle determined by $\tan^{-1}(X_C/R)$ must be less than 45°. Here,

$$V_C = \frac{X_C E}{\sqrt{R^2 + X_C^2}} = \frac{(3.18\ \text{k}\Omega)(10\ \text{V})}{\sqrt{(5\ \text{k}\Omega)^2 + (3.18\ \text{k}\Omega)^2}} = \mathbf{5.37\ V}$$

with

$$\theta_C = -\tan^{-1}\frac{R}{X_C} = -\tan^{-1}\frac{5\ \text{k}\Omega}{3.2\ \text{k}\Omega}$$
$$= -\tan^{-1}1.56 = \mathbf{-57.38°}$$

f = 10 kHz

$$X_C \cong \mathbf{1.59\ k\Omega} \quad V_C = \mathbf{3.03\ V} \quad \theta_C = \mathbf{-72.34°}$$

f = 15 kHz

$$X_C \cong \mathbf{1.06\ k\Omega} \quad V_C = \mathbf{2.07\ V} \quad \theta_C = \mathbf{-78.02°}$$

f = 20 kHz

$$X_C \cong \mathbf{795.78\ \Omega} \quad V_C = \mathbf{1.57\ V} \quad \theta_C = \mathbf{-80.96°}$$

The phasor diagrams for f = 5 kHz and f = 20 kHz appear in Fig. 15.52 to show the continuing rotation of the $\mathbf{V}_C$ vector.

Note also from Figs. 15.51 and 15.52 that the vector $\mathbf{V}_R$ and the current **I** have grown in magnitude with the reduction in the capacitive reactance.

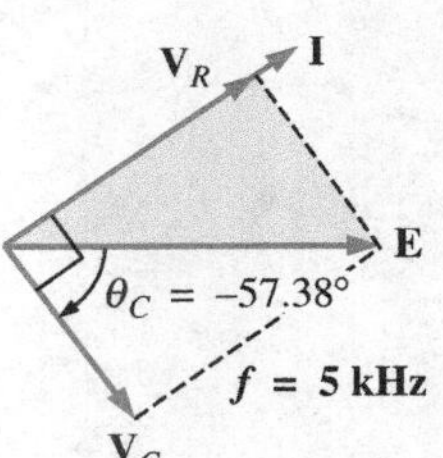

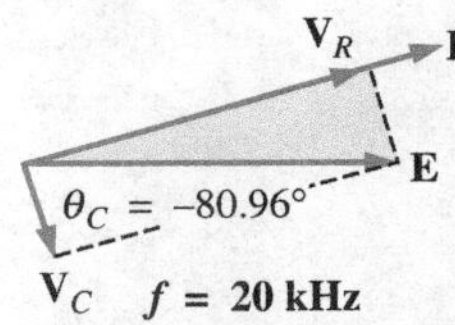

FIG. 15.52
The phasor diagram for the circuit in Fig. 15.47 for f = 5 kHz and 20 kHz.

Eventually, at very high frequencies, X_C will approach zero ohms and the short-circuit equivalent can be applied, resulting in $V_C \cong 0$ V and $\theta_C \cong -90°$ and producing the phasor diagram in Fig. 15.53. The network is then resistive, the phase angle between **I** and **E** is essentially zero degrees, and V_R and I are their maximum values.

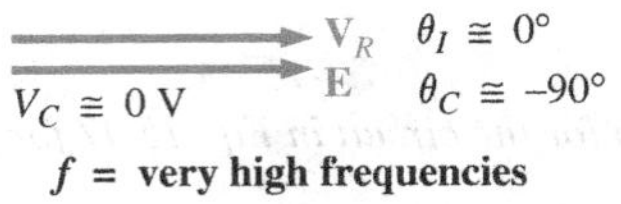

$\mathbf{V}_R$ $\theta_I \cong 0°$
$\mathbf{E}$ $\theta_C \cong -90°$
$V_C \cong 0$ V
f = **very high frequencies**

FIG. 15.53
The phasor diagram for the circuit in Fig. 15.47 at very high frequencies.

A plot of V_C versus frequency appears in Fig. 15.54. At low frequencies, $X_C >> R$, and V_C is very close to E in magnitude. As the applied frequency increases, X_C decreases in magnitude along with V_C as V_R captures more of the applied voltage. A plot of θ_C versus frequency is provided in Fig. 15.55. At low frequencies, the phase angle between $\mathbf{V}_C$ and **E** is very small since $\mathbf{V}_C \cong \mathbf{E}$. Recall that if two phasors are equal, they must have the same angle. As the applied frequency increases, the network becomes more resistive, and the phase angle between $\mathbf{V}_C$ and **E** approaches 90°. Keep in mind that, at high frequencies, **I** and **E** are approaching an in-phase situation, and the angle between $\mathbf{V}_C$ and **E** will approach that between $\mathbf{V}_C$ and **I,** which we know must be 90° ($\mathbf{I}_C$ leading $\mathbf{V}_C$).

A plot of V_R versus frequency approaches E volts from zero volts with an increase in frequency, but remember that $V_R \neq E - V_C$ due to the vector relationship. The phase angle between **I** and **E** could be plotted directly from Fig. 15.55 using Eq. (15.15).

In Chapter 21, the analysis of this section is extended to a much wider frequency range using a log axis for frequency. It will be demonstrated

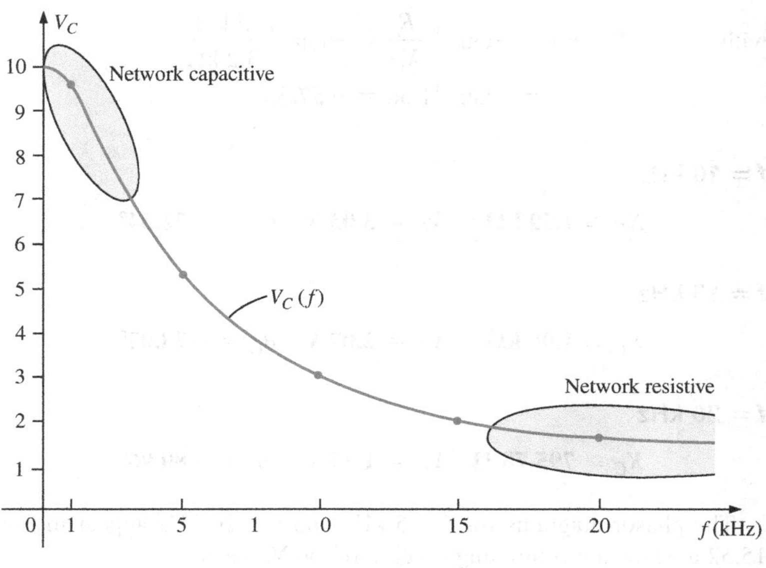

FIG. 15.54
The magnitude of the voltage V_C versus frequency for the circuit in Fig. 15.47.

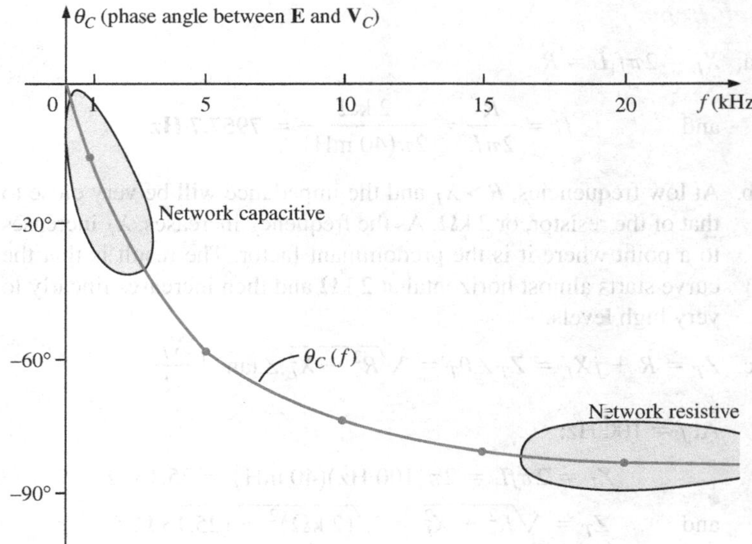

FIG. 15.55
The phase angle between **E** *and* $\mathbf{V}_C$ *versus frequency for the circuit in Fig. 15.47.*

that an *R-C* circuit such as that in Fig. 15.47 can be used as a filter to determine which frequencies will have the greatest impact on the stage to follow. From our current analysis, it is obvious that any network connected across the capacitor will receive the greatest potential level at low frequencies and be effectively "shorted out" at very high frequencies.

The analysis of a series *R-L* circuit proceeds in much the same manner, except that X_L and V_L increases with frequency and the angle between **I** and **E** approaches 90° (voltage leading the current) rather than 0°. If $\mathbf{V}_L$ is plotted versus frequency, $\mathbf{V}_L$ will approach **E**, as demonstrated in Example 15.12, and X_L will eventually attain a level at which the open-circuit equivalent is appropriate.

EXAMPLE 15.12 For the series *R-L* circuit in Fig. 15.56:

a. Determine the frequency at which $X_L = R$.
b. Develop a mental image of the change in total impedance with frequency without doing any calculations.
c. Find the total impedance at $f = 100$ Hz and 40 kHz, and compare your answer with the assumptions of part (b).
d. Plot the curve of V_L versus frequency.
e. Find the phase angle of the total impedance at $f = 40$ kHz. Can the circuit be considered inductive at this frequency? Why?

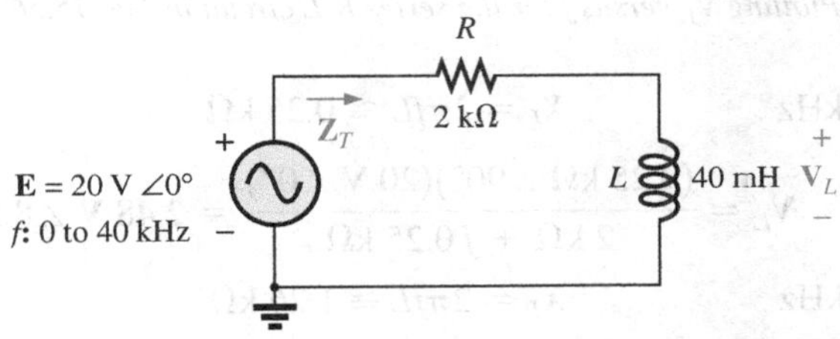

FIG. 15.56
Circuit for Example 15.12.

Solutions:

a. $X_L = 2\pi f_1 L = R$

and $$f_1 = \frac{R}{2\pi L} = \frac{2\text{ k}\Omega}{2\pi(40\text{ mH})} = \mathbf{7957.7\text{ Hz}}$$

b. At low frequencies, $R > X_L$ and the impedance will be very close to that of the resistor, or 2 kΩ. As the frequency increases, X_L increases to a point where it is the predominant factor. The result is that the curve starts almost horizontal at 2 kΩ and then increases linearly to very high levels.

c. $Z_T = R + jX_L = Z_T \angle \theta_T = \sqrt{R^2 + X_L^2} \angle \tan^{-1}\frac{X_L}{R}$

At $f = 100$ Hz:

$$X_L = 2\pi fL = 2\pi(100\text{ Hz})(40\text{ mH}) = 25.13\ \Omega$$

and $$Z_T = \sqrt{R^2 + X_L^2} = \sqrt{(2\text{ k}\Omega)^2 + (25.13\ \Omega)^2} = 2000.16\ \Omega \cong R$$

At $f = 40$ kHz:

$$X_L = 2\pi fL = 2\pi(40\text{ kHz})(40\text{ mH}) \cong 10.05\text{ k}\Omega$$

and $$Z_T = \sqrt{R^2 + X_L^2} = \sqrt{(2\text{ k}\Omega)^2 + (10.05\text{ k}\Omega)^2} = 10.25\text{ k}\Omega \cong X_L$$

Both calculations support the conclusions of part (b).

d. Applying the voltage divider rule gives

$$\mathbf{V}_L = \frac{\mathbf{Z}_L\mathbf{E}}{\mathbf{Z}_T}$$

From part (c), we know that at 100 Hz, $Z_T \cong R$, so that $V_R \cong 20$ V and $V_L \cong 0$ V. Part (c) revealed that at 40 kHz, $Z_T \cong X_L$, so that $V_L \cong 20$ V and $V_R \cong 0$ V. The result is two plot points for the curve in Fig. 15.57.

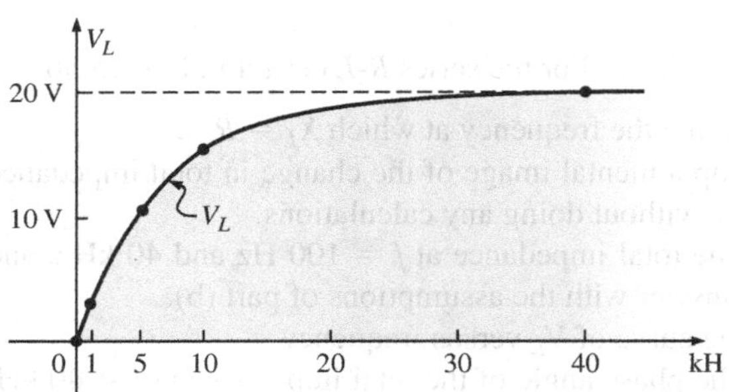

FIG. 15.57
Plotting V_L versus f for the series R-L circuit in Fig. 15.56.

At 1 kHz $$X_L = 2\pi fL \cong 0.25\text{ k}\Omega$$

and $$\mathbf{V}_L = \frac{(0.25\text{ k}\Omega \angle 90°)(20\text{ V} \angle 0°)}{2\text{ k}\Omega + j\,0.25\text{ k}\Omega} = \mathbf{2.48\text{ V} \angle 82.87°}$$

At 5 kHz $$X_L = 2\pi fL \cong 1.26\text{ k}\Omega$$

and $$\mathbf{V}_L = \frac{(1.26\text{ k}\Omega \angle 90°)(20\text{ V} \angle 0°)}{2\text{ k}\Omega + j\,1.26\text{ k}\Omega} = \mathbf{10.68\text{ V} \angle 57.79°}$$

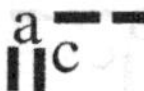

At 10 kHz $\quad X_L = 2\pi fL \cong 2.5\ \text{k}\Omega$

and $\quad \mathbf{V}_L = \dfrac{(2.5\ \text{k}\Omega \angle 90^\circ)(20\ \text{V} \angle 0^\circ)}{2.5\ \text{k}\Omega + j\,2.5\ \text{k}\Omega} = \mathbf{15.63\ V \angle 38.66^\circ}$

The complete plot appears in Fig. 15.57.

e. $\theta_T = \tan^{-1}\dfrac{X_L}{R} = \tan^{-1}\dfrac{10.05\ \text{k}\Omega}{2\ \text{k}\Omega} = \mathbf{78.75^\circ}$

The angle θ_T is closing in on the 90° of a purely inductive network. Therefore, the network can be considered quite inductive at a frequency of 40 kHz.

15.6 SUMMARY: SERIES ac CIRCUITS

The following is a review of important conclusions that can be derived from the discussion and examples of the previous sections. The list is not all-inclusive, but it does emphasize some of the conclusions that should be carried forward in the future analysis of ac systems.

For series ac circuits with reactive elements:

1. *The total impedance will be frequency dependent.*
2. *The impedance of any one element can be greater than the total impedance of the network.*
3. *The inductive and capacitive reactances are always in direct opposition on an impedance diagram.*
4. *Depending on the frequency applied, the same circuit can be either predominantly inductive or predominantly capacitive.*
5. *At lower frequencies, the capacitive elements will usually have the most impact on the total impedance, while at high frequencies, the inductive elements will usually have the most impact.*
6. *The magnitude of the voltage across any one element can be greater than the applied voltage.*
7. *The magnitude of the voltage across an element compared to the other elements of the circuit is directly related to the magnitude of its impedance; that is, the larger the impedance of an element, the larger is the magnitude of the voltage across the element.*
8. *The voltages across a coil or capacitor are always in direct opposition on a phasor diagram.*
9. *The current is always in phase with the voltage across the resistive elements, lags the voltage across all the inductive elements by 90°, and leads the voltage across all the capacitive elements by 90°.*
10. *The larger the resistive element of a circuit compared to the net reactive impedance, the closer is the power factor to unity.*

PARALLEL ac CIRCUITS

15.7 ADMITTANCE AND SUSCEPTANCE

The discussion for **parallel ac circuits** is very similar to that for dc circuits. In dc circuits, *conductance* (G) was defined as being equal to $1/R$. The total conductance of a parallel circuit was then found by adding the conductance of each branch. The total resistance R_T is simply $1/G_T$.

In ac circuits, we define **admittance** (**Y**) as being equal to $1/\mathbf{Z}$. The unit of measure for admittance as defined by the SI system is *siemens,* which has the symbol S. Admittance is a measure of how well an ac circuit will *admit,* or allow, current to flow in the circuit. The larger its value, therefore, the heavier is the current flow for the same applied potential. The total admittance of a circuit can also be found by finding the sum of the parallel admittances. The total impedance $\mathbf{Z}_T$ of the circuit is then $1/\mathbf{Y}_T$; that is, for the network in Fig. 15.58,

$$\mathbf{Y}_T = \mathbf{Y}_1 + \mathbf{Y}_2 + \mathbf{Y}_3 + \cdots + \mathbf{Y}_N \quad (15.16)$$

and

$$\mathbf{Z}_T = \frac{1}{\mathbf{Y}_T} \quad (15.17)$$

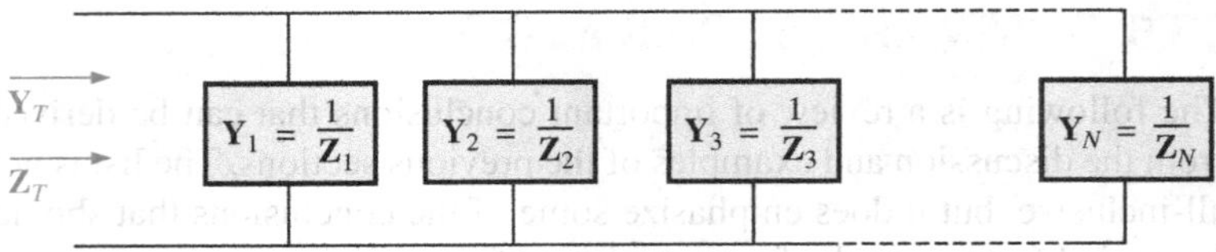

FIG. 15.58
Parallel ac network.

or, since $\mathbf{Z} = 1/\mathbf{Y}$,

$$\frac{1}{\mathbf{Z}_T} = \frac{1}{\mathbf{Z}_1} + \frac{1}{\mathbf{Z}_2} + \frac{1}{\mathbf{Z}_3} + \cdots + \frac{1}{\mathbf{Z}_N} \quad (15.18)$$

and

$$\mathbf{Z}_T = \frac{1}{\dfrac{1}{\mathbf{Z}_1} + \dfrac{1}{\mathbf{Z}_2} + \dfrac{1}{\mathbf{Z}_3} + \cdots + \dfrac{1}{\mathbf{Z}_N}} \quad (15.19)$$

matching Eq. (6.3) for dc networks.

For two impedances in parallel,

$$\frac{1}{\mathbf{Z}_T} = \frac{1}{\mathbf{Z}_1} + \frac{1}{\mathbf{Z}_2}$$

If the manipulations used in Chapter 6 to find the total resistance of two parallel resistors are now applied, the following similar equation results:

$$\mathbf{Z}_T = \frac{\mathbf{Z}_1\mathbf{Z}_2}{\mathbf{Z}_1 + \mathbf{Z}_2} \quad (15.20)$$

For N parallel equal impedances ($\mathbf{Z}_1$), the total impedance is determined by

$$\mathbf{Z}_T = \frac{\mathbf{Z}_1}{N} \quad (15.21)$$

For three parallel impedances,

$$\mathbf{Z}_T = \frac{\mathbf{Z}_1\mathbf{Z}_2\mathbf{Z}_3}{\mathbf{Z}_1\mathbf{Z}_2 + \mathbf{Z}_2\mathbf{Z}_3 + \mathbf{Z}_1\mathbf{Z}_3} \quad (15.22)$$

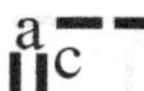

As pointed out in the introduction to this section, conductance is the reciprocal of resistance, and

$$\mathbf{Y}_R = \frac{1}{\mathbf{Z}_R} = \frac{1}{R \angle 0^\circ} = G \angle 0^\circ \tag{15.23}$$

The reciprocal of reactance ($1/X$) is called **susceptance** and is a measure of how *susceptible* an element is to the passage of current through it. Susceptance is also measured in *siemens* and is represented by the capital letter B.

For the inductor,

$$\mathbf{Y}_L = \frac{1}{\mathbf{Z}_L} = \frac{1}{X_L \angle 90^\circ} = \frac{1}{X_L} \angle -90^\circ \tag{15.24}$$

Defining

$$B_L = \frac{1}{X_L} \quad \text{(siemens, S)} \tag{15.25}$$

we have

$$\mathbf{Y}_L = B_L \angle -90^\circ \tag{15.26}$$

Note that for inductance, an increase in frequency or inductance will result in a decrease in susceptance or, correspondingly, in admittance.

For the capacitor,

$$\mathbf{Y}_C = \frac{1}{\mathbf{Z}_C} = \frac{1}{X_C \angle -90^\circ} = \frac{1}{X_C} \angle 90^\circ \tag{15.27}$$

Defining

$$B_C = \frac{1}{X_C} \quad \text{(siemens, S)} \tag{15.28}$$

we have

$$\mathbf{Y}_C = B_C \angle 90^\circ \tag{15.29}$$

For the capacitor, therefore, an increase in frequency or capacitance will result in an increase in its susceptibility.

For parallel ac circuits, the **admittance diagram** is used with the three admittances, represented as shown in Fig. 15.59.

Note in Fig. 15.59 that the conductance (like resistance) is on the positive real axis, whereas inductive and capacitive susceptances are in direct opposition on the imaginary axis.

For any configuration (series, parallel, series-parallel, and so on), the angle associated with the total admittance is the angle by which the source current leads the applied voltage. For inductive networks, θ_T is negative, whereas for capacitive networks, θ_T is positive.

For parallel ac networks, the components of the configuration and the desired quantities determine whether to use an impedance or admittance approach. If the total impedance is requested, the most direct route may be to use impedance parameters. However, sometimes using admittance parameters can also be very efficient, as demonstrated in some of the examples in the rest of the text. In general, use the approach with which you are more comfortable. Naturally, if the format of the desired quantity is spelled out, it is usually best to work with those parameters.

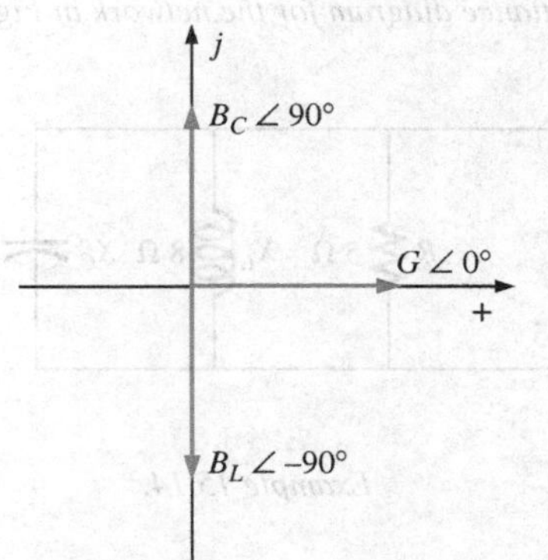

FIG. 15.59
Admittance diagram.

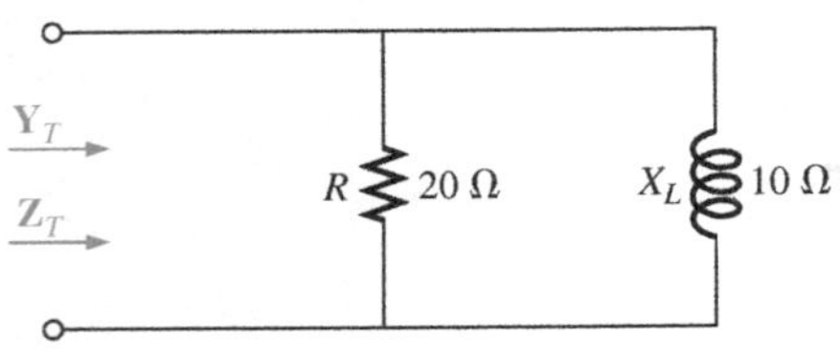

FIG. 15.60
Example 15.13.

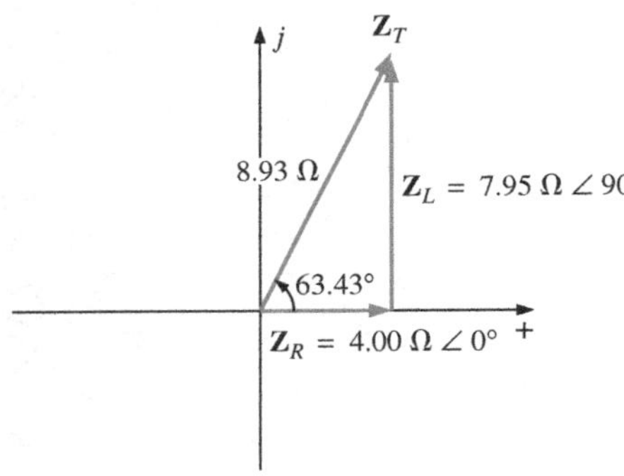

FIG. 15.61
Impedance diagram for the network in Fig. 15.60.

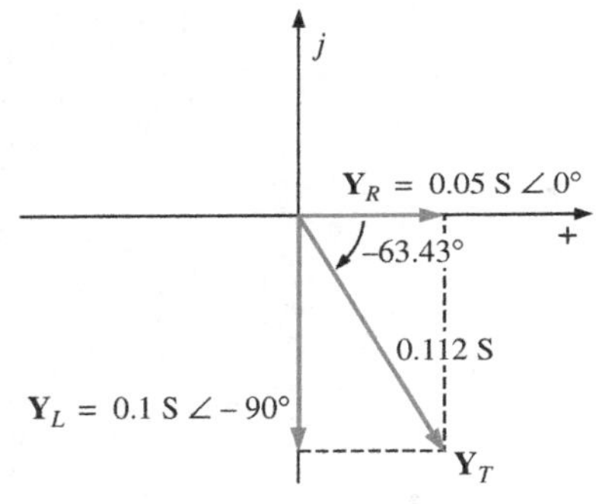

FIG. 15.62
Admittance diagram for the network in Fig. 15.60.

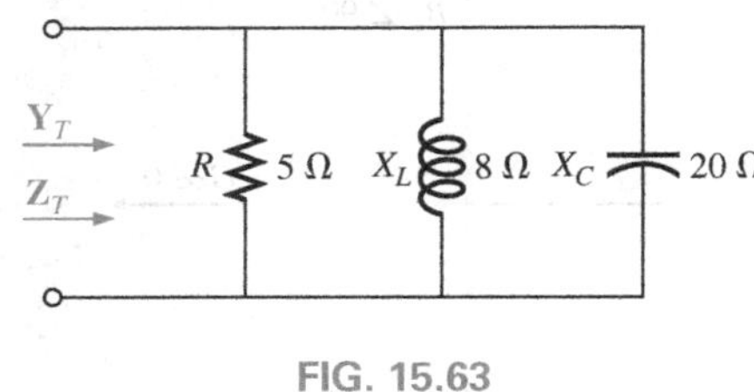

FIG. 15.63
Example 15.14.

EXAMPLE 15.13 For the network in Fig. 15.60:

a. Calculate the input impedance.
b. Draw the impedance diagram.
c. Find the admittance of each parallel branch.
d. Determine the input admittance and draw the admittance diagram.

Solutions:

a. $$\mathbf{Z}_T = \frac{\mathbf{Z}_R\mathbf{Z}_L}{\mathbf{Z}_R + \mathbf{Z}_L} = \frac{(20\ \Omega\ \angle 0°)(10\ \Omega\ \angle 90°)}{20\ \Omega + j\ 10\ \Omega}$$
$$= \frac{200\ \Omega\ \angle 90°}{22.361\ \angle 26.57°} = \mathbf{8.93\ \Omega\ \angle 63.43°}$$
$$= \mathbf{4.00\ \Omega + j\,7.95\ \Omega} = R_T + jX_{L_T}$$

b. The impedance diagram appears in Fig. 15.61.

c. $$\mathbf{Y}_R = G\ \angle 0° = \frac{1}{R}\ \angle 0° = \frac{1}{20\ \Omega}\ \angle 0° = \mathbf{0.05\ S\ \angle 0°}$$
$$= \mathbf{0.05\ S + j\,0}$$
$$\mathbf{Y}_L = B_L\ \angle -90° = \frac{1}{X_L}\ \angle -90° = \frac{1}{10\ \Omega}\ \angle -90°$$
$$= \mathbf{0.1\ S\ \angle -90° = 0 - j\,0.1\ S}$$

d. $$\mathbf{Y}_T = \mathbf{Y}_R + \mathbf{Y}_L = (0.05\ \text{S} + j0) + (0 - j0.1\ \text{S})$$
$$= \mathbf{0.05\ S - j\,0.1\ S} = G - jB_L$$

The admittance diagram appears in Fig. 15.62.

EXAMPLE 15.14 Repeat Example 15.13 for the parallel network in Fig. 15.63.

Solutions:

a. $$\mathbf{Z}_T = \frac{1}{\dfrac{1}{\mathbf{Z}_R} + \dfrac{1}{\mathbf{Z}_L} + \dfrac{1}{\mathbf{Z}_C}}$$
$$= \frac{1}{\dfrac{1}{5\ \Omega\ \angle 0°} + \dfrac{1}{8\ \Omega\ \angle 90°} + \dfrac{1}{20\ \Omega\ \angle -90°}}$$
$$= \frac{1}{0.2\ \text{S}\ \angle 0° + 0.125\ \text{S}\ \angle -90° + 0.05\ \text{S}\ \angle 90°}$$
$$= \frac{1}{0.2\ \text{S} - j\ 0.075\ \text{S}} = \frac{1}{0.2136\ \text{S}\ \angle -20.56°}$$
$$= \mathbf{4.68\ \Omega\ \angle 20.56°}$$

or

$$\mathbf{Z}_T = \frac{\mathbf{Z}_R\mathbf{Z}_L\mathbf{Z}_C}{\mathbf{Z}_R\mathbf{Z}_L + \mathbf{Z}_L\mathbf{Z}_C + \mathbf{Z}_R\mathbf{Z}_C}$$
$$= \frac{(5\ \Omega\ \angle 0°)(8\ \Omega\ \angle 90°)(20\ \Omega\ \angle -90°)}{(5\ \Omega\ \angle 0°)(8\ \Omega\ \angle 90°) + (8\ \Omega\ \angle 90°)(20\ \Omega\ \angle -90°) + (5\ \Omega\ \angle 0°)(20\ \Omega\ \angle -90°)}$$
$$= \frac{800\ \Omega\ \angle 0°}{40\ \angle 90° + 160\ \angle 0° + 100\ \angle -90°}$$

$$= \frac{800\ \Omega}{160 + j\,40 - j\,100} = \frac{800\ \Omega}{160 - j\,60}$$

$$= \frac{800\ \Omega}{170.88\ \angle -20.56^\circ}$$

$$= \mathbf{4.68\ \Omega\ \angle 20.56^\circ = 4.38\ \Omega + j\,1.64\ \Omega}$$

b. The impedance diagram appears in Fig. 15.64.

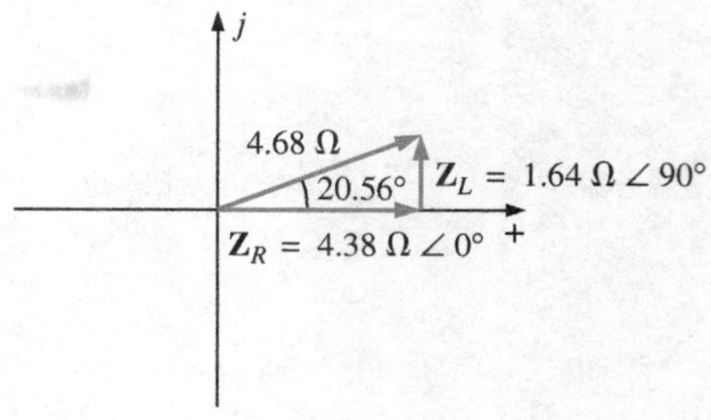

FIG. 15.64
Impedance diagram for the network in Fig. 15.63.

c. $\mathbf{Y}_R = G\ \angle 0^\circ = \frac{1}{R}\angle 0^\circ = \frac{1}{5\ \Omega}\angle 0^\circ$
$= \mathbf{0.2\ S\ \angle 0^\circ = 0.2\ S} + j\,\mathbf{0}$

$\mathbf{Y}_L = B_L\ \angle -90^\circ = \frac{1}{X_L}\angle -90^\circ = \frac{1}{8\ \Omega}\angle -90^\circ$
$= \mathbf{0.125\ S\ \angle -90^\circ = 0} - j\,\mathbf{0.125\ S}$

$\mathbf{Y}_C = B_C\ \angle 90^\circ = \frac{1}{X_C}\angle 90^\circ = \frac{1}{20\ \Omega}\angle 90^\circ$
$= \mathbf{0.050\ S\ \angle +90^\circ = 0} + j\,\mathbf{0.050\ S}$

d. $\mathbf{Y}_T = \mathbf{Y}_R + \mathbf{Y}_L + \mathbf{Y}_C$
$= (0.2\ \text{S} + j\,0) + (0 - j\,0.125\ \text{S}) + (0 + j\,0.050\ \text{S})$
$= 0.2\ \text{S} - j\,0.075\ \text{S} = \mathbf{0.214\ S\ \angle -20.56^\circ}$

The admittance diagram appears in Fig. 15.65.

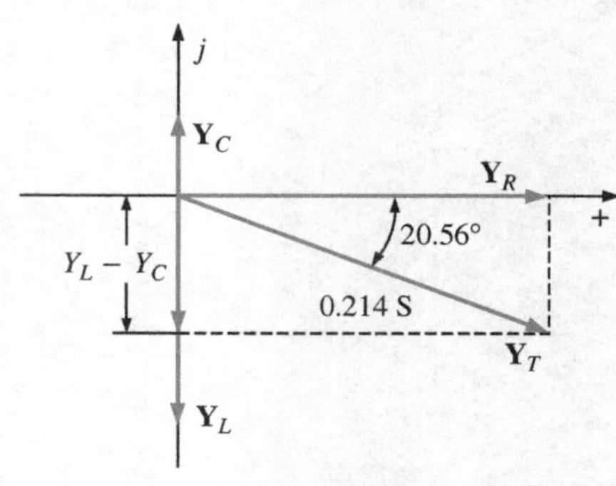

FIG. 15.65
Admittance diagram for the network in Fig. 15.63.

On many occasions, the inverse relationship $\mathbf{Y}_T = 1/\mathbf{Z}_T$ or $\mathbf{Z}_T = 1/\mathbf{Y}_T$ will require that we divide the number 1 by a complex number having a real and an imaginary part. This division, if not performed in the polar form, requires that we multiply the numerator and denominator by the conjugate of the denominator, as follows:

$$\mathbf{Y}_T = \frac{1}{\mathbf{Z}_T} = \frac{1}{4\ \Omega + j\,6\ \Omega} = \left(\frac{1}{4\ \Omega + j\,6\ \Omega}\right)\left(\frac{(4\ \Omega - j\,6\ \Omega)}{(4\ \Omega - j\,6\ \Omega)}\right)$$

$$= \frac{4 - j\,6}{4^2 + 6^2}$$

and
$$\mathbf{Y}_T = \frac{4}{52}\text{S} - j\frac{6}{52}\text{S}$$

To avoid this laborious task each time we want to find the reciprocal of a complex number in rectangular form, a format can be developed using the following complex number, which is symbolic of any impedance or admittance in the first or fourth quadrant:

$$\frac{1}{a_1 \pm jb_1} = \left(\frac{1}{a_1 \pm jb_1}\right)\left(\frac{a_1 \mp jb_1}{a_1 \mp jb_1}\right) = \frac{a_1 \mp jb_1}{a_1^2 + b_1^2}$$

or
$$\boxed{\frac{1}{a_1 \pm jb_1} = \frac{a_1}{a_1^2 + b_1^2} \mp j\frac{b_1}{a_1^2 + b_1^2}} \qquad \textbf{(15.30)}$$

Note that the denominator is simply the sum of the squares of each term. The sign is inverted between the real and imaginary parts. A few examples will develop some familiarity with the use of this equation.

EXAMPLE 15.15 Find the admittance of each set of series elements in Fig. 15.66.

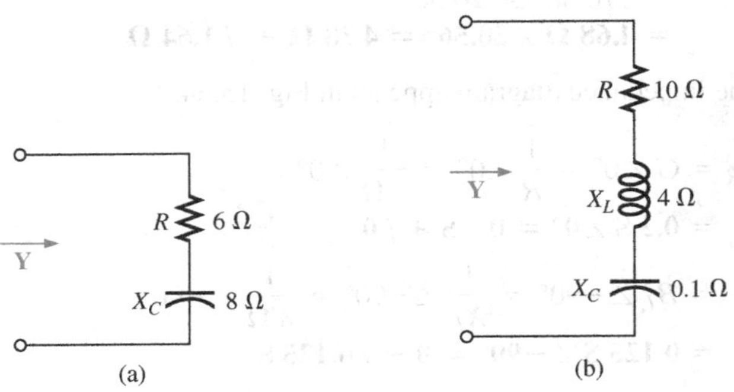

FIG. 15.66
Example 15.15.

Solutions:

a. $\mathbf{Z} = R - jX_C = 6\ \Omega - j\,8\ \Omega$

Eq. (15.30):

$$\mathbf{Y} = \frac{1}{6\ \Omega - j\,8\ \Omega} = \frac{6\ \Omega}{(6\ \Omega)^2 + (8\ \Omega)^2} + j\frac{8\ \Omega}{(6\ \Omega)^2 + (8\ \Omega)^2}$$
$$= \mathbf{\frac{6}{100}\,S} + j\,\mathbf{\frac{8}{100}\,S}$$

b. $\mathbf{Z} = 10\ \Omega + j\,4\ \Omega + (-j\,0.1\ \Omega) = 10\ \Omega + j\,3.9\ \Omega$

Eq. (15.30):

$$\mathbf{Y} = \frac{1}{\mathbf{Z}} = \frac{1}{10\ \Omega + j\,3.9\ \Omega} = \frac{10}{(10)^2 + (3.9)^2} - j\frac{3.9}{(10)^2 + (3.9)^2}$$
$$= \frac{10}{115.21} - j\frac{3.9}{115.21} = \mathbf{0.087\ S - j\,0.034\ S}$$

15.8 PARALLEL ac NETWORKS

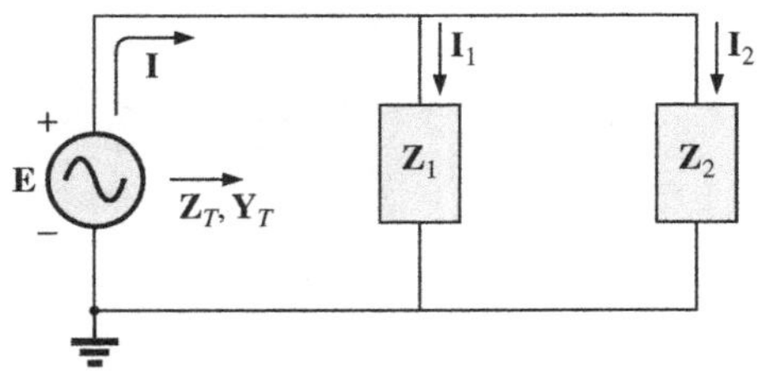

FIG. 15.67
Parallel ac network.

For the representative parallel ac network in Fig. 15.67, the total impedance or admittance is determined as described in the previous section, and the source current is determined by Ohm's law as follows:

$$\mathbf{I} = \frac{\mathbf{E}}{\mathbf{Z}_T} = \mathbf{E}\mathbf{Y}_T \qquad \textbf{(15.31)}$$

Since the voltage is the same across parallel elements, the current through each branch can then be found through another application of Ohm's law:

$$\mathbf{I}_1 = \frac{\mathbf{E}}{\mathbf{Z}_1} = \mathbf{E}\mathbf{Y}_1 \qquad \textbf{(15.32)}$$

$$\mathbf{I}_2 = \frac{\mathbf{E}}{\mathbf{Z}_2} = \mathbf{E}\mathbf{Y}_2 \tag{15.33}$$

Kirchhoff's current law can then be applied in the same manner as used for dc networks. However, keep in mind that we are now dealing with the algebraic manipulation of quantities that have both magnitude and direction. We have

$$\mathbf{I} - \mathbf{I}_1 - \mathbf{I}_2 = 0$$

or

$$\mathbf{I} = \mathbf{I}_1 + \mathbf{I}_2 \tag{15.34}$$

The power to the network can be determined by

$$P = EI \cos \theta_T \tag{15.35}$$

where θ_T is the phase angle between **E** and **I.**

Let us now look at a few examples carried out in great detail for the first exposure.

R-L

Refer to Fig. 15.68.

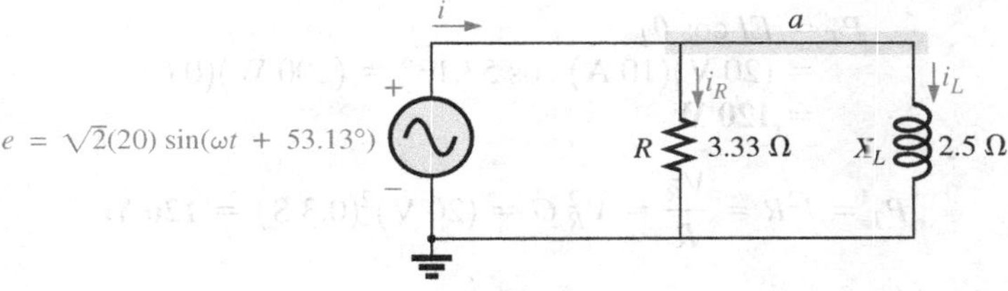

FIG. 15.68
Parallel R-L network.

Phasor Notation As shown in Fig. 15.69.

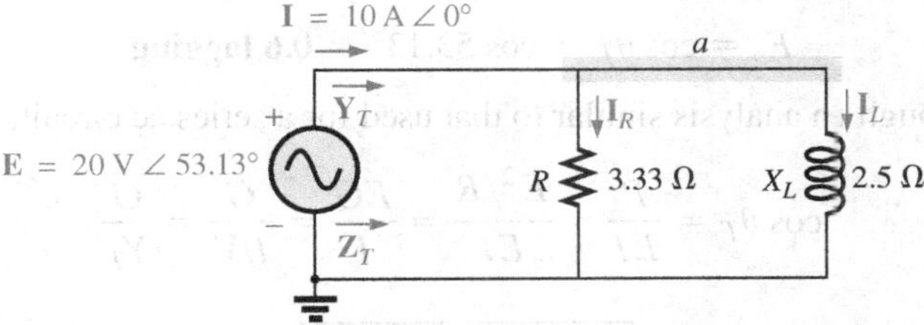

FIG. 15.69
Applying phasor notation to the network in Fig. 15.68.

$$\begin{aligned}
\mathbf{Y}_T &= \mathbf{Y}_R + \mathbf{Y}_L \\
&= G \angle 0° + B_L \angle -90° = \frac{1}{3.33\ \Omega} \angle 0° + \frac{1}{2.5\ \Omega} \angle -90° \\
&= 0.3\ \text{S} \angle 0° + 0.4\ \text{S} \angle -90° = 0.3\ \text{S} - j\,0.4\ \text{S} \\
&= \mathbf{0.5\ S \angle -53.13°}
\end{aligned}$$

$$\mathbf{Z}_T = \frac{1}{\mathbf{Y}_T} = \frac{1}{0.5\ \text{S} \angle -53.13°} = \mathbf{2\ \Omega \angle 53.13°}$$

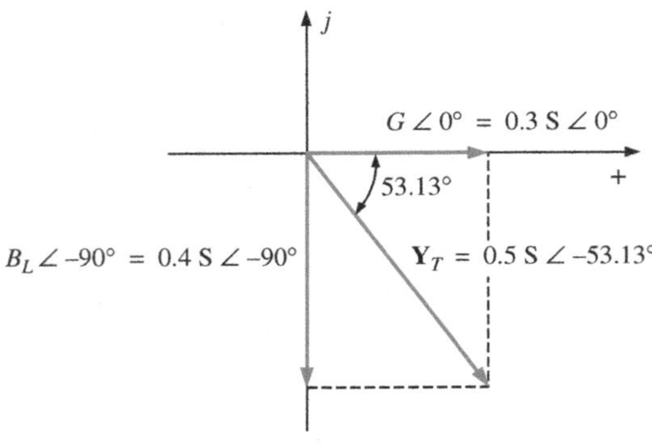

FIG. 15.70
Admittance diagram for the parallel R-L network in Fig. 15.68.

Admittance diagram: As shown in Fig. 15.70.

$$\mathbf{I} = \frac{\mathbf{E}}{\mathbf{Z}_T} = \mathbf{E}\mathbf{Y}_T = (20 \text{ V} \angle 53.13^\circ)(0.5 \text{ S} \angle -53.13^\circ) = \mathbf{10 \text{ A} \angle 0^\circ}$$

$$\mathbf{I}_R = \frac{E \angle \theta}{R \angle 0^\circ} = (E \angle \theta)(G \angle 0^\circ) = (20 \text{ V} \angle 53.13^\circ)(0.3 \text{ S} \angle 0^\circ) = \mathbf{6 \text{ A} \angle 53.13^\circ}$$

$$\mathbf{I}_L = \frac{E \angle \theta}{X_L \angle 90^\circ} = (E \angle \theta)(B_L \angle -90^\circ) = (20 \text{ V} \angle 53.13^\circ)(0.4 \text{ S} \angle -90^\circ) = \mathbf{8 \text{ A} \angle -36.87^\circ}$$

Kirchhoff's current law: At node *a*,

$$\mathbf{I} - \mathbf{I}_R - \mathbf{I}_L = 0$$

or

$$\mathbf{I} = \mathbf{I}_R + \mathbf{I}_L$$

$$10 \text{ A} \angle 0^\circ = 6 \text{ A} \angle 53.13^\circ + 8 \text{ A} \angle -36.87^\circ$$
$$10 \text{ A} \angle 0^\circ = (3.60 \text{ A} + j\,4.80 \text{ A}) + (6.40 \text{ A} - j\,4.80 \text{ A}) = 10 \text{ A} + j\,0$$

and

$$\mathbf{10 \text{ A} \angle 0^\circ = 10 \text{ A} \angle 0^\circ} \qquad \text{(checks)}$$

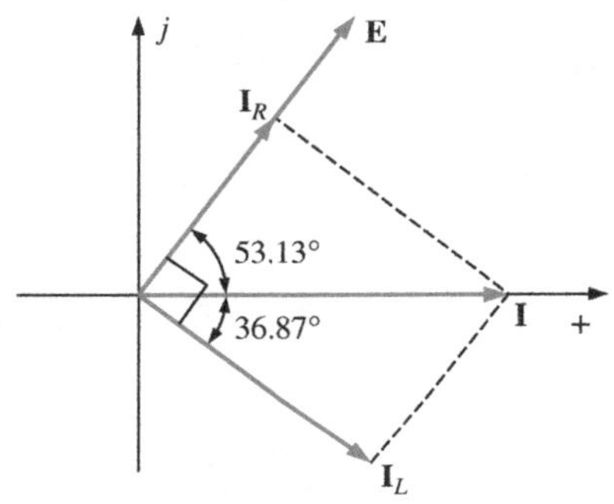

FIG. 15.71
Phasor diagram for the parallel R-L network in Fig. 15.68.

Phasor diagram: The phasor diagram in Fig. 15.71 indicates that the applied voltage **E** is in phase with the current $\mathbf{I}_R$ and leads the current $\mathbf{I}_L$ by 90°.

Power: The total power in watts delivered to the circuit is

$$P_T = EI\cos\theta_T = (20 \text{ V})(10 \text{ A})\cos 53.13^\circ = (200 \text{ W})(0.6) = \mathbf{120 \text{ W}}$$

or

$$P_T = I^2R = \frac{V_R^2}{R} = V_R^2 G = (20 \text{ V})^2(0.3 \text{ S}) = \mathbf{120 \text{ W}}$$

or, finally,

$$P_T = P_R + P_L = EI_R\cos\theta_R + EI_L\cos\theta_L = (20 \text{ V})(6 \text{ A})\cos 0^\circ + (20 \text{ V})(8 \text{ A})\cos 90^\circ = 120 \text{ W} + 0 = \mathbf{120 \text{ W}}$$

Power factor: The power factor of the circuit is

$$F_p = \cos\theta_T = \cos 53.13^\circ = \mathbf{0.6 \text{ lagging}}$$

or, through an analysis similar to that used for a series ac circuit,

$$\cos\theta_T = \frac{P}{EI} = \frac{E^2/R}{EI} = \frac{EG}{I} = \frac{G}{I/V} = \frac{G}{Y_T}$$

and

$$\boxed{F_p = \cos\theta_T = \frac{G}{Y_T}} \qquad \textbf{(15.36)}$$

where G and Y_T are the magnitudes of the total conductance and admittance of the parallel network. For this case,

$$F_p = \cos\theta_T = \frac{0.3 \text{ S}}{0.5 \text{ S}} = \mathbf{0.6 \text{ lagging}}$$

Impedance approach: The current **I** can also be found by first finding the total impedance of the network:

$$\mathbf{Z}_T = \frac{\mathbf{Z}_R\mathbf{Z}_L}{\mathbf{Z}_R + \mathbf{Z}_L} = \frac{(3.33\ \Omega\ \angle 0^\circ)(2.5\ \Omega\ \angle 90^\circ)}{3.33\ \Omega\ \angle 0^\circ + 2.5\ \Omega\ \angle 90^\circ}$$

$$= \frac{8.325\ \angle 90^\circ}{4.164\ \angle 36.87^\circ} = \mathbf{2\ \Omega\ \angle 53.13^\circ}$$

Then, using Ohm's law, we obtain

$$\mathbf{I} = \frac{\mathbf{E}}{\mathbf{Z}_T} = \frac{20\ \text{V}\ \angle 53.13^\circ}{2\ \Omega\ \angle 53.13^\circ} = \mathbf{10\ A\ \angle 0^\circ}$$

R-C

Refer to Fig. 15.72.

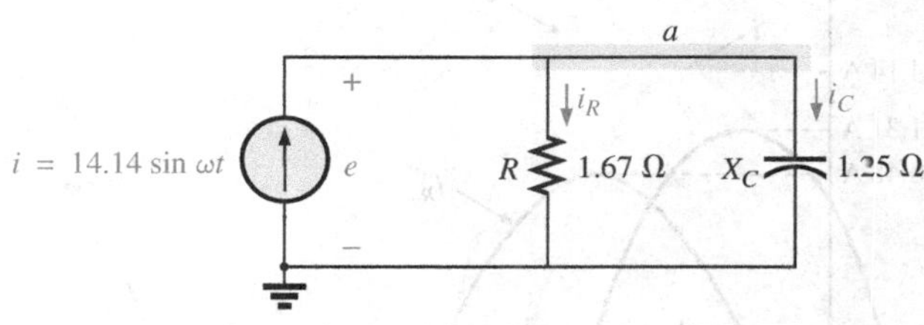

FIG. 15.72
Parallel R-C network.

Phasor Notation As shown in Fig. 15.73.

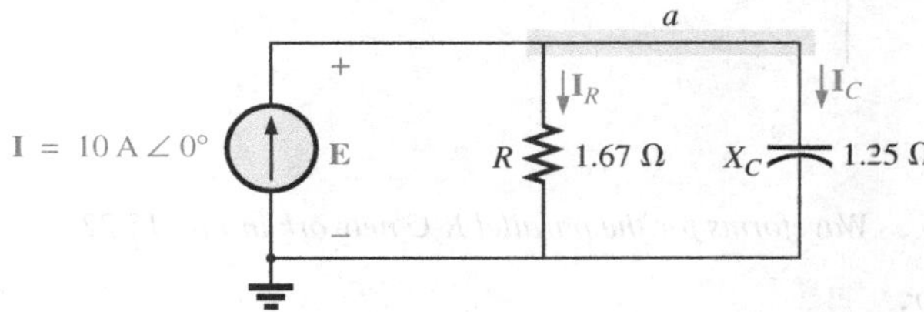

FIG. 15.73
Applying phasor notation to the network in Fig. 15.72.

$$\mathbf{Y}_T = \mathbf{Y}_R + \mathbf{Y}_C = G\ \angle 0^\circ + B_C\ \angle 90^\circ = \frac{1}{1.67\ \Omega}\ \angle 0^\circ + \frac{1}{1.25\ \Omega}\ \angle 90^\circ$$

$$= 0.6\ \text{S}\ \angle 0^\circ + 0.8\ \text{S}\ \angle 90^\circ = 0.6\ \text{S} + j\,0.8\ \text{S} = \mathbf{1.0\ S\ \angle 53.13^\circ}$$

$$\mathbf{Z}_T = \frac{1}{\mathbf{Y}_T} = \frac{1}{1.0\ \text{S}\ \angle 53.13^\circ} = \mathbf{1\ \Omega\ \angle -53.13^\circ}$$

Admittance diagram: As shown in Fig. 15.74.

$$\mathbf{E} = \mathbf{I}\mathbf{Z}_T = \frac{\mathbf{I}}{\mathbf{Y}_T} = \frac{10\ \text{A}\ \angle 0^\circ}{1\ \text{S}\ \angle 53.13^\circ} = \mathbf{10\ V\ \angle -53.13^\circ}$$

$$\mathbf{I}_R = (E\ \angle \theta)(G\ \angle 0^\circ)$$
$$= (10\ \text{V}\ \angle -53.13^\circ)(0.6\ \text{S}\ \angle 0^\circ) = \mathbf{6\ A\ \angle -53.13^\circ}$$

$$\mathbf{I}_C = (E\ \angle \theta)(B_C\ \angle 90^\circ)$$
$$= (10\ \text{V}\ \angle -53.13^\circ)(0.8\ \text{S}\ \angle 90^\circ) = \mathbf{8\ A\ \angle 36.87^\circ}$$

Kirchhoff's current law: At node *a*,

$$\mathbf{I} - \mathbf{I}_R - \mathbf{I}_C = 0$$

or

$$\mathbf{I} = \mathbf{I}_R + \mathbf{I}_C$$

which can also be verified (as for the *R-L* network) through vector algebra.

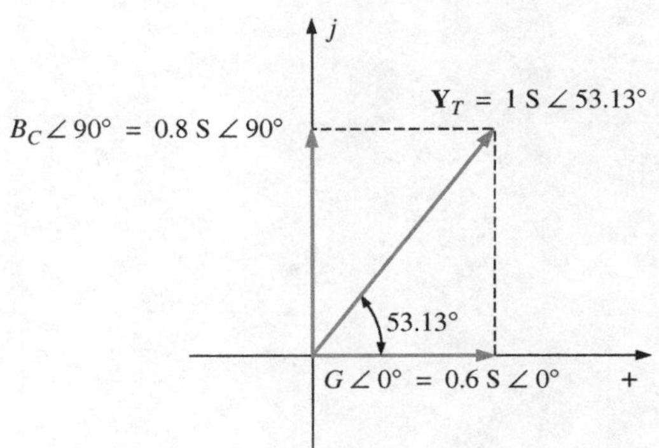

FIG. 15.74
Admittance diagram for the parallel R-C network in Fig. 15.72.

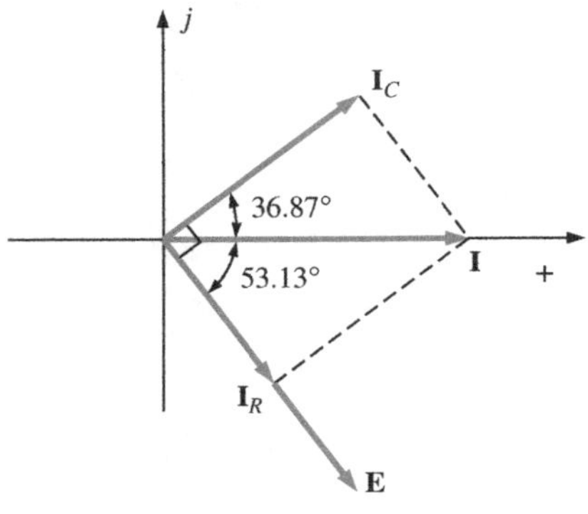

FIG. 15.75
Phasor diagram for the parallel R-C network in Fig. 15.74.

Phasor diagram: The phasor diagram in Fig. 15.75 indicates that **E** is in phase with the current through the resistor $\mathbf{I}_R$ and lags the capacitive current $\mathbf{I}_C$ by 90°.

Time domain:

$$e = \sqrt{2}(10)\sin(\omega t - 53.13^\circ) = \mathbf{14.14\sin(\omega t - 53.13^\circ)}$$
$$i_R = \sqrt{2}(6)\sin(\omega t - 53.13^\circ) = \mathbf{8.48\sin(\omega t - 53.13^\circ)}$$
$$i_C = \sqrt{2}(8)\sin(\omega t + 36.87^\circ) = \mathbf{11.31\sin(\omega t + 36.87^\circ)}$$

A plot of all of the currents and the voltage appears in Fig. 15.76. Note that e and i_R are in phase and e lags i_C by 90°.

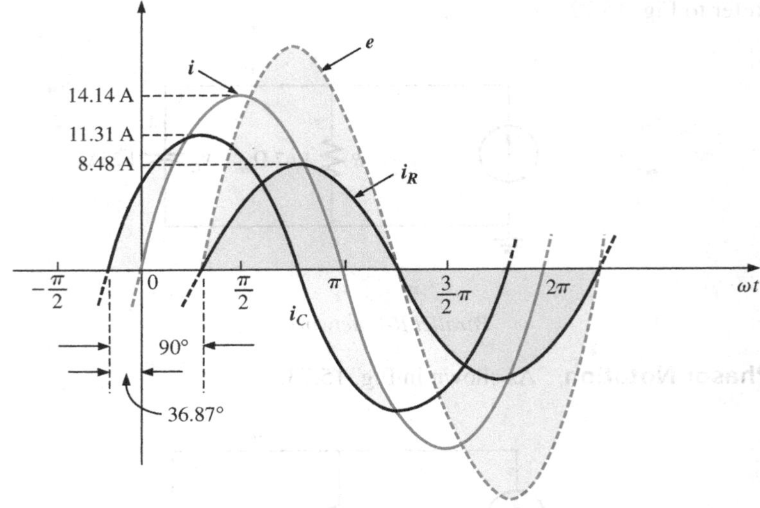

FIG. 15.76
Waveforms for the parallel R-C network in Fig. 15.72.

Power:

$$P_T = EI\cos\theta = (10\text{ V})(10\text{ A})\cos 53.13^\circ = (10)^2(0.6)$$
$$= \mathbf{60\ W}$$

or
$$P_T = E^2G = (10\text{ V})^2(0.6\text{ S}) = \mathbf{60\ W}$$

or, finally,

$$P_T = P_R + P_C = EI_R\cos\theta_R + EI_C\cos\theta_C$$
$$= (10\text{ V})(6\text{ A})\cos 0^\circ + (10\text{ V})(8\text{ A})\cos 90^\circ$$
$$= \mathbf{60\ W}$$

Power factor: The power factor of the circuit is

$$F_p = \cos 53.13^\circ = \mathbf{0.6\ leading}$$

Using Eq. (15.36), we have

$$F_p = \cos\theta_T = \frac{G}{Y_T} = \frac{0.6\text{ S}}{1.0\text{ S}} = \mathbf{0.6\ leading}$$

Impedance approach: The voltage **E** can also be found by first finding the total impedance of the circuit:

$$\mathbf{Z}_T = \frac{\mathbf{Z}_R\mathbf{Z}_C}{\mathbf{Z}_R + \mathbf{Z}_C} = \frac{(1.67\ \Omega\ \angle 0^\circ)(1.25\ \Omega\ \angle -90^\circ)}{1.67\ \Omega\ \angle 0^\circ + 1.25\ \Omega\ \angle -90^\circ}$$
$$= \frac{2.09\ \angle -90^\circ}{2.09\ \angle -36.81^\circ} = \mathbf{1\ \Omega\ \angle -53.19^\circ}$$

and then, using Ohm's law, we find

$$\mathbf{E} = \mathbf{IZ}_T = (10\text{ A }\angle 0°)(1\ \Omega\ \angle -53.19°) = \mathbf{10\ V\ \angle -53.19°}$$

R-L-C

Refer to Fig. 15.77.

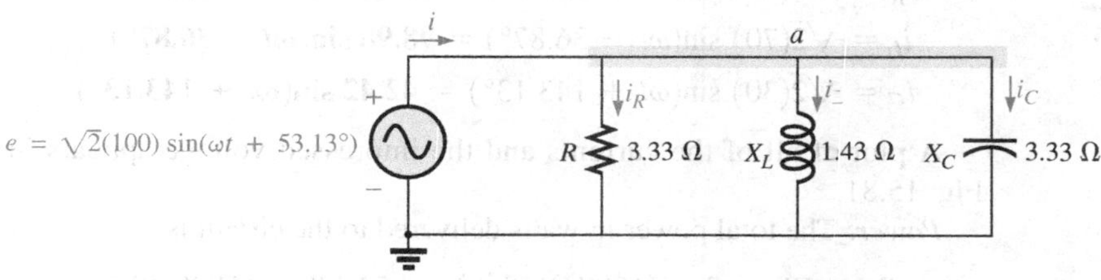

FIG. 15.77
Parallel R-L-C ac network.

Phasor notation: As shown in Fig. 15.78.

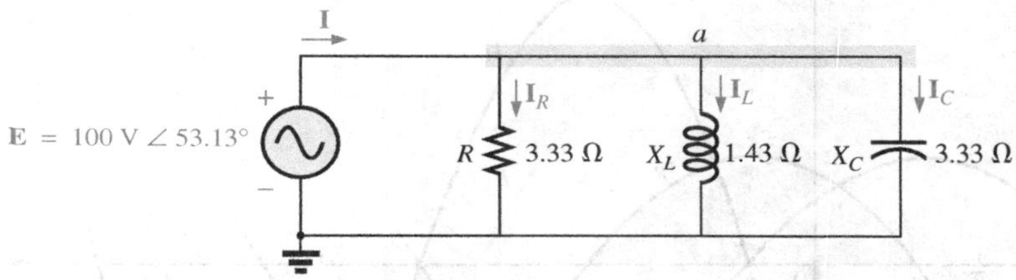

FIG. 15.78
Applying phasor notation to the network in Fig. 15.77.

$$\begin{aligned}\mathbf{Y}_T &= \mathbf{Y}_R + \mathbf{Y}_L + \mathbf{Y}_C = G\angle 0° + B_L\angle -90° + B_C\angle 90° \\ &= \frac{1}{3.33\ \Omega}\angle 0° + \frac{1}{1.43\ \Omega}\angle -90° + \frac{1}{3.33\ \Omega}\angle 90° \\ &= 0.3\text{ S}\angle 0° + 0.7\text{ S}\angle -90° + 0.3\text{ S}\angle 90° \\ &= 0.3\text{ S} - j\,0.7\text{ S} + j\,0.3\text{ S} \\ &= 0.3\text{ S} - j\,0.4\text{ S} = \mathbf{0.5\ S\ \angle -53.13°}\end{aligned}$$

$$\mathbf{Z}_T = \frac{1}{\mathbf{Y}_T} = \frac{1}{0.5\text{ S}\angle -53.13°} = \mathbf{2\ \Omega\ \angle 53.13°}$$

Admittance diagram: As shown in Fig. 15.79.

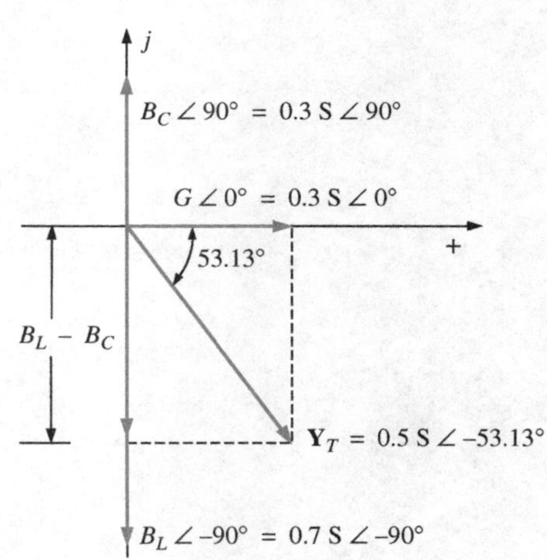

FIG. 15.79
Admittance diagram for the parallel R-L-C network in Fig. 15.77.

$$\mathbf{I} = \frac{\mathbf{E}}{\mathbf{Z}_T} = \mathbf{EY}_T = (100\text{ V}\angle 53.13°)(0.5\text{ S}\angle -53.13°) = \mathbf{50\ A\ \angle 0°}$$

$$\begin{aligned}\mathbf{I}_R &= (E\angle\theta)(G\angle 0°) \\ &= (100\text{ V}\angle 53.13°)(0.3\text{ S}\angle 0°) = \mathbf{30\ A\ \angle 53.13°} \\ \mathbf{I}_L &= (E\angle\theta)(B_L\angle -90°) \\ &= (100\text{ V}\angle 53.13°)(0.7\text{ S}\angle -90°) = \mathbf{70\ A\ \angle -36.87°} \\ \mathbf{I}_C &= (E\angle\theta)(B_C\angle 90°) \\ &= (100\text{ V}\angle 53.13°)(0.3\text{ S}\angle +90°) = \mathbf{30\ A\ \angle 143.13°}\end{aligned}$$

Kirchhoff's current law: At node *a*,

$$\mathbf{I} - \mathbf{I}_R - \mathbf{I}_L - \mathbf{I}_C = 0$$

or

$$\mathbf{I} = \mathbf{I}_R + \mathbf{I}_L + \mathbf{I}_C$$

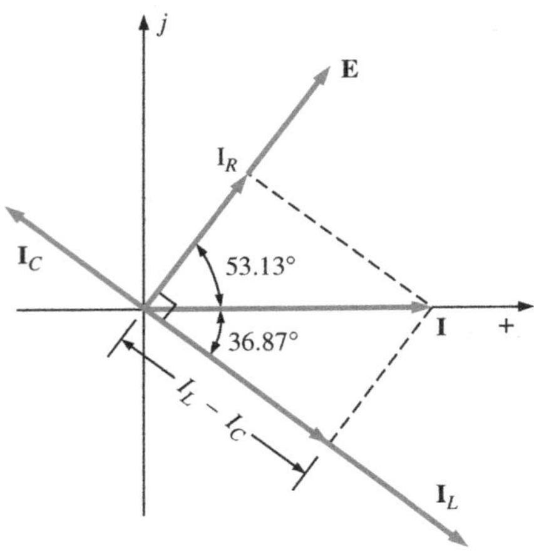

FIG. 15.80
Phasor diagram for the parallel R-L-C network in Fig. 15.77.

Phasor diagram: The phasor diagram in Fig. 15.80 indicates that the impressed voltage **E** is in phase with the current $\mathbf{I}_R$ through the resistor, leads the current $\mathbf{I}_L$ through the inductor by 90°, and lags the current $\mathbf{I}_C$ of the capacitor by 90°.

Time domain:

$$i = \sqrt{2}(50) \sin \omega t = \mathbf{70.70 \sin \omega t}$$
$$i_R = \sqrt{2}(30) \sin(\omega t + 53.13°) = \mathbf{42.42 \sin(\omega t + 53.13°)}$$
$$i_L = \sqrt{2}(70) \sin(\omega t - 36.87°) = \mathbf{98.98 \sin(\omega t - 36.87°)}$$
$$i_C = \sqrt{2}(30) \sin(\omega t + 143.13°) = \mathbf{42.42 \sin(\omega t + 143.13°)}$$

A plot of all of the currents and the impressed voltage appears in Fig. 15.81.

Power: The total power in watts delivered to the circuit is

$$P_T = EI \cos \theta = (100 \text{ V})(50 \text{ A}) \cos 53.13° = (5000)(0.6)$$
$$= \mathbf{3000 \text{ W}}$$

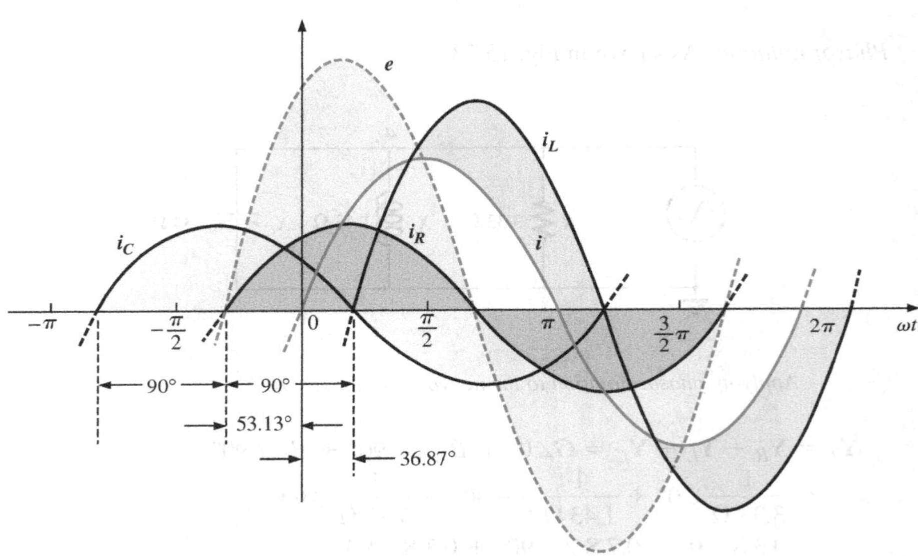

FIG. 15.81
Waveforms for the parallel R-L-C network in Fig. 15.77.

or $$P_T = E^2G = (100 \text{ V})^2(0.3 \text{ S}) = \mathbf{3000 \text{ W}}$$

or, finally,

$$\begin{aligned} P_T &= P_R + P_L + P_C \\ &= EI_R \cos \theta_R + EI_L \cos \theta_L + EI_C \cos \theta_C \\ &= (100 \text{ V})(30 \text{ A}) \cos 0° + (100 \text{ V})(70 \text{ A}) \cos 90° \\ &\qquad + (100 \text{ V})(30 \text{ A}) \cos 90° \\ &= 3000 \text{ W} + 0 + 0 \\ &= \mathbf{3000 \text{ W}} \end{aligned}$$

Power factor: The power factor of the circuit is

$$F_p = \cos \theta_T = \cos 53.13° = \mathbf{0.6 \text{ lagging}}$$

Using Eq. (15.36), we obtain

$$F_p = \cos \theta_T = \frac{G}{Y_T} = \frac{0.3 \text{ S}}{0.5 \text{ S}} = \mathbf{0.6 \text{ lagging}}$$

Impedance approach: The input current **I** can also be determined by first finding the total impedance in the following manner:

$$\mathbf{Z}_T = \frac{\mathbf{Z}_R\mathbf{Z}_L\mathbf{Z}_C}{\mathbf{Z}_R\mathbf{Z}_L + \mathbf{Z}_L\mathbf{Z}_C + \mathbf{Z}_R\mathbf{Z}_C} = \mathbf{2\ \Omega\ \angle 53.13^\circ}$$

and, applying Ohm's law, we obtain

$$\mathbf{I} = \frac{\mathbf{E}}{\mathbf{Z}_T} = \frac{100\text{ V}\ \angle 53.13^\circ}{2\ \Omega\ \angle 53.13^\circ} = \mathbf{50\ A\ \angle 0^\circ}$$

15.9 CURRENT DIVIDER RULE

The basic format for the **current divider rule** in ac circuits is exactly the same as that for dc circuits; that is, for two parallel branches with impedances $\mathbf{Z}_1$ and $\mathbf{Z}_2$ as shown in Fig. 15.82,

$$\mathbf{I}_1 = \frac{\mathbf{Z}_2\mathbf{I}_T}{\mathbf{Z}_1 + \mathbf{Z}_2} \quad \text{or} \quad \mathbf{I}_2 = \frac{\mathbf{Z}_1\mathbf{I}_T}{\mathbf{Z}_1 + \mathbf{Z}_2} \tag{15.37}$$

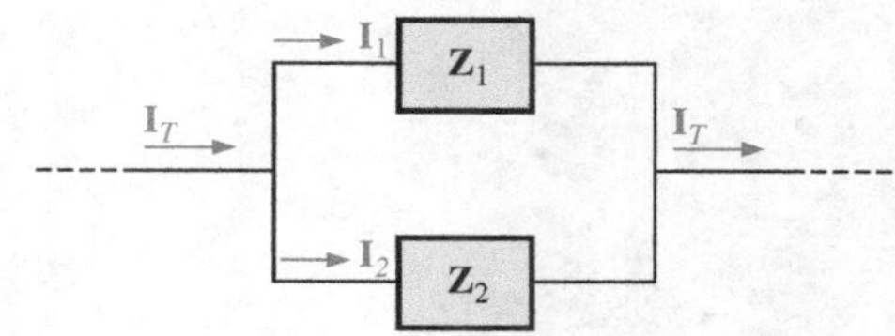

FIG. 15.82
Applying the current divider rule.

EXAMPLE 15.16 Using the current divider rule, find the current through each impedance in Fig. 15.83.

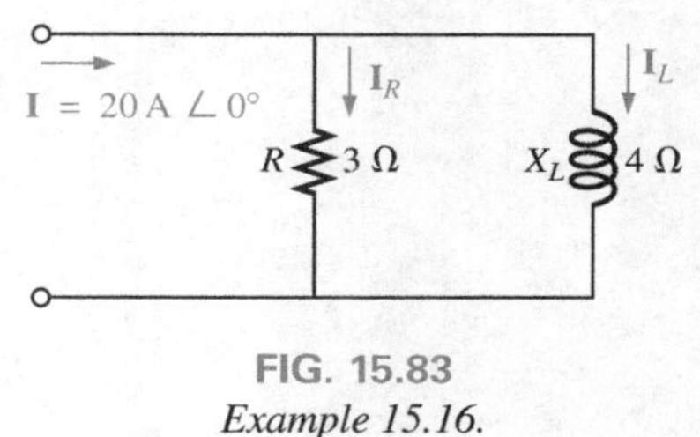

FIG. 15.83
Example 15.16.

Solution:

$$\mathbf{I}_R = \frac{\mathbf{Z}_L\mathbf{I}_T}{\mathbf{Z}_R + \mathbf{Z}_L} = \frac{(4\ \Omega\ \angle 90^\circ)(20\text{ A}\ \angle 0^\circ)}{3\ \Omega\ \angle 0^\circ + 4\ \Omega\ \angle 90^\circ} = \frac{80\text{ A}\ \angle 90^\circ}{5\ \angle 53.13^\circ}$$
$$= \mathbf{16\ A\ \angle 36.87^\circ}$$

$$\mathbf{I}_L = \frac{\mathbf{Z}_R\mathbf{I}_T}{\mathbf{Z}_R + \mathbf{Z}_L} = \frac{(3\ \Omega\ \angle 0^\circ)(20\text{ A}\ \angle 0^\circ)}{5\ \Omega\ \angle 53.13^\circ} = \frac{60\text{ A}\ \angle 0^\circ}{5\ \angle 53.13^\circ}$$
$$= \mathbf{12\ A\ \angle -53.13^\circ}$$

EXAMPLE 15.17 Using the current divider rule, find the current through each parallel branch in Fig. 15.84.

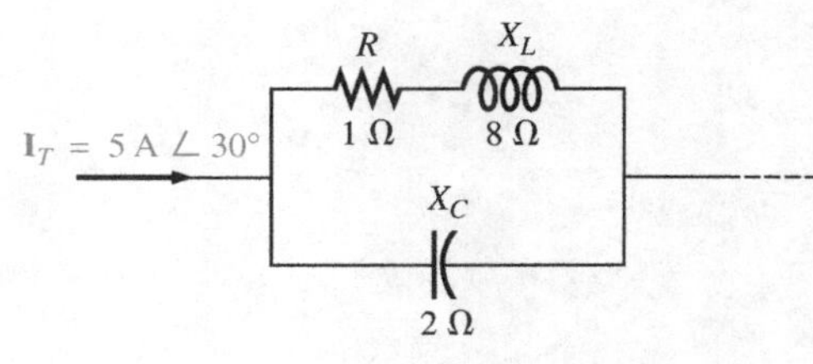

FIG. 15.84
Example 15.17.

Solution:

$$\mathbf{I}_{R\text{-}L} = \frac{\mathbf{Z}_C\mathbf{I}_T}{\mathbf{Z}_C + \mathbf{Z}_{R\text{-}L}} = \frac{(2\ \Omega\ \angle -90^\circ)(5\text{ A}\angle 30^\circ)}{-j\,2\ \Omega + 1\ \Omega + j\,8\ \Omega} = \frac{10\text{ A}\ \angle -60^\circ}{1 + j6}$$
$$= \frac{10\text{ A}\ \angle -60^\circ}{6.083\ \angle 80.54^\circ} \cong \mathbf{1.64\ A\ \angle -140.54^\circ}$$

$$\mathbf{I}_C = \frac{\mathbf{Z}_{R\text{-}L}\mathbf{I}_T}{\mathbf{Z}_{R\text{-}L} + \mathbf{Z}_C} = \frac{(1\ \Omega + j\,8\ \Omega)(5\text{ A}\ \angle 30^\circ)}{6.08\ \Omega\ \angle 80.54^\circ}$$
$$= \frac{(8.06\ \angle 82.87^\circ)(5\text{ A}\ \angle 30^\circ)}{6.08\ \angle 80.54^\circ} = \frac{40.30\text{ A}\ \angle 112.87^\circ}{6.083\ \angle 80.54^\circ}$$
$$= \mathbf{6.63\ A\ \angle 32.33^\circ}$$

15.10 FREQUENCY RESPONSE OF PARALLEL ELEMENTS

Recall that for elements in series, the total impedance is the direct sum of the impedances of each element, and the largest real or imaginary component has the most impact on the total impedance. For parallel elements, it is important to remember that *the smallest parallel resistor or the smallest parallel reactance will have the most impact on the real or imaginary component, respectively, of the total impedance.*

In Fig. 15.85, the frequency response has been included for each element of a parallel *R-L-C* combination. At very low frequencies, the impedance of the coil will be less than that of the resistor or capacitor, resulting in an inductive network in which the reactance of the inductor will have the most impact on the total impedance. As the frequency increases, the impedance of the inductor will increase, while the impedance of the capacitor will decrease. Depending on the components chosen, it is possible that the reactance of the capacitor will drop to a point where it will equal the impedance of the coil before either one reaches the resistance level.

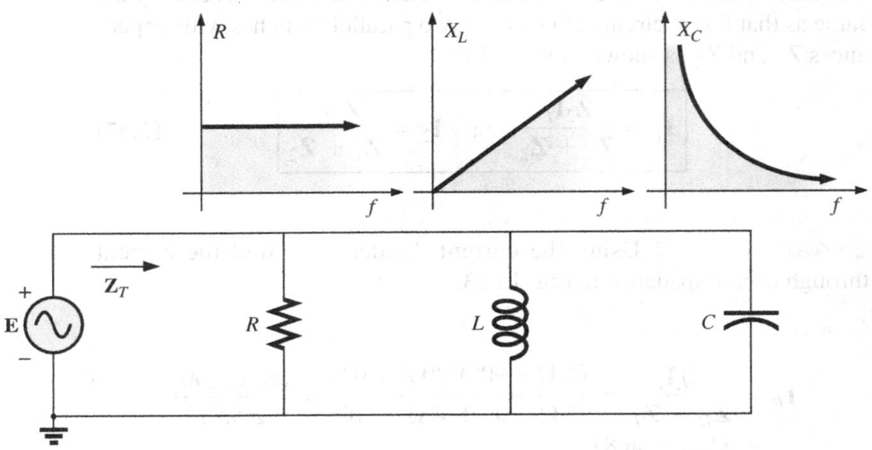

FIG. 15.85
Frequency response for parallel R-L-C elements.

Therefore, it is impossible to make too many broad statements about the effect of each element as the frequency increases. In general, however, for very low frequencies, we can assume that a parallel *R-L-C* network will be inductive as described above, and at very high frequencies it will be capacitive since X_C will drop to very low levels. In between, a point will result at which X_L will equal X_C and where X_L or X_C will equal *R*. The frequencies at which these events occur, however, depend on the elements chosen and the frequency range of interest. In general, however, keep in mind that the smaller the resistance or reactance, the greater is its impact on the total impedance of a parallel system.

Let us now note the impact of frequency on the total impedance and inductive current for the parallel *R-L* network in Fig. 15.86 for a frequency range of zero through 40 kHz.

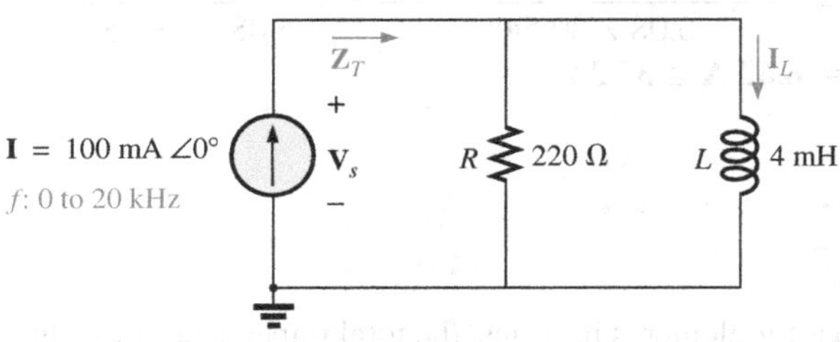

FIG. 15.86
Determining the frequency response of a parallel R-L network.

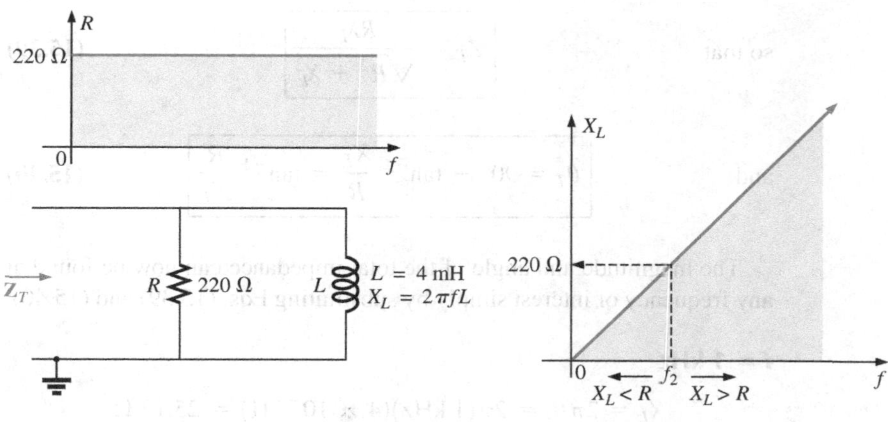

FIG. 15.87
The frequency response of the individual elements of a parallel R-L network.

$\mathbf{Z}_T$ Before getting into specifics, let us first develop a "sense" for the impact of frequency on the network in Fig. 15.86 by noting the impedance-versus-frequency curves of the individual elements, as shown in Fig. 15.87. The fact that the elements are now in parallel requires that we consider their characteristics in a different manner than occurred for the series *R-C* circuit in Section 15.5. Recall that for parallel elements, the element with the smallest impedance will have the greatest impact on the total impedance at that frequency. In Fig. 15.87, for example, X_L is very small at low frequencies compared to R, establishing X_L as the predominant factor in this frequency range. In other words, at low frequencies the network will be primarily inductive, and the angle associated with the total impedance will be close to 90°, as with a pure inductor. As the frequency increases, X_L increases until it equals the impedance of the resistor (220 Ω). The frequency at which this situation occurs can be determined in the following manner:

$$X_L = 2\pi f_2 L = R$$

and

$$f_2 = \frac{R}{2\pi L} \tag{15.38}$$

which for the network in Fig. 15.86 is

$$f_2 = \frac{R}{2\pi L} = \frac{220\ \Omega}{2\pi(4 \times 10^{-3}\ \text{H})}$$
$$\cong \mathbf{8.75\ kHz}$$

which falls within the frequency range of interest.

For frequencies less than f_2, $X_L < R$, and for frequencies greater than f_2, $X_L > R$, as shown in Fig. 15.87. A general equation for the total impedance in vector form can be developed in the following manner:

$$\mathbf{Z}_T = \frac{\mathbf{Z}_R \mathbf{Z}_L}{\mathbf{Z}_R + \mathbf{Z}_L}$$
$$= \frac{(R\angle 0^\circ)(X_L \angle 90^\circ)}{R + jX_L} = \frac{RX_L \angle 90^\circ}{\sqrt{R^2 + X_L^2}\ \angle \tan^{-1} X_L/R}$$

and

$$\mathbf{Z}_T = \frac{RX_L}{\sqrt{R^2 + X_L^2}} \underline{/90^\circ - \tan^{-1} X_L/R}$$

so that

$$Z_T = \frac{RX_L}{\sqrt{R^2 + X_L^2}} \tag{15.39}$$

and

$$\theta_T = 90° - \tan^{-1}\frac{X_L}{R} = \tan^{-1}\frac{R}{X_L} \tag{15.40}$$

The magnitude and angle of the total impedance can now be found at any frequency of interest simply by substituting Eqs. (15.39) and (15.40).

f = 1 kHz

$$X_L = 2\pi fL = 2\pi(1\text{ kHz})(4 \times 10^{-3}\text{ H}) = 25.12\ \Omega$$

and

$$Z_T = \frac{RX_L}{\sqrt{R^2 + X_L^2}} = \frac{(220\ \Omega)(25.12\ \Omega)}{\sqrt{(220\ \Omega)^2 + (25.12\ \Omega)^2}} = \mathbf{24.96\ \Omega}$$

with

$$\theta_T = \tan^{-1}\frac{R}{X_L} = \tan^{-1}\frac{220\ \Omega}{25.12\ \Omega}$$
$$= \tan^{-1}8.76 = 83.49°$$

and

$$Z_T = \mathbf{24.96\ \Omega\ \angle 83.49°}$$

This value compares very closely with $X_L = 25.12\ \Omega\ \angle 90°$, which it would be if the network were purely inductive ($R = \infty\ \Omega$). Our assumption that the network is primarily inductive at low frequencies is therefore confirmed.

Continuing, we obtain

$f =$ 5 kHz: $\mathbf{Z_T = 109.1\ \Omega\ \angle 60.23°}$
$f =$ 10 kHz: $\mathbf{Z_T = 165.5\ \Omega\ \angle 41.21°}$
$f =$ 15 kHz: $\mathbf{Z_T = 189.99\ \Omega\ \angle 30.28°}$
$f =$ 20 kHz: $\mathbf{Z_T = 201.53\ \Omega\ \angle 23.65°}$
$f =$ 30 kHz: $\mathbf{Z_T = 211.19\ \Omega\ \angle 16.27°}$
$f =$ 40 kHz: $\mathbf{Z_T = 214.91\ \Omega\ \angle 12.35°}$

At f = 40 kHz, note how closely the magnitude of Z_T has approached the resistance level of 220 Ω and how the associated angle with the total impedance is approaching zero degrees. The result is a network with terminal characteristics that are becoming more and more resistive as the frequency increases, which further confirms the earlier conclusions developed by the curves in Fig. 15.87.

Plots of Z_T versus frequency in Fig. 15.88 and θ_T in Fig. 15.89 clearly reveal the transition from an inductive network to one that has resistive characteristics. Note that the transition frequency of 8.75 kHz occurs right in the middle of the "knee" of the curves for both Z_T and θ_T.

A review of Figs. 15.49 and 15.88 reveals that a series *R-C* and a parallel *R-L* network will have an impedance level that approaches the resistance of the network at high frequencies. The capacitive circuit approaches the level from above, whereas the inductive network does the same from below. For the series *R-L* circuit and the parallel *R-C* network, the total impedance will begin at the resistance level and then display the characteristics of the reactive elements at high frequencies.

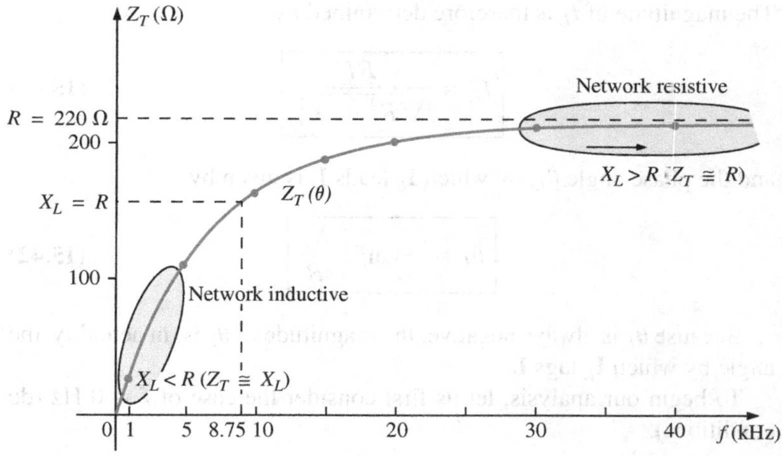

FIG. 15.88
The magnitude of the input impedance versus frequency for the network in Fig. 15.86.

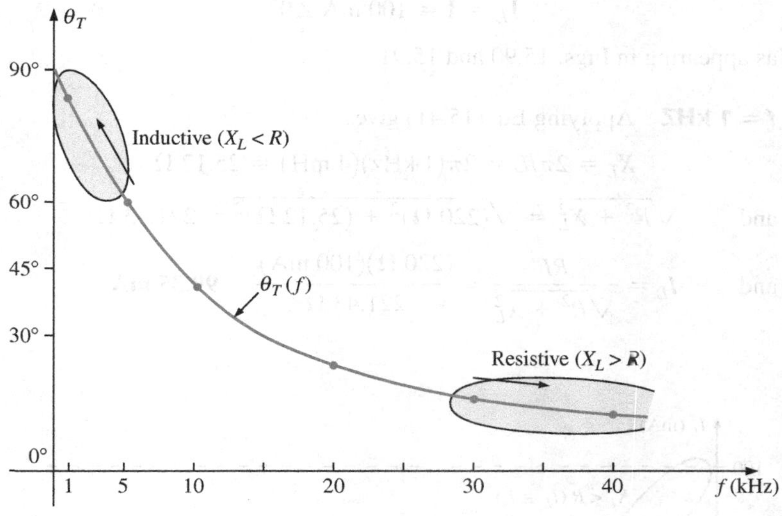

FIG. 15.89
The phase angle of the input impedance versus frequency for the network in Fig. 15.86.

$\mathbf{I}_L$ Applying the current divider rule to the network in Fig. 15.86 results in the following:

$$\mathbf{I}_L = \frac{\mathbf{Z}_R \mathbf{I}}{\mathbf{Z}_R + \mathbf{Z}_L}$$

$$= \frac{(R \angle 0°)(I \angle 0°)}{R + jX_L} = \frac{RI \angle 0°}{\sqrt{R^2 + X_L^2} \;\underline{/\tan^{-1} X_L/R}}$$

and $$\mathbf{I}_L = I_L \angle \theta_L = \frac{RI}{\sqrt{R^2 + X_L^2}} \;\underline{/-\tan^{-1} X_L/R}$$

The magnitude of I_L is therefore determined by

$$I_L = \frac{RI}{\sqrt{R^2 + X_L^2}} \qquad (15.41)$$

and the phase angle θ_L, by which $\mathbf{I}_L$ leads $\mathbf{I}$, is given by

$$\theta_L = -\tan^{-1}\frac{X_L}{R} \qquad (15.42)$$

Because θ_L is always negative, the magnitude of θ_L is, in actuality, the angle by which $\mathbf{I}_L$ lags $\mathbf{I}$.

To begin our analysis, let us first consider the case of $f = 0$ Hz (dc conditions).

$f = 0$ Hz

$$X_L = 2\pi fL = 2\pi(0\text{ Hz})L = 0\ \Omega$$

Applying the short-circuit equivalent for the inductor in Fig. 15.86 results in

$$\mathbf{I}_L = \mathbf{I} = 100\text{ mA}\angle 0°$$

as appearing in Figs. 15.90 and 15.91.

$f = 1$ kHZ Applying Eq. (15.41) gives

$$X_L = 2\pi fL = 2\pi(1\text{ kHz})(4\text{ mH}) = 25.12\ \Omega$$

and $$\sqrt{R^2 + X_L^2} = \sqrt{(220\ \Omega)^2 + (25.12\ \Omega)^2} = 221.43\ \Omega$$

and $$I_L = \frac{RI}{\sqrt{R^2 + X_L^2}} = \frac{(220\ \Omega)(100\text{ mA})}{221.43\ \Omega} = \mathbf{99.35\ mA}$$

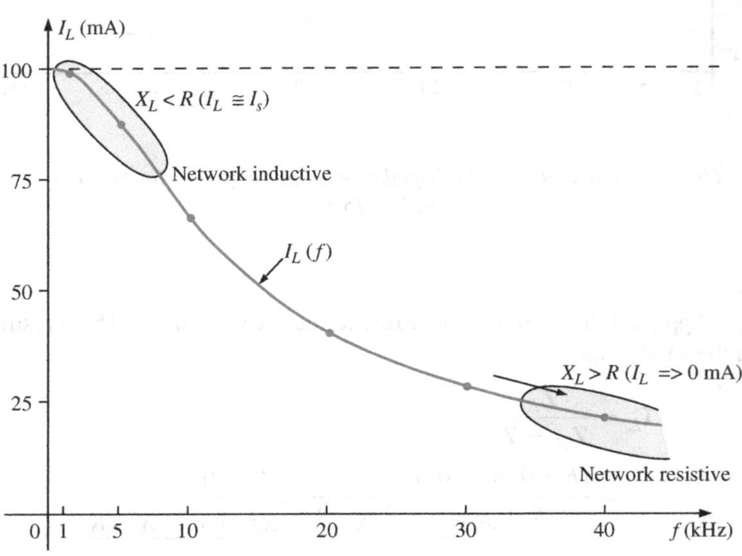

FIG. 15.90
The magnitude of the current $\mathbf{I}_L$ versus frequency for the parallel R-L network in Fig. 15.86.

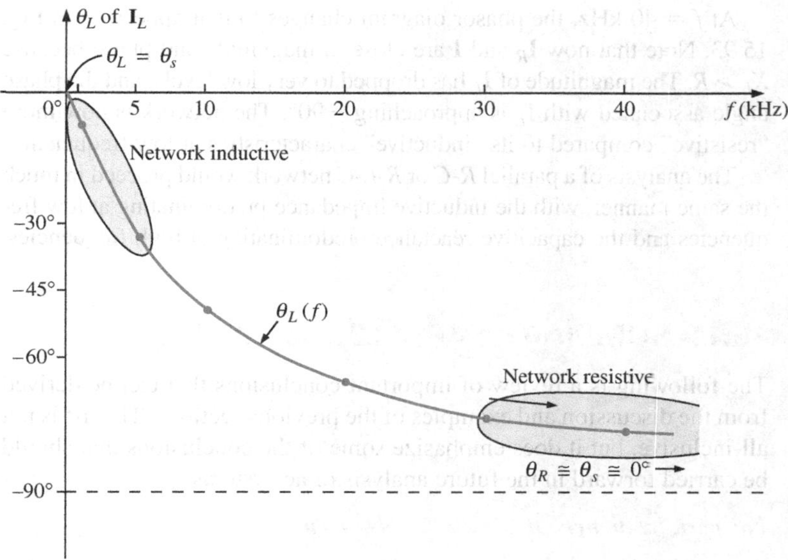

FIG. 15.91
The phase angle of the current $\mathbf{I}_L$ versus frequency for the parallel R-L network in Fig. 15.86.

with

$$\theta_L = \tan^{-1}\frac{X_L}{R} = -\tan^{-1}\frac{25.12\ \Omega}{220\ \Omega} = -\tan^{-1}0.114 = \mathbf{-6.51°}$$

and

$$\mathbf{I}_L = \mathbf{99.35\ mA\ \angle{-6.51°}}$$

The result is a current $\mathbf{I}_L$ that is still very close to the source current $\mathbf{I}$ in both magnitude and phase.

Continuing, we obtain

$$f = 5\text{ kHz:}\quad \mathbf{I}_L = \mathbf{86.84\ mA\ \angle{-29.72°}}$$
$$f = 10\text{ kHz:}\quad \mathbf{I}_L = \mathbf{65.88\ mA\ \angle{-48.79°}}$$
$$f = 15\text{ kHz:}\quad \mathbf{I}_L = \mathbf{50.43\ mA\ \angle{-59.72°}}$$
$$f = 20\text{ kHz:}\quad \mathbf{I}_L = \mathbf{40.11\ mA\ \angle{-66.35°}}$$
$$f = 30\text{ kHz:}\quad \mathbf{I}_L = \mathbf{28.02\ mA\ \angle{-73.73°}}$$
$$f = 40\text{ kHz:}\quad \mathbf{I}_L = \mathbf{21.38\ mA\ \angle{-77.65°}}$$

The plot of the magnitude of I_L versus frequency is provided in Fig. 15.90 and reveals that the current through the coil dropped from its maximum of 100 mA to almost 20 mA at 40 kHz. As the reactance of the coil increased with frequency, more of the source current chose the lower-resistance path of the resistor. The magnitude of the phase angle between $\mathbf{I}_L$ and $\mathbf{I}$ is approaching 90° with an increase in frequency, as shown in Fig. 15.91, leaving its initial value of zero degrees at $f = 0$ Hz far behind.

At $f = 1$ kHz, the phasor diagram of the network appears as shown in Fig. 15.92. First note that the magnitude and the phase angle of $\mathbf{I}_L$ are very close to those of $\mathbf{I}$. Since the voltage across a coil must lead the current through a coil by 90°, the voltage $\mathbf{V}_s$ appears as shown. The voltage across a resistor is in phase with the current through the resistor, resulting in the direction of $\mathbf{I}_R$ shown in Fig. 15.92. Of course, at this frequency $R > X_L$, and the current I_R is relatively small in magnitude.

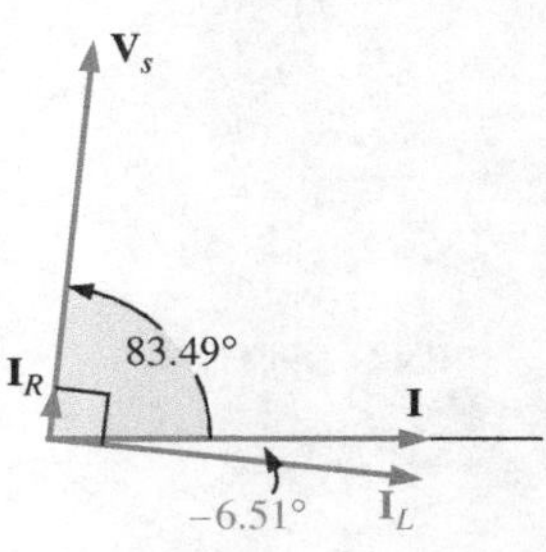

FIG. 15.92
The phasor diagram for the parallel R-L network in Fig. 15.86 at $f = 1$ kHz.

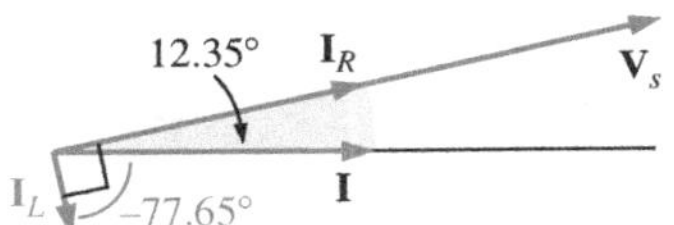

FIG. 15.93
The phasor diagram for the parallel R-L network in Fig. 15.86 at f = 40 kHz.

At $f = 40$ kHz, the phasor diagram changes to that appearing in Fig. 15.93. Note that now $\mathbf{I}_R$ and $\mathbf{I}$ are close in magnitude and phase because $X_L > R$. The magnitude of $\mathbf{I}_L$ has dropped to very low levels, and the phase angle associated with $\mathbf{I}_L$ is approaching $-90°$. The network is now more "resistive" compared to its "inductive" characteristics at low frequencies.

The analysis of a parallel *R-C* or *R-L-C* network would proceed in much the same manner, with the inductive impedance predominating at low frequencies and the capacitive reactance predominating at high frequencies.

15.11 SUMMARY: PARALLEL ac NETWORKS

The following is a review of important conclusions that can be derived from the discussion and examples of the previous sections. The list is not all-inclusive, but it does emphasize some of the conclusions that should be carried forward in the future analysis of ac systems.

For parallel ac networks with reactive elements:

1. *The total admittance or impedance will be frequency dependent.*
2. *Depending on the frequency applied, the same network can be either predominantly inductive or predominantly capacitive.*
3. *The magnitude of the current through any one branch can be greater than the source current.*
4. *The inductive and capacitive susceptances are in direct opposition on an admittance diagram.*
5. *At lower frequencies, the inductive elements will usually have the most impact on the total impedance, while at high frequencies, the capacitive elements will usually have the most impact.*
6. *The impedance of any one element can be less than the total impedance (recall that for dc circuits, the total resistance must always be less than the smallest parallel resistor).*
7. *The magnitude of the current through an element, compared to the other elements of the network, is directly related to the magnitude of its impedance; that is, the smaller the impedance of an element, the larger is the magnitude of the current through the element.*
8. *The current through a coil is always in direct opposition with the current through a capacitor on a phasor diagram.*
9. *The applied voltage is always in phase with the current through the resistive elements, leads the voltage across all the inductive elements by 90°, and lags the current through all capacitive elements by 90°.*
10. *The smaller the resistive element of a network compared to the net reactive susceptance, the closer is the power factor to unity.*

15.12 EQUIVALENT CIRCUITS

In a series ac circuit, the total impedance of two or more elements in series is often equivalent to an impedance that can be achieved with fewer elements of different values, the elements and their values being determined by the frequency applied. This is also true for parallel circuits. For the circuit in Fig. 15.94(a),

$$\mathbf{Z}_T = \frac{\mathbf{Z}_C\mathbf{Z}_L}{\mathbf{Z}_C + \mathbf{Z}_L} = \frac{(5\ \Omega\ \angle -90°)(10\ \Omega\ \angle 90°)}{5\ \Omega\ \angle -90° + 10\ \Omega\ \angle 90°} = \frac{50\ \angle 0°}{5\ \angle 90°}$$
$$= 10\ \Omega\ \angle -90°$$

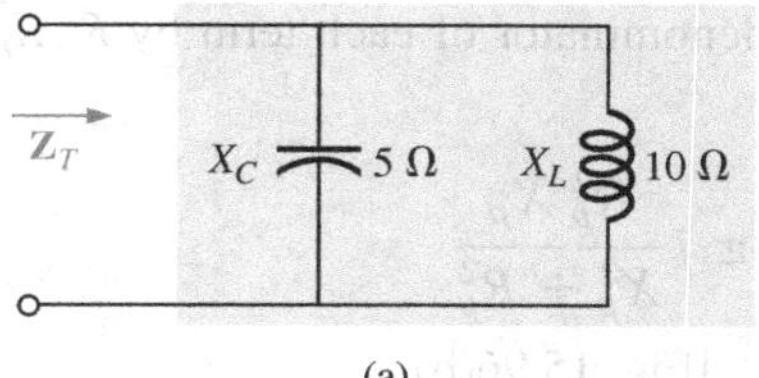

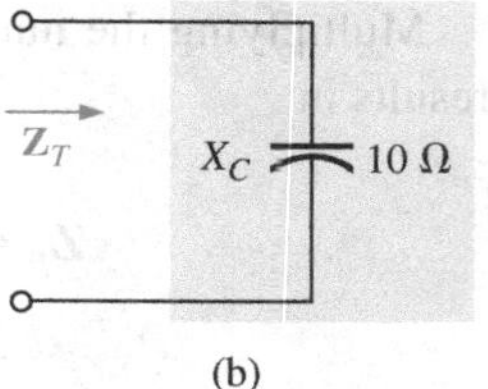

FIG. 15.94
Defining the equivalence between two networks at a specific frequency.

The total impedance at the frequency applied is equivalent to a capacitor with a reactance of 10 Ω, as shown in Fig. 15.94(b). Always keep in mind that this equivalence is true only at the applied frequency. If the frequency changes, the reactance of each element changes, and the equivalent circuit changes—perhaps from capacitive to inductive in the above example.

Another interesting development appears if the impedance of a parallel circuit, such as the one in Fig. 15.95(a), is found in rectangular form. In this case,

$$\begin{aligned}\mathbf{Z}_T &= \frac{\mathbf{Z}_L\mathbf{Z}_R}{\mathbf{Z}_L + \mathbf{Z}_R} = \frac{(4\ \Omega \angle 90^\circ)(3\ \Omega \angle 0^\circ)}{4\ \Omega \angle 90^\circ + 3\ \Omega \angle 0^\circ} \\ &= \frac{12 \angle 90^\circ}{5 \angle 53.13^\circ} = 2.40\ \Omega \angle 36.87^\circ \\ &= 1.92\ \Omega + j\,1.44\ \Omega\end{aligned}$$

which is the impedance of a series circuit with a resistor of 1.92 Ω and an inductive reactance of 1.44 Ω, as shown in Fig. 15.95(b).

The current **I** will be the same in each circuit in Fig. 15.94 or Fig. 15.95 if the same input voltage **E** is applied. For a parallel circuit of one resistive element and one reactive element, the series circuit with the same input impedance will always be composed of one resistive and one reactive element. The impedance of each element of the series circuit will be different from that of the parallel circuit, but the reactive elements will always be of the same type; that is, an *R-L* circuit and an *R-C* parallel circuit will have an equivalent *R-L* and *R-C* series circuit, respectively. The same is true when converting from a series to a parallel circuit. In the discussion to follow, keep in mind that

the term equivalent refers only to the fact that for the same applied potential, the same impedance and input current will result.

To formulate the equivalence between the series and parallel circuits, the equivalent series circuit for a resistor and reactance in parallel can be found by determining the total impedance of the circuit in rectangular form; that is, for the circuit in Fig. 15.96(a),

$$\mathbf{Y}_p = \frac{1}{R_p} + \frac{1}{\pm jX_p} = \frac{1}{R_p} \mp j\frac{1}{X_p}$$

and

$$\begin{aligned}\mathbf{Z}_p &= \frac{1}{\mathbf{Y}_p} = \frac{1}{(1/R_p) \mp j(1/X_p)} \\ &= \frac{1/R_p}{(1/R_p)^2 + (1/X_p)^2} \pm j\frac{1/X_p}{(1/R_p)^2 + (1/X_p)^2}\end{aligned}$$

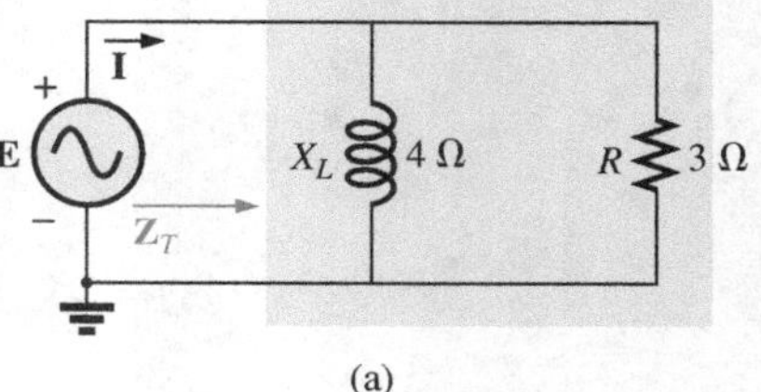

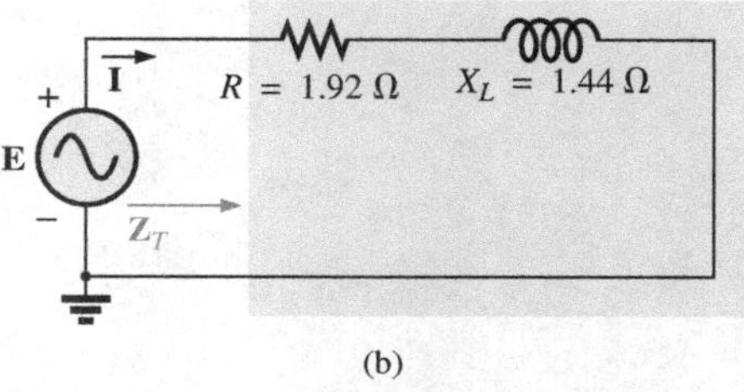

FIG. 15.95
Finding the series equivalent circuit for a parallel R-L network.

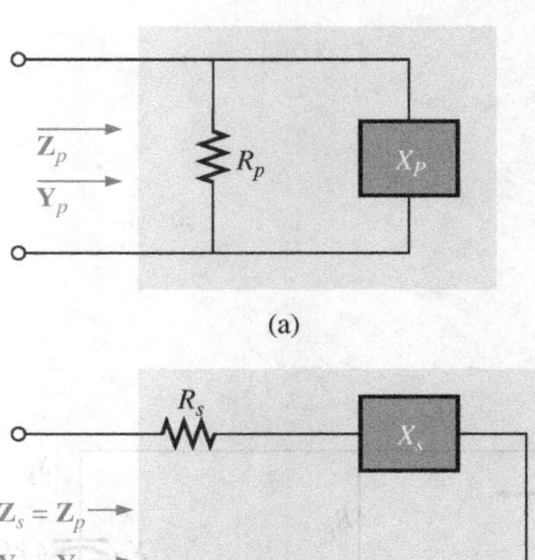

FIG. 15.96
Defining the parameters of equivalent series and parallel networks.

Multiplying the numerator and denominator of each term by $R_p^2 X_p^2$ results in

$$\mathbf{Z}_p = \frac{R_p X_p^2}{X_p^2 + R_p^2} \pm j\frac{R_p^2 X_p}{X_p^2 + R_p^2}$$
$$= R_s \pm jX_s \quad \text{[Fig. 15.96(b)]}$$

and

$$R_s = \frac{R_p X_p^2}{X_p^2 + R_p^2} \quad \textbf{(15.43)}$$

with

$$X_s = \frac{R_p^2 X_p}{X_p^2 + R_p^2} \quad \textbf{(15.44)}$$

For the network in Fig. 15.95,

$$R_s = \frac{R_p X_p^2}{X_p^2 + R_p^2} = \frac{(3\ \Omega)(4\ \Omega)^2}{(4\ \Omega)^2 + (3\ \Omega)^2} = \frac{48\ \Omega}{25} = \mathbf{1.92\ \Omega}$$

and

$$X_s = \frac{R_p^2 X_p}{X_p^2 + R_p^2} = \frac{(3\ \Omega)^2(4\ \Omega)}{(4\ \Omega)^2 + (3\ \Omega)^2} = \frac{36\ \Omega}{25} = \mathbf{1.44\ \Omega}$$

which agrees with the previous result.

The equivalent parallel circuit for a circuit with a resistor and reactance in series can be found by finding the total admittance of the system in rectangular form; that is, for the circuit in Fig. 15.96(b),

$$\mathbf{Z}_s = R_s \pm jX_s$$
$$\mathbf{Y}_s = \frac{1}{\mathbf{Z}_s} = \frac{1}{R_s \pm jX_s} = \frac{R_s}{R_s^2 + X_s^2} \mp j\frac{X_s}{R_s^2 + X_s^2}$$
$$= G_p \mp jB_p = \frac{1}{R_p} \mp j\frac{1}{X_p} \quad \text{[Fig. 15.96(a)]}$$

or

$$R_p = \frac{R_s^2 + X_s^2}{R_s} \quad \textbf{(15.45)}$$

with

$$X_p = \frac{R_s^2 + X_s^2}{X_s} \quad \textbf{(15.46)}$$

For the above example,

$$R_p = \frac{R_s^2 + X_s^2}{R_s} = \frac{(1.92\ \Omega)^2 + (1.44\ \Omega)^2}{1.92\ \Omega} = \frac{5.76\ \Omega}{1.92} = \mathbf{3.0\ \Omega}$$

and

$$X_p = \frac{R_s^2 + X_s^2}{X_s} = \frac{5.76\ \Omega}{1.44} = \mathbf{4.0\ \Omega}$$

as shown in Fig. 15.95 (a).

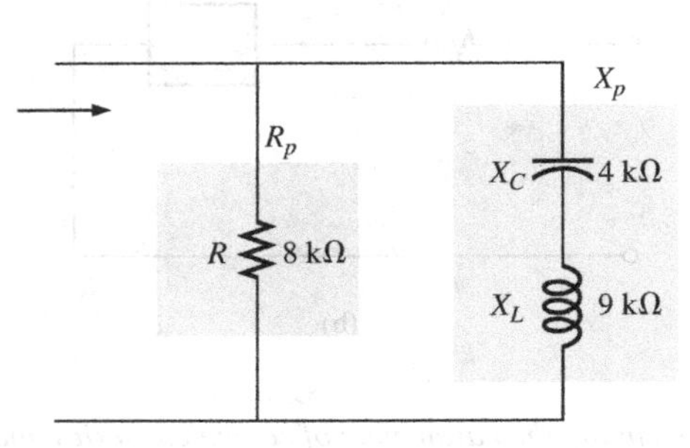

FIG. 15.97
Example 15.18.

EXAMPLE 15.18 Determine the series equivalent circuit for the network in Fig. 15.97.

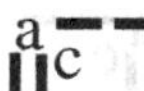

Solution:

$$R_p = 8\text{ k}\Omega$$
$$X_p(\text{resultant}) = |X_L - X_C| = |9\text{ k}\Omega - 4\text{ k}\Omega| = 5\text{ k}\Omega$$

and

$$R_s = \frac{R_p X_p^2}{X_p^2 + R_p^2} = \frac{(8\text{ k}\Omega)(5\text{ k}\Omega)^2}{(5\text{ k}\Omega)^2 + (8\text{ k}\Omega)^2} = \frac{200\text{ k}\Omega}{89} = \mathbf{2.25\text{ k}\Omega}$$

with

$$X_s = \frac{R_p^2 X_p}{X_p^2 + R_p^2} = \frac{(8\text{ k}\Omega)^2(5\text{ k}\Omega)}{(5\text{ k}\Omega)^2 + (8\text{ k}\Omega)^2} = \frac{320\text{ k}\Omega}{89} = \mathbf{3.60\text{ k}\Omega\ \ (inductive)}$$

The equivalent series circuit appears in Fig. 15.98.

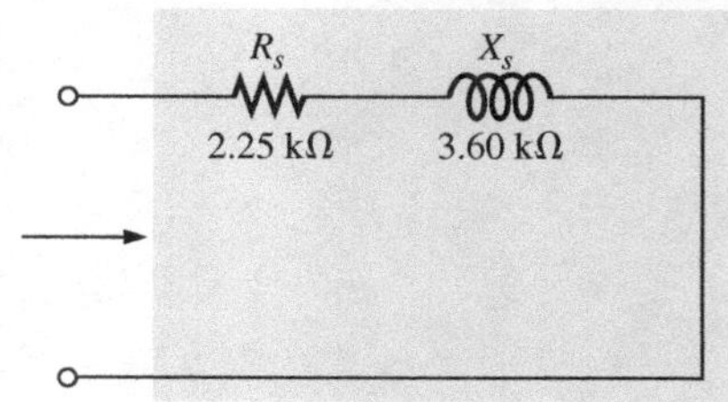

FIG. 15.98
The equivalent series circuit for the parallel network in Fig. 15.97.

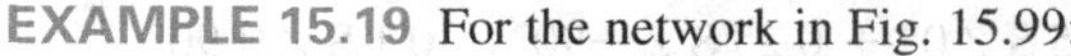

EXAMPLE 15.19 For the network in Fig. 15.99:

a. Determine $\mathbf{Y}_T$ and $\mathbf{Z}_T$.
b. Sketch the admittance diagram.
c. Find $\mathbf{E}$ and $\mathbf{I}_L$.
d. Compute the power factor of the network and the power delivered to the network.
e. Determine the equivalent series circuit as far as the terminal characteristics of the network are concerned.
f. Using the equivalent circuit developed in part (e), calculate $\mathbf{E}$, and compare it with the result of part (c).
g. Determine the power delivered to the network, and compare it with the solution of part (d).
h. Determine the equivalent parallel network from the equivalent series circuit, and calculate the total admittance $\mathbf{Y}_T$. Compare the result with the solution of part (a).

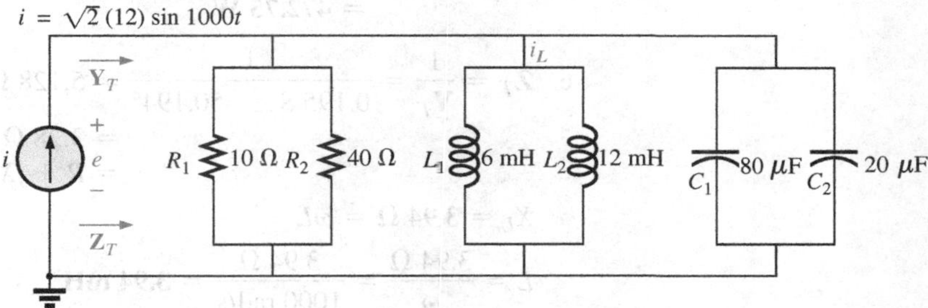

FIG. 15.99
Example 15.19.

Solutions:

a. Combining common elements and finding the reactance of the inductor and capacitor, we obtain

$$R_T = 10\ \Omega \parallel 40\ \Omega = 8\ \Omega$$
$$L_T = 6\text{ mH} \parallel 12\text{ mH} = 4\text{ mH}$$
$$C_T = 80\ \mu\text{F} + 20\ \mu\text{F} = 100\ \mu\text{F}$$
$$X_L = \omega L = (1000\text{ rad/s})(4\text{ mH}) = 4\ \Omega$$
$$X_C = \frac{1}{\omega C} = \frac{1}{(1000\text{ rad/s})(100\ \mu\text{F})} = 10\ \Omega$$

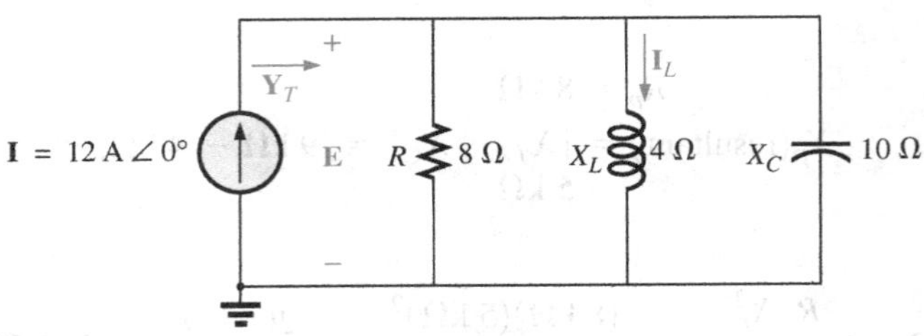

FIG. 15.100
Applying phasor notation to the network in Fig. 15.99.

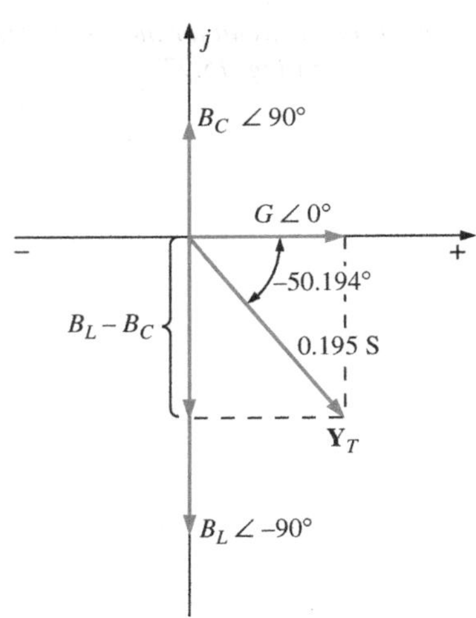

FIG. 15.101
Admittance diagram for the parallel R-L-C network in Fig. 15.99.

The network is redrawn in Fig. 15.100 with phasor notation. The total admittance is

$$\begin{aligned}\mathbf{Y}_T &= \mathbf{Y}_R + \mathbf{Y}_L + \mathbf{Y}_C \\ &= G\angle 0° + B_L\angle -90° + B_C\angle +90° \\ &= \frac{1}{8\,\Omega}\angle 0° + \frac{1}{4\,\Omega}\angle -90° + \frac{1}{10\,\Omega}\angle +90° \\ &= 0.125\text{ S}\angle 0° + 0.25\text{ S}\angle -90° + 0.1\text{ S}\angle +90° \\ &= 0.125\text{ S} - j\,0.25\text{ S} + j\,0.1\text{ S} \\ &= 0.125\text{ S} - j\,0.15\text{ S} = \mathbf{0.195\ S\ \angle -50.194°}\end{aligned}$$

$$\mathbf{Z}_T = \frac{1}{\mathbf{Y}_T} = \frac{1}{0.195\text{ S}}\angle -50.194° = \mathbf{5.13\ \Omega\ \angle 50.19°}$$

b. See Fig. 15.101.

c. $\mathbf{E} = \mathbf{I}\mathbf{Z}_T = \dfrac{\mathbf{I}}{\mathbf{Y}_T} = \dfrac{12\text{ A}\angle 0°}{0.195\text{ S}\angle -50.194°} = \mathbf{61.54\ V\ \angle 50.19°}$

$$\mathbf{I}_L = \frac{\mathbf{V}_L}{\mathbf{Z}_L} = \frac{\mathbf{E}}{\mathbf{Z}_L} = \frac{61.538\text{ V}\angle 50.194°}{4\,\Omega\angle 90°} = \mathbf{15.39\ A\ \angle -39.81°}$$

d. $F_p = \cos\theta = \dfrac{G}{Y_T} = \dfrac{0.125\text{ S}}{0.195\text{ S}} = \mathbf{0.641\ lagging}$ (**E** leads **I**)

$$\begin{aligned}P = EI\cos\theta &= (61.538\text{ V})(12\text{ A})\cos 50.194° \\ &= \mathbf{472.75\ W}\end{aligned}$$

e. $\mathbf{Z}_T = \dfrac{1}{\mathbf{Y}_T} = \dfrac{1}{0.195\text{ S}\angle -50.194°} = 5.128\,\Omega\angle +50.194°$

$$\begin{aligned}&= 3.28\,\Omega + j\,3.94\,\Omega \\ &= R + jX_L\end{aligned}$$

$$X_L = 3.94\,\Omega = \omega L$$

$$L = \frac{3.94\,\Omega}{\omega} = \frac{3.94\,\Omega}{1000\text{ rad/s}} = \mathbf{3.94\ mH}$$

The series equivalent circuit appears in Fig. 15.102.

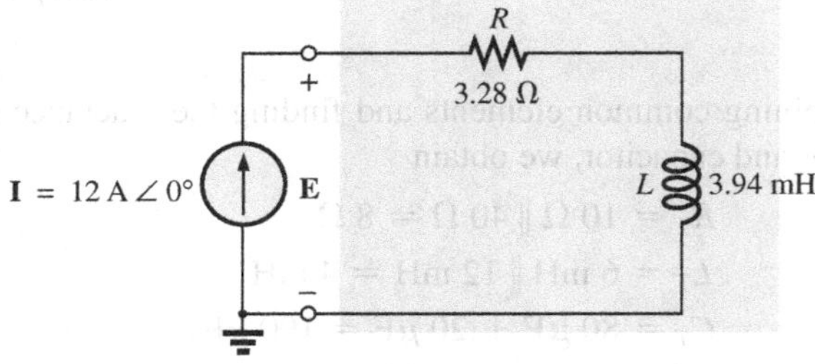

FIG. 15.102
Series equivalent circuit for the parallel R-L-C network in Fig. 15.99 with ω = 1000 rad/s.

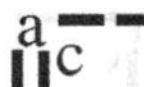

f. $\mathbf{E} = \mathbf{I}\mathbf{Z}_T = (12\text{ A}\angle 0^\circ)(5.128\ \Omega\angle 50.194^\circ)$
$= \mathbf{61.54\ V\angle 50.194^\circ}$ (as above)

g. $P = I^2R = (12\text{ A})^2(3.28\ \Omega) = \mathbf{472.32\ W}$ (as above)

h. $$R_p = \frac{R_s^2 + X_s^2}{R_s} = \frac{(3.28\ \Omega)^2 + (3.94\ \Omega)^2}{3.28\ \Omega} = \mathbf{8\ \Omega}$$

$$X_p = \frac{R_s^2 + X_s^2}{X_s} = \frac{(3.28\ \Omega)^2 + (3.94\ \Omega)^2}{3.94\ \Omega} = \mathbf{6.67\ \Omega}$$

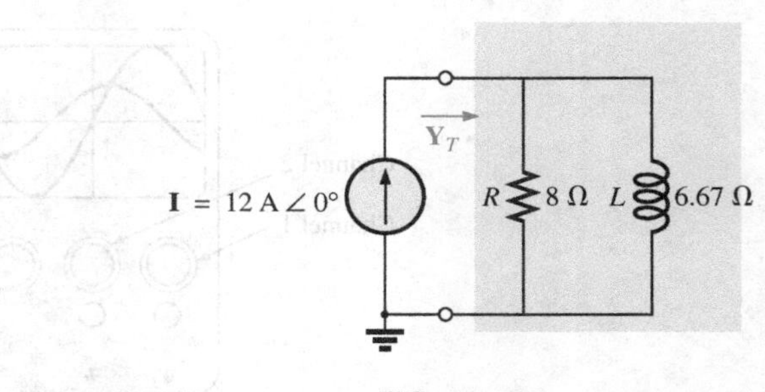

FIG. 15.103
Parallel equivalent of the circuit in Fig. 15.102.

The parallel equivalent circuit appears in Fig. 15.103.

$$\begin{aligned}\mathbf{Y}_T &= G\angle 0^\circ + B_L\angle -90^\circ = \frac{1}{8\ \Omega}\angle 0^\circ + \frac{1}{6.675\ \Omega}\angle -90^\circ \\ &= 0.125\text{ S}\angle 0^\circ + 0.15\text{ S}\angle -90^\circ \\ &= 0.125\text{ S} - j\,0.15\text{ S} = \mathbf{0.195\ S\angle 50.194^\circ} \quad \text{(as above)}\end{aligned}$$

15.13 PHASE MEASUREMENTS

Measuring the phase angle between quantities is one of the most important functions that an oscilloscope can perform. It is an operation that must be performed carefully, however, or you may obtain the incorrect result or damage the equipment. Whenever you are using the dual-trace capability of an oscilloscope, the most important thing to remember is that

both channels of a dual-trace oscilloscope must be connected to the same ground.

Measuring Z_T and θ_T

For ac parallel networks restricted to resistive loads, the total impedance can be found in the same manner as described for dc circuits: Simply remove the source and place an ohmmeter across the network terminals. However,

for parallel ac networks with reactive elements, the total impedance cannot be measured with an ohmmeter.

An experimental procedure must be defined that permits determining the magnitude and the angle of the terminal impedance.

The phase angle between the applied voltage and the resulting source current is one of the most important because (a) it is also the phase angle associated with the total impedance; (b) it provides an instant indication of whether a network is resistive or reactive; (c) it reveals whether a network is inductive or capacitive; and (d) it can be used to find the power delivered to the network.

In Fig. 15.104, a resistor has been added to the configuration between the source and the network to permit measuring the current and finding the phase angle between the applied voltage and the source current.

At the frequency of interest, the applied voltage establishes a voltage across the sensing resistor that can be displayed by one channel of the dual-trace oscilloscope. In Fig. 15.104, channel 1 is displaying the applied voltage and channel 2 the voltage across the sensing resistor. Sensitivities for each channel are chosen to establish the waveforms

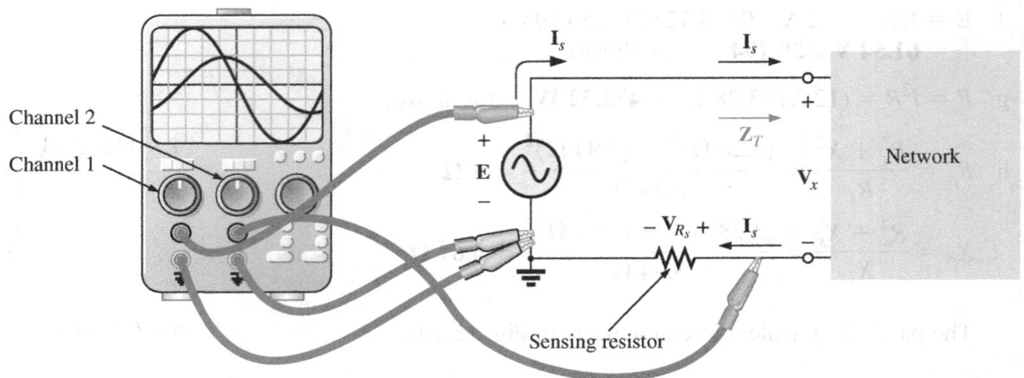

FIG. 15.104
Using an oscilloscope to measure Z_T and θ_T.

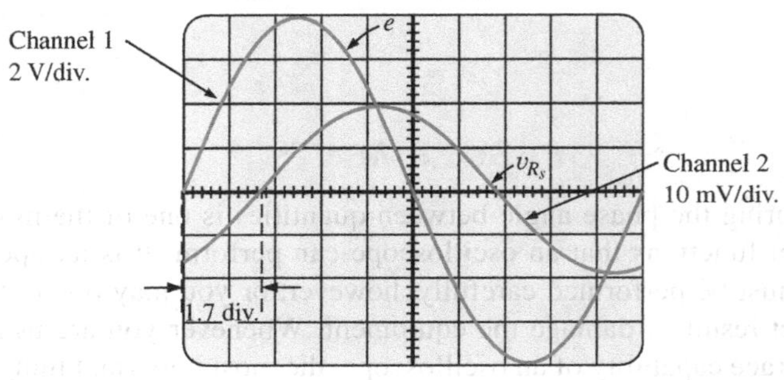

FIG. 15.105
e and v_{R_s} for the configuration in Fig. 15.104.

appearing on the screen in Fig. 15.105. As emphasized above, note that both channels have the same ground connection. In fact, the need for a common ground connection is the only reason that the sensing resistor was not connected to the positive side of the supply. Since oscilloscopes display only voltages versus time, the peak value of the source current must be found using Ohm's law. *Since the voltage across a resistor and the current through the resistor are in phase, the phase angle between the two voltages will be the same as that between the applied voltage and the resulting source current.*

Using the sensitivities, we find that the peak value of the applied voltage is

$$E_m = (4 \text{ div.})(2 \text{ V/div.}) = 8 \text{ V}$$

while the peak value of the voltage across the sensing resistor is

$$V_{R_s(\text{peak})} = (2\text{div.})(10 \text{ mV/div.}) = 20 \text{ mV}$$

Using Ohm's law, we find that the peak value of the current is

$$I_{s(\text{peak})} = \frac{V_{R_s(\text{peak})}}{R_s} = \frac{20 \text{ mV}}{10 \ \Omega} = 2 \text{ mA}$$

The sensing resistor is chosen small enough so that the voltage across the sensing resistor is small enough to permit the approximation $\mathbf{V}_x = \mathbf{E} - \mathbf{V}_{R_s} \cong \mathbf{E}$. The magnitude of the input impedance is then

$$Z_T = \frac{V_x}{I_s} \cong \frac{E}{I_s} = \frac{8 \text{ V}}{2 \text{ mA}} = \mathbf{4 \ k\Omega}$$

For the chosen horizontal sensitivity, each waveform in Fig. 15.105 has a period T defined by ten horizontal divisions, and the phase angle between the two waveforms is 1.7 divisions. Using the fact that each period of a sinusoidal waveform encompasses 360°, we can set up the following ratios to determine the phase angle θ:

$$\frac{10 \text{ div.}}{360^\circ} = \frac{1.7 \text{ div.}}{\theta}$$

and
$$\theta = \left(\frac{1.7}{10}\right)360^\circ = \mathbf{61.2^\circ}$$

In general,

$$\theta = \frac{(\text{div. for } \theta)}{(\text{div. for } T)} \times 360^\circ \qquad \textbf{(15.47)}$$

Therefore, the total impedance is

$$\mathbf{Z}_T = \mathbf{4\ k\Omega \angle 61.2^\circ} = \mathbf{1.93\ k\Omega} + \boldsymbol{j}\mathbf{3.51\ k\Omega} = R + jX_L$$

which is equivalent to the series combination of a 1.93 kΩ resistor and an inductor with a reactance of 3.51 kΩ (at the frequency of interest).

Measuring the Phase Angle between Various Voltages

In Fig. 15.106, an oscilloscope is being used to find the phase relationship between the applied voltage and the voltage across the inductor. Note again that each channel shares the same ground connection. The resulting pattern appears in Fig. 15.107 with the chosen sensitivities. This time, both channels have the same sensitivity, resulting in the following peak values for the voltages:

$$E_m = (3 \text{ div.})(2 \text{ V/div.}) = \mathbf{6\ V}$$
$$V_{L(\text{peak})} = (1.6 \text{ div.})(2 \text{ V/div.}) = \mathbf{3.2\ V}$$

The phase angle is determined using Eq. 15.45:

$$\theta = \frac{(1 \text{ div.})}{(8 \text{ div.})} \times 360^\circ$$
$$\theta = \mathbf{45^\circ}$$

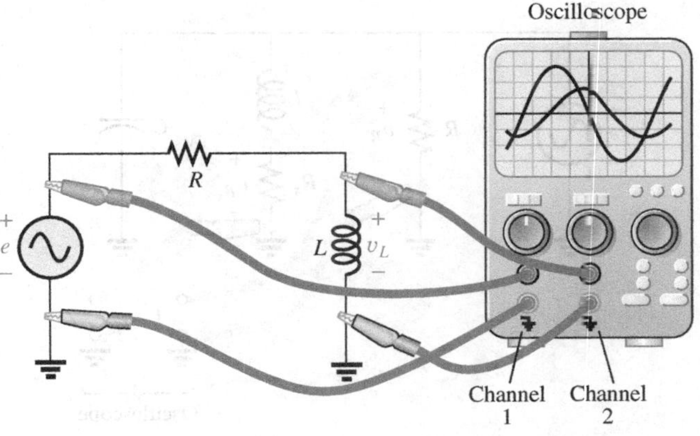

FIG. 15.106
Determining the phase relationship between e and v_L.

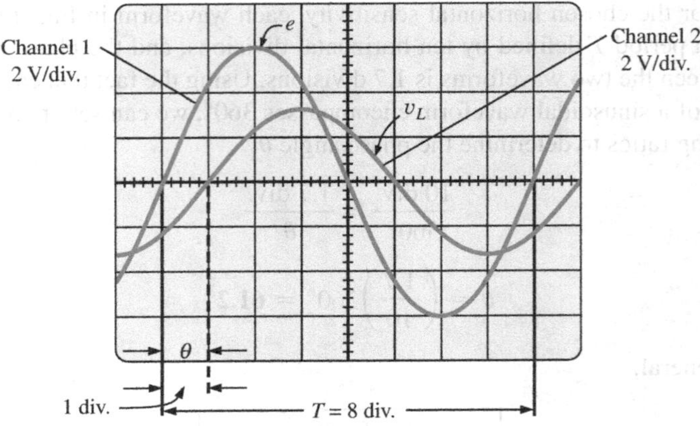

FIG. 15.107
Determining the phase angle between e and v_L for the configuration in Fig. 15.106.

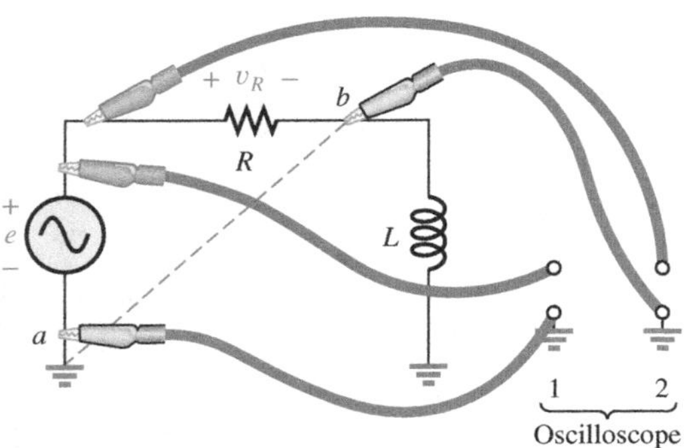

FIG. 15.108
An improper phase-measurement connection.

If the phase relationship between e and v_R is desired, the oscilloscope *cannot be connected* as shown in Fig. 15.108. *The grounds of each channel are internally connected in the oscilloscope,* forcing point b to have the same potential as point a. The result would be a direct connection between points a and b that would short out the inductive element. If the inductive element is the predominant factor in controlling the level of the current, the current in the circuit could rise to dangerous levels and damage the oscilloscope or supply. The easiest way to find the phase relationship between e and v_R would be to simply interchange the positions of the resistor and the inductor and proceed as before.

For the parallel network in Fig. 15.109, the phase relationship between two of the branch currents, i_R and i_L, can be determined using a sensing resistor, as shown in the figure. The value of the sensing resistor is chosen small enough in comparison to the value of the series inductive reactance to ensure that it will not affect the general response of the network. Channel 1 displays the voltage v_R, and channel 2 the voltage v_{R_s}. Since v_R is in phase with i_R, and v_R is in phase with i_L, the phase relationship between v_R and v_{R_s} will be the same as between i_R and i_L. The peak value of each current can be found through a simple application of Ohm's law.

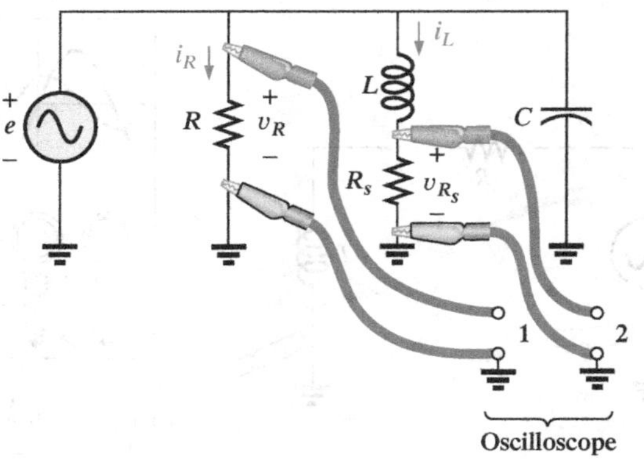

FIG. 15.109
Determining the phase relationship between i_R and i_L.

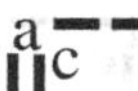

15.14 APPLICATIONS

Home Wiring

An expanded view of house wiring is provided in Fig. 15.110 to permit a discussion of the entire system. The house panel has been included with the "feed" and the important grounding mechanism. In addition, a number of typical circuits found in the home have been included to provide a sense for the manner in which the total power is distributed.

First note how the copper bars in the panel are laid out to provide both 120 V and 240 V. Between any one bar and ground is the single-phase 120 V supply. However, the bars have been arranged so that 240 V can be obtained between two vertical adjacent bars using a double-gang circuit breaker. When time permits, examine your own panel (but do not remove the cover), and note the dual circuit breaker arrangement for the 240 V supply.

For appliances such as fixtures and heaters that have a metal casing, the ground wire is connected to the metal casing to provide a direct path to ground path for a "shorting" or errant current as described in Section 6.8. For outlets that do not have a conductive casing, the ground lead is connected to a point on the outlet that distributes to all important points of the outlet.

Note the series arrangement between the thermostat and the heater but the parallel arrangement between heaters on the same circuit. In addition,

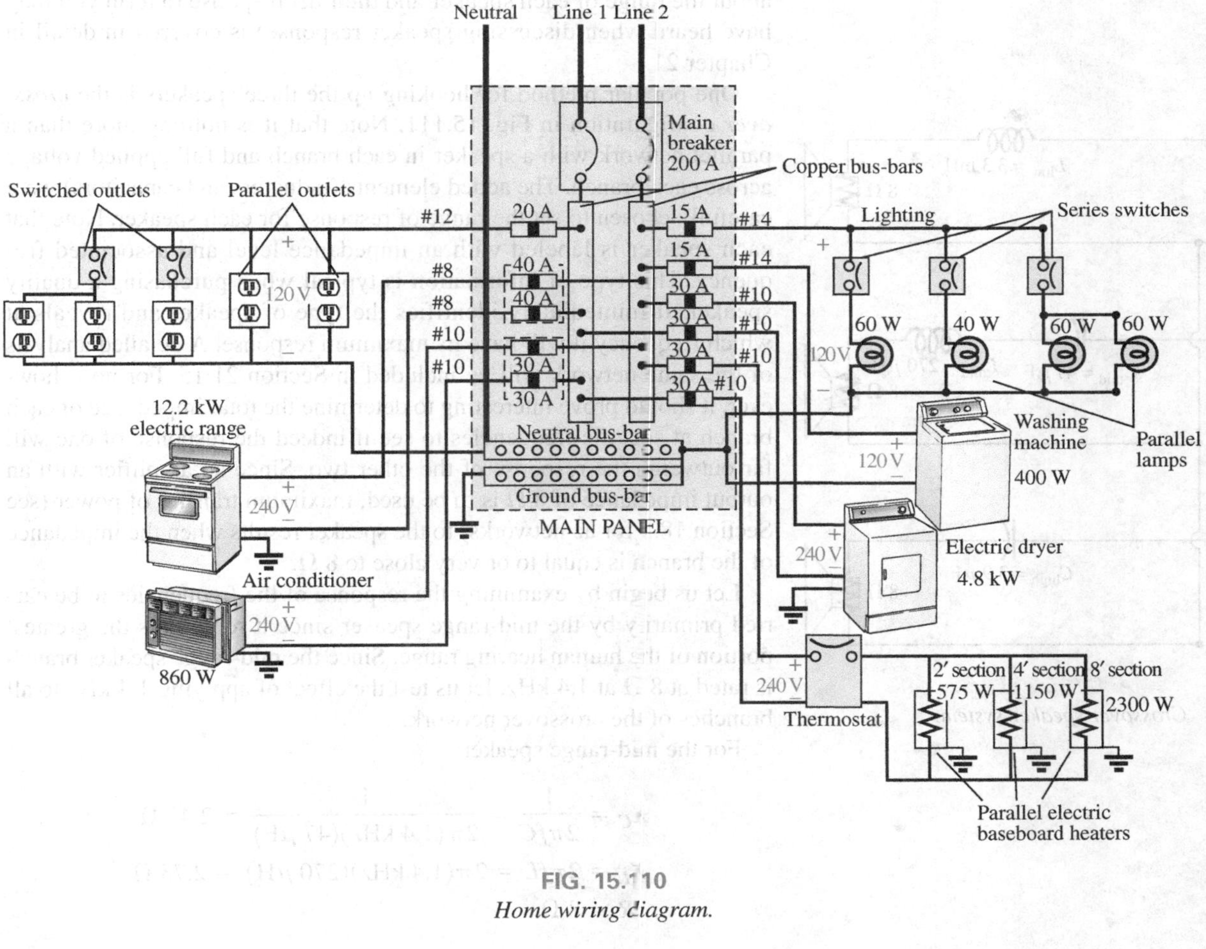

FIG. 15.110
Home wiring diagram.

note the series connection of switches to lights in the upper-right corner but the parallel connection of lights and outlets. Due to high current demand, the air conditioner, heaters, and electric stove have 30 A breakers. Keep in mind that the total current does not equal the product of the two (or 60 A) since each breaker is in a line and the same current will flow through each breaker.

In general, you now have a surface understanding of the general wiring in your home. You may not be a qualified, licensed electrician, but at least you should now be able to converse with some intelligence about the system.

Speaker Systems

The best reproduction of sound is obtained by using different speakers for the low-, mid-, and high-frequency regions. Although the typical audio range for the human ear is from about 100 Hz to 20 kHz, speakers are available from 20 Hz to 40 kHz. For the low-frequency range usually extending from about 20 Hz to 300 Hz, a speaker referred to as a *woofer* is used. Of the three speakers, it is normally the largest. The mid-range speaker is typically smaller in size and covers the range from about 100 Hz to 5 kHz. The *tweeter,* as it is normally called, is usually the smallest of the three speakers and typically covers the range from about 2 kHz to 25 kHz. There is an overlap of frequencies to ensure that frequencies aren't lost in those regions where the response of one drops off and the other takes over. A great deal more about the range of each speaker and their dB response (a term you may have heard when discussing speaker response) is covered in detail in Chapter 21.

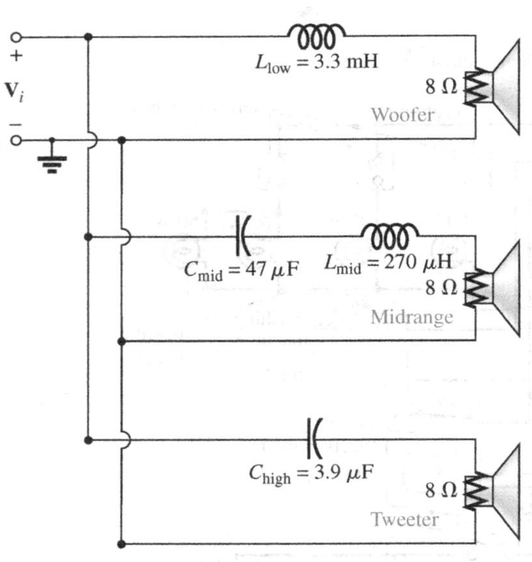

FIG. 15.111
Crossover speaker system.

One popular method for hooking up the three speakers is the *crossover* configuration in Fig. 15.111. Note that it is nothing more than a parallel network with a speaker in each branch and full applied voltage across each branch. The added elements (inductors and capacitors) were carefully chosen to set the range of response for each speaker. Note that each speaker is labeled with an impedance level and associated frequency. This type of information is typical when purchasing a quality speaker. It immediately identifies the type of speaker and reveals at which frequency it will have its maximum response. A detailed analysis of the same network will be included in Section 21.15. For now, however, it should prove interesting to determine the total impedance of each branch at specific frequencies to see if indeed the response of one will far outweigh the response of the other two. Since an amplifier with an output impedance of 8 Ω is to be used, maximum transfer of power (see Section 18.5 for ac networks) to the speaker results when the impedance of the branch is equal to or very close to 8 Ω.

Let us begin by examining the response of the frequencies to be carried primarily by the mid-range speaker since it represents the greatest portion of the human hearing range. Since the mid-range speaker branch is rated at 8 Ω at 1.4 kHz, let us test the effect of applying 1.4 kHz to all branches of the crossover network.

For the mid-range speaker:

$$X_C = \frac{1}{2\pi fC} = \frac{1}{2\pi(1.4\text{ kHz})(47\ \mu\text{F})} = 2.42\ \Omega$$

$$X_L = 2\pi fL = 2\pi(1.4\text{ kHz})(270\ \mu\text{H}) = 2.78\ \Omega$$

$$R = 8\ \Omega$$

$$\begin{aligned} \text{and} \quad Z_{\text{mid-range}} &= R + j(X_L - X_C) = 8\ \Omega + j(2.78\ \Omega - 2.42\ \Omega) \\ &= 8\ \Omega + j\,0.36\ \Omega \\ &= 8.008\ \Omega \angle -2.58° \cong 8\ \Omega \angle 0° = R \end{aligned}$$

In Fig. 15.112(a), the amplifier with the output impedance of 8 Ω has been applied across the mid-range speaker at a frequency of 1.4 kHz. Since the total reactance offered by the two series reactive elements is so small compared to the 8 Ω resistance of the speaker, we can essentially replace the series combination of the coil and capacitor by a short circuit of 0 Ω. We are then left with a situation where the load impedance is an exact match with the output impedance of the amplifier, and maximum power will be delivered to the speaker. Because of the equal series impedances, each will capture half the applied voltage or 6 V. The power to the speaker is then $V^2/R = (6\ \text{V})^2/8\ \Omega = 4.5\ \text{W}$.

At a frequency of 1.4 kHz, we would expect the woofer and tweeter to have minimum impact on the generated sound. We will now test the validity of this statement by determining the impedance of each branch at 1.4 kHz.

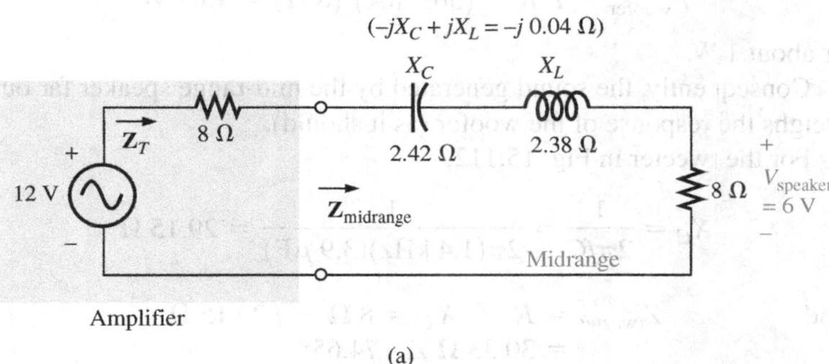

(a)

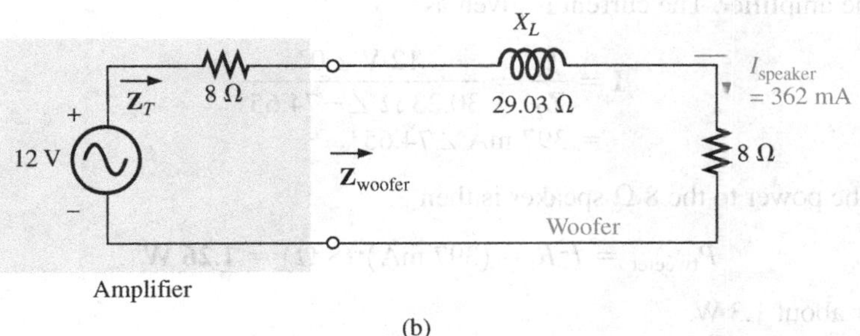

(b)

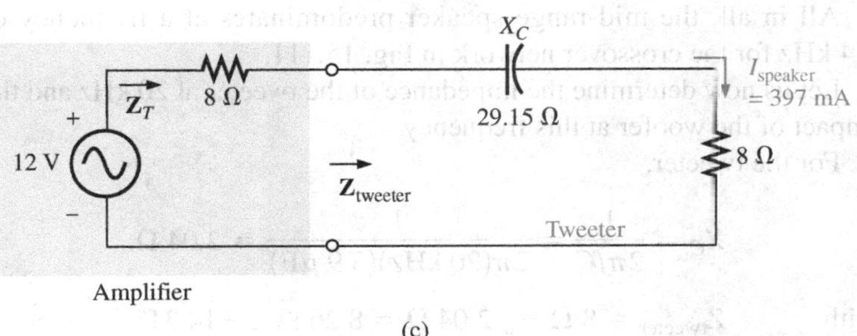

(c)

FIG. 15.112
Crossover network: (a) mid-range speaker at 1.4 kHz; (b) woofer at 1.4 kHz; (c) tweeter.

For the woofer,

$$X_L = 2\pi fL = 2\pi(1.4 \text{ kHz})(3.3 \text{ mH}) = 29.03 \ \Omega$$

and
$$\begin{aligned} Z_{\text{woofer}} &= R + jX_L = 8 \ \Omega + j\,29.03 \ \Omega \\ &= 30.11 \ \Omega \angle 74.59^\circ \end{aligned}$$

which is a poor match with the output impedance of the amplifier. The resulting network is shown in Fig. 15.112(b).

The total load on the source of 12 V is

$$\begin{aligned} Z_T &= 8 \ \Omega + 8 \ \Omega + j\,29.03 \ \Omega = 16 \ \Omega + j\,29.03 \ \Omega \\ &= 33.15 \ \Omega \angle 61.14^\circ \end{aligned}$$

and the current is

$$\begin{aligned} \mathbf{I} &= \frac{\mathbf{E}}{\mathbf{Z}_T} = \frac{12 \text{ V} \angle 0^\circ}{33.15 \ \Omega \angle 61.14^\circ} \\ &= 362 \text{ mA} \angle -61.14^\circ \end{aligned}$$

The power to the 8 Ω speaker is then

$$P_{\text{woofer}} = I^2R = (362 \text{ mA})^2(8 \ \Omega) = \mathbf{1.05 \ W}$$

or about 1 W.

Consequently, the sound generated by the mid-range speaker far outweighs the response of the woofer (as it should).

For the tweeter in Fig. 15.112,

$$X_C = \frac{1}{2\pi fC} = \frac{1}{2\pi(1.4 \text{ kHz})(3.9 \ \mu\text{F})} = 29.15 \ \Omega$$

and
$$\begin{aligned} Z_{\text{tweeter}} &= R - jX_C = 8 \ \Omega - j\,29.15 \ \Omega \\ &= 30.33 \ \Omega \angle -74.65^\circ \end{aligned}$$

which, as for the woofer, is a poor match with the output impedance of the amplifier. The current is given as

$$\begin{aligned} \mathbf{I} &= \frac{\mathbf{E}}{\mathbf{Z}_T} = \frac{12 \text{ V} \angle 0^\circ}{30.23 \ \Omega \angle -74.65^\circ} \\ &= 397 \text{ mA} \angle 74.65^\circ \end{aligned}$$

The power to the 8 Ω speaker is then

$$P_{\text{tweeter}} = I^2R = (397 \text{ mA})^2(8 \ \Omega) = \mathbf{1.26 \ W}$$

or about 1.3 W.

Consequently, the sound generated by the mid-range speaker far outweighs the response of the tweeter also.

All in all, the mid-range speaker predominates at a frequency of 1.4 kHz for the crossover network in Fig. 15.111.

Let us now determine the impedance of the tweeter at 20 kHz and the impact of the woofer at this frequency.

For the tweeter,

$$X_C = \frac{1}{2\pi fC} = \frac{1}{2\pi(20 \text{ kHz})(3.9 \ \mu\text{F})} = 2.04 \ \Omega$$

with
$$Z_{\text{tweeter}} = 8 \ \Omega - j\,2.04 \ \Omega = 8.26 \ \Omega \angle -14.31^\circ$$

Even though the magnitude of the impedance of the branch is not exactly 8 Ω, it is very close, and the speaker will receive a high level of power (actually 4.43 W).

For the woofer,

$$X_L = 2\pi fL = 2\pi(20\text{ kHz})(3.3\text{ mH}) = 414.69\ \Omega$$

with $\quad Z_{\text{woofer}} = 8\ \Omega - j\,414.69\ \Omega = 414.77\ \Omega\ \angle 88.9°$

which is a terrible match with the output impedance of the amplifier. Therefore, the speaker will receive a very low level of power (6.69 mW ≅ 0.007 W).

For all the calculations, note that the capacitive elements predominate at low frequencies and the inductive elements at high frequencies. For the low frequencies, the reactance of the coil is quite small, permitting a full transfer of power to the speaker. For the high-frequency tweeter, the reactance of the capacitor is quite small, providing a direct path for power flow to the speaker.

Phase-Shift Power Control

In Chapter 11, the internal structure of a light dimmer was examined and its basic operation described. We can now turn our attention to how the power flow to the bulb is controlled.

If the dimmer were composed of simply resistive elements, all the voltages of the network would be in phase, as shown in Fig. 15.113(a). If we assume that 20 V are required to turn on the triac in Fig. 11.68, the power will be distributed to the bulb for the period highlighted by the blue area of Fig. 15.113(a). For this situation, the bulb is close to full brightness since the applied voltage is available to the bulb for almost the entire cycle. To reduce the power to the bulb (and therefore reduce its brightness), the controlling voltage would need a lower peak voltage, as shown in Fig. 15.113(b). In fact, the waveform in Fig. 15.113(b) is such that the turn-on voltage is not reached until the peak value occurs. In this case, power is delivered to the bulb for only half the cycle, and the brightness of the bulb is reduced. The problem with using only resistive elements in a dimmer now becomes apparent: The bulb can be made no dimmer than the situation depicted by Fig. 15.113(b). Any further reduction in the controlling voltage would reduce its peak value below the trigger level, and the bulb would never turn on.

This dilemma can be resolved by using a series combination of elements such as shown in Fig. 15.114(a) from the dimmer in Fig. 11.68. Note that the controlling voltage is the voltage across the capacitor, while the full line voltage of 120 V rms, 170 V peak, is across the entire branch. To describe the behavior of the network, let us examine the case

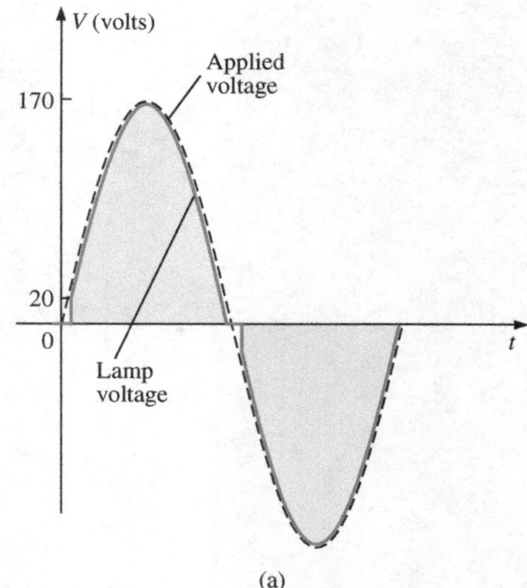

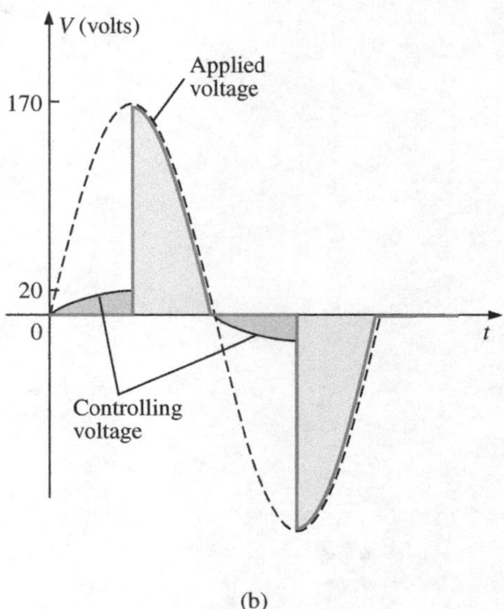

FIG. 15.113
Light dimmer: (a) with purely resistive elements; (b) half-cycle power distribution.

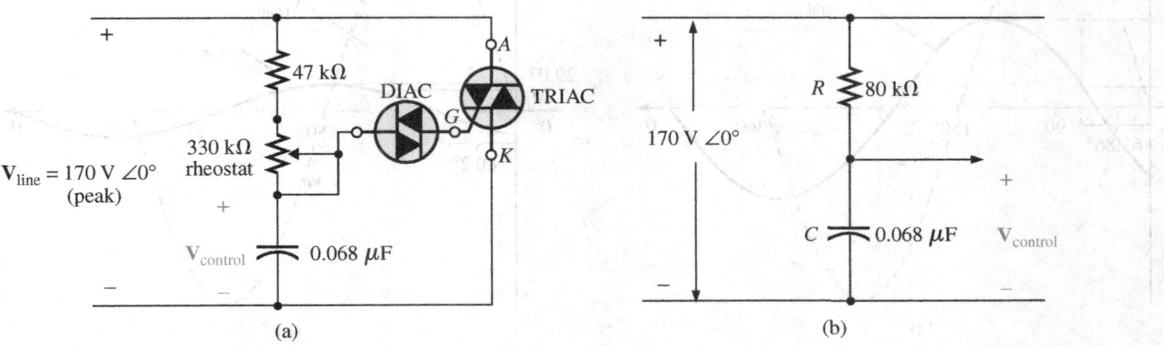

FIG. 15.114
Light dimmer: (a) from Fig. 11.68; (b) with rheostat set at 33 kΩ.

defined by setting the potentiometer (used as a rheostat) to 1/10 its maximum value, or 33 kΩ. Combining the 33 kΩ with the fixed resistance of 47 kΩ results in a total resistance of 80 kΩ and the equivalent network in Fig. 15.114(b).

At 60 Hz, the reactance of the capacitor is

$$X_C = \frac{1}{2\pi fC} = \frac{1}{2\pi(60\text{ Hz})(62\ \mu\text{F})} = 42.78\text{ k}\Omega$$

Applying the voltage divider rule gives

$$\begin{aligned}\mathbf{V}_{\text{control}} &= \frac{\mathbf{Z}_C\mathbf{V}_s}{\mathbf{Z}_R + \mathbf{Z}_C}\\ &= \frac{(42.78\text{ k}\Omega \angle -90°)(V_s \angle 0°)}{80\text{ k}\Omega - j\,42.78\text{ k}\Omega} = \frac{42.78\text{ k}\Omega\ V_s \angle -90°}{90.72\text{ k}\Omega \angle -28.14°}\\ &= 0.472\ V_s \angle -61.86°\end{aligned}$$

Using a peak value of 170 V gives

$$\begin{aligned}V_{\text{control}} &= 0.472(170\text{ V}) \angle -61.86°\\ &= 80.24\text{ V} \angle -61.86°\end{aligned}$$

producing the waveform in Fig. 15.115(a). The result is a waveform with a phase shift of 61.86° (lagging the applied line voltage) and a relatively high peak value. The high peak value results in a quick transition to the 20 V turn-on level, and power is distributed to the bulb for the major portion of the applied signal. Recall from the discussion in Chapter 11 that the response in the negative region is a replica of that achieved in the positive region. If we reduced the potentiometer resistance further, the phase angle would be reduced, and the bulb would burn brighter. The situation is now very similar to that described for the response in Fig. 15.113(a). In other words, nothing has been gained thus far by using the capacitive element in the control network. However, let us now increase the potentiometer resistance to 200 kΩ and note the effect on the controlling voltage.

That is,

$$R_T = 200\text{ k}\Omega + 47\text{ k}\Omega = 247\text{ k}\Omega$$

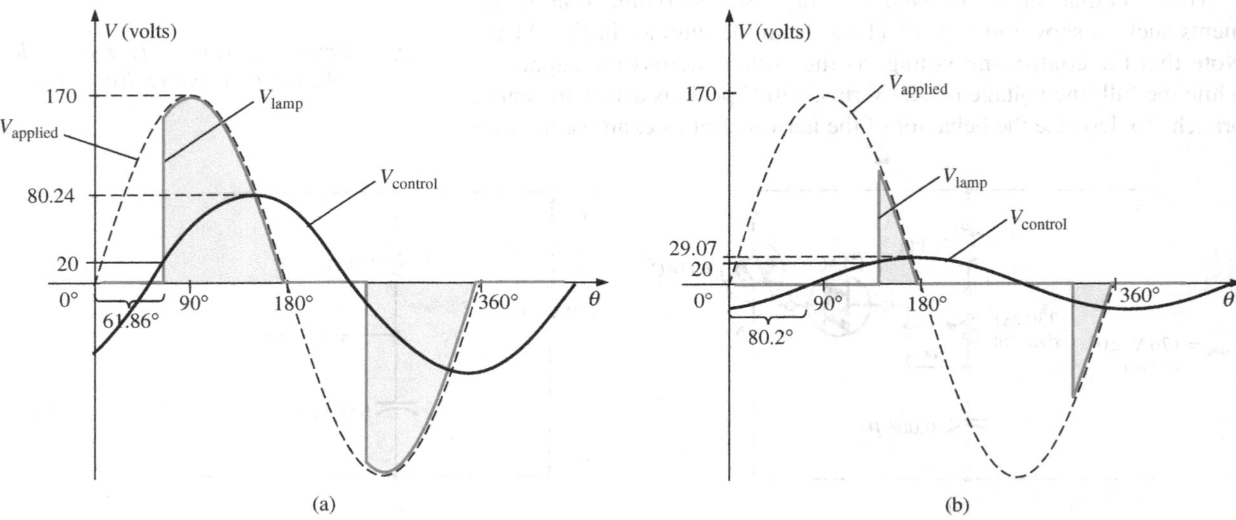

FIG. 15.115
Light dimmer in Fig. 11.68; (a) rheostat set at 33 kΩ; (b) rheostat set at 200 kΩ.

$$\mathbf{V}_{\text{control}} = \frac{\mathbf{Z}_C\mathbf{V}_s}{\mathbf{Z}_R + \mathbf{Z}_C}$$
$$= \frac{(42.78\text{ k}\Omega\angle -90°)(V_s\angle 0°)}{247\text{ k}\Omega - j\,42.78\text{ k}\Omega} = \frac{42.78\text{ k}\Omega\, V_s\angle -90°}{250.78\text{ k}\Omega\angle -9.8°}$$
$$= 0.171\,V_s\angle -80.2°$$

and using a peak value of 170 V, we have

$$\mathbf{V}_{\text{control}} = 0.171(170\text{ V})\angle -80.2°$$
$$= 29.07\text{ V}\angle -80.2°$$

The peak value has been substantially reduced to only 29.07 V, and the phase-shift angle has increased to 80.2°. The result, as depicted by Fig. 15.115(b), is that the firing potential of 20 V is not reached until near the end of the positive region of the applied voltage. Power is delivered to the bulb for only a very short period of time, causing the bulb to be quite dim, significantly dimmer than obtained from the response in Fig. 15.113(b).

A conduction angle less than 90° is therefore possible due only to the phase shift introduced by the series *R-C* combination. Thus, it is possible to construct a network of some significance with a rather simple pair of elements.

15.15 COMPUTER ANALYSIS

PSpice

Series *R-L-C* Circuit The *R-L-C* network in Fig. 15.35 is now analyzed using OrCAD Capture. Since the inductive and capacitive reactances cannot be entered onto the screen, the associated inductive and capacitive levels were first determined as follows:

$$X_L = 2\pi fL \Rightarrow L = \frac{X_L}{2\pi f} = \frac{7\ \Omega}{2\pi(1\text{ kHz})} = 1.114\text{ mH}$$

$$X_C = \frac{1}{2\pi fC} \Rightarrow C = \frac{1}{2\pi fX_C} = \frac{1}{2\pi(1\text{ kHz})3\ \Omega} = 53.05\ \mu\text{F}$$

Enter the values into the schematic as shown in Fig. 15.116. For the ac source, the sequence is **Place part** icon-**SOURCE-VSIN-OK** with **VOFF** set at 0 V, **VAMPL** set at 70.7 V (the peak value of the applied sinusoidal source in Fig. 15.35), and **FREQ** = 1 kHz. Double-click on the source symbol and the **Property Editor** appears, confirming the above choices and showing that **DF** = 0 s, **PHASE** = 0°, and **TD** = 0 s as set by the default levels. You are now ready to do an analysis of the circuit for the fixed frequency of 1 kHz.

The simulation process is initiated by first selecting the **New Simulation Profile** icon and inserting **PSpice 15-1** as the **Name** followed by **Create.** The **Simulation Settings** dialog appears and since you are continuing to plot the results against time, select the **Time Domain (Transient)** option under **Analysis type.** Since the period of each **cycle** of the applied source is 1 ms, set the **Run to time** at 5 ms so that five cycles appear. Leave the **Start saving data after** at 0 s even though there will be an oscillatory period for the reactive elements before the circuit settles down. Set the **Maximum step size** at 5 ms/1000 = 5 μs. Finally, select **OK** followed by the **Run PSpice** key. The result is a blank screen with an *x*-axis extending from 0 s to 5 ms.

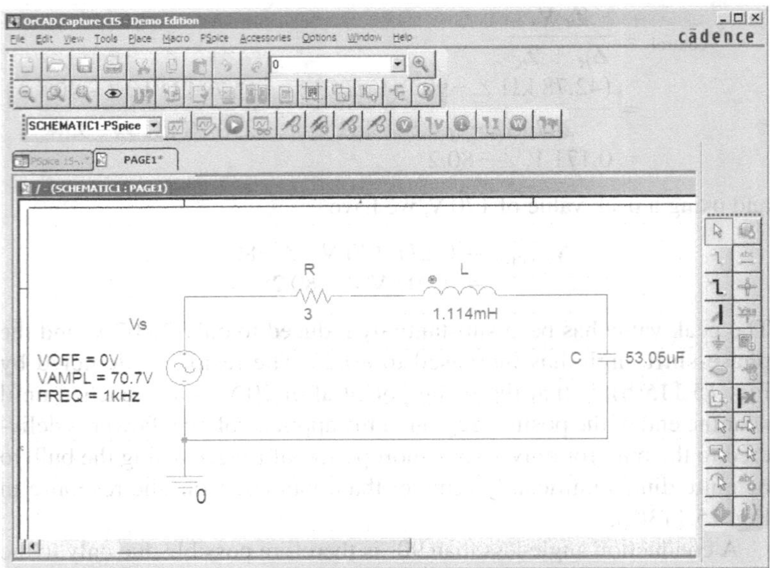

FIG. 15.116
Using PSpice to analyze a series R-L-C ac circuit.

The first quantity of interest is the current through the circuit, so select **Trace-Add-Trace** followed by **I(R)** and **OK.** The resulting plot in Fig. 15.117 clearly shows that there is a period of storing and discharging of the reactive elements before a steady-state level is established. It

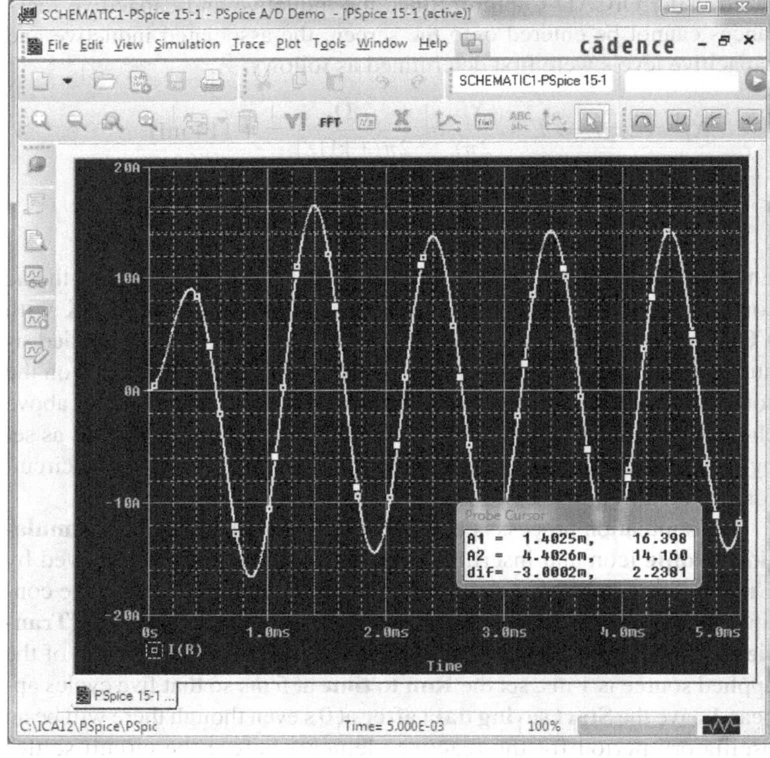

FIG. 15.117
A plot of the current for the circuit in Fig. 15.116 showing the transition from the transient state to the steady-state response.

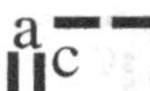

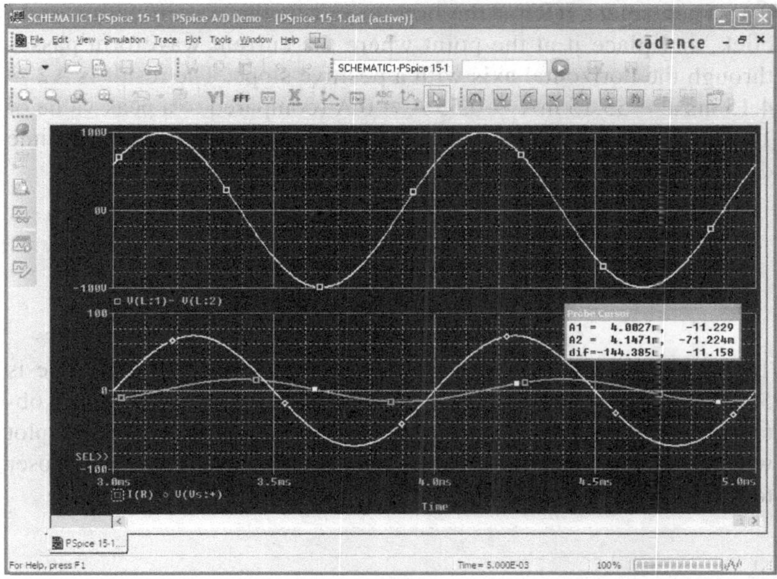

FIG. 15.118

A plot of the steady-state response ($t > 3$ ms) for v_L, v_s, and i for the circuit in Fig. 15.116.

would appear that after 3 ms, steady-state conditions have been essentially established. Select the **Toggle cursor** key, and left-click; a cursor appears that can be moved along the axis near the maximum value around 1.4 ms. In fact, the cursor reveals a maximum value of 16.4 A, which exceeds the steady-state solution by over 2 A. Right-click to establish a second cursor on the screen that can be placed near the steady-state peak around 4.4 ms. The resulting peak value is about 14.15 A, which is a match with the longhand solution for Fig. 15.35. We will therefore assume that steady-state conditions have been established for the circuit after 4 ms.

Now add the source voltage through **Trace-Add Trace-V(Vs:+)-OK** to obtain the multiple plot at the bottom of Fig. 15.118. For the voltage across the coil, the sequence **Plot-Add Plot to Window-Trace-Add Trace-V(L:1)-V(L:2)** results in the plot appearing at the top of Fig. 15.118. Take special note that the **Trace Expression** is **V(L:1)–V(L:2)** rather than just **V(L:1)** because **V(L:1)** would be the voltage from that point to ground, which would include the voltage across the capacitor. In addition, the − sign between the two comes from the **Functions or Macros** list at the right of the **Add Traces** dialog box. Finally, since we know that the waveforms are fairly steady after 3 ms, cut away the waveforms before 3 ms with **Plot-Axis Settings-X axis-User Defined-3ms to 5ms-OK** to obtain the two cycles of Fig. 15.118. Now you can clearly see that the peak value of the voltage across the coil is 100 V to match the analysis of Fig. 15.35. It is also clear that the applied voltage leads the input current by an angle that can be determined using the cursors. First activate the cursor option by selecting the cursor key (a red plot through the origin) in the second toolbar down from the menu bar. Then select **V(Vs:+)** at the bottom left of the screen with a left click, and set it at that point where the applied voltage passes through the horizontal axis with a positive slope. The result is **A1** $= 4$ ms at $-4.243\ \mu$V $\cong$

0 V. Then select **I(R)** at the bottom left of the screen with a right click, and place it at the point where the current waveform passes through the horizontal axis with a positive slope. The result is **A2** = 4.15 ms at −55.15 mA = 0.55 A ≅ 0 A (compared to a peak value of 14.14 A). At the bottom of the **Probe Cursor** dialog box, the time difference is 147.24 μs.

Now set up the ratio

$$\frac{147.24\ \mu s}{1000\ \mu s} = \frac{\theta}{360^\circ}$$

$$\theta = 52.99^\circ$$

The phase angle by which the applied voltage leads the source is 52.99°, which is very close to the theoretical solution of 53.13° obtained in Fig. 15.39. Increasing the number of data points for the plot would have increased the accuracy level and brought the results closer to 53.13°.

Multisim

We now examine the response of a network versus frequency rather than time using the network in Fig. 15.86, which now appears on the schematic in Fig. 15.119. The ac current source appears as **AC_CURRENT_SOURCE** under the **SIGNAL_CURRENT_SOURCES Family** listing. Note that the current source was given an amplitude of 1 A to establish a magnitude match between the response of the voltage across the network and the impedance of the network. That is,

$$|Z_T| = \left|\frac{V_s}{I_s}\right| = \left|\frac{V_s}{1\text{ A}}\right| = |V_s|$$

Before applying computer methods, we should develop a rough idea of what to expect so that we have something to which to compare the

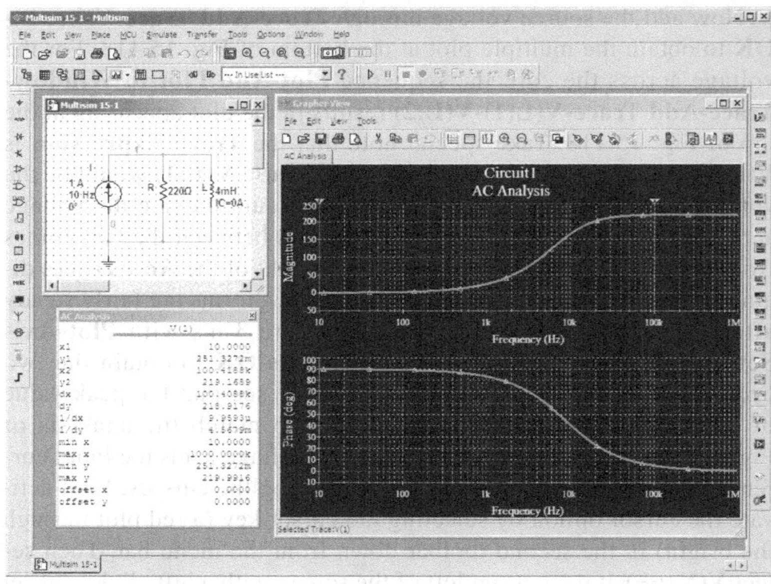

FIG. 15.119
Obtaining an impedance plot for a parallel R-L network using Multisim.

computer solution. At very high frequencies such as 1 MHz, the impedance of the inductive element will be about 25 kΩ, which when placed in parallel with the 220 Ω will look like an open circuit. The result is that as the frequency gets very high, we should expect the impedance of the network to approach the 220 Ω level of the resistor. In addition, since the network will take on resistive characteristics at very high frequencies, the angle associated with the input impedance should also approach 0°. At very low frequencies the reactance of the inductive element will be much less than the 220 Ω of the resistor, and the network will take on inductive characteristics. In fact, at, say, 10 Hz, the reactance of the inductor is only about 0.25 Ω, which is very close to a short-circuit equivalent compared to the parallel 220 Ω resistor. The result is that the impedance of the network is very close to 0 Ω at very low frequencies. Again, since the inductive effects are so strong at low frequencies, the phase angle associated with the input impedance should be very close to 90°.

Now for the computer analysis. The current source, the resistor element, and the inductor are all placed and connected using procedures described in detail in earlier chapters. However, there is one big difference this time: Since the output will be plotted versus frequency, the **AC Analysis Magnitude** in the **AC_CURRENT** dialog box for the source must be set to 1 A. In this case, the default level of **1A** matches that of the applied source, so you were set even if you failed to check the setting. In the future, however, a voltage or current source may be used that does not have an amplitude of 1, and proper entries must be made to this listing.

For the simulation, first apply the sequence **Simulate-Analyses-AC Analysis** to obtain the **AC Analysis** dialog box. Set the **Start frequency** at **10 Hz** so that you have entries at very low frequencies, and set the **Stop frequency** at **1MHz** so that you have data points at the other end of the spectrum. The **Sweep type** can remain **Decade,** but the number of points per decade will be 1000 so that you obtain a detailed plot. Set the **Vertical scale** on **Linear.** Within **Output variables, V(1).** Shifting it over to the **Selected variables for analysis** column using the **Add** keypad and then hitting the **Simulate** key results in the two plots in Fig. 15.119. Select the **Show/Hide Grid** key to place the grid on the graph, and select the **Show/Hide Cursors** key to place the **AC Analysis** dialog box appearing in Fig. 15.119. Since two graphs are present, define the one you are working on by clicking on the **Voltage** or **Phase** heading on the left side of each plot. A small red arrow appears when selected so you know which is the active plot. When setting up the cursors, be sure that you have activated the correct plot. When the red cursor is moved to 10 Hz (**x1**), you find that the voltage across the network is only 0.251 V (**y1**), resulting in an input impedance of only 0.25 Ω—quite small and matching your theoretical prediction. In addition, note that the phase angle is essentially at 90° in the other plot, confirming your other assumption above—a totally inductive network. If you set the blue cursor near 100 kHz (**x2** = 102.3 kHz), you will find that the impedance at 219.2 Ω (**y2**) is closing in on the resistance of the parallel resistor of 220 Ω, again confirming the preliminary analysis above. As noted in the bottom of the **AC Analysis** box, the maximum value of the voltage is 219.99 Ω or essentially 220 Ω at 1 MHz. Before leaving the plot, note the advantages of using a log axis when you want a response over a wide frequency range.

PROBLEMS

SECTION 15.2 Impedance and the Phasor Diagram

1. Express the impedances in Fig. 15.120 in both polar and rectangular forms.

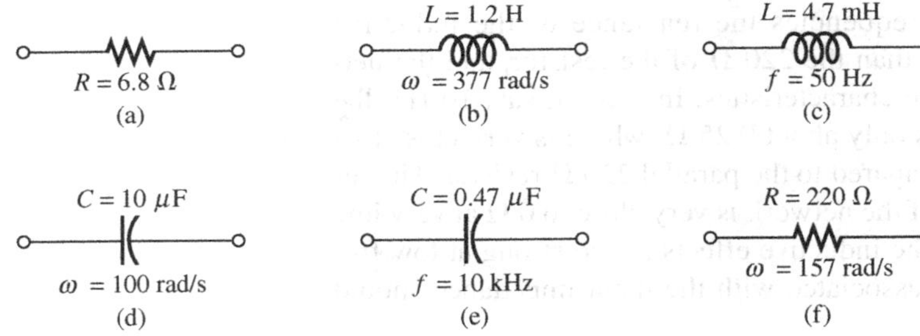

FIG. 15.120
Problem 1.

2. Find the current i for the elements in Fig. 15.121 using complex algebra. Sketch the waveforms for v and i on the same set of axes.

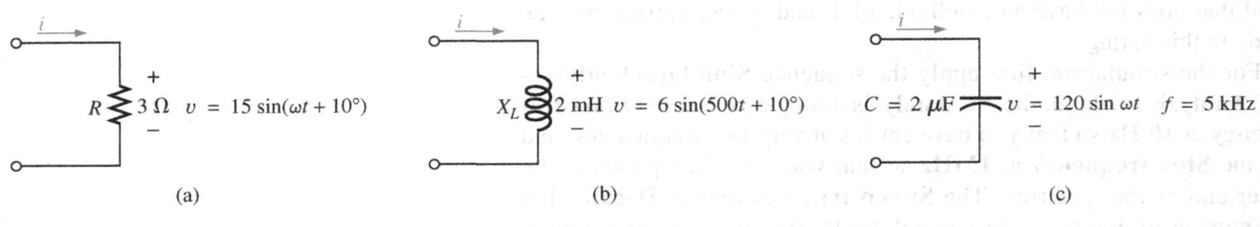

FIG. 15.121
Problem 2.

3. Find the voltage v for the elements in Fig. 15.122 using complex algebra. Sketch the waveforms of v and i on the same set of axes.

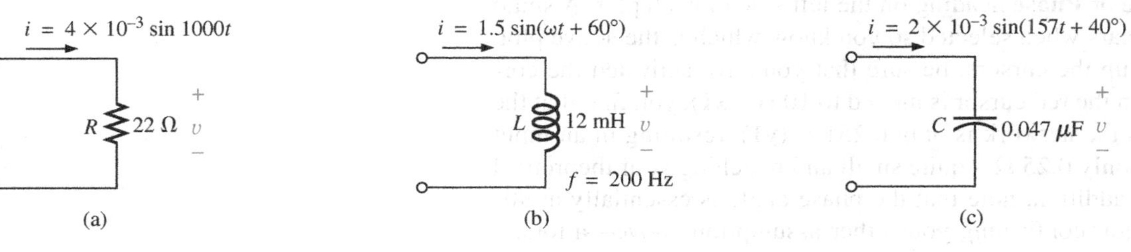

FIG. 15.122
Problem 3.

SECTION 15.3 Series Configuration

4. Calculate the total impedance of the circuits in Fig. 15.123. Express your answer in rectangular and polar forms, and draw the impedance diagram.

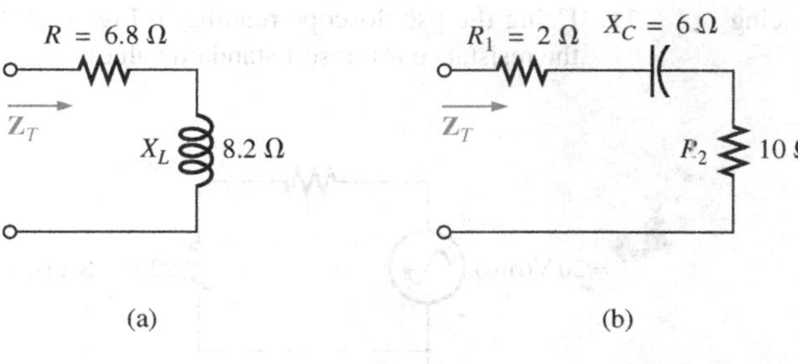

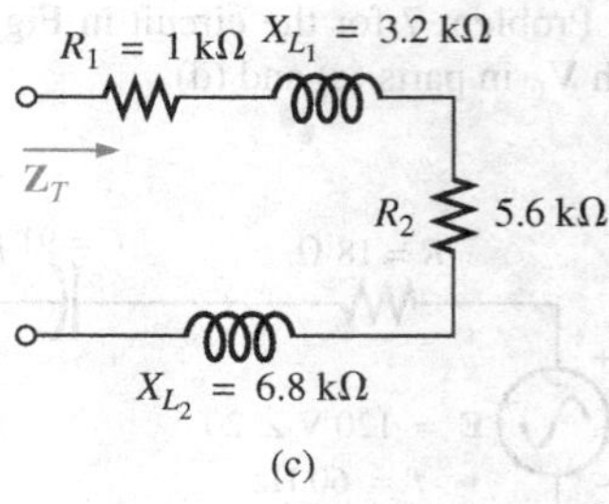

FIG. 15.123
Problem 4.

5. Calculate the total impedance of the circuits in Fig. 15.124. Express your answer in rectangular and polar forms, and draw the impedance diagram.

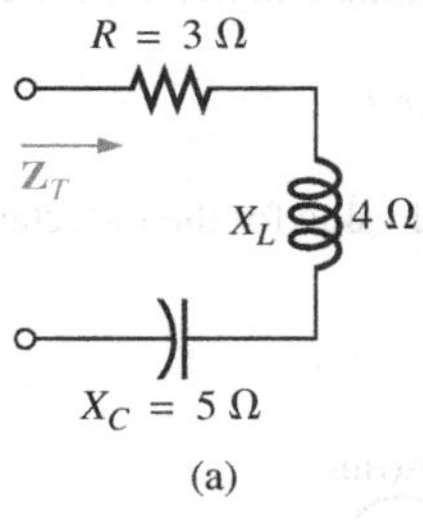

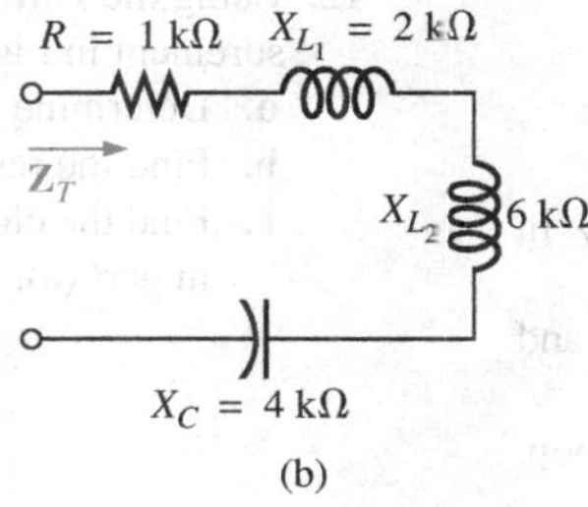

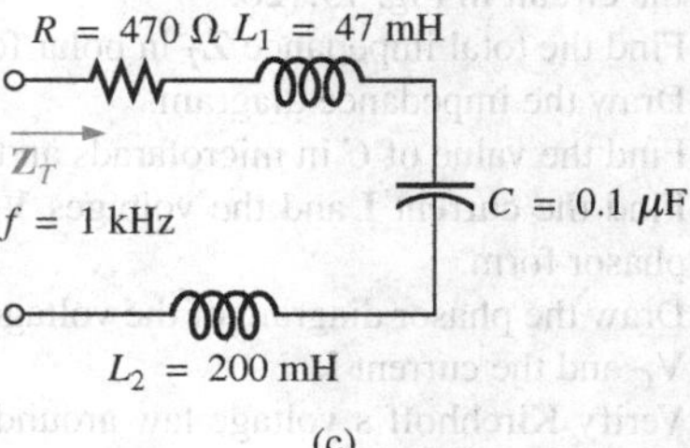

FIG. 15.124
Problem 5.

6. Find the type and impedance in ohms of the series circuit elements that must be in the closed container in Fig. 15.125 for the indicated voltages and currents to exist at the input terminals. (Find the simplest series circuit that will satisfy the indicated conditions.)

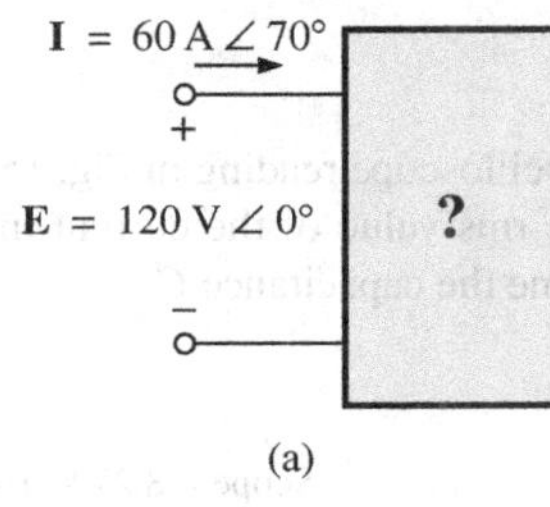

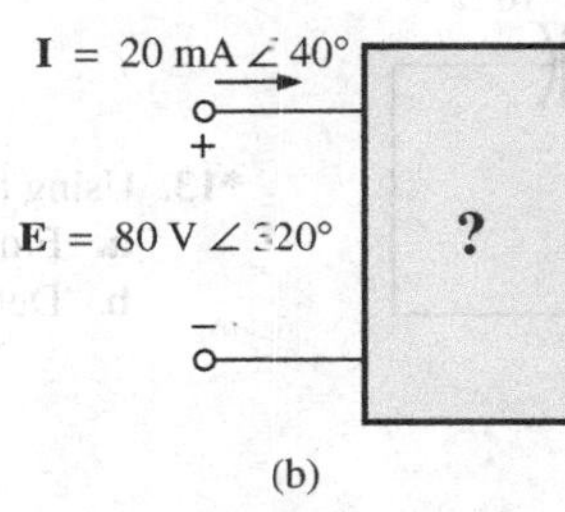

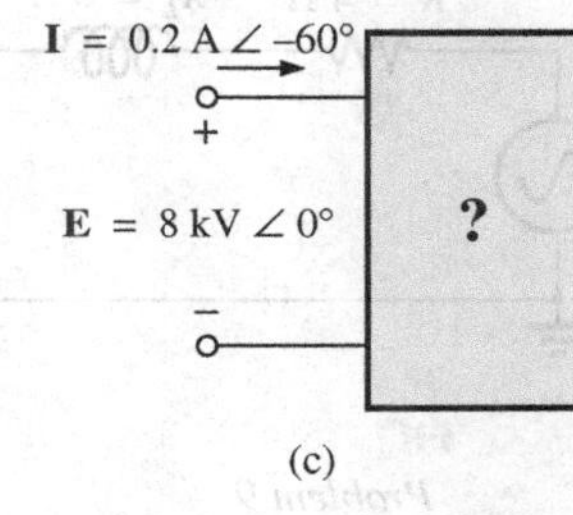

FIG. 15.125
Problems 6 and 26.

7. For the circuit in Fig. 15.126:
  a. Find the total impedance $\mathbf{Z}_T$ in polar form.
  b. Draw the impedance diagram.

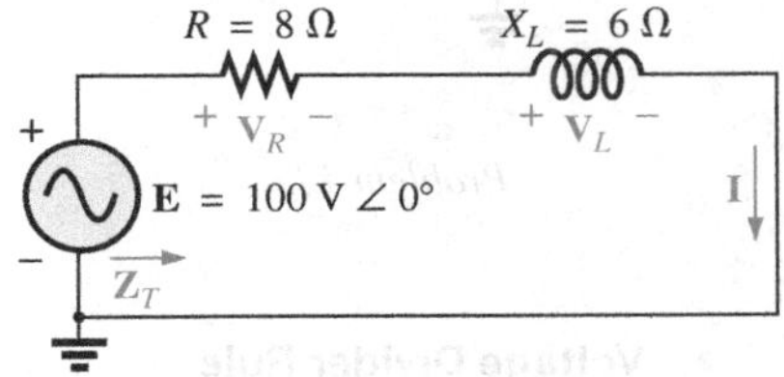

FIG. 15.126
Problems 7 and 48.

  c. Find the current **I** and the voltages $\mathbf{V}_R$ and $\mathbf{V}_L$ in phasor form.
  d. Draw the phasor diagram of the voltages **E**, $\mathbf{V}_R$, and $\mathbf{V}_L$, and the current **I**.
  e. Verify Kirchhoff's voltage law around the closed loop.
  f. Find the average power delivered to the circuit.
  g. Find the power factor of the circuit, and indicate whether it is leading or lagging.
  h. Find the sinusoidal expressions for the voltages and current if the frequency is 60 Hz.
  i. Plot the waveforms for the voltages and current on the same set of axes.

8. Repeat Problem 7 for the circuit in Fig. 15.127, replacing $\mathbf{V}_L$ with $\mathbf{V}_C$ in parts (c) and (d).

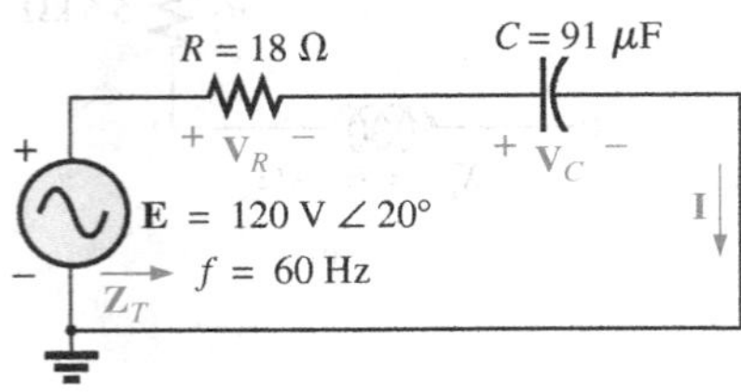

FIG. 15.127
Problem 8.

9. For the circuit in Fig. 15.128:

a. Find the total impedance $\mathbf{Z}_T$ in polar form.
b. Draw the impedance diagram.
c. Find the value of C in microfarads and L in henries.
d. Find the current **I** and the voltages $\mathbf{V}_R$, $\mathbf{V}_L$, and $\mathbf{V}_C$ in phasor form.
e. Draw the phasor diagram of the voltages **E**, $\mathbf{V}_R$, $\mathbf{V}_L$, and $\mathbf{V}_C$ and the current **I**.
f. Verify Kirchhoff's voltage law around the closed loop.
g. Find the average power delivered to the circuit.
h. Find the power factor of the circuit, and indicate whether it is leading or lagging.
i. Find the sinusoidal expressions for the voltages and current.
j. Plot the waveforms for the voltages and current on the same set of axes.

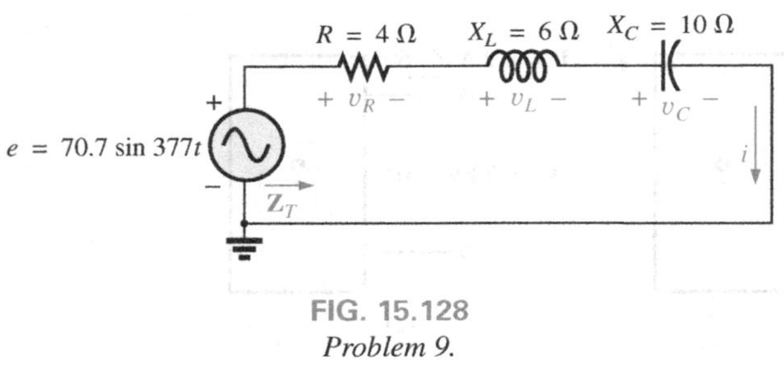

FIG. 15.128
Problem 9.

10. Repeat Problem 9 for the circuit in Fig. 15.129 except for part (c).

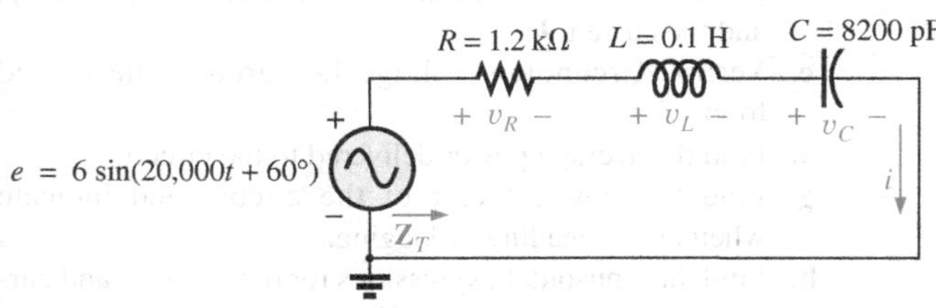

FIG. 15.129
Problem 10.

11. Using the oscilloscope reading in Fig. 15.130, determine the resistance R (closest standard value).

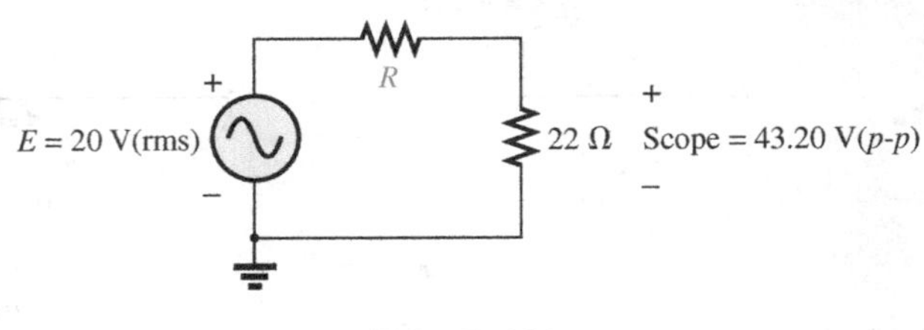

FIG. 15.130
Problem 11.

***12.** Using the DMM current reading and the oscilloscope measurement in Fig. 15.131:

a. Determine the inductance L.
b. Find the resistance R.
c. Find the closest standard value for the inductance found in part (a).

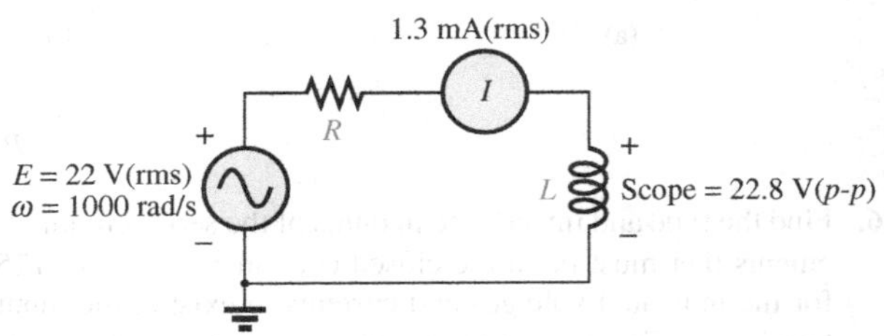

FIG. 15.131
Problem 12.

***13.** Using the oscilloscope reading in Fig. 15.132:

a. Find the rms value of the current in the series circuit.
b. Determine the capacitance C.

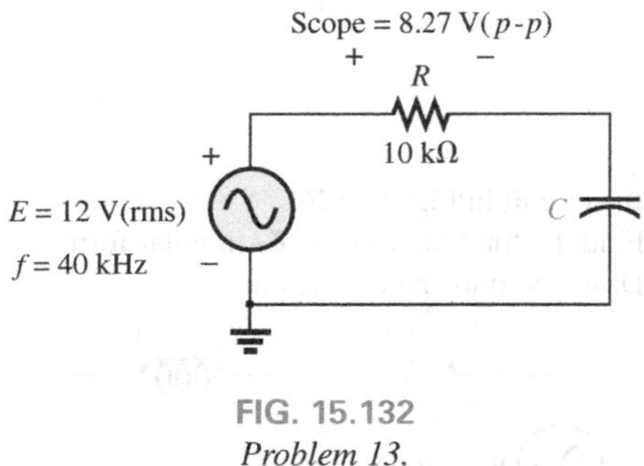

FIG. 15.132
Problem 13.

SECTION 15.4 **Voltage Divider Rule**

14. Calculate the voltages $\mathbf{V}_1$ and $\mathbf{V}_2$ for the circuits in Fig. 15.133 in phasor form using the voltage divider rule.

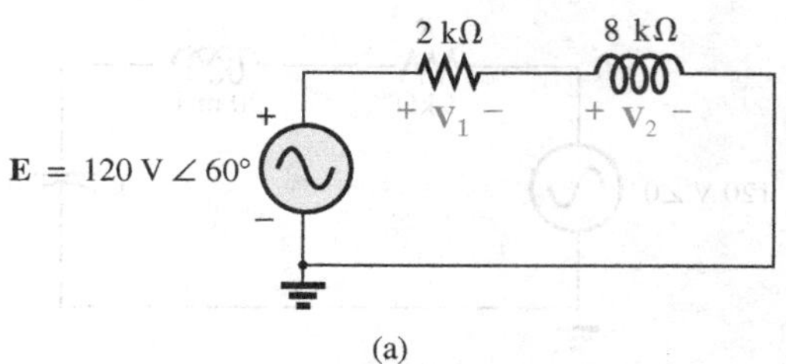

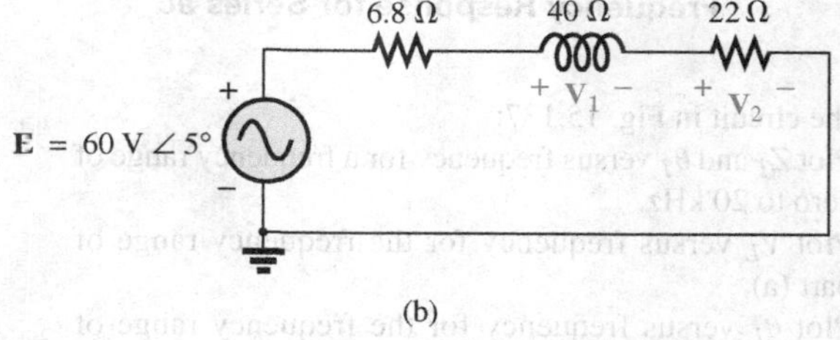

FIG. 15.133
Problem 14.

15. Calculate the voltages $\mathbf{V}_1$ and $\mathbf{V}_2$ for the circuits in Fig. 15.134 in phasor form using the voltage divider rule.

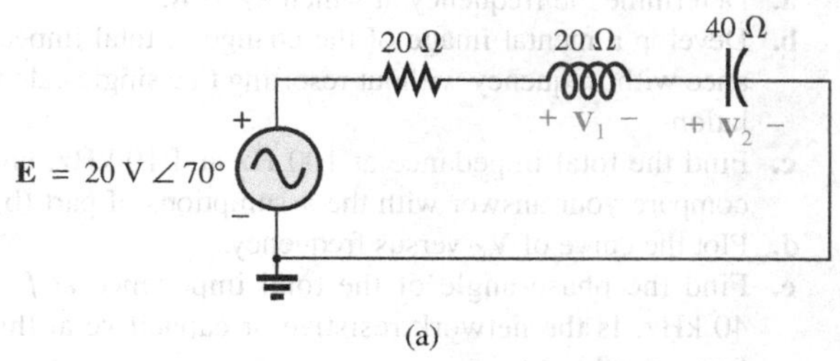

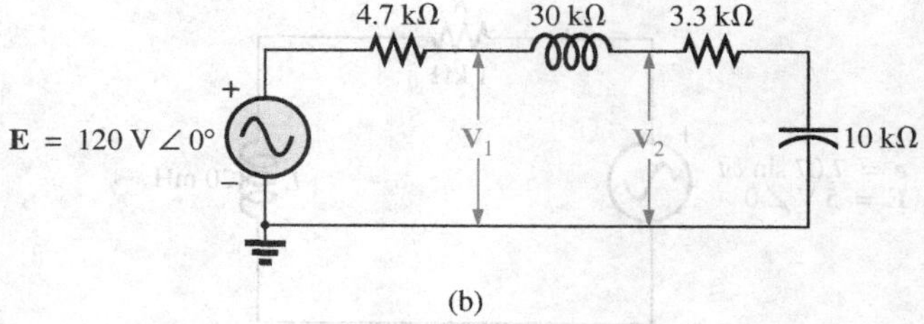

FIG. 15.134
Problem 15.

***16.** For the circuit in Fig. 15.135:

a. Determine **I**, $\mathbf{V}_R$, and $\mathbf{V}_C$ in phasor form.

b. Calculate the total power factor, and indicate whether it is leading or lagging.

c. Calculate the average power delivered to the circuit.

d. Draw the impedance diagram.

e. Draw the phasor diagram of the voltages **E**, $\mathbf{V}_R$, and $\mathbf{V}_C$, and the current **I**.

f. Find the voltages $\mathbf{V}_R$ and $\mathbf{V}_C$ using the voltage divider rule, and compare them with the results of part (a).

g. Draw the equivalent series circuit of the above as far as the total impedance and the current i are concerned.

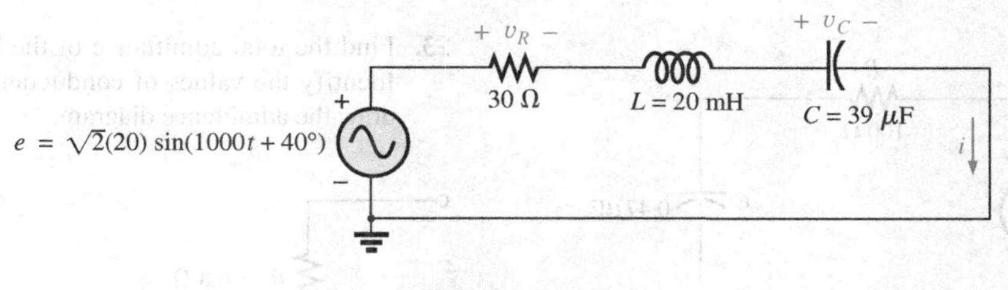

FIG. 15.135
Problems 16 and 47.

17. An electrical load has a power factor of 0.8 lagging. It dissipates 8 kW at a voltage of 200 V. Calculate the impedance of this load in rectangular coordinates.

***18.** Find the series element or elements that must be in the enclosed container in Fig. 15.136 to satisfy the following conditions:

a. Average power to circuit = 300 W.

b. Circuit has a lagging power factor.

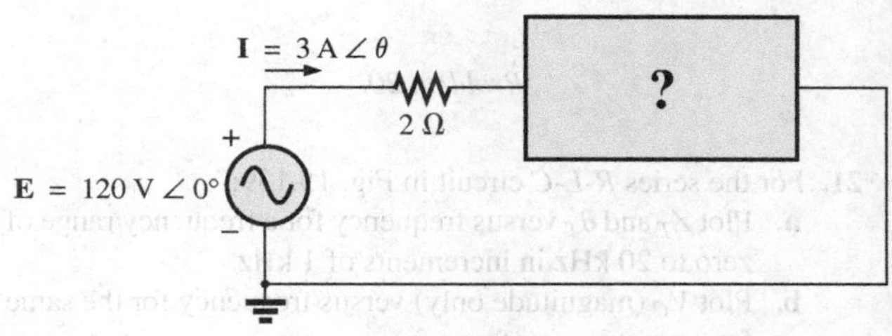

FIG. 15.136
Problem 18.

SECTION 15.5 Frequency Response for Series ac Circuits

*19. For the circuit in Fig. 15.137:
  a. Plot Z_T and θ_T versus frequency for a frequency range of zero to 20 kHz.
  b. Plot V_L versus frequency for the frequency range of part (a).
  c. Plot θ_L versus frequency for the frequency range of part (a).
  d. Plot V_R versus frequency for the frequency range of part (a).

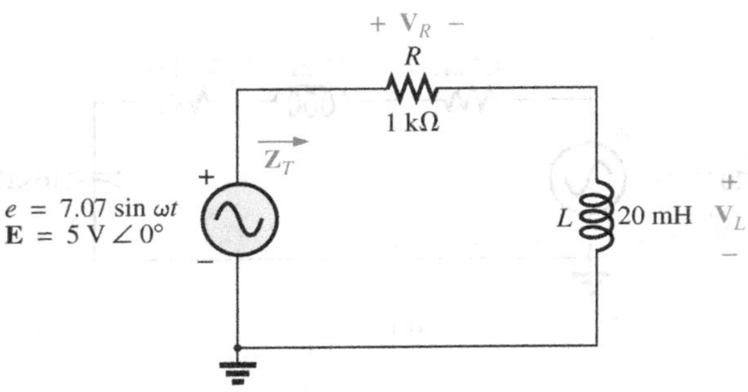

FIG. 15.137
Problem 19.

*20. For the circuit in Fig. 15.138:
  a. Plot Z_T and θ_T versus frequency for a frequency range of zero to 10 kHz.
  b. Plot V_C versus frequency for the frequency range of part (a).
  c. Plot θ_C versus frequency for the frequency range of part (a).
  d. Plot V_R versus frequency for the frequency range of part (a).

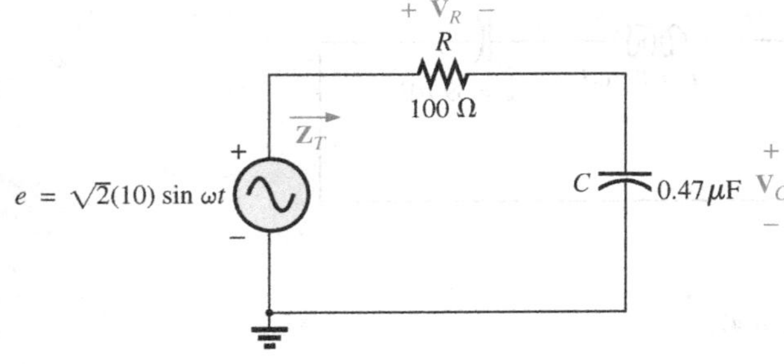

FIG. 15.138
Problem 20.

*21. For the series *R-L-C* circuit in Fig. 15.139:
  a. Plot Z_T and θ_T versus frequency for a frequency range of zero to 20 kHz in increments of 1 kHz.
  b. Plot V_C (magnitude only) versus frequency for the same frequency range of part (a).
  c. Plot I (magnitude only) versus frequency for the same frequency range of part (a).

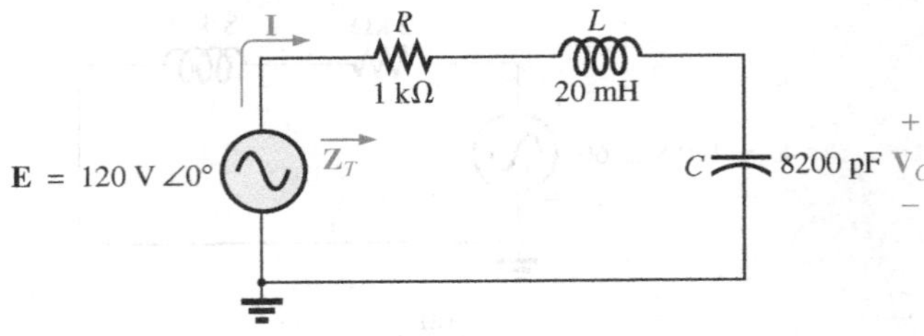

FIG. 15.139
Problem 21.

22. For the series *R-C* circuit in Fig. 15.140:
  a. Determine the frequency at which $X_C = R$.
  b. Develop a mental image of the change in total impedance with frequency without resorting to a single calculation.
  c. Find the total impedance at 100 Hz and 10 kHz, and compare your answer with the assumptions of part (b).
  d. Plot the curve of $\mathbf{V}_C$ versus frequency.
  e. Find the phase angle of the total impedance at f = 40 kHz. Is the network resistive or capacitive at this frequency?

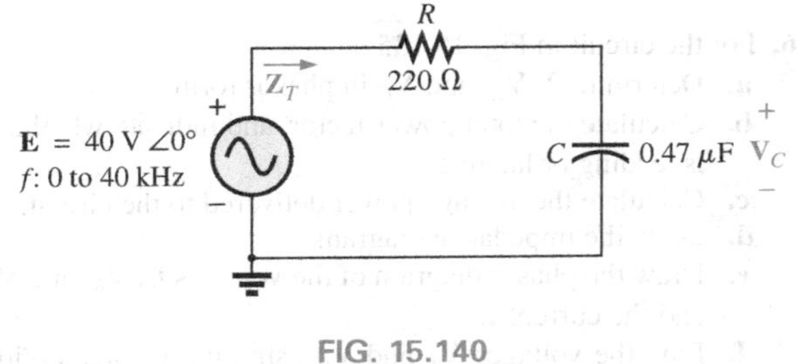

FIG. 15.140
Problem 22.

SECTION 15.7 Admittance and Susceptance

23. Find the total admittance of the branches in Fig. 15.141. Identify the values of conductance and susceptance, and draw the admittance diagram.

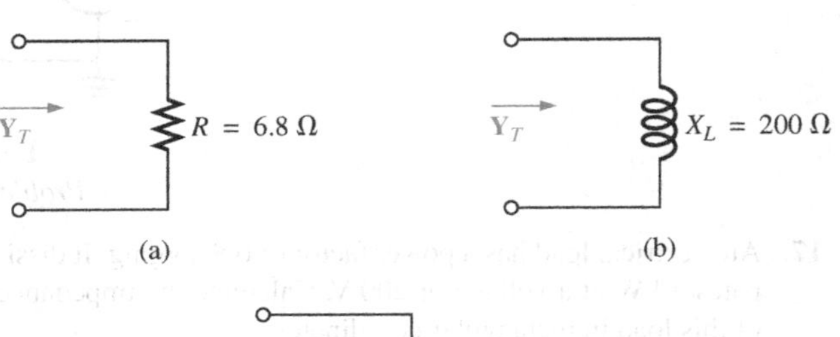

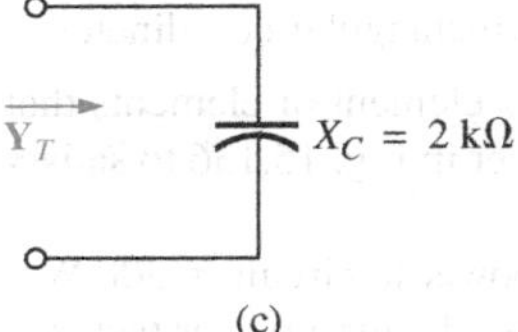

FIG. 15.141
Problem 23.

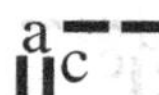

24. Find the total admittance for the networks of Fig. 15.142. Identify the values of total conductance and admittance, and draw the admittance diagram.

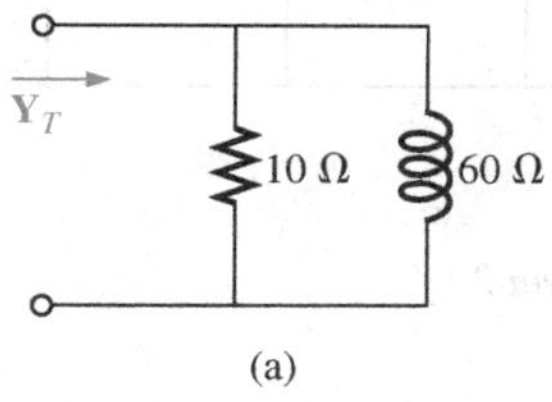

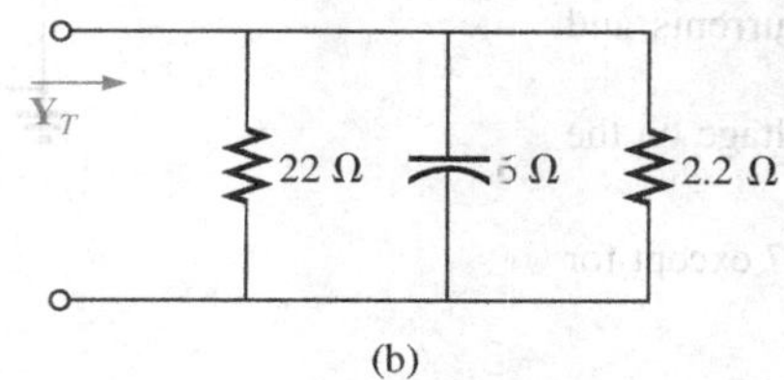

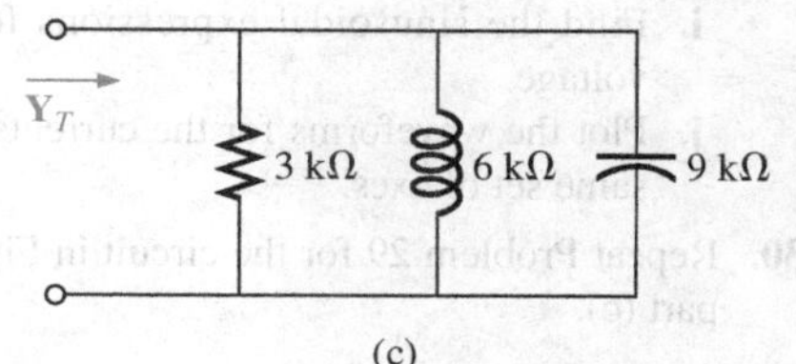

FIG. 15.142
Problem 24.

25. Find the total admittance for the networks of Fig. 15.143. Identify the values of total conductance and admittance, and draw the admittance diagram.

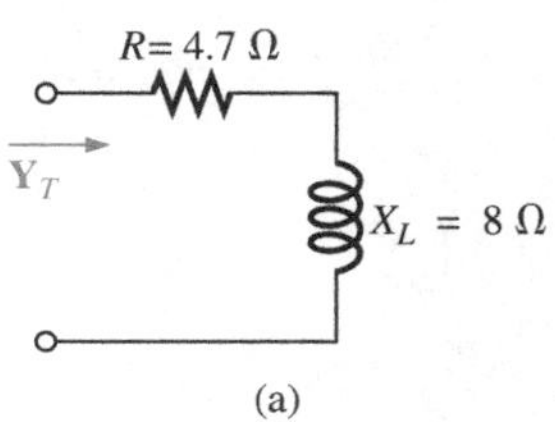

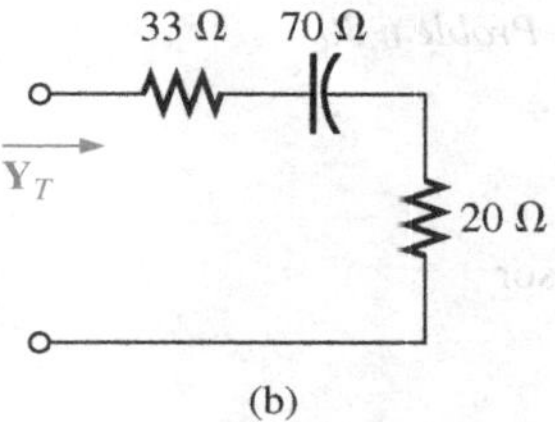

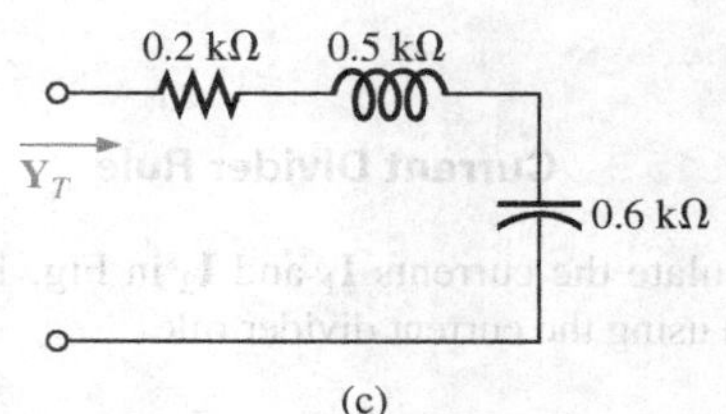

FIG. 15.143
Problem 25.

26. Repeat Problem 6 for the parallel circuit elements that must be in the closed container for the same voltage and current to exist at the input terminals. (Find the simplest parallel circuit that will satisfy the conditions indicated.)

SECTION 15.8 Parallel ac Networks

27. For the circuit in Fig. 15.144:

a. Find the total admittance $\mathbf{Y}_T$ in polar form.
b. Draw the admittance diagram.
c. Find the voltage **E** and the currents $\mathbf{I}_R$ and $\mathbf{I}_L$ in phasor form.
d. Draw the phasor diagram of the currents $\mathbf{I}_s$, $\mathbf{I}_R$, and $\mathbf{I}_L$, and the voltage **E**.
e. Verify Kirchhoff's current law at one node.
f. Find the average power delivered to the circuit.

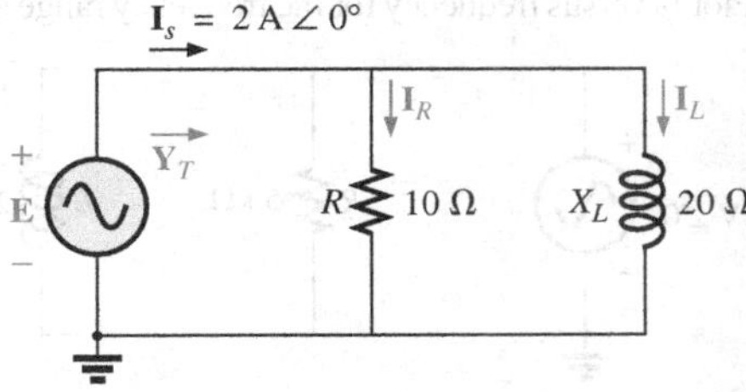

FIG. 15.144
Problem 27.

g. Find the power factor of the circuit, and indicate whether it is leading or lagging.
h. Find the sinusoidal expressions for the currents and voltage if the frequency is 60 Hz.
i. Plot the waveforms for the currents and voltage on the same set of axes.

28. Repeat Problem 27 for the circuit in Fig. 15.145, replacing $\mathbf{I}_L$ with $\mathbf{I}_C$ in parts (c) and (d).

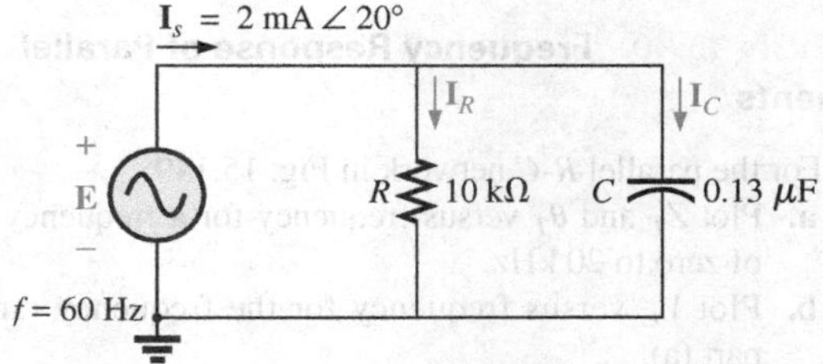

FIG. 15.145
Problem 28.

29. For the circuit in Fig. 15.146:

a. Find the total admittance and impedance in polar form.
b. Draw the admittance and impedance diagrams.
c. Find the value of C in microfarads and L in henries.
d. Find the voltage **E** and currents $\mathbf{I}_R$, $\mathbf{I}_L$, and $\mathbf{I}_C$ in phasor form.
e. Draw the phasor diagram of the currents $\mathbf{I}_s$, $\mathbf{I}_R$, $\mathbf{I}_L$, and $\mathbf{I}_C$, and the voltage **E**.

f. Verify Kirchhoff's current law at one node.
g. Find the average power delivered to the circuit.
h. Find the power factor of the circuit, and indicate whether it is leading or lagging.
i. Find the sinusoidal expressions for the currents and voltage.
j. Plot the waveforms for the currents and voltage on the same set of axes.

30. Repeat Problem 29 for the circuit in Fig. 15.147 except for part (c).

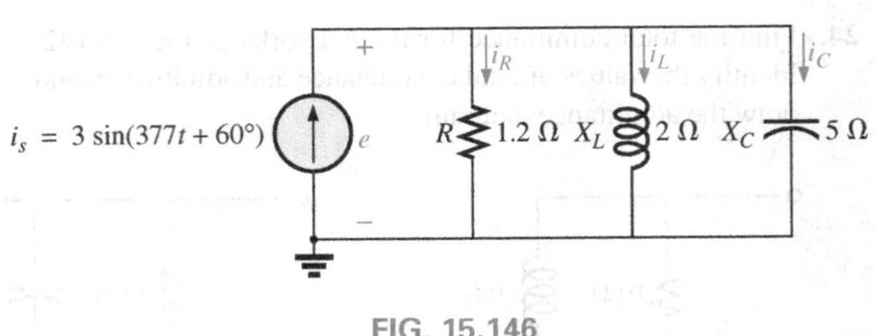

FIG. 15.146
Problem 29.

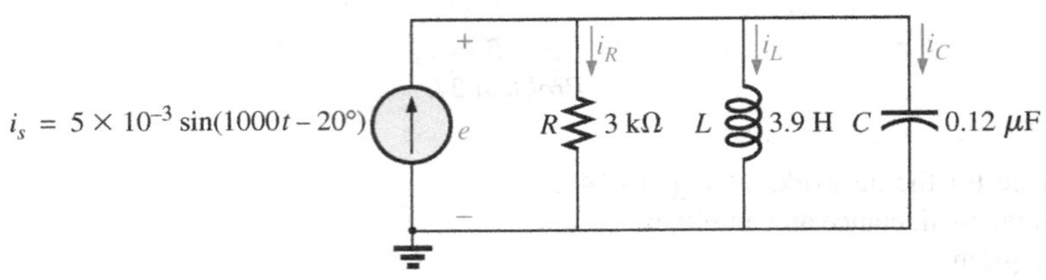

FIG. 15.147
Problem 30.

SECTION 15.9 Current Divider Rule

31. Calculate the currents $\mathbf{I}_1$ and $\mathbf{I}_2$ in Fig. 15.148 in phasor form using the current divider rule.

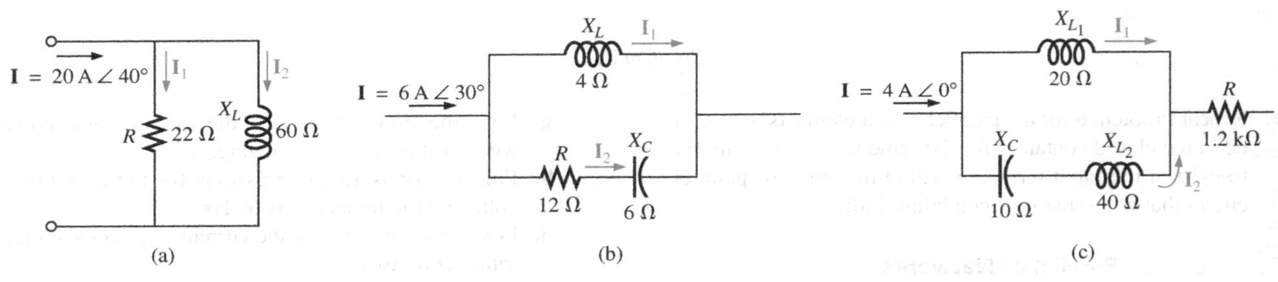

FIG. 15.148
Problem 31.

SECTION 15.10 Frequency Response of Parallel Elements

***32.** For the parallel *R-C* network in Fig. 15.149:
a. Plot Z_T and θ_T versus frequency for a frequency range of zero to 20 kHz.
b. Plot V_C versus frequency for the frequency range of part (a).
c. Plot I_R versus frequency for the frequency range of part (a).

***33.** For the parallel *R-L* network in Fig. 15.150:
a. Plot Z_T and θ_T versus frequency for a frequency range of zero to 10 kHz.
b. Plot I_L versus frequency for the frequency range of part (a).
c. Plot I_R versus frequency for the frequency range of part (a).

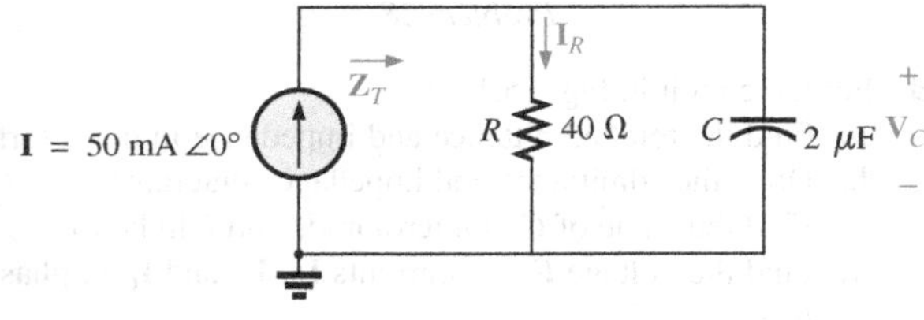

FIG. 15.149
Problems 32 and 34.

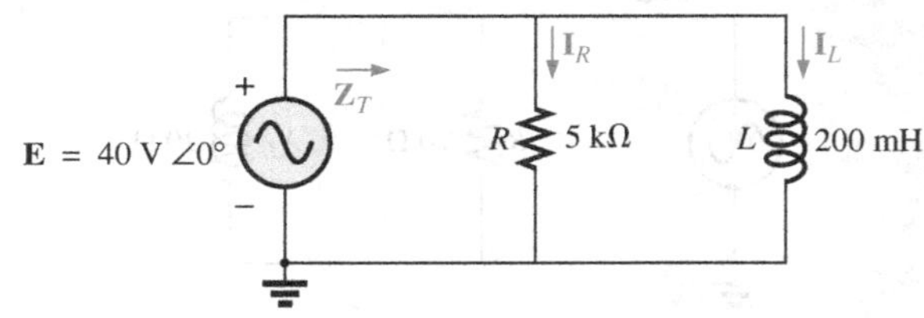

FIG. 15.150
Problems 33 and 35.

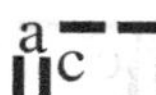

34. Plot Y_T and θ_T (of $\mathbf{Y}_T = Y_T \angle \theta_T$) for a frequency range of zero to 20 kHz for the network in Fig. 15.149.

35. Plot Y_T and θ_T (of $\mathbf{Y}_T = Y_T \angle \theta_T$) for a frequency range of zero to 10 kHz for the network in Fig. 15.150.

36. For the parallel *R-L-C* network in Fig. 15.151:

a. Plot Y_T and θ_T (of $\mathbf{Y}_T = Y_T \angle \theta_T$) for a frequency range of zero to 20 kHz.

b. Repeat part (a) for Z_T and θ_T (of $\mathbf{Z}_T = Z_T \angle \theta_T$).

c. Plot V_C versus frequency for the frequency range of part (a).

d. Plot I_L versus frequency for the frequency range of part (a).

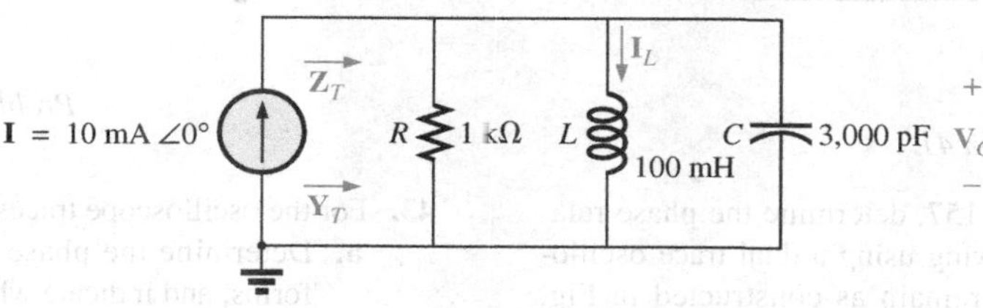

FIG. 15.151
Problem 36.

SECTION 15.12 Equivalent Circuits

37. For the series circuits in Fig. 15.152, find a parallel circuit that will have the same total impedance ($\mathbf{Z}_T$).

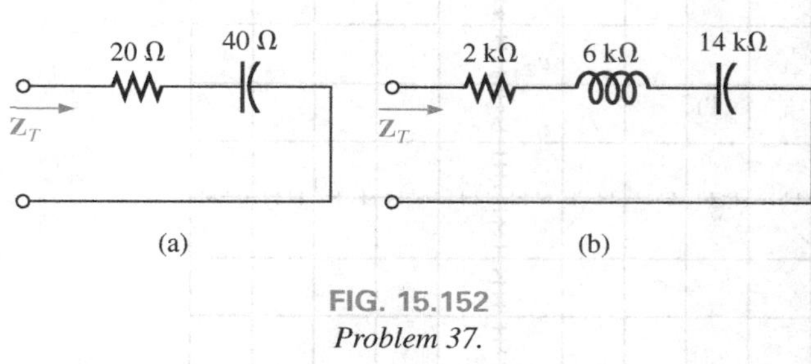

FIG. 15.152
Problem 37.

38. For the parallel circuits in Fig. 15.153, find a series circuit that will have the same total impedance.

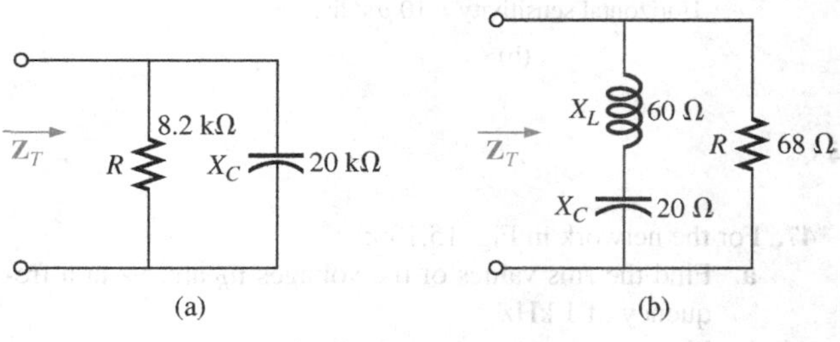

FIG. 15.153
Problem 38.

***39.** For the network in Fig. 15.154:

a. Calculate **E**, $\mathbf{I}_R$, and $\mathbf{I}_L$ in phasor form.

b. Calculate the total power factor, and indicate whether it is leading or lagging.

c. Calculate the average power delivered to the circuit.

d. Draw the admittance diagram.

e. Draw the phasor diagram of the currents $\mathbf{I}_s$, $\mathbf{I}_R$, and $\mathbf{I}_L$, and the voltage **E**.

f. Find the current $\mathbf{I}_C$ for each capacitor using only Kirchhoff's current law.

g. Find the series circuit of one resistive and reactive element that will have the same impedance as the original circuit.

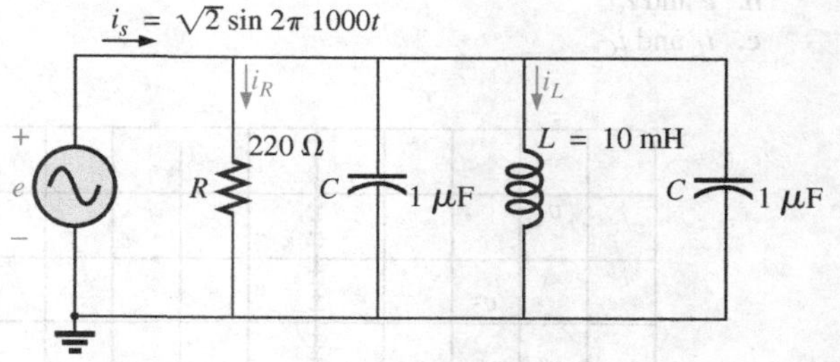

$i_s = \sqrt{2} \sin 2\pi\, 1000t$

FIG. 15.154
Problem 39.

40. Find the element or elements that must be in the closed container in Fig. 15.155 to satisfy the following conditions. (Find the simplest parallel circuit that will satisfy the indicated conditions.)

a. Average power to the circuit = 3000 W.

b. Circuit has a lagging power factor.

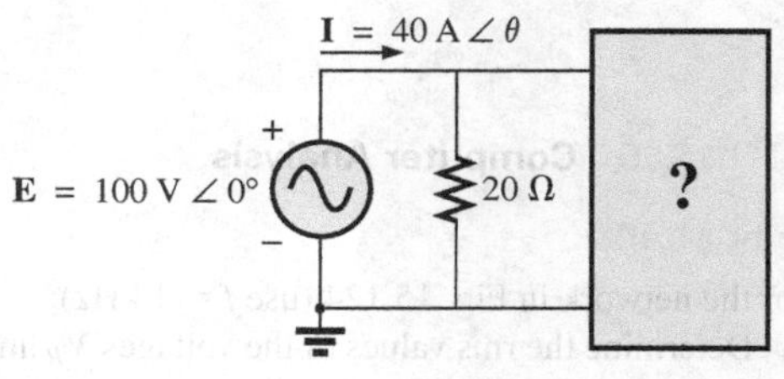

FIG. 15.155
Problem 40.

SECTION 15.13 Phase Measurements

41. For the circuit in Fig. 15.156, determine the phase relationship between the following using a dual-trace oscilloscope. The circuit can be reconstructed differently for each part, but do not use sensing resistors. Show all connections on a redrawn diagram.

a. e and v_C

b. e and i_s

c. e and v_L

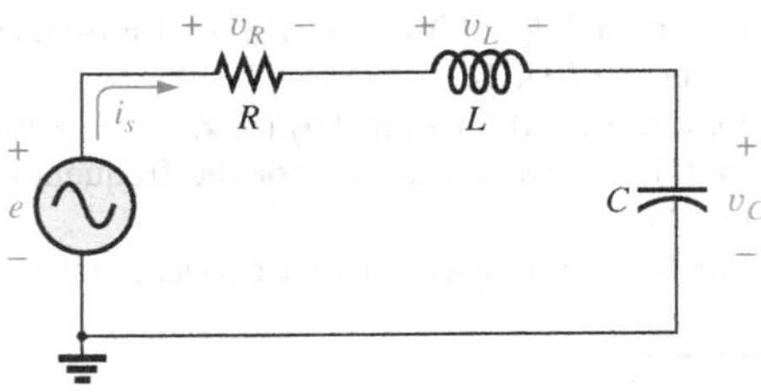

FIG. 15.156
Problem 41.

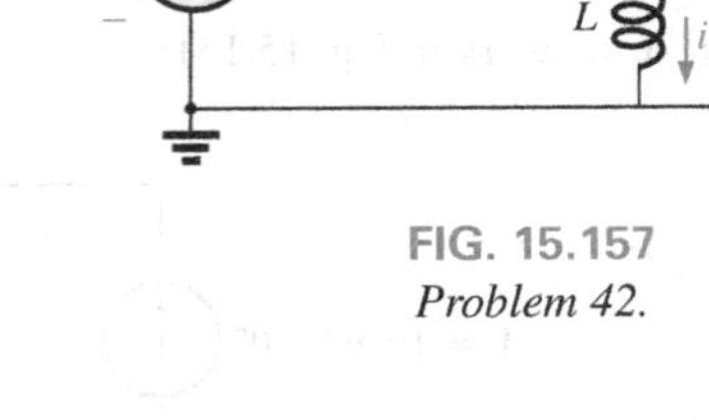

FIG. 15.157
Problem 42.

42. For the network in Fig. 15.157, determine the phase relationship between the following using a dual-trace oscilloscope. The network must remain as constructed in Fig. 15.157, but sensing resistors can be introduced. Show all connections on a redrawn diagram.
  a. e and v_{R_2}
  b. e and i_s
  c. i_L and i_C

43. For the oscilloscope traces in Fig. 15.158:
  a. Determine the phase relationship between the waveforms, and indicate which one leads or lags.
  b. Determine the peak-to-peak and rms values of each waveform.
  c. Find the frequency of each waveform.

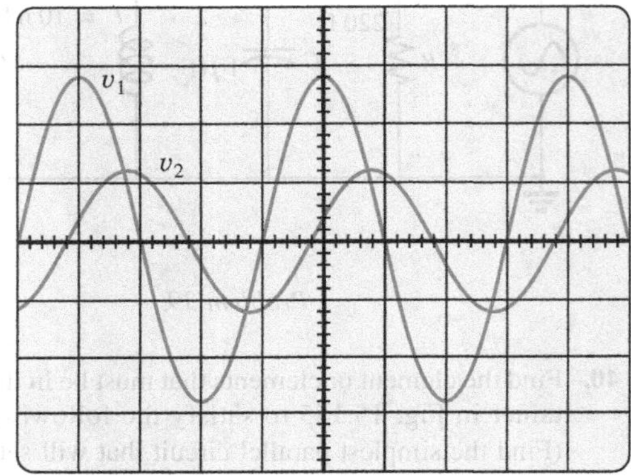

Vertical sensitivity = 0.5 V/div.
Horizontal sensitivity = 0.2 ms/div.

(a)

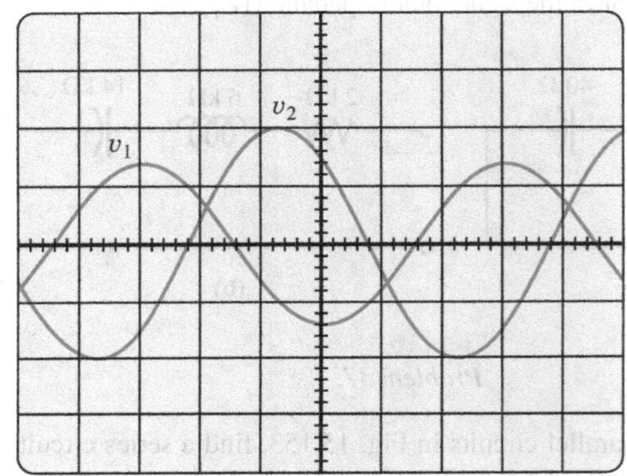

Vertical sensitivity = 2 V/div.
Horizontal sensitivity = 10 μs/div.

(b)

FIG. 15.158
Problem 43.

SECTION 15.15 Computer Analysis

PSpice or Multisim

44. For the network in Fig. 15.124 (use $f = 1$ kHz):
  a. Determine the rms values of the voltages $\mathbf{V}_R$ and $\mathbf{V}_L$ and the current **I**.
  b. Plot v_R, v_L, and i versus time on separate plots.
  c. Place e, v_R, v_L, and i on the same plot, and label accordingly.

45. For the network in Fig. 15.144:
  a. Determine the rms values of the currents $\mathbf{I}_s$, $\mathbf{I}_R$, and $\mathbf{I}_L$.
  b. Plot i_s, i_R, and i_L versus time on separate plots.
  c. Place e, i_s, i_R, and i_L on the same plot, and label accordingly.

46. For the network in Fig. 15.127:
  a. Plot the impedance of the network versus frequency from 0 to 10 kHz.
  b. Plot the current i versus frequency for the frequency range zero to 10 kHz.

***47.** For the network in Fig. 15.135:
  a. Find the rms values of the voltages v_R and v_C at a frequency of 1 kHz.
  b. Plot v_C versus frequency for the frequency range zero to 10 kHz.
  c. Plot the phase angle between e and i for the frequency range zero to 10 kHz.

GLOSSARY

Admittance A measure of how easily a network will "admit" the passage of current through that system. It is measured in siemens, abbreviated S, and is represented by the capital letter *Y*.

Admittance diagram A vector display that clearly depicts the magnitude of the admittance of the conductance, capacitive susceptance, and inductive susceptance and the magnitude and angle of the total admittance of the system.

Current divider rule A method by which the current through either of two parallel branches can be determined in an ac network without first finding the voltage across the parallel branches.

Equivalent circuits For every series ac network, there is a parallel ac network (and vice versa) that will be "equivalent" in the sense that the input current and impedance are the same.

Impedance diagram A vector display that clearly depicts the magnitude of the impedance of the resistive, reactive, and capacitive components of a network and the magnitude and angle of the total impedance of the system.

Parallel ac circuits A connection of elements in an ac network in which all the elements have two points in common. The voltage is the same across each element.

Phasor diagram A vector display that provides at a glance the magnitude and phase relationships among the various voltages and currents of a network.

Series ac configuration A connection of elements in an ac network in which no two impedances have more than one terminal in common and the current is the same through each element.

Susceptance A measure of how "susceptible" an element is to the passage of current through it. It is measured in siemens, abbreviated S, and is represented by the capital letter B.

Voltage divider rule A method through which the voltage across one element of a series of elements in an ac network can be determined without first having to find the current through the elements.

Current divider rule A method by which the current through one of two parallel branches can be determined in an ac network without first finding the voltage across the parallel branches.

Equivalent circuits For every series ac network there is a parallel ac network (and vice versa) that will be "equivalent" in the sense that the input current and impedance are the same.

Impedance diagram A vector display that clearly depicts the magnitude of the impedance of the resistive, reactive, and capacitive components of a network and the magnitude and angle of the total impedance of the system.

Parallel ac circuits A connection of elements in an ac network in which all the elements have two points in common. The voltage is the same across each element.

Phasor diagram A vector display that provides at a glance the magnitude and phase relationships among the various voltages and currents of a network.

Series ac configuration A connection of elements in an ac network in which no two impedances have more than one terminal in common and the current is the same through each element.

Susceptance A measure of how susceptible an element is to the passage of current through it. It is measured in siemens, abbreviated S, and is represented by the capital letter B.

Voltage divider rule A method through which the voltage across one element of a series of elements in an ac network can be determined without first having to find the current through the elements.

Series-Parallel ac Networks

16

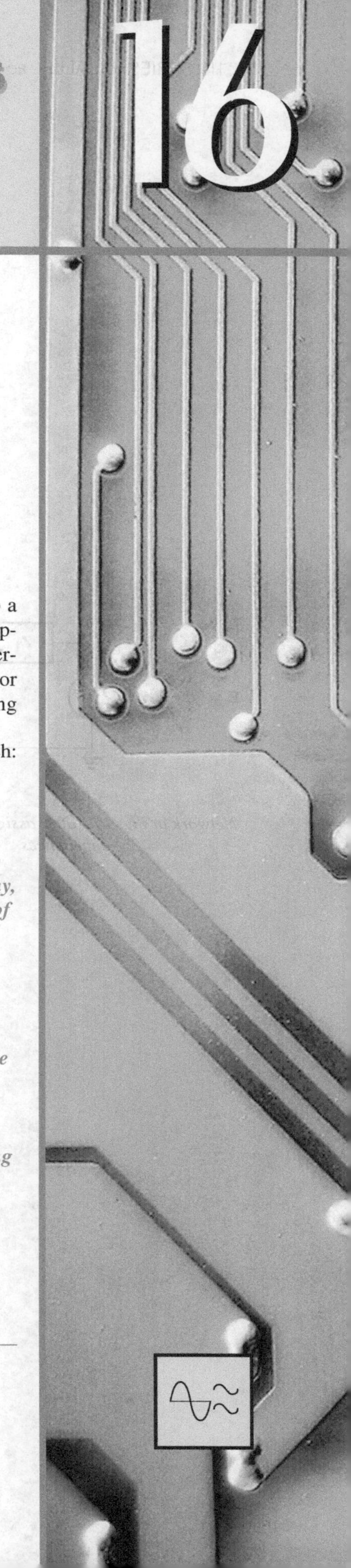

Objectives

- *Develop confidence in the analysis of series-parallel ac networks.*
- *Become proficient in the use of calculators and computer methods to support the analysis of ac series-parallel networks.*
- *Understand the importance of proper grounding in the operation of any electrical system.*

16.1 INTRODUCTION

In this chapter, we shall use the fundamental concepts of the previous chapter to develop a technique for solving **series-parallel ac networks.** A brief review of Chapter 7 may be helpful before considering these networks since the approach here is quite similar to that undertaken earlier. The circuits to be discussed have only one source of energy, either potential or current. Networks with two or more sources are considered in Chapters 17 and 18, using methods previously described for dc circuits.

In general, when working with series-parallel ac networks, consider the following approach:

1. *Redraw the network, using block impedances to combine obvious series and parallel elements, which will reduce the network to one that clearly reveals the fundamental structure of the system.*
2. *Study the problem and make a brief mental sketch of the overall approach you plan to use. Doing this may result in time- and energy-saving shortcuts. In some cases, a lengthy, drawn-out analysis may not be necessary. A single application of a fundamental law of circuit analysis may result in the desired solution.*
3. *After the overall approach has been determined, it is usually best to consider each branch involved in your method independently before tying them together in series-parallel combinations. In most cases, work back from the obvious series and parallel combinations to the source to determine the total impedance of the network. The source current can then be determined, and the path back to specific unknowns can be defined. As you progress back to the source, continually define those unknowns that have not been lost in the reduction process. It will save time when you have to work back through the network to find specific quantities.*
4. *When you have arrived at a solution, check to see that it is reasonable by considering the magnitudes of the energy source and the elements in the circuit. If not, either solve the network using another approach or check over your work very carefully. At this point, a computer solution can be an invaluable asset in the validation process.*

16.2 ILLUSTRATIVE EXAMPLES

EXAMPLE 16.1 For the network in Fig. 16.1:

a. Calculate $\mathbf{Z}_T$.
b. Determine $\mathbf{I}_s$.
c. Calculate $\mathbf{V}_R$ and $\mathbf{V}_C$.

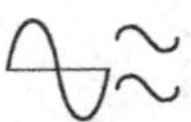

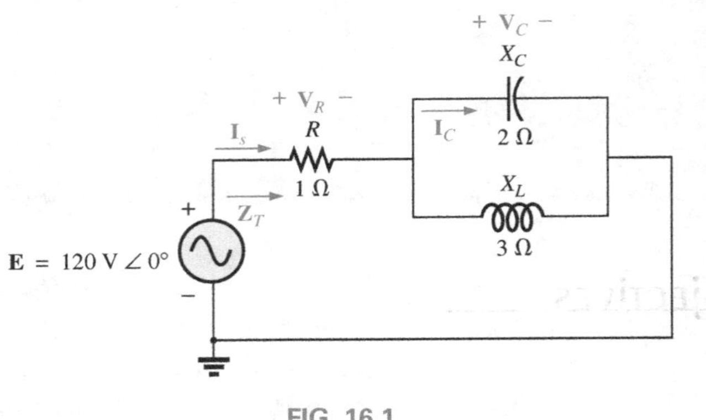

FIG. 16.1
Example 16.1.

d. Find $\mathbf{I}_C$.
e. Compute the power delivered.
f. Find F_p of the network.

Solutions:

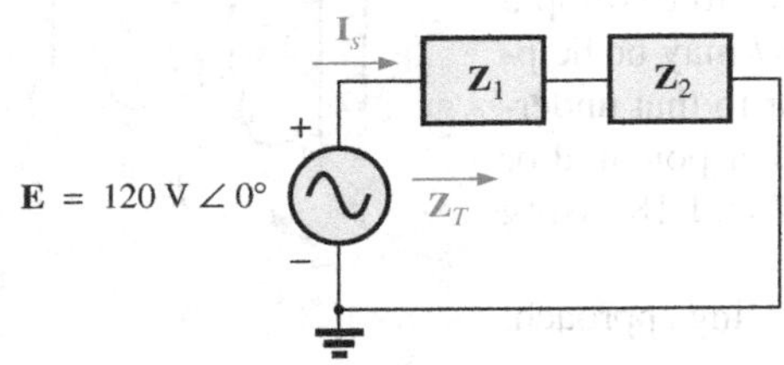

FIG. 16.2
Network in Fig. 16.1 after assigning the block impedances.

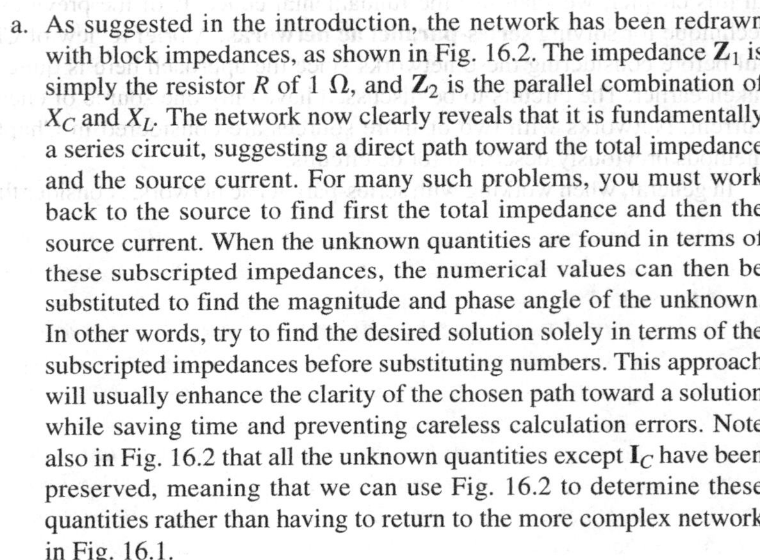

a. As suggested in the introduction, the network has been redrawn with block impedances, as shown in Fig. 16.2. The impedance $\mathbf{Z}_1$ is simply the resistor R of 1 Ω, and $\mathbf{Z}_2$ is the parallel combination of X_C and X_L. The network now clearly reveals that it is fundamentally a series circuit, suggesting a direct path toward the total impedance and the source current. For many such problems, you must work back to the source to find first the total impedance and then the source current. When the unknown quantities are found in terms of these subscripted impedances, the numerical values can then be substituted to find the magnitude and phase angle of the unknown. In other words, try to find the desired solution solely in terms of the subscripted impedances before substituting numbers. This approach will usually enhance the clarity of the chosen path toward a solution while saving time and preventing careless calculation errors. Note also in Fig. 16.2 that all the unknown quantities except $\mathbf{I}_C$ have been preserved, meaning that we can use Fig. 16.2 to determine these quantities rather than having to return to the more complex network in Fig. 16.1.

The total impedance is defined by

$$\mathbf{Z}_T = \mathbf{Z}_1 + \mathbf{Z}_2$$

with

$$\mathbf{Z}_1 = R \angle 0^\circ = 1\ \Omega \angle 0^\circ$$

$$\mathbf{Z}_2 = \mathbf{Z}_C \parallel \mathbf{Z}_L = \frac{(X_C \angle -90^\circ)(X_L \angle 90^\circ)}{-jX_C + jX_L} = \frac{(2\ \Omega \angle -90^\circ)(3\ \Omega \angle 90^\circ)}{-j\,2\ \Omega + j\,3\ \Omega}$$

$$= \frac{6\Omega \angle 0^\circ}{j\,1} = \frac{6\ \Omega \angle 0^\circ}{1 \angle 90^\circ} = 6\ \Omega \angle -90^\circ$$

and

$$\mathbf{Z}_T = \mathbf{Z}_1 + \mathbf{Z}_2 = 1\ \Omega - j\,6\ \Omega = \mathbf{6.08\ \Omega \angle -80.54^\circ}$$

b. $$\mathbf{I}_s = \frac{\mathbf{E}}{\mathbf{Z}_T} = \frac{120\ \text{V} \angle 0^\circ}{6.08\ \Omega \angle -80.54^\circ} = \mathbf{19.74\ A \angle 80.54^\circ}$$

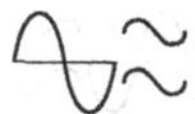

c. Referring to Fig. 16.2, we find that $\mathbf{V}_R$ and $\mathbf{V}_C$ can be found by a direct application of Ohm's law:

$$\mathbf{V}_R = \mathbf{I}_s\mathbf{Z}_1 = (19.74 \text{ A} \angle 80.54°)(1 \, \Omega \angle 0°) = \mathbf{19.74 \text{ V}\angle 80.54°}$$
$$\mathbf{V}_C = \mathbf{I}_s\mathbf{Z}_2 = (19.74 \text{ A} \angle 80.54°)(6 \, \Omega \angle -90°) = \mathbf{118.44 \text{ V}\angle -9.46°}$$

d. Now that $\mathbf{V}_C$ is known, the current $\mathbf{I}_C$ can also be found using Ohm's law:

$$\mathbf{I}_C = \frac{\mathbf{V}_C}{\mathbf{Z}_C} = \frac{118.44 \text{ V} \angle -9.46°}{2 \, \Omega \angle -90°} = \mathbf{59.22 \text{ A} \angle 80.54°}$$

e. $P_{del} = I_s^2 R = (19.74 \text{ A})^2(1 \, \Omega) = \mathbf{389.67 \text{ W}}$

f. $F_p = \cos\theta = \cos 80.54° = \mathbf{0.164 \text{ leading}}$

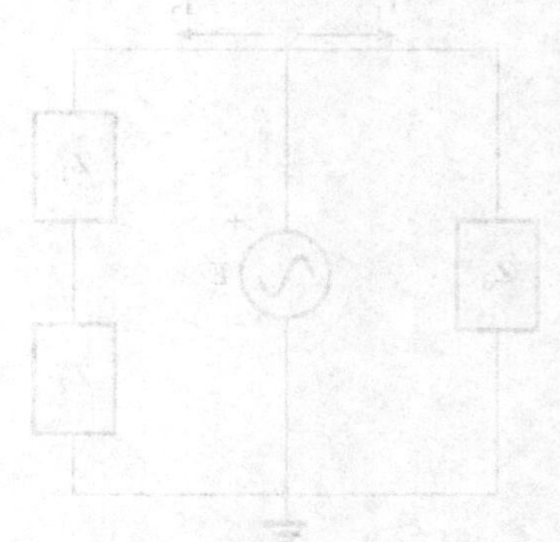

The fact that the total impedance has a negative phase angle (revealing that $\mathbf{I}_s$ leads $\mathbf{E}$) is a clear indication that the network is capacitive in nature and therefore has a leading power factor. The fact that the network is capacitive can be determined from the original network by first realizing that, for the parallel *L-C* elements, the smaller impedance predominates and results in an *R-C* network.

EXAMPLE 16.2 For the network in Fig. 16.3:

a. If **I** is 50 A ∠30°, calculate $\mathbf{I}_1$ using the current divider rule.
b. Repeat part (a) for $\mathbf{I}_2$.
c. Verify Kirchhoff's current law at one node.

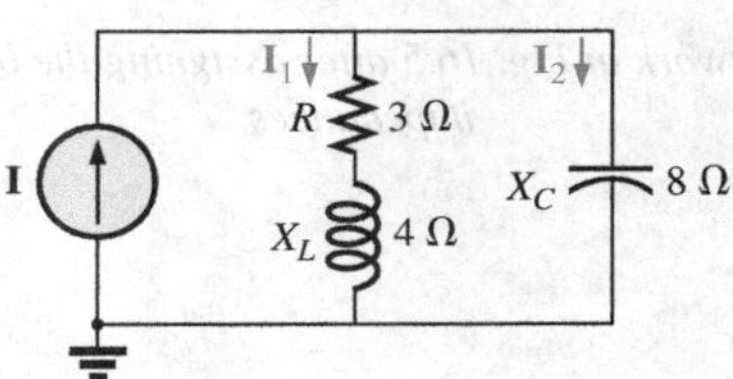

FIG. 16.3
Example 16.2.

Solutions:

a. Redrawing the circuit as in Fig. 16.4, we have

$$\mathbf{Z}_1 = R + jX_L = 3 \, \Omega + j\,4 \, \Omega = 5 \, \Omega \angle 53.13°$$
$$\mathbf{Z}_2 = -jX_C = -j\,8 \, \Omega = 8 \, \Omega \angle -90°$$

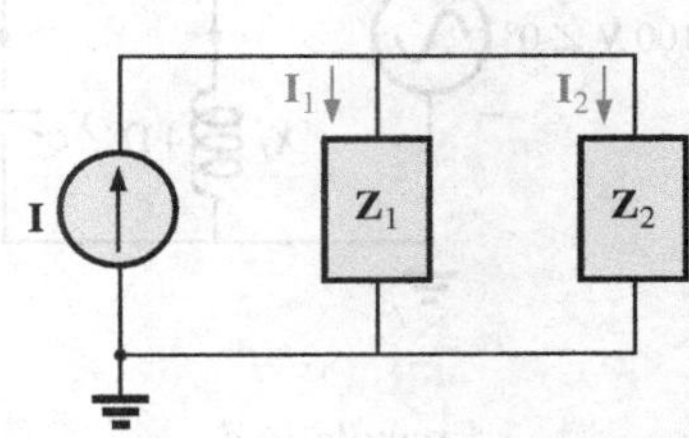

FIG. 16.4
Network in Fig. 16.3 after assigning the block impedances.

Using the current divider rule yields

$$\mathbf{I}_1 = \frac{\mathbf{Z}_2\mathbf{I}}{\mathbf{Z}_2 + \mathbf{Z}_1} = \frac{(8 \, \Omega \angle -90°)(50 \text{ A} \angle 30°)}{(-j\,8 \, \Omega) + (3 \, \Omega + j\,4 \, \Omega)} = \frac{400 \angle -60°}{3 - j4}$$
$$= \frac{400 \angle -60°}{5 \angle -53.13°} = \mathbf{80 \text{ A} \angle -6.87°}$$

b. $$\mathbf{I}_2 = \frac{\mathbf{Z}_1\mathbf{I}}{\mathbf{Z}_2 + \mathbf{Z}_1} = \frac{(5 \, \Omega \angle 53.13°)(50 \text{ A} \angle 30°)}{5 \, \Omega \angle -53.13°} = \frac{250 \angle 83.13°}{5 \angle -53.13°}$$
$$= \mathbf{50 \text{ A} \angle 136.26°}$$

c. $$\mathbf{I} = \mathbf{I}_1 + \mathbf{I}_2$$
$$50 \text{ A} \angle 30° = 80 \text{ A} \angle -6.87° + 50 \text{ A} \angle 136.26°$$
$$= (79.43 - j\,9.57) + (-36.12 + j\,34.57)$$
$$= 43.31 + j\,25.0$$
$$50 \text{ A} \angle 30° = 50 \text{ A} \angle 30° \quad \text{(checks)}$$

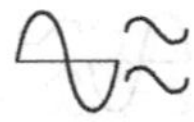

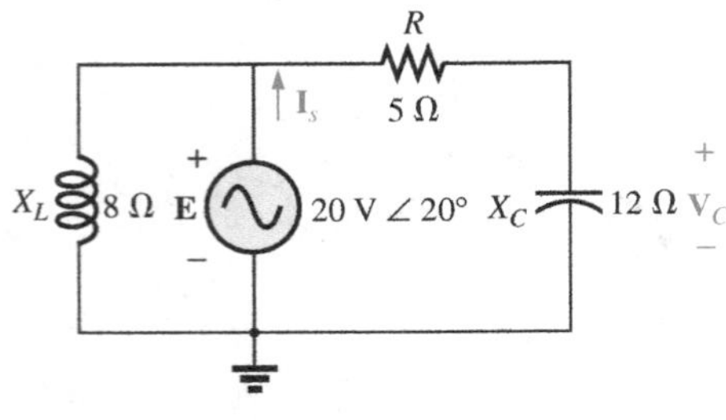

FIG. 16.5
Example 16.3.

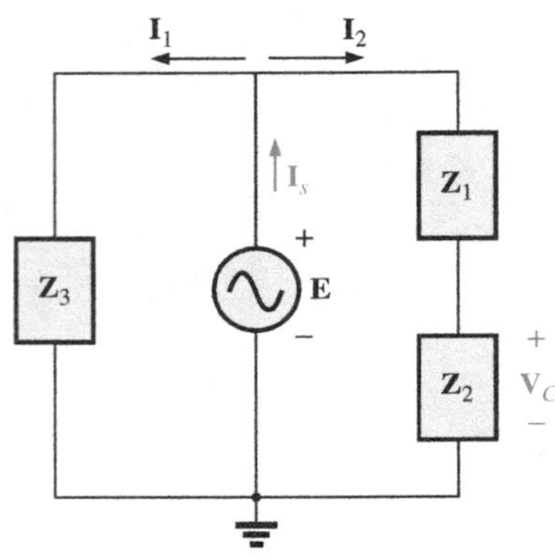

FIG. 16.6
Network in Fig. 16.5 after assigning the block impedances.

EXAMPLE 16.3 For the network in Fig. 16.5:

a. Calculate the voltage $\mathbf{V}_C$ using the voltage divider rule.
b. Calculate the current $\mathbf{I}_s$.

Solutions:

a. The network is redrawn as shown in Fig. 16.6, with

$$\mathbf{Z}_1 = 5\ \Omega = 5\ \Omega\ \angle 0^\circ$$
$$\mathbf{Z}_2 = -j\,12\ \Omega = 12\ \Omega\ \angle -90^\circ$$
$$\mathbf{Z}_3 = +j\,8\ \Omega = 8\ \Omega\ \angle 90^\circ$$

Since $\mathbf{V}_C$ is desired, we will not combine R and X_C into a single block impedance. Note also how Fig. 16.6 clearly reveals that **E** is the total voltage across the series combination of $\mathbf{Z}_1$ and $\mathbf{Z}_2$, permitting the use of the voltage divider rule to calculate $\mathbf{V}_C$. In addition, note that all the currents necessary to determine $\mathbf{I}_s$ have been preserved in Fig. 16.6, revealing that there is no need to ever return to the network of Fig. 16.5—everything is defined by Fig. 16.6.

$$\mathbf{V}_C = \frac{\mathbf{Z}_2\mathbf{E}}{\mathbf{Z}_1 + \mathbf{Z}_2} = \frac{(12\ \Omega\ \angle -90^\circ)(20\ \text{V}\ \angle 20^\circ)}{5\ \Omega - j\,12\ \Omega} = \frac{240\ \text{V}\ \angle -70^\circ}{13\ \angle -67.38^\circ}$$
$$= \mathbf{18.46\ V\ \angle -2.62^\circ}$$

b. $$\mathbf{I}_1 = \frac{\mathbf{E}}{\mathbf{Z}_3} = \frac{20\ \text{V}\ \angle 20^\circ}{8\ \Omega\ \angle 90^\circ} = 2.5\ \text{A}\ \angle -70^\circ$$

$$\mathbf{I}_2 = \frac{\mathbf{E}}{\mathbf{Z}_1 + \mathbf{Z}_2} = \frac{20\ \text{V}\ \angle 20^\circ}{13\ \Omega\ \angle -67.38^\circ} = 1.54\ \text{A}\ \angle 87.38^\circ$$

and

$$\mathbf{I}_s = \mathbf{I}_1 + \mathbf{I}_2$$
$$= 2.5\ \text{A}\ \angle -70^\circ + 1.54\ \text{A}\ \angle 87.38^\circ$$
$$= (0.86 - j\,2.35) + (0.07 + j\,1.54)$$
$$\mathbf{I}_s = 0.93 - j\,0.81 = \mathbf{1.23\ A\ \angle -41.05^\circ}$$

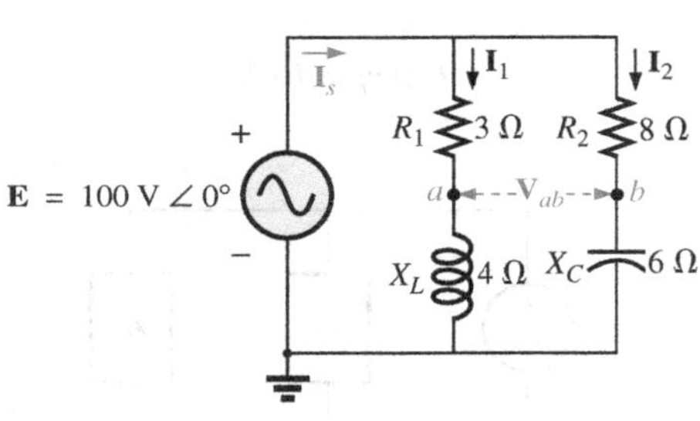

FIG. 16.7
Example 16.4.

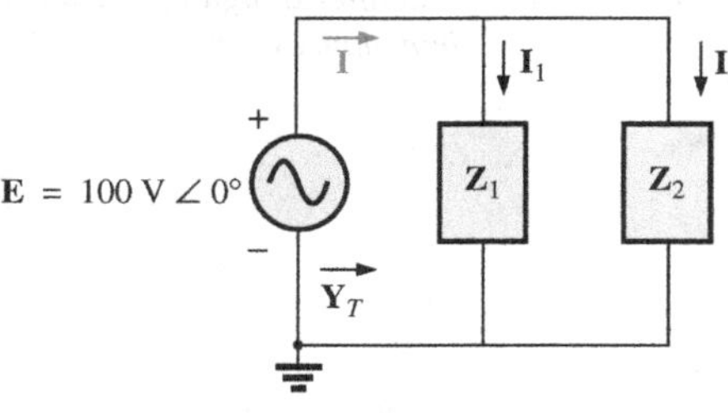

FIG. 16.8
Network in Fig. 16.7 after assigning the block impedances.

EXAMPLE 16.4 For Fig. 16.7:

a. Calculate the current $\mathbf{I}_s$.
b. Find the voltage $\mathbf{V}_{ab}$.

Solutions:

a. Redrawing the circuit as in Fig. 16.8, we obtain

$$\mathbf{Z}_1 = R_1 + jX_L = 3\ \Omega + j\,4\ \Omega = 5\ \Omega\ \angle 53.13^\circ$$
$$\mathbf{Z}_2 = R_2 - jX_C = 8\ \Omega - j\,6\ \Omega = 10\ \Omega\ \angle -36.87^\circ$$

In this case the voltage $\mathbf{V}_{ab}$ is lost in the redrawn network, but the currents $\mathbf{I}_1$ and $\mathbf{I}_2$ remain defined for future calculations necessary to determine $\mathbf{V}_{ab}$. Fig. 16.8 clearly reveals that the total impedance can be found using the equation for two parallel impedances:

$$\mathbf{Z}_T = \frac{\mathbf{Z}_1\mathbf{Z}_2}{\mathbf{Z}_1 + \mathbf{Z}_2} = \frac{(5\ \Omega\ \angle 53.13^\circ)(10\ \Omega\ \angle -36.87^\circ)}{(3\ \Omega + j\,4\ \Omega) + (8\ \Omega - j\,6\ \Omega)}$$
$$= \frac{50\ \Omega\ \angle 16.26^\circ}{11 - j\,2} = \frac{50\ \Omega\ \angle 16.26^\circ}{11.18\ \angle -10.30^\circ}$$
$$= \mathbf{4.472\ \Omega\ \angle 26.56^\circ}$$

and

$$\mathbf{I}_s = \frac{\mathbf{E}}{\mathbf{Z}_T} = \frac{100\text{ V}\angle 0^\circ}{4.472\ \Omega\ \angle 26.56^\circ} = \mathbf{22.36\ A\ \angle -26.56^\circ}$$

b. By Ohm's law,

$$\mathbf{I}_1 = \frac{\mathbf{E}}{\mathbf{Z}_1} = \frac{100\text{ V}\angle 0^\circ}{5\ \Omega\ \angle 53.13^\circ} = \mathbf{20\ A\ \angle -53.13^\circ}$$

$$\mathbf{I}_2 = \frac{\mathbf{E}}{\mathbf{Z}_2} = \frac{100\text{ V}\angle 0^\circ}{10\ \Omega\ \angle -36.87^\circ} = \mathbf{10\ A\ \angle 36.87^\circ}$$

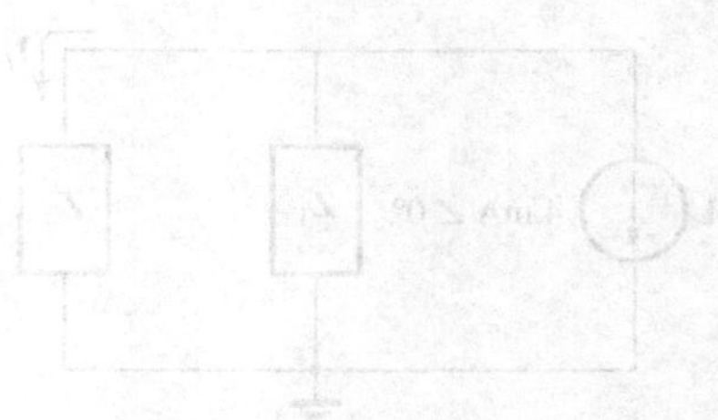

Returning to Fig. 16.7, we have

$$\mathbf{V}_{R_1} = \mathbf{I}_1\mathbf{Z}_{R_1} = (20\text{ A}\angle -53.13^\circ)(3\ \Omega\ \angle 0^\circ) = \mathbf{60\ V\ \angle -53.13^\circ}$$
$$\mathbf{V}_{R_2} = \mathbf{I}_1\mathbf{Z}_{R_2} = (10\text{ A}\angle +36.87^\circ)(8\ \Omega\ \angle 0^\circ) = \mathbf{80\ V\ \angle +36.87^\circ}$$

Instead of using the two steps just shown, we could have determined $\mathbf{V}_{R_1}$ or $\mathbf{V}_{R_2}$ in one step using the voltage divider rule:

$$\mathbf{V}_{R_1} = \frac{(3\ \Omega\ \angle 0^\circ)(100\text{ V}\angle 0^\circ)}{3\ \Omega\ \angle 0^\circ + 4\ \Omega\ \angle 90^\circ} = \frac{300\text{ V}\angle 0^\circ}{5\ \angle 53.13^\circ} = \mathbf{60\ V\ \angle -53.13^\circ}$$

To find $\mathbf{V}_{ab}$, Kirchhoff's voltage law must be applied around the loop (Fig. 16.9) consisting of the 3 Ω and 8 Ω resistors. By Kirchhoff's voltage law,

$$\mathbf{V}_{ab} + \mathbf{V}_{R_1} - \mathbf{V}_{R_2} = 0$$

or

$$\begin{aligned}\mathbf{V}_{ab} &= \mathbf{V}_{R_2} - \mathbf{V}_{R_1}\\ &= 80\text{ V}\angle 36.87^\circ - 60\text{ V}\angle -53.13^\circ\\ &= (64 + j\,48) - (36 - j\,48)\\ &= 28 + j\,96\\ \mathbf{V}_{ab} &= \mathbf{100\ V\angle 73.74^\circ}\end{aligned}$$

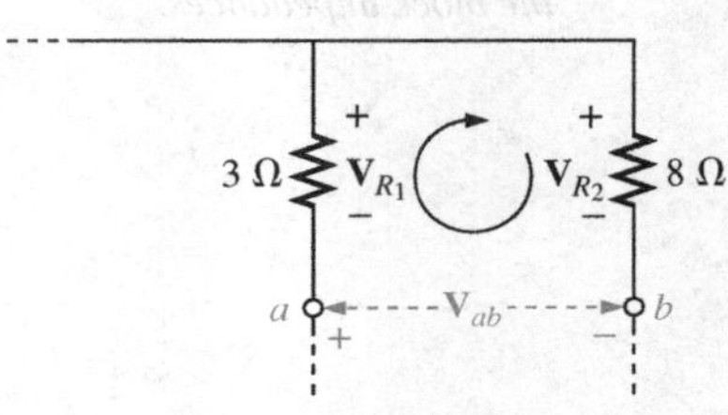

FIG. 16.9
Determining the voltage V_{ab} for the network in Fig. 16.7.

EXAMPLE 16.5 The network in Fig. 16.10 is frequently encountered in the analysis of transistor networks. The transistor equivalent circuit includes a current source **I** and an output impedance R_o. The resistor R_C is a biasing resistor to establish specific dc conditions, and the resistor R_i represents the loading of the next stage. The coupling capacitor is designed to be an open circuit for dc and to have as low an impedance as possible for the frequencies of interest to ensure that $\mathbf{V}_L$ is a maximum

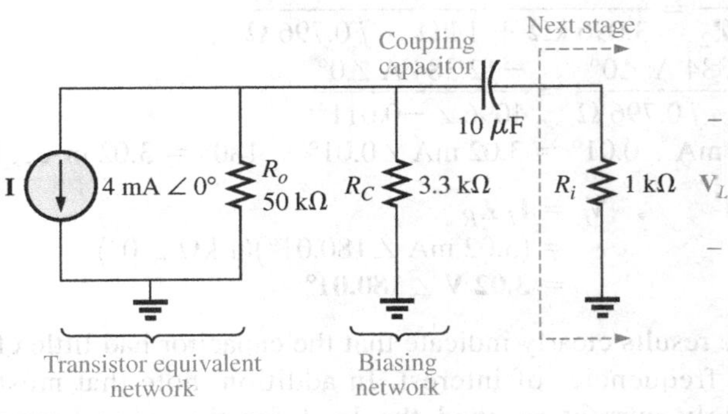

FIG. 16.10
Basic transistor amplifier.

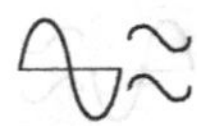

value. The frequency range of the example includes the entire audio (hearing) spectrum from 100 Hz to 20 kHz. The purpose of the example is to demonstrate that, for the full audio range, the effect of the capacitor can be ignored. It performs its function as a dc blocking agent but permits the ac to pass through with little disturbance.

a. Determine $\mathbf{V}_L$ for the network in Fig. 16.10 at a frequency of 100 Hz.
b. Repeat part (a) at a frequency of 20 kHz.
c. Compare the results of parts (a) and (b).

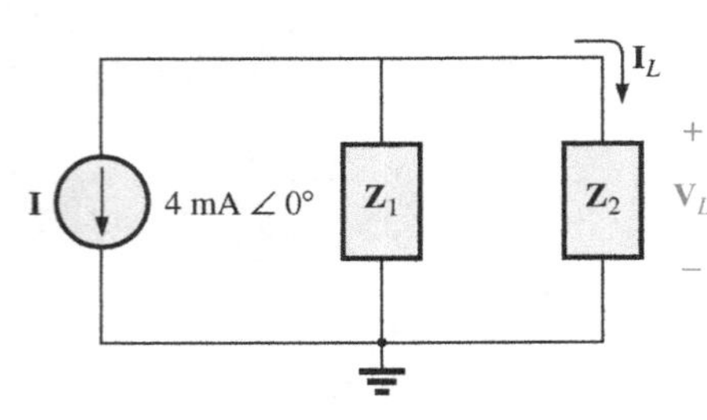

FIG. 16.11
Network in Fig. 16.10 following the assignment of the block impedances.

Solutions:

a. The network is redrawn with subscripted impedances in Fig. 16.11:

$$\mathbf{Z}_1 = 50\text{ k}\Omega \angle 0° \parallel 3.3\text{ k}\Omega \angle 0° = 3.096\text{ k}\Omega \angle 0°$$

$$\mathbf{Z}_2 = R_i - jX_C$$

$$\text{At } f = 100\text{ Hz}, X_C = \frac{1}{2\pi fC} = \frac{1}{2\pi(100\text{ Hz})(10\ \mu\text{F})} = 159.16\ \Omega$$

and
$$\mathbf{Z}_2 = 1\text{ k}\Omega - j\,159.16\ \Omega$$

Using the current divider rule gives

$$\begin{aligned}\mathbf{I}_L &= \frac{-\mathbf{Z}_1\mathbf{I}}{\mathbf{Z}_1 + \mathbf{Z}_2} = \frac{-(3.096\text{ k}\Omega \angle 0°)(4\text{ mA} \angle 0°)}{3.096\text{ k}\Omega + 1\text{ k}\Omega - j\,159.16\ \Omega}\\ &= \frac{-12.384\text{ A} \angle 0°}{4096 - j\,159.16} = \frac{-12.384\text{ A} \angle 0°}{4099 \angle -2.225°}\\ &= -3.02\text{ mA} \angle 2.23° = 3.02\text{ mA} \angle 2.23° + 180° = 3.02\text{ mA} \angle 182.23°\end{aligned}$$

and
$$\begin{aligned}\mathbf{V}_L &= \mathbf{I}_L\mathbf{Z}_R\\ &= (3.02\text{ mA} \angle 182.23°)(1\text{ k}\Omega \angle 0°)\\ &= \mathbf{3.02\ V \angle 182.23°}\end{aligned}$$

b. $$\text{At } f = 20\text{ kHz}, X_C = \frac{1}{2\pi fC} = \frac{1}{2\pi(20\text{ kHz})(10\ \mu\text{F})} = 0.796\ \Omega$$

Note the dramatic change in X_C with frequency. Obviously, the higher the frequency, the better is the short-circuit approximation for X_C for ac conditions. We have

$$\mathbf{Z}_2 = 1\text{ k}\Omega - j\,0.796\ \Omega$$

Using the current divider rule gives

$$\begin{aligned}\mathbf{I}_L &= \frac{-\mathbf{Z}_1\mathbf{I}}{\mathbf{Z}_1 + \mathbf{Z}_2} = \frac{-(3.096\text{ k}\Omega \angle 0°)(4\text{ mA} \angle 0°)}{3.096\text{ k}\Omega + 1\text{ k}\Omega - j\,0.796\ \Omega}\\ &= \frac{-12.384\text{ A} \angle 0°}{4096 - j\,0.796\ \Omega} = \frac{-12.384\text{ A} \angle 0°}{4096 \angle -0.011°}\\ &= -3.02\text{ mA} \angle 0.01° = 3.02\text{ mA} \angle 0.01° + 180° = 3.02\text{ mA} \angle 180.01°\end{aligned}$$

and
$$\begin{aligned}\mathbf{V}_L &= \mathbf{I}_L\mathbf{Z}_R\\ &= (3.02\text{ mA} \angle 180.01°)(1\text{ k}\Omega \angle 0°)\\ &= \mathbf{3.02\ V \angle 180.01°}\end{aligned}$$

c. The results clearly indicate that the capacitor had little effect on the frequencies of interest. In addition, note that most of the supply current reached the load for the typical parameters employed.

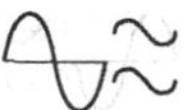

EXAMPLE 16.6 For the network in Fig. 16.12:

a. Determine the current **I**.
b. Find the voltage **V**.

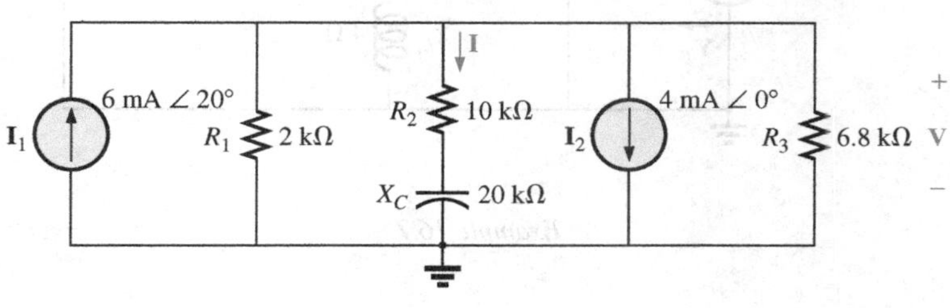

FIG. 16.12
Example 16.6.

Solutions:

a. The rules for parallel current sources are the same for dc and ac networks. That is, the equivalent current source is their sum or difference (as phasors). Therefore,

$$\begin{aligned}\mathbf{I}_T &= 6\text{ mA }\angle 20^\circ - 4\text{ mA }\angle 0^\circ \\ &= 5.638\text{ mA} + j\,2.052\text{ mA} - 4\text{ mA} \\ &= 1.638\text{ mA} + j\,2.052\text{ mA} \\ &= 2.626\text{ mA }\angle 51.402^\circ\end{aligned}$$

Redrawing the network using block impedances results in the network in Fig. 16.13, where

$$\mathbf{Z}_1 = 2\text{ k}\Omega\angle 0^\circ \parallel 6.8\text{ k}\Omega\angle 0^\circ = 1.545\text{ k}\Omega\angle 0^\circ$$

and $$\mathbf{Z}_2 = 10\text{ k}\Omega - j\,20\text{ k}\Omega = 22.361\text{ k}\Omega\angle -63.435^\circ$$

Note that **I** and **V** are still defined in Fig. 16.13.

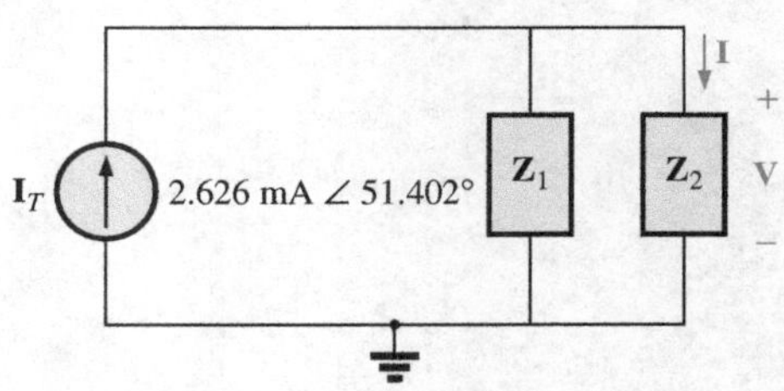

FIG. 16.13
Network in Fig. 16.12 following the assignment of the subscripted impedances.

Using the current divider rule gives

$$\begin{aligned}\mathbf{I} &= \frac{\mathbf{Z}_1\mathbf{I}_T}{\mathbf{Z}_1 + \mathbf{Z}_2} = \frac{(1.545\text{ k}\Omega\angle 0^\circ)(2.626\text{ mA}\angle 51.402^\circ)}{1.545\text{ k}\Omega + 10\text{ k}\Omega - j\,20\text{ k}\Omega} \\ &= \frac{4.057\text{ A}\angle 51.402^\circ}{11.545\times 10^3 - j\,20\times 10^3} = \frac{4.057\text{ A}\angle 51.402^\circ}{23.093\times 10^3\angle -60.004^\circ} \\ &= \mathbf{0.18\ mA\ \angle 111.41^\circ}\end{aligned}$$

b. $$\begin{aligned}\mathbf{V} &= \mathbf{I}\mathbf{Z}_2 \\ &= (0.176\text{ mA}\angle 111.406^\circ)(22.36\text{ k}\Omega\angle -63.435^\circ) \\ &= \mathbf{3.94\ V\ \angle 47.97^\circ}\end{aligned}$$

EXAMPLE 16.7 For the network in Fig. 16.14:

a. Compute **I**.
b. Find $\mathbf{I}_1$, $\mathbf{I}_2$, and $\mathbf{I}_3$.
c. Verify Kirchhoff's current law by showing that

$$\mathbf{I} = \mathbf{I}_1 + \mathbf{I}_2 + \mathbf{I}_3$$

d. Find the total impedance of the circuit.

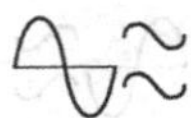

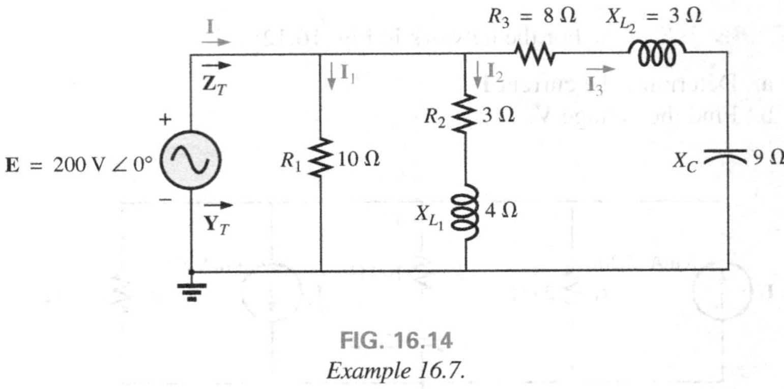

FIG. 16.14
Example 16.7.

Solutions:

a. Redrawing the circuit as in Fig. 16.15 reveals a strictly parallel network where

$$\mathbf{Z}_1 = R_1 = 10\ \Omega \angle 0°$$
$$\mathbf{Z}_2 = R_2 + jX_{L_1} = 3\ \Omega + j\,4\ \Omega$$
$$\mathbf{Z}_3 = R_3 + jX_{L_2} - jX_C = 8\ \Omega + j\ 3\ \Omega - j\ 9\ \Omega = 8\ \Omega - j\ 6\ \Omega$$

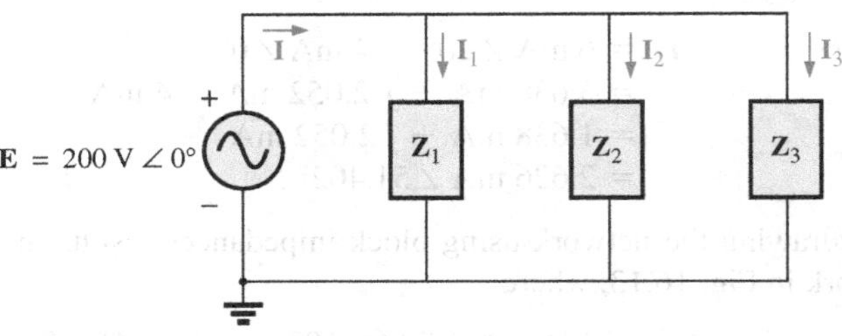

FIG. 16.15
Network in Fig. 16.14 following the assignment of the subscripted impedances.

The total admittance is

$$\begin{aligned}\mathbf{Y}_T &= \mathbf{Y}_1 + \mathbf{Y}_2 + \mathbf{Y}_3 \\ &= \frac{1}{\mathbf{Z}_1} + \frac{1}{\mathbf{Z}_2} + \frac{1}{\mathbf{Z}_3} = \frac{1}{10\ \Omega} + \frac{1}{3\ \Omega + j\,4\ \Omega} + \frac{1}{8\ \Omega - j\,6\ \Omega} \\ &= 0.1\ \text{S} + \frac{1}{5\ \Omega \angle 53.13°} + \frac{1}{10\ \Omega \angle -36.87°} \\ &= 0.1\ \text{S} + 0.2\ \text{S} \angle -53.13° + 0.1\ \text{S} \angle 36.87° \\ &= 0.1\ \text{S} + 0.12\ \text{S} - j\ 0.16\ \text{S} + 0.08\ \text{S} + j\ 0.06\ \text{S} \\ &= 0.3\ \text{S} - j\ 0.1\ \text{S} = 0.316\ \text{S} \angle -18.435°\end{aligned}$$

Calculator The above mathematical exercise presents an excellent opportunity to demonstrate the power of some of today's calculators. For the TI-89, the above operation is as shown in Fig. 16.16.

FIG. 16.16
Finding the total admittance for the network in Fig. 16.14 using the TI-89 calculator.

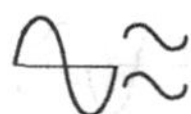

Be sure to use the negative sign for the complex number from the subtraction option and not the sign selection (−). The sign selection is used for negative angles in the polar form.

Converting to polar form requires the sequence shown in Fig. 16.17.

(• 3 − • 1 *i*) ▶ Polar ENTER ENTER **316.2E–3 ∠ –18.43E0**

FIG. 16.17
Converting the rectangular form in Fig. 16.16 to polar form.

Convert to polar form:
The current **I** is given by

$$\mathbf{I} = \mathbf{E}\mathbf{Y}_T = (200\ \text{V} \angle 0°)(0.326\ \text{S} \angle -18.435°)$$
$$= \mathbf{63.2\ A \angle -18.44°}$$

b. Since the voltage is the same across parallel branches,

$$\mathbf{I}_1 = \frac{\mathbf{E}}{\mathbf{Z}_1} = \frac{200\ \text{V} \angle 0°}{10\ \Omega \angle 0°} = \mathbf{20\ A \angle 0°}$$
$$\mathbf{I}_2 = \frac{\mathbf{E}}{\mathbf{Z}_2} = \frac{200\ \text{V} \angle 0°}{5\ \Omega \angle 53.13°} = \mathbf{40\ A \angle -53.13°}$$
$$\mathbf{I}_3 = \frac{\mathbf{E}}{\mathbf{Z}_3} = \frac{200\ \text{V} \angle 0°}{10\ \Omega \angle -36.87°} = \mathbf{20\ A \angle +36.87°}$$

c.
$$\mathbf{I} = \mathbf{I}_1 + \mathbf{I}_2 + \mathbf{I}_3$$
$$60 - j\,20 = 20 \angle 0° + 40 \angle -53.13° + 20 \angle +36.87°$$
$$= (20 + j\,0) + (24 - j\,32) + (16 + j\,12)$$
$$60 - j\,20 = 60 - j\,20 \quad \text{(checks)}$$

d.
$$\mathbf{Z}_T = \frac{1}{\mathbf{Y}_T} = \frac{1}{0.316\ \text{S} \angle -18.435°}$$
$$= \mathbf{3.17\ \Omega \angle 18.44°}$$

EXAMPLE 16.8 For the network in Fig. 16.18:

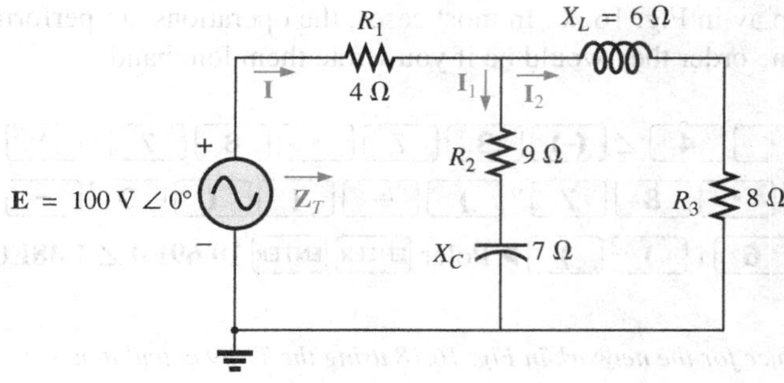

FIG. 16.18
Example 16.8.

a. Calculate the total impedance $\mathbf{Z}_T$.
b. Compute **I**.
c. Find the total power factor.
d. Calculate $\mathbf{I}_1$ and $\mathbf{I}_2$.
e. Find the average power delivered to the circuit.

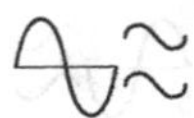

Solutions:

a. Redrawing the circuit as in Fig. 16.19, we have

$$\mathbf{Z}_1 = R_1 = 4\ \Omega \angle 0^\circ$$
$$\mathbf{Z}_2 = R_2 - jX_C = 9\ \Omega - j\,7\ \Omega = 11.40\ \Omega \angle -37.87^\circ$$
$$\mathbf{Z}_3 = R_3 + jX_L = 8\ \Omega + j\,6\ \Omega = 10\ \Omega \angle +36.87^\circ$$

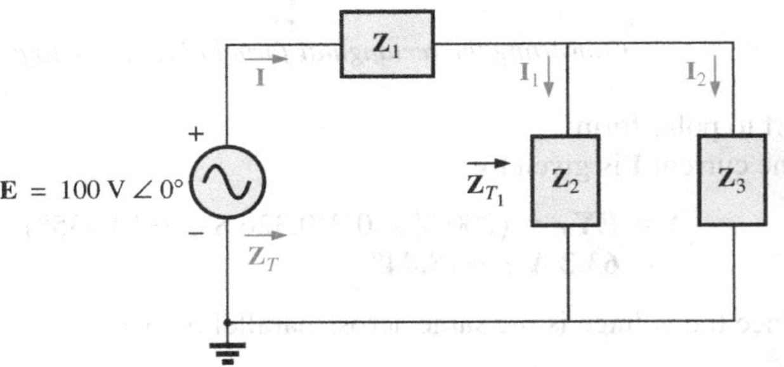

FIG. 16.19
Network in Fig. 16.18 following the assignment of the subscripted impedances.

Notice that all the desired quantities were conserved in the redrawn network. The total impedance is

$$\mathbf{Z}_T = \mathbf{Z}_1 + \mathbf{Z}_{T_1}$$
$$= \mathbf{Z}_1 + \frac{\mathbf{Z}_2\mathbf{Z}_3}{\mathbf{Z}_2 + \mathbf{Z}_3}$$
$$= 4\ \Omega + \frac{(11.4\ \Omega \angle -37.87^\circ)(10\ \Omega \angle 36.87^\circ)}{(9\ \Omega - j\,7\ \Omega) + (8\ \Omega + j\,6\ \Omega)}$$
$$= 4\ \Omega + \frac{114\ \Omega \angle -1.00^\circ}{17.03 \angle -3.37^\circ} = 4\ \Omega + 6.69\ \Omega \angle 2.37^\circ$$
$$= 4\ \Omega + 6.68\ \Omega + j\,0.28\ \Omega = 10.68\ \Omega + j\,0.28\ \Omega$$
$$\mathbf{Z}_T = \mathbf{10.68\ \Omega \angle 1.5^\circ}$$

Calculator Another opportunity to demonstrate the versatility of the calculator! For the above operation, however, you must be aware of the priority of the mathematical operations, as demonstrated in the calculator display in Fig. 16.20. In most cases, the operations are performed in the same order they would be if you wrote them longhand.

4 + (1 1 . 4 ∠ (-) 3 7 . 8 7 °) ×
(1 0 ∠ 3 6 . 8 7 °) ÷ ((9 - 7
i) + (8 + 6 i)) ▶Polar ENTER ENTER **10.69E0 ∠ 1.48E0**

FIG. 16.20
Finding the total impedance for the network in Fig. 16.18 using the TI-89 calculator.

b. $\mathbf{I} = \dfrac{\mathbf{E}}{\mathbf{Z}_T} = \dfrac{100\ \text{V} \angle 0^\circ}{10.684\ \Omega \angle 1.5^\circ} = \mathbf{9.36\ A \angle -1.5^\circ}$

c. $F_p = \cos\theta_T = \dfrac{R}{Z_T} = \dfrac{10.68\ \Omega}{10.684\ \Omega} \cong \mathbf{1}$

(essentially resistive, which is interesting, considering the complexity of the network)

d. Using the current divider rule gives

$$\mathbf{I}_2 = \frac{\mathbf{Z}_2\mathbf{I}}{\mathbf{Z}_2 + \mathbf{Z}_3} = \frac{(11.40\ \Omega\ \angle -37.87°)(9.36\ \text{A}\ \angle -1.5°)}{(9\ \Omega - j\,7\ \Omega) + (8\ \Omega + j\,6\ \Omega)}$$

$$= \frac{106.7\ \text{A}\ \angle -39.37°}{17 - j\,1} = \frac{106.7\ \text{A}\ \angle -39.37°}{17.03\ \angle -3.37°}$$

$$\mathbf{I}_2 = \mathbf{6.27\ A\ \angle -36°}$$

Applying Kirchhoff's current law (rather than another application of the current divider rule) yields

$$\mathbf{I}_1 = \mathbf{I} - \mathbf{I}_2$$

or

$$\mathbf{I} = \mathbf{I}_1 - \mathbf{I}_2$$
$$= (9.36\ \text{A}\ \angle -1.5°) - (6.27\ \text{A}\ \angle -36°)$$
$$= (9.36\ \text{A} - j\,0.25\ \text{A}) - (5.07\ \text{A} - j\,3.69\ \text{A})$$
$$\mathbf{I}_1 = 4.29\ \text{A} + j\,3.44\ \text{A} = \mathbf{5.5\ A\ \angle 38.72°}$$

e. $P_T = EI\cos\theta_T$

$$= (100\ \text{V})(9.36\ \text{A})\cos 1.5°$$
$$= (936)(0.99966)$$
$$P_T = \mathbf{935.68\ W}$$

16.3 LADDER NETWORKS

Ladder networks were discussed in some detail in Chapter 7. This section will simply apply the first method described in Section 7.6 to the general sinusoidal ac ladder network in Fig. 16.21. The current $\mathbf{I}_6$ is desired.

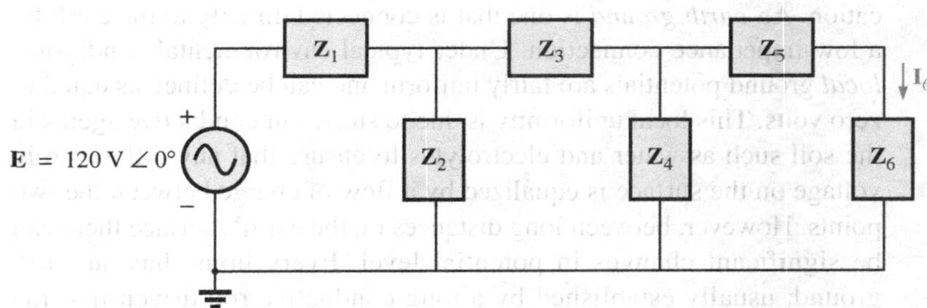

FIG. 16.21
Ladder network.

Impedances $\mathbf{Z}_T$, $\mathbf{Z}'_T$, and $\mathbf{Z}''_T$ and currents $\mathbf{I}_1$ and $\mathbf{I}_3$ are defined in Fig. 16.22. We have

$$\mathbf{Z}''_T = \mathbf{Z}_5 + \mathbf{Z}_6$$

and

$$\mathbf{Z}'_T = \mathbf{Z}_3 + \mathbf{Z}_4 \parallel \mathbf{Z}''_T$$

with

$$\mathbf{Z}_T = \mathbf{Z}_1 + \mathbf{Z}_2 \parallel \mathbf{Z}'_T$$

Then

$$\mathbf{I} = \frac{\mathbf{E}}{\mathbf{Z}_T}$$

and

$$\mathbf{I}_3 = \frac{\mathbf{Z}_2\mathbf{I}}{\mathbf{Z}_2 + \mathbf{Z}'_T}$$

with

$$\mathbf{I}_6 = \frac{\mathbf{Z}_4\mathbf{I}_3}{\mathbf{Z}_4 + \mathbf{Z}''_T}$$

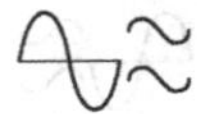

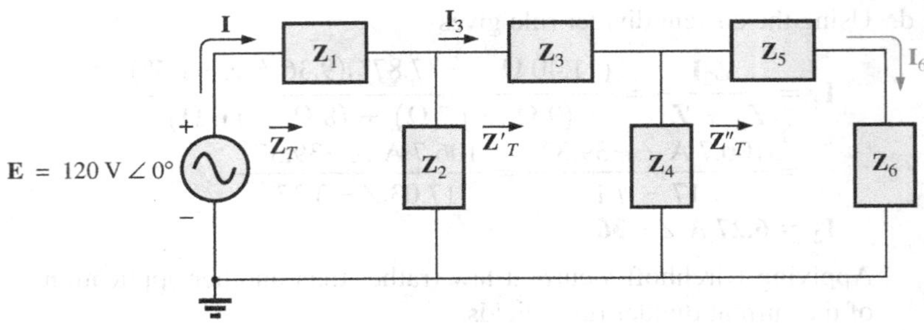

FIG. 16.22
Defining an approach to the analysis of ladder networks.

16.4 GROUNDING

Although usually treated too lightly in most introductory electrical or electronics texts, the impact of the ground connection and how it can provide a measure of safety to a design are very important topics. *Ground potential is zero volts at every point in a network that has a ground symbol.* Since all points are at the same potential, they can all be connected together, but for purposes of clarity, most are left isolated on a large schematic. On a schematic, the voltage levels provided are always with respect to ground. A system can therefore be checked quite rapidly by simply connecting the black lead of the voltmeter to the ground connection and placing the red lead at the various points where the typical operating voltage is provided. A close match normally implies that that portion of the system is operating properly.

There are various types of grounds, whose use depends on the application. An *earth ground* is one that is connected directly to the earth by a low-impedance connection. Under typical environmental conditions, *local* ground potentials are fairly uniform and can be defined as equal to zero volts. This local uniformity is due to sufficient conductive agents in the soil such as water and electrolytes to ensure that any difference in voltage on the surface is equalized by a flow of charge between the two points. However, between long distances on the earth's surface there can be significant changes in potential level. Every home has an earth ground, usually established by a long conductive rod driven into the ground and connected to the power panel. The electrical code requires a direct connection from earth ground to the cold-water pipes of a home for safety reasons. A "hot" wire touching a cold-water pipe draws sufficient current because of the low-impedance ground connection to throw the breaker. Otherwise, people in the bathroom could pick up the voltage when they touched the cold-water faucet, thereby risking bodily harm. Because water is a conductive agent, any area of the home with water, such as a bathroom or the kitchen, is of particular concern. Most electrical systems are connected to earth ground primarily for safety reasons. All the power lines in a laboratory, at industrial locations, or in the home are connected to earth ground.

A second type is referred to as a *chassis ground,* which may be *floating* or connected directly to an earth ground. A chassis ground simply stipulates that the chassis has a reference potential for all points of the network. If the chassis is not connected to earth potential (0 V), it is said to be *floating* and can have any other reference voltage for the other voltages to be compared to. For instance, if the chassis is sitting at 120 V, all measured voltages of the network will be referenced to this level. A reading of 32 V between a point in the network and the chassis ground will

therefore actually be at 152 V with respect to earth potential. Most high-voltage systems are not left floating, however, because of loss of the safety factor. For instance, if someone should touch the chassis and be standing on a suitable ground, the full 120 V would fall across that individual.

Grounding can be particularly important when working with numerous pieces of measuring equipment in the laboratory. For instance, the supply and oscilloscope in Fig. 16.23(a) are each connected directly to an earth ground through the negative terminal of each. If the oscilloscope is connected as shown in Fig. 16.23(a) to measure the voltage V_{R1}, a dangerous situation will develop. The grounds of each piece of equipment are connected together through the earth ground, and they effectively short out the resistor. Since the resistor is the primary current-controlling element in the network, the current will rise to a very high level and possibly damage the instruments or cause dangerous side effects. In this case, the supply or scope should be used in the floating mode or the resistors interchanged as shown in Fig. 16.23(b). In Fig. 16.23(b), the grounds have a common point and do not affect the structure of the network.

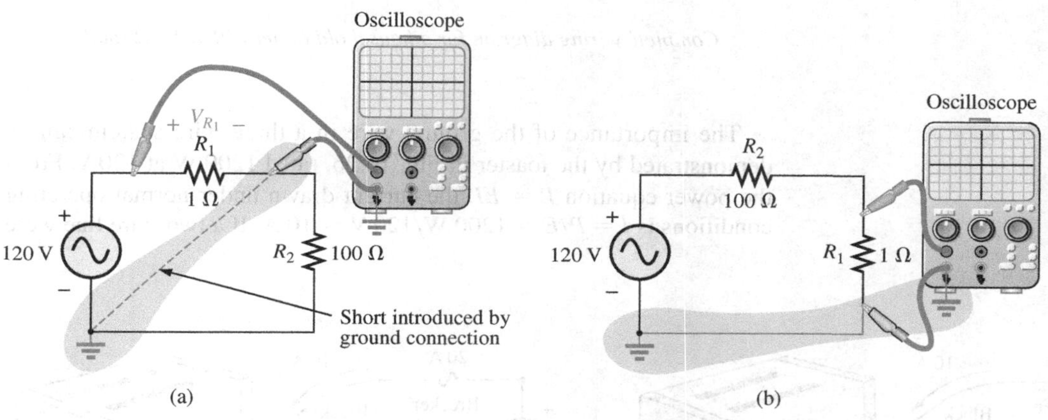

FIG. 16.23
Demonstrating the effect of the oscilloscope ground on the measurement of the voltage across resistor R_1.

The National Electrical Code requires that the "hot" (or *feeder*) line that carries current to a load be *black* and the line (called the *neutral*) that carries the current back to the supply be *white*. Three-wire conductors have a ground wire that must be *green* or *bare*, which ensures a common ground but which is not designed to carry current. The components of a three-prong extension cord and wall outlet are shown in Fig. 16.24. Note

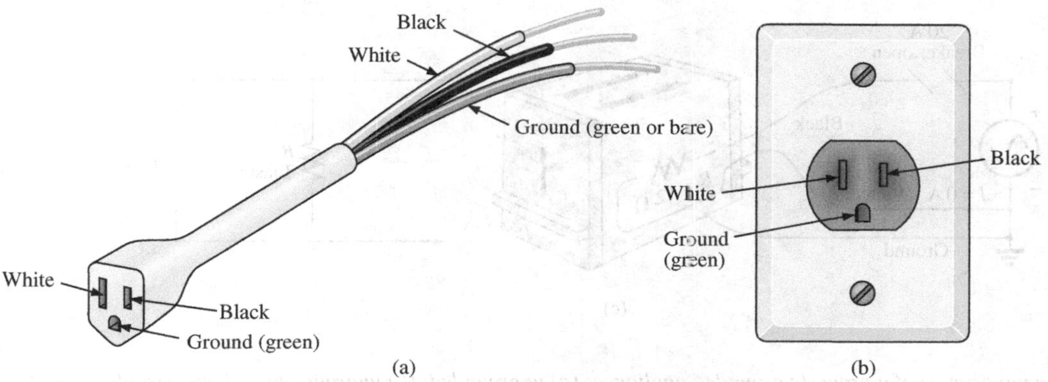

FIG. 16.24
Three-wire conductors: (a) extension cord; (b) home outlet.

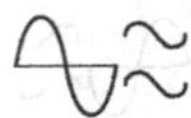

that on both fixtures, the connection to the hot lead is smaller than the return leg and that the ground connection is partially circular.

The complete wiring diagram for a household outlet is shown in Fig. 16.25. Note that the current through the ground wire is zero and that both the return wire and the ground wire are connected to an earth ground. The full current to the loads flows through the feeder and return lines.

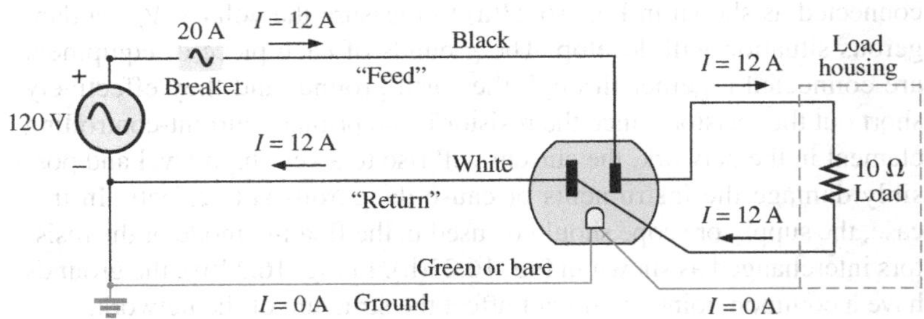

FIG. 16.25
Complete wiring diagram for a household outlet with a 10 Ω load.

The importance of the ground wire in a three-wire system can be demonstrated by the toaster in Fig. 16.26, rated 1200 W at 120 V. From the power equation $P = EI$, the current drawn under normal operating conditions is $I = P/E = 1200\text{ W}/120\text{ V} = 10\text{ A}$. If a two-wire line were

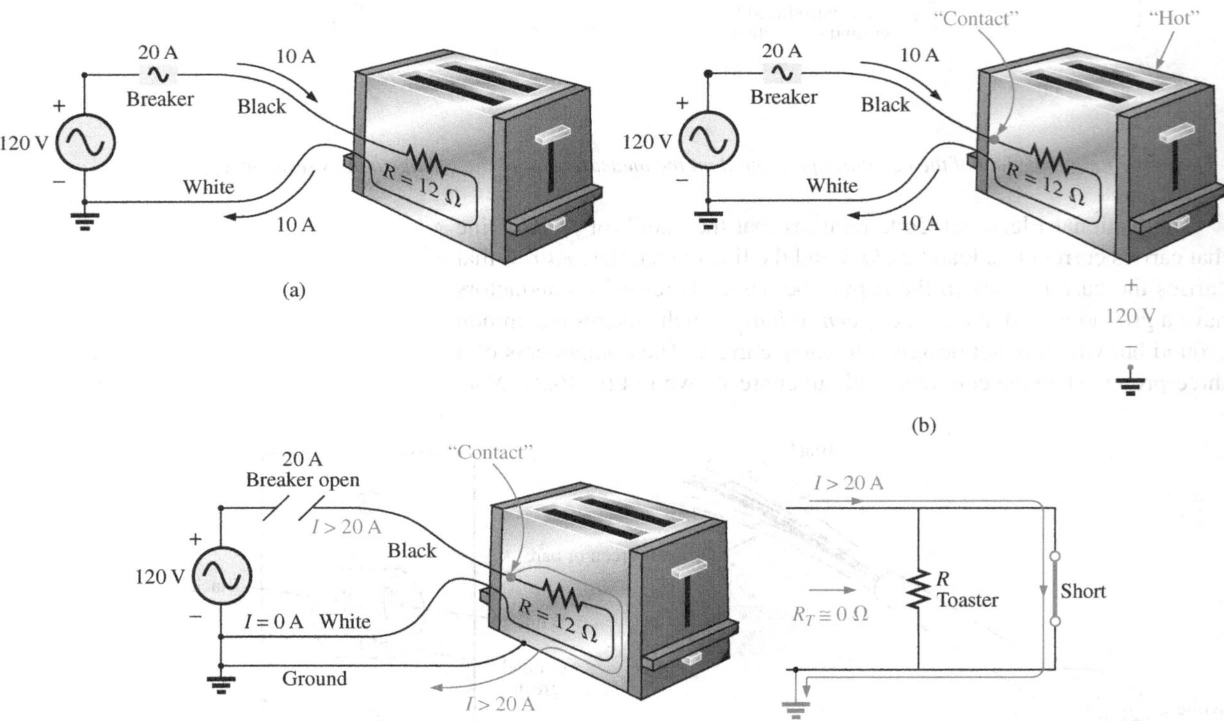

FIG. 16.26
Demonstrating the importance of a properly grounded appliance: (a) ungrounded; (b) ungrounded and undesirable contact; (c) grounded appliance with undesirable contact.

used as shown in Fig. 16.26(a), the 20 A breaker would be quite comfortable with the 10 A current, and the system would perform normally. However, if abuse to the feeder caused it to become frayed and to touch the metal housing of the toaster, the situation depicted in Fig. 16.26(b) would result. The housing would become "hot," yet the breaker would not "pop" because the current would still be the rated 10 A. A dangerous condition would exist because anyone touching the toaster would feel the full 120 V to ground. If the ground wire were attached to the chassis as shown in Fig. 16.26(c), a low-resistance path would be created between the short-circuit point and ground, and the current would jump to very high levels. The breaker would "pop," and the user would be warned that a problem exists.

Although the above discussion does not cover all possible areas of concern with proper grounding or introduce all the nuances associated with the effect of grounds on a system's performance, you should understand the importance of the impact of grounds.

16.5 APPLICATIONS

The vast majority of the applications appearing throughout the text have been of the series-parallel variety. The following are series-parallel combinations of elements and systems used to perform important everyday tasks. The ground fault circuit interrupter outlet employs series protective switches and sensing coils and a parallel control system, while the ideal equivalent circuit for the coax cable employs a series-parallel combination of inductors and capacitors.

GFCI (Ground Fault Circuit Interrupter)

The National Electric Code, the "bible" for all electrical contractors, now requires that ground fault circuit interrupter (GFCI) outlets be used in any area where water and dampness could result in serious injury, such as in bathrooms, pools, marinas, and so on. The outlet looks like any other except that it has a reset button and a test button in the center of the unit as shown in Fig. 16.27(a). The primary difference between it and an ordinary outlet is that it will shut the power off much more quickly than the breaker all the way down in the basement could. You may still feel a shock with a GFCI outlet, but the current cuts off so quickly (in a few milliseconds) that a person in normal health should not receive a serious electrical injury. Whenever in doubt about its use, remember that its cost (relatively inexpensive) is well worth the increased measure of safety.

The basic operation is best described by the network in Fig. 16.27(b). The protection circuit separates the power source from the outlet itself. Note in Fig. 16.27(b) the importance of grounding the protection circuit to the central ground of the establishment (a water pipe, ground bar, and so on, connected to the main panel). In general, the outlet will be grounded to the same connection. Basically, the network shown in Fig. 16.27(b) senses both the current entering (I_i) and the current leaving (I_o) and provides a direct connection to the outlet when they are equal. If a fault should develop such as that caused by someone touching the hot leg while standing on a wet floor, the return current will be less than the feed current (just a few milliamperes is enough). The protection circuitry senses this difference, establishes an open circuit in the line, and cuts off the power to the outlet.

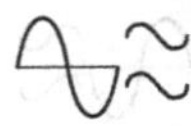

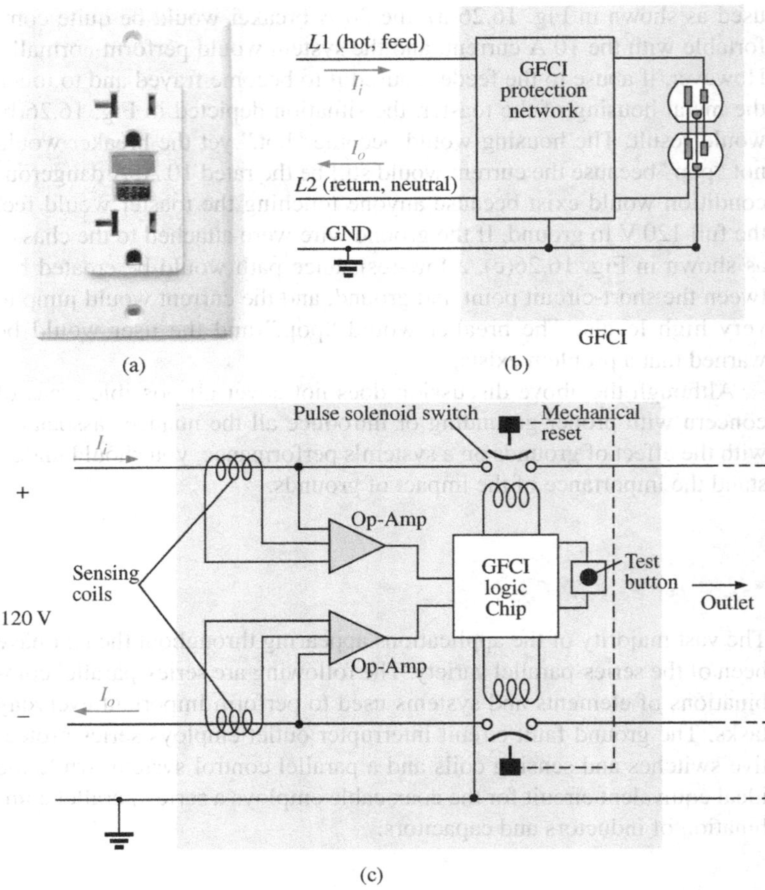

FIG. 16.27

GFCI outlet: (a) wall-mounted appearance; (b) basic operation; (c) schematic.

Fig. 16.28(a) shows the feed and return lines passing through the sensing coils. The two sensing coils are separately connected to the printed circuit board. There are two pulse control switches in the line and a return to establish an open circuit under errant conditions. The two contacts in Fig. 16.28(a) are the contacts that provide conduction to the outlet. When a fault develops, another set of similar contacts in the housing slides away, providing the desired open-circuit condition. The separation is created by the solenoid appearing in Fig. 16.28(b). When the solenoid is energized due to a fault condition, it pulls the plunger toward the solenoid, compressing the spring. At the same time, the slots in the lower plastic piece (connected directly to the plunger) shift down, causing a disconnect by moving the structure inserted in the slots. The test button is connected to the brass bar across the unit in Fig. 16.28(c) below the reset button. When pressed, it places a large resistor between the line and ground to "unbalance" the line and cause a fault condition. When the button is released, the resistor is separated from the line, and the unbalance condition is removed. The resistor is actually connected directly to one end of the bar and moves down with pressure on the bar as shown in Fig. 16.28(d). Note in Fig. 16.28(c) how the metal ground connection passes right through the entire unit and how it is connected to the ground terminal of an applied plug. Also note how it is separated from the rest of the network with the plastic housing. Although this unit appears simple on the outside and is relatively small in size, it is beautifully designed and contains a great deal of technology and innovation.

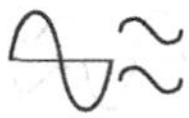

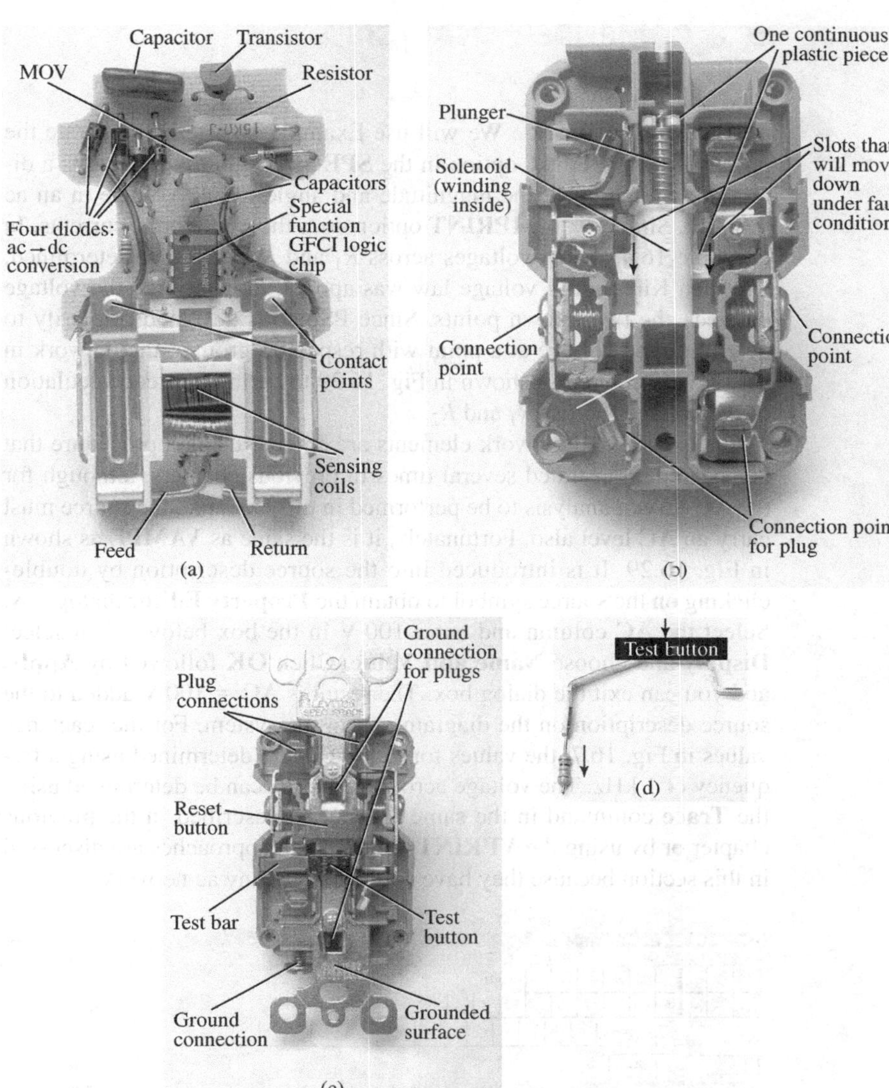

FIG. 16.28

GFCI construction: (a) sensing coils; (b) solenoid control (bottom view); (c) grounding (top view); (d) test bar.

Before leaving the subject, note the logic chip in the center of Fig. 16.28(a) and the various sizes of capacitors and resistors. Note also the four diodes in the upper left region of the circuit board used as a bridge rectifier for the ac-to-dc conversion process. The transistor is the black element with the half-circle appearance. It is part of the driver circuit for the controlling solenoid. Because of the size of the unit, there wasn't a lot of room to provide the power to quickly open the circuit. The result is the use of a pulse circuit to control the motion of the controlling solenoid. In other words, the solenoid is pulsed for a short period of time to cause the required release. If the design used a system that would hold the circuit open on a continuing basis, the power requirement would be greater and the size of the coil larger. A small coil can handle the required power pulse for a short period of time without any long-term damage.

As mentioned earlier, if unsure, install a GFCI. It provides a measure of safety—at a very reasonable cost—that should not be ignored.

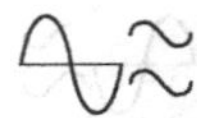

16.6 COMPUTER ANALYSIS

PSpice

ac Bridge Network We will use Example 16.4 to demonstrate the power of the **VPRINT** option in the **SPECIAL** library. It permits a direct determination of the magnitude and angle of any voltage in an ac network. Similarly, the **IPRINT** option does the same for ac currents. In Example 16.4, the ac voltages across R_1 and R_2 were first determined, and then Kirchhoff's voltage law was applied to determine the voltage between the two known points. Since PSpice is designed primarily to determine the voltage at a point with respect to ground, the network in Fig. 16.7 is entered as shown in Fig. 16.29 to permit a direct calculation of the voltages across R_1 and R_2.

The source and network elements are entered using a procedure that has been demonstrated several times in previous chapters, although for the **AC Sweep** analysis to be performed in this example, the source must carry an **AC** level also. Fortunately, it is the same as **VAMPL** as shown in Fig. 16.29. It is introduced into the source description by double-clicking on the source symbol to obtain the **Property Editor** dialog box. Select the **AC** column and enter 100 V in the box below. Then select **Display** and choose **Name and Value.** Click **OK** followed by **Apply,** and you can exit the dialog box. The result is **AC** = 100 V added to the source description on the diagram and in the system. For the reactance values in Fig. 16.7, the values for L and C were determined using a frequency of 1 kHz. The voltage across R_1 and R_2 can be determined using the **Trace** command in the same manner as described in the previous chapter or by using the **VPRINT** option. Both approaches are discussed in this section because they have application to any ac network.

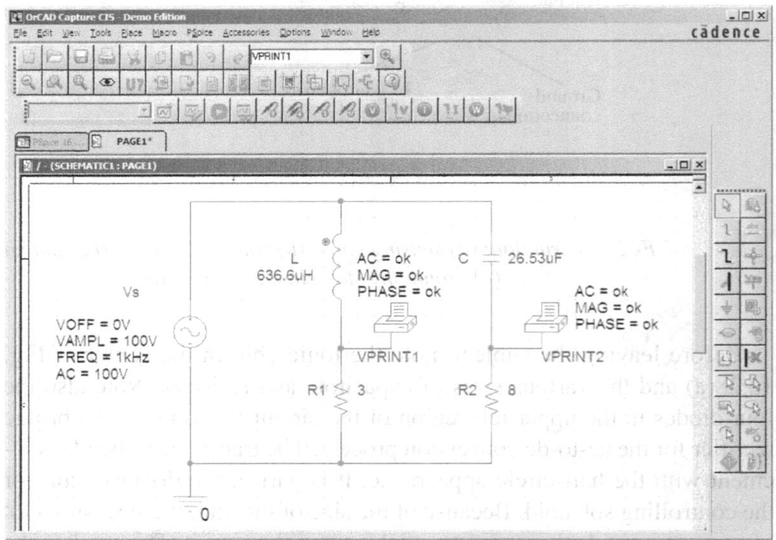

FIG. 16.29
*Determining the voltage across R_1 and R_2 using the **VPRINT** option of a PSpice analysis.*

The **VPRINT** option is under the **SPECIAL** library in the **Place Part** dialog box. Once selected, the printer symbol appears on the screen next to the cursor, and it can be placed near the point of interest. Once the printer symbol is in place, double-click on it to display the **Property Editor** dialog box. Scrolling from left to right, type the word **ok** under **AC,**

MAG, and **PHASE.** When each is active, select the **Display** key and choose the option **Name and Value** followed by **OK.** When all the entries have been made, choose **Apply** and exit the dialog box. The result appears in Fig. 16.29 for the two applications of the **VPRINT** option. If you prefer, **VPRINT1** and **VPRINT2** can be added to distinguish between the two when you review the output data. To do this, return to the **Property Editor** dialog box for each by double-clicking on the printer symbol of each and selecting **Value** and then **Display** followed by **Value Only.** Do not forget to select **Apply** after each change in the **Property Editor** dialog box. You are now ready for the simulation.

The simulation is initiated by selecting the **New Simulation Profile** icon and entering **PSpice 16-1** as the **Name.** Then select **Create** to bring up the **Simulation Settings** dialog box. This time, you want to analyze the network at 1 kHz but are not interested in plots against time. Thus, select the **AC Sweep/Noise** option under **Analysis type** in the **Analysis** section. An **AC Sweep Type** region then appears asking for the **Start Frequency.** Since you are interested in the response at only one frequency, the **Start** and **End Frequency** will be the same: 1 kHz. Since you need only one point of analysis, the **Points/Decade** will be 1. Click **OK,** and select the **Run PSpice** icon. The **SCHEMATIC1** screen appears, and the voltage across R_1 can be determined by selecting **Trace** followed by **Add Trace** and then **V(R1:1).** The result is the bottom display in Fig. 16.30 with only one plot point at 1 kHz. Since you fixed the

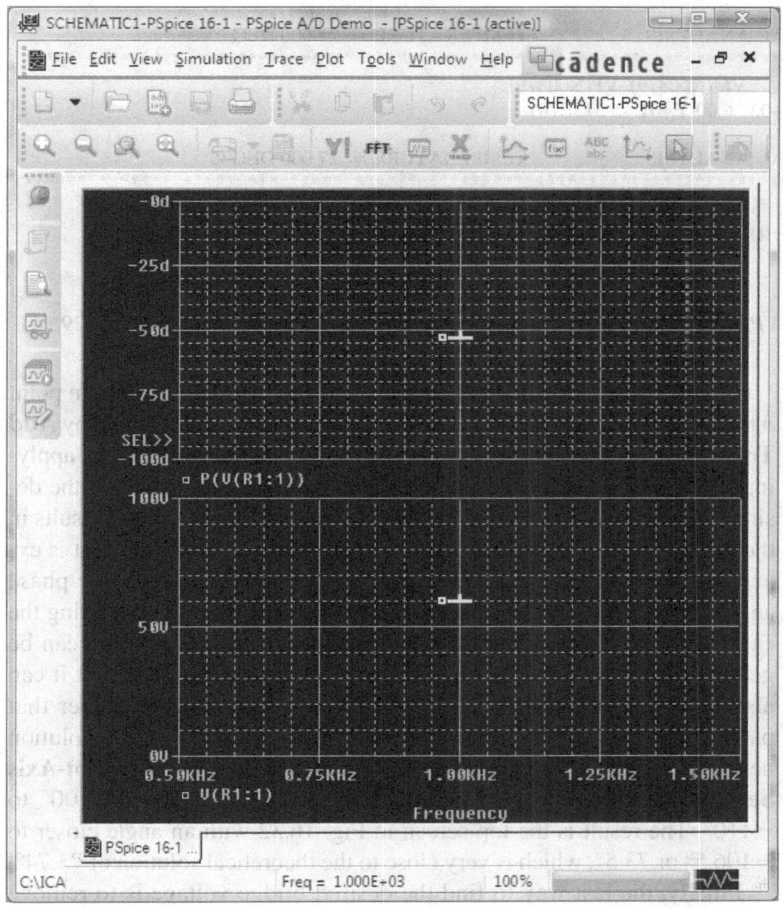

FIG. 16.30
The resulting magnitude and phase angle for the voltage V_{R_1} in Fig. 16.29.

frequency of interest at 1 kHz, this is the only frequency with a response. The magnitude of the voltage across R_1 is 60 V to match the longhand solution of Example 16.4. The phase angle associated with the voltage can be determined by the sequence **Plot-Add Plot to Window-Trace-Add Trace-P( )** from the **Functions or Macros** list and then **V(R1:1)** to obtain **P(V(R1:1))** in the **Trace Expression** box. Click **OK,** and the resulting plot shows that the phase angle is near just less than $-50°$ which is certainly a close match with the $-53.13°$ obtained in Example 16.4.

The **VPRINT** option just introduced offers another method for analyzing voltage in a network. When the **SCHEMATIC1** window appears after the simulation, exit the window using the **X,** and select **PSpice** on the top menu bar of the resulting screen. Select **View Output File** from the list that appears. You will see a long list of data about the construction of the network and the results obtained from the simulation. In Fig. 16.31, the portion of the output file listing the resulting magnitude and phase angle for the voltages defined by **VPRINT1** and **VPRINT2** is provided. Note that the voltage across R_1 defined by **VPRINT1** is 60 V at an angle of $-53.13°$. The voltage across R_2 as defined by **VPRINT2** is 80 V at an angle of 36.87°. Both are exact matches of the solutions in Example 16.4. In the future, therefore, if the **VPRINT** option is used, the results will appear in the output file.

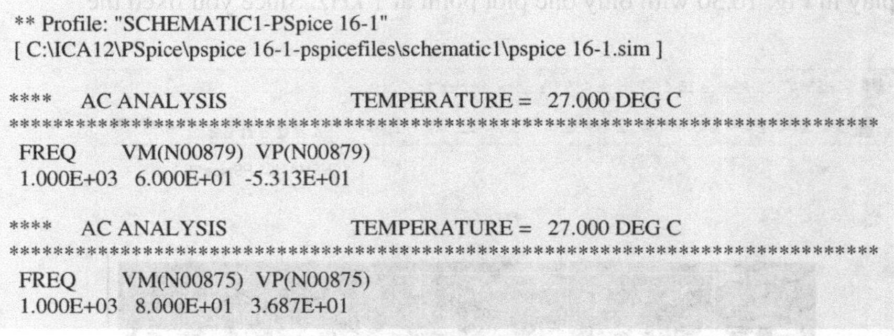

```
** Profile: "SCHEMATIC1-PSpice 16-1"
[ C:\ICA12\PSpice\pspice 16-1-pspicefiles\schematic1\pspice 16-1.sim ]

****    AC ANALYSIS                 TEMPERATURE =  27.000 DEG C
**************************************************************************

 FREQ        VM(N00879)  VP(N00879)
 1.000E+03   6.000E+01  -5.313E+01

****    AC ANALYSIS                 TEMPERATURE =  27.000 DEG C
**************************************************************************

 FREQ        VM(N00875)  VP(N00875)
 1.000E+03   8.000E+01   3.687E+01
```

FIG. 16.31

*The **VPRINT1** (V_{R_1}) and **VPRINT2** (V_{R_2}) response for the network in Fig. 16.29.*

Now you can determine the voltage across the two branches from point *a* to point *b*. Return to **SCHEMATIC1,** and select **Trace** followed by **Add Trace** to obtain the list of **Simulation Output Variables.** Then, by applying Kirchhoff's voltage law around the closed loop, you find that the desired voltage is **V(R1:1)-V(R2:1)** which when followed by **OK,** results in the plot point in the screen in the bottom of Fig. 16.32. Note that it is exactly 100 V as obtained in the longhand solution. Determine the phase angle through **Plot-Add Plot to Window-Trace-Add Trace,** creating the expression **P(V(R1:1)-V(R2:1)).** Remember that the expression can be generated using the lists of **Output variables** and **Functions,** but it can also be typed in from the keyboard. However, always remember that parentheses must be in sets—a left and a right. Click **OK,** and a solution near $-105°$ appears. A better reading can be obtained by using **Plot-Axis Settings-Y Axis-User Defined** and changing the scale to $-100°$ to $-110°$. The result is the top screen in Fig. 16.32 with an angle closer to $-106.5°$ or 73.5°, which is very close to the theoretical solution of 73.74°.

Finally, the last way to find the desired bridge voltage is to remove the **VPRINT2** option and place the ground at that point as shown in Fig. 16.33. Be absolutely sure to remove the original ground from the network. Now the voltage generated from a point above R_1 to ground

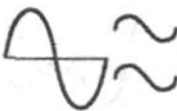

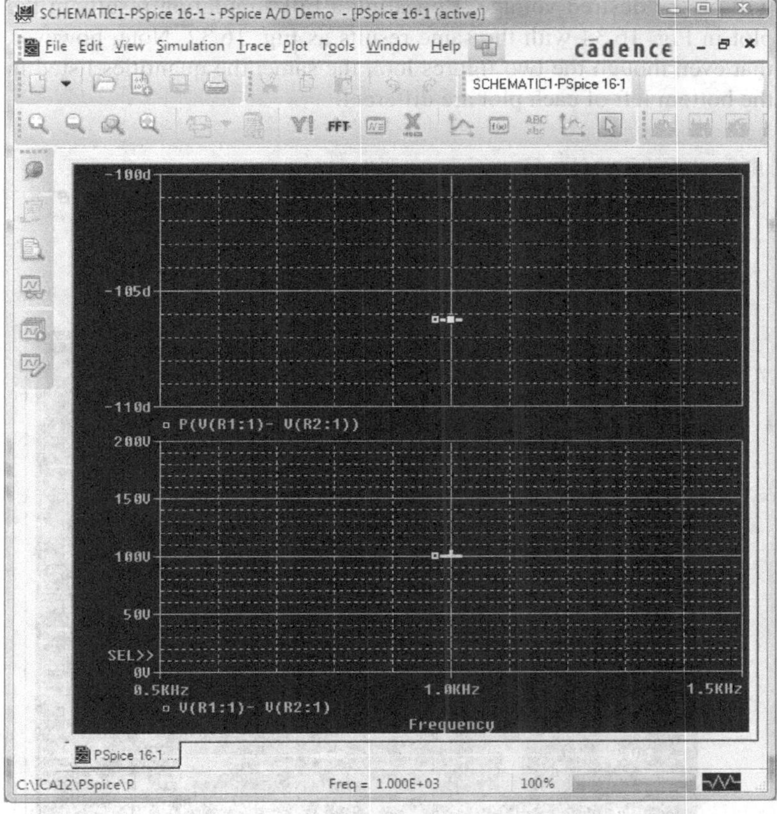

FIG. 16.32
The PSpice response for the voltage between the two points above resistors R_1 and R_2.

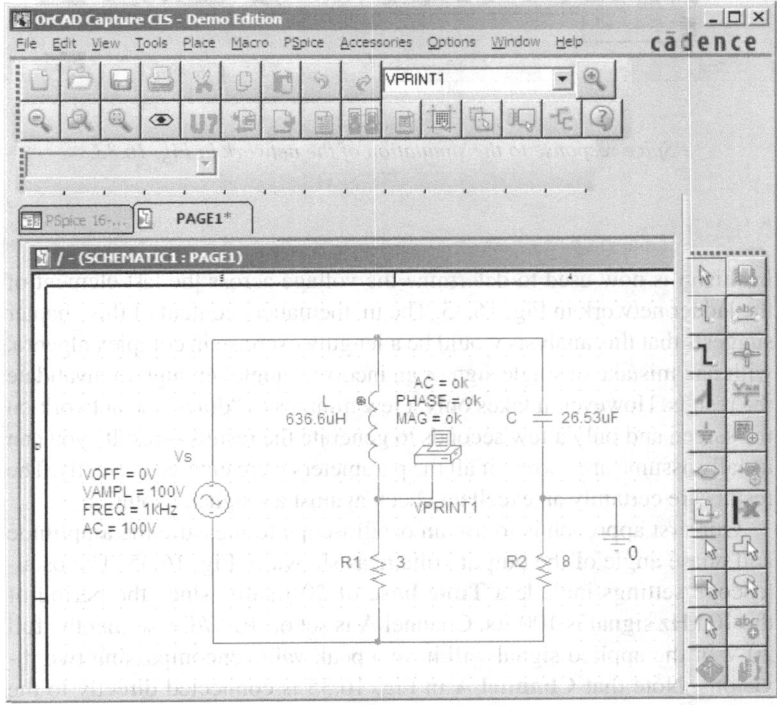

FIG. 16.33
Determining the voltage between the two points above resistors R_1 and R_2 by moving the ground connection in Fig. 16.29 to the position of ***VPRINT2****.*

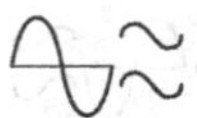

will be the desired voltage. Repeating a full simulation results in the plot in Fig. 16.34 with the same results as Fig. 16.32. Note, however, that even though the two figures look the same, the quantities listed in the bottom left of each plot are different.

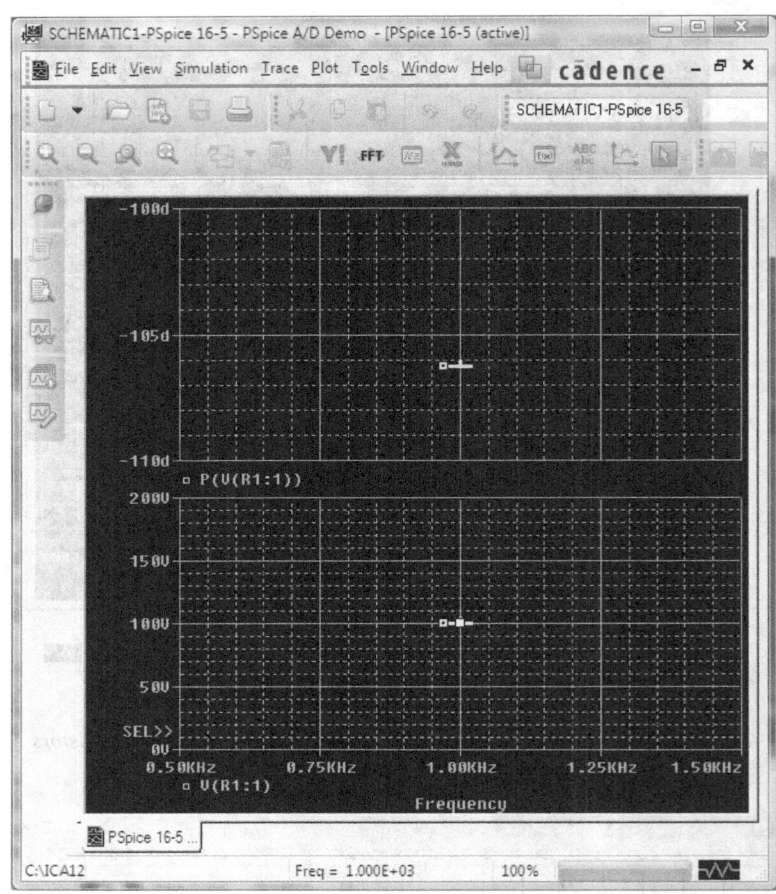

FIG. 16.34
PSpice response to the simulation of the network in Fig. 16.33.

Multisim

Multisim is now used to determine the voltage across the last element of the ladder network in Fig. 16.35. The mathematical content of this chapter suggests that this analysis would be a lengthy exercise in complex algebra, with one mistake (a single sign or an incorrect angle) enough to invalidate the results. However, it takes only a few minutes to "draw" the network on the screen and only a few seconds to generate the results—results you can usually assume are correct if all the parameters were entered correctly. The results are certainly an excellent check against a longhand solution.

Our first approach is to use an oscilloscope to measure the amplitude and phase angle of the output voltage as shown in Fig. 16.35. The oscilloscope settings include a **Time base** of 20 μs/div. since the period of the 10 kHz signal is 100 μs. Channel **A** is set on 10 V/div. so that the full 20 V of the applied signal will have a peak value encompassing two divisions. Note that **Channel A** in Fig. 16.35 is connected directly to the source **Vs** and to the **Trigger** input for synchronization. Expecting the output voltage to have a smaller amplitude resulted in a vertical sensitivity

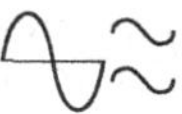

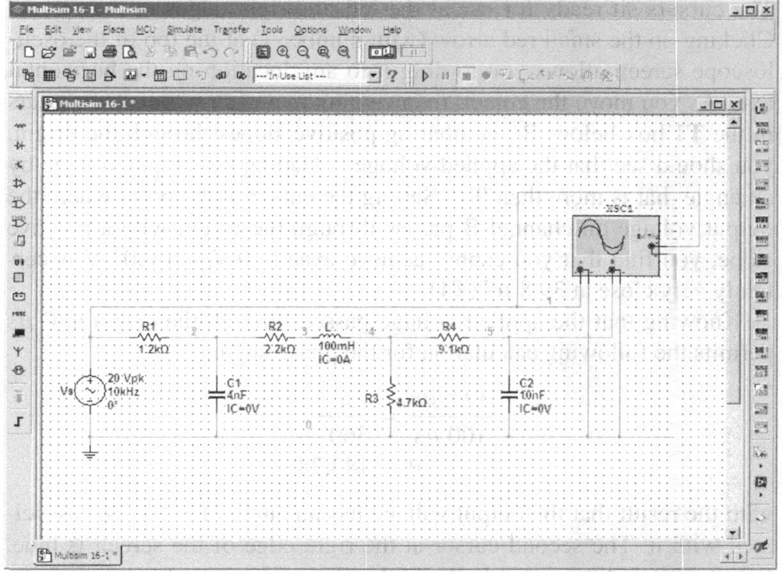

FIG. 16.35
Using the Multisim oscilloscope to determine the voltage across the capacitor C_2.

of 1 V/div. for **Channel B.** The analysis was initiated by placing the **Simulation** switch in the **1** position. It is important to realize that ***when simulation is initiated, it will take time for networks with reactive elements to settle down and for the response to reach its steady-state condition. It is therefore wise to let a system run for a while after simulation before selecting Sing. (Single) on the oscilloscope to obtain a steady waveform for analysis.***

The resulting plots in Fig. 16.36 clearly show that the applied voltage has an amplitude of 20 V and a period of 100 μs (5 div. at 20 μs/div).

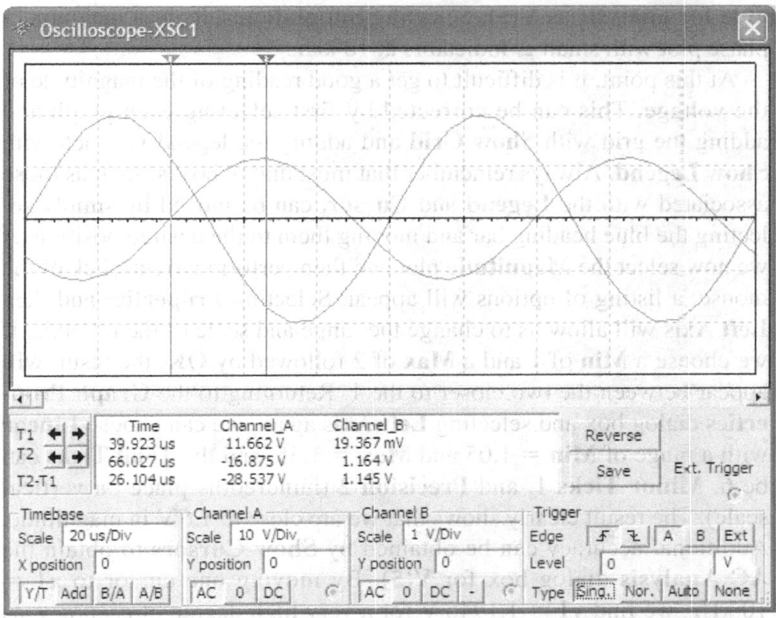

FIG. 16.36
Using Multisim to display the applied voltage and voltage across the capacitor C_2 for the network in Fig. 16.35.

The cursors sit ready for use at the left and right edges of the screen. Clicking on the small red arrow (with number 1) at the top of the oscilloscope screen allows you to drag it to any location on the horizontal axis. As you move the cursor, the magnitude of each waveform appears in the **T1** box below. By comparing positive slopes through the origin, you should see that the applied voltage is leading the output voltage by an angle that is more than 90°. Setting the cursor at the point where the output voltage on channel B passes through the origin with a positive slope, you find that you cannot achieve exactly 0 V, but 0.01 V is certainly very close at 39.8 μs **(T1)**.

Knowing that the applied voltage passed through the origin at 0 μs permits the following calculation for the phase angle:

$$\frac{39.8\ \mu\text{s}}{100\ \mu\text{s}} = \frac{\theta}{360^\circ}$$
$$\theta = 143.28^\circ$$

with the result that the output voltage has an angle of -143.28° associated with it. The second cursor at the right edge of the screen is blue. Selecting it and moving it to the peak value of the output voltage results in 1.16 V at 66.33 μs (**T2**). The result of all the above is

$$\mathbf{V}_{C_2} = 1.16\text{ V}\ \angle -143.28^\circ$$

The second approach is to use the **AC Analysis** option under the **Simulate** heading. First, realize that when you use the oscilloscope as you just did, you did not need to pass through the sequence of dialog boxes to choose the desired analysis. All that was necessary was to simulate using either the switch or the **Simulate Run** sequence—the oscilloscope was there to measure the output voltage. The **AC Analysis** approach requires that you first return to the **AC_VOLTAGE** dialog box and set the **AC Analysis** magnitude to 20 V. Then use the sequence **Simulate-Analyses-AC Analysis** to obtain the **AC Analysis** dialog box and set the **Start** and **Stop frequencies** at 10 kHz and the **Selected variable for analysis** as **V(5).** Selecting **Simulate** results in a magnitude-phase plot with small Δ indicators at 10 kHz.

At this point, it is difficult to get a good reading of the magnitude of the voltage. This can be corrected by first selecting each graph and adding the grid with **Show Grid** and adding the legend for each with **Show Legend.** Always remember that most dialog boxes, such as those associated with the **Legend** and **Cursor,** can be moved by simply selecting the blue heading bar and moving them to the desired position. If we now select the **Magnitude** plot and then perform a right-click of the mouse, a listing of options will appear. Selecting **Properties** and then **Left Axis** will allow us to change the range and scale of the left axis. If we choose a **Min** of 1 and a **Max** of 2 followed by **OK,** the result will appear between the two closer to the 1. Returning to the **Graph Properties** dialog box and selecting **Left Axis** again, we can choose **Linear** with a range of **Min** = 1.05 and **Max** = 1.35, and the **Total Ticks** can be 6, **Minor Ticks** 1, and **Precision** 2 (hundredths place on vertical scale). The result clearly shows that we are close to 1.6 V in magnitude. Additional accuracy can be obtained by **Show Cursors** to obtain the **AC Analysis** dialog box for **V(5).** By moving one cursor to **x1** = 10 kHz, we find **y1** = 1.1946 V for a very high degree of accuracy, as shown in Fig. 16.37. If we now select the **Phase(deg)** plot and use the cursor control, we find with **x1** = 10 kHz that **y1** = −142.147°, which is very close to the result obtained above.

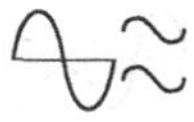

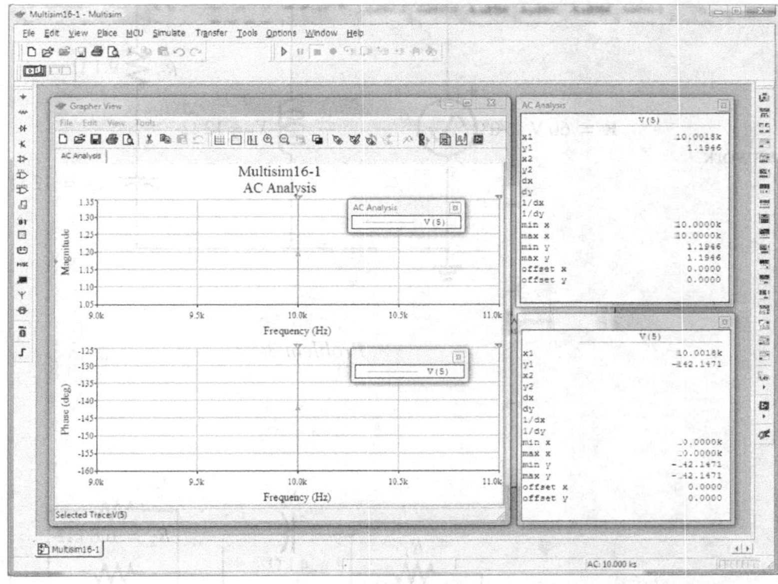

FIG. 16.37
*Using the **AC Analysis** option in Multisim to determine the magnitude and phase angle for the voltage V_{C_2} for the network in Fig. 16.35.*

In total, therefore, you have two methods to obtain an ac voltage in a network—one by instrumentation and the other through the computer methods. Both are valid, although, as expected, the computer approach has a higher level of accuracy.

PROBLEMS

SECTION 16.2 Illustrative Examples

1. For the series-parallel network in Fig. 16.38:
   a. Calculate $\mathbf{Z}_T$.
   b. Determine $\mathbf{I}_s$.
   c. Determine $\mathbf{I}_1$.
   d. Find $\mathbf{I}_2$.
   e. Find $\mathbf{V}_L$.

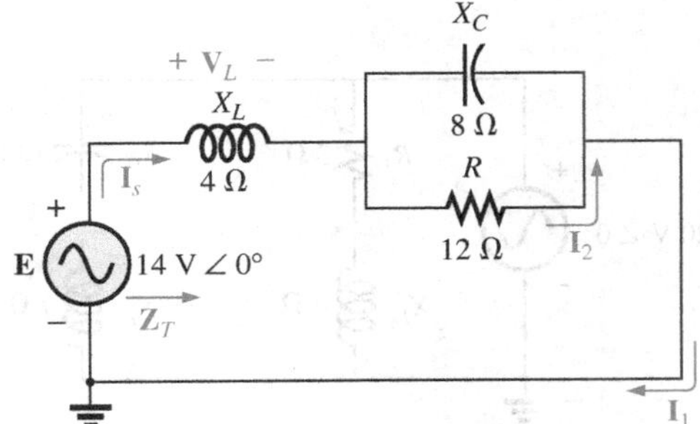

FIG. 16.38
Problem 1.

2. For the network in Fig. 16.39:
   a. Find the total impedance $\mathbf{Z}_T$.
   b. Determine the current $\mathbf{I}_s$.
   c. Calculate $\mathbf{I}_C$ using the current divider rule.
   d. Calculate $\mathbf{V}_L$ using the voltage divider rule.

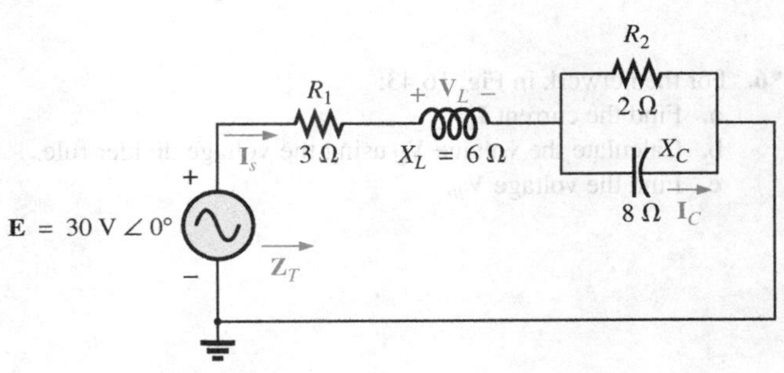

FIG. 16.39
Problems 2 and 15.

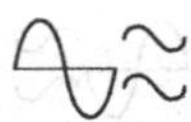

3. For the network in Fig. 16.40:
   a. Find the total impedance $\mathbf{Z}_T$.
   b. Find the current $\mathbf{I}_s$.
   c. Calculate $\mathbf{I}_2$ using the current divider rule.
   d. Calculate $\mathbf{V}_C$ using the voltage divider rule.
   e. Calculate the average power delivered to the network.

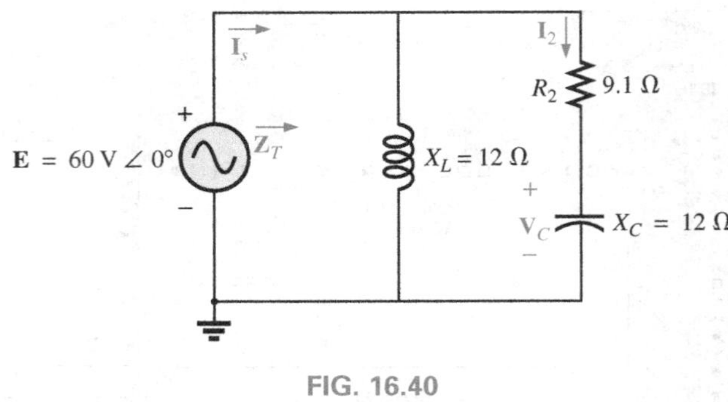

FIG. 16.40
Problem 3.

4. For the network in Fig. 16.41:
   a. Find the total impedance $\mathbf{Z}_T$.
   b. Calculate the voltage $\mathbf{V}_2$ and the current $\mathbf{I}_L$.
   c. Find the power factor of the network.

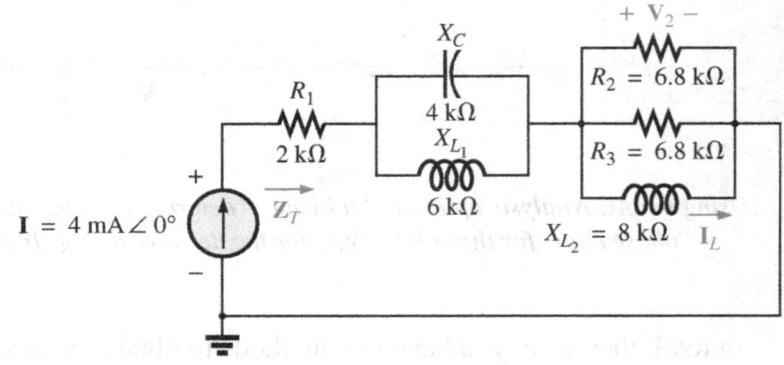

FIG. 16.41
Problem 4.

5. For the network in Fig. 16.42:
   a. Find the current $\mathbf{I}$.
   b. Find the voltage $\mathbf{V}_C$.
   c. Find the average power delivered to the network.

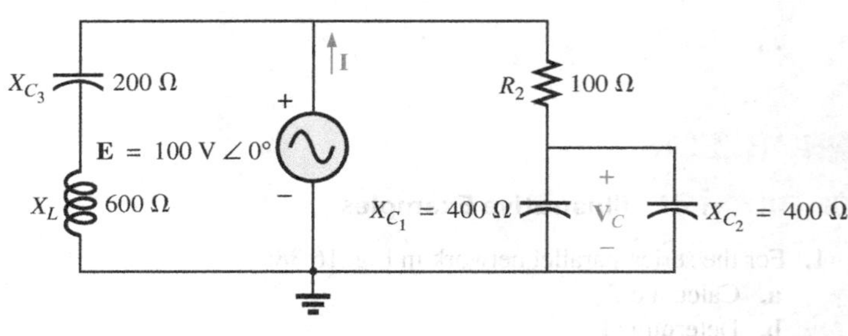

FIG. 16.42
Problem 5.

*6. For the network in Fig. 16.43:
   a. Find the current $\mathbf{I}_1$.
   b. Calculate the voltage $\mathbf{V}_C$ using the voltage divider rule.
   c. Find the voltage $\mathbf{V}_{ab}$.

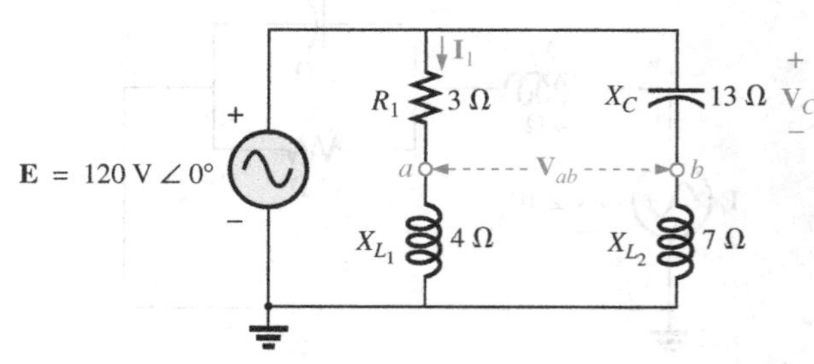

FIG. 16.43
Problem 6.

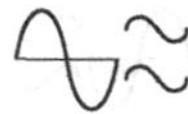

***7.** For the network in Fig. 16.44:

a. Find the current $\mathbf{I}_1$.
b. Find the voltage $\mathbf{V}_1$.
c. Calculate the average power delivered to the network.

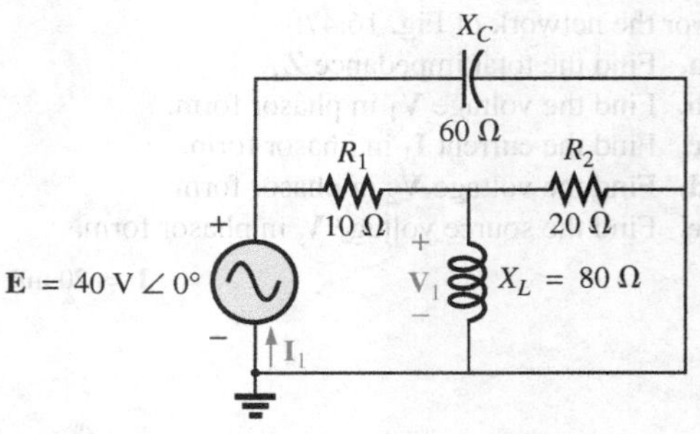

FIG. 16.44
Problems 7 and 16.

8. For the network in Fig. 16.45:

a. Find the total impedance $\mathbf{Z}_T$ and the admittance $\mathbf{Y}_T$.
b. Find the currents $\mathbf{I}_1$, $\mathbf{I}_2$, and $\mathbf{I}_3$.
c. Verify Kirchhoff's current law by showing that $\mathbf{I}_s = \mathbf{I}_1 + \mathbf{I}_2 + \mathbf{I}_3$.
d. Find the power factor of the network, and indicate whether it is leading or lagging.

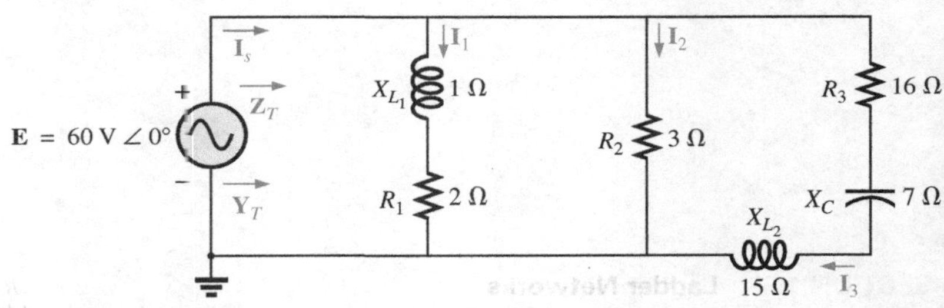

FIG. 16.45
Problem 8.

***9.** For the network in Fig. 16.46:

a. Find the total impedance $\mathbf{Z}_T$.
b. Find the source current $\mathbf{I}_s$ in phasor form.
c. Find the currents $\mathbf{I}_1$ and $\mathbf{I}_2$ in phasor form.
d. Find the voltages $\mathbf{V}_1$ and $\mathbf{V}_{ab}$ in phasor form.
e. Find the average power delivered to the network.
f. Find the power factor of the network, and indicate whether it is leading or lagging.

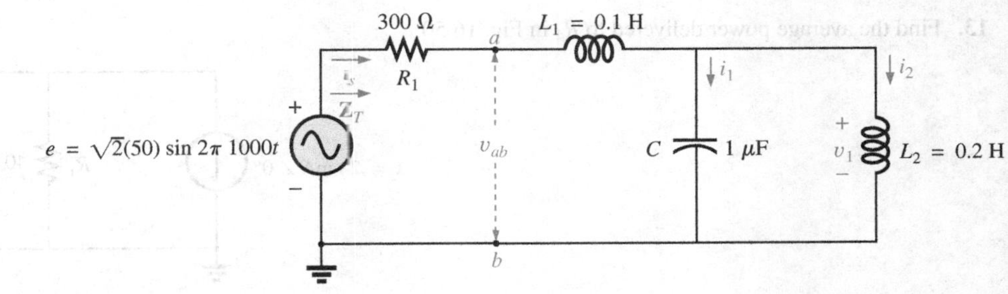

FIG. 16.46
Problem 9.

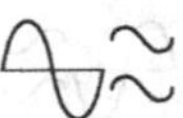

***10.** For the network of Fig. 16.47:

a. Find the total impedance $\mathbf{Z}_T$.

b. Find the voltage $\mathbf{V}_1$ in phasor form.

c. Find the current $\mathbf{I}_1$ in phasor form.

d. Find the voltage $\mathbf{V}_2$ in phasor form.

e. Find the source voltage $\mathbf{V}_s$ in phasor form.

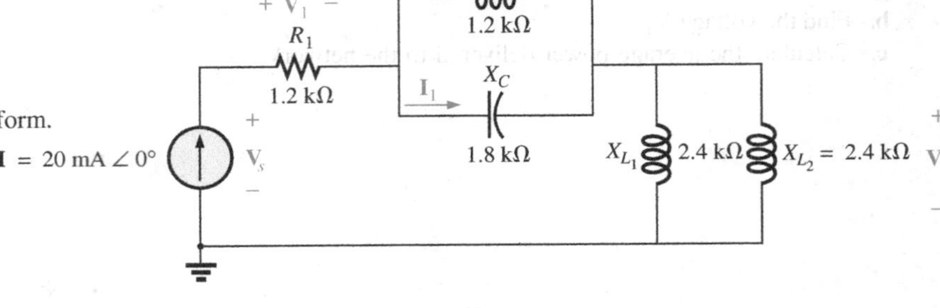

FIG. 16.47
Problem 10.

11. For the network of Fig. 16.48:

a. Find the total impedance $\mathbf{Z}_T$.

b. Find the voltage $\mathbf{V}_1$ across the 2 Ω resistor using the voltage divider rule.

c. Find the current $\mathbf{I}_1$ using Ohm's law.

d. Find the current $\mathbf{I}_s$.

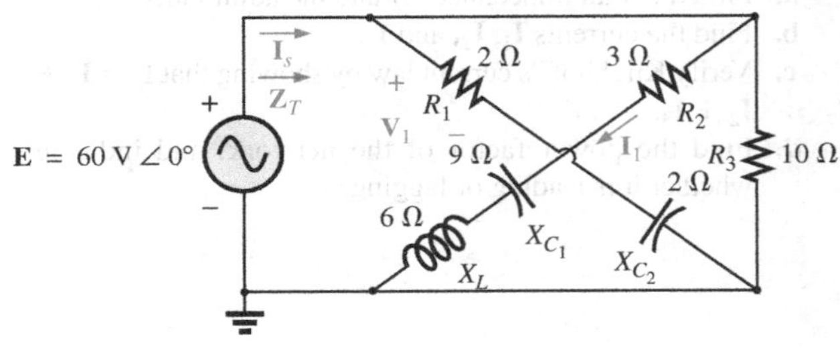

FIG. 16.48
Problems 11 and 17.

SECTION 16.3 Ladder Networks

12. Find the current $\mathbf{I}_5$ for the network in Fig. 16.49. Note the effect of one reactive element on the resulting calculations.

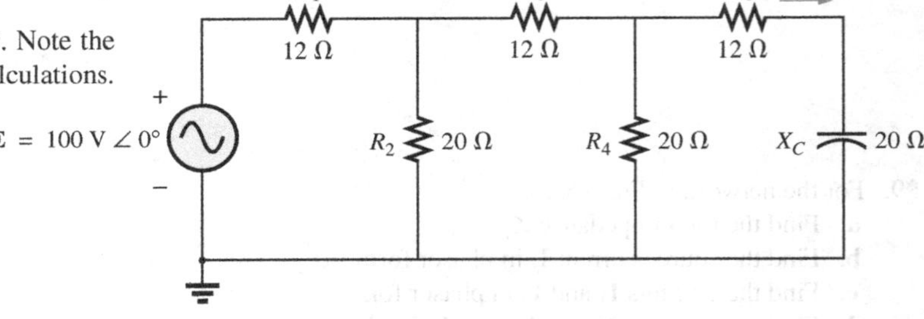

FIG. 16.49
Problem 12.

13. Find the average power delivered to R_4 in Fig. 16.50.

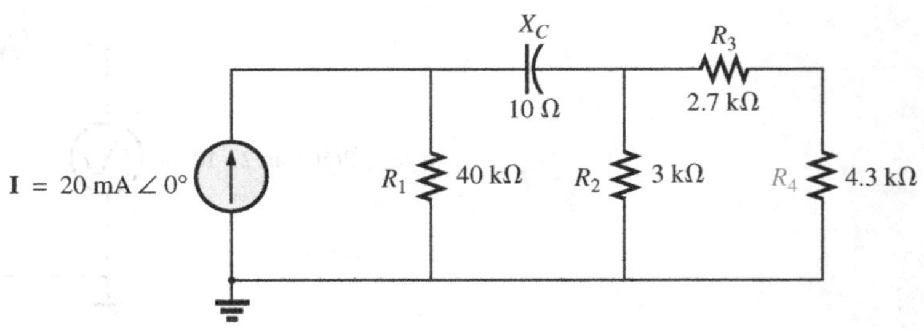

FIG. 16.50
Problem 13.

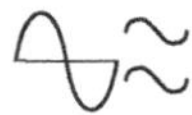

14. Find the current $\mathbf{I}_1$ for the network in Fig. 16.51.

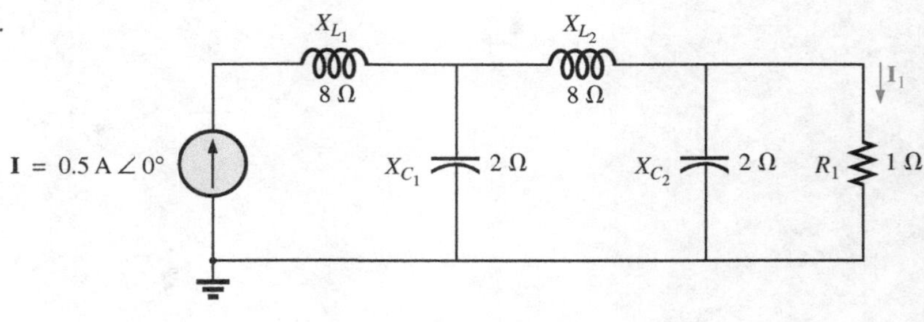

FIG. 16.51
Problems 14 and 18.

SECTION 16.6 **Computer Analysis**

PSpice or Multisim

For Problems 15 through 18, use a frequency of 1 kHz to determine the inductive and capacitive levels required for the input files. In each case, write the required input file.

***15.** Repeat Problem 2 using PSpice or Multisim.

***16.** Repeat Problem 7, parts (a) and (b), using PSpice or Multisim.

***17.** Repeat Problem 11 using PSpice or Multisim.

***18.** Repeat Problem 14 using PSpice or Multisim.

GLOSSARY

Ladder network A repetitive combination of series and parallel branches that has the appearance of a ladder.

Series-parallel ac network A combination of series and parallel branches in the same network configuration. Each branch may contain any number of elements, whose impedance is dependent on the applied frequency.

Methods of Analysis and Selected Topics (ac)

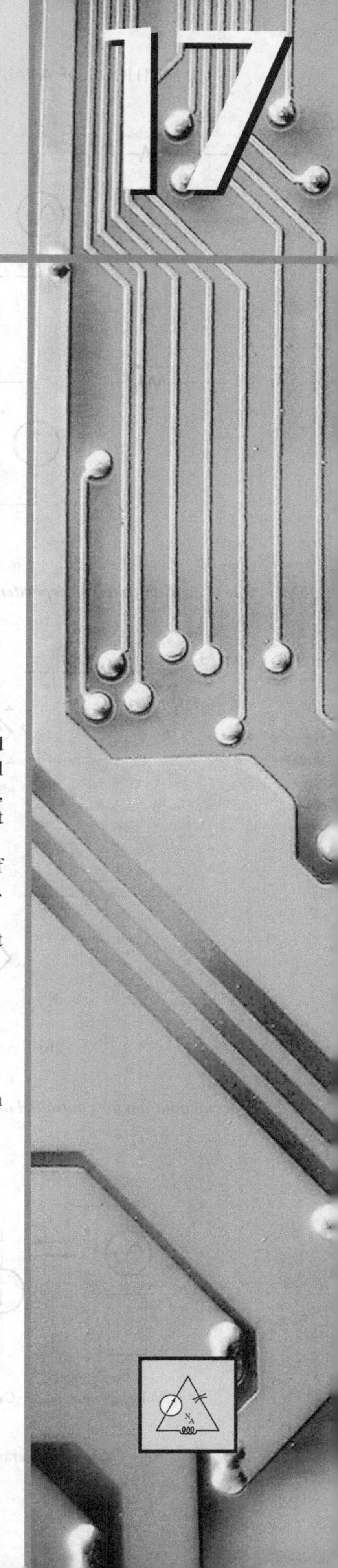

Objectives

- ***Understand the differences between independent and dependent sources and how the magnitude and angle of a controlled source is determined by the dependent variable.***
- ***Be able to convert between voltage and current sources and vice versa in the ac domain.***
- ***Become proficient in the application of mesh and nodal analysis to ac networks with independent and controlled sources.***
- ***Be able to define the relationship between the elements of an ac bridge network that will establish a balance condition.***

17.1 INTRODUCTION

For networks with two or more sources that are not in series or parallel, the methods described in the last two chapters cannot be applied. Rather, methods such as mesh analysis or nodal analysis must be used. Since these methods were discussed in detail for dc circuits in Chapter 8, this chapter considers the variations required to apply these methods to ac circuits. Dependent sources are also introduced for both mesh and nodal analysis.

The branch-current method is not discussed again because it falls within the framework of mesh analysis. In addition to the methods mentioned above, the bridge network and Δ-Y, Y-Δ conversions are also discussed for ac circuits.

Before we examine these topics, however, we must consider the subject of independent and controlled sources.

17.2 INDEPENDENT VERSUS DEPENDENT (CONTROLLED) SOURCES

In the previous chapters, each source appearing in the analysis of dc or ac networks was an **independent source,** such as E and I (or **E** and **I**) in Fig. 17.1.

The term independent specifies that the magnitude of the source is independent of the network to which it is applied and that the source displays its terminal characteristics even if completely isolated.

A dependent or controlled source is one whose magnitude is determined (or controlled) by a current or voltage of the system in which it appears.

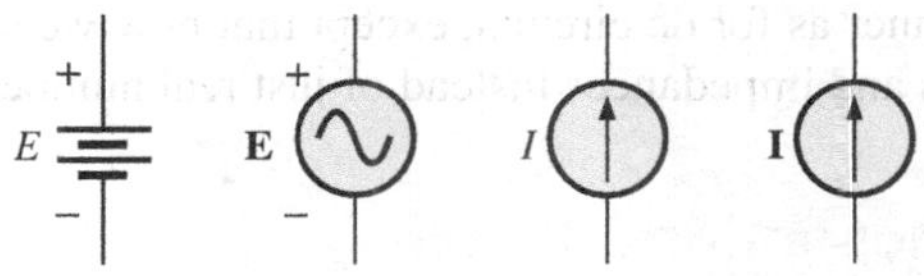

FIG. 17.1
Independent sources.

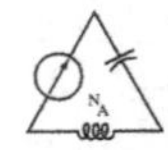

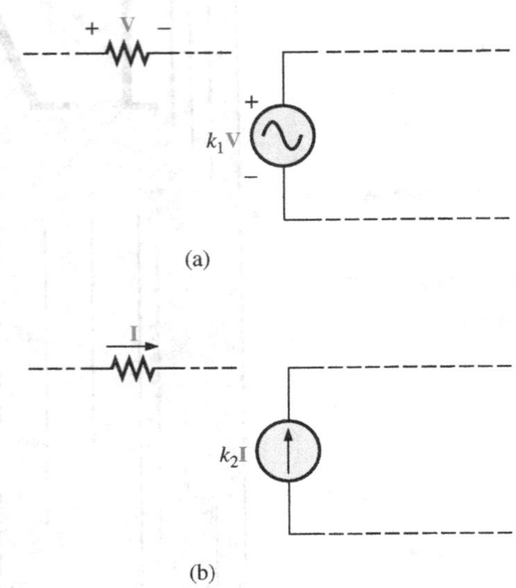

FIG. 17.2
Controlled or dependent sources.

Currently two symbols are used for controlled sources. One simply uses the independent symbol with an indication of the controlling element, as shown in Fig. 17.2. In Fig. 17.2(a), the magnitude and phase of the voltage are controlled by a voltage **V** elsewhere in the system, with the magnitude further controlled by the constant k_1. In Fig. 17.2(b), the magnitude and phase of the current source are controlled by a current **I** elsewhere in the system, with the magnitude further controlled by the constant k_2. To distinguish between the dependent and independent sources, the notation in Fig. 17.3 was introduced. In recent years, many respected publications on circuit analysis have accepted the notation in Fig. 17.3, although a number of excellent publications in the area of electronics continue to use the symbol in Fig. 17.2, especially in the circuit modeling for a variety of electronic devices such as the transistor and FET. This text uses the symbols in Fig. 17.3.

Possible combinations for controlled sources are indicated in Fig. 17.4. Note that the magnitude of current sources or voltage sources can be controlled by a voltage and a current, respectively. Unlike with the independent source, isolation such that **V** or **I** = 0 in Fig. 17.4(a) results in the short-circuit or open-circuit equivalent as indicated in Fig. 17.4(b). Note that the type of representation under these conditions is controlled by whether it is a current source or a voltage source, not by the controlling agent (**V** or **I**).

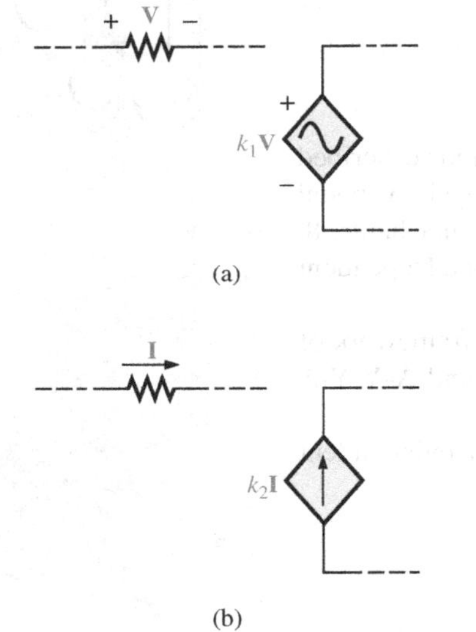

FIG. 17.3
Special notation for controlled or dependent sources.

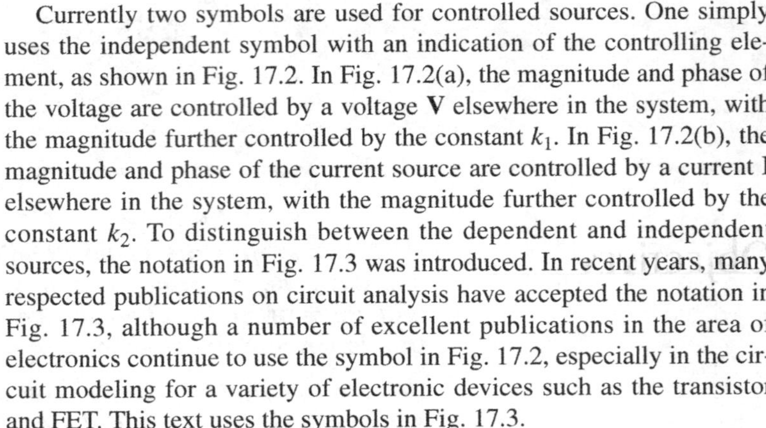

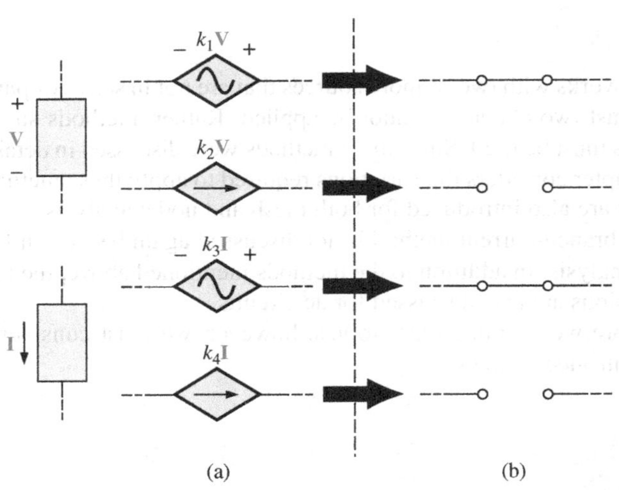

FIG. 17.4
Conditions of **V** *= 0 V and* **I** *= 0 A for a controlled source.*

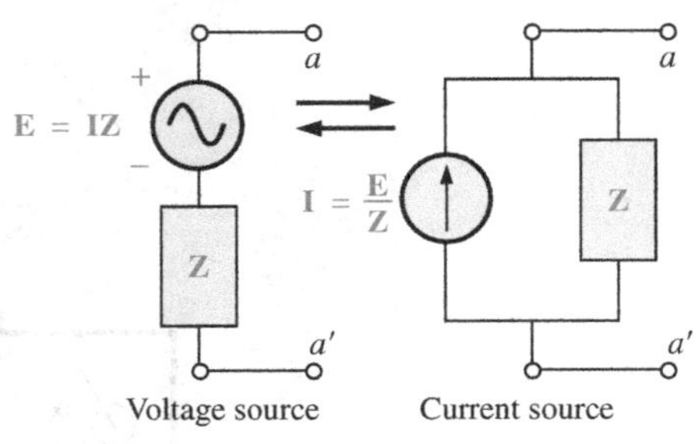

FIG. 17.5
Source Conversion.

17.3 SOURCE CONVERSIONS

When applying the methods to be discussed, it may be necessary to convert a current source to a voltage source or a voltage source to a current source. This **source conversion** can be accomplished in much the same manner as for dc circuits, except that now we shall be dealing with phasors and impedances instead of just real numbers and resistors.

Independent Sources

In general, the format for converting one type of independent source to another is as shown in Fig. 17.5.

EXAMPLE 17.1 Convert the voltage source in Fig. 17.6(a) to a current source.

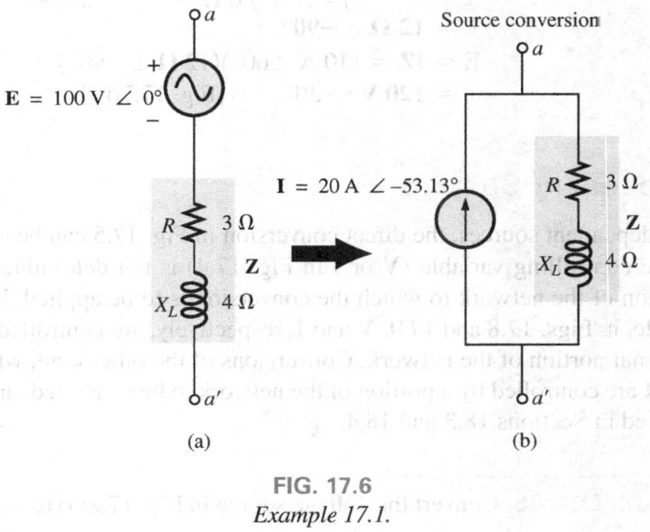

FIG. 17.6
Example 17.1.

Solution:

$$\mathbf{I} = \frac{\mathbf{E}}{\mathbf{Z}} = \frac{100\text{ V }\angle 0^\circ}{5\ \Omega\ \angle 53.13^\circ}$$
$$= \mathbf{20\ A\ \angle -53.13^\circ} \qquad \text{[Fig. 17.6(b)]}$$

EXAMPLE 17.2 Convert the current source in Fig. 17.7(a) to a voltage source.

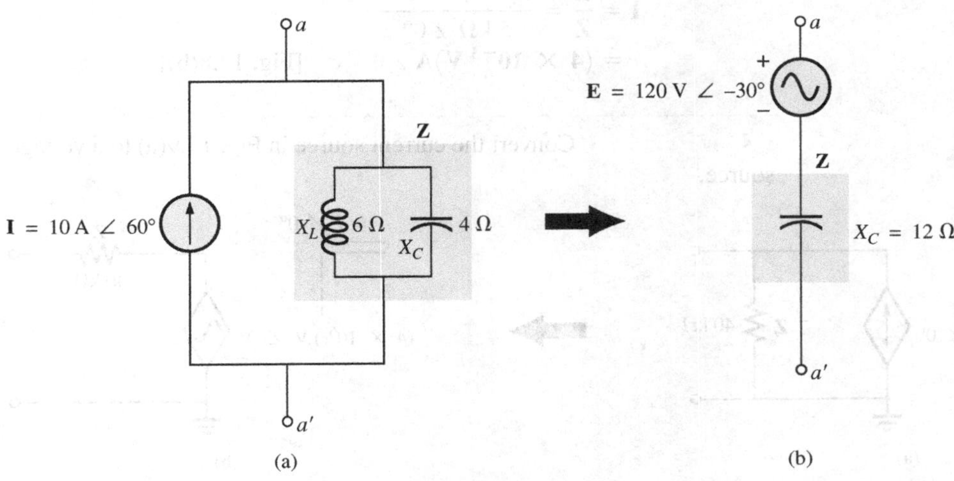

FIG. 17.7
Example 17.2.

Solution:

$$\mathbf{Z} = \frac{\mathbf{Z}_C\mathbf{Z}_L}{\mathbf{Z}_C + \mathbf{Z}_L} = \frac{(X_C \angle -90°)(X_L \angle 90°)}{-jX_C + jX_L}$$

$$= \frac{(4\ \Omega \angle -90°)(6\ \Omega \angle 90°)}{-j\,4\ \Omega + j\,6\ \Omega} = \frac{24\ \Omega \angle 0°}{2 \angle 90°}$$

$$= \mathbf{12\ \Omega \angle -90°}$$

$$\mathbf{E} = \mathbf{IZ} = (10\ \text{A} \angle 60°)(12\ \Omega \angle -90°)$$

$$= \mathbf{120\ V\angle -30°} \qquad \text{[Fig. 17.7(b)]}$$

Dependent Sources

For dependent sources, the direct conversion in Fig. 17.5 can be applied if the controlling variable (**V** or **I** in Fig. 17.4) is not determined by a portion of the network to which the conversion is to be applied. For example, in Figs. 17.8 and 17.9, **V** and **I**, respectively, are controlled by an external portion of the network. Conversions of the other kind, where **V** and **I** are controlled by a portion of the network to be converted, are considered in Sections 18.3 and 18.4.

EXAMPLE 17.3 Convert the voltage source in Fig. 17.8(a) to a current source.

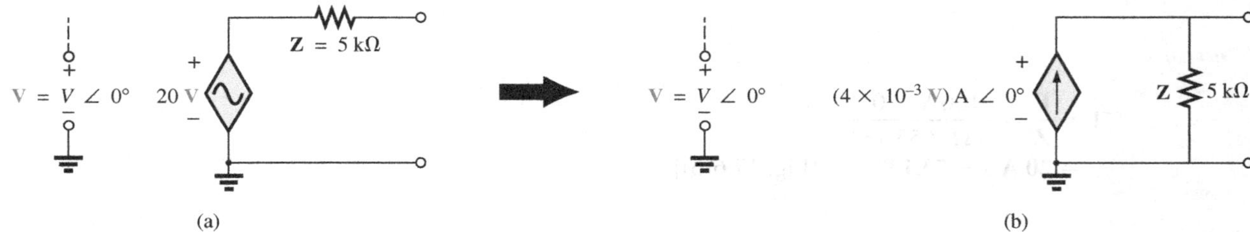

FIG. 17.8
Source conversion with a voltage-controlled voltage source.

Solution:

$$\mathbf{I} = \frac{\mathbf{E}}{\mathbf{Z}} = \frac{(20\ \mathbf{V})\ \text{V} \angle 0°}{5\ \text{k}\Omega \angle 0°}$$

$$= \mathbf{(4 \times 10^{-3}\ V)A \angle 0°} \qquad \text{[Fig. 17.8(b)]}$$

EXAMPLE 17.4 Convert the current source in Fig. 17.9(a) to a voltage source.

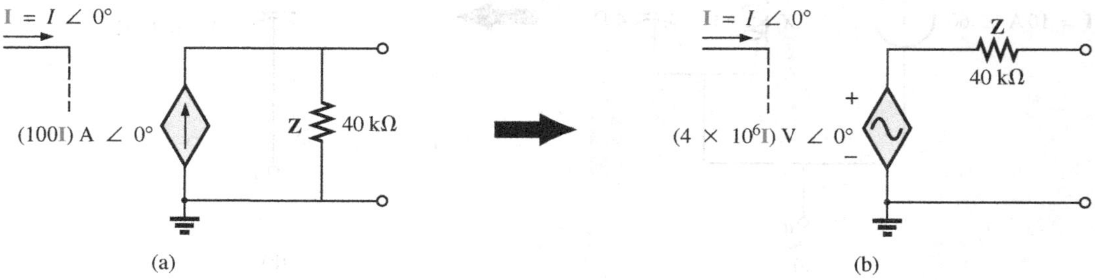

FIG. 17.9
Source conversion with a current-controlled current source.

Solution:

$$\mathbf{E} = \mathbf{IZ} = [(100\,\mathbf{I})\,\text{A}\,\angle 0°][40\,\text{k}\Omega\,\angle 0°]$$
$$= (4 \times 10^6\,\mathbf{I})\,\mathbf{V}\,\angle \mathbf{0°}$$

17.4 MESH ANALYSIS

General Approach

Independent Voltage Sources Before examining the application of the method to ac networks, the student should first review the appropriate sections on **mesh analysis** in Chapter 8 since the content of this section will be limited to the general conclusions of Chapter 8.

The general approach to mesh analysis for independent sources includes the same sequence of steps appearing in Chapter 8. In fact, throughout this section the only change from the dc coverage is to substitute impedance for resistance and admittance for conductance in the general procedure.

1. *Assign a distinct current in the clockwise direction to each independent closed loop of the network. It is not absolutely necessary to choose the clockwise direction for each loop current. However, it eliminates the need to have to choose a direction for each application. Any direction can be chosen for each loop current with no loss in accuracy as long as the remaining steps are followed properly.*
2. *Indicate the polarities within each loop for each impedance as determined by the assumed direction of loop current for that loop.*
3. *Apply Kirchhoff's voltage law around each closed loop in the clockwise direction. Again, the clockwise direction was chosen to establish uniformity and to prepare us for the format approach to follow.*
   a. *If an impedance has two or more assumed currents through it, the total current through the impedance is the assumed current of the loop in which Kirchhoff's voltage law is being applied, plus the assumed currents of the other loops passing through in the same direction, minus the assumed currents passing through in the opposite direction.*
   b. *The polarity of a voltage source is unaffected by the direction of the assigned loop currents.*
4. *Solve the resulting simultaneous linear equations for the assumed loop currents.*

The technique is applied as above for all networks with independent sources or for networks with *dependent sources where the controlling variable is not a part of the network under investigation*. If the controlling variable is part of the network being examined, a method to be described shortly must be applied.

EXAMPLE 17.5 Using the general approach to mesh analysis, find the current $\mathbf{I}_1$ in Fig. 17.10.

Solution: When applying these methods to ac circuits, it is good practice to represent the resistors and reactances (or combinations thereof) by subscripted impedances. When the total solution is found in terms of these subscripted impedances, the numerical values can be substituted to find the unknown quantities.

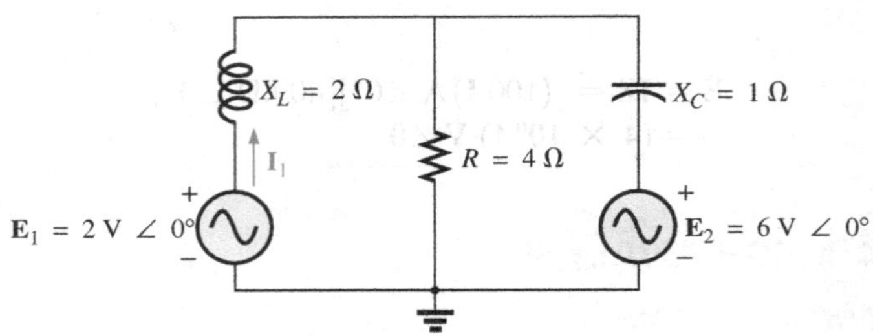

FIG. 17.10
Example 17.5.

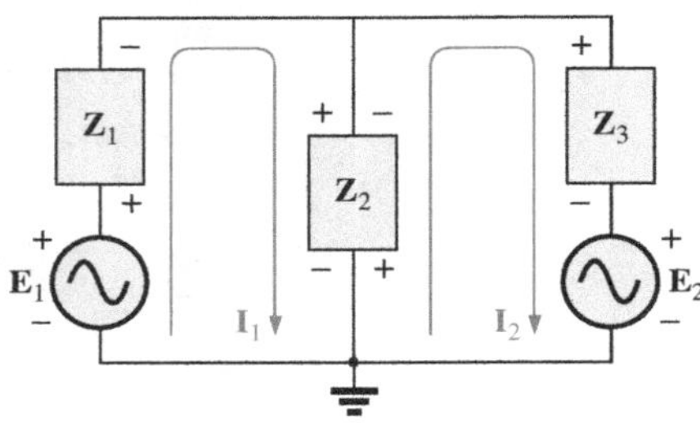

FIG. 17.11
Assigning the mesh currents and subscripted impedances for the network in Fig. 17.10.

The network is redrawn in Fig. 17.11 with subscripted impedances:

$$\mathbf{Z}_1 = +jX_L = +j\,2\ \Omega \qquad \mathbf{E}_1 = 2\ \text{V}\angle 0°$$
$$\mathbf{Z}_2 = R = 4\ \Omega \qquad \mathbf{E}_2 = 6\ \text{V}\angle 0°$$
$$\mathbf{Z}_3 = -jX_C = -j\,1\ \Omega$$

Steps 1 and 2 are as indicated in Fig. 17.11.

Step 3:

$$\begin{aligned} +\mathbf{E}_1 - \mathbf{I}_1\mathbf{Z}_1 - \mathbf{Z}_2(\mathbf{I}_1 - \mathbf{I}_2) &= 0 \\ -\mathbf{Z}_2(\mathbf{I}_2 - \mathbf{I}_1) - \mathbf{I}_2\mathbf{Z}_3 - \mathbf{E}_2 &= 0 \end{aligned}$$

or

$$\begin{aligned} \mathbf{E}_1 - \mathbf{I}_1\mathbf{Z}_1 - \mathbf{I}_1\mathbf{Z}_2 + \mathbf{I}_2\mathbf{Z}_2 &= 0 \\ -\mathbf{I}_2\mathbf{Z}_2 + \mathbf{I}_1\mathbf{Z}_2 - \mathbf{I}_2\mathbf{Z}_3 - \mathbf{E}_2 &= 0 \end{aligned}$$

so that

$$\begin{aligned} \mathbf{I}_1(\mathbf{Z}_1 + \mathbf{Z}_2) - \mathbf{I}_2\mathbf{Z}_2 &= \mathbf{E}_1 \\ \mathbf{I}_2(\mathbf{Z}_2 + \mathbf{Z}_3) - \mathbf{I}_1\mathbf{Z}_2 &= -\mathbf{E}_2 \end{aligned}$$

which are rewritten as

$$\begin{aligned} \mathbf{I}_1(\mathbf{Z}_1 + \mathbf{Z}_2) - \mathbf{I}_2\mathbf{Z}_2 \qquad\qquad &= \mathbf{E}_1 \\ -\mathbf{I}_1\mathbf{Z}_2 \qquad + \mathbf{I}_2(\mathbf{Z}_2 + \mathbf{Z}_3) &= -\mathbf{E}_2 \end{aligned}$$

Step 4: Using determinants, we obtain

$$\mathbf{I}_1 = \frac{\begin{vmatrix} \mathbf{E}_1 & -\mathbf{Z}_2 \\ -\mathbf{E}_2 & \mathbf{Z}_2 + \mathbf{Z}_3 \end{vmatrix}}{\begin{vmatrix} \mathbf{Z}_1 + \mathbf{Z}_2 & -\mathbf{Z}_2 \\ -\mathbf{Z}_2 & \mathbf{Z}_2 + \mathbf{Z}_3 \end{vmatrix}} = \frac{\mathbf{E}_1(\mathbf{Z}_2 + \mathbf{Z}_3) - \mathbf{E}_2(\mathbf{Z}_2)}{(\mathbf{Z}_1 + \mathbf{Z}_2)(\mathbf{Z}_2 + \mathbf{Z}_3) - (\mathbf{Z}_2)^2} = \frac{(\mathbf{E}_1 - \mathbf{E}_2)\mathbf{Z}_2 + \mathbf{E}_1\mathbf{Z}_3}{\mathbf{Z}_1\mathbf{Z}_2 + \mathbf{Z}_1\mathbf{Z}_3 + \mathbf{Z}_2\mathbf{Z}_3}$$

Substituting numerical values yields

$$\mathbf{I}_1 = \frac{(2\ \text{V} - 6\ \text{V})(4\ \Omega) + (2\ \text{V})(-j\,1\ \Omega)}{(+j\,2\ \Omega)(4\ \Omega) + (+j\,2\ \Omega)(-j\,2\ \Omega) + (4\ \Omega)(-j\,2\ \Omega)}$$
$$= \frac{-16 - j\,2}{j\,8 - j^2\,2 - j\,4} = \frac{-16 - j\,2}{2 + j\,4} = \frac{16.12\ \text{A}\angle -172.87°}{4.47\angle 63.43°}$$
$$= \mathbf{3.61\ A\angle -236.30°} \quad \text{or} \quad \mathbf{3.61\ A\angle 123.70°}$$

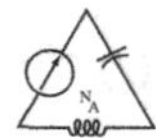

Dependent Voltage Sources For dependent voltage sources, the procedure is modified as follows:

1. Steps 1 and 2 are the same as those applied for independent voltage sources.
2. Step 3 is modified as follows: Treat each dependent source like an independent source when Kirchhoff's voltage law is applied to each independent loop. However, once the equation is written, substitute the equation for the controlling quantity to ensure that the unknowns are limited solely to the chosen mesh currents.
3. Step 4 is as before.

EXAMPLE 17.6 Write the mesh currents for the network in Fig. 17.12 having a dependent voltage source.

Solution:

Steps 1 and 2 are defined in Fig. 17.12.

Step 3:

$$\mathbf{E}_1 - \mathbf{I}_1 R_1 - R_2(\mathbf{I}_1 - \mathbf{I}_2) = 0$$
$$R_2(\mathbf{I}_2 - \mathbf{I}_1) + \mu\mathbf{V}_x - \mathbf{I}_2 R_3 = 0$$

Then substitute $\mathbf{V}_x = (\mathbf{I}_1 - \mathbf{I}_2)R_2$.

The result is two equations and two unknowns:

$$\mathbf{E}_1 - \mathbf{I}_1 R_1 - R_2(\mathbf{I} - \mathbf{I}_2) = 0$$
$$R_2(\mathbf{I}_2 - \mathbf{I}_1) + \mu R_2(\mathbf{I}_1 - \mathbf{I}_2) - \mathbf{I}_2 R_3 = 0$$

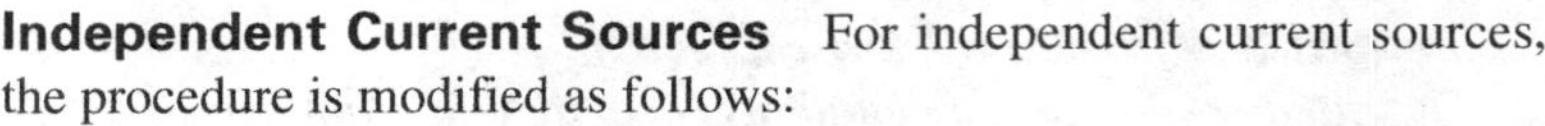

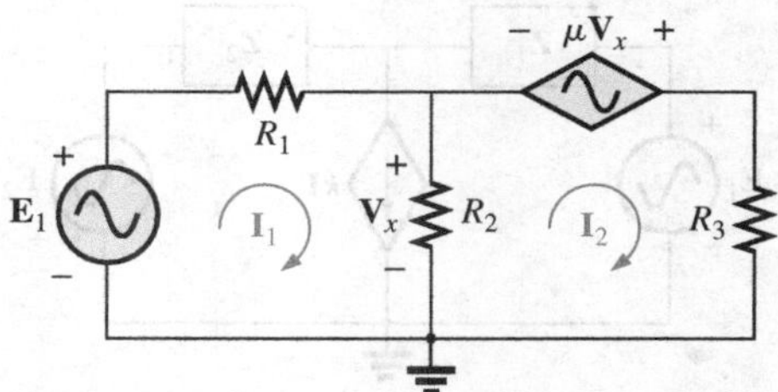

FIG. 17.12
Applying mesh analysis to a network with a voltage-controlled voltage source.

Independent Current Sources For independent current sources, the procedure is modified as follows:

1. Steps 1 and 2 are the same as those applied for independent sources.
2. Step 3 is modified as follows: Treat each current source as an open circuit (recall the *supermesh* designation in Chapter 8), and write the mesh equations for each remaining independent path. Then relate the chosen mesh currents to the dependent sources to ensure that the unknowns of the final equations are limited to the mesh currents.
3. Step 4 is as before.

EXAMPLE 17.7 Write the mesh currents for the network in Fig. 17.13 having an independent current source.

Solution:

Steps 1 and 2 are defined in Fig. 17.13.

Step 3: $\mathbf{E}_1 - \mathbf{I}_1\mathbf{Z}_1 + \mathbf{E}_2 - \mathbf{I}_2\mathbf{Z}_2 = 0$ (only remaining independent path)

with $\mathbf{I}_1 + \mathbf{I} = \mathbf{I}_2$

The result is two equations and two unknowns.

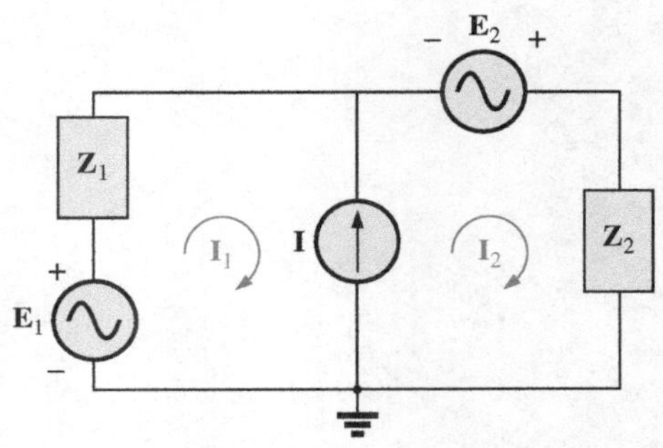

FIG. 17.13
Applying mesh analysis to a network with an independent current source.

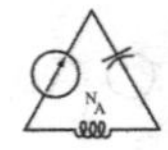

Dependent Current Sources For dependent current sources, the procedure is modified as follows:

1. Steps 1 and 2 are the same as those applied for independent sources.
2. Step 3 is modified as follows: The procedure is essentially the same as that applied for independent current sources, except now the dependent sources have to be defined in terms of the chosen mesh currents to ensure that the final equations have only mesh currents as the unknown quantities.
3. Step 4 is as before.

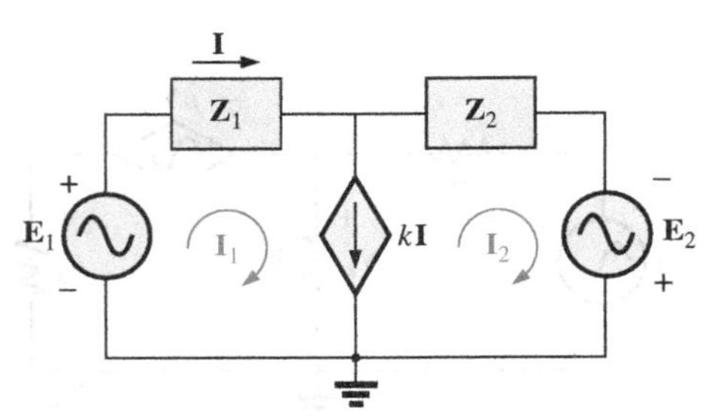

FIG. 17.14
Applying mesh analysis to a network with a current-controlled current source.

EXAMPLE 17.8 Write the mesh currents for the network in Fig. 17.14 having a dependent current source.

Solution:

Steps 1 and 2 are defined in Fig. 17.14.

Step 3: $\mathbf{E}_1 - \mathbf{I}_1\mathbf{Z}_1 - \mathbf{I}_2\mathbf{Z}_2 + \mathbf{E}_2 = 0$

and $k\mathbf{I} = \mathbf{I}_1 - \mathbf{I}_2$

Now $\mathbf{I} = \mathbf{I}_1$ so that $k\mathbf{I}_1 = \mathbf{I}_1 - \mathbf{I}_2$ or $\mathbf{I}_2 = \mathbf{I}_1(1 - k)$

The result is two e quations and two unknowns.

Format Approach

The format approach was introduced in Section 8.8. The steps for applying this method are repeated here with changes for its use in ac circuits:

1. *Assign a loop current to each independent closed loop (as in the previous section) in a clockwise direction.*
2. *The number of required equations is equal to the number of chosen independent closed loops. Column 1 of each equation is formed by summing the impedance values of those impedances through which the loop current of interest passes and multiplying the result by that loop current.*
3. *We must now consider the mutual terms that are always subtracted from the terms in the first column. It is possible to have more than one mutual term if the loop current of interest has an element in common with more than one other loop current. Each mutual term is the product of the mutual impedance and the other loop current passing through the same element.*
4. *The column to the right of the equality sign is the algebraic sum of the voltage sources through which the loop current of interest passes. Positive signs are assigned to those sources of voltage having a polarity such that the loop current passes from the negative to the positive terminal. Negative signs are assigned to those potentials for which the reverse is true.*
5. *Solve the resulting simultaneous equations for the desired loop currents.*

The technique is applied as above for all networks with independent sources or for networks with dependent sources where the controlling variable is not a part of the network under investigation. If the controlling variable is part of the network being examined, additional care must be taken when applying the above steps.

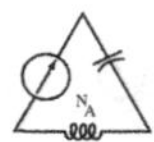

EXAMPLE 17.9 Using the format approach to mesh analysis, find the current $\mathbf{I}_2$ in Fig. 17.15.

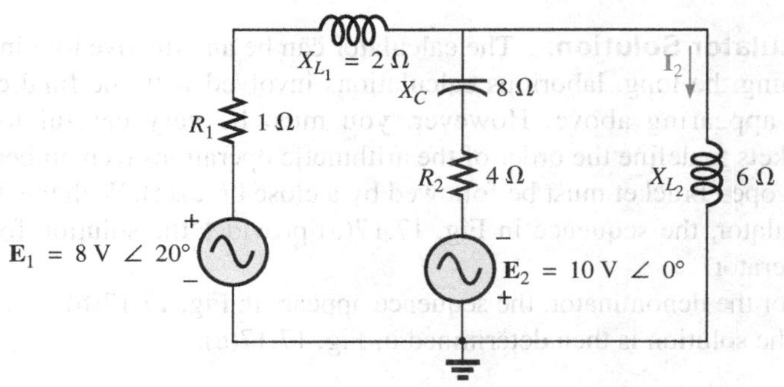

FIG. 17.15
Example 17.9.

Solution: The network is redrawn in Fig. 17.16:

$$\mathbf{Z}_1 = R_1 + jX_{L_1} = 1\ \Omega + j\,2\ \Omega \qquad \mathbf{E}_1 = 8\ \text{V}\ \angle 20°$$
$$\mathbf{Z}_2 = R_2 - jX_C = 4\ \Omega - j\,8\ \Omega \qquad \mathbf{E}_2 = 10\ \text{V}\ \angle 0°$$
$$\mathbf{Z}_3 = +jX_{L_2} = +j\,6\ \Omega$$

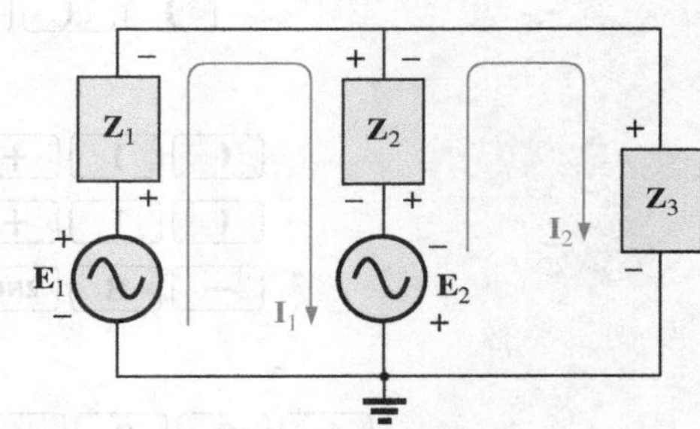

FIG. 17.16
Assigning the mesh currents and subscripted impedances for the network in Fig. 17.15.

Note the reduction in complexity of the problem with the substitution of the subscripted impedances.

Step 1 is as indicated in Fig. 17.16.

Steps 2 to 4:

$$\mathbf{I}_1(\mathbf{Z}_1 + \mathbf{Z}_2) - \mathbf{I}_2\mathbf{Z}_2 = \mathbf{E}_1 + \mathbf{E}_2$$
$$\mathbf{I}_2(\mathbf{Z}_2 + \mathbf{Z}_3) - \mathbf{I}_1\mathbf{Z}_2 = -\mathbf{E}_2$$

which are rewritten as

$$\mathbf{I}_1(\mathbf{Z}_1 + \mathbf{Z}_2) - \mathbf{I}_2\mathbf{Z}_2 = \mathbf{E}_1 + \mathbf{E}_2$$
$$-\mathbf{I}_1\mathbf{Z}_2 + \mathbf{I}_2(\mathbf{Z}_2 + \mathbf{Z}_3) = -\mathbf{E}_2$$

Step 5: Using determinants, we have

$$\mathbf{I}_2 = \frac{\begin{vmatrix} \mathbf{Z}_1 + \mathbf{Z}_2 & \mathbf{E}_1 + \mathbf{E}_2 \\ -\mathbf{Z}_2 & -\mathbf{E}_2 \end{vmatrix}}{\begin{vmatrix} \mathbf{Z}_1 + \mathbf{Z}_2 & -\mathbf{Z}_2 \\ -\mathbf{Z}_2 & \mathbf{Z}_2 + \mathbf{Z}_3 \end{vmatrix}} = \frac{-(\mathbf{Z}_1 + \mathbf{Z}_2)\mathbf{E}_2 + \mathbf{Z}_2(\mathbf{E}_1 + \mathbf{E}_2)}{(\mathbf{Z}_1 + \mathbf{Z}_2)(\mathbf{Z}_2 + \mathbf{Z}_3) - \mathbf{Z}_2^2} = \frac{\mathbf{Z}_2\mathbf{E}_1 - \mathbf{Z}_1\mathbf{E}_2}{\mathbf{Z}_1\mathbf{Z}_2 + \mathbf{Z}_1\mathbf{Z}_3 + \mathbf{Z}_2\mathbf{Z}_3}$$

Substituting numerical values yields

$$\mathbf{I}_2 = \frac{(4\ \Omega - j\,8\ \Omega)(8\ \text{V}\ \angle 20°) - (1\ \Omega + j\,2\ \Omega)(10\ \text{V}\ \angle 0°)}{(1\ \Omega + j\,2\ \Omega)(4\ \Omega - j\,8\ \Omega) + (1\ \Omega + j\,2\ \Omega)(+j\,6\ \Omega) + (4\ \Omega - j\,8\ \Omega)(+j\,6\ \Omega)}$$
$$= \frac{(4 - j\,8)(7.52 + j\,2.74) - (10 + j\,20)}{20 + (j\,6 - 12) + (j\,24 + 48)}$$

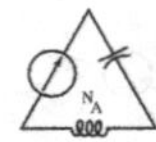

$$= \frac{(52.0 - j\,49.20) - (10 + j\,20)}{56 + j\,30} = \frac{42.0 - j\,69.20}{56 + j\,30} = \frac{80.95\text{ A }\angle -58.74^\circ}{63.53\ \angle 28.18^\circ}$$
$$= \mathbf{1.27\text{ A }\angle 286.92^\circ}$$

Calculator Solution: The calculator can be an effective tool in performing the long, laborious calculations involved with the final equation appearing above. However, you must be very careful to use brackets to define the order of the arithmetic operations (remember that each open bracket must be followed by a close bracket). With the TI-89 calculator, the sequence in Fig. 17.17(a) provides the solution for the numerator.

For the denominator, the sequence appears in Fig. 17.17(b).

The solution is then determined in Fig. 17.17(c).

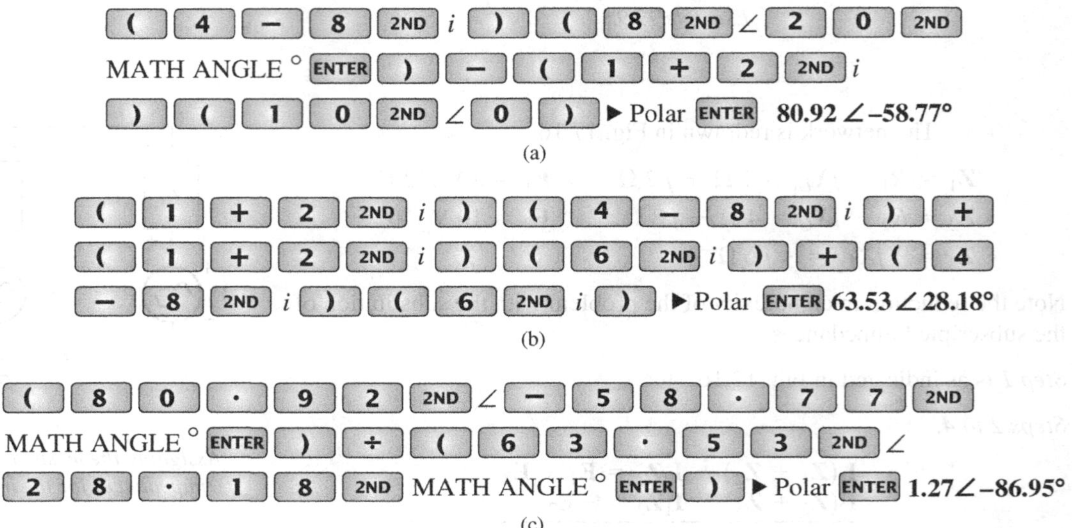

FIG. 17.17
Determining I_2 *for the network of Fig. 17.15.*

EXAMPLE 17.10 Write the mesh equations for the network in Fig. 17.18. Do not solve.

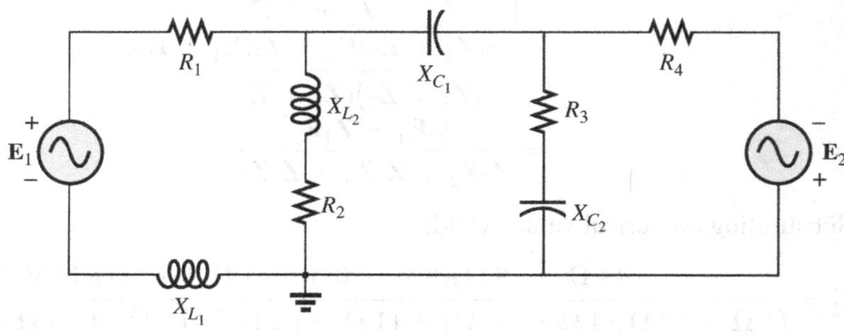

FIG. 17.18
Example 17.10.

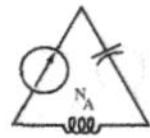

Solution: The network is redrawn in Fig. 17.19. Again note the reduced complexity and increased clarity provided by the use of subscripted impedances:

$$\begin{aligned} \mathbf{Z}_1 &= R_1 + jX_{L_1} & \mathbf{Z}_4 &= R_3 - jX_{C_2} \\ \mathbf{Z}_2 &= R_2 + jX_{L_2} & \mathbf{Z}_5 &= R_4 \\ \mathbf{Z}_3 &= jX_{C_1} \end{aligned}$$

and

$$\begin{aligned} \mathbf{I}_1(\mathbf{Z}_1 + \mathbf{Z}_2) - \mathbf{I}_2\mathbf{Z}_2 &= \mathbf{E}_1 \\ \mathbf{I}_2(\mathbf{Z}_2 + \mathbf{Z}_3 + \mathbf{Z}_4) - \mathbf{I}_1\mathbf{Z}_2 - \mathbf{I}_3\mathbf{Z}_4 &= 0 \\ \mathbf{I}_3(\mathbf{Z}_4 + \mathbf{Z}_5) - \mathbf{I}_2\mathbf{Z}_4 &= \mathbf{E}_2 \end{aligned}$$

or

$$\begin{array}{llll} \mathbf{I}_1(\mathbf{Z}_1 + \mathbf{Z}_2) & - \mathbf{I}_2(\mathbf{Z}_2) & + 0 & = \mathbf{E}_1 \\ \mathbf{I}_1\mathbf{Z}_2 & - \mathbf{I}_2(\mathbf{Z}_2 + \mathbf{Z}_3 + \mathbf{Z}_4) & - \mathbf{I}_3(\mathbf{Z}_4) & = 0 \\ 0 & - \mathbf{I}_2(\mathbf{Z}_4) & - \mathbf{I}_3(\mathbf{Z}_4 + \mathbf{Z}_5) & = \mathbf{E}_2 \end{array}$$

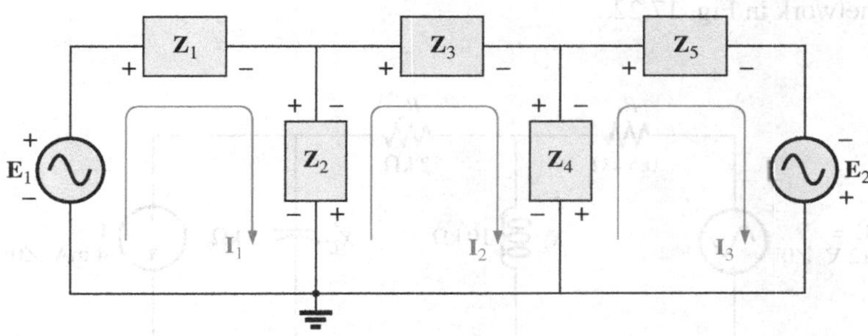

FIG. 17.19
Assigning the mesh currents and subscripted impedances for the network in Fig. 17.18.

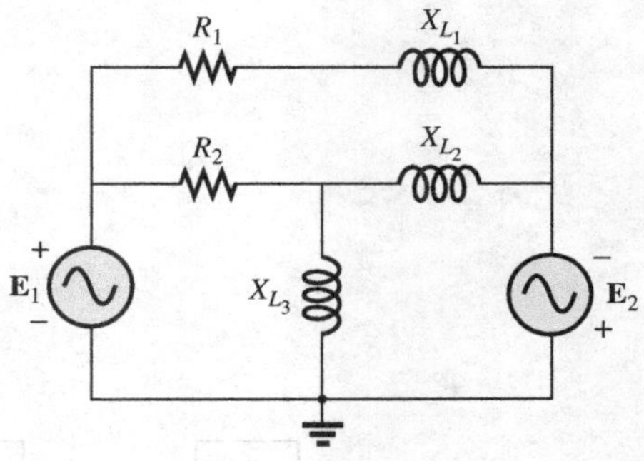

FIG. 17.20
Example 17.11.

EXAMPLE 17.11 Using the format approach, write the mesh equations for the network in Fig. 17.20.

Solution: The network is redrawn as shown in Fig. 17.21, where

$$\begin{aligned} \mathbf{Z}_1 &= R_1 + jX_{L_1} & \mathbf{Z}_3 &= jX_{L_2} \\ \mathbf{Z}_2 &= R_2 & \mathbf{Z}_4 &= jX_{L_3} \end{aligned}$$

and

$$\begin{aligned} \mathbf{I}_1(\mathbf{Z}_2 + \mathbf{Z}_4) - \mathbf{I}_2\mathbf{Z}_2 - \mathbf{I}_3\mathbf{Z}_4 &= \mathbf{E}_1 \\ \mathbf{I}_2(\mathbf{Z}_1 + \mathbf{Z}_2 + \mathbf{Z}_3) - \mathbf{I}_1\mathbf{Z}_2 - \mathbf{I}_3\mathbf{Z}_3 &= 0 \\ \mathbf{I}_3(\mathbf{Z}_3 + \mathbf{Z}_4) - \mathbf{I}_2\mathbf{Z}_3 - \mathbf{I}_1\mathbf{Z}_4 &= \mathbf{E}_2 \end{aligned}$$

or

$$\begin{array}{llll} \mathbf{I}_1(\mathbf{Z}_2 + \mathbf{Z}_4) & - \mathbf{I}_2\mathbf{Z}_2 & - \mathbf{I}_3\mathbf{Z}_4 & = \mathbf{E}_1 \\ -\mathbf{I}_1\mathbf{Z}_2 & + \mathbf{I}_2(\mathbf{Z}_1 + \mathbf{Z}_2 + \mathbf{Z}_3) & - \mathbf{I}_3\mathbf{Z}_3 & = 0 \\ -\mathbf{I}_1\mathbf{Z}_4 & - \mathbf{I}_2\mathbf{Z}_3 & + \mathbf{I}_3(\mathbf{Z}_3 + \mathbf{Z}_4) & = \mathbf{E}_2 \end{array}$$

Note the symmetry *about* the diagonal axis; that is, note the location of $-\mathbf{Z}_2$, $-\mathbf{Z}_4$, and $-\mathbf{Z}_3$ off the diagonal.

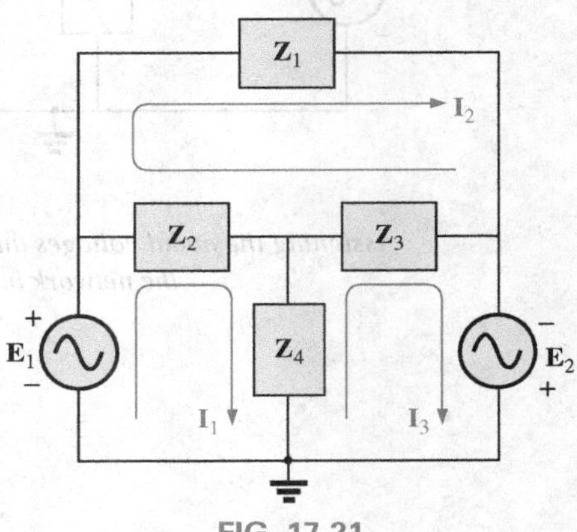

FIG. 17.21
Assigning the mesh currents and subscripted impedances for the network in Fig. 17.20.

17.5 NODAL ANALYSIS

General Approach

Independent Sources Before examining the application of the method to ac networks, a review of the appropriate sections on **nodal**

analysis in Chapter 8 is suggested since the content of this section is limited to the general conclusions of Chapter 8.

The fundamental steps are the following:

1. *Determine the number of nodes within the network.*
2. *Pick a reference node and label each remaining node with a subscripted value of voltage:* V_1, V_2*, and so on.*
3. *Apply Kirchhoff's current law at each node except the reference. Assume that all unknown currents leave the node for each application of Kirchhoff's current law.*
4. *Solve the resulting equations for the nodal voltages.*

A few examples will refresh your memory about the content of Chapter 8 and the general approach to a nodal-analysis solution.

EXAMPLE 17.12 Determine the voltage across the inductor for the network in Fig. 17.22.

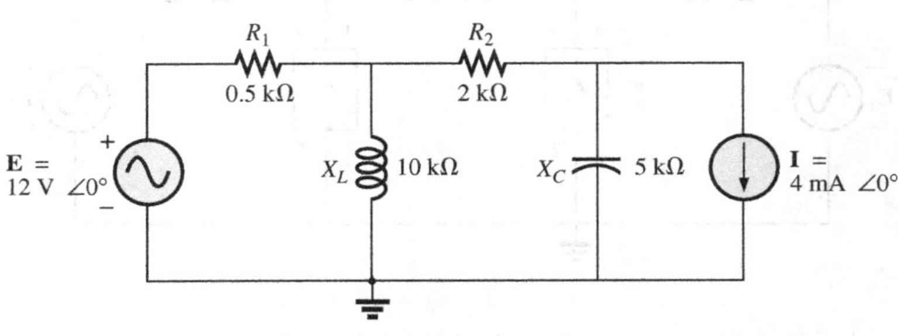

FIG. 17.22
Example 17.12.

Solution:

Steps 1 and 2 are as indicated in Fig. 17.23.

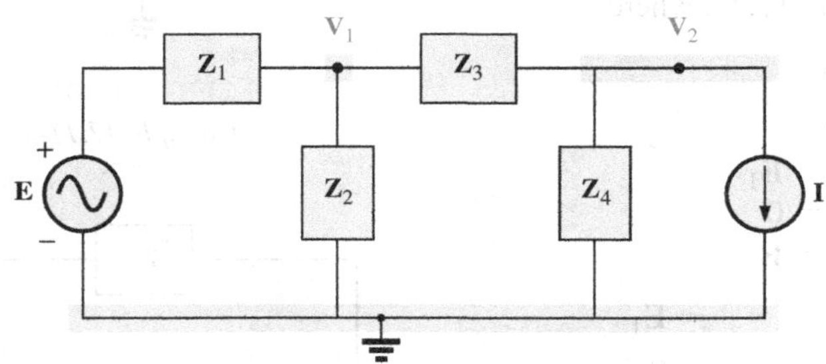

FIG. 17.23
Assigning the nodal voltages and subscripted impedances to the network in Fig. 17.22.

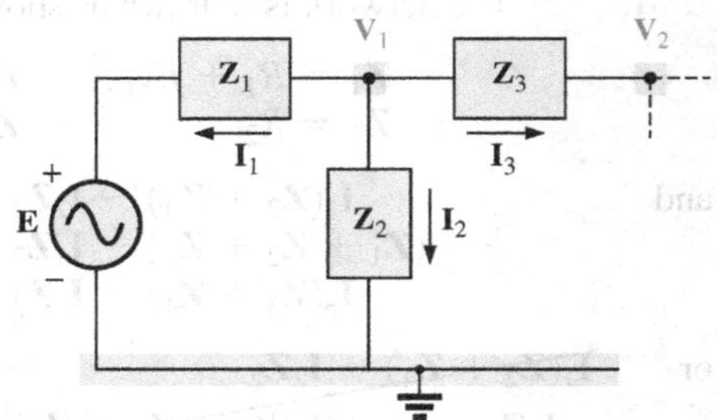

FIG. 17.24
Applying Kirchhoff's current law to the node $\mathbf{V}_1$ *in Fig. 17.23.*

Step 3: Note Fig. 17.24 for the application of Kirchhoff's current law to node $\mathbf{V}_1$:

$$\Sigma \mathbf{I}_i = \Sigma \mathbf{I}_o$$

$$0 = \mathbf{I}_1 + \mathbf{I}_2 + \mathbf{I}_3$$

$$\frac{\mathbf{V}_1 - \mathbf{E}}{\mathbf{Z}_1} + \frac{\mathbf{V}_1}{\mathbf{Z}_2} + \frac{\mathbf{V}_1 - \mathbf{V}_2}{\mathbf{Z}_3} = 0$$

Rearranging terms gives

$$\mathbf{V}_1\left[\frac{1}{\mathbf{Z}_1}+\frac{1}{\mathbf{Z}_2}+\frac{1}{\mathbf{Z}_3}\right]-\mathbf{V}_2\left[\frac{1}{\mathbf{Z}_3}\right]=\frac{\mathbf{E}_1}{\mathbf{Z}_1} \quad \textbf{(17.1)}$$

Note Fig. 17.25 for the application of Kirchoff's current law to node $\mathbf{V}_2$.

$$0 = \mathbf{I}_3 + \mathbf{I}_4 + \mathbf{I}$$

$$\frac{\mathbf{V}_2-\mathbf{V}_1}{\mathbf{Z}_3}+\frac{\mathbf{V}_2}{\mathbf{Z}_4}+\mathbf{I}=0$$

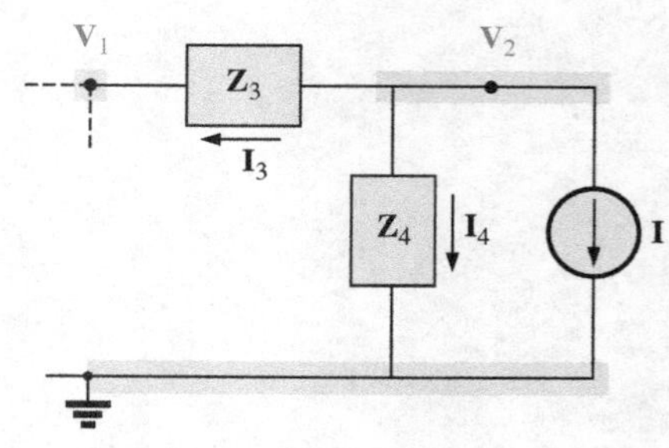

FIG. 17.25
Applying Kirchhoff's current law to the node $\mathbf{V}_2$ in Fig. 17.23.

Rearranging terms gives

$$\mathbf{V}_2\left[\frac{1}{\mathbf{Z}_3}+\frac{1}{\mathbf{Z}_4}\right]-\mathbf{V}_1\left[\frac{1}{\mathbf{Z}_3}\right]=-\mathbf{I} \quad \textbf{(17.2)}$$

Grouping equations gives

$$\begin{aligned}\mathbf{V}_1\left[\frac{1}{\mathbf{Z}_1}+\frac{1}{\mathbf{Z}_2}+\frac{1}{\mathbf{Z}_3}\right]-\mathbf{V}_2\left[\frac{1}{\mathbf{Z}_3}\right]&=\frac{\mathbf{E}}{\mathbf{Z}_1}\\ \mathbf{V}_1\left[\frac{1}{\mathbf{Z}_3}\right]-\mathbf{V}_2\left[\frac{1}{\mathbf{Z}_3}+\frac{1}{\mathbf{Z}_4}\right]&=\mathbf{I}\end{aligned}$$

$$\frac{1}{\mathbf{Z}_1}+\frac{1}{\mathbf{Z}_2}+\frac{1}{\mathbf{Z}_3}=\frac{1}{0.5\text{ k}\Omega}+\frac{1}{j\,10\text{ k}\Omega}+\frac{1}{2\text{ k}\Omega}=2.5\text{ mS}\angle-2.29^\circ$$

$$\frac{1}{\mathbf{Z}_3}+\frac{1}{\mathbf{Z}_4}=\frac{1}{2\text{ k}\Omega}+\frac{1}{-j\,5\text{ k}\Omega}=0.539\text{ mS}\angle21.80^\circ$$

and

$$\begin{aligned}\mathbf{V}_1[2.5\text{ mS}\angle-2.29^\circ]-\mathbf{V}_2[0.5\text{ mS}\angle0^\circ]&=24\text{ mA}\angle0^\circ\\ \mathbf{V}_1[0.5\text{ mS}\angle0^\circ]-\mathbf{V}_2[0.539\text{ mS}\angle21.80^\circ]&=4\text{ mA}\angle0^\circ\end{aligned}$$

with

$$\mathbf{V}_1=\frac{\begin{vmatrix}24\text{ mA}\angle0^\circ & -0.5\text{ mS}\angle0^\circ\\ 4\text{ mA}\angle0^\circ & -0.539\text{ mS}\angle21.80^\circ\end{vmatrix}}{\begin{vmatrix}2.5\text{ mS}\angle-2.29^\circ & -0.5\text{ mS}\angle0^\circ\\ 0.5\text{ mS}\angle0^\circ & -0.539\text{ mS}\angle21.80^\circ\end{vmatrix}}$$

$$=\frac{(24\text{ mA}\angle0^\circ)(-0.539\text{ mS}\angle21.80^\circ)+(0.5\text{ mS}\angle0^\circ)(4\text{ mA}\angle0^\circ)}{(2.5\text{ mS}\angle-2.29^\circ)(-0.539\text{ mS}\angle21.80^\circ)+(0.5\text{ mS}\angle0^\circ)(0.5\text{ mS}\angle0^\circ)}$$

$$=\frac{-12.94\times10^{-6}\text{ V}\angle21.80^\circ+2\times10^{-6}\text{ V}\angle0^\circ}{-1.348\times10^{-6}\angle19.51^\circ+0.25\times10^{-6}\angle0^\circ}$$

$$=\frac{-(12.01+j\,4.81)\times10^{-6}\text{ V}+2\times10^{-6}\text{ V}}{-(1.271+j\,0.45)\times10^{-6}+0.25\times10^{-6}}$$

$$=\frac{-10.01\text{ V}-j\,4.81\text{ V}}{-1.021-j\,0.45}=\frac{11.106\text{ V}\angle-154.33^\circ}{1.116\angle-156.21^\circ}$$

$$\mathbf{V}_1=\mathbf{9.95\text{ V}\angle1.88^\circ}$$

Dependent Current Sources For dependent current sources, the procedure is modified as follows:

1. Steps 1 and 2 are the same as those applied for independent sources.
2. Step 3 is modified as follows: Treat each dependent current source like an independent source when Kirchhoff's current law

is applied to each defined node. However, once the equations are established, substitute the equation for the controlling quantity to ensure that the unknowns are limited solely to the chosen nodal voltages.

3. Step 4 is as before.

EXAMPLE 17.13 Write the nodal equations for the network in Fig. 17.26 having a dependent current source.

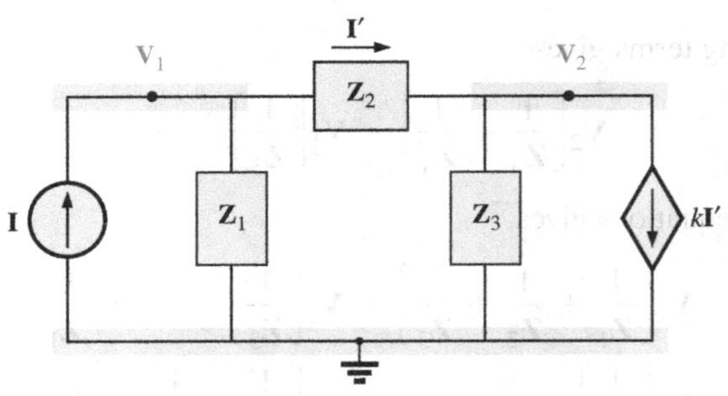

FIG. 17.26
Applying nodal analysis to a network with a current-controlled current source.

Solution:

Steps 1 and 2 are as defined in Fig. 17.26.

Step 3: At node $\mathbf{V}_1$,

$$\mathbf{I} = \mathbf{I}_1 + \mathbf{I}_2$$

$$\frac{\mathbf{V}_1}{\mathbf{Z}_1} + \frac{\mathbf{V}_1 - \mathbf{V}_2}{\mathbf{Z}_2} - \mathbf{I} = 0$$

and

$$\mathbf{V}_1\left[\frac{1}{\mathbf{Z}_1} + \frac{1}{\mathbf{Z}_2}\right] - \mathbf{V}_2\left[\frac{1}{\mathbf{Z}_2}\right] = \mathbf{I}$$

At node $\mathbf{V}_2$,

$$\mathbf{I}_2 + \mathbf{I}_3 + k\mathbf{I} = 0$$

$$\frac{\mathbf{V}_2 - \mathbf{V}_1}{\mathbf{Z}_2} + \frac{\mathbf{V}_2}{\mathbf{Z}_3} + k\left[\frac{\mathbf{V}_1 - \mathbf{V}_2}{\mathbf{Z}_2}\right] = 0$$

and

$$\mathbf{V}_1\left[\frac{1-k}{\mathbf{Z}_2}\right] - \mathbf{V}_2\left[\frac{1-k}{\mathbf{Z}_2} + \frac{1}{\mathbf{Z}_3}\right] = 0$$

resulting in two equations and two unknowns.

Independent Voltage Sources between Assigned Nodes For independent voltage sources between assigned nodes, the procedure is modified as follows:

1. Steps 1 and 2 are the same as those applied for independent sources.
2. Step 3 is modified as follows: Treat each source between defined nodes as a short circuit (recall the *supernode* classification in Chapter 8), and write the nodal equations for each remaining independent node. Then relate the chosen nodal voltages to the

independent voltage source to ensure that the unknowns of the final equations are limited solely to the nodal voltages.
3. Step 4 is as before.

EXAMPLE 17.14 Write the nodal equations for the network in Fig. 17.27 having an independent source between two assigned nodes.

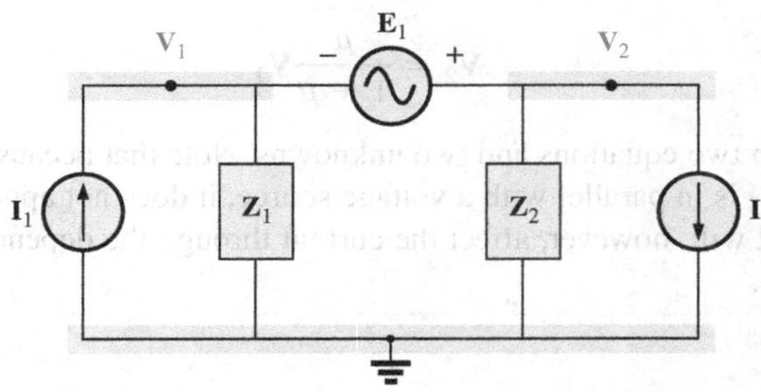

FIG. 17.27
Applying nodal analysis to a network with an independent voltage source between defined nodes.

Solution:

Steps 1 and 2 are defined in Fig. 17.27.

Steps 3: Replacing the independent source $\mathbf{E}_1$ with a short-circuit equivalent results in a supernode that generates the following equation when Kirchhoff's current law is applied to node $\mathbf{V}_1$:

$$\mathbf{I}_1 = \frac{\mathbf{V}_1}{\mathbf{Z}_1} + \frac{\mathbf{V}_2}{\mathbf{Z}_2} + \mathbf{I}_2$$

with
$$\mathbf{V}_2 - \mathbf{V}_1 = \mathbf{E}_1$$

and we have two equations and two unknowns.

Dependent Voltage Sources between Defined Nodes For dependent voltage sources between defined nodes, the procedure is modified as follows:

1. Steps 1 and 2 are the same as those applied for independent voltage sources.
2. Step 3 is modified as follows: The procedure is essentially the same as that applied for independent voltage sources, except that now the dependent sources have to be defined in terms of the chosen nodal voltages to ensure that the final equations have only nodal voltages as their unknown quantities.
3. Step 4 is as before.

EXAMPLE 17.15 Write the nodal equations for the network in Fig. 17.28 having a dependent voltage source between two defined nodes.

Solution:

Steps 1 and 2 are defined in Fig. 17.28.

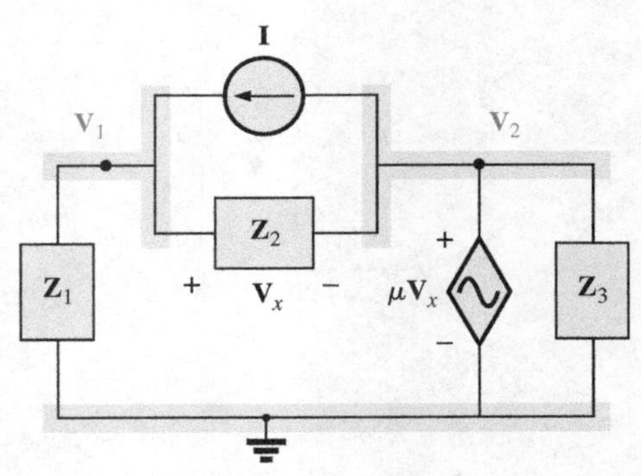

FIG. 17.28
Applying nodal analysis to a network with a voltage-controlled voltage source.

Step 3: Replacing the dependent source $\mu\mathbf{V}_x$ with a short-circuit equivalent results in the following equation when Kirchhoff's current law is applied at node $\mathbf{V}_1$:

$$\mathbf{I} = \mathbf{I}_1 + \mathbf{I}_2$$

$$\frac{\mathbf{V}_1}{\mathbf{Z}_1} + \frac{(\mathbf{V}_1 - \mathbf{V}_2)}{\mathbf{Z}_2} - \mathbf{I} = 0$$

and

$$\mathbf{V}_2 = \mu\mathbf{V}_x = \mu[\mathbf{V}_1 - \mathbf{V}_2]$$

or

$$\mathbf{V}_2 = \frac{\mu}{1 + \mu}\mathbf{V}_1$$

resulting in two equations and two unknowns. Note that because the impedance $\mathbf{Z}_3$ is in parallel with a voltage source, it does not appear in the analysis. It will, however, affect the current through the dependent voltage source.

Format Approach

A close examination of Eqs. (17.1) and (17.2) in Example 17.12 reveals that they are the same equations that would have been obtained using the format approach introduced in Chapter 8. Recall that the approach required that the voltage source first be converted to a current source, but the writing of the equations was quite direct and minimized any chances of an error due to a lost sign or missing term.

The sequence of steps required to apply the format approach is the following:

1. ***Choose a reference node and assign a subscripted voltage label to the (N − 1) remaining independent nodes of the network.***
2. ***The number of equations required for a complete solution is equal to the number of subscripted voltages (N − 1). Column 1 of each equation is formed by summing the admittances tied to the node of interest and multiplying the result by that subscripted nodal voltage.***
3. ***The mutual terms are always subtracted from the terms of the first column. It is possible to have more than one mutual term if the nodal voltage of interest has an element in common with more than one other nodal voltage. Each mutual term is the product of the mutual admittance and the other nodal voltage tied to that admittance.***
4. ***The column to the right of the equality sign is the algebraic sum of the current sources tied to the node of interest. A current source is assigned a positive sign if it supplies current to a node and a negative sign if it draws current from the node.***
5. ***Solve the resulting simultaneous equations for the desired nodal voltages. The comments offered for mesh analysis regarding independent and dependent sources apply here also.***

EXAMPLE 17.16 Using the format approach to nodal analysis, find the voltage across the 4 Ω resistor in Fig. 17.29.

Solution: Choosing nodes (Fig. 17.30) and writing the nodal equations, we have

$$\mathbf{Z}_1 = R = 4\ \Omega \qquad \mathbf{Z}_2 = jX_L = j\,5\ \Omega \qquad \mathbf{Z}_3 = -jX_C = -j\,2\ \Omega$$

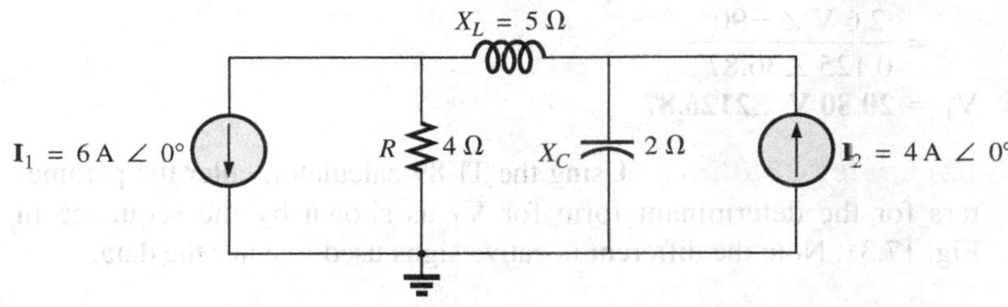

FIG. 17.29
Example 17.16.

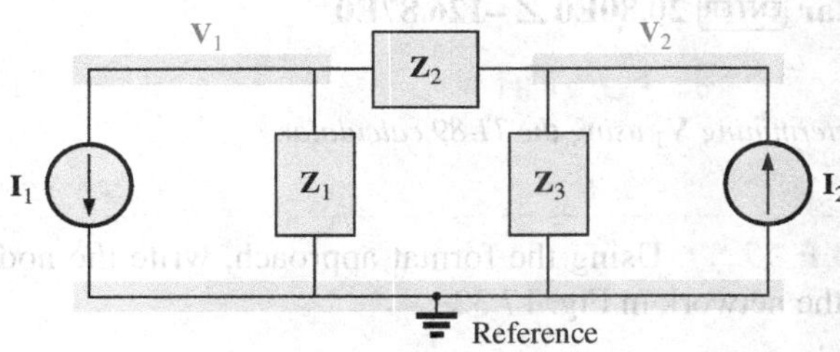

FIG. 17.30
Assigning the nodal voltages and subscripted impedances for the network in Fig. 17.29.

$$\begin{aligned}\mathbf{V}_1(\mathbf{Y}_1 + \mathbf{Y}_2) - \mathbf{V}_2(\mathbf{Y}_2) &= -\mathbf{I}_1\\ \mathbf{V}_2(\mathbf{Y}_3 + \mathbf{Y}_2) - \mathbf{V}_1(\mathbf{Y}_2) &= +\mathbf{I}_2\end{aligned}$$

or

$$\begin{aligned}\mathbf{V}_1(\mathbf{Y}_1 + \mathbf{Y}_2) - \mathbf{V}_2(\mathbf{Y}_2) \qquad\qquad &= -\mathbf{I}_1\\ -\mathbf{V}_1(\mathbf{Y}_2) \qquad + \mathbf{V}_2(\mathbf{Y}_3 + \mathbf{Y}_2) &= +\mathbf{I}_2\end{aligned}$$

$$\mathbf{Y}_1 = \frac{1}{\mathbf{Z}_1} \qquad \mathbf{Y}_2 = \frac{1}{\mathbf{Z}_2} \qquad \mathbf{Y}_3 = \frac{1}{\mathbf{Z}_3}$$

Using determinants yields

$$\begin{aligned}\mathbf{V}_1 &= \frac{\begin{vmatrix} -\mathbf{I}_1 & -\mathbf{Y}_2 \\ +\mathbf{I}_2 & \mathbf{Y}_3 + \mathbf{Y}_2 \end{vmatrix}}{\begin{vmatrix} \mathbf{Y}_1 + \mathbf{Y}_2 & -\mathbf{Y}_2 \\ -\mathbf{Y}_2 & \mathbf{Y}_3 + \mathbf{Y}_2 \end{vmatrix}} = \frac{\begin{vmatrix} -6\text{ A} & +j\,0.2\text{ S} \\ +4\text{ A} & +j\,0.3\text{ S} \end{vmatrix}}{\begin{vmatrix} 0.25\text{ S} - j\,0.2\text{ S} & +j\,0.2\text{ S} \\ +j\,0.2\text{ S} & +j\,0.3\text{ S} \end{vmatrix}}\\ &= \frac{-(\mathbf{Y}_3 + \mathbf{Y}_2)\mathbf{I}_1 + \mathbf{I}_2\mathbf{Y}_2}{(\mathbf{Y}_1 + \mathbf{Y}_2)(\mathbf{Y}_3 + \mathbf{Y}_2) - \mathbf{Y}_2^2}\\ &= \frac{-(\mathbf{Y}_3 + \mathbf{Y}_2)\mathbf{I}_1 + \mathbf{I}_2\mathbf{Y}_2}{\mathbf{Y}_1\mathbf{Y}_3 + \mathbf{Y}_2\mathbf{Y}_3 + \mathbf{Y}_1\mathbf{Y}_2}\end{aligned}$$

Substituting numerical values, we have

$$\begin{aligned}\mathbf{V}_1 &= \frac{-[(1/-j\,2\ \Omega) + (1/j\,5\ \Omega)]6\text{ A} \angle 0° + 4\text{ A} \angle 0°(1/j\,5\ \Omega)}{(1/4\ \Omega)(1/-j\,2\ \Omega) + (1/j\,5\ \Omega)(1/-j\,2\ \Omega) + (1/4\ \Omega)(1/j\,5\ \Omega)}\\ &= \frac{-(+j\,0.5 - j\,0.2)6 \angle 0° + 4 \angle 0°(-j\,0.2)}{(1/-j\,8) + (1/10) + (1/j\,20)}\\ &= \frac{(-0.3 \angle 90°)(6 \angle 0°) + (4 \angle 0°)(0.2 \angle -90°)}{j\,0.125 + 0.1 - j\,0.05}\\ &= \frac{-1.8 \angle 90° + 0.8 \angle -90°}{0.1 + j\,0.075}\end{aligned}$$

$$= \frac{2.6\text{ V}\angle -90^\circ}{0.125\angle 36.87^\circ}$$

$$\mathbf{V}_1 = \mathbf{20.80\text{ V}\angle 2126.87^\circ}$$

Calculator Solution: Using the TI-89 calculator, enter the parameters for the determinant form for $\mathbf{V}_1$ as shown by the sequence in Fig. 17.31. Note the different negative signs used to enter the data.

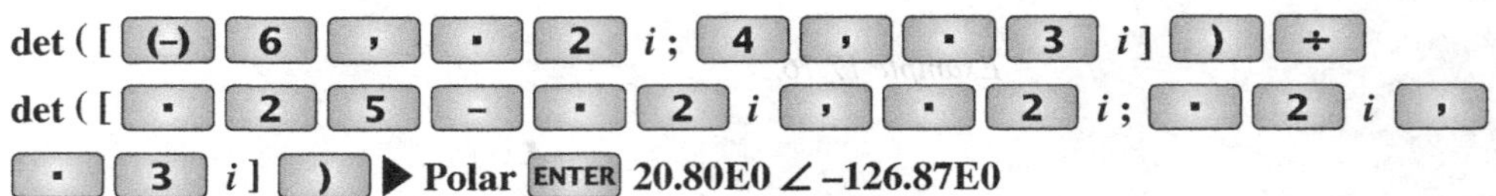

FIG. 17.31
Determining $\mathbf{V}_1$ *using the TI-89 calculator.*

EXAMPLE 17.17 Using the format approach, write the nodal equations for the network in Fig. 17.32.

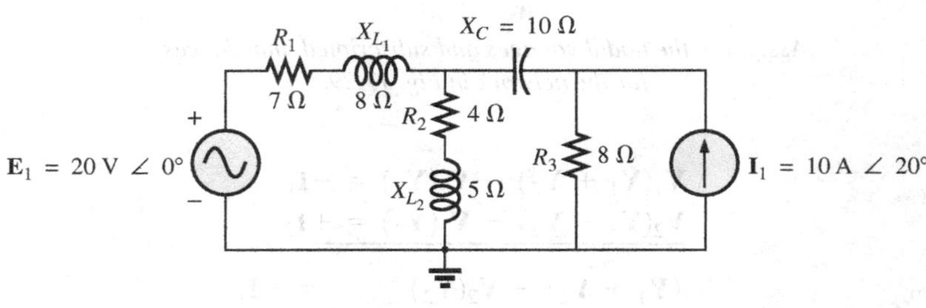

FIG. 17.32
Example 17.17.

Solution: The circuit is redrawn in Fig. 17.33, where

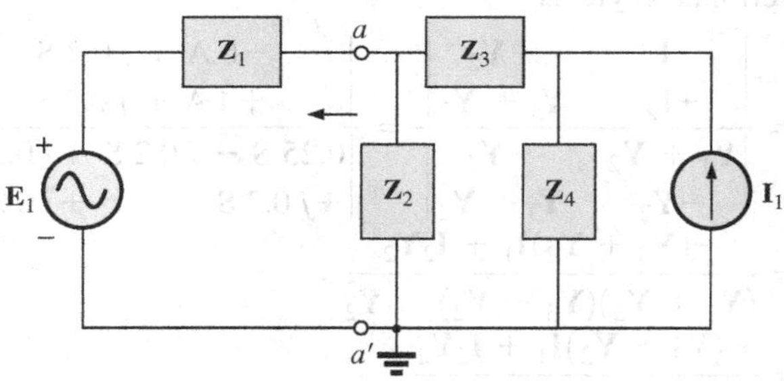

FIG. 17.33
Assigning the subscripted impedances for the network in Fig. 17.32.

$$\mathbf{Z}_1 = R_1 + jX_{L_1} = 7\ \Omega + j\,8\ \Omega \qquad \mathbf{E}_1 = 20\text{ V}\angle 0^\circ$$
$$\mathbf{Z}_2 = R_2 + jX_{L_2} = 4\ \Omega + j\,5\ \Omega \qquad \mathbf{I}_1 = 10\text{ A}\angle 20^\circ$$
$$\mathbf{Z}_3 = -jX_2 = -j\,10\ \Omega$$
$$\mathbf{Z}_4 = R_3 = 8\ \Omega$$

Converting the voltage source to a current source and choosing nodes, we obtain Fig. 17.34. Note the "neat" appearance of the network using

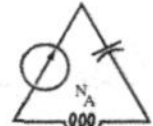

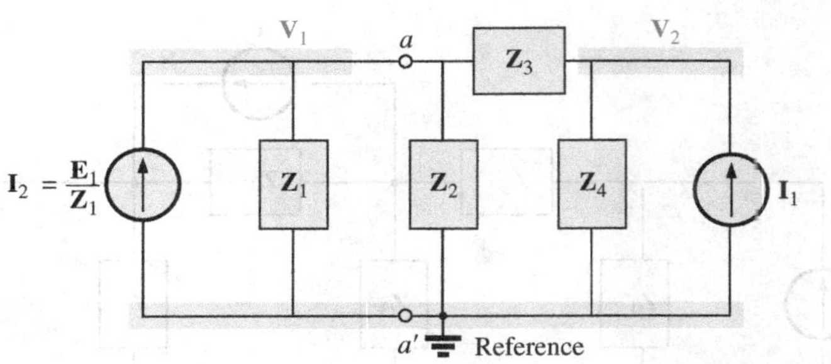

FIG. 17.34
Converting the voltage source in Fig. 17.33 to a current source and defining the nodal voltages.

the subscripted impedances. Working directly with Fig. 17.32 would be more difficult and could produce errors.

Write the nodal equations:

$$\begin{aligned} \mathbf{V}_1(\mathbf{Y}_1 + \mathbf{Y}_2 + \mathbf{Y}_3) - \mathbf{V}_2(\mathbf{Y}_3) &= +\mathbf{I}_2 \\ \mathbf{V}_2(\mathbf{Y}_3 + \mathbf{Y}_4) - \mathbf{V}_1(\mathbf{Y}_3) &= +\mathbf{I}_1 \end{aligned}$$

$$\mathbf{Y}_1 = \frac{1}{\mathbf{Z}_1} \qquad \mathbf{Y}_2 = \frac{1}{\mathbf{Z}_2} \qquad \mathbf{Y}_3 = \frac{1}{\mathbf{Z}_3} \qquad \mathbf{Y}_4 = \frac{1}{\mathbf{Z}_4}$$

which are rewritten as

$$\begin{aligned} \mathbf{V}_1(\mathbf{Y}_1 + \mathbf{Y}_2 + \mathbf{Y}_3) - \mathbf{V}_2(\mathbf{Y}_3) \quad &= +\mathbf{I}_2 \\ -\mathbf{V}_1(\mathbf{Y}_3) \quad + \mathbf{V}_2(\mathbf{Y}_3 + \mathbf{Y}_4) &= +\mathbf{I}_1 \end{aligned}$$

EXAMPLE 17.18 Write the nodal equations for the network in Fig. 17.35. Do not solve.

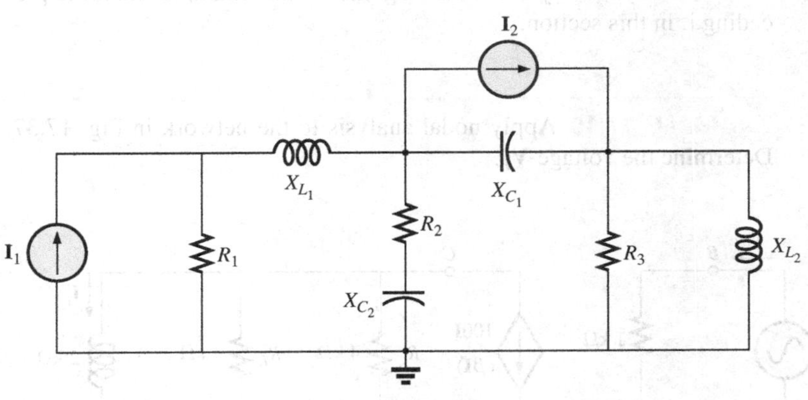

FIG. 17.35
Example 17.18.

Solution: Choose nodes (Fig. 17.36):

$$\begin{aligned} &\mathbf{Z}_1 = R_1 \qquad && \mathbf{Z}_2 = jX_{L_1} \qquad && \mathbf{Z}_3 = R_2 - jX_{C_2} \\ &\mathbf{Z}_4 = -jX_{C_1} \qquad && \mathbf{Z}_5 = R_3 \qquad && \mathbf{Z}_6 = jX_{L_2} \end{aligned}$$

and write the nodal equations:

$$\begin{aligned} \mathbf{V}_1(\mathbf{Y}_1 + \mathbf{Y}_2) - \mathbf{V}_2(\mathbf{Y}_2) &= +\mathbf{I}_1 \\ \mathbf{V}_2(\mathbf{Y}_2 + \mathbf{Y}_3 + \mathbf{Y}_4) - \mathbf{V}_1(\mathbf{Y}_2) - \mathbf{V}_3(\mathbf{Y}_4) &= -\mathbf{I}_2 \\ \mathbf{V}_3(\mathbf{Y}_4 + \mathbf{Y}_5 + \mathbf{Y}_6) - \mathbf{V}_2(\mathbf{Y}_4) &= +\mathbf{I}_2 \end{aligned}$$

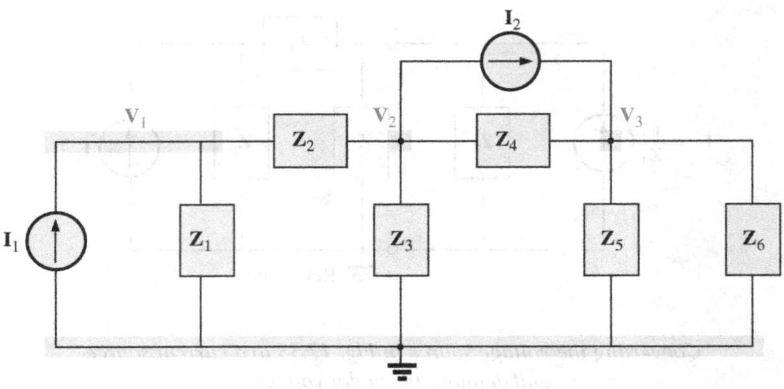

FIG. 17.36
Assigning the nodal voltages and subscripted impedances for the network in Fig. 17.35.

which are rewritten as

$$\begin{array}{lllr} \mathbf{V}_1(\mathbf{Y}_1 + \mathbf{Y}_2) & - \mathbf{V}_2(\mathbf{Y}_2) & + 0 & = +\mathbf{I}_1 \\ -\mathbf{V}_1(\mathbf{Y}_2) & + \mathbf{V}_2(\mathbf{Y}_2 + \mathbf{Y}_3 + \mathbf{Y}_4) & - \mathbf{V}_3(\mathbf{Y}_4) & = -\mathbf{I}_2 \\ 0 & - \mathbf{V}_2(\mathbf{Y}_4) & + \mathbf{V}_3(\mathbf{Y}_4 + \mathbf{Y}_5 + \mathbf{Y}_6) & = +\mathbf{I}_2 \end{array}$$

$$\mathbf{Y}_1 = \frac{1}{R_1} \qquad \mathbf{Y}_2 = \frac{1}{jX_{L_1}} \qquad \mathbf{Y}_3 = \frac{1}{R_2 - jX_{C_2}}$$

$$\mathbf{Y}_4 = \frac{1}{-jX_{C_1}} \qquad \mathbf{Y}_5 = \frac{1}{R_3} \qquad \mathbf{Y}_6 = \frac{1}{jX_{L_2}}$$

Note the symmetry about the diagonal for this example and those preceding it in this section.

EXAMPLE 17.19 Apply nodal analysis to the network in Fig. 17.37. Determine the voltage $\mathbf{V}_L$.

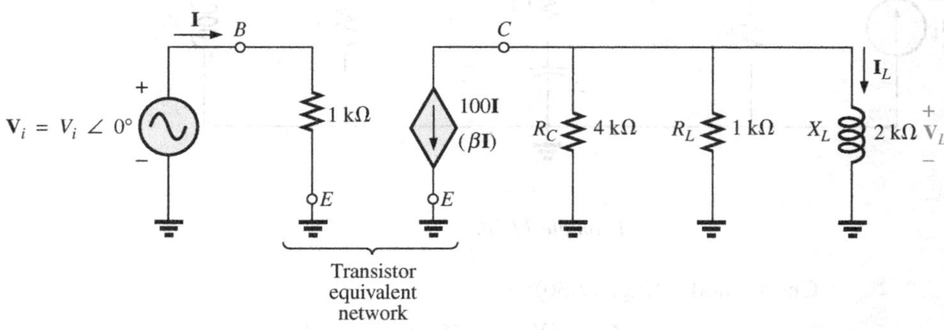

FIG. 17.37
Example 17.19

Solution: In this case, there is no need for a source conversion. The network is redrawn in Fig. 17.38 with the chosen nodal voltage and subscripted impedances.

Apply the format approach:

$$\mathbf{Y}_1 = \frac{1}{\mathbf{Z}_1} = \frac{1}{4\text{ k}\Omega} = 0.25\text{ mS }\angle 0° = G_1\angle 0°$$

$$\mathbf{Y}_2 = \frac{1}{\mathbf{Z}_2} = \frac{1}{1\text{ k}\Omega} = 1\text{ mS }\angle 0° = G_2\angle 0°$$

$$\mathbf{Y}_3 = \frac{1}{\mathbf{Z}_3} = \frac{1}{2\text{ k}\Omega\angle 90°} = 0.5\text{ mS }\angle -90° = -j\,0.5\text{ mS} = -jB_L$$

$$\mathbf{V}_1{:}(\mathbf{Y}_1 + \mathbf{Y}_2 + \mathbf{Y}_3)\mathbf{V}_1 = -100\,\mathbf{I}$$

and

$$\mathbf{V}_1 = \frac{-100\,\mathbf{I}}{\mathbf{Y}_1 + \mathbf{Y}_2 + \mathbf{Y}_3} = \frac{-100\,\mathbf{I}}{0.25\text{ mS} + 1\text{ mS} - j\,0.5\text{ mS}}$$

$$= \frac{-100\times 10^3\,\mathbf{I}}{1.25 - j\,0.5} = \frac{-100\times 10^3\,\mathbf{I}}{1.3463\angle -21.80°}$$

$$= -74.28\times 10^3\,\mathbf{I}\angle 21.80°$$

$$= -74.28\times 10^3\left(\frac{\mathbf{V}_i}{1\text{ k}\Omega}\right)\angle 21.80°$$

$$\mathbf{V}_1 = \mathbf{V}_L = \mathbf{-(74.28V_{\mathit{i}})V\angle 21.80°}$$

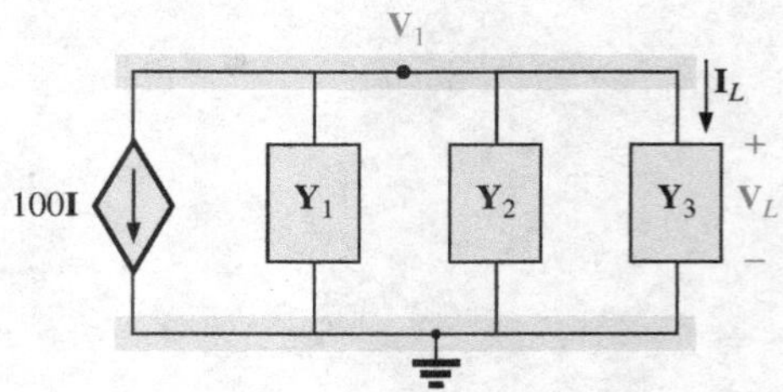

FIG. 17.38
Assigning the nodal voltage and subscripted impedances for the network in Fig. 17.37.

17.6 BRIDGE NETWORKS (ac)

The basic bridge configuration was discussed in some detail in Section 8.11 for dc networks. We now continue to examine **bridge networks** by considering those that have reactive components and a sinusoidal ac voltage or current applied.

We first analyze various familiar forms of the bridge network using mesh analysis and nodal analysis (the format approach). The balance conditions are investigated throughout the section.

Apply **mesh analysis** to the network in Fig. 17.39. The network is redrawn in Fig. 17.40, where

$$\mathbf{Z}_1 = \frac{1}{\mathbf{Y}_1} = \frac{1}{G_1 + jB_C} = \frac{G_1}{G_1^2 + B_C^2} - j\frac{B_C}{G_1^2 + B_C^2}$$

$$\mathbf{Z}_2 = R_2 \quad \mathbf{Z}_3 = R_3 \quad \mathbf{Z}_4 = R_4 + jX_L \quad \mathbf{Z}_5 = R_5$$

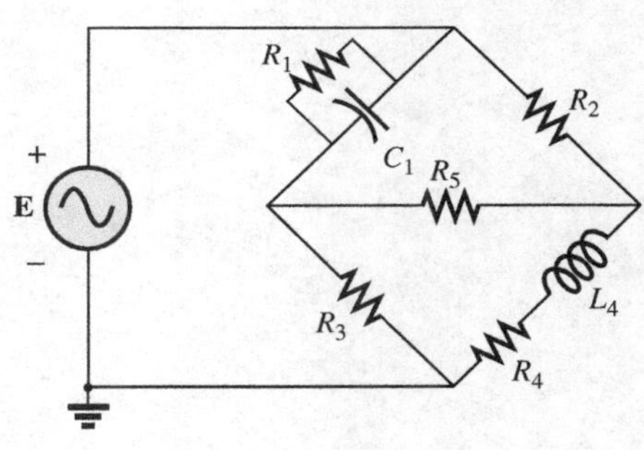

FIG. 17.39
Maxwell bridge.

Applying the format approach:

$$(\mathbf{Z}_1 + \mathbf{Z}_3)\mathbf{I}_1 - (\mathbf{Z}_1)\mathbf{I}_2 - (\mathbf{Z}_3)\mathbf{I}_3 = \mathbf{E}$$
$$(\mathbf{Z}_1 + \mathbf{Z}_2 + \mathbf{Z}_5)\mathbf{I}_2 - (\mathbf{Z}_1)\mathbf{I}_1 - (\mathbf{Z}_5)\mathbf{I}_3 = 0$$
$$(\mathbf{Z}_3 + \mathbf{Z}_4 + \mathbf{Z}_5)\mathbf{I}_3 - (\mathbf{Z}_3)\mathbf{I}_1 - (\mathbf{Z}_5)\mathbf{I}_2 = 0$$

which are rewritten as

$$\begin{aligned}
\mathbf{I}_1(\mathbf{Z}_1 + \mathbf{Z}_3) &- \mathbf{I}_2\mathbf{Z}_1 && - \mathbf{I}_3\mathbf{Z}_3 && = \mathbf{E}\\
-\mathbf{I}_1\mathbf{Z}_1 &+ \mathbf{I}_2(\mathbf{Z}_1 + \mathbf{Z}_2 + \mathbf{Z}_5) && - \mathbf{I}_3\mathbf{Z}_5 && = 0\\
-\mathbf{I}_1\mathbf{Z}_3 &- \mathbf{I}_2\mathbf{Z}_5 && + \mathbf{I}_3(\mathbf{Z}_3 + \mathbf{Z}_4 + \mathbf{Z}_5) && = 0
\end{aligned}$$

Note the symmetry about the diagonal of the above equations. For balance, $\mathbf{I}_{\mathbf{Z}_5} = 0$ A, and

$$\mathbf{I}_{\mathbf{Z}_5} = \mathbf{I}_2 - \mathbf{I}_3 = 0$$

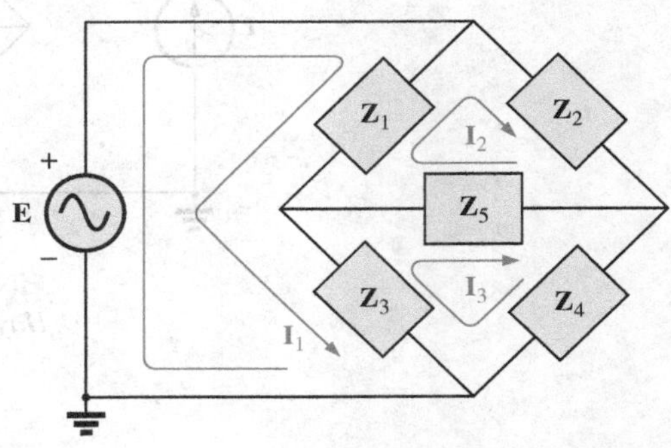

FIG. 17.40
Assigning the mesh currents and subscripted impedances for the network in Fig. 17.39.

From the above equations,

$$\mathbf{I}_2 = \frac{\begin{vmatrix} \mathbf{Z}_1 + \mathbf{Z}_3 & \mathbf{E} & -\mathbf{Z}_3 \\ -\mathbf{Z}_1 & 0 & -\mathbf{Z}_5 \\ -\mathbf{Z}_3 & 0 & (\mathbf{Z}_3 + \mathbf{Z}_4 + \mathbf{Z}_5) \end{vmatrix}}{\begin{vmatrix} \mathbf{Z}_1 + \mathbf{Z}_3 & -\mathbf{Z}_1 & -\mathbf{Z}_3 \\ -\mathbf{Z}_1 & (\mathbf{Z}_1 + \mathbf{Z}_2 + \mathbf{Z}_5) & -\mathbf{Z}_5 \\ -\mathbf{Z}_3 & -\mathbf{Z}_5 & (\mathbf{Z}_3 + \mathbf{Z}_4 + \mathbf{Z}_5) \end{vmatrix}}$$

$$= \frac{\mathbf{E}(\mathbf{Z}_1\mathbf{Z}_3 + \mathbf{Z}_1\mathbf{Z}_4 + \mathbf{Z}_1\mathbf{Z}_5 + \mathbf{Z}_3\mathbf{Z}_5)}{\Delta}$$

where Δ signifies the determinant of the denominator (or coefficients). Similarly,

$$\mathbf{I}_3 = \frac{\mathbf{E}(\mathbf{Z}_1\mathbf{Z}_3 + \mathbf{Z}_3\mathbf{Z}_2 + \mathbf{Z}_1\mathbf{Z}_5 + \mathbf{Z}_3\mathbf{Z}_5)}{\Delta}$$

and
$$\mathbf{I}_{Z_5} = \mathbf{I}_2 - \mathbf{I}_3 = \frac{\mathbf{E}(\mathbf{Z}_1\mathbf{Z}_4 - \mathbf{Z}_3\mathbf{Z}_2)}{\Delta}$$

For $\mathbf{I}_{Z_5} = 0$, the following must be satisfied (for a finite Δ not equal to zero):

$$\boxed{\mathbf{Z}_1\mathbf{Z}_4 = \mathbf{Z}_3\mathbf{Z}_2} \qquad \mathbf{I}_{Z_5} = 0 \qquad \textbf{(17.3)}$$

This condition is analyzed in greater depth later in this section.

Applying **nodal analysis** to the network in Fig. 17.41 results in the configuration in Fig. 17.42, where

$$\mathbf{Y}_1 = \frac{1}{\mathbf{Z}_1} = \frac{1}{R_1 - jX_C} \qquad \mathbf{Y}_2 = \frac{1}{\mathbf{Z}_2} = \frac{1}{R_2}$$

$$\mathbf{Y}_3 = \frac{1}{\mathbf{Z}_3} = \frac{1}{R_3} \qquad \mathbf{Y}_4 = \frac{1}{\mathbf{Z}_4} = \frac{1}{R_4 + jX_L} \qquad \mathbf{Y}_5 = \frac{1}{R_5}$$

and

$$(\mathbf{Y}_1 + \mathbf{Y}_2)\mathbf{V}_1 - (\mathbf{Y}_1)\mathbf{V}_2 - (\mathbf{Y}_2)\mathbf{V}_3 = \mathbf{I}$$
$$(\mathbf{Y}_1 + \mathbf{Y}_3 + \mathbf{Y}_5)\mathbf{V}_2 - (\mathbf{Y}_1)\mathbf{V}_1 - (\mathbf{Y}_5)\mathbf{V}_3 = 0$$
$$(\mathbf{Y}_2 + \mathbf{Y}_4 + \mathbf{Y}_5)\mathbf{V}_3 - (\mathbf{Y}_2)\mathbf{V}_1 - (\mathbf{Y}_5)\mathbf{V}_2 = 0$$

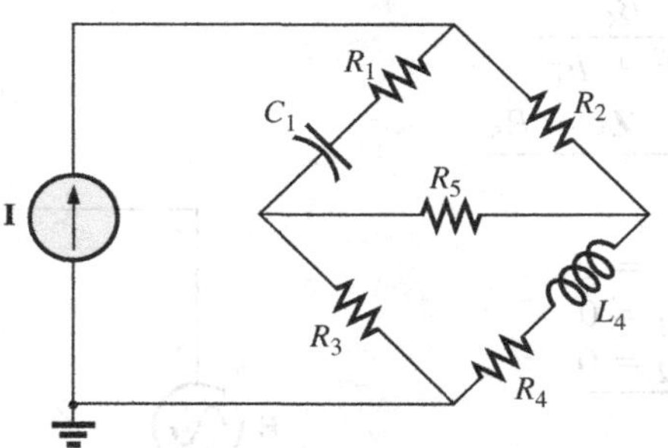

FIG. 17.41
Hay bridge.

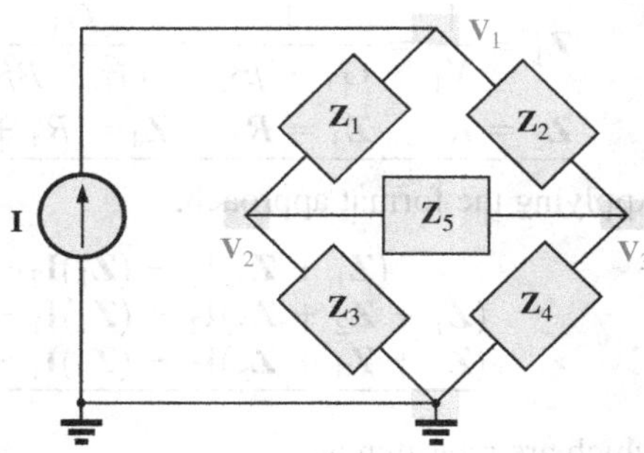

FIG. 17.42
Assigning the nodal voltages and subscripted impedances for the network in Fig. 17.41.

which are rewritten as

$$\begin{aligned} \mathbf{V}_1(\mathbf{Y}_1 + \mathbf{Y}_2) &- \mathbf{V}_2(\mathbf{Y}_1) &- \mathbf{V}_3\mathbf{Y}_2 &= \mathbf{I} \\ -\mathbf{V}_1(\mathbf{Y}_2) &+ \mathbf{V}_2(\mathbf{Y}_1 + \mathbf{Y}_3 + \mathbf{Y}_5) &- \mathbf{V}_3\mathbf{Y}_5 &= 0 \\ -\mathbf{V}_1\mathbf{Y}_2 &- \mathbf{V}_2\mathbf{Y}_5 &+ \mathbf{V}_3(\mathbf{Y}_2 + \mathbf{Y}_4 + \mathbf{Y}_5) &= 0 \end{aligned}$$

Again, note the symmetry about the diagonal axis. For balance, $\mathbf{V}_{\mathbf{Z}_5} = 0$ V, and

$$\mathbf{V}_{\mathbf{Z}_5} = \mathbf{V}_2 - \mathbf{V}_3 = 0$$

From the above equations,

$$\mathbf{V}_2 = \frac{\begin{vmatrix} \mathbf{Y}_1 + \mathbf{Y}_2 & \mathbf{I} & -\mathbf{Y}_2 \\ -\mathbf{Y}_1 & 0 & -\mathbf{Y}_5 \\ -\mathbf{Y}_2 & 0 & (\mathbf{Y}_2 + \mathbf{Y}_4 + \mathbf{Y}_5) \end{vmatrix}}{\begin{vmatrix} \mathbf{Y}_1 + \mathbf{Y}_2 & -\mathbf{Y}_1 & -\mathbf{Y}_2 \\ -\mathbf{Y}_1 & (\mathbf{Y}_1 + \mathbf{Y}_3 + \mathbf{Y}_5) & -\mathbf{Y}_5 \\ -\mathbf{Y}_2 & -\mathbf{Y}_5 & (\mathbf{Y}_2 + \mathbf{Y}_4 + \mathbf{Y}_5) \end{vmatrix}}$$

$$= \frac{\mathbf{I}(\mathbf{Y}_1\mathbf{Y}_3 + \mathbf{Y}_1\mathbf{Y}_4 + \mathbf{Y}_1\mathbf{Y}_5 + \mathbf{Y}_3\mathbf{Y}_5)}{\Delta}$$

Similarly,

$$\mathbf{V}_3 = \frac{\mathbf{I}(\mathbf{Y}_1\mathbf{Y}_3 + \mathbf{Y}_3\mathbf{Y}_2 + \mathbf{Y}_1\mathbf{Y}_5 + \mathbf{Y}_3\mathbf{Y}_5)}{\Delta}$$

Note the similarities between the above equations and those obtained for mesh analysis. Then

$$\mathbf{V}_{\mathbf{Z}_5} = \mathbf{V}_2 - \mathbf{V}_3 = \frac{\mathbf{I}(\mathbf{Y}_1\mathbf{Y}_4 - \mathbf{Y}_3\mathbf{Y}_2)}{\Delta}$$

For $\mathbf{V}_{\mathbf{Z}_5} = 0$, the following must be satisfied for a finite Δ not equal to zero:

$$\boxed{\mathbf{Y}_1\mathbf{Y}_4 = \mathbf{Y}_3\mathbf{Y}_2} \qquad \mathbf{V}_{\mathbf{Z}_5} = 0 \tag{17.4}$$

However, substituting $\mathbf{Y}_1 = 1/\mathbf{Z}_1$, $\mathbf{Y}_2 = 1/\mathbf{Z}_2$, $\mathbf{Y}_3 = 1/\mathbf{Z}_3$, and $\mathbf{Y}_4 = 1/\mathbf{Z}_4$, we have

$$\frac{1}{\mathbf{Z}_1\mathbf{Z}_4} = \frac{1}{\mathbf{Z}_3\mathbf{Z}_2}$$

or

$$\boxed{\mathbf{Z}_1\mathbf{Z}_4 = \mathbf{Z}_3\mathbf{Z}_2} \qquad \mathbf{V}_{\mathbf{Z}_5} = 0$$

corresponding with Eq. (17.3) obtained earlier.

Let us now investigate the balance criteria in more detail by considering the network in Fig. 17.43, where it is specified that **I** and **V** = 0.

Since **I** = 0,

$$\boxed{\mathbf{I}_1 = \mathbf{I}_3} \tag{17.5}$$

and

$$\boxed{\mathbf{I}_2 = \mathbf{I}_4} \tag{17.6}$$

In addition, for **V** = 0,

$$\boxed{\mathbf{I}_1\mathbf{Z}_1 = \mathbf{I}_2\mathbf{Z}_2} \tag{17.7}$$

and

$$\boxed{\mathbf{I}_3\mathbf{Z}_3 = \mathbf{I}_4\mathbf{Z}_4} \tag{17.8}$$

Substituting the preceding current relations into Eq. (17.8), we have

$$\mathbf{I}_1\mathbf{Z}_3 = \mathbf{I}_2\mathbf{Z}_4$$

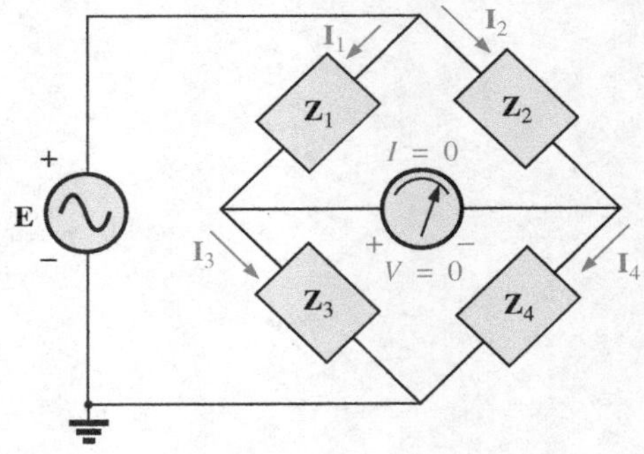

FIG. 17.43
Investigating the balance criteria for an ac bridge configuration.

and
$$\mathbf{I}_2 = \frac{\mathbf{Z}_3}{\mathbf{Z}_4}\mathbf{I}_1$$

Substituting this relationship for $\mathbf{I}_2$ into Eq. (17.7) yields

$$\mathbf{I}_1\mathbf{Z}_1 = \left(\frac{\mathbf{Z}_3}{\mathbf{Z}_4}\mathbf{I}_1\right)\mathbf{Z}_2$$

and
$$\mathbf{Z}_1\mathbf{Z}_4 = \mathbf{Z}_2\mathbf{Z}_3$$

as obtained earlier. Rearranging, we have

$$\boxed{\frac{\mathbf{Z}_1}{\mathbf{Z}_3} = \frac{\mathbf{Z}_2}{\mathbf{Z}_4}} \tag{17.9}$$

corresponding to Eq. (8.2) for dc resistive networks.

For the network in Fig. 17.41, which is referred to as a **Hay bridge** when $\mathbf{Z}_5$ is replaced by a sensitive galvanometer,

$$\mathbf{Z}_1 = R_1 - jX_C$$
$$\mathbf{Z}_2 = R_2$$
$$\mathbf{Z}_3 = R_3$$
$$\mathbf{Z}_4 = R_4 + jX_L$$

This particular network is used for measuring the resistance and inductance of coils in which the resistance is a small fraction of the reactance X_L.

Substitute into Eq. (17.9) in the following form:

$$\mathbf{Z}_2\mathbf{Z}_3 = \mathbf{Z}_4\mathbf{Z}_1$$
$$R_2R_3 = (R_4 + jX_L)(R_1 - jX_C)$$

or
$$R_2R_3 = R_1R_4 + j(R_1X_L - R_4X_C) + X_CX_L$$

so that

$$R_2R_3 + j\,0 = (R_1R_4 + X_CX_L) + j\,(R_1X_L - R_4X_C)$$

For the equations to be equal, *the real and imaginary parts must be equal.* Therefore, for a balanced Hay bridge,

$$\boxed{R_2R_3 = R_1R_4 + X_CX_L} \tag{17.10}$$

and
$$\boxed{0 = R_1X_L - R_4X_C} \tag{17.11}$$

or substituting $\quad X_L = \omega L \quad$ and $\quad X_C = \dfrac{1}{\omega C}$

we have
$$X_CX_L = \left(\frac{1}{\omega C}\right)(\omega L) = \frac{L}{C}$$

and
$$R_2R_3 = R_1R_4 + \frac{L}{C}$$

with
$$R_1\omega L = \frac{R_4}{\omega C}$$

Solving for R_4 in the last equation yields

$$R_4 = \omega^2 LCR_1$$

and substituting into the previous equation, we have

$$R_2R_3 = R_1(\omega^2 LCR_1) + \frac{L}{C}$$

Multiply through by C and factor:

$$CR_2R_3 = L(\omega^2C^2R_1^2 + 1)$$

and
$$L = \frac{CR_2R_3}{1 + \omega^2C^2R_1^2} \quad \textbf{(17.12)}$$

With additional algebra this yields:

$$R_4 = \frac{\omega^2C^2R_1R_2R_3}{1 + \omega^2C^2R_1^2} \quad \textbf{(17.13)}$$

Eqs. (17.12) and (17.13) are the balance conditions for the Hay bridge. Note that each is frequency dependent. For different frequencies, the resistive and capacitive elements must vary for a particular coil to achieve balance. For a coil placed in the Hay bridge as shown in Fig. 17.41, the resistance and inductance of the coil can be determined by Eqs. (17.12) and (17.13) when balance is achieved.

The bridge in Fig. 17.39 is referred to as a **Maxwell bridge** when $\mathbf{Z}_5$ is replaced by a sensitive galvanometer. This setup is used for inductance measurements when the resistance of the coil is large enough not to require a Hay bridge.

Applying Eq. (17.9) in the form

$$\mathbf{Z}_2\mathbf{Z}_3 = \mathbf{Z}_4\mathbf{Z}_1$$

and substituting

$$\mathbf{Z}_1 = R_1 \angle 0° \parallel X_{C_1} \angle -90° = \frac{(R_1 \angle 0°)(X_{C_1} \angle -90°)}{R_1 - jX_{C_1}}$$

$$= \frac{R_1 X_{C_1} \angle -90°}{R_1 - jX_{C_1}} = \frac{-jR_1X_{C_1}}{R_1 - jX_{C_1}}$$

$$\mathbf{Z}_2 = R_2$$

$$\mathbf{Z}_3 = R_3$$

and $\mathbf{Z}_4 = R_4 + jX_{L_4}$

we have
$$(R_2)(R_3) = (R_4 + jX_{L_4})\left(\frac{-jR_1X_{C_1}}{R_1 - jX_{C_1}}\right)$$

$$R_2R_3 = \frac{-jR_1R_4X_{C_1} + R_1X_{C_1}X_{L_4}}{R_1 - jX_{C_1}}$$

or
$$(R_2R_3)(R_1 - jX_{C_1}) = R_1X_{C_1}X_{L_4} - jR_1R_4X_{C_1}$$

and
$$R_1R_2R_3 - jR_2R_3X_{C_1} = R_1X_{C_1}X_{L_4} - jR_1R_4X_{C_1}$$

so that for balance

$$\cancel{R_1}R_2R_3 = \cancel{R_1}X_{C_1}X_{L_4}$$

$$R_2R_3 = \left(\frac{1}{\cancel{2\pi}fC_1}\right)(\cancel{2\pi f}L_4)$$

and
$$L_4 = C_1R_2R_3 \quad \textbf{(17.14)}$$

and $$R_2R_3\cancel{X_{C_1}} = R_1R_4\cancel{X_{C_1}}$$

so that

$$\boxed{R_4 = \frac{R_2R_3}{R_1}} \tag{17.15}$$

Note the absence of frequency in Eqs. (17.14) and (17.15).

One remaining popular bridge is the **capacitance comparison bridge** of Fig. 17.44. An unknown capacitance and its associated resistance can be determined using this bridge. Application of Eq. (17.9) yields the following results:

$$\boxed{C_4 = C_3\frac{R_1}{R_2}} \tag{17.16}$$

$$\boxed{R_4 = \frac{R_2R_3}{R_1}} \tag{17.17}$$

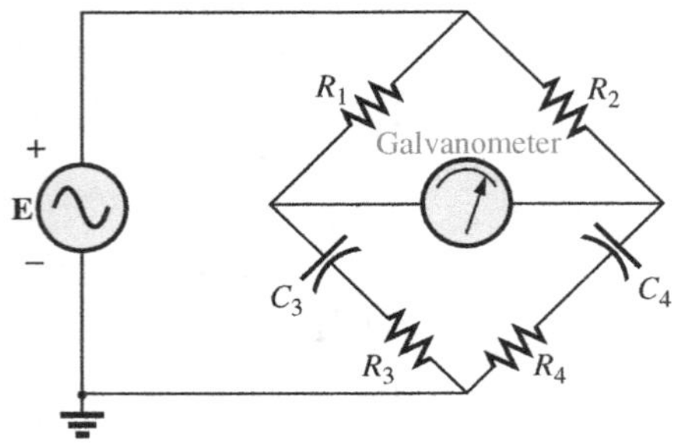

FIG. 17.44
Capacitance comparison bridge.

The derivation of these equations appears as a problem at the end of the chapter.

17.7 Δ-Y, Y-Δ CONVERSIONS

The Δ-Y, Y-Δ (or π-T, T-π as defined in Section 8.12) conversions for ac circuits are not derived here since the development corresponds exactly with that for dc circuits. Taking the **Δ-Y configuration** shown in Fig. 17.45, we find the general equations for the impedances of the Y in terms of those for the Δ:

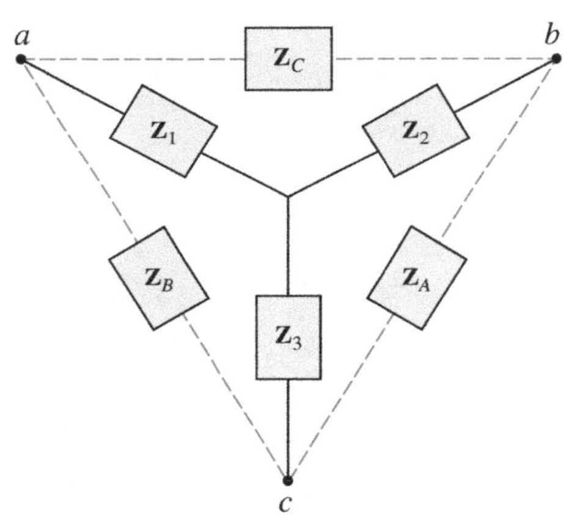

FIG. 17.45
Δ-Y configuration.

$$\boxed{\mathbf{Z}_1 = \frac{\mathbf{Z}_B\mathbf{Z}_C}{\mathbf{Z}_A + \mathbf{Z}_B + \mathbf{Z}_C}} \tag{17.18}$$

$$\boxed{\mathbf{Z}_2 = \frac{\mathbf{Z}_A\mathbf{Z}_C}{\mathbf{Z}_A + \mathbf{Z}_B + \mathbf{Z}_C}} \tag{17.19}$$

$$\boxed{\mathbf{Z}_3 = \frac{\mathbf{Z}_A\mathbf{Z}_B}{\mathbf{Z}_A + \mathbf{Z}_B + \mathbf{Z}_C}} \tag{17.20}$$

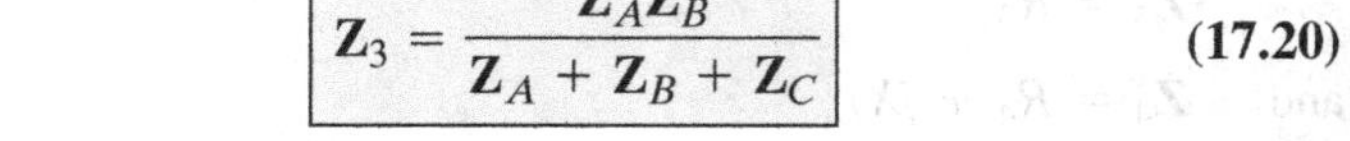

For the impedances of the Δ in terms of those for the Y, the equations are

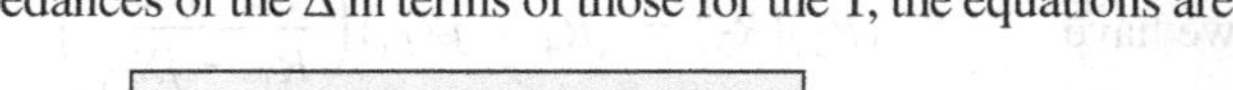

$$\boxed{\mathbf{Z}_B = \frac{\mathbf{Z}_1\mathbf{Z}_2 + \mathbf{Z}_1\mathbf{Z}_3 + \mathbf{Z}_2\mathbf{Z}_3}{\mathbf{Z}_2}} \tag{17.21}$$

$$\boxed{\mathbf{Z}_A = \frac{\mathbf{Z}_1\mathbf{Z}_2 + \mathbf{Z}_1\mathbf{Z}_3 + \mathbf{Z}_2\mathbf{Z}_3}{\mathbf{Z}_1}} \tag{17.22}$$

$$\boxed{\mathbf{Z}_C = \frac{\mathbf{Z}_1\mathbf{Z}_2 + \mathbf{Z}_1\mathbf{Z}_3 + \mathbf{Z}_2\mathbf{Z}_3}{\mathbf{Z}_3}} \tag{17.23}$$

Note that each impedance of the Y is equal to the product of the impedances in the two closest branches of the Δ, divided by the sum of the impedances in the Δ.

Further, the value of each impedance of the Δ is equal to the sum of the possible product combinations of the impedances of the Y, divided by the impedances of the Y farthest from the impedance to be determined.

Drawn in different forms (Fig. 17.46), they are also referred to as the T and π configurations.

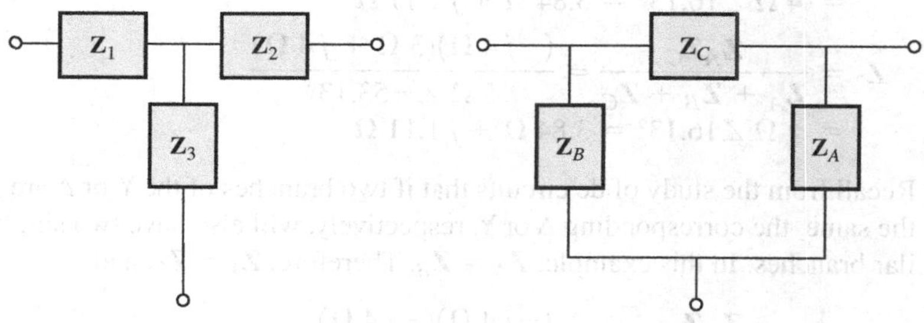

FIG. 17.46
The T and π configurations.

In the study of dc networks, we found that if all of the resistors of the Δ or Y were the same, the conversion from one to the other could be accomplished using the equation

$$R_\Delta = 3R_Y \qquad \text{or} \qquad R_Y = \frac{R_\Delta}{3}$$

For ac networks,

$$\mathbf{Z}_\Delta = 3\mathbf{Z}_Y \qquad \text{or} \qquad \mathbf{Z}_Y = \frac{\mathbf{Z}_\Delta}{3} \tag{17.24}$$

Be careful when using this simplified form. It is not sufficient for all the impedances of the Δ or Y to be of the same magnitude: *The angle associated with each must also be the same.*

EXAMPLE 17.20 Find the total impedance $\mathbf{Z}_T$ of the network in Fig. 17.47.

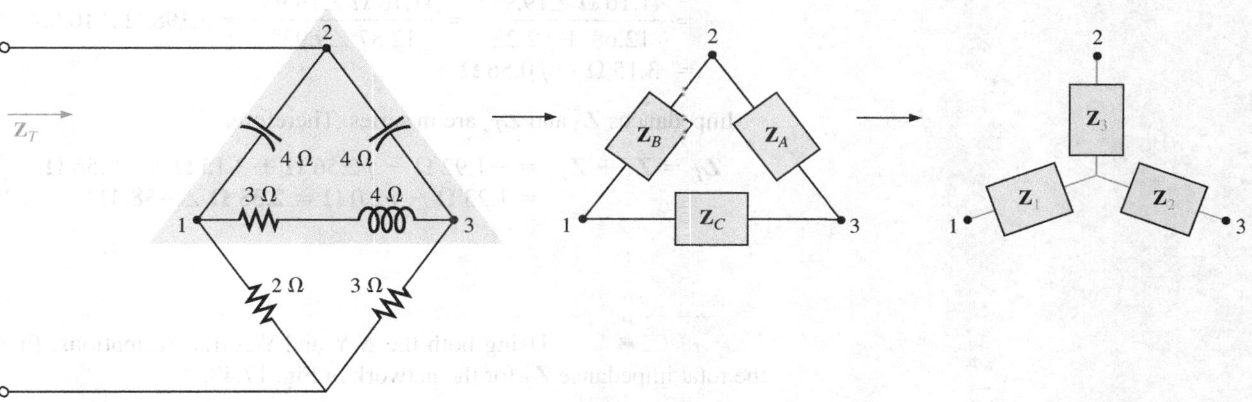

FIG. 17.47
Converting the upper Δ of a bridge configuration to a Y.

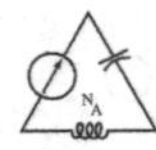

Solution:

$$\mathbf{Z}_B = -j\,4 \qquad \mathbf{Z}_A = -j\,4 \qquad \mathbf{Z}_C = 3 + j\,4$$

$$\mathbf{Z}_1 = \frac{\mathbf{Z}_B\mathbf{Z}_C}{\mathbf{Z}_A + \mathbf{Z}_B + \mathbf{Z}_C} = \frac{(-j\,4\,\Omega)(3\,\Omega + j\,4\,\Omega)}{(-j\,4\,\Omega) + (-j\,4\,\Omega) + (3\,\Omega + j\,4\,\Omega)}$$
$$= \frac{(4\,\angle -90^\circ)(5\,\angle 53.13^\circ)}{3 - j\,4} = \frac{20\,\angle -36.87^\circ}{5\,\angle -53.13^\circ}$$
$$= 4\,\Omega\,\angle 16.13^\circ = 3.84\,\Omega + j\,1.11\,\Omega$$

$$\mathbf{Z}_2 = \frac{\mathbf{Z}_A\mathbf{Z}_C}{\mathbf{Z}_A + \mathbf{Z}_B + \mathbf{Z}_C} = \frac{(-j\,4\,\Omega)(3\,\Omega + j\,4\,\Omega)}{5\,\Omega\,\angle -53.13^\circ}$$
$$= 4\,\Omega\,\angle 16.13^\circ = 3.84\,\Omega + j\,1.11\,\Omega$$

Recall from the study of dc circuits that if two branches of the Y or Δ are the same, the corresponding Δ or Y, respectively, will also have two similar branches. In this example, $\mathbf{Z}_A = \mathbf{Z}_B$. Therefore, $\mathbf{Z}_1 = \mathbf{Z}_2$, and

$$\mathbf{Z}_3 = \frac{\mathbf{Z}_A\mathbf{Z}_B}{\mathbf{Z}_A + \mathbf{Z}_B + \mathbf{Z}_C} = \frac{(-j\,4\,\Omega)(-j\,4\,\Omega)}{5\,\Omega\,\angle -53.13^\circ}$$
$$= \frac{16\,\Omega\,\angle -180^\circ}{5\,\angle -53.13^\circ} = 3.2\,\Omega\,\angle -126.87^\circ = -1.92\,\Omega - j\,2.56\,\Omega$$

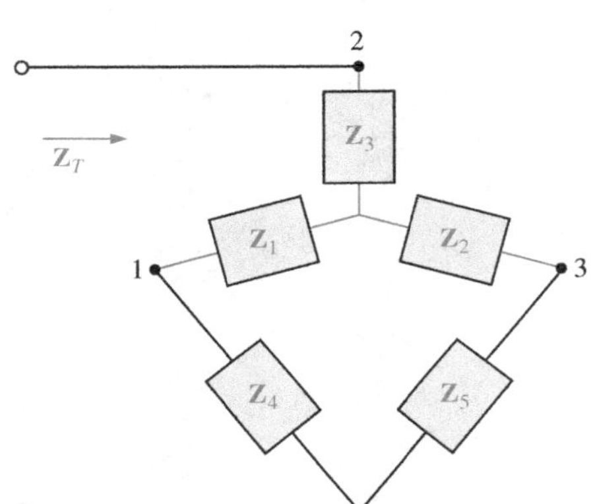

FIG. 17.48
The network in Fig. 17.47 following the substitution of the Y configuration.

Replace the Δ by the Y (Fig. 17.48):

$$\mathbf{Z}_1 = 3.84\,\Omega + j\,1.11\,\Omega \qquad \mathbf{Z}_2 = 3.84\,\Omega + j\,1.11\,\Omega$$
$$\mathbf{Z}_3 = -1.92\,\Omega - j\,2.56\,\Omega \qquad \mathbf{Z}_4 = 2\,\Omega$$
$$\mathbf{Z}_5 = 3\,\Omega$$

Impedances $\mathbf{Z}_1$ and $\mathbf{Z}_4$ are in series:

$$\mathbf{Z}_{T_1} = \mathbf{Z}_1 + \mathbf{Z}_4 = 3.84\,\Omega + j\,1.11\,\Omega + 2\,\Omega = 5.84\,\Omega + j\,1.11\,\Omega$$
$$= 5.94\,\Omega\,\angle 10.76^\circ$$

Impedances $\mathbf{Z}_2$ and $\mathbf{Z}_5$ are in series:

$$\mathbf{Z}_{T_2} = \mathbf{Z}_2 + \mathbf{Z}_5 = 3.84\,\Omega + j\,1.11\,\Omega + 3\,\Omega = 6.84\,\Omega + j\,1.11\,\Omega$$
$$= 6.93\,\Omega\,\angle 9.22^\circ$$

Impedances $\mathbf{Z}_{T_1}$ and $\mathbf{Z}_{T_2}$ are in parallel:

$$\mathbf{Z}_{T_3} = \frac{\mathbf{Z}_{T_1}\mathbf{Z}_{T_2}}{\mathbf{Z}_{T_1} + \mathbf{Z}_{T_2}} = \frac{(5.94\,\Omega\,\angle 10.76^\circ)(6.93\,\Omega\,\angle 9.22^\circ)}{5.84\,\Omega + j\,1.11\,\Omega + 6.84\,\Omega + j\,1.11\,\Omega}$$
$$= \frac{41.16\,\Omega\,\angle 19.98^\circ}{12.68 + j\,2.22} = \frac{41.16\,\Omega\,\angle 19.98^\circ}{12.87\,\angle 9.93^\circ} = 3.198\,\Omega\,\angle 10.05^\circ$$
$$= 3.15\,\Omega + j\,0.56\,\Omega$$

Impedances $\mathbf{Z}_3$ and $\mathbf{Z}_{T_3}$ are in series. Therefore,

$$\mathbf{Z}_T = \mathbf{Z}_3 + \mathbf{Z}_{T_3} = -1.92\,\Omega - j\,2.56\,\Omega + 3.15\,\Omega + j\,0.56\,\Omega$$
$$= 1.23\,\Omega - j\,2.0\,\Omega = \mathbf{2.35\,\Omega\,\angle -58.41^\circ}$$

EXAMPLE 17.21 Using both the Δ-Y and Y-Δ transformations, find the total impedance $\mathbf{Z}_T$ for the network in Fig. 17.49.

Solution: *Using the Δ-Y transformation,* we obtain Fig. 17.50. In this case, since both systems are balanced (same impedance in each branch),

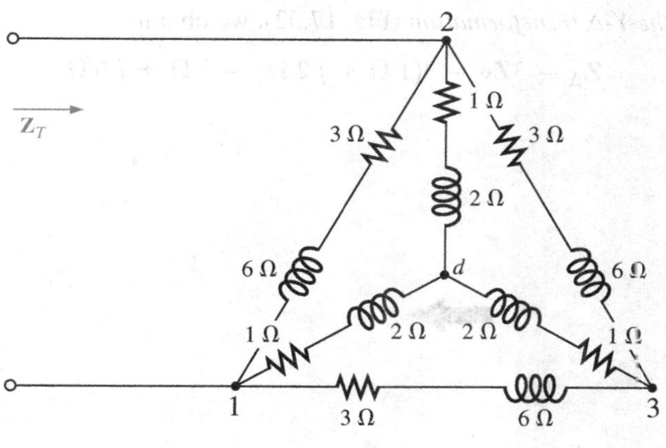

FIG. 17.49
Example 17.21.

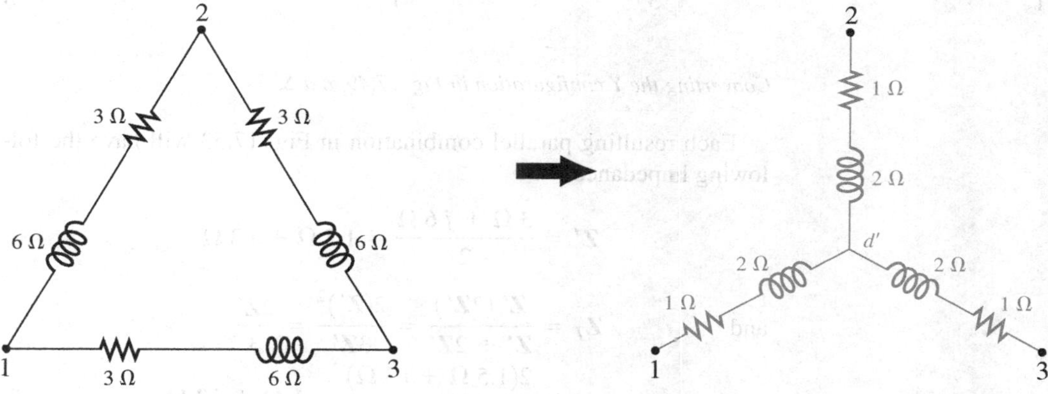

FIG. 17.50
Converting a Δ configuration to a Y configuration.

the center point d' of the transformed Δ will be the same as point d of the original Y:

$$\mathbf{Z}_Y = \frac{\mathbf{Z}_\Delta}{3} = \frac{3\ \Omega + j\,6\ \Omega}{3} = 1\ \Omega + j\,2\ \Omega$$

and (Fig. 17.51)

$$\mathbf{Z}_T = 2\left(\frac{1\ \Omega + j\,2\ \Omega}{2}\right) = \mathbf{1\ \Omega + \mathit{j}\,2\ \Omega}$$

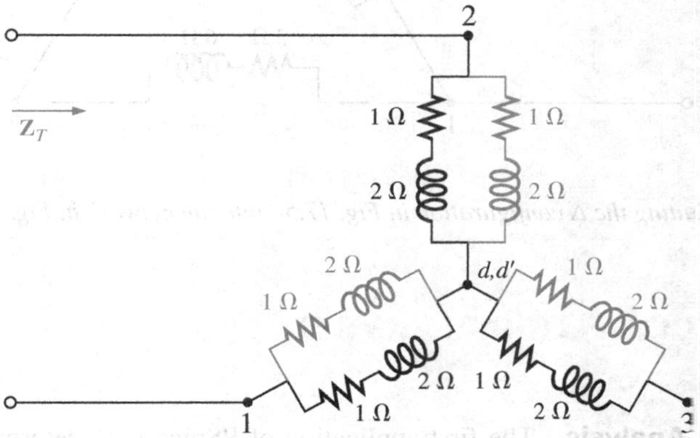

FIG. 17.51
Substituting the Y configuration in Fig. 17.50 into the network in Fig. 17.49.

Using the Y-Δ transformation (Fig. 17.52), we obtain

$$\mathbf{Z}_\Delta = 3\mathbf{Z}_Y = 3(1\ \Omega + j\,2\ \Omega) = 3\ \Omega + j\,6\ \Omega$$

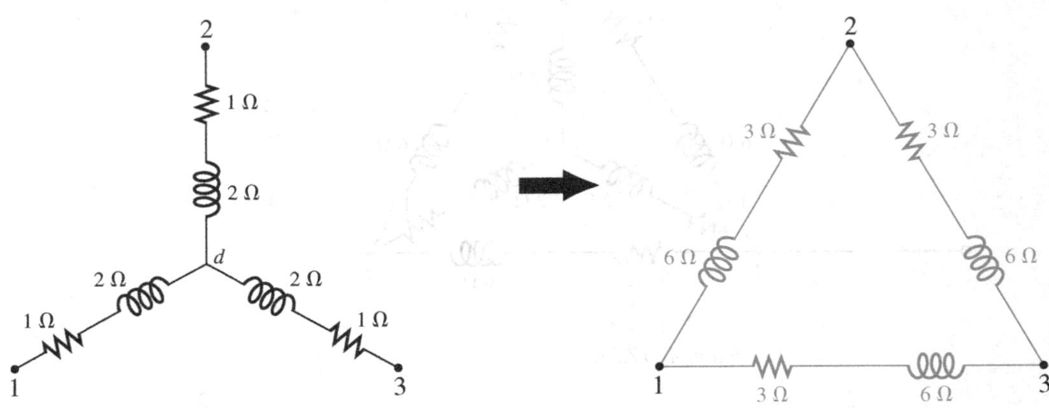

FIG. 17.52
Converting the Y configuration in Fig. 17.49 to a Δ.

Each resulting parallel combination in Fig. 17.53 will have the following impedance:

$$\mathbf{Z}' = \frac{3\ \Omega + j\,6\ \Omega}{2} = 1.5\ \Omega + j\,3\ \Omega$$

and

$$\mathbf{Z}_T = \frac{\mathbf{Z}'\,(2\mathbf{Z}')}{\mathbf{Z}' + 2\mathbf{Z}'} = \frac{2(\mathbf{Z}')^2}{3\mathbf{Z}'} = \frac{2\mathbf{Z}'}{3}$$

$$= \frac{2(1.5\ \Omega + j\,3\ \Omega)}{3} = \mathbf{1\ \Omega + \mathit{j}\,2\ \Omega}$$

which compares with the above result.

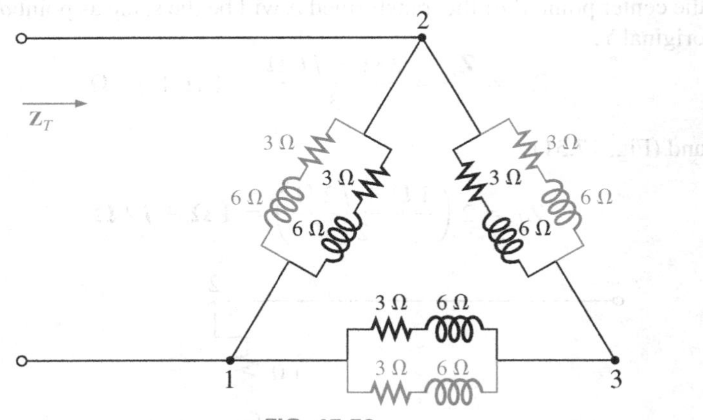

FIG. 17.53
Substituting the Δ configuration in Fig. 17.54 into the network in Fig. 17.49.

17.8 COMPUTER ANALYSIS

PSpice

Nodal Analysis The first application of PSpice is to determine the nodal voltages for the network in Example 17.16 and compare solutions. The network appears as shown in Fig. 17.54 using elements that were

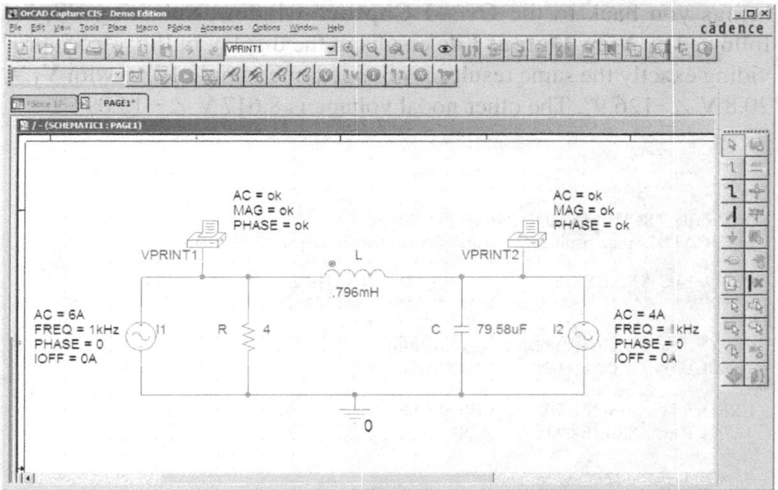

FIG. 17.54
Using PSpice to verify the results in Example 17.16.

determined from the reactance level at a frequency of 1 kHz. There is no need to continually use 1 kHz. Any frequency will do, but remember to use the chosen frequency to find the network components and when setting up the simulation.

For the current sources, choose **ISIN** so that the phase angle can be specified (even though it is 0°), although the symbol does not have the arrow used in the text material. The direction must be recognized as pointing from the + to − sign of the source. That requires that the sources $\mathbf{I}_1$ and $\mathbf{I}_2$ be set as shown in Fig. 17.54. Reverse the source $\mathbf{I}_2$ by using the **Mirror Vertically** option obtained by right-clicking the source symbol on the screen. Setting up the **ISIN** source is the same as that used with the **VSIN** source. It can be found under the **SOURCE** library, and its attributes are the same as for the **VSIN** source. For each source, set **IOFF** to 0 A; the amplitude is the peak value of the source current. The frequency will be the same for each source. Then select **VPRINT1** from the **SPECIAL** library and place it to generate the desired nodal voltages. Finally add the remaining elements to the network as shown in Fig. 17.54. For each source, double-click the symbol to generate the **Property Editor** dialog box. Set **AC** at the 6 A level for the $\mathbf{I}_1$ source and at 4 A for the $\mathbf{I}_2$ source, followed by **Display** and **Name and Value** for each. It appears as shown in Fig. 17.54. Double-clicking on each **VPRINT1** option also provides the **Property Editor,** so **OK** can be added under **AC, MAG,** and **PHASE.** For each quantity, select **Display** followed by **Name and Value** and **OK.** Then select **Value** and **VPRINT1** is displayed as **Value** only. Selecting **Apply** and leaving the dialog box results in the listing next to each source in Fig. 17.54. For **VPRINT2,** first change the listing on **Value** from **VPRINT1** to **VPRINT2** before selecting **Display** and **Apply.**

Now select the **New Simulation Profile** icon, and enter **PSpice 17-1** as the **Name** followed by **Create.** In the **Simulation Settings** dialog box, select **AC Sweep,** and set the **Start Frequency** and **End Frequency** at 1 kHz with 1 for the **Points/Decade.** Click **OK,** and select the **Run PSpice** icon; a **SCHEMATIC1** screen results. Exiting (**X**)

brings you back to the **Orcad Capture** window. Selecting **PSpice** followed by **View Output File** results in the display in Fig. 17.55, providing exactly the same results as obtained in Example 17.16 with $\mathbf{V}_1 = 20.8\text{ V}\angle-126.9°$. The other nodal voltage is $8.617\text{ V}\angle-15.09°$.

```
** Profile: "SCHEMATIC1-PSpice 17-1"
[ C:\ICA11\PSpice\pspice 17-1-pspicefiles\schematic1\pspice 17-1.sim ]

****    AC ANALYSIS                 TEMPERATURE =  27.000 DEG C
******************************************************************************
***

FREQ          VM(N01900)     VP(N01900)
1.000E+03     2.080E+01      -1.269E+02

FREQ          VM(N01908)     VP(N01908)
1.000E+03     8.617E+00      -1.509E+01
```

FIG. 17.55
Output file for the nodal voltages for the network in Fig. 17.54.

Current-Controlled Current Source (CCCS) Our interest now turns to controlled sources in the PSpice environment. Controlled sources are not particularly difficult to apply once a few important elements of their use are understood. The network in Fig. 17.14 has a current-controlled current source in the center leg of the configuration. The magnitude of the current source is k times the current through resistor R_1, where k can be greater or less than 1. The resulting schematic, appearing in Fig. 17.56, seems quite complex in the area of the controlled source, but once you understand the role of each component, the schematic is not that difficult to understand. First, since it is the only new element in the schematic, let us concentrate on the controlled source. Current-controlled current sources (CCCS) are called up under the **ANALOG** library as **F** and appear as shown in the center in Fig. 17.56. Pay attention to the direction of the current in

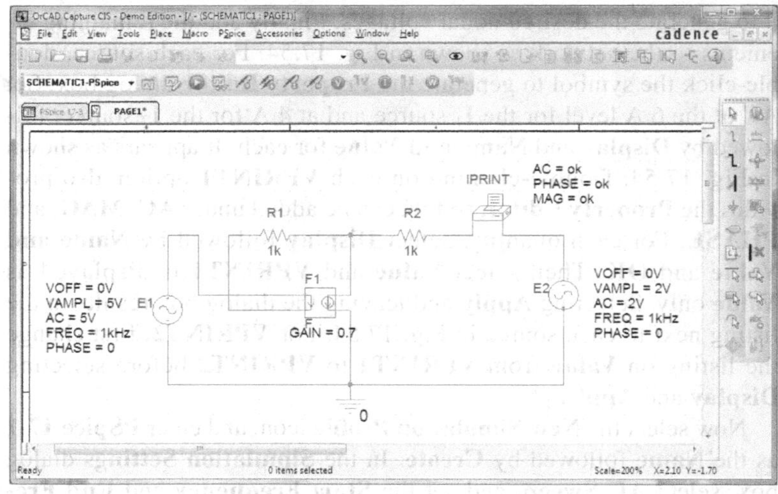

FIG. 17.56
Using PSpice to verify the results in Example 17.8.

each part of the symbol. In particular, note that the sensing current of **F** has the same direction as the defining controlling current in Fig. 17.14. In addition, note that the controlled current source also has the same direction as the source in Fig. 17.14. If you double-click on the CCCS symbol, the **Property Editor** dialog box appears with the **GAIN** (k as described above) set at 1. In this example, the gain must be set at **0.7,** so click on the region below the **GAIN** label and enter **0.7.** Then select **Display** followed by **Name and Value-OK.** Exit the **Property Editor,** and **GAIN** = **0.7** appears with the CCCS as shown in Fig. 17.56.

The other new component in this schematic is **IPRINT;** it can be found in the **SPECIAL** library. It is used to tell the program to list the current in the branch of interest in the output file. If you fail to tell the program which output data you would like, it will simply run through the simulation and list specific features of the network but will not provide any voltages or currents. In this case, the current $\mathbf{I}_2$ through the resistor R_2 is desired. Double-clicking on the **IPRINT** component results in the **Property Editor** dialog box with a number of elements that need to be defined—much like that for **VPRINT.** First enter **OK** beneath **AC** and follow with **Display-Name and Value-OK.** Repeat for **MAG** and **PHASE,** and then select **Apply** before leaving the dialog box. The **OK** tells the software program that these are the quantities that it is "ok" to generate and provide. The purpose of the **Apply** at the end of each visit to the **Property Editor** dialog box is to "apply" the changes made to the network under investigation. When you exit the **Property Editor,** the three chosen parameters appear on the schematic with the **OK** directive. You may find that the labels will appear all over the **IPRINT** symbol. No problem—just click on each, and move to a more convenient location.

The remaining components of the network should be fairly familiar, but don't forget to **Mirror Vertically** the voltage source **E2.** In addition, do not forget to call up the **Property Editor** for each source and set the level of **AC, FREQ, VAMPL,** and **VOFF** and be sure that the **PHASE** is set on the default value of 0°. The value appears with each parameter in Fig. 17.58 for each source. Always be sure to select **Apply** before leaving the **Property Editor.** After placing all the components on the screen, you must connect them with a **Place wire** selection. Normally, this is pretty straightforward. However, with controlled sources, it is often necessary to cross over wires *without* making a connection. In general, when you're placing a wire over another wire and you don't want a connection to be made, click a spot on one side of the wire to be crossed to create the temporary red square. Then cross the wire, and click again to establish another red square. If the connection is done properly, the crossed wire should not show a connection point (a small red dot). In this example, the top of the controlling current was connected first from the **E1** source. Then a wire was connected from the lower end of the sensing current to the point where a 90° turn up the page was to be made. The wire was clicked in place at this point before crossing the original wire and clicked again before making the right turn to resistor R_1. You will not find a small red dot where the wires cross.

Now for the simulation. In the **Simulation Settings** dialog box, select **AC Sweep/Noise** with a **Start** and **End Frequency** of 1 kHz. There will be **1 Point/Decade.** Click **OK,** and select the **Run Spice** key; a **SCHEMATIC1** results that should be exited to obtain the **Orcad Capture** screen. Select **PSpice** followed by **View Output File,**

and scroll down until you read **AC ANALYSIS** (see Fig. 17.57). The magnitude of the desired current is 1.615 mA with a phase angle of 0°, a perfect match with the theoretical analysis to follow. One would expect a phase angle of 0° since the network is composed solely of resistive elements.

```
** Profile: "SCHEMATIC1-PSpice 17-3"
[ C:\ICA11\PSpice\PSpice 17-3-PSpiceFiles\SCHEMATIC1\PSpice 17-3.sim ]
****     AC ANALYSIS                          TEMPERATURE = 27.000 DEG C
******************************************************************************
***
FREQ         IM(V_PRINT1)         IP(V_PRINT1)
1.000E+03    1.615E-03            0.000E+00
```

FIG. 17.57

The output file for the mesh current $\mathbf{I}_2$ in Fig. 17.14.

The equations obtained earlier using the supermesh approach were

$$\mathbf{E} - \mathbf{I}_1\mathbf{Z}_1 - \mathbf{I}_2\mathbf{Z}_2 + \mathbf{E}_2 = 0 \quad \text{or} \quad \mathbf{I}_1\mathbf{Z}_1 + \mathbf{I}_2\mathbf{Z}_2 = \mathbf{E}_1 + \mathbf{E}_2$$

and
$$k\mathbf{I} = k\mathbf{I}_1 = \mathbf{I}_1 - \mathbf{I}_2$$

resulting in
$$\mathbf{I}_1 = \frac{\mathbf{I}_2}{1-k} = \frac{\mathbf{I}_2}{1-0.7} = \frac{\mathbf{I}_2}{0.3} = 3.333\mathbf{I}_2$$

so that
$$\mathbf{I}_1(1\text{ k}\Omega) + \mathbf{I}_2(1\text{ k}\Omega) = 7\text{ V} \qquad \text{(from above)}$$

becomes
$$(3.333\mathbf{I}_2)1\text{ k}\Omega + \mathbf{I}_2(1\text{ k}\Omega) = 7\text{ V}$$

or
$$(4.333\text{ k}\Omega)\mathbf{I}_2 = 7\text{ V}$$

and
$$\mathbf{I}_2 = \frac{7\text{ V}}{4.333\text{ k}\Omega} = \mathbf{1.615\text{ mA}\angle 0^\circ}$$

confirming the computer solution.

PROBLEMS

SECTION 17.2 Independent versus Dependent (Controlled) Sources

1. Discuss, in your own words, the difference between a controlled and an independent source.

SECTION 17.3 Source Conversions

2. Convert the voltage source in Fig. 17.58 to a current source.

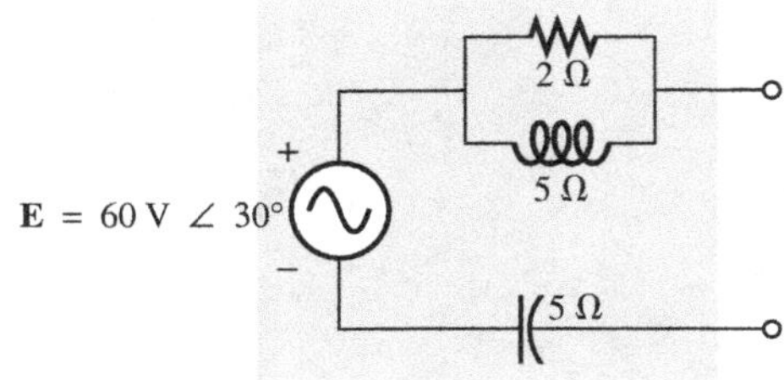

FIG. 17.58

Problem 2

3. Convert the current source in Fig. 17.59 to a voltage source.

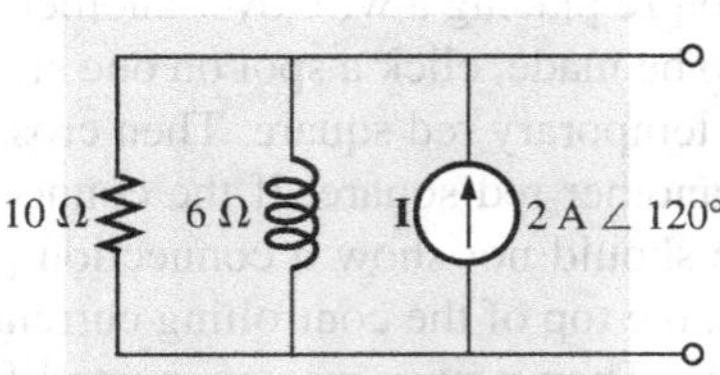

FIG. 17.59

Problem 3.

4. Convert the voltage source in Fig. 17.60(a) to a current source and the current source in Fig. 17.60(b) to a voltage source.

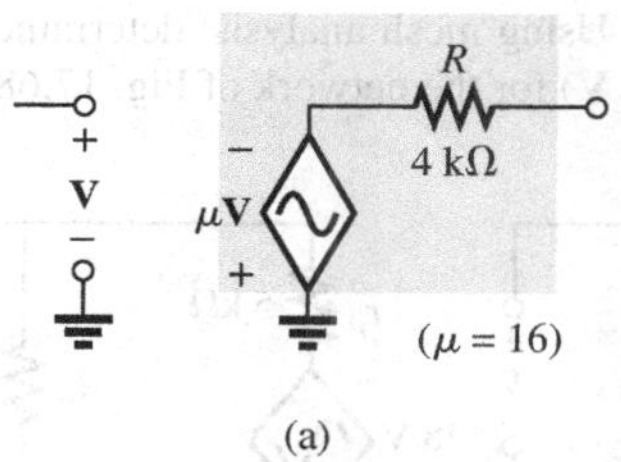

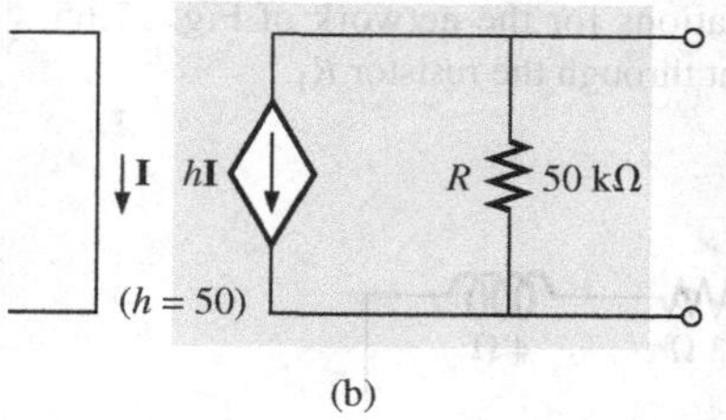

FIG. 17.60
Problem 4.

SECTION 17.4 Mesh Analysis

5. Write the mesh equations for the network of Fig. 17.61. Determine the current through the resistor *R*.

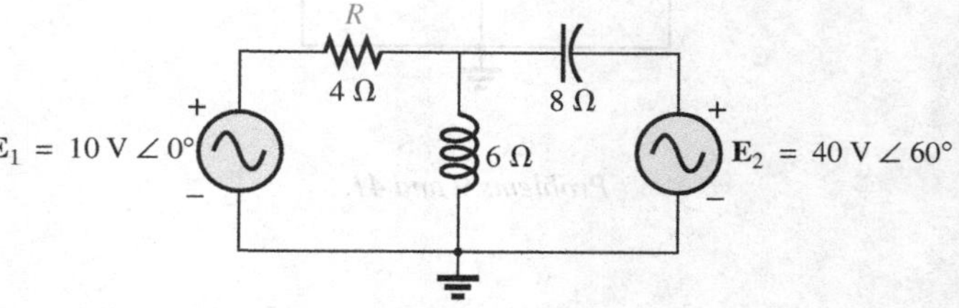

FIG. 17.61
Problems 5 and 40.

6. Write the mesh equations for the network of Fig. 17.62.

FIG. 17.62
Problem 6.

7. Write the mesh equations for the network of Fig. 17.63. Determine the current through the resistor R_1.

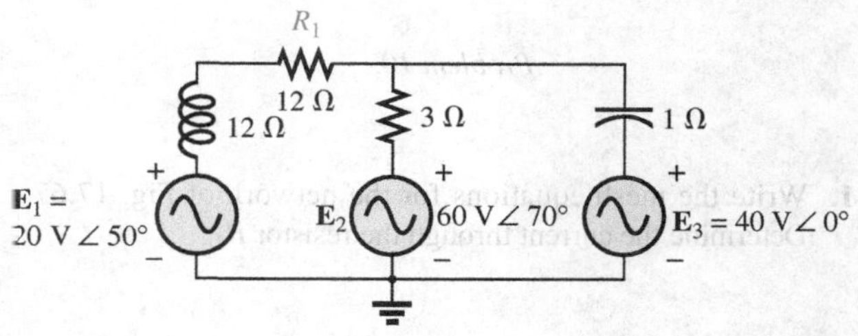

FIG. 17.63
Problems 7 and 21.

*8. Write the mesh equations for the network of Fig. 17.64. Determine the current through the resistor R_1.

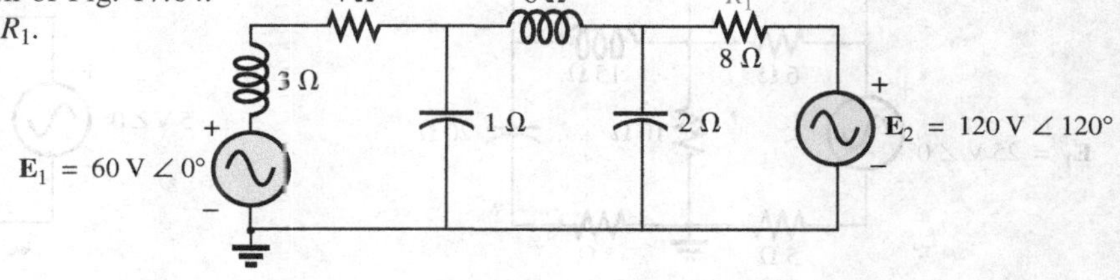

FIG. 17.64
Problem 8.

*9. Write the mesh equations for the network of Fig. 17.65. Determine the current through the resistor R_1.

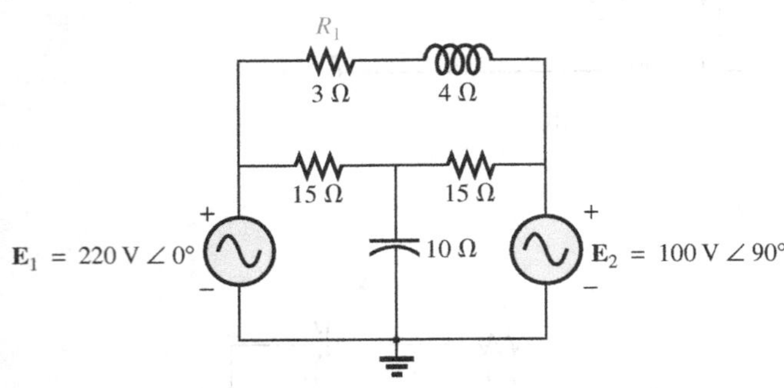

FIG. 17.65
Problems 9 and 41.

*10. Write the mesh equations for the network of Fig. 17.66. Determine the current through the resistor R_1.

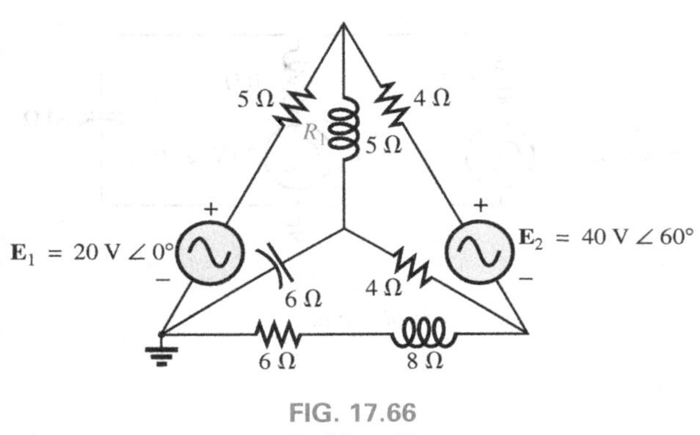

FIG. 17.66
Problem 10.

11. Write the mesh equations for the network of Fig. 17.67. Determine the current through the resistor R_1.

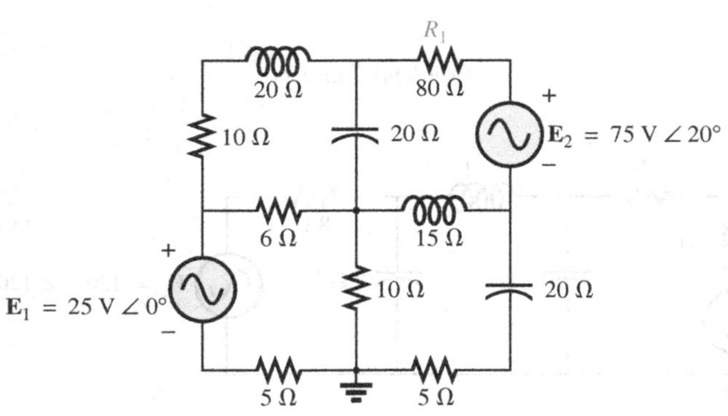

FIG. 17.67
Problems 11 and 22.

12. Using mesh analysis, determine the current $\mathbf{I}_L$ (in terms of **V**) for the network of Fig. 17.68.

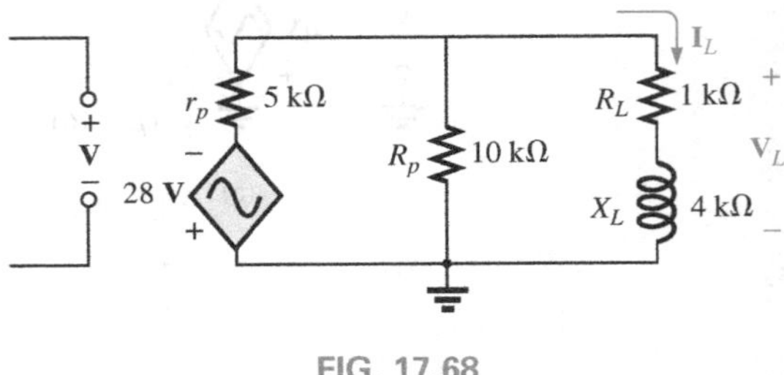

FIG. 17.68
Problem 12.

13. Using mesh analysis, determine the current $\mathbf{I}_L$ (in terms of **I**) for the network of Fig. 17.69.

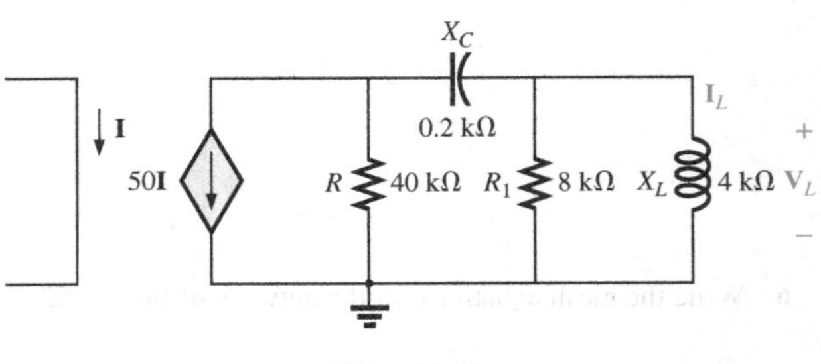

FIG. 17.69
Problem 13.

*14. Write the mesh equations for the network of Fig. 17.70, and determine the current through the 1 kΩ and 2 kΩ resistors.

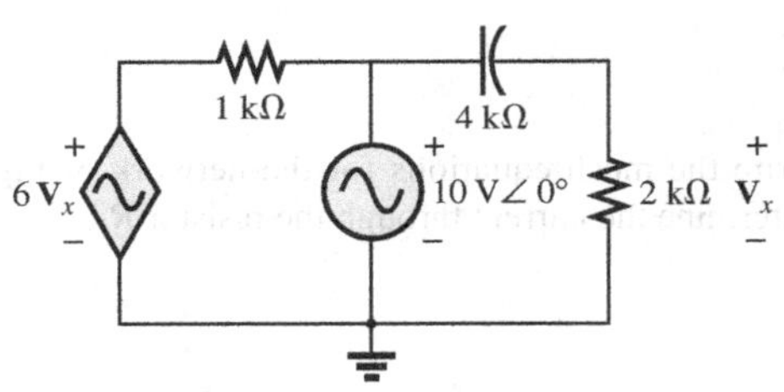

FIG. 17.70
Problems 14 and 42.

*15. Write the mesh equations for the network of Fig. 17.71, and determine the current through the 10 kΩ resistor.

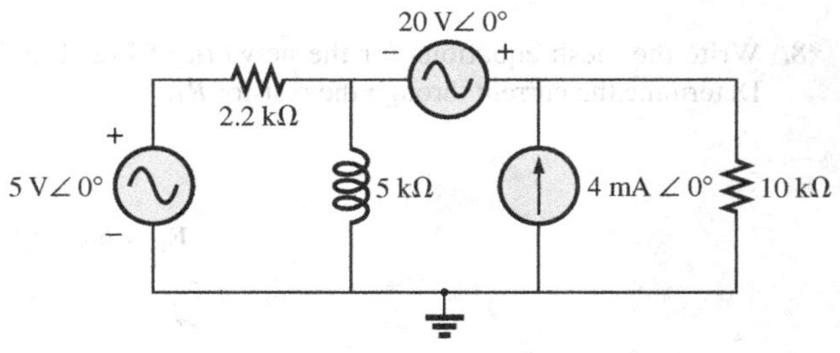

FIG. 17.71
Problems 15 and 43.

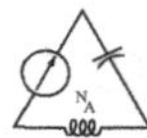

*16. Write the mesh equations for the network of Fig. 17.72, and determine the current through the inductive element.

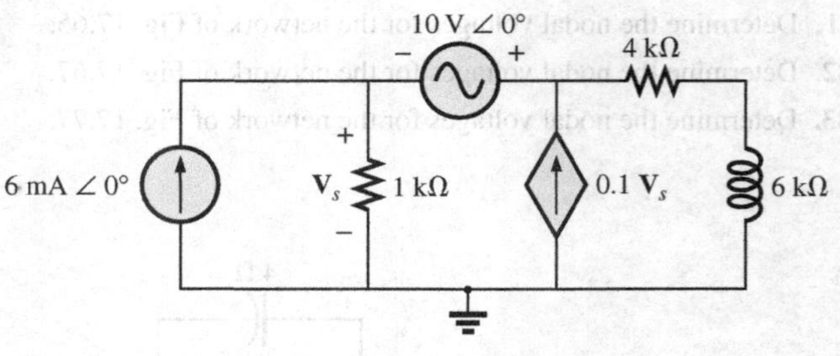

FIG. 17.72
Problems 16 and 44.

SECTION 17.5 Nodal Analysis

17. Determine the nodal voltages for the network of Fig. 17.73.

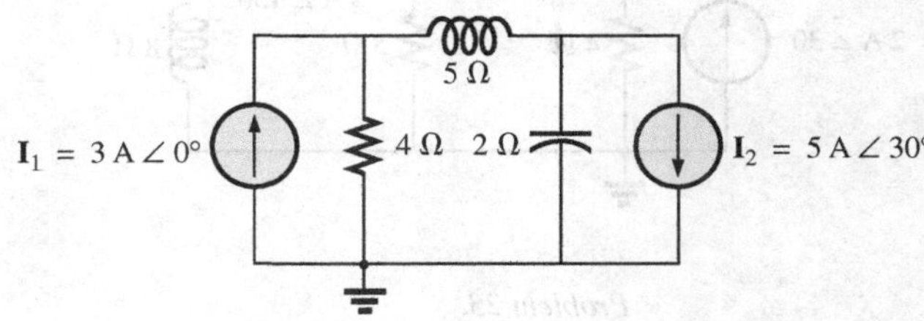

FIG. 17.73
Problem 17.

18. Determine the nodal voltages for the network of Fig. 17.74.

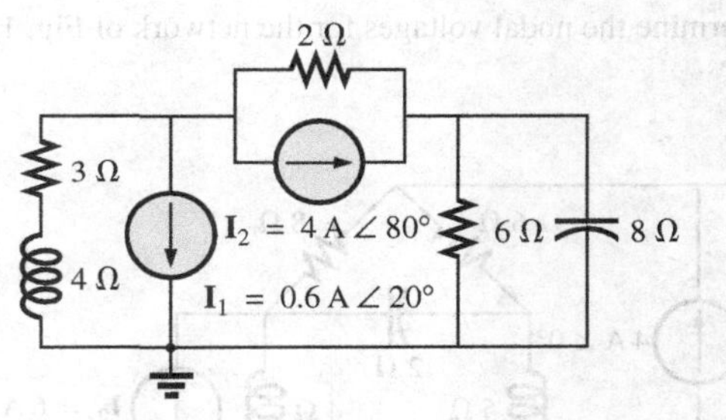

FIG. 17.74
Problems 18 and 45.

19. Determine the nodal voltages for the network of Fig. 17.75.

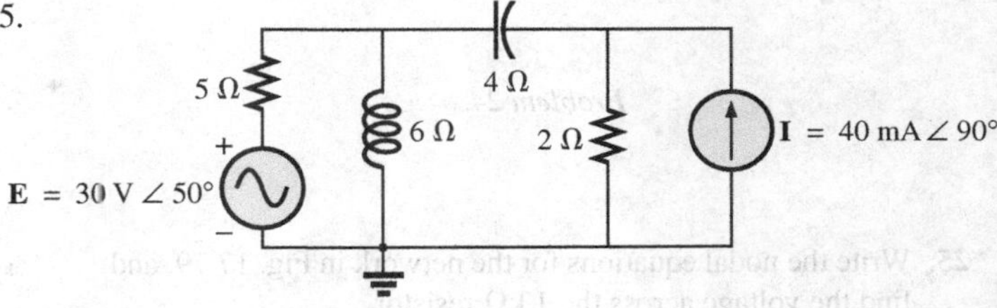

FIG. 17.75
Problem 19.

20. Determine the nodal voltages for the network of Fig. 17.76.

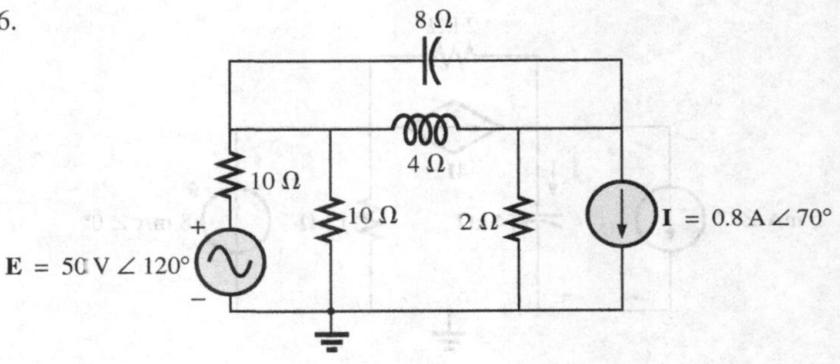

FIG. 17.76
Problem 20.

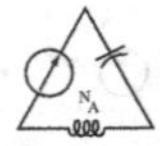

21. Determine the nodal voltages for the network of Fig. 17.65.

*22. Determine the nodal voltages for the network of Fig. 17.67.

*23. Determine the nodal voltages for the network of Fig. 17.77.

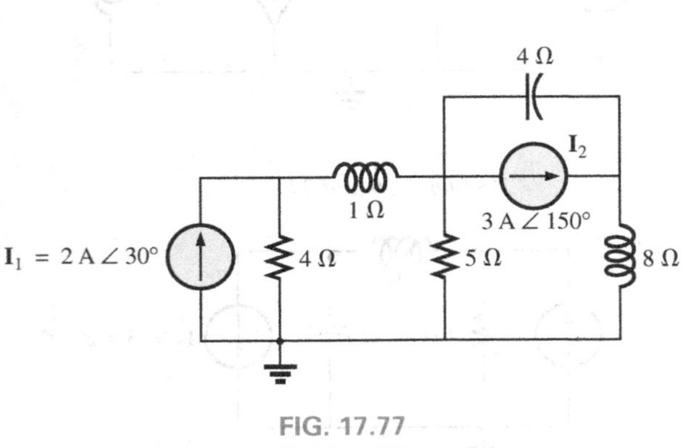

FIG. 17.77
Problem 23.

*24. Determine the nodal voltages for the network of Fig. 17.78.

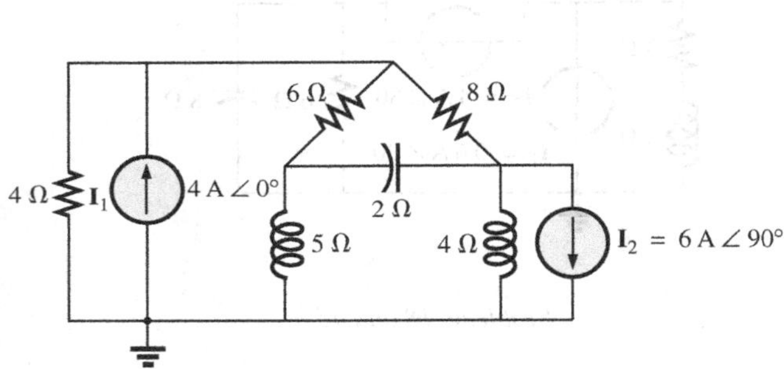

FIG. 17.78
Problem 24.

*25. Write the nodal equations for the network in Fig. 17.79, and find the voltage across the 1 kΩ resistor.

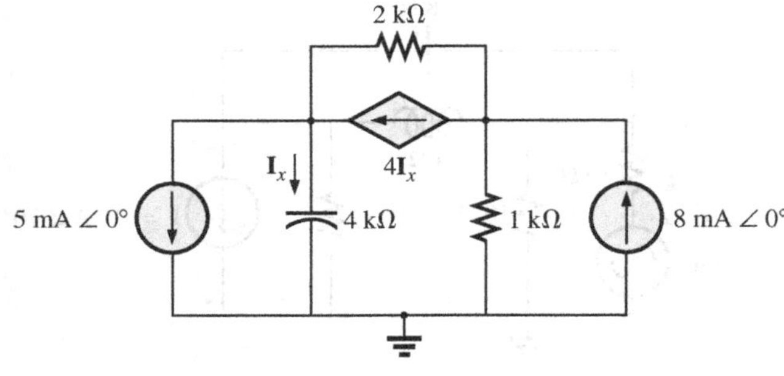

FIG. 17.79
Problems 25 and 46.

*26. Write the nodal equations for the network of Fig. 17.80, and find the voltage across the capacitive element.

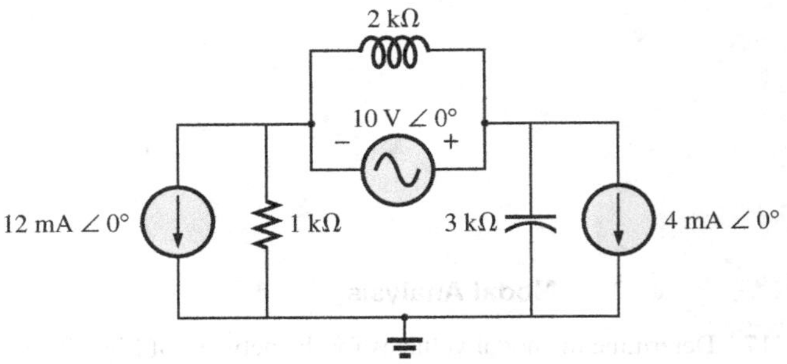

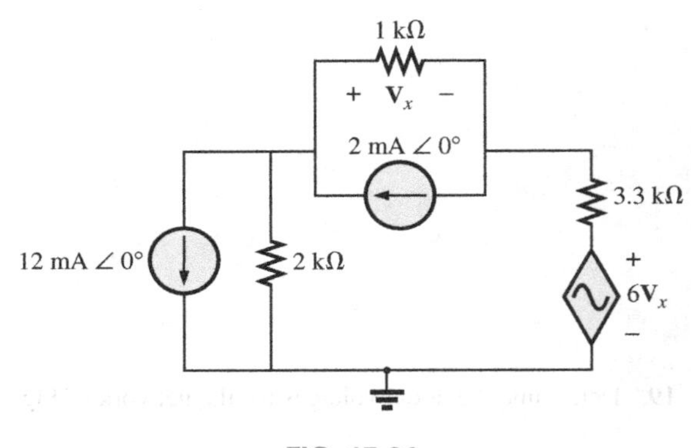

FIG. 17.80
Problems 26 and 47.

*27. Write the nodal equations for the network of Fig. 17.81, and find the voltage across the 2 kΩ resistor.

FIG. 17.81
Problems 27 and 48.

*28. Write the nodal equations for the network of Fig. 17.82, and find the voltage across the 2 kΩ resistor.

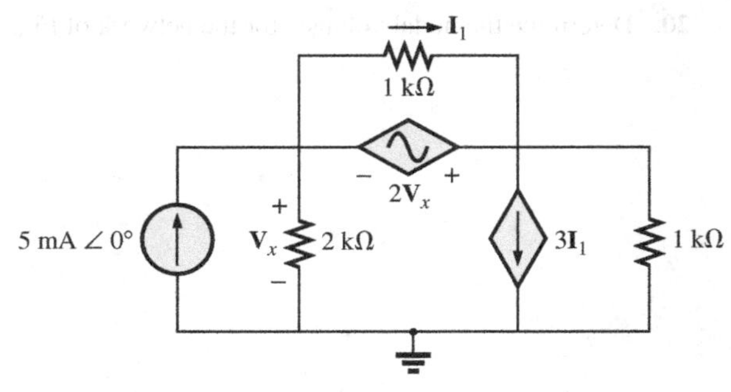

FIG. 17.82
Problems 28 and 49.

***29.** For the network of Fig. 17.83, determine the voltage $\mathbf{V}_L$ in terms of the voltage $\mathbf{E}_i$.

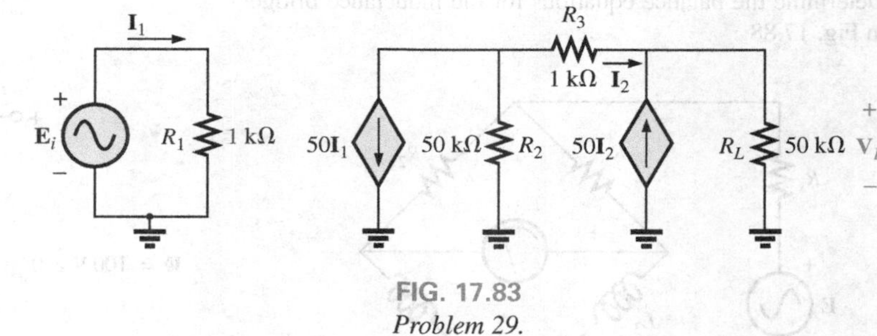

FIG. 17.83
Problem 29.

SECTION 17.6 Bridge Networks (ac)

30. For the bridge network in Fig. 17.84:
   a. Is the bridge balanced?
   b. Using mesh analysis, determine the current through the capacitive reactance.
   c. Using nodal analysis, determine the voltage across the capacitive reactance.

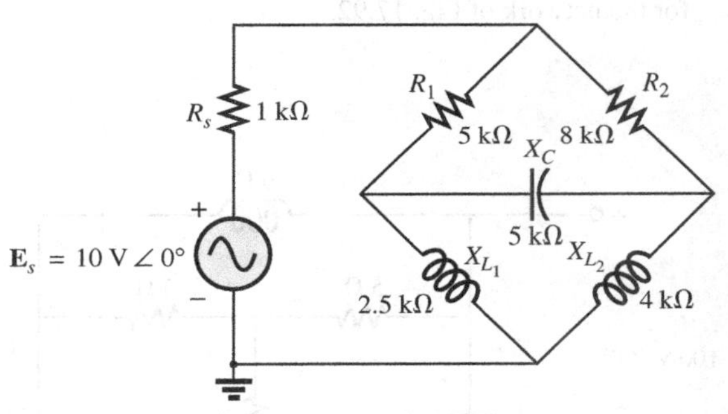

FIG. 17.84
Problem 30.

31. For the bridge network in Fig. 17.85:
   a. Is the bridge balanced?
   b. Using mesh analysis, determine the current through the capacitive reactance.
   c. Using nodal analysis, determine the voltage across the capacitive reactance.

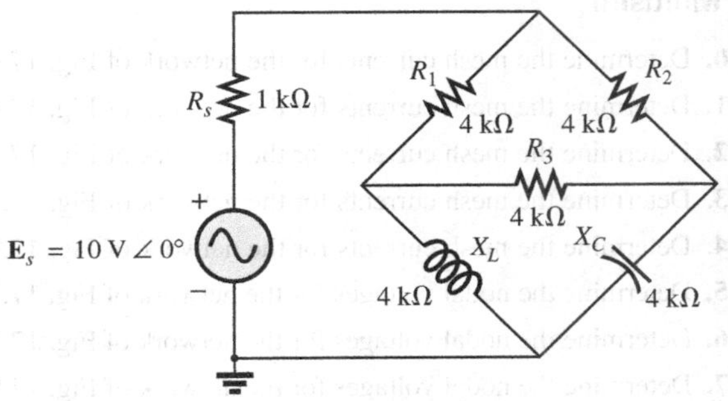

FIG. 17.85
Problem 31.

32. The Hay bridge in Fig. 17.86 is balanced. Using Eq. (17.3), determine the unknown inductance L_x and resistance R_x.

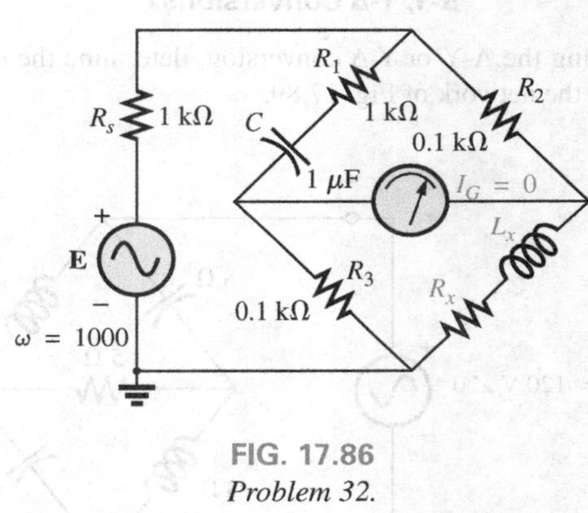

FIG. 17.86
Problem 32.

33. Determine whether the Maxwell bridge in Fig. 17.87 is balanced ($\omega = 1000$ rad/s).

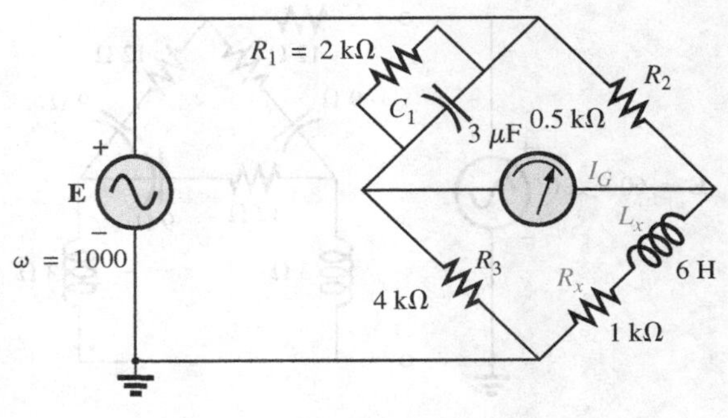

FIG. 17.87
Problem 33.

34. Derive the balance equations (17.16) and (17.17) for the capacitance comparison bridge.

35. Determine the balance equations for the inductance bridge in Fig. 17.88.

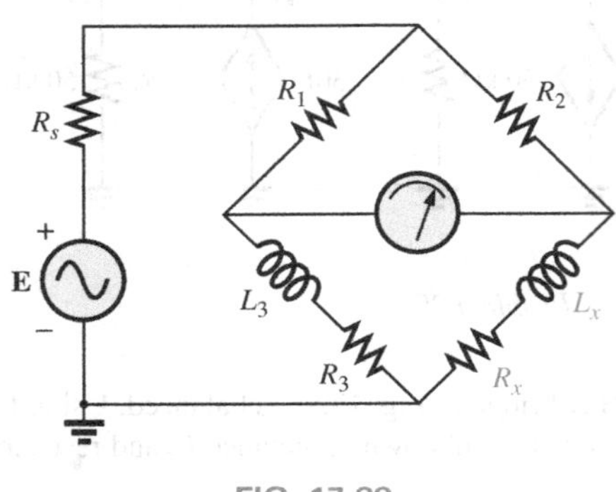

FIG. 17.88
Problem 35.

SECTION 17.7 Δ-Y, Y-Δ Conversions

36. Using the Δ-Y or Y-Δ conversion, determine the current **I** for the network of Fig. 17.89.

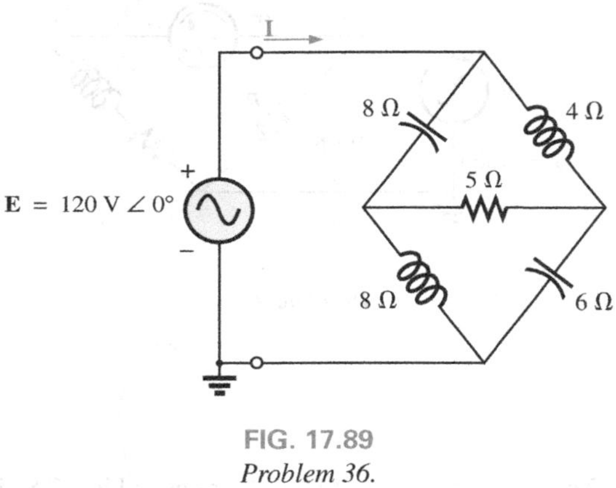

FIG. 17.89
Problem 36.

37. Using the Δ-Y or Y-Δ conversion, determine the current **I** for the network of Fig. 17.90.

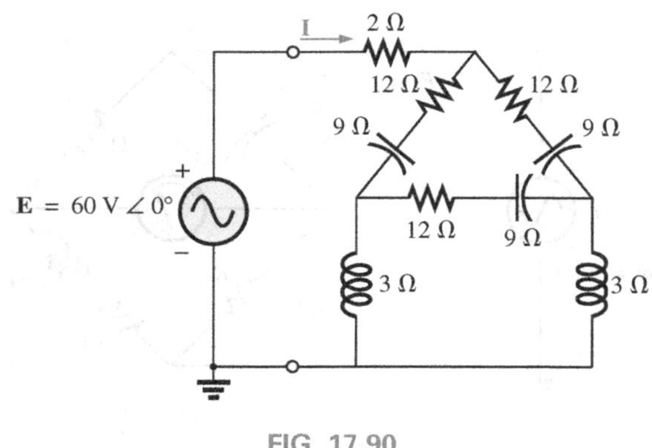

FIG. 17.90
Problem 37.

38. Using the Δ-Y or Y-Δ conversion, determine the current **I** for the network of Fig. 17.91.

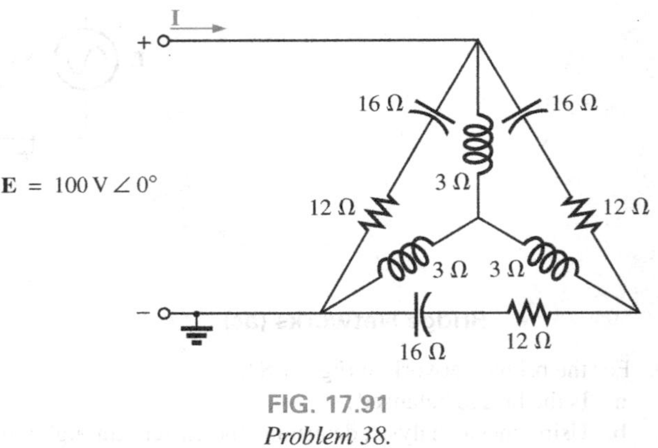

FIG. 17.91
Problem 38.

39. Using the Δ-Y or Y-Δ conversion, determine the current **I** for the network of Fig. 17.92.

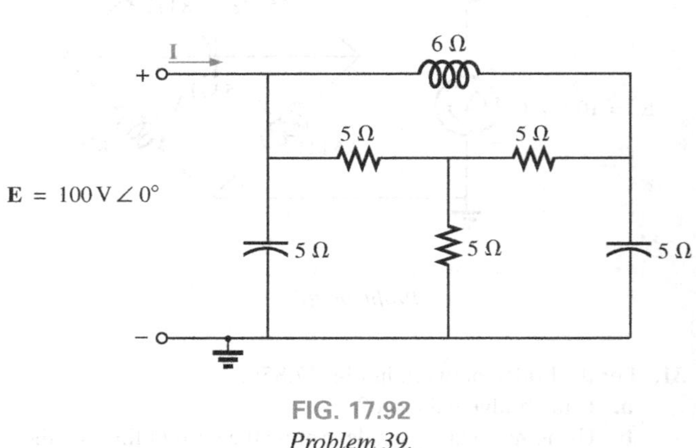

FIG. 17.92
Problem 39.

SECTION 17.8 Computer Analysis PSpice or Multisim

40. Determine the mesh currents for the network of Fig. 17.61.

41. Determine the mesh currents for the network of Fig. 17.65.

***42.** Determine the mesh currents for the network of Fig. 17.70.

***43.** Determine the mesh currents for the network of Fig. 17.71.

***44.** Determine the mesh currents for the network of Fig. 17.72.

45. Determine the nodal voltages for the network of Fig. 17.74.

***46.** Determine the nodal voltages for the network of Fig. 17.79.

***47.** Determine the nodal voltages for the network of Fig. 17.80.

***48.** Determine the nodal voltages for the network of Fig. 17.81.

***49.** Determine the nodal voltages for the network of Fig. 17.82.

GLOSSARY

Bridge network A network configuration having the appearance of a diamond in which no two branches are in series or parallel.

Capacitance comparison bridge A bridge configuration having a galvanometer in the bridge arm that is used to determine an unknown capacitance and associated resistance.

Delta (Δ) configuration A network configuration having the appearance of the capital Greek letter *delta.*

Dependent (controlled) source A source whose magnitude and/or phase angle is determined (controlled) by a current or voltage of the system in which it appears.

Hay bridge A bridge configuration used for measuring the resistance and inductance of coils in those cases where the resistance is a small fraction of the reactance of the coil.

Independent source A source whose magnitude is independent of the network to which it is applied. It displays its terminal characteristics even if completely isolated.

Maxwell bridge A bridge configuration used for inductance measurements when the resistance of the coil is large enough not to require a Hay bridge.

Mesh analysis A method through which the loop (or mesh) currents of a network can be determined. The branch currents of the network can then be determined directly from the loop currents.

Nodal analysis A method through which the nodal voltages of a network can be determined. The voltage across each element can then be determined through application of Kirchhoff's voltage law.

Source conversion The changing of a voltage source to a current source, or vice versa, which will result in the same terminal behavior of the source. In other words, the external network is unaware of the change in sources.

Wye (Y) configuration A network configuration having the appearance of the capital letter Y.

Appendixes

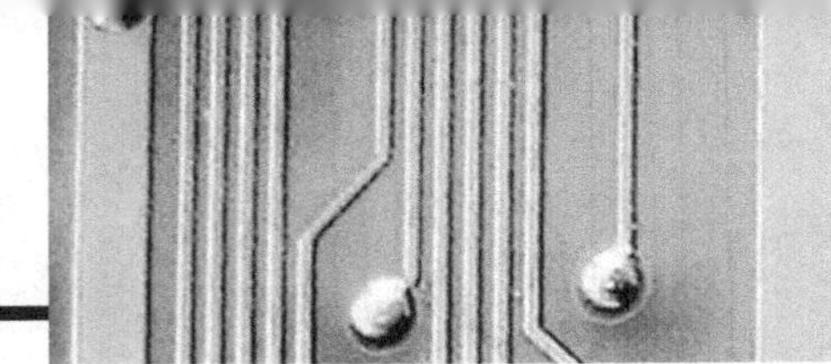

Appendix A

CONVERSION FACTORS

To Convert from	To	Multiply by
Btus	Calorie-grams	251.996
	Ergs	1.054×10^{10}
	Foot-pounds	777.649
	Hp-hours	0.000393
	Joules	1054,35
	Kilowatthours	0.000293
	Wattseconds	1054.35
Centimeters	Angstrom units	1×10^{8}
	Feet	0.0328
	Inches	0.3937
	Meters	0.01
	Miles (statute)	6.214×10^{-6}
	Millimeters	10
Circular mils	Square centimeters	5.067×10^{-6}
	Square inches	7.854×10^{-7}
Cubic inches	Cubic centimeters	16.387
	Gallons (U.S. liquid)	0.00433
Cubic meters	Cubic feet	35.315
Days	Hours	24
	Minutes	1440
	Seconds	86,400
Dynes	Gallons (U.S. liquid)	264.172
	Newtons	0.00001
	Pounds	2.248×10^{-6}
Electronvolts	Ergs	1.60209×10^{-12}
Ergs	Dyne-centimeters	1.0
	Electronvolts	6.242×10^{11}
	Foot-pounds	7.376×10^{-8}
	Joules	1×10^{-7}
	Kilowatthours	2.777×10^{-14}
Feet	Centimeters	30.48
	Meters	0.3048
Foot-candles	Lumens/square foot	1.0
	Lumens/square meter	10.764
Foot-pounds	Dyne-centimeters	1.3558×10^{7}
	Ergs	1.3558×10^{7}
	Horsepower-hours	5.050×10^{-7}
	Joules	1.3558
	Newton-meters	1.3558
Gallons (U.S. liquid)	Cubic inches	231
	Liters	3.785
	Ounces	128
	Pints	8

To Convert from	To	Multiply by
Gauss	Maxwells/square centimeter	1.0
	Lines/square centimeter	1.0
	Lines/square inch	6.4516
Gilberts	Ampere-turns	0.7958
Grams	Dynes	980.665
	Ounces	0.0353
	Pounds	0.0022
Horsepower	Btus/hour	2547.16
	Ergs/second	7.46×10^9
	Foot-pounds/second	550.221
	Joules/second	746
	Watts	746
Hours	Seconds	3600
Inches	Angstrom units	2.54×10^8
	Centimeters	2.54
	Feet	0.0833
	Meters	0.0254
Joules	Btus	0.000948
	Ergs	1×10^7
	Foot-pounds	0.7376
	Horsepower-hours	3.725×10^{-7}
	Kilowatthours	2.777×10^{-7}
	Wattseconds	1.0
Kilograms	Dynes	980,665
	Ounces	35.2
	Pounds	2.2
Lines	Maxwells	1.0
Lines/square centimeter	Gauss	1.0
Lines/square inch	Gauss	0.1550
	Webers/square inch	1×10^{-8}
Liters	Cubic centimeters	1000.028
	Cubic inches	61.025
	Gallons (U.S. liquid)	0.2642
	Ounces (U.S. liquid)	33.815
	Quarts (U.S. liquid)	1.0567
Lumens	Candle power (spher.)	0.0796
Lumens/square centimeter	Lamberts	1.0
Lumens/square foot	Foot-candles	1.0
Maxwells	Lines	1.0
	Webers	1×10^{-8}
Meters	Angstrom units	1×10^{10}
	Centimeters	100
	Feet	3.2808
	Inches	39.370
	Miles (statute)	0.000621

To Convert from	To	Multiply by
Miles (statute)	Feet	5280
	Kilometers	1.609
	Meters	1609.344
Miles/hour	Kilometers/hour	1.609344
Newton–meters	Dyne–centimeters	1×10^7
	Kilogram-meters	0.10197
Oersteds	Ampere-turns/inch	2.0212
	Ampere-turns/meter	79.577
	Gilberts/centimeter	1.0
Quarts (U.S. liquid)	Cubic centimeters	946.353
	Cubic inches	57.75
	Gallons (U.S. liquid)	0.25
	Liters	0.9463
	Pints (U.S. liquid)	2
	Ounces (U.S. liquid)	32
Radians	Degrees	57.2958
Slugs	Kilograms	14.5939
	Pounds	32.1740
Watts	Btus/hour	3.4144
	Ergs/second	1×10^7
	Horsepower	0.00134
	Joules/second	1.0
Webers	Lines	1×10^8
	Maxwells	1×10^8
Years	Days	365
	Hours	8760
	Minutes	525,600
	Seconds	3.1536×10^7

Appendix B

PSPICE AND MULTISIM

PSPICE 16.2

The PSpice software package used throughout this text is derived from programs developed at the University of California at Berkeley during the early 1970s. SPICE is an acronym for Simulation Program with Integrated Circuit Emphasis. Although a number of companies have customized SPICE for their particular use, Cadence Design Systems offers both a commercial and a demo version of OrCAD. The commercial or professional versions that engineering companies use can be quite expensive, so Cadence offers a free distribution of the enclosed demo version to provide an introduction to the power of this simulation package. This text uses the OrCAD 16.2 Demo version. Additional information can be found for the Cadence OrCAD 16.2 Release under the **OrCAD Inc.** heading of the web site at www.orcad.com. Additional hard copies can be obtained under **Cadence OrCAD Downloads,** where a request can be submitted online at the bottom of the section title **OrCAD Demo DVD.**

System requirements include:

Windows XP, Windows Vista, and Window Server 2003
Pentium 4 (32-bit) equivalent of faster
Minimum 512 MB memory (1 G or more recommended for XP and Vista)
200 MB swap space (or more)
65,000 color Windows display, with minimum 1024 × 768 (1280 × 1024) recommended

MULTISIM 10.1

Multisim is a product of Electronics Workbench, a subsidiary of National Instruments. Its web site is www.electronicsworkbench.com, and its phone number in the U.S. and Canada is 1 800-263-5552. In Europe, the web site is www.ewbeurope.com, and the phone number is 31-35-694-4444.

System requirements include:

Windows 2000/XP/NT (NT 4.0 service pack or later)
Pentium 4 or equivalent processor (600 MHz minimum)
128 MB of RAM (256 MB recommended)
100 MB of free disk space (500 MB recommended)

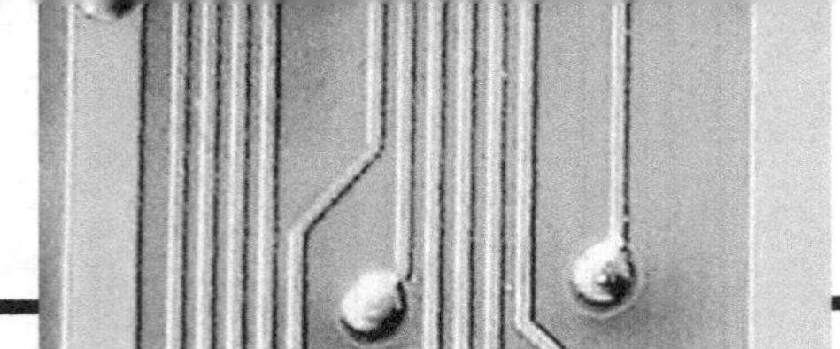

Appendix C

DETERMINANTS

Determinants are used to find the mathematical solutions for the variables in two or more simultaneous equations. Once the procedure is properly understood, solutions can be obtained with a minimum of time and effort and usually with fewer errors than when using other methods.

Consider the following equations, where x and y are the unknown variables and a_1, a_2, b_1, b_2, c_1, and c_2 are constants:

Col. 1		Col. 2		Col. 3	
a_1x	$+$	b_1y	$=$	c_1	**(C.1a)**
a_2x	$+$	b_2y	$=$	c_2	**(C.1b)**

It is certainly possible to solve for one variable in Eq. (C.1a) and substitute into Eq. (C.1b). That is, solving for x in Eq. (C.1a) gives

$$x = \frac{c_1 - b_1y}{a_1}$$

and substituting the result in Eq. (C.1b) gives

$$a_2\left(\frac{c_1 - b_1y}{a_1}\right) + b_2y = c_2$$

It is now possible to solve for y since it is the only variable remaining, and then substitute into either equation for x. This is acceptable for two equations, but it becomes a very tedious and lengthy process for three or more simultaneous equations.

Using determinants to solve for x and y requires that the following formats be established for each variable:

Col. 1 Col. 2 (for x); Col. 1 Col. 2 (for y)

$$x = \frac{\begin{vmatrix} c_1 & b_1 \\ c_2 & b_2 \end{vmatrix}}{\begin{vmatrix} a_1 & b_1 \\ a_2 & b_2 \end{vmatrix}} \qquad y = \frac{\begin{vmatrix} a_1 & c_1 \\ a_2 & c_2 \end{vmatrix}}{\begin{vmatrix} a_1 & b_1 \\ a_2 & b_2 \end{vmatrix}} \qquad \textbf{(C.2)}$$

First note that only constants appear within the vertical brackets, and that the denominator of each is the same. In fact, the denominator is simply the coefficients of x and y in the same arrangement as in Eqs. (C.1a) and (C.1b). When solving for x, replace the coefficients of x in the numerator by the constants to the right of the equal sign in Eqs. (C.1a) and (C.1b), and repeat the coefficients of the y variable. When solving for y, replace the y coefficients in the numerator by the constants to the right of the equal sign, and repeat the coefficients of x.

Each configuration in the numerator and denominator of Eq. (C.2) is referred to as a *determinant (D)*, which can be evaluated numerically in the following manner:

$$\text{Determinant} = D = \begin{vmatrix} \overset{\text{Col.}}{\underline{1}} & \overset{\text{Col.}}{\underline{2}} \\ a_1 & b_1 \\ a_2 & b_2 \end{vmatrix} = a_1b_2 - a_2b_1 \quad \text{(C.3)}$$

The expanded value is obtained by first multiplying the top left element by the bottom right and then subtracting the product of the lower left and upper right elements. This particular determinant is referred to as a *second-order* determinant since it contains two rows and two columns.

It is important to remember when using determinants that the columns of the equations, as indicated in Eqs. (C.1a) and (C.1b), must be placed in the same order within the determinant configuration. That is, since a_1 and a_2 are in column 1 of Eqs. (C.1a) and (C.1b), they must be in column 1 of the determinant. (The same is true for b_1 and b_2.)

Expanding the entire expression for x and y, we have the following:

$$x = \frac{\begin{vmatrix} c_1 & b_1 \\ c_2 & b_2 \end{vmatrix}}{\begin{vmatrix} a_1 & b_1 \\ a_2 & b_2 \end{vmatrix}} = \frac{c_1b_2 - c_2b_1}{a_1b_2 - a_2b_1} \quad \text{(C.4a)}$$

$$y = \frac{\begin{vmatrix} a_1 & c_1 \\ a_2 & c_2 \end{vmatrix}}{\begin{vmatrix} a_1 & b_1 \\ a_2 & b_2 \end{vmatrix}} = \frac{a_1c_2 - a_2c_1}{a_1b_2 - a_2b_1} \quad \text{(C.4b)}$$

EXAMPLE C.1 Evaluate the following determinants:

a. $\begin{vmatrix} 2 & 2 \\ 3 & 4 \end{vmatrix} = (2)(4) - (3)(2) = 8 - 6 = \mathbf{2}$

b. $\begin{vmatrix} 4 & -1 \\ 6 & 2 \end{vmatrix} = (4)(2) - (6)(-1) = 8 + 6 = \mathbf{14}$

c. $\begin{vmatrix} 0 & -2 \\ -2 & 4 \end{vmatrix} = (0)(4) - (-2)(-2) = 0 - 4 = \mathbf{-4}$

d. $\begin{vmatrix} 0 & 0 \\ 3 & 10 \end{vmatrix} = (0)(10) - (3)(0) = \mathbf{0}$

EXAMPLE C.2 Solve for x and y:

$$\begin{aligned} 2x + y &= 3 \\ 3x + 4y &= 2 \end{aligned}$$

Solution:

$$x = \frac{\begin{vmatrix} 3 & 1 \\ 2 & 4 \end{vmatrix}}{\begin{vmatrix} 2 & 1 \\ 3 & 4 \end{vmatrix}} = \frac{(3)(4) - (2)(1)}{(2)(4) - (3)(1)} = \frac{12 - 2}{8 - 3} = \frac{10}{5} = \mathbf{2}$$

$$y = \frac{\begin{vmatrix} 2 & 3 \\ 3 & 2 \end{vmatrix}}{5} = \frac{(2)(2) - (3)(3)}{5} = \frac{4 - 9}{5} = \frac{-5}{5} = \mathbf{-1}$$

Check:

$$\begin{aligned} 2x + y &= (2)(2) + (-1) \\ &= 4 - 1 = 3 \quad \text{(checks)} \\ 3x + 4y &= (3)(2) + (4)(-1) \\ &= 6 - 4 = 2 \quad \text{(checks)} \end{aligned}$$

EXAMPLE C.3 Solve for x and y:

$$\begin{aligned} -x + 2y &= 3 \\ 3x - 2y &= -2 \end{aligned}$$

Solution: In this example, note the effect of the minus sign and the use of parentheses to ensure that the proper sign is obtained for each product:

$$x = \frac{\begin{vmatrix} 3 & 2 \\ -2 & -2 \end{vmatrix}}{\begin{vmatrix} -1 & 2 \\ 3 & -2 \end{vmatrix}} = \frac{(3)(-2) - (-2)(2)}{(-1)(-2) - (3)(2)}$$

$$= \frac{-6 + 4}{2 - 6} = \frac{-2}{-4} = \mathbf{\frac{1}{2}}$$

$$y = \frac{\begin{vmatrix} -1 & 3 \\ 3 & -2 \end{vmatrix}}{-4} = \frac{(-1)(-2) - (3)(3)}{-4}$$

$$= \frac{2 - 9}{-4} = \frac{-7}{-4} = \mathbf{\frac{7}{4}}$$

EXAMPLE C.4 Solve for x and y:

$$\begin{aligned} x &= 3 - 4y \\ 20y &= -1 + 3x \end{aligned}$$

Solution: In this case, the equations must first be placed in the format of Eqs. (C.1a) and (C.1b):

$$\begin{aligned} x + 4y &= 3 \\ -3x + 20y &= -1 \end{aligned}$$

$$x = \frac{\begin{vmatrix} 3 & 4 \\ -1 & 20 \end{vmatrix}}{\begin{vmatrix} 1 & 4 \\ -3 & 20 \end{vmatrix}} = \frac{(3)(20) - (-1)(4)}{(1)(20) - (-3)(4)}$$

$$= \frac{60 + 4}{20 + 12} = \frac{64}{32} = \mathbf{2}$$

$$y = \frac{\begin{vmatrix} 1 & 3 \\ -3 & -1 \end{vmatrix}}{32} = \frac{(1)(-1) - (-3)(3)}{32}$$

$$= \frac{-1 + 9}{32} = \frac{8}{32} = \mathbf{\frac{1}{4}}$$

The use of determinants is not limited to the solution of two simultaneous equations; determinants can be applied to any number of simultaneous linear equations. First we examine a shorthand method that is applicable to third-order determinants only since most of the problems in the text are limited to this level of difficulty. We then investigate the general procedure for solving any number of simultaneous equations.

Consider the three following simultaneous equations

Col. 1		Col. 2		Col. 3		Col. 4
a_1x	$+$	b_1y	$+$	c_1z	$=$	d_1
a_2x	$+$	b_2y	$+$	c_2z	$=$	d_2
a_3x	$+$	b_3y	$+$	c_3z	$=$	d_3

in which x, y, and z are the variables, and $a_{1,2,3}$, $b_{1,2,3}$, $c_{1,2,3}$, and $d_{1,2,3}$ are constants.

The determinant configuration for x, y, and z can be found in a manner similar to that for two simultaneous equations. That is, to solve for x, find the determinant in the numerator by replacing column 1 with the elements to the right of the equal sign. The denominator is the determinant of the coefficients of the variables (the same applies to y and z). Again, the denominator is the same for each variable. We have

$$x = \frac{\begin{vmatrix} d_1 & b_1 & c_1 \\ d_2 & b_2 & c_2 \\ d_3 & b_3 & c_3 \end{vmatrix}}{D}, \quad y = \frac{\begin{vmatrix} a_1 & d_1 & c_1 \\ a_2 & d_2 & c_2 \\ a_3 & d_3 & c_3 \end{vmatrix}}{D}, \quad z = \frac{\begin{vmatrix} a_1 & b_1 & d_1 \\ a_2 & b_2 & d_2 \\ a_3 & b_3 & d_3 \end{vmatrix}}{D}$$

where

$$D = \begin{vmatrix} a_1 & b_1 & c_1 \\ a_2 & b_2 & c_2 \\ a_3 & b_3 & c_3 \end{vmatrix}$$

A shorthand method for evaluating the third-order determinant consists of repeating the first two columns of the determinant to the right of the determinant and then summing the products along specific diagonals as follows:

$$D = \left|\begin{matrix} a_1 & b_1 & c_1 \\ a_2 & b_2 & c_2 \\ a_3 & b_3 & c_3 \end{matrix}\right| \begin{matrix} a_1 & b_1 \\ a_2 & b_2 \\ a_3 & b_3 \end{matrix}$$

(Diagonals 4(−), 5(−), 6(−) run upward to the right; diagonals 1(+), 2(+), 3(+) run downward to the right.)

The products of the diagonals 1, 2, and 3 are positive and have the following magnitudes:

$$+a_1b_2c_3 + b_1c_2a_3 + c_1a_2b_3$$

The products of the diagonals 4, 5, and 6 are negative and have the following magnitudes:

$$-a_3b_2c_1 - b_3c_2a_1 - c_3a_2b_1$$

The total solution is the sum of the diagonals 1, 2, and 3 minus the sum of the diagonals 4, 5, and 6:

$$+(a_1b_2c_3 + b_1c_2a_3 + c_1a_2b_3) - (a_3b_2c_1 + b_3c_2c_1 + c_3a_2b_1) \quad \textbf{(C.5)}$$

Warning: This method of expansion is good only for third-order determinants! It cannot be applied to fourth- and higher-order systems.

EXAMPLE C.5 Evaluate the following determinant:

$$\begin{vmatrix} 1 & 2 & 3 \\ -2 & 1 & 0 \\ 0 & 4 & 2 \end{vmatrix} \longrightarrow \begin{vmatrix} 1 & 2 & 3 \\ -2 & 1 & 0 \\ 0 & 4 & 2 \end{vmatrix} \begin{matrix} 1 & 2 \\ -2 & 1 \\ 0 & 4 \end{matrix}$$

(with diagonals marked (−) (−) (−) above and (+) (+) (+) below)

Solution:

$$[(1)(1)(2) + (2)(0)(0) + (3)(-2)(4)] - [(0)(1)(3) + (4)(0)(1) + (2)(-2)(2)]$$
$$= (2 + 0 - 24) - (0 + 0 - 8) = (-22) - (-8)$$
$$= -22 + 8 = \mathbf{-14}$$

EXAMPLE C.6 Solve for x, y, and z:

$$1x + 0y - 2z = -1$$
$$0x + 3y + 1z = +2$$
$$1x + 2y + 3z = 0$$

Solution:

$$x = \frac{\begin{vmatrix} -1 & 0 & -2 \\ 2 & 3 & 1 \\ 0 & 2 & 3 \end{vmatrix} \begin{matrix} -1 & 0 \\ 2 & 3 \\ 0 & 2 \end{matrix}}{\begin{vmatrix} 1 & 0 & -2 \\ 0 & 3 & 1 \\ 1 & 2 & 3 \end{vmatrix} \begin{matrix} 1 & 0 \\ 0 & 3 \\ 1 & 2 \end{matrix}}$$

$$= \frac{[(-1)(3)(3) + (0)(1)(0) + (-2)(2)(2)] - [(0)(3)(-2) + (2)(1)(-1) + (3)(2)(0)]}{[(1)(3)(3) + (0)(1)(1) + (-2)(0)(2)] - [(1)(3)(-2) + (2)(1)(1) + (3)(0)(0)]}$$

$$= \frac{(-9 + 0 - 8) - (0 - 2 + 0)}{(9 + 0 + 0) - (-6 + 2 + 0)}$$

$$= \frac{-17 + 2}{9 + 4} = -\mathbf{\frac{15}{13}}$$

$$y = \frac{\begin{vmatrix} 1 & -1 & -2 \\ 0 & 2 & 1 \\ 1 & 0 & 3 \end{vmatrix} \begin{matrix} 1 & -1 \\ 0 & 2 \\ 1 & 0 \end{matrix}}{13}$$

$$= \frac{[(1)(2)(3) + (-1)(1)(1) + (-2)(0)(0)] - [(1)(2)(-2) + (0)(1)(1) + (3)(0)(-1)]}{13}$$

$$= \frac{(6 - 1 + 0) - (-4 + 0 + 0)}{13}$$

$$= \frac{5 + 4}{13} = \mathbf{\frac{9}{13}}$$

$$z = \frac{\begin{vmatrix} 1 & 0 & -1 \\ 0 & 3 & 2 \\ 1 & 2 & 0 \end{vmatrix} \begin{matrix} 1 & 0 \\ 0 & 3 \\ 1 & 2 \end{matrix}}{13}$$

$$= \frac{[(1)(3)(0) + (0)(2)(1) + (-1)(0)(2)] - [(1)(3)(-1) + (2)(2)(1) + (0)(0)(0)]}{13}$$

$$= \frac{(0 + 0 + 0) - (-3 + 4 + 0)}{13}$$

$$= \frac{0 - 1}{13} = -\frac{\mathbf{1}}{\mathbf{13}}$$

or from $0x + 3y + 1z = +2$,

$$z = 2 - 3y = 2 - 3\left(\frac{9}{13}\right) = \frac{26}{13} - \frac{27}{13} = -\frac{\mathbf{1}}{\mathbf{13}}$$

Check:

$$\left.\begin{aligned} 1x + 0y - 2z &= -1 \\ 0x + 3y + 1z &= +2 \\ 1x + 2y + 3z &= 0 \end{aligned}\right\} \begin{aligned} -\frac{15}{13} + 0 + \frac{2}{13} &= -1 \\ 0 + \frac{27}{13} + \frac{-1}{13} &= +2 \\ -\frac{15}{13} + \frac{18}{13} + \frac{-3}{13} &= 0 \end{aligned} \left.\vphantom{\begin{aligned} a \\ b \\ c \end{aligned}}\right\} \begin{aligned} -\frac{13}{13} &= -1 \checkmark \\ \frac{26}{13} &= +2 \checkmark \\ -\frac{18}{13} + \frac{18}{13} &= 0 \checkmark \end{aligned}$$

The general approach to third-order or higher determinants requires that the determinant be expanded in the following form. There is more than one expansion that will generate the correct result, but this form is typically used when the material is first introduced.

$$D = \begin{vmatrix} a_1 & b_1 & c_1 \\ a_2 & b_2 & c_2 \\ a_3 & b_3 & c_3 \end{vmatrix} = \underset{\text{Multiplying factor}}{a_1} \underbrace{\left(+ \underbrace{\begin{vmatrix} b_2 & c_2 \\ b_3 & c_3 \end{vmatrix}}_{\text{Minor}} \right)}_{\text{Cofactor}} + \underset{\text{Multiplying factor}}{b_1} \underbrace{\left(- \underbrace{\begin{vmatrix} a_2 & c_2 \\ a_3 & c_3 \end{vmatrix}}_{\text{Minor}} \right)}_{\text{Cofactor}} + \underset{\text{Multiplying factor}}{c_1} \underbrace{\left(+ \underbrace{\begin{vmatrix} a_2 & b_2 \\ a_3 & b_3 \end{vmatrix}}_{\text{Minor}} \right)}_{\text{Cofactor}}$$

This expansion was obtained by multiplying the elements of the first row of D by their corresponding cofactors. It is not a requirement that the first row be used as the multiplying factors. In fact, any *row* or *column* (not diagonals) may be used to expand a third-order determinant.

The sign of each cofactor is dictated by the position of the multiplying factors (a_1, b_1, and c_1 in this case) as in the following standard format:

$$\begin{vmatrix} + \rightarrow & - & + \\ \downarrow & & \\ - & + & - \\ & & \\ + & - & + \end{vmatrix}$$

Note that the proper sign for each element can be obtained by assigning the upper left element a positive sign and then changing signs as you move horizontally or vertically to the neighboring position.

For the determinant D, the elements would have the following signs:

$$\begin{vmatrix} a_1^{(+)} & b_1^{(-)} & c_1^{(+)} \\ a_2^{(-)} & b_2^{(+)} & c_2^{(-)} \\ a_3^{(+)} & b_3^{(-)} & c_3^{(+)} \end{vmatrix}$$

The minors associated with each multiplying factor are obtained by covering up the row and column in which the multiplying factor is located and writing a second-order determinant to include the remaining elements in the same relative positions that they have in the third-order determinant.

Consider the cofactors associated with a_1 and b_1 in the expansion of D. The sign is positive for a_1 and negative for b_1 as determined by the standard format. Following the procedure outlined above, we can find the minors of a_1 and b_1 as follows:

$$a_{1(\text{minor})} = \begin{vmatrix} \not{a}_1 & \not{b}_1 & \not{c}_1 \\ \not{a}_2 & b_2 & c_2 \\ \not{a}_3 & b_3 & c_3 \end{vmatrix} = \begin{vmatrix} b_2 & c_2 \\ b_3 & c_3 \end{vmatrix}$$

$$b_{1(\text{minor})} = \begin{vmatrix} \not{a}_1 & \not{b}_1 & \not{c}_1 \\ a_2 & \not{b}_2 & c_2 \\ a_3 & \not{b}_3 & c_3 \end{vmatrix} = \begin{vmatrix} a_2 & c_2 \\ a_3 & c_3 \end{vmatrix}$$

It was pointed out that any row or column may be used to expand the third-order determinant, and the same result will still be obtained. Using the first column of D, we obtain the expansion

$$D = \begin{vmatrix} a_1 & b_1 & c_1 \\ a_2 & b_2 & c_2 \\ a_3 & b_3 & c_3 \end{vmatrix} = a_1\left(+\begin{vmatrix} b_2 & c_2 \\ b_3 & c_3 \end{vmatrix}\right) + a_2\left(-\begin{vmatrix} b_1 & c_1 \\ b_3 & c_3 \end{vmatrix}\right) + a_3\left(+\begin{vmatrix} b_1 & c_1 \\ b_2 & c_2 \end{vmatrix}\right)$$

The proper choice of row or column can often effectively reduce the amount of work required to expand the third-order determinant. For example, in the following determinants, the first column and third row, respectively, would reduce the number of cofactors in the expansion:

$$D = \begin{vmatrix} 2 & 3 & -2 \\ 0 & 4 & 5 \\ 0 & 6 & 7 \end{vmatrix} = 2\left(+\begin{vmatrix} 4 & 5 \\ 6 & 7 \end{vmatrix}\right) + 0 + 0 = 2(28 - 30)$$
$$= \mathbf{-4}$$

$$D = \begin{vmatrix} 1 & 4 & 7 \\ 2 & 6 & 8 \\ 2 & 0 & 3 \end{vmatrix} = 2\left(+\begin{vmatrix} 4 & 7 \\ 6 & 8 \end{vmatrix}\right) + 0 + 3\left(+\begin{vmatrix} 1 & 4 \\ 2 & 6 \end{vmatrix}\right)$$
$$= 2(32 - 42) + 3(6 - 8) = 2(-10) + 3(-2)$$
$$= \mathbf{-26}$$

EXAMPLE C.7 Expand the following third-order determinants:

a. $D = \begin{vmatrix} 1 & 2 & 3 \\ 3 & 2 & 1 \\ 2 & 1 & 3 \end{vmatrix} = 1\left(+\begin{vmatrix} 2 & 1 \\ 1 & 3 \end{vmatrix}\right) + 3\left(-\begin{vmatrix} 2 & 3 \\ 1 & 3 \end{vmatrix}\right) + 2\left(+\begin{vmatrix} 2 & 3 \\ 2 & 1 \end{vmatrix}\right)$

$= 1[6 - 1] + 3[-(6 - 3)] + 2[2 - 6]$

$= 5 + 3(-3) + 2(-4)$

$= 5 - 9 - 8$

$= \mathbf{-12}$

b. $D = \begin{vmatrix} 0 & 4 & 6 \\ 2 & 0 & 5 \\ 8 & 4 & 0 \end{vmatrix} = 0 + 2\left(-\begin{vmatrix} 4 & 6 \\ 4 & 0 \end{vmatrix}\right) + 8\left(+\begin{vmatrix} 4 & 6 \\ 0 & 5 \end{vmatrix}\right)$

$= 0 + 2[-(0 - 24)] + 8[(20 - 0)]$

$= 0 + 2(24) + 8(20)$

$= 48 + 160$

$= \mathbf{208}$

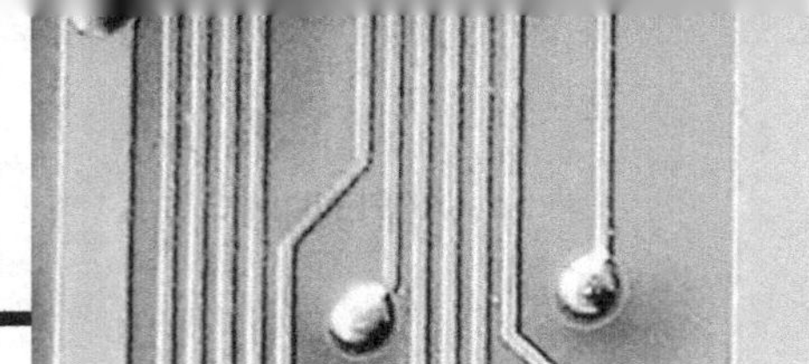

Appendix D

GREEK ALPHABET

Letter	Capital	Lowercase	Some Applications
Alpha	A	α	Area, angles, coefficients
Beta	B	β	Angles, coefficients, flux density
Gamma	Γ	γ	Specific gravity, conductivity
Delta	Δ	δ	Density, variation
Epsilon	E	ε	Base of natural logarithms
Zeta	Z	ζ	Coefficients, coordinates, impedance
Eta	H	η	Efficiency, hysteresis coefficient
Theta	Θ	θ	Phase angle, temperature
Iota	I	ι	
Kappa	K	κ	Dielectric constant, susceptibility
Lambda	Λ	λ	Wavelength
Mu	M	μ	Amplification factor, micro, permeability
Nu	N	ν	Reluctivity
Xi	Ξ	ξ	
Omicron	O	o	
Pi	Π	π	3.1416
Rho	P	ρ	Resistivity
Sigma	Σ	σ	Summation
Tau	T	τ	Time constant
Upsilon	Y	υ	
Phi	Φ	ϕ	Angles, magnetic flux
Chi	X	χ	
Psi	Ψ	ψ	Dielectric flux, phase difference
Omega	Ω	ω	Ohms, angular velocity

Appendix E

MAGNETIC PARAMETER CONVERSIONS

	SI (MKS)	CGS	English
Φ	webers (Wb) 1 Wb	maxwells $= 10^8$ maxwells	lines $= 10^8$ lines
B	T $1\text{T} = 1\ \text{Wb/m}^2$	gauss (maxwells/cm^2) $= 10^4$ gauss	lines/in.^{2} $= 6.452 \times 10^4$ lines/in.^{2}
A	$1\ \text{m}^2$	$= 10^4\ \text{cm}^2$	$= 1550$ in.^{2}
μ_o	$4\pi \times 10^{-7}$ Wb/Am	$= 1$ gauss/oersted	$= 3.20$ lines/Am
$\mathcal{F}$	NI (ampere-turns, At) 1 At	$0.4\pi NI$ (gilberts) $= 1.257$ gilberts	NI (At) 1 gilbert − 0.7958 At
H	NI/l (At/m) 1At/m	$0.4\pi NI/l$ (oersteds) $= 1.26 \times 10^{-2}$ oersted	NI/l(At/in.) $= 2.54 \times 10^{-2}$ At/in.
H_g	$7.97 \times 10^5 B_g$ (At/m)	B_g (oersteds)	$0.313B_g$ (At/in.)

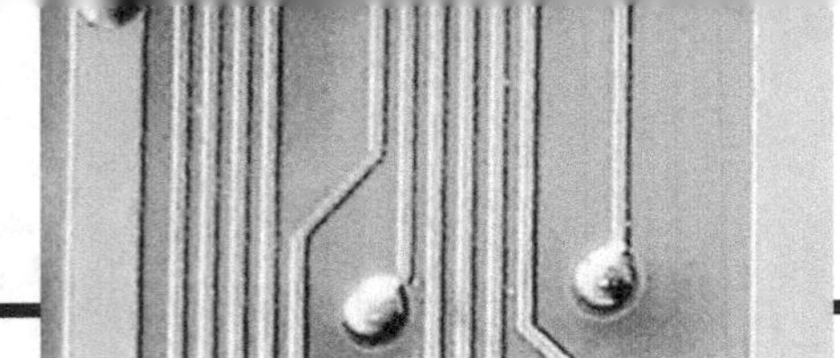

Appendix F

MAXIMUM POWER TRANSFER CONDITIONS

This appendix derives the maximum power transfer conditions for the situation where the resistive component of the load is adjustable but the load reactance is set in magnitude.*

For the circuit in Fig. F.1, the power delivered to the load is determined by

$$P = \frac{V_{R_L}^2}{R_L}$$

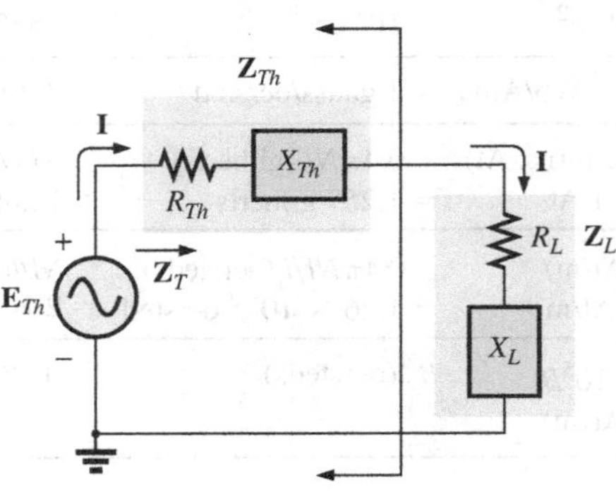

FIG. F.1

Applying the voltage divider rule gives

$$\mathbf{V}_{R_L} = \frac{R_L \mathbf{E}_{Th}}{R_L + R_{Th} + X_{Th} \angle 90° + X_L \angle 90°}$$

The magnitude of $\mathbf{V}_{R_L}$ is determined by

$$V_{R_L} = \frac{R_L E_{Th}}{\sqrt{(R_L + R_{Th})^2 + (X_{Th} + X_L)^2}}$$

and
$$V_{R_L}^2 = \frac{R_L^2 E_{Th}^2}{(R_L + R_{Th})^2 + (X_{Th} + X_L)^2}$$

with
$$P = \frac{V_{R_L}^2}{R_L} = \frac{R_L E_{Th}^2}{(R_L + R_{Th})^2 + (X_{Th} + X_L)^2}$$

Using differentiation (calculus), we find that maximum power will be transferred when $dP/dR_L = 0$. The result of the preceding operation is that

$$R_L = \sqrt{R_{Th}^2 + (X_{Th} + X_L)^2} \qquad \text{[Eq. (18.21)]}$$

The magnitude of the total impedance of the circuit is

$$Z_T = \sqrt{(R_{Th} + R_L)^2 + (X_{Th} + X_L)^2}$$

*With sincerest thanks for the input of Professor Harry J. Franz of the Beaver Campus of Pennsylvania State University.

Substituting this equation for R_L and applying a few algebraic maneuvers results in

$$Z_T = 2R_L(R_L + R_{Th})$$

and the power to the load R_L as

$$P = I^2R_L = \frac{E_{Th}^2}{Z_T^2}R_L = \frac{E_{Th}^2R_L}{2R_L(R_L + R_{Th})}$$

$$= \frac{E_{Th}^2}{4\left(\dfrac{R_L + R_{Th}}{2}\right)}$$

$$= \frac{E_{Th}^2}{4R_{av}}$$

with $R_{av} = \dfrac{R_L + R_{Th}}{2}$

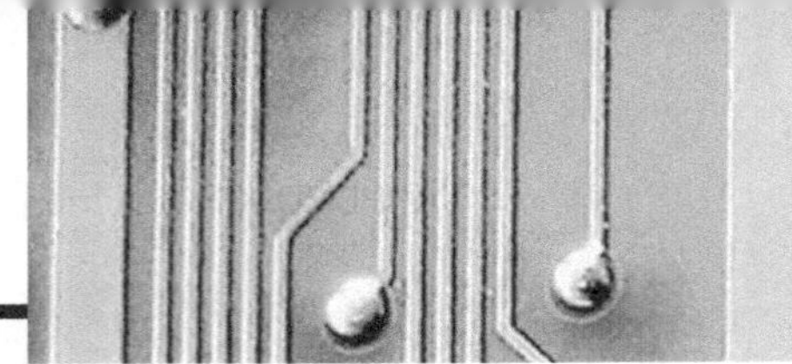

Appendix G

ANSWERS TO SELECTED ODD-NUMBERED PROBLEMS

Chapter 1

5. 29.05 mph
7. **(a)** 139.33 ft/s **(b)** 0.431 s **(c)** 40.91 mph
11. MKS, CGS = 20°C; SI, K = 293.15
13. 45.72 cm
15. **(a)** 14.6 **(b)** 56.0 **(c)** 1046.1 **(d)** 0.1 **(e)** 3.1
17. **(a)** 14.603 **(b)** 56.042 **(c)** 1046.060 **(d)** 0.063 **(e)** 3.142
19. **(a)** 15×10^3 **(b)** 5×10^{-3} **(c)** 2.4×10^6 **(d)** 60×10^3 **(e)** 4.02×10^{-4} **(f)** 2×10^{-10}
21. **(a)** 100×10^3 **(b)** 10 **(c)** 1×10^9 **(d)** 1×10^{-3} **(e)** 10 **(f)** 1×10^{24}
23. **(a)** 10×10^{-3} **(b)** 10×10^{-6} **(c)** 10×10^6 **(d)** 1×10^{-9} **(e)** 1×10^{42} **(f)** 1×10^3
25. **(a)** 1×10^6 **(b)** 10×10^{-3} **(c)** 100×10^{30} **(d)** 1×10^{-63}
27. **(a)** 1×10^{-6} **(b)** 1×10^{-5} **(c)** 1×10^{-8} **(d)** 1×10^{11}
29. Scientific: **(a)** 2.05×10^1 **(b)** 5.04×10^4 **(c)** 6.74×10^{-4} **(d)** 4.60×10^{-2}
Engineering: **(a)** 20.46×10^0 **(b)** 50.42×10^3 **(c)** 674.00×10^{-6} **(d)** 46.00×10^{-3}
31. **(a)** 0.06×10^6 **(b)** 400×10^{-6} **(c)** 0.005×10^9 **(d)** 1200×10^{-9}
33. **(a)** 90 s **(b)** 72 s **(c)** 50×10^3 μs **(d)** 160 mm **(e)** 120 ns **(f)** 4629.63 days
35. **(a)** 2.54 m **(b)** 1.22 m **(c)** 26.7 N **(d)** 0.13 lb **(e)** 4921.26 ft **(f)** 3.22 m
37. 26.82 m/s
39. 3600 quarters
41. 345.6 m
43. 44.82 min/mile
45. **(a)** 4.74×10^{-3} Btu **(b)** 7.1×10^{-4} m^3 **(c)** 1.21×10^5 s **(d)** 2113.38 pints
47. 14.4
49. 0.93
51. 3.24
53. 1.20×10^{12}

Chapter 2

3. **(a)** 1.11 μN **(b)** 0.31 N **(c)** 1138.34 kN
5. $F_2 = r_1^2 F_1 / r_2^2$
7. **(a)** 72 mN **(b)** $Q_1 = 20\ \mu C$, $Q_2 = 40\ \mu C$
9. 0.48 J
11. 8 C
13. 4.29 mA
15. 192 C
17. 3 s
19. 2.25×10^{18} electrons
21. 22.43 mA
23. 6.67 V
25. 3.34 A
27. 60.0 Ah
29. W(60 Ah) : W(40 Ah) = 1.5 : 1; I(60 Ah) : I(40 Ah) = 3600 A : 2400 A = 1.5 : 1
31. 13.89%
33. 129.6 kJ
37. **(a)** 38.1 kV **(b)** 342.9 kV

Chapter 3

1. **(a)** 500 mils **(b)** 20 mils **(c)** 250 mils **(d)** 393.7 mils **(e)** 120 mils **(f)** 39.37 mils
3. **(a)** 0.04 in. **(b)** 0.029 in. **(c)** 0.2 in. **(d)** 0.025 in. **(e)** 0.0025 in. **(f)** 0.55 in.
5. **(a)** 544 CM **(b)** 0.023 in.
7. **(a)** 942.73 CM **(b)** larger **(c)** smaller
9. **(a)** 293.82 ft **(b)** 1.47 lb **(c)** −40°F → +221°F
11. **(a)** 21.71 μΩ **(b)** 35.59 μΩ
13. 942.28 mΩ
15. **(a)** yes **(b)** A(#0) : A(#12) = 16.16 : 1, I(#0) : I(#12) = 7.5 : 1
17. **(a)** #2 **(b)** #0
19. 2.57 Ω
21. 3.69 Ω
23. **(a)** 27.85°C **(b)** −210.65°C
25. **(a)** 0.00393 **(b)** 83.61°C
27. 1.75°
29. 100.30 Ω
31. 6.5 kΩ
35. **(a)** blue, gray, black, silver **(b)** orange, orange, silver, silver **(c)** red, red, orange, silver **(d)** green, blue, green, silver
37. yes, 423 Ω to 517 Ω
39. **(a)** 0.62 kΩ **(b)** 33 kΩ **(c)** 390 Ω **(d)** 1.2 MΩ
41. **(a)** 629.72 mS **(b)** 384.11 mS
43. 500 S
49. **(a)** 21.71 μΩ **(b)** 35.59 μΩ
51. 0.15 in.
59. **(a)** 10 fc = 3 kΩ, 100 fc = 0.4 kΩ **(b)** negative **(c)** no **(d)** −321.43 Ω/fc

Chapter 4

1. 1.23 V
3. 16 kΩ
5. 72 mV
7. 54.55 Ω
9. 28.57 Ω
11. 1.2 kΩ
13. **(a)** 12.63 Ω **(b)** 8.21×10^6 J
21. 16 s
23. 2.86 s
25. 207.36 mW
27. 129.10 mA, 15.49 V
29. 120 V
31. 9.61 V
33. 32 Ω, 120 V
35. 70.71 mA, 1.42 kV
37. **(a)** 86.4 J **(b)** energy doubles, power the same
39. 59.80 kWh
41. **(a)** 120 kW **(b)** 576.92 A **(c)** 216 kWh
43. **(a)** \$2.39/day **(b)** 16 ¢/hour **(c)** 1.45 kWh **(d)** ≅ 24 **(e)** no
45. \$25.29
47. **(a)** 12 kW **(b)** 10,130 W < 12 kW (yes) **(c)** 20.26 kWh
49. 52.29 ¢
51. 84.77%
53. **(a)** 238 W **(b)** 17.36%
55. **(a)** 1657.78 W **(b)** 15.07 A **(c)** 19.38 A
57. 88%
59. 80%
61. $\eta_1 = 40\%$, $\eta_2 = 80\%$

Chapter 5

1. **(a)** E and R_1
(b) R_1 and R_2
(c) E_1, E_2 and R_1
(d) E_1 and R_1; E_2, R_3, and R_4
(e) R_3, R_4, and R_5; E and R_1
(f) R_2 and R_3

3. **(a)** 7.7 kΩ **(b)** 17.5 kΩ

5. **(a)** 99 Ω **(b)** 7.52 kΩ

7. **(a)** 1.2 kΩ **(b)** 0 Ω **(c)** ∞ Ω

9. **(a)** the most: R_3; the least: R_1
(b) R_3, $R_T = 90$ kΩ, $I_s = 0.5$ mA
(c) $V_1 = 0.6$ V, $V_2 = 3.4$ V, $V_3 = 41$ V

11. **(a)** 4 A **(b)** 36 V **(c)** 3 Ω
(d) $V_{4.7\,\Omega} = 18.8$ V, $V_{1.3\,\Omega} =$ 5.2 V, $V_{3\,\Omega} = 12$ V

13. **(a)** $R_T = 6$ kΩ, $I = 20$ mA, $V_{R_1} = 60$ V, $V_{R_2} = 20$ V, $V_{R_3} = 40$ V
(b) $P_{R_1} = 1.2$ W, $P_{R_2} = 0.4$ W, $P_{R_3} = 0.8$ W **(c)** 2.4 W
(d) 2.4 W **(e)** equal **(f)** R_1
(g) dissipated **(h)** R_1 : 2 W; $R_2 : \frac{1}{2}$ W; $R_3 = 1$ W

15. $I = 2.828$ A, $E = 90.5$ V, $R_1 = 2$ Ω, $R_2 = 29$ Ω

17. 6 Ω

19. **(a)** I(cw) = 1.17 A
(b) I(cw) = 173.91 mA

21. **(a)** $V = 2$ V **(b)** $V = 42$ V
(c) $V_1 = 8$ V, $V_2 = -4$ V

23. **(a)** $V_1 = 4$ V, $V_2 = 10$ V
(b) $V_1 = 14$ V, $V_2 = 18$ V

25. $R_2 = 100$ Ω, $R_3 = 200$ Ω

27. **(a)** 20 V **(b)** 20 V **(c)** 0.36 V

29. $V_2 = 4$ V, $V_4 = 6$ V, $I = 2$ mA, $E = 24$ V

31. **(a)** 80 Ω in series with bulb
(b) $\frac{1}{4}$ W resistor

33. $V_{R_1} = 12$ V, $V_{R_2} = 42$ V, $V_{R_3} = 6$ V

35. **(a)** $V_a = 17$ V, $V_b = 21$ V, $V_{ab} = -4$ V **(b)** $V_a = 14$ V, $V_b = 30$ V, $V_{ab} = -16$ V
(c) $V_a = 13$ V, $V_b = -8$ V, $V_{ab} = 21$ V

37. **(a)** $V_a = 20$ V, $V_b = 26$ V, $V_c = 35$ V, $V_d = -12$ V, $V_e = 0$ V **(b)** $V_{ab} = -6$ V, $V_{dc} = -47$ V, $V_{cb} = 9$ V
(c) $V_{ac} = -15$ V, $V_{db} = -38$ V

39. $R_1 = 2$ kΩ, $R_2 = 2.25$ kΩ, $R_3 = 0.75$ kΩ, $R_4 = 1.25$ kΩ

41. $V_0 = 0$ V, $V_4 = 15$ V, $V_7 = 4$ V, $V_{10} = 12$ V, $V_{23} = 12$ V, $V_{30} = -8$ V, $V_{67} = 0$ V, $V_{56} = -1$ V, $I = 3$ A↑

43. **(a)** 2 Ω **(b)** 7.14%

45. **(a)** 1.2 mA **(b)** 1.17 mA
(c) no

Chapter 6

1. **(a)** R_2 and R_3 **(b)** E and R_3
(c) R_2 and R_3 **(d)** R_2 and R_3
(e) E, R_1, R_2, R_3, and R_4 **(f)** E, R_1, R_2, R_3 **(g)** E_2, R_2, and R_3

3. **(a)** 12 Ω **(b)** 0.652 kΩ
(c) 10.81 Ω **(d)** 3 kΩ
(e) 2.62 Ω **(f)** 0.999 Ω

5. **(a)** 8 Ω **(b)** 13 kΩ **(c)** 6.8 kΩ
(d) 2.4 kΩ **(e)** 24 kΩ

7. **(a)** 1.18 Ω **(b)** ∞ Ω **(c)** 2 Ω

9. **(a)** 6 Ω **(b)** 36 V
(c) $I_s = 6$ A, $I_1 = 4.5$ A, $I_2 = 1.5$ A

11. **(a)** $I_1 = 2.4$ mA, $I_2 = 20$ mA, $I_3 = 3.53$ mA **(b)** 925.93 Ω
(c) 25.92 mA **(d)** 25.93 mA

13. **(a)** 9 Ω **(b)** 27 V

15. $E = 36$ V, $R_1 = 24$ Ω, $I_3 = 9$ A

17. **(a)** 4 Ω **(b)** 12 Ω **(c)** 10 A

19. **(a)** 761.79 Ω, $I_1 = 60$ mA, $I_2 = 12.77$ mA, $I_3 = 6$ mA
(b) $P_1 = 3.6$ W. $P_2 = 0.766$ W, $P_3 = 0.36$ W **(c)** 4.73 W
(d) 4.73 W **(e)** R_1—smallest

21. 1.56 kW

23. **(a)** 14.67 A **(b)** 256 W
(c) 14.67 A

25. **(a)** $I = 8$ A **(b)** $I_1 = 6$ mA, $I_2 = 15$ mA, $I_3 = 5$ mA

27. $R_1 = 3$ kΩ, $R_3 = 6$ kΩ, $R_T = 1.33$ kΩ, $E = 12$ V

29. **(a)** $I_1 = 64$ mA, $I_2 = 20$ mA, $I_3 = 16$ mA, $R = 3.2$ kΩ
(b) $E = 30$ V, $I_1 = 1$ A, $I_3 = 0.5$ A, $R_2 = R_3 = 60$ Ω, $P_{R_2} = 15$ W

31. **(a)** $I_1 = 3$ A, $I_2 = 4$ A
(b) $I_T = 8.5$ A, $I_1 = 6$ A

33. **(a)** 9 A **(b)** 0.9 A **(c)** 9 mA
(d) 90 μA **(e)** very little
(f) 9.1 A **(g)** 0.91 A
(h) 9.1 mA **(i)** 91 μA

35. **(a)** 6 kΩ **(b)** $I_1 = 24$ mA, $I_2 = 8$ mA

37. **(a)** $I_1 = I_2 = 3$ A, $I_L = 6$ A
(b) 36 W **(c)** 72 W **(d)** 6 A

39. $I = 3$ A, $R = 2$ Ω

41. **(a)** 6.13 V **(b)** 9 V **(c)** 9 V

43. **(a)** 16.48 V **(b)** 16.47 V
(c) 16.32 V **(d)**: (a) 13.33 V, (b) 13.25 V, (c) 11.43 V

45. 6 kΩ not connected

Chapter 7

1. **(a)** R_1, R_2, and E in series; R_3, R_4, and R_5 in parallel
(b) E and R_1 in series; R_2, R_3, and R_4 in parallel
(c) E and R_1 in series; R_2, R_3, and R_4 in parallel
(d) E_1 and R_1 in series; E_2 and R_4 in parallel
(e) E and R_1 in series; R_2 and R_3 in parallel
(f) E, R_1, R_4, and R_6 in parallel; R_2 and R_5 in parallel

3. 3.6 kΩ

5. 12 kΩ

7. **(a)** 4 Ω **(b)** $I_s = 9$ A, $I_1 = 6$ A, $I_2 = 3$ A **(c)** 6 V

9. **(a)** $V_a = 36$ V, $V_b = 60$ V, $V_c = 20$ V **(b)** $I_1 = 24$ mA, $I_2 = 35.5$ mA

11. 22.5 Ω

13. **(a)** $I = 14$ A, $I_1 = 6$ A, $I_2 = 8$ A, $I_3 = 0.8$ A

15. **(a)** $I_s = 5$ A, $I_1 = 1$ A, $I_3 = 4$ A, $I_4 = 0.5$ A **(b)** $V_a = 17$ V, $V_{bc} = 10$ V

17. **(a)** $I_E = 2$ mA, $I_C = 2$ mA
(b) $I_B = 24$ μA
(c) $V_B = 2.7$ V, $V_C = 3.6$ V
(d) $V_{CE} = 1.6$ V, $V_{BC} = -0.9$ V

19. **(a)** 1.88 Ω **(b)** $V_1 = V_4 = 32$ V
(c) 8 A **(d)** $I_s = 17$ A, $R_T = 1.88$ Ω

21. **(a)** 14 V **(b)** 9 A
(c) $V_a = -6$ V, $V_b = -20$ V

23. 30 Ω

25. **(a)** no **(b)** 6 kΩ open

27. **(a)** 5.53 Ω **(b)** 7.23 A
(c) 0.281 W

29. **(a)** 12 A **(b)** 0.5 A
(c) 0.5 A **(d)** 6 A

31. $R_1 = 0.5$ kΩ, $R_2 = 2$ kΩ, $R_3 = 4$ kΩ, $R_4 = 1$ kΩ, $R_5 = 0.6$ kΩ, $P_{R_1} = 1$ W, $P_{R_2} = 2$ W, $P_{R_3} = \frac{1}{2}$ W, $P_{R_4} = 2$ W, $P_{R_5} = 1$ W

33. **(a)** yes **(b)** $R_1 = 750$ Ω, $R_2 = 250$ Ω **(c)** $R_1 = 745$ Ω, $R_2 = 255$ Ω

35. **(a)** 1 mA **(b)** $R_{\text{shunt}} = 5$ mΩ

37. **(a)** $R_s = 300$ kΩ,
(b) 20,000 Ω/V

39. 0.05 μA

Chapter 8

1. **(a)** $I_1 = 4.8$ A, $I_2 = 1.2$ A
(b) 9.6 V

3. 31.6 V

5. $V_3 = 1.6$ V, $I_2 = 0.1$ A

7. **(a)** $I_s = 4.68$ A, $R_p = 4.7$ Ω
(b) $I_s = 4.09$ mA, $R_p = 2.2$ kΩ

9. **(a)** 18.18 A **(b)** yes, 18.18 A

11. **(a)** $I_T = 4.2$ A **(b)** 16.8 V

13. **(a)** $V_{ab} = -7$ V **(b)** 1.17 A↑

15. **(a)** $I_1(\text{cw}) = -\frac{1}{7}$ A, $I_2(\text{ccw}) = \frac{5}{7}$ A, $I_3(\text{down}) = \frac{4}{7}$ A **(b)** 4.57 V
17. $I_1(\text{cw}) = 1.45$ mA, $I_2(\text{ccw}) = 8.51$ mA, $I_3(\text{down}) = 9.96$ mA
19. **(d)** 63.69 mA
21. **(a)** $I_{E_1}(\text{ccw}) = 3.06$ A, $I_{E_2}(\text{up}) = 3.25$ A **(b)** $P_{E_2} = 39$ W, $P_{R_3} = 0.43$ W
23. **(a)** $I_1(\text{cw}) = 2.03$ mA, $I_2(\text{left}) = 0.8$ mA, $I_3 = I_4(\text{cw}) = 1.23$ mA **(b)** 4.65 V
25. **(b)** $I_1(\text{cw}) = 1.21$ mA, $I_2(\text{cw}) = -0.48$ mA, $I_3(\text{cw}) = -0.62$ mA **(c)** $I_{E_1}(\text{down}) = 1.69$ mA, $I_{E_2}(\text{up}) = 0.62$ mA
27. **(b)** $I_1(\text{cw}) = 0.03$ mA, $I_2(\text{cw}) = -0.88$ mA, $I_3(\text{cw}) = -0.97$ mA, $I_4(\text{cw}) = -0.64$ mA **(c)** 5.46 mW
29. **(a)** $I_B = 63.02\ \mu$A, $I_C = 4.42$ mA, $I_E = 4.48$ mA **(b)** $V_B = 2.98$ V, $V_E = 2.28$ V, $V_C = 10.28$ V **(c)** 70.14
31. $I_{4\Omega} = 5.53$ A, $I_{6\Omega} = 2.47$ A, $I_{8\Omega} = 0.53$ A, $I_{1\Omega} = 8.53$ A
33. **(b)** 3.25 A
35. **(b)** $I_1(\text{cw}) = 3.31$ A, $I_2(\text{cw}) = -63.69$ mA, $I_3(\text{cw}) = 0.789$ A **(c)** 3.37 A
37. **(b)** −6.44 V
39. **(b)** $I_1(\text{cw}) = 2.37$ A, $I_2(\text{cw}) = -0.20$ A, $I_3(\text{cw}) = 1.25$ A **(c)** $V_a = 4.48$ V, $V_b = 10$ V **(d)** −5.52 V
41. **(b)** $V_1 = -29.29$ V, $V_2 = -33.34$ V **(c)** 1.67 A
43. **(b)** $V_1 = -2.56$ V, $V_2 = 4.03$ V **(c)** $V_{R_1} \mp = 2.56$ V, $V_{R_2} = V_{R_5} \pm = 4.03$ V, $V_{R_4} = V_{R_3}(-+) = 6.59$ V
45. **(b)** $V_1 = 7.24$ V, $V_2 = -2.45$ V, $V_3 = 1.41$ V **(c)** $V_{5\Omega}(-+) = 3.86$ V
47. **(b)** $V_1 = -5.31$ V, $V_2 = 0.62$ V, $V_3 = 3.75$ V **(c)** 69 mA
49. $V_1 = 10.08$ V, $V_2 = 6.94$ V, $V_3 = -17.06$ V
51. **(a)** $V_1 = -10.29$ V, $V_2 = -11.43$ V **(b)** $V_{5A} \mp = 10.29$ V, $V_{3A} \mp = 11.43$ V
53. **(a)** $V_1 = -6.64$ V, $V_2 = 1.29$ V, $V_3 = 10.66$ V **(b)** 1.34 A
55. **(a)** $V_1 = -6.92$ V, $V_2 = 12$ V, $V_3 = 2.3$ V **(b)** 3.46 A
57. **(b)** 20 mA **(c)** no **(d)** no
59. **(b)** 0 A **(c)** yes **(d)** yes
61. 3.33 mA
63. 1.76 mA
65. 133.33 mA
67. 0.83 mA
69. 4.2 Ω

Chapter 9

1. **(a)** 0.1 A↑ **(b)** same **(c)** same
3. 1.25 A↓
5. 52.12 V
7. 10.66 V
9. **(a)** $R_{Th} = 4.1$ kΩ, $E_{Th} = 96$ V **(b)** 2 kΩ : 0.495 W, 100 kΩ : 85 mW
11. $R_{Th} = 2.18\ \Omega$, $E_{Th} = 9.81$ V
13. $R_{Th} = 2\ \Omega$, $E_{Th} = 60$ V
15. **(a)** $R_{Th} = 10\ \Omega$, $E_{Th} = 2$ V **(b)** 20 Ω : 66.67 mA, 50 Ω : 33.33 mA, 100 Ω : 18.18 mA
17. $R_{Th} = 4.04$ kΩ, $E_{Th} = 9.74$ V
19. **(a)** $R_{Th} = 12.5$ kΩ, $E_{Th} = 20$ V **(b)** $R_{Th} = 2.72$ kΩ, $E_{Th} = 60$ mV **(c)** $R_{Th} = 2.2$ kΩ, $E_{Th} = 16$ V
21. **(a)** $R_N = 6\ \Omega$, $I_N = 1$ A **(b)** $E_{Th} = 6$ V, $R_{Th} = 6\ \Omega$
23. $R_N = 2.18\ \Omega$, $I_N = 4.5$ A
25. $R_N = 2\ \Omega$, $I_N = 30$ A
27. $R_N = 4.04$ kΩ, $I_N = 2.41$ mA
29. **(a)** $R_N = 3\ \Omega$, $I_N = 5$ A **(b)** $V_{100\,\Omega} \mp 55.34$ V
31. **(a)** 2.18 Ω **(b)** 11.06 W
33. **(a)** 4.04 kΩ **(b)** 5.87 mW
35. 0 Ω
37. 500 Ω, $P_{max} = 1.44$ W
39. $I_L = 39.3\ \mu$A, $V_L = 220$ mV
41. $I_L = 2.25$ A, $V_L = 6.08$ V
47. **(a)** 0.36 mA **(b)** 0.36 mA **(c)** yes

Chapter 10

1. **(a)** 36×10^3 N/C **(b)** 36×10^9 N/C
3. 50 μF
5. **(a)** 16.69 V/m **(b)** 1.97 kV/m **(c)** 100 : 1
7. 348.43 pF
9. 2.66 μm
11. **(a)** 24.78 nF **(b)** 10^6 V/m **(c)** 4.96 μC
13. 25.6 kV
15. 0.35 μF
17. 470 μF, 465.3 μF–474.7 μF
19. **(a)** 0.56 s **(b)** $v_C = 20\ \text{V}(1 - e^{-t/0.56\,\text{s}})$ **(c)** $1\tau = 12.64$ V, $3\tau = 19$ V, $5\tau = 19.87$ V **(d)** $i_C = 0.2\ \text{mA}\ e^{-t/0.56\,\text{s}}$, $v_R = 20\ \text{V}e^{-t/0.56\,\text{s}}$
21. **(a)** 5.5 ms **(b)** $v_C = 100\ \text{V}(1 - e^{-t/5.5\,\text{ms}})$ **(c)** $1\tau = 63.21$ V, $3\tau = 95.02$ V, $5\tau = 99.33$ V **(d)** $i_C = 18.18\ \text{mA}\ e^{-t/5.5\,\text{ms}}$, $v_{R_2} = 60\ \text{V}\ e^{-t/5.5\,\text{ms}}$
23. **(a)** 100 μs **(b)** 4.72 V **(c)** 11.99 V
25. **(a)** 263.2 ms **(b)** $v_C = 22\ \text{V}\ (1 - e^{-t/263.2\,\text{ms}})$, $i_C = 4.68\ \text{mA}\ e^{-t/263.2\,\text{ms}}$ **(c)** 21.51 V, 0.105 mA **(d)** $v_C = 21.51\ \text{V}\ e^{-t/263.2\,\text{ms}}$, $i_C = 4.58\ \text{mA}\ e^{-t/263.2\,\text{ms}}$
27. **(a)** $v_C = 60\ \text{V}\ (1 - e^{-t/4.84\,\mu\text{s}})$, $i_C = 272.73\ \mu\text{A}\ e^{-t/4.84\,\mu\text{s}}$ **(b)** $v_C = 59.6\ \text{V}\ e^{-t/15.18\,\mu\text{s}}$, $i_C = -86.96\ \mu\text{A}\ e^{-t/15.18\,\mu\text{s}}$
29. **(a)** $v_C = 40\ \text{V} - 34\ \text{V}\ e^{-t/22.1\,\text{ms}}$ **(b)** $i_C = 7.23\ \text{mA}\ e^{-t/22.1\,\text{ms}}$
31. $v_C = -20\ \text{V} + 10\ \text{V}\ e^{-t/2.71\,\mu\text{s}}$, $i_C = -12.2\ \text{mA}\ e^{-t/2.71\,\mu\text{s}}$
33. **(a)** 55.99 mV **(b)** 139.99 mV **(c)** 2.5 ms **(d)** 8.54 ms
35. $R = 54.60$ kΩ
37. **(a)** 22.07 V **(b)** 0.81 μA **(c)** 3.58 s
39. **(a)** $v_C = -27.2\ \text{V} + 37.2\ \text{V}\ e^{-t/18.26\,\text{ms}}$, $i_C = -4.48\ \text{mA}\ e^{-t/18.26\,\text{ms}}$
41. **(a)** $v_C = 3.27\ \text{V}\ (1 - e^{-t/53.80\,\text{ms}})$, $i_C = 1.22\ \text{mA}\ e^{-t/53.80\,\text{ms}}$
43. **(a)** 19.63 V **(b)** 2.32 s **(c)** 1.15 s
45. 10 μs–20 μs: −1.18 A; 20 μs–30 μs: +7.05 A; 30 μs–40 μs: −7.05 A; 40 μs–50 μs: 0 A; 50 μs–55 μs: −4.7 A; 55 μs–60 μs: +4.7 A; 60 μs–70 μs: 0 A; 70 μs–80 μs: +4.7A; 80 μs–100 μs: −1.175 A
47. 6.67 μF
49. $V_1 = 10$ V, $Q_1 = 60\ \mu$C; $V_2 = 6.67$ V, $Q_2 = 40\ \mu$C; $V_3 = 3.33$ V, $Q_3 = 40\ \mu$C
51. $V_1 = 13.45$ V, $Q_1 = 2.96$ mC; $V_2 = 6.55$ V, $Q_2 = 2.16$ mC; $V_3 = 6.55$ V, $Q_3 = 0.786$ mC
53. 8640 pJ
55. $W_{200\,\mu F} = 9.70$ mJ, $W_{100\,\mu F} = 1.75$ mJ

Chapter 11

1. **(a)** 0.04 Wb/m² **(b)** 0.04 T **(c)** 88 At **(d)** 0.4×10^3 gauss
3. **(a)** 20.06 mH **(b)** increase ratio = μ_r
5. **(a)** 42.3 mH **(b)** 1.57 mH **(c)** 75.2 mH **(d)** 1.76 H

7. 6.0 V
9. 14 turns
11. (a) 15 μs
(b) $i_L = 1 \text{ mA}(1 - e^{-t/15\,\mu s})$
(c) $v_L = 20 \text{ V}\, e^{-t/15\,\mu s}$
$v_R = 20 \text{ V}(1 - e^{-t/15\,\mu s})$
(d) $i_L: 1\tau = 0.632$ mA,
$3\tau = 0.951$ mA,
$5\tau = 0.993$ mA; $v_L: 1\tau = 7.36$ V,
$3\tau = 0.98$ V, $5\tau = 140$ mV
13. $R = 1.2\text{ k}\Omega$, $L = 3.6$ mH
15. (a) $i_L = 9.23 \text{ mA} - 17.23 \text{ mA}\, e^{-t/30.77\,\mu s}$,
$v_L = 67.2 \text{ V}\, e^{-t/30.77\,\mu s}$
17. (a) $i_L = 2 \text{ mA} + 4 \text{ mA}\, e^{-t/19.23\,\mu s}$,
$v_L = 41.6 \text{ V}\, e^{-t/19.23\,\mu s}$
19. (a) $i_L = 6 \text{ mA}(1 - e^{-t/0.5\,\mu s})$,
$v_L = 12 \text{ V}\, e^{-t/0.5\,\mu s}$
(b) $i_L = 5.19 \text{ mA}\, e^{-t/83.3\,\mu s}$,
$v_L = -62.28 \text{ V}\, e^{-t/83.3\,\mu s}$
21. (a) $i_L = 1.3 \text{ mA}(1 - e^{-t/7.56\,\mu s})$,
$v_L = 8.09 \text{ V}\, e^{-t/7.56\,\mu s}$
(b) 0.822 mA, 2.98 V
23. (a) $i_L = -4.54 \text{ mA}(1 - e^{-t/6.67\,\mu s})$,
$v_L = -6.81 \text{ V}\, e^{-t/6.67\,\mu s}$
(b) $i_L = -3.53$ mA, $v_L = -1.52$ V
(c) $i_L = -3.53 \text{ mA}\, e^{-t/2.13\,\mu s}$,
$v_L = +16.59 \text{ V}\, e^{-t/2.13\,\mu s}$
25. (a) $i_L = 0.68 \text{ mA} + 1.32 \text{ mA}\, e^{-t/0.49\,ms}$,
$v_L = -5.43 \text{ V}\, e^{-t/0.49\text{ ms}}$
27. (a) 0.92 μs (b) 16.2 V
(c) 0.81 V
29. (a) 4.88 mA (b) 99.33 mA
(c) 13.86 ms
31. (a) 13.33 V (b) 7.98 μA
(c) 4.12 μs (d) 0.244 V
33. 0 ms–2 ms: 37.5 mV; 2 ms–6 ms: −37.5 mV; 6 ms–9 ms: +25 mV; 9 ms–13 ms: 0 V; 13 ms–14 ms: +25 mV; 14 ms–17 ms: 0 V; 17 ms–19 ms: −12.5 mV
35. 10.75 mH
37. 6.8 mH, 5.7 kΩ,
9.1 kΩ ‖ 2.45 mH
39. 25 mH, 2.2 kΩ, 18 μF
41. (a) $i_L = 3.56 \text{ mA}(1 - e^{-t/8.31\,\mu s})$,
$v_L = 4.29 \text{ V}\, e^{-t/8.31\,\mu s}$
43. $I_1 = 7$ A, $I_2 = 2$ A
45. $V_1 = 12$ V, $I_1 = 3$ A, $V_2 = -8$ V,
$I_2 = 0$ A

Chapter 12

1. Φ : CGS : 5 × 10⁴ maxwells;
English : 5×10^4 lines
B: CGS: 8 gauss; English :
51.62 lines/in.^{2}
3. (a) 0.04 T
5. 952.4×10^3 At/Wb
7. 2624.67 At/m
9. 2.13 A
11. (a) 60t (b) 13.34×10^{-4} Wb/Am
13. 2.70 A
15. 1.35 N
17. (a) 2.02 A (b) 2 N
19. 6.12 mWb
21. (a) $B = 1.5 \text{ T}(1 - e^{-H/700 \text{ At/m}})$
(b) 900 At/m: graph = 1.1 T, Eq. = 1.09 T; 1800 At/m: graph = 1.38 T, Eq. = 1.39 T; 2700 At/m: graph = 1.47 T, Eq. = 1.47 T
Excellent results
(c) $H = -700 \log_e(1 - \frac{B}{1.5\text{T}})$
(d) 1 T: graph = 750 At/m, Eq. = 769.03 At/m; 1.4 T: graph = 1920 At/m, Eq. = 1895.64 At/m
(e) 40.1 mA vs. 44 mA in Example 12.1

Chapter 13

1. (a) 10 V (b) 15 ms: −10 V, 20 ms: 0 V (c) 20 V
(d) $T = 20$ ms (e) 2
3. (a) 40 mV (b) 1.5 ms: −40 mV; 5.1 ms: −40 mV (c) 80 mV
(d) 2 ms (e) 3.5
5. (a) 1 Hz (b) 16 Hz
(c) 25 Hz (d) 40 kHz
7. 0.3 ms
9. (a) 125 mV (b) 32 μs
(c) 31.25 kHz
11. (a) 60° (b) 215° (c) 18°
(d) 108°
13. (a) 628.32 rad/s
(b) 1.57×10^3 rad/s
(c) 12.56×10^3 rad/s
(d) 25.13×10^3 rad/s
15. 2.78 ms
17. (a) 20, 60 Hz (b) 12, 120 Hz
(c) 10^6, 1591.55 Hz (d) 8, 1.6 kHz
21. 0.48 A
23. 11.54°, 168.46°
27. (a) $v = 6 \times 10^{-3} \sin(2\pi\, 2000t + 30°)$ (b)
$i = 20 \times 10^{-3} \sin(2\pi\, 60t - 60°)$
29. $v = 12 \times 10^{-3} \sin(2\pi\, 2000t + 135°)$
31. v leads i by 90°
33. in phase
35. 13.95 μs
37. ½ ms
39. 1 V
41. 2.33 V
43. (a) 0 V (b) 0 V (c) same
45. (a) 0.4 ms (b) 2.5 kHz
(c) −25 mV
47. (a) 84.85 V (b) 4.24 mA
(c) 5.66 μA
49. 1.43 V
51. $G = 0$ V, $V_{rms} = 8$ V
53. (a) $y = 2x \Rightarrow y^2 = 4x^2$
(b) 360 (c) 5.48 (d) 3.67
(e) rms ≅ 1.5 average

Chapter 14

1. —
3. (a) $3770 \cos 377t$
(b) $120 \cos(200t + 20°)$
(c) $4440.63 \cos(157t - 20°)$
(d) $200 \cos t$
5. (a) $v = 700 \sin 1000t$
(b) $v = 14.8 \sin(400t - 120°)$
7. (a) 22 mH (b) 1.2 H
9. (a) $v = 100 \sin(\omega t + 90°)$
(b) $v = 0.8 \sin(\omega t + 150°)$
(c) $v = 120 \sin(\omega t - 120°)$
11. (a) $i = 24 \sin(\omega t - 90°)$
(b) $i = 0.6 \sin(\omega t - 70°)$
13. (a) ∞ Ω (b) 530.79 Ω
(c) 15.92 Ω (d) 62.83 Ω
15. (a) 4.08 kHz (b) 34 Hz
(c) 408.09 kHz (d) 20.40 Hz
17. (a) $i = 6 \times 10^{-3} \sin(200t + 90°)$
(b) $i = 22.64 \times 10^{-6} \sin(377t + 90°)$
19. (a) $v = 1190.48 \sin(300t - 90°)$
(b) $v = 37.81 \sin(377t - 120°)$
21. (a) $X_C = 400\ \Omega$
(b) $X_L = 40\ \Omega$, $L = 254.78$ mH
(c) $R = 5\ \Omega$
23. —
25. 318.47 mH
27. 5070 pF
29. 192 W in each case
31. $i = 40 \sin(\omega t - 50°)$
33. (a) $i = 4.27 \sin(1000t - 30°)$
(b) 30 mH (c) 0 W
35. (a) $i_1 = 2.4 \sin(10^4 t + 150°)$, $i_2 = 12 \sin(10^4 t + 150°)$
(b) $i_s = 14.40 \sin(10^4 t + 150°)$
37. (a) 5.0 ∠36.87° (b) 2.83 ∠45°
(c) 12.65 ∠7.57°
(d) 1001.25 ∠2.86°
(e) 4123.11 ∠104.04°
(f) 0.894 ∠116.57°
39. (a) $4.6 + j3.86$
(b) $-6.0 + j10.39$ (c) $-j2000$
(d) $-0.006 - j0.0022$
(e) $47.97 + j1.68$
(f) $4.7 \times 10^{-4} - j1.71 \times 10^{-4}$
41. (a) $11.8 + j7.0$
(b) $151.90 + j49.90$
(c) $4.72 \times 10^{-6} + j71$
43. (a) $7.03 + j9.93$
(b) $95.7 + j22.77$
(c) 28.07 ∠−115.91°
45. (a) 8.00 ∠20°
(b) 49.68 ∠−64.0°
(c) 40×10^{-3} ∠40°

47. **(a)** $4 \angle 0°$ **(b)** $5.93 \angle -134.47°$
(c) $9.30 \angle -43.99°$
49. **(a)** $5.06 \angle 88.44°$
(b) $426 \angle 109.81°$
51. **(a)** $x = 3, y = 6$ or $x = 6, y = 3$
(b) $\theta = 30°$
53. **(a)** $14.14 \angle -180°$
(b) $4.24 \times 10^{-6} \angle 90°$
(c) $2.55 \times 10^{-6} \angle 70°$
55. $v_a = 63.25 \sin(377t + 63.43°)$
57. $v_a = 108.92 \sin(377t - 0.33°)$

Chapter 15

1. **(a)** $6.8\ \Omega \angle 0° = 6.8\ \Omega$
(b) $452.4\ \Omega \angle 90° = +j\,452.4\ \Omega$
(c) $1.48\ \Omega \angle 90° = +j\,1.48\ \Omega$
(d) $1\ \text{k}\Omega \angle -90° = -j\,1\ \text{k}\Omega$
(e) $33.86\ \Omega \angle -90° = -j\,33.86\ \Omega$ **(f)** $220\ \Omega \angle 0° = 220\ \Omega$
3. **(a)** $v = 88 \times 10^{-3} \sin 1000t$
(b) $v = 22.62 \sin(2\pi\, 200t + 150°)$
(c) $v = 270.96 \sin(157t - 50°)$
5. **(a)** $3\ \Omega - j\,1\ \Omega = 3.16\ \Omega \angle -18.43°$
(b) $1\ \text{k}\Omega + j\,4\ \text{k}\Omega = 4.12\ \text{k}\Omega \angle 75.96°$
(c) $470\ \Omega - j\,40\ \Omega = 471.7\ \Omega \angle -4.86°$
7. **(a)** $10\ \Omega \angle 36.87°$ **(b)** —
(c) $\mathbf{I} = 10\ \text{A} \angle -36.87°$, $\mathbf{V}_R = 80\ \text{V} \angle -36.87°$, $\mathbf{V}_L = 60\ \text{V} \angle 53.13°$ **(d)** —
(e) — **(f)** 800 W
(g) 0.8 lagging
(h) $v_R = 113.12 \sin(\omega t - 36.87°)$, $v_L = 84.84 \sin(\omega t + 53.13°)$, $i = 14.14 \sin(\omega t - 36.87°)$
9. **(a)** $5.66\ \Omega \angle -45°$ **(b)** —
(c) $L = 16$ mH, $C = 265\ \mu$F
(d) $\mathbf{I} = 8.83\ \text{A} \angle 45°$, $\mathbf{V}_R = 35.32\ \text{V} \angle 45°$, $\mathbf{V}_L = 52.98\ \text{V} \angle 135°$, $\mathbf{V}_C = 88.30\ \text{V} \angle -45°$ **(e)** —
(f) — **(g)** 311.8 W
(h) 0.707 leading
(i) $i = 12.49 \sin(377t + 4.5°)$, $e = 70.7 \sin 377t$, $v_R = 49.94 \sin(377t + 45°)$, $v_L = 74.91 \sin(377t + 135°)$, $v_C = 124.86 \sin(377t - 45°)$
11. $6.8\ \Omega$
13. **(a)** $292.4\ \mu$A **(b)** 100 pF
15. **(a)** $\mathbf{V}_1 = 14.14\ \text{V} \angle -155°$, $\mathbf{V}_2 = 28.29\ \text{V} \angle 25°$
(b) $\mathbf{V}_1 = 112.92$ V, $\mathbf{V}_2 = 58.66\ \text{V} \angle -139.94°$
17. $3.2\ \Omega + j\,2.4\ \Omega$
19. —
21. —
23. **(a)** $\mathbf{Y}_T = 0.147\ \text{S} \angle 0° = 0.147\ \text{S}$
(b) $\mathbf{Y}_T = 5\ \text{mS} \angle -90° = -j\,5\ \text{mS}$
(c) $\mathbf{Y}_T = 0.5\ \text{mS} \angle 90° = +j\,0.5\ \text{mS}$
25. **(a)** $54.7\ \text{mS} - j\,93.12\ \text{mS}$
(b) $6.88\ \text{mS} + j\,9.08\ \text{mS}$
(c) $4\ \text{mS} + j\,2\ \text{mS}$
27. **(a)** $111.8\ \text{mS} \angle -26.57°$ **(b)** —
(c) $\mathbf{E} = 17.89\ \text{V} \angle 26.57°$, $\mathbf{I}_R = 1.79\ \text{A} \angle 26.57°$, $\mathbf{I}_L = 0.89\ \text{A} \angle -63.43°$ **(d)** —
(e) — **(f)** 32.04 W
(g) 0.894 lagging
(h) $e = 25.30 \sin(377t + 26.57°)$, $i_R = 2.58 + \sin(377t + 26.57°)$, $i_L = 1.26 \sin(377t - 63.43°)$, $i_s = 2.83 \sin 377t$
29. **(a)** $\mathbf{Y}_T = 0.89\ \text{S} \angle -19.81°$, $\mathbf{Z}_T = 1.12\ \Omega \angle 19.81°$ **(b)** —
(c) $C = 531\ \mu$F, $L = 5.31$ mH
(d) $\mathbf{E} = 2.40\ \text{V} \angle 79.81°$, $\mathbf{I}_R = 2.00\ \text{A} \angle 79.81°$, $\mathbf{I}_L = 1.20\ \text{A} \angle -10.19°$, $\mathbf{I}_C = 0.48\ \text{A} \angle 169.81°$ **(f)** —
(g) 4.8 W **(h)** 0.941 lagging
(i) $e = 3.39 \sin(377t + 79.81°)$, $i_R = 2.83 \sin(377t + 79.81°)$, $i_L = 1.70 \sin(377t - 10.19°)$, $i_C = 0.68 \sin(377t + 169.81°)$
31. **(a)** $\mathbf{I}_1 = 18.78\ \text{A} \angle 60.14°$, $\mathbf{I}_2 = 6.88\ \text{A} \angle -29.86°$
(b) $\mathbf{I}_1 = 6.62\ \text{A} \angle 12.89°$, $\mathbf{I}_2 = 1.97\ \text{A} \angle 129.46°$
(c) $\mathbf{I}_1 = 2.4\ \text{A} \angle 0°$, $\mathbf{I}_2 = 1.6\ \text{A} \angle 0°$
33. — **35.** —
37. **(a)** $R_p = 100\ \Omega$, $X_p = 50\ \Omega$ (C)
(b) $R_p = 34\ \text{k}\Omega$, $X_p = 8.5\ \text{k}\Omega$ (L)
39. **(a)** $\mathbf{E} = 176.68\ \text{V} \angle 36.44°$, $\mathbf{I}_R = 0.803\ \text{A} \angle 36.44°$, $\mathbf{I}_L = 2.813\ \text{A} \angle -53.56°$
(b) 0.804 lagging
(c) 141.86 W
(d) — **(e)** —
(f) $1.11\ \text{A} \angle 126.43°$
(g) $142.15\ \Omega + j\,104.96\ \Omega$

Chapter 16

1. **(a)** $4\ \Omega \angle -22.65°$
(b) $3.5\ \text{A} \angle 22.65°$
(c) $3.5\ \text{A} \angle 22.65°$
(d) $1.94\ \text{A} \angle -33.66°$
(e) $14\ \text{V} \angle 112.65°$
3. **(a)** $19.86\ \Omega \angle 37.17°$
(b) $3.02\ \text{A} \angle -37.17°$
(c) $3.98\ \text{A} \angle 52.83°$
(d) $47.81\ \text{V} \angle -37.17°$
(e) 144.42 W
5. **(a)** $0.25\ \text{A} \angle 36.86°$
(b) $89.44\ \text{V} \angle -26.57°$ **(c)** 20 W
7. **(a)** $1.42\ \text{A} \angle 18.26°$
(b) $26.57\ \text{V} \angle 4.76°$ **(c)** 54.07 W
9. **(a)** $537.51\ \Omega \angle 56.07°$
(b) $93\ \text{mA} \angle -56.07°$
(c) $\mathbf{I}_1 = 106.48\ \text{mA} \angle -56.07°$, $\mathbf{I}_2 = 13.48\ \text{mA} \angle 123.93°$
(d) $\mathbf{V}_1 = 16.93\ \text{V} \angle 213.93°$, $\mathbf{V}_{ab} = 41.49\ \text{V} \angle 33.92°$
(e) 2.595 W **(f)** 0.558 lagging
11. **(a)** $1.52\ \Omega \angle -38.89°$
(b) $42.43\ \text{V} \angle 45°$
(c) $14.14\ \text{A} \angle 45°$
(d) $39.47\ \text{A} \angle 38.89°$
13. 139.71 mW

Chapter 17

1. —
3. $\mathbf{Z} = 5.15\ \Omega \angle 59.04°$, $\mathbf{E} = 10.30\ \text{V} \angle 179.04°$
5. $5.15\ \text{A} \angle -24.5°$
7. $2.55\ \text{A} \angle 132.72°$
9. $48.33\ \text{A} \angle -77.57°$
11. $0.68\ \text{A} \angle -162.9°$
13. $42.91\ \mathbf{I} \angle 149.31°$
15. $2.69\ \text{mA} \angle -174.8°$
17. $\mathbf{V}_1 = 14.68\ \text{V} \angle 68.89°$, $\mathbf{V}_2 = 12.97\ \text{V} \angle 155.88°$
19. $\mathbf{V}_1 = 19.86\ \text{V} \angle 43.8°$, $\mathbf{V}_2 = 8.94\ \text{V} \angle 106.9°$
21. $\mathbf{V}_1 = 220\ \text{V} \angle 0°$, $\mathbf{V}_2 = 96.30\ \text{V} \angle -12.32°$, $\mathbf{V}_3 = 100\ \text{V} \angle 90°$
23. $\mathbf{V}_1 = 5.74\ \text{V} \angle 122.76°$, $\mathbf{V}_2 = 4.04\ \text{V} \angle 145.03°$, $\mathbf{V}_3 = 25.94\ \text{V} \angle 78.07°$
25. $\mathbf{V}_1 = 4.37\ \text{V} \angle -128.66°$, $\mathbf{V}_2 = \mathbf{V}_{1\,\text{k}\Omega} = 2.25\ \text{V} \angle 17.63°$
27. $\mathbf{V}_1 = \mathbf{V}_{2\,\text{k}\Omega} = 10.67\ \text{V} \angle 180°$, $\mathbf{V}_2 = 6\ \text{V} \angle 180°$
29. $\mathbf{V}_L = -2451.92\mathbf{E}_i$
31. **(a)** not balanced
(b) $\mathbf{I}_{X_C} = 1.76\ \text{mA} \angle -71.54°$
(c) $\mathbf{V}_{X_C} = 7.03\ \text{V} \angle -18.46°$
33. balanced
35. $R_x = \dfrac{R_2R_3}{R_1}$, $L_x = \dfrac{R_2L_3}{R_1}$
37. $7.02\ \text{A} \angle 20.56°$
39. $36.9\ \text{A} \angle 23.87°$

Chapter 18

1. $6.09\ \text{A} \angle -32.12°$
3. $3.40\ \text{A} \angle 135.36°$
5. $v_C = 12\ \text{V} + 3.75 \sin(\omega t - 83.66°)$
7. $178.5\ \text{mA} \angle -26.57°$
9. $70.61\ \text{mA} \angle -11.31°$
11. $2.94\ \text{mA} \angle 0°$
13. $\mathbf{Z}_{\text{Th}} = 2.4\ \Omega \angle 36.87°$, $\mathbf{E}_{\text{Th}} = 80\ \text{V} \angle 36.87°$

15. $\mathbf{Z}_{Th} = 21.31\ \Omega\ \angle 32.2°$, $\mathbf{E}_{Th} = 2.13\ V\ \angle 32.2°$
17. $\mathbf{Z}_{Th} = 5.00\ \Omega\ \angle -38.66°$, $\mathbf{E}_{Th} = 77.14\ V\ \angle 50.41°$
19. **(a)** dc: $R_{Th} = 22\ \Omega$, $E_{Th} = -5\ V$; ac: $\mathbf{Z}_{Th} = 66.04\ \Omega\ \angle 57.36°$, $\mathbf{E}_{Th} = 6.21\ V\ \angle 207.36°$ **(b)** $i = -72.46\ mA + 62.36 \times 10^{-3} \sin(1000t + 173.42°)$
21. **(a)** $\mathbf{Z}_{Th} = 4.47\ k\Omega\ \angle -26.57°$, $\mathbf{E}_{Th} = 31.31\ V\ \angle -26.57°$ **(b)** 6.26 mA $\angle 63.44°$
23. $\mathbf{Z}_{Th} = 4.44\ k\Omega\ \angle -0.03°$, $\mathbf{E}_{Th} = -444.45 \times 10^3\ \mathbf{I}\ \angle 0.26°$
25. $\mathbf{Z}_{Th} = 5.10\ k\Omega\ \angle -11.31°$, $\mathbf{E}_{Th} = -50\ V\ \angle 0°$
27. $\mathbf{Z}_{Th} = -39.22\ \Omega\ \angle 0°$, $\mathbf{E}_{Th} = 20\ V\ \angle 53°$
29. $\mathbf{Z}_{Th} = 607.42\ \Omega\ \angle 0°$, $\mathbf{E}_{Th} = 1.62\ V\ \angle 0°$
31. $\mathbf{Z}_N = 21.31\ \Omega\ \angle 32.2°$, $\mathbf{I}_N = 0.1\ A\ \angle 0°$
33. $\mathbf{Z}_N = 9.66\ \Omega\ \angle 14.93°$, $\mathbf{I}_N = 2.15\ A\ \angle -42.87°$
35. **(a)** dc: $R_N = 22\ \Omega$, $I_N = -227.27$ mA; ac: $\mathbf{Z}_N = 66.04\ \Omega\ \angle 57.36°$, $\mathbf{I}_N = 94\ mA\ \angle 150°$ **(b)** $\mathbf{I} = -72.46\ mA + 62.68 \times 10^{-3} \sin(1000t + 173.22°)$
37. **(a)** $\mathbf{Z}_N = 4.47\ k\Omega\ \angle -26.57°$, $I_N = 7\ mA\ \angle 0°$ **(b)** 6.26 mA $\angle 63.44°$
39. $\mathbf{Z}_N = 4.44\ k\Omega\ \angle -0.03°$, $\mathbf{I}_N = 100\mathbf{I}\ \angle 0.29°$
41. $\mathbf{Z}_N = 25\ k\Omega\ \angle 0°$, $\mathbf{I}_N = 72\ mA\ \angle 0°$
43. $\mathbf{Z}_N = 6.65\ k\Omega\ \angle 0°$, $\mathbf{I}_N = 0.79\ mA\ \angle 0°$
45. $\mathbf{Z}_L = 1.51\ \Omega - j\,0.39\ \Omega$, $P_{max} = 1.61$ W
47. $\mathbf{Z}_L = 2.48\ \Omega + j\,5.15\ \Omega$, $P_{max} = 618.33$ W
49. $\mathbf{Z}_L = 1.38\ k\Omega - j\,5.08\ k\Omega$, $P_{max} = 50.04$ mW
51. **(a)** $\mathbf{Z}_L = 4\ k\Omega + j\,2\ k\Omega$ **(b)** 61.27 mW
53. **(a)** 7.31 nF **(b)** 2940.27 Ω **(c)** 1 mW
55. **(a)** 0.83 mA $\angle 0°$ **(b)** 0.83 mA $\angle 0°$ **(c)** the same

Chapter 19

1. **(a)** 130W **(b)** $Q_T = 0$ VAR, $S_T = 130$ VA **(c)** 0.542 A **(d)** $R_1 = 371.6\ \Omega$, $R_2 = 668.9\ \Omega$ **(e)** $I_1 = 0.348$ A, $I_2 = 0.193$ A
3. **(a)** $P_T = 400$ W, $Q_T = -400$ VAR (C), $S_T = 565.69$ VA, $F_P = 0.707$ (leading) **(b)** — **(c)** 5.66 A $\angle 135°$
5. **(a)** $P_T = 350$ W, $Q_T = -450$ VAR (C), $S_T = 570.09$ VA **(b)** 0.614 (leading) **(c)** — **(d)** 11.4 A $\angle 52.12°$
7. **(a)** $P_R = 200$ W, $P_{L,C} = 0$ W **(b)** $Q_R = 0$ VAR, $Q_C = 800$ VAR (C), $Q_L = 100$ VAR (L) **(c)** $S_R = 200$ VA, $S_C = 80$ VA, $S_L = 100$ VA **(d)** $P_T = 200$ W, $Q_T = 20$ VAR (L), $S_T = 200$ VA, $F_P = 0.995$ lagging **(e)** — **(f)** 10.05 A $\angle -5.73°$
9. **(a)** $P_L = 0$ W, $P_C = 0$ W, $P_R = 38.99$ W **(b)** $Q_L = 126.74$ VAR, $Q_C = 46.92$ VAR, $Q_R = 0$ VAR **(c)** $S_L = 126.74$ VA, $S_C = 46.92$ VA, $S_R = 38.99$ VA **(d)** $P_T = 38.99$ W, $Q_T = 79.82$ VAR (L), $S_T = 88.83$ VA, $F_P = 0.439$ (lagging) **(e)** — **(f)** 0.31 J **(g)** $W_L = 0.32$ J $W_C = 0.12$ J
11. **(a)** $\mathbf{Z}_T = 2.30\ \Omega + j\,1.73\ \Omega$ **(b)** 4000 W
13. **(a)** $P_T = 900$ W. $Q_T = 0$ VAR, $S_T = 900$ VA, $F_F = 1$ **(b)** 9 A $\angle 0°$ **(c)** — **(d)** load 1: $X_C = 20\ \Omega$; load 2: $R = 2.83\ \Omega$; load 3: $R = 5.66\ \Omega$; $X_L = 4.72\ \Omega$
15. **(a)** $P_T = 1100$ W, $Q_T = 2366.26$ VAR (C), $S_T = 2609.44$ VA, $F_P = 0.422$ (leading) **(b)** 521.89 V $\angle -65.07°$ **(c)** load 1: $R = 1743.38\ \Omega$, $X_C = 1307.53\ \Omega$; load 2: $R = 43.59\ \Omega$, $X_C = 99.88\ \Omega$
17. **(a)** 7.81 kVA **(b)** 0.640 (lagging) **(c)** 65.08 A **(d)** 1105 μF **(e)** 41.67 A
19. **(a)** 128.14 W **(b)** *a-b*: 42.69 W, *b-c*: 64.03 W, *a-c*: 106.72 W, *a-d*: 106.72 W, *c-d*: 0 W, *d-e*: 0 W, *f-e*: 21.34 W
21. **(a)** $R = 5\ \Omega$, $L = 132.03$ mH **(b)** $R = 10\ \Omega$ **(c)** $R = 15\ \Omega$, $L = 262.39$ mH

Chapter 20

1. **(a)** $\omega_s = 250$ rad/s, $f = 39.79$ Hz **(b)** $\omega_s = 3496.50$ rad/s, $f_s = 556.49$ Hz
3. **(a)** 2 kΩ **(b)** 120 mA **(c)** $V_R = 12$ V, $V_L = 240$ V, $V_C = 240$ V **(d)** 20 **(e)** $L = 63.7$ mH, $C = 15{,}920$ pF **(f)** 250 Hz **(g)** $f_1 = 4.88$ kHz, $f_2 = 5.13$ kHz
5. **(a)** 400 Hz **(b)** $f_1 = 5.8$ kHz, $f_2 = 6.2$ kHz **(c)** $X_L = X_C = 45\ \Omega$ **(d)** 375 mW
7. **(a)** 10 **(b)** 20 Ω **(c)** 1.5 mH, 3.98 μF **(d)** $f_1 = 1.9$ kHz, $f_2 = 2.1$ kHz
9. **(a)** $R = 10\ \Omega$, $L = 13.26$ mH, $C = 27.07$ μF, $f_1 = 8.34$ kHz, $f_2 = 8.46$ kHz
11. **(a)** 1 MHz **(b)** 160 kHz **(c)** $R = 720\ \Omega$, $L = 0.716$ mH, $C = 35.38$ pF **(d)** 56.23 Ω
13. **(a)** 159.16 kHz **(b)** 4 V **(c)** 40 mA **(d)** 20
15. **(a)** 1.027 MHz **(b)** 114.1 V **(c)** 13.69 W **(d)** 669 mW
17. $R = 91$ kΩ (closest to 93.33 kΩ), $C = 240$ pF (closest to 250 pF)
19. **(a)** $f_s = 7.12$ kHz, $f_p = 6.65$ kHz, $f_m = 7.01$ kHz **(b)** $X_L = 20.88\ \Omega$, $X_C = 23.94\ \Omega$ **(c)** 55.56 Ω **(d)** $Q_p = 2.32$, BW = 2.87 kHz **(e)** $I_C = 92.73$ mA, $I_L = 99.28$ mA **(f)** 2.22 V
21. **(a)** 3558.81 Hz **(b)** 138.2 V **(c)** 691 mW **(d)** 575.86 Hz
23. **(a)** 98.54 Ω **(b)** 8.21 **(c)** 8.05 kHz **(d)** 4.83 V **(e)** $f_1 = 7.55$ kHz, $f_2 = 8.55$ kHz
25. $R_s = 2.79$ kΩ, $C = 31{,}660$ pF
27. **(a)** 251.65 kHz **(b)** 4.44 kΩ **(c)** 14.05 **(d)** 17.91 kHz **(e)** $f_s = 251.65$ kHz, $Z_{T_p} = 49.94\ \Omega$, $Q_p = 2.04$, BW = 95.55 kHz **(f)** $f_s = 251.65$ kHz, $Z_{T_p} = 13.33$ kΩ, $Q_p = 21.08$, BW = 11.94 kHz **(g)** Network: $L/C = 100 \times 10^3$; part (e): $L/C = 1 \times 10^3$; part (f): $L/C = 400 \times 10^3$ **(h)** As the L/C ratio increased, BW decreased and V_p increased.

Chapter 21

1. **(a)** left: 1.54 kHz, right: 5.62 kHz **(b)** bottom: 0.22 V, top: 0.52 V
3. **(a)** 1000 **(b)** 10^{12} **(c)** 1.59 **(d)** 1.1 **(e)** 10^{10} **(f)** 1513.56 **(g)** 10.02 **(h)** 1,258,925.41

5. 1.68
7. −0.30
9. **(a)** 1.85
(b) 18.45 dB
11. 13.01
13. 38.49
15. 24.08 dB_s
17. —
19. **(a)** $f_c = 3617.16$ Hz; $f = f_c: A_v = 0.707$; $f = 0.1f_c: A_v = 0.995$; $f = 0.5f_c: A_v = 0.894$; $f = 2f_c: A_v = 0.447$; $f = 10f_c: A_v = 0.0995$
(b) $f = f_c: \theta = -45°$; $f = 0.1f_c: \theta = -5.71°$; $f = 0.5f_c: \theta = -26.57°$; $f = 2f_c: \theta = -63.43°$; $f = 10f_c: \theta = -84.29°$
21. $C = 0.265\ \mu F$
23. **(a)** $f_c = 3.62$ kHz; $f = f_c: A_v = 0.707$; $f = 2f_c: A_v = 0.894$; $f = 0.5f_c: A_v = 0.447$; $f = 10f_c: A_v = 0.995$; $f = \frac{1}{10}f_c: A_v = 0.0995$
(b) $f = f_c: \theta = 45°$; $f = 2f_c: \theta = 26.57°$; $f = 0.5f_c: \theta = 63.43°$; $f = 10f_c: \theta = 5.71°$; $f = \frac{1}{10}f_c: \theta = 84.29°$
25. $R = 795.77\ \Omega$, $R_{standard} = 750\ \Omega + 47\ \Omega = 797\ \Omega$
27. **(a)** low-pass section: $f_{c_1} = 795.77$ Hz; high-pass section: $f_{c_2} = 1.94$ kHz; $f = f_{c_1}: V_o = 0.654V_i$; $f = f_{c_2}: V_o = 0.64V_i$; At $f = f_{c_1} + \frac{BW}{2} = 1.37$ kHz; $V_o = 0.706V_i$
(b) BW defined at $0.5V_i$; $f = 500$ Hz: $V_o = 0.515V_i$; $f = 4$ kHz: $V_o = 0.429V_i$; from plot BW ≅ 2.9 kHz with $f_{center} = 1.93$ kHz
29. **(a)** $f_s = 98.1$ kHz
(b) $Q_s = 16.84$, BW = 5.83 kHz
(c) $f = f_s: A_v = 0.93$; $f_1 = 95.19$ kHz, $f_2 = 101.02$ kHz; $f = f_1: V_o = 0.64$ V; $f = f_2: V_b = 0.66$ V
(d) $f = f_s: V_{o_{max}} = 0.93$ V; $f_1 = 95.19$ kHz, $V_o = 0.66$ V; $f_2 = 101.02$ kHz: $V_o = 0.66$ V
31. **(a)** $Q_s = 12.5$
(b) BW = 400 Hz, $f_1 = 4.8$ kHz, $f_2 = 5.2$ kHz
(c) $f = f_s$, $V_o = 25$ mV,
(d) $f = f_s: V_o = 25$ mV
33. **(a)** $f_p = 726.44$ kHz (band-stop); f(band-pass) = 2.01 MHz
35. **(a), (b)** $f_c = 7.20$ kHz
(c) $f = 0.5f_c: A_{v_{dB}} = -7$ dB; $f = 2f_c: A_{v_{dB}} = -0.969$ dB; $f = \frac{1}{10}f_c: A_{v_{dB}} = -20.04$ dB; $f = 10f_c: A_{v_{dB}} = -0.043$ dB
(d) $f = 0.5f_c: A_v = 0.447$; $f = 2f_c: A_v = 0.894$;
(d) $f = 0.5f_c: \theta = 63.43°$; $f = f_c: \theta = 45°$; $f = 2f_c: \theta = 26.57°$
37. **(a), (b)** $f_c = 13.26$ kHz
(c) $f = 0.5f_c: A_{v_{dB}} = -0.97$ dB; $f = 2f_c: A_{v_{dB}} = -6.99$ dB; $f = \frac{1}{10}f_c: A_{v_{dB}} = -0.04$ dB; $f = 10f_c: A_{v_{dB}} = -20.04$ dB
(d) $f = 0.5f_c: A_v = 0.894$; $f = 2f_c: A_v = 0.447$
(e) $f = 0.5f_c: \theta = -26.57°$, $f = f_c: \theta = -45°$, $f = 2f_c: \theta = -63.43°$
39. **(a)** $f_1 = 642.01$ Hz, $f_c = 457.47$ Hz, vertical shift = −2.94 dB **(b)** $f = f_1: \theta = 45°$; $f = f_c: \theta = 54.52°$; $f = f_1/2: \theta = 63.44°$; $f = \frac{1}{10}f_1: \theta = 84.29°$; $f = 2f_1: \theta = 26.57°$; $f = 10f_1: \theta = 5.71°$
41. **(a)** $f_1 = 19.41$ kHz, $f_c = 1.92$ kHz, vertical shift = −20 dB
(b) $f = f_c = f_1: \theta = -39.29°$; $f = 10$ kHz: $\theta = -51.88°$
43. **(a)** $f_1 = 945.66$ Hz, $f_c = 7.59$ kHz, vertical shift = −18.08 dB
(b) $f = f_1: \theta = 37.89°$; $f = f_c: \theta = 37.89°$; $f = 4$ kHz: $\theta = 48.96°$
45. **(a)** $f_1 = 180$ Hz, $f_2 = 18$ kHz, BW = 17,820 Hz, $f = 180$ Hz: $A_{v_{dB}} = -2.99$ dB ≅ −3 dB, $f = 18$ kHz: $A_{v_{dB}} = -3.105$ dB ≅ −3 dB
(b) $f = f_1: \theta = 90°$; $f = 1.8$ kHz: $\theta = 0.12° \cong 0°$; $f = 18$ kHz: $\theta = -90°$
47. $A_v = -120/[(1 - j\,50/f)(1 - j\,200/f)(1 + jf/36\text{ kHz})]$
49. $A_v = 1/(1 + jf/2000)$, $f_c = 2$ kHz
51. $A_{v_{dB}} = 20\log_{10}\sqrt{1 + (f_1/1000)^2} + 20\log_{10}\sqrt{1 + (f_2/2000)^2} + 40\log_{10}1/\sqrt{1 + (f_3/3000)^2}$; $f = 1$ kHz: $A_{v_{dB}} = 3.06$ dB, $f = 2$ kHz: $A_{v_{dB}} = 6.81$ dB, $f = 3$ kHz: $A_{v_{dB}} = 9.1$ dB, 0 dB slope for asymptote at 13.06 dB for $f >> f_3$
53. **(a)** woofer, 400 Hz: $A_v = 0.673$; tweeter, 5 kHz: $A_v = 0.678$
(b) woofer, 3 kHz: $A_v = 0.015$; tweeter, 3 kHz: $A_v = 0.337$
(c) midrange, 3 kHz: $A_v = 0.998$

Chapter 22

1. **(a)** 50 mH
(b) $e_p = 1.6$ V, $e_s = 5.12$ V
(c) $e_p = 15$ V, $e_s = 12$ V
3. **(a)** 355.56 mH
(b) $e_p = 24$ V, $e_s = 0.6$ V
(c) $e_p = 15$ V, $e_s = 12$ V
5. **(a)** 5 V **(b)** 625.59 μWb
7. 120 Hz
9. 30 Ω
11. 12,000 turns
13. **(a)** 3 **(b)** 2.78 W
15. **(a)** 364.55 Ω ∠86.86°
(b) 329.17 mA ∠−86.86°
(c) $\mathbf{V}_{R_e} = 6.58$ V ∠−86.86°, $\mathbf{V}_{X_e} = 14.48$ V ∠3.14°, $\mathbf{V}_{X_L} = 105.33$ V ∠3.14°
17. —
19. 3.2 H
21. $\mathbf{I}_1(\mathbf{Z}_{R_1} + \mathbf{Z}_{L_1}) + \mathbf{I}_2(\mathbf{Z}_m) = \mathbf{E}_1$; $\mathbf{I}_1(\mathbf{Z}_m) + \mathbf{I}_2(\mathbf{Z}_{L_2} + \mathbf{Z}_{R_L}) = 0$
23. **(a)** 20 **(b)** 83.33 A
(c) 4.17 A
(d) $I_s = 4.17$ A, $I_p = 83.33$ A
25. **(a)** $\mathbf{V}_L = 25$ V ∠0°
(b) $\mathbf{I}_L = 5$ A ∠0°
(b) $\mathbf{Z}_L = 80$ Ω ∠0°
(c) $\mathbf{Z}_{1/2} = 20$ Ω ∠0°
27. **(a)** $\mathbf{E}_Z = 40$ V ∠0°, $\mathbf{I}_2 = 3.33$ A ∠60°, $\mathbf{E}_3 = 30$ V ∠60°, $\mathbf{I}_3 = 3$ A ∠60°
(b) $R_1 = 64.52$ Ω
29. $[\mathbf{Z}_1 + \mathbf{Z}_{L_1}]\mathbf{I}_1 - \mathbf{Z}_{M_{12}}\mathbf{I}_2 + \mathbf{Z}_{M_{13}} = \mathbf{E}_1$; $\mathbf{Z}_{M_{12}}\mathbf{I}_1 - [\mathbf{Z}_2 + \mathbf{Z}_3 + \mathbf{Z}_{L_2}]\mathbf{I}_2 + \mathbf{Z}_2\mathbf{I}_3 = 0$; $\mathbf{Z}_{M_{13}}\mathbf{I}_1 + \mathbf{Z}_2\mathbf{I}_2 + [\mathbf{Z}_2 + \mathbf{Z}_4 + \mathbf{Z}_{L_3}]\mathbf{I}_3 = 0$

Chapter 23

1. **(a)** 120.1 V **(b)** 120.1 V
(c) 12.01 A **(d)** 12.01 A
3. **(a)** 120.1 V **(b)** 120.1 V
(c) 16.98 A **(d)** 16.98 A
5. **(a)** $\theta_2 = -120°$, $\theta_3 = +120°$
(b) $\mathbf{V}_{an} = 120$ V ∠0°, $\mathbf{V}_{bn} = 120$ V ∠−120°, $\mathbf{V}_{cn} = 120$ V ∠120°
(c) $\mathbf{I}_{an} = 8$ A ∠−53.13°, $\mathbf{I}_{bn} = 8$ A ∠−173.13°, $\mathbf{I}_{cn} = 8$ A ∠66.87° **(e)** 8 A
(f) 207.85 V
7. $V_\phi = 127.0$ V, $I_\phi = 8.98$ A, $I_L = 8.98$ A
9. **(a)** $\mathbf{E}_{AN} = 12.7$ kV ∠−30°, $\mathbf{E}_{BN} = 12.7$ kV ∠−150°, $\mathbf{E}_{CN} = 12.7$ kV ∠90°
(b–c) $\mathbf{I}_{an} = \mathbf{I}_{Aa} = 11.29$ A ∠−97.54°, $\mathbf{I}_{bn} = \mathbf{I}_{Bb} = 11.29$ A ∠−217.54°, $\mathbf{I}_{cn} = \mathbf{I}_{Cc} = 11.29$ A ∠22.46°

(d) $\mathbf{V}_{an} = 12.16$ kV $\angle -29.34°$, $\mathbf{V}_{bn} = 12.16$ kV $\angle -149.34°$, $\mathbf{V}_{cn} = 12.16$ kV $\angle -90.66°$

11. (a) 120.1 V (b) 208 V (c) 13.36 A (d) 23.15 A

13. (a) $\theta_2 = -120°$, $\theta_3 = +120°$ (b) $\mathbf{V}_{ab} = 208$ V $\angle 0°$, $\mathbf{V}_{bc} = 208$ V $\angle -120°$, $\mathbf{V}_{ca} = 208$ V $\angle 120°$ (d) $\mathbf{I}_{ab} = 9.46$ A $\angle 0°$, $\mathbf{I}_{bc} = 9.46$ A $\angle -120°$, $\mathbf{I}_{ca} = 9.46$ A $\angle 120°$ (e) 16.38 A (f) 120.1 V

15. (a) $\theta_2 = -120°$, $\theta_3 = +120°$ (b) $\mathbf{V}_{ab} = 208$ V $\angle 0°$, $\mathbf{V}_{bc} = 208$ V $\angle -120°$, $\mathbf{V}_{ca} = 208$ V $\angle 120°$ (d) $\mathbf{I}_{ab} = 86.67$ A $\angle -36.87°$, $\mathbf{I}_{bc} = 16.67$ A $\angle -156.87°$, $\mathbf{I}_{ca} = 86.67$ A $\angle 83.13°$ (e) 150.11 A (f) 120.1 V

17. (a) $\mathbf{I}_{ab} = 15.33$ A $\angle -73.30°$, $\mathbf{I}_{bc} = 15.33$ A $\angle -193.30°$, $\mathbf{I}_{ca} = 15.33$ A $\angle 46.7°$ (b) $\mathbf{I}_{Aa} = 26.55$ A $\angle -103.30°$, $\mathbf{I}_{Bb} = 26.55$ A $\angle 136.70°$, $\mathbf{I}_{Cc} = 26.55$ A $\angle 16.70°$ (c) $\mathbf{E}_{AB} = 17.01$ kV $\angle -0.59°$, $\mathbf{E}_{BC} = 17.01$ kV $\angle -120.59°$, $\mathbf{E}_{CA} = 17.01$ kV $\angle 119.41°$

19. (a) 208 V (b) 120.09 V (c) 7.08 A (d) 7.08 A

21. $V_\phi = 69.28$ V, $I_\phi = 2.89$ A, $I_L = 2.89$ A

23. $V_\phi = 69.28$ V, $I_\phi = 5.77$ A, $I_L = 5.77$ A

25. (a) 440 V (b) 440 V (c) 29.33 A (d) 50.8 A

27. (a) $\theta_2 = -120°$, $\theta_3 = +120°$ (b) $\mathbf{V}_{ab} = 100$ V $\angle 0°$, $\mathbf{V}_{bc} = 100$ V $\angle -120°$, $\mathbf{V}_{ca} = 100$ V $\angle 120°$ (d) $\mathbf{I}_{ab} = 5$ A $\angle 0°$, $\mathbf{I}_{bc} = 5$ A $\angle -120°$, $\mathbf{I}_{ca} = 5$ A $\angle 120°$ (e) 8.66 A

29. (a) $\theta_2 = -120°$, $\theta_3 = +120°$ (b) $\mathbf{V}_{ab} = 100$ V $\angle 0°$, $\mathbf{V}_{bc} = 100$ V $\angle -120°$, $\mathbf{V}_{ca} = 100$ V $\angle 120°$ (d) $\mathbf{I}_{ab} = 7.07$ A $\angle 45°$, $\mathbf{I}_{bc} = 7.07$ A $\angle -75°$, $\mathbf{I}_{ca} = 7.07$ A $\angle 165°$ (e) 12.25 A

31. $P_T = 2160$ W, $Q_T = 0$ VAR, $S_T = 2160$ VA, $F_p = 1$

33. $P_T = 7210.67$ W, $Q_T = 7210.67$ VAR (C), $S_T = 10{,}197.42$ VA, $F_P = 0.707$ (leading)

35. $P_T = 7.26$ kW, $Q_T = 7.26$ kVAR (L), $S_T = 10.27$ kVA, $F_p = 0.707$ (lagging)

37. $P_T = 287.93$ W, $Q_T = 575.86$ VAR (L), $S_T = 643.83$ VA, $F_p = 0.447$ (lagging)

39. $P_T = 900$ W, $Q_T = 1200$ VAR (L), $S_T = 1500$ VA, $F_p = 0.6$ (lagging)

41. $12.98\ \Omega - j\,7.31\ \Omega$

43. (a) 9,237.6 V (b) 80 A (c) 1276.8 kW (d) 0.576 lagging (e) $\mathbf{I}_{Aa} = 80$ A $\angle -54.83°$ (f) $\mathbf{V}_{an} = 7773.45$ V $\angle -4.87°$ (g) $62.52\ \Omega + j\,74.38\ \Omega$ (h) System: 0.576 lagging; Load: 0.643 lagging (i) 93.98%

45. (b) $P_T = 5899.64$ W, $P_{\text{meter}} = 1966.55$ W

49. (a) 120.09 V (b) $\mathbf{I}_{an} = 8.49$ A, $\mathbf{I}_{bn} = 7.08$ A, $\mathbf{I}_{cn} = 42.47$ A (c) $P_T = 4.93$ kW, $Q_T = 4.93$ kVAR (L), $S_T = 6.97$ kVA, $F_p = 0.707$ (lagging) (d) $\mathbf{I}_{an} = 8.49$ A $\angle -75°$, $\mathbf{I}_{bn} = 7.08$ A $\angle -195°$, $\mathbf{I}_{cn} = 42.47$ A $\angle 45°$ (e) 35.09 A $\angle -43.00°$

Chapter 24

1. (a) positive-going (b) 2 V (c) 0.2 ms (d) 6 V (e) 6.5% (f) 625 Hz (g) 12.5%

3. (a) positive-going (b) 10 mV (c) 3.2 ms (d) 20 mV (e) 6.9%

5. $V_2 = 13.58$ mV

7. (a) 120 μs (b) 8.33 kHz (c) maximum: 440 mV minimum: 80 mV

9. prf = 125 kHz Duty cycle = 62.5%

11. (a) 8 μs (b) 2 μs (c) 125 kHz (d) 0 V (e) 3.46 mV

13. 18.88 mV

15. 117 mV

17. $v_C = 4\text{ V}(1 - e^{-t/20\text{ ms}})$

19. $i_C = -8\text{ mA } e^{-t}$

21. $i_C = 4\text{ mA } e^{-t/0.2\text{ ms}}$

23. $0 \rightarrow \frac{T}{2}$: $v_C = 20$ V, $\frac{T}{2} \rightarrow T$: $v_C = 20\text{ V}e^{-t/0.2\text{ ms}}$, $T \rightarrow \frac{3}{2}T$: $v_C = 20\text{ V}(1 - e^{-t/0.2\text{ ms}})$, $\frac{3}{2}T \rightarrow 2T$: $v_C = 20\text{ V}e^{-t/02\text{ ms}}$

25. $\mathbf{V}_{\text{scope}} = 10\text{ V}\angle 0°$, $\theta_{\mathbf{Z}_s} = \theta_{\mathbf{Z}_p} = -59.5°$

Chapter 25

1. (I): (a) no (b) no (c) yes (d) no (e) yes (II): (a) yes (b) yes (c) yes (d) yes (e) no (III): (a) yes (b) yes (c) no (d) yes (e) yes (IV): (a) no (b) no (c) yes (d) yes (e) yes

7. (a) 19.04 V (b) 4.53 A

9. 71.87 W

11. (a) $2 + 2.08\sin(400t - 33.69°) + 0.5\sin(800t - 53.13°)$ (b) 2.51 A (c) $24 + 24.96\sin(400t - 33.69°) + 6\sin(800t - 53.13°)$ (d) 30.09 V (e) $16.64\sin(400t + 56.31°) + 8\sin(800t + 36.87°)$ (f) 13.06 V (g) 75.48 W

13. (a) $1.2\sin(400t + 53.13°)$ (b) 0.85 A (c) $18\sin(400t + 53.13°)$ (d) 12.73 V (e) $18 + 23.98\sin(400t - 36.87°)$ (f) 24.73 V (g) 10.79 W

15. $2.26 \times 10^{-3}\sin(377t + 93.66°) + 1.92 \times 10^{-3}\sin(754t + 1.64°)$

17. $30 + 30.27\sin(20t + 7.59°) + 0.5\sin(40t - 30°)$

Chapter 26

1. $Z_i = 986.84\ \Omega$

3. (a) $I_{i_1} = 10\ \mu$A (b) $Z_{i_2} = 4.5$ kΩ (c) $E_{i_3} = 6.9$ V

5. $Z_o = 44.59$ kΩ

7. $Z_o = 10$ kΩ

9. (a) $\mathbf{A}_v = -392.98$ (b) $\mathbf{A}_{v_T} = -320.21$

11. (a) $\mathbf{A}_{v_{NL}} = -2398.8$ (b) $\mathbf{E}_i = 50$ mV (c) $\mathbf{Z}_i = 1$ kΩ

13. (a) $\mathbf{A}_G = 6.067 \times 10^4$ (b) $\mathbf{A}_{G_T} = 4.94 \times 10^4$

15. (a) $\mathbf{A}_{v_T} = 1500$ (b) $\mathbf{A}_{i_T} = 187.5$ (c) $\mathbf{A}_{i_1} = 15$, $\mathbf{A}_{i_2} = 12.5$ (d) $\mathbf{A}_{i_T} = 187.5$

17. (a) $\mathbf{z}_{11} = (\mathbf{Z}_1\mathbf{Z}_2 + \mathbf{Z}_1\mathbf{Z}_3)/(\mathbf{Z}_1 + \mathbf{Z}_2 + \mathbf{Z}_3)$, $\mathbf{z}_{12} = \mathbf{Z}_1\mathbf{Z}_3/(\mathbf{Z}_1 + \mathbf{Z}_2 + \mathbf{Z}_3)$, $\mathbf{z}_{21} = \mathbf{z}_{12}$, $\mathbf{z}_{22} = (\mathbf{Z}_1\mathbf{Z}_3 + \mathbf{Z}_2\mathbf{Z}_3)/(\mathbf{Z}_1 + \mathbf{Z}_2 + \mathbf{Z}_3)$

19. (a) $\mathbf{y}_{11} = (\mathbf{Y}_1\mathbf{Y}_2 + \mathbf{Y}_1\mathbf{Y}_3)/(\mathbf{Y}_1 + \mathbf{Y}_2 + \mathbf{Y}_3)$, $\mathbf{y}_{12} = -\mathbf{Y}_1\mathbf{Y}_2/(\mathbf{Y}_1 + \mathbf{Y}_2 + \mathbf{Y}_3)$, $\mathbf{y}_{21} = \mathbf{y}_{12}$, $\mathbf{y}_{22} = (\mathbf{Y}_1\mathbf{Y}_2 + \mathbf{Y}_2\mathbf{Y}_3)/(\mathbf{Y}_1 + \mathbf{Y}_2 + \mathbf{Y}_3)$

21. $\mathbf{h}_{11} = \mathbf{Z}_1\mathbf{Z}_2/(\mathbf{Z}_1 + \mathbf{Z}_2)$, $\mathbf{h}_{21} = -\mathbf{Z}_1/(\mathbf{Z}_1 + \mathbf{Z}_2)$, $\mathbf{h}_{12} = \mathbf{Z}_1/(\mathbf{Z}_1 + \mathbf{Z}_2)$, $\mathbf{h}_{22} = (\mathbf{Z}_1 + \mathbf{Z}_2 + \mathbf{Z}_3)/(\mathbf{Z}_1\mathbf{Z}_3 + \mathbf{Z}_2\mathbf{Z}_3)$

23. $\mathbf{h}_{11} = (\mathbf{Y}_1 + \mathbf{Y}_2 + \mathbf{Y}_3)/(\mathbf{Y}_1\mathbf{Y}_2 + \mathbf{Y}_1\mathbf{Y}_3)$, $\mathbf{h}_{21} = -\mathbf{Y}_2/(\mathbf{Y}_2 + \mathbf{Y}_3)$, $\mathbf{h}_{12} = \mathbf{Y}_2/(\mathbf{Y}_2 + \mathbf{Y}_3)$, $\mathbf{h}_{22} = \mathbf{Y}_2\mathbf{Y}_3/(\mathbf{Y}_2 + \mathbf{Y}_3)$

25. (a) 47.62 (b) −99

27. $\mathbf{Z}_i = 9{,}219.5\ \Omega \angle -139.4°$, $\mathbf{Z}_o = 29.07$ k$\Omega \angle -86.05°$

29. $\mathbf{h}_{11} = 2.5$ kΩ, $\mathbf{h}_{12} = 0.5$, $\mathbf{h}_{21} = -0.75$, $\mathbf{h}_{22} = 0.25$ mS

Index